SECOND EDITION

HANDBOOK OF THERMOPLASTICS

SECOND EDITION

HANDBOOK OF THERMOPLASTICS

EDITED BY

Olagoke Olabisi
Kolapo Adewale

CRC Press
Taylor & Francis Group
Boca Raton London New York

CRC Press is an imprint of the
Taylor & Francis Group, an **informa** business

CRC Press
Taylor & Francis Group
6000 Broken Sound Parkway NW, Suite 300
Boca Raton, FL 33487-2742

© 2016 by Taylor & Francis Group, LLC
CRC Press is an imprint of Taylor & Francis Group, an Informa business

No claim to original U.S. Government works

Printed on acid-free paper
Version Date: 20151007

International Standard Book Number-13: 978-1-4665-7722-0 (Hardback)

This book contains information obtained from authentic and highly regarded sources. Reasonable efforts have been made to publish reliable data and information, but the author and publisher cannot assume responsibility for the validity of all materials or the consequences of their use. The authors and publishers have attempted to trace the copyright holders of all material reproduced in this publication and apologize to copyright holders if permission to publish in this form has not been obtained. If any copyright material has not been acknowledged please write and let us know so we may rectify in any future reprint.

Except as permitted under U.S. Copyright Law, no part of this book may be reprinted, reproduced, transmitted, or utilized in any form by any electronic, mechanical, or other means, now known or hereafter invented, including photocopying, microfilming, and recording, or in any information storage or retrieval system, without written permission from the publishers.

For permission to photocopy or use material electronically from this work, please access www.copyright.com (http://www.copyright.com/) or contact the Copyright Clearance Center, Inc. (CCC), 222 Rosewood Drive, Danvers, MA 01923, 978-750-8400. CCC is a not-for-profit organization that provides licenses and registration for a variety of users. For organizations that have been granted a photocopy license by the CCC, a separate system of payment has been arranged.

Trademark Notice: Product or corporate names may be trademarks or registered trademarks, and are used only for identification and explanation without intent to infringe.

Visit the Taylor & Francis Web site at
http://www.taylorandfrancis.com

and the CRC Press Web site at
http://www.crcpress.com

To our children—Lara Olabisi, Simi Olabisi, Wande Olabisi, Toyosi Olabisi, Banji Olabisi, Lanre Olabisi, Dr. Ronke Olabisi, Toke Olabisi, and Bunmi Olabisi; Tomi Adewale, Funmi Adewale, and Isaac Adewale

Contents

Preface xi
Editors xiii
Contributors xv

Chapter 1 Polyolefins 1

Olagoke Olabisi

Chapter 2 Vinyl Alcohol Polymers 53

Kereilemang Khana Mokwena Nthoiwa, Carlos A. Diaz, and Yash Chaudhari

Chapter 3 Polyvinyl Butyral 89

Christian Carrot, Amine Bendaoud, and Caroline Pillon

Chapter 4 Polyacrylonitrile 139

Johannis C. Simitzis and Spyridon K. Soulis

Chapter 5 Polyacrylates 169

Joke Vandenbergh and Thomas Junkers

Chapter 6 Polyacetals 193

Kinga Pielichowska

Chapter 7 Polyethers 251

Rebecca Klein and Frederik R. Wurm

Chapter 8 Aromatic Polyamides 285

Miriam Trigo, Jesús L. Pablos, Félix C. García, and José M. García

Chapter 9 Thermoplastic Polyesters 319

Ganesh Kannan, Sarah E. Grieshaber, and Wei Zhao

Chapter 10 Polycarbonates 347

Pierre Moulinié

Chapter 11 Liquid Crystalline Polymers 369

Xiao Li and Haifeng Yu

Chapter 12	Thermoplastic Polyurethanes	387
	Koh-Hei Nitta and Mizue Kuriyagawa	
Chapter 13	Fluoroplastics	397
	Thierry A. Blanchet	
Chapter 14	Polyarylethersulfones	419
	M. Jamal El-Hibri and Shari W. Axelrad	
Chapter 15	Ketone-Based Thermoplastics	461
	Yundong Wang and Mukerrem Cakmak	
Chapter 16	Polyimides	491
	Hooman Abbasi, Marcelo Antunes, and José Ignacio Velasco	
Chapter 17	Polyphenylquinoxalines	533
	Jin-Gang Liu	
Chapter 18	Aromatic Polyhydrazides and Their Corresponding Polyoxadiazoles	571
	Emilia Di Pace, Paola Laurienzo, Mario Malinconico, and Maria Grazia Volpe	
Chapter 19	Polybenzimidazoles	617
	Yan Wang, Tingxu Yang, Kayley Fishel, Brian C. Benicewicz, and Tai-Shung Chung	
Chapter 20	Conductive Thermoplastics	669
	Louis M. Leung	
Chapter 21	Advanced Thermoplastic Composites	693
	Salvatore Iannace, Gianfranco Carotenuto, Luigi Sorrentino, Mariano Palomba, and Luigi Nicolais	
Chapter 22	Natural Fiber Thermoplastic Composites	727
	Norma E. Marcovich, María M. Reboredo, and Mirta I. Aranguren	
Chapter 23	Material Selection, Design, and Application	753
	Eungkyu Kim, Patrick Lee, Joe Dooley, and Mark A. Barger	

Contents

Chapter 24 Laser Processing of Thermoplastic Composites 805
Peter Jaeschke and Verena Wippo

Chapter 25 Bioplastics 835
Caisa Johansson

Chapter 26 Thermoplastic Additives: Flame Retardants 877
Kolapo Peluola Adewale

Chapter 27 Recycling of Thermoplastics 919
Jesús María García-Martínez and Emilia P. Collar

Chapter 28 Environment Health and Safety: Regulatory and Legislative Issues 941
Emilia P. Collar and Jesús María García-Martínez

Index 957

Preface

The global thermoplastics market, representing more than 10% of the global chemical industry, remains the fastest-growing segment of the world economy. This growth is driven by several forces, among which are (1) the widening sphere as well as the demanding requirements of emerging thermoplastics applications; (2) the need for conservation of global natural resources, recycling, and environmental protection; (3) competitive basic, mission-oriented, and applied R&D, including 3-D printing; (4) the revolutionary and evolutionary scientific and technological innovations by scientists and engineers making fundamental breaks from the past; and (5) megatrends such as globalization, urbanization, energy demand, and climate change. Thermoplastics account for about 90% of all polymers produced worldwide.

Tailor-made materials with controlled microstructures continue to emerge not only in polyolefins and other commodity thermoplastics but also in bioplastics, polar thermoplastics, thermoplastics elastomers, synthetic water-soluble thermoplastics, high-performance thermoplastics, high-temperature thermoplastics, specialty thermoplastics for superfunction membranes, conductive thermoplastics, polymeric nonlinear optical (NLO) materials, liquid crystalline polymers, natural fiber thermoplastic composites, advanced thermoplastics composites, as well as nanocomposites. Thermoplastics are generally characterized by higher impact strength, easier processability, and superior adaptability to complex mold designs than thermosets. Components made of metals are typically replaced by thermoplastics and thermoplastics containing composites and nanocomposites.

The first edition of the *Handbook of Thermoplastics* was published in 1997. The purpose of this second edition is to revise and update the first edition incorporating developments and advances in thermoplastics with regard to new materials, additives, materials selection, design, processing, and applications. Compared to the first edition, some chapters have been deleted and new chapters added. For each of the thermoplastics contained in the first edition, the second edition includes the latest advances. A feature of the new edition is the discussion of bio-derived thermoplastics.

The *Handbook of Thermoplastics, Second Edition* is an authoritative source for a worldwide audience in industry, academia, government, and nongovernment organizations. It is indispensable for experienced and practicing professionals, namely, polymer, plastics, chemical, industrial, mechanical, electrical and electronics, design, manufacturing, automotive, aerospace, and bioengineers; materials, organic, physical, and biological scientists; polymer, organic, physical, and industrial chemists; and upper-level undergraduate and graduate students in these disciplines. It provides comprehensive, up-to-date coverage for each thermoplastic, including

- History, development, and commercialization milestones
- Polymer formation mechanisms and process technologies
- Structural and phase characteristics as they affect use properties
- Blends, alloys, copolymers, composites, and their commercial relevance
- Processing, performance properties, and applications
- Any other issue that relates to current and prospective developments in science, technology, environmental impact, and commercial viability

These points were regarded as guidelines, and every contributor was urged to choose a specific format that would make the handbook a timeless reference. A thorny element for each author, and indeed a most important issue that will continue to challenge the thermoplastics industry, relates to environmental waste. The industry continues to develop products having a long life and capable of withstanding aggressive conditions. The rapid proliferation of thermoplastics in an impressive array of applications is evidence of the success of the industry. This very success implies that many thermoplastics, discarded after they have fulfilled their purposes, will pose formidable disposal challenges. This issue is underscored by the inclusion of the last two chapters on recycling, environment, health, safety, regulatory, and legislative issues

The *Handbook of Thermoplastics, Second Edition* is composed of 28 chapters prepared by 65 internationally recognized authorities from 14 countries. It contains more than 4600 bibliographic citations plus over 650 tables, drawings, and photographs. Each chapter includes full references at the end, was edited, reviewed, and revised where necessary, but the authors are responsible for the content. Although no attempt was made to rigorously group the chapters into subsections, there are some subtle groupings as well as overlaps. This is inspired by the reality of the changing thermoplastics industry with its overlapping product families, flexible output, and thermoplastics applications. A pragmatic approach was taken, based on the conventional wisdom embodied in the broad classification of thermoplastics in terms of their applications loosely superimposed on their general spectrum of performance properties, namely, *commodity*, *transitional*, *engineering*, *high performance*, and *high temperature*. The final outcome, representing a cohesive treatment premised on this particular perspective, illustrates the phenomenal progress and the still evolving array of thermoplastics.

It is hoped that this single-volume collective work will serve its intended purposes, contributing to the dialogue on questions that will continue to arise: What should be the priorities and targets for future development and future investments in the thermoplastics industry? What are the prospective developmental patterns? What will be the available opportunities and the prevailing threats? What are the possible strategic approaches? What, in short, are the new sets of thermoplastics products that are likely to be produced and what technologies are likely to be used in the decades ahead?

The efforts of the contributors in preparing and revising their manuscripts for the handbook are deeply appreciated. We acknowledge the support of Corrpro Companies, Inc. Houston, Texas and Hanwha Advanced Materials America, Opelika, Alabama. As always, we are indebted to our spouses, Dr. Juliet Enakeme Olabisi and Dr. Joyce Titilola Adewale, for their considerable understanding.

Editors

Olagoke Olabisi, PhD, is the director of Internal Corrosion Engineering at Corrpro, Houston, Texas, USA. A coauthor, editor, or coeditor of four books, the author or a coauthor of more than 50 professional papers and book chapters, and a holder of nine international patents, he is a fellow of the Nigerian Society of Chemical Engineers and has been a member of the Polymer Processing Society, the American Institute of Chemical Engineers, the American Chemical Society, NACE International, and the Association of Consulting Chemists and Chemical Engineers. Dr. Olabisi earned his BSc (1969) in chemical engineering from Purdue University, West Lafayette, Indiana, his MSc (1971) in chemical engineering from the University of California, Berkeley, California, and his PhD degree (1973) in macromolecular science and engineering from Case Western Reserve University, Cleveland, Ohio.

Kolapo Peluola Adewale, PhD, is the process manager of Hanwha Advanced Materials America, in Opelika, Alabama, USA. He is a holder of 16 international patents and the author or a coauthor of 14 professional papers. He is a member of the Society of Plastic Engineers and the American Institute of Chemical Engineers. Dr. Adewale earned his BSc degree (1981) and his MSc (1984) in chemical engineering from the University of Lagos, Lagos, Nigeria, and his PhD (1995) in polymer science and engineering from the University of Akron, Akron, Ohio.

Contributors

Hooman Abbasi
Department of Materials Science and Metallurgy
Technical University of Catalonia
UPC·BarcelonaTech
Terrassa, Spain

Kolapo Peluola Adewale
Hanwha Advanced Materials America
Opelika, Alabama

Marcelo Antunes
Department of Materials Science and Metallurgy
Technical University of Catalonia
UPC·BarcelonaTech
Terrassa, Spain

Mirta I. Aranguren
Institute for Materials Science and Technology (INTEMA)
Facultad de Ingeniería
Universidad Nacional de Mar del Plata-CONICET
Mar del Plata, Argentina

Shari W. Axelrad
Solvay Specialty Polymers, LLC
Alpharetta, Georgia

Mark A. Barger
Dow Chemical Company
Midland, Michigan

Amine Bendaoud
Université de Lyon
CNRS, UMR5223, Ingénierie des Matériaux Polymères
Université de Saint-Etienne, Jean Monnet
Saint-Etienne, France

Brian C. Benicewicz
Department of Chemistry and Biochemistry
University of South Carolina
Columbia, South Carolina

Thierry A. Blanchet
Rensselaer Polytechnic Institute
Troy, New York

Mukerrem Cakmak
University of Akron
Akron, Ohio

Gianfranco Carotenuto
Istituto per i Polimeri
Compositi e Biomateriali
Consiglio Nazionale delle Ricerche
Portici, Italy

Christian Carrot
Université de Lyon
CNRS, UMR5223, Ingénierie des Matériaux Polymères
Université de Saint-Etienne, Jean Monnet
Saint-Etienne, France

Yash Chaudhari
Department of Packaging Science
Rochester Institute of Technology
Rochester, New York

Tai-Shung Chung
Department of Chemical and Biomolecular Engineering
National University of Singapore
Singapore

Emilia P. Collar
Polymer Engineering Group (GIP)
Institute of Polymer Science and Technology (ICTP)
Consejo Superior de Investigaciones Científicas (CSIC)
Madrid, Spain

Carlos A. Diaz
Department of Packaging Science
Rochester Institute of Technology
Rochester, New York

Emilia Di Pace
Institute of for Polymers, Composites
 and Biomaterials
CNR
Pozzuoli (Naples), Italy

Joe Dooley
JDooley Consulting LLC
Charlestown, Indiana

M. Jamal El-Hibri
Solvay Specialty Polymers, LLC
Alpharetta, Georgia

Kayley Fishel
Department of Chemistry and Biochemistry
University of South Carolina
Columbia, South Carolina

Félix C. García
Departamento de Química
Facultad de Ciencias
Universidad de Burgos
Burgos, Spain

José M. García
Departamento de Química
Facultad de Ciencias
Universidad de Burgos
Burgos, Spain

Jesús María García-Martínez
Polymer Engineering Group (GIP)
Institute of Polymer Science and Technology
 (ICTP)
Consejo Superior de Investigaciones Científicas
 (CSIC)
Madrid, Spain

Sarah E. Grieshaber
SABIC Innovative Plastics
Pittsfield, Massachusetts

Salvatore Iannace
Istituto per i Polimeri
Compositi e Biomateriali
Consiglio Nazionale delle Ricerche
Portici, Italy

Peter Jaeschke
Laser Zentrum Hannover e.V.
Hannover, Germany

Caisa Johansson
Faculty of Health, Science and Technology
Department of Engineering and Chemical
 Sciences
Karlstad University
Karlstad, Sweden

Thomas Junkers
Polymer Reaction Design
Institute for Materials Research (IMO-IMOMEC)
Hasselt University
Diepenbeek, Belgium

Ganesh Kannan
SABIC Americas, Inc.
Sugar Land, Texas

Eungkyu Kim
Dow Chemical Company
Midland, Michigan

Rebecca Klein
Institute of Organic Chemistry
Johannes Gutenberg-University Mainz
and
Graduate School "Material Science in Mainz"
Mainz, Germany

Mizue Kuriyagawa
Department of Materials and Chemical
 Engineering
Kanazawa University
Ishikawa, Japan

Paola Laurienzo
Institute for Polymers, Composites
 and Biomaterials
CNR
Pozzuoli (Naples), Italy

Patrick Lee
College of Engineering and Mathematical
 Sciences
The University of Vermont
Burlington, Vermont

Contributors

Louis M. Leung
Department of Chemistry
Hong Kong Baptist University
Kowloon, Hong Kong

Xiao Li
Department of Materials Science
 and Engineering
College of Engineering
Peking University
Beijing, China

Jin-Gang Liu
Laboratory of Advanced Polymer Materials
Institute of Chemistry
Chinese Academy of Sciences
Beijing, China

Mario Malinconico
Institute for Polymers, Composites
 and Biomaterials
CNR
Pozzuoli (Naples), Italy

Norma E. Marcovich
Institute for Materials Science and Technology
 (INTEMA)
Facultad de Ingeniería
Universidad Nacional de Mar del
 Plata-CONICET
Mar del Plata, Argentina

Pierre Moulinié
Covestro, LLC
Pittsburgh, Pennsylvania

Luigi Nicolais
Istituto per i Polimeri
Compositi e Biomateriali
Consiglio Nazionale delle Ricerche
Portici, Italy

Koh-Hei Nitta
Department of Materials and Chemical
 Engineering
Kanazawa University
Ishikawa, Japan

Kereilemang Khana Mokwena Nthoiwa
Department of Packaging Science
Rochester Institute of Technology
Rochester, New York

Olagoke Olabisi
Corrpro Companies, Inc.
Houston, Texas

Jesús L. Pablos
Departamento de Química
Facultad de Ciencias
Universidad de Burgos
Burgos, Spain

Mariano Palomba
Istituto per i Polimeri
Compositi e Biomateriali
Consiglio Nazionale delle Ricerche
Portici, Italy

Kinga Pielichowska
AGH University of Science and Technology
Faculty of Materials Science and Ceramics
Department of Biomaterials
Kraków, Poland

Caroline Pillon
Université de Lyon
CNRS, UMR5223, Ingénierie des Matériaux
 Polymères
Université de Saint-Etienne, Jean Monnet
Saint-Etienne, France

María M. Reboredo
Institute for Materials Science and Technology
 (INTEMA)
Facultad de Ingeniería
Universidad Nacional de Mar del
 Plata-CONICET
Mar del Plata, Argentina

Johannis C. Simitzis
National Technical University of Athens
School of Chemical Engineering
Athens, Greece

Luigi Sorrentino
Istituto per i Polimeri
Compositi e Biomateriali
Consiglio Nazionale delle Ricerche
Portici, Italy

Spyridon K. Soulis
National Technical University of Athens
School of Chemical Engineering
Athens, Greece

Miriam Trigo
Departamento de Química
Facultad de Ciencias
Universidad de Burgos
Burgos, Spain

Joke Vandenbergh
Polymer Reaction Design
Institute for Materials Research
   (IMO-IMOMEC)
Hasselt University
Diepenbeek, Belgium

José Ignacio Velasco
Department of Materials Science and
   Metallurgy
Technical University of Catalonia
UPC·BarcelonaTech
Terrassa, Spain

Maria Grazia Volpe
Institute of Food Science
CNR
Avellino, Italy

Yan Wang
School of Chemistry and Chemical
   Engineering
Huazhong University of Science and
   Technology
Wuhan, China

Yundong Wang
Teknor Apex Company
Pawtucket, Rhode Island

Verena Wippo
Laser Zentrum Hannover e.V.
Hannover, Germany

Frederik R. Wurm
Max Planck Institute for Polymer Research
Mainz, Germany

Tingxu Yang
Department of Chemical and Biomolecular
   Engineering
National University of Singapore
Singapore

Haifeng Yu
Department of Materials Science and
   Engineering
College of Engineering
Peking University
Beijing, China

Wei Zhao
SABIC Innovative Plastics
Pittsfield, Massachusetts

1 Polyolefins*

Olagoke Olabisi

CONTENTS

1.1 Introduction and Historical Background 2
1.2 Catalyst Systems for Olefin Polymerization 3
    1.2.1 Chromium-Based Catalysts 3
        1.2.1.1 Phillips Catalysts 3
        1.2.1.2 Organochromium Catalysts 4
    1.2.2 Ziegler–Natta Catalyst Systems 5
    1.2.3 SSC Systems 9
        1.2.3.1 Metallocene SSC 9
        1.2.3.2 Post-Metallocene SSC 12
1.3 Production Technology 13
    1.3.1 Free Radical Polymerization Processes 13
        1.3.1.1 High-Pressure Autoclave Reactor Process 13
        1.3.1.2 Tubular Reactor Process 13
    1.3.2 Polymerization Processes for Polyethylenes 14
        1.3.2.1 High-Pressure Processes 14
        1.3.2.2 Low-Pressure Liquid Slurry Processes 15
        1.3.2.3 Low- and Medium-Pressure Solution Processes 16
        1.3.2.4 Low-Pressure Gas-Phase Processes 17
    1.3.3 Polymerization Processes for PP 18
        1.3.3.1 Low-Pressure Liquid Pool Slurry Phase Processes 19
        1.3.3.2 Low-Pressure Modular Gas-Phase Reactor Processes 19
    1.3.4 Polymerization Processes for Other Polyolefins 19
    1.3.5 Process Technologies for SSCs 19
1.4 Polyolefin Structure–Property Relationships 20
    1.4.1 Polyethylenes 22
        1.4.1.1 Branched LDPE 22
        1.4.1.2 Linear LDPEs 23
        1.4.1.3 High-Density Polyethylenes 25
        1.4.1.4 Ultrahigh-Molecular-Weight Polyethylenes 26
    1.4.2 Polypropylene 26
    1.4.3 Poly(butene-1) 27
    1.4.4 Poly(4-methylpentene-1) 28
    1.4.5 Polyolefin Elastomers 28
    1.4.6 Polyolefin Blends and Copolymers 28
        1.4.6.1 Polyolefin Blends 28
        1.4.6.2 Polyolefin Copolymers 30
    1.4.7 Polyolefins from SSCs 32
        1.4.7.1 SSC Polyethylenes 33
        1.4.7.2 SSC PPs 35

---

* Based in part on the first-edition chapters on polyolefins.

	1.4.8	Poly(cycloolefins) from SSCs	39
		1.4.8.1 Poly(cycloolefins) by Vinyl Polymerization	39
		1.4.8.2 Poly(cycloolefins) by ROMP	39
	1.4.9	Cycloolefin Copolymers from SSCs	40
1.5	Polyolefin Composites and Nanocomposites		41
	1.5.1	Conventional Polyolefin Composites	41
	1.5.2	Polyolefin Nanocomposites	42
		1.5.2.1 Nanocomposite Formation by Physical Methods	42
		1.5.2.2 Nanocomposite Formation by Chemical Reaction	43
1.6	Processing Methods for Polyolefins		44
References			45

1.1 INTRODUCTION AND HISTORICAL BACKGROUND

The first known polyolefin, polyethylene, was discovered by two research scientists at the Imperial Chemical Industries (ICI) in 1933. See Ref. [1] polymerized ethylene using less than 0.2% oxygen as a free radical polymerization initiator at 200°C and pressures of 0.1–0.3 GN/m^2. The first commercial free radical polymerization plant, using peroxides and/or peroxyesters, was in operation in September 1939. By the early 1940s, low-density polyethylene (LDPE) production was already based on two high-pressure technologies, namely, autoclave reactor and tubular reactor, yielding two significantly different product streams, one for extrusion coatings and the other for film production. The ICI's free radical polymerization process involves the following four reaction steps: initiation, propagation, termination, and chain transfer. Chain transfer incorporates disproportionation, hydrogen abstraction, scission reactions, and intermolecular as well as intramolecular hydrogen transfer.

Although the original LDPE process did not involve the use of catalysts, its autoclave reactor and tubular reactor technologies are still in use as they have been adapted for a variety of catalysts to make a variety of polyolefins. Generally, polyolefins consist of LDPE, linear low-density polyethylene (LLDPE), very-low-density polyethylene (VLDPE), ultra-low-density polyethylene (ULDPE), medium-density polyethylene (MDPE), high-density polyethylene (HDPE), ethylene–octene copolymers, polypropylene (PP), stereo-block PP, olefin block copolymers, propylene–butane copolymers, propylene-based elastomers, polyolefin plastomers, poly(α-olefin)s, and ethylene–propylene–ethylidenenorbornene (EPDM). Polyolefins are known for their low energy demand during polymerization and melt processing.

In the catalyzed production of polyolefins, the most crucial differences are evident in the microstructure of the polymers resulting from each of the catalysts. Taking polyethylene as an example, titanium-based catalyst normally yields narrow-molecular-weight linear polyethylene, whereas the vanadium- and chromium-based catalysts yield intermediate-molecular-weight linear polyethylene or broad-molecular-weight distributions. On the other hand, the modern metallocene and nonmetallocene single-site catalysts (SSCs) are able to produce narrow-molecular-weight linear polyethylene with long chain branching.

Polyolefins have established themselves among the most widely used commodity polymers during the last eight decades. They now represent more than 50% of global production capacity for all commodity plastics. The global plastics market was 1.5 million tons in 1945 and 245 million tons in 2006 [2,3]. The global thermoplastics market, representing approximately 10% of the global chemical industry [4,5], was about 90 million tons in 1995, 60% of which was accounted for by polyolefins [6]. In 2007, the global polyolefin market consisted of 65 million tons for polyethylene with 6% projected growth rate, 40 million tons for PP with 8% projected growth rate, and 5 million tons for other olefins polymers. It is noteworthy that the global consumption of plastics is 50% more than steel consumption in volume. Today, the cost of polyolefins is going down on account of the natural gas boom from shale gas [7].

Polyolefins find applications in a variety of commercial space including construction, agriculture, transportation, appliances, electronics, and communication. Polyolefins are mostly used in packaging applications and *use and throw* products. Other applications include durables and consumable products such as plastic pipes, wire and cable coatings, shoe soles, storage containers, garbage bags, industrial packaging films, food packaging films, rigid food containers, paper coating, etc. Polyolefins also find applications in insulation, furniture, textiles, banknotes, and many more. Exceptional properties are the driving forces in developing new polyolefins, broadening the property envelope and expanding their boundaries toward the areas traditionally occupied by more sophisticated, expensive, and sometimes hazardous materials. Polyolefins are economically and ecologically attractive materials; they possess an extraordinary recycling capacity. Sometimes referred to as *solid crude oil*, they offer advantages over other plastics in recycling. They may be degraded by catalytic hydrogenation, cracked, or used as a source for incineration.

This chapter will focus on all polyolefins including polyolefin nanocomposites. In Section 1.2, four types of polyolefin catalysts will be discussed: (1) chromium-based catalysts; (2) Ziegler–Natta catalysts; (3) metallocene SSC; and (4) postmetallocene SSC. All four categories are important for polyethylenes, but the last three categories of catalysts are far more relevant for PPs.

1.2 CATALYST SYSTEMS FOR OLEFIN POLYMERIZATION

Much like the uncatalyzed free radical initiated polymerization, the ionic polymerization processes of the transition metal halides and the transition metal oxide catalysts also involve initiation, propagation, and termination steps. The catalysts could be homogeneous or heterogeneous. A typical heterogeneous olefin polymerization catalyst system may consist of (1) a support, (2) a surface-modifying reductant, (3) a catalyst precursor, and (4) a cocatalyst that activates the catalyst. The order of addition of components has an effect on the overall nature of the catalyst system. The support normally has to be pretreated either by physical dehydroxylation (calcination), chemical dehydroxylation, thermal degassing, or surface modification using a reductant [8]. The factors affecting the overall performance of a supported catalyst include (1) dispersion of the catalyst precursor, (2) transformation characteristics during support pretreatment, (3) interaction of the catalyst precursor with the support, (4) possible agglomeration of the catalyst precursor, and (5) catalyst impurities and poisoning. An important element of catalyst design is the prevention of dangerous runaway reactions, particularly in gas-phase polymerization where explosion could be especially devastating.

1.2.1 CHROMIUM-BASED CATALYSTS

The first solution phase process for the production of linear HDPE, involving the use of transition metal oxide catalysts at 100–250°C and pressures of 3–5 MN/m^2 [9,10], was carried out in 1950–1952. Independently, Standard Oil of Indiana and Phillips Petroleum Company used molybdenum oxide and chromium oxide catalysts, respectively, to produce HDPE (after vaporizing the solvent). By the 1960s, catalyst development efforts enabled the low-temperature production of linear HDPE solid using a slurry phase reactor with an inert solvent. High-activity catalysts, developed by the middle of the 1960's finally enabled the introduction of gas-phase ethylene polymerization. Today, several variants of these processes are in operation in different parts of the world using modern catalysts. PP production followed a similar trend except that it almost always lags behind polyethylene.

1.2.1.1 Phillips Catalysts

The chromium oxide ethylene polymerization catalyst was a product of serendipity at the Phillips Petroleum Company in 1950 when it was discovered by He, Lanning, Hogan, and Banks [10]. Chromium oxide was supposed to aid the conversion of refinery stack gases into motor fuel; however, it converted the ethylene in the stack gases into polyethylenes. The chromium oxide catalyst consists of a refractory support and an oxide of Cr(II), Cr(III), Cr(VI), or any inorganic chromium compound that could be calcined to chromium oxide, such as chromic nitrate, chromium sulfate, ammonium chromate, chromium carbonate, chromyl chloride, and t-butyl chromate [11–14].

Since the early days, the Phillips catalyst has been modified using a variety of inorganic and organic compounds including boron trichloride, boron phosphate, boron ester, isopropyl borate, trimethyl borate, ammonium tetrafluoroborate, ammonium hexafluorosilicate, α,ω-aliphatic diene, isoprene/nickel oxide, isoprene/nickel nitrate, isoprene/nickel acetate, isoprene/nickel chloride, titanium/tetraisopropyl titanate, magnesium ethoxide/tetraisopropyl titanate, dibutyl magnesium/tetraisopropyl titanate, trialkyl dialkyl phosphate-titanate, or benzene. The supports that have been used include silica, alumina (could be fluorided or phosphated), silica-alumina, zirconia, zirconia-silica cogel, thoria, germania, or mixtures thereof. Triisobutyl aluminum, 1,5-hexadiene/triethyl aluminum, 1,7-octadieneand/triethyl aluminum, and triethyl borane have been used as cocatalysts for chromium oxide catalysts [11–14].

The SiO_2-supported chromium oxide catalyst could be activated at 750–1500°C in a stream of nonreducing moisture-free gas containing oxygen or in vacuo at temperatures between 400°C and 900°C. Photoreduction with a mercury lamp light source, in the presence of carbon monoxide, was also found to be just as effective, and it could be done at temperatures as low as 200°C. Generally, activation with carbon monoxide, as with triethyl borane, reduces the oxidation state of chromium. On the other hand, molybdenum oxide (or cobalt molybdate) on γ-alumina titania or zirconia support needs a reducing gas such as hydrogen or carbon monoxide.

The effects of metal oxide loading on reactivity and kinetic profile have been studied. Ethylene polymerization is presumed to occur through the reaction of an adsorbed monomer with an adjacent monomer or a polymer similarly adsorbed on the solid surface. In the case of $Cr/AlPO_4$ catalyst, the P/Al ratio has an effect on the catalyst activity as well as the tendency toward bimodality of the polymer molecular-weight distribution. In general, however, chromium-based catalysts yield linear HDPE with intermediate- or broad-molecular-weight distribution. Perhaps, the single outstanding issue with chromium catalysts is the fact that the active site structures remain a matter of controversy until today.

1.2.1.2 Organochromium Catalysts

The organochromium family of catalysts, due to Union Carbide (UCC is now part of Dow Chemical Company), consists of closed ring bis(cyclopentadienyl)chromate (chromocene) [15] and/or bis(triphenylsilyl)chromate [16]. The latter is the reaction product of triphenylsilanol and chromium trioxide, which when supported on silica-alumina is very active particularly for the gas-phase ethylene polymerization process. The former, when supported on high-surface area silica, liberates a cyclopentadienyl (Cp) group and polymerizes ethylene via a coordinated anionic mechanism. Other organochromium catalysts with π-bonded ligands include open ring chromocene [17], namely, dimethylpentadienyl chromate $[Cr(DMPD)_2]$, mixed open/closed ring chromocene, namely, Cr(Cp)DMPD [18], biscumene [Cr(0)] [19], $Cr(neopentyl)_4$, bismesitylene [Cr(O)] [20], as well as those with σ-bonded ligands such as $Cr_4(trimethylsilyl)_8$, $Cr(trimethylsilyl)_4$, chromium acetyl acetonate, chromium acetate, chromium stearate, $Cr(benzene)_2$, $Cr(octate)_3$, as well as organophosphoryl chromium compounds. The supports normally used are silica, aluminophosphate, alumina, and fluorided and phosphated alumina.

The organochromium catalysts are not as thermally stable as the chromium oxide family of catalysts, but a calcined chromium oxide catalyst could be modified by impregnation with organochromium compound in order to form a highly active mixed Cr–Cr catalyst. This is exemplified by the modification of hexavalent t-butyl chromate with zero-valent organochromium dicumene Cr(0) as well as that of divalent Cr(II) or hexavalent Cr(VI) oxide with the divalent alkyl chromium $Cr_4(trimethylsilyl)_8$ [19]. Similarly, organochromium esters could form bimetallic complexes with metal chlorides. The hydrated chromium acetate $[Cr(CH_3COO)_3 \cdot H_2O]$/magnesium chloride and anhydride chromium stearate $[Cr(C_{17}H_{35}COO)_3]$/magnesium chloride systems have been extensively investigated [21,22] with $AlEt_2Cl$, $Al_2Et_3Cl_3$, $AlEtCl_2$, or $AlEt_3$ as the cocatalyst. The effects of chromium loading on reactivity and kinetic profile have also been studied for organochromium catalysts [23].

Polyolefins

1.2.2 Ziegler–Natta Catalyst Systems

The other type of ionic polymerization process for the production of linear HDPE became a reality in 1953 when Karl Ziegler discovered [24] the first-generation transition metal halide catalyst by combining the salts or oxides of the periodic table groups IV-B and V-B metals with organometallic aluminum alkyls cocatalyst. Guilio Natta's major contribution [25] was the use of the Ziegler catalyst, namely, $TiCl_4$–$AlEt_3$, for the isospecific polymerization of propylene in 1954, and the resulting family of catalysts is collectively called the Ziegler–Natta catalysts. Stereoregularity is an important practical property in the polymerization of vinyl monomers, CH_2==CHR, which is capable of yielding polymers that are atactic, characterized by a random arrangement of R; isotactic, characterized by an arrangement of R uniformly on one side of the polymer backbone; and syndiotactic, characterized by an arrangement of R on the alternate side of the polymer backbone plane.

The second-generation $MgCl_2$ and/or donor-supported Ziegler–Natta catalyst system, which were at least 100 times more active, led to the development of the low-pressure polymerization processes for polyolefins and synthetic elastomers. This revolutionary development resulted in the simplified gas-phase low-pressure polymerization plant operation without the need for the removal of residual trace catalyst from the polymer, making nonpelletized LLDPE, ULDPE, or VLDPE, PP, and its copolymers. Simonazzi and Giannini [26] provided an impressive array of the accomplishments in the science, engineering, and technology of the Ziegler–Natta catalysis. Although the review sought to highlight the significant role of Montell (now called Basell), it provides an insight into the worldwide efforts related to the simplification of the polyolefin process technologies in terms of economics, versatility, safety, and environmental efficiency.

Basically, both the heterogeneous transition metal halide and oxide catalyst systems are characterized by the following common features: (1) a solid surface for monomer adsorption; (2) a transition metal that is easily converted from one to the other of its several valence states; and (3) a propensity for the formation of organometallic compound with another organometallic compound or a monomer. However, stereospecific polymerization of butene-1 or propylene (small, nonpolar, volatile monomers) requires the presence of a strong complexing active center adsorbed on a solid surface [27–30].

The Ziegler–Natta [24,25] catalyst system consists of two components, namely, the transition metal compound customarily called the catalyst and the alkylaluminum compound customarily called the cocatalyst. Some typical examples of these compounds are presented in Table 1.1. The reactions of the various catalysts and cocatalysts have been studied extensively, and the product derived from the reaction between, for example, $TiCl_4$ and $AlEt_3$ is known to consist of a partial colloidal mixture of the titanium halides at various oxidation states [26]. No complex compound was found that includes the two metal atoms such as titanium and aluminum [27–30]. The preferred Ziegler–Natta titanium catalyst compounds are the high-surface-area violet crystalline forms of $TiCl_3$, and the commercially utilized titanium trichlorides are normally activated by hydrogen or by organometallic compounds such as organoaluminum, organozinc, or organomagnesium compounds. Complete reduction of $TiCl_4$ to $TiCl_3$ could be accomplished with $EtAlCl_2$ or Et_2AlCl at 1:1 or 2:1 ratios, respectively [31–34].

The activity and yield of the catalyst largely depend on the nature of the cocatalyst (activator) and on the catalyst/cocatalyst ratio. The effects of additional organic adjuncts attached to the aluminum cocatalyst underscore the fact that the activity of a catalyst system depends strongly on the cocatalyst type [35–37]. Dual functional titanium catalysts and benzyl derivatives of titanium, which are active in the absence of aluminum trialkyl, also exist [38–40].

Selected Ziegler–Natta catalysts, based on zirconium and vanadium, are as follows: (1) $Zr(OC_3H_7)_4$ and $Zr[OCH(CH_2CH_3)_2]_4$; (2) $Zr(OC_4H_9)_2Cl_2$, $Zr(OC_6H_{13})_2Cl_2$, and $Zr(OC_8H_{17})_2Cl_2$; (3) VCl_3 [40], VCl_4 [41,42], and $VCl_3(THF)_3$ [42,43]; (4) $VOCl_3$ [42,44–46], $VO(OBu)_3$ [47], and $VO(OC_2H_5)_3$; (5) vanadyl acetate [48]; and (6) mixtures. Unlike titanium or the other transition metal catalysts, vanadium catalysts need promoters such as chloroform [43,49], Freon-11 [42], dichloromethane or methylene dichloride [41–43], trichlorofluoromethane [42,43,49], 1,1,1-trichloroethane [42,49,50], hexachloropropane, heptachloropropane, or octachloropropane [51]. Because of the structural and chemical homogeneity of its

TABLE 1.1
Examples of Two-Component Ziegler–Natta Catalyst Systems

Transition Metal Salt	Organometallic Compounds
$TiCl_4$	Et_3Al
$Zr(OC_3H_7)_4$	Et_3Al
$Zr[OCHEt_2]_4$	Et_3Al
VCl_3	Et_2AlCl
$V(acac)_3$[a]	Et_2AlCl
$Cr(acac)_3$[a]	Et_2AlCl
$CoCl_2$ 2 pyridine	Et_2AlCl
$Zr(OC_3H_7)_4$	Et_2AlCl
$Zr[OCHEt_2]_4$	Et_2AlCl
$TiCl_4$	BuLi
$TiCl_3$	Bu_2Mg
Cp_2[b]$TiCl_2$	$EtAlCl_2$
$TiCl_4$, VCl_3, or $TiCl_3$	Et_2AlCl
$TiCl_4$, VCl_3, or $TiCl_3$	$(i\text{-}C_4H_9)_xAl_y(C_5H_{10})_z$[c]
$TiCl_4$, VCl_3, or $TiCl_3$	$Et_3Al_2Cl_3$
$TiCl_4$, VCl_3, or $TiCl_3$	Et_3Al
$TiCl_4$, VCl_3, or $TiCl_3$	$(i\text{-}C_4H_9)_3Al$
$TiCl_4$, VCl_3, or $TiCl_3$	$(i\text{-}C_4H_9)_2AlH$
	DEAC
	Isoprenyl
	EASC[d]
	TEAL
	TIBAL
	DIBAL-

Source: Simonazzi, T. and U. Giannini, *Gazz. Chim. Ital.* 124: 533, 1994; SRI International, Polyolefin Markets and Resin Characteristics, Vol. 2, Project No. 3948, SRI, Menlo Park, CA, 1983; SRI International, Polyolefins Production and Conversion Economics, Vol. 3, Project No. 3948, SRI, Menlo Park, CA, 1983; SRI International, Polyolefin Production Technology, Vol. 4, Project No. 3948, SRI, Menlo Park, CA, 1983.

[a] acac, acetylacetonate anion.
[b] Cp_2, cyclopentadienyl.
[c] Where $z \approx 2x$, made by reacting TIBAL or DIBAL-H with isoprene.
[d] Ethyl aluminum sesquichloride; $Zr(OC_3H_7)_4$ and $Zr[OCH(CH_2CH_3)_2]_4$ react with Et_3Al and Et_2AlCl.

active center, the homogeneous vanadium-based catalysts are traditionally used for the production of ethylene–propylene rubber (EPR) copolymer and ethylene–propylene–diene monomer (EPDM) terpolymers. The preferred cocatalyst is halogenated aluminum alkyls, and the preferred promoters include ethyl trichloroacetate, *n*-butyl perchlorocrotonate, and benzotrichloride. In the production of LLDPE, silica-supported vanadium catalysts are particularly active in the presence of halocarbon promoters resulting in a higher α-olefin comonomer incorporation rate and better comonomer distribution along the polymer chain. However, the vanadium-based catalysts are less capable of controlling the molecular-weight distribution, yielding intermediate- or broad-molecular-weight distribution compared with those based on titanium, zirconium, or hafnium. Calcium carbonate-mixed silica support could also be used for the vanadium-based catalysts.

Several methods exist for the preparation of the varieties of supported Ziegler–Natta catalysts. Some of these are impregnation, milling, comilling [32], or solution methods. Cocrystallization using low-valency transition metal carbonyls [52] such as $Mn_2(CO)_{10}$, $Mn(CO)_5Cl$, $V(CO)_6$, and $Fe(CO)_8$ results in solid solutions, such as $FeCl_2 \cdot 2TiCl_3$ and $MnCl_2 \cdot 2TiCl_3$, which are known to be quite active. Dialkyl magnesium compounds have also been used as reducing agents including the following: dimethyl magnesium, diethyl magnesium, di-n-butyl magnesium, n-butyl-s-butyl magnesium, ethyl-n-butyl magnesium, ethyl-n-hexyl magnesium, dihexyl magnesium, and butyloctyl magnesium [32,33,36,52]. Metal chloride reducing agents, such as $SiCl_4$ [50] and BCl_3 [37,53], have also been used.

Generally, the most active catalyst is based on titanium, and the high-activity, high-yield $MgCl_2$-supported titanium chloride catalyst is produced either by dry comilling of $MgCl_2$ and titanium halides or by cocondensing $MgCl_2$ vapor with the vaporized toluene/$TiCl_4$ or heptane/$TiCl_4$ or diisopropylbenzene/$TiCl_4$ substrates [35] or by solution. The solubility of $MgCl_2$ in the electron donor solvent, such as tetrahydrofuran (THF), increases in the presence of the reducing Lewis acid such as aluminum chloride, ethyl aluminum, and boron trichloride. This is a good technique for activating the magnesium halide-based titanium or vanadium catalysts [53]. However, the catalyst reactivity and stereospecificity of the $MgCl_2$-supported titanium chloride is related to the structure of α-$TiCl_3$, γ-$TiCl_3$, and δ-$TiCl_3$ vis-à-vis that of the $MgCl_2$ support [26]. The crystalline layer structure of the violet $TiCl_3$ is similar to that of $MgCl_2$, and dry comilling of the two results in favorable epitaxial placement of the active dimeric titanium chloride on the (100) lateral planes of $MgCl_2$ exposing a larger number of stereospecific sites, and hence the increased propagation rate. While the lateral (100) surfaces are known to be stereospecific, the (110) planes are known to be aspecific.

In addition, the chemical nature and porosity of $MgCl_2$ are said to play more effective roles than the specific surface area [34,54]. Indeed, complexes containing titanium and magnesium bonded by double-chloride bridges have been observed, exposing the titanium atoms on the catalyst surface where they are more accessible [26]. Silica, silica–alumina, modified or unmodified, as well as MgO supports have been used with mixed results [31–36,53,55–57]. Catalyst modifiers such as $NdCl_3$, $BaCl_2$, $ZnCl_2$, $ZnEt_2$, and Grignard reagents (C_6H_5MgCl) have been used. Magnesium alkoxide modifiers that have been used include magnesium methoxide and magnesium ethoxide [31,32,55,58–61]. For the $MgCl_2$/TiX_4/$Al(iBu)_3$ system, the nonchloride ligands impart decreased activities, although the resulting polyolefins might have improved properties [62]. With nonchloride ligands, the activity of the titanium-based catalysts increases with decreasing electron-releasing capability of the ligand [63] in the following order: $Ti(OC_6H_5)_4 > Ti(O(CH_2)_3CH_3)_4 > Ti(N(C_2H_5)_2)_4$. This is further illustrated by another study where the catalyst activity is in accordance with the following order [64]: $TiCl_4 > TiCl_2(OBu)_2 > TiCl(OBu)_3$.

The high-yield, high-stereospecificity $MgCl_2$ donor-supported titanium chloride catalyst was first developed by using a Lewis base modifier like the esters of aromatic monocarboxylic or phthalic acids or alkylphthalate as the internal donor, which is added in the preparation of supported catalysts. This fourth-generation Ziegler–Natta catalyst system, first developed in the 1980s, is still extensively used in polyolefin manufacturing. Other internal donors, developed beyond the year 2000, include 1,3-dione, isocyanate, 1,3-diether, malonic ester, succinate, 1,3-diol ester, glutaric acid ester, diamine, 1,4-diol, 1,5-diol ester, phthalate esters, and cycloalkyl esters.

Further developments led to the use of an additional modifier as the external donor such as bifunctional Lewis base, which is added during the olefin polymerization process. The bifunctional Lewis base is essentially a cocatalyst, such as polyalkoxysilane, along with the aluminum trialkyl, which is a strong Lewis acid. The use of the internal and external donors led to the possibility of controlling the morphology of the catalyst granules. The catalyst morphology (size, shape, and porosity) is important, and there are significant differences in the polymerization rate pattern between granular versus powdered catalyst or between spherical powder versus granular powder [37,65,66]. Indeed, the morphology of the resulting polymer particle replicates that of the catalyst, which essentially acts as a template for the polymer growth, justifying the trend toward spherical catalyst particles [26,65,66]. The polymer particle size distribution is also similar to that of the catalyst. This similarity has made it possible to control the polymer granule size, its porosity, and its stereoregularity. Additionally, the polymer molecular-weight distribution could be similarly controlled by changing the structure of the external donor.

The organic silane external donors that have generated most interest in terms of stereospecificity include dicyclopentyl dimethoxy silane (DCPDMS), diisopropyl dimethoxy silane (DIPDMS), methyl cyclohexyl dimethoxy silane (CHMDMS), and diisobutyl dimethoxy silane (DIBDMS), in that order. The excellent variety of internal and external donors in use consolidated the position of the fourth-generation Ziegler–Natta catalyst system as one of the most widely used in polyolefin manufacturing.

A further advance, made in 1989 by Himont, made it possible to use a single donor, such as 1,3 diethers, eliminating the need for the external donor. The development was considered to be the fifth-generation Ziegler–Natta catalyst systems. The mechanism of the Lewis base donors is presumed to be related to the selective poisoning and/or modification of the aspecific catalyst sites through complexation with the base. The 1,3 diether is characterized by the desired oxygen–oxygen distance, which is crucial for chelating to the tetracoordinate magnesium atoms located on the (110) aspecific plane of $MgCl_2$.

This line of catalyst development made it possible to exploit the living polymerization capability of each catalyst granule whereby each granule contains a *living polymer* whose growth could be spontaneously continued by the addition of the same or different sets of monomers resulting in block or multimonomer copolymers. Thus, not only does the technology facilitate the synthesis of homopolymers whose molecular weight ranges from very low to extremely high, but the technology also facilitates the *in situ* manufacture of heterophase polyolefin copolymers, blends, and alloys, incorporating nonolefinic comonomers with each catalyst granule serving as a self-contained reactor. The traditional and high-yield $MgCl_2$-supported Ziegler–Natta catalyst systems for PP polymerization appear in Table 1.2 and those for polyethylenes appear in Table 1.3. Catalyst preparation effort sometimes results in process innovation exemplified by Borealis development of an emulsion process. Focused effort by BASF and SINOPEC BRICI also yielded unique Ziegler–Natta catalyst systems. Supported or unsupported Ziegler–Natta catalyst systems are also used in olefin oligomerization [67].

Ziegler–Natta catalyst research effort continues with multiple purposes. The first is focused on catalyst morphology and particle size distribution with the goal of controlling the resulting polyolefin particle size distribution, fine content, and bulk density. The second is to increase the catalyst activity while decreasing its surface area and porosity. The third is devoted to catalyst improvement

TABLE 1.2
Traditional and High-Yield $MgCl_2$-Supported Ziegler–Natta Catalyst Systems for Propylene Polymerization

Catalyst System	Activity [kg PP/(mol Ti) MPa h]	Isotactic Index[a] (%)
$TiCl_3$–Et_2AlCl	76	90–95
$MgCl_2$-supported $TiCl_4$–Et_3Al	9000	30–50
$MgCl_2$-supported $TiCl_4/LB_1$[b]–Et_3Al	7000	50–60
$MgCl_2$-supported $TiCl_4/LB_1$[b]–Et_3Al/LB_2[c]	6000	92–95
$MgCl_2$-supported $TiCl_4/LB_3$[d]–Et_3Al/LB_4[e]	15,000	98–99
$MgCl_2$-supported $TiCl_4/LB_5$[f]–Et_3Al	20,000	97–99

Source: Simonazzi, T. and U. Giannini, *Gazz. Chim. Ital. 124*: 533, 1994.
[a] Weight percent of polymer insoluble in boiling *n*-heptane.
[b] LB_1, ethyl benzoate.
[c] LB_2, methyl 4-methylbenzoate.
[d] LB_3, diisobutyl phthalate.
[e] LB_4, dicyclopentyldimethoxysilane.
[f] LB_5, 2,2-diisobutyl-1,3-dimethoxypropane.

TABLE 1.3
Traditional and High-Yield MgCl$_2$-Supported Ziegler–Natta Catalyst Systems for Ethylene Polymerization

Catalyst System	Activity [kg polym/(mol Ti) MPa h]
TiCl$_3$–Et$_2$AlCl	320
TiCl$_3$Et$_3$Al	710
MgCl$_2$-supported TiCl$_4$–Et$_3$Al	17,000

Source: Simonazzi, T. and U. Giannini, *Gazz. Chim. Ital.* 124: 533, 1994.

in terms of stereospecificity. It was such an effort that resulted in evolutionary catalysts, which enabled the formation of diversified molecular structures during polymerization. Such catalysts enabled reproducible molecular and morphological structures for homopolymers, copolymers, and blends, and the control of molecular weight, molecular-weight distribution, monomeric units' configuration, chain configuration, and size and shape of polymer particles. These were the periodic table groups IV catalysts [68–70] now christened metallocene SSC systems.

1.2.3 SSC Systems

The key distinctions between Ziegler–Natta catalyst and SSC relate to the following characteristics. Ziegler–Natta catalysts possess multiple reaction sites, use simple aluminum alkyls as cocatalysts, use internal and external electron donors, and produce polyolefins with broad-molecular-weight distribution, nonuniform chain lengths, high bulk density, and high soluble content. On the other hand, SSCs are single sited, use alkyl aluminoxane and bulky anions as cocatalysts, use no internal or external electron donors, and yield polyolefins with narrow-molecular-weight distribution, uniform chain lengths, low bulk density, and low soluble content. SSCs are characterized by higher activity, outstanding ability to incorporate sterically demanding comonomers, and the ability to vary the comonomer distribution (alternating, random, or block) over the entire polymer chain backbone. From an ecological viewpoint, the total ash arising from the incineration of SSC polymer is only a fraction of what is produced by an equivalent Ziegler–Natta polymer.

1.2.3.1 Metallocene SSC

The typical chemical structure of a Group IV metallocene catalyst is characterized by a well-defined organometallic molecular complex constrained mostly by tetrahedral geometry. Metallocene SSC consists of a transition metal atom that is sterically hindered in that it is sandwiched between π-carbocyclic ancillary ligands such as Cp, fluorenyl, indenyl, or other substituted structures, and it is sometimes referred to as a half-sandwich titanium amide system. When the two π-carbocyclic ancillary ligands on either side of the transition metal are unbridged, the metallocene is nonstereorigid and is characterized by C_{2v} symmetry. When the two ligands are bridged, the metallocene is stereorigid, and it is called ansa-metallocene, which could be characterized by C_1, C_2, or C_s symmetry. Although the π-ligands are most common, halides and σ-homoleptic hydrocarbyls are the other two classes of ligands of the Group IV metallocene SSCs. In addition, the bridging moieties on ansa-metallocenes have electronic and steric implications. Variations within ligands and/or bridging moieties result in variations of catalyst stability, catalytic activity, kinetic profile, polymer stereoregularity, monomer/comonomer incorporation capability, molecular-weight characteristics, and polymer microstructure.

The first homogeneous metallocene catalyst was discovered in 1957 by Natta et al. [25], who replaced the chloride ligand of the Ziegler–Natta transition metal catalyst with bis(cyclopentadienyl) titanium compounds together with aluminum alkyls for ethylene polymerization. A breakthrough occurred when Kaminsky and his coworkers noticed that the addition of water to trialkyl aluminum in a molar ratio of 1:1 significantly improved the catalyst activity [68]. Thus, by the 1980s and 1990s, homogeneous catalysts based on group IVA metallocenes and an aluminoxane, especially methylaluminoxane (MAO), as cocatalysts gained widespread industrial and scientific interest [68–75]. The catalyst components, as well as the active species generated from them, are soluble in hydrocarbons; thus they were originally referred to as homogeneous Ziegler–Natta catalysts but later called post Ziegler–Natta catalysts. These are the periodic table groups IV [68–70] metallocene SSC systems based primarily on three key transition metals, namely, zirconium (Zr), titanium (Ti), and hafnium (Hf), and are now referred to as SSC systems.

1.2.3.1.1 MAO Cocatalyst

MAO cocatalyst is a mixture of oligomers formed by the controlled reaction of trimethylaluminum and water under elimination of methane. Its structure was a riddle until Barron [76–78] and Sinn [79] discovered that association/dissociation phenomena, condensation and cleavage by trimethylaluminum, characterize the dynamic behavior of MAO in solution. MAO is an amorphous, pyrophoric solid that is soluble in toluene but insoluble in hexane. The average number of aluminum units in the cluster of MAO varies between 10 and 20; higher-molecular-weight MAO compound is essentially insoluble. MAO is oligomeric and its degree of oligomerization, n, strongly influences the metallocene catalyst activity. Depending on the metallocene type and composition, the most effective range for MAO is $3 < n < 50$. The catalyst stability, catalytic activity, kinetic profile, polymer stereoregularity, monomer/comonomer incorporation capability, molecular-weight characteristics, and polymer microstructure are affected not only by the amount of MAO but also by the metallocene/MAO ratios. In large-scale commercial processes, heterogenization has enabled the reduction of the amount of metallocene/MAO ratios such that the ratio of aluminum to transition metal is low. Silica, silica–alumina, resinuous materials, and mixtures have been used as catalyst supports.

MAO is crucial in the formation of the metallocenium active species. The generation of the active species involves alkylation of the metallocene by the MAO and abstraction of a methyl group from the dimethyl metallocene. The resulting active species is a 14-valence electron cationic alkylmetallocenium ion [80] formed by dissociation of the metallocene–aluminoxane complex. Because the catalyst adjunct, [MAO Me]$^-$ anion, is essentially noncoordinating or weakly coordinating, the incoming olefin monomer coordinates instead with the cationic alkylmetallocenium ion, via the metal–carbon bond, into the previous monomer. Thus, the growing chain migrates through the 4-center transitional state regenerating the vacant coordination site for the next incoming monomer.

The Lewis acid catalyst adjunct, the [MAO Me]$^-$ anion, may be replaced by another Lewis acid such as Me$_2$AlF [15], [B(C$_6$H$_5$)$_4$]$^-$ [16], [C$_2$B$_9$H$_{12}$]$^{2-}$ [81], [B(C$_6$F$_5$)$_3$] [82], [Ph$_3$C][B(C$_6$F$_5$)$_4$] [83,84], [R$_2$R'NH][B(C$_6$F$_5$)$_4$] [85], or [Ph$_3$C][B(C$_6$F$_4$(SiR$_3$)] [86], in combination with trialkylaluminum (TMA) compounds as alkylating agent and dichlorometallocenes or simply in combination with dialkylmetallocenes. In most of these substituted systems, two problems arise. Because of the highly unsaturated character of the active species, a scavenger is needed, usually TMA. Secondly, almost all the Lewis acids interact somewhat with the active species and may not be regarded as noncoordinating [87]. The only systems that seem to fulfill the requirements as good cocatalysts are those based on [Ph$_3$C][B(C$_6$F$_5$)$_4$], [R$_3$NH][B(C$_6$F$_5$)4], or [Ph$_3$C][B(C$_6$F$_4$(SiR$_3$)].

1.2.3.1.2 Importance of Metallocene SSC Symmetry on Tacticity

Owing to the fact that the polymer chain migrates during insertion, the symmetry of the metallocene SSC is of fundamental importance to the tacticity of the polymer produced. When a chiral metallocene active center is combined with a prochiral monomer such as propylene or hexane-1 monomer, a diastereotopic transition state is formed. This makes possible two sets of activation energies for monomer insertion. Stereospecificity may arise either from the chiral β-carbon atom at the terminal monomer unit of the growing chain (chain end control) or from the chiral catalyst site (enantiomorphic site control). Therefore, the microstructure of the polymer produced depends on the mechanism of stereocontrol as well as on the nature of the metallocene used.

The first chiral-bridged zirconocene synthesized in 1984 by Brintzinger and used as an isospecific polymerization catalyst by Kaminsky was racemic ethylenebis(4,5,6,7-tetrahydro-1-indenyl)zirconium dichloride [88–90]. Ewen [91] showed that the analogous ethylenebis(1-indenyl)titanium dichloride (a mixture of the meso form and racemate) produces a mixture of isotactic PP (iPP) and atactic PP (aPP). The chiral titanocene as well as the zirconocene were shown to work by enantiomorphic site control. With titanocene, the achiral mesostructure causes the formation of atactic polymer. Also, achiral metallocenes like $C_{p2}ZrCl_2$ or $[Me_2Si(Flu)_2]ZrCl_2$ produce aPP. The polymerization is not stereo- but regioselective due to the bent structure of the tetrahedral active complex favoring 1, 2 insertions. The nonbridged metallocene catalyst $(2-Ph-Ind)_2ZrCl_2$ is a mixture of the meso form and racemate. The racemate produces iPP, whereas the meso form produces aPP.

Using C_s-symmetrical metallocenes, syndiotacticity is due to enantiotopic vacancies formed by the chain migratory insertion [92–94] during polymerization. If, however, the π-carbocyclic ancillary ligand used in the catalyst has substituents at key positions, isospecific polymerization of prochiral monomer units could occur owing to steric hindrance. Chiral C_2-symmetric metallocenes, like bridged bis(indenyl) compounds, possess homotopic coordination sites, which favor identical orientation of the approaching prochiral monomer resulting in isospecific polymerization. For achiral C_{2v} symmetric metallocenes, the polymer tacticity is determined by the chain-end controlled mechanism, and atactic polyolefins are formed because the configuration of the asymmetric center of the last-inserted prochiral monomer unit occasionally changes. Hemi-isotactic polymerization of prochiral monomer units normally results from the use of C_1-symmetrical metallocene SSC. That is, every other methyl group has isotactic placement; the remaining methyl groups are randomly placed.

1.2.3.1.3 Significance of Metallocene SSC in Polyolefins

The metallocene SSC has the capability of producing polyolefins with terminal unsaturation, which are used for building functionalities in the polymer. Considering only the olefins, the metallocene SSC possesses an extraordinary versatility for polymerizing a variety of monomers that include simple olefins, diolefins, cyclic olefins, and cyclic diolefins. Higher-molecular-weight polyolefins are enhanced with Group IV bis(2-R-indenyl) *ansa*-metallocenes [73]. Metallocenes living homopolymerization and copolymerization capability is exceptional, and the number of copolymer compositional permutations made possible is large. Mixed metallocene SSC made possible the development of new types of polymer blends (PBs) that are made *in situ*. These mixed metallocenes differ in terms of transition metals, C_1, C_2, C_{2v}, or C_s symmetry, π-ligands, halides ligands, σ-homoleptic hydrocarbyl ligands, bridging moieties, Lewis acid catalyst modifiers, or scavengers.

The first commercial utilization of Kaminsky-type metallocene SSC, a conventional biscyclopentadienyl catalyst, was in 1991 by Exxon (now ExxonMobil) using high-pressure autoclaves and gas-phase reactors (Exxpol technology) to produce EXACT® polymers. This was followed in 1992 by Dow using an ansa-cyclopentadienylamido constrained geometry SST to produce AFFINITY™ polyolefin plastomers, ENGAGE™ polyolefin elastomers (POEs), ELITE™ enhanced polyethylene, and NORDEL™-IP EPDM. Later developments included the slurry process INSITE™ technology as well as the gas-phase technologies for LLDPE and HDPE.

Commercial production of syndiotactic PP (sPP) utilized a silica-supported metallocene SSC. By 2003, ExxonMobil introduced propylene–ethylene copolymers Vistamaxx™, and Mitsui introduced propylene–butylene copolymers Tafmer® XM both based on metallocene SSC. Attention was also focused on the process of catalyst support preparation that enabled significant reduction of the MAO cocatalyst content while increasing the metallocene catalyst activity. One such method involved spray-drying of clay and SiO_2 reactant to form microspherical particles with high porosity, which was used as a catalyst activator during pre-polymerization and a catalyst support during polymerization processes. This is exemplified by IOLA, which was developed by W. R. Grace & Co.

1.2.3.2 Post-Metallocene SSC

Group IV metallocenes are well-defined organometallic molecular complexes characterized mostly by tetrahedral geometry. It is the precise catalyst structure that facilitated incremental improvements in science and technology that eventually lead to the development of the second-generation postmetallocene SSC systems [95]. These new-generation SSC systems are based on a variety of transition metals including Group IV and other ligands in addition to the confining Cp-based ancillary ligands generally used with Group IV–metallocene SSC systems.

Aside from the tetrahedral geometry that mostly characterizes the metallocene SSC, the new-generation SSC systems are characterized by a variety of coordination geometries including square planar, trigonal bipyramidal, octahedral, as well as tetrahedral geometries. While metallocene SSC enables significant control of polyolefin composition distribution and long-chain branching, the postmetallocene SSCs, in addition, enable significant control of polyolefin purity, tacticity distribution, and unique comonomer incorporation. The key impact of postmetallocene SSC technology is in enabling the effective control of stereo-/regio-defects on polymer backbone during polyolefin polymerization as well as comonomer *blockiness* for olefin block copolymers.

The novel well-defined structures of the postmetallocene SSC enable unusual polymerization reactions including (1) polymerization with exceedingly fast monomer insertion [96]; (2) extensive branch formation through chain walking [97]; (3) extensive syndiospecific propylene polymerization using C_2-symmetric catalyst complexes [98]; (4) chain-end functionalization with vinyl groups and aluminum alkyls [7]; (5) copolymerization with vinyl functionalized polar olefins [97,99]; (6) polymerization of ultrahigh-molecular-weight polymers [100] with α-diimine Ni and Pd complexes bearing sterically demanding ortho-substituted N-aryl groups [97]; and (7) adaptable olefin living polymerization mechanisms [101]. Indeed, an emerging concept posits that the unusual living polymerization mechanisms are occasioned by noncovalent attractive, rather than steric, intramolecular interactions in postmetallocene SSCs [102]. The unique features of postmetallocene SSC have resulted in exceptional polymers that were hardly possible with previous chromium-based catalysts, Ziegler–Natta catalysts, or metallocene SSC systems.

Postmetallocene catalyst systems include Group III to Group XIII catalysts. The ligands include (1) Cp and other carbon-donor ligands; (2) chelating amides and related ligands; (3) chelating alkoxides, aryloxides, and related ligands; (4) chelating phosphorus-based ligands; (5) non-Cp-based amide, amine, and phenoxy ligands; (6) neutral bis(imino)pyridine; (7) neutral α-diimine, neutral nitrogen-based, and related ligands; (8) anionic ligands; and (9) monoanionic ligands [95]. Recent developments have demonstrated that, when fluorine-containing ancillary ligands are used, noncovalent interactions control polymerization reactions rather than steric influences [102]. That is, postmetallocene SSCs are predominantly based on alkyl or aryl ligands containing nitrogen, oxygen, sulfur, or phosphorus atoms and are usually devoid of Cp, indenyl, or fluorine-based ligands. A variety of metal–ligand combinations are available involving such metals as nickel, iron, copper, zinc, and titanium. In spite of the tremendous success of late transition metal systems, an overriding metal–ligand combination rule remains elusive.

1.2.3.2.1 Postmetallocene SSC Polyolefins

Dow Chemical Company was the first to commercialize postmetallocene SSC polyolefins with the use of metal–ligand combinations of zirconium or hafnium with pyridyl amine-based ligands. Under the trade name of VERSIFY, propylene–ethylene plastomers and elastomers are characterized by narrow-molecular-weight and unique composition distribution. The second postmetallocene SSC polyolefins were the INFUSE™ ethylene–octene block copolymers. These were produced using chain shuttling technology, whereby two independent postmetallocene catalyst systems with differing monomer selectivity characteristics are involved along with a diethyl zinc chain shuttling agent [103]. Perhaps the best known postmetallocene SSC systems used for the production of PP are (1) a diimine nickel/palladium system, (2) a pyridine diimine iron/cobalt system, and (3) a salicylaldiminato titanium/zirconium system. The industrial impact of high-performance postmetallocene SSCs is yet to be fully realized.

Polyolefins

1.3 PRODUCTION TECHNOLOGY

The first commercial polyolefin process was introduced by ICI [1,104] in 1939. Since then, large capacity innovative technologies, with process optimizations, have continued to fuel the expansion of polyolefin market worldwide. The contemporary commercial olefin polymerization processes include (1) loop reactors (Spheripol, Borstar, Phillips, ExxonMobil, Sinopec ST, Hypol-II), (2) autoclave reactors (Hypol-I, Exxon Sumitomo, Rexene), (3) fluidized bed reactors (Unipol, Sumitomo, Catalloy), and (4) stirred bed reactors (Innovene, Horizone, Novolen). Other industrial participants include Chevron Phillips, Hostalen, LyondellBasell (Lupotech), and Dow (Unipol).

As a result of the competitive nature of polyolefin business, there have been a number of expensive patent litigations exemplified by DuPont (LLDPE), Montecatini/Basell (PE), Exxon, Phillips, Mobil and Dow (SSC catalysts), and British Petroleum and UCC (now part of Dow Chemical Company) for the polyolefin process. The historical perspectives of polyolefin process developments are discussed next.

1.3.1 Free Radical Polymerization Processes

Branched LDPE was first commercialized [1] in 1939 by ICI using its high-pressure technology, based on tubular or stirred autoclave reactors at temperatures of 200–300°C and pressures of 0.1–0.3 GN/m^2. Although the high-pressure technology was developed for free radical polymerization, subsequent developments of tubular or stirred autoclave reactors have enabled the use of chromium-based catalysts, Ziegler–Natta catalysts, metallocene SSCs, and postmetallocene SSCs.

1.3.1.1 High-Pressure Autoclave Reactor Process

The high-pressure autoclave reactor process operates at pressures of about 0.2 GN/m^2 at temperatures of 150–315°C. The commercial high-pressure autoclave technology available for the production of branched LDPE generally uses oxygen or peroxide initiators. They were originally proprietary to the following companies: ICI, Cities Services/ICI, CdF Chimie, Dow Chemical, DuPont, El Paso, Gulf, National Distillers/ICI, and Sumitomo. The LDPE is produced at 15–20% conversion, and the density range falls within 0.915–0.925 gm/cm^3. A telogen, such as butane, is normally added to the feed stream for molecular-weight control. The LDPEs are generally used for injection molding, extrusion coating, heavy-duty film, and wire and cable covering. The autoclave reactor is also used in the production of ethylene–vinyl acetate (EVA) copolymers.

The control of the LDPE density, chain branching, crystallinity, molecular weight, and its distribution require a judicious adjustment of pressure, temperature profile, initiator type and concentration, and telogen type and concentration. For example, low temperatures are used for the production of injection molding grades requiring broad-molecular-weight distribution. Uniform high temperature and high initiator concentration are used for the production of polymers with narrow-molecular-weight distribution. Long-chain branching (LCB) is favored by high temperature, low pressure, and high initiator concentration. An increase in temperature also leads to an increase in chain transfer reaction (decrease in molecular weight), whereas an increase in pressure leads to an increase in chain growth reaction resulting in an increase in density. Initiator and telogen type and concentration could affect the LDPE chain length by a factor of 2–4. Density, which depends on the extent of chain branching vis-à-vis the polymer molecular weight, is also sensitive to temperature and pressure to a much lesser extent.

1.3.1.2 Tubular Reactor Process

The flow charts for the tubular and autoclave reactor processes are similar except for the reactor section where the heavy-walled reactor tubing replaces the autoclave. The reactor tubing is 1000–2000 m long and has 25–50 mm internal diameter, and the process operates at pressures of about 0.3 GN/m^2 with temperatures ranging between 170°C and 330°C. The commercial high-pressure tubular technology available for the production of branched LDPE generally uses oxygen or peroxide initiators. They were originally proprietary to the following companies: ICI, ANIC, Arco, ATO Chimie, BASF, Exxon, El Paso, Imhausen, Distillers/ICI, Stemicarbon, Sumitomo, UCC (now part of Dow Chemical Company), and VEB-Leuna Werke.

Molten LDPE is produced at a conversion of 25–35%, which is higher than that of the autoclave processes principally because of the easier method of heat removal through cooling jackets. As with the autoclave reactor, polymer density, chain branching, crystallinity, and molecular weight and its distribution are controlled during the polymerization process. A telogen is normally added to the feed stream for molecular-weight control, and the density of the resulting LDPE ranges between 0.918 and 0.93 g/cm^3. The polymers find applications in tough, stiff clarity film and packaging, as industrial liners, heavy-duty bags, shrink film, lamination film, wire, and cable. The tubular reactor is also used in the production of EVA co- and terpolymers characterized by high-environmental-stress cracking resistance useful in frozen food packaging and other high-clarity and high-gloss applications. Certain specific operating conditions of the tubular reactor process could be related to the following structure–property characteristics:

- The higher operating pressure results in higher reaction propagation rate, higher density, lower degree of branches, higher molecular weight, and stiffer polymers.
- The plug flow character leads to elongated as opposed to the near-spherical structure of the polymer made in the autoclave.
- The variability of the temperatures and pressures somewhat results in the production of polymers characterized by a relatively wide-molecular-weight distribution.

1.3.2 Polymerization Processes for Polyethylenes

The advent of the Phillips and Ziegler catalysts gave birth to the low-pressure olefin polymerization processes operating at pressures of about 2–10 MN/m^2 and temperatures of about 100°C. In most commercial polymerization processes, the supported catalyst system is conveyed to the reactor either in the form of a powdery solid or a Bingham fluid composition [105,106]. While pressure is a key variable for density control in the high-pressure branched LDPE polymerization, the amount and type of α-olefin comonomer in the feed composition is the key density-controlling variable for the low-pressure LLDPE processes. In addition, the average molecular weight of the polymer is responsive to the polymerization temperature, whereas the molecular-weight distribution is responsive to the catalyst system. Hydrogen is used as a molecular-weight modifier.

1.3.2.1 High-Pressure Processes

Although the high-pressure autoclave and tubular processes were developed for branched LDPE, catalyst developments resulted in several adapted high-pressure technologies for linear MDPE, linear HDPE, and LLDPE. The available high-pressure technologies, with their corresponding catalyst systems, were proprietary to the following companies [104,107–109]:

1.3.2.1.1 High-Pressure Autoclave Reactor
- ARCO technology using a Ziegler catalyst system, produces LLDPE.
- Bayer technology using a silyl ester catalyst system, produces linear MDPE.
- Dow Chemical technology using a Ziegler catalyst system, produces linear MDPE, HDPE, and LLDPE.
- CdF Chimie technology using a Ziegler catalyst system, produces linear MDPE, HDPE, and LLDPE.

Polyolefins

1.3.2.1.2 High-pressure tubular reactor

- ATO Chimie technology using a Ziegler catalyst system, produces LLDPE.
- Dow Chemical technology using a Ziegler catalyst system, produces linear MDPE, linear HDPE, and LLDPE.
- El Paso/Montedison technology using a Ziegler catalyst system, produces LLDPE.
- Imhausen technology using a Ziegler catalyst system, produces linear MDPE, linear HDPE, and LLDPE.
- Mitsubishi Petrochemical technology using a Ziegler catalyst system, produces linear MDPE, linear HDPE, and LLDPE.

1.3.2.2 Low-Pressure Liquid Slurry Processes

The first low-pressure linear HDPE slurry technology to utilize the Ziegler catalyst was commercialized in 1955 by Hoechst. The process is based on a stirred-tank reactor containing a heavy hydrocarbon diluent with a continuous ethylene feed. The reactor is operated at a pressure of 0.8 MN/m^2, a temperature of 85°C, and an average residence time of 2.7 h. The concentration of the solid polymer particles in the reactor discharge stream ranges between 13 and 45 wt%, depending on the particular polymer grade, and the overall ethylene conversion is generally 98 wt% [104,107–110]. The Ziegler catalyst system consists of a reaction product of magnesium tetrachloride, titanium tetrachloride, and titanium tetraisopropylate, with aluminum isopropylate as the cocatalyst.

The Mitsubishi continuous stirred-tank reactor process, utilizing either a chromium oxide or titanium–vanadium catalyst system, operates at a pressure of 3.5 MN/m^2, a temperature of 80–90°C, and an average residence time of 2 h with an overall ethylene conversion of 95 wt%. The Montedison stirred-tank reactor technology is essentially similar to the Mitsubishi process [107–109]. Another stirred-tank reactor technology was developed by R.G.C. Jenkins and Company [111], and a multiple cascade reactor technology was developed by BP Chemicals Ltd. [112].

The liquid pool slurry process technology for PP, developed by El Paso Polyolefins Company, was also used for linear polyethylene provided that isobutane or propane is the liquid hydrocarbon diluent for the ethylene monomer feed and with appropriately modified catalyst [104,107–109]. It is based on a jacket reactor with a transition metal catalyst system that was proprietary to Mitsui Petrochemical/Montedison. It operates at a pressure of 2.6 MN/m^2, temperature of 60°C, and an average residence time of 1–2 h. Polymerization takes place in the presence of the diluent, and the average spherical particle size of the polymer formed is about 1200 µm. The concentration of the solid polymer particles in the product stream is 30–43 wt%.

The first low-pressure linear polyethylene slurry loop reactor technology, using supported chromium oxide catalyst in a light hydrocarbon diluent, was commercialized in 1961 by Phillips Petroleum Company [11]. Isobutane was used as the diluent in a continuous-path double-loop reactor operating at a pressure of 3.5 MN/m^2, a temperature range of 85–110°C, and an average residence time of 1.5 h. A modified chromium/titanium catalyst system enabled the production of a broad range of polymer products, which, as with the original Phillips catalyst, do not have to be de-ashed but are stabilized and pelletized. The polymer is formed inside the pores of the catalyst that it eventually pulverizes, distributing the microscopic catalyst particles uniformly. The concentration of the solid polymer particles in the reactor discharge stream ranged between 18 and 50 wt%, depending on the particular polymer grade. The overall ethylene conversion is generally about 98 wt%. Another process, developed by Phillips Petroleum Company [113], used chromium-based catalyst on high silica–titania cogel support with a cocatalyst of triethyl borane or diethyl aluminum ethoxide. The polymers produced are characterized by a range of density between 0.915 and 0.965 g/cm^3.

The Solvay heavy hydrocarbon diluent slurry technology, based also on loop reactors, used a supported Ziegler–Natta catalyst system consisting normally of magnesium ethylate, titanium butylate, and butyl aluminum dichloride, with aluminum as the cocatalyst. The hydrocarbon diluent was n-hexane, and the continuous-path double-loop reactor operated at a pressure of 3 MN/m^2, a temperature of 85°C, and an average residence time of 2.5 h. The concentration of the solid polymer particles in the reactor discharge stream ranged between 16 and 50 wt%, depending on the particular polymer grade. The overall ethylene conversion is about 97 wt%. The liquid slurry technology available for the production of linear polyethylenes and the corresponding catalyst systems were proprietary to the following companies [107–110].

1.3.2.2.1 Slurry Phase Heavy-Diluent Stirred-Tank Reactor
- Amoco technology using a proprietary catalyst system
- Asahi Chemical technology using a Ziegler catalyst system
- Hercules technology using a Ziegler catalyst system
- Exxon technology using a Ziegler catalyst system
- Chisso technology using a Ti/Mg/V catalyst system
- Dow Chemical technology using a Ziegler catalyst system
- DuPont technology using a Ziegler catalyst system
- Hoechst technology using a Ziegler catalyst system
- ICI technology using a Ziegler catalyst system
- Hüls technology using a Ziegler catalyst system
- Idemitsu technology using a Ziegler catalyst system
- Mitsubishi Chemical technology using Ti, Cr, V, and Ti/V catalyst systems
- Mitsubishi Petrochemical technology using a Ziegler catalyst system
- Mitsui Petrochemical technology using a Ziegler catalyst system
- Montedison technology using a Ziegler catalyst system
- Shell technology using a Ziegler catalyst system
- Sumitomo technology using a Ziegler catalyst system
- Stamicarbon technology using a Ziegler catalyst system

1.3.2.2.2 Slurry Phase Light-Diluent Stirred-Tank Reactor
- Sumitomo technology using a Ziegler catalyst system
- Mitsui Toatsu technology using a Ziegler catalyst system

1.3.2.2.3 Slurry Phase Light-Diluent Loop Reactor
- Chemplex technology using Cr/Sn/Al/Ti catalyst systems
- Solvay technology using a Ziegler catalyst system
- Phillips technology using Ziegler Cr/Ti, Cr/P, Cr, Ti/Mg, V, and Ti/V catalyst systems

1.3.2.2.4 Slurry Phase Heavy-Diluent Loop Reactor
- National Distillers/ICI technology using a Ziegler catalyst system
- Solvay technology using a Ziegler catalyst system

1.3.2.2.5 Slurry Phase Liquid Pool Reactor
- El Paso technology using a Ziegler catalyst system
- Montell (now called Basell) technology using Ti/Mg Ziegler–Natta catalyst systems
- Montedison technology using a Ziegler catalyst system

1.3.2.3 Low- and Medium-Pressure Solution Processes

Solution polymerization processes predate, but were superseded, by the slurry processes [107–110]. As opposed to the slurry processes, polymerization took place in a solvent at high pressure and above the melting point of the polymer. The polymer formed dissolves in the solvent resulting in a homogeneous single-phase liquid product stream that is subsequently devolatilized. The major advantage of the solution processes is the wide operating temperature range limited only by the polymer solubility and the polymer degradation temperatures. This enables a more efficient control of the polymer molecular-weight distribution. The need for the energy-intensive solvent vaporization is its major disadvantage. Improved solution processes were used by Dow, DuPont of Canada, Mitsui, Phillips, Stamicarbon/DSM, and Eastman Kodak.

The DuPont solution process technology, based on a stirred-tank reactor system, used soluble Ziegler–Natta catalyst system such as titanium or vanadium tetrachloride with triisobutyl aluminum as the cocatalyst in a solvent such as cyclohexane. It is an adiabatic process operating at a pressure of 10 MN/m^2, a temperature of 200°C, and an average residence time of 2 min with an overall ethylene conversion of about 88 wt%. Another process due to DuPont [112] operates at a temperature range of 105–320°C and a pressure of 1.7 MN/m^2. The Stamicarbon technology is an adiabatic low-pressure stirred-reactor process operating at a pressure of 3 MN/m^2, temperatures of 130–175°C, and an average residence time of 5 min with an overall ethylene conversion of 95 wt%. The Dow technology is a cooled low-pressure twin stirred-reactor process based on a conventional soluble Ziegler catalyst. The reactors operate at a pressure range of about 1.9–2.6 MN/m^2, a temperature of 160°C, and an overall residence time of 30 min with an overall ethylene conversion of 94 wt%. The solution phase technology available for the production of linear polyethylenes and the corresponding catalyst systems were proprietary to the following companies [107–110].

1.3.2.3.1 Solution Phase Medium-Pressure Adiabatic Reactor
- Amoco technology using a Mo oxide catalyst system
- DuPont technology using a Ziegler catalyst system
- Mitsui Petrochemical technology using a Mo oxide catalyst system
- Eastman technology using a Ziegler catalyst system
- Phillips technology using Cr and Cr/Ti catalyst systems
- Stamicarbon technology using a Ziegler catalyst system

1.3.2.3.2 Solution Phase Low-Pressure Cooled Reactor
- Dow technology using a Ziegler catalyst system

1.3.2.4 Low-Pressure Gas-Phase Processes

The gas-phase, fluidized bed ethylene polymerization technology, first described by UCC (now part of Dow Chemical Company) in 1957, was commercialized in 1968. The UCC low-pressure Unipol process, introduced in 1975, was based on silica-supported titanium-modified chromium oxide catalysts [114–116], although high-activity Mg-Ti catalysts and a variety of process modifications were later developed by UCC [116–120] and British Petroleum [121–124]. The reactor was operated at a pressure of 2 MN/m^2, a fluidized bed temperature of 75–100°C, depending on the particular polymer grade, and an average residence time of 3–5 h with an overall ethylene conversion of 97 wt%. The reactor is operated at 85–100°C for HDPE or 75–100°C for LLDPE. The polyethylene product range is broad, and the particle size could be as high as 15–20 times the size of the original catalyst particle. The product withdrawal rate is used in ensuring a constant fluidized bed volume. UCC licensed the Unipol process in the United States, Japan, Europe, Australia, and Saudi Arabia. It remains one of the single largest technologies for LLDPE production. The process is characterized by simplicity, reaction uniformity, and suitability for large-scale production.

The BP Chimie/Napthachimie process is similar to the Unipol technology except for the use of a proprietary highly active titanium/alkylaluminum catalyst system. It is operated at a pressure range of 0.5–3.3 MN/m^2 and a fluidized bed temperature of 60–100°C. The vertical continuous stirred (mechanical) bed gas-phase polyolefin technology developed by BASF came on stream in 1976 as Novolen process [107–110,125]. It is operated at a pressure of 3.4 MN/m^2, a temperature of 100–110°C, depending on the particular polymer grade, and an average residence time of 4 h with either a magnesium-supported two-component Ziegler catalyst or a silica-supported modified chromium oxide catalyst. The horizontal continuous stirred-bed gas-phase polyolefin technology, developed by Amoco [126–128], is based on a compartmentalized cylindrical vessel stirred by a series of axially mounted longitudinal paddles. It is operated at a pressure of 2 MN/m^2, a temperature of 82–88°C, depending on the particular polymer grade, and an average residence time of 4.3 h with either a magnesium-supported two-component Ziegler catalyst or a silica-supported Phillips catalyst.

The Spherilene gas-phase technology developed by Montell (now called Basell) [6,26] is a hybrid process for the production of linear polyethylenes, in a spherical granular shape, whose molecular weight ranges from very low to very high. The process consists of a loop reactor followed by a fluidized bed gas-phase reactor, and it is equipped with a stripping unit so effective that the final polymer contains no monomer residue. It is characterized by high productivity, rapid and low-cost changeover of polymer grades, and excellent product quality.

The gas-phase technology available for the production of linear polyethylenes and the corresponding catalyst systems were proprietary to the following companies [107–110].

1.3.2.4.1   Gas-Phase Fluidized Bed Reactor
- UCC technology using proprietary Cr, Cr/Ti, and Ti/Mg catalyst systems
- Cities Services/ICI technology using a proprietary Ziegler catalyst system
- ICI technology using a proprietary Ziegler catalyst system
- Mitsubishi Petrochemical technology using a proprietary Ziegler catalyst system
- Mitsui Petrochemical technology using a proprietary Ziegler catalyst system
- Sumitomo technology using a Ziegler catalyst system
- Nippon Oil technology using a proprietary Ziegler catalyst system
- Shell technology using a proprietary Ziegler catalyst system
- BP Chimie/Napthachimie technology using a proprietary Ziegler catalyst system

1.3.2.4.2   Gas-Phase Horizontal Stirred-Bed Reactor
- Amoco-Chisso technology using proprietary Ti/Mg and Cr catalyst systems
- Amoco technology using a proprietary Ti/Mg catalyst system

1.3.2.4.3   Gas-Phase Vertical Stirred-Bed Reactor
- BASF technology using proprietary Cr/Ti/Mg/Sn/Al catalyst systems
- Amoco-Chisso technology using proprietary Ti/Mg and Cr catalyst systems

1.3.2.4.4   Modular Gas-Phase Fluidized Bed Reactor
- Montell (now called Basell) technology using a proprietary Ti/Mg catalyst system

1.3.2.4.5   Hybrid Slurry-Phase Loop Reactor/Gas-Phase Fluidized Bed Reactor
- Montell (now called Basell) technology using proprietary Ti/Mg catalyst systems

The olefins gas-phase fluidized bed technology used presently can be traced to the earlier processes, namely, Unipol, Sumitomo, and Catalloy.

1.3.3   Polymerization Processes for PP

Practically all of the above processes have been used in the production of PP. The special process technologies originally developed for PP and propylene copolymers included variants of the liquid pool slurry process; the Eastman Kodak solution process; the Novolen gas-phase process jointly developed by BASF, ICI, and Quantum; as well as the Catalloy modular gas-phase process of Montell (now called Basell). Contemporary and emerging catalyst developments have resulted in adapted PP process technologies. The autoclave reactor technology used presently for liquid bulk PP processes is due to Exxon Sumitomo, Rexene, and Hypol-I. The loop reactor technology used presently for liquid bulk PP processes is due to Spheripol, Borstar, Phillips, Exxon Mobil, ST (Sinopec), and Hypol-II. The gas-phase fluidized bed PP technologies used presently are due to Unipol, Sumitomo, and Catalloy.

1.3.3.1 Low-Pressure Liquid Pool Slurry Phase Processes

The liquid pool slurry process technology for PP, developed by El Paso Polyolefins Company [109,110], uses liquid propylene as the diluent as well as the feed. It is based on a jacket reactor with a highly active, highly stereospecific transition metal Ziegler catalyst system that is proprietary to Mitsui Petrochemical/Montedison. It operates at a pressure of 2.6 MN/m^2, a temperature of 60°C, and an average residence time of 1.3 h, and the concentration of the solid polymer particles in the product stream is 43 wt%. The removal of aPP or catalyst residue is not required because of the high productivity and stereospecificity of the catalyst.

The Spheripol process, developed by Montell (now called Basell) [6,26], is based on a proprietary high-yield, high-stereospecificity donor/MgCl$_2$-supported titanium-based Ziegler–Natta catalyst system. The process operates without wastewater, with no solid waste, with minimal steam requirement, and with significantly low power requirement. A significant percentage of the global production of PP is based on the Spheripol as well as the Spherizone process, developed in 2003 by Montell (now called Basell). The Spherizone process consists of a multizone circulating reactor (MZCR) with two regions, namely, Riser and Downcomer. The key advantage of Spherizone, compared with the multireactor technology, is the much shorter residence time that enables uniformity of polymer particles [129–131].

1.3.3.2 Low-Pressure Modular Gas-Phase Reactor Processes

The Catalloy gas-phase technology, developed by Montell (now called Basell) [6,26,129–131], consists of three mutually independent gas-phase reactors in series and is also known as reactor particle technology. It is based on a high-yield, highly stereospecific, high-surface-area catalyst system whose morphology ranges from a dense spherical shape to a sponge-like structure. The process operates with no by-product, no liquid waste, and no solid waste. It is capable of producing (1) PP with high polydispersity and high stereoregularity, (2) random copolymers containing up to 15 wt% comonomers, and (3) multiphase alloys containing up to 70 wt% multimonomer copolymers.

1.3.4 POLYMERIZATION PROCESSES FOR OTHER POLYOLEFINS

Ultrahigh-molecular-weight polyethylene (UHMWPE) is produced in the slurry-phase heavy-diluent stirred-tank reactor by Hercules and Hoechst based on the proprietary Ziegler catalyst system. Polybutene-1 (PB-1) and copolymers are made in a slurry-phase light-diluent stirred-tank reactor by Hüls and in a solution-phase medium-pressure adiabatic reactor by Hüls and Shell using the Ziegler catalyst system. In addition, PB-1 is made by Mitsui Petrochemical, which also manufactures poly(4-methylpentene-1) based on a stereospecific Ziegler–Natta catalyst. The tubular high-pressure process is used in the production of EVA co- and terpolymers.

1.3.5 PROCESS TECHNOLOGIES FOR SSCs

Contemporary and emerging catalyst developments have resulted in several adapted technologies for producing SSC polymers. New technologies with improved flexibility were developed based on new catalyst design. The processes, earlier developed for large-volume polyolefins, have been adapted to enable the exploitation of the unique advantages of contemporary and emerging SSC developments. Thus, different process technologies ranging from solution-, slurry-, and gas-phase processes have found applications using a variety of SSCs.

1.4 POLYOLEFIN STRUCTURE–PROPERTY RELATIONSHIPS

The choice of a catalyst, polymerization process, and the appropriate control of reactor conditions determine the polyolefin molecular weight, molecular-weight distribution, density, and other properties [107–109]. From the free radical initiated noncatalyzed polymerization technology of the 1930s, to Ziegler–Natta and chromium catalyst technologies of the 1950s, to the metallocene SSC catalyst technologies of the 1990s, and the postmetallocene SSC catalyst technologies of the present, polyolefins continue to evolve capturing significant market share in applications that were not previously accessible. To underscore the impact of SSC in the polyolefin industry, Section 1.4.7 is devoted to the discussion of polyolefins from SSCs.

A classification of polyolefins and their corresponding densities appears in Table 1.4. A list of possible polyolefin structural characteristics and properties is presented in Table 1.5. Density is generally a reflection of polyolefin linearity. The higher the density, the higher the following polymer characteristics and performance properties: (1) chain linearity, (2) stiffness, (3) tensile strength, (4) tear strength, (5) softening temperature, and (6) brittleness [101]. On the other hand, polyolefin failure properties such as impact strength, flexural strength, and environmental stress crack resistance (ESCR) decrease as the polyolefin density increases. A simplified outline of polyolefin catalyst/technology–structure–property interdependence [26] is presented in Figure 1.1.

Polyolefins are used in several different applications primarily because of their wide range of resin characteristics and end-product properties. These properties, particularly processability, physical, and mechanical properties, are highly dependent on the average molecular weights and molecular-weight distribution. For example, injection molding requires resins with low melt viscosity and elasticity, whereas blow molding requires resins characterized by high molecular weights, molecular-weight distribution, and high melt viscosity and elasticity. The molecular-weight characteristics are also the determinants of molded part warpage, propensity for failure, and ESCR, albeit not in the same direction. For example, as the high-molecular-weight species in polyolefin resin increase, the positive salutary effects on the ESCR properties would be counterbalanced by the negative effects of part warpage, surface roughness, and opacity.

The use temperatures of the different polyolefins are limited by the glass transition temperature (T_g) at the low end and the crystalline melting point (T_m) at the upper end. However, the degree of mechanical property retention is better related to either the heat deflection temperatures (HDT or DTUL) or the Vicat softening points. Of all polyolefins, only the poly(4-methylpentene-1)

TABLE 1.4
Classification and Density of Polyolefins

Polyethylene Type	Macromolecular Classification	Density Range (g/cm^3)
LDPE	Homopolymer	0.910–0.925
MDPE	Homopolymer	0.926–0.940
LLDPE	Copolymer	0.910–0.940
VLDPE	Copolymer	0.890–0.915
HDPE	Copolymer	0.941–0.959
HDPE	Homopolymer	0.960 and higher
HMWPE	Homopolymer	0.947–0.955
UHMWPE	Homopolymer	0.940
Polypropylene	Homopolymer	0.904–0.906
Ethylene–propylene copolymer	Copolymer	0.904–0.907
PB-1	Homopolymer	0.910
Poly(4-methyl pentene-1)	Homopolymer	0.830

TABLE 1.5
Possible Polyolefin Structural Characteristics and Properties

Structural Characteristics	Physical Properties	Mechanical Properties
Molecular weights	Density	Tensile strength
Molecular-weight	Average particle size	Impact strength (tensile, Izod, dart)
Distribution	Particle density	Flexural strength
Polydispersity	Surface texture, gloss	Tear strength
SCB	Film clarity, haze	Tensile strength, at yield
LCB	Thermal expansion	Ultimate tensile strength
Unsaturation	Specific heat	Tensile modulus
Crystallinity	Thermal conductivity	Flexural modulus
Morphology	Electrical conductivity	Stiffness, rigidity
Stereoregularity	Glass transition temperature, T_g	Toughness, ductility
Randomness	Melting temperature, T_m	Brittleness
Chain linearity	Crystallization temperature, T_c	Notch sensitivity
Processing Properties	Softening temperature	Elongation at yield
Shear viscosity	Heat distortion temperature	Elongation at break
Extensional viscosity	Permeability (gas, water vapor)	Orientation factor
Intrinsic viscosity	**Degradation Properties**	Creep resistance
Formability, malleability	Thermal degradation	Strain hardening
Melt flow (index, rate)	Oxidative degradation	Resilience
Melt strength	Photodegradation	Hardness
P-V-T relation	Biodegradation	Compressive strength
	Shear degradation	Shear yield strength
	Environmental stress cracking	Shear ultimate strength
		Puncture resistance
		Lubricity
		Abrasion resistance

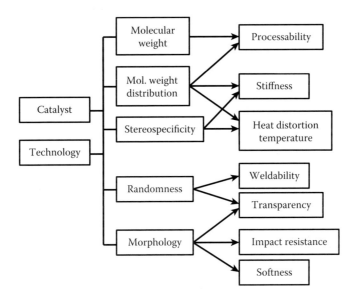

FIGURE 1.1 Catalyst/technology–structure–property relationship of polyolefins.

copolymer has a Vicat softening point (179°C), which is higher than that of PP (145–150°C), and a melting point of 240°C compared to 165–179°C for PP. The specific heat of polyolefins is in the following order polyethylenes > PP > PB-1, but PP has about the lowest thermal conductivity and thermal expansion coefficient. Although polyolefins generally have good ESCR (with PP being better than PE), all polyolefins require a degree of stabilization against all forms of degradation.

1.4.1 Polyethylenes

The density of conventional polyethylenes, commercially manufactured with the Ziegler–Natta or chromium catalyst technologies, ranges between 0.890 and 0.980 g/cm^3. The density of polymethylene is 0.980 g/cm^3 and that of the amorphous phase of polyethylene is 0.850 g/cm^3. The density of conventional polyethylene depends on its relative content of LCB and short-chain branching (SCB) [132]. The degree and type of chain branching strongly influence the molecular-weight distribution, degree of crystallinity, lamellar morphology, density, and rheology of polyethylenes. The LCB may vary from 20 to thousands of carbons, and the SCB may have 5–10 carbons depending on whether it is copolymerized with α-olefin. SCB could be attached either to the main polymer backbone or to the LCB.

SCB is expressed as methyl groups per 1000 carbon atoms and is measured by infrared absorption techniques. Empirical relations exist for comparing the degree of LCB of different polymers [133]. LCB is normally irrelevant for polymer properties below the melting point; however, it is relevant in determining the rheological and processing properties, which are highly dependent on the molecular-weight distribution. LDPE has LCB and, consequently, has much broader-molecular-weight distributions [132] compared to the other polyolefins. Because of their relatively broad-molecular-weight distributions, polyethylenes have a broad melting range.

An industrially important empirical flow parameter for linear polyethylenes is the melt flow index. A melt index is influenced by molecular weight, its distribution, chain branching, branch length, and many other molecular structural parameters. In spite of the absence of a definitive correlation, there is a reasonable consensus that a melt flow index is directly related to polyolefin part clarity and mold shrinkage, but inversely related to melt viscosity, impact strength, notch sensitivity, creep resistance, heat resistance, toughness, melt strength, and average molecular weight. The apparent relationship is utilized in many ways including the down-gauging of molded parts, extruded films, and profiles. Nonetheless, it has been shown that the molecular weights of two LDPEs having similar melt indexes but different LCBs could differ by a factor of 2. That is, a melt index is a poor indicator of the molecular structure for LDPE. The comparative processing and performance property profiles of LDPE, HDPE, and LLDPE appear in Table 1.6 [134].

All polyethylenes are non-Newtonian, but LDPE is less so than HDPE, LLDPE, PP, and PB-1, in that order, and could therefore be more easily compounded with colorants and additives. Its melt viscosity is far more temperature-sensitive and shear-sensitive, and it has an enhanced extensional viscosity. On the other hand, the more non-Newtonian polyolefins have a more plug-like velocity profile with a corresponding higher pumping efficiency. Their viscosities are less temperature- or shear-sensitive, and they have considerably less extensional viscosities.

1.4.1.1 Branched LDPE

Branched LDPE is not closely packed as it contains a substantial amount of LCB; it looks much like a tree. It has a density of 0.910–0.930 g/cm^3, 45–60% crystallinity, and respectable flexibility. LDPE is characterized by a broad-molecular-weight distribution with a polydispersity index as high as 20. It has a broad molecular weight ranging between 17,000 and 30,000, but skewed toward the higher-molecular-weight components. Although its melting temperature is normally centered around 110°C, its actual range depends on the degree and randomness of branching [135].

LDPE, made in a highly back-mixed autoclave reactor, tends to branch equally in all directions giving rise to somewhat spherical molecules. On the other hand, commercial products of a tubular

TABLE 1.6
Processing and Performance Properties of LDPE, HDPE, and LLDPE

Property	LDPE	HDPE	LLDPE Relative to LDPE	LLDPE Relative to HDPE
Tensile strength (MN/m^2)	6.9–15.9	21.4–38	Higher	Lower
Elongation (%)	90–650	50–800	Higher	Higher
Impact strength (J/12.7 mm)	No break	1.02–8.15	Better	Similar
Environmental stress cracking resistance	–	–	Better	Same
Heat distortion temperature (°C)	40–50	60–82	15°C higher	Lower
Stiffness (4.5 MN/m^2)	1.18–2.42	5.53–10.4	Higher	Lower
Warpage			Less	Similar
Processability	Excellent	Good		Easier
Haze (%)	40	–	Worse	Better
Gloss (45° %)	83	–	Worse	Better
Clarity	Near transparent to opaque	Translucent to opaque	Worse	Better
Melt strength	–	–	Lower	Lower
Softening point range (°C) Permeability (ml cm^{-2} g^{-1} mil^{-1} cm) H_g^{-1} at 25°C × 10^{-8}	85–87	120–130	Narrower	Narrower
(a) H$_2$O vapor	420	55	Better	Worse
(b) CO$_2$	60	13	Better	Worse

Source: Mukherjee, A. K. et al., *Popular Plastics*: 15 (October), 1985.

reactor contain fewer and longer branches on the main backbone. LDPE also contains SCB consisting of ethyl and butyl groups whose frequency distribution is narrow. For a given melt index, higher LCB implies higher mechanical properties, broader-molecular-weight distribution, higher flow resistance, and enhanced possibility of entanglement with concomitant effects on performance properties of the finished products. The degree of LCB, rather than a melt index, is the indicator of LDPE molecular structure and properties. It is because of the enhanced melt strength and elasticity imparted by the high degree of LCB that LDPE is a preferred material for blown film, shrink film, and extrusion coatings. The negative aspect of its high tolerance for extension is that its melt strength could be exceeded during blown film processing unless adequate allowance is made in die design.

The individual crystallite dimensions in LDPE films are about 10–30 nm; an agglomerate of the crystallites results in a spherulite whose dimension is larger than the wavelength of light. Consequently, higher film clarity is obtained if the formation of aggregates is minimized and the spherulite size is reduced. The translucence of LDPE-blown film is due partly to the film surface unevenness and partly to the presence of spherulites whose size is a function of the degree of branching and the material thermal history. Because of this, melt elasticity, viscosity, temperature, and their effects on surface texture are, in the final analysis, the primary determinants of film clarity in commercial LDPE. The morphology of LDPE solid depends on the relative magnitude of the rates of crystallite nucleation and spherulite growth, both of which are significant [136].

1.4.1.2 Linear LDPEs

Conventional linear LDPE is a copolymer of ethylene and a linear α-olefin comonomer such as propylene, butene-1, pentene-1, hexene-1, heptene-1, octene-1, decene-1, tetradecene-1, or methyl-4-pentene-1 [133,134,136–138]. Commercially available conventional LLDPE, made with conventional catalyst technology, normally contains 8–12% of butene-1, hexene-1, or octene-1, and it has a nonrandom broad comonomer distribution with density ranges between 0.900 and 0.945 g/cm³. With SSC technology, the polyolefin produced is VLDPE, which is structurally similar to LLDPE [139]. VLDPE has a density between 0.890 and 0.915 g/cm³. There is an inverse relationship between comonomer content and LLDPE density and between the molecular weight and density. The SCBs of the low-molecular-weight fractions are two to four times those for the high-molecular-weight fractions. This heterogeneity is catalyst-specific, and it depends on the production technology used [140]. A homogeneous LLDPE with random and narrow comonomer sequence distribution as well as narrow-molecular-weight distribution is characterized by lower density, lower melting point, higher impact strength, significantly lower film haze, and improved film properties in the machine and transverse directions. This type of LLDPE essentially belongs in Section 1.4.7.

In spite of the shortcomings of conventional LLDPE, it still has much improved properties over LDPE of the same density, molecular weight, and molecular-weight distribution. The primary advantages of LLDPE, arising from its backbone linearity and the presence of SCB, are as follows: higher tensile strength, impact strength, toughness, stiffness, film gloss, puncture resistance, tear strength, ESCR, and permeability of water vapor and carbon dioxide. Up to a point, the extent of the property improvement is directly related to the amount and chain length of the oligomeric comonomer [132,141]. For equally the same reasons, its optical properties, melt strength, and rheological properties are not necessarily optimum. For the same density, LLDPE is about 5% more crystalline and has a melting point that is about 14°C higher than that of LDPE [142].

The α-olefin comonomer forms the SCBs on the linear ethylene chain backbone of LLDPE disrupting the crystallization process and limiting the crystalline size structure, thus increasing the percentage of the amorphous phase at the expense of the crystalline phase and lowering the density [107,136]. In conventional LLDPE, the SCB frequency distribution is uneven, and, combined with the highly linear character of the polymer, the nonuniformity increases the relative magnitude of the rate of crystallite nucleation and that of spherulite growth resulting in higher haze values. The haziness of LLDPE-blown film is due partly to the film surface unevenness and partly to the presence of spherulites whose size is a function of the nonuniformity of SCB and the material thermal history. Higher density LLDPE has lower SCB and higher nonuniform frequency distribution with a corresponding larger spherulite and higher haze. Because of this, melt elasticity, viscosity, temperature, and their effects on surface texture are not as important in determining film clarity as they are for LDPE.

LLDPE is more non-Newtonian than LDPE, but its higher melt viscosity is not as temperature-sensitive or as shear-sensitive as LDPE. Because of its higher non-Newtonian character, its velocity profile in a channel is more plug-like with a corresponding higher pumping efficiency. But the absence of LCB and the narrower polydispersity imply a lower melt strength and a lower shear thinning behavior, which, when combined with the higher viscosity, makes LLDPE more prone to melt fracture given the higher pressure and screw torque required for commercial extrusion. Because of the absence of LCB and the narrower polydispersity, LLDPE melt strain-hardens: it is soft in extension but stiff in shear. It therefore has to be processed at a higher melt temperature with a wider die gap to permit the use of a lower pressure and a higher drawdown ratio in blown films [107]. For the same melt index and density, LLDPE-cast film properties are superior to those of LDPE; consequently, a higher melt index LLDPE could easily yield a lower film thickness and a higher productivity.

Generally, LLDPE has penetrated the market normally dominated by other polyethylenes principally because of the high optical quality of its film and its corresponding performance properties even at rather low film thickness. The best LLDPE would be characterized by uniform intermolecular- and intramolecular comonomer SCB distribution with a variety of comonomers such as betene-1, hexene-1, octene-1, and 4-methylpentene-1. The super random copolymer of Dow Chemical, made in the proprietary solution process, is based on octene-1, and it has a high compositional uniformity and film performance. On the other hand, the LLDPEs produced in the gas-phase process using supported titanium chloride catalysts are characterized by nonuniform composition distribution

Polyolefins

and high hexane extractables, and so are the LLDPEs made with the high-pressure retrofitted LDPE processes based on butene-1. The estimated annual growth rate for LLDPE is 7–8%.

1.4.1.3 High-Density Polyethylenes

HDPE is closely packed with a density of 0.960–0.980 g/cm^3, a crystallinity as high as 95%, and a melting point as high as 138.5°C, with the highest values corresponding to those of polymethylene. Commercially available conventional HDPE normally contains 1–3 wt% of butene-1, hexene-1, or octene-1. It has a few SCBs with a polydispersity index in the range 5–15, although its molecular-weight distribution is generally sharper than that of a corresponding LDPE. The situation is of course different for the HDPE produced with SSC technology discussed in Section 1.4.7. The homopolymer MDPE has a density ranging between 0.926 and 0.940. Quenched HDPE could have a density as low as 0.945 g/cm^3 because of the reduced crystallinity [135,136]. However, the morphology of HDPE solid is determined by the relative magnitude of the rates of crystallite nucleation and spherulite growth, both of which are so high that it is nearly impossible to quench HDPE fast enough to obtain spherulites that are small enough to give high optical quality. Film clarity could only be improved through the use of heterogeneous nucleating agents.

Nonetheless, the longer the molecular weight of HDPE is, the more the crystallization process is inhibited, limiting the crystalline size structure and increasing the percentage of the amorphous phase at the expense of the crystalline phase resulting in reduced density. The increased amorphous phase enhances the impact properties and lowers the yield properties, hardness, as well as stiffness up to a point. As the molecular weight of HDPE increases, the ultimate tensile properties and elongation increase up to the point where chain entanglement becomes important. There is an inverse relationship between the melt index and the molecular weight, but the intrinsic viscosity is related to the viscosity average molecular weight as expressed by the Mark–Houwink equation. Generally, HDPE molecular weight has an effect on the following properties: density, tensile strength, tensile elongation, impact strength, toughness, stiffness, and permeation characteristics. Up to a point, the HDPE molecular weight is inversely related to its hardness and flexural stiffness, whereas it is directly related to Vicat softening point and Izod impact strength. On balance, however, it is the molecular-weight distribution that determines the end-product physical, mechanical, rheological, and processing properties.

HDPE is non-Newtonian and its velocity profile in a channel is plug-like with a high pumping efficiency; this is truer for the polymer with the narrow-molecular-weight distribution. The melt viscosity of HDPE with a narrow-molecular-weight distribution is not particularly temperature-sensitive or shear-sensitive, making it more amenable to higher drawdowns when used as fibers and monofilaments. The HDPE with a broader-molecular-weight distribution has a higher melt index, higher melt strength, and an enhanced shear thinning ability, which facilitate processing in blow molding of large containers, extrusion of parts with large cross-sectional areas, and blow extrusion of thin crisp films used as replacements for paper. On the other hand, for a given melt index, the HDPE with a narrower-molecular-weight distribution has higher impact strength. Generally, HDPE has excellent low-temperature flexibility, and its brittleness temperature is almost independent of its melt index. The high-molecular-weight HDPE [weight average molecular weight (M_w) of 3–4 × 10^5] with fractional melt indices is used in the manufacture of carrier bags, garbage bags, drums, and pipes, whereas the medium-molecular-weight varieties ($M_w \approx 2.5 \times 10^5$) find applications as wrapping paper, grease-proof paper for meat, and flower packaging.

The specialty-grade HDPE remains the high-molecular-weight HDPE with the best performance–processability balance for blow molding and extrusion applications such as thin film, pipes, and heavy drums. A judicious choice of catalyst and polymerization process enables the production of HDPE with the best combination of structural molecular parameters vis-à-vis the required properties. For this premium grade of HDPE, the stirred-tank slurry processes are best suited. The commodity large-volume HDPE is dominated by the cost-effective gas-phase processes that have essentially surpassed the most licensed light diluent slurry loop reactor process. HDPE includes the narrow-molecular-weight injection molding grades as well as the medium-molecular-weight extrusion grade for films and pipes. The Unipol gas-phase process is able to produce bimodal HDPE grades, and the polymerization technology from Mitsui Petrochemical produces what has been called "superpolyethylene." Of all the available processes, the new Spherilene process developed by Montell (now called Basell) is the most versatile as it is able to produce HDPE with molecular weights ranging from very low to very high [6,26]. It could therefore supply both the specialty and the commodity market.

1.4.1.4 Ultrahigh-Molecular-Weight Polyethylenes

The UHMWPE [142] with M_w in excess of 3.1×10^6 and a density of 0.94 g/cm³ is structurally similar to HDPE with folded zigzag crystalline conformations having fold spacings between 10 and 50 nm. The chemical properties, electrical properties, tensile impact strength, and elastic modulus are similar except that UHMWPE is characterized by exceptional self-lubricating properties and low-temperature performance properties, and its abrasion resistance is superior to that of abrasion-resistant steel. It normally decomposes before it melts, and some of its processing techniques mimic those of powder metallurgy.

The UHMWPE, prepared under exceedingly high pressures between 0.2 and 0.7 GN, christened RT UHMWPE by DuPont, is tough, malleable, permanently deformable without breaking, and processable by solid-phase extrusion or other solid phase-forming technique [142]. Its impact strength and elastic modulus are comparable to those of polycarbonate. The RT UHMWPE has extended chain crystal structure fold spacings of about 1000 nm compared to about 250 nm for the extended chain crystal structure of conventional HDPE, produced under high-pressure crystallization, which is brittle at ordinary temperatures. The RT UHMWPE finds application as machined parts for prosthetic devices, small gears, automotive parts (that ordinarily would have required ABS or polyacetal), and many other applications requiring its outstanding combination of properties. Like other polyethylenes, its upper use temperature is below 100°C.

1.4.2 POLYPROPYLENE

There are basically three forms of PP homopolymers: crystalline, amorphous, and elastomeric [6,26,107,143,144]. The SCB of PP interferes with its polymer crystallization, limiting the size and formation of its crystallites, increasing the proportion of amorphous polymer, and decreasing the density. The syndiotactic crystalline form has a calculated density of 0.90 g/cm³. The isotactic crystalline form has a measured density of 0.90–0.91 g/cm³, a degree of crystallinity 60–80%, T_m of 165–179°C, and Vicat softening point of 145–150°C, and it is capable of significant supercooling. Aside from being dependent on the method of measurement, the measured crystallinity depends on the molecular weight, the molecular-weight distribution, crystallization conditions, and annealing. The atactic amorphous PP is soft and gummy. It has a measured density of 0.86–0.89 g/cm³, its mechanical properties are significantly lower than those of the crystalline forms, and it finds application primarily as hot melt adhesives and bitumen binders.

The crystalline form of PP depends on the relative magnitude of the rates of crystallite nucleation and spherulite growth, both of which are determined by the cooling rate as well as the relative nearness of the crystallization temperature (T_c) to the crystalline melting point (T_m). When T_c is slightly below T_m and the cooling rate is slow, spherulite growth predominates and the resulting material is highly crystalline. For this material, the higher the crystallinity, the higher the density, tensile yield strength, and elastic modulus, and conversely, the lower the impact strength, toughness, and elongation. With high cooling rates, T_c decreases fast, nucleation becomes predominant, smaller spherulites are formed, and the degree of crystallinity is diminished. For this material, impact strength, toughness, and elongation are high, but density, tensile yield strength, and elastic modulus are low. If the cooling rate is fast enough, small-enough spherulites could be formed to obtain high optical quality [145]. Film clarity could be further improved through the use of heterogeneous nucleating agents [146], which would also enhance the crystallinity, density, tensile yield strength, and elastic modulus.

The weight-average molecular weight of PPs strongly influences the physical, mechanical, rheological, and processing characteristics. PP melt is non-Newtonian at high shear rates but approaches Newtonian behavior when the shear rates are very low. Its shear rate dependence is a function of the polydispersity index. PP with broad-molecular-weight characteristics has a better processability, particularly for injection molding and extrusion applications, where PP competes with amorphous polymers such as impact modified polystyrene and polycarbonates. Because of its crystalline nature, the significant shrinkage of PP is a disadvantage relative to its amorphous competition, but its high T_m and the concomitant high heat distortion temperature compensate adequately. Broad-molecular-weight distribution PP could also be obtained by blending PP of different polydispersity indices, and those with good-enough melt strength find applications in thermoforming, pipe extrusion, blown film, and blow molding.

There is an inverse relationship between the melt flow index and the molecular weight, but the intrinsic viscosity is related to the viscosity average molecular weight through the Mark–Houwink equation. Commercial PP could have a narrow or a broad range of polydispersity index between 5 and 15. The narrow polydispersity index PPs with a high melt flow index are used in injection molding of thin-walled containers, cast film production, and fiber spinning. They are characterized by high strength, toughness, ductility, elasticity, resilience, and orientation. Oriented PP films find applications in food packaging, graphics, and solar control; the film tapes are used as primary carpet backing fabric, sacks, ropes, twine, and nets for seedlings [147]. PPs with controlled rheology and narrow polydispersity index (3.0) are obtained by the controlled chemical or thermal degradation of high molecular grades (visbreaking). This grade of PP could also be produced by the Montell (now called Basell) technology using donor-assisted Ti/Mg catalyst systems [6,26].

Other PPs made with conventional catalyst technologies include (1) high-modulus homopolymers, random and block copolymers for solid-state pressure-forming applications requiring superior clarity, stiffness, impact strength, and heat distortion; and (2) high-melt-strength PPs with enhanced elongation viscosity for application as extrusion coating, foaming, blow molding, melt-phase thermoforming, and a host of other applications that previously eluded PPs [6,26].

1.4.3 Poly(butene-1)

There are primarily three crystalline polymorphs of PB-1 classified as follows [148–153]: (1) the hexagonal or rhombohedral polymorph form I (T_m = 136°C) and form I' (T_m = 98–103°C); (2) tetragonal form II (T_m = 124°C); and (3) orthorhombic form III (T_m = 106°C). Although each polymorph has different properties with form I having the highest density, crystallinity, hardness, rigidity, stiffness, and tensile yield strength, the commercially available PB-1 is 98–99.5% isotactic and 50–55% crystalline, and its weight-average molecular weight is $2.5–6.5 \times 10^5$.

Structurally, PB-1 is similar to PP except that it has ethyl chain branches rather than methyl. The commercially available PB-1 exists first in the form II polymorph structure, which has 11 monomer units in 3 turns or 40 monomer units in 11 turns of its helical molecular structure, before it transforms irreversibly to the form I polymorph with 3 monomer units per turn in its helix. The phase transformation occurs within seconds, days, or weeks depending on the prevailing pressure, temperature, state of strain, molecular weight, tacticity, catalyst residue, nucleating agents, additive content, and comonomer composition and distribution. The addition of about 10% of PP has been known to shorten the transformation half-time by as much as 70%.

The physical properties of PB-1 cut across all other polyolefins, and its general characteristics are a cross between polyethylenes and PP. Its T_g is −25°C, which is in between that of polyethylenes (−60°C) and PP (≈0°C); its T_m is much like that of HDPE, i.e., 124–136°C, and it has a measured density of 0.91 g/cm^3, which is essentially identical to that of PP. Its tensile yield strength is about the same as that of MDPE; its flexibility is much like that of LDPE having a density of 0.92 g/cm^3, and its high-temperature mechanical property retention is much like that of HDPE.

The non-Newtonian character of PB-1 is much like that of LDPE, and it could therefore be easily compounded with colorants, additives, and high filler loading. The relative magnitude of its specific heat compares as follows: polyethylenes > PP > PB-1. The tear strength of its film is about six times that of LDPE and about three times that of LLDPE [154,155]. Compared to PP and PEs, PB-1 has good electrical properties, moisture barrier properties, an excellent ESCR, as well as an outstanding abrasion, impact, and creep resistance. While polyethylenes and PPs do not strain-harden, PB-1 strain-hardens as its chain orientation and the corresponding material state of stress are a strong function of strain.

In the melt, PB-1 has excellent melt strength, and it finds applications in pipe extrusion for hot water pipes, blown film, laboratory/medical ware, food/meat packaging, agricultural packaging, compression wraps, and hot-fill containers. The blown film is characterized by high strength, toughness, ductility, elasticity, resilience, and orientation potential.

1.4.4 Poly(4-methylpentene-1)

Poly(4-methylpentene-1) is principally isotactic. It has a tetragonal crystalline form whose helical molecular conformation has seven monomer units in two turns of its helix. It has a high T_m (235–250°C) and a low density of about 0.83 g/cm^3—the lowest density of all polyolefins. Its helical conformation is presumed to be retained even within the amorphous phase. Consequently, its crystalline and amorphous phases have a similar density, and this is responsible for its exceptional transparency with a light transmission value of 90–93% [156]. Its monomer, unlike that of other polyolefins, is not a by-product of steam-cracking operations; it results from propylene dimerization. The monomer confers on poly(4-methylpentene-1) a crowded side chain branching, which in turn confers a rigid backbone to the stereoregular variety. The commercial poly(4-methylpentene-1) copolymer variety, with a trade name of TPX, has the best high-temperature mechanical property retention of all of the polyolefins (up to 205°C) [157]. For this reason, it is capable of hundreds of sterilization cycles with high-pressure steam (150°C) in laboratory and medical applications. Because of the bulky and crowded side groups, poly(4-methylpentene-1) has a significant free volume that is responsible for its high water vapor permeability, high gas permeability, high oxygen uptake, and a high susceptibility to photooxidation.

The physical properties of poly(4-methylpentene-1) are generally similar to those of PP: their tensile strength, flexural modulus, hardness, and impact strength are similar. However, poly(4-methylpentene-1) is transparent while PP is translucent at best. Poly(4-methylpentene-1) is thixotropic and could easily be compounded with colorants, additives, and fillers. Its melt viscosity is far more temperature-sensitive than that of all the other polyolefins and could be processed in injection molding, blow molding, and extrusion. Its market includes laboratory/medical ware, lighting, automotive, appliances, electronics, and electrical parts. It has superior electrical properties.

1.4.5 Polyolefin Elastomers

Commercially available POEs include the following: (1) poly(transisoprene); (2) poly(chloroprene); (3) poly(1,2 butadiene); (4) poly(styrene-co-butadiene); (5) nitrile rubber; (6) butyl rubber; (7) EPR; (8) EPDM rubber; and others. The estimated global consumption of polyolefin elastomers in 1995 is 8 million tons [6]. The subject of thermoplastic elastomers is very broad and it is the basis of several books. Because of this, only EPR and EPDM will be discussed briefly along with propylene copolymers in Section 1.4.6.2.2.

1.4.6 Polyolefin Blends and Copolymers

1.4.6.1 Polyolefin Blends

PBs are mixtures of structurally different homopolymers, copolymers, terpolymers, and the like. The copolymers, terpolymers, etc., may be random, alternating, graft, block, star-like, or comb-like, as long as the constituent materials exist at the polymeric level. The raison d'être of PBs is the cost/

performance ratio. An expensive polymer, whose property spectrum is much higher than is needed for a new application, may be blended with an appropriate inexpensive polymer whose property spectrum is such that the resulting polyblend has an attractive cost/performance ratio. Thus, the standard of performance demanded by the new application is satisfied by a mixture of commercially available polymers without the need to develop a new polymer or to invest in a new plant. The situation is even more attractive if, by a judicious choice of compatibilizers, a significant degree of synergy, rather than additivity, is achieved in the polyblends and a new material is essentially created [158–161].

PBs could be classified in terms of the technology of manufacture. For polyolefins, mechanical PB is probably the most relevant and is made by melt-blending the polymers either on an open roll, an extruder, or any other suitable intensive mixer at a processing temperature well above the T_m. Depending on the state of thermal stability of the polymers being mixed, the high-intensity processing shear could initiate degradation, resulting in free radicals that could cause polymeric reactions and possible cross-linking of the constituent polymers. This may or may not be desirable.

PBs could also be classified in terms of the polymer–polymer phase behavior. Most of the commercially important polyolefin PBs presented in Table 1.7 are multiphase and would be characterized

TABLE 1.7
Commercially Important Polyolefin Blends

HMWPP/PP

PP/LDPE

PP/HDPE

PP/EPDM

PP/PIB/LDPE

PP/EPDM/HDPE

PP-EP/HDPE

PP/EPR or EBR/LDPE or EVA

HDPE/EPR or EBR/LDPE or EVA

PB-1/EPR or EBR/LDPE or EVA

LDPE/LLDPE

LDPE/LLDPE/EVA

HMW HDPE/HDPE

HMW MDPE/MDPE

HMW HDPE/LLDPE

PB-1/PP

PB-1/HDPE

Polyolefins/S-EB-S/engineering thermoplastics

PP/S-EB-S/PBT

PP/S-EB-S/polyamides

Polyolefins/S-EP-S/engineering thermoplastics

Polyolefin/S-EP-S/polyphenylene ether

Polyolefin/S-EP-S/thermoplastic polyurethane

Source: SRI International, Polyolefin Markets and Resin Characteristics, Vol. 2, Project No. 3948, SRI, Menlo Park, CA, 1983; SRI International, Polyolefins Production and Conversion Economics, Vol. 3, Project No. 3948, SRI, Menlo Park, CA, 1983; SRI International, Polyolefin Production Technology, Vol. 4, Project No. 3948, SRI, Menlo Park, CA, 1983; Mai, K. and J. Xu, Toughening of thermoplastics. In *Handbook of Thermoplastics,* ed. O. Olabisi, Marcel Dekker, New York, 1997; Chang, F. C., Compatibilized thermoplastics blends. In *Handbook of Thermoplastics,* ed. O. Olabisi, Marcel Dekker, New York, 1997; Hudson, R. L., Polypropylene blend compositions, U.S. Patent 4,296,022, 1981.

by low ductility and elongation [107–109,161–164]. This is why the blends must be compatibilized to ensure a very effective interfacial stress transfer between the constituent polymers. Depending on the morphology of the mixture, the resulting PB could, in some cases, be referred to as polyolefin alloys.

Generally, a block copolymer could operate as a dispersing agent (emulsifier), reducing the interfacial tension and the particle size of the dispersed phase. It could significantly contribute to the dispersion of the constituent polymers and improve elongation, tensile strength, and impact strength of the PB. In effect, an A-co-B nonreactive compatibilizer could increase the compatibility of a normally immiscible mixture of A and B homopolymers. For example, the impact properties of the blends of poly(ethylene-b-propylene)/HDPE, poly(ethylene-b-propylene) EP (rubber), and PP/EP (rubber) are enhanced because the ethylene or the propylene components of the copolymer provide the needed compatibilization effects. Pendant reactive chemical groups, namely, the A–X reactive compatibilizers and ionomer-reactive compatibilizers, have also been used to enhance polymer compatibility. PB-1 is compatible with PP and HDPE, and the PB-1/PP films find applications where the required toughness, creep, ESCR, memory, and heat sealability properties would have precluded the use of either PP or PB-1. HDPE enhances the heat distortion temperature and the T_m of PB-1 [153].

1.4.6.2 Polyolefin Copolymers

Inasmuch as most polyolefin blends are incompatible, copolymerization is one method of creating new commercially viable polyolefin materials such as random, block, and graft copolymers from the wide range of olefin monomers. LLDPE is a copolymer of ethylene and a-olefins. Other commercially important polyolefin copolymers are as follows: (1) those based on ethylene such as EVA copolymer; (2) those based on propylene such as ethylene–propylene (EP) copolymer; (3) those based on butene-1 such as copolymers of butene-1 with ethylene, propylene, pentene-1, 3-methylbutene-1, 4-methylpentene-1, and/or octene-1; and (4) those based on 4-methylpentene-1 such as the copolymers of 4-methylpentene-1 with pentene-1 and hexene-1.

1.4.6.2.1 EVA Copolymers

Ethylene-vinyl acetate (EVA) copolymers a random statistical copolymer that normally contains up to 40% vinyl acetate, is usually made in a free radical process at pressures higher than 0.1 GN/m². The vinyl acetate–ethylene (VAE) copolymers containing above 40% and up to 100 wt% vinyl acetate are made either in a medium-pressure solution-phase process or in a low-pressure emulsion polymerization process. It is the EVA rather than the VAE that is generally predominant and is considered a variant of LDPE with lower crystallinity but higher density. When the vinyl acetate content approaches 50 wt%, the crystallinity decreases to zero, as the polymer density approaches 1.0 g/cm³. The property advantages of EVA could be appreciated by comparing the stretch wrap film, containing 9–12 wt% vinyl acetate, with LDPE or plasticized PVC. The EVA has superior stretchability, puncture resistance, retained wrap force, and tear resistance—all of the properties that are *sine qua non* for stretch-wrap films [163,164]. EVAs containing 2–40% vinyl acetate find applications as high-clarity films for packaging, heavy-duty stretch-wrap films, agricultural films, injection molding, profile extrusion, shoe soles, and hot melt adhesives.

1.4.6.2.2 PP Copolymers

The conventional poly(ethylene-co-propylene) or EP copolymer is available primarily in two forms, namely, random or block copolymers. The *raison d'être* for EP copolymer is to increase the low-temperature impact properties as well as the amorphous phase of the crystalline PP. Random EP copolymers have good melt strength and are used in thermoforming and for blow molding of bottles requiring high clarity. The thin film of the single-phase random copolymer is transparent as the material has a significantly reduced crystallinity. The random copolymerization of propylene with as much as 20 wt% α-olefins such as decene-1 or longer yields a tough polymer whose density is 0.896 g/cm³ with significantly improved low-temperature properties and polymer rheology.

On the other hand, the film of the block copolymer is essentially opaque, reflecting the presence of the two types of crystallite that render the EP block copolymer multiphase. In comparison with a PP homopolymer, an EP block copolymer exhibits lower density, lower brittleness temperature, higher impact strength, higher toughness, higher elongation, lower notch sensitivity, plus the characteristic hinge break mode of the PP homopolymer [6,26,107,143,144]. The ESCR of an EP block copolymer is outstanding, and the effect of the weight-average molecular weight is similar to that of the PP homopolymer. It finds applications where its property profile makes it superior to PP.

The $MgCl_2$ catalyst technology was the first to produce, in one reactor, a tough, high-impact, multiphase PP alloy containing a blend of iPP homopolymer as the continuous phase and EP rubber as a uniformly dispersed phase [6,26]. Ordinarily, this sort of material would have had to be made by blending the component polymers in an intensive mixer such as an extruder. In addition, random copolymers containing up to 15% comonomers as well as multiphase alloys containing up to 70% multimonomer are produced with the same catalyst technology. The ability to combine olefinic and nonolefinic monomers has made it possible to produce polymers that combine the desirable properties of olefins with the desirable characteristics of amorphous engineering thermoplastics [6,26].

EPR copolymer containing 25–70 wt% propylene is amorphous and elastomeric. When made in the presence of a small amount of nonconjugated diene, such as 1,4-hexadiene, dicyclopentadiene, or 5-ethylidene-2-norbornene, EPDM rubber is obtained. Because the only remaining double bond is pendant to the backbone, EPDM, like natural rubber, is less sensitive to oxygen and ozone. EPDM is characterized by excellent low-temperature performance properties, good dielectric properties, and excellent resistance to chemicals. It is commercially produced in solution and suspension processes. Its unique structural molecular parameters include the possible presence of random/alternating/block comonomer distributions, variable block sequence lengths, a measure of tail-to-tail enchainment, and absence or presence of crystallinity and of gels. There are observable relationships between these structural parameters and the characteristic elastomeric performance properties such as cure ability, processability, wear resistance, elasticity, and mechanical properties.

1.4.6.2.3 Poly(butene-1) Copolymers

Copolymers [148] of butene-1 with ethylene, propylene, pentene-1, 3-methylbutene-1, 4-methylpentene-1, and octene-1 have been investigated. Of these, 3-methylbutene-1, 4-methylpentene-1, and octene-1 are easily incorporated into the PB-1 lattice structure, and they significantly retard the phase transformation of polymorph form II to that of polymorph form I. On the contrary, ethylene and propylene are not well incorporated into the PB-1 lattice structure, and, even at very low quantities, they significantly accelerate the phase transformation to such an extent that the form I polymorph precipitates directly from the melt. Although poly(ethylene-co-butene-1) with more than 20 mol% ethylene comonomer is virtually amorphous with a T_g less than −40°C, the random copolymer, poly(propylene-co-pentene-1) retains a small measure of crystallinity, and its 50:50 block copolymer actually contains both the PP crystalline structure and the form I structure of PB-1 with the respective melting points. The poly(1-pentene-co-butene-1) exists in the form I polymorph with up to 84 mol% pentene-1 comonomer. The injection molding processability of PB-1 is enhanced by the incorporation of ethylene or propylene comonomer, and, at very low comonomer concentrations, propylene has been known to increase the isotacticity of the polymer.

1.4.6.2.4 Poly(4-methylpentene-1) Copolymers

The commercially available variety of poly(4-methylpentene-1), manufactured by Mitsui Petrochemical, is christened TPX and contains pentene-1 or hexene-1 comonomer. It is more crystalline than the homopolymer with a melting range of 230–240°C, a Vicat softening temperature of 179°C, and a heat distortion temperature of 100°C. The corresponding values for the homopolymer are 100°C, 245°C, and 58°C, respectively [156,157]. TPX has higher elongation, impact, and softening temperature than the homopolymer; it has a higher melting and softening temperature than PP. Like its homopolymer, its film is exceptionally transparent, whereas the PP film is translucent at best. It could be processed in

TABLE 1.8
Commercialization of Metallocene-Based Products in 1995

Company	Capacity (kt/y)	Location
Ethylene/α-olefin copolymers		
Dow	112	USA
	56	Spain
Exxon	100	USA
	15	USA
Mitsubishi Yuka	100	Japan
Nippon Petrochemical	100	Japan
Ube Industries	20	Japan
Total	**453**	
Isotactic polypropylene		
BASF	12	Germany
Chisso	20	Japan
Exxon	100	USA
Hoechst	100	Germany
Mitsui	75	Japan
Total	**307**	

Source: Catalyst Consultants.

injection molding, blow molding, and extrusion processes. Its major applications are identical to those of the homopolymer; it is also used in polyolefin composites.

1.4.7 Polyolefins from SSCs

The development and commercialization of polyolefins from SSCs began in the early 1990s [102,165–167]. As shown in Table 1.8, the commercialization of metallocene-based products in 1995 included ethylene/α-olefin copolymers and iPP. SSC polyolefins are generally characterized by lower density and other unique features that improve processability, physical, and mechanical performance. Such features include narrow-molecular-weight distributions, long-chain branching, narrow composition distributions, and blocky structure. Compared to earlier generation technologies, SSC polyolefins are relatively simple and the catalysts are more predictable, enabling the designing of tailor-made polyolefin products through the use of molecular architecture approach and economic models.

SSC systems enable the synthesis of modern polyolefins with tailor-made microstructures, morphology control, and controlled isotactic/syndiotactic, hemi-isotactic, and isotactic–atactic properties. PBs and block copolymers are also generated within the process of polymerization. SSC polyolefin manufacturing has now attained a high level of technological optimization resulting in low energy consumption processes and enhanced polymer properties. The global market size of SSC polymers is modest, but they find high-performance applications in food and specialty packaging, wire and cable applications (low-voltage flexible cable insulation/high-voltage jacketing), hoses, tubes, weather stripping, gaskets, foams, sealants, carpet fiber, elastic fiber and film, belts, footwear, as well as apparel applications such as swimwear, wrinkle-free shirts, woven fabrics, denims, and other durable products. In addition, thermoplastic polyolefins (TPOs), produced using SSC systems, have replaced engineering plastics in interior and exterior automotive applications such as bumper fascia, claddings, air dams, and instrument panels on account of superior cost/performance scenario.

TABLE 1.9
Catalyst Systems and the Resulting Polyolefin Microstructures

Property	Ti Catalyst	V Catalysts	Metallocene
Molecular-weight distribution	Narrow 4–6	Medium to broad	Very narrow ($M_w/M_n = 2$)
α-Olefin incorporation	Moderate	High	Low to very high
Composition distribution	Heterogeneous	Homogeneous	Homogeneous
α-Olefin blocks	Isotactic	Atactic to isotactic	Syndiotactic to atactic to isotactic
Polymer unsaturation	Low	Very low	0.1–1 vinyl group per chain

1.4.7.1 SSC Polyethylenes

Metallocenes have been used most successfully in the production of LLDPEs since the production technology was commercially operational in 1991. LLDPE is a copolymer of ethylene and α-olefins. While the copolymerization of ethylene and α-olefins using conventional Ziegler–Natta catalysts results in an isotactic arrangement of the α-olefin blocks, with SSCs, the polymer stereochemistry varies over the whole range of possible microstructures involving syndiotactic to atactic to isotactic. Table 1.9 presents the comparison of catalyst systems in terms of polymer microstructures produced.

Spherilene process was introduced by Himont whose successor company was Montell and the last successor company is Basell. The process was a low-cost gas-phase technology for polyethylene using a spherical $MgCl_2$-supported catalyst to control the polymer morphology on a level similar to that of PP gas-phase processes [131]. The catalyst and cocatalyst are reacted in a prepolymerization loop in a diluent slurry to achieve morphology and mileage control. The polyethylene produced has a spherical morphology and a high bulk density. In addition, a combination of two gas-phase reactors in series allows the synthesis of PBs within the production process. The two gas-phase reactors, separated by a hook loop, are used for the production of the polymer, which is stabilized and bagged downstream. Pelletization is optional but not necessary due to the spherical form of the polymer. The polyethylene products, as well as the blends, have properties that are similar to the competitive polymers produced by different technologies.

Another gas-phase technology that improved upon the capacity of older plants was introduced by Exxon [168]. It exploits the following advantages of SSC catalysts: (1) good comonomer incorporation, (2) homogeneous composition, and (3) high activity of the copolymerization. In gas-phase polymerization, the reactor capacity is mainly determined by the heat removal capability of the gas stream circulated. Thus, capacity is dependent on the recycle gas composition and the physical properties of the reactant to determine the following: (1) gas-phase density (the higher the density becomes, the more gas is circulated), (2) gas dew point (control is necessary to keep the fluidized bed free of liquid), and (3) reactor temperature (higher temperature allows higher dew points, increasing the heat removal capacity). Hence, by injecting the reactants as condensed phase, it is possible to utilize their heat of vaporization to increase the plant capacity. By controlling the density of the fluidized bed, it is possible to increase the amount of liquid above 10 wt% per cycle without the formation of hot spots, lumps, or instabilities of the reactor-supercondensed mode technology. The capacity of the plant, as well as the product range, may be improved further by using appropriate postmetallocene SSC catalysts.

The better comonomer incorporation allows lower comonomer concentrations, and the better response to molecular-weight regulation by hydrogen leads to a decrease in hydrogen concentration by a factor of 100. Thus, a more condensing agent may be added, improving the polymer density as well as the heat removal capacity. The narrow-molecular-weight distribution and the uniform composition of the copolymer allow higher temperatures because of the absence of low-molecular-weight, high-comonomer-content fraction that could otherwise become sticky at elevated temperatures. It is possible to increase the capacity of a plant by a factor of up to 2 with this technology and at the same time broaden the product envelope significantly by incorporating additional comonomer and utilizing comonomers like hexene and octene for the gas-phase processes. Other development by the Westlake Company enabled efficient control of comonomer distribution. Through adaptation of Sclairtech technology by NOVA Chemicals, bimodal SSC LLDPEs became a reality using dual reactors.

Generally, polyethylenes produced by metallocene catalysts are characterized by narrow-molecular-weight distributions (M_w/M_n = 2–2.5) reflecting the single-site character of the catalyst. They are highly linear and usually contain a vinyl end group in each polymer chain. The molecular weight may be controlled by the choice of the SSC, by the experimental conditions (monomer/SSC ratio, temperature), or by addition of small amounts of hydrogen. Zirconium metallocene catalysts are more active than their hafnium or titanium analogs. Indeed, hafnocenes have lower numbers of active species, and rapid decomposition decreases the activity of titanium metallocene systems [169]. The more effective incorporation of comonomers results in lower consumption of comonomer and better recovery efficiency. The combination of two SSCs enables the possibility of combining the toughness of high-molecular-weight polymers with the processability of lower-molecular-weight polymers.

The uniformity of active species in SSC enables the uniformity of the physical properties of polyolefins. Advantages over conventional Ziegler–Natta catalysts include (1) higher activity, (2) uniform comonomer incorporation, (3) narrow-molecular-weight distribution, (4) high ability to incorporate sterically demanding comonomers, and (5) the possibility of controlling comonomer distribution in the polymer chain from alternating or random or blocky structures through a judicious selection of SSCs. The higher activity is reflected in the reduced amount of ash and metal found after incineration of the polymer. With conventional catalysts, the amount of ash and metal is much higher (30 ppm Al, 7–8 ppm Ti, 330 ppm total ash) than a metallocene catalyst (<20 ppm Al, <0.5 ppm Zr, <20 ppm total ash) [170], making SSC quite attractive from an ecological view point.

The absence of low-molecular-weight high-comonomer-content molecules in LLDPE results in low extractables, enabling applications in food packaging and medical sectors. In the conventional LLDPE, the high-molecular-weight low-comonomer-content fraction has a crystallization temperature that is significantly higher than that of the fractions containing more comonomer units, and these latter chains form the nuclei for crystallization. This heterogeneous nucleation leads to the formation of thick lamellae with minimum tie molecules. On the contrary, the nucleation in the copolymers produced by SSC is almost homogeneous, due to the narrow comonomer distribution, resulting in thinner lamellae and a significant number of tie molecules [171–173]. As a result of the homogeneous nucleation, the melting point of SSC LLDPE, which decreases as the density decreases, is lower than that of the conventional LLDPE. The lower melting point allows lower heat seal temperatures, whereas the increased number of tie molecules enhances the dart impact strength. Another advantage of homogeneous composition in SSC polyethylenes is the higher clarity of the products. In conventional LLDPE, the crystallites formed by the low-monomer-content chain scatter light and give rise to haze.

The major drawback of the SSC LLDPE is its viscous nature occasioned by the absence of LCB and low-molecular-weight fraction. This may be overcome by (1) using mixtures of SSCs to produce bimodal distributed polymers, (2) blending of copolymers of different molecular-weight distributions, or (3) using SSC that allows a desired amount of LCB.

For the copolymerization of ethylene and higher α-olefins such as octene-1, Dow [174] and Exxon introduced half-sandwich metallocenes, namely, the constrained geometry catalysts. Using the constrained geometry catalysts, the incorporation of octene, plus the vinyl-terminated polymers formed during the polymerization, leads to a product containing a limited amount of long-chain branches. LCB in this case differs from LCB in LLDPE for the following reasons: (1) the long-chain branches are fewer and longer; (2) the level of LCB is controlled by the catalyst used; (3) the long-chain branches are almost linear [175]. Without affecting the polymer properties, the low level of LCB significantly improves processability to a level that is superior to that of conventional PE in spite of the narrow-molecular-weight distribution. The polymers extrude about 2.5 times faster than conventional LLDPEs, and their melt tensions, which are correlatable to melt strength and bubble stability, are about twice as high. This is important for film blowing applications.

Copolymers that contain up to 20 wt% α-olefin, christened plastomers, are flexible thermoplastics that match the property profiles of conventional LLDPEs and VLDPEs at much lower densities. Copolymers containing more than 20 wt% of comonomers are called polyolefin elastomers (POEs). Their flexibility, clarity, and tensile strength enable them to replace other competitive flexible thermoplastics such as polyvinyl chloride (PVC), EVA, EMA, styrene block copolymers, as well as conventional EPR and EPDM [176].

1.4.7.2 SSC PPs

The fact that LLDPE production was ahead of that of PP may be attributed to the difficulties that SSC catalysts had in the beginning in competing with high mileage Ziegler–Natta catalysts. Nonetheless, the properties of SSC PP in terms of isospecificity, molecular mass, as well as tailor-made microstructure represent a significant advantage for SSC PP. Highly stereoselective metallocene SSCs produce highly crystalline and stiff PPs. These polymers exhibit a stiffness that is 25–30% above that of conventional PP, essentially equivalent to reinforced PPs [177]. Packages made from these PPs have reduced wall thicknesses, are easier to recycle, and show enhanced impact strength, heat resistance, lower density, and lifetime stability.

While iPP could be made in appreciable amount by both Ziegler–Natta and SSC, sPP can only be made in appreciable amount by SSC. The tailoring of the polymer microstructure by the choice of appropriate SSC has enhanced the market penetration of iPP, sPP, as well as the high-molecular-weight aPP. The term *stereospecificity* does not refer to the extractable aPP, which contributes to the fact that conventional iPP always has a melting point of 160–165°C.

1.4.7.2.1 Isotactic PP

The properties and the melting point of iPPs prepared by SSCs are determined by the amount of irregularities (stereo- and regio-errors) randomly distributed along the polymer chain. The most important feature of SSC iPP is the low amount of extractables, which makes it possible for the PPs to be used for food wrapping and other applications even at cooking temperature. Depending on their substitution pattern, SSC PPs have melting points between 132°C and 165°C. This is presented in Table 1.10, which illustrates a broad range of product properties obtained in propylene polymerization experiments with various metallocene/MAO catalysts. The table compares the catalyst productivity, PP molecular weight, melting point, and isotacticity.

Metallocene SSC, supported on 1,3,2,4-dimethyl benzylidene sorbitol (DMBS), an effective nucleating agent, is used in producing iPP grades with granular morphology and fine spherulites [178]. The state-of-the-art metallocenes are supported on silica treated with MAO [179,180]. They operate at low Al/Zr ratios and are used in gas-phase processes [181]. With these free-floating catalyst powders, polymer particle morphology control is possible and could match conventional systems with particle diameters of several hundred millimeters with narrow particle size distribution and bulk densities above 0.45 g/cm^3. Table 1.11 presents a comparison of iPPs prepared by metallocene SSC and high-mileage $MgCl_2$-supported Ziegler–Natta catalyst systems in a bulk polymerization at 70°C.

The low melting points obtained with some SSCs even at high pentad isotacticities are occasioned by the 2,1 and 1,3 mis-insertions [182,183]. Low-melting-point polymers with conventional catalysts are obtained by copolymerization of propylene with small amounts of ethylene. Table 1.12 compares the properties of low-melting-point Ziegler–Natta propylene copolymers, metallocene propylene copolymers, and homopolymers. The data illustrate the enhanced stiffness and transparency of the polymers produced using metallocene SSC.

The molecular-weight distribution of SSC iPP (M_w/M_n = 2–2.5) is lower than that of the conventional grade obtained by peroxide degradation (M_w/M_n = 3–4). Its processing performance by thin-wall molding or fiber spinning is good. Compared with the conventional iPP grades, metallocene

TABLE 1.10
Metallocene Range of Catalyst Systems and the Resulting Polypropylene Isotacticity[a]

Metallocene	Productivity [kg PP/(mmol Zr*h)]	$M_w \times 10^{-3}$ (g/mol)	m.p. (°C)	Isotacticity (% mmmm)
[En(Ind)$_2$]ZrCl$_2$	188	24	132	78.5
[Me$_2$Si(Ind)$_2$]ZrCl$_2$	190	36	137	81.7
[Me$_2$Si(IndH$_4$)$_2$]ZrCl$_2$	48	24	141	84.5
[Me$_2$Si(2Me–Ind)$_2$]ZrCl$_2$	99	195	145	88.5
[Me$_2$Si(2Me–4iPr–Ind)$_2$]ZrCl$_2$	245	213	150	88.6
[Me$_2$Si(2,4Me$_2$–Cp)$_2$]ZrCl$_2$	97	31	149	89.2
[Me$_2$Si(2Me–4tBu–Cp)$_2$]ZrCl$_2$	10	19	155	94.3
[Me$_2$Si(2Me–4,5BenzInd)$_2$]ZrCl$_2$	403	330	146	88.7
[Me$_2$Si(2Me–4Ph–Ind)$_2$]ZrCl$_2$	755	729	157	95.2
[Me$_2$Ge(2Me–4Ph–Ind)$_2$]ZrCl$_2$	750	1135	158	–
[Me$_2$Si(2Me–4Naph–Ind)$_2$]ZrCl$_2$	875	920	161	99.1

Source: Antberg, M. et al., *Makromol. Chem. Macromol. Symp.* 48/49: 333, 1991; Spaleck, W. et al., *Organometallics 13*: 954, 1994.

[a] Conditions: Bulk polymerization in liquid propylene at 70°C, Al/Zr ratio 15,000.

TABLE 1.11
Comparison of Isotactic Polypropylenes by SSC and Ziegler–Natta Catalysts at 70°C

Property	(I)	(II)	(III)	(IV)
Melting point (°C)	139	151	160	162
M_w/M_n	2.2	2.3	2.5	5.8
Tensile modulus (N/mm^2)	1060	1440	1620	1190
Hardness (N/mm^2)	59	78	86	76
Impact resistance Izod (mJ/mm^2)	128	86	100	103
Light transmission (% 1 mm plate)	56	44	35	34
Melt flow rate (°/min)	2	2	2	2

Source: Reprinted from *Catalyst Design for Tailor-Made Polyolefins*, K. Soga and M. Terano (eds.), Shiomura, T. Kohno, M. Inoue, N. et al., Syndiotactic polypropylene, p. 327, Copyright 1994, with permission from Elsevier and Kodansha Ltd., Tokyo.

Note: MAO catalyst systems with Al/Zr = 15,000: (I) [En(IndH$_4$)$_2$]ZrCl$_2$; (II) [Me$_2$Si(4,5Benzind)$_2$]ZrCl$_2$; (III) [Me$_2$Si(4,6iPrInd)$_2$]ZrCl$_2$; Ziegler–Natta catalyst system: (IV) MgCl$_2$ supported TiCl$_4$–Et$_3$Al.

products possess enhanced mechanical strength, which could be further improved by tailoring the molecular-weight distribution.

Metallocenes iPP waxes are used as pigment dispersants, toner, or lacquer surfaces [184]. With an appropriate choice of metallocene, it is possible to prepare iPP with molecular weights ranging between 10,000 and 70,000 g/mol and with melting points ranging between 140°C and 160°C. With conventional catalysts, hydrogen or polymer degradation is used to control the molecular weight while the melting point is controlled by the addition of a comonomer. Table 1.13 compares iPP waxes prepared by metallocene SSC to waxes produced by Ziegler–Natta catalysts using molecular-weight regulation by hydrogen, visbreaking of Ziegler–Natta random EP copolymers, or through

TABLE 1.12
Comparison of Low-Melting-Point Ziegler–Natta Ethylene–Propylene Copolymers, Metallocene Ethylene–Propylene Copolymers, and Propylene Homopolymers

Property	Ziegler–Natta Copolymer	Metallocene Copolymer	Metallocene Homopolymer
Melting point (°C)	141	140	142
Tensile modulus (N/mm^2)	620	940	1120
Hardness (N/mm^2)	41	59	65
Impact resistance Izod (mJ/mm^2)	23.1	11.3	7.3
Light transmission (% 1 mm plate)	57	65	48
Extractables (% hexane, 69°C)	7.9	1.1	0.7

Source: Soga, K. et al., *Makromol. Chem. Rapid Commun.* 8: 305, 1987.

TABLE 1.13
Isotactic Polypropylene Waxes Prepared by Metallocene SSC I and II, Compared to Waxes Produced by Ziegler–Natta Catalyst Using Methods III, IV, and V

Property	(I)	(II)	(III)	(IV)	(V)
$M_w \times 10^{-3}$ (g/mol)	68	50	40	44	36
M_w/M_n	1.8	2.0	3.8	2.2	1.9
C_2/C_4 (%)	–	–	–	4.0/2.4	–
m.p. (°C)	163	133	159	133	155
Crystallinity (%)	69	50	60	30	59
Isotacticity (mmmm %)	96	85	91	80	91
Double bonds/chain	0.5–1	1	0	4	5
Mis-insertions/1000°C	1.7 1.3	4.7 1.3		19 C_2	n.d.
	0.2 2.1	0.3 2.1	0.3 2.1	7 C_4	
Melt viscosity (200°C) (mm^2/s)	2800	2521	900	1684	1040
Hardness (bar)	2000	874	1800	423	1870
Yellowness index	0.5	0.8	1	5–6	1–2
Dropping point (°C)	168	147	160	149	169
Congealing temp. (°C)	n.d.	113	117	102	124

Source: Reprinted from *Catalyst Design for Tailor-Made Polyolefins*, K. Soga and M. Terano (eds.), Shiomura, T. Kohno, M. Inoue, N. et al., Syndiotactic polypropylene, p. 327, Copyright 1994, with permission from Elsevier and Kodansha Ltd., Tokyo.

Note: (I) [Me$_2$Si(2Me–4tBuCp)$_2$]ZrCl$_2$/MAO; (II) [Me$_2$Si(IndH$_4$)$_2$]ZrCl$_2$/MAO; (III) Ziegler–Natta propylene polymerization with molecular-weight regulation by hydrogen; (IV) Visbreaking of Ziegler–Natta random ethylene–propylene copolymers; and (V) Ziegler–Natta propylene homopolymers.

Ziegler–Natta propylene homopolymerization. SSCs offer several property combinations not accessible with conventional systems. For example, the vinyl end groups of the metallocene products are utilized for functionalization, whereas with conventional catalysts, only saturated end groups are formed due to the high amount of hydrogen used for molecular-weight regulation. With conventional catalysts, the reactor normally has to be run under non-optimal conditions (high temperature, high hydrogen pressure, and lower productivity) to obtain the desired product. There is also the problem of a decreased output rate, which is necessitated by the difficulty of condensing the hydrogen/

TABLE 1.14
Comparison of Isotactic and Syndiotactic Polypropylenes of the Same Stereoregularities Prepared by Metallocene Catalysts

Property	[Me$_2$Si(Ind)$_2$]ZrCl$_2$	[Me$_2$C(Flu)(CP)]ZrCl$_2$
Tacticity	Iso	Syndio
mmmm%/rrrr%	83.1	83.6
Niso/nsyn	33	25
m.p. (°C)	138.4	133.2
Crystallinity DSC (%)	41.6	27.2
MFI 230/5	16.4	21.1
Density	0.899	0.885

propylene feed with its large amount of hydrogen coupled with the low heat removal capacity of these mixtures. If run at optimum conditions, polymer degradation with expensive peroxides must be used to control the molecular weight. Metallocenes avoid these difficulties and thereby enable process simplification.

1.4.7.2.2 Syndiotactic PP

Using C$_s$-symmetrical metallocenes, the production of sPP is possible due to enantiotopic vacancies formed by the chain migratory insertion [92–94] during polymerization. As illustrated in Table 1.14, sPPs, produced by SSC metallocene catalysts, show a higher level of irregularities than iPP. With the same degree of tacticity, the syndiotactic polymer exhibits a lower melting point, lower density, lower crystallinity, and lower crystallization rate [185,186]. The smaller crystal size in sPP enables a higher clarity of the material but is also responsible for its inferior gas barrier properties, limiting its utility in food packaging applications. On the other hand, its strong resistance to radiation allows medical applications. Other advantages of sPP are the higher viscous and elastic modulus at high shear rates and its outstanding impact strength, which disappears at low temperatures. The combination of flexibility, clarity, and tensile strength as well as the low heat seal temperatures enables sPPs to replace PVC, EVA, and LLDPE in films, foils, and extruder products [175].

1.4.7.2.3 Reactor Blends and Copolymers of Stereospecific PP

In the 1980s, reactor granules of spherical morphology were first introduced with processes such as the Unipol with conventional Ziegler–Natta catalysts. These products do not need to be melt-extruded [187,188]. The advent of SSC catalysts facilitated the use of two reactors in series, for the production of reactor blends of PP, by transferring the propylene homopolymer-containing active catalyst to a second reactor where an EPR phase is produced. Combination of the products of different composition in the two reactors allows the synthesis of a broad range of products with medium to "super-high" impact strength. By replacing the PP homopolymer of these blends with a random copolymer matrix, the flexibility and tensile strength at low temperatures are enhanced although the melting point and stiffness are decreased. Blending of these copolymers with LLDPE results in soft PPs (less than 100 MPa tensile-modulus) and enables sPPs to replace thermoplastic elastomers like EVA, plasticized PVC, and SEBS [189].

The excellent performance of SSC copolymers offers improvements in impact properties. Among the wide variety of properties of impact copolymers, it is safe to say that the stiffness of the material is determined by the matrix material, whereas the impact resistance depends largely on the elastomeric EPR phase. While conventional catalysts have some inhomogeneities in the EPR phase due to crystalline ethylene-rich sequences, the more homogeneous comonomer distribution obtained with metallocene catalysts results in a totally amorphous EPR phase [190]. For applications demanding broader-molecular-weight distributions, two or more SSCs may be combined to give a tailor-made molecular-weight distribution.

1.4.8 Poly(cycloolefins) from SSCs

Strained cyclic olefins like cyclobutene, cyclopentene, and norbornene can be used as monomers and comonomers in a wide variety of polymers. Generally, they can be polymerized by double-bond opening (vinyl polymerization) or by ring-opening metathesis polymerization (ROMP).

1.4.8.1 Poly(cycloolefins) by Vinyl Polymerization

Homopolymerization of cyclic olefins by double-bond opening is achieved by several transition metal catalysts, such as palladium catalysts as well as metallocene catalysts. The polymers feature two chiral centers per monomer unit and therefore are ditactic. While polymers produced by achiral palladium catalysts seem to be atactic using chiral metallocene catalysts, highly tactic, crystalline materials could be produced featuring extraordinarily high melting points (in some cases above the decomposition temperature) and extreme chemical resistance.

The microstructures of these polymers have been investigated using oligomers as models. Norbornene was shown to polymerize via cis-exo insertion [191,192], whereas in the case of cyclopentene, quite unusual cis- and trans-1,3 insertions are observed [192–195].

1.4.8.2 Poly(cycloolefins) by ROMP

The concept of catalyzed metathesis polymerization of cyclic olefin parallels the catalyzed double-bond opening ethylene polymerization occasioned by the 1953 Karl Ziegler's discovery [24] of the first-generation transition metal halide catalyst with an organometallic aluminum alkyl cocatalyst, which earned Ziegler and Natta the 1963 Nobel Prize in Chemistry. It was, however, Guilio Natta, in 1966, that successfully undertook the polymerization of cycloheptene, cyclooctene, and cyclododecene using a catalyst consisting of a combination of tungsten hexachloride with either triethylaluminum or diethylaluminum chloride. Thus, the ring-opening polymerization of cyclic alkenes to polyalkenemers was born, even though the underlying carbene catalyzed metathesis mechanism was not understood at the time. Several generations of ROMP catalysts were later developed leading to the revelation of carbene catalyzed metathesis mechanism and the award of the 2005 Nobel Prize in Chemistry to Dr. Yves Chauvin at the Institut Français du Pétrole, Rueil-Malmaison, France; Professor Robert H. Grubbs, California Institute of Technology, Pasadena, California; and Professor Richard R. Schrock, Massachusetts Institute of Technology, Cambridge, Massachusetts [196].

Catalyzed ROMP of cyclic olefins is driven by the ring strain in monomers such as cyclobutene, cyclopentene, dicyclopentadiene, and norbornene. First, a transition metal (ruthenium, molybdenum, tungsten) carbene catalyst complex is formed. The carbene catalyst complex attacks the double bond in the monomer ring structure forming a highly strained metallocyclic intermediate that opens up, yielding a double-bonded unit with the transition metal as well as a terminal double bond. This constitutes the beginning of living ring-opening catalyzed metathesis polymerization. Variations within ligands, bi- and tricyclic rings, and ring substituting moieties and/or solvents result in variations of catalyst stability, catalytic activity, kinetic profile, catalyst regio- and stereoselectivity, monomer/comonomer incorporation capability, and polymer molecular-weight characteristics and microstructure.

Important industrial ROMP polymers include Vestanamer, a trans-polyoctenamer (from cyclooctene), polynorbornene (Norsorex), and polydicyclopentadiene (PDCPD, a side reaction of norbornene polymerization). These polymers could also be functionalized for further reaction with a variety of nonpolar monomers or oligomers.

1.4.9 Cycloolefin Copolymers from SSCs

The homopolymers of cycloolefins like norbornene or tetracyclododecene are barely processable due to their high glass transition temperatures and their insolubility in common organic solvents. Functionalization enables copolymerizing of ROMP polyalkenemers with a variety of monomers or oligomers resulting in copolymers with impressive mechanical, thermal, rheological, and crystallization properties. Also, the metallocyclic intermediates enable the production of block copolymers by changing feed compositions.

Cycloolefin copolymers (COCs) of cyclic olefins with ethylene or α-olefins represent a new class of thermoplastic amorphous polyolefins [197–199]. Early attempts to produce such copolymers were made using a heterogeneous two-component Ziegler–Natta catalyst system, namely, $TiCl_4$/$AlEt_2Cl$ [24,25]. In the 1980s, vanadium catalysts were used for the copolymerization, but real progress was made by utilizing metallocene SSC. Metallocenes are about 10 times more active than vanadium systems, and, by judicious choice of the metallocenes, the comonomer distribution is varied from statistical to alternating. Statistical copolymers are amorphous if more than 15 mol% of cycloolefin is incorporated in the polymer chain. The glass transition temperature can be varied over a wide range by appropriate choice of the cycloolefin, with the right amount of the cycloolefin incorporated in the polymer backbone.

As for the ethylene/norbornene copolymerization, it is possible to produce copolymers with molecular-weight distributions of M_w/M_n = 1.1–1.4 by controlling the polymerization conditions [200]. This *pseudo-living polymerization* enables the production of block copolymers by changing the feed composition. Statistical copolymers are transparent due to their amorphous character; they are colorless and show a high optical anisotropy. Because of their high carbon/hydrogen ratio, these polymers have a high refractive index of 1.53 as illustrated in Table 1.15 for an ethylene/norbornene copolymer at 50 mol% incorporation. Their stability against hydrolysis and chemical degradation in combination with their stiffness and very good processability makes them potential materials for optical applications in compact disks, lenses, and optical fibers [201].

The ethylene/norbornene alternating copolymer has a glass transition temperature of 130°C and a melting point of 295°C. Thermoplastic processing is therefore possible at 300–330°C. Its melting point as well as its crystallinity may be influenced by the choice of the metallocene and polymerization conditions. Compared to the statistical copolymers, the alternating structures are characterized

TABLE 1.15
Properties of a Random Ethylene Norbornene Copolymer Containing 52 mol% of Norbornene

Mechanical Properties	
Density (g/cm³)	1.02
Glass transition temperature (°C)	150
Tensile modulus, ISO 527 (MPa)	3100
Tensile strength, ISO 527 (MPa)	66
Elongation at break, ISO 527 (%)	2–3
Optical Properties	
Clarity	White, clear
Anisotropy	Very low
Refractive index	1.53

Source: Land, H.-T. and D. Niederberg, *Kunststoffe/plast Europe* 85(8): 13, 1995; Land, H.-T., New cyclic-olefin copolymers from metallocenes. In *Proceedings of the International Congress on Metallocene Polymers Metallocenes '95*, 217, Brussels Schotland Business Research Inc., 1995.

by better heat resistance and are unaffected by nonpolar solvents. The diameter of the crystallites is about 0.05–1 mm; thus, these copolymers are transparent. Similar alternating structures could be obtained by the ring-opening polymerization of multicyclic polyolefins followed by hydrogenation of the unsaturated polymer.

1.5 POLYOLEFIN COMPOSITES AND NANOCOMPOSITES

The birth of the composite, as a material containing polymers plus particulate fillers and fiber reinforcements, could be traced to the first truly synthetic plastic, a phenolic resin, christened *Bakelite*, after Dr. Leo Hendrick Baekeland [202] who founded the Bakelite Corporation, which later became the Union Carbide Corporation (UCC). The pressure vessel used to commercialize Bakelite, called the Bakelizer, was designated the first National Historic Chemical Landmark in November 1993 and is housed at the Smithsonian Institution's National Museum of American History in Washington, DC. In the 1910s, the addition of wood flour to Bakelite gave rise to the first synthetic composite. By the 1950s, synthetic composite had included the glass fiber-reinforced plastics, and this proceeded the era, in the late 1960s, of intense interest in composites beginning with the development of the high-performance carbon and aramid fibers.

1.5.1 CONVENTIONAL POLYOLEFIN COMPOSITES

Fibers that could be used in polyolefin composites include acicular particulates, alumina, aramid, boron, ceramic, glass, carbon, polyolefins, modern rigid rod polymer fibers, and knitted reinforcements. The desired features of fibers are as follows: high aspect ratio, chemical stability, thermal stability, low cost, low health hazard, minimum grain size, minimum porosity, minimum surface flaws, minimum surface roughness, high specific strength, high specific stiffness, and high toughness. No single fiber is characterized by the best combination of all of these properties, and a significant body of knowledge exists pertaining to the dependence of the polyolefins–fiber composite performance on the interplay among the individual fiber component, matrix, interface, nature of damage, and failure mechanisms. The key applications of polymer–fiber composites are in simple primary structures, and macromechanics has been used to elucidate the paramount issues dictating the design requirements for polymer–fiber composites in structural applications [203].

Particulate fillers that have been used as polyolefin additives or extenders are as follows: wood powder, glass spheres, hollow silicates, calcium silicates (wollastonites, $CaSiO_3$), silica minerals (diatomaceous earth, kaolin clays), magnesium silicates (talc, mica, asbestos), calcium sulfate whiskers, barium sulfate, calcium carbonate, and carbon black. Fillers or fibers are used either to lower the cost or to improve the physical properties of polymeric materials. In general, however, fibers have the strongest effects on polyolefin properties followed by plate-like and particulate fillers, in that order.

Flexural modulus, tensile modulus, stiffness, abrasion resistance, antistatic behavior, and heat distortion temperature are almost always improved, and the coefficient of thermal expansion is considerably reduced by the addition of fibers or fillers. The same cannot be said for impact strength (toughness) and elongation to break (ductility) unless an appropriate interfacial agent is used. Toughness and ductility are low in filled polymers because of strain magnification, as rigid inclusions constitute stress concentration flaws and all of the strains are imposed on the diminished quantity of ductile matrix. This is why polyolefin composites are weak and brittle if the interfacial agent is not effective. Coupling agents improve adhesion and thereby enhance the composite toughness with little or no increase in ductility, resulting in a strong but brittle material. On the contrary, a decoupling agent decreases adhesion but facilitates microcavitation, which significantly increases ductility with little or no increase in toughness, resulting in a ductile but weak material.

The most desired interfacial agent is that which significantly increases both toughness and ductility through enhanced particle-to-matrix adhesion and the formation of a discernible tough interface. This implies diminished interface stress concentration and uniform microcavitation so as to obtain an overall reduced stress even if a moderate triaxial state of stress exists throughout the composite. Synergism is an ultimate goal. The search for such reinforcement promoters [204,205] constitutes a dynamic ongoing research area in industry, universities, and research institutes. So far, there is a loose consensus that shear banding, crazing, and/or microcavitation are the essential mechanisms for toughness and ductility, as they are largely responsible for creating desired new surfaces that are capable of absorbing large amounts of energy during deformation.

In preparing polyolefin composites, the interfacial agents are first deposited on the fillers and/or fibers before the melt compounding and/or interfacial reaction is effected. The interfacial agents that have been used with polyolefins include organosilanes, organotitanates, other organometallic compounds, cross-linking agents, and polymerization catalysts. Composite materials could be processed in one or more of the following: autoclave molding, filament winding, hand lay, injection molding, and pultrusion. The various aspects of the methods are well documented.

Under the influence of adverse environmental agents, the ultimate properties of polyolefin composites are susceptible to degradation. Such an adverse environment could be something as innocuous as moisture. The effect of adsorbed moisture is to degrade the matrix-dependent properties with a resultant effect on the load-bearing performance. In a less friendly environment, plasticization and possibly environmental stress cracking could occur.

1.5.2 Polyolefin Nanocomposites

Polyolefin nanocomposites are multiphase systems, with filler particle dimensions in the range of 1–100 nm, consisting principally of polyethylene, PP, copolymers, or blends intimately mixed with nanoscale fillers. Relevant nanoparticles are nanoclays, nanotubes, nanofibers, and nanopowders; specific examples include carbon nanofibers, carbon nanotubes, graphite (graphene), nanometals, nanometal oxides, and nanoscale inorganic fillers and fibers. The types of clay used include mainly montmorillonite (MMT), hectorite, saponite, as well as organoclays. The performance of polyolefin nanocomposites depends on the morphology of the nanocomposites. For clay nanocomposites, the comprehensive properties depend strongly on the degree to which deagglomeration, dispersion, intercalation, and exfoliation of clay platelets are achieved within the polyolefin matrix.

Polyolefin nanocomposites with deagglomerated and well-dispersed nanoclay, but no expansion of the interlayer spacing of the nanoclay platelets, are normally characterized as *immiscible* [206]. If the intensity of mixing is such that the interlayer spacing of the platelets is expanded but the characteristic diffraction peak of the nanoclay remains observable (with clay loading > 2%), the polyolefin nanocomposites are normally characterized as *intercalated*. If, on the other hand, the deagglomerated nanoclay is well dispersed and well intercalated, and the interlayer spacing of the platelets is so expanded that the characteristic diffraction peak of the nanoclay is not observable, the resulting polyolefin nanocomposites are characterized as *exfoliated*, provided that the clay loading is greater than 2%. As an example, the interlayer spacing of MMT, in a highly exfoliated polyolefin/MMT nanocomposite, is higher than 3.27 nm [207–212]. Polyolefin nanocomposite formation process could be roughly delineated into physical method and chemical reaction process.

1.5.2.1 Nanocomposite Formation by Physical Methods

Physical process may include solution method, latex method, and melt processing. Because of its capacity to achieve significant exfoliation of clay platelets within the polymer matrix, the commercial physical method of choice is melt processing involving formulation compounding and fabrication. In this regard, exfoliation depends on the nanocomposite loading, the surfactant used to form the organoclay, the effectiveness of nanofiller surface treatment, the positive interaction or affinity of polyolefin–filler surfaces, the polyolefin melt flow, and the intensity of the dispersive energy of the processing method. Aside from the degree of exfoliation, appropriate orientation of nanoplatelet/nanofiber is also affected by the type of processing used. In general, twin screw extrusion compounding is the commercial physical method of choice for producing polyolefin nanocomposites.

With this method, polyolefin nanocomposites have penetrated the following markets: automobile, packaging, and fire-retardant. Nanoplatelet/nanofiber ability to reduce flammability as well as the maximum heat release rate during combustion [208] reduces the amount of flame-retardant additives that need to be incorporated. Indeed, TPO nanocomposites have been replacing conventional composites in automotive applications [209] since the first thermoplastic polyamide nanocomposites [209] were commercially introduced as timing belt cover in 1991 by Toyota Motor.

For polyolefin nanocomposites in injection or extrusion blow molding, melt strength, rather than melt viscosity, is the key controlling resin characteristic. Consequently, nanoclay-filled TPO nanocomposites, used as a matrix, have found applications in film-blowing and other blow-molding operations on account of enhanced *melt strength*. Polyolefin nanocomposites are also being used as a matrix for conventional fillers in special applications requiring tailor-made property sets where the melt flow index of TPO nanocomposites is appropriately adjusted to be identical to that of the original polyolefin matrix used in such applications.

Ordinarily, polyethylenes and PPs, particularly sPP, possess inferior gas barrier properties, compared to PB-1, limiting their applications in food packaging. However, polyolefin nanocomposites have enhanced gas barrier properties on account of the tortuous path created by the nanofillers. Because of this, the barrier properties of these polyolefins are being enhanced with the addition of exfoliated nanoclay platelets having an appropriate aspect ratio to alter the diffusion path of penetrant molecules. Because of the excellent surface finish characteristics of nanocomposites, the addition of the exfoliated nanoclay platelets does not impair the smoothness and transparency of the resultant polyolefin thin film. Consequently, single-site polyolefin nanocomposites, with enhanced stiffness, tensile strength, gas barrier properties, and tensile and dynamic storage modulus, are assuming a preeminent role in food-packaging applications.

1.5.2.2 Nanocomposite Formation by Chemical Reaction

Nanocomposite chemical reaction process is the *in situ* polymerization or copolymerization of olefins with nanoclay. This formation process could involve the use of any of three types of nanoclay (MMT, hectorite, or saponite) and any of four types of catalysts, namely, Ziegler–Natta catalysts, metallocene catalysts, nonmetallocene catalysts, and late transition metal catalysts [210]. Beginning with immobilizing the precatalysts onto the clay, the nanocomposite chemical reaction formation method ends with olefin polymerization within the interlayer of the nanoclay platelets. The layered structure of the polymerizing system has an effect on the chain transfer and the termination steps, thereby affecting the molecular weight and other microstructure of the polymers or copolymers.

The heat of olefin polymerization and propagation has positive effects on the intercalation and exfoliation of nanoclay in the polyolefin matrix. By doing so, the nanoclay interlayer spacing is so expanded that complete exfoliation of nanoclay is accomplished [211]. The nanoclay serves as a catalyst support or an adjunct, which remains with the final nanocomposite. For polyethylene nanocomposites, both the crystalline and the amorphous phases of the matrix are known to be affected with concomitant effect on the γ, β, and α transitions [212]. Increasing amount of nanofiller is known to produce increasing percentage of crystallinity and higher melting temperature, suggesting a heightened heterogeneous nucleating activity of the polymer in the presence of nanofillers.

Generally, completely exfoliated polyolefin nanocomposites are characterized by excellent surface finish and high modulus-to-weight, strength-to-weight, and surface-to-volume ratios, which could result in weight reduction of up to 40% compared to conventional polyolefin composites. With chemical reaction formation process, polyolefin nanocomposites having less than 5 wt% nanoclay are characterized by complete exfoliation. Such nanocomposites are generally superior to those formed through physical methods particularly for impact properties, elongation at rupture, and tensile and flexural properties. They possess high heat distortion temperature, enhanced glass transition temperature, high on-set decomposition temperature, outstanding mechanical properties, excellent gas barrier properties, respectable flame-retardant properties, and effective dyeability. Other properties affected by the higher degree of exfoliation include crystallinity, improved thermal oxidative stability, reduced thermal expansion coefficient, improved melt flow, and enhanced hardness.

Selecting the most appropriate formation method is the key to realizing the variety of property advantages inherent in polyolefin nanocomposites. Regardless of the method of formation, the comprehensive properties of polyolefin nanocomposites are superior to those of conventional polyolefin composites reinforced with microsized conventional glass fiber and/or other fillers.

1.6 PROCESSING METHODS FOR POLYOLEFINS

The key processing methods for polyolefins are injection molding, compression molding, rotational molding, blow molding, structural foam molding [213], structural web molding [214,215], extrusion, blown film extrusion, and cast film extrusion [132]. Aside from the independent processing variables of time, temperature, and pressure, the choice of an appropriate processing machine for a given polyolefin product is critical. Available machines include single screw extruder, twin screw extruder, gear pumps, Buss-kneaders, Readco mixers, and Farrel continuous mixers. Each of these machines finds appropriate applications in compounding polyolefins with particulate fillers and/or fiber reinforcements to produce composites or nanocomposites.

In injection molding, the polyolefin granules or pellets are placed in a hopper that continuously feeds the heated barrel of an extruder. The polymer is melted and the molten material is injected, under high pressure, into a relatively cold mold where the material solidifies replicating the shape of the mold cavity. It is essential for the melt viscosity to be sufficiently low to ensure that the mold cavity is filled in a minimum possible cycle time. Injection molding is a cyclic process.

In rotational molding (rotomolding or rotoforming), finely ground polyolefin powders are heated inside a rotating mold where the polymer melts and uniformly coats the inner surface of the mold. The mold is cooled in a special chamber just prior to part removal. The process is used for the production of large complex polyolefin parts such as large containers, storage tanks, water tanks, and portable sanitary facilities. Rotational molding is also a cyclic process.

The injection or extrusion blow molding makes hollow parts through the formation of a parison that is expanded, with a gas, against a mold cavity. Smaller containers (<1 L) are produced with injection blow molding, whereas extrusion blow molding is suitable for larger containers and for containers with handles. Melt strength, rather than melt viscosity, is the key controlling resin characteristic, and large containers require high-molecular-weight polyolefins with broad-molecular-weight distribution that are easier to process and less likely to exhibit parison sag. Blow molding is a cyclic process.

In extrusion forming, polyolefin granules or pellets are placed into a hopper that continuously feeds the heated barrel of an extruder. The polymer is plasticated and melted, and the molten material is pumped through a die of roughly the same shape as the final product such as sheets, pipes, films, and wire-and-cable coatings. The extruded product is drawn by some type of takeoff equipment, sized, and cooled until solidified. Extrusion is a continuous process.

In blown film extrusion, molten polyolefin is extruded through a circular die whereby the die mandrel introduces an internal air pressure that expands the extruded tube from 1.5 to 2.5 times the die diameter. Melt strength, rather than melt viscosity, is the key controlling resin characteristic; die swell and melt fracture are undesirable. Blown film extrusion is a continuous process.

In cast film extrusion process, molten polyolefin is extruded, as a thin sheet on a mirror-surfaced chill roll, through a large die whose size is equal to the width of the film to be cast. The extruded thin sheet is then drawn down by other rolls. Cast film extrusion is a continuous process. Cast films are used as diaper liners, pallet stretch wrap, household cling wrap, and overwrap.

REFERENCES

1. Bett, K. E., Crossland, B., Ford, H. and Gardner, A. K. 1983. Review of the engineering developments in the high pressure polyethylene process 1933–1983, *Proceedings of the Golden Jubilee Conference, Polyethylenes 1933–1983*, Plastics and Rubber Institute, London.
2. Realising the value of recycled plastics. 2010. *Market Situation Report*, WRAP, Banbury, UK.
3. Plastics Europe—The compelling facts about plastics. 2008. Association of Plastic Manufacturers, Brussels, Belgium.
4. Olabisi, O. 1997. Conventional polyolefins. In *Handbook of Thermoplastics*, ed. O. Olabisi, pp. 1–38, Marcel Dekker, New York.
5. Treat, J. E. Global value management opportunities in thermoplastics. In *Thermoplastics Beyond the Year 2000: A Paradigm*, eds. O. Olabisi and A. G. Maadhah, KFUPM Press, Dhahran, Saudi Arabia.
6. Albizzati, E. 1995. New polymeric materials technologies, *Proceedings of the Expert-Group Meeting on Techno-Economic Aspects of the Commercial Application of New Materials in the ESCWA Region*, organized by the UNIDO and ESCWA at Al-Ain, October 1–3.
7. Gas works. 2012. *The Economist*, July 14, available at http://www.economist.com/sites/default/files/20120714_natural_gas.pdf.
8. Johnson, R. N. 1975. Catalyst modified with certain strong reducing agents and silane compounds and use in polymerization of olefins, U.S. Patent 3,879,368.
9. Peters, E. F. 1954. Polymerization of conditioned olefin charging stocks with molybdenum catalyst, U.S. Patent 2,692,259.
10. Hogan, J. P. and Banks, R. L. 1958. Polymers and production thereof, U.S. Patent 2,825,721.
11. Hogan, J. P. 1983. Catalysis of the Phillips Petroleum Company polyethylene process. In *Applied Industrial Catalysis*, Vol. 1, ed. B. E. Leach, p. 149, Academic Press, New York.
12. McDaniel, M. P. and Johnson, M. M. 1986. A comparison of Cr/SiO_2 and $Cr/AlPO_4$ polymerization catalysts, I. Kinetics, *J. Catal. 101*: 446.
13. McDaniel, M. P. and Johnson, M. M. 1987. Comparison of Cr/SiO_2 and $Cr/AlPO_4$ polymerization catalysts: 2. Chain transfer, *Macromolecules 20*: 773.
14. Woo, T. W. and Woo, S. I. 1990. Ethylene polymerization with Phillips catalyst co-catalyzed with Al(i-Bu)3, *J. Catal. 123*: 215.
15. Karol, F. J. 1972. Chromocene catalysts for ethylene polymerization: Scope of the polymerization, *J. Polym. Sci. 10 A-1*: 2621.
16. Karol, F. J. 1972. Ethylene polymerization with supported bis-triphenylsilyl chromate, *J. Polym. Sci. 10 A-1*: 2609.
17. Smith, P. D. and McDaniel, M. P. 1989. Ethylene polymerization catalysts from supported organo-transition metal complexes. I. Pentadienyl derivatives of Ti, V, and Cr, *J. Polym. Sci. A Polym. Chem. 27*: 2695.
18. Freeman, J. W., Wilson, D. R. and Ernst, R. D. 1987. Ethylene polymerization over organochromium catalyst: A comparison between closed and open pentadienyl ligands, *J. Polym. Sci. A Polym. Chem. 25*: 2063.
19. Benham, E. A., Smith, P. D., Hsieh, E. T. and McDaniel, M. P. 1988. Mixed organo/oxide chromium polymerization catalysts, *J. Macromol. Sci. Chem. A25*: 259.
20. McDaniel, M. P. 1988. Controlling polymer properties with the Phillips chromium catalysts, *Ind. Eng. Chem. Res. 27*: 1559.
21. Soga, K., Chen, S. I., Shiono, T. and Doi, Y. 1985. Preparation of highly active Cr-catalysts for ethylene polymerization, *Polymer 26*: 1891.
22. Soga, K., Chen, S. I., Doi, Y. and Shiono, T. 1986. Polymerization of ethylene and propylene with $Cr(C_{17}H_{35}COO)3/AlEt_2Cl$/metal chloride catalysts, *Macromolecules 19*: 2893.
23. Smith, P. D. and McDaniel, M. P. 1990. Ethylene polymerization catalysts from supported organotransition metal complex: II. Chromium alkyls, *J. Polym. Sci. A Polym. Chem. 28*: 3587.
24. Ziegler, K., Holzkamp, E., Martin, H. and Breil, H. 1955. The Mulheim low pressure polyethylene process, *Agew. Chem. 67*: 541.
25. Natta, G., Pino, P., Mazzanti, G. and Giannini, U. 1957. A crystallizable organometallic complex containing titanium and aluminum, *J. Am. Chem. Soc. 79*: 2975.
26. Simonazzi, T. and Giannini, U. 1994. Forty years of development in Ziegler–Natta catalysis: From innovations to industrial realities, *Gazz. Chim. Ital. 124*: 533.
27. Natta, G. and Danusso, F. eds. 1967. *Stereoregular Polymers and Stereospecific Polymerizations*, Vols. 1 and 2, Pergamon Press, Oxford.

28. Hoffmann, J. D., Davis, G. T. and Lauritzen, J. I. 1976. The rate of crystallization of linear polymers with chain folding. In *Treatise on Solid State Chemistry*, Vol. 3, ed. N. B. Hannay, Plenum Press, New York.
29. Burfield, D. R. and Quirk, R. P. 1983. *Transition Metal Catalyzed Polymerization: Alkenes and Dienes*, Harwood Academic, New York.
30. Kissin, Y. V. 1985. *Isospecific Polymerization of Olefins with Heterogeneous Ziegler–Natta Catalysts*, Springer-Verlag, New York.
31. Lin, S., Wang, H., Zhang, Q., Lu, Z. and Lu, Y. 1985. Ethylene polymerization with modified supported catalysts, *Proceedings of the International Symposium on Future Aspects of Olefin Polymerization*, Tokyo, Japan, pp. 91–107.
32. Zakharov, V. A., Makhtarulin, S. I., Perkovets, D. V., Moroz, E. M., Mikenas, T. B. and Bukatov, G. D. 1986. Structure, composition and activity of supported titanium-magnesium catalysts for ethylene polymerization. In *Studies in Surface Science and Catalysis: Catalytic Polymerization of Olefins*, Vol. 25, eds. T. Keii and K. Soga, pp. 71–88, Elsevier, New York.
33. Salajka, S., Kratochvila, J., Hamrik, O., Kazda, A. and Gheorghiu, M. 1990. One phase supported titanium-based catalysts for polymerization of ethylene and its copolymerization with 1-alkene. I. Preparation and properties of catalysts, *J. Polym. Sci. A Polym. Chem.* 28: 1651.
34. Ivanchev, S. S., Baulin, A. A. and Rodionov, A. G. 1980. Promotion by supports of the reactivity of propagating species of Ziegler supported catalytic systems for the polymerization and copolymerization of olefins, *J. Polym. Sci. A Polym. Chem.* 18: 2045.
35. Muñoz-Escalona, A., Gallardo, J. A., Hernandez, J. G. and Albornoz, L. A. 1988. Morphology of SiO_2 supported Ziegler-Natta catalysts and their produced polyethylenes. In *Transition Metal Catalyzed Polymerizations: Ziegler–Natta and Metathesis Polymerization* eds. R. P. Quirk and R. E. Hoff, p. 512, Cambridge University Press, Cambridge, UK.
36. Ellestad, O. H. 1985. Polymerization of ethylene and propylene by $SiO2/TiCl_4$/Etx $AlCl_3$–x catalysts, *J. Mol. Catal.* 33: 289.
37. Muñoz-Escalona, A., Garcia, H. and Albornoz, A. 1987. Homo and copolymerization of ethylene with highly active catalysts based on TiCl4 and Grignard compounds, *J. Appl. Polym. Sci.* 34: 977.
38. Beach, D. L. and Kissin, Y. V. 1984. Dual functional catalysts for ethylene polymerization to branched polyethylene. I. Evaluation of catalysts systems, *J. Polym. Sci. Polym. Chem.* 22: 3027.
39. Kissin, Y. V. and Beach, D. L. 1986. Dual functional catalysis for ethylene polymerization to branched polyethylene. II. Kinetics of ethylene polymerization with a mixed homogeneous-heterogeneous Ziegler-Natta catalyst system, *J. Polym. Sci. Polym. Chem.* 24: 1069.
40. Kissin, Y. V. and Beach, D. L. 1986. A novel multifunctional catalytic route for branched polyethylene synthesis. In *Studies in Surface Science and Catalysis: Catalytic Polymerization of Olefins*, eds. T. Keii and K. Soga, p. 443, Elsevier, New York.
41. Martin, J. L. 1985. Olefin polymerization, U.S. Patent 4,507,449.
42. Best, S. A. 1986. Polymerization catalyst, production and use, U.S. Patent 4,579,835.
43. Kao, S.-C., Cann, K. J., Karol, F. J., Marcinkowsky, A. E., Godde, M. G. and Theobald, E. H. 1989. Catalyst for regulating the molecular weight distribution of ethylene polymers, U.S. Patent 4,845,067.
44. Roling, P. V., Veazey, R. L. and Aylward, D. E. 1986. A process for polymerizing a monomer charge, Eur. Patent 0,196,830.
45. Veazy, R. L. and Pennington, B. T. 1986. A process for polymerizing a monomer charge, Eur. Patent 0,197,685.
46. Aylward, D. E. 1986. A process for polymerizing a monomer charge, Eur. Patent 0,197,690.
47. Best, S. A. 1986. Polymerization catalyst, production and use, U.S. Patent 4,579,834.
48. Desmond, M. J., Benton, K. C. and Weinert, R. J. 1984. Catalysts for the polymerization of ethylene, U.S. Patent 4,482,639.
49. Zoeckler, M. T. and Karol, F. J. (UCC). 1988. Ethylene polymerization catalyst, Eur. Patent 0,285,137.
50. Spitz, R., Pasquet, V. and Guyot, A. 1987. Linear low density polyethylene prepared in gas phase with bisupported SiO_2/$MgCl_2$ Ziegler–Natta catalysts. In *Transition Metals and Organometallics as Catalysts for Olefin Polymerization*, eds. W. Kaminsky and H. Sinn, p. 405, Springer-Verlag, New York.
51. Cann, K. J., Karol, F. J., Lee, H. H. and Marcinkowsky, A. E. 1990. Ethylene polymerization catalyst, Eur. Patent 0,361,520 A2.
52. Gavens, P. D., Bottrill, M., Kelland, J. W. and McMeeking, J. 1982. Ziegler–Natta catalysis. In *Comprehensive Organometallic Chemistry: The Synthesis, Reaction, and Structures of Organometallic Compounds*, eds. G. Wilkinson, F.G.A. Stone, and E.W. Abel, p. 476, Pergamon Press, Oxford.

53. Simon, A. and Grobler, A. 1980. Some contributions to the characterization of active sites in Mg-supported Ziegler–Natta catalysts, *J. Polym. Sci. Polym. Chem. 18*: 3111.
54. Muñoz-Escalona, A., Hernandez, J. G. and Gallardo, J. A. 1985. Design of supported Ziegler–Natta catalysts using SiO_2 as carrier, *Proceedings of the International Symposium on Future Aspects of Olefin Polymerization*, Tokyo, pp. 123–146.
55. Muñoz-Escalona, A., Martin, A. and Hidalgo, J. 1981. Influence of the carrier characteristics on the catalytic activity of supported $TiCl_4/Al(C_2H_5)2Cl$ Ziegler catalyst in the ethylene polymerization, *Eur. Polym. J. 17*: 367.
56. Muñoz-Escalona, A., Alarcon, C., Albornoz, L. A., Fuentes, A. and Squera, J. A. 1987. Morphological characterization of Ziegler–Natta catalysts and nascent polymers. In *Transition Metals and Organometallics as Catalysts for Olefin Polymerization*, eds. W. Kaminsky and H. Sinn, p. 417, Springer-Verlag, New York.
57. Fanelli, A. J., Burlew, J. V. and Marsh, G. B. 1989. The polymerization of ethylene over TiCl4 supported on alumina aerogels: Low pressure results, *J. Catal. 116*: 318.
58. Damyanov, D., Velikova, M. and Petkov, L. 1979. Catalytic activity of a supported catalyst for ethylene polymerization obtained by modification of vulcasil with titanium tetrachloride vapor, *Eur. Polym. J. 51*: 233.
59. Greco, A., Bertolini, G., Bruzzone, M. and Cesca, S. 1979. Novel binary chlorides containing $TiCl_3$ as components of coordination catalysts for ethylene polymerization. II. Behavior of the resulting catalyst systems, *J. Appl. Polym. Sci. 23*: 1333.
60. Kim, I. I., Kim, J. H. and Woo, S. I. 1990. Kinetic study of ethylene polymerization by highly active silica supported TiCl4/MgCl2 catalysts, *J. Appl. Polym. Sci. 39*: 837.
61. Ittel, S. D., Mulhaupt, R. and Klabunde, U. 1986. Vapor synthesis of high-activity catalysts for olefin polymerization, *J. Polym. Sci. A Polym. Chem. 24*: 3447.
62. Czaja, K. and Szczegot, K. 1986. Physico-chemical characterization of polyethylenes obtained in the presence of different catalysts containing alkoxy ligands. *Polymery-Tworzywa Wieloczasteczkowe 51*: 402.
63. Zucchini, U., Cuffiani, I. and Pennini, G. 1984. Ethylene polymerization by high yield co-milled catalysts. I. Influence of titanium ligands on the activity of the catalyst, *Makromol. Chem. Rapid Commun. 5*: 567.
64. Etherton, B. P. 1987. Polymerization catalyst, production and use, Eur. Patent 0,240,254.
65. Zucchini, U., Saggese, G. A., Cuffiani, I. and Foschini, G. 1988. Progress in tailor-made Ziegler–Natta catalysts for ethylene polymerization. In *Transition Metal Catalyzed Polymerization: Ziegler–Natta and Metathesis Polymerization*, eds. R. P. Quirk and R. E. Hoff, p. 450, Cambridge University Press, Cambridge, UK.
66. Galli, P., Cecchin, G. and Simonazzi, T. 1988. *Proceedings of the 32nd IUPAC Symposium on Frontiers in Macromolecular Science*, Blackwell Scientific, Kyoto, August 1–5, 1989.
67. Olabisi, O., Abdillahi, M. M., Saeed, M. R., Zahoor, M. A. and Al-Sherehy, F. 2001. Catalyst and process for ethylene oligomerization, U.S. Patent 6,184,428 B1.
68. Galli, P. 1994. The breakthrough in catalysis and processes for olefin polymerization: Innovative structures and a strategy in the materials area for the twenty-first century, *Prog. Polym. Sci. 19*: 959.
69. Arndt, M. 1997. New polyolefins. In *Handbook of Thermoplastics*, ed. O. Olabisi, pp. 39–56, Marcel Dekker, New York.
70. Olabisi, O., Atiqullah, M. and Kaminsky, W. 1997. Group 4 metallocenes: Supported and unsupported, *J.M.S.-Revs. Macromol. Chem. Phy. C37(3)*: 519.
71. Sinn, H. and Kaminsky, W. 1980. Ziegler–Natta catalysis, *Adv. Organomet. Chem. 18*: 99.
72. Gupta, V. K., Satish, S. and Bhardwaj, I. S. 1994. Metallocene complexes of group 4 elements in the polymerization of monoolefins, *J.M.S.-Revs. Macromol. Chem. Phys. C34(3)*: 439.
73. Brintzinger, H. H., Fischer, D., Mühlhaupt, R., Rieger, B. and Waymouth, R. 1995. Stereospezifische Olefinpolymerisation mit chiralen Metallocenkatalysatoren, *Angew. Chem. 107*: 1255.
74. Reddy, S. S. and Sivaram, S. 1994. Homogeneous metallocene–methylaluminoxane catalyst systems for ethylene polymerization, *Prog. Polym. Sci. 19*: 309.
75. Huang, J. and Rempel, G. L. 1994. Ziegler–Natta-catalysts for olefin polymerization: Mechanistic insights from metallocene systems, *Prog. Polym. Sci. 19*: 459.
76. Gibson, V. C. and Spitzmesser, S. K. 2003. Advances in non-metallocene olefin polymerization catalysis. *Chem. Rev. 103*: 283.
77. Mason, M. R., Smith, J. M., Bott, S. G. and Barron, A. R. 1993. Hydrolysis of tertBu3 Al: The first structural characterization of aluminoxanes [(R2Al)2O] and (RAlO)n, *J. Am. Chem. Soc. 115*: 4871.

78. Harlan, C. J., Mason, M. R. and Barron, A. R. 1994. tert-Butylaluminum hydroxides and oxides: Structural relationship between alkylalumoxanes and alumina gels, *Organometallics 13*: 2857.
79. Harlan, C. J., Bott, S. G. and Barron, A. R. 1995. Three-coordinate aluminum is not a prerequisite for the catalytic activity in the zirconocene-alumoxane polymerization of ethylene, *J. Am. Chem. Soc. 117*: 6465.
80. Bliemeister, J., Hagendorf, W., Harder, A. et al. 1994. The role of MAO-activators. In *Ziegler Catalysts*, eds. G. Fink, R. Mühlhaupt, and H. H. Brintzinger, p. 57, Springer-Verlag, Berlin.
81. Jordan, R. F. 1991. Chemistry of cationic dicyclopentadienyl group 4 metal–alkyl complexes, *Adv. Organomet. Chem. 32*: 325.
82. Hlatky, G. G., Turner, H. W. and Eckmann, R. R. 1992. Metallacarboranes as labile anions for ionic zirconocene olefin polymerization catalysts, *Organometallics 11*: 1413 (and references therein).
83. Yang, X., Stern, C. L. and Marks, T. J. 1994. Cationic zirconocene olefin polymerization catalysts based on the organo–Lewis acid B(C_6F_5)$_3$. A synthetic, structural, solution dynamics, and polymerization catalytic study, *J. Am. Chem. Soc. 116*: 10015 (and references therein).
84. Ewen, J. A. and Elder, M. J. 1993. Syntheses and models for stereospecific metallocenes, *Makromol. Chem. Macromol. Symp. 66*: 179.
85. Chien, J. C. W., Tsai, W.-M. and Rausch, M. D. 1991. Isospecific polymerization of propylene catalyzed by rac-En(Ind)2ZrMe+ cation, *J. Am. Chem. Soc. 113*: 8570.
86. Hlatky, G. G., Turner, H. W. and Eckmann, R. R. 1989. Ionic, base-free zirconocene catalysts for ethylene polymerization, *J. Am. Chem. Soc. 111*: 2728.
87. Jia, L., Yang, X., Ishihara, A. and Marks, T. J. 1995. Protected (fluoroaryl)borates as effective counteranions for cationic metallocene polymerization catalysts, *Organometallics 14*: 3135.
88. Bachmann, M. and Lancaster, S. J. 1992. Base-free cationic 14-electron alkyls of Ti, Zr, and Hf as polymerization catalysts: A comparison, *J. Organomet. Chem. 434(1)*: C1–C5.
89. Kaminsky, W., Külper, K., Brintzinger, H. H. and Wild, F. R. W. P. 1985. Polymerization von Propen und Buten mit einem chiralen Zirconocen und MAO als Cokatalysator, *Angew. Chem. Int. Ed. Engl. 24*: 507.
90. Kaminsky, W. 1986. Stereoselektive Polymerization von Olefinen mit homogenen, chiralen Ziegler-Natta-Katalysatoren, *Angew. Makromol. Chem. 145(146)*: 149.
91. Wild, F. R. W. P., Wasiucionek, M., Huttner, G. and Brintzinger, H. H. 1985. Synthesis and crystal structure of a chiral ansa-zirconocene derivatives with ethylene-bridged tetrahydroindenyl ligands, *J. Organomet. Chem. 288*: 63.
92. Ewen, J. A. 1984. Mechanisms of stereochemical control in propylene polymerization with soluble group 4B metallocene/MAO catalysts, *J. Am. Chem. Soc. 106*: 6355.
93. Ewen, J. A., Jones, R. L., Razavi, A. and Ferrara, J. D. 1988. Syndiospecific propylene polymerizations with group 4 metallocenes, *J. Am. Chem. Soc. 110*: 6255.
94. Gwen, J. A., Elder, M. J., Jones, L. et al. 1991. Metallocene/PP structural relationships: Indications on polymerization and stereochemical control mechanism, *Makromol. Chem. Macromol. Symp. 48/49*: 253.
95. Shiomura, T., Kohno, M., Inoue, N. et al. 1994. Syndiotactic polypropylene. In *Catalyst Design for Tailor-Made Polyolefins*, eds. K. Soga and M. Terano, p. 327, Kodansha Ltd., Tokyo, and Elsevier, Amsterdam.
96. Arai, N. and Ohkuma, T. 2012. Design of molecular catalysts for achievement of high turnover number in homogeneous hydrogenation, *Chem. Rec. 12*: 284–289.
97. Ittel, S. D., Johnson, L. K. and Brookhart, M. 2000. Late-metal catalysts for ethylene homo- and copolymerization, *Chem. Rev. 100*: 1169–1203.
98. Tian, J. and Coates, G. W. 2000. Development of a diversity-based approach for the discovery of stereoselective polymerization catalysts: Identification of a catalyst for the synthesis of syndiotactic polypropylene, *Angew. Chem., Int. Ed. 39*: 3626–3629.
99. Nakamura, A., Ito, S. and Nozaki, K. 2009. Coordination-insertion copolymerization of fundamental polar monomers, *Chem. Rev. 109*: 5215–5244.
100. Nakayama, Y., Saito, J., Bando, H. and Fujita, T. 2006. MgCl2/R ' nAl(OR)(3-n): An excellent activator/support for transition-metal complexes for olefin polymerization *Chem.–Eur. J. 12*: 7546–7556.
101. Sakuma, A., Weiser, M.-S. and Fujita, T. 2007. Living olefin polymerization and block copolymer formation with FI catalysts, *Polym. J. 39*: 193.
102. Iwashita, A., Chan, M. C. W., Makio, H. and Fujita, T. 2014. Attractive interactions in olefin polymerization mediated by post-metallocene catalysts with fluorine-containing ancillary ligands, *Catal. Sci. Technol., 4*: 599–610. doi:10.1039/c3cy00671a.

103. Arriola, D. J., Carnahan, E. M., Hustad, P. D., Kuhlman, R. L. and Wenzel, T. T. 2006. Catalytic production of olefin block copolymers via chain shuttling polymerization. *Science 312*: 714–719.
104. Choi, K. Y. and Ray, W. H. 1985. Recent development in transition metal catalyzed olefin polymerization. A survey I: Ethylene polymerization, *J. Macromol. Sci. Rev. Macromol. Chem. Phys. C 25*: 1.
105. Lynch, T. J. and Rowatt, R. J. 1980. Stereospecific olefin polymerization process, U.S. Patent 4,182,817.
106. Dumain, A. and Raufast, C. 1985. Device and process for introducing a powder with catalytic activity into a fluidized bed polymerization reactor, Eur. Patent 0,157,584.
107. SRI International. 1983. *Polyolefin Markets and Resin Characteristics*, Vol. 2, Project No. 3948, SRI, Menlo Park, CA.
108. SRI International. 1983. *Polyolefins Production and Conversion Economics*, Vol. 3, Project No. 3948, SRI, Menlo Park, CA.
109. SRI International. 1983. *Polyolefin Production Technology*, Vol. 4, Project No. 3948, SRI, Menlo Park, CA.
110. Short, J. N. 1983. Low pressure ethylene polymerization processes. In *Transition Metal Catalyzed Polymerization*, ed. R. P. Qurik, p. 651, Harwood Academic, New York.
111. Tsubaki, K., Morinaga, H., Matsuo, Y. and Iwabuchu, T. 1979. Two stage polymerization of ethylene employing multicomponent catalyst system, U.K. Patent 2,020,672.
112. Speakman, J. G. and Dows, D. W. 1981. Ethylene polymerization process, U.K. Patent 1,601,861.
113. Bogg, E. A. 1989. Process for olefin polymerization, Eur. Patent 0,307,907.
114. Levine, I. J. and Karol, F. J. 1977. Preparation of low and medium density ethylene polymer in fluid bed reactor, U.S. Patent 4,011,382.
115. Miller, A. R. 1977. Fluidized bed reactor, U.S. Patent 4,003,712.
116. Staub, R. B. 1983. The Unipol process: Technology to serve the world's polyethylene markets, *Proceedings of the Golden Jubilee Conference: Polyethylenes 1933–1983*, London, pp. B5.4.1–B5.4.14.
117. Goeke, G. L., Wagner, B. E. and Karol, F. J. 1981. Impregnated polymerization catalyst, process for preparing, and use for ethylene copolymerization, U.S. Patent 4,302,565.
118. Karol, F. J., Goeke, G. L., Wagner, B. E., Fraser, W. A., Jorgensen, R. J. and Friis, N. 1981. Preparation of ethylene copolymers in fluid bed reactor, U.S. Patent 4,302,566.
119. Brown, G. L., Warner, D. F. and Byon, J. H. 1981. Exothermic polymerization in a vertical fluid bed reactor system contacting cooling means therein and apparatus therefor, U.S. Patent 4,255,542.
120. Karol, F. J. and Jacobson, F. I. 1986. Catalysis and the Unipol process, studies in surface science and catalysis. In *Catalytic Polymerization of Olefins*, Vol. 25, eds. T. Keii and K. Soga, pp. 323–337, Elsevier, New York.
121. Bailly, J. C. A. and Speakman, J. G. 1986. Process for the polymerization of ethylene or the copolymerization of ethylene and alpha olefins in a fluidized bed in the presence of a chromium-based catalyst, Eur. Patent 0,175,532.
122. Durand, D. and Morterol, F. R. 1986. Process for starting up the polymerization of ethylene or copolymerization of ethylene and at least one other alpha-olefin in the gas phase in the presence of a catalyst based on chromium oxide, Eur. Patent 0,179,666.
123. Dumain, A. and Raufast, C. 1989. Process for gas phase polymerization of olefins in a fluidized bed reactor, Eur. Patent 0,301,872.
124. Dumain, A., Havas, L. and Engel, J. 1990. Gas phase alpha-olefins polymerization process in the presence of an activity reactor, Eur. Patent 0,359,444.
125. Trieschmann, H.-G., Ambil, K.-H., Rau, W. and Wisseroth, K. 1977. Method of removing heat from polymerization reactions of monomers in the gas phase, U.S. Patent 4,012,573.
126. Jezl, J. L., Peters, E. F. and Shepard, J. W. 1976. Process for the vapor phase polymerization of monomers in a horizontal, quench-cooled, stirred bed reactor using essentially total off-gas recycle and melt finishing, U.S. Patent 3,965,083.
127. Peters, E. F., Spangler, M. J., Michaels, G. O. and Jezl, J. L. 1976. Vapor phase reactor off-gas recycle system for use in the vapor state polymerization of monomers, U.S. Patent 3,971,768.
128. Jezl, J. L. and Peters, E. F. 1978. Horizontal reactor for the vapor phase polymerization of monomers, U.S. Patent 4,129,701.
129. Montell Technology Company-US5733987. 1992. PCT Patent, 92/21706.
130. Basell Poliolefine Italia S.R.L. [IT/IT]. 2008. PCT Patent, 2008/015113.
131. Covezzi, M. 1995. The Spherilene process: Linear polyethylenes, *Macromol. Symp. 89*: 577.
132. Faucher, J. A. and Reding, F. P. 1965. Relationship between structure and fundamental properties. In *Crystalline Olefin Polymers*, eds. R. A. V. Raff and K. W. Doak, p. 677, InterScience, New York.

133. Dighton, G. L. 1989. Polyethylene. In *Alpha Olefins Applications Handbook*, eds. G. R. Lappin and J. D. Sauer, pp. 63–97, Marcel Dekker, New York.
134. Karol, F. J. and Wu, C. 1975. Ethylene polymerization studies with supported cyclopentadienyl, arene, and allyl chromium catalysts, *J. Polym. Sci. Chem. Ed. 13*: 7.
135. Mukherjee, A. K., Dhara, S. K. and Sharma, P. K. 1985. A new ethylene polymer: linear low density polymer (LLDPE), *Popular Plastics* 15 (October).
136. Mandelkern, L. 1964. *Crystallization of Polymers*, McGraw-Hill, New York.
137. Mitsui Petrochemical Industries. 1974. Non-elastic ethylene copolymers and their preparations, British Patent 1,355,245.
138. Mitsui Toatsu Chemicals. 1976. Ethylene polymerization catalyst, French Patent 2,292,717.
139. Rundlof, L. C. 1985–1986. Polyethylene and ethylene copolymers: Very low density polyethylene. In *Modern Plastics Encyclopedia*, ed. J.E. Homans, p. 59, McGraw-Hill, New York.
140. DuPont. 1970. Homogeneous partly crystalline ethylene copolymers, British Patent 1,209,825.
141. DuPont. 1978. Hydrocarbon interpolymer compositions, U.S. Patent 4,076,698.
142. SRI International. 1985. *Plastic Films*, p. 95, Report No. 159, SRI, Menlo Park, CA.
143. Frank, H. P. 1968. *Polypropylene*, Gordon and Breach, New York.
144. van der Ven, S. 1990. *Polypropylene and Other Polyolefins: Polymerization and Characterization*, Elsevier, Amsterdam.
145. Beck, N. H. 1965. DTA study of heterogeneous nucleation of crystallization in polypropylene, *J. Appl. Polym. Sci. 9*: 2131.
146. Voeks, J. F. 1968. Modification of crystalline structure of crystallizable high polymers, U.S. Patent 3,367,926.
147. Gray, D. J. 1977. Manufacture of glass-clear polypropylene film, *Plast. Rubber Process 2*: 60.
148. Rudin, I. D. 1968. *Poly(1-butene)—Its Preparation and Properties*, Gordon and Breach, New York.
149. Turner-Jones, A. 1963. Poly-1-butylene type II crystalline form, *J. Polym. Sci. B, 8*: 455.
150. Lindergren, R. C. 1970. Polybutylene: Properties of a packaging material, *Polym. Eng. Sci. 10*: 163.
151. Icenogle, R. D. and Klingensmith, G. B. 1987. Characterization of the stereostructure of three poly-1-butenes: Discrimination between intramolecular and intermolecular distributions of defects in stereoregularity, *Macromolecules 20*: 2788.
152. Foglia, A. J. 1969. Polybutylene, its chemistry, properties and applications, *J. Polym. Symp. 11*: 1.
153. Eastman Kodak. 1973. Poly-1-butene resins, U.S. Patent 3,733,373.
154. Macgregor, P. W. 1981. Polybutylene: Prospects for the Eighties, *Soc. Plastics Engineers, Regional Technical Conference*, Processes, Materials, Applications, Houston, TX.
155. Rohn, C. L. 1974. Predicting the properties of poly(butylene-1) blown films, *J. Polym. Sci. Polym. Symp. 46*: 161.
156. Isaacson, R. B. 1964. Properties of semi-crystalline polyolefins: poly(4-methyl-1-pentene), *J. Appl. Polym. Sci. 8*: 2789.
157. Raine, H. C. 1969. TPX—The development of a new plastic, *Appl. Polym. Symp. 11*: 39.
158. Olabisi, O., Robeson, L. M. and Shaw, M. T. 1979. *Polymer–Polymer Miscibility*, Academic Press, New York.
159. Olabisi, O. 1982. Polyblends. In *Kirk-Othmer Encyclopedia of Chemical Technology*, 3rd ed., Vol. 18, p. 443, eds. H.F. Mark et al. John Wiley & Sons, New York.
160. Olabisi, O. 1995. Developments in polymer blends and related technologies: prospects for applications in the ESCWA region, *Proceedings of the Expert-Group Meeting on Techno-Economic Aspects of the Commercial Application of New Materials in the ESCWA Region*, Organized by the UNIDO and ESCWA at Al-Ain, October 1–3.
161. Mai, K. and Xu, J. 1997. Toughening of thermoplastics. In *Handbook of Thermoplastics*, ed. O. Olabisi, Marcel Dekker, New York.
162. Chang, F. C. 1997. Compatibilized thermoplastics blends. In *Handbook of Thermoplastics*, ed. O. Olabisi, Marcel Dekker, New York.
163. Hudson, R. L. 1981. Polypropylene blend compositions, U.S. Patent 4,296,022.
164. Narain, H. 1979. Ethylene-vinyl acetate (EVA) copolymers: Preparation, properties and applications, *J. Sci. Ind. Res. 38*: 25.
165. Imanishi, Y. and Naga, N. 2001. Developments in olefin polymerizations with transition metal catalyst, *Prog. Polym. Sci. 26*: 1147–1198.
166. Chun, P. S. and Swogger, K. W. 2008. Olefin polymer technologies—History and recent progress at the Dow Chemical Company, *Polym. Sci. 33*: 749–819.
167. Qiao, J., Guo, M., Wang, L. et al. 2011. Recent advances in polyolefin technology, *Polym Chem. 2*: 161.

168. Hemmer, J. L. 1994. Recent advances in Exxpol Technology, *Proceedings of the Polyethene World Congress*, Maack Business Services.
169. Kaminsky, W. 1994. Zirconocene catalysts for olefin polymerization, *Catal. Today 20*: 257.
170. Akimoto, A. 1995. New metallocene catalyst for high-temperature polymerization, *Proceedings of the International Congress on Metallocene Polymers Metallocenes '95*, Schotland Business Research Inc., Brussels, p. 439.
171. Kashiwa, N. 1994. Feature of metallocene-catalyzed polyolefins. In *Catalyst Design for Tailor-Made Polyolefins*, eds. K. Soga and M. Terano, p. 381, Kodansha Ltd., Tokyo, and Elsevier, Amsterdam.
172. Kashiwa, N. and Todo, A. 1993. The distinguished features of metallocene-based polyolefins, *Proceedings of the Metallocenes Conference MetCon '93*, Catalysts Consultants Inc., Houston, TX, p. 235.
173. Hosoda, S., Uemura, A., Shigematsu, Y., Yamamoto, I. and Kojima, K. 1994. Structure and properties of ethylene/α-olefin copolymers polymerized with homogenous and heterogeneous catalysts. In *Catalyst Design for Tailor-Made Polyolefins*, eds. K. Soga and M. Terano, p. 365, Kodansha Ltd., Tokyo, and Elsevier, Amsterdam.
174. Stevens, J. C. 1994. Insite catalyst structure/activity relationships for olefin polymerization. In *Catalyst Design for Tailor-Made Polyolefins*, eds. K. Soga and M. Terano, p. 277, Kodansha Ltd., Tokyo, and Elsevier, Amsterdam.
175. Clayfield, T. E. 1995. Tailor-made plastics creations, *Kunststoffe/plast Europe 85(8)*: 1038.
176. Swogger, K. W. 1994. Novel molecular structure opens up new applications for insite based polymers. In *Catalyst Design for Tailor-Made Polyolefins*, eds. K. Soga and M. Terano, p. 284, Kodansha Ltd., Tokyo, and Elsevier, Amsterdam.
177. Langius, L. J. M. 1995. Hochkristallines Polyolefin. *Kunststoffe 85*: 1122.
178. Yi, Q., Wen, X., Dong, J. and Han, C. C. 2007. In-reactor compounding metallocenic isotactic poly(propylene)/nucleation agent compositions by employing the nucleation agent as a catalyst support, *Macromol. React. Eng. 1*: 307–312.
179. Kaminsky, W. and Renner, F. 1993. High melting polypropenes by silica-supported zirconocene catalysts, *Makromol. Chem. Rapid Commun. 14*: 239.
180. Soga, K., Kim, H. J. and Shiono, T. 1994. Polymerization of propene with highly isospecific SiO_2-supported zirconocene catalysts activated by common alkylaluminiums, *Macromol. Chem. Phys. 195*: 3347 (and references therein).
181. Hungenberg, K. D., Kerth, J., Langhauser, F. and Müller, P. 1994. Progress in gas phase polymerization of propylene with supported TiCl4 and metallocene catalysts. In *Catalyst Design for Tailor-Made Polyolefins*, eds. K. Soga and M. Terano, p. 373, Kodansha Ltd., Tokyo, and Elsevier, Amsterdam.
182. Rieger, B., Mu, X., Mallin, D. T., Rausch, M. D. and Chien, J. C. W. 1990. Degree of stereochemical control of rac-[En(Ind)$_2$ZrCl$_2$/MAO] catalyst and properties of anisotactic PPs, *Macromolecules 23*: 3559.
183. Soga, K., Shiono, T., Takemura, S. and Kaminsky, W. 1987. Isotactic polymerization of propene with En(H4Ind)$_2$ZrCl$_2$ combined with MAO, *Makromol. Chem. Rapid Commun. 8*: 305.
184. Hungenberg, K. D., Kerth, J., Langhauser, F., Marczinke, B. and Schlund, R. 1994. Gas phase polymerization of olefins with Ziegler–Natta and metallocene/aluminoxane catalysts: A comparison. In *Ziegler Catalysts*, eds. G. Fink, R. Mühlhaupt, and H. H. Brintzinger, p. 363, Springer-Verlag, Berlin.
185. Antberg, M., Dolle, V., Haftka, S. et al. 1991. Stereospecific polymerizations with metallocene catalysts: Products and technological aspects, *Makromol. Chem. Macromol. Symp. 48/49*: 333.
186. Spaleck, W., Küber, F., Winter, A. et al., 1994. The influence of aromatic substituents on the polymerization behavior of bridged zirconocene catalysts, *Organometallics 13*: 954.
187. Karol, F. J. 1995. Catalysis and the Unipol process in the 1990s, *Macromol. Symp. 89*: 563.
188. Schwager, H. 1992. Polypropylen-Reaktorblends, *Kunststoffe 82*: 499.
189. Langhauser, F., Fischer, D. and Seelert, S. 1995. Metallocene catalysts and the BASF Novolen process: Strong partners for propylene impact copolymer production, *Proceedings of the International Congress on Metallocene Polymers Metallocenes '95*, Schotland Business Research Inc., Brussels, p. 243.
190. Seiler, E. 1995. Polypropylene still has scope for innovation-even in its established fields of application, *Kunststoffe/plast Europe 85(8)*: 30.
191. Arndt, M., Engehausen, R., Kaminsky, W. and Zoumis, K. 1995. Hydrooligomerization of cycloolefins—A view of the microstructure of polynorbornene, *J. Mol. Catal. A Chem. 101*: 171.
192. Arndt, M. and Kaminsky, W. 1995. Microstructure of poly(cycloolefins) produced by metallocene/methylaluminoxane (MAO) catalysts, *Macromol. Symp. 97*: 225.
193. Collins, S. and Kelly, W. M. 1992. The microstructure of polycyclopentene produced by polymerization of cyclopentene with homogeneous Ziegler–Natta catalysts, *Macromolecules 25*: 233.

194. Kelly, W. M., Taylor, N. J. and Collins, S. 1994. Polymerization of cyclopentene using metallocene catalysts: Polymer tacticity and properties, *Macromolecules 27*: 4477.
195. Arndt, M. and Kaminsky, W. 1995. Polymerization with metallocene catalysts. Hydroligomerization and NMR investigations concerning the microstructure of poly(cyclopentenes), *Macromol. Symp. 95*: 167.
196. The Royal Swedish Academy of Sciences. 2005. Advanced Information on the Nobel Prize in Chemistry, available at http://www.nobelprize.org/nobel_prizes/chemistry/laureates/2005/advanced-chemistryprize 2005.pdf (retrieved March 25, 2015).
197. Kaminsky, W. and Noll, A. 1993. Copolymerization of norbornene and ethene with homogenous zirconocenes/methylaluminoxane catalysts, *Polym. Bull. 31*: 175.
198. Arndt, M., Kaminsky, W. and Schupfner, G. U. 1995. Comparison of different metallocenes for olefin polymerization, *Proceedings of the International Congress on Metallocene Polymers Metallocenes '95*, Schotland Business Research Inc., Brussels, p. 403.
199. Chedron, H., Brekner, M.-J. and Osan, F. 1994. Cycloolefin Copolymere: Eine neue Klasse transparenter Thermoplaste, *Angew. Makromol. Chem. 223*: 121.
200. Land, H.-T. and Niederberg, D. 1995. CDs to be produced from COC in future? *Kunststoffe/plast Europe 85(8)*: 13.
201. Land, H.-T. 1995. New cyclic-olefin copolymers from metallocenes, *Proceedings of the International Congress on Metallocene Polymers Metallocenes '95*, Schotland Business Research Inc., Brussels, p. 217.
202. Baekeland, L. H. 1909. Condensation products of phenols and formaldehyde, U.S. Patent 939,966.
203. Jones, F. R. ed. 1992. *Handbook of Polymer-Fibre Composites*, Longman, London.
204. Ancker, F. H., Ashcraft, A. C. and Wagner, E. R. 1983. Synergistic reinforcement promoter systems for filled polymers, U.S. Patent 4,409,342.
205. Ancker, F. H. 1989. Method of preparing mixtures of incompatible hydrocarbon polymers, U.S. Patent 4,873,116.
206. Paul, D. R. and Robeson, L. M. 2008. Polymer nanotechnology: Nanocomposites, *Polymer 49*: 3187–3204.
207. Bergman, J. S., Chen, H., Giannelis, E. P., Thomas, M. G. and Coates, G. W. 1999. Synthesis and characterization of polyolefin-silicate nanocomposites: A catalyst intercalation and *in situ* polymerization approach, *Chem. Commun. 92*: 2179–2180.
208. Morgan, A. B. 2006. Flame retarded polymer layered silicate nanocomposites: A review of commercial and open literature systems, *Polym. Adv. Technol. 17*: 206–217.
209. Kim, D. H., Fasulo, P. D., Rodgers, W. R. and Paul, D. R. 2007. Structure and properties of polypropylene-based nanocomposites: Effect of PP-g-MA to organoclay ratio, *Polymer 48*: 5308.
210. He, F.-A., Zhang, L.-M., Jiang, H.-L., Chen, H.-L., Wu, Q. and Wang, H.-H. 2007. A new strategy to prepare polyethylene nanocomposites by using late-transition metal catalyst supported on AlEt3-activated organoclay, *Comp. Sci. Technol. 67*: 1727–1733.
211. Qian, J. and Guo, C.-Y. 2010. Polyolefin nanocomposites from olefin polymerization between clay layers, *Open Macromol. J. 4*: 1–14.
212. Alexandre, M., Dubois, P., Sun, T., Garces, J. M. and Jerome, R. 2002. Polyethylene-layered silicate nanocomposites prepared by the polymerizing-filling technique: Synthesis and mechanical properties, *Polymer 43*: 2123–2132.
213. Olabisi, O. 1981. Structural foam molding process, U.S. Patent 4,255,368.
214. Olabisi, O. 1981. Process for molding structural web articles, U.S. Patent 4,247,515.
215. Olabisi, O. 1983. Structural web molding, *Plastics Eng. 38(10)*: 24–28.

2 Vinyl Alcohol Polymers*

*Kereilemang Khana Mokwena Nthoiwa,
Carlos A. Diaz, and Yash Chaudhari*

CONTENTS

- 2.1 Introduction 54
- 2.2 Historical Development and Commercialization 55
- 2.3 Global Production and Consumption... 55
- 2.4 Polymerization and Processing Technologies 55
  - 2.4.1 Polymerization of Vinyl Acetate to Polyvinyl Acetate................................... 56
  - 2.4.2 Hydrolysis of Polyvinyl Acetate to PVOH.. 56
  - 2.4.3 Other Methods of Producing PVOH.. 57
  - 2.4.4 Degree of Hydrolysis 57
  - 2.4.5 Degree of Polymerization and Molecular Weight of PVOH 59
- 2.5 Molecular Structure of PVOH 61
  - 2.5.1 Chain Configuration 61
  - 2.5.2 Branching.. 62
  - 2.5.3 End Groups 62
  - 2.5.4 Tacticity 63
- 2.6 Solid State Properties of PVOH.. 65
  - 2.6.1 Crystal Structure of PVOH... 65
  - 2.6.2 Thermal Properties of PVOH 66
  - 2.6.3 Surface Properties of PVOH... 67
  - 2.6.4 Barrier, Mechanical, and Optical Properties of PVOH Films......................... 67
- 2.7 Solution Properties of PVOH.. 69
  - 2.7.1 Solubility of PVOH in Water 69
  - 2.7.2 Other Solvents and Diluents 70
  - 2.7.3 Precipitation of PVOH from Aqueous Solution by Inorganic Salts 70
  - 2.7.4 Viscosity of PVOH Solutions... 71
  - 2.7.5 Gelation in PVOH Solutions 72
- 2.8 Modification of PVOH.. 73
  - 2.8.1 Copolymers... 73
    - 2.8.1.1 Ethylene Vinyl Alcohol Copolymers 73
    - 2.8.1.2 Vinyl Alcohol–Maleic Acid Copolymers 74
    - 2.8.1.3 Cationic Copolymers 74
    - 2.8.1.4 Block and Graft Copolymers 74
    - 2.8.1.5 Others... 75
  - 2.8.2 Chemically Modified PVOH 75
    - 2.8.2.1 Esterification 75
    - 2.8.2.2 Acetalization 76
    - 2.8.2.3 Cross-Linking 76
    - 2.8.2.4 Others... 76

---

* Based in parts on the first-edition chapter on vinyl alcohol polymers.

2.9 Complexes and Blends 77
    2.9.1 Metal Complexes 77
    2.9.2 PVOH–Iodine Complexes..77
    2.9.3 Blends, Composites, and Nanocomposites 77
2.10 Applications...78
    2.10.1 Water-Soluble Films 78
    2.10.2 Packaging Films and Coatings 78
    2.10.3 Blended PVOH Films 78
    2.10.4 Polarizer Films 79
    2.10.5 Molded Products..79
    2.10.6 Fibers 79
    2.10.7 Textile Sizes...80
    2.10.8 Stabilizers/Protective Colloids 80
    2.10.9 Adhesives...80
    2.10.10 Binders...81
    2.10.11 Membranes and Gels 81
    2.10.12 Production of Polyvinyl Butyral..82
References..82

2.1 INTRODUCTION

Polyvinyl alcohol (PVOH) is a synthetic polymer made from the partial or full hydrolysis of polyvinyl acetate (PVA) to remove acetate groups. Addition polymerization of the vinyl alcohol monomer is not possible, because this species does not exist in the free state (Tang and Alavi 2011). PVOH is a hydrophilic polymer that dissolves in water over a wide range of temperatures (Goodship and Jacobs 2009; Hassan and Peppas 2000; Kadajji and Betageri 2011). This polymer is atactic and has the ability to crystallize, as opposed to PVA. Table 2.1 summarizes properties of PVOH in the different forms in which it is available commercially. Properties depend on factors such as molecular weight, degree of hydrolysis, water content, and the presence of additives (i.e., plasticizers) and, to some extent, on the method of manufacture. PVOH is nontoxic with a high barrier to oxygen and aromas, and it is resistant to oil, grease, and organic solvents. There are many different grades of PVOH that are produced by varying the molecular structure during the manufacturing process,

TABLE 2.1
Representative Properties of PVOH

Property	Typical Value	Remarks
Color	Clear, white to yellow	
Density (g/cm^3)	1.19–1.31	
Melting temperature (°C)	180–240	Increases with degree of hydrolysis
Boiling point (°C)	228	
Degradation temperature (°C)	180	Unplasticized PVOH
Glass transition temperature (°C)	75–85	Dry, increases with degree of hydrolysis
Degree of crystallinity (%)	23.4	Partially hydrolyzed
	46.5	Fully hydrolyzed

Source: Goodship, V. and D. Jacobs, *Polyvinyl Alcohol: Materials, Processing and Applications (Rapra Review Reports)*, Smithers Rapra Press: Shrewsbury, Shropshire, England, 2009; Marten, F.L., Vinyl Alcohol Polymers, in *Kirk-Othmer Encyclopedia of Chemical Technology*, 5th ed., edited by Arza Seidel and Mickey Bickford, 591–627, Hoboken, NJ: John Wiley & Sons, 2004; Kumaki, Y. et al., Enhanced Polyvinyl Alcohol as a Barrier Paper Coating for Food Packaging, in *TAPPI PaperCon.*, Nashville, Tennessee, 2014.

resulting in a diverse array of commercial uses in medical (e.g., hydrogels for tissue scaffolds, lenses, corneal prostheses, drug delivery devices), packaging, paper, textiles, and other applications. The first practical use of PVOH in large quantities was for warp sizing of rayon and other synthetic fibers. It has been used as an emulsifier or a stabilizer in emulsion polymerization and as a thickening agent for aqueous dispersions (Sakurada 1985).

2.2 HISTORICAL DEVELOPMENT AND COMMERCIALIZATION

The discovery and chemistry of PVOH is closely linked to that of PVA. In 1912, Fritz Klatte discovered that vinyl acetate (VAc) is formed as a by-product during the manufacture of ethylidenediacetate from acetylene and acetic acid (Sakurada 1985). Later in 1924, W.O. Herrmann and W. Haehnel at Wacker Chemie in Germany added an alkali to a clear alcoholic solution of PVA to saponify the polymeric ester. Based on their experience with the saponification of monomeric VAc, Hermann and Haehnel expected to obtain a resinous precipitate, but instead, they found an ivory-colored water-soluble PVOH (Sakurada 1985; Sakurada and Okaya 1998). PVOH was also studied independently by H. Staudinger and his coworkers, and in 1926, they reported the reversible change between PVOH and PVA via esterification and saponification as strong evidence of the theory of the macromolecule.

2.3 GLOBAL PRODUCTION AND CONSUMPTION

In the early days, production of both PVA and alcohol was carried out in Japan, Germany, the United States, France, and the United Kingdom. Today, the market for PVOH is global and covers the major geographic regions, including Asia-Pacific (dominated by China and Japan), North America, South America, and Eastern and Western Europe. Worldwide consumption of PVOH has been estimated at 1,000,000 tons annually as of 2010 (Magdum et al. 2013). Demand is forecasted to grow at an average annual rate of around 3.5% from 2012 to 2017 (IHS Chemical 2013).

Kuraray Co. is the largest single manufacturer of PVOH, accounting for 16% of all world capacity and offering over 50 different grades of this material (Goodship and Jacobs 2009). Kuraray acquired the Germany-based Clariant GmbH in 2001 and the US-based MonoSol, LLC, in 2012 to expand its VAc business in the global market. As of 2012, Kuraray had production locations in Japan, Singapore, and Germany, with a total production capacity of about 235,000 metric tons per year for PVOH resins under the trade names of Mowiol, Mowiflex, and POVAL. Kuraray also produces PVOH films under trademarks Mowiflex TC for common thermoplastic processes such as blown-film extrusion and injection molding, and EXCEVAL for water-soluble films. MonoSol, LLC, is the largest supplier of water soluble polymer films, compounds, and solutions under several trademarks with production facilities in the United States and the United Kingdom. Other major manufacturers of resins and water-soluble and polarizer films include Sekisui Chemicals Co. (SELVO), DuPont (ELVANOL), Nippon Gohsei (GOHSENOL, BOVLON), Chan Chun Petrochemical Co., Aicello Chemicals (SOLUBLON), Jiangmen Proudly Water Soluble Plastic Co. Ltd., and Changzhou Water Soluble Co. Ltd.

2.4 POLYMERIZATION AND PROCESSING TECHNOLOGIES

The basic repeating unit (constitutional unit) of PVOH is vinyl alcohol. However, the vinyl alcohol monomer is unstable, and it cannot be isolated or obtained in high concentrations. It is well known that VAc is obtained by the addition of acetic acid to acetylene. It is, therefore, quite natural to assume that vinyl alcohol would be formed by the reaction of water with acetylene. This reaction, however, produces acetaldehyde and not vinyl alcohol. It is very probable that vinyl alcohol is formed at some point during the reaction, but it changes instantaneously to acetaldehyde (Sakurada and Okaya 1998). Because of this unstable nature of the vinyl alcohol monomer, the most common

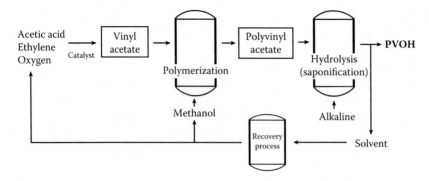

FIGURE 2.1 Overview of PVOH polymerization process.

method to synthesize PVOH is through a two-step process involving free-radical polymerization of VAc followed by a hydrolysis process to convert the acetate functional groups into hydroxyl groups. Figure 2.1 shows a general process scheme for commercial production of PVOH starting from the preparation of VAc.

2.4.1 Polymerization of Vinyl Acetate to Polyvinyl Acetate

The polymerization of VAc to PVA takes place via free-radical chain polymerization in an organic solvent such as methanol (Marten and Zvanut 1992a). The polymerization process includes the typical steps of initiation, chain propagation, possible chain transfer and reinitiation, and finally, termination, either by disproportionation or combination. Details of free-radical polymerization of PVOH have been extensively described by various authors, including Marten and Zvanut (1992a) and Sakurada and Okaya (1998). The polymerization is usually initiated by chemical initiators, by ultraviolet (UV) light, or by ionizing radiation (Tesoro 1986). Solution polymerization is mostly used to produce PVA for PVOH manufacture, and the most common initiators in solution polymerization are peroxides and azo compounds (Marten and Zvanut 1992a). Emulsion polymerization can be utilized to produce PVOH products for applications such as adhesives. The common initiators for emulsion polymerization are hydroperoxide, persulfate, and redox systems.

2.4.2 Hydrolysis of Polyvinyl Acetate to PVOH

PVOH is obtained from hydrolysis of PVA by replacing acetate groups with hydroxyl groups in the polymer chain using either a base or acid catalyst. Hermann and Haehnel, who were the first to produce PVOH from PVA, obtained it by adding an ethanol solution containing potassium hydroxide to a solution of PVA at room temperature (Sakurada 1985). The alkaline hydrolysis or alcoholysis process, which is better described as transesterification or saponification rather than a simple hydrolysis, is represented in the reaction scheme shown in Figure 2.2.

In industry, methanol is used as the solvent for the hydrolysis of PVA. Methanol acts as a chain-transfer agent and, together with the type and quality of initiator, enables the molar masses to be adjusted to various values. It also serves to remove the heat produced in polymerization by evaporative cooling. Methyl acetate is obtained after saponification, as shown in Figure 2.2. For economic reasons, the methanol must be recovered and recycled to the process. A small amount of acid or base is added in catalytic amounts to the methanol solvent to promote an ester exchange reaction between the PVA and methanol, converting the acetate groups of the PVA to hydroxyl groups to generate a reaction to form PVOH (Marten and Zvanut 1992b). Alkaline catalysts such as sodium or potassium hydroxide/methoxide/ethoxide, calcium hydroxide, and sodium carbonate, and acid catalysts such as sulfuric or hydrochloric acid can be used.

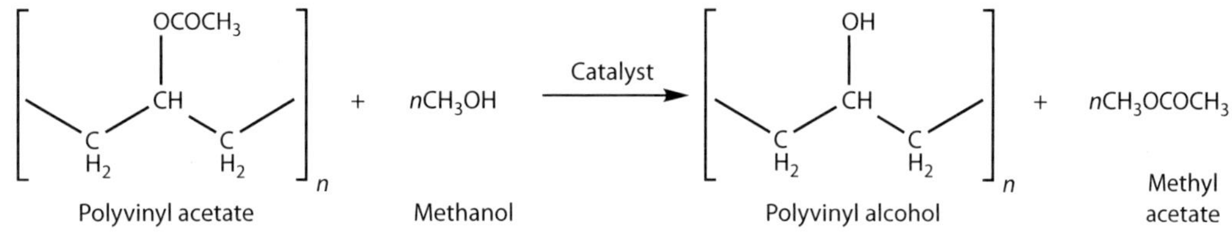

Vinyl Alcohol Polymers

FIGURE 2.2 Transesterification of polyvinyl acetate to fully hydrolyzed PVOH.

2.4.3 OTHER METHODS OF PRODUCING PVOH

While industrial production of PVOH mainly follows the polymerization of PVA, there have been some developments for synthesizing the polymer from other monomers, resulting in PVOH of varying tacticities and properties. Some of the monomers that have been used for synthesis of PVOH include vinyl formate, vinyl benzoate, vinyl ethers, *t*-butyl vinyl ether, vinyl pivalate, and benzyl ether. Yamada and coworkers (1998) carried out the polymerization of bulky vinyl esters, including vinyl trifluoroacetate, vinyl pivalate, vinyl diphenylacetate, and vinyl 2,2-bis(trifluoromethyl) propionate to produce PVOH. In another investigation, the saponification of polymers obtained from vinyl ester polymerization using fluoroalcohols as solvents produced syndiotactic PVOH (Nagara et al. 2001). Using toluene as a solvent at −78°C, Ohgi and Sato (1999) obtained an isotactic PVOH from poly(*tert*-butyl vinyl ether) polymerized with $BF_3 \cdot OEt_2$ catalyst and other derivatives. The hydrolysis was made with HBr at 0°C in methylene chloride or toluene. Heterotactic PVOH with similar syndiotactic fractions to the commercial one was synthesized from poly(vinyl dimethylphenylsilyl ether) hydrolysis by Palos and coworkers (2005). Lin and coworkers (2006) produced ultrahigh-molecular-weight PVOH using a two-stage photoconcentrated emulsion polymerization of VAc. However, to our knowledge, none of these alternative paths are used commercially.

2.4.4 DEGREE OF HYDROLYSIS

During transesterification of PVA to PVOH, depending on the preparation conditions, it is possible to prepare products with differing acetyl group contents. The degree of hydrolysis in the finished PVOH is determined by the molar percentage of acetate groups in the PVA feedstock that are replaced by hydroxyl groups. The residual acetyl group content can be adjusted by varying the concentration of the catalyst. Commercial PVOH grades are generally classified as fully hydrolyzed, intermediate, or partially hydrolyzed, as shown in Table 2.2.

TABLE 2.2
Degree of Hydrolysis of Commercial PVOH Grades

Grade	Degree of Hydrolysis (mol.%)	Residual Acetyl Groups (mol.%)
Fully hydrolyzed	98–99	0.6–1.5
Intermediate	93–97	2.7–6.9
Partially hydrolyzed	85–90	10–15

Source: Data from Clariant GmbH, Mowiol Polyvinyl Alcohol, 1999, available at http://www.cbm.uam.es/mkfactory.esdomain/webs/cbmso/images/TextBox/mowiol.pdf; Goodship, V. and D. Jacobs, *Polyvinyl Alcohol: Materials, Processing and Applications (Rapra Review Reports)*, Smithers Rapra Press: Shrewsbury, Shropshire, England, 2009.

Figure 2.3 represents a scheme for preparation of partially and fully hydrolyzed PVOH from saponification of PVA and from reacetylation of PVOH. Fully hydrolyzed grades of PVOH contain less than 1.5 mol.% acetyl groups and have an abundance of hydroxyl groups, which contributes to strong intramolecular and intermolecular hydrogen bonding (Marten 2003). Partially hydrolyzed PVOH contains about 10–15 mol.% acetyl groups. The distribution of residual acetyl groups in partially hydrolyzed PVOH is critically affected by the hydrolysis procedure adopted. Partially hydrolyzed PVOH, which was prepared by saponification of PVA, has the VAc and vinyl alcohol units distributed in a block structure arrangement along a polymer chain, i.e., step 1 in Figure 2.3 (Erbil 2000; Hayashi 1996; Matsuzawa 1997). The "blockiness" of the copolymer can be increased by adding a proportion of aggregation agents of PVOH such as benzene, methyl acetate, and acetone to the alkaline hydrolysis solvent. Partially acetylated PVOH, on the other hand, results in a random distribution of the acetyl groups, i.e., step 3 in Figure 2.3.

The sequence of the VAc units in the copolymers can be quantitatively evaluated using infrared (IR) spectroscopy, melting point, and iodine color reaction (Hayashi 1996). In the ^{13}C nuclear magnetic resonance (NMR) spectra of copolymers, the block characters and the mean run lengths of the VAc and vinyl alcohol units in a copolymer are quantitatively calculated from the three methylene carbon lines assigned to the three diad sequences *OH–OH*, *OH–OAc*, and *OAc–OAc* (Van der Velden and Beulen 1982; Wu and Sheer 1977). The number representing the average

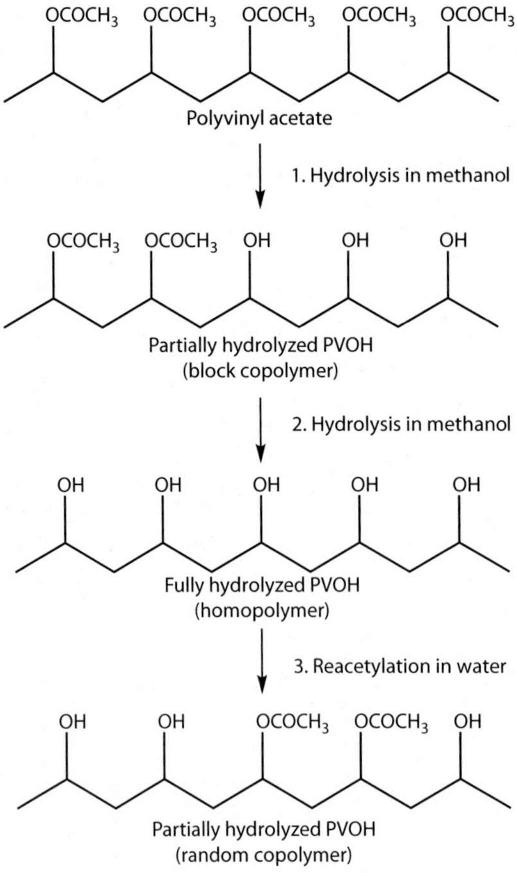

FIGURE 2.3 Scheme for preparation of fully and partially hydrolyzed PVOH.

sequence length of the VAc unit, L_{OAc}, can be determined using ^{13}C NMR spectra and the following equation:

$$L_{OAc} = \frac{2(OAc)}{(OH-OAc)} \quad (2.1)$$

where *OAc* is the mole fraction of the VAc unit and *OH–OAc* the diad fraction in the sum of the area due to *OH–OH*, *OH–OAc*, and *OAc–OAc* diads estimated from the absorption due to methylene carbons (Bugada and Rudin 1984). The iodine color reaction is a simple method to distinguish the distribution of VAc units in a partially hydrolyzed PVOH. An aqueous iodine–iodine solution has no response upon an aqueous dilute solution of PVOH and reacetylated PVOH, but it shows a red color reaction with that of partially hydrolyzed PVOH, as well as that with PVA (Hayashi 1996).

2.4.5 Degree of Polymerization and Molecular Weight of PVOH

The degree of polymerization (DP) of commercial PVOH ranges widely from below 300 to over 4000, depending on manufacturers and end-use applications. The DP for PVOH can be calculated from the molecular weight (M_W) by taking into account the relative amounts of acetate monomers in partially hydrolyzed PVOH according to the following equation (Clariant GmbH 1999):

$$DP = \frac{M_W}{(86 - 0.42 \cdot \text{degree of hydrolysis})} \quad (2.2)$$

The average DP of a polymer is estimated by dividing the rate of polymerization by the rate of production of the stable polymer molecule. Stable polymer molecules of PVOH are formed by termination, monomer transfer, and solvent transfer. According to Sakurada and Okaya (1998), the rate of polymerization is directly proportional to the concentration of monomer and to the square root of the concentration of the initiator. The average DP of PVOH is influenced by the conditions during polymerization, including the type and concentration of the solvent and initiator (Sakurada 1985). The DP of PVOH is lower than that of the parent VAc and was shown to be constant at about 2000 from 0% to 80% conversion, which is the same as the initial DP of the VAc (Sakurada and Okaya 1998). The drop of the DP of PVA by saponification is due to the cutting of branches at the acetyl groups of PVA (Wheeler et al. 1952). Sakurada and Okaya explained that the number of branches by transfer increases in proportion to conversion; hence, the DP of PVA increases with conversion. However, in the saponification of PVA to PVOH, all acetyl groups, including the budding points of the branches, are converted to hydroxyl groups during deacetylation, and the branches are separated from the trunks (Sakurada and Okaya 1998).

The average molecular weight and its distribution depend on the method and conditions of the polymerization of PVA. Chain-transfer reactions control the molecular weight of precursors generally obtained by radical polymerization of VAc. Branching by chain transfer was found to increase the proportion of both high- and low-molecular-weight components in the system. Bulk and suspension polymerization broaden the distribution of molecular weight of both PVA and PVOH by branching. Table 2.3 shows four PVOH grades based on their nominal number average molecular weights (M_n), DP, and aqueous solution viscosity of 4 wt.% solution at 20°C measured with a Brookfield viscometer.

Absolute molecular weights of PVOH can be determined from dilute aqueous solutions using osmotic pressure and light-scattering techniques. A widely used indirect method is the measurement of the limiting viscosity or the intrinsic viscosity [η], which yields an average molecular weight. To obtain molecular weight from the limiting viscosity number, the Mark–Houwink equation, which is

TABLE 2.3
DP, Number Average Molecular Weight, and Viscosity of Commercial PVOH Grades

Grades	DP	M_n	Solution Viscosity (cPs)
Low	550	22,000–27,000	4–7
Intermediate	900	35,000–40,000	13–18
Medium	1500	75,000–82,000	26–30
High	2200	89,000–100,000	48–65

Source: Data from Marten, F.L., Vinyl Alcohol Polymers, in *Encyclopedia of Polymer Science and Technology*, vol. 8, 399–437, Hoboken, NJ: John Wiley & Sons, 2003; Chang Chun Petrochemical, PVA Polyvinyl Alcohol, 2009, available at http://www.perrychem.com/files/PVA_English_2009_.pdf; Erbil, H.Y., *Vinyl Acetate Emulsion Polymerization and Copolymerization with Acrylic Monomers*. Boca Raton, FL: CRC Press, 2000; Nagy, D.J., Size Exclusion of Polyvinyl Alcohol and Polyvinyl Acetate, in *Handbook of Size Exclusion Chromatography and Related Techniques*, edited by Chi-San Wu, 269–295, New York: Marcel Dekker, 2003.

specific for polymer solvents, is commonly used. Several Sakurada–Houwink–Mark–Kuhn viscosity equations were presented using the absolute molecular weights (Brandrup and Immergut 1989). The molecular weight of poorly fractionated PVOH determined by the osmotic pressure technique is presented by following equation:

$$[\eta] = 6.66 \times 10^{-4} M^{0.64} = 7.50 \times 10^{-3} DP^{0.64} \text{ (dl/g, 30°C)} \tag{2.3}$$

where M is the viscosity average molecular weight. On the other hand, the equation for the homogeneous polymers is given as follows (Matsuzawa 1997):

$$[\eta] = 4.53 \times 10^{-4} M^{0.64} \text{ (dl/g, 30°C)} \tag{2.4}$$

The intrinsic viscosity obtained by extrapolating reduced viscosity to the infinite dilution is nearly equal to $(\ln \eta_{rel})/C$, where η_{rel} is the relative viscosity and C the concentration (Matsuzawa 1997). Since the aqueous solutions of syndiotactic PVOH are not stable at temperatures below 100°C, the molecular weight is normally determined for the reacetylated PVA. Nagy used size-exclusion chromatography (SEC) to determine Mark–Houwink equation for fully hydrolyzed PVOH and found constants comparable to those obtained by universal calibration viscometry and those from low-angle laser light scattering (Nagy 1993).

PVOH prepared by free-radical polymerization of VAc generally has broad molecular weight distribution with a typical polydispersity index of approximately 2 (Forder et al. 1995; Nagy 2003). Molecular weight distribution can be determined using several techniques such as osmometry, light scattering of aqueous solutions, viscometry, gel permeation, or SEC. The molecular weight distribution of PVOH has been obtained by fractional precipitation using a water–propanol or water–acetone system (Matsuzawa 1997). Aqueous SEC coupled with low-angle laser light scattering was used to characterize the molecular weight distributions of PVOH using the columns packed with hydrophilic gel particles and 0.05 N $NaNO_3$ solution as a mobile phase (Nagy 1986).

Both molecular weight and degree of hydrolysis greatly influence the overall behavior of PVOH and sometimes determine the specific end-use applications. Figure 2.4 shows the general trends of the effects of molecular weight (DP) and degree of hydrolysis on several properties of PVOH. Some of these properties are discussed in detail in subsequent sections. In summary, at a constant degree of hydrolysis, an increase of the molecular weight increases the solution viscosity in water; increases the tensile strength, water and solvent resistance, and adhesive strength of the resultant PVOH film; and decreases the ease of solvation in water. At constant molecular weight, the increase

Vinyl Alcohol Polymers

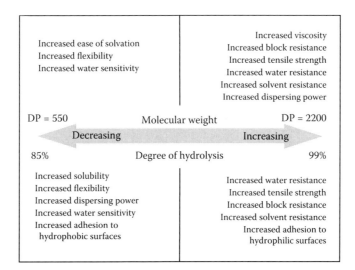

FIGURE 2.4 Effect of molecular weight and degree of hydrolysis on the properties of PVOH. (Adapted from SEKISUI, Selvol™ Polyvinyl Alcohol in Emulsion Polymerization, 2011, available at http://www.sekisui-sc.com/products/selvol/index.html; Marten, F.L., Vinyl Alcohol Polymers, in *Encyclopedia of Polymer Science and Technology*, vol. 8, 399–437, Hoboken, NJ: John Wiley & Sons, 2003.)

in the degree of hydrolysis increases water and solvent resistance, crystallinity, tensile and adhesive strength, and the ability to adhere to hydrophilic surfaces of the resultant PVOH films. Properties such as water solubility are more sensitive to the degree of hydrolysis, whereas solution viscosity is more sensitive to molecular weight (or DP).

2.5 MOLECULAR STRUCTURE OF PVOH

2.5.1 Chain Configuration

It has been shown that industrial polymerization processes for PVOH predominantly produce polymer chains bearing the substituents on alternate carbon atoms with successive monomer units oriented in the same direction, thus creating a head-to-tail or the 1,3-glycol structure (Sakurada and Okaya 1998). This type of structural arrangement reflects the selectivity of monomer addition to the free-radical chain polymerization. Studies by Marvel and Denoon Jr. (1938) indeed showed that PVOH has a 1,3-glycol chain structure as has been suggested originally by Staudinger and coworkers, and Hermann and Haehnel. Despite the predominance of the 1,3-glycol structure, evidence available in the literature indicates that the head-to-head or tail-to-tail structures are also possible, in which a pair of substituents alternate regularly on consecutive carbon atoms. These arrangements yield a 1,2-glycol structure. Commercial PVOH product usually contains about 1–2% of head-to-head or tail-to-tail configurations with 1,2-glycol units, an amount that is considered to have insignificant influence on the physical properties of PVOH (Murahashi 1967; Sakurada and Okaya 1998). PVOH from PVA prepared at 60°C showed 1.5 mol.% of 1,2-glycol linkages, a value determined from the decrease in the molecular weight after cleavage of PVOH by periodic acid (Matsuzawa 1997). Syndiotactic PVOH derived from polymerization of poly(vinyl trifluoroacetate) at 60°C has about 1.0 mol.% of 1,2-glycol linkages. Formation of 1,2-glycol linkages has been attributed to occasional abnormal addition of monomer in the chain-growth phase of polymerization (Flory and Leutner 1948). The most important factor that affects the content of 1,2-glycol linkages is the temperature of polymerization, which has been found to be independent of molecular weight, polymerization solvent, or method of polymerization (Marten 2003). It is also possible that the 1,2-glycol linkage is formed by the coupling reaction in the termination, although the contribution of stable polymer molecules produced by termination to the amount of 1,2-glycol linkages is small.

2.5.2 Branching

During polymerization of many monomers, side reactions such as a transfer to polymer can occur, inadvertently leading to branch points. Branching can also be introduced deliberately during polymerization with the choice of the proper initiator or the addition of a polyfunctional monomer or agent. Both long-chain and short-chain branches have been reported in PVOH. The possible mechanisms for the formation of branches include by chain transfer to the acetoxy group and through chain transfer to the main chain with branches during polymerization (Marten and Zvanut 1992b). Branches generated through the acetoxy group are hydrolyzable, and these result in an alcohol group in the main chain and a methyl acetate ester end group on the branch. Those connected to the main chain are nonhydrolyzable. The number of short-chain branches in PVOH is identical to the molar percentage of short-chain branches generated during the formation of PVA. Short-chain branches are formed during polymerization of VAc when the growing radical "backbites" through a six-membered ring intermediate, yielding a four-carbon atom short side chain. Short-chain branches determined by NMR for commercial PVOHs are in the range of 0.6–7 per 1000 carbon atoms (Matsuzawa 1997). The extent of long-chain branches was shown by an SEC coupled with a low-angle light-scattering detector system to be 0–2 per 1000 carbon atoms for commercial PVOHs. The branching points are considered to be methine carbons in the main chains.

2.5.3 End Groups

Various kinds of end groups exist in PVOH, which can greatly influence the color and heat stability of the polymer. These groups are incorporated during the polymerization process by chain-transfer, initiation, or termination reactions (Marten and Zvanut 1992a). The PVOH molecules obtained by bulk polymerization have one carboxyl group at each end. During polymerization, chain transfer to aldehyde introduces the carbonyl groups in the following manner (Matsuzawa 1997):

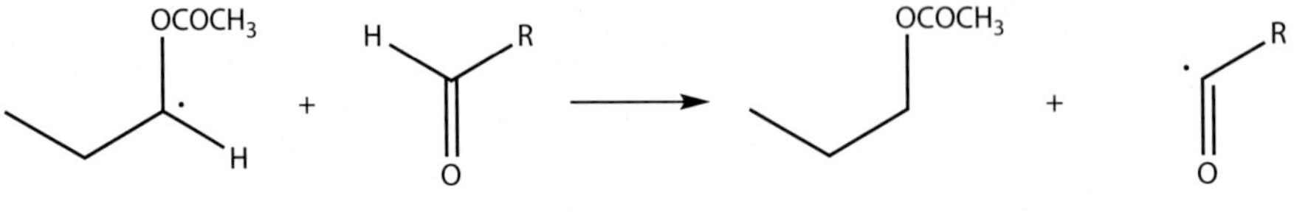

The carbonyl groups lead to the formation of double bonds during heating. Ketone groups are incorporated by chain transfer to aldehydes during the polymerization of VAc. The formation of acid end groups or corresponding methyl esters is a result of long-chain branching. These groups have an influence on the properties of PVOH as an emulsifier. Hydrophobic end groups are normally due to the incorporation of initiator residues. However, end groups incorporated in this fashion are normally limited in number and length, with some influence mainly in low-molecular-weight polymers (Noro 1973). Chain transfer to the terminal acetyl methyl group of PVA would lead to an end group that is VAc ester, and alcoholysis would convert it to a carboxylic acid end group and/or a carboxylic acid sodium salt end group. The chain-transfer reaction is an abstraction of the hydrogen on the acetyl group of VAc (Amiya and Uetsuki 1985). The growing radical is transferred to the VAc monomer, which then reinitiates the polymerization. The reaction results in an unsaturated end group, which is still polymerizable. The inclusion of this terminal end group in a growing chain leads to a trifunctional branch point with the incorporation of the entire polymer molecule as a long-chain branch. Aldehyde end groups are due to termination by disproportionation, which leads to a saturated and an unsaturated end group (Noro 1973). PVOHs with hydrophobic end groups are prepared from PVA, and the end groups result from the chain-transfer reaction in polymerization. The

Vinyl Alcohol Polymers

alkylthiols, fluorocarbons containing thiol groups, and a thiol containing silanol compound are used as chain-transfer agents (Matsuzawa 1997). Since the chain-transfer constant to *n*-dodecyl mercaptan in the polymerization of VAc at 60°C is 22, low-molecular-weight PVOH with an *n*-dodecyl group at the end can easily be prepared. The end group structure is as follows:

2.5.4 Tacticity

The three stereo regularities of isotactic, syndiotactic, and atactic (or heterotactic) can be obtained depending on the type of vinyl ester precursor monomer used to produce the PVOH. The conventional polymerization of VAc and subsequent hydrolysis of PVA results mainly in atactic configuration with substituents randomly oriented on either side of the polymer backbone. Various vinyl monomer precursors are used to obtain a variety of stereo regularities, including vinyl pivalate, *t*-butyl vinyl ether, benzyl vinyl ether, vinyl trifluoroacetate, vinyl formate, and vinyl trimethylsil ether (Choi et al. 2000; Forder et al. 1995; Fukae et al. 1990, 1997; Lyoo et al. 2001; Yamamoto et al. 1992). In general, the vinyl ether monomers yield isotactic-rich polymers, while the vinyl esters yield syndiotactic-rich polymers (Lyoo et al. 2003). High-molecular-weight PVOH with high syndiotacticity was successfully prepared from photoemulsion polymerization and saponification of vinyl pivalate.

Various NMR techniques are used to identify the contents of isotactic, syndiotactic, and heterotactic triads. The ratios of syndiotactic, isotactic, and heterotactic triads were determined as 39.3%, 11.3%, and 49.4%, respectively, from NMR spectroscopy (Fukae et al. 1990). DeMember and coworkers (1972) used ^{1}H NMR to estimate the contents of isotactic, syndiotactic, and heterotactic triads (mm, rr, mr) in d_6-dimethyl sulfoxide (d_6-DMSO). Figure 2.5 illustrates the spectra of isotactic (i), syndiotactic (s), and atactic (a) PVOHs. Peaks due to methine protons corresponding to mm, mr, and rr triads are recognized at δ = 4.7, 4.3, and 3.9 ppm, respectively. The content of pentads is estimated from the split absorptions due to methine carbons in ^{13}C NMR spectra using d_6-DMSO as

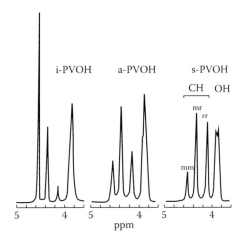

FIGURE 2.5 ^{1}H NMR spectrum of i-PVOH (r% = 20), a-PVOH (r% = 55), and s-PVOH (r% = 66). (Adapted from Matsuzawa, S., Vinyl Alcohol Polymers, in *Handbook of Thermoplastics*, edited by Olagoke Olabisi, 269–289, New York: Marcel Dekker, 1997.)

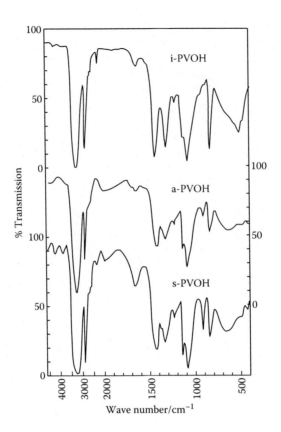

FIGURE 2.6 IR spectrum of i-PVOH (r% = 20), a-PVOH (r% = 55), and s-PVOH (r% = 66). (Adapted from Matsuzawa, S., Vinyl Alcohol Polymers, in *Handbook of Thermoplastics*, edited by Olagoke Olabisi, 269–289, New York: Marcel Dekker, 1997.)

a solvent (Ovenall 1984). The diad contents are estimated from IR spectra. Figure 2.6 illustrates the IR spectra of i-PVOH, s-PVOH, and a-PVOHs, where the absorptions at 916 and 849 cm^{-1} are due to the syndiotactic and the isotactic diads (r, m), respectively. The content of the racemic diad (r) is estimated from the following equation:

$$r\% = 72.4(D_{916}/D_{869})^{0.43} \tag{2.5}$$

where D is the optical density and equal to 20%, 55%, and 60% for i-PVOH, a-PVOH, and s-PVOH, respectively (Murahashi et al. 1966).

The effect of the tacticity of PVOH on its physical properties is known to be significant. An increase in the syndiotacticity of PVOH has been reported to affect the physical properties, such as solubility in solvents, melting temperature, heat resistance, tensile strength, and the modulus (Choi et al. 2000; Fukae et al. 1990, 1997; Sakurada 1985; Yamamoto et al. 1992). A melting point of 248°C was observed for syndiotactic PVOH obtained from vinyl pivalate on differential scanning calorimetry (DSC) thermograms as compared to 229°C for high-molecular-weight PVOH derived from VAc and 231°C for the common commercial free-radical polymerization (Fukae et al. 1990). It was noted that the melting point becomes higher as syndiotacticity increases. The heat of fusion estimated from endothermic peaks on DSC thermograms was approximately the same (Fukae et al. 1997).

2.6 SOLID STATE PROPERTIES OF PVOH

2.6.1 CRYSTAL STRUCTURE OF PVOH

The highly crystalline structure of PVOH has been noted in the early years of its commercial production. The PVA from which PVOH is derived has not been synthesized in crystalline form, because the presence of acetate side groups does not facilitate crystallization of the polymer. In 1935, Halle and Fuller presented an x-ray fiber diagram for a stretched PVOH film suggesting the crystalline nature of the material with a fiber period estimated to be 2.5 Å (0.25 nm) (Matsuzawa 1997). In 1948, Bunn suggested that the random steric pattern of substituents made PVA noncrystallizable but that PVOH, with the smaller hydroxyl groups, was crystalline regardless of stereoregularity. Bunn proposed a monoclinic unit cell of PVOH with a = 0.781 nm, b = 0.252 nm (chain axis), c = 0.551 nm, and β = 91°. Figure 2.7 shows the c and b projections (Matsuzawa 1997). The lattice model indicated an atactic structure in which hydroxyl groups are randomly arranged in left- and right-hand positions (Bunn 1948). The molecular repeat distance of 2.52 Å in PVOH indicates that the chain has a simple plane zigzag configuration.

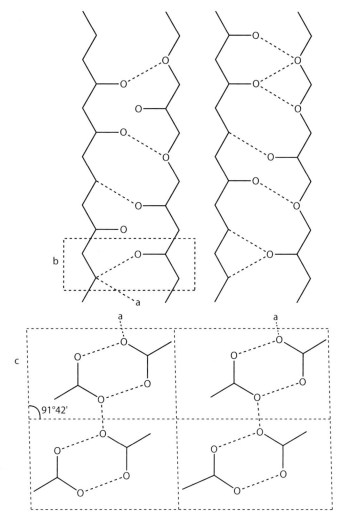

FIGURE 2.7 Crystal model of PVOH. (Adapted from Matsuzawa, S., Vinyl Alcohol Polymers, in *Handbook of Thermoplastics*, edited by Olagoke Olabisi, 269–289, New York: Marcel Dekker, 1997.)

2.6.2 THERMAL PROPERTIES OF PVOH

On heating, PVOH undergoes phase transitions such as glass transition and melting and eventually decomposes. Literature values of these transitions vary depending on the degree of hydrolysis, DP, tacticity, water content, etc., but could also be due to the methods used to measure them. Reported values of glass transition temperature of dry PVOH lie between 70°C and 90°C, depending on the thermal history. PVOH samples that have no thermal history above 130°C have a glass transition temperature around 70°C, while those samples treated at 230°C showed values between 85°C and 90°C (Sakurada and Okaya 1998). The glass transition temperature of wet PVOH decreases with the concentration of water according to the following Fox relation:

$$\frac{1}{T_g} = \frac{1-C}{T_{gp}} + \frac{C}{T_{gw}} \qquad (2.6)$$

where T_{gp} and T_{gw} are the glass transition temperatures of dry PVOH and pure water, respectively; C is the concentration of water.

Rault and coworkers (1995) performed DSC studies and noted that the glass transition temperature of PVOH varies with water content in the range between 0% and 30%. Above a concentration of 30%, it was found that water started to crystallize upon cooling and the glass transition temperature did not decrease but remained constant and equal to the melting temperature of ice. The glass transition temperature of dry PVOH was also found to be lowered by the presence of residual acetyl groups up to 12 mol.% (Rault et al. 1995); however, the effect of residual groups is not as important as that of the presence of water. The methods used to determine the glass transition temperature contribute to the differences of values reported in literature. For example, the glass transition temperature obtained from dilatometry is higher than values estimated from the refractive index–temperature relation, whereas values from differential scanning calorimetry are lower (Rault et al. 1995).

A range of values for melting point of PVOH has been reported for fully hydrolyzed PVOH. PVOH derived from VAc has a melting point of 223°C, whereas PVOHs with the isotactic diads of 26 and 86 have melting points of about 285°C and 240°C, respectively (Murahashi 1966). Differential thermal analysis methods are frequently used to detect the melting point of PVOH. Tubbs (1965) reported a melting point of 228°C using differential thermal analysis, while a melting temperature of 230°C was reported for PVOH with crystallinity of 65% (Holland and Hay 2001). Pocsan and coworkers (1997) identified a peak at 140–220°C on DSC curves as the temperature range where melting of partially hydrolyzed PVOH occurred. The melting point of PVOH has been shown to depend on the degree of hydrolysis, polymer composition, distribution of chain units, length of crystallized units, and also the method of preparation of the polymers (Tubbs 1966). PVOH derived from partial saponification of PVA with sodium hydroxide was reported to have high melting points as compared to that synthesized by acid alcoholysis (Pocsan et al. 1997). The melting point decreases with increasing plasticizer contents such as residual acetyl and glycerol, and the heat of fusion is 1.6 kcal/mol per PVOH segment (Tubbs 1965).

Direct experimental determination of the melting point can be complicated by the inherent thermal instability of PVOH as decomposition occurs very close to the melting point. To circumvent this problem, measurements are usually performed on the polymer in the presence of a diluent, and the melting point calculated from an expression that relates the melting point depression caused by the volume fraction of the diluent as follows (Tubbs and Wu 1992):

$$\frac{1}{T_m} - \frac{1}{T_m^0} = \left(\frac{RV_\mu}{\Delta H_\mu V_1}\right)[(1-V_2) - x_1(1=V_2)^2] \qquad (2.7)$$

where T_m is the melting point of the polymer–diluent system, T_m^0 is the melting point of the pure polymer, V_μ is the molar volume of a crystalline segment, V_1 is the molar volume of the diluent, and V_2 is the molar volume of the polymer, R is the gas constant, ΔH_μ is the heat of fusion of a crystalline segment, and x_1 is the polymer–diluent interaction parameter. With respect to tacticity, the melting temperature of PVOH generally follows the order syndiotactic (230–267°C) > atactic (228–240°C) > isotactic (212–235°C).

2.6.3 Surface Properties of PVOH

Surface properties of PVOH such as adhesion strongly depend on the degree of hydrolysis. The surface tension of aqueous solutions of PVOH is lower than that of pure water, and it decreases with time. Because of the presence of hydroxyl groups, PVOH has the capability of hydrogen bonding with its solvents and decreases their surface tension similar to other surface-active agents (Bhattacharya and Ray 2004). Surface tension also decreases with increasing residual acetyl group content at the same concentration, but the lowering is greater in partially hydrolyzed PVOH with the acetyl groups in block type than that with randomly distributed acetyl groups (Matsuzawa 1997). The former is useful as a protective colloid. The rate of decrease of the surface tension of aqueous solutions for syndiotacticity-rich PVOH is greater than that for commercial PVOH (atactic) (Matsuzawa et al. 1980); PVOH has a tendency to migrate from bulk to the surface, forming a thin film on the solution surface after a certain time. The equilibrium energies of wetting (contact angle × surface tension) of a 3% PVOH (DP = 1700, fully hydrolyzed) aqueous solution at 20°C to Teflon, polyethylene, and polyamide 6 are −15.1, −2.4, and 44.5×10^{-3} Nm^{-1}, respectively. In the case of partially hydrolyzed (88 mol.%) PVOH, corresponding energies are −4.4, 7.0, and 37.4×10^{-3} Nm^{-1}, respectively. The hydrophobic polymers are wettable for the partially hydrolyzed PVOH, whereas for hydrophilic ones, the situation is different (Matsuzawa 1997). Zuo and coworkers (2013) found that residual VAc units may segregate to the surface of PVOH films, affecting the adsorption of solvent molecules such as ethanol. Upon immersion of PVOH films in ethanol, surface tension decreased (20° increase of water contact angle) for highly hydrolyzed samples (i.e., 97.7% and 99%), whereas the surface tension increased in samples with a degree of hydrolysis less than 96%, changing the hydrophilic and hydrophobic nature (wettability) of the films.

Adhesion of PVOH to various substrates depends primarily upon the number of acetyl groups and the fiber substrate to which it is being applied. For example, on hydrophilic fibers such as cotton and rayon, fully hydrolyzed and intermediate grades of PVOH exhibit good adhesion. This is because these grades will have a higher number of hydroxyl groups still remaining on the polymer chain. Consequently, the size will have better specific adhesion because of the potential for hydrogen bonding at the size–fiber interface. On the other hand, hydrophobic fibers exhibit poor wetting characteristics toward water-based sizing agents, which results in a lack of adhesion to the fully hydrolyzed and intermediate grades.

2.6.4 Barrier, Mechanical, and Optical Properties of PVOH Films

Generally, PVOH is not processable unless modified with plasticizers. For this reason, solution casting is the method of choice. In some special cases, PVOH films can be produced by common converting processes such as cast-film, blown-film, and injection-molding processes. PVOH films are prepared by casting about 10% solution containing plasticizer through a slit onto a rotating drier subsequent to evaporation. The films are also prepared by the extrusion of concentrated solutions. Glycols such as glycerol, sorbitol, urea, etc. are used as plasticizers (Matsuzawa 1997). Films produced from PVOH have an excellent barrier to gases such as oxygen, nitrogen, and carbon dioxide at very low relative humidity (RH). The oxygen permeability of biaxially oriented PVOH film (BOVLON, Nippon Gohsei) at 20°C and 0% RH is <0.0015 cc mil/100 in.^{2} day (Table 2.4). In comparison, oxygen permeability of ethylene vinyl alcohol (EVOH) with 29 mol.% ethylene (Soarnol,

TABLE 2.4
Film Properties of PVOH

Property	Value	Remarks
Oxygen permeability (cc mil/100 in.^{2} day)	0.0015	20°C, 0% RH
	0.04	25°C, 75% RH
Water vapor permeability (g mil/100 in.^{2} day)	4.6	40°C, 90% RH
Tensile strength (Pa)	2.5×10^4	MD and TD
Tear strength (mN/μm)	4.5	
Elongation (%)	50	MD
	45	TD
Transparency (%)	91.3	
Haze (%)	1.3	

Source: Data obtained from Nippon Gohsei, Polyvinyl Alcohol (PVOH) Film: Bovlon, 2014, available at http://www.nichigo.co.jp/pvohfilm/english/bovlon/property.html.

Note: MD, machine direction; TD, transverse direction.

Nippon Gohsei) at the same temperature and RH conditions is <0.005 cc mil/100 in.^{2} day, that of high-barrier polyvinylidene chloride (PVDC) is 0.15 cc mil/100 in.^{2} day, and that of polyamide 6 is 3.9 cc mil/100 in.^{2} day (Nippon Gohsei 2012). However, the gas barrier properties of PVOH films severely deteriorate under conditions of high humidity. As shown in Table 2.4, the oxygen permeability of the film increased 24 times when RH was increased from 0% to 75%.

Mechanical properties also have a strong dependence on RH. Increasing RH decreases the tensile strength and Young's modulus, but increases the elongation and tear strength. Figure 2.8 shows the effect of RH on the tensile strength of biaxially oriented PVOH film. At low RH, PVOH films are hard and brittle, but they become soft and tough at high RH. The plasticizing effect of water vapor makes the films more flexible, and this dependence is greater for partially hydrolyzed grades than for fully hydrolyzed ones. The tear strength at high RH is greater for partially hydrolyzed grades than fully hydrolyzed PVOH grades, owing to their higher equilibrium moisture content. In general, the dependence of the mechanical properties on RH is greater with partially hydrolyzed grades than with fully hydrolyzed grades (Toyoshima 1973).

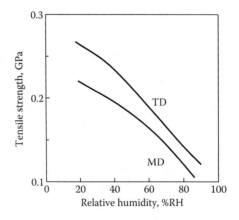

FIGURE 2.8 Tensile strength of biaxially stretched PVOH film (14 μm) as a function of relative humidity at 20°C. MD, machine direction; TD, transverse direction. (Adapted from Matsuzawa, S., Vinyl Alcohol Polymers, in *Handbook of Thermoplastics*, edited by Olagoke Olabisi, 269–289, New York: Marcel Dekker, 1997.)

Orientation and heat treatment increase tensile strength because of the molecular alignment, development of crystallinity, and limited cross-linking. Unoriented commercial PVOH films have tensile strengths of about 100 and 54 MPa at 0% and 65% RH, respectively, whereas biaxially stretched commercial PVOH films have a tensile strength of 210 MPa at 65% RH (Matsuzawa 1997). Notomi and coworkers (1976) reported tensile strengths of over 116 MPa, excellent dimensional stability, and impact resistance for biaxially oriented, fully hydrolyzed PVOH films at low temperatures. Nonstretched PVOH films are brittle and have low impact resistance. In this case, plasticizers such as polyhydric alcohol (e.g., glycerin, diethylene glycol, dipropylene glycol) or urea are added at comparatively large quantities (e.g., 10–15 wt.%). Poor dimensional stability and leaching are some of the limitations of plasticized films. In fully hydrolyzed films, tensile modulus increases sharply, while tensile strength increases almost linearly with heat treatment temperature. Tear strength and elongation decrease markedly at heat treatment temperatures above 160°C (Toyoshima 1973). PVOH film has a light transmittance (white light source) of 60–66%, which is higher than that of polyethylene, and has good adhesion to ink (Matsuzawa 1997). PVOH films also have no electrostatic charge and have high oil and organic solvent resistance.

2.7 SOLUTION PROPERTIES OF PVOH

2.7.1 Solubility of PVOH in Water

Water is the most common and important solvent for PVOH. The solubility of PVOH in water is greatly affected by the degrees of hydrolysis and polymerization, the effect of the former being especially significant. The strong intramolecular and intermolecular hydrogen bonding in fully hydrolyzed PVOH reduces its solubility in water (Marten 2003), requiring temperatures greater than 70°C to dissolve. On the other hand, the presence of residual acetate groups in partially hydrolyzed PVOH weakens the hydrogen bonds and allows solubility at lower temperatures. The hydrophobic nature of the acetate groups results in a negative heat of solution, which increases as the number of acetate groups is increased (Marten 2003, 2004). The solubility of PVOH as a function of degree of hydrolysis is shown in Figure 2.9.

Another way to increase the dissolution temperature is by increasing the degree of crystallinity. This can be accomplished via heat treatment. The dissolution temperature of PVOH increases with increasing crystal content (Matsuzawa 1997). In the crystallites, the increase in the amount of intermolecular hydrogen bonds heightens the dissolution temperature. Figure 2.10 shows the dissolution temperature plotted against heat treatment temperature. The PVOH heated at 210°C dissolves at

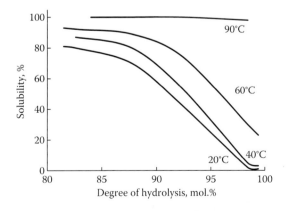

FIGURE 2.9 Solubility in water versus degree of hydrolysis at a dissolving time of 15 min. (Data obtained from Clariant GmbH, Mowiol Polyvinyl Alcohol, 1999, available at http://www.cbm.uam.es/mkfactory.esdomain/webs/cbmso/images/TextBox/mowiol.pdf.)

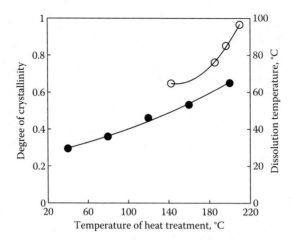

FIGURE 2.10 Degree of crystallinity and dissolution temperature against temperature of heat treatment. (Adapted from Matsuzawa, S., Vinyl Alcohol Polymers, in *Handbook of Thermoplastics*, edited by Olagoke Olabisi, 269–289, New York: Marcel Dekker, 1997.)

95°C, but aggregates may still be present in the solution. Aggregates decrease with increasing dissolution temperature and time (Matsuzawa 1997).

2.7.2 Other Solvents and Diluents

In addition to water, a number of other highly polar and hydrophilic solvents or solvent mixtures can dissolve PVOH, including dimethyl sulfoxide (DMSO), ethylene glycol, and *N*-methyl pyrrolidone, acetamide, dimethylformamide (Marten 2004; Tacx et al. 2000; Wright et al. 2013). The Huggins constant K_H is a relative measure of the interaction between polymer and solvent and can be used to provide information on the suitability of a solvent or solvent mixtures for a polymer. Thermodynamically, the lower the value of K_H is, the greater is the dissolving power of a solvent. In a good solvent, there is a relatively extended chain conformation, whereas in a poor solvent, the polymer chains collapse with intramolecular aggregation occurring (Wright et al. 2013). K_H for water was reported to be 0.5 or above (Tacx et al. 2000; Wright et al. 2013).

2.7.3 Precipitation of PVOH from Aqueous Solution by Inorganic Salts

PVOH can be precipitated from aqueous solution by inorganic salts and modified through the interaction of hydroxyl groups with inorganic salts. However, the large amount of salt required makes it difficult to carry out the salting process on an industrial scale (Yamagata and Banno 1978). The coagulation of aqueous PVOH solution in inorganic salt solutions depends on the kinds of inorganic salt. The minimum concentrations of sodium sulfate, zinc sulfate, and sodium chloride to coagulate a 5% PVOH solution are 0.70, 1.0, and 3.1 N, respectively (Matsuzawa 1997). Yamaura and Naitoh (2002) reported that the degree of crystallinity of PVOH increased with the addition of NaCl and it is possible to prepare high-modulus fiber spun from a PVOH solution of sodium chloride. Kubo and coworkers (2009) reported that the crystallites of PVOH could be destroyed by the addition of $Mg(NO_3)_2 \cdot 6(H_2O)$ and that $PVOH/Mg(NO_3)_2 \cdot 6(H_2O)$ blend films with low crystallinity were very soft and rubber-like. Jiang and coworkers (2012) employed a plasticizer consisting of magnesium chloride hexahydrate to modify the properties of PVA film and reported that the plasticizer was highly compatible with PVOH.

Vinyl Alcohol Polymers

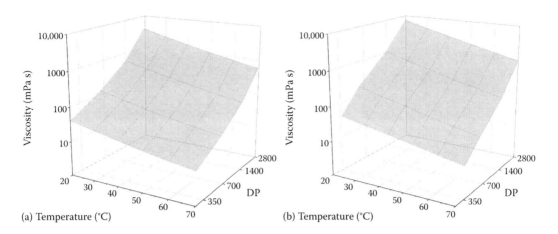

FIGURE 2.11 Viscosity of PVOH solutions as a function of temperature and degree of polymerization (DP) (15 wt.% in water). (a) Partially hydrolyzed PVOH, DH: 87.7%; (b) fully hydrolyzed PVOH, DH: 98.4%. (Data obtained from Clariant GmbH, Mowiol Polyvinyl Alcohol, 1999, available at http://www.cbm.uam.es/mkfactory.esdomain/webs/cbmso/images/TextBox/mowiol.pdf.)

2.7.4 Viscosity of PVOH Solutions

The viscosity behavior of PVOH solutions depends on the polymer concentration, degree of hydrolysis, molecular weight (DP), temperature, stereoregularity of polymer chains, and thermal history (Alves et al. 2011; Briscoe et al. 2000; Choi et al. 2001; Gao et al. 2010a,b; Lămătic et al. 2009; Onogi et al. 1967). Figure 2.11 shows the effect of temperature and DP on viscosity of a 15 wt.% aqueous solution of PVOH. In general, for any given DP, fully hydrolyzed PVOH produces solutions with a higher viscosity than do partially hydrolyzed ones. As the temperature increases, the viscosity decreases and obeys the Arrhenius equation (El-Hefian et al. 2010); DP has the opposite effect. DP and concentration have a stronger effect on viscosity than that of the degree of hydrolysis and temperature.

The flow behavior of polymer solutions is usually represented by the zero-shear viscosity obtained from shear experiments at relatively low shear rates. The viscosity of PVOH solution at zero shear rate in the concentration range 0–80% is represented as follows:

$$\eta = \eta_0 \{1 + (A'/k) \exp(B/T) DP^\alpha (1 - \gamma C) C\} \quad (2.8)$$

where η_0 is the viscosity of water, T the absolute temperature, and C the concentration in g/100 mL, $A' = 1.6 \times 10^{-3}$, $k = 14$, $B = 495 \pm 20$, $\alpha = 0.64$, and $\gamma = 0.0045$. This equation reduces to the following equation (Baker's equation) in the concentration range of 0–20% and the temperature range of 20–40°C:

$$\eta = \eta_0 \{1 + ([\eta]/k)C\}^k \quad (2.9)$$

The flow of concentrated solutions is non-Newtonian, and the velocity gradient (dv/dr) for a capillary is expressed as follows:

$$-dv/dr = \tau/\eta^* + (m/\eta^*)\tau^n \quad (2.10)$$

where η^* is the zero-shear viscosity, τ the shear stress, and m and n the constants. The value of n varies with concentration and the DP. The n value of the PVOH of DP = 1020 varies from 1.51 to 2.51 between the concentrations of 39.2 and 64.9 g/100 mL at 80°C (Matsuzawa 1997).

When the zero-shear-rate viscosity is plotted against molecular weight or concentration, the plots show an inflection point at a critical molecular weight or concentration showing a marked increase in viscosity of PVOH solutions (Onogi et al. 1967). It was also suggested that the structures of PVOH solutions before and after the critical concentration differ (Bercea et al. 2013). The critical concentration is said to correspond to the transition from nonentangled chains to entangled polymer chains. Below the critical concentration, individual PVOH molecules are said to be dispersed in water in the form of thread-filled spheres containing plenty of water, and above it, entanglement occurs between molecular spheres. Above the critical concentration, the behavior of the dissolved polymer depends on the interactions of the polymer with the solvent molecules, the intramolecular interactions between polymer segments, and also the intermolecular interactions between different neighboring polymer chains (Onogi et al. 1967). Intermolecular entanglements predominate on the overall dynamics of polymer chains. Once the chains are long enough, flow becomes much more difficult because forces applied to one polymer chain are transmitted to and distributed among many other chains. Bercea and coworkers (2013) found that below the critical value, PVOH solutions exhibited Newtonian behavior over a range of shear rates from 2×10^{-4} s^{-1} to 10^3 s^{-1}. The transition from Newtonian to non-Newtonian behavior at the different shear rates was dependent on the concentration, i.e., at shear rates higher than 1 s^{-1} for a concentration of 5 g/dL, 0.3 s^{-1} for 10 g/dL, and 0.03 s^{-1} for 15 g/dL.

2.7.5 Gelation in PVOH Solutions

Aqueous solutions of PVOH undergo ageing that is reflected by an increase in viscosity over time, and at sufficiently high polymer concentrations, gel formation occurs. Figure 2.12 shows the increase in viscosity with time for concentrated aqueous solutions of PVOH with formation of thermoreversible opaque gels at temperatures below 30°C. The rate of gelation increases with decreasing temperature and increasing concentration but decreases with increasing residual acetyl group content (Matsuzawa 1997). Mixtures of DMSO/water with DMSO contents of 40–80% are poorer solvents than water, and PVOH solutions of the mixtures are known to form gels easily (Matsuzawa 1997). The gelation process is time and temperature dependent, and the properties of gels formed depend on the ratio of DMSO to water. Higher gelation rates were observed for PVOH solutions containing mixtures of DMSO and water compared with aqueous solutions of only PVOH in water (Ohkura et al. 1992). A critical concentration where gelation occurs has been identified by several authors. The maximum rate of gelation is observed for a solution of the mixture with DMSO content of 60% by volume (Matsuzawa 1997). Hoshino and coworkers (1996) reported gel formation in PVOH

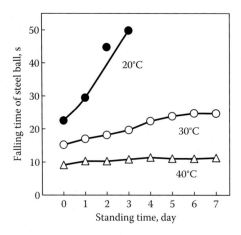

FIGURE 2.12 Change of viscosity of aqueous PVOH solutions with standing time, DP = 1500, 13%. ●, 20°C; ○, 30°C; △, 40°C. (Adapted from Matsuzawa, S., Vinyl Alcohol Polymers, in *Handbook of Thermoplastics*, edited by Olagoke Olabisi, 269–289, New York: Marcel Dekker, 1997.)

solutions with DMSO composition range of 20–80% by weight. A critical concentration of 60–70% was identified, below which a clear gel was formed, which eventually became turbid. Above this critical concentration, the solution becomes turbid before the formation of gels.

2.8 MODIFICATION OF PVOH

PVOH can be modified for a variety of reasons, including enhancing water insolubility, creation of hydrogels, adding specific reactivity to the polymer, manufacturing high-barrier films, etc. Copolymerization, cross-linking, heat treatment, chemical modification, blending, and compounding with different fillers and nanofillers are among the approaches to modify PVOH. The different modification approaches are discussed below.

2.8.1 Copolymers

Most copolymers of vinyl alcohol with another monomer are prepared by the hydrolysis (saponification) of the copolymers with VAc. Many copolymers containing ethylene, acrylic acid, maleic acid, maleic anhydride, polyethylene glycol, poly(e-caprolactone) (Zaleska et al. 2009), PVA (Hartley 1959), allyl sulfonic acid, quaternary aminopropylmethacrylamide, ethylisopropyl acrylamide, vinyl versatate/maleic acid, vinylpyridine (Li et al. 1998), N-vinyl succinimide, vinyl trialkoxysilane, N-alkoxymethacrylamide, styrene (Lu et al. 1998), etc. have been studied, and some of them are produced industrially (Finch 1992).

2.8.1.1 Ethylene Vinyl Alcohol Copolymers

EVOH was first prepared in 1966 from ethylene–VAc copolymer (Matsuzawa 1997; Teiichiro et al. 1968). The copolymer suppressed the hydrophilic nature of PVOH while retaining high-performance barrier properties at normal humidity. EVOH quickly found industrial appeal as a high oxygen barrier film, and it is currently a widely used food packaging material in both multi-layer films and molded rigid containers. Similar to the production of PVOH, EVOH entails the copolymerization of ethylene and VAc followed by hydrolysis. Solubilized ethylene is mixed with the solution containing VAc and fed to the polymerization vessel. Random copolymerization occurs with VAc content in the polymer being slightly higher than in the feed. The relative amounts of ethylene and vinyl alcohol in EVOH copolymers affect almost all the properties. Therefore, EVOH processing and end-use applications are based on the copolymer composition. EVOH resins are commercially available in a range of compositions, most commonly encompassing vinyl alcohol contents of about 52–76 mol.% (Mokwena and Tang 2012). In general, copolymers of higher vinyl alcohol content have properties resembling those of PVOH. Similarly, those with higher ethylene contents resemble properties of PE. For instance, EVOH copolymers with lower ethylene contents have better gas barrier properties. Data from Kuraray (2014) show that EVOH of 27 mol.% ethylene offers a 10 times higher oxygen barrier than that of 44 mol.% ethylene. Zhang and coworkers (2001) reported oxygen transmission rates four to six times higher for an EVOH copolymer with 44 mol.% when compared to a film containing 32 mol.% ethylene. Copolymer composition also affects water absorption and transmission of EVOH copolymers. Copolymers with low ethylene content absorb more water than those with higher ethylene content (Cava et al. 2006; Zhang et al. 1999).

EVOH copolymers are crystalline, with different crystalline systems depending on the composition. For example, the crystal system of a copolymer with contents of ethylene below 40 mol.% is monoclinic. The relationship between density (d) and the crystallinity (α) is expressed by the following equation for the copolymer with 40 mol.% ethylene (Matsuzawa 1997):

$$1/d = -0.0483\alpha + 0.8923 \qquad (2.11)$$

The melting point (T_m) decreases with increasing ethylene content up to 80 mol.% ethylene. The glass transition temperature (T_g) decreases with increasing ethylene content. For example, a copolymer with 38 mol.% ethylene has a T_m of 172°C and a T_g of 53°C. The tensile strength, Young's modulus, and elongation at break of the copolymer with ethylene content of 44 mol.% at a temperature of 20°C and RH of 65% are 52 MPa, 2.0 GPa, and 280%, respectively (values from Kuraray EVAL).

The major limitation for EVOH is its inherent sensitivity in moisture, which poses a negative effect on its oxygen barrier and other properties. When exposed to humid conditions, EVOH absorbs significant amounts of moisture, as reported by several authors (Aucejo et al. 1999; Lagaron et al. 2001; Mokwena et al. 2011; Zhang et al. 1999). Because of this moisture sensitivity, EVOH in food packaging is invariably in the form of multilayer structures, protected from moisture by water barrier materials such as polyethylene and polypropylene (PP). Flexible pouches for food that undergo retort processing are especially susceptible to deterioration of the oxygen barrier due to the extreme temperatures and moisture conditions encountered. Significant increases in oxygen transmission of multilayer films were reported by several authors. A tenfold increase was reported by Mokwena et al. (2009) for polyethylene terephthalate (PET)/EVOH/PP film after retort at 121°C for 10 min, while Lopez-Rubio et al. (2005) reported over 400 times increase for PP/EVOH/PP film after retort at 121°C for 20 min.

2.8.1.2 Vinyl Alcohol–Maleic Acid Copolymers

The VAc copolymer with dibasic maleic acid can be prepared by the copolymerization of both monomers in methanol. Maleic acid improves water solubility at low contents. The copolymers containing 1–6% maleic acid with vinyl alcohol are soluble in water in a range of degrees of hydrolysis of 50–100 mol.% compared to PVOH, which is soluble at high degrees of hydrolysis, 82–100 mol.%. The copolymers containing a small amount of hydrophobic vinyl versatate and hydrophilic maleic acid are surface active (Matsuzawa 1997). When maleic anhydride is used in the copolymerization, only the copolymers with degrees of hydrolysis above 95 mol.% are soluble in water, due to intermolecular ester formation. Paper coated with partially hydrolyzed vinyl alcohol–maleic acid copolymers has rather high resistance to air permeability (Matsuzawa 1997). Graft copolymers of acrylic acid onto PVOH have been synthesized with the aim of removing lead ions from aqueous solutions (Chowdhury et al. 2009).

2.8.1.3 Cationic Copolymers

Attaching cationic groups onto PVOH imparts ionic bonding, providing stronger interaction than hydrogen bonding alone. The copolymer with 6 mol.% quaternary trimethylaminopropyl methacrylamide (QAPM) is useful for the aggregation of pulp fibers because of its adhesion to the negatively charged pulp fibers. Vinyl alcohol–N-alkyl acrylamide copolymers are thermally cross-linked to form hot water-resistant films (Matsuzawa 1997). Sang and Xiao (2008) synthesized cationically modified PVOH by copolymerization of VAc and diallyldimethyl ammonium chloride, followed by alkaline hydrolysis. Single- and dual-component flocculation systems were evaluated as alternatives to traditional synthetic dual-polymer systems associated with environmental pollution.

2.8.1.4 Block and Graft Copolymers

Synthesis of well-defined block copolymers of PVOH is challenging because the radical polymerization of VAc is difficult to control (Tong et al. 2009). Block copolymers of PVOH with the polymers and copolymers derived from acrylic acid, methyl acrylate, acrylamide, acrylonitrile, ethylene oxide, etc. have been investigated. PVOH–polyacrylic acid (PAA) block polymers are prepared by the polymerization of acrylic acid in the presence of PVOH with a thiol end group in water, using a water-soluble radical initiator. PVOH–PAA (100:20) block copolymers are compatible with gelatin, and the films made from the copolymers have good extensibility (Matsuzawa 1997). Recently, well-defined block copolymers have been made with the reversible addition–fragmentation chain-transfer (RAFT) polymerization of VAc with xanthates (Stenzel 2008; Tong et al. 2009) and cobalt-mediated radical polymerization of VAc (Debuigne et al. 2005).

Tezuka and coworkers (1989) prepared PVOH (hydrophilic hard segment)–graft–polytetrahydrofuran (hydrophilic soft segment). The alcholate of 3-hydroxypropyldimethylvinylsilane initiates the polymerization of tetrahydrofuran leaving the vinyl groups, which initiate the polymerization of

VAc, followed by hydrolysis (Matsuzawa 1997). Styrene–divinylbenzene copolymer prepared with a tetraethylthiuram disulfide initiator has a thiuram group, which initiates the block polymerization of VAc. PVOH–polystyrene block copolymer containing sulfonic acid as a hydrolysis catalyst is prepared after sulfonation and saponification (Arai et al. 1980). PVOH–PVAc graft polymers are prepared in the emulsion polymerization of vinyl acetate in the presence of PVOH as a protective colloid. The graft polymerization starts by the abstraction of hydrogens from carbon atoms adjacent to the hydroxyl group with radicals (Matsuzawa 1997).

The flexible polymer polyethylene glycol (PEG) serves as a functional branch chain in PVOH–graft–PEG copolymers, which find uses in controlled drug release, membranes, and coatings (Muschert et al. 2009). The grafted copolymer is soluble in water independently of pH and is insoluble in most organic solvents (Heuschmid et al. 2013).

2.8.1.5 Others

Vinyl alcohol–acrylic acid copolymers form lactone rings when the vinyl acetate–methyl acrylate copolymer is hydrolyzed by alkali followed by neutralization. The copolymer prepared from vinyl acetate and acrylic acid is easily soluble in water (Finch 1992; Matsuzawa 1997). Vinyl alcohol–alkyleneoxyacrylate (ALOAC) copolymer with ALOAC of 5.5 mol.% has a melting point of 184°C and improved extrusion and injection-molding processability (Marten 1986). Vinyl alcohol–vinylanethylsulfone copolymers are amphilitic; they are soluble in water and methanol (Imai et al. 1985a). Vinyl acetate–vinyl methanesulfonic copolymers (MVSs) react with 2-methyl-2-oxalidone (OXZ) to form PVOH–poly(OXZ) graft copolymers (Tezuka et al. 1987). Vinyl alcohol–n-butyldimethylsiloxyldimethylvinylsilane copolymers give films characterized by hydrophobic surfaces (Tezuka et al. 1986). Silane-modified PVOH provides higher strengths and better attachment to mineral substances such as pigments, fillers, and glass (Hashemzadeh and Fickert 2008).

2.8.2 Chemically Modified PVOH

The functionality of the hydroxyl group in PVOH could be modified with esterification, etherification, and acetalization. Incorporation of phosphonic acid groups gives ionic conduction ability to PVOH (Takada et al. 2002). Hydrophobic and alkyl groups result in high surface and interfacial activities. Transformation of hydroxyl groups into carbonyl groups improves flame retardance (Liu and Chiu 2003; Zaikov and Lomakin 1997). PVOH etherified with N,N-dialkylalkoxymethylamine is used as a cationic polymer with quaternization (Matsuzawa 1997). PVOH has also been modified with oxidation, phosphorylation (Liu and Chiu 2003), and cross-linking.

2.8.2.1 Esterification

The PVOH esters with inorganic acids such as sulfuric, nitric, phosphoric, boric, and titanic acid were prepared before 1962, and boric acid has been used as a gelling agent. The cross-linking reaction occurs in alkaline solution, and the following structure for the cross-linking was presented (Matsuzawa 1997; Shibayama et al. 1988):

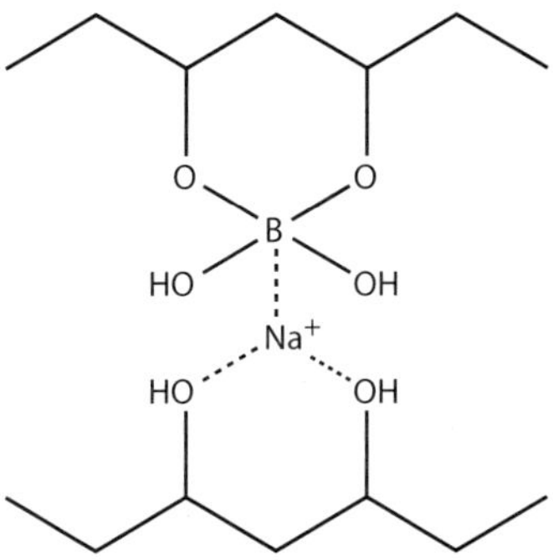

The PVOH esters with organic acids are prepared with the use of acid anhydrides, acylchlorides, and ester exchange. Acetylation, acetoacetylation, and cinnamylation had been known before 1959. Acetoacetylation is carried out with ester exchange of hydroxyl groups in PVOH and acetoacetate (Matsuzawa 1997). Titantrichloride is used as a gelling agent. PVOH containing acetoacetyl groups has hot water resistance and an increased waterproof strength with heat treatment. PVOH containing cinnamate groups is cross-linked with light with the following reaction scheme (Robertson et al. 1959):

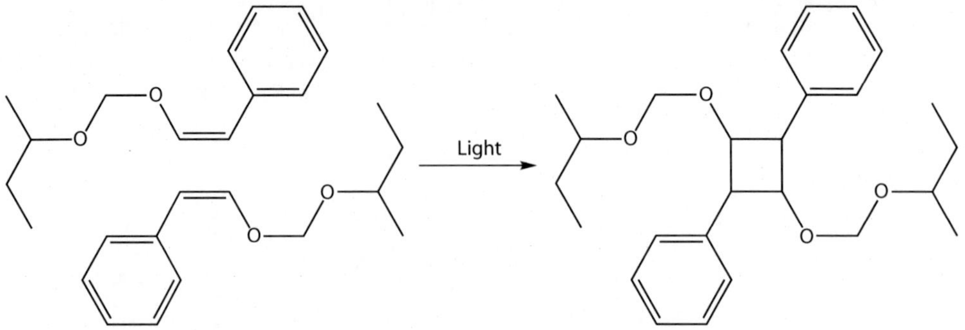

2.8.2.2 Acetalization

Acetalization by a monoaldehyde such as formaldehyde is a practical method to produce a PVOH structure that is insoluble in water. In this case, insolubility is obtained by substituting hydrophobic groups for the hydroxyl group (Kumeta et al. 2003). The aldehydes with two aldehyde groups in a molecule such as glutaric aldehyde are used as cross-linking agents. The acetals made with the use of acetoacetaldehyde are water soluble even at 40–50 mol.%. The formalization of PVOH is used for the preparation of vinylon fibers and sponges. Reacting PVOH with butyraldehyde in the presence of an acid catalyst produces polyvinyl butyral (PVB) and is extensively used in laminated safety glass, paint, adhesives, and binders (Fernández et al. 2006).

2.8.2.3 Cross-Linking

Cross-linking of PVOH is used to produce hydrogels, which have been employed in biomedical applications such as contact lenses, ophthalmic materials, tendon repair, and drug delivery (Schmedlen et al. 2002). Methods to produce hydrogels include physical, by freeze-thaw processes, chemical, and radiation. Intermolecular cross-linking by acetalization is a practical method used as previously discussed. Other chemical cross-linking with the use of aldehydes, carboxylic acids, inorganic acids, polyacrolein monoisocyanates and diisocyanates, divinyl sulfates, glycidyl methacrylate, etc. have been studied (Matsuzawa 1997). Schmedlen and coworkers (2002) investigated photoactive PVOH hydrogels for use as tissue engineering scaffolds. PVOH in aqueous solution could be cross-linked by γ irradiation in the absence of oxygen (Danno 1958), and PVOH treated with fiber-reactive dye could be cross-linked with UV light in vacuum (Matsuzawa 1997). The PVOH bearing 4-hydroxylstyryl quinolinium groups could be cross-linked by visible light (Ichimura 1982). Water resistance can be achieved by a physical cross-linking network among small crystals of PVOH formed by heat treatment. When heat-treated with drawing, PVOH fibers become insoluble even in boiling water. However, the insolubility of PVOH does not last for extended periods of time (Kumeta et al. 2003). A PVOH with an aldehyde group at both ends can be prepared through the oxidative cleavage of 1,2-glycol bonds in PVOH with periodic acid. The modified PVOH can be used as a cross-linking agent. The PVOH (86% hydrolyzed)–cupric acetate film prepared at room temperature becomes insoluble in water by heating it at 90°C (Matsuzawa 1997).

2.8.2.4 Others

When PVOH is reacted with methyl vinyl sulfoxide using sodium hydroxide as a catalyst, Michael addition occurs to form $-CH_2-CH(OCH_2CH_2SOCH_3)-$. The resulting polymer becomes soluble with increasing sulfoxide content (Imai et al. 1985b). Tributyl tin monochloride reacts with hydroxyl groups of PVOH to form $-CH_2-CH(SnC_4H_5)-$, a polymer that is thermally stable (Carraher and Piersma 1973). Hydroxyl groups in PVOH are converted to carbonyl groups by Oppenauer oxidation (Kun and Cassidy 1960).

2.9 COMPLEXES AND BLENDS

2.9.1 METAL COMPLEXES

Most of the transition metal ions such as Cu^{2+}, Fe^{3+}, Ni^{2+}, and Zn^{2+} can form stable complexes with electron-rich compounds, e.g., amine (NH_2), hydroxyl (OH), and thiol (SH) groups (Feng et al. 2012). Cu(II)–PVOH complex was found in 1972 to be formed at pH above 7.0 and to have a stability constant of $9.7 \times 10^{15} (NO_3^-)$. An intermolecular chelate complex model was presented by Hojo and coworkers (1974). Zn, Co, Ni, and Fe(III) complexes were also studied. Fe(III)/Cu(II) complexes accelerate the decomposition of hydrogen peroxide (Matsuzawa 1997). A cluster model in which a cluster is surrounded with a PVOH segment was presented by Narisawa (Matsuzawa 1997; Narisawa et al. 1989). The copper–PVOH–lactic acid complex is linear and considered to be a precursor of a superconductor Ba–Y–Cu oxide filament (Tomita and Goto 1994). $NiCl_2$-doped PVOH film is resistant to γ-ray degradation and heat treatment (El-Shahawy 1994; Khaled et al. 1993; Matsuzawa 1997).

2.9.2 PVOH–IODINE COMPLEXES

PVOH was found to be colored green with iodine by Staudinger in 1927. The complex was also found to turn blue in the presence of boric acid, and the stretched complex films produce polarized light (Land 1948; Matsuzawa 1997). PVOH–iodine complexes are widely used for polarizing films in practical applications (Yang and Horii 2008).

2.9.3 BLENDS, COMPOSITES, AND NANOCOMPOSITES

Recent environmental regulations and concerns have prompted renewed efforts to develop products and processes compatible with the environment. PVOH-based blends and nanocomposites show a versatile range of properties at acceptable cost and biodegradation rate and can be employed in a wide range of applications. PVOH, though not always miscible, is generally compatible with polymers having polar groups. Miscibility depends on the interaction of hydroxyl groups of PVOH with the side groups of a second polymer. PVOH–starch blends have haze due to microphase separation. The polymers are compatible but not miscible. The size of the microdomain decreases when the PVOH containing carboxyl groups is used, leading to a decrease in haze. PVOH and poly(N-vinyl-2-pyrrolidone) are miscible, and the blend forms transparent films (Matsuzawa 1997). PVOH and polyacrylic acid form a polymer complex (Vázquez-Torres et al. 1993). Polyaniline is dispersed as microirregular particles in PVOH showing no interaction between them (Chen and Fang 1991). The above blends except PVOH/polyaniline blend are prepared using water as a common solvent. PVOH/cellulose blends are prepared with solvents such as a dimethylformamide (DMF)/N_2O_4 mixture and concentrated phosphoric acid, while PVOH/chitosan blends are prepared with 15% acetic acid (Matsuzawa 1997). Polyaniline particles are prepared with the polymerization of aniline in PVOH (Chen and Fang 1991). Electroactive polyaniline–PVOH composite films were prepared by Mirmohseni and Wallace (2003).

Nanocomposites, prepared using PVOH as a matrix, have attracted much attention in the past 15 years. Layered silicate, silica, cadmium sulfide nanoparticles, and carbon nanotubes are among the studied nanoreinforcements. The preparation methods are usually solution cast or *in situ* polymerization. Naturally organic nanofillers, such as cellulose nanocrystals, can enhance processability, biocompatibility, and biodegradability compared to inorganic ones (Roohani et al. 2008; Tang and Alavi 2011). Recently, fabrication of PVOH/grapheme oxide nanocomposites has been reported (Bao et al. 2011; Liang et al. 2009).

2.10 APPLICATIONS

PVOH has been one of the most versatile polymers available for many different applications in a wide range of industries, from packaging, paper and paperboard, textiles, adhesives, chemical industry, building products, and cosmetics, to ceramics. Its wide application across industries arises from the ease of influencing the properties of the polymer by controlling the DP or molecular weight and the degree of hydrolysis of the acetate groups.

2.10.1 Water-Soluble Films

Water-soluble PVOH is commonly used in packaging unit dosage quantities of products such as dyes, laundry detergents, bleaches, agricultural chemicals, disinfectants, cleaning chemicals, etc. The benefits of these premeasured doses include reducing exposure of users to potentially harmful chemicals; delivery of accurate dosages of chemicals that may be difficult to measure; and the disposal of used packages. The water-soluble pouches and sachets dissolve as they release their contents into cold or warm aqueous systems without the user having to open the packaging. They are also used for water-soluble laundry bags, seed tapes, and sanitary pads (Finch 1992). Partially hydrolyzed PVOHs with a degree of hydrolysis of around 90 mol.% are soluble in cold water because the residual acetate groups act as wedges preventing close association of neighboring chains and hydrogen bonding interaction, thus facilitating water penetration and dissolution (Goswami et al. 2004).

2.10.2 Packaging Films and Coatings

Fully hydrolyzed PVOH is typically used as films and coatings in food packaging applications for its excellent oxygen barrier properties, especially at low RH. Since the oxygen barrier of PVOH is severely compromised at higher RH, PVOH films for packaging moist products are therefore likely to be used as part of multilayer structures where the PVOH layer is protected by moisture barrier materials such as polypropylene and polyethylene. PVOH films are optically clear, have good adhesion to inks, and are tough with high tensile and abrasion resistance properties. PVOH is approved for use as a surface coating for paper and paperboard food contact materials under Title 21 of the Code of Federal Regulations (CFR), part 170, with the following food types: low-moisture fats and oils, dry solids with a surface containing no free fat or oil; and dry solids with a surface containing free fat or oil. PVOH coatings can be applied in typical converting processes to substrates such as polyethylene, polypropylene, polyethylene terephthalate, or paper for packaging a variety of dry and snack food applications (Rolando 2000).

2.10.3 Blended PVOH Films

There is a lot of research on blending PVOH with a variety of synthetic and natural polymers with the aim of improving processing. PVOH is usually included in blends for its film-forming properties, good gas barrier, adhesion/binding ability, mechanical properties, water solubility, biodegradability, etc., depending on the proposed applications. Blends of PVOH with polysaccharides have been studied as biomaterials for possible uses in biomedical applications (Silva et al. 2013). Blown films of PVOH–starch have been prepared from the mixtures of PVOH–starch–glycerol–water systems. The films have a good oxygen barrier and are biodegradable (Matsuzawa

1997). Moraes and coworkers (2009) studied thermal and rheological characterizations of film-forming blends of gelatin and PVOH. The blend films with cellulose, hydroxyl propyl methyl cellulose, chitosan, and gelatin have been studied in order to prepare ultrafiltration films, tablet-coating materials, artificial skins, and strong gelatin films, respectively (Matsuzawa 1997). Cast films from blends of starch/PVOH/nanosilicone dioxide were found to have improved water resistance and mechanical properties (Tang et al. 2008). The blend films with polyacrylic acid and polyamide 6 have higher thermal stability than PVOH and higher oxygen barrier than polyamide 6, respectively. Both blend films with polyaniline (Chen and Fang 1991) and polypyrrole (Matsuzawa 1997) are electroconducting.

2.10.4 POLARIZER FILMS

Polarizing films are widely used in a variety of applications, including high-precision optical devices, photography filters, antiglare sunglasses, and liquid crystal displays (LCDs). PVOH film is the most commonly employed polarizing film for LCD panels. Polarization is the property of absorbing light along the orientation of a film, i.e., dichroism, which comes from uniaxially oriented elongated shaped pigments in the uniaxially oriented polymer film (Dirix et al. 1995). In polarized films, PVOH film is oriented by tensile drawing at temperatures close to, but below, the melting temperature of the polymer. Dichroic dye (iodine) molecules are introduced in the oriented films either by chemical modification of the polymer or by incorporation of the dyes in the films prior to or after drawing (Dirix et al. 1995). The dichroic iodine molecules appear to adopt the orientation of the polymer, and the films exhibit the desirable characteristics of polarizers: i.e., the absorption of light is dependent on the polarization direction of the incident radiation. Generally, the stained PVOH film is laminated to triacetate cellulose to provide physical rigidity and humidity protection.

2.10.5 MOLDED PRODUCTS

PVOH can be made extrudable by using mixtures such as PVOH (33%), water (50%), plasticizers, fillers, and pigments subsequent to drying and elongation (Matsuzawa 1997). These are used for belts and canvases. Rollers are manufactured by wrapping the sheet prepared from molding compound containing PVOH (30–40%), plasticizer (40%), and water (or water–methanol mixture [30–20%]) around a roll core subsequent to vulcanization, used for printing (Matsuzawa 1997). Recent efforts aim at making injection-moldable PVOH compounds fully biodegradable in water. Semicrystalline grades can be soluble in cold water (down to 10°C). Applications include containers, soluble cores for lost-core molding, and single-use medical applications.

2.10.6 FIBERS

Hermann and coworkers, early on, discovered that PVOH could be made into strong, elastic, and degradable fibers that can be used in surgical sutures. A fiber called Synthofil was manufactured in a pilot study as early as 1935. The general process to make the fiber involves forcing a PVOH solution through spinnerets into a coagulation bath of sodium sulfate solution, after which the wet fiber is stretched, washed, and then dried. The dried fiber is then heat-treated to increase the crystallinity and then treated with formaldehyde to increase its water resistance (Lindemann 1989). Characteristics of PVOH fibers are high strength, low elongation, and high modulus. This fiber is also very hygroscopic, has good dimensional stability under dry heat, has good chemical resistance to chemicals (e.g., alkali), and has high durability against UV rays. The main fields of application are in the rubber industry, agriculture, fishing, nonwovens, paper, plastic reinforcement, cement reinforcement, and others.

2.10.7 TEXTILE SIZES

PVOH has been one of the most versatile size materials available for warp sizing formulations since it was commercialized as a textile warp sizing agent around 1965 (Goswami et al. 2004). A study published in 1956 discussed the use of PVOH as warp size for cotton and cotton blends (Abrams et al. 1956). The authors found that a high-viscosity, fully hydrolyzed PVOH in combination with glycerin (and optionally with starch) gave the lowest shedding values. Although the benefits of PVOH as a warp size were readily recognized very early, its high price (i.e., 65–85 cents/lb. compared to 6–8 cents/lb. for starch in the 1930s and 1940s) limited its extensive use in textiles (Lindemann 1989). In the 1960s, the problem of stream pollution, partially caused by discharging starch wastes, lead to renewed interest in PVOH for warp sizes for cotton and cotton blends. The versatility of PVOH is attributed to the fact that it can be manufactured into a variety of modified forms and can be utilized alone or in combination with other size materials such as starch, polyester resins, acrylic copolymers, and carboxymethyl cellulose (CMC), or as a binder with numerous fiber substrates such as glass, acrylic, polyester, cellulose acetate, polyamide, and numerous styles (Goswami et al. 2004). The most important properties of a size necessary for textile processing and weaving are tensile strength, extensibility, recovery from extension, abrasion resistance, and ease of removal (Goswami et al. 2004). The tensile strength of the PVOH film increases as the chain length (DP or molecular weight) of each grade increases and as the percentage of acetyl removal (degree of hydrolysis) increases. Partially hydrolyzed grades are 20–25% weaker than the fully hydrolyzed grades (Goswami et al. 2004). PVOH can be added to weaker sizes such as starch to improve the performance of these size products. One of the added advantages of PVOH is its compatibility with most of the other film formers employed in sizing.

2.10.8 STABILIZERS/PROTECTIVE COLLOIDS

PVOH is employed industrially as a protective colloid in emulsion or suspension polymerizations of vinyl monomers, particularly vinyl acetate and vinyl chloride (Hayashi 1996). PVOH fulfills the dual function of emulsifying the monomer and stabilizing the polymer particles formed during the process. Partially hydrolyzed grades of PVOH increase particle stability and prevent agglomeration of polymer particles as they are formed. The hydrophilic polymers adsorb to the surface of the hydrophobic polymer particles, convert their surfaces from hydrophobic to hydrophilic, and sterically stabilize the polymer particles. To enable the hydrophobic polymers to effectively adsorb to the hydrophobic polymer particles, the stabilizer should be copolymers of hydrophilic and hydrophobic monomers rather than hydrophilic homopolymers. Grades of partially hydrolyzed PVOH with approximately 10–20% residual vinyl acetate units are most useful as stabilizers for polymerization. Stability increases with acetyl content and with the blockiness of the distribution of the acetate groups. Increasing the blockier distribution of the vinyl acetate units in a copolymer produces smaller particles and higher viscosity of the latex, as well as an increase in the content of vinyl acetate units in a copolymer.

2.10.9 ADHESIVES

The excellent adhesion of PVOH to materials such as paper, cardboard and cellophane, textiles, leather, porous ceramic surfaces, etc. makes it useful as an adhesive and coating for these materials. PVOH adhesives exhibit rapid tack for quick setting. PVOH can be used as a humectant to retard the loss of water due to its hydrophilic nature (Rolando 1998). Because of their water sensitivity, partially hydrolyzed grades are used in remoistenable adhesives (Matsuzawa 1997), while fully hydrolyzed grades are used in formulating quick-setting, water-resistant adhesives, including paper-laminating adhesives for use in the manufacture of solid fiberboard, linerboard, and spiral-wound tubes and cans. PVOH is also used as a modifier for adhesives based on resin dispersions. PVOH-based adhesives are used for packaging (paperboard and fiberboard laminating), corrugated boards, tube winding, etc. (Rolando 1998).

2.10.10 Binders

PVOH has been studied for use as concrete reinforcement material due to its good mechanical properties (e.g., high strength, high modulus) and also due to its ability to bond well with a cementitious matrix (Zheng and Feldman 1995). PVOH fibers have a positive effect on the bending strength of their composites, due to a good interfacial bond between the fibers and cement matrix. The good interfacial bond is attributed to a noncircular cross-section of the fibers and to hydrogen bonds between the fibers and the cement matrix. The formation of the associated microstructure has been attributed to the effect of PVOH on the nucleation of CH and C–S–H at the fiber surface (Chu et al. 1993). PVOH fiber–reinforced high-performance cementitious materials have the potential to be used for external walls, for radiation-shielding walls, and in the construction industry as a high-strength, high-impact, and lightweight concrete. They are also used as binders of color developers in thermosensitive, pressure-sensitive, and electrosensitive papers; metal fibers in papers; inorganic fibers and wood-based particles in boards; electrode-forming substances; components of battery separators; phosphors in television tube screens; and fertilizers. Other applications include temporary binders for ceramic powders, sand and metal powders before heat treatment in ceramics, sand molds, and high-performance metals (Matsuzawa 1997). Fully or partially hydrolyzed or modified PVOHs are used as pigment coatings for paper and paperboard and for coating of release, security, and photographic papers. Papers coated with Polyvinyl nitrate (PVN)–poly(N-vinyl-2-pyrrolidone) blend are used for jet printing. The emulsifying, binding, film-forming, and thickening properties of PVOH have all found use in cosmetic preparations. Microcapsules of perfumes, pharmaceuticals, dyes, etc., have also been prepared with the use of PVOH (Matsuzawa 1997). Cleansing creams, shaving creams, and facial masks are also formulated using PVOH.

2.10.11 Membranes and Gels

PVOH hydrogels and membranes have been developed for biomedical applications such as contact lenses, artificial pancreases, hemodialysis, and synthetic vitreous humor, as well as for implantable medical materials to replace cartilage and meniscus tissues (Baker et al. 2012). Various kinds of PVOH membranes have been prepared through cross-linking. The membrane cross-linked with oxalic acid is used for reverse osmosis; that cross-linked with a C_3–C_8 aldehyde is used for pervaporation (selective vaporization permeation of a component in a liquid mixture through a membrane); that cross-linked with glutaraldehyde and swollen with salt solution is used for the perm selectivity of oxygen over nitrogen; that cross-linked by photochemical reaction of PVOH bearing aromatic azido groups is used for the immobilization of β-glycosidase (Matsuzawa 1997); that with quaternary ammonium groups is used for electrical conductivity (Nakanishi et al. 1991); the cross-linked PVOH–chitosan blend membrane is used for controlled drug delivery; the cross-linked PVOH–polystyrene sulfonic acid blend membrane is used for esterification and pervaporation in 1-propanol–propionic acid mixture (David et al. 1992); and the sulfoxide-modified PVOH membrane is used for the permselectivity of sulfur dioxide over nitrogen. PVOH gels carrying enzymes and pharmaceuticals as well as those used as biomedical materials have been investigated (Matsuzawa 1997).

Hydrogels have received significant attention because of their exceptional promise in biomedical applications. The structures contain covalent bonds produced by the simple reaction of one or more comonomers, physical cross-links from entanglements, association bonds such as hydrogen bonds or strong van der Waals' interactions between chains, or crystallites bringing together two or more macromolecular chains (Muppalaneni and Omidian 2013; Peppas 1987). Properties such as high water content, elastic nature when swollen, biocompatibility, and the ability to absorb water and aqueous solutions make PVOH hydrogels attractive for biomedical applications such as tissue replacement material, soft contact lens materials, and articular cartilages (Baker et al. 2012; Hassan and Peppas 2000). The cross-links, either physical or chemical, provide the structural stability the hydrogel needs after it swells in the presence of water or biological fluids. The degree of crosslinking dictates the amount of fluid uptake and, thus, the physical, chemical, and diffusional properties of the polymer, and ultimately, its biological properties. In pharmaceutical applications, PVOH hydrogels are favored because of their biocompatibility, film forming, and swelling properties (Muppalaneni and Omidian 2013). Commercial PVOH hydrogels are found in tablet, ophthalmic, transdermal, and implant dosage forms.

2.10.12 Production of Polyvinyl Butyral

PVOH is used as a chemical intermediate in the reaction with aldehydes in the presence of small amounts of mineral acid to yield PVB. With its excellent adhesive and film-forming properties, its strong binding power, and its outstanding optical transparency, PVB is used in applications including interlayers for safety glasses in automotives, paints, lacquers, and varnishes (e.g., primers for metals and anticorrosion paints); printing inks; temporary binders; and adhesives.

REFERENCES

Abrams, E., C.W. Rougeux, and J.N. Coker. 1956. "Polyvinyl Alcohol as a Warp Size for Various Staple Yarns." *Textile Research Journal* 26 (11): 875–880.

Alves, M.H., B.E.B. Jensen, A.A.A. Smith, and A.N. Zelikin. 2011. "Polyvinyl Alcohol Physical Hydrogel: New Vista on a Long Serving Biomaterial." *Macromolecular Bioscience* 11 (10): 1293–1313.

Amiya, S. and M. Uetsuki. 1985. "Determination of 1, 2-Glycol Linkages in Polyvinyl by 400 MHz ^{1}H-NMR and ^{13}C-NMR Alcohol)." *Analytical Sciences* 1: 91–92.

Arai, K., Y. Ogiwara, and C. Kuwabara. 1980. "Preparation of Copoly(vinyl Alcohol–styrenesulfonic Acid) Resin and Its Catalytic Activity on Hydrolysis of Carbohydrates. II. Two-Step Polymerization Using Tetraethylthiuram Disulfide as an Initiator." *Journal of Applied Polymer Science* 25 (12): 2935–2941.

Aucejo, S., C. Marco, and R. Gavara. 1999. "Water Effect on the Morphology of EVOH Copolymers." *Journal of Applied Polymer Science* 74 (5): 1201–1206.

Baker, M.I., S.P. Walsh, Z. Schwartz, and B.D. Boyan. 2012. "A Review of Polyvinyl Alcohol and Its Uses in Cartilage and Orthopedic Applications." *Journal of Biomedical Materials Research. Part B, Applied Biomaterials* 100 (5): 1451–1457.

Bao, C., Y. Guo, L. Song, and Y. Hu. 2011. "Poly(vinyl Alcohol) Nanocomposites Based on Graphene and Graphite Oxide: A Comparative Investigation of Property and Mechanism." *Journal of Materials Chemistry* 21 (36): 13942.

Bercea, M., S. Morariu, and D. Rusu. 2013. "*In Situ* Gelation of Aqueous Solutions of Entangled Poly(vinyl Alcohol)." *Soft Matter* 9 (4): 1244–1253.

Bhattacharya, A. and P. Ray. 2004. "Studies on Surface Tension of Poly(vinyl Alcohol): Effect of Concentration, Temperature, and Addition of Chaotropic Agents." *Journal of Applied Polymer Science* 93 (1): 122–130.

Brandrup, J. and E.H. Immergut. 1989. *Polymer Handbook*, 3rd ed. New York: Wiley-Interscience.

Briscoe, B., P. Luckham, and S. Zhu. 2000. "The Effects of Hydrogen Bonding upon the Viscosity of Aqueous Poly(vinyl Alcohol) Solutions." *Polymer* 41 (10): 3851–3860.

Bugada, D.C. and A. Rudin. 1984. "Characterization of Poly(vinyl Alcohol–acetate) by ^{13}C N.m.r. and Thermal Analyses." *Polymer* 25 (12): 1759–1766.

Bunn, C.W. 1948. "Crystal Structure of Polyvinyl Alcohol." *Nature* 161: 929–930.

Carraher, C.E. and J.D. Piersma. 1973. "Modification of Poly(vinyl Alcohol) through Reaction with Tin Reactants." *Die Angewandte Makromolekulare Chemie* 28 (1): 153–160.

Cava, D., C. Sammon, and J.M. Lagaron. 2006. "Water Diffusion and Sorption-Induced Swelling as a Function of Temperature and Ethylene Content in Ethylene–Vinyl Alcohol Copolymers as Determined by Attenuated Total Reflection Fourier Transform Infrared Spectroscopy." *Applied Spectroscopy* 60 (12): 1392–1398.

Chang Chun Petrochemical. 2009. "PVA Polyvinyl Alcohol." Available at http://www.perrychem.com/files/PVA_English_2009_.pdf.

Chen, S.A. and W.G. Fang. 1991. "Electrically Conductive Polyaniline–Poly(vinyl Alcohol) Composite Films: Physical Properties and Morphological Structures." *Macromolecules* 24 (6): 1242–1248.

Choi, J.H., W.S. Lyoo, H.D. Ghim, and S.-W. Ko. 2000. "High Molecular Weight Syndiotacticity-Rich Poly(vinyl Alcohol) Gel in Aging Process." *Colloid & Polymer Science* 278 (12): 1198–1204.

Choi, J.H., S.-W. Ko, B.C. Kim, J. Blackwell, and W.S. Lyoo. 2001. "Phase Behavior and Physical Gelation of High Molecular Weight Syndiotactic Poly(vinyl Alcohol) Solution." *Macromolecules* 34 (9): 2964–2972.

Chowdhury, P., A. Mukherjee, B. Singha, and S.K. Pandit. 2009. "Synthesis of Graft Copolymer of Poly(acrylic Acid) and Poly(vinyl Alcohol) in the Presence of Methylene Bisacrylamide Crosslinker and Investigation of Its Efficiency in Removing Lead Ion from Aqueous Solution." *Journal of Macromolecular Science, Part A* 46 (5): 547–553.

Chu, T.-J., R.E. Robertson, H. Najm, and A.E. Naaman. 1993. "Effects of Poly(Vinyl Alcohol) on Fiber Cement Interfaces. Part II: Microstructures." *Advanced Cement Based Materials* 1 (3): 122–130.

Clariant GmbH. 1999. "Mowiol Polyvinyl Alcohol." Available at http://www.cbm.uam.es/mkfactory.esdomain/webs/cbmso/images/TextBox/mowiol.pdf.

Danno, A. 1958. "Gel Formation of Aqueous Solution of Polyvinyl Alcohol Irradiated by Gamma Rays from Cobalt-60." *Journal of the Physical Society of Japan* 13 (7): 722–727.

David, M.O., Q.T. Nguyen, and J. Néel. 1992. "Pervaporation Membranes Endowed with Catalytic Properties, Based on Polymer Blends." *Journal of Membrane Science* 73 (2–3): 129–141.

Debuigne, A., J.-R. Caille, N. Willet, and R. Jérôme. 2005. "Synthesis of Poly(vinyl Acetate) and Poly(vinyl Alcohol) Containing Block Copolymers by Combination of Cobalt-Mediated Radical Polymerization and ATRP." *Macromolecules* 38 (23): 9488–9496.

DeMember, J.R., H.C. Haas, and R.L. MacDonald. 1972. "NMR of Vinyl Polymers. I. A New and Sensitive Probe for Configuration in Poly(vinyl Alcohol) Polymer Chains." *Journal of Polymer Science Part B: Polymer Letters* 10 (5): 385–389.

Dirix, Y., T.A. Tervoort, and C. Bastiaansen. 1995. "Optical Properties of Oriented Polymer/Dye Polarizers." *Macromolecules* 28: 486–491.

El-Hefian, E.A., M.M. Nasef, and A.H. Yahaya. 2010. "Preparation and Characterization of Chitosan/Polyvinyl Alcohol Blends—A Rheological Study." *E-Journal of Chemistry* 7 (s1): S349–S357.

El-Shahawy, M.A. 1994. "Structural State of Ni(II) in Heat-Treated PVA–NiCl$_2$ Composites." *Polymer Degradation and Stability* 43 (1): 75–79.

Erbil, H.Y. 2000. *Vinyl Acetate Emulsion Polymerization and Copolymerization with Acrylic Monomers*. Boca Raton, FL: CRC Press.

Feng, Q., B. Tang, Q. Wei, D. Hou, S. Bi, and A. Wei. 2012. "Preparation of a Cu(II)–PVA/PA6 Composite Nanofibrous Membrane for Enzyme Immobilization." *International Journal of Molecular Sciences* 13 (10): 12734–12746.

Fernández, M.D., M.J. Fernández, and P. Hoces. 2006. "Synthesis of Poly(vinyl Butyral)s in Homogeneous Phase and Their Thermal Properties." *Journal of Applied Polymer Science* 102 (5): 5007–5017.

Finch, C.A. 1992. *Polyvinyl Alcohol—Developments*, 2nd ed. New York: Wiley.

Flory, P.J. and F.S. Leutner. 1948. "Occurrence of head-to-head arrangements of structural units in polyvinyl alcohol." *Journal of Polymer Science. Part A. Polymer Chemistry* 3 (6): 880–890.

Forder, C., S.P. Armes, and N.C. Billingham. 1995. "Synthesis of Polyvinyl Alcohols with Narrow Molecular Weight Distribution from Poly(benzyl Vinyl Ether) Precursors." *Polymer Bulletin* 35: 291–297.

Fukae, R., T. Yamamoto, O. Sangen, T. Saso, T. Kako, and M. Kamachi. 1990. "Dynamic Mechanical Behaviors of Poly(vinyl Alcohol) Film with High Syndiotacticity." *Polymer Journal* 22 (7): 636–637.

Fukae, R., T. Yamamoto, Y. Fujita, N. Kawatsuki, O. Sangen, and M. Kamachi. 1997. "Poly(vinyl Alcohol) with High Diad-Syndiotacticity and High Mellting Point." *Polymer Journal* 29 (3): 293–295.

Gao, H., J. He, R. Yang, and L. Yang. 2010a. "Characteristic Rheological Features of High Concentration PVA Solutions in Water with Different Degrees of Polymerization." *Journal of Applied Polymer Science* 116: 2734–2741.

Gao, H.-W., R.-J. Yang, J.-Y. He, and L. Yang. 2010b. "Rheological Behaviors of PVA/H$_2$O Solutions of High-Polymer Concentration." *Journal of Applied Polymer Science* 116: 1459–1466.

Goodship, V. and D. Jacobs. 2009. *Polyvinyl Alcohol: Materials, Processing and Applications (Rapra Review Reports)*. Smithers Rapra Press: Shrewsbury, Shropshire, England.

Goswami, B.C., R.D. Anandjiwala, and D. Hall. 2004. *Textile Sizing*. New York: Marcel Dekker.

Hartley, F.D. 1959. "Graft Copolymer Formation during the Polymerization of Vinyl Acetate in the Presence of Polyvinyl Alcohol." *Journal of Polymer Science* 34 (127): 397–417.

Hashemzadeh, A. and K.E. Fickert. 2008. "Silane-Modified Polyvinyl Alcohols." US Patent 20080281035 A1.

Hassan, C.M. and N.A. Peppas. 2000. "Structure and Applications of Poly(vinyl Alcohol) Hydrogels Produced by Conventional Crosslinking or by Freezing/Thawing Methods." In *Biopolymers—PVA Hydrogels, Anionic Polymerisation Nanocomposites SE-2*, edited by J.Y. Chang, Vol. 153, 37–65. Berlin, Heidelberg: Springer.

Hayashi, S. 1996. "Protective Colloids." In *Polymeric Materials Encyclopedia*, edited by J.C. Salamone, Vol. 2, 1316–1321. Boca Raton, FL: CRC Press.

Heuschmid, F.F., P. Schuster, and B. Lauer. 2013. "Nonclinical Toxicity of the Grafted Copolymer Excipient PEG–PVA." *Food and Chemical Toxicology* 51 Suppl 1: S1–S2.

Hojo, N., H. Shirai, and S. Hayashi. 1974. "Complex Formation between Poly(vinyl Alcohol) and Metallic Ions in Aqueous Solution." *Journal of Polymer Science: Polymer Symposia* 47 (1): 299–307.

Holland, B.J. and J.N. Hay. 2001. "The Thermal Degradation of Poly(vinyl Alcohol)." *Polymer* 42 (16): 6775–6783.

Hoshino, H., S. Okada, H. Urakawa, and K. Kajiwara. 1996. "Gelation of Poly(vinyl Alcohol) in Dimethyl Sulfoxide/water Solvent." *Polymer Bulletin* 37 (2): 237–244.

Ichimura, K. 1982. "Preparation of Water-Soluble Photoresist Derived from Poly(vinyl Alcohol)." *Journal of Polymer Science: Polymer Chemistry Edition* 20 (6): 1411–1417.

IHS Chemical. 2013. "Polyvinyl Alcohols." Available at http://www.ihs.com/products/chemical/planning/ceh/polyvinyl-alcohols.aspx.

Imai, K., T. Shiomi, Y. Tezuka, and K. Takahashi. 1985a. "Syntheses of Vinyl Sulfoxide/Vinyl Acetate-Type Copolymers." *Journal of Macromolecular Science: Part A—Chemistry* 22 (10): 1347–1358.

Imai, K., T. Shiomi, Y. Tezuka, K. Takahashi, and M. Satoh. 1985b. "Michael Addition Reaction of Vinyl Sulfoxide with Poly(Vinyl Alcohol)." *Journal of Macromolecular Science: Part A—Chemistry* 22 (10): 1359–1369.

Jiang, X., X. Zhang, D. Ye, X. Zhang, and H. Dai. 2012. "Modification of Polyvinyl Alcohol Films by the Addition of Magnesium Chloride Hexahydrate." *Polymer Engineering and Science* 52 (7): 1565–1570.

Kadajji, V.G. and G.V. Betageri. 2011. "Water Soluble Polymers for Pharmaceutical Applications." *Polymers* 3 (4): 1972–2009.

Khaled, M.A., F. Sharaf, M.S. Risk, and M.M. El-Ockr. 1993. "Effect of γ-Irradiation on the Refractive Indices and Optical Absorption of Poly(vinyl Alcohol) Doped with NiCl2 and CrCl3." *Polymer Degradation and Stability* 40 (3): 385–388.

Kubo, J., N. Rahman, N. Takahashi, T. Kawai, G. Matsuba, K. Nishida, Y. Kanaya, and M. Yamamoto. 2009. "Improvement of Poly(vinyl Alcohol) Properties by the Addition of Magnesium Nitrate." *Journal of Applied Polymer Science* 112 (3): 1647–1652.

Kumaki, Y., M. Kawagoe, S. Takada, P. Garcia, and L. Neufeld. 2014. "Enhanced Polyvinyl Alcohol as a Barrier Paper Coating for Food Packaging." In *TAPPI PaperCon*. Nashville, Tennessee.

Kumeta, K., I. Nagashima, K. Matsui, and K. Mizoguchi. 2003. "Crosslinking Reaction of Poly(vinyl Alcohol) with Poly(acrylic Acid) (PAA) by Heat Treatment: Effect of Neutralization of PAA." *Journal of Applied Polymer Science* 90 (9): 2420–2427.

Kun, K.A. and H.G. Cassidy. 1960. "Electron Exchange Polymers. XIV. Steric Hindrance in the Chemical Modification of High Polymers." *Journal of Polymer Science* 44 (144): 383–389.

Kuraray Co. 2014. "Technical Bulletin No. 110. Gas Barrier Properties of EVAL Resins." Available at http://www.eval-americas.com/media/36916/tb_no_110.pdf.

Lagaron, J.M., A.K. Powell, and G. Bonner. 2001. "Permeation of Water, Methanol, Fuel and Alcohol-Containing Fuels in High-Barrier Ethylene–Vinyl Alcohol Copolymer." *Polymer Testing* 20 (5): 569–577.

Lămătic, I.-E., M. Bercea, and S. Morariu. 2009. "Intrinsic Viscosity of Aqueous Polyvinyl Alcohol Solutions." *Revue Roumaine de Chimie* 54 (11–12): 981–986.

Land, E.H. 1948. "Treatment of Polarizing Polyvinyl Alcohol–Iodine Sorption Complex Image with Boric Acid." US Patent 2445581.

Li, B., L. Niu, W. Kou, Q. Deng, G. Cheng, and S. Dong. 1998. "Synthesis of a Self-Gelatinizable Grafting Copolymer of Poly(vinyl Alcohol) for Construction of an Amperometric Peroxidase Electrode." *Analytical Biochemistry* 256 (1): 130–132.

Liang, J., Y. Huang, L. Zhang, Y. Wang, Y. Ma, T. Guo, and Y. Chen. 2009. "Molecular-Level Dispersion of Graphene into Poly(vinyl Alcohol) and Effective Reinforcement of Their Nanocomposites." *Advanced Functional Materials* 19 (14): 2297–2302.

Lin, C.-A., C.-R. Wu, and H.-C. Tsai. 2006. "Synthesis of Ultra-High-Molecular-Weight PVA Using Two-Stage Photo-Concentrated Emulsion Polymerization of Vinyl Acetate." *Designed Monomers & Polymers* 9 (3): 305–315.

Lindemann, M.K. 1989. "Vinyl Acetate and the Textile Industry." *Textile Chemist and Colorist* 21 (1): 21–28.

Liu, Y.-L. and Y.-C. Chiu. 2003. "Novel Approach to the Chemical Modification of Poly(vinyl Alcohol): Phosphorylation." *Journal of Polymer Science Part A: Polymer Chemistry* 41 (8): 1107–1113.

Lopez-Rubio, A., P. Hernandez-Munoz, E. Gimenez, T. Yamamoto, R. Gavara, and J.M. Lagarón. 2005. "Gas Barrier Changes and Morphological Alterations Induced by Retorting in Ethylene Vinyl Alcohol-Based Food Packaging Structures." *Journal of Applied Polymer Science* 96 (6): 2192–2202.

Lu, Z., X. Huang, and J. Huang. 1998. "Synthesis and Characterization of Amphiphilic Diblock Copolymer of Polystyrene and Polyvinyl Alcohol Using Ethanolamine–Benzophenone as Photochemical Binary Initiation System." *Journal of Polymer Science Part A: Polymer Chemistry* 36 (1): 109–115.

Lyoo, W.S., S.S. Han, J.H. Choi, H.D. Ghim, S.W. Yoo, J. Lee, S.I. Hong, and W.S. Ha. 2001. "Preparation of High Molecular Weight Poly(vinyl Alcohol) with High Yield Using Low-Temperature Solution Polymerization of Vinyl Acetate." *Journal of Applied Polymer Science* 80: 1003–1012.

Lyoo, W.S., J.H. Yeum, B.C. Ji, H.D. Ghim, S.S. Kim, J.H. Kim, J.Y. Lee, and J. Lee. 2003. "Preparation of Water-Soluble Syndiotacticity-Rich High Molecular Weight Poly(Vinyl Alcohol) Microfibrillar Fibers Using Copolymerization of Vinyl Pivalate and Vinyl Acetate and Saponification." *Journal of Applied Polymer Science* 88: 31482–31487.

Magdum, S.S., G.P. Minde, and V. Kalyanraman. 2013. "Rapid Determination of Indirect COD and Polyvinyl Alcohol from Textile Desizing Wastewater." *Pollution Research* 32 (3): 515–519.

Marten, F.L. 1986. "Copolymers of Vinyl Alcohol and Poly(alkyleneoxy)acrylates." US Patent 4618648.

———. 2003. "Vinyl Alcohol Polymers." In *Encyclopedia of Polymer Science and Technology*, Vol. 8, 399–437. Hoboken, NJ: John Wiley & Sons.

———. 2004. "Vinyl Alcohol Polymers." In *Kirk-Othmer Encyclopedia of Chemical Technology*, 5th ed., edited by A. Seidel and M. Bickford, 591–627. Hoboken, NJ: John Wiley & Sons.

Marten, F.L. and C.W. Zvanut. 1992a. "Manufacture of Polyvinyl Acetate for Polyvinyl Alcohol." In *Polyvinyl Alcohol—Developments*, edited by C.A. Finch, 31–56. Chichester, West Sussex, England.

———. 1992b. "Hydrolysis of Polyvinyl Acetate to Polyvinyl Alcohol." In *Polyvinyl Alcohol—Developments*, edited by C.A. Finch, 57–76. Chichester, West Sussex, England.

Marvel, C.S. and C.E. Denoon Jr. 1938. "The Structure of Vinyl Polymers. II. Polyvinyl Alcohol." *Journal of the American Chemical Society* 60 (5): 1045–1051.

Matsuzawa, S. 1997. "Vinyl Alcohol Polymers." In *Handbook of Thermoplastics*, edited by O. Olabisi, 269–289. New York: Marcel Dekker.

Matsuzawa, S., K. Yamaura, N. Yoshimoto, I. Hortkawa, and M. Kuroiwa. 1980. "Adsorption of Stereoregular Poly(vinyl Alcohols) at Air–Water Interface." *Colloid and Polymer Science* 258 (2): 131–135.

Mirmohseni, A. and G.G. Wallace. 2003. "Preparation and Characterization of Processable Electroactive Polyaniline–Polyvinyl Alcohol Composite." *Polymer* 44 (12): 3523–3528.

Mokwena, K.K. and J. Tang. 2012. "Ethylene Vinyl Alcohol: A Review of Barrier Properties for Packaging Shelf Stable Foods." *Critical Reviews in Food Science and Nutrition* 52 (7): 640–650.

Mokwena, K.K., J. Tang, C.P. Dunne, T.C.S. Yang, and E. Chow. 2009. "Oxygen Transmission of Multilayer EVOH Films after Microwave Sterilization." *Journal of Food Engineering* 92 (3): 291–296.

Mokwena, K.K., J. Tang, and M.-P. Laborie. 2011. "Water Absorption and Oxygen Barrier Characteristics of Ethylene Vinyl Alcohol Films." *Journal of Food Engineering* 105 (3): 436–443.

Moraes, I.C.F., R.A. Carvalho, A.M.Q.B. Bittante, J. Solorza-Feria, and P.J.A. Sobral. 2009. "Film Forming Solutions Based on Gelatin and Poly(vinyl Alcohol) Blends: Thermal and Rheological Characterizations." *Journal of Food Engineering* 95 (4): 588–596.

Muppalaneni, S. and H. Omidian. 2013. "Polyvinyl Alcohol in Medicine and Pharmacy: A Perspective." *Journal of Developing Drugs* 2 (3): 1–2.

Murahashi, S. 1966. "PVA—Selected Topics on Its Synthesis." *Macromolecular Chemistry* 3: 435.

———. 1967. "Poly(vinyl Alcohol)—Selected Topics on Its Synthesis." *Pure and Applied Chemistry* 15: 435–452.

Murahashi, S., S. Nozakura, M. Sumi, H. Yuki, and K. Hatada. 1966. "NMR Study of the Tacticity of Poly(vinyl Acetate)." *Journal of Polymer Science Part B: Polymer Letters* 4 (1): 65–69.

Muschert, S., F. Siepmann, B. Leclercq, B. Carlin, and J. Siepmann. 2009. "Drug Release Mechanisms from Ethylcellulose: PVA–PEG Graft Copolymer-Coated Pellets." *European Journal of Pharmaceutics and Biopharmaceutics* 72 (1): 130–137.

Nagara, Y., T. Nakano, Y. Okamoto, Y. Gotoh, and M. Nagura. 2001. "Properties of Highly Syndiotactic Poly(vinyl Alcohol)." *Polymer* 42 (24): 9679–9686.

Nagy, D.J. 1986. "Molecular Weight Determination of Poly(vinyl Alcohol) Using Aqueous Size Exclusion Chromatography/Low-Angle Laser Light Scattering." *Journal of Polymer Science Part C: Polymer Letters* 24 (2): 87–93.

———. 1993. "A Mark–Houwink Equation for Poly(vinyl Alcohol) from SEC–Viscometry." *Journal of Liquid Chromatography* 16 (14): 3041–3058.

———. 2003. "Size Exclusion of Polyvinyl Alcohol and Polyvinyl Acetate." In *Handbook of Size Exclusion Chromatography and Related Techniques*, edited by C.-S. Wu, 269–295. New York: Marcel Dekker.
Nakanishi, T., T. Akiyama, T. Seike, M. Satoh, and J. Komiyama. 1991. "Electrical Conductivity of Cationic Poly(vinyl Alcohol) Membranes Having Quaternary Ammonium Groups." *Journal of Membrane Science* 64 (3): 247–254.
Narisawa, M., K. Ono, and K. Murakami. 1989. "Interaction and Structure of PVA–Cu(II) Complex: 1. Binding of a Hydrophobic Dye toward PVA–Cu(II) Complex." *Polymer* 30 (8): 1540–1545.
Nippon Gohsei. 2012. "OTR Barrier Properties—Soarnol (EVOH)." Available at http://www.soarnol.com/eng/s_data/s_data24.html.
———. 2014. "Polyvinyl Alcohol (PVOH) Film: Bovlon." Available at http://www.nichigo.co.jp/pvohfilm/english/bovlon/property.html.
Noro, K. 1973. "Manufacture of Polyvinyl Acetate for Polyvinyl Alcohol." In *Polyvinyl Alcohol—Properties and Applications*, edited by C.A. Finch, 67–89. New York: John Wiley & Sons.
Notomi, R., T. Shigeyoshi, and M. Sugiyama. 1976. "Preparation of Biaxially Oriented Polyvinyl Alcohol Film." US Patent 3985849.
Ohgi, H. and T. Sato. 1999. "Highly Isotactic Poly(vinyl Alcohol). 2. Preparation and Characterization of Isotactic Poly(vinyl Alcohol)." *Macromolecules* 32 (8): 2403–2409.
Ohkura, M., T. Kanaya, and K. Keisuke. 1992. "Gels of Polyvinyl Alcohol from Dimethylsulphoxide/water Solutions." *Polymer* 33 (17): 3686–3690.
Okaya, T. 1992. "General Properties of Polyvinyl Alcohol in Relation to Its Applications." In *Polyvinyl Alcohol—Developments*, edited by C.A. Finch, 2nd ed., 1–29. Chichester, West Sussex, England.
Onogi, S., S. Kimura, T. Kato, T. Masuda, and N. Miyanaga. 1967. "Effects of Molecular Weight and Concentration on Flow Properties of Concentrated Polymer Solutions." *Journal of Polymer Science, Part C* 15 (1): 381–406.
Ovenall, D.W. 1984. "Microstructure of poly(vinyl alcohol) by 100-MHz carbon 13 NMR." *Macromolecules* 17 (8): 1458–1464.
Palos, I., G. Cadenas-Pliego, S.Y. Knjazhanski, E.J. Jiménez-Regalado, E.G. De Casas, and V.H. Ponce-Ibarra. 2005. "Poly(vinyl Alcohol) Obtained by Hydrolysis of Poly(vinyl Silyl Ethers) and Poly(vinyl Ethers) Synthesized with Indenyltitanium Trichloride." *Polymer Degradation and Stability* 90 (2): 264–271.
Peppas, N.A. 1987. *Hydrogels in Medicine and Pharmacy*. Boca Raton, FL: CRC Press.
Pocsan, I., S. Serban, G. Hubca, M. Dimonie, and H. Iovu. 1997. "The Distribution of Sequences of Partially Hydrolyzed Polyvinylic Alcohol." *European Polymer Journal* 33 (10–12): 1805–1807.
Rault, J., R. Gref, Z.H. Ping, Q.T. Nguyen, and J. Néel. 1995. "Glass Transition Temperature Regulation Effect in a Poly(vinyl Alcohol)–Water System." *Polymer* 36 (8): 1655–1661.
Robertson, E.M., W.P. van Deusen, and L.M. Minsk. 1959. "Photosensitive Polymers. II. Sensitization of Poly(vinyl Cinnamate)." *Journal of Applied Polymer Science* 2 (6): 308–311.
Rolando T.E. 1998. Solvent Free Adhesives. *Rapra Review Reports*, Report 101, 9 (5), Rapra Technology Ltd.
———. 2000. "Flexible Packaging—Adhesives, Coatings and Processes." *Rapra Review Reports*, Report 122, 11 (2), Rapra Technology Ltd.
Roohani, M., Y. Habibi, N.M. Belgacem, G. Ebrahim, A.N. Karimi, and A. Dufresne. 2008. "Cellulose Whiskers Reinforced Polyvinyl Alcohol Copolymers Nanocomposites." *European Polymer Journal* 44 (8): 2489–2498.
Sakurada, I. 1985. *Polyvinyl Alcohol Fibers*, New York: Marcel Dekker.
Sakurada, I. and T. Okaya. 1998. "Vinyl Fibers." In *Handbook of Fiber Chemistry*, edited by M. Lewin and E.M. Pearce, 2nd ed. New York: Marcel Dekker.
Sang, Y. and H. Xiao. 2008. "Clay Flocculation Improved by Cationic Poly(vinyl Alcohol)/anionic Polymer Dual-Component System." *Journal of Colloid and Interface Science* 326 (2): 420–425.
Schmedlen, R.H., K.S. Masters, and J.L. West. 2002. "Photocrosslinkable Polyvinyl Alcohol Hydrogels That Can Be Modified with Cell Adhesion Peptides for Use in Tissue Engineering." *Biomaterials* 23 (22): 4325–4332.
SEKISUI. 2011. "Selvol™ Polyvinyl Alcohol in Emulsion Polymerization." Available at http://www.sekisui-sc.com/products/selvol/index.html.
Shibayama, M., M. Sato, Y. Kimura, H. Fujiwara, and S. Nomura. 1988. "11B N.m.r. Study on the Reaction of Poly(vinyl Alcohol) with Boric Acid." *Polymer* 29 (2): 336–340.
Silva, F.E.F., M.C.B. Di-Medeiros, K.A. Batista, and K.F. Fernandes. 2013. "PVA/Polysaccharides Blended Films: Mechanical Properties." *Journal of Materials,* vol. 2013, Article ID 413578, 6 p. 2013.
Stenzel, M.H. 2008. "RAFT Polymerization: An Avenue to Functional Polymeric Micelles for Drug Delivery." *Chemical Communications (Cambridge, England)* (30): 3486–3503.

Tacx, J.C.J.F., H.M. Schoffeleers, A.G.M. Brands, and L. Teuwen. 2000. "Dissolution Behavior and Solution Properties of Polyvinylalcohol as Determined by Viscometry and Light Scattering in DMSO, Ethyleneglycol and Water." *Polymer* 41 (3): 947–957.

Takada, N., T. Koyama, M. Suzuki, M. Kimura, K. Hanabusa, H. Shirai, and S. Miyata. 2002. "Ionic Conduction of Novel Polymer Composite Films Based on Partially Phosphorylated Poly(vinyl Alcohol)." *Polymer* 43 (7): 2031–2037.

Tang, S., P. Zou, H. Xiong, and H. Tang. 2008. "Effect of Nano-SiO_2 on the Performance of Starch/Polyvinyl Alcohol Blend Films." *Carbohydrate Polymers* 72 (3): 521–526.

Tang, X. and S. Alavi. 2011. "Recent Advances in Starch, Polyvinyl Alcohol Based Polymer Blends, Nanocomposites and Their Biodegradability." *Carbohydrate Polymers* 85 (1): 7–16.

Teiichiro, C., H. Kazuo, and H. Katsuaki. 1968. "Process for Making Improved Films from Saponified Ethylene–Vinyl Acetate Copolymers." US Patent 3419654.

Tesoro, G. 1986. "Polyvinyl Alcohol Fibers, by Ichiro Sakurada, Marcel Dekker, New York, 1985, 449 pp." *Journal of Polymer Science Part C: Polymer Letters* 24 (9): 485–486.

Tezuka, Y., S. Tanaka, and K. Imai. 1986. "Preparation and Saponification of Vinyl siloxane–Vinyl Acetate Copolymers." *Polymer* 27 (1): 123–128.

Tezuka, Y., Y. Horie, and K. Imai. 1987. "Preparation and Reactions of Vinyl Sulphonate–Vinyl Acetate Copolymers." *Polymer* 28 (6): 1025–1029.

Tezuka, Y., A. Okabayashi, and K. Imai. 1989. "Synthesis of Poly(vinyl Alcohol)–Graft–Poly(tetrahydrofuran)." *Die Makromolekulare Chemie* 190 (4): 753–762.

Tomita, H. and T. Goto. 1994. "Solutions Spinning of the High Tc Oxide Superconductor. II. Effect of Organic Acid on the Complex Formation of PVA with the Copper (II) Ion." *Journal of Applied Polymer Science* 51 (6): 1151–1157.

Tong, Y.-Y., Y.-Q. Dong, F.-S. Du, and Z.-C. Li. 2009. "Block Copolymers of Poly(ethylene Oxide) and Poly(vinyl Alcohol) Synthesized by the RAFT Methodology." *Journal of Polymer Science Part A: Polymer Chemistry* 47 (7): 1901–1910.

Toyoshima, K. 1973. "Properties of Polyvinyl Alcohol Films." In *Polyvinyl Alcohol: Properties and Applications*, edited by C.A. Finch, 339–388. New York: John Wiley & Sons.

Tubbs, R.K. 1965. "Melting Point and Heat of Fusion of Poly(vinyl Alcohol)." *Journal of Polymer Science Part A: General Papers* 3 (12): 4181–4189.

Tubbs, R.K. 1966. "Sequence Distribution of Partially Hydrolyzed Poly(vinyl Acetate)." *Journal of Polymer Science Part A-1: Polymer Chemistry* 4 (3): 623–629.

Tubbs, R.K. and T.K. Wu. 1992. "Thermal Properties of Polyvinyl Alcohol." In *Polyvinyl Alcohol—Developments*, edited by C.A. Finch, 2nd ed., 167–181. New York: John Wiley & Sons.

Van der Velden, G. and J. Beulen. 1982. "300-MHz H NMR and 25-MHz ^{13}C NMR Investigations of Sequence Distributions in Vinyl Alcohol–Vinyl Acetate Copolymers." *Macromolecules* 15: 1071–1075.

Vázquez-Torres, H., J.V. Cauich-Rodríguez, and C.A. Cruz-Ramos. 1993. "Poly(vinyl Alcohol)/poly(acrylic Acid) Blends: Miscibility Studies by DSC and Characterization of Their Thermally Induced Hydrogels." *Journal of Applied Polymer Science* 50 (5): 777–792.

Wheeler, O.L., S.L. Ernst, and R.N. Crozier. 1952. "Molecular Weight Degradation of Polyvinyl Acetate on Hydrolysis." *Journal of Polymer Science* 8 (4): 409–423.

Wright, E.J., G.P. Andrews, C.P. McCoy, and D.S. Jones. 2013. "The Effect of Dilute Solution Properties on Poly(vinyl Alcohol) Films." *Journal of the Mechanical Behavior of Biomedical Materials* 28: 222–231.

Wu, T.K. and M.L. Sheer. 1977. "Carbon-13 NMR Determination of Pentad Tacticity of Poly(vinyl Alcohol)." *Macromolecules* 10 (3): 529–531.

Yamada, K., T. Nakano, and Y. Okamoto. 1998. "Stereospecific Free Radical Polymerization of Vinyl Esters Using Fluoroalcohols as Solvents." *Macromolecules* 31 (22): 7598–7605.

Yamagata, T. and S. Banno. 1978. "Process for Separating Polyvinyl Alcohol from Its Solution." US Patent 4078129A.

Yamamoto, T., R. Fukae, T. Saso, O. Sangen, M. Kamachi, T. Sato, and Y. Fukunishi. 1992. "Synthesis of High Molecular Weight Polyvinyl Alcohol of Various Tactic Contents through Photo-Emulsion Copolymerization of Vinyl Acetate and Vinyl Pivalate." *Polymer Journal* 24 (1): 115–119.

Yamaura, K. and M. Naitoh. 2002. "Preparation of High Performance Films from Polyvinyl alcohol/NaCl/H2O Systems." *Journal of Materials Science* 37: 705–708.

Yang, H. and F. Horii. 2008. "Investigation of the Structure of Poly(vinyl Alcohol)–Iodine Complex Hydrogels Prepared from the Concentrated Polymer Solutions." *Polymer* 49 (3): 785–791.

Zaikov, G.E. and S.M. Lomakin. 1997. "Innovative Type of Low Flammability Varnish Based on Poly(vinyl Alcohol)." *Polymer Degradation and Stability* 57 (3): 279–282.

Zaleska, I.M., M. Kitagawa, S. Sugihara, and I. Ikeda. 2009. "Synthesis of Biocompatible and Biodegradable Block Copolymers of Polyvinyl Alcohol–Block–Poly(ε-Caprolactone) Using Metal-Free Living Cationic Polymerization." *Journal of Polymer Science Part A: Polymer Chemistry* 47 (19): 5169–5179.

Zhang, Z., I.J. Britt, and M.A. Tung. 1999. "Water Absorption in EVOH Films and Its Influence on Glass Transition Temperature." *Journal of Polymer Science Part B: Polymer Physics* 37: 691–699.

Zhang, Z., I.J. Britt, and M.A. Tung. 2001. "Permeation of Oxygen and Water Vapor through EVOH Films as Influenced by Relative Humidity." *Journal of Applied Polymer Science* 82 (8): 1866–1872.

Zheng, Z. and D. Feldman. 1995. "Synthetic Fiber-Reinforced Concrete." *Progress in Polymer Science* 20 (94): 185–210.

Zuo, B., Y. Hu, X. Lu, S. Zhang, H. Fan, and X. Wang. 2013. "Surface Properties of Poly(vinyl Alcohol) Films Dominated by Spontaneous Adsorption of Ethanol and Governed by Hydrogen Bonding." *Journal of Physical Chemistry C* 117 (7): 3396–3406.

3 Polyvinyl Butyral

Christian Carrot, Amine Bendaoud, and Caroline Pillon

CONTENTS

- 3.1 Historical Development and Commercialization 90
  - 3.1.1 History 90
  - 3.1.2 Economic Data 90
- 3.2 Polymer Formation 91
  - 3.2.1 Polymerization 91
  - 3.2.2 Postpolymerization Chemistry 96
- 3.3 Structural and Phase Characteristics 100
  - 3.3.1 IR Spectroscopy 100
  - 3.3.2 UV Spectroscopy 100
  - 3.3.3 ^{13}C NMR Spectroscopy 100
  - 3.3.4 ^{1}H NMR Spectroscopy 100
  - 3.3.5 Dosing of Acetate 103
  - 3.3.6 Glass Transition Temperature 104
  - 3.3.7 Solvents, Solution Properties, and Molecular Weight Characterization 104
  - 3.3.8 Thermal Degradation (High Temperature Conditions in Nitrogen) 106
- 3.4 Polymer Engineering Properties 106
  - 3.4.1 Density 106
  - 3.4.2 Surface Properties 106
  - 3.4.3 Adhesion Properties 109
  - 3.4.4 Electrical Properties 110
  - 3.4.5 Mechanical Properties 111
  - 3.4.6 Photodegradation and Photoaging 112
  - 3.4.7 Thermo-Oxidation 114
- 3.5 Processing Technologies 115
  - 3.5.1 Thermal Properties for Processing 115
  - 3.5.2 Viscoelastic and Rheological Properties 115
  - 3.5.3 Extrusion of Films 116
- 3.6 Additive Effect on Properties and Applications 116
  - 3.6.1 Plasticizers 116
  - 3.6.2 Antioxidants 117
  - 3.6.3 Light and UV Stabilizers 118
  - 3.6.4 Flame Retardants 118
- 3.7 Blends and Composites 118
  - 3.7.1 Blends 118
  - 3.7.2 Composites and Nanocomposites 120
- 3.8 Applications 122
  - 3.8.1 Safety Glasses 122
  - 3.8.2 Module Encapsulation in PV Application 126
  - 3.8.3 Miscellaneous Applications 128
  - 3.8.4 Recycling of PVB 129
- References 130

3.1 HISTORICAL DEVELOPMENT AND COMMERCIALIZATION

3.1.1 History

The development of polyvinyl butyral (PVB) has been primarily driven by the car industry and its demand for safety glass. The discovery of the toughening of glass by lamination with a synthetic layer dates back to the beginning of the twentieth century and was the result of the clumsiness of Edouard Benedictus, a French scientist, in his laboratory. He accidentally knocked over a glass flask of cellulose nitrate reagent on the floor and observed that the star-crazed flask did not break. The laminated glass was discovered and given the name *Triplex* by Benedictus. The process of manufacture of laminated glass was set up 30 years later, but until 1935, the only interlayer to be used was the cellulose nitrate (named *Pyralin*). It was the only transparent, clear, and sufficiently strong material available to be bonded to glass with a natural resin (Canada balsam).

The discovery of poly(vinyl alcohol) (PVA) in Germany by Haehnel and Herrmann in 1924 (see Ref. [1] for a complete story), subsequent patenting of its acetalization [2,3], and large development efforts by a number of companies during the 1930s and 1940s lead among others to the patenting of PVB by the chemical company Union Carbide and Carbon Corporation in 1935 [4]. The new thermoplastic polymer was successful in replacing the cellulose nitrate in laminated safety glass for automobiles. Though the commercialization of various vinyl acetal polymers had already begun during the early 1930s, PVB prepared from butyraldehyde (BA) was found to be superior to other vinyl acetal polymers owing to its better cold temperature characteristics and toughness over a wide range of temperatures at lower cost.

Since then, the leading manufacturers of PVB have changed their offer of PVB for laminated glass to meet different types of lamination process (degassing or vacuum calendering and vacuum processes with or without autoclave) and the requirements of growth and productivity of automotive (from 1960), building (from 1980), and PV (from 2000) industries. Also, PVB has found other applications as a binder and coating material owing to its adhesion properties related to bonding capability, which in turn is related to the chemistry of the monomeric units. Nevertheless, laminated products remain the leading application, though the PVBs used today are no longer comparable to those of the first generation because their chemical composition has been adjusted and physical performances strongly improved, especially following technological processing innovations. Specialty-grade PVBs (PVB acoustic, solar PVB, structural PVB, PVB PV) meet the emerging needs of different markets. In particular, the manufacturers of PVBs have ignored the PV market in the 1980s, letting ethylene-vinyl acetate (EVA) become the leading material in this area, because of its superior properties for permeability and flexibility without additives [5]. Since then, the market crisis in the automotive and building industries has renewed the development effort toward selected grades meeting the requirements of the solar module industry.

3.1.2 Economic Data

The total world capacity of PVB is not published, but it is estimated to be over 200,000 tons. The main producer of PVB worldwide is Solutia, the former Monsanto, with trademark Saflex. There are two main producers, Solutia and DuPont, both reluctant to reveal their capacity figures. It is estimated that Solutia controls 50% of the world supply of PVB, DuPont controls 30%, and smaller companies such as Sekisui in Japan and the former Hüls plant at Troisdorf control the remaining 20% (see Table 3.1) [6].

In 2011, 89% of PVB was used for the production of construction and automotive industry safety glass, 4% was used in solar PV materials, and the remaining 7% was used in paints, glues, dyes, and other materials [7]. Its main use is in car windshields, which is about 90% in the United States but only 60% in Europe; its application in construction glazing is relatively more important in Europe. In 2008, the global PVB resin consumption was 270,000 tons, with a market size of about US$2200 million. The demand for this product was estimated to increase at an average growth rate of 6%,

TABLE 3.1
PVB Producers

Manufacturer	Trade Name	Patent	Base
Solutia[a]	Butvar	1938	USA
DuPont	Butacite	1937	USA
Kuraray[b]	Trosifol	1953	Japan
	Mowital		
	Pioloform		
Sekisui	Silec	1958	Japan
Hindustan Inks & Resins Ltd (HIRL)	–	–	India
Formosa Chemicals & Fiber Corp. (FCFC)	–	–	Taiwan

[a] Monsanto, with trademark Saflex, was at the origin of PVB for interlayer films in glass panes. Solutia was formed from the demerger from Monsanto, in September 1997. Solutia has been completely bought by Eastman Kodak in 2012.

[b] Formerly held by HT Troplast and sold to Kuraray in 2005 and also by Clariant and sold in 2002.

leading to a 2015 consumption that will amount to 410,000 tons. Solutia, Sekisui, DuPont, and Kuraray occupy 96% of the market share. The automotive industry is the largest ultimate customer of PVB films since safety glass is mainly used for automotive windshields where industry safety standards are very strict. According to statistics [7], in 2009, the automobile industry PVB films represented a market size of US$818 million with a 6.5% annual growth rate. There is a potential in its use in side windows and rear windows because the movable side window must be very hard, and at present, the majority of movable glass materials used are relatively fragile and not safe enough.

The construction industry is the second largest user of PVB films as an interlayer material embedded in safety glass, being mainly used in office buildings. In 2009, the global construction industry consumption of PVB films amounted to 89,000 tons, with US$614 million in sales revenue. At present, many factors promote the PVB membrane application in the building trade, such as regulation for residential buildings to use safe glass and better noise reduction.

The demand of the PV market is increasing since PVB membrane is used in the PV modules as a sealing material. The PV cell panel is characterized by long service life, generally up to 20 years, and sealing is a means to prolong the life of modules. At present, the PV cell panel sealants are dominated by EVA, with PVB as a thermoplastic material being the second candidate.

3.2 POLYMER FORMATION

3.2.1 Polymerization

Poly(vinyl acetate), which is polymerized from vinyl acetate by suspension or solution polymerization, is the starting material for the production of PVB. First, PVA is obtained from poly(vinyl acetate) by means of transesterification (alcoholysis) usually with methanol and base catalysis (Scheme 3.1).

Hydrolysis of poly(vinyl acetate) produces PVA with predominantly 1,3-glycol units and also 1,2-glycol units that come from head-to-head configuration of vinyl acetate monomers and, depending on the conditions of polymerization, also branches. Second, the well-known reaction between aldehydes and alcohols, where the nucleophilic addition of an alcohol to the carboxylic group of an aldehyde produces a hemiacetal, and unstable hemiacetals further react with another alcohol to yield a stable acetal ring, is used to form acetal under conditions of acid catalysis. Finally, in the net reaction, one molecule of aldehyde reacts with two molecules of alcohol (Scheme 3.2).

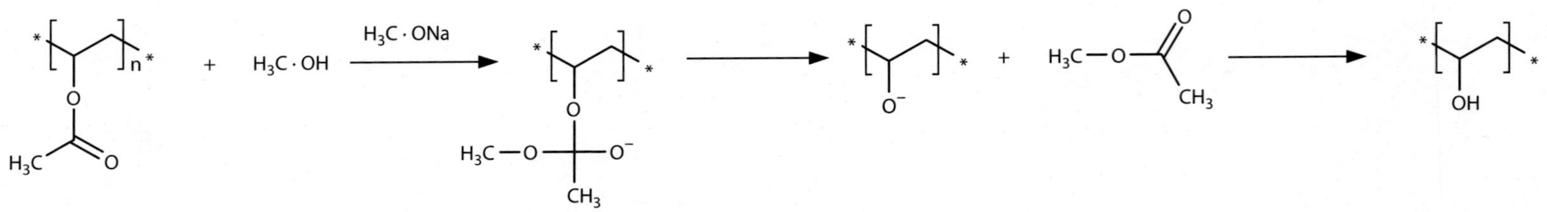

SCHEME 3.1 Transesterification of poly(vinyl acetate).

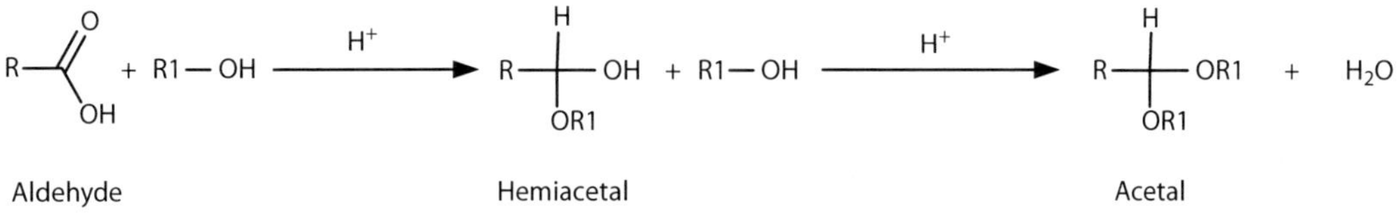

Polyvinyl Butyral

SCHEME 3.2 Conversion from aldehyde to acetal.

PVA reacts with an aldehyde primarily to form six-membered rings between adjacent, intramolecular hydroxyl groups; to a lesser extent, intermolecular acetals are also formed. This side reaction can lead to branched and eventually cross-linked polymer. The reaction can be used for the making of any vinyl acetal polymer and proceeds on the various units in intramolecular reactions (Schemes 3.3 and 3.4) and, to a lesser extent, in intermolecular reactions (Scheme 3.5).

As a particular case, the reaction of an aqueous PVA solution with butanal (BA) in the presence of small amounts of mineral acid, acting as a catalyst, yields PVB. The hydroxyl groups of PVA react with BA to form 1,3-dioxane rings (acetal rings), and partial or even almost complete exchange of the original functional group for a new one can occur (Scheme 3.6). Detailed mechanisms are given in Ref. [8].

Since not all hydroxyl groups of the PVA react with aldehyde, PVB invariably contains a certain percentage of hydroxyl groups. Moreover, a small percentage of the acetyl groups always remain in the polymer chain from the upstream transesterification during which polyvinyl acetate is converted to PVA (generally less than 2.5 mol%). Therefore, the final product has the character of a terpolymer of vinyl butyral, vinyl alcohol (VA), and vinyl acetate (Scheme 3.7).

From statistical reasons, mainly because the cyclization reaction requires the presence of two adjacent hydroxyl groups on the PVA chain, Flory [9] calculated the theoretical upper limit of the acetalization reaction of PVAs assuming no vinyl acetate unit. The amount of isolated and unreacted

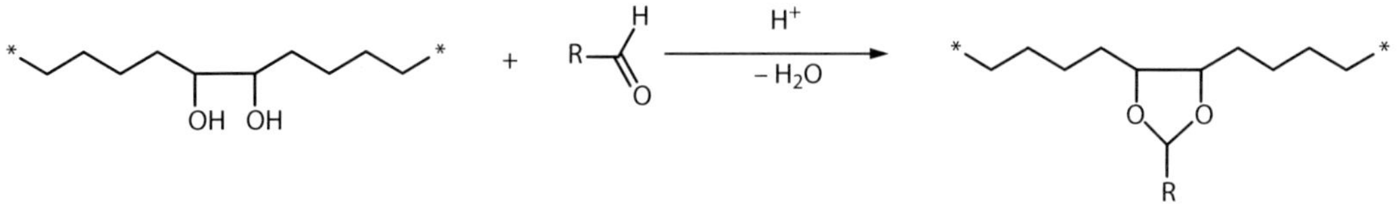

SCHEME 3.3 Intramolecular acetalization of the 1,3-glycol group.

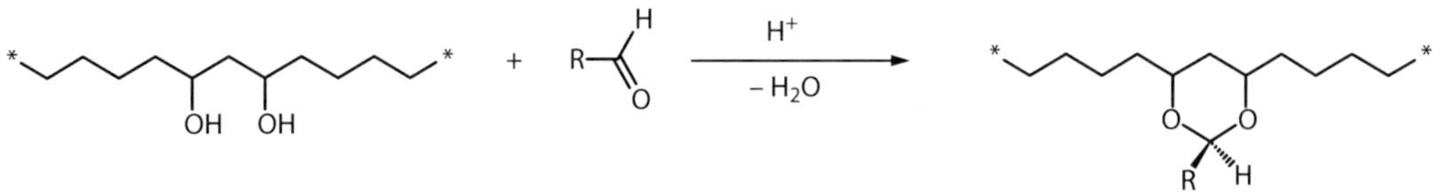

SCHEME 3.4 Intramolecular acetalization of the 1,2-glycol group.

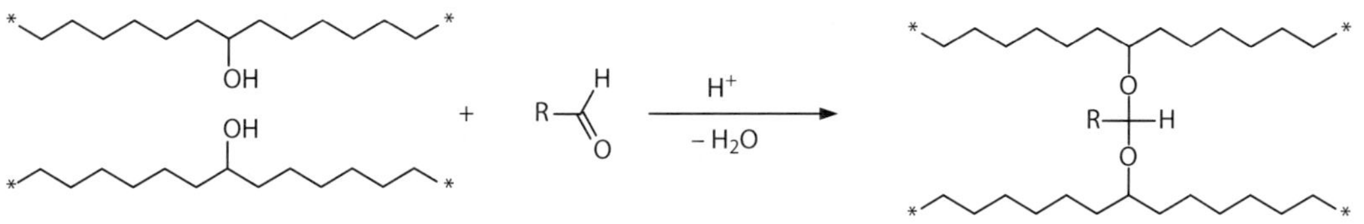

SCHEME 3.5 Intermolecular acetalization of the 1,3-glycol group.

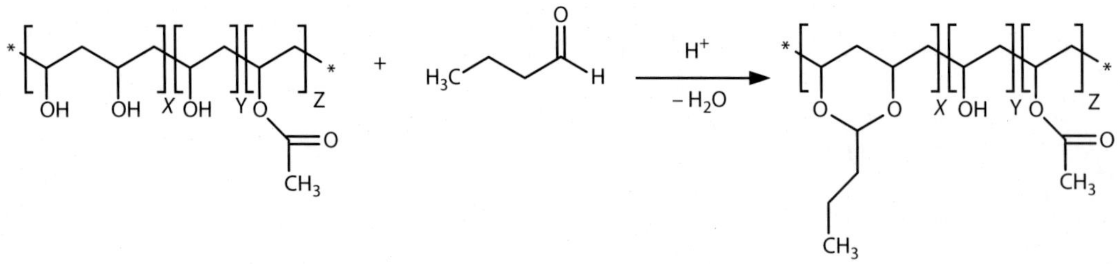

SCHEME 3.6 Acetalization of PVA in the case of BA.

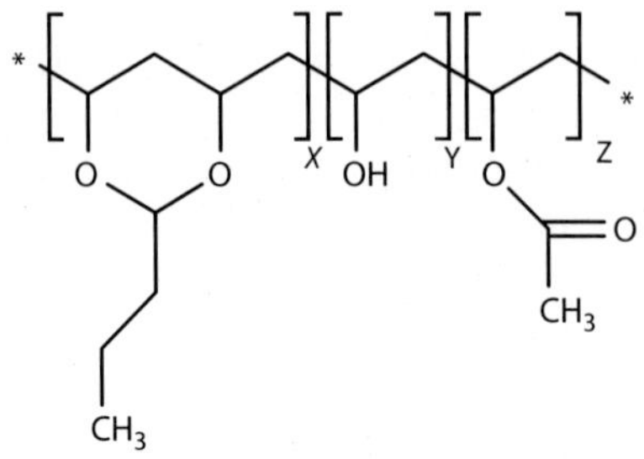

SCHEME 3.7 Terpolymer structure of PVB.

hydroxyl units should be 0.184 mol% for 1,3- and 1,2-glycol containing PVA and 0.135 mol% for pure 1,3-glycol containing PVA if acetalization is irreversible. For acetalization with BA, always assuming no vinyl acetate unit, this corresponds to an unreacted VA content between 12.25 and 8.84 wt% in PVB. Attempts have been made to synthesize polyvinyl acetals with a degree of acetalization exceeding the predicted values [10]. Raghavendrachar and Chanda [11] suggested a possibility of achieving degrees of acetalization above 86.5% while keeping the assumption of irreversibility, but the predicted value has not been significantly exceeded. Indeed, Flory initially assumed that acetalization was an irreversible reaction, but the question of the reversibility of formation of acetal rings was left open. Reversibility of acetalization and interchange can theoretically lead to higher levels of acetalization and can be taken into account in statistical calculations [12]. However, in practice, intermolecular acetalization limits higher levels of acetalization in commercial products to about 82 mol%. Consequently, PVBs always contain at least 7% by weight residual VA units. It is standard to use PVBs with VA contents of 9% to approximately 30% by weight (Table 3.2).

Since the residual acetate content is kept as low as possible (0–2.5 mol%), the balance between residual hydroxyl groups and acetal rings depends on the reaction conditions, and this balance provides distinct properties to the final polymer. The properties are determined mainly by three

TABLE 3.2
Typical Monomer Contents for PVB

Manufacturer	Trade Name	VA Content (wt%)	VAc Content (wt%)	Reference
Solutia	Butvar	11–20	0–2	[13]
DuPont	Butacite			
Kuraray	Mowital	11–27	0–2.5	[14]
Kuraray	Pioloform	14–18	0–2.5	[14]
Sekisui	S-Lec			

factors: the degree of acetalization, the degree of polymerization, and the used aldehydes, as they determine the polarity and molecular weight of the polymer. For the PVB for safety glass, the reaction of acetalization is stopped at about 75% conversion of the hydroxyl groups so that the residual or unreacted hydroxyl groups provide the required strength and adhesion to glass, since a good bond between the components of the laminate is imperative. Assuming no remaining acetate, the molar composition in VA (m in mol% VA) is deduced from the weight fraction of VA, w in wt% VA, by the following equation:

$$m = 100 \frac{142w}{44(100-w)+(142w)} \qquad (3.1)$$

Special grades of PVB were also obtained by the use of both butanal and ethanal [14] or furfural [15,16], leading to the slightly different structures (Schemes 3.8 and 3.9). Ionomeric PVB can be synthesized by using the sodium salt of o-benzaldehyde sulfonic acid in a preliminary step, and then going on the polymerization in the conventional way. Ionomeric PVB containing 3 and 5 mol% of ionomer groups has been used [17]. Grafting of polydimethylsiloxane, polyoxyethylene [18], and also long branches of PVB [19] has been described.

High acetalization reaction can be conducted in homogeneous solution processes. The acetals of PVA are generally prepared by the action of aldehydes on PVA in an aqueous medium. However, because these acetals are nonsoluble in water, the contact time of the reactants is not very long. Theoretically, 100% acetalization of PVA should be possible if the reaction is carried out homogeneously in a non-aqueous nonpolar solvent when hydrogen bonding is the least. It was reported [20] that various acetalizations of PVA, including conversion to PVB, could be carried out until 90–95 mol% in a mixture of DMF/benzene (4:1 v/v) using a catalytic concentration of ethyl nitrate dimethyl sulfoxide (1.4×10^{-3} mol/mol PVA); 93% acetalization was also claimed using N-methyl-2-pyrrolidone (NMP) as a solvent for PVA and PVB [21]. The simultaneous methanolysis and butyralization of poly(vinyl acetate) was also conducted via the reaction of methanol and BA with

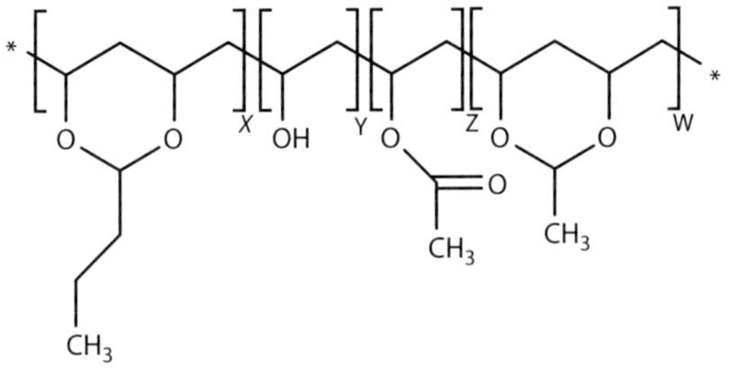

SCHEME 3.8 Poly(vinyl butyral-co-ethanal).

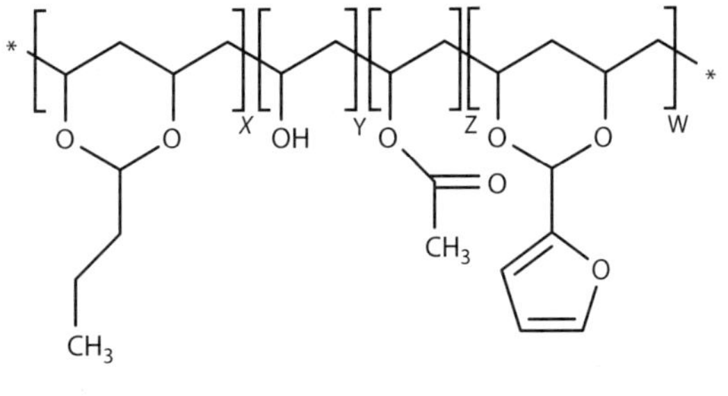

SCHEME 3.9 Poly(vinyl butyral-co-furfural).

polyvinyl acetate under batch conditions at temperatures from 70°C to 90°C at the vapor pressure of the reactants [22].

Finally, from the chemical engineering point of view, for the preparation of poly(vinyl acetal), five different processes are known [23–25]:

1. *Precipitation.* An aqueous solution of PVA reacts with the aldehyde until the homogeneous phase reaction changes to a heterogeneous reaction when the acetal precipitates.
2. *Dissolution.* PVA powder is suspended in a suitable nonsolvent that dissolves the aldehyde and the final product. The reaction starts out heterogeneously and is completed homogeneously.
3. *Homogeneous reaction.* The reaction starts in a water solution of PVA. A solvent for the acetal, which can mix with water, is added continuously to prevent precipitation.
4. *Heterogeneous reaction.* PVA in film or fiber form reacts with the aldehyde.
5. *Direct conversion of poly(vinyl acetate) to poly(vinyl acetal).* This is also a homogeneous reaction. The poly(vinyl acetate) is dissolved in a suitable solvent and is hydrolyzed by a strong mineral acid and acetalized at the same time.

A detailed list of patents on the ingredients and polymerization of PVB can be found in Ref. [3]. Other details on synthesis can be found in Ref. [26] and in various papers on the subject [18,27,28]. Hydrolysis and acetalization can be conducted either sequentially or concurrently. In practice, only two basic processes with variants are used commercially today, and they are described hereafter. More detailed schemes of the polymerization arrangements, especially concerning the solvent process using sequential hydrolysis and acetalization and the aqueous process, for chemical engineering are described in Ref. [3].

The solution process is carried out in an organic solvent. To limit the undesirable poly(vinyl acetate) content, simultaneous hydrolysis and acetalization cannot be used. Poly(vinyl acetate) is first saponified by transesterification with ethanol and a mineral acid catalyst to produce PVA that precipitates and therefore can be separated from remaining ethanol and ethyl acetate by centrifugation. PVA is then acetalized in a separate operation, after being slurried with ethanol, and heated with BA and the acid catalyst. As the acetalization reaction proceeds in the solvent process, the slurry of PVA dissolves as PVB in the reaction mixture. Upon completion of the acetalization reaction, PVB is precipitated into water, and aqueous neutralization of the acid catalysis occurs. Neutralization, with sodium hydroxide or potassium hydroxide, and washing stages are necessary to remove the acid catalyst and to improve resin thermal stability [29]. By-products can be recovered. Resin has very low levels of gel from intermolecular acetalization.

The aqueous process is carried out in water, where PVA dissolves easily. The acetal is precipitated during the reaction of the aldehyde with an aqueous solution of PVA containing acetic acid and a mineral acid. Acidity is neutralized after completion of the reaction. Precipitated resin in the aqueous acetalization shows higher levels of intermolecular acetalization, which can be avoided by emulsifiers ammonium thiocyanate or urea that improve solubility in water.

To increase the average molecular weight and polydispersity, small quantities of various dialdehyde or trialdehyde can be added during the acetalization to promote a low amount of cross-linking. Other cross-linking agents are cited next.

3.2.2 Postpolymerization Chemistry

PVB is a thermoplastic material that is soluble in a large number of organic solvents, but it can be cross-linked with various species owing to OH groups in order to improve solvent, chemical, and thermal resistances. The cross-linking capacity depends on the number of OH groups available. Phenolic (Scheme 3.10), melamine (Scheme 3.11), epoxy (Scheme 3.12), isocyanate (Scheme 3.13), aldehyde (Scheme 3.14), and formaldehyde hydrate can be used [30]. Urea and melamine as well as polyisocyanate enable curing at room temperature with acid catalysts.

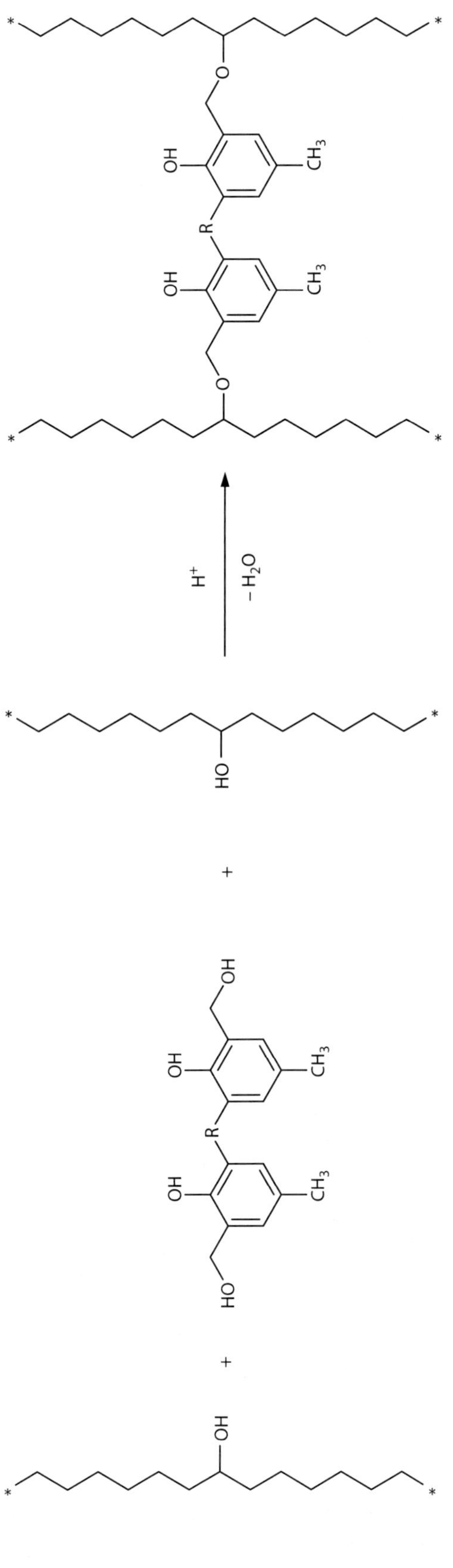

Polyvinyl Butyral

SCHEME 3.10 Reaction of PVB with phenolic compounds.

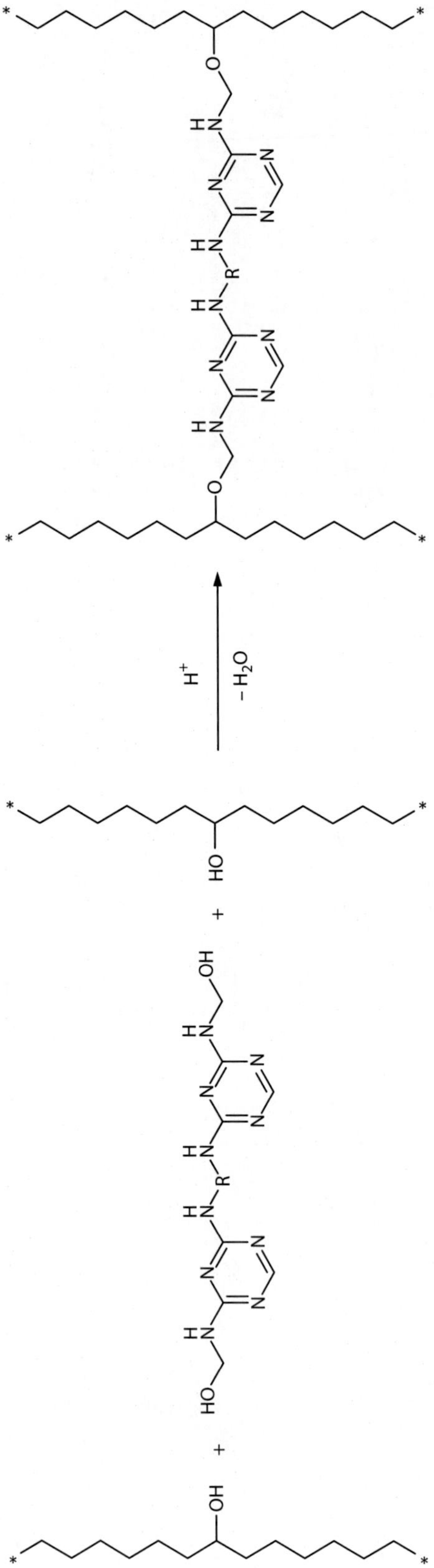

SCHEME 3.11 Reaction of PVB with melamine.

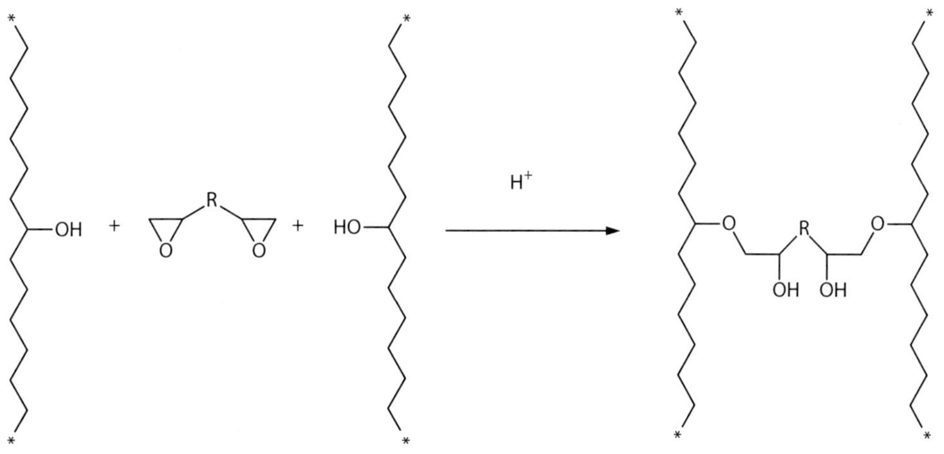

SCHEME 3.12 Reaction of PVB with epoxy compounds.

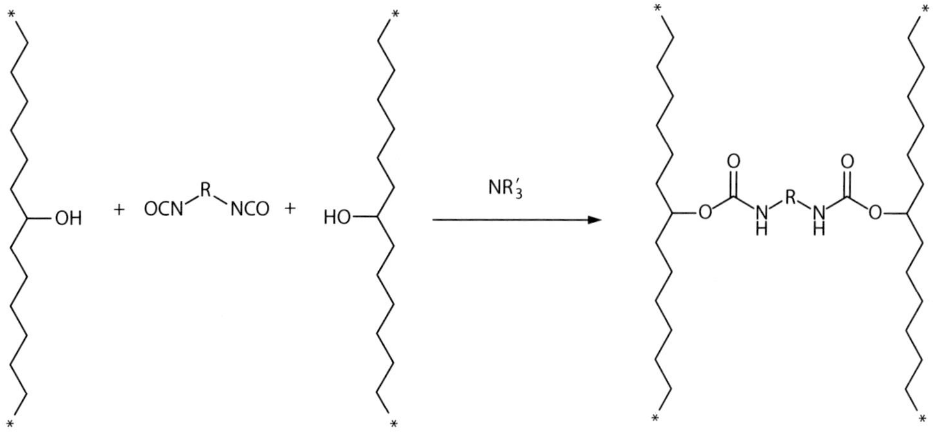

SCHEME 3.13 Reaction of PVB with isocyanates.

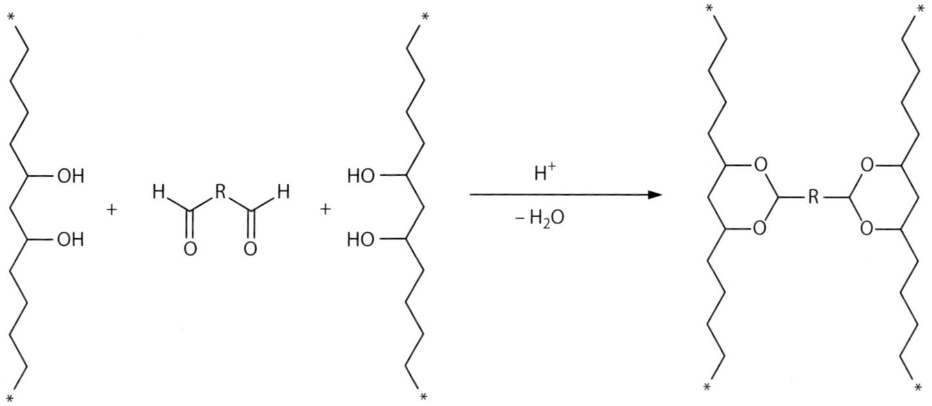

SCHEME 3.14 Reaction of PVB with aldehydes.

3.3 STRUCTURAL AND PHASE CHARACTERISTICS

Finally, PVB is a random terpolymer composed of vinyl butyral, VA, and vinyl acetate monomeric units. Structure and properties are strongly dependent on the hydroxyl content. For example, it controls the adhesion to surfaces, improves properties of thermoset resin, influences cross-linking behavior, and affects miscibility and morphology of blends. Therefore, much analytical effort has been exerted to develop reliable tools for its characterization [31–33]. PVB is not a hazardous product. PVB used for the process is a colorless and odorless powder. Users still need to protect themselves from the dust, which can irritate throat, eyes, and respiratory system and cause coughing. Under air combustion, PVB will produce carbon dioxide. Butavar and Mowital are nontoxic and suitable for food packaging according to US Food and Drug Administration 21 CFR.

3.3.1 IR Spectroscopy

Fourier transform infrared (FTIR) spectra show typical peaks attributed to PVA, with OH stretching vibration bands at 3490 cm^{-1}, plus absorptions arising from the presence of acetalic functions (in the region between 1050 and 1150 cm^{-1}). The band at 1740 cm^{-1} appears because of the acetate group. Figure 3.1 and Table 3.3 give the assignation of the various peaks [21,34].

3.3.2 UV Spectroscopy

Figure 3.2 exhibits the UV absorption spectra of PVB. The UV–vis spectral analysis shows a strong absorption band at around UV at 415 nm and a small shoulder at 390 nm [35]. The absorption band observed for PVB is expected for hydroxyl-containing groups related to n to π^* electronic transition characteristic of oxygen lone pair of electrons due to the OH groups, or of the acetyl C=O and C–O–C cyclic electrons present in the polymer backbone, all functional groups involved in intramolecular and intermolecular hydrogen bonding.

3.3.3 ^{13}C NMR Spectroscopy

Due to the terpolymer nature of PVB, ^1H and ^{13}C nuclear magnetic resonance (NMR) spectroscopy is useful to determine the PVB structure [36]. Since the acetalization degree is reflected in the solubility of PVB in various solvents, spectra of PVB must be obtained in dimethyl sulfoxide (DMSO)-d6 for low acetalization and in $CDCl_3$ for high acetalization rates. Deuterated methanol (CD_3OD) was also used [37]. ^{13}C NMR spectrum of PVB (90 mol% VB) in deuterated chloroform ($CDCl_3$) is given in Figure 3.3 with tetramethylsilane (TMS) as the internal standard. The ^{13}C NMR assignments of peaks and chemical shifts are given in Table 3.4 in correspondence with the figure and the racemic (r) and meso (m) stereostructures [21,26,36]. Indeed, as in any vinyl monomeric unit with asymmetric carbon, acetal ring conformations are due to the rotational isomeric states of meso and racemic diads of parent PVA (Scheme 3.15), also responsible for the meso and racemic peaks of the remaining VA [38]. Of course, the degree of acetalization and the racemic stereostructure fraction of acetal rings are correlated.

3.3.4 ^1H NMR Spectroscopy

Figure 3.4 gives an example of ^1H NMR spectrum in DMSO-d6 of PVB with 21–27 wt% VB using TMS as the internal standard. The ^1H NMR assignments of peaks and chemical shifts are given in Table 3.5. In addition to confirmation of structures, it enables the quantification of the extent of acetalization by the following equation:

Polyvinyl Butyral

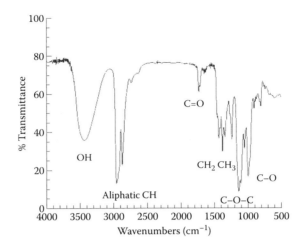

FIGURE 3.1 FTIR spectrum of PVB with 24–27 wt% VA.

TABLE 3.3
IR Assignation of Functional Groups

Wave (cm^{-1})	Groups
3490	OH (VA)
2850 and 3000	CH$_3$, CH$_2$, CH
1400 and 1280	Alkynes
1740	C=O (VAc)
1200	Ester
1100	C–O–C (VB)
960	Acetal (VB)

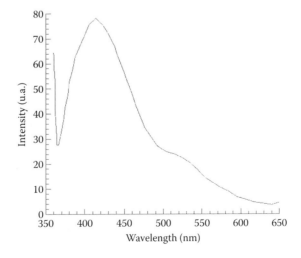

FIGURE 3.2 UV absorption spectrum of PVB with 18–20 wt% VA. (From Kalu, E.E. et al., *Int. J. Electrochem. Sci.*, 7, 5297–5313, 2012. © Copyright ESG. Adapted and reproduced with permission.)

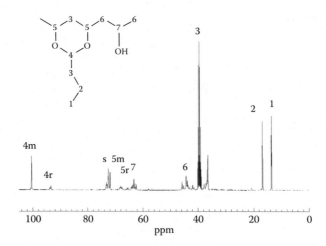

FIGURE 3.3 ^{13}C NMR spectrum of PVB (90 mol% VB) in CDCl$_3$ with TMS as standard. (Adapted from Fernandez, M. et al., *J Appl Polym Sci* 102, 5007–5017, 2006.)

TABLE 3.4
Assignments of ^{13}C NMR Peaks

Number	Structural Unit	Chemical Shift (ppm)
1	Butyral CH$_3$	13.7–18.1–18.2
2	Butyral CH$_2$	21.9–22.0
3	Butyral CH$_2$	41.1–41.6
		46.6–46.8
4m	Meso-butyral O–CH–O	105.5–105.6
4r	Racemic butyral O–CH–O	98.5–98.6
5m	Meso butyral CH	76.4–78.8
5r	Racemic butyral CH	72.4–72.5
6	Vinyl alcohol–CH$_2$	41.1–41.6
		49.3–49.4
7	Vinyl alcohol–CH	68.0–69.0

Source: Fernandez, M. et al., *J Appl Polym Sci* 102, 5007–5017, 2006; Dhaliwal, A. and Hay, J., *Thermochim Acta* 391, 245–255, 2002.

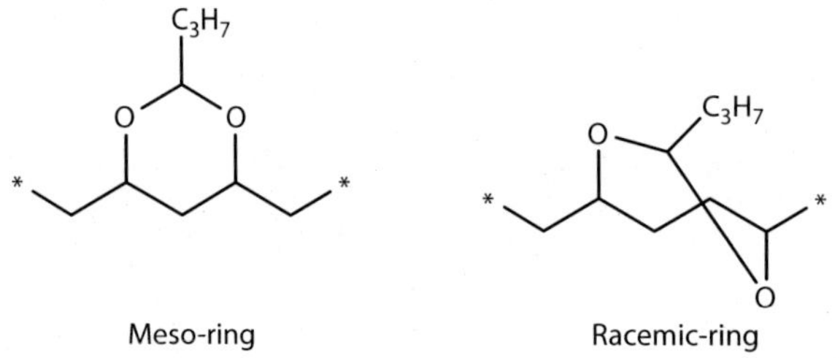

SCHEME 3.15 Meso and racemic stereostructures of PVB.

Polyvinyl Butyral

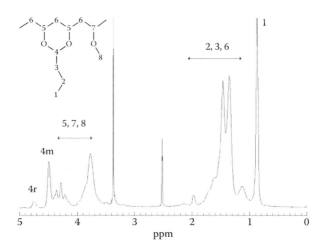

FIGURE 3.4 ^{1}H NMR spectrum of PVB (24–27 wt% VA) in DMSO with TMS as standard.

TABLE 3.5
Assignments of ^{1}H NMR Peaks

Number	Structural Unit	Chemical Shift (ppm)
1	Butyral methyl CH$_3$	0.95
2, 3, 6	Butyral and alcohol methylene CH$_2$	1.2–1.8
4m	Meso-butyral dioxymethine O–CH–O	4.6
4r	Racemic butyral dioxymethine O–CH–O	4.8
5, 7, 8	Methine butyral CH and vinyl OH	3.6–4.4

Source: Fernandez, M. et al., *J Appl Polym Sci* 102, 5007–5017, 2006.

$$VB = \frac{2}{\left(\dfrac{3A_{CH_2}}{A_{CH_3}}\right) - 6} \quad (3.2)$$

The dioxymethine protons 4m and 4r could be utilized to determine the fraction of the stereostructure of butyral rings. All resonances in the range 15–50 ppm correspond to methylene and those from 60 to 110 ppm to methines. Other finer interpretations can be found in Ref. [37], which use both proton and carbon NMR techniques coupled to two-dimensional J-resolved spectroscopy for better assignment of the peaks.

3.3.5 Dosing of Acetate

According to Ref. [39], the quantity of acetate can be dosed by a simple method. Iodine–iodide solution placed on a PVB sample changes its color in a few seconds. PVB with less than 15% acetate content gives a yellow color, 15–27% produces green, 27–40% gives a bluish-green color, and over 40% produces blue.

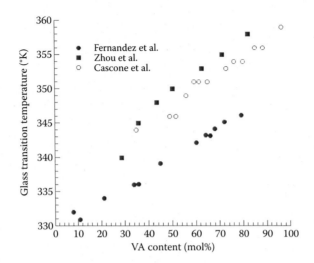

FIGURE 3.5 Variation of the glass transition temperature of PVB with the VA content. (Data from Zhou, Z. et al., *Turk J Chem*, 21, 229–238, 1997; Cascone, E. et al., *J Appl Polym Sci.*, 82, 2934–2946, 2001; Fernandez, M. et al., *J Appl Polym Sci.*, 102, 5007–5017.)

3.3.6 Glass Transition Temperature

Generally, PVB is amorphous, only polymers with a very high VA content (higher than 63.3 wt% VA) are crystallizable, and the melting temperature was found to be between 171°C and 218°C, approaching that of PVA [40]. In all other cases, PVB is amorphous and displays an easily observable glass transition. The effect of the vinyl content on the glass transition of PVB has been studied in detail by varying the VA content [8,21,40], and the glass transition temperature was found to increase upon the VA molar content (Figure 3.5). Butyral groups have a plasticizing effect, and the increase in the content of hydroxyl groups gives rise to an increase in hydrogen bonding, which leads to a significant departure from group additivity. The T_g of PVB with low VA content (10 mol%) is nearly 330 K, that of PVB with 50 mol% is 350 K, whereas that of a PVB with 80 mol% is 358 K. Higher values were reported on unplasticized Butvar PVB (18 wt% VA) at 80°C [41] and by Dhaliwal (74 K) [36].

3.3.7 Solvents, Solution Properties, and Molecular Weight Characterization

The useful relations among molecular weight, intrinsic viscosity [η], and specific η_{sp} or relative viscosity η_r are given in the following (in which all constants are solvent- and temperature-dependent):

$$\text{the Mark–Houwink relation: } [\eta] = KM^a \quad (3.3)$$

$$\text{the Huggins and Kramer relation: } \frac{\eta_{sp}}{C} = [\eta] + K_H[\eta]^2 C \quad (3.4)$$

$$\text{the Kramer relation: } \frac{\ln(\eta_r)}{C} = [\eta] - K_K[\eta]^2 C \quad (3.5)$$

For PVB, the Mark–Houwink exponent (*a*) has been found to be 0.72 in tetrahydrofuran (THF) at 25°C. The *K* parameter depends on the hydroxyl content being 2.89 × 10⁻⁴ mL/g for 20 wt% VA and slightly less (2.52 × 10⁻⁴ mL/g) for 10 wt% VA [42]. Results in glycol ether [43] showed values of $a = 0.72$, $K_H = 0.41$ g/dL, and $K_K = -0.10$ g/dL for PVB with 22 wt% VA. Virial and refraction indices increment dn/dc of PVB with 43 mol% VA can be found in Refs. [19,44,45] according to Table 3.6. For methanol and methyl isobutyl ketone, most mixtures are better solvents for PVB than either individual solvent that leads to polymer aggregation.

Most of the industrial PVBs have weight average molecular weights in the range 40,000–250,000 g/mol [13,14]. The molecular weight of PVB can be determined by size-exclusion chromatography (SEC). THF can be used as a solvent to measure the molecular weight of PVB by SEC [46], but *hexafluoroisopropanol* (HFIP) is required for PVB with VA content higher than 20% [47]. Indeed, HFIP eliminates aggregates and gives better accuracy of molecular weight averages as demonstrated by low-angle laser light scattering (LALLS). Normal-phase gradient polymer elution chromatography (NP-GPEC) was used for determination of molecular weight distribution (MWD) and VA content using dimethylacetamide (DMAc) with 0.5% lithium chloride (LiCl) as the solvent [31].

The solubility parameters have been given by theoretical analysis using the solubility parameter concept and blend miscibility [48]. The dispersion, polar, and hydrogen bonding contributions and the total solubility parameter of the repeating unit δ_D, δ_P, δ_H, and δ have been found to be 7.72, 2.90, 3.26, and 8.87 MPa$^{1/2}$, respectively, for VB and 7.62, 6.80, 11.54, and 15.41 MPa$^{1/2}$ for VA. Overall, different values were found by using the Hoftyzer–Van Krevelen method of structural groups or other calculation methods (see Table 3.7) [49–51].

TABLE 3.6
Polymer Solvent Data

Solvent	Temperature (°C)	A_2 (mL mol/g^2)	dn/dc
Methanol (MeOH)	25	−0.5	0.1562
Methanol (MeOH)	50	–	0.1676
Methyl isobutyl ketone (MIBK)	44	–	0.1015
MIBK/MeOH (3:1 vol)	25	0.00079	0.1120–0.1145
MIBK/MeOH (1:1 vol)	25	0.00036	0.1284
MIBK/MeOH (9:1 vol)	25	0.00015	0.1062
Acetic acid	25	0.00104	0.1173
Acetic acid	80		0.1302
DMAc–0.5% LiCl	35	0.00139	0.0570

Source: Data from Striegel, A.M., *Polym Int* 53, 1806–1812, 2004; Paul, C. and Cotts, P., *Macromolecules* 19, 692–699, 1986; Paul, C. and Cotts, P., *Macromolecules* 20, 1986–1991, 1987.

TABLE 3.7
Solubility Parameters

Name	VA (wt%)	δ_D	δ_P	δ_H	Reference
Mowital B60H	18–21	15.5	6.5	10.4	[50]
Mowital B30T	24–27	19.1	9.5	12.2	[51]
Mowital B30H	18–21	18.6	12.9	10.3	[49]
Mowital B60H	18–21	20.2	11.2	13.3	[49]
Butvar B76	11.5–13.5	18.6	4.4	13.0	[49]
Butvar B-90, B-98	18–20	21.7	7.9	14.6	
Butvar B-72, B-74	17.5–20	21.2	8.7	14.0	
Butvar B-76, B-79	11–13.5	17.7	7.2	12.6	

PVB is soluble in the following solvents: acetic acid, acetone, methanol, ethanol, 2-propanol, butanol, 2-butoxyethanol, cyclohexanone, benzyl alcohol, 1-methoxy-propanol-2, butyl glycol, *n*-butyl acetate, ethyl acetate, *N,N*-dimethylacetamide, *N,N*-dimethylformamide, *N,N*-dimethylsulfoxide, NMP, and THF, but solubility is strongly dependent on the composition and especially the VA content. PVB is also resistant to weak and strong acids, and weak and strong bases. Table 3.8 shows the solubility of commercial PVB (Mowital, Pioloform, and Butvar) in various solvents. The VA content is clearly the important parameter, whereas molecular weight has a secondary effect.

3.3.8 Thermal Degradation (High Temperature Conditions in Nitrogen)

The thermo-oxidation and thermal degradation of PVB at high temperatures in various atmospheres have been studied by many authors. Indeed, with PVB being used as a binder for ceramics processing, its elimination during the process of sintering is a critical issue for the quality of the final product. Under nitrogen, PVB starts to decompose at 280°C (ramp of 5°C/min) and under air at 200°C. As it relates to the processing properties, thermal degradation in air will be addressed in Section 3.4.7. The main products of degradation are BA, 2-butanal, butanoic acid, acetic acid, 2,5-dihydrofuran, and butanol. Traces of other chemicals such as 1-phenyl ethanone, 3,5-diphenyl, 1,2,4-trioxolane, and 2,4-hexadiene are also detectable. The degradation of PVB takes place in three steps: First, water from the VA copolymer is eliminated between 315°C and 350°C with formation of a small amount of butanal. The second step is the degradation of side groups and main chain scission. In the second step, 85% of the polymer is degraded between 350°C and 445°C. Finally, cross-linked and cyclic compounds are formed above 445°C [52]. Hydroxyl groups have a dramatic effect on the stability of PVB [8,21]. The decomposition of PVB takes place in one step when the percentage of VA unit is less than 80%, and PVB starts to decompose at 280°C with a weight loss between 80% and 90% at 480°C. For higher VA content, decomposition of PVB takes place in two steps with maximum at 316°C and at 382°C and weight loss of 40% and 50%, respectively (Figure 3.6). As has already been indicated in the synthesis part, the addition of bases such as hydroxides increases the stability of PVB, and the starting decomposition temperature can be delayed until 330°C in air [29].

3.4 POLYMER ENGINEERING PROPERTIES

3.4.1 Density

The density of PVB ranges from 1.083 g/cm^3 for PVB with a VA content of 11 wt% (28.5 mol%) to 1.100 g/cm^3 at a VA content of 19 wt% (43.1 mol%) [8,13]. The variation of the specific volume with temperature was given by Wilski for similar VA contents [53] with a value of variation with temperature of 2.52×10^{-6} cm^3/g/°C below the glass transition and 6.91×10^{-6} cm^3/g/°C above it (Figure 3.7).

3.4.2 Surface Properties

Surface tension of PVB was evaluated by the pendant drop method [54] and found to be 36.4 ± 0.2 mN/m at 240°C with linear temperature dependence of −0.6 mN/m between 240°C and 260°C. The value of 38 mN/m was calculated with dispersion and polar components of 23.0 and 15.0 mN/m, respectively [51]. Values of the critical surface tension by the Zisman method given at room temperatures (24–25 mN/m [55,56]) for various VA contents show very little influence of OH groups.

TABLE 3.8
Solubility of Commercial PVB in Various Solvents, 10% Solution unless Indicated

Polymer	Butvar B76 B79	Mowital B30HH	Mowital B60HH	Butvar B90 B98	Mowital B16H B20H	Mowital B30H	Mowital B45H	Mowital B60H	Mowital B75H[a]	Butvar B72[a] B74[a]	Mowital BX860[b]	Mowital B45M	Mowital B30T B60T	Pioloform BL16[c]
VA wt%	11.5–13.5	11–14	12–16	18–20	18–21	18–21	18–21	18–21	18–21	17.5–20	18–21	21–24	24–27	14–18
Methanol	SW	PS	PS	S	S	S	S	S	S	S	PS	PS	S	S
Ethanol	S	S	S	S	S	S	S	S	S	S	S	S	S	S
n-Propanol	S	S	S	S	S	S	S	S	S	S	S	S	S	S
n-Butanol	S	S	S	S	S	S	S	S	S	S	S	S	S	S
Diacetone alcohol	S	S	S	S	S	S	S	S	S	PS	S	S	S	S
Tetrahydrofuran	S	S	S	S	S	S	S	S	S	S	S	S	S	S
Methyl cellulose	S	S	S	S	S	S	S	S	S	S	S	S	S	S
Ethyl cellulose	S	S	S	S	S	S	S	S	S	S	S	S	S	S
Butyl cellulose	S	S	S	S	S	S	S	S	S	S	S	S	S	S
Methyl acetate	S	S	S	PS	PS	PS	PS	PS	I	I	PS	PS	I	S
Ethyl acetate	S	S	S	PS	PS	S	PS	PS	S	I	PS	PS	I	S
Butyl acetate	S	S	S	PS	PS	PS	PS	PS	I	I	PS	PS	I	S
Acetone	S	S	S	SW	S	S	S	PS	I	I	PS	PS	PS	S
Methyl ethyl ketone	S	S	S	PS	S	S	S	PS	I	SW	PS	PS	S	S

(*Continued*)

TABLE 3.8 (CONTINUED)
Solubility of Commercial PVB in Various Solvents, 10% Solution unless Indicated

Polymer	Butvar B76 B79	Mowital B30HH	Mowital B60HH	Butvar B90 B98	Mowital B16H B20H	Mowital B30H	Mowital B45H	Mowital B60H	Mowital B75H[a]	Butvar B72[a] B74[a]	Mowital BX860[b]	Mowital B45M	Mowital B30T B60T	Pioloform BL16[c]
Methyl isobutyl ketone	S	S	S	I	PS	S	PS	I	I	I	PS	PS	I	S
Cyclohexanone	S	S	S	S	S	S	S	S	S	S	PS	PS	S	S
Toluene	PS	P	PS	PS	PS	PS	PS	PS	I	I	I	I	I	PS
Xylene	PS	P	PS	PS	PS	PS	PS	I	I	I	I	I	I	PS
Acetic acid	S	S	S	S	S	S	S	S	S	S	S	S	S	S
Dimethyl sulfoxide	S	S	S	S	S	S	S	S	S	S	S	S	S	S

Source: Kuraray, *Polyvinylbutyral of Superior Quality*. Hattersheim am Main, Germany: Kuraray Europe GmbH, 2013; Solutia, *Butvar Polyvinyl Butyral Properties and Uses. On-line Technical Bulletin N°. 2008084E.* St. Louis, Mo: Solutia, Inc., 2008.

Note: I: insoluble; PS: partially soluble; S: soluble; SW: swelling.

[a] 5% solution.
[b] 14–18 wt% acetate.
[c] With ethanal.

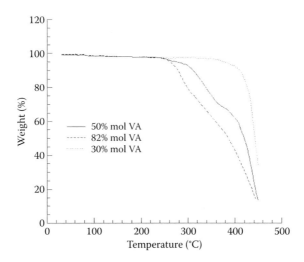

FIGURE 3.6 Weight loss in ATG for PVB with various VA content. (From Zhou, Z. et al., *Turk J Chem*, 21, 229–238, 1997. © Copyright Tübitak. Adapted and reproduced with permission.)

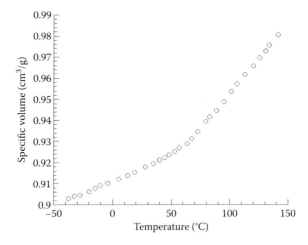

FIGURE 3.7 Specific volume of PVB: variation with temperature. (From Wilski, V. H.: Die spezifische Wärme des Polyvinylbutyrals. *Angew Makromol Chem*. 1969. 6. 101–108. Copyright Wiley-VCH Verlag GmbH & Co. KGaA. Reproduced with permission.)

3.4.3 Adhesion Properties

The hydroxyl groups of PVB are also responsible for its outstanding adhesion to many substrates, the most important for technological applications being its affinity to glass. Both chemical bonding and hydrogen bonding with glass silanols are involved (Figure 3.8). Figure 3.9 shows the effect of the hydroxyl content on the work of adhesion measured on glass beads [57]. According to measurements on interlayers for laminated glass [58], the activity of hydroxyl depends on the pH and on the water content in PVB sheets. Alkali modification decreases the adhesion of PVB to glass, whereas acidic modification increases it and provides higher stability of adhesion regarding the water content in PVB sheet. Therefore, adhesion depends on the amount of organic acid added and on the provenance of PVB and its plasticizers. Improvements of adhesion by silane coupling have been explored. The mono-alkoxy silanes have weak impact on adhesion when compared to untreated

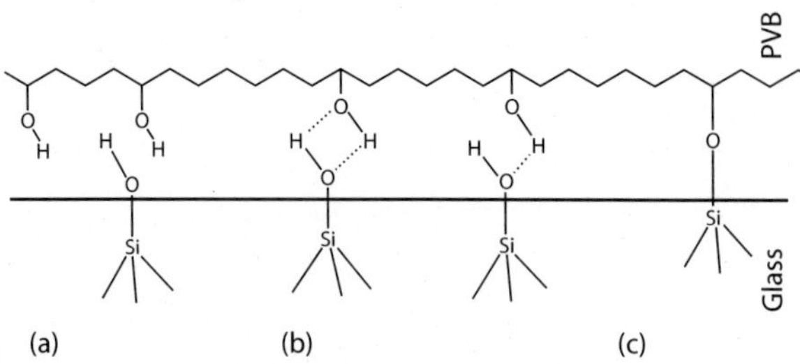

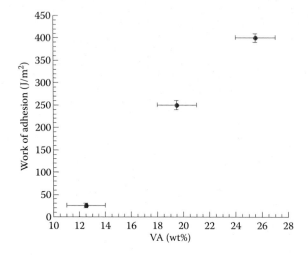

FIGURE 3.8 Adhesion to glass owing to interactions of hydroxyls with silanol groups. (a) Interacting groups. (b) Hydrogen bonding. (c) Covalent bonding.

FIGURE 3.9 Effect of VA content on adhesion. (Data from Nguyen, F.N. and Berg, J.C., *J. Adhes. Sci. Technol*, 18. 1011–1026.)

glass due to the poor attachment of the mono-alkoxy silane to the mineral surface. Di- and tri-alkoxy silanes enhance adhesion strength due to stability of their bonds to the glass surface and formation of interphases with the polymer [59]. Aminosilanes form acid–base moieties between the hydroxyl groups of PVB and amino groups of aminosilane, and the adhesion strength is controlled by the length of the amino functional group, suggesting that the adhesion promotion is controlled by compatibility and penetration of the silane organo functional group [60].

3.4.4 Electrical Properties

In PVB, two major relaxation processes—one above room temperature, designated the β or secondary relaxation, and the other below room temperature, designated the α or primary relaxation—have been detected by dielectric measurements [61–63] and confirm results by thermally stimulated depolarization current spectroscopy [64,65] and mechanical measurements [30] (cf. also Section 3.4.5). Typical dispersion data are shown in Figure 3.10. The α process is attributed to the rotation of dipoles from one quasi-stable position to another. It is possible only at higher temperatures at which the diffusional motion of segments occurs, and it is strongly dependent on the molecular structure of the polymer. In PVB, the activation energy for this relaxation was found to be 63 kJ/mol. It is not a single relaxation process, but rather it has a distribution of relaxations. The β process is associated with the deorientation of the aligned dipoles involving the acetate/hydroxyl groups when large-scale

Polyvinyl Butyral

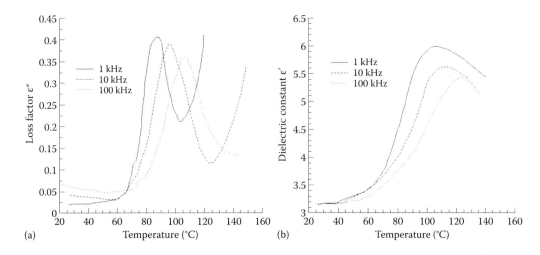

FIGURE 3.10 Dielectric properties of PVB (7 wt% VA). (a) Loss factor. (b) Dielectric constant. (From Funt, B., *Can J Chem*, 30, 84–91, 1952, © Canadian Science Publishing or its licensors. Adapted and reproduced with permission.)

TABLE 3.9
Dielectrical Properties of Butvar PVB Resins

Frequency (Hz)	B-72, B-74 ε'	B-72, B-74 ε''	B-76, B-79 ε'	B-76, B-79 ε''	B-90 ε'	B-90 ε''	B-98 ε'	B-98 ε''
50	3.2	0.0064	2.7	0.0050	3.2	0.0066	3.3	0.0064
10^3	3.0	0.0062	2.6	0.0039	3.0	0.0059	3.0	0.0061
10^6	2.8	0.0270	2.6	0.0130	2.8	0.0220	2.8	0.0230
10^7	2.7	0.0310	2.5	0.0150	2.7	0.0230	2.8	0.0240

Source: Data from Solutia, *Butvar Polyvinyl Butyral Properties and Uses. On-line Technical Bulletin N°. 2008084E.* St. Louis, Mo: Solutia, Inc., 2008.

conformational rearrangement of the main chain is frozen. The activation energy of this relaxation is lower (33 kJ/mol in PVB), and it becomes quite difficult to observe by conventional dielectric spectroscopy with high-molecular-weight samples. The β relaxation is of Debye type with a symmetrical distribution of relaxation times. Thermally stimulated depolarization (TSD) current studies also reveal a relaxation process at high temperatures that has been attributed to the release of trapped space charges. Comparisons of the dielectric behavior and particularly of the α dispersion of various polyacetals reflect the internal plasticization with increasing size of the substituent group (formal to butyral) and the shifting of dispersion to lower temperatures [66]. Typical values of dielectric constant and dissipation factor at room temperature are given for Butvar PVB in Table 3.9 [13].

3.4.5 Mechanical Properties

Dynamic mechanical spectroscopy on neat PVB around the glass transition usually displays the α relaxation process [30,63] with a usual decrease of three decades of the storage modulus from the glassy plateau at about $G' = 1$ GPa until a rubbery plateau at $G' = 1.2$ MPa in shear (Figure 3.11). Consistent values in tensile tests, $E' = 2$ GPa in the glass state and a rubbery plateau at $E' = 5$ MPa, were reported in Ref. [63].

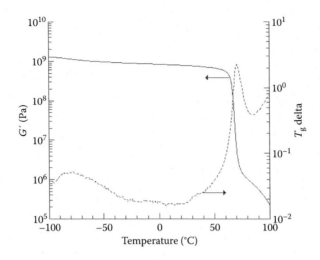

FIGURE 3.11 Shear dynamic properties of PVB (24–27 wt% VA) at 1 rad/s.

TABLE 3.10
Mechanical Properties of Butvar PVB Resins

	ASTM Standard	Butvar B72–B74	Butvar B76–B79	Butvar B90–98
VA content (wt%)		17.5–20	11.5–13.5	18–20
Yield stress (MPa)	D638-58T	47–54	40–47	43–50
Breaking strength (MPa)	D638-58T	48–55	32–39	39–46
Elongation at yield (%)	D638-58T	8	8	8
Elongation at break (%)	D638-58T	70–75	110	100–110
Tensile modulus (GPa)	D638-58T	2.3–2.4	1.9–2.0	2.1–2.2
Flexural strength at yield (MPa)	D790-59T	83–90	72–79	76–83
Hardness, Rockwell M	D785-51	115	100	110–115
Hardness, Rockwell E	D785-51	20	5	20
Impact strength notched Izod	D256-56	0.587	0.477	0.374–0.481

Source: Data from Solutia, *Butvar Polyvinyl Butyral Properties and Uses. On-line Technical Bulletin N°. 2008084E.* St. Louis, Mo: Solutia, Inc., 2008.

Typical tensile properties of various commercial unplasticized PVBs at room temperature are summarized in Table 3.10.

3.4.6 Photodegradation and Photoaging

For more than 50 years, PVB has been employed in the repair and preservation of museum objects as an adhesive, an impregnant, a consolidant, or a surface coating. Therefore, the issue of its photochemical stability under moderate conditions of exposure has been widely addressed. The studies have shown that the region of short wavelengths is the most detrimental from the point of view of PVB photooxidation. When covering the sheets with glass used for making windshields, the intensities of shorter wavelengths are suppressed, but still relatively long wavelength light remains capable of inducing the photooxidation of PVB, and an additional protection of PVB material in safety glass, by UV stabilizers, may be necessary. The use of PVB in the solar module industry has renewed the scientific interest for a better understanding of the mechanisms involved and on the manner in which PVB may chain-break, cross-link, or discolor upon exposure principally to near-ultraviolet and visible radiation at various temperatures in quite severe conditions [67]. Although the structure of PVB should not theoretically allow absorption of light from the near-UV region, this polymer is readily degraded by these wavelengths. Commercial PVB, however, contains ketonic carbonyl groups already present in the starting PVA. The tertiary hydrogens present in the PVB structure represent the most likely sites of radical attack, which, in the presence of oxygen, lead eventually to the formation of hydroperoxides. The latter will easily be split, by the action of either light or heat, to give oxygen-centered radicals, which subsequently yield molecular fragments. The process of photooxidation is autocatalytic, which is the key role being played by ketones and hydroperoxide intermediates [68]. It was suggested that the weakest C–H bond is that in the acetal cycle between oxygens [69] followed by that in the VA segment. Influence of the remaining acetate has also been pointed out but only in the case of unusually high acetate contents [63]. In the course of photochemical aging under visible, middle, and near-ultraviolet wavelengths, PVB passes through the usual sequence of events in photoaging (Figure 3.12). At first, during a period of induction, the measured concentration of peroxide rises, passes through a maximum, and then declines to a notably low level. A change in weight follows the same trend, and when the loss reaches 2–3%, brittleness and discoloration occur. The initial gain and subsequent loss in weight are closely associated with the rise and fall of peroxide concentration. The decline in the peroxide content may be attributed to photolysis of hydroperoxides; the larger the butyral group content, the more intense the formation of carbonyl. The detailed mechanism for the deterioration of PVB comprises chain-breaking, growth of carbonyl, and release of low-molecular-weight fragments predominated by BA (butanal) and butyric acid. The release of aldehyde is far more important because it arises simply as a reversal of the initial reaction between aldehyde and PVA involved in the synthesis of the polymer. Nevertheless, both aldehyde and acid can contribute to the decomposition of hydroperoxides that may remove peroxides from the solid polymer after prolonged photooxidation. Volatile peroxides formed by scission reactions in hydroperoxide-containing segments near the ends of the polymer chain or chains, or by recombination of hydroxyl and alkoxy radicals formed by the homolysis of solid-state hydroperoxides, have also been observed [70]. However, in a material that has had some exposure to ultraviolet, PVB surprisingly develops no noticeable yellowness when heated moderately for a substantial length of time [68]. Apparently, the formation of appreciable color, which

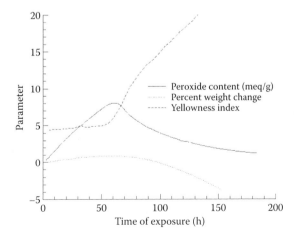

FIGURE 3.12 Photo-oxidation of PVB (11–14 wt% VA) under UVB exposure. (Reprinted from Feller, R. L. et al., *Polym Degrad Stab.*, 92, 920–931. Copyright 2007, with permission from Elsevier.)

is essentially a thermally induced process, tends to occur at a point at which the marked decline in peroxide concentration has taken place and on a polymer having the lower degree of butyral substitution. A further reason why discoloration of PVB has not been a marked feature in many photochemical exposures is that yellowing is observed when the polymer is in the dark, whereas bleaching occurs when it is exposed to light due to the disruption of the bonds in chromophores. Finally, at moderate temperatures below the glass transition temperature, PVB tends to chain-break when exposed to visible and near-ultraviolet wavelengths, and at higher temperatures, cross-linking is expected.

3.4.7 Thermo-Oxidation

Due to its composition, PVB is very sensitive to the degradation and migration of plasticizers. The thermal oxidation of PVB in mild conditions such as those encountered during processing (150–180°C) has been reviewed essentially in view of the stabilization of the polymers and of long-term aging and recycling [71] since it can have an effect on the optical and mechanical properties of the interlayer of safety glasses [39]. Almost all the functional groups characteristic of the polymer being investigated are susceptible to the thermal degradation process [72]. In the case of high acetate contents, the initial point of degradation is the acetate group. In standard PVB, the degradation of the cyclic butyral moiety is important when the temperature reaches 100°C with the production of butanoic acid, butanol, and butanal [52,73], and the reduction of the melt viscosity indicates chain scission rather than cross-linking (Scheme 3.16). Plasticization by water or by added plasticizer has a favorable effect since it decreases the viscosity of the polymer and self-heating during the processing step [74]. In plasticized PVB, the relative thermal stability of commercial PVB samples as measured by thermogravimetry, from mass loss against temperature plots [36], shows that weight loss of 20% and 70% occurs in two distinct regions between 200–300°C and 300–500°C. Below 260°C, the change in mass can be attributed to plasticizer loss by evaporation in the case of dibutyl sebacate. Nevertheless, the loss of plasticizer at common processing temperatures up to 200°C is not significant. PVB decomposes by side group elimination above 260°C. BA, butenal, water, and acetic acid are the major components from these elimination reactions, but benzene and other aromatic products are observed from the decomposition of polyene residues. Instability of PVB can also be attributed to acid materials, or other substances that are easily oxidized into acids, which attack

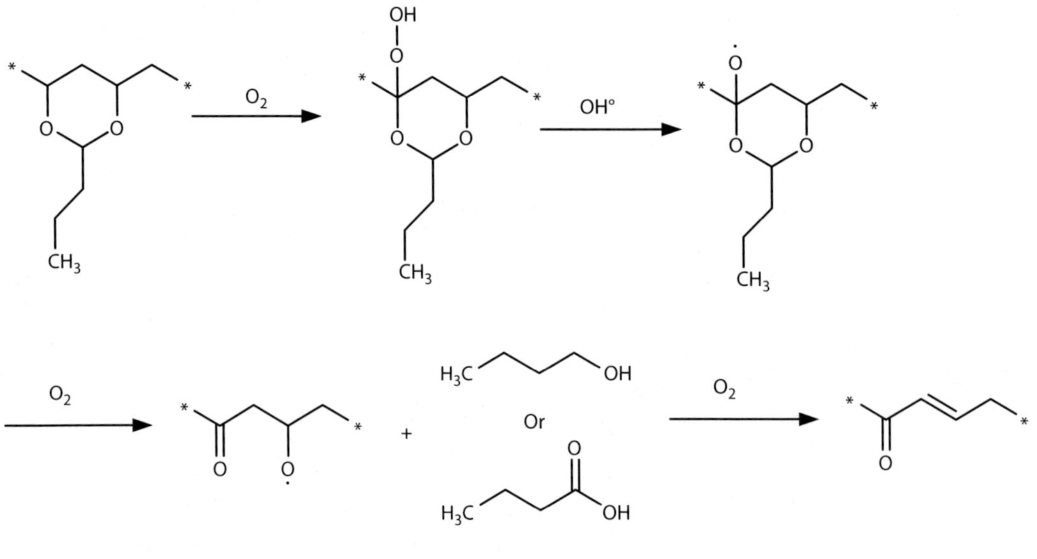

SCHEME 3.16 Degradation pathways during thermo-oxidation of PVB.

the 1,3-dioxane ring and then result in its opening. That is why these acid materials are removed or neutralized by purification or addition of bases like inorganic bases, such as sodium hydroxide, calcium hydroxide, and barium hydroxide, or organic bases, like hexamine, at the end of the synthesis process [29].

3.5 PROCESSING TECHNOLOGIES

3.5.1 THERMAL PROPERTIES FOR PROCESSING

Since thermal, physical, and rheological data are requested in software packages that are useful and nowadays widely used for the modeling of processing techniques and design, a few useful values are proposed hereafter. The heat capacity of unplasticized PVB with 18–22 wt% VA [53] at 20°C is 1.36 J (g K)$^{-1}$, and the thermal conductivity [41] is 0.236 W (m K)$^{-1}$. At the glass transition, the variation of the heat capacity is 0.577 J (g K)$^{-1}$.

3.5.2 VISCOELASTIC AND RHEOLOGICAL PROPERTIES

The linear viscoelastic properties of PVB with 18 wt% VA [75] show a high level of rubbery plateau modulus (1.3 MPa) and a low molecular weight between entanglements M_e (2670 g/mol). Consequently, it displays a rubbery state in a wide temperature range. The onset stress of shark-skin failure is around 0.18 MPa, which is almost identical to that of a conventional polyethylene. This high onset shear stress of the shark-skin failure enables high output rate at extrusion processing. The correlation with roughness during lamination is addressed in Ref. [76]. Other rheological data on a PVB with weight average molecular weight of 143,000 g/mol and glass transition at 15–20°C, both in shear and extension (extensional rate from 0.01 to 0.1 s^{-1}), can be found in Ref. [77] with a value of 1 MPa for the rubbery plateau. Coefficients of William-Landel and Ferry for time–temperature superposition are also given at 105°C ($C_1 = 53$ and $C_2 = 256$°C). Extensional measurements at 105°C show the classical behavior with slight extension thickening behavior at high shear rates (>0.1 s^{-1}) and steady plateau below this value (Figure 3.13). Results on miscible blends with polyethylene glycol (PEG) show that the viscosity fits the Arrhenius model very well in the homogeneous region with an activation energy of 30 kJ/mol [78]. Addition of ionomeric PVB was claimed to improve the flow properties [17].

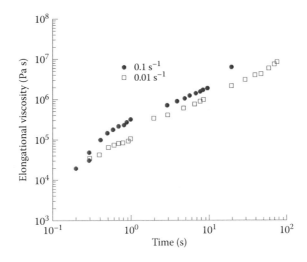

FIGURE 3.13 Extensional viscosity of PVB at 105°C. (From Juang, Y.-J. et al., *Polym Eng Sci*, 41, 275–292. © Copyright 2001 Society of Plastics Engineers. Adapted and reproduced with permission.)

3.5.3 EXTRUSION OF FILMS

The extrusion of films is a step that is commonly used for PVB films utilized in laminated safety glasses. The extrusion process of the resin into a film is complex and delicate. The resin, the plasticizer, and the additives are stored in silos and further mixed with appropriate amounts of additives. After homogenization, the mixture is transferred by gravity in the hopper of an extruder. Under the effect of temperature (>200°C) and of the extrusion screw rotation, the mixture melts into a viscous liquid. With PVB, the resin is partially melted before plasticizer addition, and the position of injection of plasticizer is carefully controlled; delayed injection can induce yellowness and surging, while premature injection can cause plugging in the feeding zone. Water must be extracted from the polymer by degassing under vacuum. Filtration of the melt before it flows into the die head of the extruder is necessary to remove mixture impurities such as unmelted solid particles and other contaminants. The die controls the shape and the surface roughness of the PVB interlayer, and its lips are often water-cooled. In close proximity to the exit of the die, the polymer film is cast onto a roll with a specially prepared surface that imparts the desired surface characteristics to one side of the molten polymer. The PVB sheet is mostly manufactured at thicknesses 0.38 mm (for architectural industry) and 0.76 mm (for automotive and architectural use). Special applications require thicknesses of 1.14 and 1.52 mm. The width of the PVB sheet can be up to 3.5 m. Manufactured PVB sheet is rolled and either separated by thin patterned polyethylene sheet or rolled undercooled (15°C) to avoid blocking because the PVB sheet must not be stuck for subsequent use [79].

3.6 ADDITIVE EFFECT ON PROPERTIES AND APPLICATIONS

3.6.1 PLASTICIZERS

The PVB resins are commonly used with appropriate additives such as antioxidants, anti-UV stabilizers, or inorganic pigments for colored PVB. The choice and amount of plasticizer used in the formulation of the PVB are essential especially concerning the mechanical and rheological properties. The plasticizer must be compatible and soluble with the resin, and this miscibility is greatly influenced by the VA content with some partitioning between rich and depleted zones within the polymer [41]. A priori calculation with the solubility parameters can be used for this purpose [48]. Phthalates have been used, but regulations have restricted the use of these additives; therefore, this class of plasticizers is not addressed in this chapter. An exhaustive list of phthalate-free plasticizers and of the limits for their use is given in Ref. [80]. The best known plasticizers are tetraethylene glycol di-n-heptanoate and triethyleneglycol di-(2-ethyl hexanoate) (3GO or TEGH) [81] and dibutyl sebacate for architectural PVB or automotive [82]; dihexyl adipate (DHA), dioctyl adipate (DOA), hexyl cyclohexyl adipate, or mixtures of heptyl and nonyl adipates for laminated glass in aeronautics and acrylics; whereas PEG is used in cathodes for fuel cells, biphenyl or its derivatives in electrographic photoconductors, and diglycidyl ether of bisphenol A as resin plasticizers in aqueous emulsions. The plasticizer content varies between 15 wt% (structural PVB) and 30 wt% (acoustic PVB). The lightly cross-linked PVB resins can be plasticized with 45 wt%, but 30 wt% is a classical amount for safety glass use with PVBs containing 17–25 wt% VA. Oil-modified sebacid alkyds and mixtures of phosphates, adipates, and polymeric plasticizers are also used in coatings. Plasticizers and PVB can be premixed in a mixer, possibly with an extrusion step in the melt process or within a common solvent in the casting process. PEGs with various molecular weights have also been investigated as plasticizers and for viscosity reduction. Lower-molecular-weight PEG (PEG 400) gives better results, comparable to the efficiency of phthalate plasticizers [83].

The glass transition temperature depends on the percentage and type of plasticizer; the higher the increase in the plasticizer content, the greater the decrease in the glass transition temperature. In a limited range of plasticizer amount, the variation of the glass temperature of the mix follows

Polyvinyl Butyral

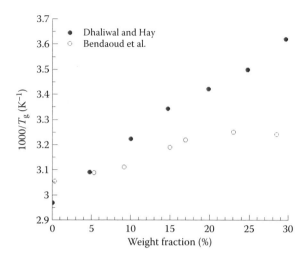

FIGURE 3.14 Variation of the glass transition temperature with the amount of plasticizer (PVB 22–24 wt% VA) and DBS. (Data from Dhaliwal, A. and Hay, J., *Thermochim Acta*, 391, 245–255, 2002; Bendaoud, A. et al., *Macromol Mater Eng*, 298, 1259–1268, 2013.)

the Fox law [36]; however, a limit of miscibility is also observed [51]. Figure 3.14 shows the results for dibutyl sebacate (DBS) in PVB containing 24 wt% VA. In addition to reducing the stiffness of polymers, plasticizers increase strain softening [84] and absorption of impact forces in the case of laminated glass, control adhesion to glass, reduce residual stress in the photoconductive layer of solar modules, affect the processing temperature, sometimes generating undesirable bubbles, and improve film-forming properties or processability especially in high extrusion throughput processes. In glass PVB sheet, plasticizers must not reduce light transmittance and adhesion to glass and must not increase yellowness. The analysis of the plasticizers can be performed by extraction from the polymer either by using organic solvents that are nonsolvents of the polymer such as *n*-heptane [36], or supercritical fluids, ethylene [85], or ethyl acetate modified supercritical CO_2 [81], followed by FTIR spectroscopy or chromatography [36].

3.6.2 Antioxidants

One of the reasons for deterioration of the optical properties of PVB could be the oxidation, which occurs during processing or use of the polymer. To prevent it, thermo- and photostabilizers are added to the polymer. However, these additives, which are soluble at high temperatures in the polymer melt, may become insoluble at low temperatures in the polymer, giving rise to phase separation and clouding. Therefore, as for plasticizers, the limit of solubility is important, and it is also connected to the type and amount of stabilizer; since plasticizers increase the mobility of polymer chains, they affect the solubility and diffusion of additives. Upon small additions of a plasticizer, antioxidant dissolves better than predicted by a sum of its solubilities in polymers and plasticizers. Larger additions of the plasticizer cause a strong polymer–plasticizer interaction, which decreases the antioxidant dissolution. The solubility of antioxidant in DHA and triethylene glycol is lower than its solubility in dibutyl sebacate [82]; 0.1–0.2 wt% of hindered phenols are classically used among various species: methyl ether of 3,5-di-tert-butyl-4-hydroxyphenyl propionic acid, ester of 3,5-di-tert-butyl-4-hydroxyphenyl propionic acid and ethyleneglycol, ester of 3,5-di-tert-butyl-4-hydroxyphenyl propionic acid and pentaeritrite, dimethyl-(3,5-di-tert-butyl-4-hydroxy-benzyl) phosphonate, diphenyl-(3,5-di-tert-butyl-4-hydroxybenzyl) phosphonate, phenyl ether of 2,2-methylene-bis-(4-methyl-6-tert-butylphenyl) phosphoric acid, and tris(2,4-di-tert-butylphenyl) phosphite [86].

3.6.3 Light and UV Stabilizers

UV stabilization of PVB may be particularly desirable for long-term strong exposure such as those required for the encapsulation of PV modules. High transparency is also required in the efficient wavelength that yields the highest quantum efficiency for the specific PV materials. The effects of UV radiation can be mitigated by the inclusion of a UV absorber such as a benzotriazole (0.2–0.35 wt%). Finally, combination with antioxidants such as a hindered amine light stabilizer (HALS; 0.1–0.2 wt%) and possibly a phenolic phosphonite can be used to decompose peroxide radicals that may form due to thermal or UV exposure [86]. HALS is not consumed as opposed to the phenolic phosphonite, which is oxidized to produce phosphate and phenols. Benzophenone has shown incompatibility with phenyl phosphonite because they had a significant tendency to form chromophores at least in other polymer systems with acetate groups [87].

3.6.4 Flame Retardants

Alumina trihydrate (ATH) performed best as a hydrated filler-type flame retardant of plasticized PVB especially in combination with low-level addition of red phosphorus that significantly improved limiting oxygen index (LOI) values. The presence of hydroxyl groups on the polymer backbone had a beneficial effect. Magnesium hydroxide addition had a limited effect on the LOI of plasticized PVB, and addition of red phosphorus made little difference because of the mismatch between the decomposition temperature of $Mg(OH)_2$ and the temperature at which the PVB plasticizer vaporizes [88].

3.7 BLENDS AND COMPOSITES

3.7.1 Blends

PVB has been used in various polymer blends as a minor component especially to modify the properties of the matrix polymer. The number of patents and publications in this area is rapidly increasing, essentially because large amounts of recycled PVB of quite good quality and purity become available from the recycling of laminated glass particularly from the automotive industry. Therefore, the polymers are generally blended in the melt using the conventional extrusion blending technique or directly on the injection molding machine. Otherwise, virgin PVB can be found in polymer blends for membranes made by film casting in a common solvent or for coating materials. These later types of blends will be mentioned but not addressed in detail since it is not the aim of the present book and the focus on thermoplastics was preferred. Generally, as in the case of most of high-molecular-weight polymer blends, miscibility is seldom observed between PVB and other polymer species. In particular, due to the terpolymer nature of PVB and to strong variations in the composition, even homologous blends of PVB with variation of chemical compositions can be immiscible [89]. However, ionomeric PVB was used as a viscosity modifier in conventional PVB [17]. PVB is often expected to toughen the semicrystalline polymer matrices especially when plasticized PVB is used. Compatibility can be achieved by hydrogen bonding or interchain reaction between the hydroxyl groups of PVB and the reactive chain ends of polyamides or polyesters; otherwise, the use of compatibilizing agents has been suggested. For example, PVB has been widely reported to be a toughening agent for polyamide 6. Miscibility of PVB and polyamide 6 directly relates to the overall VA content in the blend as demonstrated in both binary and ternary blends of PVB, PVA, and nylon-6. Miscibility window is observed in the range of 55–70 wt% of VA in PVB for binary blends [90] and approximately 40–60 wt% of total VA in the ternary blends [91]. Methylene and amide groups of nylon-6 were observed to interact favorably with VA in the amorphous phase. For lower VA contents, using plasticized PVB from recycled laminated automotive windshields, super-toughened material can be obtained for 40 wt% PVB in the blend [92]. Virgin material can also be used as a toughening additive [93]. Combination with styrene–ethylene/butylene–styrene copolymer (SEBS) as an impact modifier has been used, and the best impact properties are obtained for PVB contents in the range 20–35 wt% [94,95]. Addition of a maleated SEBS (SEBS-g-MA) is not necessary in PA6 but improves the mechanical properties in polyamide 6-6 [96]. Attempts to use PVB as a compatibilizer between polyamide 6 and polypropylene have failed [97]. On the other hand, the use of compatibilizers in blends of PVB with olefinic polymers such as polypropylene is necessary to overcome the poor affinity between the hydrophobic and hydrophilic species. PP–PVB blends show very little improvement of mechanical properties. However, adding PP as the minor phase and obtaining thermoplastic elastomers after dynamic cross-linking and phase inversion in the melt can be a promising route [93]. Maleic anhydride grafted polypropylene (PP-g-MA) and reactive compatibilization have been used for other blends [40]; but the effect of the compatibilizer remains limited, and in composites, with mica as a filler, little toughening was observed [98]. Polyvinylidene fluoride is a crystalline technical polymer whose hydrophobicity can be modulated by addition of PVB for membrane applications [99]. It is fully miscible with PVB until 30 wt% of PVB [100,101]. Polyesters also belong to a class of crystalline polymers that has been addressed for blending with PVB. In coatings or for improvement of the weathering resistance [102] because of incompatibility, layering of the phase can provide a protective surface layer. Blending in the melt starting from cyclic oligomers and inducing miscibility by grafting during the polymerization have been reported [103]. Biodegradable polyesters such as polycaprolactone (PCL) have been tested in tape casting, green tapes, and ceramics to change the degradability of the substrate [104,105] and also because PVB changes the crystallization behavior of PCL. In particular, addition of a few percent of amorphous PVB strikingly reduces the nucleation frequency of the PCL crystals; therefore, it reduces the nucleation density and allows the spherulites of PCL to grow on the order of centimeters. The morphological changes also include a marked enhancement in the regularity of the lamellar organization in the banded spherulite; the band spacings of PCL spherulites decrease with increasing PVB content, indicating a shortening of the period of lamellar twisting, giving rise to interesting electrical properties in presence of conductive fillers (carbon black [CB]) as, for example, a positive temperature coefficient [97,106–109]. Other biodegradable polyesters coming from biological and agricultural resources have been investigated, especially poly(lactic acid) [110] and poly(3-hydroxybutyrate) [111], the latter showing partial miscibility in PVB with 24–36 wt% VA after melt-blending. Cellulose acetate is miscible with a fraction of PVB lower than 20 wt% and can be used for pervaporation in a membrane field [112]. Other nonpolyester biodegradable polymers from agricultural resources such as thermoplastic starch (TPS) can be blended with PVB by melt extrusion [113]. Polyether of the poly(ethylene-glycol) family can be easily melt-blended to bring some plasticizing effect [83] or viscosity modification [78]. PVB microporous membranes have been prepared via thermally induced phase separation because the PVB/PEG systems show an upper critical solution temperature (UCST) behavior and remain homogeneous at 150°C and 160°C while phase separation occurs at 140°C. Control of the morphology can be tuned by using an emulsifier such as an ethoxylated–propoxylated block copolymer [114–116] or by ternary blends with polystyrene (PS), because PS has better compatibility with PVB than PEG [117]. Combination with poly(ethylene-co-vinyl alcohol) (EVOH) generates both UCST and LCST behaviors. Cycloaliphatic polyether thermoplastic polyurethanes (TPUs) are melt-blended with PVB for improvement of resistance of the PVB interlayer [118], and poly(ethersulfone) (PES) can be used in blends obtained by casting for membranes for ultrafiltration [119].

Though it is expected to toughen rigid semicrystalline thermoplastic, especially when using plasticized PVB owing to its rubbery behavior at room temperature, PVB is also used in blends with amorphous materials. However, in the latter, it may be expected that significant transfer of plasticizers occurs within the amorphous phases, especially when the glass transition of the host polymer is low. Polymethyl methacrylate (PMMA) is not miscible with PVB, but a UCST-type phase behavior was observed in certain blends obtained in the melt. The miscibility range depends on the copolymer composition and on the molecular weight of PMMA with a minimum at 22–33 wt% VA. This is mainly attributed to a copolymer effect in which the unfavorable interaction between PVB and PMMA is minimized at a certain copolymer composition due to the repulsive interaction between

VA and VB units within the PVB chains [120]. The morphology of films obtained by spin-coating has been observed by AFM [121]. Recycled scraps of PVB were also used in blends with excellent optical properties and toughness owing to compatibilization in the melt by the incorporation of a copolymer of methyl methacrylate–butadiene–styrene (MBS) [122]. Blends with polyvinyl chloride (PVC) have been thoroughly studied because of the large amounts of recycled PVB of excellent quality that become available on the market. The cheap recycled and plasticized material can be used as a filler material that may provide interesting additional properties such as toughening [93]. Attention must be paid to the migration and compatibility of the plasticizer of each phase, since PVB can accept only a limited amount of plasticizers in comparison to PVC [51]. PVB was reported to act as a thermal stabilizer of PVC owing to radical substitution of chlorine atoms on PVC chain by PVB, interrupting the dehydrochlorination of PVC and HCl scavenging by acetic acid or butyral moieties [72,123]. Miscibility has been observed for amounts of PVB lower than 10 wt% [124]. Blends with phenolic resins (phenol-formaldehyde) without chemical reaction with the OH groups (but with strong hydrogen bonding) and without catalyst enable the preparation of mesoporous carbon fibers [125,126]. Finally, miscellaneous blends with technical polymers such as polyvinyl-pyrrolydone (PVP) [127,128] or polyaniline and sulfonated PS [129] are reported for membrane or humidity sensors. A variety of polymer blends with PVB are shown in Table 3.11.

3.7.2 Composites and Nanocomposites

This section will address mostly the particulate composites; indeed laminated glass can be considered as an evident laminated composite material based on the sandwich of glass and PVB, but this will be developed in Section 3.8. Apart from this case, most of the composites with a PVB matrix are nanocomposites. Because of its wide use in the green tape industry for ceramics and sol-gel applications, as a binder, PVB has been used in a great variety of mixtures involving various minerals. In this section also, only the composites and nanocomposites with functional applications other than binding will be addressed. The use of PVB as a binder will also be dealt with in Section 3.8. In fact, most of the work also refers to nanocomposites, and only a few conventional composites are mentioned. Among these, polyamide fibers in combination with silica [130,131] have been used in solvent-casting to generate high mechanical and impact resistance. Leather fibers (30–70 wt%) were mixed with PVB by extrusion and act as a reinforcing material with good distribution and adhesion between matrix and fibers [132]. Glass beads were melt-blended with PVB [59], mostly as a demonstrator for the use of silanes regarding adhesion to glass. Silica aerogel can be mixed at any amount with PVB to lower the thermal conductivity [133].

The other examples are nanocomposites, though unusually large amounts of filler were sometimes used. For example, silica was used at 15 wt% in PVB to prepare nanofibers by electrospinning and to enable modulation of the diameter of the fibers [134]. Nanocomposites with Fe_3O_4 nanoparticles at contents up to 14.5 vol% were synthesized, spray-dried, and further processed in injection molding, showing well-distributed nanoparticles and reduction of glass transition temperature, magnetic losses, coercivity, and saturation [135]. Barium titanate nanopowder was introduced in PVB in the range 60–90 wt% to change volume resistivity, dielectric constant, and dissipation factor and their variation with temperature [136]. High dielectric constant and low dielectric loss have also been obtained by addition of ferroelectric ceramic lead zirconate titanates (PZTs) at 85 vol% amounts [137]. Nanocomposites with clays of the montmorillonite type (cloisite with sodium cations) were obtained in the melt with various intercalants such as 3-aminopropyltriethoxysilane (APS), trimethyl chlorosilane (TMCS), and octyltrimethoxysilane (OTMS) and also using commercially modified montmorillonite (Cloisite 30B) [138]. All nanocomposites until 3 wt% of nanoclay reached a mix of intercalated and exfoliated structures with an increase in the breaking strength and significant changes in the light transmission. Nanocomposites of PVB and titanium oxide (TiO_2) were prepared by the sol-gel process in film casting [139]. Pure PVB became stiff and brittle by the addition of TiO_2, with more than a twofold increase in the Young's modulus with 2 wt% of TiO_2.

TABLE 3.11
Blends with PVB

Blend	Targeted Properties	Process	Reference
PVB-PA6-PVA	Miscibility	Melt	[91]
PVB-PA6	Miscibility–toughening–tensile properties	Melt-virgin or recycled PVB	[90,92,93]
PVB-PA6-SEBS	Tensile and impact properties	Melt-recycled PVB	[94,95]
PVB-PA66-SEBS	Tensile and impact properties	Melt-recycled PVB	[96]
PVB-PA6-PP	Tensile and impact properties	Melt-recycled PVB	[95]
PVB-PP	Tensile and impact properties	Melt	[93]
PVB-PP	Tensile and impact properties	Melt, TPE, dynamic cross-linking	[93]
PVB-PP-g-MA	Crystallinity and tensile properties	Melt	[40]
PVB-PP-PP-g-MA-mica	Tensile and impact properties	Melt	[98]
PVB-PVDF	Tensile properties, surface properties, hydrophobicity, permeation	Casting	[100,101,140]
PVB-PBT	Dielectric properties, weathering resistance	Melt, casting	[102,103]
PVB-PCL	Degradability	Casting	[104,105]
PVB-PCL-CB	Electric properties	Casting	[97,106–109]
PVB-PLA	Tensile properties	Casting	[110]
PVB-P3HB	Miscibility	Melt	[111]
PVB-cellulose acetate	Pervoration membrane	Casting	[112]
PVB-TPS	Tensile properties	Melt	[113]
PVB-PEG	Plasticizer, viscosity modifier	Melt	[78,83]
PVB-PEG-P(EG-b-PG-b EG)	Permeation, hydrophobicity	Melt	[114–116]
PVB-PS-PEG	Membrane, irradiation, tensile properties	Casting	[117]
PVB-TPU	Impact properties	Melt	[118]
PVB-PES	Ultrafiltration membrane	Casting	[119]
PVB-PMMA	Morphology	Melt-spin coating	[120,121]
PVB-PMMA-MBS	Optical properties—toughness	Melt, recycled PVB	[122]
PVB-PVC	Toughness, tensile properties, thermal stabilizer, permeation, hydrophobicity, surface properties	Melt or casting, Recycled or virgin PVB	[51,72,93,123,124]
PVB-phenol formaldehyde	Carbon fibers	Melt-pyrolysis	[125,126]
PVB-PVP	Ultrafiltration	Spin-coating	[127,128]
PVB-PANI-SO$_3$PS	Humidity sensors	Casting	[129]

Thermal and abrasion resistances were also improved. The solubility and the degree of swelling in ethanol became drastically lower by mixing a small amount of TiO_2, and nanocomposites containing 5 wt% were ethanol-selective in permeation. Sol-gel generation of nanoparticles is generally more efficient to generate enhanced properties both with silica or titanium dioxide even when prepared by melt compounding from colloidal sol in comparison to simple mixing of solid particles. Introduction of these nanocomposites within Kevlar/PVB composites showed very high mechanical properties with 5–7 wt% of nanofiller [130] and a glass temperature decrease. Functionalized mesoporous silica added at amounts until 1.5 wt% in PVB composite films obtained by solution casting films yields thermodependence of the conductivity. Doping of silica was obtained by tripentaerythritol [141]. PVB–Al(NO$_3$)$_3$ sol has been reported as a precursor of alumina fibers in composites with PVB [142] owing to the reaction of aluminum nitrate nonahydrate with vinyl acetates and VAs. Carbonaceous materials have been also investigated using various allotropic forms. CB is used in combination with PCL to obtain a positive temperature coefficient of conductivity [97,106–109]; graphene provides improvements of the thermal and mechanical properties inducing toughness and flexibility for cast composites [34]; and oriented graphite modulates the thermal conductivity of composite sheets prepared by tape casting [143].

A wide development of functional nanocomposites is driven by the improvement and generation of new properties in the glass industry particularly in energy-saving glass since nanoparticles enable to keep transparency in the visible spectrum while still strongly interacting with the UV spectrum. Antimony-doped tin oxide (ATO) nanoparticles were employed to fabricate ATO/PVB nanocomposites. UV–Vis–NIR spectra show that the addition of ATO nanoparticles can significantly enhance the thermal insulating efficiency of ATO/PVB nanocomposites. With the increase in the ATO content, the thermal insulating efficiency is increased. UV can be almost fully absorbed. Visible transmittance is over 72.7%, and haze is below 2% when the ATO content is in the range of 0.1–0.5 wt%. The thermal insulating tests indicate a 1–3°C temperature reduction [139].

3.8 APPLICATIONS

PVB fits the checklist for an enormous number of applications, such as safety glasses for architecture and automotive, increasingly for encapsulation of solar modules in PV applications, and as a binder in various coatings, paintings, enamels, adhesives, and inks. All these applications are related to its tremendous adhesion properties on various materials through covalent and hydrogen bonding by the hydroxyl groups.

3.8.1 Safety Glasses

PVB is an essential constituent of lightfast, highly transparent, elastic, and tear-proof films. The manufacture of PVB films for laminated safety glass is one of the most important applications of PVB. The outstanding properties of the PVB films include toughness, good adhesion to glass, good light resistance, and excellent transparency. The films are responsible for the special safety advantages of laminated safety glass. Laminated safety glass is a combination of two or more annealed or tempered glass sheets linked together with one or more films of PVB. Therefore, laminated glass is a composite material combining the properties of the glass and those of the polymer (glass adhesion, elasticity, impact resistance). The synthetic layer plays two roles. On one hand, it dissipates part of the impact energy by viscoplastic deformation and avoids projection of fragments because the glass splinters also adhere to the film minimizing the risk of injury. On the other hand, after impact, it will keep a residual strength ensuring the maintenance of the glazing. The interlayers can improve the mechanical performance of a glass and also, acoustically by attenuation of given frequencies, provide protection from UV and heat control while retaining transparency. Grades of PVB with a high molecular weight are combined with plasticizers to produce safety glass films in a thermoplastic extrusion process. The PVB film is pressed between two glass panels to create fracture-resistant and shatter-proof laminated safety glass used mainly in the building and automotive industries. Though other synthetic materials, resins, or gels can be used [144], PVB is most commonly used for laminated glazing applications because of its high compatibility with glass, due to its high transparency and its remarkable adhesive properties, and finally due to its viscoelastic properties guaranteeing a large deformation before failure and a maximum damping at low frequencies at room temperature [145–147].

The manufacture of laminated glass is performed through several steps including washing, assembly, degassing, autoclaving, and cleaning [148]. After precutting and possible shaping, in a

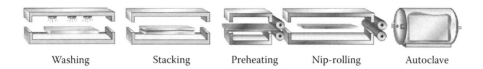

Washing   Stacking   Preheating   Nip-rolling   Autoclave

FIGURE 3.15 Process of lamination.

washing step, glass products are washed to remove all traces of contamination and dried thoroughly. The assembly is performed in a closed chamber to avoid dust at a relative humidity of 30% and temperature of 18–20°C to avoid premature joining of glass and PVB film. Glass and PVB are stacked depending on the desired composition. Trimming of the laminate can be performed before entry in a preheating furnace. Degassing is the next operation and is critical to eliminate the air trapped between the PVB and glass and to get the proper clarity [76]. Sealing the edges avoids entrance of air during the final step of autoclaving (Figure 3.15). This is done by calendering in a preheating oven with temperature conditions depending on the composition and production speed. If the temperature of the preheating stage is too low, laminated glass cannot seal, while, on the contrary, too high temperature will cause seal edge early, which may stop the exhaust of air trapping large air bubbles in the laminated glass. For curved glass, such as automobile windshields, curvature is accomplished by differentially stretching the sheet heated at about 85–100°C over a tapered shaping drum. Rolls of sheet or precut interlayer blanks are usually stored or shipped refrigerated at 3–10°C, or shipped at ambient conditions with an interleaved thin sheet of plastic such as polyethylene to prevent blocking. The laminated glass is disposed into a vacuum bag or a vacuum rubber ring is put around the glass. Low pressure conditions (0.5–0.8 bar) are used to exhaust air out from the laminated glass before the sealing of the edges. The exhausting has two periods: one with cold air below 25°C during about 15 min, and the second in hot exhausting conditions with a temperature of laminated glass surface between 80°C and 110°C during about 30 min. Comparing the surface transparency of laminated glass through both production ways, the vacuum exhausting method is better than hot rolling exhausting. The final adhesion of the glass and PVB is carried out at a pressure of 12 bars and at temperature of 135–145°C in the autoclaving step, which allows a flow of PVB that fits perfectly onto the glass surface and creates adhesion. Temperature reduction is performed until 45°C under pressure; otherwise the edge of laminated glass will show bubbles. The cycle times depend on the composition of the laminated glass. If possible, a second calendering can be necessary to eliminate the excess of PVB by creep. Checking of humidity, adherence, and impact strength is required to the European standard EN ISO 12543-4. Manufacturing process also allows addition of dyes to be used as a decorative element. For laminated safety glass, the strength of the adhesive bond between the glass and the interlayer must be carefully controlled [58]. If adhesion is too high, a projectile can easily penetrate the laminate because cracks made in the glass propagate through the interlayer. If adhesion is too low, glass retention during an impact will be reduced, even though the interlayer is not penetrated. Alkanoate salts as well as moisture can act to reduce adhesion by competing with resin hydroxyls for bonding sites on the glass surface [3,58].

Typically, safety glass is composed of two or more glass sheets connected to each other by sheets of PVB. The number of leaves can reach up to 18 sheets, the thinnest being two glasses of 2 mm combined with an interlayer of 0.38 mm of PVB. Usually, safety laminated glasses are separated in various types depending on the level of performance and use: protection against injuries, against fall of people, against falling objects, against vandalism and burglary, against gunfire, and against explosions, and also enhanced protection. The composition of the safety glass is often referred to as xxx.y, with xxx indicating the glass panel and y the number of PVB sheets of 0.38 mm thickness. For example, 444.8 means three glass panels of 4 mm each separated by 8 layers of PVB with individual thickness of 0.38 mm. The thickness can be easily calculated by the sum of individual thickness and therefore is the specific weight (in kg/m^2) with a glass density of 2.5 g/cm^3 and PVB

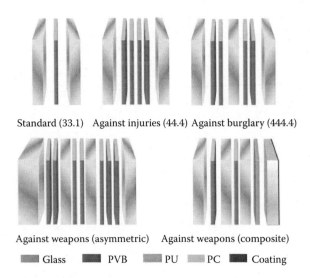

FIGURE 3.16 Various configurations of laminated glass.

density of 1 g/cm³. Double- or triple-laminated glasses are used for separation, stairwell, roof, balcony, shop display, and protection against burglary, vandalism, and riot including explosion (Figure 3.16) [149].

In architecture, for protection of, or simple security to protect, people from accidental impact against a glass, the safety glass must withstand shocks of a person falling against a window. If the glass cracks, PVB will hold the glass pieces together and prevent their dangerous dispersion. This uses two glass panels with a minimum of one PVB layer, and it is used for the glass in doors, windows, shower screens, bus shelters, and roof glazing. For protection against injuries with risk of fall, for example, in a tall building, the glass pieces must remain in place and protect the fall. At minimum, two glass panels are laminated with two PVB layers; typical compositions range from 33.2 to 44.6. They are used for balcony railings, banisters, spandrels, balustrades, internal partitions, lift partitions, and façade glazing to the ground level. These compositions are also used for protection from simple burglary that does not use heavy objects. Glasses developed to be used against organized burglary are resistant to destruction of a window by means of a hammer or an ax to make an opening of 400 × 400 mm. Typical compositions are 444.8 to 444.12. 'Depending on breaking strength requirements, they can be used in private residences, commercial office blocks, urban event locations such as high-risk shop windows. Security against weapons requires the use of asymmetrical laminated glasses such as 3663.6 or a combination with other polymers. PVB composites are layered with a film of poly(ethylene terephthalate), acrylic, or polycarbonate, or metal fabric in addition to glass [150]. The application ranges from counters and glazing in banks, vehicles transporting money, and embassies. A grid of copper that is extremely thin (0.08 mm in diameter) and barely visible can also be introduced between the PVB layers in order to trigger an alarm in case of cutting the thread or to provide radio-frequency shielding from an external source. In addition to security issues, the PVB interlayer has also a significant effect on other functional properties. It enables the blocking of 99% of the UV rays (below 380 nm) of the sun preventing the fading of colors of fabric, furniture, and flooring over time, especially upon addition of particles (Figure 3.17) [151,152] or by use of diffractive holographic elements [153]. Colored films can also be used for visual improvements. Due to its viscoelastic properties and low T_g especially upon addition of a plasticizer, PVB provides also the damping of vibration and sound. To increase damping effectively and to obtain good sound attenuation, laminated glass is used with a viscoelastic plastic interlayer that changes vibration energy into heat energy. Special grades known as acoustic PVB are used in multilayer assemblies. These grades have a low glass transition temperature between 16°C and

Polyvinyl Butyral

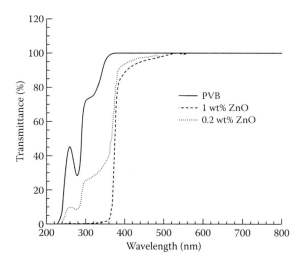

FIGURE 3.17 UV shielding by ZnO nanocomposites. (From Zeng, X.-F. et al., Fabrication of highly transparent ZnO/PVB nanocomposite films with novel UV-shielding properties, 2010. In Nanoelectronics Conference (INEC), 3rd International, IEEE, 2010, 777–778, © Copyright IEEE. Adapted and reproduced with permission.)

18°C, and therefore the relaxation modulus at 20°C at 3 s (which is considered to be representative of a typical wind load) is 10 to 100 times less than that of conventional PVB (Figure 3.18) [154–158]. For example, the use of acoustic PVB yields 10 dB of noise reduction for a 6 mm laminated glass in the 1000–3000 Hz band (Figure 3.19), which is the noise transparency range that lets the most irritating noise through inducing a 50% reduction of perceived sounds [159].

Looking at the history of automotive glazing, laminated glass with a PVB interlayer was first introduced in the United States in 1935 and regulated by law in 1938. In 1964, high penetration resistance laminate was developed for passenger safety. New functions are now required for automotive glazing such as comfort, security, and environmental issues in addition to safety. Modifying

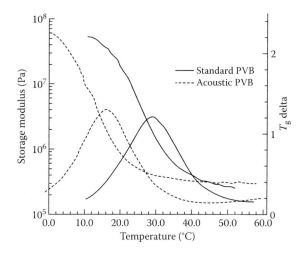

FIGURE 3.18 Storage modulus and damping of conventional PVB (Saflex RB11) and acoustic PVB (Saflex AC41). (From Savineau, G., Intercalaires pour verres feuilletés, 2013. In Techniques de l'Ingenieur, N4404, 1–11. Paris: Editions Techniques de l'Ingenieur. © Copyright Techniques de l'Ingenieur. Adapted and reproduced with permission.)

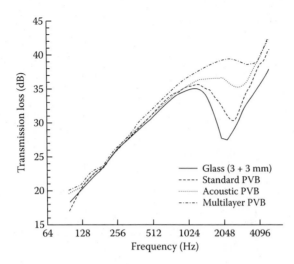

FIGURE 3.19 Noise reduction for various laminated glass. (With kind permission from Springer Science+Business Media: *AutoTechnology*, Functional laminated glazing with advanced PVB technology, 4, 2004, 62–65, Yoshioka, T. © Copyright Springer.)

the PVB interlayer is a practical way of adding new functions. Noise reduction is targeted and can be further improved by the use of multilayer films, keeping a high mechanical resistance. Indeed, the acoustic interlayer is inferior to the normal PVB interlayer in terms of mechanical strength and handling because of low elasticity or softness required for damping at ambient temperature. For example, three-layer make-up has a low elasticity PVB layer as a core for maximum acoustic performance and normal PVB resin as outer layers for mechanical strength and handling ease. Damping is particularly increased in the 100–300 Hz band associated to engine noise [160]. Solar control PVB for windshields has also been developed with very high absorbance in UV and IR and 70% transmittance in visible light [142], decreasing air temperature in the vehicle by 1°C in a midday average weather condition in Colorado [161]. An interlayer with gradient color bands is also made for automotive windshields. The band is printed on the interlayer surface or tinted by coextrusion with pigmented resin.

3.8.2 Module Encapsulation in PV Application

Since the 1980s, the experience in lamination and high level of control of these technologies has driven intensive developments toward the use of PVB films for solar module encapsulation, and nowadays, PVB is increasingly assuming a leading role among alternative encapsulation materials for solar cells as EVA copolymers. Reliable and long-term protection of solar cells from external influences is required for solar modules. Transparent plastic encapsulation materials contribute to the durability of PV modules and to the long-term generation of electricity from sunlight. The first manufacturer worldwide to develop a special PVB film for use in solar modules was Kuraray Europe GmbH with Trosifol trademark. Solar modules using PVB films have been successfully undergoing outdoor weathering tests since the beginning of the 1980s. To ensure module durability and long-term power generation, cell encapsulation materials have to display certain important features such as mechanical protection of the cell, protection from weathering, electrical insulation, external impact resistance, barrier to oxygen and water vapor, prevention of cell corrosion, adapted adhesion to other module components (glass, cell, backing film, contact), and obviously high transparency and high UV protection (to prevent degradation of the plastic). There are few plastics that display all of these properties reliably in practice. Classical solar modules using crystalline Si wafers are encapsulated between two layers of PVB (Figure 3.20) and glass. In thin-film PV

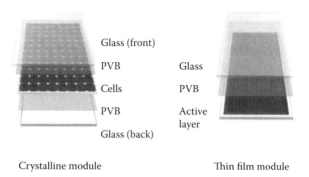

FIGURE 3.20 Various types of solar module assemblies.

modules, the semiconductor photoactive layer is applied directly on a glass substrate and separated from the front glass by a PVB film. The few polymer materials suitable as encapsulation products and industrially used today are mainly cross-linked elastomers and thermoplastic materials. The cross-linked elastomer products are EVA, TPU and acrylate encapsulation resins, and two-component silicones that cross-link on exposure to heat and/or UV light to form a rubber-elastic material and thus firmly embed the solar cell. Thermoplastic products melt without cross-linking on exposure to heat, thus retaining their original chemical composition. PVB is the most important encapsulation alternative, particularly for thin-film solar modules, and competes with TPU, ionoplastics, modified polyolefins, and thermoplastic silicones (TSI). The advantage of PVB for PV application is the high transparency [162] of this material in the effective wavelength window of solar modules above 450 nm (Figure 3.21) [163]. However, PV modules must also stand for corrosion and aging, and the major causes are humidity and UV radiation. The encapsulant is important as it stores and disperses water on the one hand and decreases the transparency due to UV radiation on the other hand. The thermoplastic foils (TPU and TSI) show the highest water vapor transmission rate, whereas the ionomer foils show the lowest. The saturation concentration of PVB is the highest, followed by TPU and EVA, while silicone and ionomer store practically no water [164]. Ionomer and PVB offer higher protection from environmental exposure for thin-film modules than

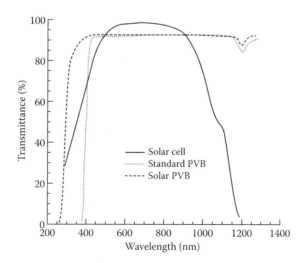

FIGURE 3.21 Transparency of PVB in the range of maximum yield of silicium solar devices. (From Kuraray, Trosifol—Solar Module Optimisation with PVB Film—Solar Technology in Perfection, Hattersheim am Main, Germany: Kuraray Europe GmbH. Adapted and reproduced with permission.)

EVA, with diffusion being higher in PVB but solubility being higher in EVA. Adhesion of PVB to glass remains a key advantage [165]. The performance and reliability of a PV module can also be affected by the degradation behavior of the polymeric components, especially in hard weathering conditions caused mainly by UV irradiation, temperature, and humidity. High irradiation doses, large temperature gradients between night and day, and high humidity at high temperature may easily impair the components with various kinds of degradation processes, especially discoloration, that lead to transmission losses and delamination. Corrosion of the metallization is less critical with PVB than with EVA that may release acetic acid, but PVB suffers from sensitivity to hydrolysis.

Depending on the type of PV module, PVB can be processed in single-stage vacuum laminators or in the two-stage process using de-airing in a bag or in a calendar with subsequent autoclaving. In vacuum laminators, films are de-aired, pressed, and cooled; the cycle time is 8–20 min for glass/back sheet and 8–25 min for glass/glass modules, depending on glass thickness. On the same principle as laminated safety glass, glass/glass modules can be de-aired in bags made of plastic or rubber and in a vacuum of less than 100 mbar in a hot-air oven. The maximum temperature is about 135°C. The modules are finally laminated in an autoclave at elevated temperature and pressure. Roll laminator permits high machining speeds, even for large-format modules, with short cycle time (1 min).

3.8.3 Miscellaneous Applications

Many coatings, paintings, enamels, adhesives, and inks are based on PVB due to its excellent adhesive properties in connection to the hydroxyl content and to the flexibility that can be brought to the thermosetting materials. The hydroxyl groups enable good wetting of most substrates and provide reactive sites for chemical combination, self-crosslinking, or cross-linking with phenolics, epoxies, isocyanates, melamines, or ureas. It is used both in powdery and liquid coats.

PVB resins may be used in stoving or wire enamels to overcoat magnet wires so that coils made from that wire can be easily bonded with heat (hot air bonding, oven, resistance, or electric) at moderate temperatures, 110–140°C, or by solvent activation with ethanol or methanol. Such wires are used in stepping motors for quartz watches, sound application coils, sensors, and RFID cards.

PVB is used alone or in combination with a wide variety of resins to give functional surface coating compositions. Films that may be air-dried, baked, or cured at room temperature can be obtained by dip-coating for improved corrosion resistance [102,166,167]. In protective coatings for metal, PVB is used as wash primers to obtain effective corrosion inhibition due to good adhesion to metal and fast drying. It improves coating uniformity, minimizes cratering, improves adhesion, and increases coating toughness and flexibility. It is widely used on metal surfaces: in storage tanks, ships, airplanes, bridges, and dam locks. Wash primer is very effective for inhibition of corrosion in a single treatment. Combination of PVB, chrome pigment, phosphoric acid, and metal complex yields the phosphatizing of the metal surface and acts as a corrosion-inhibiting pigment in an adhering binder also supplying chromate ions to repair pin holes in the phosphate film. Adhesion is improved by stoving at temperatures up to about 140°C. Because of concerns about the toxicity of chromium compounds, systems using zinc borate, phosphate, or molybdates are increasingly used.

Coating metal foil (aluminum, brass, tin, lead, iron) with PVB solution increases moisture-proofness and printability. Application is made by brush, spray, dip, and fluidized bed for drum and can linings, and curing enables to meet the extractability requirements of the US Food and Drug Administration for indirect food additive uses.

PVB is used as a component of wash coats and sealers for wood providing holdout, intercoat adhesion, moisture resistance, flexibility, toughness, and impact resistance to the coating. The wood substrate is protected against discoloration for polyester and polyurethane coatings. Combination with phenolic resins gives excellent barriers to bleeding of terpenes from wood.

PVB combined with nitrocellulose and plasticizer has been used in paint for leather. But generally it is also suitable to coat textiles to give water and stain resistance without affecting appearance, feel, drape, and color of the fabric. PVB can be applied as a transparent film to practically all common textile materials such as cotton, wool, silk, nylon, viscose rayon, and other synthetics.

PVB and aluminum isopropanol were deposed by dip coating on carbon fiber and used as a precursor of a hybrid coating of aluminum oxide and amorphous carbon that improves thermal stability of carbon fiber and its adhesion toward aluminum matrix [168].

Mixtures of PVB and phenolic resin are used as structural adhesives for bonding metal, glass, leather, wood, cloth, paper, and other materials. After the solvent has been completely removed at room temperature, the surfaces are heated and pressed together with curing time and temperature at 160°C for 15–30 min. In printed-circuit board adhesives where peel strength, blister resistance, and dielectric properties are required, such mixtures are used in joining the prepregs of phenolic laminate with copper foil. PVB can also be a base for hot melt adhesives providing a tough, clear film with adhesive strength. PVB binds the glass beads used in retroreflective films for license plates, decals, and road signs.

PVB is used in printing inks because it is soluble in mild solvents such as alcohols and esters, and it improves flexibility, adhesion, and toughness of inks used for gravure, screen, ink jet printing, and flexography. PVB is also soluble in aromatic and fast-drying solvents for high-speed and high-quality printing. Actually, it is used in many printing technology as conductive pastes, printer ribbons, pen inks, and offset printing plates. It also serves as a secondary binder in toners to increase viscosity, to improve film integrity over the fuser roll, and to minimize the level of fine particles [14,162].

Finally, owing to its outstanding binding power and elasticity, PVB is suitable for the manufacture of ceramic slurry and green tape with alumina [169,170], ferrites [171,172], barium titanate [173,174], silica zirconium, titanium, chromium, cerium, yttrium, and nickel [175–177]. Since it combusts with virtually no residue during sintering, it is an ideal temporary binder for the manufacture of high-performance ceramics provided that suitable conditions are used. This was demonstrated by many degradation studies in presence of various inorganics such as SiGe or $FeSi_2$ [178,179], silica, mullite, α-alumina, and γ-alumina [180], glass ceramics and silver [181]. These tapes are used in the manufacture of multilayer electronic devices [182], and even in fuel cells [183], solid polymer electrolytes in dye-sensitized solar cells [184,185], and electrochromic devices (using lithium iodide, lithium thiocyanate, lithium perchlorate, and lithium triflate as dopant salts) [186] or sensors such as radiation dosimeters [187]. Nanofibers with polyaniline were used as humidity sensors [129].

3.8.4 Recycling of PVB

In the last few years, interest in recycling of PVB waste has been raised and it has driven a lot of investigations. Indeed, the total amount of produced PVB sheets including all industrial applications was estimated to be about 120,000 tons/year [79], and at present, PVB is much less recycled than glass, though the price could be half that of virgin PVB. The main issues for the recycling of laminated PVB are the removal of glass from the polymer, the prodegrading life cycle of PVB submitted to UV radiation and heat, and the unavoidable reprocessing step. Recycled PVB is separated into three categories. Category 1 includes transparent PVB (98% of PVB, free of any impurity), category 2 refers to colorless PVB (2% of impurities and 96% of PVB), while category 3 represents colored PVB (70% of PVB and 4% of impurities). PVB is recovered from laminated glass by mechanical peeling. More sophisticated and effective techniques [188] use mechanical cracking in a first stage, followed by chemical separation in alkali water solution at 60°C and final mechanical brushing of glass residues. Separation is efficient, though saponification of plasticizers can be expected. The only problem of this technique is that the saponification can leach plasticizers, and it changes PVB properties if excess of alkali (>0.5%) is used. A method for separation and discoloration of colored PVB sheets using ethyl acetate and decolorization reagent was proposed by Wang et al. [189]. Recovery of PVB with properties close to those of the initial plasticized PVB is possible, including chemical properties, amount of plasticizer, strain at break, tensile strength, fracture energy, and even optical transmittance, provided that the processing temperature is lower than

260°C and that low amounts of alkaline reagent are used in the separation step [36]. Loss of plasticizers can be expected but the amount can be tuned. Low processing temperatures (below 150°C) were suggested for dry PVB, while PVB with moderate moisture contents (8 wt%) can stand higher temperatures because of a plasticizing effect [71]. Recycled PVB can be used for laminated glass and also as toughening agents in blends with other thermoplastics (see Section 3.7.1).

REFERENCES

1. Lindeman, M. K. 1989. Vinyl acetate and the textile industry. *Text Chem Color* 21, 21–28.
2. Herrmann, W. O. and Haehnel, W. 1927. Uber den Poly-vinylalkohol. *Ber Deut Chem Ges (A and B Ser)* 60, 1658–1663.
3. Wade, B. 2004. Vinyl acetal polymers. In *Encyclopedia of Polymer Science and Technology*, 3rd Ed., Vol. 8, ed. H. F. Mark, 381–399. New York: John Wiley & Sons.
4. Perkins, G. A. 1940. Process for making polyvinyl acetal resins. U.S. Patent 2,194,613 A.
5. Green, M. A. 2005. Silicon photovoltaic modules: A brief history of the first 50 years. *Prog Photovoltaics: Res Appl* 13, 447–455.
6. Pardos, F. 2004. *Plastic Films: Situation and Outlook: A Rapra Market Report*. Shawbury: Rapra Technology Limited.
7. Frost and Sullivan. 2010. *Global PVB Film Market. Frost and Sullivan Report*. Mountain View, CA: Frost and Sullivan.
8. Zhou, Z., David, D., MacKnight, W. and Karasz, F. 1997. Synthesis characterization and miscibility of polyvinyl butyrals of varying vinyl alcohol contents. *Turk J Chem* 21, 229–238.
9. Flory, P. J. 1939. Intramolecular reaction between neighboring substituents of vinyl polymers. *J Am Chem Soc* 61, 1518–1521.
10. Noma, K., Koh, T. and Tsuneda, T. 1949. Limits of acetalization of polyvinyl alcohol (in Japanese). *Kobunshi Kagaku* 6, 439–443.
11. Raghavendrachar, P. and Chanda, M. 1983. Neighbouring group effect on the kinetics of acetalisation of poly (vinyl alcohol). *Eur Polym J* 19, 391–397.
12. Flory, P. J. 1950. Statistical mechanics of reversible reactions of neighboring pairs of substituents in linear polymers. *J Am Chem Soc* 72, 5052–5055.
13. Solutia. 2008. *Butvar Polyvinyl Butyral Properties and Uses. On-line Technical Bulletin N°. 2008084E*. St. Louis, MO: Solutia, Inc.
14. Kuraray. 2013. *Polyvinylbutyral of Superior Quality*. Hattersheim am Main, Germany: Kuraray Europe GmbH.
15. Kurskii, Y. A., Beloded, L., Kobyakova, N., Kurskaya, V., Ludanova, K., Pashkina, E., Ryabinina, M. Y. and Rodionov, A. 2011. A GLC and NMR analysis of the composition of polyvinyl butyral furfural acetal. *Russ J Appl Chem* 84, 1914–1919.
16. Kurskii, Y. A., Kobyakova, N. K., Pashkina, E. P., Rodionov, A. G., Ryabinina, M. Y., Shavyrin, A. S. and Faerman, V. I. 2014. The composition and structure of polyvinyl butyral furfural: NMR study. *Polym Sci Ser A* 56, 40–47.
17. DasGupta, A. M., David, D. J. and Misra, A. 1991. Modification of polyvinyl butyral rheological properties via blending with ionomeric polyvinyl butyral. *Polym Bull* 25, 657–660.
18. Toncheva, V., Ivanova, S. and Velichkova, R. 1992. Modified poly (vinyl acetals). *Eur Polym J* 28, 191–198.
19. Striegel, A. M. 2004. Mid-chain grafting in PVB-graft-PVB. *Polym Int* 53, 1806–1812.
20. Chetri, P. and Dass, N. 2001. Preparation of poly (vinyl butyral) with high acetalization rate. *J Appl Polym Sci* 81, 1182–1186.
21. Fernandez, M., Fernandez, M. and Hoces, P. 2006. Synthesis of poly (vinyl butyral)s in homogeneous phase and their thermal properties. *J Appl Polym Sci* 102, 5007–5017.
22. O'Neill, M. L., Newman, D., Beckman, E. J. and Wilkinson, S. P. 1999. Solvent-free generation of poly (vinyl acetals) directly from poly (vinyl acetate). *Polym Eng Sci* 39, 862–871.
23. Blomstrom, T. P. 1989. Vinyl acetal polymers. In *Encyclopedia of Polymer Science and Engineering*, 2nd Ed., Vol. 17, ed. H. F. Mark, 136–167. New York: John Wiley & Sons.
24. Marten, F. L. 1989. Vinyl alcohol polymers. In *Encyclopedia of Polymer Science and Engineering*, 2nd Ed., Vol. 17, ed. H. F. Mark, 167–198. New York: John Wiley & Sons.
25. Marten, F. L. 2004. Vinyl alcohol polymers. In *Encyclopedia of Polymer Science and Technology*, 3rd Ed., Vol. 8, ed. H. F. Mark, 400–443. New York: John Wiley & Sons.

26. Sundararajan, P. R. 1999. Poly(vinyl butyral). In *Polymer Data Handbook*, Chap. 10, ed. J. E. Mark, 381–399. New York: Oxford University Press.
27. Fitzhugh, A. F. and Crozier, R. N. 1952. Relation of composition of polyvinyl acetals to their physical properties. I. Acetals of saturated aliphatic aldehydes. *J Polym Sci, Part A: Polym Chem* 8, 225–241.
28. Hajian, M., Koohmareh, G. A. and Rastgoo, M. 2010. Investigation of factors affecting synthesis of polyvinyl butyral by Taguchi method. *J Appl Polym Sci* 115, 3592–3597.
29. Liu, R., He, B. and Chen, X. 2008. Degradation of poly (vinyl butyral) and its stabilization by bases. *Polym Degrad Stab* 93, 846–853.
30. Takahashi, Y. 1961. Viscoelastic properties of the phenolic resin-polyvinyl butyral system. *J Appl Polym Sci* 5, 468–477.
31. Striegel, A. M. 2002. Determining the vinyl alcohol distribution in poly (vinyl butyral) using normal-phase gradient polymer elution chromatography. *J Chromatogr A* 971, 151–158.
32. Blomstrom, T. P. 2005. Polyvinyl acetal resins. In Coatings Technology Handbook, 3rd Ed., Chap. 60, ed. A. A. Tracton 60.1–60.11. Boca Raton, FL: CRC Press.
33. Corroyer, E., Brochier-Salon, M.-C., Chaussy, D., Wery, S. and Belgacem, M. N. 2013. Characterization of commercial polyvinylbutyrals. *Int J Polym Anal Charact* 18, 346–357.
34. Hajian, M., Reisi, M. R., Koohmareh, G. A. and Jam, A. R. Z. 2012. Preparation and characterization of polyvinylbutyral/graphene nanocomposite. *J Polym Res* 19, 1–7.
35. Kalu, E. E., Daniel, M. and Bockstaller, M. R. 2012. Synthesis, characterization, electrocatalytic and catalytic activity of thermally generated polymer-stabilized metal nanoparticles. *Int J Electrochem Sci* 7, 5297–5313.
36. Dhaliwal, A. and Hay, J. 2002. The characterization of polyvinyl butyral by thermal analysis. *Thermochim Acta* 391, 245–255.
37. Bruch, M. D. and Bonesteel, J. A. K. 1986. Interpretation of the proton NMR spectrum of poly (vinyl butyral) by two-dimensional NMR. *Macromolecules* 19, 1622–1627.
38. Berger, P. A., Remsen, E. E., Leo, G. C. and David, D. J. 1991. Characterization of acetal ring conformations in poly (vinyl acetal) resins using two-dimensional nuclear magnetic resonance spectroscopy. *Macromolecules* 24, 2189–2193.
39. El-Din, N. M. S. and Sabaa, M. W. 1995. Thermal degradation of poly (vinyl butyral) laminated safety glass. *Polym Degrad Stab* 47, 283–288.
40. Cascone, E., David, D., Di Lorenzo, M., Karasz, F., Macknight, W., Martuscelli, E. and Raimo, M. 2001. Blends of polypropylene with poly (vinyl butyral). *J Appl Polym Sci* 82, 2934–2946.
41. Schaefer, J., Garbow, J. R., Stejskal, E. and Lefelar, J. 1987. Plasticization of poly (butyral-covinyl alcohol). *Macromolecules* 20, 1271–1278.
42. Cotts, P. M., O. A. C. 1985. Dilute solutions of poly(vinylbutyral): Characterization of aggregated and non-aggregated solutions. In *Microdomains in Polymer Solutions*, ed. P. Dubin, 101–121. New York: Plenum Press.
43. Marani, D., H. J. W. M. 2013. Use of intrinsic viscosity for evaluation of polymer-solvent affinity. *Ann T Nord Rheol Soc* 21, 255–261.
44. Paul, C. and Cotts, P. 1986. Effects of aggregation and solvent quality on the viscosity of semidilute poly (vinyl butyral) solutions. *Macromolecules* 19, 692–699.
45. Paul, C. and Cotts, P. 1987. Static and dynamic light scattering from poly (vinyl butyral) solutions: Effects of aggregation and solvent quality. *Macromolecules* 20, 1986–1991.
46. Mrkvickova, L., Danhelka, J. and Pokorny, S. 1984. Characterization of commercial polyvinylbutyral by gel permeation chromatography. *J Appl Polym Sci* 29, 803–808.
47. Remsen, E. E. 1991. Determination of molecular weight for poly (vinyl butyral) using size exclusion chromatography/low-angle laser light scattering (SEC/LALLS) in hexafluoroisopropanol. *J Appl Polym Sci* 42, 503–510.
48. David, D. and Sincock, T. 1992. Estimation of miscibility of polymer blends using the solubility parameter concept. *Polymer* 33, 4505–4514.
49. Hansen, C. M. 2007. Appendix A.2. In *Hansen Solubility Parameters: A User's Handbook*, 2nd Ed., ed. C. M. Hansen, 495. Boca Raton, FL: CRC Press.
50. Lubasova, D. and Martinova, L. 2011. Controlled morphology of porous polyvinyl butyral nanofibers. *J Nanomater* 2011, 1–6.
51. Bendaoud, A., Carrot, C., Charbonnier, J. and Pillon, C. 2013. Blends of plasticized polyvinyl butyral and polyvinyl chloride: Morphology analysis in view of recycling. *Macromol Mater Eng* 298, 1259–1268.
52. Liau, L. C., Yang, T. C. and Viswanath, D. S. 1996. Reaction pathways and kinetic analysis of PVB thermal degradation using TG/FT-IR. *Appl Spectrosc* 50, 1058–1065.

53. Wilski, V. H. 1969. Die spezifische Wärme des Polyvinylbutyrals. *Angew Makromol Chem* 6, 101–108.
54. Morais, D., Valera, T. S. and Demarquette, N. R. 2006. Evaluation of the surface tension of poly (vinyl butyral) using the pendant drop method. *Macromol Symp* 245, 208–214.
55. Newman, S. 1967. The effect of composition on the critical surface tension of polyvinyl butyral. *J Colloid Interface Sci* 25, 341–345.
56. Wu, S. 1968. Estimation of the critical surface tension for polymers from molecular constitution by a modified Hildebrand-Scott equation. *J Phys Chem* 72, 3332–3334.
57. Nguyen, F. N. and Berg, J. C. 2004. The effect of vinyl alcohol content on adhesion performance in poly (vinyl butyral)/glass systems. *J Adhes Sci Technol* 18, 1011–1026.
58. Tupy, M., Merinska, D., Svoboda, P., Kalendova, A., Klasek, A. and Zvonicek, J. 2012. Effect of water and acid—Base reactants on adhesive properties of various plasticized poly (vinyl butyral) sheets. *J Appl Polym Sci* 127, 3474–3484.
59. Miller, A. and Berg, J. 2003. Effect of silane coupling agent adsorbate structure on adhesion performance with a polymeric matrix. *Composites Part A* 34, 327–332.
60. Harding, P. and Berg, J. 1998. The adhesion promotion mechanism of organofunctional silanes. *J Appl Polym Sci* 67, 1025–1033.
61. Funt, B. 1952. Dielectric dispersion in solid polyvinyl butyral. *Can J Chem* 30, 84–91.
62. Mehendru, P., Kumar, N., Arora, V. and Gupta, N. 1982. Dielectric relaxation studies in polyvinyl butyral. *J Chem Phys* 77, 4232–4235.
63. Saad, G. R., El-Shafee, E. and Sabaa, M. 1995. Dielectric and mechanical properties in the photodegradation of poly (vinyl butyral) films. *Polym Degrad Stab* 47, 209–215.
64. Jain, K., Kumar, N. and Mehendru, P. 1979. Electrical and dielectric properties of polyvinyl butyral I. Studies of charge storage mechanism. *J Electrochem Soc* 126, 1958–1963.
65. Mehendru, P., Jain, K. and Kumar, N. 1980. Electrical and dielectric properties of polyvinylbutyral II: The effect of the molecular weight on the charge storage mechanism. *Thin Solid Films* 70, 7–10.
66. Funt, B. and Sutherland, T. 1952. Dielectric properties of polyvinyl acetals. *Can J Chem* 30, 940–947.
67. Feller, R. L., Curran, M., Colaluca, V., Bogaard, J. and Bailie, C. 2007. Photochemical deterioration of poly (vinylbutyral) in the range of wavelengths from middle ultraviolet to the visible. *Polym Degrad Stab* 92, 920–931.
68. Reinohl, V., Sedlar, J. and Navratil, M. 1981. Photo-oxidation of poly (vinylbutyral). *Polym Photochem* 1, 165–175.
69. Mikhailik, O., Seropegina, Y. N., Melnikov, M. Y. and Fock, N. 1981. The kinetics and mechanism of photo-ageing of polyvinylbutyral. *Eur Polym J* 17, 1011–1019.
70. Butler, C. H. and Whitmore, P. M. 2014. Evolution of peroxide species during the photooxidation of poly (vinyl butyral). *J Appl Polym Sci* 131, doi:10.1002/app.39753.
71. Merinska, D., Tupy, M., Kaspsarkova, V., Popelkova, J., Zvonicek, J., Pistek, D. and Svoboda, P. 2009. Degradation of plasticized PVB during reprocessing by kneading. *Macromol Symp* 286, 107–115.
72. Etienne, S., Becker, C., Ruch, D., Germain, A. and Calberg, C. 2010. Synergetic effect of poly (vinyl butyral) and calcium carbonate on thermal stability of poly (vinyl chloride) nanocomposites investigated by TG-FTIR-MS. *J Therm Anal Calorim* 100, 667–677.
73. Liau, L. C., Yang, T. C. and Viswanath, D. S. 1996. Mechanism of degradation of poly (vinyl butyral) using thermogravimetry/fourier transform infrared spectrometry. *Polym Eng Sci* 36, 2589–2600.
74. Monakhova, T., Mar'in, A. and Shlyapnikov, Y. A. 1993. Thermal oxidation of highly plasticized poly (vinyl butyral). *Polym Sci* 35, 1271–1273.
75. Arayachukiat, S., Siriprumpoonthum, M., Nobukawa, S. and Yamaguchi, M. 2014. Viscoelastic properties and extrusion processability of poly (vinyl butyral). *J Appl Polym Sci* 131, doi:10.1002/app.40337.
76. Juang, Y.-J., Bruer, D., Lee, L. J., Koelling, K. W., Srinivasan, N., Drummond, C. H. and Wong, B. C. 2001. A method for assessing the effect of polymer sheeting rheology, surface pattern, and processing conditions on glass lamination. *J Appl Polym Sci* 80, 521–528.
77. Juang, Y.-J., Lee, L. J. and Koelling, K. W. 2001. Rheological analysis of polyvinyl butyral near the glass transition temperature. *Polym Eng Sci* 41, 275–292.
78. Qiu, Y.-R. and Ouyang, W. 2012. Rheological behavior of poly (vinyl butyral)/polyethylene glycol binary systems. *Mater Sci Eng, C* 32, 167–171.
79. Tupy, M., Merinska, D. and Kasparkova, V. 2012. PVB sheet recycling and degradation. In *Material Recycling—Trends and Perspectives*, Chap. 5, ed. D. Achilias, 133–150. Rijeka, Czech Republic: InTech.
80. Wypych, G. 2012. Plasticizers use and selection for specific polymer. In *Handbook of Plasticizers*, 2nd Ed., Chap. 11, ed. G. Wypych, 381–399. Toronto: ChemTech Publishing.

81. Ude, M., Ashraf-Khorassani, M. and Taylor, L. 2002. Supercritical fluid extraction of plasticizers in poly (vinyl butyral) (PVB) and analysis by supercritical fluid chromatography. *Chromatographia* 55, 743–748.
82. Marin, A., Tatarenko, L. A. and Shlyapnikov, Y. A. 1998. Solubility of antioxidants in poly (vinyl butyral). *Polym Degrad Stab* 62, 507–511.
83. Kim, D.-H., Lim, K.-Y., Paik, U. and Jung, Y.-G. 2004. Effects of chemical structure and molecular weight of plasticizer on physical properties of green tape in $BaTiO_3$/PVB system. *J Eur Ceram Soc* 24, 733–738.
84. Ellis, B. and Lim, B. 1984. Stress-softening and strain-hardening of plasticized poly (vinyl butyral). *J Mater Sci Lett* 3, 620–622.
85. Shende, R. V., Kline, M. and Lombardo, S. J. 2004. Effects of supercritical extraction on the plasticization of poly (vinyl butyral) and dioctyl phthalate films. *J Supercrit Fluids* 28, 113–120.
86. Hintersteiner, I., Sternbauer, L., Beissmann, S., Buchberger, W. W. and Wallner, G. M. 2014. Determination of stabilisers in polymeric materials used as encapsulants in photovoltaic modules. *Polym Test* 33, 172–178.
87. Kempe, M. 2011. Overview of scientific issues involved in selection of polymers for PV applications. In *Photovoltaic Specialists Conference (PVSC), 2011 37th IEEE*, pp. 000085–000090. New York: IEEE.
88. Burns, M., Wagenknecht, U., Kretzschmar, B. and Focke, W. W. 2008. Effect of hydrated fillers and red phosphorus on the limiting oxygen index of poly (ethylene-co-vinyl actate)-poly (vinyl butyral) and low density polyethylene-poly (ethylene-co-vinyl alcohol) blends. *J Vinyl Addit Technol* 14, 113–119.
89. Kitaura, T., Fadzlina, W. N., Ohmukai, Y., Maruyama, T. and Matsuyama, H. 2013. Preparation and characterization of several types of polyvinyl butyral hollow fiber membranes by thermally induced phase separation. *J Appl Polym Sci* 127, 4072–4078.
90. Jeong, H., Rooney, M., David, D., MacKnight, W., Karasz, F. and Kajiyama, T. 2000. Miscibility of polyvinyl butyral/nylon 6 blends. *Polymer* 41, 6003–6013.
91. Jeong, H., Rooney, M., David, D., MacKnight, W., Karasz, F. and Kajiyama, T. 2000. Miscibility and characterization of the ternary crystalline system: Poly (vinyl butyral)/poly (vinyl alcohol)/nylon6. *Polymer* 41, 6671–6678.
92. Valera, T. S. and Demarquette, N. R. 2008. Polymer toughening using residue of recycled windshields: PVB film as impact modifier. *Eur Polym J* 44, 755–768.
93. Hofmann, G. H. and Lee, W. 2006. Modification of polymers by using poly (vinyl butyral)-based additives. *J Vinyl Addit Technol* 12, 33–36.
94. Cha, Y.-J., Lee, C.-H. and Choe, S. 1997. The influence of scrap poly (vinyl butyral) film on the morphology and mechanical properties of nylon 6 toughened with SEBS-g-MA. *J Ind Eng Chem* 3, 257–262.
95. Cha, Y., Lee, C. and Choe, S. 1998. Morphology and mechanical properties of nylon 6 toughened with waste poly (vinyl butyral) film. *J Appl Polym Sci* 67, 1531–1540.
96. Lai, S.-M. and Lin, Y.-C. 2006. Mechanical properties of nylon 6, 6/polyvinyl butyral blends. *Polym-Plast Technol Eng* 45, 421–428.
97. Lee, C.-H., Cha, Y. J. and Choe, S. 1998. Effects of recycled poly (vinyl butyral) film on the morphology and mechanical properties of the blends containing polypropylene and nylon 6. *J Ind Eng Chem* 4, 161–169.
98. Jarvela, P. A., Shucai, L. and Jarvela, P. K. 1997. Dynamic mechanical and mechanical properties of polypropylene/poly (vinyl butyral)/mica composites. *J Appl Polym Sci* 65, 2003–2011.
99. Yan, L. and Wang, J. 2011. Development of a new polymer membrane-PVB/PVDF blended membrane. *Desalination* 281, 455–461.
100. Kuleznev, V., Mel'nikova, O. and Klykova, V. 1978. Dependence of modulus and viscosity upon composition for mixtures of polymers. Effects of phase composition and properties of phases. *Eur Polym J* 14, 455–461.
101. Singh, Y. and Singh, R. 1983. Compatibility studies on solutions of polymer blends by viscometric and ultrasonic techniques. *Eur Polym J* 19, 535–541.
102. Takeshita, Y., Kamisho, T., Sakata, S., Sawada, T., Watanuki, Y., Nishio, R. and Ueda, T. 2013. Dual layer structural thermoplastic polyester powder coating film and its weathering resistance. *J Appl Polym Sci* 128, 1732–1739.
103. Tripathy, A. R., Chen, W., Kukureka, S. N. and MacKnight, W. J. 2003. Novel poly (butylene terephthalate)/poly (vinyl butyral) blends prepared by in situ polymerization of cyclic poly (butylene terephthalate) oligomers. *Polymer* 44, 1835–1842.
104. Rohindra, D. R. and Khurma, J. R. 2003. Biodegradation study of poly (e-caprolactone)/poly (vinyl butyral) blends. *South Pac J Nat Sci* 21, 47–49.

105. Rohindra, D., Sharma, P. and Khurma, J. 2005. Soil and microbial degradation study of poly (e-caprolactone)/polycaprolactone)/poly (vinyl butyral) blends. *Macromol S* 224, 323–332.
106. Lee, J.-C., Ajima, K. N., Ikehara, T. and Nishi, T. 1997. Conductive-filler-filled poly (e-caprolactone)/poly (vinyl butyral) blends. II. Electric properties (positive temperature coefficient phenomenon). *J Appl Polym Sci* 65, 409–416.
107. Lee, J.-C., Nakajima, K., Ikehara, T. and Nishi, T. 1997. Conductive-filler-filled poly (e-caprolactone)/poly (vinyl butyral) blends. I. Crystallization behavior and morphology. *J Appl Polym Sci* 64, 797–802.
108. Nozue, Y., Kurita, R., Hirano, S., Kawasaki, N., Ueno, S., Iida, A., Nishi, T. and Amemiya, Y. 2003. Spatial distribution of lamella structure in PCL/PVB band spherulite investigated with microbeam small and wide-angle X-ray scattering. *Polymer* 44, 6397–6405.
109. Wu, T.-M. and Cheng, J.-C. 2003. Morphology and electrical properties of carbon-black-filled poly (e-caprolactone)/poly (vinyl butyral) nanocomposites. *J Appl Polym Sci* 88, 1022–1031.
110. Khurma, J. R., Rohindra, D. R. and Devi, R. 2005. Miscibility study of solution cast blends of poly (lactic acid) and poly (vinyl butyral). *South Pac J Nat Sci* 23, 22–25.
111. Chen, W., David, D. J., MacKnight, W. J. and Karasz, F. E. 2001. Miscibility and morphology of blends of poly (3-hydroxybutyrate) and poly (vinyl butyral). *Polymer* 42, 8407–8414.
112. Qian, J.-W., Chen, H.-L., Zhang, L., Qin, S. and Wang, M. 2002. Effect of compatibility of cellulose acetate/poly (vinyl butyral) blends on pervaporation behavior of their membranes for methanol/methyl tert-butyl ether mixture. *J Appl Polym Sci* 83, 2434–2439.
113. Sita, C., Burns, M., Hassler, R. and Focke, W. 2006. Tensile properties of thermoplastic starch-PVB blends. *J Appl Polym Sci* 101, 1751–1755.
114. Qiu, Y. R., Rahman, N. A. and Matsuyama, H. 2008. Preparation of hydrophilic poly (vinyl butyral)/Pluronic F127 blend hollow fiber membrane via thermally induced phase separation. *Sep Purif Technol* 61, 1–8.
115. Qiu, Y. R. and Matsuyama, H. 2010. Preparation and characterization of poly (vinyl butyral) hollow fiber membrane via thermally induced phase separation with diluent polyethylene glycol 200. *Desalination* 257, 117–123.
116. Ouyang, W. and Qiu, Y.-R. 2011. Rheological properties of polyvinyl butyral/Pluronic F127/PEG200 blend systems. *J Cent South Univ Technol* 18, 1891–1896.
117. Nizam El-Din, H. M. and El-Naggar, A. W. M. 2008. Effect of g-irradiation on the physical properties and dyeability of poly (vinyl butyral) blends with polystyrene and poly (ethylene glycol). *Polym Compos* 29, 597–605.
118. Sincock, T. and David, D. 1992. Miscibility and properties of poly (vinyl butyral) and thermoplastic polyurethane blends. *Polymer* 33, 4515–4521.
119. Shen, F., Lu, X., Bian, X. and Shi, L. 2005. Preparation and hydrophilicity study of poly (vinyl butyral)-based ultrafiltration membranes. *J Membr Sci* 265, 74–84.
120. Chen, W., David, D. J., MacKnight, W. J. and Karasz, F. E. 2001. Miscibility and phase behavior in blends of poly (vinyl butyral) and poly (methyl methacrylate). *Macromolecules* 34, 4277–4284.
121. Almeida, A., Gliemann, H., Schimmel, T. and Petri, D. 2005. Characterization of PMMA/PVB blend films by means of AFM. *Microsc Microanal* 11, 122–125.
122. Choi, H. K., Lee, Y. M., Yoon, J. H. and Choi, S. 1996. The effect of MBS on the compatibility of scrap PVB/PMMA blends. *J Kor Inst Rubber Tech* 31, 23–32.
123. Mohamed, N. A. and Sabaa, M. W. 1999. Thermal degradation behaviour of poly (vinyl chloride)-poly (vinyl butyral) blends. *Eur Polym J* 35, 1731–1737.
124. Peng, Y. and Sui, Y. 2006. Compatibility research on PVC/PVB blended membranes. *Desalination* 196, 13–21.
125. Ozaki, J., Endo, N., Ohizumi, W., Igarashi, K., Nakahara, M., Oya, A., Yoshida, S. and Iizuka, T. 1997. Novel preparation method for the production of mesoporous carbon fiber from a polymer blend. *Carbon* 35, 1031–1033.
126. Ozaki, J.-I., Ohizumi, W. and Oya, A. 2000. A TG-MS study of poly (vinyl butyral)/phenol-formaldehyde resin blend fiber. *Carbon* 38, 1515–1519.
127. Zhang, P., Wang, Y., Xu, Z. and Yang, H. 2011. Preparation of poly (vinyl butyral) hollow fiber ultrafiltration membrane via wet-spinning method using PVP as additive. *Desalination* 278, 186–193.
128. Lang, W.-Z., Shen, J.-P., Zhang, Y.-X., Yu, Y.-H., Guo, Y.-J. and Liu, C.-X. 2013. Preparation and characterizations of charged poly (vinyl butyral) hollow fiber ultrafiltration membranes with perfluorosulfonic acid as additive. *J Membr Sci* 430, 1–10.
129. Lin, Q., Li, Y. and Yang, M. 2012. Polyaniline nanofiber humidity sensor prepared by electrospinning. *Sens Actuators, B* 161, 967–972.

130. Torki, A., Zivkovic, I., Radmilovic, V., Stojanovic, D., Radojevic, V., Uskokovic, P. and Aleksi, R. 2010. Dynamic mechanical properties of nanocomposites with poly (vinyl butyral) matrix. *Int J Mod Phys B* 24, 805–812.
131. Torki, A., Stojanovic, D., Zivkovic, I., Marinkovic, A., Skapin, S., Uskokovic, P. and Aleksic, R. 2012. The viscoelastic properties of modified thermoplastic impregnated multiaxial aramid fabrics. *Polym Compos* 33, 158–168.
132. Ambrosio, J., Lucas, A., Otaguro, H. and Costa, L. 2011. Preparation and characterization of poly (vinyl butyral)-leather fiber composites. *Polym Compos* 32, 776–785.
133. Kim, G. and Hyun, S. 2003. Effect of mixing on thermal and mechanical properties of aerogel-PVB composites. *J Mater Sci* 38, 1961–1966.
134. Chen, L.-J., Liao, J.-D., Lin, S.-J., Chuang, Y.-J. and Fu, Y.-S. 2009. Synthesis and characterization of PVB/silica nanofibers by electrospinning process. *Polymer* 50, 3516–3521.
135. Kirchberg, S., Rudolph, M., Ziegmann, G. and Peuker, U. 2012. Nanocomposites based on technical polymers and sterically functionalized soft magnetic magnetite nanoparticles: Synthesis, processing, and characterization. *J Nanomater* 2012, 20.
136. Joshi, N., Rakshit, P., Grewal, G., Shrinet, V. and Pratap, A. 2013. Temperature dependence dielectric properties of modified barium titanate-PVB composites. In *Proceeding of International Conference on Recent Trends in Applied Physics and Material Science: Ram 2013*, Vol. 1536, pp. 691–692. Melville, NY: AIP Publishing.
137. Dong, L., Xiong, C., Quan, H. and Zhao, G. 2006. Polyvinyl-butyral/lead zirconate titanates composites with high dielectric constant and low dielectric loss. *Scripta Materialia* 55, 835–837.
138. Pistek, D., Merinska, D., Dujkova, Z. and Tupy, M. 2010. The mechanical and optical properties of the PVB nanocomposites. In *Proceedings of the 3rd WSEAS International Conference on Advances in Sensors, Signals and Materials*, pp. 26–29. World Scientific and Engineering Academy and Society (WSEAS).
139. Nakane, K., Kurita, T., Ogihara, T. and Ogata, N. 2004. Properties of poly (vinyl butyral)/TiO$_2$ nanocomposites formed by sol-gel process. *Composites Part B* 35, 219–222.
140. Yan, W. 2013. Fabrication and thermal insulating properties of ATO/PVB nanocomposites for energy saving glass. *J Wuhan Univ Technol Mater Sci Ed* 28, 912–915.
141. Roy, A. S., Gupta, S., Seethamraju, S., Madras, G. and Ramamurthy, P. C. 2014. Impedance spectroscopy of novel hybrid composite films of polyvinylbutyral (PVB)/functionalized mesoporous silica. *Composites Part B* 58, 134–139.
142. Zhang, Y., Ding, Y., Li, Y., Gao, J. and Yang, J. 2009. Synthesis and characterization of polyvinyl butyral-Al (NO3) 3 composite sol used for alumina based fibers. *J Sol-Gel Sci Technol* 49, 385–390.
143. Zhou, S. X., Zhu, Y., Du, H. D., Li, B. H. and Kang, F. Y. 2012. Preparation of oriented graphite/polymer composite sheets with high thermal conductivities by tape casting. *New Carbon Mater* 27, 241–249.
144. Alsaed, O. and Jalham, I. 2012. Polyvinyl butyral (PVB) and ethyl vinyl acetate (EVA) as a binding material for laminated glass. *Jordan J Mech Ind Eng* 6, 127–133.
145. Bennison, S. J., Jagota, A. and Smith, C. A. 1999. Fracture of glass/poly (vinyl butyral) (Butacite®) laminates in biaxial flexure. *J Am Ceram Soc* 82, 1761–1770.
146. Duser, A. V., Jagota, A. and Bennison, S. J. 1999. Analysis of glass/polyvinyl butyral laminates subjected to uniform pressure. *J Eng Mech* 125, 435–442.
147. Foraboschi, P. 2007. Behavior and failure strength of laminated glass beams. *J Eng Mech* 133, 1290–1301.
148. Shin, P., 2012. How to make PVB laminated glass with PVB film polyvinyl butyral. Available at http://www.glasssummit.com/how-to-make-pvb-laminated-glass-with-pvb-film-polyvinyl-butyral.
149. AGCGlassEurope. 2013. *Toughened and Laminated Safety Glass. Technical Information*. Louvain-La-Neuve, Belgium: AGC Glass Europe.
150. Hooper, P., Sukhram, R., Blackman, B. and Dear, J. 2012. On the blast resistance of laminated glass. *Int J Solids Struct* 49, 899–918.
151. Qingshan, L., Wenjie, G. and Pengsheng, M. 2008. Application of poly (vinyl butyral) nanocomposites in environment design. *Adv Nat Sci* 1, 81–88.
152. Zeng, X.-F., Li, X., Tao, X., Shen, Z.-G. and Chen, J.-F. 2010. Fabrication of highly transparent ZnO/PVB nanocomposite films with novel UV-shielding properties. In *Nanoelectronics Conference (INEC), 2010, 3rd International*, pp. 777–778. New York: IEEE.
153. Matusevich, V., Tolstik, E., Kowarschik, R., Egorova, E., Matusevich, Y. I. and Krul, L. 2013. New holographic polymeric composition based on plexiglass, polyvinyl butyral, and phenanthrenquinone. *Opt Commun* 295, 79–83.
154. Lilly, J. G. 2004. Recent advances in acoustical glazing. *Sound Vib* 38, 8–13.

155. D'Haene, P. and Savineau, G. 2007. Mechanical properties of laminated safety glass—FEM study. In *Proceedings of the Glass Processing Days 2007, GPD 2007*, pp. 593–598.
156. Sanz-Ablanedo, E., Lamela, M., Rodriguez-Perez, J. R. and Arias, P. 2010. Modelizacion y contraste experimental del comportamiento mecánico del vidrio laminado estructural. *Mater Constr* 60, 131–141.
157. Froli, M. and Lani, L. 2011. Adhesion, creep and relaxation properties of PVB in laminated safety glass. In *Proceedings of the Glass Processing Days 2011, GPD 2011*, pp. 1–4.
158. Savineau, G. 2013. Intercalaires pour verres feuilletés. In *Techniques de l'ingénieur Sciences et technologies du verre*, N4404, 1–11. Paris: Editions T.I.
159. Solutia. 2008. *Saflex SilentGlass Technology Product Bulletin, Americas*. St. Louis, MO: Solutia, Inc.
160. Yoshioka, T. 2004. Functional laminated glazing with advanced PVB technology. *AutoTechnology* 4, 62–65.
161. Rugh, J., Chaney, L., Ramroth, L., Venson, T. and Rose, M. 2013. Impact of solar control PVB glass on vehicle interior temperatures, air-conditioning capacity, fuel consumption, and vehicle range. In *SAE 2013 World Congress and Exhibition*, 2013-01-0553. SAE Technical Paper.
162. Kempe, M. D. 2010. Ultraviolet light test and evaluation methods for encapsulants of photovoltaic modules. *Sol Energy Mater Sol Cells* 94, 246–253.
163. Kuraray 2012. *Trosifol—Solar Module Optimisation with PVB Film—Solar Technology in Perfection*. Hattersheim am Main, Germany: Kuraray Europe GmbH.
164. Peike, C., Hülsmann, P., Blüml, M., Schmid, P., Weiss, K.-A. and Köhl, M. 2012. Impact of permeation properties and backsheet-encapsulant interactions on the reliability of PV modules. *ISRN Renew Energy* 2012, 1–5.
165. Kim, N. and Han, C. 2013. Experimental characterization and simulation of water vapor diffusion through various encapsulants used in PV modules. *Sol Energy Mater Sol Cells* 116, 68–75.
166. Barranco, V., Carmona, N., Villegas, M. A. and Galvan, J. C. 2010. Tailored sol-gel coatings as environmentally friendly pre-treatments for corrosion protection. *ECS Trans* 24, 277–290.
167. David, G., Ortega, E., Chougrani, K., Manseri, A. and Boutevin, B. 2011. Grafting of phosphonate groups onto PVA by acetalyzation. Evaluation of the anti-corrosive properties for the acetalyzed PVA coatings. *React Funct Polym* 71, 599–606.
168. Peng, P., Li, X., Yuan, G., She, W., Cao, F., Yang, D., Zhuo, Y., Liao, J., Yang, S. and Yue, M. 2001. Aluminum oxide/amorphous carbon coatings on carbon fibers, prepared by pyrolysis of an organic–inorganic hybrid precursor. *Mater Lett* 47, 171–177.
169. Jean, J.-H. and Wang, H.-R. 2001. Organic distributions in dried alumina green tape. *J Am Ceram Soc* 84, 267–272.
170. Mukherjee, A., Khan, R., Bera, B. and Maiti, H. S. 2008. I: Dispersibility of robust alumina particles in non-aqueous solution. *Ceram Int* 34, 523–529.
171. Jean, J.-H., Yeh, S.-F. and Chen, C.-J. 1997. Adsorption of poly (vinyl butyral) in nonaqueous ferrite suspensions. *J Mater Res* 12, 1062–1068.
172. Hsiang, H.-I., Chen, C.-C. and Tsai, J.-Y. 2005. Dispersion of nonaqueous Co_2Z ferrite powders with titanate coupling agent and poly (vinyl butyral). *Appl Surf Sci* 245, 252–259.
173. Cho, Y.-S., Yeo, J.-G., Jung, Y.-G., Choi, S.-C., Kim, J. and Paik, U. 2003. Effect of molecular mass of poly (vinyl butyral) and lamination pressure on the pore evolution and microstructure of $BaTiO_3$ laminates. *Mater Sci Eng, A* 362, 174–180.
174. Yun, J. W. and Lombardo, S. J. 2008. Permeability of laminated green ceramic tapes as a function of binder loading. *J Am Ceram Soc* 91, 1553–1558.
175. Masia, S., Calvert, P. D., Rhine, W. E. and Bowen, H. K. 1989. Effect of oxides on binder burnout during ceramics processing. *J Mater Sci* 24, 1907–1912.
176. Jaw, K.-S., Hsu, C.-K. and Lee, J.-S. 2001. The thermal decomposition behaviors of stearic acid, paraffin wax and polyvinyl butyral. *Thermochim Acta* 367, 165–168.
177. Liau, L. C. and Chien, Y. 2006. Kinetic investigation of $ZrO2$, $Y2O3$, and Ni on poly (vinyl butyral) thermal degradation using nonlinear heating functions. *J Appl Polym Sci* 102, 2552–2559.
178. Salam, L. A., Matthews, R. D. and Robertson, H. 2000. Pyrolysis of polyvinyl butyral (PVB) binder in thermoelectric green tapes. *J Eur Ceram Soc* 20, 1375–1383.
179. Salam, L. A., Matthews, R. D. and Robertson, H. 2000. Optimisation of thermoelectric green tape characteristics made by the tape casting method. *Mater Chem Phys* 62, 263–272.
180. Nair, A. and White, R. L. 1996. Effects of inorganic oxides on polymer binder burnout. I. Poly (vinyl butyral). *J Appl Polym Sci* 60, 1901–1909.

181. Liau, L. C., Liau, J. and Chen, Y. 2004. Study of the composition effect of glass ceramic and silver on poly (vinyl butyral) thermal degradation with thermogravimetric analysis. *J Appl Polym Sci* 93, 2142–2149.
182. El-Sherbiny, M. and El-Rehim, N. 2001. Spectroscopic and dielectric behavior of pure and nickel-doped polyvinyl butyral films. *Polym Test* 20, 371–378.
183. Seo, J., Kuk, S. and Kim, K. 1997. Thermal decomposition of PVB (polyvinyl butyral) binder in the matrix and electrolyte of molten carbonate fuel cells. *J Power Sources* 69, 61–68.
184. Gopal, S., Ramchandran, R. and Agnihotry, R. S. 1997. Polyvinyl butyral based solid polymeric electrolytes: Preliminary studies. *Solar Energy Mater Solar Cells* 45, 17–25.
185. Chen, K.-F., Liu, C.-H., Huang, H.-K., Tsai, C.-H. and Chen, F.-R. 2013. Polyvinyl butyral-based thin film polymeric electrolyte for dye-sensitized solar cell with long-term stability. *Int J Electrochem Sci* 8, 3524–3539.
186. Gopal, S., Agnihotry, S. and Gupta, V. 1996. Ionic conductivity in poly (vinyl butyral) based polymeric electrolytes: Effect of solvents and salts. *Solar Energy Mater Solar Cells* 44, 237–250.
187. Abdel-Fattah, A., Beshir, W., Hegazy, E.-S. A. and Ezz El-Din, H. 2001. Photo-luminescence of Risø B3 and PVB films for application in radiation dosimetry. *Radiat Phys Chem* 62, 423–428.
188. Tupy, M., Mokrejs, P., Merinska, D., Svoboda, P. and Zvonicek, J. 2014. Windshield recycling focused on effective separation of PVB sheet. *J Appl Polym Sci* 131, doi: 10.1002/app.39879.
189. Wang, H., Xie, C., Yu, W. and Fu, J. 2012. Efficient combined method of selective dissolution and evaporation for recycling waste polyvinylbutyral films. *Plast Rubber Compos* 41, 8–12.

4 Polyacrylonitrile*

Johannis C. Simitzis and Spyridon K. Soulis

CONTENTS

4.1 History, Development, and Commercialization Milestones 139
4.2 Polymer Formation and Process Technologies 141
    4.2.1 Acrylonitrile 141
        4.2.1.1 Physical Properties of Acrylonitrile 141
        4.2.1.2 Chemical Properties of Acrylonitrile 141
        4.2.1.3 Manufacture of Acrylonitrile 142
        4.2.1.4 Health and Safety Factors 142
    4.2.2 Polymerization of Acrylonitrile 143
        4.2.2.1 Homopolymerization 143
        4.2.2.2 Copolymerization 144
4.3 Structural and Phase Characteristics and Their Effects on the Properties of Polyacrylonitrile 144
    4.3.1 Structure 144
    4.3.2 Phase Characteristics 146
        4.3.2.1 Glass and Other Transitions 146
        4.3.2.2 Heat Treatment 146
    4.3.3 Physical Properties 147
        4.3.3.1 Solubility and Molecular Weight Relations 147
        4.3.3.2 Other Physical Properties 148
    4.3.4 Chemical Properties 148
4.4 Thermoplastic Blends, Alloys, Copolymers, and Composites, and Their Commercial Relevance 149
    4.4.1 Copolymers of Acrylonitrile–Butadiene (Nitrile Rubber, NBR) 149
    4.4.2 Copolymers of Acrylonitrile in High-Barrier Applications 149
    4.4.3 Copolymers of Acrylonitrile in Optically Active Applications 150
    4.4.4 Copolymer of Acrylonitrile in Medical Applications 150
    4.4.5 Other Copolymers of Acrylonitrile and Blends 150
4.5 Polyacrylonitrile Processing, Performance Properties, and Applications 150
4.6 New Technologies Based on Polyacrylonitrile 152
4.7 Conclusion 155
References 155

4.1 HISTORY, DEVELOPMENT, AND COMMERCIALIZATION MILESTONES

Acrylonitrile (vinyl cyanide), CH_2=CHCN, which is mainly produced by the reaction of propylene and ammonia, is one of the most important monomers of the polymer industry. Its polymer, polyacrylonitrile (PAN), is important for the production of synthetic fibers. Acrylonitrile in combination with other monomers (mainly butadiene) is also used for the preparation of synthetic rubbers [1–4].

PAN has been known since the late 1920s, based on the German patent literature. In the beginning, acrylonitrile was used only in the preparation of copolymers (mainly with butadiene to form

---

* Based in parts on the first-edition chapter on polyacrylonitrile.

Buna-N rubber). The first commercial applications of acrylonitrile polymers were developed in Germany during World War II for use in the fabrication of oil-resistant rubbers [1,4]. PAN could not be spun because of its insolubility in common solvents and because its higher melting point compared to its thermal decomposition point precluded its melt spinning. PAN fiber preparation became feasible during World War II with the provision of suitable spinning solvents by I.G. Farbenindustrie in Germany and DuPont Co. in the United States [1,2].

Acrylic fiber was commercialized by DuPont under the trade name Orlon in the late 1940s. The industrial production of acrylic fiber was followed by other companies using different spinning solvents or other comonomers with acrylonitrile or other spinning techniques. According to the Federal Trade Commission, in 1960, an *acrylic fiber* was defined as "a manufactured fiber in which the fiber-forming substance is any long-chain synthetic polymer composed of at least 85% by weight of acrylonitrile units as shown."

$$-[CH_2 - CH(CN)]-$$

A *modacrylic fiber*, however, was defined as "a manufactured fiber composed of less than 85% but at least 35% by weight of acrylonitrile units" [2,5,6]. For modacrylic fibers, other comonomers such as vinyl chloride, vinylidene chloride, and vinyl bromide are commercially used to improve the flame resistance of the fiber due to the halogen atom present [1–3].

The worldwide production of acrylic and modacrylic fibers increased between 1960 and 1980 because of such advantages of acrylic fibers as their woollike appearance and feel, and their favorable economic position owing to the relatively low cost of manufacturing acrylonitrile. In the decade 1984–1994, the production growth rate was lower due primarily to the worldwide recession that affected all man-made fibers [7–10]. The world production of acrylonitrile in 1991 was 2,385,800 metric tons [7]. Taking as 100 the production of that year, the corresponding index in the year 2000 was about 130, showing the highest value during these 10 years; the index in 2001 was 110 [11]. Then, the world acrylonitrile production increased yearly until reaching the record value of 5200 million tons in 2007 (though in 2008, the global economic difficulties caused an abrupt decrease of almost 14%) [12].

Industries in many countries produce acrylic (and modacrylic) fibers. Seven such companies are in Japan; five each in the United States and Germany; four in India; three each in Italy and Mexico; two each in France, the United Kingdom, Spain, Brazil, Korea, and Taiwan; and one each in Ireland, Bulgaria, Romania, Greece, Poland, Hungary, the former Yugoslavia, Israel, Turkey, Iran, Argentina, Peru, and Bangladesh [2]. Acrylic fibers have found application in the textile industry for the manufacture of sweaters, socks, craft yarns, blankets, and carpets [1,10,13–15]. The world fiber production share in 1980 was 35% polyester, 22% nylon, 14% acrylic, 7% olefin, and 22% cellulosic. The individual fiber production patterns in 2000 was 56% polyester, 12% nylon, 8% acrylic, 18% olefin, and 6% cellulosic. In 2000, the global production for acrylic (and modacrylic) fibers was 2.7 million metric tons [16].

Acrylonitrile homopolymer, i.e., PAN, has found little application because of its difficult processability (PAN is not a typical thermoplastic polymer). Even in PAN used in the production of synthetic fibers, small amounts (e.g., 1%) of comonomers are added in the polymerization reactor to improve some properties of the final fibers, like dye uptake, stability, etc. The copolymers of acrylonitrile with other monomers combine the unusual and desirable properties of acrylonitrile unit with a variation of melt processability. The content of acrylonitrile in polymers of general applications ranges up to ca. 50%, while for barrier applications (especially for oxygen and carbon dioxide), it ranges up to approximately 75% [1,4].

Since the beginning of acrylic fiber industrial manufacture (ca. 1960), acrylic fibers represented more than 70% of the whole consumption of acrylonitrile in Europe, Latin America, and the Far

East. In the United States, this consumption gradually decreased from 50% to about 30%. Worldwide acrylonitrile producers include six companies in Japan; four in the United States; three each in Germany and Italy; two in the United Kingdom; and one each in France, Austria, Netherlands, Russia, Spain, Romania, Bulgaria, Poland, Mexico, Brazil, India, Taiwan, South Korea, North Korea, and China [4]. Historically, the prices of acrylonitrile mainly depended on the industrial process of its manufacture (e.g., the introduction and broad licensing of the Sohio low-cost propylene ammoxidation process in 1963) and the oil price. In approximate terms, the price of acrylonitrile is lower than $1/kg (ca. $0.9/kg), while the price of acrylic fibers is around $2/kg and that of synthetic rubbers (styrene acrylonitrile [SAN]) is lower than $2/kg (around $1.9/kg) [4,8,17,18].

Acrylonitrile is also used as the raw material for the production of acrylamide. The latter is polymerized to polyacrylamide, which is used for water and sewage treatments, textile treatments, etc. Other applications of acrylonitrile polymers include antioxidants, adhesives and binders, dyes, emulsifying agents, surface coatings, etc. In addition, PAN fibers are the precursor materials for the manufacture of carbon fibers by controlled pyrolysis. Carbon fiber-reinforced plastics are extremely stiff and low in density. PAN membranes have been used in various separations including mass transfer via biocatalysis. Acrylonitrile is also useful in the modification of other polymers, e.g., acrylonitrile grafted starch, which has very high water absorbency [2–4,19,20].

4.2 POLYMER FORMATION AND PROCESS TECHNOLOGIES

4.2.1 ACRYLONITRILE

4.2.1.1 Physical Properties of Acrylonitrile

Pure acrylonitrile is a colorless liquid with a penetrating odor of peach pits. Acrylonitrile is miscible with most polar and nonpolar organic solvents, like methanol, ethanol, acetone, benzene, toluene, xylene, (petroleum) ether, carbon tetrachloride, etc. [3,4]. The solubility, at 20°C, of water in acrylonitrile is 3.1 wt.%, and that of acrylonitrile in water is 7.3 wt.%. The boiling point (°C) of acrylonitrile is 77.3 (760 mm Hg), 64.7 (450 mm Hg), and 8.7 (50 mm Hg). The density (g/mL) of acrylonitrile is 0.806 (20°C) and 0.784 (41°C). Acrylonitrile forms azeotrope with many solvents, e.g., methanol (boiling point [BP] = 61.4°C, wt.% = 39), benzene (BP = 73.3°C, wt.% = 47), and water (BP = 71°C, wt.% = 88) [3,4,21,22]. The ultraviolet spectra and the Raman spectra appear in Ref. [23]; other important physical properties of acrylonitrile are described in Refs. [4,20].

4.2.1.2 Chemical Properties of Acrylonitrile

Acrylonitrile is a very reactive substance. This behavior is due to the presence of both a double bond and an electron-accepting nitrile group. Generally, acrylonitrile may react by the nitrile group or the double bond. Both types of reactions can also occur simultaneously [3,4,24]. The polymerization of the carbon–carbon double bond is the most important reaction and is detailed below.

The reactions of the *nitrile group* include hydration/hydrolysis, alcoholysis, reactions with olefins and alcohols, and reactions with aldehydes and methylol compounds. The nitrile group is partially hydrolyzed in concentrated 85% sulfuric acid to produce acrylamide sulfate, which then, by neutralization, yields acrylamide. In dilute acid or alkali conditions, acrylic acid is produced by complete hydrolysis. Acrylonitrile reacts with primary alcohols in the presence of sulfuric acid as a catalyst to yield acrylic esters. The nitrile group of acrylonitrile reacts with aldehydes in the presence of sulfuric acid as a catalyst, e.g., with formaldehyde, the reaction yields N,N'-methylenebisacrylamide [24].

The reactions of the *double bond* include hydrogenation, halogenation, hydroformylation, hydrodimerization, Diels–Alder reactions, and reactions with azo compounds. Acrylonitrile can be hydrogenated with a metal catalyst to yield propionitrile, which, by further hydrogenation, yields propylamine. Halogenation of acrylonitrile at low temperatures takes place slowly to yield 2,3-dihalopropionitriles, whereas at elevated temperatures, 3-substituted isomers are formed. Acrylonitrile reacts with a mixture of hydrogen and carbon monoxide (oxo synthesis), with cobalt octacarbonyl as a catalyst, to yield β-cyanopropionaldehyde. It can be reductively dimerized by a chemical or electrochemical method to give adiponitrile. Acrylonitrile with diazomethane compounds form pyrazolines, which are easily converted to cyclopropanes [24].

An important type of reaction is the cyanoethylation reactions (Michael-type additions) [25]. Most compounds possessing a labile hydrogen atom can be added to the double bond of acrylonitrile to form cyanoethyl groups, e.g., propionitriles:

$$CH_2=CHCN + RH \rightarrow RCH_2CH_2CN$$

The cyanoethylation of some natural polymers, which possess labile hydrogen atoms, like cellulose, lead to the formation of cyanoethyl derivatives, which are of some industrial importance [25].

4.2.1.3 Manufacture of Acrylonitrile

Acrylonitrile was manufactured in the past according to the older processes by synthetic reactions with C_2 compounds (e.g., acetylene, ethylene oxide, or acetaldehyde) and HCN. Today, virtually all acrylonitrile throughout the world is produced from propene (propylene) using Standard Oil of Ohio's (Sohio's) ammoxidation process and other similar modified processes. *Ammoxidation* is the catalytic oxidative reaction of activated methyl groups (as in propene) with NH_3 to form a nitrile group [12]. The first commercial plant started in 1960 using bismuth molybdate catalyst ($Bi_2O_3 \cdot MoO_3$). Approximately stoichiometric amounts of propene and NH_3 are reacted with a slight excess of air and added H_2O in a fluidized bed at 450°C and 1.5 bar [11]. The main reaction is as follows [4,13]:

$$H_2C=CHCH_3 + NH_3 + 1.5O_2 \rightarrow H_2C=CHCN + 3H_2O$$

$$(\Delta H = -120 \text{ kcal/mol})$$

The development of this process resulted in a drastic reduction in the price of acrylonitrile and an increase in high-volume applications of acrylonitrile in fibers and plastics.

4.2.1.4 Health and Safety Factors

Acrylonitrile is a neurotoxin with toxic liquid and vapor. It is flammable and explosive; extreme care is required for the handling of this substance. Acrylonitrile toxicity is mainly due to its molecule rather than to the release of cyanide ions [26]. It could enter the body through inhalation, ingestion, or skin contact and should not exceed 2 ppm in all workplaces. The explosive limit of its mixture with air is approximately 3.05–17.0% by volume, and the flash point is approximately −5°C (the ignition temperature is 481.0°C) [3,4,26].

Generally, acrylonitrile can be managed satisfactorily in both laboratory and industrial scale by applying the necessary safety precautions (e.g., ventilation, protective clothing, washing immediately with large amounts of water in cases of contact with the skin). During storage, acrylonitrile can be polymerized by peroxides formed from impurities. Lead, copper, magnesium, and their alloys catalyze chemical changes in acrylonitrile and should not be used in the manufacture of storage containers [3,4].

The risk of cancer mortality and incidence among 2559 employees exposed to acrylonitrile at two production plants of PAN (Orlon/DuPont USA) in 1944–1991 was studied [27]. Overall mortality was lower than expected in a comparison with the US population and all DuPont employees: 454 deaths, with standardized mortality ratios (SMRs) of 69 and 91, respectively. All the cancer death ratios were lower than expected in a similar comparison. The SMR values for specific sites did not differ significantly from the expected values. The cancer morbidity patterns were similarly unremarkable.

Polymers and copolymers of acrylonitrile are generally risk-free because the amount of the residual acrylonitrile in finished resins or products is very low (approximately 1 ppm in acrylic fibers). For food-contact applications, specific regulations must be applied [3,4].

4.2.2 POLYMERIZATION OF ACRYLONITRILE

4.2.2.1 Homopolymerization

4.2.2.1.1 Polymerization Mechanism

Pure acrylonitrile does not polymerize in the absence of initiators or light. It polymerizes readily in the presence of free radicals or anionic initiators and by irradiation with light (wavelength less than 2900 A, free-radical mechanism) [3,4]. The free-radical polymerization of acrylonitrile can be induced with conventional initiators (benzoyl peroxide, 2,2′-azobis[isobutyronitrile] [AIBN], etc.) at temperatures below 100°C [28,29] or with other initiators (e.g., pentavalent vanadyl ion–thiourea combination) [30] or redox systems (e.g., $Na_2S_2O_5$, $K_2S_2O_8$, and $FeSO_4$) in aqueous media at low temperatures (20°C and lower) [31]. Polymerization could also be induced by electropolymerization [32–34], light [35,36], and radiation (by x-rays and γ-rays) [37]. The use of a magnetic field during the radical polymerization of acrylonitrile with an AIBN initiator increases the rate of polymerization and alters the structure and properties of the PAN [38]. Oxygen is a very strong inhibitor of the polymerization of acrylonitrile. In the presence of oxygen, peroxides are formed until the oxygen is exhausted. Then the polymerization of acrylonitrile may ultimately proceed to an explosion because of the thermal decomposition of the peroxides [3,4].

On the other hand, acrylonitrile is one of the most reactive monomers for anionic initiators [39–42]. The latter include different types such as alfin catalysts [43], alkoxides [44], butyllithium [45], etc. In these systems, the active center is the $-CH_2CHCN^-$ anion, and the propagation may be readily terminated by proton donors. Therefore, these polymerizations are carried out in aprotic solvents, e.g., straight-chain ethers, cyclic ethers (like tetrahydrofuran), and hydrocarbons (toluene or hexane). The anionic polymerization of acrylonitrile is less important compared to the free-radical polymerization, because yellow products are often obtained, cyanoethylation as a side reaction takes place, and "living" species are difficult to observe owing to the insolubility of the polymer in the usual solvents [3]. To avoid such side reactions, living/controlled free-radical polymerization can be used, especially by using atom transfer radical polymerization (ATRP; with appropriate choice of initiator, catalyst, and solvent), leading to PAN with very low polydispersities [46].

4.2.2.1.2 Polymerization Processes

PAN is not soluble in its monomer, and therefore, the polymer is precipitated during bulk polymerization. Consequently, acrylonitrile cannot be polymerized into useful shapes by casting. Autoacceleration is observed during the bulk polymerization of acrylonitrile, and the rate of polymerization is continuously increased up to at least 20% of conversion [47]. As polymerization proceeds, the viscosity increases, and it is more difficult to remove the heat of polymerization in a batch bulk process. Consequently, the polymerization could get out of control. One way to overcome these problems is to use a continuous bulk process whose conversion in the effluent, after the reactor, is 54.6% [4,48]. Generally, the polymerization of the acrylonitrile can be carried out as a bulk, solution, suspension, slurry, or emulsion process.

The solution process is used to prepare acrylic polymers, which, after the reactor, are directly solution spun to fiber. Dimethylformamide (DMF) is commonly used as the solvent of this continuous slurry process where the monomer is isolated as small suspended droplets in an aqueous medium. The polymerization mechanism is similar to bulk polymerization, and rather high conversion can be achieved, for example, 90% within a residence time in the reactor of 1.69 h [4]. Furthermore, side reactions, leading to structural defects, chain scission, and development of color, have been reported [49]. As far the emulsion process is concerned, a redox system is used to carry out the polymerization at low temperatures, yielding a polymer with better color than that from polymerization at higher temperatures with the common initiator systems [4,50]. Although continuous polymerization of acrylonitrile is the typical method used in the industry, in the laboratory, polymerization in bulk or in solution may be carried out in a flask or in a suitable laboratory reactor. Such polymerizations of acrylonitrile are described in Refs. [51,52].

4.2.2.2 Copolymerization

Acrylonitrile can be readily copolymerized with electron-donor monomers, and it has more than 800 copolymers. Copolymerization depends on the Q (resonance term) and ε (+,–; polarity term) values of the acrylonitrile ($Q = 0.60$, $\varepsilon = 1.20$) and of the other monomer [3,53,54]. Generally, the monomer units in acrylonitrile copolymers are randomly arranged, but by using special techniques, alternating and block copolymers can be obtained. The reactivity ratios of acrylonitrile (M_1) and of the other monomer (M_2) are given in detail in the literature [3,53,55]. For example, the reactivity ratios of the free-radical copolymerization of acrylonitrile (r_1) with various monomers (r_2) are as follows: acrylamide, $r_1 = 0.875$ and $r_2 = 1.357$ at 30°C; acrylic acid, $r_1 = 0.35$ and $r_2 = 1.15$ at 50°C; butyl acrylate, $r_1 = 1.003$ and $r_2 = 1.005$ at 60°C; isobutylene, $r_1 = 1.02$ and $r_2 = 0$ at 60°C; methyl acrylate, $r_1 = 1.5$ and $r_2 = 0.84$ at 50°C, or $r_1 = 0.84$ and $r_2 = 0.83$ at 65°C) [3,56]; styrene, $r_1 = 0.07$ and $r_2 = 0.37$ at 50°C; vinyl acetate, $r_1 = 4.05$ and $r_2 = 0.06$ at 60°C; and vinyl chloride, $r_1 = 3.7$ and $r_2 = 0.07$ at 50°C. Copolymerization of acrylonitrile with a basic or acidic monomer is the best way to introduce dye sites onto the polymer chain because PAN itself does not accept dyes readily [3].

The copolymerization of acrylonitrile is carried out as a bulk, suspension, slurry, or emulsion process. Generally, alternating copolymers are formed by using a strong acceptor monomer (e.g., vinylidene cyanide) together with a strong donor monomer (e.g., styrene). Although acrylonitrile is an electron acceptor, in order to achieve alternating copolymers, a special technique is applied. Acrylonitrile forms complexes readily with charge-transfer agents (e.g., organoaluminum, metallic halides). These complexes are strong electron acceptors, and they interact with strong donor monomers to form alternating copolymers; for example, an alternating copolymer of acrylonitrile and vinyl acetate is formed in the presence of $ZnCl_2$ or Ziegler–Natta catalyst [57,58], and other alternating copolymers of acrylonitrile are similarly formed [59,60]. In contrast, block copolymers of acrylonitrile could be formed by chemical techniques [61] and other techniques, such as ultrasonics and radiation. Diblock (AB) or triblock (ABA) copolymers of acrylonitrile can be formed with other monomers such as acrylic acid, methyl acrylate, methyl methacrylate, styrene, vinyl acetate, and vinyl chloride [4]. Graft copolymers of acrylonitrile [62,63] such as acrylonitrile grafted onto a copolymer of vinylpyridine and acrylic acid [3] have been prepared by grafting techniques (e.g., high-energy radiation).

4.3 STRUCTURAL AND PHASE CHARACTERISTICS AND THEIR EFFECTS ON THE PROPERTIES OF POLYACRYLONITRILE

4.3.1 STRUCTURE

The main characteristic of the PAN molecule is its strongly polar *nitrile groups*. The monomer units have head-to-tail linkage with nitrile groups on alternate carbon atoms at very close distance. The CN groups have the following different possibilities to interact with their surroundings [64]: (1) The high dipole moment of CN groups causes strong attraction or repulsion (depending on the orientation) toward other molecules or groups also possessing a high dipole moment. (2) The lone-pair orbital of nitrogen can participate in hydrogen bonding (e.g., with water, in an electron donor–acceptor (EDA) complex formed with Lewis acids). (3) Electrons in the π-orbitals of the nitrile triple bond can interact with transition-metal ions. The PAN molecule is assumed to have a rigid, irregularly helical conformation with the nitrile groups assuming varying angles relative to the helical axis because of intramolecular repulsion between adjacent CN groups and intermolecular attraction between macromolecules [64].

There are many publications about the structure of PAN, which, to date, is still unresolved because of significant discrepancies. The morphology of PAN is difficult to resolve, and PAN belongs to the category of neither conventionally amorphous nor semicrystalline polymers. It is suggested that

PAN consists of the following three forms: (1) one phase combining some of the properties of both crystalline and amorphous phases; (2) two different amorphous phases with various types of secondary intermolecular forces and one crystalline phase; and (3) three phases, namely amorphous, paracrystalline, and crystalline [65–72]. However, PAN can be obtained in crystalline forms. The hexagonal or the orthorhombic structure for the syndiotactic PAN and the tetrahedral unit cell for the isotactic PAN have been described [73–81]. Two-dimensional order between adjacent macromolecules is attributed to PAN fibers [65,77]. PAN may contain both amorphous and ordered phases, but highly drawn PAN fibers contain only the hexagonal mesophase [82]. Recently, this hexagonal mesophase structure was confirmed by high resolution transmission electron microscopy (HRTEM) measurements of PAN fibers [83–85]. The ordered structures were not uniformly distributed in the whole fiber diameter, but they tend to be mainly concentrated in the outer fiber layers, with the inner layers having more amorphous areas and voids [86].

The chain conformation of PAN is determined by wide-angle x-ray scattering (WAXS), nuclear magnetic resonance (NMR), etc. [70,71,80,81,87–95]. There is not a single unique description in the literature about the tacticity of PAN. It has been described as completely atactic [96–100], as isotactic partially crystalline [101], as 80% syndiotactic [102], and as isotactic and syndiotactic in a ratio of 50:50 [91]. These discrepancies can be interpreted in terms of the influence of the preparation conditions on the crystallinity and the tacticity, so that different degrees of stereoregularity are achieved [103,104]. PANs prepared with radical initiators are approximately 75% syndiotactic, and those prepared by anionic polymerization, usually at low temperatures, are more isotactic [92,105,106]. Isotactic PAN can be prepared by using a zeolite as a host with radical initiators [107]. PAN fibers formed by solid-state coextrusion of a film of nascent PAN under controlled conditions reveal an x-ray long-period scattering, depending on the preparation conditions, including the initial morphology and the drawing conditions [108].

The Fourier transform infrared (FTIR) spectra of PAN homopolymer (from commercial and laboratory production) and a copolymer of acrylonitrile with methyl acrylate (from industrial production) are represented in Figure 4.1. The spectra band and the corresponding groups have been described in the literature [109–115]. The change of the bands of the polymer after heat treatment can be correlated to the formation reaction (see Section 4.6).

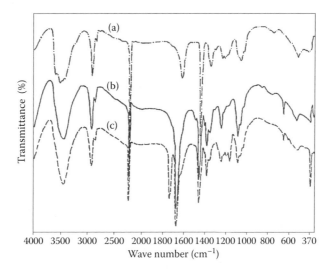

FIGURE 4.1 FTIR spectra of polyacrylonitrile. (a) PAN homopolymer, laboratory prepared (M_v = 120,000). (b) PAN homopolymer, commercial Dralon T (M_v = 90,000). (c) PAN copolymer with 5% methyl acrylate, commercial Dralon N (M_v = 90,000). (The characteristic bands are discussed in Section 4.6.)

The polar nature of PAN provides the characteristic properties of the polymer, namely, hardness and rigidity, and resistance to most chemicals and solvents.

4.3.2 Phase Characteristics

4.3.2.1 Glass and Other Transitions

Many transitions of PAN, described in the literature, fall between 39°C and 180°C depending on the method used (e.g., birefringence, dynamic mechanical measurements); sample form (film or fiber); thermal treatment; and relative humidity [64,116,117]. Most authors agree on two major transitions [64,118–123]: one at about 80–100°C, assuming that the thermal movement (onset of backbone mobility) overcomes the weak dispersion forces, and the other near 140°C, related to a loosening of the intermolecular dipole–dipole interactions. Values between 85°C and 106°C are identified with the glass temperature. By mechanical measurements, the secondary transition for the amorphous phase is determined at 79°C, and the dipole–dipole interaction for the quasicrystalline phase, at 99°C (considered as its glass transition temperature). The main transition for the amorphous phase at 157°C is usually attributed to concerted motions of the pendent CN groups and is very sensitive to modifications. For example, it disappears by stretching for acrylic fibers [124]. By heating unoriented PAN, two glass transitions are observed at ~100°C and ~150°C. Upon further heating at moderate rates (e.g., 10°C/min), a degradation exotherm is seen at ~340°C. But if a very fast heating rate (100°C/min) is used, a melting endotherm can be seen at ~320°C before the degradation exotherm. By heating highly oriented PAN, a single glass transition is seen at ~100°C, whereas the melting/degradation still occurs at ~340°C [125]. Copolymers have lower glass temperature.

Under common conditions, PAN is not stable up to its melting point. Therefore, its melting point can be determined only by extremely high heating rates, e.g., 40°C/min by differential thermal analysis (DTA) [64] or over 1000°C/min by immersion into a bath of molten metal [67–69]. Under such conditions, the melting point of PAN was determined to be in the range of 320°C ± 5°C [76,103,126]. Crystallization has been performed by using fractionated PAN, and several morphological states (ovals, spherulites, rectangular single crystals, twinned crystals) have been observed at various temperatures [127–129].

Dried gel films of isotactic PAN from solution were initially drawn by solid-state coextrusion (first-stage draw), followed by a further tensile draw at 100–200°C (second-stage draw). The temperature for an optimum second-stage draw of isotactic PAN is 130–140°C, which is significantly lower than that of atactic PAN (160–180°C) [130]. Microvoids are observed in the pristine fibers and are also formed in the temperature range of 190–470°C [131]. By heating PAN, a first lamellar mesophase shows up at about 200–300°C, and a second lamellar mesophase appears in an inert atmosphere at temperatures higher than about 350°C [132].

4.3.2.2 Heat Treatment

PAN heated at temperatures in the range of 150°C develops progressively intense colors. During the oxidative thermal treatment of PAN at high temperatures, several reactions take place. The most considerable reaction is the polymerization of the nitrile groups mainly at the temperature range of 200–300°C to form the so-called polyimine ladder polymer [133–137]. This cyclization reaction is exothermic, as measured by differential scanning calorimenrty (DSC), presented in Figure 4.2. After the oxidative thermal treatment of PAN, the polymer becomes insoluble. However, acrylic polymers that have been heat-treated at 200–300°C for 10 min or more under inert conditions in the dark are soluble in strong acid–water solutions [138]. For flameproofing of PAN fiber, oxygen must be present during the heat treatment, while heat treatment under inert conditions does not result in a flameproof material [114,138–147]. An overview of the flame retardation of acrylic fibers is given

Polyacrylonitrile

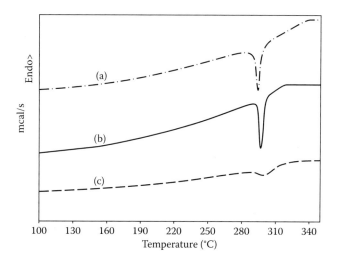

FIGURE 4.2 DSC of polyacrylonitrile. (a) PAN homopolymer, laboratory prepared (M_v = 120,000). (b) PAN homopolymer, commercial Dralon T (M_v = 90,000). (c) PAN copolymer with 5% methyl acrylate, commercial Dralon N (M_v = 90,000; scanning rate: 10°C/min, under nitrogen flow).

by Ref. [148]. High-performance carbon fibers can be fabricated by a controlled high-temperature degradation or processing of PAN fibers (see Section 4.6).

4.3.3 Physical Properties

4.3.3.1 Solubility and Molecular Weight Relations

4.3.3.1.1 Solubility

PAN is a relatively insoluble polymer, and its solvents are generally polar substances, such as N,N-dimethylformamide, N,N-dimethylacetamide, dimethylsulfone, dimethyl sulfoxide (DMSO), ethylene carbonate, propylene carbonate, and nitrophenols. Other solvents involve concentrated aqueous solutions of very soluble salts, such as LiBr, $ZnCl_2$, NaCNS, $NaClO_4$, and quaternary ammonium salts. Other types of PAN solvents are concentrated sulfuric acid and nitric acid [149]. Various PAN solutions, such as a highly concentrated PAN–propylene carbonate solution, have been studied [150–155]. The Hildebrand parameter of PAN, δ_2 (J/cm^3)$^{1/2}$, is 31.5 (experimental) or ~26.0 (estimated) [156].

4.3.3.1.2 Molecular Weight Relations

The constants for the [η]–M relationships expressed in terms of M_n or M_w are tabulated in Refs. [143–145,150,151,157,158] according to the method used (osmotic pressure, light scattering, etc.); the solvent; the temperature; and the molecular weight range of PAN. The following relationship has been determined for the viscosity-average molecular weight M_v of PAN in an N,N-dimethylformamide solvent at 25°C:

$$[\eta] = 2.43 \times 10^{-4} M_v^{0.75}$$

where [η] is the intrinsic viscosity (in dL/g).

This relation can be applied over the range of the degree of polymerization: 150–1320 [3,159].

4.3.3.2 Other Physical Properties

Some physical properties of PAN are as follows [149,160–162]:

- Density: 1.17 g/cm^3
- Tensile strength: σ = 4000 kp/cm^2
- Elongation at break: ε = 8–20%
- Maximum temperature for permanent use: 200°C
- Maximum moisture uptake: 2.5–3%

Some thermodynamic and thermal data for PAN are as follows [156]:

- Heat of polymerization: –72.4 ± 2.2 kJ/mol
- Heat capacity, c_p: 13.77 J/mol K at T = 50 K, 86.18 J/mol K at T = 370 K
- Specific heat of combustion, Δh_c: 30.6 kJ/g (experimental), 31.5 kJ/g (calculated)
- Thermal decomposition temperature: 250–310°C
- Exothermal decomposition range: 238–299°C
- Thermal conductivity (disks from powder), κ: 0.440 mW/cm K (at 20 K), 1.600 mW/cm K (at 100 K)
- Volumetric coefficient of thermal expansion $(1/V)(dV/dT)_p$: 2.8 × 10^{-4} – 3.8 × 10^{-4} K^{-1} (above T_g), 1.4 × 10^{-4} – 1.6 × 10^{-4} K^{-1} (below T_g)

PAN can achieve very high, persistent electrical polarization. This arises from the strong dipole moment of the nitrile groups and the quasicrystalline structure of PAN [163].

PAN has a remarkable barrier property to oxygen and carbon dioxide, but it has high permeability to helium. In addition, PAN has high permeability and high sorption to water vapor due to the high polarity of PAN. This is the only disadvantage for barrier applications of PAN [164–166].

4.3.4 Chemical Properties

PAN has a high resistance to the common chemicals, but the nitrile groups and α-hydrogens react with some reagents. Hydrolysis with hot aqueous alkali solutions leads to the formation of poly(acrylic acid), and hydration of PAN with concentrated sulfuric acid leads to a solution [167]. Acidic hydrolysis of PAN with 65% HNO$_3$ at 21°C leads to a multiblock copolymer containing acrylonitrile and acrylamide units [168]. Hydroxylamine reacts with PAN to produce amidoximes and hydroxamic acids [167]. PAN can be grafted with vinyl acetate in an emulsion [169], and poly(5-vinyltetrazole) can be prepared by a polymer-analogous conversion of PAN [170,171]. Irradiation of PAN induces free radicals, which can lead to other reactions, such as cross-linking or grafting in the presence of a monomer [172,173]. PAN reacts with poly(vinyl alcohol) in DMSO without any catalyst to form conjugated carbon–nitrogen sequences [170]. PAN can be modified with several primary amines, such as ethanolamine, ethylenediamine, and ethylamine, and with aqueous inorganic acid or alkali [174–182]. PAN can form complexes with iodine [183,184]. The surface of PAN (containing 10% acetate groups) as fibers and finely ground powder has been modified by enzymatic treatment using a nitrile hydratase as enzyme, which selectively converted the pendant nitrile groups into the corresponding amides [185].

PAN (fibers) has poor dyeability because of the high degree of order and the orientation of molecular nitrile groups. Consequently, there is a lack of segmental mobility, which makes the penetration of the dye molecule difficult. The use of small amounts of some comonomers in acrylonitrile leads to the improvement of the dyeability of the final acrylic fibers [186–188]. The ability of the structurally modified or unmodified PAN to adsorb dyes from aqueous solutions is interpreted in terms of

EDA interactions between the main groups of the polymer and the dye [175–178]. A sorbent for the extraction of uranium was prepared by converting the nitrile groups of PAN to amidoxime groups [189].

4.4 THERMOPLASTIC BLENDS, ALLOYS, COPOLYMERS, AND COMPOSITES, AND THEIR COMMERCIAL RELEVANCE

Acrylonitrile homopolymer (PAN) is not processable as a thermoplastic mass, because it degrades before melting. Even small amounts of comonomers are incorporated with acrylonitrile in order to improve some properties of acrylic fibers (e.g., dye uptake). Heat treatment of acrylic fibers under appropriate conditions can improve their dimensional stability and their resistance to permanent deformation. Heating causes structural changes of the fibers, affecting their physical and dyeing properties [190–192].

PAN or its copolymers are used as fibers, sorbents, and stationary phases of gas chromatography (e.g., cross-linked porous copolymers of acrylonitrile and divinylbenzene). Chemical modification of these products with amines leads to a group of specific sorbents, amino exchange resins, and carriers for biologically active substances [193].

The usual copolymers of acrylonitrile with other monomers are melt processable. The acrylonitrile content varies up to 50% and 75% for general applications and barrier applications, respectively. The main use of PAN is in acrylic fibers (see details in Section 4.5). Nitrile polymers (copolymers of acrylonitrile with some monomers) belong to the high-barrier polymers, which are used in packaging, especially for food and beverages, to replace glass and metal containers [194]. Two important classes of acrylonitrile copolymers are acrylonitrile–butadiene–styrene (ABS) and SAN. Some industrially important comonomers of acrylonitrile include styrene, vinyl chloride, acrylic acid, vinyl acetate, isobutylene, and acrylate. Some other copolymers of acrylonitrile are reported below.

4.4.1 Copolymers of Acrylonitrile–Butadiene (Nitrile Rubber, NBR)

The synthetic rubber based on butadiene and acrylonitrile was first described in 1930. After vulcanization, the rubber has an excellent resistance to oil and petrol. Pilot plant production of this copolymer under the name Buna N was started in Germany in 1934. The acrylonitrile content has the most profound effect on the properties of the vulcanized NBR, influencing its general resistance to aliphatic hydrocarbons. In addition, high acrylonitrile content improves the tensile strength, abrasion resistance, hardness, processing, and compatibility with plastics. However, low-temperature flexibility, resilience, and elasticity are decreased [195]. The physical properties of NBRs are good after compounding with carbon black, and they may be processed to conductive rubbers [196]. The largest applications of NBRs are in the engineering industries for oil seals, O-rings, gaskets, fuel, and oil hose [195,196]. NBRs may be reinforced by incorporation of phenolic resins, whereas blends of NBR and polyvinyl chloride (PVC) have better resistance to ozone and weathering in comparison to NBR itself [195].

4.4.2 Copolymers of Acrylonitrile in High-Barrier Applications

Commercial high-barrier resins are copolymers of acrylonitrile of different types, such as Lopac (70% acrylonitrile and 30% styrene), Barex (90% copolymer made up of 74% acrylonitrile and 26% methyl acrylate plus 10% butadiene rubber graft), and Cycopac (90% copolymer made up of 74% acrylonitrile and 26% styrene, plus 10% butadiene rubber graft) [194]. Generally, acrylonitrile provides high tensile strength, stiffness, chemical resistance, and a gas barrier; methyl acrylate improves the processability and the stability; and the elastomer provides the toughness. Another type of acrylonitrile copolymer is those with vinylidene chloride. Poly(vinylidene chloride) is the best high-barrier polymer with the lowest permeation rates with respect to oxygen, carbon dioxide, and water. All these materials find their main application in packaging as gas barriers [194].

4.4.3 Copolymers of Acrylonitrile in Optically Active Applications

Optically active compounds are capable of rotating linear polarized light by a specific angle from its plane of vibration. Such an optically active polymer is the copolymer of acrylonitrile with benzofuran, which is prepared in the presence of optically active aluminum compounds as complexing agents of acrylonitrile. This copolymer is an alternating copolymer whereby the alternating dyad contributes to the optical activity [197,198]. Highly soluble (60)fullerene-based PAN derivatives have been synthesized that have nonlinear optical (NLO) properties. NLO materials draw interest for potential applications in many fields of photonics, optical communication, and laser technology [199].

4.4.4 Copolymer of Acrylonitrile in Medical Applications

The copolymer of acrylonitrile with 2-dimethylaminoethyl methacrylate has blood compatibility, and it can have medical applications as adsorbent coatings for hypoperfusion or dialysis membranes [4,200–204].

4.4.5 Other Copolymers of Acrylonitrile and Blends

Other copolymers of acrylonitrile are used in reverse-osmosis membranes, as composite membranes, due to the high stability of the acrylonitrile unit in solvents and in activated membranes for biocatalysis [205]. Partially cyanoethylated poly(vinyl alcohol) in its copolymer with formaldehyde and hydroquinone improves the redox capacities because of the polar nitrile groups present (electron-transfer polymers) [206]. Graft copolymers of acrylonitrile impart hydrophilic features to starch, resulting in higher water absorption of the starch [207,208]; grafting onto cotton improves the moisture retention of cotton [209], and grafting onto non-wood-based cellulosic leads to the formation of superabsorbents [210].

PAN, as a constituent in polyblend fibers, was first examined in detail as early as 1956. These fibers were a heterogeneous bicomponent mixture of PAN with cellulose, cellulose acetate, and silk. Today, there are some commercially available polyblend fibers of PAN such as Chinon K-6 consisting of 70% graft of PAN on 30% casein. Other polyblend (or alloy) fibers are formed by intimately mixing two or more polymers before delivery to the spinneret [211]. Blends of PAN with cellulose [212] and solutions of PAN blends [212,213] have been studied. In addition, interpenetrating polymer networks (IPNs) based on PAN and other polymers, such as poly(zinc acrylate) [214] or natural rubber [215], have been synthesized.

4.5 POLYACRYLONITRILE PROCESSING, PERFORMANCE PROPERTIES, AND APPLICATIONS

PAN homopolymer cannot be melt processed. It can only be processed by special techniques, but its copolymers may be processable as rubbers, fibers, and plastics.

Acrylonitrile–butadiene copolymer (or generally diene monomers) is processed in equipment normally used for *rubber processing*, which includes the addition of other compounds (e.g., carbon black) and vulcanization [195,196].

Clear, inherent, bulky articles of PAN cannot be prepared by conventional techniques. PAN cannot be melt-cast into sheets [216]; however, after the modification of the inherent polymer particle structure, PAN can be compression-molded with conventional equipment. The change in structure is accomplished by mechanical and thermal treatment using a differential-speed rubber mill heated to 120–190°C. The effect of the shear accompanied by a molecular weight decrease of PAN results in a highly ordered structure with planar orientations. In this form, PAN can be compression-molded into tough, clear articles with retention of the laminar structure [217]. Compression-molded clear articles of PAN can also be prepared by using molding powder consisting of preformed PAN, acrylonitrile, and initiator [216]. Generally, to produce polymers with good molding characteristics, copolymers of acrylonitrile are used, and the processing of these copolymers is similar to the conventional processing of (thermo)plastics. Melt-spun high-acrylonitrile fibers based on its copolymers (including also termonomers, e.g., acrylonitrile, methyl acrylate, and itaconic acid) [218] are currently under development by some companies [219].

The solution fiber spinning of PAN (homopolymer or copolymer) is carried out by two fundamentally different methods, namely, wet spinning and dry spinning. There is also a hybrid method, the dry jet-wet spinning, which has the characteristics of both methods. In all of the methods, the PAN or its copolymer is dissolved in a solvent such as DMF or dimethylacetamide to form a highly viscous solution, which is called dope. The spinning dope is then deaerated, filtrated, and heated to around 100°C. This solution is extruded through a spinneret having tens of thousands of holes with diameters ca. 0.05–9.38 mm. The choice of a particular diameter depends on the desired fiber diameter (i.e., the tex or denier of the fiber) for a defined application [220–225].

In conventional wet spinning, the spinneret or jet is immersed in a bath containing a mixture of solvent and nonsolvent where the polymer is coagulated and the initial shape of the fiber is formed [162,224–226]. The composition of the coagulation bath has a profound effect on the morphology (and the properties) of the as-spun fibers. When the bath is DMSO poor, the fibers have a kidney-like cross section, may develop skin–core structure, and are less crystalline; PAN fibers with better characteristics are produced using a coagulation bath with high DMSO concentration (70 wt.%) [227]. From the study of the ternary system PAN–DMF–H_2O, it was found that the presence of a small amount of water in the spinning dope leads to coagulation of the PAN fibers through spinodal decomposition. The fibers thus produced contain fewer voids and have enhanced mechanical properties [228]. For the hybrid method, the spinneret is suspended approximately 0.6–5 cm above the spin bath. This has the advantage of increased spinning speed and improved fiber properties [2,229]. In all cases, the coagulated filaments are withdrawn from the spin bath by using a set of rolls that permit the stretching of the filaments, as well as the washing and further stretching of the resultant fibers. The initial gelation and coagulation of the extruded filament are very important for the properties of the end fiber. The choice of the spin bath conditions, the solvent and its concentration, as well as the temperature is very important [220,230]. The gel density and fibril size could be increased by the choice of the spin bath temperatures, and this also leads to an improvement of the mechanical properties of the fiber (higher strength and modulus of elasticity, etc.) [231–235]. The changes in the structure of spun acrylic fibers during processing have also been described [236–241]. The modeling and simulation of the coagulation process of PAN wet spinning have been described in Ref. [242].

In a conventional dry spinning, the spinning dope is extruded in a 300- to 900-hole spinneret assembly at high temperatures along with a stream of heated gas into a vertical chamber [2,220,221]. The filament formation takes place through the evaporation of the solvent, and the filament is stretched (14 times its initial length for dry spinning, 22 times for wet spinning) [243,244]. The structure of the filament strongly depends on the art of the solvent removal from the extruded filament. In dry or wet spinning, an external skin is formed on the initial filament that behaves like a pipe filled with a viscous liquid. If the rate of diffusion of the solvent from the filament to the bath is equal to the rate of diffusion in opposite direction, then the cross section of the filament is round. This occurs in wet spinning, where the coagulation liquid (e.g., 70% DMF and 30% water) can also diffuse into the filament. However, there is no opposite diffusion in dry spinning, and therefore, the cross section is not round but has a dog bone shape [2,220,221]. It has also been recorded that acrylonitrile that is polymerized in a solvent such as DMF can be directly spun into fiber [245,246]. Besides the wet and dry spinning of PAN that are industrially applied for the production of acrylic fibers, gel spinning and electrospinning are more recently researched. High orientation is achieved by the gel spinning of PAN, which is limited by the chain entanglements; the latter affect the elasticity and play an essential role in fiber formation and development of fiber strength [247]. The electrostatic fiber production or electrospinning offers unique capabilities for producing novel synthetic fibers of unusually small diameter and good mechanical performance (nanofibers), and fabrics with controllable pore structure and high surface area [248]. In the last decade, the production of nanofibers based on PAN has gained considerable interest; nanofibers have the advantage of very high surface-to-volume ratio, and they can be applied in many fields such as catalysis or filtration [249]. In continuation of the use of PAN as carbon fiber precursor, there is considerable interest in manufacturing carbon nanofibers (CNFs) from PAN nanofibers. It has been reported that centimeter-long PAN-based CNFs have been produced with Young's modulus up to 200 GPa and tensile strength up to 3.5 GPa [250–252]. The increase in crystallinity of PAN nanofibers led to an increase in the heat released during the nitrile cyclization reactions [253].

PAN can also be processed in other ways for some other applications. For example, chemically activated PAN membranes can be prepared from a 15% solution of PAN that is precipitated in water. The water is then replaced with ethanol, which is also replaced with ether before drying [169]. Such membranes can also be made by other techniques [254–261]. PAN adsorbents, based on acrylic fiber wastes, can be prepared by the use of an appropriate chemical modification or low-temperature pyrolysis [262,263]. Acrylic fibers also find applications as asbestos substitutes (e.g., in brakes); in industrial filters, battery separators, and synthetic paper; and as precursor for carbon fibers (see Section 4.6). Other nontextile uses of PAN include tubular fibers for reverse osmosis, gas separation, ion-exchange dialysis, and ultrafiltration. Hollow fibers made from copolymers of acrylonitrile with vinyl acetate absorb liquid monomers, which can then be polymerized *in situ* [2]. Solutions of PAN in DMF sprayed into supercritical fluid carbon dioxide form hollow fibers and highly oriented microfibrils (<1 μm diameter) [264]. Hollow fibers prepared from a copolymer of acrylonitrile can be used in membranes [265,266]. Acrylic fibers are also used as reinforcing fibers in molding resins of acrylonitrile–styrene copolymer. Electrically conducting fibers based on acrylonitrile copolymer treated in a solution of copper and other substances have also been prepared [2]. A PAN-based gel-like electrolyte is one of the candidates for the electrolyte of rechargeable lithium batteries [267]. The modification of acrylic and other fibers for accomplishing antistatic properties is discussed in the review in Ref. [268].

4.6 NEW TECHNOLOGIES BASED ON POLYACRYLONITRILE

When the important development of PAN as a textile fiber started around the 1960s, the problems of discoloration in use were studied and solved through a better understanding of the degradation mechanism of the discoloration process [269]. Some years later, the study of the thermal degradation of PAN became very interesting and necessary to enable the production of carbon fibers with optimum properties.

Carbon fibers had already been prepared by Thomas Edison in 1879 by the pyrolysis of cellulosic fibers yielding very low tensile strength. The preparation of carbon fibers with improved properties began in 1959 and 1961 by using as a precursor rayon and PAN, respectively [270]. The need for high-performance composite materials in the mid-1960s for aeronautic and aerospace applications led to the intensive research on and development of strong carbon fibers [271]. The specific Young's modulus of high-modulus carbon fibers exceeds that of glass fibers by up to an order of magnitude. Consequently, advanced composites or carbon fiber-reinforced polymers achieved the same or even better mechanical properties compared to metallic materials [272]. Today, PAN is the main precursor material used for the manufacture of commercial carbon fibers [272,273]. Thus, research for the improvement of carbon fibers includes also research on the precursor materials, i.e., the homopolymer (PAN) and the acrylic copolymers as well as their spinning process. Furthermore, the last years many papers about carbon fibers have been published and great part of them concerns the thermal degradation and the behavior of PAN. Such papers contribute also to the investigation of the structure of PAN (and its copolymers), mainly in the form of fibers; its phase characteristics; and its physical and chemical properties.

The conversion of acrylic precursor fibers to carbon fibers involves three processing steps: (1) stabilization, (2) carbonization, and (3) graphitization. The last step takes place only for the manufacture of high-modulus carbon fibers [271–280].

The PAN precursor has been modified to improve its properties by incorporation of a suitable acidic comonomer during polymerization, which increases its hydrophilicity and catalyzes the cyclization of nitrile groups during the stabilization step [281]. The effects of various solvents employed as media for polymerization have been interpreted from the results of the thermal and structural (x-ray diffraction) methods [282]. Wet- or dry-spun acrylic fibers can be modified during postspinning accompanied commonly by stretching: (1) The presence of certain plasticizers (e.g., DMF, CuCl) for reducing the diameter of PAN has not only the advantage of uniformity during the thermal stabilization but also that of fewer defects per unit volume for the corresponding carbon fibers, and hence, their better mechanical properties [283–285]. (2) Through coatings with certain resins, the exotherm of the cyclization reaction is suppressed. (3) Certain reagents like Lewis acids, organic bases, inorganic compounds, etc. are used, which act as catalysts for the cyclization reaction [285].

The first step in the production of carbon fibers is the *stabilization* of the fibers. Commonly, this step is a heat treatment of PAN (homopolymer or copolymer) fibers in the presence of air (or oxygen) up to around 300°C. This treatment can also take place in two substeps: one up to around 240°C (i.e., at temperatures lower than the onset of the exothermic peak due to the cyclization reaction of the nitrile groups [see Figure 4.2]) followed by another at up to around 300°C [271,274–279,286–291]. Treatment in other gas media (e.g., ozone, NO_2, SO_2, HCl) without or with oxygen [292–295] and treatment in reactive liquid media (e.g., nitrobenzene, solution of Lewis acids) without or followed by oxidation in air [296,297] have been proposed. However, air (or oxygen) heat treatment is used mostly for the stabilization. It has been demonstrated by *in situ* NMR investigation into the thermal degradation of PAN why air is the preferred medium for stabilization of PAN [298]. It is very important to avoid the shrinkage of the fibers that occurs during this treatment, or, more preferably, it is necessary to further stretch the fibers in order to increase the orientation of the macromolecules [106,271,272,299–303]. During the first processing step, many reactions take place. Generally, the most important reactions are as follows: (1) the cyclization reaction of the nitrile that is accompanied by heat evolution (exothermic reaction) and (2) the dehydrogenation reactions leading to the formation of C=C in the macromolecule backbone. In addition to these reactions, the role of oxygen is very important (see below) [269,271,290–306].

The progress of these reactions can be followed mainly with the aid of FTIR and DSC (see also Figures 4.1 and 4.2). The progress of the dehydrogenation reaction can be followed by monitoring the decrease of the CH_2 band (2940 cm^{-1}). The increase of the cyclization reaction results in the decrease of the CN band (nitrile group, 2240 cm^{-1}). The increase of the oxidation reactions results in the increase of various bands, e.g., OH and/or NH band (3330–3230 cm^{-1}) or C=O band (1740 cm^{-1}), which shows a small increase during the oxidation. The appearance of a band at 1600 cm^{-1} is characteristic of the conjugated double bonds (C=C), due to dehydrogenation, or C=N– group, due to cyclization, with an aromatic ring. For a copolymer with methyl acrylate (MA), the C=O band (1740 cm^{-1}) and the C–O– band of the ester group (1170 cm^{-1}) are attributed to MA [111–114,279,307,308]. Based on FTIR spectra [309,310] or wide-angle x-ray diffraction (WAXD) [311], the aromatization index, conversion index, or stabilization index has been determined in order to follow the stabilization process.

Oxygen is incorporated in the macromolecule through the substitution of hydrogen forming allylic hydroperoxide groups, which then decompose to give carbonyl. Hydrogen-bonded interaction between carbonyl groups and secondary amine groups in neighboring macromolecules contributes to the physical stabilization of the polymer structure. Furthermore, oxygen initiates the cyclization reaction at lower temperatures, thus reducing the intensity of the exotherm. All of these reactions result in the formation of short ladder segments in the macromolecule. Scission reactions take place at temperatures higher than 250°C, reflecting the breakdown of uncondensed macromolecules [269,271]. By proper plotting of the treatment temperature versus treatment time of PAN fibers, three regions can be distinguished: (1) no chemical reactions (only physical phenomena), (2) oxidation and/or dehydrogenation reactions, and (3) domination of cyclization reactions [312]. Thus, it is useful to choose suitable heating programs for the oxidative treatment of PAN fibers [313]. This novel methodological approach has been extensively exploited by several research groups [314–316].

Other aspects of the stabilization process comprise improvement of the production equipment, e.g., a new oxidation oven [317]; new techniques (for instance, a fiber-on-fiber abrasion test in oxidized PAN [318]); electron beam-induced cyclization, which does not need any external or internal catalysts and leads to reduced cyclization times [319]; and near-edge x-ray absorption fine structure (NEXAFS) spectra of heat-treated PAN [320]. The dielectric behavior of PAN in its pristine stage displays a strong secondary relaxation and a weak relaxation having lower activation energy in the cyclized stage [321]. Stabilized fibers show also a reducing ability like activated carbon fibers (ACFs), i.e., they can be oxidized by an oxidant such as Au^{+3}, Fe^{+3}, etc. [322].

The *carbonization* process takes place between 400°C and 1000°C in an inert gas (e.g., N_2) or in a vacuum, and it is accompanied by many reactions. Variable amounts of H_2O, CO, and CO_2 are released during the carbonization depending on the extent of the oxidation. NH_3, HCN, and N_2 are basic scission products formed during this treatment. Small amounts of H_2 and CH_4 (the amount of the latter is higher for the copolymers of acrylonitrile with methyl acrylate) are also released [111,271,323]. The chemical state of nitrogen during the pyrolysis of PAN fibers below 1000°C differs between the surface, having more tertiary nitrogen –N<, and the core of the fibers [324]; moreover, various nitrogen functionalities are formed [325]. Generally, this step is characterized by the removal of the hetero atoms from the ladder macromolecules and the development of a graphite-like structure [269,271,279,326–328]. The effects of the properties of initial PAN fibers [329], their microstructure [330], the common processing [331], or new processing with high magnetic fields in the carbonization step [332] on the modulus and the strength of the carbon fibers produced have been investigated in order to improve their mechanical properties.

The *graphitization* process involves the treatment of the carbon fibers formed, and it takes place at temperatures above 2000°C up to 3000°C (commonly 2500°C) in the presence of argon or in vacuum. The graphite-like structure of the fibers is improved, resulting in higher modulus of the fibers. Carbon fibers that have been graphitized are also called graphite fibers; although their structure is not the graphite structure, it is similar [326].

The carbon fibers are divided according to their values of modulus of elasticity (M) and tensile strength (T) and are designated as follows: (1) high modulus–high tensile (HMT), M = 400–450 GPa, T = 1700–2500 MPa; (2) high modulus (HM), M = 300–700 GPa, T = 2000–2500 MPa; (3) high tensile (HT), M = 200–250 GPa, T = 2500–3200 MPa; and (4) low modulus (LM), M = 50–150 GPa, T = 500 MPa [273]. HMT, HM, and HT carbon fibers are used in the production of composite materials for the manufacture of a large number of articles, ranging from aerospace vehicle parts to sporting goods and prosthetics. On the other hand, the industrial production of low-modulus carbon fibers has increased during recent years. These fibers are much cheaper, and they have their own areas of practical application, such as heat-insulating screens, filtering, sorption layers, and coatings [272,273].

Other perspectives of PAN application involve combination with other materials, such as a vermiculite–PAN intercalation compound [333], PAN-coated silica used as support for copper catalyst in the dehydrogenation reaction of methanol [334], concrete reinforced by PAN fibers [335], and PAN fibers used as reinforcing materials or for producing filters or friction coatings [336]. Carbon films have been prepared from composite materials of PAN and vapor-grown carbon fiber (VGCF) [337]. ACFs produced from acrylic textile fibers have significant advantages over the more traditional powder or granular forms; they have high surface area, high adsorption capacity, and very high rates of adsorption from the gas or liquid phase [338–341]. Hydrogen can be adsorbed on carbon materials; the adsorption amount for activated carbon reaches values of ~0.5 wt.% hydrogen at ambient temperature up to 5 wt.% under cryogenic conditions, whereas nanostructured carbon materials allow 8 wt.% hydrogen adsorption at ambient temperature [342]. Nanofibers could revolutionize hydrogen storage for fuel cell applications [343]. Nanocomposites have been synthesized from PAN and CdS by γ irradiation [344]. Nanotechnology is regarded as a key technology for the twenty-first century, and carbon nanotubes will find a steadily growing application field (e.g., space travel, field-emission displays, transistors with extremely fast switching rates). Nanotubes can be manipulated only in composites, and it is necessary to find suitable matrix systems [345]. The preparation of unique carbons by using various templates (well-characterized inorganic solids) has been described [346,347]. One-, two-, and three-dimensional carbon can be prepared. A monomer, like acrylonitrile, can be intercalated into, e.g., montmorillonite, and then polymerized; the composite produced is subjected to carbonization at 700°C, the resultant clay–carbon composite is washed with HF and HCl solutions to dissolve the clay framework, and the insoluble fraction is a two-dimensional carbon [346]. The ladder polymer obtained by pyrolysis of PAN has third-order NLO properties with potential applications as communication devices, optical computing devices, etc. [348,349]. PAN used in composites with superconducting ceramic powder (e.g., YBCO powder), after carbonization, provides both the elastic and the transport current characteristics required [350]. Polymer electrolytes based on PAN [351] as well as carbon anodes for a lithium secondary battery based on PAN have been used [352].

4.7 CONCLUSION

PAN is an important polymer that is characterized by hardness, rigidity, and resistance to most chemicals and solvents. Thermally treated PAN under common conditions decomposes before melting, and as a result, melt processing as a homopolymer is not possible. Thus PAN is processed as a copolymer or as a solution in the manufacture of acrylic fibers. The latter is the most important application of PAN. Important copolymers are ABS, SAN, and NBR. Recently, PAN has become the most important precursor for the manufacture of carbon fibers, which are fundamental materials for advanced composites. There are also many possibilities for the development of new materials based on PAN because of its relatively low production cost, the reactivity of its nitrile group, and its ability to be combined with other monomers and polymers.

REFERENCES

1. ABC-Chemie. 1979. *Polyacrylonitrile/Polyacrylonitrile Fiber*, Vol. 2, pp. 1099–1101. Verlag Harri Deutsch, Thun, Frankfurt/Main.
2. Bach, H.C. and Knorr, R.S. 1985. Acrylic fibers. In *Encyclopedia of Polymer Science and Engineering*, eds. H.F. Mark, N.M. Bikales, C.G. Overberger, G. Menges and J.I. Kroschwitz, Vol. 1, pp. 334–388. John Wiley & Sons, New York.
3. Bamford, C.H. and Eastmond, G.C. 1964. Acrylonitrile polymers. In *Encyclopedia of Polymer Science and Technology*, eds. H.F. Mark, N.G. Gaylord and N.M. Bikales, Vol. 1, pp. 374–425. Interscience Publishers/John Wiley & Sons, New York.
4. Peng, F.M. 1985. Acrylonitrile polymers. In *Encyclopedia of Polymer Science and Engineering*, eds. H.F. Mark, N.M. Bikales, C.G. Overberger, G. Menges and J.I. Kroschwitz, Vol. 1, pp. 426–470. John Wiley & Sons, New York.
5. Gwinne, J.W. et al., 1960. *Rules and Regulations under the Textiles Fiber Products Identification Act*, p. 4. U.S. Federal Trade Commission, Washington, DC.
6. Technical Committee 38, 2013. ISO-2076. Textiles-*Man-made Fibres Generic Names*. International Organization for Standardization, Switzerland.
7. UN Statistics Division. 1993. *Industrial Statistics Yearbook 1991*, Vol. II, pp. 3511–3543. United Nations, New York.
8. Anon., 1994. Growth continues in chemical production/synthetic fibers. *Chem. Eng. News* 34 (July 4).
9. Anon., 1990. Growth slackens appreciably as 1980s end/synthetic fibers. *Chem. Eng. News* 40 (June 18).

10. Greek, B.F. 1984. Slumping synthetic fibers may get help from import limits. *Chem. Eng. News* 11 (September 3).
11. Anon., 2002. Production: Down but not out. *Chem. Eng. News* 60 (June 24).
12. Garmston, S. 2009. Acrylonitrile and derivatives, world supply/demand report 2009. PCI Acrylonitrile Ltd., Guilford.
13. Weissermel, K. and Arpe, H.-J. 1993. *Industrial Organic Chemistry*, pp. 299–307. VCH Verlagsgesellschaft GmbH, Weinheim, New York.
14. Greek, B.F. 1986. Restraints on imported apparel lift synthetic fiber output. *Chem. Eng. News* 9 (November 17).
15. Koslowski, H.J. 1986. Der Synthesefasermarkt der Welt. *Kunststoffe* 76:1158.
16. Fiberfacts. Available at http://www.fibersource.com/f-info/fiber%20production.htm.
17. Anon., 1992. Acrylonitrile. *Chem. Ind. Newsletter* 4 (March–April).
18. Anon., 1987. Synthetic fibres. *Technol Newsletter* 28:4 (January 5).
19. Adams, D.F. 1985. Applications in Aerospace, especially in the USA. In *Carbon Fibres and Their Composites*, ed. E. Fitzer, pp. 205–228. Springer-Verlag, Berlin.
20. Logemann, H. 1961. Polymerisation ungesaettigter Nitrile, insbesondere des Acrylonitrils. In *Methoden der Organischen Chemie*, Band XIV/1, ed. Houben-Weyl, pp. 973–1009. Georg-Thieme Verlag, Stuttgart.
21. Horsley, L.H. 1947. Table of azeotropes and nonazeotropes. *Anal. Chem.* 19:508.
22. Anon., 1959. *The Chemistry of Acrylonitrile*, 2nd ed., pp. 2–10. American Cyanamid Co., New York.
23. Fukuyama, T. and Kuchitsu, K. 1970. Structure of acrylonitrile as determined by electron diffraction and spectroscopy. *J. Mol. Struct.* 5(1–2):131.
24. Stueben, K. 1970. Acrylonitrile and related compounds. In *High Polymers*, eds. E.C. Leonard, Vol. XXIV, Part 1, pp. 31–51. Wiley-Interscience, New York.
25. Bikales, N.M. 1966. Cyanoethylation, In *Encyclopedia of Polymer Science and Technology*, eds. H.F. Mark, N.G. Gaylord and N.M. Bikales, Vol. 4, pp. 533–562. Interscience Publishers/John Wiley & Sons, New York.
26. Roth, L. 1988. *Krebs erzeugende Stoffe*, pp. 88, 89, 105, 106. Wissenschaftliche Verlagsgesellschaft GmbH, Stuttgart.
27. Wood, S.M., Buffler, P.A., Burau, K. and Krivanek, N. 1998. Mortality and morbidity of workers exposed to acrylonitrile in fiber production. *Scand. J. Work Environ. Health* 24:54.
28. Vasishtha, R. and Srivastava, A.K. 1991. Polymerization of methacrylic acid and acrylonitrile by p-nitrobenzyl triphenyl phosphonium ylide. *Polym. Eng. Sci.* 31(8):567.
29. Chiang, R. and Friedlander, H.N. 1966. Influence of catalyst depletion or deactivation on polymerization kinetics: III. Solution polymerization of acrylonitrile in N,N-dimethylformamide at $-30°C$. *J. Polym. Sci. (A-1)* 4:2857.
30. Wu, J.-Y. and Yang, C.-X. 1990. Polymerization of acrylonitrile initiated by vanadium(V)–thiourea reaction. *J. Polym. Sci.* 28:1289.
31. Ebdon, J.R., Huckerby, T.N. and Hunter, T.C. 1994. Free-radical aqueous slurry polymerizations of acrylonitrile: 1. End-groups and other minor structures in polyacrylonitriles initiated by ammonium persulfate/sodium metabisulfate. *Polymer* 35(2):250.
32. Teng, F.S. and Mahalingem, R. 1986. Rate parameters in polyacrylonitrile film formation trough electropolymerization. *Polym. Commun.* 27:342.
33. Balasubramanian, T.R. and Mahadevan, V. 1988. Electro-initiated polymerization of acrylonitrile in aqueous acetic acid containing $Mn(OAc)_2.4H_2O$ and $CNCH_2COOH$. *J. Polym. Sci. (A)* 26:301.
34. Noel, S., Newton, P., Bodin, C. et al. 1994. Tribological behavior of heat-treated thin films of electropolymerized polyacrylonitrile. *Surf. Interface Anal.* 22(1–12):393.
35. Barton, J., Capek, I. and Hrdlovic, P. 1975. Photoinitiation. II. Kinetics of the acrylonitrile polymerization photoinitiated by aromatic hydrocarbons. *J. Polym. Sci., Polym. Chem. Ed.* 13:2671.
36. Kapur, G.S. and Brar, A.S. 1993. Microstructure determination of poly-acrylonitrile prepared by photopolymerization using uranyl ion by 13C-NMR spectroscopy. *J. Polym. Mater.* 10(1):37.
37. Batty, N.S. and Guthrie, J.T. 1980. Aspects of the kinetics of the radiation-induced polymerization of acrylonitrile in dimethylformamide. *J. Appl. Polym. Sci.* 25(11):2539.
38. Bag, D.S. and Maiti, S. 1998. Polymerization under magnetic field—II. Radical polymerization of acrylonitrile, styrene and methyl methacrylate. *Polymer* 39(3):525.
39. Chiang, R., Rhodes, J.H. and Evans, R.A. 1966. Solution polymerization of acrylonitrile catalyzed by sodium triethylthioisopropoxyaluminate: A polyacrylonitrile with high structural regularity. *J. Polym. Sci., Part A-1* 4:3089.

40. Feit, B.-A., Mirelman, D. and Zilkha, A. 1965. Anionic heterogeneous polymerization of acrylonitrile by butyllithium. I. *J. Appl. Polym. Sci.* 9:2459.
41. Feit, B.-A., Mirelman, D., Katz, Y. and Zilkha, A. 1965. Anionic heterogeneous polymerization of acrylonitrile by butyllithium. II. Effect of Lewis bases on the molecular weight. *J. Appl. Polym. Sci.* 9:2475.
42. Pai Verneker, V.R. and Shaha, B. 1989. Metallic lithium-initiated bulk polymerization of acrylonitrile. *Polym. Commun.* 30:92.
43. Furukawa, J., Tsuruta, T. and Morimoto, K. 1957. Vinyl polymerization by mixed ionic catalysts containing alkylsodium as component. *Kogyo Kagaku Zasshi* 60:1402.
44. Zilkha, A., Feit, B.A. and Frankel, M. 1959. Alkoxides as initiators of anionic polymerization of vinyl monomers. I. Polymerization of acrylonitrile by use sodium alkoxides. *J. Chem. Soc.* 1959:928.
45. Frankel, M., Ottolenghi, A., Albeck, M. and Zilkha, A. 1959. Anionic polymerization of vinyl monomers with butyl–lithium. *J. Chem. Soc.* 1959:3858.
46. Matyjaszewski, K., Mu Jo, S., Paik, H.J. and Gaynor, S.G. 1997. Synthesis of well-defined polyacrylonitrile by atom transfer radical polymerization. *Macromolecules* 30:6398.
47. Billmeyer, Jr., F.W. 1971. *Textbook of Polymer Science*, pp. 358, 414. Wiley-Interscience/John Wiley & Sons, New York.
48. Patron, L., Moretti, A., Tedesco, R. and Pasqualetto, R. 1974. Bulk polymerization of acrylonitrile. *DOS* 2,400,043 (July 4).
49. Kricheldorf, H.R., Nuyken, O. and Swift, G. 2005. *Handbook of Polymer Synthesis*, 2nd ed., pp. 299–306. Marcel Dekker, New York.
50. Brubaker, M.M. 1949. Rapid polymerization process. U.S. Patent 2,462,354 (February 22).
51. Braun, D., Cherdron, H. and Kern, W. 1971. *Praktikum der makromolekularen organischen Chemie*, pp. 142, 143, 158–160, 214–216. Dr. Alfred Huethig Verlag, Heidelberg.
52. McCaffery, E.M. 1970. *Laboratory Preparation for Macromolecular Chemistry*, pp. 89–99. McGraw-Hill, New York.
53. Young, L.J. 1961. Copolymerization parameters. *J. Polym. Sci.* 54:411.
54. Vollmert, B. 1988. *Grundriss der Makromolekularen Chemie*, Vol. I, pp. 156, 157. E. Vollmert Verlag, Karlsruhe.
55. Greenlay, R.Z. 1999. Free Radical Copolymerization Reactivity Ratios. In Polymer Handbook (4th ed.), eds. J. Brandrup, E.H. Immergut and E.A. Grulke, pp. II-181, 194–200. John Wiley & Sons, New York.
56. Li, B., Yuan, H. and Pan, Z. 1989. Monomer reactivity ratios of acrylonitrile with methyl acrylate in continuous aqueous copolymerization at high conversion. In *Polymer Reaction Engineering*, eds. K.H. Reichert and W. Geiseler, pp. 222–229. VCH Verlagsgesellschaft, Weinheim.
57. Chen, C.S.H. 1976. 1:1 Alternating copolymer of acrylonitrile and vinyl acetate. *J. Polym. Sci., Polym. Chem. Ed.* 14:2109.
58. Sharma, Y.N., Gandhi, V.G. and Bhardwaj, I.S. 1980. Investigations on the kinetics of copolymerization of vinyl acetate with acrylonitrile by $Co(acac)_3$–$Al(C_2H_5)_3$ catalyst system. *J. Polym. Sci., Polym. Chem. Ed.* 18:59.
59. Wang, H., Chu, G., Srisiri, W., Padias, A.B. and Hall, Jr., H.K. 1994. Free radical copolymerization of electron-rich vinyl monomers with acrylonitrile complexed to zinc chloride. *Acta Polym.* 45:26.
60. Chaturverdi, B. and Srivastava, A.K. 1994. Copolymerization of acrylonitrile with chromium acrylate initiated by styrene–arsenic sulfide complex. *Polymer* 35(3):642.
61. Craubner, H. 1980. Macromolecular N-nitroso-acrylamines as initiators for polyreactions. I. Synthesis of block copolymers from polycondensates and olefinic monomers. *J. Polym. Sci., Polym. Chem. Ed.* 18:2011.
62. Chapiro, A. 1957. Swelling of graft copolymers of acrylonitrile on polyethylene in dimethylformamide. *J. Polym. Sci.* 23:377.
63. El'Garf, S.A., Konkin, A.A. and Rogovin, Z.A. 1966. Synthesis of poly-acrylonitrile graft copolymers. *Vysokomol. soyed.* 8(1):42.
64. Henrici-Olive, G. and Olive, S. 1979. Molecular interactions and macroscopic properties of polyacrylonitrile and model substances. *Adv. Polym. Sci.* 32:125.
65. Bohn, C.R., Schaefgen, J.R. and Statton, W.O. 1961. Laterally ordered polymers: Polyacrylonitrile and poly(vinyl trifluoroacetate). *J. Polym. Sci.* 55:531.
66. Andrews, R.D., Miyachi, K. and Doshi, R.S. 1974. Iodine swelling of poly-acrylonitrile. Effect of orientation and evidence for a three-phase structure. *J. Macromol. Sci., Phys. (B)* 9(2):281.
67. Hinrichsen, G. and Orth, H. 1971. Direct evidence of colloidal structure on drawn polyacrylonitrile. *J. Polym. Sci. (B)* 9:529.
68. Hinrichsen, G. 1972. Structural changes of drawn polyacrylonitrile during annealing. *J. Polym. Sci. (C)* 38:303.

69. Kenyon, A.S. and McC.J. Rayford. 1979. Mechanical relaxation processes in polyacrylonitrile polymers and copolymers. *J. Appl. Polym. Sci.* 23(3):717.
70. Grobelny, J., Tekely, P. and Turska, E. 1981. A broad-line nuclear magnetic resonance investigation of polyacrylonitrile phase structure and chain conformation. *Polymer* 22:1649.
71. Grobelny, J., Sokol, M. and Turska, E. 1984. A study of conformation, configuration and phase structure of polyacrylonitrile and their mutual dependence by means of WAXS and ^{1}H BL-n.m.r. *Polymer* 25:1415.
72. Xu, X., Xiao, W. and Chen, K. 1994. Morphology and crystallinity of particles formed from dilute solutions of poly(vinyl alcohol-b-acrylonitrile), poly(vinyl alcohol) and polyacrylonitrile. *Eur. Polym. J.* 30(12):1439.
73. Craig, J.P., Knudsen, J.P. and Holland, V.F. 1962. Characterization of acrylic fiber structure. *Text. Res. J.* 32(6):435.
74. Hinrichsen, G. and Orth, H. 1971. Zur Struktur verstreckter Folien und Faeden sowie aus verduennten Loesungen hergestellter Einkristalle aus Poly-acrylonitril. *Kolloid-Z. u. Z. Polymere* 247:844.
75. Miller, R.L. 1975. Crystallographic data for various polymers. In *Polymer Handbook*, eds. J. Brandrup and E.H. Immergut, pp. III-1, 12. John Wiley & Sons, New York.
76. Krigbaum, W.R. and Tokita, N. 1960. Melting point depression study of polyacrylonitrile. *J. Polym. Sci.* 43:467.
77. Klement, J.J. and Geil, P.H. 1968. Growth and drawing of polyacrylonitrile crystals grown from solution. *J. Polym. Sci. (A-2)* 6:1381.
78. Allen, R.A., Ward, I.M. and Bashir, Z. 1994. The variation of the d-spacings with stress in the hexagonal polymorph of polyacrylonitrile. *Polymer* 35(19):4035.
79. Allen, R.A., Ward, I.M. and Bashir, Z. 1994. An investigation into the possibility of measuring an "X-ray modulus" and new evidence for hexagonal packing in polyacrylonitrile. *Polymer* 35(10):2063.
80. Allen, R.A., Ward, I.M. and Bashir, Z. 1994. An investigation into the possibility of measuring an "X-ray modulus" and new evidence for hexagonal packing in polyacrylonitrile. *Polymer* 35(10):2063.
81. Allen, R.A., Ward, I.M. and Bashir, Z. 1994. The variation of the d spacings with stress in the hexagonal polymorph of polyacrylonitrile. *Polymer* 35(19):4035.
82. Bashir, Z. 2001. The hexagonal mesophase in atactic polyacrylonitrile: A new interpretation of the phase transitions in the polymer. *J. Macromol. Sci.-Phys.* B40(1):41.
83. Yu, M.J., Wang, C.G., Bai, Y.J., Zhu, B., Ji, M. and Xu, Y. 2008. Microstructural evolution in polyacrylonitrile fibers during oxidative stabilization. *J. Polym. Sci. B: Polym. Phys.* 46:759.
84. Bai, Y.J., Wang, C.G., Lun, N., Wang, Y.X., Yu, M.J. and Zhu, B. 2006. HRTEM microstructures of PAN precursor fibers. *Carbon* 44:1773.
85. He, D.X., Wang, C.G., Bai, Y.J., Lun, N., Zhu, B. and Wang, Y.X. 2007. Microstructural evolution during thermal stabilization of PAN fibers. *J. Mater. Sci.* 42:7402.
86. Ge, H., Liu, H., Chen, J. and Wang, C. 2008. The skin–core structure of poly(acrylonitrile-itaconic acid) precursor fibers in wet-spinning. *J. Appl. Polym. Sci.* 108:947.
87. McMahon, P.E. 1967. Wideline NMR studies of polyacrylonitrile. *J. Polym. Sci. (A-2)* 5:271.
88. Ganster, J., Fink, H.-P. and Zenke, I. 1991. Chain conformation of polyacrylonitrile: A comparison of model scattering and radial distribution functions with experimental wide-angle x-ray scattering results. *Polymer* 32(9):1566.
89. Minagawa, M., Ute, K., Kitayama, T. and Hatada, K. 1994. Determination of stereoregularity of gamma.-irradiation canal polymerized poly-acrylonitrile by 1H 2D J-Resolved NMR spectroscopy. *Macromolecules* 27(13):3669.
90. Hunter, T.C. 1994. Microstructural characterization of polyacrylonitrile by NMR and FTIR. In *Proceedings of the 3rd Euro-American Conference on Functional Polymers and Biopolymers, Macromolecules '92, Macromolecular Reports—International Rapid Publication Supplement to the Journal of Macromolecular Science—Pure and Applied Chemistry*, 31 Suppl. 6–7, pp. 1061–1068.
91. Svegliado, G., Talamini, G. and Vidotto, G. 1967. Stereoregularity of polyacrylonitrile determined by NMR. *J. Polym. Sci. (A-1)* 5:2875.
92. Yamadera, R. and Murano, M. 1967. Studies on tacticity of polyacrylonitrile. I. High-resolution nuclear magnetic resonance spectra of poly-acrylonitrile. *J. Polym. Sci. (A-1)* 5:1059.
93. Minagawa, M., Takasu, T., Shinozaki, S., Yoshii, F. and Morishita, N. 1995. 13C n.m.r. and g.p.c.–low angle laser light scattering measurements on polyacrylonitrile prepared by urea clathrate polymerization in the solid state for the optimization of tacticity. *Polymer* 36(12):2343.
94. Kaji, H. and Schmidt-Rohr, K. 2001. Conformation and dynamics of atactic poly(acrylonitrile). 2. Torsion angle distributions in meso dyads from two-dimensional solid-state double-quantum 13C NMR. *Macromolecules* 34:7368.

95. Katsuraya, K., Hatanaka, K., Matsuzaki, K. and Minagawa, M. 2001. Assignment of finely resolved ^{13}C NMR spectra of polyacrylonitrile. *Polymer* 42:6323.
96. Dunn, P. and Ennis, B.C. 1970. Thermal analysis of polyacrylonitrile. Part I. The melting of polyacrylonitrile. *J. Appl. Polym. Sci.* 14:1795.
97. Arcus, C.L. 1955. The stereoisomerization of addition polymers. I. The stereochemistry of addition and configuration of maximum order. *J. Chem. Soc.* 2801.
98. Liang, C.Y. and Krimm, S. 1958. Infrared spectra of high polymers. VII. Polyacrylonitrile. *J. Polym. Sci.* 31:513.
99. Hu, X., Johnson, D.J. and Tomka, J.G. 1995. Molecular modelling of the structure of polyacrylonitrile fibres. *J. Text. Inst.* 86(2):322.
100. Rizzo, P., Auriemma, F., Guerra, G., Petraccone, V. and Corradini, P. 1996. Conformational disorder in the pseudohexagonal form of atactic polyacrylonitrile. *Macromolecules* 29:8852.
101. Kotake, Y., Yoshihara, T., Sato, H., Yamada, N. and Yasushi, J. 1967. On the structure of crystalline polymethacrylonitrile. *J. Polym. Sci. (B)* 5(2):163.
102. Kolt'tsov, A.I., Kamalov, S. and Vol'kenshtein, M.V. 1967. Anisotropy of nuclear magnetic resonance spectra of polyacrylonitrile fibres. *Vysokomol. soyed. (A9)* 1:131.
103. Chiang, R. 1965. Dissolution and crystallization temperatures of high polymers. II. New method of characterization of polyacrylonitrile. *J. Polym. Sci. (A)* 3:2019.
104. Talamini, G. and Vidotto, G. 1964. Heterogeneous block polymerization. *Chim. Ind. (Milan)* 46(4):371.
105. Nakano, Y., Hisatani, K. and Kamide, K. 1995. Synthesis of ultra-high molecular weight polyacrylonitrile with highly isotactic content (mm > 0.60) using dialkylmagnesium/polyhydric alcohol system as catalyst. *Polym. Int.* 36(1):87.
106. Nakano, Y., Hisatani, K. and Kamide, K. 1994. Synthesis of highly isotactic (mm > 0.70) polyacrylonitrile by anionic polymerization using diethylberyllium as a main initiator. *Polym. Int.* 35(2):207.
107. Jung, K.T., Hwang, D.K., Shul, Y.G., Han, H.S. and Lee, W.S. 2002. The preparation of isotactic polyacrylonitrile using zeolite. *Mater. Lett.* 53:180.
108. Sawai, D. and Kanamoto, T. 2000. X-ray long period of poly(acrylonitrile) fibers prepared by solid-state coextrusion of nascent powder. *Polymer J.* 32(10):895.
109. Koenig, J.L., Cornell, S.W. and Witenhafer, D.E. 1967. Infrared technique for the measurement of structural changes during the orientation process in polymers. *J. Polym. Sci. (A-2)* 5:301.
110. Minagawa, M., Miyano, K., Takahashi, M. and Yoshii, F. 1988. Infrared characteristic absorption bands of highly isotactic poly(acrylonitrile). *Macromolecules* 21:2387.
111. Simitzis, J. 1975. Einfluss von Copolymergehalt und Verstreckung von PAN-Fasern auf das Pyrolyseverhalten und die Eigenschaften daraus hergestellter Kohlenstoff-Fasern. Ph.D. Thesis, University of Karlsruhe, Germany.
112. Simitzis, J. 1976. Kinetik des Schrumpfprozesses bei der Herstellung von Kohlenstoff-Fasern aus PAN-Fasern. *Chemiker-Zeitung* 100:416.
113. Simitzis, J. 1977. Einfluss von MA-Copolymergehalt der PAN-Fasern auf ihr Pyrolyseverhalten waehrend der Oxidation. *Coll. Polym. Sci.* 255:1074.
114. Peebles, Jr., L.H. 1976. Acrylonitrile polymers, degradation. In *Encyclopedia of Polymer Science and Technology*, eds. H.F. Mark, N.G. Gaylord and N.M. Bikales, Supplement Vol. 1, pp. 1–25. Interscience Publishers/John Wiley & Sons, New York.
115. Minagawa, M., Taira, T., Kondo, K., Yamamoto, S., Sato, E. and Yoshii, F. 2000. Conformation effect and FT-IR diffuse reflection spectroscopy of stereoregular isotactic poly(acrylonitrile) prepared by urea clathrate polymerization. *Macromolecules* 33:4526.
116. Hori, T., Zhang, H. S., Shimizu, T. and Zollinger, H. 1988. Change of water states in acrylic fibers and their glass transition temperatures by DSC measurements. *Text. Res. J.* 58:227.
117. Raja, R.A., Raju, B.B. and Varadarajan, T.S. 1994. Application of fluorescence probe technique for determination of glass-transition temperature of polymers: Studies in polyacrylonitrile (PAN). *J. Appl. Polym. Sci.* 54(6):827.
118. Kimmel, R.M. and Andrews, R.D. 1965. Birefringence effects in acrylonitrile polymers. II. The nature of the 140°C transition. *J. Appl. Phys.* 36(10):3063.
119. Andrews, R.D. and Okuyama, H. 1968. Rheo-optical behavior of poly-acrylonitrile: Creep and creep recovery. *J. Appl. Phys.* 39(11):4909.
120. Cotten, G.R. and Schneider, W.C. 1963. Fractionation of acrylonitrile copolymers. I. Development of a method. *J. Appl. Polym. Sci.* 7(4):1243.
121. Hayakawa, R., Nishi, T., Arisawa, K. and Wada, Y. 1967. Dielectric relaxation in the paracrystalline phase in polyacrylonitrile. *J. Polym. Sci. (A-2)* 5:165.

122. Rizzo, P., Guerra, G. and Auriemma, F. 1996. Thermal transitions of polyacrylonitrile fibers. *Macromolecules* 29:1830.
123. Hu, X.-P. and Hsieh, Y.-L. 1997. Structure of acrylic fibres prior to cyclization. *Polymer* 38(6):1491.
124. Okajima, S., Ikeda, M. and Takeuchi, A.J. 1968. A new transition point of polyacrylonitrile. *J. Polym. Sci. (A-1)* 6:1925.
125. Bashir, Z. 2001. The hexagonal mesophase in atactic polyacrylonitrile: A new interpretation of the phase transitions in the polymer. *J. Macromol. Sci.-Phys.* B40(1):41.
126. Hinrichsen, G. 1971. Untersuchungen zum Schmelzen von Polyacrylnitril. *Ang. Makrom. Chem.* 20:121.
127. Gohil, R.M., Patel, K.C. and Patel, R.D. 1976. Crystallization of poly-acrylonitrile: Growth mechanisms for various growth features. *Coll. Polym. Sci.* 254(10):859.
128. Boucher, E.A., Langdon, D.J. and Manning, R.J. 1972. Effect of heat treatment on the morphology and structure of crystals, powders, and fibers of polyacrylonitrile and saran. *J. Polym. Sci. (A-2)* 10:1285.
129. Min, B.G., Son, T.W. and Jo, W.H. 1993. Direct formation of highly oriented, fibrillary structure from polyacrylonitrile melt. *Polym. Prepr. (Am. Chem. Soc., Div. Polym. Chem.)* 34(2):825.
130. Sawai, D., Yamane, A., Kameda, T. et al. 1999. Uniaxial drawing of isotactic poly(acrylonitrile): Development of oriented structure and tensile properties. *Macromolecules* 32:5622.
131. Thunemann, A.F. and Ruland, W. 2000. Microvoids in polyacrylonitrile fibers: A small-angle x-ray scattering study. *Macromolecules* 33:1848.
132. Thunemann, A.F. 2000. Lamellar mesophases in polyacrylonitrile: A Synchrotron small-angle x-ray scattering study. *Macromolecules* 33:2626.
133. Grassie, N. and Hay, J.N. 1962. Thermal coloration and insolubilization in polyacrylonitrile. *J. Polym. Sci.* 56:189.
134. Grassie, N. and Hay, J.N. 1963. Thermal degradation of poly(methyl vinyl ketone) and its copolymers with acrylonitrile. *Makromol. Chem.* 64:82.
135. Friedlander, H.N., Peebles, L.H., Brandrup, J. and Kirby, J.R. 1968. The chromophore of polyacrylonitrile, VI. Mechanism of color formation in polyacrylonitrile. *Macromolecules* 1(1):79.
136. Bashir, Z., Manns, G., Service, D.M., Bott, D.C., Herbert, I.R., Ibbett, R.N. and Church, S.P. 1991. Investigation of base induced cyclization and methine proton abstraction in polyacrylonitrile solutions. *Polymer* 32(10):1826.
137. Mailhot, B. and Gardette, J.-L. 1994. Mechanism of thermolysis, thermo-oxidation and photooxidation of polyacrylonitrile. *Polym. Degrad. Stab.* 44(2):223.
138. Peebles, Jr., L.H. 1976. Acrylonitrile polymers, degradation. In *Encyclopedia of Polymer Science and Technology*, eds. H.F. Mark, N.G. Gaylord and N.M. Bikales, Supplement Vol. 1, pp. 1–25. Interscience Publishers/John Wiley & Sons, New York.
139. Vosburgh, W.G. 1990. The heat treatment of orlon acrylic fiber to render it fireproof. *Text. Res. J.* 30:882.
140. Zhang, J., Horrocks, A.R. and Hall, M.E. 1994. The flammability of poly-acrylonitrile and its copolymers. IV. The flame retardant mechanism of ammonium polyphosphate. *Fire Mater.* 18(5):307.
141. Ko, T.-H. 1994. Preparation of flame resistant fibers from poly-acrylonitrile fibers. *Mater. Chem. Phys.* 38(3):289.
142. Zhang, J., Hall, M.E. and Horrocks, A.R. 1993. The flammability of poly-acrylonitrile and its copolymers. I. The flammability assessment using pressed powdered polymer samples. *J. Fire Sci.* 11(5):443.
143. Hall, M.E., Horrocks, A.R. and Zhang, J. 1994. The flammability of poly-acrylonitrile and its copolymers. *Polym. Degrad. Stab.* 44(3):379.
144. Hall, M.E., Zhang, J. and Horrocks, A.R. 1994. Flammability of poly-acrylonitrile and its copolymers. III. Effect of flame retardants. *Fire Mater.* 18(4):231.
145. Surianarayanan, M., Rao, S.P., Vijayaraghavan, R. and Raghavan, K.V. 1998. Thermal behavior of acrylonitrile polymerization and polyacrylonitrile decomposition. *J. Hazard. Mater.* 62:187.
146. Xu, J.Z., Tian, C.M., Ma, Z.G., Gao, M., Guo, H.Z. and Yao, Z.H. 2001. Study on the thermal behavior and flammability of the modified polyacrylonitrile fibers. *J. Thermal Anal. Calorim.* 63:501.
147. Huang, T.J., Hanyon, W.J. and Chapman, M.R. 2001. Fire retardant and heat resistant yarns and fabrics made therefrom. U.S. Patent 6,287,686 (September 11).
148. Bajaj, P., Agrawal, A.K., Dhand, A., Kasturia, N. and Hansraj, N. 2000. Flame retardation of acrylic fibers: An overview. *J.M.S.-Rev. Macromol. Chem. Phys.* C40(4):309.
149. Fester, W. 1975. Physical constants of poly(acrylonitrile). In *Polymer Handbook*, eds. J. Brandrup and E.H. Immergut, pp. V-37–V-40. John Wiley & Sons, New York.
150. Akki, R., Desai, P. and Abhiraman, A.S. 1994. Morphological implications of phase transitions in polymer solutions: Study of polyacrylonitrile-based solutions. *J. Appl. Polym. Sci.* 54(9):1263.

151. Loan, S., Grigorescu, G., Loan, C. and Simionescu, B.C. 1994. Excluded volume effect in polyacrylonitrile solutions. *Polym. Bull. (Berlin)* 33:119.
152. Bashir, Z. 1992. Thermoreversible gelation and plasticization of polyacrylonitrile. *Polymer* 33:4304.
153. Minagawa, M., Takasu, T., Morita, T., Shirai, H., Fujikura, Y. and Kameda, Y. 1996. The steric effect of solvent molecules in the dissolution of polyacrylonitrile from five different N,N-dimethylformamide derivatives as studied using Raman spectroscopy. *Polymer* 37(3):463.
154. Morariu, S., Bercea, M., Ioan, C., Ioan, S. and Simionescu, B.C. 1999. Conformational characteristics of oligo- and polyacrylonitrile in dilute solution. *Eur. Polym. J.* 35:377.
155. Bercea, M., Morariu, S., Ioan, C., Ioan, S. and Simionescu, B.C. 1999. Viscometric study of extremely dilute polyacrylonitrile solutions. *Eur. Polym. J.* 35:2019.
156. Korte, S. 1999. Physical constants of poly(acrylonitrile). In *Polymer Handbook*, eds. J. Brandrup, E.H. Immergut and E.A. Grulke, pp. V59–V66. John Wiley & Sons, New York.
157. Chiang, R. and Stauffer, J.C. 1967. Association of polyacrylonitrile prepared by low-temperature, solution polymerization with an organometallic catalyst. *J. Polym. Sci. (A-2)* 5:101.
158. Krigbaum, W.R. and Kotliar, A.M. 1958. The molecular weight of poly-acrylonitrile. *J. Polym. Sci.* 32:323.
159. Cleland, R.L. and Stockmayer, W.H. 1955. Intrinsic viscosity–molecular weight relation for polyacrylonitrile. *J. Polym. Sci.* 17:473.
160. Davis, C.W. and Shapiro, P. 1964. Acrylic fibers. *Encycl. Polym. Sci. Technol.* 1:342.
161. Beevers, R.B. 1968. The physical properties of polyacrylonitrile and its copolymers. *Macromol. Rev.* 3:113.
162. Marzolph, H. 1962. Aufbau und Eigenschaften von Polyacrylnitril-Faeden. *Angew. Chem.* 74(16):628.
163. Stupp, S.I. and Carr, S.H. 1979. Electric field-induced structure in poly(acrylonitrile). *Colloid Polym. Sci.* 257(9):913.
164. Allen, S.M., Fujii, M., Stannett, V., Hopfenberg, H.B. and Williams, J.L. 1977. The barrier properties of polyacrylonitrile. *J. Membrane Sci.* 2(2):153.
165. Huvard, G.S., Stannett, V.T., Koros, W.J. and Hopfenberg, H.B. 1980. The pressure dependence of carbon dioxide sorption and permeation in poly(acrylonitrile). *J. Membrane Sci.* 6(2):185.
166. Ranade, G., Stannett, V. and Koros, W.J. 1980. Temperature dependence and energetics of the equilibrium sorption of water vapor in glassy polyacrylonitrile. *J. Appl. Polym. Sci.* 25(10):2179.
167. Schouteden. F.L.M. 1957. Polyacrylamidoximes. *Makromol. Chem.* 24(1):25.
168. Krentsel, L.B., Kudryavtsev, Y.V., Rebrov, A.I., Litmanovich, A.D. and Plate, N.A. 2001. Acidic hydrolysis of polyacrylonitrile: Effect of neighboring groups. *Macromolecules* 34:5607.
169. Hayes, R.A. 1953. Polymeric chain-transfer reactions. Polymerization of some vinyl monomers in the presence of vinyl polymers. *J. Polym. Sci.* 11:531.
170. Hu, C.-M. and Chiang, W.-Y. 1990. Studies of reactions with polymers. V. The reaction mechanism and resultant structure of PVA with PAN in DMSO without any catalyst. *J. Polym. Sci. (A)* 28:353.
171. Gaponik, P.N., Ivashkevich, O.A., Karavai, V.P. et al. 1994. Polymers and copolymers based on vinyl tetrazoles. 1. Synthesis of poly(5-vinyltetrazole) by polymer-analogous conversion of polyacrylonitrile. *Angew. Makromol. Chem.* 219:77.
172. Simitzis, J. 1979. Einfluss der γ-Bestrahlung auf das Pyrolyseverhalten der Polyacrylnitril-(PAN-) Fasern. *Atomkernenerg./Kerntech.* 33:52.
173. Simitzis, J. 1981. Untersuchung des Pyrolyseverhaltens und der Eigenschaften der mit γ-Strahlen vorbehalten Polyacrylnitril-(PAN-)Fasern. *Atomkernenerg./Kerntech.* 38:205.
174. Chiang, W.-Y. and Hu, C.-M. 1990. Studies of reactions with polymers. VI. The modification of PAN with primary amines. *J. Polym. Sci. (A)* 28:1623.
175. Simitzis, J. 1994. Modified polyacrylonitrile for adsorption applications. *Acta Polym.* 45:104.
176. Simitzis, J. 1995. Diffusion-limited sorption of dyes on modified acrylics and acrylic copolymers. *Polymer* 36:1017.
177. Simitzis, J. 1995. Thermally modified acrylic polymers as sorbents. *Polym. Int.* 36:279.
178. Simitzis, J. 1995. Characterization of acrylic polymers by dye-adsorption. *Int. J. Polym. Anal. Charact.* 1:1.
179. Zhang, B.W., Fischer, K., Bieniek, D. and Kettrup, A. 1994. Synthesis of carboxyl group containing hydrazine-modified polyacrylonitrile fibers and application for the removal of heavy metals. *React. Polym.* 24(1):49.
180. Batty, N.S. and Guthrie, J.T. 1978. Degradation of polyacrylonitrile in solution by alkali. *Polymer* 19:1145.

181. Bajaj, P., Chavan, R.B. and Manjeet, B. 1985. Saponification kinetics of acrylonitrile terpolymer and polyacrylonitrile. *J. Macromol. Sci.-Chem.* A22(9):1219.
182. Puntigam, H.R. and Voelker, T. 1967. *Acryl- und Methacrylverbindungen*, pp. 72, 73. Springer-Verlag, Berlin.
183. Cho, H.H., Kim, H.S. and Choi, S.C. 1993. Structural study of polyacrylonitrile–iodine complex. Microstructural feature of polyacrylonitrile–iodine complex. *Han'guk Somyu Konghakhoechi* 30(5):370.
184. Kim, H.S. and Choi, S.C. 1994. Crystalline structure of polyacrylonitrile–iodine complex. *J. Appl. Polym. Sci.* 53(11):1403.
185. Battistel, E., Morra, M. and Marinetti, M. 2001. Enzymatic surface modification of acrylonitrile fibers. *Appl. Surface Sci.* 117:32.
186. Bell, J.P. and Murayama, T. 1968. Relation between dynamic mechanical properties and dye diffusion behavior in acrylic fibers. *J. Appl. Polym. Sci.* 12:1795.
187. Peebles, Jr., L.H., Thompson, Jr., R.B., Kirby, J.R. and Gibson, M.E. 1972. Basic dyeability and acid content of high-conversion polyacrylonitrile. *J. Appl. Polym. Sci.* 16:3341.
188. Bajaj, P. and Munukutla, S.K. 1990. Effect of spinning dope additives on dyeing behavior of acrylic fibers. *Text. Res. J.* 60:113.
189. Katragadda, S., Gesser, H.D. and Chow, A. 1997. The extraction of uranium by amidoximated orlon. *Talanta* 45:257.
190. Rohner, R.M. and Zollinger, H. 1986. Porosity versus segment mobility in dye diffusion kinetics—A differential treatment: Dyeing of acrylic fibers. *Text. Res. J.* 56:1.
191. Gupta, A.K. and Maiti, A.K. 1982. Effect of heat treatment on the structure and mechanical properties of polyacrylonitrile fibers. *J. Appl. Polym. Sci.* 27:2409.
192. Saechtling, H. 1986. *Kunststoff Taschenbuch*, p. 287. Carl Hanser Verlag, Munich.
193. Kolarz, B.N., Wojaczynska, M., Trochimczuk, A. and Luczynski, J. 1988. Porous terpolymers: Poly(acrylonitrile-co-ethyl/butyl/acrylate-co-divinylbenzene). *Polymer* 29:1137.
194. Nemphos, S.P., Salame, M. and Steingiser, S. 1976. Barrier polymers. In *Encyclopedia of Polymer Science and Technology*, eds. H.F. Mark and N.M. Bikales, Supplement Vol. 1, pp. 65–95. Interscience Publishers/John Wiley & Sons, New York.
195. Blow, C.M. 1971. *Rubber Technology and Manufacture*, pp. 31, 119–125. Plastics & Rubber Institute/Newnes-Butterworths, London, Boston.
196. Norman, R.H. 1970. *Conductive Rubbers and Plastics*, pp. 59, 63, 64. Applied Science, London.
197. Schulz, R.C. 1968. Optical active polymers. In *Encyclopedia of Polymer Science and Technology*, eds. H.F. Mark, N.G. Gaylord and N.M. Bikales, Vol. 9, pp. 507–524. Interscience Publishers/John Wiley & Sons, New York.
198. Kobayashi, E., Furukawa, J. and Nagata, S. 1979. Studies on the optical active alternating copolymer of benzofuran or 1,3-cyclooctadiene with acrylic monomers, *J. Polym. Sci., Polym. Chem. Ed.* 17:2093.
199. Xiao, L., Chen, Y., Cai, R. and Huang, Z.-E. 1999. Synthesis and characterization of [60]fullerene-based nonlinear optical polyacrylonitrile derivatives. *J. Mater. Sci. Lett.* 18:833.
200. Zheng, H., Xue, H., Zhang, Y. and Shen, Z. 2002. A glucose biosensor based on microporous polyacrylonitrile synthesized by single rare-earth catalyst. *Biosens. Bioelectron.* 17(6–7):541.
201. Stefanovic, V., Vlahovic, P., Kostic, S. and Mitic-Zlatkovic, M. 1998. In vitro blood compatibility evaluation of cuprophan and polyacrylonitrile membranes. *Nephron* 79(3):350.
202. Salmon, J., Cardigan, R., Mackie, I., Cohen, S.L., Machin, S. and Singer, M. 1997. Continuous venovenous haemofiltration using polyacrylonitrile filters does not activate contact system and intrinsic coagulation pathways. *Intensive Care Med.* 23(1):38.
203. Guastoni, C., Tetta, C., Hoenich, N.A. et al. 1996. Mechanisms and kinetics of the synthesis and release of platelet-activating factor (PAF) by polyacrylonitrile membranes. *Clin. Nephrol.* 46(2):132.
204. Gasche, Y., Pascual, M., Suter, P.M., Favre, H., Chevrolet, J.C. and Schifferli, J.A. 1996. Complement depletion during haemofiltration with polyacrylonitrile membranes. *Nephrol. Dial. Transpl.* 11(1):117.
205. Tiersch, B., Hicke, H.-G., Becker, M. and Paul, D. 1994. Elektronen-mikroskopische Charakterisierung von chemisch aktivierten PAN-Membranen. In *5. Berliner Polymeren-Tage*, October 5–7, p. 113. Potsdam, Berliner Verband fuer Polymerforschung e.V.
206. Cassidy, H.G. and Kun, K.A. 1966. Electron-transfer polymers. In *Encyclopedia of Polymer Science and Technology*, eds. H.F. Mark, N.G. Gaylord and N.M. Bikales, Vol. 5, pp. 693–738. Interscience Publishers/John Wiley & Sons, New York.
207. Turner, J.E., Shen, M. and Lin, C.C. 1980. Hydrophilic behavior of HSPAN/PVA membrane. *J. Appl. Polym. Sci.* 25(7):1287.
208. Stout, E.I., Trimnell, D., Doane, W.M. and Russell, C.R. 1977. Graft copolymers of starch–polyacrylonitrile prepared by ferrous ion–hydrogen peroxide initiation. *J. Appl. Polym. Sci.* 21(9):2565.

209. Hirai, A., Tsuji, W., Kitamaru, R. and Hosono, M. 1976. Structure of the decrystallized cotton in fabrics prepared by alkali and acrylonitrile treatments. *J. Appl. Polym. Sci.* 20(12):3365.
210. Lokhande, H.T. and Varadarajan, P.V. 1993. A new approach in the production of non-wood-based cellulosic super-absorbents through the polyacrylonitrile (PAN) grafting method. *Bioresour. Technol.* 45(3):161.
211. Hersch, S.P. 1985. Polyblend fibers. In *High Technology Fibers*, eds. M. Lewin and J. Preston, Part A, pp. 1–50. Marcel Dekker, New York.
212. Marsano, E., Bianchi, E., Conio, G. and Tealdi, A. 1994. Cellulose–polyacrylonitrile blends: 3. Polymer–polymer interactions in dimethylacetamide/LiCl solution determined by viscometry. *Polymer* 35(16):3565.
213. Somanathan, N., Senthil, P., Viswanathan, S. and Arumugam, V. 1994. Compatibility studies of solution-blended polyacrylonitrile/poly(*n*-butyl methacrylate). *J. Appl. Polym. Sci.* 54(10):1537.
214. Gupta, N. and Srivastava, A.K. 1994. Study of the morphology properties of interpenetrating polymer networks of poly(zinc acrylate) and polyacrylonitrile. *Polymer* 35(17):3769.
215. Harun, M.G. and Tong, C.C. 1994. Interpenetrating polymer networks from natural rubber and polyacrylonitrile. *Macromol. Rep., A31, No Suppl.* 1–2:35.
216. Griffith, R.K. 1968. Molding of polyacrylonitrile. *J. Appl. Polym. Sci.* 12:1939.
217. Griffith, R.K. and Zorska, R. 1969. Structural modification and molding of polyacrylonitrile. *J. Appl. Polym. Sci.* 13:1159.
218. Rangarajan, P., Bhanu, V.A., Godshall, D., Wilkes, G.L., McGrath, J.E. and Baird, D.G. 2002. Dynamic oscillatory shear properties of potentially melt processable high acrylonitrile terpolymers. *Polymer* 43:2699.
219. Davidson, J.A., Jung, H.-T., Hudson, S.D. and Percec, S. 2000. Investigation of molecular orientation in melt-spun high acrylonitrile fibers. *Polymer* 41:3357.
220. Groebe, V. and Meyer, K. 1959. Acrylic fibers. XXII. Influence of precipitant on fiber formation in the wet spinning of polyacrylonitrile. *Faserforsch. u. Textiltechnik* 10:214.
221. Carter, M.E. 1971. *Essential Fiber Chemistry*, p. 115. Marcel Dekker, New York.
222. Takahashi, M.J. and Watanabe, M. 1961. Acrylic fiber. XX. The effect of solvent content in the coagulation bath on the filament formation. *J. Soc. Textile Cellulose Ind. Japan* 17:243.
223. Sunden, O. Sonnerskog, S.H., Sunden, N.B. and Larrson, H.E. 1961. Wet-spinning of fibers containing polyacrylonitrile. U.S. Patent 2,967,085 and 2,967,086 (January 3).
224. Bruson, H.A. 1958. Copolymers of acrylonitrile and vinyl halohydrins. U.S. Patent 2,826,566 (March 11).
225. Ehlers, F.A. and Tomela, D.R. 1958. Acrylonitrile polymer dissolved in dimethyl sulfoxide. U.S. Patent 2,858,290 (October 28).
226. Stanton, G.W., Lefferdink, T. B. and Spence, T. C. 1957. Controlled coagulation of salt-spun polyacrylonitrile. U.S. Patent 2,790,700 (April 30).
227. Peng, G.Q., Zhang, X.H., Wen, Y.F., Yang, Y.G. and Liu, L. 2008. Effect of coagulation bath DMSO concentration on the structure and properties of polyacrylonitrile (PAN) nascent fibers during wet-spinning. *J. Macromol. Sci. B* 47:1130.
228. Tan, L., Pan, D. and Pan, N. 2008. Thermodynamic study of a water–dimethylformamide–polyacrylonitrile ternary system. *J. Appl. Polym. Sci.* 110:3439.
229. Zwick, M. 1967. Spinning of fibers from polymer solutions undergoing phase separation. I. Practical considerations and experimental study. *Appl. Polym. Symp.* 6:109.
230. Knudsen, J.P. 1963. The influence of coagulation variables on the structure and physical properties of an acrylic fiber. *Text. Res. J.* 33:13.
231. Takahashi, M. 1961. Acrylic fiber XXVI. Effects of spinning conditions on the fine structure of filaments. *Chem. High Polymers (Tokyo)* 18:163.
232. Statton, W.O. 1962. Microvoids in fibres as studied by small angle scattering of x-rays. *J. Polym. Sci.* 58:905.
233. Paul, D.R. and McPeters, A.L. 1977. Effect of spin orientation on drawing of wet-spun fibers. *J. Appl. Polym. Sci.* 21:1699.
234. Bajaj, P. and Kumari, M.S. 1989. Physiomechanical properties of fibers from blends of acrylonitrile terpolymer and its hydrolyzed products. *Text. Res. J.* 59:191.
235. Kulshreshtha, A.K., Garg, V.N. and Sharma, Y.N. 1986. Plastic deformation, crazing and fracture morphology of acrylic fibers. *Text. Res. J.* 56:484.
236. Craig, J.P., Knudsen, J.P. and Holland, V.F. 1962. Characterization of acrylic fiber structure. *Text. Res. J.* 32(6):435.
237. Bell, J.P. and Dumbleton, J.H. 1971. Changes in the structure of wet-spun acrylic fibers during processing. *Text. Res. J.* 41:196.

238. Sotton, M. and Vialard, A.M. 1971. Contribution to the study of the porous structure of polyacrylonitrile (PAN) fibers. *Text. Res. J.* 41:834.
239. Law, S.J. and Mukhopadhyay, S.K. 1997. The construction of a phase diagram for a ternary system used for the wet spinning of acrylic fibers based on a linearized cloud point curve correlation. *J. Appl. Polym. Sci.* 65:2131.
240. Law, S.J. and Mukhopadhyay, S.K. 1998. Compositional changes during the wet spinning of acrylic fibers from aqueous sodium thiocyanate solvent. *J. Appl. Polym. Sci.* 69:1459.
241. Law, S.J. 2000. Formation/structure/property relationships of wet-spun acrylic fibers. *Chem. Fibers Int.* 50(2):65.
242. Oh, S.C., Wang, Y.S. and Yeo, Y.K. 1996. Modelling and simulation of the coagulation process of poly(acrylonitrile) wet-spinning. *Ind. Eng. Chem. Res.* 35:4796.
243. Diemunsch, J., J. Chabert and A. Banderet. 1965. Contribution à l'Etude de la Sorption des Fibers Textiles. *I.T.F. Bull.* 117:201.
244. Moreton, R. 1971. The spinning of polyacrylonitrile fibers for the production of carbon fibers. In *Intern. Conference in Carbon Fibers*, London, Paper No. 12.
245. Melacini, P., Patron, L., Doria, G. and Tedesco, R. 1975. Continuous preparation of spinning solutions of acrylic polymers. *DOS* 2,428,093 (January 16).
246. Ellwood, P. 1969. Illuminating is the word for acrylic fiber method. *Chem. Eng.* 76(18):90.
247. Rroughton, R.M., Gowayed, Y., Wang, W., Li, Y., Guo, L. and Yang, S.S. 2001. Development of chain orientation during the deformation of semiliquid polymers. *National Textile Center, U.S.A., Annual Report* 1 (November). Available at http://www.eng.auburn.edu/department/te/faculty/Broughton.
248. Warner, S.B., Buer, A., Ugbolue, S.C., Rutledge, G.C. and Shin, M.Y. 1998. A fundamental investigation of the formation and properties of electrospun fibers. *National Textile Center, U.S.A., Annual Report* 83 (November).
249. Nataraj, S.K., Yang, K.S. and Aminabhavi, T.M. 2012. Polyacrylontirle-based nanofibers—A review. *Prog. Polym. Sci.* 37:487.
250. Arshad, S.N., Naraghi, M. and Chasiotis, I. 2011. Strong carbon nanofibers from electrospun polyacrylonitrile. *Carbon* 49:1710.
251. Gu, S.Y. Ren, J. and Wu, Q.L. 2005. Preparation and structures of electrospun PAN nanofibers as a precursor of carbon nanofibers. *Synth. Met.* 155:157.
252. Zhou, Z., Lai, C., Zhang, L. et al. 2009. Development of carbon nanofibers from aligned electrospun polyacrylonitrile nanofiber bundles and characterization of their microstructural, electrical, and mechanical properties. *Polymer* 50:2999.
253. Hou, X., Yang, X., Zhang, L., Waclawik, E. and Wu, S. 2010. Stretching-induced crystallinity and orientation to improve the mechanical properties of electrospun PAN nanocomposites. *Mater. Design* 31:1726.
254. Oechel, A., Ulbricht, M., Tomaschewski, G. and Hicke, H.G. 1994. Photo-modification of ultrafiltration membranes. 3. Photochemically induced gas-phase grafting of acrylic monomers onto polyacrylonitrile ultrafiltration membranes. *J. Inf. Rec. Mater.* 21(5–6):633.
255. Fritzsche, A.K., Arevalo, A.R., Moore, M.D. and O'Hara, C. 1993. The surface structure and morphology of polyacrylonitrile membranes by atomic force microscopy. *Adv. Filtr. Sep. Technol.* 7:481.
256. Chung, T.S., Kafchinski, E.R., Spak, M., Bembry-Ross, B. and Gle, C. Wensley. 1994. A coated microporous polyacrylonitrile hollow fiber as high-performance composite membrane for fluid separation. Patent No. 5324430 U.S. (940628), Cont.-in-part of U.S. Ser. No 993,931, p. 6.
257. Kobayashi, T., Miyamoto, T., Nagai, T. and Fujii, N. 1994. Polyacrylonitrile ultrafiltration membranes containing negatively charged groups for permeation and separation of dextran and dextran sulfate. *J. Appl. Polym. Sci.* 52(10):1519.
258. Kim, I.C., Yun, H.G. and Lee, K.H. 2002. Preparation of asymmetric polyacrylonitrile membrane with small pore size by phase inversion and post-treatment process. *J. Membrane Sci.* 199(1–2):75.
259. Kobayashi, T., Wang, H.Y. and Fujii, N. 1998. Molecular imprint membranes of polyacrylonitrile copolymers with different acrylic acid segments. *Anal. Chim. Acta* 365(1–3):81.
260. Germic, L., Ebert, K., Bouma, R.H.B., Borneman, Z., Mulder, M.H.V. and Strathmann, H. 1997. Characterization of polyacrylonitrile ultrafiltration membranes. *J. Membrane Sci.* 132(1):131.
261. Ulbricht, M. and Oechel, A. 1996. Photo-bromination and photo-induced graft polymerization as a two-step approach for surface modification of polyacrylonitrile ultrafiltration membranes. *Eur. Polym. J.* 32(9):1045.
262. Simitzis, J. 1995. Modification of wastes of PAN-fibers for adsorption applications. *Angew. Makromol. Chem.* 228:13.

263. Simitzis, J., Terlemesian, E. and Mladenov, I. 1995. Utilization of waste of PAN-fibers as adsorbents by chemical and thermal modification. *Eur. Polym. J.* 31(12):1261.
264. Luna-Barcenas, G., Kanakia, S.K., Sanchez, I.C. and Johnston, K.P. 1995. Semicrystalline microfibrils and hollow fibres by precipitation with a compressed-fluid antisolvent. *Polymer* 36(16):3173.
265. Zhang, Y., Shi, L., Li, R., Li, H., Han, S. and He, B. 1999. Preparation of acrylonitrile–methyl methacrylate–sodium sulfonate acrylate tricopolymer hollow fiber membrane. *Polym. Adv. Technol.* 10(1–2):86.
266. Rivereau, A.S., Darquy, S., Chaillous, L. et al. 1997. Reversal of diabetes in non-obese diabetic mice by xenografts of porcine islets entrapped in hollow fibres composed of polyacrylonitrile–sodium methallylsulphonate copolymer. *Diabetes Metab.* 23(3):205.
267. Sotomura, T., Adachi, K., Taguchi, M., Iwaku, M., Tatsuma, T. and Oyama N. 1999. Developing stable, low impedance interface between metallic lithium anode and polyacrylonitrile-based polymer gel electrolyte by preliminary voltage cycling. *J. Power Sources* 81–82:192.
268. Bajaj, P., Gupta, A.P. and Ojha, N. 2000. Antistatic and hydrophilic synthetic fibers: A critique. *J.M.S.-Rev. Macromol. Chem. Phys.* C40(2–3):105.
269. Grassie, N. and Scott, G. 1985. *Polymer Degradation & Stabilisation*, pp. 49–54. Cambridge University Press, Cambridge.
270. Shindo, A. 1961. Studies on graphite fibres. *Rep. Government Industr. Res. Inst., Osaka*, Nr. 317.
271. Ehrburger, P. and Donnet, J.B. 1985. Carbon and graphite fibers. In *High Technology Fibers*, eds. M. Lewin and J. Preston, pp. 169–220. Marcel Dekker, New York.
272. Fitzer, E. 1985. Technical status and future prospects of carbon fibres and their application in composites with polymer matrix (CFRPs). In *Carbon Fibres and their Composites*, ed. E. Fitzer, pp. 3–45. Springer-Verlag, Berlin.
273. Ermolenko, I.N., Lyubliner, I.P. and Gulko, N.V 1990. *Chemically Modified Carbon Fibers and their Application*. VCH-Verlagsgesellschaft GmbH, Weinheim.
274. Watt, W., Phillips, L.N. and Johnson, W. 1969. High-strength high-modulus carbon fibres. *The Engineer* 221:815 (May 27).
275. Grassie, N. and McGuchan, R. 1971. Pyrolysis of polyacrylonitrile and related polymers-III. Thermal analysis of preheated polymers. *Eur. Pol. J.* 7:1357.
276. Turner, W.M. and Johnson, F.C. 1969. The pyrolysis of acrylic fiber in inert atmosphere. I. Reactions up to 400°C. *J. Appl. Polym. Sci.* 13:2073.
277. Grassie, N. and McGuchan, R. 1972. Pyrolysis of polyacrylonitrile and related polymers-V. Thermal analysis of α-substituted acrylonitrile polymers. *Eur. Pol. J.* 8:243.
278. Grassie, N. and McGuchan, R. 1970. Pyrolysis of polyacrylonitrile and related polymers—I. Thermal analysis of polyacrylonitrile. *Eur. Pol. J.* 6:1277.
279. Henrici-Olive, G. and Olive, S. 1983. The chemistry of carbon fiber formation from polyacrylonitrile. *Adv. Polym. Sci.* 51:1.
280. Gupta, A.K., Paliwal, D.K. and Bajaj, P. 1991. Acrylic precursors for carbon fibers. *J.M.S.-Rev. Macromol. Chem. Phys.* C31(1):1.
281. Devasia, R., Reghunadhan-Nair, C.P. and Ninan, K.N. 2002. Solvent and kinetic penultimate unit effects in the copolymerization of acrylonitrile with itaconic acid. *Eur. Polym. J.* 38:2003.
282. Sanchez-Soto, P.J., Aviles, M.A., del Rio, J.C., Gines, J.M., Pascual, J. and Perez-Rodriguez, J.L. 2001. Thermal study of the effect of several solvents on polymerization of acrylonitrile and their subsequent pyrolysis. *J. Anal. Appl. Pyrolysis* 58–59:155.
283. Chen, J.C. and Harrison, I.R. 2002. Modification of polyacrylonitrile (PAN) carbon fiber precursor via post-spinning plasticization and stretching in dimethyl formamide (DMF). *Carbon* 40:25.
284. Mittal, J., Mathur, R.B., Bahl, O.P. and Inagaki, M. 1998. Post spinning treatment of PAN fibers using succinic acid to produce high performance carbon fibers. *Carbon* 36(7–8):893.
285. Mittal, J., Mathur, R.B. and Bahl, O.P. 1997. Post spinning modification of PAN fibres—A review. *Carbon* 35(12):1713.
286. Ko, T.-H., Chiranairadul, P. and Lin, C.-H. 1991. The influence of continuous stabilization on the properties of stabilized fibers and the final activated carbon fibers. Part I. *Polym. Eng. Sci.* 31(19):1618.
287. Kim, J., Kim, Y.C., Ahn, W. and Kim, C.Y. 1993. Reaction mechanisms of polyacrylonitrile on thermal treatment. *Polym. Eng. Sci.* 33(22):1452.
288. Bailey, J.E. and Clarke, A.J. 1971. Carbon fibre formation—The oxidation treatment. *Nature* 234:529.
289. Kirby, J.R., Brandrup, J. and Peebles, Jr., L.H. 1968. On the chromophore of polyacrylonitrile. II. The presence of ketonic groups in polyacrylonitrile. *Macromolecules* 1(1):53.
290. Hay, J.N. 1968. Thermal reactions of polyacrylonitrile. *J. Polym. Sci. (A-1)* 6:2127.

291. Watt, W. and Green, J. 1971. The pyrolysis of polyacrylonitrile. In *Intern. Conf. on Carbon Fibres*, London, The Plastics Institute, Paper No. 4.
292. Huxley, R.V. 1971. Carbon filaments. Brit. Patent 1 224 884 (1967/1971), Courtaulds Ltd., England.
293. Morita, K., Miyaji, H. and Ono, K. 1970. High strength carbon fibers. *DOS* 1,958,361 (September 10). Toray Industries Inc., Tokyo.
294. Shindo, A. 1971. On the carbonization of acrylic fibres in hydrochloric acid vapour. In *Intern. Conf. on Carbon Fibres*, London, The Plastics Institute, Paper No. 3.
295. Lee, J.K., Shim, H.J., Lim, J.C. et al. 1997. Influence of tension during oxidative stabilization on SO_2 adsorption characteristics of polyacrylonitrile (PAN) based activated carbon fibers. *Carbon* 35(6):837.
296. Mueller, D.J. 1973. Zur Herstellung von Kohlenstoff-Fasern aus Polyacrylnitril uber die katalytische Zyklisierung des Polymeren. Ph.D. Thesis, University of Karlsruhe, Germany.
297. Overhoff, D. 1970. (Sigri Elektrographit GmbH), Carbon and graphite fibers from polyacrylonitrile. *DOS* 1,929,849 (December 17).
298. Martin, S.C., Liggat, J.J. and Snape, C.E. 2001. *In situ* NMR investigation into the thermal degradation and stabilisation of PAN. *Polym. Degrad. Stabil.* 74:407.
299. Tsai, J.-S. 1994. Orientation change for polyacrylonitrile precursor during oxidation. *J. Mater. Sci. Lett.* 13(16):1162.
300. Wang, P.H., Liu, J. and Li, R.Y. 1994. Physical modification of polyacrylonitrile precursor fiber: Its effect on mechanical properties. *J. Appl. Polym. Sci.* 52(12):1667.
301. Wang, P.H. 1998. Aspects on prestretching of PAN precursor: Shrinkage and thermal behavior. *J. Appl. Polym. Sci.* 67:1185.
302. Kim, H.S., Shioya, M. and Takaku, A. 1999. Kinetic studies on hot-stretching of polyacrylonitrile-based carbon fibers by using internal resistance heating. Part I. Changes in resistivity and strain. *J. Mater. Sci.* 34:329.
303. Kim, H.S., Shioya, M. and Takaku, A. 1999. Kinetic studies on hot-stretching of polyacrylonitrile-based carbon fibers by using internal resistance heating. Part II. Changes in structure and mechanical properties. *J. Mater. Sci.* 34:3307.
304. Xue, T.J., McKinney, M.A. and Wilkie, C.A. 1997. The thermal degradation of polyacrylonitrile. *Polym. Degrad. Stabil.* 58:193.
305. Herrera, M., Wilhelm, M., Matuschek, G. and Kettrup, A. 2001. Thermoanalytical and pyrolysis studies of nitrogen containing polymers. *J. Anal. Appl. Pyrolysis* 58–59:173.
306. Bajaj, P., Sreekumar, T.V. and Sen, K. 2001. Thermal behavior of acrylonitrile copolymers having methacrylic and itaconic acid comonomers. *Polymer* 42:1707.
307. Memetea, L.T., Billingham, N.C. and Then, E.T.H. 1995. Hydroperoxides in polyacrylonitrile and their role in carbon-fibre formation. *Polym. Degrad. Stabil.* 47:189.
308. Pandey, G.C., Rao, K.V. and Kumar, A. 1995. Quantitative application of DRIFT spectroscopy: Determination of composition of special acrylic fibers. *Polym. Test.* 14:489.
309. Ogawa, H. and Saito, K. 1995. Oxidation behavior of polyacrylonitrile fibers evaluated by new stabilization index. *Carbon* 33(6):783.
310. Tsai, J.S. 1997. Comparison of batch and continuous oxidation processes for producing carbon fibre based on PAN fibre. *J. Mater. Sci. Lett.* 16:361.
311. Zhu, Y., Wilding, M.A. and Mukhopadhyay, S.K. 1996. Estimation, using infrared spectroscopy, of the cyclization of poly(acrylonitrile) during the stabilization stage of carbon fibre production. *J. Mater. Sci.* 31:3831.
312. Soulis, S. and Simitzis, J. 2005. Thermomechanical behaviour of poly[acrylonitrile-co-(methyl acrylate)] fibres oxidatively treated at temperatures up to 180°C. *Polym. Int.* 54:1474.
313. Simitzis, J. and Soulis, S. 2008. Correlation of chemical shrinkage of polyacrylonitrile fibres with kinetics of cyclization. *Polym. Int.* 57:99.
314. Fazlitidinova, A.G., Tyumwntev, V.A., Podkopayev, S.A. and Shveikin, G.P. 2010. Changes in polyacrylonitrile fiber fine structure during thermal stabilization. *J. Mater. Sci.* 45:3998–4005.
315. Yongping, H., Tongqing, S., Haojing, W. and Dong, W. 2008. A new method for the kinetic study of cyclization reaction during stabilization of polyacrylonitrile fibers. *J. Mater. Sci.* 43:4910–4914.
316. Xiao, S., Cao, W., Wang, B., Xu, L. and Chen, B. 2003. Mechanism and kinetics of oxidation during the thermal stabilization of polyacrylonitrile fibers. *J. Appl. Polym. Sci.* 127: 3198.
317. Rogers, J.H., Albus, E.T., Sprague, P.S. and Wimberger, R.J. 2000. Oxidation oven. U.S. Patent 6,027,337 (February 22).
318. Zhu, Y., Wilding, M.A. and Mukhopadhyay, S.K. 1996. Fibre-on-fibre abrasion in oxidised polyacrylonitrile. *J. Text. Inst.* 87(3):417.

319. Dietrich, J., Hirt, P. and Herlinger, H. 1996. Electron-beam-induced cyclisation to obtain C-fibre precursors from polyacrylonitrile homopolymers. *Eur. Polym. J.* 32:617.
320. Kikuma, J., Warwick, T., Shin, H.-J., Zhang, J. and Tonner, B.P. 1998. Chemical state analysis of heat-treated polyacrylonitrile fiber using soft X-ray spectromicroscopy. *J. Electron Spectrosc. Relat. Phenom.* 94:271.
321. Thunemann, A.F. 2000. Dielectric relaxation of polyacrylonitrile in its pristine and cyclized stage. *Macromolecules* 33:1790.
322. Wang, P., Hong, K. and Zhu, Q. 1996. The reduction property of thermally treated polyacrylonitrile fibres. *Polymer* 37(24):5533.
323. Nielsen, M., Jurasek, P., Hayashi, J. and Furimsky, E. 1995. Formation of toxic gases during pyrolysis of polyacrylonitrile and nylons. *J. Anal. Appl. Pyrolysis* 35:43.
324. Mittal, J., Konno, H., Inagakia, M. and Bahl, O.P. 1998. Denitrogenation behavior and tensile strength increase during carbonization of stabilized PAN fibers. *Carbon* 36(9):1327.
325. Pels, J.R., Kapteinjn, F., Moulijn, J.A., Zhu, Q. and Thomas, K.M. 1995. Evolution of nitrogen functionalities in carbonaceous materials during pyrolysis. *Carbon* 33(11):1641.
326. Jenkins, G.M. and Kawamura, K. 1976. *Polymeric Carbons–Carbon Fibre, Glass and Char*, pp. 52–82. Cambridge University Press, Cambridge.
327. Setnescu, R., Jipa, S., Setnescu, T., Kappel, W., Kobayashi, S. and Osawa, Z. 1999. IR and X-ray characterization of the ferromagnetic phase of pyrolysed polyacrylonitrile. *Carbon* 37:1.
328. Monthioux, M., Bahl, O.P., Mathur, R.B., Dhami, T.L., Dwivedi, H.O. and Sharma, S.P. 2000. Controlling PAN-based carbon fibre texture via surface energetics. *Carbon* 38:475.
329. Serkov, A.T. and Zlatoustova, L.A. 2000. Strength of a carbon fibre as a function of the physicomechanical properties of the initial polyacrylonitrile fibre. *Fibre Chem.* 32(4):279.
330. Huang, Y. and Young, R.J. 1995. Effect of fibre microstructure upon the modulus of PAN- and pitch-based carbon fibres. *Carbon* 33(2):97.
331. Edie, D.D. 1998. The effect of processing on the structure and properties of carbon fibers. *Carbon* 36(4):345.
332. Sung, M.G., Sassa, K., Tagawa, T. et al. 2002. Application of a high magnetic field in the carbonization process to increase the strength of carbon fibers. *Carbon* 40:2013.
333. Aviles, M.A., Sanchex-Soto, P.J., Justo, A. and Perez-Rodriguez, J.L. 1994. Compositional variation of Sialon phase produced after carbothermal reduction and nitridation of a vermiculite–polyacrylonitrile intercalation compound. *Mater. Res. Bull.* 29(10):1085.
334. Liu, Z.-T., Lu, D.-S. and Guo, Z.-Y. 1994. Polyacrylonitrile coated silica as support for copper catalyst in methanol dehydrogenation to methyl formate. *Appl. Catal., A* 118(2):163.
335. Haehne, H., Techen, H. and Woerner, J.-D. 1993. Eigenschaften von mit Polyacrylnitril-Fasern verstaerktem Beton (Properties of concrete reinforced with polyacrylonitrile fibers). *Beton- und Stahlbetonbau* 88(1):5.
336. Neuert, R. 2001. (Acordis Kehlheim GmbH), High-strength high-modulus polyacrylonitrile fibers, method for their production and use. U.S. Patent 6,228,966 (May 8).
337. Zhu, D., Xu, C., Nakura, N. and Matsuo, M. 2002. Study of carbon films from PAN/VGCF composites by gelation/crystallization from solution. *Carbon* 40:363.
338. Wang, P.H., Liu, J., Zho, J. and Xu, C.Y. 1997. Effect of heat treatment on surface properties of polyacrylonitrile-based activated carbon fibres. *J. Mater. Sci. Lett.* 16:187.
339. Carrott, P.J.M., Nabais, J.M.V., Ribeiro Carrott, M.M.L. and Pajares, J.A. 2001. Preparation of activated carbon fibres from acrylic textile fibres. *Carbon* 39:1543.
340. Laszlo, K., Tombacz, E. and Josepovits, K. 2001. Effect of activation on the surface chemistry of carbons from polymer precursors. *Carbon* 39:1217.
341. Severini, F., Formaro, L., Pegoraro, M. and Posca, L. 2002. Chemical modification of carbon fiber surfaces. *Carbon* 40:735.
342. Strobel, R., Jorissen, L., Schliermann, T. et al. 1999. Hydrogen adsorption on carbon materials. *J. Power Sources* 84:221.
343. Browning, D., Mepsted, C., Giwa, C., Green, K., Lakeman, B. and Barnes, P. 1999. Carbon: The key to power. *Chem. Ind.* 21:839.
344. Qiao, Z., Xie, Y., Xu, J., Zhu, Y. and Qian, Y. 2000. Synthesis of CdS/polyacrylonitrile nanocomposites by γ-irradiation. *Mater. Res. Bull.* 35:1355.
345. Lahr, B. and Sandler, J. 2000. Carbon nanotubes. *Kunststoffe, Plast Europe* 90(1):42.
346. Tomita, A. 2000. Preparation of unique carbons by using various templates. In *1st World Conference on Carbon, Eurocarbon 2000*, July 9–13, Vol. I, pp. 7–8, Berlin.

347. Li, Z. and Jaroniec, M. 2001. Silica gel-templated mesoporous carbons prepared from mesophase pitch and polyacrylonitrile. *Carbon* 39:2077.
348. Yan, J., Wu, J., Zhu, H., Zhang, X., Sun, D., Li, F. and Sun, M. 1995. Third-order nonlinear optical property of a heterocyclic ladder polymer. *Opt. Commun.* 116:425.
349. Pospisil, J., Samoc, M. and Zieba, J. 1998. Third-order nonlinear optical properties of a ladder polymer obtained by pyrolysis of polyacrylonitrile. *Eur. Polym. J.* 34:899.
350. Schlesinger, Y. and Mogilko, E. 1994. Characterization of superconducting YBCO/polyacrylonitrile composites. *Physics C* 232:37.
351. Huang, B., Wang, Z., Li, G., Huang, H., Xue, R., Chen, L. and Wang, F. 1996. Lithium ion conduction in polymer electrolytes based on PAN. *Solid State Ionics* 85:79.
352. Wu, Y., Fang, S. and Jiang, Y. 1998. Carbon anodes for a lithium secondary battery based on polyacrylonitrile. *J. Power Sources* 75(2):201.

5 Polyacrylates*

Joke Vandenbergh and Thomas Junkers

CONTENTS

5.1 Introduction 169
    5.1.1 Acrylics Industries 169
    5.1.2 Historical Milestones 170
5.2 Chemistry 170
    5.2.1 Monomers 170
        5.2.1.1 Synthesis 170
        5.2.1.2 Storage 172
    5.2.2 Polymerizations 173
        5.2.2.1 Free-Radical Polymerization 173
        5.2.2.2 Living Polymerizations 176
        5.2.2.3 Stereospecific Polymerization 177
5.3 Properties of Polyacrylates 178
    5.3.1 Thermal Degradation of Polymethacrylates 178
    5.3.2 Flame Retardancy of Acrylics 180
    5.3.3 Optical Properties 181
        5.3.3.1 Scattering Studies 181
    5.3.4 Mechanical Properties 182
5.4 Commercial Formulations and Processing 182
    5.4.1 PMMA and Other Polymethacrylates 182
        5.4.1.1 Types and Production of PMMA 182
        5.4.1.2 Applications of PMMA 184
        5.4.1.3 PMMA Blends 185
        5.4.1.4 Other Types of Polymethacrylates 187
    5.4.2 PAA and Its Polyacrylate Esters 187
References 188

5.1 INTRODUCTION

5.1.1 ACRYLICS INDUSTRIES

Acrylic materials have been studied for over 160 years and constitute (meth)acrylic acid and their ester derivatives, (meth)acrylates (see Figure 5.1). Today's acrylics industry can be divided in two distinct multibillion-dollar markets: on one hand, the polyacrylic acid (PAA) and its ester derivative polyacrylate (PAc) market and, on the other hand, the poly(methyl methacrylate) (PMMA) market. For many decades, PMMA was the predominant (meth)acrylic ester produced worldwide. A recent US market report predicts that the global PMMA market is expected to be worth US$9.7 billion by 2017, growing at a compound annual growth rate (CAGR) of 6.5% from 2012 till 2017.[1] However, due to growing demand for superabsorbent polymers and water treatment polymers in emerging countries, the acrylic acid (AA) market is even expected to grow to US$14 billion by 2018,[2] thereby overtaking the PMMA market. The superabsorbent polymer market (PAA) is expected to grow at a

---

* Based in parts on the first-edition chapter on polyacrylates.

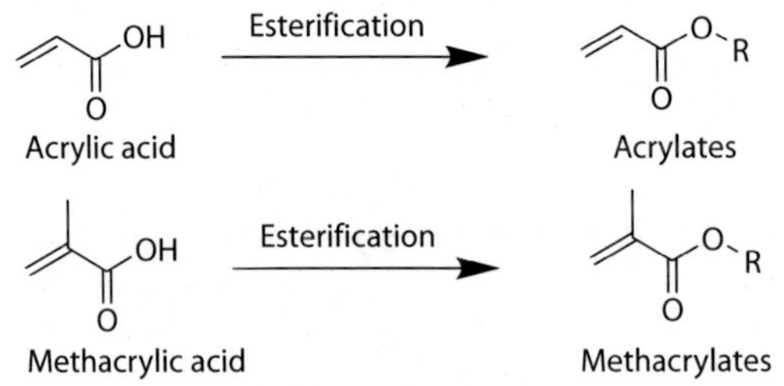

FIGURE 5.1 (Meth)acrylic acid and its (meth)acrylate derivatives.

CAGR of 5.2% from 2013 till 2018, while the acrylate markets, especially 2-ethylhexyl esters, are expected to grow at a CAGR of 4.6% from 2013 till 2018. Major players in the (P)MMA market are Mitsubishi Rayon (Japan), Arkema SA (France), LG MMA (South Korea), Chi Mei Corp. (Taiwan), Sumimoto Chemical Company Ltd. (Japan), Evonik Industries (Germany), BASF (Germany), Dow Chemical Company (USA), AkzoNobel (Netherlands), Quinn Plastics (UK), and Cytec Industries (USA). Regarding the PAA and Pac market, key manufacturers are Nippon Shokubai Company Ltd. (Japan), Arkema SA, Evonik Industries, BASF, and Dow Chemical Company.

5.1.2 Historical Milestones

The earliest reports on AA and the polymerization of this material date back from the end of the nineteenth century. As early as 1901, the German chemist Dr. Otto Röhm had finalized his PhD dissertation on the "polymerization products of acrylic acid." In 1907, he founded the Röhm & Haas Company, together with businessman Otto Haas. Although the commercial activities of the company initially were concentrated on the production of enzymes for leather treatment, a research team was founded in 1910 to study and develop the synthesis and polymerization of acrylic acid and, later, methacrylic acids (MAAs). In 1915, Röhm secured a patent on the use of polyacrylic esters as a binder component in industrial paints and lacquers. In 1918, after World War I, the German scientist Dr. Walter Bauer joined the research team of Röhm & Haas. Under joint management of Röhm and Bauer, the research team developed a commercial method for the polymerization of methyl methacrylate (MMA) in 1933. The resulting polymer, a transparent hard shatterproof acrylic glass, was introduced onto the market under the brand name Plexiglas. Due to its vitreous properties, lightweight, and increased strength compared to ordinary glass, it became widely used in the armament industry for the fabrication of aircraft cockpit windows. After World War II, new civilian applications were developed for PMMA, such as illuminated advertising, glass roofing, and façade design. Furthermore, acrylic resins found their way into dental applications and the paint and textile industry. Other companies such as ICI and DuPont also became involved in acrylic production, and a range of thermoplastic products was developed. Ever since, acrylic polymers have been widespread within innumerable everyday and specialty applications, still proving their strengths and benefits to date.

5.2 CHEMISTRY

5.2.1 Monomers

5.2.1.1 Synthesis

The most widely used industrial production method for AA consists of the catalytic oxidation of propene, a by-product stemming from oil refineries (see Figure 5.2).[3] In a first oxidation, propene is converted into acrolein, which, in a second step, is oxidized to form AA. In an alternative process,

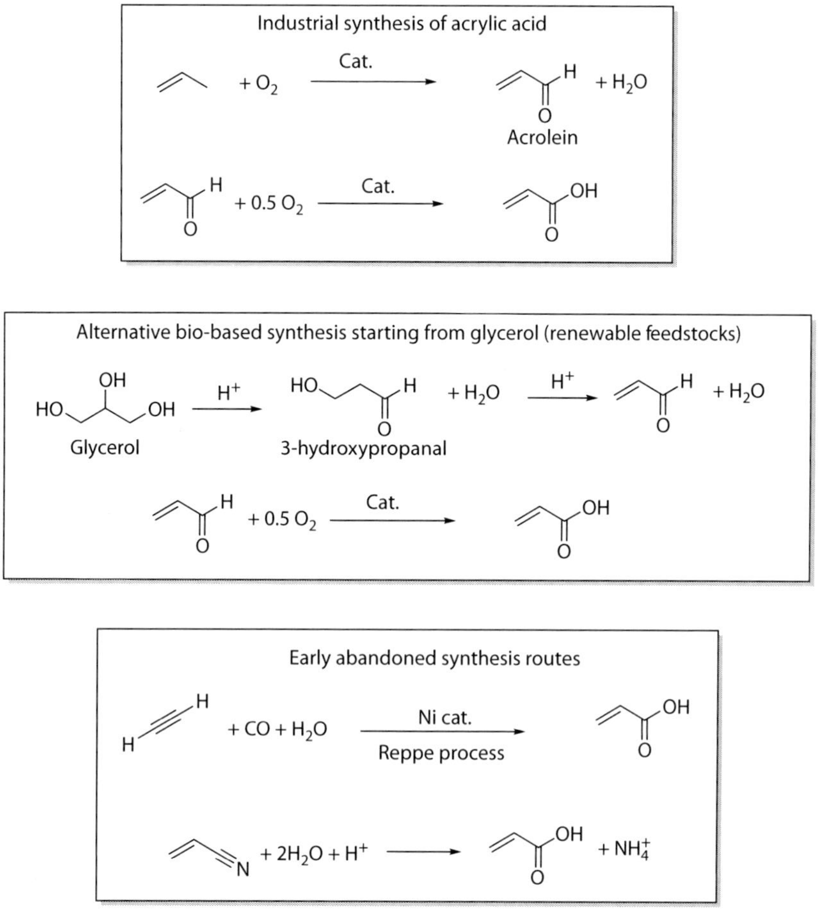

FIGURE 5.2 Synthesis routes toward acrylic acid.

developed by the Dow Chemical Company and OPX Biotechnologies Inc., AA is produced from renewable feedstocks by dehydrating glycerol into 3-hydroxypropanal and further dehydration into acrolein. Finally, the obtained acrolein is again oxidized into AA.

Some older synthesis methods are, for example, the hydrolysis of acrylonitrile; however, this method also generates ammonium by-products and was therefore abandoned. Another early synthesis method consists of the hydrocarboxylation of acetylene by a Reppe process.[4] However, due to the need for poisonous nickel carbonyl catalysts and high pressures of carbon monoxide, nowadays, this route also is almost never applied.

For the synthesis of MAA, a number of industrial processes have been developed (Figure 5.3).[5] The most widely applied process is a C_4-based route whereby acetone cyanohydrin is converted into methacrylamide sulfate by using excesses of sulfuric acid. In a second step, the methacrylamide sulfate is hydrolyzed into MAA or methanolyzed into MMA directly. This procedure requires huge amounts of sulfuric acid, and ammonium sulfate is formed as a by-product, which, in turn, creates large waste streams that require expensive recovery and recycle plants.

Alternative C_4 routes consist of the catalytic oxidation of isobutene or *t*-butanol into methacrolein and subsequent oxidation into MAA. Furthermore, a C_2-based route is developed involving the hydroformylation of ethylene into propanaldehyde followed by condensation with formaldehyde to give methacrolein that, in a final step, is oxidized into MAA. In a final C_3

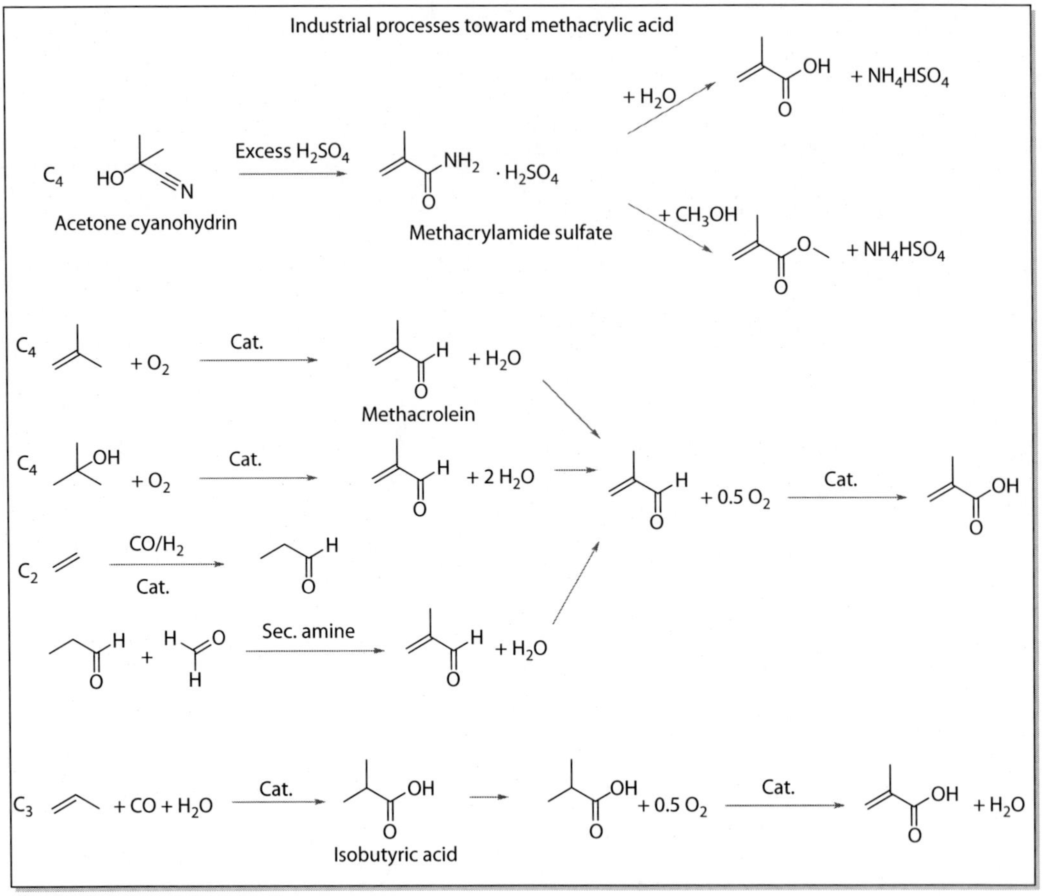

FIGURE 5.3 Industrial synthesis routes toward MAA.

route, propene is reacted with carbon monoxide, forming isobutyric acid, which is oxidatively dehydrogenated into MAA. In general, the choice for a certain synthesis route is based on the economics of production, which are partly driven by recovery costs and market prices for intermediates and by-products.

Both AA and MAA can undergo esterification reactions by reacting with an appropriate alcohol to form corresponding acrylates or (meth)acrylates. Furthermore, higher methacrylates can be synthesized by transesterification of MMA.

5.2.1.2 Storage

Since acrylic monomers are susceptible to premature polymerization, they are transported and stored with a stabilizer that inhibits polymerization. Mostly, hydroquinone monomethyl ether is used in a concentration range of 10–100 ppm. The stabilizer is only effective in the presence of low amounts of oxygen; therefore, flasks and drums are only filled 97% with an acrylic monomer, allowing adequate oxygenation of the inhibitor. In order to increase the shelf life of both the monomer and the inhibitor, acrylic monomers have to be stored in a cool place. AA and MAA preferentially have to be stored between 15°C and 25°C in order to prevent freezing of the monomers. (Meth)acrylates, on the other hand, can safely be stored in a freezer. Furthermore, the monomers have to be shielded from light by storing them in a dark place.

5.2.2 Polymerizations

5.2.2.1 Free-Radical Polymerization

5.2.2.1.1 Kinetic Overview

The most important mode of polymerizations toward (meth)acrylates is free-radical polymerization (FRP). Thereby, reactions are usually started by initiators that decompose into radicals upon thermal or photochemical activation. These radicals then add to a first monomer unit, after which the generated radical undergoes chain propagation and termination reactions. The rate of polymerization is thus dependent on the rate of initiation, propagation, and termination, following Equation 5.1:

$$R_P = \frac{k_p \cdot (2f \cdot k_d)^{0.5}}{k_t^{0.5}} \cdot c_I^{0.5} \cdot c_M \tag{5.1}$$

where R_p stands for the overall rate of polymerization, k_p the propagation rate coefficient, f the initiator efficiency, k_d the initiator decomposition rate coefficient, k_t the termination rate coefficient, c_I the initiator concentration, and c_M the monomer concentration. The average degree of polymerization $\overline{DP_n}$ is likewise dependent on the same parameters found in Equation 5.2:

$$\overline{DP_n} = \frac{k_p}{(2f \cdot k_d)^{0.5} \cdot k_t^{0.5}} \cdot c_I^{-0.5} \cdot c_M \tag{5.2}$$

The faster a monomer propagates, the faster is the polymerization, and the higher is the molecular weight of the residual polymer under otherwise constant conditions. Higher termination rate coefficients result in the opposite: slower polymerizations and lower molecular weights. The nature of the termination reaction is complex due to the diffusion control of the reaction. k_t is hence a function of temperature, pressure, and the choice of solvent, but also of bulk viscosity, chain length, and polymer concentration. Even though advances have been made to resolve the single dependencies, usually, average termination rates are used for modeling and simple predictions. k_t for methacrylates and acrylates is typically within an order of magnitude to 10^8 L mol^{-1} s^{-1} and decreases with sterical hindrance for dilute solutions. Due to viscosity increases during polymerization, the rate of termination may change by several orders of magnitude over the period of reaction (see Section 5.2.2.1.3). k_p shows distinct family behavior and is significantly higher for acrylates when compared to the methacrylate counterpart. Also, when increasing the size of the ester, propagation rates increase significantly, which may be explained by shielding of the polar ester group. Typical values and activation energies for a selection of acrylates as benchmarked and recommended by the International Union of Pure and Applied Chemistry (IUPAC) are given in Table 5.1.

5.2.2.1.2 Backbiting

Methacrylates are well described by a simple radical polymerization mechanism. Next to initiation, propagation, and termination, the only significant side reactions are weak transfer-to-monomer or transfer-to-solvent reactions. Acrylates, conversely, exhibit a further, very significant side reaction, namely, transfer to polymer.[11] In transfer-to-polymer reactions, the growing chain-end macroradical transfers its radical center in an H-shift reaction to a remote location on its own or any surrounding backbone. The high driving force for this reaction stems from the added radical stability that results from that process. While growing acrylate macroradicals are secondary in nature, the transfer products are tertiary radicals. Transfer to polymer can be subdivided into three distinct reaction pathways, which are random intermolecular transfer, random intramolecular transfer, and nonrandom intramolecular transfer. In the nonrandom route, often referred to as *backbiting*, the H-shift

TABLE 5.1
Propagation Rate Coefficients at 25°C and Their Corresponding Arrhenius Parameters for Selected (Meth)acrylates as Benchmarked by the IUPAC

Monomer	k_p (25°C), L mol^{-1} s^{-1}	A, L mol^{-1} s^{-1}	EA, kJ/mol
Methyl acrylate[6]	13,100	1.41×10^7	17.3
Butyl acrylate[7]	16,400	2.24×10^7	17.9
Methyl methacrylate[8]	323	2.67×10^6	22.4
Ethyl methacrylate[9]	323	4.07×10^6	23.4
Butyl methacrylate[7]	370	3.80×10^6	22.9
Dodecyl methacrylate[7]	525	2.51×10^6	21.0
Cyclohexyl methacrylate[10]	589	6.31×10^6	23.0
Glycidil methacrylate[8]	600	6.17×10^6	22.9
Benzyl methacyrylate[8]	657	6.76×10^6	22.9
Isobornyl methacrylate[8]	553	6.17×10^6	23.1

reaction occurs via a six-membered transition ring structure (see Figure 5.4). In this way, the radical can travel along the backbone of the chain but is mostly found in close proximity to the chain end.[12]

Backbiting and random transfer events are very significant reactions since they lower the reactivity of the macroradicals and hence lead to a reduced polymerization rate. At the same time, they also influence the structure of the polymer. Backbiting results in short-chain branches in the polymer after addition and further growth of the tertiary midchain radicals; random transfer events likewise result in long-chain branching. Backbiting is a particularly fast reaction, whereby addition of monomers to the transfer product is slow (in the order of magnitude of the corresponding methacrylate propagation rate coefficients). Typically, in an acrylate polymerization at 60°C, up to 80% of the radicals in the polymerization are of tertiary nature.[13]

While backbiting (and other transfer events) thus have an influence on branching levels of the product as well as on the overall polymerization rate (midchain radicals are more persistent compared to chain-end radicals and hence slow down overall propagation rates), also, secondary reactions can occur at elevated temperatures (usually noticeable from 80°C upward). Midchain radicals can undergo β-scission reactions, which lead to chain scission and formation of unsaturated polymer species. If midchain radical concentration is high enough (which is certainly the case in high-temperature polymerization), formation of these unsaturated species can become the main reaction pathway. In such cases, the unsaturated polymers can act as macromonomers and take part in chain addition, effectively establishing an addition–fragmentation chain-transfer equilibrium (see also Section 5.2.2.2.3). The consequence of such a high-temperature kinetic scenario is that products are mostly of low molecular weight and feature unsaturated end groups.

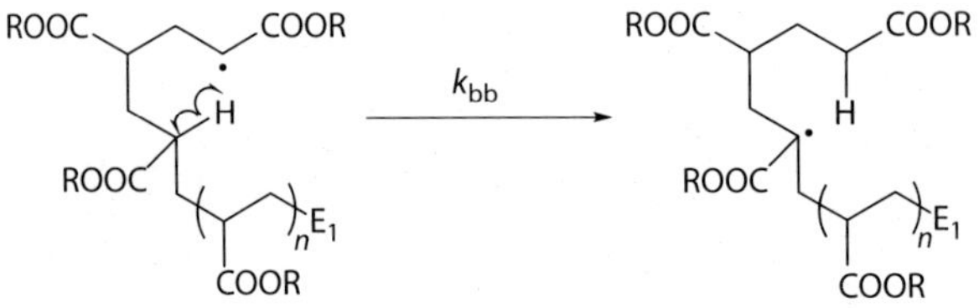

FIGURE 5.4 Backbiting reaction via a six-membered ring transition structure during acrylate polymerization.

Polyacrylates

5.2.2.1.3 Practical Aspects

The diffusion-control kinetics of the termination reaction has major ramifications for medium- to high-conversion polymerization. The polymerization undergoes an autoacceleration as the viscosity of the polymerizing medium increases, as termination becomes inhibited by the lowered diffusion of radicals. This phenomenon is known as the Trommsdorff effect (alternatively the Norrish effect or the gel effect). The Trommsdorff effect can cause problems when polymerizing in bulk as heat removal becomes problematic and can cause not only deleterious effects on the polymer properties but also a considerable safety hazard. The heat of polymerization for a number of acrylate monomers is given in Table 5.2.[14] The Trommsdorff effect is more pronounced for methacrylates and typically plays only a minor role in acrylate polymerization. One method of quantifying the strength of the Trommsdorff effect is to use the gel effect index, γ, which is given by

$$\gamma = R_p/R_{p0} - 1 \tag{5.3}$$

where R_p and R_{p0} are the rates of polymerization in the absence (theoretical only) and presence of the Trommsdorff effect.[15] For MMA, the gel effect starts to take place at 14% conversion (average degree of polymerization = 3580), and the γ value has a maximum at 73% conversion.[16] These specific numbers will vary with molecular weight.

Another important aspect of the polymerization process is the shrinkage in volume that occurs on the transformation from monomer to polymer. For the bulk polymerization of MMA, the shrinkage is given by

$$v = v_M(1 + \varepsilon x) \tag{5.4}$$

where v_M and v denote the specific volumes of the pure monomer and reaction mixture, respectively; x is the degree of conversion; and ε is the contraction constant. The contraction constant for MMA is given by[17]

$$-e = 0.232911 + 0.000516T \ (T \text{ in } °C) \tag{5.5}$$

and the specific volume of MMA is given by

$$V_m = 1.025934 \times 0.001494T \text{ cm}^3/\text{g} \ (T \text{ in } °C) \tag{5.6}$$

TABLE 5.2
Heat of Polymerization for Some Methacrylate Monomers

Monomer	Heat of Polymerization (kJ/mol)
Methacrylic acid	56.5
Methyl methacrylate	57.7
Ethyl methacrylate	57.7
n-Butyl methacrylate	56.5
2-Hydroxyethyl methacrylate	49.8
2-Hydroxypropyl methacrylate	50.6

Source: Data from Nemec, J. W. and Kirch, L. S., Methacrylic acid and derivatives, *Encyclopaedia of Polymer Science and Engineering* (H. F. Mark, ed.), Wiley-Interscience, New York, 1985.

Thus, the shrinkage at 60°C for fully polymerized MMA is approximately 25.6%. The shrinkage decreases with increasing size of the ester side chain.

5.2.2.2 Living Polymerizations

Although free-radical polymerization is a very robust technique to generate high-molar-mass bulk polymers, the method only allows for limited control over the molecular weight, dispersity, and end groups of the polymers formed. Since certain specialty polyacrylic applications require control over these properties and, sometimes, complex polymer architectures such as block copolymers are desired, a number of living polymerization techniques have been developed. In a living polymerization, the polymer chain ends remain active, even after all monomers are consumed.

5.2.2.2.1 Anionic Polymerization

Anionic polymerization of acrylic monomers employs organometallic initiators based on lithium or higher alkali metals to control the polymerization.[18] Often, very low temperatures and specific solvents are required in order to obtain a living polymerization process. Most reports on anionic polymerization of (meth)acrylics were published in the early 1990s. However, since then, the anionic polymerization technique has been encompassed by the development of the more versatile controlled radical polymerization (CRP) techniques, and while still in use, anionic polymerization has become a niche technique for synthesis of very specialized materials.

5.2.2.2.2 Group Transfer Polymerization

A special type of ionic polymerization, named group transfer polymerization (GTP) was developed by DuPont in the 1980s as a specific method for the polymerization of acrylic monomers.[19] In GTP, an organosilicon compound is used to initiate polymerization in solution in the presence of a nucleophile or Lewis acid catalyst. GTP is operative at ambient temperatures and allows tight control over molecular weight. Nevertheless, GTP is only used to a minor extent in industrial processes, due to the development of the CRP techniques in the mid-1990s. However, a very recent review reports on the improvement of GTP in many aspects due to the introduction of newly developed initiators and Lewis base/acid organocatalysts such as N-heterocyclic carbenes.[20] Due to these recent developments, GTP retains its potential as an acrylic polymerization technique.

5.2.2.2.3 Controlled Radical Polymerization

In the mid-1990s, a number of CRP techniques were developed, which allowed control over molecular weight, dispersity, end groups, and molecular architecture (e.g., block, star, comb, graft copolymers). Among these techniques, the most important methods are atom transfer radical polymerization (ATRP),[21,22] nitroxide-mediated polymerization (NMP),[23,24] and reversible addition–fragmentation chain-transfer polymerization (RAFT).[25] Furthermore, catalytic chain-transfer polymerization (CCTP)[26] and cobalt-mediated radical polymerization (CMRP)[27] are valuable techniques for acrylic polymerizations that also have to be mentioned. For more elaborate information on the specifics and differences between these CRP techniques, the reader is referred to the papers mentioned in the reference section.

ATRP is based on a catalytic process using a transition-metal complex (mostly copper), stabilized by specific ligands and initiated by specific initiators.[28,29] ATRP is suitable for polymerization of acrylates and methacrylates. The technique is, in most cases, not suited for polymerization of AA and MAA, due to the interference between the monomers' acid group and the ligand applied to stabilize the active copper complex. However, poly(meth)acrylic acid can be obtained by first polymerizing *tert*-butyl (meth)acrylate (or other bulky esters) by ATRP, followed by hydrolysis of the tertiary ester groups.

NMP is a method that makes use of an alkoxyamine initiator, which is thermally initiated to mediate the polymerization reaction. No metal catalyst is applied, which makes NMP a very simple and straightforward method to produce living polymers.[30] In the case of acrylic monomers,

TABLE 5.3
Overview of Which CRP Methods Are Suitable to Produce Certain Acrylic Monomers

	Acrylic Acid	Methacrylic Acid	Acrylates	Methacrylates
ATRP	No	No	Yes	Yes
NMP	Yes	No	Yes	No
RAFT	Yes	Yes	Yes	Yes
CCTP	No	Yes	No	Yes
CMRP	Yes	No	Yes	No

sometimes, a small amount of free nitroxide radicals is added to the alkoxyamine to improve control. NMP is suitable for both AA and acrylates but unsuitable for MAA and methacrylates since these monomers exhibit too-high activation/deactivation equilibrium constants and are generally prone to disproportionation reactions. Controlled polymerization of MMA can only be achieved by applying an SG1-based alkoxyamine initiator with a high dissociation rate constant and by adding a small amount of styrene (10%) as a comonomer to mediate the reaction.

While ATRP and NMP are based on a persistent radical effect to induce livingness, RAFT uses a degenerative chain-transfer mechanism to establish control. To this end, specific thiocarbonylthio compounds (RAFT agents) have been developed that are able to mediate polymerization. Nowadays, mostly trithiocarbonate-based and dithiobenzoate-based RAFT agents are employed. For each monomer type to be polymerized, some specific RAFT agents are accessible that provide the best control. As such, RAFT polymerization is the most versatile route toward monomer compatibility; all acrylic monomers can in principle be synthesized via RAFT.[31,32]

CCTP[33] and CMRP[34] are both CRP techniques that make use of cobalt complexes to mediate polymerization, where CCTP belongs to nonliving techniques. The difference between the two techniques lies in the nature of control. CCTP is based on a catalytic chain-transfer process and only works with monomers that have an α-methyl group with hydrogens prone to abstraction by the cobalt complex. As such, CCTP is suited for MAA and methacrylates. CMRP, on the other hand, is based on a reversible termination via the persistent radical effect and is only suited for monomers that are not prone to hydrogen abstraction, in this case, AA and acrylates.

In general, the different CRP techniques have established a variety of synthetic possibilities toward the design of novel polyacrylic materials with unprecedented properties. In Table 5.3, an overview is given on which CRP methods are most suited to polymerize each type of acrylic monomer. Finally, it must be mentioned that to date, only a limited number of commercially available products are yet derived from CRP, but numbers of products are strongly increasing in recent years due to the nowadays increased maturity of the methodologies.[35] Regardless, CRP can definitely be considered as one of the most revolutionary methods toward future next-generation specialty polymer materials.

5.2.2.3 Stereospecific Polymerization

Radical processes do not directly allow control of the stereospecificity of the resulting polymers, and materials made by FRP are, in the majority, atactic. Yet, by choosing suitable reaction conditions, influence on the tacticity can be gained, especially via living polymerization methods. Hatada and coworkers published extensively on the synthesis of stereospecific methacrylates using anionic initiators.[36–38] The T_g of PMMA increases with both the syndiotacticity of the polymer and increasing molecular weight. Hatada et al.[37] estimate the T_g of 100% s-PMMA to be 141°C at infinite molecular weight. Ute et al. found that over the molecular weight range of 13-mer to 50-mer, plots of T_g versus reciprocal molecular weight for stereospecific PMMA were linear.[39] Stereospecific poly(hydroxyethyl methacrylate) (PHEMA) cannot be synthesized in the same way, because of the hydroxyl functionality, which reacts with nucleophilic centers; therefore s-PHEMA

(84% s-triad tacticity) was synthesized by the photopolymerization of hydroxyethyl methacrylate (HEMA) in methanol or ethanol at −40°C with a conventional azo initiator, azobis(methyl isobutyrate) (AMIB).[40] Highly isotactic PHEMA has been synthesized by blocking the hydroxyl group by forming the benzoate ester. An anionic polymerization was then initiated using a dialkyl copper lithium catalyst at −10°C. The resulting poly(benzoxyethyl methacrylate) was hydrolyzed to yield i-PHEMA (80% i-triad tacticity). α-Substituted acrylates have also successfully been synthesized with stereoregularity using lithiated anionic initiators.[41] In the same study, it was shown that the Li$^+$ ion is responsible for the stereocontrol. Addition of lithium salts to radical polymerizations did not induce stereospecificity, yet zinc salts have been shown to have a distinct effect.

Other pathways toward stereospecific polymerization of (meth)acrylates are given via metallocene-mediated polymerization, whereby tacticity is controlled by the choice of the organometallic ligands employed. Isotactic and syndiotactic polymers can be achieved via this route.[42,43]

5.3 PROPERTIES OF POLYACRYLATES

A range of PMMA resins are produced by the primary suppliers, with the PMMA product in the form of sheet product (cast or extruded) or pellets/chips for injection molding. The difference between the grades will mainly center around color, hardness, and melt flow. Some typical properties of an acrylic resin produced for injection molding applications are given in Table 5.4.[44] Some further physical property data for acrylic homopolymers are reported in Tables 5.5 and 5.6.[45,46]

5.3.1 Thermal Degradation of Polymethacrylates

Polymethacrylates have the potential for degradation via a nonradical unzipping mechanism at elevated temperatures and under ultraviolet (UV) irradiation.[47] This can be a disadvantage when trying to maintain a high thermal stability of the materials in applications such as coatings. As a result, MMA is often copolymerized with a small amount of methyl acrylate, which acts as a chain

TABLE 5.4
Physical Properties of Various Unmodified PMMA Resins

Property	Flow Grade[a]			Test Method
	Grade 5	Grade 6	Grade 8	
Heat deflection temp. under load, annealed at 1.82 MPa, °C (°F)	74 (165)	80 (176)	90–102 (194–216)	ASTM D-648
Continuous use temp., °C (°F)	60–74 (140–165)	63–80 (145–176)	71–102 (160–216)	No load
Melt flow rate, g/10 min	24	17	2–8	ASTM D-1238 condition 1
Refractive index number	1.49	1.49	1.49	ASTM D-542
Specific gravity	1.18	1.18	1.19	ASTM D-792
Light transmission, %	92	92	92	ASTM D-1003
Tensile strength, MPa	61	66	70	ASTM D-638
Flexural modulus, GPa	3.036	3.036	3.105	ASTM D-790
Impact strength, J/m	12.3	12.3	12.3	ASTM D-256 notched Izod
Rockwell hardness, M scale	84	89	97	ASTM D-785
Water absorption, %	0.3	0.3	0.3	ASTM D-570 wt. gain, 24 h

Source: Data from Cassidy, R. T., Acrylics, *Engineering Materials Handbook*, Vol. 2 (C. A. Dostal, ed.), ASM International, Materials Park, OH, 1988, p. 103.

[a] Flow grade per ASTM D-788.

TABLE 5.5
Physical Properties of Members of the Polymethacrylate Family

Homopolymer	Abbreviation	CAS Registry Number	T_g, °C	Density at 20°C, g/cm^3	Solubility Parameter, (J/cm^3)$^{1/2}$	Refractive Index, (n_D^{30})
Poly(methyl methacrylate)	PMMA	[9011-14-7]	105	1.190	18.6	1.490
Poly(ethyl methacrylate)	PEMA	[9003-42-3]	65	1.119	18.3	1.485
Poly(n-propyl methacrylate)	PnPrMA	[25609-74-9]	35	1.085	18.0	1.484
Poly(isopropyl methacrylate)	PiPrMA	[26655-94-7]	81	1.033	—	1.552
Poly(n-butyl methacrylate)	PnBMA	[9003-63-8]	20	1.055	17.8	1.483
Poly(secbutyl methacrylate)	PsBMA	[29356-88-5]	60	1.052	—	1.480
Poly(isobutyl methacrylate)	PiBMA	[9011-15-8]	53	1.045	16.8	1.477
Poly(t-butyl methacrylate)	PtBMA	[25213-39-2]	107	1.022	17.0	1.4638
Poly(n-hexyl methacrylate)	PHMA	[25087-17-6]	−5	1.007[25]	17.6	1.4813
Poly(lauryl methacrylate)	PLMA	[25719-52-2]	−65	0.929[25]	16.8	1.474
Poly(2-hydroxyethyl methacrylate)	PHEMA	[25249-16-5]	90–101	1.274	27.5	1.5119

Source: Based on data in Stickler, M. and Rhein, T., Polymethacrylates, *Ullmann's Encyclopaedia of Industrial Chemistry*, Vol. A21, Wiley-VCH, Weinheim, Germany, 1992.

TABLE 5.6
Physical Properties of Members of the Polyacrylate Family

Homopolymer	Abbreviation	CAS Registry Number	T_g, °C	Density at 25°C, g/cm³	Refractive Index $\left(n_D^{25}\right)$
Poly(methyl acrylate)	PMA	[9003-21-8]	22	1.20	1.479
Poly(ethyl acrylate)	PEA	[9003-32-1]	−8	1.13	1.464
Poly(n-propyl acrylate)	PnPrA	[24979-82-6]	−25	1.10	1.462
Poly(n-butyl acrylate)	PnBA	[9003-49-0]	−43	1.06	1.474
Poly(secbutyl acrylate)	PsBA	[30347-35-4]	−6	1.06	1.462
Poly(isobutyl acrylate)	PiBA	[26335-74-0]	−17	1.05	1.464
Poly(t-butyl acrylate)	PtBA	[25232-27-3]	55	1.03	1.468
Poly(n-hexyl acrylate)	PHA	[27103-47-5]	−51	0.98	1.468
Poly(lauryl acrylate)	PLA	[26246-92-4]	−17	0.99	1.468
Poly(2-ethyl hexyl acrylate)	PEHA	[9003-77-4]	−58	0.99	1.433

Source: Based on data in Penzel, E., Polyacrylates, *Ullmann's Encyclopaedia of Industrial Chemistry*, Vol. A21, Wiley-VCH, Weinheim, Germany, 1992, p. 157.

stopper preventing severe degradation.[48] Several authors have shown that the nature of the chain end in PMMA has a significant effect on the onset of degradation. Especially, the presence of a vinyl end group (stemming from termination by disproportionation) may be held responsible for the onset of degradation. Saturated end-groups and head-to-head termination products from bimolecular termination also may cause unzipping of chains, even if slower.

Yet, it appears that careful selection of polymerization conditions is essential for controlling the termination chemistry, and terminating a large number of chains via a chain-transfer mechanism might enhance thermal stability of the thermoplastic.

Degradation via unzipping can also be exploited in recycling acrylic monomers from waste thermoplastic product. There are a number of processes that have potential to recover the monomer; melting vessels and heated screw extruders do not generally yield sufficient monomer (50–80% yield). A molten lead bath can be utilized, but the associated environmental concerns with lead provide a strong disincentive for this process. Kaminsky and Franck suggested the use of an indirectly heated fluidized bed process at a temperature of 450°C.[49] They report a monomer recovery level of 97%, with the exact yield dependent on the pyrolysis temperature. The main component in the liquid besides MMA is methyl acrylate.

5.3.2 Flame Retardancy of Acrylics

One of the primary limitations to the use of acrylics as materials is their fire performance. Several approaches can be taken to improve matters. If clarity is not important, then fillers such as alumina trihydrate (ATH) can be used. In the event of fire, ATH undergoes an endothermic reaction, producing water, which retards the spread of fire. An alternative approach is to encourage char formation in the polymer via the use of phosphorus reagents. Clouet et al. have conducted a number of studies on phosphorus-containing acrylates and methacrylates located at chain ends, as pendant groups and as blocks in the middle of polymer chains.[50,51] The limiting oxygen index (LOI) values of the polymers seemed to show that the flame-retardant efficiency of the phosphorus was dependent on both the nature of the phosphorus compound and its position in the polymer chain. Polymers with phosphorus at chain ends and in the middle of the chain exhibited better flame-retardant properties.

TABLE 5.7
Optical Properties of PMMA

Refractive Index Parameter	Values
n_D^{20} (589 nm)	1.491
n_C^{20} (656.3 nm)	1.488
n_F^{20} (486.1 nm)	1.496
Abbe value	61.4
Rate of change in index with temperature, $dn/dt \times 10^{-5}/°C$	−12.5
Luminous transmittance at 3.175 mm, %	92

Source: Data from Wolpert, H. D., Optical properties, *Engineering Materials Handbook*, Vol. 2 (C. A. Dostal, ed.), ASM International, Materials Park, OH, 1988, p. 481.

5.3.3 OPTICAL PROPERTIES

PMMA is one of the most widely used optical plastics. Some of the optical properties of PMMA are listed in Table 5.7. The popularity of PMMA stems from its ease of molding, ease of fabrication, and high optical transmission (92%). Water absorption has some effect on optical properties as samples immersed in water for 10 days show a 0.2% increase in linear dimensions. Of the 8% transmission loss, 7.5% is due to surface reflection; therefore, antireflection coatings are often used where transmission is critical. Magnesium fluoride with an index of refraction of 1.38 in thin films is widely used for a single quarter-wavelength coating on acrylic. A single element with two surfaces, when coated, will have reflection losses reduced from 8% to 3% at 460 nm, 2.5% at 530 nm, and 3% at 600 nm.[52] PMMA core optical fibers are now in general use despite the drawback that their attenuations are usually in the region of 100 dB/km, which makes them inadequate for long-range links. In theory, it is possible to improve the performance of PMMA, as shown by Koike et al., who prepared PMMA where the scattering loss was reduced to 13 dB/km.[53] They found that the isotropic scattering loss varied between 9 and 400 dB/km depending on the conditions of polymerization or heat treatment, whereas the anisotropic scattering loss was almost unchanged in the range of 4–6 dB/km. The origin of the scattering losses was ascribed to isotropic strain inhomogeneities caused by shrinkage and the polymerization exotherm during polymerization, thus indicating that careful control over the Trommsdorff effect is crucial when preparing PMMA for optical applications. The use of doped PMMA for nonlinear optical applications has been described in the literature.[54] The dopant is often an azobenzene-containing moiety that can be simply trapped in the PMMA matrix, or methacrylates containing azobenzene side groups can be used as comonomers.[55] High-refractive-index acrylic polymers can be synthesized via copolymerization with carbazole-containing acrylates and methacrylates.[56]

5.3.3.1 Scattering Studies

Lovell and Windle reported an extensive wide angle x-ray scattering (WAXS) study on atactic and highly syndiotactic PMMA.[57] This work indicated that the conformation of PMMA is effectively all-trans regardless of the stereochemistry of the chains. Precise values of backbone rotation angles (10°, 10°, −10°, −10°) and bond angles (110° and 128°) indicate that the chain consists of curved segments, with a regular conformation persisting over 16 to 20 backbone bonds. Further, it was suggested that this persistent curvature is the reason why syndiotactic PMMA (in the absence of solvent) cannot be induced to crystallize despite the stereoregularity of the individual units.

WAXS has also been applied to PHEMA.[58] The polymeric PHEMA chains adopt a similar conformation to PMMA. Like PMMA, PHEMA has a curved backbone that is thought to result from a relief of severe steric crowding between the methyl and acrylate groups. Similar to PMMA, this curvature is suggested to be responsible for the prevention of crystallization even in relatively stereoregular PHEMA.

Light-scattering studies on glassy PMMA found that the polymer scattered excess light beyond that expected for a typical supercooled liquid.[59] They attributed this to quasicrystalline regions, frozen strains, low-molecular-weight species, or foreign contaminants such as dust. The heterogeneities causing the excess scattering were in the order of 1000 Å or greater.

5.3.4 Mechanical Properties

For a full description of the fracture behavior of glassy thermoplastic polymers such as PMMA, the reader is referred to Kinloch and Young.[60] Bowden conducted a study on shear yielding and has shown the existence of broad, diffuse shear bands for PMMA samples under compression.[61] Narisawa et al. studied notched bars of PMMA subjected to three-point bending over a range of low-strain rates from 10^{-2} to $10^{-4}\,\text{s}^{-1}$;[62] all tests were conducted under plane-strain conditions. They found that internal crazes initiated at the intersection of shear bands. According to Kinloch and Young, the fracture mechanism that emerges is as follows: (1) strain softening in plane-strain deformation causing strain inhomogeneities, (2) intersection of shear bands causing a stress concentration, (3) an internal craze initiating at the maximum stress concentration, and (4) craze breakdown and crack initiation and growth.

The nature of crazing in PMMA has been deduced from optical interference patterns[63] and has been compared with the predicted craze shape from the Dugdale model,[60] which assumes a craze surface stress, σ_{cs}, of 80 MPa. The susceptibility of PMMA to fracture at relatively low stresses via a crazing mechanism upon exposure to a variety of liquid environments has been reported by Graham et al.[64] All of the liquids, especially alcohols, caused a drastic decrease in the value of σ_{cs} compared to the value in air. The value of the crack-opening displacement at the crack tip in crack growth, δ_{tc}, has been shown to be dependent on molecular weight; the value of δ_{tc} at 20°C increased from 1.2 to 2.7 μm as M_w increased from 1.1×10^5 to 9×10^6 g/mol. Some aspects of product design need to take account of the material properties of acrylics. It should be noted that acrylics exhibit room temperature creep, and while the initial strength of acrylics is high, long-term stress loading can result in stress crazing. Consequently, design stresses must be limited to 10.35 MPa. The high-impact grades have good creep resistance under moderate stress levels. Continuous stress loadings at room temperature should be limited to 3.45 MPa.[44]

5.4 COMMERCIAL FORMULATIONS AND PROCESSING

5.4.1 PMMA and Other Polymethacrylates

5.4.1.1 Types and Production of PMMA[65,66]

A general overview of the different industrial processing methods toward PMMA is depicted in Figure 5.5.

5.4.1.1.1 Cast Sheet

Cast sheet PMMA is produced directly from MMA monomer syrup by two methods: cell cast or continuous cast. Cell casting is a batch process where an MMA syrup is introduced between two sheets of high-quality glass, which are sealed. The MMA syrup inside the sealed glass cells is then polymerized in specially developed hot-air ovens or water baths. The residence time can vary from several hours to days depending on the thickness of the sheet. A postpolymerization process at higher temperatures (110–120°C above the T_g of PMMA) is used to complete the final 10–15% of conversion. This process is fairly time- and labor-intensive but produces a very high-quality mirror-like surface finish. In addition to the batch process of cell casting, a PMMA cast sheet can also be

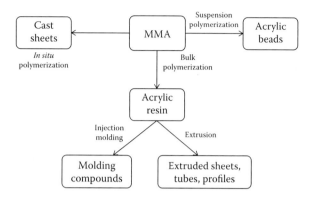

FIGURE 5.5 Industrial processing formulations toward PMMA.

produced by a continuous process whereby the MMA syrup is polymerized between two highly polished stainless steel bands.

Cast sheet PMMA (better known under the names Plexiglas, Perspex, Lucite, etc.) has a very high molecular weight. Therefore, it is rigid; craze resistant; and easy to handle, cut, and shape. Moreover, cast sheet PMMA has excellent optical properties.[67] The optical-grade cast sheet PMMA is crystal clear and exhibits 92% light transmission, which is even higher than normal glass. Cast sheets are obtainable in a wide range of sizes, thicknesses, and colors.

5.4.1.1.2 PMMA Resin

PMMA resin is mainly produced from polymerizing MMA in bulk or in solution, employing a series of continuous-flow stirred-tank reactors (CFSTRs). For continuous bulk polymerization, reaction temperatures between 130°C and 170°C are applied, and the monomer slurry is polymerized till conversion ranges of 40–80%. On the other hand, continuous solution polymerization is carried out at lower temperatures around 70–80°C, and after polymerization in CFSTR, the residual solvent is evaporated in a devolatilizer. The crude PMMA resin can then be processed via extrusion and/or injection molding.[68] In an extrusion process, the PMMA pellets are melted and conveyed in a long horizontal chamber containing one or more rotating screws. The shear force of the screws in combination with the heated chamber melts the resin, which, at the end of the extruder, is transported through a special formed die, which shapes the plastic in its final form. By this technique, various PMMA tubes, rods, profiles, and sheets can be produced. However, some PMMA applications require a very specific shape, which cannot be provided by extrusion. In this case, injection molding is applied whereby plastic pellets are molted (for instance in an extruder) and subsequently injected in a special mold under pressure. Typically, the plastic is injected at temperatures between 195°C and 245°C into a cooler mold having a temperature between 50°C and 95°C. After the mold is completely filled with the molten plastic, the polymer is cooled down, forming a plastic part in the exact shape of the mold. This technique is, for instance, used for the manufacture of automobile parts, electronic pieces, domestic goods, etc.

5.4.1.1.3 PMMA Spherical Beads

Acrylic beads are spherical solid PMMA beads that are available in diameter ranges from 50 to 1000 μm. Acrylic beads are synthesized by suspension polymerization, a heterogeneous polymerization technique that utilizes certain stabilizers such as gelatin, poly(vinyl alcohol), alkali salts of MAA, or inorganic compounds. The particle size can be controlled by selecting the appropriate stabilizer and stirring conditions. The polymer is isolated as beads by filtration and dried. Most properties of acrylic beads are the same as for acrylic resin. However, due to their spherical shape and small size, acrylic beads exhibit increased surface areas, faster dissolving in solvents, specific light-diffusion or light-scattering properties, and general processing advantages.

5.4.1.1.4 Modified PMMA

Since pure PMMA sometimes does not exhibit the property standards necessary for certain applications, the production formulations are often modified by adding (co)monomers, additives, or fillers. For instance, during processing of PMMA, small amounts of methyl acrylate are added as a (co) monomer in the polymerization. This enhances the thermal stability of the PMMA by decreasing the tendency to depolymerize (unzipping) during heat processing. Furthermore, in order to improve impact strength of the PMMA, it is often copolymerized with butyl acrylate. Likewise, addition of MAA increases the glass transition temperature of the polymer for higher-temperature uses such as lighting applications. Also, plasticizers can be added to modify glass transition, impact strength, or processing properties. In order to protect the PMMA from UV light, or to give it a certain color, dyes can be added during polymerization. Finally, fillers can be applied to alter the final material properties or to make end products cheaper (improve cost effectiveness).

5.4.1.2 Applications of PMMA[1]

5.4.1.2.1 Signs and Displays

Optical-grade PMMA sheets exhibit excellent clarity and transparency and are furthermore extremely robust and weather resistant. Therefore, they are widely used for illuminated signage, display cabinets, and display windows. Since PMMA also filters damaging UV light and can be colored in any possible shade in the spectrum, these sheets are also applied in museum casings, poster holders, and advertisement boards and panels.

5.4.1.2.2 Construction and Architecture

Due to its tremendous impact and UV resistance combined with excellent optical clarity, PMMA is the material of first choice for window and door profiles, canopies, and panels. It can also easily be molten into specially designed three-dimensional shapes and applied in balustrades, car ports, and façade design. Furthermore, because PMMA is so extremely strong and resistant to abrasive action, it is used for the world's most impressive aquaria and marine centers to provide the best possible viewing experience. Next, PMMA facilitates light transmission and provides good heat insulation, making it the ideal material for building green houses and verandas. Finally, PMMA sheets are applied as sound barriers alongside noisy highways.

5.4.1.2.3 Automotive and Transportation

PMMA sheets are applied in car windows, motorcycle windshields, and number plate covers. Acrylic resin is colored and injection-molded into exterior front, rear, and indicator car-light covers and for interior light covers. Because acrylic sheets are also salt resistant, they are used in marine transportation, for instance, for the windows of a ship. Furthermore, PMMA can resist very low subzero temperatures and high pressure differences, making it ideal for use in aviation purposes, such as plane and helicopter windows, canopies, and windscreens.

5.4.1.2.4 Lighting

Since PMMA sheets exhibit excellent transparency, superior light transmission, and good heat insulation, they are extensively used in lighting applications in displays, lit shelving, and illuminated signs. They can be obtained in any possible color, ideally for use in disco lights. They are also the perfect material for designing LED lights.

5.4.1.2.5 Sanitary

PMMA is not only extremely robust; it is, additionally, resistant to repeated exposure to chemicals and soap. It can be injection-molded into every shape desirable. Therefore, it is widely applied as

Polyacrylates

the perfect material for bathtubs, washstands, Jacuzzis, and steam baths. Furthermore, PMMA cast sheets are used in shower cabin doors, in shower or bathtub panels, and for other purposes in washroom areas of leisure complexes and fitness areas.

5.4.1.2.6 Electronics

The excellent optical clarity, high light transmission, and scratch resistance of PMMA makes it the perfect material for use in screens for LCD/LED TVs, PC monitors, laptops, cell phones, and electronic equipment displays. It is also used in infrared transmitter/receiver windows. Due to its UV and heat resistance, PMMA sheets are also used as cover material for solar panels where their excellent light transmission allows for very high energy conversion efficiencies.

5.4.1.2.7 Medical and Health

As mentioned before, PMMA is a durable and easy-to-clean material; therefore, it is extensively used for storage cabinets in hospitals, doctor's practices, and research labs. For the same reason, transparent PMMA sheets are used to fabricate incubators for newborns. Besides these applications, PMMA is also applied as dental cavity fillings and bone cement due to its superior biocompatibility, reliability, and durability. Finally, acrylic resins are used for the manufacture of all kinds of disposable medical diagnostics, such as blood cuvettes or drug testing devices.

5.4.1.2.8 Furniture and Design

As a final application range, PMMA is used by designers and furniture fabricants to make specially designed chairs, tables, kitchen cabinets, and others. The material is easy to obtain in any shape, color, and special finishes. Finally, PMMA resin or sheets can be used in an unlimited number of household applications, ranging from salad bowls to table mats.

5.4.1.3 PMMA Blends

In order to design polymer materials with optimal properties, blending is an ideal technique. Blends exhibit combinations of desired properties of the individual polymers and, in some cases, even behave superiorly over the individual homopolymers. Therefore, in this section, some of the most well-studied blends of PMMA with other polymer materials are described.

5.4.1.3.1 PMMA/Poly(Vinylidene Fluoride)

The mixing behavior of blends of PMMA with poly(vinylidene fluoride) (PVDF) has been widely studied with numerous techniques.[69–72] While PMMA is an amorphous material, PVDF is a semicrystalline polymer. Small amounts of PVDF additives in PMMA act as plasticizer, while low amounts of PMMA additives in PVDF improve the processability of the resulting polymer blend. PMMA/PVDF blends are used for weatherable film applications and architectural coatings,[73] as well as nanophotonic applications[74] and polymeric membranes in battery technology.[75] Furthermore, upon using supercritical CO_2, PMMA/PVDF foams can be prepared.[76] Atactic and syndiotactic PMMA/PVDF blends are miscible in the melt over the whole compositional range below the lower critical solution temperature (LCST), while isotactic PMMA/PVDF blends show phase separation. Specific hydrogen bond interactions between the carbonyl groups of PMMA and the acidic hydrogens in PVDF are responsible for a negative enthalpy of mixing and the resulting molecular compatibility between the two polymers.[77] A number of Flory–Huggins interaction parameters have been determined using melting point depression techniques,[69,70] among others.

5.4.1.3.2 PMMA/Poly(Ethylene Oxide)

Also, blends between PMMA and poly(ethylene oxide) (PEO) have been widely studied. The blends usually consist of amorphous regions (both PMMA and PEO) and semicrystalline regions (PEO). The crystallization of PEO in the PMMA/PEO blend is dependent on molecular weight and on the tacticity of the PMMA.[78,79] Atactic and syndiotactic PMMA/PEO blends are miscible,[80,81] and single T_g values are reported. For blends with isotactic PMMA, two T_g values were determined, and a two-phase system exists.[82] Small angle x-ray scattering (SAXS) and small angle neutral scattering (SANS) studies demonstrate that amorphous a-PMMA or s-PMMA is incorporated in between the crystalline PEO lamellae, while i-PMMA is rejected to interfibrillar PEO spherulite regions.[83] The specific interaction between PMMA and PEO is considered as weak van der Waals forces, although some studies also suggest more specific interactions between the carbonyl group of PMMA and the oxygen atom of PEO.[84]

Among other applications, PMMA/PEO blends are applied as polymer electrolytes.[85] Furthermore, a recent study reports on the use of PMMA/PEO blends as phase change materials for thermal energy storage applications.[86]

5.4.1.3.3 PMMA/PC

PMMA and polycarbonate (PC) are two transparent polymeric materials that are widely applied in various optical applications.[87] By making PMMA/PC blends, the excellent optical clarity and surface hardness of PMMA can be combined with the superior toughness and very high glass transition temperature of PC.[88] However, it is far from trivial to obtain homogeneous blends from these two materials.[89,90] A number of studies have indicated that blend miscibility depends strongly on the solvent casting conditions and on the tacticity of the PMMA.[91–93] Furthermore, PMMA/PC mixtures display an LCST. Since the T_g of the mixture lies very close to the LCST, phase separation occurs when the mixture is heated above T_g, giving rise to immiscible blends.[94] The phase separation in PMMA/PC blends has been studied by numerous techniques, such as differential scanning calorimetry (DSC), thermogravimetric analysis (TGA), Fourier transformed infrared spectroscopy (FT-IR), solid-state nuclear magnetic resonance (NMR), X-ray photoelectron spectroscopy (XPS), and time-of-flight secondary ion mass spectrometry (SIMS).[95–97] Regarding applications, PMMA/PC blends have been investigated for use in gas separation membranes as well as in nanofibers.[98,99]

5.4.1.3.4 PMMA/Poly(Vinyl Chloride)

Blends of PMMA with poly(vinyl chloride) (PVC) are the most applied mixtures in commercial applications. PMMA acts as an impact modifier for PVC and is therefore highly interesting to use as a mixing component in modified PVC. PMMA/PVC blends have been used in industrial or consumer goods; as medical, electrical, or chemical engineering equipment; for food or beverage containers; for power tool housings; in aircraft components; etc.[100]

PVC/PMMA blends can be miscible over the whole composition range[101] however, the miscibility is dependent on many factors, such as temperature (miscible below LCST), blend production method, molar mass, and PMMA tacticity.[102,103] The polymers are compatible because of the specific hydrogen bonding interaction between the carbonyl groups of PMMA and the acidic hydrogens of PVC.[104]

TABLE 5.8
Overview of the Most Cited Articles on (Meth)acrylic Blends in the Period 2010–2014

Acrylic Polymer	Polymer 2	Ref.	# Citations
Ethylene/acrylate copolymer	Polyamide	105	44
Polysulfone-graft-PEG methyl ether methacrylate	Polyethersulfone	106	36
PMMA	PS-b-PMMA	107	35
PMMA	Poly(4-vinylphenol-b-styrene)	108	34
PMMA (europium doped)	Ionic liquid [C6mim][Tf2N]	109	33
Poly(vinyl pyrrolidone) (PVP)-b-PMMA-b-PVP	Polyethersulfone	110	32
Poly(benzyl methacrylate)	Poly(Butyl acrylate) (PBuA)	111	32
Ethylene/methyl acrylate copolymer	Copolyamide (PA6/12)	112	27
Ethylene/BuA/glycidyl methacrylate terpolymer	Polylactide (PLA)	113	26
Ethylene/acrylate copolymer	PLA	114	19

5.4.1.3.5 Other Blends

Besides the blends of PMMA with all of the above mentioned polymers, also, a great number of other mixtures of (meth)acrylic polymers with other polymers have been investigated. Since listing all the possible blend combinations would be endless work, we decided to create Table 5.8, which contains the most cited recent references on acrylic polymer blends (period 2010–2014), in order to give the reader an impression on the current state of the art in acrylic blend research.

5.4.1.4 Other Types of Polymethacrylates

Since (meth)acrylates are based on (meth)acrylic acid, a large array of monomers can be synthesized simply by esterification with (functional) alcohols. Thus, depending on application, (meth)acrylates with very specific structure and functionality are thus often tailor-made. Typical functional monomers contain polyethylene glycol chain segments or urethane moieties on the ester side chain. Most commonly used, however, are simpler monomers such as alkyl esters and hydroxyethyl (meth)acrylate.

5.4.1.4.1 PHEMA

PHEMA is a polar methacrylate that has the ability to swell in water to form a hydrogel.[115] It is biocompatible, inert, and permeable for oxygen. Therefore, the polymer is used in the fabrication of soft contact lenses, as an embedding medium for microscopy studies, and as a coating layer for implants and prostheses.

5.4.1.4.2 Polymethacrylates with Long Hydrophobic Alkyl Side Chains

Tert-butyl methacrylate has a bulky hydrophobic side chain and is used as comonomer in acrylic resins for coating applications. Cyclohexyl methacrylate is a cyclic hydrophobic monomer that is applied in resins to improve chemical resistance and weather ability. Behenyl methacrylate, lauryl methacrylate, isodecyl methacrylate, isotridecyl methacrylate, and stearyl methacrylate are monomers with longer hydrophobic side chains carrying between 10 and 22 carbon atoms. They are used as (co)monomers in acrylic resins to enhance flexibility, hydrophobicity, adhesive properties, weather ability, and impact resistance of the resulting materials. They are used in coatings and adhesives,[116] cosmetics, inks and pigment dispersions, and plastic processing, and for mining and oil or gas extractions.

5.4.1.4.3 Multifunctional Methacrylates

1,3-butylene glycol dimethacrylate and mono-, diethylene, triethylene, or polyethylene glycol dimethacrylate are difunctional hydrophilic methacrylate monomers that are used as cross-linkers in various acrylic resins. The have low viscosity and low shrinkage. They promote adhesion, flexibility, impact resistance, and abrasion resistance in the final products.

Besides these dimethacrylates, dimethyl, diethyl, and *tert*-butylaminoethyl methacrylate are monofunctional amine methacrylates that have dual methacrylic and amine reactivity and therefore exhibit excellent adhesion to metallic and plastic substrates. By quaternization of the amine functionality to ammonium salts, water solubility can be generated.

The multifunctional methacrylates described above are used in inks, paints, coatings, adhesives, cosmetics, the oil and mining industry, and additives for concrete in road construction.

5.4.2 PAA AND ITS POLYACRYLATE ESTERS

PAA and its ester derivatives are used for a broad range of applications.[117] Among others, PAA, polymethyl acrylate (PMA), polyethyl acrylate (PEA), polybutyl acrylate (PBuA), and polyethylhexyl acrylate (PEHA) are the most widely used elastomers, which find applications as dispersants and stabilizers in acrylic paints and lacquers for surface coatings.[116] Furthermore, they are used in various adhesives and sealants such as superglues (polycyanoacrylates)[118] and as additives for various plastics (for instance, as plasticizer in PMMA; see Section 5.4.1.1.4). PMA and PEA are also used as fiber modifiers in textile manufacture, and copolymers of acrylates with, for instance, 2-chloroethyl vinyl ether are applied as synthetic rubbers.[119] In general, they are widely applied in industrial processing, high-performance formulations, and building and construction industries.

A second major market for PAA derivatives is superabsorbent polymers (SAPs).[120] Salts of PAA such as sodium polyacrylate and, to a lesser extent, ammonium, potassium, or lithium polyacrylate have the potential to absorb 200 to 300 times their own mass in water. Therefore, these polymers are widely used in applications such as baby diapers, sanitary napkins, and protective underwear. Furthermore, they are used to control water spills, as artificial snow for motion pictures, in agriculture as water retention agents, and to keep moisture from penetrating into underground power cables.

Finally, the AA derivatives are intensively used in various water treatment applications.[121] For instance, copolymers of acrylamide and AA are used as anionic flocculants to purify water reservoirs or water circulation systems. Furthermore, AA polymers are used as scale inhibitors to avoid precipitation of calcium carbonate and other low-solubility salts from cooling or heating water circuits.

REFERENCES

1. Polymethyl Methacrylate (PMMA) Market by Application, Grade & Form—Global Trends & Forecasts (2012–2017), MarketsandMarkets, Dallas, TX, July 2012, Report Code: AD1177.
2. Acrylic Acid & Its Derivatives Market—By Derivative Types (Esters/Acrylates—Methyl, Ethyl, Butyl, 2-EH; Polymers—Elastomers, SAP, Water Treatment Polymers; Other Derivatives), Applications & Geography—Global Trends & Forecast to 2018. MarketsandMarkets, Dallas, TX, June 2013, Report Code: CH1148.
3. Ohara, T.; Sato, T.; Shimizu, N.; Prescher, G.; Schwind, H.; Weiberg, O.; Marten, K.; Greim. H.; Acrylic acid and derivatives; *Ullmann's Encyclopedia of Industrial Chemistry*, Wiley-VCH, Weinheim, Germany, 2011, doi:10.1002/14356007.a01_161.pub3.
4. Reppe, W.; Friederich, H. H.; Lautenschlager, H.; Laib, H.; Acrylic acid, DE 949654 19560927 patent, 1956.
5. Bauer, W.; Methacrylic acid and derivatives; *Ullmann's Encyclopedia of Industrial Chemistry*, Wiley-VCH, Weinheim, Germany, 2011, doi:10.1002/14356007.a16_441.pub2.
6. Barner-Kowollik, C.; Beuermann, S.; Buback, M.; Castignolles, P.; Charleux, B.; Coote, M. L.; Hutchinson, R. A.; Junkers, T.; Lacík, I.; Russell, G. T.; Stach, M.; van Herk, A. M.; Critically evaluated rate coefficients in radical polymerization—7. Secondary-radical propagation rate coefficients for methyl acrylate in bulk; *Polym. Chem.*, **2014**, *5*, 204–212.
7. Asua, J. M.; Beuermann, S.; Buback, M.; Castignolles, P.; Charleux, B.; Gilbert, R. G.; Hutchinson, R. A.; Leiza, J. R.; Nikitin, A. N.; Vairon, J. P.; van Herk, A. M.; Critically evaluated rate coefficients for free-radical polymerization, 5. Propagation rate coefficients for butyl acrylate; *Macromol. Chem. Phys.*, **2004**, *205*, 2151–2160.
8. Beuermann, S.; Buback, M.; Davis, T. P.; Gilbert, R. G.; Hutchinson, R. A.; Olaj, O. F.; Russell, G. T.; Schweer, J.; van Herk, A. M.; Critically evaluated rate coefficients for free-radical polymerization, 2. Propagation rate coefficients for methyl methacrylate; *Macromol. Chem. Phys.*, **1997**, *198*, 1545.
9. Beuermann, S.; Buback, M.; Davis, T. P.; Gilbert, R. G.; Hutchinson, R. A.; Kajiwara, A.; Klumperman, B.; Russell, G. T.; Critically evaluated rate coefficients for free-radical polymerization—3. Propagation rate coefficients for alkyl methacrylates; *Macromol. Chem. Phys.*, **2000**, *201*, 1355.
10. Beuermann, S.; Buback, M.; Davis, T. P.; García, N.; Gilbert, R. G.; Hutchinson, R. A.; Kajiwara, A.; Kamachi, M.; Lacík, I.; Russell, G. T.; Critically evaluated rate coefficients for free-radical polymerization, 4, Propagation rate coefficients for methacrylates with cyclic ester groups; *Macromol. Chem. Phys.*, **2003**, *204*, 1338.
11. Junkers, T.; Barner-Kowollik, C.; The role of mid-chain radicals in acrylate free radical polymerization: Branching and scission; *J. Polym. Sci. Polym. Chem.*, **2008**, *46*, 7585–7605.
12. Vandenbergh, J.; Junkers, T.; Macromonomers from AGET activation of poly(n-butyl acrylate) precursors: Radical transfer pathways and midchain radical migration; *Macromolecules*, **2012**, *45*, 6850–6856.
13. Willemse, R. X. E.; van Herk, A. M.; Panchenko, E.; Junkers, T.; Buback, M.; PLP-ESR monitoring of mid-chain radicals in n-butyl acrylate polymerization; *Macromolecules*, **2005**, *38*, 5098–5103.

14. Nemec J. W.; Kirch, L. S.; Methacrylic acid and derivatives; *Encyclopaedia of Polymer Science and Engineering* (H. F. Mark, ed.), Wiley-Interscience, New York, 1985.
15. O'Driscoll, K. F.; Dionisio, J. M.; Mahabadi, H. K.; The temperature dependence of the gel effect in free-radical vinyl polymerization; *ACS Symp. Ser.*, **1979**, *104*, 361.
16. Huang, J.; The kinetics of styrene/methyl methacrylate free radical copolymerization. MSc thesis, University of Waterloo, Waterloo, Canada, 1988.
17. Eastmond, G. C.; Free-radical polymerization: An international study, *Makromol. Chem. Macromol. Symp.*, **1987**, *10/11*, 71.
18. Varshney, S. K.; Hautekeer, J. P.; Fayt, R.; Jérôme, R.; Teyssié, P.; Anionic polymerization of (meth) acrylic monomers. Effect of lithium salts as ligands for the living polymerization of methyl methacrylate using monofunctional initiators; *Macromolecules*, **1990**, *23*, 2618–2622.
19. Webster, O. W.; Hertler, W. R.; Sogah, D. Y.; Farnham, W. B.; Rajan-Babu, T. V.; Group transfer polymerization. A new concept for addition polymerization with organosilicon initiators; *J. Am. Chem. Soc.*, **1983**, *105*, 5706–5708.
20. Fuchise, K.; Chen, Y.; Satoh, T.; Kakuchi, T.; Recent progress in organocatalic group transfer polymerization; *Polym. Chem.*, **2013**, *4*, 4278–4291.
21. Wang, J.; Matyjaszewski, K.; "Controlled"/"living" radical polymerization. Atom transfer radical polymerization in the presence of transition-metal complexes; *J. Am. Chem. Soc.*, **1995**, *117*, 5614–5615.
22. Kato, M.; Kamigaito, M.; Sawamoto, M.; Higashimura, T.; Polymerization of methyl methacrylate with the carbon tetrachloride/dichlorotris-(triphenylphosphine)ruthenium(ii)/methylaluminum bis(2,6-di-tert-butylphenoxide) initiating system: Possibility of living radical polymerization; *Macromolecules*, **1995**, *28*, 1721–1723.
23. Hawker, C. J.; Frechet, J. M. J.; Grubbs, R. B.; Dao, J.; Preparation of hyperbranched and star polymers by a "living" self-condensing free radical polymerization; *J. Am. Chem. Soc.*, **1995**, *117*, 10763–10764.
24. Moad, G.; Rizzardo, E.; Alkoxyamine-initiated living radical polymerization: Factors affecting alkoxyamine homolysis rates; *Macromolecules*, **1995**, *28*, 8722–8728.
25. Chiefari, J.; Chong, Y. K.; Ercole, F.; Krstina, J.; Jeffery, J.; Le, T. P. T.; Mayadunne, R. T. A.; Meijs, G. F.; Moad, C. L.; Moad, G.; Rizzardo, E.; Thang, S. H.; Living free-radical polymerization by reversible addition fragmentation chain transfer: The RAFT process; *Macromolecules*, **1998**, *31*, 5559–5562.
26. Wayland, B. B.; Basickes, L.; Mukerjee, S.; Wei, M.; Fryd, M.; Living radical polymerization of acrylates initiated and controlled by organocobalt porphyrin complexes; *Macromolecules*, **1997**, *30*, 8109–8112.
27. Davis, T. P.; Haddleton, D. M.; Maloney, D. R.; Cobalt mediated free radical polymerization of acrylic monomers; *Trends Polym. Sci.*, **1995**, *3*, 365–373.
28. Matyjaszewski, K.; Atom transfer radical polymerization (ATRP): Current status and future perspectives; *Macromolecules*, **2012**, *45*, 4015–4039.
29. Braunecker, W. A.; Matyjaszewski, K.; Controlled/living radical polymerization: Features, developments and perspectives; *Prog. Polym. Sci.*, **2007**, *32*, 93–146.
30. Nicolas, J.; Guillaneuf, Y.; Lefay, C.; Bertin, D.; Gigmes, D.; Charleux, B.; Nitroxide-mediated polymerization; *Prog. Polym. Sci.*, **2013**, *38*, 63–235.
31. Lowe, A. B.; McCormick, C. L.; Reversible addition–fragmentation chain transfer (RAFT) radical polymerization and the synthesis of water-soluble (co)polymers under homogeneous conditions in organic and aqueous media; *Prog. Polym. Sci.*, **2007**, *32*, 283–351.
32. Moad, G.; Rizzardo, E.; Thang, S. H.; Radical addition–fragmentation chemistry in polymer synthesis; *Polymer*, **2008**, *49*, 1079–1131.
33. Heuts, J. P. A.; Roberts, G. E.; Biasutti, J. D.; Catalytic chain transfer polymerization: An overview; *Aust. J. Chem.*, **2002**, *55*, 381–398.
34. Debuigne, A.; Poli, R.; Jérôme, C.; Jérôme, R.; Detrembleur, C.; Overview of cobalt-mediated radical polymerization: Roots, state of the art and future prospects; *Prog. Polym. Sci.*, **2009**, *34*, 211–239.
35. Destarac, M.; Controlled radical polymerization: Industrial stakes, obstacles and achievements; *Macromol. React. Eng.*, **2010**, *4*, 165–179.
36. Kitayama, T.; Ute, K.; Hatada, K.; Synthesis of stereoregular polymers and copolymers of MMA by living polymerization and their characterization by NMR spectroscopy; *Polym. J.*, **1990**, *23*, 5.
37. Kitayama, T.; Masuda, E.; Yamaguchi, M.; Nishimura, T.; Hatada, K.; Syndio specific polymerization of methacrylates by *t*-phosphine-triethylaluminum; *Polym. J.*, **1992**, *24*, 817.
38. Hatada, K.; Ute, K.; Tanaka, K.; Okamoto, Y.; Kitayama, T.; Living and highly isotactic polymerization of MMA by *t*-butyl MgBr in toluene; *Polym. J.*, **1986**, *18*, 1037.
39. Ute, K.; Miyatake, N.; Hatada, K.; Glass transition temperature and melting temperature of uniform isotactic and syndiotactic poly(methyl methacrylate)s from 13mer to 50mer; *Polymer*, **1995**, *36*, 1415.

40. Gregonis, D. E.; Russell, G. A.; Andrade, J. D.; deVisser, A. C.; Preparation and properties of stereoregular PHEMA polymers and hydrogels; *Polymer*, **1978**, *19*, 1279.
41. Okamoto, Y.; Habaue, S.; Uno, T.; Baraki, H.; Stereospecific polymerization of α-substituted acrylates; *Macromol. Symp.*, **2000**, *157*, 209–216.
42. Collins, S.; Ward, D. G.; Suddaby, K. H.; Group-transfer polymerization using metallocene catalysts: Propagation mechanisms and control of polymer stereochemistry; *Macromolecules*, **1994**, *27*, 7222.
43. Caporaso, L.; Gracia-Budria, J.; Cavallo, L.; Stereospecificity in metallocene catalyzed acrylate polymerizations: The chiral orientation of the growing chain selects its own chain end enantioface; *J. Am. Chem. Soc.*, **2006**, *128*, 16649–16654.
44. Cassidy, R. T.; Acrylics; *Engineering Materials Handbook*, Vol. 2 (C. A. Dostal, ed.), ASM International, Materials Park, OH, 1988, p. 103.
45. Stickler, M.; Rhein, T.; Polymethacrylates; *Ullmann's Encyclopaedia of Industrial Chemistry*, Vol. A21, Wiley-VCH, Weinheim, Germany, 1992.
46. Penzel, E.; Polyacrylates; *Ullmann's Encyclopaedia of Industrial Chemistry*, Vol. A21, Wiley-VCH, Weinheim, Germany, 1992, p. 157.
47. Bennet, F.; Hart-Smith, G.; Gruendling, T.; Davis, T. P.; Barker, P. J.; Barner-Kowollik, C.; Degradation of poly(methyl methacrylate) model compounds under extreme environmental conditions; *Macromol. Chem. Phys.*, **2010**, *211*, 1083–1097.
48. Grassie, N.; Torrance, B. J. D.; Thermal degradation of copolymers of methyl methacrylate and methyl acrylate. I. Products and general characteristics of the reaction; *J. Polym. Sci. A*, **1968**, *6*, 3303.
49. Kaminsky, W.; Franck, J.; Monomer recovery by pyrolysis of poly(methyl methacrylate) (PMMA); *J. Anal. Appl. Pyrol.*, **1991**, *19*, 311.
50. Clouet, G.; Knipper, M.; Brossas, J.; Thermal degradation of phosphonated PMMA, *Polym. Degr. Stab.*, **1987**, *17*, 151.
51. Reghunadhan Nair, C. P.; Clouet, G.; Guilbert, Y.; Flame and thermal resistance of phosphorus functionalized PMMA and P(STY); *Polym. Degr. Stab.*, **1989**, *26*, 305.
52. Wolpert, H. D.; Optical properties; *Engineering Materials Handbook*, Vol. 2 (C. A. Dostal, ed.), ASM International, Materials Park, OH, 1988, p. 481.
53. Koike, Y.; Tanio, N.; Ohtsuka, Y.; Light scattering and heterogeneities in low-loss poly(methyl methacrylate) glasses; *Macromolecules*, **1989**, *22*, 1367.
54. Kuzyk, M. G.; Paek, U. C.; Dirk, C. W.; Guest-host polymer fibres for nonlinear optics; *Appl. Phys. Lett.*, **1991**, *59*, 902.
55. Eckl, M.; Muller, H.; Strohriegl, P.; Nonlinear optically active polymethacrylates with high glass transition temperatures; *Macromol. Chem. Phys.*, **1995**, *196*, 315.
56. Davis, T. P.; Gallagher, M. J.; Ranasinghe, M. G.; Zammit, M. D.; Synthesis of high refractive index acrylic copolymers; *J. Mat. Chem.*, **1994**, *4*, 1359.
57. Lovell, R.; Windle, A. H.; Determination of the local conformation of PMMA from WAXS; *Polymer*, **1981**, *22*, 175.
58. Mitchell, G. R.; Brown, D. J.; Windle, A. H.; Wide-angle scattering for swollen and glassy poly(2-hydroxyethyl methacrylate); *Makromol. Chem.*, **1983**, *184*, 1937.
59. Judd, R. E.; Crist, B.; Light scattering studies of structure in glassy poly(methyl methacrylate); *J. Polym. Sci. Polym. Lett. Ed.*, **1980**, *18*, 717.
60. Kinloch, A. J.; Young, R. J.; *Fracture Behaviour of Polymers*, Elsevier, New York, 1990.
61. Bowden, P. B.; Formation of microshear bands in polystyrene and poly(methyl methacrylate); *Phil. Mag.*, **1970**, *22*, 455.
62. Narisawa, I.; Ishikawa, M.; Ogawa, H.; Notch brittleness of ductile glassy polymers under plane strain; *J. Mat. Sci.*, **1980**, *15*, 2059.
63. Weidmann, G. W.; Doll, W.; Some results of optical interference measurements of critical displacements at the crack tip; *Int. J. Fract.*, **1978**, *14*, R189.
64. Graham, I. D.; Williams, J. G.; Zichy, E. L.; Craze kinetics for poly(methyl methacrylate) in liquids; *Polymer*, **1976**, *17*, 439.
65. Lynch, J.; Acrylic cast sheet and its manufacture; *PCT Int. Appl.*, WO 2003061937 patent, 2003.
66. Albrecht, K..; Stickler, M.; Rhein, T.; Polymethacrylates; *Ullmann's Encyclopedia of Industrial Chemistry*, Wiley-VCH, Weinheim, Germany, 2013, doi:10.1002/14356007.a21_473.pub2.
67. Kroschwitz, J. I.; Optical properties; *Concise Encyclopedia of Polymer Science and Engineering*; John Wiley & Sons, New York, **1998**, pp. 683–684.
68. Elias, H.-G.; Acrylic polymers; *Macromolecules Vol. 2: Industrial Polymers and Syntheses*; Wiley-VCH, Weinheim, Germany, **2007**, pp. 294–297.

69. Canalda, J. C.; Hoffmann, T.; Martinez-Salazar, J.; On the melting behavior of polymer single crystals in a mixture with a compatible polymer. 1. Poly(vinylidene fluoride)/poly(methyl methacrylate) blends; *Polymer*, **1995**, *36*, 981–985.
70. Nishi, T.; Wang, T. T.; Melting point depression and kinetic effects of cooling on crystallization in poly(vinylidene fluoride)–poly(methyl methacrylate) mixtures; *Macromolecules*, **1975**, *8*, 909–915.
71. Okabi, Y.; Murakami, H.; Osaka, N.; Saito, H.; Inoue, T.; Morphology development and exclusion of noncrystalline polymer during crystallization in PVDF/PMMA blends; *Polymer*, **2010**, *51*, 1494–1500.
72. Ma, W.; Zhang, J.; Wang, X.; Wang, S.; Effect of PMMA on crystallization behavior and hydrophilicity of poly(vinylidene fluoride)/poly(methyl methacrylate) blend prepared in semi-dilute solutions; *Appl. Surf. Sci.*, **2007**, *253*, 8377–8388.
73. Robeson, L. M.; Applications of polymer blends: Emphasis on recent advances; *Polym. Eng. Sci.*, **1984**, *24*, 587–597.
74. Quan, S.-L.; Lee, H.-S.; Lee, E.-H.; Park, K.-D.; Lee, S. G.; Chin, I.-J.; Ultrafine PMMA (QDs)/PVDF core-shell fibers for nanophotonic applications; *Microelectron. Eng.*, **2010**, *87*, 1308–1311.
75. Rajendran, S.; Kannan, R.; Mahendran, O.; An electrochemical investigation of PMMA/PVDF blend based polymer electrolytes; *Mater. Lett.*, **2001**, *49*, 172–179.
76. Walker, T. A.; Melnichenko, Y. B.; Wignall, G. D.; Spontak, R. J.; Phase behavior of poly(methyl methacrylate)/poly(vinylidene fluoride) blends with and without high-pressure CO_2; *Macromolecules*, **2003**, *36*, 4245–4249.
77. Leonard, C.; Halary, J. L.; Monnerie, L.; Hydrogen bonding in PMMA-fluorinated polymer blends: FTIR investigations using ester model molecules; *Polymer*, **1985**, *26*, 1507–1513.
78. John, E.; Ree, T.; Crystallization of poly(ethylene oxide) in binary polymer blends: Influence of tacticity of poly(methyl methacrylate); *J. Polym. Sci., Part A: Polym. Chem.*, **1990**, *28*, 385–398.
79. Martuscelli, E.; Pracella, M.; Yue, W. P.; Influence of composition and molecular mass on the morphology, crystallization and melting behavior of poly(ethylene oxide)/poly(methyl methacrylate) blends; *Polymer*, **1984**, *25*, 1097–1106.
80. Zawanda, J. A.; Ylitalo, C. M.; Fuller, G. G.; Colby, R. H.; Long, T. E.; Component relaxation dynamics in a miscible polymer blend: Poly(ethylene oxide)/poly(methyl methacrylate); *Macromolecules*, **1992**, *25*, 2896–2902.
81. Lodge, T. P.; Wood, E. R.; Haley, J. C.; Two calorimetric glass transitions do not necessarily immiscibility: The case of PEO/PMMA; *J. Polym. Sci. Part B: Polym. Phys.*, **2006**, *44*, 756–763.
82. Arai, F.; Takeshita, H.; Dobashi, M.; Takenaka, K.; Miya, M.; Shiomi, T.; Effects of liquid–liquid phase separation of crystallization of poly(ethylene glycol) in blends with isotactic poly(methyl methacrylate); *Polymer*, **2012**, *53*, 851–856.
83. Cimmino, S.; Di Pace, E.; Martuscelli, E.; Silvestre, C.; Evaluation of the equilibrium melting temperature and structure analysis of poly(ethylene oxide)/poly(methyl methacrylate) blends; *Makromol. Chem.*, **1990**, *191*, 2447–2454.
84. Ramano Rao, G.; Castiglioni, C.; Gussoni, M.; Zerbi, G.; Martuscelli, E.; Probing the structure of polymer blends by vibrational spectroscopy: The case of poly(ethylene oxide) and poly(methyl methacrylate) blends; *Polymer*, **1985**, *26*, 811–820.
85. Osman, Z.; Ansor, N. M.; Chew, K. W.; Kamarulzaman, N.; Infrared and conductivity studies on blends of PMMA/PEO based polymer electrolytes; *Ionics*, **2005**, *11*, 431–435.
86. Sari, A.; Alkan, C.; Karaipekli, A.; Uzun, O.; Poly(ethylene glycol)/poly(methyl methacrylate) blends as novel form-stable phase change materials for thermal energy storage; *J. Appl. Polym. Sci.*, **2010**, *116*, 929–933.
87. Ide, F.; *Transparent Plastics of Optical Access Generation*, CMC Publications, Tokyo, 2005, p. 37.
88. Li, Y.; Shimizu, H.; Fabrication of nanostructured polycarbonate/poly(methyl methacrylate) blends with improved optical and mechanical properties by high-shear processing; *Polym. Eng. Sci.*, **2011**, *51*, 1437–1445.
89. Marin, N.; Favis, B. D.; Co-continuous morphology development in partially miscible PMMA/PC blends; *Polymer*, **2002**, *43*, 4723–4731.
90. Kyu, T.; Saldanha, J. M.; Miscible blends of polycarbonate and poly(methyl methacrylate); *J. Polym. Sci., Part C: Polym. Lett.*, **1988**, *26*, 33–40.
91. Saldanha, J. M.; Kyu, T.; Influence of solvent casting on the evolution of phase morphology of PC/PMMA blends; *Macromolecules*, **1987**, *20*, 2840–2847.
92. Kyu, T.; Ko, C.-C.; Lim, D.-S.; Smith, S. D.; Noda, I.; Miscibility studies on blends of polycarbonate with syndiotactic poly(methyl methacrylate); *J. Polym. Sci., Part B: Polym. Phys.*, **1993**, *31*, 1641–1648.
93. Kyu, T.; Lim, D. S.; Phase decomposition in blends of polycarbonate and isotactic poly(methyl methacrylate); *Macromolecules*, **1991**, *24*, 3645–3650.

94. Nishimoto, M.; Kekkula, H.; Paul, D. R.; Role of slow phase separation in assessing the equilibrium phase behavior of PC–PMMA blends; *Polymer*, **1991**, *32*, 272–278.
95. Asano, A.; Takegoshi, K.; Kikichi, K.; Solid-state NMR study of miscibility and phase separation of polymer blend: Polycarbonate/poly(methyl methacrylate); *Polym. J.*, **1992**, *24*, 555–562.
96. Lhoest, J. B.; Bertrand, P.; Weng, L. T.; Dewez, J. L.; Combined time-of-flight secondary ion mass spectrometry and x-ray photoelectron spectroscopy study of the surface segregation of poly(methyl methacrylate) (PMMA) in bisphenol A polycarbonate PMMA blends; *Macromolecules*, **1995**, *28*, 4631–4637.
97. Singh, A. K.; Mishra, R. K.; Prakash, R.; Maiti, P.; Singh, A. K.; Pandey, D.; Specific interactions in partially miscible polycarbonate (PC)/poly(methyl methacrylate) (PMMA) blends; *Chem. Phys. Lett.*, **2010**, *486*, 32–36.
98. Wei, M.; Kang, B.; Sung, C.; Mead, J.; Core-sheath structure in electrospun nanofibers from polymer blends; *Macromol. Mater. Eng.*, **2006**, *291*, 1307–1314.
99. Lai, J.-Y.; Huang, S.-J.; Huang, S.-L.; Shyu, S. S.; Poly(methyl methacrylate)/polycarbonate membrane for gas separation; *Sep. Sci. Tech.*, **1995**, *30*, 461–476.
100. Utracki, L. A.; *Polymer Blends Handbook*, Springer, New York, 2003, p. 41.
101. Zhou, C.; Gao, Y.; Qi, X.; Tan, Z.; Sun, S.; Zhang, H.; Phase separation of impact modified PVC/PMMA blends under melt blending conditions, *J. Vinyl Addit. Technol.*, **2013**, *19*, 11–17.
102. Ramesh, S.; Yahaya, A. H.; Arof, A. K.; Miscibility studies of PVC blends (PVC/PMMA and PVC/PEO) based polymer electrolytes; *Solid State Ionics*, **2002**, *148*, 483–486.
103. Schurer, J. W.; de Boer, A.; Challa, G.; Influence of the tacticity of poly(methyl methacrylate) on the compatibility with poly(vinyl chloride); *Polymer*, **1975**, *16*, 201–204.
104. Kogler, G.; Mirau, P. A.; Two-dimensional NMR studies of intermolecular interactions in poly(vinyl chloride)/poly(methyl methacrylate) mixtures; *Macromolecules*, **1992**, *25*, 598–604.
105. Baudouin, A.-C.; Devaux, J.; Bailly, C.; Localization of carbon nanotubes at the interface in blends of polyamide and ethylene-acrylate copolymer; *Polymer*, **2010**, *51*, 1341–1354.
106. Yi, Z.; Zhu, L.-P.; Xu, Y.-Y.; Zhao, Y.-F.; Ma, X.-T.; Zhu, B.-K.; Polysulfone-based amphiphilic polymer for hydrophilicity and fouling-resistant modification of polyethersulfone membranes; *J. Membr. Sci.*, **2010**, *365*, 25–33.
107. Li, X.; Gao, J.; Xue, L.; Han, Y.; Porous polymer films with gradient-refractive-index structure for broadband and omnidirectional antireflection coatings; *Adv. Funct. Mater.*, **2010**, *20*, 259–265.
108. Chen, S. C.; Kuo, S.-W.; Jeng, U. S.; Su, C.-J.; Chang, F.-C.; On modulating the phase behavior of block copolymer/homopolymer blends via hydrogen bonding; *Macromolecules*, **2010**, *43*, 1083–1092.
109. Lunstroot, K.; Driesen, K.; Nockemann, P.; Viau, L.; Mutin, P. H.; Vioux, A.; Binnemans, K.; Ionic liquid as plasticizer for europium(iii)-doped luminescent pol(methyl methacrylate) films; *Phys. Chem. Chem. Phys.*, **2010**, *12*, 1879–1885.
110. Ran, F.; Nie, S.; Zhao, W.; Li, J.; Su, B.; Sun, S.; Zhao, C.; Biocompatibility of modified polyethersulfone membranes by blending an amphiphilic triblock co-polymer of poly(vinyl pyrrolidone)-b-poly(methyl methacrylate)-b-poly(vinyl pyrrolidone); *Acta Biomater.*, **2011**, *7*, 3370–3381.
111. Feldman, K. E.; Kade, M. J.; Meijer, E. W.; Hawker, C. J.; Kramer, E. J.; Phase behavior of complementary multiply hydrogen bonded end-functional polymer blends; *Macromolecules*, **2010**, *43*, 5121–5127.
112. Baudouin, A. C.; Bailly, C.; Devaux, J.; Interface localization of carbon nanotubes in blends of two copolymers; *Polym. Degrad. Stab.*, **2010**, *95*, 389–398.
113. Liu, H.; Song, W.; Chen, F.; Guo, L.; Zhang, J.; Interaction of microstructure and interfacial adhesion on impact performance of polylactide (PLA) ternary blends; *Macromolecules*, **2011**, *44*, 1513–1522.
114. Afrifah, K. A.; Matuana, L. M.; Impact modification of polylactide with a biodegradable ethylene/acrylate copolymer; *Macromol. Mater. Eng.*, **2010**, *295*, 802–811.
115. Holly, F. J.; Refojo, M. F.; Wettability of hydrogels I. Poly(2-hydroxyethyl methacrylate), *J. Biomed. Mater. Res.*, **1975**, *9*, 315–326.
116. Paul, S.; *Surface Coatings Science and Technology*, 2nd Ed., Wiley, New York, 1995.
117. Penzel, E.; Polyacrylates; *Ullmann's Encyclopedia of Industrial Chemistry*, Wiley-VCH, Weinheim, Germany, 2000, doi:10.1002/14356007.a21_157.
118. Stepanski, H.; Acrylate chemistry for adhesives. Principles, properties and possible uses; *Adhaesion-Kleben&Dichten*, **2013**, *57*, 26–33.
119. Mendelsohn, M. A.; Minter, H. F.; Acrylic elastomers. I. Properties of some modified acrylic rubbers; *Rubber Age (New York)*, **1964**, *95*, 403–406.
120. Buchholz, F. L.; Recent advances in superabsorbent polyacrylates; *Trends Polym. Sci.*, **1994**, *2*, 277–281.
121. Godlewski, I. T.; Schuck, J. J.; Libutti, B. L.; Polymer for use in water treatment; US 4029577 A 19770614, US patent, 1997.

6 Polyacetals*

Kinga Pielichowska

CONTENTS

6.1 Introduction 193
6.2 History 194
6.3 Raw Materials for Polyacetals 194
    6.3.1 Monomers 194
        6.3.1.1 Formaldehyde.. 194
        6.3.1.2 Trioxane 196
        6.3.1.3 Comonomers 196
    6.3.2 Reaction Media 197
    6.3.3 Polymerization Initiators 197
    6.3.4 End-Group Capping 199
    6.3.5 POM Additives 199
6.4 Polymerization Mechanisms for Polyacetals 201
6.5 Characteristics of Polyacetals 208
6.6 Properties of Polyacetals 210
6.7 Polyacetals Processing 214
    6.7.1 Melt Processing.. 214
        6.7.1.1 Injection Molding.. 214
        6.7.1.2 Melt Extrusion... 219
    6.7.2 Solid-State Processing 220
    6.7.3 Other Methods 220
6.8 Polyacetal Blends... 223
6.9 Polyacetal Composites and Nanocomposites 226
    6.9.1 Polyacetal Composites 226
    6.9.2 Polyacetal Nanocomposites 226
6.10 Polyacetal Applications 232
6.11 Polyacetals Recycling 233
6.12 Polyacetals Market 233
6.13 Developments and New Resins 234
6.14 Conclusions 235
References 235

6.1 INTRODUCTION

Polyoxymethylene (POM) is one of the most important engineering (thermo) plastics (ENPLAs) with the main chain composed of the $-CH_2-O-$ unit. The term *polyoxymethylene* (also known as either acetal, polyacetal, or acetal resins) is used to describe high-molecular-weight polymers and copolymers of formaldehyde (Masamoto 1993). POM is the first member of the polyether series; however, it has very different properties than the rest of polyethers. Acetal homopolymer refers to

---

* Based in parts on the first-edition chapter on polyacetal.

resin containing solely the carbon–oxygen backbone, while for the copolymer, the oxymethylene structure is occasionally interrupted by a comonomer unit (Lüftl et al. 2006; Pielichowska 2012a). Polyacetal is the first commercial ENPLA that has been produced on industrial scale by DuPont since 1960 under the trade name Delrin® (Masamoto 1993). POM possesses excellent properties, such as high mechanical strength, good abrasion resistance, and exceptional dimensional stability, and has a wide range of applications in various industries (Goossens and Groeninckx 2006; Yu et al. 2011). For instance, it is commonly used in the food industry, as paintball markers, to make frame sliders and knee pucks for motorcycle riders, as well as for slide gloves that are used when longboarding (Srivastava et al. 2011).

6.2 HISTORY

First observation on formaldehyde polymers was reported by Butlerov in 1859, but it was erroneously designated as a dimer. He found that formaldehyde could gradually form a white, resinous solid, which would rapidly decompose at 150°C and could easily be dissolved in a dilute acid or base (Chiang and Huang 1997). Over 60 years later, Staudinger (1960) confirmed earlier Delépine (1897) suggestions that this material is a high-molecular-weight substance. Staudinger et al. also found that end-capping prevents hemiacetal degradation of POM (Staudinger and Gaule 1916; Staudinger et al. 1929). In 1922, Hammick and Boeree described the formation of white polymer by sublimation of trioxane; however, formaldehyde polymers with sufficient molecular weight to be commercially significant were described only in 1942 in patent of E. I. du Pont de Nemours & Co. (Austin and Frank 1942). In early 1950s, DuPont performed intensive studies on preparation and characterization of properties of formaldehyde polymers. They discovered that tough solid polymers were readily obtained from high-purity formaldehyde using ionic initiators. These polymers were stabilized by replacing the hydroxyl groups on the polymer chain ends with ester groups. As a result, polymers with excellent tensile, impact, and compression strengths, good abrasion, and wear resistance were obtained—these properties allowed replacing metals in numerous applications (Masamoto 1993; Koch and Lindvig 1959; Linton and Goodman 1959; Schweitzer, Macdonald, and Punderson 1959). In 1956, DuPont registered the name of polyacetal as *Delrin* and constructed a commercial plant in 1957 with a capacity of 6000 tons/year at Parkersburg in West Virginia. By 1960, the plant capacity had increased to 12,000 tons/year in order to cater to the emerging new applications (Chiang and Huang 1997). Next, in 1962, Celanese developed and introduced to production acetal resin synthesized by the copolymerization of trioxane and cyclic ethers, such as ethylene oxide (Walling, Brown, and Bartz 1962), under the trade name Ceclon. Since the introduction of polyacetal resin on the market by DuPont and Celanese, many companies strived to develop new technology. Twenty years after, companies such as BASF, Asahi Kasei, and Mitsubishi Gas Chemical had also started the production of POM. In Table 6.1, the largest producers of POM are presented.

6.3 RAW MATERIALS FOR POLYACETALS

6.3.1 MONOMERS

6.3.1.1 Formaldehyde

Formaldehyde with the general formula HCHO and molar mass 30.03 g/mol (CAS No. 50-00-0) is the simplest aliphatic aldehyde. Its synonyms are as follows: methanal, methyl aldehyde, methylene oxide, methylene glycol, oxymethylene, oxymethane, paraform, and formalin (solution of 37% by weight). Physical properties of formaldehyde are summarized in Table 6.2.

TABLE 6.1
Polyacetal Producers

Resin Trade Name	Producer's Name	Year of Production Starting	Homopolymer or Copolymer	Location
Delrin	Du Pont	1960	Homopolymer	United States
Ceclon	Celanese	1962	Copolymer	United States
Tenac	Ashai Chemical	1972	Homopolymer and copolymer	Japan
Hostaform	Ticona/Celanese	1961	Copolymer	Germany
Duracon	Polyplastics	1962	Copolymer	Japan
Ultraform	Ultraform (joint venture of BASF and Degussa)	1970	Copolymer	United States
Lupital	Mitsubishi Gas Chemical		Copolymer	Japan
Tepcon	TEPCO	1992	Copolymer	Taiwan
Kepital	KEPCO	1988	Copolymer	South Korea
Lucel	LG Chem.		Copolymer	South Korea
Tarnoform	ZA Tarnów-Mościce	1990s	Copolymer	Poland

TABLE 6.2
Properties of Formaldehyde

Property	Information
Color	Colorless
Physical state	Gas
Odor	Pungent, suffocating, highly irritating
Odor threshold	50 ppm
Water	0.5–1.0 ppm
Air	
Melting point (°C)	−92
Boiling point (°C)	−19.5
d (air = 1.000)	1.067
d (−20°C)	0.815
Vapor pressure at 25°C	Gas: vapor pressure > bp, 3883 mmHg
Polymerization	Polymerizes (readily in water)
Henry's law constant at 25°C	3.27×10^{-7} atm·m^3/mol
Incompatibilities	Reacts with alkalies, acids, and oxidizers
Autoignition temp. (°C)	300
Flash point (°C)	60°C
Explosive limits (% by vol. in air)	7–72
Solubility	Very soluble up to 55 wt.% in H$_2$O, also in organic solvents (alcohols, ethers, acetone, benzene)

Source: Adapted from ATSDR, Toxicological profile for formaldehyde, 1999.

Formaldehyde can present a moderate health hazard injuring the eyes (1 ppm concentration can cause burning in the eyes), skin, and the respiratory system. It is mutagenic, teratogenic, and probably carcinogenic to humans. Humans exposed to 1–2 ppm of formaldehyde in air can exhibit the symptoms of itching eyes, burning nose, dry and sore throat, sneezing, coughing, headache, feeling thirsty, and disturbed sleep. Inhalation of high concentration of formaldehyde can lead to death.

Exposure to 700 ppm for 2 h was fatal to mice; cats died after an 8 h exposure (Patnaik 2007). Conventionally, formaldehyde is produced on an industrial scale by methanol oxidation:

$$CH_3OH + 1/2 O_2 \xrightarrow[320-370°C]{Fe_2(MoO_4)_3-MoO_3} HCHO + H_2O$$

In this method, excess air is used with metal oxide as a catalyst, and the cracking reaction temperature is 320–370°C. Conversion is up to 98.4% at a rather high reaction rate. The metal oxide catalyst used is not easily poisoned and is normally used continuously for more than a year (Chiang and Huang 1997). Another method is the excess methanol method that uses silver as a catalyst in cracking methanol at 600–650°C.

$$CH_3OH \xrightarrow[500-650°C]{Ag} HCHO + H_2 \uparrow$$

Conversion in this method is about 80%. The reaction rate is moderate, but the catalyst is easily poisoned in a short time. Methanol oxidation process (Masamoto, Iwaisako et al. 1993; Masamoto, Matsuzaki, and Morishita 1993; Masamoto, Matsuzaki et al. 1993) consists of methylal formation and methylal oxidation.

$$2CH_3OH + HCHO \xrightarrow{H^+} CH_3OCH_2OCH_3 + H_2O$$

$$CH_3OCH_2OCH_3 + O_2 \xrightarrow{Fe_2O_3MoO_3X} 3HCHO + H_2O$$

This process requires less energy for evaporation of water. The yield of formaldehyde is high, and the need for the recovery of diluted aqueous formaldehyde could be eliminated (Chiang and Huang 1997). Formaldehyde monomer can be also obtained by heating a mineral oil slurry containing 10–40% dry paraformaldehyde to 115–140°C or by adding paraformaldehyde-di(2-ethylhexyl) phthalate to a heated generator (Miller 1949). Another method is the thermal decomposition of cyclohexyl hemiformal (Goodman and Sherwood 1961), α-POM (Austin and Frank 1942), and trioxane (Sandler 1974).

6.3.1.2 Trioxane

1,3,5-Trioxane (CAS No. 110-88-3, molar mass 90.08 g/mol) is a cyclic triether, also known as 1,3,5-trioxacyclohexane or trioxymethylene, and is a white stable solid used as a monomer for polyacetals. Trioxane forms spontaneously when formaldehyde gas is heated or when its 2% sulfuric acid solution is heated followed by extraction with chloroform. Trioxane used for polymerization should be purified by fractional distillation or recrystallization in methylene chloride or petroleum ether. The properties of trioxane are given in Table 6.3.

Trioxane during heating undergoes thermal decomposition to formaldehyde. Exposure to trioxane causes irritation to the eyes, irritation to the airways, disturbances to the nervous system, and, in chronic contact, possible allergies to the skin (Consortium 2000).

6.3.1.3 Comonomers

Components that can copolymerize with trioxane should undergo cationic (co)polymerization, i.e., cyclic acetals, cyclic ethers, lactones, or cyclic esters (Kern et al. 1966). As it was described in patent literature (Eckardt et al. 1999), suitable comonomers for the preparation of the POM copolymers with 1,3,5-trioxane are cyclic acetals (preferably formals, having 5 to 11, especially 5 to 8, ring members, in particular cyclic formals of aliphatic or cycloaliphatic diols having 2 to 8, preferably

TABLE 6.3
Trioxane Properties

Property	Information
Melting point (°C)	60–63
Boiling point (°C)	114–116
Odor	Ethanol like
d (65°C)	1.17
Ignition point (°C)	410°C
Flash point (°C)	45
Explosive limits (% by vol. in air)	3.6–28.7
Solubility	Freely soluble in water

Source: RSC, 1,3,5-Trioxane, 2014. Available at http://www.chemspider.com/Chemical-Structure.7790.html (July 23, 2014).

2, 3, or 4, carbon atoms, whose carbon chain may be interrupted by an oxygen atom at intervals of 2 carbon atoms) and cyclic ethers having 3 to 5, preferably 3, ring members. The most popular configuration consists of 2–6% of ethylene oxide or 1,3-dioxolane. Comonomers used in acetal resin synthesis are presented in Table 6.4.

6.3.2 Reaction Media

During polyacetal synthesis, different inert reaction media can be used, such as propane, pentane, cyclohexane, hexane, benzene, heptane, toluene, xylene, decahydronaphthalene, carbon tetrachloride, or C3–C10 inert hydrocarbons. Polar solvents such as methylene chloride, ethylene dichloride, water, and nitrobenzene are also used (Chiang and Huang 1997; Sandler 1974).

6.3.3 Polymerization Initiators

Formaldehyde polymerization:

- *Anionic initiators*: amines, phosphines, ammonium and sulfonium salts, amides, amidines, etc.
- *Cationic initiators*: sulfuric or phosphoric acid, Lewis acids (e.g., $SnCl_4$, $SnBr_4$, $BF_3 \cdot OET_2$), stable cations (acetyl perchlorate), organometallic initiators (calcium stearate, stannous acrylate, dibutyltin dilaurate)

For commercial production of polyacetal homopolymer, only anionic initiators are used (Masamoto 1993).

Trioxane polymerization and copolymerization:

- *Anionic initiators*: amides, amidines, tertiary amines, ammonium salts, cyclic nitrogen-containing compounds, arsine, phosphine, stibine, alkoxides, and metal alkyls (Chiang and Huang 1997)
- *Cationic initiators*: BF_3, $BF_3 \cdot OEt_2$, $BF_3 \cdot Bu_2O$, $SnCl_4$, $TiCl_4$, $ZrCl_4$, thionyl chloride, phosphorus trichloride, SbF_3, CH_3COClO_4, $FeCl_3$, $SBCl_5$, H_2SO_4, salts with anion of either hexachloroarsenate, hexafluoroarsenate, hexachlorophosphate, or hexafluorophosphate, molybdenyl acetylacetonate, derivatives of trifluoromethanesulfonic acid, $HClO_4$ (Bartz 1960; Burg, Schlaf, and Cherdron 1971; Chen 1971, 1976a,b; Fischer et al. 1967; Jaacks and Kern 1963; Kennedy 1966; Penczek et al. 1973; Sandler 1974; Schneider 1957)

TABLE 6.4
Comonomers Used in Polyoxymethylene Copolymer Synthesis

Comonomer Group	Comonomer	Chemical Structure	Ref.
Cyclic formals	1,3-Dioxolane		Weissermel et al. 1967b
	1,3-Dioxane		Weissermel et al. 1967b
	1,3-Dioxepane		Weissermel et al. 1967b
	1,3,6-Trioxocane		Weissermel et al. 1967b
	1,3,5-Trioxacycloheptane		Shieh, Lay, and Chen 2003
Epoxy compounds and cyclic ethers	Trimethylene oxide		Berardinelli and Hudgin 1961
	Tetrahydrofuran		Berardinelli and Hudgin 1961
	Ethylene oxide		McAndrew 1969; Weissermel et al. 1967a
	Propylene oxide		Weissermel et al. 1967a; McAndrew 1969
	Glycidyl phenyl ether		Makabe, Okita, and Yamamoto 1992; McAndrew 1969
	Glycidyl naphthyl ether		Makabe, Okita, and Yamamoto 1991; Makabe, Okita, and Yamamoto 1992
	Styrene oxide		McAndrew 1969; Makabe, Okita, and Yamamoto 1992; Makabe, Okita, and Yamamoto 1991

(Continued)

TABLE 6.4 (CONTINUED)
Comonomers Used in Polyoxymethylene Copolymer Synthesis

Comonomer Group	Comonomer	Chemical Structure	Ref.
	Butanediol diglycidyl ether		McAndrew 1969
Lactone	γ-Butyrolactone		Defazio and Kray 1962; Mantell 1966
	β-Propiolactone		Mantell 1966; Kern et al. 1966
	δ-Valerolactone		Mantell 1966
	ε-Caprolactone		Kiss 1967; Mantell 1966
Cyclic carbonate	Ethylene carbonate		Kutsuma and Yamase 1969
Vinyl derivatives	Styrene		Kern et al. 1966; Hohr, Cherdron, and Kern 1962

6.3.4 END-GROUP CAPPING

POM is usually end-capped using acetic anhydride (Masamoto 1993; Masamoto and Matsuzaki 1994) or long-chain alkyl groups (Masamoto et al. 2000).

6.3.5 POM ADDITIVES

- *Antioxidants and heat stabilizers*: As the stabilizers for POM phenolic compounds having complex structure, dicyandiamides, aromatic amines, hydrazines, ureas, sulfur compounds, and polyamides are usually used. These stabilizers have never been used alone but have been usually used in combination with two kinds of stabilizers consisting of phenolic compounds, as an antioxidant, and nitrogen-containing compounds such as polyamides or dicyandiamines, as a heat stabilizer (Fukuda, Ishida, and Matsuoka 1975).

    The hindered phenolic antioxidants such as pentaerythritol tetrakis[3-(3,5-di-t-butyl-4-hydroxyphenyl)propionate], 2,2′-methylene bis(4-methyl-6-t-butylphenol), 1,3,5-trimethyl-2,4,6-tris(3,5-di-t-butyl-4-hydroxybenzyl)benzene, 4,4′-methylenebis(2,6-di-t-butylphenol), 4,4′-butylidenebis(6-t-butyl-3-methylphenol), 1,6-hexanediol bis[3-(3,5-di-t-butyl-4-hydroxyphenyl)propionate], 2,2′-thiodiethylbis[3-(3,5-di-t-butyl-4-hydroxyphenyl)-propionate], n-octadecyl 3-(4′-hydroxy-3′,5′-di-t-butylphenyl)-propionate, di-stearyl 3,5-di-t-butyl-4-hydroxybenzyl-phosphonate, triethylene glycol bis[3-(3-t-butyl-5-methyl-4-hydroxyphenyl) propionate], 2-t-butyl-6-(3-t-butyl-5-methyl-2-hydroxybenzyl)-4-methylphenyl acrylate, and N,N′-hexamethylenebis(3,5-di-t-butyl-4-hydroxyhydrocinnamamide) have been used as POM antioxidants (Sakurai et al. 1985; Sugio et al. 1982; Sugiyama 1992; Walter, Schuette, and Goerrissen 1989). As heat stabilizers alkaline earth metal silicates or alkaline earth metal glycerophosphates, alkaline earth metal compounds, calcium, barium, and strontium, or the salts of these metals with saturated or unsaturated carboxylic acids, the hydroxides of magnesium and metal-containing compounds selected from the group consisting of the hydroxides, inorganic acid salts, carboxylic acid salts, and alkoxides of an alkali metal have been used (Roos and Wolters 1973; Schuette et al. 1983; Sugio et al. 1982; Walter, Schuette, and Goerrissen 1989).

- *Formaldehyde or formic acid-trapping agents*: Formaldehyde reactive nitrogen-containing compounds (i.e., dicyandiamide, amino-substituted triazines, copolycondensates of amino-substituted triazine and formaldehyde), formaldehyde reactive nitrogen-containing polymers (i.e., polyamide resins, polymers obtained by polymerizing acrylamide and its derivative or acrylamide and its derivative with other vinyl monomers in the presence of a metal alcoholate, polymers obtained by polymerizing acrylamide and its derivative or acrylamide and its derivative with other vinyl monomers in the presence of a radical polymerization catalyst, polymers containing nitrogen groups such as amine, amide, urea, urethane), and hydroxides, inorganic acid salts, carboxylates, or alkoxides of alkali metals or alkaline earth metals (Horio and Yoshinaga 2003; Sakurai et al. 1985; Schuette et al. 1983).
- *Lubricants*: Aliphatic esters of polyethylene wax, polyoxyalkyleneglycols, polyamide, poly(vinylidene difluoride), perfluoropolyether, modified polyolefins, such as a maleic anhydride modified polyolefins, solid lubricant polytetrafluoroethylene (PTFE)/MoS_2 (Chang and Lin 2012, 2013; Horio and Yoshinaga 2003; Kurz and Schleith 2003; Sun, Yang, and Li 2008a).
- *Molding assistant*: Polyethylene glycol 6000, stearamide (Sterzel et al. 1992).
- *UV-stabilizers*: 2-2(Hydroxy-5-t-octylphenyl)benzotriazole, bis(2,2,6,6-tetramethyl-4-piperidinyl)sebacate (Mulholland 1990), a mixture of 2-[2-hydroxy-3-tert.-butyl-5 (2-carbo-n-octyloxyethyl)-phenyl]-5-chlorobenztriazole and 2-[2-hydroxy-3-tert.-butyl-5-[2-carbo-(2-ethylhexyl)-oxyethyl]-phenyl]-5-chlorobenztriazole (Rody and Slongo 1989), carbon blacks, 2,4-bis-(2′-hydroxyphenyl)-6-alkyl-S-triazines, 4-hydroxybenzophenones, 1,3-bis-(2′-hydroxybenzoyl), benzene derivatives oxalic acid diamides, such as oxalic acid dianelide, 2-(2′-hydroxyphenyl)-benzotriazoles (Colombo, Neri, and Riva 1999; Nun, Schauhoff, and Dorn 1997).
- *Impact modifiers*: Thermoplastic polyurethanes (TPUs; Grossman and Lutz 2000; Wagner and Roehr 1965), 2-octylacrylate (Bergeret-Richaud et al. 2012), acrylic rubber composites composed of two or more acrylic rubber components having a different glass transition temperature with a vinyl monomer and a compound with a sulfonic acid group or a sulfuric acid group (Ito and Nakamura 2001), two-phase mixture of polybutadiene and styrene/acrylonitrile (ABS) (Burg and Cherdron 1971, 1972), polybutadiene (DiEdwardo and Kusumgar 1984; Memon and Weese 1995), graft copolymer prepared on an acrylic ester/butadiene basis (Bronstert, Schmidt, and Schuette 1971), polydiene/polyalkylene oxide block polymer (Schmidt et al. 1975), polysiloxanes or silicone rubbers (Schmitt and Sterzel 1978), polyacrylate, polymethacrylate, polyacrylonitrile, polymethacrylonitrile (Burg et al. 1991).
- *Nucleating agents*: Talc, combination of talc and POM terpolymer (Zierer, Ziegler, and Kurz 2004), silane-coated talc, polymeric nucleating material comprising ethylene/methyl acrylate copolymer and aliphatic amide wax (Kassal, Mori, and Shinohara 2001), boron nitride (Collins 1966), silica, polyimides, copper phthalocyanine, calcium carbonate,

diatomite, dolomite and encapsulated nucleant in polyolefins (Vaidya 1994), a branched or cross-linked acetal copolymer or terpolymer, a melamine-formaldehyde resin (Kassal 1999).

- *Colorants*: Carbon black (Van De Walle and Fourcade 1968), inorganic pigments such as titanium dioxide, ultramarine blue, cobalt blue and others, and organic pigments and dyes such as phthalocyanines, anthraquinones, and others, either individually or as a mixture or together with polymer-soluble dyes (Disch et al. 2001), Li(Co, Zn, Cd)Ti$_3$O as blue-green pigments (Arno and Herbert 1975), iron oxides (Ayala and Joyce 1990).
- *Flame retardants*: Amine, ammonium or amidine phosphate (Bruner and Busse 1969), aluminum hydroxide (Hull, Witkowski, and Hollingbery 2011), aluminum hydroxide/melamine/novolac resin synergistic system (Wang, Liu, and Wang 2010), MMT (Kiliaris and Papaspyrides 2010), magnesium hydroxide with melamine, linear novolac resin, and triphenyl phosphate (Liu, Wang, and Wang 2012).
- *Antistatic agents*: Polyetheresteramide (Jeong, Kim, and Lee 2007), esters of mono-, di-, tri-, or tetra-glycerol with fatty acid, alkanesulfonates (Bergner 1993), polyalkylene polyols or metal salt-dissolved polyalkylene polyols (Komatsu, Igarashi, and Erami 1995), conductive fillers such as carbon black, carbon fiber, carbon nanotubes, stainless-steel fiber masterbatch (Richaud 2014).
- *Fillers*: CaCO$_3$, MgCO$_3$, BaSO$_4$, SiO$_2$, Al(OH)$_3$, glass beads, dolomite, carbon black, graphite, wood flour, talc, mica, clays, wollastonite, cellulose fibers, glass fibers, carbon fibers and nanofibers, aramid fibers, kenaf fibers, PET fibers (Abdullah, Dan-Mallam, and Yusoff 2013; Hsu, Hwang, and Ting 2010; Kawaguchi et al. 2010; Richaud 2014; Siengchin, Psarras, and Karger-Kocsis 2010; Xu et al. 2012; Zhao and Ye 2009).

6.4 POLYMERIZATION MECHANISMS FOR POLYACETALS

POM can be obtained by polymerization of formaldehyde or trioxane using the ionic polymerization method. Cationic polymerization of trioxane is performed via ring opening reactions, whereby formaldehyde can be polymerized either anionically or cationically (see Schemes 6.1 and 6.2; Weissermel et al. 1967b).

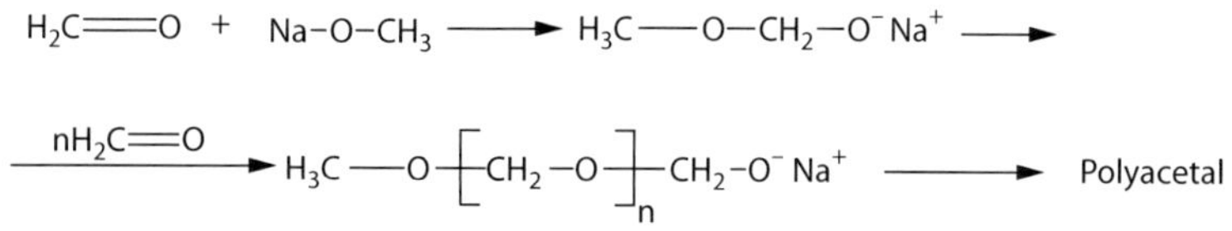

SCHEME 6.1 Anionic polymerization of formaldehyde.

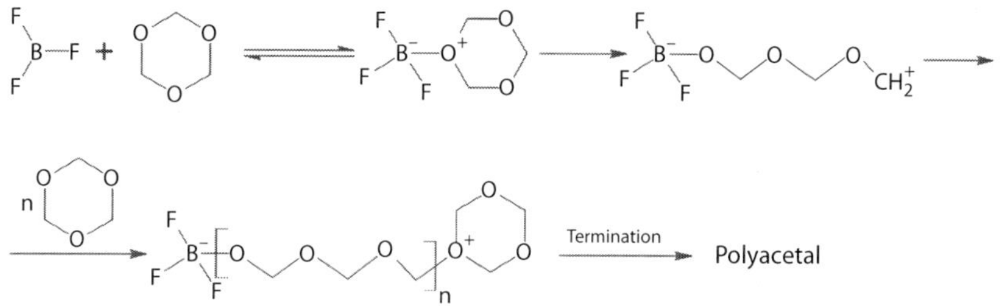

SCHEME 6.2 Cationic polymerization of 1,3,5-trioxane.

Polyacetal can be produced by formaldehyde polymerization in aqueous solutions, by polymerization of anhydrous gaseous formaldehyde, as well as by polymerization in bulk and solution of monomeric liquid or gaseous formaldehyde according to the reaction shown in the following (Sandler and Karo 1992; Spence 1933):

$$nCH_2 = O \rightarrow [-CH_2-O-]_n$$

Trioxane can be polymerized in the solid state, in the melt, in the gas phase, in suspension, and in solution with a POM precipitation stage. POM does not dissolve in the reaction mixture, and thus, after the induction period, phase separation accompanied by crystallization takes place. The phase separation and crystallization of POM during its polymerization lead to complicated reaction mechanism and kinetics (Shieh and Chen 1999). Kern and Jaacks (1960) demonstrated that the cationic polymerization of 1,3,5-trioxane in methylene dichloride with boron trifluoride is initiated by a direct interaction of BF_3 with trioxane without a cocatalyst such as water, and that it proceeds through a zwitterion or by a free-ion mechanism as it is presented in Scheme 6.3.

On the other hand, Miki et al. (1966) shown that the polymerization followed the ion-pair mechanism with a step-by-step insertion of a monomer unit, and the additive (cocatalyst) participates in the initiation of the polymerization reaction as it is displayed in Scheme 6.4.

In the work by Leese and Baumber (1965), it was observed that, during an induction period, soluble low molecular weight polymer is produced, which eventually forms crystal nuclei, and these nuclei initiate an accelerating reaction whose rate depends on the area of the crystal surface available for polymer deposition or formation. A progressive increase in the size of the original crystals was suggested rather than constant formation of fresh nuclei. Berlin and Yenikolopyan (1969) proposed that trioxane polymerization can be regarded as an equilibrium process, with the polymerization active centers either dissolved or positioned on the solid surfaces of the polymer. The following reversible processes were postulated to occur during polymerization of trioxane: (1) crystallization and solution of the active ends; (2) addition of the monomer and depolymerization of dissolved active ends; and (3) the same reactions involving an active center positioned on the solid surface. This heterogeneous reaction is characterized by reversible polymerization–depolymerization equilibrium

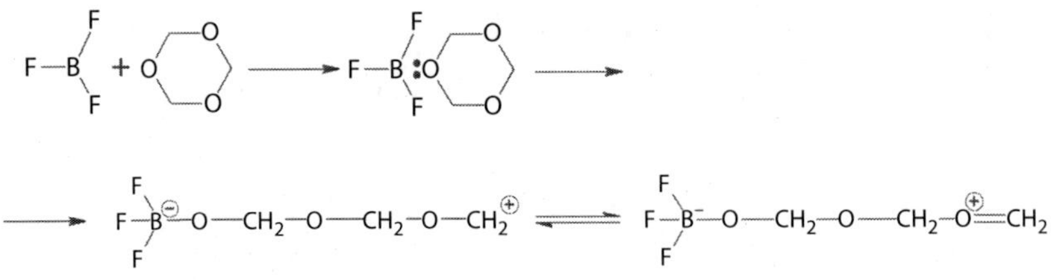

SCHEME 6.3 Initiation of polymerization process of 1,3,5-trioxane by boron trifluoride. (Adapted from Miki, T. et al., *Bulletin of the Chemical Society of Japan* 39(11):2480–2485, 1966; Kern, W., and V. Jaacks, *Journal of Polymer Science* 48(150):399–404, 1960.)

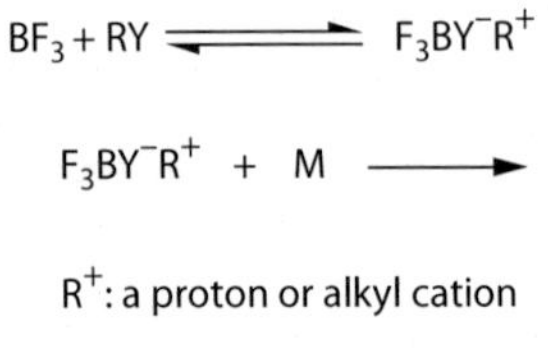

$$BF_3 + RY \rightleftharpoons F_3BY^-R^+$$

$$F_3BY^-R^+ + M \rightarrow RM^+...BF_3Y^-$$

R^+: a proton or alkyl cation

SCHEME 6.4 Initiation of polymerization process of 1,3,5-trioxane by boron trifluoride. (Adapted from Miki, T. et al., *Bulletin of the Chemical Society of Japan* 39(11):2480–2485, 1966.)

between the dissolved monomer and the solid polymer (Rodríguez-Baeza 1991). Currently, there is a general agreement to consider that the reaction is initiated by the Brönsted acid resulting, in very low concentration, from the equilibrated reaction of BF_3/OR_2 with residual water (coinitiation) (Sharavanan et al. 2009) as it is presented in Scheme 6.5.

Iguchi et al. (1969) observed extended chain crystals (ECCs) of polymers formed in the course of cationic polymerization of trioxane. They revealed that long, needle-like nuclei formed in the beginning of the polymerization grew into feather-like platelets with twin habit, and the degree of crystallinity reached a very high value due to full extension of molecular chains and their regular alignment in the direction of the crystal growth. Rodríguez-Baeza and Catalán Saravia (1985) studied the effect of the initiator concentration upon the kinetics of polymerization, induction times, and morphology and on the crystallization behavior of POM. The mechanism that is responsible for the POM homopolymer crystal growth, when perchloric acid was used as the initiator, was described by Wegner et al. (1980) and Dröscher and Wegner (1979). They postulated that the kinetics of the polymerization reaction is controlled by those parameters, which are typical for the mechanism of crystal growth, and the polymerization starts after a certain nucleation period of time has passed. The growth of the individual crystals with time is governed by characteristic dislocations, which control the lateral growth as well as the thickening of the crystals. Large amounts of the initiator are adsorbed onto the surfaces of the growing crystals, catalyzing the insertion of monomers into the already formed polymer chain. The POM crystal is schematically shown in Figure 6.1, and microphotographs of the hexagonal pyramidal form of POM crystals are presented in Figure 6.2.

The lateral growth of the hexagonal base lamella is due to the addition of monomer or oligomer to cationically active chain ends, while the thickness growth of the crystals occurs due to a spiral growth; the surface of the crystals at the beginning of the reaction shows regular steps of spiral growth, which is propagated from the center of the crystals (Rodríguez-Baeza and Catalán Saravia 1985). The spiral front proceeds during the polymerization reaction as it is schematically drawn in Figure 6.3.

Dröscher and Wegner (1979) postulated that the reversible polymerization step leads only to a stable product if it occurs near to the kink position of the growing spiral. The reaction proceeds until the enlarged fold reaches the upper surface and a new kink position is built. Additionally, for the polymerization reaction itself, a transacetalization reaction by which the monomers or the always present cyclic oligomers (Jaacks 1969) are incorporated into the chain fold (Plesch and Westerma 1968, 1969) can occur. Besides, an addition onto a cationically active chain end (Weissermel et al. 1967a) can be considered—as a result of both reactions, extended polymer chains are formed. Schematic formation of extended and folded POM crystals is presented in Figures 6.4 and 6.5.

A similar mechanism is observed in the copolymerization of trioxane with various cyclic formals. Thus, crystalline copolymers are obtained with the co-units incorporated into the POM crystals, which behave like solid solutions (Dröscher and Wegner 1979). Additionally, some polyacetals, e.g., poly-(1,3-dioxolane), may be prepared either as almost exclusively cyclic or as predominantly

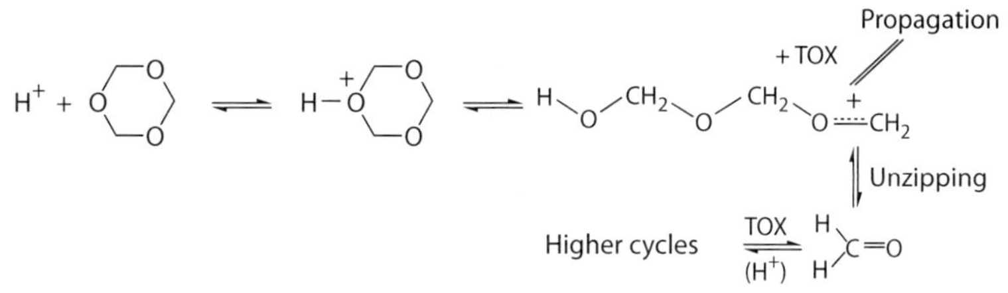

SCHEME 6.5 Polymerization of 1,3,5-trioxane with a protonic initiator. (Reprinted with permission from Sharavanan, K., E. Ortega, M. Moreau, C. Lorthioir, F. Laupretre, P. Desbois, M. Klatt, and J. P. Vairon, *Macromolecules*, 8702–8710. Copyright 2009 American Chemical Society.)

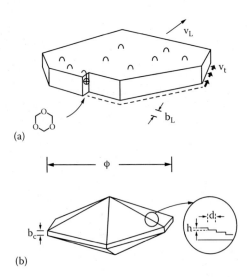

FIGURE 6.1 (a) Schematic representation of a POM crystal. (b) Geometric model of a POM crystal, which grows on the basis of a screw dislocation to an ideal bipyramidal form. (Wegner, G., M. Rodriguez-Baeza, A. Lücke, and G. Lieser: Kinetik und Mechanismus der kationischen Polymerisation von Trioxan: Ein katalysierter Kristallwachstumsprozeß. *Die Makromolekulare Chemie* 1980, 181(8), 1763–1790. Copyright Wiley-VCH Verlag GmbH & Co. KGaA. Reproduced with permission.)

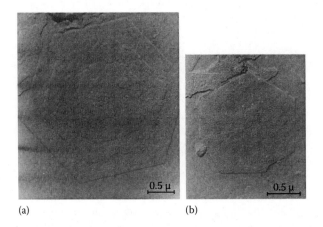

FIGURE 6.2 Electron micrograph (replica) of POM crystals from the homopolymerization of trioxane with $HClO_4$ in CH_2Cl_2: $HClO_4$,10-5 M [trioxane], 2.5 M. (Reprinted with permission from Dröscher, M. and G. Wegner, *Industrial & Engineering Chemistry Product Research and Development*, 259–263. Copyright 1979 American Chemical Society.)

linear macromolecules, depending on the starting monomer concentration (Szymanski, Kubisa, and Penczek 1983).

In case of copolymerization of trioxane with comonomers (Sharavanan et al. 2009), the initiation reactions are governed by the relative reactivity of trioxane and, for example, 1,3-dioxolane, and led to trioxane-cyclic oxonium and dioxolane-cyclic oxonium as in reactions 2 and 3, respectively. Because the reactivity of dioxolane is higher than that of trioxane (the basicity of dioxolane with pKa = 7.55 is higher than that of trioxane with pKa = 10.00; Yamashita et al. 1966), reaction 3 prevails over reaction 2. Additionally, the trioxane-cyclic oxonium can convert to formaldehyde as in reaction 4 (Allcock 1967; Withey and Whalley 1963), which then reacts with dioxolane-cyclic

Polyacetals

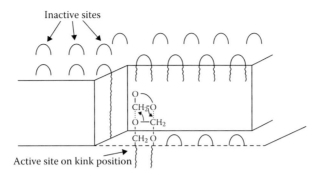

FIGURE 6.3 Schematic representation of the polymerization reaction near the kink position of a POM crystal. (With kind permission from Springer Science+Business Media: *Industrial & Engineering Chemistry Product Research and Development*, Kinetics of a simultaneous polymerization and crystallization: Cationic polymerization and copolymerization of trioxane, 18(4), 1979, 259–263, Dröscher, M. and G. Wegner.)

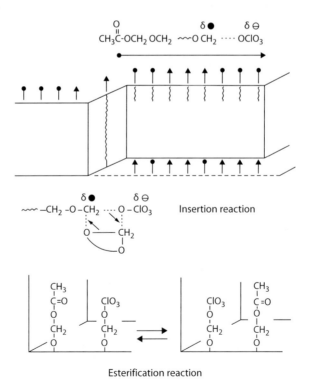

FIGURE 6.4 Mechanism of the thickness growth. The crystal surface contained acetyl ester end groups and growing perchlorate ester end groups. Folded chains are not present. (With kind permission from Springer Science+Business Media: *Frontiers in Polymer Science*, edited by W. Wilke, Effect of the initiator acetyl perchlorate upon the morphology, kinetics and mechanisms of crystal growth of poly(oxymethylene), 1985, 49–58, Rodríguez-Baeza, M., and R. E. Catalán Saravia.)

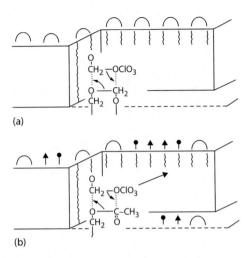

FIGURE 6.5 Crystal growth by participation of folded chains. (a) Insertion reaction. (b) Liberation of the initiator. (With kind permission from Springer Science+Business Media: *Frontiers in Polymer Science*, edited by W. Wilke, Effect of the initiator acetyl perchlorate upon the morphology, kinetics and mechanisms of crystal growth of poly(oxymethylene), 1985, 49–58, Rodríguez-Baeza, M. and R. E. Catalán Saravia.)

oxonium to form the 1,3,5-trioxepane by-product (Scheme 6.6) (Allcock 1967; Collins et al. 1979, 1981; Shieh, Lay, and Chen 2003; Wamser 1951; Withey and Whalley 1963; Yamashita et al. 1966).

During propagation, the trioxene-cyclic oxonium from reaction 2 can react with formaldehyde to produce tetroxocane—reaction 6. Next, trioxane appears to be considerably consumed for the copolymerization as in reactions 7 or 8. The propagation of copolymer chains may also proceed through the reactions of the trioxane-cyclic oxonium and formaldehyde—reactions 9 and 10 in Scheme 6.7 (Shieh, Lay, and Chen 2003).

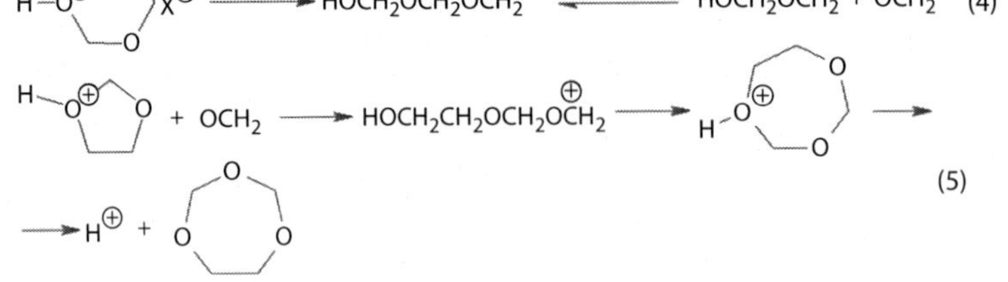

SCHEME 6.6 Initiation reactions of POM copolymer. (From Shieh, Y. T. et al., *Journal of Polymer Research* 10(3):151–160, 2003; Collins, G. L. et al., *Journal of Polymer Science: Polymer Chemistry Edition* 19(7):1597–1607, 1981; Wamser, C. A., *Journal of the American Chemical Society* 73(1):409–416, 1951; Collins, G. L. et al., *Journal of Polymer Science: Polymer Letters Edition* 17(10):667–671, 1979; Yamashita, Y. et al., *Journal of Polymer Science Part A-1: Polymer Chemistry* 4(9):2121–2135, 1966; Withey, R. J. and E. Whalley, *Transactions of the Faraday Society* 59(0):901–906, 1963; Allcock, H. R., *Heteroatom Ring Systems and Polymers*: Academic Press, 1967.)

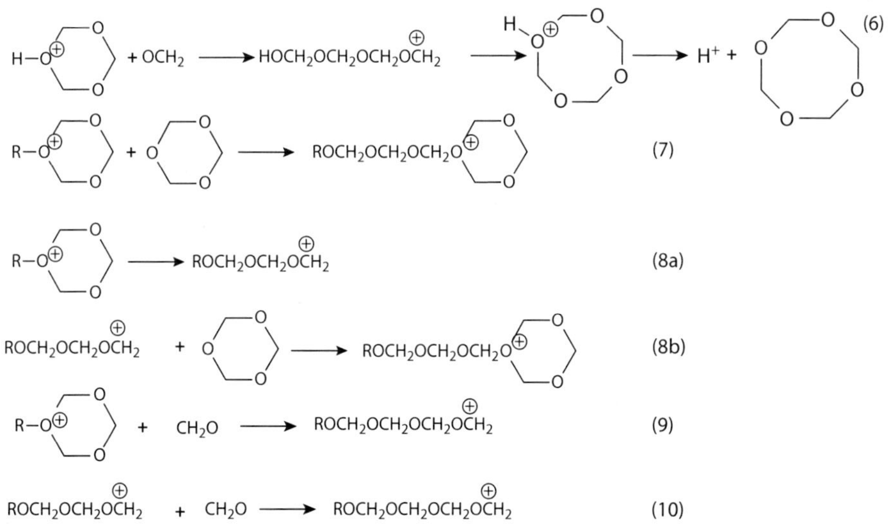

Polyacetals

SCHEME 6.7 Propagation reactions of POM copolymer. (From Shieh, Y. T. et al., *Journal of Polymer Research* 10(3):151–160, 2003.)

The oxocarbenium cations are less stable than cyclic oxonium cations (Enikolopyan et al. 1967) and can be stabilized by complexing with oxygen-containing cyclic formals, oligomers, or polymers to form oxonium intermediate species. The stabilized oxocarbenium cations can participate in various chain transfer reactions, e.g., hydride transfer reactions (Shieh, Yeh, and Chen 1999; Vorobena et al. 1974), transacetalization reactions (Cherdron 1972; Weissermel et al. 1964), and back-biting reactions (Miki, Higashimura, and Okamura 1967), to form methoxy groups in the chain ends, random copolymers, and tetroxocane by-products, respectively (Shieh, Lay, and Chen 2003).

The influence of the reaction conditions, such as initiator and comonomer concentration on the transfer to residual water and the product formation during the short induction period, was examined during the bulk copolymerization of 1,3,5-trioxane with 1,3-dioxepane initiated by perchloric acid hydrate(s) (Sharavanan et al. 2009), and a consistent description of the global polyacetal formation process was proposed (see Figure 6.6).

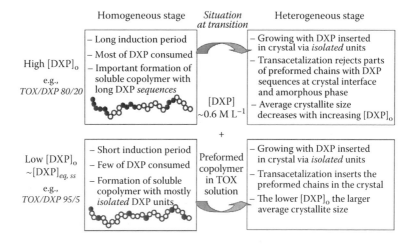

FIGURE 6.6 Simplified overview of the 1,3,5-trioxane and 1,3-dioxepane (TOX-DXP) copolymerization process in bulk. (Reprinted with permission from Sharavanan, K., E. Ortega, M. Moreau, C. Lorthioir, F. Laupretre, P. Desbois, M. Klatt, and J. P. Vairon, *Macromolecules*, 8702–8710. Copyright 2009 American Chemical Society.)

6.5 CHARACTERISTICS OF POLYACETALS

From a fit to the measured heat capacity data, it has been reported that the microstructure of POM comprises crystalline, amorphous, and rigid amorphous phase, and the relative fractions of these phases are strongly influenced by sample preparation conditions (Kumaraswamy et al. 2012; Suzuki, Grebowicz, and Wunderlich 1985). POM crystallizes in two crystallographic forms: as a stable hexagonal POM (h-POM) consisting of 9/5 (or 29/16) helical conformation and metastable orthorhombic POM (o-POM) consisting of 2/1 helical molecules (Carazzolo 1963; Carazzolo and Mammi 1963; Kobayashi 1993; Morishita et al. 1989; Tadokoro et al. 1960; Uchida and Tadokoro 1967). An orthorhombic unit cell contains two polymer chains, each containing two monomeric units, while the hexagonal form contains one chain of nine monomeric units per unit cell (Raimo 2014; Zamboni and Zerbi 1964). h-POM can be obtained through the crystallization from the melt or dilute solutions of POM. The single crystals of h-POM obtained in dilute solution show the typical hexagonal morphology with folded chain crystals (FCCs). Cationic polymerization of trioxane using boron trifluoride and water as catalyst yields feather-shaped platelets and needle-like (polymer whisker) single hexagonal crystals consisting of ECCs (see Figure 6.7; Iguchi 1973; Iguchi, Kanetsun H, and Kawai 1969; Morishita et al. 1989).

Orthorhombic POM was found in a polymerization system of alkaline aqueous formaldehydes (Carazzolo and Mammi 1963). Moth-shaped platelets and rod-like o-POM single crystals consisting of ECC were obtained as by-products in a cationic polymerization of trioxane, too (Kobayashi, Morishita, and Shimomura 1989). The metastable o-POM single crystals transform irreversibly to the stable h-POM single crystals at 69°C with an endothermic effect of 0.6 kJ/mol per -CH_2O- unit on heating, while the morphology, the fiber axis orientation, and the extended-chain structure of the starting orthorhombic crystals remain unchanged throughout the thermal transition (Kobayashi et al. 1987). The hexagonal form has unit cell dimensions $a = b = 4.45$ Å and $c = 17.3$ Å; a and b axes are in the same plane perpendicular to the c-axis, which is parallel to the molecular chain (Carazzolo 1963; Zhao and Ye 2011b). The orthorhombic form has unit cell dimensions $a = 4.77$ Å, $b = 7.65$ Å, and $c = 17.80$ Å. The orthorhombic form can be transformed to hexagonal form by an expansion along the a-axis and a contraction along the b-axis (Preedy and Wheeler 1972; Raimo 2014). It was found that the conformation around the C–O bonds of POM helices is predominantly of the same sign of gauche type (Raimo 2014). At room temperature in SAXS results, two peaks L_1 at 0.45 nm^{-1} and L_2 at 0.90 nm^{-1}, corresponding to long periods of 14 and 7 nm, respectively, could be observed. In the works by Hama and Tashiro (2003a,b,c), this effect was explained by using a

FIGURE 6.7 Scanning electron micrographs of needle-like POM crystals. (Iguchi, M. and I. Murase: "Shish kebab" structures formed on needle-like polyoxymethylene crystals. *Journal of Polymer Science Part A: Polymer Chemistry.* 1975. 13. 1461–1465. Copyright Wiley-VCH Verlag GmbH & Co. KGaA. Reproduced with permission.)

lamellar insertion model, where new thinner lamellae are inserted between the thicker lamellae formed earlier as shown in Figure 6.8.

Generally, by crystallization from dilute solutions, single crystals of various shapes, e.g., rhomb, hexagon, or truncated figures, are formed. POM crystallizes as a thin single crystal of FCC morphology when the dilute solution is slowly cooled to room temperature (Figure 6.9a), but, as shown in Figure 6.9, it forms a whisker or a single crystal of fully ECC morphology when the ring-opening polymerization of trioxane is performed with a cationic catalyst (Iguchi 1983; Tashiro et al. 2004). However, during crystallization from the melt or saturated solutions, spherulitic morphology can be obtained as can be seen in Figure 6.10. The fastest growth rate of crystalline polyacetal spherulites occurs at 90°C (Chiang and Huang 1997).

In the POM copolymers, comonomer units can be arranged in the amorphous or in the crystalline phase, depending on the chemical structure and composition. The incorporation of comonomer units leads to considerable decrease in degree of crystallinity and melting point (Pielichowska 2012a,b). Kobayashi et al. (1993) found that in POM copolymers containing 3–30 wt.% of tetramethylene oxide units, with an increase in the comonomer content, the degree of crystallinity decreased. Tetramethylene oxide units were mainly located in the amorphous phase and at the amorphous/crystalline interface. Additionally, incorporation of tetramethylene oxide units in the interfacial regions or within crystals leads to a distortion of the POM hexagonal cell with an increase in the cell volume (Kobayashi et al. 1993; Raimo 2014).

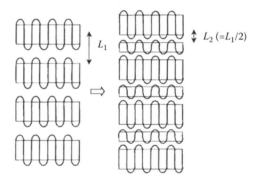

FIGURE 6.8 An insertion model of new lamellae in between the already existing lamellar stacking structure. (Reprinted from *Polymer*, 44(7), Hama, H., and K. Tashiro, Structural changes in non-isothermal crystallization process of melt-cooled polyoxymethylene[II] evolution of lamellar stacking structure derived from SAXS and WAXS data analysis, 2159–2168, Copyright 2003, with permission from Elsevier.)

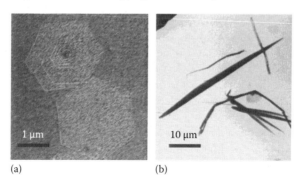

FIGURE 6.9 Electron microscopy images of (a) FCC and (b) whisker or ECC of POM. (Reprinted with permission from Tashiro, K., T. Kamae, H. Asanaga, and T. Oikawa, *Macromolecules*, 826–830. Copyright 2004 American Chemical Society.)

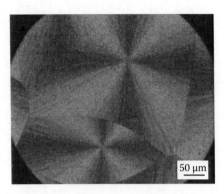

FIGURE 6.10 Polarizing optical micrographs of POM. (Ding, Q., W.-L. Dai, and P. Zhang: The effect of polyvinylidene fluoride on nonisothermal crystallization behavior of polyoxymethylene. *Polymer Engineering & Science*. 2007. 47. 2034–2040. Copyright Wiley-VCH Verlag GmbH & Co. KGaA. Reproduced with permission.)

Structure analysis of POM has been carried out by Uchida and Tadokoro (1967) and then by other authors (Aich and Hagele 1985); structural parameters are collected in Table 6.5.

Tashiro et al. (2004) investigated the molecular structure of POM whisker single crystal, and the averaged structural parameters were as follows: bond length CO = 1.42 Å, bond angles OCO = 109.0° and COC =114.8°, and torsional angle COCO = 77.8°. Molecular chain structure is displayed in Figure 6.11.

The Fourier transform infrared (FTIR) spectrum shows the absorption band of the carbonyl of acetoxy group at 1755 cm^{-1}, methylene of oxymethylene unit at 1470 cm^{-1}, and some other characteristic absorption bands at 938, 970, 980, 1028, 1125, and 2900 cm^{-1}. The absorption bands of the metastable conformations appear at 850, 920, 970, 1030, 1150, and 1165 cm^{-1} (Chiang and Huang 1997). Differences in FTIR spectra for POM homopolymer and copolymer are presented in Figure 6.12.

Due to presence of ECC and FCC morphologies, remarkable differences in the infrared spectra can be observed (Figure 6.13).

In the nuclear magnetic resonance (NMR) spectrum of the acetoxy group (CH_3COO-), proton appears at 2.1 ppm, and the proton of the oxymethylene unit —(CH_2O)— is located at 5.0 ppm (Chiang and Lo 1988).

6.6 PROPERTIES OF POLYACETALS

POM homopolymer is characterized by a high degree of crystallinity 50–80%, with an equilibrium melting temperature of 184°C and heat of fusion of 326 J/g (Pielichowska et al. 2013; Suzuki, Grebowicz, and Wunderlich 1985). POM has a melting point around 170–180°C with deflection point at 97°C. High crystallinity makes POM a material with high strength, stiffness, toughness, good creep resistance, fatigue endurance, good solvent resistance, and a low coefficient of friction. The water absorption of polyacetal is <0.2% giving the polymer excellent weight retention characteristics (Chiang and Huang 1997). POM shows a relatively narrow melting peak in differential scanning calorimetry (DSC) analysis; however, in a study by Sauer et al. (2000), it was suggested that the melting and recrystallization of POM crystals occurs over the broad temperature range of 60–160°C.

POM shows five relaxation transitions: ε-transition below −259°C and δ-transition around −223°C, which have not yet been assigned; γ-transition extending from −120°C to −70°C comprising a doublet; β-transition near 0°C, which has been identified as the glass transition; and a strong α-transition observed over a broad range of temperatures from 50°C to 150°C (Kumaraswamy et al. 2012; Miki 1970; Papir and Baer 1970, 1971). The latter transition was attributed to motions of amorphous chain segments constrained to various degrees by the extent of their association with crystalline lamellae. There is an increase in motion as temperature is increased and as crystalline constraints *soften* due

TABLE 6.5
Summary of Crystalline Structures of POM Obtained by X-Ray Diffraction

Inter/Intramolecular Parameters	Carazzolo (Carazzolo and Mammi 1963, Carazzolo 1963)		Uchida (Uchida and Tadokoro 1967)	Andrews (Andrews and Martin 1973)	Takahashi (Takahashi and Tadokoro 1979)	Gramlich (Gramlich 1977)	
	Hexagonal	Orthorhombic	Hexagonal	Hexagonal	Hexagonal	Orthorhombic	Orthohexagonal
Lattice Constants (nm)							
a	0.447	0.477	0.447	0.448	0.447	0.477	0.447
b	0.447	0.765	0.447	0.448	0.447	0.765	0.774
c	1.73;5.559(1)	0.356	1.739;5.602	0.173	1.739;5.602	0.356	7.34
Proposed helix type	2*9/5; 2*29/16	2*2/1	2*9/5; 2*29/16		2*9/5; 2*29/16	2*2/1	2*38/21
Helical pitch	0.349	0.35	–	0.346	–	0.356	0.350
Radius of helix (nm)	–	0.079	0.068	0.072	–	–	–
Bond Length (nm)							
C–O	0.143	–	0.1421;0.1418	0.1430	0.1433;0.1429	0.1410	0.1405
C–H	–	–	–	–	0.106	0.098	0.105
Valence Angle (°)							
COC	–	112.68	112.40;113.00	106.88	112.3;112.9	1114.1	114
OCO	–	112.68	110.82;11.1.37	–	109.9;110.4	112.8	112
Dihedral Angle (°)							
CH_2–O	102.62;103.65	–	101.78;102.55	106.3	102.4;103.0	115.6	102

Source: With kind permission from Springer Science+Business Media: *Progress in Colloid and Polymer Science*, Atomistic calculation of chain conformations and crystal-structures of polyoxymethylene, 71, 1985, 86–95, Aich, R., and P. C. Hagele.

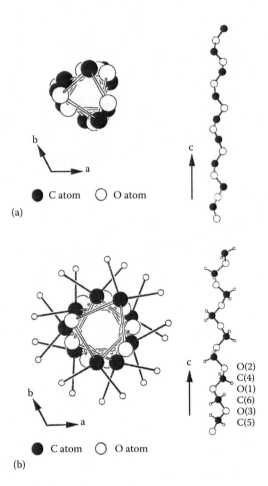

FIGURE 6.11 Molecular structure of POM obtained (a) by applying the direct method to the electron diffraction data of the whisker and (b) by the refinement of structure (a) by combining with the x-ray diffraction data taken for γ-ray polymerized sample. (Reprinted with permission from Tashiro, K., T. Kamae, H. Asanaga, and T. Oikawa, *Macromolecules*, 826–830. Copyright 2004 American Chemical Society.)

to melting of thin lamellae. The energy of these motions was estimated, and this was shown to coincide with the POM α-transition (Bershtein et al. 2002; Hagemeyer, Schmidt-Rohr, and Spiess 1989; Kentgens, Deboer, and Veeman 1987; Kentgens et al. 1985; Kumaraswamy et al. 2012). β-Relaxation process was ascribed to intrachain molecular motions or motions of long molecular segments in the disordered phase (Sauer et al. 1997; Siengchin et al. 2008). Some early dielectric results suggested that the transition at the temperature –70°C is the glass transition because of its *cooperative* character (Read and Williams 1961). However, because a strong γ-transition is observed in single crystal mats of POM and it also occurs at the same temperature in melt crystallized POM, it was suggested that it is not the glass transition but transition occurring in defected regions within the crystal lamellae (Takayanagi 1963). Hojfors et al. (1977) showed that it is an amorphous local relaxation in the noncrystalline phase, whereas the low-temperature shoulder (γ′-transition) on the main γ-transition, seen most clearly at low frequencies, is due to relaxation of dislocations or defects in the crystals (Sauer et al. 1997).

POM homopolymer displays a very poor thermal stability. POM undergoes predominant depolymerization during thermal aging at processing temperatures, typically above 150°C, and thermal aging is noticeably accelerated in the presence of oxygen (Fayolle et al. 2008; Kern and Cherdron 1960). It has the lowest limiting oxygen index (LOI = 15) and very high oxygen content (53%)

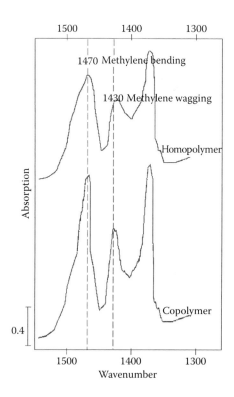

FIGURE 6.12 FTIR spectra of POM homopolymer versus POM copolymer with the characteristic absorption bands corresponding to CH$_2$ wagging and bending. (Reprinted from Ramirez N. V. et al., *Polymer-Plastics Technology and Engineering*, 48:470–477, 2009. With permission from Taylor & Francis.)

compared to the other polymers, and is therefore extremely flammable (Harashina, Tajima, and Itoh 2006; Hilado 1990; Wang, Liu, and Wang 2010). POM burns violently in blooming blue flame, producing a large number of flaming molten drops, which make the fire easily spread. Because of lack of potential reactive functional groups, it is hard for polyacetal itself to participate in the char-forming reaction during combustion (Wang, Liu, and Wang 2010). POM is considered as the most difficult thermoplastic polymer to be flame retarded (Sun et al. 2010).

POM chains are sensitive to acidic or basic compounds (protonic or Lewis acids such as formic acid, hydrogen halide, and antimony halide) and easily catalytically degrade in the presence of such compounds to produce highly flammable formaldehyde monomers (Sun et al. 2010; Wang, Liu, and Wang 2010). At an elevated temperature, POM easily decomposes in an *unzipping* reaction to yield formaldehyde that can be easily oxidized to formic acid, which accelerates the polymer decomposition (see Scheme 6.8; Archodoulaki, Lüftl, and Seidler 2007; Bruner and Busse 1969; Harashina, Tajima, and Itoh 2006; Liu et al. 2008; Sun et al. 2010).

Hydroxyl-terminated POMs undergo thermal and thermo-oxidative degradation through acidolysis and hydrolysis starting at the chain ends (Kern et al. 1961). In order to increase the thermal stability, the hydroxyl groups at the chain ends have to be removed by acetylation (end-capping; Lüftl et al. 2008). Another method of improvement of POM thermal stability is to copolymerize trioxane with comonomers such as ethylene oxide or dioxolane to produce POM copolymer, which consists of oxymethylene segments with oxyethylene units in the main chain (Walling, Brown, and Bartz 1962)—depolymerization of oxymethylene segments can be stopped at the first oxyethylene unit. POM is also affected by irradiation that is damaging POM resins and deteriorating its physical and engineering properties (Kassem, Bassiouni, and El-Muraikhi 2002; Subramanyam and Subramanyam 1987; Zahran 1998). POM exhibits very good mechanical properties, like high

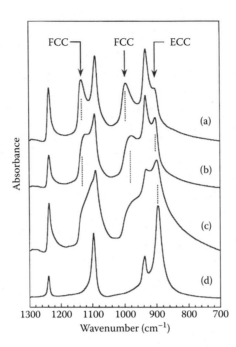

FIGURE 6.13 Infrared spectra of POM samples prepared under various conditions. (a) Solution-grown lamellar mat. (b) Film cast from HFIP solution. (c) Film cooled from the melt. (d) Whisker. (Reprinted from *Polymer* 44(10), Hama, H., and K. Tashiro, Structural changes in non-isothermal crystallization process of melt-cooled polyoxymethylene. [I] Detection of infrared bands characteristic of folded and extended chain crystal morphologies and extraction of a lamellar stacking model, 3107–3116, Copyright 2003, with permission from Elsevier.)

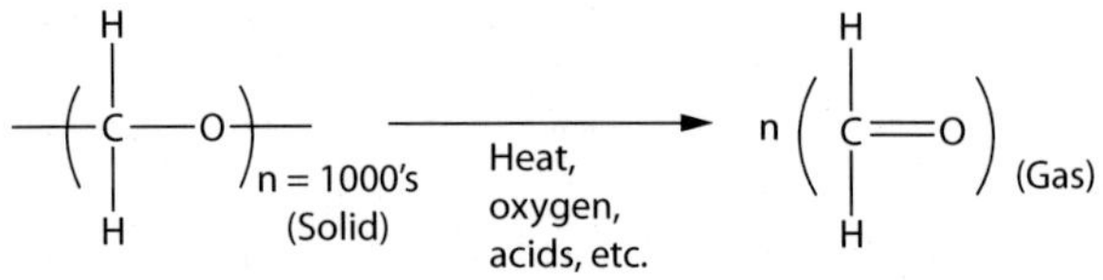

SCHEME 6.8 Methods of degradation of pure POM. (With kind permission from Springer Science+Business Media: *Journal of Materials Science*, Investigation on thermo-stabilization effect and nonisothermal degradation kinetics of the new compound additives on polyoxymethylene, 44, 2009, 1251–1257, Shi, J., B. Jing, X. Zou, H. Luo, and W. Dai.)

tensile strength, flexural modulus, and deflection temperature. Additionally, it possesses high creep and fatigue resistance, as well as resistance to friction and wear (Sun et al. 2010). POM characteristics include low friction coefficient due to the flexibility of linear molecular chains and good wear-resistant properties because of its high crystallinity and high bond energy (Sun, Yang, and Li 2008a). Because of excellent friction and wear property, dimensional stability, and chemical stability, POM is widely used to produce self-lubricating mechanical parts (Hu et al. 2007). Selected properties of different POM grades are summarized in Table 6.6.

6.7 POLYACETALS PROCESSING

6.7.1 MELT PROCESSING

6.7.1.1 Injection Molding

POM can be processed by means of typical processing methods like extrusion or injection molding. Injection molding is a two-step cyclical process consisting of melt generation by a rotating screw,

TABLE 6.6
Selected Mechanical, Thermal, and Electrical Properties of POM

POM	Test Method ISO	Delrin, DuPont (DuPont 2000)	Ultraform, BASF (BASF 2007a)	Hostaform C 9021, Ticona (Ticona 2007)	Tenac 3010, Homopolymer, Asahi Kasei Chemicals Corporation (Asahi 2010)	Duracon, Polyplastics Co. (Standard Grades) (Polyplastics Co. 2014)	Lucel, LG Chem (Injection Molding POM Grades) (Chem 2009)	Tepcon, Polyplastics Co. (Standard Grades) (Polyplastics 2013)	KEP Kepital, Korea Engineering Plastics Co., Acetal Homopolymers (Plastics 2014)	KEP Kepital, Korea Engineering Plastics Co., Acetal Copolymer (Plastics 2014)	Tarnoform, Grupa Azoty S.A. (Standard Grades) (Azoty 2007)
Density (g/cm^3)	ISO 1183	1.42	1.54–1.58	1.41	1.42	1.41	1.41	1.41	1.41	1.41	1.41
Water absorption, saturation in water at 23°C (%)	DIN 53 495/1L	ASTM D570	–	0.65	0.2	–	–	–	0.22 24 h at 73°C	0.22 24 h at 73°C	0.7
Moisture absorption, saturation under standard climatic conditions 23°C/50% r.h. (%)	–	0.9–1.4 ASTM D570	–	0.20	–	–	–	–	–	–	0.2
Melt flow rate MFR 190/2.16 (g/10 min)	190°C, 2.16 kg, ISO 1133	1–17 ASTM D123	2.6	8 (cm^3/10 min)	2.8	–	9–45	–	30–45 ASTM D1238	9–27	2.5–48
Tensile modulus (MPa)	ISO 527-1,2	–	2600	3100	–	2500–2800	–	–	–	–	2350–3000
Tensile strength (MPa)	ISO 527-1,2	67–73 ASTM D638	62–63	64	71	59–63	620–635 kg/cm^2 ASTM D638	59–63	60–62.2 ASTM D638	59–63	60–68

(*Continued*)

TABLE 6.6 (CONTINUED)
Selected Mechanical, Thermal, and Electrical Properties of POM

POM	Test Method ISO	Delrin, DuPont (DuPont 2000)	Ultraform, BASF (BASF 2007a)	Hostaform C 9021, Ticona (Ticona 2007)	Tenac 3010, Homopolymer, Asahi Kasei Chemicals Corporation (Asahi 2010)	Duracon, Polyplastics Co. (Standard Grades) (Polyplastics Co. 2014)	Lucel, LG Chem (Injection Molding POM Grades) (Chem 2009)	Tepcon, Polyplastics Co. (Standard Grades) (Polyplastics Co. 2013)	KEP kepital, Korea Engineering Plastics Co., Acetal Homopolymers (Plastics 2014)	KEP kepital, Korea Engineering Plastics Co., Acetal Copolymer (Plastics 2014)	Tarnoform, Grupa Azoty S.A. (Standard Grades) (Azoty 2007)
Tensile strain at break ($v = 50$ mm/min) (MPa)	—	66–73 ASTM D638	—	—	—	27–40	—	27–40	—	—	—
Elongation at yield (%)	23°C, ISO 527-1,2	8–25 ASTM D638	10–11	—	—	—	—	—	—	—	7–14
Elongation at break (%)	23°C, ISO 527	15–80 ASTM D638	31–32	—	65	—	50–65% ASTM D638	—	45–65	50–60	9–65
Tensile creep modulus, 1000 h, elongation ≤ 0.5% (MPa)	23°C, ISO 899-1	—	1300	—	1300	—	—	—	—	—	—
Charpy impact strength (kJ/m²)	23°C, ISO 179/1eU	—	250C	180P	—	—	—	—	—	—	—
Charpy notched impact strength (kJ/m²)	23°C, ISO 179/1eA	—	6	6.5	13	5.0–8.0	—	5.0–8.0	—	—	4–8
Izod notched impact strength (kJ/m²)	23°C, ISO 180/1A	53–107 ASTM D256 [J/m]	6.5	—	108	—	5.5–7.0 kg·cm/cm ASTM D256	—	0.49–0.70 J/cm ASTM D256	0.53–0.64 ASTM D256	5–9

(Continued)

TABLE 6.6 (CONTINUED)
Selected Mechanical, Thermal, and Electrical Properties of POM

POM	Test Method ISO	Delrin, DuPont (DuPont 2000)	Ultraform, BASF (BASF 2007a)	Hostaform C 9021, Ticona (Ticona 2007)	Tenac 3010, Homopolymer, Asahi Kasei Chemicals Corporation (Asahi 2010)	Duracon, Polyplastics Co. (Standard Grades) (Polyplastics Co. 2014)	Lucel, LG Chem (Injection Molding POM Grades) (Chem 2009)	Tepcon, Polyplastics Co. (Standard Grades) (Polyplastics Co. 2013)	KEP Kepital, Korea Engineering Plastics Co., Acetal Homopolymers (Plastics 2014)	KEP Kepital, Korea Engineering Plastics Co., Acetal Copolymer (Plastics 2014)	Tarnoform, Grupa Azoty S.A. (Standard Grades) (Azoty 2007)
Ball indentation hardness H 358/30 (MPa)	23°C, ISO 358N, 30S, 2039-1	–	125–130	–	–	–	–	–	–	–	135–150
Heat deflection temp. 1.8 MPa (°C)	ISO 75-2/A	123–132 ASTM D648	95	–	96	90–100	18.6 kg 110°C 4.6 kg 160°C ASTM D648 (Unannealed)	90–100	110 Unannealed; ASTM D648	110 Unannealed; ASTM D648	110–120
Melting point, DSC (°C)	DSC, ISO 3146	178 ASTM D3418	166	166	–	–	–	–	165	165	167
Vicat softening temperature VST/B/50 (°C)	ISO 306	–	150	150	–	–	–	–	–	–	150–160
Coeff. of linear thermal expansion, flow direction ($1 \times 10^{-5}\,K^{-1}$)	30–60°C, ASTM D696	1.10–1.26 ASTM E831	12 (DIN 53 752)	1.1 (ISO 11359-2)	–	11–12	–	11–13	–	–	11
Specific heat (J/g K)	20°C	–	–	–	–	–	–	–	–	–	1.48
Thermal conductivity (W/m·K)	–	0.28–0.36	–	–	–	–	–	–	–	–	–

(Continued)

TABLE 6.6 (CONTINUED)
Selected Mechanical, Thermal, and Electrical Properties of POM

POM	Test Method ISO	Delrin, DuPont (DuPont 2000)	Ultraform, BASF (BASF 2007a)	Hostaform C 9021, Ticona (Ticona 2007)	Tenac 3010, Homopolymer, Asahi Kasei Chemicals Corporation (Asahi 2010)	Duracon, Polyplastics Co. (Standard Grades) (Polyplastics Co. 2014)	Lucel, LG Chem (Injection Molding POM Grades) (Chem 2009)	Tepcon, Polyplastics Co. (Standard Grades) (Polyplastics Co. 2013)	KEP Kepital, Korea Engineering Plastics Co., Acetal Homopolymers (Plastics 2014)	KEP Kepital, Korea Engineering Plastics Co., Acetal Copolymer (Plastics 2014)	Tarnoform, Grupa Azoty S.A. (Standard Grades) (Azoty 2007)
Max. service temperature, up to a few hours (°C)	—	—	100	—	—	—	—	—	—	—	100
Dielectric constant at 100 Hz/1 MHz	IEC 60250	3.6–3.7 (1 MHz) ASTM D150	3.8/3.8	—	—	—	—	—	3.7 (1 MHz) ASTM D150	—	—
Dissipation factor at 100 Hz/1 MHz	IEC 60250	0.005–0×006 (1 MHz) ASTM D150	0.001/0.005	0.0020/0.0050	—	—	$3.8 \cdot 10^{-4}$ ASTM D150, 1MHz	—	0.007 (1 MHz) ASTM D150	—	10×10^{-4} 85×10^{-4}
Volume resistivity (Ω-cm)	IEC 60093	1×10^{14}–1×10^{15} ASTM D25	10^{13}	10^{12}	—	10^{14}	10^{14} ASTM D257	—	10^{14} ASTM D257	—	—
Surface resistivity (Ω)	IEC 60093	2×10^{14}–$6 \cdot 10^{15}$ ASTM D257	10^{13}	10^{14}	—	10^{16}	10^{16} ASTM D257	—	10^{16} ASTM D257	—	10^{14}
Dielectric strength K20/K20 (kV/mm)	IEC 60243-1	16.8–18.0 ASTM D149	40	35	—	19	24 ASTM D149	19	19.0 ASTM D149	—	25
Comparative tracking index CTI, test solution A	IEC 60112	—	CTI 600	CTI 600	—	—	—	CTI 600<	—	—	—
Comparative tracking index CTI M, test solution B	IEC 60112	—	CTI 600 M	—	—	—	—	—	—	—	—

TABLE 6.7
Typical Injection Molding Parameters for Several POM Grades

		Celcon (Ticona 2000)	Hostaform (Ticona 2006)	Ultraform (BASF 2007a)	Delrin (DuPont 2006)	Tarnoform (Azoty 2007)
Cylinder temperature (°C)	Rear	182	170–180	180–220	180–190	175
	Center	188	180–210	180–220	190–200	185
	Front	193	190–210	180–220	200–215	195
	Nozzle	198	190–210	205–220	190	200
Melt temperature (°C)		182–199	190–210	180–220	205–220	180–220
Mold surface temperature (°C)		82–121	80–120	60–120	50–90	60–120
First stage injection pressure (MPa)		76–138	60–120			60–120
Second (hold) injection pressure (MPa)		76–138	60–120		60–110	60–180
First stage injection fill time (s)		2–5	0.5–1		0.5–5	
Back pressure (MPa)		0–0.3			<1	0–1
Screw peripheral speed (m/s)		20–40 rpm	0.1–0.3	Up to 0.3	0.15–0.3	0.1–0.4
Overall cycle (s)		15–60 (depending on wall thickness)			35–60	35–45
Drying		82°C for 3 h		100–110°C for 3 h	80°C for 2–4 h	100–120°C for 2–3 h

Source: Pielichowska, K.: Polyoxymethylene processing. In *Polyoxymethylene Handbook: Structure, Properties, Applications and Their Nanocomposites.* 107–151. 2014. Copyright Wiley-VCH Verlag GmbH & Co. KGaA. Reproduced with permission.

then filling of the mold with molten polymer by the forward ramming of the screw (called a reciprocating screw), followed by a very short packing stage necessary to pack more polymer in the mold to offset the shrinkage after cooling and solidification. The material is held in the mold under high pressure until it has solidified sufficiently to allow ejection (Agazzi et al. 2013; Vlachopoulos and Strutt 2003). Typical injection molding parameters for different POM grades are presented in Table 6.7.

Since the 1980s, microsystem technology has grown in importance, and a rapidly growing interest in microinjection molding (μIM) is observed, which is related to the production of small plastic parts for electronics, biomedical, and other industries, and for the fabrication of components for complex microelectromechanical systems (Attia, Marson, and Alcock 2009; Drummer and Vetter 2011). POM has been widely used in μIM experiments because of its processing characteristics, such as low viscosity, fast molding, and good processing stability for deposit-free molding (Chu et al. 2010; Ong, Zhang, and Woo 2006).

6.7.1.2 Melt Extrusion

Extrusion processes are utilized widely to manufacture on industrial scale sheets, filaments, profiles and pipes, wire coatings, and cable coverings (Chung 2000). Extrusion consists of the following sequence of steps: (1) heating and melting the polymer, (2) pumping the polymer to the shaping unit, (3) forming the melt into the required shape and dimensions, and (4) cooling and solidifying (Vlachopoulos and Wagner 2001). The quality of the extruded polymer is highly dependent upon the homogeneity of the molten polymer being fed into the die, which should ideally be supplied at a constant pressure, temperature, and throughput (Vera-Sorroche et al. 2013). Recommended conditions of POM extrusion process are presented in Table 6.8 (Pielichowska 2014b).

TABLE 6.8
Recommended Extrusion Parameters for Several POM Grades

		Celcon Acetal Copolymer (Ticona 2000)	Ultraform (BASF 2007a)	Delrin (DuPont 2006)	Tarnoform (Azoty 2007)
Guide values for the screw geometry (extrusion)	Overall length	24	20–25		20–25
	Feed section		8		8–9
	Compression section		3–5		3–5
	Metering section		9–12		9–10
Melt temperature (°C)		180–205		205–225	175–185
Screw speed (Zhao et al.)		33	42		40
Die temperature (°C)		230	175		

Source: Pielichowska, K.: Polyoxymethylene processing. In *Polyoxymethylene Handbook: Structure, Properties, Applications and Their Nanocomposites*, 107–151, 2014. Copyright Wiley-VCH Verlag GmbH & Co. KGaA. Reproduced with permission.

6.7.2 Solid-State Processing

Solid-state extrusion is a technique that was first applied to solid-phase forming of metals in the 1960s (Pugh and Low 1964–1965). Since then, numerous works have been reported on the hydrostatic extrusion of both semicrystalline and amorphous polymers (Hope and Ward 1981; Mohanraj et al. 2006; Nakayama and Kanetsuna 1975; Schaffner, Vinson, and Jungnickel 1991). Hydrostatic extrusion was successfully applied to many thermoplastics such as polyethylene (PE), polytetrafluoroethylene (PTFE), polypropylene (PP), poly(ethylene terephthalate) (PET), polyoxymethylene (POM), polyamides (PA), and poly(methyl methacrylate) (PMMA) (Dröscher 1982; Kim and Porter 1988). The hydrostatic extrusion behavior of POM was investigated by Coates and Ward (1978) for a range of different POM molecular weight grades; extruded materials with axial Young's moduli reaching values as high as 24 GPa were obtained. Also Kastelic et al. (1983) revealed that significant gains in POM physical properties, especially in tensile and impact strengths, can be achieved for extruded samples. Higher tensile strength of the hydrostatically extruded POM was attributed to the higher extent of voiding present in the die-drawn samples as detected by SAXS and density measurements (Mohanraj et al. 2006).

6.7.3 Other Methods

POM can be processed by using other methods such as extrusion assisted by supercritical carbon dioxide, blow molding, melt blowing, compression molding, rolling, sintering, or spinning (Pielichowska 2014b). Technical details of the extrusion assisted by supercritical carbon dioxide of POM were presented in a Meyer and Kinslow patent (Meyer and Kinslow 1991). There are still very little data concerning blow molding of POM. The reason could be that POM resins have generally poor melt tension due to limitation of the polymer molecular weight, and difficulties have been encountered during the blow molding process (Murao et al. 1993). It was discovered that during crystallization from a stressed polymer melt, no randomly nucleated spherulitic structures but instead a highly oriented *row-nucleated* morphology can be observed. As a result, during crystallization of POM films from stressed melts (film blowing), the surface is covered with protruding lamellar edges highly oriented in the extrusion direction (see Figure 6.14; Garber and Clark 1970).

FIGURE 6.14 The surface of Delrin blown film is completely covered with lamellar edges highly oriented normal to the extrusion direction. The lamellae are twisted into bundles aligned in the transverse direction. Magnification ×17,000. (Reprinted from Garber, C. A., and E. S. Clark, *Journal of Macromolecular Science, Part B* 4 (3):499–517, 1970. With permission from Taylor & Francis.)

Production method of a hollow, sheet- or rod-shaped molding from a POM copolymer was described by Murao et al. in patents EP 0568308 A2 (Murao et al. 1993) and US 5458839 A (Murao et al. 1995). POM/polyurethane composites containing 5–40 wt.% of polyurethane fabricated by blow molding method were characterized by significantly increased melt strength, decreased rate of crystallization, and increased and more consistent die swell as compared with pure POM (Flexman 1984). In the patent US 2011/0309558 A1 (Wilhelm, Klenz, and Binkowski 2011), process of the production of POM articles using blow molding method was described, whereas in another patent, a description of the method for the manufacturing of a container or parts of a container for fuel or compressed gases from high-impact-resistant POM, thermoplastic elastomer, and a coupling agent was given (Latz, Lowell, and Ziegler 2012). Recently, useful reports on commercially available POM grades for processing by blow molding method have been published (BASF 2008). The fundamentals of melt blowing method have been elaborated in the 1950s and since then a variety of polymers have been processed by this method, but a very limited number of reports are available for melt blowing of POM. Recently, it was shown that POM can be attenuated to microfibers, but production of well-bonded webs requires special efforts to tune the process conditions. If they are properly fixed, the obtained products exhibit excellent air permeability, good resilience, and low friction properties (Zhao 2005).

There are also few data concerning POM processing using compression molding method. It was revealed that processing POM by hot compaction and compression molding methods leads to a significant increase in the degree of crystallinity for the sintered samples, as compared with the compression-molded ones. A nearly twofold increase in the Young's modulus as a result of a greater continuity of the stiff crystalline phase in the sintered material, as compared with the compression-molded one, was observed (Al Jebawi, Sixou, and Seguela 2007; Al Jebawi et al. 2006).

The properties of POM processed by rolling were investigated by Gezovich and Geil (1971). Tensile test results of rolled POM indicated that the yield stress increases along the roll direction with a decrease in the yield stress perpendicular to the roll direction. Gray and McCrum (1971) investigated the differences between the textures produced by cold and hot rolling of POM. They found that POM rolling at a temperature close to the melting point produced a [0001] texture, that is, with the most densely packed planes [0001] aligned parallel to the rolling plane. In the work by Preedy and Wheeler (1972), the textures observed were interpreted in terms of a stress-induced martensitic transformation of the hexagonal phase to an orthorhombic phase. Along this research line, two alternative models for the development of texture in rolled POM were described by Miles and Mills (1975). The first proposed that slip systems in the crystalline phase are the main deformation factor, whereas the second approach was based on Wilchinsky's model (Wilchinsky 1964) of rigid crystals rotating in a deforming amorphous matrix. After analysis, the authors concluded that the former approach is in better agreement with experimental data.

The pressure-assisted sintering of POM powder in the solid state (hot compaction) is an alternative way to the melt compression or injection molding processes. Generally, the sintered POM materials show very high stiffness and fairly high stress at break. Native POM powder was processed with powders obtained from grinding of melt-extruded pellets by sintering method by Al Jebawi et al. (2006). They observed a significant increase in crystallinity and a strong loss of ductility; however, the sintered samples have been easily machined out into strips for mechanical testing. The authors postulated that during sintering, molecular interpenetration occurs over very limited depth at the contact surface of crystals from adjacent particles via chain reptation through the intermediate amorphous layer. Chain entanglements are kept excluded from the crystal, and they involve the switching of crystal stems in close-neighbor positions (Al Jebawi, Sixou, and Seguela 2007).

POM can be processed to produce fibers by using the spinning method, too. POM fibers are used in the production of conveyer belts, driving belts, ropes, sails, fishing nets and trawls, tarpaulins, fire hoses, petrol- and oil-resistant hoses, electrical insulations and filter materials, industrial fabrics, and other articles for which the operational temperature does not exceed 120–130°C (Egorov, Grzhimalovskii, and Yudin 1970). Generally, POM satisfies the main requirements for a fiber-forming polymer, i.e., it has sufficiently high molecular weight and elongated macromolecules without branches or transverse chemical bonds, and it is able to undergo transition to the viscous-liquid state without decomposition (Egorov, Grzhimalovskii, and Yudin 1970). However, in-fibril voids may occur at a high crystallization rate, and thus fibers are easily cut during melt spinning and stretching processes; as a result, long fibers having proper strength cannot be obtained (Samon et al. 2001). Nevertheless, it was found that quenching of the yarn (when it is still in the molten state or a few degrees below melting point) retards and prevents crystallization, and the yarn does not reach a high crystallinity (SPA 1966). Fabrication method of POM fibers with high strength and high modulus of elasticity was described in the patent by Kikutani and Okawa (2004), whereas production of POM fibers by melt spinning, involving a drawing operation, was described in another patent (Koji and Tomonobu 2001). By applying a POM copolymer with a melt flow rate in the range 0.3–30 ml/min, POM fibers can be also obtained (Zierer et al. 2008). Pressurized drawing process is applied for commercial production of high-performance POM fibers (Kikutani 2002).

Recently, nanofibers produced by electrospinning have attracted significant interest; however, due to POM poor solubility in common solvents, some technical problems were encountered. However, Kongkhlang and coworkers (Kongkhlang, Kotaki et al. 2008; Kongkhlang, Tashiro et al. 2008) reported on successful electrospinning of POM nanofibers using hexafluoroisopropanol (HFIP) as a solvent (see Figure 6.15).

Different orientation processes, such as solid-state extrusion, die drawing, and roll drawing, may be applied to improve tensile modulus and tensile strength of POM. POM can form highly crystalline fibers, which exhibit marked elastic behavior under specific conditions of stress-induced crystallization (Quynn and Brody 1971). Ultrahigh modulus POM can be also obtained in a two-stage drawing procedure—after an initial draw ratio of approximately 7, the sample is slowly drawn at 150°C to its final draw ratio of >15 (Clark and Scott 1974). In the work by Takasa et al. (2006a) on the press drawing process, the draw ratio reached 6.0 and the tensile modulus was 4.5 GPa, whereby the degree of crystallinity of polyacetal decreased from 70% to 65%. After applying the press drawing followed by subsequent simultaneous biaxial drawing, the tensile modulus reached 11 GPa when the draw ratio was 14, and the degree of crystallinity increased from 71% to 76–80% (Takasa, Miyashita, and Takeda 2006b). Highly oriented POM/multiwalled carbon nanotube (MWCNT) composites with the tensile strength and modulus reaching 900 MPa and 12 GPa, respectively, without remarkable drop of the elongation at break, have been produced by solid hot stretching technology (Zhao and Ye 2011b,c). The crystalline orientation and tensile modulus of the oriented POM from die-drawing and hydrostatic extrusion at similar draw ratios were comparable, but the

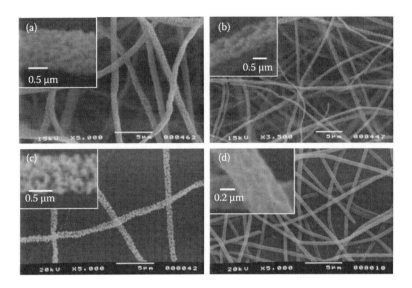

FIGURE 6.15 Scanning electron micrographs of the electrospun POM nanofibers from an HFIP solution with copolymer contents of (a) 0, (b) 1.5, (c) 4.4, and (d) 13 wt.% (electrostatic field strength = 15 kV/10 cm and relative humidity = 75%). (Reprinted from Kongkhlang, T., M. Kotaki, Y. Kousaka, T. Umemura, D. Nakaya, and S. Chirachanchai, *Macromolecules*, 4746–4752. Copyright 2008 American Chemical Society.)

strengths were different at higher draw ratios—the strength of the hydrostatically extruded samples was higher than that of the die-drawn samples (Mohanraj et al. 2006).

Products made from POM often require some machining after they have been processed, but no special machines or processes are required for machining POM parts. The machines that are normally used in the plastics and metal industries with high-performance superspeed steel tools or hard metal tools can be used. It is recommended to use the highest possible cutting speed and optimum chip removal, and apply very sharp tools and not too high clamping pressures. To cool the workpiece during machining, compressed air is to be used. Usually for finishing molded articles, for bevelling, smoothing, burring, and fitting the edges and corners of sheets of plastic, files are applied. For POM, the coarsest files are required. Milled curved-tooth files with coarse, single-cut, shear-type teeth are particularly suitable. Polyacetal can be also semifinished by wet sanding by using conventional belt and disc sanding equipment. The ashing operation removes deep scratches, parting lines, and harsh surface imperfections (Berins 1991; Waack 2014).

For POM, assembling different methods such as hot-plate welding, friction welding, hot riveting, and ultrasonic riveting can be used. For joining of moldings made of POM, adhesive systems based on epoxy resins, polyurethanes, or methacrylates may be applied, whereas in hot-plate welding method, the surfaces to be joined are brought up to temperature by light contact with a PTFE-coated hot plate (220–240°C) for 5–30 s depending on the shape and POM viscosity, and then are welded together under pressure. Assembling using friction welding method requires frictional speeds in the range 100–300 m/min at contact pressures of 0.2–0.5 N/mm^2. POM surfaces can also be decorated after proper pretreatment using mechanical pretreatment (roughening), acid etching, or primers, or physical pretreatment by painting, vacuum metallizing, electroplating, hot stamping, or laser marking (Pielichowska 2014b; Ticona 2006).

6.8 POLYACETAL BLENDS

Polymer blends can be completely or partially miscible, or fully immiscible. Generally, only a few miscible blends have been identified, while most polymer mixtures form immiscible blends, in which there is a distinct interphase (Chiang and Huang 1997; He, Zhu, and Inoue 2004).

POM and novolac resin are miscible at low temperatures due to hydrogen bonds between the –OH group in novolac and ether oxygen in POM (Wang, Li, and Jin 2000). In miscible crystalline/crystalline blends of POM with poly(ethylene oxide) (PEO), it was observed that with increasing PEO content in the blend, the crystallization temperature of POM decreased and the multiple crystalline morphologies including two kinds of interfibrillar or interlamellar structures were formed. In the PEO-rich blend, POM dispersed into PEO spherulites, producing different interfibrillar or interlamellar structures (see Figure 6.16; Bai and Wang 2012; Liu et al. 2012; Liu, Bai, and Wang 2012).

Other miscible blends containing POM are those with polyvinylphenol (PVPh) and terpenephenol (TPh) (Egawa et al. 1996; French, Machado, and Linvien 1992). Miscibility was attributed to the hydrogen-bonding interactions between the phenolic OH group of PVPh and the ether oxygen of POM, and it was suggested that mixing is preferentially induced in the noncrystalline phase (Matsumoto et al. 1997). Similar miscibility explanation was postulated for TPh (Egawa et al. 1996). In contrast, POM/phenoxy blends are composed of two almost pure phases that suggest their immiscibility. However, mechanical properties of the blends were those of a compatible blend (Erro, Gaztelumendi, and Nazabal 1996). POM/epoxy resin blends were obtained by curing at temperatures above (180°C) and below (150°C and 145°C) the melting temperature of POM (T_m = 168°C), and it was revealed that in blends, both phase separation and crystallization occur. The increase in the molar mass of the epoxy resin during the cure reaction leads to phase separation (reaction-induced phase separation [RIPS]) and results in a thermoplastic-rich phase and an epoxy-rich phase in the blend. Authors observed that crystallization after curing at 180°C can only be induced by cooling, while curing at 150°C results in RIPS, preceding isothermal crystallization at the cure temperature (see Figure 6.17; Goossens, Goderis, and Groeninckx 2006; Goossens and Groeninckx 2006, 2007).

To improve the wear resistance or to reduce the friction coefficient, POM is blended with PTFE (Hu 2000), ultrahigh-molecular-weight polyethylene (UHMWPE) (Ratnagiri 2009), high-density

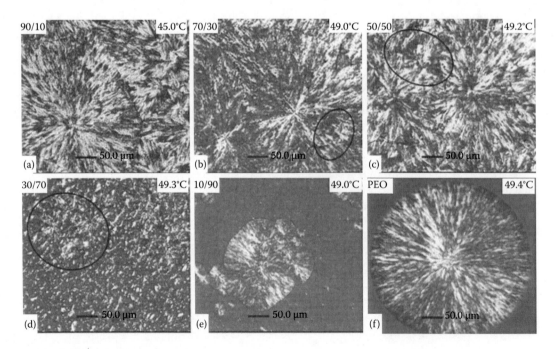

FIGURE 6.16 Polarized light micrographs of neat PEO and PEO of POM/PEO blends during non-isothermal crystallization of PEO at a cooling rate of 1°C/min. The shown temperature was the crystallization temperature of PEO in the samples. (With kind permission from Springer Science+Business Media: *Journal of Polymer Research*, Effect of blend composition on crystallization behavior of polyoxymethylene/poly(ethylene oxide) crystalline/crystalline blends, 19, 2012, 9787, Liu X., S. Bai, M. Nie, and Q. Wang.)

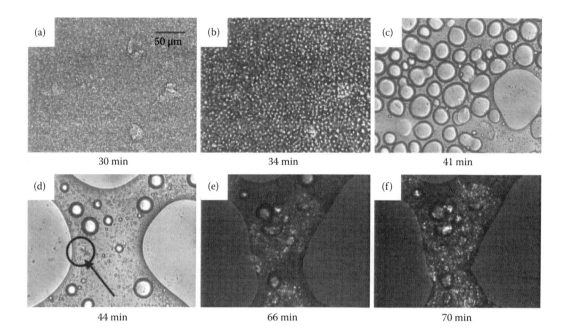

FIGURE 6.17 OM pictures of a blend with 20 wt.% POM, cured for 2 h at 150°C. RIPS is followed by isothermal crystallization inside the POM-rich phase. The arrow indicates the start of a growing spherulite. (Reprinted from Goossens, S., B. Goderis, and G. Groeninckx, *Macromolecules*, 2953–2963. Copyright 2006 American Chemical Society.)

polyethylene (HDPE) (Mamunya et al. 2007; Su et al. 2009), polyetheretherketone (PEEK) and/or polyetherimide (PEI) (Suzuki and Nagahama 1987), silicone polymer (Takayama et al. 1993), and epoxy-novolac (Chiang and Huang 1997; Murasawa et al. 1994). Addition of linear low-density polyethylene (LLDPE) and ethylene acrylic acid (EAA) into POM leads to improvements in the friction and wear properties (Chen, Cao, and Li 2006a,b; Utracki 1998).

Blends with TPUs (Chang and Yang 1990; Chiang and Lo 1988; Gao, Qu, and Fu 2004; Kumar, Arindam et al. 1993; Kumar, Mahesh et al. 1993; Kumar, Neelakantan, and Subramanian 1993, 1995; Mehrabzadeh and Rezaie 2002; Peng et al. 2009; Tang et al. 2013), polyurethane rubber (Pecorini, Hertzberg, and Manson 1990), and other types of polyurethanes (PUs) (Chiang and Huang 1989), nitrile rubber (NBR) (Zhang et al. 2000), ionomers of ethylene–methacrylic acid copolymer ionized with sodium cation (EMA-Na) and zinc cation (EMAZn) with methyl methacrylate–styrene–butadiene terpolymer (MBS) (Wang and Cui 2005), and ethylenepropylene diene (EPDM) (Chiang and Huang 1993; Uthaman, Majeed, and Pandurangan 2006; Uthaman, Pandurangan, and Majeed 2007) were prepared to enhance the impact strength of POM (Chiang and Huang 1989). Numerous other blends of POM have been investigated such as POM/poly(styrene–butadiene–styrene) (Wang 2011), POM with homopolymer or copolymer of acrylic or methacrylic acid and fluoropolymer (Katsumata 1992), POM/PS (Romankevich et al. 1982), or POM/polyamide (Page 2000). Electrically conductive materials could be made by blending polyacetal and polyaniline (Terlemezyan, Ivanova, and Tacheva 1993) or polyindene (PIn) (Cabuk, Sari, and Unal 2010).

In POM/PP blends, it was revealed that the impact and flexural modulus was higher than that of PP itself—it suggested that POM and PP may be at least partially compatible (Soundararajan and Shit 2001). The ethylene vinyl alcohol (EVOH) copolymers have been demonstrated to be effective compatibilizers of PP–POM blends (Huang et al. 2003). POM with polycarbonate (PC) forms immiscible blends, and incorporation of PC significantly increased the modulus and yield strength of POM (Chang, Yang, and Wu 1991). POM)/poly(ε-caprolactone) (PCL) reactive blend prepared via a chain–transfer reaction was characterized by a microscopically phase-separated morphology in which the diameter of the PCL microphase was below 100 nm, which could not be achieved through simple blending in classical POM/PCL blends (Kawaguchi et al. 2008). It was revealed that among poly(ethylene glycol-*co*-cyclohexane-1,4-dimethanol terephthalate) (PETG)/POM blends, the elongation at break, energy to tensile failure, notched Izod impact strength, and fracture toughness were the highest for 50/50 POM/PETG composition, but the mechanical properties of the 50/50 blend were still inferior to the neat PETG and POM (Lam, Bakar, and Ishak 2005; Lam et al. 2004).

6.9 POLYACETAL COMPOSITES AND NANOCOMPOSITES

6.9.1 Polyacetal Composites

Different reinforcements have been incorporated into POM to increase its mechanical properties (see Table 6.9).

On the other hand, selected polymer matrices can be reinforced with POM whisker crystals. Good reinforcement effect was obtained with matrix resins such as epoxide and unsaturated polyester, as the modulus of the whisker is 1011 N/m^2, i.e., almost equivalent to the ideal crystal modulus of POM (Komatsu 1996; Iguchi et al. 1982).

6.9.2 Polyacetal Nanocomposites

Polymer nanocomposites often exhibit enhancements in strength and stiffness with loss of elongation at break, and show better mechanical, thermal, and electrical properties. POM nanocomposites can be prepared using different methods, and selection of proper preparation route depends on the type and size of nanofillers. Usually, in the first step, compounding process is necessary to disperse nanofillers in the polymer matrix, and in the second step, nanocomposite is shaped through, e.g., injection molding method. In Table 6.10, methods of nanoparticle dispersion in the POM matrix were summarized (Pielichowska 2014a).

To avoid agglomeration effects, it is recommended to use twin screw extruders with proper length-to-diameter (L/D) ratio because the shearing forces in single screw extruders are not always strong enough to break up large agglomerates, resulting in poor filler dispersion in the polymer matrix. Due to low heat resistance of POM, the length of screws should be limited as much as possible (Pielichowska 2014a).

Depending on the type, pretreatment, and incorporation way of nanoadditives, different effects on POM properties can be observed. Hence, Zeng et al. (2007) investigated the POM/CNT nanocomposites, and they revealed that carboxylic acid functional groups on MWCNT surfaces can cause a drastic decrease in thermal stability of the POM matrix due to acidolysis and hydrolysis reactions, but also showed a nucleation effect on POM, leading to an increase in crystallization temperature and crystallization rate. In the case of POM nanocomposites with CNTs purified by graphitization, such a decrease in the thermal stability was not observed. Additionally, incorporation of graphitized MWCNTs leads to significant improvement of electrical properties, but they did not induce nucleation and did not increase the crystallization rate of POM (Zeng et al. 2010; Zhao and Ye 2011a). Moreover, incorporation of MWCNTs leads to increase in the storage and loss modulus in the whole temperature range (Zhao and Ye 2011a). For highly oriented POM/MWCNT composites, the tensile strength, modulus, and thermal conductivity of the drawn composites were enhanced by 620%, 870%, and 180%, respectively (Zhao and Ye 2011b). In POM nanocomposites with carbon nanofibers (CNFs), the stiffness, strength, resistance to creep, and stress relaxation behaviors of POM were all improved, while the ductility reduced by CNF incorporation (Siengchin, Psarras, and Karger-Kocsis 2010).

For POM-based nanocomposites with montmorillonite (MMT), it was found that the addition of MMT, either Na–MMT or organo–MMT, may accelerate the overall nonisothermal crystallization process of POM (Xu, Ge, and He 2001). The introduction of MMT into POM matrix can

TABLE 6.9
POM/Filler Composites

Filler	Comments	Ref.
Glass fiber (GF)	Improved mechanical properties. POM/GF composites are now broadly used in automobiles, electronics, communication devices, and consumer products.	Barker and Price 1970; Hardy and Wagner 1969; Hardy 1971; Kastelic et al. 1983; Hope, Richardson, and Ward 1982; Curtis, Hope, and Ward 1982; Hine, Duckett, and Ward 1993; Hine et al. 1996, 1997; Zachariev, Rudolph, and Ivers 2004, 2010; Kawaguchi, Masuda, and Tajima 2008; Hashemi, Gilbride, and Hodgkinson 1996; Hsu, Hwang, and Ting 2010
Glass beads (GB)	Flexural strengths and impact strengths showed negative deviation from the rule of mixtures.	Hashemi, Elmes, and Sandford 1997
Polyamide 6 (PA6)	After incorporation of 1–4 wt.% of PA6, the impact strength and elongation were improved markedly, with retention of good tensile strength.	Li, Xie, and Yang 2006
Graphite	The thermal conductivity of POM was improved remarkably from 0.36 W/(m·K) for neat POM to 1.15 W/(m·K) in POM/graphite (30 wt.%) graphite, with similar tensile strength and notched Charpy impact.	Zhao and Ye 2009
Cellulose fibers	The modulus, deflection temperature under load, flexural strength, and thermal conductivity were enhanced with increasing CelF content, and the composites had an advantage of specific modulus compared to glass fiber filled POM.	Bledzki, Mamun, and Feldmann 2012; Kawaguchi et al. 2010
Fique fibers	The mechanical properties of treated fiber (by alkalization, using formaldehyde, glycidyl methacrylate, and an isocyanate compound) composites were higher than those of the untreated ones due to the enhancement of the fiber-matrix adhesion tendency of fique fibers.	Ganan and Mondragon 2005
Kenaf/PET	Increasing the PET fiber content significantly improved the composite resistance to environmental degradation.	Abdullah, Dan-Mallam, and Yusoff 2013
Woven kenaf fiber	The tensile strength of the woven POM/kenaf composite increased from 67 to 75 MPa.	Dan-Mallam, Abdullah, and Yusoff 2014
PU/Carbon Nanofiber	CNF reinforcement increased the stiffness, resistance to creep, and tensile strength, reduced the elongation at break, and improved the thermo-oxidative stability of POM.	Siengchin, Psarras, and Karger-Kocsis 2010

improve mechanical properties, thermal stability, and enhanced char formation (Leszczynska and Pielichowski 2012; Pielichowski and Leszczynska 2006). It was also revealed that POM/MMT systems show different behavior compared with other polymer/clay composites, depending on the type of surfactant or an excessive use of MMT (Sun, Ye, and Zhao 2007). Improvement in the tensile strength, elongation at break at lower filler loadings (3 wt.%), and thermal stability was observed for POM filled with organomodified fluorohectorite (OFH) (Jose and Alagar 2011). It was revealed that the crystalline structure of POM is not affected upon the addition of different POSS molecules, but the thermal stability of POM under inert atmosphere considerably increased with the addition of any of POSS grades tested. The best results were found for POM composites with aminopropylisobutyl–POSS—the onset of degradation temperature increased by more than 50°C

TABLE 6.10
Methods of Nanoparticle Dispersion in POM Matrix

Nanoparticle Type	Nanoparticle Content (wt.%)	Preparation Method	Equipment	Conditions	Ref.
MWCNTs	0.5	Conventional melt mixing	Internal two-roller mixer	Preheating up to 190°C at a rotation speed of 30 rpm, and after polymer melting, MWCNTs were added for 10 min at 190°C at 90 rpm.	Zeng et al. 2007, 2010
MWCNTs	0–20; 0–5; 0–12 vol.%	Solution–evaporation method with sonication and next melt mixing	Twin screw extruder	MWNTs were dispersed in phenylcarbinol using sonication, and POM powder was dissolved in the dispersion system at 150°C; then stabilizers were added.	Zhao and Ye 2010, 2011a,c
MMT and Na-MMT	5	Melt mixing	Roller mill	Temperature: 175–180°C, 10 min.	Xu, Ge, and He 2001
Organo-modified MMT	0–5	Melt mixing	Internal mixer	Temperature: 180°C, 15 min, rotor speed 60 rpm.	Pielichowski and Leszczynska 2006; Leszczynska and Pielichowski 2012
Alkyl ammonium/MMT	0–5	Melt mixing	Twin screw extruder	Temperature: 180–220°C, rotor speed of 45 rpm.	Sun, Ye, and Zhao 2007
		In situ intercalative polymerization/melt processing	Laboratory equipment/twin screw extruder	1,3,5-trioxane and 1,3-dioxolane were mixed together with MMT and a catalyst; the system was in situ polymerized at 65°C; next compounded with POM and other stabilizers; temperature: 180–220°C, rotor speed of 45 rpm.	
		Solvent intercalation/melt processing	Laboratory equipment/twin screw extruder	MMT was dispersed in phenylcarbinol using sonication, and then the POM powder was dissolved in the dispersion system at 150°C; dry POM/MMT system was compounded with POM and other stabilizers; temperature: 180–220°C, rotor speed of 45 rpm.	

(Continued)

TABLE 6.10 (CONTINUED)
Methods of Nanoparticle Dispersion in POM Matrix

Nanoparticle Type	Nanoparticle Content (wt.%)	Preparation Method	Equipment	Conditions	Ref.
Bentonite-based organo-modified clays	1	Solid-state pan milling intercalation/melt processing	Laboratory equipment/twin screw extruder	The mixture of POM powder and MMT was milled by three-dimensional grinding discs; next it was compounded with POM and other processing aids; temperature: 180–220°C, rotor speed of 45 rpm.	Kongkhlang, Kousaka et al. 2008
Organo-modified fluorohectorite (OHF)	0–5	Melt mixing	Co-rotating twin screw extruder	Temperature: 210°C, screw speed 100 rpm.	Jose and Alagar 2011
		Melt mixing	Twin screw extruder	POMs were placed in a twin screw extruder that was preheated up to 190°C; next the weighed OFH clay was added into the molten POM matrix, and then the POM-OFH mixture was mixed; temperature: 190°C, 10 min, rotor speed 90 rpm.	
MMT surface modified with trimethyl stearyl ammonium	0–8	Melt mixing	Twin screw extruder	Temperature: 150–200°C, screw speed 50 rpm.	Wacharawichanant, Sahapaibounkit, and Saeueng 2012
Glycidylethyl-POSS, aminopropylisobutyl-POSS, poly(ethylene glycol)-POSS	0–10	Melt mixing	Twin screw corotating microcompounding extruder	Temperature: 190°C, 5 min, screw speed 100 rpm.	Sanchez-Soto et al. 2010
Glycidyl-POSS, glycidylisobutyl-POSS, octaisobutyl-POSS, trisilanophenyl-POSS	0–5	Melt mixing	Twin screw corotating Brabender internal mixer	Temperature: 190°C, 6 min, screw speed of 60 rpm.	Illescas et al. 2010

(*Continued*)

TABLE 6.10 (CONTINUED)
Methods of Nanoparticle Dispersion in POM Matrix

Nanoparticle Type	Nanoparticle Content (wt.%)	Preparation Method	Equipment	Conditions	Ref.
Glycidyl-POSS, glycidylisobutyl-POSS, aminopropylisobutyl-POSS, poly(ethylene glycol)-POSS	2.5	Melt mixing	Twin screw corotating Brabender internal mixer	Temperature: 190°C, 6 min, screw rate 60 rpm.	Ramirez, Sanchez-Soto, and Illescas 2011
Monosilanolisobutyl-POSS	0–10	Melt mixing	Twin screw corotating microcompounding extruder	Temperature: 190°C, 5 min, screw speed of 100 rpm.	Illescas et al. 2011
Dimethylsiloxy-POSS, ethyl epoxycyclohexy-POSS	0–1	Melt mixing	Single screw extruder, equipped with a zone of intensive mixing	Temperature: 180°C, L/D ratio of the screw 34, screw speed 10 rpm.	Czarnecka-Komorowska et al. 2012
HAp	0.5	Melt mixing	Twin screw extruder	The temperature profiles of the barrel: 190–195–200–200–200°C, rotary speed 50 rpm, L/D ratio of the screw 25.	Pielichowska 2008
HAp	0–10	Melt mixing	Twin screw extruder	The temperature profiles of the barrel: 212–220–220–215°C, rotary speed 50, L/D ratio of the screw 25. The temperature profiles of the barrel: for T200 185–190–190–190°C, for T300 185–190–190–185°C, and for T411 190–190–190–185°C, rotary speed 50 rpm.	Pielichowska, Szczygielska, and Spasowka 2012; Pielichowska et al. 2013 Pielichowska 2012b,c

(Continued)

TABLE 6.10 (CONTINUED)
Methods of Nanoparticle Dispersion in POM Matrix

Nanoparticle Type	Nanoparticle Content (wt.%)	Preparation Method	Equipment	Conditions	Ref.
Stearic acid modified nano-HAp	0–5	Melt mixing	Twin screw extruder	POM pellets were preheated up to 190°C in an extruder, the weighed n-SHA was added into the molten POM matrix, and then the POM/n-SHA mixtures were mixed further; temperature 190°C, 10 min, screw speed 90 rpm.	Jose and Alagar 2012
ZnO	0–8	Melt mixing	Twin screw extruder	Temperature: 170–200°C, screw speed 50 rpm.	Wacharawichanant et al. 2008
ZnO	0–12	Melt mixing	Twin screw extruder	Temperature: 170–210°C, screw speed 30 rpm.	Wacharawichanant 2008
MoS_2	2	In situ polymerization	—	The cationic ring-opening copolymerization of trioxane and dioxolane (5.0 wt. %) proceeded at 65°C in solution	Wang et al. 2008
Al_2O_3	0–12	Melt mixing	Twin screw extruder	Temperature: 160–170°C, screw speed 25 rpm.	Sun, Yang, and Li 2008b
Al_2O_3	0–4	Melt mixing	Twin screw extruder	Temperature: 150–200°C, screw speed 50 rpm.	Wacharawichanant, Sahapaibounkit, and Saeueng 2011
Boehmite alumina	3	Melt mixing	Twin screw extruder	Temperature: 150–190°C, rotor speed 200 rpm.	Siengchin 2013
$BaSO_4$	0–3	Melt mixing	Extruder	Temperature: 190–190–190–185°C, screw speed 25 rpm.	Romero-Ibarra et al. 2012
SiO_2	0–5	Melt mixing	Twin screw extruder	Temperature: 170–180°C, screw speed 80 rpm.	Xu et al. 2012; Guo et al. 2014

Source: Pielichowska, K., Polyoxymethylene nanocomposites. In *Manufacturing of Nanocomposites with Engineering Plastics*, edited by V. Mittal, 2014. With permission from Woodhead Publishing, Cambridge, UK.

(Ramirez, Sanchez-Soto, and Illescas 2011). In case of POM/hydroxyapatite (HAp) nanocomposites, it was found that for both POM homopolymer (Pielichowska 2012a) and POM copolymer (Pielichowska, Szczygielska, and Spasowka 2012), the degree of crystallinity of POM increases with an increase in HAp content of up to 0.5–1.0% and next decreases with an increase in the HAp content (Pielichowska et al. 2013), while Young's modulus increases and elongation at break and tensile strength decrease with an increase in the HAp content. Thermogravimetric analysis results indicated that with an increase in the HAp content, the thermal stability of the POM matrix decreases significantly depending upon the POM grade and molecular weight (Pielichowska 2012c). POM nanocomposites with nano-HAp modified with stearic acid exhibited improvements in tensile strength and elongation at break at lower filler content and decreased with higher filler loading, while Young's modulus showed an increase with respect to the loading (Jose and Alagar 2012).

Nanometric ZrO_2 particles exert a heterogeneous nucleating effect leading to formation of thinner POM spherulites, but they have a little influence on the friction coefficient. The lowest wear appeared at the 1.0 wt.% content of ZrO_2 nanoparticles (Wang et al. 2007). Addition of molybdenum disulfide (MoS_2) decreases the thermal stability of the POM matrix, but larger friction reduction and wear resistance, especially under high load, were observed (Hu et al. 2010; Hu, Wang et al. 2009; Wang et al. 2008). For POM/ZnO nanocomposites, a decrease in tensile strength and an increase in Young's modulus with increasing filler content were observed. Additionally, the thermal stability of POM/ZnO nanocomposites increased with an increase in filler content (Wacharawichanant 2008; Wacharawichanant et al. 2008). It was also found that corundum nanoparticles were more effective in enhancing the tribological properties of POM nanocomposites in oil-lubricated conditions than in dry sliding conditions (Sun, Yang, and Li 2008b). These nanoparticles also have some influence on the tensile strength and Young's modulus (Wacharawichanant, Sahapaibounkit, and Saeueng 2011). In another work, spherical nanoparticles of barium sulfate ($BaSO_4$) were introduced into the POM matrix, and the radio-opacity to x-ray tests revealed an increase in the optical contrast as the $BaSO_4$ concentration increases (Romero-Ibarra et al. 2012). In POM-based composites containing reactable nano-SiO_2 (RNS), it was observed that the addition of RNS can enhance the tensile strength and Young's modulus of composites, and the impact strength showed a significant improvement at low contents of RNS. However, with the increase in the RNS content, the impact strength of composites shows a gradual decrease (Xu et al. 2012).

6.10 POLYACETAL APPLICATIONS

Currently, the major use of POM is in electrical and electronics sectors producing capacitors, insulators, and other components. Promising growth rates for POM are expected in the transportation industry, as well as in the food packaging sector (GPM 2014). The most common POM applications include

- Automotive constructions (door handles, gears, wipers, speedometers, fans, dashboards, carburettor parts, moldings, safety belts, seat adjustment, fuel pumps, window lifters)
- Electrical engineering (gears for washing machines, cabinets for radios, housing for drills)
- Mechanical engineering (gears, brushes, pumps, impellers, gaskets, extruder screws)
- Packaging (tanks, aerosol cans)
- Toys and fancy goods (buckles for belts, zippers, hangers)
- Sports and leisure (tennis racquet handles, ski bindings, windsurfing boards)
- Domestic articles (aerosol bottles, parts for lawn mowers and sowing machines, fasteners, kitchen utensils, electronic brushes, shavers, CD recorders)
- Plumbing (pipes, tanks)
- Biomedical applications (functional and mechanical components for use in medical devices such as insulin, pens, inhalers, or atomizing systems; device components such as spring elements, snapfits, gearwheels, bearings, sleeves, valves or nozzles; tribological systems such as cylinders, screws, sliders, or nuts; clamps for inner tubes, clips, or plug-in

connectors; pharmaceutical closures; handles for surgical instruments; transmission and connecting elements for conveyor chains and belts, hip and knee prostheses, occluder disc of the Björk–Shiley Delrin (BSD) heart valve, in dentistry as a substitute for acrylic resins, and metals in many prosthetic applications (BASF 2013; McKellop, Milligan, and Rostlund 1996; Moore et al. 1998; Thompson 2001; Thompson, Northmore-Ball, and Tanner 2001; Zilberman 2005)

- Musical instruments (mouthpieces for flutes, bagpipes) (Tokarz, Pawlowski, and Kedzierski 2014)

Automotive, electrical, and electronic segments are the major users of POM with automotive constituting 32.4% of the global POM market in 2012. The other segments—medical devices, pharmaceutics, sport goods, etc.—shall add to the global demand for POM in the near future (PR Newswire 2014a).

6.11 POLYACETALS RECYCLING

POM grades offered by the manufacturers extend from extremely free-flowing products for the injection molding of thin-walled parts through high-viscosity grades for extrusion. POM types include impact-modified, high-stiffness, creep resistance, weather resistance, high-sliding-resistant, special lubricant, low warpage, high-impact, flexible, filled, and reinforced polymers in a wide range of sorts—this diversity may cause some problems during recycling by recompounding and remelting (Brandrup 1996). Additionally, recompounding and remelting have rather limited application for POM because it leads to extensive POM decomposition (Fleischer, Muck, and Reuschel 1992) with formaldehyde release. Another option is mechanical recycling, which allows plastic wastes to be used as raw materials for other products (Aznar et al. 2006), and primary recycling, better known as re-extrusion. In primary recycling, scraps and industrial or single-polymer plastic wastes and parts are reintroduced to the extrusion cycle together with pristine polymer to produce final articles (Al-Salem 2009). Unfortunately, this method also has some limitations—in technological practice, a 3:1 ratio of virgin to reground resin is commonly applied. For POM, further degradation was observed during the multiple processing, whereas investigations of melt-flow index and molar mass show effects correlated to late-term changes of the molar mass distribution during degradation processes initiated by thermomechanical and thermo-oxidative stresses as well as by the exposure to ultraviolet irradiation (weathering; Archodoulaki et al. 2007).

Chemical recycling technologies such as pyrolysis, gasification, solvolysis, or acidolysis can be used in which plastic wastes are converted into different products through several chemical processes. POM undergoes disintegration in aqueous solutions of acids, and the products are formaldehyde and methylene diol/trioxane. As a result, formaldehyde monomer solution can be reacted to form trioxane, and virgin POM can be obtained again (Blair 1994; Brandrup 1996).

6.12 POLYACETALS MARKET

The POM market value was $2.2 billion in 2012 and is estimated to reach about $3.3 billion in 2018, growing at a compound annual growth rate (CAGR) of 6.8% (GPM 2014). The Asia-Pacific region dominated the POM market, accounting for 45.6% of the POM market revenue in 2012, and is expected to have the highest market revenue in 2018 followed by the Western European market. With major market share for POM, the former market is expected to be the second fastest growing market with an estimated CAGR of 5.3%, by volume, from 2013 to 2019. The Western European market is expected to grow at a rate of 6.2% in revenue terms and 4.1% in consumption terms from 2013 to 2018. The rest of the world (RoW) including Latin America, the Middle East, and Africa is anticipated to show substantial growth due to increasing demand for consumer and automotive products and is expected to be the fastest growing market in the future, with a CAGR of 8.6% from

2013 to 2018. North America accounted for 19.8% of the total POM market share in 2012 with expected lower growth in POM demand connected to decreasing market for electrical products such as switchboards due to decline in the construction industry (PR Newswire 2014a,b).

POM is highly concentrated, with the top six players (BASF SE, E.I. DuPont de Nemours and Company, Polyplastics, Formosa Plastics Corporation, Celanese Corporation, Asahi Kasei Chemicals Corporation and A. Schulman) accounting for more than 58% of the total POM global market in 2012 (PR Newswire 2014a).

POM is commercially available as homopolymer under trade names Delrin from du Pont (market share 15–16%) and Tenac from Asahi Chemical, as well as in copolymer form under trade names Hostaform and Celcon from Celanese (market share 20–21%), Duracon and Tepcon from Polyplastics (market share 17–18%), Kepital from Korea Engineering Plastics (market share 7–9%), Ultraform from BASF (market share 6–8%), Lupital from Mitsubishi Gas Chemical, Lucel from LG Chem., and Tarnoform from Grupa AZOTY SA (http://www.adzpom.com/CompanyEn.htm). Polyacetal resins are available as basic or standard grades, grades with improved slip/wear properties (modified with additives e.g., MoS_2, graphite, PTFE, chemical lubricants, or mineral fillers), reinforced grades (with glass fibers, glass spheres, carbon fibers), high-impact or toughened grades (modified with rubber or TPUs), and as special grades (e.g., for contact with fuels, electrical conductive grades, medical grades, low emission grades) (Lüftl and Visakh 2014).

Concerning POM pricing, the standard-grade POM pellet price is about US$900–1800 per ton, whereas special grades can cost more than US$100,000 per ton. The recycled polyacetal materials with purity of 92–96% are available at a price from US$800–1000 per ton, and the price depends on POM quality and the amount of contaminants. The price range of semifinished parts, such as sheets, rods, etc., made from standard grades is US$1000–6000 per ton, whereby semifinished parts produced from special POM grades may be purchased from US$4000 to more than US$100,000 per ton (Lüftl and Visakh 2014).

6.13 DEVELOPMENTS AND NEW RESINS

Major POM producers are successively introducing novel polyacetal grades in the market. For instance, BASF introduced in 2007 two new POM grades, Ultraform S2320 003 PRO and W2320 003 PRO, showing high resistance to chemicals, good sliding friction properties, and high capacity of bearing mechanical loads. They are particularly suitable for functional components such as valves, metering systems, springs, slide, and gear elements (BASF 2007b, 2013). Another development from this company is Ultraform E3120 BM designed for hollow objects with demanding requirements, fabricated by extrusion blow molding (BASF 2008).

The new high-strength POM copolymer grade was recently brought to the market by Ticona—it combines the mechanical properties traditionally associated with high-strength POM homopolymer with the thermal and chemical (hot water and caustic) stability originating from traditional POM copolymers (Gronner 2011). Additionally, Ticona has introduced 11 grades of Celcon MT (Medical Technology) acetal copolymer and four grades of Celanex MT PBT for medical uses. These polyacetals, compliant with FDA regulations, are designed as components of drug-delivery systems, medical devices, and packaging.

Celanese Corporation celebrated 50 years of the Celcon/Hostaform POM copolymer product line by expanding its S series of impact-modified POM to include two new extreme toughened XT grades for applications that require impact strength and flexibility in demanding environments (IMA 2014). New low-emission DuPont Delrin 300TE acetal resin (POM) is the first impact-modified POM from DuPont to meet the automotive industry's demanding requirements for the use of plastics in vehicle interiors (DuPont 2014a). The very latest grades of Delrin are low-emission homopolymer grades, medium-high viscosity acetals, grades with food contact compliance, and grades for healthcare applications (DuPont 2014b).

6.14 CONCLUSIONS

Excellent properties of POM, such as exceptional dimensional stability, high mechanical strength and good abrasion resistance, as well as resistance to friction and wear make this polymer a suitable candidate used to produce various mechanical parts, including self-lubricating components. POM characteristics arise from the flexibility of the linear macrochains and high crystallinity as well as high bond energy. However, at elevated temperatures, POM undergoes degradation in an *unzipping* reaction with evolution of formaldehyde, which can be easily oxidized to formic acid, which promotes autocatalytic POM decomposition. To increase the thermal stability, acetylation reaction for removing the hydroxyl groups at the chain ends or copolymerization of trioxane with ethylene oxide or dioxolane, which stops the depolymerization reactions, is applied on industrial scale.

Among well-established POM processing methods, microsystem technologies, especially microinjection molding for production of small parts, and complex microelectromechanical systems for, e.g., electronics and biomedical sectors are developing rapidly. This is mostly due to POM processing characteristics including low viscosity, fast molding, and good processing stability for deposit-free molding.

Novel POM-based materials showing improved wear resistance or reduced friction coefficient comprise blends with polytetrafluoroethylene, ultrahigh-molecular-weight polyethylene, or PEEK. New class of materials are POM nanocomposites, which exhibit major enhancements in strength and stiffness with loss of elongation at break, and show better mechanical, thermal, and electrical properties. POM nanocomposites can be obtained by various methods, e.g., melt processing, and the choice of preparation route depends on the type and size of nanofillers. Importantly, proper dispersion of the nanofiller in the polymer matrix is the crucial step to avoid detrimental agglomeration effects. Then, the nanocomposite is shaped through, e.g., injection molding.

POM grades include impact-modified, high-stiffness, creep resistant, weather resistant, high-sliding-resistant, special lubricant, low warpage, high-impact, flexible, filled, and reinforced polymers in a wide range of sorts—although this diversity makes it possible to offer a selected POM grade for a given application, key POM producers are successively introducing new polyacetal grades in the market and are expanding the production facilities, such as Al Jubail or Industrial Park Hoechst chemical complexes. These developments correspond well with the anticipated growth of POM market value from $2.2 billion in 2012 to about $3.3 billion in 2018 at a CAGR of 6.8%.

REFERENCES

Abdullah, M. Z., Y. Dan-Mallam, and P. S. M. M. Yusoff. 2013. Effect of environmental degradation on mechanical properties of kenaf/polyethylene terephtalate fiber reinforced polyoxymethylene hybrid composite. *Advances in Materials Science and Engineering*, 8. doi:10.1155/2013/671481.

Agazzi, A., V. Sobotka, R. LeGoff, and Y. Jarny. 2013. Optimal cooling design in injection moulding process—A new approach based on morphological surfaces. *Applied Thermal Engineering* 52 (1):170–178.

Aich, R. and P. C. Hagele. 1985. Atomistic calculation of chain conformations and crystal-structures of polyoxymethylene. *Progress in Colloid and Polymer Science* 71:86–95.

Al Jebawi, K., B. Sixou, R. Seguela, G. Vigier, and C. Chervin. 2006. Hot compaction of polyoxymethylene, part 1: Processing and mechanical evaluation. *Journal of Applied Polymer Science* 102 (2):1274–1284.

Al Jebawi, K., B. Sixou, and R. Seguela. 2007. Hot compaction of polyoxymethylene. II. Structural characterization. *Journal of Applied Polymer Science* 106 (2):757–764.

Allcock, H. R. 1967. *Heteroatom Ring Systems and Polymers*. New York: Academic Press.

Al-Salem, S. M. 2009. Establishing an integrated databank for plastic manufacturers and converters in Kuwait. *Waste Management* 29 (1):479–484.

Andrews, E. H. and G. E. Martin. 1973. Structure of polyoxymethylene produced by radiation-induced, solid-state polymerization of trioxane. *Journal of Materials Science* 8 (9):1315–1324.

Archodoulaki, V. M., S. Lüftl, T. Koch, and S. Seidler. 2007. Property changes in polyoxymethylene (POM) resulting from processing, ageing and recycling. *Polymer Degradation and Stability* 92 (12):2181–2189.

Archodoulaki, V. M., S. Lüftl, and S. Seidler. 2007. Degradation behavior of polyoxymethylene: Influence of different stabilizer packages. *Journal of Applied Polymer Science* 105 (6):3679–3688.

Arno, B. and M. Herbert. 1975. Inorganic pigments, patent US 3876441 A.

Asahi. 2010. Tenac™ 3010 Homopolymer, Homopolymer Asahi Kasei Chemicals Corporation, http://catalog.ides.com/Datasheet.aspx?I=26119&E=10394.

ATSDR. 1999. Toxicological profile for formaldehyde, http://www.atsdr.cdc.gov/toxprofiles/tp111.pdf.

Attia, U. M., S. Marson, and J. R. Alcock. 2009. Micro-injection moulding of polymer microfluidic devices. *Microfluidics and Nanofluidics* 7 (1):1–28.

Austin, P. R. and C. E. Frank. 1942. Polymers of formaldehyde, patent US 2,296,249.

Ayala, J. A. and G. A. Joyce. 1990. Iron oxides used in coloring plastics, US patent 4952617 A.

Aznar, M. P., M. A. Caballero, J. A. Sancho, and E. Francés. 2006. Plastic waste elimination by co-gasification with coal and biomass in fluidized bed with air in pilot plant. *Fuel Processing Technology* 87 (5):409–420.

Azoty. 2007. Tarnoform®—Nowoczesne tworzywo poliacetalowe, Grupa AZOTY S.A.

Bai, S. B. and Q. Wang. 2012. Effect of injection speed on phase morphology, crystallization behavior, and mechanical properties of polyoxymethylene/poly(ethylene oxide) crystalline/crystalline blend in injection processing. *Polymer Engineering and Science* 52 (9):1938–1944.

Barker, S. J. and M. B. Price. 1970. *Polyacetals*. London: Iliffe for the Plastics Institute.

Bartz, K. W. 1960. Catalytic polymerization of trioxane, US patent 2947728 A.

BASF. 2007a. Product range, Ultraform—Polyoxymethylene (POM), http://www.plasticsportal.net/wa/plasticsEU~tr_TR/function/conversions:/publish/common/upload/engineering_plastics/Ultraform_range_chart.pdf.

BASF. 2007b. Product: Ultraform® PRO, Application: Healthcare & Diagnostics, http://www.plasticsportal.net/wa/plasticsEU~ru_RU/function/conversions:/publish/common/upload/application_examples/ultraform/Ultraform_healthcare_diagnostics.pdf.

BASF. 2008. Hollow but strong, news release, http://www.plasticsportal.net/wa/plasticsEU~en_GB/portal/show/common/plasticsportal_news/2008/08_167?doc_lang=pl_PL.

BASF. 2013. Engineering plastics for medical solutions, http://www2.basf.us//PLASTICSWEB/displayanyfile?id=0901a5e1802dc06c.

Berardinelli, F. M. and D. E. Hudgin. 1961. Trioxane polymer stabilization, US patent 2989509 A.

Bergeret-Richaud, M., P. Dargelos, S. Girois, and R. Pirri. 2012. New impact modifier and impact modified thermoplastic composition, WO patent 2012038441 A1.

Bergner, K. D. 1993. Plastic molding compound treated with an antistatic agent, patent US 5200446 A.

Berins, M. 1991. *Plastics Engineering Handbook of the Society of the Plastics Industry, Plastics Engineering*. New York: Van Nostrand Reinhold.

Berlin, A. A. and N. S. Yenikolopyan. 1969. Thermodynamics and kinetics of polymerization in the presence of a solid polymer. *Polymer Science U.S.S.R.* 11 (12):3038–3050.

Bershtein, V. A., L. M. Egorova, V. M. Egorov, N. N. Peschanskaya, P. N. Yakushev, M. Y. Keating, E. A. Flexman, R. J. Kassal, and K. P. Schodt. 2002. Segmental dynamics in poly(oxymethylene) as studied via combined differential scanning calorimetry/creep rate spectroscopy approach. *Thermochimica Acta* 391 (1–2):227–243.

Blair, L. M. 1994. Aqueous process for recycling acetal polymer and moldings thereof, WO patent 1993024439A1.

Bledzki, A. K., A. A. Mamun, and M. Feldmann. 2012. Polyoxymethylene composites with natural and cellulose fibres: Toughness and heat deflection temperature. *Composites Science and Technology* 72 (15):1870–1874.

Brandrup, J. 1996. *Recycling and Recovery of Plastics*. Munich: Hanser Publishers.

Bronstert, K. D., F. D. Schmidt, and W. D. Schuette. 1971. Thermoplastische Formmassen hoher Schlagzaehigkeit Thermoplastic molding compounds of high Impact strength, DE patent 1964156 A1.

Bruner, W. M. and W. F. Busse. 1969. Flame retardant polyoxymethylene composition, patent US 3485793 A.

Burg, K. and Cherdron, H. 1971. Thermoplastische Formmassen auf Polyacetal-Basis A thermoplastic molding material based on polyacetal, DE patent 1931392 A1.

Burg, K. and Cherdron H. 1972. Polyoxymethylenes containing a butadiene polymer and a vinyl aromatic hydrocarbon or a methyl methacrylate polymer, US patent 3642940 A.

Burg, K., H. Schlaf, and H. Cherdron. 1971. Einfluß verschiedener Initiatoren auf die Homo- und Copolymerisation des Trioxans. *Die Makromolekulare Chemie* 145 (1):247–258.

Burg, K., H. Cherdron, F. Kloos, and H. Schlaf. 1991. Mixed with shell-core polymer of a crosslinked polydiene base; low temperature impact strength, US patent 5039741 A.

Butlerov, A. M. 1859. Ueber einige Derivate des Jodmethylens. *Annalen der Chemie und Pharmacie* 111:242–252.

Cabuk, T. Z., B. Sari, and H. I. Unal. 2010. Preparation of novel polyindene/polyoxymethylene blends and investigation of their properties. *Journal of Applied Polymer Science* 117 (6):3659–3664.

Carazzolo, G. A. 1963. Structure of normal crystal form of polyoxymethylene. *Journal of Polymer Science Part A-General Papers* 1 (5):1573.

Carazzolo, G. A. and M. Mammi. 1963. Crystal structure of a new form of polyoxymethylene. *Journal of Polymer Science Part A-General Papers* 1 (3):965–983.

Chang, F. C. and M. Y. Yang. 1990. Mechanical fracture-behavior of polyacetal and thermoplastic polyurethane elastomer toughened polyacetal. *Polymer Engineering and Science* 30 (9):543–552.

Chang, J. P. and C. E. Lin. 2012. Durable polyoxymethylene composition, patent US 20120238680 A1.

Chang, J. P. and C. E. Lin. 2013. Durable polyoxymethylene composition, patent EP 2371901 B1.

Chang, F. C., M. Y. Yang, and J. S. Wu. 1991. Blends of polycarbonate and polyacetal. *Polymer* 32 (8):1394–1400.

Chem, L. G. 2009. Lucel, http://www.lgchem.com.tr/lg-product-page.php?country=1&productid=7.

Chen, C. S. H. 1971. Process for the preparation of oxymethylene copolymers, US patent 3597397 A.

Chen, C. S. H. 1976a. Copolymerization of trioxane with cyclic acetals—Effect of catalyst on polymerization and monomer reactivity. *Journal of Polymer Science Part A-Polymer Chemistry* 14 (1):143–151.

Chen, C. S. H. 1976b. Effect of gegen ion in cationic polymerization and copolymerization of trioxane. *Journal of Polymer Science: Polymer Chemistry Edition* 14 (1):129–142.

Chen, J. Y., Y. Cao, and H. L. Li. 2006a. An investigation on wear mechanism of POM/LLDPE blends. *Journal of Applied Polymer Science* 101 (1):48–53.

Chen, J. Y., Y. Cao, and H. L. Li. 2006b. Investigation of the friction and wear behaviors of polyoxymethylene/linear low-density polyethylene/ethylene-acrylic-acid blends. *Wear* 260 (11–12):1342–1348.

Cherdron, H. 1972. New trioxane copolymers. *Journal of Macromolecular Science: Part A—Chemistry* 6 (6):1077–1088.

Chiang, W. Y. and C. Y. Huang. 1989. The effect of the soft segment of polyurethane on copolymer-type polyacetal polyurethane blends. *Journal of Applied Polymer Science* 38 (5):951–968.

Chiang, W. Y. and C. Y. Huang. 1993. Properties of copolymer-type polyacetal ethylene propylene diene terpolymer blends. *Journal of Applied Polymer Science* 47 (1):105–112.

Chiang, W.-Y. and C.-Y. Huang. 1997. Polyacetal. In *Handbook of Thermoplastics*, edited by O. Olabisi. New York: Marcel Dekker.

Chiang, W. Y. and M. S. Lo. 1988. Properties of copolymer-type polyacetal polyurethane blends. *Journal of Applied Polymer Science* 36 (7):1685–1700.

Chu, J. S., M. R. Kamal, S. Derdouri, and A. Hrymak. 2010. Characterization of the microinjection molding process. *Polymer Engineering and Science* 50 (6):1214–1225.

Chung, C. I. 2000. *Extrusion of Polymers: Theory and Practice*. Munich: Hanser Publishers.

Clark, E. S. and L. S. Scott. 1974. Superdrawn crystalline polymers: A new class of high-strength fiber. *Polymer Engineering & Science* 14 (10):682–686.

Coates, P. D. and I. M. Ward. 1978. Hydrostatic extrusion of polyoxymethylene. *Journal of Polymer Science Part B-Polymer Physics* 16 (11):2031–2047.

Collins, I. O. D. 1966. Boron nitride incorporated in polymer products, US patent 3261800 A.

Collins, G. L., R. K. Greene, F. M. Berardinelli, and W. V. Garruto. 1979. Observations on the initiation of trioxane with boron trifluoride dibutyl etherate. *Journal of Polymer Science: Polymer Letters Edition* 17 (10):667–671.

Collins, G. L., R. K. Greene, F. M. Berardinelli, and W. H. Ray. 1981. Fundamental considerations on the mechanism of copolymerization of trioxane and ethylene oxide initiated with boron trifluoride dibutyl etherate. *Journal of Polymer Science: Polymer Chemistry Edition* 19 (7):1597–1607.

Colombo, R., C. Neri, and R. M. Riva. 1999. 2-(2′-hydroxyphenyl)benzotriazoles used as u.v. stabilisers, WO patent 1999023093 A1.

Consortium, Trioxane Manufacturers. 2000. 1,3,5-trioxane, http://www.epa.gov/hpv/pubs/summaries/triox/c12863a.pdf.

Curtis, A. C., P. S. Hope, and I. M. Ward. 1982. Modulus development in oriented short-glass-fiber-reinforced polymer composites. *Polymer Composites* 3 (3):138–145.

Czarnecka-Komorowska, D., T. Sterzynski, H. Maciejewski, and M. Dutkiewicz. 2012. The effect of polyhedral oligomeric silsesquioxane (POSS) on morphology and mechanical properties of polyoxymethylene (POM). *Composites Theory and Practice* 12 (4):232–236.

Dan-Mallam, Y., M. Z. Abdullah, and P. S. M. M. Yusoff. 2014. The effect of hybridization on mechanical properties of woven kenaf fiber reinforced polyoxymethylene composite. *Polymer Composites* 35 (10):1900–1910.

Defazio, C. A. and R. J. Kray. 1962. Trioxane copolymers, US patent 3026299 A.

Delépine, M. 1897. Sur une nouvelle méthode de préparation des amines primaires, Comptes rendus de l'Académie des sciences. *Compt. Rend. Acad. Sci.* 124:292–295.

DiEdwardo, A. H. and R. M. Kusumgar. 1984. Oxymethylene polymer modified with 1,2-polybutadiene exhibiting enhanced tensile impact properties, patent US 4424307 A.

Ding, Q., W.-L. Dai, and P. Zhang. 2007. The effect of polyvinylidene fluoride on nonisothermal crystallization behavior of polyoxymethylene. *Polymer Engineering & Science* 47:2034–2040.

Disch, S., P. Eckardt, M. Hoffmockel, K. F. Mück, and G. Reuschel. 2001. Low formaldehyde emission, patent US 6306940 B1.

Dröscher, M. 1982. Solid-state extrusion of semicrystalline copolymers. In *Advances in Polymer Science, Synthesis and Degradation Rheology and Extrusion*, 47, 119–138. Berlin, Heidelberg: Springer.

Dröscher, M. and G. Wegner. 1979. Kinetics of a simultaneous polymerization and crystallization: Cationic polymerization and copolymerization of trioxane. *Industrial & Engineering Chemistry Product Research and Development* 18 (4):259–263.

Drummer, D. and K. Vetter. 2011. Expansion-injection-molding (EIM) by cavity near melt compression—About the process characteristic. *CIRP Journal of Manufacturing Science and Technology* 4 (4):376–381.

DuPont. 2000. DuPont Engineering Polymers, Delrin® acetal resin—Product and Properties Guide.

DuPont. 2006. DuPont™ Delrin® acetal resin Molding Guide.

DuPont. 2014a. DuPont™ Delrin® POM Extends Scope of Use of Plastics in Vehicle Interiors, http://www.dupont.com/industries/automotive/safety-systems-interior/articles/delrin-automotive-interiors.html.

DuPont. 2014b. Latest Grades of Delrin® Acetal, http://www.dupont.com/products-and-services/plastics-polymers-resins/thermoplastics/Articles/grades-of-delrin.html (accessed August 5, 2014).

Eckardt, P., M. Hoffmockel, K. F. Muck, G. Reuschel, S. Verma, and M. G. Yearwood. 1999. Process for the preparation of polyacetal copolymers, patent US 5962623 A.

Egawa, Y., S. Imanishi, A. Matsumoto, and F. Horii. 1996. Solid-state C-13 NMR study on miscibility of polyoxymethylene/terpenephenol blends. *Polymer* 37 (25):5569–5575.

Egorov, B. A., A. S. Grzhimalovskii, and A. V. Yudin. 1970. Polyformaldehyde fibres. *Fibre Chemistry* 1 (4):403–408.

Enikolopyan, N. S., V. J. Irzhak, I. P. Kravchuk, O. A. Plechova, G. V. Rakova, L. M. Romanov, and G. P. Savushkina. 1967. Some peculiarities of cationic polymerization of trioxane. *Journal of Polymer Science Part C: Polymer Symposia* 16 (4):2453–2462.

Erro, R., M. Gaztelumendi, and J. Nazabal. 1996. Transitions and mechanical properties in polyoxymethylene/phenoxy blends. *Journal of Polymer Science Part B-Polymer Physics* 34 (6):1055–1062.

Fayolle, B., J. Verdu, M. Bastard, and D. Piccoz. 2008. Thermooxidative ageing of polyoxymethylene, Part 1: Chemical aspects. *Journal of Applied Polymer Science* 107 (3):1783–1792.

Fischer, E., K. Kullmar, K. Weissermel, and M. Reiher. 1967. Polyacetals and a process of preparing them, patent US 3316217 A.

Fleischer, D., K. F. Muck, and G. Reuschel. 1992. Recycling of polyacetal. *Kunststoffe-German Plastics* 82 (9):763–766.

Flexman, E. A. 1984. Blow moulding of polyoxymethylene/polyurethane compositions, patent EP 0121407 A2.

French, R. N., J. M. Machado, and D. Linvien. 1992. Miscible polyacetal poly(vinyl phenol) blends. 1. Predictions based on low-molecular-weight analogs. *Polymer* 33 (4):755–759.

Fukuda, H., S. Ishida, and T. Matsuoka. 1975. Polyacetal composition, patent US 3903197 A.

Ganan, P. and I. Mondragon. 2005. Effect of fiber treatments on mechanical behavior of short fique fiber-reinforced polyacetal composites. *Journal of Composite Materials* 39 (7):633–646.

Gao, X. L., C. Qu, and Q. Fu. 2004. Toughening mechanism in polyoxymethylene/thermoplastic polyurethane blends. *Polymer International* 53 (11):1666–1671.

Garber, C. A. and E. S. Clark. 1970. Morphology and deformation behavior of "row-nucleated" polyoxymethylene film. *Journal of Macromolecular Science, Part B* 4 (3):499–517.

Gezovich, D. M. and P. H. Geil. 1971. Deformation of polyoxymethylene by rolling. *Journal of Materials Science* 6 (6):509.

Goodman, J. H. H. and J. L. T. Sherwood. 1961. Process of polymerizing formaldehyde in the presence of onium compounds as initiators, patent US 2994687 A.

Goossens, S. and G. Groeninckx. 2006. Mutual influence between reaction-induced phase separation and isothermal crystallization in POM/epoxy resin blends. *Macromolecules* 39 (23):8049–8059.

Goossens, S. and G. Groeninckx. 2007. High melting thermoplastic/epoxy resin blends: Influence of the curing reaction on the crystallization and melting behavior. *Journal of Polymer Science Part B-Polymer Physics* 45 (17):2456–2469.

Goossens, S., B. Goderis, and G. Groeninckx. 2006. Reaction-induced phase separation in crystallizable micro- and nanostructured high melting thermoplastic/epoxy resin blends. *Macromolecules* 39 (8):2953–2963.

GPM. 2014. Global Polyoxymethylene Market 2014–2019—Applications Analysis: Electrical & Electronics, Consumer, Industrial, Transportation, and Others, http://uk.reuters.com/article/2014/05/08/research-and-markets-idUKnBw085905a+100+BSW20140508.

Gramlich, V. 1977. ECM-4, 4th Europ Cryst Meeting, Oxford, Abstract Book, 531.

Gray, R. W. and N. G. McCrum. 1971. Orientation in rolled polyoxymethylene. *Nature-Physical Science* 234 (49):117–118.

Gronner, R. 2011. *A New High Strength POM Copolymer*. ANTEC, Brookfield, http://www.4spe.org/Resources/resource.aspx?ItemNumber=10148.

Grossman, R. F. and J. T. Lutz. 2000. *Polymer Modifiers and Additives, Plastics Engineering*. Taylor & Francis.

Guo, L. P., X. M. Xu, Y. D. Zhang, and Z. J. Zhang. 2014. Effect of functionalized nanosilica on properties of polyoxymethylene-matrix nanocomposites. *Polymer Composites* 35 (1):127–136.

Hagemeyer, A., K. Schmidt-Rohr, and H. W. Spiess. 1989. Two-dimensional nuclear magnetic resonance experiments for studying molecular order and dynamics in static and in rotating solids. In *Advances in Magnetic and Optical Resonance*, edited by W. S. Warren, 85–130. San Diego, CA: Academic Press.

Hama, H. and K. Tashiro. 2003a. Structural changes in isothermal crystallization process of polyoxymethylene investigated by time-resolved FTIR, SAXS and WAXS measurements. *Polymer* 44 (22):6973–6988.

Hama, H. and K. Tashiro. 2003b. Structural changes in non-isothermal crystallization process of melt-cooled polyoxymethylene. [I] Detection of infrared bands characteristic of folded and extended chain crystal morphologies and extraction of a lamellar stacking model. *Polymer* 44 (10):3107–3116.

Hama, H. and K. Tashiro. 2003c. Structural changes in non-isothermal crystallization process of melt-cooled polyoxymethylene[II] evolution of lamellar stacking structure derived from SAXS and WAXS data analysis. *Polymer* 44 (7):2159–2168.

Hammick, D. L. and A. R. Boeree. 1922. Preparation of [small alpha]-trioxymethylene and a new polymeride of formaldehyde. *Journal of the Chemical Society, Transactions* 121:2738–2740.

Harashina, H., Y. Tajima, and T. Itoh. 2006. Synergistic effect of red phosphorus, novolac and melamine ternary combination on flame retardancy of poly(oxymethylene). *Polymer Degradation and Stability* 91 (9):1996–2002.

Hardy, G. F. 1971. Fracture toughness of glass fiber-reinforced acetal polymer. *Journal of Applied Polymer Science* 15 (4):853–866.

Hardy, G. F. and H. L. Wagner. 1969. Tensile behavior of glass fiber-reinforced acetal polymer. *Journal of Applied Polymer Science* 13 (5):961–975.

Hashemi, S., M. T. Gilbride, and J. Hodgkinson. 1996. Mechanical property relationships in glass-filled polyoxymethylene. *Journal of Materials Science* 31 (19):5017–5025.

Hashemi, S., P. Elmes, and S. Sandford. 1997. Hybrid effects on mechanical properties of polyoxymethylene. *Polymer Engineering and Science* 37 (1):45–58.

He, Y., B. Zhu, and Y. Inoue. 2004. Hydrogen bonds in polymer blends. *Progress in Polymer Science* 29 (10):1021–1051.

Hilado, C. J. 1990. *Flammability Handbook for Plastics*. Lancaster, PA: Technomic Pub. Co.

Hine, P. J., R. A. Duckett, and I. M. Ward. 1993. The fracture-behavior of short glass-fiber-reinforced polyoxymethylene. *Composites* 24 (8):643–649.

Hine, P. J., N. Davidson, R. A. Duckett, A. R. Clarke, and I. M. Ward. 1996. Hydrostatically extruded glass-fiber-reinforced polyoxymethylene. 1. The development of fiber and matrix orientation. *Polymer Composites* 17 (5):720–729.

Hine, P. J., S. Wire, R. A. Duckett, and I. M. Ward. 1997. Hydrostatically extruded glass fiber reinforced polyoxymethylene. 2. Modeling the elastic properties. *Polymer Composites* 18 (5):634–641.

Hohr, L., H. Cherdron, and W. Kern. 1962. Über den Angriff des Carboniumions am Monomeren Styrol bei der Kationischen Copolymerisation von Trioxanund Styrol. 19. Über Polyoxymethylene. *Makromolekulare Chemie* 52:59–69.

Hojfors, R. J. V., E. Baer, and P. H. Geil. 1977. Dynamic-mechanical study of molecular motions in solid polyoxymethylene copolymers from 4 to 315-degrees-K. *Journal of Macromolecular Science-Physics* B13 (3):323–348.

Hope, P. S. and I. M. Ward. 1981. An activated rate theory approach to the hydrostatic extrusion of polymers. *Journal of Materials Science* 16 (6):1511–1521.

Hope, P. S., A. Richardson, and I. M. Ward. 1982. The hydrostatic extrusion and die-drawing of glass-fiber-reinforced polyoxymethylene. *Polymer Engineering and Science* 22 (5):307–313.

Horio, M. and Y. Yoshinaga. 2003. Polyoxymethylene resin composition, patent US 6602953 B1.

Hsu, C. I., J. R. Hwang, and P. H. Ting. 2010. Influence of process conditions on tensile properties polyoxymethylene composite. *Journal of Reinforced Plastics and Composites* 29 (16):2402–2412.

Hu, K. H., J. Wang, S. Schraube, Y. F. Xu, X. G. Hu, and R. Stengler. 2009. Tribological properties of MoS_2 nano-balls as filler in polyoxymethylene-based composite layer of three-layer self-lubrication bearing materials. *Wear* 266 (11–12):1198–1207.

Hu, P., T. Shang, M. Jiang, and M. Chen. 2007. Investigation of the thermal decomposition properties of polyoxymethylene. *Journal of Wuhan University of Technology-Materials Science Edition* 22 (1):171–173.

Hu, X. G. 2000. Friction and wear behaviors of toughened polyoxymethylene blend under water lubrication. *Polymer-Plastics Technology and Engineering* 39 (1):137–150.

Hu, X., K. Hu, Y. Xu, and R. Stengler. 2010. Synthesis of nano-MoS_2 particles and its role in the self-lubrication of polyacetal-based composite. In *Advanced Tribology*, Proceedings of CIST2008 & ITS-IFToMM2008, edited by J. Luo, Y. Meng, T. Shao, and Q. Zhao, 491–492. Beijing: Tsinghua University Press; Berlin Heidelberg: Springer-Verlag.

Huang, J. M., H. J. Cheng, J. S. Wu, and F. C. Chang. 2003. Blends of poly(propylene) and polyacetal compatibilized by ethylene vinyl alcohol copolymers. *Journal of Applied Polymer Science* 89 (6):1471–1477.

Hull, T. R., A. Witkowski, and L. Hollingbery. 2011. Fire retardant action of mineral fillers. *Polymer Degradation and Stability* 96 (8):1462–1469.

Iguchi, M. 1973. Growth of needle-like crystals of polyoxymethylene during polymerisation. *British Polymer Journal* 5 (3):195–198.

Iguchi, M. 1983. Generation of orthorhombic polyoxymethylene in a cationic polymerization system of trioxane. *Polymer* 24 (7):915–920.

Iguchi, M., H. Kanetsun, and T. Kawai. 1969. Growth of polyoxymethylene crystals in course of polymerization of trioxane in solution. *Makromolekulare Chemie* 128:63–82.

Iguchi, M. and I. Murase. 1975. "Shish kebab" structures formed on needle-like polyoxymethylene crystals. *Journal of Polymer Science Part A: Polymer Chemistry* 13: 1461–1465.

Iguchi, M., T. Suehiro, Y. Watanabe, Y. Nishi, and M. Uryu. 1982. Composite-materials reinforced with polyoxymethylene whiskers. *Journal of Materials Science* 17 (6):1632–1638.

Illescas, S., A. Arostegui, D. A. Schiraldi, M. Sanchez-Soto, and J. I. Velasco. 2010. The role of polyhedral oligomeric silsesquioxane on the thermo-mechanical properties of polyoxymethylene copolymer based nanocomposites. *Journal of Nanoscience and Nanotechnology* 10 (2):1349–1360.

Illescas, S., M. Sanchez-Soto, H. Milliman, D. A. Schiraldi, and A. Arostegui. 2011. The morphology and properties of melt-mixed polyoxymethylene/monosilanolisobutyl-POSS composites. *High Performance Polymers* 23 (6):457–467.

IMA. 2014. New POM grades for tougher applications, http://www.injectionmouldingasia.com/feb2014/news1.html.

Ito, M. and K. Nakamura. 2001. Impact modifier, process for producing the same, and thermoplastic resin composition containing the same, patent WO 2001081465 A1.

Jaacks, V. 1969. Cationic copolymerization of trioxane and 1,3-dioxolane. In *Addition and Condensation Polymerization Processes*, edited by N. A. J. Platzer, 371–386. Washington, DC: ACS.

Jaacks, V. and W. Kern. 1963. Initiatoren für die Polymerisation des Trioxans. 20. Über Polyoxymethylene. *Makromolekulare Chemie* 62:1–17.

Jeong, C. R., T. K. Kim, and S. E. Lee. 2007. Permanent anti-static polyoxymethylene resin composition, patent WO 2007119972 A1.

Jose, A. J. and M. Alagar. 2011. Development and characterization of organoclay-filled polyoxymethylene nanocomposites for high performance applications. *Polymer Composites* 32 (9):1315–1324.

Jose, A. J. and M. Alagar. 2012. Fabrication of bioactive polyoxymethylene nanocomposites for bone tissue replacement. *Macromolecular Symposia* 320 (1):24–37.

Kassal, R. J. 1999. Polyoxymethylene resin compositions having improved molding characteristics, patent WO 1999035191 A1.

Kassal, R. J., H. Mori, and K. Shinohara. 2001. Polyoxymethylene resin compositions having improved molding characteristics, patent WO 2001023473 A1.

Kassem, M. E., M. E. Bassiouni, and M. El-Muraikhi. 2002. The effect of gamma-irradiation on the optical properties of polyoxymethylene compacts. *Journal of Materials Science-Materials in Electronics* 13 (12):717–719.

Kastelic, J., A. Buckley, P. Hope, and I. Ward. 1983. Hydrostatic extrusion of glass reinforced and unreinforced Celcon® POM. In *Structure-Property Relationships of Polymeric Solids*, edited by A. Hiltner, 175–191. New York: Plenum Press.

Katsumata, T. 1992. Weather-resistant polyacetal resin compositions and molded articles thereof, patent US 5086095 A.

Kawaguchi, K., E. Masuda, and Y. Tajima. 2008. Tensile behavior of glass-fiber-filled polyacetal: Influence of the functional groups of polymer matrices. *Journal of Applied Polymer Science* 107 (1):667–673.

Kawaguchi, K., H. Nakao, E. Masuda, and Y. Tajima. 2008. Morphology and nonisothermal crystallization of a Polyacetal/Poly(epsilon-caprolactone) reactive blend prepared via a chain-transfer reaction. *Journal of Applied Polymer Science* 107 (2):1269–1279.

Kawaguchi, K., K. Mizuguchi, K. Suzuki, H. Sakamoto, and T. Oguni. 2010. Mechanical and physical characteristics of cellulose-fiber-filled polyacetal composites. *Journal of Applied Polymer Science* 118 (4):1910–1920.

Kennedy, C. D., W. R. Sorenson, and G. G. McClaffin. 1966. *Am. Chem. Soc. Polymer Preprints* 7:667.

Kentgens, A. P. M., A. F. De Jong, E. De Boer, and W. S. Veeman. 1985. A 2D-exchange NMR study of very slow molecular motions in crystalline poly(oxymethylene). *Macromolecules* 18 (5):1045–1048.

Kentgens, A. P. M., E. Deboer, and W. S. Veeman. 1987. Ultraslow molecular motions in crystalline polyoxymethylene—A complete elucidation using two-dimensional solid-state NMR. *Journal of Chemical Physics* 87 (12):6859–6866.

Kern, W. and H. Cherdron. 1960. Der abbau von polyoxymethylenen. Poloxymethylene. 14. Mitteilung. *Die Makromolekulare Chemie* 40 (1):101–117.

Kern, W. and V. Jaacks. 1960. Some kinetic effects in the polymerization of 1,3,5-trioxane. *Journal of Polymer Science* 48 (150):399–404.

Kern, W., H. Baader, H. Cherdron, L. Hohr, H. Deibig, A. Giefer, V. Jaacks, and A. Wildenau. 1961. Polyoxymethylene. *Angewandte Chemie-International Edition* 73 (6):177–186.

Kern, W., H. Deibig, A. Giefer, and V. Jaacks. 1966. Polymerization and copolymerization of trioxan. *Pure Appl. Chem.* 12 (1–4):371–386.

Kikutani, T. 2002. Formation and structure of high mechanical performance fibers. II. Flexible polymers. *Journal of Applied Polymer Science* 83 (3):559–571.

Kikutani, T. and H. Okawa. 2004. Polyoxymethylene fiber and method for production thereof, patent US 6818294 B2.

Kiliaris, P. and C. D. Papaspyrides. 2010. Polymer/layered silicate (clay) nanocomposites: An overview of flame retardancy. *Progress in Polymer Science* 35 (7):902–958.

Kim, B. S. and R. S. Porter. 1988. Uniaxial drawing of poly(ethyleneoxide)/poly(methyl methacrylate) blends by solid-state extrusion. *Journal of Polymer Science Part B: Polymer Physics* 26 (12):2499–2508.

Kiss, K. D. 1967. Oxymethylene-epsilon caprolactam copolymers and method of preparation, US patent 3299005 A.

Kobayashi, M. 1993. Polymorphism and morphologies of linear and macrocyclic poly(oxymethylene). In *Crystallization of Polymers*, edited by M. Dosière, 283–299. Amsterdam: Springer.

Kobayashi, M., H. Morishita, M. Shimomura, and M. Iguchi. 1987. Vibrational spectroscopic study on the solid-state phase transition of poly(oxymethylene) single crystals from the orthorhombic to the trigonal phase. *Macromolecules* 20 (10):2453–2456.

Kobayashi, M., H. Morishita, and M. Shimomura. 1989. Pressure-induced phase transition of poly(oxymethylene) from the trigonal to the orthorhombic phase: Effect of morphological structure. *Macromolecules* 22 (9):3726–3730.

Kobayashi, M., T. Adachi, Y. Matsumoto, H. Morishita, T. Takahashi, K. Ute, and K. Hatada. 1993. Polarized raman and infrared studies of single-crystals of orthorhombic modification of polyoxymethylene and its linear oligomer—Crystal-structures and vibrational assignments. *Journal of Raman Spectroscopy* 24 (8):533–538.

Koch, T. A. and P. E. Lindvig. 1959. Molecular structure of high molecular weight acetal resins. *Journal of Applied Polymer Science* 1 (2):164–168.

Koji, I. and N. Tomonobu. 2001. Production of polyoxymethylene fiber, patent JP1999000362724.

Komatsu, T. 1996. Effects of surface treatments of superdrawn polyoxymethylene fibers on adhesion to epoxy resins. *Journal of Applied Polymer Science* 59 (7):1137–1143.

Komatsu, Y., H. Igarashi, and T. Erami. 1995. Antistatic agent for polyacetal resins, patent US 5478877 A.

Kongkhlang, T., M. Kotaki, Y. Kousaka, T. Umemura, D. Nakaya, and S. Chirachanchai. 2008. Electrospun polyoxymethylene: Spinning conditions and its consequent nanoporous nanofiber. *Macromolecules* 41 (13):4746–4752.

Kongkhlang, T., Y. Kousaka, T. Umemura, D. Nakaya, W. Thuamthong, Y. Pattamamongkolchai, and S. Chirachanchai. 2008. Role of primary amine in polyoxymethylene (POM)/bentonite nanocomposite formation. *Polymer* 49 (6):1676–1684.

Kongkhlang, T., K. Tashiro, M. Kotaki, and S. Chirachanchai. 2008. Electrospinning as a new technique to control the crystal morphology and molecular orientation of polyoxymethylene nanofibers. *Journal of the American Chemical Society* 130 (46):15460–15466.

Kumar, G., M. Arindam, N. R. Neelakantan, and N. Subramanian. 1993. Stress-relaxation behavior of polyacetal-thermoplastic polyurethane elastomer blends. *Journal of Applied Polymer Science* 50 (12):2209–2216.

Kumar, G., L. Mahesh, N. R. Neelakantan, and N. Subramanian. 1993. Studies on thermal-stability and behavior of polyacetal and thermoplastic polyurethane elastomer blends. *Polymer International* 31 (3):283–289.

Kumar, G., N. R. Neelakantan, and N. Subramanian. 1993. Mechanical behavior of polyacetal and thermoplastic polyurethane elastomer toughened polyacetal. *Polymer-Plastics Technology and Engineering* 32 (1–2):33–51.

Kumar, G., N. R. Neelakantan, and N. Subramanian. 1995. Polyacetal and thermoplastic polyurethane elastomer toughened polyacetal—Crystallinity and fracture-mechanics. *Journal of Materials Science* 30 (6):1480–1486.

Kumaraswamy, G., N. S. Surve, R. Mathew, A. Rana, S. K. Jha, N. N. Bulakh, A. A. Nisal, T. G. Ajithkumar, P. R. Rajamohanan, and R. Ratnagiri. 2012. Lamellar melting, not crystal motion, results in softening of polyoxymethylene on heating. *Macromolecules* 45 (15):5967–5978.

Kurz, K. and O. Schleith. 2003. Polyoxymethylene moulding compound containing a lubricant, the use thereof and moulded bodies produced therefrom, patent US 20030171470 A1.

Kutsuma, T. and Y. Yamase. 1969. Process for the production of trioxane copolymers, patent US 3428590 A.

Lam, K. L., A. A. Bakar, Z. A. M. Ishak, and J. Karger-Kocsis. 2004. Amorphous copolyester/polyoxymethylene blends: Thermal, mechanical and morphological properties. *Kautschuk Gummi Kunststoffe* 57 (11):570.

Lam, K. L., A. A. Bakar, and Z. A. M. Ishak. 2005. Effects of compatibilizer and testing speed on the mechanical and morphology behaviors of co-continuous amorphous copolyester—Polyoxymethylene blends. *Polymer Engineering and Science* 45 (5):710–719.

Latz, G., L. Lowell, and U. Ziegler. 2012. High impact resistant polyoxymethylene for extrusion blow molding, patent WO 2012130988 A1.

Leese, L. and M. W. Baumber. 1965. Kinetics and mechanism of trioxan polymerization. *Polymer* 6 (5):269–286.

Leszczynska, A. and K. Pielichowski. 2012. The mechanical and thermal properties of polyoxymethylene (POM)/organically modified montmorillonite (OMMT) engineering nanocomposites modified with thermoplastic polyurethane (TPU) compatibilizer. In *Polymer Composite Materials: From Macro, Micro to Nanoscale*, edited by A. Boudenne, 201–209. Stafa-Zurich: Trans Tech Publications Ltd.

Li, Y. L., T. X. Xie, and G. S. Yang. 2006. Studies on novel composites of polyoxymethylene/polyamide 6. *Journal of Applied Polymer Science* 99 (1):335–339.

Linton, W. H. and H. H. Goodman. 1959. Physical properties of high molecular weight acetal resins. *Journal of Applied Polymer Science* 1 (2):179–184.

Liu, Y. A., M. F. Liu, D. Y. Xie, and Q. Wang. 2008. Thermoplastic polyurethane-encapsulated melamine phosphate flame retardant polyoxymethylene. *Polymer-Plastics Technology and Engineering* 47 (3):330–334.

Liu, X., S. B. Bai, M. Nie, and Q. Wang. 2012. Effect of blend composition on crystallization behavior of polyoxymethylene/poly(ethylene oxide) crystalline/crystalline blends. *Journal of Polymer Research* 19:9787.

Liu, X., S. B. Bai, and Q. Wang. 2012. Influence of shearing on crystallization behavior of polyoxymethylene/poly(ethylene oxide) crystalline/crystalline blends. *Journal of Macromolecular Science Part B-Physics* 51 (4):642–653.

Liu, Y., Z. Wang, and Q. Wang. 2012. Effects of magnesium hydroxide and its synergistic systems on the flame retardance of polyformaldehyde. *Journal of Applied Polymer Science* 125 (2):968–974.

Lüftl, S. and P. M. Visakh. 2014. Polyoxymethylene: State of art, new challenges and opportunities. In *Polyoxymethylene Handbook: Structure, Properties, Applications and Their Nanocomposites*, edited by S. Lüftl, P. M. Visakh, and S. Chandran, 1–19. Hoboken, NJ: John Wiley & Sons; Salem, MA: Scrivener Publishing.

Lüftl, S., V. M. Archodoulaki, and S. Seidler. 2006. Thermal-oxidative induced degradation behaviour of polyoxymethylene (POM) copolymer detected by TGA/MS. *Polymer Degradation and Stability* 91 (3):464–471.

Lüftl, S., V. M. Archodoulaki, T. Koch, and S. Seidler. 2008. Effects of the additive package on the thermal properties of a commercial polyoxymethylene homopolymer. *Journal of Vinyl & Additive Technology* 14 (1):21–27.

Makabe, Y. T. H. R., S. Okita, and Y. Yamamoto. 1991. Polyoxymethylene multi-copolymer and its resin composition, patent EP 0412783 A2.

Makabe, Y., S. Okita, and Y. Yamamoto. 1992. Copolymer of trioxane, another cyclic ether and one of glycidyl phenyl ether, styrene oxide, glycidyl naphthyl ether, mechanical strength, patent US 5079330 A.

Mamunya, Y. P., Y. V. Muzychenko, E. V. Lebedev, G. Boiteux, G. Seytre, C. Boullanger, and P. Pissis. 2007. PTC effect and structure of polymer composites based on polyethylene/polyoxym ethylene blend filled with dispersed iron. *Polymer Engineering and Science* 47 (1):34–42.

Mantell, R. M. 1966. Production of trioxane polymers with cationic exchange material as catalyst, US patent 3252940 A.

Masamoto, J. 1993. Modern polyacetals. *Progress in Polymer Science* 18 (1):1–84.

Masamoto, J. and K. Matsuzaki. 1994. End capping during polymerization of formaldehyde for manufacturing acetal homopolymer. *Polymer-Plastics Technology and Engineering* 33 (2):221–232.

Masamoto, J., T. Iwaisako, M. Chohno, M. Kawamura, J. Ohtake, and K. Matsuzaki. 1993. Development of a new advanced process for manufacturing polyacetal resins. 1. Development of a new process for manufacturing highly concentrated aqueous formaldehyde solution by methylal oxidation. *Journal of Applied Polymer Science* 50 (8):1299–1305.

Masamoto, J., K. Matsuzaki, T. Iwaisako, K. Yoshida, K. Kagawa, and H. Nagahara. 1993. Development of a new advanced process for manufacturing polyacetal resins. 3. End-capping during polymerization for manufacturing acetal homopolymer and copolymer. *Journal of Applied Polymer Science* 50 (8):1317–1329.

Masamoto, J., K. Matsuzaki, and H. Morishita. 1993. Development of a new advanced process for manufacturing polyacetal resins. 2. Vapor-liquid-equilibrium of formaldehyde water-system. *Journal of Applied Polymer Science* 50 (8):1307–1315.

Masamoto, J., K. Yajima, S. Sakurai, S. Aida, M. Ueda, and S. Nomura. 2000. Microphase separation in polyoxymethylene end-capped with a long-chain alkyl. *Polymer* 41 (19):7283–7287.

Matsumoto, A., Y. Egawa, T. Matsumoto, and F. Horii. 1997. Miscibility of polyoxymethylene blends as revealed by high-resolution solid-state C-13-NMR spectroscopy. *Polymers for Advanced Technologies* 8 (4):250–256.

McAndrew, F. B. 1969. Copolymerization of trioxane with an epoxy-containing comonomer in the presence of an aldehyde, patent US 3445433 A.

McKellop, H. A., H. L. Milligan, and T. Rostlund. 1996. Long term biostability of polyacetal (Delrin®) implants. *Journal of Heart Valve Disease* 5:S238–S242.

Mehrabzadeh, M. and D. Rezaie. 2002. Impact modification of polyacetal by thermoplastic elastomer polyurethane. *Journal of Applied Polymer Science* 84 (14):2573–2582.

Memon, N. A. and R. H. Weese. 1995. Stabilized modifier and impact modified thermoplastics, patent US 5451624 A.

Meyer, B. H. and J. C. Kinslow. 1991. Purification, impregnation and foaming of polymer particles with carbon dioxide, patent US 5049328 A.

Miki, K. 1970. Dynamic mechanical properties of polyoxymethylene, IV. *Polymer Journal* 1 (4):432–441.

Miki, T., T. Higashimura, and S. Okamura. 1966. Kinetic studies of the solution polymerization of trioxane catalyzed by $BF_3 \cdot O(C_2H_5)_2$. VI. The catalytic mechanism of boron trifluoride coordination complexes in ethylene dichloride. *Bulletin of the Chemical Society of Japan* 39 (11):2480–2485.

Miki, T., T. Higashimura, and S. Okamura. 1967. Rates of polymer formation and monomer consumption in the solution polymerization of trioxane catalyzed by $BF_3 \cdot O(C_2H_5)_2$. *Journal of Polymer Science Part A-1: Polymer Chemistry* 5 (1):95–106.

Miles, M. J. and N. J. Mills. 1975. Theories for development of rolling textures in polyoxymethylene. *Journal of Materials Science* 10 (12):2092–2111.

Miller, J. A. R. 1949. Generation of monomeric formaldehyde gas from formaldehyde polymers, patent US 2460592 A.

Mohanraj, J., M. J. Bonner, D. C. Barton, and I. M. Ward. 2006. Physical and mechanical characterization of oriented polyoxymethylene produced by die-drawing and hydrostatic extrusion. *Polymer* 47 (16):5897–5908.

Moore, D. J., M. A. R. Freeman, P. A. Revell, G. W. Bradley, and M. Tuke. 1998. Can a total knee replacement prosthesis be made entirely of polymers? *The Journal of Arthroplasty* 13 (4):388–395.

Morishita, H., M. Kobayashi, H. Kuwahara, M. Shimomura, and M. Iguchi. 1989. On the morphology of rodlike trigonal polyoxymethylene single crystals. *Science Bulletin of the Faculty of Education, Nagasaki University* 41:13–19.

Mulholland, B. M. 1990. UV-light stabilized polyoxymethylene molding compositions, patent EP 0368635 A1.

Murao, T., K. Yamamoto, G. Reuschel, and D. Fleischer. 1993. Blow or extrusion molding product of polyoxymethylene resin and process for producing the same, patent EP 0568308 A2.

Murao, T., K. Yamamoto, G. Reuschel, and D. Fleischer. 1995. Method of blow or extrusion molding polyoxymethylene resin, patent US 5458839 A.

Murasawa, K., K. Shioda, Z. Tsuru, K. Nomura, and K. Matsuo. 1994. Polyacetal resin compositions for antifriction abrasion-resistant composites. *Chemical Abstract* 120:32310p.

Nakayama, K. and H. Kanetsuna. 1975. Hydrostatic extrusion of solid polymers. *Journal of Materials Science* 10 (7):1105–1118.

Nun, E., S. Schauhoff, and K. Dorn. 1997. Polyoxymethylene with improved resistance to zinc and/or copper ions, process for the production thereof and use thereof, US patent 5641830 A.

Ong, N. S., H. Zhang, and W. H. Woo. 2006. Plastic injection molding of high-aspect ratio micro-rods. *Materials and Manufacturing Processes* 21 (8):824–831.

Page, I. B. 2000. *Polyamides as Engineering Thermoplastic Materials*. Shawbury, UK: Rapra Technology Limited.

Papir, Y. S. and E. Baer. 1970. Effect of molecular organization on cryogenic relaxation behavior of polyoxymethylene. *Bulletin of the American Physical Society* 15 (3):352.

Papir, Y. S. and E. Baer. 1971. Internal friction of polyoxymethylene from 4.2 degrees to 300 degrees K. *Materials Science and Engineering* 8 (6):310–322.

Patnaik, P. 2007. *A Comprehensive Guide to the Hazardous Properties of Chemical Substances*. Hoboken, NJ: John Wiley & Sons.

Pecorini, T. J., R. W. Hertzberg, and J. A. Manson. 1990. Structure property relations in an injection-molded, rubber-toughened, semicrystalline polyoxymethylene. *Journal of Materials Science* 25 (7):3385–3395.

Penczek, S., J. Fejgin, P. Kubisa, K. Matyjaszewski, and M. Tomaszewicz. 1973. New highly efficient initiators for the copolymerization of 1,3-dioxolane with 1,3,5-trioxane based on the derivatives of trifluoromethanesulfonic acid. *Die Makromolekulare Chemie* 172 (1):243–247.

Peng, P., Y. Z. Chen, Y. F. Gao, J. Yu, and Z. X. Guo. 2009. Phase morphology and mechanical properties of the electrospun polyoxymethylene/polyurethane blend fiber mats. *Journal of Polymer Science Part B-Polymer Physics* 47 (19):1853–1859.

Pielichowska, K. 2008. Preparation of polyoxymethylene/hydroxyapatite nanocomposites by melt processing. *International Journal of Material Forming* 1:941–944.

Pielichowska, K. 2012a. Polyoxymethylene-homopolymer/hydroxyapatite nanocomposites for biomedical applications. *Journal of Applied Polymer Science* 123 (4):2234–2243.

Pielichowska, K. 2012b. The influence of molecular weight on the properties of polyacetal/hydroxyapatite nanocomposites. Part 1. Microstructural analysis and phase transition studies. *Journal of Polymer Research* 19:9775.

Pielichowska, K. 2012c. The influence of molecular weight on the properties of polyacetal/hydroxyapatite nanocomposites. Part 2. In vitro assessment. *Journal of Polymer Research* 19 (2):9788

Pielichowska, K. 2014a. Polyoxymethylene nanocomposites. In *Manufacturing of Nanocomposites with Engineering Plastics*, edited by V. Mittal. Cambridge, UK: Woodhead Publishing.

Pielichowska, K. 2014b. Polyoxymethylene processing. In *Polyoxymethylene Handbook: Structure, Properties, Applications and Their Nanocomposites*, edited by S. Lüftl, P. M. Visakh, and S. Chandran, 107–151. Hoboken, NJ: John Wiley & Sons; Salem, MA: Scrivener Publishing.

Pielichowska, K., A. Szczygielska, and E. Spasowka. 2012. Preparation and characterization of polyoxymethylene-copolymer/hydroxyapatite nanocomposites for long-term bone implants. *Polymers for Advanced Technologies* 23 (8):1141–1150.

Pielichowska, K., E. Dryzek, Z. Olejniczak, E. Pamula, and J. Pagacz. 2013. A study on the melting and crystallization of polyoxymethylene-copolymer/hydroxyapatite nanocomposites. *Polymers for Advanced Technologies* 24 (3):318–330.

Pielichowski, K. and A. Leszczynska. 2006. Polyoxymethylene-based nanocomposites with montmorillonite: An introductory study. *Polimery* 51 (2):143–149.

Plastics, Korea Engineering. 2014. KEP Kepital®, http://www.kepital.com/en/product/kepital.php.

Plesch, P. H. and P. Westermann. 1968. Polymerization of 1,3-dioxolane. I. Structure of polymer and thermodynamics of its formation. *Journal of Polymer Science Part C-Polymer Symposium* (16PC):3837.

Plesch, P. H. and P. Westermann. 1969. Polymerization of 1,3-dioxepan. I. Structure of polymer and thermodynamics of its formation and a note on 1,3-dioxan. *Polymer* 10 (2):105–111.

Polyplastics Co., Ltd. 2013. TEPCON® POM Grade Catalog, https://www.polyplastics.com/Gidb/TopSelectBrandAction.do?brandSelected=1.2&_LOCALE=ENGLISH.

Polyplastics Co., Ltd. 2014. DURACON® POM Grade Catalog, https://www.polyplastics.com/Gidb/GradeListDownloadAction.do;jsessionid=abc1a5a12ecc6dd8d5151a0144d5?brandSelected=1.1.

Preedy, J. E. and E. J. Wheeler. 1972. Phase-transformations in cold-rolled polyoxymethylene. *Nature-Physical Science* 236 (65):60–61.

PR Newswire. 2014a. Global polyoxymethylene market is expected to reach USD 3.39 billion by 2019: Transparency market research. In *bc-Transparency-Market-R*: Y.

PR Newswire. 2014b. Polyoxymethylene market worth $3.3 billion by 2018. In *bc-MarketsandMarkets*: Y.

Pugh, H. L. I. D. and A. H. Low. 1964–1965. The hydrostatic extrusion of difficult metals. *Journal of the Institute of Metals* 93: 201–217.

Quynn, R. G. and H. Brody. 1971. Elastic "hard" fibers. I. *Journal of Macromolecular Science, Part B* 5 (4):721–738.

Raimo, M. 2014. Structure and morphology of polyoxymethylene. In *Polyoxymethylene Handbook: Structure, Properties, Applications and Their Nanocomposites*, edited by S. Lüftl, P. M. Visakh, and S. Chandran, 163–191. Hoboken, NJ: John Wiley & Sons; Salem, MA: Scrivener Publishing.

Ramirez, N. V., M. Sanchez-Soto, S. Illescas, and A. Gordillo. 2009. Thermal degradation of polyoxymethylene evaluated with FTIR and spectrophotometry. *Polymer-Plastics Technology and Engineering* 48:470–477.

Ramirez, N. V., M. Sanchez-Soto, and S. Illescas. 2011. Enhancement of POM thermooxidation resistance through POSS nanoparticles. *Polymer Composites* 32 (10):1584–1592.

Ratnagiri, R. 2009. Polyacetal-ultrahigh molecular weight polyethylene blends, patent WO 2009015300 A1.

Read, B. E. and G. Williams. 1961. The dielectric and dynamic mechanical properties of polyoxymethylene (Delrin). *Polymer* 2 (3):239–255.

Richaud, E. 2014. Polyoxymethylene additives. In *Polyoxymethylene Handbook: Structure, Properties, Applications and Their Nanocomposites*, edited by S. Lüftl, P. M. Visakh, and S. Chandran, 53–105. Hoboken, NJ: John Wiley & Sons; Salem, MA: Scrivener Publishing.

Rodríguez-Baeza, M. 1991. Poly(oxymethylene) crystals grown in the polymerization of trioxane initiated with perchloric acid in nitrobenzene. *Polymer Bulletin* 26 (5):521–528.

Rodríguez-Baeza, M. and R. E. Catalán Saravia. 1985. Effect of the initiator acetyl perchlorate upon the morphology, kinetics and mechanisms of crystal growth of poly(oxymethylene). In *Frontiers in Polymer Science*, edited by W. Wilke, 49–58. Darmstadt: Steinkopff Verlag.

Rody, J. and M. Slongo. 1989. 2-(2-Hydroxyphenyl)-benztriazoles, their use as UV-absorbers and their preparation, patent US 4853471 A.

Romankevich, O. V., T. I. Zhila, S. Y. Zabello, N. A. Sklyar, and S. Y. Frenkel. 1982. Rheological properties of the melts of polyoxymethylene atactic polystyrene blends. *Vysokomolekulyarnye Soedineniya Seriya A* 24 (11):2282–2290.

Romero-Ibarra, I. C., E. Bonilla-Blancas, A. Sanchez-Solis, and O. Manero. 2012. Influence of X-ray opaque $BaSO_4$ nanoparticles on the mechanical, thermal and rheological properties of polyoxymethylene nanocomposites. *Journal of Polymer Engineering* 32 (4–5):319–326.

Roos, G. and E. Wolters. 1973. Stabilized polyoxymethylene molding compositions containing a mixture of an alkaline earth metal salt and an ester of (alkylhydroxyphenyl)-carboxylic acids with a polyol, patent US 3743614 A.

RSC. 2014. 1,3,5-Trioxane, http://www.chemspider.com/Chemical-Structure.7790.html (accessed July 23, 2014).

Sakurai, M., J. Miyawaki, T. Umemura, T. Kawata, S. Kiboshi, and A. Shibata. 1985. Acetal resin composition, patent US 4506053 A.

Samon, J. M., J. M. Schultz, B. S. Hsiao, S. Khot, and H. R. Johnson. 2001. Structure development during the melt spinning of poly(oxymethylene) fiber. *Polymer* 42 (4):1547–1559.

Sanchez-Soto, M., S. Illescas, H. Milliman, D. A. Schiraldi, and A. Arostegui. 2010. Morphology and thermomechanical properties of melt-mixed polyoxymethylene/polyhedral oligomeric silsesquioxane nanocomposites. *Macromolecular Materials and Engineering* 295 (9):846–858.

Sandler, S. 1974. *Polymer Syntheses*. San Diego, CA: Academic Press.

Sandler, S. R. and W. Karo. 1992. *Polymer Syntheses, Organic Chemistry: A Series of Monographs*. New York: Academic Press.

Sauer, B. B., P. Avakian, E. A. Flexman, M. Keating, B. S. Hsiao, and R. K. Verma. 1997. A.C. dielectric and TSC studies of constrained amorphous motions in flexible polymers including poly(oxymethylene) and miscible blends. *Journal of Polymer Science Part B: Polymer Physics* 35 (13):2121–2132.

Sauer, B. B., R. S. Mclean, J. D. Londono, and B. S. Hsiao. 2000. Morphological changes during crystallization and melting of polyoxymethylene studied by synchrotron X-ray scattering and modulated differential scanning calorimetry. *Journal of Macromolecular Science-Physics* B39 (4):519–543.

Schaffner, F., J. Vinson, and B. J. Jungnickel. 1991. Hydrostatic extrusion rate relaxation in polyethylene, polyoxymethylene, and their blends. *Angewandte Makromolekulare Chemie* 185:137–145.

Schmidt, F. D. C. D., G. D. C. D. Fahrbach, W. D. C. D. Schenk, and E. Seiler. 1975. Thermoplastische formmassen hoher schlagfestigkeit, patent DE 2408487 A1.

Schmitt, B. D. C. D. and H. J. D. C. D. Sterzel. 1978. Schlagzaehe thermoplastische formmassen, patent DE 2659357 A1.
Schneider, A. K. 1957. Catalytic process for polymerizing trioxane to tough, high molecular weight polyoxymethylene, US patent 2795571 A.
Schuette, W., A. Hilt, M. Walter, and K. Boehlke. 1983. Polyacetals having an improved maximum sustained-use temperature, patent US 4386178 A.
Schweitzer, C. E., R. N. Macdonald, and J. O. Punderson. 1959. Thermally stable high molecular weight polyoxymethylenes. *Journal of Applied Polymer Science* 1 (2):158–163.
Sharavanan, K., E. Ortega, M. Moreau, C. Lorthioir, F. Laupretre, P. Desbois, M. Klatt, and J. P. Vairon. 2009. Cationic copolymerization of 1,3,5-trioxane with 1,3-dioxepane: A comprehensive approach to the polyacetal process. *Macromolecules* 42 (22):8702–8710.
Shi, J., B. Jing, X. Zou, H. Luo, and W. Dai. 2009. Investigation on thermo-stabilization effect and nonisothermal degradation kinetics of the new compound additives on polyoxymethylene. *Journal of Materials Science* 44:1251–1257.
Shieh, Y.-T. and S.-A. Chen. 1999. Kinetics and mechanism of the cationic polymerization of trioxane. I. Crystallization during polymerization. *Journal of Polymer Science Part A: Polymer Chemistry* 37 (4):483–492.
Shieh, Y.-T., M.-J. Yeh, and S.-A. Chen. 1999. Kinetics and mechanism of the cationic polymerization of trioxane. II. Consideration of hydride transfer. *Journal of Polymer Science Part A: Polymer Chemistry* 37 (22):4198–4204.
Shieh, Y. T., M. L. Lay, and S. A. Chen. 2003. Kinetics and mechanism of the cationic polymerization of trioxane. III. Copolymerization with cyclic formals. *Journal of Polymer Research* 10 (3):151–160.
Siengchin, S. 2013. Dynamic mechanic and creep behaviors of polyoxymethylene/boehmite alumina nanocomposites produced by water-mediated compounding: Effect of particle size. *Journal of Thermoplastic Composite Materials* 26 (7):863–877.
Siengchin, S., J. Karger-Kocsis, G. C. Psarras, and R. Thomann. 2008. Polyoxymethylene/polyurethane/alumina ternary composites: Structure, mechanical, thermal and dielectric properties. *Journal of Applied Polymer Science* 110 (3):1613–1623.
Siengchin, S., G. C. Psarras, and J. Karger-Kocsis. 2010. POM/PU/Carbon nanofiber composites produced by water-mediated melt compounding: Structure, thermomechanical and dielectrical properties. *Journal of Applied Polymer Science* 117 (3):1804–1812.
Soundararajan, S. and S. C. Shit. 2001. Studies on properties of poly olefins: Poly propylene copolymer (PPcp) blends with poly oxy methylenes (POM). *Polymer Testing* 20 (3):313–316.
SPA, Montedison. 1966. Process for the spinning of crystalline polyoxymethylene polymers, GB 1095750 A.
Spence, R. 1933. The polymerisation of gaseous formaldehyde. *Journal of the Chemical Society (Resumed)* 279:1193–1197.
Srivastava, S., S. J. La'Verne, I. A. Khan, P. Ali, and V. D. Gupta. 2011. Phonons and heat capacity of polyoxymethylene. *Journal of Applied Polymer Science* 122 (2):1376–1381.
Staudinger, H. 1960. *Die hochmolecularen organischen Verbindungen*. Berlin, New York: Springer-Verlag.
Staudinger, H. and A. Gaule. 1916. Vergleich der Stickstoff-Abspaltung bei verschiedenen aliphatischen Diazoverbindungen. *Berichte der deutschen chemischen Gesellschaft* 49 (2):1897–1918.
Staudinger, H., R. Singer, H. Johner, M. Lüthy, W. Kern, D. Russidis, and O. Schweitzer. 1929. Über hochpolymere Verbindungen. Über die Konstitution der Polyoxymethylene. *European Journal of Organic Chemistry* 474:145–275.
Sterzel, H. J., J. H. H. T. Maat, J. Ebenhoech, and M. Meyer. 1992. Polyoxymethylene homo- and copolymer, removing binder, patent US 5145900 A.
Su, R., J. X. Su, K. Wang, C. Y. Yang, Q. Zhang, and Q. Fu. 2009. Shear-induced change of phase morphology and tensile property in injection-molded bars of high-density polyethylene/polyoxymethylene blends. *European Polymer Journal* 45 (3):747–756.
Subramanyam, H. N. and S. V. Subramanyam. 1987. Thermal-expansion of irradiated polyoxymethylene. *European Polymer Journal* 23 (3):207–211.
Sugio, A., A. Amemiya, M. Kimura, Y. Otuki, and K. Kawaguchi. 1982. Stabilized oxymethylene copolymer composition, patent US 4342680 A.
Sugiyama, N. 1992. Polyoxymethylene compositions containing amine polymer having pendant—NH.sub.2 functional groups, patent US 5128405 A.
Sun, T. J., L. Ye, and X. W. Zhao. 2007. Thermostabilising and nucleating effect of montmorillonite on polyoxymethylene. *Plastics Rubber and Composites* 36 (7–8):350–359.

Sun, L. H., Z. G. Yang, and X. H. Li. 2008a. Mechanical and tribological properties of polyoxymethylene modified with nanoparticles and solid lubricants. *Polymer Engineering and Science* 48 (9):1824–1832.

Sun, L.-H., Z.-G. Yang, and X.-H. Li. 2008b. Study on the friction and wear behavior of POM/Al2O3 nanocomposites. *Wear* 264 (7–8):693–700.

Sun, S. Y., Y. D. He, X. D. Wang, and D. Z. Wu. 2010. Flammability characteristics and performance of halogen-free flame-retarded polyoxymethylene based on phosphorus-nitrogen synergistic effects. *Journal of Applied Polymer Science* 118 (1):611–622.

Suzuki, K. and M. Nagahama. 1987, Japanese Patent 004,748, Jan 10, 1987.

Suzuki, H., J. Grebowicz, and B. Wunderlich. 1985. Heat capacity of semicrystalline, linear poly(oxymethylene) and poly(oxyethylene). *Die Makromolekulare Chemie* 186 (5):1109–1119.

Szymanski, R., P. Kubisa, and S. Penczek. 1983. Mechanism of cyclic acetal polymerization. End of a controversy? *Macromolecules* 16 (6):1000–1008.

Tadokoro, H., T. Yasumoto, S. Murahashi, and I. Nitta. 1960. Molecular configuration of polyoxymethylene. *Journal of Polymer Science* 44 (143):266–269.

Takahashi, Y. and H. Tadokoro. 1979. Least-squares refinement of molecular-structure of polyoxymethylene. *Journal of Polymer Science Part B-Polymer Physics* 17 (1):123–130.

Takasa, K., N. Miyashita, and K. Takeda. 2006a. Micro-structure and characteristics of highly oriented polyoxymethylene obtained by press and biaxial drawing. *Journal of Applied Polymer Science* 99 (3):835–844.

Takasa, K., N. Miyashita, and K. Takeda. 2006b. Young modulus and degree of crystallization of highly-elongated polyoxymethylene. *Journal of Applied Polymer Science* 101 (2):1223–1227.

Takayama, K., T. Endo, O. Kanoto, and N. Matsunaga. 1993. Self-lubricated polyacetal molding compositions and molded articles formed thereof, patent US 5177123 A.

Takayanagi, M. 1963. Viscoelastic properties of crystalline polymers. *Memoirs of the Faculty of Science Kyushu University* 23:41.

Tang, W. H., H. H. Wang, J. Tang, and H. L. Yuan. 2013. Polyoxymethylene/thermoplastic polyurethane blends compatibilized with multifunctional chain extender. *Journal of Applied Polymer Science* 127 (4):3033–3039.

Tashiro, K., T. Kamae, H. Asanaga, and T. Oikawa. 2004. Structural analysis of polyoxymethylene whisker single crystal by the electron diffraction method. *Macromolecules* 37 (3):826–830.

Terlemezyan, L., B. Ivanova, and S. Tacheva. 1993. Electrically conductive polyaniline polyoxymethylene blends. *European Polymer Journal* 29 (7):1019–1023.

Thompson, M. S. 2001. The design of a novel hip resurfacing prosthesis, PhD thesis. London: University of London.

Thompson, M. S., M. D. Northmore-Ball, and K. E. Tanner. 2001. Tensile mechanical properties of polyacetal after one and six months' immersion in Ringer's solution. *Journal of Materials Science-Materials in Medicine* 12 (10–12):883–887.

Ticona. 2000. Processing Celcon® acetal copolymer, http://www.hipolymers.com.ar/pdfs/celcon/procesamiento/Processing%20Celcon%20acetal%20copolymer.pdf.

Ticona. 2006. Hostaform® Polyoxymethylene Copolymer (POM), http://www.eurotecsrl.info/pdf/Hostaform%20brochure.pdf.

Ticona. 2007. Hostaform C 9021/POM/unfilled, http://www.b2bpolymers.com/TDS/Ticona_Hostaform_C9021.pdf.

Tokarz, L., S. Pawlowski, and M. Kedzierski. 2014. Polyoxymethylene applications. In *Polyoxymethylene Handbook: Structure, Properties, Applications and Their Nanocomposites*, edited by S. Lüftl, P. M. Visakh, and S. Chandran, 153–161. Hoboken, NJ: John Wiley & Sons; Salem, MA: Scrivener Publishing.

Uchida, T. and H. Tadokoro. 1967. Structural studies of polyethers. 4. Structure analysis of polyoxymethylene molecule by 3-dimensional fourier syntheses. *Journal of Polymer Science Part A-2-Polymer Physics* 5 (1pa2):63–81.

Uthaman, N., A. Majeed, and A. Pandurangan. 2006. Impact modification of polyoxymethylene (POM). *E-Polymers* 6:438–446.

Uthaman, R. N., A. Pandurangan, and S. S. M. A. Majeed. 2007. Mechanical, thermal, and morphological characteristics of compatibilized and dynamically vulcanized polyoxymethylene/ethylene propylene diene terpolymer blends. *Polymer Engineering and Science* 47 (6):934–942.

Utracki, L. A. 1998. Polyoxymethylene (acetal resins). In *Commercial Polymer Blends*, 399–406. London: Springer, Chapman & Hall.

Vaidya, S. R. 1994. Polyoxymethylene compositions containing at least one encapsulated nucleant, patent US 5298537 A.

Van De Walle, T. and R. Fourcade. 1968. Black acetal resin compositions, patent US 3397170 A.

Vera-Sorroche, J., A. Kelly, E. Brown, P. Coates, N. Karnachi, E. Harkin-Jones, K. Li, and J. Deng. 2013. Thermal optimisation of polymer extrusion using in-process monitoring techniques. *Applied Thermal Engineering* 53 (2):405–413.

Vlachopoulos, J. and D. Strutt. 2003. Polymer processing. *Materials Science and Technology* 19 (9):1161–1169.

Vlachopoulos, J. and J. R. Wagner. 2001. *The SPE Guide on Extrusion Technology and Troubleshooting*, SPE Division Series. Brookfield, CT: Society of Plastics Engineers.

Vorobena, G. A., G. M. Trofimova, A. A. Berlin, and N. S. Yenikolopyan. 1974. A thermodynamic approach to regulating macromolecular and molecular structures during polymer synthesis. *Polymer Science U.S.S.R.* A16(7):1729–1736.

Waack, G. 2014. Machining nylon and acetal parts from stock shapes, *The Plastics Distributor and Fabricator*, May/June, http://www.plasticsmag.com/features.asp?fIssue=May/Jun-06.

Wacharawichanant, S. 2008. Effect of zinc oxide on morphology and mechanical properties of polyoxymethylene/zinc oxide composites. *Macromolecular Symposia* 264:54–58.

Wacharawichanant, S., S. Thongyai, A. Phutthaphan, and C. Eiamsam-Ang. 2008. Effect of particle sizes of zinc oxide on mechanical, thermal and morphological properties of polyoxymethylene/zinc oxide nanocomposites. *Polymer Testing* 27 (8):971–976.

Wacharawichanant, S., P. Sahapaibounkit, and U. Saeueng. 2011. Study on mechanical and morphological properties of polyoxymethylene/Al_2O_3 nanocomposites. *TIChE International Conference 2011*, Songkhla, Thailand.

Wacharawichanant, S., P. Sahapaibounkit, and U. Saeueng. 2012. Effect of clay on the properties of polyoxymethylene/clay nanocomposites. *Advanced Materials Research* 488–489:82–86.

Wagner, K. and H. Roehr. 1965. Zu Formkoerpern wiederholt thermoplastisch zu verarbeitende Formmassen, patent DE 1193240 B.

Walling, C. T., F. Brown, and K. W. Bartz. 1962. Copolymers, patent US 3027352 A.

Walter, M., W. Schuette, and H. Goerrissen. 1989. Polyoxymethylene molding materials having improved thermal stability, their preparation and their use, patent US 4837400 A.

Wamser, C. A. 1951. Equilibria in the system boron trifluoride—Water at 25°. *Journal of the American Chemical Society* 73 (1):409–416.

Wang, Q. A. 2011. Thermal stability of polyoxymethylene and its blends with poly(ethylene-methylacrylate) or poly(styrene-butadiene-styrene). *Journal of Applied Polymer Science* 121 (1):376–388.

Wang, X. D. and X. Cui. 2005. Effect of ionomers on mechanical properties, morphology, and rheology of polyoxymethylene and its blends with methyl methacrylate-styrene-butadiene copolymer. *European Polymer Journal* 41 (4):871–880.

Wang, X. D., H. Q. Li, and R. G. Jin. 2000. Study on the miscibility and phase behavior of polyoxymethylene with Novolak. *Journal of Materials Science & Technology* 16 (4):427–430.

Wang, J., X. G. Hu, M. Tian, and R. Stengler. 2007. Study on mechanical and tribological property of nanometer ZrO_2-filled polyoxymethylene composites. *Polymer-Plastics Technology and Engineering* 46 (5):469–473.

Wang, J., K. H. Hu, Y. F. Xu, and X. G. Hu. 2008. Structural, thermal, and tribological properties of intercalated polyoxymethylene/molybdenum disulfide nanocomposites. *Journal of Applied Polymer Science* 110 (1):91–96.

Wang, Z. Y., Y. A. Liu, and Q. Wang. 2010. Flame retardant polyoxymethylene with aluminium hydroxide/melamine/novolac resin synergistic system. *Polymer Degradation and Stability* 95 (6):945–954.

Wegner, G., M. Rodriguez-Baeza, A. Lücke, and G. Lieser. 1980. Kinetik und Mechanismus der kationischen Polymerisation von Trioxan: Ein katalysierter Kristallwachstumsprozeß. *Die Makromolekulare Chemie* 181 (8):1763–1790.

Weissermel, K., E. Fischer, K. Gutweiler, and H. D. Hermann. 1964. Zur Copolymerisation des Trioxans. *Kunststoffe* 54:410–415.

Weissermel, K., E. Fischer, K. Gutweiler, H. D. Hermann, and H. Cherdron. 1967a. Polymerisation von Trioxan. *Angewandte Chemie* 79 (11):512–520.

Weissermel, K., E. Fischer, K. Gutweiler, H. D. Hermann, and H. Cherdron. 1967b. Polymerization of trioxane. *Angewandte Chemie International Edition in English* 6 (6):526–533.

Wilchinsky, Z. W. 1964. Orientation in crystalline polymers related to deformation. *Polymer* 5:271–281.

Wilhelm, W., R. Klenz, and D. Binkowski. 2011. Method for producing molded parts from a polyoxymethylene polymer, patent US 20110309558 A1.

Withey, R. J. and E. Whalley. 1963. Pressure effect and mechanism in acid catalysis. Part 11.—Depolymerization of paraldehyde and trioxane. *Transactions of the Faraday Society* 59:901–906.

Xu, W., M. Ge, and P. He. 2001. Nonisothermal crystallization kinetics of polyoxymethylene/montmorillonite nanocomposite. *Journal of Applied Polymer Science* 82 (9):2281–2289.

Xu, X. M., L. P. Guo, Y. D. Zhang, and Z. J. Zhang. 2012. Mechanical and thermal properties of reactable nano-SiO2/polyoxymethylene composites. In *Advanced Engineering Materials II, Pts 1–3*, edited by C. X. Cui, Y. L. Li, and Z. H. Yuan, 103–109. Stafa-Zurich: Trans Tech Publications Ltd.

Yamashita, Y., T. Tsuda, M. Okada, and S. Iwatsuki. 1966. Correlation of cationic copolymerization parameters of cyclic ethers, formals, and esters. *Journal of Polymer Science Part A-1: Polymer Chemistry* 4 (9PA1):2121–2135.

Yu, N., L. H. He, Y. Y. Ren, and Q. Xu. 2011. High-crystallization polyoxymethylene modification on carbon nanotubes with assistance of supercritical carbon dioxide: Molecular interactions and their thermal stability. *Polymer* 52 (2):472–480.

Zachariev, G., H. V. Rudolph, and H. Ivers. 2004. Damage accumulation in glass fibre reinforced polyoxymethylene under short-term loading. *Composites Part A: Applied Science and Manufacturing* 35 (10):1119–1123.

Zachariev, G., H.-V. Rudolph and H. Ivers. 2010. Residual strength of GFR/POM as a function of damage. *Journal of Physics: Conference Series* 240 (1):012166.

Zahran, R. R. 1998. Effect of gamma-irradiation on the ultrasonic and structural properties of polyoxymethylene. *Materials Letters* 37 (1–2):83–89.

Zamboni, V. and G. Zerbi. 1964. Vibrational spectrum of new crystalline modification of polyoxymethylene. *Journal of Polymer Science Part C-Polymer Symposium* 7 (1):153–161.

Zeng, Y., Z. Ying, J. H. Du, and H. M. Cheng. 2007. Effects of carbon nanotubes on processing stability of polyoxymethylene in melt-mixing process. *Journal of Physical Chemistry C* 111 (37):13945–13950.

Zeng, Y., P. F. Liu, J. H. Du, L. Zhao, P. M. Ajayan, and H. M. Cheng. 2010. Increasing the electrical conductivity of carbon nanotube/polymer composites by using weak nanotube-polymer interactions. *Carbon* 48 (12):3551–3558.

Zhang, X. F., Y. Zhang, Z. L. Peng, X. Y. Shang, and Y. X. Zhang. 2000. Dynamically vulcanized nitrile rubber/polyoxymethylene thermoplastic elastomers. *Journal of Applied Polymer Science* 77 (12):2641–2645.

Zhao, R. 2005. Melt blowing polyoxymethylene copolymer. *International Nonwovens Journal* 14:19–24.

Zhao, X. W. and L. Ye. 2009. Study on the thermal conductive polyoxymethylene/graphite composites. *Journal of Applied Polymer Science* 111 (2):759–767.

Zhao, X. W. and L. Ye. 2010. Preparation, structure, and property of polyoxymethylene/carbon nanotubes thermal conducive composites. *Journal of Polymer Science Part B-Polymer Physics* 48 (8):905–912.

Zhao, X. W. and L. Ye. 2011a. Structure and mechanical properties of polyoxymethylene/multi-walled carbon nanotube composites. *Composites Part B-Engineering* 42 (4):926–933.

Zhao, X. W. and L. Ye. 2011b. Structure and properties of highly oriented polyoxymethylene produced by hot stretching. *Materials Science and Engineering a-Structural Materials Properties Microstructure and Processing* 528 (13–14):4585–4591.

Zhao, X. W. and L. Ye. 2011c. Structure and properties of highly oriented polyoxymethylene/multi-walled carbon nanotube composites produced by hot stretching. *Composites Science and Technology* 71 (10):1367–1372.

Zhao, Y., W. Zhang, L. P. Liao, W. J. Li, and Y. Xin. 2011. Microencapsulation of epoxy resins for self-healing material. In *Manufacturing Processes and Systems, Pts 1–2*, edited by X. H. Liu, Z. Y. Jiang, and J. T. Han, 1031–1035. Stafa-Zurich: Trans Tech Publications Ltd.

Zierer, D., U. Ziegler, and K. Kurz. 2004. Nucleated polyacetal molding materials having increased crystallization speed, their use and shaped molded bodies produced therefrom, patent US 20040030094 A1.

Zierer, D., R. Bernstein, J. Schweitzer, and K. Kurz. 2008. Polyoxymethylene fibers, production thereof and use thereof, patent US 7410696 B2.

Zilberman, U. 2005. Formaldehyde from POM brackets. *American Journal of Orthodontics and Dentofacial Orthopedics* 128:147–148.

7 Polyethers*

Rebecca Klein and Frederik R. Wurm

CONTENTS

7.1	Introduction	252
7.2	Polyepoxides	253
	7.2.1 Polymerization	253
	7.2.1.1 Anionic Ring-Opening Polymerization	253
	7.2.1.2 Cationic Ring-Opening Polymerization	254
	7.2.1.3 Monomer-Activated Anionic Polymerization	255
7.3	Poly(Ethylene Glycol)s	256
	7.3.1 History	256
	7.3.2 Properties	256
	7.3.3 Applications	257
7.4	Multifunctional Poly(Ethylene Glycol)	259
7.5	Poly(Propylene Glycol)	261
	7.5.1 Properties	262
	7.5.2 Pluronic	262
7.6	Polyglycerol	263
	7.6.1 Hyperbranched Polyglycerol	264
	7.6.1.1 History	264
	7.6.1.2 Properties	264
	7.6.1.3 Applications	265
	7.6.2 Linear Polyglycerol	265
	7.6.2.1 History	266
	7.6.2.2 Properties	266
	7.6.2.3 Applications	266
7.7	Polyoxymethylene	267
	7.7.1 History	267
	7.7.2 Polymerization	267
	7.7.2.1 Anionic Polymerization of Formaldehyde	267
	7.7.2.2 Cationic Ring-Opening Polymerization of Trioxane	268
	7.7.2.3 Properties	269
	7.7.2.4 Applications	269
7.8	Polyoxetanes	269
	7.8.1 History	270
	7.8.2 Polymerization	270
	7.8.3 Properties	270
	7.8.4 Applications	271
7.9	Polytetrahydrofuran	272
	7.9.1 History	272
	7.9.2 Polymerization	272
	7.9.3 Properties	272
	7.9.4 Applications	272

---

* Based in parts on the first-edition chapter on polyethers.

7.10 Polyethers from Renewable Resources 273
  7.10.1 Polysaccharides 273
    7.10.1.1 Cellulose 273
    7.10.1.2 Starch 274
    7.10.1.3 Hydroxyethyl Starch 275
    7.10.1.4 Lignin 275
  7.10.2 Furan-Based Polymers 277
7.11 Poly(Phenylene Ether) 278
  7.11.1 History 278
  7.11.2 Polymerization 279
  7.11.3 Properties 279
  7.11.4 Applications 279
References 279

7.1 INTRODUCTION

Ethers are chemical compounds with an oxygen bridge between two organic (alkyl or aryl) residues. The word *ether* originates from Greek and means "higher air" or "air of fire." These compounds are very common in natural products, e.g., they can be found in the glycosidic linkages of polysaccharides but also in other natural materials and drugs. The ether bond is relatively stable and inert, which renders these materials interesting, e.g., as solvents. The C–O–C bond is angulated (ca. 112°) and polar, while the length of the C–O bond spans ca. 143 pm (Figure 7.1). The solubility in water decreases with increasing length of the alkyl chain due to the increasing sterical hindrance resulting in polarity. The physical properties strongly vary from the corresponding alcohols, and the ethers show lower melting and boiling points due to the loss of active hydrogen bonding. As ethers can accept hydrogen bonds from water, they interact also with water and are found to be hygroscopic and often rather "wet" solvents (diethylether can dissolve up to ca. 8% of water).

There are several routes to synthesize ethers: the most prominent pathway is the Williamson ether synthesis, which was first published in 1850.[1] The ether linkage is generated by a nucleophilic substitution of an alkali alkoxide on an alkylating reagent (typically a haloalkane). Another way is the acid-catalyzed condensation of two alcohols resulting in an ether bond. As this reaction is reversible, ethers can find only limited use under acidic conditions.

Cyclic ethers like epoxides are accessible via catalytic oxidation or epoxidation with peroxy acids. The easy accessibility of these compounds and the high stability of the ether bonds (in the presence of radicals, for example) together with their polarity render them very interesting building blocks for polyethers, an old class of materials already investigated by Staudinger during the beginnings of the field of the macromolecular chemistry.

Crown ethers are another interesting class of cyclic, oligomeric ethers. The structure, consisting typically of $-CH_2-CH_2-O-$ repeating units, is reminiscent of a crown, which explains the name. Depending on the number of repeating units, these compounds form complexes with different metal cations, e.g., 18-crown-6 is 100 times more selective for potassium as sodium.

In this chapter, different polyethers (Figure 7.2) and their history, polymerization, applications, and properties will be discussed in detail.

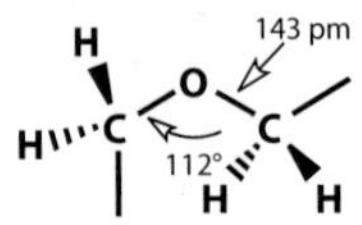

FIGURE 7.1 General structure of an ether with bond angle and bond length.

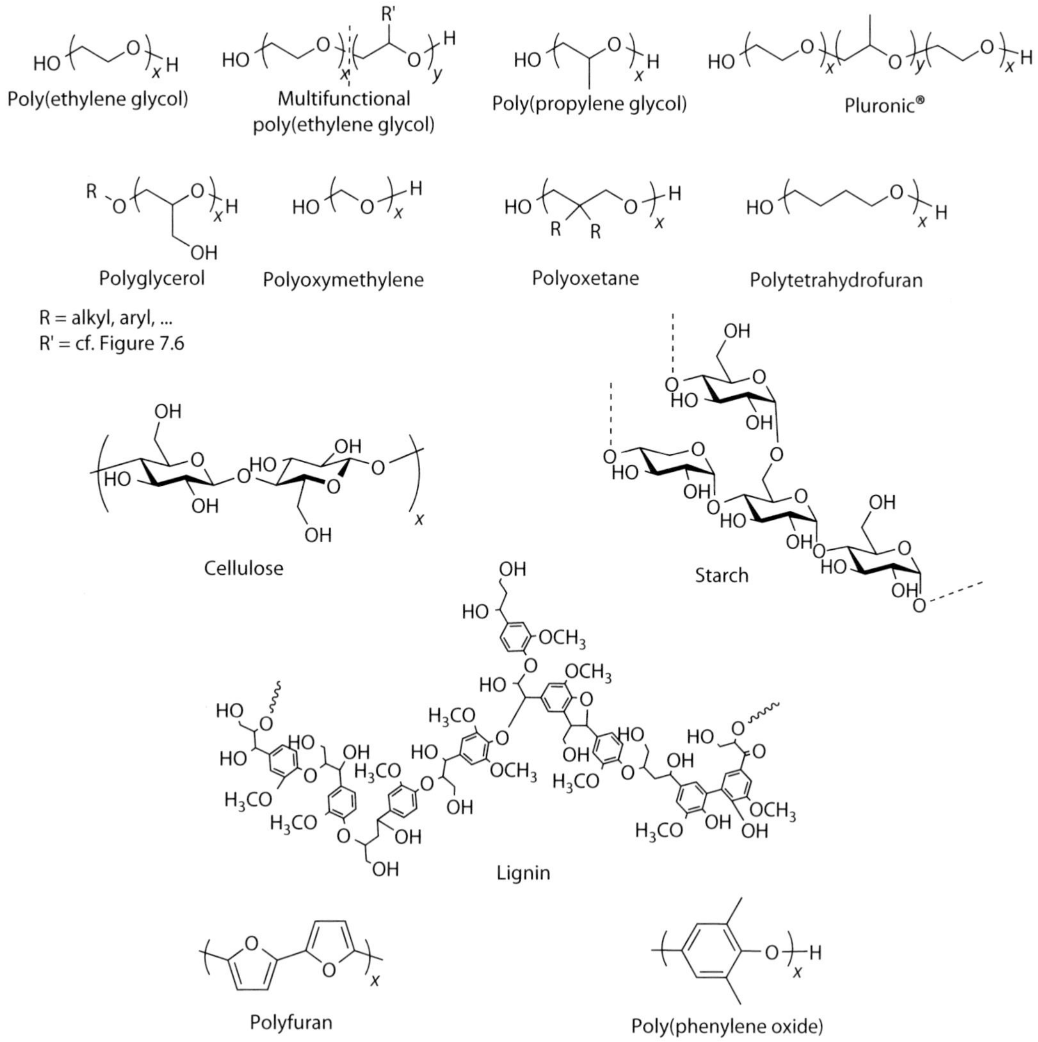

Polyethers

FIGURE 7.2 Chemical structures of some discussed polyethers.

7.2 POLYEPOXIDES

Epoxides or oxiranes are the smallest cyclic ethers that can be produced (Figure 7.7). They are typically used as monomers for ring-opening polymerization (ROP) due to their high ring strain (112 kJ/mol). Polyethers derived from epoxides are a very important class of materials and are produced worldwide in several tons per year for various commodities or high-performance applications.[2] The polymers are used as precursors for polyurethanes and surfactants but also in biomedical, cosmetic, or nutritional applications and in more specialized areas like lithium-ion batteries. The development, synthesis, and applications of these materials are discussed in detail in the following sections.

7.2.1 POLYMERIZATION

7.2.1.1 Anionic Ring-Opening Polymerization

The anionic ring-opening polymerization (AROP) of epoxides was first described by Staudinger[3] in 1933 and Flory[4] in 1940. It is the most important and widely used polymerization method in

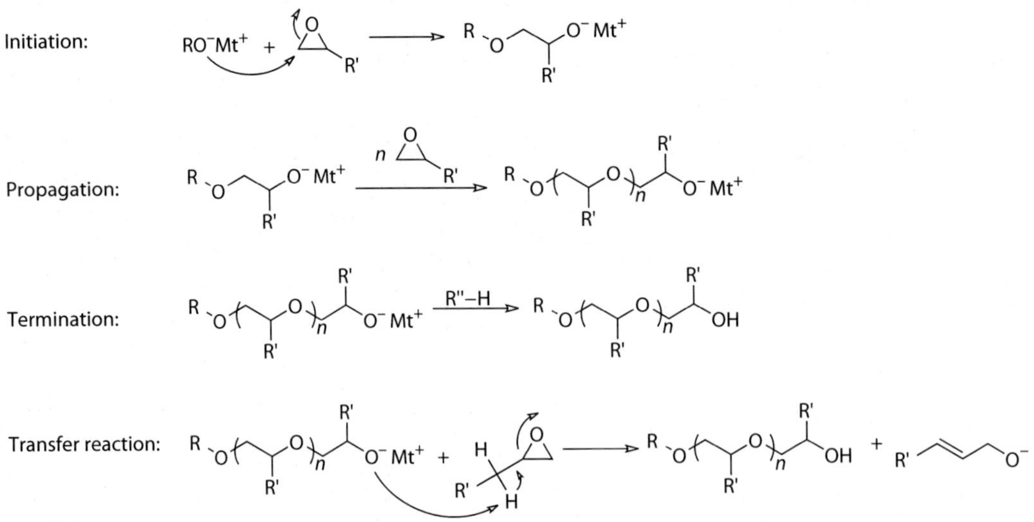

FIGURE 7.3 Mechanism of the anionic ring-opening polymerization (AROP) of (functional) epoxides.

academia as well as in industry due to fast reaction kinetics and the excellent control over molecular weight and molecular weight distribution. The initiation is performed through basic alkali metal derivatives like hydrides, alkyls, aryls, and amides and mostly alkoxides of sodium, potassium, and cesium.[2,5] In the case of lithium alkoxides, typically, no polymerization is observed,[6] which can be explained with the hard and soft acid and base (HSAB) concept.[7] The interaction between the comparably hard oxygen atom and the hard lithium cation is very strong, and no monomer insertion occurs. Only very slow oligomerization over a period of several weeks has been observed.[8] Using this concept, the increasing polymerization rate constants with increasing size of the counterion ($Na^+ < K^+ < Cs^+$) can be explained. Therefore, the choice of solvent and counterion is of great importance. Commonly, potassium is used because of the efficient polymerization results and lower costs compared to cesium. However, cesium exhibits higher polymerization rates and reduces transfer reactions in the case of substituted oxiranes like propylene oxide (PO) (Figure 7.3).[9] Furthermore, crown ethers, cryptates, or phosphazene bases can increase the polymerization rate through complexation of the cation, thus increasing the nucleophilicity of the free alkoxide anion and thereby leading to a more reactive chain end.[10,11] The initiation step is a nucleophilic substitution (S_N2) of the alkoxide and the oxirane, resulting in the formation of a new alkoxide, which is able to attack a new monomer, thereby leading to the propagation of the polymer chain. In the AROP, typically, only monosubstituted monomers are applied, which can only be attacked at the sterically less hindered side.

After complete monomer consumption, termination is achieved by the addition of an acidic, hydrogen-carrying compound, mostly water or alcohols, leading to hydroxyl functionalized chain ends (Figure 7.3).

7.2.1.2 Cationic Ring-Opening Polymerization

The cationic ring-opening polymerization (CROP) of epoxides is a rarely used method due to the predominating formation of cyclic products, which can be explained through intramolecular chain-transfer reactions (backbiting).[5] In the late 1980s, it was shown that the amount of cyclic oligomers is drastically reduced in the presence of compounds containing hydroxyl groups.[12] In this case, the so-called activated monomer (AM) mechanism introduced by Szwarc[13] is more likely than the conventional active chain-end (ACE) mechanism.

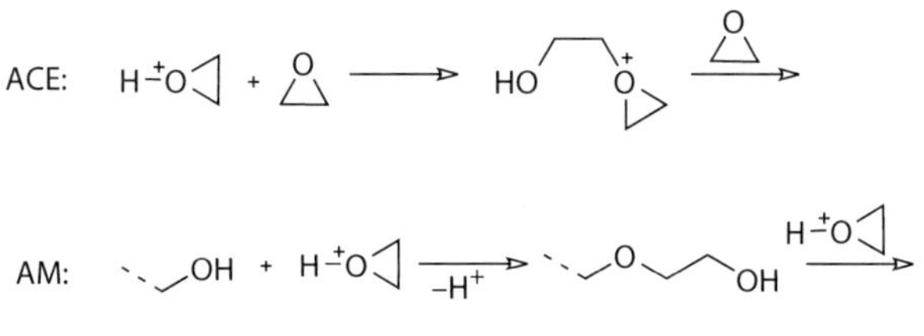

Polyethers

FIGURE 7.4 Active chain end (ACE) versus activated monomer (AM) mechanism.

In the case of the AM mechanism, the active centers are located on the monomer, and the polymer chain is neutral, thereby strongly reducing the backbiting process (Figure 7.4). Both mechanisms are competing, and in order to have the AM mechanism prevail, the monomer concentration has to be kept low, which can be conducted via slow monomer addition (SMA). The propagation constant for cationic polymerizations is much higher (10^4–10^6 mol^{-1} L s^{-1}) in comparison to anionic polymerizations (1–10^4 mol^{-1} L s^{-1}). Therefore, the polymerization is much faster, and sufficient heat dissipation is necessary to control molecular weight and molecular weight distribution.[14] Despite these problems, the CROP of oxiranes exhibits features of a living process; however, it is limited in terms of the molar mass of the resulting polymers and therefore rarely applied in industry. However, one application is the synthesis of polyepichlorohydrine (PECH).

7.2.1.3 Monomer-Activated Anionic Polymerization

Some epoxide derivatives like PO or glycidyl ethers undergo side reactions during the harsh basic conditions of the AROP. Undesired proton abstractions from the methyl group in the case of PO or the methylene group in the case of glycidyl ethers lead to the formation of unsaturated species (Figure 7.3).[9,15] These side reactions during the oxyanionic polymerization limit the molecular weights of the obtained polyethers to lower than 10,000 g/mol in the case of PO.[16] The use of the monomer-activated anionic polymerization protocol proceeds under milder (typically less basicity and lower temperatures) conditions, thereby reducing the side reactions and leading to high-molecular-weight polymers (M_n up to 150,000 g/mol). This strategy was developed by Deffieux et al. and employs a trialkylaluminum catalyst (e.g., i-Bu$_3$Al) and a tetraalkylammonium salt as the respective inititator.[16,17] The coordination of the trialkylaluminum to the tetraalkylammonium salt initiator takes place, and a 1:1 ate complex is formed. An excess of the Lewis acid (i.e., the trialkylaluminum catalyst), especially for monomers with additional coordination sites, with respect to the initiator is required to activate the monomer, which is then attacked by the nucleophilic ate complex (Figure 7.5). Some of the monomers polymerized with this technique are ethylene oxide (EO),

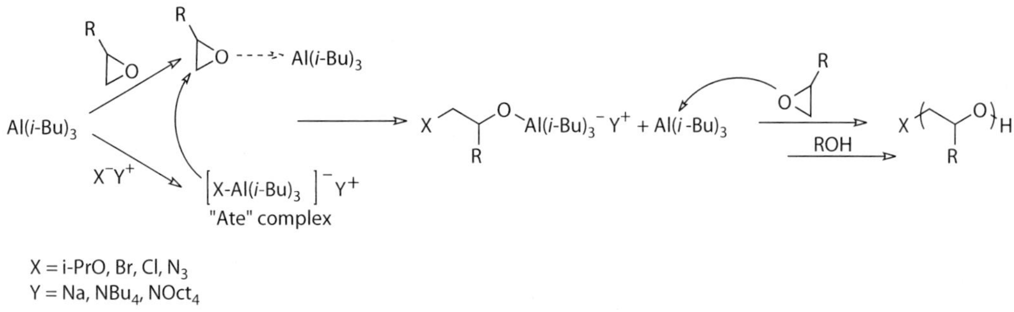

FIGURE 7.5 Mechanism of the monomer-activated anionic polymerization.

PO,[16] ethoxyethyl glycidyl ether (EEGE),[18] epichlorohydrine (ECH),[19] glycidyl methyl ether (GME),[20] allyl glycidyl ether (AGE),[21] glycidyl methacrylate,[22] and various hydrophobic epoxides.[23]

More systems for activated epoxide polymerizations are reviewed in an excellent article by Carlotti and coworkers.[2] All systems have in common that they reduce side reactions and increase molar masses and kinetics; however, the catalyst is present in a rather large amount and needs to be removed from the polymerization mixture afterwards, which turns out to be quite challenging for some systems.[20]

7.3 POLY(ETHYLENE GLYCOL)S

Poly(ethylene glycol) (PEG), also known as poly(ethylene oxide) (PEO) for molecular weights above 20,000 g/mol, is a polymer based on $-CH_2-CH_2-O-$ repeating units. Depending on the molecular weight, it occurs as a viscous liquid or semicrystalline solid (melting area ca. 60°C, glass transition ca. −60°C, see below) highly flexible and water-soluble polyether. It exhibits very low immunogenicity, antigenicity, and toxicity and is therefore widely used in medicine, pharmacy, cosmetics, and food applications.[24,25]

7.3.1 History

The first synthesis of PEG goes back to 1859, when Wurtz reported the polymerization of EO in the presence of water, ethylene glycol, and acetic acid.[25] In 1929, Staudinger used activated alumina[26] and, later, oxides of calcium,[27] strontium, or zinc as catalysts to obtain very-high-molecular-weight polymers.[3] The commercialization of PEG followed during the 1930s through base catalysis and initiation with alcoholates.[28] In the late 1970s, Abuchowski et al.[29,30] reported on pioneering works for the covalent conjugation of PEG (PEGylation), which improved pharmacokinetic properties of protein- and peptide-based drugs. These works paved the way for the success story of PEG; today, there are several Food and Drug Administration (FDA)-approved, PEGylated drugs in the clinics.[31] Furthermore, multifunctional PEGs (*mf*-PEGs) (cf. Section 2.2.4), which were first reported in the mid-1990s,[32,33] open up tremendous possibilities for further developments in this area but also many more applications.

7.3.2 Properties

The physical characteristics of PEG strongly depend on its molecular weight. PEGs with molecular weights lower than 400 g/mol are nonvolatile liquids, whereas PEG 600 exhibits a melting area between 17°C and 22°C, resulting in a liquid at room temperature and a paste-like texture at low temperatures. PEGs with molecular weights higher than 2000 g/mol are solid substances like flocs or powders. Rigidity and melting point (T_m) increase with molecular weight until the T_m constitutes a constant value of 60°C for higher molecular weights. Therefore, physical mixing of PEGs with different molecular weights enables the adjustment of the T_m, e.g., in a physiological interesting range.[25] In the same way as the melting point increases with increasing molecular weight, the degree of crystallinity increases, too. The tendency to form crystalline phases with increasing chain length is due to the lower segmental mobility and more convenient geometrical alignment. The degree of crystallization (X_c) can be calculated via the heat of melting of the polymers using Equation 7.1, where ΔH_m^0 is the heat of melting of 100% crystalline PEG (196.8 J/g)[34] and ΔH_m is the heat of melting of the investigated polymer.

$$X_c = \frac{\Delta H_m}{\Delta H_m^0} \quad (7.1)$$

Pielichowski et al. found that X_c increases from 85.7% for PEG 1000 to 96.4% for PEG 35,000.[34] The glass transition temperature (T_g) of PEG exhibits as −57°C.

Despite the high degree of crystallinity, a very interesting and unique property of PEG is its excellent water solubility. PEGs with low molecular weight are soluble in water in every concentration, whereas the solubility decreases with increasing molecular weight. However, a 50% homogeneous solution of PEG with M_n = 35,000 g/mol can be obtained at room temperature. This property is intriguing as all other aliphatic polyethers except poly(vinyl methyl ether) are completely insoluble in water.[25] The outstanding water solubility was attributed to the distance between the repeating oxygen atoms in the polymer chain, which accords with the distance of oxygen in liquid water. This enables the formation of an expanded network of hydrogen bonds between the PEG chain and the water molecules, which was shown via theoretical works and molecular simulations.[35] PEG is a thermoresponsive polymer, and its solubility in water is reduced with increasing temperature. Only below the lower critical solution temperature (LCST), which is around 100°C for PEG, the hydrophilic surface of the polymer can build up the described hydrate shell, and the polymer chain is stretched. Above the critical demixing temperature (cloud point temperature), the polymer is insoluble and precipitates.[36]

The second very important property of PEG is the remarkably low toxicity, concerning both the acute as well as chronic toxicity by oral or intravenous uptake. The World Health Organization (WHO) assigned the estimated acceptable daily uptake at up to 10 mg/kg.[37] The uptake of PEG in the gastrointestinal tract decreases with increasing molecular weight. Animal studies have shown that PEG 5000 is not absorbed from the intestine over 5 h, whereas 2% of PEG 2000 is absorbed. After acute oral and dermal exposure, PEG is practically nontoxic, with animal LD_{50} values greater than 2 g/kg and 10 g/kg, respectively.[37] However, PEG has a laxative effect and is used as laxative, mainly in the United States. Besides the very low toxicity, PEG is nonimmunogenic and nonantigenic. Recent studies, however, reveal that the efficiency of PEGylated drugs can be reduced over time as antibodies against PEG can be generated in vivo.

7.3.3 APPLICATIONS

The excellent water solubility of PEG renders it highly interesting as a nonionic, surface-active agent (i.e., surfactant). In these nonionic surfactants, the PEG chain is attached to lipophilic blocks, like long-chained alkyl residues (fatty acids or fatty alcohols [ca. C_{12}–C_{16}]), via ether or ester bonds. The most important nonionic surfactants are PEG–monoalkyl ethers, PEG–monoalkylphenyl ethers (alkylphenolethoxylates), PEG–fatty acid esters, PEG–glyceryl fatty acid ester, PEG–sorbitan fatty acid ester, and poloxamers, which are block copolymers of poly(propylene oxide) (PPO) and PEG (cf. Section 2.4) (Figure 7.6). The chain length of the hydrophilic and hydrophobic blocks exhibits a strong influence on the properties of the surfactant and allows the adjustment of the properties.

These PEG-based surfactants are of great importance in cosmetics. Especially, in skin care products, the surfactants are employed to obtain a stable mixture of water and fats, e.g., for lipid-enriched cream. Furthermore, these surfactants can be found in shower gel or soap to remove dirt particles. Copolymers of PEG and silicones are applied in hair-care products and improve combability and hair straightening. Unmodified PEG, especially with relatively low molecular weight (M_n = ~500 g/mol), is used as a biocompatible solvent or humectants in creams. Higher-molecular-weight (M_n > 2000 g/mol) PEGs enable the adjustment of the viscosity and the stabilization of emulsions.[25]

Additionally, these surfactants can be found in food to stabilize water-in-oil emulsions like butter and mayonnaise, or oil-in-water emulsions like milk and cream and help to solubilize instant powder.

In the late 1970s, Abuchowski and coworkers introduced the concept of the covalent protein conjugation to PEG (PEGylation, conducted typically by methoxy-PEG [mPEG] carrying a single protein-reactive end group).[29,30] Since then, some conjugates have been approved by the FDA and used as therapeutics. The PEGylated drugs increase protein half-life as they reduce kidney clearance and increase blood circulation time due to steric shielding against degradation. Furthermore, they are known to exhibit a so-called stealth effect, i.e., they avoid phagocytosis and removal from

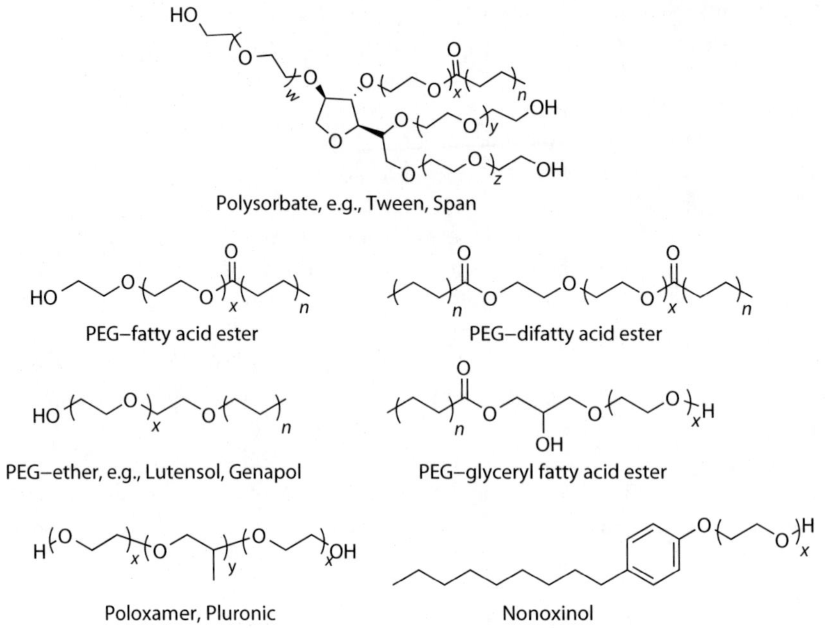

FIGURE 7.6 PEG surfactants applied in cosmetics and pharmaceutical products.

the bloodstream. They exhibit low opsonization rates and reduce immunological responses to proteins by perturbance of antibody binding or breakdown of the biomolecule by enzymes.[24] The same benefits of PEG are also used in liposomal formulations, which are stabilized by PEGylation to generate so-called stealth liposomes.[38,39] All these factors lead to less frequent dosing due to increased plasma half-lives of the polymer–drug conjugates and thus improvement for the patients. Currently, 10 FDA-approved PEG–protein conjugates and one PEG–aptamer conjugate for different diseases are on the market,[40] with several more under current investigation (Table 7.1).

TABLE 7.1
Marketed PEGylated Drugs

Brand	Active Substance	Approval	Disease
Adagen	Adenosine deaminase	1990	Severe combined immunodeficiency
Oncaspar	Asparginase	1994	Acute lymphoblastic leukemia
PEG-Intron	Interferon α-2b	2000	Hepatitis C
Pegasys	Interferon α-2a	2001	Hepatitis C
Neulasta	G-CSF	2002	Chemotherapy-induced neutropenia
Somavert	GHA	2003	Acromegaly
Macugen	Anti-VEGF aptamer	2004	Wet age-related macular degeneration
Mircera	EPO	2007	Renal anemia after chronic kidney disease
Cimzia	Anti-tumor necrosis factor Fab'	2009	Crohn's disease
Krystexxa	Uricase	2010	Chronic gout refractory
Omontys	Erythropoiesis-stimulating agent	2012	Renal anemia after chronic kidney disease

Source: Dingels, C. et al., *Chem. unserer Zeit*, 45, 2011; Alconcel, S.N.S. et al., *Polym. Chem.*, 2, 2011; Pfister, D. and Morbidelli, M., *J. Control. Release*, 180, 2014.

Note: EPO, erythropoietin; G-CSF, granulocyte-colony stimulating factor; GHA, growth hormone receptor antagonist; VEGF, vascular endothelial growth factor.

Another interesting field for the use of PEG are lithium-ion batteries: the ion conductivity properties of PEG were first described by Fenton et al. in 1973.[41] The oxygen atoms along the backbone can interact with Li ions and are therefore used as solid electrolytes. The efficient storage of electrical energy is a crucial aspect for future technologies especially with respect to renewable energy sources and the connected challenge of energy storage. Therefore, safe, fast, robust, and effective energy storage devices are an essential requirement for modern accumulators. Li-ion batteries are a very prominent example for effective energy storage; however, the applied electrolytes are aprotic organic solvents, which are inflammable and can show leakage and short lifetimes. An advancement of these liquid electrolytes is the use of a leakproofed and flame-resistant polymer, e.g., PEG and its derivatives. The PEG segments act as multivalent ligands to promote the ion conductivity (similar to crown ethers acting as binding metal cations) as they allow the migration of Li ions between both electrodes. To date, however, the obtained Li-ion conductivities cannot keep up with the liquid electrolytes, as the crystallization of PEG hinders the ion movement. Recently, the Li-ion conductivities of branched PEG have been investigated to overcome the high crystallinity.[42] At the moment, a compromise is applied where the liquid electrolyte is embedded in a swollen polymer matrix, at least reducing the risk of leakage.[43]

Furthermore, PEG can be found in textiles, ceramics, lubricants, metalworking, urethanes, rubber chemicals, and electronics and is the key player among aliphatic polyethers. It is also applied in TentaGel resins, where PEG chains are grafted onto cross-linked polystyrene (PS) particles.

7.4 MULTIFUNCTIONAL POLY(ETHYLENE GLYCOL)

PEG is *the* polymer in various biomedical applications due to its excellent solubility in aqueous and organic media; it shows no or little immunogenicity, antigenicity, or toxicity and is highly flexible, as discussed in detail before.[44,45] However, as a linear polymer, it has a maximum of two functional (end) groups and lacks further functional groups along the backbone, limiting its loading capacity. The copolymerization of EO with functional EO derivatives, first employed in the mid-1990s,[32,33] overcomes the limited loading capacities and has received increased attention in recent years. The AROP of EO with appropriate comonomers enables the synthesis of highly controlled polymers with an adjustable number of functional groups at the ethylene glycol backbone, which is desirable for not only biomedical but also a lot of other applications, such as lubricants or adhesives. Figure 7.7 depicts some functional epoxides that were used for the synthesis of *mf*-PEGs, focusing on the one-pot copolymerization of EO and the respective oxirane to produce statistical PEG-based copolymers. Indeed, these monomers can also be used for the synthesis of block copolymers, which will not be discussed in this work.[46–48]

The synthesis of these monomers—if not commercially available—usually starts from ECH or glycidol and an alcohol or amine carrying the functional group in a one- or two-step procedure. Starting materials are easily accessible and common in chemical industry, which renders these materials interesting for different applications. The functional epoxide is obtained via intramolecular nucleophilic substitution ($S_N i$) when ECH and the respective alcohol or amine are used as the starting materials.[49–55] Another route is the protection of the hydroxyl group of glycidol with ethyl vinyl ether under acidic catalysis.[56]

The random copolymerization of EO and functional epoxides (Figure 7.7) enables further modification of the PEG backbone. The introduction of hydroxyl groups is important because alcohols show a versatile reactivity, and as glycerol derivatives, these structures are typically nontoxic. As alcohols can undergo nucleophilic attack and can initiate the AROP themselves, these groups have to be protected. This can be realized by acetals or ketals such as the ethoxyethyl protective group.[56] The random copolymerization of EO with EEGE[33] enabled the synthesis of multi-hydroxyl-functional PEG, which can be used for bioconjugation,[57,58] to obtain graft structures[59] or highly branched polyethers.[60] The copolymerization with isopropylidene glyceryl glycidyl ether (IGG) and following dilute acidic deprotection allows the introduction of two vicinal hydroxyl groups

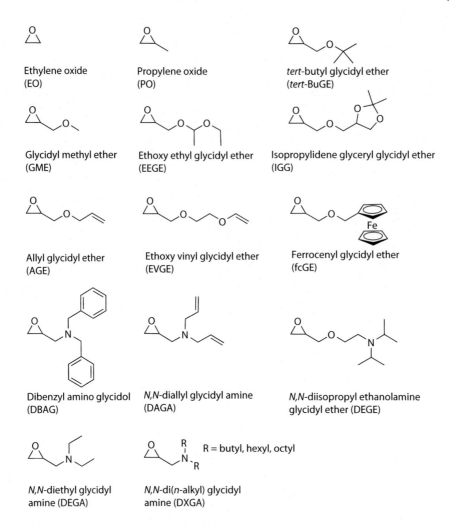

FIGURE 7.7 Overview of different (functional) epoxides.

per monomer unit.[50,61] Furthermore, the 1,2-diol can easily form cyclic acetals with aldehydes or ketones and enables the pH-controlled, reversible attachment or release of small compounds.

The introduction of allyl or vinyl moieties via AGE or ethoxy vinyl glycidyl ether (EVGE)[51] enables further reactions, e.g., through thiol-ene addition to attach biomolecules[62] or through hydrosilylation, enabling the synthesis of complex architectures.[63]

The copolymerization of EO with ferrocenyl glycidyl ether (fcGE)[54] leads to redox-responsive and thermoresponsive copolymers where the iron center can be oxidized reversibly. The introduction of amine moieties in the PEG backbone is of high interest with respect to biomedical applications and was performed in several works. N,N-dibenzyl glycidyl amine (DBAG)[49] and N,N-diallyl glycidyl amine (DAGA)[52] carry protected amines, which can be deprotected after polymerization and enable the introduction of free amine moieties. The use of N,N-diethyl glycidyl amine (DEGA),[53] N,N-di(n-alkyl) glycidyl amine (DXGA),[55] and N,N-diisopropyl ethanolamine glycidyl ether (DEGE)[64] allows the synthesis of cationic polyelectrolytes, as the amine groups can be quaternized.

Under certain reaction conditions, the copolymerization of glycidyl derivatives with EO is perfectly random, but also, gradient copolymers, especially for the amino derivatives, have been reported.[52–54]

One very interesting feature of PEG and the presented copolymers is their temperature-dependent water solubility. As these polyethers are often applied in biomedical applications, the water solubility

TABLE 7.2
Overview of Thermal Properties of Different *mf*-PEGs

Comonomer	co/EO, %	T_c, °C	T_g, °C	T_m, °C	Ref.
Homo-PEG	–	100	−57	66	20,49
GME	31 to 78	55 to 98	−69 to −60	10 to 38	20
EEGE	3 to 67	–	−71 to −64	−2 to 45	57
IGG	9 to 53	47 to 81	−59 to −49	13 to 31	61,65
AGE	9 to 82	40 to 75	−78 to −62	−16 to 57	65,66
EVGE	5 to 88	9 to 83	−60 to −55	16 to 34	51,65
DBAG	2 to 15	22 to 71	−54 to −40	20 to 46	49,65
DAGA	3 to 24	29 to 94	−69 to −56	−5 to 53	52
DEGE	10 to 40		−66 to −57		64
DEGA	4 to 29	55 to 97	−65 to −71	−10 to 40	53
DButGA	4 to 25	37 to 100[a]	−73 to −66	15 to 45	55
DHexGA	4 to 21	52 to 81[a]	−78 to −71	12 to 46	55
DOctGA	4 to 9	77 to 90[a]	−75 to −70	43 to 49	55
FcGE	3 to 28	7 to 82	−59 to −49	8 to 43	54

Note: T_c, cloud point temperature; T_g, glass transition temperature; T_m, melting temperature.
[a] Measured at pH = 9.

is an important issue and an additional handle for tuning material properties. PEG itself shows an LCST in water of approximately 100°C. The LCST describes the critical temperature below which a polymer is miscible for all compositions. Table 7.2 shows the thermal properties of the different copolymers, including the cloud point temperature (T_c), the glass transition temperature (T_g), and the melting temperature (T_m). Through variation and the amount of the comonomer along the PEG backbone, the T_c can be tuned in a broad temperature range (values from 7°C up to 98°C have been reported). The influence of the comonomers on the T_g of the materials is rather low and varies from −78°C to −40°C, whereas the T_m is much more strongly influenced, as the comonomer units act as defects in the PEG crystal, and deviations between −16°C and 57°C and variable degree of crystallinity have been reported. The PEG homopolymer shows a T_g of −57°C[20] and a T_m of 66°C.[49] The published data of all copolymers are summarized in Table 7.2.

As shown above, various possibilities for the introduction of functional groups along the polyepoxide backbone have been developed and allow adjusting of the properties of the resulting polymer. Other functional groups are rarely investigated today, and this area still has much room for further development and the search for new functional epoxide derivatives. Many functional PEGs, e.g., α,ω-(hetero)telechelics, are commercially available from Iris Biotech or Rapp Polymere; they can be synthesized using functional initiators or terminating agents and postpolymerization modifications.[67]

7.5 POLY(PROPYLENE GLYCOL)

Poly(propylene glycol) (PPG), also called PPO for higher molecular weights, is an amorphous, water-insoluble polymer. It can be synthesized in a stereoirregular, atactic, stereoregular, isotactic, or syndiotactic manner but is mainly used in its atactic form. The additional methyl group per monomer unit compared to PEG breaks the crystalline structure as well as the hydrophilic behavior. PPG is synthesized via ROP of PO using a double metal cyanide catalyst.[2,68] In 2003, 6.6 million tons of PO were produced, and about 60% of this was converted to PPG.[69] The main application is the use as surfactant in combination with PEG, called Pluronic (cf. Section 7.5.2), or in Jeffamine (available from Huntsman), where primary amino groups are attached to the end of PO or EO homopolymer and copolymers, and other applications are rare. However, it is used as a rheology modifier in polyurethane foams, in drug formulations, and as reference compound in mass spectrometry.

7.5.1 Properties

PPG is an amorphous material and is a liquid in the molecular weight range of 200–6000 g/mol in the case of the atactic PPG. Its T_g lies in the range of −72°C to −67°C.[70] The degradation temperature (T_{deg}) depends on the stereoregularity of the polymer and is in the range of 265°C to 365°C.[71] PPGs are soluble in most organic solvents; however, the solubility in water is temperature dependent. Polymers with molecular weights below 700 g/mol are water soluble below 18°C,[72] whereas higher-molecular-weight polymers are water insoluble.

In the isotactic PPO, the PO units have the same configuration; however, the material is optically inactive as some molecules have the opposite configuration. The thermal properties slightly differ, with a T_g of −75°C and a T_m of 78°C.[73] Due to the regular structure, these polymers exhibit higher crystallinity compared to the atactic polymers.

7.5.2 Pluronic

The combination of PEG and PPG polymers leads to amphiphilic structures. ABA triblock copolymers (Figure 7.8) of these polymers have been commercially available since the 1950s under the name poloxamer or the trademark Pluronic.[74]

They are synthesized via sequential polymerization of PO and EO in the presence of alkaline catalysts like sodium or potassium hydroxide. First, the PPG block is synthesized as the middle block, and both obtained hydroxyl end groups subsequently are used to ring-open EO, resulting in two PEG end blocks. Through variation of the total molecular weights of both blocks as well as the block sequence, the hydrophilic–lipophilic balance (HLB) can be adjusted over a broad range to stabilize dispersions (Table 7.3). Commercially available copolymers typically exhibit molecular weights in the range of 2000–20,000 g/mol and PEG contents in the range of 20–80 wt.%.[74] The nomenclature follows a system where the letter describes the viscosity state at room temperature, like solid (flake), F; pasty, P; or liquid, L. The last number multiplied by 10 gives the relative mass% of the EO units, and the number(s) above give the relative molecular weight of the PPG block.

A minimum PPG block size is necessary to obtain phase segregation, whereas the need for water solubility in many commercial applications also sets an upper limit for the PPG block size. Pluronic shows the typical surfactant properties, like the ability to interact with hydrophobic surfaces and biological membranes due to their amphiphilic character. Above a certain critical micelle concentration (CMC), the polymers self-assemble into micelles in aqueous solution to minimize the contact of the hydrophobic segments with the polar aqueous environment. Typically, the copolymers that are used for drug delivery show CMCs in the range of 5×10^{-3} to 1% at 37°C.[75] The diameters of the micelles vary from ca. 10–100 nm.[76] The structure of these micelles can be spherical, rodlike, or lamellar depending on the PEG and PPG block length, the concentration of the polymers, and the temperature.[77] The core of the micelle consists of the hydrophobic PPG segments, which are separated from the hydrophilic aqueous environment through the PEG shell. The number of block copolymers that form one micelle is called the *aggregation number* and varies from several to over 100 for spherical micelles.[75] The core itself can serve as a pool to incorporate various hydrophobic

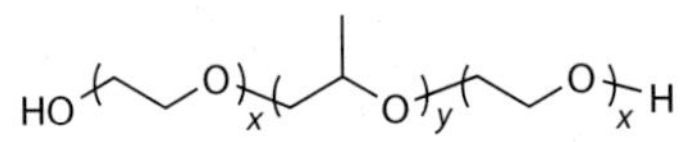

FIGURE 7.8 Structure of Pluronic block copolymers containing two PEG blocks and one PPG block.

TABLE 7.3
Physicochemical Properties of Pluronic Block Copolymers

Copolymer	MW[a]	No. of EO Units[b]	No. of PO Units[b]	HLB[c]	Cloud Point in 1% Aqueous Solution,[c] °C
L101	3800	8.6	59.0	1	15
L61	2000	4.6	31.0	3	24
L92	3650	16.6	50.3	6	26
P103	4950	33.8	59.7	9	86
P84	4200	38.2	43.5	14	74
L35	1900	21.6	16.4	19	73
F87	7700	122.5	39.8	24	>100
F68	8400	152.7	29.0	29	>100

Source: Kabanov, A.V. et al., *J. Control. Release*, 82, 2002.
Note: MW, molecular weight.
[a] Average molecular weights from manufacturer (BASF, Wyandotte, MI).
[b] Calculated using average molecular weights.
[c] Determined by manufacturer.

substances or to increase the miscibility of substances with different hydrophilicity, whereas the PEG shell leads to a stabilization in the dispersed state. Therefore, the polymers are mainly used in industrial applications, e.g., as lubricant, defoamer, humectants, and plasticizer; cosmetics; and pharmaceuticals. It was shown that up to 20–30 wt.% therapeutic or diagnostic reagents can be incorporated into the core of the Pluronic micelles.[78]

As pointed out before, the water solubility of both PPG and, especially, PEG is strongly temperature dependent. Therefore, also, the micelle formation of Pluronic strongly correlates with temperature. Below room temperature, both blocks are hydrated, and the block copolymer exhibits relatively high solubility in water. With increasing temperature, the PPG block dehydrates and therefore becomes insoluble, which leads to micelle formation. The temperature where micelle formation occurs is called critical micelle temperature (CMT) and is in the range from ca. 25°C to 40°C, i.e., below or near body temperature, for most Pluronic copolymers.[75]

It was shown that the incorporation of drugs into Pluronic block copolymer micelles results in an increased solubility, metabolic stability, and circulation time for the drug.[79] Furthermore, these micelles can be used as biological response modifiers.[79] The key player for the biological activity of these materials is their ability to incorporate into cell membranes and the subsequent translocation into the cells. This allows the polymer to influence various cellular functions like mitochondrial respiration, ATP synthesis, apoptotic signal transduction, and gene expression.[79] Through these effects, it was shown that the interaction of Pluronic micelles with multidrug-resistant cancer cells resulted in a sensitization of these cells, leading to an increase in cytotoxic activity of the incorporated drugs by two or three orders of magnitude.[80,81] Additionally, the polymers inhibit the drug efflux systems in the blood–brain barrier as well as in the small intestine, leading to enhanced transport of drugs to the brain.[75]

All these findings render Pluronic a highly interesting class of surfactants with multiple possible applications in the important field of drug delivery systems.[79] However, most of the material is applied in industrial applications as surfactant, lubricant, defoamer, etc.

7.6 POLYGLYCEROL

Polyglycerols (PGs; also known as *polyglycidols* when synthesized from glycidol, i.e., 2,3-epoxy-1-propanol) are polymers that derive from the natural metabolite glycerol. Glycerol is common in the human body, e.g., as the backbone of triglycerides and phospholipids,[82] and is currently also used as a renewable raw material as it is generated as a side product during biodiesel production.[83] That is one reason why glycerol-based polymers are attracting increased attention. Due to the three alcohol groups in glycerol, besides linear polymers, branched polymers also can be generated with perfect (dendritic) or statistical (hyperbranched) architecture. In the following, the properties and applications of two glycerol-based polymers, namely, hyperbranched (*hb*) as well as linear (*lin*) PG, will be discussed.

7.6.1 Hyperbranched Polyglycerol

7.6.1.1 History

Hyperbranched, i.e., statistically branched, PG (*hb*PG) can be synthesized via ring-opening multi-branching polymerization (ROMBP) of glycidol, a cyclic latent AB_2 monomer, which was already studied by several groups in the 1960s.[84] In 1985, Vandenberg published the characterization of the branching of PG, which was prepared by anionic polymerization.[85] After the introduction of the term "hyperbranched (*hb*) polymers" for a random dendritic branching in polymers by Webster and Kim[86] in the 1990s, cationic[87,88] and anionic[89,90] polymerization techniques were used for the synthesis of *hb*PG from glycidol by Penczek, Kubis, and Dworak. The SMA technique by Frey et al. further improved control over hyperbranching polymerization and enabled the synthesis of *hb*PG with moderate to narrow molecular weight distributions (M_w/M_n = 1.3–1.6).[89] However, molecular weights were limited ($M_n \leq$ 6000 g/mol). The SMA technique circumvents the self-initiation of glycidol and undesired cyclization reactions, as the concentration of the monomer is kept low, and thereby, the reaction with the multifunctional hyperbranched polymer is favored. The fast cation-exchange equilibrium ensures that all hydroxyl groups are potentially active propagation sites, resulting in a branched polymer structure (degree of branching [DB] = 0.53–0.59) with narrow distribution. The DB is a value to measure the perfection of a hyperbranched structure.[91] For a linear structure, the DB is 0, and for a perfect dendrimer, 1, whereas hyperbranched polymers exhibit values in between. The DB is usually determined from nuclear magnetic resonance (NMR) spectra and calculated with Equation 7.2, whereas D and L are the integrals of the NMR resonances of the dendritic and linear units, respectively.[92]

$$\mathrm{DB} = \frac{2D}{2D + L} \tag{7.2}$$

To overcome the limit of the molecular weight, Frey et al. developed a two-step protocol using a low-molecular-weight PG macroinitiator (M_n = 500 and 1000 g/mol), which is deprotonated and serves as an initiator for the ROMBP of glycidol.[93] With this method, *hb*PGs with molecular weights up to 24,000 g/mol, moderate to narrow molecular weight distributions (M_w/M_n = 1.3–1.8), and a DB in the range of 0.60–0.63, which approximates the theoretical limit for SMA conditions (DB = 0.66), were obtained.

Very-high-molecular-weight (up to 700,000 g/mol) *hb*PGs with narrow molecular weight distributions (M_w/M_n = 1.1–1.4) can be obtained through the addition of dioxane as an emulsifying agent during the ROMBP of glycidol.[94] The mechanism is not fully understood yet; however, Brooks et al. postulate faster cation exchange in the presence of dioxane leading to the low molecular weight disperties.[94]

7.6.1.2 Properties

*Hb*PG is a biocompatible polymer comparable to PEG with a flexible and very stable polyether backbone.[95] Compared to PEG, it has higher oxidative stability, and the multiple hydroxyl end groups allow manifold further modifications, e.g., with drugs or cross-linkers. Additionally, these

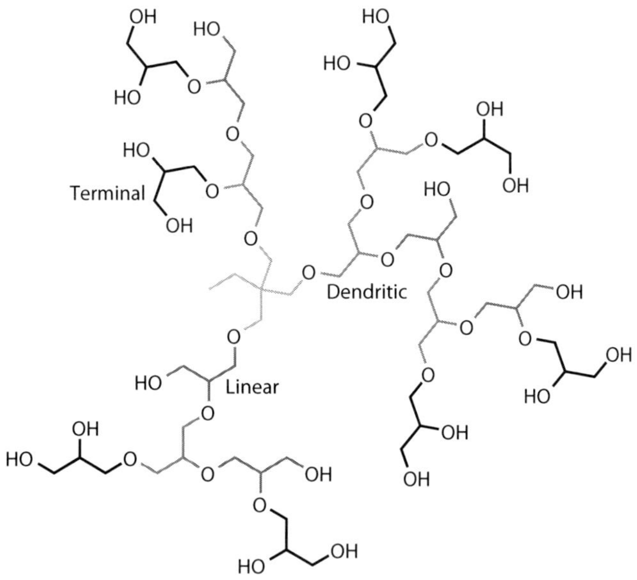

Polyethers

FIGURE 7.9 Chemical structure of hyperbranched polyglycerol (*hb*PG) with different repeating units.

hydroxyl functionalities provide excellent water solubility of the polymers, which is important in terms of biomedical applications. Due to the hyperbranched structure, *hb*PG exhibits low intrinsic viscosities and high densities. During the synthesis, different units, namely, dendritic (D), linear (L), and terminal (T) units, arise, as depicted in Figure 7.9. The DB and, thereby, the intrinsic viscosity can be influenced through the copolymerization of glycidol with AB monomers that result only in linear units, e.g., AGE or phenylglycidyl ether (PGE)[96] or EO.[42] In the case of PGE, the DB was varied in the range from 9% to 58%. Furthermore, the solubility as well as the thermal properties of the polymers can be adjusted through the copolymerization. For *hb*PG homopolymers, the T_g is ca. −20°C,[96] depending on the molecular weight and the number of hydroxyl groups.

7.6.1.3 Applications

The excellent biocompatibility and water solubility as well as the one-step synthesis render *hb*PG also interesting for industrial applications, especially in medicine and pharmacology. Oligoglycerols and their fatty acid esters are used on a large scale in cosmetics or cleaning agents; as surfactants; and in the food industry, pharmaceuticals, polymer additives, and electronics.[97] *Hb*PG is also an attractive candidate in more sophisticated material applications, e.g., in targeted drug, dye, and gene delivery; in protein-resistant nonfouling surfaces in medical devices and biosensors; or in hydrogels.[97,98]

Furthermore, *hb*PG can be used as soluble catalyst support in organic synthesis or in biomineralization processes.[99–102]

7.6.2 Linear Polyglycerol

Linear PGs (*lin*PG) (Figure 7.10) are, to date, hardly considered in industry, in contrast to *hb*PG, as they are typically synthesized via AROP from protected glycidol derivatives, like trimethylsilyl glycidyl ether (TMSGE), EEGE, *tert*-butyl glycidyl ether (*t*BGE), and AGE, and require, in addition to monomer synthesis and purification, a subsequent postpolymerization deprotection step. In academia, EEGE is the most favorite monomer due to the facile removal of the acetal protecting group under mild acidic conditions.

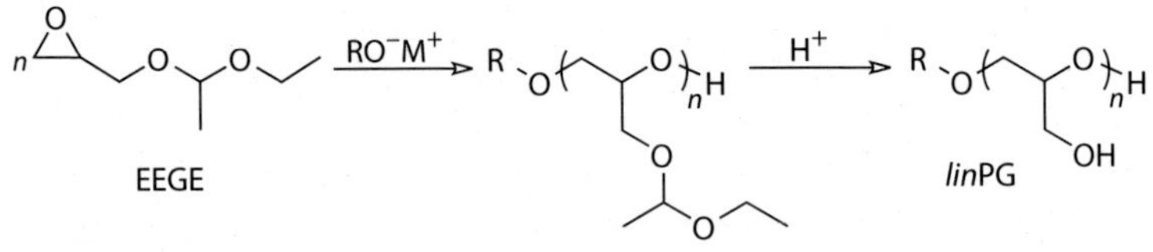

FIGURE 7.10 Synthesis of linear polyglycerol (*lin*PG) with EEGE as monomer.

7.6.2.1 History

After the first attempts in 1986, the controlled synthesis of *lin*PGs with relatively high molecular weights was reported in 1994. Taton et al. synthesized polymers with molecular weights up to 30,000 g/mol and moderate molecular weight dispersities (M_w/M_n = 1.38–1.89) via AROP of EEGE and cesium hydroxide as the inititator.[33] Later, Dworak et al. produced a series of *lin*PGs with low molecular weight dispersities (M_w/M_n < 1.20) with potassium or cesium alkoxides as the respective initiators.[103,104] Using "graft-on-graft" strategies with repeating cycles of deprotonation, polymerization, and acidic deprotection, highly branched polymers with high molecular weights (M_n > 1,800,000 g/mol) have been reported.[105,106] Since then, different functional initiators[18,107–112] as well as termination agents[113] have been applied for the preparation of well-defined polymers with exactly one functional moiety at the chain end.[113–115]

7.6.2.2 Properties

The atactic nature of *lin*PG after the AROP results in viscous and amorphous materials, which show high water solubility and exhibit biocompatibilities similar to or higher than PEG.[95] Haag et al. reported that molecular weights of at least 1000 g/mol are necessary to provide biorepellent properties.[112] Furthermore, a slightly lower chain flexibility compared to PEG was observed.[116] The thermal properties of the polymers are not investigated systematically; however, T_g ranging from −8°C to −27°C have been mentioned.[117] In comparison to PEG, a slightly higher thermal and oxidative stability has been claimed;[118] however, detailed degradation studies are missing. In literature, one example is reported with an onset of degradation at temperatures exceeding 250°C.[119] It is further expected that biodegraded products are less harmful as they are glycerol derivatives, which are highly abundant in nature. An overview of the thermal properties of the epoxide-derived polyethers described in this chapter can be found in Table 7.4.

7.6.2.3 Applications

To date, only the low-molecular-weight oligomers diglycerol and triglycerol are commercially available while higher-molecular-weight *lin*PG is not available on an industrial scale. However, the described properties render these materials interesting alternatives for biomedical applications,

TABLE 7.4
Overview of Thermal Properties and Water Solubility of Epoxide-Derived Polyethers

Polymer	T_g, °C	T_m, °C	T_{deg}, °C	Solubility in Water	Ref.
PEG	−57	66	345–410	✓	20,36,120
PPO	−70	60	265–365	✗	70,71
*hb*PG	−20	–	307–400	✓	96,121
*lin*PG	−8 to −27	–	>250	✓	117,119

Note: T_{deg}, degradation temperature; T_g, glass transition temperature; T_m, melting temperature.

e.g., in intravenous applications,[95] in drug delivery and controlled release systems, as antifouling coatings, or in bioconjugation. A review on the current research was published recently visualizing the promising possibilities of these polymers.[117]

7.7 POLYOXYMETHYLENE

Polyoxymethylene (POM), also called polyacetal, is a high-molecular-weight polymer and copolymer of formaldehyde based on $-CH_2O-$ repeating units. Due to the polyacetal structure, the polymer is degraded by acids or elevated temperatures. Therefore, it is often copolymerized with cyclic ethers, e.g., EO or dioxolane, to interrupt the acetal structure, thus leading to higher stability. POM is one of the high-performance engineering polymers due to its excellent mechanical properties, which will be further discussed in the following sections.

7.7.1 History

The synthesis of POM was first described by Butlerov[122] in 1859 and further investigated by Staudinger in 1925.[123] However, it took almost 100 years before DuPont started further investigations and the commercialization of POM in the 1950s. Staudinger, Trautz, and Ufer;[124] Walker;[125] and Carruthers and Norrish[126] observed that rigid and solid polymers based on formaldehyde units can be obtained from formaldehyde and ionic initiators. Different initiators of multiple nucleophiles and electrophiles like ordinary acids and bases; amines; Lewis acids; organometallic compounds; phosphines; and carbonyl complexes of iron; cobalt; and nickel were ,employed. These polymers exhibited fast hydrolytic degradation due to their pure acetal structure and their potential at that time was questionable. Stabilization of the terminal hydroxyl groups, i.e., hemiacetals, was achieved through esterification, as already studied by Staudinger,[123] resulting in polymers with excellent tensile, impact, and compression strengths and low abrasion. The patenting of this product was done in 1956, and commercial production began in 1960.[127] In 1962, Celanese started the production of a POM copolymer by copolymerizing trioxane with cyclic ethers like EO,[128] which further improved the thermal stability of the polymer and led to a rapid worldwide expansion of the acetal resin.

7.7.2 Polymerization

7.7.2.1 Anionic Polymerization of Formaldehyde

Staudinger first reported on the anionic polymerization of formaldehyde;[123] the main challenge during this process is the production of very pure and water-free formaldehyde, which is rather laborious and cost intensive. Two plausible mechanisms are discussed for the anionic polymerization of formaldehyde.[129] The first describes the nucleophilic addition of the anionic chain end at the carbonyl double bond of the monomer (Figure 7.11).

In the second route, the crystalline polymer grows stepwise in aqueous or alcoholic solution through an equilibrium reaction with the solvated monomer. In this case, the crystallization is the driving force for the polymerization. Initiators can be amines, phosphines, ammonium or phosphonium salts, amides, or amidines.[130] The ion pair, which is obtained by the reaction with water or other protic contaminations in very low concentration, acts as an active initiator (Figure 7.12).

Impurities in formaldehyde (e.g., water) act as chain-transfer agents, and the obtained polymer chains carry hydroxyl end groups, i.e., unstable hemiacetals. Therefore, the chain ends have to be acetylated with acetic anhydride[131] to prevent spontaneous depolymerization of the polymer backbone.

$$\sim\!\!CH_2O^- + CH_2=O \longrightarrow \sim\!\!CH_2OCH_2O^-$$

FIGURE 7.11 Anionic polymerization of formaldehyde via addition at the double bond.

Initiation: $R_3NH^+OH^- + CH_2=O \longrightarrow HOCH_2O^-R_3NH^+$

Propagation: $HO(CH_2O)_nCH_2O^-R_3NH^+ + CH_2=O \longrightarrow HO(CH_2O)_{n+1}CH_2O^-R_3NH^+$

Transfer: $HO(CH_2O)_nCH_2O^-R_3NH^+ + H_2O \longrightarrow HO(CH_2O)_nCH_2OH + R_3NH^+OH^-$

Reinitiation: $R_3NH^+OH^- + CH_2=O \longrightarrow HOCH_2O^- + R_3NH^+$

FIGURE 7.12 Mechanism for the anionic polymerization of formaldehyde in aqueous or alcoholic solution.

The cationic polymerization of formaldehyde is also possible, but it is more complicated than the anionic polymerization, as the abstraction of a hydride ion from formaldehyde is energetically unfavorable and the positive charge cannot be stabilized by formaldehyde. Therefore, this method is rarely used. Lewis acids, e.g., $SnCl_4$ or $SnBr_4$, can be used as initiators producing POM with similar molecular weight distributions and end groups as the anionic mechanism.[132]

7.7.2.2 Cationic Ring-Opening Polymerization of Trioxane

The CROP of trioxane is the general pathway to POM: initiators can be protonic acids such as sulfuric acid, trichloroacetic acid or trifluoromethanesulfonic acid, solvo acids, and metal halides such as BF_3, $FeCl_3$, and $SnCl_4$.[133] The polymerization can be performed in bulk, in melt, in the gas phase, in suspension, or in solution; however, the different methods can lead to different products. During the CROP of trioxane, a characteristic induction period is observable where only monomeric formaldehyde and a few oligomers but no macromolecular products are formed. In this period, the carbenium ion releases formaldehyde as long as a temperature-controlled equilibration concentration is reached, after which the polymerization starts.[134] The mechanism of the polymerization was described by Meerwein et al.[135] In the first step, the addition of the initiator cation at an oxygen of the trioxane ring leads to the formation of an oxonium ion. In the following step, the cleavage of the carbon–oxygen bond opens the ring and results in a resonance-stabilized cation, which is reactive enough to attack another monomer, leading to chain propagation (Figure 7.13). The reaction rate strongly depends on the choice of catalyst, catalyst concentration, polymerization temperature, and solvent, which influence the chain length as well as the occurrence of side reactions such as transfer reactions, transacetalization, and hydride shift.[136]

The polymer is both temperature and acid labile and decomposes slowly under formaldehyde release due to the acetal backbone: this process is called unzipping (T_{deg} under N_2 = 318°C).[137] To prevent the unzipping, trioxane is copolymerized with cyclic ethers, e.g., EO, 1,3-dioxolane, 1,3-dioxepane, or similar structures. In this case, the unzipping process stops as soon as it reaches the carbon–carbon bond.[136] The copolymerization influences the crystallinity of the chain, resulting in a lower melting temperature (T_m) and reduced mechanical properties but also improved processability.[138]

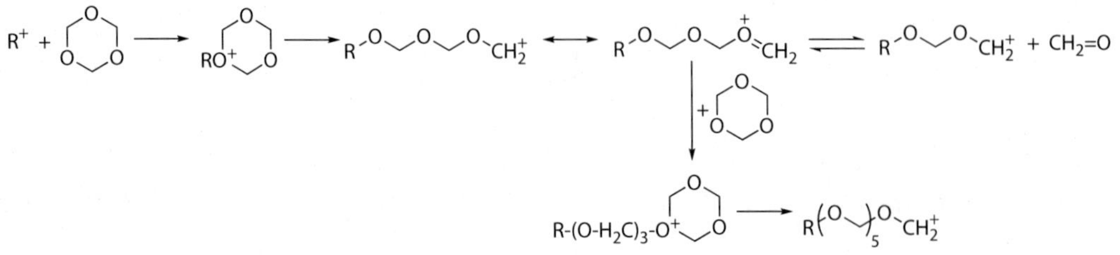

FIGURE 7.13 Mechanism of the cationic ring-opening polymerization of trioxane. (From Meerwein, H. et al., *Angew. Chem.*, 72, 1960.)

TABLE 7.5
Properties of POM Homopolymers and Copolymers

Property	POM-H	POM-C	Ref.
Density at 20°C, g/cm^3	1.42	1.41	136,141
T_m, °C	175–185	165	142
T_g, °C	−85	−60	140
Degree of crystallinity, %	65	41	139,143
Heat of fusion J/g	326	135	143
Tensile strength, MPa	70	61	141
Elongation, %	25	40–75	141
Elasticity modulus	3600	2829	141,144

7.7.2.3 Properties

The high crystallinity of POM provides excellent mechanical properties like high tensile strength, stiffness and toughness, low friction, and high fatigue resistance.[139] However, the high crystallinity leads to the insolubility of POM in most organic solvents. At room temperature, it is soluble in hexafluoroisopropanol, and at elevated temperatures (120°C), in polar solvents like dimethylformamide (DMF) or dimethyl sulfoxide (DMSO). Additionally to the mechanical stability, POM shows low moisture absorption and high scratch resistance. It is resistant against strong bases, almost all organic solvents, and diluted acids (pH > 4). However, concentrated acids, hydrofluoric acids, and oxidizers attack the acetal structure, resulting in rapid depolymerization and degradation.[140] The properties of POM homopolymers (POM-Hs) and POM copolymers (POM-Cs), consisting of trioxane and dioxolane, slightly differ and are summarized in Table 7.5.

7.7.2.4 Applications

POM can be processed by injection, blow or foam molding, or extrusion, where the temperatures should be between 180°C and 220°C. If the temperatures exceed 250°C, degradation can occur. The service temperature is about 120°C.[140] The polymer is used in applications that were traditionally confined to metals due to their high mechanical stability, low friction and wear resistance, and long-term characteristics such as resistance against chemicals and hot water. The materials are cost competitive with metals due to the low density and, therefore, the cost per gram. POM is produced in the scale of about 1 million tons per year and can be found in a broad spectrum of applications, ranging from the automotive industry (e.g., mechanical or electrical engineering or vehicle manufacturing) to medical devices (e.g., insulin pens) and garments (e.g., zippers). Further modification with glass fibers, minerals, conducting fillers, lubricants, or other additives enlarges the scope of applications.

7.8 POLYOXETANES

Polyoxetanes are polymers that are obtained through the CROP of four-membered cyclic compounds with three carbons and one oxygen, known as oxetanes. Structures of oxetane monomers with different side chains are shown in Figure 7.14. The properties of the polymers strongly depend on these side chains of the oxetane ring, and the polymers range from amorphous liquids to highly crystalline solids. Furthermore, the variation of the side chains allows the synthesis of linear or hyperbranched structures.

In comparison to oxiranes (112 kJ/mol), oxetanes (106 kJ/mol) exhibit a lower ring strain and are less reactive,[145] which has to be taken into account during the polymerization process.

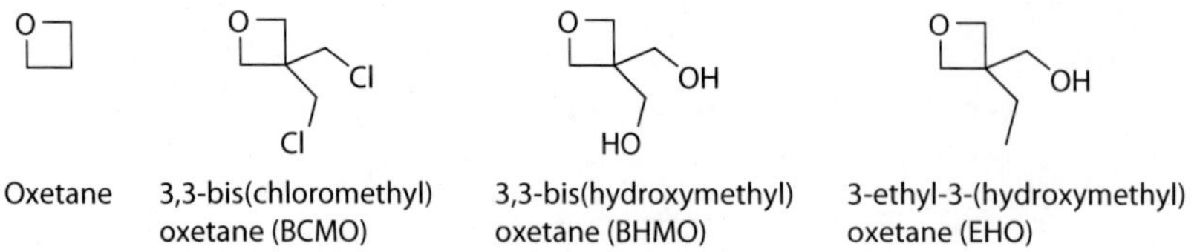

Oxetane   3,3-bis(chloromethyl) oxetane (BCMO)   3,3-bis(hydroxymethyl) oxetane (BHMO)   3-ethyl-3-(hydroxymethyl) oxetane (EHO)

FIGURE 7.14 Structure of oxetanes carrying different side chains.

7.8.1 History

The first reported polymerization of an oxetane was that of 3,3-bis(chloromethyl)oxetane (BCMO).[146] In 1955 and 1956, detailed investigations of the polymerization of different 3,3-disubstituted oxetanes[147] and the homopolymerization of oxetane itself[148] followed. In the 1950s, the polymer PBCMO was commercialized under the trade name Penton or Pentaplast. The polymers can be synthesized with molecular weights up to 350,000 g/mol and exhibit excellent chemical resistance against aggressive media, as they are stable even in concentrated H_2SO_4 at 120°C. PBCMO shows comparable thermal and mechanical properties as nylon 6 and was applied for adhesives or coatings.[149] However, due to the high costs, the production was stopped 15 years later.

Another very stable polyoxetane is linear poly(bis(3,3-hydroxymethyl)oxetane) (PBHMO), which is obtained by the CROP of 3,3-bis(hydroxymethyl)oxetane (BHMO). In 1989, Vandenberg compared the limited solubility in organic solvents and water with cellulose.[150] The polymer exhibits a very high T_m of 314°C and is applied in films or fibers.

7.8.2 Polymerization

Oxetanes exhibit a higher basicity (oxetane, $pK_a = -2.02$; oxirane, $pK_a = -3.7$)[151] and lower ring strain (oxetane, 106 kJ/mol versus oxirane, 113 kJ/mol)[152] than oxiranes, which leads to a more facile polymerization following the cationic mechanism, in contrast to oxiranes, which are polymerized predominantly via an anionic mechanism.[27] However, insertion polymerization, anionic polymerization,[153] and free-radical ROP[154] are reported in some cases. As initiators for the CROP, different electrophilic reagents can be applied, e.g., acids like trifluoromethane sulfonic acid, Lewis acids like $BF_3 \cdot OEt_2$ or $SnCl_4$,[147] aluminum alkyl-based initiators,[155] or preformed trialkyl oxonium-ion salts. When using $BF_3 \cdot OEt_2$ as the initiator, a cocatalysator, like water, ethanol, or hydroxyl-terminated polymers, is necessary; otherwise, no polymerization occurs.[147] Polymerization can be conducted in bulk or in solvents like dichloromethane, toluene, or DMSO, but additionally, reaction temperature and reaction time strongly influence the molecular weight and molecular weight distribution of the polymers. The reaction times vary between 3 h and several days, and the reaction temperature, between −50°C and 130°C in the case of 3-ethyl-3-(hydroxymethyl)oxetane (EHO).[156]

In the case of hydroxyl-functional oxetanes, the rate of monomer addition influences the DB of the resulting polymers, as both the ACE and the AM mechanism compete. If the ACE mechanism prevails, mainly branched chains will be obtained, whereas the AM mechanism results in linear chains. Therefore, if the monomer is slowly added to a solution of the initiator, the ACE mechanism is preferred, and mainly linear chains are obtained.[157] In the case of poly(3-ethyl-3-(hydroxymethyl) oxetane) (PEHO), different DBs (9–45%)[158–161] are reported, depending on the reaction conditions like temperature and monomer addition techniques (Figure 7.15).

7.8.3 Properties

The properties of the polymers strongly depend on the symmetry, sterics, and polarity of the side chains of the oxetane monomers and vary from totally amorphous liquids to highly crystalline solids. The melting temperatures of polyoxetanes with different functional groups are summarized in

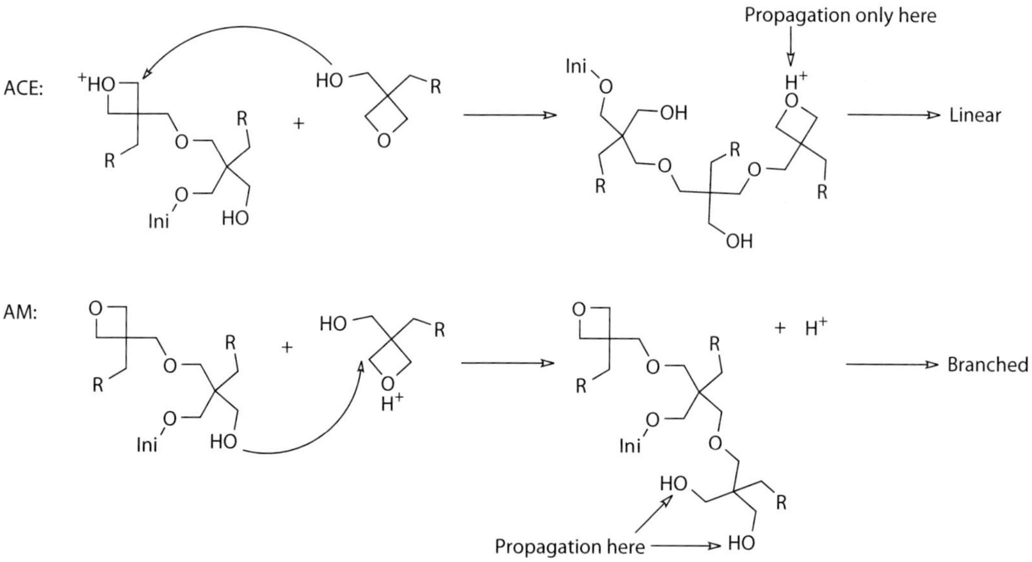

FIGURE 7.15 ACE versus AM mechanism for the cationic ring-opening polymerization of hydroxyl-functional oxetanes leading to linear or branched polyoxetanes.

TABLE 7.6
Melting Temperatures of Polyoxetanes with Different Functionalities

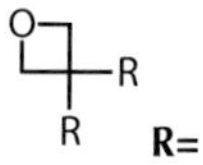

R=	T_m, °C
H	35
Methyl	47
Fluoromethyl	135
Chloromethyl	180
Bromomethyl	220
Iodomethyl	290
Cyanomethyl	175

Table 7.6 and range from 35°C for unsubstituted oxetane polymers to 290°C for 3,3-bis(iodomethyl) oxetane polymers. The copolymerization of different monomers allows the adjustment of the T_m over a broad temperature range.[162] In Table 7.6, only symmetric oxetanes are mentioned; however, asymmetric oxetanes also are accessible.[151]

The polymers are resistant to common organic solvents and are only soluble in concentrated sulfuric acid or pyridine. The copolymerization with different functional oxetanes or oxiranes enables the adjustment of the solubility and opens up new possible applications.[151,163,164]

7.8.4 APPLICATIONS

PBCMO was the only oxetane-based polymer that was commercially available in the 1950s and 1960s. The polymer shows a relatively high heat distortion temperature (99°C) and a low water absorption, which renders it interesting for items that require sterilization, e.g., in biomedical applications. Furthermore, the polymer is stable against common solvents and only attacked by strong acids.[165] Applications were mainly in the chemical processing and precision injection-molded industrial parts like valves, precision gears, corrosion-free coatings, films, and adhesives.

7.9 POLYTETRAHYDROFURAN

Polytetrahydrofuran (PTHF), also known as poly(tetramethylene ether) glycol (PTMEG), consists of –$CH_2CH_2CH_2CH_2O$– repeating units and can be obtained by the CROP of tetrahydrofuran (THF). The commercially available products usually exhibit molecular weights of 1000 and 2000 g/mol and are used in the production of polyurethanes and polyesters due to their excellent elastomeric properties.

7.9.1 History

The first reports on the polymerization of THF were published in the late 1930s by Meerwein et al.;[149] however, intensive studies were performed mainly after World War II, summarized in several reviews and books.[14,166] High-molecular-weight (M_n = 2000–3000 g/mol) PTHF has excellent elastomeric properties; however, its production is rather cost intensive. Therefore, since 1960, only low-molecular-weight PTHF (200–2000 g/mol) is commercially available and used as a chain segment in polyurethanes or polyesters.

7.9.2 Polymerization

PTHF is synthesized via the CROP of THF using the common initiators like strong protonic acids (H_2SO_4, HSO_3F, or $HClO_4$) or Lewis acids (BF_3, PF_5, $SbCl_5$). The polymerization proceeds with a living character, and therefore, termination can be achieved by adding, e.g., water or amines. The termination is necessary to obtain stable polymers; otherwise, depolymerization during the drying state may occur. However, adjusting of the molecular weights is not possible with this method, and undesired high-molecular-weight compounds can be formed. To prevent these undesired high-molecular-weight species, transfer agents can be added: first, they allow the control of molecular weight, and second, functional groups can directly be introduced at the chain end. For example, the addition of acetic anhydride as a transfer agent directly leads to acetate end groups that can be hydrolyzed after the polymerization, thereby giving a bishydroxy functional compound with adjusted molecular weight.[149,165]

7.9.3 Properties

PTHF is a semicrystalline polymer that is soluble in many polar organic solvents like chloroform and ethyl acetate, and unpolar organic solvents, e.g., benzene and toluene. The appearance changes depending on the molecular weight: low-molecular-weight materials are sticky, viscous oils at room temperature, and with increasing molecular weight, the texture changes over waxes to tough materials. In the same order, the T_g and T_m changes. For very low molecular weights (M_n = 250 g/mol), the polymer exhibits a T_m of −15°C and a T_g of −98°C, whereas polymers with a molecular weight of M_n = 2000 g/mol exhibit a T_m of 33°C[167] and a T_g of −84°C.[168] As the polymers solidify at room temperature, it is important to preheat them before transfer, e.g., in tank trucks. Furthermore, PTHF is hygroscopic and has to be stored in closed tanks under a nitrogen atmosphere. Under elevated temperatures or the presence of oxygen, the polymers undergo degradation, and antioxidants like amines or pyrocatechols must be added.[169]

7.9.4 Applications

PTHF is mainly applied as an elastomeric compound in polyurethanes or polyesters. The polyurethanes with PTHF soft segments have outstanding properties. They exhibit excellent hydrolytic stability and elastomeric properties, high abrasion resistance, and high water vapor permeability, which render these materials highly interesting for the textile industry. The PTHF-based polyurethanes are known as Spandex or Elastan fibers, which are processed in swimsuits, underwear, and sports clothes. Furthermore, thermoplastic polyurethane elastomers (TPU) are applied in tires, automotive parts, cable sheathing, roof and floor coatings, and colorless films.

7.10 POLYETHERS FROM RENEWABLE RESOURCES

Today, the market for materials derived from natural and renewable resources is increasing tremendously due to climate change, environmental degradation, and resource scarcity. Therefore, scientists and engineers developed new techniques and production processes to isolate renewable raw materials and turn them into commercially interesting products. However, also, the very first polymers used in everyday life were based on natural products like cellulose.

In the following sections, some polyether-based renewable materials will be discussed in detail. Polysaccharides for example are polyacetals, which can be regarded formally as diethers of geminal diols.

7.10.1 Polysaccharides

7.10.1.1 Cellulose

Cellulose is the most abundant organic polymer, with an annual biomass production of about 1.5×10^{12} tons.[170] The molecular structure consists of repeating β-D-glucopyranose molecules that are linked covalently via β-1,4-glycosidic bonds, i.e., an acetal between the equatorial hydroxyl group of C_4 and the C_1 carbon atom, as depicted in Figure 7.16.

Every second glucopyranose molecule is rotated 180° on the plane, resulting in the typical long, linear chain of cellulose. Cellulose can be found in wood, cotton, and other plant fibers but also in some bacteria and algae. Depending on the origin of the cellulose, material properties and chain length can vary. The degree of polymerization (DP) varies between 300 and 1700 for wood pulp and 800 and 10,000 for other plant fibers and bacterial cellulose, whereas regenerate fibers have a DP of 250–500.[170] The structure of cellulose directly suggests the characteristic properties like hydrophilicity (many OH groups), degradability under acidic conditions (acetal linkages), and high chemical functionality. Additionally, the possibility to build up an extensive network based on hydrogen bonds leads to partially crystalline fiber structures. Due to these crystalline domains, cellulose is insoluble in water and most organic solvents; however, mixtures of dimethylacetamide and lithium chloride or DMSO and tetrabutylammonium fluoride dissolve cellulose. Nevertheless, commercially available products are produced in bulk or in the swollen state. In that case, the activation steps like breaking of the hydrogen bonds and the swelling process are very important to achieve quantitative functionalization of the hydroxyl groups.

Cellulose is mainly used as paperboard or paper; however, since the groundbreaking work of the Hyatt Manufacturing Company in 1870, where the first thermoplastic polymer, a material derived from nitrocellulose and camphor named celluloid, was described, cellulose appeared in many other industrial applications.[171] Celluloid is still applied today, e.g., in table tennis balls, in music instruments, and as film materials in the photography and music industry. The development of the viscose

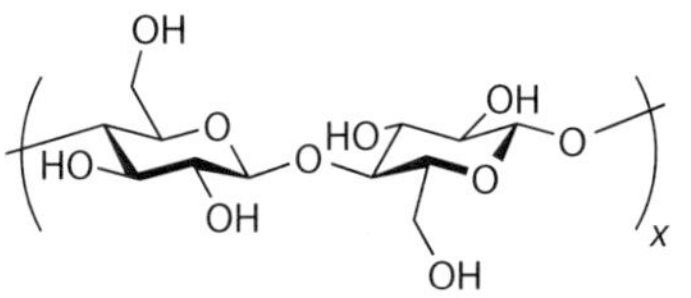

FIGURE 7.16 Chemical structure of cellulose.

TABLE 7.7
Mechanical Properties of Bacterial Cellulose in Comparison to Poly(Propylene) and Poly(Ethylene Terephthalate)

Material	Young's Modulus, GPa	Tensile Strength, MPa	Elongation, %
Bacterial cellulose	15–35	200–300	1.5–2.0
Polypropylene	1.0–1.5	30–40	100–600
Poly(ethylene terephthalate)	3–4	50–70	50–300

Source: Yamanaka, S. and Watanabe, K., Applications of Bacterial Cellulose, in *Cellulosic Polymers—Blends, and Composites*, edited by Gilbert, R.D., 207–215, Hanser Publishers, Munich, New York, Cincinnati, 1994.

process[170] (where cellulose is transformed into cellulose xanthogenate and afterwards, a solution of the xanthogenate is spun into aqueous sodium hydroxide, leading to viscose fibers) enabled the industrial production of rayon (viscose fiber), which applied in textiles and cellophane and is very important for food packaging. Today (in 2003, about 3.2 million tons),[170] cellulose esters are applied in composites and laminates as binders and fillers due to their good mechanical and optical properties.[172] Furthermore, they are used in membranes or other separation media, e.g., water supply, food and beverage processing, and medicine or bioscience research. Applications in manifold filters or superabsorbers can additionally be found.[170,172]

A new emerging research area focuses on the synthesis of cellulose by laboratory bacterial cultures. These enable the adjustment of the molecular weight, the molecular weight distribution, and the supramolecular structure, which determine the properties of the cellulose.[173] The bacterial-derived cellulose shows different characteristics from the plant-derived cellulose, e.g., the degree of polymerization and crystallinity are very high, with 2000 to 8000 repeating units and crystallinity values of 60–90%. The cellulose shows high purity and an extremely high water content of 90% or more.[170] Due to the very high crystallinity of the bacterial cellulose, these materials exhibit excellent mechanical properties, which are summarized in Table 7.7 and illustrate the outstanding possibilities for further development.[174]

7.10.1.2 Starch

Starch is another polysaccharide and serves as the natural energy storage in plants. In contrast to cellulose, the single glucose molecules are attached via α-1,4-glycosidic bonds (Figure 7.17) and therefore can easily hydrolyze. In addition, every 15 to 30 glucose units, the monomers are also connected via α-1,6-glycosidic bonds, leading to a branched structure. The linear structures are called amylose, whereas the branched structure is called amylopectin and makes up about 75–80% of the starch of plants.[175]

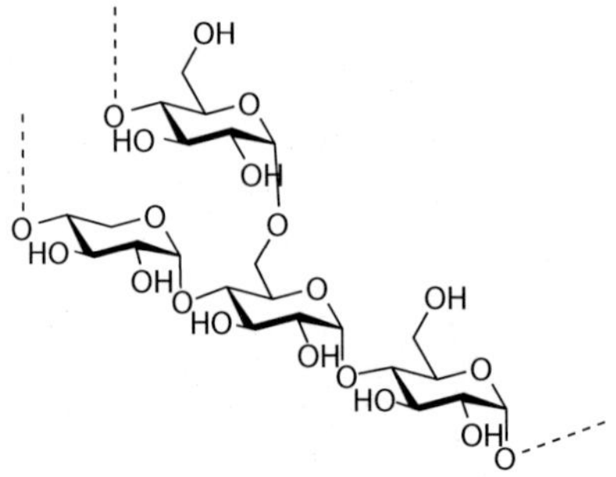

FIGURE 7.17 Chemical structure of starch.

The branched structure enables the fast degradation of starch starting at different branches and thereby releases more energy in a shorter time frame. Furthermore, these branching points lead to efficiently packed parallel left-handed double helices.[176] As already discussed for cellulose, the composition and structure of starch granules of plants differ, leading to different properties and functions of the starch. Therefore, starch is often modified chemically, physically, enzymatically, or genetically to open up new possible applications, reviewed by Kaur et al.[177] Starch is insoluble in water but swells upon heating in aqueous environments, leading to an increase in the viscosity. Therefore, the most important application of starch is the use as thickener or binder in the food industry, e.g., in soups, puddings, sauces, salad dressings, and noodles. Other industrial applications are in papermaking as a stabilizer, in clothing to stiffen collars or shirts, or in the construction industry as adhesive or glue.[176] Recent applications are the use as delivery vehicles for active proteins and biodegradable films.[178] The new developments illustrate the possibility of starch-based materials, which are not costly, are widely available, and show great diversity for water-based systems and processes.

7.10.1.3 Hydroxyethyl Starch

Hydroxyethyl starch (HES) is a man-made starch derivative with hydroxyethyl groups at carbon position 2, 3, or 6 and is applied as a blood plasma substitute. Amylopectin derived from potato or corn is functionalized through hydroxyethylation using EO in the presence of an alkaline catalyst[179] to ensure water solubility and to prevent enzymatic degradation through amylase. Different molecular weights (70–450 kDa) and degrees of hydroxyethyl substitution (0.4–0.7) influence the metabolic rate in vivo. The higher the degree of substitution and the higher the C_2/C_6 ratio, the slower is the metabolization in vivo.[180] After severe blood loss due to a trauma or surgery, HES is administered to regulate the oncotic pressure. However, HES can cause anaphylactic reactions like hypersensitivity or bradycardia, or renal failures. Additionally, high-molecular-weight HES can accumulate in the body after repeated infusion. Therefore, in 2013, the European Medicines Agency and the American FDA suggested abandoning the use of HES for patients at risk, e.g., patients with sepsis or burns.[181]

This example shows the possibilities new materials offer but also the problems that can appear if possible consequences and effects are not investigated sufficiently.

7.10.1.4 Lignin

Lignin (Figure 7.18) is one of the most abundant renewable substances besides cellulose and can be extracted from plants, representing about 15–30% of their total mass.[182] It is a nonpreferred by-product of the paper industry, produced in millions of tons; however, only 1–2% of lignin is applied in industry.[183] One major problem with the application of lignin is the difficult isolation from plants and the varying structure depending on its origin, composition, and broad molecular weight distribution. Raw lignin is typically insoluble in most solvents and needs to be treated or depolymerized before use.

The principal function of lignin is the stabilization of plant tissues, enabling growth above ground level. The structure of lignin is based on three basic phenol derivatives, named monolignols, that make up all types of lignins found in nature but vary in their total composition depending on the type of plant. These compounds are *p*-coumaryl alcohol, coniferyl alcohol, and sinapyl alcohol (Figure 7.19), which are connected via aromatic and aliphatic ether bonds to build up a hyper-branched polyether.

These bonds are formed during the peroxidase-mediated dehydrogenation of the monolignol units.[182] Due to the phenolic and aliphatic hydroxyl groups, many further modifications are possible, like alkylation, dealkylation, oxyalkylation, amination, carboxylation, acylation, halogenations, oxidation and reduction, sulfonation, silylation, and phosphorylation.[184]

Different processes are used to extract lignin from wood, which influence the properties of the obtained lignin. In the Kraft process, wood is treated with Na_2S/NaOH solution at elevated

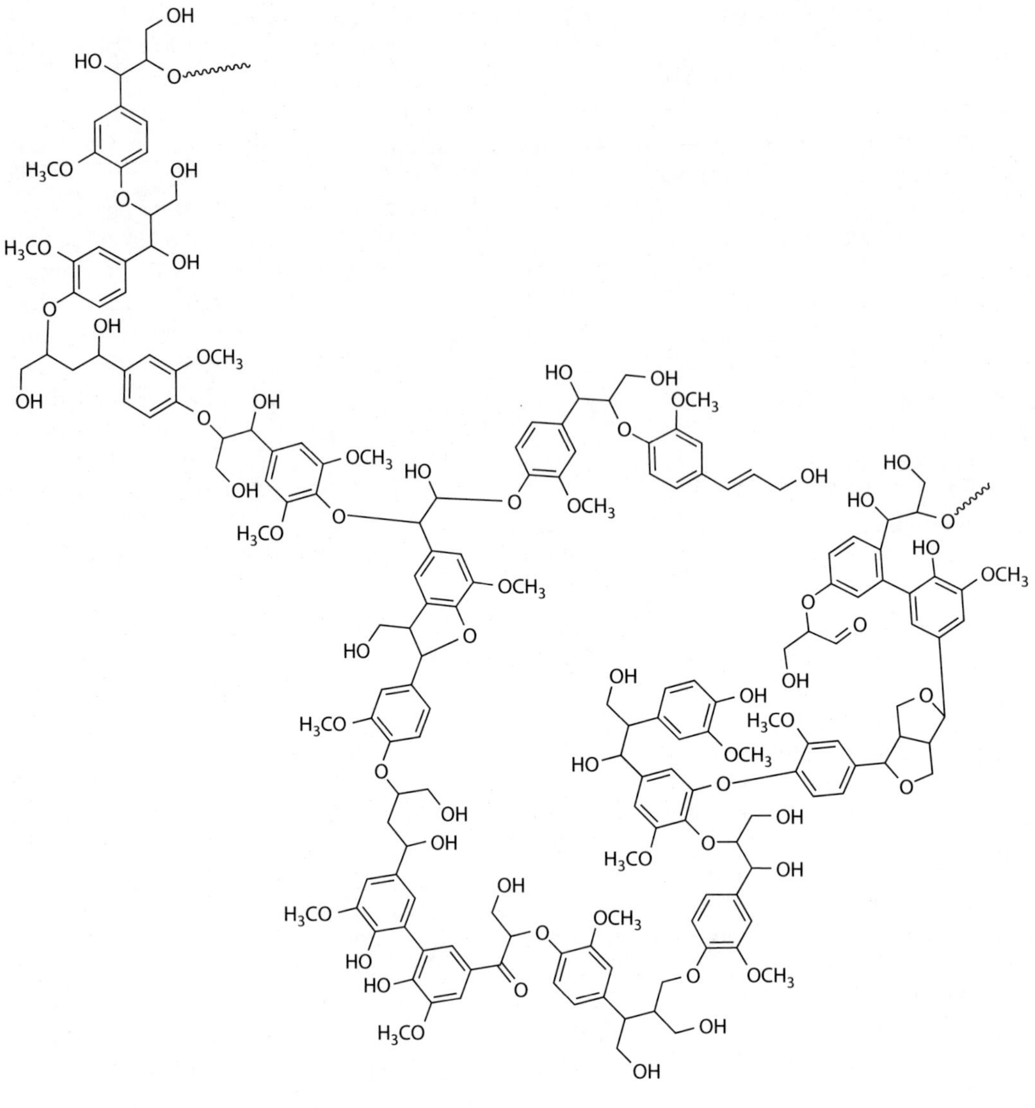

FIGURE 7.18 Possible structure of a lignin polymer.

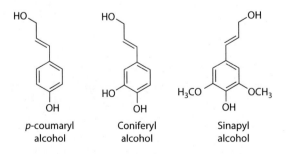

FIGURE 7.19 Three monolignols for the synthesis of lignin.

Polyethers

temperatures (155–175°C) for several hours, which allows the separation of the solid cellulose and the black liquid lignin. Kraft lignin has lower molecular weight than the original lignin and is a hydrophobic, water-insoluble material.[182] The traditional sulfite process delivers lignin sulfonates using calcium, magnesium, or ammonium salts of sulfurous acids (sulfites or bisulfites). The extracted lignin carries both hydrophobic as well as hydrophilic properties and is water soluble. The molecular weights are higher than in the Kraft process as the procedure is less aggressive.[182,185] Furthermore, eight different organic solvent-based procedures are used. In this case, lignin is extracted from biomass using a mixture of organic solvent and water at elevated temperatures and pressure. The produced organosolv lignin is less modified than Kraft lignin and is mostly sulfur-free, avoiding the characteristic odor of Kraft lignin and ligninsulfonates.[182] Another process uses steam at high temperatures (180–200°C) and pressures and subsequent decompression in the presence of chemicals to extract lignin from wood. This steam-exploded lignin is water soluble and exhibits small amounts of carbohydrate and wood-extract impurities.[182]

Applications of lignin are rare at the moment, and most of the extracted material is directly burned at the factories to recover energy.[186] Only 2% of the material is applied in industrial or agricultural uses (e.g., as binders for bricks, ceramics, or road dust; as heavy-metal chelating agents; as emulsifiers; or as adhesives for chipboards).[182,183,185,187] A recent study uses lignin nanocarriers as a potential drug carrier for agricultural applications.[183]

Lignin is an underestimated and unconsidered raw material at the moment. But the enormous unused reservoirs demand for new technologies to separate lignin from cellulose and to provide more homogeneous materials, which would make lignin more interesting for commercial applications.

7.10.2 Furan-Based Polymers

Furfural, the precursor for furan derivatives, is a renewable material deriving from hemicelluloses. About 200,000 tons are produced annually through heating of hemicelluloses in sulfuric acid at fairly high temperatures, leading to first a hydrolysis and then a cyclodehydration.[188] The obtained furfural can be transformed to many furfuryl-based compounds like furfuryl alcohol, furfuryl amine, furoic acid, or furan and THF (Figure 7.20).

Furfuryl-based polymers can carry the hetero cycle in the main chain or in the side chains. Poly(furfuryl alcohol) (PFA) is the most prominent furfural-derived polymer, carries the furfuryl group in the main chain, and is polymerized via acid-catalyzed polycondensation leading to

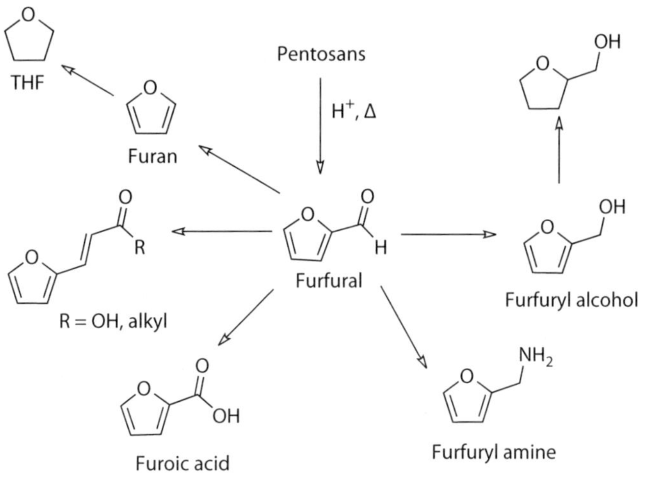

FIGURE 7.20 Furfuryl-based chemicals derived from renewable pentosans. (From Lichtenthaler, F.W., *Acc. Chem. Res.*, 35, 2002.)

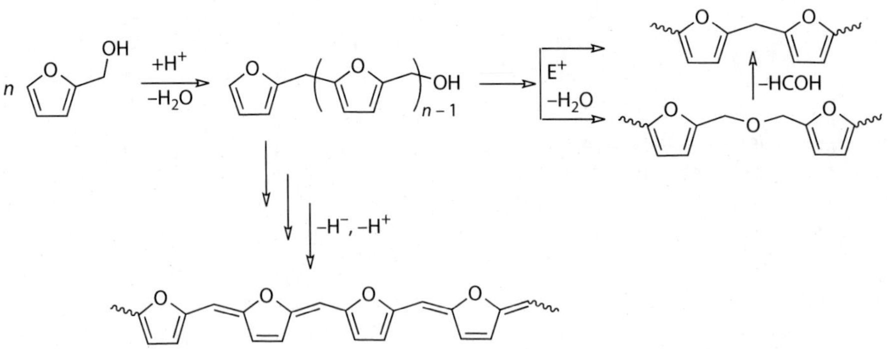

FIGURE 7.21 Polycondensation of furfuryl alcohol and formation of conjugated sequences.

black cross-linked materials. Choura et al. investigated the mechanism of the polycondensation and explained the black color through the formation of conjugated sequences along the chain after the loss of hydride ions followed by the deprotonation of the formed carbenium ions. Furthermore, these structures can undergo cross-linking reactions, explaining the insolubility of the arising resins (Figure 7.21).[189]

PFA is used as an adhesive and binder.[190] Additionally, it is used as nanoporous carbons, glassy carbons, and polymer nanocomposites due to compatibility with many organic polymers and inorganic materials and the high carbon yield when pyrolized. These carbons are applied in a broad range of applications, e.g., as adsorbents, separation membranes, catalysts, and electrodes of fuel cells or lithium-ion batteries.[191] Furthermore, PFA is used as noncorrosive coating in industrial floors, tanks, and reaction vessels due to the excellent resistance against strong acids, bases, and many organic solvents as well as the high thermal stability and fire resistance.[192]

7.11 POLY(PHENYLENE ETHER)

Poly(phenylene ether) (PPE), also known as poly(phenylene oxide) (PPO) or poly(oxy-2,6-dimethyl-1,4-phenylene) (Figure 7.22), is a high-temperature thermoplastic. It is mainly used in blends with PS due to the difficult processing of the homopolymer.

7.11.1 History

Hay and coworkers discovered PPE in 1965, when they observed an exothermic reaction together with an increase in viscosity during the oxidation of 2,6-dimethylphenol at room temperature.[193] Commercialization by General Electric followed in 1960. In those days, the major priority was the development of an economical process for the synthesis of the monomer 2,6-dimethylphenol as this was only available as a side product from coal tar distillation.[193] The breakthrough of PPE was the introduction of NORYL resins in 1966. These resins are homogeneous blends of PPE with PS in different ratios, which enable the adjustment of the polymer properties, resulting in easier processability and additionally lowering the price of the material.[194] Later, heterogeneous blends of PPE with nylon were introduced.[195]

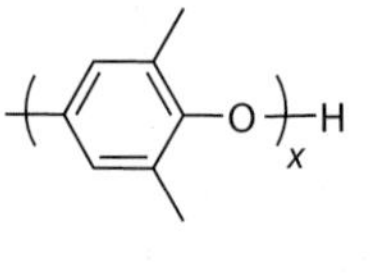

FIGURE 7.22 Chemical structure of poly(phenylene ether).

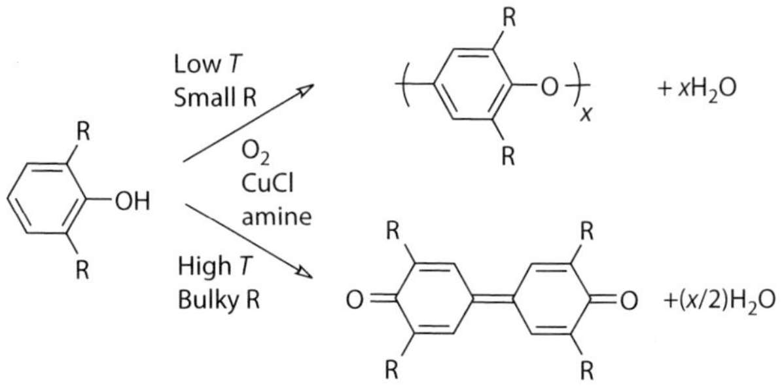

Polyethers

FIGURE 7.23 Possible reactions of 2,6-disubstituted phenols under polymerization conditions.

7.11.2 Polymerization

Aromatic polyethers in general can be obtained by oxidative coupling polymerization of 2,6-disubstituted phenols.[196,197] Oxygen is passed through a solution of the phenol, an amine, and a catalytic amount of copper(I) salt in an organic solvent, like toluene or methanol at 25–50°C. A side reaction is the formation of diphenoquinones, which is more prominent in the case of bulky substituents (Figure 7.23).[198]

A major problem during the polymerization is the formation of water as a coproduct, which hydrolyzes and deactivates the catalyst. Therefore, N,N-di-*tert*-butylethylene diamine (DBEDA) is used, which forms a very hydrolytically stable complex with copper and gives a very active catalyst.[199] Another possibility to synthesize aromatic polyethers is the Ullmann condensation with 4-bromophenols; however, this process is too expensive for industrial purposes.[193] Further functionalization of the polymers is possible via metallation with butyl lithium; halogenations on either the nucleus or the methyl group; or electrophilic substitution reactions with halogens, sulfuric acid, etc.[193,200]

7.11.3 Properties

PPE is an amorphous material with special thermal properties as the T_g is 208°C. Therefore, PPE has to be melt-processed at very high temperatures, which leads to oxidative degradation reactions. As PPE is miscible with PS in all ratios, almost all material is blended with PS, commercially available as NORYL resins. These blends improve the processability (injection molding or extrusion) of the material and reduce costs. NORYL resins exhibit excellent resistance to water, acids, and bases; show high heat and flame resistance; and have dimensional stability.[73] Through blending with other materials like poly(diphenyl siloxane), polycarbonate, elemental sulfur, or nylon, a broad spectrum of properties can be covered.[193]

7.11.4 Applications

PPE/PS blends are applied in electronic areas, like computers or keyboard frames; in large articles, like panels, or hospital and office furniture; and in automotive applications, like instrument panels, consoles, or wheel covers. As the materials show very low water absorption and are resistant against acids and bases, they are also used in medicine, e.g., for sterilizable instruments.[162]

REFERENCES

1. Williamson, A. *Philos. Mag. Ser. 3*, **1850**, *37*, (251), 350–356.
2. Brocas, A.-L.; Mantzaridis, C.; Tunc, D.; Carlotti, S. *Prog. Polym. Sci.*, **2013**, *38*, (6), 845–873.
3. Staudinger, H.; Lohmann, H. *Justus Liebigs Ann. Chem.*, **1933**, *505*, (1), 41–51.

4. Flory, P. J. *J. Am. Chem. Soc.*, **1940**, *62*, (6), 1561–1565.
5. Penczek, S.; Cyprik, M.; Duda, A.; Kubisa, P.; Slomkowski, S. *Prog. Polym. Sci.*, **2007**, *32*, (2), 247–282.
6. Tonhauser, C.; Frey, H. *Macromol. Rapid. Commun.*, **2010**, *31*, (22), 1938–1947.
7. Pearson, R. G. *J. Am. Chem. Soc.*, **1963**, *85*, (22), 3533–3539.
8. Hsieh, H. L.; Quirk, R. P. 1996. *Anionic Polymerization: Principles and Practical Applications.* New York: Marcel Dekker.
9. de Lucas, A.; Rodríguez, L.; Pérez-Collado, M.; Sánchez, P.; Rodríguez, J. F. *Polym. Int.*, **2002**, *51*, (10), 1066–1071.
10. Boileau, S.; Deffieux, A.; Lassalle, D.; Menezes, F.; Vidal, B. *Tetrahedron Lett.*, **1978**, *19*, (20), 1767–1770.
11. Deffieux, A.; Boileau, S. *Polymer*, **1977**, *18*, (10), 1047–1050.
12. Goethals, E. J. 1984. Cationic polymerization and related processes. *Proceedings of 6th Internat. Symposium held in Ghent, Belgium, August 30–September 2, 1983.* London: Academic Press.
13. Szwarc, M. *Adv. Polym. Sci.*, **1965**, *4*, 1–65.
14. Kubisa, P. 1996. Cationic polymerization of heterocyclics. In *Cationic Polymerizations*, Matyjaszewski, K., Ed., 437–553. New York: Marcel Dekker.
15. Hans, M.; Keul, H.; Moeller, M. *Polymer*, **2009**, *50*, (5), 1103–1108.
16. Labbé, A.; Carlotti, S.; Billouard, C.; Desbois, P.; Deffieux, A. *Macromolecules*, **2007**, *40*, (22), 7842–7847.
17. Carlotti, S.; Billouard, C.; Gautriaud, E.; Desbois, P.; Deffieux, A. *Macromol. Symp.*, **2005**, *226*, (1), 61–68.
18. Gervais, M.; Brocas, A.-L.; Cendejas, G.; Deffieux, A.; Carlotti, S. *Macromolecules*, **2010**, *43*, (4), 1778–1784.
19. Meyer, J.; Keul, H.; Möller, M. *Macromolecules*, **2011**, *44*, (11), 4082–4091.
20. Müller, S. S.; Moers, C.; Frey, H. *Macromolecules*, **2014**, *47*, (16), 5492–5500.
21. Brocas, A.-L.; Cendejas, G.; Caillol, S.; Deffieux, A.; Carlotti, S. *J. Polym. Sci., Part A: Polym. Chem.*, **2011**, *49*, (12), 2677–2684.
22. Labbé, A.; Brocas, A.-L.; Ibarboure, E.; Ishizone, T.; Hirao, A.; Deffieux, A.; Carlotti, S. *Macromolecules*, **2011**, *44*, (16), 6356–6364.
23. Gervais, M.; Brocas, A.-L.; Deffieux, A.; Ibarboure, E.; Carlotti, S. *Pure Appl. Chem.*, **2012**, *84*, (10), 2103–2111.
24. Pelegri-O'Day, E. M.; Lin, E.-W.; Maynard, H. D. *J. Am. Chem. Soc.*, **2014**, *136*, (41), 14323–14332.
25. Dingels, C.; Schömer, M.; Frey, H. *Chem. unserer Zeit*, **2011**, *45*, (5), 338–349.
26. Staudinger, H.; Schweitzer, O. *Ber. dtsch. Chem. Ges. A/B*, **1929**, *62*, (8), 2395–2405.
27. Matyjaszewski, K.; Möller, M., Eds. 2012. *Polymer Science: A Comprehensive Reference.* Oxford, UK: Elsevier.
28. McClelland, C. P.; Bateman, R. L. *Chem. Eng. News*, **1945**, *23*, (3), 247–251.
29. Abuchowski, A.; van Es, T.; Palczuk, N. C.; Davis, F. F. *J. Biol. Chem.*, **1977**, *252*, 3578–3581.
30. Abuchowski, A.; McCoy, J. R.; Palczuk, N. C., van Es, T.; Davis, F. F. *J. Biol. Chem.*, **1977**, *252*, 3582–3586.
31. Alconcel, S. N. S.; Baas, A. S.; Maynard, H. D. *Polym. Chem.*, **2011**, *2*, (7), 1442.
32. Koyama, Y.; Umehara, M.; Mizuno, A.; Itaba, M.; Yasukouchi, T.; Natsume, K.; Suginaka, A. *Bioconjug. Chem.*, **1996**, *7*, (3), 298–301.
33. Taton, D.; Le Borgne, A.; Sepulchre, M.; Spassky, N. *Macromol. Chem. Phys.*, **1994**, *195*, (1), 139–148.
34. Pielichowski, K.; Flejtuch, K. *Polym. Adv. Technol.*, **2002**, *13*, (10–12), 690–696.
35. Kjellander, R.; Florin, E. *J. Chem. Soc., Faraday Trans. 1*, **1981**, *77*, (9), 2053.
36. Saeki, S.; Kuwahara, N.; Nakata, M.; Kaneko, M. *Polymer*, **1976**, *17*, (8), 685–689.
37. Fruijtier-Pölloth, C. *Toxicology*, **2005**, *214*, (1–2), 1–38.
38. Immordino, M. L.; Dosip, F.; Cattel, L. *Int. J. Nanomed.*, **2006**, *1*, (3), 297–315.
39. Müller, S. S.; Dingels, C.; Hofmann, A. M.; Frey, H. 2013. Polyether-based lipids synthesized with an epoxide construction kit: multivalent architectures for functional liposomes. In *Tailored Polymer Architectures for Pharmaceutical and Biomedical Applications*, Scholz, C., Kressler, J., Eds., 11–25. Washington, DC: American Chemical Society.
40. Pfister, D.; Morbidelli, M. *J. Control. Release*, **2014**, *180*, 134–149.
41. Fenton, D. E.; Parker, J. M.; Wright, P. V. *Polymer*, **1973**, *14*, (11), 589.
42. Wilms, D.; Schömer, M.; Wurm, F.; Hermanns, M. I.; Kirkpatrick, C. J.; Frey, H. *Macromol. Rapid. Commun.*, **2010**, *31*, (20), 1811–1815.
43. Agrawal, R. C.; Pandey, G. P. *J. Phys. D: Appl. Phys.*, **2008**, *41*, (22), 223001.
44. Obermeier, B.; Wurm, F.; Mangold, C.; Frey, H. *Angew. Chem. Int. Ed. Engl.*, **2011**, *50*, (35), 7988–7997.
45. Knop, K.; Hoogenboom, R.; Fischer, D.; Schubert, U. S. *Angew. Chem.*, **2010**, *122*, (36), 6430–6452.

46. Lee, J. S.; Feijen, J. *J. Control. Release*, **2012**, *161*, (2), 473–483.
47. Locatelli, E.; Comes Franchini, M. *J. Nanopart. Res.*, **2012**, *14*, (12), 1e17.
48. Wilms, V. S.; Frey, H. *Polym. Int.*, **2013**, *62*, (6), 849–859.
49. Obermeier, B.; Wurm, F.; Frey, H. *Macromolecules*, **2010**, *43*, (5), 2244–2251.
50. Wurm, F.; Nieberle, J.; Frey, H. *Macromolecules*, **2008**, *41*, (6), 1909–1911.
51. Mangold, C.; Dingels, C.; Obermeier, B.; Frey, H.; Wurm, F. *Macromolecules*, **2011**, *44*, (16), 6326–6334.
52. Reuss, V. S.; Obermeier, B.; Dingels, C.; Frey, H. *Macromolecules*, **2012**, *45*, (11), 4581–4589.
53. Reuss, V. S.; Werre, M.; Frey, H. *Macromol. Rapid Commun.*, **2012**, *33*, (18), 1556–1561.
54. Tonhauser, C.; Alkan, A.; Schömer, M.; Dingels, C.; Ritz, S.; Mailänder, V.; Frey, H.; Wurm, F. R. *Macromolecules*, **2013**, *46*, (3), 647–655.
55. Herzberger, J.; Kurzbach, D.; Werre, M.; Fischer, K.; Hinderberger, D.; Frey, H. *Macromolecules*, **2014**, *47*, 7679–7690, 141103162334009.
56. Fitton, A. O.; Hill, J.; Jane, D. E.; Millar, R. *Synthesis*, **1987**, *1987*, (12), 1140–1142.
57. Mangold, C.; Wurm, F.; Obermeier, B.; Frey, H. *Macromol. Rapid. Commun.*, **2010**, *31*, (3), 258–264.
58. Li, Z.; Chau, Y. *Bioconjug. Chem.*, **2009**, *20*, (4), 780–789.
59. Huang, J.; Li, Z.; Xu, X.; Ren, Y.; Huang, J. *J. Polym. Sci., Part A: Polym. Chem.*, **2006**, *44*, (11), 3684–3691.
60. Dimitrov, P.; Hasan, E.; Rangelov, S.; Trzebicka, B.; Dworak, A.; Tsvetanov, C. *Polymer*, **2002**, *43*, (25), 7171–7178.
61. Mangold, C.; Wurm, F.; Obermeier, B.; Frey, H. *Macromolecules*, **2010**, *43*, (20), 8511–8518.
62. Yoshihara, C.; Shew, C.-Y.; Ito, T.; Koyama, Y. *Biophys. J.*, **2010**, *98*, (7), 1257–1266.
63. Wurm, F.; Schüle, H.; Frey, H. *Macromolecules*, **2008**, *41*, (24), 9602–9611.
64. Lee, A.; Lundberg, P.; Klinger, D.; Lee, B. F.; Hawker, C. J.; Lynd, N. A. *Polym. Chem.*, **2013**, *4*, (24), 5735.
65. Mangold, C.; Obermeier, B.; Wurm, F.; Frey, H. *Macromol. Rapid. Commun.*, **2011**, *32*, (23), 1930–1934.
66. Obermeier, B.; Frey, H. *Bioconjug. Chem.*, **2011**, *22*, (3), 436–444.
67. Thompson, M. S.; Vadala, T. P.; Vadala, M. L.; Lin, Y.; Riffle, J. S. *Polymer*, **2008**, *49*, (2), 345–373.
68. Kim, I.; Ahn, J.-T.; Ha, C. S.; Yang, C. S.; Park, I. *Polymer*, **2003**, *44*, (11), 3417–3428.
69. Peretti, K. L.; Ajiro, H.; Cohen, C. T.; Lobkovsky, E. B.; Coates, G. W. *J. Am. Chem. Soc.*, **2005**, *127*, (33), 11566–11567.
70. Lai, J.; Trick, G. S. *J. Polym. Sci. A-1 Polym. Chem.*, **1970**, *8*, (9), 2339–2350.
71. Madorsicy, S. L.; Straus, S. *J. Polym. Sci.*, **1959**, *36*, (130), 183–194.
72. Mortensen, K.; Schwahn, D.; Janssen, S. *Phys. Rev. Lett.*, **1993**, *71*, (11), 1728–1731.
73. Tsvetanov, C. B. 1997. Polyethers. In *Handbook of Thermoplastics*, Olabisi, O., Adewale, K., Eds., 575–598. New York: Marcel Dekker.
74. Alexandridis, P. *Curr. Opin. Colloid Interface Sci.*, **1997**, *2*, 478–489.
75. Kabanov, A. V.; Batrakova, E. V.; Alakhov, V. Y. *J. Control. Release*, **2002**, *82*, 189–212.
76. Kabanov, A. V.; Nazarova, I. R.; Astafieva, I. V.; Batrakova, E. V.; Alakhov, V. Y.; Yaroslavov, A. A.; Kabanov, V. A. *Macromolecules*, **1995**, *28*, (7), 2303–2314.
77. Nagarajan, R. *Colloids Surf., B*, **1999**, *16*, (1–4), 55–72.
78. Kozlov, M. Y.; Melik-Nubarov, N. S.; Batrakova, E. V.; Kabanov, A. V. *Macromolecules*, **2000**, *33*, (9), 3305–3313.
79. Batrakova, E. V.; Kabanov, A. V. *J. Control. Release*, **2008**, *130*, (2), 98–106.
80. Batrakova, E. V.; Li, S.; Elmquist, W. F.; Miller, D. W.; Alakhov, V. Y.; Kabanov, A. V. *Br. J. Cancer*, **2001**, *85*, (12), 1987–1997.
81. Venne, A.; Li, S.; Mandeville, R.; Kabanov, A.; Alakhov, V. *Cancer Res*, **1996**, *15*, (56), 3626–3629.
82. Zhang, H.; Grinstaff, M. W. *Macromol. Rapid. Commun.*, **2014**, *35*, (22), 1906–1924.
83. Yang, F.; Hanna, M. A.; Sun, R. *Biotechnol. Biofuels*, **2012**, *5*, 13.
84. Sandler, S. R.; Berg, F. R. *J. Polym. Sci. A-1 Polym. Chem.*, **1966**, *4*, (5), 1253–1259.
85. Vandenberg, E. J. *J. Polym. Sci. Polym. Chem. Ed.*, **1985**, *23*, (4), 915–949.
86. Kim, Y. H.; Webster, O. W. *J. Am. Chem. Soc.*, **1990**, *112*, (11), 4592–4593.
87. Tokar, R.; Kubisa, P.; Penczek, S.; Dworak, A. *Macromolecules*, **1994**, *27*, (2), 320–322.
88. Dworak, A.; Walach, W.; Trzebicka, B. *Macromol. Chem. Phys.*, **1995**, *196*, (6), 1963–1970.
89. Sunder, A.; Hanselmann, R.; Frey, H.; Mülhaupt, R. *Macromolecules*, **1999**, *32*, (13), 4240–4246.
90. Sunder, A.; Frey, H.; Mülhaupt, R. *Macromol. Symp.*, **2000**, *153*, (1), 187–196.
91. Hawker, C. J.; Lee, R.; Frechet, J. M. J. *J. Am. Chem. Soc.*, **1991**, *113*, (12), 4583–4588.
92. Hölter, D.; Burgath, A.; Frey, H. *Acta Polym.*, **1997**, *48*, (1), 30–35.
93. Wilms, D.; Wurm, F.; Nieberle, J.; Böhm, P.; Kemmer-Jonas, U.; Frey, H. *Macromolecules*, **2009**, *42*, (9), 3230–3236.

94. Kainthan, R. K.; Muliawan, E. B.; Hatzikiriakos, S. G.; Brooks, D. E. *Macromolecules*, **2006**, *39*, (22), 7708–7717.
95. Kainthan, R. K.; Janzen, J.; Levin, E.; Devine, D. V.; Brooks, D. E. *Biomacromolecules*, **2006**, *7*, (3), 703–709.
96. Sunder, A.; Türk, H.; Haag, R.; Frey, H. *Macromolecules*, **2000**, *33*, (21), 7682–7692.
97. Calderón, M.; Quadir, M. A.; Sharma, S. K.; Haag, R. *Adv. Mater.*, **2010**, *22*, (2), 190–218.
98. Wilms, D.; Stiriba, S.-E.; Frey, H. *Acc. Chem. Res.*, **2010**, *43*, (1), 129–141.
99. Sunder, A.; Krämer, M.; Hanselmann, R.; Mülhaupt, R.; Frey, H. *Angew. Chem. Int. Ed.*, **1999**, *38*, (23), 3552–3555.
100. Krämer, M.; Stumbé, J.-F.; Türk, H.; Krause, S.; Komp, A.; Delineau, L.; Prokhorova, S.; Kautz, H.; Haag, R. *Angew. Chem. Int. Ed.*, **2002**, *41*, (22), 4252–4256.
101. Mecking, S.; Thomann, R.; Frey, H.; Sunder, A. *Macromolecules*, **2000**, *33*, (11), 3958–3960.
102. Balz, M.; Barriau, E.; Istratov, V.; Frey, H.; Tremel, W. *Langmuir*, **2005**, *21*, (9), 3987–3991.
103. Dworak, A.; Panchev, I.; Trzebicka, B.; Walach, W. *Polym. Bull.*, **1998**, *40*, (4–5), 461–468.
104. Dworak, A.; Panchev, I.; Trzebicka, B.; Walach, W. *Macromol. Symp.*, **2000**, *153*, (1), 233–242.
105. Walach, W.; Kowalczuk, A.; Trzebicka, B.; Dworak, A. *Macromol. Rapid. Commun.*, **2001**, *22*, (15), 1272–1277.
106. Hans, M.; Gasteier, P.; Keul, H.; Moeller, M. *Macromolecules*, **2006**, *39*, (9), 3184–3193.
107. Wurm, F.; Dingels, C.; Frey, H.; Klok, H.-A. *Biomacromolecules*, **2012**, *13*, (4), 1161–1171.
108. Hofmann, A. M.; Wurm, F.; Frey, H. *Macromolecules*, **2011**, *44*, (12), 4648–4657.
109. Moers, C.; Nuhn, L.; Wissel, M.; Stangenberg, R.; Mondeshki, M.; Berger-Nicoletti, E.; Thomas, A.; Schaeffel, D.; Koynov, K.; Klapper, M.; Zentel, R.; Frey, H. *Macromolecules*, **2013**, *46*, (24), 9544–9553.
110. Thomas, A.; Bauer, H.; Schilmann, A.-M.; Fischer, K.; Tremel, W.; Frey, H. *Macromolecules*, **2014**, *47*, (14), 4557–4566.
111. Gunkel, G.; Weinhart, M.; Becherer, T.; Haag, R.; Huck, W. T. S. *Biomacromolecules*, **2011**, *12*, (11), 4169–4172.
112. Weinhart, M.; Grunwald, I.; Wyszogrodzka, M.; Gaetjen, L.; Hartwig, A.; Haag, R. *Chem. Asian J.*, **2010**, *5*, (9), 1992–2000.
113. Thomas, A.; Wolf, F. K.; Frey, H. *Macromol. Rapid. Commun.*, **2011**, *32*, (23), 1910–1915.
114. Thomas, A.; Niederer, K.; Wurm, F.; Frey, H. *Polym. Chem.*, **2013**, *5*, (3), 899.
115. Haamann, D.; Keul, H.; Klee, D.; Möller, M. *Macromolecules*, **2010**, *43*, (15), 6295–6301.
116. Weber, M.; Bujotzek, A.; Andrae, K.; Weinhart, M.; Haag, R. *Mol. Simul.*, **2011**, *37*, (11), 899–906.
117. Thomas, A.; Müller, S. S.; Frey, H. *Biomacromolecules*, **2014**, *15*, 1935–1954.
118. Siegers, C.; Biesalski, M.; Haag, R. *Chem. Eur. J.*, **2004**, *10*, (11), 2831–2838.
119. Atkinson, J. L.; Vyazovkin, S. *Macromol. Chem. Phys.*, **2011**, *212*, (19), 2103–2113.
120. Calahorra, E.; Cortazar, M.; Guzmán, G. M. *J. Polym. Sci. B Polym. Lett. Ed.*, **1985**, *23*, (5), 257–260.
121. Klein, R.; Schüll, C.; Berger-Nicoletti, E.; Haubs, M.; Kurz, K.; Frey, H. *Macromolecules*, **2013**, *46*, (22), 8845–8852.
122. Butlerow, A. *Ann. Chem. Pharm.*, **1859**, *111*, (2), 242–252.
123. Staudinger, H.; Lüthy, M. *Helv. Chim. Acta*, **1925**, *8*, (1), 41–64.
124. Trautz, M.; Ufer, E. *J. Prakt. Chem.*, **1926**, *113*, 105–136.
125. Walker, F. *J. Am. Chem. Soc.*, **1933**, *55*, (7), 2821–2826.
126. Carruthers, J. E.; Norrish, R. G. W. *Trans. Faraday Soc.*, **1936**, *32*, 195–208.
127. Masamoto, J. *Prog. Polym. Sci.*, **1993**, *18*, (1), 1–84.
128. Walling, C.; Brown, F.; Bartz, K. *Copolymers*, March 27, 1962.
129. Brown, N. *J. Macromol. Sci. Part A: Chem.*, **1967**, *1*, (2), 209–230.
130. Künzel, V. E.; Giefer, A.; Kern, W. *Makromol. Chem.*, **1966**, *96*, (1), 17–29.
131. Masamoto, J. 1994. Polyacetal resin composition. JP19920281188 19921020, October 5.
132. Vogl, O. *J. Macromol. Sci.: Part A—Chem.*, **1975**, *9*, (5), 663–685.
133. Jaacks, V. V.; Kern, W. *Makromol. Chem.*, **1963**, *62*, (1), 1–17.
134. Kern, W.; Jaacks, V. *J. Polym. Sci.*, **1960**, *48*, (150), 399–404.
135. Meerwein, H.; Delfs, D.; Morschel, H. *Angew. Chem.*, **1960**, *72*, (24), 927–934.
136. Weissermel, K.; Fischer, E.; Gutweiler, K.; Hermann, H. D.; Cherdron, H. *Angew. Chem. Int. Ed. Engl.*, **1967**, *6*, (6), 526–533.
137. Mück, K.-F. 1999. Physical constants of poly(oxymethylene). In *Polymer Handbook*, Brandrup, J.; Immergut, E. H.; Grulke, E. A., Eds., V/97–V/112. New York: Wiley.
138. Wilski, H. *Makromol. Chem.*, **1971**, *150*, (1), 209–222.
139. Zhao, R. *INJ Summer*, **2005**, *14*, (2), 19–24.

140. Kaiser, W. 2011. *Kunststoffchemie für Ingenieure: Von der Synthese bis zur Anwendung.* München: Hanser.
141. DeLassus, P. T.; Whiteman, N. F. 1999. Physical and mechanical properties of some important polymers. In *Polymer Handbook*, Brandrup, J., Immergut, E. H., Grulke, E. A., Eds., V/159–V/169. New York: Wiley.
142. Vieweg, R. 1975. *Kunststoff Handbuch, Grundlagen.* München: Carl Hanser Verlag.
143. Kumaraswamy, G.; Surve, N. S.; Mathew, R.; Rana, A.; Jha, S. K.; Bulakh, N. N.; Nisal, A. A.; Ajithkumar, T. G.; Rajamohanan, P. R.; Ratnagiri, R. *Macromolecules*, **2012**, *45*, (15), 5967–5978.
144. Samyn, P.; Driessche, I.; Schoukens, G. *J. Polym. Res.*, **2007**, *14*, (5), 411–422.
145. Wolk, J. L.; Sprecher, M.; Basch, H.; Hoz, S. *Org. Biomol. Chem.*, **2004**, *2*, (7), 1065–1069.
146. Farthing, A. C.; Reynolds, R. J. William. *J. Polym. Sci.*, **1954**, *12*, (1), 503–507.
147. Farthing, A. C. *J. Chem. Soc.*, **1955**, 3648.
148. Rose, J. B. *J. Chem. Soc.*, **1956**, 542.
149. Vairon, J.-P.; Spassky, N. 1996. Industrial cationic polymerizations: An overview. In *Cationic Polymerizations*, Matyjaszewski, K., Ed., 683–750. New York: Marcel Dekker.
150. Vandenberg, E. J.; Mullis, J. C.; Juvet, R. S. *J. Polym. Sci., Part A: Polym. Chem.*, **1989**, *27*, (9), 3083–3112.
151. Christ, E.-M.; Müller, S. S.; Berger-Nicoletti, E.; Frey, H. *J. Polym. Sci. Part A: Polym. Chem.*, **2014**, *52*, (19), 2850–2859.
152. Burkhard, J. A.; Wuitschik, G.; Rogers-Evans, M.; Müller, K.; Carreira, E. M. *Angew. Chem.*, **2010**, *122*, (48), 9236–9251.
153. Hirano, T.; Nakayama, S.; Tsuruta, T. *Makromol. Chem.*, **1975**, *176*, (6), 1897–1900.
154. Bailey, W. J. *Polym. J.*, **1985**, *17*, (1), 85–95.
155. Kambara, S.; Hatano, M. *J. Polym. Sci.*, **1958**, *27*, (115), 584–586.
156. Bednarek, M.; Biedron, T.; Helinski, J.; Kaluzynski, K.; Kubisa, P.; Penczek, S. *Macromol. Rapid Commun.*, **1999**, *20*, (7), 369–372.
157. Magnusson, H.; Malmström, E.; Hult, A. *Macromol. Rapid Commun.*, **1999**, *20*, (8), 453–457.
158. Bednarek, M. *Polym. Int.*, **2003**, *52*, (10), 1595–1599.
159. Mai, Y.; Zhou, Y.; Yan, D.; Lu, H. *Macromolecules*, **2003**, *36*, (25), 9667–9669.
160. Zhu, Q.; Wu, J.; Tu, C.; Shi, Y.; He, L.; Wang, R.; Zhu, X.; Yan, D. *J. Phys. Chem. B*, **2009**, *113*, (17), 5777–5780.
161. Yan, D.; Zhou, Y.; Hou, J. *Science*, **2004**, *303*, (5654), 65–67.
162. Olabisi, O.; Adewale, K., Eds. 1997. *Handbook of Thermoplastics: Polyethers.* New York: Marcel Dekker.
163. Xia, Y.; Wang, Y.; Wang, Y.; Wang, D.; Deng, H.; Zhuang, Y.; Yan, D.; Zhu, B.; Zhu, X. *Macromol. Chem. Phys.*, **2011**, *212*, (10), 1056–1062.
164. Garrido, L.; Riande, E.; Guzmán, J. *Makromol. Chem., Rapid Commun.*, **1983**, *4*, (11), 725–729.
165. Pruckmayr, G.; Dreyfuss, P.; Dreyfuss, M. P. 2000. Polyethers, tetrahydrofuran and oxetane polymers. In *Kirk-Othmer Encyclopedia of Chemical Technology*, Pruckmayr, G., Dreyfuss, P., Dreyfuss, M. P., Eds. Hoboken, NJ: John Wiley & Sons.
166. Dreyfuss, P. 1981. *Poly(tetrahydrofuran).* New York: Gordon and Breach.
167. Trick, G. S.; Ryan, J. M. *J. Polym. Sci., C Polym. Symp.*, **1967**, *18*, (1), 93–103.
168. Wetton, R. E.; Williams, G. *Trans. Faraday Soc.*, **1965**, *61*, 2132.
169. Davis, A.; Golden, J. H. *Makromol. Chem.*, **1965**, *81*, (1), 38–50.
170. Klemm, D.; Heublein, B.; Fink, H.-P.; Bohn, A. *Angew. Chem. Int. Ed. Engl.*, **2005**, *44*, (22), 3358–3393.
171. Balser, K.; Hoppe, L.; Eichler, T.; Wendel, M.; Astheimer, A.-J. 1986. Cellulose esters. In *Ullmann's Encyclopedia of Industrial Chemistry*, Gerhartz, W., Yamamoto, Y. S., Campbell, F. T., Pfefferkorn, R., Rounsaville, J. F., Eds., 419–459. Weinheim, Germany: Weinheim VCH.
172. Edgar, K. J.; Buchanan, C. M.; Debenham, J. S.; Rundquist, P. A.; Seiler, B. D.; Shelton, M. C.; Tindall, D. *Prog. Polym. Sci.*, **2001**, *26*, (9), 1605–1688.
173. Klemm, D.; Schumann, D.; Udhardt, U.; Marsch, S. *Prog. Polym. Sci.*, **2001**, *26*, (9), 1561–1603.
174. Yamanaka, S.; Watanabe, K. 1994. Applications of bacterial cellulose. In *Cellulosic Polymers—Blends, and Composites*, Gilbert, R. D., Ed., 207–215. Munich: Hanser Publishers.
175. Fardet, A. *Nutr. Res. Rev.*, **2010**, *23*, (1), 65–134.
176. Jobling, S. *Curr. Opin. Plant Biol.*, **2004**, *7*, (2), 210–218.
177. Kaur, B.; Ariffin, F.; Bhat, R.; Karim, A. A. *Food Hydrocolloids*, **2012**, *26*, (2), 398–404.
178. Rindlav-Westling, Å.; Stading, M.; Gatenholm, P. *Biomacromolecules*, **2002**, *3*, (1), 84–91.
179. Warren, B. B.; Durieux, M. E. *Anesth. Analg.*, **1997**, *84*, 206–212.
180. Treib, J.; Baron, J.-F.; Grauer, M. T.; Strauss, R. G. *Intensive Care Med.*, **1999**, *25*, (3), 258–268.

181. Food and Drug Administration. 2013. *FDA Safety Communication: Boxed Warning on Increased Mortality and Severe Renal Injury, and Additional Warning on Risk of Bleeding, for Use of Hydroxyethyl Starch Solutions in Some Settings*, http://www.fda.gov/BiologicsBloodVaccines/SafetyAvailability/ucm 358271.htm.
182. Calvo-Flores, F. G.; Dobado, J. A. *ChemSusChem*, **2010**, *3*, (11), 1227–1235.
183. Yiamsawas, D.; Baier, G.; Thines, E.; Landfester, K.; Wurm, F. R. *RSC Adv.*, **2014**, *4*, (23), 11661.
184. Wang, J.; Manley, R.; Feldman, D. *Prog. Polym. Sci.*, **1992**, *17*, (4), 611–646.
185. Lora, J. H.; Glasser, W. G. *J. Polym. Environ.*, **2002**, *10*, (1/2), 39–48.
186. Thielemans, W.; Can, E.; Morye, S. S.; Wool, R. P. *J. Appl. Polym. Sci.*, **2002**, *83*, (2), 323–331.
187. Stewart, D. *Ind. Crops Prod.*, **2008**, *27*, (2), 202–207.
188. Lichtenthaler, F. W. *Acc. Chem. Res.*, **2002**, *35*, (9), 728–737.
189. Choura, M.; Belgacem, N. M.; Gandini, A. *Macromolecules*, **1996**, *29*, (11), 3839–3850.
190. Magalhães, W. L. E.; da Silva, R. R. *J. Appl. Polym. Sci.*, **2004**, *91*, (3), 1763–1769.
191. Wang, H.; Yao, J. *Ind. Eng. Chem. Res.*, **2006**, *45*, (19), 6393–6404.
192. McKillip, W. J. 1981. Furan and derivatives. In *Kirk-Othmer Encyclopedia of Chemical Technology*, Kirk, R. E., Othmer, D. F., Seidel, A., Eds., 501–527. Hoboken, NJ: Wiley-Interscience.
193. Hay, A. S. *J. Polym. Sci., Part A: Polym. Chem.*, **1998**, *36*, (4), 505–517.
194. Giragosian, N. H. 1978. New product pricing. In *Successful Product and Business Development*, Giragosian, N. H., Ed., 143–156. New York: Marcel Dekker.
195. Hay, A. *Prog. Polym. Sci.*, **1999**, *24*, (1), 45–80.
196. Hay, A. S.; Blanchard, H. S.; Endres, G. F.; Eustance, J. W. *J. Am. Chem. Soc.*, **1959**, *81*, (23), 6335–6336.
197. Jayakannan, M.; Ramakrishnan, S. *Macromol. Rapid Commun.*, **2001**, *22*, (18), 1463.
198. Percec, V.; Hill, D. H. 1996. Step-growth electrophilic oligomerization and polymerization reactions. In *Cationic Polymerizations*, Matyjaszewski, K., Ed., 555–682. New York: Marcel Dekker.
199. Mobley, D. P. *J. Polym. Sci. Polym. Chem. Ed.*, **1984**, *22*, (11), 3203–3215.
200. Líška, J.; Borsig, E. *J. Macromol. Sci., Polym. Rev.*, **1995**, *35*, (3), 517–529.

8 Aromatic Polyamides*

Miriam Trigo, Jesús L. Pablos, Félix C. García, and José M. García

CONTENTS

8.1 Introduction 285
8.2 Synthesis 287
    8.2.1 Monomer Synthesis 288
    8.2.2 Low-Temperature Solution Methods 288
    8.2.3 High-Temperature Solution Methods...289
    8.2.4 Alternative Polymerization Methods 290
8.3 Polyamides with Controlled Structures 290
    8.3.1 Chain-Growth Polycondensation 290
    8.3.2 Constitutional Isomerism..294
    8.3.3 Spherical Aromatic Polyamides 296
8.4 Properties..297
8.5 Chemical Stability 305
8.6 Processing..305
    8.6.1 Wet Spinning 307
        8.6.1.1 Wet Spinning of MPIA 307
        8.6.1.2 Wet Spinning of PPTA...307
        8.6.1.3 Wet Spinning of ODA/PPTA 308
    8.6.2 Dry Spinning 308
    8.6.3 Dry-Jet Wet Spinning 308
    8.6.4 Film Casting of PPTA 308
8.7 Global Market View 309
8.8 Applications...309
    8.8.1 Expanding Applications of Aromatic Polyamides 310
        8.8.1.1 Aromatic Polyamides with Unique Properties 310
        8.8.1.2 Aromatic Polyamides as Biomaterials and for Use in Medical Applications 311
        8.8.1.3 Cross-Linked Aromatic Polyamides: A Further Step in High-Performance Materials.. 312
        8.8.1.4 Future Developments 312
8.9 Special Additives...312
8.10 Environmental Impact and Recycling 313
References...313

8.1 INTRODUCTION

Poly(amide)s, or polyamides, are polymers containing the amide group –CO–NH– in their repeating unit.[1] These polymers are widespread and can be found both in nature, i.e., in proteins, polypeptides, silk, and wool, and in synthetic compounds, including aliphatic polyamides, or nylons, and aromatic polyamides, commonly referred to as aramids.

---

* Based in parts on the first-edition chapter on polyamides.

Both aromatic (aramids) and aliphatic (nylons) polyamides are considered to be engineering materials. However, aramids broadly outperform the mechanical properties and the thermal and chemical resistance of nylons, and the differences are a consequence of the nature of the aromatic backbone of aramids. According to the definition given by the US Federal Trade Commission (FTC), wholly aromatic polyamides are synthetic polyamides in which at least 85% of the amide groups are bound directly to two aromatic rings.* Because of their chemical structure, wholly aromatic polyamides exhibit a number of outstanding characteristics, such as high chemical and thermal resistance, excellent mechanical properties, and low flammability, qualifying them to be classified as high-performance materials.[1,2]* These materials are useful in advanced technologies, where they are increasingly used to replace ceramics and metals.[3-6]

Poly(p-phenylene terephthalamide) (PPTA) and poly(m-phenylene isophthalamide) (MPIA) are the best-known commercial aramids. They are also the simplest and were the first to be synthesized. In solution, they can be transformed into flame- and cut-resistant, high-tensile-strength synthetic fibers. The transformed materials are widely used in technological applications, such as protective and sport fabrics; bulletproof body armor; coatings, fillers, or advanced composites in the aerospace and armament industries; industrial filters; and electrical insulation.

However, because of their chemical structure, their solubility in most common organic solvents is very poor, and their transition temperatures are higher than their decomposition temperatures. Thus, they are difficult to process, limiting their use in some applications.

As a result, basic and applied research is performed with two objectives: (1) to overcome the problems associated with the processing difficulties and poor solubility without changing their high-performance properties and (2) to expand the scope of their use as high-performance materials to new or promising fields of materials research, such as reverse osmosis (RO), gas- or ion-exchange membranes, nanocomposites with advanced thermal and mechanical performance, optically active materials, and electroluminescent or photoluminescent materials.

In the late 1950s, nylon and polyester were the most technologically advanced synthetic materials. To achieve maximum tenacity, the polymer chains had to be extended and assume almost perfect crystalline packing, which required mechanically drawing the fiber after melt spinning. The tenacity and elastic modulus of polyester and nylons were far from the theoretically possible ones. In 1965, DuPont discovered an all-*para*-oriented aramid, poly(p-benzamide) (PBA), marketed under the name Fiber B. Its production lasted only a few years, most likely due to economic reasons.[7,8] This fiber was replaced by PPTA, which was marketed under the trade name Kevlar, in 1971. The polycondensation of PPTA was initially performed at low temperature in a solution of terephthaloyl dichloride (TPC) and p-phenylenediamine (PPD) in hexamethylphosphoramide (HMPA). Later, the solvent was replaced by N-methyl-2-pyrrolidone (NMP) with $CaCl_2$. To process PPTA, it must be transformed into fibers by spinning from lyotropic solutions in concentrated sulfuric acid at high temperatures due to its insolubility in organic solvents.

At the same time, DuPont produced an alternative to the all-*para*-aramid fiber, MPIA, which was originally called HT-1 and became commercially available in 1967 under the trade name Nomex.[9] The structure of this fiber is less linear than that of PPTA, resulting in a decrease in its cohesive energy and tendency to crystallize, but it still exhibits good mechanical properties and has a higher thermal resistance to fire.

A very strong *para*-aramid fiber was developed and produced by Teijin Limited and became available under the trade name Technora in 1987. This fiber was obtained by copolymerizing TPC, PPD, and 3,4-oxydianiline (ODA), producing a polymer, ODA/PPTA, with enhanced solubility. Due to the asymmetry of the ODA monomer, this material is less ordered and has a lower cohesive energy than PPTA and MPIA. It is still a high-performance fiber, however, with good fatigue resistance, high tensile strength, long-term dimensional stability, and excellent resistance to corrosion, heat, and chemicals.

---

* Rules and regulations under the Textile Fiber Products Identification Act (http://www.ftc.gov/os/statutes/textile/rr-textl.pdf), Part 303.7 (Generic names and definitions for manufactured fibers), US Federal Trade Commission (FTC).

Aromatic Polyamides

8.2 SYNTHESIS

Scheme 8.1 shows the monomers used in the syntheses of these aramids. These monomers are now widely available at medium cost. Scheme 8.2 presents the structures of commercial aromatic polyamides that are currently relevant, and Table 8.1 lists information for commercial aramids, including the companies producing them, their type, and their brand names. The first two companies, DuPont and Teijin, have a dominant position in the aramid market.

Aromatic polyamides are commonly prepared according to the following two methods: (1) reaction between diacid chlorides and diamines at low temperatures and (2) direct polycondensation of aromatic diacids with diamines in solution at high temperatures (Yamazaki–Higashi method). In both cases, polar aprotic solvents, such as HMPA, NMP, N,N-dimethylformamide (DMF), or

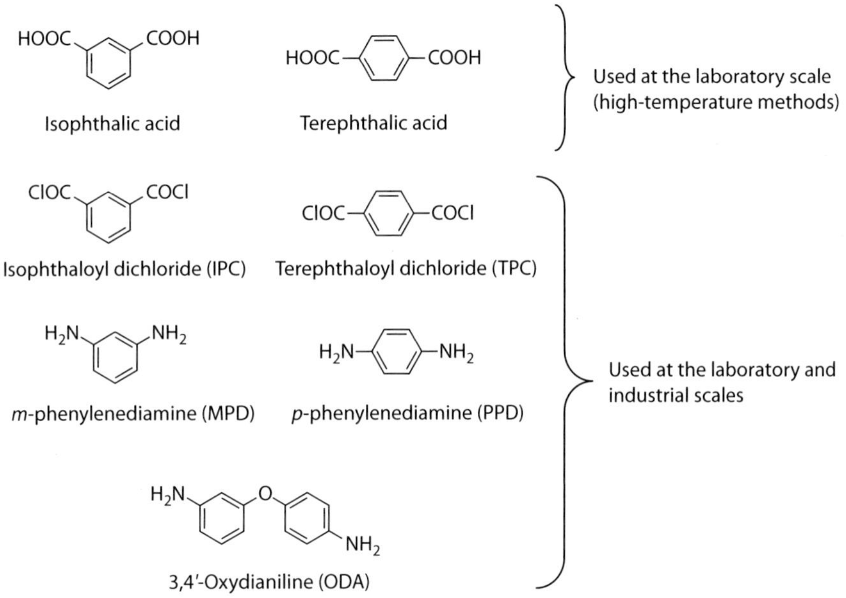

SCHEME 8.1 Monomers used in the synthesis of aramids.

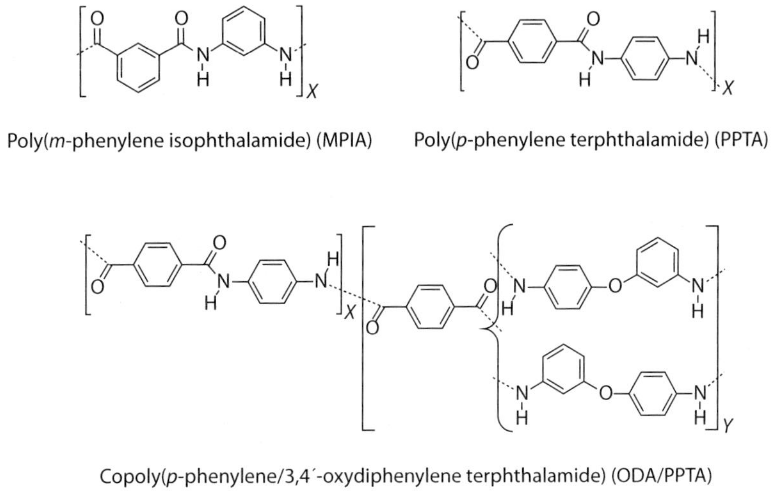

SCHEME 8.2 Commercially important aromatic polyamides.

TABLE 8.1
Company, Aramid Type, and Brand Names of Commercial Aramids

Company	Aramid Type	Brand Name
DuPont	*Meta*-aramid	Nomex
	Para-aramid	Kevlar
Teijin	*Meta*-aramid	Teijinconex
	Para-aramid	Twaron
	Copolymer ODA/PPTA	Technora
Kolon Industries	*Para*-aramid	Heracron
Hyosung	*Para*-aramid	Alkex
SRO Group	*Meta*-aramid	X-Fiper
Yantai	*Meta*-aramid	Newstar
Spandex Co.	*Para*-aramid	Taparan
Woongjin	*Meta*-aramid	Arawin

N,N-dimethylacetamide (DMA), are used. To increase the solubility by decreasing the strength of the interchain hydrogen bonds, LiCl and $CaCl_2$ salts or a combination of these salts is added to the mixture, particularly in the synthesis of the *para*-aramid PPTA. Important factors in the polymerization process include the monomer stoichiometry, purity of the solvent and monomers, stirring intensity, anhydrous conditions, monomer concentration, and temperature.[10] Scheme 8.3 shows, as an example, the preparation of MPIA at low (a) and high (b) temperatures using triphenyl phosphite (TPP) as the condensation promoter.[11]

8.2.1 Monomer Synthesis

Commonly, polymerization reactions should be performed using pure solvents and monomers under anhydrous conditions. At the laboratory scale, the solvents are dried with phosphorous pentoxide, for example, and distilled twice. All salts added to the polymerization reaction should be carefully dried to maintain the anhydrous conditions. Isophthalic acid and terephthalic acid are purified by crystallization in water. Isophthaloyl dichloride (IPC) and TPC are prepared by reacting their acids with $SOCl_2$ and then crystallizing them from dry heptane. *Meta*-phenylenediamine (MPD), PPD, and 3,4′-oxydianiline are purified by double vacuum sublimation.

8.2.2 Low-Temperature Solution Methods

At the laboratory scale, the low-temperature solution method is generally preferred when the diacid chloride can be easily obtained from the aromatic diacid.

From a commercial viewpoint, low-temperature solution methods are used to condense PPD and TPC in NMP with $CaCl_2$ or MPD and IPC in DMA as the solubility promoter for the synthesis of

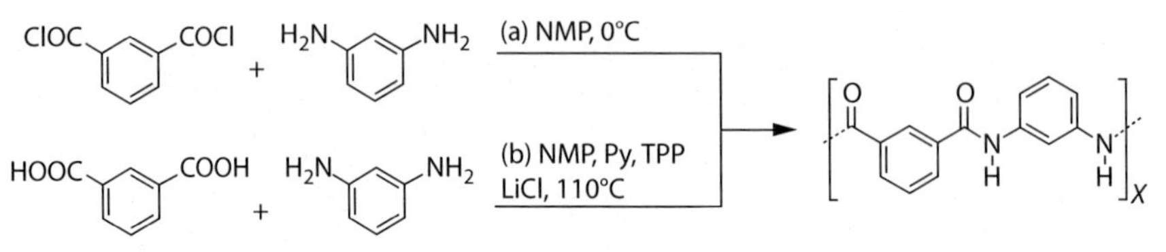

SCHEME 8.3 Preparation of MPIA at (a) low and (b) high temperatures.

PPTA and MPIA, respectively. ODA/PPTA is prepared commercially by the polycondensation of PPD and ODA (50% each) with isophthaloyl dichloride in NMP without solubility promoters, and the evolved HCl is neutralized with $Ca(OH)_2$. The main drawback of this method is that extremely high monomer purity is needed to obtain high-molecular-weight materials.

The number average molecular weight (M_n) typically obtained for condensation polymers, and thus aromatic polyamides, is 10×10^3 to 30×10^3 g/mol. This method produces polymers with a polydispersity between approximately 2 for lower-molecular-weight polymers and 3 for polymers with weight average molecular weights greater than 35×10^3 g/mol.[10]

At the laboratory scale, the synthesis procedure can be modified by silylating the diamines, which increases the reactivity of the amino group. Silylated diamines are moisture sensitive, and to avoid the need for isolation and purification steps, this procedure is performed *in situ*.[12]

Some of the earliest reports of aramid synthesis employed a two-phase polycondensation system at room temperature, namely, interfacial polymerization.[13] In this method, the diamine and diacid chloride are dissolved in water and a solvent with limited water solubility, respectively. The aqueous solution usually contains a surfactant and a base to neutralize the HCl generated. Upon vigorous stirring, a polymer precipitate is formed. The reaction occurs within seconds in the organic solvent at the interface. This technique has some drawbacks that make it difficult to develop on a commercial scale: the stoichiometry at the interface is difficult to control because the instantaneous concentration is diffusion dependent and has only a limited dependence on the monomer concentrations, and the precipitation of the growing polymer chains usually produces aramids with a broad molecular weight distribution unsuitable for fibers or film-forming materials.[14] Nevertheless, aramids with properties similar to those prepared via solution polycondensation methods can be prepared by optimizing the solvent type and volume, monomer concentration, and stirring rate. Films can be cast from solutions with desirable chemical properties. In addition, water-soluble dihydrochloride derivatives of water-insoluble diamines can be used as monomers.[15]

8.2.3 HIGH-TEMPERATURE SOLUTION METHODS

If the diacid chloride cannot be obtained from the corresponding aromatic diacid or it is of low quality, meaning that side reactions are likely to occur or it is heat or moisture sensitive, aromatic diamines and diacids can be reacted via direct polycondensation. This method was developed by Yamazaki and Higasi in the mid-1970s,[16] but it is not currently used commercially. In this phosphorylation method, diacids and diamines are directly condensed at 110°C in NMP in the presence of condensation-promoting agents, such as TPP and pyridine (Py), and salts, such as LiCl, $CaCl_2$, or a mixture of the two. In addition to the requirement of high-purity monomers, another drawback of this method is that side reactions can occur due to the high-temperature conditions. Therefore, model compounds are often prepared first to verify the absence of side reactions under the polymerization conditions when monomers with sensitive functional groups are used.

Recently, a new technique for polymer synthesis involving the microwave-assisted polycondensation of aromatic diacids and diamines under Yamazaki conditions has been developed. Microwave radiation (MW) is a widely used, nonconventional energy source capable of promoting chemical reactions in an extremely fast and unconventional way. The MW system replaces conventional heating and temperature control systems, reducing the reaction time from approximately 4 h to 2 min while maintaining the inherent viscosities obtained with the conventional methods and suppressing the side reactions.[17–20]

MW has also proved to be effective for the polycondensation of aromatic or aliphatic diisocyanates and diacids to yield semiaromatic polyamides.[21] Inherent viscosities between 0.51 and 0.2 dL/g are obtained from the reaction, which is completed in less than 5 min.

Recent efforts have also been directed toward the development of efficient and cleaner polycondensation methods by decreasing the use of traditional toxic and volatile organic solvents. To avoid the use of harmful solvents such as pyridine or NMP in the Yamazaki polyamidation method, or

DMF, DMA, and NMP in low-temperature polycondensation, environmentally friendly aramids have been prepared using ionic liquids.[22] Ionic liquids are salts that exist in the liquid phase at room temperature and are capable of dissolving aromatic polyamides due to their high polarity, high dielectric constant, high thermal stability, and low vapor pressure. Accordingly, environmentally friendly polyamides have been prepared with TPP to promote the direct condensation of diacids and diamines and the reaction of diacid dichlorides and diamines at low temperatures.[21–33]

In short, both conventional and high-temperature methods using TPP as a condensation promoter or MW-assisted polycondensation to reduce reaction times can be employed in a polycondensation procedure because comparable inherent viscosities and molecular weights are obtained for polymers with identical chemical structures.

8.2.4 Alternative Polymerization Methods

It is beyond the scope of this chapter to cover all the synthetic methods provided by organic chemistry to obtain the aromatic or semiaromatic amide bonds used to prepare polyamides.

These methods include, among others, the reaction of diacids and diisocyanates to synthesize wholly and partially aromatic polyamides, direct polycondensation using thionyl chloride as the activating agent, condensation of diacids with the formamidinium salts of aromatic diamines, use of diamines and CS_2, reaction of aromatic diacid phenyl esters with amines, reaction of diacid halides and activated diamines through *in situ* silylation, palladium-catalyzed carbonylation–polycondensation from dihaloaryl compounds and aromatic diamines, etc. Some of these methods are described by Fink,[5] Vollbracht,[6] Gaymans,[7] Sekiguchi and Coutin,[34] Kubota,[35] and Rabani.[36]

8.3 POLYAMIDES WITH CONTROLLED STRUCTURES

Polycondensation reactions generally follow a step-growth pattern in which condensation reactions between molecules of different degrees of polymerization occur, making it difficult to control the molecular weight of the polymer. The average degree of polymerization obeys Flory's statistical treatment and Carothers' equation, which give a theoretical molecular weight polydispersity of 2 at high conversion, and is based on the assumption that all monomer and polymer functional groups are equally reactive. When monomers with unequal reactivity resulting from different functional groups on AB monomers, the asymmetry of bifunctional monomers (AA+BB), or the use of A_xB_y monomers with defined structures are employed, the polymers that are produced do not obey these conventional rules. As a consequence, it is difficult to synthesize polymers with a controlled molecular weight distribution in this manner.[1] On the other hand, living polymerization allows the molecular weight to be controlled and leads to products with a narrow molecular weight distribution. This type of polymerization is possible if polycondensation proceeds via a chain-growth mechanism instead of step-growth polycondensation. This change in the polycondensation mechanism is possible because nature already uses a chain-growth polycondensation process to synthesize monodisperse biopolymers such as DNA and polypeptides.[37]

8.3.1 Chain-Growth Polycondensation

Experimental results in which some high-molecular-weight condensation polymers were obtained with low conversions paved the way for chain-growth polycondensation methods. These methods involve suppressing the reaction between monomers by selectively activating the polymer-propagating terminal, forcing the monomer to react only with the initiator and polymer end group. Yokoyama et al. published a detailed review of chain-growth polycondensation that discusses its theory, structures, and applications.[37,38]

One way to obtain polymers with controlled structures is to selectively activate polymer end groups by changing the substituent effects between the monomer functional groups. For example, in an AB monomer, the substituent effect of the B site decreases the reactivity of the A site, thereby suppressing the step-growth reaction between the monomers. The substituent effect changes when the monomer reacts with the reactive site of an initiator. If the newly formed bond increases the reactivity of the polymer end group, the next monomer only reacts with the propagating end of the polymer. However, this polymerization cannot be completely controlled, because of the low solubility of the resulting polymer, side reactions, or insufficient inhibition of step-growth polycondensation. Yet, the condensation polymerization can be completely controlled by changing the resonance or inductive effects of the substituents at reaction conditions.

For example, the polycondensation of phenyl-4-octylaminobenzoate in the presence of the initiator phenyl-4-nitrobenzoate and a base (a combination of N-octyl-N-triethylsilylaniline, CsF, and 18-crown-6) under mild conditions yields a polymer (I) with a well-defined structure (Scheme 8.4).

The molecular weight of the polyamide can be controlled by the feed ratio of the monomer to initiator up to $M_n = 22,000$, and a polydispersity of $(M_w/M_n) \leq 1.1$ can be achieved. In this polymerization, the reactivity is controlled by the difference in the resonance effects of the aminyl anion and polymer amide linkage as follows (Scheme 8.5): the deprotonation of the AB monomer (2) in the presence of a base yields the aminyl anion (4), which is capable of deactivating the phenyl ester moiety of (2) through the resonance effect (re). Specifically, the amide anion decreases the electrophilicity of the phenyl ester moiety at the para position because it is strongly electron donating

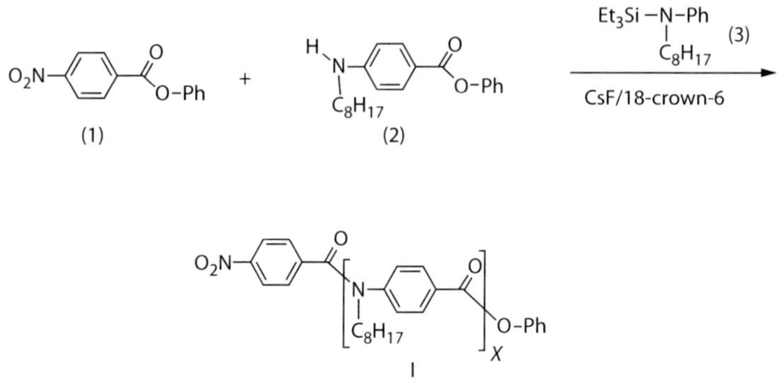

SCHEME 8.4 Polycondensation of phenyl-4-octylaminobenzene.

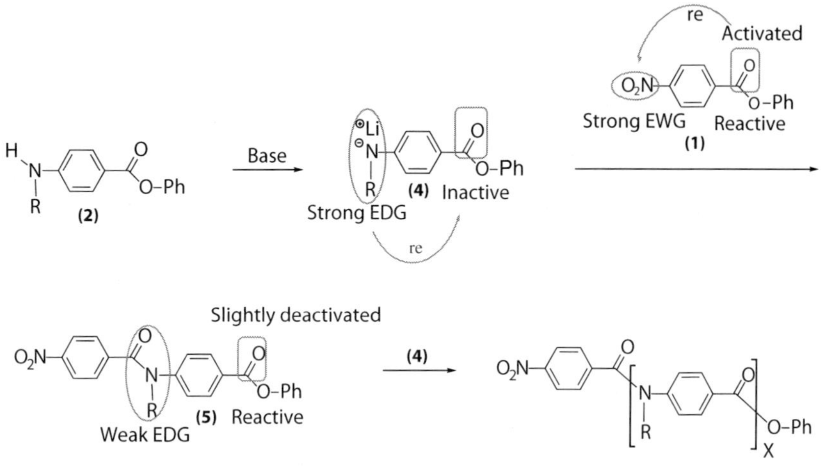

SCHEME 8.5 Control mechanism based on resonance effects in polycondensation reactions.

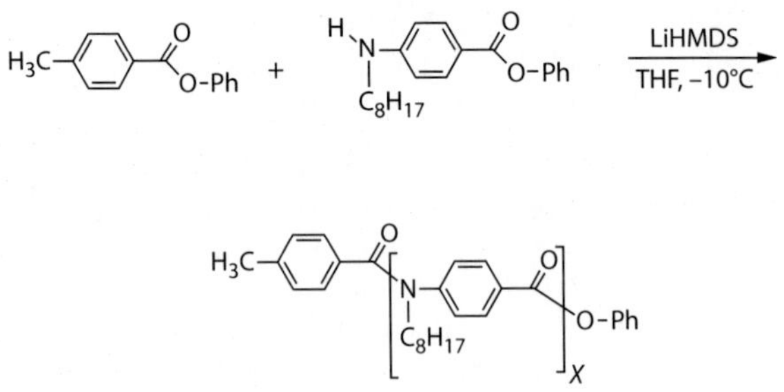

SCHEME 8.6 Polycondensation of phenyl-4-octylaminobenzoate using LiHMDS as the base and phenyl-4-methylbenzoate as the initiator.

(electron-donating group [*EDG*]). At the same time, the phenyl ester moiety is strongly activated by the initiator (1) due to the presence of an electron-withdrawing group (EWG), the nitro group, at the *para* position. The initial reaction of the aminyl anion (4) with the initiator (1) produces a new amide group (5). This amide linkage is weakly electron donating and thus facilitates the chain growth because the phenyl ester of this chain reacts with the monomer with the aminyl anion.[39]

Unfortunately, this reaction requires unusual bases, CsF and 18-crown-6, and it is necessary to separate the synthesized polyamide from other by-products. The procedure can thus be improved by using a commercially available base, lithium hexamethyldisilazide (LiHMDS), and phenyl-4-methylbenzoate as the initiator (Scheme 8.6).[40]

Following the same procedure, well-defined poly(*m*-benzamide)s can be obtained (Scheme 8.7). In this case, the key substituent effects between the amino group and ester moiety of the monomer are inductive effects (*ies*) instead of resonance effects. It is believed that even if inductive effects are weaker than resonance effects, the negative charge on the amide anion formed by deprotonation deactivates the ester moiety enough to avoid the reaction between the aminyl anions.[41–43]

The large cohesive energy of aromatic polyamides due to their density, effectiveness, and highly directional interchain amide–amide hydrogen bonds results in the outstanding mechanical and thermal properties of these materials. *N*-alkyl polyamides are highly soluble because they lack amide–amide hydrogen bonds. Soluble aramids (N–H) can be obtained by protecting the amino group with an alkyl group (*N*-alkyl, such as 4-octyloxybenzyl, OOB). Then, the benzyl groups can be easily removed by treating the polyamide with the proper reagent at mild conditions.[44,45] To obtain a polymer with a specific molecular weight and polydispersity, the polymerization should start from

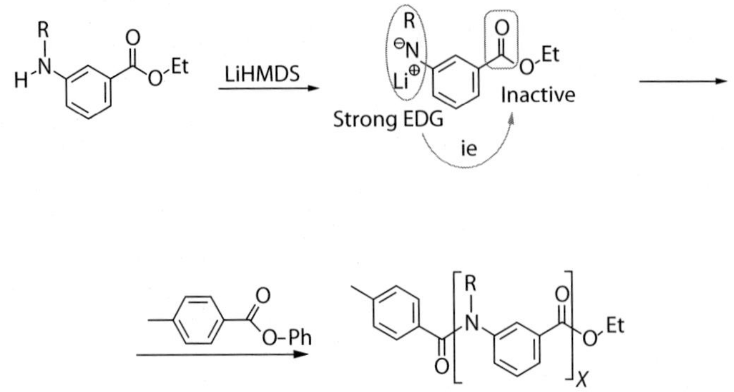

SCHEME 8.7 Control mechanism in polycondensation reactions based on inductive effects.

an initiator unit and continue via chain-growth polymerization, also called living polymerization. This subsequent addition of a suitable monomer enables the preparation of either *para–para* or *meta–para* (Scheme 8.8) condensation block copolymers and offers a new method for designing materials with *a la carte* properties.[41,44,45] The diblock copolymer IV has a narrow molecular weight distribution, is soluble in organic solvents, and is self-assembled in tetrahydrofuran (THF), yielding intriguing micrometer-sized bundles and aggregates of flake-like structures. These structures most likely result from interchain hydrogen bonding of the *p*-benzamide blocks. It has been suggested that these block copolymers and star polymers with well-defined PBA units could be a suitable starting point for the generation of aramid nanoarchitectures.

It is also possible to obtain copolymers and chain-growth polymers. Polymer I reacts with poly(ethylene glycol) (PEG), NaH, and monomethyl ether to yield VII, a PEG and polyamide block copolymer (Scheme 8.9).[44,45] A triblock copolymer was prepared in a similar reaction between PEG and polymer I.[46] Recent research also explored the synthesis of well-defined diblock copolymers of aromatic polyamides and aromatic polyethers.[47]

A block copolymer of polyamide and polytetrahydrofuran [poly(THF)] was also prepared via a similar procedure using an amino-protecting group (Boc, *tert*-butoxycarbonyl). When

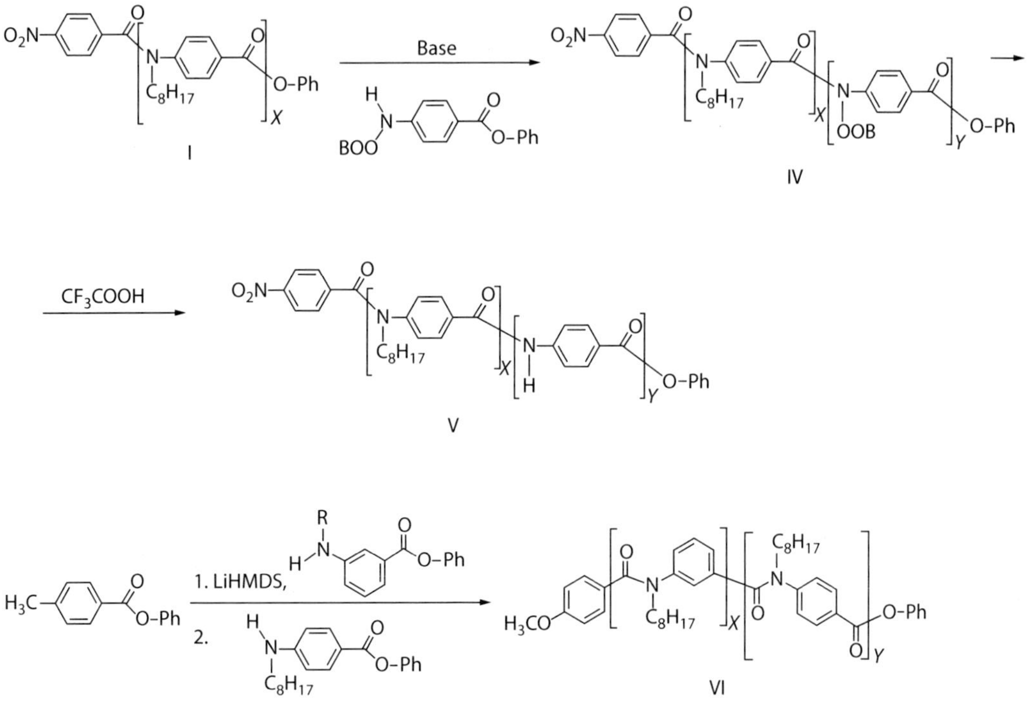

SCHEME 8.8 Preparation of condensation block copolymers.

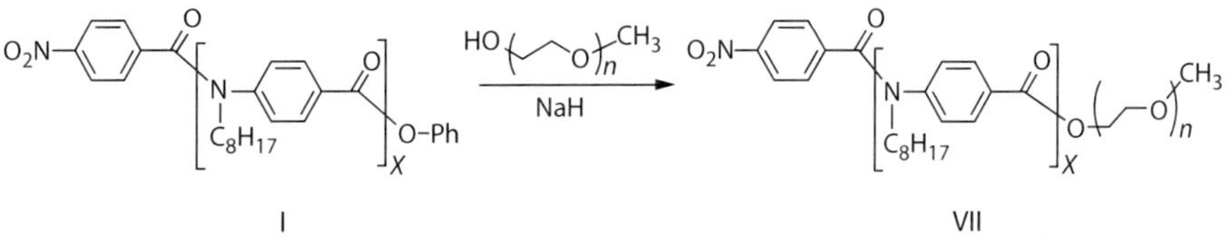

SCHEME 8.9 Synthesis of a PEG and polyamide block copolymer.

trifluoromethanesulfonic anhydride was used as an initiator for the THF polymerization, a poly(THF)–polyamide–poly(THF) triblock copolymer was obtained.[48]

8.3.2 CONSTITUTIONAL ISOMERISM

The unequal reactivity of monomers can be due to either asymmetry or induced asymmetry. In the former, the reaction rates are different because the monomer functional groups are not chemically equivalent. In the latter, the reaction of one functional group alters the reactivity of the second functional group; hence, the monomer functional groups are equally reactive until one of them reacts.

Only the isomerism associated with monomer asymmetry is considered here. When substituents are introduced on the diacid or diamine monomers in a way that breaks the symmetry of the monomeric units, constitutional isomerism of the polymer results.

Ueda summarized the practical and theoretical aspects of the constitutional isomerism in one-step condensation polymerizations.[49] While the relationship between the polyamide structure and properties attributed to constitutional isomerism is not well known, it is assumed that the cohesive energy and crystallinity are strongly affected by the molecular order. Thus, if the structural regularity could be controlled, the thermal, mechanical, and solubility properties would be at least partially controlled.[50–52]

Some of the previously mentioned work has described the synthesis of perfectly ordered structures in both several synthetic steps and time-consuming multistep synthetic procedures to compare the characteristics of the resulting polymers.

When nonsymmetric (XabX) and symmetric (YccY) monomers are polycondensed, two ordered structures, head to tail (H-T) and head to head/tail to tail (H-H/T-T), can be obtained together with an infinite set of disordered polymers. The three general cases are shown in Figure 8.1.

The influence of constitutional isomerism on the physical properties of polyamides was first studied by Pino et al.,[53] who investigated the theoretical aspects of the structural regularity of polyamides. These researchers defined a probability parameter, s, as the probability of two adjacent nonsymmetric units pointing in the same direction in a chain. This parameter was used to quantify the overall structural regularity of the polymers.

When XabX is reacted with YccY, the shortest structural elements in the polymer are –acca–, –accb–, –bcca–, and –bccb–, where –accb– and –bcca– are indistinguishable. The probability of an –accb– pattern is given by

$$s = \frac{[\text{accb}]}{\left([\text{acca}] + [\text{accb}] + [\text{bccb}]\right)} \quad (8.1)$$

where [accb] includes [accb] and [bcca].

In terms of the kinetics, it was concluded that differences in the reactivity of the functional groups in a nonsymmetric monomer are not significant enough to produce $s = 0$ condensation

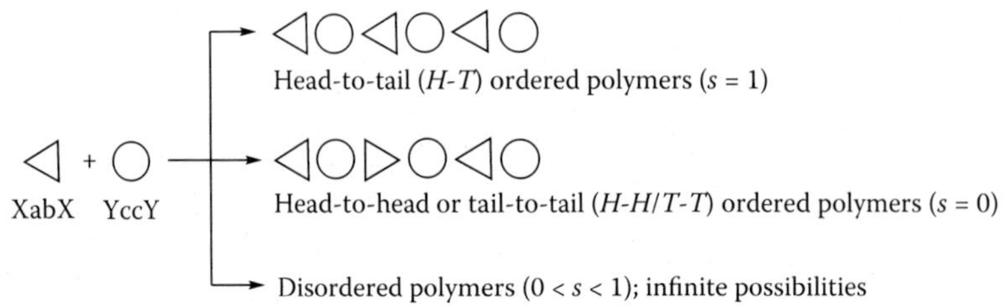

FIGURE 8.1 Three general cases of polycondensed symmetric (YccY) and nonsymmetric (XabX) monomers.

polymers,[54] and as a consequence, the immediate mixing of two nonsymmetric monomers yields random polymers. To synthesize *H-H* or *T-T* polymers with $s = 0$, the symmetric YccY monomer must be slowly added to the nonsymmetric XabX monomer. Only XbaccabX would be present in the media after half of the YccY is added, and only –bccb– units would be formed when the remaining YccY is added. Thus, the resulting polymer would have only –acca– and –bccb– units, and the *s* value would be 0.

Following this approach, the polymerization of 1,3-diamine-4-chlorobenzene (Scheme 8.10, VIII) was performed, and the sequence distribution was analyzed.[51] These polyamides were prepared in a one-stage polycondensation reaction of nonsymmetric monomers, and the constitutional isomerism was controlled (to the degree that the reaction kinetics were limited) by the reaction temperature, mode of monomer addition, and monomer feed rates. The constitutional order, which was determined by ^{1}H and ^{13}C nuclear magnetic resonance (NMR) spectroscopy, varied with *s* values between 0.28 and 0.5. A later paper from the same authors evaluated the properties of aramids prepared from 1,3-diamine-4-halobenzene and compared them to those of polyamides derived from symmetrically dihalogenated MPDs, specifically 1,3-diamine-4,6-dihalobenzene, where the halogen was F, Cl, or Br.[52] The study revealed that the presence of a nonsymmetric monohalogen-substituted structural unit enhances the solubility and influences the thermal properties due to the inherent disorder in the *H-H*, *H-T*, and *T-T* sequence distribution.

The synthesis, characterization, and constitutional isomerism of new aromatic polyamides containing pendant aryloxy groups based on asymmetrically substituted MPDs (Scheme 8.10, IX) was recently studied by Pal et al.[55] This work revealed that under normal polymerization conditions, all the polymers showed a preference for *H-H/T-T* sequences with an *s* value between 0.35 and 0.37, T_g in the range of 237–254°C, and good solubility in polar aprotic solvents from which films could be cast.

Poly(amide-ester) was prepared by the direct polycondensation of a symmetric monomer (IPC) and nonsymmetric monomer (4-aminophenethyl alcohol) to investigate ordered *H-H/T-T* aliphatic–aromatic polyamides.[50] In this case, IPC was slowly added to 4-aminophenethyl alcohol following

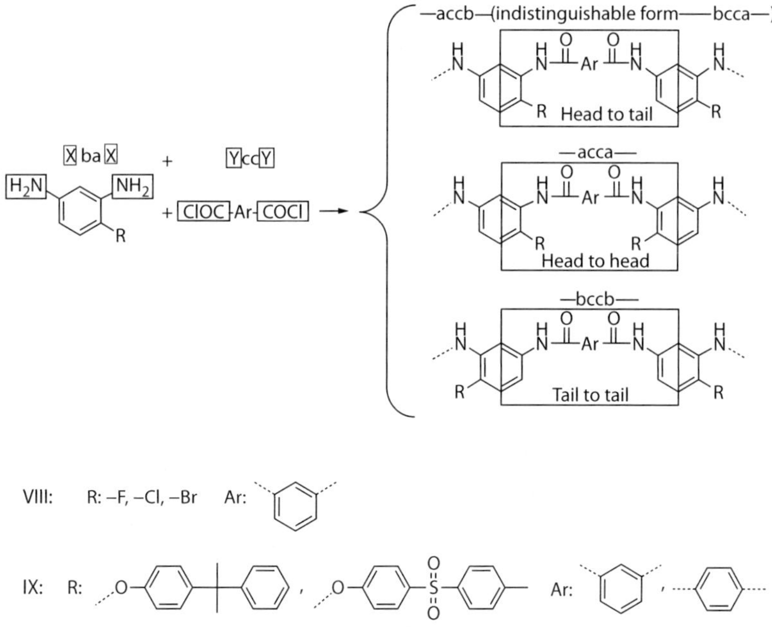

SCHEME 8.10 Aromatic polyamides containing pendant aryloxy groups based on asymmetrically substituted *meta*-phenylenediamines.

the standard low-temperature polycondensation method, yielding the ordered *H-H*/*T-T* poly(amide-ester). *H-H*/*T-T* and random poly(amide-ester)s were also prepared to compare the polymers. The microstructure of the polymers was studied by ^{1}H and ^{13}C NMR spectroscopy. The influence of the structure on the physical properties and the ability to control the structure was demonstrated by the fact that the ordered *H-H*/*T-T* polymer was semicrystalline while the random poly(amide-ester) was amorphous. Additionally, ordered *H-H*/*T-T* semiaromatic polyamides were prepared by the slow addition of isophthalic acid (a symmetric monomer) to *N*-phenylethylenediamine (a nonsymmetric monomer) in a direct polycondensation.[56] On the other hand, to obtain *H-H*/*T-T*– and *H-T*-ordered polyamides, a multistep procedure was followed. By simultaneously mixing the diamine and IPC in NMP, a random polyamide, which had a higher solubility, was obtained.

A discussion of constitutional isomerism in terms of constitutional order, theory, and polyamide structures identified over the last decade can be found in the reviews of Ueda[49] and Gentle and Suter.[57]

8.3.3 Spherical Aromatic Polyamides

Aromatic polyamides are important high-performance polymers due to their outstanding mechanical, chemical, and thermal properties that derive from their regular, linear structures and interchain hydrogen bonds. Their structures also make them highly insoluble, intractable (difficult to transform), and resistant to melting. Modifying their structure by introducing asymmetric or bulky pendant units increases their solubility and makes them easier to melt, especially when the formation of amide–amide linkages is inhibited. One way to suppress the formation of these bonds is to modify the polymer architecture. The introduction of a dendritic structure (sphere-like geometry) prevents chain entanglement and produces new processable high-performance materials. Ideally, the molecular weight of these polymers is known, and they are spherical, defect-free, and monodisperse. This polymer structure gives rise to unique properties, such as low solution viscosity, good solubility, and poor mechanical properties.[58] Because of their properties, these polymers are suitable materials for applications in which they encapsulate molecules, e.g., catalysis, chromatography, self-assembly, and medical and biological applications. Spherical polymers can be prepared by tedious multiple-step procedures involving either divergent or convergent growth.[59] The former is based on the reaction of previously formed dendrons with a central core molecule, resulting in growth of the dendritic molecules in one direction from the core. The latter is based on the slow addition of low-molecular-weight building blocks to a polyfunctional core molecule, leading to a radial and globally symmetric growth.[60] However, another class of branched polymers, the hyperbranched polymers, are easier to prepare and are usually synthesized in a one-pot reaction under conventional polymerization procedures, making them of significant industrial interest. In contrast to dendrimers, hyperbranched polymers are neither monodisperse nor defect-free. Moreover, while perfect dendrimers are only composed of dendritic (*D*) and terminal (*T*) structural units, hyperbranched polymers also have linear (*L*) units, and their structure depends on the structure of their monomer unit.

The structure can be characterized empirically by the ratio of the dendritic, linear, and terminal units, which describes the degree of branching (DB) or most probable $\overline{\text{DB}}$ and the average number of branches (ANB):[61,62]

$$\text{DB} = \frac{2D}{2D+L} = \frac{D+T-N}{D+L+T-N} \tag{8.2}$$

where *D*, *T*, and *L* are the total number of dendritic, terminal, and linear units, respectively, and *N* is the number of hyperbranched polymer molecules. DB = 1 for a perfect dendrimer, DB = 0 for a linear polymer, and DB < 1 for hyperbranched polymers.

This equation is limited to AB_2 monomers. For the general case of hyperbranched polymers based on AB_x ($x > 2$) monomers, the following equation is used:

Aromatic Polyamides

$$\overline{\text{DB}} = \left(\frac{x-1}{x}\right)^{x-1} \qquad (8.3)$$

where x is the multiplicity of the monomer. Later, it was proposed that to determine the structure of hyperbranched polymers, the number of branches deviating from the linear direction per nonterminal monomer unit should be considered. For an AB_2 system, the average number of branches is defined as

$$\text{ANB} = \frac{D}{D+L} \qquad (8.4)$$

A more detailed overview of the synthesis, properties, and applications of dendritic and hyperbranched aromatic polyamides can be found in the work of Scholl et al.[63] and Jikei and Kakimoto.[58]

8.4 PROPERTIES

The research and development of aramid fibers started in the 1960s with the introduction of Kevlar and Nomex by the DuPont Company and continues to grow today. A description of the research history of principal aramid fibers is given in Table 8.2.

Aromatic polyamides are used as high-performance materials due to their good solvent, thermal (high transition and degradation temperatures), mechanical (high unit tensile strength and modulus), heat, flame, and chemical resistance, which can be attributed to their fully aromatic chemical structure containing amide linkages. As it has been previously pointed out, the properties of aramids are closely linked to their aromatic structure and *meta* and *para* substitutions.

TABLE 8.2
Research History of Principal Aramid Fibers

Year	Material	Developer/Producer	Base Polymer
1965	Fiber B	DuPont Co., USA	
1967	Nomex	DuPont Co., USA	MPIA
1971	Kevlar	DuPont Co., USA	PPTA
1972	Teijinconex	Teijin Ltd., Japan, and DuPont Co., USA	MPIA
1982	Twaron (originally Fiber X and Arenka)	AKZO, Netherlands/Teijin Ltd., Japan	PPTA
1987	Technora	Teijin Ltd., Japan	ODA/PPTA

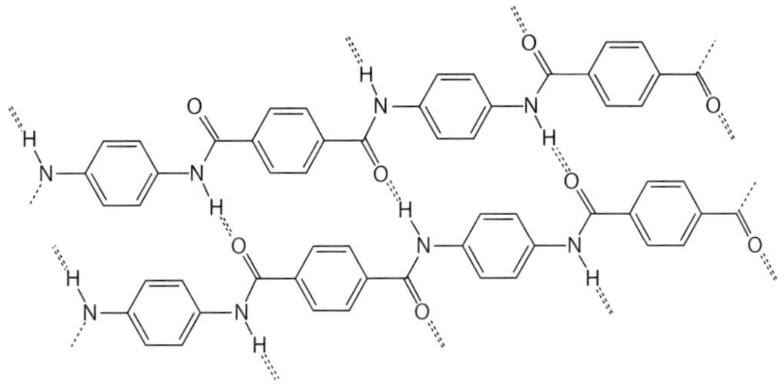

SCHEME 8.11 Schematic representation of the highly directional interchain hydrogen bonds formed in the PPTA structure.

Aromatic polyamides with all-*para* substitutions (PPTA)[9] are highly crystalline materials that exhibit large-scale organization and large cohesive energies. They consist of stiff rodlike macromolecular chains that interact with each other via strong and highly directional hydrogen bonds (Scheme 8.11), leading to compact intermolecular packing that favors the formation of crystalline domains.[11] Another partially *para*-substituted copolyamide resulting from the copolymerization of TPC, PPD, and ODA has enhanced solubility (ODA/PPTA).

Compared to *para*-oriented aramids, aromatic polyamides with all-*meta* orientations in the phenylene ring (MPIA) are significantly less crystalline (exhibit a moderate tendency to crystallize under suitable conditions, i.e., heat drawing) and have less linear structures, which are associated with a decrease in the cohesive energy and tendency to crystallize. These aromatic polyamides still exhibit high thermal and mechanical resistance.

The physical, thermal, and tensile properties of aramid fibers are summarized in Tables 8.3 through 8.5.[10,11,64] Aramid fibers exhibit unique properties due to their significantly higher tensile strengths and moduli than other organic fibers. Their thermal and mechanical properties depend on their aromatic structures and amide linkages.

Some of the most important properties of aromatic polyamides are their very good thermal properties (Table 8.4) due to the high bond dissociation energies of C–C and C–N bonds, which result in decomposition temperatures that exceed 400°C and 500°C for *meta* and *para* aromatic polyamides, respectively.[65,66] Because of their hydrogen bonding and chain rigidity, their glass transition temperatures (T_g) are very high, with values of 272–275°C for MPIA and 295°C for PPTA (low crystallinity, measured by Aharoni).[67] These structural characteristics are also responsible for their

TABLE 8.3
Physical Properties of Commercial Aramid Fibers

Physical Properties	Polymer		
	MPIA	PPTA	ODA/PPTA
Density, g/cm^3	1.37–1.38	1.44	1.39
Water uptake at 65% RH, %	5.2	3.9	4.0

Source: Data from Garcia, J.M. et al., Aromatic polyamides (aramids), in *Handbook of Engineering and Specialty Thermoplastics*, vol. 6, edited by Thomas, S. and Vishak, P.M., 141–181, John Wiley & Sons, Hoboken, NJ, and Scrivener Publishing, Salem, MA, 2011.
Note: RH, relative humidity.

TABLE 8.4
Thermal Properties of Aromatic Polyamides

Thermal Properties	Polymer		
	MPIA	PPTA	ODA/PPTA
T_g, °C	275	–	–
T_m, °C	365 (decompose)	>500 (decompose)	>500 (decompose)
T_d, °C, in N$_2$	400–430	520–540	500
T_d, °C, in air	420	430–480	–

Source: Data from Garcia, J.M. et al., Aromatic polyamides (aramids), in *Handbook of Engineering and Specialty Thermoplastics*, vol. 6, edited by Thomas, S. and Vishak, P.M., 141–181, John Wiley & Sons, Hoboken, NJ, and Scrivener Publishing, Salem, MA, 2011.

TABLE 8.5
Tensile Properties of Aramid Fibers

Tensile Properties	Polymer		
	MPIA	PPTA	ODA/PPTA
Strength, GPa	0.59–0.86	2.9–3.0	3.4
Modulus, GPa	7.9–12.1	70–112	72
Elongation, %	20–45	2.4–3.6	4.6
Crystallinity, %	Highly crystalline (68–95)	Highly crystalline	Highly oriented/less crystalline
Flammability (LOI)	28–32	29	25

Sources: Data from Gallini, J., Polyamides aromatic, in *Encyclopedia of Polymer Science and Technology*, vol. 3, edited by Mark, H.F. and Kroschwitz, J.I., 558–584, John Wiley & Sons, New York, 1989; Garcia, J.M. et al., Aromatic polyamides (aramids), in *Handbook of Engineering and Specialty Thermoplastics*, vol. 6, edited by Thomas, S. and Vishak, P.M., 141–181, John Wiley & Sons, Hoboken, NJ, and Scrivener Publishing, Salem, MA, 2011.

limited solubility and good chemical resistance (high energy absorption, corrosion resistance, good dielectric behavior, nonmagnetic properties, and good fatigue resistance).

Aramid fibers have a specific strength and stiffness that are greater than those of glass fibers and exhibit very good tensile properties at temperatures over 400°C. They are also able to absorb large amounts of energy during fracture. The macroscopic mechanical properties of a polymer are determined by the molecular structure of the chain, which results in chain alignment in one direction with a narrow orientation distribution around the filament axis. Tensile properties are typically measured in a stress–strain test after drying at 140°C. They are comparable to those of cast metals, and aramids are stronger than steel wire and stiffer than glass, making them suitable replacements for metal (Table 8.5). For example, the strength and modulus of Kevlar aramid fibers (PPTA) are twice and nine times larger, respectively, than those of high-strength nylon.[68] Another set of properties that are becoming increasingly important in industry is related to impact behavior, including behavior due to collisions, explosions, and bullet impacts. This type of behavior is measured using standard test methods for testing the impact strength under ordinary conditions.[69,70]

The principal properties of commercial fibers are listed in tables to provide an overview of the following commercial aramid products: Kevlar (Tables 8.6 and 8.7); Twaron (PPTA), Technora (ODA/PPTA), and Teijinconex (Tables 8.8 and 8.9); and Nomex (MPIA) (Table 8.10).[10,71,72]*

Kevlar is chemically stable under a wide range of exposure conditions, including over long periods of exposure and at elevated temperatures. However, several strong aqueous acids, bases, and sodium hypochlorite can degrade these materials.*

A comparison of Kevlar to other commercial materials reveals that Kevlar yarn has a break tenacity more than five times that of steel wire and twice that of nylon, polyester, or glass fibers (Figure 8.2, left). At 150°C, Kevlar has a modulus more than three times that of Dacron and nylon at room temperature and two times that of steel wire (Figure 8.2, right).

Twaron (PPTA) is another high-performance commercial *para*-aramid fiber with high strength (excellent strength-to-weight properties), modulus, and dimensional stability; excellent heat, cut, and chemical resistance; no melting point; low flammability; and insulator properties.

Technora is a *para-/meta*-aramid fiber with high tensile strength (this material is much stronger than steel on a weight basis); good fatigue resistance; long-term dimensional stability; and excellent

---

* Technical Guide for NOMEX® Brand Fiber, H-52703, 1999; Technical Guide for KEVLAR® Aramid Fiber, H-77848, 2000; Forging the Future with KEVLAR®, H-51009, E. I. du Pont de Nemours & Co., Inc., Wilmington, Del., 1997. Teijinconex®, Technical Information, CN02/91.2, Teijinconex®, Aromatic Polyamide Fiber; High Tenacity Aramid Fiber; Technora®, Teijin.com/advanced fibres, Teijin Ltd., Japan. Twaron®, Product Information: Yarns, Fibers and Pulp, Twaron® in Brake Linings and Clutch Facings, Akzo Nobel, the Netherlands.

TABLE 8.6
Principal Properties of Kevlar Aramid Yarns and Fibers

Density, g/cm³	Tensile Strength at Elevated Temperatures, GPa		Strength, GPa	Elongation at Break, %	Modulus, GPa
	200°C	250°C			
1.44–1.47	2.2–2.6	2–2.4	2.9–3	2.4–3.6	71–112

Glass Transition Temperature, °C	Max. Temperature Range for Use in Air, °C	Decomposition Temperature in Air, °C	Heat of Combustion, J/Kg	Breaking Tenacity, GPa	Flammability, LOI, %
345–360	427–482	427–482	35×10^6	29.2–30	29

Source: Data from Ozawa, S. and Matsuda, K., Aramid copolymer fibers, in *Handbook of Fiber Science and Technology, Vol. III, High Technology Fibers Part B*, edited by Lewin, M. and Preston, J., 1:1–34, Marcel Dekker, New York, 1989; Technical Guide for NOMEX® Brand Fiber, H-52703, 1999; Technical Guide for KEVLAR® Aramid Fiber, H-77848, 2000; Forgoing the Future with KEVLAR®, H-51009, E. I. du Pont de Nemours & Co., Inc., Wilmington, Del., 1997. Teijinconex®, Technical Information, CN02/91.2, Teijinconex®, Aromatic Polyamide Fiber; High Tenacity Aramid Fiber; Technora®, Teijin.com/advanced fibres, Teijin Ltd., Japan. Twaron®, Product Information: Yarns, Fibers and Pulp, Twaron® in Brake Linings and Clutch Facings, Akzo Nobel, the Netherlands.

TABLE 8.7
Properties of Different Types of Kevlar

Fiber	Characteristic	Crystalline Modulus, GPa	Loop Elongation, %	Modulus, GPa	Tensile Strength, GPa	Extension to Break, %	Density, g/cm³
Kevlar 29	Regular	153	2.1	70	2.9	4.0	1.44
Kevlar 49	High modulus	156	1.3	135	2.9	2.8	1.45
Kevlar 100	Colored	156	2.3	60	2.9	3.9	1.44
Kevlar 119	High durability	156	2.7	55	3.1	4.4	1.44
Kevlar 129	High strength	156	–	99	3.4	3.3	1.45
Kevlar 149	Ultrahigh modulus	156	0.6	143	2.3	1.5	1.47
Kevlar 981	–	–	–	–	120	3.5	2.8	–

Source: Data from Yang, H.H., *Kevlar Aramid Fibers*, E. I. du Pont Fibers, John Wiley & Sons, Hoboken, NJ, 1993; Jassal, M. and Gosh, S., *Indian J. Fiber Text. Res.*, 27, 2002.

resistance to corrosion, heat, chemicals, and saltwater. At high temperatures, Technora retains more than a half of its room-temperature tensile strength and modulus (Figure 8.3).

Teijinconex is a *meta*-linked aromatic polyamide fiber with excellent heat resistance and low flammability; it does not catch fire when exposed to direct flame or heat, and the material neither burns nor melts, which means it cannot stick to skin. It is lightweight, comfortable to wear, strong, and soft, and it offers excellent resistance to long-term heat exposure. It is also self-lubricating and has good abrasion resistance.

Twaron yarns are very strong; their tensile strength is two to three times higher than that of high-strength polyester and polyamide yarns and five times higher than that of steel (on a weight basis), as observed in their stress–strain curves (Figure 8.4).

TABLE 8.8
Properties of Twaron, Technora, and Teijinconex Fibers

Fiber	Twaron	Technora	Teijinconex
Density, g/cm^3	1.44–1.45	1.39	1.38
Tensile strength, GPa	2.4–3.6	3.4	0.62–0.69
Tenacity, MPa	18.7–26.5	26.5	5.1–7.8
Modulus, GPa	70–110	72	7.9–9.8
Elongation at break, %	2.2–4.4	4.6	35–45
Moisture, wt.%	3.2–5	1.9	5–5.5
Glass transition temperature, °C	–	–	280
Decomposition or melting temperature, °C	500	500	400
LOI	29–37	25	29–32

Source: Data from Gallini, J., Polyamides aromatic, in *Encyclopedia of Polymer Science and Technology*, vol. 3, edited by Mark, H.F. and Kroschwitz, J.I., 558–584, John Wiley & Sons, New York, 1989; Technical Guide for NOMEX® Brand Fiber, H-52703, 1999; Technical Guide for KEVLAR® Aramid Fiber, H-77848, 2000; Forgoing the Future with KEVLAR®, H-51009, E. I. du Pont de Nemours & Co., Inc., Wilmington, Del., 1997. Teijinconex®, Technical Information, CN02/91.2, Teijinconex®, Aromatic Polyamide Fiber; High Tenacity Aramid Fiber; Technora®, Teijin.com/advanced fibres, Teijin Ltd., Japan. Twaron®, Product Information: Yarns, Fibers and Pulp, Twaron® in Brake Linings and Clutch Facings, Akzo Nobel, the Netherlands.

TABLE 8.9
Comparison of the Properties of Teijinconex to Those of Other Natural (Not Colored) Organic and Conventional Synthetic Fibers

	Unit	Teijinconex SF, reg	Teijinconex SF, HT	Nylon66 SF	Polyester SF	Cotton SF	Glass –
Tensile strength	cN/dtex	4.4–4.9	5.3–6.2	3.9–6.6	4.1–5.7	2.6–4.9	8.5
	MPa	590–690	740–830	440–750	580–800	390–600	2900–4000
Elongation	%	35–45	20–30	25–60	20–50	6–10	2–4
Young's modulus	cN/dtex	57–71	83–88	8–26	22–62	60–80	270
	GPa	7.8–9.8	11.7–12.7	1.0–3.0	3.0–8.5	5.8–11.0	80
Moisture content	%	5.0–5.5	5.0–5.5	3.5–5.0	0.4–0.5	7.0	–
Specific gravity	–	1.38	1.38	1.14	1.38	1.50–1.54	2.5

Source: Data from Technical Guide for NOMEX® Brand Fiber, H-52703, 1999; Technical Guide for KEVLAR® Aramid Fiber, H-77848, 2000; Forgoing the Future with KEVLAR®, H-51009, E. I. du Pont de Nemours & Co., Inc., Wilmington, Del., 1997. Teijinconex®, Technical Information, CN02/91.2, Teijinconex®, Aromatic Polyamide Fiber; High Tenacity Aramid Fiber; Technora®, Teijin.com/advanced fibres, Teijin Ltd., Japan. Twaron®, Product Information: Yarns, Fibers and Pulp, Twaron® in Brake Linings and Clutch Facings, Akzo Nobel, the Netherlands.

Note: HT, high-tenacity type; reg, regular type; SF, staple fiber.

TABLE 8.10
Properties of Different Types of Nomex

Fiber	Nomex 430	Nomex 450	Nomex 455/462
Density, g/cm^3	1.38	1.37	–
Tenacity, GPa	0.63	0.37	0.33
Elongation, %	30.5–31	22	21
Initial modulus, GPa	10.8–11.9	–	–
Loop tenacity, GPa	0.5–0.52	–	–
Heat of combustion, J/kg	28.1×10^6	28.1×10^6	28.1×10^6
Specific heat at 25°C, cal/g °C	0.30	0.30	0.26

Source: Data from Technical Guide for NOMEX® Brand Fiber, H-52703, 1999; Technical Guide for KEVLAR® Aramid Fiber, H-77848, 2000; Forgoing the Future with KEVLAR®, H-51009, E. I. du Pont de Nemours & Co., Inc., Wilmington, Del., 1997. Teijinconex®, Technical Information, CN02/91.2, Teijinconex®, Aromatic Polyamide Fiber; High Tenacity Aramid Fiber; Technora®, Teijin.com/advanced fibres, Teijin Ltd., Japan. Twaron®, Product Information: Yarns, Fibers and Pulp, Twaron® in Brake Linings and Clutch Facings, Akzo Nobel, the Netherlands.

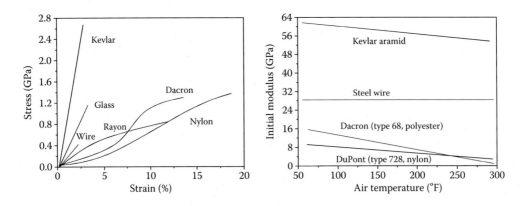

FIGURE 8.2 A comparison of Kevlar to other commercial materials.

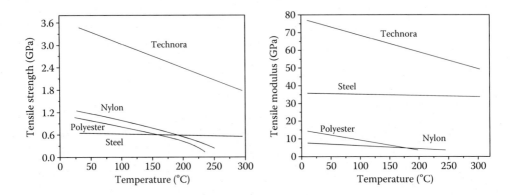

FIGURE 8.3 Technora is a *para-/meta*-aramid fiber with high tensile strength and modulus.

Aromatic Polyamides

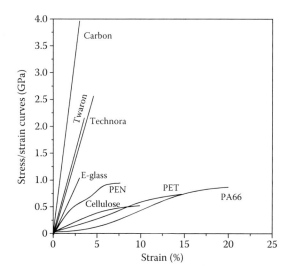

FIGURE 8.4 Twaron yarns are very strong, with a tensile strength higher than that of high-strength polyester and polyamide yarns and five times higher than that of steel (on a weight basis).

Teijinconex also exhibits performance properties superior to those of natural organic and conventional synthetic fibers such as polyester and nylon (Table 8.9).*

Finally, Nomex is used in both 100% Nomex materials and blends of Nomex and Kevlar.[†] Nomex 430, a 100% Nomex material, is a highly crystalline natural (not colored) filament yarn with high strength and chemical resistance. It is a natural staple fiber with higher crystallinity and strength than type 455 and 462 fibers and is considered to be a 100% *meta*-aramid staple. Nomex 455 (or Nomex III) and 462 (or Nomex IIIA) are blends with Kevlar that exhibit higher-performance thermal protective properties (Table 8.10). Type 455 has a lower crystallinity and is lower in strength than type 450. Nomex 462 is a blend with Kevlar and a static dissipative fiber, P-140. The properties of Nomex 455 and Nomex 462 are similar except for the static dissipative properties of Nomex IIIA.

In general, Nomex fibers exhibit very good resistance to many chemicals, including most hydrocarbons and organic solvents, which gives these materials excellent durability and wear life. While Nomex also shows resistance to alkalis at room temperature, it is degraded by strong alkalis at high temperatures. Nevertheless, data obtained in a laboratory pulse-jet unit demonstrate that Nomex is more durable than polyester when treated with or without acid at temperatures from 107°C to 177°C.[†]

Optically, aromatic polyamides are sensitive to ultraviolet (UV) light.[73] Materials that are exposed to indoor light or sunlight have to be protected by means of coating or with additives. The critical wavelength region is 300–450 nm for PPTA.

Another interesting property of *para*-aramids is the dependence of the viscosity on the polymer concentration in H_2SO_4 solutions (Figure 8.5).[74] These polymer solutions exhibit liquid crystalline behavior. Thus, at polymer concentrations below 12%, the polymer solution is isotropic, and the viscosity increases sharply with increasing polymer concentration. Above concentrations of 12%,

---

* Technical Guide for NOMEX® Brand Fiber, H-52703, 1999; Technical Guide for KEVLAR® Aramid Fiber, H-77848, 2000; Forgoing the Future with KEVLAR®, H-51009, E. I. du Pont de Nemours & Co., Inc., Wilmington, Del., 1997. Teijinconex®, Technical Information, CN02/91.2, Teijinconex®, Aromatic Polyamide Fiber; High Tenacity Aramid Fiber; Technora®, Teijin.com/advanced fibres, Teijin Ltd., Japan. Twaron®, Product Information: Yarns, Fibers and Pulp, Twaron® in Brake Linings and Clutch Facings, Akzo Nobel, the Netherlands.

† Technical Guide for NOMEX® Brand Fiber, H-52703, 1999; Technical Guide for KEVLAR® Aramid Fiber, H-77848, 2000; Forgoing the Future with KEVLAR®, H-51009, E. I. du Pont de Nemours & Co., Inc., Wilmington, Del., 1997. Teijinconex®, Technical Information, CN02/91.2, Teijinconex®, Aromatic Polyamide Fiber; High Tenacity Aramid Fiber; Technora®, Teijin.com/advanced fibres, Teijin Ltd., Japan. Twaron®, Product Information: Yarns, Fibers and Pulp, Twaron® in Brake Linings and Clutch Facings, Akzo Nobel, the Netherlands.

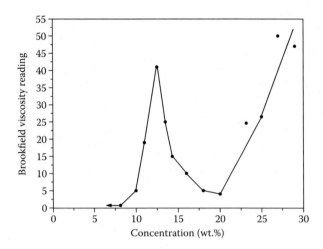

FIGURE 8.5 An interesting property of *para*-aramids is the unusual dependence of the viscosity on the polymer concentration in H_2SO_4 solutions.

the viscosity decreases as the concentration increases because the solution becomes anisotropic. At a concentration of approximately 19%, the solution viscosity increases again, and above 20%, the solution becomes completely liquid crystalline.[71] This behavior is especially important in fiber production via wet spinning. Solutions of MPIA and ODA/PPTA are isotropic and exhibit a conventional concentration–viscosity relationship.

The intrinsic viscosity $[\eta]$ of a polymer is related with its weight average molecular weight (M_w) by the Mark–Houwink equation, $[\eta] = kM_w^{\alpha}$, where the values of the constants k and α for polymer solutions in 96% sulfuric acid are 0.013 and 0.84, respectively, for MPIA, and 0.008 and 1.09, respectively, for PPTA.[10] For quality control of the molecular weight at the laboratory scale, the inherent viscosity (η_{inh}) is usually measured for practical reasons, i.e., η_{inh} can be obtained in a single measurement in minutes by measuring the relative viscosity (η_{rel}) and using the equation $\eta_{inh} = \ln(\eta_{rel})/C$. Its value at a given aramid concentration (C), usually 0.5 g/dL, and temperature with a given solvent are key parameters for ensuring that a minimum molecular weight is achieved. Thus, it is generally accepted that a polymer prepared at the laboratory scale has a reasonably high molecular weight if $\eta_{inh} \geq 0.5$ dL/g (solvent: DMA, temperature = 25 ± 0.1°C). On the other hand, inherent viscosities greater than 1.7 dL/g and between 4 and 7 dL/g (solvent: 96% H_2SO_4, temperature = 30 ± 0.1°C) are generally expected for commercial MPIA and PPTA fibers, respectively. The organic solvents used in aramid solutions play an important role in the polymer properties, and special care is required when analyzing and comparing any data. For instance, MPIA solutions in H_2SO_4 have a higher viscosity, normal stresses, and storage modulus than DMA/LiCl solutions at a given concentration and temperature.[75]

One of the special properties of aromatic polyamides is their heat resistance and flammability (flame retardancy and chemical resistance) as measured by the most common flammability parameter for fibers, the limiting oxygen index (LOI). LOI is defined as the minimum oxygen concentration that will support sustained combustion of the material. The LOI for *meta*-aramids is between 30% and 32% vol., with weight loss at 450°C and use temperature of 370°C. The *para*-aramids' LOI is between 28% and 30% vol., with char yields above 450°C. The LOI of the copolymer *para*-aramid Technora is 25% vol., with an ignition point of 650°C.[76] Nomex shrinks away from a high heat source or flame, and a thick char, which acts as a thermal barrier, is produced by burning.

In summary, a comparison of the properties of aramids and other synthetic yarns demonstrates the excellent advantages of aramids; such as very high tenacity; high elasticity modulus; low creep; and good heat, fatigue, and chemical resistance.

New research fields are advancing the development and applications of high-performance materials with improved properties. For example, dendrimers and hyperbranched polymers exhibit interesting and unique properties, such as good solubility, low viscosity, and encapsulation abilities, because of their spherical architecture and despite their rigid aromatic amide and can therefore be employed in various fields. End-group modification of hyperbranched polymers affects their properties, such as the solubility and T_g; thus, dendritic and hyperbranched aromatic polyamides have been developed in recent years to produce new processable high-performance materials with good solubility and low solution viscosity.[58,63] Poly(amide imide)s (PAIs) are also described as high-performance materials with properties between those of aliphatic polyamides and aramids (T_g = 280°C). From an industrial perspective, the lower performance of PAIs compared to that of aromatic polyamides is offset by their facile preparation. Because aromatic polyamides have special properties such as high salt rejection, water permeability, and fouling tolerance, these materials are used in RO membranes.[2]

8.5 CHEMICAL STABILITY

Another fundamental property of aromatic polyamides is their chemical resistance to most hydrocarbons and organic solvents, for example, under different exposure conditions. This property is important, from the transformation of these materials into finished goods to their final application. PPTA is only soluble in strong acids (sulfuric, hydrofluoric, or methanesulfonic acid), while MPIA and ODA/PPTA are soluble in amide solvents (NMP, DMA, DMF) and dimethyl sulfoxide (DMSO). Their solubilities can be increased by adding salts (LiCl and $CaCl_2$). The solutions are stable under conventional conditions. However, the amide linkages in aramids are disrupted after long periods of exposure to strong aqueous acids and bases, leading to their degradation and loss of strength. The resistance of aramid fibers and yarn to solvents and acid/base solutions is given in Table 8.11.[10]*

8.6 PROCESSING

The applications of aromatic polyamides are based on the forms, such as fibers (continuous multifilament yarn or staple fibers), pulp, and films, of the polymers that can be prepared from their solutions.

Polymer processing converts polymers into fibrous forms by fiber growth or extrusion. Commercial aramids have very high melting points, and their transformation using conventional processing techniques, such as extrusion or injection, is not possible. To overcome this difficulty, films and fibers are prepared from the polymer solutions using spinning methods, which handle a large number of fibers, minimizing the cost of commercial processes. There are three principal spinning methods, wet, dry, and dry-jet wet spinning, and solution casting of films is also possible.

In general, the spinning solution is initially heated to 80°C to reach a suitable processing temperature, which is necessary for the production of fibers from highly concentrated solutions in 100% sulfuric acid. The concentration limit of the polymer in the spinning solution is 20 wt.%.[77] When concentrations above this limit are used, the undissolved material affects the spinnability, and the resulting material has inferior mechanical properties. The polymer solutions are then extruded through a spinneret to form fibers. The presence of a small air gap allows for the reorientation and elongational stretching of the polymer chains. Finally, the highly directional orientation of the fibers

---

* Technical Guide for NOMEX® Brand Fiber, H-52703, 1999; Technical Guide for KEVLAR® Aramid Fiber, H-77848, 2000; Forgoing the Future with KEVLAR®, H-51009, E. I. du Pont de Nemours & Co., Inc., Wilmington, Del., 1997. Teijinconex®, Technical Information, CN02/91.2, Teijinconex®, Aromatic Polyamide Fiber; High Tenacity Aramid Fiber; Technora®, Teijin.com/advanced fibres, Teijin Ltd., Japan. Twaron®, Product Information: Yarns, Fibers and Pulp, Twaron® in Brake Linings and Clutch Facings, Akzo Nobel, the Netherlands.

TABLE 8.11
Chemical Resistance of Aramid Yarns in Terms of Percentage Strength Loss Values of MPIA and PPTA

Chemical		Concentration, %	Temperature, °C	Time, h	Strength Loss, %	
					MPIA	PPTA
Acids	Hydrochloric	10	21	1000	41–80	
		37	71	10	81–100	
		37	21	24		–
		10	21	100		41–80
	Nitric	1	21	100	11–20	11–20
		10	21	100	21–40	41–80
		70	21	100	41–80	
		70	21	24	41–80	
	Sulfuric	10	21	100	–	–
		70	21	100	–	21–40
	Acetic	100	99	100	–	
		100	21	24		–
		40	21	1000		11–20
	Formic	40	21	1000	–	
		90	21	100	–	–
Bases	Ammonium hydroxide	28	21	1000	–	
	Sodium hydroxide	10	21	1000	–	41–80
		10	99	100	81–100	81–100
		40	21	100	–	
		40	21	1000	11–20	
Oxidizing and reducing agents	Sodium hypochlorite	0.01, pH 10	21	1000	–	
		0.01, pH 10	71	100	11–20	
		0.04, pH 11	71	100	11–20	
		0.1	21	1000		81–100
	Hydrogen peroxide	0.4, pH 7	71	100	–	
Organic solvents	Acetone, benzene, carbon tetrachloride, ethyl ether, ethyl alcohol, methylene chloride	100	21	1000	–	–
Salt solutions	Sodium chloride	10	99	100	–	–
		10	121	100	11–20	41–80
	Sodium carbonate	5	121	100	11–20	
	Ferric chloride	3	99	100	21–40	41–80

Source: Data from Gallini, J., Polyamides aromatic, in *Encyclopedia of Polymer Science and Technology*, vol. 3, edited by Mark, H.F. and Kroschwitz, J.I., 558–584, John Wiley & Sons, New York, 1989; Garcia, J.M. et al., Aromatic polyamides (aramids), in *Handbook of Engineering and Specialty Thermoplastics*, vol. 6, edited by Thomas, S. and Vishak, P.M., 141–181, John Wiley & Sons, Hoboken, NJ, and Scrivener Publishing, Salem, MA, 2011.

Aromatic Polyamides

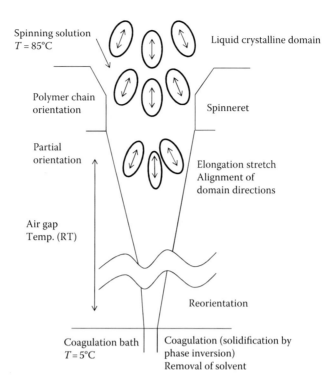

FIGURE 8.6 Aramid production through a dry-jet wet spinning process. RT, room temperature.

is achieved via coagulation with cold water, which extracts the solvent from the polymer, resulting in the formation of the polymer fiber (Figure 8.6).

8.6.1 Wet Spinning

Polymer solutions are converted into fibers by dissolving the dry aromatic polyamide in an organic solvent and subsequently precipitating it in a coagulating bath, where the solvent is diluted. This process gives polymer threads with specific structures.

8.6.1.1 Wet Spinning of MPIA

This process involves dissolving the dry polymer at low temperatures and then heating the dispersion to approximately 100°C to form a clear solution. After that, the MPIA solution is wet-spun into an aqueous solution containing a high concentration of an inorganic salt. The coagulated fiber is washed and then drawn and posttreated to obtain fibers with excellent mechanical properties.

8.6.1.2 Wet Spinning of PPTA

For *para*-polyamides (PPTA), the polymer is dissolved in 98–100% sulfuric acid at a concentration greater than 18%,[78] and the solution is subsequently pumped through a spinneret into an aqueous coagulating bath to produce highly oriented fibers. The fibers formed during this spinning process are highly oriented, although high-temperature drawing, which is commonly used for other polyamide fibers, is not employed, and they are very stiff (tensile moduli of 50–75 GPa).

8.6.1.3 Wet Spinning of ODA/PPTA

Fibers of ODA/PPTA are obtained from an NMP solution containing $CaCl_2$ that is filtered and deaerated.[79,80] The solution is then pumped through a spinneret into a hot water/$CaCl_2$ bath. The resulting fibers have tenacities in the range of 2.6–3.3 GPa.

8.6.2 Dry Spinning

In dry spinning, the polymer solution is also forced through a spinneret, but the solvent is evaporated in a warm air current to produce fibers that are nearly solvent-free. For MPIA fibers, a DMF/$LiCl_2$ solution is moved through an air column maintained at 225°C, resulting in fibers with a tenacity of 0.6 GPa and an elongation of 30%.[81]

8.6.3 Dry-Jet Wet Spinning

The dry-jet wet spinning process (Figure 8.6) was invented by Blade.[82] In this process, an anisotropic solution of aramid polymer is extruded through spinneret holes to orient the liquid crystalline domains along the direction of flow through an air gap and overcome the deorientation process at the capillary exit. Then, the filaments are washed and neutralized in a coagulation bath.[71]

8.6.4 Film Casting of PPTA

Film casting is used to produce biaxially oriented PPTA films.[83] PPTA must first be dissolved in sulfuric acid in a concentration that produces a liquid crystalline state (it is possible to condense the material directly in an NMP solution and cast the solution, which produces a transparent film without having to use concentrated sulfuric acid).[84] Then, the viscous solution is extruded through a die to a belt, where it is subjected to high-humidity warm air, causing the solution to become isotropic. The casting solution is heated and then placed in a coagulating liquid, such as dilute sulfuric acid. The resulting film is washed with water to remove the acid and stretched biaxially to orient the polymer. Finally, it is dried and heat-treated.

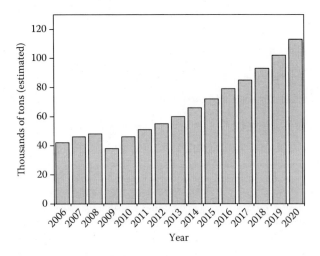

FIGURE 8.7 The annual global *para*-aramid market and outlook.

Aromatic Polyamides

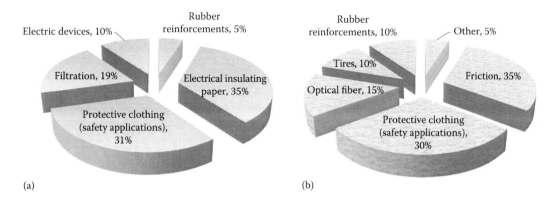

FIGURE 8.8 Market breakdown for *meta*-aramid (a) and *para*-aramid (b).

8.7 GLOBAL MARKET VIEW

The global aramid market is dominated by two companies, DuPont and Teijin, because of the difficulty in accessing the technology and extremely well-protected intellectual property. The world aramid demand peaked in 2008 and dropped 20% to a minimum in 2009. It subsequently recovered to prerecession levels and has an expected annual growth of approximately 7–9% (Figure 8.7).

The commercial applications of aramids will be described in the next section. The annual increase in demand and present global market breakdown (Figure 8.8) reflect the increasing need for lightweight materials in many end products, which would help to reduce energy consumption and carbon footprints and provide greater security and safety.[85]

8.8 APPLICATIONS

Aromatic polyamides became commercially available with the marketing launch of Nomex fibers in the 1960s; Kevlar fibers in 1971; and another *para*-aramid, Twaron, which is similar to Kevlar, and the copolyamide Technora in the 1980s. Aromatic polyamides opened up new horizons in materials research for improving thermal and mechanical properties.

Most commonly, aromatic polyamides are employed in advanced aromatic composites or high-performance fiber-derived products (fibers, filaments, yarns, fabrics, etc.) because of the thermal, chemical, and mechanical resistance of their highly oriented, crystalline fibers.

In the textile market, MPIA fibers are important in fire protective clothing for firefighters and operators in chemical factories and are used as fire-blocking materials in other places with a real fire danger, such as aircraft upholstery and hospitals, because of its inherent flame, abrasion, wear, and chemical resistance.

According to the International Organization for Standardization (ISO), the unit used to express the fiber fineness is the *tex* (1 dtex = 1 g/1000 m). Although the fiber thickness ranges from 1 to 100 μm, it is difficult to measure, and the tex is used to statistically validate the fineness of a fiber (fiber density). Another commonly used unit is the *denier* (1 denier = 1 g/9000 m).[85]

Because of their chemical resistance and good insulating characteristics, mixtures of MPIA fibrids and floc[86] can be used in paper, another important market, which includes electrical insulation in motors and generators.

PPTA and ODA/PPTA fibers are used in a wide range of applications where high strength and abrasion resistance are important. These fibers are used in protective clothing for police and military personnel because they are high-strength, lightweight, ballistic-resistant materials.[87,88] PPTA is also employed in cut-resistant fabrics that are used to make gloves and chainsaw chaps. High-strength

TABLE 8.12
Specific Examples of Uses of Aromatic Polyamides

Field of Application	Examples of Use
Friction materials	Brake pads
Optical applications	Optical fiber cables
Protective applications	Firefighting, ballistics, cut protection
Reinforcing fibers	Tires, pipes
Sporting equipment	Sail cloths

composites containing PPTA are used in the aircraft and transportation industries, military and police vehicles, and gaskets for applications where it is necessary to seal adjacent surfaces to prevent gas leakage.

Other markets of interest include ropes and cables based on PPTA fibers. These fibers can provide tension reinforcement for fiber-optic cables and can be used as rubber reinforcement fibers in tires, conferring stability and durability; and rubber transmission belts and hoses that require high temperature, pressure, and chemical resistance.

PPTA films have a higher tensile modulus than polyester and are therefore more desirable in industry. The combination of their high tensile modulus and high-temperature stability makes PPTA films suitable for magnetic recording media, acoustic diaphragms, and electronic applications.

It is important to note that thin-film membranes have recently received tremendous attention for a wide range of applications, such as desalination, purification of chemical and biological products, wastewater treatment, chemical sensors, gas separation, etc.[89–92] Due to the wide temperature range over which aromatic polyamides can be employed, these materials could be used in cryogenic fuel tanks, which would test their helium gas permeability.[93] Some specific uses are given in Table 8.12.

Aromatic hyperbranched polyamides are also useful for a variety of applications as supports for the preparation of magnetic nanoparticles, rheology modifiers of linear polyamides, and as supports for protein immobilizers.[63]

8.8.1 Expanding Applications of Aromatic Polyamides

8.8.1.1 Aromatic Polyamides with Unique Properties

The design, synthesis, and preparation of chiral polymers (optically active polymers [OAPs]), optically active polyamides in particular, is a currently relevant research field because of their applications in chromatographic techniques, nonlinear optical devices, and chiral liquid crystalline ferroelectrics.[94–96] The simplest approach generally involves the polymerization of chiral monomers. Pendant polyisophthalamides with a lateral L-isoleucine core group were synthesized by Mallakpour et al.[21,97] These polymers are optically active. Furthermore, the presence of amino acids in the polymer architecture makes them environmentally friendly, and they could potentially be used as the chiral stationary phase in gas chromatography (GC) to separate racemic mixtures. In another work, OAPs, specifically polyisophthalamides having a pendant L-alanine or methionine and phthalimide groups, were polymerized in ionic liquids. The resulting polymers had decomposition temperatures greater than 300°C and an LOI greater than 35 (according to the van Krevelen and Hoftyzer equation[98]), making them self-extinguishing materials.

Luminescent and electrochromic polyamides emit photoluminescence (PL) or electroluminescence (EL). Light-emitting phenomena are characteristic of luminescence materials that are exposed to an electric current or absorb photons.[99,100] The properties of these aromatic polyamides are interesting for the production of organic light-emitting diodes (OLEDs), or more specifically,

TABLE 8.13
Examples of Uses of Polyamide Membrane Technology

Use of gas membrane technology
Separation of carbon dioxide and water from natural gas
Separation of nitrogen and sulfur oxides from industrial gas streams and polluted atmosphere
Enrichment of synthesis gas
Recovery of hydrogen from ammonia synthesis
Separation of nitrogen and oxygen from air
Other uses in the petrochemical industry

polymer light-emitting diodes (PLEDs), or luminescent converters (LUCOs).[101,102] Electrochromism (EC) is a luminescent phenomenon in which the optical properties of a material change reversibly when it is oxidized or reduced. This property is useful for mirrors and intelligent earth-tone chameleon materials.[103]

Particularly, the preparation of blue light-emitting polymers is of great interest because blue light can be converted to green or red light using the proper dyes, while the reverse transformation is not possible. Hence, much work has been focused on introducing 1,3,4-oxadiazole rings into the polymer backbone or as a pendant structure.[104,105] These rings are electron withdrawing, which facilitates the injection and transport of electrons.

Aromatic polyamides, which form the active layer in membrane technologies,[106–108] have been used in RO and nanofiltration membranes in water purification systems for many years because of their high salt rejection, contamination tolerance, and water permeability.[109] These techniques are mainly used to produce drinking water because the membranes are designed to prevent the passage of solutes and allow only water to pass through the material. They have direct applications in wastewater treatment, seawater desalination, and dialysis. Polyamide membranes can also be used in other applications, such as biochemical, biomedical, and semiconductor devices, that require purified, high-quality water. Several investigations have attempted to improve the chlorine resistance, salt retention, and flux of polyamide membranes.[110–113]

Novel permselective polymers with improved performance for the separation and purification of gases and vapor mixtures are also being developed. These applications have high economic and environmental impact, and polyamide membranes would make the processes that use gas membrane technologies more environmentally friendly. Further research and development of these membranes would increase the membrane performance and produce materials with superior processability and gas permeability. Some examples of these processes are summarized in Table 8.13.

8.8.1.2 Aromatic Polyamides as Biomaterials and for Use in Medical Applications

It is important to note that *para*-aramids can be used in medical applications, replacing metal alloys in artificial implants and prostheses and preventing problems associated with scattered radiation in diagnostic examinations. Recent works have reported the use of polyamides in artificial hair and wigs because they mimic these materials due to the amide linkage.[114] In other work, Ahamad[115] explored the formation of complexes between polyamide dendrimers and Pd(II) or Pt(II) (metallodendrimers) and studied their antibacterial activity against gram-positive and gram-negative bacteria. The cytotoxicity assays of the dendrimers and metallodendrimers showed that the second-generation dendrimers were active against all the pathogenic bacterial strains.

Another possible biomedical application is using aramids to study the interactions between DNA and artificial binders to understand processes and mechanisms in the context of gene therapy.[116]

8.8.1.3 Cross-Linked Aromatic Polyamides: A Further Step in High-Performance Materials

The properties of aramids are typically improved by introducing new monomers, chemically modifying the structure of existing polymers, and cross-linking. The first method has been extensively exploited in basic research and has mainly involved introducing flexible bonds into the main chain and bulky side groups to reduce the cohesive energy and rigidity and improve the solubility. This process usually leads to an undesirable loss in the thermal or mechanical resistance. The only example of commercial success with this strategy is the ODA/PPTA copolymer (Technora).* The second method, the chemical modification of aramids, is not practical, due to the insolubility of the polymers in conventional organic solvents. Hence, cross-linking is usually used to improve the mechanical and thermal resistance of organic polymers, but this strategy has been used infrequently for aramids due to economic and technical difficulties. Thus, a few examples of the thermal cross-linking of functional aromatic polyamides[117–121] or of aramids using cross-linking agents have been described.[122]

Recently, the introduction of a reactive azide group into the aramid structure has been shown to yield functional polymers that can be easily scaled up to current fiber production facilities (i.e., dry, wet, or dry-jet wet spinning) and have enhanced mechanical, thermal, and electrical isolation properties of the commercial aramids.[123]

8.8.1.4 Future Developments

Because some of their special and unique properties, such as their high transition temperatures and poor solubility in common organic solvents, limit their applications due to processing difficulties, current research on aromatic polyamides aims to improve their processability, mechanical properties, and solubility for use in new applications, such as biodegradable materials, molecular recognition systems, cross-linked fibers, fluorescent and intelligent yarns, etc.[124,125]

8.9 SPECIAL ADDITIVES

To improve the poor resistance of aramid fibers to UV light and prevent color change from UV exposure, particularly in the textile industry, UV absorbers or light screeners are incorporated into aramid fibers or used during fabric manufacturing with these materials. Encapsulation of the aramid fibers by meshing with other fibers is also used to protect aramid fibers in the 300–450 nm range.

Due to the high crystallinity of these fibers and protective barrier formed by the external fibers that shield the internal fibers, the normal textile dye process is not possible. Filament yarns with low crystallinity are therefore employed when dyed fibers are needed. Textile pigment printing with an insoluble coloring material that adheres to the fabric by a resin binder, which might be a UV stabilizer, is another dyeing option.[126] These resin binders are acrylic copolymer binders or modified nitrile polymer binders that increase the photostability of the aramid fibers.[127]

For both aromatic and aliphatic polyamides, the electrical charge at the polymer surface must be dissipated for certain applications due to the risk of sparking upon discharge. To increase the electrical conductivity, sulfonated polyaniline can be added.[128,129] Another method for producing conductive aramids is electroless plating, which results in a very electrically conductive fiber. In this process, the fibers are impregnated with metal complexes, which are subsequently activated by reduction with hydrogen and immersed into an electroless plating solution.[130]

---

* Technical Guide for NOMEX® Brand Fiber, H-52703, 1999; Technical Guide for KEVLAR® Aramid Fiber, H-77848, 2000; Forging the Future with KEVLAR®, H-51009, E. I. du Pont de Nemours & Co., Inc., Wilmington, Del., 1997. Teijinconex®, Technical Information, CN02/91.2, Teijinconex®, Aromatic Polyamide Fiber; High Tenacity Aramid Fiber; Technora®, Teijin.com/advanced fibres, Teijin Ltd., Japan. Twaron®, Product Information: Yarns, Fibers and Pulp, Twaron® in Brake Linings and Clutch Facings, Akzo Nobel, the Netherlands.

8.10 ENVIRONMENTAL IMPACT AND RECYCLING

The reduction in CO_2 emissions to the atmosphere is currently one of the principal goals of the transportation industry (automobile and aircraft industries), and the use of aramid-reinforced composites can decrease the weight of cars and airplanes to reduce fuel consumption and, consequently, CO_2 emissions.

High-performance aramid materials have been increasingly used to replace asbestos as a friction material in devices such as vehicle brakes and clutch components because aramid materials are safer than asbestos[131] during manipulation, transformation, and recovery.

The LOI values of commercial aramids are higher than 25, indicating that these materials, especially MPIA, are self-extinguishing[132] and therefore reduce the tendency to propagate flames after the ignition source is removed. The gases produced by burning aramids are similar to those produced by wool and contain a small amount of toxic gases.

The recyclability of aromatic polyamide fibers depends on how the fibers were incorporated into the product. Polyamide fabrics in textile applications are easily recyclable into quality products that can be reused in gloves, boat ropes, brake pads, etc. The quality of the waste materials is a matter of research because it is a key parameter to produce high-quality recycled goods. The recovery, recycle, or reuse of aramids processed as composite materials, e.g., in resins or rubbers, is a challenging matter. Chemical methods related with matrix degradation are envisaged for this purpose.[133]

REFERENCES

1. J.M. García, F.C. García, F. Serna, J.L. de la Peña, High-performance aromatic polyamides, *Prog. Polym. Sci.*, 35:623–686, 2010.
2. M. Trigo-López, P. Estévez, N. San-José, A. Gómez-Valdemoro, F.C. García, F. Sena, J.L. de la Peña, J.M. García, Recent patents on aromatic polyamides, *Recent Patents Mater. Sci.*, 2:190–208, 2009.
3. P.E. Cassidy, *Thermally Stable Polymers, Synthesis and Properties*, Marcel Dekker, New York, 1980.
4. H.H. Yang, *Aromatic High-Strength Fibers*, Wiley Interscience, New York, 1989.
5. J.K. Fink, *High Performance Polymers*, William Andrew, New York, 2008.
6. L. Vollbracht, Aromatic polyamides, in G. Allen, B. Bevington, G.V. Eastmond, A. Ledwith, S. Russo, P. Sigwald (eds.), *Comprehensive Polymer Science*, Vol. 5, 373–383, Pergamon Press, Oxford, 1989.
7. R.J. Gaymans, Polyamides, in M.E. Rogers, T.E. Long (eds.), *Synthetic Methods in Step-Growth Polymers*, Vol. 3, 135–196, John Wiley & Sons, Hoboken, NJ, 2003.
8. H.G. Elias, *Macromolecules, Vol. 2, Industrial Polymers and Synthesis*, Wiley-VCH, Weinheim, 2007.
9. M.S. Reisch, High-performance fibers find expanding military, industrial uses, *Chem. Eng. News*, 65:9–14, 1987.
10. J. Gallini, Polyamides aromatic, in H.F. Mark, J.I. Kroschwitz (eds.), *Encyclopedia of Polymer Science and Technology*, Vol. 3, 558–584, John Wiley & Sons, New York, 1989.
11. J.M. Garcia, F.C. Garcia, F. Serna, J.L. de la Peña, Aromatic polyamides (aramids), in S. Thomas, P.M. Vishak (eds.), *Handbook of Engineering and Specialty Thermoplastics*, Vol. 6, 141–181, John Wiley & Sons, Hoboken, NJ, and Scrivener Publishing, Salem, MA, 2011.
12. Y. Imai, Y. Oishi, Novel synthetic methods for condensation polymers using silylated nucleophilic monomers, *Prog. Polym. Sci.*, 14:173–193, 1989.
13. P.W. Morgan, *Condensation Polymers*, Interscience Publishers, New York, 1965.
14. J. Preston, Aromatic polyamides, in H.F. Mark, N.M. Bikales, C.G. Overberger, G. Menges (eds.), *Encyclopedia of Polymer Science and Engineering*, Vol. 11, 381–389, John Wiley & Sons, New York, 1988.
15. J.M. García, J.C. Álvarez, J.G. de la Campa, J. de Abajo, Synthesis and characterization of aliphatic–aromatic poly(ether amide)s, *Macromol. Chem. Phys.*, 198:727–737, 1997.
16. N. Yamazaki, F. Higasi, J. Kawataba, Studies on reactions of the *N*-phosphonium salts of pyridines. XI. Preparation of the polypeptides and polyamides by means of triaryl phosphites in pyridine, *J. Polym. Sci. Polym. Chem. Ed.*, 12:2149–2154, 1974.
17. S. Mallakpour, Z. Rafiee, Microwave-induced synthesis of new optically active and soluble polyamides containing pendant 4-(2-phthalimidiylpropanoylamino)benzoylamino-groups, *Amino Acids*, 37:665–672, 2009.

18. S. Mallakpour, H. Seyedjamali, One-pot polyamidation reaction of optically active aromatic diacid containing methionine and phthalimide moieties with aromatic diamines under microwave irradiation and traditional heating, *Eur. Polym. J.*, 44:3615–3619, 2008.
19. S. Mallakpour, Z. Rafiee, Expeditious synthesis of novel aromatic polyamides from 5-[3-phenyl-2-(9,10-dihydro-9,10-ethanoanthracene-11,12-dicarboximido)propanoylamino]isophthalic acid and various diamines using microwave-assisted polycondensation, *React. Funct. Polym.*, 69:252–258, 2009.
20. P. Carretero, R. Sandin, R. Mercier, A.E. Lozano, J.G. de la Campa, J. de Abajo, Microwave-induced synthesis of aromatic polyamides by the phosphorylation reaction, *Aust. J. Chem.*, 62:250–253, 2009.
21. S. Mallakpour, M. Taghavi, A facile microwave-assisted synthesis of novel optically active polyamides derived from 5-(3-methyl-2-phthalimidylpentanoylamino)isophtalic acid and different diisocyanates, *Eur. Polym. J.*, 44:87–97, 2008.
22. S. Mallakpour, M. Dinari, High performance polymers in ionic liquids: A review on prospects for green polymer chemistry. Part I, *Polyamides Iran. Polym. J.*, 19:983–1004, 2010.
23. Y.S. Vygodskii, E.I. Lozinskaya, A.S. Shaplov, Ionic liquids as novel reaction media for the synthesis of condensation polymers, *Macromol. Rapid. Commun.*, 23:676–680, 2002.
24. S. Mallakpour, Z. Rafiee, Safe and fast polyamidation of 5-(4-(2-phthalimidiylpropanoylamino) benzoylamino)isophtalic acid with aromatic diamines in ionic liquid under microwave irradiation, *Polymer*, 49:3007–3013, 2008.
25. S. Mallakpour, Z. Rafiee, Use of ionic liquid and microwave irradiation as a convenient, rapid and eco-friendly method for synthesis of novel optically active and thermally stable aromatic polyamides containing N-phthaloyl-L-alanine pendent group, *Polym. Degrad. Stab.*, 93:753–759, 2008.
26. S. Mallakpour, M. Dinari, Preparation of thermally stable and optically active organosoluble aromatic polyamides containing L-leucine amino acid under green conditions, *Polym. Bull.*, 63:623–635, 2009.
27. S. Mallakpour, M. Taghavi, Direct polyamidation in green media: Studies on thermal degradation of novel organosoluble and optically active flame retardant polyamides, *React. Funct. Polym.*, 69:206–215, 2009.
28. S. Mallakpour, S. Meratian, Preparation and characterization of thermostable chiral extended polyamides bearing N-phthaloyl-L-leucine pendent architectures in green media, *J. Appl. Polym. Sci.*, 111:1209–1215, 2009.
29. S. Mallakpour, H. Seyedjamali, Ionic liquid catalyzed synthesis of organosoluble wholly aromatic optically active polyamides, *Polym. Bull.*, 62:605–614, 2009.
30. S. Mallakpour, M. Dinari, Soluble new optically active polyamides derived from 5-(4-methyl-2-phthalimidylpentanoylamino)-isophthalic acid and different diisocyanates under microwave irradiation in molten ionic liquid, *J. Appl. Polym. Sci.*, 112:244–253, 2009.
31. S. Mallakpour, A. Zadehnazari, Synthesis of optically active and thermally stable polyamides with bulky aromatic side chain in an ionic liquid (tetrabutylammonium bromide), *High Perform. Polym.*, 22:567–580, 2010.
32. S. Mallakpour, M. Dinari, Environmentally friendly methodology for preparation of amino acid containing polyamides, *J. Polym. Environ.*, 18:705–713, 2010.
33. S. Mallakpour, H. Seyedjamali, Ionic liquid as a green media for rapid synthesis of optically active organosoluble polyamides, *Des. Monomers Polym.*, 13:377–386, 2010.
34. H. Sekiguchi, B. Coutin, Polyamides, in H.R. Kricheldor (ed.), *Handbook of Polymer Synthesis, Part A*, Vol. 14, 807–939, Marcel Dekker, New York, 1992.
35. Y. Kubota, S. Nakada, Y. Sug, New heat-resistant and soluble aramids synthesized by palladium-catalyzed carbonylation polycondensation, *Mater. Trans.*, 43:326–331, 2002.
36. G. Rabani, A. Kraft, Synthesis of poly(ether-esteramide) elastomers by a palladium-catalyzed polycondensation of aromatic diiodides with telechelic diamines and carbon monoxide, *Macromol. Rapid Commun.*, 23:375–379, 2002.
37. T. Yokozawa, A. Yokoyama, Chain-growth polycondensation: The living polymerization process in polycondensation, *Prog. Polym. Sci.*, 32:147–172, 2007.
38. A. Yokoyama, T. Yokozawa, Converting step-growth to chain growth condensation polymerization, *Macromolecules*, 40:4093–4101, 2007.
39. T. Yokozawa, T. Asai, R. Sugi, A. Ishigooka, S. Hiraoka, Chain-growth polycondensation for nonbiological polyamides of defined architecture, *J. Am. Chem. Soc.*, 122:8313–8324, 2000.
40. T. Yokozawa, D. Muroya, R. Sugi, A. Yokoyama, Convenient method of chain-growth polycondensation for well-defined aromatic polyamides, *Macromol. Rapid Commun.*, 26:979–981, 2005.

41. R. Sugi, A. Yokoyama, T. Furuyama, M. Uchiyama, T. Yokozawa, Inductive effect-assisted chain-growth polycondensation. Synthetic development from *para*- to *meta*-substituted aromatic polyamides with low polydispersities, *J. Am. Chem. Soc.*, 127:10172–10173, 2005.
42. T. Ohishi, R. Sugi, A. Yokoyama, T. Yokozawa, A variety of poly(*m*-benzamide)s with low polydispersities from inductive effect-assisted chain-growth polycondensation, *J. Polym. Sci. A: Polym. Chem.*, 44:4990–5003, 2006.
43. R. Sugi, T. Ohishi, A. Yokoyama, T. Yokozawa, Novel water-soluble poly(*m*-benzamide)s: Precision synthesis and thermosensitivity in aqueous solution, *Macromol. Rapid. Commun.*, 27:716–721, 2006.
44. T. Yokozawa, M. Ogawa, A. Sekino, R. Sugi, A. Yokoyama, Chain-growth polycondensation for well-defined aramide, Synthesis of unprecedented block copolymer containing aramide with low polydispersity, *J. Am. Chem. Soc.*, 124:15158–15159, 2002.
45. T. Yokozawa, M. Ogawa, A. Sekino, R. Sugi, A. Yokoyama, Synthesis of well-defined poly(*p*-benzamide) from chain-growth polycondensation and its application to block copolymers, *Macromol. Symp.*, 199:187–195, 2003.
46. R. Sugi, Y. Hitaka, A. Sekino, A. Yokoyama, T. Yokozawa, Bidirectional propagation of chain-growth polycondensation: Its application to poly(ethylene glycol)–aromatic polyamide–poly(ethylene glycol) triblock copolymer with low polydispersity, *J. Polym. Sci. A: Polym. Chem.*, 41:1341–1346, 2003.
47. A. Yokoyama, T. Masukawa, Y. Yamazaki, T. Yokozawa, Successive chain-growth condensation polymerization for the synthesis of well-defined diblock copolymers of aromatic polyamide and aromatic polyether, *Macromol. Rapid Commun.*, 30:24–28, 2009.
48. R. Sugi, A. Yokoyama, T. Yokozawa, Synthesis of well-defined telechelic aromatic polyamides by chain-growth polycondensation: Application to the synthesis of block copolymers of polyamide and poly(tetrahydrofuran), *Macromol. Rapid Commun.*, 24:1085–1090, 2003.
49. M. Ueda, Sequence control in one-step condensation polymerization, *Prog. Polym. Sci.*, 24:699–730, 1999.
50. L. Li, K. Yonetake, O. Haba, T. Endo, M. Ueda, Constitutional isomerism. IV. Synthesis and characterization of poly(amide-ester)s from isophthaloyl chloride and 4-aminophenethyl alcohol, *Polym. J.*, 33:364–370, 2001.
51. J.M. García, F.C. García, F. Serna, Constitutional isomerism in polyamides derived from isophthaloyl chloride and 1,3-diamine-4-chlorobenzene, *J. Polym. Sci. A: Polym. Chem.*, 41:1202–1215, 2003.
52. F. Serna, F. García, J.L. de la Peña, V. Calderón, J.M. García, Properties, characterization and preparation of halogenated aromatic polyamides, *J. Polym. Res.*, 14:341–350, 2007.
53. P. Pino, G.P. Lorenzi, U.W. Suter, P. Casartelli, I. Steinman, F.J. Bonner, Control of structure isomerism in polyamides, *Macromolecules*, 11:624–626, 1978.
54. U.W. Suter, P. Pino, Structural isomerism in polycondensates. 2. Aspects for monomers with independent functional groups, *Macromolecules*, 17:2248–2255, 1984.
55. R.R. Pal, P.S. Patil, M.M. Salunkhe, N.N. Maldar, P.P. Wadgaonkar, Synthesis, characterization and constitutional isomerism study of new aromatic polyamides containing pendant groups based on asymmetrically substituted *meta*-phenylene diamines, *Eur. Polym. Sci.*, 45:953–959, 2009.
56. L. Li, O. Haba, T. Endo, M. Ueda, Constitutional isomerism: Synthesis and characterization of ordered polyamides from isophthalic acid and *N*-phenylethylenediamine by direct polymerization, *High Perform. Polym.*, 13:217–232, 2001.
57. F.T. Gentle, W.W. Suter, Constitutional regularity in linear condensation polymers, in F. Allens, J.C. Bevington (eds.), *Comprehensive Polymer Science*, Vol. 5, 97–115, Pergamon, Oxford, 1989.
58. M. Jikei, M. Kakimoto, Dendritic aromatic polyamides and polyimides, *J. Polym. Sci. A: Polym. Chem.*, 42:1293–1309, 2004.
59. J.M.J. Fréchet, Functional polymers and dendrimers—Reactivity, molecular architecture, and interfacial energy, *Science*, 263:1710–1715, 1994.
60. C.J. Hawker, J.M.J. Fréchet, Preparation of polymers with controlled molecular architecture—A new convergent approach to dendritic macromolecules, *J. Am. Chem. Soc.*, 112:1738–1747, 1990.
61. C.J. Hawker, R. Lee, J.D.J. Fréchet, One-step synthesis of hyperbranched dendritic polyesters, *J. Am. Chem. Soc.*, 113:4583–4588, 1991.
62. D. Hölter, A. Burgath, H. Frey, Degree of branching in hyperbranched polymers, *Acta Polym.*, 48:30–35, 1997.
63. M. Scholl, Z. Kadlecova, H.-A. Klok, Dendritic and hyperbranched polyamides, *Prog. Polym. Sci.*, 34:24–61, 2009.

64. S. Ozawa, K. Matsuda, Aramid copolymer fibers, in M. Lewin, J. Preston (eds.), *Handbook of Fiber Science and Technology, Vol. III, High Technology Fibers Part B*, Vol. 1, 1–34, Marcel Dekker, New York, 1989.
65. B.A.K. Chaudhuri, Y. Min, E.M. Pearce, Thermal properties of wholly aromatic polyamides, *J. Polym. Sci., Polym. Chem. Ed.*, 18:2949–2958, 1980.
66. R. Takatsuka, R.K. Uno, F. Toda, Y. Iwakura, Study on wholly aromatic polyamides containing methyl-substituted phenylene linkage, *J. Polym. Sci., Polym. Chem. Ed.*, 15:1905–1915, 1977.
67. S.M. Aharoni, S.A. Curran, N.S. Murthy, Poly(*p*-phenyleneterephthalamide) and poly(*m*-phenylene-isophthalamide): Positional isomers with partial miscibility, *Macromolecules*, 25:4431–4436, 1992.
68. A.E. Zachariades, R.S. Porter, *The Strength and Stiffness of Polymers*, Marcel Dekker, New York, 1983.
69. ASTM Standard ASTM D256-06a, Test methods for determining the Izod pendulum impact resistance of plastics, ASTM International, West Conshohocken, PA, 2007.
70. ISO Standard 179, Plastics—Determination of Charpy impact properties—Part 1: Non-instrumented impact test, International Organization for Standardization, Geneva, Switzerland, 2005.
71. H.H. Yang, *Kevlar Aramid Fiber*, John Wiley & Sons, Chichester, 1993.
72. M. Jassal, S. Gosh, Aramid fibers, an overview, *Indian J. Fiber Text. Res.*, 27:290–306, 2002.
73. H. Zhang, J. Zhang, J. Chen, X. Hao, S. Wang, X. Feng, Y. Guo, Effects of solar UV irradiation on the tensile properties and structure of PPTA fiber, *Polym. Degrad. Stabil.*, 91:2761–2767, 2006.
74. T.I. Bair, P.W. Morgan, Optically anisotropic spinning dopes of polycarbonamides, US Patent 3673143, assigned to E. I. du Pont de Nemours and Company (Wilmington, DE), 1972.
75. H. Aoki, J.L. White, J.F. Fellers, A rheological and optical properties investigation of aliphatic (nylon 66, PγBLG) and aromatic (Kelvar®, Nomex®) polyamide solutions, *J. Appl. Polym. Sci.*, 23:2293–2314, 1979.
76. S. Bourbigot, X. Flambard, Heat resistance and flammability of high performance fibers: A review, *Fire Mater.*, 26:155–168, 2002.
77. S. Rebouillat, Aramids, in J.W.S. Hearle (ed.), *High-Performance Fibers*, Vol. 2, 31–68, Woodhead Publishing, Cambridge, 2000.
78. H. Blades, Certificate of correction, US Patent 3767756, assigned to E. I. du Pont de Nemours and Company (Wilmington, DE), 1973.
79. S. Ozawa, K. Matsuda, *Handbook of Fiber Science and Technology, Vol. III: High Technology Fibers, Part B*, Marcel Dekker, New York, 1989.
80. S. Ozawa, Y. Nakagawa, K. Matsuda, T. Nishihara, H. Yunoki, Novel aromatic copolyamides prepared from 3,4′ diphenylene type diamines, and shaped articles therefrom, US Patent 4075172, assigned to Teijin Ltd., 1978.
81. S.L. Kwolek, P.W. Morgan, W.R. Sorenson, Process of making wholly aromatic polyamides, US Patent 3063966, assigned to E. I. du Pont de Nemours & Co., 1962.
82. H. Blades, High strength polyamide fibres and films, US Patent 3869429, assigned to E. I. du Pont de Nemours and Company (Wilmington, DE), 1975.
83. T. Imanishi, S. Muraoka, Transparent poly-*p*-phenylene-terephtalamide film, US Patent 4752643, assigned to Asahi Chemical Industry Co, Ltd., 1988.
84. T. Yamada, M. Nakatani, Aromatic polyamide film, method for producing the same, and magnetic recording medium and solar cell using the same, US Patent 5853907, assigned to Asahi Kasei Kogyo Kabushiki Kaisha (Osaka, Japan), 1998.
85. P.A. Koch, *Groses Textil-Lexikon*, Deutsche Verlags-Anstalt, Stuttgart, 1966.
86. G.C. Gross, Synthetic paper structures of aromatic polyamides, US Patent 3756908, assigned to E. I. du Pont de Nemours and Company (Wilmington, DE), 1973.
87. D.C. Prevorsek, G.A. Harpell, D. Wertz, Ballistic resistant fabric articles, US Patent 5677029, assigned to AlliedSignal Inc. (Morristown, NJ), 1997.
88. Chitrangad, Aramid ballistic structure, US Patent 6030683, assigned to E. I. du Pont de Nemours and Company (Wilmington, DE), 2000.
89. D.J. Mohan, L. Kullova, A study on the relationship between preparation condition and properties/performance of polyamide TFC membrane by IR, DSC, TGA, and SEM techniques, *Desalin. Water Treat.*, 51:586–596, 2013.
90. S. Yu, M. Liu, X. Liu, C. Gao, Performance enhancement in interfacially synthesized thin-film composite polyamide–urethane reverse osmosis membrane for seawater desalination, *J. Membr. Sci.*, 342:313–320, 2009.
91. R.S. Harisha, K.M. Hosamani, R.S. Keri, S.K. Nataraj, T.M. Aminabhavi, Arsenic removal from drinking water using thin film composite nanofiltration membrane, *Desalination*, 252:75–80, 2010.

92. D. Rana, Y. Kim, T. Matsuura, H.A. Arafat, Development of antifouling thin-film-composite membranes for seawater desalination, *J. Membr. Sci.*, 367:110–118, 2011.
93. M. Bubacz, A. Beyle, D. Hui, C.C. Ibeh, Helium permeability of coated aramid papers, *Composites Part B*, 39:50–56, 2008.
94. S. Itsuno, Chiral polymer synthesis by means of repeated asymmetric reaction, *Prog. Polym. Sci.*, 30:540–558, 2005.
95. Y. Okamoto, Chiral polymers, *Prog. Polym. Sci.*, 25:159–162, 2000.
96. M. Nakagawa, Y. Ikeuchi, M. Yoshikawa, H. Yoshida, S. Sakurai, Optical resolution of racemic amino acid derivatives with chiral polyamides bearing glutamyl residue as a diacid component, *J. Appl. Polym. Sci.*, 123:857–865, 2012.
97. S. Mallakpour, M. Dinari, Microwave step-growth polymerization of 5-(4-methyl-2-phthalimidylpentanoyl-amino)isophthalic acid with different diisocyanates, *Polym. Adv. Technol.*, 19:1334–1342, 2008.
98. D.W. Van Krevelen, P.J. Hoftyzer, *Properties of Polymers, Their Estimation and Correlation with Chemical Structure*, Elsevier, New York, 1976.
99. D.Y. Kim, H.N. Cho, C.Y. Ki, Blue light emitting polymers, *Prog. Polym. Sci.*, 25:1089–1139, 2000.
100. L. Akcelrud, Electroluminescent polymers, *Prog. Polym. Sci.*, 28:875–962, 2003.
101. P. Estevez, H. El-Kaoutit, F.C. Garcia, F. Serna, J. de la Peña, J.M. Garcia, Chemical modification of the pendant structure of wholly aromatic polyamides: Toward functional high-performance materials with tuned chromogenic and fluorogenic behavior, *J. Polym. Sci. Part A: Polym. Chem.*, 48:4823–4833, 2010.
102. J.L. Barrio-Manso, P. Calvo, F.C. García, J.L. Pablos, T. Torroba, J.M. García, Functional fluorescent aramids: Aromatic polyamides containing a dipicolinic acid derivative as luminescent converters and sensory materials for the fluorescence detection and quantification of Cr(VI), Fe(III) and Cu(II), *Polym. Chem.*, 4:4256–4264, 2013.
103. C.W. Chang, G.S. Liou, Novel anodic electrochromic aromatic polyamides with multi-stage oxidative coloring based on N,N,N',N'-tetraphenyl-p-phenylenediamine derivatives, *J. Mater. Chem.*, 18:5638–5646, 2008.
104. M. Bruma, E. Hamciuc, B. Schulz, T. Köpnick, Y. Kaminorz, J. Robison, Aromatic polymers with side oxadiazole rings as luminescent materials in LEDs, *Macromol. Symp.*, 199:511–521, 2003.
105. M. Bruma, E. Hamciuc, B. Schulz, T. Kopnick, Y. Kaminorz, J. Robison, Synthesis and study of new polyamides with side oxadiazole rings, *J. Appl. Polym. Sci.*, 87:714–721, 2002.
106. Y.H. La, J. Diep, R. Al-Rasheed, D. Miller, L. Krupp, G.M. Geise, A. Vora, B. Davis, M. Nassar, B.D. Freeman, M. McNeil, G. Dubois, Enhanced desalination performance of polyamide bi-layer membranes prepared by sequential interfacial polymerization, *J. Membr. Sci.*, 437:33–39, 2013.
107. G. Han, S. Zhang, X. Li, T.S. Chung, High performance thin film composite pressure retarded osmosis (PRO) membranes for renewable salinity-gradient energy generation, *J. Membr. Sci.*, 440:108–121, 2013.
108. Y.H. Na, J. Diep, R. Al-Rasheed, B. Davis, G. Geise, A. Vora, M. Nassar, G. Dubois, Novel polyamide bilayer membranes with improved desalination performance, *Technical Proceedings of the 2012 NSTI Nanotechnology Conference and Expo, NSTI-Nanotech*, 687–669, 2012.
109. J.E. Cadotte, Interfacially synthesized reverse osmosis membrane, US Patent 4277344, assigned to FilmTec Corporation, 1981.
110. S. Konagaya, H. Kuzumoto, O. Watanabe, New reverse osmosis membrane materials with higher resistance to chlorine, *J. Appl. Polym. Sci.*, 75:1357–1364, 2000.
111. Y. Jin, Z. Su, Effects of polymerization conditions on hydrophilic groups in aromatic polyamide thin films, *J. Membr. Sci.*, 330:175–179, 2009.
112. H.A. Shawky, Performance of aromatic polyamide RO membranes synthesized by interfacial polycondensation process in a water–tetrahydrofuran system, *J. Membr. Sci.*, 339:209–214, 2009.
113. N.A. Mohamed, A.O.H. Al-Dossary, Structure–property relationships for novel wholly aromatic polyamide-hydrazides containing various proportions of *para*-phenylene and *meta*-phenylene units. III. Preparation and properties of semipermeable membranes for water desalination by reverse osmosis separation performance, *Eur. Polym. J.*, 39:1653–1667, 2003.
114. Y. Shirakashi, T. Watanabe, O. Asakura, A. Irikura, M. Watanabe, H. Kojima, N. Imai, Artificial hair, wig using the same, and method of making artificial hair, US Patent 2008314402A1, assigned to M. Yoshimura, C. Yoshimura, LP Washington, DC (US), 2008.
115. T. Ahamad, S.F. Mapolie, S.M. Alshehri, Synthesis and characterization of polyamide metallodendrimers and their anti-bacterial and anti-tumor activities, *Med. Chem. Res.*, 21:2023–2031, 2012.
116. M.S. Peters, M. Li, T. Schrader, Interactions of calix[n]arenes with nucleic acids, *Nat. Prod. Commun.*, 7:409–417, 2012.

117. L.J. Markoski, K.A. Walker, G.A. Deeter, G.E. Spilman, D.C. Martin, J.S. Moore, Cross-linkable copolymers of poly(*p*-phenyleneterephthalamide), *Chem. Mater.*, 5:248–250, 1993.
118. L.S. Tan, N. Venkatasubramanian, Aromatic polyamides containing keto-benzocyclobutene pendant, *J. Polym. Sci. Part A: Polym. Chem.*, 34:3539–3549, 1996.
119. J. Bos, W. Hendrikus, J. Nijenhuis, Crosslinked aramid polymer, WO2009130244A3, assigned to Teijin Aramid B.V, 2008.
120. Y.J. Kim, I.S. Chung, I. In, S.Y. Kim, Soluble rigid rod-like polyimides and polyamides containing curable pendent groups, *Polymer*, 46:3992–4004, 2005.
121. W. Seeeny, Improvements in compressive properties of high modulus fibres by crosslinking, *J. Polym. Sci. Part A: Polym. Chem.*, 30:1111–1122, 1992.
122. J. Bos, S.C. Noordewier, Crosslinkable aramid polymers, WO2008028605, assigned to Teijin Aramid B.V, 2008.
123. M. Trigo-López, J.L. Barrio-Manso, F. Serna, F.C. García, J.M. García, Crosslinked aromatic polyamides: A further step in high-performance materials, *Macromol. Chem. Phys.*, 214:2223–2231, 2013.
124. L. Wang, Y. Wang, L. Ren, Synthesis and characterization of novel biodegradable aromatic–aliphatic poly(ester amide)s containing ethylene oxide moieties, *J. Appl. Polym. Sci.*, 109:1310–1318, 2008.
125. A. Gómez-Valdemoro, V. Calderón, N. San-José, The extraction of environmentally polluting cations from aqueous media with novel polyamides containing cation- and anion-selective host units, *J. Polym. Sci. Part. A: Polym. Chem.*, 47:670–681, 2009.
126. G.M. Kent, K.L. Johnson, D.R. Gadoury, R.L. Mumford, Ultraviolet stability of aramid and aramid-blend fabrics by pigment dyeing or printing, US Patent 6451070B1, assigned to BASF Corporation (Mount Olive, NJ), 2002.
127. Y. Xing, X. Ding, UV photo-stabilization of tetrabutyl titanate for aramid fibers via sol-gel surface modification, *J. Appl. Polym. Sci.*, 103:3113–3119, 2007.
128. C.H. Hsu, Electrically conductive fibers, WO Patent 1997022740, assigned to E. I. Du Pont de Nemours & Company (US), 1997.
129. J.D. Hartzler, Electrically-conductive para-aramid pulp, US Patent 6436236B1, assigned to E. I. du Pont de Nemours & Company (Wilmington, DE), 2002.
130. X. Zhao, K. Hirogaki, I. Tabata, S. Okubayashi, T. Hori, A new method of producing conductive aramid fibers using supercritical carbon dioxide, *Surf. Coat. Technol.*, 201:628–636, 2006.
131. G.L. Hearn, J.A. Amner, P.J. Walker, Method and apparatus to identify asbestos and aramid, US Patent 6144208, assigned to Ford Global Technologies, Inc. (Dearborn, MI), 2000.
132. A.R. Horrocks, M. Tunc, L. Cegielka, The burning behavior of textiles and its assessment by oxygen-index methods, *Text. Prog.*, 18:1–186, 1989.
133. L. Yuyan, L. Li, M. Linghui, The experimental research on recycling of aramid fibers by solvent method, *J. Reinf. Plast. Compos.*, 28:2211–2220, 2009.

9 Thermoplastic Polyesters*

Ganesh Kannan, Sarah E. Grieshaber, and Wei Zhao

CONTENTS

9.1 Poly(Ethylene Terephthalate) 320
    9.1.1 History 320
    9.1.2 PET Polymerization Process 320
        9.1.2.1 Antimony 320
        9.1.2.2 Germanium 320
        9.1.2.3 Other PET Catalysts 320
        9.1.2.4 PET Synthesis 321
        9.1.2.5 Solid-State Polymerization 321
    9.1.3 Properties and Applications of PET 322
9.2 Poly(Butylene Terephthalate) 324
    9.2.1 History 324
    9.2.2 Raw Materials 324
    9.2.3 PBT Polymerization Process 327
    9.2.4 Green Chemistry 329
    9.2.5 PBT Crystallization 329
    9.2.6 PBT Properties and Applications 329
9.3 Polyesters Based on Cyclohexanedimethanol 331
    9.3.1 History of CHDM-Based Polyesters 331
    9.3.2 Polymerization Process 332
    9.3.3 Properties and Applications 334
9.4 Naphthalate Polyesters 336
    9.4.1 History 336
    9.4.2 Polymerization Process 337
    9.4.3 Thermal Properties of Naphthalate Polyesters 337
    9.4.4 Gas Permeability 338
        9.4.4.1 Gas Barrier Properties of PEN/PET Blends 339
    9.4.5 Chemical Resistance 339
    9.4.6 Optical (UV) Properties 339
    9.4.7 Fluorescence of Naphthalate Polyesters 340
    9.4.8 Photodegradation 340
    9.4.9 Mechanical Properties 340
    9.4.10 Naphthalate-Based Elastomers 340
    9.4.11 Applications 341
References 342

---

* Based in parts on the first-edition chapter on thermoplastic polyesters.

9.1 POLY(ETHYLENE TEREPHTHALATE)

9.1.1 History

Calico Printer's Association Limited initiated research in 1940 to develop polyesters for synthetic fiber applications. In 1941, John Rex Whinfield and James Tennant Dickson of the Calico Printer's Association patented poly(ethylene terephthalate) (PET) [1] based on the early research of Wallace Carothers on condensation polymerization of aliphatic acids such as adipic acid with ethylene glycol (EG), but Carothers had not studied the condensation of EG with aromatic acids such as terephthalic acid (TPA) to produce PET. Whinfield and Dickson, along with inventors W.K. Birtwhistle and C.G. Ritchiethey, also created the first PET polyester fibers called Terylene in 1941 [2] (first manufactured by Imperial Chemical Industries [ICI]). When DuPont learned about the new fibers from ICI (which acquired Calico industries in 1947), it negotiated and purchased patent rights and then proceeded further to commercialization.

9.1.2 PET Polymerization Process

Until the mid-1960s, dimethyl terephthalate (DMT) was the preferred feedstock for PET production, mainly due to its higher purity than its acid form, TPA. When Amoco developed a high-purity TPA process, the acid form gained acceptance and has since become the primary feedstock. The acid process completely eliminated the additional step of converting TPA to dimethyl ester, which added further cost, and this was also an important criterion for the worldwide acceptance of TPA-based PET. Owing to the commercial importance of PET, large volumes of work on various catalysts used for PET polymerization have been reported, and a detailed discussion on the catalysts follows.

The metal catalysts based on antimony and germanium have dominated the industrial production process from the time of invention. Since PET has gained commercial importance from 1950 onward, new research has been focused on improving the process, especially the catalysts, both inorganic and organic. Although a plethora of new catalysts were discovered, replacement of antimony and germanium as industrial-scale catalysts has not succeeded to date. So far, no negative data have been obtained due to the presence of antimony in PET sold for food applications.

9.1.2.1 Antimony
Among the various antimony compounds such as trioxide [3], triacetate [4], and glycolate [5], the most widely used industrial version is the trioxide. The other two catalysts do not show any significant difference from PET synthesized with antimony trioxide.

9.1.2.2 Germanium
Germanium dioxide-catalyzed [6] PET shows higher optical clarity, resulting in its use in film applications. It is also used by Japanese producers for manufacturing bottle-grade PET. Because of its tendency to support oxidative degradation, germanium catalysts are used together with stabilizers based on phosphorus.

9.1.2.3 Other PET Catalysts
Due to the commercial importance of PET, an abundance of patents focusing on either new catalyst or combinations of existing catalysts have been reported, a few of which are described. Examples of mixture catalysts include zinc acetate/antimony acetate/phosphate ester [7], manganese benzoate/antimony triacetate/phosphate ester [7], manganese benzoate/acetyltriisopropyl titanate with or without cobalt acetate [7], antimony triacetate/zinc diacetate/titanium glycolate [8], and manganese acetate/cobalt acetate/antimony trioxide/titanium isopropoxide [9]. Examples of other novel compounds include triethanolamine titanium chelate [10]; titanium glycolate [8]; cyanides of calcium, cadmium, or zinc (for PET prepolymer) [11]; cobalt acetate [12]; stannous formate [13]; acetylacetonates of manganese, cobalt, cadmium, zinc, or magnesium [14]; phenylstibonic acid [15]; *p*-chlorophenylstibinic acid [15];

Thermoplastic Polyesters

cobalt tungstate [16]; lead oxide [17]; lead fluoride [18]; trifluoroacetates of manganese, magnesium, or calcium [19]; and perfluorobutyrates of barium, tin, or calcium [19]. Regarding metal-free organic catalysts for PET synthesis, an example can be found using N-heterocyclic carbene as a catalyst [20]. The above-mentioned examples are by no means exhaustive, as the literature has been flooded with several catalysts, and our effort is to provide a brief understanding of various catalyst approaches that have been reported.

The majority of efforts in developing new catalysts are focused on finding a substitute for antimony. However, until a regulatory demand for antimony-free PET is put into effect, antimony will continue to be the main catalyst used commercially.

9.1.2.4 PET Synthesis

PET is synthesized from EG as the diol and either an acid (TPA) or an ester (DMT) (Scheme 9.1). Unlike typical condensation polymerization, which requires equal stoichiometric (1:1) amounts of acid and alcohol, EG is typically charged in excess to the acid [7]. During the first stage of polymerization, the intermediate bis-(2-hydroxyethyl)-terephthalate (BHET) monomer is produced and further polymerized under reduced pressure with heat to produce PET resin [21], as shown in Scheme 9.1.

The final molecular weight is influenced by factors such as time, temperature, and vacuum. Typically, the polymerization temperature is increased gradually in the range of 200–290°C, and the melt polymerization time can be in the range of 2–5 h. DMT reacts faster than TPA due to its lower melting point (142°C). Although DMT was initially used commercially, TPA is currently the monomer of choice for PET production.

9.1.2.5 Solid-State Polymerization

As mentioned earlier, one of the factors that influence the molecular weight is the polymerization time. In order to get very high molecular weights, the polymerization time has to be significantly longer. As the polymer melt resides for a longer time at a very high temperature, the polymer often develops a brown color due to thermal degradation. Commercially, the requirement is not only higher molecular weight but also no color formation. Films and fibers are made by melt-phase polymerization until intrinsic viscosities (IVs) of 0.5–0.7 dL/g are reached. Above 0.7 dL/g (soft drink and beverage bottles, tire cord filaments, and industrial fibers), the solid-state polymerization

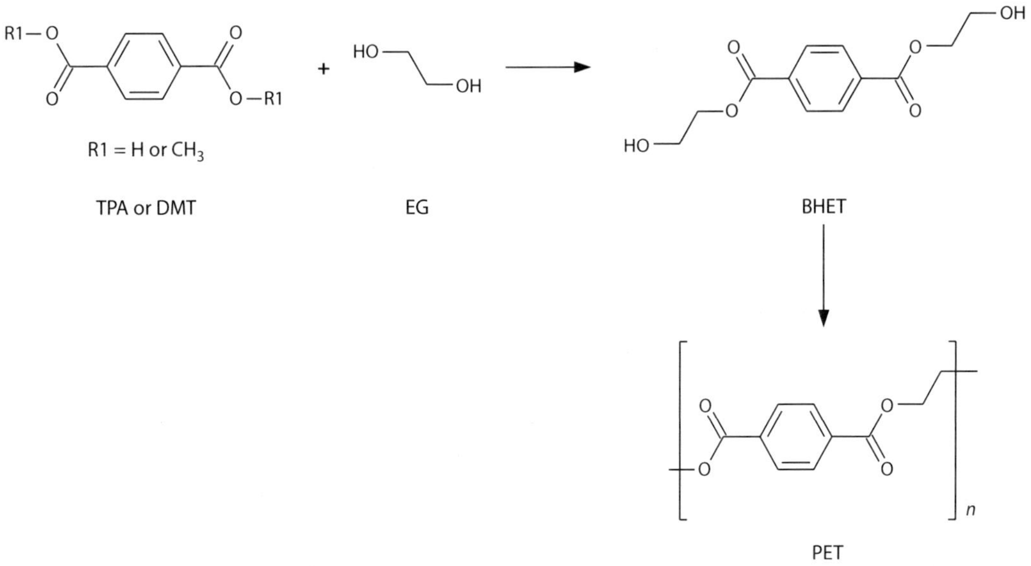

SCHEME 9.1 PET manufacturing route.

(SSP) procedure is used by heating a low-molecular-weight PET below its melting point (245°C) but above its glass transition temperature (80°C) [22]. Between the melting point and glass transition, with heating, the polymerization continues, albeit at a slower rate, but ensures that no color bodies are formed. The ideal temperature range for SSP is 200–220°C. Vacuum or inert gas inside the SSP chamber ensures the removal of EG, water, and acetaldehyde [22].

The main mechanism for SSP is the diffusion of end groups (acid [or ester] and alcohol) and joining of the two chains with the elimination of by-products such as water (or methanol). Since the crystalline phase is highly ordered and possible movements are negligible, most of the reaction occurs in the amorphous domain [23]. Additionally, since the end group produces defects in the crystallites, most of the end groups are essentially in the amorphous phase, resulting in polymerization driven by the amorphous state [23]. Similar to melt polymerization, removal of water (or methanol) after the condensation by inert gas or vacuum drives the reaction forward.

9.1.3 Properties and Applications of PET

Fully annealed PET has a melting point of 280°C, but commercial PET has lower crystallinity, and the melting point is between 255°C and 265°C [24]. The reason for the low crystallinity of commercial PET is both defects in the backbone generated by chemical impurities (diethylene glycol produced from EG) during polymerization becoming a part of the PET backbone [24], or deliberately added chemicals such as isophthalic acid (IPA) that improve processability and lower the crystallization rate. The addition of IPA to PET improves the melt strength that is required for bottle applications [25]. Typical properties of PET collected from data sheets are given in Table 9.1.

Due to the short $-CH_2-CH_2-$ spacer between the aromatic rings, PET crystallizes slowly, but the crystallization rate and, subsequently, the crystallinity in PET can be strongly influenced by the molecular weight. Low molecular weight allows the chain to align easily compared to higher molecular weight, resulting in higher crystallinity for the low-molecular-weight grades. A combination of molecular weight and crystallinity influences the final mechanical properties of a particular grade. For example, yield strain differs with both crystallinity as well as molecular weight [26], as shown in Table 9.2.

PET is flammable and requires flame-retardant additives for applications needing flame-retardant performance. The limiting oxygen index (LOI) for PET is approximately 21 [27], which means oxygen from air can support combustion. PET does not produce char while burning, which leads to complete combustion. Additives to make flame-retardant PET products are mostly halogen based

TABLE 9.1
Typical Properties of PET

Property	Unit	Value
Glass transition temperature	°C	80–82
Melting temperature	°C	245–265
Heat of fusion	J/g	104
Coefficient of thermal expansion	$10^{-6}\ K^{-1}$	20–80
Heat deflection temperature (0.45 MPa load)	°C	115
Specific gravity (amorphous)	–	1.33
Specific gravity (crystalline)	–	1.39
Tensile strength	MPa	55–75
Tensile modulus	GPa	2–4
Elongation at break	%	50–165
Izod impact strength (notched, 23°C)	J/m	13–35

TABLE 9.2
Yield Strain (%) at 23°C [26] for Amorphous and Crystalline PET Grades of Different Molecular Weights

	Yield Strain (%)	
IV of PET	Amorphous PET	PET with 40% Crystallinity
0.62	6.7	8.9
0.68	5.7	7.9
0.75	6.0	8.3
0.86	5.7	7.9

Source: Dixon, E.R. and Jackson, J.B., *J. Mater. Sci.*, 3:464, 1968.

such as brominated polystyrene [28] along with antimony (sodium antimonate) as a synergist, or phosphorous-based additives [28]. For phosphorus compounds, although phosphates provide good flame retardancy, the PET molecular weight decreases due to the hydrolysis by phosphoric acid, which is generated from the phosphate. On the other hand, red phosphorous provides excellent flame resistance but generates highly toxic phosphine [29] while burning.

Similar to most of the commercial polymers, PET is an insulator, which allows for applications such as electrical connectors.

The crystallinity of PET results in good resistance to most acids, alkalis, alcohols, greases, and oils. Since the barrier to gases is generally related to the free volume of the polymer, crystalline polymers provide excellent barrier properties because of the close packing of molecular chains. PET has been approved for food contact applications and is generally considered very safe to use. All the soda bottles in the world are made from PET as well as most of the drinking water bottles.

PET does not have weathering performance at the level of aliphatic polymers such as polyethylene and aliphatic nylons, but its weatherability is acceptable for most short- to medium-term applications.

Optically, PET has very high clarity and transmittance, depending on the processing conditions used. The slow crystallization rate of PET means that any process with fast cooling results in a clear product (bottles, films, sheets, etc.). Processes with a longer residence time or annealing at high temperatures cause PET to crystallize, which results in opaque products.

Among the commercial polyesters, PET is the most commonly used material not only due to its excellent properties but also because it has the lowest cost compared to other engineering thermoplastic polyesters. Similar to nylons and polycarbonates (PCs), drying of polyester (including PET) pellets in an oven at an ambient temperature for several hours is important to prevent polymer degradation by moisture during processing. Several different grades (differentiated by IV) of PET are made by the industry for various processing methods and final products. A few of the most common applications are also listed in Table 9.3.

TABLE 9.3
Grades of PET/Processing Type/Applications

Grade (IV dL/g)	Processing	Applications
0.68	Injection molding	Cups, connectors
0.76	Blow molding, film extrusion, fiber spinning	Bottles for mineral water, containers, cast film, fiber
0.8	Blow molding	General bottle applications
0.84	Blow molding, thermoforming	Bottles for carbonated drinks, juices, oils, cosmetics and food
0.9	Blow molding	4–5 gal. bottles, detergent bottles, jars, deep-draw cups

The advantage of low cost combined with crystallinity and higher thermal stability will continue to be the main reason for PET to be considered as the first material of choice for any new application and uniquely positioned to expand into new markets in the future. New inventions in processing and molding technologies will open new vistas for PET that have not yet been explored.

9.2 POLY(BUTYLENE TEREPHTHALATE)

In addition to PET, poly(butylene terephthalate) (PBT) shares the designation of the most commercially significant thermoplastic polyester. According to a BCC market report, the global market of PET in 2012 was around 270,000 t, while the market size of PBT in 2012 was about 950,000 t [30]. The markets for both PET and PBT are projected to grow continuously in the coming years. Unlike PET, for which the major applications are fibers, films, and bottles, the main application for PBT is injection-molded parts, for which most utilize glass-filled or mineral-filled PBT. The butylene group in the PBT backbone provides the polymer with higher chain mobility than the ethylene group in the PET backbone and results in a faster crystallization speed. Faster crystallization makes PBT more suitable than PET for injection-molding processes [31].

9.2.1 History

The history of PBT dates back to the 1940s and is mingled with the history of PET. The theory of the stepwise growth process was documented by Carothers in the late 1920s, but it is difficult to unambiguously determine the inventor of PBT. Its commercialization began in the late 1960s with the first introduction of Calenex 3300, a 30% glass fiber-reinforced PBT injection-molding grade, by Ticona (now Celanese group) [32]. PBT neat resins, filled PBT grades, and blends of PBT are often found in automotive applications, home appliances, electrical and electronic devices, and other engineering thermoplastic parts due to their high flow characteristics, chemical resistance, thermal and mechanical properties, dimensional stability, and colorability.

9.2.2 Raw Materials

The production of PBT generally starts from two raw materials: 1,4-butanediol (BDO) and DMT or TPA. BDO can be synthesized from various raw materials using different processes. For example, Reppe chemistry (shown in Scheme 9.2) is a process that uses acetylene and formaldehyde to produce 1,4-butynediol, which is then hydrogenated to obtain BDO. Reppe chemistry was invented in the 1930s [33] and is still a widely used route to produce BDO in industry. The Reppe reaction takes place using a catalyst based on copper oxide with bismuth oxide added [34]. And the synthesis in a

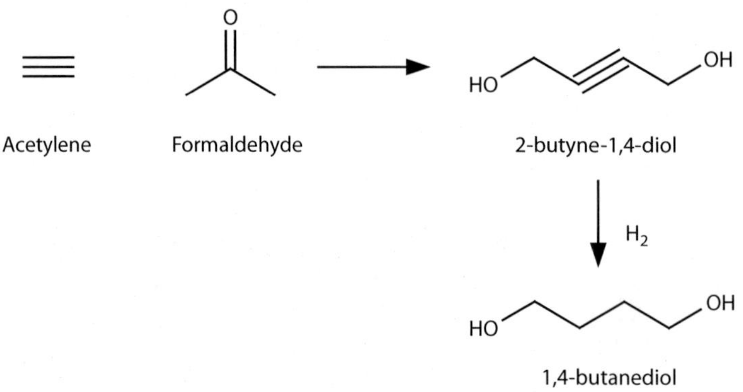

SCHEME 9.2 Reppe chemistry to produce BDO.

fluid bed reactor system happens at 70–100°C, 1–6 bar [35–37]. The pH of the reaction is controlled at 5–8 by the addition of alkali [35–37]. The hydrogenation of 1,4-butynediol can be completed with a nickel or palladium catalyst.

Other feedstocks for the synthesis of BDO include propylene oxide (Scheme 9.3) [34,38,39], butadiene (Scheme 9.4) [34,38,40], and maleic anhydride (Scheme 9.5) [34,38,41].

Recently, many companies have been developing bio-based routes for chemical synthesis due to the increasing price of oil and a broader awareness of sustainability, and consequently, bio-based BDO is gaining momentum in industry. Genomatica has developed and patented a one-step process to produce BDO through fermentation using non-naturally occurring microbial organisms [42], which is called the GENO BDO process. In late 2012, Genomatica and DuPont Tate & Lyle Bio Products Company, LLC, successfully produced over 2000 t of BDO through direct fermentation in a 5-week campaign, which was announced as "the first successful commercial-scale production

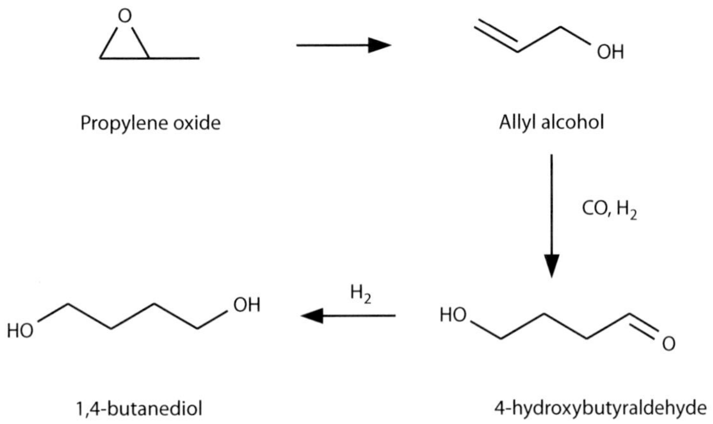

SCHEME 9.3 BDO synthesis from propylene oxide.

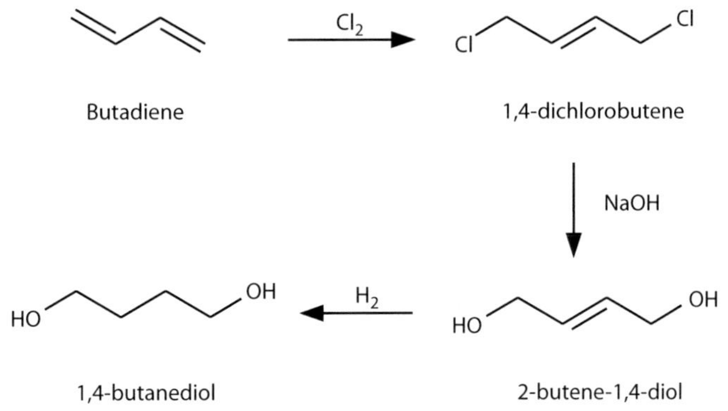

SCHEME 9.4 BDO production from butadiene.

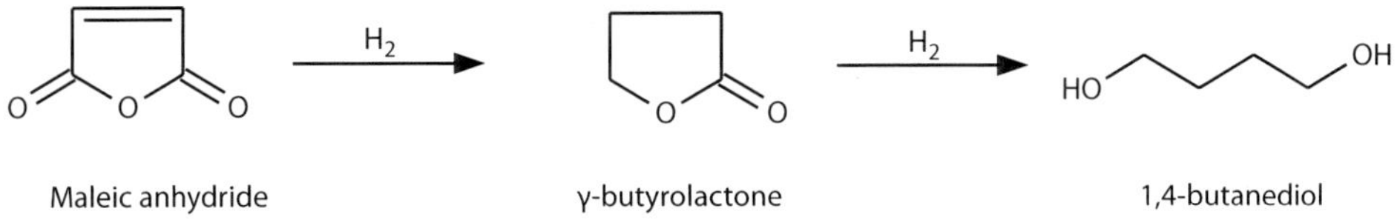

SCHEME 9.5 BDO production from maleic anhydride.

of BDO using a bio-based manufacturing process" [43]. Recently, Toray has successfully produced bio-based PBT from Genomatica's bio-based BDO. The bio-based PBT has equivalent physical properties to the petroleum-based PBT. DSM also announced their approval of bio-based BDO from the GENO BDO process in their Arnitel block-copolyester product line, where the hard segment is PBT.

BioAmber is another player in the bio-based BDO market. BioAmber started as a producer of bio-based succinic acid (SA) and later exclusively licensed DuPont's hydrogenation catalyst technology to produce bio-based BDO and tetrahydrofuran (THF) (from bio-based BDO). BioAmber has also built an exclusive partnership with Evonik to optimize DuPont's catalyst [44].

While progress has been made in the production of BDO from biorenewable resources, TPA and DMT production processes still rely on fossil-based feedstock. Both TPA and DMT manufacturing start with p-xylene as feedstock. The majority of TPA and DMT production worldwide is based on a cobalt/manganese/bromine catalyst developed by Scientific Design for oxidation of p-xylene [45]. In the process, acetic acid, which serves as the solvent, is charged into the reactor along with p-xylene and a catalyst. Excess compressed air is also charged into the reactor and reacts to oxidize the p-xylene. In this oxidation process, three intermediates are produced: p-tolualdehyde, p-toluic acid, and 4-formyl-benzoic acid, as shown in Scheme 9.6. The TPA produced precipitates from the acetic acid solvent due to its low solubility. During precipitation, a significant amount of 4-formyl-benzoic acid is trapped in the precipitate and prohibits further oxidation to TPA due to the similarity of TPA and 4-formyl-benzoic acid. TPA produced by this process contains such high amounts of residual 4-formyl-benzoic acid that it cannot be used to produce polyesters directly.

There are several processes to purify crude TPA and remove the residual 4-formyl-benzoic acid. One process was developed by Amoco Chemical [46] and accounts for the majority of the share of TPA feedstock for polyesters. The process is based on hydrogenation of 4-formyl-benzoic acid to p-toluic acid, followed by a series of crystallization steps of TPA [47]. The p-toluic acid is more soluble than TPA and remains in the liquid phase, while purified TPA crystals are precipitated out. Other processes include multistage oxidation, where after the oxidation process described previously, the slurry is transferred to another reactor and the temperature of the system is further increased. At the higher temperature, the TPA becomes more soluble in the slurry. The trapped 4-formyl-benzoic acid is then released and oxidized into TPA. Hydrolysis of DMT is another process used to produce a very small percentage of the TPA feedstock [48].

As in the production of TPA, the production of DMT also starts with p-xylene [48]. The p-xylene is first oxidized into p-toluic acid. Methanol is then added for the esterification reaction to form

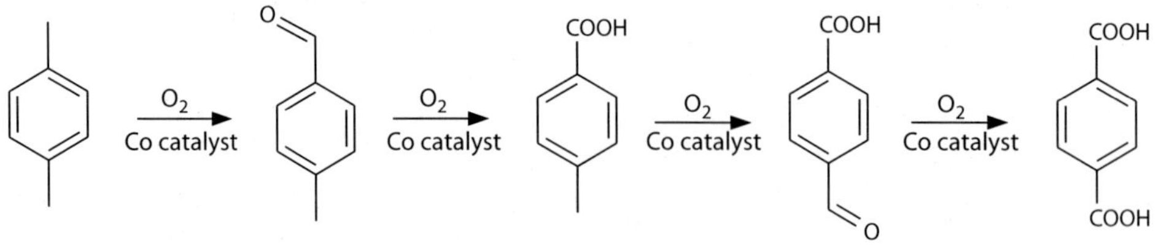

SCHEME 9.6 Oxidation process to produce TPA.

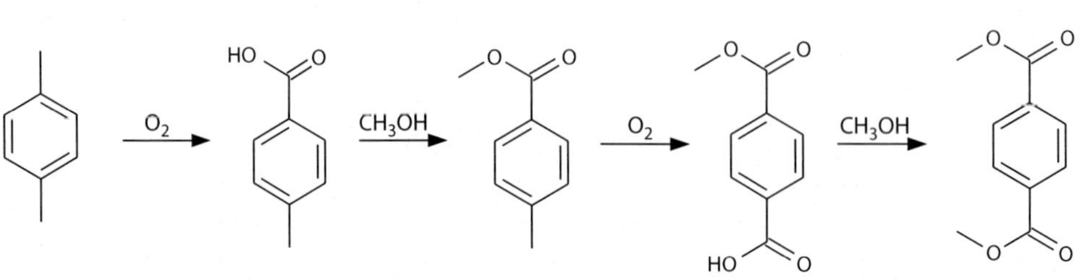

SCHEME 9.7 DMT manufacturing process.

methyl *p*-toluate. In the following step, methyl *p*-toluate is oxidized to produce monomethyl terephthalate. Finally, methanol is again introduced to react with monomethyl terephthalate and produce DMT, as in Scheme 9.7. DMT can be purified easily by crystallization and distillation. In addition to this process, DMT can also be produced by direct esterification of crude TPA with methanol.

9.2.3 PBT Polymerization Process

The synthesis chemistry of PBT has not changed significantly throughout the years, although there have been advances in processing and catalyst technology. Traditionally, the production of PBT resin started with DMT and BDO [49]. The first step is the transesterification reaction between DMT and BDO to produce the diester, bis(4-hydroxybutyl-terephthalate), or ester oligomers with hydroxybutyl ester end groups. Scheme 9.8 shows the diester structure. The reaction takes place at a temperature between 150°C and about 210°C, at atmospheric or slightly lowered pressure. During this first step, methanol is removed from the system as a by-product. The end of the first step is indicated by complete removal of methanol.

The second step is the polycondensation reaction of the diester or ester oligomers, during which the molecular weight of PBT is increased as the polycondensation reaction progresses (Scheme 9.9). The polycondensation takes place at temperatures close to 260°C and at a reduced pressure below 100 Pa [49]. In this step, BDO is constantly removed from the system as a by-product to advance the reaction. As the PBT molecular weight builds up, the melt viscosity increases significantly, which limits the rate of BDO removal. Hence, applied effective agitation is required to facilitate the BDO removal. The second step is completed when the desired molecular weight of PBT is achieved. The typical reaction time for the polycondensation step is 4 to 5 h. In industry, a common practice is to produce two grades of resin with high and low molecular weights. Any molecular weight between the two for a desired product or application can then be achieved by blending the low-molecular-weight and high-molecular-weight resins.

Due to the advancement of TPA purification technology and the lower price of TPA as a raw material, a new process using TPA and BDO as raw chemicals (shown in Scheme 9.10) has become less costly compared to the traditional process using DMT and BDO. Instead of transesterification between DMT and BDO to produce the diester or ester oligomers, the new process employs a direct

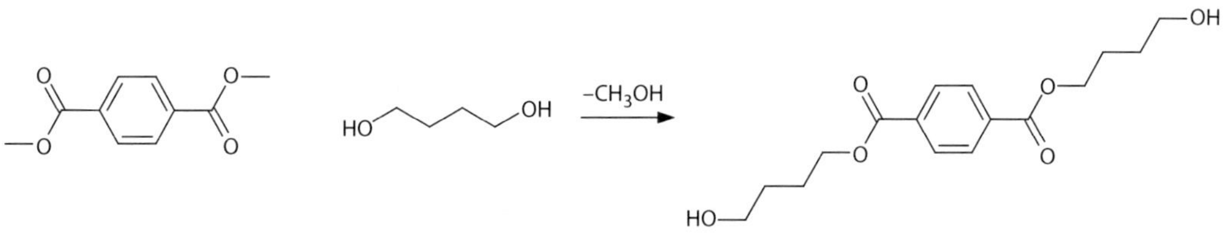

SCHEME 9.8 Transesterification of DMT and BDO.

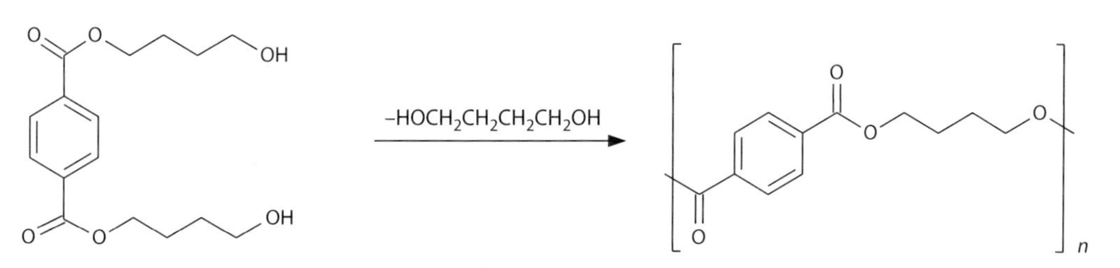

SCHEME 9.9 Polycondensation step to produce PBT.

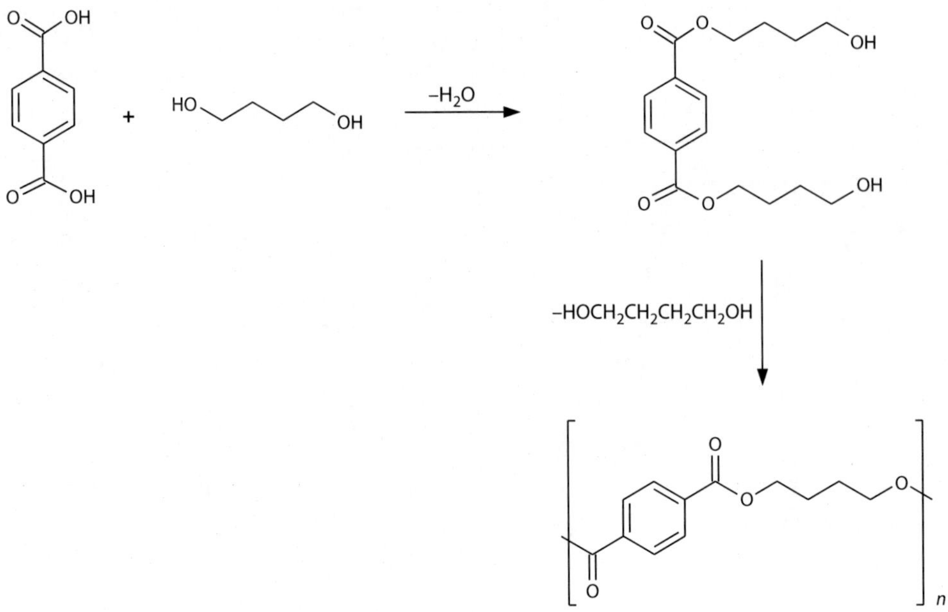

SCHEME 9.10 PBT from TPA and BDO.

esterification reaction of TPA with an excess amount of BDO [50–52]. During the reaction, water is removed from the reaction vessel. The second step of the process is to produce PBT from bis(4-hydroxybutyl-terephthalate). The ester oligomer formed is very similar for both routes, with either TPA or DMT as the starting raw material.

Industrial catalysts for PBT production are primarily titanium alkoxides, which can catalyze both direct esterification reactions and transesterification reactions [50,52]. In contrast to PET production, in PBT production, the same catalyst is used for both reactions due to its high efficiency. After PBT production, the catalyst remains in the resin and maintains its activity. Even at very low levels, these active catalysts can lead to other transesterification reactions when blended with other polymers such as PET and PC, but weak Lewis acids can be added into the blend to deactivate the catalysts [53,54].

The major side reaction in the production of PBT is the generation of THF [55–57]. The TPA-based process produces as much as twice the amount of THF by-product as that generated by the DMT-based process. Thus, another key point in making the TPA-based process more successful is to reduce THF formation during the reaction. THF forms in the reaction through two routes (Scheme 9.11) [58,59]. One is the acid-catalyzed dehydration of BDO, which is believed to cause

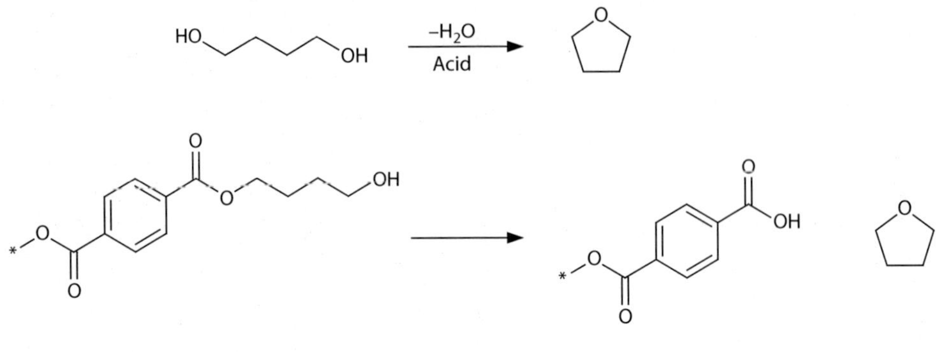

SCHEME 9.11 THF by-product formation.

the major amount of THF formation in the TPA-based process. The other route is backbiting of the hydroxybutyl ester end group to form BDO.

9.2.4 GREEN CHEMISTRY

Environmental awareness and the concept of sustainability have not only touched the raw materials aspect of PBT production, as previously mentioned for bio-based BDO, but have also promoted PBT production from postconsumer recycle (PCR) PET. PET is widely used as a packaging material for soft drinks, food, cosmetics, cooking oils, and many more products, and PET is the largest recycled plastic material in Europe. According to PETCORE Europe, in 2012, an equivalent of more than 60 billion PET bottles were recycled, which represents more than 52% of all postconsumer PET bottles [60].

One representative case of PBT production from PCR PET is IQ VALOX, produced by the Innovative Plastics business of SABIC. The process starts with PCR PET collection, sorting, cleaning, and grinding to produce PCR PET flakes. The PCR PET flakes are heated together with EG to break down the PET polymer chain and form bis-hydroxyethyl terephthalate (BHET) [61,62]. BDO is then added to make bis-hydroxybutyl terephthalate (BHBT) through a transesterification reaction. The next step is a polycondensation reaction to polymerize BHBT at elevated temperature and reduced pressure.

The process to make PBT from PCR PET is a chemical recycle process that degrades the PCR PET polymer chains to monomers or oligomers, and builds up the new PBT polymer chains from these monomers and oligomers. Unlike physical recycle processes by which the material performance will be deteriorated due to degradation and contamination, the PBT produced through a chemical recycle process from PCR PET is expected to have the same properties as PBT produced from virgin raw materials.

9.2.5 PBT CRYSTALLIZATION

PBT has very good chemical resistance, low moisture absorption, and relatively high strength and stiffness due to the existence of its crystalline phase. Typically, PBT has about 35% crystallinity [63]. The glass transition temperature of PBT is approximately 45°C [49] depending on the crystallinity, and the melting temperature is approximately 225°C [64]. The crystallinity, glass transition temperature, and melting temperature of PBT depend on the thermal history of the material. For instance, by annealing, the crystallinity of PBT can reach 60% [65–67]. Annealing of PBT for long periods of time can also shift the melting temperature and glass transition temperature toward higher values due to perfection of the ordered schemes. Various sizes of PBT crystals usually exist, and therefore, the differential scanning calorimetry (DSC) curve often shows multiple melting peaks. Understanding the thermal history of the PBT material is key to deciphering the DSC results.

PBT crystallizes in two different forms, the alpha form and the beta form, and both have a triclinic unit cell. For the alpha crystal, $a = 4.83$ Å, $b = 5.96$ Å, and $c = 11.62$ Å; $\alpha = 99.9°$, $\beta = 115.2°$, and $\gamma = 111.3°$ [68,69]. For beta crystals, $a = 4.95$ Å, $b = 5.67$ Å, and $c = 12.95$ Å; $\alpha = 101.7°$, $\beta = 121.8.2°$, and $\gamma = 99.9°$ [70]. The beta crystal usually forms under stress.

9.2.6 PBT PROPERTIES AND APPLICATIONS

PBT is an engineering thermoplastic that has applications in the automotive, electrical, electronics, health care, and building and construction markets, mainly for injection-molded parts. Although PBT can also be used to make fiber and film products, the applications of PBT in the form of fibers and films are far less than those for injection-molded parts. For fiber and film applications, PET plays the biggest role in the polyester family. PBT is limited in its capability to expand in this market due to its relatively high price along with its fast crystallization speed that results in a translucent rather than a transparent film.

In most injection-molding applications, PBT resin needs to be reinforced and enhanced with inorganic fillers to gain high modulus and strength. Among inorganic fillers, glass fiber is the most popular because it can achieve the desired properties at a relatively low cost. Other fillers for PBT include talc, clay, silica, mica, wollastonite, glass beads, and carbon fibers. Table 9.4 lists the properties of PBT and comparisons of PBT with glass fiber-filled PBT [71]. PBT resins are also formulated to various levels of flame resistance with flame-retardant and synergistic additives. Flame-retardant products are required for many applications such as electrical equipment, electronic devices, and home appliances. Typical flame-retardant ratings of PBT seen in the market are UL 94 V0 at 0.8, 1.6, and 3.2 mm, with slight differences in the vicinity of these thicknesses from different suppliers. A UL 94 5VA rating is also possible for PBT with a thickness of 2.0 mm or above [72].

Traditional flame retardants used for PBT are bromine-containing compounds or polymers [72], such as brominated carbonate oligomers, decabromodiphenyl ether (decaBDE), brominated PC, and brominated polystyrene. Antimony oxide can be added into the formulation for its synergistic effects with the primary flame-retardant additive.

However, with regulations on halogen-containing plastics becoming more stringent, it is often desirable to remove all halogen-containing flame-retardant agents from the PBT formulation for many applications. Some of the most successful alternatives are phosphorous-based flame retardants. The simplest phosphorous-based flame retardant for PBT is red phosphorous, which is also supplied in a coating form. The coating can help to reduce the toxic phosphine emitted during the processing [72].

A combination of nitrogen-based chemicals together with phosphorous-based flame retardants can also be used to flame-retard PBT. For example, PBT formulated with a combination of melamine pyrophosphate and resorcinol diphosphate can achieve a UL 94 rating of V0 at 0.8 mm [73].

In addition to PBT resin and reinforced PBT resin in various applications, PBT is also blended with other engineering thermoplastics to achieve unique combinations of properties. Blends of PBT with commercial significance include PBT/PET blends, PBT/PC blends, and impact-modified PBT.

PBT/PET blends are usually reinforced by glass fibers or minerals, similar to reinforced PBT [74]. Some physical and mechanical properties of glass fiber-reinforced PBT/PET blends are shown

TABLE 9.4
Typical Properties of PBT and Glass Fiber-Filled PBT Resins

Properties	Unit	Standard	PBT Resin	PBT+15% GF	PBT+20% GF	PBT+30% GF
Specific gravity	–	ASTM D792	1.31	1.40	1.44	1.53
Tensile strength	MPa	ASTM D638	55	90	115	125
Tensile elongation	%	ASTM D638	150	4	4	4
Flexural strength	MPa	ASTM D790	80	145	175	190
Flexural modulus	GPa	ASTM D790	2	5	6	8
Heat deflection temperature (1.82 MPa)	°C	ASTM D648	60	205	207	210
Izod impact strength (notched, room temperature)	kgf cm/cm	ASTM D256	5	5	7	10

Note: GF, glass fiber.

TABLE 9.5
Typical Properties of Glass Fiber-Reinforced PBT/PET Blends

Properties	Unit	Standard	PBT/PET+15% GF	PBT/PET+30% GF	PBT/PET+40% GF
Specific gravity	–	ASTM D792	1.43	1.54	1.64
Tensile stress at break	kgf/cm^2	ASTM D638	910	1070	1480
Tensile strain at yield	%	ASTM D638	3	3	2
Tensile modulus	kgf/cm^2	ASTM D638	50,900	81,500	130,000
Flexural modulus	kgf/cm^2	ASTM D790	45,800	71,300	108,900
Izod impact strength (notched, 23°C)	cm kgf/cm	ASTM D256	3	8	10
Instrumented impact total energy (23°C)	cm kgf	ASTM D3763	61	71	80
Heat deflection temperature (1.82 MPa, 3.2 mm, unannealed)	°C	ASTM D648	160	195	193

in Table 9.5. Although PBT and PET are miscible in the molten state as well as in the amorphous phase, they crystallize separately to form individual crystalline domains. Compared to glass fiber-reinforced PBT, glass fiber-reinforced PBT/PET blends have better surface appearance and gloss. The mechanical properties of the reinforced PBT/PET blends are dictated mainly by the glass fiber content. Another advantage of PBT/PET blends is their low cost due to the relatively low price of PET. In PBT/PET blends, transesterification may happen during processing [75], and this can be undesirable if it happens to the extent that crystallinity is reduced and fast crystallization is hindered.

PBT/PC blends combine the advantages of amorphous PC with crystalline PBT. Amorphous PC provides dimensional stability, good impact properties, and paint adhesion, while crystalline PBT provides good chemical resistance and heat performance. The remaining catalyst in the PBT resin causes transesterification to occur when PBT and PC are compounded [76]. To prohibit the transesterification reaction, a quencher is often added into the formulation to deactivate the catalyst [77]. The quencher is usually a phosphate salt such as zinc phosphate or mono zinc phosphate.

Impact-modified PBT blends are blends of PBT with impact modifiers [78–80], including methacrylate–butadiene–styrene (MBS), acrylonitrile–butadiene–styrene (ABS), and styrene–ethylene–butadiene–styrene (SEBS). The rubbery butadiene phase of the copolymers provides good impact properties. In some cases, these blends can maintain high impact properties at reduced temperatures.

9.3 POLYESTERS BASED ON CYCLOHEXANEDIMETHANOL

The commercial success of PET and PBT along with the quickly growing market of polyester fibers inspired further industrial research into new monomers for polyester synthesis during the 1950s. One of the most important and successful of these monomers was the high-heat diol 1,4-cyclohexanedimethanol (CHDM) developed by Tennessee Eastman Co., a division of Eastman Kodak Co. [81].

9.3.1 History of CHDM-Based Polyesters

Linear polyesters synthesized from cis- or trans-CHDM and TPA were patented in 1959 and became known as poly(1,4-cyclohexylenedimethylene terephthalate) (PCT) [82,83]. PCT is a crystalline polyester with a high melting point along with better weatherability, higher glass transition temperature, and higher heat distortion temperature than PET, and was reported be useful in forming fibers and films [81]. PCT was first introduced by Tennessee Eastman as Kodel fiber, and a second line of Kodel fibers made from PET was later introduced by Eastman and then discontinued in the 1990s [84].

The original patents also disclosed the discovery of amorphous CHDM copolyesters, synthesized by polymerizing both EG and CHDM with TPA, but the first commercialization of an amorphous copolyester containing CHDM did not occur until many years later [81]. The amorphous poly(ethylene terephthalate-co-1,4-cyclohexylenedimethylene terephthalate) copolyesters are known as glycol-modified PCT (PETG) when the CHDM content is less than 50% and glycol-modified PCT (PCTG) when the CHDM content is greater than 50%. PETG was introduced by Eastman in 1977 as Kodar PETG 6763 and contained 70% EG and 30% CHDM [81]. The Eastman line of PETG and PCTG products has since been expanded to include several resins for injection-molding and blow-molding applications. Eastman later introduced PCT as an injection-molding resin in 1987 under the name Ektar, which included glass-reinforced PCT, but it is now called Thermx [85] and is owned by Celanese.

CHDM can also be polymerized with dimethyl trans-1,4-cyclohexanedicarboxylate (DMCD) or 1,4-cyclohexanedicarboxylic acid (CHDA) to yield completely aliphatic, cyclic, partially crystalline copolyesters known as poly(cyclohexylene dimethylene cyclohexanedicarboxylate) (PCCD). PCCD copolymers of trans-CHDA and CHDM that can form films and fibers were reported in 1959 [86]. However, they were not commercialized until many years later, most likely because of the exact stoichiometry needed to obtain high molecular weight and difficulty in crystallizing from the melt [87].

9.3.2 Polymerization Process

The two major manufacturers of CHDM are currently Eastman Chemical in the United States and SK Chemical in Korea [88]. CHDM is a considerably more expensive diol monomer than EG but provides added value in properties such as higher heat and toughness compared to PET. The CHDM monomer is synthesized by catalytic hydrogenation of dimethylene terephthalate (DMT) in two steps, as shown in Scheme 9.12 [81]. DMT is first converted to dimethyl 1,4-cyclohexanedicarboxylate (DMCD), followed by conversion of DMCD to the CHDM diol [81]. The copper chromite catalyst is often used in industrial processes and yields a 70:30 trans/cis mixture of CHDM. The trans/cis ratio of the monomer isomers varies depending on the catalyst type and reaction conditions, and the ratio affects the properties of PCT and other polymers derived from CHDM [81].

Polyesters derived from CHDM are usually synthesized via a melt-phase polycondensation process with a metal catalyst in two steps, as shown in Scheme 9.13. Similar to other polyesters such as PET and PBT, the first step is an esterification with an excess of diols to form oligomers. This is followed by a second polycondensation step under high vacuum and elevated temperature to build

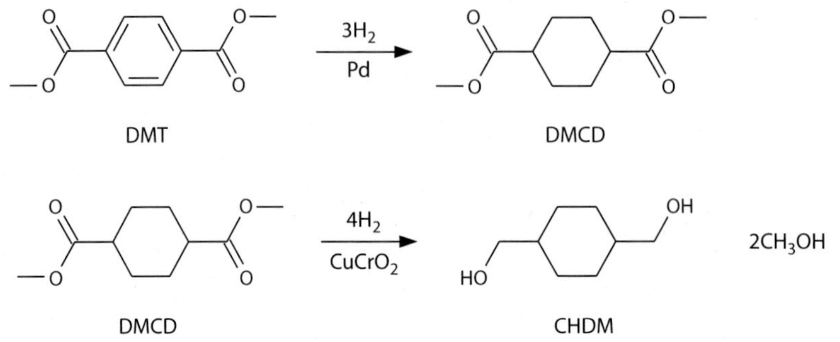

SCHEME 9.12 CHDM monomer synthesis by catalytic hydrogenation.

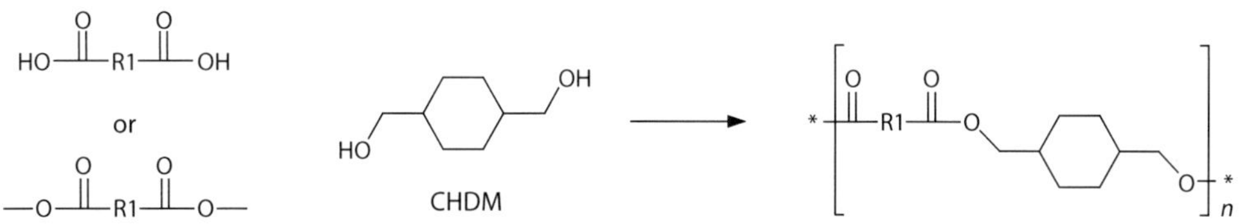

Thermoplastic Polyesters

SCHEME 9.13 Polyester synthesis by polymerization of CHDM with diesters or diacids.

molecular weight [84]. The catalyst for the melt polymerization process can affect the molecular weight and color of the polymer formed. For PETG/PCTG synthesis, titanium catalysts are very active, but the copolymers tend to have a yellow color due to side reactions that also occur quickly [89]. Some catalysts such as antimony and cobalt yield polymers with a gray color due to reduction of the metals [90]. Combinations of manganese, zinc, titanium, and optional germanium catalysts with a phosphorus inhibitor and cobalt toning agent have been shown to reduce the yellow color of PETG/PCTG, and the material can be used in forming sheets [89,90]. Improvements in color and clarity have also been demonstrated by optimizing the process conditions to create a more active esterification product that requires lower catalyst amounts and reduced temperature for polycondensation, along with higher amounts of phosphorus stabilizer [91].

The main types of CHDM polyesters are shown in Table 9.6. PCT is the crystalline polyester generated from CHDM and DMT or TPA. PETG and PCTG are amorphous copolyesters synthesized by the addition of EG in amounts above or below 50%, respectively, and acid-modified PCT (PCTA) is a crystalline copolyester synthesized by the addition of other diacids such as IPA at low levels. PCCD is produced from CHDM and DMCD to generate completely aliphatic copolyesters. The thermal properties of each polyester are shown in Table 9.7, and mechanical properties are listed in Table 9.8.

TABLE 9.6
CHDM-Derived Polyesters and Copolyesters

Diester/Diacid	Diol	Polymer
DMT/TPA	CHDM	PCT 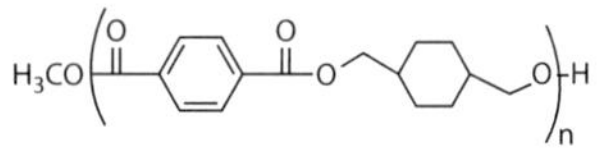
TPA + IPA	CHDM	PCTA 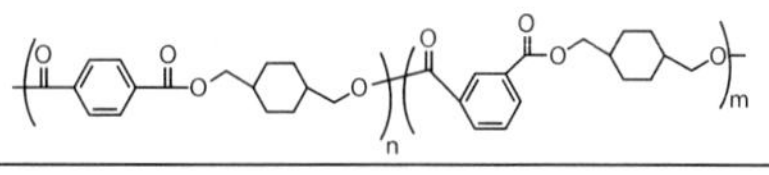
DMT/TPA	CHDM (less than 50%) + EG	PETG 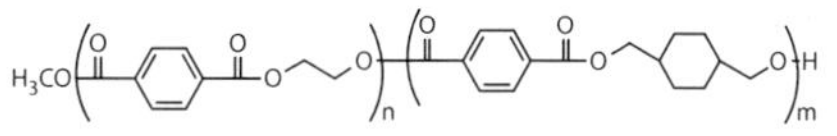
DMT/TPA	CHDM (greater than 50%) + EG	PCTG 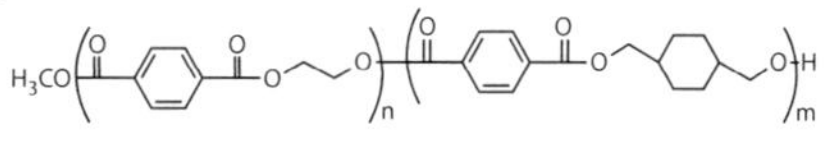
DMCD	CHDM	PCCD 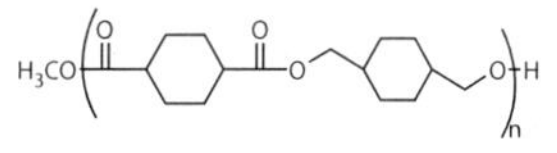

TABLE 9.7
Thermal Properties of CHDM-Derived Polyesters and Copolyesters

Polymer	T_g, °C	T_m, °C
PCT	60–95	250–290
	80–90	285–290
PCTA	88	230–280
PETG	81	–
PCTG	84	255
PCCD	40–70	205–235

TABLE 9.8
Mechanical Properties of CHDM-Derived Polyesters and Copolyesters

Polyester	PCT	PCTA	PETG	PCTG	PCCD
Specific gravity	1.2	1.2	1.27	1.23	1.13
Tensile stress at yield, MPa		47	50	45	13
Tensile stress at break, MPa	40–65	51	28	52	23
Elongation at yield, %	–	5	4.3	5	38
Elongation at break, %	170–350	300	110	330	400
Flexural modulus, MPa	1600–1900	1800	2100	1900	150
Izod impact strength, notched at 23°C, J/m	80–100	80	101	NB	–
Izod impact strength, notched at −40°C, J/m	–	40	37	64	40

Note: NB, no break.

9.3.3 Properties and Applications

CHDM polyesters generally have high strength, impact, transparency, and chemical resistance. CHDM is a unique monomer in that polymerization with the trans isomer yields PCT with a higher melting point of 315–320°C, while the cis isomer yields PCT with a melting point of 260–267°C [82]. Mixtures of isomers yield PCT polyesters with melting points of about 250–290°C [42]. The glass transition temperature of the polymer is also dependent on the trans/cis ratio of CHDM isomers. The T_g is about 60°C for PCT synthesized from cis-CHDM and 95°C for PCT from trans-CHDM. Methods have been developed to synthesize and isolate the higher-melting-point trans isomer [92], as well as to improve control over the trans/cis ratio [93]. Typical commercial PCT with a 70:30 trans/cis CHDM ratio crystallizes slower than PBT, at a rate similar to PET, although the crystallization rate is also dependent on the trans/cis ratio [84].

PCT was first used as a textile fiber [85]. Today, PCT is commonly used in connector applications, and the heat distortion temperature and stiffness can be enhanced further by the addition of glass filler. The heat deflection temperature (HDT) of 30% glass-filled PCT is typically 250–265°C, which is higher than that of glass-filled PET (220–230°C) or PBT (200–210°C) [94] but lower than liquid crystal polymers (288°C) [88].

Although its thermal properties are useful in many applications, the processing window of PCT is narrow due to its high melting point, which nearly approaches the decomposition temperature. To lower the melting point, PCT can be modified with low levels of other diacids, such as IPA to PCTA, a crystalline copolyester with good resistance to hydrolysis [85]. The IPA content in the copolyester is usually maintained at less than 25% to prevent difficulty in crystallization and the need to use artificial nucleation techniques [95]. PCTA is available as Durastar from Eastman Chemical, and these copolyesters have a lower melting point and wider processing window than PCT but retain the transparency, toughness, and chemical and hydrolysis resistance [85]. PCTA has good hydrolytic stability and low moisture uptake due to its hydrophobic backbone and slower hydrolysis than copolyesters containing EG [81], and it requires less drying prior to processing. Most other CHDM-containing copolyesters must be dried carefully before processing to prevent loss of molecular weight and mechanical properties, although methods have been developed to reduce drying time and minimize hydrolytic degradation [96].

CHDM can also be incorporated into glycol-containing polyesters such as PET to produce amorphous polyesters, and the mole percent of CHDM in the copolyester influences the crystallization rate and melting temperature of the copolyester [81]. The mole percent of CHDM reduces the crystallinity and slows the crystallization rate of PET. An amorphous window where the polymers do not crystallize occurs around a ratio of about 70:30 EG/CHDM. As the level of CHDM is increased further, the melting point again becomes present and begins to increase along with the crystallization rate [96].

PETG copolyesters, as mentioned above, contain less than 50% CHDM and are useful in applications for containers and blow-molded bottles. They are often used in packaging and are produced as Spectar by Eastman Chemical Company [81]. PETG has been copolymerized with other monomers such as 2,6-naphthalene dicarboxylic acid (NDA) (0–40%) and SA (0–40%) to modify the T_g and shrinkage behavior [97]. At levels of CHDM greater than 50%, the copolyesters are known as PCTG. Due to the higher levels of CHDM, PCTG has higher impact strength, toughness, glass transition temperature, and heat distortion temperature but lower modulus than PETG. Both PETG and PCTG copolyesters have excellent transparency and are resistant to many organic solvents as well as γ-ray sterilization, rendering them useful in applications for kitchen and other household appliances or medical devices [81], along with automotive and electronic applications [98].

PCCD is a useful copolyester material for applications where greater ultraviolet (UV) stability and weatherability are required. Compared to PETG and PCTG, PCCD has better resistance to photooxidation as a result of its completely aliphatic nature. PCCD has a high melting point (225–235°C), but its lower crystallization temperature (152–171°C) and T_g (40–70°C) inhibit its usefulness in some applications [99]. The process conditions and stoichiometry have been optimized to enhance crystallinity [100], and various attempts have been made to increase the crystallinity and raise the T_g by incorporation of other aliphatic monomers such as bicyclo[2.2.2]octyl (DMCD-2) [101]. Amides such as 1,6-bis(4-carbomethoxycyclohexylcarboxamido) hexane have also been incorporated [99] and successfully raised T_c to 180–190°C without significantly affecting T_m. Incorporation of 2,2'-(sulfonylbis(4,1-phenyleneoxy))-bis(ethanol) is a method of raising the T_g of PCCD while maintaining low melt viscosity [102]. Higher amounts of trans-cyclohexyl diester isomer also raise the crystallization temperature, and attempts have been made to control reaction conditions by minimizing catalyst, temperature, and reaction time to maximize the trans/cis isomer ratio and lead to higher crystallization temperatures [87].

The widespread use of bisphenol A (BPA)-containing polymers and health concerns about BPA exposure has inspired development of several alternative, BPA-free copolyesters. In 2008, Eastman introduced Tritan, a copolyester derived from DMT, CHDM, and 2,2,4,4-tetramethyl-1,3-cyclobutanediol (TMCD) [103]. Its monomers and extracts have tested negative for estrogenic or androgenic activity [103]. Like PCTG/PETG and PCCD, Tritan is a tough, durable, transparent polyester with good impact and chemical resistance, and it is used for many similar applications such as bottles, housewares, small appliances, and medical devices [104].

ECOZEN is another BPA-free, bio-based copolyester of EG, TPA, CHDM, and isosorbide introduced by SK Chemicals in 2012 [105]. The isosorbide biomonomer is corn based and provides higher heat resistance than PETG [105]. ECOZEN has a heat distortion temperature of 80–110°C, with good chemical resistance and transparency, and has found applications in LED lighting, consumer electronics, housewares, and appliances [106].

Unlike other polyesters such as PET and PBT, many CHDM-derived polyesters are miscible with PC over a wide compositional range and can be utilized to enhance the properties of PC blends. Blends of CHDM polyesters with PC are fully transparent and maintain good mechanical properties such as impact, tensile, and flexural strength, but with the added benefits of improved flow and hydrolysis resistance compared to traditional PC blends [107,108]. The polyesters also provide enhanced chemical resistance to a variety of solvents. Compared to PC alone, CHDM polyester/PC blends have improved resistance to ionizing radiation, demonstrating significantly reduced yellowing after exposure to various doses of gamma and electron beam radiation often used for sterilization of medical devices [109].

SABIC's Innovative Plastics business has developed the XYLEX product line based on transparent PC/amorphous polyester blends. In addition to clarity, XYLEX resins provide high impact strength; ductility; chemical resistance to solvents including cleaning fluids, automotive fluids, food and beverages, and alcohols; good weatherability; and high flow in selected grades [110]. XYLEX is useful in applications such as eyewear, health care, consumer electronics, housewares, and automotive [110,111].

9.4 NAPHTHALATE POLYESTERS

Concurrent with the invention of PET and PBT in the 1940s, other aromatic acid components were being explored as alternatives to TPA (or DMT). This resulted in the discovery and subsequent commercialization of naphthalate-based polyester poly(ethylene naphthalate) (PEN).

The aromatic monomer, namely, naphthalene dicarboxylic acid (NDA), is synthesized by the oxidation of 2,6-dialkyl napthalene by cobalt/manganese or other catalysts with a promoter such as bromine [112]. Two minor impurities in the reaction include trimellitic acid and 2-formyl naphthoic acid (obtained from the partial oxidation of 2,6-dialkyl naphthalene) [112]. Additional products such as bromonapthalene dicarboxylic acid are also observed in very small amounts [113]. In 1995, BP Amoco commercialized 2,6-naphthalene dicarboxylic acid dimethyl ester (NDC), produced from o-xylene and 1,4-butadiene, and NDC now sells up to 60 million lb./year. Other companies that followed Amoco in NDC production are Mitsubishi and Teijin of Japan. Even today, the high cost of NDC causes a niche market response to NDC-based polymers compared to commodity polymers like PET.

9.4.1 History

Prior to the piloting of PEN by Teijin in 1970, research was focused on an economical commercial route to manufacture NDC and NDA. The first attempts to make NDA (or NDC) mimicked the oxidation process of p-xylene to TPA. Instead of p-xylene, alkyl naphthalenes were oxidized to NDA [112]. Lillwitz [114] reviewed various other approaches for NDA production such as diiodonaphthalene, transcarboxylation, alkyl aromatic acylation, alkyl aromatic base-catalyzed condensations, and various aromatic alkylations.

The route chosen by BP Amoco for the production of NDA (or NDC) is the oxidation of 2,6-dimethyl naphthalene, derived from the alkenylation of o-xylene with 1,3-butadiene followed by subsequent cyclization, dehydrogenation, and isomerization (Scheme 9.14). Today, the cost of NDA remains expensive due to difficulties in the alkenylation of o-xylene with 1,3-butadiene. The stumbling block associated with alkenylation of o-xylene with 1,3-butadiene is primarily attributed to the very strong basic catalyst used in the reaction. Common bases, such as alkali metal oxides and alkali metal-exchanged zeolites, do not perform as well as the strongly basic sodium–potassium eutectic alloy (22 wt.% sodium and 78 wt.% potassium) [114]. It is paramount to invent novel a catalyst or process conditions that would drastically bring down the cost of NDA. Else, the market for NDA-based polymers will continue to be niche, in spite of being studied for over 60 years.

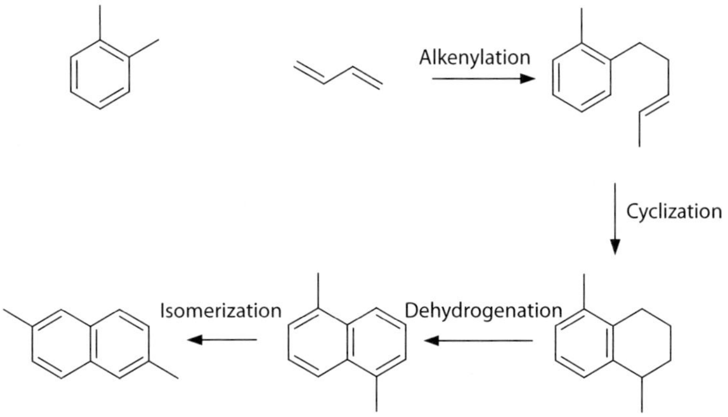

Thermoplastic Polyesters

SCHEME 9.14 NDA process.

9.4.2 Polymerization Process

Similar to other aromatic polyesters, the aromatic diacid (NDA or NDC) is polymerized with an aliphatic diol, mainly EG [115], propylene glycol, or butylene glycol, in the presence of a transesterification catalyst to produce PEN, poly(propylene naphthalate) (PTN) [116], or poly(butylene naphthalate) (PBN) [116], respectively. Compared to PEN, both PTN and PBN (commercialized by Teijin) are more expensive, and hence, most commercial interest generally involves studying PEN for various applications. Typical PEN polymerization processes are similar to those of PET and PBT, using excess diol to diacid (2:1 ratio) in the first stage to produce oligomers in the presence of a nitrogen atmosphere at a temperature range of 150–240°C, followed by further polymerization under vacuum at 290°C [115]. For PBN and PTN, the final polymerization temperature is lower, between 240°C and 255°C [116]. Due to the crystalline nature, SSP can also be used for polyesters such as PEN [117]. SSP not only reduces the color of the final polymer but also reduces degradation, which is important in developing useful applications involving both clarity and performance.

9.4.3 Thermal Properties of Naphthalate Polyesters

Increasing the flexibility and length of the spacer from the shorter $-(CH_2)_2-$ in the case of EG used for PEN to the longer $-(CH_2)_4-$ in BDO used for PBN results in lowering of both T_g and T_m (Table 9.9) [118]. PBN has a significantly higher melting temperature compared to PTN, but PTN only has a moderate increase in glass transition temperature over PBN. Faster crystallization leads to a decrease in molding cycle time, which is very important for molding a greater number of parts in a shorter amount of time. PBT is currently the material of choice for faster molding cycles, but PBN can crystallize even faster than PBT. Compared to a crystallization time of 3 s for PBT, a mere 1 s is required for PBN.

TABLE 9.9
Thermal (T_g and T_m) Data of Naphthalate Polyesters

Property, °C	PEN	PTN	PBN
T_g	122	87	82
T_m	270	207	243

Source: Naphthalates. 2001. Engineering resin applications. BP. Available at http://wenku.baidu.com/view/c56094of05087632311212dc.html.

TABLE 9.10
Thermal Properties of Random Copolyesters Made from PPN and PTN: T_m (First Heating), ΔH_m (First Heating), and T_g (Second Heating)

Composition (Tere-Ester/Naph-Ester)	T_m, °C	ΔH_m, %	T_g, °C
100/0	227	77	47
92/8	221	70	50
85/15	212	62	52
70/30	192	52	57
65/35	185	48	59
55/45	169	37	65
45/55	150	35	67
35/65	169	29	71
25/75	176	34	75
10/90	199	42	82
0/100	207	47	87

Source: Lorenzetti, C., Finelli, L., Lotti, N., Vannini, M., Gazzano, M., Berti, C., and Munari, A. *Polymer*, 46:4041, 2005.

In any crystalline polymer system, introducing defects by copolymerization will result in poor crystal lattice packing, and consequently, the melting point and percentage of crystallinity will decrease. Poly(1,3-propylene naphthalate) (PPN) was transesterified under melt with another polyester, namely, poly(1,3-propylene terephthalate) (PPT) [119], in the presence of a titanium tetrabutoxide catalyst at 260°C. The product was a random copolymer that displayed a reduction in both percentage of crystallinity as well as melting point, depending on the amount of comonomer in the system (Table 9.10). A gradual increase in T_g was also observed (Table 9.10).

Similarly, both PET/PEN random copolymers [120] as well as PET/PEN blends [120] had a lower melting point compared to the individual polymers due to entropy affecting the crystallization phenomena. At about 15% comonomer concentration (PET or PEN), the random copolymer essentially became amorphous [120], while the blends still showed a melting point [120]. Further, the blends showed higher T_m compared to copolymers of the same composition [120]. The difference in melting points of the same composition in a copolymer or blend is due to more homopolymer linkages in the case of the blend, leading to higher crystallinity and, hence, a higher melting point. Low- and high-naphthalate-content PET copolymers were commercialized by Shell company under the brand name HiPERTUF, targeted for packaging applications such as beer, food, carbonated drinks, hot fill, cosmetics, and pharmaceuticals.

Unlike melting temperature, the glass transition temperature of PEN/PET blends linearly increases with the ratio of PEN to PET [120], so that the T_g of new commercial products can easily be tailored by blending various amounts of PEN with PET. Practical applications include hot-filled, pasteurized, and washable containers, and outdoor glazing.

Similar to PET [121], PEN also exhibits a triclinic crystalline morphology (alpha and beta form) [122] (Table 9.11). Below 200°C, PEN crystallizes in the alpha form. At much higher temperatures, it crystallizes in the beta form.

9.4.4 Gas Permeability

PEN and PET are both slow-crystallizing polymers, meaning clear products such as films and bottles can be made. A major requirement for any bottle packaging is to provide a good barrier to gases,

TABLE 9.11
Crystal Unit Cell Data of PET and PEN

Polymer	a, nm	b, nm	c, nm	Alpha	Beta	Gamma
PET	0.49	0.594	1.075	98.5	118	112
PEN, alpha	0.651	0.575	1.32	81.33	144	100
PEN, beta	0.926	1.559	1.273	121.6	95.57	122.52

Source: Zhang, X. 2014. *Fundamentals of Fiber Science.* Lancaster, PA: DEStech; Ito, M. and Kikutani, T. 2002. PEN: Scheme and properties. In *Handbook of Thermoplastic Polymers: Homopolymers, Copolymers, Blends and Composites*, ed. S. Fakirov, 463–482. Weinheim: Wiley-VCH.

TABLE 9.12
Gas Transmission, cc mil/100 in.^{2} day

Polymer	CO_2, Amorphous	CO_2, Oriented	O_2, Amorphous	O_2, Oriented	H_2O, Amorphous
PET	69	31	10.9	6.1	4.31
PEN	15	6	4.4	1.6	2.12

Source: Naphthalates. 2001. General properties of naphthalate-containing polymers. BP. Available at http://wenku.baidu.com/view/0a24810bbb68a98271fefadf.

and PEN showed significantly superior barrier performance compared to PET [120] (Table 9.12). The crystal orientation leads to closer packing of polymer chains, resulting in improved barrier performance over amorphous polymers. PTN is even more superior to PEN as a barrier for CO_2, and oxygen [118].

9.4.4.1 Gas Barrier Properties of PEN/PET Blends

Since PET is significantly cheaper than PEN, for all commercial practical purposes, adding more PET to blends with PEN significantly reduces the cost of the final product. Compared to the linear behavior for gas barrier properties of PET/PEN copolymers, actual data on the blends were significantly lower, indicating that the blends provided an improved barrier. This is attributed to the morphology of the PET and PEN domains, which directly influences the barrier [120].

9.4.5 CHEMICAL RESISTANCE

PEN is significantly superior to PET in terms of chemical, oxidative, and hydrolytic stability [120]. Improved chemical resistance is useful in applications such as packaging of pharmaceuticals, cosmetics, and chemicals (Table 9.13).

9.4.6 OPTICAL (UV) PROPERTIES

Unlike DMT, which does not absorb UV or visible light [120], NDC absorbs UV below 400 nm [120]. When polymerized, these monomers produce either PET or PEN, respectively. UV absorption occurs in PEN, while PET does not absorb UV light [120]. The UV absorption property causes PEN and other naphthalate polyesters/copolyesters to display improved UV barrier performance, resulting in greater retention of mechanical strength [120].

TABLE 9.13
Chemical Resistance Comparison between PET and PEN (Measured by Retention of Elongation at Break, %)

Polymer	10% HCl, 5 Weeks	10% NaOH, 2 Weeks	Ammonia Gas, 10 Weeks	Thermal Aging, 2 Weeks at 130°C, 100% Relative Humidity
PET	0	0	0	20
PEN	60	50	96	50

Source: Naphthalates. 2001. General properties of naphthalate-containing polymers. BP. Available at http://wenku.baidu.com/view/0a24810bbb68a98271fefadf.

9.4.7 Fluorescence of Naphthalate Polyesters

Fluorescence is defined as the emission of light that was absorbed at a different wavelength. Due to the UV absorption property of naphthalate, polyesters containing naphthalate groups exhibit fluorescence [123]. PEN fluoresces a blue color when excited [123]. Among blends such as PET/PEN, the extent of fluorescence is directly related to the amount of napthalate (PEN) present in the composition. With lower amounts of PEN, the blue color may not be visible. The fluorescence property can be used as an identification marker for naphthalate-containing compositions [123].

9.4.8 Photodegradation

The formation of excited state dimers in naphthalate systems when exposed to sunlight or other sources such as UV results in the appearance of color (yellow) [124]. PEN is superior to PET during weathering, for this reason presence even a small amount of naphthalate improves the weathering performance of PET [118].

9.4.9 Mechanical Properties

Since PEN crystallizes slowly, Table 9.14 reflects the mechanical properties of amorphous PEN [118]. Compared to PET, generally superior performance in most of the properties is displayed by PEN [120]. Table 9.15 shows the mechanical properties against the ratio of PEN/PET for 50 µm films and fibers [120]. PET shows better elongation compared to PEN. On the other hand, significant shrinkage is observed for PET, which could be due to the higher crystallization rate of PET over PEN.

9.4.10 Naphthalate-Based Elastomers

It is well known that fast-crystallizing polymers such as PBT can be copolymerized with a flexible diol such as poly(ethylene glycol) to result in the formation of elastomers. Industrially, the most well-known elastomer is called Hytrel. Produced by DuPont, Hytrel is a copolymer of PBT having poly(tetramethylene oxide) flexible linkages. Since PBN crystallizes similarly to PBT, elastomers involving a poly(tetramethylene oxide) glycol soft linkage were also produced from PBN, and Table 9.16 lists the property comparison [118]. Higher melt temperatures combined with superior tensile strength and flex modulus were observed for PBN elastomers.

TABLE 9.14
Mechanical Properties of Amorphous PEN

Tensile strength yield, MPa	81.4
Elongation at yield, %	7.2
Ultimate tensile strength, MPa	47.6
Elongation at break, %	100
Flex modulus, MPa	2440
Tensile modulus, MPa	2400
Notch Izod, J/m	26.7
Tensile impact, J/m	0.25
Rockwell hardness, R scale	123

Source: Naphthalates. 2001. Engineering resin applications. BP. Available at http://wenku.baidu.com/view/c560 94cf05087632311212dc.html.

TABLE 9.15
Mechanical Property Comparison between PEN and PET

Property	PEN/PET Ratio
Young's modulus, MD, MPa	1.31
Young's modulus, TD, MPa	1.27
Tensile strength, MD, MPa	1.39
Tensile strength, TD, MPa	1.36
Stress at 5% Elongation, MD, MPa	0.57
Stress at 5% Elongation, TD, MPa	0.7
Elongation at break, MD, %	0.57
Elongation at break TD, %	0.7
Thermal shrinkage at 150°C, MD %	0.46
Thermal shrinkage at 150°C, TD %	0.30
Fiber tenacity, g/den	1.1
Fiber modulus, g/den	3.27
Fiber elongation at break, %	0.57
Fiber boiling water shrinkage, %	0.2
Fiber dry heat shrinkage at 177°C, %	0.5

Source: Naphthalates. 2001. General properties of naphthalate-containing polymers. BP. Available at http://wenku.baidu.com/view/0a24810bbb68a98271fefadf.

Note: MD, machine direction; TD, transverse direction.

9.4.11 Applications

The main applications of naphthalate polyesters are in the area of films, fibers, and containers. Films, such as the Teonex brand of PEN films produced by Teijin, are sold for high-density data storage tapes, integrated circuit cards, automotives, capacitors, electrical insulation, batteries, and opticals. Fibers, such as Performance Fibers A-701, are sold for tire and industrial applications. Container applications include food, beer, carbonated drinks, hot-fill juices, cosmetics, and pharmaceuticals, using materials such as Shell's HiPERTUF PET/PEN copolymer.

TABLE 9.16
Elastomer Comparison Having Hard Blocks of PBT or PBN

Property	PBN Elastomer	PBT Elastomer
Specific gravity,	1.21	1.12
Shore D hardness,	55	47
Melt temperature, °C	225	182
Vicat temperature, °C	190	127
Tensile strength at yield, MPa	16.4	7.8
Elongation at yield, %	20	50
Tensile strength at break, MPa	26.3	18.8
Tensile elasticity, MPa	34.8	45.1
Flex modulus, MPa	139.4	70.6
Taber abrasion, mg/1000	16	12
Volume resistivity, omega, cm	1.50×10^{15}	1.80×10^{12}

Source: Naphthalates. 2001. Engineering resin applications. BP. Available at http://wenku.baidu.com/view/c56094cf05087632311212dc.html.

REFERENCES

1. Carraher, Jr., C.E. 2014. *Polymer Chemistry*. Boca Raton, FL: Taylor & Francis.
2. Whinfield, J.R. 1946. Chemistry of terylene. *Nature*, 158:930–931.
3. Billica, H.R. 1951. Production of polyethylene terephthalate using antimony trioxide as polymerization catalyst. US Patent 2647885A.
4. Thomas, R.R. 1966. Preparation of antimony triacetate. US Patent 3415860A.
5. Julliard, G.C., Therese, J.M., Perry, M., Wayne, W.A., and Cheuk, Y. 1999. Preparation of polyesters employing antimony catalysts and acidic phosphorus compounds. WO Patent 2001014452A1.
6. Jen, Z.C. 2001. Manufacturing method for decreasing the cyclic oligomer content in polyester. US Patent 6392005B1.
7. Finley, M.E., Nicholas, R.C., and Ronald, T.A. 1974. Process and catalyst-inhibitor system for preparing synthetic linear polyester. US Patent 3907754A.
8. Bashir, Z., Munif, A.M., Rao, M.V., and Padmanabhan, S. 2010. Process for making polyethylene terephthalate. WO Patent 2011020619A1.
9. Cholod, M.S. and Shah, N.M. 1982. Catalyst system for a polyethylene terephthalate polycondensation. US Patent 4356299A.
10. Bander, J.A., Lazarus, S.D., and Twilley, I.C. 1980. Catalytic process for preparation of polyesters. US Patent 4260735A.
11. John, P.A. and Mary, J.S. 1967. Preparation of polyethylene terephthalate using cyanide esterification catalyst. US Patent 3451973A.
12. Charles, Jr., H.H. 1951. Production of polyethylene terephthalate with cobaltous acetate as catalyst. US Patent 2641592A.
13. Patrick, H.H. 1955. Stannous formate catalyst for preparing polyethylene terephthalate. US Patent 2892815A.
14. James, B.B., William, K.E., and Julian, K.L. 1955. Catalytic production of polyethylene terephthalate. US Patent 2857363A.
15. Joseph, G.N. and John, J.V. 1969. Stibinic and stibonic catalysts for polyethylene terephthalate. US Patent 3642702A.
16. Siggel, E., Koepp, H.M., and Rein, W. 1966. Process for the production of polyethylene terephthalate. US Patent 3398124A.
17. Emmette, F.I. 1948. Production of polyethylene terephthalate. US Patent 2534028A.
18. John, A.P. and Mary, J.S. 1967. Process of preparing polyethylene terephthalate using lead fluoride as transesterification catalyst and as condensation catalyst. US Patent 3457239A.

19. Hartmann, A., Forster, P.F., and Otto, V.O.H. 1965. Process for manufacturing polyethylene terephthalate employing metal salts of halogenated haliphatic acids. US Patent 3385830A.
20. Nyce, G.W., Lamboy, J.A., Connor, E.F., Waymouth, R.M., and Hedrick, J.L. 2002. Expanding the Catalytic Activity of Nucleophilic N-Heterocyclic Carbenes for Transesterification Reactions. *Org. Lett.*, 4(21):3587–3590.
21. Bartolome, L., Imran, M., Cho, B.G., Al-Masry, W.A., and Kim, D.H. 2012. Recent developments in the chemical recycling of PET. In *Material Recycling—Trends and Perspectives*, ed. Achilias, D. Available at http://www.intechopen.com/books/material-recycling-trends-and-perspectives/recent-developments-in-the-chemical-recycling-of-pet.
22. Papaspyrides, C.D. and Vouyouika, S.M. 2009. *Solid State Polymerization*. Hoboken, NJ: John Wiley & Sons.
23. Duh, B. 2006. Effects of crystallinity on solid-state polymerization of poly(ethylene terephthalate). *J. Appl. Polym. Sci.*, 102(1):623–632.
24. Fakirov, S. 1997. Polyethylene terephthalate. In *Handbook of Thermoplastics*, ed. Olabisi, O., 449–464. New York: Marcel Dekker.
25. Scheirs, J. 2000. *Compositional and Failure Analysis of Polymers: A Practical Approach*. New York: John Wiley & Sons.
26. Dixon, E.R. and Jackson, J.B. 1968. The inter-relation of some mechanical properties with molecular weight and crystallinity in poly(ethylene terephthalate). *J. Mater. Sci.*, 3:464.
27. Tewarson, A. 2003. Flammability of polymers. In *Plastics and the Environment*, ed. Andrady, A.L., 403–489. Hoboken, NJ: John Wiley & Sons.
28. Schiers, J. 2003. Additives for the modification of poly(ethylene terephthalate) to produce engineering-grade polymer. In *Modern Polyesters*, eds. Schiers, J. and Long, T.E., 495–540. Chichester: John Wiley & Sons.
29. Largman, T. 1981. Phosphine suppressants for polymeric compositions including red phosphorus as a flame retardant. US Patent 4356282A.
30. Schlechter, M. 2013. *Engineering Resins, Polymer Alloys and Blends: Global Markets*. BCC Research, Wellesley, MA.
31. Gallucci, R. and Patel, B. 2003. Poly(butylene terephthalate). In *Modern Polyesters: Chemistry and Technology of Polyesters and Copolyesters*, eds. Scheirs, J. and Long, T.E. Chichester: John Wiley & Sons.
32. Available at http://www.celanese.com/engineered-materials/products/Celanex--PBT/What-is-PBT.aspx.
33. Copenhaver, J.W. and Bigelow, M.H. 1949. *Acetylene and Carbon Monoxide Chemistry*. New York: Reinhold.
34. Haas, T., Jaeger, B., Weber, R., Mitchell, S.F., and King, C.F. 2005. New diol processes:1,3-propanediol and 1,4-butanediol. *Appl. Catal. A: Gen.*, 280(1):83–88.
35. Dehler, J., Hoffmann, H., Joschek, H.I., Reiss, W., Schnur, R., and Winderl, S. 1978. Manufacture of butynediol. U.S. Patent No. 4,093,668. Washington, DC: U.S. Patent and Trademark Office.
36. Hoffmann, H., Reiss, W., Schroeder, W., and Winderl, S. 1978. Manufacture of butynediol. U.S. Patent No. 4,067,914. Washington, DC: U.S. Patent and Trademark Office.
37. Hoffmann, H., Reiss, W., Schnur, R., Winderl, S., and Zehner, P. 1976. Process for the manufacture of butynediol. U.S. Patent No. 3,957,888. Washington, DC: U.S. Patent and Trademark Office.
38. Sampat, B.J. 2011. 1,4-Butanediol: A techno-commercial profile. *Chemical Weekly*. Available at http://www.chemicalweekly.com/Profiles/1,4-Butanediol.pdf.
39. Mueller, H. 1995. Tetrahydrofuran. In *Ullmann's Encyclopedia of Industrial Chemistry*. Weinheim: Wiley-VCH.
40. Kouba, J.K. and Snyder, R.B. 1987. Coproduction of butanediol and tetrahydrofuran and their subsequent separation from the reaction product mixture. U.S. Patent No. 4,656,297. Washington, DC: U.S. Patent and Trademark Office.
41. Felthouse, T.R. 1998. Growing maleic anhydride through its uses and processes. *Chemist*, 75(5):22–27.
42. Burk, M.J., Van Dien, S.J., Burgard, A.P., and Niu, W. Compositions and methods for the biosynthesis of 1,4-butanediol and its precursors. US Patent US8067214 B2.
43. Available at http://www.genomatica.com/news/press-releases/successful-commercial-production-of-5-million-pounds-of-bdo/.
44. Available at http://www.bio-amber.com/products/en/products/bdo_1_4_butanediol.
45. Saffer, A. and Barker, R.S. 1958. Preparation of aromatic polycarboxylic acids. US Patent 2833816 A.
46. Meyer, D. 1967. Fiber-grade terephthalic acid by catalytic hydrogen treatment of dissolved impure terephthalic acid. US Patent 3584039.

47. Fisher, J. 1973. Terephthalic acid recovery by continuous flash crystallization. US Patent 3931305.
48. Sheehan, R. 1995. Terephthalic acid, dimethyl terephthalate, and isophthalic acid. In *Ullmann's Encyclopedia of Industrial Chemistry*. Weinheim: Wiley-VCH.
49. Antic, V.V. and Pergal, M.V. 2011. Poly(butylene terephthalate)—Synthesis, properties, applications. In *Handbook of Engineering and Specialty Thermoplastics*, eds. Thomas, S. and Visakh, P.M. Beverly, MA: Scrivener Publishing.
50. Hoeschele, G.K. and McGrik, R.H. 1992. Process for preparing polyesters using a catalyst mixture of tetraalkyl titanates and zirconates. EP0472179 A2.
51. Kawaguchi, K., Nakane, T., and Hijikata, K. 1994. Process for the preparation of polyester resin. EP0578464 A1.
52. Griebler, W.-D., Hirschberg, E., Hirthe, B., Schmidt, W., and Thiele, U. 2003. Titanium containing catalyst and process for production of polyester. EP0736560 B1.
53. Lei, C. and Chen, D. 2008. Effect of di-n-dodecyl phosphate on the transesterification reaction in a poly(butylene terephthalate)/polycarbonate blend. *J. Appl. Polym. Sci.*, 109(2):1099–1104.
54. Van Bennekom, A.C.M., Van Den Berg, D., Bussinkt, J., and Gaymans, R.J. 1997. Blends of amide modified polybutylene terephthalate and polycarbonate: Phase separation and morphology. *Polymer*, 38(20):5041–5049.
55. Komatsu, Y., Matano, K., and Suzuki, S. 1998. Manufacture of white poly(butylene terephthalate) with low amount of THF generation. JP10243740.
56. Ramaraju, D., Ravi, G.R., Starkey, K., Donovan, M., and Agarwal, P. 2007. Process of making polyesters. US20070197738 A1.
57. Chang, S.J. and Tsai, H.B. 1992. Effect of salts on the formation of THF in preparation of PBT by TPA process. *J. Appl. Polym. Sci.*, 45(2):371–373.
58. Pilati, F., Manaresi, P., Fortunato, B., Munari, A., and Passalacqua, V. 1981. Formation of poly(butylene terephthalate): Growing reactions studied by model molecules. *Polymer*, 22(6):799–803.
59. Pilati, F., Manaresi, P., Fortunato, B., Munari, A., and Passalacqua, V. 1981. Formation of poly(butylene terephthalate): Secondary reactions studied by model molecules. *Polymer*, 22(11):1566–1570.
60. Petcore Europe press release. 2014. More than 60 billion PET bottles recycled 2012! Available at http://www.petcore.org/news/more-60-billion-pet-bottles-recycled-2012.
61. Agarwal, P., Cohoon, K., and Dhawan, S. et al. 2010. Use scrap PET to make PBT thermoplastic molding compositions involves depolymerizing PET in presence of alcohol; butanediol used to make polybutylene terephthalate can come from biomass (i.e. renewable monomer); energy savings/reduced carbon dioxide emissions by use of waste PET. US Patent 7799836 B2.
62. Agarwal, P., Cohoon, K., and Dhawan, S. et al. 2012. Process for making polybutylene terephthalate (PBT) from polyethylene terephthalate (PET). US Patent 8088834 B2.
63. Cheng, S.Z.D., Pan, R., and Wunderlich, B. 1988. Thermal analysis of poly(butylene terephthalate) for heat capacity, rigid-amorphous content, and transition behavior. *Macromol. Chem.*, 189(10):2443–2458.
64. Smith, J.G., Kibler, C.J., and Sublett, B.J. 1966. Preparation and properties of poly(methylene terephthalate). *J. Polym. Sci., Part A*, 4(7):1851–1859.
65. Jaquiss, D.B.G., Borman, W.F.H., and Campbel, R.W. 1982. Polyesters, thermoplastics. In *Encyclopedia of Chemical Technology*, eds. Mark, H.F., Othmer, D.F., Overberger, C.G., and Seaborg, G.T., 549–574. New York: John Wiley & Sons.
66. Chang, E.P., Kirsten, R.O., and Slagowski, E.L. 1978. The effect of additives on the crystallization of poly(butylene terephthalate). *Polym. Eng. Sci.*, 18(12):932–936.
67. Illers, K.H. 1980. Heat of fusion and specific volume of poly(ethylene terephthalate) and poly(butylene terephthalate). *Colloid. Polym. Sci.*, 258(2):117–124.
68. Mencick, Z. 1975. The crystal structure of poly(tetramethylene terephthalate). *J. Polym. Sci.: Polym. Phys. Ed.*, 13(11):2173–2181.
69. Stambaugh, B., Koenig, J.L., and Lando, J.B. 1979. X-ray investigation of the structure of poly-(tetramethylene terephthalate). *J. Polym. Sci.: Polym. Phys. Ed.*, 17(6):1053–1062.
70. Yokouchi, M., Sakakibara, Y., Chatani, Y., Tadokoro, T., Tanaka, T., and Yoda, K. 1976. Structures of two crystalline forms of poly(butylene terephthalate) and reversible transition between them by mechanical deformation. *Macromolecules*, 9(2):266–273.
71. CCP PBT datasheet. Available at http://www.ccp.com.tw/.
72. Weil, E.D. and Levchik, S.V. 2004. Commercial flame retardancy of thermoplastic polyesters—A review. *J. Fire Sci.*, 22(4):229–350.
73. Penn, R. 1998. Flame retarded poly(butylene terephthalate) composition. US Patent 5,814,690.

74. Vu-Khanh, T., Denault, J., Habib, P., and Low, A. 1991. The effect of injection molding on the mechanical behavior of long-fiber reinforced PBT/PET blends. *Compos. Sci. Technol.*, 40(4):423–435.
75. Montaudo, G., Puglisi, C., and Samperi, F. 1998. Mechanism of exchange in PBT/PC and PET/PC blends. Composition of the copolymer formed in the melt mixing process. *Macromolecules*, 31(3):650–661.
76. Porter, R.S. and Wang, L.H. 1992. Compatibility and transesterification in binary polymer blends. *Polymer*, 33(10):2019–2030.
77. Clark, A.H., Courson, R.D., Gallucci, R.R., and Walsh, E.B. 1995. High density polyester–polycarbonate molding composition. US Patent 5,441,997.
78. Wu, J., Mai, Y.W., and Cotterell, B. 1993. Fracture toughness and fracture mechanisms of PBT/PC/IM blend. *J. Mater. Sci.*, 28(12):3373–3384.
79. Wang, X.H., Zhang, H.X., Wang, Z.G., and Jiang, B.Z. 1997. Toughening of poly(butylene terephthalate) with epoxidized ethylene propylene diene rubber. *Polymer*, 38(7):1569–1572.
80. Mohd Ishak, Z.A., Ishiaku, U.S., and Karger-Kocsis, J. 1999. Hygrothermal aging and fracture behavior of styrene-acrylonitrille/acrylate based core-shell rubber toughened poly(butylene terephthalate). *J. Appl. Polym. Sci.*, 74(10):2470–2481.
81. Turner, S.R. 2004. Development of amorphous copolyesters based on 1,4-cyclohexane-dimethanol. *J. Polym. Sci., Part A: Polym. Chem.*, 42(23):5847–5852.
82. Kibler, C.J., Bell, A., and Smith, J.G. 1959. Linear polyesters and polyester-amides from 1,4-cyclohexanedimethanol. US Patent 2,901,466.
83. Kibler, C.J., Bell, A., and Smith, J.G. 1962. Linear polyester of 1,4-cyclohexianedimethanol and aminocarboxylic acids. US Patent 3,033,826.
84. East, A.J. 2006. Polyesters, thermoplastic. In *Kirk-Othmer Encyclopedia of Chemical Technology*, eds. A. Seidel and M. Bickford. 31–95. New York: John Wiley & Sons.
85. Turner, S.R., Seymour, R.W., and Dombroski, J.R. 2004. Amorphous and crystalline polyesters based on 1,4-cyclohexanedimethanol. In *Modern Polyesters: Chemistry and Technology of Polyesters and Copolyesters*, eds. Scheirs, J. and Long, T.E. Chichester: John Wiley & Sons.
86. Caldwell, J.R. and Gilkey, R. 1959. Fiber-forming polyesters from trans-1,4-cyclohexanedicarboxylic compounds and 1,1-cyclohexane dimethanol. US Patent 2,891,930.
87. Brunelle, D.J. and Jang, T. 2006. Optimization of poly(1,4-cyclohexylidene cyclohexane-1,4-dicarboxylate) (PCCD) preparation for increased crystallinity. *Polymer*, 47(11):4094–4104.
88. Leaversuch, R.D. 2004. Thermoplastic polyesters: It's a good time to know them better. *Plastics Technology*. Available at http://www.ptonline.com/articles/thermoplastic-polyesters-it's-a-good-time-to-know-them-better.
89. Yau, C.C. and Cherry, C. 1995. Copolyester of cyclohexanenedimethanol and process for producing such polyester. US Patent 5,385,773.
90. Yau, C.C. and Cherry, C. 1994. Copolyester of cyclohexanedimethanol and process for producing such polyester. US Patent 5,340,907.
91. Adams, V.S., Hataway, J.E., and Roberts, K.A. 1997. Process for preparing copolyesters of terephthalic acid ethylene glycol and 1,4-cyclohexanedimethanol exhibiting a neutral hue high clarity and increased brightness. US Patent 5,681,918.
92. Hasek, R.H. and Knowles, M.B. 1959. Preparation of trans-1,4-cyclohexanedimethanol. US Patent 2,917,549.
93. Rathmell, C., Spratt, R.C., and Tuck, M.W.M. 1995. Preparation of cyclohexanedimethanol with a particular ratio. US Patent 5,395,987.
94. Sastri, V.R. 2010. Engineering thermoplastics: Acrylics, polycarbonates, polyurethanes, polyacetals, polyesters, and polyamides. In *Plastics in Medical Devices: Properties, Requirements, and Applications*, 121–173. Amsterdam: Elsevier.
95. Schulken, R.M., Boy, R.E., and Cox, R.H. 1964. Differential thermal analysis of linear polyesters. *J. Polym. Sci., Part C*, 6(1):17–25.
96. Dickerson, J.P., Brink, A.E., Oshinski, A.J., and Seo, K.S. 1997. Copolyesters based on 1,4-cyclohexanedimethanol having improved stability. US Patent 5,656,715.
97. Tsai, Y., Fan, C.H., Hung, C.Y., and Tsai, F.J. 2008. Amorphous copolyesters based on 1,3/1,4-cyclohexanedimethanol: Synthesis, characterization and properties. *J. Appl. Polym. Sci.*, 109(4):2598–2604.
98. Turner, S.R., Seymour, R.W., and Smith, T.W. 2001. Cyclohexanedimethanol polyesters. In *Encyclopedia of Polymer Science and Technology*. Hoboken, NJ: John Wiley & Sons.
99. Brunelle, D.J. 1999. Polyesteramides of high crystallinity and reagents for their preparation. US Patent 5,939,519.

100. Patel, B.R., Smith, G.F., and Banach, T.E. 1999. Crystalline polyester resins and process for their preparation. US Patent 5,986,040.
101. Liu, Y. and Turner, S.R. 2010. Synthesis and properties of cyclic diester based aliphatic copolyesters. *J. Polym. Sci., Part A: Polym. Chem.*, 48(10):2162–2169.
102. Turner, S.R. and Sublett, B.J. 2000. Low melt viscosity amorphous copolyesters with enhanced glass transition temperatures. US Patent 6,120,889.
103. Kline, T.R. and Ruhter, M.C. 2012. Alternatives analysis report for bisphenol-A in infant formula cans and baby food jar lids. Published by Maine Department of Environmental Protection. Available at http://www.maine.gov/dep/safechem/documents/AAR-Report-December2012.pdf (accessed July 21, 2015).
104. Available at http://www.eastman.com/Brands/Eastman_Tritan/Pages/Overview.aspx.
105. Laird, K. 2012. Green matter: A dash of isosorbide makes all the difference. *Plastics Today*. Available at http://www.plasticstoday.com/blogs/green-matter-dash-isosorbide-makes-all-difference-112920121.
106. Available at http://www.skchemicals.com/korean/product/ecozen/sub/sub_overview.asp?menum=11.
107. Scott, S.W. 1983. Thermoplastic molding composition. US Patent 4,391,954.
108. Scott, S.W. 1995. Thermoplastic molding composition. US Patent 5,478,896.
109. Allen, R.B. and Avakian, R.W. 1988. Method for enhancing ionizing radiation. US Patent 4,778,656.
110. Innovative Plastics: RESISTANCE+DURABILITY: Chemical Resistance Performance Testing for Healthcare Materials. Available at http://www.pod-sabic-ip.com/KBAM/Reflection/Assets/Thumbnail/22863_3.pdf (accessed July 21, 2015).
111. VERSATILITY + DURABILITY: XYLEX™ PC/POLYESTER RESIN. Available at http://www.pod-sabic-ip.com/KBAM/Reflection/Assets/Thumbnail/7129_15.pdf (accessed October 12, 2014).
112. Yamashita, G. and Yamamoto, K. 1972. Process for the preparation of 2,6-naphthalenedicarboxylic acid. US Patent 3870754.
113. McMahon, R.F., Greene, Jr., J.D., and Peterson, D.A. 1997. Process for preparing 2,6-naphthalenedicarboxylic acid. US Patent 6114575.
114. Lillwitz, L.D. 2001. Production of dimethyl-2,6-naphthalenedicarboxylate: Precursor to polyethylene naphthalate. *Appl. Catal. A.*, 221:337.
115. Lee, K.K., Cho, B.H., and Kim, Y.W. 1993. Process for preparing polyethylene naphthalate. US Patent 5294695.
116. Kulkarni, S. and Raj, B. 2007. Naphthalate based polyester resin compositions. US Patent 20070232763 A1.
117. Kwon, I.-H., Bang, Y.-H., Lee, J., Lee, D.-J., and Lee, I.-H. 2003. High tenacity polyethylene-2,6-naphthalate fibers. European Patent 1510604 B1.
118. Naphthalates. 2001. Engineering resin applications. BP. Available at http://wenku.baidu.com/view/c56094cf05087632311212dc.html.
119. Lorenzetti, C., Finelli, L., Lotti, N., Vannini, M., Gazzano, M., Berti, C., and Munari, A. 2005. Synthesis and characterization of poly(propylene terephthalate/2,6-naphthalate) random copolyesters. *Polymer*, 46:4041.
120. Naphthalates. 2001. General properties of naphthalate-containing polymers. BP. Available at http://wenku.baidu.com/view/0a24810bbb68a98271fefadf.
121. Zhang, X. 2014. *Fundamentals of Fiber Science*. Lancaster, PA: DEStech.
122. Ito, M. and Kikutani, T. 2002. PEN: Scheme and properties. In *Handbook of Thermoplastic Polymers: Homopolymers, Copolymers, Blends and Composites*, ed. S. Fakirov, 463–482. Weinheim: Wiley-VCH.
123. Santhanagopalakrishnan, P. and Kulkarni, S.T. 2008. Polyester resins. In *Polyesters and Polyamides*, eds. Deopura, B.L., Alagirusamy, R., Joshi, M., and Gupta, B., 1–40. Boca Raton, FL: CRC Press.
124. Marien, A., Hagemann, J., and Weber, B. 1998. Polyalkylene naphthalene film comprising specific UV-absorber. US patent 5824465.

10 Polycarbonates*

Pierre Moulinié

CONTENTS

10.1 Introduction 348
    10.1.1 Aliphatic and Bio-Based Polycarbonates 348
    10.1.2 Aromatic Polycarbonates 350
10.2 Synthesis of Polycarbonates 350
    10.2.1 Interfacial Reaction 351
    10.2.2 Melt Reaction 351
        10.2.2.1 Preparation of Diphenyl Carbonate 352
10.3 Properties of Polycarbonate and Its Copolymers 352
    10.3.1 BPA Polycarbonate 352
    10.3.2 High-Heat (Co)Polycarbonates 353
    10.3.3 PC Containing Biphenyl Segments 354
    10.3.4 Polyarylate Block Copolymers Including Resorcinol 355
    10.3.5 Branched PC 355
    10.3.6 Copolymers with Polydimethylsiloxane 356
10.4 Characterization of Polycarbonate 357
    10.4.1 Analytical Techniques Used to Characterize Polycarbonate 357
    10.4.2 Thermal Analysis and Stability Studies 357
    10.4.3 Rheology 358
    10.4.4 Relaxation Studies 358
10.5 Modification of Polycarbonates 359
    10.5.1 Stabilization 359
    10.5.2 Flame-Retardant Polycarbonates 359
    10.5.3 Impact Modification 360
    10.5.4 Blends and Alloys 361
        10.5.4.1 PC/ABS 361
        10.5.4.2 PC/Polyesters 362
    10.5.5 Filled Polycarbonates 363
10.6 Processing of Polycarbonates 364
    10.6.1 Injection Molding of Polycarbonate and Its Resins 364
    10.6.2 Extrusion, Thermoforming, and Blow Molding of Polycarbonate 365
10.7 Summary and Outlook 365
Acknowledgments 365
References 365

---

* Based in parts on the first-edition chapter on polycarbonates.

10.1 INTRODUCTION

The term *polycarbonate* applies to several types of polymeric materials, which contain carbonate linkages (Scheme 10.1). Although their chemistry has similarities to polyesters, polycarbonates are easier to prepare with phenolic building blocks, which yield higher thermal stability, higher rigidity, and higher glass transitions. Several types of polycarbonates are possible:

- Carbonate-based polyols used for polyurethanes, which are not covered in this chapter.
- Aliphatic polycarbonates: prepared with similar synthesis methods as aromatic polycarbonates but based on aliphatic building blocks. Though these were not common, they appear to be gaining in significance.
- Aromatic polycarbonates: the most common type of polycarbonates.

Though commercially available polycarbonates are overwhelmingly aromatic, there has been progress in the past 10 years toward introduction of aliphatic polycarbonates. Both aliphatic and aromatic polycarbonates will be discussed in this chapter. A general description of the monomers and polymerization pathways is given first. In subsequent sections, material characteristics, recent discoveries, and polycarbonate blends with various ingredients are also discussed.

10.1.1 ALIPHATIC AND BIO-BASED POLYCARBONATES

Because aromatic polycarbonates have higher thermal stability and superior engineering properties, development of aliphatic polycarbonates has been comparatively slow. Nevertheless, there have been noteworthy developments, which are briefly covered in this section. Engineering thermoplastics based on isosorbide and 2,2,4,4-tetramethylcyclobutanediol have been recently described. Both building blocks are relatively rigid and bulky, which helps raise the glass transition and rigidity.

Copolycarbonates based on isosorbide have been recently commercialized by Mitsubishi Chemical under the trade name Durabio and by Teijin under the trade name Planext. The structure of isosorbide is shown in Scheme 10.2. Note that other stereoisomers are possible (iditol and mannitol, among others). Though isosorbide-based polycarbonates were first reported some time ago [1], recent interest likely stems from increased availability of high-purity isosorbide and increasing interest in polymers derived from bio-based feedstocks. Isosorbide's straightforward isolation from sugars and its ability to be useful for preparation of polycarbonates or polyesters have driven much of the interest in this class of resin materials [2,3]. Table 10.1 summarizes the T_g values for some of the copolymer compositions recently disclosed with isosorbide.

Literature values for polycarbonates based solely on isosorbide gave a relatively wide range of glass transition temperatures, although all references described different synthetic routes for

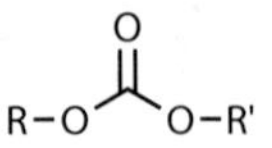

SCHEME 10.1 Carbonate linkage.

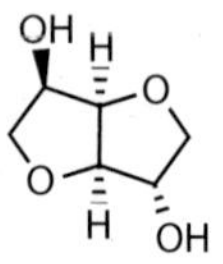

SCHEME 10.2 Isosorbide (D-glucitol, 1,3:3,6-dianhydro-).

TABLE 10.1
Carbonate-Type (Co)Polymers with Isosorbide, the Relative Amounts of Each (Co)Monomer, and the Resulting Glass Transition Temperatures

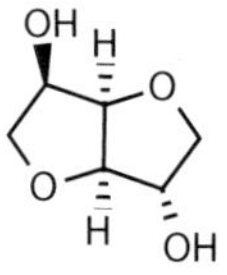

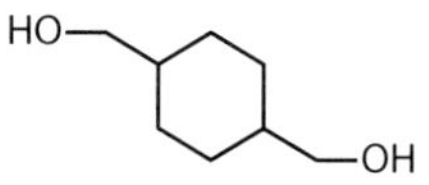

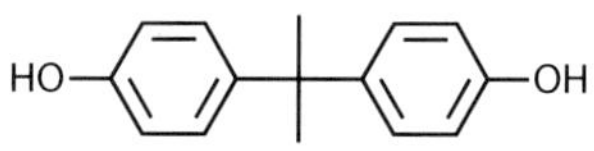

Isosorbide (%)	1,4-cyclohexanedimethanol (%)	Bisphenol-A (%)	T_g (°C)	Reference
100	–	–	148–160	[4,5]
68	32	–	124	[5]
37	63	–	89	[5]
50	–	50	152	[6]

polymerization. Diphenyl carbonate (DPC) is commonly used when polymerizing via melt reaction (described in Section 10.2.2), though the use of dimethyl carbonate has also been described [4]. All polycarbonates produced with isosorbide were reported to be amorphous and transparent. The disclosure by Fuji indicated that the homopolymer with isosorbide had much lower impact strength than the copolymers they described [5]. While inclusion of 1,4-cyclohexanedimethanol in the structure lowered the heat resistance of the resulting copolymers, it helped improve the impact performance. Preparation of bisphenol-A (BPA)–isosorbide copolymers has also been described, resulting in glass transition temperatures around 152°C [6]. Additional features of such copolymers include increased hardness and lower refractive index.

Due to the monomers' specific three-dimensional structure, it can also impart some chirality to the resulting polycarbonate chain and yield polymers capable of rotating polarized light [7]. Previous work has shown that aliphatic polycarbonates containing methylene chain structures can biodegrade [8]. Inclusion of isosorbide with methylene-based chains also exhibited some degree of biodegradability [9]. Those polymers were more flexible and more easily hydrolyzed than those shown in Table 10.1. Thus, engineering resins in Table 10.1 are not expected to be biodegradable.

Acar and Brunelle described the use of 2,2,4,4-tetramethylcyclobutanediol, whose structure is shown in Scheme 10.3, as a comonomer for aliphatic polycarbonates [10], prepared by melt reaction with DPC. This monomer is prepared via a ketene prepared from isobutyric acid and hydrogenation of the resulting dione [10]. Polycarbonates prepared with this monomer were amorphous with glass transition temperatures between 120°C and 124°C. Though high glass transitions were possible when this monomer was used in copolycarbonates, the authors noted a change in cis/trans ratio after processing at higher temperatures.

This monomer has been used in copolyesters to raise the glass transition temperature for polyesters based on terephthalic acid, ethylene glycol, and 1,4-cyclohexanedimethanol [10]. Recent work suggests that the cis isomer of the monomer is critical to high-impact performance of the polyesters derived from it [10].

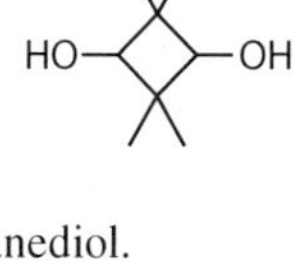

SCHEME 10.3 2,2,4,4-tetramethylcyclobutanediol.

TABLE 10.2
Polycarbonate Producers and Trademarks or Names for Their Resins

Producer	Trade Name	Producer	Trade Name
Chi Mei (Taiwan)	Wonderlite	Mitsubishi (Japan)	Novarex
Covestro (Germany)	Makrolon	Sabic Innovative Plastics (Saudi Arabia)	Lexan, Sabic PC
Honam (Korea)	Hopelex	Samsung Cheil (Korea)	Infino
Idemitsu (Japan)	Tarflon	Samyang (Korea)	Trirex
Kazanorgsintez (Russia)	PC	Teijin (Japan)	Panlite
LG Chem (Korea)	Lupoy	Trinseo (USA)	Calibre

10.1.2 Aromatic Polycarbonates

Aromatic polycarbonates are the largest portion of this class of materials. Most often, the generic term *polycarbonate* refers to BPA polycarbonate (PC). For clarity, we will use *PC* to refer to polycarbonate with BP as a building block.

PC was discovered independently in Germany by Dr. Herman Schnell of Bayer and in the United States by Daniel Fox of General Electric. The earliest patent for PC was filed with the German Patent Office on October 16, 1953 [11]. A 1956 publication by Schnell summarizes the glass transitions and crystallization behaviors of various polycarbonates prepared with various bisphenols [12]. A review by Brunelle also summarizes characteristics of polycarbonates prepared with other aromatic building blocks [13].

PC's unique combination of transparency, rigidity, ductility, and heat resistance made the polymer successful in many applications. There has been a steady increase in the number of PC producers over the years, and companies with significant PC production are listed in Table 10.2. PC is now a global business, with all regions having significant manufacturing capabilities. The number of producers has increased mostly in Asia over the past 10 years, and global demand in 2015 has been estimated to exceed 4000 kt [14]. PC's clarity, impact strength, heat resistance, and processability continue to make it a resin of choice for demanding applications such as automotive lighting, including new high-intensity LED lighting, and architectural roofing.

Section 10.2 discusses the synthetic routes employed to prepare PC, though it should be noted that these reaction schemes are often useful for other aromatic or aliphatic polycarbonates. Some structure–property relationships in polycarbonates with various monomers, and copolymers, are discussed in Section 10.3. Important insights from some characterization techniques are summarized in Section 10.4. Transparent and opaque blends of polycarbonate, which continue to be successful in automotive components or electronic devices, are covered in Section 10.5.

10.2 SYNTHESIS OF POLYCARBONATES

Although this section highlights the synthetic routes with BPA polycarbonate (PC), the reactions can use other dihydroxyl monomers. Some reaction conditions may need to account for the polymer's semicrystalline behavior, if the monomer's structure facilitates crystallization (i.e., temperatures and solvents).

The structure of linear PC is shown in Scheme 10.4.

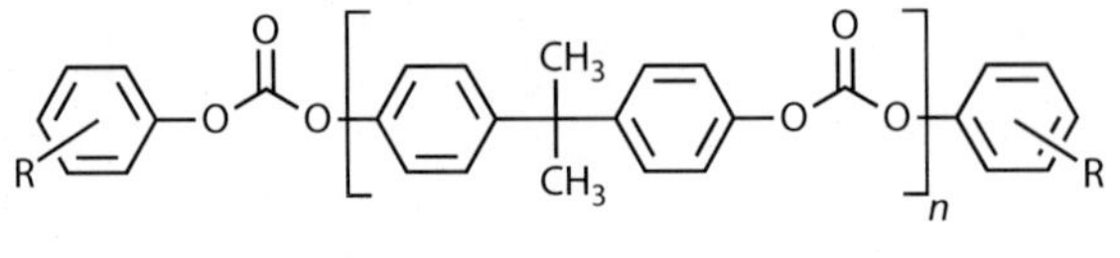

SCHEME 10.4 Bisphenol-A polycarbonate.

The two most common synthetic routes to PC are interfacial and melt reactions. Both interfacial and melt-reaction pathways use BPA as a monomer, which is synthesized by reacting acetone and phenol via an acid-catalyzed reaction and recrystallized to a high purity with a suitable solvent [15]. It should be noted that BPA was discovered many decades earlier than PC, as a patent application filed in 1930 already described an improved route to its synthesis [16].

Other pathways have been sought from the interfacial and melt reactions over the years. Hallgren disclosed a direct synthetic route to PC that employed BPA, carbon monoxide, and organometallic catalysts derived from elements in group VIIIB. Though this is of interest as a route without using phosgene and making use of trapped atmospheric CO_2, molecular weights and yields were relatively low [17]. Another commonly reported pathway is reacting chloroformates with phenols, although chloroformates are often prepared by phosgenation.

In the next two sections, the interfacial and melt-reaction processes to prepare PC are described. Both have their advantages, and it should be pointed out that both pathways are capable of yielding high-purity PC capable of meeting stringent quality requirements, such as for food-contact applications.*

10.2.1 Interfacial Reaction

As is the case for preparation of an ester by reacting an alcohol with an acid halide, PC can be prepared by reaction of phosgene and BPA in a two-phase scheme. In this method, an aqueous solution containing a bisphenolate of BPA is reacted with phosgene. An organic solvent capable of dissolving polycarbonate is also part of this mixture, for example, dichloromethane. The reaction occurs at the interface, and the mechanism involves ionic species, which can reside in the aqueous phase, while the high-molecular-weight product resides in the organic phase. The molecular weight of the product in the organic phase continues to grow as the reaction proceeds at the aqueous–organic interface. Preparation of esters and carbonates by interfacial reaction also requires an amine catalyst, specifically a trialkylamine or pyridine. The generalized scheme is shown in Scheme 10.5.

A benefit to this approach is that the catalysts and chlorides form salts in the aqueous phase. Once the aqueous phase is removed and the organic phase is adequately washed, the resulting product does not contain residual catalysts. As with condensation-type polymerization reactions, monofunctional monomers are useful to terminate the polymer chains and control the molecular weight of the product. Thus, BPA PC made by an interfacial reaction is especially stable due to the product not containing reactive end groups or residual catalysts (which is the case for polyesters). A comparison of chain terminator on polycarbonate properties is included in a patent application filed by Okamoto in 1988 [18]. PC terminated with cumylphenol gave lower birefringence than PC terminated with phenol or *t*-butylphenol. With careful reaction conditions (most often at high dilution and by not using chain terminators), cyclic polycarbonates can be prepared, where both reacting chain ends can react and form a carbonate linkage. Kricheldorf and coworkers successfully isolated and characterized cyclic PC [19]. Conditions used to help promote the formation of cyclics are believed to be particularly useful when scaling up reactions to understand the effect that dilution can have on their occurrence.

10.2.2 Melt Reaction

Use of the so-called melt-reaction route to PC makes use of diphenyl carbonate (DPC), which can be reacted with BPA by a condensation reaction (Scheme 10.6).

The reaction also benefits from the use of catalysts, though recently published work shows that extremely low concentrations should be used [20]. High temperatures capable of keeping the PC product in a viscous melt state (though polycarbonate is amorphous) combined with continuous

---

* See 21. Code of Federal Regulations §177.1580—Polycarbonate Resins.

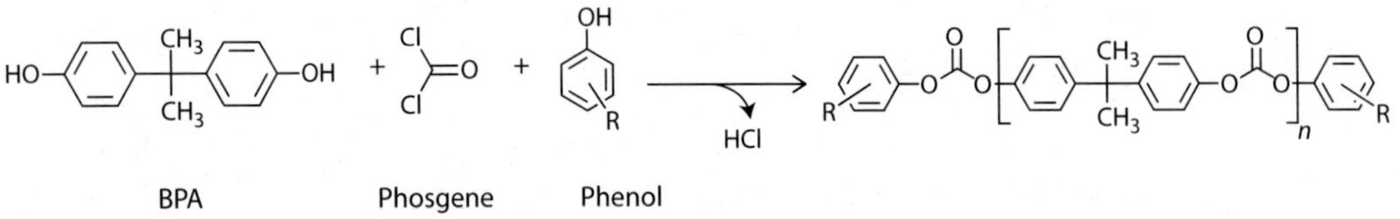

SCHEME 10.5 Interfacial reaction scheme for PC.

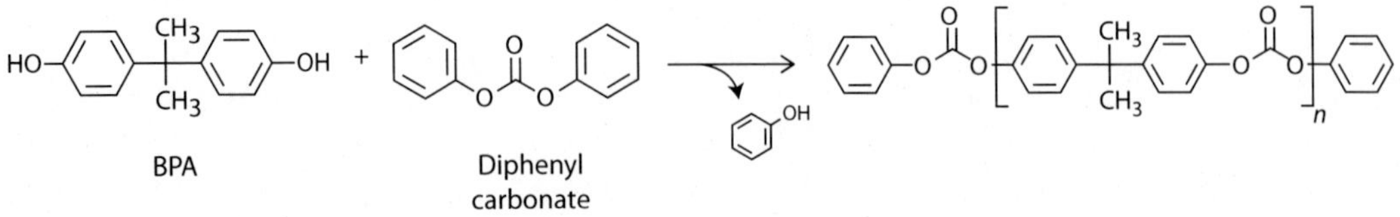

SCHEME 10.6 Melt-reaction scheme for PC.

stirring and effective removal of phenol by reduced pressure are necessary to build the molecular weight of the product. Although there exists significant patent literature on typical reaction conditions and optimization from production settings, Hsu and coworkers have studied and modeled reaction conditions for the process in a 70 L reactor system [21]. Melt polymerization has attracted significant interest as a route to PC that does not involve phosgene or solvents, and most of the recently introduced production capacities use this process. Recent work, however, has suggested that the higher residence time at elevated temperature may cause a higher incidence of thermally induced rearrangement products compared to interfacial processes [22]. Recent work with supercritical carbon dioxide revealed that, much like with solid-state polycondensation of polyesters, it can act as a plasticizer and improve PC's solid-state polymerization reaction kinetics with PC prepolymers. This route would permit molecular weight increases at lower temperatures [23].

10.2.2.1 Preparation of Diphenyl Carbonate

Though there continues to be much work to find alternate routes to prepare PC, most recent developments have described novel pathways for the preparation of DPC. In a path similar to the interfacial process for the preparation of polycarbonate, DPC can be synthesized by reacting an aqueous solution of a phenolate (derived from phenol) with phosgene [24]. Researchers at Asahi Chemical Industry disclosed a process for preparation of DPC by reacting phenol with carbon monoxide, which involved conversion of dimethyl carbonate into DPC [25]. Asahi Kasei also described preparation of DPC from ethylene oxide, CO_2, and phenol [26]. More recently, a scheme involving carbon dioxide with phenol and propylene oxide was disclosed by researchers at Shell, where diethyl carbonate and diisopropyl carbonate were used in an intermediate step toward DPC [27].

10.3 PROPERTIES OF POLYCARBONATE AND ITS COPOLYMERS

10.3.1 BPA Polycarbonate

Generally, PC based on BPA is an amorphous polymer with a glass transition temperature near 145°C. Crystallization of PC has been reported under special conditions such as with supercritical CO_2 [28], or with specific solvents like acetone [29]. Melting points from about 210°C to 290°C have been reported in the literature [30]. In this section, however, we will focus on properties of amorphous PC.

Amorphous PC has excellent impact strength and can withstand usage temperatures well above 100°C. Table 10.3 summarizes some of the engineering properties of the polymer, which are

TABLE 10.3
Engineering Properties of Makrolon 3208

Property	Standard	Unit	Value
Tensile modulus	ISO 527-1, -2	MPa	2350
Notched Izod impact, 25°C	ISO 180A	kJ/m^2	65, partial breaks
Heat distortion temp. under load (HDTUL)	ISO 75-1, -2	°C	
0.45 MPa			137
1.80 MPa			124
Flammability (UL 94 at 1.5 mm)	UL 94	Rating	V2

Source: Makrolon 3208 product data sheet, Covestro LLC.

common to PC grades of various molecular weights, according to International Organization for Standardization (ISO) test standards and SI units. More extensive engineering data are available from PC suppliers, who issue detailed data sheets for the grades they supply.

The melt volume rate (MVR) at 300°C/1.2 kg (ISO) typically ranges from 5 to 35 cm^3/10 min, depending on the grade.* Grades for optical discs require even higher flow, as microscopic features need to be consistently replicated on the mold surface with fast production rates (with an extremely high MVR of 16.5 cm^3/10 min at 2.16 kg/250°C). Though V2 is not the most stringent rating for UL 94 testing, PC's mainly aromatic character helps slow burning. Flame-retardant-type PC is discussed in greater detail in Section 10.5.2.

Polycarbonate possesses excellent stability at melt-processing conditions. The processes described in Section 10.2 yields polymer chains that do not have active functional end groups. Blends of various-molecular-weight polycarbonates have been shown to not undergo transesterification-type reactions and maintained consistent mechanical properties, although some brittleness was observed at low temperature for low-molecular-weight PC [31]. The notched Izod impact strength and heat distortion temperature under load (HDTUL) are well above values for other transparent polymers like polyesters, polystyrene (PS), or poly(methyl methacrylate) (PMMA). Its high modulus is also useful in designing parts with thinner walls and even helps assure rapid cycle times during injection molding [32].

Physical properties beyond the conventional data sheet properties of BPA PC have also been extensively studied. Polycarbonate benefits from a high modulus and T_g, which correlate to excellent stress relaxation properties [33]. Modeling of the stress relaxation and creep behavior of PC has been reported, where additional effects from conditioning or prior deformation were also found [34]. Mechanical properties have also been used for modeling purposes to ultimately predict the useful life of parts made from polycarbonate, which are discussed in greater detail in Section 10.4 [35,36].

Although PC is known to absorb modest amounts of moisture under normal atmospheric conditions, its properties generally remain unchanged with changes in humidity. Ito et al. extensively studied the effect of absorbed moisture on PC, observed an increase in T_g at higher absorbed moisture levels, and suggested that H$_2$O acted as an antiplasticizer in PC [37].

10.3.2 High-Heat (Co)Polycarbonates

Replacing BPA with more bulky or sterically hindered monomers raises the glass transition temperature beyond PC's 145°C. High-heat copolycarbonates have been commercialized by Covestro (Apec) and Sabic IP (Lexan XHT) and fulfill specialty requirements where very high temperatures are encountered, such as high-intensity lighting or firefighting equipment.

---

* See for example, datasheets for Makrolon 2258, 2458 and 3208 from Covetsro.

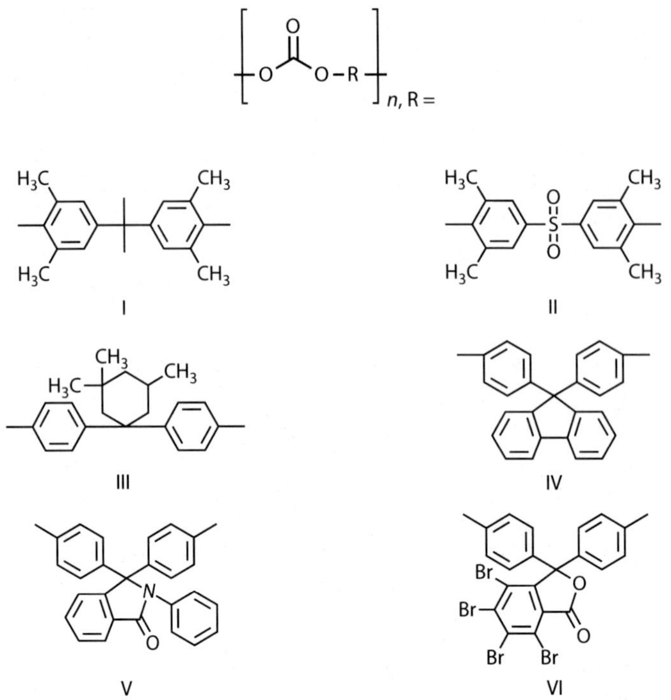

FIGURE 10.1 Structures of monomers for producing high-temperature polycarbonates.

Figure 10.1 illustrates the structures of some building blocks useful in yielding polycarbonates with glass transitions well above the T_g of 145°C for PC. As an example that increasing bulkiness and decreasing mobility raise T_g, a polycarbonate based on tetramethyl–BPA (structure I) has a glass transition temperature of 207°C [38]. This polymer has many properties superior to PC and maintains respectable ductility, though not as high as PC. Structure II, where the bulkiness of four methyl groups and an –SO$_2$– replaces BPA's isopropyl segment, yields a T_g of 260°C [39]. Paul and Bier described how a polycarbonate made with a monomer derived from 3,3,5-trimethylcyclohexane-1-one (structure III) has a glass transition as high as 238°C [40]. The flammability characteristics of polycarbonate based on fluorenone (BPF-PC, structure IV), a high-heat transparent material with a T_g of 275°C, was studied by Kambour et al. [39]. Mahood et al., with General Electric (now Sabic Innovative Plastics), described polymers based on 2,3-dihydro-3,3-bis(4-hydroxyphenyl)-2-phenyl-1H-isoindol-1-one (structure V) with T_g values of 250°C or higher [41]. The polymers described by Mahood also were shown in earlier work by Lin and Pearce to have increased both stability and limiting oxygen index (LOI) [42]. A tetrabromophenolphthalein-based polycarbonate (structure VI) yielded a T_g of 290°C and LOI levels between 65 and 75 [43]. Additional bisphenols and polycarbonates have been summarized in a review article compiled by Schmidhauser and Sybert [44].

There have also been efforts to produce bio-based polycarbonates with high-heat characteristics. For example, terpene diphenol (TPD), derived from terpene (which can be obtained from conifers), was used to synthesize various polycarbonates whose glass transitions went up to 214°C, depending on TPD content, where it boosted the glass transition normally observed from PC [43].

10.3.3 PC Containing Biphenyl Segments

Incorporation of biphenyl segments in polymers has been investigated to improve the impact performance of PC and also as a monomer for carbonate-based liquid crystalline materials [45]. The linear nature of this monomer affords better chain packing, while still allowing significant rotations of the phenyl rings. Thus, the ability of such segments in PC-type polymers to improve mechanical properties has been attributed to their high aspect ratio [46]. So-called tetraaryl–BPA (TABPA) was studied by Boyles and coworkers for the polymers' mechanical properties. When copolymerized with BPA, or hexafluoro-BPA, polycarbonates with glass transitions ranging from 150°C up to 175°C were obtained [47]. Significantly enhanced toughness was observed for one of their polymers where the bisaryl was copolymerized with BPA [48]. Similarly, the incorporation of 4,4′-dihydroxydiphenol into the PC chain significantly increased the ductility of BPA PC [47]. As little as 30 mol.% randomly incorporated into the polycarbonate backbone significantly improved the low-temperature notched-impact performance. A higher glass-transition temperature (163°C versus 155°C) and higher LOI index (35% versus 28%) were also observed. Though other polymers are known to crystallize with such monomers, 30 mol.% with BPA is completely amorphous and transparent. Blends with copolycarbonates based on 4,4′-dihydroxydiphenol also had excellent impact properties [49].

10.3.4 Polyarylate Block Copolymers Including Resorcinol

Though new block copolymers are continually introduced, PC–polyarylate copolymers have proven useful in further combining properties of PC with a polyester.

While the use of various additives can improve of polycarbonates' resistance to ultraviolet (UV) radiation (see Section 10.5.1), it is useful to prepare block copolymers from polycarbonates and polyesters such that the outstanding UV resistance of polyarylates that include resorcinol in their structure could be combined with PC. The resulting materials can significantly improve the UV resistance of PC-based materials. The UV resistance of these block copolymers was found to be superior to what was achievable with UV additives, and greater protection was afforded with higher polyarylate content in the blocks [50]. For example, so-called Lexan SLX copolymers commercialized by Sabic reportedly combine these features and offer excellent UV resistance thanks to the polyarylate, which becomes a self-protecting material after undergoing a UV-induced Fries rearrangement reaction [51,52]. These copolymers have been incorporated into transparent and opaque weatherable materials [53]. The environmental stress-cracking performance (ESC) of this class of commercial resins was deemed to be superior to PC's [54]. The self-protecting nature of polyarylates stems from the formation of o-hydroxybenzophenone, which has a significantly broader absorption spectrum where the irradiated polymer becomes more opaque to UV radiation from 355 to 385 nm. The o-hydroxybenzophenone moieties are structurally similar to commercially available UV blockers (for example, Chimassorb 81 [55]).

10.3.5 Branched PC

Branched polycarbonates are commercially available and can be prepared by using specialty building blocks that are incorporated directly into the molecular backbone. Monomers that contain more than two reactive phenolic sites allow for branching directly during polymerization. The use of 1,1,1-tris-p-hydroxyphenylethane (THPE, Figure 10.2, structure VII) has been described in research

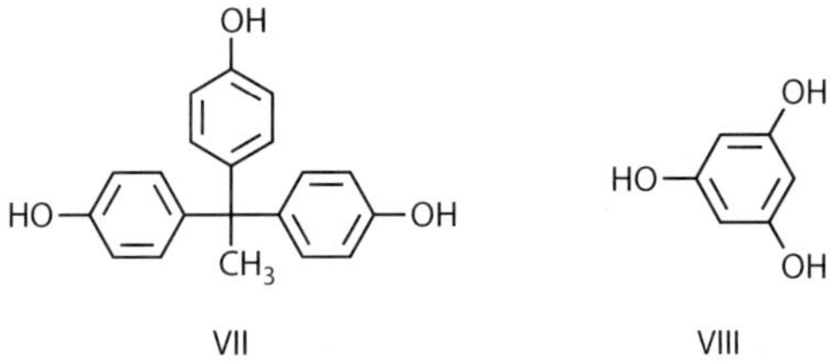

FIGURE 10.2 Structures useful for preparation of branched polycarbonate.

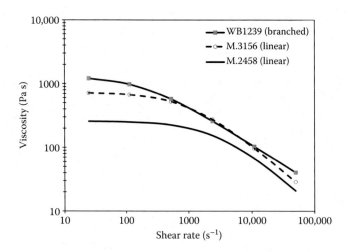

FIGURE 10.3 Shear–viscosity curves of linear and branched polycarbonate. (From CAMPUS® data for Makrolon WB1239 [branched], Makrolon 3156 [linear], and Makrolon 2458 [linear] at 300°C.)

articles [56] and in patent literature. The use of trihydroxybenzene (Figure 10.2, structure VIII, also referred to as *phloroglucinol*) as a branching agent for PC has been described in the literature [57], where increased occurrence of cross-linked microgels was observed [58], as well as a slight effect on the polymers' glass transition temperature [59].

Branched PC can be prepared by the interfacial process, or melt process [60]. The degree of long-chain branching can be examined by either gel-permeation chromatography (GPC; described briefly below) or solution viscosity (described in Section 10.4.1) [59]. While the impact resistance of a branched polycarbonate is not significantly different from BPA PC, the rheological properties are significantly changed [61]. Zhai and coworkers reported slightly lower T_g values for branched PC and slightly higher free-volume fractions [62]. Figure 10.3 compares the shear–viscosity curves for a linear and a branched polycarbonate. Branched materials are known to exhibit high viscosities at low shear and undergo more shear thinning at high shear.

The greater degree of molecular entanglement for branched materials also increases strain hardening that can occur during processing, which makes these polymers particularly suitable for parts made by thermoforming or blow molding [63]. So-called hyperbranched PCs have also been described in the literature [64,65], though they have proven more informative in investigating relaxation mechanisms in the PC backbone [66] than being used as engineering materials.

Combination of branched PC (prepared with THPE) with siloxane block copolymers was described to modify rheological and surface characteristics of the resulting material (i.e., lower wettability) [67].

More recently, the use of specialty chain terminators that can subsequently react with each other and incorporate branched sites after heating or melt processing has also been described [68–70]. Marks and coworkers incorporated 4-hydroxybenzocyclobutene, *m*-ethynylphenol, or methacrylate in interfacially polymerized PC and induced branching and cross-linking at 300°C. The benefit to this approach is the ability to achieve higher degrees of branching than is feasible with trifunctional monomers. This has been touted as being effective at further improving the flame retardancy of PC [71]. Morizur recently reported on the relative behaviors of different end groups used as branching agents [72].

10.3.6 Copolymers with Polydimethylsiloxane

Polydimethylsiloxane, or PDMS, is recognized for its stability and flexibility and is commonly used in medical applications. PDMS is also found in core–shell impact modifiers, which will be described in Section 10.5.3. As such, PDMS has been of particular interest for many resins, including PC, to boost impact performance. Simple PDMS–PC blends have been described in the literature and show excellent impact-strength [73]. Block copolymers of PC and PDMS have been known for some time already, and their preparation is straightforward from suitably capped PDMS precursors [74]. Sabic Innovative Plastics commercialized copolymers of PC and PDMS under the trade name Lexan EXL, which are promoted as materials with superior impact strength.

Various comonomers (i.e., for the polycarbonate segments) have also been investigated in producing PC–siloxane copolymers [75]. These copolymers do not always exhibit improved impact strength, as impact performance depends on the right balance of rigidity and flexibility in the polymer backbone. Zhou and Osby described blending PDMS with PC by melt kneading and with a twin-screw extrusion and obtained notched-impact performance that was comparable to a commercial Lexan EXL grade by incorporation of low levels of PDMS in PC [73]. They also provided evidence that PC and PDMS transesterified to form a copolymer during melt mixing. Additional approaches to prepare PC–PDMS copolymers in a kneader were described by Korn et al., and impact strength superior to high-molecular-weight PC was also observed [76].

10.4 CHARACTERIZATION OF POLYCARBONATE

10.4.1 Analytical Techniques Used to Characterize Polycarbonate

The molecular weight of polycarbonate is conveniently determined by GPC, also referred to as *size-exclusion chromatography* in solvents such as methylene chloride or tetrahydrofuran (THF). Systems equipped with UV, light-scattering, or index-of-refraction detection units are suitable. Experiments used to model hydrolytic stability based on GPC experiments were reported by Zinbo and Golovoy [77,78].

Though PS standards are most often used for molecular weight calibration of GPC systems, polycarbonate standards have been made commercially available. The absolute determination of PC's molecular weight can be done with nuclear magnetic resonance spectroscopy (NMR), where the relative abundance of chain end groups is compared to repeating units. Ito and coworkers reported their method to calculate the molecular weight of PC using ^{1}H NMR for PC terminated with *t*-butylphenol [79].

Fourier-transform infrared (FTIR) spectroscopy, or even Raman, is suitable to recognize polycarbonate. Many spectra have been reported in the literature. Although these methods are also useful for examining additives in PC, the detection limit may be somewhat limited.

Both ^{13}C and ^{1}H NMR are also useful in identifying polycarbonate. Though solid-state NMR methods are now fairly common, PC is soluble in $CDCl_3$, a common solvent used for NMR. This method is also capable of characterization of chain terminators, comonomers, or even isomers of BPA. As will be discussed in Section 10.4.4, solid-state NMR was combined with low-temperature relaxation studies to help identify specific motions responsible for PC relaxation modes observed in dynamic mechanical analysis (DMA).

10.4.2 Thermal Analysis and Stability Studies

BPA PC has been well studied by thermogravimetry (TGA) and differential scanning calorimetry (DSC). The degradation mechanism has been studied by combining these methods with other analytical techniques such as FTIR, gas- or liquid-chromatography (GC or LC) and mass spectrometry (MS). Thus TGA-FTIR, TGA-GC/MS and TGA-LC/MS have been used to carefully study PC decomposition products [80]. Degradation was found to follow chain scission of the isopropylidene linkage, the weakest link in the polymer chain. Interestingly, oxygen was also found to contribute to branching reactions. Derivation of kinetic equations using results from TGA at various heating rates has also been studied for BPA PC [81,82]. Isothermal TGA experiments can be used to model degradation kinetics, though several articles note inconsistencies from the models derived from isothermal TGA experiments [83,84]. Though such models have been used to predict the useful life of PC in an application, such models are complicated by unavoidable differences between the experiments and conditions encountered during actual usage. For example, TGA experiments are based on recording mass losses after degradation occurs, where chain scission and oxidation do not necessitate a mass loss. Furthermore, Pickett presented work that shows that commercially available polymers have been improved over several decades and have considerably better stability than what was found in earlier stability studies [38].

10.4.3 Rheology

Information on the rheology of linear and branched PC is crucial to processors, and these characteristics have been extensively studied and modeled [85,86]. These have also become more prevalent than viscosity data obtained from solutions in glass capillaries (Ubbelohde viscometer). Furthermore, rheological measures such as melt viscosity and MVR (measured by ASTM or ISO standards) are provided by the major PC suppliers for most of their resins. Though MVR is not a complete measure of a polymers' viscoelastic properties, it is still commonly used by processors to quickly estimate a material's processability. Melt viscosity is typically measured from 50 to 20,000 s^{-1} at temperatures of 300°C for pure PC. Lower temperatures (between 250°C and 270°C) are typically used to obtain rheological data for blends, due to their higher flowability.

10.4.4 Relaxation Studies

The relaxation of PC has been well studied as this has been shown to play an important role in explaining the high ductility of PC. Dynamic mechanical analysis (DMA) is particularly useful for relaxation studies as it is possible to carefully examine the deformation physics at various frequencies and temperatures. Low temperatures are needed to carefully examine γ-transitions that occur well below −50°C. A typical DMA plot for PC is shown in Figure 10.4.

The storage modulus shows that PC retains its rigidity over a very wide temperature range until it approaches the glass transition. The glass transition is taken as either the peak of the loss-modulus curve or the peak at tan δ. Much work has been done to study the sub-T_g relaxations in PC to understand how they relate to the toughness of the material. Yee and Smith published an excellent comparison of relaxation studies with DMA to examine the sub-T_g transitions of PC analogues with

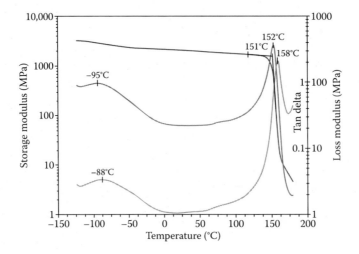

FIGURE 10.4 Plot of storage, loss moduli, and tan δ for PC.

slight differences in repeating units [87]. Substitutions on BPA's phenyl rings, or at the isopropylidene segment, afforded systematic adjustments to PC's chain packing and the overall mobility of the polymer chains. The assignment of so-called γ-relaxations at low temperature to rotation of the phenylene rings was corroborated by investigations done by NMR.

10.5 MODIFICATION OF POLYCARBONATES

In addition to meeting many requirements by themselves, PC blends have proven useful for meeting more specialized design and engineering requirements. The following sections provide a brief overview of components often blended with PC to tailor the properties to specific applications (e.g., flammability ratings).

10.5.1 STABILIZATION

Although polycarbonate's chemistry imparts excellent stability, the use of various stabilizers has been shown to be beneficial to further improving stability against UV light or stability at high temperatures. Stabilizers based on benzotriazole have proven useful for stabilization of PC against UV light. These stabilizers help trap UV-generated radicals that are transferred from the polymer chain. So-called UV blockers such as o-hydroxybenzophenones are also helpful in protection from UV degradation by absorbing much of the radiation in that spectrum [57]. The major PC suppliers offer most of their products in versions that include UV stabilizers for applications intended for outdoor use. Techniques based on adding the stabilizer to the surface of a molded part by means of infusion have proven helpful also [88,89]. As mentioned in Section 10.3, PC exposed to UV will undergo Fries rearrangement reactions. Work by Pickett suggests that the degradation products from UV exposure catalyze faster degradation by UV [90]. Thus, for outdoor applications, UV stabilization is critical.

Medical devices are often sterilized by radiation, either with γ-rays from a ^{60}Co source or from an electron beam. This method is particularly useful as devices can be sterilized within their packaging material. PC is suitable for sterilization by radiation and has been shown to retain its impact strength after sterilization [91]. After irradiation, PC acquires an intense yellow/green color, which fades over several days. This color change has been extensively studied, and an excellent review has been written by Factor [92]. Several additives and chemistries have been described that are capable of accepting radicals from the PC chain and reducing the color shift after sterilization [93,94].

Stabilizers based on phosphorous acid esters are known to be effective in improving the thermal stability of polycarbonate. Since the reactions the stabilizers participate in can sometimes be different (i.e., thermo-oxidative versus hydrolysis), the combination of stabilizers has also proven to be helpful [95].

10.5.2 FLAME-RETARDANT POLYCARBONATES

The flammability characteristics of polycarbonate have been extensively studied, and various reviews are available [96]. Various standards for flammability performance exist for electrical devices, vehicles, aircraft, mass transit, and construction. The most common flammability standards used for comparison of plastics are UL 94 ratings and LOI testing. Descriptions of several standards and these tests can be found elsewhere [97]. An LOI of 25% has been reported for PC, which is significantly lower (i.e., less combustible) than more aliphatic materials. Though general-purpose polycarbonate is able to meet a UL 94 rating of V2 at 1.5 mm thickness, flame-retardant grades are able to meet more stringent V0 requirements. Even more stringent 5VA ratings are known at higher thicknesses [98]. Levchik and Weil reviewed many of the flame-retardants used for polycarbonate and polycarbonate blends [99].

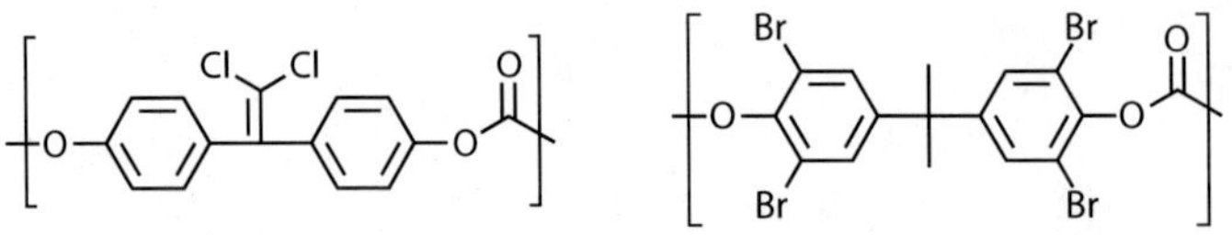

Bisphenol-C PC,
$T_g = 168°C$ [100]

Tetrabromobisphenol-A PC,
$T_g = 170–202°C$ [101]

FIGURE 10.5 Structures of common halogenated building blocks used to make polycarbonates useful as flame-retardants. (From Factor, A. and Orlando, C., *J. Polym. Sci.: Polym. Chem. Ed.* 18, 579–592, 1980; Chung, J. Y. and Paul, W. G., *58th Annu. Tech. Conf.—Soc. Plast. Eng.* 2125–2129, 2000.)

Figure 10.5 illustrates the structure of common halogenated building blocks used to make polycarbonates, which are also useful for improving the flame retardancy of PC. Inclusion of halogen atoms in the polymer backbone significantly increases the flame retardancy of the material. For example, PC blended with polycarbonates based on bisphenol-C (Figure 10.5, left) were shown to increase the LOI value from 25% for PC up to levels beyond 50% [100]. Though halogens are known to easily produce radicals that can scavenge oxygen during combustion, work with bisphenol-C polycarbonate suggests they also help promote cross-linking of the polymer, which can suppress dripping and help the formation of char [102]. Similarly, flame retardants that include bromine in their chemical structure are common owing to bromine's slightly lower propensity to form radials than chlorine (due to lower electronegativity) and its bulky structure, which helps its polymer have high-heat properties and assures that mixtures with PC have high heat resistance.

The polycarbonate based on tetrabromobisphenol-A (Figure 10.5, right) is usually supplied in oligomeric form, and variation of T_g at relatively low molecular weights should be expected. Work by Chung and Paul suggests that this polymer is miscible with PC [101].

Phosphorus-based flame retardants are widely used as nonhalogen methods of achieving necessary UL performance. Phosphates have the important disadvantage of significantly plasticizing many polymers and lowering the heat resistance of the resin [103]. Calcium phosphinate salts have been shown to be a promising means to overcome this limitation in various polymers [104] and achieve V0 listings, even with PC/polyester blends [105]. More recently, phosphonate type chemistry has been used to improve flame-retardant characteristics of polycarbonate [106]. Like with phosphate-based flame-retardant additives, these help generate char during burning, which helps insulate underlying material and stop the propagating flame. A benefit to phosphonates, however, is that they are less effective at plasticization and can improve the thermal properties of flame-retardant blends. Polymeric phosphonates using BPA as a comonomer are amorphous and transparent.

Silicone-based flame retardants are also recognized as being effective for improving the ratings for polycarbonate. Siloxane copolycarbonates, discussed earlier in Section 10.3.6, have higher LOI values compared to PC. Tang and coworkers combined silicone and phosphorus chemistry by synthesizing novel oligophosphonium phosphates based on a siloxane backbone. When blended with polycarbonate, these novel additives were shown to be effective also in increasing LOI and reducing the heat of combustion, though the samples were not optically clear [107].

Multiwalled carbon nanotubes (MWCNTs) were also reported to be able to lower the tendency of dripping during combustion. At low levels, the improved barrier performance imparted by MWCNT reduced oxygen migration into the burning material and lowered the peak heat-release rates recorded during cone calorimetry [108].

10.5.3 Impact Modification

PC already benefits from outstanding impact strength yet is blended with elastomers to either increase flexibility or provide additional strength against impact. Impact modification has been shown to be particularly useful when the quality of polycarbonate feedstock varies, such as with recycled PC [109].

Blending elastomers with PC is a challenge, due to the different nature of the elastomers and PC—where PC is often more polar than elastomers based on olefins or siloxane. Adding core–shell elastomers to PC (i.e., impact modification) is a convenient method to impart more ductility to PC [110]. The core of these additives is elastomeric, while the grafted shell increases its' compatibility with PC and aids dispersion in the surrounding matrix. Improved performance of impact-modified PC was also observed when the blends were further annealed [111]. For the case of PC blended with polyolefins, such as polyethylene, a compatibilizer may be necessary to help with droplet breakup. Polyolefins copolymerized with acrylic monomers (methyl methacrylate or butyl acrylate) are known to have better compatibility with PC. Glycidyl-methacrylate (GMA)- or maleic anhydride (MA)-functionalized copolymers are useful for grafting with the PC portion. Most often, such blends are characterized by exhibiting low gloss due to relatively large cross-linked, particles at the surface, although some advanced processing methods have been shown to overcome this and yield a high gloss [112,113]. Thermoplastic urethane elastomers (so-called TPU resins) are also known to work as impact modifiers for PC, either alone or in combination with other impact modifiers [114,115].

10.5.4 BLENDS AND ALLOYS

Polycarbonate blends have been very successful commercially as they provide the right balance between material performance and productivity. These are also more common than impact-modified PC. From polycarbonate/polyester blends suitable for automotive exteriors to flame-retardant PC/acrylonitrile–butadiene–styrene (ABS) housings for consumer electronics, PC has been an important contributor due to the right property profile it brings [116]. In most cases, PC is the continuous phase of the blend, having a higher propensity to encapsulate other polymers [117]. The major polycarbonate suppliers offer PC alloys, with PC/ABS being more common that PC/polyesters. Over the years, more sophisticated composites have been introduced by careful selection of the blend components and blending processes. For example, electrically conductive pellets were described by Cofer et al. for specialized applications requiring electrical conductivity [118]. Fundamental studies on thermally conductive PC composites have been reported recently, and commercial grades have also been introduced (e.g., Makrolon TC grades) [119].

10.5.4.1 PC/ABS

Polycarbonate blends with acrylonitrile–butadiene–styrene copolymers (PC/ABS) are commercially significant and available from Covestro (Bayblend), Sabic IP (Cycoloy), Trinseo (Emerge), or Cheil (Infino). PC and ABS form immiscible blends, and the material properties can be tailored by carefully formulating the PC and ABS components. PC's toughness and high heat resistance together with ABS' ductility and processability provide an excellent combination as an engineering resin. Greco and Sorrentino provided an excellent overview of fundamental research done into PC/ABS blends [120]. Table 10.4 compares properties for various commercially available PC/ABS grades.

TABLE 10.4
Engineering Properties of Various Bayblend PC/ABS Grades

Property	Test Standard	Unit	Bayblend T50XF	Bayblend T65XF	Bayblend T85XF
Melt volume rate	ISO 1133	cm^3/10 min	19	18	19
Tensile modulus (25°C)	ISO 527	MPa	2100	2400	2300
Notched Izod impact strength (25°C)	ISO 180	kJ/m^2	45	45	35
Vicat softening temperature (50 N, 120°C/h)	ISO 306	°C	115	120	130

Source: Data sheets provided by Covestro LLC for Bayblend T50XF, T65XF, and T85XF.

Higher heat resistance usually results from the use of higher proportions of polycarbonate in the blend. It is important to note that without the presence of an elastomeric phase in the blend (i.e., simple PC/poly(styrene-acrylonitrile), or PC/SAN), the material would be brittle, with very low notched-impact strength [121]. PC/ABS blends usually form a well-dispersed ABS phase. However, undesirable coalescence of the phases has been reported [122], and some work has been done to develop suitable compatibilizers for SAN and PC to overcome this [123,124]. Recently, addition of PMMA was shown to improve PC/ABS blends via a compatibilizing effect [125]. PMMA is generally recognized to also be immiscible with PC and forms blends with higher opacity than PC/ABS due to the greater differences in refractive indices of PC and PMMA.

Properties of PC/ABS can also be optimized with careful consideration of attributes in the polycarbonate, ABS, or styrene–acrylonitrile copolymer. For example, the chemical resistance of PC/ABS has been shown to be superior when using a branched polycarbonate [126]. Furthermore, the hydrolysis resistance of PC/ABS has been shown to largely rely on the composition of the ABS [127,128].

10.5.4.2 PC/Polyesters

Principally due to polyesters' ability to crystallize and, thereby, impart improved chemical resistance, PC alloys with polyesters are suitable for applications where higher chemical resistance than PC or PC/ABS is required. It follows that the higher crystalline behavior of polybutylene terephthalate (PBT) versus polyethylene terephthalate (PET) gives PC/PBT higher chemical resistance than PC/PET. It should be noted, however, that the T_g of PET (67–81°C) is higher than that of PBT (50–60°C), and therefore, PC/PET blends offer superior heat resistance. PET also has a higher crystalline melting point than PBT (260°C versus 226°C). Some summaries of PC/polyesters have been published elsewhere [129]. PC/polyester alloys have been commercialized as both general-purpose and flame-retardant resins under the trade names Makroblend (Covestro), Xenoy (Sabic Innovative Plastics), and Emerge (Trinseo).

Unlike polycarbonates, which are essentially free of residual catalysts and have capped molecular chains, polyesters have active carboxylic acid or ester chain ends and residual organometallic catalysts [130]. As such, these polymers can undergo transreactions when blended together [131]. Montaudo and coworkers did extensive studies on transreactions between polycarbonate and polyesters and the reaction mechanism (reviewed recently in a PC review article their group coauthored) [132].

As with PC/ABS, PC/polyester blends benefit from the inclusion of impact-modifiers tailored to an optimized morphology [133]. It is noteworthy that PBT and PC show surprisingly good compatibility with evidence of some miscibility, which is believed to be key to the excellent impact properties exhibited by PC/PBT alloys [134,135].

Poly(trimethylene terephthalate), or PTT, has been introduced commercially in the past 10 years, mostly due to the availability of 1,3-propanediol from bio-based pathways. As a polyester, PTT's properties are intermediate between PET and PBT. Blends of PTT with PC also have been studied and shown to undergo transesterification [136]. Studies with DMA (described earlier in Section 10.4) showed the blends to be partially miscible [137].

PCTG, a copolyester polymerized from terephthalic acid, ethylene glycol, and cyclohexanedimethanol (Eastar, commercially available from Eastman Chemical, or Skygreen, from SK Chemical), is known to produce miscible blends with PC that are optically clear [138]. This blend has been shown to have advantages in terms of improved chemical resistance or radiation resistance [139,140]. Such specialty resins are commercially available (Eastalloy from Eastman Chemical or Makroblend from Covestro).

Another novel PC/polyester blend system is PC blends with poly(L-lactic acid) (PLA). These resins have been introduced to reduce petroleum consumption in engineering resins and take advantage of resins produced with chemical building blocks derived from plant sources. PLA is a semicrystalline resin and is now widely used in single-use packaging applications and available from Natureworks (Ingeo). Despite PLA having a relatively low glass transition temperature (55–60°C), its crystallinity and melting point were high enough to make suitable impact-modified PC/PLA blends [141].

Polycarbonates

10.5.5 Filled Polycarbonates

The use of fillers is a convenient method to raise the modulus beyond levels reported in Table 10.2. In general, higher stiffness is achievable with glass or carbon fibers. Reinforcement with fibers, however, is more directionally dependent due to their tendency to orient along flow lines in an injection-molded part. The modulus as a function of filler content is shown for commercially available glass fiber- and carbon fiber-reinforced PC in Figure 10.6. Carbon fiber reinforcement at 30% or higher is also indicated to be electrically conductive.

Another important benefit to filled grades is a significant decrease in the coefficient of linear thermal expansion (CLTE), which can be particularly important for composite parts made from a multishot injection-molding process. Incorporation of fillers invariably reduces the toughness of resins, and much work has been done to carefully optimize the resin to maintain ductility.

In general, carbon fiber reinforcement achieves higher rigidity than glass fiber. Mineral reinforcement is not common in PC, though widely available for PC/ABS or PC/polyester resins.

To improve the toughness of reinforced resins, research has concentrated on nano-sized fillers, which can also interact with the matrix at the molecular level. For example, clay-based nanofillers, which, when fully exfoliated, impart stiffness and toughness to nylon-based resins, were also the subject of much work with polycarbonate-based resins. Clay-based fillers require special treatments to promote intercalation and exfoliation of the clay layers, along with significant shear during blending. In many cases, the increased modulus from reinforcement comes at the expense of lower thermal stability and embrittlement of the material [142].

PC reinforced with MWCNTs was also investigated for unique mechanical properties and the potential for materials that could offer electrical conductivity as well. These fillers, when mixed with PC in the melt, showed viscosity increases well beyond what was normally observed for carbon fibers, due to the high aspect ratio of MWCNT and interactions with PC [143]. Similarly, PC–MWCNT composites exhibited lower percolation thresholds for electrical conductivity. The same research group also noted that there is variation between industrially available MWCNT in terms of their dispersion and agglomeration [144], as well as processing effects [145].

Composites with expanded graphite have also been recently described, where thermal conductivity is sufficiently high to use PC composites as heat sinks for LED lighting [146]. Like other carbon-based fillers, graphite also imparts electrical conductivity, and this feature was used to prepare strain-sensing composites [147].

Polyhedral oligomeric silsesquioxane (POSS), a cage-like nanofiller with particle sizes around 1–3 nm have attracted interest as materials that could increase a polymer's modulus and help retain ductility. The backbone of the cage structure is SiO_2 and $R–SiO_2$. Zhao and Schiraldi studied POSS

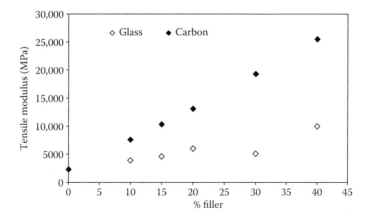

FIGURE 10.6 Modulus versus filler content for glass fiber- and carbon fiber-reinforced PC.

particles with various substituent groups in PC. Interestingly, trisilanolphenyl–POSS had excellent compatibility in PC and resulted in optically clear samples at up to 5% reinforcement, though with reduced T_g, likely due to plasticization effects [148].

Inorganic whiskers have also been investigated as a means to improve properties observed with fiber reinforcement. Aluminum borate and potassium titanate whiskers were investigated as fillers for PC. Though significant improvement in stiffness were observed, thermal analysis (DSC and TGA) revealed that the fibers induced thermal degradation of the PC [149]. Recent work with alumina nanorods in PC included synthetic grafting of PC to the reinforcement, which significantly helped dispersion and improved mechanical properties, compared to ungrafted reinforcement. Films of these alumina-grafted PC composites, which included a masterbatch preparation step, showed some degree of optical transparency due to excellent dispersion of the rods in the PC matrix [150].

10.6 PROCESSING OF POLYCARBONATES

As resin suppliers have extensive literature to design parts and process PC resins available from their websites, only a brief summary is given in the sections that follow.

10.6.1 INJECTION MOLDING OF POLYCARBONATE AND ITS RESINS

Parts made from polycarbonates and blends are most often made by injection molding. While optimal filling speeds, cooling times, and temperatures are unique for every situation, ranges for injection-molding temperatures and shrinkages for various commercially available PC resins are summarized in Table 10.5.

The melt and mold temperatures generally follow the same trends as the materials' softening temperature. Molding shrinkage is higher for PC/PBT and PC/PET due to crystallization of the polyesters. Though PC and high-heat PC should have similar shrinkages due to their structural similarities, these are cooled from higher melt temperatures, which increases the degree of shrinkage. It should be noted that the values in Table 10.5 are general guidelines and that some instances may require higher temperatures than those provided. Processing information for aliphatic polycarbonates was not yet available, but their lower thermal stability and heat resistance compared to aromatic polycarbonates makes it reasonable to expect their processing temperatures to more closely

TABLE 10.5
Typical Settings for Injection-Molding Various Polycarbonate Resins

Resin	Melt Temperature, °C	Mold Temperature, °C	Molding Shrinkage, %
PC	280–320	80–100	0.5–0.8
High-heat PC	310–340	100–150	0.8–0.9
Filled PC	310–330	80–130	0.3–0.5
PC/ABS	240–280	70–100	0.5–0.7
PC/PBT	250–270	60–80	0.8–1.0
PC/PET	260–280	60–80	0.6–0.8

Source: BMS AG publication, Injection Molding of High-Quality Molded Parts: Processing Data and Advice, MS 00062579, April 2013; Thermoplastic Resins Property Guide, Covestro LLC 20628, December 2011.

resemble settings for PC blends, as their temperature ranges are below PC. Copolycarbonates' temperature settings will depend on the resulting viscosity and heat resistance of the resin.

Drying polycarbonates and blends is crucial to producing defect-free parts. Though there are many configurations possible for adequate drying, the main target is a moisture level of 0.02% or less. Lower moisture levels (i.e., 0.01% or less) are highly recommended for PC/polyesters, as polyester hydrolysis can promote transesterification of the resins [134,150,151]. Inadequate drying can often be recognized by bubbling of melt purges or high occurrences of splay [152].

Since most molders produce parts to aesthetics, a convenient method to quickly ascertain whether degradation is a factor in the injection-molding process is to measure MVR on resin taken from molded parts and comparing it to the suppliers' data.

10.6.2 EXTRUSION, THERMOFORMING, AND BLOW MOLDING OF POLYCARBONATE

For extrusion, thermoforming, and blow molding, a resin that has relatively high viscosity at low shear is preferable. Though high-molecular-weight linear resins are known to be suitable in some circumstances, the use of a branched PC resin is often preferable. As with injection molding, literature is available from resin suppliers with suggestions on processing resins with extrusion, thermoforming, and blow molding [153–155]. These techniques are seldom applied to specialty copolycarbonates. Extrusion and thermoforming of PC blends, especially PC/ABS, has become more common [156].

10.7 SUMMARY AND OUTLOOK

As plastics manufacturing gets increasingly sophisticated, the versatility of polycarbonate to solve problems in engineering will continue to be exploited. The literature base summarizing studies with polycarbonates continues to grow and increases at a steady rate. We can therefore continue to expect innovations with polycarbonates that will expand the versatility of this engineering material. We can also anticipate many developments in alternate synthesis routes to polycarbonate, and copolymers, which will likely enable the emergence of even more copolycarbonates, both aromatic and aliphatic.

ACKNOWLEDGMENTS

Permission by Covestro, LLC, to publish this chapter is gratefully acknowledged. Sincere thanks to all friends, coworkers, and colleagues who have helped with our understanding of polycarbonate and the significant progress and innovation witnessed over the years.

REFERENCES

1. H. Medem, M. Schreckenberg, R. Dhein, W. Nouvertné, H. Rudolph, EP 25937, filed April 1, 1981.
2. D.-J. Kim, W.-J. Yoon, J.-R. Kim, Y.-J. Lee, S.-Y. Hwang, US2013/0295306, filed January 30, 2012.
3. F. Fenouillot, A. Rousseau, G. Colomines, R. Saint-Loup, J.-P. Pascault, *Progr. Polym. Sci.* **35** (2010) 578–622.
4. Q. Li, W. Zhu, C. Li, G. Guan, D. Zhang, Y. Xiao, L. Zheng, *J. Polym. Sci., Part A: Polym. Chem.* **51** (2012) 1387–1397.
5. M. Fuji, M. Akita, EP 2033981, filed June 14, 2007.
6. D. Dhara, A. A. G. Shaikh, G. Chatterjee, C. Seetharaman, US2005/0143554, filed May 28, 2004.
7. P. Schuhmacher, H. R. Kricheldorf, S.-J. Sun, DE 19631658, filed February 12, 1998.
8. T. Artham, M. Doble, *Macromol. Biosci.* **8** (2008) 14–24.
9. M. Yokoe, K. Aoi, M. Okada, *J. Appl. Polym. Sci.* **98** (2005) 1679–1687.
10. A. E. Acar, D. Brunelle, *ACS Symp. Ser.* **898** (2005) 216–228.
11. H. Schnell, L. Bottenbruch, H. Krimm, DE 971,790, filed October 16, 1953.
12. H. Schnell, *Angew. Chem.* **68** (1956) 633–660.

13. D. J. Brunelle, "Polycarbonates" in *Encyclopedia of Polymer Science and Technology*, A. Seidel, ed. John Wiley & Sons, Hoboken, NJ, 2006.
14. PR Newswire, November 30, 2011 citing Chemical Market Associates completion of World ABS and PC analysis.
15. K. Buser, W. Schafer, GB 902,350, filed September 13, 1960.
16. S. Kohn, E. Schub, US 1,978,949, filed August 27, 1930.
17. H. Hallgren, US 4,096,168, filed October 12, 1976.
18. M. Okamoto, US 4,997,904, filed August 11, 1988.
19. H. R. Kricheldorf, G. Schwartz, S. Böhme, C.-L. Schultz, R. Wehrmann, *Macromol. Chem. Phys.* **204** (2003) 1398–1405.
20. J.-P. Hsu, J.-J. Wong, *Ind. Eng. Chem. Res.* **45** (2006) 2672–2676.
21. J.-P. Hsu, J.-J. Wong, S. Tseng, *J. Appl. Polym. Sci.* **108** (2008) 694–704.
22. A. C. Hagenaars, J.-J. Pesce, C. Bailly, B. A. Wolf, *Polymer* **42** (2001) 7653–7661.
23. C. Shi, J. DeSimone, D. J. Kiskerow, G. W. Roberts, *Macromolecules* **34** (2001) 7744–7750.
24. J. H. Pearson, S. Tryon, US 2,335,441, filed August 2, 1940.
25. S. Fukuoka, M. Tojo, M. Kawamura, JP 04235951, filed January 11, 1991.
26. S. Fukuoka, M. Tojo, H. Hachiya, M. Aminaka, K. Hasegawa, *Polym. J.* **39** (2007) 91–114.
27. E. Van der Heide, T. M. Nisbet, G. G. Vaporciyan, C. Leonardus, M. Vrouwenvelder, US 7,732, 629, filed January 21, 2008.
28. E. Beckman, R. S. Porter, *J. Polym. Sci., Part B: Polym. Phys.* **25** (1987) 1511–1517.
29. R. P. Sheldon, P. R. Blakey, *Nature*, **195** (1962) 172–173.
30. J. Lu, D. Xi, R. Huang, L. Li, *J. Macromol. Sci., Part B: Phys.* **50** (2011) 1018–1030.
31. K. Cheah, W. D. Cook, *Polym. Eng. Sci.* **43** (2003) 1727–1739.
32. I. Menego, M. Yeager, P. Moulinié, *71st Annu. Tech. Conf.—Soc. Plast. Eng.* (2013) 1522–1526.
33. M. G. Khametova, *Chem. Petrol. Eng.* **48** (2013) 558–562.
34. F. Khan, E. Krempl, *Polym. Eng. Sci.* **44** (2004) 1783–1791.
35. T. A. P. Engels, L. C. A. van Breemen, L. E. Govaert, H. E. H. Meijer, *Polymer* **52** (2011) 1811–1819.
36. J. E. Pickett, D. J. Coyle, *Polym. Degr. Stab.* **98** (2013) 1311–1320.
37. E. Ito, Y. Kobayashi, *J. Appl. Polym. Sci.* **22** (1978) 1143–1149.
38. V. Serini, D. Freitag, H. Vernalaken, *Angew. Makromol. Chem.* **55** (1976) 175.
39. R. P. Kambour, W. V. Ligon, R. R. Russell, *J. Polym. Sci: Polym. Lett.* **16** (1978) 327–333.
40. W. G. Paul, P. N. Bier, "New high heat polycarbonates: Structure, properties" in *Applications in Application of High Temperature Polymers*, R. R. Luise, ed. CRC Press, Clearwater, FL, 1997.
41. J. A. Mahood, E. E. Gurel, US 7,358,321, filed November 29, 2005.
42. M. S. Lin, E. M. Pearce, *J. Polym. Sci.: Polym. Chem. Ed.* **19** (1981) 2659–2670.
43. Y. Xin, H. Uyama, *J. Polym. Res.* **19** (2012) 1–7.
44. J. Schmidhauser, P. D. Sybert, *J. Macromol. Sci.—Polym. Rev.* **C41** (2001) 325–367.
45. H. R. Kricheldorf, S.-J. Sun, A. Gerken, T.-C. Chang, *Macromolecules* **29** (1996) 8077–8082.
46. D. A. Boyles, T. S. Filipova, J. T. Bendler, G. Longbrake, J. Reams, *Macromolecules* **38** (2005) 3622–3629.
47. A. Karbach, D. Drechsler, C.-L. Schultz, U. Wollborn, M. Moethrath, M. Erkelenz, J. Y. Chung, J. Mason, *ACS Symp. Ser.* **898** (2005) 96–111.
48. D. A. Boyles, P. Dehmer, J. Bendler, T. Filipova, *SAMPE Conf. Proc.* **51** (2006) 284/1–10.
49. J. Y. Chung, J. P. Mason, M. Erkelenz, R. Wehrmann, *63rd Annu. Tech. Conf.—Soc. Plast. Eng.* (2005) 2080–2084.
50. M. Diepens, P. Gijsman, *Polym. Degr. Stab.* **94** (2009) 1808–1813.
51. V. V. Korshak, S. V. Vinogradova, S. A. Siling, S. R. Rafikov, Z. Y. A. Fomina, V. V. Rode, *J. Polym. Sci.: Part A-1* **7** (1969) 157–172.
52. S. M. Cohen, R. H. Young, A. H. Markhart, *J. Polym. Sci.: Part A-1* **9** (1971) 3263–3299.
53. P. Sybert, S. Klei, D. Rosendale, J. Di, D. Shen, *63rd Annu. Tech. Conf.—SPE* (2005) 2523–2527.
54. X. Li, *Polym. Degr. Stab.* **90** (2005) 44–52.
55. F. A. Cangelosi, L. H. Davis, R. L. Gray, J. A. Stretanski, D. J. Jakiela, "UV stabilizers" in *Encyclopedia of Polymer Science and Technology*, (4th ed.), H. F. Mark, ed. pp. 452–493. John Wiley & Sons, Hoboken, NJ, 2014.
56. S. P. Kim, J.-S. Lee, S.-H. Kim, B.-H. Lee, S. H. Kim, W.-G. Kim, *J. Ind. Eng. Chem.* **5** (1999) 268–273.
57. Z. Dobkowski, J. Brzeziński, *Eur. Polym. J.* **17** (1981) 537–540.
58. Z. Czlonkowska-Kohutnicka, Z. Dobkowski, *Eur. Polym. J.* **18** (1982) 911–915.
59. Z. Dobkowski, *Eur. Polym. J.* **18** (1982) 563–567.

60. S. Konrad, H. W. Heuer, K.-H. Koehler, C. Muennich, R. Wehrmann, US 8,487,025, filed April 7, 2011.
61. M.-Y. Lyu, J. S. Lee, Y. Pae, *J. Appl. Polym. Sci.* **80** (2001) 1814–1824.
62. W. Zhai, J. Yu, W. Ma, J. He, *Macromolecules* **40** (2007) 73–80.
63. F. Stadler, A. Nishioka, J. Stange, K. Koyama, H. Muenstedt, *Rheol. Acta* **46** (2007) 1003–1012.
64. D. H. Bolton, K. L. Wooley, *J. Polym. Chem.: Part A: Polym. Chem.* **40** (2002) 823–835.
65. M. Miyasaka, T. Takazoe, H. Kudo, T. Nishikubo, *Polym. J.* **42** (2010) 852–859.
66. D. H. Bolton, J. M. Goetz, D. Gan, J. A. Byers, B. Poliks, K. L. Wooley, J. Schaefer, *Macromolecules* **36** (2003) 2368–2373.
67. Md. M. Islam, D.-W. Seo, H.-H. Jang, Y.-D. Lim, K. M. Shin, W.-G. Kim, *Macromol. Res.* **19** (2011) 1278–1286.
68. M. J. Marks, J. Newton, D. C. Scott, S. E. Bales, *Macromolecules* **31** (1998) 8781–8788.
69. M. J. Marks, J. Newton, *J. Polym. Chem.: Part A: Polym. Chem.* **38** (2000) 2340–2351.
70. M. J. Marks, J. Newton, S. E. Bales, *J. Polym. Chem.: Part A: Polym. Chem.* **38** (2000) 2352–2358.
71. Y. Xu, L. Wen, Z. Chen, Y. Niu, J.-F. Morizur, US2014/0079934, filed September 14, 2012.
72. J.-F. Morizur, *71st Annu. Tech. Conf.—Soc. Plast. Eng.* (2013) 112–114.
73. W. Zhou, J. Osby, *Polymer* **51** (2010) 1990–1999.
74. H. A. M. van Aert, L. Nelissen, P. J. Lemstra, D. J. Brunelle, *Polymer* **42** (2001) 1781–1788.
75. W.-C. Shih, C.-C. M. Ma, J.-C. Yang, J.-T. Gu, L.-D., Tsai, *J. Appl. Polym. Sci.* **75** (2000) 545–552.
76. M. R. Korn, E. H. Jonsson, R. J. Kumpf, *Polym. Prepr.* **38** (1997) 464–465.
77. A. Golovoy, M. Zinbo, *Polym. Eng. Sci.* **29** (1989) 1733.
78. M. Zinbo, A. Golovoy, *Polym. Eng. Sci.* **32** (1992) 786–791.
79. Y. Ito, H. Ogasawara, Y. Ishida, H. Ohtani, S. Tsuge, *Polym. J.* **28** (1996) 1090–1095.
80. B. N. Jang, C. A. Wilkie, *Thermochim. Acta.* **426** (2005) 73–84.
81. J. E. Robertson, T. C. Wara, *Polym. Mat. Sci. Eng.* **81** (1999) 138–139.
82. H. Polli, L. A. M. Pontes, A. S. Araujo, *J. Therm. Anal. Calorim.* **79** (2005) 383–387.
83. B. Janković, *J. Polym. Res.* **16** (2009) 213–230.
84. I. Blanco, L. Abate, M. L. Antonelli, F. A. Bottino, *Polym. Degr. Stab.* **98** (2013) 2291–2296.
85. Z. Dobkowski, *Eur. Polym. J.* **18** (1982) 1051–1059.
86. J. H. Park, J. C. Hyun, W. N. Kim, S. R. Kim, S. C. Ryu, *Macromol. Res.* **10** (2002) 135–139.
87. A. F. Yee, S. A. Smith, *Macromolecules* **14** (1981) 54–64.
88. J. Buekers, P. Bier, U. Grigo, EP 677547, filed April 3, 1995.
89. R. A. Pyles, R. L. Archey, D. M. Derikart, US 7504054, filed December 6, 2004.
90. J. E. Pickett, *Polym. Degr. Stab.* **96** (2011) 2253–2265.
91. P. Moulinié, J. J. Charles, *69th Annu. Tech. Conf.—Soc. Plast. Eng.* (2011) 1100–1104.
92. A. Factor, "Degradation of bisphenol A polycarbonate by light and γ-ray irradiation" in *Handbook of Polycarbonate Science and Technology*, D. G. LeGrand, J. T. Bendler, eds. Marcel Dekker, New York, 2000.
93. V. M. Nace, EP 228525, filed October 31, 1985.
94. B. Elbert, E. Jahnke, EP 2568004, filed September 8, 2011.
95. J. Eiffler, R. J. Lee, US 5,362,783, filed June 8, 1993.
96. R. E. Lyon, M. L. Janssens, "Polymer flammability," Report DOT/FAA/AR-05/14 (May 2005).
97. M. M. Hirschler, "Regulations, codes, and standards relevant to fire issues in the United States" in *Fire Retardancy of Polymeric Materials*, Second Ed., C. A. Wilkie, A. B. Morgan, eds. pp. 587–670. CRC Press, Boca Raton, FL, 2010.
98. M. Rogunova, N. Sunderland, J. P. Mason, H. Franssen, B. Krauter, US2011/0071241, filed September 23, 2009.
99. S. V. Levchik, E. D. Weil, *J. Fire Sci.* **24** (2006) 137–151.
100. A. Factor, C. Orlando, *J. Polym. Sci.: Polym. Chem. Ed.* **18** (1980) 579–592.
101. J. Y. Chung, W. G. Paul, *58th Annu. Tech. Conf.—Soc. Plast. Eng.* (2000) 2125–2129.
102. S. I. Stoliarov, P. R. Westmoreland, *Polymer* **44** (2003) 5469–5475.
103. J. Green, *Plast. Eng.* **62** (2001) 173–271.
104. F. Brandstetter, V. Muench, H. Naarmann, K. Penzien, EP 58379, filed August 25, 1982.
105. T. Eckel, V. Buchholz, E. Wenz, DE 102007061762.
106. J.-P. Lens, L. Kagumba, M. A. Lebel, *PMSE Prepr.* **243** (2012) PMSE-140.
107. Z. Tang, Y. Li, Y. J. Zhang, P. Jian, *Polym. Degr. Stab.* **97** (2012) 638–644.
108. B. Schartel, U. Braun, U. Knoll, M. Bartholmai, H. Goering, D. Neubert, P. Pötschke, *Polym. Eng. Sci.* **48** (2008) 149–158.
109. S. Sun, Y. He, X. Wang, D. Wu, *J. Appl. Polym. Sci.* **116** (2010) 2451–2464.

110. H. Xu, S. Tang, L. Yang, W. Hou, *J. Polym. Sci.: Part B: Polym. Phys.* **48** (2010) 1970–1977.
111. T. A. P. Engels, B. A. G. Schrauwen, L. E. Govaert, H. E. H. Meijer, *Macromol. Mater. Eng.* **294** (2009) 114–121.
112. R. R. Gallucci, P. J. Hans, J. M. R. Janssen, M. R. Pixton, W. D. Mordecai, US 5,814,712, filed April 25, 1996.
113. P. Moulinié, EP 2,035,499, filed June 8, 2007.
114. W. Nouvertné, F. J. Gielen, P. Tacke, U. Grigo, B. Quiring, C. Lindner, EP 337206, filed March 30, 1989.
115. D. E. Henton, D. M. Naeger, F. M. Plaver, US 5,219,933, filed March 25, 1991.
116. B. Hager, D. Wittmann, E. Wenz, *66th Annu. Tech. Conf.—Soc. Plast. Eng.* (2008) 1301–1305.
117. S. Y. Hobbs, M. E. J. Dekkers, V. H. Watkins, *Polymer* **29** (1988) 1598–1602.
118. C. G. Cofer, D. E. McCoy, US2002/0108699, filed August 15, 2002.
119. E. H. Weber, M. L. Clingerman, J. A. King, *J. Appl. Polym. Sci.* **88** (2003) 112–122.
120. R. Greco, A. Sorentino, *Adv. Polym. Tech.* **13** (1994) 249–258.
121. G. Weber, J. Schoeps, *Angew. Makromol. Chem.* **136** (1985) 45–64.
122. G. Wildes, H. Keskkula, D. R. Paul, *Polymer* **40** (1999) 5609–5621.
123. G. S. Wildes, T. Harada, H. Keskkula, D. R. Paul, V. Janarthanan, A. R. Padwa, *Polymer* **40** (1999) 3069–3082.
124. D. H. Bolton, P. Moulinie, D. M. Derikart, N. Köhncke, US 6,670,420.
125. J. C. Lim, J.-K. Park, *J. Appl. Polym. Sci.* **95** (2005) 689–699.
126. A. Seidel, T. Eckel, H. Warth, US 7,186,787.
127. J. L. Derudder, A. A. Volkers, U.S. App. 20060004154.
128. E. Avtomonov, A. Seidel, T. Eckel, H. Eichenauer, U.S. App. 20090239991.
129. J. M. R. C. A. Santos, J. Guthrie, *J. Mat. Chem.* **16** (2006) 237–245.
130. W. Michaeli, H. Seidel, *Kunstst.* **99** (2009) 40–44.
131. L. Yan, R. Hufen, P. Moulinié, A. Karbach, S. Konrad, C.-L. Schultz, *65th Annu. Tech. Conf.—Soc. Plast. Eng.* (2007) 2355–2359.
132. F. Samperi, M. S. Montaudo, G. Montaudo, "Polycarbonates" in *Handbook and Engineering and Specialty Thermoplastics*, T. Sabu, P. M. Visakh, eds. pp. 493–528. John Wiley & Sons, Hoboken, NJ, 2011.
133. M. E. J. Dekkers, S. Y. Hobbs, V. H. Watkins, *J. Mat. Sci.* **23** (1988) 1225–1230.
134. P. Moulinié, S. Hobeika, R. Hufen, *66th Annu. Tech. Conf.—Soc. Plast. Eng.* (2008) 2087–2091.
135. S. Y. Hobbs, M. E. J. Dekkers, V. H. Watkins, *J. Mat. Sci.* **23** (1988) 1219–1224.
136. L.-T. Lee, E. M. Woo, *Coll. Polym. Sci.* **282** (2004) 1308–1315.
137. A. Yavari, A. Asadinezhad, S. H. Jafari, H. A. Khonakdhar, F. Boehme, R. Haessler, *Eur. Polym. J.* **41** (2005) 2880–2886.
138. L. U. Kim, J. H. Lee, C. K. Kim, *PMSE Prepr.* **95** (2006) 693–695.
139. R. W. Fonseca, P. H. T. Vollenberg, R. C. Cook, US2005/0143532, filed November 5, 2004.
140. J. P. Mason, US 5,491,179, filed November 23, 1994.
141. J. Y. J. Chung, J. P. Mason, US2010/0041831, filed August 15, 2008.
142. K. Nevalainen, J. Vuorinen, V. Villman, R. Suihkonen, P. Järvelä, J. Sundelin, T. Lepistö, *Polym. Eng. Sci.* **49** (2009) 631–640.
143. P. Pötschke, T. D. Fornes, D. R. Paul, *Polymer* **43** (2002) 3247–3255.
144. S. Pegel, P. Pötschke, G. Petzold, I. Alig, S. M. Dudkin, D. Lellinger, *Polymer* **49** (2008) 974–984.
145. G. Kasaliwal, A. Göldel, P. Pötschke, *J. Appl. Polym. Sci.* **112** (2009) 3494–3509.
146. X. Li, M. Sagal, US2012/0319031, filed June 15, 2011.
147. S.-H. Hwang, H. W. Park, Y.-B. Park, M.-K. Um, J.-H. Byun, S. Kwon, *Compos. Sci. Technol.* **89** (2013) 1–9.
148. Y. Zhao, D. Schiraldi, *Polymer* **46** (2005) 11640–11647.
149. S. C. Tjong, W. Jiang, *J. Appl. Polym. Sci.* **73** (1999) 2247–2253.
150. H. R. Hakimelahi, L. Hu, B. B. Rupp, M. R. Coleman, *Polymer* **51** (2010) 2494–2502.
151. Bayer MaterialScience AG Product Literature, "The injection molding of high-quality molded parts—Preparing the material: Drying," Publication PCS-1141e (July 17, 2008).
152. Bayer MaterialScience AG Product Literature, "Injection molding—Defects, causes, remedies," Publication MS00056040 (March 2013).
153. Bayer MaterialScience AG Brochure, "Makrolon® WB1239 water in good shape. Technical information source," Publication MS00041660 (October 2008).
154. Bayer Corporation Publication, "A processing guide for thermoforming," Publication KU-F7011 (May 1999).
155. "Lexan® sheet—Processing guide, forming, fabricating, finishing and decorating solid un-coated and coated sheet," Publication SABIC-SFS-6212 (2009).
156. E. Wenz, P. Moulinie, T. Eckel, V. Buchholz, D. Wittmann, B. Hager, F. J. Zaganiacz, US2007/0060678, filed September 14, 2005.

11 Liquid Crystalline Polymers*

Xiao Li and Haifeng Yu

CONTENTS

11.1 Introduction 369
11.2 Molecular Design of LCPs 370
    11.2.1 MCLCPs 370
    11.2.2 SCLCPs 371
11.3 Characterization Methods and Synthesis of LCPs 372
    11.3.1 POM 372
    11.3.2 DSC 373
    11.3.3 XRD 374
    11.3.4 Synthesis of LCPs 375
11.4 Physical and Chemical Properties 375
    11.4.1 Viscosity 375
    11.4.2 Mechanical Properties 375
11.5 LCBCs 376
11.6 Photoresponsive LCPs 377
11.7 SLCPs 379
    11.7.1 Hydrogen-Bonded SLCPs 379
    11.7.2 Halogen-Bonded SLCPs 380
    11.7.3 Other SLCPs 382
11.8 LCEs 382
11.9 Conclusions 384
References 384

11.1 INTRODUCTION

Liquid crystalline polymers (LCPs) are the combination of two independent disciplines, liquid crystals (LCs) and polymers. In 1888, Austrian botanist Friedrich Reinitzer observed two melting temperatures when he studied the compound of cholesteryl benzoate, which gave rise to an intermediate state in addition to crystalline and isotropic states and opened the gate of research on LCs. LCs and polymers were discovered at around the same time, but polymers have become one of the most important materials in human life. However, LCs saw little usage until the early 1970s. Since then, they have attracted wide interest around the world and been applied, especially in the display industry, to such consumer items as portable TVs, notebooks, smart cell phones, and desktop computers [1].

    The first LCP discovered by humans was the tobacco mosaic virus in solution in the 1940s. In 1964, the first commercialized LCP was invented by a female scientist working in DuPont, with the famous trade name Kevlar, which is composed of aromatic polyamides. The Kevlar fiber is extremely strong with a high modulus, is relatively lightweight, and has high thermal resistivity and chemical resistivity. Now it is widely applied in the fields of aerospace, shipbuilding, friction materials, and so on. Although Kevlar has been successfully implemented in many products and the study of LCPs covers a variety of fields, unfortunately, most LCPs are still under research in the laboratory.

---

* Based in parts on the first-edition chapter on liquid crystalline polymers.

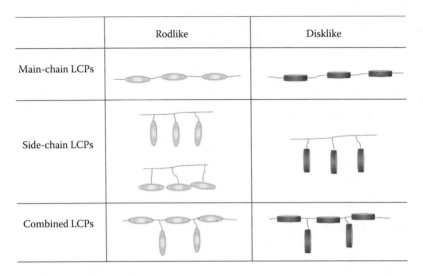

FIGURE 11.1 Classification of liquid crystalline polymers.

LCPs can be divided into natural LCPs, including biomacromolecules such as cellulose and its derivatives, polypeptides, DNA and RNA, and synthetic LCPs. Synthetic LCPs are polymerized from LC monomers and exhibit the properties of both LCs and polymers. Generally, LCPs contain mesogens that are rodlike or disklike. In terms of molecular structure, LCPs are mainly divided into main-chain LCPs (MCLCPs), side-chain LCPs (SCLCPs), and combined LCPs; the classification is depicted in Figure 11.1. Furthermore, to realize the multiple functions of LCPs, liquid crystalline block copolymers (LCBCs), supramolecular LCPs (SLCPs), liquid crystalline elastomers (LCEs) and stimuli-responsive LCPs were synthesized in the laboratory and recently have become the hot spots in the field of LCPs.

LCPs show improved processability and mechanical properties compared with low-molecular-weight LC molecules; therefore, they can be used as high-modulus materials and high-strength composites. At the same time, LCPs keep the LC's characteristics as low mass molecules, such as the anisotropic physical properties and the low viscosity. These features render LCPs not only a kind of structural material but also a novel kind of smart material, which covers diverse application fields, including biomaterials, aeronautical materials, nanotechnology and photonic devices, and so on. In short, the research on LCPs is of great significance.

In this chapter, a general description of LCPs, including properties, synthesis, characterization, and research progress, will be presented.

11.2 MOLECULAR DESIGN OF LCPs

11.2.1 MCLCPs

The principles of the molecular design of MCLCPs have been satisfactorily clarified during the last decades. To form lyotropic LCPs, the rules should be as follows: (1) they should contain a certain length of rigid-rod structures, and (2) the solubility of LCPs in the solvent should be higher than the critical concentration. For thermotropic LCPs, there are two essential principles: (1) the polymer melting point must be low enough to grant a reasonably wide processing temperature range, and (2) the macromolecules must nevertheless possess sufficient anisotropy.

So far, two kinds of LCPs, aromatic polyamides and benzazole polymers (PBZ), have been synthesized and shown to display the lyotropic LC behaviors. Aromatic polyamides are typical rigid-chain LCPs and occupy important positions in lyotropic LCPs. In 1968, S. Kwolek, working

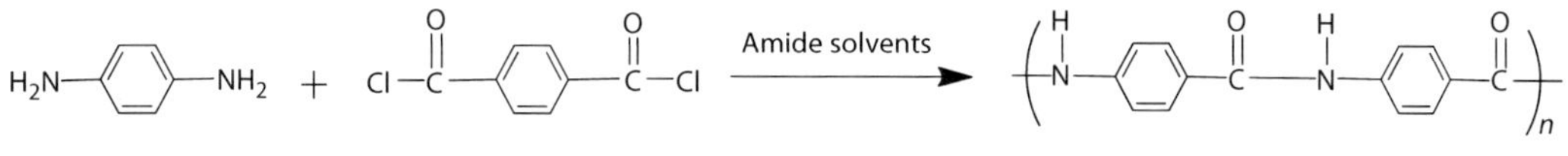

FIGURE 11.2 The synthesis of poly(4-4′-phenylene terephthalamide).

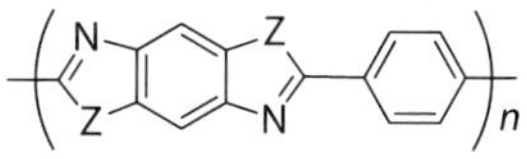

SCHEME 11.1 Molecular structures of benzazole polymers.

at DuPont de Nemours discovered, that aromatic polyamides of the poly(4-benzamide) (PBA) type can be obtained as mesophasic, spinnable solutions in strongly interacting solvents such as concentrated sulfuric acid or N,N-dialkylamides with added LiCl [2]. Another important lyotropic LCP is poly(4-4′-phenylene terephthalamide) (PPTA), which has become popular as a high-modulus, high-strength fiber with the trade name Kevlar. PPTA can be easily synthesized by polycondensation, as shown in Figure 11.2.

PBZ has recently been used to produce high-modulus fibers, too. Different from aromatic polyamide, the main chain of PBZ is fully composed of cyclic structures. The molecular structure is demonstrated in Scheme 11.1.

When Z is a different atom, this structure represents a different molecule [3].

When Z = S, it is poly(*p*-phenylenebenzobisthiazole), PBT.
When Z = O, it is poly(*p*-phenylenebenzobissoxazole), PBO.
When Z = N, it is poly(*p*-phenylenebenzobisimidazole), PBI.

In addition to synthetic LCPs, several natural macromolecules (like cellulose and its derivatives), DNA, RNA, and polypeptides also show lyotropic LCs [4].

Thermotropic MCLCPs are mainly composed of polyesters. However, as temperature goes up, polyesters usually decompose before they reach their melting points. Thus, modifications have to be made to lower their melting temperatures so that the polymers can be kept in the LC state before decomposition. The Nobel laureate de Gennes predicted that the addition of a flexible segment between the rigid units should afford semiflexible polymers exhibiting thermotropic LC features [5]. Therefore, the most common strategies to reduce the melting point are (1) the introduction of aliphatic spacers linking the mesogenic groups and (2) the synthesis of random copolyesters containing appropriately modified comonomers.

When aliphatic spacers are introduced into LC materials, the longer the spacer length is, the lower the isotropic temperature would be. Besides, an odd–even effect is observed in the thermal properties of LCPs. Some of the polymers even show different phases; when n is an odd number, it is in the nematic phase, and if n turns even, the polymer shows smectic characteristics [6].

11.2.2 SCLCPs

The long-range order of MCLCPs lies in the orientation of the whole polymer chain. However, the order of SCLCPs mainly depends on the alignment of mesogenic units instead of the polymer main chain. For example, some of the SCLCPs show photoelectric properties as the low-molecular-weight LC molecules do. Therefore, SCLCPs can hardly be used as high-modulus fibers, but they are quite meaningful for functional materials.

Classified by the linking ways of main-chain and mesogen units, SCLCPs can be divided into two parts, end-on and side-on SCLCPs. We can obtain SCLCPs in two ways: (1) introduction of

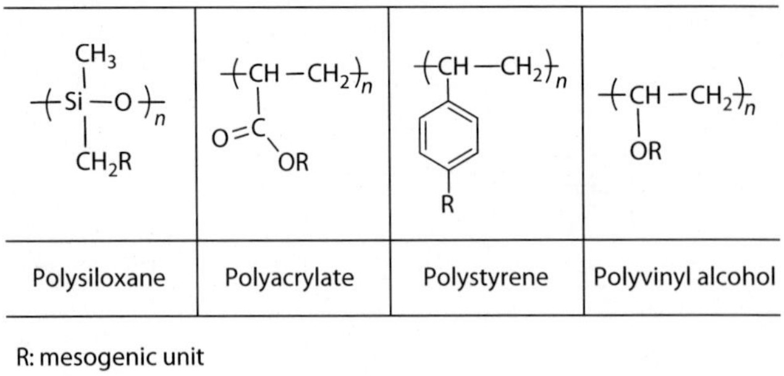

Polysiloxane	Polyacrylate	Polystyrene	Polyvinyl alcohol
$\!-\!(Si\!-\!O)_n\!-\!$ with CH_3 and CH_2R	$\!-\!(CH\!-\!CH_2)_n\!-\!$ with $O=C\!-\!OR$	$\!-\!(CH\!-\!CH_2)_n\!-\!$ with phenyl-R	$\!-\!(CH\!-\!CH_2)_n\!-\!$ with OR

R: mesogenic unit

FIGURE 11.3 Main-chain polymers of side-chain liquid crystalline polymers.

side-chain mesogens onto the main-chain polymer via chemical reaction and (2) polymerization of monomers with mesogens on the side.

Three core parts influence the performance of SCLCPs: the main chain, side-chain mesogen units, and the spacer linking them together. Usually, four kinds of monomers are used as the main-chain polymer: acrylics, vinyl alcohol, siloxane, and styrene. The molecular structures of the four kinds of polymers are shown in Figure 11.3. Better flexibility of the main-chain polymer leads to lower glass transition temperature (T_g) and a wider range of LC temperatures. Recently, polysiloxane has attracted much research interest for its good flexibility. As the movement and orientation of the mesogens are constrained by main-chain polymers, a flexible spacer has to be introduced to avoid the influence of main-chain polymers on mesogens. As the spacer between the side chain and main chain becomes longer, the ordering of LCPs increases; for example, polymers take on the smectic phase instead of the nematic phase, the T_g is decreased, and the phase transition temperature becomes higher.

In 1987, famous Chinese chemist Q.F. Zhou proposed the concept of mesogen-jacketed LCPs (MJLCPs), as also called rigid SCLCPs, which are special side-on SCLCPs with very short spacers or without spacers [7]. Actually, in terms of structure, MJLCPs are similar to SCLCPs. However, they resemble MCLCPs in terms of rigidity and LC characteristics, since there is no spacer between the main chain and the rigid side chain, which renders the main chain to be straight. This means that the spacer between main-chain polymers and side-chain mesogen units seems unnecessary. The establishment of this model is of great significance to prepare SCLCPs with the properties of MCLCPs.

11.3 CHARACTERIZATION METHODS AND SYNTHESIS OF LCPs

The unique properties of LCPs are of great importance in scientific research. The most common characterization methods for LCPs are polarizing optical microscopy (POM), differential scanning calorimetry (DSC), and x-ray diffraction (XRD). Besides, nuclear magnetic resonance (NMR), Fourier-transform infrared (FTIR), and ultraviolet (UV)-Vis spectroscopy, florescence spectroscopy can be used to explore the molecular structure and the LC behaviors of LCPs. The three most common methods are discussed below in detail.

11.3.1 POM

To make sure that a sample shows an LC phase, the easiest and most direct method is POM. Through POM, one can find the textures of the sample and then speculate on the phase of LCPs. LC textures form because defects exist in the LCPs, and perfect orientation can only show a colored bright field. The usual textures of different phases are listed in Table 11.1.

For the nematic LC phase, the schlieren texture is often observed in thinner samples, while threaded textures usually appear in thicker ones. As the ordering of LCP increases, identifying the

Liquid Crystalline Polymers

TABLE 11.1
The Usual Textures of Different LCP Phases

Liquid Crystal Phase	Possible Textures
Nematic	Threaded, schlieren, droplets
Smectic	Fan texture, broken focal conic, bâtonnets
Cholesteric	Fingerprint, planar texture, focal conic

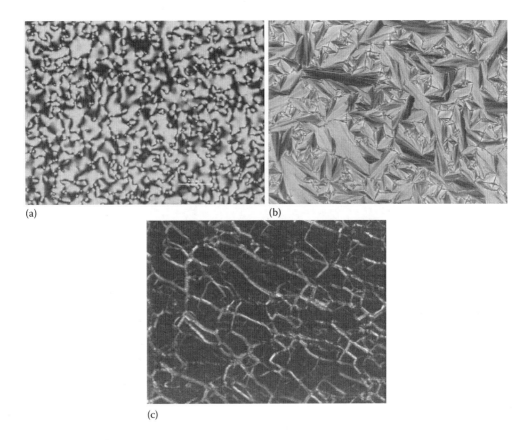

FIGURE 11.4 Typical textures of liquid crystal materials. (a) Schlieren texture of nematic phase; (b) focal conic texture of smectic A phase; (c) planar texture of cholesteric phase.

LC phase becomes more difficult. For instance, POM textures cannot provide enough evidence for the smectic LC state; thus, XRD detection is often needed to precisely determine the phase of LC materials. Figure 11.4 shows some typical POM textures of LCPs.

11.3.2 DSC

DSC curves indicate the thermal behavior of LCPs, such as the phase transition temperature and the enthalpy change during the transitions. Sometimes, it can be confusing, as the second-time DSC curve of the same sample frequently differs from the first one, resulting from thermal history. For example, heating or cooling at a rapid rate may lead to incomplete transition of different phases. Yu synthesized one diblock copolymer, PEO-b-PM$_6$ABOC$_2$, which shows a nematic LC phase [8].

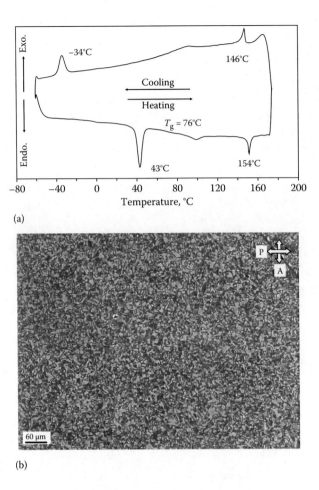

FIGURE 11.5 Characterization of the diblock copolymer PEO-*b*-PM$_6$ABOC$_2$ showing nematic LC phase. (a) DSC curve; (b) POM image at 120°C.

Figure 11.5a is the DSC curve of this polymer. Upon heating, an obvious T_g appears at 76°C, and two peaks at 43°C and 154°C represent phase transition temperatures, corresponding to the melting point of polyethylene oxide (PEO) and the LC-to-isotropic phase transition temperature. On cooling, the clearing temperature appears at 146°C, which is lower than 154°C due to the overcooling effect. Although the transition temperature is clear on the DSC curve, we cannot decide which phase transition occurs at this temperature. Thus, POM is helpful to confirm the exact phase, just as Figure 11.5b shows.

11.3.3 XRD

XRD is an important tool in crystal engineering. From the XRD results, we can get the degree of crystallization, crystal lattice, crystal size, molecular orientation, and so on. With this information, the exact phase of LCPs can be determined, and it is especially effective in identifying smectic LC phases. The principle of XRD is widely known as Bragg's law, which is simply depicted in Figure 11.6. For smectic LCPs, XRD gives the diffraction angle, and the layer distance can be calculated according to Bragg's law. Comparing the calculated results with the molecular length, we can acquire the tilt angle of the LC molecules and, thereby, the LC phases. In addition, small-angle x-ray scattering (SAXS) and wide-angle x-ray scattering (WAXS) are now frequently used to get more precise results.

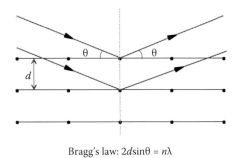

Bragg's law: $2d\sin\theta = n\lambda$

FIGURE 11.6 Illustration of Bragg's law.

11.3.4 Synthesis of LCPs

Benefiting from the development of materials chemistry and physics, diverse LCPs have been designed and synthesized. Several methods such as anionic, cationic, and free-radical polymerization have been used to synthesize LCPs. Besides, LCBCs can be made through atom transfer radical polymerization (ATRP), reversible addition–fragmentation chain transfer (RAFT), and nitroxide-mediated polymerization (NMP). In addition, LCs can be obtained by supramolecular self-assembly through hydrogen bonds or halogen bonds [9].

11.4 PHYSICAL AND CHEMICAL PROPERTIES

11.4.1 Viscosity

The biggest difference between LCPs and general polymers is that LCPs are highly ordered and have high fluidity. Compared with general polymers of the same molecular weight, LCPs show much lower viscosity at their LC temperatures. In addition, at the transition temperature from isotropic phase to LC phase, the viscosity decreases obviously. Viscosity fluctuates when molecular conformation, such as molecular weight, molecular length, molecular polarization, number of π electrons, and conjugation extent, changes. Commonly, as these factors go up, the viscosity increases. Besides, the viscosity is also affected by some external factors, including the concentration of lyotropic LCPs, the ambient temperature, and the shearing rate. Specially, the viscosity decreases with increasing shear rate, which is known as the shear thinning effect and is advantageous for applications. For instance, the filling of molded plastic products is more convenient, and the molded shape is sharper and better.

11.4.2 Mechanical Properties

Among all the mechanical properties of LCPs, elasticity is the most attractive one. The same as the low-molecular-weight LC, three kinds of deformation exist in LCPs: splaying, twisting, and bending, with the elastic constants of K_{11}, K_{22}, and K_{33}, respectively. For small molecular LCs, the elastic constants depend on the chemical structure of the molecule. However, when it comes to polymers, the molecular chain length and the order parameter could also contribute to them. Take K_{33}, for example, which increases rapidly with a higher order parameter.

In the development of composite materials, the compounding of LCPs with other materials has aroused much interest. The addition of a small amount of LCPs into other polymers improves the fluidity and processability of these materials. Besides, compounding LCPs with other low-cost polymers can reduce the costs and overcome some disadvantages of anisotropic materials at the same time. For example, the mixture of nylon and LCPs possesses a low coefficient of thermal expansion.

11.5 LCBCs

Generally, LCBCs are the integration of LCPs and other amorphous or non-LC polymers. Different from random copolymers, the polymerization degree of LCBCs is better controlled by living radical polymerization, such as ATRP, RAFT, and so on. Usually, the component polymers of LCBCs are immiscible; thus, the microphase separation often occurs in their bulk films. The phase-segregated morphology varies from spheres and cylinders to lamellae phases with the increase of the volume ratio of the LC component; the morphology of microphase separation of liquid crystalline diblock copolymers is depicted in Figure 11.7 [10]. Furthermore, the interaction between the microphase separation and LC ordering, known as supramolecular cooperative motion (SMCM), is of great importance in the formation of the periodically ordered nanostructure, as shown in Figure 11.8.

It has been reported that the LCBC composed of hydrophilic PEO and hydrophobic polymethacrylate containing an azobenzene moiety in the side chain can easily form PEO cylinders in thin films [11]. Experiments have also been performed to prove the stability of microphase separation. For example, when one PEO-based LCBC, shown in Scheme 11.2, is heated above the isotropic temperature, the cylinder phase turns into spheres [12]. Upon cooling below the clearing point, the spherical PEO turns back into cylinders. In addition, this order–order phase transition is repeatable.

However, long-range ordered nanostructures of the LCBCs cannot be formed without any treatment. Therefore, several methods, including thermal annealing, photoalignment, mechanical rubbing, and electric or magnetic field, have been developed to obtain the macroscopically aligned nanostructures. Upon thermal annealing, the PEO cylinders show an ordered alignment perpendicular to the substrate with hexagonal packing. Besides, mechanical rubbing methods can be used to fabricate parallel

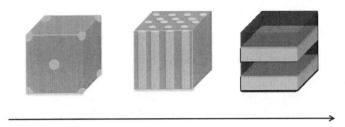

Increasing volume ratio of liquid crystalline components

FIGURE 11.7 Microphase separation of liquid crystalline block copolymers.

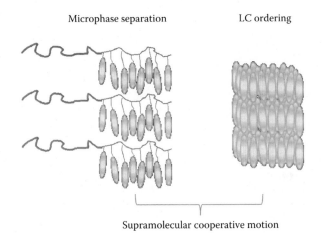

Supramolecular cooperative motion

FIGURE 11.8 Supramolecular cooperative motion in liquid crystalline block copolymers.

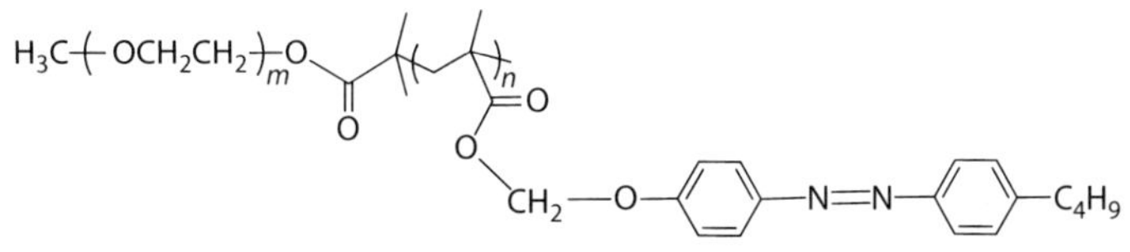

Liquid Crystalline Polymers

[Structure: H₃C-(OCH₂CH₂)ₘ-O-C(=O)-[CH₂-C]ₙ-C(=O)-O-CH₂-O-C₆H₄-N=N-C₆H₄-C₄H₉]

SCHEME 11.2 Molecular structure of PEO-based liquid crystalline block copolymers.

patterning in the PEO-based LCBC, and all of the cylinders are oriented along the rubbing direction [13]. The introduction of light-responsive moieties such as azobenzenes into the LCBCs provides the microphase separation with a photo-controllable feature because azobenzenes can act as both chromophores and mesogens [14]. Under the exposure of linearly polarized light, PEO nanocylinders can be aligned perpendicularly to the polarization direction, which is easily obtained in arbitrary areas.

The regularly aligned nanostructures can be used as nanotemplates to prepare a well-ordered array of all kinds of nanoparticles, such as Ag nanoparticles [15], Au nanoparticles [16], and SiO₂ nanorods [17]. All of these indicate the potential application of LCBCs in nanotechnology and functional materials.

11.6 PHOTORESPONSIVE LCPs

Recently, the fabrication of smart functional materials with controlled properties by external stimuli has attracted much attention [18]. Among these external stimuli, light is especially advantageous for its easily obtained supply, its being noncontact, and its reversible method [19]. Recently, several molecular structures have been used as photoresponsive moieties, including azobenzenes, spiropyrans, stiff stilbene, dithienylethene, and so on. Photoresponsive properties usually contain photoisomerization, photochemical phase transitions, photoalignment, and photoinduced cooperative motion. Azobenzene is becoming one of the most commonly adopted chromophores in photoresponsive materials for its reversible photoisomerization, and it can act as both a mesogen and a photoresponsive group [9]. As shown in Figure 11.9a, the trans form of azobenzene is a rigid rodlike structure, which easily shows the LC phase, while the cis form is isotropic. This trans/cis isomerization could lead to a photochemical phase transition, for example, from an ordered structure to a disordered structure, as depicted in Figure 11.9b.

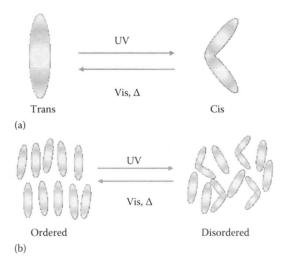

FIGURE 11.9 Photoresponsive properties of azobenzene LCs. (a) Photoisomerization and (b) the photochemical phase transitions.

Another important feature of photoresponsive LCPs is photoalignment. According to Weigert's effect, azobenzenes with their transition moments parallel to the polarization direction of linearly polarized light can be easily activated to the excited states and then follow the trans/cis isomerization. However, if the azobenzene molecules are perpendicular to the polarization direction, the molecules become inert, as shown in Figure 11.10a [20]. Thus, when linearly polarized light is irradiated to azobenzene molecules, the molecules will be aligned perpendicularly to the polarization direction of the linearly polarized light. Based on this effect, shown in Figure 11.10b, the photo-triggered molecular cooperative motion is introduced into LC copolymers. When the photoinert mesogens exist in copolymers with photoresponsive mesogens, they can be realigned even though they do not absorb the light. First, the photoresponsive mesogens are aligned following Weigert's effect, and then the photoinert mesogens realign along the direction of the ordered photoresponsive ones. Usually, only a small amount of chromophores (e.g., 1%) can drive the whole system to an ordered state upon irradiation of linearly polarized light. Molecular cooperative motion is quite useful in light-driven actuators and devices since there is a magnification of the energy input [21–25].

Recently, a newly developed approach to control the LC alignment has been reported by Wang and coworkers [26]. In this work, highly aligned carbon nanotubes (CNTs) are used to drive the cross-linking LCPs to align along the length of the CNTs. The use of CNTs can improve the mechanical properties and electric conductivity. In addition, single wall carbon nanotubes (SWNTs) can effectively absorb heat and then transform light into thermal energy [27].

Photoresponsive LCPs have been applied in diverse fields, such as photo-actuators and photonic devices, nanotemplates and nanostructures, light-responsive micelles, and so on. When the LCPs are amphiphilic, the hydrophobic block and the hydrophilic block could self-assemble into micellar aggregates of various morphologies, including star micelles, crew-cut micelles, rods, and vesicles [28], These micellar aggregates have wide applications in various fields, including catalysis, templates, logic gates, sensors, and so on. But one of the most popular applications should be their potential use for controlled drug delivery [29]. So far, azobenzene is the most popular motif in photoresponsive LCPs, as the apolar trans isomer can be converted to the polar cis one upon UV light irradiation and this is reversible upon irradiation of visible light [30]. Zhao and coworkers have

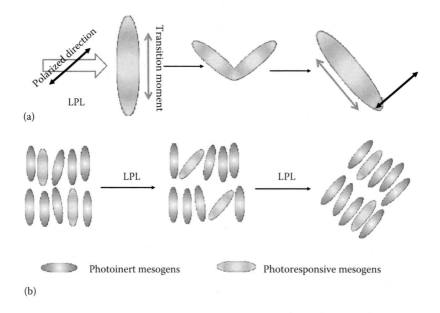

FIGURE 11.10 Photoresponsive properties of azobenzene-based LCPs. (a) Photoalignment and (b) molecular cooperative motion.

designed a micellar system that can be disrupted by UV irradiation and reform after exposure to visible light based on the amphiphilic LCPs [28,31–33].

Another important application for photoresponsive LCPs is holographic grating. Usually, the phase separation of polymer blends seriously scatters visible light, leading to a decrease in diffraction efficiency, which is one of the most important parameters of holographic gratings. Decreasing the size of phase separation to the nanoscale can eliminate the scattering effect, thus increasing diffraction efficiency. It has been reported that the surface modulation of holographic grating can be enhanced by controlling the nanoscale microphase separation process [34]. Yu et al. synthesized a series of PMMA-based diblock copolymers and diblock random copolymers with azobenzene as the mesogen in side chain. All of them showed photoinduced alignment and were subjected to record holographic gratings, although they show different behavior, as the content of mesogenic block varies. When the mesogenic content in block copolymers becomes lower, the photoinduced mass transportation is prohibited, leading to lower surface relief than homopolymers [35].

11.7 SLCPs

11.7.1 Hydrogen-Bonded SLCPs

Supramolecular chemistry has become a hot spot in scientific research since 1987, when the Nobel Prize in Chemistry was awarded to Donald J. Cram, Jean-Marie Lehn, and Charles J. Pedersen in recognition of their outstanding work in this area. As Lehn described, supramolecular chemistry is the chemistry of the intermolecular bond, covering the structures and functions of the entities formed by the association of two or more chemical species [36]. Different from covalent bonds, the intermolecular bonds include weaker and reversible bonds, including hydrogen bonding, halogen bonding, π–π interaction, metal coordination, and ionic bonding [37]. As a result, diverse SLCPs have been developed through these interactions.

In 1989, Kato et al. first reported supramolecular LCs obtained by the interaction of hydrogen bonding. He constructed one novel LC system by intermolecular bonding between two independent components, a pyridine ring and a carboxylic moiety. The dimers show two LC phases, nematic and smectic phases [38]. After that, disklike LC supramolecules, trimers and bifunctional hydrogen-bonded LCs, were achieved, respectively. Paleos reviewed the molecular design and synthesis of hydrogen-bonded LCPs before 2000 in detail [39].

Following the first supramolecularly hydrogen-bonded LC between pyridines and carboxylic moieties, diverse functional materials based on SLCPs have been explored. Zhao and coworkers first synthesized one homopolymer and block copolymer bearing azopyridine moieties in the side chain [40]. Scheme 11.3 shows the molecular structure of these polymers [41]. Upon addition of carboxylic acids, these amorphous polymers can easily self-assemble into an LC phase through hydrogen bonding, and a cholesteric LC phase is accessible if a chiral acid is added. With the photoactive property of azobenzenes and the self-assembled properties of pyridine, azopyridine polymers could realize some novel photoresponsive functions.

Yu et al. fabricated a novel kind of SLCP microparticle with one amorphous azopyridyl polymer and a dicarboxylic acid of different alkyl chain length and then explored their morphology, their photoresponsive characteristics, and the stability of the hydrogen bond under photo-irradiation. The morphology of nanoparticles can be controlled by the ratio of the hydrogen-bond donor to acceptor, as shown in Figure 11.11. Having both the photoresponsive LC feature and intriguing morphologies, these SLCP microparticles are promising for optoelectronic applications [42,43]. Furthermore, hydrogen bonding can be used in multiresponsive organogels and the fabrication of gratings. Chen and coworkers synthesized one carboxylic azobenzene polymer that could form a multiresponsive gel in dimethyl sulfoxide (DMSO) through hydrogen bonding and H aggregations [44]. Figure 11.12 sketches the multiresponsive behavior of the supramolecular organogels.

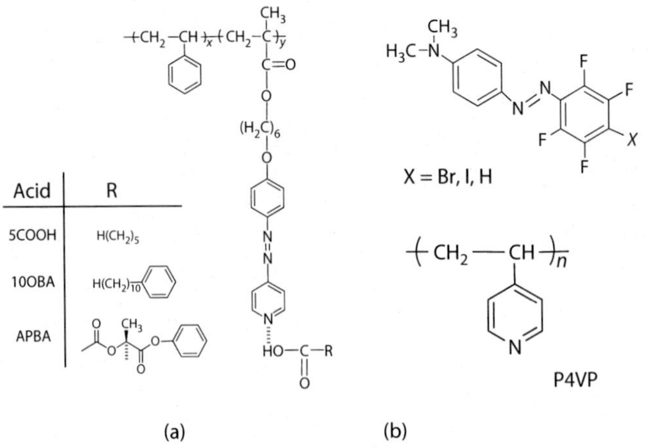

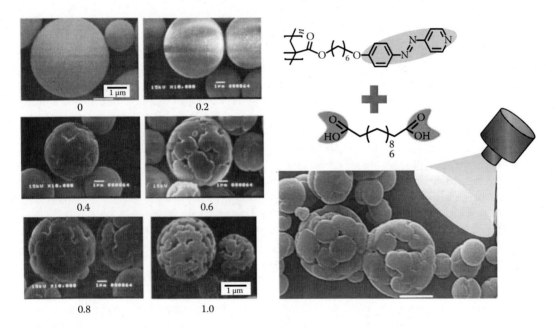

SCHEME 11.3 Typical examples of supramolecular liquid crystalline polymers.

FIGURE 11.11 Microparticles fabricated with supramolecular LCPs showing controlled morphologies and photoresponsive behaviors.

Seki et al. reported one example of surface-relief grating formation in a hydrogen-bonded SLCP system. It is well known that azobenzenes are very important in the mass transport process upon holographic recording, but they are not advantageous when it comes to applications, as they strongly absorb light. Taking advantage of the reversible property of the hydrogen bond, the azobenzenes can be easily removed from the surface-relief grating, and at the same time, the morphology of the surface-relief grating is unaffected [45].

11.7.2 Halogen-Bonded SLCPs

Analogous to hydrogen bonding, halogen bonding is one of the noncovalent interactions, where halogen atoms (I, Br, Cl, F) function as electrophilic species [46]. Although hydrogen bonding is the

Liquid Crystalline Polymers

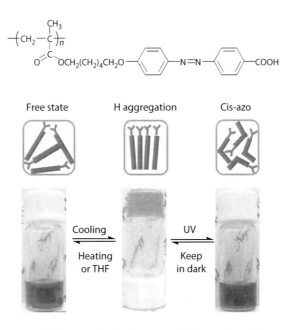

FIGURE 11.12 Multiresponsive hydrogen-bonded organogel of one carboxylic azo polymer.

most frequently used noncovalent interaction to fabricate supramolecularly self-assembled systems, halogen bonding demonstrates superior properties over hydrogen bonding. As a result of the strong electric-withdrawing property of halogen atoms, especially the iodine atom, halogen bonding is more directional and stronger than hydrogen bonding. Besides, the interaction strength can be well tuned by choosing different halogen atoms [47,48]. The strongest interaction of halogen bonding is found to be the interaction of N-I [49]. Above all, halogen bonding is a promising noncovalent interaction in the construction of organic self-assembled systems.

Bruce first reported halogen-bonded LCs formed between an alkoxystibazole and various substituted phenols [50]. Although neither of the components shows mesogenic properties, the complexes of iodopentafluorobenzene and iodine with a range of 4-alkoxystilbazoles showed LC feature [51]. Recently, Yu et al. have successfully obtained bromine-bonded LCs with high mesophase stability [52], which further developed the halogen-bonded LCs. In addition, reversible photoinduced phase transition behaviors were first reported for the halogen-bonded LC systems, as shown in Figure 11.13.

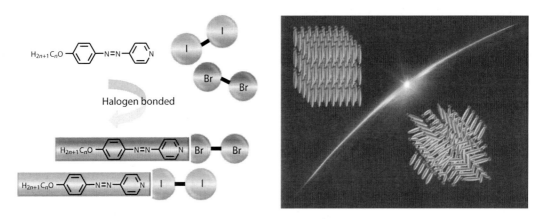

FIGURE 11.13 Fabrication of halogen-bonded LCs and their reversible photoinduced phase transition behaviors.

Pierangelo and coworkers synthesized almost 100 kinds of halogen-bonded complexes and explored the LC characteristic and thermal behaviors of these complexes [53].

Though halogen bonding is the least exploited among all kinds of noncovalent interactions, its applications have been widely reported, including crystal engineering [46,54], microelectric devices, holographic storage, and biological systems. Arri and coworkers fabricated surface-relief grating using halogen-bonded SLCPs, and the molecular structure is shown in Scheme 11.3b. Compared with hydrogen-bonded systems, halogen-bonded systems show better performance in forming surface-relief grating. It was reported that both the diffraction efficiency and the modulation depth obtained were better in halogen-bonded complexes [55].

11.7.3 Other SLCPs

Metallomesogens are metal complexes exhibiting mesogenic properties. Though research on photoactive LCs has been popular recently, few reports about metallomesogens that combine metal ions and organic ligands have been found [56]. Bruce introduced Ag into organic mesogen ligands, forming ionic mesophases that are stable at room light [57]. Besides, the alkoxystilbazole complexes of different silver salts showed diverse mesophases, including nematic, smectic, columnar, and even cubic phases [58]. The LC and photophysical properties of some of the azopyridine containing silver mesogens were explored in detail [59].

Pd pincer surfactant is also reported to form mesomorphic supramolecular complexes with pyridine polymers. Davidi and coworkers fabricated mesomorphic comb polymers containing poly (2-vinyl pyridine) (P2VP) and Pd–SCS pincer, and explored the formation mechanism as well as the dynamic behavior of the interaction [60,61]. Furthermore, complexes of Pd–SCS pincer and P2VP-b-PS copolymers could form hierarchical structures due to selective separation, which is promising in the construction of hierarchical arrays of inorganic nanoclusters [62].

11.8 LCEs

By cross-linking, LCEs combine LCs' ordering property and the mechanical properties of elastomers. If a chromophore is induced into LCEs, a shape and volume change in LCEs can be produced thermally or photochemically, which could make it possible to convert the light energy directly into mechanical work [63]. De Gennes had predicted this possibility to induce a large deformation of LCE by phase transition earlier [5,64].

Basically, LCEs show thermoelastic properties: when LCEs are heated to the isotropic phase, they contract; on the contrary, they expand upon cooling below the phase transition temperature. Take azobenzenes, for example; the distance between the 4- and 4′-carbons in benzene rings is 9.0 Å in trans form and 5.5 Å in cis form. Eisenbach first reported the contraction of cross-linked amorphous polymers containing azobenzenes [65]. Then the contraction ratio was significantly enhanced by reversible photochemical phase transition. Finkelmann succeeded in obtaining a reversible contraction of 20% in azo-containing cross-linking polymers [66]. Furthermore, the films of these LCEs were fabricated and showed rapid contraction of 18% upon UV irradiation [67].

However, the deformation of LCEs is not limited to 2-D contraction and expansion; 3-D movements of LCEs have provoked the interest of scientists. So far, several unique ways of 3-D movement have been reported, including oscillating, twisting, swimming, and rotation, which can be used for robotic arms, and actuators thus provide a new approach to convert light energy directly into mechanical energy [68,69]. Ikeda et al. first reported photoresponsive LCE films containing azobenzenes showing the bending and unbending behavior upon UV irradiation and proposed the bimetal mechanism of this bending performance [70]. As the volume contraction of the azobenzene molecule is induced only on the surface area upon UV irradiation, the film will bend toward the incident light. Intriguingly, a further experiment was carried out and found that the bending direction is affected by the mesogenic alignment in LCE films. For example, the homeotropically aligned

Liquid Crystalline Polymers 383

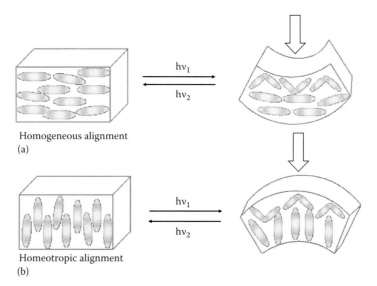

FIGURE 11.14 Photoinduced deformation of LCEs with mesogens in (a) homogeneous and (b) homeotropic alignment.

LCE films bend away from the incident light, as depicted in Figure 11.14 [71]. Based on this bending and unbending movement, diverse movements of LCEs are also achieved. Yamada and coworkers first reported light-driven motors with laminated films composed of LCEs, which could roll into cycles. The motor, which is simply homemade and driven totally by light without a battery or gears, can convert the light into mechanical force durably [68]. What is more, novel LCE films have been prepared that could move like inchworms or robotic arms upon photoirradiation [69].

Recently, Yu successfully fabricated polymer-dispersed liquid crystal (PDLC)-like films with wrinkled LCP microparticles as dispersed phases and polyvinyl alcohol (PVA) as continuous phases, showing photoinduced deformation behaviors like LCEs, as shown in Figure 11.15 [72]. Although the photomechanical details have not been clarified yet, it is expected that almost all the general polymer materials can be used for fabricating these PDLC-like films with photomechanical features. This demonstrates that such kinds of soft materials coupling light and mechanical responses together are theoretically unlimited. Numerous applications, from nanoscale to macroscopic, are

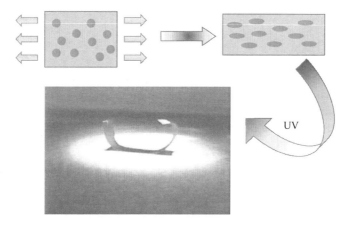

FIGURE 11.15 Photomechanical response of PDLC-like films fabricated with wrinkled LCP microparticles coupling with mechanical stretching.

possible, especially where an efficient power supply to mechanical motions is battery-free and noncontact.

Although the movements of these LCE films are charming and promising in application, there are still problems to be solved, such as responsive time, photodriven mechanism, and deformation types [73]. Thus, further research should be carried out to satisfy the requirements of practical applications.

11.9 CONCLUSIONS

LCPs, combining the properties of both polymers and LCs, are promising for industrial processes and functional materials. Through molecular design, we can easily change the molecular structures according to actual conditions. Therefore, diverse LCPs have been invented and developed in laboratories all over the world, demonstrating the potential of LCPs in future applications. The research on LCPs is a prolific field in scientific endeavors, and it expected that more functions and further applications of LCPs will become a reality in the near future.

REFERENCES

1. X. Wang and Q. F. Zhou, *Liquid crystalline polymers*, World Scientific, Singapore, 2004.
2. S. L. Kwolek, P. W. Morgan and J. R. Schaefgen, Liquid crystalline polymers, in *Encyclopedia of Polymer Science and Technology*, Vol. 9 (H. Mark, ed.), Wiley, New York, 1987.
3. Q. J. Zhang, *An introduction to liquid crystalline polymers*, Press of University of Science and Technology of China, Beijing, 2013.
4. H. Gao, *Liquid crystal chemistry*, Tsinghua University Press, China, 2011.
5. P. G. De Gennes, Physique moléculaire, *C. R. Acad. Sci. B*, 281: 101 (1975).
6. W. R. Krigbaum, J. Watanabe and T. Ishikawa, Investigation of the mesophase properties of polymers based on 4,4′-dihydroxybiphenyl, *Macromolecules*, 16: 1271 (1983).
7. X. F. Chen, Z. Shen, X. H. Wan, X. H. Fan, E. Q. Chen, Y. Ma and Q. F. Zhou, Mesogen-jacketed liquid crystalline polymers, *Chem. Soc. Rev.*, 39: 3072 (2010).
8. H. F. Yu, T. Kobayashi and G. Hu, Macroscopic control of microphase separation in an amphiphilic nematic liquid-crystalline diblock copolymer, *Polymer*, 52: 1154 (2011).
9. H. F. Yu, Photoresponsive liquid crystalline block copolymers: From photonics to nanotechnology, *Prog. Polym. Sci.*, 39: 781 (2014).
10. H. F. Yu, T. Kobayashi and H. Yang, Liquid-crystalline ordering helps block copolymer self-assembly, *Adv. Mater.*, 23: 3337 (2011).
11. S. A. Boyer, J. P. E. Grolier, H. Yoshida and T. Iyoda, Effect of interface on thermodynamic behavior of liquid crystalline type amphiphilic diblock copolymers, *J. Polym. Sci. Part B: Polym. Phys.*, 45: 1354 (2007).
12. H. F. Yu, J. Li, T. Ikeda and T. Iyoda, Macroscopic parallel nanocylinder array fabrication using a simple rubbing technique, *Adv. Mater.*, 18: 2213 (2006).
13. H. F. Yu, J. Li, T. Iyoda and T. Ikeda, Control of regular nanostructures self-assembled in an amphiphilic diblock liquid-crystalline copolymer, *Mol. Cryst. Liq. Cryst.*, 478: 271 (2007).
14. H. F. Yu, T. Iyoda and T. Ikeda, Photoinduced alignment of nanocylinders by supramolecular cooperative motions, *J. Am. Chem. Soc.*, 128: 11010 (2006).
15. J. Li, K. Kamata, S. Watanabe and T. Iyoda, Template- and vacuum-ultraviolet-assisted fabrication of Ag-nanoparticle array on flexible and rigid substrates, *Adv. Mater.*, 19: 1267 (2007).
16. S. Watanabe, R. Fujiwara, M. Hada, Y. Okazaki and T. Iyoda, Specific recognition of nanophase-separated surfaces of amphiphilic block copolymers by hydrophilic and hydrophobic gold nanoparticles, *Angew. Chem. Int. Ed.*, 46: 1120 (2007).
17. H. F. Yu and T. Kobayashi, Fabrication of stable nanocylinder arrays in highly birefringent films of an amphiphilic liquid-crystalline diblock copolymer, *ACS Appl. Mater. Interfaces*, 1: 2755 (2009).
18. J. F. Xu, Y. Z. Chen, D. Wu, L. Z. Wu, C. H. Tung and Q. Z. Yang, Photoresponsive hydrogen-bonded supramolecular polymers based on a stiff stilbene unit, *Angew. Chem. Int. Ed.*, 52: 9738 (2013).
19. H. F. Yu, Recent advances in photoresponsive liquid-crystalline polymers containing azobenzene chromophores, *J. Mater. Chem. C*, 2: 3047 (2014).
20. H. F. Yu, Y. Naka, A. Shishido, T. Iyoda and T. Ikeda, Effect of recording time on grating formation and enhancement in an amphiphilic diblock liquid-crystalline copolymer, *Mol. Crys. Liq. Crys.*, 498: 29 (2009).

21. T. Ikeda, Photomodulation of liquid crystal orientations for photonic applications, *J. Mater. Chem.*, 13: 2037 (2003).
22. T. Ikeda and O. Tsutsumi, Optical switching and image storage by means of azobenzene liquid-crystal films, *Science*, 268: 1873 (1995).
23. T. Seki, Dynamic photoresponsive functions in organized layer systems comprised of azobenzene-containing polymers, *Polym. J.*, 36: 435 (2004).
24. Y. Zhao and J. He, Azobenzene-containing block copolymers: The interplay of light and morphology enables new functions, *Soft Matter*, 5: 2686 (2009).
25. N. Kawatsuki, Photoalignment and photoinduced molecular reorientation of photosensitive materials, *Chem. Lett.*, 40: 548 (2011).
26. H. Wang and F. Glorius, Mild rhodium (III)-catalyzed C–H activation and intermolecular annulation with allenes, *Angew. Chem. Int. Ed.*, 51: 7318 (2012).
27. X. Sun, W. Wang, L. Qiu, W. Guo, Y. Yu and H. Peng, Unusual reversible photomechanical actuation in polymer/nanotube composites, *Angew. Chem. Int. Ed.*, 51: 8520 (2012).
28. G. Wang, X. Tong and Y. Zhao, Preparation of azobenzene-containing amphiphilic diblock copolymers for light-responsive micellar aggregates, *Macromolecules*, 37: 8911 (2004).
29. I. W. Hamley, Nanotechnology with soft materials, *Angew. Chem. Int. Ed.*, 42: 1692 (2003).
30. O. Hamley, E. Poggi, J. F. Gohy and C. A. Fustin, Functionalized stimuli-responsive nanocages from photocleavable block copolymers, *Macromolecules*, 47: 183 (2013).
31. X. Tong, G. Wang, A. Soldera and Y. Zhao, How can azobenzene block copolymer vesicles be dissociated and reformed by light, *J. Phys. Chem. B*, 109: 20281 (2005).
32. F. D. Jochum and P. Theato, Thermo- and light responsive micellation of azobenzene containing block copolymers, *Chem. Commun.*, 46: 6717 (2010).
33. H. Wu, J. Dong, C. Li, Y. Liu, N. Feng, L. Xu and G. Wang, Multi-responsive nitrobenzene-based amphiphilic random copolymer assemblies, *Chem. Commun.*, 49: 3516 (2013).
34. H. F. Yu, K. Okano, A. Shishido, T. Ikeda, K. Kamata, M. Komura and T. Iyoda, Enhancement of surface-relief gratings recorded in amphiphilic liquid-crystalline diblock copolymer by nanoscale phase separation, *Adv. Mater.*, 17: 2184 (2005).
35. H. F. Yu, Y. Naka, A. Shishido and T. Ikeda, Well-defined liquid-crystalline diblock copolymers with an azobenzene moiety: Synthesis, photoinduced alignment and their holographic properties, *Macromolecules*, 41: 7959 (2008).
36. J. M. Lehn, Perspectives in supramolecular chemistry—From molecular recognition towards molecular information processing and self-organization, *Angew. Che. Int. Ed.*, 29: 1304 (1990).
37. G. V. Oshovsky, D. N. Reinhoudt and W. Verboom, Supramolecular chemistry in water, *Angew. Chem. Int. Ed.*, 46: 2366 (2007).
38. T. Kato and J. M. Frechet, A new approach to mesophase stabilization through hydrogen bonding molecular interactions in binary mixtures, *J. Am. Chem. Soc.*, 111: 8533 (1989).
39. T. Felekis, L. Tziveleka, D. Tsiourvas and C. M. Paleos, Liquid crystals derived from hydrogen-bonded supramolecular complexes of pyridinylated hyperbranched polyglycerol and cholesterol-based carboxylic acids, *Macromolecules*, 38: 1705 (2005).
40. L. Cui and Y. Zhao, Azopyridine side chain polymers: An efficient way to prepare photoactive liquid crystalline materials through self-assembly, *Chem. Mater.*, 16: 2076 (2004).
41. L. Cui, S. Dahmane, X. Tong, L. Zhu and Y. Zhao, Using self-assembly to prepare multifunctional diblock copolymers containing azopyridine moiety, *Macromolecules*, 38: 2076 (2005).
42. H. Liu, T. Kobayashi and H. F. Yu, Easy fabrication and morphology control of supramolecular liquid-crystalline polymer microparticles, *Macromol. Rapid Commun.*, 32: 378 (2011).
43. H. F. Yu, H. Liu and T. Kobayashi, Fabrication and photoresponse of supramolecular liquid-crystalline microparticles, *ACS Appl. Mater. Interfaces*, 3: 1333 (2011).
44. D. Chen, H. Liu, T. Kobayasi and H. F. Yu, Multiresponsive reversible gels based on a carboxylic azo polymer, *J. Mater. Chem.*, 20: 3610 (2010).
45. N. Zettsu, T. Ogasawara, N. Mizoshita, S. Nagano and T. Seki, Photo-triggered surface relief grating formation in supramolecular liquid crystalline polymer systems with detachable azobenzene unit, *Adv. Mater.*, 20: 516 (2008).
46. P. Metrangolo, F. Meyer, T. Pilati, G. Resnati and G. Terraneo, Halogen bonding in supramolecular chemistry, *Angew. Chem. Int. Ed.*, 47: 6114 (2008).
47. P. Metrangolo, H. Neukirch, T. Pilati and G. Resnati, Halogen bonding based recognition processes: A world parallel to hydrogen bonding, *Acc. Chem. Res.*, 38: 386 (2005).

48. A. Priimagi, G. Cavallo, P. Metrangolo and G. Resnati, The halogen bond in the design of functional supramolecular materials: Recent advances, *Acc. Chem. Res.*, 46: 2686 (2013).
49. V. Amico, S. V. Meille, E. Corradi, M. T. Messina and G. Resnati, Perfluorocarbon–hydrocarbon self-assembling: 1D infinite chain formation driven by nitrogen–iodine interactions, *J. Am. Chem. Soc.*, 120: 8261 (1998).
50. H. L. Nguyen, P. N. Horton, M. B. Hursthouse, A. C. Legon and D. W. Bruce, Halogen bonding: A new interaction for liquid crystal formation, *J. Am. Chem. Soc.*, 126: 16 (2004).
51. L. J. McAllister, C. Präsang, J. P. Wong, W. Thatcher, R. J. Whitwood, A. C. Donnio, B. P. O'Brien, P. B. Karadakov and D. W. Bruce, Halogen-bonded liquid crystals of 4-alkoxystilbazoles with molecular iodine: A very short halogen bond and unusual mesophase stability, *Chem. Commun.*, 49: 3946 (2013).
52. Y. Chen, H. F. Yu, L. Zhang, H. Yang and Y. Lu, Photoresponsive liquid crystals based on halogen bonding of azopyridines, *Chem. Commun.*, 50: 9647 (2014).
53. D. W. Bruce, P. Metrangolo, F. Meyer, T. Pilati, C. Präsang, G. Resnati and A. C. Whitwood, Structure–function relationships in liquid-crystalline halogen-bonded complexes, *Chem.-Eur. J.*, 16: 9511 (2010).
54. P. Metrangolo, Y. Carcenac, M. Lahtinen, T. Pilati, K. Rissanen, A. Vij and G. Resnati, Nonporous organic solids capable of dynamically resolving mixtures of diiodoperfluoroalkanes, *Science*, 323: 1461 (2009).
55. A. Priimagi, G. Cavallo, A. Forni, M. Gorynsztejn-Leben, M. Kaivola, P. Metrangolo, R. Milani, A. Shishido, T. Pilati, G. Resnati and G. Terraneo, Halogen bonding versus hydrogen bonding in driving self-assembly and performance of light-responsive supramolecular polymers, *Adv. Func. Mater.*, 22: 2572 (2012).
56. I. Aiello, M. Ghedini, F. Neve and D. Pucci, Synthesis and mesogenic properties of rodlike bis(alkylphenylazo)-substituted N,N'-salicylidenediaminato nickel (II), copper (II), and oxovanadium (IV) complexes, *Chem. Mater.*, 9: 2107 (1997).
57. D. W. Bruce, D. A. Dunmur, E. Lalinde, P. M. Maitlis and P. Styring, Novel types of ionic thermotropic liquid crystals, *Nature*, 323: 791 (1986).
58. D. W. Bruce, Calamitics, cubics, and columnars liquid-crystalline complexes of silver (I), *Acc. Chem. Res.*, 33: 831 (2000).
59. P. Sudhadevi Antharjanam, V. A. Mallia and S. Das, Novel azopyridine-containing silver mesogens: Synthesis, liquid-crystalline, and photophysical properties, *Chem. Mater.*, 14: 2687 (2002).
60. I. Davidi, A. Semionov, D. Eisenberg, G. Goobes and R. Shenhar, Mesomorphic behavior induced by stacking interactions between poly(2-vinyl pyridine) and palladium pincer surfactants in the solid state, *Soft Matter*, 8: 7393 (2012).
61. I. Davidi, D. Hermida-Merino, K. Keinan-Adamsky, G. Portale, G. Goobes and R. Shenhar, Dynamic behavior of supramolecular comb polymers consisting of poly(2-vinyl pyridine) and palladium-pincer surfactants in the solid state, *Chemistry*, 20: 6951 (2014).
62. I. Davidi, D. Patra, D. Hermida-Merino, G. Portale, V. M. Rotello, U. Raviv and R. Shenhar, Hierarchical structures of polystyrene-block-poly(2-vinylpyridine)/palladium–pincer surfactants: Effect of weak surfactant–polymer interactions on the morphological behavior, *Macromolecules*, 47: 5774 (2014).
63. H. F. Yu and T. Ikeda, Photocontrollable liquid-crystalline actuators, *Adv. Mater.*, 23: 2149 (2011).
64. P. G. De Gennes, M. Hébert and R. Kant, Artificial muscles based on nematic gels, *Macromol. Symp.*, 113: 39 (1997).
65. C. D. Eisenbach, Isomerization of aromatic azo chromophores in poly(ethyl acrylate) networks and photomechanical effect, *Polymer*, 21: 1175 (1980).
66. H. Wermter and H. Finkelmann, Liquid crystalline elastomers as artificial muscles, *e-Polymer*, 013: 1 (2001).
67. M. H. Li, P. Keller, B. Li, X. Wang and M. Brunet, Light-driven side-on nematic elastomer actuators, *Adv. Mater.*, 15: 569 (2003).
68. M. Yamada, M. Kondo, J.-I. Mamiya, Y. Yu, M. Kinoshita, C. J. Barrett and T. Ikeda, Photomobile polymer materials: Towards light-driven plastic motors, *Angew. Chem. Int. Ed.*, 47: 4986 (2008).
69. M. Yamada, M. Kondo, R. Miyasato, Y. Naka, J.-I. Mamiya, M. Kinoshita, A. Shishido, Y. Yu, C. J. Barrett and T. Ikeda, Photomobile polymer materials—Various three-dimensional movements, *J. Mater. Chem.*, 19: 60 (2008).
70. T. Ikeda, M. Nakano, Y. Yu, O. Tsutsumi and A. Kanazawa, Anisotropic bending and unbending behavior of azobenzene liquid-crystalline gels by light exposure, *Adv. Mater.*, 15: 201 (2003).
71. Y. Yu and T. Ikeda, Soft actuators based on liquid-crystalline elastomers, *Angew. Chem. Int. Ed.*, 45: 5416 (2006).
72. H. F. Yu, C. Dong, W. Zhou, T. Kobayashi and T. Yang, Photomechanical behaviors of hybrid films of liquid-crystalline polymer microparticles, *Small*, 7: 3039 (2011).
73. L. Yu, Z. Cheng, Z. Dong, Y. Zhang and H. F. Yu, Photomechanical response of polymer-dispersed liquid crystals/graphene oxide nanocomposites, *J. Mater. Chem. C*, 2: 8501 (2014).

12 Thermoplastic Polyurethanes*

Koh-Hei Nitta and Mizue Kuriyagawa

CONTENTS

12.1 Introduction 387
12.2 Structure and Morphology 387
    12.2.1 Chemical Structure 387
    12.2.2 Hydrogen Bond 390
    12.2.3 Phase Separation 390
12.3 Dynamic-Mechanical Properties 391
12.4 Mechanical Properties 392
    12.4.1 Effect of Segmented Structure of TPUs on Stress–Strain Properties 392
    12.4.2 Effect of Hard-Segment Content on Stress–Strain Properties 393
References 394

12.1 INTRODUCTION

Polyurethanes (PUs) were first discovered by Bayer and coworkers in 1937 [1–3] as a competitive response to the work on nylon by scientists from DuPont de Nemours. Because of the versatility of PU chemistry, this class of polymer can be used to produce an extremely broad spectrum of materials ranging from flexible foams to thermoplastic elastomers [4–9]. Thermoplastic polyurethanes (TPUs) were developed in the 1950s as a new type of PU in Germany by Bayer-Fabenfabriken and in the United States by B.F. Goodrich. The Alliance for the Polyurethane Industry (API) describes TPUs as "bridging the gap between rubber and plastics" because TPUs offer the mechanical properties of rubber but can be processed as thermoplastics [10]. The TPUs consist of linear soft segments (long-chain diols) and linear hard segments (diisocyanates and chain extenders). The wide applicability of TPU materials arises from their high modulus, drawability, high strength, and good toughness. In general, phase separation occurs in most TPUs because of the intrinsic incompatibility between the hard and soft segments. The hard segments, composed of polar materials such as urethane and/or urea, can form hydrogen bonds between carbonyls and amino groups and thus tend to cluster or aggregate into ordered hard domains, whereas the soft segments form amorphous domains. Improving the microscale phase separation of hard and soft segments during preparation and processing can influence the mechanical properties of the material. The domain size of the hard segments is on the order of 10 nm, which is similar to the domain size in the crystalline lamellae of typical thermoplastics such as polyethylene and polypropylene.

Figure 12.1 shows the segmented multiblock structure of TPU. The dispersed domains of hard segments are particularly important because they act as cross-linkers, conferring high elasticity upon the material. The plasticity and viscosity of the material arise because the intermolecular interaction between hard segments is based on hydrogen bonding.

12.2 STRUCTURE AND MORPHOLOGY

12.2.1 CHEMICAL STRUCTURE

In general, the soft segments are prepared from polyols, while the hard segments consist of diisocyanates and short polyols as chain extenders. The structure of TPUs can be controlled by the

---

* Based in parts on the first-edition chapter on thermoplastic polyurethanes.

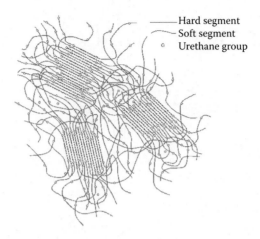

FIGURE 12.1 Illustration of molecular morphology of a typical TPU.

differences in reactivity of each component. TPUs are typically synthesized by either the prepolymer technique or the one-shot method. In the prepolymer method, diisocyanate and long-chain diol are mixed in the relative amounts required to obtain the specified concentrations of soft and hard segments, and the reaction is carried out until all hydroxyl groups are reacted [5]. In the case of the one-shot method, the reaction components (polyols, isocyanates, and chain extenders) are simultaneously mixed and reacted at once. The one-shot method yields a narrower distribution of hard-segment lengths, compared to the prepolymer procedure. Details of the chemical reactions are covered in several excellent references [11–14]. Some prepolymers and chain extenders are exemplified.

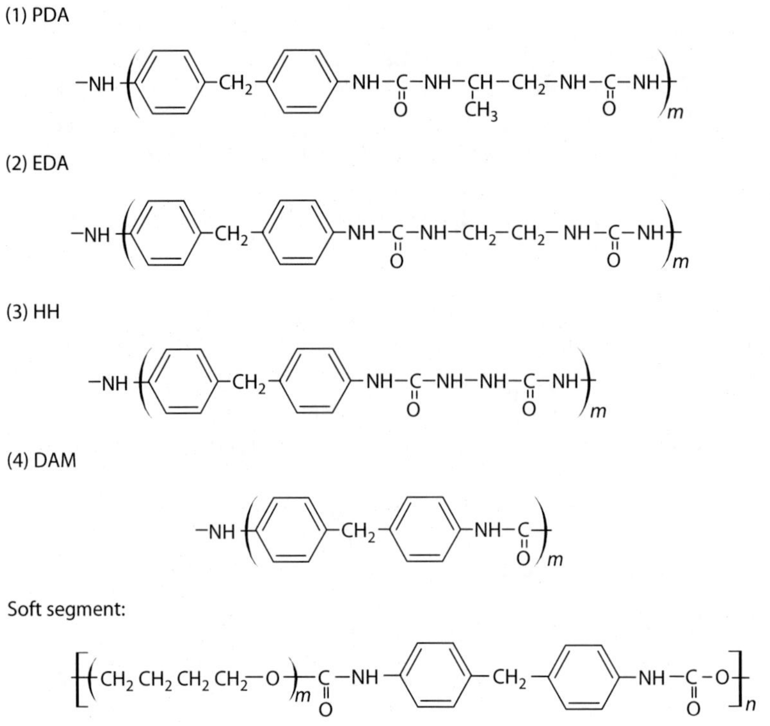

(1) PDA

(2) EDA

(3) HH

(4) DAM

Soft segment:

SCHEME 12.1 PDA, EDA, HH, DAM based hard segments.

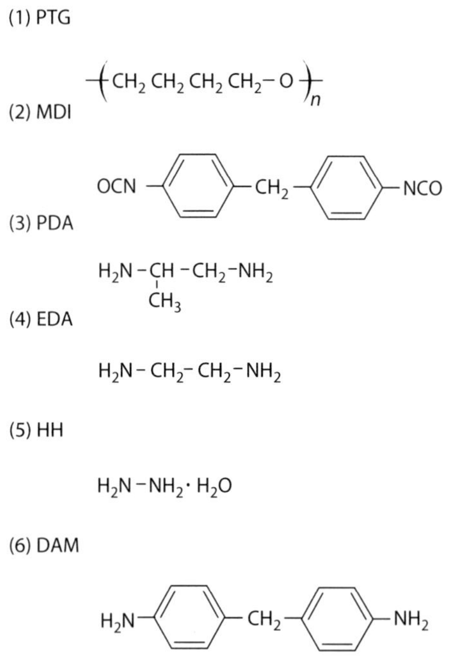

(1) PTG

$-(CH_2\ CH_2\ CH_2\ CH_2-O)_n$

(2) MDI

OCN—⟨C$_6$H$_4$⟩—CH$_2$—⟨C$_6$H$_4$⟩—NCO

(3) PDA

H$_2$N–CH–CH$_2$–NH$_2$
       |
       CH$_3$

(4) EDA

H$_2$N–CH$_2$–CH$_2$–NH$_2$

(5) HH

H$_2$N–NH$_2$·H$_2$O

(6) DAM

H$_2$N—⟨C$_6$H$_4$⟩—CH$_2$—⟨C$_6$H$_4$⟩—NH$_2$

SCHEME 12.2 (1) PTG: poly(tetramethylene glycol); (2) MDI: 4,4'-diphenylmethane diisocyanate; (3) PDA: propylene diamine; (4) EDA: ethylene diamine; (5) HH: hydra hydrate; (6) DAM: diamino diphenyl methane.

Hard segments are shown in Scheme 12.1. In addition, typical primary structures of segmented PU using these prepolymers PTG and MDI and chain extenders PDA, EDA, HH, DAM are shown in Scheme 12.2.

TPUs contain a high concentration of polar groups, i.e., the urethane groups resulting from isocyanate–hydrogen reactions, as well as esters, urea, and other moieties. The relatively strong interaction between these polar groups plays a central role in determining the mechanical properties of the thermoplastics. Table 12.1 shows the empirical cohesive energies and specific volumes of each polar group.

The cohesive energy of polar urethane and urea is much greater than that of other groups. The hydrogen bonding and dipole–dipole interactions between hard segments are sufficiently strong to provide a network of physical cross-links between flexible soft segments. Consequently, the TPU materials exhibit the mechanical properties of covalently cross-linked elastomers. At room temperature, the strong

TABLE 12.1
Cohesive Energy and Specific Volume of Polar Groups Found in TPUs

Group	Cohesive energy/kJ mol^{-1}	Specific volume/cm^3 mol^{-1}
Urethane (–NHCOO–)	36.5	43.5
Urea (–NHCONH–)	35.5	36.2
Phenylene (–C$_6$H$_4$–)	16.3	83.9
Methylene (–CH$_2$–)	2.8	21.8
Ether (–O–)	4.2	7.3
Ester (–COO–)	12.1	28.9
Ketone (–CO–)	11.1	21.6

Source: Saunders, H. and Frisch, K.C., *Polyurethanes: Chemistry and Technology*, Part I and II, InterScience Publishers, New York, 1962.

interaction between the hard segments dominates the modulus, hardness, and toughness, but at elevated temperatures, their elastomeric properties decrease depending on their ability to remain associated.

The soft segments of TPU materials typically comprise aliphatic polyethers or aliphatic polyesters. These polymers have a glass transition temperature below room temperature. They are amorphous, and their optimum molecular weight (which can range from 600 to 3000) depends on the mechanical and thermal properties required of the final TPU.

12.2.2 Hydrogen Bond

The primary structure of PU indicates that carbonyl-to-amino hydrogen bonds occur between urea or urethane groups. The hydrogen bond, formed between a proton donor and a proton acceptor, is the strongest secondary chemical bond (Table 12.1).

Infrared spectroscopy can be used to confirm the existence of hydrogen bonds. The –NH group in the urethane bond is a strong proton donor, while the carbonyl group (C=O) and alkoxy (OR) groups are both proton acceptors. The N–H group participating in the hydrogen bond displays a stretching absorption band at 3400–3300 cm^{-1}. The band corresponding to the hydrogen bonds between the urethane N–H and the urethane C=O appears at 3355–3300 cm^{-1}. The band corresponding to the hydrogen bond between the N–H and the ether appears at 3310–3290 cm^{-1}. The nonbonded N–H group exhibits a band at about 3450 cm^{-1} whose intensity decreases during the urethane reaction, indicating that almost all of the N–H groups in TPU are hydrogen bonded. TPU exhibits two C=O stretching vibration bands at 1730–1690 cm^{-1}; of these, one at 1720–1670 cm^{-1} is assigned to hydrogen-bonded C=O groups, and the second at around 1740–1730 cm^{-1} is assigned to free C=O groups. Simple carbonates absorb at 1739 cm^{-1}. In addition, N–H bending and C–N stretching bands appear at 1575–1530 cm^{-1}, and a C–O stretching band appears at 1250–1220 cm^{-1}. The urea bond exhibits an N–H stretching band at 3400–3300 cm^{-1} and a C=O stretching band at 1660 cm^{-1}. The allophanate and biuret groups also exhibit N–H stretching and C=O stretching bands, at 1750–1710 cm^{-1} and 1708–1653 cm^{-1}, respectively. These are overlapped by the C=O stretching bands of the urea and urethane groups.

12.2.3 Phase Separation

The degree of phase separation in TPUs increases with the chain length of the soft segments [15]. In general, polyether-type TPUs show a higher degree of phase separation than the polyester-type TPUs (MDI based) because polyethers are less compatible with MDI. The DAM hard segments show a well-defined crystalline structure, with a monoclinic unit cell whose dimensions are $a = 0.472$ nm, $b = 1.13$ nm, $c = 1.16$ nm, and $\gamma = 116.5°$ [16]. The packing of MDI–butanediol hard segments shows a triclinic unit cell with the following approximate dimensions: $a = 0.52$ nm, $b = 0.48$ nm, $c = 3.50$ nm, $\alpha = 115°$, $\beta = 121°$, and $\gamma = 116.5°$ [17].

The phase separation of various TPU materials has also been investigated by dynamic-mechanical analysis, electron microscopy [18–21], small-angle x-ray scattering [22–33], electron spin resonance [34,35], differential scanning calorimetry [36–43], and atomic force microscopy [44,45]. The hard segments form domains, whose size is on the order of 5–100 nm and which act as physical crosslinkers. The morphology of the TPU also depends upon the primary structure of the soft segments, which influences the aggregation of the hard segments.

The length and other properties of the chain extender components affect both the T_g of the soft segments and the crystallizability of the hard segments [46–49]. The segregation of the hard segments can be reduced by changing the chain extender in TPU from 1,4-butanediol to 1,5-pentanediol or to 1,3-butanediol. TPU with 1,3-butanediol as the chain extender is not able to crystallize, because of its methyl side groups, which improve the mixing of the hard segments with the soft-segment matrix, whereas the TPU having 1,4-butanediol as the chain extender has the highest degree of crystallinity.

12.3 DYNAMIC-MECHANICAL PROPERTIES

The miscibility of the hard and soft segments in TPU depends on the differences in their glass-to-rubber transition temperature (T_g) [50]. The glass-to-rubber transition occurs at the onset of micro-Brownian segmental motion, which has been identified by dynamic-mechanical spectra as a primary dispersion or relaxation mechanism. The compatibility between the soft and hard components can be readily confirmed from dynamic-mechanical measurements. The loss modulus E'' spectrum of a two-component system with phase-separated morphology typically shows double peaks, each of which is assigned to the T_g of one component. In contrast, the spectrum of a two-component system where both components are miscible, having no clear phase-separated morphology, shows only a single broad peak whose position is between that of the two original T_g peaks of the pure components.

Figure 12.2 shows the dynamic-mechanical spectra of various elastic segmented TPUs, differing in hard-segment content. The TPUs were prepared from MDI for the hard segment, with 1.4-BD as the chain extender. Polypropylene glycol (PPG) was used as the soft segment. In the temperature range from 0°C to 100°C, the loss modulus E'' exhibits a single peak, which is assigned to the T_g of the soft-segment component. The peak shifts to higher temperatures upon increasing the content of hard segments. The region beyond the T_g peaks expresses the plateau modulus E', corresponding to rubberlike properties and extending up to the flow region that corresponds to relaxation of the hard segments. This behavior indicates that the segregated hard segments act as the cross-linkers. The molecular weight between cross-links M_c can be estimated from the plateau modulus using the relation $E' = 3nRT$, where n is the effective cross-linking density and R is the gas constant. These values are listed in Table 12.2.

To quantitatively characterize the molecular mobility of soft segments, the apparent activation energy for micro-Brownian motion of soft segments was determined from the frequency dependence of T_g, on the assumption that the glass transition relaxation progresses in accordance with the Arrhenius equation. An example of the frequency dependence of dynamic-mechanical spectra is shown in Figure 12.3. The slope of the Arrhenius plot, where the logarithmic values of frequency are plotted against the inverse of temperature, gives the apparent activation energy of micro-Brownian motion of soft segments. These values are listed in Table 12.2.

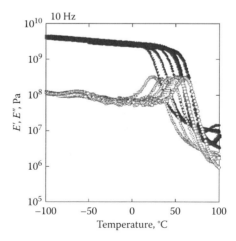

FIGURE 12.2 Dynamic-mechanical spectra of PU materials, each labeled with a numeral that denotes the hard segment content in vol.%. ▽, PU49; △, PU52; ◇, PU55; □, PU58; and ○, PU61. Open symbols indicate E', and closed symbols indicate E''.

TABLE 12.2
Glass Transition Temperature and Molecular Weight between Domains for PUs

	$T_g/°C$	M_c	Activation Energy, kJ
PU49	22	700	700
PU52	33	880	880
PU55	42	1010	1010
PU58	51	1310	1310
PU61	61	2300	2300

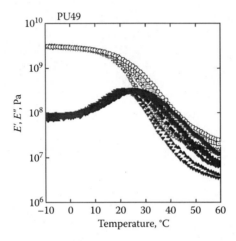

FIGURE 12.3 Dynamic-mechanical spectra of PU49 measured at different frequencies: ▽, 1 Hz; △, 3 Hz; ◇, 10 Hz; □, 30 Hz; and ○, 100 Hz. Open symbols indicate E', and closed symbols indicate E''.

12.4 MECHANICAL PROPERTIES

12.4.1 EFFECT OF SEGMENTED STRUCTURE OF TPUs ON STRESS–STRAIN PROPERTIES

The thermoplastic properties in segmented PU materials depend strongly on their two-component phase-separated morphology. The domains of hard segments in TPU act as junction points, while the flexible soft segments that link adjacent hard domains exhibit an entropic mechanical response. The soft segments play an important role in determining the modulus, strength, and hardness of TPUs [1,5] in the temperature range between the T_g of both components. Hydrogen bond dissociation for the hard segments occurs above the T_g [1], which is usually in the region of 80–120°C [28,29]. Therefore, raising the temperature above the T_g loosens cohesion within the hard segments, causing the segmented PU materials to exhibit thermoplasticity.

The chemical structures of both the diisocyanate and chain extender affect the stress–strain properties, while the type of chain extender influences the tensile strength. Using 1,4-BD as the chain extender gives the highest tensile strength [1].

In general, the longer the soft segment, the higher the elongation at break [46]. TPUs based on polyesters have higher tensile strength than those based on polyether and polybutadiene. Flexible segments with very low molecular weight exhibit poor elasticity but high stress tolerance, whereas the high-molecular-weight segments exhibit high elasticity with lower stress tolerance. TPUs based on branched polyesters with side chains have lower tensile strength during uniaxial extension, compared to TPUs based on linear polyesters. This is because the side chains hinder crystallization within the material.

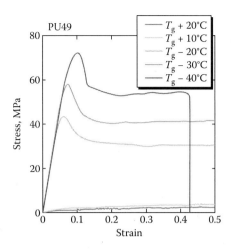

FIGURE 12.4 Stress–strain curves in the yielding region of PU49 measured near the T_g of the soft segments.

Figure 12.4 shows the stress–strain curve for PU49, measured near the T_g of the soft segments. Below the T_g of the soft segments, the stress is almost proportional to the applied strain in the initial strain region. The Young's modulus can be conventionally estimated from the slope of the line. After this elastic region, the material shows a clear yield point, located at the maximum of the nominal stress–strain curve. The yield point is associated with the onset of *temporal* plastic deformation. Beyond the yield point, a concave contraction initiates on the specimen and coalesces into a well-defined neck. The yield stress and strain increase linearly as the temperature decreases. At a temperature of $T_g - 30°C$, the PU48 breaks at the necked region. However, at $T_g - 20°C$ and $T_g - 10°C$, the neck expands to the whole of the specimen, and a strain-hardening region appears, after which the stress steeply increases again. Further extension causes the specimen to fracture. The overall behavior is similar to the plastic behavior of semicrystalline polymer solids.

Above T_g, the stress level drastically drops, the tensile behavior is transformed to rubberlike behavior, and the sample specimens recover their original size and shape after the stress is released.

12.4.2 Effect of Hard-Segment Content on Stress–Strain Properties

Here, we exemplify the mechanical contribution of the hard-segment content using the PU samples in Table 12.2. As shown in Figure 12.5, the Young's modulus and the yield stress slightly decrease

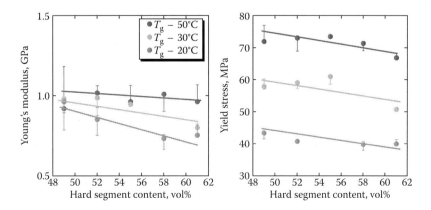

FIGURE 12.5 Young's modulus and yield stress plotted against the hard-segment content of PU samples. Data are from Table 12.2, below the T_g of the soft segments.

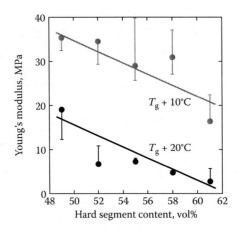

FIGURE 12.6 Young's modulus and yield stress plotted against the hard-segment content of PU samples. Data are from Table 12.2, above the T_g of the soft segments.

as the content of hard segments increases. This is much different from the mechanical properties of typical semicrystalline polymers such as polyethylene and polypropylene, which show a strong positive dependence of the Young's modulus and yield stress on the crystallinity.

The stress–strain curves measured above the T_g of the soft segments exhibit antisigmoidal behavior, a typical tensile response for rubberlike materials. The stress level and Young's modulus decrease as the content of hard segments increases, as well as with increasing temperature (Figure 12.6). The mechanical response of rubbery materials originates from entropic elasticity and the concentration of junction points. As temperature increases, entropic elasticity increases. A larger number of junction points per molecule increases the overall concentration of junction points. These results suggest that the looseness of interaction between hard segments dominates the macroscopic mechanical response of these TPU materials at elevated temperatures.

REFERENCES

1. Hepburn, C. 1992. *Polyurethane Elastomers*, Elsevier Applied Science, Amsterdam.
2. Legge, N. R., Holden, G., Schroeder, H. E., Eds. 1987. *Thermoplastic Elastomers: A Comprehensive Review*, Hanser, Munich.
3. Woods, G. 1990. *The ICI Polyurethanes Book*, 2nd ed., Wiley, Chichester.
4. Ng, N. H., Appegrezza, A. E. Seymour, R. W., Cooper, S. L. 1973. Effect of segment size and polydispersity on the properties of polyurethane block polymers. *Polymer* 14: 225.
5. Petrovic, Z. S., Ferguson, J. 1991. Polyurethane elastomers. *Prog. Polym. Sci.* 16: 695–836.
6. Koberstein, J. T., Russell, T. P. 1986. Simultaneous SAXS–DSC study of multiple endothermic behavior in polyether-based polyurethane block copolymers. *Macromolecules* 19: 714–720.
7. Koverstein, J. T., Stein, R. S. 1983. Small-angle x-ray scattering studies of microdomain structure in segmented polyurethane elastomers. *J. Polym. Sci. Polym. Part A-2 Polym. Phys.* 21: 1439–1472.
8. Furukawa, M., Komiyama, M., Yokoyama, T. 1996. Characterization of polyurethane network elastomers. *Angew. Makro. Chem.* 240: 205–211.
9. Furukawa, M., Hamada, Y., Kojio, K. 2003. Aggregation structure and mechanical properties of functionally graded polyurethane elastomers. *J. Polym. Sci. Polym. Phys.* 41: 2355–2363.
10. Qi, H. I., Boyce, M. C. 2003. Stress–strain behavior of thermoplastic polyurethane. *Mech. Mater.* 37: 817–839.
11. Saunders, H., Frisch, K. C. 1962. *Polyurethanes: Chemistry and Technology*, Part I and II, InterScience Publishers, New York.
12. Buist, J. M., Gudgeon, H. 1969. *Advances in Polyurethane Technology*, MacLaren and Sons, London.
13. Woods, G. 1987. *The ICI Polyurethane Book*, John Wiley & Sons, New York.

14. Kimura, I., Ishihara, H., Ono, H., Yoshihara, N., Nomura, S., Kawai, H. 1974. Morphology and deformation mechanism of segmented poly(urethaneureas) in relation to spherulitic crystalline textures. *Macromolecules* 7: 355–363.
15. Hepburn, C. 1992. *Polyurethane Elastomers*, 2nd Ed., Elsevier, London, p. 51.
16. Ishihara, H., Kimura, I., Yoshihara, N. 1983. Studies on segmented polyurethane—urea elastomers: Structure of segmented polyurethane—urea based on poly(tetramethylene glycol), 4,4'-diphenylmethane diisocyanate, and 4,4'-diaminodiphenylmethane. *J. Macromol. Sci.-Phys.* B22: 713–733.
17. Blackwell, J., Gardner, K. H. 1979. Structure of the hard segments in polyurethane elastomers. *Polymer* 20: 13–17.
18. Wang, C. S., Kenney, D. J. 1995. Effect of hard segments on morphology and properties of thermoplastic polyurethanes. *J. Elastomers Plastics* 27: 182–199.
19. Chen-Tsai, C. H. Y., Thomas, E. L., Macknight, W. J., Schenider, N. S. 1986. Structure and morphology of segmented polyurethanes. 3. Electron microscopy and small angle X-ray scattering studies of amorphous random segmented polyurethanes. *Polymer* 27: 659–666.
20. Serrano, M., Macknight, W. J., Schenider, N. S. 1987. Transport-morphology relationships in segmented polybutadiene polyurethane. 1. Experimental results. *Polymer* 28: 1667–1673.
21. Joseph, M. D., Savina, M. R., Harris, R. F. 1992. Effects on properties of varying the cis/trans isomer distribution in polyurethane elastomers made with 1,4-cyclohexane diisocyanate. *J. Appl. Polym. Sci.* 44: 1125–1133.
22. Bonart, R., Muller, E. H. 1974. Phase separation in urethane elastomers as judged by low-angle x-ray scattering. I. Fundamentals. *J. Macromol. Sci. Phys.* B10: 177–189.
23. Li, Y., Gao, T., Chu, B. 1992. Synchrotron SAXS studies of the phase separation kinetics in a segmented polyurethane. *Macromolecules* 25: 1737–1742.
24. Visser, S. A., Cooper, S. L. 1991. Analysis of small-angle x-ray scattering data for model polyurethane ionomers: Evaluation of hard-sphere models. *Macromolecules* 24: 2584–2593.
25. Leung, L. M., Koberstein, J. T. 1985. Small-angle scattering analysis of hard-microdomain structure and microphase mixing in polyurethane elastomers. *J. Polym. Sci. Polym. Phys. Ed.* 23: 1883–1913.
26. Tang, W., Farris, R. J., Macknight, W. J., Eisenbach, C. D. 1994. Segmented polyurethane elastomers with liquid crystalline hard segments. 1. Synthesis and phase behavior. *Macromolecules* 27: 2814–2812.
27. Smith, T. L. 1974. Tensile strength of polyurethane and other elastomeric block copolymers. *J. Polym. Sci. Polym. Phys. B* 12: 1825–1848.
28. Briber, R. M., Thomas, E. L. 1983. Investigation of two crystal forms in MDI/BDO-based polyurethanes. *J. Macromol. Sci. Phys.* B22: 509–528.
29. Okamoto, D. T., O'Connel, E. M., Cooper, S. L., Root, T. W. 1993. Solid-state ^{13}C nuclear magnetic resonance characterization of MDI-based polyurethanes. *J. Polym. Sci. Polym. Phys.* 31: 1163–1177.
30. Smith, T. L. 1986. In *Encyclopedia of Materials Science and Engineering* (M. B. Bever, ed.), Pergamon Press, Oxford, p. 1341.
31. Petrovic, Z. S., Budinski-Simendic, J. 1985. Study of the effect of soft-segment length and concentration on properties of polyetherurethanes. I. The effect on physical and morphological properties. *Rubber Chem. Technol.* 58: 685–700.
32. Petrovic, Z. S., Budinski-Simendic, J. 1985. Study of the effect of soft-segment length and concentration on properties of polyetherurethanes. II. The effect on mechanical properties. *Rubber Chem. Technol.* 58: 701–712.
33. Pechhold, E., Pruckmayr, G. 1982. Polytetramethylene ether glycol: Effect of concentration, molecular weight and distribution on properties of MDI/BDO-based polyurethanes. *Rubber Chem. Technol.* 55: 76–87.
34. Kumler, P. L., Boyer, R. F. 1976. ESR studies of polymer transition. I. *Macromolecules* 9: 903–910.
35. Chen, W. P., Schlick, S. 1990. Study of phase separation in polyurethanes using paramagnetic labels: Effect of soft-segment molecular weight and temperature. *Polymer* 31: 308–314.
36. Lee, H. S., Wang, Y. K., Hsu, S. L. 1987. Spectroscopic analysis of phase separation behavior of model polyurethanes. *Macromolecules* 20: 2089–2095.
37. Schneider, N. S., Paik Sung, C. S. 1977. Transition behavior and phase segregation in TDI polyurethanes. *Polym. Eng. Sci.* 17: 73–80.
38. Harrell Jr., L. L. 1969. Segmented polyurethanes: Properties as a function of segment size and distribution. *Macromolecules* 2: 607–612.
39. Li, C., Yu, X., Speckhard, T. A., Cooper, S. L. 1988. Synthesis and properties of polycyanoethylmethyl–siloxane polyurea urethane elastomers: A study of segmental compatibility. *J. Polym. Sci. Polym. Phys. Ed.* 26: 315–337.

40. Hesketh, T. R., Van Bogart, J. W. C., Cooper, S. L. 1980. Differential scanning calorimetry analysis of morphological changes in segmented elastomers. *Polym. Eng. Sci.* 20: 190–197.
41. Yoon, S. C., Ratner, B. D., Ivan, B., Kennedy, J. P. 1994. Surface and bulk structure of segmented poly(ether urethanes) with perfluoro chain extenders. 5. Incorporation of polyisobutylene macroglycols. *Macromolecules* 27: 1548–1554.
42. Kakati, D. K., Gosain, R., George, M. H. 1994. New polyurethane ionomers containing phosphonate groups. *Polymer* 35: 398–402.
43. Boufi, S., Belgacem, M. N., Quillerou, J. Gandini, A. 1993. Urethanes and polyurethanes bearing furan moieties. 4. Synthesis, kinetics, and characterization of linear polymers. *Macromolecules* 26: 6706–6717.
44. McLean, R. S., Sauer, B. B. 1997. Tapping-mode AFM studies using phase detection for resolution of nanophases in segmented polyurethanes and other block copolymers. *Macromolecules* 30 (26): 8314–8317.
45. Kojio, K., Kugumiya, S., Uchiba, Y., Nishino, Y., Furukawa, M. 2009. The microphase-separated structure of polyurethane bulk and thin films. *Polymer Journal* 41: 118–124.
46. Brunette, C. M., Hsu, S. L., Rossman, M., Macknight, W. J., Schnieder, N. S. 1981. Thermal and mechanical properties of linear segmented polyurethanes with butadiene soft segments. *Polym. Eng. Sci.* 21: 668–674.
47. Leung, L. M., Koberstein, J. T. 1986. DSC annealing study of microphase separation and multiple endothermic behavior in polyether-based polyurethane block copolymers. *Macromolecules* 19: 706–713.
48. Petrovic, Z. S., Javni, I. 1989. Effect of soft-segment length and concentration on phase separation in segmented polyurethane. *J. Polym. Sci. Polym. Phys.* 27: 545–560.
49. Ferguson, J., Patsavoullis, D. 1972. Chemical structure–physical property relationships in polyurethane elastomeric fibers: Property variation in polymers containing high hard segment concentration. *Eur. Polym. J.* 8: 385–396.
50. Furukawa, M., Hamada, Y., Kojio, K. 2003. Aggregation structure and mechanical properties of functionally graded polyurethane elastomers. *J. Polym. Sci. Polym. Phys.* 41: 2355–2363.

13 Fluoroplastics*

Thierry A. Blanchet

CONTENTS

13.1 Introduction 397
13.2 Polytetrafluoroethylene 398
    13.2.1 Polymerization 398
    13.2.2 Structure 399
        13.2.2.1 Molecular Structure 399
        13.2.2.2 Crystal Structure 400
    13.2.3 Properties 401
        13.2.3.1 Thermal Properties 401
        13.2.3.2 Chemical Properties 402
        13.2.3.3 Electrical Properties 402
        13.2.3.4 Mechanical Properties 403
    13.2.4 Processing 404
        13.2.4.1 Molding 404
        13.2.4.2 Ram and Paste Extrusion 405
        13.2.4.3 Coatings and Composites 406
    13.2.5 Tribological Bearing Behavior 406
        13.2.5.1 Friction 406
        13.2.5.2 Wear 408
        13.2.5.3 Wear-Resistant Composites 408
        13.2.5.4 Radiation-Modified Tribological Behavior 409
    13.2.6 Electronics 410
13.3 Other Fluoropolymers 411
    13.3.1 Polychlorotrifluoroethylene 411
    13.3.2 Fluorinated Ethylene Propylene 411
    13.3.3 Polyvinylidene Fluoride 412
    13.3.4 Polyvinyl Fluoride 412
    13.3.5 Ethylene Chlorotrifluoroethylene 413
    13.3.6 Ethylene Tetrafluoroethylene 413
    13.3.7 Perfluoroalkoxy 414
13.4 Other Continuing Fluoropolymer Developments 415
References 415

13.1 INTRODUCTION

The chance discovery of polytetrafluoroethylene (PTFE) by Roy Plunkett in 1938 occurred amidst research on new routes to synthesize novel refrigerants, one of which used the monomer tetrafluoroethylene (TFE) gas as a starting point. Upon finding a cylinder of supposedly compressed TFE gas to display no pressure, Plunkett had the cylinder cut open to find a white, waxy solid that had formed via polymerization of the TFE. This unusual solid seemed extremely inert, appearing to be insoluble in all solvents and to have an extremely high melt temperature [1]. Since that time, PTFE and the

---

* Based in parts on the first-edition chapter on polytetrafluoroethylene.

numerous other fluoroplastics derived from it have become commercially important thermoplastics with broad application in the bearings, electronics, and chemical process industries. Utilization of such fluoroplastics often results from their unique combinations of exceptional properties: high electrical resistivity and low dielectric constant, high melt temperature, chemical inertness, and unusual surface properties, especially leading to the extremely low friction that they may impart to dry sliding bearing contacts. The following is a broad review of the structure, properties, processing, and application of PTFE, as well as a summary of many of the other fluoroplastic variations and TFE copolymers that have since been developed, comparing and contrasting each to the basic PTFE homopolymer.

13.2 POLYTETRAFLUOROETHYLENE

Upon its discovery of PTFE, the DuPont company quickly thereafter developed production processes for the monomer TFE and subsequent polymerization, with a pilot plant in Arlington, NJ, directed to the World War II effort. The unusual combination of properties of this new polymer led to the consideration of PTFE as a gasket material in the Manhattan Project's development of the atomic bomb [1]. The commercial availability of PTFE was impeded for roughly 10 years after its discovery as a result of difficulties in reacting the TFE monomer safely, as it has a wide explosive range in air (14–43%) [2]. In 1944, such a TFE semiworks explosion killed two people and injured several others [3]. Small-scale commercial production of PTFE by DuPont was delayed until 1947 and then scaled up in 1950 at a new facility in Parkersburg, WV [1]. Also in 1947, Imperial Chemical Industries built its first production plant in Western Europe [4]. PTFE polymer has since become available from several manufacturers under a variety of trade names, such as DuPont (Teflon), ICI (Fluon), Ausimont (Halon), Hoechst (Hostaflon), and Daikin Kogyo (Polyflon) [2].

13.2.1 Polymerization

TFE, the building block for PTFE, was first synthesized from tetrafluoromethane in an electric arc furnace in 1933. Since that time, other synthesis routes for the monomer have been developed. One such route involves the pyrolysis of difluorochloromethane, $CHClF_2$, which yields the TFE monomer $CF_2=CF_2$ as well as the hydrogen chloride by-product HCl. The difluorochloromethane reactant is produced through the partial fluorination of chloroform ($CHCl_3$) with hydrogen fluoride (HF) in the presence of an antimony trifluoride catalyst, a reaction that also yields HCl as a by-product. Furthermore, the chloroform reactant is produced via the reaction of methane (CH_4) with chlorine (Cl_2), which once again leads to an HCl by-product. HF used in chloroform fluorination is produced from fluorspar (calcium fluoride) and sulfuric acid, additionally yielding the by-product $CaSO_4$. Highly toxic perfluoroisobutylene, $CF_2=C(CF_3)_2$, is also produced in small quantities during TFE synthesis [2].

Because of the corrosive nature of the numerous products and reactants (particularly HF and HCl), platinum-lined nickel reactors may be used. The TFE monomer subsequently must be stored with added terpenes to inhibit premature polymerization. Equipment hot spots must be avoided, as they may initiate violent TFE disproportionation to carbon and carbon tetrafluoride. The complexity of the synthesis of the monomer, the need for corrosion-resistant reactor materials, the burden of large amounts of HCl waste by-product, and the hazards associated with handling the monomer all contribute to the inherently high expense of the materials produced from TFE [2], which may vary in recent years from ~$9/kg for mechanical grade PTFE upwards for specially grades [5].

PTFE, $-(CF_2CF_2)_n-$, is produced via the free radical-initiated addition polymerization of TFE gas under pressure in an aqueous medium [1]. The polymerization reaction is extremely exothermic. Three forms of PTFE may be produced, depending on polymerization conditions. If a vigorously agitated aqueous medium with little or no dispersing agent is employed, the polymer partially coagulates to form a precipitated resin (granular resin). The irregularly shaped resin is subsequently dried and ground to a powder of the desired particle size. If instead, polymerization is carried out in an aqueous medium containing adequate dispersing agent, to disrupt coagulation, under only mild agitation (aqueous dispersion polymerization), smaller colloidal particles (~0.2 μm) are formed. This aqueous dispersion is then typically concentrated to roughly 60–65% solid by weight. A third, fine powder (coagulated dispersion) resin form may also be produced through the use of an initiator and emulsifying agents when the aqueous medium is stirred gently [2].

13.2.2 Structure

13.2.2.1 Molecular Structure

PTFE is capable of polymerization to extremely high molecular weight, on the order of $M_n = 10^6$–10^7 [2], or perhaps even in excess of 10^8 [6]. As PTFE does not dissolve in common solvents, methods other than usual solution light-scattering techniques must be used to quantify molecular weight. One such technique is based on the dependency of heat of crystallization, ΔH_c as measured by differential scanning calorimetry, upon number-averaged molecular weight M_n [6]. Specific-gravity techniques may also be employed because the crystallinity and resultant density of polymer cooled slowly from the melt will be greater for PTFE of lower molecular weight [2].

The polymerization of TFE leads to an extremely linear molecule, with a smooth profile lacking branches or bulky side groups. Branching would require the rupture of strong C–F bonds, which is unlikely during the polymerization of TFE [7]. Only under extremely high pressures (~5000 atm at ambient temperatures) does the molecule adopt a planar zigzag conformation (phase III) similar to that of linear polyethylene [8–10].

Under more typical pressure conditions, the zigzag must twist about its length and assume a helical conformation to accommodate the large fluorine atoms [11]. A slight opening of the C–C bond angle to about 116° must also occur, as shown in Figure 13.1 [11,12]. Below 19°C (phase II), a helix

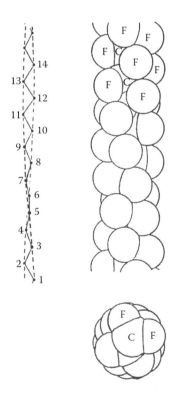

FIGURE 13.1 Helical (twisted zigzag) structure of PTFE molecule (side views and end view). Below 19°C, 13 CF$_2$ groups are required for the helix to complete a 180° twist. (From Bunn, C.W. and Howells, R.R., *Nature*, 174, 1954.)

exists with a 13.8° angular rotation about each C–C backbone bond, such that 13 CF_2 groups are required to complete a 180° twist of the helix and thus compose the helix repeat unit [11,13]. The C–C twist angle may instead be thought of as its supplement, 166.2°, and the conformation termed 13_6 to denote that 13 CF_2 groups are required to complete six full twists of the helix [14]. Above 19°C, the helix conformation is instead 15_7, requiring 15 CF_2 groups to compose a 180° twist [13,15,16].

13.2.2.2 Crystal Structure

The large fluorine atoms and resultant helical chain conformation cause the linear PTFE molecules to be rodlike cylinders, fully extended and rigid. The linearity and smooth profile of the molecule permit crystallization into a banded structure [17,18], as shown in Figures 13.2 and 13.3. Band lengths may vary between 10 and 100 μm, with the width of the bands ranging from 0.2 to 1 μm, and these dimensions are dependent on the rate of cooling following thermal processing [18]. Regularly spaced striations are noted on the band surface, perpendicular to the length of the band. These striations result from the crystalline "slices," which are stacked much like a deck of cards, separated by thin disordered regions of amorphous polymer. The crystalline slice thickness is roughly equal to the spacing of the observed striations, which is typically 20–30 nm [19]. The molecular and crystalline structure of PTFE leads to its unusual frictional behavior, which will be discussed in detail in a later section.

Crystallinity of virgin as-polymerized resin may be as high as 92–98%, due to the linearity of the PTFE molecule [2]. It has been observed that the linear PTFE molecules lie perpendicular to the band length, parallel to the observed striations, within the plane of the crystalline slices [17]. At temperatures below 19°C (phase II), these molecules arrange themselves in a triclinic pseudohexagonal lattice with a spacing between neighboring molecule chain axes of roughly 0.554 nm [11]. As temperature is increased to above 19°C (phase IV), the change to a hexagonal lattice accompanies the transition in helix conformation, with an increased spacing between molecule axes of 0.566 nm [11,15,20]. Upon further increases in temperature, a transition to phase I is experienced at 30°C [16]. Phase I still consists of a 15_7 helical conformation, but the extent of disorder regarding the rotational orientation of molecules about their long axes is increased. Correspondingly, the spacing between molecular axes is further increased [14]. Another transition may also occur at 150°C, due to translational motion and disorder along the chain axis [21]. The sharp transition from a triclinic to a hexagonal unit cell at 19°C corresponds to a 1.3% increase in volume [22]. The volume change over the range from $T < 19°C$ to $T > 30°C$ may be as large as 1.8%. As this is over a range of temperatures that may be experienced by a manufactured component under roughly ambient conditions, the volume change accompanying these transitions can be a significant design concern [23].

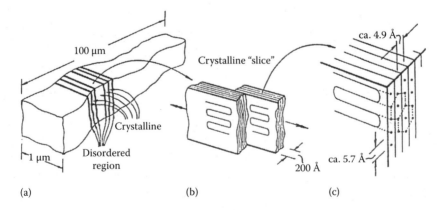

FIGURE 13.2 Crystalline structure of PTFE: (a) band; (b) crystalline slices, which may slip during deformation via shear in the disordered regions that separate them; (c) hexagonal array of PTFE molecules within crystalline slice. (From Makinson, K.R. and Tabor, D., *Proc. Roy. Soc.*, 281, 1964, p. 58, Fig. 1a.)

FIGURE 13.3 Fracture surface of PTFE displaying artifacts of banded structure. Several bands are highlighted by white arrows. The processing of this sample produced a band width of approximately 0.3 µm. Parallel striations within these bands result from the stacking of crystalline slices. (From Tanaka, K., Effects of Various Fillers on the Friction and Wear of PTFE-Based Composites, *Friction and Wear of Polymer Composites*, edited by Friedrich, K., 137, Elsevier, Amsterdam, 1986.)

As a result of the molecular and crystalline structure, the density of PTFE is high for a polymer, typically ranging from 2.1 to 2.3 g/cm^3 (this upper limit corresponding to near-perfectly crystalline PTFE). The exact value will depend on the degree of crystallinity of the polymer, which may be as high as 98% upon polymerization [7], and may vary from less than 40% [24] to 70% [18] depending on subsequent processing conditions. Density will also be affected by the considerable volume changes, which accommodate transitions in the helical conformation of the molecule at 19°C and 30°C [23], as well as porosity and void content.

13.2.3 Properties

13.2.3.1 Thermal Properties

Outside of the range of temperatures where molecular conformation transitions predominate, the coefficient of thermal expansion is roughly ~10^{-4}°C^{-1}, the specific heat ~1.0 kJ/kg °C, and the thermal conductivity ~0.25 W/m °C [25,26]. The first melting of the virgin, as-polymerized PTFE occurs at 342°C, transforming it from a white solid to a transparent gel. This first melting process is irreversible, with a corresponding irreversible reduction in crystallinity. Upon subsequent solidification, melting will thereafter occur at 327°C instead. This is the temperature most commonly cited as the melting point of PTFE [2].

Due to its extremely high molecular weight, PTFE does not truly liquefy, and therefore, the shape of the melt is stable. Even at 380°C, the gel will have a viscosity in the 10^{10}–10^{12} Pa s range [27]. There is likely to be limited molecular movement in the absence of stress, and the original grouping of molecules in the initial solid may roughly be preserved in the melt and reappear upon cooling [17]. Melting is accompanied by a roughly 30% increase in volume [2]. This unique melt behavior for a thermoplastic necessitates specialized processing techniques (to be discussed later).

In vacuum, PTFE does not begin to thermally degrade until roughly 440°C, with degradation proceeding at a maximum rate near 540°C. The polymer "unzips," yielding the TFE monomer. (In fact, heating PTFE to ~600°C in vacuum is the preferred and conventional method for those working with TFE to obtain laboratory-scale amounts of the monomer [2].) In addition to temperature, the rate of decomposition depends on the type of PTFE resin, time, and, to a lesser extent, pressure and nature of its environment. Under an ambient atmosphere, steady-state fractional weight loss via thermal degradation of the granular resin will proceed at a rate of only ~10^{-9}%/h at 260°C, increasing to ~2 × 10^{-6}%/h at 315°C, and reaching ~10^{-3}%/h by 370°C. Initial degradation rates will be considerably higher (e.g., 10^{-4}%/h at 260°C) but will reduce to these steady-state values upon continual exposure. The fine powder coagulated dispersion resin form decomposes at a significantly greater rate [26]. Maximum service temperatures for PTFE do not often exceed 260°C, so that thermal degradation may not often pose great concern [4].

13.2.3.2 Chemical Properties

PTFE is nearly inert, with high resistance to chemical attack. This may result from its strong interatomic bonding and from the shielding of the carbon backbone by the sheath provided by the fluorine atoms [2]. It is not dissolved in common solvents, due in part to its high molecular weight, though dissolution may be accomplished by only a few rare chemicals at higher temperatures. It can be attacked by alkali metals, fluorine and strong fluorinating agents, sodium hydroxide, and pure oxygen, but only at elevated temperatures and pressures [1,2,4]. Water, common acids, bases, aqueous solutions, and even organic solvents are not absorbed by PTFE to any substantial extent. Measured weight changes in such environments upon exposure for various combinations of duration and temperature are small enough to be considered experimentally insignificant. One exception is carbon tetrachloride, which causes a 2.5% increase in PTFE weight after 8 h at 100°C, or a 3.7% increase at 200°C. (Another exception is fluorocarbon oils, which may permeate PTFE more readily, due to their similar chemical composition. PTFE may be dissolved in perfluorinated kerosene at 300°C.) The permeability of gases and vapors into PTFE is also generally low, particularly in the instance of highly crystalline PTFE [2]. As a result of its inert nature and resistance to permeation, PTFE is commonly utilized in the chemical process industry as a barrier resin for coatings, linings, valves, gaskets, and seals. Additionally, PTFE does not absorb electromagnetic radiation in the visible or ultraviolet (UV) range, and thus, light does not induce photochemical reactions or degradation. As a result, the polymer has a very high resistance to weathering and ages well, thus seeing broad outdoor application [1,4]. PTFE has an opaque white appearance and an index of refraction of 1.325 [28].

PTFE is attacked by electron irradiation and similarly by x-ray and γ radiation as these may, in turn, serve as internal sources of secondary electron production within the polymer [29]. Irradiation leads to the formation of primary and secondary radicals through cleavage of C–C and C–F bonds, respectively. In air, the primary and secondary radicals tend to react with O_2 at higher doses, favoring chain scission instead of branching or cross-linking reactions. Such scission results in a decrease in molecular weight, with PTFE fragments having acid fluoride end groups, –COF. In the presence of water vapor, acid fluoride end groups may, in turn, lead to the formation of carboxylic acid end groups, –COOH. Scission resulting from secondary radicals may also proceed by the formation of a primary radical and another chain fragment with an olefin end group –CF=CF_2. The sequence of such scission-related reactions, and Fourier-transform infrared (FTIR) absorbance bands of the resultant end groups, has been discussed by Fisher and Corelli [30]. Scission and the corresponding reduction in molecular weight are also evidenced by the reduction of melt viscosity with increasing radiation dose in air [31]. If irradiation is instead performed in vacuum, in the absence of O_2 and H_2O, the free radicals may be "trapped" and stable for a considerable lifetime. Cross-linking is favored over scission in vacuum, as deduced from various x-ray photoelectron spectroscopy (XPS) studies [29,32–34]. In addition to scission, branching and cross-linking reactions, radiation-induced crystallization and increased density are also observed in PTFE [30,35,36].

13.2.3.3 Electrical Properties

PTFE possesses excellent insulating characteristics. Its volume resistivity is ~1.9 × 10^{19} Ω cm (most other polymers fall in the 10^8–10^{17} Ω cm range), with surface resistivity exceeding 10^{17} Ω cm [25]. Due to the nonpolar nature of the PTFE molecule and the corresponding small polarization experienced in electric fields, due solely to the displacement of its electrons and nuclei, its dielectric constant is very small, roughly 2.1 [28]. The dielectric strength varies considerably and is dependent on specimen thickness [25]. For a 2 mm thickness, a dielectric strength of 23,600 V/mm has been measured [2]. Due to its chemical inertness, PTFE resins are resistant to tracking, a phenomenon whereby localized pyrolysis of the polymer may lead to the formation of carbon-conducting paths, resulting in dielectric failure [28]. In alternating-current fields, the small polarization that occurs, due solely to electron and nuclei displacement, responds to the field without lag, and thus, the dielectric loss is small, often as low as 10^{-5}. This value may change with the frequency of the alternating field, particularly for the sintered granular resin, where it may peak at 0.0003 at frequencies above 10^8 Hz [28]. This is still considerably better than that measured for polar polymers such as polyvinyl chloride (PVC). Dielectric constant and loss vary little over the useful range of temperature for PTFE, −40°C to 260°C [25]. Finally, PTFE finds itself at the negative end of the triboelectric series, with a negative electrostatic charge being acquired upon contact with most other polymers [28].

13.2.3.4 Mechanical Properties

The mechanical properties of PTFE are somewhat difficult to quantify as a result of its cold-flow behavior. Under a sustained load, PTFE will creep. Under an imposed strain, stress relaxation will occur. For example, at room temperature under a constant tensile stress of 6.8 MPa, PTFE that has cold-flowed to a strain of 5% soon after loading will continue to elongate, reaching roughly 13% strain by 50 h and 17% strain by 100 h. As a result, an apparent modulus is often discussed for PTFE, with the denominator consisting of the sum of the initial elastic strain and strain resulting from creep. Apparent modulus is therefore a function of time. Some delayed elastic recovery will occur upon unloading of PTFE. Cold flow has a significant effect on the handling of PTFE, e.g., when utilized as a gasket material between bolted flange faces. As stress relaxation will result in a decrease in bolt pressure and corresponding leak, bolts will require retightening following the initial period of most rapid creep. Use of thinner gaskets is also helpful [23]. The tendency to creep can be reduced through addition of inorganic fillers.

At room temperature, the tensile modulus of elasticity, based on initial elastic strain, is measured to be roughly 400 MPa [7], with a Poisson ratio of 0.46. Elastic response begins to deviate from linearity at strains of only a few percent [23]. During tensile testing, PTFE will begin to cold-draw at ~10–13 MPa. The tensile strength of PTFE may vary from ~17 to 31 MPa, whereas elongation at break may range from 300% to 450% [7]. Hardness as measured on the durometer D scale may vary from 50 to 65 [26], also being measured as 98 on the durometer A scale, 58 on the Rockwell R scale [25], and 4 kg/mm^2 on the Vickers scale [37]. Impact strength at room temperature is measured by the Izod test to be roughly ~110–160 J/m [26].

Stress–strain curves, as expected, are drastically affected by temperature; however, even at 260°C, a tensile strength of roughly 6.5 MPa is maintained. Tensile strengths at other temperatures have been measured: 8 MPa at 200°C, 16 MPa at 100°C, 33 MPa at −56°C, and 100 MPa at −195°C. At −56°C, elongation at break is still greater than 30%, being reduced to 6% as temperature is dropped to −195°C [23]. The impact strength of PTFE increases with temperature, exceeding 320 J/m at 77°C [26]. The Poisson ratio at 100°C has been measured as 0.36 [23].

Fillers such as bronze or graphite particles or chopped glass fibers, used to reduce the cold-flow deformation of PTFE and especially to improve its wear resistance (to be discussed later in detail), do not have a large effect on the tensile strength of PTFE. Fillers may provide significant improvements in impact strength, although the elongation capabilities of the polymer are often somewhat reduced [26]. Irradiation processing has also been shown to drastically increase the creep resistance of PTFE, although elongation properties are considerably compromised [35]. However, when irradiated in air, the tensile strength is reduced roughly by a factor of 2 for doses between 2 and 70 Mrad due to scission and reduced molecular weight [35]. If, instead, it is irradiated in vacuum (so as to prevent scission reactions between radicals and oxygen, and favor cross-linking) and at higher temperature (330°C), large increases in tensile strength may be achieved while retaining elongations in the 400% range [38].

13.2.4 Processing

13.2.4.1 Molding

Since the low flow rate of the extremely viscous gel-like PTFE melt is impractical for processing techniques ordinarily used for polymers, other techniques resembling powder metallurgy methods for ceramics and metals are often adopted. Such techniques typically consist of a room-temperature pressing of dry powder (usually granular resin powder) in a mold to the desired shape, followed by sintering of the preform above the melt temperature, which allows the particles to coalesce into a dense homogeneous structure.

Recommended pressures for preforming range from 16 to 31.5 MPa, depending on the grade (size and shape) of the particles being used. Once the desired pressure is reached, a dwell time of 2–3 min/kg of PTFE is recommended for preforming. For some applications, lower preform pressures may be acceptable, though they are usually accompanied by increased porosity. The compression ratio (volume of the unpressed powder divided by the volume of the resultant preform) is usually between 2.7 and 4.4 and is strongly dependent on the shape, size, and density of the initial unpressed powder. Such large compression ratios do not pose difficulty for the preforming of relatively thin articles but will require a long stroke of the press for thicker or longer moldings. Compaction should take place slowly to allow time for air to escape and prevent porosity. For thin moldings, presses may be closed as rapidly as 150 mm/min; however, closure may need to be as gradual as 5 mm/min for thicker, longer moldings [39]. As void content is primarily determined during cold forming, poor press technique and porosity can have a considerably negative effect on tensile strength, elongation at break, and flex fatigue life, as well as dielectric strength [23].

For convenience, preform moldings are normally compacted from the top only. Due to particle/mold friction, such a technique will result in a pressure gradient through the preform, with the lower sections of the molding being subjected to the lower pressures. This difference in pressures may not be of concern in thinner preforms, but for thicker moldings, it is advisable to apply double-end pressing. In this case, the midpoint will experience the lower pressures. As a general guide, the maximum satisfactory length for a cylindrical preform is roughly eight times its diameter. For a tubular perform, the length is limited to 16 times the wall thickness due to these particle/mold friction effects. For longer preforms, successive charging of the powder into the mold and subsequent pressing may be used, though it is suggested that in the preliminary stages, low pressure should be employed (~1 MPa) to avoid the formation of distinct weak interfaces between successive powder charges. Higher final pressure may be applied after all charges have been successively compressed [39].

When pressing is complete, the pressure should be reduced slowly to allow gradual elastic recovery of the preform. As the preform will have a radial compressive stress while in the mold, it should be ejected from the mold with one smooth continuous stroke to avoid the circumferential cracking that would otherwise result from start–stop ejection. As preforms will contain some residual stresses upon ejection, it is advisable to allow a relaxation period of 12–24 h at room temperature prior to the sintering operations [39].

Sintering is usually preformed as a batch process at a temperature between 360°C and 380°C in an electrically heated oven of uniform temperature, normally achieved with a circulating air system. For preforms with section thicknesses <100 mm, a sintering temperature of 380°C is employed, whereas sintering is performed at 360°C for those with thicknesses in excess of 175 mm. The recommended dwell time at the sintering temperature also varies with minimum section thickness, varying linearly from 3 h for a 50 mm thickness to 13 h for a 250 mm thickness. Some sintering will occur at any temperature in excess of the melting point; however, sintering at lower temperatures approaching 327°C will require significantly longer dwell times. Use of temperatures in excess of 390°C to speed sintering is not recommended, as PTFE degradation in air will become significant [39]. Molecular weight is determined primarily by that of the initial powder; however, sintering to temperatures in excess of 390°C may result in decomposition and a reduction of molecular weight [23]. Even at lower temperatures, venting should be provided for toxic decomposition products [39].

In order to minimize stresses in the preform that result from nonuniformities in temperature (especially for preforms of greater section thickness), heating should be gradual. Recommended heating rates should be roughly 100°C/h for section thicknesses of 50 mm, decreasing to roughly 10°C/h for a preform with a section thickness of 200 mm. As a volume increase of 25% occurs upon melting, a dwell period may also be included at 310°C, so that temperature may equilibrate and become more uniform before passing through the melting point [39]. The cooling rate is similarly important because in thicker moldings, fast cooling results in high thermal stresses, which may lead to cracking or distortion. The cooling rate dictates the degree of crystallinity [24] and the size of the crystalline bands [18] that result upon solidification, ranging from 37% to 64% and from 0.2 to ~1.0 μm, respectively, with the higher values corresponding to slower rates of cooling. The lower crystallinity retained by quenching or fast cooling leads to higher tensile strengths and elongations at break, and better flex life. However, slow cooling will produce a material with greater creep resistance, lower permeability to gases and solvents, and lower residual stress and distortion [39]. The cooling rate will depend on the desired final properties. In slow cooling, a dwell period at 300°C may be included to reduce residual stresses. Stress-free moldings can be attained by an annealing period at 250°C. This may be included in the cooling schedule, or a separate annealing operation may be utilized. Very thin sections that withstand more rapid heating and cooling rates may be introduced using a conveyor to transport them into and out of ovens preheated to 380°C in a continuous process [39].

PTFE is typically free-sintered, although in some instances, pressure-cooling techniques are employed wherein the preform is not removed from the mold prior to sintering. Pressure cooling can reduce the tendency of moldings to distort upon cooling; however, the process is more expensive than free sintering and induces residual stresses in the molding that may require annealing upon removal. Free sintering of thin sections may require a means of stabilization, such as being pinned down by a lightly loaded plate, to prevent distortion during cooling [39]. Another process, coining, involves heating a sintered preform to its melt temperature, transferring it to a mold of desired shape, rapidly pressing the heated preform at low pressure, and cooling within the mold to retain the desired shape. Sintered bodies may also be machined to final desired shapes. A tape or sheet may be skived from large cylindrical sintered billets [2], or produced by a calendering process in which powder or a preform is fed between rolls that compress the charge into a thin solid, which is subsequently passed through sintering ovens [7].

13.2.4.2 Ram and Paste Extrusion

As mentioned, the previously described press/sinter molding fabrication techniques are not appropriate for the production of continuous moldings. Screw extrusion is difficult due to the high melt viscosity of PTFE as well as the tendency of the unmelted powder to be compacted by the screw, although some presintered powder grades reduce screw compaction difficulties. Pipes, rods, and other long moldings may, however, be produced by some grades of granular powder resin (free-flowing and presintered powders) via reciprocating ram extrusion. This extrusion process consists of the sequential charging of PTFE powder and preforming by a reciprocating ram, pressing the continuous product down a heated tube where melting and coalescence occur. As a result of thermal expansion and related tube friction, considerable forces are necessary for ram extrusion of granular resin PTFE powder [1,2].

Fine powder (coagulated dispersion) resins may be employed for "paste" or lubricated extrusion using a ram extruder. The powder is mixed with 15–25% hydrocarbon lubricant such as kerosene, white oil, or naphtha, with the resultant blend appearing much like the powder alone. The blend is forced through a finishing die by a constant-rate ram. The blend may be preformed at low pressure (2.5 MPa) prior to extrusion or used directly as a powder. The lubricant allows the blend to be readily compressed and easily extruded through low to medium reduction ratios for the production of thin-walled tubing and hose, wire insulation, ribbon, and tape. Inadequate lubricant concentration results in high extrusion pressures, whereas excessive lubricant prevents coalescence of particle powders during extrusion. Shear stresses on particles may result in fibrillation and alignment along the extrusion direction, increasing longitudinal strength. The continuous extrudate is subsequently passed through a multistage oven. The lubricant is vaporized at a temperature between 93°C and 302°C, and the residual PTFE is subsequently sintered typically at 370°C. The operation, which is essentially semicontinuous, is interrupted occasionally to allow for replacement of polymer charges [1,2].

13.2.4.3 Coatings and Composites

Aqueous dispersions of PTFE may be employed to form coatings on surfaces, or to form films or fibers. These dispersions usually contain smaller PTFE particles, with diameters ranging from 0.1 to 0.3 μm, at a concentration between 30% and 60%. The dispersion also contains a surfactant wetting agent, at between 2.5% and 10%. Spraying, flow coating, or dip coating, as used with cookware and bakeware, is followed by drying via infrared lamp or forced convection ovens below 100°C. Drying is then followed by baking at 260–316°C to remove the wetting agent, and subsequently by sintering at 360°C or more. Thicker coatings may be acquired with multiple dips. Such coatings provide chemical resistance to the underlying substrate material. Freestanding thin films may be formed through casting of the aqueous dispersion upon a smooth surface, such as a polished stainless steel belt. Subsequent drying, baking, and sintering of the film is followed by stripping from the belt. Thicker films may be formed by repeating these steps prior to stripping. Fiber may be formed by forcing the aqueous dispersion through a spinneret and into a coagulating bath. Fibers are subsequently heated, sintered, and hot-drawn to develop strength [1,2].

Aqueous dispersions may also be used for the fabrication of composite materials. This may be done via the impregnation of fibrous or porous structures, such as asbestos, or fiberglass fabric. The fabric, for example, would be passed in a continuous process from a feed spool, through a dip tank containing the dispersion, into a multistage tower oven to dry, bake off the wetting agent, and sinter the PTFE, and would subsequently be collected on a take-up spool. Such architectural fabrics, which take advantage of both the strength of the glass fibers and the weatherability of the PTFE matrix, are widely used today as roofing on sports stadiums and other large arenas. Co-coagulation may be used to form composite materials containing a second component, which is particulate. A dispersion of this second component could be mixed with that of PTFE and the two components coagulated together. The resulting coagulation can be processed by the previously described molding or lubricated extrusion techniques [1]. Particulate fillers, including discontinuous fibers, can instead be introduced into PTFE matrices by mixing with granular powders followed by pressing and sintering, although care must be taken to avoid settling of either of the components. These second components must be able to withstand the thermal conditions required for the subsequent sintering of PTFE powders.

13.2.5 Tribological Bearing Behavior

13.2.5.1 Friction

It has been known for quite some time that PTFE polymer exhibits self-lubricating behavior. In dry sliding contact with smooth counterfaces, the friction coefficient, μ, will often be less than 0.1, particularly at lower speeds and higher temperatures [19,40–42]. Coefficients of dry sliding friction as low as $\mu = 0.056$ have been claimed under some conditions [1]. Additionally, static coefficients of friction may be lower than kinetic coefficients of friction, such that the application of PTFE may prevent stick–slip motion instabilities [2]. The self-lubricating behavior of PTFE, which results from the formation of low-shear-strength PTFE "running" films on the polymer surface as well as "transfer" films on the counterface, has led to the broad application of PTFE as a solid lubricant in a variety of bearing systems where fluid film lubrication is precluded.

Makinson and Tabor [19] were among the first to attempt a structure-based explanation for the tribological behavior of PTFE. During low-speed (1 mm/s) sliding contact with a smooth glass counterface at room temperature, PTFE was found to form a transfer film on the glass with a thickness averaging 20 nm as measured by interference microscopy. This dimension is coincident with that of the crystalline slice thickness noted in the banded structure [17], suggesting a connection between PTFE structure and surface film formation. Speerschneider and Li [18] suggested that the deformation of PTFE occurs via the relative sliding of neighboring crystalline slices, with slip being localized within the noncrystalline regions separating them. In cases where the band axis was parallel to the tensile load axis, slice rotation would first occur to enable deformation via slip between slices [18]. Adopting this deformation model, Makinson and Tabor [19] suggested that the sliding displacement of PTFE is similarly accommodated by slip within the disordered region, transferring a layer of PTFE of thickness corresponding to the slice thickness (20 nm) to the counterface, not unlike a pack of cards being spread out on a table. At low speeds, this film formation process is considered to be the self-lubrication mechanism, and Makinson and Tabor [19] measured resultant friction coefficients as low as $\mu = 0.07$.

Tanaka et al. [40] invoked a similar model to explain the transfer wear of PTFE under low-speed conditions against smooth glass counterfaces. Transfer films were described as fibrous, with fibers running along the sliding direction. A transfer "fiber" is formed by the serial connection of crystalline slices that originate from slip within the amorphous regions of bands located at each area of real contact on the PTFE surface. The fiber thickness would therefore be coincident with the slice thickness (roughly 20–30 nm), whereas the fiber width would coincide with the width of the band.

Makinson and Tabor [19] and Tanaka et al. [40] each noted alignment of PTFE molecules comprising the surface films along the sliding direction, deduced from electron diffraction. As the model of the banded structure suggests that linear PTFE molecules lie in the plane of the crystalline slice, it was concluded that a preferred orientation of the transferred crystalline slices must therefore exist in which the slices are rotated such that the molecule chains align with the sliding direction.

Makinson and Tabor [19] do mention that an alternate explanation for the diffraction data indicating molecular orientation along the sliding direction could instead involve the extension of individual molecules. Based on the directionality and length of splits that would develop in aged transfer films, Steijn [43] arrived at this same conclusion, as did Pooley and Tabor [44] in a later study. Observing the curled edges of such splits, Pooley and Tabor [44] also measured a PTFE transfer-film thickness of only 2 nm. As this is less than the crystalline slice thickness by an order of magnitude, it was suggested that the development of thin, oriented PTFE transfer films could indeed result from the drawing and extension of individual molecular chains along the sliding direction. This drawing process is enabled by the linearity of the straight-chain PTFE molecule and the degree of lateral mobility that exists among neighboring chains.

This observation is also in agreement with those made using a number of analytical techniques that have been developed over the past several decades. Pepper [45] used Auger electron spectroscopy (AES) to estimate PTFE transfer-film thickness, formed during sliding contact with an atomically clean metal under vacuum, to be only two to four atomic layers. Wheeler [46] employed XPS also to measure atomic-scale transfer-film thickness produced under similar sliding conditions. Furthermore, Blanchet et al. [47] performed angle-resolved XPS on such thin transfer films and found them to consist of strands that fractionally cover the metal counterface to an increasing degree as subsequent sliding passes of the PTFE are made. Lauer et al. [48] used infrared emission spectroscopy and ellipsometry to estimate the PTFE transfer-film thickness formed during sliding contact with steel in air to be monomolecular and to deduce molecular orientation. Yang et al. [49] employed nuclear reaction analysis to measure similarly thin PTFE transfer films formed on steel counterfaces during the initial traversals by a PTFE pin. Dietz et al. [50] successfully used atomic force microscopy (AFM) to image molecular structure and orientation in transfer films formed atop a glass substrate after a PTFE bar has been dragged across it. In all of the above instances, the film thickness is at least an order of magnitude less than the crystalline slice thickness, indicating that the transfer-film formation process must involve the extension, drawing, and orientation of individual PTFE molecules.

The extension, drawing, and orientation of individual PTFE molecules, as well as the potential slip localized within the amorphous region separating crystalline slices, are viscoelastic deformation processes. The force necessary for such processes (manifesting itself as friction) depends on the relative magnitudes of the strain rate imposed (related to the sliding speed, V) and the reciprocal of the relaxation time of the viscoelastic polymer (which has an Arrhenius temperature dependence for semicrystalline polymers) [51]. Thus, friction is observed to increase with increasing speed and decreasing temperature. The friction coefficient of dry sliding systems has been found to be proportional to the product of a speed-related factor, V^n, and a temperature-related factor, $\exp\{\Delta H/(RT)\}$, where the speed exponent n has been estimated to be 0.26–0.4 [40,42]. The activation energy, ΔH, was originally thought to be in the 20–40 kcal/mol range [52] but has since been measured to be in the 7–10 kcal/mol range [40–42,53]. The friction coefficient has also been observed to decrease with increasing contact pressure.

13.2.5.2 Wear

Despite its low-friction properties, PTFE undergoes severe rates of sliding wear at most typical engineering speeds. At very low speeds (less than ~10 mm/s at room temperature), wear may actually proceed at a mild rate (volume lost per distance slid per unit normal load) of ~10^{-5} mm^3/Nm. At these speeds, mild transfer wear involves the cyclic removal and replenishment of the thin running and transfer films that afford PTFE its self-lubricating behavior. As speed is increased or temperature decreased, a sharp transition occurs as the wear rate becomes severe, jumping by roughly two orders of magnitude to the ~10^{-3} mm^3/Nm range [40–42]. This transition is accompanied by the generation of large platelike debris with a thickness of several micrometers and dimensions of several hundred micrometers in the plane of the debris plate [42,54]. The friction coefficient also increases with increasing speed or decreasing temperature, and the transition to severe wear tends to always occur at speed/temperature combinations that yield a friction coefficient slightly above 0.10, inducing this severe wear mechanism [42]. Thus, unfilled PTFE alone is generally not an appropriate material for bearing components in higher-speed sliding contacts.

13.2.5.3 Wear-Resistant Composites

It has been shown that the unacceptably high rate of transfer wear of PTFE can be drastically reduced, typically by a couple orders of magnitude, by the inclusion of any of a very broad variety of particulate microfillers [55–58]. Many mechanisms have been proposed over the years to explain this wear reduction, and these can be roughly broken down into four categories: prevention of surface film formation, increased transfer-film adhesion, preferential load support by a wear-resistant filler, and control of debris size.

It had been suggested that the transfer wear process of unfilled PTFE involves the formation of surface films through the destruction of the banded structure as slip occurs in the amorphous regions separating neighboring crystalline slices. Tanaka et al. [59–62] proposed that microfillers reduce wear by preventing this destruction of the banded structure. This mechanism suggests that surface films are no longer formed, which counters microscopic observation of sliding surfaces as well as the low coefficients still provided by some composites that depend on the formation of running and transfer films. Briscoe et al. [63] instead suggested that transfer films still form on the sliding counterfaces but that fillers increase the adhesion of these transfer films such that the removal step of the cyclic film deposition/removal transfer wear process becomes limiting. Much speculation has been made regarding the role of the filler in this increased transfer-film adhesion, varying from direct PTFE/counterface chemical bonding that may result from localized stress, temperature, or oxide removal induced by the filler, to bonding of the filler itself to the counterface with the PTFE film mechanically interlocked with these adherent filler particles [63–69]. However, the occurrence of these reactions of the counterface with the relatively inert PTFE has been questioned [46,47,70,71]. As transfer films formed by PTFE are in layers, with deposition occurring atop previously deposited films [49], the occurrence of chemical bonding between the first transfer layer and the counterface, or lack thereof, may have little effect on transfer wear behavior [70].

Lancaster [56] suggested that wear-resistant fillers reduce wear simply because they support a fraction of the normal load, thus shielding the PTFE matrix. Such a contribution to wear resistance can be demonstrated by simple rule-of-mixtures approaches, with various contact pressures being assumed for the filler particulates and matrix material [72,73]. Ricklin [74] suggested that, in addition to preferential load support, the wear reduction mechanism of particulate microfillers might have a larger contribution from the control of the size and shape of the debris formed. Unlike the production of large platelike debris by the unfilled PTFE, the microfillers seem to disrupt this wear particle formation and instead yield finer debris, with a resultant lower rate of wear for the composite material [42,47,54]. This mechanism is aided by an enrichment of the surface region that occurs upon sliding, as the more wear-resistant fillers tend to accumulate [59,60,75,76].

PTFE composites are now used in many applications where conventional fluid lubricants are inappropriate for reasons of evaporation, contamination, pumping, or maintenance. Such applications include textile or food processing equipment, business paper handling machines [55], machine tool guideways [76,77], bridge support bearings that allow thermal deck expansions, vacuum systems, space mechanisms [78], and even retainers for rolling element bearings in the liquid oxygen turbopumps of missiles [79] or the Space Shuttle [80].

Many hard fillers employed in these composites, although reducing the wear of PTFE, may also induce an abrasive wear of the mating counterface as well as sacrifice some of the originally desirable low-friction behavior of the matrix polymer [5,42,79,81]. Polymeric fillers (instead of hard inorganic fillers), such as polyoxybenzoate [81], polyphenylene sulfide [81], and fluorinated ethylene propylene (FEP) [82] have also been shown to reduce the wear of PTFE without inducing such abrasive wear of the mating counterface.

As opposed to the PTFE microcomposites discussed above, where microfiller particles may enable wear that is reduced typically by a couple orders of magnitude, more recent investigations on nanoparticle fillers have revealed a greater ability to reduce PTFE wear by up to four or even five orders of magnitude. Rather than being more general across a broad range of filler materials, as seems to be the case for microcomposite performance, such more highly effective nanofiller wear reduction performance seems to be much more limited, not only to particular filler materials, such as alumina and carbon, but furthermore, to only certain of their phases [83,84].

13.2.5.4 Radiation-Modified Tribological Behavior

Briscoe and Ni [35] demonstrated that the irradiation of unfilled PTFE to doses of 20–100 Mrad can greatly enhance wear resistance, in some cases by nearly two orders of magnitude. Unfortunately, as irradiation alters the linear molecular profile that originally afforded PTFE its self-lubricity, the observed doubling of the kinetic coefficient of friction would not be unexpected [35]. However, although they did not report on wear, McLaren and Tabor [37] found instead that at sliding speeds of 0.1 and 1 m/s, the friction coefficient of irradiated PTFE had maxima at a dose of 5 Mrad and then decreased with increasing dose, settling at a value similar to that of the unirradiated PTFE. It has since been shown that PTFE irradiated in air to doses in excess of 5 Mrad will have both a friction coefficient and a wear rate that drop with increasing dose. At a dose of 30 Mrad, the friction is similar to that of unirradiated PTFE, whereas the wear rate has been reduced by a factor of roughly 2000 from that of the unirradiated PTFE [85]. In the 2.5–5 Mrad dose range, Fisher and Corelli [30] found that PTFE irradiated in air experienced branching and cross-linking, whereas higher doses heavily favored scission and the formation of lower-molecular-weight chain fragments, which may explain the return to lower friction behavior.

The discussion of wear and friction of irradiated PTFE and PTFE composites thus far has assumed that mating counterfaces are smooth, as is normally the case for engineering surfaces. However, if the mating counterfaces are rough and abrasive to the polymer, irradiation of PTFE or compounding of it with fillers into composites may actually increase abrasive wear rate by compromising elongation and ductility [56,86]. In sliding against rough counterfaces, the unfilled unmodified polymer will be superior as its wear rate is relatively independent of surface roughness over the range R_a = 0.02–1.0 µm [87]. This probably results from the transfer behavior of PTFE, which considerably modifies and smoothens rougher counterfaces.

Irradiation may also be used for the production of PTFE micropowders or waxes. Scission of PTFE with a molecular weight originally on the order of $M_n = 10^7$ or more by heavy doses of electron or γ irradiation yields a decrease in M_n to values of $2.5–25 \times 10^4$. Micropowders may also be formed by thermal or shear degradation. These low-molecular-weight micropowders, which are commonly added to plastics, coatings, finishes [2], lubricants [88], and composite coatings [89], are widely used in industry to provide low-friction, nonstick properties.

13.2.6 Electronics

As previously mentioned, the exceptional electrical resistivity and dielectric strength of PTFE and its resistance to tracking, weathering, chemical attack, and melting have led to its utilization as an insulator in hook-ups, computer back panels, interconnects, and various wiring, spaghetti tube electrical sleeving, and cabling. Due to its low dielectric constant and dielectric loss, PTFE is a particularly attractive insulation for high-frequency AC applications, where the alternating polarization of polymers such as PVC would lead to significant dielectric heating and potential thermal breakdown [7]. PTFE's low dielectric constant also implies that there will be less electrical interference and cross talk between neighboring circuits. As a result, PTFE is a popular substrate for high-frequency printed circuit boards, where the trend is toward faster circuits on smaller boards with copper circuit lines packed more closely [90].

Typically, such circuit boards are produced by a multistep process in which a thick copper film is mechanically bonded to the PTFE substrate. The copper is covered with a photoresist that is subsequently patterned. The unwanted copper is removed chemically, leaving the desired circuits. This process is not optimal, because it is subtractive and results in troublesome chemical wastes, and it involves thick copper films so that fine, tightly packed circuit lines are difficult to produce [90]. Furthermore, stress release that occurs upon etching of the thicker copper film results in warpage that makes subsequent fabrication difficult unless weighty metal backers have been attached to the circuit board substrate to resist deflection [91]. An additive process yielding thinner microstrip lines would be more satisfactory.

Several researchers have studied the effect of radiation on the adhesion of metals to PTFE surfaces [92,93], as a potential means for writing microstrip circuit lines directly on PTFE. Wheeler and Pepper [92] showed that x-ray irradiation, which could be masked to form a pattern on the substrate, produced chemical changes that improved the adhesion of subsequently evaporated nickel films. Jiang et al. [93] showed a similar effect upon evaporated copper films using a pulsed UV excimer laser, as a result of chemical and topographical changes in the PTFE substrate. Assuming that the evaporated metal, adhering less strongly to unmodified regions of the PTFE, could be preferentially removed mechanically by adhesive tape, for example, thin microstrip lines composing the desired printed circuit would remain.

Rye [33,94] developed an alternative whereby evaporated copper microstrip circuit lines would only remain on regions not irradiated. Regions of PTFE irradiated in vacuum, either by x-ray or electron beam, cross-link and therefore are not chemically treated by subsequent chemical etching by sodium in liquid ammonia or by complexes of sodium and naphthalene. During subsequent chemical vapor deposition with Cu(I) β-diketonate precursors, copper metal deposition occurs only on etched (unirradiated) regions, yielding the desired printed circuit. Or electroless copper deposition may also be employed, whereby copper from a metal-ion solution only adheres strongly to etched regions. The subsequent application and peeling of adhesive tape would remove copper only from the irradiated regions. These more straightforward additive processes would enable the formation of thinner, narrower (50 µm) circuit lines that are more closely packed than lines formed by conventional subtractive processes. As this technique is further refined, it has the potential to evolve from a means of connecting microelectronic devices on printed PTFE circuit boards to a means of actual fabrication of microelectronic devices themselves on PTFE substrates [88].

13.3 OTHER FLUOROPOLYMERS

Sperati [1] described the main advantage of PTFE as often being the material that can perform tasks under demanding conditions that no other materials can tackle. Its main disadvantage is that it is expensive, not only as a result of the extreme care that must be taken during polymerization but also due to the fabrication techniques that must be employed that are so unconventional for a thermoplastic. In attempts to address some of PTFE's shortcomings, most notably its lack of melt processability, other fluoropolymers have since been developed and commercialized in response. While addressing a shortcoming, each such fluoropolymer alternative often compromises a desirable attribute held by PTFE, spurring continued development of further varieties. The resulting fluoropolymers are often broadly categorized by whether they are fully perfluoropolymers or only partially fluorinated, or by whether they are homopolymers or are copolymerized. For further reference, several thorough reviews on the topic of fluoropolymers have recently been conducted [95,96]. The comparably brief overview of these alternate fluoropolymers that follows is sequenced according to the timeline of their commercialization. For each fluoropolymer, commercial suppliers are stated, and the approximate values of various properties stated for these fluoropolymers are typical among those more exhaustively listed on the supplier websites of technical specifications.

13.3.1 POLYCHLOROTRIFLUOROETHYLENE

Though its initial commercialization as Kel-F 81 by the M.W. Kellogg Company did not occur until 1953, the pre–World War II discovery of polychlorotrifluoroethylene (PCTFE) by the I.G. Farben Company, Germany, in 1934 actually predates by a few years that of PTFE. It is a partially fluorinated homopolymer, with chlorine replacing a fluorine in the monomer, $-(CF_2CFCl)_n-$. The larger chlorine disrupts the helical symmetry of the molecules as well as their close packing, though still maintaining semicrystallinity upon cooling with a microstructure that is spherulitic rather than banded. This disruption of the smooth molecular profile results in a reduction of melt point from 327°C for PTFE to approximately 212°C, enabling melt processability with melt viscosity drastically reduced to less than 10^1 Pa s as compared to >10^{10} Pa s for PTFE. Thermal degradation of PCTFE via chain scission may begin to occur significantly at temperatures as low as 250°C, however, not far beyond its melt point; thus, precise control of temperature must be exercised during such melt processing.

PCTFE's tensile strength of 34–39 MPa is increased from the 17–31 MPa range for PTFE, as is its hardness, in the 80–90 range Shore D as compared to 50–65 for PTFE, and its resistance to cold-flow deformation. PCTFE is also stiffer, with its tensile modulus increased to approximately 1.4 GPa, from 0.4 GPa for PTFE. The ductility of PCTFE is somewhat reduced, down to the 100–250% range for elongation at break, less than the typical values of 300–450% for PTFE. It is also relatively expensive. However, the primary trade-off accompanying PCTFE's improved melt processability is its diminished maximum operating temperature, reduced to the 132–180°C range from a typical value of 260°C for PTFE.

PCTFE has excellent resistance to chemical attack approaching that of PTFE, as well as excellent resistance to water absorption of PTFE, and thus is frequently applied in a film form as a protective barrier coating. Kel-F 81 PCTFE continued to be offered by the 3M Company from the 1950s until being discontinued in 1996, yet it remains commercially available from various other suppliers such as Daikin (Neoflon), although still often being commonly referred to as Kel-F.

13.3.2 FLUORINATED ETHYLENE PROPYLENE

Invented and commercially introduced by DuPont in 1960, FEP copolymerizes TFE, $-(CF_2CF_2)-$, with typically 5% up to no more than 15% hexafluoropropylene (HFP), $-(CF_2CFCF_3)-$. As with the chlorine of PCTFE, the bulky trifluoromethyl $-CF_3$ group of the hexafluoropropylene monomer disrupts the smooth molecular profile and close packing of PTFE, reducing the melt point of this semicrystalline and fully fluorinated polymer to 260°C and thus making it the first melt-processable perfluoropolymer. Its melt viscosity is reduced from >10^{10} Pa s for PTFE down to the range of 10^4–10^5 Pa s, and since it is fully fluorinated, it better maintains the corrosion and chemical resistance of PTFE. Though its 200°C maximum operating temperature is superior to PCTFE, the reduced melt point associated with melt processability still renders this maximum operating temperature lower than the 260°C value typical of PTFE.

The strength, hardness, and ductility of FEP do not differ greatly from PTFE, falling within a comparable range, while values of elastic modulus may be slightly (~15%) higher. While FEP is tougher than PTFE, resisting break during Izod impact testing, it has a lesser fatigue resistance as represented by folding endurance, down from >10^6 cycles for PTFE to a range of 5–80 × 10^3 cycles for FEP. The disruption of the smooth linear molecular profile by the bulky trifluoromethyl side groups and the reduced mobility by which the molecules may draw by one another result in sliding friction that is increased twofold to threefold relative to those of PTFE but, by the same mechanism, allow FEP to better resist cold-flow and creep deformation.

FEP maintains many of the electrical properties of PTFE such as high volume resistivity, while having an improved dielectric strength that is approximately doubled to 53 V/mm, with wire and cable insulation thus being a common application. FEP is commercially available through various suppliers, including DuPont under the name Teflon FEP.

13.3.3 Polyvinylidene Fluoride

Though not introduced into the commercial marketplace until another 13 years later by Pennwalt Corporation, the polymerization of the partially fluorinated polyvinylidene fluoride (PVDF), –$(CH_2CF_2)_n$–, was first claimed in 1948. This replacement of two fluorines by hydrogen within the repeating unit of PTFE also renders the fluoropolymer melt processable, with a melt viscosity reduced to the range of 200–17,000 Pa s depending on the processing temperature, with its 177°C melt point being the lowest among melt-processable fluoropolymers. These melt characteristics make PVDF attractive for processing of wire and cable insulation, with its electrical resistivity of ~5 × 10^{14} Ω cm and high dielectric strength of ~65 kV/mm, as well as for extruding tubing and piping, with it comprising an ~20% share of worldwide fluoropolymer production, second only to PTFE itself at about 70%. PVDF is relatively inexpensive among fluoropolymers and is commercially available through several producers, such as Arkema (Kynar) and Solvay (Hylar, Solef).

PVDF's practical low melt temperature is, in turn, accompanied by a reduced maximum use temperature of approximately just 150°C. The tensile strength of PVDF can be considerably higher than PTFE, from ~31 MPa, which is the upper end of the range for PTFE, up to as high as ~50 MPa, with hardness being increased from Shore D values of 50–65 for PTFE up to 75–80. It is also twofold to fourfold stiffer than PTFE, with a Young's tensile modulus in the range of 1–2 GPa, though its ductility in some instances is greatly reduced, with elongations at break of 250% to as low as only 50%.

With the replacement of half the fluorine by hydrogen, the specific gravity of PVDF drops from a value of 2.2 more typical of perfluorinated polymers down to 1.7. While generally maintaining the helical molecular conformation of PTFE and solidifying into a semicrystalline state, under special conditions of stretched tension from room temperature to approximately 50°C, the smaller hydrogen may allow the molecule to still accommodate the remaining larger fluorine while also reverting to a planar zigzag conformation. This phase is unusual in that it is piezoelectric, an order of magnitude more so than observed for any other polymer. With greater compliance than piezoelectric materials that are typically ceramic, PVDF is a promising sensor material for health and usage monitoring of structural systems.

13.3.4 Polyvinyl Fluoride

DuPont polymerized polyvinyl fluoride (PVF), –$(CH_2CHF)_n$–, in 1949, developing and commercially introducing it by the name Tedlar in 1961. With the replacement of fluorine by smaller hydrogen beyond that of PVDF, the molecular conformation becomes the planar zigzag of polyethylene rather than the helical conformation of PTFE, with specific gravity again reduced to 1.7. As with PVDF, the melt point is also greatly reduced, down to 190–200°C, also with a reduced maximum use temperature of 107°C. While it is a thermoplastic, it is not melt-processed, as the 204°C degradation temperature at which it begins to evolve hazardous HF gas is so close to the melt point and presents an impractically narrow melt-processing temperature range. While possessing broad chemical resistance, at elevated temperatures above 100°C there are a few highly polar solvents into which PVF can be dissolved, and it is by such solutions that PVF can be processed into film and coating forms by approaches such spraying, dipping, or casting onto substrates, or extrusion as plastisol to other forms such as self-standing films, which may subsequently be adhered to other surfaces or laminated, with the solvent being subsequently driven off. Given its resistance to weathering and staining, PVF sees application in film form upon fabrics and metallic architectural sidings, and also upon aircraft cabin interiors, given its low flammability. PVF additionally is commonly used in thin-film form as a mold release agent in molding processes.

13.3.5 Ethylene Chlorotrifluoroethylene

With the discovery that ethylene could be copolymerized well with fluorinated monomers, an alternating copolymer of chlorotrifluoroethylene, $-(CF_2CFCl)-$, with ethylene, $-(CH_2CH_2)-$, at a 1:1 ratio was the first of such to be introduced commercially, by Ausimont in 1970, and is currently commercially available in several grades as Halar through Solvay. As with PVF, such reduced fluorination via smaller hydrogen allows ethylene chlorotrifluoroethylene (ECTFE) to adopt an extended planar zigzag molecular conformation, with the solid being semicrystalline and specific gravity reduced to 1.7 from the 2.1–2.2 values of PCTFE and PTFE. Its melt point of 242°C and resistance to thermal degradation up to temperatures of 380°C provide practical melt processability, with melt viscosities ranging from $\sim 10^3$–10^5 Pa s over the various available grades. ECTFE melts, however, can be corrosive to typical screw and barrel extrusion components; thus, corrosion-resistant extruder materials are required.

As with PVDF, which similarly is only partially fluorinated with high hydrogen content, ECTFE has high tensile strength of ~55 MPa, hardness of ~75 on the Shore D scale, and tensile modulus of elasticity of ~1.6 GPa, relative to those of PTFE (17–31 MPa, 50–65, and 0.4 GPa, respectively). ECTFE, however, is not as compromised as PVDF in its ductility, maintaining an elongation at break of ~250%, closer to that of PTFE. Like PVDF, useful mechanical properties are maintained by this melt-processable ECTFE fluoropolymer through a maximum use temperature of about 150°C, above which it becomes prone to a slow brittle stress cracking failure mechanism under sustained loading.

ECTFE has the highest dielectric strength among fluoropolymers, approaching 80 kV/mm with a resistivity $\sim 10^{16}$ Ω cm making it an attractive wire and cable insulation material, particularly in applications potentially benefitting from its excellent low flammability. As a protective coating, ECTFE is exceptional, as it possesses the greatest resistance of all fluoropolymers to the permeation of smaller molecules that may otherwise lead to corrosion of underlying substrates. In addition to melt processing, ECTFE also lends itself well to coating from fine powders via electrostatic approaches.

13.3.6 Ethylene Tetrafluoroethylene

Following ECTFE, DuPont likewise copolymerized ethylene, $-(CH_2CH_2)-$, with alternating TFE, $-(CF_2CF_2)-$, and commercialized this ethylene tetrafluoroethylene (ETFE) in the early 1970s under the name Tefzel. As with other fluoropolymers of only partial fluorination and with high hydrogen content, ETFE adopts an extended planar zigzag molecular conformation within a semicrystalline structure, with a reduced 1.76 specific gravity. Though melt point, 267°C, is somewhat higher than

ECTFE, it is still readily melt-processable as thermal degradation does not become great until above ~380°C, providing a wide window, 325–345°C, of typical processing temperatures for which ETFE's melt viscosity is within a practical range of 700–10,000 Pa s. As with other highly hydrogenated fluoropolymers, ETFE possesses higher strength and stiffness than perfluoropolymers, with a tensile strength 40–46 MPa, hardness of 72 Shore D, and a tensile modulus of elasticity ~827 MPa, while maintaining ductility with elongations at break in the 150–300% range, which, like ECTFE, more nearly approaches the higher values of perfluoropolymers, in contrast to values as low as 50% for the comparably hydrogen-containing PVDF. As a result, ETFE possesses high toughness, fully resisting break during Izod impact testing, as also observed for FEP. ETFE maintains such useful mechanical properties through a maximum service temperature of 150°C, equivalent to the ECTFE copolymer. As compared to their perfluorinated counterparts, the hydrogen content of such partially fluorinated polymers results in much higher values of the friction coefficient, with ETFE typically in the range of 0.3–0.4.

ETFE possesses a very high dielectric strength, nearly equal to the 80 kV/mm of ECTFE, and even higher resistivity, >10^{17} Ω cm, with its melt processability rendering it a useful wire and cable insulation material, especially in nuclear applications due to its ability to withstand ionizing radiation, with a rate of cross-linking that balances out the rate of scission degradation. Purposefully providing such radiation cross-linking following ETFE's melt processing can expand its maximum service temperature, up to 240°C. ETFE is similarly resistant to UV radiation and weather, having demonstrated Florida aging exposures of 15 years while remaining unaffected as well as accelerated aging exposures simulating twice that duration. Combined with the nonstick self-cleaning surface characteristics of fluoropolymers, ETFE in sheet form has seen extensive application as a flexible roofing material for sports arenas and other building structures. In addition to DuPont, ETFE is commercially available through several other producers, for example Asahi Glass (Fluon ETFE) and Daikin (Neoflon ETFE).

13.3.7 Perfluoroalkoxy

DuPont invented perfluoroalkoxy (PFA) and commercially introduced it in 1972. It is a perfluorinated copolymer of TFE, –(CF_2CF_2)–, with –(CF_2CFOR)–. In the case of PFA propyl vinyl ether (PPVE) monomer, the alkoxy links a fluorinated propyl R side group –$CF_2CF_2CF_3$, while in the case of PFA methyl vinyl ether (PMVE) monomer with the alkoxy instead linking a fluorinated methyl R side group –CF_3, the resulting copolymer may be distinguished as methyl perfluoroalkoxy (MFA). These copolymers possess melt points that may range from 253°C to as high as 315°C, approaching that of PTFE, yet the inclusion of the PFA groups at a typical 1% level relative to TFE provides sufficient disruptively bulky side groups to possess low melt viscosities in the range of 4000–30,000 Pa s, practical for melt processing. PFA is resistant to thermal degradation and can be melt-processed at temperatures up to 445°C, though processing temperatures in the range of 340–395°C are more typical.

As a perfluoropolymer, the specific gravity returns to higher values of about 2.15. The mechanical properties of tensile strength 25 MPa, hardness of about 60 Shore D, and tensile modulus in the range of 440–480 MPa, well below the increased levels observed in hydrogen-containing PVDF and ETFE that are only partially fluorinated, return to levels comparable to the other PTFE and FEP perfluoropolymers. The primary advantage offered by PFA is the increase in maximum service temperature from 200°C for the similar FEP copolymer back up to the 260°C value for PTFE, while maintaining FEP's melt processability. While not as fatigue resistant as PTFE, the folding endurance of up to 500 kilocycles for PFA is far less negatively compromised than FEP, which is characterized by a maximum endurance of 80 kilocycles. Also, as compared to these other perfluoropolymers, PFA offers the advantage of a higher dielectric strength, ~80 kV/mm, comparable to those offered by the partially fluorinated alternatives and four times that of PTFE. Coupled with the excellent chemical resistance common to perfluoropolymers, such properties make PFA an attractive selection for wire and cable insulation, tubing, and plastic labware applications. In addition to

DuPont's Teflon PFA, other commercially available PFA copolymers include Daikin's Neoflon and Solvay's Hyflon PFA.

13.4 OTHER CONTINUING FLUOROPOLYMER DEVELOPMENTS

With the ability to copolymerize FEP, ETFE, ECTFE, and PFA, certainly, combinations of fluoropolymer properties may be further tailored via terpolymers, and THV (TFE/HFP/VDF) and HTE (HFP/TFE/ethylene) serve as a couple of the more common examples of such further developments. And whereas the most prominent industrial fluoropolymers surveyed above are all semicrystalline, in 1985, DuPont introduced a Teflon AF amorphous fluoropolymer, formed by copolymerizing TFE with perflurodimethyldioxole (PDD) with side groups sufficiently bulky to inhibit crystallization. Asahi Glass coincidentally developed an amorphous homopolymer, Cytop, using perfluorobutenylvinylether (PBVE), while in 1997, Solvay similarly introduced Hyflon AD, another amorphous copolymer. These amorphous fluoropolymers newly offer optical clarity and also solubility in select solvents, thus additionally enabling thin-film coating processes. Additionally, with the commercial development of fuel cells over recent decades, extensive use as a membrane material has been made of DuPont's Nafion, a copolymer of TFE and perfluoroalkylvinylether whose conversion of end groups to sulfonic acid introduces useful ionic properties, placing fluoropolymers centrally among the further development of ionomer materials [97]. With its unique properties and such evolving new applications, PTFE and its family of related fluoroplastics will continue to affect our lives in positive ways well into its second century.

REFERENCES

1. C. A. Sperati, Fluorocarbon polymers, polytetrafluoroethylene (PTFE), *Handbook of Plastic Materials and Technology* (I. I. Rubin, ed.), John Wiley & Sons, New York, 1990, p. 117.
2. S. V. Gangal, Polytetrafluoroethylene homopolymer of tetrafluoroethylene, *Encyclopedia of Polymer Science and Engineering*, Vol. 16 (H. F. Mark and J. I. Kroschwitz, eds.), John Wiley & Sons, New York, 1989, p. 577.
3. H. S. Eleuterio, Fluoropolymers: An explorer's vantage point, *J. Macromol. Sci. Chem. A28*: 897 (1991).
4. Fluon Fluoropolymers Polytetrafluoroethylene, ICI Literature Reference FB/200/GB/1/05 89, 1989.
5. S. Ebnesajjad, Introduction to fluoropolymers, *Applied Plastics Engineering Handbook* (M. Kutz, ed.), Elsevier, Amsterdam, 2011, p. 49.
6. B. C. Arkles and M. J. Schireson, The molecular weight of PTFE wear debris, *Wear 39*: 177 (1976).
7. F. W. Billmeyer, *Textbook of Polymer Science*, John Wiley & Sons, New York, 1984, p. 398.
8. R. G. Brown, Vibrational spectra of polytetrafluoroethylene: Effects of temperature and pressure, *J. Chem. Phys. 40*: 2900 (1964).
9. R. I. Beecroft and C. A. Swenson, Behavior of polytetrafluoroethylene (Teflon) under high pressures, *J. Appl. Phys. 30*: 1793 (1959).
10. T. W. Bates and W. H. Stockmayer, Conformational energies of perfluoroalkanes. III. Properties of polytetrafluoroethylene, *Macromolecules 1*: 17 (1968).
11. C. W. Bunn and R. R. Howells, Structures of molecules and crystals and fluorocarbons, *Nature 174*: 549 (1954).
12. N. G. McCrum, C. P. Buckley, and C. B. Bucknall, *Principles of Polymer Engineering*, Oxford University Press, New York, 1992, p. 36.
13. R. E. Moynihan, The molecular structure of perfluorocarbon polymers. Infrared studies on polytetrafluoroethylene, *J. Am. Chem. Soc. 81*: 1045 (1959).
14. S. K. Biswas and K. Vijayan, Friction and wear of PTFE: A review, *Wear 158*: 193 (1992).
15. H. A. Rigby and C. W. Bunn, A room-temperature transition in polytetrafluoroethylene, *Nature 164*: 583 (1949).
16. F. A. Quinn, D. E. Roberts, and R. N. Work, Volume–temperature relationships for the room temperature transition in Teflon, *J. Appl. Phys. 22*: 1085 (1951).
17. C. W. Bunn, A. J. Cobbold, and R. P. Palmer, The fine structure of polytetrafluoroethylene, *J. Polym. Sci. 28*: 365 (1958).
18. C. J. Speerschneider and C. H. Li, Some observations on the structure of polytetrafluoroethylene, *J. Appl. Phys. 33*: 1871 (1962).

19. K. R. Makinson and D. Tabor, The friction and wear of polytetrafluoroethylene, *Proc. Roy. Soc. 281*: 49 (1964).
20. P. Marx and M. Dole, Specific heat of synthetic high polymers. V. A study of the order–disorder transition in polytetrafluoroethylene, *J. Am. Chem. Soc. 77*: 4771 (1955).
21. T. Yamamoto and T. Hara, X-ray diffraction study of crystal transformation and molecular disorder in poly (tetrafluoroethylene), *Polymer 23*: 521 (1982).
22. N. G. McCrum, An internal friction study of polytetrafluoroethylene, *J. Polym. Sci. 34*: 355 (1959).
23. L. H. Gillepsie, D. O. Saxton, and F. M. Chapman, New design data for FEP and PTFE. 1. Strength and deformation, *Machine Design*, January 21, 1960, p. 126.
24. T. Hu and N. S. Eiss, The effects of molecular weight and crystallinity on wear of polytetrafluoroethylene, *Proceedings of the International Conference on the Wear of Materials*, 1983, p. 636.
25. L. H. Gillepsie, D. O. Saxton, and F. M. Chapman, New design data for FEP and PTFE. 2. Thermal, wear and electrical properties, *Machine Design*, February 18, 1960, p. 156.
26. Teflon Fluorocarbon Resin Mechanical Design Data, E. I. du Pont de Nemours and Co., Inc.
27. D. V. Rosato, *Plastics Encyclopedia and Dictionary*, Oxford University Press, Oxford, 1993, p. 571.
28. C. M. Hall, *Polymer Materials*, John Wiley & Sons, New York, 1981, p. 92.
29. D. R. Wheeler and S. V. Pepper, X-ray photoelectron and mass spectroscopic study of electron irradiation and thermal stability of polytetrafluoroethylene, *J. Vac. Sci. Technol. A 8*: 4046 (1990).
30. W. K. Fisher and J. C. Corelli, Effect of ionizing radiation on the chemical composition, crystalline content and structure, and flow properties of polytetrafluoroethylene, *J. Polym. Sci. Polym. Chem. 19*: 2465 (1981).
31. W. K. Fisher and J. C. Corelli, Flow properties of various irradiated resins of PTFE (Teflon), *J. Appl. Polym. Sci. 27*: 3769 (1982).
32. J. A. Kelber, J. W. Rogers, and S. J. Ward, Effects of low-energy electron bombardment on the surface structure and adhesive properties of polytetrafluoroethylene, *J. Mat. Res. 1*: 717 (1986).
33. R. Rye, Spectroscopic evidence for radiation-induced crosslinking of poly(tetrafluoroethylene), *J. Polym. Sci. B Polym. Phys. 31*: 357 (1993).
34. D. R. Wheeler and S. V. Pepper, Effect of X-ray flux on polytetrafluoroethylene in X-ray photoelectron spectroscopy, *J. Vac. Sci. Technol. 20*: 226 (1982).
35. B. J. Briscoe and Z. Ni, The friction and wear of γ-irradiated polytetrafluoroethylene, *Wear 100*: 221 (1984).
36. X. Zhong, L. Yu, W. Zhao, J. Sun, and Y. Zhang, Radiation-induced crystal defects in PTFE, *Polym. Degrad. Stab. 40*: 97 (1993).
37. K. G. McLaren and D. Tabor, The friction and deformation properties of irradiated polytetrafluoroethylene (PTFE), *Wear 8*: 3 (1965).
38. J. Sun, Y. Zhang, and X. Zhong, Radiation crosslinking of polytetrafluoroethylene, *Polym. Commun. 35*: 2881 (1994).
39. ICI Polytetrafluoroethylene processing directions.
40. K. Tanaka, Y. Uchiyama, and S. Toyooka, The mechanism of wear of polytetrafluoroethylene, *Wear 23*: 153 (1973).
41. T. A. Blanchet, F. E. Kennedy, and X. Tian, The mild/severe wear transition of PTFE polymer in oscillatory sliding contacts, *Proceedings of the International Conference on the Wear of Materials*, Orlando, FL, 1991, p. 689.
42. T. A. Blanchet and F. E. Kennedy, Sliding wear mechanism of polytetrafluoroethylene (PTFE) and PTFE composites, *Wear 153*: 229 (1992).
43. R. P. Steijn, The sliding surface of polytetrafluoroethylene: An investigation with the electron microscope, *Wear 12*: 193 (1968).
44. C. M. Pooley and D. Tabor, Friction and molecular structure: The behavior of some thermoplastics, *Proc. Roy. Soc. 329*: 251 (1972).
45. S. V. Pepper, Auger analysis of films formed on metals in sliding contact with halogenated polymers, *J. Appl. Phys. 45*: 2947 (1974).
46. D. R. Wheeler, The transfer of polytetrafluoroethylene studied by X-ray photoelectron spectroscopy, *Wear 66*: 355 (1981).
47. T. A. Blanchet, F. E. Kennedy, and D. T. Jayne, XPS analysis of the effect of fillers on PTFE transfer film development in sliding contacts, *Tribol. Trans. 36*: 535 (1993).
48. J. L. Lauer, B. G. Bunting, and W. R. Jones, Investigation of frictional transfer films of PTFE by infrared emission spectroscopy and phase-locked ellipsometry, *Tribol. Trans. 31*: 282 (1988).
49. E. L. Yang, J. P. Hirvonen, and R. O. Toivanen, Effect of temperature on the transfer film formation in sliding contact of PTFE with stainless steel, *Wear 146*: 367 (1991).

50. P. Dietz, P. K. Hansma, K. J. Ihn, F. Motamedi, and P. Smith, Molecular structure and thickness of poly(tetrafluoroethylene) films measured by atomic force microscopy, *J. Mat. Sci. 28*: 1372 (1993).
51. K. G. McLaren and D. Tabor, Visco-elastic properties and the friction of solids, *Nature 197*: 856 (1963).
52. K. C. Ludema and D. Tabor, The friction and visco-elastic properties of polymeric solids, *Wear 9*: 329 (1966).
53. R. P. Steijn, Sliding experiments with polytetrafluoroethylene, *ASLE Trans. 11*: 235 (1968).
54. S. Bahadur and D. Tabor, The wear of filled polytetrafluoroethylene, *Wear 98*: 1 (1984).
55. S. Ricklin and R. R. Miller, Filled Teflon for dry bearings, *Mat. Meth. 40*: 112 (1954).
56. J. K. Lancaster, The effect of carbon fibre reinforcement on the friction and wear of polymers, *Br. J. Appl. Phys. 1*: 549 (1968).
57. Fluorocomp Composites, LNP Corporation, Bulletin 106-686, 1986.
58. T. A. Blanchet and F. E. Kennedy, Effects of oscillatory speed and mutual overlap on the tribological behavior of PTFE and selected PTFE-based self-lubricating composites, *Tribol. Trans. 34*: 327 (1991).
59. K. Tanaka, Y. Uchiyama, S. Ueda, and T. Shimizu, Friction and wear of PTFE-based composites, *Proceedings of the JSLE-ASLE International Lubrication Conference '75, Tokyo* (T. Sakurai, ed.), Elsevier, Amsterdam, 1976, p. 110.
60. K. Tanaka, Friction and wear of glass and carbon fiber-filled thermoplastic polymers, *J. Lub. Technol. 99:* 408 (1977).
61. K. Tanaka and S. Kawakami, Effect of various fillers on the friction and wear of polytetrafluoroethylene-based composites, *Wear 79*: 221 (1982).
62. K. Tanaka, Effects of various fillers on the friction and wear of PTFE-based composites, *Friction and Wear of Polymer Composites* (K. Friedrich, ed.), Elsevier, Amsterdam, 1986, p. 137.
63. B. J. Briscoe, M. D. Steward, and A. J. Groszek, The effect of carbon aspect ratio on the friction and wear of PTFE, *Wear 42*: 99 (1977).
64. B. J. Briscoe, A. K. Pogosian, and D. Tabor, The friction and wear of high density polythene: The action of lead oxide and copper oxide fillers, *Wear 27*: 19 (1974).
65. W. A. Brainard and D. H. Buckley, Adhesion and friction of PTFE in contact with metals as studied by Auger spectroscopy, field ion and scanning electron microscopy, *Wear 26*: 75 (1973).
66. P. Cadman and G. M. Gossedge, Studies of polytetrafluoroethylene transfer layers produced by rubbing in ultrahigh vacuum using a relatively simple apparatus, *Wear 51*: 57 (1978).
67. P. Cadman and G. M. Gossedge, The chemical interaction of metals with polytetrafluoroethylene, *J. Mat. Sci. 14*: 2672 (1979).
68. P. Cadman and G. M. Gossedge, The chemical nature of metal–polytetrafluoroethylene tribological interactions as studied by X-ray photoelectron spectroscopy, *Wear 54*: 211 (1979).
69. J. Gao and H. Dang, Molecular structure variations in friction of stainless steel/PTFE and its composite, *J. Appl. Polym. Sci. 36*: 36 (1988).
70. D. Gong, B. Zhang, Q. Xue, and H. Wang, Effect of tribochemical reaction of polytetrafluoroethylene transferred film with substrates on its wear behavior, *Wear 137*: 267 (1990).
71. D. Gong, Q. Xue, and H. Wang, ESCA study on tribochemical characteristics of filled PTFE, *Wear 148*: 161 (1991).
72. N. Axen and S. Jacobson, A model for the abrasive wear resistance of multiphase materials, *Wear 174*: 187 (1994).
73. T. A. Blanchet, A model for polymer composite wear behavior, including preferential load support and surface accumulation of filler particulates, *Tribol. Trans. 38*: 821 (1995).
74. S. Ricklin, Review of design parameters for filled PTFE bearing materials, *Lub. Eng. 33*: 487 (1977).
75. M. N. Gardos, Theory and practice of self-lubricated oscillatory for high-vacuum applications, *Lub. Eng. 37*: 641 (1981).
76. E. Jisheng and D. T. Gawne, Tribological performance of bronze-filled PTFE facings for machine tool slideways, *Wear 176*: 195 (1994).
77. Anonymous, *Industrial Lubrication and Tribology*, May/June 1990, p. 14.
78. R. A. Rowntree, E. W. Roberts, and J. M. Todd, Tribological design: The spacecraft industry, *Tribological Design of Machine Elements* (D. Dowson et al., eds.), Elsevier, Amsterdam, 1989, p. 389.
79. M. Nosaka, M. Oike, M. Kikuchi, K. Kamijo, and M. Tajiri, Self-lubricating performance of ball bearings for the LE-7 liquid oxygen rocket-turbopump, *Lub. Eng. 49*: 677 (1993).
80. T. J. Chase, Wear modes active in angular contact ball bearings in liquid oxygen environment of the space shuttle, *Lub. Eng. 49*: 313 (1993).
81. B. C. Arkles, J. Theberge, and M. J. Schireson, Wear behavior of thermoplastic polymer-filled PTFE composites, *Lub. Eng. 33*: 33 (1977).

82. M. Hong and S. Pyun, Effect of fluorinated ethylene propylene copolymer on the wear behavior of polytetrafluoroethylene, *Wear 143*: 87 (1991).
83. D. L. Burris and W. G. Sawyer, Improved wear resistance in alumina–PTFE nanocomposites with irregular shaped nanoparticles, *Wear 260*: 915 (2006).
84. S. S. Kandanur, M. A. Schrameyer, K. F. Jung, M. E. Makowiec, S. Bhargava, and T. A. Blanchet, Effect of activated carbon and various other nanoparticle fillers on PTFE wear, *Trib. Trans. 57*: 821 (2014).
85. T. A. Blanchet and Y. Peng, Wear resistant polytetrafluoroethylene via electron irradiation, *Lub. Eng. 52*: 489 (1996).
86. B. J. Briscoe and P. D. Evans, The influence of asperity deformation on the abrasive wear of γ-irradiated poly(tetrafluoroethylene), *Proceedings of the International Conference on the Wear of Materials*, 1989, p. 449.
87. K. Tanaka and T. Nagai, Effect of roughness on the friction and wear of polytetrafluoroethylene and polyethylene, *Proceedings of the International Conference on the Wear of Materials*, 1985, p. 397.
88. L. Tocci, Slick 50 tackles industry—And its detractors, *Lubes-n-Greases 1*: 1 (1995).
89. E. A. Rosset, S. Mischler, and D. Landolt, Surface chemistry effects on friction of Ni–P/PTFE composite coatings, *Dissipative Processes in Tribology* (D. Dowson et al., eds.), Elsevier, Amsterdam, 1994, p. 329.
90. R. Dagani, New simple process forms copper circuits directly on Teflon, *Chem. Eng. News 70(1)*: 24 (1992).
91. D. J. Doyle, A low cost alternative to reinforcing PTFE based printed circuit boards, *Proceedings of IEEE MTT-S International Microwave Symposium*, Vol. 3, San Diego, CA, 1994, p. 1149.
92. D. R. Wheeler and S. V. Pepper, Summary abstract: improved adhesion of Ni films on X-ray damaged polytetrafluoroethylene, *J. Vac. Sci. Technol. 20*: 442 (1982).
93. W. Jiang, M. G. Norton, and J. T. Dickinson, Surface modification of polytetrafluoroethylene and the deposition of copper films, *Mat. Res. Symp. Proc. 304*: 97 (1993).
94. R. Rye, Electron irradiation of poly(tetrafluoroethylene): Effect on adhesion and comparison with X-rays, *Langmuir 6*: 338 (1990).
95. S. Ebnesajjad and P. R. Khaladkar, *Fluoropolymers Applications in Chemical Processing Industries*, William Andrew Publishing, Norwich, NY, 2005.
96. J. G. Drobny, *Fluoroplastics*, Smithers Rapra Review Reports, Shrewsbury, Great Britain, 2005.
97. H. Teng, Overview of the development of the fluoropolymer industry, *Applied Sciences 2*: 496 (2012).

14 Polyarylethersulfones*

M. Jamal El-Hibri and Shari W. Axelrad

CONTENTS

14.1 Overview 420
14.2 Synthesis 421
    14.2.1 Nucleophilic Displacement Route 421
    14.2.2 Other Synthesis Routes 424
14.3 Processing 427
    14.3.1 Rheology 427
    14.3.2 Drying 428
    14.3.3 Injection Molding 428
    14.3.4 Extrusion and Thermoforming 428
    14.3.5 Solution Processing 429
    14.3.6 Secondary Fabrication Operations 429
    14.3.7 Effect of Processing on Properties 429
14.4 Properties 430
    14.4.1 Physical, Thermal, and Mechanical 430
    14.4.2 Electrical 435
    14.4.3 Flammability Behavior 435
    14.4.4 Radiation Resistance 435
    14.4.5 Chemical Resistance and Hydrolytic Stability 436
    14.4.6 Solubility 437
14.5 Blends and Composites 438
    14.5.1 Blends 438
        14.5.1.1 Blends with Polyaryletherketones and Polyphenylenesulfide ... 439
        14.5.1.2 Blends with Aramids 439
        14.5.1.3 Miscible Blends 439
        14.5.1.4 Blends with Other Polysulfones 440
        14.5.1.5 Blends with Liquid Crystalline Polymers 440
        14.5.1.6 Other Blends 440
    14.5.2 Composites 441
14.6 Health, Safety, and Environmental Aspects 442
    14.6.1 Health 442
    14.6.2 Safety 442
    14.6.3 Sustainability Aspects 442
14.7 End-Use Applications 443
    14.7.1 Food Service and Food Processing 443
    14.7.2 Plumbing Components 443
    14.7.3 Membranes 444
    14.7.4 Medical Devices 447
    14.7.5 Commercial Aircraft Interiors 448
    14.7.6 Toughening of Advanced Composites 449

---

* Based in parts on the first-edition chapter on polyarylethersulfones.

14.7.7 Structural Foams	450
14.7.8 High-Performance Coatings	450
14.7.9 Electrical and Electronic Components	451
14.8 Economic Aspects	451
References	452

14.1 OVERVIEW

Polyarylethersulfones are a family of aromatic amorphous thermoplastics that possess unique high-performance properties as engineering materials. As the name implies, the backbone repeat units of a polyarylethersulfone are primarily composed of aryl groups, aromatic ether linkages, and aromatic sulfone groups. A key structural feature of these polymers that distinguishes them from other aromatic polymers is the presence of the *para*-linked diarylsulfone grouping (Scheme 14.1) as part of the main backbone repeat unit.

The polymers are transparent, mechanically, tough and rigid, and exhibit high glass transition temperatures and excellent thermal-oxidative resistance. They are also known for their resistance against hydrolysis and against acidic and basic attack, as well as for their easy processability and great melt thermal stability that allows melt processing at temperatures up to 400°C. This combination of features coupled with ease of fabrication has won polyarylethersulfones a wide and increasingly diversified range of applications.

There are three polyarylethersulfones that have grown to be of commercial interest today. These are known by the following common generic designations and acronyms:

- Polysulfone (PSU)
- Polyethersulfone (PES)
- Polyphenylsulfone (PPSU)

PSU has the following repeat unit structure (Scheme 14.2).

Its glass transition temperature is 185°C. It was the first member of this class of polymers to become commercial, having been first introduced in 1965 by Union Carbide Corporation as Bakelite polysulfone. PSU is the best known and most commercially utilized polymer in this family today. It is available from two commercial sources.

PES was first introduced commercially by Imperial Chemical Industries, Ltd. It has a T_g of 220°C and the following repeat unit structure (Scheme 14.3).

This polymer is available from three major commercial suppliers today.

PPSU is distinguished by the presence of the biphenyl linkages in its backbone, its repeat unit structure being as follows (Scheme 14.4).

It is a somewhat unique member of the polyarylethersulfone family. While having the same T_g as that of PES, this polymer exhibits supertough characteristics, which are similar to those well

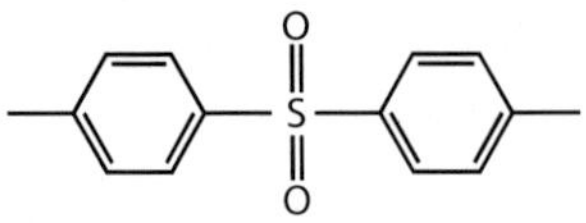

SCHEME 14.1 Para-linked diarylsulfone structure.

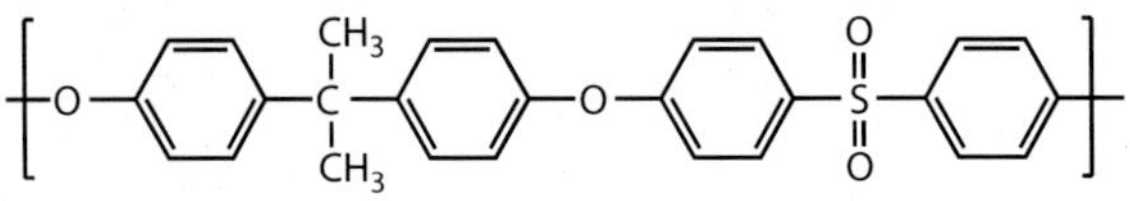

SCHEME 14.2 PSU repeat unit structure.

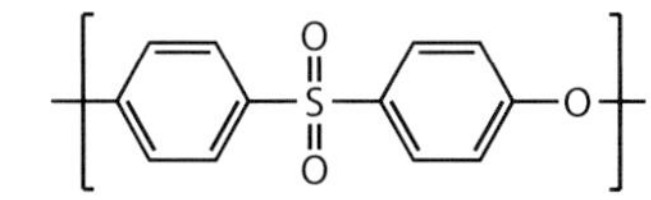

SCHEME 14.3 PES repeat unit structure.

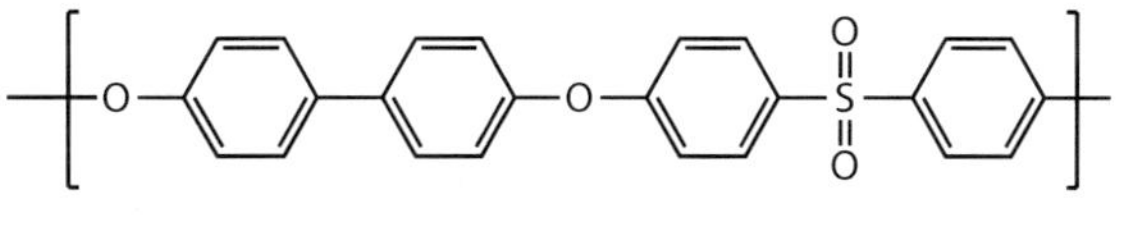

SCHEME 14.4 PPSU repeat unit structure.

known for bisphenol A polycarbonate (PC). This polymer continues to be the most notched impact-resistant amorphous plastic in its temperature class that is available commercially. It possesses other distinguishing attributes that will be discussed in depth in a later section. The resin was first briefly introduced commercially by Union Carbide in 1977 and then withdrawn. It was later fully commercialized by Amoco Corporation in 1990 as Radel R PPSU, and the polymer is available today from two commercial suppliers.

Before going further, it is important to point out that the polymer designations *polysulfone*, *polyethersulfone*, and *polyphenylsulfone* are simply convenient common names that have become accepted over time. They were not conceived to accurately represent the chemical makeup of each of the above three polymers. The shorthand notations *PSU* (also referred to as *PSF* in the polymer literature), *PES* (also referred to as *PESU*), and *PPSU* (also referred to as *PPSF* in some references) are common notations for these polymers that have become accepted by industry associations and by other international standardization bodies like ASTM and International Organization for Standardization (ISO). In this work, the word *polysulfones* refers to the polymer class as a whole. In addition to the most descriptive term *polyarylethersulfones*, other accepted common names for polymers of this family include *polyethersulfones*, *polyarylsulfones*, and *sulfone polymers*.

The two main routes for polysulfone synthesis are as follows: (1) the aromatic nucleophilic displacement route and (2) the electrophilic Friedel–Crafts synthesis. These two methods were discovered almost simultaneously on two continents in the early to mid-1960s. Today, the nucleophilic displacement route is used exclusively for commercial production of these polymers. This synthesis involves reacting equimolar amounts of 4,4′-dihalodiphenylsulfone (industrially, 4-4′-dichlorodiphenylsulfone [DCDPS] is always used) with the appropriate bisphenol. In the case of PSU, bis(4-hydroxyphenyl)-2,2-propane (bisphenol A) is used, while 4,4′-dihydroxydiphenylsulfone (bisphenol S) is needed to make PES. PPSU synthesis, on the other hand, requires 4,4′-dihydroxydiphenyl (biphenol) as the bisphenol. The structures of these and other bisphenols are shown in Table 14.1.

14.2 SYNTHESIS

As stated above, the most important method by far for the synthesis of polysulfones is that known as the aromatic nucleophilic displacement (or substitution) polycondensation reaction. The emphasis of the first portion of this section will thus focus on a description of this synthesis scheme, especially as it relates to the synthesis of PSU, PES, and PPSU. The second subsection will briefly touch on other methods that have been used and reported in the literature for the synthesis of polysulfones.

14.2.1 Nucleophilic Displacement Route

The first successful application of nucleophilic displacement chemistry to the synthesis of polysulfones emerged from the pioneering efforts of Drs. R.N. Johnson and A.G. Farnham of Union Carbide Corporation in the mid-1960s [1–3]. Their discovery of this polymerization route paved the way for

TABLE 14.1
Glass Transition Temperatures[a] of Polysulfones Produced from the Polycondensation of 4,4′-Dichlorodiphenylsulfone with Various Bisphenols

Bisphenol	Structure	T_g (°C)[a]
4,4′-dihydroxydiphenyl oxide	HO–C₆H₄–O–C₆H₄–OH	170
4,4′-dihydroxydiphenyl sulfide	HO–C₆H₄–S–C₆H₄–OH	175
4,4′-dihydroxydiphenyl methane	HO–C₆H₄–CH₂–C₆H₄–OH	180
Bis(4-hydroxyphenyl)-2,2-propane	HO–C₆H₄–C(CH₃)₂–C₆H₄–OH	185
Hydroquinone	HO–C₆H₄–OH	200
Bis(4-hydroxyphenyl)-2,2-perfluoropropane	HO–C₆H₄–C(CF₃)₂–C₆H₄–OH	205
4,4′-dihydroxybenzophenone	HO–C₆H₄–C(=O)–C₆H₄–OH	205
4,4′-dihydroxydiphenyl sulfone	HO–C₆H₄–SO₂–C₆H₄–OH	220
4,4′-dihydroxydiphenyl	HO–C₆H₄–C₆H₄–OH	220
Bis(4-hydroxyphenyl)-1,4-benzene	HO–C₆H₄–C₆H₄–C₆H₄–OH	250
Bis(1-hydroxyphenyl-4-sulfonyl)-4,4′-diphenyl	HO–C₆H₄–SO₂–C₆H₄–C₆H₄–SO₂–C₆H₄–OH	265

Source: Reprinted from Kwiatkowski, G.T. et al., *Makromol. Chem., Macromol. Symp.*, 54/55, 1992, copyright 1992, with permission from John Wiley & Sons.

[a] Glass transition values reported rounded to nearest 5°C and are DSC onset values.

the development of a large family of high-performance polyarylether resins, which includes polyarylethersulfones, polyaryletherketones, and polyetherimides (PEIs). The essence of this polycondensation scheme is the reaction of an aromatic dihydric phenol such as bisphenol A with an aromatic dihalo compound in which the halogens are activated by an electron-withdrawing group in the presence of a base.

The synthesis of polysulfone can be accomplished by a two-step process, the first of which involves the *in situ* conversion of bisphenol A to its disodium salt of this bisphenol using sodium hydroxide. The reaction is carried out in a polar aprotic solvent such as dimethyl sulfoxide (DMSO) according to Equation 14.1:

Polyarylethersulfones

$$\text{HO-}\bigcirc\text{-C(CH}_3\text{)}_2\text{-}\bigcirc\text{-OH} + 2\text{NaOH} \longrightarrow \text{NaO-}\bigcirc\text{-C(CH}_3\text{)}_2\text{-}\bigcirc\text{-ONa} + 2\text{H}_2\text{O} \quad (14.1)$$

The water that is generated from this reaction must be removed continuously and thoroughly to preclude the possibility of any hydrolysis of the 4,4′-dichlorodiphenylsulfone (DCDPS) monomer that is added for the second step in the reaction, as per Equation 14.2:

$$\text{NaO-}\bigcirc\text{-C(CH}_3\text{)}_2\text{-}\bigcirc\text{-ONa} + \text{Cl-}\bigcirc\text{-SO}_2\text{-}\bigcirc\text{-Cl} \longrightarrow$$

$$\left[\text{O-}\bigcirc\text{-C(CH}_3\text{)}_2\text{-}\bigcirc\text{-O-}\bigcirc\text{-SO}_2\text{-}\bigcirc\right]_n + 2\text{NaCl} \quad (14.2)$$

The reaction rate for polymerizations of this type is governed by a number of factors, including the basicity of the dialkali salt of the bisphenol and its solubility in the reaction medium. The halogen type on the dihalo compound also has an effect, the reactivity order being F > Cl > Br > I [4]. Although the difluoro monomer offers faster reaction kinetics, the much more attractive economics of the dichloro monomer make it by far the industrially preferred choice. Another factor is the purity of the DCDPS monomer. Chlorines in the *meta* position do not undergo nucleophilic displacement to any significant extent. Therefore, if there is any appreciable 3,4′ isomer content in the DCDPS, this may prevent the attainment of sufficiently high molecular weights as the 3,4′ species behaves much like a monofunctional endcapping agent.

Molecular weight control in the nucleophilic synthesis of polysulfones may be achieved by one or more of the following means: (1) the addition of a monofunctional analog of one of the monomers and/or (2) the use of unbalanced stoichiometry as governed by Carothers' principle of functionality for condensation polymers.

PSU can also be produced by using a weak base such as an alkali metal carbonate in a dipolar aprotic solvent. Dipolar aprotic solvents that could be used for these polymerizations include DMSO, N-methyl-2-pyrrolidone (NMP), dimethyl acetamide (DMAc), sulfolane, and diphenylsulfone. PES and PPSU are also typically produced using this method. One of the advantages of this polymerization route is that no separate step is required for the formation of a dialkali metal salt of the bisphenol. This polymerization scheme is exemplified by the reaction of 4,4′-dihydroxydiphenylsulfone (bisphenol S) with DCDPS in the presence of an alkali metal (M) carbonate to produce PES [5], as can be seen in Equation 14.3,

$$\text{HO-}\bigcirc\text{-SO}_2\text{-}\bigcirc\text{-OH} + \text{Cl-}\bigcirc\text{-SO}_2\text{-}\bigcirc\text{-Cl} \xrightarrow[\text{Solvent}]{M_2CO_3}$$

$$\left[\text{O-}\bigcirc\text{-SO}_2\text{-}\bigcirc\text{-O-}\bigcirc\text{-SO}_2\text{-}\bigcirc\right]_n + 2\text{MCl} + \text{H}_2\text{O} + \text{CO}_2 \quad (14.3)$$

while the completely analogous polymerization of PPSU based on 4,4′-dihydroxydiphenyl (bisphenol) is described by Equation 14.4

$$\text{Cl}-\overset{\overset{O}{\|}}{\underset{\underset{O}{\|}}{S}}-\!\!\bigcirc\!\!-O-\!\!\bigcirc\!\!-\overset{\overset{O}{\|}}{\underset{\underset{O}{\|}}{S}}-\text{Cl} \;+\; \bigcirc\!\!-O-\!\!\bigcirc \;\xrightarrow[\text{Catalyst}]{-2\text{HCl}}$$

$$\left[\!\!\bigcirc\!\!-\overset{\overset{O}{\|}}{\underset{\underset{O}{\|}}{S}}-\!\!\bigcirc\!\!-O\right]_n$$

(14.4)

Alkali metal carbonates such as sodium carbonate and potassium carbonate can be used in these reactions. Following polymerization and termination, the polysulfone is ready for recovery.

An alternate form of the nucleophilic displacement polymerization of PES described by Equation 14.3 involves the partial hydrolysis of DCDPS to achieve identical equivalents of chlorine and phenolic hydroxyl functionality. This step is followed by the A-B type polycondensation of this difunctional polymer in the presence of a molar equivalent of potassium hydroxide [6].

In a variation on the base-catalyzed nucleophilic displacement chemistry known as the Ullmann synthesis, polysulfones and other polyarylethers are synthesized from aromatic dihydroxy compounds, which are reacted with aromatic dibromo compounds in the presence of cuprous chloride as a catalyst. The advantage of the method is that it does not require the dibromo compound to be activated by an electron-withdrawing group (e.g., sulfone). Hence, a more diverse variety of backbone structures is possible. Details of this synthesis method are described in a US patent [7].

As stated above, one of the traditional requirements of nucleophilic polycondensation reactions is a dipolar aprotic solvent to solvate both the dihydric phenol salt and the polymer. A phase transfer catalysis process was developed recently by Savariar that circumvents this requirement [8,9]. The method successfully relies on 2.2.2-Cryptand as a catalyst for the synthesis of polysulfone in chlorobenzene [8]. Less reactive crown ethers can also be used as catalysts, but those require dichlorobenzene as a solvent because of the higher temperatures needed [9]. High-molecular-weight PPSU was shown to be achievable via this route using dichlorobenzene as a solvent. The method was not fully successful with PES, and only low-molecular-weight polymer was achievable. Cross-linked polystyrene-bound 2.2.2-Cryptand was reported to be an effective catalyst for these polymerizations, which allows effective recovery and reuse of the catalyst.

By starting with more than one bisphenol during a nucleophilic displacement synthesis, random copolymers and multipolymers can be readily produced. Such copolymers typically take on physical characteristics that vary in proportion to the volume or weight fraction of the constituent homopolymer repeat units in the copolymer backbone. Copolymerization of sulfone polymers using multiple bisphenols usually results in relatively random sequencing of the representative repeat units. This is by virtue of the ether–ether interchange reactions that occur from the displacement action of the phenate nucleophile at the aromatic ether linkages activated by sulfone groups. The result is a copolymer with relatively random repeat unit sequencing along the main backbone [10–12].

14.2.2 Other Synthesis Routes

Besides aromatic nucleophilic displacement chemistry, several alternate synthesis routes exist for the preparation of polysulfones. For various reasons, none of these routes has gained commercial importance. This section briefly introduces these alternate synthesis routes with reference to pertinent literature sources. All of the synthesis methods mentioned are documented in the journal literature or in published patents.

At a time contemporaneous to the development of the nucleophilic displacement polycondensation route, ICI Ltd. researchers in the United Kingdom developed the electrophilic Friedel–Crafts synthesis methods for PES [5,13–15]. One approach was based on the reaction of bis(4-chlorosulfonylphenyl)ether with diphenylether, per Equation 14.5:

$$\text{HO-}\bigcirc\text{-}\bigcirc\text{-OH} + \text{Cl-}\bigcirc\text{-SO}_2\text{-}\bigcirc\text{-Cl} \xrightarrow[\text{Solvent}]{M_2CO_3}$$

$$\left[\text{-O-}\bigcirc\text{-}\bigcirc\text{-O-}\bigcirc\text{-SO}_2\text{-}\bigcirc\text{-}\right]_n + 2MCl + H_2O + CO_2 \quad (14.5)$$

A similar polymerization was also carried out using 4-chlorosulfonyldiphenylether as a single monomer, as shown in Equation 14.6:

$$\bigcirc\text{-O-}\bigcirc\text{-SO}_2\text{-Cl} \xrightarrow[\text{Catalyst}]{-HCl} \left[\bigcirc\text{-SO}_2\text{-}\bigcirc\text{-O-}\right]_n \quad (14.6)$$

A high-PES polymer was shown to be consistently achievable by this electrophilic route; however, the backbone structure was not always linear. The single monomer route above has been reported to be the preferred method of the two as it was found to afford structures that were virtually free from branching and that were exclusively *para* linked—both necessary backbone features for mechanically tough PES. Other polysulfones were found to be amenable to synthesis using the same chemistry with different starting materials. The Friedel–Crafts synthesis route was used by 3M Corporation in the late 1960s for the manufacture of a high-temperature polysulfone that was sold for a brief period under the trade name Astrel 360. The polymer was believed to have been based on the electrophilic polycondensation of 4-chlorosulfonyldiphenyl with 4-chlorosulfonyldiphenylether to produce a copolymer having a T_g of 265°C, as shown in Equation 14.7:

$$\bigcirc\text{-O-}\bigcirc\text{-SO}_2\text{-Cl} + \bigcirc\text{-}\bigcirc\text{-SO}_2\text{-Cl} \xrightarrow[\text{Catalyst}]{-HCl}$$

$$\text{Copolymer of } \left[\bigcirc\text{-O-}\bigcirc\text{-SO}_2\right] + \left[\bigcirc\text{-}\bigcirc\text{-SO}_2\right] \quad (14.7)$$

A novel synthesis of polysulfones based on the nickel-catalyzed coupling of diarylchlorides was reported on by Colon et al. [16–19]. Because of the nature of the coupling reaction, this synthesis always produces a phenyl–phenyl linkage in the polymer repeat unit. As such, the synthesis is particularly suited for the preparation of PPSU and other biphenyl-containing polysulfones. The complete synthesis is actually a two-step process. The first step involves the low-molecular-weight nucleophilic condensation of DCDPS with 4-chlorophenol to produce a diaryl chloride intermediate, which undergoes self-coupling in the presence of zero-valent nickel, triphenylphosphine, and an excess of zinc. The reaction occurs under mild conditions with temperatures typically being in the 60–80°C range. Zinc chloride is recovered as a by-product of the reaction. The reaction sequence can be depicted as in Equation 14.8:

$$(14.8)$$

References 16 and 17 deal with the synthesis of the diaryl halide intermediate, whereas References 18 and 19 discuss the polymerization step with particular emphasis on PPSU synthesis.

Oxidative coupling via the Scholl reaction has also been successfully utilized to synthesize a polysulfone [20]. A high polymer was obtained upon treating 4,4'-di(1-naphthoxy)diphenylsulfone and 4,4'-di(1-naphthoxy)benzophenone with ferric chloride. The reaction requires equimolar quantities of Lewis acid and is restricted to systems such as naphthoxy-based monomers, which can undergo the Scholl reaction.

Decarboxylation of aromatic PCs containing electron-withdrawing groups (e.g., sulfone) in their repeat unit was reported in an early reference to be a viable route for polysulfone synthesis [21]. Application of reactions of this type has been the subject of research interest [22,23].

Polysulfones can also be prepared using melt polymerization [24]. More expensive 4,4'-fluorodiphenylsulfone is used instead of less reactive DCDPS. The bisphenols are typically converted into a silyated form before the polymerization (Equation 14.9). Fluoride ions (CsF) are used in catalytic amounts to initiate the polymerization. The byproduct of the polymerization, trifluorosilylfluoride, being a gas, is easily removed. These polymerizations do not use expensive solvents and their purification is simpler as no metal salts are formed.

$$(14.9)$$

Polyarylethersulfones

14.3 PROCESSING

Polysulfones are completely linear polymers that may be processed by all thermoplastics processing methods, with negligible chemical and physical change at temperatures up to 400°C. The most important method for processing polysulfones is injection molding. However, they may be extruded into sheets, film, tubing, rods, slabs, and other shapes, or extruded onto wire as electrical insulation coatings. Blow molding may also be applied. Extruded sheets may be thermoformed by vacuum, pressure, and plug-assist methods. Solution casting is also mentioned in the technical literature. The ease of processing has contributed to the broad acceptance of polysulfones for many engineering uses.

14.3.1 Rheology

The melt rheology behavior of polysulfones is characterized by its relatively weak response to shear compared with most aliphatic backbone polymers. The resistance to shear thinning is due to the rigidity of the aromatic backbone, which limits polymer chain orientation during shearing. The rheological behavior of polysulfones is, in many respects, analogous to that of bisphenol A PC. In fact, the flow behavior of PSU is almost identical to that of PC when one measures the PSU viscosity at a temperature 50–60°C higher than that of PC, thus compensating for the difference in glass transition of the two materials. This similarity is illustrated in Figure 14.1, which juxtaposes apparent viscosity–shear rate plots for PSU and PC along with curves for polystyrene and low-density polyethylene. This comparison is made over the same shear rate range but using a temperature appropriate for each polymer. The rheology characteristics of PES and PPSU are fundamentally similar to those of PSU, although melt temperatures required for these two polymers to match PSU viscosities are usually 20–40°C higher. The melt viscosity values for each of these resins will vary depending on the molecular weight/melt flow grade under consideration. The viscosity differences can be as high as an order of magnitude between the low- and high-viscosity grades. PSU, PES, and PPSU are all supplied commercially in three viscosity grades—high, medium (general purpose), and low—to meet the processing needs of different applications. Figure 14.1 shows the viscosity of polysulfones superimposed on viscosity curves for PC, polyethylene, and polystyrene to illustrate

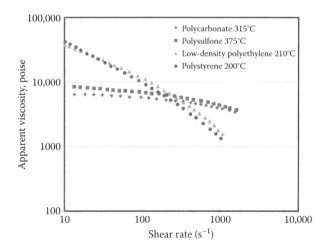

FIGURE 14.1 Comparison of the melt viscosity dependence on shear rate for four polymers at typical processing temperatures: Udel P-1700 PSU at 375°C, polycarbonate at 315°C, polystyrene at 200°C, and low-density polyethylene at 210°C.

the key difference of low shear-thinning behavior seen in aromatic amorphous polymers like polysulfones and PCs in comparison to aliphatic polymers.

14.3.2 Drying

Before subjecting a polysulfone to melt-fabrication operations, the resin should be dried to remove sorbed moisture. While water does not degrade or chemically react with polysulfones during melt processing, it can cause foaming of the melt, resulting in surface defects like bubbles, streaking, and splay marks on fabricated parts. To prevent this from occurring, the resin should be dried to a target moisture content of less than 500 ppm. This is easily achieved by drying in either a conventional forced-air oven or a desiccated hopper dryer unit at temperatures in the range of 135–165°C for 3–4 h. The drying temperature can be increased up to 180°C for PES and PPSU if there is a need to cut down on drying time. If an oven is used for drying, the resin should ideally be placed in trays to a depth of no greater than 2 in.

14.3.3 Injection Molding

Injection molding is the technique used most widely for melt fabrication of polysulfones. Reciprocating-screw injection-molding machines equipped with a screw having a 20:1 length/diameter (L/D) ratio and a 2.5:1 compression ratio are typical. Straight-through nozzles are the preferred type, and nozzle shutoff valving is generally not desired, due to shear heating and dead-spot considerations. Sprues should be short and relatively thick, and tapered, runners full round or trapezoidal. Gate size and location are extremely important for obtaining defect-free parts. Gates should be located so that resin flow is from thick to thin sections, and they should be relatively large. Melt temperatures used for injection-molding polysulfones can vary widely depending on the resin, grade, and geometry of the part being molded. Typical melt temperatures are in the range of 340–370°C for PSU, 350–390°C for PES, and 360–400°C for PPSU. Mold temperatures are typically 100–140°C for PSU and 130–170°C for PES and PPSU. Mold temperatures on the high side of the above-stated ranges are preferred whenever possible to minimize molded-in stresses in the parts. Excessively high injection speeds coupled with low melt temperatures are to be avoided also for the same reason. Optimum injection speeds are usually in the medium to medium-high range. Screw speeds should be maintained low (40–80 rpm) whenever possible to minimize shear heating and darkening of the polymer. The injection pressures needed are usually high, but screw back pressures should be kept at a minimum (2–4 bars typically) to minimize shear-induced heating and discoloration.

14.3.4 Extrusion and Thermoforming

PSU, PES, and PPSU can be readily extruded into a variety of shapes using conventional single-screw extruders. A general-purpose screw having a 20:1 L/D and 2.5 compression ratio is suitable for most purposes. Barrel zone temperatures typically range from 280°C to 340°C for PSU and from 300°C to 360°C for PES and PPSU. Die temperature settings are commonly in the range 320–360°C depending on the type of product, throughput rate, allowable pressure drops, as well as other factors. For sheet and film extrusion operations, oil-heated rollers are necessary ahead of the take-up system to ensure good surface quality, uniformity, and gloss as well as low residual stresses. An extruded sheet is often postfabricated in a subsequent thermoforming process to produce the finished article. Polysulfones readily lend themselves to a variety of thermoforming types, including pressure, vacuum, and plug assisted. It is important that the sheet be dried to below 0.05% prior to thermoforming; otherwise, bubbling or blistering can occur during the thermoforming step, causing defects in the finished part. A typical forming temperature for PSU is 250°C, whereas 290°C is more appropriate for PES and PPSU. These temperatures allow sufficient softening while providing resistance to sagging during the heat-up period. The higher melt viscosity grades are naturally preferred for thermoforming fabrication.

One important consideration in the melt processing of sulfones is proper metallurgy and good practice. Sulfidation is the reaction of unprotected steel with sulfur liberated during the melt processing. This process results in the creation of iron sulfide on the inside of a machine screw and barrel assembly or within a hot tooling. It is evidenced by a thin, black, brittle film of iron sulfide that can be sloughed off into the melt, causing defects. This behavior can be limited by avoiding excessively high melt temperatures, residence times, and back pressure; by selecting proper metallurgy; and by maintaining proper equipment hygiene.

14.3.5 Solution Processing

Articles made from sulfone polymers are also fabricated using various solution processing techniques. Membranes, coatings, films, and fibers are prepared by dissolution of polymer in a suitable solvent, forming, and evaporation of the solvent, or by casting into a nonsolvent, thereby achieving a porous structure (e.g., of a membrane) by coagulation. Sulfone polymers are soluble in various aprotic polar solvents such as dimethylformamide (DMF), DMAc, and NMP. PSU offers the broadest solubility, and PPSU the least. The use of powdered grades of polymer facilitates the dissolution process [25].

14.3.6 Secondary Fabrication Operations

Most of the commonly used secondary part finishing operations for thermoplastics are possible for polysulfones. These operations include annealing, metallizing, machining, and joining, to mention just a few. Annealing is sometimes employed to remove residual molded-in stresses from injection-molded parts. PSU can be annealed effectively at temperatures of 160–170°C. Annealing times required in an air oven are 4 h at 160°C or 1 h at 170°C. For PES and PPSU, the useful temperature range for annealing purposes is 190–200°C. Among the benefits of annealing are enhanced environmental resistance, increased stiffness, reduced creep, and enhanced resistance to dimensional changes when the part is exposed to elevated temperatures. The downside to annealing is a reduction in impact resistance and ductility. Annealing of injection-molded sulfone polymer parts can be important if the polymer is anticipated to be used under extreme temperature conditions and dimensional stability is critical. Metallizing of polysulfones can be achieved via vacuum metallizing techniques. Polysulfones can be readily machined using ordinary metalworking tools. The good toughness of these polymers allows deep smooth cuts without chipping or shattering. However, machining oils should be avoided, if possible, in order to preclude the possibility of environmental stress cracking (ESC) during machining. Joining of polysulfones is possible by a variety of techniques. These include heat staking, ultrasonic welding, and laser welding. Due to their amorphous nature and solubility, polysulfones can also be solvent bonded, although this method has practical limitations, not to mention the environmental shortcomings. Epoxy-type adhesives are also useful for bonding polysulfones to themselves or to other materials.

14.3.7 Effect of Processing on Properties

Several works have been reported on processing of polysulfones with respect to the effects of processing on the structure and properties of the polymers. Residual stresses in injection-molded PSU have been analyzed by Siegmann et al. [26]. Studies on the effects of successive processing cycles on the structure and properties of PES and PSU [27–29] have shown that the chemical structure and the mechanical properties do not significantly change as a consequence of reprocessing with the processing conditions used. However, a progressive darkening of the polymer is observed upon reprocessing [27,28]. Results from these studies confirm the stability of polysulfones in processing conditions as well as the possibility of reprocessing them without important changes in their structure and properties, with the exception of the color change.

14.4 PROPERTIES

14.4.1 Physical, Thermal, and Mechanical

The basic physical and thermal properties of PSU, PES, and PPSU are listed in Table 14.2. Being completely amorphous, polysulfones are transparent in their natural form. They generally have a slight yellow/amber hue. This coloration is not inherent in the polymer but, rather, is a consequence of exposure to high temperatures during resin manufacture and subsequent fabrication steps. Significant improvements have been made over the years in manufacturing practices to minimize

TABLE 14.2
Physical and Thermal Properties of Polysulfone (PSU), Polyethersulfone (PES), and Polyphenylsulfone (PPSU)

Methods	PSU	PES	PPSU
Color	Light yellow	Light amber	Light amber
Clarity	Transparent	Transparent	Transparent
Haze,[a] %	<7	<7	<7
Light transmittance,[b] %	80	70	70
Refractive index	1.63	1.67	1.67
Density, g/cm^3	1.24	1.37	1.29
Glass transition temperature,[c] °C	185	220	220
Heat deflection temperature,[d] °C	174	204	207
Continuous service temperature,[e] °C	160	180	180
Coefficient of linear thermal expansion, °C^{-1}	5.1×10^{-5}	5.5×10^{-5}	5.5×10^{-5}
Specific heat at 23°C, J/g K	1.00	1.12	1.17
Thermal conductivity, W/m K	0.26	0.18	0.35
Water absorption,[f] %			
In 24 h	0.22	0.61	0.37
At equilibrium	0.62	2.1	1.1
Mold shrinkage, cm/cm	0.005	0.006	0.006
Temperature at 10% TGA wt. loss,[g] °C			
In nitrogen	512	547	550
In air	507	515	541

Source: Reprinted from *Kirk-Othmer Encyclopedia of Chemical Technology,* 4th ed., vol. 19, with permission from John Wiley & Sons.

[a] As measured on 3.1-mm-thick specimens (ASTM D1003).
[b] Typical values; varies with color; measured on 3.1-mm-thick specimens.
[c] Onset value as measured by differential scanning calorimetry.
[d] As measured on 3.1-mm-thick ASTM specimens under a load of 1.82 MPa (ASTM D648).
[e] Practical maximum long-term use temperatures for PSU and PES based on UL 746 thermal rating data; value for PPSU is estimated.
[f] According to ASTM D570 and measured from a bone dry state.
[g] Thermogravimetric analysis conducted at a heating rate of 10°C/min and 20 mL/min gas (nitrogen or air) flow rate.

color development. Of the three polysulfones commercially available today, PSU offers the lowest color and highest degree of optical clarity. Recent developments in manufacturing technology have resulted in near-water-white clarity for PSU [30]. The light transmittance values given in Table 14.2 are typical for resins undergoing moderate levels of thermal exposure during melt fabrication. The refractive indices of these polymers are relatively high thanks to the aromaticity of the backbone structure.

The thermal performance features of polysulfones are manifested in the following three key characteristics: (1) high glass transition temperatures, (2) long-term thermal-oxidative endurance, and (3) a very-high-temperature melt thermal stability. Polysulfones in general also have a low coefficient of thermal expansion and low mold shrinkage. The basis for the high thermal and oxidative resistance attributes is rooted in the mostly aromatic (PSU) or fully aromatic (PES, PPSU) backbone structures that are resonant with high bond dissociation energies. The bond dissociation energies of aryl ether, aryl sulfone, and aryl quaternary carbon are also impressive, although these bonds are not as strong as the aromatic C–C and C–H bonds. Last but not least, the sulfur atom in these polymers is in its highest oxidation state, thereby precluding further oxidation.

The short-term high-temperature structural capability of amorphous thermoplastics, including polysulfones, is governed by the T_g of the polymer. The maximum practical-use temperature is defined by the 1.82 MPa stress heat deflection temperature (HDT). This temperature limit is 174°C for PSU, and 204°C and 207°C for PES and PPSU, respectively. The high HDTs of polysulfones are complemented by a high degree of thermal-oxidation resistance, as stated above. This allows prolonged high-temperature utilization of these resins in the temperature range of 150–190°C, depending on the polymer, formulation, and use conditions. The thermal and oxidative stabilities of PSU, PES, and PPSU can be appreciated from the thermogravimetric data in Table 14.2 and Figure 14.2. As a result of this stability, polysulfones can be easily melt-fabricated at temperatures up to 400°C without any adverse side effects on resin integrity.

The dynamic-mechanical properties of polysulfones have been studied by several authors and are characterized by a major low-temperature γ transition at around −100°C; a low-intensity broad β peak near 80°C also appears for quenched samples. The transition at −100°C is believed to be at least in part a result of the 180° flips of the phenyl rings about the ether bonds [31]; this mobility of

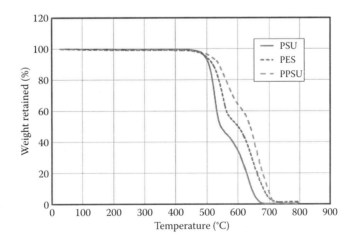

FIGURE 14.2 Thermogravimetric curves for PSU, PES, and PPSU, generated under a heating rate of 10°C/min in air.

backbone microstructure appears critical to the toughness properties of the polymer. Another mechanism that is postulated to be operative is a concerted motion of the sulfone group with complexed water [32]. The absorption of moisture increases the magnitude of this transition but is not essential for its existence. Paul et al. [33] analyzed the dynamic-mechanical behavior of polysulfones made from a range of bisphenols and correlated the low-temperature transition phenomena with polysulfone backbone structure. Robeson et al. [32] compared the dynamic-mechanical spectra of different polysulfones and analyzed the origin of the different transitions observed. Ting [34] analyzed the dynamic-mechanical behavior of PSU, PES, and PPSU. The incorporation of additional groups in the backbone produced additional low-temperature relaxation peaks in the loss curve.

The structural origin of the secondary relaxation processes in PSU [35] and the effect of annealing on the observed γ and β relaxation peaks have been analyzed, as well as the behavior of PSU in the glass transition region [36] and its dynamic-mechanical behavior [37] compared with that of three structurally related polymers. Finally, the effects of thermal history on some of the transitions [38], the enthalpy relaxations of PSU [39], and the dynamic-mechanical behavior of PSU and other thermoplastics deformed in uniaxial tension below their T_g [40] have also been studied.

Polysulfones are rigid and tough polymers with adequate mechanical characteristics for demanding engineering uses even without reinforcement. Their strength and stiffness at room temperature are high in comparison with most aliphatic backbone amorphous thermoplastics. The polymers exhibit ductile yielding over a wide range of temperatures and deformation rates. High unnotched impact resistance and retention of ductility at low temperatures have been tied to the low-temperature γ transition. Brittle mode failure in polysulfones is usually observed when −100°C is approached or traversed.

Room-temperature mechanical properties of PSU, PES, and PPSU are presented in Table 14.3. Among the three polymers, PPSU exhibits the highest elastic limit during tensile deformation. PES, on the other hand, offers the highest tensile strength of the three polysulfones. Other tensile, flexural, and compressive properties are quite comparable for the three polymers.

The high notched Izod impact of PPSU stands out as a distinguishing feature for this resin compared to both PSU and PES. The big difference in notched impact properties seen between PPSU on the one hand and PES on the other underscores the important dependence of key mechanical properties such as impact on subtle microstructural features. The dependence of impact properties of polysulfones on various backbone structural features was discussed by Attwood et al. [41]. One of the key factors noted is the importance of an all-*para*-linked backbone structure. This enhances the occurrence of chain entanglements that are important to mechanical toughness characteristics.

The deformation behavior of PES was studied by Plummer and Donald [42]. Crazing became increasingly likely as the temperature was raised. The transition temperature was found to depend on the strain rate and on the molecular weight of the polymer. The effects of stress aging on deformation of thin films of PES have also been studied [43]. Ting [34] studied the fracture behavior of PSU, PES, and PPSU. It was found that the fracture energy, G_{IC}, follows the order PPSU > PSU > PES. This is shown in Figure 14.3. The postyielding behavior of PSU [44], a ductile-to-brittle transition in PES [45], and the relationship between structure and deformation behavior of PSU and PES [46] have also been investigated.

The mechanical properties shown in Table 14.3 should be viewed as typical values for PSU, PES, and PPSU under short-term and idealized loading conditions. For material selection purposes, it is important to take into consideration end-use aspects, which might include long-term loads and creep, cyclic stresses (leading to fatigue), as well as temperature and environmental effects. During continual or long-term intermittent operation, polysulfones retain a high proportion of their initial properties upon exposure to high temperatures. The creep resistance is also maintained even at high temperatures [47]. The static and flexural fatigue behavior of PSU, among other high-performance polymers, has been analyzed by Trotignon et al. [48]. While

TABLE 14.3
Typical Room-Temperature Mechanical Properties of Polysulfone (PSU), Polyethersulfone (PES), and Polyphenylsulfone (PPSU)

Property	ASTM Test Method	PSU	PES	PPSU
Tensile[a] (yield) strength, MPa[b]	D638	70.3	83.0	72.0
Tensile modulus, GPa[c]	D638	2.48	2.60	2.30
Elongation at yield, %	D638	5.7	6.5	7.2
Elongation at break, %	D638	75	40	90
Flexural strength, MPa	D790	106	111	91
Flexural modulus, GPa	D790	2.69	2.90	2.40
Compressive strength, MPa	D695	96	100	99
Compressive modulus, GPa	D695	2.58	2.68	1.73
Shear (yield) strength, MPa	D732	41.4	50	62
Notched Izod impact, J/m[d]	D256	69	85	694
Unnotched Izod impact	D256	NB	NB	NB
Tensile impact, kJ/m[2e]	D1822	420	340	400
Poisson ratio, at 0.5% strain	–	0.37	0.39	0.42
Rockwell hardness	D785	M69	M88	M86
Abrasion resistance,[f] mg/1000 cycles	D1044	20	19	20

Source: Reprinted from *Kirk-Othmer Encyclopedia of Chemical Technology,* 4th ed., vol. 19, with permission from John Wiley & Sons.

Note: NB, no break.

[a] Tensile, flexural, and impact properties based on 3.1-mm-thick ASTM specimens.
[b] To convert MPa to psi, multiply by 145.
[c] To convert GPa to psi, multiply by 145,000.
[d] To convert J/m to ft. lbf/in., divide by 53.38.
[e] To convert kJ/m2 to ft. lbf/in.2, divide by 2.10.
[f] Taber abrasion test using CS-17 wheel and 1000 g load for 1000 cycles.

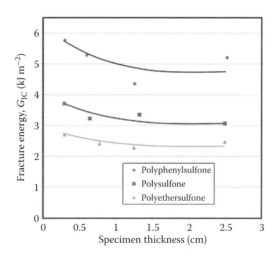

FIGURE 14.3 Effect of chemical structure and specimen thickness on the fracture energy of polysulfones. (Reprinted from Aitken, C.L. et al., *Macromolecules,* 25, 1992, copyright 1981, with permission from Chapman & Hall, Inc.)

adequate for many engineering uses, fatigue resistance of sulfone polymers is not as good as it is for semicrystalline engineering resins. Glass fiber or other reinforcement significantly enhances the fatigue resistance of polysulfones.

The effect of temperature on tensile stress–strain behavior of PSU is shown in Figure 14.4. The resin exhibits useful long-term structural capabilities up to a temperature of 160°C. PES and PPSU extend that temperature limit to about 180°C. The mechanical properties of polysulfones are maintained at low temperatures. Thus, PSU has a notched Izod impact strength of 64 J/m at −40°C, as compared to 69 J/m at 22°C. PSU, PES, and PPSU components fail in a ductile manner even at cryogenic temperatures.

The resistance of PSU to long-term thermal aging as manifested by the retention of mechanical properties is depicted in Figure 14.5. Analogous behavior is observed for PES and PPSU.

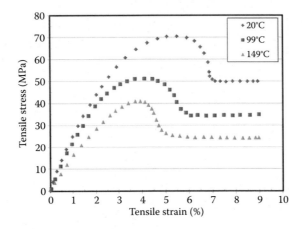

FIGURE 14.4 Tensile stress–strain curves of PSU showing yielding behavior at three temperatures. (Reprinted from *Kirk-Othmer Encyclopedia of Chemical Technology*, 4th ed., vol. 19, with permission from John Wiley & Sons.)

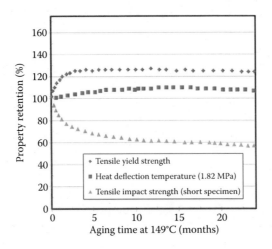

FIGURE 14.5 Effect of long-term thermal aging of PSU at 149°C in air on tensile strength, heat deflection temperature, and tensile impact strength. (Reprinted from *Kirk-Othmer Encyclopedia of Chemical Technology*, 4th ed., vol. 19, by permission from John Wiley & Sons.)

TABLE 14.4
Electrical Properties of Polysulfone (PSU), Polyethersulfone (PES), and Polyphenylsulfone (PPSU)

Property	Test Method	PSU	PES	PPSU
Dielectric strength, 3.2 mm,[a] kV/mm	ASTM D49	16.6	15.5	14.6
Volume resistivity, ohm cm	ASTM D257	7×10^{16}	9×10^{16}	9×10^{15}
Dielectric constant	ASTM D150			
At 60 Hz		3.18	3.65	3.44
At 10^3 Hz		3.17	3.65	3.45
At 10^6 Hz		3.19	3.52	3.45
Dissipation factor	ASTM D150			
At 60 Hz		0.0008	0.0019	0.0006
At 10^3 Hz		0.0008	0.0023	
At 10^6 Hz		0.0051	0.0048	0.0076

Source: Reprinted from *Kirk-Othmer Encyclopedia of Chemical Technology,* 4th ed., vol. 19, with permission from John Wiley & Sons.
[a] Thickness.

14.4.2 Electrical

Polysulfones exhibit good electrical-insulative properties, which include high dielectric strengths, low dissipation factors, and low dielectric constants. These capabilities are retained over wide temperature and frequency ranges. Selected electrical properties for PSU, PES, and PPSU are shown in Table 14.4. The favorable electrical properties of polysulfones in combination with the high-temperature capabilities and inherent flame retardancy characteristics make these polymers candidates for many electrical and electronic uses, as discussed in Section 14.7. Although the electrical breakdown of PES has been studied [49], few works have been published dealing with the electrical properties of polysulfones or with their flammability behavior, which is introduced in the next section.

14.4.3 Flammability Behavior

Polysulfones exhibit good inherent flame retardancy characteristics. PES and PPSU are both inherently flame retardant down to very low thicknesses. While still good in comparison to most other thermoplastics, PSU is not as resistant to burning in its natural form as PES and PPSU. This is believed to be due to the aliphatic isopropylidene moiety present in the backbone. There are commercially available formulated PSU grades, however, which are certified UL 94 V0 at low thicknesses, as the resin can be flame-retarded relatively easily through the addition of traditional flame retardants. All polysulfones demonstrate high oxygen indices and low flame spread and smoke release indices, particularly the wholly aromatic ones such as PES and PPSU. Because of their excellent flammability properties, the sulfones are used in a variety of aerospace applications. Vertical burn, smoke density, toxic gas emission, and heat release are critical properties, particularly for aerospace interior applications. All of the sulfones will pass Federal Aviation Administration (FAA) requirements for smoke density and toxic gas emission. PES and PPSU offer vertical burn performance, but only select grades of PPSU will pass the heat release testing. Flammability properties for PSU, PES, and PPSU can be found in Table 14.5.

14.4.4 Radiation Resistance

Polysulfones are resistant to many frequency ranges in the electromagnetic spectrum. This includes microwave, visible, and infrared. Although chemical reactions may take place under γ radiation [50],

TABLE 14.5
Flammability Behavior of Polysulfone (PSU), Polyethersulfone (PES), and Polyphenylsulfone (PPSU)

	Test Method	Unit	PSU	PES	PPSU
General Flammability Properties					
Flammability rating	UL 94	–	V0 at 6.1 mm[a]	V0 at 0.8 mm[a]	V0 at 0.8 mm[a]
Limiting oxygen index	ASTM D86	%	30.0	38.0	38.0
Smoke density	ASTM E662	–	90 at 1.5 mm[a]	35 at 6.2 mm[a]	30 at 6.2 mm[a]
Self-ignition temp	ASTM D1929	°C	621	502	–
Aerospace Industry–Specific Properties					
60 s vertical burn	DMS 1510				
Burning time		s	30	3	0
Burning length		in	3.6	2.8	<3
Longest burn particle		s	7	1	0
National Bureau of Standards smoke density	ASTM F814/E662				
D_s at 1.5 min		–	1	0	1
D_s at 4 min		–	40	3	1
D_{max}		–	40	3	2
D_{max} time		s	3.98	3.92	3.88
Toxicity – Flaming	BSS 7239				
HCN	ATS 1000	ppm	1	1	<1
CO	ABD0031	ppm	75	30	<100
$NO + NO_2$		ppm	1	1	<1
SO_2		ppm	<1	<1	<1
HF		ppm	<1	<1	<1
HCl		ppm	50	15	<1

[a] Thickness.

x-ray [51], and electron radiation [52–54], the resistance of polysulfones to x-ray, electron beam radiation [55], and γ radiation [56] is good under the limits of most practical applications. Polysulfones do not fare well in response to ultraviolet (UV) radiation [57–60], however. In their neat form, polysulfones absorb heavily in this region of the spectrum, with attendant losses in resin molecular integrity at and directly beneath the surface exposed to the UV source. As a consequence, the resin loses its strength and impact characteristics to a significant degree. Polysulfones are therefore not recommended for use in their natural form where they would be directly exposed to sunlight for prolonged periods of time. Pigmentation and reinforcement reduce the vulnerability of the resin to UV-induced mechanical property losses, and the addition of carbon black to these polymers can counteract these effects almost completely. Painting or coating is also an option whenever feasible as a means of blocking UV light from reaching the polysulfone surface.

14.4.5 Chemical Resistance and Hydrolytic Stability

One of the known weaknesses of amorphous plastics is their susceptibility to environmental attack when exposed to certain chemical environments under stress. The phenomenon of ESC of plastics, while well known and documented, is still not fully understood. Nonetheless, it is generally accepted that the chemical penetrates into the free volume of the amorphous phase, thereby weakening the secondary intermolecular forces between polymer chains and segments. The resistance of a polymer to ESC by a given chemical is directly related to the difference between the solubility parameters of the polymer and the chemical in question.

Because of their highly polar and aromatic character, polysulfones exhibit better ESC resistance characteristics than most other amorphous polymers. Their high-solubility parameters relative to aliphatic hydrocarbons afford good resistance against such environments. They can therefore be used effectively in a wide range of environments where they need to come in contact with oils and greases even at elevated temperatures. The resins also exhibit excellent resistance against hydrogen-bonding chemicals and solvents such as alcohols, glycols, and aliphatic amines. Common ketones, esters, and aromatic and chlorinated hydrocarbons represent the closest match of solubility parameters to those of sulfone polymers. Consequently, these chemicals can be problematic stress cracking agents for polysulfones if the polymer is under stress. The addition of glass fiber or other reinforcements to polysulfones can usually counteract most ESC resistance limitations of the neat resins in such environments, thus making their use possible. The cases where this approach will not work are those where the chemical in question has solvating power over the polymer. Examples of such solvents can be found in the next section, which is devoted to solubility of sulfone polymers.

Different researchers have studied the interactions of liquids and gases with polysulfones and their effects on polymer properties. Pearce et al. [61] obtained crystalline gels from PES/methylene chloride solutions, and Nazábal et al. [62] studied the effects of different fluid environments on the mechanical properties of PES. The effects of thermal and hydrolytic aging on PES, among other polymers [63], have also been analyzed.

Studies on the effect of sorbed carbon dioxide on the dynamic-mechanical behavior of PSU [64,65] showed that it acts as a strong plasticizer for PSU; the antiplasticization by different substances on PSU was analyzed by Maeda and Paul [66]. The effects of antiplasticization on gas sorption and transport were also studied. Free-volume interpretations were proposed for the experimental results [67].

A qualitative rating for the environmental resistance of PSU, PES, and PPSU to various organic and other environments can be found in Table 14.6. As can be seen from the table, the order of resistance for the three polymers generally follows the order PPSU > PES > PSU. Apart from the resistance to various organic environments, the resistance of polysulfones against hydrolytic or environmental attack by various aggressive aqueous media is outstanding. The polymers are highly resistant to hot aqueous environments including boiling water, high-pressure steam, mineral acids and bases, inorganic salt solutions, and ionic detergents. This is usually a key reason that polysulfones are selected over PCs, polyesters, polyamides, and polyimides in applications involving aggressive aqueous media [68]. The order of chemical resistance to organic media for the three polysulfones follows the order shown above; the order of resistance to hot water and steam environments follows a different order, PPSU > PSU > PES. The reason why this is so is the much higher affinity of PES to water relative to the other two polymers. The sulfone moiety is the hygroscopic group in sulfone polymers and, as such, is responsible for most of the water uptake by the polymer. The equilibrium water uptake of the three polymers is shown in Table 14.2 as 0.62%, 2.1%, and 1.1% for PSU, PES, and PPSU, respectively. The much higher water uptake of PES in the presence of boiling water or high-pressure steam results in accelerated hygrothermal annealing, which results in embrittlement in PES sooner than it will occur in PSU or PPSU under the same exposure conditions. A recent study determined through molecular simulations that the sulfone (SO_2) moiety is the group predominantly responsible for water affinity and water absorption of these polymers [69]. Due to the supertough nature of PPSU, it is the most resistant among the three polymers to hygrothermal annealing. Hence, the resistance order of the three polymers in terms of resistance to boiling water and steam environments is PPSU > PSU > PES [70].

14.4.6 Solubility

Because of their highly aromatic and polar character, polysulfones are not soluble in a large number of solvents. A handful of polar solvents will dissolve the vast majority of polysulfones (including

TABLE 14.6
Resistance[a,b] of Unreinforced Polysulfone (PSU), Polyethersulfone (PES), and Polyphenylsulfone (PPSU) Resins to Various Environments[c]

Environment	Rating		
	PSU	PES	PPSU
Hot water/steam	1	1	1
Aliphatic hydrocarbons	1	1	1
Alcohols and glycols	2	2	2
Electrolyte solutions	1	1	1
Nonoxidizing acids	3	2	1
Bases	3	2	1
Ionic surfactants	1	1	1
Nonionic surfactants	6	4	3
Carbon tetrachloride	8	5	2
Esters	9	8	6
Ketones	10	8	7
Aromatic hydrocarbons	10	9	7
Chlorinated hydrocarbons	10	8	8

Source: Reprinted from *Kirk-Othmer Encyclopedia of Chemical Technology,* 4th ed., vol. 19, with permission from John Wiley & Sons.

[a] Rating codes range from 1 (excellent) through 5 (good) to 10 (very poor).
[b] The above data are general indications and are for comparative purposes only; actual resistance will depend on many factors including stress, temperature, concentration, and exposure duration.
[c] Stress cracking resistance is usually enhanced substantially with the addition of reinforcing agents.

PSU, PES, and PPSU) at slightly above room temperature. These solvents include NMP, DMAc, pyridine, and aniline. 1,1,2-Trichloroethane and 1,1,2,2-tetrachloroethane are also good solvents for polysulfones but are usually avoided because of their potentially harmful health effects.

PSU exhibits solubility in a wider range of solvents than does PES or PPSU. In addition to solubility in the polar solvents cited above, PSU is also readily soluble in cyclic ethers like tetrahydrofuran (THF) and 1,4-dioxane, and in a wide range of chlorinated solvents like chloroform, dichloromethane, and chlorobenzene. The enhanced solubility qualities of PSU have been a key attribute in the development of PSU for applications requiring solution-based fabrication operations such as the manufacture of asymmetric hollow fiber membranes. The enhanced solubility of PSU relative to both PES and PPSU is believed to be due in part to the lower solubility parameter of PSU, which is, in turn, due to the presence of the aliphatic gem dimethyl group in its backbone. PES and PPSU undergo solvent-induced crystallization in a number of the solvents that dissolve PSU at room temperature, such as dichloromethane and THF.

14.5 BLENDS AND COMPOSITES

14.5.1 Blends

Blending of two or more polymers to achieve specific property combinations based on the properties of the parent polymers has continued to be a very attractive materials development approach in the plastics industry for several decades. Polysulfones offer both an opportunity as well as a challenge in this area.

Most polymer blends comprising polysulfones are high-performance blends. Among them, those with a semicrystalline component, liquid crystalline polymers (LCPs), or thermosetting resins, as well as the miscible blends merit separate discussion. The polysulfone blends used in membrane applications are included in Section 14.7.3 membranes. The ones with thermosetting resins are discussed in Section 14.7, devoted to epoxy toughening by polysulfones. The rest of the blends are discussed in Section 14.5.1.6.

By the term *compatibility*, we imply the presence of useful blend performance characteristics and stable phase morphology for practical considerations. Miscibility refers to complete solubility on a molecular level to achieve a one-phase system.

Reasonably compatible blends of polysulfones with other polymers include those formed with polycarbonates [71], polyarylates [72], polyimides [73], PEIs [74,75], polyaryletherketones [76–84], and polyphenylenesulfides [85–93].

14.5.1.1 Blends with Polyaryletherketones and Polyphenylenesulfide

The blends of PES, and semicrystalline polyaryletherketones, mainly polyether ether ketone (PEEK), have been a focus of attention in the patent and scientific literature [76–84] in recent years. The miscibility and mechanical behavior of PES/PEEK blends have been studied [76–79]. Their phase behavior depends on the blending method and on the processing conditions. Blends obtained from solution are reported to be miscible [76]; however, when obtained from the melt state, they are found to be either partially miscible [77] or practically immiscible [78,79]. Melt-mixed PES/PEEK blends show mechanical compatibility despite their almost complete phase separation [77–79].

PES/polyphenylene sulfide (PPS) [85,86] and PSU/PPS blends [87–93] have also been studied. The presence of polysulfones gives rise to clear toughness improvements. The phase behavior of PES/PPS blends has been studied by differential scanning calorimetry (DSC), solid-state nuclear magnetic resonance (NMR) [85], and dynamic-mechanical analysis [86]. Evidence of partial miscibility was found in solution blends by NMR [85]; however, in the same study, "mechanical blends" appeared completely immiscible. Blends of PSU and PPS obtained either by solution [90] or by melt mixing [87,88,91–93] were phase-separated regardless of blend composition. The mechanical properties of the blends [88,91,92] as well as the melting behavior of PPS in its blends with PSU [93] have also been reported.

14.5.1.2 Blends with Aramids

Blends of polysulfones, mainly PES, and aromatic polyamides (aramids) have received special attention in the literature [94,95]. In most cases, the studies dealt with the miscibility, crystallization behavior, and mechanical properties. Aramid crystallization was induced and accelerated by PES [95]. Polymerization blending [94] of PES and an aramid gave rise to miscible blends over the entire composition range, with improved mechanical properties. Solution-blended PES and other aramids phase-separated, giving rise to films low in modulus or strength, although tough and ductile in some compositions [94].

14.5.1.3 Miscible Blends

The research and successful development of new polymer blends that are miscible could assure the availability of new mechanically compatible materials. Besides, and as a consequence of their compatibility, the mechanical, chemical, and thermal properties, as well as other characteristics of the blends, are usually close to those predicted by the additivity rule. Furthermore, synergistic property enhancement can occur, giving rise to economically attractive materials for some applications.

The arylether and sulfone groupings are, for the most part, chemically inert. As such, this stability against chemical reaction precludes the possibility of interchain reactions that could facilitate miscibility through the creation of polymer–polymer block structures, which generally leads to compatibilization and/or miscibility. Blends of polysulfones with other polymers must therefore rely exclusively on favorable intermolecular interactions for the achievement of blend compatibility or miscibility.

Thermodynamic miscibility of melt-processable polysulfone blends with other polymers is extremely rare. Some examples of PES miscibility with phenoxy (Ph) [89,90,96,97], polyimides [73,98–102], poly(ethylene oxide) (PEO) [103–105], poly(ethyl oxazoline) [106], and polyhydroxyether of phenolphthalein [107] are discussed in the literature. A lower critical solution temperature (LCST) was reported in most of the blends. Only one of these blend systems [99] has been reported to form melt-processable single-phase mixtures. The transport and sorption properties of the PES/Ph blend [108,109] have also been examined. Blends of PPSU and PEI were examined and reported to be miscible [110]. Also, blends of PPSU with some polyphenylenes exhibit miscibility behavior when cast into films from solution [111].

14.5.1.4 Blends with Other Polysulfones

None of the binary mixtures comprising PSU, PES, and PPSU forms a miscible blend. However, such blends do form compatible mixtures with a finely dispersed stable phase morphology. Blends of PSU and PES and their properties are discussed by Harris and Robeson in a US patent [112]. Blend miscibility behavior of polysulfones having high T_g (>230°C) with other polysulfones such as PES and PPSU has also been reported [113,114]. Properties of PPSU blends with either PSU or PES over the full range of compositions are described by El-Hibri et al. [115–117]. Over a certain range of compositions, both the PPSU/PSU and PPSU/PES systems exhibit supertoughness when the PPSU volume fraction in the blend is maintained above about 65%.

14.5.1.5 Blends with Liquid Crystalline Polymers

LCPs have found application as high-modulus fibers and films with unique properties, due to the formation of ordered solutions (lyotropic LCPs) or melts (thermotropic LCPs) [118]. Moreover, when blended with thermoplastics, they give low melt viscosities and appear very elongated, producing materials with similar structure and some superior properties to those of traditional fiber-reinforced analogs.

Studies of blends of polysulfones and LCPs focus essentially on the characterization of morphology [119–123], phase behavior [120–122,124,125], rheology, and mechanical properties [120–123], as well as crystallization [126] and compatibilization [122].

In most cases, LCPs are immiscible with polysulfones. However, analysis of PES [122] or PSU [120] blends with the Vectra A LCP resin found a slight decrease in the T_g of the polysulfone.

Blends of LCPs with polysulfones, as with other thermoplastics, generally bring considerable reinforcement in tensile and flexural strength and modulus of the polysulfone, with the counterbalance of a reduction in ductility [120–122]. The increase in tensile strength or modulus is due [120] to the very high molecular orientation and fibrillation of the LCP.

14.5.1.6 Other Blends

Melt blends of polysulfones with olefinic, vinyl, or styrenic polymers generally offer little promise due to the big disparity between the melt, thermal, and oxidative stabilities of these polymers vis-à-vis polysulfones. These so-called commodity plastics do not possess the requisite stability for melt processing in the 340°C+ temperature regime, which is required to properly melt-fabricate polysulfones. This precludes viable systems even if such blends show signs of compatibility. In addition to the thermal processing limitations, however, there is also usually very poor compatibility behavior between polysulfones and these polymer types. Callaghan and Paul [127] published a major study in which the interaction energies between polysulfones and other several such polymers were discussed as they relate to blend miscibility and compatibility questions.

Attempts at blending elastomeric polymers with PSU to achieve high notched impact resistance in the polysulfone have been generally unsuccessful. This has been due to the poor interphase compatibility between PSU and commercially available rubbers and elastomers that are used today in polymer impact modification technology, not to mention the limited thermal stability. Two partially successful attempts based on a polyurethane in one case [128] and on polydimethylsiloxane in the other [129] are reported in US patents.

The effects of repeated molding (e.g., regrind) operations on the degradation and mechanical properties of a PSU/PC commercial blend have been evaluated and compared with those of its two components [130]. Polybenzimidazole (PBI) is immiscible with PSU; however, PSU/PBI alloys exhibit some interesting and synergistic properties [131]. Moreover, mechanically compatible PES/PEI blends [75] and immiscible PSU/Ph blends [132] have been obtained.

14.5.2 Composites

Polysulfones are very suitable for the preparation of specialty functional compounds. They are thermally stable, amenable to compounding, and chemically compatible with a number of fillers and additives commonly used in polymer modification.

Formulations containing chopped glass and carbon fiber of loadings up to 40% can be found. The tensile and flexural properties at room and high temperatures, as well as resistance to stress cracking by aggressive chemical environments or UV radiation [133], can be enhanced by the addition of fibrous reinforcements such as chopped glass fiber. For example, 30% glass-reinforced PES shows a flexural modulus of 8.0 GPa at 180°C, as compared to 8.4 GPa at 20°C. These enhancements do come at the expense of a loss in ductility and an increase in melt viscosity. Typical mechanical properties of glass-reinforced PSU and PES at room temperature are given in Table 14.7. Carbon fiber also imparts further increases in mechanical properties. Comparisons of the effects of the addition of carbon fiber to various polymers on mechanical properties and failure mechanisms have been studied extensively [134,135]. Carbon fibers, graphite fibers, stainless steel fibers, and graphite can also be incorporated to modify electrical properties in some cases to provide electrostatic dissipation or conductivity. Glass bead-reinforced and mineral-reinforced formulations are also available and are desirable when isotropic reinforcement is important.

Other reinforcements are also used to improve various properties. Formulations containing polytetrafluoroethylene (PTFE) are found on the market. PTFE is added to provide lubricity by lowering the coefficient of friction at loadings up to 15%. In some cases PTFE, is used in conjunction with graphite and minerals to improve overall wear resistance. Mineral fillers also can improve dielectric strength and reduce thermal expansion. Commercially available high-temperature flame retardants are also used along with other inorganic fillers such as titanium dioxide pigment to improve the heat release and flame propagation properties.

TABLE 14.7
Typical Physical and Mechanical Properties of Glass-Reinforced (GR) Polysulfone and Polyethersulfone

Property	Method	Polysulfone			Polyethersulfone		
		10% GR	20% GR	30% GR	10% GR	20% GR	30% GR
Tensile strength, MPa	D638	77.9	96.5	108	86	105	126
Tensile modulus, GPa	D638	3.65	5.17	7.38	3.8	5.7	8.6
Tensile elongation, %	D638	4.1	3.2	2.0	5.8	3.2	1.9
Flexural strength, MPa	D790	128	148	154	145	162	179
Flexural modulus, GPa	D790	3.79	5.52	7.58	4.1	5.2	8.1
Izod impact, J/m	D256	64	69	74	48	59	75
Tensile impact, kJ/m^2	D1822	101	114	109	59	65	71
Heat deflection temp., °C	D648	179	180	181	211	214	216
Coefficient of linear thermal expansion, mm/mm °C	D696	3.6	2.5	2.0	3.6	3.1	3.1
Specific gravity	D1505	1.33	1.40	1.49	1.43	1.51	1.58
Mold shrinkage, mm/mm	–	0.004	0.003	0.002	0.005	0.004	0.003

Source: Reprinted from *Kirk-Othmer Encyclopedia of Chemical Technology,* 4th ed., vol. 19, with permission from John Wiley & Sons.

14.6 HEALTH, SAFETY, AND ENVIRONMENTAL ASPECTS

14.6.1 Health

Aromatic polysulfones are widely recognized for being chemically inert polymers. PSU has been successfully used for over three decades in applications where safety is of utmost importance. Examples of such applications include food service, potable water plumbing, dairy processing components, and medical devices.

Many PSU, PES, and PPSU grades comply with the requirements of several governmental and/or regulatory agencies, including but not limited to the following: US Food and Drug Administration (FDA), European Commission Directive 2002/72/EC, Underwriters Laboratories (UL), National Sanitation Foundation (NSF), International Standard 51: Food Equipment Materials, and Standard 61: Drinking Water System Components and 3A Dairy Certification. In addition, there are international water contact standards such as the French ACS, the United Kingdom's BS6920, and the German KTW and W270. Some specific grades have been tested for biocompatibility and may comply with ISO 10993 and/or USP Class VI. Regulatory approvals will vary by company and by specific grades; therefore, it is always best to contact the supplier and request a letter assuring compliance of specific individual grades when such compliance is needed for the end use.

14.6.2 Safety

The high thermal stability and oxidative resistance of the backbone structures in polysulfones translate to the virtual absence of any organic volatile releases when the polymers are melt-fabricated within their recommended melt-processing ranges of up to 380°C. At temperatures above 380°C, trace amounts sulfur dioxide, methane, and other organic compounds begin to evolve. As is required with the melt fabrication of other plastic materials, good ventilation of the molding or extrusion area is recommended when melt-fabricating polysulfones into finished parts. Moreover, where different resins are processed, the complete purging of molding or extrusion and tooling equipment of any residual presence of thermally sensitive resins—such as vinyls and polyacetals—must be ensured. Such polymers can undergo violent degradation reactions when exposed to temperatures above 300°C. Attention to fabrication equipment hygiene is therefore an important safety practice in connection with melt processing of polysulfones and, indeed, all other high-temperature thermoplastics.

14.6.3 Sustainability Aspects

There are numerous applications in which the use of sulfone polymers has been proven advantageous from a polymer property standpoint, including consumer goods, medical, plumbing, electrical/electronic, aerospace, and industrial, to name a few. In recent years, however, the increasing worldwide concern about product sustainability fueled the requirement for a better understanding of these product's environmental profiles, which has been made possible through life cycle assessment (LCA) studies. These studies can produce a list containing the quantity of emissions to the environment as well as the amount of energy, water, and material consumed along the entire manufacturing stage of the products, including raw materials and transportation. These are then translated into several standard environmental impact categories, such as global warming potential, energy resources, water use, toxicity, etc. Full LCA analyses of products made with high-performance sulfone polymers in many applications have shown their contribution to sustainability and product ecodesign by providing positive environmental impacts compared to alternative materials and/or processes. For instance, water purification membrane processes using polysulfone-based filtration membranes have beneficial environmental impacts compared to conventional water treatment processes [136]. In addition, the use of sulfone polymers in membrane-based pretreatment systems allows for less fouling and higher fluxes in water purification operations, consequently decreasing their overall footprint and the energy required per cubic meter of freshwater produced [137]. Polysulfones may also be used in the manufacturing of proton-exchange membranes for fuel cells while this technology is characterized as an environmentally responsible alternative to conventional energy systems [138]. In the healthcare industry, sulfone polymers can be used in medical devices exposed to multiple cycles of cleaning and sterilization, while in some cases, reusable devices have shown reduced environmental impacts over the life of the products [139]. In the plumbing industry, a recent LCA study involving data on plastic pipe systems manufactured by companies encompassing more than 50% of the European market has shown that cross-linked polyethylene (PEX) pipe systems using PPSUs as fittings have a lower environmental impact than those made from copper [140]. Finally, the lighter weight of products made with sulfone polymers in structural aircraft applications offers significant weight reduction opportunities, leading to lower fuel consumption and resulting in less CO_2 emissions. In fact, it has been reported that the implementation of weight reduction initiatives in aircraft has saved more than 10 million gal. of jet fuel and removed 221 million lb. of CO_2 emissions per year [141].

The high thermal-oxidative stability of polysulfones allows these polymers to be reprocessed without significant loss to properties. Tensile strength, modulus, strain to break as well as notched Izod and tensile impact measurements taken from PPSU reprocessed up to nine times show no significant changes. Conventional use of regrind involves mixing 75% virgin resin with 25% clean regrind. This ability allows for improving yield by reprocessing scrap parts, start-up resin, and injection-molding runners. The stability of the sulfone polymers also allows—at least in principle—for complete reclamation of an article, thus facilitating cradle-to-grave planning.

14.7 END-USE APPLICATIONS

Polysulfones are used in a wide and ever-expanding range of applications that take advantage of their unique combination of properties. The following segments of this section delve into some detail in the description of various current and emerging application areas of polysulfones.

Because of the shared fundamental commonalities of PSU, PES, and PPSU, these three polymers could be used interchangeably in many applications. Whenever possible, however, PSU is selected in consideration of its lower cost. Its higher clarity and lower color also make it more attractive for some uses. PES is selected over PSU in uses where higher levels of heat resistance (HDT), inherent flame retardancy, or environmental resistance are needed. PPSU represents a step improvement over the other two polymers in impact resistance, hydrolytic stability, and overall ability to retain ductility and mechanical integrity in the face of harsh thermal and environmental conditions. Due to its higher cost, however, it is used only when both PSU and PES fail to meet the requirements of a particular end use.

14.7.1 Food Service and Food Processing

The combination of outstanding hydrolytic stability, high-temperature resistance, and chemical resistance to most household and industrial cleaners have made polysulfones a natural fit in food and hot beverage service and processing applications. Such applications include coffee pots and coffee maker components, institutional food serving trays, and steam-heated food service trays. Food processing applications include automated milking machine equipment and conveyor belt components used in food processing. Durable microwavable bowls for consumers have also been made and sold extensively for many years out of sulfone polymers, thanks to the excellent microwave compatibility, impact resistance, transparency, staining resistance, and colorability; good overall aesthetics; and of course, compliance with regulatory food contact approvals.

14.7.2 Plumbing Components

Uses in functional household plumbing also take advantage of the excellent hydrolytic stability as well as the resistance to dissolved chlorine, which is used today by many municipalities to prevent bacterial growth. Plumbing components that have utilized PSU and PPSU over the years (not

FIGURE 14.6 Cutaway section of Soterna thermal energy boiler, which utilizes PPSU as material of construction of its boiler pipes. (Photo courtesy of Soterna S. Coop, Navarra, Spain.)

much PES is used in plumbing) include faucet waterways and mixing valves, distribution manifolds, hot-water heater components, and plumbing fittings. Since the 1990s PPSU revolutionized the plumbing industry by providing a viable and robust thermoplastic alternative to brass fittings for PEX pipes. The ability of PPSU to be used in pressure fittings was facilitated by the robust toughness of the material and the ability to withstand long-term hydrostatic stress at the elevated temperatures of hot-water service. In a recent alternative energy application, PPSU was successfully employed as the material of construction for heat-exchanger tubes on a solar water heater (Figure 14.6), replacing stainless steel in this type of end use. Key selection criteria included long-term resistance to hot chlorinated water as well as the long-term hydrostatic hoop stress endurance at elevated temperatures now proven for PPSU. Plumbing uses of polysulfones extend to the chemical processing arena, where polysulfones find uses in corrosion-resistant pipes, fittings, flow meters, and pump components, as well as in filter modules, support plates, and tower packing. Glass-reinforced grades are used where exposure to severe chemical environments is anticipated.

14.7.3 Membranes

One of the most important applications of aromatic polysulfones, especially PSU and PES, is their use in membranes. This is enabled by their ease of formation into semiporous membrane structures, including flat sheets and hollow fibers, as well as by their excellent mechanical strength, thermal stability, and chemical resistance. They are commonly produced using nonsolvent-induced phase separation (sometimes called diffusion-induced phase separation), wherein a highly viscous polymer solution, or membrane dope, is extruded into an appropriate cross section (flat sheet or hollow fiber) and then immersed in a nonsolvent for the polymer [142–144]. This initiates a phase separation into polymer-rich domains (which form the membrane structure) and polymer lean domains (which form the internal pores). In addition, the nonsolvent extracts the remaining solvent, leaving only the porous polymer membrane. A wide variety of membrane architectures can be manufactured including so-called spongelike and fingerlike structures. The membranes may also be asymmetric with porous or nonporous surfaces or symmetric. Membrane structures are typically observed using scanning electron microscopy (SEM), and three of the most common morphologies are depicted in Figure 14.7. Commonly, other polymers are blended into the membrane dope to facilitate formation of the desired membrane architecture as well as to increase the hydrophilicity of the membranes. The most common polymers used in these blends are poly(N-vinyl-2-pyrrolidone) (PVP) [145] and polyethylene glycol (PEG) [146].

Polyarylethersulfones

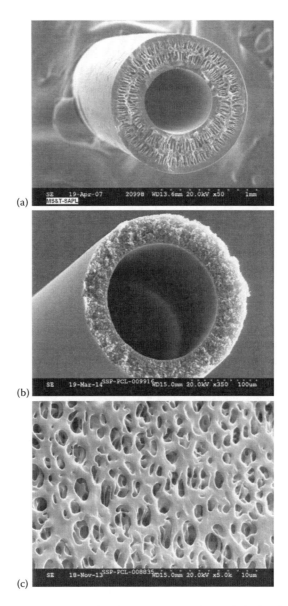

FIGURE 14.7 Representative images of the three most typical morphologies found in polyarylethersulfone porous membranes. (a) Fingerlike, (b) spongelike, and (c) pores. (Images courtesy of Solvay Specialty Polymers.)

Aromatic polysulfones find use in virtually all types of membranes including gas separations, reverse osmosis (RO), forward osmosis (FO), nanofiltration (NF), ultrafiltration (UF), and microfiltration (MF). A special and extremely important application is hemodialysis, which enables life-sustaining support of patients with poor kidney function [147,148]. Finally, sulfonated polysulfones have been considered for use in fuel-cell membranes.

Gas separations require nonporous membranes, but they are commonly made as asymmetric hollow fibers with a nonporous top layer (referred to as a dense skin) and a porous support layer. Gas separations are different compared to liquid separations in that the permeability for a given gas (and thus selectivity between two gases) is a material property of the polymer and thus does not depend on the membrane morphology (neglecting defects that may occur in the nonporous separation layer and small contributions to the resistance from a support layer, if present). Polysulfones have among the best combinations of permeability and O_2/N_2, selectivity enabling their application in oxygen and nitrogen generation [149].

Virtually all RO membranes manufactured today are thin-film composites, comprised of a cross-linked polyamide layer that is interfacially polymerized onto a porous polysulfone support membrane [150,151]. This composite structure is further supported by a nonwoven mat typically comprised of poly(ethylene terephthalate), permitting their use at the high pressures required to desalinate seawater. RO membranes are cast into a continuous flat sheet, which is then assembled into spiral-wound modules, and this is currently the most economical method to produce potable water from seawater. Recently, researchers have begun to incorporate inorganic nanoparticles, such as zeolite A [152,153], into the RO membranes to achieve better performance. FO membranes may also be based on thin-film composite technology similar to RO membranes, but large-scale application of FO is difficult because of the need for a "draw solution." Nevertheless, FO is being considered for water purification and power production [154]. Power production is possible because when salt transfers from the high-salinity side of the membrane to the low-salinity side, the pressure of the low-salinity side increases, and it may be possible to harness this effect to produce energy in locations where rivers meet the sea [155]. NF membranes are similar to RO membranes, except that the polyamide layer is optimized for higher flux but with lower salt rejection. NF can be used to desalinate so-called brackish water, which is water having lower salt concentrations than saltwater but with salinity too high for drinking.

UF and MF membranes reject materials on the basis of size (or molecular weight), and thus, the separation ability depends on both the material and the membrane morphology. The boundary between UF and MF is generally considered to be a size of about 0.1 µm, corresponding to a molecular weight cutoff of around 100,000 Da, with UF rejecting the smaller molecules. They can be manufactured in both a flat sheet and hollow fiber form. Researchers have also proposed the addition of inorganic particles such as zeolite A [156] or silver nanoparticles [157] to improve water flux and/or reduce membrane fouling. UF and MF applications are extremely diverse, for example, potable water production (either to directly produce potable water or as pretreatment for RO), wastewater treatment, biopharmaceutical production (especially proteins and enzymes), and beer and juice concentration and clarification. The excellent chemical resistance of PSU and PES enables their use in UF and MF applications since they must be periodically cleaned with agents such as acids, bases, and chlorine (bleach) to control membrane fouling [158]. Fouling can be caused by precipitation of inorganic substances, accumulation of organic substances such as proteins, or even growth of microbial communities. UF and MF membranes operate with a pressure difference between the two sides of the membrane.

While hemodialysis membranes are morphologically similar to UF membranes, they operate with a concentration difference between the two sides of the membrane, although the separation is still based on size. Almost all of the hemodialysis membranes today are produced with PSU or PES due to their better biocompatibility compared to other materials that have been used (mostly cellulose derivatives). All hemodialysis membranes are hollow fibers with the blood from the patient flowing through the fiber lumen and a "dialysate" saline solution flowing outside the fibers. Hemodialysis membranes are designed to mimic a human kidney, which removes uremic toxins (e.g., urea, phosphorous, β_2-microglobulin) while retaining valuable larger proteins such as albumin [159].

Many attempts have been made to modify sulfone polymers to allow for the production of improved membranes by either using monomers with pendant functional groups or post modification [160–162]. Sulfonation [163,164] is a technique that is commonly used to prepare more hydrophilic membranes, which are believed to permit higher water fluxes with reduced fouling. Halomethylation and lithiation can also be practiced to prepare reactive sites on the polysulfone backbone, and these sites can be further reacted [165] to prepare functionalized membranes. In practice, neither technique is practiced on an industrial scale, because of the health and safety aspects associated with these reactions. Graft copolymers have also been prepared to try to improve the hydrophilicity and reduce fouling [166]. Many attempts have been made to modify polysulfones for hemodialysis to improve the biocompatibility and reduce the need for anticoagulation therapies during dialysis, for example, by attaching vitamin E [167] or heparin-like substances [168] to the surface. Surface modification of the membranes after production can also be done by using appropriate high-energy techniques such as plasma-mediated grafting or electron beam-mediated grafting [169].

Sulfonated polysulfones have been proposed for use in proton-exchange membranes and direct methanol fuel cells because their fixed charges allow for high conductivity [170]. The current state-of-the-art materials for fuel cells are based on poly(perfluorosulfonic acids). Sulfonated polysulfones have an advantage of lower methanol crossover and, potentially, the ability to work under lower relative humidity conditions or higher operational temperatures, but their oxidative stability is not as good. Roy et al. illustrate a number of approaches to design a polymer architecture with the best possible performance [170]. For example, polymerization of random and block copolymers of hydrophilic (sulfonated polysulfone either from sulfonated monomers or via post modification) and hydrophobic (unsulfonated polysulfone and/or poly[perfluoro-]) segments can control the water swelling, ion-exchange capacity, methanol permeability, etc. This remains an active area of research.

14.7.4 Medical Devices

Medical applications are a major area of use for polysulfones, as the resins offer a full range of sterilizability options. The resins adequately withstand most current means of sterilization, including heat, steam, boiling water, γ radiation, ethylene oxide, and many chemical disinfectants. One major area of application for polysulfones, especially PPSU, is in cases and trays that carry and organize surgical and other medical and dental instruments. These containers must withstand hundreds (sometimes thousands) of steam autoclave sterilization cycles as well as have excellent impact resistance and biocompatibility with the ISO 10993 standard. To date, PPSU has been the only unreinforced thermoplastic able to replace stainless steel in this type of application. Medical and dental instrument handles are another big area for PPSU for the same reasons. In addition to the functional capabilities of sulfone polymers, the transparency of these polymers and their ability to be colored in a broad range of bright and vibrant colors have also contributed to their adoption in hospital settings, where pleasant appearance and aesthetics are often important.

In addition to the now well-established use of polysulfones in conventional medical devices and other components of medical instrumentation and equipment, since 2007, PSU and PPSU also have been offered on the market in grades that are for implantable medical devices that go into the human body for durations longer than 30 days. Such grades are produced in compliance with the ISO 13485 standard and under the relevant aspects of current good manufacturing practices (GMP). These grades are available under the trade names Eviva PSU and Veriva PPSU. The interest in polysulfones for implant applications dates back to the early 1980s, where PSU was used for the first time as the material of construction for the groundbreaking Jarvik 7 artificial heart, a form of which is still in use to this day [171]. Since then, a number of other PSU implant applications have been developed and validated. Typically, these applications take advantage of the ability of PSU to hold very tight dimensional tolerances from injection-molding processes. The ability to use ultrasonic welding for seamless integration of multiple parts is also a consideration. The excellent dielectric properties also make the material a good candidate because of its magnetic resonance imaging (MRI) compatibility. Above all, however, the purity, biocompatibility, and chemical inertness in the human body are chief selection criteria. The implant applications where PSU is used are ones in which there is no load-bearing function to the implanted device. Past studies have shown that polysulfones are not suitable in load-bearing implant applications in contact with biological fluids containing lipids or phospholipids, such as bone marrow, due to susceptibility to ESC [172]. Examples of successful PSU implant applications include pacemaker battery housings and infusion ports being used today for drug delivery in chemotherapy and other treatments (Figure 14.8).

FIGURE 14.8 Implantable drug infusion system available from Plan 1 Health utilizes two ultrasonically welded Eviva PSU parts for its housing. (Photo courtesy of Plan 1 Health, SRL, Amaro, Italy.)

14.7.5 Commercial Aircraft Interiors

Sulfone-based polymers are used in a variety of aerospace interior applications, where low heat release, flame propagation, smoke generation, and toxic gas emission are critical performance criteria. Flame-retardant materials result in improving passenger survival in an emergency by increasing the time passengers have to exit an aircraft. PPSU exhibits superior flammability performance. It also offers excellent impact resistance and compatibility with cleaners and other chemicals used in the industry, thereby enabling the design of durable interior parts. PPSU is used in a number of applications, including passenger service units, air return grills, window frames, cockpit interiors, and exit doors. More recently, PPSU has also been fabricated into airline catering trolleys [173] and serving dishes. The trolley shown in Figure 14.9 offers weight savings, damage tolerance, and recyclability advantages over a conventional aluminum or composite trolley. The use of high-performance materials such as polysulfones allows manufacturers to build lighter aircraft, which yield large savings in fuel economy.

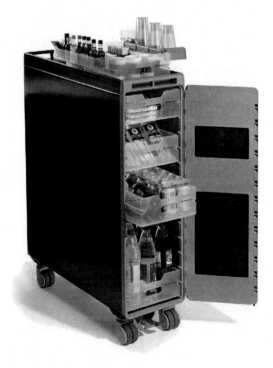

FIGURE 14.9 Fully recyclable airline catering trolley made from extruded PPSU profiles and injection-molded PPSU components. (Photo courtesy of Aerocat B.V.)

14.7.6 Toughening of Advanced Composites

Polysulfones are effective tougheners for advanced carbon fiber–epoxy thermoset composites that are replacing some heavier metal structural components on new-generation commercial airplanes to significantly reduce fuel consumption and simplify manufacturing [174–177]. The epoxies most often used in this demanding application are trifunctional and tetrafunctional epoxies, which, when cured, give networks with high cross-link density with exceptional high-temperature and chemical resistance.

By themselves, cured epoxies and other thermosets are brittle and most often require the addition of well-dispersed fillers, such as a rubber, to improve the impact strength, or toughness of the composite. Adding rubbers, however, reduces both the overall mechanical properties and thermal resistance of the thermoset. Certain high-performance thermoplastics like PES and PEI with high glass transition temperatures (T_g = ~220°C), however, can be used with the advanced multifunctional epoxies to increase epoxy fracture toughness (as measured by the critical stress parameter, K_{IC}, or the fracture energy, G_{IC}), by a factor of at least two to three without significant losses in mechanical and thermal properties. Suitable thermoplastic tougheners must be soluble in the epoxy and then phase-separate during the curing step. PPSU is insoluble in epoxies and is not currently used as a toughener. PSU is soluble in epoxies and can be used as an epoxy toughener, but it does not have as high a T_g as PES and is not typically used in advanced epoxies for aerospace applications.

In a typical molding/curing procedure, about 10–20 wt.% finely ground PES powder is first dissolved in the epoxy at elevated temperatures to give a clear, viscous solution. The hardener, often 4,4′-diaminodiphenylsulfone (DDS), is then added, and the mixture is infused into a carbon fiber matrix (~60 wt.% C) to give a prepreg. After the prepreg is formed into the desired shape, it is then cured at about 180°C for several hours using either an autoclave or newer out-of-autoclave processes [178,179]. More recently, PES electrospun nanofibers were claimed to boost toughness at lower loadings than powders [180].

During the curing step, as the epoxy begins to polymerize, a PES thermoplastic-rich phase separates by spinodal decomposition to give a co-continuous two-phase morphology, as shown in Figure 14.10, believed to be required for suitable toughening [181,182].

Often, PES for toughening is supplied with reactive end groups, such as –OH, or –NH$_2$ (e.g., Solvay Virantage reactive PES [rPES] grades). It is widely believed that these reactive groups aid in interfacial bonding with the epoxy to give optimal toughening. There are, however, some contradictory studies indicating that the reactive groups may not be necessary to afford effective toughening

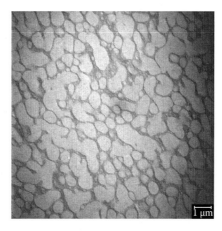

FIGURE 14.10 Transmission electron microscopy (TEM) image of co-continuous two-phase morphology of cured epoxy-rich (light) and PES-rich (dark) phases. (Image courtesy of Solvay Specialty Polymers.)

for some formulations and curing conditions [177]. At a given loading, the higher the average molecular weight of the PES toughener, the higher the degree of toughening. With increasing molecular weight, however, the epoxy/DDS/PES precure blend becomes quite viscous and difficult to process. Composite molders continuously search for the optimum balance of properties and processability by adjusting the PES molecular weight, using mixtures of epoxies, and controlling the curing conditions.

14.7.7 Structural Foams

Aromatic polysulfones, mainly PPSU and PES, have been used as high-performance foam material [183,184] due to their excellent strength, thermal stability, and chemical resistance. When converted into foam form, additional attributes such as density reduction, resiliency, thermal insulation, and acoustic insulation can be enhanced.

The foaming of polysulfone polymers has been practiced and studied widely in academic institutions through combinations of physical and chemical foaming via batch, injection, and extrusion processes [185]. However, the majority of commercially available polysulfone foams are prepared by physical foaming via a melt extrusion process to yield low-medium-density foams (densities of 40–130 kg/m^3).

Due to foam cellular structure, characteristics of the cell morphology are absolutely critical in determining the foam's final properties for a given polymer [186–190]. Some key morphological parameters are cell size, cell size distribution, cell density, open cell content, cell-wall uniformity, etc. Various techniques have been utilized to characterize cell morphology [190], but the main technique for cells < 1 mm is SEM. An example SEM photograph of a PPSU foam is shown in Figure 14.11.

Currently, there are two major commercial suppliers of sulfone polymer foams on a worldwide basis. These are Solvay Specialty Polymers and DIAB International. Solvay Specialty Polymers manufactures and supplies a PPSU type of polymer foam. DIAB manufactures and supplies a PES type of polymer foam. Both can be used in aircraft, rail, and marine applications in neat form or as a foam core for composite sandwich applications, as exemplified in Figure 14.12. As described previously in Section 14.4 for the neat PPSU and PES polymers, the relative ranking of physical, thermal, and mechanical properties translates to the foams, as well.

14.7.8 High-Performance Coatings

High-performance coatings may be formulated with PES resin. Due to its properties and FDA food contact status [191], the resin is used commercially as a binder resin in nonstick fluoropolymer

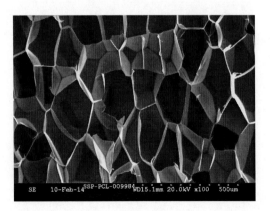

FIGURE 14.11 SEM micrograph showing the closed cell morphology of a PPSU structural foam with a density of 60 kg/m^3. (Image courtesy of Solvay Specialty Polymers.)

FIGURE 14.12 A sandwich-construction structural panel comprised of a 60-kg/m^3-density PPSU foam core and continuous carbon fiber–epoxy top and bottom skin layers. (Photo courtesy of Solvay Specialty Polymers.)

bakeware coatings, industrial primers, and rice cooker coatings [192,193]. The polymer backbone of aromatic and sulfone groups provides excellent thermal stability and very good chemical resistance to strong acids and strong bases along with good hydrolytic stability. The polar sulfone groups in the backbone PES provide very good adhesion to metal substrates. PES has the greatest weight percentage of sulfone per repeat unit among commercial sulfone-based polymers at 28%. The polymer end groups may be functional phenolic groups, such as in Veradel 3600RP, which can react with suitable electrophiles and also provides for interaction with substrates.

PES is typically formulated into a solvent-based coating. Solvents that may be used to dissolve PES are *N*-methyl pyrrolidinone (NMP), NMP/methylene chloride, NMP/xylene, DMF, and DMAC. Small particles of PES powders may be obtained by grinding or precipitating small particles. These may be incorporated into waterborne formulations [194]. The coatings are applied by typical methodologies such as a spray, dip, roller, or curtain and then baked at elevated temperatures (200–400°C) to evaporate solvents and form a cohesive film.

14.7.9 Electrical and Electronic Components

Electrical and electronics applications take advantage of the excellent dielectric properties of polysulfones at elevated temperatures as well as their inherent flame retardancy, which continues to gain in importance in light of the need to move away from flame-retardant additives, many of which pose environmental challenges. A good example of an end use in this domain that has been in industrial use since the early 1990s is wire insulation coating for high-voltage transformer magnet wire. The insulation coating is applied via direct melt extrusion on top of the magnet wire, thus eliminating the need for solvent-based coatings of thermoset varnishes and enamels, which, besides being less sound environmentally, usually do not have the thermal aging resistance or the electrical properties offered by PPSU. Examples of other electrical/electronic uses include printed circuit boards, circuit breaker components, lighting sockets, automotive fuses, and electronic and fiber-optic connectors.

14.8 ECONOMIC ASPECTS

There are currently three major commercial suppliers for polysulfones on a worldwide basis. These are Solvay Specialty Polymers, BASF Corporation, and Sumitomo Chemical Co. Each of these firms manufactures and supplies a PES-type polymer. Solvay and BASF also supply PSU and PPSU. Table 14.8 gives the trade names associated with each of the polysulfones that are manufactured by the three major suppliers.

Sulfone polymers are commercially available in various forms such as powders, pellets, sheets, films, stock shapes, fibers, and foams.

TABLE 14.8
Suppliers and Trade Names of Commercially Available Polysulfone (PSU), Polyethersulfone (PES), and Polyphenylsulfone (PPSU) Resins[a]

Supplier	Polyarylethersulfone Type	Trade Name
Solvay Specialty Polymers	PSU	Udel
	PES	Veradel
	PPSU	Radel
BASF Corp.	PSU	Ultrason S
	PES	Ultrason E
	PPSU	Ultrason P
Sumitomo Chemical Co.	PES	Sumikaexcel

[a] As of June 2014.

REFERENCES

1. R. A. Clendinning, A. G. Farnham, and R. N. Johnson, The development of polysulfone and other polyarylethers, in *High Performance Polymers: Their Origin and Development* (R. B. Seymour and G. S. Kirshenbaum eds.), Elsevier, New York, 1986, pp. 149–158.
2. R. N. Johnson, A. G. Farnham, R. A. Clendinning, W. F. Hale, and C. N. Merriam, Poly(arylethers) by nucleophilic aromatic substitution. I. Synthesis and properties, *J. Polym. Sci., Part A-1 5*: 2375 (1967).
3. R. N. Johnson and A. G. Farnham, Polyarylene polyethers, U.S. Pat. 4,108,837 (1978) to Union Carbide Corporation.
4. R. J. Cotter, *Engineering Plastics, a Handbook of Polyarylethers*, Gordon and Breach Publishers, Basel, Switzerland, 1995, p. 4.
5. J. B. Rose, Discovery and development of the Victrex™ polyarylethersulfones, in *High Performance Polymers: Their Origin and Development* (R. B. Seymour and G. S. Kirshenbaum eds.), Elsevier, New York, 1986, pp. 169–185.
6. D. A. Barr and J. B. Rose, Aromatic polymers from dihalo-benzenoid compounds and alkali metal hydroxide, Br. Pat. 1,153,035 (1965) to Imperial Chemical Industries, P.L.C.
7. A. G. Farnham and R. N. Johnson, Polyarylene polyethers, U.S. Pat. 3,332,909 (1967) to Union Carbide Corporation.
8. S. Savariar, Process for preparation of poly(aryl ether) polymers by macro bicyclic catalysts, U.S. Pat. 5,239,043 (1993) to Amoco Corporation.
9. S. Savariar, Process for preparation of poly(aryl ether) polymers by macro monocyclic catalysts, U.S. Pat. 5,235,019 (1993) to Amoco Corporation.
10. T. E. Atwood, A. B. Newton, and J. B. Rose, Kinetic investigation of the synthesis of a polyethersulfone, *Brit. Polym. J. 4*: 391 (1972).
11. A. Bunn, Monomer sequence determinations of aryl ether sulfone copolymers as determined by ^{13}C n.m.r., *Brit. Polym. J. 20*: 307 (1988).
12. I. Fukawa, T. Tanabe, and H. Hachiya, Trans-etherification of aromatic polyetherketone and aromatic polyethersulfone, *Polym. J. 24*: 173 (1992).
13. B. E. Jennings, M. E. B. Jones, and J. B. Rose, Synthesis of poly(arylene sulfones) and poly(arylene ketones) by reactions involving substitution at aromatic nuclei, *J. Polym. Sci. Part C Polym. Lett. 1*: 715 (1967).
14. M. E. B. Jones, Polysulphones and method of preparation, U.S. Pat. 4,008,203 (1977) to Imperial Chemical Industries, P.L.C.
15. J. B. Rose, Preparation and properties of poly(arylene ether sulphones), *Polymer 15*: 456 (1974).
16. I. Colon and D. R. Kelsey, Coupling of aryl chlorides by nickel and reducing metals, *J. Org. Chem. 51*: 2627 (1986).
17. M. Q. Zhang, J. R. Xu, H. M. Zeng, K. Friedrich, Q. Huo, Z. Y. Zhang, and F. C. Yun, Fractal approach to the critical filler volume faction of an electrically conductive polymer composite, *J. Mater. Sci. 30*: 4226 (1995).

18. I. Colon, Integrated process for the preparation of substantially linear high molecular weight thermoplastic polymers from aryl polyhalide monomers, U.S. Pat. 4,400,499 (1983) to Union Carbide Corporation.
19. G. T. Kwiatkowski, I. Colon, M. J. El-Hibri, and M. Matzner, Aromatic biphenylene polymers: Synthesis via nickel coupling of aryl dichlorides, *Makromol. Chem., Macromol. Symp. 54/55*: 199 (1992).
20. V. Perec and H. Nava, Synthesis of aromatic polyethers Scholl Reaction. I. Poly(1,1'-dinaphthyl ether phenyl sulfone)s and poly(1,1'-dinaphthyl ether phenyl ketone)s, *J. Polym. Sci. Part A Polym. Chem. 26*: 783 (1988).
21. H. Witt, H. Holtschmidt, and E. Muller, Thermische Cyclisierung von o-chlorarylisothiocyanaten, *Angew. Chem. 82*: 79 (1970).
22. P. J. Mayska and E. Tresper, Process for the preparation of polyether-sulphones, U.S. Pat. 4,775,738 (1988) to Bayer A.G.
23. M. J. Mullins, S. P. Crain, E. P. Woo, D. J. Murray, and S. E. Bales, Poly(aryl ether)–poly(aryl carbonate) block copolymers and their preparation, U.S. Pat. 4,994,533 (1991) to Dow Chemical Co.
24. H. R. Kricheldorf and G. Nier, New polymer synthesis, IX, Synthesis of poly(ether sulfone) from sylilated diphenols or hydroxybenzoic acids, *J. Polym. Sci. Polym. Chem. Ed., 21*: 2283 (1983).
25. "Solution processing guide for polymer membranes," Solvay Specialty Polymers, 2011.
26. A. Siegmann, S. Kenig, and A. Buchman, Residual stresses in injection-molded amorphous polymers, *Polym. Eng. Sci. 27*: 1069 (1987).
27. K. Qi and R. Huang, Effect of processing on the structure and properties of PES, *Polym. Plast. Technol. Eng. 33*: 121 (1994).
28. A. Arzak, J. I. Eguiazábal, and J. Nazábal, El Reprocesado por Inyección en la Poli(étersulfona), *Fetraplast 5*: 16 (1992).
29. P. Sanchez, P. M. Remiro, and J. Nazabal, Influence of reprocessing on the mechanical properties of a commercial polysulfone/polycarbonate blend, *Polym. Eng. Sci. 32*: 861 (1992).
30. M. J. El-Hibri, Polysulfone compositions exhibiting very low color and high light transmittance properties and articles made therefrom, U.S. Pat. 7,423,110 (2008) to Solvay Advanced Polymers, LLC.
31. J. J. Dumais, A. L. Cholli, L. W. Jelinski, J. L. Hedrick, and J. E. McGrath, Molecular basis of the β-transition in poly(arylene ether sulfones), *Macromolecules 19*: 1884 (1986).
32. L. M. Robeson, A. G. Farnham, and J. E. McGrath, Synthesis and dynamic mechanical characteristics of poly(aryl ethers), *Appl. Polym. Symp. 26*: 373 (1975).
33. C. L. Aitken, J. S. McHattie, and D. R. Paul, Dynamic mechanical behavior of polysulfones, *Macromolecules 25*: 2910 (1992).
34. R. Y. Ting, Fracture energy and surface topology in the cracking of high performance sulphone polymers, *J. Mat. Sci. 16*: 3059 (1981).
35. J. R. Fried, A. Letton, and W. J. Welsh, Secondary relaxation processes in bisphenol-A polysulphone, *Polymer 31*: 1032 (1990).
36. E. Macho, J. M. Alberdi, A. Alegría, and J. Colmenero, Dynamic mechanical behaviour of a polysulfone in the glass transition region, *Makromol. Chem., Macromol. Symp. 20/21*: 451 (1988).
37. A. Alegría, E. Macho, and J. Colmenero, Dynamic mechanical study of four amorphous polymers around and above the glass transition. Breakdown of the time–temperature superposition principle in the frame of the coupling model, *Macromolecules 24*: 5196 (1991).
38. L. C. E. Struik, Effect of thermal history on secondary relaxation processes in amorphous polymers, *Polymer 28*: 57 (1987).
39. T. Hatakeyama, H. Yoshida, S. Hirose, and H. Hatakeyama, Enthalpy relaxations of polyethers having phenylene groups in the main chain, *Thermochim. Acta 163*: 175 (1990).
40. G. W. Adams and R. J. Farris, Latent energy of deformation of amorphous polymers: 2. Thermomechanical and dynamic mechanical properties, *Polymer 30*: 1829 (1989).
41. T. E. Attwood, M. B. Cinderey, and J. B. Rose, Effects of repeat unit structure on the toughness of poly(aryl ether sulphone)s, *Polymer 34*: 1322 (1993).
42. C. J. G. Plummer and A. M. Donald, The deformation behavior of polyethersulfone and polycarbonate, *J. Polym. Sci. Part B Polym. Phys. 27*: 325 (1989).
43. C. J. G. Plummer and A. M. Donald, The effect of stress ageing on deformation in thin films of polystyrene and poly(ether sulphone), *Polymer 32*: 3322 (1991).
44. G. W. Adams and R. J. Farris, Latent energy of deformation of amorphous polymers: 1. Deformation calorimetry, *Polymer 30*: 1824 (1989).
45. C. J. G. Plummer and A. M. Donald, The ductile–brittle transition in macroscopic tensile tests on polyethersulfone, *J. Appl. Polym. Sci. 41*: 1197 (1990).

46. H.-T. Chiu and D.-S. Hwung, The relationship between structure and deformation behavior of sulfone polymers, *Polym. Eng. Sci. 35*: 499 (1995).
47. K. V. Gotham and S. Turner, Poly(ether sulphone) as an engineering material, *Polymer 15*: 665 (1974).
48. J. P. Trotignon, J. Verdu, C. Martin, and E. Morel, Fatigue behavior of some temperature-resistant polymers. *J. Mater. Sci. 28*: 2207 (1993).
49. M. Hikita, T. Hirose, Y. Ito, T. Mizutani, and M. Ieda, Investigation of electrical breakdown of polymeric insulating materials using a technique of pre-breakdown current measurements, *J. Phys. D: Appl. Phys. 23*: 1515 (1990).
50. E.-S. A. Hegazy, T. Sasuga, M. Nishii, and T. Seguchi, Irradiation effects on aromatic polymers: 1. Gas evolution by gamma irradiation, *Polymer 33*: 2897 (1992).
51. G. Marletta, S. Pignataro, A. Tóth, I. Bertóti, T. Székely, and B. Keszler, X-ray, electron, and ion beam induced modifications of poly(ether sulfone), *Macromolecules 24*: 99 (1991).
52. D. J. T. Hill, D. A. Lewis, and J. H. O'Donnell, Accelerated failure of bisphenol-A polysulfone during electron beam irradiation under an applied stress, *J. Appl. Polym. Sci. 44*: 115 (1992).
53. T. Sasuga, N. Hayakawa, and K. Yoshida, Electron beam irradiation effects on mechanical relaxation of aromatic polysulphones, *Polymer 28*: 236 (1987).
54. E.-S. A. Hegazy, T. Sasuga, M. Nishii, and T. Seguchi, Irradiation effects on aromatic polymers: 2. Gas evolution during electron-beam irradiation, *Polymer 33*: 2904 (1992).
55. A. Davis, M. H. Gleaves, J. H. Golden, and M. B. Huglin, The electron irradiation stability of polysulphone, *Makromol. Chem. 129*: 63 (1969).
56. J. R. Brown and J. H. O'Donnell, The radiation stability of an aromatic polysulfone, *J. Polym. Sci. Part C Polym. Lett. 8*: 121 (1970).
57. S.-I. Kuroda, A. Nagura, K. Horie, and I. Mita, Degradation of aromatic polymers. III. Crosslinking and chain scission during photodegradation of polysulphones, *Eur. Polym. J. 25*: 621 (1989).
58. S.-I. Kuroda, I. Mita, K. Obata, and S. Tanaka, Degradation of aromatic polymers: Part IV—Effect of temperature and light intensity on the photodegradation of polyethersulfone, *Polym. Deg. Stabil. 27*: 257 (1990).
59. T. Yamashita, H. Tomitaka, T. Kudo, K. Horie, and I. Mita, Degradation of sulfur-containing aromatic polymers: Photodegradation of polyethersulfone and polysulfone, *Polym. Deg. Stabil. 39*: 47 (1993).
60. A. Rivaton and J. L. Gardette, Photodegradation of polyethersulfone and polysulfones, *Polym. Deg. Stab. 66*: 385 (1999).
61. S. Makhija, K. Levon, and E. M. Pearce, Gelation/crystallization of polyethersulfone, *Polym. Prepr. 32*: 473 (1991).
62. A. Arzak, J. I. Eguiazábal, and J. Nazábal, Effect of a fluid environment on the properties of poly(ether sulfone), *J. Macromol. Sci.-Phys. B33*: 259 (1994).
63. M. Edge, N. S. Allen, J. H. He, M. Derham, and Y. Shinagawa, Physical aspects of the thermal and hydrolytic ageing of polyester, polysulphone and polycarbonate films, *Polym. Deg. Stabil. 44*: 193 (1994).
64. J. R. Fried, H.-C. Liu, and C. Zhang, Effect of sorbed carbon dioxide on the dynamic mechanical properties of glassy polymers, *J. Polym. Sci. Part C Polym. Lett. 27*: 385 (1989).
65. J. R. Fried, C. Zhang, and H.-C. Liu, Effect on sorbed carbon dioxide on the impact strength of amorphous glassy polymers, *J. Polym. Sci. Part C Polym. Lett. 28*: 7 (1990).
66. Y. Maeda and D. R. Paul, Effect of antiplasticization on gas sorption and transport. I. Polysulfone, *J. Polym. Sci. Part B Polym. Phys. 25*: 957 (1987).
67. Y. Maeda and D. R. Paul, Effect of antiplasticization on gas sorption and transport. III. Free volume interpretation, *J. Polym. Sci. Part B Polym. Phys. 25*: 1005 (1987).
68. L. M. Robeson and S. T. Crisafulli, Microcavity formation in engineering polymers exposed to hot water, *J. Appl. Poly. Sci. 28*: 2925 (1983).
69. G. Marque, J. Verdu, V. Prunier and D. Brown, A molecular dynamics simulation study of three polysulfones in dry and hydrated states, *J. Polym. Sci: Part B: Polym. Phys. 48*: 2312 (2010).
70. "Sterilization compatibility overview high-performance medical-grade plastics," Solvay Specialty Polymers Product Literature, 2013.
71. B. P. Barth, Mixtures of polycarbonates and polyarylene polyethers, U.S. Pat. 3,365,517 (1968) to Union Carbide Corporation.
72. E. Nield, Thermoplastic blends of aromatic polysulfones and thermoplastic polyesters, U.S. Pat. 3,742,087 (1973) to ICI, Ltd.
73. J. E. Harris and G. T. Brooks, Miscible blends of imide containing polymers with poly(aryl sulfones), U.S. Pat. 5,037,902 (1991) to Amoco Corporation.

74. L. M. Robeson, M. Matzner, and L. M. Maresca, Blends of poly(aryl ether) resins and polyetherimide resins, U.S. Pat. 4,293,670 (1982) to Union Carbide Corporation.
75. G. Qipeng, Q. Lingwei, D. Mengxian, and F. Zhiliu, Tensile properties of blends of poly(ether sulphone) with a poly(ether imide), *Eur. Polym. J. 28*: 1045 (1992).
76. X. Yu, Y. Zheng, Z. Wu, X. Tang, and B. Jiang, Study on the compatibility of the blend of poly(aryl ether ether ketone) with poly(aryl ether sulfone), *J. Appl. Polym. Sci. 41*: 2649 (1990).
77. T. M. Malik, Thermal and mechanical characterization of partially miscible blends of poly(ether ether ketone) and polyethersulfone, *J. Appl. Polym. Sci. 46*: 303 (1992).
78. A. Arzak, J. I. Eguiazábal, and J. Nazabal, Phase behaviour and mechanical properties of poly(ether ether ketone)–poly(ether sulphone) blends, *J. Mater. Sci. 26*: 5939 (1991).
79. A. Arzak, J. I. Eguiazábal, and J. Nazabal, Compatibility in immiscible poly(ether ether ketone)/poly(ether sulphone) blends, *J. Appl. Polym. Sci. 58*: 653 (1995).
80. J. E. Harris and L. M. Robeson, Blends of a biphenyl containing poly(aryl ether sulfone) and a poly(aryl ether ketone), U.S. Pat. 4,804,724 (1989) to Amoco Corporation.
81. J. E. Harris and L. M. Robeson, Blends of a biphenyl containing poly(aryl ether sulfone) and a poly(aryl ether ketone), U.S. Pat. 4,713,426 (1987) to Amoco Corporation.
82. T. Tsutsumi, Y. Goto, M. Amano, and T. Takahashi, Polyetherketone compositions, Jpn. Kokai Tokyo Koho JP 02 01,759 (1990) to Mitsui Toatsu Chemicals Inc.
83. P. Ittemann and G. Heinz, Heat resistant molding compositions containing aromatic polyetherpolysulfones and polyether–polyketones with good interphase adhesion, Ger. Offen. DE 3,807,296 (1989) to BASF A.G.
84. R. D. Birch and K. J. Artus, Manufacture of tubular membranes from aromatic polyether–ketone–aromatic polyether sulfone blends, Eur. Pat. Appl. EP 417,908 (1991) to Imperial Chemical Industries, PLC.
85. X. Zhang and Y. Wang, Investigation on domain structure of poly(phenylene sulphide) and poly(ether sulphone) blends by solid-state nuclear magnetic resonance methods, *Polymer 30*: 1867 (1989).
86. C. Yang, M. Zhang, H. Zeng, and J. Zhang, Study on internal friction characteristics of poly(phenylene sulfide)/poly(ethersulfone) blends and composites reinforced by carbon fiber, *Polym. J. 24*: 339 (1992).
87. M.-F. Cheung, A. Golovoy, H. K. Plummer, and H. van Oene, Polysulphone and poly(phenylene sulphide) blends: 1. Thermal characterization and phase morphology, *Polymer 31*: 2299 (1990).
88. M.-F. Cheung, A. Golovoy, and H. van Oene, Polysulphone and poly(phenylene sulphide) blends: 2. Mechanical behaviour, *Polymer 31*: 2307 (1990).
89. H. Zeng, G. He, and G. Yang, The dynamic mechanical behavior of blends of poly(phenylene sulfide) with acetylene-terminated sulfone. *Angew. Makromol. Chem. 143*: 25 (1986).
90. H. Zeng and K. Mai, Dynamic mechanical behaviour and phase separation of mixtures of polysulfone with bis[4-(4-ethynylphenoxy)phenyl]-sulfone or poly(phenylene sulfide), *Makromol. Chem. 187*: 1787 (1986).
91. S. Akhtar and J. L. White, Characteristics of binary and ternary blends of poly(p-phenylene sulfide) with poly(bis phenol A) sulfone and polyetherimide, *Polym. Eng. Sci. 31*: 84 (1991).
92. M.-F. Cheung and H. K. Plummer, Tensile fracture morphology of polysulfone–poly(phenylene sulfide) blends, *Polym. Bull. 26*: 349 (1991).
93. K. Mai, M. Zhang, H. Zeng, and S. Qi, Double melting phenomena of polyphenylene sulfide and its blends, *J. Appl. Polym. Sci. 51*: 57 (1994).
94. H. Hayashi, S. Nakata, M. Kakimoto, and Y. Imai, Polymerization blending for compatible poly(ether sulfone)/aramid blend based on polycondensation of an N-silylated aromatic diamine with an aromatic diacid chloride in poly(ether sulfone) solution, *J. Appl. Polym. Sci. 49*: 1241 (1993).
95. S. Nakata, M. Kakimoto, and Y. Imai, Miscibility, crystallization behaviour and mechanical properties of binary blends of aramids and poly(ether sulphone)s, *Polymer 33*: 3873 (1992).
96. H. Saito, D. Tsutsumi, and T. Inoue, Temperature dependence of the Flory interaction parameter in a single-phase mixture of poly(hydroxy ether of bisphenol-A) and poly(ether sulfone), *Polym. J. 22*: 128 (1990).
97. V. B. Singh and D. J. Walsh, The miscibility of polyethersulfone with phenoxy resin, *J. Macromol. Sci.-Phys. B25*: 65 (1986).
98. K. Jeremic, F. E. Karasz, and W. J. MacKnight, Influence of solvent and temperature on the phase behavior of poly(aryl sulfone)/polyimide blends, *N. Polym. Mat. 3*: 163 (1992).
99. M. J. El-Hibri, J. E. Harris, and J. L. Melquist, Blends of polyether sulfones and polyimides, U.S. Pat. 5,191,035 (1993) to Amoco Corporation.
100. Y.-J. Cha, E.-T. Kim, T.-K. Ahn, and S. Choe, Mechanical and morphological phase behavior in miscible polyethersulfone, and polyimide blends, *Polym. J. 26*: 1227 (1994).

101. K. Liang, J. Grebowicz, E. Valles, F. E. Karasz, and W. J. MacKnight, Thermal and rheological properties of miscible polyethersulfone/polyimide blends, *J. Polym. Sci. Part B Polym. Phys. 30*: 465 (1992).
102. J. Grobelny, D. M. Rice, F. E. Karasz, and W. J. MacKnight, High resolution solid state ^{13}C nuclear magnetic resonance study of poly(ether sulfone)/polyimide blends, *Polym. Commun. 31*: 86 (1990).
103. D. J. Walsh and V. B. Singh, The phase behaviour of a poly(ether sulfone) with poly(ethylene oxide), *Makromol. Chem. 185*: 1979 (1984).
104. D. J. Walsh, S. Rostami, and V. B. Singh, The thermodynamics of polyether sulfonepoly(ethylene oxide) mixtures, *Makromol. Chem. 186*: 145 (1985).
105. W. Guo and J. S. Higgins, Miscibility and kinetics of phase separation in blends of poly(ethylene oxide) and poly(ether sulphone), *Polymer 32*: 2115 (1991).
106. H. Nakamura, J. Maruta, T. Ohnaga, and T. Inoue, Phase separation in a mixture of poly(ether sulphone) and poly(ethyl oxazoline): Observation in real and reciprocal spaces, *Polymer 31*: 303 (1990).
107. G. Qipeng, Miscibility of poly(hydroxyether of phenolphthalein) with poly(ether sulphone), *Eur. Polym. J. 28*: 1395 (1992).
108. B. T. Swinyard, P. S. Pagoo, J. A. Barrie, and R. Ash. The transport and sorption of water in polyethersulphone, polysulphone, and polyethersulphone/phenoxy blends, *J. Appl. Polym. Sci. 41*: 2479 (1990).
109. M. J. Reimers and T. A. Barbari, Gas sorption and diffusion in hydrogen-bonded polymers. II. Polyethersulfone/polyhydroxyether blends. *J. Polym. Sci. Part B Polym. Phys. 32*: 131 (1994).
110. J. Ramiro, J. I. Eguiazábal, and J. Nazábal, New miscible poly(ether imide)/poly(phenyl sulfone) blends, *Macromol. Mater. Eng. 291*: 707 (2006).
111. P. J. Jones, L. C. Paslay, and S. E. Morgan, Effects of chain conformation on miscibility, morphology, and mechanical properties of solution blended substituted polyphenylene and polyphenylsulfone, *Polymer 51*: 738 (2010).
112. J. E. Harris and L. M. Robeson, Composition for making circuit board substrates and electrical connectors, U.S. Pat. 4,743,645 (1988) to Amoco Corporation.
113. S. M. Andrews, Synthesis, characterization, and blends of high temperature poly(arylether sulfone)s, *J. Polym. Sci. Part A Polym. Chem. 30*: 221 (1992).
114. J. E. Harris and L. M. Robeson, Miscible blends of poly(aryl ether sulfones), U.S. Pat. 4,804,723 (1989) to Amoco Corporation.
115. B. L. Dickinson, M. J. El-Hibri, and M. E. Sauers, Poly(aryl ether sulfone) compositions comprising poly(phenylene ether sulfone), U.S. Pat. 5,086,130 (1992) to Amoco Corporation.
116. M. J. El-Hibri, B. L. Dickinson, and M. E. Sauers, Poly(aryl ether sulfone) compositions, U.S. Pat. 5,164,466 (1992) to Amoco Corporation.
117. M. J. El-Hibri and B. L. Dickinson, Properties of polyphenylsulfone blends with polysulfone and polyethersulfone, *Soc. Plast. Eng. ANTEC '93 Conference Proceedings*, New Orleans, LA, 1993, pp. 202–204.
118. L. C. Sawyer and M. Jaffe, Structure-property relationships in liquid crystalline polymers, in *High Performance Polymers* (E. Baer and A. Moet, eds.), Hanser, New York, 1991, p. 56.
119. V. G. Kulichikhin, O. V. Vasil'Eva, I. A. Litvinov, E. M. Antipov, I. L. Parsamyan, and N. A. Platé, Rheology and morphology of polymer blends containing liquid-crystalline component in melt and solid state, *J. Appl. Polym. Sci. 42*: 363 (1991).
120. S. M. Hong, B. C. Kim, K. U. Kim, and I. J. Chung, Rheology and physical properties of polysulfone in-situ reinforced with a thermotropic liquid-crystalline polyester, *Polym. J. 23*: 1347 (1991).
121. A. Golovoy, M. Kozlowski, and M. Narkis, Characterization of thermotropic liquid crystalline polyester/polysulfone blends, *Polym. Eng. Sci. 32*: 854 (1992).
122. F. Yazaki, A. Kohara, and R. Yosomiya, Polymer blends of polyethersulfone with all aromatic liquid crystalline co-polyester, *Polym. Eng. Sci. 34*: 1129 (1994).
123. J. Q. Zheng and T. Kyu, Phase transformations in a thermotropic liquid crystalline copolyester and its blends with polyether sulfone, *Polym. Eng. Sci. 32*: 1004 (1992).
124. B. S. Hsiao, R. S. Stein, N. Weeks, and R. Gaudiana, Light scattering of thermotropic polyester and polysulfone mixtures in solution, *Macromolecules 24*: 1299 (1991).
125. S. Cohen-Addad, R. S. Stein, and P. Esnault, Liquid crystalline polyester and polysulphone mixtures: Observation of phase separation and aggregation, *Polymer 32*: 2319 (1991).
126. J. Xu, W. Xian, and H. Zeng, Polyethersulphone induced crystallization with a liquid crystal polymer, *Polym. Commun. 32*: 336 (1991).
127. T. A. Callaghan and D. R. Paul, Estimation of interaction energies by the critical molecular weight method: 2. Blends with polysulfones, *J. Polym. Sci. Part B Polym. Phys. 32*: 1847 (1994).

128. R. Lauchlan and G. Shaw, Polyarylene polyether resin–polyurethane blend of high impact, tensile and flexural strengths, Br. Pat. 1,436,014 (1976) to Uniroyal, Inc.
129. E. G. Hendricks, Mixtures of organopolysiloxanes and polyarylene polyethers, U.S. Pat. 3,423,479 (1969) to Union Carbide Corporation.
130. P. Sanchez, P. M. Remiro, and J. Nazabal, Influence of reprocessing on the mechanical properties of a commercial polysulfone/polycarbonate blend, *Polym. Eng. Sci. 32*: 861 (1992).
131. T.-S. Chung, M. Glick, and E. J. Powers, Polybenzimidazole and polysulfone blends, *Polym. Eng. Sci. 33*: 1042 (1993).
132. B. T. Swinyard, J. A. Barrie, and D. J. Walsh, Phase behavior of polysulphone with poly(ethylene oxide) and phenoxy polymers, *Polym. Commun. 28*: 331 (1987).
133. M. M. Qayyum and J. R. White, Weathering of glass-fiber-reinforced polystyrene and poly(ethersulfone), *Polym. Compos. 11*: 24 (1990).
134. R. Weiss and W. Huettner, High performance carbon fiber-reinforced polysulfone, *Mat. Sci. Monogr. 41*: 415 (1987).
135. G. M. Lin and J. K. L. Lai, Fracture mechanism in short fiber-reinforced thermoplastic resin composites, *J. Mat. Sci. 28*: 5240 (1993).
136. A. Bonton, C. Bouchard, B. Barbeau, and S. Jedrzejak, Comparative life cycle assessment of water treatment plants, *Desalination 284*: 42 (2012).
137. W. Y. W. Ning, How is the environmental footprint of RO membranes being reduced through energy and biofouling improvements, WaterWorld.mht, last updated in June 2014.
138. I. Dincer, Hydrogen and fuel cell technologies for sustainable future, *Jordan J. Mech. Ind. Eng. 2*: 1 (2008).
139. Moving (back) to reusables in the OR: Greening the OR implementation module, http://www.GreeningtheOR.org, last updated in June 2014.
140. Cross-linked polyethylene (PEX) pipe systems vs copper environmental impact comparison, *The European Plastic Pipes and Fittings Association*, http://www.teppfa.eu, last updated in June 2014.
141. FuelSmart, http://hub.aa.com/en/nr/media-kit/operations/fuelsmart, last updated in June 2014.
142. W. W. Y. Lau, M. D. Guiver, and T. Matsuura, Phase separation in polysulfone/solvent/water and polyethersulfone/solvent/water systems, *J. Membr. Sci. 1991*: 219 (1991).
143. J. Wijmans, J. Kant, M. H. V. Mulder, and C. A. Smolders, Phase separation phenomena in solutions of polysulfone in mixtures of a solvent and a nonsolvent: Relationship with membrane formation, *Polymer 26*: 1539 (1985).
144. I. Cabasso, E. Klein, and J. K. Smith, Polysulfone hollow fibers. I. Spinning and properties, *J. Appl. Polym. Sci. 20*: 2377 (1976).
145. H. Matsuyama, T. Maki, M. Teramoto, and K. Kobayashi, Effect of PVP additive on porous polysulfone membrane formation by immersion precipitation method, *Sep. Sci. Technol. 38*: 3449 (2003).
146. A. Idris, N. Mat Zain, and M. Y. Noordin, Synthesis, characterization, and performance of asymmetric polyethersulfone (PES) ultrafiltration membranes with polyethylene glycol of different molecular weights as additives, *Desalination 207*: 324 (2007).
147. S. K. Bowry, E. Gatti, and J. Vienken, Contribution of polysulfone membranes to the success of convective dialysis therapies, *Contrib. Nephrol. 173*: 110 (2011).
148. J. Vienken, Von der Schiessbaumwolle zum Polysulfon: Eine kleine Geschiste der Dialysemembran (From gun cotton to polysulfones: A small history of dialysis membranes), *Nieren- und Hochdruckkrankheiten 32*: 263 (2003).
149. T. A. Barbari, W. J. Koros, and D. R. Paul, Gas transport in polymers based on bisphenol A, *J. Polym. Sci. Part B: Polym. Phys. 26*: 709 (1988).
150. A. K. Ghosh and E. M. V. Hoek, Impacts of support membrane structure and chemistry on polyamide–polysulfone interfacial composite membranes, *J. Membr. Sci. 336*: 140 (2009).
151. R. J. Peterson, Composite reverse osmosis and nanofiltration membranes, *J. Membr. Sci. 83*: 81 (1993).
152. B.-H. Jeong et al., Interfacial polymerization of thin film nanocomposites: A new concept for reverse osmosis membranes, *J. Membr. Sci. 294*: 1 (2007).
153. C. Kong, T. Shintani, and T. Tsuru, "Pre-seeding"-assisted synthesis of a high performance polyamide–zeolite nanocomposite membrane for water purification, *New J. Chem. 2010*: 2101 (2010).
154. N. Y. Yip, A. Tiraferri, W. A. Phillip, J. D. Schiffman, and M. Elimelech, High performance thin film composite forward osmosis membrane, *Environ. Sci. Technol. 44*: 3812 (2010).
155. K. Gerstandt, K. V. Peinemann, S. E. Skilhagen, T. Thorsen, and T. Holt, Membrane processes in energy supply for an osmotic power plant, *Desalination 224*: 64 (2008).
156. F. Liu, B.-R. Ma, D. Zhou, Y. Xiang, and L. Xue, Breaking through tradeoff of polysulfone ultrafiltration membranes by zeolite 4A, *Microporous Mesoporous Mater. 186*: 113 (2014).

157. K. Zudrow, L. Brunet, S. Mahendra, D. Li, A. Zhang, Q. Li, and P. J. Alvarez, Polysulfone ultrafiltration membranes impregnated with silver nanoparticles show improved biofouling resistance and virus removal, *Water Res. 43*: 715 (2009).
158. B. Pellegrin, R. Prulho, A. Rivaton, S. Thérias, J.-L. Gardette, E. Gaudichet-Maurin, and C. Causserand, Multi-scale analysis of hypochlorite induced PES/PVP ultrafiltration membranes degradation, *J. Membr. Sci. 447*: 287 (2013).
159. A. K. Cheung and J. K. Leypoldt, The hemodialysis membranes: A historical perspective, current state, and future prospect, *Semin. Nephrol. 17*: 196 (1997).
160. L. Breitbach, E. Hinke, and E. Staude, Heterogeneous functionalizing of polysulfone membranes, *Angew. Makromol. Chem. 184*: 183 (1991).
161. C. Dizman, M. A. Tasdelen, and Y. Yagci, Recent advances in the preparation of functionalized polysulfones, *Polym. Int. 62*: 991 (2013).
162. C. Zhao, J. Xue, F. Ran, and S. Sun, Modification of polyethersulfone membranes—A review, *Prog. Mater. Sci. 58*: 76 (2013).
163. A. Noshay and L. M. Robeson, Sulfonated polysulfone, *J. Appl. Polym. Sci. 20*: 1885 (1976).
164. C. H. Lee et al., Disulfonated poly(aryleneethersulfone) random copolymer thin film composite membrane fabricated using a benign solvent for reverse osmosis applications, *J. Membr. Sci. 389*: 363 (2012).
165. M. Yoshikawa, K. Tsubouchi, M. D. Guiver, and G. P. Robertson, Modified polysulfone membranes III: Pervaporation separation fo benzene–cyclohexane mixtures thoruhg carboxylated polysulfone membranes, *J. Appl. Polym. Sci. 74*: 407 (1999).
166. Y.-Y. Xu, Z. Yi, L.-P. Zhu, J.-Y. Wang, J.-L. Shi, and B.-K. Zhu, Synthesis of amphiphilic PSF-g-PEGMA copolymer by ATRP and its hydrophilic modification of PES membranes, *Polym. Mater. Sci. Eng. 100*: 93 (2009).
167. G. J. Dahe, R. S. Teotia, S. S. Kadam, and J. R. Bellare, The biocompatibility and separation performance of antioxidative polysulfone/vitamin E TPGS composite hollow fiber membranes, *Biomaterials 32*: 352 (2011).
168. X. Huang, D. Guduru, Z. Xu, J. Vienken, and T. Groth, Immobilization of heparin on polysulfone surface for selective adsorption of low-density lipoprotein (LDL), *Acta Biomater. 6*: 1099 (2010).
169. A. Schulze, B. Marquardt, S. Kaczmarek, R. Schubert, A. Prager, and M. R. Buchmeiser, Electron beam based functionalization of polyethersulfone membranes, *Macromol. Rapid Commun. 31*: 467 (2010).
170. A. Roy, M. A. Hickner, X. Yu, Y. Li, T. E. Glass, and J. E. McGrath, Influence of chemical composition and sequence length on the transport properties of proton exchange membranes, *J. Polym. Sci. Part B: Polym. Phys. 44*: 2226 (2006).
171. J. G. Copeland et al., Cardiac replacement with a total artificial heart as a bridge to transplantation, *New Engl. J. Med. 351*: 859 (2004).
172. C. M. Asgian, L. N. Gilbertson, E. E. Blessing and R. D. Crowninshield, Environmentally induced fracture of polysulfones in lipids, *Transactions: Fifteenth Annual Meeting of the Society of Biomaterials*, Birmingham, AL, 1989, p. 17.
173. R. Knoppers, Developing a full-polymer aircraft catering trolley from PPSU, *High Performance Plastics Conference*, Cologne, Germany, February 2011.
174. J. H. Hodgkin, G. P. Simon, and R. J. Varley, Thermoplastic toughening of epoxy resins: A critical review, *Polym. Adv. Technol. 9*: 3 (1998).
175. A. J. MacKinnon, S. D. Jenkins, P. T. McGrail, and R. A. Pethrick, Cure and physical properties of thermoplastic modified epoxy resins based on polyethersulfone, *J. Appl. Polym. Sci. 58*: 2345 (1995).
176. C. B. Bucknall and I. K. Partridge, Addition of polyethersulphone to epoxy resins, *Br. Polym. J. 15*: 71 (1983).
177. J. C. Hedrick, N. M. Patel, and J. E. McGrath, Toughening of epoxy resin networks with functionalized engineering thermoplastics, in *Toughened Plastics 1: Science and Engineering* (C. K. Riew and J. K. Gillham, eds.), Advances in Chemistry Series 233, ACS, Washington, DC, 1993, p. 293.
178. J. Zhang, Q. Guo, and B. Fox, Structural and material properties of a rapidly cured thermoplastic-toughened epoxy system, *J. Appl. Polym. Sci.* 113: 485 (2009).
179. S. L. Agius, K. J. C. Magniez, and B. L. Fox, Fracture behaviour of a rapidly cured polyethersulfone toughened carbon fibre/epoxy composite, *Compos. Struct. 92*: 2119 (2010).
180. G. Li, P. Li, Y. Yu, X. Jia, S. Zhang, X. Yang, and S. Ryu, Novel carbon fiber/epoxy composite toughened by eletrospun polysulfone nanofibers, *Mater. Lett. 62*: 511 (2008).
181. T. Inoue, Reaction-induced phase decomposition in polymer blends, *Prog. Polym. Sci. 20*: 119 (1995).
182. Y. W. S. Zhang, F. Chen, and C. C. Han, Dynamically asymmetric phase separation and morphological structure formation in the epoxy/polysulfone blends, *Macromolecules 44*: 7465 (2011).

183. "Radel R-1050 PPSU Foam," *Solvay Specialty Polymers Technical Bulletin—Ver 2.3*, 2014.
184. M. Drewniak, The effect of environmental conditions on the performance of PES based foams, *SPE Foams Conference Proceedings*, 2013.
185. H. Sun and J. E. Mark, Preparation, characterization, and mechanical properties of some microcellular polysulfone foams, *J. Appl. Polym. Sci.* 86: 1692 (2002).
186. L. G. Gibson and M. F. Ashby, *Cellular Solids: Structure and Properties*, Pergamon, New York, 1988.
187. S. T. Lee and D. Scholz, eds., *Polymeric Foams: Technology and Developments in Regulation, Process, and Products*, CRC Press, Boca Raton, FL, 2009.
188. S. T. Lee and N. S. Ramesh, eds., *Polymeric Foams: Mechanisms and Materials*, CRC Press, Boca Raton, FL, 2004.
189. R. Gendron, ed., *Thermoplastic Foam Processing: Principles and Development*, CRC Press, Boca Raton, FL, 2005.
190. J. L. Throne, *Thermoplastic Foam Extrusion: An Introduction*, Carl Hanser Verlag, Munich, 2004.
191. United States Federal Code of Regulations No.: 21CFR177.2440.
192. L. W. McKeen, *Fluorinated Coatings and Finishes Handbook*, William Andrew Publishing, New York, 2006, p. 49.
193. M. Hagiwara, K. Kiwa, T. Ogita, and L. G. P. J. D'Haenens, Process for coating metal surfaces with a fluororesin using a primer, US Pat. 5,626,907 (1997) to E. I. DuPont De Nemours and Company.
194. N. Tomihashi, K. Ogita, and H. Sanemasa, Water-based primer composition for fluororesin coating, US Pat. 6,333,372 (2001) to Daikin Industries, Ltd.

15 Ketone-Based Thermoplastics*

Yundong Wang and Mukerrem Cakmak

CONTENTS

15.1 Introduction 461
15.2 Synthesis 463
15.3 Structure 463
    15.3.1 Crystal Structures 464
        15.3.1.1 Influence of Chemical Architecture 466
15.4 Properties 467
    15.4.1 Thermal and Optical Properties 467
    15.4.2 Crystallization Behavior 469
        15.4.2.1 Effects of Heat Treatment 470
    15.4.3 Mechanical Properties 471
    15.4.4 Chemical Resistance 473
    15.4.5 Radiation Resistance 473
    15.4.6 Fire and Smoke 475
    15.4.7 Electrical Properties 475
15.5 Blends, Compounds, and Composites 475
    15.5.1 Blends 475
    15.5.2 Compounds and Composites 477
15.6 Processing 478
    15.6.1 Effect of Solid-State Deformation ($T_p < T_g$) 478
    15.6.2 Effect of Rubbery Region Processing ($T_g < T_p < T_{cc}$) 479
    15.6.3 Effect of Melt Processing ($T_p > T_m$) 480
        15.6.3.1 Melt Spinning 480
        15.6.3.2 Injection Molding 482
15.7 Applications 483
References 484

15.1 INTRODUCTION

Polyaryl ether ketones, or PAEKs, are highly aromatic, mostly semicrystalline thermoplastics. The letters *E* and *K* refer to the sequence of ether and ketone units in their structures. Examples of this new class of high-performance thermoplastics are PEEEK, PEEK, PEEKEK, PEK, PEEKK, PEKEKK, PEKK, and PEKKK, with their chemical repeat units shown in Figure 15.1. Because of their relatively stiff aromatic polymer backbones, PAEKs exhibit some of the highest thermal transitions found in commercially available polymers today. As a result, these polymers can be continuously used at temperatures as high as 240°C in the case of polyether ether ketone (PEEK) [1]. They possess a good combination of properties, including toughness and high mechanical strength over a broad temperature range, thermal stability, solvent and chemical resistance, very low smoke emission in flammability tests, and good electrical properties. They have been shown to resist basic as well as moderately acidic environments.

---

* Based in parts on the first-edition chapters on ketone-based thermoplastics.

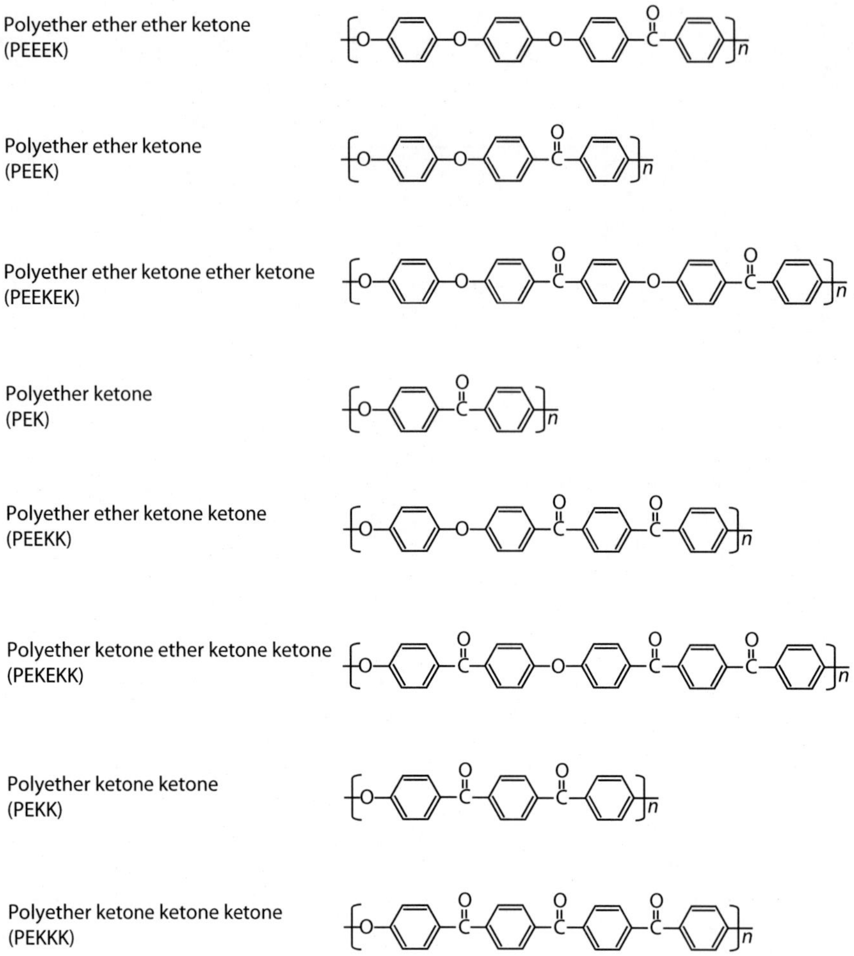

Polyether ether ether ketone (PEEEK)	
Polyether ether ketone (PEEK)	
Polyether ether ketone ether ketone (PEEKEK)	
Polyether ketone (PEK)	
Polyether ether ketone ketone (PEEKK)	
Polyether ketone ether ketone ketone (PEKEKK)	
Polyether ketone ketone (PEKK)	
Polyether ketone ketone ketone (PEKKK)	

FIGURE 15.1 Examples of PAEK unit structures.

The most successful PAEK is PEEK, first produced commercially by Imperial Chemical Industries (ICI) under the trade name VICTREX PEEK. The VICTREX PEEK brand and products are now owned and marketed by Victrex Polymer Solutions with more than 4000 tons of annual capacity. Several other companies have been making PEEK at smaller capacities or are active in developing PEEK technologies in different regions of the world. Another well-recognized PAEK is polyether ketone (PEK), first commercialized by Raychem Corporation under the trade name STILAN (expired, no longer registered) and later by Victrex Polymer Solutions under the trade name VICTREX PEEK HT. Other PAEKs include polyether ketone ketone (PEKK), polyether ether ketone ketone (PEEKK), and polyether ketone ether ketone ketone (PEKEKK), some of which are no longer available in the marketplace.

After more than 50 years of research and development, PAEKs find use in almost all industrial sectors, including aerospace, transportation or automotive, oil and gas, electronic devices, semiconductors and displays, food processing, medical and biomedical, and consumer. They are converted to usable products using such processes as wire coating, powder coating, injection molding, extrusion, compression molding, and machining, and as a matrix in fiber-reinforced or nanoparticle-filled composites.

There have been numerous patents citing a very large number of applications using PEEK and its chemically close relatives. A few examples of these applications are flexible printed circuit boards

Ketone-Based Thermoplastics 463

[2], magnetic tapes with good heat resistance [2], heat-resistant laminates with polyimides [3], automotive engine parts, and composites for a variety of aerospace applications. More detailed examples will be given in Section 15.7.

15.2 SYNTHESIS

The first synthesis of aromatic PEEK was reported by Bonner [4]. Many other researchers and companies have devoted their efforts to synthesizing PAEKs using different routes and to improving the polymerization yield, cost, and molecular weight of the PAEK polymers [5–22]. The two main routes for making PAEKs are nucleophilic routes and electrophilic routes. Other refined routes include soluble precursor routes [13,23–27], carbon–carbon coupling routes [28–30], and ring-opening polymerization routes [31–34]. The semicrystalline PEEK is commercially synthesized by the reaction of 4,4′-difluorobenzophenone or 1,4-bis(4-fluorobenzoyl)benzene with hydroquinone in the presence of alkali carbonates [15]. This is shown in Figure 15.2.

Inasmuch as the fluorinated monomers are expensive, there have been attempts to replace them with chlorinated versions. This resulted in too slow a reaction rate to be commercially viable. Later, Hoffmann et al. [15] reported that the use of a new catalyst, N-alkyl(4-N',N'-dialkylamino)pyridinium, in these reactions not only achieves high reaction rates but also allows for the synthesis of high-molecular-weight PAEKs.

PEK, PEKEKK, and PEKK can be made using electrophilic routes. Both Raychem and BASF have made PEK and PEKEKK using this route in the past. PEKK is made commercially from diphenylether and phthaloyl chloride using an electrophilic route [35]. This route is taken since it is difficult to make PEKK using nucleophilic routes due to the complexity of the monomer involved. The crystallinity of PEKK is determined by the ratio of terphthaloyl to isophthaloyl chloride. The isophthaloyl chloride units disrupt the crystallization process and reduce the melting point of PEKK when its content is increased. PEKK is also made by Polymics with many grades having variable glass transition temperatures and melting points. Electrophilic routes can also be taken to make PEEK as described in patents from Gharda and Victrex [36,37].

Synthesis of branched polyaryl ketones was reported by Brugel [38], who used the trifunctional comonomers such as 1,3,5-benzene tricarboxylic acid chloride to produce branches along the polymer backbone. In a more recent study, linear and hyperbranched polyether ketones containing flexible oxyethylene spacers were synthesized by Jeol et al. [39]. Over the past 50 years, there have been a very large number of papers published and numerous patents filed in this area, not all of which will be discussed here. Additional information on the chemistry of PAEKs can be found in a review article [40], a thesis [41], and books or book chapters [42–45].

15.3 STRUCTURE

Having a relatively rigid chain backbone, PAEK can be transformed to an amorphous state by rapid quenching from the melt under reasonable laboratory conditions, such as ice water quenching.

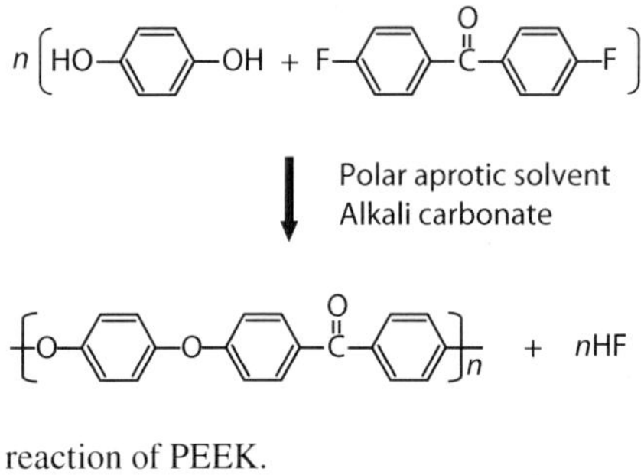

FIGURE 15.2 Polymerization reaction of PEEK.

When heated from the amorphous state, it is known to crystallize at temperatures above the glass transition temperature. For PEEK, this crystallization occurs at around 170–180°C. This type of crystallization, which occurs by heating from the amorphous state, is commonly called *cold crystallization*. Amorphous PEEK is also known to be crystallized by the action of certain solvents such as methylene chloride. PEEK is also readily crystallized upon cooling from melt at moderate cooling rates [46].

15.3.1 CRYSTAL STRUCTURES

PEEK, being similar in chemical structure to poly(phenylene oxide) [47], crystallizes predominantly in the orthorhombic Pbcn space group with $a = 7.75$ Å, $b = 5.86$ Å, and $c = 10$ Å unit cell parameters [48]. In this structure, two chains pass through the unit cell, one at the center and one at the four corners. The consecutive phenyl group planes along the PEEK chain form a 37° angle with each other (Figure 15.3). PEK also crystallizes into the orthorhombic Pbcn space group with $a = 7.65$ Å, $b = 5.97$ Å, and $c = 10.1$ Å unit cell parameters [49].

For PAEKs, replacing ether with ketone linkages was found to stiffen the polymer chain without altering the unit cell parameters significantly. This implies that the ether and ketone linkages are crystallographically equivalent. A number of other researchers also studied the crystal structure of PEEK [48,50–56,58,68] and other PAEKs [49,68]. The lattice parameters, as determined by various investigators, are summarized in Table 15.1.

Thermal history was also found to affect the unit cell parameters of PEEK by Wakelyn [51] and Hay et al. [57]. Hay et al. indicate that a-, b-, and c-axis dimensions decrease with an increase in annealing temperature. This was attributed to the reduction of disorder in the lateral packing of the chains resulting from variations of the torsional angle of the phenyl planes about the c-axis. Presumably, as a result of these changes in the unit cell dimensions, crystalline densities in the range of 1.382–1.415 have been estimated by various investigators. The density of the amorphous PEEK was reported to be in the 1.263–1.264 range [46,58].

Assuming the additivity of bond polarizabilities, the intrinsic refractive indices and birefringence have been determined by Shimizu et al. [52] based on their own unit cell parameters and

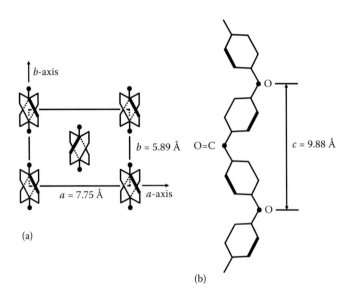

FIGURE 15.3 Arrangement of PEEK molecules in the unit cell: (a) projection along c-axis; (b) projection along a-axis. (Redrawn from Dawson, P.C. and Blundell, D.J. *Polymer*, 21, 1980. With permission.)

TABLE 15.1
Unit Cell Parameters of PAEKs by Various Investigators

PAEKs	Crystalline Form	Reference	a, Å	b, Å	c, Å	p_c, g/cm^3
PEEK	Form 1	Dawson [48]	7.75	5.86	10.0	1.40
PEEK[a]	Form 1	Rueda et al. [50]	7.75	5.89	9.883	1.415
PEEK	Form 1	Wakelyn [51]	7.88	5.94	10.16	1.341
PEEK[a]	Form 1	Shimizu et al. [52]	7.80	5.92	10.0	1.382
PEEK[a]	Form 1	Fratini et al. [53]	7.83	5.94	9.86	1.392
PEEK[a]	Form 1	Blundell [68]	7.76	5.89	9.95	1.404
PEEK	Form 1	Hay et al. [49]	7.78	5.92	10.56	1.378
PEEK	Form 1	Kumar et al. [54]	7.79	5.91	10.00	1.384
PEEK	Form 1	Voice et al. [55]	7.97	5.93	9.80	1.380
PEEK[a]	Form 1	Liu et al. [56]	7.80	5.92	10.05	
PEEK[a]	Form 2	Liu et al. [56]	4.75	10.60	10.86	
PEEKEK[a]	Form 1	Blundell [68]	7.79	5.94	9.96	
PEK[a]	Form 1	Blundell [68]	7.76	6.00	10.01	
PEEKK[a]	Form 1	Blundell [68]	7.80	6.01	10.01	
PEKEKK[a]	Form 1	Blundell [68]	7.74	6.04	10.05	
PEKK(T)	Form 1	Gardner et al. [63]	7.69	6.06	10.16	
PEKK(T)	Form 1	Ho et al. [67]	7.67	6.06	10.08	
PEKK(T)	Form 2	Ho et al. [67]	4.17	11.34	10.08	
PEKK(T)	Form 2	Gardner et al. [63]	3.93	5.75	10.16	
PEKK(T)[a]	Form 2	Blundell [68]	4.17	11.34	10.08	1.395

[a] Indicates that uniaxially oriented samples were used to determine the crystal structure.

by Cakmak [59] based on the parameters obtained by Fratini et al. [53], as well as by several other investigators [54,55,60]. The results are presented in Table 15.2. The refractive index is the highest along the chain axis and the lowest along the a-axis.

Elastic moduli of PEEK and PEK along their chain direction in the crystalline regions were measured by Nishino et al. [61] using x-ray diffraction techniques. These are 71 GPa for PEEK and 57 GPa for PEK. The moduli measured in the directions transverse to the chain axes result in 5.9 GPa for (110), 4.3 GPa for (200), and 5.0 for (020) planes of PEEK. These are about one order of magnitude smaller as compared to those values along the chain axis. Shimizu et al. [62] report 69 GPa for crystalline regions and 63 GPa for amorphous regions.

TABLE 15.2
Optical Properties of PEEK by Various Investigators

Property	Cakmak [59]	Shimizu et al. [52]	Kumar et al. [54]	Voice et al. [55]	Karacan [60]
n_a	1.5247	1.507	1.48	1.514	1.505
n_b	1.7450	1.732	1.77	1.729	1.715
n_c	1.9448	1.936	1.97	1.924	1.964
$\Delta n_{ca}^0 (n_c - n_a)$	0.42	0.429			
$\Delta n_{cb}^0 (n_c - n_b)$	0.1998	0.204			
$\Delta n_{ba}^0 (n_b - n_a)$	0.2203	0.225			
$\Delta n_c^0 (= n_c - (n_a + n_b)/2)$	0.310	0.321	0.345	0.302	0.354
Δn_{am}^0	0.281	–			

15.3.1.1 Influence of Chemical Architecture

When the percentage of ketone linkages is increased (polyphenylene oxide [PPO] = 0%, PEEK = 33.5%, PEK = 50%, PEKK = 67%), the b- and c-axes increase in size, and the a-axis decreases [63]. PAEKs containing a larger percentage of ketone linkages were found to be polymorphic depending on the technique used for crystallization. If PEKK containing all *para* linkages in its backbone is crystallized from the melt, it exhibits a crystal structure similar to that exhibited by PEEK (called form 1). A new crystalline form (form 2) is observed when the amorphous PEKK is subjected to solvent crystallization or cold crystallization [64,65]. The formation of form 2 crystals was attributed to the low chain mobilities during solvent and cold crystallization [66]. This form 2 was found to be thermodynamically less stable than form 1 and, as a result, possesses lower melting temperature. In a later work, Gardner and coworkers showed that the crystallization into form 2 is favored if the intrachain linkage stiffness is increased. This was accomplished by changing the 1,3-substituted (iso) moiety with a 1,4-substituted (tere) moiety. All of their results are summarized in Figure 15.4.

Electron diffraction studies carried out by Ho et al. [67] revealed that form 2 has unit cell dimensions of $a = 4.2$ Å, $b = 11.3$ Å, and $c = 10.1$ Å, and it is consistent with the two-chain orthorhombic unit cell proposed by Blundell et al. [68], as shown in Figure 15.5, when projected along the c-axis onto the ab plane.

PEKEKK [69], polyether biphenyl ether ketone ketone containing metapehnyl links (PEDEKmK) [70], and PEEKK [71] were also observed to exhibit polymorphism. For many years, PEEK and PEK were believed to be monomorphic. However, Liu et al. [56] later reported that PEEK exhibits strain-induced polymorphism. Form 2 was found to coexist with the dominant form 1 structure in uniaxial-oriented PEEK. The form 2 structure has a unit cell dimension of $a = 4.75$ Å, $b = 10.60$ Å, and $c = 10.86$ Å.

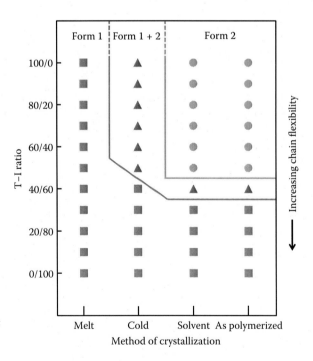

FIGURE 15.4 PEKK phase diagram showing the influence of method of crystallization and the chain flexibility on the formation of crystal form 1 and form 2. (Redrawn from Gardner, K.H. et al. *Polymer*, 35, 1994. With permission.)

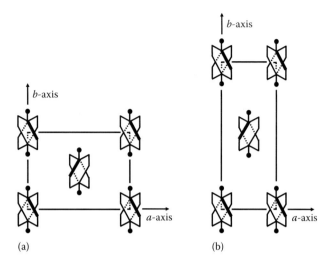

FIGURE 15.5 (a) Form 1 and (b) form 2 crystal structures for PEKK projected along the *c*-axis. (Redrawn from Blundell, D.J. and Newton, A.B., *Polymer*, 32, 1991. With permission.)

15.4 PROPERTIES

PAEKs possess a good combination of properties, including toughness and high mechanical strength over a broad temperature range, thermal stability, solvent and chemical resistance, very low smoke emission in flammability tests, and good electrical properties. They were shown to resist basic as well as moderately acidic environments.

15.4.1 Thermal and Optical Properties

PEEK exhibits a glass transition temperature of about 145°C. Upon heating from the amorphous state, it exhibits a cold crystallization temperature of about 170–180°C and a melting temperature of about 335°C [46]. Both the glass transition temperature and the melting temperatures increase linearly with an increase of the percentage of ketone linkages, as shown in Figure 15.6. A more recent study conducted by Dosière et al. [72] compared the structure and thermal properties of a poly(aryl ether ketone ether ketone naphthyl ketone) (PEKEKNK) with those of poly(aryl ether ketone ether ketone ketone) (PEKEKK). It was found that the presence of the naphthyl group increases the rigidity of the chain even further. The glass transition temperature of PEKEKNK is 20°C, but the melting temperature is only 2°C higher than that of PEKEKK.

Optical properties of PEEK are given in Table 15.2. Blundell and Osborn [46] determined various thermal and structural parameters for PEEK. These are given in Table 15.3.

Dynamic mechanical studies indicate additional transitions, such as the β transition, occurring below the glass transition region. The transition termed γ occurs around −100°C, and this is associated with the molecular motion of water bound to the main chain [73]. Electron beam irradiation causes the appearance of a new transition called β′ situated in the 40–100°C range.

Under a nitrogen atmosphere, PEEK shows good thermal stability at temperatures as high as 550–580°C, beyond which significant weight loss occurs [81], with the decomposition products being phenol and benzoquinone [74]. Exposure of PEEK to temperatures in the 350–380°C range in air for extended periods results in the increase of melt viscosity, primarily as a result of cross-linking [75]. In a later study, Tsai et al. [76] reported that the onset point of thermal decomposition for PEEK is about 450°C. Abate et al. [77] have investigated the thermal kinetic stability of model

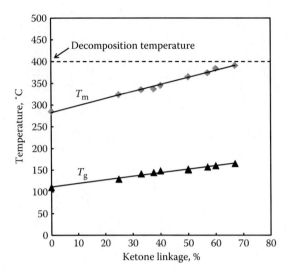

FIGURE 15.6 Thermal transition temperature versus percent keto content. (Redrawn from Gardner, K.H. et al., *Polymer*, 33, 1992. With permission.)

TABLE 15.3
Thermal and Structural Parameters of PEEK

Property	PEEK
T_g, °C	144
T_m, °C	335
T_{cc}, °C (cold crystallization)	180
T_{cm}, °C (crystallization from melt)	283
T_m, °C (equilibrium melting point)	395
S_e, erg/cm² (end-surface free energy)	49
S_s, erg/cm² (side-surface free energy)	38
ΔH_f, kJ/kg	130
ΔS_f, kJ/kg °C	0.33

Source: Data from Blundell, D.J. and Osborn, B.N., *Polymer*, 24, 1983. With permission.

polymers having sulfone, ketone, and ether groups and found that the stability of the polymers is the best with sulfone, followed by ketone and then by ether. This trend correlates well with the chain rigidity.

For amorphous and unfilled PAEKs, the heat deflection temperature (HDT) depends on the glass transition temperatures of the polymers, which correlate well with the ketone linkage in percentage or rigidity of the polymer chain. When crystallized, HDT can be increased substantially to a point close to the melting point of PAEKs with the addition of glass fibers or other reinforcing fillers.

One of the advantages with PAEKs is their high continuous use temperature (CUT) or relative temperature index (RTI) as defined by Underwriters Laboratory or UL. Due to their excellent thermal stability, PAEKs have CUT or RTI values in the range of 180–260°C, which is much higher than some of the other engineering plastics having higher HDT values but poorer CUT values.

To understand the effect of molecular weight and percentage of ketone linkage on thermal properties of PAEKs, three different types of unfilled PAEKs from Victrex plc are compared in Table 15.4.

TABLE 15.4
Thermal Properties of Various Commercially Available PAEKs

Thermal Properties	Condition	Test Method	PEEK 150G	PEEK 450G	PEEK 650G	PEK G22	PEK G45	PEKEKK G45
Melting point, °C		ISO 11357	343	343	343	373	373	387
Glass transition, °C	Onset	ISO 11357	143	143	143	152	152	162
Specific heat capacity, kJ kg^{-1} °C^{-1}	23°C	DSC	2.2	2.2	2.2	2.2	2.2	
Coefficient of thermal expansion, ppm K^{-1}	Along flow below T_g	ISO 11359	50	45	45	45	45	45
	Average below T_g		55	55	65	55	55	55
	Along flow above T_g		120	120	125	75	75	105
	Average above T_g		140	140	160	130	130	125
Heat deflection temperature, °C	1.8 MPa	ISO 75A-f	156	152	152	163	163	172
Thermal conductivity, W m^{-1} K^{-1}	23°C	ISO 22007-4	0.29	0.29	0.29	0.29	0.29	0.29
Relative thermal index	Electrical	UL 746B	260	260				
	Mechanical w/o impact		240	240				
	Mechanical w/impact		180	180				

Source: Data from Victrex, data sheets, revision July 2012, available at http://www.victrex.com/en/datasheets/datasheets.php.

All data shown in this table are typical values taken from the data sheets on the website of Victrex plc [78]. For PEEK, the effect of molecular weight is presented by comparing three different Victrex grades having increased molecular weight in the order of 150G < 450G < 650G. It is clear that molecular weight does not seem to have much effect on thermal properties. However, with increased chain rigidity, PEK and PEKEKK show increased melting and glass transition temperatures compared to PEEK.

15.4.2 Crystallization Behavior

Due to the relatively slow crystallization character, PEEK can be quenched into an amorphous state, and crystallization can be studied from the glassy as well as the molten state. Blundell and Osborn [46] were the first to report the thermal crystallization behavior of PEEK. Their studies indicate that PEEK shows a maximum rate of crystallization at about 230°C, which is roughly halfway between the glass transition temperature and the melting temperature. The crystal growth habit was found to be positively birefringent spherulites. In a later study, Kumar et al. [75] showed that the PEEK spherulites grown between room temperature and 320°C are negatively birefringent, with the *b* crystallographic axes aligned along the radius of the spherulites.

Lovinger and Davis [79] melt-crystallized spherulites from ultrathin films and found that these spherulites grow with the *b* crystallographic axis along the radial direction. These crystals grew

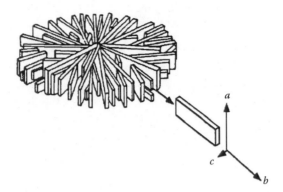

FIGURE 15.7 Thin-film spherulites of PEEK with unusual cylindrical symmetry. (From Lovinger, A.J. and Davis, D.D., *J. Appl. Phys.*, 58, 1985. With permission.)

with the lamellae on edge (*bc* crystallographic plane parallel to the substrate surface), with the *b*-axis along the radial direction, the *c*-axis in the tangential direction, and the *a*-axis making up the cylinder axis of this crystal, shown schematically in Figure 15.7. These results also indicate that these spherulites are negatively birefringent because $n_c \gg n_b$, as presented in Table 15.2. The refractive index is the highest along the chain axis and the lowest along the *a*-axis. This result is also in agreement with that of Kumar et al.

Spear-shaped single crystals of PEEK were grown from dilute solutions of PEEK in α-chloronaphthalene or benzophenone by Lovinger and Davis [80]. Electron diffraction studies of these spear-like crystals revealed that the *b* crystallographic axes lie in the long axis of the crystals, whereas the *a*-axis is transverse, and the *c*-axis is normal to the lamella plane.

Others also have studied the bulk crystallization behavior of unfilled PEEK as well as in the presence of other materials such as carbon fibers [81] under isothermal [82–85] as well as nonisothermal [82,86] conditions. The carbon fiber surfaces preferentially nucleate crystallization, and as a result, epitaxial growth of PEEK crystals occurs on these surfaces upon cooling from the melt. These studies also found that, in general, PEEK exhibits high nucleation densities. PEEK crystals persist at temperatures as high as the equilibrium melting temperature (~390°C) [82]. Therefore, "self-seeding" occurs if the crystallization experiments are carried out without specimen premelting at this temperature.

Crystallization of amorphous PEEK by heating to temperatures between T_g and T_m results in a poorly defined structure where random lamellar aggregates and small spherulites are observed [79]. This poorly defined structure is a consequence of low chain mobility present during crystallization at low temperatures. Amorphous PEEK can also be crystallized when exposed to solvents such as methylene chloride [87].

Crystallization kinetics of PAEKs was studied by Vasconcelos et al. using differential scanning calorimetry (DSC) [88]. The study was performed to obtain the relationship between the kinetics, content, nucleation, and geometry of the crystalline phases, according to the parameters of the Avrami and Kissinger models. It was found that PEEK has a higher crystallinity than PEKK. Another study on the crystallization and melting behavior of PEEK was conducted by Sandler et al. using a time-resolved x-ray technique, as reported in Reference [89]. In addition, a number of studies were recently conducted to understand the effects of supercritical carbon dioxide fluid [90], molecular weight [91], and shear [92] on the crystallization behavior of PEEK.

15.4.2.1 Effects of Heat Treatment

DSC scans of annealed samples indicate a minor melting endotherm at about 10°C above the annealing temperature [63] in addition to the primary melting peak located at 335°C. One can also

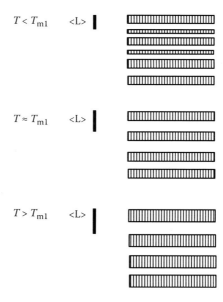

FIGURE 15.8 Schematic representation of double-melting mechanism. Thinner lamellae melt at the lower endothermic peak, and subsequently, the remaining lamellae are thickened. L represents long period. (Redraw from Hsiao, B.S. et al., *Polymer*, 34, 1993. With permission.)

generate multiple endotherms in the DSC curves by step cooling from the melt. This behavior has also been observed by others in PEEK and other polymers [93–95]. In the past, this behavior was attributed to two possible mechanisms: In the first mechanism, the two endotherms represent the melting of two separate morphological features in the structure. In the second mechanism, these are as a result of melting and recrystallization of the original structure.

According to the small angle x-ray scattering (SAXS) and wide angle x-ray scattering (WAXS) studies of Blundell [96], both endotherms are due to a single crystal structure at room temperature. The lower-melting endotherm represents crystals existing prior to the DSC scan. Between the two endotherms, constant melting and recrystallization occur. The higher-melting endotherm signifies the point where the net difference between the rates of melting and recrystallization reaches a maximum. Lee and Porter [97] also agreed with Blundell's conclusions.

Based on a comparison of surface replicas of isothermally melt-crystallized samples and those postannealed between the two endotherms, Bassett et al. [93] concluded that the "secondary" low-temperature melting endotherm is due to the material crystallized between the lamellae. However, their studies did not rule out the recrystallization and melting hypothesis. Blundell [96], using SAXS results, concluded that postannealing, for as little as 2 min between the two endotherms, was sufficient to bring about reorganization leading to lamellar thickening. A few years later, Hsiao and coworkers [98], using online SAXS measurements, showed that the lower-melting endotherm corresponds to the melting of preexisting thinner lamellae and the higher-melting endotherm corresponds to the melting of lamellae that have been thickened as a result of partial recrystallization of the material that melted at lower temperatures. This behavior is shown schematically in Figure 15.8.

15.4.3 Mechanical Properties

Mechanical properties of PAEKs are influenced by their chemical architecture and molecular weight. In addition, other features such as crystallinity, crystalline morphology, molecular orientation, and

TABLE 15.5
Mechanical Properties of Various Commercially Available PAEKs

Mechanical Properties	Condition	Test Method	PEEK 150G	PEEK 450G	PEEK 650G	PEK G22	PEK G45	PEKEKK G45
Tensile strength, MPa	Yield, 23°C	ISO 527	110	100	95	115	115	115
Tensile elongation, %	Break, 23°C	ISO 527	25	45	45	20	30	20
Tensile modulus, GPa	23°C	ISO 527	3.7	3.7	3.5	3.7	3.7	4.3
Flexural strength, MPa	At 3.5% strain, 23°C	ISO 178	130	125	115	140	140	135
	At yield, 23°C		175	165	155	185	185	180
	125°C		90	85	85	110	110	110
	175°C		19	18	16	32	32	36
	275°C		13	13	9	16	16	21
Flexural modulus, GPa	23°C	ISO 178	4.3	4.1	4.0	4.2	4.2	4.1
Compressive strength, MPa	23°C	ISO 604	130	125	120	140	140	145
	120°C		80	70	65	90	90	90
	200°C					30	30	35
Charpy impact strength, kJ m^{-2}	Notched, 23°C	ISO 179/1eA	4.0	7.0	9.5	3.8		4.0
	Unnotched, 23°C	ISO 179/U	No break	No break	No break	No break		No break
Izod impact strength, kJ m^{-2}	Notched, 23°C	ISO 180/A	5.0	7.5	8.0	5.0	7.0	6.0
	Unnotched, 23°C	ISO 180/U	No break	No break	No break	No break	No break	No break

Source: Data from Victrex, data sheets, revision July 2012, available at http://www.victrex.com/en/datasheets/datasheets.php.

structural gradients as determined by the process and processing conditions also affect the mechanical properties of the PAEK products or parts being tested. For this reason, Table 15.5 is prepared from data shown on the website of Victrex plc [78] for three different unfilled PAEKs under the same test methods, with specimens having the same geometry prepared under appropriate conditions, defined by the manufacture of these resins.

For PEEK, the effect of molecular weight is presented by comparing three different Victrex grades having increased molecular weight in the order of 150G < 450G < 650G. In general, elongation at break and impact strength are improved with the increase in molecular weight, while tensile strength at yield and flexural strength show the opposite trend. In comparison to PEEK, PEK and PEKEKK seem to show improved tensile strength, flexural strength, and compressive strength at all temperatures tested.

Although the mechanical properties of PAEKs are not viewed as exceptional at room temperature compared to some other engineering polymers, PAEKs do show superior performance in property retention at elevated temperatures, especially against more aggressive environments, due to their excellent thermal stability. In addition, PAEKs possess excellent wear and fatigue resistance, even at relatively high temperatures.

In a more recent study, mechanical properties of PEEK 450G have been extensively investigated by Rae et al. [99]. The compressive properties were measured at strain rates between 1×10^{-4} and 3000 s^{-1} and temperatures between $-85°C$ and 200°C. The tensile properties were measured between the strain rates of 2.7×10^{-5} and 1.9×10^{-2} s^{-1} and at temperatures between $-50°C$

and 150°C. The Taylor impact properties were investigated as a function of velocity, and various large-strain compression tests were undertaken to explain the results. The fracture toughness was investigated as a function of temperature and compared with previous literature. As with all semicrystalline polymers, the mechanical response is a strong function of the strain rate and testing temperature. The phenomenon of darkening observed in Taylor impacted samples is shown to be due to reduced crystallinity brought about by a large compressive strain. For samples deformed to large compressive strains using a variety of techniques and strain rates, the measured Vickers hardness was found to decrease in accordance with reduced crystallinity measured by other techniques.

15.4.4 Chemical Resistance

The chemical resistance of PAEKs is reviewed by Cotter [42], Pritchard [100], Fink [43], and Kemmish [44]. For commercially available PAEK products and compounds, Reference [101] provides valuable assessment of their resistance to common industrial chemicals and fluids. Chemical resistance of PEEK is shown in Table 15.6, according to Reference [102], where a more complete list can be found.

PAEKs are known to resist swelling of most common chemicals, due to their semicrystalline nature. Since crystallinity is affected by the process, care should be taken to maximize the degree of crystallinity for PAEKs to ensure that adequate chemical resistance is achieved. In many cases, this can be done by postannealing.

PAEKs have excellent resistance to hydrolysis due to a lack of hydrolysable groups such as esters, amides, and imides. They are also resistant to common industrial fluids used in automotive, oil and gas, building and construction, appliance, and semiconductor industries. However, PAEKs are also prone to attack by halogens that are capable of halogenating the aromatic rings. Strong acids such as highly concentrated sulfuric acid and nitric acid may also dissolve PAEKs. At temperatures close to its melting point, PEEK can be dissolved in some aromatic solvents, such as aromatic esters and ketones. Many solvents, such as methylene chloride and chloroform, can also soften amorphous PEEK and reduce its glass transition temperature to below room temperature causing, it to go through a solvent-induced crystallization process. This is already discussed in Section 15.4.2. Beck et al. [103] conducted a systematic study to understand the solubility of PEEK in 107 different organic compounds. The results of their study can be found in a table in Reference [103].

15.4.5 Radiation Resistance

PAEKs are known to resist γ-rays, electrons, protons, and neutrons because of their ability to delocalize the bonding electrons in the aromatic rings. This makes it relatively difficult for high-energy particles to break the polymer chain or generate a free radical at or adjacent to the aromatic ring. The radiation resistance of PEEK has been investigated for potential applications, such as in nuclear power plants and satellites in outer space [104]. A good review was given for medical and biomedical applications in Reference [105]. The environmental effect was reported in a number of studies [106–111]. The presence of oxygen was found to enhance the degradation process.

Recently, Rajak and Malviya [112] studied the aging effect of γ irradiation on physicomechanical and thermal properties of PAEKs. These properties are important for medical and food packaging applications. It was found that PAEKs remain stable up to 45 kGY and up to the age of 30 days of irradiation. A slight variance (<9%) was observed in degradation temperature and flexural modulus. The overall effect on physicomechanical and thermal properties of PAEKs during irradiation of up to 45 kGy was insignificant due to simultaneous cross-linking and chain scission.

PAEKs have limited resistance to ultraviolet (UV) radiation due to strong absorption of UV radiation by the aromatic moieties. This is demonstrated by a study conducted by Nakamura et al. [113] for PEEK sheets under tensile loads. Chemical analysis based on Fourier-transform infrared spectrometry (FTIR) and x-ray photoelectron spectroscopy (XPS) showed photochemical scission

TABLE 15.6
Chemical Resistance of PEEK

Chemical	Resistance at 20°C	Resistance at 60°C	Resistance at 100°C
Acetic acid (glac./anh.)	R	R	NR
Acetone	R	R	R
Alcohols	R	R	R
Aliphatic esters	R	R	R
Aluminum chloride	R	R	R
Aluminum sulfate	R	R	R
Ammonium chloride	R	R	R
Amyl acetate	R	R	R
Aniline	R	R	R
Antimony trichloride	R	R	R
Aqua regia	R	R	ND
Aromatic solvents	R	R	R
Benzene	R	R	R
Benzoic acid	R	R	R
Boric acid	R	R	R
Bromide (K) solution	R	R	R
Bromine	NR	NR	NR
Bromine liquid, tech.	NR	NR	NR
Bromine water, saturated aqueous	R	R	R
Carbon disulfide	R	R	R
Carbon tetrachloride	R	R	R
Chlorine, wet	R	R	R
Chlorobenzene	R	R	R
Copper salts (most)	R	R	R
Cyclohexane	R	R	R
Ether	R	R	R
Glycerine	R	R	R
Glycols	R	R	R
Hydrazine	R	R	R
Hydrochloric acid (conc.)	R	R	ND
Hydrogen peroxide (30–90%)	R	R	R
Hypochlorites (Na 12–14%)	NR	NR	NR
Nitric acid (50%)	R	R	R
Nitric acid (90%)	NR	NR	NR
Oils, diesel	R	R	R
Phosphoric acid (95%)	R	R	R
Phenol	R	R	ND
Sea water	R	R	R
Silicone fluids	R	R	R
Sulfuric acid (70%)	R	R	R
Sulfuric acid (95%)	NR	NR	NR
Sulfuric acid, fuming	NR	NR	NR
Trichlorethylene	R	R	R
Zinc chloride	R	R	R

Source: Data from K-Mac Plastics, K-Mac Plastics data sheets, K-Mac Plastics, Wyoming, MI, available at http://k-mac-plastics.net, accessed January 11, 2014.

Note: ND, no data; NR: not recommended; R, resistant.

TABLE 15.7
Typical Fire and Smoke Properties and Toxicity Information for PEEK

Fire, Smoke, and Toxicity	Condition	Test Method	PEEK 450G
Glow wire test, °C	2 mm thickness	IEC 60695-2-12	960
Limiting oxygen index, % O_2	0.4 mm thickness	ISO 4589	24
	3.2 mm thickness		35
Toxicity index	CO content	NES 713	0.074
	CO_2 content		0.15
	Total gases		0.22

Source: Data from Victrex, data sheets, revision July 2012, available at http://www.victrex.com/en/datasheets/datasheets.php.

caused by UV exposure. Thermal properties, measured by DSC, indicated that a cross-linking reaction occurred during the radiation tests. Tensile properties of PEEK sheets after UV radiation clearly showed a tendency to embrittlement affected not only by cross-linking but also by the orientation of molecular chains resulting from the temperature rise of the specimens. Furthermore, the applied tensile stress during exposure accelerated molecular scission and disturbed the cross-linking effects of the tensile properties. To improve the UV stability of PAEKs, one can incorporate a certain level of UV absorbers and stabilizers. However, most of these additives are not stable enough to survive the high processing temperatures of PAEKs. Adding carbon black with small primary particle size to PAEKs can potentially improve the UV resistance of the finished products or parts if coloring them black is cosmetically acceptable.

15.4.6 Fire and Smoke

PAEKs are high-performance polymers with unique fire performance. PAEKs do not support combustion in air and have low smoke emission. They burn only in concentrated oxygen environments with a limiting oxygen index (LOI) of around 35%. When PAEKs burn, they release only nontoxic gases such as water and carbon dioxide. This unique property sets PAEKs apart from many other high-performance polymers that give off toxic gases during combustion. Table 15.7 shows the typical fire and smoke properties and toxicity information for Victrex PEEK 450G [78].

15.4.7 Electrical Properties

PAEKs are good electrical insulators, having an excellent dielectric constant and dissipation factors. They can be used over a wide temperature range in different environments. They offer better insulation options with low smoke and low toxicity in fire situations. However, PAEKs have a relatively low comparative tracking index, as shown in Table 15.8 [78]. This may indicate that PAEKs are somewhat susceptible to electrical breakdown on the surface, possibly due to degradation of PAEKs.

15.5 BLENDS, COMPOUNDS, AND COMPOSITES

15.5.1 Blends

Based on the glass transition temperature observations, PEEK was shown to be miscible with polyetherimide (PEI) in the entire composition range [114,115] by melt mixing as well as solution mixing [116] techniques [117,118]. The noncrystallizable PEI chains having high T_g reduces the

TABLE 15.8
Typical Electrical Properties of PAEKs

Electrical Properties	Condition	Test Method	PEEK 450G	PEK G22	PEKEKK G45
Dielectric strength, kV mm^{-1}	2 mm thickness	IEC 60243-1	23	23	23
	50 µm thickness		190		
Comparative tracking index, V		IEC 60112	150	150	150
Loss tangent	23°C, 1 MHz	IEC 60250	0.004		0.004
Dielectric constant	23°C, 1 kHz	IEC 60250	3.1		3.0
	23°C, 50 Hz		3.0		
	200°C, 50 Hz		4.5		
Volume resistivity, Ω cm	23°C	IEC 60093	10^{16}	10^{16}	10^{16}
	125°C		10^{15}	10^{15}	
	275°C		10^{9}	10^{9}	

Source: Data from Victrex, data sheets, revision July 2012, available at http://www.victrex.com/en/datasheets/datasheets.php.

crystallizability [114,117] of PEEK, and the melting point disappears in the DSC scans of the blends containing more than 50% PEI. It appears that during crystallization, the PEI chains are expelled into the regions between the branches of the spherulites instead of being pushed out at the boundaries defined by the spherulite growth fronts. Chen and Porter [116] attributed this behavior to the kinetic competition between the growth rate of the crystallizable component and the diffusion rate of the noncrystallizable component d, first proposed by Keith and Padden [119], where $d = D/G$, D being the diffusion coefficient of the noncrystallizable component and G the growth rate of the crystallizable component. In PEEK/PEI blends, the T_g of PEI is high (230°C) as compared to that of PEEK (145°C). As a result, PEI has a rather small diffusion rate, and it is entrapped within the body of the growing spherulite in the interlamellar regions during crystallization.

In a separate study [120], PEEK and polyamideimide (PAI; condensation product of trimellitic anhydride and a 7:3 molar ratio of 4,4'-oxydianiline to *m*-phenylenediamine) was found to be immiscible. However, sulfonation or sulfamination of PEEK brought about miscibility at all composition levels. This was attributed to the formation of electron donor–acceptor complexes between sulfonated/sulfaminated phenylene rings of PEEK and *N*-phenylene groups of polyimides. Sulfonated PEEK was also found to compatibilize the PEI and PAI blends, which are normally immiscible [121].

The miscibility of PEEK, PEKK, and PEK with a new thermoplastic polyimide (N-TPI) composed of 4,4'-bis(3-aminophenoxy)biphenyl and pyromellitic dianhydride was studied by Sauer and Hsiao [122]. PEEK is immiscible with N-TPI, and the miscibility of PAEKs with N-TPI is increased with the increase in the fraction of ketone linkages in the PAEKs. Charge-transfer complexation of polyimide was offered as a possible explanation of miscibility in these blends. Harris and Robeson [123] also studied the miscibility of PEEK with other PAEKs and related the miscibility to the ketone content in the other PAEKs, as shown in Table 15.9. PEEK having a 33% ketone content is in the midrange.

There were also other reports of using PEEK in PEEK/PEI/liquid crystalline polymer (LCP) ternary blends as a matrix together with PEI [124,125]. In these blends, the LCP remains a distinct phase and forms microfibrils in the PEI/PEEK matrix. This was reported to help improve the mechanical properties. In later studies, Bicakci and Cakmak [126,127] investigated the ternary diagrams of PEEK, PEI, and polyethylenenaphthalate (PEN). PEI was found to cause increased miscibility between PEEK and PEN, which are normally not miscible with each other. Ternary compositions containing more than 40% PEI were found to be completely miscible based on the observed glass transition temperatures.

TABLE 15.9
Miscibility of PEEK with Other PAEKs

PAEKs	Ketone Content, %	Miscibility with PEEK
PPE	0	I
PEEEK	25	M
PEK	50	M
PEEKK	50	M
P(E)0.43(K)0.57[a]	57	M
PEKK	67	I

Source: Data from Harris, J. and Robeson L.M., *J. Polym. Sci. Phys.*, 25, 1987. With permission.
Note: I, immiscible and not isomorphic; M, miscible and isomorphic.
[a] Random copolymers.

There have been many industrial attempts to blend PAEKs with other polymers to reduce the total system cost or to enhance the performance of PAEKs. This is reflected in the numerous patents filed in the past three decades by various companies, including Union Carbide, ICI, Solvay, Mitsubishi Plastics, General Electric, Cantor Coburn LLP, Kureha, Ticona, Hoechst Celanese, PBI Performance Products, DuPont, Victrex, etc. Not all these companies are currently in the business making PAEKs and their blends with other polymers.

Addition of polyaryl ether sulfones (PAESs) to PAEKs can potentially reduce the cost and improve the dimensional stability of PAEKs [128,129]. But this may lead to sacrifices in chemical resistance due to a lack of crystallinity in PAES. PAESs are not miscible with PAEKs but are partially compatible. With proper melt mixing steps, finely dispersed PAES morphology can be developed. Blending PEI into PAEKs can increase the glass transition temperature of the blends. This may potentially offer an option for their use in flexible printed circuit board applications [130]. For various reasons, PAEKs have been also blended with other polymers such as Extem [131], polyimide/siloxane copolymers [132] for improved flexibility, polyphenylene sulfone (PPS) [133–135], polybenzimidazole (PBI) [136–140], fluoropolymers [141], thermotropic liquid crystalline polyesters (TLCP) [142], polyphenylenes [143], etc.

15.5.2 COMPOUNDS AND COMPOSITES

Many unfilled virgin PAEKs come in the form of powders or granules. They can be processed in their original forms by the conventional plastic processing techniques. However, for many demanding applications, further improvements in certain properties are needed. This is normally achieved by adding glass fiber, carbon fiber, graphite, fillers, nanoplatelets, fluoropolymers, and other functional additives to virgin PAEKs through a melt compounding step. PAEK compounds are typically produced by PAEK manufacturers and by some customer compounders. Table 15.10 shows some examples of PAEK polymers, unfilled PAEKs, and PAEK compounds from different manufacturers.

Adding glass fiber or carbon fiber to PAEKs increases the mechanical strength, stiffness, and modulus of the final compounds. As expected, carbon fiber offers a better reinforcement effect and lower density compared to glass fiber. However, the initial unfilled PAEK resins need to have reduced viscosity or lower molecular weight to be able to incorporate a high level of glass fiber or carbon fiber. Lubricants can also be added to PAEKs in addition to carbon fiber to further improve the wear resistance to PAEKs. Typical lubricating additives include polytetrafluoroethylene (PTFE), graphite, fluoropolymer synthetic oils, etc. For electronic applications, improved conductive performance or static dissipation is needed. This can be achieved by adding conductive fillers to PAEKs, such as milled carbon fiber, conductive carbon black, metal fibers, carbon nanotubes, etc.

TABLE 15.10
Examples of Commercially Available PAEK Products

Trade Name	Product Offering	Manufacturer
VICTREX PEEK	PEEK	Victrex Polymer Solutions [78]
VICTREX HT	PEK	Victrex Polymer Solutions [78]
VICTREX ST	PEKEKK	Victrex Polymer Solutions [78]
APTIV Films	PEEK films	Victrex Polymer Solutions [78]
VICOTE Coatings	PEEK coatings, as powder or aqueous dispersions	Victrex Polymer Solutions [78]
VICTREX Pipes	PEEK pipe stockings	Victrex Polymer Solutions [78]
PEEK-OPTIMA	PEEK	Invibio Biomaterial Solutions [144]
Vestakeep	PEEK, PEEK compounds, stock shapes	Degussa Industries (Evonik) [145]
G-PAEK	PEK, PEK compounds, PEK film, PEK filament, and PEK stock shapes	Gharda Chemicals Ltd. [146], Gharda Plastics [147]
GAPEKK	PEKK and PEKK compounds	Gharda Plastics [147]
G-COAT	PEK surface coatings	Gharda Plastics [147]
Ketron	PEEK and PEEK compounds	Quadrant Engineering Plastic Products [148]
TECAPEEK	PEEK and PEKEKK compounds and stock shapes	Ensinger Inc. [149]
AvaSpire	PAEK unfilled and compounds	Solvay Plastics [150]
Ketaspire	PEEK unfilled and compounds	Solvay Plastics [150]
RTP Compounds	PEEK, PEK, PEKEKK, and PEKK compounds	RTP [151]
LARPEEK	PEEK unfilled and compounds	LATI S.p.A. [152]

PAEKs can be made into composites and laminates for even more superior mechanical strength and toughness [153–155]. Continuous glass fiber or carbon fiber can be used to reinforce the PAEKs through proper wetting. The fracture toughness of the composite depends on the interfacial interaction between the fiber and the resin and on the matrix morphology [156]. There are a number of proprietary processes used commercially by companies making these composites. A detailed review can be found in Reference [44].

15.6 PROCESSING

PAEKs can be processed using most conventional thermoplastic processing techniques, including extrusion, injection molding, wire coating, melt spinning, spin welding, machining, laser sintering, etc. PAEKs are available commercially in the form of powders, granules, and compounds. PAEK products are also available in films, fibers, stock shapes, and foams.

The behavior of PEEK during processing under deformation may be divided into the following: (1) melt processing at temperature $T_p > 335°C$ (such as melt spinning and injection molding); (2) processing in the rubbery region between the T_g (145°C) and the T_{cc} (170–180°C) (such as uniaxial and biaxial stretching; and (3) solid-state processing at temperature $T_p < T_g$ (such as solid-state extrusion and cold rolling). Note that T_p is the processing temperature and T_{cc} is the cold crystallization temperature. All of these processes have been applied to PEEK in the past. The following sections summarize the behavior of PEEK processed under each of these conditions.

15.6.1 Effect of Solid-State Deformation ($T_p < T_g$)

Deformation of polymers at low temperatures affords large preferential chain orientations. Several techniques have been developed to apply deformation fields on PEEK from amorphous precursors.

The mode of deformation depends on the starting material being amorphous or crystalline. Plummer and Kausch [157] found that at temperatures below the glass transition, the deformation is governed by simple shear if the starting PEEK is amorphous. The strain-induced crystallization also accompanies the deformation of PEEK below the glass transition temperature [158]. In the semicrystalline PEEK, the deformation is characterized by both shear and crazing. Plummer and Kausch [157] also discovered that if they raise the temperature of deformation, the crazing diminishes and disappears completely above T_g, leaving shear as the dominant deformation mechanism. Zone drawing and zone annealing were shown to increase the dynamic modulus of PEEK films up to 13 GPa [159].

One of the negative aspects of deformation below the glass transition temperature is that microvoids are created in the structure, which can act as failure initiators. In order to reduce this dilational problem, other processes incorporating compressive forces during deformation have been developed. These are solid-state extrusion and die drawing techniques. In the solid-state extrusion [160], a solid billet is forced through a die, where substantial deformation occurs. PEEK deformed by this process could attain tensile moduli of about 6.5 GPa and tensile strengths of about 600 MPa.

In die drawing, the billet is pulled through a narrow die and is free to neck down in the deformation field [161]. In a bending test, PEEK billets, drawn to a draw ratio of 4, attained a modulus of 11 GPa with this process. This process was also shown to impart significant orientation of chains along the deformation direction. This polymer was also processed using cold rolling by Wu and Schultz [162], who reported significant enhancements of the chain orientation along the rolling direction and associated increases in properties.

15.6.2 Effect of Rubbery Region Processing ($T_g < T_p < T_{cc}$)

Polymer films of high strength and thermal stability are useful for applications involving electrical insulation of motors, transformers, and condensers [163]. Processing of crystallizable polymers in the temperature range between the T_g and T_{cc} from their amorphous precursors is called rubbery state processing and results in high preferential orientation levels. If present, stress-induced crystallization helps to produce films with uniform thickness. For this purpose, PEEK films are cast using a sheet-casting die onto a casting roll stabilized at temperatures of 35–50°C in order to obtain essentially amorphous films. The use of electrical pinning devices on the casting system helps in obtaining films of uniform thickness. The temperatures of these films are raised to a range between 150°C and 165°C, where thermally activated crystallization rates are relatively low. If the temperature is raised beyond 165°C, the films prematurely crystallize, essentially locking in the structure and preventing the uniform deformation. If the thermal crystallization occurs prior to deformation, attempts to stretch the films below the melt temperature usually accompany neck formation or premature fracture.

Typically, the films are processed by a two-step biaxial stretching process. In the first stage of this process, the temperature of the films, stabilized in the 150–165°C range [164], is stretched by passing the films through a sequence of drawing rolls. The difference in the rotation speed of these rolls provides the stretching action. This preoriented film is then heated to 180–200°C and stretched in the transverse direction using a tenter frame stretcher. The films are then exposed to temperatures of 245–270°C, where the crystallization rates are the maximum for PEEK, in order to improve the crystallinity and produce films with good thermal stability. During this heat setting process, the films are maintained in the grips of the machine in order to preserve the orientation developed during the stretching stages while increasing crystallinity. PEEK films with a secant modulus as high as 4 GPa and breaking stress of about 300 MPa were reported in films subjected to a 1.5 × 3.4 stretching sequence.

In a later publication, the importance of stress hardening during deformation in the rubbery region on the final thickness uniformity and surface roughness was reported [165]. This study indicates that during the stretching of the PEEK films, they at first become nonuniform in thickness, and as soon as the strain hardening is substantially developed, the uniformity of the thickness is reestablished. This effect is sometimes called the self-leveling effect and is a direct result of strain hardening due to stress-induced crystallization.

The structural development in biaxially stretched films of PEEK was studied extensively by Simhambhatla [166] using WAXS pole figure and SAXS techniques. When stretched biaxially from the amorphous state, the a-axes orient in the normal direction, and this is enhanced when the equal biaxial condition is attained. Sequential stretching brings about a bimodal orientation, indicated in the WAXS pole figures of (110) planes. One of these populations is oriented in the first stretching direction, and the second population is in the second stretching direction. This bimodal orientation is caused by the development of oriented structures in the first stretching direction, and subsequent stretching normal to the first direction causes reorientation of a portion of this first population in the second direction. This second population is further increased in size by the additional oriented crystallization of polymer chains that remained amorphous during the first stretching step. More recently, Daver et al. [167] studied the influence of rubbery-stage large uniaxial deformation followed by relaxation and retraction stages on the mechano-optical behavior of cast amorphous PEEK film. At the early stage of stretching, the birefringence remains linearly proportional to stress, and PEEK remains amorphous. Beyond a critical stress level, a deviation from linearity is observed where a small fraction of the material exhibits extreme preferred orientation, with substantial transitional disorder resembling nematic-like order. The onset of this deviation also coincides with the onset of strain hardening. Further stretching beyond this transition leads to stress-induced crystallization.

Structural development in amorphous blends of PEEK and PEI was investigated by Bicakci and Cakmak [168] under uniaxial deformation in the rubbery state. Prior to the critical draw ratio, PEEK/PEI blends remain amorphous. Subsequent to the onset of strain hardening, the stress rises rapidly as a result of strain-induced crystallization of the chain highly oriented in the direction of stretching. During this stage, a highly structured physical network is formed where oriented crystallites are believed to act as nodes in the network structure. Wide-angle x-ray scattering data indicate that chains in these nodes are highly oriented, although the three-dimensional crystalline lattice is not well established, due to poor axial registry of these highly oriented chains. During subsequent heat setting at high temperatures, when the PEEK/PEI blends are drawn to levels below the onset of strain hardening, a large relaxation process occurs, followed by fast crystallization and long-term slow structural rearrangement stages [169]. When the PEEK/PEI films are predrawn to levels beyond a critical structural level (crystallinity and orientation), the initial relaxation stage disappears. This may indicate that beyond a critical structural order, a long-range physical network is formed, which results in the development of an oriented crystalline phase during heat setting. The addition of noncrystallizable PEI chains was found to retard the formation of this network structure, leading to lower orientation levels.

Bicakci and Cakmak [170] also investigated the uniaxial deformation behavior of ternary blends of PEN with PEI and PEEK, with the aims to increase the glass transition temperature and reduce neck formation during deformation of PEN in the rubbery state. The addition of PEI increases the glass transition temperature and eliminates the neck formation but decreases the crystallizability. PEEK is added to compensate for the decreased crystallizability of PEN. A similar study was conducted by Zhou and Cakmak [171] for ternary blends of PEN with PEI and PEEK under biaxial deformation, with findings similar to those found under uniaxial deformation.

15.6.3 Effect of Melt Processing ($T_p > T_m$)

The effect of shear flow in the molten PEEK was found to accelerate the crystallization process [172]. This occurs as a result of extension of polymer chains, leading to the formation of numerous nuclei.

15.6.3.1 Melt Spinning

Since the flow field in melt spinning is mostly extensional, except near the die exit, where the transition from shear to extensional flow occurs, the crystallization behavior is accelerated due to the increased efficiency of chain orientation in this field.

In this process, the PEEK is heated to the 380–400°C [173] range before it is pumped through the spinnerets to form fibers. PEEK fibers are produced at take-up speeds of 250 [173], 380 [174], and up to 950 [62] m/min, which are typically well below the speeds used to produce fibers such as polyester and nylon with low orientation levels. This is partly as a result of the very strong dependency of orientation (birefringence) on the take-up speeds in this polymer, which causes rapid increases in spin-line tension with the increase in take-up speeds. The fibers produced with speeds of up to about 150 m/min [173,174] show increasing preferential orientation of the chain axes along the fiber axis, as reflected in the measured birefringences, without any sign of formation of crystalline order.

When the birefringence levels reach 0.08–0.09 [173,174], the fibers produced begin to exhibit crystalline order, as shown in Figure 15.9. Subsequent drawing steps at temperatures of 160–200°C result in significant increases in the orientation of chains in both crystalline and amorphous regions. Herman's orientation factors as high as 0.9–0.95 are easily achieved in spun–drawn fibers [174,62]. This increase in orientation is reflected in the increases in modulus and tensile strength and decreases in elongation to break values. The maximum value for modulus was about 9 GPa and for tensile strength about 0.8–1.5 GPa [174]. The melt-spun and drawn fibers of PEEKEK containing slightly higher ketone linkages in the polymer backbone exhibit moduli as high as 12–13 GPa and tensile strengths of about 0.7 GPa, as reported by Hsiung and Cakmak [175].

Waddon et al. [176] observed unusual orientation effects in fibers that were gently drawn at temperatures about 30°C above the melting point. For fibers drawn at 356°C (type 1), the b-axis was oriented in the draw direction and acted like an axis of cylindrical symmetry. SAXS maxima were observed in the equatorial region for these samples. For fibers drawn at 368°C (type 2), smeared (020) maxima were observed in the meridional direction (as were the SAXS maxima).

The authors [176] explained these observations on the basis of a uniplanar fanlike arrangement of lamellae for PEEK, first proposed by Lovinger and Davis [177]. At lower temperatures, type 1 orientation is developed because of the presence of more residual nuclei, which form lamellae impinged together into sheaves with the long axis of the lamellae oriented in the flow direction. At higher temperatures, the preexisting nuclei are reduced in number, thereby reducing impingement and fostering the formation of "lamellar cartwheels" (spherulites are formed at still higher temperatures) representing the type 2 orientation. These observations were confirmed directly by scanning electron micrographs and indirectly by birefringence measurements.

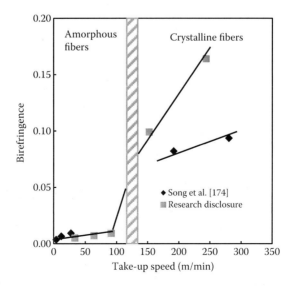

FIGURE 15.9 Birefringence as a function of take-up speed. (Data from Spinning and Drawing of Polyetheretherketone (PEEK), *Res. Discl. No. 21602*, 104, 1982; Song, S.S. et al. *Sen-i Gakkaishi*, 45, 1989.)

15.6.3.2 Injection Molding

About 35–40% of the plastics produced today are converted to usable products using an injection-molding process. In this process, the polymer pellets are melted and rapidly injected into a cavity whose temperature is kept well below the melting point of the polymer. As a result, the molten polymer experiences substantial melt deformation while being subjected to severe thermal gradients before it goes through solidification to take the shape of the cavity. This complex thermal deformation history results in fairly intricate structural variations in polymers such as PEEK [178] and other PAEKs [179–181].

As we have observed in the previous section, PEEK remains amorphous when subjected to cooling even in air, which is a poor conductor of heat. When it is injected into steel molds that are kept at temperatures below the glass transition temperatures, PEEK melt undergoes severe loss of heat near the skin and vitrifies into glass. This polymer, however, was found to have a highly crystallized sublayer appearing optically as dark regions facing the broad surface of the dumbbell-shaped sample in the cross-sectional cuts.

These dark regions were found to be highly oriented and formed as a result of shear-induced crystallization that took place during the filling stage. The interior of these samples, however, was unexpectedly found to vitrify into glass. The crystallinity profiles of a series of samples molded at different mold temperatures are shown in Figure 15.10. The crystallinity of the interior gradually increases with the increase in mold temperature, and above about 150°C, PEEK crystallinity gradients become very similar to those of fast-crystallizing polymers such as polyethylene and polypropylene. In the latter polymers, the crystallinity is generally low at the skin as a result of the rapid cooling of the polymer in these regions, and typically, it increases gradually toward the core, where cooling rates are much lower as a result of the larger insulation effect due to the thickness of the material between the interior and the skin regions. This behavior was only observed in the PEEK parts molded at temperatures where the thermally activated crystallization rates are substantial. This range is 170–240°C for PEEK.

The WAXS pole figure studies of the injection-molded PEEKs revealed that chain axes near the surface are oriented along the flow direction and the a crystallographic axes are oriented normal

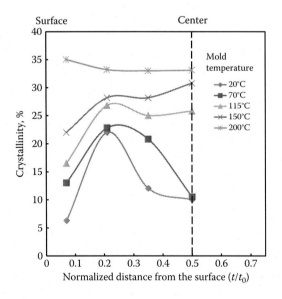

FIGURE 15.10 Crystallinity distribution on the normal direction of PEEK samples molded at mold temperatures ranging between 20°C and 200°C with 23.3 cm^3/s injection flow rate. (From Hsiung, C.M. and Cakmak, M., *J. Appl. Polym. Sci.*, 47, 1993. With permission.)

to the surface of the part indicating (001) (200) uniplanar axial texture. It appears that the polymer chains orient in the shear flow in order to pose minimum resistance to the flow. As a result, the polymer chains that contain highly planar molecules flexibly connected along the main chain are oriented with their broad surfaces parallel to the shear (1)–neutral (3) plane under the action of shear [80,177]. For instance, this was observed in a polymer containing significant planar groups such as naphthalene polyethylene naphthalate [182]; these groups orient parallel to the broad surfaces of the samples, which essentially form a 1 (flow direction)–3 (transverse direction) surface. In the case of PEEK, this direction coincides with the a-axis (being the minor axis of the ellipsoidal tube envelope encapsulating the PEEK chains).

The formation of a three-layer structure has been described in terms of a semiempirical model that links the stress effects with crystallization [183,184]. In this model, the crystallization induction time envelope (induction time-versus-temperature curve) is empirically related to the stress field that the polymer experiences. The induction time-versus-temperature curves are known for quiescent (unstressed) conditions at high and low temperatures. The polymer takes longer to start crystallizing, and the minimum time in this curve generally corresponds to a temperature halfway between the glass transition temperature and the melting temperature (for PEEK, this is approximately 230°C). When the melt is stressed, this envelope moves to shorter times and becomes broader in the temperature scale.

In order to take the nonisothermal character of this process into account, this model assumes that the polymer melt experiences small isothermal steps as it cools during the flow and experiences a series of these envelopes. The stress that a given fluid element experiences continually changes during the flow and subsequent equilibration stage.

15.7 APPLICATIONS

PAEKs are among the best high-performance polymers available in the marketplace. Compared to other polymers such as PPS, PAEKs exhibit better chemical and solvent resistance due to their crystallizability and ordered structures when processed properly, as described in the previous section. However, PAEKs are quite expensive and used only in niche applications where improved functional performance is required over other conventional materials.

PAEKs have been successfully used in aerospace applications, replacing metal, conventional thermoset composites, and other polymers because of its light weight, processing flexibility, reduced manufacturing cost, flame retardancy, low smoke generation, and improved functional performance in harsh environments.

In automotive applications, PAEKs are used to substitute metal in gears, seals, and supporting rings. Their use facilitates weight reduction; lower fuel consumption; lower CO_2 emissions; and reductions in noise, vibration, and harshness. For many years, PAEKs have been specified and utilized in safety-critical applications such as anti-lock braking systems (ABS), vacuum pumps, and turbocharger waste-gate controls. Many PAEK parts can be made without the need for secondary operations such as machining. These parts offer very good dry-running properties and are being used in systems with little or no oil lubrication.

In electronic applications, PAEKs offer high stiffness; high strength-to-weight ratio; high wear, chemical, and creep resistance; dimensional stability; low outgassing; radiotransparency; low moisture absorption; and high temperature resistance. Parts made with PAEKs for mobile devices can be thinner and lighter. Other successful applications include switches, circuit boards, connectors, sensors, audio speakers, parts for printer and copiers, etc.

In medical applications, PAEKs have been used in all three classes of applications: short-term and limited-use applications, and prolonged implantable (<30 days) and permanent implantable applications. These medical-grade PAEKs are manufactured under strict guidelines and regulations by a number of companies such as Invibio Biomaterial Solutions (a division of Victrex plc), Solvay, Oxford Performance Materials, and Evonik. Examples of PAEK medical applications include spinal devices, orthopedic devices, and trauma devices. Other specialty implants, including craniomaxillofacial (CMF), neuro, cardio, and pharmaceutical devices, are being rapidly developed. Among all sectors, medical devices are becoming more and more attractive for PAEKs, with more patents filed and technical papers published recently.

PAEKs have also found successes in energy, industrial, and semiconductor sectors. With more development underway, PAEKs will enjoy continuous growth in these sectors for years to come.

REFERENCES

1. H. J. Dillon, Ketone based resins: PEEK, *Modern Plastics Encyclopedia*, 48 (1988).
2. M. Kamura, H. Tabuse, and N. Kuramoto, Jap. Patent 63004934 A2 (88/4934) (1988).
3. K. Katayama, Jap. Patent 62148260 A2 (87/148260) (1987).
4. W. H. Bonner, U. S. Patent 3065205 (1962).
5. I. Goodman, J. E. McIntyre, and W. Russell, British Patent 971227 (1964).
6. R. N. Johnson, A. G. Farnham, R. A. Clendinning, W. F. Hale, and C. N. Merriam, Poly(aryl ethers) by nucleophilic aromatic substitution. I. Synthesis and properties, *J. Polym. Sci., Part A-1* 5: 2375 (1967).
7. Y. Iwakura, K. Uno, and T. Takiguchi, Syntheses of aromatic polyketones and aromatic polyamide, *J. Polym. Sci., Part A-1* 6: 3345 (1968).
8. J. B. Rose and P. A. Staniland, U. S. Patent 4320224 (1982).
9. J. B. Rose, European Patent 63874 (1982).
10. V. Jansons and H. C. Gors, PCT Int. Publication, WO8403891 (1984).
11. D. K. Mohanty, Y. Sachdeva, J. L. Hedrick, J. F. Wolfe, and J. E. McGrath, Synthesis and transformations of tetramethyl Bis A polyaryl ethers, *Am. Chem. Soc. Div. Polym. Chem. Polym. Prepr. 25*: 19 (1984).
12. J. Devaux, D. Delimoy, D. Daoust, R. Legras, J. P. Mercier, C. Straszielle, and E. Nield, On the molecular weight determination of a poly(aryl-ether-ether-ketone) (PEEK). *Polymer 26*: 1994 (1985).
13. D. K. Mohanty, R. C. Lowery, G. D. Lyle, and J. E. McGrath, Ketimine modifications as a route to novel amorphous and derived semicrystalline poly(arylene ether ketone) homo- and copolymers, *Int. SAMPE Symp. Exp. 32*: 408 (1987).
14. V. Jansons, H. C. Gors, S. Moore, R. H. Reamey, and P. Becker, U. S. Patent 4,698,393 (1987).
15. U. Hoffmann, M. Klapner, and K. Mullen, Phase-transfer catalyzed synthesis of Poly(ether ketones)s, *Polym. Bull. 30*: 481 (1993).
16. H. S. Sterzel, U. S. Patent 4645819 (1987).
17. M. J. Mullins and E. P. Woo, Synthesis and properties of poly(aromatic ketones), *J. Macromol. Sci. Rev. Macromol. Chem. C27*: 313 (1987).
18. M. Ueda and M. Sato, Synthesis of aromatic poly(ether ketones), *Macromolecules 20*: 2675 (1987).
19. T. E. Attwood, P. C. Dawson, J. L. Freeman, L. R. J. Hoy, J. B. Rose, and P. A. Staniland, Synthesis and properties of polyaryl ether ketones, *Polymer 22*: 1096 (1981).
20. H. R. Kricheldorf and G. Bier, New polymer syntheses. II. Preparation of aromatic PEK's from silylated bisphenols, *Polymer 25*: 1151 (1984).
21. M. Tokai, Y. Sakaguchi, and Y. Kato, Preparation and properties of poly(ether ketone)s derived from 2,6-naphthalenedicarboxylic acid chloride, *High Perform. Polym. 7*: 267 (1995).
22. J. Xie, W. Y. Peng, G. Li, and J. M. Jiang, Synthesis of poly(aryl ether ketone)s from new bisphenol monomers, *J. Fiber Bioeng. Informatics 3*: 142 (2010).
23. D. K. Mohanty, T. S. Lin, T. C. Ward, and J. E. McGrath, Novel synthesis of thermoplastic semicrystalline poly(arylene ether ketone), *Int. SAMPE Symp. Exp. 31*: 945 (1986).
24. D. R. Kelsey, L. M. Robeson, and R. A. Clendinning, Defect-free, crystalline aromatic poly(etherketones): A synthetic strategy based on acetal monomers, *Macromolecules 20*: 1204 (1987).
25. W. Risse and D. Y. Sogah, Synthesis of soluble high molecular weight poly(aryl ether ketones) containing bulky substituents, *Macromolecules 23*: 4029 (1990).
26. R. W. Phillips, V. V. Sheares, E. T. Samulski, and J. M. DeSimone, Isomeric poly(benzophenone)s: Synthesis of highly crystalline poly(4,4'-benzophenone) and amorphous poly(2,5-benzophenone), a soluble poly(*p*-phenylene) derivative, *Macromolecules 27*: 2354 (1994).
27. A. Pandya, J. Yang, and H. W. Gibson, A new polyketone synthesis involving nucleophilic substitution via carbanions derived from bis(α-aminonitriles). 1. Semicrystalline poly(arylene ketone sulfones), *Macromolecules 27*: 1367 (1994).

28. M. Ueda and F. Ichikawa, Synthesis of aromatic poly(ether ketone)s by nickel-catalyzed coupling polymerization of aromatic dichlorides, *Macromolecules 23*: 926 (1990).

29. G. T. Kwiatkowski and I. Colon, High molecular weight aromatic biphenylene polymers by nickel coupling of aryl dichlorides, in *Contemporary Topics in Polymer Science*, vol. 7, J. C. Salamone and J. S. Riffle, Eds., Plenum Press, New York, pp. 57–74 (1992).

30. G. A. Deeter and J. S. Moore, A new polymerization reaction for the synthesis of aromatic polyketones, *Macromolecules 26*: 2535 (1993).

31. H. M. Colquhoun, C. C. Dudman, M. Thomas, C. A. O'Mahoney, and D. J. J. Williams, Synthesis, structure, and ring-opening polymerisation of strained macrocyclic biaryls: A new route to high-performance materials, *Chem. Soc., Chem. Commun. (4)*: 336 (1990).

32. M. Chen, F. Fronczek, and H. W. Gibson, Concise synthesis and characterization of 30-membered macrocyclic monomer for poly(ether ether ketone), *Macromol. Chem. Phys. 197*: 4069 (1996).

33. M. Chen and H. W. Gibson, Large-sized macrocyclic monomeric precursors of poly(ether ether ketone): Synthesis and polymerization, *Macromolecules 29*: 5502 (1996).

34. M. Chen and H. W. Gibson, Novel macrocycle by friedel–crafts acylation cyclization, *Macromolecule 30*: 2516 (1997).

35. E. G. Brugel, European Patent 0,225,144 (1986).

36. K. Gharda, P. Trivedi, V. Iyer, U. Vakil, and S. Limaye, European Patent 1170318 B1 (2001).

37. D. Kemmish and B. Wilson, U. S. Patent 6909015 B2 (2001).

38. E. G. Brugel, U. S. Patent 4720537 (1988).

39. I. Y. Jeon, L. S. Tan, and J. B. Baek, Synthesis of linear and hyperbranched poly(etherketone)s containing flexible oxyethylene spacers, *J. Polym. Sci.: Part A: Polym. Chem. 45*: 5112 (2007).

40. V. L. Rao, Polyether ketones, *J. Macromol. Sci. Polym. Rev. C35*: 661 (1995).

41. J. L. Yang, Part I: Synthesis of aromatic polyketones via soluble precursors derived from bis(alpha-aminonitrile)s, Ph.D. Dissertation, Virginia Polytechnic Institute and State University, Blacksburg, VA, USA, 1998.

42. R. J. Cotter, *Engineering Plastics: A Handbook of Polyarylethers*, Gordon and Breach Science Publishers, Basel, Switzerland (1995).

43. J. K. Fink, Poly(aryl ether ketone)s, in *High Performance Polymers*, William Andrew, Norwich, NY, pp. 209–236 (2008).

44. D. Kemmish, *Update on the Technology and Application of Polyaryletherketones*, iSmithers, Shawbury, Shrewsbury, Shropshire, UK (2010).

45. J. Wang, Polyether ether ketone, in *Handbook of Engineering and Specialty Thermoplastics, Volume 3: Polyethers and Polyesters*, S. Thomas and P. M. Visakh, Eds., Wiley-Scrivener, Salem, MA, USA (2011).

46. D. J. Blundell and B. N. Osborn, The morphology of poly(aryl ether ether ketone), *Polymer 24*: 953 (1983).

47. J. Boone and E. P. Magre, Crystal structure of poly *p*-phenylene oxide, *Makromol. Chem. 126*: 130 (1969).

48. P. C. Dawson and D. J. Blundell, X-ray data for poly aryl ether ketones, *Polymer 21*: 577 (1980).

49. J. Hay, D. Kemmish, J. Langford, and A. Rae, The structure of crystalline PEEK, *Polym. Commun. 25*: 175 (1984).

50. D. R. Rueda, F. Ania, A. Richardson, and I. M. Ward, X-ray diffraction study of die drawn poly(aryl ether ketone) (PEEK), *Polym. Commun. 24*: 258 (1983).

51. N. T. Wakelyn, On the structure of PEEK, *Polym. Commun. 25*: 306 (1984).

52. J. Shimizu, T. Kikutani, Y. Ookashi, and A. Takaku, Melt spinning of PEEK, *Sen-I Gakaishi 41*: 59 (1985).

53. A. V. Fratini, E. M. Cross, R. B. Whitaker, and W. W. Adams, Refinement of the structure of PEEK fiber in an orthorhombic unit cell, *Polymer 27*: 861 (1986).

54. S. Kumar, D. P. Anderson, and W. W. Adams, Crystallization and morphology of polyarylether ether ketone, *Polymer 27*: 329 (1986).

55. A. M. Voice, D. I. Bower, and I. M. Ward, Molecular orientation in uniaxially drawn poly(aryl ether ether ketone): 1. Refractive index and X-ray measurements, *Polymer 34*: 1154 (1993).

56. T. X. Liu, S. G. Wang, Z. S. Mo, and H. F. Zhang, Crystal structure and drawing-induced polymorphism in poly(aryl ether ether ketone), *J. Appl. Polym. Sci. 73*: 237 (1999).

57. J. N. Hay, J. I. Langford, and J. R. Lloyd, Variation of unit cell parameters of aromatic polymers with crystallization temperature, *Polymer 30*: 489 (1989).

58. J. N. Hay, D. J. Kemmish, J. I. Langford, and A. I. M. Rae, The structure of crystallization PEEK, *Polym. Commun.* 25: 175 (1984).
59. M. Cakmak, Intrinsic birefringence of poly ether ether ketone, *J. Polym. Sci. Polym. Lett.* 27: 119 (1989).
60. I. Karacan, X-ray diffraction studies of poly(aryl ether ether ketone) fibers with different degrees of crystallinity and orientation, *Fibers Polym.* 6: 206 (2005).
61. T. Nishino, K. Tada, and K. Nakamae, Elastic modulus of crystalline regions of PEEK, PEK and poly *p*-phenylene sulfide, *Polymer 33*: 736 (1992).
62. J. Shimizu, T. Kikutani, Y. Ookoshi, and A. Takaku, Melt spinning of PEEK and the structure and properties of resulting fibers, *Seni-I Gakkaishi 43*: 507 (1987).
63. K. H. Gardner, B. S. Hsiao, R. R. Matheson, and B. A. Wood, Structure, crystallization and morphology of poly(aryl ether ether ketone ketone), *Polymer 33*: 2483 (1992).
64. R. R. Matheson, Y. T. Chia, P. Avakian, and K. H. Gardner, Solvent induced crystalline of thermoplastics, *Polym. Prep.* 29: 468 (1988).
65. P. Avakian, K. H. Gardner, and R. R. Matheson, A comment on crystallization of PEKK and PEEK resins, *J. Polym. Sci. Lett.* 28: 243 (1990).
66. K. H. Gardner, B. S. Hsiao, and K. L. Faron, Polymorphism in poly(aryletherketone)s, *Polymer 35*: 2290 (1994).
67. R. M. Ho, S. Z. D. Cheng, B. S. Hsiao, and K. H. Gardner, Crystal morphology and phase identifications in poly(aryl ether ketone)'s and their copolymers. 1. Polymorphism of PEKK, *Macromolecules 27*: 2136 (1994).
68. D. J. Blundell and A. B. Newton, Variations in the crystal lattice of PEEK and related *para* substituted aromatic polymers. 2. Effect of sequence and proportion of ether and ketone links, *Polymer 32*: 308 (1991).
69. D. R. Rueda, M. C. García-Gutíerrez, F. Ania, M. C. Zolotukhin, and F. J. Baltá Calleja, Crystallization kinetics and polymorphism in aromatic polyketones (PEKEKK) with different molecular weight, *Macromolecules 31*: 8201 (1998).
70. S. E. Wang, J. Z. Wang, T. X. Liu, Z. S. Mo, H. F. Zhang, D. C. Yang, and Z. W. Wu, The crystal structure and drawing-induced polymorphism in poly(aryl ether ketone)s, 2. Poly(ether ether ketone ketone), PEEKK, *Macromol. Chem. Phys. 198*: 969 (1997).
71. S. E. Wang, T. X. Liu, Z. S. Mo, J. Z. Wang, F. Xu, and Z. W. Wu, The crystal structure and drawing induced polymorphism in poly(aryl ether ketone)s, 3. Crystallization during hot-drawing of poly(ether ether ketone ketone), *Macromol. Rapid Commun. 18*: 83 (1997).
72. M. Dosière, D. Villers, M. G. Zolotukhin, and M. H. J. Koch, Comparison of the structure and thermal properties of a poly(aryl ether ketone ether ketone naphthyl ketone) with those of poly(aryl ether ketone ether ketone ketone), *e-Polymers* no. 130 (2007).
73. T. Sasuga and M. Hagiwara, Molecular motions of noncrystalline poly(aryl ether ether ketone) PEEK and influence of electron beam irradiation, *Polymer 26*: 501 (1985).
74. R. B. Prime and J. C. Seferis, Thermooxidative decomposition of poly ether ether ketone, *J. Polym. Sci. Lett.* 24: 641 (1986).
75. S. Kumar, D. P. Anderson, and W. W. Adams, Crystallization and morphology of polyarylether ether ketone, *Polymer 27*: 329 (1986).
76. C. J. Tsai, L. H. Perng, and Y. C. Ling, A study of thermal degradation of poly(aryl-ether-ether-ketone) using stepwise pyrolysis/gas chromatography/mass spectrometry, *Rapid Commun. Mass Spectrom. 11*: 1987 (1997).
77. L. Abate, A. Pollicino, A. Recca, I. Blanco, and A. Orestano, Kinetics of the isothermal degradation of model polymers containing ether, ketone and sulfone groups, *Polym. Degrad. Stabil. 87*: 271 (2005).
78. Available at http://www.victrex.com/en/datasheets/datasheets.php, data sheets revision July 2012.
79. A. J. Lovinger and D. D. Davis, Electron microscopic investigation of the morphology of melt crystallized polyarylether ketone, *J. Appl. Phys. 58*: 2843 (1985).
80. A. J. Lovinger and D. D. Davis, Solution crystallization of poly(ether ether ketone), *Macromolecules 19*: 1861 (1986).
81. Y. Lee and R. S. Porter, Crystallization of poly(ether ether ketone) (PEEK) in carbon fiber composites, *Polym. Eng. Sci. 26*: 633 (1986).
82. Y. Lee and R. S. Porter, Effects of thermal history on crystallization of poly ether ether ketone) (PEEK), *Macromolecules 21*: 2770 (1988).

83. C. N. Velisaris and J. C. Seferis, Crystallization kinetics of polyether ether ketone (PEEK) matrices, *Polym. Eng. Sci. 26*: 1574 (1986).
84. P. Cebe and S. D. Hong, Crystallization behavior of poly ether ether ketone, *Polymer 27*: 1183 (1986).
85. H. X. Nguyen and H. Ishida, Molecular analysis of the crystallization behavior of poly(aryl ether ether ketone), *J. Polym. Sci. Phys. 24*: 1079 (1986).
86. P. Cebe, Nonisothermal crystallization of poly ether ether ketone aromatic polymer composite, *Polym. Comp. 9*: 271 (1988).
87. E. J. Stober, J. C. Seferis, and J. D. Keenan, Characterization and exposure of poly ether ether ketone (PEEK) to fluid environments, *Polymer 25*: 1845 (1984).
88. G. C. Vasconcelos, R. L. Mazur, E. C. Botelho, M. C. Rezende, and M. L. Costa, Evaluation of crystallization kinetics of poly (ether-ketone-ketone) and poly (ether-ether-ketone) by DSC, *J. Aerosp. Technol. Manag., São José dos Campos 2*: 155 (2010).
89. J. Sandler, A. H. Windle, P. Werner, V. Altsta, D. T. M. V. Es, and M. S. P. Shaffer, Time-resolved x-ray studies of poly(aryl ether ether ketone) crystallization and melting behavior, part 2: Melting, *J. Mater. Sci. 38*: 2135 (2003).
90. D. Wang, H. Gao, W. Jiang, and Z. H. Jiang, Effect of supercritical carbon dioxide on the crystallization behavior of poly(ether ether ketone), *J. Polym. Sci.: Part B: Polym. Phys. 45*: 2927 (2007).
91. Q. S. Lu, Z. G. Yang, X. H. Li, and S. L. Jin, Synthesis, morphology, and melting behavior of poly(ether ether ketone) of different molecular weights, *J. Appl. Polym. Sci. 114*: 2060 (2009).
92. G. Y. Zhao, Y. F. Men, Z. H. Wu, X. L. Jig, and W. Jiang, Effect of shear on the crystallization of the poly(ether ether ketone), *J. Polym. Sci.: Part B: Polym. Phys. 48*: 220 (2010).
93. D. C. Bassett, R. H. Olley, and I. A. M. Al Raheil, On crystallization phenomena in PEEK, *Polymer 29*: 1745 (1988).
94. Y. Lee and R. S. Porter, Effects of thermal history on crystallization of PEEK, *Macromolecules 21*: 2770 (1988).
95. P. H. Holdsworth and A. Turner-Jones, The melting behavior of heat crystallized PET, *Polymer 12*: 195 (1971).
96. D. J. Blundell, On the interpretation of multiple melting peaks in PEEK, *Polymer 28*: 2248 (1987).
97. Y. Lee and R. S. Porter, Double melting behavior of PEEK, *Macromolecules 20*: 1336 (1987).
98. B. S. Hsiao, K. H. Gardner, D. Q. Wu, and B. Chu, Time resolved X-ray study of poly aryl ether ether ketone) crystallization and melting behavior. 2. Melting, *Polymer 34*: 3996 (1993).
99. P. J. Rae, E. N. Brown, and E. B. Orler, The mechanical properties of poly(ether-ether-ketone) (PEEK) with emphasis on the large compressive strain response, *Polymer 48*: 598 (2007).
100. G. Pritchard, Anti-corrosion polymers: PEEK, PEKK and other polyaryls, *Rapra Review Report* No. 80, Rapra, Technology, Shawbury, Shrewsbury, UK (1995).
101. W. Woishnis, *Chemical Resistance of Plastics and Elastomers*, 4th ed., Elsevier, William Andrew, New York (2007).
102. K-Mac Plastics Data sheets, K-Mac Plastics, Wyoming, MI. Available at http://k-mac-plastics.net, accessed January 11, 2014.
103. H. N. Beck, R. A. Lundgard, and R. D. Mahoney, U. S. Patent 5064580 (1991).
104. J. G. Funk and G. F. Sykes Jr., Space radiation effects on Poly(aryl-ether-ketone) thin films and composites, *SAMPE Quart. 19(3)*: 19 (1988).
105. S. Kurtz, *PEEK Biomaterials Handbook*, Elsevier, William Andrew, New York (November 2011).
106. O. Yoda, The radiation effect on non-crystalline poly(aryl-ether-ketone) as revealed by x-ray diffraction and thermal analysis, *Polym. Commun. 25*: 238 (1984).
107. T. Sasuga and M. Hagiwara, Mechanical relaxation of crystalline poly(aryl-ether-ether-ketone) (PEEK) and influence of electron beam irradiation, *Polymer 27*: 821 (1986).
108. T. Sasuga and M. Hagiwara, Radiation deterioration of several aromatic polymers under oxidative conditions, *Polymer 28*: 1915 (1987).
109. E. A. Hegazy, T. Sasuga, M. Nishii, and T. Seguchi, Irradiation effects on aromatic polymers: 1. Gas evolution by gamma irradiation, *Polymer 33*: 2897 (1992).
110. E. A. Hagazy, T. Sasuga, M. Nishii, and T. Seguchi, Irradiation effects on aromatic polymers: 2. Gas evolution during electron-beam irradiation, *Polymer 33*: 2904 (1992).
111. E. A. Hagazy, T. Sasuga, and T. Seguchi, Irradiation effects on aromatic polymers: 3. Changes in thermal properties by gamma irradiation, *Polymer 33*: 2911 (1992).
112. N. L. Rajak and M. Malviya, Polyaryletherketone: Ageing effect on physicomechanical and thermal properties, *Can. Chem. Trans. 1*: 202 (2013).

113. H. Nakamura, T. Nakamura, T. Noguchi, and K. Imagawa, Photodegradation of PEEK sheets under tensile stress, *Polym. Degrad. Stabil. 91*: 740 (2006).
114. J. E. Harris and L. M. Robeson, Miscible blends of PAEK's and poly ether imides, *J. Appl. Polym. Sci. 35*: 1877 (1988).
115. F. M. Froix, M. Petersen, N. G. R. Farrar, and M. C. Hathaway, European Patent 0163464 A1 (1985).
116. H. L. Chen and R. S. Porter, Phase and crystallization behavior of solution blended poly ether ether ketone and poly ether imide, *Polym. Eng. Sci. 32*: 1870 (1992).
117. G. Grevecoeur and G. Groenincx, Binary blends of PEEK and PEI miscibility, crystallization behavior and semicrystalline morphology, *Macromolecules 24*: 1190 (1991).
118. S. D. Hudson, D. D. Davis, and A. J. Lovinger, Semi-crystalline morphology of PEEK/PEI blends, *Bull. Am. Phys. Soc. 36*: 632 (1991).
119. H. D. Keith and F. J. Padden, Spherulitic crystallization from melt. I. Fractionation and impurity segregation and their influence on crystalline morphology, *J. Appl. Phys. 35*: 1270 (1964).
120. R. J. Karcha and R. S. Porter, Miscible blends of modified polyarylether ketones, *J. Polym. Sci. Phys. 31*: 821 (1993).
121. R. J. Karcha and R. S. Porter, Ternary blends of sulphonated PEEK and two aromatic polyimides, *Polymer 33*: 4866 (1992).
122. B. B. Sauer and B. S. Hsiao, Miscibility of three different poly(aryl ether ketones) with a high melting thermoplastic polyimide, *Polymer 34*: 3315 (1993).
123. J. Harris and L. M. Robeson, Isomorphic behavior of poly(aryl ether ketone) blends, *J. Polym. Sci. Phys. 25*: 311 (1987).
124. R. E. S. Bretas, D. Collias, and D. G. Baird, Dynamic rheological properties of poly etherimide/PEEK/liquid crystalline polymer ternary blends, *Polym. Eng. Sci. 34*: 1492 (1994).
125. G. Crevecoeur and G. Groeninckx, Melt spinning of in-situ composites of a thermotropic liquid crystalline polyester in a miscible matrix of PEEK and PEI, *Polym. Camp. 13*: 244 (1992).
126. S. Bicakci and M. Cakmak, *Proceedings of Polymer Processing Society Regional Meeting*, Akron, Ohio, 1995.
127. S. Bicakci and M. Cakmak, Phase behavior of ternary blends of poly(ethylene naphthalate), poly(ether imide) and poly(ether ether ketone), *Polymer 39*: 4001 (1998).
128. L. Robeson and J. Harris, U. S. Patent 4624997 (1984).
129. S. Weinberg, S. Shorrock, and M. El Hibri, U. S. Patent 2009048379 (2009).
130. K. Taniguchi and S. Yamada, WO02057343 (2001).
131. A. Aneja, R. Gollucci, R. Odle, and K. Sheth, U. S. Patent 2007197739 (2007).
132. G. Haralur, G. Kailasam, and K. Sheth, U. S. Patent 2009234060 (2009).
133. N. Nishihata and M. Tada, European Patent 1416015A1 (2003).
134. M. Ajbani, A. Auerback, and K. Feng, WO2009128825 A1 (2009).
135. M. El-Hibri, E. Ryan, N. Harry, R. Empaynado, and B. Stern, WO2008129059 A1 (2008).
136. L. Disano, B. Ward, and E. Alvarez, European Patent 0392855B1 (1990).
137. T. Andres, E. Alvarez, R. Hughes, W. Cooper, and C. Wang, U. S. Patent 5391605 (1993).
138. R. Hughes, U. S. Patent 5844036 (1997).
139. B. Dawkins, M. Gruender, G. Copeland, and T. Hsu, WO2008097709A1 (2008).
140. B. Dawkins, M. Gruender, G. Copeland, and J. Zucker, WO2008097675 (2008).
141. J. Lahijani, U. S. Patent 2001016625 (2001).
142. B. Wilson and D. Flath, GB2424890 (2005).
143. N. Maljkvic, R. Chavers, and M. El Hibri, WO2007101847A2 (2007).
144. Available at http://invibio.com, accessed January 21, 2014.
145. Available at http://industrial.vestakeep.com, accessed January 22, 2014.
146. Available at http://www.gharda.com, accessed January 22, 2014
147. Available at http://www.ghardaplastics.com, accessed January 22, 2014.
148. Available at http://www.quadrantplastics.com, accessed January 21, 2014.
149. Available at http://www.ensinger-inc.com, accessed January 21, 2014.
150. Available at http://www.solvayplastics.com, accessed January 22, 2014
151. Available at http://www.rtpcompany.com, accessed January 22, 2014
152. Available at http://www.lati.com, accessed January 22, 2014
153. J. Karger-Kocsis and K. Friedrich, Temperature and strain-rate effects on the fracture toughness of poly(ether ether ketone) and its short glass-fibre reinforced composite, *Polymer 27*: 1753 (1986).
154. K. Lee and Y. Weitsman, Effects of nonuniform crystallinity on stress distributions in cross-ply graphite/PEEK (APC-2) laminates, *J. Compos. Mater. 25*: 1143 (1991).

155. Q. F. Cheng, Z. P. Fang, Y. H. Xu, and X. S. Yi, Morphological and spatial effects on toughness and impact damage resistance of PAEK-toughened BMI and graphite fiber composite laminates, *Chin. J. Aeronaut. 22*: 87 (2009).
156. T. Q. Li, M. Q. Zhang, K. Zhang, and H. M. Zeng, The dependence of the fracture toughness of thermoplastic composite laminates on interfacial interaction, *Compos. Sci. Technol. 60*: 465 (2000).
157. C. J. G. Plummer and H. H. Kausch, Deformation of thin films of PEEK, *Polymer 34*: 309 (1993).
158. R. Russo, Effect of temperature on the drawing behavior of PEEK, *J. Appl. Polym. Sci. 46*: 2177 (1992).
159. T. Kunugi, A. Mizushima, and T. Hayakawa, Preparation of high modulus and high strength PEEK film by zone annealing, *Polym. Commun. 27*: 175 (1986).
160. Y. Lee, J. M. Lefebre, and R. S. Porter, Uniaxial draw of poly(aryl ether ether ketone) by solid state extrusion, *J. Polym. Sci. Phys. 26*: 795 (1988).
161. A. Richardson, F. Ania, D. R. Rueda, I. M. Ward, and F. J. Balta Calleja, The production and properties of poly(aryl ether ketone) (PEEK) rods oriented by drawing through a conical die, *Polym. Eng. Sci. 25*: 355 (1985).
162. G. Wu and J. M. Schultz, Fracture behavior of oriented PEEK, *Polym. Eng. Sci. 29*: 405 (1989).
163. F. Nobuo and H. Haruo, European Patent 0059077 A1 (1982).
164. B. P. Griffin and I. D. Luscombe, Production of oriented films of aromatic polyether ketones, *Res. Disci. No. 21601*: 103 (1982).
165. M. Cakmak and M. Simhambhatla, Dynamics of uni- and biaxial deformation and its effects on the thickness uniformity and surface roughness of poly ether ether ketone films, *Polym. Eng. Sci. 35*: 1562 (1995).
166. M. Simhambhatla, Fundamental Studies on Biaxial Stretching, Plasma Treatment and Vacuum Metallization of Poly (ether ether ketone) (PEEK) films, Masters Thesis, The University of Akron, Akron, OH, USA (1991).
167. F. Daver, A. Blake, and M. Cakmak, Stages of structural ordering leading to stress induced crystallization of PEEK films: A mechano-optical study on deformation, relaxation and retraction, *Macromolecules 42*: 2626 (2009).
168. S. Bicakci and M. Cakmak, Development of structural hierarchy during uniaxial drawing of PEEK/PEI blends from amorphous precursors, *Polymer 43*: 149 (2002).
169. S. Bicakci and M. Cakmak, Kinetics of rapid structural changes during heat setting of preoriented PEEK/PEI blends films as followed by spectral birefringence technique, *Polymer 43*: 2737 (2002).
170. S. Bicakci and M. Cakmak, Uniaxial deformation behavior of ternary blends of poly(ethylene naphthalate), poly(ether imide) and poly(ether ether ketone) films, *Polymer 39*: 5405 (1998).
171. X. X. Zhou and M. Cakmak, Influence of composition and annealing on the structure development in biaxially stretched PEN/PEI/PEEK ternary blends, *Polymer 47*: 6362 (2006).
172. M. Chien and A. W. Weiss, Strain induced crystallization behavior of poly(ether ether ketone) (PEEK), *Polym. Eng. Sci. 28*: 6 (1988).
173. Spinning and drawing of polyetheretherketone (PEEK), *Res. Discl. No. 21602*: 104 (1982).
174. S. S. Song, J. L. White, and M. Cakmak, Structure development in the melt spinning and drawing of poly ether ether ketone fibers, *Sen-i Gakkaishi 45*: 243 (1989).
175. C. M. Hsiung and M. Cakmak, Effect of melt spinning and cold drawing on structure and properties of poly(aryl ether ketone) PAEK fibers, *Polym. Eng. Sci. 31*: 172 (1991).
176. A. J. Waddon, A. Keller, and D. J. Blundell, Six axis type orientations in melt drawn fibers of PEEK, *Polym. Bull. 19*: 297 (1988).
177. A. J. Lovinger and D. D. Davis, Electron microscopic investigation of the morphology of a melt crystallized poly aryl ketone, *J. Appl. Phys. 58*: 2843 (1985).
178. C. M. Hsiung, M. Cakmak, and J. L. White, Crystallization phenomena in the injection molding of polyether ether ketone and its influence on mechanical properties, *Polym. Eng. Sci. 30*: 967 (1990).
179. C. M. Hsiung and M. Cakmak, Effect of processing conditions on the structural gradients developed in injection molded poly aryl ether ketone parts. I. Characterization by micro beam X-ray diffraction technique, *J. Appl. Polym. Sci. 47*: 125 (1993).
180. C. M. Hsiung and M. Cakmak, Effect of processing conditions on the crystallinity, orientation gradients and mechanical properties of injection molded poly aryl ether ketone. II. Large dumbbell parts, *J. Appl. Polym. Sci. 47*: 149 (1993).
181. Y. Ulcer, M. Cakmak, and C. M. Hsiung, Effect of processing conditions on the structure development in the welding region of injection molded poly aryl ether ketone. III, *J. Appl. Polym. Sci. 55*: 1241 (1995).

182. Y. Ulcer and M. Cakmak, Hierarchical structural gradients in injection molded poly(ethylene 2,6 naphthalene-dicarboxylate) parts, *Polymer 35*: 5651 (1994).
183. C. M. Hsiung, M. Cakmak, and Y. Ulcer, A structure oriented model to simulate the shear induced crystallization in injection molding polymers: A Lagrangian approach, *Polymer 37*: 4555 (1996).
184. C. M. Hsiung and M. Cakmak, Computer simulations of crystallinity gradients developed in injection molding of slowly crystallizing polymers, *Polym. Eng. Sci. 31*: 1372 (1991).

16 Polyimides*

Hooman Abbasi, Marcelo Antunes, and José Ignacio Velasco

CONTENTS

16.1 Introduction 491
16.2 Polymer Synthesis 493
    16.2.1 Polyimides by Polycondensation 493
        16.2.1.1 Polyimides via Precursors from Dianhydrides and Their Derivatives 493
        16.2.1.2 Polyimides from Dianhydrides and Diisocyanates or Derivatives 495
        16.2.1.3 Polyimides from Dianhydrides and Disilylated Diamines 495
        16.2.1.4 Polyimides from Diimides and Their Derivatives 496
        16.2.1.5 Miscellaneous Synthesis 496
    16.2.2 Polyaddition 497
        16.2.2.1 Diels–Alder Reaction 497
        16.2.2.2 Michael Addition 498
16.3 Design and Synthesis of Organic-Soluble and/or Melt-Processable (Thermoplastic)
     Polyimides 498
    16.3.1 Aromatic Polyimides 498
    16.3.2 Aliphatic–Aromatic Polyimides 500
16.4 Characterization 501
16.5 Structures and Properties 502
    16.5.1 Polyimides in Solution 502
    16.5.2 Polyimides in the Solid State 503
    16.5.3 LC Properties 504
    16.5.4 Polymer Alloys 505
    16.5.5 Commercial TPIs 507
16.6 Functional Polyimides 509
    16.6.1 Gas Permeability and Permselectivity 509
    16.6.2 Photosensitive Polyimides 510
    16.6.3 NLO Polyimides 511
    16.6.4 Electrical Insulating and Electrolyte Polyimides 511
References 512

16.1 INTRODUCTION

Polyimides are polymers containing an imide linkage as either an open-chain structure or as a heterocyclic unit in the polymer backbone. Although a few articles about polyimides with an open-chain imide linkage have been reported [1], a vast number of papers have been published about the synthesis and applications of polyimides with heterocyclic units, ever since the first report concerning polyimides was published by Bogert and Renshaw [2] in 1908. However, it was only in the early 1960s that polyimides were successfully introduced as commercial polymeric materials by E.I. du Pont de Nemours [3–7]. In general, polyimides are aromatic polymers composed of five-membered aromatic imide units and aromatic rings in the repeating unit, being ranked among

---

* Based in parts on the first edition chapter on polyimides.

the most heat-resistant polymers, with outstanding mechanical, electrical, and chemical properties, hence being widely used as high-temperature plastics, adhesives, dielectrics, photoresists, nonlinear optical (NLO) materials, membrane materials for separation, and Langmuir–Blodgett (LB) films, among others. Furthermore, polyimides are used in a wide range of applications, including aerospace, defense, and optoelectronics; they are also used in liquid crystal alignments, composites, electroluminescent devices, electrochromic materials, polymer electrolyte fuel cells, polymer memories, fiber optics, etc. [8]. Nevertheless, conventional polyimides show poor solubility in most organic solvents and display poor thermoplasticity. Because of this, they are impossible to fabricate using conventional processing procedures and must be processed in the form of their precursor, polyamic acid, which is thermally and chemically converted to the corresponding polyimides.

Currently, a new class of polyimides having modified structures, so-called thermoplastic polyimides (TPIs), have attracted a great deal of attention as high-performance materials, much because they are soluble in organic solvents and processable by means of conventional processing techniques just like any other thermoplastic. They exhibit good physical properties comparable to traditional polyimides and have potential applications as high-temperature films, composite matrices containing a variety of fillers (e.g., carbon nanotubes, clay, or graphene), cable insulators, adhesives, etc. [9–13]. TPIs can be processed either in the form of the precursor (polyamic acid) or in the form of conventional polyimide, though generally they can be processed in the fully imidized form. They are easily fabricated in the molten state by injection molding, extrusion, and compression molding; they can be drawn into oriented films in the melt or shaped to cast films or spun fibers from polymer solutions.

Commercial high-performance TPI LARC-TPI (Figure 16.1a) was developed by NASA Langley Research Center in the late 1970s and is on the market under license by Mitsui Toatsu Chemical Inc. Modified LARC-ITPI (Figure 16.1b), an isomer of LARC-TPI, and semicrystalline LARC-CPI, which has improved melt flow and is processable by injection molding, was also developed. Mitsui Toatsu Chemical also developed other semicrystalline TPI, NEW-TPI (Figure 16.2), in the late 1980s. The NEW-TPI is supplied as pellets (AURUM) and films (Regulus). TPIs have a high content of metalinkages or flexible atoms (C=O and O) between the benzene rings in the polymer backbones. Their glass transition temperature (T_g) is below 265°C. Amorphous TPI (XU-218), which contains bulky groups such as indane derivatives together with the aromatic rings, is also marketed by Ciba Geigy. This polyimide has a much higher T_g (320°C).

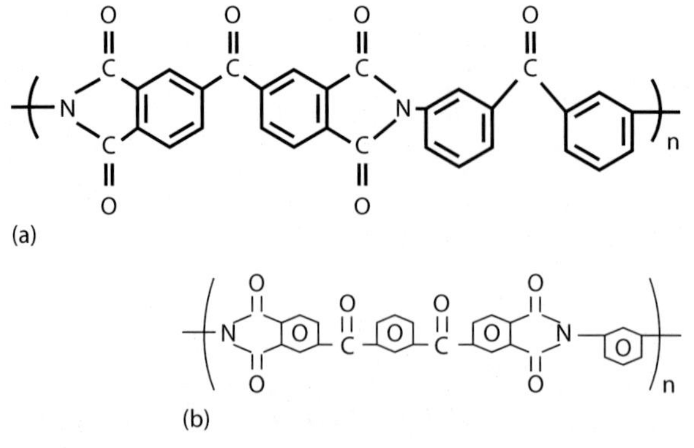

FIGURE 16.1 Chemical structures of (a) LARC-TPI (From Hara, S. and H. Inata, Mixtures of polyamide-imides, containing 3,4′-diaminodiphenylether as diamine component, and of a plasticizer, EP0210851 A2, 1987) and (b) LARC-ITPI (From Hergenrother, P.M., High performance polymer development, National Aeronautics and Space Administration, Technology 2000, pp. 149–155, 1991).

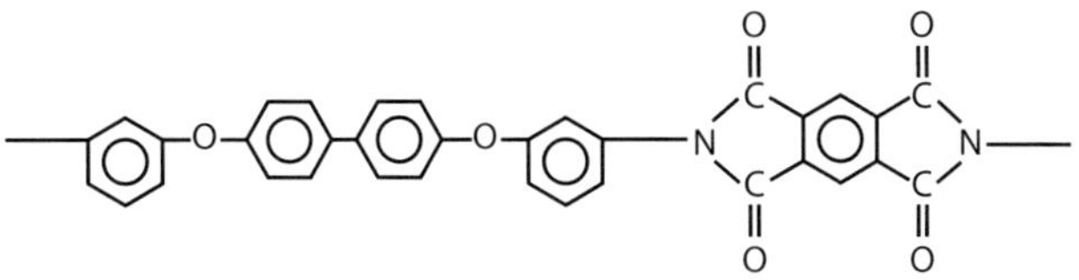

FIGURE 16.2 Chemical structure of NEW-TPI. (From Srinivas, S. et al., *Macromolecules* 30(4):1012–1022, 1997.)

16.2 POLYMER SYNTHESIS

Generally speaking, thermoplastic (organic-soluble and/or melt-processable) polyimides are synthesized by polycondensation or polyaddition.

16.2.1 POLYIMIDES BY POLYCONDENSATION

16.2.1.1 Polyimides via Precursors from Dianhydrides and Their Derivatives

Polycondensation of dianhydrides or their derivatives with diamines has been successfully employed for preparing TPIs.

16.2.1.1.1 Polyimides from Dianhydrides and Diamines

This is the most popular technique for the preparation of conventional polyimides and the synthesis of TPIs, which are derived by thermal or chemical cyclodehydration or polyamic acid precursors from dianhydrides and diamines by one- and two-step methods. Usually, chemical dehydration gives polyimides with better solubilities in organic solvents than thermal dehydration, as the side reaction is suppressed under mild reaction conditions [14]. However, polyimides produced using thermal dehydration present better thermal properties, i.e., higher T_g and decomposition temperature, than that of polyimides obtained from chemical dehydration, which can be explained according to morphological polymer changes [15].

Kinetic study of the cyclodehydration for model compounds derived from pyromellitic dianhydride or *bridged* anhydrides, whose central substituent is an electron donor or an electron acceptor, and aromatic amine indicates that the reaction paths are very different, mainly depending on the type of central substituent experimental conditions (isotherms, liquid, or solid state) and the type of central substituents [16,17].

When preparing organic-soluble polyimides in a two-step reaction, the intermediate polyamic acid can be converted to the corresponding polyimide by thermal treatment in nonpolar aprotic solvents such as *N*-methyl-2-pyrrolidone (NMP) and *N,N*-dimethylacetamide (DMAc), while aromatic polyimides having phenyl groups and trifluoromethyl groups at 2-2′-position of the diphenyl ether moieties can be prepared from bis(4-amino-2-biphenyl)ether and bis(4-amino-2-trifluoromethylphenyl)ether, respectively. The phenyl groups influence the rotational flexibility of the diphenyl ether units and the intermolecular interaction among the polyimide chains [18]. Polyimidization proceeds in hot organic solvents according to a second-order reaction, where the carboxylic group (COOH) in polyamic acids comparatively acts between two polymer chains as catalyst. Cosolvents like *o*-dichlorobenzene facilitate the azeotropic removal of the by-product, water, which is produced by the cyclodehydration of the intermediate polyamic acids. Thermal and microwave imidization of cast or spin-coated films of polyamic acids also gives the corresponding polyimides. In addition, semiaromatic polyimides with 69–96% degree of imidization prepared using 4,4′-oxydianiline, 4,4′-diamino-3,3′dihydroxybiphenyl, and bicyclo[2.2.2]-oct-7-ene-2,3,5,6-tetra carboxylic dianhydride (BCDA) in NMP, NMP/cyclohexanone, and NMP/γ-butyrolactone (BCDA-based polyimides) showed good organic solubility derived from both the hydrogen-bonded interaction of the phenolic hydroxyl groups and the π–π interaction of the phenyl rings [19–24].

Organic-soluble polyimides can also be prepared from dianhydrides and diamines by high-temperature solution polycondensation in a one-step reaction. Good solvents for this technique are aromatic solvents with high boiling points, such as nitrobenzene, o-dichlorobenzene, benzonitrile, α-chloronaphthalene, phenol, p-chlorophenol, NMP, salicylic acid, and cresols [20–41].

A novel aromatic diamine, 2,2′,6,6′-tetraphenyl-4,4′-oxydianiline (4PhODA, 4), has been synthesized by oxidation, bromination, Suzuki coupling, and reduction of 4,4′-oxydianiline (4,4′-ODA). Highly phenylated polyimides have been prepared from diamine 4 and six commercially available aromatic dianhydrides by one-step reaction. These highly phenylated polyimides present excellent solubility in organic solvents such as NMP and DMAc, even at room temperature, and in N,N-dimethyl formamide (DMF), chloroform, and m-cresol at 60°C. Films obtained by casting from their NMP or m-cresol solutions present a combination of transparency, high flexibility, and toughness, with glass transition temperatures varying from 240°C to as high as 298°C. The enhanced solubility combined with good thermal stability could allow these phenylated polyimides to be further functionalized by various aromatic electrophilic substitutions on four phenyl rings for polyelectrolyte applications [20].

Benzoic acid and p-hydroxybenzoic acid [29,35,36], as well as isoquinoline [30,34,37], promote the imidization, acting as catalysts. A mixture of benzoic acid and isoquinoline has an effect in m-cresol and naphthalene. It is important to remove the by-product completely in order to get high-molecular-weight polyimides by this technique. Polyphosphoric acid, which is used for the preparation of aromatic completely cyclized polyimides [38], is suitable for the synthesis of polyimides as a condensation reagent via the high-temperature solution polycondensation [39]. Polyimides with high molecular weight can be prepared by melt polycondensation using extrusion equipment [40,41].

16.2.1.1.2 Polyimides from Diesters and Diamines

Polyimides with high thermal stability are industrially produced using the acylation of diamines by semiesters of tetracarboxylic acid. The semiesters of bis-o-phthalic acids are commonly obtained by the dissolution of the corresponding aromatic dianhydrides in boiling alcohols. When the alcoholic solutions are treated later on with diamines, the H-complexes, in which the functional groups of the semiester and those of amine are bound by hydrogen bonds, will spontaneously form. The heat treatment of the H-complexes results in the elimination of water and alcohol to form imide cycles (acylation of amines by semiesters) [36,42]. Due to their higher cost, polyamic acid esters have not been favored, although it has been found that the use of phosphoramides such as diphenyl(2.3-dihydro-2-thioxo-3-enzoxazoyl)phosphonate (DDTBP) in combination with a base catalyst will couple and chemically imidize a mixture of diamine and tetracarboxylic diacid diester [43]. Aromatic polyimides containing bis(phenoxy)naphthalene units have been synthesized from 2,3-bis(4-aminophenoxy)naphthalene (BAPON) and various aromatic tetracarboxylic dianhydrides by the usual two-step procedure, which includes ring-opening polyaddition to give polyamic acids, followed by cyclodehydration to polyimides. This technique has an advantage over the traditional two-step method inasmuch as the presence of a small amount of moisture in the reaction system presents no problem; alcoholysis does not disturb the production of polyamic acids. In addition, the diester–dicarboxylic acids have better solubility in the solvents for polymerization; hence, the solution polycondensation of aromatic diester–dicarboxylic acids with aromatic diamines by the one-step method is successfully applied to the preparation of controlled or uncontrolled molecular weight soluble polyimides. Model reaction suggests that polycondensation proceeds by the regeneration of the dianhydrides from the corresponding diester–dicarboxylic acids [44]. The imidization behavior of isomeric poly(amic ethyl ester) of the polyimides, based on pyromellitimide subjected to isothermal temperature treatments, depends strongly on the spin casting solvents [45]. Polymerization conditions and catalysts affect polymer's viscosity. The most effective catalyst is a 1:1 mixture of benzoic acid added at the beginning of the reaction and isoquinoline added after a few hours. In the reaction system, amine catalysts act effectively in the following order: isoquinoline > quinoline > imidazole [46].

High-molecular-weight aliphatic polyimides can be prepared by the two-step melt polycondensation of the 1:1 salt from diamine and diester–dicarboxylic acids [3]. This technique is only valid for TPIs. Thermal polycondensation of the 1:1 salt of diester–dicarboxylic acids and oxydianiline conducted under reduced pressure or under high pressure gives crystalline polyimides, whose melting temperatures are much higher [47]. Aliphatic–aromatic polyimides are also prepared by the melt- or solid-state polycondensation of 1:1 salts from aliphatic diamines and aromatic dianhydrides under high pressure [48]. Polycondensation proceeds rapidly in the one-step reaction with the elimination of water resulting in high crystalline polyimides, whose T_g steps decrease with aliphatic chain lengths.

Diester–dicarbonyl chloride can be used for the preparation of polyamic esters instead of the diester–dicarboxylic acids. It reacts with the diamines to give polyamide esters by the traditional Schotten–Baumann reaction, as acid chloride is subject to nucleophilic attack by diamines. The advantage of the reaction is that the intermediate polyamic acid alkyl esters are not easily hydrolyzed. The intermediates can be readily imidized on heating to produce the corresponding polyimides with the release of alcohol [13,49]. In addition, direct polycondensation of diester–dicarboxylic acids with diamines using phosphorous compounds as condensation reagents gives polyimides [50,51]. Amines such as dodecylamine and phenecylamine catalytically promote the imidization of polyamide esters at high temperatures [52].

16.2.1.1.3 Polyimides from Tetracarboxylic Acids and Diamines

High-molecular-weight aliphatic–aromatic polyimides are obtained by polycondensation of the salts prepared from tetracarboxylic acids and aliphatic diamines at high temperature and high pressure. In this reaction, the intermediate polyamic acids are not detected during polycondensation, as it appears that the imidization and formation of polyamic acids take place at the same time. This means that the imidization rate is very fast and, at high temperature, the tetracarboxylic acids are dehydrated to yield dianhydrides [53,54].

16.2.1.2 Polyimides from Dianhydrides and Diisocyanates or Derivatives

It is known that phthalic anhydride reacts with aromatic and aliphatic isocyanates to give *N*-aryl- and/or *N*-alkylphthalimides. This reaction can be used to prepare homo- and copolyimides from dianhydrides and diisocyanates [55–58]. Hydrolysis of the diisocyanate forms a diamine, which is used to prepare the polyamic acid and then converted into a polyimide by general imidization techniques [59]. By using this reaction, the resulting thermoplastic urethane-modified polyimides provide better retention of stress–strain properties at an elevated temperature. Such polymers are synthesized from 2,4-TDI, poly(oxytetramethylene)glycol, and pyromellitic dianhydride [60]. Polyimides from diisocyanates containing oxyethylene and polymethylene chains [61] are semicrystalline and are insoluble in organic solvents. Polyimides with flexible alkoxy and trioxaoctyl or tetraoxaundecyl side chains are synthesized from diisocyanates blocked by imidazole, which have improved organic solubilities [62,63]. Whatever the origin of the effectiveness, microwave-assisted synthesis seems particularly interesting for polymerization. Microwave radiation can be used in step growth polymerization for the direct synthesis of aromatic polyimides from the reaction of aromatic diisocyanates and dianhydrides. Polyimides obtained via this method have shown superior inherent viscosities and higher yields when compared to polyimides obtained by the conventional solution method [64,65].

16.2.1.3 Polyimides from Dianhydrides and Disilylated Diamines

N-Silylated diamines react with dianhydrides at room temperature to give intermediate poly(amide trimethylsilyl ester)s, which are converted to the corresponding polyimides by subsequent imidization with the elimination of trimethylsilanol on heating. Korshak et al. [66] have reported the preparation of aliphatic–aromatic polyimides with block-like sequences using this method. Hydroxy-containing aromatic polyimides prepared from silylated 2,4-diaminophenol and tetracarboxylic dianhydrides have poorer solubility in organic solvents and higher T_g values than analogous polyimides obtained from *m*-phenylene diamine, primarily because of the presence of intermolecular hydrogen bonding [67]. The imidization rate of the silylated polymers is a little faster than that of the polyamic acid precursors using the traditional method [66]. The use of silylated amines has several drawbacks, such as the necessity to synthesize and purify activated monomers, which are difficult to isolate because of their sensitivity to moisture. Silylated amines are also more expensive than diamines. To avoid such problems, studies have shown that using in situ silylated diamines that were produced by adding chloro(trimethyl)silane (CTMS) or other silylating agents to the diamine solutions can be effective. When sterically hindered amines or amines with strong electron-withdrawing groups are used, silylation can improve the low reactivity of the diamines [8,68].

16.2.1.4 Polyimides from Diimides and Their Derivatives

Direct polyesterification using organic phosphorous compounds as condensation reagents, transesterification, or the Schotten–Baumann reaction of dicarboxylic acid derivatives of trimellitimide, pyromellitimide, and bisimides with aromatic dials (having flexible linkages), aliphatic diols, or polyethylene glycols lead to the production of organic-soluble or melt-processable poly(ester-imide) s and poly(anhydride-co-imide)s [69–75]. A new generation of poly(urethane-imide-imide)s thermoplastic elastomers have been synthesized via the reaction of NCO-terminated polyurethane with 2,2′-pyromellitdiimidodisuccinic anhydride chain extender [76]. In copolymers containing compatible monomers, there is a linear relationship between T_g and monomer fractions. T_g values of copolymers depend on the differing size and chemical nature of the comonomers and its dependence with the intra- and intermolecular structure of copolymer chains [74,75,77]. Friedel–Crafts reaction of monomers having imide rings in the P_2O_5–CH_3SO_3H mixture results in soluble and meltable semicrystalline homo- and copoly(phenylene ether imide ketone)s [78]. A new kind of aromatic diamine monomer containing pyridine unit, 2,6-Bis(3-aminobenzoyl)pyridine (BABP), was synthesized by the Friedel–Crafts acylation of benzene with 2,6-pyridinedicarbonyl chloride to form 2,6-dibenzoylpyridine (DBPY), the nitrification of DBPY with nitric acid (99%) to form dinitro compound (BNBP), and the deoxidization of BNBP using $SnCl_2$ in ethanol, being successively used [79].

Polyimides are also synthesized by utilizing aromatic bis(*o*-iodoester)s or tetraiodoaromatic compounds and primary diamines in the presence of carbon monoxide and a palladium catalyst [80–83]. Polyimides from the bis(*o*-iodoester) are fully imidized, but tetraiodoaromatic compounds give polymers with incompletely imidized structures and/or branching due to side reaction. The synthetic scope of this polymerization is limited to polyimides that are soluble in dipolar aprotic solvents. Diamines containing bulky units, such as trimethylindane, or with no steric hindrance and higher basicity are normally used for the synthesis of organic-soluble high-molecular-weight polyimides. Model reaction suggests the presence of amido linkage due to incomplete decyclization. The conventional polycondensation from dianhydrides and diamines is a more preferred technique for polyimide synthesis [83,84].

Novel star-shaped imide compounds containing electron-donating triphenylamine and/or electron-withdrawing bis(trifluoromethyl)phenyl side group have been synthesized via a two-step process. The imide compounds obtained were characterized by nuclear magnetic resonance (NMR), Fourier transform infrared (FTIR), differential scanning calorimetry (DSC), TGA, melting point analyzer, EA, and solubility measurements. In addition, their optical and electrical properties were evaluated by fluorescence spectroscopy, UV–vis spectroscopy, and cyclic voltammetry (CV), exhibiting deep blue emission (443 nm), along with high T_m (382°C) and relatively high T_g (148°C) [85].

16.2.1.5 Miscellaneous Synthesis

The chemical dehydration of polyamic acids to polyimides by acetic anhydride and pyridine results in the formation of isoimide structure together with imide structures in the polymer backbone. However, the isoimide unit is easily converted to imide without elimination of volatile compounds by thermal isomerization. The polyisoimide has better solubility, lower T_g, and lower melt flow property than the corresponding polyimide [86]. The use of thermal isomerization enables the formation of TPIs [86] that could find applications as high-temperature adhesives [87], positive-working photosensitive precursors [88], and other functional materials [89]. Various polyisoimides prepared using dehydrating agents could easily be converted to the corresponding polyimides by thermal treatment (>250°C).

The addition of inorganic materials such as polyhedral oligomeric silsesquioxane (POSS) into polyimides could enable the formation of high-performance materials. In this sense, polyimide nanocomposites have been prepared using tethered POSS in their side chains [90–92], and linear polyimides have been obtained from a double decker-shaped silsesquinoxane dianhydride [93]. Other miscellaneous structures with specific properties have also been prepared based on various monomers, such as heterocycle-containing monomers [94–97], carbazole-containing monomers [98], perylene-containing monomers [99,100], and nonlinear optical monomers [101,102].

16.2.2 Polyaddition

16.2.2.1 Diels–Alder Reaction

By a proper choice of monomers and reaction conditions, the Diels–Alder reaction of a bisdiene and a bisdienophile has often been utilized for the preparation of linear high-molecular-weight polyimides, avoiding the formation of by-products [103].

The Diels–Alder reaction is a [4+2] reaction that is capable of providing simple, efficient, and clean procedures to generate new bonds by inter- or intramolecular coupling and represents one of the most useful synthetic methods in organic chemistry. In this reaction, a dienophile is typically added to a conjugated diene to give a cyclic product called *adduct*. One interesting feature of this reaction is its thermal reversibility, which implies that its equilibrium can be easily displaced toward the reagents by heating (retro-Diels–Alder) [104–106].

Diels–Alder polymers degrade and form monomeric units via the retro-Diels–Alder reaction at high temperatures. Soluble polyimides have been prepared by the Diels–Alder reaction of bismaleimide with various bisdienes [biscyclopentadienes, bis(aminediene)s, bisfurans, and bis(2-pyrone)s] [107–114]. The reaction between hexafluoroacetone azine and dinaamides also yields polymers [115]. Oligomers are obtained from disorbonylamides and dimaleoylamides. High-molecular-weight soluble poly(hydrophthalimide)s are synthesized by solution polycondensation in a refluxing chloronaphthalene and are then converted to aromatic polyimides by dehydrogenation. Full aromatization is difficult to perform quantitatively. Bisfurans containing siloxane linkage could be used to make high-molecular-weight aromatized polyimides by the Diels–Alder reaction followed by aromatization in acetic anhydride [116]. Particularly, a novel polymerization of mesoionic sydnone and bismaleimide has been used for making a new class of polyimides with good thermal stability and solubility [117]. Synthesis of thermoplastic poly(imidosilane)s via disiloxane equilibration reactions has been reported [118,119]. Perfectly alternating copolymers are prepared by this method.

Recently, polyimides have been prepared by Chi and coworkers using in situ Diels–Alder reaction of 1,4-bis[4-(methyloxy)phenyloxy]-2,3,6,7-tetrakis-(bromomethyl)benzene (MPBB) with four arylenebismaleimides [120]. Furthermore, a novel thermoplastic–thermosetting merged polyimide system has been developed via Diels–Alder intermolecular polymerization of bisfuran, namely, 2,5-bis(furan-2-ylmethylcarbamoyl) terephthalic acid A with a series of bismaleimides. The obtained Diels–Alder adducts were aromatized and imidized, i.e., cyclized, through carboxylic and amide groups in order to generate thermoplastic–thermosetting merged polyimides. Bisfuran A was prepared by the condensation of pyromellitic dianhydride with furan-2-ylmethanamine and characterized by elemental, spectral, thermal, and LCMS analyses [121]. Poly(ester-urethane-imide)s have also been prepared by Diels–Alder polyaddition of 1,6-hexamethylene-bis(2-furanylmethylcarbamate) with various bismaleimides containing ester groups in the backbone. The Diels–Alder reaction was carried out in *m*-cresol at 110°C, followed by thermal and chemical aromatization of tetrahydrophthalimide intermediates. The monomers and polymers were characterized by IR,

^{1}H NMR spectroscopy, and elemental analysis. Thermal properties of the polymers were investigated by differential scanning calorimetry and dynamic thermogravimetric analysis [122].

16.2.2.2 Michael Addition

Linear poly(bismaleimidediamine)s with crown ether and polydimethylsiloxane chain or substituted fluorine atom has been synthesized from bismaleimides by means of Michael addition [123] in *m*-cresol [124–126]. Insoluble thermoset polyimides have also been obtained by this method using diamines and bismaleimides [127].

16.3 DESIGN AND SYNTHESIS OF ORGANIC-SOLUBLE AND/OR MELT-PROCESSABLE (THERMOPLASTIC) POLYIMIDES

Organic-soluble and/or melt-processable (thermoplastic) polyimides are prepared from dianhydrides or their derivatives and diamines by using the same polycondensation techniques as those used for conventional polyimides. Various difunctional monomers containing imide structures would also yield TPIs by general polycondensation, improving their solubility and processability without sacrificing their outstanding physical properties. The following structural modifications could be affected in order to decrease interchain interaction or reduce polymer's backbone stiffness: introduction of angular or flexible linkages into the backbone and introduction of kinked linkages (meta- and orthocatenation) or unsymmetrical and cardo (loop) structures. Among flexible linkages, ether linkages are the most popular ones, although large polar or nonpolar pendant bulky groups are also commonly placed along the polymer backbone [44,128–147]. Also common is the disruption of the symmetry or regularity of the polymer chains by means of copolymerization [148].

Polyimides are commonly classified into two categories: fully aromatic polyimides, which are made up of five-membered aromatic imide rings, benzene rings together with flexible linkages such as SO_2, O, C=O, CH_2, $C(CF_3)_2$, as well as meta- or orthocatenation between the aromatic rings in the repeating units [149–174]; and aliphatic–aromatic polyimides having flexible segments such as $(CH_2)_n$, $(CH_2O)_n$, $[Si(CH_3)_2O]_n$, $[Si(CH_3)_2]_n$, or $(CF_2)_n$ and aromatic or aliphatic imide rings in the backbone.

16.3.1 Aromatic Polyimides

The organic-soluble and/or melt-processable (thermoplastic) fully aromatic polyimides have been prepared using the dianhydrides and diamines listed in Table 16.1. Some monomers are known from an early date and are commercially available. Flexible linkages such as ether, CH_2, C=O, SO_2, $C(CF_3)Ph$, and $C(CF_3)_2$ have been introduced into polyimide backbones in order to inhibit aggregation due to intermolecular interaction between polymer chains, to lower phase transition temperatures, and to get organic-soluble polyimides. Nevertheless, glass transition temperature reduction has been reported in theses polyimides in comparison with those containing rigid backbones [175,176].

In general, fluorine-containing polymers are known to show several attractive properties such as low dielectric constant, low refractive index, low water absorption, good solubility in organic solvents, low thermal expansion, and optical transparency. Fluorine-containing organic-soluble and melt-processable fully aromatic polyimides [177–189] are prepared from fluorinated or fluorine-containing dianhydrides or its diesters and diamines in a one- or two-step reaction. Incorporation of 6F and CF_3 groups in the polymer backbone improves the solubility of polyimides and reduces their melting temperature, favoring processability. Polyimides with 3F groups demonstrate good thermo-oxidative stability [174]. Polyimides prepared from diamine containing the CF_3 group are amorphous due to the suppression of the coplanar structure. In polyimides made from fluorinated dianhydrides, T_g decreases with increasing fluorine content, being characterized by lower T_g values than the analogous fluorine-free or trimethyl-substituted polymers. The trimethyl group, substituted

TABLE 16.1
Diamines and Dianhydrides Commonly Used for Polyimide Preparation

on the biphenyl, has a disturbing effect on the coplanarity of the rigid biphenyl ring. These polyimides can be used as materials for microelectronics and aerospace engineering.

Dianhydrides and diamines with bent structures or units [175,176,190–203] are used in the preparation of organic-soluble and melt-processable rigid rod-like aromatic polyimides using the one- or two-step method. Some of them can be used to make transparent and flexible films having excellent tensile properties. These polyimides have relatively high T_g values and good thermal stability despite their kinked linkages. Diphenylprehnitric anhydride gives organic-soluble polyimides with a bend caused by the prehnitimide structure in a one-step polycondensation. These polyimides have very high T_g values, excellent thermal stabilities, and proper mechanical properties [204]. Polyimides containing binaphthyl display extremely high T_g values (above 430°C) [205]. On the other hand, polyimides having cyclic side cardo groups (fluorene, phthalide, phthalimide) obtained via the two-step method or made by thermal or chemical imidization also have good thermal stability and excellent solubility in organic solvents such as NMP, m-cresol, and o-chlorophenol.

The other approach used to improve the processability of polyimides considers the introduction of bulky pendant groups such as phenyl or phenoxy along the polymer backbone [44,132–134]. Organic-soluble rod-like polyimides with pendant groups are synthesized using mono- and diphenylated pyromellitic dianhydride in phenolic solvents with isoquinoline as catalyst by the one-step method [135]. Polyimides containing tetraphenylthiophene units are prepared by the traditional two-step method [136,137]. A series of new noncoplanar polyimides based on biphenyldiamine containing a naphthalene group have been successfully prepared by the polycondensation method using various tetracarboxylic dianhydrides and aromatic dicarboxylic acids [132].

The introduction of pendant phenyl groups increases T_g [135,137]. Soluble polyimides containing perylene units with tetraphenoxy or 4-*tert*-butylphenoxy groups and a heterocyclic fused ring, such as 9,9-disubstituted xanthene, in the backbone, have successfully been prepared [139,140]. They exhibit good mechanical properties, lower dielectric constant, lower moisture absorption, and low coefficient of thermal expansion (CTE), demonstrating potential applications in electronics. For instance, 2,2-dichloro-4,4′,5,5′-benzophenone tetracarboxylic dianhydride can be used to make fully imidized soluble polyimides containing pendant chlorine atoms, which are inherently more soluble than the corresponding chlorine-free benzophenone-based polyimides due to the increased noncoplanarity and/or the solubilizing character of chlorine moieties [141]. The incorporation of rigid and 3D bulky triptycene structure into the polyimide backbone is effective in improving solubility and thermal properties. Polyimides containing this structure are soluble in DMAc, DMF, and dimethylsulfoxide (DMSO), showing higher T_g values (around 400°C) than analogous polyamides [142]. Polyimides having pendant perfluoroalkyl groups prepared by chemical imidization are soluble in chloroform and tetrachloroethane, and hence are good candidates for electronic applications [143]. High T_g polyimides containing pendant phenolic hydroxy groups are synthesized via the one-step solution polycondensation and are soluble in amide solvents and dilute aqueous bases [147]. The phenolic hydroxy groups could serve as reactive sites for the introduction of pendant functional groups. The reaction of di(hydroxyalkyl) compounds and diimides under Mitsunobu conditions facilitates the preparation of polyimides by the one-step process. Reaction conditions provide an alternative and convenient method for the design of NLO-functionalized polyimides [206–208].

Cardo copolyimides displaying different microstructures, determined by the synthesis method, have been prepared. Thermal cyclodehydration and one-step high-temperature polycondensation lead to random copolyimides whose impact strength is higher than that of block copolyimides formed by catalytic imidization of the intermediate polymers [66]. Imide-aryl ether copolymers containing a 1,2,4-triazole ring and prepared by the two-step method [209] display good mechanical properties. Random copolyimides have been synthesized using diamino functional room temperature ionic liquid (RTIL) and have been characterized. Incorporation of $[C_{12}(DAPIM)_2][NTf_2]_2$ within the 6FDA-MDA backbone led to reduced thermal stability, glass transition temperature, d-spacing, fractional free volume, and increments in density. Gas permeability, solubility, and diffusivity of the random copolyimides decreased, and permeability selectivity increased compared with pure 6FDA-MDA polyimide [210]. Transparent fluorine-containing copolyimides exhibit significantly improved physical properties [211]. Block copolyimides containing fluorenyl cardo moieties (FR) in the polyimide blocks and poly(ethylene oxide) (PEO) sequences have been prepared using the two-step method. The combination of both moieties led to novel copolyimides with good solubility in a variety of solvents, which could easily be processed into films. Copolyimides with a proportion of PEO from 30 to 56 wt% resulted amorphous, showing proper mechanical properties and less water absorption than other PEO-containing polyimides [212].

16.3.2 Aliphatic–Aromatic Polyimides

Polyimides composed of alicyclic rings such as bicyclooctene, bicycloheptane, decahydronaphthalene, and cyclobutane have been developed by the traditional two-step imidization of alicyclic dianhydrides and various aromatic diamines [213–217]. Polyimides with bicyclooctene and tricyclododecane rings decompose at around 360°C as a result of retro-Diels–Alder reaction, but the ones containing bicycloheptane and decahydronaphthalene are not retrodegradable, possessing both solubility in organic solvents as well as excellent thermal stability. Thermal decomposition occurs in two steps. Alicyclic bicyclooctene polyimides have UV cutoff wavelengths between 300 and 400 nm and are good candidates for colorless polyimides much like fluorinated polyimides. As for the alicyclic polyimides based on a cyclobutane ring, the degree of imidization and density depends on temperature, with the fully imidized polyimides with the highest densities being obtained at 350°C. Polyimides containing an m-linkage in the diamine moieties have a higher density and a higher ordered structure than those with p-linkage. The imidization rate is low compared to the aromatic polyamic acids [218].

Alicyclic polyimides can be synthesized from an alicyclic dianhydride, 1,8-dimethylbicyclo[2.2.2]oct-7-ene-2,3,5,6-tetracarboxylic dianhydride (DMEA), and several multialkyl-substituted 4,4′-diaminodiphenylmethane compounds, including 3,3′-dimethyl-4,4′-diaminodiphenyl methane (DMDA), 3,3′,5,5′-tetramethyl-4,4′-diaminodiphenyl methane (TMDA), and 3,3′,5,5′-tetraethyl-4,4′-diaminodiphenylmethane (TEDA). Comparatively, alicyclic polyimides exhibit better solubility and transparency but worse thermal stability and mechanical properties than aromatic polyimides. Moreover, they present good solubility in organic solvents such as NMP, DMF, DMAc, chloroform, tetrahydrofuran, and *m*-cresol. In addition, alicyclic polyimides show a thermal decomposition temperature around 450°C, which is lower than common aromatic polyimides [219].

Polycondensation of aromatic dianhydrides and diamines with flexible spacers between benzene rings or alicyclic diamines results in aliphatic–aromatic TPIs [220–225]. The investigations concerning aliphatic–aromatic polyimides are focused on the production of thermoplastic polymers with extremely low T_g's (around 22°C) and melting temperatures below 300°C [220], typical of many other semicrystalline thermoplastics. The intermolecular charge transfer complex is generally formed between the imide rings and the amine moieties of aromatic polyimides [225], with the reason being the coloration and the high dielectric constants of aromatic polyimides. Polyimides based on alicyclic diamines restrain the formation of intermolecular charge-transfer complexes without reducing thermal stability. The high T_g values of these polyimides are almost similar to those of analogous aromatic polyimides and even higher in some cases with the use of adamantyl and dicyclohexyl units, presumably because of a proper balance among stiffness, polarity, and free volume of their backbone units without having to incorporate expensive fluorine or silicon compounds into their structures [226,227]. In polyimides containing benzophenone units, the C=O group photocrosslinks with the alicyclic units [224]. Polyimides with anhydride groups are prepared from alkylene dioxy derivatives [228]. TPIs composed of polydimethylsiloxane segments are soluble in organic solvents [229,230], characterized by sub-ambient T_g values, and are phase-separated. The thermal stability of these polyimides is comparable to that of aromatic polyethers [229].

16.4 CHARACTERIZATION

During the preparation of polyimides by thermal or chemical dehydration, various chemical reactions, equilibria for polyimide formation, and/or dehydration take place. FT-IR, Raman and UV spectroscopies, as well as ^{1}H, ^{13}C, ^{15}N, and ^{19}F NMR spectroscopies are available tools for determining the chemical species in the reaction systems and for understanding the reaction processes.

Imidization monitoring by FT-IR spectroscopy in polyimides having complex structures comprising multiple aromatic rings such as 6F linkage-containing polymers indicates that in the later stages of imidization, the measurements are either insensitive to changes of IR spectra or in conflict with the observation by other methods, although FT-IR is able to give at least semiquantitative data of the imidization. One major factor for the poor agreement with other data is that the imide bands found at 1780 cm^{-1} (symmetrical carbonyl) and 730 cm^{-1} (imide deformation) overlap with those of terminal anhydride groups formed by dissociation of the polyamic acids during imidization [231]. Another problem is that the above-mentioned imide bands are also subject to the dichloric effect. FT-IR spectrum data also illustrate the occurrence of side reactions other than the imide formation during imidization, i.e., isoimide and anhydride formation, and the formation of interchain imide linkage. The imide band at 1370 cm^{-1} does not overlap with anhydride peaks and shows little effect resulting from anisotropy [231,232].

The typical doublet bands of C=O group stretching bands in the imide group are found at 1718 cm^{-1} (asymmetric stretch) and 1783 cm^{-1} (symmetric stretch) [233–235]. Other characteristic peaks of imide groups shown here include C–N stretching at 1354 cm^{-1}, transverse stretching of C–N–C groups at 1087 cm^{-1}, and out-of-plane bending of C–N–C groups at 714 cm^{-1} [233]. It has been identified and discussed that the intensity of the characteristic imide ring band found at 1363–1381 cm^{-1} varies depending on temperature [236].

NMR spectroscopy is a powerful tool for getting information during imidization. Imide conversion is calculated from the ^{1}H NMR signal intensities of COOH protons, whose measurement shows that the imidization starts at around 70°C and finishes at around 200°C [237]. Absence of amide and acid proton in ^{1}H NMR spectra may be indicative of full imidization [238]. ^{13}C NMR spectroscopy is used for monitoring thermal and chemical imidization in polyamic acids as well as model diphthalic dianhydride compounds under various conditions [239,240]. Proportions of isomeric units for conventional polyamic acids can be estimated from the relative proportions of the same carbon with different isomeric connectivities. In this case, it can be roughly estimated because the degree of nuclear Overhauser units seems to be almost equal in all isomeric moieties [241]. Depolymerization and polycondensation processes can be measured using ^{13}C NMR spectrum analysis [239], which is also supported by other data. Solid-state ^{13}C NMR spectroscopy has been used for investigating polyimide structures [242]. The relationship between chemical shifts and reactivities of diamines and dianhydrides can be investigated by ^{15}N, ^{1}H, and ^{13}C NMR spectra. Particularly, ^{15}N NMR spectroscopy is a useful tool for investigating diamine reactivities. Chemical shifts are capable of being used to indicate the electron properties of monomers [243,244]. ^{19}F NMR study is advantageous for amic acid and imide chemistry, as the signals for the monomers, namely, amic acids and isoimide and imide compounds, are easily identified. Amic acid equilibria and imide interchange reactions can be quantitatively identified. In many cases, ^{19}F NMR spectrum data agree with that of ^{13}C NMR spectra, particularly for carbon–fluorine coupling constants [245]. However, this approach is limited to fluorine-containing polyimides. ^{19}F NMR is not affected by oxidation and hydrogenation, as expected from ^{1}H NMR. Nonetheless, sharp single peaks in this spectra indicate the high purity of monomers, whereas ^{31}P NMR peak shifted up and down field can indicate oxidation and hydrogenation, respectively [246].

Soluble polyimides prepared by solution imidization have successfully been analyzed by SEC (also known as GPC or GFC) for determining the molecular properties of synthetic and natural macromolecules in solution. NMP with 0.06 M LiBr is a good solvent for a standard mobile phase for polyimides in the presence of a small amount of electrolyte (LiBr or H_3PO_4 in NMP), which is able to suppress the exclusion effect of NMP. Tetrahydrofuran (THF), chloroform, DMAc, and DMF are also proper solvents [247,248]. Cross-linked polystyrene (PS) beads are used as stationary phase in the SEC column. Average molecular weights of the polyimides are calculated using universal SEC calibration with PS standard [249]. Agreement of theoretical average molecular weights with the data measured in various solvents confirms that the universal SEC calibration concept is valid for semiflexible polyimides. The weight-average molecular weights are consistent with that estimated using low-angle laser light scattering. Mark–Houwink–Sakurada constants a and K values calculated from molecular weight-intrinsic viscosity data were 0.66 and 0.037, respectively. The a value indicates that the polymer has a stiff chain. Unperturbed chain dimension of the polyimide is obtained by the Stockmayer–Fixmann extrapolation method. The value of $r_0/M^{1/2}$ was 8.6×10^{-9} cm·mol$^{1/2}$·g$^{1/2}$ [250,251]. The values of a and K measured in 0.1 M LiBr/DMAc for the thermally imidized soluble polyimide with a 6F linkage were about 0.6 and 1.14×10^{-3}, respectively [252].

16.5 STRUCTURES AND PROPERTIES

16.5.1 POLYIMIDES IN SOLUTION

The solubility parameter (SP) calculated using the Hansen SP [253] for polyimides having benzophenone moiety was 12.96 cal$^{1/2}$/cm$^{3/2}$, while that obtained using the equilibrium swelling method was between 12.96 and 13.1 cal$^{1/2}$/cm$^{3/2}$ [254]. Hansen SP is composed of three partial SPs accounting for dispersion forces, polar forces, and hydrogen bonding [255]. Conformation of organic-soluble rigid rod-like polyimides in solution can be analyzed by two independent methods that show that the polymer has rather low persistent lengths of 13 nm, although being formed by rigid

Polyimides

rod and fully linear building blocks. The bending fluctuation of the bonds is responsible for the somewhat high flexibility of the polymer chain [256]. An organic-soluble rigid rod polyimide demonstrates thermoreversible gelation due to the thermally induced phase separation and subsequent liquid crystalline (LC) transitions in *m*-cresol [257,258]. Uniform submicrometer particles of high-temperature polyimides are prepared by precipitation from a homogeneous solution and a mixture of water and NMP using a new pre-preg process developed for high-temperature polymers. Narrow size distribution particles with an average diameter of 0.2–0.5 μm have been reported [259].

16.5.2 Polyimides in the Solid State

Crystal structure, morphology, and phase transition of organic-soluble and TPIs and oligomers have been investigated by FT-IR, x-ray diffraction, electron diffraction, and transmission electron microscopy (TEM) [260–268].

Studies of the crystallization process of TPIs at the molecular level by FT-IR show that it involves intramolecular rotations, with H-bonding interactions acting as physical cross-linking points, leading to the benzimidazole/imide *mixed layer* packing (MLP) in which the imide ring of one chain and the amine phenyl ring of another adopt a coplanar *sandwich* conformation [269]. The solution-precipitated morphology of a sulfonated aromatic polyimide prepared in the melt depends on crystallization conditions. Sheaf-like crystals are formed when it is held in DMAc solution for an extended period. A similar crystalline morphology is seen in stiff polymers like polyetheretherketone (PEEK) and PEK. If the solvent is evaporated in moist air before crystallization, the result is a full amorphous polymer [268]. The crystal unit cell parameters of some TPIs are listed in Table 16.2

TABLE 16.2
Crystal Unit Cell Parameters of Various Polyimides

Repeat unit	a (nm)	b (nm)	c (nm)	α, β, γ (°)
	0.85	0.56	1.23	90
	0.84	0.56	1.66	90
	0.84	0.55	2.09	90
	Cell I 0.836	0.563	3.303	90
	Cell II 0.593	0.470	3.300	90
	0.823	0.558	2.209	90
	0.84	0.565	5.33	90
	0.789	0.629	2.51	90

[265]. Quantitative thermal analysis of a family of semicrystalline polyimides with ethylene glycol sequences shows that the heat capacity (23–640 K) increases above the T_g [270]. Mechanical properties of aliphatic–aromatic polyimides having benzophenone moiety are comparable to other thermoplastic engineering plastics, though the toughness is about twice as high as that of the commercial ones. The crystal unit cell parameter of the polymer is $a = 0.960$ nm, $b = 0.582$ nm, $c = 2.46$ nm, and $\gamma = 81.1°$. Investigation of relaxation and molecular motion by dynamic mechanical and dielectric analyses indicates the presence of α, β, and γ relaxation processes [174]. β relaxation found in various polyimides having flexible linkages is related to the rotation of rigid segments [271]. The presence of flexible aliphatic groups does not deteriorate the mechanical stability of polyimide ionomers, at least up to 50 mol% composition, compared to fully aromatic polyimides. Another study reveals that copolyimides containing a suitable ratio of adamantyl moieties together with flexible aliphatic siloxane groups display improved thermal and mechanical stabilities [272,273].

Besides interactions between polar imide groups, aromatic structures are usually believed to be the primary reason for the high thermal stability and mechanical properties of polyimides, hence the use of aromatic monomers (dianhydride and diamine) dominating over the use of aliphatic ones. Only a small portion of studies have considered the use of aliphatic monomers to synthesize semi-aromatic or aliphatic PIs, while most of them still had alicyclic structures on the polymer backbones and similar solubility to conventional fully aromatic PIs [274,275].

The mechanical properties of polyimides are influenced by many factors, such as chemical structure, viscosity, molecular weight, preparation procedure, heating history, sample preparation, and the method of property determination. Thus, it is likely that differences in mechanical properties are concealed by large experimental uncertainties. In general, polyimides exhibit modulus values of 1.5–3.0 GPa and tensile strengths of 70–100 MPa. However, the elongation at break may range from 2% to 15%, depending on chemical structure. Polyimides containing flexible linkage units such as ether and isopropylidene in the main chain exhibit higher elongations. In addition, noncoplanar, asymmetrical, and amorphous polyimides usually show higher elongations. It is a general rule but not absolute that polyimides with high mechanical modulus display lower elongation [8,276].

16.5.3 LC Properties

Polyimides showing thermotropic LC properties are classified in two groups: aliphatic–aromatic polyimides [277–298] and fully aromatic polyimides [299–309]. Most of the former are composed of asymmetrical imide rings, trimellitimide or phthalimide, and aliphatic oligothioether or oligoether spacers. The poly(ester-imide)s, having a regular sequence of flexible spacer and aromatic group, have shown smectic mesophases due to the great difference in polarity between the aliphatic spacer and aromatic imide [310]. A series of poly(ester-imide)s consisting of trimellitimide ring and flexible spacers have been synthesized by Kricheldorf et al., and the relationship between polymer structures and LC properties has been evaluated [277–285,311].

The oligoether spacers prevent the formation of layered structures. The LC properties strongly depend on the presence of aromatic ester groups. Polyimides derived from 4,4′-dihydroxybiphenyl show LC phases. Rheological studies of aliphatic–aromatic poly(ester-imide)s illustrate that the viscosity in the nematic state is considerably lower than that corresponding to the scaling law $\eta \propto M^{3.5}$. The viscosity curves display three regions typical of LC polyesters [285]. The imide dicarboxylic acid is a poor mesogen. Aliphatic–aromatic polyimides having symmetrical imide structures [286–298], pyromellitimide, naphthalenetetracarboxylic diimide, and perylenetetracarboxylic diimide demonstrate no LC properties without the presence of aromatic ring. Despite the fact that in polyimides containing biphenyltetracarboxylic diimide the mesogenic character of biphenyltetracarboxylic diimide is low [294], the LC properties depend on the structures of polymer repeating units. On the other hand, model compounds of biphenyltetracarboxylic diimide present LC phases [312]. Polyimides devoid of ester linkage based on biphenyltetracarboxylic diimide and aliphatic chains have no LC melts [313].

Fully aromatic poly(ester-imide)s composed of trimellitimide unit and aromatic rings with pendant substituted groups and bent linkages are reported to show thermotropic LC nematic phase. Aromatic poly(ester-imide)s based on symmetrical pyromellitimide, biphenyltetracarboxylic diimide, benzophenonetetracarboxylic diimide, diphenyl ether dicarboxylic diimide, diphenylsulfonetetracarboxylic diimide, and hexafluoroisopropylidenetetracarboxylic diimide have been prepared. A thermotropic LC character is found for the poly(ester-imide)s that are based on biphenyl and diphenyl ether dicarboxylic diimide [303,304]. This is explained by the assumption that interchain interaction (donor–acceptor and dipole–dipole) between temporary coplanar chain segments, rather than geometric factors like chain stiffness, linear conformation, or aspect ratio [303,305], plays a significant role in the stabilization of nematic phase. Fully aromatic polyimides having neither ester linkage nor carbonyl linkage have also been prepared, showing LC phase and improved melt processability [308].

A systematic series of nine all-aromatic diamines built around para-, meta-, and ortho-substituted aryl ether units, with either 2, 3, or 4 ether units per monomer, have been synthesized. These *flexible* diamines were polymerized using 3,3′,4,4′-Biphenyltetracarboxylic dianhydride (BPDA) and 3,3′,4,4′-oxydiphthalic dianhydride (ODPA) in order to investigate the possibility of forming all-aromatic poly(ether imide)s with LC properties. The general observation is that the thermal transition temperatures (T_g and T_m) drop rapidly when the number of aryl ether units increases (para > ortho > meta) but seems to level off at three [314]. Polyimides with methylene or ethylene oxide sequences in the diamine moiety are readily prepared from biphenyl dianhydride and diamines with terminal aminophenoxy units. Such polymers with three to six ethylene oxide units exhibit liquid crystallinity at elevated temperatures and rigid crystalline behavior at lower temperatures. However, the temperature regimes for LC behavior for the polymers with sequences of three and four ethylene oxide units are close to the thermal decomposition temperatures. In contrast, polyimides containing five or six ethylene oxide units exhibit liquid-crystalline phases that are stable over the temperature range of 50°C or more. The LC morphology was smectic A in all cases. Enthalpies of transition between the crystalline and liquid-crystalline phases and LC and isotropic liquid phases have been determined. The former were in the order of 20 $J \cdot g^{-1}$ and the latter about 10 $J \cdot g^{-1}$ [315]. Novel siloxane-containing LC polyimides with methyl, chloro, and fluoro substituents on mesogenic units have been developed from siloxane-containing diamines with pyromellitic dianhydride (PMDA) or BPDA, and their thermotropic LC behavior was examined. Among these, chloro and fluoro substituents were effective for the formation of LC phases, particularly when substituted away from the center of the mesogenic unit. The isotropization temperature is not much affected, though crystal-LC transition temperatures significantly decreased. On the other hand, the methyl substituent tends to interrupt liquid crystallization as well as crystallization. Thus, the fluoro-substituted polyimide derived from BPDA exhibited the lowest crystalline–LC transition temperature ($T_{cr} - l_c$ = 134°C) among all polyimides, showing a wide liquid crystal temperature up to 238°C. X-ray diffraction measurements conducted on the oriented mesophases of fibrous polyimides showed that they formed SmA and SmC as high- and low-temperature mesophases, respectively [316].

16.5.4 Polymer Alloys

In general, polymer alloys are classified into three categories: polymer blends, block-graft copolymers, and interpenetrating polymer networks (IPNs). Among these, polymer blends provide the most cost-effective technique in terms of polymer properties modification. There are two types of polyimide blends: (1) polyimide/polyimide blends having different molecular structures and (2) blends of polyimides with high-performance polymers.

In flexible polyimide/polyimide blends and fluorine-containing polyimide blends [317], some blend systems are miscible on a molecular level, but polymer blend of rod-like polyimide and flexible fluorine-containing polyimides demonstrate phase separation [318,319]. Intermolecular charge–transfer interaction plays a role in the miscibility of the blends. The miscibility has been investigated by means of DSC, SEM [320], DMA [321,322], FT-IR [319], ^{13}C CPMAS NMR [283], and fluorescence spectroscopy [317,321]. DSC and DMA measurements have shown that the blends have one single T_g, which means that the systems are miscible [317,322,323]. It is shown that the x-ray diffraction method is convenient in determining the miscibility of polyimide blends [324]. The miscibility of polyamic acid/polyimide blends depends on the imidization method. Thermal imidization of the blends retains miscibility, but chemical imidization leads to phase separation [325]. While moisture diffusion in compatible polyimide blends is faster, it is much lower than that in pure semiflexible polyimide based on fluorine dianhydride [319].

Some examples of miscible blend systems of high-performance polymers and polyimides are polybenzimidazole (PBI)/polyimide blends [326–328], polyethersulfone/polyimide blends [329,330], PBI/polyimidesulfone [331], aramid/polyimide [332], and LCP/polyimide blends [320,333–337]. PBI and PI are good candidates for precursor blends for CNFs because both polymers are thermally and chemically stable and electrospin well from organic solvents such as NMP or DMAc. There have been numerous reports on blending PBI and PI for the development of high-performance polymers, showing that PBI and PI are miscible on the molecular level because of hydrogen bonding between N–H groups of PBI and carbonyl groups of PI [328].

Systematic investigation by FT-IR spectroscopy proves the existence of hydrogen bonding interaction. Composition-dependent frequency shifts in the NH stretching bands of PBI and in the C=O stretching bands of polyimide are observed. Aramid and polyimide blends are miscible over the entire composition range, and single T_g values are detected. This is explained by the contribution of the Coleman–Panter disorientational entropy change [332]. The miscibility of the polymer blends of three different poly(aryl ether ketone)s (PAEKs) and TPI (AURUM) correlates with the percentage of ketone linkages in PAEK. The higher the percentage of ketone linkages, the higher the degree of miscibility [338] for poly(ether imide)/thermotropic copolyester (Rodrun) blends with LC polymer contents of 5%, 20%, and 40%. Previous extrusion or different injection speeds changed neither the morphology nor the mechanical properties of the blends, which improved significantly only at higher temperatures. The lack of significant increase in properties was due to the very low viscosity ratio-induced lack of fibrillation and overall poor mixing level of the blends, which was seen even in blends injected after previous extrusion [333]. The morphology of the system depends on the concentration of the compatibilizer, and the impact strengths are higher than that of the noncompatibilized system [337]. Compatibility of PEI/LCP blend systems in the presence of polyphosphazene was determined by considering the variation in glass transition temperature. Enhanced thermal stability and tensile properties point toward the superior interfacial adhesion [334].

Block copolymers based on polyimides represent a new class of polymeric materials with potential high-performance applications. There are two types of block copolymers arising from polyimides, characterized by either the *hard–hard* segments or the *hard–soft* segments. Hard–hard block copolymers (e.g., LC polyamide–polyimide [339], poly(ether ketone)–polyimide, polyimide–polyimide [340], poly(ethersulfone)–polyimide [341,342], LC poly(ester-imide)s [343], poly(aryl ether benzoxazole)–polyimide [344], and poly(imide siloxane) [345]) are prepared by the usual polycondensation method from presynthesized telechelic polymers or oligomers having terminal functional groups [346]. Poly(siloxane-imide)s [347] are also prepared by varying the average molecular weight of flexible siloxane diamine. Poly(imide siloxane)s of different soft and hard block lengths have been prepared by reacting two different diamines via two-pot solution imidization technique [348]. The mechanical properties of many block copolymers are greatly improved by the introduction of polyimides in the backbone. Multiphase morphologies are observed. In the LC block copolymers, the formation of LC phase depends on the block composition, block length, and coupling agent. The formation also requires the proper balance between rigid blocks and spacers [343].

Polyimide–polydimethylsiloxane [349–351], polyimide–polyformal [352], polyimide–poly(propylene oxide), or polyimide–poly(methyl methacrylate) block copolymers, consisting of hard and soft segments, have been prepared by the polycondensation or transimidization method [350–353].

However, series of polyimide–polydimethylsiloxane have been prepared using in situ sol–gel reaction of diethoxydimethylsilane (DEDMS) and imidization of poly(amide acid) (PAA) with higher corresponding toughness [354]. They are characterized by microphase-separated morphology and far greater flexibility than polyimide homopolymers. Polyimide–poly(propylene oxide) block copolymers could produce stable nanofoams upon thermal decomposition of the propylene oxide coblocks, which are expected to have low dielectric constants and heat-resistant properties [353].

Graft copolyimides containing poly(dimethylsiloxane)s and fluorinated alkoxy groups in the side chains are prepared by thermal and chemical dehydration two-step reactions. In addition, the graft copolymer with the poly(dimethylsiloxane) segments in the side chain is a good compatibilizer for the polymer blends of poly(dimethylsiloxane) and polyimides. It also leads to glass transition temperatures as low as −120°C and very flexible backbone, good thermal and oxidative stability, high gas permeability, excellent dielectric properties, and hydrophobic low energy surfaces with excellent water repellency, physiological inertness, and biocompatibility. Although polydimethylsiloxane is the most widely used backbone, possibility of attaching a variety of inert or reactive substituents (R) on the tetravalent silicon atom in the siloxane ($-R_2Si-O-$) backbone, together with the ease of reactive oligomer preparation through acid or base catalyzed equilibration reactions, also plays critical roles that make silicones attractive intermediates in polymer synthesis. Very low SP of polydimethylsiloxane (around 15.5 $(J/cm^3)^{1/2}$) allows the synthesis of microphase separated block and segmented copolymers with interesting structure–morphology–property behavior. Another advantage offered by silicone polymer chemistry is the possibility of using different polymerization techniques for copolymer synthesis, which includes step-growth (condensation or addition), chain-growth (free radical, ATRP, and anionic), and ring-opening polymerization methods. Cyclic silicone monomers, reactive intermediates, and functionally terminated oligomers are commercially available from various sources. Flexible silicone polymer chemistry offers unique advantages for the preparation of a wide range of tailor-designed copolymers with interesting combinations of properties, which makes them useful for applications such as biomaterials, foul release coatings, and gas separation membranes [348,355–358].

Polymer blends are often thermodynamically unstable. During chain extension, phase separation occurs, resulting in a substantial reduction of mechanical properties. IPN or semi-IPN techniques have been considered as possible approaches to solve the problem. The miscible blends of thermosetting and TPIs could be converted to IPN by various treatments [359–363]. IPNs have good processability because of the thermoplastic polymer and high-temperature properties arising from the thermosetting one. Improvements in fracture toughness, thermal properties, and synergy in processability have been reported.

16.5.5 Commercial TPIs

AURUM TPI resin, developed by Mitsui Inc. [364], EXTEM XH from SABIC Co. (invented originally by GE), and Vespel TPIs from DuPont Co. are recently developed polyimides that have received increasing attention in recent years as melt-processable, high-performance, semicrystalline thermoplastic polymers.

EXTEM XH has stable characteristics and good processability at high temperatures, showing potential for gas separation operated in harsh environments [365].

TPIs have superior mechanical properties, high-temperature stability, good solvent resistance, and good melt processability. Incorporation of flexible units and linkages in the backbone increases the molecular mobility, lowers the T_g, and improves processability. These TPIs can be processed by injection-molding or extrusion, and can be pelletized. On heating, TPI forms folded chain crystals whereby two ether bonds in the repeating unit contribute to chain folding. The supermolecular structures do not develop below the T_g, and that of solution-cast TPI specimens depends on the molecular chain length and orientation in the original as-cast film [366,367]. The x-ray structure analysis shows the unit cell dimensions to be $a = 7.89$ Å, $b = 6.29$ Å, c (fiber axis) $= 25.11$ Å, and β = 90.0°. For instance, the unit cell of AURUM was found to contain two polymer chains with 1/1 helical symmetry. The molecular conformation of TPI showed four degrees of freedom in terms of ether linkages, accounting for the comparatively lower melting temperature of TPI. The overall chain conformation of TPI was straight and close to the fully extended conformation [368]. Vespel is the trademark of this durable high-performance polyimide manufactured by DuPont. This rather expensive polymer is mostly used in aerospace, semiconductor, and transportation technologies. It combines heat resistance, lubricity, dimensional stability, and chemical and creep resistance. This polymer has a T_g of approximately 289°C, and thermal degradation of this TPI depicts a degradation stage at 579.6°C according to the degradation of an imide segment [369,370].

The elastic modulus (E_1) of TPI's crystalline region in the direction parallel to the chain axis has been found to be around 55 GPa, while in the perpendicular direction, it was found to be 3.2–3.5 GPa. This implies that no special interaction exists within the NEW-TPI [371]. Isothermal cold crystallization of the NEW-TPI shows that the Avrami exponent is 3.5 for $T_c < 330°C$ and decreases as T_c increases. Results indicate that the crystallization can be modeled according to heterogeneous nucleation and 3D crystal growth. Time-resolved small angle X-ray scattering (SAXS) shows that both crystallinity and crystal perfection increase during crystallization. The crystals formed by isothermal cold crystallization are very small, with lamellar thickness ranging from about 1.5 to 2 times the monomer repeating unit [372]. Kinetics of nonisothermal cold crystallization are considerably slow in comparison to that of other high-performance thermoplastics such as poly(phenylene sulfide) (PPS) and PEEK, and TPI has a much narrower processing window [373]. The dielectric and dynamic mechanical relaxation behaviors have been investigated to characterize the T_g relaxation and to explore the amorphous phase behavior of TPI. It is suggested that the NEW-TPI has a very small amount of tightly bound or rigid amorphous material [374]. The crystallinity and the dynamic modulus increase markedly by zone drawing [375], resulting in oriented crystallites [376]. The zone drawing also enhances the dynamic modulus for the blend of NEW-TPI/Xydar [376]. TPI has good electron radiation resistance at high temperatures, and the irradiation improves tensile strength and Young's modulus at high temperature owing to cross-linking [377]. TPI also has good flame resistance with a typical limiting oxygen index (LOI) value of 47.0.

LARC-CPI has been developed for high-performance applications such as structural adhesives and composites. It is important to understand the crystalline behavior and the morphological feature of semicrystalline LARC-CPI, inasmuch as crystallinity affects material properties such as stiffness, impact resistance, tensile strength, elongation at break, solvent and environmental resistances, optical properties, as well as processing. Dynamic glassy modulus of LARC-CPI increases from about 3.0 GPa (undrawn) to 9.5 GPa ($\lambda = 3.8$) by applying zone annealing. The zone-drawn films are highly oriented, and the crystal lattice parameters are $a = 8.0 \pm 0.2$, $b = 5.9 \pm 0.2$, and $c = 36.5 \pm 0.3$ Å. The SEM and TEM data lend support to the Avrami exponent of about 2 obtained by crystallization kinetics using DSC. The thickness of the lamellae forming shear-like stack is 100 Å. The heat of fusion for a fully crystalline LARC-CPI sample is estimated to be 125 J/g [378]. Polyimides having isomeric chemical structures also display a semicrystalline-like behavior. Three distinct crystalline morphologies, namely ellipsoid, cubic, and needle-like, which are embedded in an amorphous matrix, have been observed [379]. The CTEs of the crystalline lattice of the NEW-TPI and LARC-CPI have been compared, and the linear CTEs relating the unit cell parameters with temperature (T) to their values at 0°C appear in Table 16.3.

TABLE 16.3
CTE of the Crystalline Lattice of the NEW-TPI and LARC-CPI

NEW-TPI	LARC-CPI
$a = 7.82(1 + 93 \times 10^{-6} T)$	$a = 8.06(1 + 117 \times 10^{-6} T)$
$b = 6.36(1 + 60 \times 10^{-6} T)$	$b = 6.12(1 + 75 \times 10^{-6} T)$

LARC-CPI has a larger crystal volume expansion coefficient than NEW-TPI, meaning that the former crystal will contract more after processing [380]—NEW-TPI: $V \le 1258(1 + 153 \times 10^{-6}\,T)$ and LARC-CPI: $V \le 1855(1 + 192 \times 10^{-6}\,T)$.

The TPI, LARC-TPI, originally developed by NASA Langley Research Center, if thermally imidized becomes completely amorphous. Mitsui Toatsu developed a chemically imidized material that has a crystalline transition at about 272°C. This temperature increases with time on annealing because of the presence of residual solvents, which induce LARC-TPI's crystallization [381,382]. LARC-ITPI is an excellent high-temperature matrix material for selected future aerospace applications where solvent resistance is not a key requirement [383]. The structures of LARC-TPI and LARC-ITPI are presented in Figure 16.2.

16.6 FUNCTIONAL POLYIMIDES

16.6.1 Gas Permeability and Permselectivity

In general, polymeric membranes having high permeability tend to have a low permselectivity [384–403]. On the other hand, it is generally accepted that high T_g polymers lead to higher gas permselectivity for a pair of gases. Hence, high T_g of the polyimides is one of the important criteria to achieve a better separation performance, besides their high thermal stability and mechanical strength [404]. One of the effective design principles targets intrachain rigidity and neglects the interchain rigidity necessary for suppression of CO_2-induced plasticization, which often reduces selectivity in high-pressure mixed-gas CO_2/CH_4 separations [405,406]. For instance, it is noteworthy to mention that rigid aromatic dianhydrides 6FDA [2,2-bis(3,4-dicarboxyphenyl)hexafluoropropane] and PMDA [benzene-1,2,4,5-tetracarboxylic dianhydride]-based polyimides exhibit excellent ideal separation performances closer to the latest upper boundary limit drawn by Robeson [407]. There are several approaches that have been adopted to reduce the hydrogen bonding and to restrict close packing. It is reported in several literatures that the introduction of bulky, packing-disruptive, propeller-shaped triphenylamine (TPA) groups into the polyimide backbone not only increases the solubility and processability of the polymers by reducing the close packing and hydrogen bonding but also helps in maintaining their high thermal and mechanical property, whereas studies showed that TPA-containing polymers displayed interesting gas separation behavior [408–411].

For many gas separation applications (biogas and fuel gas) involving CO_2 capture, the gas mixture is often moist or sometimes even saturated with water vapor. Moreover, water typically swells the polymer matrix, increasing polymer chain mobility and thus reducing gas diffusion resistance. The immobilized water retained in the matrix forms transport passageways for gas permeation through the membrane. The permeability of CO_2 in water has been found to be as high as 1983 Barrer, which is three to six orders of magnitude higher than in the polymer matrix. Therefore, the presence of water vapor is expected to increase both solubility and diffusivity of the gas in the membranes. However, the hydrophobic aromatic backbone of polyimide restricts the amount of water in the matrix, thus limiting permeability increase under wet conditions. Composite membranes comprising a polymer bulk phase (continuous phase) and a filler phase (dispersed phase) provide a promising solution for the above-mentioned problems. The use of appropriate fillers with multifunctional properties provides the possibility to enhance membrane design [412]. Membranes based on the high-temperature thermal rearrangement (TR) of α-hydroxy-polyimides produce exceptional gas separation properties, with permselectivity for CO_2/CH_4 often surpassing that of other polymeric membrane types. This phenomenon is believed to be due to the conversion of the α-hydroxy-polyimide into polybenzoxazole (PBO), while at the same time producing binodal cavity sizes and increased fractional free volume within the membrane. In addition, the thermal and chemical resistance of PBOs makes them attractive for membrane gas separation. A disadvantage of the TR process is the fact that the resulting membrane has reduced mechanical strength because of increased rigidity [413–417]. Finally, the interdependence of chemical structures and gas transport properties provides an opportunity of specific tailoring of the molecular structure to obtain desired separation performance for specific applications. Therefore, selection of an appropriate monomer pair is crucial for the preparation of membranes with high selectivity for a specific application [418–422]. Many structures such as hyperbranched PI [423], idan structure [424], brominated PI [425], noncoplanar structure [426], and polyimides with bulky units [427], among others, proved to suit this application.

16.6.2 Photosensitive Polyimides

Most photosensitive polyimides (PSPIs) are developed from photosensitized precursor polyamic acids. However, these polymers have two main disadvantages: on the one hand, they require thermal treatment for imidization, promoting shrinkage; on the other hand, they have limited shelf-life stability. Currently, there are two types of common PSPIs: positive-acting resists [428–430] and negative-acting resists [431]. Most of the commercially standard negative-tone PSPI resists are based on poly(amic acid)s, in which cross-linking groups are introduced at the side chains, such as reactive methacrylates or ion linkages by acid amine. These PSPIs have different photosensitive mechanisms to form negative-type patterns using organic solvents as developers [431–433]. Block copolyimides having hydroxyl groups have been prepared with addition of the ester based on 2,3,4-trihydroxybenzophenone with 1,2-naphthoquinonediazide-5-sulfonic acid p-cresol ester (PC5) or 1,2-naphthoquinonediazide-5-sulfonic acid (NT200) as photoreactive compound [434,435]. Other studies have concentrated on the synthesis of positive-tone PSPIs in aqueous base using diazonaphthoquinone (DNQ) or DNQ derivative compounds [436–439].

Negative-type PSPIs usually contain side-chain methacryloyl or acryloyl cross-linking groups and a photosensitizer. The negative tone photoinitiator-free PSPI that incorporated the photosensitive 4,4-bis[(4-amino)thiophenyl] benzophenone (BATPB) into its backbone and methacryloyl or acryloyl groups into its side chains showed upon UV irradiation, with the BAPTB structure in the polyimide chain undergoing photolysis to produce several types of radicals that can initiate polymerization of the methacryloyl or acryloyl groups to form the cross-linked system [440]. Negative-type PSPI with side-chain acryloyl groups show good absorbance by i-line. As the polyimide side chain was converted to particular non-acryloyl groups, the polyimide demonstrated less absorbance at 635 nm [432].

Positive-working PSPIs having protective groups based on photoinduced acidolysis are prepared by the polycondensation of fluorinated dianhydride and diamines, followed by the protection of the hydroxy group on benzene ring in the backbone [441,442]. Polyimides act as positive-working photoreactive materials in the presence of a photoacid generator such as p-nitrobenzyl-9,10-diethoxyanthracene-2-sulfonate and offer an excellent pattern with good profile and high sensitivity. Polyimides having side-chain 1,2-naphthoquinone (NQD) groups have unique lithographic behavior. Depending on the content of NQD, either the positive-working mode could be observed with an aqueous base developer or the negative-working mode could be achieved with an organic solvent developer [443]. The fluorine-containing aromatic polyimides with phenolic OH group provide an aqueous-base-developable positive-working photoresist system when formulated with a DNQ sensitizer. When they are modified by the reaction of the soluble polyimides with methacryloyl chloride, they become negative-working [444]. Polyimides with pendant methylthiomethyl group based on benzophenone moiety are photosensitive, but are not thermally stable and are negative-working photoresist materials [445,446]. The pendant groups might increase the photosensitivity of benzophenone-containing polyimide system [447]. Polyisoimide containing D4SB as a photosensitive compound is a positive-type PSPI precursor, where D4SB acts as a dissolution controller [88].

Polyimides containing disilane units in the main chain are organic-soluble and photosensitive. These polyimides are obtained using the two-step method from diamines having the silane linkage and dianhydrides, or by the polycondensation of diaminosilanes and bisphenols containing imide rings. The resulting polymers are sensitive to UV irradiation, and their inherent viscosities decrease on exposure to UV irradiation, being candidates for positive-working resistant materials [448–450].

16.6.3 NLO Polyimides

Second-order NLO polyimides having side-chain chromophores with high-temperature stability are prepared by polycondensation or polyaddition in order to enhance the second harmonic generation (SHG) thermal/temporal stability. This is because polyimides display a great structural versatility as well as an impressive thermal stability and optical transparency. Polyimides prepared by the conventional two-step method followed by chemical imidization have T_g values of 160–188°C [451] and are endowed with relatively large d_{33} values (between 35.15 and 45.20 pm/V at 532 nm). Various cross-linkable or high glass transition temperature polymers have been developed to enhance the stability of dipole orientation at elevated temperatures. Therefore, NLO polymers with T_g's as high as 291°C have been prepared by covalent bonding of a chromophore to the backbone of hydroxyl polyimides via a Mitsunobu reaction. Poled films of the polymers were measured for their SHG effect and showed d_{33} values up to 20 pm/V. Some of the films show a stable NLO response even to 90% of remaining NLO intensity at 125°C during 200 h [452]. To obtain NLO polyimides exhibiting excellent thermal stability and large electro-optical (EO) coefficients, a series of novel hyperbranched polyimides from a tris(ether anhydride) [1,1,1-tris[4-(3,4-dicarboxyphenoxy)phenyl] ethane trianhydride (III)] and a difunctional chromophore [(2,4-diamino-4′-(4-nitrophenyldiazenyl) azobenzene) (DNDA)] were synthesized through ring-opening polyaddition at room temperature to form poly(amic acid)s, followed by curing process. The presence of the imide rings in the hyperbranched NLO polymers imparted them with excellent thermal stability. Moreover, their highly hyperbranched architectures resulted in site-isolation effects, which restricted the aggregation of the chromophore units. All of these hyperbranched polymers exhibited larger EO coefficients (14.6–17.2 pm/V) and better temporal stability at 120°C than those of linear NLO polymers. In addition, waveguide properties for hyperbranched NLO polymers (3.4–4.6 dB cm^{-1} at 1310 nm) were also observed. Relative to the analogous linear NLO polyimide, the hyperbranched NLO polyimides exhibited superior temporal stability, larger EO coefficients, and lower optical losses [453].

The SHG signal is stable enough to withstand a high temperature environment for a long period [454–457]. The polyimide reveals only minor decay in SHG efficiency over the first 24 h at 85°C. Long-term stability of the NLO property is observed at 170°C after 150 h. NLO polyimide synthesized by posttricyanovinylation of the parent polymer (T_g: 158°C) with tetracyanoethylene (TCNE) has a good thermal stability. A polyimide containing 32% m/m NLO shows a d_{33} value of 16 pm/V. Poled side-chain polyimide treated at 90°C for over 1000 h displays an EO activity with negligible change during heating [457,458].

16.6.4 Electrical Insulating and Electrolyte Polyimides

Owing to their excellent electrical resistivity (1.4 × 10^{17} Ω·cm at 23°C for Kapton polyimide), PIs are suitable materials for electrical insulating purposes. Prepared conventional PI films (CPI) with different combinations of monomers depict that PMDA-DDE has the lowest dielectric constant and dissipation factor among the prepared CPI films. PMDA-based films have shown a higher insulation than BTDA-based ones. It is clear that BTDA has the more polarizable groups than PMDA [459]. The decrease in conductivity has also been achieved using nanofillers such as silica and alumina [460,461] or other nano-inorganic fillers [462].

Despite drawbacks such as high cost, low conductivity at dry condition, and high methanol permeability, polymeric membranes are established as the most promising membranes in fuel cells as electrolytes. Although perfluorinated ionomer membranes (Nafion) are highly proton-conductive and chemically and physically stable at moderate temperatures, their high gas permeability, high cost, and environmental inadaptability limit their application in fuel cell production. In proton conductive aliphatic/aromatic polyimide ionomers, the aliphatic segments introduced both in the main and side chains could effectively improve the hydrolytic stability of polyimide ionomers without sacrificing other properties such as proton conductivity and oxidative and mechanical stability

[272]. Sulfonated polyimides have also been extensively investigated [463–466]. For instance, sulfonated polybenzimidazoles exhibit superior performance when compared to Nafion (currently the most used membrane in fuel cells) at higher temperatures [467,468].

REFERENCES

1. Kurita, K. et al., Synthesis and properties of an open-chain polyimide from benzyloxyamine and terephthaloyl chloride. *Journal of Polymer Science Part A: Polymer Chemistry*, 1989. **27**(13): pp. 4297–4303.
2. Bogert, M.T. and R.R. Renshaw, 4-amino-0-phthalic acid and some of its derivatives. 1. *Journal of the American Chemical Society*, 1908. **30**(7): pp. 1135–1144.
3. Edwards, W.M. and R.I. Maxwell, Polyimides of pyromellitic acid. 1955, US Patent 2,710,853.
4. Gresham, W.F. and M.A. Naylor Jr., Novel polyimides. 1956, US Patent 2,731,447.
5. Prince, M. and J. Hornyak, High pressure reactions. IV. Poly(pyromellitamic acid)s. *Journal of Polymer Science: Polymer Letters Edition*, 1966. **4**: pp. 601–604.
6. E. I. du Pont de Nemours & Co., Polyimides of pyromellitic acid. 1956, Br. Patent 762,152.
7. E. I. du Pont de Nemours & Co., Polyimides of pyromellitic acid. 1958, Ger. Patent 1,031,510.
8. Liaw, D.-J. et al., Advanced polyimide materials: Syntheses, physical properties and applications. *Progress in Polymer Science*, 2012. **37**(7): pp. 907–974.
9. Saeed, M.B. and M.-S. Zhan, Adhesive strength of nano-size particles filled thermoplastic polyimides. Part-I: Multi-walled carbon nano-tubes (MWNT)–polyimide composite films. *International Journal of Adhesion and Adhesives*, 2007. **27**(4): pp. 306–318.
10. Saeed, M.B. and M.-S. Zhan, Adhesive strength of nano-size particles filled thermoplastic polyimides. Part-II: Aluminum nitride (AlN) nano-powder–polyimide composite films. *International Journal of Adhesion and Adhesives*, 2007. **27**(4): pp. 319–329.
11. Samyn, P. et al., The sliding behaviour of sintered and thermoplastic polyimides investigated by thermal and Raman spectroscopic measurements. *Wear*, 2008. **264**(9–10): pp. 869–876.
12. Wilson, D., H.D. Stenzenberger, and P.M. Hergenrother, *Polyimides*. 1990, Munich: Springer.
13. Abadie, M.J. and B. Sillion, Polyimides and other high-temperature polymers. *Proceedings of the 2nd European Technical Symposium on Polyimides and High-Temperature Polymers (STEPI 2), Montpellier, France, June 4–7, 1991*. 1991, Elsevier Science Ltd.
14. Yang, C.P. and J.H. Lin, Syntheses and properties of aromatic polyamides and polyimides derived from 9,9-bis [4-(p-aminophenoxy) phenyl] fluorene. *Journal of Polymer Science Part A: Polymer Chemistry*, 1993. **31**(8): pp. 2153–2163.
15. Liou, G.-S. et al., Synthesis and properties of new aromatic poly(amine-imide)s derived from N,N′-bis(4-aminophenyl)-N,N′-diphenyl-1,4-phenylenediamine. *Journal of Polymer Science Part A: Polymer Chemistry*, 2002. **40**(21): pp. 3815–3822.
16. Grenier-Loustalot, M.-F., F. Joubert, and P. Grenier, Mechanisms and kinetics of the polymerization of thermoplastic polyimides. I, Study of the pyromellitic anhydride/aromatic amine system. *Journal of Polymer Science. Part A. Polymer Chemistry*, 1991. **29**(11): pp. 1649–1660.
17. Grenier-Loustalot, M.-F., F. Joubert, and P. Grenier, Mechanisms and kinetics of polymerization of thermoplastic polyimides. II. Study of "bridged" dianhydride/aromatic amine systems. *Journal of Polymer Science A Polymer Chemistry*, 1993. **31**: pp. 3049–3063.
18. Morikawa, A. et al., Synthesis and characterization of polyimide from 4,4′-diamino diphenyl ether having substituents at 2,2′-position. *Journal of Photopolymer Science and Technology*, 2013. **26**(3): pp. 367–372.
19. Morikawa, A., F. Miyata, and J. Nishimura, Synthesis and properties of polyimides from 1,4-bis(4-amino-2-phenylphenoxy) benzene and 4,4′-bis(4-amino-2-phenylphenoxy) biphenyl. *High Performance Polymers*, 2012. **24**(8): pp. 783–792.
20. Ogata, Y. et al., Aggregation behavior of organic-soluble semi-aromatic polyimides in N-methyl pyrrolidone systems. *Journal of Polymer Science Part B: Polymer Physics*, 2012. **50**(18): pp. 1312–1320.
21. Chen, J.-C. et al., Highly phenylated polyimides containing 4,4′-diphenylether moiety. *Reactive and Functional Polymers*, 2014. **78**: pp. 23–31.
22. Tsuda, Y., Y. Matsuda, and T. Matsuda, Soluble polyimides bearing long-chain alkyl groups on their side chain via polymer reaction. *International Journal of Polymer Science*, 2012. Article ID 972541, 10 p.
23. Hasanain, F. and Z.Y. Wang, New one-step synthesis of polyimides in salicylic acid. *Polymer*, 2008. **49**(4): pp. 831–835.

24. Kuznetsov, A.A. et al., New alternating copolyimides by high temperature synthesis in benzoic acid medium. *High Performance Polymers*, 2004. **16**(1): pp. 89–100.
25. Feiring, A., B. Auman, and E. Wonchoba, Synthesis and properties of fluorinated polyimides from novel 2,2′-bis(fluoroalkoxy) benzidines. *Macromolecules*, 1993. **26**(11): pp. 2779–2784.
26. McGrath, J. et al., Synthesis and blend behavior of high performance homo-and segmented thermoplastic polyimides, in *Makromolekulare Chemie. Macromolecular Symposia*. 1991, Wiley Online Library.
27. Kishanprasad, V. and P. Gedam, Polyamic acids: Thermal and microwave imidization and film properties. *Journal of Applied Polymer Science*, 1993. **50**(3): pp. 419–429.
28. Kim, Y. et al., Kinetic and mechanistic investigations of the formation of polyimides under homogeneous conditions. *Macromolecules*, 1993. **26**(6): pp. 1344–1358.
29. Kaneda, T. et al., High-strength–high-modulus polyimide fibers I. One-step synthesis of spinnable polyimides. *Journal of Applied Polymer Science*, 1986. **32**(1): pp. 3133–3149.
30. Harris, F.W. and S.L.-C. Hsu, Synthesis and characterization of polyimides based on 3,6-diphenylpyromellitic dianhydride. *High Performance Polymers*, 1989. **1**(1): pp. 3–16.
31. Buzin, P.V. et al., New AB polyetherimides obtained by direct polycyclocondensation of aminophenoxy phthalic acids. *High Performance Polymers*, 2004. **16**(4): pp. 505–514.
32. Kuznetsov, A.A. et al., High temperature polyimide synthesis in "active" medium: Reactivity leveling of the high and the low basic diamines. *High Performance Polymers*, 2007. **19**(5): pp. 711–721.
33. Kuznetsov, A.A., A.Y. Tsegel'skaya, and P.V. Buzin, One-pot high-temperature synthesis of polyimides in molten benzoic acid: Kinetics of reactions modeling stages of polycondensation and cyclization. *Polymer Science Series A*, 2007. **49**(11): pp. 1157–1164.
34. Myung, B.Y., C.J. Ahn, and T.H. Yoon, Synthesis and characterization of polyimides from novel 1-(3′,5′-bis(trifluoromethyl)benzene) pyromelliticdianhydride (6FPPMDA). *Polymer*, 2004. **45**(10): pp. 3185–3193.
35. Kuznetsov, A.A., One-pot polyimide synthesis in carboxylic acid medium. *High Performance Polymers*, 2000. **12**(3): pp. 445–460.
36. Sroog, C.E., Polyimides. *Progress in Polymer Science*, 1991. **16**(4): pp. 561–694.
37. Ho Choi, K. et al., Synthesis and characterization of new alkali-soluble polyimides and preparation of alternating multilayer nano-films therefrom. *Polymer*, 2004. **45**(5): pp. 1517–1524.
38. Jin, L. et al., Homogenous one-pot synthesis of polyimides in polyphosphoric acid. *European Polymer Journal*, 2009. **45**(10): pp. 2805–2811.
39. Sakaguchi, Y. and Y. Kato, Synthesis of polyimide and poly(imide-benzoxazole) in polyphosphoric acid. *Journal of Polymer Science Part A: Polymer Chemistry*, 1993. **31**(4): pp. 1029–1033.
40. Silvi, N. et al., Method for preparing polyimide and polyimide prepared thereby. 2006, Patent WO2005037890 A1.
41. Xanthos, M., *Reactive extrusion: Principles and practice*. 1992, Munich: Hanser Publishers, pp. 82–84.
42. Artem'eva, V.N. et al., Investigation of polyimide formation from the complexes of the diester of benzophenonetetracarboxylic acid with diamines 7. Role of theo-carboxylic group of acid diester in the formation and imidization of H-complexes. *Russian Chemical Bulletin*, 1995. **44**(6): pp. 1021–1026.
43. Naiini, A. et al., Process for making polyimides from diamines and tetracarboxylic diacid diester. 1998, US Patent 5,789,525.
44. Yang, C.P. and W.T. Chen, Synthesis and properties of novel aromatic polyimides of 2,3-bis(4-aminophenoxy)naphthalene. *Macromolecules*, 1993. **26**(18): pp. 4865–4871.
45. Stoffel, N.C. et al., Solvent and isomer effects on the imidization of pyromellitic dianhydrideoxydianiline-based poly(amic ethyl ester)s. *Polymer*, 1993. **34**(21): pp. 4524–4530.
46. Sęk, D., P. Pijet, and A. Wanic, Investigation of polyimides containing naphthalene units: 1. Monomer structure and reaction conditions. *Polymer*, 1992. **33**(1): pp. 190–193.
47. Kumagai, Y. et al., A facile solid-state synthesis of crystalline oxydianiline-based polypyromellitimides via salt monomers. *Polymer*, 1995. **36**(14): pp. 2827–2833.
48. Itoya, K. et al., High-pressure synthesis of aliphatic polyimides via salt monomers composed of aliphatic diamines and oxydiphthalic acid. *Macromolecules*, 1994. **27**(15): pp. 4101–4105.
49. Becker, K.H. and H.W. Schmidt, Para-linked aromatic poly(amic ethyl esters): Precursors to rodlike aromatic polyimides. 1. Synthesis and imidization study. *Macromolecules*, 1992. **25**(25): pp. 6784–6790.
50. Houlihan, F. et al., Synthesis and characterization of the tert-butyl ester of the oxydianiline/pyromellitic dianhydride polyamic acid. *Macromolecules*, 1989. **22**(12): pp. 4477–4483.
51. Ueda, M. and H. Mori, Synthesis of polyamic acid di-tert-butyl esters by direct polycondensation of di-tert-butyl esters of tetracarboxylic acids with diamines using diphenyl (2,3-dihydro-2-thioxo-3-benzoxazolyl) phosphonate. *Die Makromolekulare Chemie*, 1993. **194**(2): pp. 511–521.

52. Volksen, W. et al., Base-catalyzed cyclization of ortho-aromatic amide alkyl esters: A novel approach to chemical imidization. *Polymers for Microelectronics*, 1994. **537**: pp. 403–416.
53. Imai, Y. and K. Itoya, High pressure polymerization and simultaneous processing for high temperature polymers. *Kino Zairyo (Function and Materials) (Japan)*, 1992. **12**(12): pp. 32–39.
54. Itoya, K. et al., Synthesis of polyimides by high-pressure polycondensation of aromatic tetracarboxylic acids with aliphatic diamines. *Polymer Preprints, Japan*, 1992. **41**: pp. 355–360.
55. Takekoshi, T., Synthesis of Polyimides, in *Kirk-Othmer encyclopedia of chemical technology*, Othmer, K., ed. Vol. 19. 1996, New York: John Wiley & Sons, pp. 813–837.
56. Verbicky, J.W., Polyimides, in *Encyclopedia of polymer science & technology*, Mark, H., Bikales, N., Overberger, C., Menges, G., Kroschwitz, J., eds. Vol. 12. 2004, New York: John Wiley & Sons, pp. 364–366.
57. Ghosh, M. and A. Barbalata, *Polyimides: Fundamentals and applications*. 1996, New York: Marcel Dekker.
58. Feldman, D. and A. Barbalata, *Synthetic polymers: Technology, properties, applications*. 1996, New York: Chapman and Hall.
59. Pan, J.Q. et al., Synthesis and properties of new copolymers containing hindered amine. *Journal of Applied Polymer Science*, 1996. **61**(8): pp. 1405–1412.
60. Sendijarević, A.A. et al., Synthesis and properties of urethane-modified polyimides. *Journal of Polymer Science Part A: Polymer Chemistry*, 1990. **28**(13): pp. 3603–3615.
61. Avadhani, C., P. Wadgaonkar, and S. Vernekar, Synthesis and characterization of oxyethylene containing diisocyanates and polyimides therefrom. *Journal of Polymer Science Part A: Polymer Chemistry*, 1990. **28**(7): pp. 1681–1691.
62. Wenzel, M., M. Ballauff, and G. Wegner, Rigid rod polymers with flexible side chains, 4. Synthesis, structure and phase behaviour of polyimides prepared from pyromellitic anhydride and 2,5-di-n-alkoxy-1,4-phenylene diisocyantes. *Die Makromolekulare Chemie*, 1987. **188**(12): pp. 2865–2873.
63. Helmer-Metzmann, F. et al., Rigid rod polymers with flexible side chains, 7. Synthesis and phase behaviour of polyimides prepared from pyromellitic anhydride and N,N′-bis(1-imidazolylcarbonyl)-2,5-bis (alkoxyalkoxy-1,4-phenylenediamines). *Die Makromolekulare Chemie*, 1989. **190**(5): pp. 985–994.
64. Yeganeh, H., B. Tamami, and I. Ghazi, A novel direct method for preparation of aromatic polyimides via microwave-assisted polycondensation of aromatic dianhydrides and diisocyanates. *European Polymer Journal*, 2004. **40**(9): pp. 2059–2064.
65. Brunel, R. et al., Assisted microwave synthesis of high molecular weight poly(aryletherketone)s. *High Performance Polymers*, 2008. **20**(2): pp. 185–207.
66. Korshak, V. et al., Synthesis and investigation of the properties of cardo-copolyimides with different microstructure. *Die Makromolekulare Chemie*, 1983. **184**(2): pp. 235–252.
67. Oishi, Y. et al., Synthesis and properties of hydroxyl-containing aromatic polyimides from trimethylsilylated 2,4-diaminophenol and aromatic tetracarboxylic dianhydrides. *Journal of Polymer Science Part A: Polymer Chemistry*, 1993. **31**(1): pp. 293–296.
68. Munoz, D.M. et al., Experimental and theoretical study of an improved activated polycondensation method for aromatic polyimides. *Macromolecules*, 2007. **40**(23): pp. 8225–8232.
69. Kishanprasad, V.S. and P.H. Gedam, Synthesis and characterization of polyester-imides from imidodicarboxylic acid monomers and ethylene glycol. *Journal of Applied Polymer Science*, 1993. **48**(7): pp. 1151–1162.
70. de Abajo, J. and J.G. de la Campa, Processable aromatic polyimides, in *Progress in polyimide chemistry I*, Kricheldorf, H.R., eds. 1999, Berlin, Heidelberg: Springer, pp. 23–59.
71. Venkatesan, D. and M. Srinivasan, Synthesis and characterization of polyesterimides containing ether linkages. *Journal of Macromolecular Science, Part A: Pure and Applied Chemistry*, 1993. **30**(11): pp. 801–814.
72. Orzeszko, A. and K. Mirowski, Synthesis and structural studies of new copolyimides. *Die Makromolekulare Chemie*, 1990. **191**(3): pp. 701–707.
73. Li, C.H., C.C. Chen, and K.M. Chen, Studies on the synthesis and properties of copolyesterimide. *Journal of Applied Polymer Science*, 1994. **52**(12): pp. 1751–1757.
74. Staubli, A. et al., Characterization of hydrolytically degradable amino acid containing poly (anhydride-co-imides). *Macromolecules*, 1991. **24**(9): pp. 2283–2290.
75. Staubli, A., E. Mathiowitz, and R. Langer, Sequence distribution and its effect on glass transition temperatures of poly(anhydride-co-imides) containing asymmetric monomers. *Macromolecules*, 1991. **24**(9): pp. 2291–2298.

76. Yeganeh, H. and M.A. Shamekhi, Poly(urethane-imide-imide), a new generation of thermoplastic polyurethane elastomers with enhanced thermal stability. *Polymer*, 2004. **45**(2): pp. 359–365.
77. Fernández-García, M. et al., Free radical copolymerization of 2-hydroxyethyl methacrylate with butyl methacrylate: Determination of monomer reactivity ratios and glass transition temperatures. *Polymer*, 2000. **41**(22): pp. 8001–8008.
78. Parthiban, A. and N. Sundaram, Preparation and characterization of Co-and homopoly(phenylene etherimideketone)s. *Journal of Polymer Science Part A: Polymer Chemistry*, 1993. **31**(5): pp. 1233–1241.
79. Zhang, S. et al., Study on synthesis and characterization of novel polyimides derived from 2,6-bis(3-aminobenzoyl) pyridine. *European Polymer Journal*, 2005. **41**(5): pp. 1097–1107.
80. Perry, R.J. and S.R. Turner, Palladium catalyzed routes to polyimides, in *Makromolekulare Chemie. Macromolecular Symposia*. 1992, Wiley Online Library.
81. Yoneyama, M., T. Nasa, and K. Arai, Novel synthesis of linear polyimides by palladium-catalyzed carbonylation polycondensation of aromatic diamides, aromatic dibromides, and carbon monoxide. *Polymer Journal*, 1998. **30**(9): pp. 697–701.
82. Perry, R.J., R.S. Turner, and R.W. Blevins, Preparation of aromatic polyimides from CO, primary diamine and bis (halo aromatic carboxylic acid ester). 1993, US Patent 5,216,118.
83. Perry, R.J., S.R. Turner, and R.W. Blevins, Palladium-catalyzed formation of poly(imide-amides). 1. Reactions with diiodo imides and diamines. *Macromolecules*, 1994. **27**(15): pp. 4058–4062.
84. Perry, R.J. et al., Synthesis of polyimides via the palladium-catalyzed carbonylation of bis (o-iodo esters) and diamines. *Macromolecules*, 1995. **28**(10): pp. 3509–3515.
85. Jeon, E. and T.-H. Yoon, Synthesis and characterization of star-shaped imide compounds. *Rapid Communication in Photoscience*, 2012. **1**: pp. 19–20.
86. Reinecke, M. and H. Ritter, Renewable resources, 2. Poly-Diels-Alder additions with disorboylamides as bisdienes and a dimaleoylamide as bisdienophile. *Macromolecular Chemistry and Physics*, 1994. **195**(7): pp. 2445–2455.
87. Mochizuki, A., T. Teranishi, and M. Ueda, Preparation and adhesion properties of polyisoimide as a high-temperature adhesive. *Polymer*, 1994. **35**(18): pp. 4022–4027.
88. Mochizuki, A. et al., Positive-working alkaline-developable photosensitive polyimide precursor based on polyisoimide using diazonaphthoquinone as a dissolution inhibitor. *Polymer*, 1995. **36**(11): pp. 2153–2158.
89. Mochizuki, A., T. Teranishi, and M. Ueda, Preparation and properties of polyisoimide as a polyimide-precursor. *Polymer Journal*, 1994. **26**(3): pp. 315–323.
90. Leu, C.-M., Y.-T. Chang, and K.-H. Wei, Polyimide-side-chain tethered polyhedral oligomeric silsesquioxane nanocomposites for low-dielectric film applications. *Chemistry of Materials*, 2003. **15**(19): pp. 3721–3727.
91. Leu, C.-M. et al., Synthesis and dielectric properties of polyimide-chain-end tethered polyhedral oligomeric silsesquioxane nanocomposites. *Chemistry of Materials*, 2003. **15**(11): pp. 2261–2265.
92. Leu, C.-M., Y.-T. Chang, and K.-H. Wei, Synthesis and dielectric properties of polyimide-tethered polyhedral oligomeric silsesquioxane (POSS) nanocomposites via POSS-diamine. *Macromolecules*, 2003. **36**(24): pp. 9122–9127.
93. Wu, S. et al., Synthesis and characterization of semiaromatic polyimides containing POSS in main chain derived from double-decker-shaped silsesquioxane. *Macromolecules*, 2007. **40**(16): pp. 5698–5705.
94. Hamciuc, C. et al., Silicon-containing heterocyclic polymers and thin films made therefrom. *Journal of Applied Polymer Science*, 2006. **102**(3): pp. 3062–3068.
95. Leng, W. et al., Synthesis and characterization of nonlinear optical side-chain polyimides containing the benzothiazole chromophores. *Macromolecules*, 2001. **34**(14): pp. 4774–4779.
96. Leng, W. et al., Synthesis of nonlinear optical polyimides containing benzothiazole moiety and their electro-optical and thermal properties. *Polymer*, 2001. **42**(22): pp. 9253–9259.
97. Hsiao, S.-H. and L.-M. Chang, Synthesis and properties of poly(imide-hydrazide)s and poly(amide-imide-hydrazide)s from N-[p-(or m-)carboxyphenyl]trimellitimide and aromatic dihydrazides or p-aminobenzhydrazide via the phosphorylation reaction. *Journal of Polymer Science Part A: Polymer Chemistry*, 2000. **38**(9): pp. 1599–1608.
98. Liou, G.-S., S.-H. Hsiao, and H.-W. Chen, Novel high-Tg poly(amine-imide)s bearing pendent N-phenylcarbazole units: Synthesis and photophysical, electrochemical and electrochromic properties. *Journal of Materials Chemistry*, 2006. **16**(19): pp. 1831–1842.
99. Niu, H. et al., New perylene polyimides containing p-n diblocks for sensitization in TiO_2 solar cells. *Polymers for Advanced Technologies*, 2004. **15**(12): pp. 701–707.

100. Huang, W. et al., Synthesis and characterization of highly soluble fluorescent main chain copolyimides containing perylene units. *European Polymer Journal*, 2003. **39**(6): pp. 1099–1104.
101. Chen, S. et al., Synthesis and properties of novel side-chain-sulfonated polyimides from bis[4-(4-aminophenoxy)-2-(3-sulfobenzoyl)]phenyl sulfone. *Polymer*, 2006. **47**(8): pp. 2660–2669.
102. Luo, J. et al., A side-chain dendronized nonlinear optical polyimide with large and thermally stable electrooptic activity. *Macromolecules*, 2004. **37**(2): pp. 248–250.
103. Bailey, W.J., *Diels-Alder polymerization. Step growth polymerization.* Vol. 279. 1972, New York: Marcel Dekker.
104. Carruthers, W., *Cycloaddition reactions in organic synthesis.* 1990, Oxford: Pergamon Press.
105. Fringuelli, F. and A. Taticchi, *Dienes in the Diels-Alder reaction.* 1990, New York: John Wiley & Sons.
106. Gheneim, R., C. Perez-Berumen, and A. Gandini, Diels–Alder reactions with novel polymeric dienes and dienophiles: Synthesis of reversibly cross-linked elastomers. *Macromolecules*, 2002. **35**(19): pp. 7246–7253.
107. Smith, J. and R. Ottenbrite, Preparation of polyimides utilizing the Diels–Alder reaction, II: Bis [N-(butadienyl-2-methyl)] amino arylenes with bismaleimides. *Polymers for Advanced Technologies*, 1992. **3**(7): pp. 373–381.
108. Kuramoto, N., K. Hayashi, and K. Nagai, Thermoreversible reaction of Diels–Alder polymer composed of difurufuryl adipate with bismaleimidodiphenylmethane. *Journal of Polymer Science Part A: Polymer Chemistry*, 1994. **32**(13): pp. 2501–2504.
109. Diakoumakos, C.D. and J.A. Mikroyannidis, Polyimides derived from Diels–Alder polymerization of furfuryl-substituted maleamic acids or from the reaction of bismaleamic with bisfurfurylpyromellitamic acids. *Journal of Polymer Science Part A: Polymer Chemistry*, 1992. **30**(12): pp. 2559–2567.
110. Mikroyannidis, J.A., Furyl-maleimide in situ generated ab-monomers: Synthesis, characterization, and Diels-Alder polymerization. *Journal of Polymer Science Part A: Polymer Chemistry*, 1992. **30**(9): pp. 2017–2024.
111. Alhakimi, G. and E. Klemm, Synthesis of a tetra (maleimide) as intermediate compound in a linear Diels–Alder polyaddition of bismaleimides with bis(2-pyrone)s. *Journal of Polymer Science Part A: Polymer Chemistry*, 1995. **33**(5): pp. 767–770.
112. Alhakimi, G., E. Klemm, and H. Görls, Synthesis of new polyimides by Diels–Alder reaction of bis (2-pyrone)s with bismaleimides. *Journal of Polymer Science Part A: Polymer Chemistry*, 1995. **33**(7): pp. 1133–1142.
113. Meador, M.A.B., D.C. Malarik, and M.A. Olshavsky, Processing studies and thermal stability of addition polymers from 1,4,5,8-tetrahydro-1,4; 5,8-diepoxyanthracene and Bis-dienes. *Journal of Polymer Science Part A: Polymer Chemistry*, 1992. **30**(2): pp. 305–312.
114. He, X., V.R. Sastri, and G.C. Tesoro, 1,4-Bis(5-methylfurfuryl)benzene: Polymerization with siloxane containing dimaleimides. *Die Makromolekulare Chemie, Rapid Communications*, 1988. **9**(3): pp. 191–194.
115. Nuyken, O., G. Maier, and K. Burger, Polymerization by [3 + 2] cycloaddition, 3. Polyimides. *Die Makromolekulare Chemie*, 1989. **190**(8): pp. 1953–1965.
116. Meador, M.A.B. et al., Evidence for thermal dehydration occurring in Diels–Alder addition polymers. *Macromolecules*, 1989. **22**(11): pp. 4385–4387.
117. Sun, K.K., Novel polymerization reaction: Double cycloaddition of sydnone and bismaleimide. *Macromolecules*, 1987. **20**(4): pp. 726–729.
118. Swint, S. and M.A. Buese, Synthesis of poly (imidosiloxanes) via disiloxane equilibration reactions. *Macromolecules*, 1990. **23**(21): pp. 4514–4518.
119. Cella, J.A. et al., Siloxane equilibration during the condensation reactions of organosilicon functional amines and anhydrides. *Macromolecules*, 1992. **25**(23): pp. 6355–6360.
120. Chi, J. H. et al., Synthesis of new alicyclic polyimides by Diels–Alder polymerization. *Journal of Applied Polymer Science*, 2007. **106**(6): pp. 3823–3832.
121. Patel, Y.S. and H.S. Patel, Thermoplastic-thermosetting merged polyimides via furan-maleimide Diels–Alder polymerization. *Arabian Journal of Chemistry*, 2013, doi:10.1016/j.arabjc.2013.04.010.
122. Gaina, V. and C. Gaina, Synthesis and characterization of poly(ester-urethane-imide)s by Diels–Alder polyaddition. *Polymer-Plastics Technology and Engineering*, 2002. **41**(3): pp. 523–540.
123. Bergmann, E.D., D. Ginsburg, and R. Pappo, The Michael Reaction. *Organic reactions*. 1959. **10**: pp. 179–555.
124. Kurmanaliev, M., E.E. Ergozhin, and I.K. Izteleuova, New poly(bismaleimide-diamine)s containing crown-ether groups in the main chain. *Die Makromolekulare Chemie*, 1993. **194**(10): pp. 2655–2661.
125. Laurienzo, P. et al., Synthesis and characterization of new siloxane-modified addition polyimides. *Macromolecular Chemistry and Physics*, 1994. **195**(9): pp. 3057–3065.

126. Misra, A.C. and G. Tesoro, Synthesis and properties of octafluoro-benzidine bis-maleimide and of its reaction products with fluorinated diamines. *Polymer*, 1992. **33**(5): pp. 1083–1089.
127. Di Bella, S. et al., Film polymerization—A new route to the synthesis of insoluble polyimides containing functional nickel(II) Schiff base units in the main chain. *European Journal of Inorganic Chemistry*, 2004. **2004**(13): pp. 2701–2705.
128. Hsiao, S.-H. and Y.-T. Chou, Synthesis and electrochromic properties of aromatic polyimides bearing pendent triphenylamine units. *Polymer*, 2014. **55**(10): pp. 2411–2421.
129. Yang, C.-P. and Y.-Y. Su, Colorless polyimides from 2,3,3′,4′-biphenyltetracarboxylic dianhydride (α-BPDA) and various aromatic bis(ether amine)s bearing pendent trifluoromethyl groups. *Polymer*, 2005. **46**(15): pp. 5797–5807.
130. Kim, Y.-H., H.-S. Kim, and S.-K. Kwon, Synthesis and characterization of highly soluble and oxygen permeable new polyimides based on twisted biphenyl dianhydride and spirobifluorene diamine. *Macromolecules*, 2005. **38**(19): pp. 7950–7956.
131. Kim, Y.J. et al., Soluble rigid rod-like polyimides and polyamides containing curable pendent groups. *Polymer*, 2005. **46**(12): pp. 3992–4004.
132. Liaw, D.-J. et al., High thermal stability and rigid rod of novel organosoluble polyimides and polyamides based on bulky and noncoplanar naphthalene-biphenyldiamine. *Macromolecules*, 2005. **38**(9): pp. 4024–4029.
133. Mikroyannidis, J.A., Aromatic polyamides and polyimides with benzoxazole or benzothiazole pendent groups prepared from 5-(2-benzoxazole)- or 5-(2-benzothiazole)-1,3-phenylenediamine. *Macromolecules*, 1995. **28**(15): pp. 5177–5183.
134. Akutsu, F. et al., Preparation of polyimides from 4,5-bis(4-aminophenyl)-2-phenylimidazole and aromatic tetracarboxylic acid dianhydrides. *Macromolecular Rapid Communications*, 1994. **15**(5): pp. 411–415.
135. Giesa, R. et al., Synthesis and thermal properties of aryl-substituted rod-like polyimides. *Journal of Polymer Science Part A: Polymer Chemistry*, 1993. **31**(1): pp. 141–151.
136. Jeong, H.-J. et al., Synthesis and characterization of novel aromatic polyimides from 3,4-bis(4-aminophenyl)-2,5-diphenylthiophene and aromatic tetracarboxylic dianhydrides. *Polymer Journal*, 1994. **26**(3): pp. 373–377.
137. Moy, T. et al., Synthesis of soluble, thermosetting polyimides derived from 1,4-phenylenebis(phenylmaleic anhydride) and aromatic diamines. *Journal of Polymer Science Part A: Polymer Chemistry*, 1994. **32**(12): pp. 2377–2385.
138. Harris, F. and Y. Sakaguchi, Soluble aromatic polyimides derived from new phenylated diamines (retroactive coverage). *Polymeric Materials: Science and Engineering*, 1989. **60**: pp. 187–191.
139. Trofimenko, S. and B.C. Auman, Polyimides based on 9,9-disubstituted xanthene dianhydrides. *Macromolecules*, 1994. **27**(5): pp. 1136–1146.
140. Dotcheva, D., M. Klapper, and K. Müllen, Soluble polyimides containing perylene units. *Macromolecular Chemistry and Physics*, 1994. **195**(6): pp. 1905–1911.
141. Falcigno, P.A., S. Jasne, and M. King, Fully imidized soluble polyimides based on a novel dianhydride: 2,2′-dichloro-4,4′,5,5′-benzophenone tetracarboxylic dianhydride. *Journal of Polymer Science Part A: Polymer Chemistry*, 1992. **30**(7): pp. 1433–1441.
142. Kasashima, Y. et al., Synthesis and properties of aromatic polyamides and polyimides from 9,10-dihydro-9,10-o-benzenoanthracene-1,4-diamine. *Polymer Journal*, 1994. **26**(10): pp. 1179–1185.
143. Auman, B.C. et al., Synthesis of a new fluoroalkylated diamine, 5-[1H,1H-2-bis(trifluoromethyl)-heptafluoropentyl]-1,3-phenylenediamine, and polyimides prepared therefrom. *Polymer*, 1995. **36**(3): pp. 651–656.
144. Yuki, Y., Ü. Tunca, and H. Kunisada, New comb-like aromatic polyamides and polyimides containing 1,3,5-triazine rings in their side chains. *Polymer Journal*, 1990. **22**(10): pp. 945–950.
145. Lin, J.-K. et al., Synthesis and characterization of new comb-like aromatic polymers based on 2,4-bis(amino-N-octadecylanilino)-6-substituted-1,3,5-triazines. *Polymer Journal*, 1990. **22**(1): pp. 47–55.
146. Lin, J.-K. et al., Synthesis and characterization of new comb-like aromatic polyamides, polyimides, and polyureas containing 1,3,5-triazine rings in their side chains. *Polymer Journal*, 1989. **21**(9): pp. 709–717.
147. Moy, T. and J. McGrath, Synthesis of hydroxyl-containing polyimides derived from 4,6-diaminoresorcinol dihydrochloride and aromatic tetracarboxylic dianhydrides. *Journal of Polymer Science Part A: Polymer Chemistry*, 1994. **32**(10): pp. 1903–1908.
148. Hasegawa, M. et al., Colorless polyimides with low coefficient of thermal expansion derived from alkyl-substituted cyclobutanetetracarboxylic dianhydrides. *Polymer International*, 2014. **63**(3): pp. 486–500.

149. Ghassemi, H. and A.S. Hay, Red pigmentary polyimides from N,N′-diamino-3,4,9,10-perylenetetracarboxylic acid bisimide. *Macromolecules*, 1994. **27**(15): pp. 4410–4412.
150. Hergenrother, P.M. and S.J. Havens, Polyimides containing carbonyl and ether connecting groups. II. *Journal of Polymer Science Part A: Polymer Chemistry*, 1989. **27**(4): pp. 1161–1174.
151. Hergenrother, P., M. Beltz, and S. Havens, Polyimides containing carbonyl and ether connecting groups. III. *Journal of Polymer Science Part A: Polymer Chemistry*, 1991. **29**(10): pp. 1483–1489.
152. Boston, H.G., A.K. St Clair, and J.R. Pratt, Polyimides derived from a methylene-bridged dianhydride. *Journal of Applied Polymer Science*, 1992. **46**(2): pp. 243–253.
153. Yang, C.P., S.H. Hsiao, and C.C. Jang, Synthesis and properties of polyimides derived from 1,6-bis (4-aminophenoxy) naphthalene and aromatic tetracarboxylic dianhydrides. *Journal of Polymer Science Part A: Polymer Chemistry*, 1995. **33**(9): pp. 1487–1493.
154. de la Campa, J.G. et al., Polyimides from 3,4:3″,4″-m-terphenyltetracarboxylic dianhydride. Synthesis and characterization. *Macromolecular Rapid Communications*, 1994. **15**(5): pp. 417–424.
155. Hergenrother, P.M. and S.J. Havens, Polyimides containing quinoxaline and benzimidazole units. *Macromolecules*, 1994. **27**(17): pp. 4659–4664.
156. Chao, H.-I. and E. Barren, Synthesis and evaluation of polyimides derived from spirobisindane dietheranhydride. *Polymer Preprints (USA)*, 1992. **33**(1): pp. 1024–1025.
157. Kasashima, Y. et al., Preparation and properties of polyamides and polyimides from 4,4″-diamino-*o*-terphenyl. *Polymer*, 1995. **36**(3): pp. 645–650.
158. Moy, T., C. DePorter, and J. McGrath, Synthesis of soluble polyimides and functionalized imide oligomers via solution imidization of aromatic diester-diacids and aromatic diamines. *Polymer*, 1993. **34**(4): pp. 819–824.
159. Yang, C.P. and W.T. Chen, Synthesis and properties of polyimides derived from 4,4′-(2,7-naphthylenedioxy) dianiline and aromatic tetracarboxylic dianhydrides. *Die Makromolekulare Chemie*, 1993. **194**(11): pp. 3061–3069.
160. Akutsu, F. et al., Preparation of polyimides from 2,3-bis(4-aminophenyl) quinoxalines and aromatic tetracarboxylic dianhydrides. *Die Makromolekulare Chemie, Rapid Communications*, 1990. **11**(12): pp. 673–677.
161. Rao, V.L. and J. Bijimol, Polyimides derived from ether ketone diamines. *Die Makromolekulare Chemie*, 1991. **192**(5): pp. 1025–1032.
162. Havens, S. and P. Hergenrother, Polyimides containing carbonyl and ether connecting groups. IV. *Journal of Polymer Science Part A: Polymer Chemistry*, 1992. **30**(6): pp. 1209–1212.
163. Preston, J. and J. Carson Jr., Processable heat-resistant resins based on novel monomers containing *o*-phenylene rings: 2. Ordered benzoxazole-and benzthiazole-imide copolymers. *Polymer*, 1993. **34**(4): pp. 830–834.
164. Hougham, G., G. Tesoro, and J. Shaw, Synthesis and properties of highly fluorinated polyimides. *Macromolecules*, 1994. **27**(13): pp. 3642–3649.
165. Husk, G.R., P.E. Cassidy, and K.L. Gebert, Synthesis and characterization of a series of polyimides derived from 4,4′-[2,2,2-trifluoro-1-(trifluoromethyl)ethylidene]bis[1,3-isobenzofurandione]. *Macromolecules*, 1988. **21**(5): pp. 1234–1238.
166. Marek Jr., M. et al., New soluble polyimides prepared from 4,4′-(alkylenediyldioxy) dianilines. *Polymer*, 1994. **35**(22): pp. 4881–4888.
167. Matsuura, T. et al., Polyimides derived from 2,2′-bis(trifluoromethyl)-4,4′-diaminobiphenyl. 2. Synthesis and characterization of polyimides prepared from fluorinated benzenetetracarboxylic dianhydrides. *Macromolecules*, 1992. **25**(13): pp. 3540–3545.
168. Stoakley, D.M., A.K. St Clair, and C.I. Croall, Low dielectric, fluorinated polyimide copolymers. *Journal of Applied Polymer Science*, 1994. **51**(8): pp. 1479–1483.
169. Park, J. et al., Synthesis and characterization of soluble alternating aromatic copolyimides. *Macromolecules*, 1994. **27**(13): pp. 3459–3463.
170. Ando, S., T. Matsuura, and S. Sasaki, Perfluorinated polyimide synthesis. *Macromolecules*, 1992. **25**(21): pp. 5858–5860.
171. Brink, M. et al., Synthesis and characterization of a novel '3F'-based fluorinated monomer for fluorine-containing polyimides. *Polymer*, 1994. **35**(23): pp. 5018–5023.
172. Scola, D.A., Synthesis and thermooxidative stability of poly[1,4-phenylene-4,4′-(2,2,2-trifluoro-1-phenylethylidene) bisphthalimide] and other fluorinated polyimides. *Journal of Polymer Science Part A: Polymer Chemistry*, 1993. **31**(8): pp. 1997–2008.
173. Takeichi, T., S. Ogura, and Y. Takayama, Soluble polyimides that contain curable internal acetylene groups in the backbone. *Journal of Polymer Science Part A: Polymer Chemistry*, 1994. **32**(3): pp. 579–585.

174. Rogers, M. et al., Semicrystalline and amorphous fluorine-containing polyimides. *Polymer*, 1993. **34**(4): pp. 849–855.
175. Ma, T. et al., Synthesis and characterization of novel polyimides derived from 4-phenyl-2,6-bis [3-(4-aminophenoxy)-phenyl]-pyridine diamine and aromatic dianhydrides. *Polymer Degradation and Stability*, 2010. **95**(7): pp. 1244–1250.
176. Bu, Q. et al., Preparation and properties of thermally stable polyimides derived from asymmetric trifluoromethylated aromatic diamines and various dianhydrides. *Polymer Degradation and Stability*, 2011. **96**(10): pp. 1911–1918.
177. Yang, C.-P., R.-S. Chen, and K.-H. Chen, Organosoluble and light-colored fluorinated polyimides based on 2,2-bis[4-(4-amino-2-trifluoromethylphenoxy)phenyl]propane and aromatic dianhydrides. *Journal of Applied Polymer Science*, 2005. **95**(4): pp. 922–935.
178. Yang, C.-P., S.-H. Hsiao, and K.-L. Wu, Organosoluble and light-colored fluorinated polyimides derived from 2,3-bis(4-amino-2-trifluoromethylphenoxy)naphthalene and aromatic dianhydrides. *Polymer*, 2003. **44**(23): pp. 7067–7078.
179. Yang, C.-P., S.-H. Hsiao, and M.-F. Hsu, Organosoluble and light-colored fluorinated polyimides from 4,4′-bis(4-amino-2-trifluoromethylphenoxy)biphenyl and aromatic dianhydrides. *Journal of Polymer Science Part A: Polymer Chemistry*, 2002. **40**(4): pp. 524–534.
180. Xie, K. et al., Synthesis and characterization of soluble fluorine-containing polyimides based on 1,4-bis(4-amino-2-trifluoromethylphenoxy)benzene. *Journal of Polymer Science Part A: Polymer Chemistry*, 2001. **39**(15): pp. 2581–2590.
181. Chung, C.-L. and S.-H. Hsiao, Novel organosoluble fluorinated polyimides derived from 1,6-bis (4-amino-2-trifluoromethylphenoxy)naphthalene and aromatic dianhydrides. *Polymer*, 2008. **49**(10): pp. 2476–2485.
182. Dhara, M.G. and S. Banerjee, Fluorinated high-performance polymers: Poly(arylene ether)s and aromatic polyimides containing trifluoromethyl groups. *Progress in Polymer Science*, 2010. **35**(8): pp. 1022–1077.
183. Ge, Z., L. Fan, and S. Yang, Synthesis and characterization of novel fluorinated polyimides derived from 1,1′-bis(4-aminophenyl)-1-(3-trifluoromethylphenyl)-2,2,2-trifluoroethane and aromatic dianhydrides. *European Polymer Journal*, 2008. **44**(4): pp. 1252–1260.
184. Zhao, X. et al., Novel polyfluorinated polyimides derived from α,α-bis(4-amino-3,5-difluorophenyl) phenylmethane and aromatic dianhydrides: Synthesis and characterization. *European Polymer Journal*, 2008. **44**(3): pp. 808–820.
185. Wang, C. et al., Novel fluorinated polyimides derived from 9,9-bis(4-amino-3,5-difluorophenyl)fluorene and aromatic dianhydrides. *Polymer Degradation and Stability*, 2009. **94**(10): pp. 1746–1753.
186. Tao, L. et al., Synthesis and characterization of highly optical transparent and low dielectric constant fluorinated polyimides. *Polymer*, 2009. **50**(25): pp. 6009–6018.
187. Hsiao, S.-H. et al., Synthesis and characterization of novel fluorinated polyimides derived from 1,3-bis(4-amino-2-trifluoromethylphenoxy)naphthalene and aromatic dianhydrides. *European Polymer Journal*, 2010. **46**(9): pp. 1878–1890.
188. Lu, Y. et al., Organosoluble and light-colored fluorinated semialicyclic polyimide derived from 1,2,3,4-cyclobutanetetracarboxylic dianhydride. *Journal of Applied Polymer Science*, 2012. **125**(2): pp. 1371–1376.
189. Meador, M.A.B. et al., Dielectric and other properties of polyimide aerogels containing fluorinated blocks. *ACS Applied Materials & Interfaces*, 2014. **6**(9): pp. 6062–6068.
190. Hsiao, S.-H. and Y.-J. Chen, Structure–property study of polyimides derived from PMDA and BPDA dianhydrides with structurally different diamines. *European Polymer Journal*, 2002. **38**(4): pp. 815–828.
191. Sahadeva Reddy, D. et al., Synthesis and characterization of soluble poly(ether imide)s based on 2,2′-bis(4-aminophenoxy)-9,9′-spirobifluorene. *Polymer*, 2003. **44**(3): pp. 557–563.
192. Wang, C.-S. and T.-S. Leu, Synthesis and characterization of polyimides containing naphthalene pendant group and flexible ether linkages. *Polymer*, 2000. **41**(10): pp. 3581–3591.
193. Shang, Y. et al., Synthesis and characterization of novel fluorinated polyimides derived from 4-phenyl-2,6-bis[4-(4′-amino-2′-trifluoromethyl-phenoxy)phenyl]pyridine and dianhydrides. *European Polymer Journal*, 2006. **42**(5): pp. 981–989.
194. Guo, X. et al., Synthesis and properties of novel sulfonated polyimides from 2,2′-bis(4-aminophenoxy) biphenyl-5,5′-disulfonic acid. *Journal of Polymer Science Part A: Polymer Chemistry*, 2004. **42**(6): pp. 1432–1440.

195. Yang, F. et al., Characterizations and thermal stability of soluble polyimide derived from novel unsymmetrical diamine monomers. *Polymer Degradation and Stability*, 2010. **95**(9): pp. 1950–1958.
196. Yu, X. et al., Synthesis and properties of thermoplastic polyimides with ether and ketone moieties. *Journal of Polymer Science Part A: Polymer Chemistry*, 2010. **48**(13): pp. 2878–2884.
197. Zhang, G. et al., Synthesis and characterization of a new type of sulfonated poly(ether ether ketone ketone)s for proton exchange membranes. *Journal of Applied Polymer Science*, 2010. **116**(3): pp. 1515–1523.
198. Chen, J.-C. et al., Synthesis and properties of organosoluble polyimides derived from 2,2′-dibromo- and 2,2′,6,6′-tetrabromo-4,4′-oxydianilines. *Journal of Applied Polymer Science*, 2010. **117**(2): pp. 1144–1155.
199. Yang, F. et al., Synthesis and properties of novel organosoluble polyimides derived from bis[3-(4-amino-2-trifluoromethylphenoxy) phenyl] ether. *European Polymer Journal*, 2009. **45**(7): pp. 2053–2059.
200. Wu, Z. et al., Novel soluble and optically active polyimides containing axially asymmetric 9,9′-spirobifluorene units: Synthesis, thermal, optical and chiral properties. *Polymer*, 2012. **53**(25): pp. 5706–5716.
201. Zhang, S. et al., Synthesis and properties of novel polyimides derived from 2,6-bis(4-aminophenoxy-4′-benzoyl)pyridine with some of dianhydride monomers. *Polymer*, 2005. **46**(25): pp. 11986–11993.
202. Wang, X. et al., Synthesis and characterization of novel polyimides derived from pyridine-bridged aromatic dianhydride and various diamines. *European Polymer Journal*, 2006. **42**(6): pp. 1229–1239.
203. Ha, Y. et al., Microstructure and properties of rigid rod-like polyimide/flexible coil-like poly(amide-imide) molecular composite films. *Macromolecular Research*, 2010. **18**(1): pp. 14–21.
204. Wang, Z.Y. and Y. Qi, Synthesis and properties of poly(aryl prehnitimide)s. *Macromolecules*, 1995. **28**(12): pp. 4207–4212.
205. Gao, J.P. and Z.Y. Wang, Synthesis and properties of polyimides from 4,4′-binaphthyl-1,1′,8,8′-tetracarboxylic dianhydride. *Journal of Polymer Science Part A: Polymer Chemistry*, 1995. **33**(10): pp. 1627–1635.
206. Gubbelmans, E. et al., Chromophore-functionalised polymides with high-poling stabilities of the nonlinear optical effect at elevated temperature. *Polymer*, 2002. **43**(5): pp. 1581–1585.
207. He, M. et al., Synthesis and nonlinear optical properties of soluble fluorinated polyimides containing hetarylazo chromophores with large hyperpolarizability. *Polymer*, 2009. **50**(16): pp. 3924–3931.
208. Yoon, C.-B. and H.-K. Shim, Facile synthesis of new NLO-functionalized polyimides via Mitsunobu reaction. *Journal of Materials Chemistry*, 1999. **9**(10): pp. 2339–2344.
209. Carter, K.R., R.D. Miller, and J.L. Hedrick, Synthesis and properties of imide-aryl ether 1,2,4-triazole random copolymers. *Polymer*, 1993. **34**(4): pp. 843–848.
210. Li, P. and M.R. Coleman, Synthesis of room temperature ionic liquids based random copolyimides for gas separation applications. *European Polymer Journal*, 2013. **49**(2): pp. 482–491.
211. Shiang, W.R. and E.P. Woo, Soluble copolyimides with high modulus and low moisture absorption. *Journal of Polymer Science Part A: Polymer Chemistry*, 1993. **31**(8): pp. 2081–2091.
212. Maya, E.M. et al., Fluorenyl cardo copolyimides containing poly(ethylene oxide) segments: Synthesis, characterization, and evaluation of properties. *Journal of Polymer Science Part A: Polymer Chemistry*, 2008. **46**(24): pp. 8170–8178.
213. Chun, B.-W., Preparation and characterization of organic-soluble optically transparent polyimides from alicyclic dianhydride, bicyclo[2.2.2]-oct-7-ene-2,3,5,6-tetracarboxylic dianhydride. *Polymer*, 1994. **35**(19): pp. 4203–4208.
214. Yamada, M. et al., Soluble polyimides with polyalicyclic structure. 2. Polyimides from bicyclo[2.2.1]heptane-2-exo-3-exo-5-exo-6-exo-tetracarboxylic2,3:5,6-dianhydride. *Macromolecules*, 1993. **26**(18): pp. 4961–4963.
215. Itamura, S. et al., Soluble polyimides with polyalicyclic structure. 1. Polyimides from bicyclo[2.2.2]oct-7-ene-2-exo,3-exo,5-exo,6-exo-tetracarboxylic2,3:5,6-dianhydrides. *Macromolecules*, 1993. **26**(14): pp. 3490–3493.
216. Kusama, M., T. Matsumoto, and T. Kurosaki, Soluble polyimides with polyalicyclic structure. 3. Polyimides from (4arH,8acH)-decahydro-1t,4t:5c,8c-dimethanonaphthalene-2t,3t,6c,7c-tetracarboxylic 2,3:6,7-dianhydride. *Macromolecules*, 1994. **27**(5): pp. 1117–1123.
217. Su, C.Y. and Y.D. Lee, Synthesis and characterization of polyimides based on novel tricyclo[6,2,2,02,7] dianhydride. *Journal of Polymer Science Part A: Polymer Chemistry*, 1990. **28**(12): pp. 3347–3362.
218. Tsujita, Y. et al., Characterization of alicyclic polyimides based on cyclobutane ring dianhydride and aromatic diamine. *Journal of Applied Polymer Science*, 1994. **54**(9): pp. 1297–1304.
219. Liu, J.-G. et al., Synthesis and characterization of highly organo-soluble polyimides based on alicyclic 1,8-dimethyl-bicyclo[2.2.2]oct-7-ene-2,3,5,6-tetracarboxylic dianhydride and aromatic diamines. *Chinese Journal of Polymer Science*, 2004. **22**(6): pp. 511–519.
220. Tjugito, S. and W.A. Feld, Polyimides derived from 1,2-bis(4-aminophenoxy) propane. *Journal of Polymer Science Part A: Polymer Chemistry*, 1989. **27**(3): pp. 963–970.

221. Patil, D., Synthesis and properties of polyamides, polyimides, and a polyamide-imide based on trimethylene glycol-di-p-aminobenzoate. *Journal of Macromolecular Science-Chemistry*, 1990. **27**(3): pp. 331–337.
222. Acevedo, M. and F. Harris, Polyimides derived from 2-methyl-2-propyl-1,3-bis(4-aminophenoxy) propane and 2,2-dimethyl-1,3-bis(4-aminophenoxy) propane. *Polymer*, 1994. **35**(20): pp. 4456–4461.
223. Feld, W.A. and T.B. Le, Polyimides containing nonaromatic nitrogen linkages. *Journal of Polymer Science Part A: Polymer Chemistry*, 1992. **30**(6): pp. 1099–1102.
224. Jin, Q., T. Yamashita, and K. Horie, Polyimides with alicyclic diamines. II. Hydrogen abstraction and photocrosslinking reactions of benzophenone-type polyimides. *Journal of Polymer Science Part A: Polymer Chemistry*, 1994. **32**(3): pp. 503–511.
225. Jin, Q. et al., Polyimides with alicyclic diamines. I. Syntheses and thermal properties. *Journal of Polymer Science Part A: Polymer Chemistry*, 1993. **31**(9): pp. 2345–2351.
226. Mathews, A., I. Kim, and C.-S. Ha, Synthesis, characterization, and properties of fully aliphatic polyimides and their derivatives for microelectronics and optoelectronics applications. *Macromolecular Research*, 2007. **15**(2): pp. 114–128.
227. Mathews, A.S., I. Kim, and C.S. Ha, Fully aliphatic polyimides from adamantane-based diamines for enhanced thermal stability, solubility, transparency, and low dielectric constant. *Journal of Applied Polymer Science*, 2006. **102**(4): pp. 3316–3326.
228. Teshirogi, T., Polyimides from 4,4′-(alkylene-α, ω-dioxy) bis(phenylsuccinic anhydride)s and bis(phenylglutaric anhydride)s. *Journal of Polymer Science Part A: Polymer Chemistry*, 1989. **27**(2): pp. 653–660.
229. Moon, Y.D. and Y.M. Lee, Preparation and thermal kinetics of poly(imide-siloxanes). *Journal of Applied Polymer Science*, 1993. **50**(8): pp. 1461–1473.
230. Fitzgerald, J.J., S.E. Tunney, and M.R. Landry, Synthesis and characterization of a fluorinated poly(imide-siloxane) copolymer: A study of physical properties and morphology. *Polymer*, 1993. **34**(9): pp. 1823–1832.
231. Pryde, C., FTIR studies of polyimides. II. Factors affecting quantitative measurement. *Journal of Polymer Science Part A: Polymer Chemistry*, 1993. **31**(4): pp. 1045–1052.
232. Pryde, C., IR studies of polyimides. I. Effects of chemical and physical changes during cure. *Journal of Polymer Science Part A: Polymer Chemistry*, 1989. **27**(2): pp. 711–724.
233. Qiao, X. and T.-S. Chung, Diamine modification of P84 polyimide membranes for pervaporation dehydration of isopropanol. *AIChE Journal*, 2006. **52**(10): pp. 3462–3472.
234. Chen, H., Y. Xiao, and T.-S. Chung, Synthesis and characterization of poly(ethylene oxide) containing copolyimides for hydrogen purification. *Polymer*, 2010. **51**(18): pp. 4077–4086.
235. Albrecht, W. et al., Amination of poly(ether imide) membranes using di- and multivalent amines. *Macromolecular Chemistry and Physics*, 2003. **204**(3): pp. 510–521.
236. Georgiev, A. et al., Investigation of solid state imidization reactions of the vapour deposited azo-polyimide thin films by FTIR spectroscopy. *Journal of Molecular Structure*, 2014. **1074**: pp. 100–106.
237. Matsuura, T. et al., Polyimide derived from 2,2′-bis(trifluoromethyl)-4,4′-diaminobiphenyl. 1. Synthesis and characterization of polyimides prepared with 2,2′-bis(3,4-dicarboxyphenyl) hexafluoropropane dianhydride or pyromellitic dianhydride. *Macromolecules*, 1991. **24**(18): pp. 5001–5005.
238. Choi, H. et al., Soluble polyimides from unsymmetrical diamine containing benzimidazole ring and trifluoromethyl pendent group. *Polymer*, 2008. **49**(11): pp. 2644–2649.
239. Ando, S., T. Matsuura, and S. Nishi, 13C n.m.r. analysis of fluorinated polyimides and poly (amic acid)s. *Polymer*, 1992. **33**(14): pp. 2934–2939.
240. Korshak, V. et al., Investigation of the cyclization of poly(amic acid)s by 13C NMR spectroscopy. *Die Makromolekulare Chemie, Rapid Communications*, 1984. **5**(10): pp. 695–700.
241. Denisov, V. et al., The isomeric composition of poly(acid) amides according to ^{13}C-NMR spectral data. *Polymer Science USSR*, 1979. **21**(7): pp. 1644–1650.
242. Wang, Y. et al., Effect of pre-imidization on the aggregation structure and properties of polyimide films. *Polymer*, 2012. **53**(19): pp. 4157–4163.
243. Damaceanu, M.-D. et al., Highly transparent and hydrophobic fluorinated polyimide films with ortho-kink structure. *European Polymer Journal*, 2014. **50**: pp. 200–213.
244. Ando, S., T. Matsuura, and S. Sasaki, 15N-, 1H-, and 13C-NMR chemical shifts and electronic properties of aromatic diamines and dianhydrides. *Journal of Polymer Science Part A: Polymer Chemistry*, 1992. **30**(11): pp. 2285–2293.
245. Smith, C.D. et al., ^{19}F nmr investigation of aromatic amic acid and imide model compounds. 1. *Polymer*, 1993. **34**(23): pp. 4852–4862.

246. Jeong, K.U., J.J. Kim, and T.H. Yoon, Synthesis and characterization of novel polyimides containing fluorine and phosphine oxide moieties. *Polymer*, 2001. **42**(14): pp. 6019–6030.
247. Halim, A. et al., Synthesis and self-assembly of polyimide/poly(dimethylsiloxane) brush triblock copolymers. *Polymer*, 2013. **54**(2): pp. 520–529.
248. Miyatake, K. et al., Durability of sulfonated polyimide membrane in humidity cycling for fuel cell applications. *Journal of Power Sources*, 2012. **204**: pp. 74–78.
249. Liu, C.-P. et al., Facile chemical method of etching polyimide films for failure analysis (FA) applications and its etching mechanism studies. *Microelectronics Reliability*, 2014. **54**(5): pp. 911–920.
250. Konáš, M. et al., Molecular weight characterization of soluble high performance polyimides. 1. Polymer-solvent-stationary phase interactions in size exclusion chromatography. *Journal of Polymer Science Part B: Polymer Physics*, 1995. **33**(10): pp. 1429–1439.
251. Konáš, M. et al., Molecular weight characterization of soluble high performance polyimides. 2. Validity of universal sec calibration and absolute molecular weight calculation. *Journal of Polymer Science Part B: Polymer Physics*, 1995. **33**(10): pp. 1441–1448.
252. Young, P.R. et al., Characterization of a thermally imidized soluble polyimide film. *Journal of Polymer Science Part A: Polymer Chemistry*, 1990. **28**(11): pp. 3107–3122.
253. Hansen, C.M., 50 Years with solubility parameters—Past and future. *Progress in Organic Coatings*, 2004. **51**(1): pp. 77–84.
254. Lee, H.R. and Y.D. Lee, Solubility behavior of an organic soluble polyimide. *Journal of Applied Polymer Science*, 1990. **40**(11–12): pp. 2087–2099.
255. Soroko, I., M.P. Lopes, and A. Livingston, The effect of membrane formation parameters on performance of polyimide membranes for organic solvent nanofiltration (OSN): Part A. Effect of polymer/solvent/non-solvent system choice. *Journal of Membrane Science*, 2011. **381**(1–2): pp. 152–162.
256. Schmitz, L. and M. Ballauff, Characterization of a stiff-chain polyimide in solution. *Polymer*, 1995. **36**(4): pp. 879–882.
257. Kyu, T. et al., Structure evolution and kinetics of thermoreversible gelation in a rigid-rod polyimide/m-cresol system. *Macromolecules*, 1994. **27**(7): pp. 1861–1868.
258. Cheng, S.Z. et al., Gel/sol and liquid-crystalline transitions in solution of a rigid-rod polyimide. *Macromolecules*, 1991. **24**(8): pp. 1883–1889.
259. Lin, T. et al., Preparation of submicrometre polyimide particles by precipitation from solution. *Polymer*, 1993. **34**(4): pp. 772–777.
260. Jia, H. et al., Preparation and characterization of polyimide magnetic hollow nanospheres. *Materials Letters*, 2012. **68**: pp. 86–89.
261. Le, N.L., Y. Wang, and T.-S. Chung, Synthesis, cross-linking modifications of 6FDA-NDA/DABA polyimide membranes for ethanol dehydration via pervaporation. *Journal of Membrane Science*, 2012. **415–416**: pp. 109–121.
262. Patterson, D.A. et al., Membrane characterisation by SEM, TEM and ESEM: The implications of dry and wetted microstructure on mass transfer through integrally skinned polyimide nanofiltration membranes. *Separation and Purification Technology*, 2009. **66**(1): pp. 90–97.
263. Tamai, S. et al., Synthesis and characterization of thermally stable semicrystalline polyimide based on 3,4′-oxydianiline and 3,3′,4,4′-biphenyltetracarboxylic dianhydride. *Polymer*, 2001. **42**(6): pp. 2373–2378.
264. O'Mahoney, C.A. et al., Single-crystal x-ray studies of aromatic oligomers: Conformation and packing in isomeric pyromellitimide-ether-sulfones. *Macromolecules*, 1991. **24**(24): pp. 6527–6530.
265. Liu, J. et al., Crystal structure, morphology, and phase transitions in aromatic polyimide oligomers. 3. Poly(1,4-phenyleneoxy-1,3-phenylene pyromellitimide). *Journal of Polymer Science Part B: Polymer Physics*, 1994. **32**(16): pp. 2705–2713.
266. Liu, J. et al., Crystal structure, morphology, and phase transitions in aromatic polyimide oligomers. 1. Poly(4,4′-oxydiphenylene pyromellitimide). *Macromolecules*, 1994. **27**(4): pp. 989–996.
267. Liu, J. et al., Crystal structure, morphology and phase transitions in aromatic polyimide oligomers: 2. Poly(1,4-phenylene-oxy-1,4-phenylene-oxy-1,4-phenylene pyromellitimide). *Polymer*, 1994. **35**(19): pp. 4048–4056.
268. Waddon, A. and F. Karasz, Crystalline and amorphous morphologies of an aromatic polyimide formed on precipitation from solution. *Polymer*, 1992. **33**(18): pp. 3783–3789.
269. Luo, L. et al., The evolution of macromolecular packing and sudden crystallization in rigid-rod polyimide via effect of multiple H-bonding on charge transfer (CT) interactions. *Polymer*, 2014. **55**(16): pp. 4258–4269.
270. Cheng, S.Z. et al., Glass transition and crystal melting in semicrystalline polyimides. *Journal of Polymer Science Part B: Polymer Physics*, 1990. **28**(5): pp. 655–674.

271. Sun, Z. et al., Beta relaxation in polyimides. *Polymer*, 1992. **33**(22): pp. 4728–4731.
272. Asano, N. et al., Aliphatic/Aromatic polyimide ionomers as a proton conductive membrane for fuel cell applications. *Journal of the American Chemical Society*, 2006. **128**(5): pp. 1762–1769.
273. Mathews, A.S., I. Kim, and C.-S. Ha, Synthesis and characterization of novel fully aliphatic polyimidosiloxanes based on alicyclic or adamantyl diamines. *Journal of Polymer Science Part A: Polymer Chemistry*, 2006. **44**(18): pp. 5254–5270.
274. Chung, C.-M. et al., Preparation of porous thin films of a partially aliphatic polyimide. *Journal of Applied Polymer Science*, 2006. **101**(1): pp. 532–538.
275. Li, B., T. Liu, and W.-H. Zhong, High modulus aliphatic polyimide from 1,3-diaminopropane and ethylenediaminetetraacetic dianhydride: Water soluble to self-patterning. *Polymer*, 2011. **52**(22): pp. 5186–5192.
276. Liaw, D.-J., Synthesis and characterization of new highly soluble organic polyimides, in *Macromolecular nanostructured materials*. Ueyama, N., Harada, A., eds., 2004, Munich: Springer, pp. 80–100.
277. de Abajo, J. et al., Liquid crystal polyimides: 17. Thermotropic poly(ester imide)s based on trimellitimide and diamino oligoether spacers. *Polymer*, 1994. **35**(25): pp. 5577–5585.
278. de Abajo, J. et al., Liquid-crystalline polyimides, 1. Thermotropic poly (ester-imide) s derived from 4-hydroxyphthalic acid, 4-aminophenol and various aliphatic α,ω-diacids. *Die Makromolekulare Chemie*, 1990. **191**(3): pp. 537–547.
279. Pardey, R. et al., Monotropic liquid crystal behavior in two poly(ester imides) with even and odd flexible spacers. *Macromolecules*, 1992. **25**(19): pp. 5060–5068.
280. Adduci, J.M., J.V. Facinelli, and R.W. Lenz, Synthesis and characterization of random semi-flexible thermotropic liquid crystalline poly(ester-imide)s. *Journal of Polymer Science Part A: Polymer Chemistry*, 1994. **32**(15): pp. 2931–2936.
281. Kricheldorf, H.R. and M. Berghahn, LC-Polyimides. XIX. Chiral and thermotropic poly(esterimide)s based on N-(4′-caboxyphenyl) trimellitimide and novel chiral spacers. *Journal of Polymer Science Part A: Polymer Chemistry*, 1995. **33**(3): pp. 427–439.
282. Kricheldorf, H.R. et al., Layer structures. 2. Influence of spacers on chain packing and phase transitions of poly(ester-imide)s derived from N-(4-carboxyphenyl) trimellitimide. *Macromolecules*, 1994. **27**(9): pp. 2540–2547.
283. Kricheldorf, H.R. and R. Pakull, New polymer syntheses. 20. Liquid-crystalline poly(ester imides) derived from trimellitic acid,α,ω-diaminoalkanes, and 4,4′-dihydroxybiphenyl. *Macromolecules*, 1988. **21**(3): pp. 551–557.
284. Pardey, R. et al., Liquid crystal transition and crystallization kinetics in poly(ester imide)s. *Macromolecules*, 1994. **27**(20): pp. 5794–5802.
285. de Abajo, J. et al., Rheological features of thermotropic and isotropic poly(ester imide)s. *Polymer*, 1995. **36**(8): pp. 1683–1687.
286. Kricheldorf, H.R., R. Pakull, and S. Buchner, New polymer syntheses. 24. Liquid crystal poly(ester imides) derived from benzophenonetetracarboxylic dianhydride and ω-amino acids. *Macromolecules*, 1988. **21**(7): pp. 1929–1935.
287. Kricheldorf, H.R., R. Pakull, and G. Schwarz, LC-polyimides, 10. Poly(ester-imide)s derived from pyromellitic dianhydride and ω-amino acids and various diphenols or α,ω-diols. *Die Makromolekulare Chemie*, 1993. **194**(4): pp. 1209–1224.
288. Inoue, T. et al., First observation of a thermotropic liquid crystal in a simple polyimide derived from 1,11-diaminoundecane and 4,4″-terphenyltetracarboxylic acid. *Macromolecules*, 1995. **28**(18): pp. 6368–6370.
289. Sun, S.J. and T.C. Chang, Study on the thermotropic liquid crystalline polycarbonates. IV. Synthesis and properties of liquid crystalline poly(imide-carbonate)s. *Journal of Polymer Science Part A: Polymer Chemistry*, 1994. **32**(16): pp. 3039–3046.
290. Karayannidis, G., D. Stamelos, and D. Bikiaris, Poly(ester-imide)s derived from 3,4,9,10-perylenetetracarboxylic acid, ω-aminododecanoic acid and α,ω-alkanediols. *Die Makromolekulare Chemie*, 1993. **194**(10): pp. 2789–2796.
291. Sato, M. et al., Thermotropic liquid crystalline aromatic-aliphatic polyimides—5. Preparation and properties of homo- and copoly(imide-carbonate)s based on benzophenonetetracarboxylic diimide and biphenyl units. *European Polymer Journal*, 1996. **32**(5): pp. 639–645.
292. Sato, M., S. Ujiie, and T. Hirata, Thermotropic liquid-crystalline aromatic-aliphatic polyimides, 7. Poly(ester-imide)s based on 3,4:3″,4″-p-terphenyldicarboximide. *Macromolecular Chemistry and Physics*, 1996. **197**(9): pp. 2765–2774.
293. Sato, M. et al., Synthesis and properties of new aromatic-aliphatic homo-and copoly(ester-imide)s based on 3,4,3′,4′-diphenylsulphone tetracarboxdiimide. *High Performance Polymers*, 1995. **7**(3): pp. 347–355.

294. Hirata, T., M. Sato, and K.I. Mukaida, Thermotropic liquid-crystalline aromatic-aliphatic polyimides, 3. Poly(imide-carbonate)s composed of 3,4:3′,4′-biphenyldicarboximide. *Macromolecular Chemistry and Physics*, 1994. **195**(5): pp. 1611–1622.
295. Hirata, T., M. Sato, and K.I. Mukaida, Thermotropic liquid-crystalline aromatic-aliphatic polyimides, 4. Poly(imide-urethane-carbonate)s based on pyromellitdiimide and 3,4:3′,4′-biphenyldicarboximide. *Macromolecular Chemistry and Physics*, 1994. **195**(6): pp. 2267–2277.
296. Sato, M., T. Hirata, and K.I. Mukaida, Thermotropic liquid-crystalline aromatic-aliphatic polyimides, 6. Poly(ester-imide)s based on 3,4:3′,4′-biphenyldicarboximide. *Macromolecular Rapid Communications*, 1994. **15**(3): pp. 203–209.
297. Hirata, T., M. Sato, and K.I. Mukaida, Thermotropic liquid-crystalline aromatic-aliphatic polyimides, 2. Effect of aliphatic spacer lengths and polymer compositions on the liquid-crystalline properties of poly(imide-carbonate)s. *Die Makromolekulare Chemie*, 1993. **194**(10): pp. 2861–2874.
298. Sato, M., T. Hirata, and K.i. Mukaida, Thermotropic liquid-crystalline aromatic-aliphatic polyimides, 1. Novel poly(imide-carbonate)s containing pyromellitimide ring. *Die Makromolekulare Chemie*, 1992. **193**(7): pp. 1729–1737.
299. Kricheldorf, H.R., A. Domschke, and S. Böhme, LC-polyimides—8. Thermotropic poly(ester-imide)s from n-(4-carboxyphenyl)-4-nitrophthalimide. *European Polymer Journal*, 1992. **28**(10): pp. 1253–1258.
300. de Abajo, J. et al., Lc-polyimides—6. Thermotropic poly(ester-imide)s derived from N-(4-carboxyphenyl) trimellitimide and various diphenols. *European Polymer Journal*, 1992. **28**(3): pp. 261–265.
301. Kricheldorf, H.R., F. Ruhser, and G. Schwarz, New polymer syntheses. 48. Thermotropic copolyesters of 4′-hydroxybiphenyl-4-carboxylic acid and 4-hydroxybenzoic acid or 3-chloro-4-hydroxybenzoic acid. *Macromolecules*, 1991. **24**(18): pp. 4990–4996.
302. Kricheldorf, H.R., A. Domschke, and G. Schwarz, Liquid-crystalline polyimides. 3. Fully aromatic liquid-crystalline poly(ester imide)s derived from N-(4-carboxyphenyl) trimellitimide and substituted hydroquinones. *Macromolecules*, 1991. **24**(5): pp. 1011–1016.
303. Kricheldorf, H.R. et al., Liquid-crystalline polyimides. 12. Fully aromatic thermotropic poly(ester-imide)s derived from diphenylether-3,3′,4,4′-tetracarboxylic imide. *Journal of Macromolecular Science, Part A: Pure and Applied Chemistry*, 1995. **32**(2): pp. 311–330.
304. Kricheldorf, H.R., V. Linzer, and C. Bruhn, Liquid-crystalline polyimides. 11. Thermotropic aromatic poly(ester-imide)s based on biphenyl-3,3′,4,4′-tetracarboxylic imide. *Journal of Macromolecular Science—Pure and Applied Chemistry*, 1994. **31**(9): pp. 1315–1328.
305. Kricheldorf, H.R. et al., Liquid-crystalline polyimides. 15. Role of conformation and donor-acceptor interactions for the nematic order of poly(ester imides). *Macromolecules*, 1993. **26**(19): pp. 5161–5168.
306. Krichhdori, H.R. et al., LC-Polyimides. IX. Poly(ester-imide)s of N-(4′-carboxyphenyl) trimellitimide containing carborane 1, 7-dicarboxylic acid, or isophthalic acid. *Journal of Polymer Science Part A: Polymer Chemistry*, 1993. **31**(1): pp. 279–282.
307. Kircheldorff, H. and N. Probst, Chiral thermotropic copoly(ester-imide)s based on isosorbide and N-(4-carboxyphenyl)-trimellitimide. *Macromolecular Rapid Communications*, 1995. **16**: pp. 231–237.
308. Asanuma, T. et al., Synthesis of thermotropic liquid crystal polyimide and its properties. *Journal of Polymer Science Part A: Polymer Chemistry*, 1994. **32**(11): pp. 2111–2118.
309. Kricheldorf, H.R. and R. Huner, Liquid-crystalline poly(imide)s 2. Thermotropic fully aromatic poly(ester-imide)s derived from 4-hydroxyphthalic acid and various substituted terephthalic acids. *Die Makromolekulare Chemie, Rapid Communications*, 1990. **11**(5): pp. 211–215.
310. Kim, S.O., T.K. Kim, and I.J. Chung, Synthesis and rheological investigation on phase behaviors of semiflexible type amorphous liquid crystalline poly(ester-imide)s. *Polymer*, 2000. **41**(12): pp. 4709–4717.
311. Sapich, B. et al., Synthesis, dielectric, and photochemical study of liquid crystalline main chain poly(ester imide)s containing cinnamoyl moieties. *Macromolecules*, 2001. **34**(16): pp. 5694–5701.
312. Eiselt, P., S. Denzinger, and H.-W. Schmidt, Liquid crystalline model compounds based on 3,3′,4,4′-biphenyltetracarboxylic dianhydride (BPDA). *Liquid Crystals*, 1995. **18**(2): pp. 257–262.
313. Kricheldorf, H.R. and V. Linzer, Liquid crystalline polyimides: 18. Thermotropic polyimides based on biphenyl-3,3′,4,4′-tetracarboxylic anhydride. *Polymer*, 1995. **36**(9): pp. 1893–1902.
314. Dingemans, T.J. et al., Poly(ether imide)s from diamines with para-, meta-, and ortho-arylene substitutions: Synthesis, characterization, and liquid crystalline properties. *Macromolecules*, 2008. **41**(7): pp. 2474–2483.
315. Costa, G. et al., Segmented polyimides with poly(ethylene oxide) blocks exhibiting liquid crystallinity. *Macromolecules*, 2008. **41**(3): pp. 1034–1040.
316. Shoji, Y. et al., Synthesis and liquid crystalline behavior of laterally substituted polyimides with siloxane linkages. *Macromolecules*, 2010. **43**(21): pp. 8950–8956.

317. Hasegawa, M., J. Ishii, and Y. Shindo, Polyimide/Polyimide blend miscibility probed by perylenetetracarboxydiimide fluorescence. *Macromolecules*, 1999. **32**(19): pp. 6111–6119.
318. Liang, K. et al., Thermal and rheological properties of miscible polyethersulfone/polyimide blends. *Journal of Polymer Science Part B: Polymer Physics*, 1992. **30**(5): pp. 465–476.
319. Jou, J.-H. and P.-T. Huang, Compatibility effect on moisture diffusion in polyimide blends. *Polymer*, 1992. **33**(6): pp. 1218–1222.
320. Blizard, K. and R. Haghighat, Processing and properties of polyimide melt blends containing a thermotropic liquid crystalline polymer. *Polymer Engineering & Science*, 1993. **33**(13): pp. 799–807.
321. García, M.G., J. Marchese, and N.A. Ochoa, Aliphatic–aromatic polyimide blends for H2 separation. *International Journal of Hydrogen Energy*, 2010. **35**(17): pp. 8983–8992.
322. Tang, H. et al., Study on structure-property relationship of thermosetting polyimide/thermoplastic polyimide blends. *Macromolecular Rapid Communications*, 1994. **15**(9): pp. 677–682.
323. Zhang, P. et al., Study on structure-property relationship of polyimide blends, 2. Structure and miscibility of polyimide PBPI-E/PTI-E blends. *Die Makromolekulare Chemie*, 1993. **194**(7): pp. 1871–1877.
324. Jou, J.-H. and P.-T. Huang, X-ray diffraction study of polyimide blends compatibility. *Polymer Journal*, 1990. **22**(10): pp. 909–918.
325. Makhija, S., E. Pearce, and T. Kwei, Kinetics of imidization of poly (amic acid) in miscible and immiscible polymer blends. *Journal of Applied Polymer Science*, 1992. **44**(5): pp. 917–925.
326. Jung, K.-H. and J.P. Ferraris, Preparation and electrochemical properties of carbon nanofibers derived from polybenzimidazole/polyimide precursor blends. *Carbon*, 2012. **50**(14): pp. 5309–5315.
327. Chung, T.-S., W.F. Guo, and Y. Liu, Enhanced Matrimid membranes for pervaporation by homogenous blends with polybenzimidazole (PBI). *Journal of Membrane Science*, 2006. **271**(1–2): pp. 221–231.
328. Ahn, T.-K., M. Kim, and S. Choe, Hydrogen-bonding strength in the blends of polybenzimidazole with BTDA-and DSDA-based polyimides. *Macromolecules*, 1997. **30**(11): pp. 3369–3374.
329. Zhan, Z. et al., High performance zeolite NaA membranes synthesized on the inner surface of zeolite/PES–PI blend composite hollow fibers. *Journal of Membrane Science*, 2014. **471**: pp. 299–307.
330. Mansourpanah, Y. et al., Changing the performance and morphology of polyethersulfone/polyimide blend nanofiltration membranes using trimethylamine. *Desalination*, 2010. **256**(1–3): pp. 101–107.
331. Lee, D.S. and G. Quin, Miscibility of polybenzimidazole/polyimidesulfone and related copolymers blends. *Polymer Journal*, 1989. **21**(9): pp. 751–762.
332. Nakata, S., M.-A. Kakimoto, and Y. Imai, Infrared spectroscopic and thermal investigations of a new miscible blend system composed of 3, 4'-oxydianiline derived polyisophthalamide and 4,4'-oxydianiline/3,3', 4,4'-biphenyltetracarboxylic dianhydride based polyimide. *Polymer Journal*, 1993. **25**(6): pp. 569–576.
333. Bastida, S., J.I. Eguiazábal, and J. Nazábal, Effects of mixing procedure on the morphology and properties of poly(ether imide)/thermotropic copolyester blends. *European Polymer Journal*, 1999. **35**(9): pp. 1661–1669.
334. Bose, S. et al., Synthesis and effect of polyphosphazenes on the thermal, mechanical and morphological properties of poly(etherimide)/thermotropic liquid crystalline polymer blend. *Materials & Design*, 2010. **31**(3): pp. 1148–1155.
335. Rath, T. et al., Mechanical, morphological and thermal properties of in situ ternary composites based on poly(ether imide), silicone rubber and liquid crystalline polymer. *Materials Science and Engineering: A*, 2008. **490**(1–2): pp. 198–207.
336. Seo, Y. et al., Compatibilizing effect of a poly(ester imide) on the properties of the blends of poly(ether imide) and a thermotropic liquid crystalline polymer: 1. Compatibilizer synthesis and thermal and rheological properties of the *in situ* composite system. *Polymer*, 1995. **36**(3): pp. 515–523.
337. Seo, Y. et al., Compatibilizing effect of a poly(ester imide) on the properties of the blends of poly(ether imide) and a thermotropic liquid crystalline polymer: 2. Morphology and mechanical properties of the *in situ* composite system. *Polymer*, 1995. **36**(3): pp. 525–534.
338. Sauer, B.B. and B.S. Hsiao, Miscibility of three different poly(aryl ether ketones) with a high melting thermoplastic polyimide. *Polymer*, 1993. **34**(15): pp. 3315–3318.
339. Chang, K.Y. and Y.D. Lee, Molecular composite. I. Novel block copolymers of liquid-crystalline polyamide and amorphous polyimide: Synthesis and characterization. *Journal of Polymer Science Part A: Polymer Chemistry*, 1993. **31**(11): pp. 2775–2784.
340. Oishi, Y. et al., Preparation and properties of molecular composite films of block copolyimides based on rigid rod and semi-flexible segments. *Polymer Journal*, 1989. **21**(10): pp. 771–780.
341. Oishi, Y. et al., Synthesis and properties of new block copolymers based on poly(ether sulfone) and polyimides. *Journal of Polymer Science Part A: Polymer Chemistry*, 1993. **31**(4): pp. 933–938.

342. Chang, K.Y., H.M. Chang, and Y.D. Lee, Molecular composite. II. Novel block copolymer and semi-interpenetrating polymer network of rigid polyamide and flexible polyimide. *Journal of Polymer Science Part A: Polymer Chemistry*, 1994. **32**(14): pp. 2629–2639.
343. Eck, T. and H.F. Gruber, Thermostable block copolymers, 1. Thermotropic poly(ester-imide) block copolymers. *Macromolecular Chemistry and Physics*, 1994. **195**(11): pp. 3541–3565.
344. Hedrick, J. et al., Imide–aryl ether benzoxazole block copolymers. *Journal of Polymer Science Part A: Polymer Chemistry*, 1990. **28**(9): pp. 2255–2268.
345. Ghosh, A. et al., New fluorinated poly(imide siloxane) random and block copolymers with variation of siloxane loading. *Journal of Macromolecular Science, Part A: Pure and Applied Chemistry*, 2010. **47**(7): pp. 671–680.
346. Fukami, A. et al., Dynamic mechanical properties of quinazolone–imide block copolymer. *Journal of Applied Polymer Science*, 1991. **42**(11): pp. 3065–3071.
347. Jiang, X. et al., New fluorinated siloxane-imide block copolymer membranes for application in organophilic pervaporation. *Desalination*, 2011. **265**(1–3): pp. 74–80.
348. Ghosh, A. et al., Synthesis and characterization of fluorinated poly(imide siloxane) block copolymers. *European Polymer Journal*, 2009. **45**(5): pp. 1561–1569.
349. Hamciuc, E. et al., Polyimide–polydimethylsiloxane copolymers containing nitrile groups. *European Polymer Journal*, 2009. **45**(1): pp. 182–190.
350. Rogers, M. et al., Perfectly alternating segmented polyimide-polydimethyl siloxane copolymers via transimidization. *Journal of Polymer Science Part A: Polymer Chemistry*, 1994. **32**(14): pp. 2663–2675.
351. Nakata, S. et al., Synthesis and properties of new block copolymers based on polydimethylsiloxane and tetraphenylethylene-containing polyimide. *Journal of Polymer Science Part A: Polymer Chemistry*, 1993. **31**(13): pp. 3425–3432.
352. Grade, M.M. and J.W. Verbicky, Polyimide–polyformal block copolymers. *Journal of Polymer Science Part A: Polymer Chemistry*, 1989. **27**(5): pp. 1467–1480.
353. Hedrick, J. et al., High temperature nanofoams derived from rigid and semi-rigid polyimides. *Polymer*, 1995. **36**(14): pp. 2685–2697.
354. Takeichi, T. et al., Effect of in situ-formed polydimethylsiloxane on the properties of polyimide hybrids. *Reactive and Functional Polymers*, 2010. **70**(10): pp. 755–760.
355. Yilgör, E. and I. Yilgör, Silicone containing copolymers: Synthesis, properties and applications. *Progress in Polymer Science*, 2014. **39**(6): pp. 1165–1195.
356. Ghosh, A. et al., Synthesis, characterization, and properties of new siloxane grafted copolyimides. *Journal of Applied Polymer Science*, 2012. **123**(5): pp. 2959–2967.
357. Ghosh, A. et al., New semifluorinated siloxane-grafted copolyimides: Synthesis and comparison with their linear analogs. *Macromolecular Materials and Engineering*, 2011. **296**(5): pp. 391–400.
358. Pei, X., G. Chen, and X. Fang, Preparation and characterization of poly(imide siloxane) block copolymers based on diphenylthioether dianhydride isomer mixtures. *High Performance Polymers*, 2011. **23**(8): pp. 625–632.
359. Garg, P., R.P. Singh, and V. Choudhary, Selective polydimethylsiloxane/polyimide blended IPN pervaporation membrane for methanol/toluene azeotrope separation. *Separation and Purification Technology*, 2011. **76**(3): pp. 407–418.
360. Cha, J.-R. and M.-S. Gong, Preparation of epoxy/polyelectrolyte IPNs for flexible polyimide-based humidity sensors and their properties. *Sensors and Actuators B: Chemical*, 2013. **178**: pp. 656–662.
361. Wang, X.Y. et al., Thermosetting polyimide resin matrix composites with interpenetrating polymer networks for precision foil resistor chips based on special mechanical performance requirements. *Applied Surface Science*, 2014. **299**: pp. 73–80.
362. Jiang, B. et al., Proton conducting membranes based on semi-interpenetrating polymer network of fluorine-containing polyimide and perfluorosulfonic acid polymer via click chemistry. *Electrochimica Acta*, 2014. **132**: pp. 457–464.
363. Pan, H. et al., Semi-interpenetrating polymer networks based-on end-group crosslinked fluorine-containing polyimide via click chemistry. *Electrochimica Acta*, 2013. **89**: pp. 577–584.
364. Margolis, J.M., *Engineering plastics handbook*. 2006, New York: McGraw-Hill.
365. Peng, N. et al., Evolution of ultra-thin dense-selective layer from single-layer to dual-layer hollow fibers using novel Extem® polyetherimide for gas separation. *Journal of Membrane Science*, 2010. **360**(1–2): pp. 48–57.
366. Takahashi, T. et al., Supermolecular structure of a new, wholly aromatic thermoplastic polyimide. *Journal of Macromolecular Science—Physics*, 1994. **33**(1): pp. 63–74.

367. Ma, S.P. et al., Morphology of solution-cast thin films of wholly aromatic thermoplastic polyimides with various molecular weights. *Polymer*, 1994. **35**(26): pp. 5618–5625.
368. Goodwin, A.A. et al., Absorption of low molecular weight penetrants by a thermoplastic polyimide. *Polymer*, 2000. **41**(19): pp. 7263–7271.
369. Khoee, S. et al., An investigation into the improvement of adhesive strength of polyimides by incorporation of elastomeric nanoparticles. *Journal of Colloid and Interface Science*, 2009. **336**(2): pp. 872–878.
370. Foroutan, M. and S. Khoee, Thermal degradation kinetics of poly(imide-silica) hybrid films. I. Nanocomposites prepared from HFPA and bis(4-amino phenyl) ether by sol–gel process. *Journal of Applied Polymer Science*, 2007. **104**(5): pp. 3228–3235.
371. Okuyama, K., H. Sakaitani, and H. Arikawa, X-ray structure analysis of a thermoplastic polyimide. *Macromolecules*, 1992. **25**(26): pp. 7261–7267.
372. Huo, P.P., J.B. Friler, and P. Cebe, NEW-TPI thermoplastic polyimide: Thermal analysis and small-angle X-ray scattering. *Polymer*, 1993. **34**(21): pp. 4387–4398.
373. Friler, J.B. and P. Cebe, Development of crystallinity in NEW-TPI polyimide. *Polymer Engineering & Science*, 1993. **33**(10): pp. 587–597.
374. Huo, P.P. and P. Cebe, NEW-TPI thermoplastic polyimide: Dielectric and dynamic mechanical relaxation. *Polymer*, 1993. **34**(4): pp. 696–704.
375. Teverovsky, J.B. et al., Effect of zone annealing on LARC-CPI thermoplastic polyimide. *Journal of Applied Polymer Science*, 1994. **54**(4): pp. 497–505.
376. Aihara, Y. and P. Cebe, Zone drawn NEW-TPI thermoplastic polyimide and its blends with Xydar liquid crystalline polymer. *Polymer Engineering & Science*, 1994. **34**(16): pp. 1275–1286.
377. Hirade, T. et al., Radiation effect of aromatic thermoplastic polyimide (new-TPI). *Polymer*, 1991. **32**(14): pp. 2499–2504.
378. Muellerleile, J.T. et al., Crystallization behaviour and morphological features of LARC-CPI. *Polymer*, 1993. **34**(4): pp. 789–806.
379. Gautreaux, C., J. Pratt, and T. St Clair, A study of crystalline transitions in a thermoplastic polyimide. *Journal of Polymer Science Part B: Polymer Physics*, 1992. **30**(1): pp. 71–82.
380. Brillhart, M.V. and P. Cebe, Thermal expansion of the crystal lattice of novel thermoplastic polyimides. *Journal of Polymer Science Part B: Polymer Physics*, 1995. **33**(6): pp. 927–936.
381. Wang, J. et al., Solvent-induced crystallization of aromatic polyimide. *Polymer*, 1989. **30**(4): pp. 718–721.
382. Hou, T., N. Wakelyn, and T. St Clair, Investigation of crystalline changes in LaRC-TPI powders. *Journal of Applied Polymer Science*, 1988. **36**(8): pp. 1731–1739.
383. Hou, T. et al., IM7/LARC ™-ITPI polyimide composites. *Polymer*, 1994. **35**(23): pp. 4956–4969.
384. Tanaka, K. et al., Effects of trifluoromethyl side groups on gas permeability and permselectivity in polyimides. *Polymer Journal*, 1994. **26**(10): pp. 1186–1189.
385. Tanaka, K., H. Kita, and K.I. Okamoto, Sorption of carbon dioxide in fluorinated polyimides. *Journal of Polymer Science Part B: Polymer Physics*, 1993. **31**(9): pp. 1127–1133.
386. Tanaka, K. et al., Permeability and permselectivity of gases in fluorinated and non-fluorinated polyimides. *Polymer*, 1992. **33**(3): pp. 585–592.
387. Tanaka, K. et al., Effect of methyl substituents on permeability and permselectivity of gases in polyimides prepared from methyl-substituted phenylenediamines. *Journal of Polymer Science Part B: Polymer Physics*, 1992. **30**(8): pp. 907–914.
388. Tanaka, K. et al., Gas permeability and permselectivity in homo-and copolyimides from 3,3',4, 4'-biphenyltetracarboxylic dianhydride and 3,3'- and 4,4'-diaminodiphenylsulfones. *Polymer Journal*, 1990. **22**(5): pp. 381–385.
389. Tanaka, K. et al., The effect of morphology on gas permeability and permselectivity in polyimide based on 3,3',4,4'-biphenyltetracarboxylic dianhydride and 4,4'-oxydianiline. *Polymer Journal*, 1989. **21**(2): pp. 127–135.
390. Chun, B.W., Packing density and gas permeability of copolyimides containing bicyclo[2,2,2]oct-7-ene-2,3,5,6-tetracarboxylic dianhydride constituent. *Journal of Polymer Science Part B: Polymer Physics*, 1995. **33**(4): pp. 731–737.
391. Costello, L. and W. Koros, Thermally stable polyimide isomers for membrane-based gas separations at elevated temperatures. *Journal of Polymer Science Part B: Polymer Physics*, 1995. **33**(1): pp. 135–146.
392. Zoia, G. et al., Permeability relationships of polyimide copolymers. *Journal of Polymer Science Part B: Polymer Physics*, 1994. **32**(1): pp. 53–58.
393. Coleman, M. and W. Koros, The transport properties of polyimide isomers containing hexafluoroisopropylidene in the diamine residue. *Journal of Polymer Science Part B: Polymer Physics*, 1994. **32**(11): pp. 1915–1926.

394. Stern, S., Y. Liu, and W. Feld, Structure/permeability relationships of polyimides with branched or extended diamine moieties. *Journal of Polymer Science Part B: Polymer Physics*, 1993. **31**(8): pp. 939–951.
395. Stern, S. et al., Structure/permeability relationships of polyimide membranes. Applications to the separation of gas mixtures. *Journal of Polymer Science Part B: Polymer Physics*, 1989. **27**(9): pp. 1887–1909.
396. Yamamoto, H. et al., Structure/permeability relationships of polyimide membranes. II. *Journal of Polymer Science Part B: Polymer Physics*, 1990. **28**(12): pp. 2291–2304.
397. Hirose, T. et al., The solubility of carbon dioxide and methane in polyimides at elevated pressures. *Journal of Polymer Science Part B: Polymer Physics*, 1991. **29**(3): pp. 341–347.
398. Okamoto, K.I. et al., Sorption and diffusion of water vapor in polyimide films. *Journal of Polymer Science Part B: Polymer Physics*, 1992. **30**(11): pp. 1223–1231.
399. Moylan, C.R., M.E. Best, and M. Ree, Solubility of water in polyimides: Quartz crystal microbalance measurements. *Journal of Polymer Science Part B: Polymer Physics*, 1991. **29**(1): pp. 87–92.
400. Lokhandwala, K., S. Nadakatti, and S. Stern, Solubility and transport of water vapor in some 6FDA-based polyimides. *Journal of Polymer Science Part B: Polymer Physics*, 1995. **33**(6): pp. 965–975.
401. Moe, M. et al., Effects of film history on gas transport in a fluorinated aromatic polyimide. *Journal of Applied Polymer Science*, 1988. **36**(8): pp. 1833–1846.
402. Okamoto, K.-I. et al., Gas permeability and permselectivity of polyimides prepared from 4,4′-diaminotriphenylamine. *Polymer Journal*, 1992. **24**(5): pp. 451–457.
403. Hachisuka, H. et al., Gas transport properties of annealed polyimide films. *Journal of Polymer Science Part B: Polymer Physics*, 1991. **29**(1): pp. 11–16.
404. Kumar Sen, S. and S. Banerjee, High Tg, processable fluorinated polyimides containing benzoisoindoledione unit and evaluation of their gas transport properties. *RSC Advances*, 2012. **2**(15): pp. 6274–6289.
405. Swaidan, R. et al., Effects of hydroxyl-functionalization and sub-Tg thermal annealing on high pressure pure- and mixed-gas CO_2/CH_4 separation by polyimide membranes based on 6FDA and triptycene-containing dianhydrides. *Journal of Membrane Science*, 2015. **475**: pp. 571–581.
406. Guiver, M.D. and Y.M. Lee, Polymer rigidity improves microporous membranes. *Science*, 2013. **339**(6117): pp. 284–285.
407. Robeson, L.M., The upper bound revisited. *Journal of Membrane Science*, 2008. **320**(1): pp. 390–400.
408. Bera, D. et al., Highly gas permeable aromatic polyamides containing adamantane substituted triphenylamine. *Journal of Membrane Science*, 2015. **474**: pp. 20–31.
409. Hu, Y.-C. et al., Novel triphenylamine-containing ambipolar polyimides with pendant anthraquinone moiety for polymeric memory device, electrochromic and gas separation applications. *Journal of Materials Chemistry*, 2012. **22**(38): pp. 20394–20402.
410. Chang, C.-W. et al., Novel organosoluble aromatic polyimides bearing pendant methoxy-substituted triphenylamine moieties: Synthesis, electrochromic, and gas separation properties. *Journal of Polymer Science Part A: Polymer Chemistry*, 2008. **46**(24): pp. 7937–7949.
411. Yen, H.-J. et al., Triphenylamine-based polyimides with trimethyl substituents for gas separation membrane and electrochromic applications. *Journal of Polymer Science Part A: Polymer Chemistry*, 2011. **49**(16): pp. 3637–3646.
412. Li, X. et al., Constructing CO_2 transport passageways in Matrimid® membranes using nanohydrogels for efficient carbon capture. *Journal of Membrane Science*, 2015. **474**: pp. 156–166.
413. Park, H.B. et al., Thermally rearranged (TR) polymer membranes for CO_2 separation. *Journal of Membrane Science*, 2010. **359**(1–2): pp. 11–24.
414. Jiang, Y. et al., Cavity size, sorption and transport characteristics of thermally rearranged (TR) polymers. *Polymer*, 2011. **52**(10): pp. 2244–2254.
415. Ribeiro, C.P. et al., Aromatic polyimide and polybenzoxazole membranes for the fractionation of aromatic/aliphatic hydrocarbons by pervaporation. *Journal of Membrane Science*, 2012. **390–391**: pp. 182–193.
416. Calle, M., A.E. Lozano, and Y.M. Lee, Formation of thermally rearranged (TR) polybenzoxazoles: Effect of synthesis routes and polymer form. *European Polymer Journal*, 2012. **48**(7): pp. 1313–1322.
417. Scholes, C.A. et al., Thermal rearranged poly(benzoxazole)/polyimide blended membranes for CO_2 separation. *Separation and Purification Technology*, 2014. **124**: pp. 134–140.
418. Wind, J.D., D.R. Paul, and W.J. Koros, Natural gas permeation in polyimide membranes. *Journal of Membrane Science*, 2004. **228**(2): pp. 227–236.
419. Tanaka, K. et al., Gas permeation and separation properties of sulfonated polyimide membranes. *Polymer*, 2006. **47**(12): pp. 4370–4377.

420. Er, O.O. et al., Copolyimide membranes for gas separation. *Desalination*, 2006. **200**(1–3): pp. 259–261.
421. Wang, Y.-C. et al., Sorption and transport properties of gases in aromatic polyimide membranes. *Journal of Membrane Science*, 2005. **248**(1–2): pp. 15–25.
422. Velioğlu, S., M.G. Ahunbay, and S.B. Tantekin-Ersolmaz, Investigation of CO_2-induced plasticization in fluorinated polyimide membranes via molecular simulation. *Journal of Membrane Science*, 2012. **417–418**: pp. 217–227.
423. Suzuki, T., Y. Yamada, and J. Sakai, Gas transport properties of ODPA-TAPOB hyperbranched polyimide-silica hybrid membranes. *High Performance Polymers*, 2006. **18**(5): pp. 655–664.
424. Xiao, Y. et al., Structure and properties relationships for aromatic polyimides and their derived carbon membranes: Experimental and simulation approaches. *The Journal of Physical Chemistry B*, 2005. **109**(40): pp. 18741–18748.
425. Xiao, Y. et al., Effects of brominating Matrimid polyimide on the physical and gas transport properties of derived carbon membranes. *Macromolecules*, 2005. **38**(24): pp. 10042–10049.
426. Chenar, M.P. et al., Gas permeation properties of commercial polyphenylene oxide and Cardo-type polyimide hollow fiber membranes. *Separation and Purification Technology*, 2006. **51**(3): pp. 359–366.
427. Huang, S.-H. et al., Gas separation properties of aromatic poly (amide-imide) membranes. *European Polymer Journal*, 2006. **42**(1): pp. 140–148.
428. Hasegawa, M. and J. Nakano, Colorless polyimides derived from cycloaliphatic tetracarboxylic dianhydrides with controlled steric structures (4). Applications to positive-type photosensitive polyimide systems with controlled extents of imidization. *Journal of Photopolymer Science and Technology*, 2009. **22**(3): pp. 411–415.
429. Hasegawa, M. and A. Tominaga, Fluorene-containing poly(ester imide)s and their application to positive-type photosensitive heat-resistant materials. *Macromolecular Materials and Engineering*, 2011. **296**(11): pp. 1002–1017.
430. Jin, X. and H. Ishii, A novel positive-type photosensitive polyimide based on soluble block copolyimide showing low dielectric constant with a low-temperature curing process. *Journal of Applied Polymer Science*, 2006. **100**(5): pp. 4240–4246.
431. Higashihara, T. et al., Recent progress in negative-working photosensitive and thermally stable polymers. *Reactive and Functional Polymers*, 2013. **73**(2): pp. 303–315.
432. Yoshida, M. et al., High-modulus negative photosensitive polyimide for i-line. *Journal of Photopolymer Science and Technology*, 2014. **27**(2): pp. 207–210.
433. Li, T.-L. and S.L.-C. Hsu, Preparation and properties of thermally conductive photosensitive polyimide/boron nitride nanocomposites. *Journal of Applied Polymer Science*, 2011. **121**(2): pp. 916–922.
434. Jin, X. and H. Ishii, A novel positive-type photosensitive polyimide based on soluble block copolyimide showing low dielectric constant with a low-temperature curing process. *Journal of Applied Polymer Science*, 2006. **100**(5): pp. 4240–4246.
435. Jin, X. and H. Ishii, A novel positive-type photosensitive polyimide having excellent transparency based on soluble block copolyimide with hydroxyl group and diazonaphthoquintone. *Journal of Applied Polymer Science*, 2005. **96**(5): pp. 1619–1624.
436. Lien-Chung Hsu, S., H.-T. Chen, and S.-J. Tsai, Novel positive-working and aqueous-base-developable photosensitive poly(imide benzoxazole) precursor. *Journal of Polymer Science Part A: Polymer Chemistry*, 2004. **42**(23): pp. 5990–5998.
437. Hsu, S.L.-C. et al., Synthesis and characterization of a positive-working, aqueous-base-developable photosensitive polyimide precursor. *Journal of Applied Polymer Science*, 2002. **86**(2): pp. 352–358.
438. Morita, K. et al., New positive-type photosensitive polyimide having sulfo groups. *Polymer*, 2003. **44**(20): pp. 6235–6239.
439. Watanabe, Y. et al., A new positive-type photosensitive alkaline-developable alicyclic polyimide based on poly(amic acid silylester) as a polyimide precursor and diazonaphthoquinone as a photosensitive compound. *Chemistry of Materials*, 2002. **14**(4): pp. 1762–1766.
440. Jiang, X. et al., A novel negative photoinitiator-free photosensitive polyimide. *Polymer*, 2006. **47**(9): pp. 2942–2945.
441. Omote, T., K.I. Koseki, and T. Yamaoka, Fluorine-containing photoreactive polymide 5: A novel positive-type polyimide based on photoinduced acidolysis. *Journal of Polymer Science Part C: Polymer Letters*, 1990. **28**(2): pp. 59–64.
442. Omote, T., K.I. Koseki, and T. Yamaoka, Fluorine-containing photoreactive polyimide. 6. Synthesis and properties of a novel photoreactive polyimide based on photo-induced acidolysis and the kinetics for its acidolysis. *Macromolecules*, 1990. **23**(22): pp. 4788–4795.

443. Omote, T. et al., Fluorine-containing photoreactive polyimide. 7. Photochemical reaction of pendant 1, 2-naphthoquinone diazide moieties in novel photoreactive polyimides. *Macromolecules*, 1990. **23**(22): pp. 4796–4802.
444. Lee, H.R. and Y.D. Lee, Synthesis, characterization, and kinetics studies of organic soluble photosensitive copolyimide. *Journal of Polymer Science Part A: Polymer Chemistry*, 1989. **27**(5): pp. 1481–1497.
445. Ho, B.C., Y.S. Lin, and Y.D. Lee, Synthesis and characteristics of organic soluble photoactive polyimides. *Journal of Applied Polymer Science*, 1994. **53**(11): pp. 1513–1524.
446. Chiang, W.Y. and W.P. Mei, Synthesis and characterization of photosensitive polyimides from methylthiomethyl-substituted 4,4'-diaminodiphenylmethanes and 3,3',4,4'-benzophenonetetracarboxylic dianhydride. *Journal of Polymer Science Part A: Polymer Chemistry*, 1993. **31**(5): pp. 1195–1201.
447. Lee, H.R., T.A. Yu, and Y.D. Lee, Characterization and dissolution studies of a benzophenone-containing organic-soluble polyimide. *Macromolecules*, 1990. **23**(2): pp. 502–509.
448. Padmanaban, M., M.-A. Kakimoto, and Y. Imai, Preparation and properties of new disilane-containing polyamide and polyimides from diaminodisilanes and bisphenol compounds. *Polymer Journal*, 1990. **22**(7): pp. 587–592.
449. Padmanaban, M. et al., Synthesis and characterization of a new soluble photosensitive polyimide from bis(4-aminophenyl) tetramethyldisilane and 4,4'-sulfonyldiphthalic dianhydride. *Die Makromolekulare Chemie, Rapid Communications*, 1990. **11**(1): pp. 15–18.
450. Padmanaban, M. et al., Preparation and properties of new photosensitive polyimides from bis (4-aminophenyl) tetramethyldisilane and various aromatic tetracarboxylic dianhydrides. *Journal of Polymer Science Part A: Polymer Chemistry*, 1990. **28**(12): pp. 3261–3269.
451. Tambe, S.M. et al., Synthesis and characterization of thermally stable second-order nonlinear optical side-chain polyimides containing thiazole and benzothiazole push–pull chromophores. *Optical Materials*, 2009. **31**(6): pp. 817–825.
452. Van den Broeck, K. et al., High glass transition chromophore functionalised polyimides for second-order nonlinear optical applications. *Polymer*, 2001. **42**(8): pp. 3315–3322.
453. Chen, Y.-C. et al., Thermally stable hyperbranched nonlinear optical polyimides using an "A2+B3" approach. *Materials Chemistry and Physics*, 2011. **127**(1–2): pp. 107–113.
454. Lin, J.T. et al., Poled polymeric nonlinear optical materials. Exceptional second harmonic generation temporal stability of a chromophore-functionalized polyimide. *Chemistry of Materials*, 1992. **4**(6): pp. 1148–1150.
455. Yu, D. and L. Yu, Design and synthesis of functionalized polyimides for second-order nonlinear optics. *Macromolecules*, 1994. **27**(23): pp. 6718–6721.
456. Miller, R. et al., Donor-embedded nonlinear optical side chain polyimides containing no flexible tether: materials of exceptional thermal stability for electrooptic applications. *Macromolecules*, 1995. **28**(14): pp. 4970–4974.
457. Jen, A.-Y. et al., Thermally stable nonlinear optical polyimides: Synthesis and electro-optic properties. *Journal of the Chemical Society. Chemical Communications*, 1994. **8**: pp. 965–966.
458. Becker, M.W. et al., Large and stable nonlinear optical effects observed for a polyimide covalently incorporating a nonlinear optical chromophore. *Chemistry of Materials*, 1994. **6**(2): pp. 104–106.
459. Deligöz, H. et al., Electrical properties of conventional polyimide films: Effects of chemical structure and water uptake. *Journal of Applied Polymer Science*, 2006. **100**(1): pp. 810–818.
460. Tanaka, T., G.C. Montanari, and R. Mulhaupt, Polymer nanocomposites as dielectrics and electrical insulation-perspectives for processing technologies, material characterization and future applications. *Dielectrics and Electrical Insulation, IEEE Transactions on*, 2004. **11**(5): pp. 763–784.
461. Yang, C., P.C. Irwin, and K. Younsi, The future of nanodielectrics in the electrical power industry. *Dielectrics and Electrical Insulation, IEEE Transactions on*, 2004. **11**(5): pp. 797–807.
462. Yang, C. and P.C. Irwin. The electrical conduction in polyimide nanocomposites, in *Electrical Insulation and Dielectric Phenomena, 2003. Annual Report. Conference on*. 2003.
463. Guo, X. et al., Novel sulfonated polyimides as polyelectrolytes for fuel cell application. 2. Synthesis and proton conductivity of polyimides from 9,9-bis(4-aminophenyl)fluorene-2,7-disulfonic acid. *Macromolecules*, 2002. **35**(17): pp. 6707–6713.
464. Genies, C. et al., Soluble sulfonated naphthalenic polyimides as materials for proton exchange membranes. *Polymer*, 2001. **42**(2): pp. 359–373.
465. Genies, C. et al., Stability study of sulfonated phthalic and naphthalenic polyimide structures in aqueous medium. *Polymer*, 2001. **42**(12): pp. 5097–5105.

466. Besse, S. et al., Sulfonated polyimides for fuel cell electrode membrane assemblies (EMA). *Journal of New Materials for Electrochemical Systems*, 2002. **5**: pp. 109–112.
467. Asensio, J., S. Borros, and P. Gomez-Romero, Proton-conducting polymers based on benzimidazoles and sulfonated benzimidazoles. *Journal of Polymer Science Part A: Polymer Chemistry*, 2002. **40**(21): pp. 3703–3710.
468. Bae, J.-M. et al., Properties of selected sulfonated polymers as proton-conducting electrolytes for polymer electrolyte fuel cells. *Solid State Ionics*, 2002. **147**(1): pp. 189–194.

17 Polyphenylquinoxalines*

Jin-Gang Liu

CONTENTS

17.1 Introduction and Scope 533
17.2 Historical Development of Polyphenylquinoxalines 535
17.3 Polymerization and Processing Technologies of Polyphenylquinoxalines 537
    17.3.1 Polymerization Chemistry 537
        17.3.1.1 Monomer Synthesis Chemistry 537
        17.3.1.2 PPQ Synthesis Chemistry 543
    17.3.2 PPQ Processing Technologies 554
        17.3.2.1 PPQ Films 555
        17.3.2.2 PPQ Moldings and Composites 555
        17.3.2.3 PPQ Foams 559
17.4 Applications of Polyphenylquinoxalines 561
    17.4.1 Energy Applications 561
    17.4.2 Optical Applications 563
    17.4.3 Microelectronic Applications 564
17.5 Conclusions 564
Acknowledgments 564
References 565

17.1 INTRODUCTION AND SCOPE

Polyphenylquinoxalines (PPQs), first developed by Hergenrother in 1967, represent a class of high-performance thermoplastic heteroaromatic polymers [1]. They are characterized by the phenyl-substituted quinoxaline ring in the structures. The special molecular structure endows the polymers with many favorable properties, such as high thermal and thermo-oxidative stability resulting from their highly conjugated backbones; good thermoplasticity because of the linear molecular skeletons; good solubility in organic solvents and relatively low dielectric constants caused by the bulky molecular packing; excellent hydrolytic resistance and low moisture absorption due to their relatively low contents of polar components; and so on [2]. Thus, PPQs have found various potential applications as functional or structural components in high-tech fields, including as matrixes for high-temperature composites used in extreme environments [3], interlayer dielectrics (ILDs) or insulation coatings for microelectronic fabrication [4], structural adhesives for high-temperature bonding [5], proton-exchange membranes (PEMs) for fuel cells [6–9], ultrafiltration membranes for water or gas treatments [10], electron transport materials (ETMs) for organic light-emitting diodes (OLEDs) [11–14], and nonlinear optical materials for optoelectronic devices [15].

    Although PPQs have been extensively studied as an important functional polymer for many decades, they have not achieved wide commercialization like the other heteroaromatic polymers invented at the same time, such as polyimides (PIs). On one hand, this is mainly due to the relatively high cost and very limited commercial availability of the starting monomers, bis(α-diketone) and bis(o-diamine) monomers [16,17]. On the other hand, the functionality of conventional PPQs needs to be further expanded so as to meet the demands of high-tech fields. Considering the main obstacles

---

* Based in parts on the first-edition chapter on polyphenylquinoxalines.

hindering the rapid development of PPQs, several basic research and practical application issues are now being performed. They include (1) development of more convenient, effective, and cost-saving preparation methods for both monomers and PPQs; (2) enrichment of commercially available bis(o-diamine) and bis(α-diketone) monomers; (3) functionalization of conventional PPQs; and (4) exploration of environmentally friendly and low-cost processing techniques for PPQs. In fact, the investigations mentioned above have never ceased since the birth of PPQs. Several research groups worldwide have continuously carried out the basic and applied research for the polymers. The progress of their work on PPQs has been intermittently reviewed in the literature. For instance, in 1976, the inventor of PPQs, Hergenrother, reviewed the synthesis, processing characteristics, thermal properties, and applications of standard and ether-containing PPQ polymers as adhesive and composites [18]. In 1983, Korshak and Rusanov reviewed the research and development of PPQs [19]. One year later, Krongauz summarized the developing status and prospects of PPQs, emphasizing the synthesis procedures for the polymers [20]. In 1986, Hergenrother reviewed the progress of high-performance thermoplastics research and modification at the NASA Langley Research Center (LaRC), USA [21]. Three thermoplastic polymers—PIs, poly(arylene ethers), and PPQs—were involved in the review. For PPQs, improving processability, mechanical properties, and service temperatures was successfully achieved by copolymerization or cross-linking procedures at LaRC, NASA. In 1989, Bruma reviewed the thermostable polymers containing phenylquinoxaline and other heterocyclic units [22]. In 1990, Béland reviewed the high-performance thermoplastic resins and their composites in his monograph [23]. In this book, the neat resin and composite properties, processing techniques, and applications of amorphous PPQs in aircraft were introduced. In the author's opinion, PPQs offered high chemical and thermal stability coupled with high mechanical properties and could produce high-quality reinforced laminates. However, PPQ is not expected to be available as a matrix for reinforced composite materials, due to its high viscosity, low modulus, and high cost. In 1997, Bruma summarized the research and application status of PPQs in the first edition of the current handbook [24]. In 1998, Lu, at the Institute of Chemistry, Chinese Academy of Sciences (ICCAS), China, reviewed the progress of PPQs in a monograph [25]. Lu et al. established the PPQ research in China and developed novel PPQs for ultrafiltration membranes, gas separation membranes, and conductive films. Lu and coworkers achieved a kilogram-scale production of 1,4-bis(phenylglyoxalyl)benzene, one of the most important monomers for PPQ synthesis, via a newly established low-toxic route. In 1999, Rabilloud, at the Institute of French Petroleum (IFP), France, published the first monograph in the literature on polyquinoxalines (PQs) and PPQs [26]. In this book, the synthesis chemistry, processing technology, and applications of PPQs were presented in detail. Especially, the detailed properties of the only commercially available PPQ, IP200 PPQ produced in IFP-Cemote, France, were introduced. In 2003, Hergenrother reviewed the use, design, synthesis, and properties of high-performance/high-temperature polymers [27]. In this review, the poly(arylene ethers) polymers containing phenylquinoxaline rings, PPQ polymers derived from the self-polymerizable monomers, and PPQ oligomers terminated with phenylethynyl (C_6H_5–C≡C–) groups were summarized. In 2004 Bruma et al. summarized the polymers containing preformed phenylquinoxaline rings [28]. In 2012, Hergenrother et al. reviewed the PQs and PPQ matrix resins and composites [29].

Based on the previous reviews, this chapter will review the latest progress on the research and development of PPQs. The scope of this review mainly covers the following aspects: (1) brief historical development and commercialization of PPQs; (2) progress of PPQ synthesis chemistry (monomer synthesis and polymer preparation); (3) newly improved polymerization and processing technologies for PPQs (films, composites, and foams); and (4) new functionalization and applications of PPQs in high-tech fields. Particular attention is paid to monomer developments for PPQs in the current chapter, which is different from the previous reviews on PPQs. It is undeniable that monomer development is a critical factor in PPQs' functionalization and commercialization. Low-cost and high-functional monomers will make the production of PPQs more economical and practical. Thus, research and development of corresponding monomers, including bis(α-diketones) and bis(o-diamines), are emphatically introduced in the review. To the best of our knowledge, there is little literature concerning the comprehensive development of monomers for PPQs. In addition, considering that the properties of PPQ-based composites and adhesives have been extensively reviewed in the literatures, they are only briefly reviewed in the current chapter. Conversely, the latest applications of PPQs in energy, optical, and microelectronic fields are discussed in detail.

17.2 HISTORICAL DEVELOPMENT OF POLYPHENYLQUINOXALINES

PPQs have been investigated as a high-performance polymer for many decades, although their development did not progress quickly enough in the past half century. Figure 17.1 shows the numbers of publications (including papers, patents, proceedings, etc.) on PPQ research from the birth of the polymer in 1967 to the present day. The numbers were searched and counted with the SciFinder database (American Chemical Society). Less than 1000 publications were found using the key word *polyphenylquinoxaline*. Indeed, limited PPQ compositions have achieved large-scale manufacturing and wide commercialization up to now. However, in recent years, with the ever-increasing demands of special functional polymers for high-tech fields, the PPQ industry is facing promising development opportunities.

Figure 17.2 briefly summarizes the histories of PPQ research and development from the 1960s to the present day. The intensive investigations in high-temperature polymers in the 1960s have led to the development of numerous classes of thermal-stable polymers, such as PQs, PIs, polybenzimidazoles (PBIs), polybenzoxazoles (PBOs), polypyrrolones (PPYs), "ladder" polymers and so on. A large proportion of the high-temperature polymers gradually died out in the following years due to the absence of good processability and mechanical properties, precluding their use in practical applications. Conversely, several heteroaromatic polymers survive to this day, and some of them have achieved commercialization and are widely used in modern industry. PPQ is one of such polymers. PPQs were originally developed in order to improve the poor thermo-oxidative stability of linear PQs at elevated temperatures. Experimental results proved the effectiveness of this molecular design. Introduction of phenyl substituents in PQs indeed led to higher decomposition temperatures in both air and inert atmospheres and better thermo-oxidative stability when subjected to 316°C and 371°C isothermal aging than the unsubstituted PQs. However, the more powerful driving force that promotes the development of PPQs is their favorable combination of processability and physical and chemical properties. These properties are highly desired not only for the conventional mechanical and electrical industry, but also in many newly rising high-tech industries, such as the microelectronic, optoelectronic, and clean-energy industries. For instance, PPQs have been proven to be some of the best hydrolytic-resistant high-temperature polymers, and they possess excellent hydrolysis resistance to various (neutral, acidic,

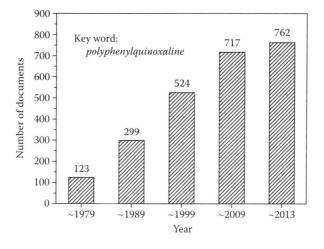

FIGURE 17.1 Publication numbers on PPQs listed in SciFinder database (August 2013).

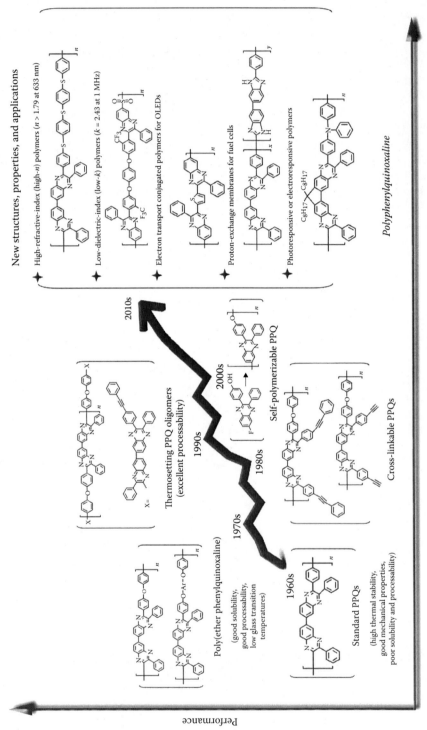

FIGURE 17.2 Brief developing histories for PPQs.

or alkaline) waters, not only at room temperature but also at elevated temperatures. PPQ films could maintain integral shapes and flexibility even after boiling in distilled water at 250°C (vapor pressure of water, 3.9 MPa) for 60 h [25]. Under the same conditions, PI films would have been completely hydrolyzed, leaving black residues. As another example, commercially available IP200 PPQ films, are not degraded when aged for 1 month in 50% sodium hydroxide solution up to 150°C or 40% sodium hydroxide solution up to 160°C. They are also stable in boiling water up to 1 year [26]. This unique characteristic makes PPQ a good candidate in various applications with high-humidity and high-temperature circumstances. Another attractive performance advantage for PPQs is that they are usually soluble in organic solvents (m-cresol, chloroform, etc.) in the form of precyclized macromolecules. Thus, PPQ films or coatings could be obtained at a relatively low temperature (below 220°C), which is very beneficial for many temperature-sensitive applications. The bulky phenyl-substituted quinoxaline rings in PPQs are not very polar, so the dielectric constants of PPQs are usually lower than the other high-temperature polymers. This feature makes PPQs a good low-dielectric-constant (low-k) material as the ILD for ultralarge integrated circuit (ULSI) fabrications [30].

The special characteristics of PPQs motivate scientists and engineers to continuously investigate the structure–property relationships of the polymers. As can be seen from Figure 17.2, the favorable and attractive properties of PPQs have greatly accelerated their development in the following 30 years after their invention in the 1960s. For example, ether-containing PPQs were developed in the 1970s in order to improve the processability of the conventional nonether PPQs [31]. Cross-linkable and thermosetting PPQ polymers or oligomers containing ethynyl or phenylethynyl side chains or end cappers were developed in the 1980s in order to improve the solvent resistance and dimensional stability or decrease the high-temperature thermoplasticity of linear PPQs at elevated temperatures [18]. In the 2000s, a novel and convenient synthesis procedure for PPQs, that is, the self-polymerization route of quinoxaline-containing monomers, was expanded, although the idea had been established at the end of the 1980s [32]. This route offered the advantage of having an inherent equal stoichiometry; thus, PPQs with high molecular weights can be easily prepared. Since the beginning of the twenty-first century, research and development on PPQs has begun to exhibit a growing trend. Various PPQs with new structures, new properties, and new applications have been developed. PPQs have been extensively investigated as high-performance components for optoelectronic, microelectronics, and energy fields.

Although PPQs show an attractive combination of properties, only a few materials have been commercialized in recent years. For example, several linear and cross-linked PPQs (401 PPQ resins) were commercially supplied by Whittaker Corp. [33,34]. The most important material is IP200 PPQ commercialized in the 1980s by Cemota Corp. [26]. The typical properties of IP200 PPQ are summarized in Table 17.1. It can be clearly seen from the properties of IP200 PPQ films that the polymers possess excellent combined properties, which are highly desirable for high-tech fields. PPQ appears to be the best hydrolytic-resistant high-temperature polymer, whose resistance to neutral and aqueous base solutions is superior to any other counterpart polymers, such as PQs and PIs. Furthermore, the superiorities of PPQs are constantly being revealed in practice. Thus, it can be anticipated that PPQs might find more applications in modern industry. As long as the cost of PPQs can be decreased to an acceptable level, their wide commercialization will become a reality in the near future.

17.3 POLYMERIZATION AND PROCESSING TECHNOLOGIES OF POLYPHENYLQUINOXALINES

17.3.1 POLYMERIZATION CHEMISTRY

17.3.1.1 Monomer Synthesis Chemistry

As mentioned above, the high cost and very limited commercially available of the starting monomers, including bis(α-diketone) and bis(o-diamine) compounds, have been among the main obstacles hindering the functionalization and commercialization of PPQs. The high cost of the monomers for

TABLE 17.1
Properties of IP200 PPQ Film

	IP200 PPQ
Solubility[a]	Soluble in phenol, cresol, xylenols, chlorophenols, hexafluoro-2-propanol. Partially soluble in chloroform, 1,1,2,2-tetrachloroethane. Not soluble in NMP, DMAc, DMF, DMSO, γ-butyrolactone (γ-BL), alcohols, ketones, aromatic hydrocarbons, pyridine.
Chemical resistance	Excellent resistance to neutral and aqueous base solutions. Medium resistance to aqueous acid solutions (not soluble in ammonium fluoride; slightly soluble in concentrated hydrofluoric acid; soluble in concentrated sulfuric acid, hydrochloric acid, and methane sulfonic acid).
Water absorption (10 μm), %	0.4 (25% RH); 1.2 (50% RH); 2.8 (95% RH)
T_{onset} (in N_2), °C	560
T_g, °C	365
Tensile strength, MPa	118.5
Tensile modulus, GPa	2.30
Elongation at break, %	10.0
Dielectric constant (1 kHz)	2.40 (35°C); 2.30 (252.5°C)
Dissipation factor (tanδ, 1 kHz)	0.0003 (35°C); 0.0005 (252.5°C)
Volume resistivity, Ω cm	3×10^{17} (0% RH); 1×10^{16} (60% RH); 9×10^{13} (95% RH)
Surface resistivity, Ω	1.1×10^{16} (0% RH); 1×10^{15} (60% RH); 2×10^{13} (95% RH)

Source: Rabilloud, G., *High-Performance Polymers. 2. Polyquinoxalines and Polyimides*, Editions Technip, Paris, 1999.
Note: DMAc, *N,N*-dimethylacetamide; DMF, *N,N*-dimethylformamide; DMSO, dimethyl sulfoxide; NMP, *N*-methyl-2-pyrrolidone; RH, relative humidity; T_g, glass transition temperature; T_{onset}, onset of thermal degradation temperature.
[a] Solubility was measured by stirring 1 g of IP200 PPQ film in 100 g of solvent at 25°C for 24 h.

PPQs is mainly due to the synthetic and purification difficulties for both bis(α-diketone) and bis(*o*-diamine) compounds. Figure 17.3 summarizes the common bis(α-diketone) and bis(*o*-diamine) monomers reported in the literature. Few of them have achieved large-scale production up till now. However, after many decades of research, the synthesis procedures for the monomers have been well established in the literature.

17.3.1.1.1 Synthesis of Bis(α-Diketones)

Basically, there are two main ways for synthesis of bis(α-diketones), including (1) an oxidative route of α-methylene ketones, acyloins, or α-phenylacetylene compounds and (2) nucleophilic substitution reactions of nitrobenzil or its derivatives with diols. The former route is illustrated in Figures 17.4 and 17.5. In the route shown in Figure 17.4, 1,4-bis(phenylglyoxaloyl)benzene(I), the most important bis(α-diketone) monomer for PPQ, is synthesized via the oxidization of various intermediates [35,36]. These intermediates involved bis(phenylketo)xylylene, bis(α-methylene ketone), or bis(α-dimethylene) compounds derived from all kind of staring materials. The intermediates were oxidized with various oxidizing agents, including selenium dioxide (SeO_2) and dimethyl sulfoxide (DMSO)/hydrobromic acid (HBr) systems. By this procedure, bis(α-diketone) monomers containing *meta*-terphenyl moieties (shown as II and III in Figure 17.4) [37] and trifunctional tris-benzil (shown as IV in Figure 17.4) [38] have also been synthesized. In the above reactions, SeO_2 has been proven to be one of the most effective oxidizing agents for bis(α-diketone) synthesis; however, it also has been proven to be a highly toxic substance to human health and environments. This greatly limits its application. Comparatively, DMSO/HBr is a promising oxidizing agent. However, the high cost of the starting materials makes this route very expensive. The route presented in

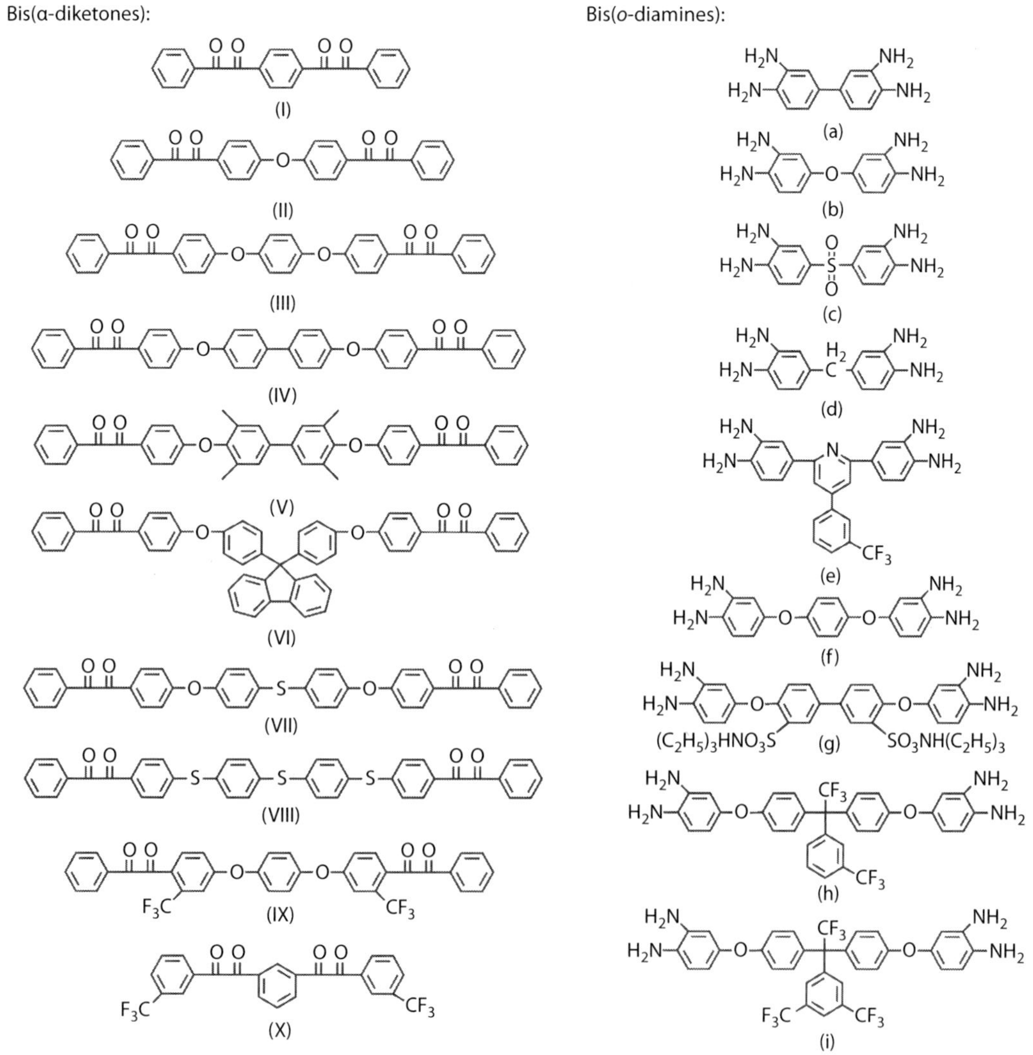

Polyphenylquinoxalines

FIGURE 17.3 Chemical structures of typical monomers for PPQ synthesis. (I) 1,4-bis(phenylglyoxaloyl)benzene; (II) 4,4′-oxydibenzil; (III) 1,4-bis(4-benzilyloxy)benzene; (IV) 4,4′-bis(4-benzilyloxy)biphenyl; (V) 4,4′-bis(4-benzilyloxy)-3,3′,5,5′-tetramethylbiphenyl; (VI) 9,9-bis[(4-benzilyloxy)phenyl]fluorene; (VII) 4,4′-thiobis[(4-phenyleneoxy)benzil]; (VIII) 4,4′-thiobis[(4-phenylenesulfanyl)benzil]; (IX) 1,4-bis(4-phenylglyoxaloyl-3-trifluoromethylphenoxy)benzene; (X) 1,3-bis[(3-trifluoromethylphenyl)glyoxaloyl]benzene. (a) 3,3′-diaminobenzidine; (b) 3,3′,4,4′-tetraaminodiphenylether; (c) 3,3′,4,4′-tetraaminodiphenylsulfone; (d) 3,3′,4,4′-tetraaminodiphenylmethane; (e) 2,6-bis(3,4-diaminophenyl)-4-(3′-trifluoromethylphenyl)pyridine; (f) 1,4-bis(3,4-diaminophenoxy)benzene; (g) di(triethylammonium)-4,4′-bis(3,4-diaminophenoxy)biphenyl-3,3′-disulfonate; (h) 1,1-bis[4-(3′,4′-diaminophenoxy)phenyl]-1-(3″-trifluoromethylphenyl)-2,2,2-trifluoroethane; (i) 1,1-bis[4-(3′,4′-diaminophenoxy)phenyl]-1-[3″,5″-bis(trifluoromethyl)phenyl]-2,2,2-trifluoroethane.

Figure 17.5 uses another class of important intermediates, that is, bis(α-phenylethynyl) compounds [39,40]. The intermediates are synthesized via a palladium-catalyzed coupling reaction of halo-substituted benzenes and aryl alkynes with cuprous iodide (CuI) as the catalyst. The triple bonds in the intermediates are then oxidized with potassium permanganate ($KMnO_4$) affording the target bis(α-diketones). This route avoids the use of highly toxic SeO_2 as the oxidizing agent. However, the high cost of palladium reagents makes it difficult for this route to be expanded for mass production of bis(α-diketones).

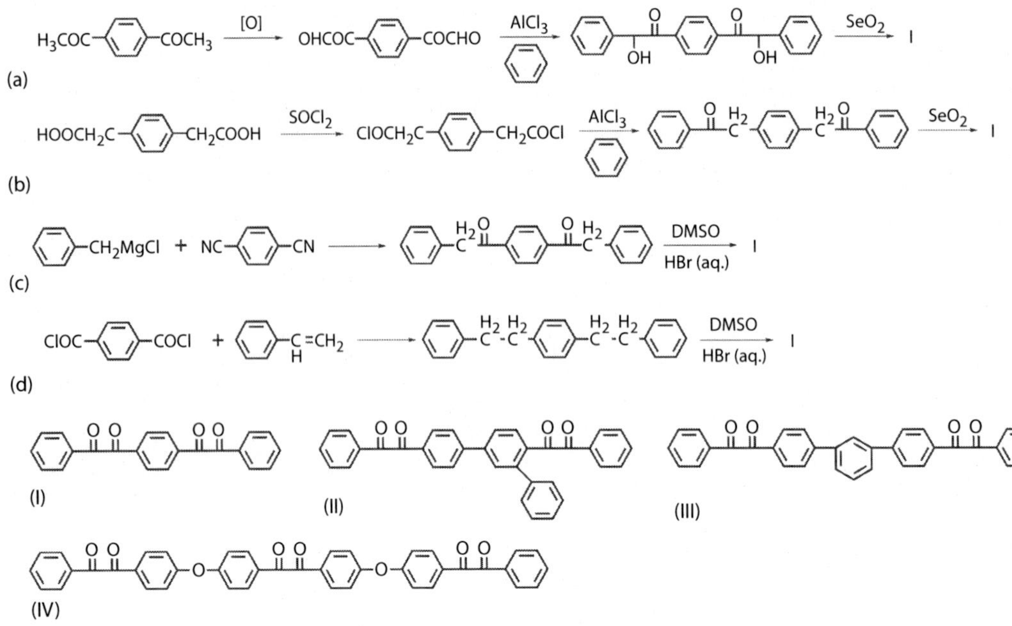

FIGURE 17.4 Synthesis of bis(α-diketones) via oxidative route.

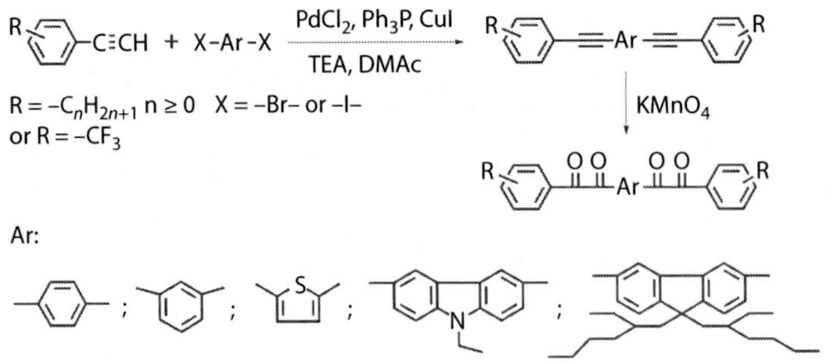

FIGURE 17.5 Synthesis of bis(α-diketones) via oxidative reactions of bis(α-diphenylacetylene). Note: TEA, triethylamine.

Although the above route is operative, it suffers from high cost, low yields, long operating periods, and poor mass production. In addition, the hazard of the highly toxic SeO_2 oxidative agent makes this route a less-than-convenient procedure. In 1974, Heath et al. reported a procedure for the synthesis of bisether-linked bis(α-diketone) monomers via the nucleophilic aromatic nitro displacement reactions of 4-nitrobenzil and aromatic diols, as presented in Figure 17.6 [41]. This procedure is thought to be one of the most promising cost-effective routes to synthesize bis(α-diketone) monomers due to the easily available starting materials [42]. 4-Nitrobenzil can be easily produced from the industrial-grade benzoin via a two-step procedure, including a first acetylation followed by a nitration reaction. Since various diols and dithiols have been commercialized in modern industry, this route provides an effective method achieving the diversification of the bis(α-diketone) monomers. It is to be noted that these bis(α-diketone) monomers contain two flexible ether linkages in their structures. The PPQs derived from the ether-containing bis(α-diketones) usually exhibit relatively lower glass transition temperatures (T_gs) in comparison with their nonether analogues due to the rotational flexibility along the biphenyl ether units in their backbones. This reduction

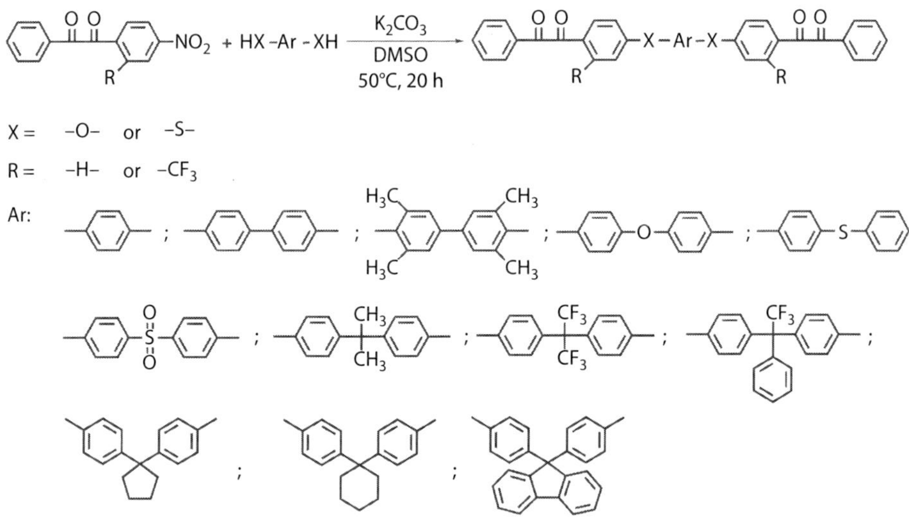

Polyphenylquinoxalines

FIGURE 17.6 Synthesis of bis(α-diketones) via nucleophilic displacement reactions.

of T_g values for ether PPQs can be remedied by several molecular designs. For instance, introduction of bulky substituents at the 2- and 2′-positions of the biphenyl ether moiety in PPQs could effectively prevent the internal rotation along the biphenyl ether units, thus increasing the thermal stability of the molecular chains of PPQs at elevated temperatures [43]. In addition, introduction of bulky substituents, such as fluorene groups, in PPQs could also decrease the rotational flexibility of the molecular chains along the biphenyl ether due to the steric effects of the substituents [44]. Lastly, introduction of heat-cross-linkable groups, such as phenylethynyl substituents is also helpful at increasing the T_g values of ether PPQs [45].

By this route, a series of bis(α-diketone) monomers containing ether or thioether bonds have been developed, and their typical chemical structures and characterization are tabulated in Table 17.2. These bis(α-diketone) compounds can be easily prepared with high yields and purified by recrystallization. In addition, this route can be further expanded by using the substituted 4-nitrobenzil as the starting material. For example, a series of fluorinated bis(α-diketone) monomers were developed in our laboratory very recently [46]. 4-Nitro-2-trifluoromethylbenzil (NTFB) was first successfully synthesized in our laboratory. The nitro in NTFB was activated by both the electron-withdrawing benzil and trifluoromethyl groups. Thus, the nitro can easily undergo displacement reactions with various diols to afford novel fluoro-containing bis(α-diketone) monomers. The obtained bis(α-diketone) monomers with polymerizable grade purities can serve as valuable monomers for functional PPQ synthesis. Figure 17.7 shows the ^{1}H nuclear magnetic resonance (NMR) spectrum of a typical fluoro-containing bis(α-diketone) monomer, 1,4-bis(4-phenyl-glyoxaloyl-3-trifluoromethyl-phenoxy)benzene, developed in our laboratory.

17.3.1.1.2 Synthesis of Bis(o-Diamines)

For bis(o-diamine) monomers, the amounts of commercially available species are much more than those of bis(α-diketones). This is mainly thanks to the commercialization of another important high-temperature thermoplastic polymer, PBI, which uses bis(o-diamines) as the starting monomers [47]. In addition, some other heteroaromatic polymers, such as PPY, also use bis(-diamines) as the monomers [48]. In recent years, with the commercialization of PBIs, some of the bis(o-diamine) monomers, including 3,3′-diaminobenzidine [49], 3,3′,4,4′-tetraaminodiphenylether [50], 3,3′,4,4′-tetraaminodiphenyl-methane [51], and 3,3′,4,4′-tetraaminodiphenylsulfone [52], achieved mass production. On the other hand, the synthesis procedures for bis(o-diamine) compounds have been well established in the literature.

TABLE 17.2
Chemical Structures and Characterization of Bis(α-Diketone) Monomers

X	R	Ar	Melting Point, °C	¹H NMR, ppm	Ref.
–O–	–H	–C₆H₄– (para)	159.0	7.99–7.96; 7.68–7.65; 7.54–7.50; 7.13; 7.06–7.04	[41,43]
–O–	–H	biphenyl	167.7	7.99–7.96; 7.68–7.65; 7.60–7.59; 7.54–7.50; 7.17–7.15; 7.08–7.07	[41,43]
–O–	–H	tetramethylbiphenyl	201.7	7.94–7.90; 7.79–7.76; 7.63–7.59; 7.53; 6.99–6.97; 2.11	[43]
–O–	–H	isopropylidene-diphenyl	Wax	7.65–6.47; 1.62	[41]
–O–	–H	diphenyl sulfide	127.5	7.99–7.97; 7.70–7.66; 7.51–7.55; 7.42–7.39; 7.08–7.04	[42]
–O–	–H	fluorene-based	212.8	7.98–7.93; 7.82–7.80; 7.69–7.65; 7.54–7.50; 7.44–7.40; 7.34–7.31; 7.26–7.24; 7.03–7.01; 6.96–6.94	[44]
–O–	–H	diphenylacetylene-biphenyl	ND	8.2–7.0	[45]
–S–	–H	diphenyl sulfide	158.9	7.91–7.89; 7.86–7.84; 7.81–7.77; 7.64–7.60; 7.57–7.55; 7.47–7.45; 7.37–7.35	[42]
–O–	–CF₃	–C₆H₄–	172.2	7.98–7.96; 7.89–7.87; 7.83–7.79; 7.67–7.61; 7.38; 7.32–7.29	[46]
–O–	–CF₃	biphenyl	Wax	8.16–7.98; 7.97–7.96; 7.84–7.80; 7.67–7.64; 7.35–7.34	[46]

Note: ND, not detected.

The common routes for bis(*o*-diamine) synthesis are summarized in Figure 17.8, and the typical monomers developed from these routes are tabulated in Table 17.3. Routes (I) and (II) have general applicability and are often utilized to prepare aromatic bis(*o*-diamine) monomers. Route (I) usually produces bis(*o*-diamine) compounds with a single linkage, such as ether, thioether, sulfone, and so on. According to the characteristics of the linkages, different reaction conditions might be used. For example, when the single linkages were –O–, –S–, or –CH$_2$–, concentrated nitric acid (concentration, ~70%) would be suitable for the nitration reaction in step 2 [50,51]. However, when the linkages were electron-withdrawing –SO$_2$– or –CO– groups, fuming nitric acid had to be used to facilitate the reaction [52]. This difference is mainly attributed to the different inductive effects of the linkages on the nitration reaction.

Route (II) was developed in recent years to produce novel bis(*o*-diamine) compounds with bisether linkages in the structures [53,54]. Similar to the route for preparing bis(α-diketones) containing bisether

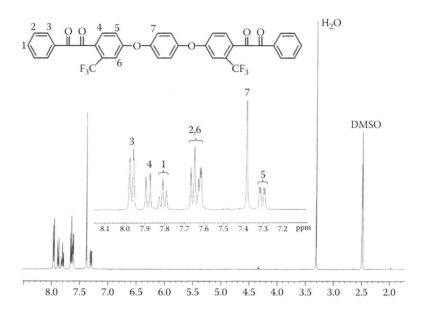

FIGURE 17.7 ^{1}H NMR spectrum of fluorinated bis(α-diketone) monomer (DMSO-d_6, 400 MHz).

groups, route (II) also provides an effective pathway for developing functional bis(o-diamine) monomers. As shown in Table 17.3, various bis(o-diamine) monomers have been produced by this route. For example, in our laboratory, several fluoro-containing aromatic bis(o-diamine) monomers, including 1,1-bis[4-(3′,4′-diaminophenoxy)phenyl]-1-(3″-trifluoro -methylphenyl)-2,2,2-trifluoroethane (*h* in Figure 17.3) and 1,1-bis[4-(3′,4′-diamino-phenoxy)phenyl]-1-[3″,5″-bis(trifluoromethyl)phenyl]-2,2,2-trifluoroethane (*i* in Figure 17.3), were synthesized by this route in order to develop novel PPQs with ultralow dielectric constants [55]. The typical ^{1}H–^{13}C heteronuclear single quantum correlation (HSQC) spectrum of bis(o-diamine) monomer *i* is shown in Figure 17.9.

Routes (III) and (IV) in Figure 17.8 are not common pathways for developing bis(o-diamine) monomers. However, they are usually utilized to develop some important bis(o-diamine) monomers for specific applications. For instance, using bis(o-diamine) monomer with hexafluoroisopropylidene linkage, 2,2′-bis(3,4-diaminophenyl)hexafluoropropane can be synthesized form route (III) [56], whose cost is much lower than that of the traditional procedure using expensive 2,2′-bis(4-aminophenyl)hexafluoropropane as the crude material [57]. Route (IV) suggests a convenient pathway developed in our laboratory to synthesize rigid-rod pyridine-bridged bis(o-diamine) monomers [58]. This series of bis(o-diamine) compounds are important monomers for functional PPQs with enhanced solubility, processability, and tensile modulus, while their thermal stability at elevated temperatures was maintained as much as possible.

17.3.1.2 PPQ Synthesis Chemistry

Generally, PPQs can be prepared by three routes, including (1) phenylquinoxaline ring formation polymerization reactions from bis(α-diketone) and bis(o-diamine) monomers; (2) polymerization via self-polymerizable AB or AB$_2$ monomers; and (3) polymerization from monomers containing preformed phenylquinoxaline rings. These three routes have their own advantages and disadvantages and have all been adopted in the literature to prepare PPQ polymers.

17.3.1.2.1 Polymerization from Bis(α-Diketone) and Bis(o-Diamine) Monomers

From the entry for *polyphenylquinoxaline* in the dictionary published recently [59], the most attractive synthesis procedure for PPQs is by copolycondensation of an aromatic bis(o-diamine) and a

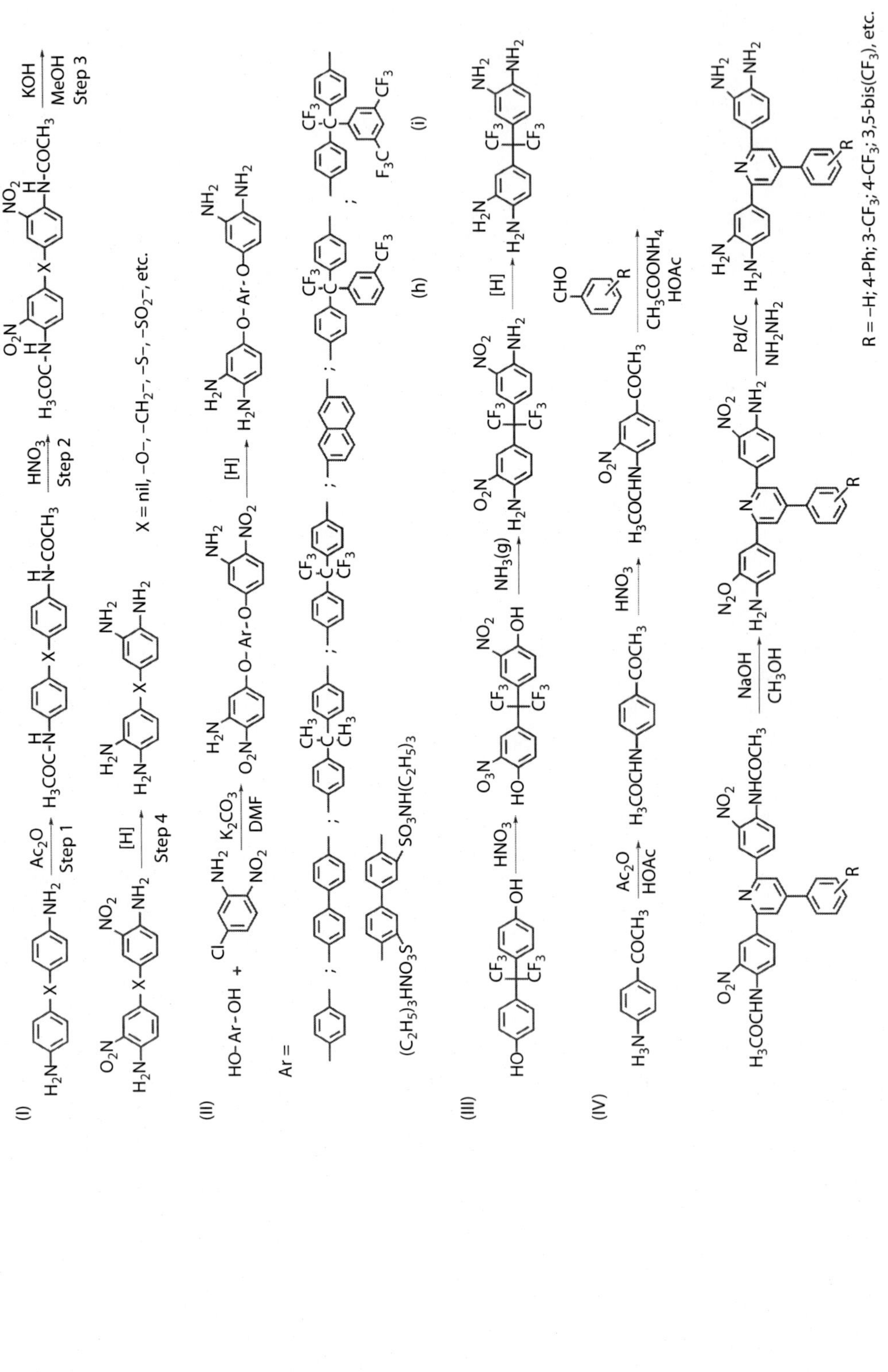

FIGURE 17.8 Synthesis procedures for bis(α-diamine) monomers.

TABLE 17.3
Chemical Structures and Characterization of Bis(o-Diamine) Monomers

Structure: H$_2$N–C$_6$H$_3$(NH$_2$)–X–C$_6$H$_3$(NH$_2$)–NH$_2$

X	Melting Point, °C	^{1}H NMR, ppm	Ref.
Nil	176–177	ND	[49]
–O–	150–151	ND	[50]
–CH$_2$–	232–236	9.82; 7.49–7.48; 7.12–7.10; 3.81; 2.02	[51]
–SO$_2$–	174	6.89–6.45; 5.20; 4.79	[52]
–O–C$_6$H$_4$–O–	226.5	6.81; 6.49–6.45; 6.23–6.22; 6.17–6.12; 4.60; 4.24	[53]
–O–C$_6$H$_4$–C$_6$H$_4$–O–	204.0	7.54–7.50; 6.93–6.90; 6.53–6.50; 6.29–6.28; 6.14–6.10; 4.50	[53]
–O–C$_6$H$_4$–C(CH$_3$)$_2$–C$_6$H$_4$–O–	144.5	8.03–7.99; 7.49; 7.36–7.33; 7.11–7.08; 6.41–6.40; 6.30–6.26; 1.70	[53]
–O–(naphthyl)–O–	135.5	7.81–7.78; 7.05–7.01; 6.98–6.97; 6.53–6.50; 6.28–6.27; 6.16–6.12; 4.66	[53]
–O–C$_6$H$_3$(SO$_3$NH(C$_2$H$_5$)$_3$)–C$_6$H$_3$(SO$_3$NH(C$_2$H$_5$)$_3$)–O–	ND	7.94–7.93; 7.41–7.37; 6.70–6.67; 6.52–6.49; 6.26–6.25; 6.14–6.10; 4.51; 3.12–3.05; 1.09–1.05	[54]
–O–C$_6$H$_4$–C(CF$_3$)(3-CF$_3$-C$_6$H$_4$)–C$_6$H$_4$–O–	ND	7.82–7.81; 7.69–7.67; 7.43–7.42; 7.31; 6.98–6.96; 6.93–6.91; 6.54–6.52; 6.30; 6.17–6.15; 4.67; 4.38	[55]
–O–C$_6$H$_4$–C(CF$_3$)(3,5-(CF$_3$)$_2$-C$_6$H$_3$)–C$_6$H$_4$–O–	ND	8.28; 7.56; 6.99–6.97; 6.93–6.92; 6.52–6.50; 6.29–6.28; 6.15–6.13; 4.66; 4.38	[55]

Note: ND, not detected.

bis(α-diketone) monomer in an appropriate solvent such as a mixture of *m*-cresol and xylene. This standard route provides a flexible molecular design for PPQ synthesis. One can choose proper bis(*o*-diamine) or bis(α-diketone) monomers to develop PPQ polymers with desired properties. This common pathway is presented in Figure 17.10. The variation of –X– and –Y– linkages will produce PPQ polymers with various properties. The reaction conditions and mechanism for this route have been well studied, and many important conclusions have been established. For example, for polymerization solvents, *m*-cresol has been proven to be the most effective, which not only serves as a good solvent for PPQ but also helps to form the preferred cis conformer of the benzil through hydrogen bonding. In addition, it has been proven that most of the PPQs can be prepared at ambient temperature when the polymerization of bis(*o*-diamine) or bis(α-diketone) monomers is conducted in phenolic solvents [26]. By this standard procedure, hundreds of PPQs with various functionalities

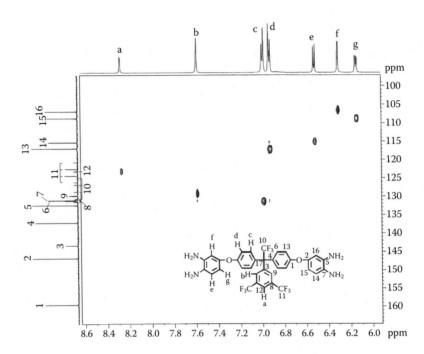

FIGURE 17.9 $^1H-^{13}C$ HSQC spectrum of the fluorinated bis(o-diamine) monomer (DMSO-d_6, 400 MHz).

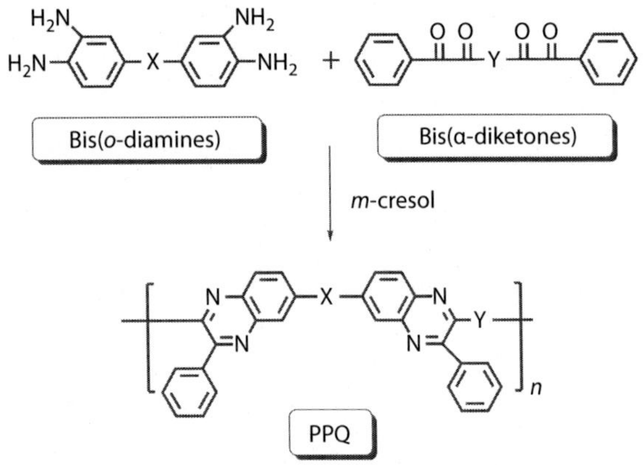

FIGURE 17.10 Synthesis of PPQs via polymerization of bis(α-diketone) and bis(o-diamine) monomers.

have been developed in the literature. One can find the detailed structure–property relationships of these PPQs in the published review articles [18–29]. Here, in Table 17.4, 20 representative PPQs, PPQ-1 to PPQ-20, are presented.

It can be seen that, according to the chemical structures of the starting bis(o-diamine) or bis(α-diketone) monomers, PPQs with quite different inherent viscosities, solubilities, thermal stabilities, and refractive indices might be obtained. For instance, PPQs derived from 3,3′-diaminobenzidine (X = nil) usually exhibit higher inherent viscosities than those from other bis(o-diamine) monomers due to the highest reactivity of the monomer. For the rigid PPQ systems (PPQ-1, -2, -4, and -5) [11,26,60], the glass transition temperatures (T_g) can be as high as 370°C (PPQ-5); however, for the flexible ether-containing PPQs, the value decreases to 215°C (PPQ-17). It should be worth noting that two ether-containing PPQs, PPQ-9 (T_g, 325°C) and PPQ-12 (T_g, 300°C), exhibited high T_g values. This is mainly because both of the *ortho*-substituted methyl groups and the bulky fluorene

TABLE 17.4
Chemical Structures and Properties of PPQs

No.	X	Y	Solubility	$[\eta]_{inh}$, dL/g	T_g, °C	$T_{10\%}$, °C	n	Ref.
PPQ-1	Nil	tolyl	MC, CF	2.62	365	ND	ND	[26]
PPQ-2	Nil	dimethylthiophene	MC, FA	1.09	>300	ND	ND	[11]
PPQ-3	Nil	bis(tolyloxy)phenyl	MC, CF, DMAc	0.81	290	ND	ND	[31]
PPQ-4	Nil	bis(tolyl)aminophenyl	MC, CF	0.81	308	577	ND	[60]
PPQ-5	Nil	dimethyl-ethylcarbazole	MC, CF	0.81	370	580	ND	[39]
PPQ-6	Nil	dibutylfluorene	MC, CF	ND	208	415	ND	[40]
PPQ-7	Nil	bis(tolyloxy)phenyl	MC, CF, NMP	0.45	280	566	ND	[43]
PPQ-8	Nil	bis(tolyloxyphenyl)	MC, CF, NMP	0.68	286	577	1.7739	[43]

(Continued)

TABLE 17.4 (CONTINUED)
Chemical Structures and Properties of PPQs

No.	X	Y	$[\eta]_{inh}$, dL/g	Solubility	T_g, °C	$T_{10\%}$, °C	n	Ref.
PPQ-9	Nil		0.83	MC, CF, NMP	325	497	1.7332	[43]
PPQ-10	Nil		0.85	MC, CF, NMP	248	570	1.7665	[42]
PPQ-11	Nil		0.83	MC, CF, NMP	235	552	1.7953	[42]
PPQ-12	Nil		0.84	MC, CF, NMP	300	579	ND	[44]
PPQ-13	–O–		0.58	MC, CF, NMP	243	546	ND	[43]
PPQ-14	–O–		0.77	MC, CF, NMP	255	569	1.7355	[43]
PPQ-15	–O–		0.95	MC, CF, NMP	287	487	1.6998	[43]

(Continued)

TABLE 17.4 (CONTINUED)
Chemical Structures and Properties of PPQs

No.	X	Y	[η]$_{inh}$, dL/g	Solubility	T_g, °C	$T_{10\%}$, °C	n	Ref.
PPQ-16	–O–		0.82	MC, CF, NMP	221	562	1.7277	[42]
PPQ-17	–O–		0.79	MC, CF, NMP	215	554	1.7590	[42]
PPQ-18	–O–		0.77	MC, CF, NMP	281	566	ND	[44]
PPQ-19	–S(=O)$_2$–		0.77	MC, CF, NMP	244	526	1.7295	[42]
PPQ-20	–S(=O)$_2$–		0.73	MC, CF, NMP	230	523	1.7606	[42]

Note: [η]$_{inh}$, inherent viscosity; CF, chloroform; DMAc, N,N-dimethylacetamide; FA, formic acid; MC, m-cresol; n, refractive index at the wavelength of 633 nm; ND, not detected; NMP, N-methyl-2-pyrrolidione; $T_{10\%}$, temperature at 10% weight loss; T_g, glass transition temperature.

groups can restrict the free rotation of ether bonds along the phenyl rings, thus increasing the glass transition temperatures. With respect to the solubility, introduction of ether or thioether linkages in PPQs enhanced their solubility in organic solvents. Nonether PPQs (PPQ-1 to PPQ-6) were only soluble in *m*-cresol and chloroform; however, those containing flexible (thio)ether linkages (PPQ-7 to PPQ-20) were also soluble in *N*-methyl-2-pyrrolidione (NMP). This enhancement of solubility is mainly due to the flexible ether or thioether groups, which hinder the close packaging of the PPQ molecular chains and allow the diffusion of solvent molecules into the PPQ chains. As we know, *m*-cresol and chloroform both pose potential hazards to human health and environments. However, NMP has been widely used in microelectronics and other high-tech fields, and the good solubility of the ether PPQs in NMP greatly expands their applications in modern industries. The fabrication of PPQ films in environmentally friendly solvents, such as NMP, will be discussed later in the chapter. Refractive indices (n) of PPQ polymers have been investigated in our laboratory very recently [42], though before this, very little attention has been paid to the optical properties of PPQ films. From the viewpoint of structure characteristics, PPQ possesses high contents of aromatic components with high molar refractions. Thus, the polymer has intrinsic high refractive index values. Considering that high-refractive-index (high-n) polymers have currently attracted much attention in fabricating high-performance optoelectronic devices [61–63], it would be worth investigating the effects of the molecular structures of PPQs on their refractive indices. As shown in Table 17.4, PPQ-8 showed a high n value of 1.7739 at a wavelength of 633 nm. Introduction of sulfur groups endowed the PPQ with much higher n values due to the high molar refraction of the sulfur-containing groups. PPQ-11 revealed a value of 1.7953, which is among the highest values reported in the literature. Meanwhile, the birefringences of PPQ-11 were as low as 0.0004. Thus, the polymer is a good candidate for advanced optoelectronic devices.

17.3.1.2.2 Polymerization from Self-Polymerizable Monomers

An alternate route for PPQ synthesis is through the polymerization of self-polymerizable monomers. Such monomers essentially include two types of compounds. One is the compounds containing both *o*-diamino and diketone moieties, such as the monomer 3,4-diaminobenzil. Another one is the monomers containing preformed phenylquinoxaline rings. Figures 17.11 through 17.13 summarize three different scenarios for this kind of procedure. Fang reported a high-performance PPQ polymer (PPQ-21) derived from the self-polymerizable monomer, 3,4-diaminobenzil (3), via

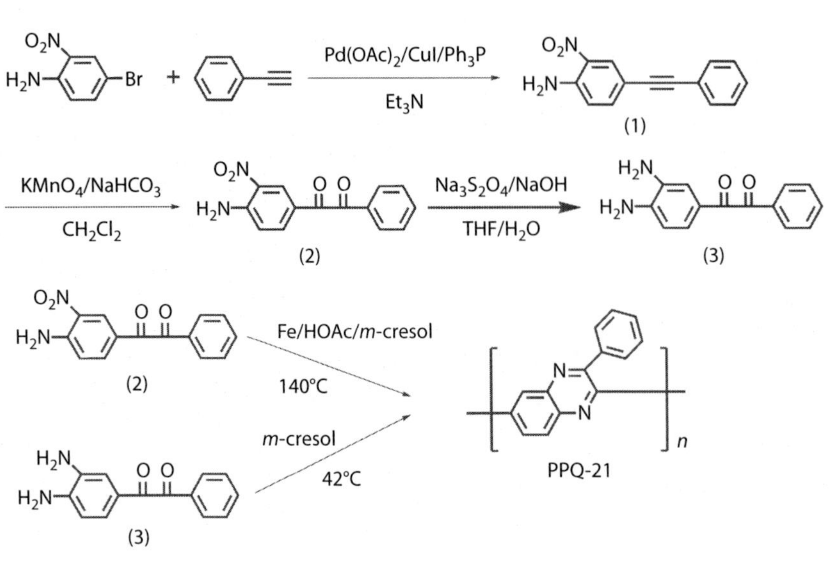

FIGURE 17.11 Synthesis of poly(3-phenylquinoxaline) via self-polymerization route.

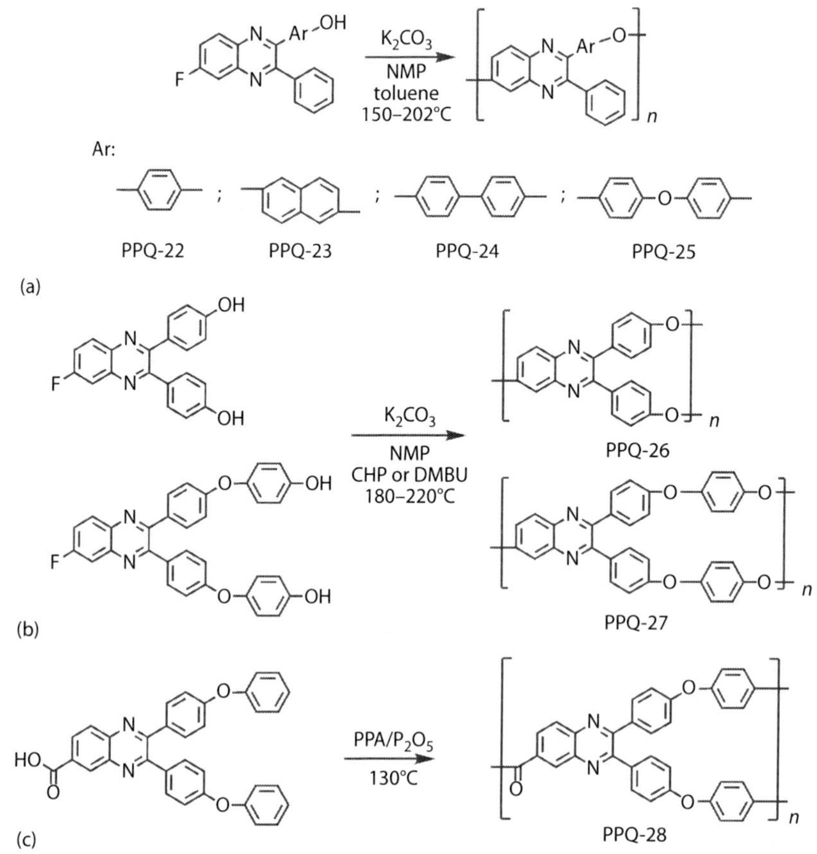

FIGURE 17.12 Synthesis of poly(arylene ether phenylquinoxaline)s via self-polymerization route.

a standard polycondensation procedure, or from 3-nitro-4-aminobenzil (2) via a reductive polycyclization procedure, as shown in Figure 17.11 [64]. The latter procedure produced high-molecular-weight PPQs, which exhibited enhanced or similar chemical and physical properties with PPQ-1 [derived from 1,4-bis(phenylglyoxalyl)benzene and 3,3′-diaminobenzidine], shown in Table 17.4. For example, PPQ-21 exhibited a higher T_g value (354°C versus 345°C), a comparable weight loss in isothermal aging performed at 371°C in air (5.3% after 33 h versus 6.9% after 25 h), and a similar coefficient of thermal expansion (CTE) value in the range of 50–320°C (35 versus 33 ppm/°C). In addition, the two polymers exhibited similar solubility in phenolic solvents and halogenated hydrocarbons. The PPQ-21 insulation layer with a low dielectric constant of 2.44 at a frequency of 100 kHz to 10 MHz could be easily obtained by spin-coating its m-cresol/xylene solution on a silicon wafer, followed by thermal curing at elevated temperatures. The superiority of this procedure is that it can achieve an inherent equal molar stoichiometry of the reactive ketone and amino moieties and can thus produce PPQs with high molecular weight. The disadvantage of this route is the very limited availability of the starting self-polymerizable monomers, restricting the modification and functionalization of conventional PPQs.

Harris et al. developed another procedure for synthesizing PPQs via self-polymerizable quinoxaline monomers, as shown in Figure 17.12a [65–68]. The route is through the aromatic nucleophilic substitution reactions of an AB monomer containing a preformed quinoxaline ring. Thus, the obtained polymers are actually poly(arylene ether phenylquinoxaline)s, a series of polymers combining the merits of both poly(arylene ether)s and PPQs. The typical properties of the polymers are tabulated in Table 17.5. It can be obviously observed that the inherent equal molar stoichiometry of the fluoro and

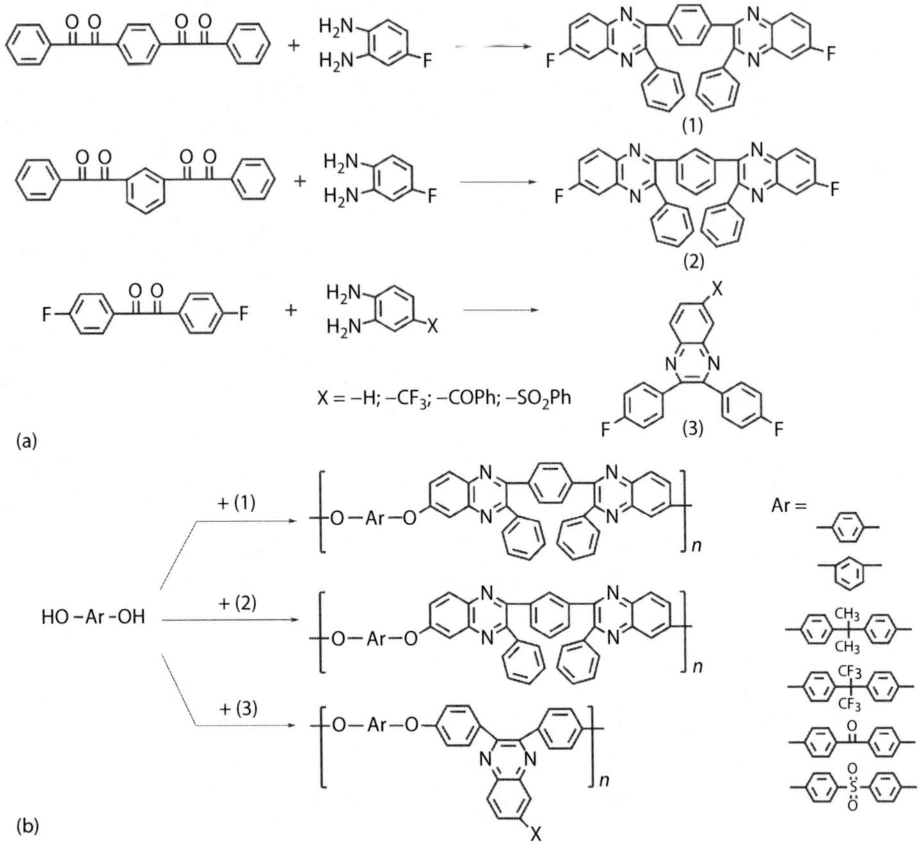

FIGURE 17.13 Synthesis of PPQs from difluoro compounds containing preformed quinoxaline rings. (a) Fluoro-substituted monomers; (b) PPQs.

phenol moieties endowed the derived PPQ-22 to PPQ-25 polymers with high inherent viscosities and high tensile properties. The flexible and tough films exhibited tensile strengths higher than 90 MPa and elongations at break higher than 88%. PPQ-22 possessed the highest tensile strength (109 MPa), whereas PPQ-25 had the highest elongation value (100%). This high mechanical property is due to the presence of the arylene ether linkages in the polymers. Meanwhile, the polymers maintained the intrinsic characteristics of PPQs, including good thermal stability and good solubility in *m*-cresol, chloroform, and even NMP. The good combination of excellent tensile properties along with high thermal stability makes this series of polymers very attractive for high-performance applications.

Besides the linear poly(arylene ether phenylquinoxaline)s derived from AB monomer containing preformed quinoxaline rings, a series of hyperbranched PPQs have also been developed via AB_2 monomers, as shown in Figure 17.12b and c [69]. For example, Srinivasan et al. successfully prepared hyperbranched PPQs containing poly(arylene ether)s from the self-polymerizable AB_2 monomers containing a single aryl fluoride and two phenolic hydroxyl groups [70]. The fluoro elements were activated by the fused pyrazine ring; thus, the nucleophilic substituent reaction proceeded smoothly. For the polymerization, various reaction conditions, including solvents [NMP; *N*-cyclohexyl-2-pyrrolidone, (CHP); or *N,N'*-dimethylpropyleneurea (DMPU)], reaction temperature, and polycondensation time, were optimized in order to obtain high-molecular-weight PPQs. The obtained PPQ-26 and PPQ-27 polymers are both soluble in NMP. In order to endow the hyperbranched PPQs with low melting viscosities, phenylethynyl-terminated hyperbranched PPQs were developed [71]. The obtained resins possessed good melting flowability, and the cured PPQ thermosets exhibited good thermo-oxidative stability (5% weight loss temperature, 560–568°C). Another hyperbranched

TABLE 17.5
Chemical Structures and Properties of PPQs via Self-Polymerizable Monomers

No.	Ar	$[\eta]_{inh}$, dL/g	Solubility	T_g, °C	$T_{5\%}$, °C	σ, MPa	E, GPa	ε, %
PPQ-22	phenylene	1.64	MC, CF, NMP	251	542	109	3.7	93
PPQ-23	naphthylene	2.51	MC, CF, NMP	278	552	104	2.6	83
PPQ-24	biphenylene	1.97	MC, CF, NMP	287	530	95	3.0	93
PPQ-25	diphenyl ether	1.41	MC, CF, NMP	206	535	92	3.3	100

Source: Klein, D.J. et al., *Macromolecules*, 34, 2001.

Note: $[\eta]_{inh}$, inherent viscosity—determined in *m*-cresol at 30°C; ε, elongation at break; σ, tensile strength; CF, chloroform; E, tensile modules; MC, *m*-cresol; NMP, *N*-methyl-2-pyrrolidione; $T_{5\%}$, temperature at 5% weight loss determined by thermogravimetric analysis (TGA) with a heating rate of 10°C/min; T_g, glass transition temperatures determined by differential scanning calorimetry (DSC) with a heating rate of 10°C/min.

polymer, poly(phenylquinoxaline-ether-ketone), was prepared from a self-polymerizable AB$_2$ monomer, 2,3-bis(4-phenoxyphenyl)quinoxaline-6-carboxylic acid, via the Friedel–Crafts reaction in a polyphosphoric acid (PPA)/phosphorus pentoxide (P$_2$O$_5$) medium at 130°C, as shown in Figure 17.12c [72]. The derived PPQ-28 film was thermally stable, with 5% weight loss temperature over 500°C in both air and helium atmospheres. The film showed ultraviolet absorption maxima at 365–370 nm and emission maxima at 433–446 nm. By a similar procedure, hyperbranched polymers containing alternating quinoxaline and benzoxazole repeat units have also been prepared [73].

17.3.1.2.3 Polymerization from Monomers Containing Preformed Phenylquinoxaline Rings
The route for producing PPQs via monomers containing preformed quinoxaline rings is shown in Figures 17.13 and 17.14. According to the different characteristics of the monomers, this route can be subdivided into different situations. For example, if the monomers were fluoro-substituted bisquinoxalines, as shown in Figure 17.13a, poly(aryl ether phenylquinoxaline)s would be obtained by nucleophilic aromatic substitution reactions of these monomers with various bisphenoxides in NMP (Figure 17.13b) [74–76]. The obtained high-molecular-weight polymers usually possessed comparable thermal stability to the standard PPQs.

Another situation is that the monomers containing quinoxaline rings were diamine compounds synthesized according to the procedures shown in Figure 17.14a. Then, the monomers were polymerized with various reactive monomers (diacid, diacid chlorides, dianhydride, etc.), affording all kinds of PPQs containing ether, ester, sulfide, sulfone, amide, imide, hexafluoroisopropylidene, silylene, and other flexible linkages (Figure 17.14b). Bruma et al. developed a series of such PPQs via this route and reviewed their structure–property relationships [77–80]. The PPQ polymers containing the linkages mentioned above usually combine the favorable properties of several high-temperature polymers. Thus, they usually possess good comprehensive properties. This molecular design is very beneficial and effective for expanding the application fields of conventional PPQs, because a single PPQ polymer could hardly meet all the property requirements of one specific application.

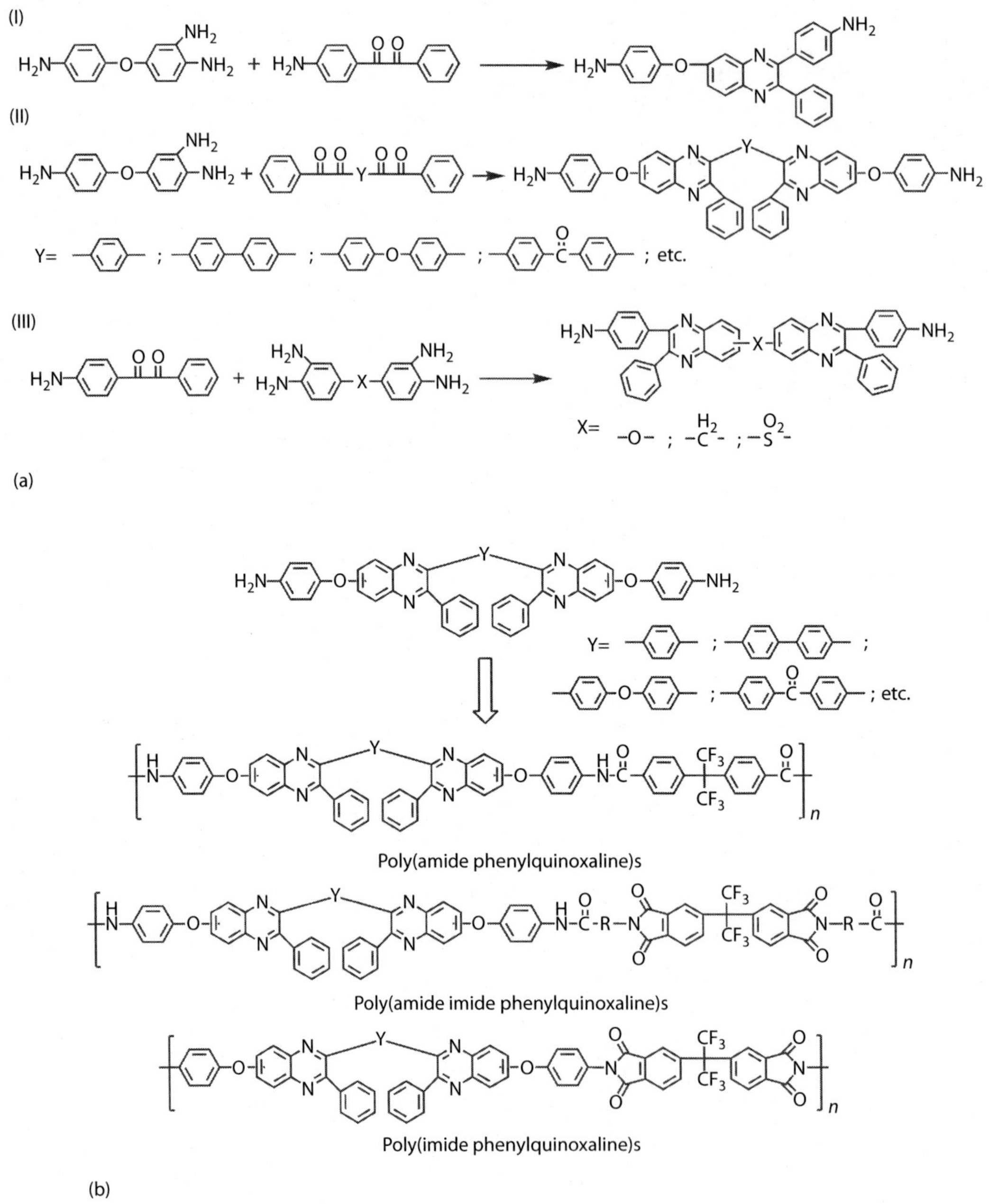

FIGURE 17.14 Synthesis of PPQs from diamine containing preformed quinoxaline rings. (a) Diamines; (b) PPQs.

17.3.2 PPQ Processing Technologies

The thermoplastic and amorphous nature of PPQs along with their good solubility makes them processable using a variety of techniques, such as melt and solution procedures. PPQ can be fabricated into various forms, including powder, varnish, film, foam, adhesive, and composite. In solution form, PPQs can be used directly for prepreg and adhesive-tape formulations, film casting, etc. The polymer can be isolated from solution and then compression molded. Below, some processing techniques of PPQ films, composites, and foams are discussed.

17.3.2.1 PPQ Films

PPQ films can be readily prepared by casting the as-prepared PPQ solutions, or the solutions prepared by dissolving the solid PPQ resin in solvents. In PPQ film fabrications, *m*-cresol, chloroform, 1,1,2,2-tetrachloroethane, and mixtures of *m*-cresol and xylene are usually used as the solvents. These solvents, more or less, will do harm to the operator's health and the ambient environment. Because of the current interest in the environmental impact of toxic solvents used in PPQ processing, research aimed at lowering the toxicity of the solvents is of prime importance. Improvement of the solubility of PPQs in good solvents will lead to a low impact on the environment. As mentioned above, bisether-linked PPQs usually are soluble in NMP, which is a solvent commonly used for polymer processing in the microelectronic and display industry. Thus, processing of ether-containing PPQ films via their NMP solutions might be a green procedure.

Figure 17.15 shows an experimental course for preparing a PPQ film. The PPQ resin was first dissolved in NMP at a solid content of 10–20 wt.% to afford a PPQ solution. The solution was purified by filtration through a 0.45 μm Teflon or sintered ceramic filter to remove any contaminates that might affect the quality of the film. Then, the solution was cast onto a clean glass, and the PPQ film was obtained by thermally baking the solution with a stepwise heating procedure up to 250°C. Figure 17.16 illustrates the appearances of PPQ resins and the formed PPQ-7 and PPQ-13 films (shown in Table 17.4) [43]. Flexible and tough PPQ films could be obtained by this experimental procedure.

A project at the ICCAS focused on achieving continuous production of PPQ film at a small to medium scale in order to meet some specific applications. PPQ derived from 1,4-bis(4-benzilyloxy) benzene and 3,3'-diaminobenzidine (PPQ-7 in Table 17.4) was chosen as the resin due to its relatively low cost, and NMP was the solvent. The continuous apparatus was similar to that used for other high-temperature films, especially PI films, as shown in Figure 17.17. PPQ resin was dissolved in NMP, and the obtained PPQ solution was cast on the preheated steel belt via a nozzle in the flow-casting equipment. Then, the solvent-containing PPQ wet films were peeled off the steel belt and transferred into the continuous curing furnace. The film was then stretched from the longitudinal and transversal directions so as induce the orientation of molecular chains in PPQ. The cured PPQ film with a very low content of solvent was wrapped and cut into the desired width. The thickness of the PPQ film was controlled to be 12.5–50 μm. PPQ films with improved tensile properties were obtained, demonstrating experimentally that continuous production of the PPQ film is feasible.

17.3.2.2 PPQ Moldings and Composites

PPQs have been recognized to be one of the most important thermoplastic matrixes for high-temperature composites [81]. The T_g values of PPQs are in the range of 240–370°C, while the thermal decomposition temperatures of PPQs are often higher than 500°C. The process temperature window

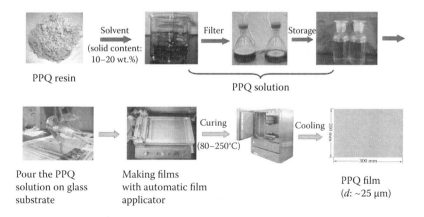

FIGURE 17.15 Processing steps for PPQ films in the laboratory.

FIGURE 17.16 Appearances of PPQ resins and films. (a) PPQ-7; (b) PPQ-13.

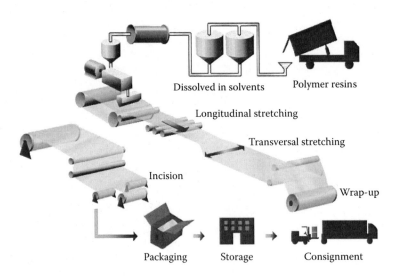

FIGURE 17.17 Suggested continuous production of PPQ films.

for PPQs is expected to be substantially wide, so the conventional thermoforming processes, such as injection molding, compression molding, and extrusion molding, might be applicable for the common PPQs. Very recently, Hergenrother et al. reviewed PQs and PPQ matrix resins and their composites [27], summarizing both research and development of PPQ moldings and composites.

Most of the applications of PPQ composites make use of their unusual thermal and thermo-oxidative stability. However, high glass transition temperatures and high melting viscosities of conventional PPQs make the systems difficult to fabricate. It has been well recognized that aromatic ether linkages inserted in aromatic polymer main chains usually provide them a significantly lower energy of internal rotation. Such a structural modification usually leads to lower T_g and crystalline melting temperatures as well as significant improvements in processing characteristics of the polymers without greatly sacrificing their thermal stability. On the other hand, by virtue of the successful application of phenylethynyl-terminated imide oligomer technology [82], introduction of a crosslinkable phenylethynyl end capper into the PPQ oligomers might be another effective procedure, achieving a good compromise between good processability and high thermal stability [83–85].

Combining the two molecular design pathways mentioned above, a series of phenylethynyl-terminated PPQs (PEPPQs) with bulky fluorene linkages in the main chain were recently developed in our laboratory, as shown in Figure 17.18 [86]. PPQ-29 to PPQ-32, with various designed molecular weights ranging from 2,500 to 20,000 g/mol, were synthesized from 9,9-bis(4-benzilyloxyphenyl) fluorene, 3,3'-diaminebenzine, and 4-phenylethynylbenzil (PEBZ) end capper. The rheological properties of the PPQs were investigated in order to evaluate the processability of the polymers. The results are shown in Figure 17.19 and Table 17.6. PPQ-29 and PPQ-30 reached the lowest viscosity at about 350°C with the values of 21 Pa s and 568 Pa s, respectively. However, under the same conditions, PPQ-31 and PPQ-32 exhibited much higher viscosities. When the temperatures are higher than 350°C, the melt viscosities of PPQs increase incrementally due to the cross-linking of phenylethynyl groups. This result indicates that PPQ-29 and PPQ-30 are readily processable under suitable pressures and temperatures. Considering the low melt viscosities and relatively narrow processing windows of current PPQs, they might be processed via the compression-molding procedure. Then, PPQ-29 and PPQ-30 thermosets were fabricated by thermally curing the corresponding PPQ resins according to the thermal molding process, including the temperature schedule and the applied pressure depicted in Figure 17.20. The parameters (300°C/30 min + 350°C/20 min + 370°C/2 h/1.5 MPa) were optimized and determined according to the rheological behavior of the PPQ resins. By this procedure, opaque deep-brown PPQ-29 and PPQ-30 pure resin molding plates (300 × 200 × 2.5 mm) were successfully prepared, whose thermal

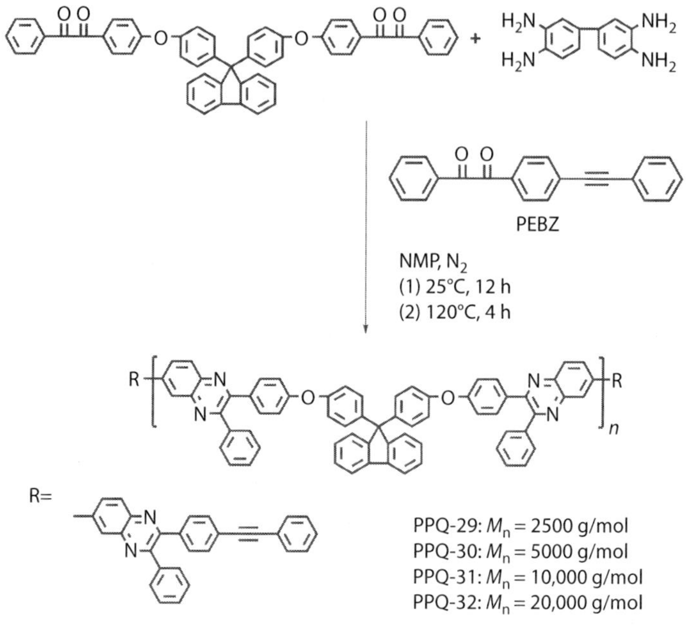

FIGURE 17.18 Synthesis of phenylethynyl-terminated PPQ oligomers.

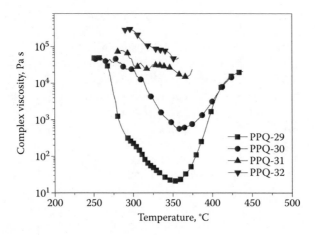

FIGURE 17.19 Rheological behaviors of PEPPQ oligomers.

TABLE 17.6
Inherent Viscosities and Rheological Properties of PEPPQs

PPQ	Inherent Viscosity, dL/g	Complex Viscosity, Pa s			Lowest Viscosity, Pa s
		275°C	325°C	375°C	
PPQ-29	0.14	4983	49	68	21 at 351°C
PPQ-30	0.19	36,960	3700	764	568 at 359°C
PPQ-31	0.28	57,330	31,790	23,740	14,940 at 364°C
PPQ-32	0.51	179,000	92,240	ND	ND

Source: Hergenrother, P.M., *Macromolecules*, 14, 1981.
Note: Inherent viscosity: measured at 30°C with a 0.5 g/dL solution in NMP. ND, not detected.

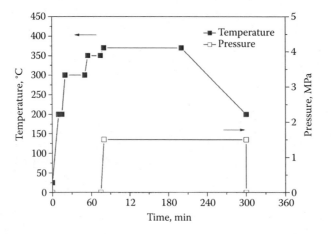

FIGURE 17.20 Thermal processing parameters for PEPPQ oligomers.

TABLE 17.7
Thermal and Mechanical Properties of Cured PPQ Thermosets

PPQ	T_g, °C	$T_{5\%}$, °C	R_{w750}, %	T_s, MPa	T_m, GPa	E_b, %	F_s, MPa	F_m, GPa
PPQ-29	338	558	67	75	1.9	4.0	142	2.8
PPQ-30	325	564	66	88	1.8	7.6	155	2.9

Note: E_b, elongation at break; F_m, flexural modulus; F_s, flexural strength; R_{w750}, residual weight ratios at 750°C in nitrogen; $T_{5\%}$, temperature at 5% weight loss in nitrogen; T_g, glass transition temperatures measured by DMA; T_m, tensile modulus; T_s, tensile strength.

and mechanical properties are presented in Table 17.7. The T_g values (tanδ peak in dynamic mechanical analysis [DMA] measurements) of the cured PPQs were 338°C (PPQ-29) and 325°C (PPQ-30), indicating the good thermal stability of the ether-containing PPQs. The mechanical properties of PPQ-29 are a bit inferior to PPQ-30, which is due to the higher cross-linking density in PPQ-29. PPQ-30 exhibited good comprehensive properties, comparable with the PEPPPQ resins derived from 4,4′-oxydibenzil, 3,3′-diaminobenzine, and PEBZ reported in the literature [87]. In summary, the synergic effects of phenylethynyl terminals and bulky fluorene moieties endow the PPQs with a combination of good properties, making them good candidates for high-temperature structural composites or adhesives.

17.3.2.3 PPQ Foams

In recent years, high-temperature-resistant polymer foams have attracted increasing attention due to their extraordinary properties and their existing and potential applications in wide a variety of technological areas. For example, in microelectronic fabrications, high-temperature polymer ILDs with low dielectric constants (usually abbreviated as low-k polymers, where k refers to the dielectric constant) and dielectric dissipation factors have been highly desired due to their ability to achieve a more rapid signal transport speed and a lower signal cross talk [30]. According to a specification by the International Technology Roadmap for Semiconductors (ITRS), by 2016, a k value below 2.0 at 1 GHz and a dielectric loss below 0.003 have to be met for the applied ILDs [88]. The common methodologies for reducing the dielectric constants of high-temperature polymer ILDs, including introduction of groups with low molar polarizability (fluorinated groups, etc.) or with large molar volumes, could only achieve a lowest k value of around 2.5. In recent years, it was found that incorporation of air voids ($k = 1.0$) into high-temperature polymers, that is, developing polymer foams or aerogels, might be a promising method for achieving a k value lower than 2.5. However, the attempts to reduce the k values of PIs below 2.5 seem to be more challenging because the conventional methods could only achieve very limited air loading.

Labadie et al. studied the effects of nano pores on the dielectric constant of PPQ foam [89]. First, they prepared triblock copolymers with PPQ as the thermal-stable main chain and poly(propylene oxide) (PO) as the thermally labile block, as illustrated in Figure 17.21. The polymers were dissolved into tetrachloromethane to afford polymer solutions. The solvent in the solutions was then evaporated to form the polymer films at 150°C under nitrogen. The films were subsequently thermally baked at 275°C in air for 9 h to obtain the tough and ductile films with no apparent decrease in mechanical properties relative to the PPQ homopolymer. The decomposed PO segments produced pores of nanometer size scale (8–10 nm) in the films. The dielectric constants of the PPQ foamed films were as low as 2.3 (45°C, 2 MHz), which is apparently lower than that of the pristine PPQ films (k, 2.7).

Recently, Merlet et al. reported new processing techniques for nanocellular PPQ foams [90,91]. In their procedures, porous structures were obtained by the *in situ* generation of foaming agents (CO_2 and isobutene) during the thermal treatment of PPQ-containing thermolabile groups

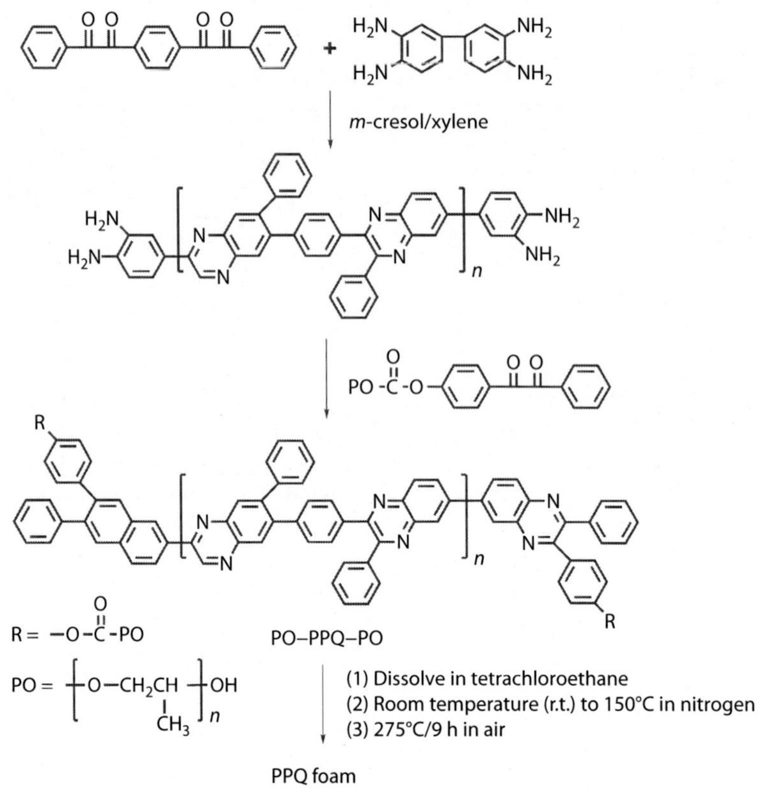

FIGURE 17.21 Preparation of PPQ foams.

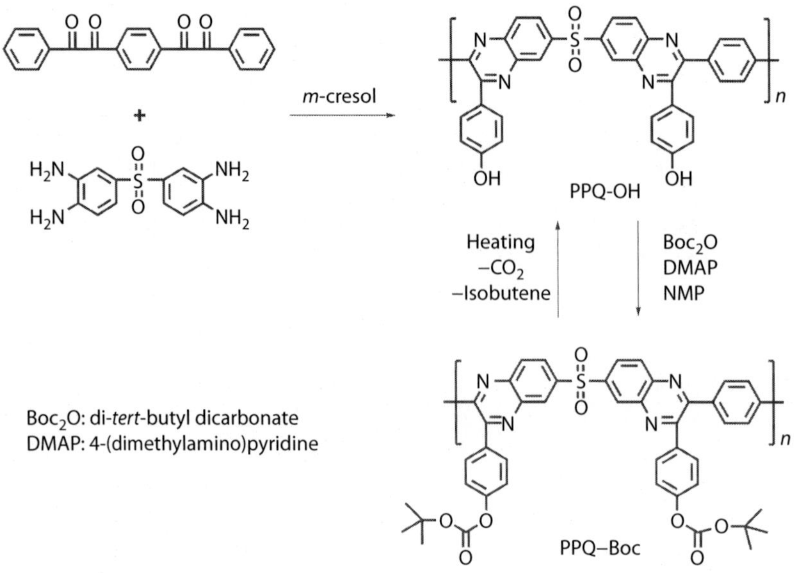

FIGURE 17.22 Synthesis of PPQ foams by *in situ* generation of foaming agents.

(*tert*-butyloxycarbonyl, etc.). For this purpose, a novel *tert*-butyloxycarbonyl functionalized PPQ film was first synthesized according to the procedure shown in Figure 17.22. The prepared PPQ films were then placed in a convection oven under a nitrogen atmosphere. By adjusting the appropriate foaming temperature and time, porous PPQ materials with a wide range of pore sizes (from nanoscale to macroscale) were successfully prepared. By this procedure, PPQ foams with dielectric constants as low as 2.4 were obtained. Good combined properties make these PPQ foams good candidates for microelectronic fabrications.

17.4 APPLICATIONS OF POLYPHENYLQUINOXALINES

The application of PPQs has been summarized in a monograph [26]. PPQs have been applied as components for adhesives; composites (structural compositions, antifriction self-lubricating materials, etc.); membranes (for alkaline batteries, ultrafiltration, or gas separation); wire insulation; film capacitors; and electronics. Since then, the application of PPQs has expanded and deepened. The most promising application areas for PPQs include the energy industry, optical fields, and microelectronic packaging.

17.4.1 Energy Applications

In all of the applications for PPQs in the energy industry, those in clean and efficient energy systems, such as high-performance battery fabrications, seem to be the most promising. In this area, PPQs have mainly been used as conductive materials for battery electrodes, separators for alkaline batteries, materials for battery electrolyzers [92], and PEMs for fuel cells.

For battery electrodes, in the 2000s, researchers at NEC Tokin Corporation, Japan, patented a series of proton-conductive polymer secondary batteries using PPQ as the negative electrode [93,94]. The proton secondary battery containing conductive polymers is a new type of product after the lithium secondary battery. The batteries are characterized by a high working voltage, stable discharge curve, small self-discharge, long cyclic life, and no pollution. Such batteries are usually constructed by arranging the positive electrodes and negative electrodes piled up via separators, and electrolyte solutions (usually aqueous sulfuric acid) are then filled therein. PPQ-1 with a π-conjugated structure (structure shown in Table 17.4) was used as the active material of negative electrodes in the batteries. PPQ-1 and carbon powder (conductive adjuvant) mixed with a weight ratio of 3:1 were completely stirred by a homogenizer and were injected into a mold of the desired size to form a negative electrode having the desired density and thickness by a hot press. The batteries fabricated with PPQ/C composites as the negative electrode, indole trimer or polyindole/C as the positive electrode, porous polyolefin membranes as the separator, and sulfuric acid/hydrogen sulfate as the electrolytic solution exhibited both high capacity and cycle characteristics. In 2013, researches at Silver-H Plus Technology Co. Ltd., Taiwan patented all-plastic secondary batteries using nitrogen-containing conductive polymers, including polyindole, polyaniline, or PPQs as the electrode materials, and polyolefin or polyethersulfone (PES) as the separators [95]. The plastic proton batteries can be charged 100,000 times and will compete with rechargeable lithium batteries.

For battery separators, early in the 1970s, researchers from the US Navy investigated the possibility of using PPQ/polymer blends as the battery separators [96]. A separator is a porous membrane placed between electrodes of opposite polarity, permeable to ionic flow but preventing electric contact of the electrodes [97]. The separator plays a key role in all kinds of batteries. An ideal separator for a battery should fulfill the following requirements: (1) good electronic insulator; (2) readily wetted by electrolyte (minimal electrolyte resistance); (3) chemical resistance to degradation by electrolyte, impurities, and electrode reactants and products; (4) mechanical and dimensional stability; (5) effective in preventing migration of particles or colloidal or soluble species between the two electrodes; and (6) uniform in thickness and other properties. Common separator materials include cellulosic or modified cellulosic materials, microporous polyolefins, polyamide, poly(tetrafluoroethylene) membranes, and so on. Excellent thermal and hydrolysis resistance along with the good mechanical properties of PPQ films also make them suitable as separators for batteries. Angres et al. investigated the behaviors of PPQ/cellulose acetate extruded membranes as separators for alkaline silver–zinc batteries [34]. The membranes were produced by dissolving partially acetylated cellulose and PPQ resin in chloroform and then casting the resulting solution on a plate, allowing the chloroform to evaporate until cloudiness was observed. The plate was then immersed in an aqueous methanol solution, affording the free-standing membrane. The prepared membranes exhibited good antioxidation and anticorrosion properties in strong alkaline electrolytes. When saturated with potassium hydroxide solution (45 wt.%), the PPQ/cellulose membrane had a low electrical resistance of 40–70 Ω cm, which was beneficial for maximum diffusion of electrolytes. Favorable properties of the PPQ membranes under battery circumstances made them excellent candidates as separators for alkaline batteries.

Besides the potential and existing applications of PPQs as electrodes or separators for batteries, the polymers have also been investigated as a class of promising high-temperature membranes for PEM fuel cells (PEMFCs) due to their excellent hydrolytic resistance and thermal stability [9]. PEMFCs have been identified as one of the most promising environmentally friendly energy systems [98]. The PEM is one of the key technical challenges for PEMFCs. Currently, the main obstacle limiting the application of PPQ membranes in PEMFCs is their short lifetime. For instance, sulfonated PPQ (SPPQ) membranes have been used as ionomer membranes ("BAM-1G") in the direct methanol fuel cell products produced by Ballard Power Systems Inc., Canada [99]. Although the BAM-1G membrane exhibited comparable fuel cell performance to that of the current Nafion 117 perfluorosulfonic acid membrane (a product of DuPont, USA), its lifetime was only 350 h [100]. By contrast, a lifetime of over 60,000 hours under fuel cell conditions has been achieved with commercial Nafion membranes [9]. In order to improve the lifetime of PPQ PEMs, considerable basic and applied research has been done. Kopitzke et al. at the Florida Solar Energy Center systemically investigated the sulfonated mechanism and thermal stability of PPQs [101,102]. First, they successfully achieved the sulfonation of solvent-cast PPQ films (PPQ-3 in Table 17.4) via a "soak-and-bake" method [101]. Sulfonation was thought to mainly occur on the benzene rings *ortho* to the electron-donating ether linkage. In addition, they evaluated the thermal stability of several aromatic polymers, including polyetheretherketone (PEEK), PES, PPQ, PBI, and PI before and after sulfonation [102]. It was observed that after sulfonation, the thermal stability of most of the polymers significantly decreases under inert conditions but slightly deteriorates in a saturated water vapor environment. Among the polymers, SPPQ exhibited exceptional stability under both wet and dry conditions, which might be due to the occurrence of some cross-linking reaction during sulfonation of PPQ. Rusanov et al. developed SPPQs via two different routes: sulfonation of PPQs with a sulfuric acid–oleum mixture and direct polymerization from monomers at 125°C using a sulfuric acid–oleum mixture as a solvent, as shown in Figure 17.23a [6]. Both procedures provided flexible and tough SPPQs membranes with tensile strengths higher than 70 MPa. The proton conductivity value of the SPPQs membranes was comparable with those of Nafion 117 membrane. Seel and Benicewicz reported the PPQ/PBI composite membranes synthesized via the PPA process for high-temperature fuel cell systems, as shown in Figure 17.23b [103]. A PBI component was introduced into the systems in order to improve the high-temperature dimensional stability of pure PPQ membranes. The experimental results indicated that an advantageous balance of properties was found for a PPQ/PBI copolymer of 58 mol.% PPQ and 42 mol.% PBI. The composite membrane was found to have a phosphoric acid doping level of 39.2 mol phosphoric acid per mole of polymer repeat unit and a proton conductivity of 0.24 S/cm at 180°C. In addition, the composite membrane exhibited excellent long-term stability in a fuel cell circumstance during a 2900 h lifetime performance test.

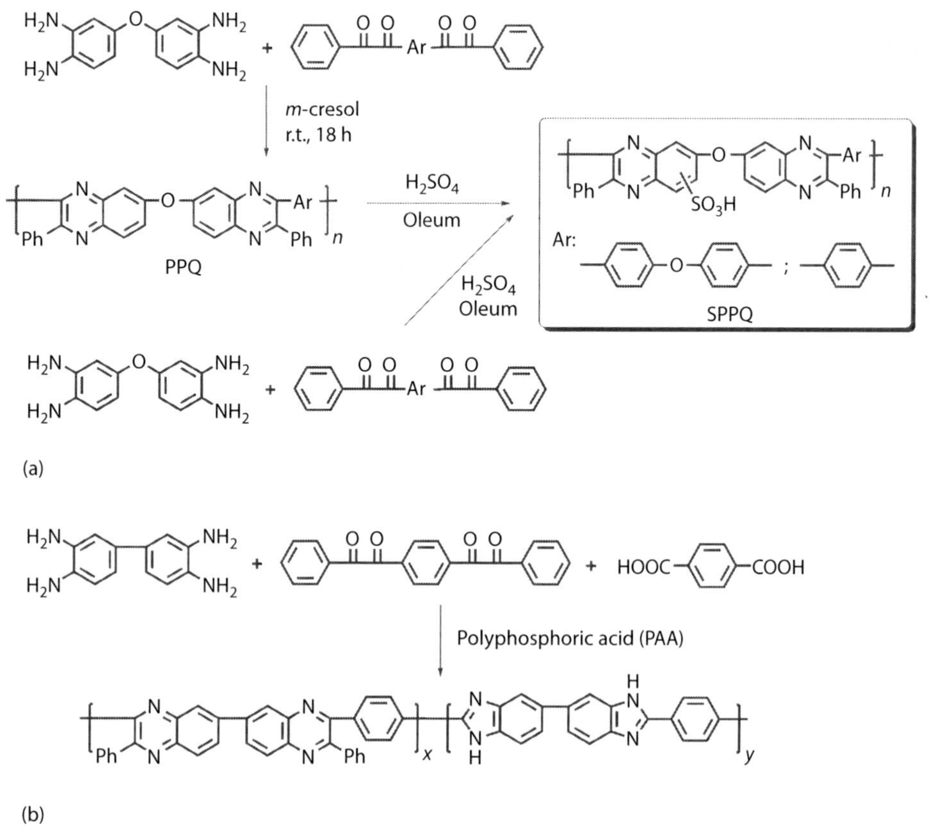

Polyphenylquinoxalines

FIGURE 17.23 (a) Synthesis of sulfonated PPQs via two different routes; (b) phosphoric acid-doped PPQ/PBI copolymer.

17.4.2 OPTICAL APPLICATIONS

As a well-known class of π-electron-deficient polymers, PPQs possess many desired properties as components for optical applications. Up to now, PPQs have been extensively investigated as ETMs for OLEDs, high-refractive-index polymers for optical lenses, and nonlinear optical materials for optoelectronic devices.

For applications as ETMs for OLEDs, PPQs have attracted much attention in recent years [14,104,105]. For instance, O'Brien et al. reported the use of PPQ (PPQ-1 in Table 17.4) as an electron transport conjugated polymer for polymer light-emitting diodes [11]. A tenfold increase in external quantum efficiency (EME) value (0.35%) was obtained in the poly(phenylenevinylene)-based (PPV-based) OLEDs with a brightness up to 250 cd/m^2. Similarly, Cui et al. reported a PPQ containing thiophene segments (PPQ-2 in Table 17.4), which had an electron affinity (EA) value of 3.0 eV and an ionization potential (IP) value of 5.4 eV [12]. The higher EA and IP values for the PPQ endowed it with better electron transport properties, with an EME value of 0.4% and a brightness over 700 cd/m^2, which was a 40-fold enhancement over the single-layer diodes. The PPQ was thought to be one of the best ETMs for PPV-based OLEDs. Zhan et al. developed a series of conjugated polymers containing oxadiazole, quinoline, and quinoxaline moieties in the main chain based on fluorene [40]. The obtained PPQ (PPQ-6 in Table 17.4) exhibited a lowest unoccupied molecular orbital (LUMO) level of −3.37 eV among the polymer series, indicating good electron transport ability. In addition, PPQs containing carbazole [39], benzothiophene [106], and dibenzothiophene [107] also exhibited good electroluminescent characteristics.

Besides the applications of PPQs as ETMs for OLEDs, the other PPQs with special optical characteristics, such as PPQs with ultrahigh refractive indices (PPQ-11 in Table 17.4) [42], electrochromic PPQs containing triarylamino segments (PPQ-4 in Table 17.4) [60], and PPQs with nonlinear optical susceptibility [108] have also been developed and reported in the literature.

17.4.3 Microelectronic Applications

The good film-forming ability, excellent thermal stability, good adhesion to various substrates, good dimensional stability at elevated temperatures, and low dielectric constants and dissipation factors of PPQs make this class of polymers ideal for investigation as ILDs in wafer fabrication, passivation coatings, and dielectric layers in high-density interconnecting systems. Early in the 1980s, PPQs were studied for use in microelectronic fields [109]. In several review articles, PPQs have been identified as a kind of low-dielectric-constant (low-k), high-temperature polymer with potential applications as ILDs for microelectronics [30,110]. The k values of conventional PPQs are usually in the range of 2.8–3.0. By introduction of air into PPQs (PPQ foams), the k values can be reduced to 2.3–2.4, as mentioned before. PPQs have also been recognized as candidate dielectrics in bump-on-polymer (BOP) structures for wafer-level packaging [111].

Pireaux et al. investigated the physicochemical surface properties of IP200 PPQ and PPQ/copper interfaces [112]. In multichip module (MCM) interconnection systems fabrication, aluminum and copper are often used. Aluminum usually shows good chemical inertness with respect to PPQ dielectric. However, the resistance losses of aluminum lines are approximately 50% higher than those of copper [26]. In contrast, copper has a low resistivity and can be applied by either sputtering or plating procedures, but it is usually incompatible with PPQ. Copper oxide easily forms on the surface of PPQ dielectric, which will act as a catalyst initiating the degradation of the dielectric during the subsequent thermal treatment. For example, when the thermal treatment included 30 min at 350°C in nitrogen, followed by 5 min in oxygen at that temperature, the PPQ/copper interface was severely damaged. The depth of copper diffusion in the polymer film was more than 0.1 μm. An effective remedy preventing the diffusion of copper in PPQ dielectric might be sputtering a passivation layer on the copper surface. The passivation layer can be chromium, nickel, or titanium, all of which are inert to the PPQ layer.

17.5 CONCLUSIONS

As a fine chemical with small quantities but very high added value, PPQ is attracting much attention from both the materials manufacturers and the engineers in high-tech fields. Although the applications of PPQs have concentrated on some very specific areas (areas where there are several competitive polymers), some of the characteristics of PPQs, such as their excellent thermoplasticity and long-term hydrolysis resistance, are the best among high-temperature polymers. With the continuous discovery of new functions for the polymers and reduction in their manufacturing cost, PPQs will expand into new markets and new applications. In the near future, PPQ, the "old-age" polymer, will regain new vitality and serve a number of high-tech applications.

ACKNOWLEDGMENTS

Financial support from the National Natural Science Foundation of China (no. 51173188) and National Basic Research Program of China (2014CB643605) are gratefully acknowledged.

REFERENCES

1. Hergenrother, P. M. and Levine, H. H. 1967. Phenyl-substituted polyquinoxalines. *J Polym Sci, Part A1* 5: 1453–66.
2. Hergenrother, P. M. 1971. Linear polyquinoxalines. *J Macromol Sci, Part C: Polym Rev* C6: 1–28.
3. Sung, N. H. and McGarry, F. J. 1976. The mechanical and thermal properties of graphite fiber reinforced polyphenylquinoxaline and polyimide composites. *Polym Eng Sci* 16: 426–36.
4. Webster, J. R. 1998. *Thin film polymer dielectrics for high-voltage applications under severe environments*. PhD diss., Virginia Polytechnic Institute and State University, Blacksburg, VA, USA.
5. Kim, B. S., Korleski, J. E., Zhang, Y., Klein, D. J. and Harris, F. W. 1999. Development of a new poly(phenylquinoxaline) for adhesive and composite applications. *Polymer* 40: 4553–62.
6. Rusanov, A. L., Belomoina, N. M., Bulycheva, E. G., Yanul, A. A., Likhatchev, D. Y., Dobrovolskii, Y. A., Iojoiu, C., Sanchez, J. Y., Voytekunas, V. Y. and Abadie, M. J. M. 2008. Preparation and characterization of sulfonated polyphenylquinoxalines. *High Perform Polym* 20: 627–41.
7. Rusanov, A. L., Likhatchev, D., Kostoglodov, P. V., Müllen, K. and Klapper, M. 2005. Proton-exchanging electrolyte membranes based on aromatic condensation polymers. *Adv Polym Sci* 179: 83–134.
8. Hickner, M. A., Ghassemi, H., Kim, Y. S., Einsla, B. R. and McGrath, J. E. 2004. Alternative polymer systems for proton exchange membranes (PEMs). *Chem Rev* 104: 4587–612.
9. Li, Q. F., He, R. H., Jensen, J. O. and Njerrum, N. J. 2003. Approaches and recent development of polymer electrolyte membranes for fuel cells operating above 100°C. *Chem Mater* 15: 4896–915.
10. Wrasidlo, W. J. and Spiegelman, S. 1979. Ultrafiltration membranes based on heteroaromatic polymers. *US Patent* 4259251.
11. O'Brien, D., Weaver, M. S., Lidzey, D. G. and Bradley, D. D. C. 1996. Use of poly(phenylquinoxaline) as an electron transport material in polymer light-emitting diodes. *Appl Phys Lett* 69: 881–3.
12. Cui, Y. T., Zhang, X. J. and Jenekhe, S. A. 1999. Thiophene-linked polyphenylquinoxaline: A new electron transport conjugated polymer for electroluminescent devices. *Macromolecules* 32: 3824–6.
13. O'Brien, D., Bleyer, A. and Bradley, D. D. C. 1996. Electroluminescence applications of a poly(phenylquinoxaline). *Synth Met* 76: 105–8.
14. Kulkarni, A. P., Tonzola, C. J., Babel, A. and Jenekhe, S. A. 2004. Electron transport materials for organic light-emitting diodes. *Chem Mater* 16: 4556–73.
15. Zhan, X. W., Liu, Y. Q., Zhu, D. B., Xu, G., Liu, X. C. and Ye, P. X. 2003. Highly efficient, thermally stable and optically transparent third-order nonlinear optical copolymers consisting of fluorene and quinoxaline/quinoline units. *Appl Phys A* 77: 375–8.
16. Unroe, M. R., Reinhardt, B. A. and Arnold, F. E. 1987. Phenylquinoxaline resin monomers. *US Patent* 4683309.
17. Elce, E. and Hay, A. S. 1996. A new synthesis of bisbenzils and novel poly(phenylquinoxaline)s therefrom. *Polymer* 37: 1745–9.
18. Hergenrother, P. M. 1976. Polyphenylquinoxalines—High performance thermoplastics. *Polym Eng Sci* 16: 303–8.
19. Korshak, V. V. and Rusanov, A. L. 1983. Phenyl-substituted polyheteroarylenes. *Russ Chem Rev* 52: 459–68.
20. Krongauz, Y. S. 1984. Current status and prospects of development of polyphenylquinoxalines. Review. *Polym Sci USSR* 26: 245–62.
21. Hergenrother, P. M. 1986. High performance thermoplastics. *Angew Makromol Chem* 145: 323–41.
22. Burma, M., Sava, I., Simionescu, C. I., Belomoina, N. M., Krongauz, E. S. and Korshak, V. V. 1989. Thermostable polymers containing phenylquinoxaline and other heterocyclic units. *J Macromol Sci: Part A Chem* 26: 969–88.
23. Béland, S. 1990. *High performance thermoplastic resins and their composites*. Park Ridge, NJ: Noyes.
24. Bruma, M. 1997. Polyphenylquinoxalines. In *Handbook of Thermoplastics*, ed. Olabisi, O., 771–98. New York: Marcel Dekker.
25. Lu, F. C. 1998. Some heterocyclic polymers and polysiloxanes. *J Macromol Sci—Rev Macromol Chem Phys* C38: 143–205.
26. Rabilloud, G. 1999. *High-performance polymers. 2. Polyquinoxalines and polyimides*. Paris: Editions Technip.
27. Hergenrother, P. M. 2003. The use, design, synthesis, and properties of high performance/high temperature polymers: An overview. *High Perform Polym* 15: 3–45.

28. Bruma, M., Hamciuc, E., Sava, I. and Belomoina, N. M. 2004. Polymers containing phenylquinoxaline rings. *Russ Chem Bull, Int Ed* 53: 1813–23.
29. Hergenrother, P. M. and Connell, J. W. 2012. Polyquinoxaline matrix resins and composites. In *Wiley encyclopedia of composites*, Second Edition, eds. Nicolais, L. and Borzacchiello, A., 1–17. New York: John Wiley & Sons.
30. Maier, G. 2001. Low dielectric constant polymers for microelectronics. *Prog Polym Sci* 26: 3–65.
31. St Clair, A. K. and Johnston, N. J. 1977. Ether polyphenylquinoxalines. II. Polymer synthesis and properties. *J Polym Sci, Polym Chem* 15: 3009–21.
32. Klein, D. J., Modarelli, D. A. and Harris, F. W. 2001. Synthesis of poly(phenylquinoxaline)s via self-polymerizable quinoxaline monomers. *Macromolecules* 34: 2427–37.
33. Pike, R. A. 1975. Oxidatively stable modified polyphenylquinoxaline resin. *US Patent* 3909481.
34. Angres, I., Duffy, J. V. and Matesky, S. J. 1979. Polymeric membranes which contain polyphenylquinoxalines and which are useful as battery separators. *US Patent* 4158649.
35. Wrasidlo, W. and Augl, J. M. 1969. Phenylated polyquinoxalines from bis(phenylglyoxaloyl)benzene. *J Polym Sci, Part A-1: Polym Chem* 7: 3393–405.
36. Jandke, M., Strohriegl, P., Berleb, S., Werner, E. and Brütting, W. 1998. Phenylquinoxaline polymers and low molar mass glasses as electron-transport materials in organic light-emitting diodes. *Macromolecules* 31: 6434–43.
37. Rabilloud, G. and Sillion, B. 1978. New polyphenylquinoxalines linked by *m*-terphenyl flexibilizing groups. *J Polym Sci: Polym Chem Ed* 16: 2093–111.
38. Rafter, R. T. and Harrison, E. S. 1976. Tris-benzil crosslinked polyphenylquinoxalines. *Polym Eng Sci* 16: 318–22.
39. Keshtov, M. L., Mal'tsev, E. I., Lypenko, D. A., Brusentseva, M. A., Sosnovyi, M. A., Vasnev, V. A., Peregudov, A. S., Vannikov, A. V. and Khokhlov, A. R. 2006. New carbazole-containing polyphenylquinoxalines: Synthesis, photophysical, and electroluminescent properties. *Polym Sci Ser B* 48: 1135–46.
40. Zhan, X. W., Liu, Y. Q., Wu, X., Wang, S. and Zhu, D. B. 2002. New series of blue-emitting and electron-transporting copolymers based on fluorene. *Macromolecules* 35: 2529–37.
41. Heath, D. R. and Wirth, J. G. 1974. Polyetherquinoxalines. *US Patent* 3852244.
42. Li, C., Li, Z., Liu, J. G., Zhao, X. J., Yang, H. X. and Yang, S. Y. 2010. Synthesis and characterization of organo-soluble thioether-bridged polyphenylquinoxalines with ultra-high refractive indices and low birefringences. *Polymer* 51: 3851–8.
43. Li, C., Li, Z., Liu, J. G., Yang, H. X. and Yang, S. Y. 2010. Multi-methyl-substituted polyphenylquinoxalines with high solubility and high glass transition temperatures: Synthesis and characterization. *J Macromol Sci, Part A: Pure Appl Chem* 47: 248–53.
44. Li, C., Li, Z., Liu, J. G., Yang, H. X. and Yang, S. Y. 2010. Fluorene-bridged polyphenylquinoxalines with high solubility and good thermal stability: Synthesis and properties. *Chin Polym Sci* 28: 971–80.
45. Lindley, P. M. and Reinhardt, B. A. 1991. Intramolecular cyclization of pendant phenylethynyl groups as a route to solvent resistance in polyphenylquinoxalines. *J Polym Sci, Part A: Polym Chem* 29: 1061–71.
46. Li, C. 2012. *Molecular design, synthesis and properties of functional polyphenylquinoxalines*. PhD diss., Institute of Chemistry, Chinese Academy of Sciences, Beijing, China.
47. Foster, R. T. and Marvel, C. S. 1965. Polybenzimidazoles. IV. Polybenzimidazoles containing aryl ether linkages. *J Polym Sci, Part A* 3: 417–21.
48. Liu, J. G., Wang, L. F., Yang, H. X., Li, H. S., Li, Y. F., Fan, L. and Yang, S. Y. 2004. Synthesis and characterization of new polybenzimidazopyrrolones derived from pyridine-bridged aromatic tetraamines and dianhydrides. *J Polym Sci, Part A: Polym Chem* 42: 1845–56.
49. Vogel, H. and Marvel, C. S. 1961. Polybenzimidazoles, new thermally stable polymers. *J Polym Sci* 50: 511–39.
50. Relles, H. M., Orlando, C. M., Heath, D. R., Schluenz, R. W., Manello, J. S. and Hoff, S. 1977. Diether polyphenylquinoxalines. Monomers via nitro dispacement–^{13}C-NMR analysis of monomers and polymers. *J Polym Sci: Polym Chem* 15: 2441–50.
51. Chen, H. Y., Cronin, J. A. and Archer, R. D. 1995. Synthesis and characterization of two new Schiff-bases and their soluble linear cerium (IV) polymers. *Inorg Chem* 34: 2306–15.
52. Stille, J. K. and Arnold, F. E. 1966. Polyquinoxaline. III. *J Polym Sci, Part A-1* 4: 551–62.
53. Wang, J. H., Li, N. W., Zhang, F., Zhang, S. B. and Liu, J. 2009. Synthesis and properties of soluble poly[bis(benzimidazobenzisoquinolinones)] based on novel aromatic tetraamine monomers. *Polymer* 50: 810–16.

54. Li, N. W., Cui, Z. M., Li, S. H., Zhang, S. B. and Xie, W. 2009. Synthesis and properties of water soluble poly[bis(benzimidazobenzisoquinolinone)] ionomers for proton exchange membranes fuel cells. *J Membr Sci* 326: 420–8.
55. Tao, L. M. 2010. *Synthesis and characterization of fluorinated heteroaromatic polymers*. PhD diss., Institute of Chemistry, Chinese Academy of Sciences, Beijing, China.
56. Koros, W. J. and Walker, D. R. B. 1993. Polyamides and polypyrrolones for fluid separation membranes. *US Patent* 5262056.
57. Vora, R. H., Paul, C. S. and Menczel, J. D. 1991. Novel polybenzimidazolone polymers based hexafluoro aromatic tetraamines. *US Patent* 5075419.
58. Liu, J. G., Wang, L. F., Fan, L. and Yang, S. Y. 2005. Synthesis and characterization of polybenzimidazopyrrolones based on pyridine-bridged aromatic tetraamines and dianhydrides. In *Polyimides and other high temperature polymers*, Vol. 3, ed. Mittal, K. L., 155–73. Boca Raton, FL: Taylor & Francis.
59. Gooch, J. W. 2011. *Encyclopedic dictionary of polymers*, Second Edition, 569. Springer: New York.
60. Keshtov, M. L., Buzin, M. I., Petrovskii, P. V., Makhaeva, E. E., Kochurov, V. S., Marochkin, D. V. and Khokhlov, A. R. 2011. Synthesis, photophysical, and electrochromic properties of new triarylamino-containing polyphenylquinoxalines. *Polym Sci Ser B* 53: 257–66.
61. Liu, J. G., Nakamura, Y., Ogura, T., Shibasaki, Y., Ando, S. and Ueda, M. 2008. Optically transparent sulfur-containing polyimide–TiO_2 nanocomposite films with high refractive index and negative pattern formation from poly(amic acid)–TiO_2 nanocomposite film. *Chem Mater* 20: 273–81.
62. Liu, J. G. and Ueda, M. 2009. High refractive index polymers: Fundamental research and practical applications. *J Mater Chem* 19: 8907–19.
63. Liu, J. G., Nakamura, Y., Ogura, T., Shibasaki, Y., Ando, S. and Ueda, M. 2007. High refractive index polyimides derived from 2,7-bis(4-aminophenylenesulfanyl)thianthrene and aromatic dianhydrides. *Macromolecules* 40: 4614–20.
64. Fang, T. 1991. Poly(3-phenylquinoxaline): A new thermally stable polymer. *Macromolecules* 24: 444–9.
65. Klein, D. J., Korleski, J. E. and Harris, F. W. 2001. Synthesis of polyphenylquinoxaline copolymers via aromatic nucleophilic substitution reactions of an A-B quinoxaline monomer. *J Polym Sci, Part A: Polym Chem* 39: 2037–42.
66. Polce, M. J., Klein, D. J., Harris, F. W., Modarelli, D. A. and Wesdemiotis, C. 2001. Structural characterization of quinoxaline homopolymers and quinoxaline/ether sulfone copolymers by matrix-assisted laser desorption ionization mass spectrometry. *Anal Chem* 73: 1948–58.
67. Baek, J. B. and Harris, F. W. 2005. Synthesis and polymerization of new self-polymerizable quinoxaline monomers. *J Polym Sci, Part A: Polym Chem* 43: 78–91.
68. Baek, J. B. and Harris, F. W. 2005. Development of an improved synthetic route to an AB phenylquinoxaline monomer. *J Polym Sci, Part A: Polym Chem* 43: 801–4.
69. Baek, J. B. and Harris, F. W. 2005. Fluorine- and hydroxyl-terminated hyperbranched poly(phenylquinoxalines) (PPQs) from copolymerization of self-polymerizable AB and AB_2, BA, and BA_2 monomers. *Macromolecules* 38: 1131–40.
70. Srinivasan, S., Twieg, R., Hedrick, J. L. and Hawker, C. J. 1996. Heterocycle-activated aromatic nucleophilic substitution of AB_2 poly(aryl ether phenylquinoxaline) monomers. 3. *Macromolecules* 29: 8543–5.
71. Baek, J. B. and Harris, F. W. 2004. Synthesis and thermal behavior of phenylethynyl-terminated linear and hyperbranched polyphenylquinoxalines. *J Polym Sci, Part A: Polym Chem* 42: 6318–30.
72. Baek, J. B. and Tan, L. S. 2006. Hyperbranched poly(phenylquinoxaline-ether-ketone) synthesis in poly(phosphoric acid)/P_2O_5 medium: Optimization and some interesting observations. *Macromolecules* 39: 2794–803.
73. Baek, J. B., Simko, S. R. and Tan, L. S. 2006. Synthesis and chain-end modification of a novel hyperbranched polymer containing alternating quinoxaline and benzoxazole repeat units. *Macromolecules* 39: 7959–66.
74. Hedrick, J. and Labadie, J. W. 1990. Poly(aryl ether phenylquinoxalines). *Macromolecules* 23: 1561–8.
75. Hedrick, J., Twieg, R., Matray, T. and Carter, K. 1993. Heterocycle-activated aromatic nucleophilic substitution: Poly(aryl ether phenylquinoxalines). 2. *Macromolecules* 26: 4833–9.
76. Labadie, J. W., Hedrick, J. L. and Boyer, S. K. 1992. Poly(aryl ether–phenylquinoxaline)s: Self-polymerization of an AB monomer via quinoxaline-activated ether synthesis. *J Polym Sci, Part A: Polym Chem* 30: 519–23.

77. Bruma, M., Schulz, B., Kopnick, T., Dietel, R., Stiller, B., Mercer, F. and Reddy, V. N. 1998. Investigation of thin films made from silicon-containing polyphenylquinoxaline-amides. *High Perform Polym* 10: 207–15.
78. Bruma, M. and Schulz, B. 2001. Silicon-containing aromatic polymers. *J Macromol Sci, Part C: Polym Rev* 41: 1–40.
79. Bruma, M. 2007. High performance polymers containing phenylquinoxaline and silicon in the main chain. *Rev Roumaine Chim* 52: 309–18.
80. Bruma, M. 2008. Polyphenylquinoxalines containing hexafluoroisopropylidene groups. *Rev Roumaine Chim* 53: 5–13.
81. Abadie, M. J. M., Voytekunas, V. Y. and Rusanov, A. L. 2006. State of the art organic matrices for high performance composites: A review. *Iran Polym J* 15: 65–77.
82. Connell, J. W., Smith, Jr., J. G. and Hergenrother, P. M. 2000. Oligomers and polymers containing phenylethynyl groups. *J Macromol Sci, Part C: Polym Rev* 40: 207–30.
83. Hergenrother, P. M. 1981. Poly(phenylquinoxalines) containing phenylethynyl groups. *Macromolecules* 14: 898–904.
84. Hergenrother, P. M. 1983. Polyphenylquinoxalines containing pendant phenylethynyl groups: Preliminary mechanical properties. *J Appl Polym Sci* 28: 355–66.
85. Ooi, I. H., Hergenrother, P. M. and Harris, F. W. 2000. Synthesis and properties of phenylethynyl-terminated, star-branched, phenylquinoxaline oligomers. *Polymer* 41: 5095–107.
86. Li, C., Liu, J. G., Ji, M., Yang, H. X. and Yang, S. Y. 2011. Phenylethynyl-terminated polyphenylquinoxaline engineering plastics with high thermal stability and good processability. Paper presented at the 5th International Symposium on Engineering Plastics, Kunming, China.
87. Connell, J. W., Smith, Jr., J. G. and Hergenrother, P. M. 1997. Chemistry and adhesive properties of phenylethynyl-terminated phenylquinoxaline oligomers. *J Adhesion* 60: 15–26.
88. Kohl, P. A. 2011. Low-dielectric constant insulators for future integrated circuits and packages. *Annu Rev Chem Biomol Eng* 2: 379–401.
89. Labadie, J. W., Hedrick, J. L. Wakharkar, V. and Hofer, D. C. 1992. Nanopore foams of high temperature polymers. *IEEE Trans Compon Hybr Manufact Technol* 15: 925–30.
90. Merlet, S., Marestin, C., Schiets, F., Romeyer, O. and Mercier, R. 2007. Preparation and characterization of nanocellular poly(phenylquinoxaline) foams. A new approach to nanoporous high performance polymers. *Macromolecules* 40: 2070–8.
91. Merlet, S., Marestin, C., Romeyer, O. and Mercier, R. 2008. "Self-foaming" poly(phenylquinoxaline)s for the designing of macro and nanoporous materials. *Macromolecules* 41: 4205–15.
92. Renaud, R. and Leroy, R. L. 1982. Separator materials for use in alkaline water electrolysers. *Int J Hydrogen Energy* 7: 155–66.
93. Nobuta, T., Nishiyama, T., Kamisuki, H., Harada, G., Kurosaki, M., Nakagawa, Y., Yoshida, S. and Mitani, M. 2004. Secondary battery of proton conductive polymer. *US Patent* 6800395.
94. Kamisuki, H., Nishiyama, T., Harada, G., Yoshida, S., Kurosaki, M., Nakagawa, Y., Nobuta, T. and Mitani, M. 2005. Secondary battery of proton conductive polymer. *US Patent* 6899974.
95. Chuang, C. C., Hsu, C. Y. and Perng, L. H. 2013. Plastics electrode material and secondary cell using the material. *US Patent* 8377546.
96. William, P., Kilroy, W. P. and Laughlin, L. 1979. Diffusion measurements on a new battery separator membrane polyphenylquinoxaline–cellulose acetate): A comparison with standard separator materials. *Mol Cryst Liq Cryst* 54: 201–5.
97. Arora, P. and Zhang, Z. M. 2004. Battery separators. *Chem Rev* 104: 4419–62.
98. Sopian, K. and Daud, W. R. W. 2006. Challenges and future developments in proton exchange membrane fuel cells. *Renew Energ* 31: 719–27.
99. Apanel, G. and Johnson, E. 2004. Direct methanol fuel cells—Ready to go commercial? *Fuel Cells Bull* 11: 12–17.
100. Steck, A. and Stone, C. 1997. Development of BAM membrane for fuel cell applications. In *Proceedings of the 2nd International Symposium on New Materials for Fuel-Cell and Modern Battery Systems*, eds. Savadogo, O. and Roberge, P. R., 792–807. Ecole Polytechnique de Montreal, Montreal, Canada.
101. Kopitzke, R. W., Linkous, C. A. and Nelson, G. L. 1998. Sulfonation of a poly(phenylquinoxaline) film. *J Polym Sci Part A: Polym Chem* 36: 1197–9.
102. Kopitzke, R. W., Linkous, C. A. and Nelson, G. L. 2000. Thermal stability of high temperature polymers and their sulfonated derivatives under inert and saturated vapor conditions. *Polym Degrad Stab* 67: 335–44.

103. Seel, D. C. and Benicewicz, B. C. 2012. Polyphenylquinoxaline-based proton exchange membranes synthesized via the PAA process for high temperature fuel cell systems. *J Membr Sci* 405–6: 57–67.
104. Hughes, G. and Bryce, M. R. 2005. Electron-transporting materials for organic electroluminescent and electrophosphorescent devices. *J Mater Chem* 15: 94–107.
105. Weaver, M. S., Lidzey, D. G., Fisher, T. A., Pate, M. A., O'Brien, D., Bleyer, A., Tajbakhsh, A., Bradley, D. D. C., Skolnick, M. S. and Hill, G. 1996. Recent progress in polymers for electroluminescence: Microcavity devices and electron transport polymers. *Thin Solid Films* 273: 39–47.
106. Keshtov, M. L., Mal'tsev, E. I., Lypenko, D. A., Brusentseva, M. A., Sosnovyi, M. A., Il'ina, M. N., Vasnev, V. A. et al. 2008. Synthesis and photophysical properties of polyphenylquinoxalines with thiophene and benzothiophene units in side chains. *Polym Sci Ser A* 50: 18–24.
107. Keshtov, M. L., Mal'tsev, E. I., Lypenko, D. A., Brusentseva, M. A., Sosnovyi, M. A., Vasnev, V. A., Peregudov, A. S., Vannikov, A. V. and Khokhlov, A. R. 2007. Synthesis and photophysical properties of polyphenylquinoxalines with thiophene and dibenzothiophene units in the backbone. *Polym Sci Ser B* 49: 75–9.
108. Yan, J., Wu, J. Y., Zhu, H. Y., Zhang, X. T., Sun, D., Hu, Y. M., Li, F. M. and Sun, M. 1995. Excited-state enhancement of the third-order nonlinear optical susceptibility of nonether polyphenylquinoxaline. *Opt Lett* 20: 255–7.
109. Labadie, J. W. and Hedrick, J. L. 1989. Recent advances in high temperature polymers for microelectronic applications. *SAMPE J* 25: 18–23.
110. Treichel, H., Ruhl, G., Ansmann, P., Wurl, R., Muller, C. and Dietlmeier, M. 1998. Low dielectric constant materials for interlayer dielectric. *Microelectron Eng* 40: 1–19.
111. Reche, J. J. H. and Kim, D. H. 2003. Wafer level packaging having bump-on-polymer structure. *Microelectron Reliab* 43: 879–94.
112. Pireaux, J J., Gregoire, C., Bellard, L., Cros, A., Torres, J., Palleau, J., Templier, F. and Lazare, S. 1994. Growth, physical properties, and adhesion of copper–polyphenylquinoxaline interfaces. *J Appl Phys* 76: 1244–55.

18 Aromatic Polyhydrazides and Their Corresponding Polyoxadiazoles*

Emilia Di Pace, Paola Laurienzo, Mario Malinconico, and Maria Grazia Volpe

CONTENTS

18.1 Introduction and Historical Background 572
    18.1.1 Polyhydrazides and Polyoxadiazoles as Thermal-Resistant Polymers 572
    18.1.2 Synthesis 572
        18.1.2.1 One-Step Synthesis 572
        18.1.2.2 Two-Step Synthesis 573
    18.1.3 Polyoxadiazoles as Electroactive Polymers 573
18.2 Polyhydrazides and Polyoxadiazoles for High-Temperature Applications 575
    18.2.1 Aromatic Polyetheraroyl Hydrazides Based on 4-Oxybenzoyl Units and Their Corresponding Polyoxadiazoles 575
        18.2.1.1 Synthesis of Monomers and Polymers 575
        18.2.1.2 Polymerization Procedure 577
        18.2.1.3 Thermogravimetric Analysis 577
        18.2.1.4 Thermal Analysis 579
        18.2.1.5 X-Ray Diffraction Analysis 580
        18.2.1.6 Studies on the Cyclization Reaction by Infrared Analysis 581
        18.2.1.7 Thermal Analysis 582
        18.2.1.8 Optical Microscopy 584
        18.2.1.9 X-Ray Diffraction Analysis 585
        18.2.1.10 Conclusions 588
    18.2.2 Aromatic Poly(Ether-Alkyl) Hydrazides Based on 4,4′-Dihydroxybiphenyl Units and Their Cyclization to Polyoxadiazoles 588
        18.2.2.1 Synthesis and Characterization of Poly(Ether-Alkyl) Hydrazides 588
        18.2.2.2 Thermogravimetric Analysis 591
        18.2.2.3 Thermal Analysis 592
        18.2.2.4 X-Ray Diffraction Analysis 593
        18.2.2.5 Conversion of Polyhydrazides to Polyoxadiazoles 594
        18.2.2.6 Thermal Analysis 595
        18.2.2.7 X-Ray Diffraction Analysis 599
        18.2.2.8 Optical Microscopy 600
        18.2.2.9 Conclusions 603
18.3 Polyhydrazides and Polyoxadiazoles for Electro-Optical Applications 603
    18.3.1 Alkoxy-Substituted Poly(p-Phenylene-1,3,4-Oxadiazole) 603
    18.3.2 Polyoxadiazoles as Electroluminescent Materials 608

---

* Based in parts on the first-edition chapter on aromatic polyhydrazides and their corresponding polyoxadiazoles.

18.4 Polyhydrazides and Polyoxadiazoles for High Specialty Applications 610
  18.4.1 Polyoxadiazoles for High-Performance Membranes 610
    18.4.1.1 Fluorinated Polyoxadiazoles 610
    18.4.1.2 Azobenzene-Containing Polyhydrazides and Polyoxadiazoles 611
    18.4.1.3 Sulfonated Polyoxadiazoles 611
  18.4.2 Examples of Polyhydrazides and Polyoxadiazoles Containing Different
          Functional Moieties in the Backbone 612
    18.4.2.1 Polyhydrazides and Polyoxadiazoles Containing Carbohydrate Moieties 612
    18.4.2.2 Polyhydrazides and Polyoxadiazoles Containing 1,3-Imidazolidine-
              2,4,5-Trione Rings 612
  18.4.3 Polyhydrazides with Shape-Persistent Properties 613
Acknowledgments 614
References 614

18.1 INTRODUCTION AND HISTORICAL BACKGROUND

18.1.1 POLYHYDRAZIDES AND POLYOXADIAZOLES AS THERMAL-RESISTANT POLYMERS

Since the 1970s, interest has grown enormously in heat-resistant, rigid-chain polymers for application in extreme environments. Kevlar [1] is the most successful example, followed later by poly(aryl ether ketone)s such as poly(ether ether ketone) (PEEK) [2]. In order to improve the processability of such aromatic polymers, it is common practice to decrease their melting points either by the introduction of defects or by the insertion of aliphatic units along the chains. Moreover, such polymers are receiving growing attention as molecular reinforcement in classical thermoplastics. With such an approach, blends are prepared by mixing the rigid polymer with the flexible one in a common solvent at a critical concentration, followed by coagulation in a nonsolvent, normally under spinning conditions. The mechanical properties of these systems depend mainly on the state of dispersion of the rigid-rod component in the flexible coil matrix [3–6].

Aromatic polyoxadiazoles are interesting candidates for application in extreme conditions. Poly(p-phenylene-1,3,4-oxadiazoles) (PODZs) couple structural characteristics analogous to those of poly(p-phenylene terephtalamide) (PPTA) and poly(buthylene terephthalate) (PBT) and ease of fabrication starting from monomers that are commercially available. PODZs were discovered in the 1960s and developed in the 1970s [7–10]. Their fibers can be spun from sulfuric acid even though the solutions do not display lyotropic behavior. Nevertheless, the resulting fibers are strong and thermally resistant.

18.1.2 SYNTHESIS

The first attempts at the synthesis of PODZ involved the direct reaction between bis-tetrazoles and diacyl chlorides [11]. The polymers obtained were characterized by low M_w; they were infusible and insoluble. Different methods for PODZ synthesis have been described in the literature since then [12–17]. The more successful procedures for high-M_w PODZ are discussed below. Both the methods described here present more advantages, such as higher molecular weight and lower residual hydrazide content.

18.1.2.1 One-Step Synthesis

The synthesis followed by Iwakura et al. [18] consists of direct polycondensation of a diacid and hydrazide in oleum or polyphosphoric acid (PPA), according to the following scheme:

This method does not allow the isolation of the intermediate polyhydrazide (PHZ), which forms and converts immediately to polyoxadiazoles. So, only insoluble polymers are obtained. However, Gomes et al. [19] showed that high reaction temperatures (160°C) favor the solubility of the resulting polymers when the direct PODZ syntheses in PPA solutions are performed.

18.1.2.2 Two-Step Synthesis

This method consists of the low-temperature condensation of a dihydrazide and a diacyl chloride, which leads to the formation of an intermediate PHZ, according to the following scheme:

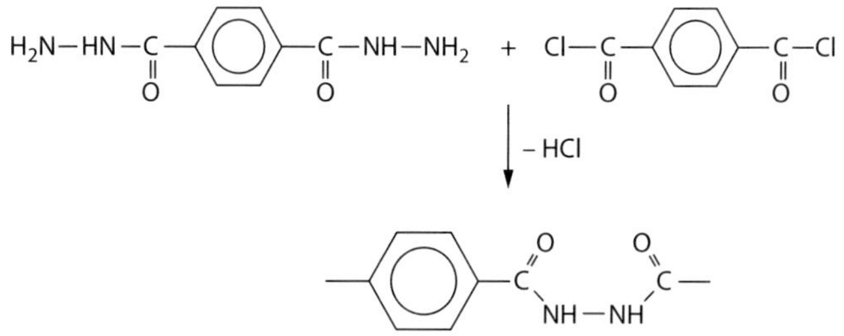

The reaction is carried out at low temperature, normally between 0°C and 15°C. Subsequently, the PHZ is isolated in film or fiber form and heated at very high temperatures (not less than 280°C), where it undergoes intramolecular cyclodehydration according to the following scheme [14]:

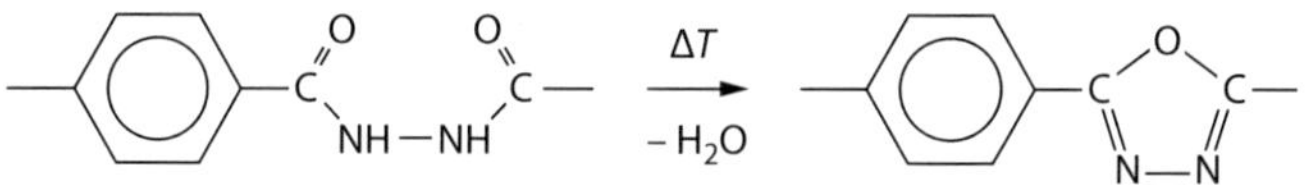

Thermal cyclodehydration is accompanied by simultaneous degradation of hydrazide groups. However, Cassidy and Fawcett [20] reported that the degradation reaction can be minimized if the reaction is accomplished 25–50°C below the cyclodehydration transition temperature. Shrinkage of the films due to ring closure of hydrazide groups to 1,3,4-oxadiazole occurs during cyclization, making it difficult to produce flexible PODZ films. However, it has been shown that it is possible to flexibilize PODZ films, optimizing the reaction conditions of PHZ cyclodehydration by an accurate control of parameters (temperature and time of cyclization) [19]. Particularly, in order to obtain PODZ film useful as membranes, the cyclization temperature must be lower than the T_g of the initial PHZ.

In an alternative to thermal treatment, cyclodehydration can be chemically activated by dehydrating agents, such as PPA or $POCl_3$.

The relative tractability of the precursor PHZs, i.e., solubility and sometimes moldability, renders them very attractive for several applications, for example, preparation of PODZ membranes for gas separation, or incorporation by solution or melt-blending methods into polar thermoplastic matrices, like polyamides or polyesters, to improve properties such as modulus and thermal resistance.

18.1.3 Polyoxadiazoles as Electroactive Polymers

The discovery that certain polymeric and organic materials, if opportunely treated, can conduct electricity almost as efficiently as copper [21] sparked a general interest in electronic and optoelectronic devices made entirely from plastics or from simple organic materials. Among these, organic crystals composed of three (anthracene) to five (pentacene) linked rings of carbon atoms and fullerene systems (C_{60}) have been studied for their properties of superconductivity [22]. As far as polymers are concerned, electrically active polymers can be divided into two main groups: filled polymers and molecularly (inherently) conductive polymers. The first group can be viewed as pseudoactive as far as polymers are concerned, since the electrical properties come from the addition of particles of electrically conductive materials, usually carbon, aluminum, or steel. In this case, the polymer matrix functions simply as a forming and protective medium in which the particles provide the electrical properties [23]. The molecularly conductive polymers, instead, have the characteristic of containing alternate double or triple bonds or aromatic units in the main chain, which can delocalize electrons. In contrast, full saturated polymers have been classified as insulators and employed for their good properties as insulating materials.

Conducting polymers have been found to be promising candidates for a wide range of applications such as electrochromic devices, sensors, light-emitting diodes, etc. The conductive polymers show values of inherent electrical conductivity between 10^{-5} and 10^{-15} Ω^{-1} cm^{-1}; in the range of between about 10 and 10^{-12} Ω^{-1} cm^{-1}, they are classified as semiconductors [24] (the valence band is completely full and the conduction band is completely empty at the absolute zero of temperature, but there is an energy gap between the valence and conduction bands of no more than about 2 eV). Higher values, up to 10^3 Ω^{-1} cm^{-1} (which is typical of a metallic conductor), can be measured when the polymers are doped with charge-transferring elements or compounds (extrinsic conductivity). These consist of either electron donors (alkaline metals) or electron acceptors (halogens, AsF_5).

Polyacetylene is the most widely investigated electroactive polymer, but a large number of other conjugated polymers of high interest are now also established [25–32]. Of these, great efforts are devoted to polypyrrole, poly(*p*-phenylene), polythiophene, poly(*p*-phenylenevinylene), and more recently, polyaniline. Poly(*p*-phenylene sulfide) is an example of a nonconjugated chain that produces conductivity when doped, in the semiconductor range. A large number of potential applications have been foreseen for these polymers, for instance, regulating photovoltaics of transduction of the solar energy in electric energy or less sophisticated applications such as low-cost diodes, thermistors, chemical sensors, etc.

One of the main goals of the research on conductive polymers is to increase the long-term stability and the ability to reproduce them. As for polyacetylene, the major limit is represented by its instability due to susceptibility to the degradative action of oxygen. The material exposed to air is quickly oxidized, losing the characteristics of crystallinity, metallic brightness, flexibility (it becomes brittle), and, if doped, electrical conductivity. Concerning the aromatic polymers, the major limit is represented by the insolubility and infusibility that make them hardly processable.

At the end of the 1970s, it was shown that, upon dopage with selected agents, PODZ develops high electron conductivity, although it is insulating and flame-retardant in the undoped state. In comparison to polyacetylene, PODZ introduces excellent thermal and oxidative resistance (thermal degradation: $450°C < T < 500°C$) [33,34]. Furthermore, the highly conjugated chain with an electron population characterized by high mobility makes PODZ appealing for application in the electro-optical field. Moreover, oxadiazole moieties act as electron transporting, due to their electron deficiency, and have been incorporated in low-molecular-weight compounds and in polymer backbones and side chains to obtain photoluminescent materials for applications in the field of organic light-emitting diodes (OLEDs). Also, for PODZ, the consideration holds on its intractability: it is infusible, and it is soluble only in strong mineral acids such as sulfuric acid and PPA.

Recently, numerous studies have been focused on the possibility to improve the solubility and the processability of polymers introducing flexible side ramifications like alkylic [35,36] or alkoxylic [37–45] chains. It is interesting to note that the peculiar structure of the main chain of such polymers is not altered by the insertion of ramifications. In general, however, not only are such "interferences" limited to improve solubility and lower the melting and glass transition temperatures, but they also influence other properties such as mechanical properties and thermal and thermo-oxidative stability. Substitution of *para*-phenylene units for *meta*-phenylene units is also

Aromatic Polyhydrazides and Their Corresponding Polyoxadiazoles

a reported strategy to disturb the crystalline structure while retaining the conjugated aromatic backbone [46,47].

This chapter will review the recent achievements in the field of modified PHZs and polyoxadiazoles for applications as thermostable and/or electroactive polymers.

18.2 POLYHYDRAZIDES AND POLYOXADIAZOLES FOR HIGH-TEMPERATURE APPLICATIONS

18.2.1 AROMATIC POLYETHERAROYL HYDRAZIDES BASED ON 4-OXYBENZOYL UNITS AND THEIR CORRESPONDING POLYOXADIAZOLES

The synthesis of liquid crystalline (LC) PHZs with mesophases stable below the temperature of thermal cyclization to the oxadiazole is highly desirable to further enhance their mechanical properties. 4-Oxybenzoyl units are structural elements known to impart thermotropic behavior to copolyesters [48]. The use of suitable spacers, such as $(CH_2)_n$, into the rigid backbone of 4-oxybenzoyl units, contributes to the stability of the nematic phases at a temperature well below the decomposition temperature of the polymer [49]. In the present section, the synthesis and physical–chemical characterization of new polyetheraroyl hydrazides incorporating 4-oxybenzoyl units and methylenic spacers are reported [50].

18.2.1.1 Synthesis of Monomers and Polymers

The polyetheraroyl hydrazides (PEHZ) investigated had the following basic repeating unit:

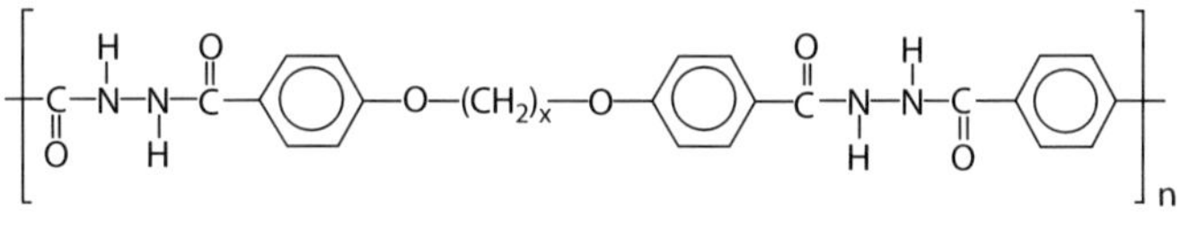

where x ranges from 2 to 12. They were obtained by reacting the corresponding aroyl dichlorides and terephthaloyl dihydrazide as follows:

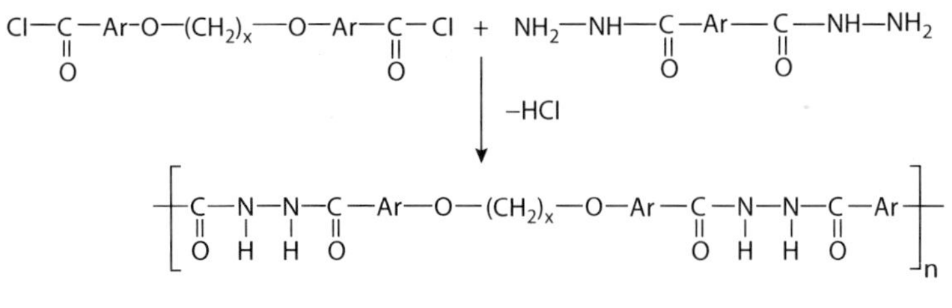

The α,ω-bis(4-chloroformylphenoxy)alkanes were reported in the literature [49] as units able to impart LC behavior to aromatic polyesters. Their synthesis starts with the preparation of the diester from a dibromoalkane and 4-ethylhydroxybenzoate:

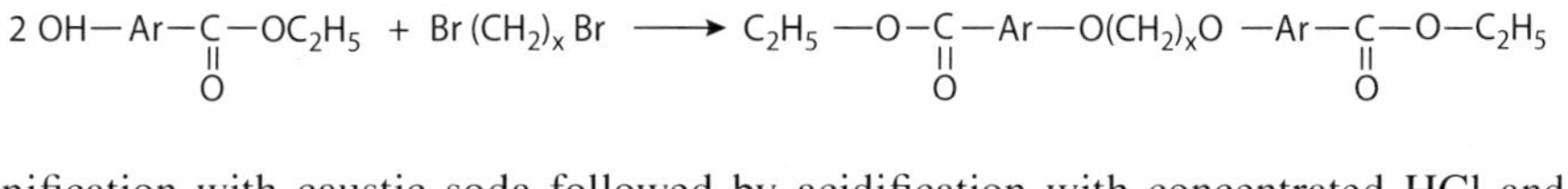

Saponification with caustic soda followed by acidification with concentrated HCl and subsequent reaction with thionyl chloride led to the corresponding aroyl dichloride with high purity. Differential scanning calorimetry (DSC) and Fourier-transform infrared (FTIR) characterization was performed on the diester and diacid intermediates. As examples, the DSC trace and FTIR spectrum of diester and diacid intermediates corresponding to $x = 8$ are shown in Figures 18.1 and 18.2,

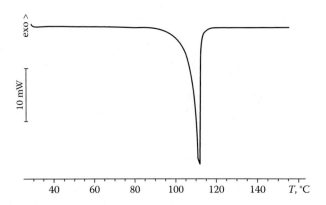

FIGURE 18.1 DSC trace of α,ω-bis(4-carboethoxyphenoxy)octane.

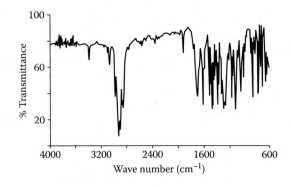

FIGURE 18.2 FTIR spectrum of α,ω-bis(4-carboxyphenoxy)octane.

respectively. The terephthaloyl dihydrazide was simply obtained by reacting 2 mol of hydrazine hydrate with 1 mol of dimethylterephthalate [7].

Sample codes and inherent viscosity of the polymers are given in Table 18.1. The number in the code stands for the length of the methylene sequence in the repeating unit. The viscosities seem to be independent of the length of the methylene chains. It was impossible to correlate η_{inh} with the viscosity average molecular weight as no report in the literature was found for K and a values of PHZs similar, at least, to ours.

TABLE 18.1
Sample Codes and Inherent Viscosity of the Polymers

Sample Codes	Alkane Length	η_{inh}[a]
PEHZ2	$(CH_2)_2$	0.55
PEHZ4	$(CH_2)_4$	1.14
PEHZ7	$(CH_2)_7$	0.57
PEHZ8	$(CH_2)_8$	0.95
PEHZ10	$(CH_2)_{10}$	0.61
PEHZ12	$(CH_2)_{12}$	0.65

[a] $\eta_{inh} = \ln(t/t_0)c$, where $c = 0.2$ g/dL in hexamethylphosphoramide (HMPA) at 30°C.

18.2.1.2 Polymerization Procedure

Low-temperature solution polymerization (0–10°C) under nitrogen was used for the preparation of the polyetheraroylhydrazide polymers. In a three-necked flask equipped with an efficient mechanical stirrer, a suspension of terephthalic dihydrazide (TDH) in dimethylacetamide (DMA) containing 5% LiCl was prepared in a nitrogen atmosphere. When a fine suspension was obtained, equimolar amounts of α,ω-bis(4-chloroformylphenoxy)alkane were added. The bath temperature was maintained around 0°C by means of a water/ice bath.

At the early stage of the reaction, the mixture became a transparent solution, and the viscosity increased. After 1 h, the solution became opalescent, and after a few minutes, it became a stirrable gel-like mixture. The resulting mixture was left under stirring at room temperature (RT) for 12 h before pouring it into water. After repeated washing with distilled water until neutral, the powder was dried under vacuum at 90°C. The polymerization yield is practically quantitative. At the end of the polymerization, the concentration of polymer in DMAc/LiCl was about 5%.

18.2.1.3 Thermogravimetric Analysis

Thermogravimetric (TG) experiments were run under N_2 atmosphere. The traces of the most spaced samples, i.e., PEHZ2, PEHZ8, and PEHZ12, are reported in Figures 18.3 to 18.5, respectively. Two peaks were detected for the samples whose attribution could be done on the basis of literature data. In fact, it is well known that all PHZs undergo a thermal cyclization (T_{cycl}) before thermal degradation (T_{deg}), according to the reaction scheme reported in Section 18.1. This reaction, also well documented by infrared (IR) spectroscopic analysis, gave rise to weight loss (evolved water) in TG experiments before degradation. The values of T_{cycl} and T_{deg}, corresponding to all samples,

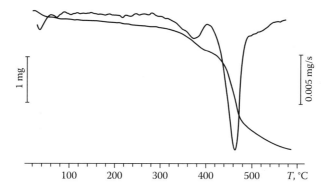

FIGURE 18.3 TGA trace of PEHZ2 (under N_2).

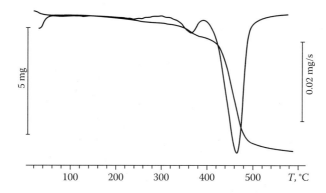

FIGURE 18.4 TGA trace of PEHZ8 (under N_2).

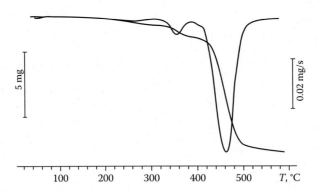

FIGURE 18.5 TGA trace of PEHZ12 (under N_2).

are reported in Table 18.2, whereas T_{cycl} data are included in Figure 18.6, together with T_g and T_m values obtained by DSC experiments. It is evident that the increase in the number of methylenes in the repeating monomeric unit caused a reduction in the cyclization temperature, while the degradation temperature showed a plateau for more than four methylene units. The regular decrease in the T_{cycl} values was a consequence of the enhanced mobility of the backbone of PEHZs with increasing number of CH_2; in fact, this would lead to an easier arrangement of the hydrazide linkage to give the oxadiazole ring.

TABLE 18.2
TGA Analysis of the Prepared Polymers

Sample	T_{cycl}, °C	T_{deg}, °C
PEHZ2	378	465
PEHZ4	369	448
PEHZ7	369	448
PEHZ8	362	458
PEHZ10	361	461
PEHZ12	356	461

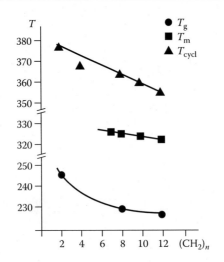

FIGURE 18.6 T_{cycl} (▲), °C as obtained by TGA experiments; T_g (●), °C and T_m (■), °C as obtained by DSC experiments of PEHZs as a function of the number of methylene units.

18.2.1.4 Thermal Analysis

All thermal data of PEHZs are collected in Table 18.3. According to the table, there was a diminution of melting temperatures, increasing the number of methylene units, at least for the even series of samples. The recorded melting temperatures (T_m) are given in Figure 18.6. Typical DSC traces in heating experiments are shown in Figures 18.7 to 18.9 for PEHZ2, PEHZ8, and PEHZ12 samples in heating experiments. Some relevant aspects can be summarized as follows:

- Multiple peaks were observed for samples PEHZ8 and PEHZ12, whereas a single broad endotherm was obtained for PEHZ2.
- According to the TG experiments, the higher temperature peak for PEHZ8 and PEHZ12 was associated with thermal cyclization; thus, the two peaks recorded at lower temperatures could be associated with the melting processes.

TABLE 18.3
DSC Values of T_m, T_{cycl}, T_g of Prepared Polyhydrazides

Sample	T_m,[a] °C	T_{cycl}, °C	T_g,[b] °C
PEHZ2	–	378	246
PEHZ4	–	365	–
PEHZ7	327–346	360	–
PEHZ8	326–347	357	230
PEHZ10	325–335	354	–
PEHZ12	323–333	350	227

[a] For crystalline polyhydrazides, two melting peaks are always detected.
[b] Glass transitions (T_g) of PEHZ4, PEHZ7, and PEHZ10 are not clearly detected.

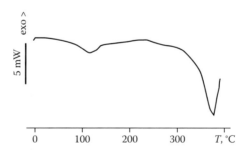

FIGURE 18.7 DSC trace (I run) of PEHZ2.

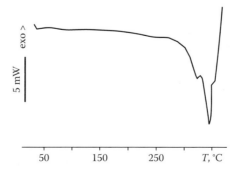

FIGURE 18.8 DSC trace (I run) of PEHZ8.

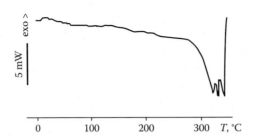

FIGURE 18.9 DSC trace (I run) of PEHZ12.

- Only one broad peak was detected for the PEHZ2 sample, centered at the cyclization temperature observed in TG experiments. This suggested that PEHZ2 was essentially amorphous.
- The glass transition temperatures were hardly recorded. For those samples where the T_g was clearly detectable, the effect of the T_g diminution caused by the increasing number of CH_2 was very evident (see also Figure 18.6).

18.2.1.5 X-Ray Diffraction Analysis

Wide-angle x-ray scattering (WAXS) powder spectra of samples of PEHZ2, PEHZ8, and PEHZ12 are reported in Figures 18.10 to 18.12. The highly crystalline nature of samples with a higher number of methylenes in the repeating unit (8 and 12) was evident from the spectra, whereas the sample with two CH_2 groups in the repeating unit was essentially amorphous.

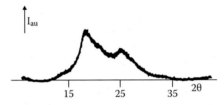

FIGURE 18.10 WAXS spectrum of PEHZ2.

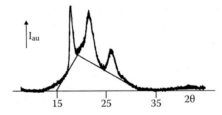

FIGURE 18.11 WAXS spectrum of PEHZ8.

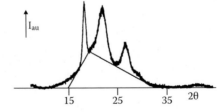

FIGURE 18.12 WAXS spectrum of PEHZ12.

TABLE 18.4
WAXS Data of PEHZs

Sample	$2\theta°$	d, Å	$1/A^a$	X_c
PEHZ2	18.632	4.88	0.588	0
	25.433	3.50	0.555	
PEHZ8	18.423	4.82	1.111	0.42
	22.069	4.04	0.454	
	26.711	3.35	0.714	
PEHZ12	18.474	4.82	1.250	0.41
	22.241	4.02	0.555	
	26.312	3.40	0.555	

[a] Reciprocal of half-height width (A) of the reflection (per degree of reflection angle $2\theta°$).

For the sake of comparison, an index of crystallinity (X_c) was determined by using the method described in Figures 18.11 and 18.12 for the separation of the crystalline and amorphous contributions. The calculated X_c, together with values of width at half-height ($1/A$) and Bragg distance (d), is reported in Table 18.4; the d values were calculated by applying Bragg's law, $\lambda = 2d \sin \theta$. The crystalline state of PEHZ8 and PEHZ12 and the amorphous state of PEHZ2 agreed well with the results of DSC analysis.

18.2.1.6 Studies on the Cyclization Reaction by Infrared Analysis

The dehydration of PHZs to polyoxadiazoles (see Section 18.1) occurs at elevated temperatures (300–360°C), and it is possible to follow it by IR spectroscopy, DSC, and TG analysis (TGA). Table 18.5 summarizes the data on cyclization temperatures of PEHZn to polyetheraroyloxadiazoles (PEODZn). The FTIR spectra of PEHZ8 and PEODZ8 are reported in Figure 18.13. The characteristic frequencies of the PHZ are the N–H stretching frequency around 3200 cm^{-1} (not reported in Figure 18.1), the carbonyl stretching at 1650 cm^{-1}, and the C–N stretching frequency at 1270 cm^{-1} (indicated by arrows). In addition, aromatic PHZs show the characteristic phenyl ring vibration at 1510 cm^{-1}. After cyclization, the polyoxadiazoles no longer show the peaks relative to N–H, C=O, and C–N stretching. Instead, the characteristic frequency of the oxadiazole ring at 963 cm^{-1} is observed.

TABLE 18.5
Codes and Cyclization Temperatures of PEHZn

Codes	Alkane Length	T_{cycl},[a] °C
PEHZ2	$(CH_2)_2$	378
PEHZ4	$(CH_2)_4$	369
PEHZ7	$(CH_2)_7$	365
PEHZ8	$(CH_2)_8$	362
PEHZ10	$(CH_2)_{10}$	361
PEHZ12	$(CH_2)_{12}$	356

[a] Obtained in dynamic thermogravimetric experiments under N_2.

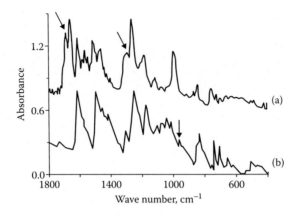

FIGURE 18.13 Infrared spectra of (a) PEHZ8 and (b) PEODZ8.

18.2.1.7 Thermal Analysis

Thermal characterization is carried out on PEODZn samples that have been cyclized directly inside the DSC pan [51]. For the sake of comparison and clarity, only traces relative to the second run of PEODZ2, PEODZ8, and PEODZ12 (i.e., after the cyclization run from PHZs) are reported in Figure 18.14.

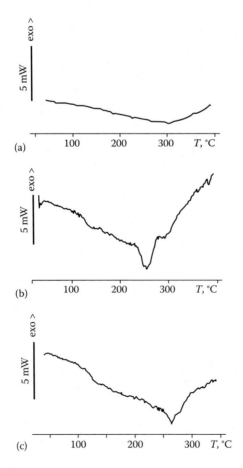

FIGURE 18.14 DSC trace (II run) of (a) PEODZ2, (b) PEODZ8, and (c) PEODZ12.

TABLE 18.6
Thermal Data of PEODZn (°C)[a]

Codes	T_g	T_{m1}	T_{m2}
PEODZ2	183		
PEODZ4	165		
PEODZ7	136	274	–
PEODZ8	133	260	300
PEODZ10	126	280	–
PEODZ12	123	280	

[a] Values refer to heating runs after cyclization.

The analyzed samples show different thermal behaviors. In fact, PEODZ2 and PEODZ4 do not display any detectable melting peak but show a clear crystallization exotherm (T_c), whereas PEODZ8 and PEODZ12 (as well as PEODZ7 and PEODZ10) exhibit clearly defined melting endotherms.

As can be seen from reported DSC traces, a well-defined T_g is detected for all samples after cyclodehydration, whose position regularly decreases by increasing the number of methylenes, as a consequence of the increased mobility of the main chain. All thermal data are summarized in Table 18.6, in which the trend in the T_g value is evident. Less regular is the behavior of T_m, which displays double peaks for the PEODZ8 sample. The crystallization kinetics are shown for PEODZ8 and PEODZ12 samples.

18.2.1.7.1 Bulk Crystallization

The isothermal crystallization process was investigated according to the following procedure. The sample, sealed in a DSC pan, was heated at the cyclization temperature and kept there for 5 min. The temperature was then rapidly lowered to the chosen level and the sample allowed to crystallize. The heat evolved during the isothermal crystallization was recorded as a function of the crystallization time. The half-time of crystallization plotted as function of crystallization temperature for the two samples under investigation is reported in Figure 18.15. All crystallizations are very fast in the temperature interval investigated. The data show a smaller increase in the half-time of crystallization for PEODZ12 compared with PEODZ8. That probably is due to the fact that the PEODZ8 has more nuclei than the PEODZ12, as reported in Section 18.2.1.8. The longer methylene sequences of PEODZ12 compared with PEODZ8 may at least partially contribute to the decrease of the crystallization rate from the former to the latter.

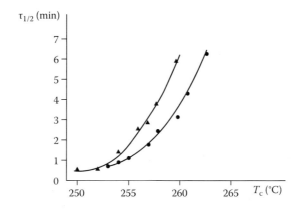

FIGURE 18.15 Half-time of crystallization $\tau_{1/2}$ as a function of T_c for PEODZ8 (●) and PEODZ12 (▲).

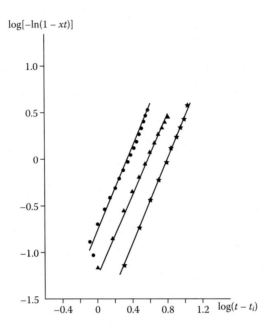

FIGURE 18.16 Avrami plots for PEODZ8 at different crystallization temperatures: $T_c = 263°C$ (★), $T_c = 261°C$ (▲), and $T_c = 260°C$ (●).

The Avrami equation [52] is used to analyze the data of the bulk kinetics of crystallization:

$$\ln[-\ln(1 - X_t)] = \ln K_n + n \ln(t - t_i)$$

where X_t is the weight fraction of material crystallized at time t, n is the Avrami index, K_n is the overall kinetic rate constant, and t_i is the induction time of nucleation. The weight fraction, X_t, of the material crystallized at time t was calculated from the ratio of the heat generated at time t and the total heat corresponding to the completion of the crystallization. The experimental data fit the Avrami equation well, as shown in Figure 18.16. The values of the Avrami index, contrary to the theoretical predictions, are nonintegral for both samples, and they range between 2 and 2.5, indicating mixed crystallization and nucleation mechanisms.

18.2.1.8 Optical Microscopy

The morphology of PEODZ8 and PEODZ12 samples as a function of temperature has been followed using an optical microscope equipped with an automated hot stage. Films used for the optical microscopy were evaporated directly on a glass slide from hot HMPA (initial concentration $c = 0.5$ g/dL). It must be pointed out that, at this initial stage, the polymer is still in the polyhydrazidic structure. The following thermal cycle was used: from RT to cyclization temperature at the heating rate of 20°C/min, followed by an isothermal 5 min hold at that temperature in order to complete the cyclization of the samples. Finally, the temperature was lowered at 20°C/min to RT; several experiments were carried out in order to determine an optimal isothermal crystallization temperature. Only for PEODZ12 was the experiment performed successfully, as shown in Figure 18.17.

The dimensions of such spherulites are in the range of 100–150 μm and they melted at 305°C. By quenching the PEODZ12 sample from the cyclization temperature directly to RT, the texture appears to be microspherulitic, as shown in Figure 18.18.

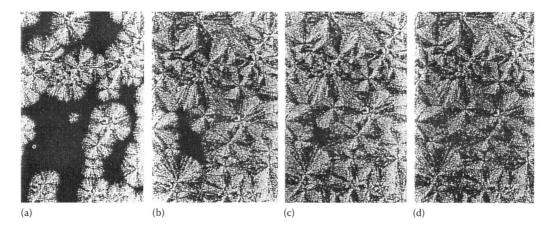

FIGURE 18.17 Optical micrographs of PEODZ12 at $T_c = 275°C$ at different times: (a) 1 min, (b) 2 min, (c) 3 min, and (d) 4 min.

FIGURE 18.18 Optical micrographs of PEODZ12 quenched from 360°C at room temperature.

During cyclization, the birefringence of the PEODZ8 sample, due to the persistence of microspherulites, never disappears completely and increases upon cooling, as shown in Figure 18.19; the PEODZ8, then, undergoes partial melting at 310°C. Such behavior could be attributed, at least in part, to a complete cyclization of PEHZ to PEODZ at the chosen temperature (360°C). Therefore, the persistence of microspherulites should be due to the residual PEHZ molecules.

It is also found that the increase by four methylene units going from PEODZ8 to PEODZ12 strongly reduces the nucleation density of PEODZs at all crystallization temperatures.

18.2.1.9 X-Ray Diffraction Analysis

WAXS powder spectra of PEODZ2, PEODZ8, and PEODZ12 are reported in Figure 18.20. It is evident that the three samples show well-defined crystalline peaks. PEODZ2, which is amorphous before the cyclization reaction, shows a WAXS pattern characterized by sharp crystalline peaks. It must be noted that such a polymer has no detectable melting peak in DSC analysis, probably because melting and decomposition are superimposed for very low values of $(CH_2)_x$. PEODZ8 and PEODZ12 show typical patterns of semicrystalline polymers. The scattering angles (2θ) for the three samples as obtained by WAXS analysis are summarized in Table 18.7. The crystallinity index, X_c, has been determined according to the Hermans–Weidinger method [53]. The Bragg distance, d, is obtained from Bragg's law:

$$\lambda = 2d \sin \theta$$

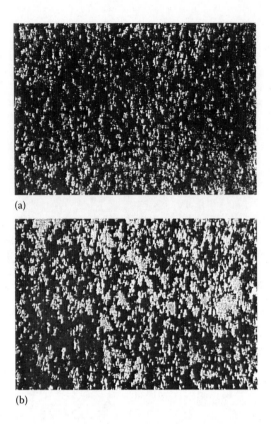

FIGURE 18.19 Optical micrographs of PEODZ8 at different temperatures: (a) 360°C and (b) 310°C.

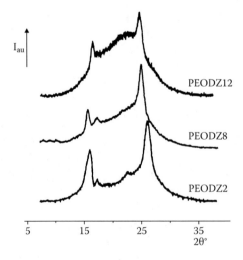

FIGURE 18.20 WAXS powder spectra (CuKα, Ni-filtered radiation) of PEODZ2, PEODZ8, and PEODZ12.

TABLE 18.7
Scattering Angle 2θ, Bragg Distance d, and Other WAXS Structural Parameters for PEODZ2, PEODZ8, and PEODZ12

Sample	2θ	d, Å	1/A[a]	X_c[b]
PEODZ2	15.856	6.60	0.769	30.79
	25.90	3.44	0.588	
PEODZ8	1570	5.65	1.111	26.41
	25.15	3.54	0.952	
PEODZ12	16.50	5.37	0.769	8.17
	24.70	3.60	0.625	

[a] Reciprocal of half-height width (A) of the reflection (per degree of reflection angle 2θ).
[b] Crystallinity index according to the Hermans–Weidinger method.

The calculated X_c values together with values of width at half-height (1/A) and Bragg distance (d) are also reported in Table 18.7.

It is possible to follow the cyclization reaction with the use of WAXS. In Figure 18.21 are reported the diffractograms corresponding to three different stages of the reaction (5, 10, and 15 min) for the PEODZ8 sample. It is evident that this progressive disappearance of PEHZ reflections was substituted by the reflections of PEODZ.

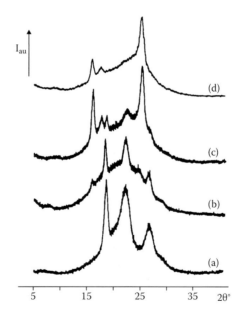

FIGURE 18.21 WAXS powder spectra (CuKα, Ni-filtered radiation) of cyclization of PEHZ8 sample at three different times: (a) 5 min, (b) 10 min, (c) 15 min, and (d) PEODZ8.

18.2.1.10 Conclusions

The polyetheraroyl hydrazides synthesized and described herein are crystalline polymers with high melting temperature for seven or more CH_2 units. Thermal cyclization to polyoxadiazole derivatives was obtained at a temperature above the melting point. The polyetheraroyl oxadiazoles obtained show interesting thermal properties. Their moldability coupled with good thermal resistance renders them valuable for application as engineering plastics. In fact, as the number of methylenes in the main chain increases, the melting temperature increases within the range 260–280°C. The degradation temperatures are very high as a consequence of the introduction of aliphatic spacers even though they are below the thermal degradation of fully aromatic polyoxadiazoles. For very short CH_2 sequences, melting is superimposed on degradation. The crystallization is rapid, and a spherulitic texture develops, whose shape depends on the length of the spacers.

18.2.2 Aromatic Poly(Ether-Alkyl) Hydrazides Based on 4,4′-Dihydroxybiphenyl Units and Their Cyclization to Polyoxadiazoles

It is well established that the combination of biphenol and alkane spacers leads to thermotropic polyesters and polyesterimides [54,55]. It was found that the number of methylenes in the repeating unit is critical for the formation of a stable mesophase. Furthermore, the insertion of a methylene sequence may impart moldability and gives rise to a clear glass transition in the polyoxadiazoles below the decomposition temperature. The synthesis and characterization of this class of PHZs and corresponding polyoxadiazoles are reported hereafter.

18.2.2.1 Synthesis and Characterization of Poly(Ether-Alkyl) Hydrazides

Poly(ether-alkyl) hydrazides have been prepared with the following basic repeating unit:

$$\left[-(CH_2)_n-O-\text{C}_6\text{H}_4-\text{C}_6\text{H}_4-O-(CH_2)_n-\overset{O}{\underset{\|}{C}}-\underset{H}{\overset{}{N}}-\underset{H}{\overset{}{N}}-\overset{O}{\underset{\|}{C}}-\text{C}_6\text{H}_4-\overset{O}{\underset{\|}{C}}-\underset{H}{\overset{}{N}}-\underset{H}{\overset{}{N}}-\overset{O}{\underset{\|}{C}}- \right]$$

with $n = 3$, 4, and 5. They are obtained by the previously described low-temperature polycondensation method, well known to give high-molecular-weight PHZs [8] between the dichloride and TDH:

$$x\text{ClC}(CH_2)_nO-\text{C}_6\text{H}_4-\text{C}_6\text{H}_4-O(CH_2)_n\text{CCl} + x\text{NH}_2\text{NHC}-\text{C}_6\text{H}_4-\text{CNHNH}_2 \rightarrow (P) + (X-1)\text{HCL}$$

TDH was prepared starting from hydrazine hydrate and dimethylterephthalate, according to the standard procedures [7]. Dichlorides were obtained from the corresponding diethyl esters starting from biphenol and the ethyl ester of an ω-bromoalkylcarboxylic acid, through the Williamson reaction [56]:

$$2\text{Br}(CH_2)_n\text{COEt} + \text{HO}-\text{C}_6\text{H}_4-\text{C}_6\text{H}_4-\text{OH} \xrightarrow{\text{NaOCH}_3} \text{EtOC}(CH_2)_n-O-\text{C}_6\text{H}_4-\text{C}_6\text{H}_4-O-(CH_2)_n\text{COEt} + 2\text{NaBr}$$

The reaction occurred in two steps. In the first step, there was formation of monoester. The conditions to shift the equilibrium toward diester formation were very critical; an excess of both sodium methoxide and bromoester improved the yield. Sodium methoxide was chosen as it is most effective in promoting the formation of biphenate salt, an intermediate of the reaction. The products were recovered by pouring the reaction mixture into water and purified by repeated recrystallizations from diluted methanol solution.

The ether esters were characterized by DSC, ^{1}H nuclear magnetic resonance (NMR), elemental analysis (Table 18.8), and IR spectroscopy. Characteristic absorption bands due to aromatic ring stretching (1602 cm^{-1}), C=O stretching (1738 cm^{-1}), and –O–C stretching (1256 cm^{-1}) were observed. The IR spectrum of the sample with $n = 4$ is given in Figure 18.22a. No significant changes were found in the spectra of the samples with different values of n.

Subsequent saponification with potassium hydroxide in ethanol and reaction of the diacid with thionyl chloride were carried out according to standard procedures. Also, the diacids were characterized by DSC, ^{1}H NMR (Table 18.9), and IR spectroscopy. Figure 18.22b shows the spectrum of the sample with $n = 4$ as an example. Additional bands due to C=O and –OH of the acid groups (1702 and 3070 cm^{-1}, respectively) were observed, whereas the ester band at 1738 cm^{-1} was absent.

The polycondensation reaction of PEHZ was performed following the same experimental procedure previously described for polyetheraroyl hydrazides in Section 18.2.1.2.

The sample codes, η_{inh}, and elemental analysis of polymer are reported in Table 18.10. The obtained viscosities were found to be independent of the length of CH$_2$ sequences and were not very high. This may be at least in part due to gelation polymerization, which can kinetically limit

TABLE 18.8
Melting Points (mp), Proton Chemical Shifts (δ_H), and Elemental Analysis of Ether Esters of General Formula (H$_5$C$_2$OOC(CH$_2$)$_n$–O–)$_2$ with $n = 3$ (Ia), 4 (Ib), and 5 (Ic)

Sample	mp, °C	δ_H(CDCl$_3$), ppm	Elemental Analysis[a]	
			C, %	H, %
Ia	94.5	1.27 (6H, t)	69.6	7.3
		1.84 (4H, m)	(69.56)	(7.25)
		2.34 (4H, t)		
		4.08 (4H, t)		
		4.11 (4H, q)		
		6.69–7.62 (8H, m)		
Ib	89.0	1.25 (6H, t)	70.4	7.7
		1.85 (8H, m)	(70.59)	(7.69)
		2.32 (4H, t)		
		4.09 (4H, t)		
		4.14 (4H, q)		
		6.71–7.65 (8H, m)		
Ic	93.5	1.28 (6H, t)	71.5	8.1
		1.85 (12H, m)	(71.49)	(8.08)
		2.38 (4H, t)		
		4.05 (4H, t)		
		4.11 (4H, q)		
		6.87–7.45 (8H, m)		

[a] Figures in parentheses indicate calculated values.

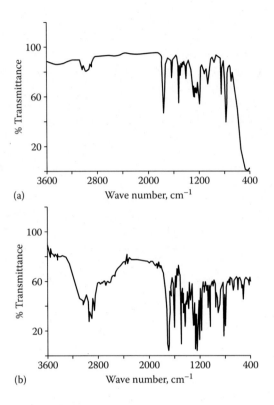

FIGURE 18.22 (a) FTIR spectrum of Ib sample; (b) FTIR spectrum of IIb sample.

TABLE 18.9
Melting Points (mp) and Proton Chemical Shifts (δ_H) of Intermediate Ether Acids of General Formula $[HOOC(CH_2)_n-O-]_2$ with n = 3 (IIa), 4 (IIb), and 5 (IIc)

Sample	mp, °C	δ_H(DMSO), ppm
IIa	252.5	1.70 (4H, m)
		2.28 (4H, t)
		3.99 (4H, t)
		6.96–7.56 (8H, m)
		12.11 (2H, s)
IIb	209.9	1.71 (8H, m)
		2.30 (4H, t)
		3.97 (4H, t)
		6.98–7.56 (8H, m)
		12.10 (2H, s)
IIc	201.0	1.68 (12H, tm)
		2.31 (4H, t)
		3.97 (4H, t)
		6.95–7.61 (8H, m)
		12.11 (2H, s)

Aromatic Polyhydrazides and Their Corresponding Polyoxadiazoles

TABLE 18.10
Codes, Number of Methylenic Groups in the Repeating Unit (n), η_{inh}, and Elemental Analysis of the Polyhydrazides

Codes	n	η_{inh},[a] dL/g	Elemental Analysis,[b] %		
			C	H	N
PEHZB	3	0.47	65.1	5.3	10.8
			(65.12)	(5.43)	(10.85)
PEHZV	4	0.35	66.1	5.7	10.3
			(66.18)	(5.88)	(10.29)
PEHZH	5	0.48	67.2	6.3	9.7
			(67.13)	(6.29)	(9.79)

Note: B, butyrrate; H, hexanoate; V, valerate.
[a] $T = 30°C$, $c = 0.20$ g/dL in HMPA.
[b] Figures in parentheses indicate calculated values.

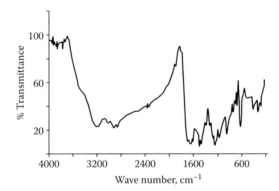

FIGURE 18.23 FTIR spectrum of PEHZV.

molecular weight. No correlation could be found between η_{inh} and M_w, as values of α and K are not reported in the literature.

IR spectroscopic analysis was performed on polymers. The PHZs showed N–H stretching at around 3200 cm^{-1}, carbonyl stretching at 1650 cm^{-1}, and C–N stretching at 1270 cm^{-1} (see Figure 18.23).

18.2.2.2 Thermogravimetric Analysis

Figure 18.24 shows the TG traces in N_2 for all samples. A weight loss before degradation of about 6% was found in every sample. This was attributed to the evolution of water as a consequence of thermal cyclization to the corresponding polyoxadiazoles [57].

Cyclization temperatures (T_{cycl}) are reported in Table 18.11, together with those of degradation.

T_{cycl} decreases with an increasing number of (CH_2) in the repeating unit derived from the enhanced mobility of the polymeric chain, which allowed easier rearrangement of the hydrazidic group to give the oxadiazolic ring. In the previous section, we reported the thermal behavior of similar polyetheraroyl hydrazides.

The cyclization temperatures were dependent on the number of methylene units in the chain and were about 20° higher than the observed values of PHZs reported in the present section, for the

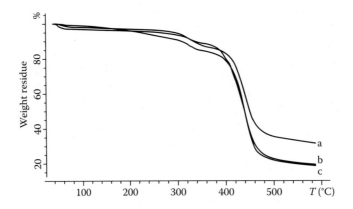

FIGURE 18.24 TG traces of (a) PEHZB, (b) PEHZV, and (c) PEHZH. Heating rate = 20°C/min.

TABLE 18.11
Cyclization and Degradation Temperatures of the PEHZ Samples

Sample	T_{cycl}, °C	T_{deg}, °C
PEHZB	340	440
PEHZV	324	444
PEHZH	320	440

Note: PEHZ, poly(ether-alkyl) hydrazide.

same number of methylene units. This is due to the fact that in the present case, the hydrazide group is directly linked to an aliphatic chain.

The degradation temperature was not influenced by the number of methylene sequences, according to Table 18.11. This is not surprising, because only short methylene sequences have been investigated. According to the literature on aromatic polyimides, the insertion of aliphatic spacers of a length of less than six methylene units does not depress thermal degradation behavior [58].

18.2.2.3 Thermal Analysis

DSC traces for "as-polymerized" PEHZB, PEHZV, and PEHZH (i.e., coagulated and repeatedly washed in water and dried in vacuo at 100°C) with no further thermal treatment are reported in Figures 18.25 to 18.27, respectively. Thermal data are summarized in Table 18.12.

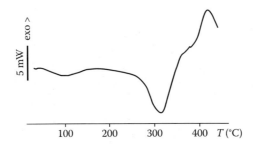

FIGURE 18.25 DSC curve of PEHZB. Heating rate = 20°C/min.

Aromatic Polyhydrazides and Their Corresponding Polyoxadiazoles

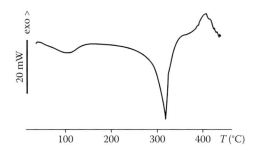

FIGURE 18.26 DSC curve of PEHZV. Heating rate = 20°C/min.

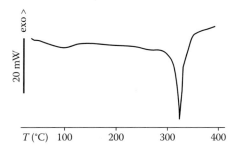

FIGURE 18.27 DSC curve of PEHZH. Heating rate = 20°C/min.

TABLE 18.12
Thermal Analysis Data for the PEHZ Samples

Sample	T_{endo}, °C	T_g, °C
PEHZB	312	–
PEHZV	315	–
PEHZH	321	240

The broad endotherm at about 100°C was due to the evolution of absorbed water (not present in the second run). All samples showed an endotherm just above 300°C. Such an endotherm was very broad for PEHZB, sharp with a shoulder on the right side in PEHZH, and resolved into two peaks for PEHZV. According to TGA (see previous paragraph), the PHZs underwent thermal cyclization in a temperature range rather corresponding to that of the DSC endotherm. The broad peaks and shoulders may thus be related to the overlapping of melting and cyclization. A clear T_g (240°C) was detected only in the case of the PEHZH sample, i.e., the sample with a higher number of methylenes in the repeating unit. This suggests that the T_g of the other samples is higher than 240°C and may overlap with the melting and cyclization processes.

The rigid biphenolic unit in the repeating unit followed by a flexible spacer (in this case, a short aliphatic chain) is reported to be able to impart LC properties to polyesters [54,55]. Nevertheless, it is reported that the number of methylene units able to give mesophase behavior in rigid polymers such as polyesters or polyesterimides is very critical, and sometimes, the mesophase behavior starts to appear only for n above 6. In our case, we found no evidence of a mesophase behavior for all cases investigated.

18.2.2.4 X-Ray Diffraction Analysis

WAXS powder spectra of PEHZ samples are reported in Figure 18.28. It is evident that the only sample that showed well-defined crystalline peaks was PEHZH. Samples of PEHZB and PEHZV

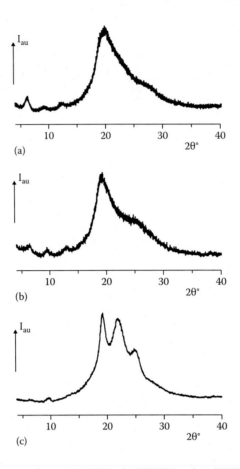

FIGURE 18.28 WAXS powder spectra: (a) PEHZB, (b) PEHZV, and (c) PEHZH. Scanning rate = 0.5°/min.

TABLE 18.13
WAXS Parameters for PEHZH Sample

Sample	X_c, %	2θ, degree	d, Å	$1/A$, per degree
PEHZH	16.87	19.20	4.62	0.50
		21.80	4.08	0.26
		24.80	3.59	0.34

showed WAXS diffractograms characterized by a rather broad halo with a maximum centered at $2\theta = 20.0°$ and $19.2°$ and a shoulder at $2\theta = 26.5$ and $24.0°$, respectively. These halos were mainly due to diffuse scattering from amorphous regions. This was probably related to the higher number of methylenes in the repeating unit of the PEHZH sample.

WAXS structural parameters for the PEHZH sample are summarized in Table 18.13. The crystallinity index, X_c, was determined according to the Hermans–Weidinger method [53]. The interplanar spacing, d, was obtained from Bragg's law: $\lambda = 2d \sin \theta$. The reciprocal of the width at half-height (A) of reflection has been reported as an index of the dimension and perfection of the crystallites.

18.2.2.5 Conversion of Polyhydrazides to Polyoxadiazoles

Cyclodehydration reactions were carried out under vacuum at temperatures of 320–340°C, using a thermostatted air bath. Samples were checked periodically during thermal cyclization by IR

Aromatic Polyhydrazides and Their Corresponding Polyoxadiazoles

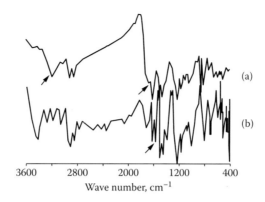

FIGURE 18.29 FTIR spectra of (a) PEHZH and (b) PODZH. (The poor resolution of spectrum [b] is due to the difficulty of grinding the sample to very fine powder.)

spectroscopy in order to follow the progress of the reaction over time. The polyoxadiazoles obtained had the following general formula:

with m values of 3, 4, and 5; they were coded PODZB, PODZV, and PODZH, respectively. (Letters B, V, and H are derived from the corresponding PHZs.)

The proper temperature of cyclodehydration has been obtained from TGA experiments. The progress of reaction was followed by IR spectroscopy.

The IR spectrum of the PODZH sample is reported in Figure 18.29 as an example, together with that of its corresponding polyhydrazide (PEHZH). Arrows mark the more significant bands, which change upon cyclization, according to the literature [59]: 3200 and 1650 cm^{-1}, which disappear, are attributed to N–H and C=O stretching, respectively; at 1560 cm^{-1}, a new band due to C=N stretching was present in the spectrum of PODZH. The band at 970 cm^{-1}, attributed in the literature to the =C–O–C= bending in whole aromatic polyoxadiazoles, was not clearly detected; this may be due to the overlapping of CH_2 bending in the same range of wave numbers. Moreover, some new minor bands appeared, which were in agreement with literature on polyoxadiazole IR analysis (710, 760, 850, and 1080–1068 cm^{-1}). IR analysis showed that the conversion was complete within 10 min for all of the samples.

18.2.2.6 Thermal Analysis

A thermal analysis was conducted on samples that were cyclized directly into the DSC plans. Experiments of cooling from the cyclization temperature to RT and subsequent heating showed crystallization and melting peaks for all the PODZs. Thermograms of experiments performed at a heating rate of 20°C/min are reported in Figures 18.30 to 18.32. A multiple heating cycle revealed that there were no significant changes in crystallization or melting behavior. All samples showed two peaks of crystallization on cooling and a single, somewhat broad melting endotherm on heating. Scans performed at higher and lower heating rates (40°C/min and 5°C/min), aimed at finding the evidence of the eventual occurrence of two distinct melting peaks that might overlap at 20°C/min, showed that the variation of the heating rate never led to multiple melting peaks.

It is worth noting that melting always occurred well before the start of the degradation process, confirming that the introduction of short aliphatic segments within the structure of fully aromatic,

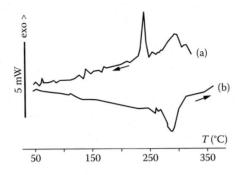

FIGURE 18.30 DSC curves of PODZB. (a) Cooling; (b) heating.

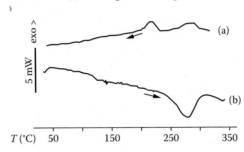

FIGURE 18.31 DSC curves of PODZV. (a) Cooling; (b) heating.

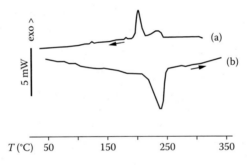

FIGURE 18.32 DSC curves of PODZH. (a) Cooling; (b) heating.

rigid chains of PODZs leads to the onset of a processability window [49]. Furthermore, a clearly detectable T_g at about 100°C (taken during heating) was present in all samples.

Thermal parameters are summarized in Table 18.14. The enthalpy of fusion increases with an increasing number of methylenes, as a consequence of improved ability of the chains to organize into a crystalline lattice. T_m, T_c, and T_g decrease with increasing CH_2. This trend is more pronounced

TABLE 18.14
Glass Transition, Crystallization, Melting, and Degradation Temperatures (°C) and Enthalpy of Fusion (J/g) of PODZs

Sample	T_g	T_c (I)	T_c (II)	T_m	T_{degr}	ΔH
PODZB	122	294	243	288	440	20.1
PODZV	120	286	219	275	440	24.7
PODZH	100	237	205	239	440	34.0

going from the PODZV ($m = 4$) to the PODZH ($m = 5$) sample, suggesting the existence of an even–odd effect [60]. Unfortunately, the number of samples investigated was not large enough to allow a more precise analysis. On the contrary, the degradation temperatures, as obtained from TGA measurements, were dependent on m. The values of T_{degr} showed that the polyoxadiazoles retain good heat resistance properties.

18.2.2.6.1 Bulk Crystallization

Typical crystallization isotherms obtained by plotting the degree of crystallization at time t, X_t, versus time during an isothermal crystallization from the melt, are reported in Figure 18.33a using PODZB as an example. From such curves, the half-time of crystallization, $\tau_{1/2}$, defined as the time required to develop half of the final crystallinity, was obtained. The variation of $\tau_{1/2}$ with the crystallization temperature is reported in Figure 18.33b for all of the samples; it is evident that the $\tau_{1/2}$ falls in different temperature ranges, according to thermal analysis.

Plotting the melting temperatures versus the corresponding isothermal crystallization temperatures and applying the relation of Hoffmann [61], the equilibrium melting temperatures, $T°_m$, were calculated to be 336°C, 300°C, and 265°C for PODZB, PODZV, and PODZH, respectively. In Figure 18.33c, the $\tau_{1/2}$ versus the undercooling ($\Delta T = T°_m - T_c$) is reported. The ΔT values, too, were different for the three samples, indicating different rates and different mechanisms of crystal growth. From the experimental data, we also calculated the "morphological factor," γ [62], which was 1.80 for PODZV, 1.75 for PODZH, and 1.35 for PODZB. As the γ factor is correlated to the chain flexibility, the values reflect the lower value of m for PODZB. From this thermodynamic parameter, it could be supposed that the kinetics of crystallization of PODZB is the slowest (e.g.,

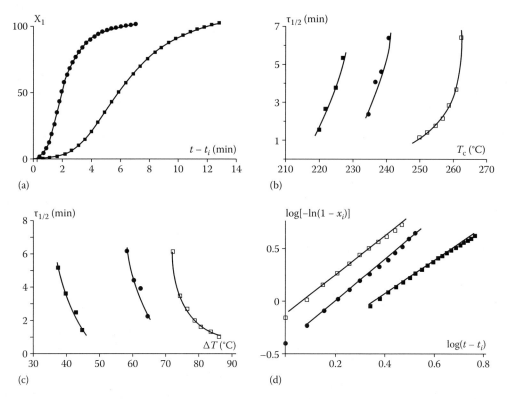

FIGURE 18.33 (a) Degree of crystallization versus time of PODZB: $T_c = 255°C$ (●), $T_c = 261°C$ (■). (b) Half-time of crystallization versus T_c: PODZB (■), PODZV (●), PODZH (□). (c) Half-time of crystallization versus T: PODZB (□), PODZV (●), PODZH (■). (d) Avrami plot of PODZB: $T_c = 250°C$ (□), $T_c = 252°C$ (●), $T_c = 255°C$ (■).

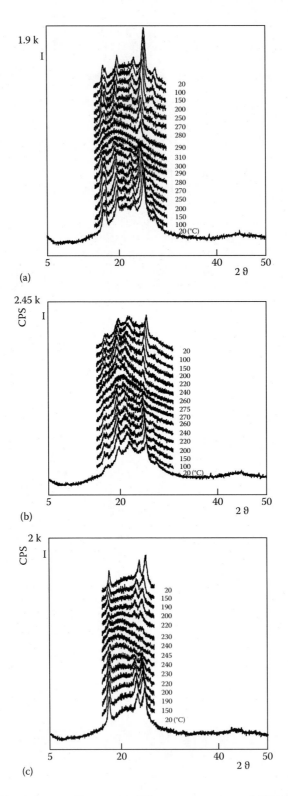

FIGURE 18.34 (a) Wide-angle x-ray diagrams at different temperatures of PODZB. (b) Wide-angle x-ray diagrams at different temperatures of PODZV. (c) Wide-angle x-ray diagrams at different temperatures of PODZH.

see also Figure 18.33c, which shows that the ΔT necessary to reach a given $\tau_{1/2}$ is higher for PODZB). On the contrary, Figure 18.33b shows that $\tau_{1/2}$ of PODZB remains very low within a broader range of T_c, indicating a faster overall kinetic with respect to the other samples. As the overall bulk crystallization comprises the nucleation rate plus the crystal growth rate, we can invoke a higher nucleation rate for PODZB. This hypothesis is supported by the morphological analysis (see Section 18.2.2.8).

The bulk kinetics of crystallization were analyzed using the Avrami treatment [52] for the kinetics of phase changes. The experimental data appear to fit the Avrami equation for any T_c (plots of $\log[-\ln(1 - X_t)]$ versus $\log[t - t_i]$ are reported in Figure 18.33d for PODZB). The values of n determined from the slopes of the above-mentioned straight lines were in all cases, noninteger and ranged between 2 and 3. Assuming that the nucleation is heterogeneous, this result could indicate that the growth of the crystals can occur in two and/or three dimensions. Studies are in progress on this subject in order to analyze the nucleation process (rate and mechanism) of the materials.

18.2.2.7 X-Ray Diffraction Analysis

X-ray measurements (WAXS) were performed at different temperatures, ranging from RT to T_m, and vice versa, to study the melting and crystallization processes. Figure 18.34a through c shows the spectra for PODZB, PODZV, and PODZH, respectively. Spectra at T higher than RT were recorded in a narrower range of 2θ to reduce the acquisition time. The crystallinity degrees (X_c) obtained at RT (e.g., Figure 18.35) are reported in Table 18.15. All of the polymers have a low X_c; from observation of the spectra, it can be noted that the reflections fall in the same range of diffraction (16–28° 2θ), but their number decreases with increasing CH_2 as a consequence of structural changes in the elemental cell of the polymers, which also will reflect on the crystal texture, as confirmed by morphological studies (see below). As no correlation was found with d spacing calculated for fully extended chains, it is supposed that the macromolecules pack in a very irregular way.

All of the polymers completed the crystallization process only at low temperatures. This could account for the two crystallization peaks found in the DSC thermograms: the first broad peak can be attributed to an initial crystallization very close to T_m, which leads to the formation of very irregular

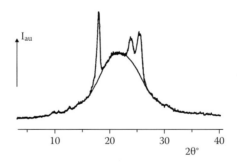

FIGURE 18.35 Wide-angle x-ray diagram of PODZH.

TABLE 18.15
Crystallinity Degrees Obtained from X-Ray Diagrams at Room Temperature

Sample	X_c, %
PODZB	17
PODZV	18
PODZH	13

and small crystals, whereas the exothermic peak at lower temperature is attributed to the growth of more perfect crystals.

The more significant variations of d spacing (in both heating and cooling experiments) with temperature for PODZB, PODZV, and PODZH were reported in Figures 18.36 to 18.38 (the d spacings were calculated by applying Bragg's law to the more significant reflections). As can be seen, all variations were linear with temperature. The thermal linear expansion coefficients, α, for each d were simply calculated from the coefficients of the straight lines [62,63] and reported alongside each line. Good agreement was found between those calculated in heating and cooling experiments.

The reflections in the low range of 2θ, however, which correspond to d spacing of 5–5.3 Å, hardly changed with temperature. They were supposed to represent distances between (001) crystallographic planes (if c is the chain axis), which are more firmly held. This might confirm the above hypothesis on a nonextended chain conformation.

18.2.2.8 Optical Microscopy

The morphological characteristics of the samples are dependent on the molecular structure and temperature. Films used for optical microscopy were obtained by solution casting of the more soluble PHZ precursors from hexamethylenephosphoramide and were then cyclized directly into the hot stage of the microscope.

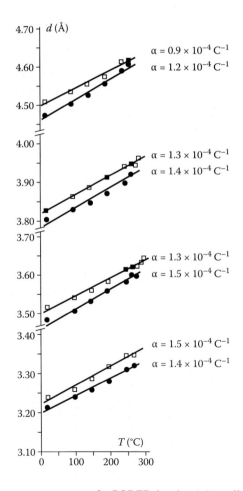

FIGURE 18.36 d spacings versus temperature for PODZB: heating (□), cooling (●).

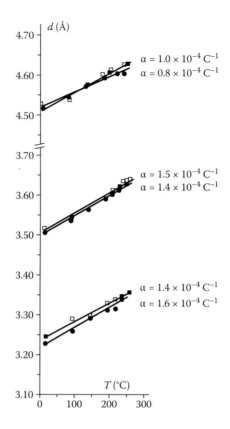

FIGURE 18.37 d spacings versus temperature for PODZV: heating (□), cooling (●).

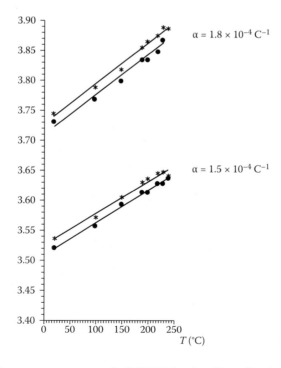

FIGURE 18.38 d spacing versus temperature for PODZH: heating (*), cooling (●).

The starting polyhyrazides have different cyclization temperatures (T_{cycl}), namely, 340°C for PODZB, 325°C for PODZV, and 320°C for PODZH. At their own cyclization temperature, no birefringence was observed for PODZV and PODZH samples, whereas the PODZB film showed bright spots. After cyclization has occurred, the temperature was slowly decreased from T_{cycl} to T_c; micrographs were taken at selected times during isothermal crystallization. Microspherulites developed for the PODZB sample (Figure 18.39a), whereas PODZV and PODZH showed a macrospherulitic morphology (Figure 18.39b and c). The spherulites have about the same dimension (38 ± 3 μm) but different shape, as one can easily see. In particular, the PODZH sample showed a dendrite structure. The birefringency observed at T_{cycl} for the PODZB sample indicates that the crystals of the starting material were not completely destroyed when the cyclization reaction occurred. Thus, the original microcrystalline structure of the PHZ might act as a nucleating agent, giving rise to spherulites of lower dimension with respect to the other investigated samples.

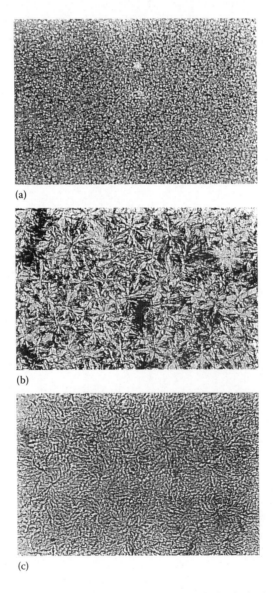

FIGURE 18.39 (a) Optical micrograph of PODZB at T_c = 255°C. (b) Optical micrograph of PODZV at T_c = 239°C. (c) Optical micrograph of PODZH at T_c = 255°C.

18.2.2.9 Conclusions

New poly(ether-alkyl) hydrazides, whose synthesis and properties are described in this section, showed different thermal behavior according to the number of methylene units along the chain. In fact, the degree of crystallinity of as-polymerized polymers increased with the number of CH_2 units, whereas the glass transition temperature T_g and temperature of cyclization to polyoxadiazoles decreased in the same direction. A detailed investigation of the melting process was severely limited by the superposition of cyclization.

The poly(ether-alkyl) oxadiazoles derived through the thermal cyclization of the PHZs were crystalline, with melting and glass transition temperatures that decreased with an increasing number of methylene spacers between aromatic units. The T_m values were comparable with those of thermoplastic polycondensates of wide application (e.g., polyesters and polyamides), whereas the T_g values were higher due to the high chain stiffness of this class of polymers. Structural and morphological analyses have shown that a different number of CH_2 units leads to differences in both the elemental cell and the crystal texture. The thermal and structural properties render these polyoxadiazoles very attractive as reinforcing agents in thermoplastic (i.e., polyester) or thermostable (i.e., polyimide) matrix blends.

18.3 POLYHYDRAZIDES AND POLYOXADIAZOLES FOR ELECTRO-OPTICAL APPLICATIONS

Conjugated polymers are very attractive on account of their unique electrical and optical properties [64–67]. Conjugated polymers are characterized by delocalization of the electrons along the backbone, which consists of aromatic rings, in some cases alternated with double bonds. PODZs, which contain alternating *p*-phenylene units and 1,3,4-oxadiazole rings, are particularly appealing in the field.

The aromatic, conjugated structure, in turn, brings insolubility and infusibility to the polymers, resulting in poor processability. Therefore, the synthesis of soluble and stable total conjugated polymers is an important goal and represents an ongoing challenge for chemists. A number of general techniques have been developed in order to improve processability, including block copolymerization [68], increase of chain flexibility by introduction of flexible points in the main chain [69,70], side-chain substitution on the aromatic ring [71], use of processable precursor polymers [72], and formation of polymer blends [73,74]. It has been demonstrated that long alkylic side chains act as an internal plasticizer [75–80], hence helping in improving processability by lowering T_g and/or T_m, and simultaneously impart solubility. The synthesis and characterization of new poly(1,3,4-oxadiazole)s containing flexibilizing ether linkages in the backbone and pentadecyl side chains is reported as an example of combination of two different strategies [81]. Flexible films have been also obtained by inserting dimethylsilane units in the backbone of copoly(hydrazide-imide) copolymers and their copoly(oxadiazole-imide) derivatives [82]. Silicon-containing aromatic polymers have attracted much scientific and technological interest for the production of optoelectronic materials, due to the ability of silicon, when placed among aromatic neighbors, to give a σ–π conjugation and thus support the transport of electrons along the macromolecular chain [83,84]. Hamciuc et al. [82] reported that the T_g of the oxadiazole-imide copolymers containing dimethylsilane was well below the T of degradation, generating a good window of processability [84].

Among the various methods, the insertion of flexible side chains is the more adequate strategy in order to impart processability and solubility while preserving the chain conjugation, and it has been applied to a series of homopolyoxadiazoles and copolyoxadiazoles for applications in the optoelectronic fields.

18.3.1 ALKOXY-SUBSTITUTED POLY(*p*-PHENYLENE-1,3,4-OXADIAZOLE)

Along with improving solubility in organic solvents, another important consequence of the introduction of large and flexible substituents on aromatic PODZ is expected to be the constitutional disorder introduced by the side chains, which can lower the crystallinity of PODZ and induce the formation of mesophases, an important property for electro-optical applications of the material. Alkoxy-substituted PODZ, having the general repeating unit:

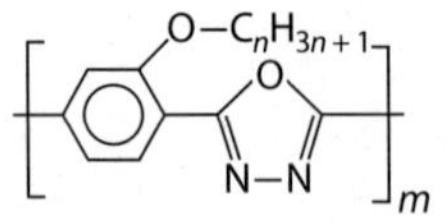

have been reported [47,85]. The length of the side chains, n, varied from 2 to 10. PODZn were obtained through cyclodehydration of corresponding initial polyhydrazides (PHZn). PHZn were prepared by polycondensation of alkoxy-substituted dichlorides and dihydrazides, whose synthesis is described for the first time in Ref. [85] as well. As the alkoxylic ramifications are sensitive to acid hydrolysis and, furthermore, they are not very stable at temperatures close to 300°C, PHZn were cyclized using POCl$_3$ as a dehydrating solvent. The complete route of synthesis of polyoxadiazoles (PODn) through the PHZn precursors is depicted in:

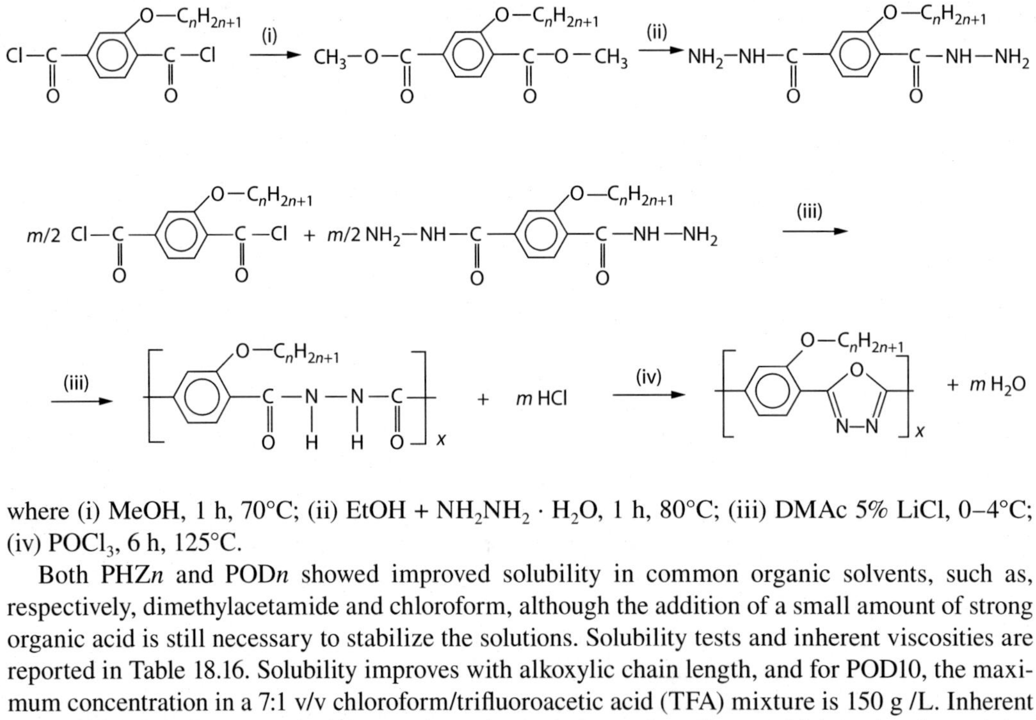

where (i) MeOH, 1 h, 70°C; (ii) EtOH + NH$_2$NH$_2$ · H$_2$O, 1 h, 80°C; (iii) DMAc 5% LiCl, 0–4°C; (iv) POCl$_3$, 6 h, 125°C.

Both PHZn and PODn showed improved solubility in common organic solvents, such as, respectively, dimethylacetamide and chloroform, although the addition of a small amount of strong organic acid is still necessary to stabilize the solutions. Solubility tests and inherent viscosities are reported in Table 18.16. Solubility improves with alkoxylic chain length, and for POD10, the maximum concentration in a 7:1 v/v chloroform/trifluoroacetic acid (TFA) mixture is 150 g /L. Inherent viscosities of PHZn are typical for condensation polymers of medium to high molecular weights (0.9–1.4) and are not significantly influenced by side-chain length. For PODn, inherent viscosity is always reduced, due to the different solvent and chemical structure. The glass transition temperature, T_g, lies in the range 165–230°C and depends on the side-chain length. The PODn show good thermal stability in nitrogen up to 270°C (Table 18.17). Dynamico-mechanical analysis (data not reported) showed that PODn are characterized by moderate chain flexibility and highly flexible lateral substituents.

The introduction of side chains with a fixed, repeated separation induced a regular structure of the polymer backbone. X-ray diffraction data revealed a structural organization in molecular sheets with a comblike arrangement of the polymeric chains, as depicted in Figure 18.40. This arrangement allows the segregation between the flexible side chains and the rigid main chains that are incompatible with each other. A typical diffractogram is shown in Figure 18.41. The diffraction peak at 26° is equal for all the samples and corresponds to a d spacing of 3.4–3.5 Å: for the proposed layered packing model, this is the spacing between adjacent chains (layer thickness), and

TABLE 18.16
Inherent Viscosity (η_{inh}) and Solubility of PHZ*n* and POD*n* Polymer

Code[a]	η_{inh},[b] dL/g	Solubility					
		DMSO	CHCl$_3$	TFA	CHCl$_3$/TFA	Trichloroethane/TFA	CHCl$_3$/CH$_3$SO$_3$H
PHZ2	0.91	++	--	++	++	++	++
		++	--	++	++	++	++
PHZ5	1.37	++	--	++	++	++	++
		++	--	++	++	++	++
PHZ8	1.31						
PHZ10	1.35						
POD2	0.71	--	--	--	++	--	--
		--	--	--	++	+-	+-
POD5	0.89	--	--	--	++	+-	+-
POD8	0.92	--	--	--	++	+-	+-
POD10	0.96						

Source: Reprinted with permission from Gillo, M. et al., *Chem. Mater.*, 14, 1539–1547. Copyright (2002) American Chemical Society.

Note: ++, soluble; +–, partially soluble; ––, unsoluble.

[a] 2, 5, 8, and 10 refer to the number of C atoms of the alkoxylic side chain.

[b] Solvent: DMSO (PHZ*n*); CHCl$_3$/TFA, volume ratio of 2:1 (POD2), 3:1 (POD5), 4:1 (POD8), and 7:1 (POD10).

TABLE 18.17
Glass Transition (T_g) and Degradation (T_d) Temperatures of PHZ*n* and POD*n*

Sample	T_g (DSC), °C	T_d (TGA), °C
PHZ2	240	300
PHZ5	234	300
PHZ8	202	280
PHZ10	190	250
POD2	231	340
POD5	205	340
POD8	173	320
POD10	165	300

Source: Reprinted with permission from Gillo, M. et al., *Chem. Mater.*, 14, 1539–1547. Copyright (2002) American Chemical Society.

further, it coincides with the typical distance observed between aromatic planes in aromatic systems (e.g., graphite). The intense diffraction peak at a low angle shifts to lower values, increasing the side-chain length. The corresponding *d* spacing, which increases linearly with the side-chain length (Figure 18.42), represents the width of the macromolecular layer. The increment of *d* corresponds to 1.20 Å for methylene. This result accounts for the zigzag planar conformation adopted by the aliphatic segment of the side group in the plane of the molecular layer.

Electron conductivity (σ) of POD8 was measured, and a value of $(1.8 \pm 0.2)10^{-12}$ S/cm was obtained, which is higher than that of terephthalic POD (10^{-16} S/cm). Optical properties of POD*n* were evaluated, and the absorbance spectrum of a POD8 film in total light in the range 200–1000 nm

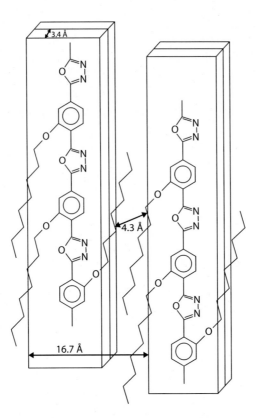

FIGURE 18.40 Representation of the molecular layer packing of POD8. (Reprinted from Capitani, D. et al., *J. Polym. Sci., Polym. Chem. Ed.*, 41, 2003.)

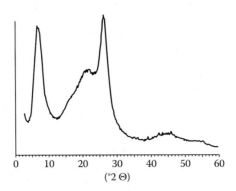

FIGURE 18.41 X-ray powder pattern of POD5. (Reprinted with permission from Gillo, M. et al., *Chem. Mater.*, 14, 1539–1547. Copyright (2002) American Chemical Society.)

is shown in Figure 18.43 as an example. The main feature of this curve is the very high absorbance in the range of 200–406 nm and a high constant transmission from 406 up to 2500 nm, suggesting the possibility to employ such a material as an ultraviolet (UV) optical filter or as a coating for shielding elements. Another relevant feature is the high transmittivity in the near-infrared (NIR) region, which is, as is well known, the typical telecommunication band (1300–1500 nm). The calculated value of the refractive index for POD8 is 1.568 ± 0.002. These optical characteristics make POD*n* very attractive for preparing planar waveguides, useful in integrated optics.

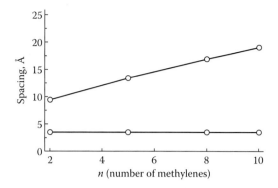

FIGURE 18.42 *d* spacing versus the number of methylenes (*n*) of the alkoxy side chain for POD*n*. (Reprinted with permission from Gillo, M. et al., *Chem. Mater.*, 14, 1539–1547. Copyright (2002) American Chemical Society.)

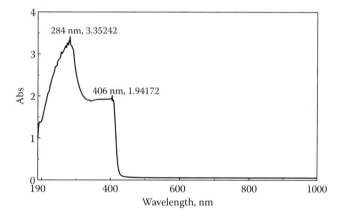

FIGURE 18.43 Total absorbance spectrum of POD8 in the range 200–1000 nm. (Reprinted with permission from Gillo, M. et al., *Chem. Mater.*, 14, 1539–1547. Copyright (2002) American Chemical Society.)

Further progress in solubility and film flexibility was obtained by Capitani et al. [47] by introducing small percentages of comonomers, such as fumaric or methyl-fumaric and *meta*-alkoxy-substituted phenylene, randomly distributed along the macromolecular chain:

Fumaryl unit (*fu*)   Mesaconic unit (*me*)   Octanoxy isophthaloyl unit (*iso*)

The introduction of "defects," such as *trans* double bonds (simple or methyl-substituted) or *meta*-substituted phenylene rings, disturbs the regular enchainment. An improvement in solubility while retaining the structural organization and the optical and electro-optical properties of interest was achieved. In the same work, a study of the conversion of precursor PHZs to polyoxadiazole by ^{15}N solid-state NMR was performed. The analysis showed that the conversion is incomplete. This is a serious drawback, and optimization of cyclodehydration conditions is needed in order to increase the effective conjugated length in the final polymers.

18.3.2 Polyoxadiazoles as Electroluminescent Materials

Electron-deficient oxadiazole units have been found to be efficient in promoting electron transport when they are incorporated into conjugated polymer main chains or attached as side groups, and are of interest for the fabrication of OLEDs. Usually, for fabricating an OLED, alternating thin layers of molecules with electron-transporting properties and molecules with hole-transporting properties are coupled. Polymers can be easily processed by spin coating, a fast and versatile technique for fabrication of optical devices, provided that they are soluble in opportune solvents [86]. Furthermore, it is possible to produce monolayer devices by introducing along the polymer chain both electron- and hole-transporting segments.

An example of monolayer OLED based on a low-molecular-weight compound containing oxadiazole moieties was reported by Bugatti et al. [87]. The authors combined in a single molecule alkoxy-substituted oxadiazole rings as electron-transporting moieties and carbazole as hole-transporting moieties. The introduction of alkoxy side chain has been used also in this case as a powerful strategy to improve solubility in common organic solvents.

The insertion of flexible aliphatic chains along and side-attached to the backbone, beyond moderating the melting temperature and improving the solubility, may induce the formation of a LC phase. The LC order is particularly interesting when optical active systems are required to be anisotropic at the macroscopic level. Examples of PHZ and PODZ with LC properties are reported in the literature. De Prisco et al. [88,89] synthesized a new class of copoly(ester-hydrazide)s (PEHs) containing low-molecular-weight polyester blocks bridged through terephthalic hydrazide units. The polyester precursors show LC properties, and the corresponding PEH retains the thermotropic behavior. A mesophase with a wide range of stability in the melt to promote the appearing of a nematic phase is formed. PEH showed good thermal stability together with easier processability, guaranteed by the thermotropic polyester block.

Acierno et al. [90] synthesized segmented polyethers containing alkoxy-substituted oxadiazoles characterized by the formation of a nematic phase (Figure 18.44). Taking advantage of the nematic phase, oriented fibers that emit polarized photoluminescence were obtained by melt spinning (Figure 18.45)

Copoly(aryl-ether)s containing alternated dihexyloxy-2,5-distyrilbenzene (DSB) or -2,5-distyrilethoxybenzene as emitting segments and different oxadiazole moieties as electron-transporting segments were reported by Chen et al. [91]. Ether spacers were employed to isolate the segments by confining the π-conjugation range, to allow the copolymers to emit at a definite λ. The resulting copolymers were thermally stable until 400°C and soluble in common organic solvents, such as chloroform and tetrahydrofuran, allowing fabrication by spin coating of double-layer LED devices.

FIGURE 18.44 Nematic optical texture of a poly(ether-oxadiazole) sample (crossed polarizers; 20× magnification). (Reprinted with permission from Capitani, D. et al., *Polym. J.*, 33, 575–583. Copyright (2001) SPSJ.)

FIGURE 18.45 Photoluminescence of a poly(ether-oxadiazole) fiber sample irradiated at 370 nm in the dark. (Reprinted with permission from Acierno, D. et al., *Macromolecules*, 36, 6410–6415. Copyright (2003) American Chemical Society.)

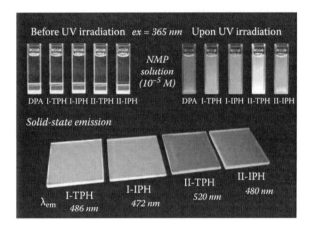

FIGURE 18.46 Images showing the appearance of polymer solutions and solid films of polyhydrazides (I-TPH, I-IPH) and polyoxadiazoles (II-TPH, II-IPH) bearing pyrenylamine moieties before and after exposure to a standard laboratory UV lamp (excited at 365 nm). 9,10-diphenylanthracene (DPA) was used as the standard. (Kung, Y-C, *Polym. Chem.*, 2, 1720–1727, 2011. Reproduced by permission of The Royal Society of Chemistry.)

Recently, new aromatic PHZs bearing pyrenylamine moieties were reported by Kung et al. [92]. Pyrene is one of the most useful fluorogenic units for fluorescent sensors, due to its ability to form an excimer state. The PHZs were prepared from a newly synthesized dicarboxylic acid containing pyrenylamine [93,94] and terephthalic or isophthalic dihydrazide via the Yamazaki–Higashi phosphorylation reaction [95,96]. Hydrazide polymers and their corresponding polyoxadiazoles, obtained by thermal cyclodehydration in the solid state, showed a medium to high fluorescence emission in the blue to yellowish-green region (Figure 18.46). The polymer films revealed interesting multicolored electrochromic behaviors, with color changes from a pale yellow neutral state to pale green oxidized states. Additionally, the polyoxadiazoles exhibited strong color changes from pale yellow to orange, red, or deep blue, due to the formation of radical anions of the oxadiazole and pyrene units as a consequence of reduction processes.

Polyoxadiazoles containing 3,4-alkylendioxythiophenes were synthesized by Ojha et al. [97] following the precursor PHZ route. The PHZ was synthesized starting from 3,4-alkylendioxythiophene dihydrazide and 3,4-ethylenedioxythiophene dichloride or terephthalic dichloride:

(Adapted with permission from Ojha, U.P. et al., *Synt. Metals*, 132, 2003.)

3,4-ethylenedioxythiophene (DEOT) is used as rich electron block in electroluminescent polymers. The optical and electrochemical properties indicate that the polymers can be good light-emitted materials. Nevertheless, the problems regarding the poor solubility and processability of the polyoxadiazoles hold, and further modification by insertion of flexible groups in the chain or as lateral substituents is still necessary.

18.4 POLYHYDRAZIDES AND POLYOXADIAZOLES FOR HIGH SPECIALTY APPLICATIONS

18.4.1 Polyoxadiazoles for High-Performance Membranes

The high basicity of the heterocyclic rings has made polyoxadiazoles, as well as other polymers with heterocyclic units (e.g., benzimidazole, benzazole), interesting polymers for use in proton-conductive membranes, after doping with acids [98]. Membranes of polyoxadiazole doped with phosphoric acid have been recently reported for fuel cell application [99]. Due to its heat and solvent resistance, polyoxadiazole also has been used for high-performance membranes for gas separation [100] and ultrafiltration [101].

18.4.1.1 Fluorinated Polyoxadiazoles

Polyoxadiazoles containing trifluoromethyl (CF_3) groups as lateral substituents [102] maintain the good thermal stability, and on the other hand, show improved film flexibility. A series of polymers named Oxad have been recently proposed for applications as high-temperature polymer electrolyte membranes for fuel cells:

Oxad-6F-D (two aromatic rings in *meta* position)

Oxad-6F-E (all aromatic rings in *para* position)

18.4.1.2 Azobenzene-Containing Polyhydrazides and Polyoxadiazoles

Azobenzene-based polymers are interesting due to their photoresponsive behavior, which can be utilized in optical switches, optical data recording, or optical information storage [103–105]. An aromatic azo group is a chromophoric system, so polymers containing azo groups in the main chain are intrinsically colored; furthermore, when inserted in the main chain, azo groups impart rigidity to the chain, so azo-polymers are of special interest as materials with liquid-crystal [106,107] or nonlinear optical properties [108]. A series of wholly aromatic, self-colored azopolyamide-hydrazides and corresponding oxadiazoles have been recently reported by Mohamed et al. [46]; the resulting polymers show increased solubility, chain flexibility, and hydrophilicity, and their physicochemical characteristics can be modulated through the comonomer ratio for a variety of applications. The polymers exhibit a great affinity for water sorption, and their films are useful, for example, as reverse-osmosis membranes in water-pollution control and as semipermeable membranes for desalination of artificial and natural seawater [109–111].

18.4.1.3 Sulfonated Polyoxadiazoles

For applications in membranes, the introduction of ionic groups (sulfonic, phosphonic, carboxylic, etc.) gives additional properties to the polymer, which enable their application in fuel cells, electrodialysis, and ultrafitration. Furthermore, ionic groups improve solubility, which is essential for use as coating or membranes. Sulfonation of PODZ has been described by Vetter et al. [112]. Sulfonated polyoxadiazoles containing $C(CH_3)_2$ or $C(CF_3)_2$ groups and ether linkages along the main chain have been synthesized:

$X = C(CH_3)_2$
$X = C(CF_3)_2$

The condensation between an oxadiazole-containing bisfluoride monomer and bisphenols was chosen as the route of synthesis [113,114] to avoid side reactions and to overcome the problem of an incomplete cyclization of hydrazide to oxadiazole. Sulfonation was then performed with sulfuric acid to different extents. The presence of ether linkage and/or $C(CF_3)_2$ groups renders the films more flexible. Polymers are thermally stable and water soluble at a sulfonation degree of around 1.0 but water insoluble at higher sulfonation degrees. Good solubility in dimethylformamide (DMF) or sulfuric acid enabled the preparation of films, so sulfonated PODZs are good candidates as materials for membrane formation.

18.4.2 Examples of Polyhydrazides and Polyoxadiazoles Containing Different Functional Moieties in the Backbone

18.4.2.1 Polyhydrazides and Polyoxadiazoles Containing Carbohydrate Moieties

Recently, PHZs containing carbohydrate moieties (namely, 2,3,4,5-tetra-acetyl-galactose) as spacers have been synthesized with the low-temperature polycondensation method by Nasr et al. [115]. The authors introduced also different pendant groups, such as 4-nitrobenzoyl or octanoxy-substituted aromatic rings:

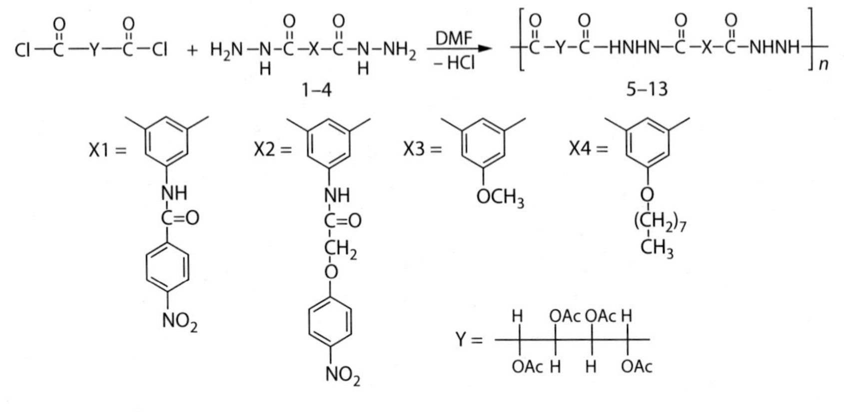

(Adapted from Nasr, M.A.M. et al., *React. Funct. Polym.*, 65, 2005.)

The obtained PHZs were promptly soluble at RT in apolar protic solvents, such as DMF and dimethylsilfoxide (DMSO), and unsoluble or only partially soluble in halogenated solvents. The PHZs were further modified by a deacetylation reaction of carbohydrate units, and contrary to acetylated PHZs, the deacetylated polymers were soluble also in concentrated sulfuric acid, due to the strong hydrogen interactions between free hydroxyl groups. The relative polyoxadiazoles obtained by cyclodehydration showed reduced solubility compared to PHZs and were characterized by good film flexibility.

18.4.2.2 Polyhydrazides and Polyoxadiazoles Containing 1,3-Imidazolidine-2,4,5-Trione Rings

The polymers containing 1,3-imidazolidine-2,4,5-trione rings on the macromolecular backbone are known as poly(parabanic acid)s. This kind of polymer can be obtained starting from diisocyanates: by a polyaddition reaction of diisocyanates with hydrogen cyanide followed by hydrolysis [116] or by a polyaddition–polycondensation reaction between diisocyanates and bisesteroxalamides [117]. Poly(parabanic acid)s were also obtained by a cyclocondensation reaction with oxalyl chloride on polyureas [118,119]:

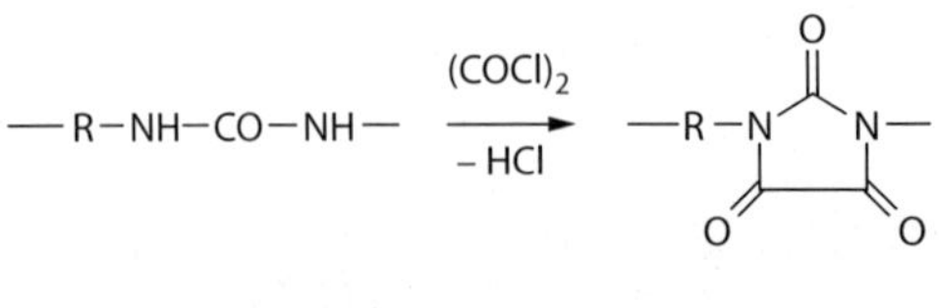

Parabanic polymers showed an electrical conductivity of $0.38–4.0 \times 10^{-16}\ \Omega^{-1}\ cm^{-1}$, which means that all polymers have good electrical insulator properties.

Caraculacu et al. [120] described a series of parabanic PHZ copolymers based on 1,3-bis(4-chloroformyl-phenylene) parabanic acid (DCPP). The copolymers were prepared by a polycondensation reaction of dihydrazides or aromatic dihydrazines and the dichloride derivative of DCPP:

Aromatic Polyhydrazides and Their Corresponding Polyoxadiazoles

The corresponding polyoxadiazoles were obtained by thermally or chemically activated cyclodehydration. Polymers that contain parabanic structures were generally soluble in polar solvents and presented high thermal stability and glass transition temperatures up to 200°C.

The PHZs can also be transformed in polychelates with metal salts producing diverse colored precipitates. PHZs based on dihydrazides can be further chemically transformed by using different reactive compounds such as aliphatic or aromatic monoisocyanates that can be attached to NH hydrazide groups on polymers.

18.4.3 Polyhydrazides with Shape-Persistent Properties

Side-chain introduction has been tuned in order to impart shape-persistent characteristics to PHZs. Zhou et al. [121] synthesized a series of novel aromatic PHZs with zigzag structures and hydrophobic, hydrophilic, or amphiphilic side chains, using the Yamazaki polymerization conditions. Suitable side chains provide solubility in organic solvents. Hydrogen bonds between the RO of side chains and HN of hydrazide form intramolecular six-membered nanocycles, similar to oligomeric foldamers [122–130], which increase the planarity of the backbones. As a consequence, the macromolecules exhibit stacking behavior, and the resulting structural organization of macromolecular chains in solution is very stable. Further, these novel PHZs are able to self-assemble in solution into vesicles (micelles) or fibers, forming organogels in solvents of different polarities (Figure 18.47).

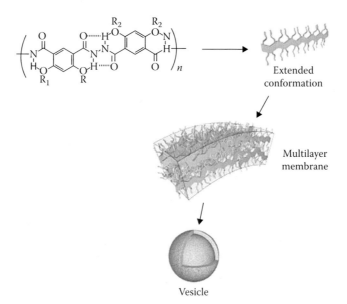

FIGURE 18.47 Scheme of self-assembling in solution of aromatic hydrazide-based polymers. (Adapted from Zhou, C. et al., *Macromol. Chem. Phys.* 211, 2012.)

ACKNOWLEDGMENTS

The authors are indebted to Luigi Calandrelli, Luciano Amelino, and Nunzio Lanzetta for x-ray, thermal, and spectroscopic data, respectively.

REFERENCES

1. Dupont de Nemours, E. I. 1974. U.S. Patent 3,817,941.
2. Goodmann, I., McIntire, J.E., and Russell, W. 1964. British Patent 971,227.
3. Wang, W.F.H., Wiff, D.R., Banner, C.L., and Helminiak, T.E. 1983. *J. Macromol. Sci. Part B* 22:231–257.
4. Wickliff, S.M., Malone, M.F., and Ferris, R.J. 1987. *J. Appl. Polym. Sci.* 34:931–944.
5. Takayanagi, M., Ogata, J., Morikawa, M., and Kai, T. 1980. *J. Macromol. Sci. Part B* 17:591–615.
6. Takayanagi, M. 1984. Polymer composites of rigid and flexible molecules. In *Interrelation between Processing Structure and Properties of Polymeric Materials*, eds. Seferis, J.C. and Theocaris, P.S. Amsterdam: Elsevier.
7. Campbell, T.W., Foldi, V.S., and Farago, J. 1959. *J. Appl. Polym. Sci.* 2:155–162.
8. Frazer, A.H., Sweeny, W., and Wallenberger, T. 1964. *J. Polym. Sci. Part A* 2:1157–1169.
9. Frazer, A.H. and Sarasohn, I.M. 1966. *J. Polym. Sci. Part A* 4:S.1649–S.1664.
10. Huisgen, R. 1981. *Chem. Heterocyclic Comp.* 17:419–433.
11. Ahire, C.J. and Marvel, C.S. 1961. *Makromol. Chem.* 44:388–397.
12. Cotter, R.J. and Matzner. M. 1972. Ring-Forming Polymerizations, In *Hetherocyclic rings*, vol. 13B, eds. Blomquist, A.F. and Wasserman, H. New York: Academic Press.
13. Frazer, A.H. and Wallenberger, F.T. 1964. *J. Polym. Sci. Part A* 2:1137–1145.
14. Frazer, A.H. and Wallenberger, F.T. 1964. *J. Polym. Sci. Part A* 2:1147–1156.
15. Frazer, A.H., Sweeny, W., and Wallenberger, F.T. 1964. *J. Polym. Sci. Part A* 2:S.1171–S.1179.
16. Abd-Alla, M.A. and Aly, K.I. 1992. *J. Macromol. Sci. Part A* 29:185–192.
17. Hensema, E.R., Boom, J.P., Mulder, M.H.V. et al. 1994. *J. Polym. Sci. Part A* 32:513–525.
18. Iwakura, Y., Uno, K., and Hara, S. 1965. *J. Polym. Sci. Part A* 3:45–54.
19. Gomes, D.F., Borges, C.P., and Pinto, J.C. 2001. *Polymer* 42:851–865.
20. Cassidy, P.E. and Fawcett, N.C. 1979. *J. Macromol. Sci.-Part C* 17:209–266.
21. Chiang, C.K., Fincher, C.R., Park, Y.W. et al. 1977. *Phys. Rev. Lett.* 39:1098–1101.
22. Phillips, P. 2000. *Nature* 406:687–688.
23. Salaneck, W.R., Clark, D.T., and Samuelson, E.J. 1991. *Science and Applications of Conducting Polymers*. Bristol: Adam Hilger.
24. Hide, F., Dìaz-Garcìa, M.A., Schwartz, B.J., Andersson, M.R., Pei, Q., and Heeger, A.H. 1996. *Science* 273:1833–1836.
25. Mort, J. and Pfister, G. 1982. *Electronic Properties of Polymer*. New York: John Wiley & Sons.
26. Kryzewski, M. 1980. *Semiconducting Polymers*. Warsaw: PWN-Polish Scientific Publishers.
27. Gau, S.C., Milliken, J., Pron, A., McDiarmid, A.G., and Heeger, A.J. 1979. *J. C. S. Chem. Comm.* 15:662–663.
28. Feldblum, A., Park, Y.W., Heeger, A.J. et al. 1981. *J. Polym. Sci.: Polym. Phys. Ed.* 19:173–179.
29. Shacklette, L.W., Eckhardt, H., Chance, R.R., Ivory, D.M., Miller, G.G., and Baughman, R.H. 1981. Highly conducting poly(*p*-phenylene) via solid-state polymerization of oligomers. In *Conductive Polymers, Polym. Sci. Technol.*, Vol. 15, pp. 115–123, ed. Seymour, R.B. New York: Plenum Press.
30. Skotheim, T.A., Elsenbaumer, R.L., and Reynolds, J.R. 1998. *Handbook of Conducting Polymers*. New York: Marcel Dekker.
31. Shacklette, L.W., Elsenbaumer, R.L., Chance, R.R., Eckhardt, H., Frommer, J.E., and Baughman, R.H. 1981. *J. Chem. Phys.* 75:1919–1927.
32. Kobayashi, M., Chen, J., Chung, T.C., Moraes, F., Heeger, A.J., and Wudl, F. 1984. *Synth. Met.* 9:77–86.
33. Yamamoto, T., Sanechika, K., and Yamamoto, A. 1980. *J. Polym. Sci. Polym. Lett. Ed.* 18:9–12.
34. Tsuitsui, T., Fukuta, Y., Hara, T., and Saito, S. 1987. *Polym. J.* 19:719–725.
35. Calandrelli, L., Immirzi, B., Kummerlowe, C., Malinconico, M., and Riva, F. 1998. Thermal resistant polyoxadiazoles and its blends: Synthesis and characterization. *Recent Res. Develop. Polym. Sci.* 2:569–597.
36. Stern, R., Ballauff, M., Lieser, G., and Wegner, G. 1991. *Polymer* 11:2096–2105.
37. Huang, W., Yu, W.L., Meng, H., Pei, J., and Li, S.F.Y. 1998. *Chem. Mater.* 10:3340–3345.
38. Caruso, U., Pragliola, S., Roviello, A., Sirigu, A., and Iannelli, P. 1995. *Macromolecules* 28:6089–6094.

39. Centore, R., Roviello, A., Sirigu, A., and Kricheldorf, H.R. 1994. *Macromol. Chem. Phys.* 195:3009–3016.
40. Kricheldorf, H.R. and Engelhardt, J. 1990. *Makromol. Chem.* 191:2017–2026.
41. Ballauff, M. and Schmidt, G.M. 1987. *Makromol. Chem., Rapid Commun.* 8:93–97.
42. Lee, K.S., Kim, H.M., Rhee, J.M., and Lee, S.M. 1991. *Makromol. Chem.* 192:1033–1040.
43. Bao, Z., Chen, Y., Cai, R., and Yu, L. 1993. *Macromolecules* 26:5281–5286.
44. Rodriguez-Parada, J.M., Duran, R., and Wegner, G. 1989. *Macromolecules* 22:2507–2516.
45. Harkness, B.R. and Watanabe, J. 1991. *Macromolecules* 24:6759–6753.
46. Mohamed, N.A., Sammour, M.H., and Elshafai, A.M. 2012. *Molecules* 17:13969–13988.
47. Capitani, D., Laurienzo, P., Malinconico, M., Proietti, N., and Roviello, A. 2003. *J. Polym. Sci., Polym. Chem. Ed.* 41:3916–3928.
48. Dicke, H.R. and Lenz, R.W. 1983. *J. Polym. Sci. Polym. Chem. Ed.* 21:2581–2588.
49. Griffin, A.C. and Havens, S.J. 1981. *J. Polym. Sci. Polym. Phys. Ed.* 19:951–969.
50. Lanzetta, N., Malinconico, M., Martuscelli, E., and Volpe, M.G. 1993. *J. Polym. Sci. Polym. Chem. Ed.* 31:1315–1322.
51. Calandrelli, L., Di Pace, E., Laurienzo, P., Malinconico, M., Martuscelli, E., and Volpe, M.G. 1994. *J. Polym. Sci. Polym. Phys. Ed.* 32:1248.
52. Avrami, M.J. 1939. *J. Chem. Phys.* 7:1103–1112.
53. Hermans, P.H. and Weidinger, A. 1961. *Makromol. Chem.* 50:98–115.
54. Kricheldorf, H.R., Pakull, R., and Buchmner, S. 1989. *J. Polym. Sci., Part A* 27:431–446.
55. Kricheldorf, H.R., Pakull, R., and Buchmner, S. 1988. *Macromolecules* 21:1929–1935.
56. Laurienzo, P., Malinconico, M., Martuscelli, E., Perenze, N., and Volpe, M.G. 1993. *Polym. J.* 25:227–236.
57. Di Pace, E., Fichera, A.M., Laurienzo, P., Malinconico, M., Martuscelli, E., Perenze, N., and Volpe, M.G. 1994. *J. Polym. Sci. B Polym. Phys.* 32:1643–1651.
58. Dolui, S.K. and Maiti, S. 1986. *Angew. Chem.* 141:31–47.
59. Varma, I.K. and Geetha, C.K. 1975. *J. Appl. Polym. Sci.* 19:2869–2878.
60. Lenz, R.W. 1985. *Polym. J.* 17:105–115.
61. Hoffman, J.D. 1964. *SPE Trans.* 4:315–362.
62. Flory, P.J. 1953. *Principles of Polymer Chemistry*. Ithaca, NY: Cornell University Press.
63. Brown, N.E., Swapp, S.M., Bennett, C.L., and Navrotsky, A. 1993. *J. Appl. Cryst.* 26:77–81.
64. Hatano, M., Kambara, S., and Okamoto, S. 1961. *J. Polym. Sci.* 51:S26–S29.
65. Shirakawa, H., Louis, E.J., MacDiarmid, A.G., Chiang, C.K., and Heeger, A.J. 1977. *J. Chem. Soc. Chem. Commun.* 16:578–580.
66. Chiang, C.K., Fincher, C.R., Park, Y.W. et al. 1977. *Phys. Rev. Lett.* 39:1098–1101.
67. Chiang, C.K., Druy, M.A., Gau, S.C. et al. 1978. *J. Am. Chem. Soc.* 100:1013–1015.
68. Diaz, A.F., Kanazawa, K.K., and Gardini, G.P. 1979. *J. Chem. Soc., Chem. Commun.* 635–636.
69. Diaz, A.F. and Logan, A. 1980. *J. Electroanal. Chem.* 111:111–114.
70. Tourillon, G. and Garnier, F. 1982. *J. Electroanal. Chem.* 135:173–178.
71. Capistran, J.D., Gagnon, D.R., Antoun, S., Lenz, R.W., and Karasz, F.E. 1984. *Polym. Prepr.* 25:282–283.
72. Aldissi, M., Hou, M., and Farrell, J. 1987. *Synth. Met.* 17:229–234.
73. Aldissi, M. 1987. *Polym. Plast. Technol. Eng.* 26:45–49.
74. Maglio, G., Palumbo, R., Tortora, M., Trifuoggi, M., and Varricchio, G. 1998. *Polymer* 39:6407–6413.
75. More, A.S., Patil, A.S., and Wadgaonkar, P.P. 2010. *Polym. Degrad. Stabil.* 95:837–844.
76. More, A.S., Sane, P.S., Patil, A.S., and Wadgaonkar, P.P. 2010. *Polym. Degrad. Stabil.* 95:1727–1735.
77. More, A.S., Pasale, S.K., and Wadgaonkar, P.P. 2010. *Eur. Polym. J.* 46:557–567.
78. Sadavarte, N.V., Avadhani, C.V., Naik, P.V., and Wadgaonkar, P.P. 2010. *Eur. Polym. J.* 46:1307–1315.
79. Sadavarte, N.V., Halhalli, M.R., Avadhani, C.V., and Wadgaonkar, P.P. 2009. *Eur. Polym. Mater.* 45:582–589.
80. More, A.S., Naik, P.V., Kumbhar, K.P., and Wadgaonkar, P.P. 2010. *Polym. Int.* 59:1408–1414.
81. More, A.S., Menon, S.K., and Wadgaonkar, P.P. 2012. *J. Appl. Polym. Sci.* 124:1281–1289.
82. Hamciuc, C., Hamciuc, E., and Brumă, M. 2006. *Rev. Roumaine Chim.* 51:773–780.
83. Janietz, S. and Schulz, B. 1996. *Eur. Polym. J.* 32:465–474.
84. Schulz, B., Kaminorz, Y., and Brehmer, L. 1997. *Synth. Met.* 84:449–450.
85. Gillo, M., Laurienzo, P., Iannelli, P. et al. 2002. *Chem. Mater.* 14:1539–1547.
86. Taylor, F. 2001. *Metal Finish.* 99/1:16–21.
87. Bugatti, V., Concilio, S., Iannelli, P. et al. 2006. *Synth. Met.* 156:13–20.
88. Capitani, D., De Prisco, N., Laurienzo, P., Malinconico, M., Proietti, N., and Roviello, A. 2001. *Polym. J.* 33:575–583.

89. De Prisco, N., Laurienzo, P., Malinconico, M., Roviello, A., and Volpe, M.G. 1999. *Mol. Cryst. Liq. Cryst.* 336:223–228.
90. Acierno, D., Amendola, E., Bellone, S. et al. 2003. *Macromolecules* 36:6410–6415.
91. Hsieh, B.Y., Yeh, K.M., and Chen, Y. 2005. *J. Polym. Sci., Polym. Chem.* 43:5009–5022.
92. Kung, Y.C. and Hsiao, S.H. 2011. *Polym. Chem.* 2:1720–1727.
93. Winnik, F.M. 1993. *Chem. Rev.* 93:587–614.
94. Kawano, S.I., Yang, C., Ribas, M., Baluschev, S., Baumgarten, M., and Mullen, K. 2008. *Macromolecules* 41:7933–7937.
95. Preston, J. and Hoffbert, Jr., W.L. 1978. *J. Polym. Sci., Polym. Symp.* 65:13–27.
96. Higashi, F. and Kokubo, N. 1980. *J. Polym. Sci., Polym. Chem. Ed.* 18:1639–1642.
97. Ojha, U.P., Krishnamoorthy, K., and Kumar, A. 2003. *Synth. Met.* 132:279–283.
98. Savinell, R.F. and Litt, M.H. 1996. US Patent 5,525,436.
99. Zaidi, S.M.J., Chen, S.F., Mikhailenko, S.D., and Kaliaguine, S.J. 2000. *New Mater. Electrochem. Syst.* 3:27–33.
100. Hensema, H.R., Mulder, M., and Smolders, C.A. 1991. *Bull. Soc. Chem. Belg.* 100:120–129.
101. Leibnitz, E., Eisold, C., and Paul, D. 1993. *Angew. Makromol. Chem.* 210:197–205.
102. Hajduk, B., Jarka, P., Weszka, J. et al. 2010. *J. Achiev. Mater. Manuf. Eng.* 40:7–14.
103. Ho, C.H., Yang, K.N., and Lee, S.N. 2001. *J. Polym. Sci. Part A* 39:2296–2307.
104. Cui, L., Tong, X., Yan, X., Liu, G., and Zhao, Y. 2004. *Macromolecules* 37:7097–7103.
105. Kim, T.D., Lee, G.U., Lee, K.S., and Kim, O.K. 2000. *Polymer* 41:5237–5245.
106. Tuo, X., Chen, Z., Wu, L., Wang, X., and Liu, D. 2000. *Polym. Prepr.* 41:1405–1406.
107. Meng, X., Natansohn, A., Barrett, C., and Rochon, P. 1996. *Macromolecules* 29:946–952.
108. He, Y., Wang, H., Tuo, X., Deng, W., and Wang, X. 2004. *Opt. Mater.* 26:89–93.
109. McKinney, Jr., R. and Rhodes, J.H. 1971. *Macromolecules* 4:633–637.
110. Dvornic, P.R. 1986. *J. Polym. Sci. Part A* 24:1133–1160.
111. Mohamed, N.A. and Al-Dossary, A.O.H. 2003. *Eur. Polym. J.* 39:1653–1667.
112. Vetter, S. and Nunes, S.P. 2004. *React. Funct. Polym.* 61:171–182.
113. Tsai, C.J. and Chen, Y. 2002. *J. Polym. Sci. Polym. Chem.* 46:293.
114. Hamciuc, C., Hamciuc, E., Bruma, M., Klapper, M., Pakula, T., and Demeter, A. 2001. *Polymer* 42:5955–5961.
115. Nasr, M.A.M., Kassem, A.A., Madkour, A.E., and Ali Ahmed, M.Z.M. 2005. *React. Funct. Polym.* 65:219–228.
116. Patton, T.L. 1971. *Polym. Prepr.* 12:163.
117. Reese, J. and Kraft, K. 1970. *Ger. Offen* 1,920,845 (*Chem. Abstr.* 74:23719h).
118. Caraculacu, G., Scorţanu, E., and Caraculacu, A.A. 1983. *Eur. Polym.* 19:143.
119. Scorţanu, E., Nicolaescu, I., Caraculacu, G., Diaconu, I., and Caraculacu, A. 1998. *Eur. Polym. J.* 34:1265–1272.
120. Caraculacu, A.A., Scorţanu, E., and Hitruc, G.E. 2001. *Eur. Polym. J.* 37:2491–2497.
121. Zhou, C., Cai, W., Wang, G.T., Zhao, X., and Li, Z.T. 2010. *Macromol. Chem. Phys.* 211:2090–2101.
122. Hou, J.L., Shao, X.B., Chen, G.J., Zhou, Y.X., Jiang, X.K., and Li, Z.T. 2004. *J. Am. Chem. Soc.* 126:12386–12394.
123. Li, C., Wang, G.T., Yi, H.P., Jiang, X.K., Li, Z.T., and Wang, R.X. 2007. *Org. Lett.* 9:1797.
124. Gong, B. 2001. *Chem. Eur. J.* 7:4336–4342.
125. Huc, I. 2004. *Eur. J. Org. Chem.* 17–29.
126. Zhao, X. and Li, Z.T. 2010. *Chem. Commun.* 46:1601–1616.
127. Cai, W., Wang, G.T., Xu, Y.X., Jiang, X.K., and Li, Z.T. 2008. *J. Am. Chem. Soc.* 130:6936–6937.
128. Cai, W., Wang, G.T., Du, P., Wang, R.X., Jiang, X.K., and Li, Z.T. 2008. *J. Am. Chem. Soc.* 130:13450–13459.
129. You, L.Y., Jiang, X.K., and Li, Z.T. 2009. *Tetrahedron* 65:9494–9504.
130. Ferguson, J.S., Yamato, K., Liu, R., He, L., Zeng, X.C., and Gong, B. 2009. *Angew. Chem. Int. Ed.* 48:3150–3154.

19 Polybenzimidazoles*

Yan Wang, Tingxu Yang, Kayley Fishel, Brian C. Benicewicz, and Tai-Shung Chung

CONTENTS

19.1 Introduction 618
19.2 Synthesis 619
    19.2.1 General Route 619
    19.2.2 Specific Case for PBI 619
    19.2.3 Product Requirements and Catalyst Effects 621
        19.2.3.1 Requirements 621
        19.2.3.2 Catalyst Effects 622
19.3 PBI Fiber Formation 622
    19.3.1 Dope Preparation 622
    19.3.2 Dry-Spinning Process 623
    19.3.3 Hot Drawing 624
    19.3.4 Sulfonation and Stabilization 624
19.4 PBI Blend Fibers 624
    19.4.1 PBI/PI Blends 624
    19.4.2 PBI/PAr Blends 627
    19.4.3 PBI/HMA 631
    19.4.4 PBI/PSF Blends 635
    19.4.5 PBI/PAI Blends 636
    19.4.6 PBI/PVPy Blends 637
19.5 Molded PBI Parts 637
19.6 PBI Matrix Resins and Composites 640
19.7 PBI Applications in High-Temperature Polymer Electrolyte Membrane Fuel Cells 640
    19.7.1 Low versus High Operational Temperature for PEM Fuel Cells 640
        19.7.1.1 PBIs for High-Temperature PEM Fuel Cells 641
        19.7.1.2 Preparation of PBI–Acid Membranes 641
    19.7.2 Impact of Chemistry on Fuel Cell Performance 643
        19.7.2.1 Meta-PBI 643
        19.7.2.2 Para-PBI 645
        19.7.2.3 AB-PBI 648
19.8 PBI-Based Membranes for Pervaporation Separation 649
    19.8.1 PBI-Based Membranes with Various Modifications 650
        19.8.1.1 Cross-Linking 650
        19.8.1.2 Sulfonation 651
        19.8.1.3 Polymer Blending 652
        19.8.1.4 Mixed-Matrix Membranes 652
    19.8.2 PBI Hollow Fiber Membranes 653

---

* Based in parts on the first-edition chapter on polybenzimidazoles.

19.9	PBI-Based Membranes for Gas Separation	654
	19.9.1 Monomer-Level Optimization	654
	19.9.2 N-Substitution Modification	657
	19.9.3 Chemical Cross-Linking	657
	19.9.4 Polymer Blending	657
	19.9.5 Mixed-Matrix Membranes	657
19.10	Current PBI Products	659
Acknowledgments		660
Abbreviations		660
References		662

19.1 INTRODUCTION

Vogel and Marvel [1] synthesized the first aromatic polybenzimidazoles (PBIs) in 1961. Because this polymer showed exceptional thermal and oxidative stability, National Aeronautics and Space Administration (NASA) and Air Force Material Laboratory sponsored DuPont and Hoechst Celanese to undertake fundamental research works on this material in the early stages of the development for aerospace and defense applications. Before the 1980s, major applications of PBI were in fire-blocking applications, thermal protective clothing, and reverse osmosis (RO) membranes. Its applications became diverse by the 1990s; molded PBI parts and microporous membranes were developed. The former exhibited superior performance as sealing elements in high-temperature corrosive environments. Hoechst Celanese commercialized PBI in 1983. Today, PBI Performance Products Inc. (http://www.pbiproducts.com/) is the major producer of high-performance PBI materials.

Many authors have reviewed the general chemistry and historical development of various PBI polymers [2–8]. Lee et al. [2] and Frazer [3] summarized the early chemistry work on PBIs before 1968 and their applications as adhesives in composites. Neuse [4] extended Lee et al.'s work and included molded PBI parts and PBI membranes. Powers and Serad [5] and Buckley et al. [6] gave in-depth historical reviews on PBI syntheses, properties, and applications of PBI fibers. They provided detailed information about the needed purities of monomers and flow charts for a large-scale production of PBI resins and fibers. Cassidy [7] and Critchley et al. [8] studied polycondensation reactions of various PBIs and emphasized the use of PBI for aerospace composites and adhesives. They also investigated PBI thermal degradation. Jaffe et al. [9] reviewed the previous work on miscible PBI blends, whereas Choe [10] summarized the effects of various catalysts on PBI polymerization. Since the early twentieth century, PBI has been extensively used in high-temperature fuel cell applications, and several reviews have summarized its applications in this aspect [11–15]. A comprehensive review of PBI on its historical development and future R&D was given by Chung [16] in 1997. A review on its synthesis, properties, processing, and applications was also conducted by Dang et al. [17] recently.

In short, PBIs are a class of high-performance heterocyclic polymers, typically synthesized from a condensation reaction of aromatic bis-o-diamines and dicarboxylates. Benzimidazole moiety is the repeating unit in the polymer molecular backbone. Poly(2,2′-(m-phenylene)-5,5′-bibenzimidazole) is the only commercially available polybenzimidazole. This composition was chosen due to the fact that it had an extremely high glass transition temperature (T_g) of 425–435°C and a superior flame resistance. Since it did not burn and could not produce much smoke, its early marked focus was for the defense industry as high-value-added fire-blocking materials and thermal protective clothing. PBI was the acronym for this composition. This material cost about $40/lb. in 1988 [18].

One of the objectives of PBI alloys is to tailor existing expensive material to a new/unique set of property/performance/price specifications through the combinations of low-cost or high-performance materials. Various miscible or partially miscible blends based on Hoechst Celanese PBI and commercially available polyimides (PI), polyamideimide (PAI), poly(4-vinyl pyridine) (PVPy), polyarylate (PAr), and high-modulus aramid (HMA) have been discovered [19–41]. Blend

Polybenzimidazoles

miscibility was evidenced in the form of infrared (IR) spectra, single T_g values, and well-defined single tan δ relaxations. The intermolecular interactions involving the >NH and carbonyl groups are the major driving forces for the miscibility. Immiscible blends such as PBI and polysulfone (PSF) were also reported [41–43]. Lithium chloride (LiCl) plays an important role on dope stability and process window of these miscible and nonmiscible blends.

Model and Lee [44,45] and Model et al. [46] investigated the uniqueness of PBI flat membranes, hollow fibers, as well as fabrication process, whereas Hughes et al. [47] summarized the overall processes, from PBI powder formation, cold molding, sintering, to their performance. Brooks et al. [48] reported the water absorption phenomenon of PBI materials.

19.2 SYNTHESIS

19.2.1 GENERAL ROUTE

In fact, Brinker and Robinson [49] were the first inventors of aliphatic PBIs, whereas Vogel and Marvel [1,50] modified their approach and synthesized the first aromatic PBIs. Since then, many methods for synthesizing PBIs have been invented [2–8]. Experimental data indicate that aromatic PBIs have thermal properties that are remarkably better than their aliphatic ones, as illustrated by Table 19.1 for several types of PBIs [3,5].

The general route to synthesize PBIs from aromatic tetraamines (bis-o-diamines) and dicarboxylates is illustrated in Figure 19.1 [5]. Generalized monomers are shown on the left-hand side, whereas the resultant PBIs are on the right-hand side. R_1 and R_2 are either aromatic or aliphatic and could contain other groups, such as ether, ketone, sulfone, etc. The phenyl groups shown in Figure 19.1 could be replaced by naphthyl groups. Polymerization could take place in either melt or solution.

19.2.2 SPECIFIC CASE FOR PBI

Poly[2,2′-(m-phenylene)-5,5′-bibenzimidazole] is the dominant commercially available PBI and has received the most attention. Two approaches have been developed to synthesize this polymer: one is a conventional two-stage reaction, and the other is a single-stage reaction. Figure 19.2 illustrates

TABLE 19.1
Structure and Stability of Several Polybenzimidazoles

Tetraamine	Acid	MP(°C)	Weight Loss (%) in N_2	Weight Loss in Air (%)[a]
	3,4-Diaminobenzoic	>600	0.4	–
Biphenyl	Terephthalic	>600	0	–
Benzene	Terephthalic	>600	1.0	–
Biphenyl	Isophthalic	>600	0.4	5.2
Benzene	Isophthalic	>600	0.3	–
Diphenylether	Isophthalic	>400	–	–
Biphenyl	Phthalic	>500	0.4	7.0
Biphenyl	4,4′-Oxydibenzoic	>400	–	–
Biphenyl	Biphenyl-4,4′ diacid	>600	0.8	–
Biphenyl	Biphenyl-2,2′ diacid	>430	8.0	–

Source: A. H. Frazer, *Polymer Reviews*, vol. 17, Interscience, New York, 1968, p. 138; Reprinted from *High Performance Polymers: Their Origin and Development* (R. B. Seymour and G. S. Kirshenbaum, eds.), E. D. Powers, and G. A. Serad, History and development polybenzimidazoles, 355, Copyright 1986, with permission from Elsevier.

[a] Weight loss after 1 h at 500°C, and after 1 h at 400°C and 450°C.

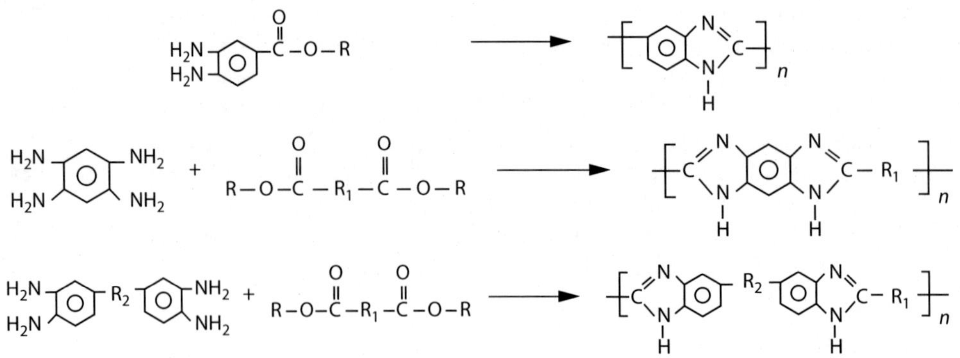

FIGURE 19.1 Polybenzimidazoles synthesized from different monomers. (Reprinted from *High Performance Polymers: Their Origin and Development* (R. B. Seymour and G. S. Kirshenbaum, eds.), E. D. Powers, and G. A. Serad, History and development polybenzimidazoles, p. 355, Copyright 1986, with permission from Elsevier.)

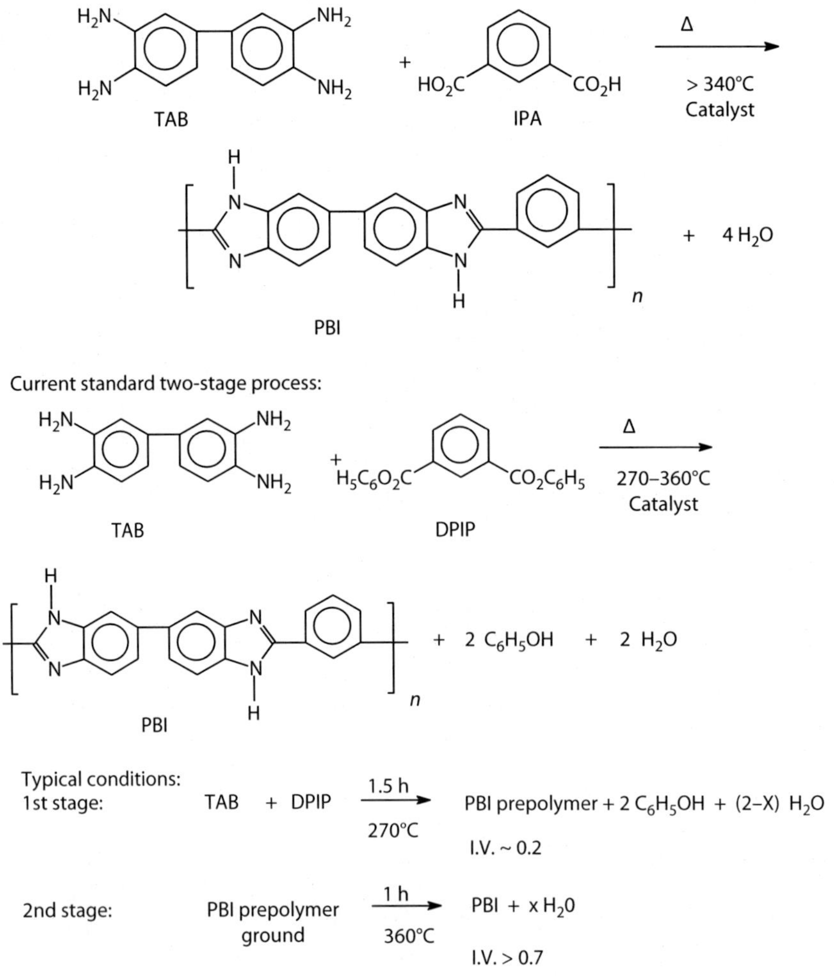

FIGURE 19.2 Two routes to synthesize PBI. (E. W. Choe: Catalysts for the preparation of polybenzimidazoles. *J. Appl. Polym. Sci.* 1994. 53. 497. Copyright Wiley-VCH Verlag GmbH & Co. KGaA. Reproduced with permission.)

TABLE 19.2
Major Differences between Single-Stage and Two-Stage Polymerizations

	Single-Stage Polymerization	Two-Stage Polymerization
Major Monomers	TAB and Isophthalic Acid (IPA)	TAB and Diphenylisophthalate (DPIP)
Reaction temperature and time	400°C for 1 h	1st stage: 270°C for 1.5 h 2nd stage: 360°C for 1.0 h
By-products	Water	Phenol and water
Antifoaming agent	No	Yes
Catalysts	Needed	Optional
Cost	Medium	High

Source: E. W. Choe, *J. Appl. Polym. Sci.* 53: 497, 1994.

TABLE 19.3
Preferred Monomer Suppliers and Properties

	TAB	DPIP	IPA
Producer	Hoechst	Amoco	Burdick and Jackson
Purity (%)	96.7	99.9+	99.5
Melting point (°C)	177.0	339	136

Source: E. W. Choe, *J. Appl. Polym. Sci.* 53: 497, 1994.

the details of these two reaction schemes [10]. The major differences between these two approaches are summarized in Table 19.2.

Polymerization of this PBI takes place in either melt or solution. The reaction kinetics and mechanisms are well known [5,6], and there is no need to repeat them here. However, it is important to point out that polymerization should take place under anaerobic conditions due to the fact that aromatic amines tend to form oxides. PBI made under nitrogen environment seems to have a better quality (less gel) than that made in vacuum [5,6,10]. In the case of a two-stage polymerization scheme, a voluminous foam is created owing to the entrainment of gaseous by-products. This foam can be significantly reduced by increasing the internal pressure of the reactor to 2.1–4.2 MPa (300–600 psi) or by adding an antifoaming agent. Foaming is not a severe phenomenon in the single-stage polymerization.

Buckley et al. [6] described the specifications of monomers for PBI polymerization. In short, extreme purity is required. Table 19.3 lists the monomer suppliers and properties used by Choe [10]. Triaminobiphenyl ($\approx$2.9%) and other extraneous impurities (0.2%) are major impurities in 3,3′,4,4′-tetraaminobiphenyl (TAB). The former may slightly influence chain growth during polymerization, whereas the latter may induce branching and gelation [6]. These impurities must be reduced to trace amounts if possible.

19.2.3 PRODUCT REQUIREMENTS AND CATALYST EFFECTS

19.2.3.1 Requirements

Before discussing the effects of catalysts on PBI resins, we need to define the required quality for a fiber-grade PBI. Generally, a synthesized PBI is useful for fiber spinning if it meets the following three criteria [5,6,10,51]:

1. Inherent viscosity (IV) is between 0.7 and 0.75 dL g^{-1}.
2. The plugging value is greater than 0.5 g/cm^2.
3. The insoluble matter is less than 1%.

IV is measured in a concentration of 0.4% PBI in 100 mL of 97% sulfuric acid. Since IV is a simple but indirect measurement of molecular weight, a polymer with a reasonably high IV usually yields a fiber with better mechanical properties. The plugging value is defined as the theoretical weight in grams of polymer that would pass through 1 cm^2 of a filter in infinite time in order to plug it. This is a measurement of solution filterability and is obtained by plotting the weight of polymer, which passes through a sheet of Gelman type A glass paper (Gelman Science Inc., Ann Arbor, MI) under 1 atm pressure versus time before becoming blocked. The bigger the plugging value, the better the solution filterability. The PBI concentration normally used in this measurement is about 5–7% in 97.0 ± 0.1% sulfuric acid. A detailed technique for measuring the plugging value appears elsewhere [6,10,51]. The insoluble matter is measured in N,N-dimethylacetamide (DMAc) solution containing 2% LiCl. The insoluble materials may result from the impurities of the monomers.

19.2.3.2 Catalyst Effects

19.2.3.2.1 Single-Stage Reaction

Choe [10] briefly reviewed previous work on the effects of catalysts on various PBIs [52–59] and conducted an extensive study on PBI syntheses. More than 15 catalysts were used to investigate the effects of a catalyst's chemistry on PBI's IV and plugging value. The following are considered to be good catalysts because they produce PBIs with an IV greater than 0.7 dL g^{-1}, a plugging value greater than 0.5, and an insoluble matter in DMAc less than 1%: dichlorophenylphosphine, chlorodiphenylphosphine, triphenyl phosphite, diphenylphosphine oxide, diphenylchlorophosphate, triphenyl phosphate, dimethoxyphenylphosphine, dibutoxyphenylphosphine, o-phenylphosphorochloridite, phenyl N-phenylphosphoramidochloridate, and dichlorodimethylsilane.

Choe [10] also reported the effects of catalyst concentration on PBI polymerization and showed that IV increases with an increase in catalyst concentration reaching almost a plateau at 1% concentration. Polymerization temperature also plays an important role on PBI properties. If 1% dichlorophenylphosphine ($C_6H_5PCl_2$) is used as the catalyst, a temperature of 390°C is required to yield a satisfactory product. A temperature of 400°C would be needed if one were to use 1% triphenyl phosphite ($C_6H_5O)_3P$ as the catalyst.

19.2.3.2.2 Two-Stage Reaction

Choe's data suggest that phosphorus-containing catalysts usually produce satisfactory products in the two-stage polymerization of PBI resins. Buckley et al. [6] had described the detailed process conditions and reactor vessels (sizes and reactor materials) for the two-stage polymerization reaction.

19.3 PBI FIBER FORMATION

19.3.1 Dope Preparation

Principally, PBI fibers can be prepared through dry-spinning, wet-spinning, and dry-jet wet-spinning processes, but dry spinning is the preferred method. A few solvents, such as sulfuric acid, dimethylformamide, dimethylsulfoxide, and DMAc, would dissolve PBI, but DMAc is preferred for dry-spinning PBI fibers [5,6,10,51,60]. Usually, spinning dopes contain 25–26 wt% of PBI and have a viscosity around 2000–3000 poise measured at room temperature. In order to prevent dopes from gelling or phasing out, 1–5 wt% LiCl or zinc chloride ($ZnCl_2$) (based on the weight of DMAc) is added to the spinning dopes. LiCl is preferred because it can be leached out easily during the subsequent washing stage. Table 19.4 summarizes the typical properties of PBI and sulfonated PBI fibers [61].

TABLE 19.4
Properties of PBI and Sulfonated PBI Fibers

Properties	Unstabilized PBI	Stabilized PBI
Denier per filament (dpf)	1.5	1.5
Tenacity (g/d)	3.1	2.7
Modulus (g/d)	90	45
Break elongation (%)	90	30

Source: Reprinted from *High Performance Polymers: Their Origin and Development* (R. B. Seymour and G. S. Kirshenbaum, eds.), E. D. Powers, and G. A. Serad, History and development polybenzimidazoles, 355, Copyright 1986, with permission from Elsevier; A. Buckley, D. Stuetz, and G. A. Serad: Polybenzimidazoles. *Encyclopedia of Polymer Science and Engineering* (J. I. Kroschwitz, ed.). 572. 1987. Copyright Wiley-VCH Verlag GmbH & Co. KGaA. Reproduced with permission; A. B. Conciatori et al., *J. Appl. Polym. Sci.* 19: 49, 1967; Hoechst Celanese PBI brochure, P.O. Box 32414, Charlotte, NC 28232-9973.
Note: Data obtained from the PBI brochure. g/d × 8.83 = cN/tex.

19.3.2 Dry-Spinning Process

The dry-spinning process consists of three elements: a spinneret set, a circulated dry column, and a take-up unit. A spinneret may have 50–1000 holes with a diameter of about 75–100 μm. The take-up speed varies from 150 to 500 m/min. The dope is filtered and metered by a gear pump to the spinneret at 70–110°C, and the jet face temperature is about 100–150°C. Fiber is formed when most of the DMAc is vaporized. Generally, this is carried out during spinning by the circulating N_2 in a dry column, which is about 6.6 m long at a temperature of about 200–220°C, as illustrated in Figure 19.3. The residual DMAc in the as-spun fiber is removed by washing with water. The as-spun fiber property is weak; its tenacity is about 0.11–15 N/tex (1.3–1.7 g/day), modulus about 2.6–4.4 N/tex (30–50 g/day), and elongation at break of about 100–120%.

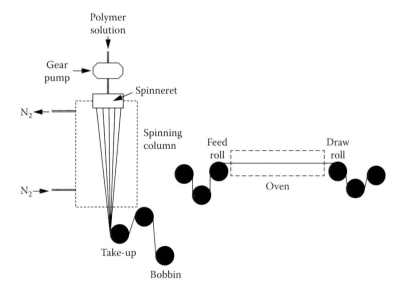

FIGURE 19.3 Typical drying-spinning and hot-draw devices.

19.3.3 Hot Drawing

In order to improve fiber properties, the as-spun fiber could be hot-drawn by passing the dried filaments through a heat muffle furnace at approximately 400–440°C dependent on the speed. Skewed rolls before and after the muffle furnace accurately maintain the filaments at different speeds, as illustrated in Figure 19.3. The spin-line tension causes the filaments to elongate and the polymeric structure within the fibers become possibly better organized.

When the draw ratio varies from 1 to 4, the initial modulus may increase from 3.3 to 11 N/tex (from 36 to 122 g/day), whereas elongation at break may drop from 110% to 18%.

19.3.4 Sulfonation and Stabilization

In order to prevent PBI fabrics from shrinkage during burning, the drawn PBI yarn is acid-treated and stabilized to form a salt with the imidazole ring structure [5,6,10,51]. Sulfonation is conducted by dipping a hot-drawn PBI yarn in a 2% sulfuric acid bath for 2 h at 50°C. Stabilization of a sulfuric acid-treated fiber is carried out by again passing it through a heat muffle furnace of approximately 380–440°C depending on the process speed. The flame shrinkage of a PBI fiber is reduced from >50% to <10% after treatment.

19.4 PBI BLEND FIBERS

Although PBI has good mechanical properties, it is difficult to be fabricated into large parts because of its high glass transition temperature (T_g ~425–435°C). Its moisture regain is high, and its thermo-oxidative stability at temperatures above approximately 260°C is not as good as that of some high-performance PIs. Blending may overcome the above shortcomings and extend its application range into the area of functional polymers. Luckily, PBI possesses both donor and acceptor hydrogen-bonding sites, which are capable of participating in specific interactions and thus favorable to form miscible blends with some other polymers [9].

19.4.1 PBI/PI Blends

PBI/PI blends have received most attention because the blends synergize their strengths and overcome their individual shortcomings [27–31,62–66]. A research team was formed in the early 1980s to overcome the weaknesses of PBI through blending with a variety of PIs and to fundamentally understand the phase nature of the blends. This multisector research team includes the University of Massachusetts, Virginia Polytechnic and State University, Lockheed Aeronautical Systems, General Electric Aircraft Engine Business Group, and Hoechst Celanese. They report that PBI was miscible with a broad range of PIs, including ether imides, fluoro-containing imides, and others [9,19–31,64–66]. Two blends were chosen as examples in this review. They were the 85:15 blend of PBI/PEI and the 10:90 PBI blend with a copolyimide containing 37.5 mol% of 4,4'-hexafluoroisopropylidenediphthalic anhydride (6FDA). The PEI used in this study was Ultem 1000 produced by General Electric. Table 19.5 illustrates the glass transition temperatures and other key parameters of the blend components. PBI has outstanding compressive properties that are essentially rendered onto the blends as illustrated in Tables 19.5 and 19.6 [9]. In addition, PBI/PEI and PBI/6FCoPI (i.e., 6FDA-polyimide) blends have better chemical resistance. For example, both PEI and 6FCoPI are attacked and softened by common solvents such as methylene chloride and acetone, whereas PBI and the two blends appear to be insoluble, even after 2-year exposure as thin films. Mechanical properties of PBI/GFCoPI blends are listed in Table 19.7.

Although PBI/PI blends show a single T_g over the entire composition range on a first heating by dynamic mechanical analysis (DMA), this single phase is metastable over much of the

TABLE 19.5
Structure and Properties of Candidate High-Temperature Matrix Polymers

Properties	PBI	PEI	6FCoPI
T_g (°C)	420	220	340
Tensile Properties			
Strength (MPa)	100	108	97.2
Elongation (%)	1.8	33	4.4
Modulus (GPa)	5.68	3.18	3.36
Flexural Properties			
Strength (MPa)	100	143	153
Modulus (GPa)	6.32	3.39	3.83
Compressive Properties			
Strength (MPa)	397	150	183
Modulus (GPa)	6.46	3.3	3.71

Source: M. Jaffe et al., *Adv. Polym. Sci.* 117: 297, 1994.

TABLE 19.6
Mechanical Properties of 85:15 PBI/PEI Blends

Tensile Properties	
Strength (MPa)	158
Elongation (%)	3.4
Modulus (GPa)	5.34
Flexural Properties	
Strength (MPa)	248
Modulus (GPa)	5.73
Compressive Properties	
Strength (MPa)	300
Modulus (GPa)	5.18

Source: M. Jaffe et al., *Adv. Polym. Sci.* 117: 297, 1994.

composition range in most cases [9,19–22,25,65]. Fourier transform IR spectroscopy (FTIR) data indicate that the origin of the miscible behavior in this system is a strong hydrogen bond interaction between the imidazole hydrogen and the carbonyl of the PI. Experimental data indicate that during the long annealing times above T_g often associated with the molding cycles for these materials, the carbonyl IR absorbance shift induced by the hydrogen bonding disappears. This indicates that phase separation has taken place in the blend, confirming the metastable nature of the observed miscibility.

Figure 19.4 illustrates the weight loss of the 85:15 PBI/PEI blend as a function of time during exposure in air at 315°C [9]. The rate of degradation is faster than that of either PBI or PEI. This is probably due to the fact that PBI accelerates the degradation of the PEI. In other words, the overall rate of degradation is exacerbated by the strong intimate interaction present within the blend.

TABLE 19.7
Mechanical Properties of 10:90 PBI/6FCoPI Blends

Tensile Properties	
Strength (MPa)	103
Elongation (%)	4.8
Modulus (GPa)	3.36
Flexural Properties	
Strength (MPa)	156
Modulus (GPa)	3.90
Compressive Properties	
Strength (MPa)	187
Modulus (GPa)	3.81

Source: M. Jaffe et al., *Adv. Polym. Sci.* 117: 297, 1994.

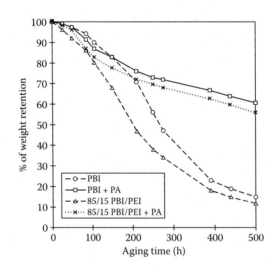

FIGURE 19.4 Effect of PA treatment on the thermooxidative stability of PBI and PBI/PEI blend films aging at 315°C. (With kind permission from Springer Science+Business Media: *Adv. Polym. Sci.* High performance polymer blends, 117, 1994, 297, M. Jaffe, P. Chen, E. W. Choe, T. S. Chung, and S. Makhija.)

Protonation of the imidazole ring by phosphoric acid (PA) improves the thermooxidative stability of PBI, as well as the blend.

Figure 19.5 illustrates the weight loss of the 10:90 PBI/6FCoPI blend as a function of time during exposure in air at 315°C. The rate of degradation of PBI is significantly reduced by the addition of 6FCoPL. These excellent thermal properties indicate that the 10:90 PBI/6FCoPI is a matrix candidate for use at temperatures up to 315°C.

A series of asymmetric hollow fiber membranes spun from PBI/PI blend solutions have been developed by Chung's group at the National University of Singapore recently [27,29,30,62,63]. Consistent with the aforementioned studies, the miscibility and molecular interactions between PBI and PEI were confirmed by various characterizations. An increase in PBI percentage in the spinning solutions could result in membranes with a tighter morphology, lesser finger-like voids, and significantly lower gas permeance. However, the gas separation [62] and pervaporation [27] performance of the resultant membranes were significantly improved.

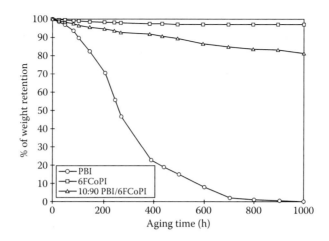

FIGURE 19.5 Thermooxidative stability of PBI and 6FCoPI blend films aged in air at 315°C. (With kind permission from Springer Science+Business Media: *Adv. Polym. Sci.* High performance polymer blends, 117, 1994, 297, M. Jaffe, P. Chen, E. W. Choe, T. S. Chung, and S. Makhija.)

19.4.2 PBI/PAr Blends

PAr is a group of wholly aromatic polyesters derived from aromatic dicarboxylic acids and diphenols or their derivatives. They are amorphous in nature with a much lower cost than that of PBL. PAr and PBI have many common polar organic solvents (e.g., 1-methyl-2-pyrrolidinone [NMP], DMAc, dimethylsulfoxide [DMSO], etc.). NMP solutions containing 10 wt% of PBI and PAr were usually homogeneous and had no insolubles. After being kept at room temperature for a number of days, PBI-rich dopes precipitate, but these phased-out solids could be easily redissolved with a mild heating (e.g., 100°C for 20 min). Based on the haze level, the stability of the PBI/PAr/NMP increases with an increase in the relative PAr concentrations [33,34,36].

Figure 19.6 illustrates that the IV of the solution blends exceeds the rule of mixtures at a concentration of 0.5%. This result suggests that PBI and PAr exhibit strong interactions in a dilute solution such that the resulting hydrodynamic size of the blends is greater than the calculated averages based on each component. Corroborating evidence of a PBI–PAr interaction was observed by FTIR. Based on the carbonyl stretching of a pure PAr film, it was found that the signal of an 80:20 PBI/PAr film showed a dramatic downfield shift (i.e., from 1741 to 1730 cm^{-1}).

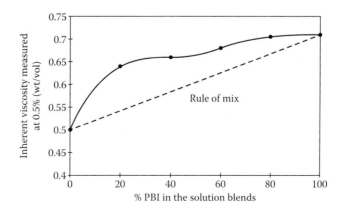

FIGURE 19.6 Relationship between the PBI/PAr blend IVs and compositions. (From T. S. Chung and P. N. Chen, Sr., *J. Appl. Polym. Sci.* 40: 1209, 1990; T. S. Chung and P. N. Chen Sr., *Polym. Eng. Sci.* 30: 1, 1990.)

This shift (see Figure 19.7) indicates the existence of intermolecular H bonding between PBI and PAr in the film blend. Similar to the case of PBI/PEI and PBI/6FCoPI, blending PAr with PBI improves the solvent resistance of PAr. For example, PAr is soluble in methylene chloride and tetrahydrofuran, whereas PBI is insoluble in these solvents. An 80:20 PAr/PBI film kept its physical integrity after soaking in methylene chloride for 30 min, whereas a pure PAr film would completely dissolve within 10 s.

Table 19.8 summarizes the as-spun fiber tensile properties and illustrates the minimum LiCl effects on fiber properties. These fibers were further drawn at elevated temperatures at various draw ratios, and the drawn fibers became stronger, as illustrated in Table 19.9. Drawing at 400°C at a ratio of 3.0 gave the best tensile modulus and strength. Table 19.10 provides a comparison of tensile properties among PBI/PAr, PBI/PSF, PBI/PEI, and PBI fibers. This comparison was based on the highest modulus and tenacity of PBI/PAr, PBI/PSF, and PBI/PEI fibers obtained from the previous literature. Dry-spun PBI/PAr has a very impressive tensile modulus and strength and can be suitable for engineering and aerospace applications. The thermal stability of PAr in the film blends is dramatically improved with the presence of PBI, and the degree of improvement increased with the

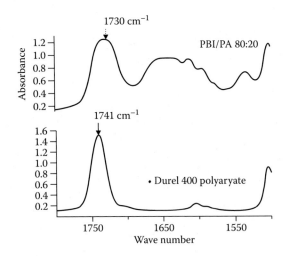

FIGURE 19.7 FTIR confirmation of the existence of intermolecular hydrogen bonding between PBI and PAr. (From T. S. Chung and P. N. Chen, Sr., *J. Appl. Polym. Sci.* 40: 1209, 1990; T. S. Chung and P. N. Chen Sr., *Polym. Eng. Sci.* 30: 1, 1990.)

TABLE 19.8
As-Spun 80:20 PBI/PAr and 80:20 PBI/PSF Fiber Properties

Sample	Spinning Method	Denier (dpf)	Initial Modulus (g/d)	Tenacity (g/d)	Elongation at Break (%)
80:20 PBI/PAr (LiCl)	Dry-spun	3.690	48.4	1.38	73.9
80:20 PBI/PAr (no LiCl)	Dry-spun	5.271	42.6	1.53	87.3
80:20 PBI/PSF	Wet-spun	30	30.2	0.97	4.0
80:20 PBI/PSF (LiCl)	Dry-spun	4.46	43.5	1.417	82.5
80:20 PBI/PSF (no LiCl)	Dry-spun	5.818	32.2	1.374	106.0

Note: g/d × 8.83 = cN/tex.

Polybenzimidazoles

TABLE 19.9
Properties of (80:20) PBI/PAr Fibers (No LiCl)

Draw Ratio	Temp. (°C)	Denier (dpf)	Initial Modulus (g/d)	Tenacity (g/d)	Elongation at Break (%)
2	400	1.919	77.69	3.118	29.50
2.5	400	1.915	91.86	3.313	17.92
3.0	400	1.501	141.15	4.611	6.75
2	420	1.668	66.50	2.922	33.75
3	420	1.453	80.74	3.378	19.75
4	420	0.951	113.86	4.014	11.61

Note: g/d × 8.83 = cN/tex.

TABLE 19.10
Property Comparison among Hot-Drawn PBI/PSF, PBI/PAr, PBI/Ultem, and PBI Fibers

Sample	Denier (dpf)	Draw Ratio	Initial Modulus (g/d)	Tenacity (g/d)	Elongation at Break (%)
PBI (LiCl)	1.7	2.0 at 440°C	91.0	4.0	31.9
PBI (no LiCl)	1.24	2.0 at 440°C	87.0	4.0	29.7
80:20 PBI/PAr (no LiCl)	1.501	3.0 at 400°C	141.15	4.611	6.75
80:20 PBI/PSF wet-spun (LiCl)	796	3.5 at 400°C	110.9	2.8	3.08
80:20 PBI/PSF (LiCl)	1.98	4.0 at 420°C	112.1	3.6	7.3
80:20 PBI/PSF (no LiCl)	1.00	4.5 at 420°C	134.1	4.84	5.9
75:25 PBI/Ultem	0.96	3.0 at 420°C	112.0	4.32	9.9

Note: g/d × 8.83 = cN/tex.

increase in the PBI concentration (see Figure 19.8). Using the weight loss at 550°C as the reference, it is interesting to notice that the thermal stability of the blends was almost linearly proportional to the relative concentration of PBI (see Figure 19.9).

Table 19.11 summarizes the mechanical properties of hot-drawn and then acid-treated/stabilized fibers. Both initial modulus and tenacity dropped due to the post-treatment, whereas their elongation

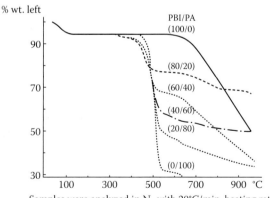

FIGURE 19.8 TGA study of PBI/PAr film blends. (From T. S. Chung and P. N. Chen, Sr., *J. Appl. Polym. Sci.* 40: 1209, 1990; T. S. Chung and P. N. Chen Sr., *Polym. Eng. Sci.* 30: 1, 1990.)

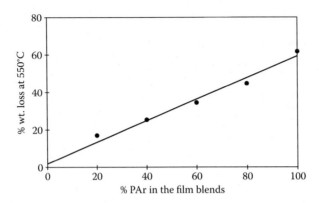

FIGURE 19.9 Correction between weight loss and PBI concentration in PBI/PAr films at 550°C. (From T. S. Chung and P. N. Chen, Sr., *J. Appl. Polym. Sci.* 40: 1209, 1990; T. S. Chung and P. N. Chen Sr., *Polym. Eng. Sci.* 30: 1, 1990.)

TABLE 19.11
Tensile Properties of Sulfonated and Stabilized PBI, PBI/PAr, and PBI/PSF Fibers

Sample	Denier (dpf)	Initial Modulus (g/d)	Tenacity (g/d)	Elongation at Break (%)
PBI (LiCl)	1.5	45.0	2.7	30
80:20 PBI/PAr (no LiCl)	1.274	80.1	3.1	13.4
80:20 PBI/PSF (no LiCl)	2.526	58.7	2.6	19.1
80:20 PBI/PSF (LiCl)	1.337	71.5	2.24	5.9

Note: g/d × 8.83 = cN/tex.

increased. By hot-drawing an as-spun fiber 1–2 weeks after it had been spun, washed, and dried, the resultant fiber properties would drop significantly. Table 19.12 provides examples. This phenomenon is probably due to the effect of the vaporization of residual DMAc solvent on the phase stability of a compatible blend. Therefore, hot-drawing *immediately after* dry-spinning is the key to preparing high-modulus and high-tenacity fibers. Phase separation was evident in the transmission electron microscopy (TEM) micrographs (Figure 19.10) of an as-spun fiber (without LiCl), which was left at room temperature for 2 months. In contrast, the as-spun fibers with LiCl have a uniform texture with no evident phase separation as illustrated in Figure 19.11, and this structure is similar to the textures observed on the control PBI fibers. Figure 19.12 illustrates the FTIR spectra of 80:20 PBI/PAr films before and after the sulfonation and stabilization processes.

TABLE 19.12
Aging Effect on Tensile Properties of Hot-Drawn Fibers

Sample	Process Condition	Denier (dpf)	Initial Modulus (g/d)	Tenacity (g/d)	Elongation at Break (%)
80:20 PBI/PAr (no LiCl)	DIAS	1.501	141.1	4.6	6.75
80:20 PBI/PAr (no LiCl)	DL	1.600	101.2	3.7	7.01
80:20 PBI/PAr (LiCl)	DL	1.141	68.3	2.17	14.23
80:20 PBI/PSF (no LiCl)	DIAS	1.698	80.5	4.04	18.5
80:20 PBI/PSF (no LiCl)	DL	1.812	91.8	3.77	16.7

Note: g/d × 8.83 = cN/tex. DIAS: draw immediately after spun; DL: draw later.

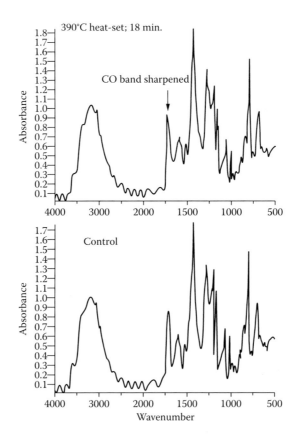

FIGURE 19.10 FTIR spectra for sulfonated 80:20 PBI/PAr (without LiCl) blends. (From T. S. Chung and P. N. Chen, Sr., *J. Appl. Polym. Sci.* 40: 1209, 1990; T. S. Chung and P. N. Chen Sr., *Polym. Eng. Sci.* 30: 1, 1990.)

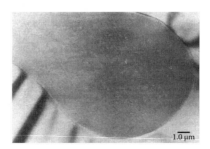

FIGURE 19.11 TEM micrographs of OsO_4 stained 80:20 PBI/PAr (without LiCl) blend fibers. (From T. S. Chung and P. N. Chen, Sr., *J. Appl. Polym. Sci.* 40: 1209, 1990; T. S. Chung and P. N. Chen Sr., *Polym. Eng. Sci.* 30: 1, 1990.)

19.4.3 PBI/HMA

Blending flexible coil polymers with rigid rod-like polymers has also generated interest. The resultant composites may have better fracture and impact toughness as well as thermal stability and flammability resistance. Two approaches of dispersing a rod-like polymer in a matrix were attempted. One involved synthesizing block copolymers, and the other involved blending solutions of two polymers. Takayanagi et al. [67] used the first approach to blend wholly aromatic polyamide such as poly(1,4-benzamide) (PBA), poly(1,4-phenyleneterephthalamide) (PPTA), and their block copolymers with nylon-6 or nylon-66. They found microfibrils of PPTA dispersed in a fractured

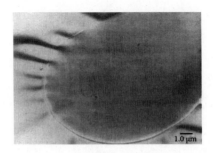

FIGURE 19.12 TEM micrographs of OsO$_4$ stained 80:20 PBI/PAr (with LiCl) blend fibers. (From T. S. Chung and P. N. Chen, Sr., *J. Appl. Polym. Sci.* 40: 1209, 1990; T. S. Chung and P. N. Chen Sr., *Polym. Eng. Sci.* 30: 1, 1990.)

surface of the polymer composites. Wright–Patterson Air Force Laboratory was the driving force for the second approach and is also the leader for rod-like/random coil chain composites. They used very expensive materials, such as poly(*p*-phenylenebenzobisthiazole) (PBZT), as reinforcing elements [68–73]. Arnold and Arnold [73] reviewed on their work.

High-modulus polyaramides (HMAs) used here were synthesized from a low-temperature solution condensation reaction using terephthalic dichloride, *p*-phenylenediamine (25 mol%), 3,3′-dimethylbenzidine (37.5 mol%), and 1,4-bis-(4′-aminophenoxy)benzene (37.5 mol%) [35]. They have tensile properties similar to those of Teijin's HM-50 and DuPont's Kevlar. Figure 19.13 illustrates the TGA of neat polymers and their blends. The decomposition temperature of a 50:50 blend was in the range 400–450°C, i.e., higher than that of the neat HMA polymer, which demonstrated that PBI protected HMA from thermal degradation. The molecular interaction between the HMA and PBI was detected by observing the frequency of the amide–carbonyl band. In pure HMA, this band appeared at 1657 and was shifted to 1655 in an as-spun fiber, to 1647 in a heat-treated and drawn fiber [35]. The molecular interaction was found to be stronger on processing at a higher temperature as the bond energy of the carbonyl was weakest in this blend. The reduction in the carbonyl bond order would result from an interaction of the C=O with N–H. Figure 19.14 illustrates the TMA curves of HMA, PBI, and 50:50 HMA/PBI blend fibers. For establishing miscibility in a blend system, one would expect a single tan δ peak intermediate in position to the two-component material peaks for a miscible blend system. This figure indicates that the 50:50 blend has a single tan δ at 380°C, which is between the peak of HMA (257.5°C) and PBI (425°C).

The 50:50 HMA/PBI blend fibers were prepared from both dry-spun and wet-spun processes. The wet-spun fiber properties were slightly inferior to those of dry-spun ones, as shown in Table 19.13.

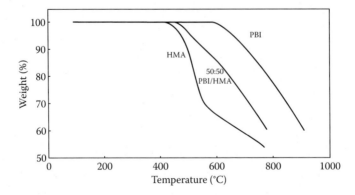

FIGURE 19.13 TGA curves of neat HMA, PBI, and 50:50 PBI/HMA. (From T. S. Chung and F. K. Herold, *Polym. Eng. Sci.* 31:1950, 1991; T. S. Chung et al., Fibers from blends of PBI with polyaramide and polyarylate, *Proceedings of Fiber Producer Conf.* (May) 1991, Greenville, SC. 3A-2.)

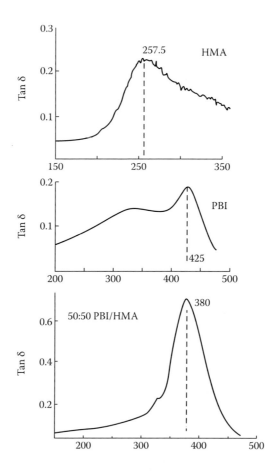

FIGURE 19.14 TMA curves of PBI, HMA, and 50:50 PBI/HMA fibers. (From T. S. Chung and F. K. Herold, *Polym. Eng. Sci.* 31:1950, 1991; T. S. Chung et al., Fibers from blends of PBI with polyaramide and polyarylate, *Proceedings of Fiber Producer Conf.* (May) 1991, Greenville, SC. 3A-2.)

TABLE 19.13
50:50 HMA/PBI Fiber Properties (Total Solid Content = 11% in DMAc)

Sample	Draw Ratio at 400°C	Denier (dpt)	Initial Modulus (g/d)	Tenacity (g/d)	Elongation at Break (%)
Wet-spun fiber		9.8	73	1.78	23.46
Wet-spun and then hot-drawn	1.5	5.75	272	8.06	3.9
Wet-spun and then hot-drawn	2	4.54	269.8	6.93	3.12
Dry-spun fiber		3.60	81.4	3.781	55.2
Dry-spun and then hot-drawn	2.5	1.406	252.1	9.65	10.38
Dry-spun and then hot-drawn	3.5	0.952	288.6	10.26	7.6
Dry-spun and then hot-drawn	4.5	0.783	302	10.83	7.3

Note: g/d × 8.83 = cN/tex.

TABLE 19.14
20:80 HMA/PBI Fiber Properties (Total Solid Content = 17.8% in DMAc)

Sample	Draw Ratio at 400°C	Denier (dpf)	Initial Modulus (g/d)	Tenacity (g/d)	Elongation at Break (%)
			A: Wet-Spun		
As-spun		24.500	41.1	1.080	4.03
Hot-drawn	3	7.820	139.4	4.834	11.17
Hot-drawn	5	4.653	165.1	6.336	10.13
Hot-drawn	7	3.347	181.0	7.330	10.43
			B: Dry-Spun		
As-spun		4.933	49.96	1.740	80.48
Hot-drawn	2.5	2.021	137.91	5.031	18.44
Hot-drawn	3.5	1.554	170.64	6.400	10.87
Hot-drawn	4.5	1.121	184.76	5.710	6.47

Note: g/d × 8.83 = cN/tex.

Table 19.14 provides the tensile properties of dry-spun and wet-spun 20:80 HMA/PBI fibers. The best conditions for obtaining a higher tensile modulus and strength for hot-drawing dry- and wet-spun fibers were different. In the case of the wet-spun fibers, the coagulation bath also had a significant effect on the ultimate fiber properties. A mixture of ethylene glycol (EG)/DMAc solvent provided a better coagulation process and yielded a higher tensile modulus and strength fiber than that of a water bath [35].

Similar to PBI/PAr blend fibers, as-spun PBI/HMA fibers seemed to become brittle after standing in the laboratory for a few weeks, probably due to the evaporation of residual DMAc solvent. Therefore, the hot-drawing process should take place immediately after dry spinning; otherwise fiber properties will decay.

The mechanical properties of PBI/HMA blend fibers follow the molecular composite theory developed by Halpin and Tsai [74]. The relationship of the composite modulus, E_{11}, to the individual moduli of the components is as follows:

$$E_{11}/E_m = (1 + \alpha b V_f)/(1 - b V_f) \qquad (19.1)$$

where

$$b = (E_f/E_m - 1)/(E_f/E_m + \alpha) \qquad (19.2)$$

where V_f is the volume fraction of fiber, and E_t and E_m are the moduli of the fiber and matrix, respectively. As the aspect ratio, α, becomes large, E_{11} approaches the limiting upper bound given by the following:

$$E_{11} = E_f V_f + E_m V_m \qquad (19.3)$$

where V_m is the volume fraction of the matrix. This equation predicts that the composite modulus and tensile strength follow a linear *rule of mixtures* of the fiber and matrix properties. Since the density of PBI is very close to that of HMA, the above can be further simplified as

$$E_{11} = E_f V_f + E_m V_m \qquad (19.4)$$

where W is the weight fraction of the constituents, and subscripts f and m refer to fiber and matrix, respectively. Figures 19.15 and 19.16 present the comparisons of the tensile modulus and strength of the hot-drawn fibers obtained by experiments and predicted by the Halpin and Tsai equation

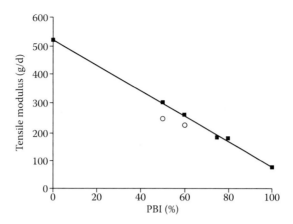

FIGURE 19.15 Tensile modulus of HMA/PBI as a function of PBI content. ■, Hot-drawn immediately after dry-spun; ○, drawn later. (From T. S. Chung and F. K. Herold, *Polym. Eng. Sci.* 31:1950, 1991; T. S. Chung et al., Fibers from blends of PBI with polyaramide and polyarylate, *Proceedings of Fiber Producer Conf.* (May) 1991, Greenville, SC. 3A-2.)

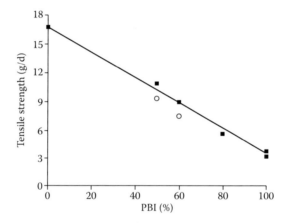

FIGURE 19.16 Tensile strength of HMA/PBI as a function PBI content. ■, Hot drawn immediately after dry-spun; ○, hot drawn later. (From T. S. Chung and F. K. Herold, *Polym. Eng. Sci.* 31:1950, 1991; T. S. Chung et al., Fibers from blends of PBI with polyaramide and polyarylate, *Proceedings of Fiber Producer Conf.* (May) 1991, Greenville, SC. 3A-2.)

(straight lines). The agreement is good, and this implies that the PBI/HMA blend is miscible at the molecular level. However, the agreement becomes poor if the as-spun fibers were aged and later hot-drawn, as illustrated in these two figures.

Recently, PBI/Kevlar (poly *p*-phenylene terephthalamide) blend staple fibers were studied by Arrieta et al. [37,38] and Genc et al. [75] on their moisture sorption characteristics and thermal aging properties. The study showed that exposure to elevated temperatures (190–320°C) resulted in a rapid decrease in tensile breaking force retention for a fabric made of a 60:40 wt% blend of Kevlar and PBI fibers [37]. X-ray diffraction, Raman, and differential thermal analyses (DTA) were carried out to evaluate the effect of thermal aging on the material's crystallinity and transition temperatures [38].

19.4.4 PBI/PSF Blends

PSF has a better hydrolytic stability than polyesters and polycarbonates. Its thermal stability is also very impressive. Thermal gravimetric analysis shows that Amoco's Udel P1700 is stable in air up to 450°C (840°F). Therefore, PBI/PSF blends were prepared to yield a low-cost, high-performance PBI variant.

Spinning dopes with different PBI/PSF ratios were prepared with a polymer concentration of 25.6 wt% and LiCl of 2 wt%. Preliminary fiber properties indicate that the 80:20 PBI/PSF blend has the best tensile properties [41,42]. In addition, no phase separation was observed in the 80:20 PBI/PSF blend dope over a few months. Phase separation slowly occurred in the 60:40 PBI/PSF sample after a few weeks. The 20:80 PBI/PSF dopes phase-separate after a few days. Dopes were also prepared without the addition of LiCl. Their formulation was 27 wt% of 80:20 PBI/PSF in DMAc.

The presence of LiCl in a spinning dope plays an important role in determining the dope viscosity, process window, and fiber properties. The addition of LiCl resulted in a lower dope viscosity due to some interaction among LiCl, PBI, and PSF [42]. As a consequence, the process windows for fiber spinning are quite different for dopes with and without LiCl. Dopes without LiCl have a broad process window for fiber spinning, and the resultant fibers have superior properties.

Table 19.8 summarizes the as-spun PBI/PSF fiber tensile properties and shows that LiCl has minimum effects on as-spun fiber properties. LiCl effects appear on hot-drawn fibers. Table 19.10 lists the best tensile modulus and strength of hot-drawn 80:20 PBI/PSF fibers and compares them to those of PBI/PAr, PBI/Ultem, and PBI fibers. Dry-spun 80:20 PBI/PSF fibers (without LiCl) and PBI/PAr (without LiCl) have the best and almost the same tensile modulus and strength. Their properties are at least comparable to those of PBI/Ultem fibers. Compared to standard PBI fibers, these two blends have higher tensile moduli and strengths than those of PBI fibers, whereas the elongations at break of the blends are inferior to those of the PBI.

FTIR spectra indicate a slight shift of the 1244 ether band of 4 cm^{-1} to lower frequency in the PBI/PSF blends, which would suggest a slight interaction between the NH and the O. The magnitude of this molecular interaction is very small compared to that for PBI/HMA (10 cm^{-1} wavelength shift) and for PBI/PAr (20 cm^{-1} wavelength shift). SEM pictures indicate that PBI/PSF blends are not miscible because they phase-separate.

The mechanical properties of acid-treated and stabilized 80:20 PBI/PSF fibers are presented in Table 19.11. Both 80:20 PBI/PSF with and without LiCl fibers have better moduli than that of stabilized PBI fibers, whereas the PBI has the highest elongation at break. This is due to the fact that these PBI/PSF fibers have been hot-drawn at high draw ratios. The 80:20 PBI/PSF fiber with LiCl appeared to have PBI-like limiting oxygen indices (LOIs) and flame shrinkage behavior [41].

The PBI/PSF fiber does not show aging phenomenon if hot-drawn later, as illustrated in Table 19.12. There is no major difference in properties for PBI/PSF fibers drawn at different periods, whereas tensile modulus and strength properties drop significantly for PBI/PAr and PBI/HMA fibers. This phenomenon is probably due to the effect of the vaporization of residual DMAc solvent on the phase stability of a compatible blend. As PBI/PSF blends are not compatible, the effect of DMAc on the phase stability is minor.

PBI/PSF blend hollow fiber membranes were also wet-spun from highly concentrated immiscible blend solutions by Chung et al. [43]. A sharply defined unit-step morphological change was found in the middle of the membrane cross section. A yellowish halo with a wavelength of 580–595 nm was observed, and its formation was possibly caused by a physical phenomenon, but not by the phase separation of PBI and PSF.

19.4.5 PBI/PAI Blends

Generally, PAIs are prepared by the condensation polymerization of a trifunctional acid anhydride (e.g., trimellitic anhydride [TMA]) with an aromatic diamine (e.g., 4,4′-methylene- or 4,4′-oxydianiline, i.e., MDA or ODA). These polymers are characterized by excellent high-temperature properties with T_g values typically above 270°C and continuous service temperatures of about 230°C. Since PAI has an imide structure on its polymer chains, hydrogen bonding could be formed between the carbonyl group of PAI's PI units and the NH group of PBI. PBI and Torlon 4000T were discovered to be miscible at a molecular level over the whole composition range as confirmed by microscopy, differential scanning calorimeter (DSC), FTIR, and DMA [32]. FTIR spectra show the existence of hydrogen-bonding interactions in the polymer blends, whereas DSC and DMA studies confirmed the existence of a single glass transition in each blend.

19.4.6 PBI/PVPy Blends

Poly(2,2'-(m-phenylene)-5,5'-bibenzimidazole) is miscible with PVPy over the entire composition range [40]. Single T_g values, intermediate between the two pure polymers, were observed. FTIR data indicate that the –NH group of PBI forms a hydrogen bond with the N of PVPy. However, the thermal stability of these blends is not significantly high but similar to the thermal stability of pure PVPy.

19.5 MOLDED PBI PARTS

Three-dimensional PBI parts were developed from the melt-derived PBI powders by a hot compression molding (hcm) process in the 1980s [18,76,77]. The original process was very complicated and time consuming. For example, PBI powders were first cold-pressed in a billet mold, and then the clamped mold was transferred to an oven for heating at a temperature above the T_g of PBI (425–435°C) for a period of 5–10 h. After cooling, the billet was cut and machined to the desired shape. This process was significantly simplified by Hughes et al. [47], who discovered that the use of porous PBI powders with reasonable moisture content was the key. In other words, porous PBI powders with reasonable moisture content could be cold-compacted at room temperature, while melt-derived PBI powders made directly from the solid state reactor (the second reaction in a two-stage PBI polymerization) could not. Figure 19.17 illustrates the morphology of melt-derived PBI powders and porous PBI powders. The porous PBI powders are prepared by spraying a 12 wt% PBI/DMAc solution into a water mist, as illustrated in Figure 19.18. The spraying nozzle is 100 µm with a pressure of 100 psi, and the PBI/DMAc dope viscosity is about 700 cp.

Two stages are needed for fabrication of PBI parts: cold-compact porous PBI powders, followed by their placement in a graphite powder bed under a pressure of 1–3 kpsi at 425–500°C for a few hours, as shown in Figure 19.19. This process is referred to as powder-assisted hot isostatic pressing (HIP) [47]. Table 19.15 summarizes the results and compares the mechanical properties of PBI parts made from the different approaches. Clearly, samples made from the new approach (df as defined in the table) yield the best properties.

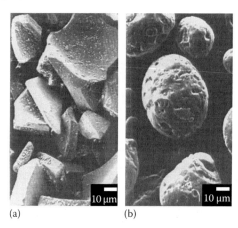

(a)   (b)

FIGURE 19.17 Morphology of melt-derived PBI powders (a) and spray precipitation porous PBI powders (b). (O. R. Hughes, P. N. Chen, W. M. Cooper, L. P. DiSano, E. Alvarez, and T. E. Andres: PBI powder processing to performance parts. *J. Appl. Polym. Sci.* 1994. 53. 485. Copyright Wiley-VCH Verlag GmbH & Co. KGaA. Reproduced with permission.)

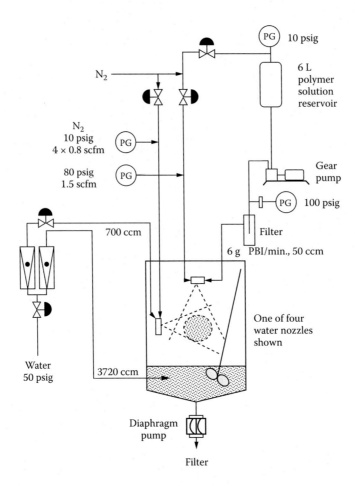

FIGURE 19.18 Schematic diagram of the spray precipitation reactor. (O. R. Hughes, P. N. Chen, W. M. Cooper, L. P. DiSano, E. Alvarez, and T. E. Andres: PBI powder processing to performance parts. *J. Appl. Polym. Sci.* 1994. 53. 485. Copyright Wiley-VCH Verlag GmbH & Co. KGaA. Reproduced with permission.)

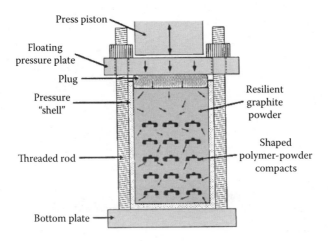

FIGURE 19.19 Pressure vessel and clamp assembly for powder-assisted HIP process. (O. R. Hughes, P. N. Chen, W. M. Cooper, L. P. DiSano, E. Alvarez, and T. E. Andres: PBI powder processing to performance parts. *J. Appl. Polym. Sci.* 1994. 53. 485. Copyright Wiley-VCH Verlag GmbH & Co. KGaA. Reproduced with permission.)

TABLE 19.15
Molded PBI Materials: Process Conditions and Properties

PBI Powder Type	Filler	Molding Conditions					Tensile Properties				Compression Strength (kpsi)	Izod Impact Strength (Unnotched) (ft-lb/in.)
		Process	Temp. (°C)	Press. (kpsi)	Time (h)	Density (g/cm³)	Strength (kpsi)	Elong. (%)	Modulus (mpsi)			
Melt-derived	None	hcm	462	3.00	0.50	1.278	22.9	3.0	0.850	57	6.7
	None	hcm	463	3.00	4.00	1.276	27.1	3.8	0.830	–	9
Spray-precipitated	None	hcm	461	3.00	0.50	1.285	26.5	3.5	0.857	–	9.8
	None	hcm	460	2.00	2.00	1.280	27.8	3.7	0.870	60	–
	None	hcm	463	3.00	4.00	1.277	31.3	4.4	0.836	–	15.2
Spray-precipitated	None	df	463	1.00	4.00[a]	1.300	32	5.0	0.950	–	–

Note: hcm = Hot compression molded to panel shape, machined to final test shape. df = direct formed: cold compacted and HIP'd to billet shape, machine to final test shape.

[a] Approximate time and temperature; actual conditions vary with vessel size.

19.6 PBI MATRIX RESINS AND COMPOSITES

The use of PBI as a matrix resin for continuous fiber composites has received a lot of attention since the 1960s and 1970s because of its high performance at elevated temperatures [78–83]. The amount of research in these areas has been gradually decreased because of (1) the availability of many high-performance polymers, such as polyetherketone (PEEK), liquid crystalline polymers (LCPs), PIs, and polyphenylene sulfide (PPS) in the 1980s and 1990s, and (2) PBI's high water absorption characteristics. For example, PBI can absorb 15–18% water by weight, and this water may degrade the mechanical properties and bonding strengths of a PBI composite. Water absorption phenomenon of PBI was investigated by Brooks et al. [48], and molecular modeling of hydrated PBI was undertaken by Iwamoto [84].

PBI-based mixed-matrix fibers with different nanofillers are developed in recent years, including carbon fibers [79], carbon nanotubes [80], polyhedral oligomeric silsesquioxane (POSS) nanoparticles [81], nanofibers [82], and zeolitic imidazolate frameworks (ZIFs) [83]. It was reported that the incorporation of a small amount (0.5 wt%) of POSS nanoparticles into the PBI dope has significant influence on both the morphology and the separation performance of the forward osmosis dual-layer hollow fiber membranes [81]. The in-plane thermal conductivity of the carbon nanotube/PBI composites could be increased by a factor of 50 when incorporating 1.94 wt% carbon nanotubes into the composite nanofibers [80]. Hollow fiber membranes fabricated from the blend solution of PBI and synthesized nanoporous ZIF-8 nanoparticles for gas separation showed an impressive enhancement in H_2 permeability as high as a hundred times without any significant reduction in H_2/CO_2 selectivity [83]. The details are discussed in Section 19.9.5.

19.7 PBI APPLICATIONS IN HIGH-TEMPERATURE POLYMER ELECTROLYTE MEMBRANE FUEL CELLS

Through the evolution of technology and the rapid increase in global population, the demand for energy is ever increasing. According to the National Petroleum Council, the world's energy demand increased by approximately 50% over the last 25 years and is expected to do so again by 2030 [85]. Due to the ever-rising demand for energy and the finite resources available, it is no surprise that alternative energy sources and methods of clean energy production are needed. British Petroleum and Royal Dutch Shell, two of the world's largest oil companies, estimate that by 2050, one-third of the world's energy will need to be produced from alternative energy sources as a result of the growing population and the limited resources available [86]. A clean energy conversion device that has gained worldwide attention because of its potential use in stationary and mobile devices is the polymer electrolyte membrane (PEM) fuel cell, also known as the proton exchange membrane fuel cell [87]. Due to the excellent production of energy and the lack of harmful by-products, PEM fuel cells have received much attention as an alternative to combustion-based energy production [88]. PEM fuel cells produce energy at the heart of the cell, the membrane electrode assembly (MEA), which consists of a polymer matrix doped with electrolyte and sandwiched between two gas diffusion layers coated with metal catalyst. When fueled by hydrogen, the catalyst at the anode side splits the hydrogen into protons and electrons. The protons diffuse across the polymer membrane, and the electrons travel around the membrane through an external circuit. The protons and electrons react with the oxidant (typically oxygen or air) at the metal catalyst layer on the cathode side and produce water, thereby completing the electrochemical cycle. Hydrogen gas is the most commonly used fuel source for PEM fuel cells, although other fuel sources such as methanol, ethanol, and methane have been studied [89].

19.7.1 Low versus High Operational Temperature for PEM Fuel Cells

While PEM fuel cells have received much attention as promising energy conversion devices, limitations are associated with the availability and development of materials that would allow for the cost of manufacturing and production of these devices to compete with existing energy sources [88].

One possible way to overcome this limitation is to increase the efficiency of these PEM fuel cells. By operating at elevated temperatures (>100°C), the efficiency of the fuel cell is greatly increased and offers many advantages over lower operational temperatures [90]. PEM fuel cells that use low boiling point dopants such as water typically operate around 80°C to ensure that the electrolyte is not vaporized [89]. When water is used as the electrolyte, most commonly used with a perfluorosulfonic acid membrane, additional components to prevent membrane dehydration are required. To ensure that heat generated by the cell does not vaporize the electrolyte, large heat exchangers are necessary. Another additional component is the dual-phase water system needed to regulate humidity [89]. If the humidity is too low, the membrane will dry out causing conductivity to dramatically decrease; however, if the humidity is too high, water condenses and the gas diffusion electrodes are flooded [89,91]. Under fully humidified conditions, a perfluorosulfonic acid membrane (e.g., Nafion) doped with water can reach a conductivity as high as 0.1 S cm^{-1} [91]. An additional problem associated with low operational temperatures is the poisoning of the catalyst from impurities in the fuel source, mainly carbon monoxide and hydrogen sulfide [91]. The tolerance of the catalyst to impurities, which has been found to be temperature dependent, is much less problematic at elevated temperatures due to reversible binding of the impurity to the catalyst. Therefore, when operating at low temperatures, an extremely pure fuel source is required [89]. In order for a PEM fuel cell to operate at high temperatures (120–200°C), the electrolyte must have a high-enough boiling point to withstand operational temperature, and the polymer membrane must be thermally stable. Typically inorganic acids with low vapor pressures, such as PA and sulfuric acid, are used as electrolytes for high-temperature PEM fuel cells. There are many benefits to operating at these increased temperatures. Since these polymers and electrolytes can withstand high temperatures, smaller heat exchangers can be used [89] and the fuel cells do not require a humidification process. As previously stated, the tolerance to gas impurities is greatly increased at these temperatures due to reversible binding. Not only is the catalyst less likely to be poisoned from the gas impurities, but also fuel sources with larger amounts of impurities can still be effectively used, thereby lowering reformation costs [89]. Additionally, operating at higher temperatures has been shown to increase the electrode kinetics, thus increasing performance and efficiency overall.

19.7.1.1 PBIs for High-Temperature PEM Fuel Cells

PBIs have attracted much attention as PEMs in high-temperature fuel cells due to their thermal and chemical resistances as well as their ability to form proton-conducting membranes. The basic sites along the PBI backbone allow for an acid–base complex to form with PA and sulfuric acid [90]. When doped with PA, this complex demonstrates high proton conductivity, low gas permeability, and a stable operational life of over two years. Though PBI doped with PA is conductive, it is important to note that PBI without an electrolyte has a negligible conductivity [90]. It is often beneficial to control the amount of PA dopant in the membrane. Excess PA can block oxygen diffusion into the catalyst layer, and insufficient PA can lead to incomplete contact between the PBI/PA complex and the catalyst layer, thereby decreasing efficiency [91]. PBIs imbibed with acid are promising candidates for low-cost, high-performance PEM materials and are readily producible [92].

19.7.1.2 Preparation of PBI–Acid Membranes

The morphology of a polymer membrane, which is greatly influenced by polymer processing, can directly affect the properties of a PBI membrane. The method through which the PBI is imbibed impacts the amount of PA that can be contained within the membrane, thereby affecting conductivity. In the past, three preparation methods have been explored for imbibing PBI with PA.

19.7.1.2.1 Conventional Acid Imbibing

The conventional and most common method of doping PBI with PA consists of a multistep process, which is illustrated in Figure 19.20 [93,94]. The PBI is formed by two-stage melt/solid polymerization, and the powder form is dissolved in an organic solvent such as dimethylacetamide (DMAc)

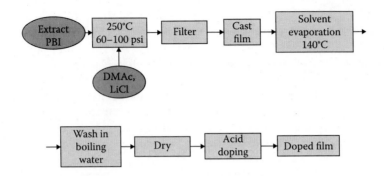

FIGURE 19.20 Step-by-step method of conventional PA imbibing of polybenzimidazole.

with LiCl for a stabilizer under pressure at 60–100 psi and at 250°C [89]. Any cross-linked PBI is removed by filtering, and the non-cross-linked polymer is then cast from solution and dried under vacuum to remove DMAc. The film is then washed in boiling distilled water to remove the LiCl and any remaining DMAc. Finally, the polymer is dried and placed in a PA bath for doping [89,92].

19.7.1.2.2 Porous PBI

In 2004, another method of imbibing PBI with PA was reported. In this method, a porous PBI film is created by casting a solution containing both PBI and a porogen (phthalates or phosphates) onto an untreated glass plate [95]. Once cast, the film is dried at 110°C for 2 h, and the porogen is removed by soaking the film in methanol. After drying a second time, the porous membrane is immersed in a concentrated PA solution for 4 days, and the membrane is blotted dry [95]. Figure 19.21 demonstrates the step-by-step process of synthesizing and imbibing the porous PBI membrane with PA.

19.7.1.2.3 Sol–Gel Process

The sol–gel process, also termed the polyphosphoric acid (PPA) process, was reported in 2005 by Xiao et al. [93]. Through this method, PBI is polymerized in PPA, which acts as both the polymerization solvent and the polycondensation reagent. Polymerization takes place at 190–220°C for 16–24 h and is then cast while hot directly onto untreated glass plates [93]. Once cast, the PPA is hydrolyzed under controlled conditions. Since both PBI and PPA are hygroscopic, moisture from the

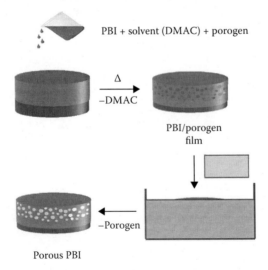

FIGURE 19.21 Synthesis of porous polybenzimidazole. (Reprinted with permission from D. Mecerreyes et al., *Chem. Mater.* 16: 604. Copyright 2004 American Chemical Society.)

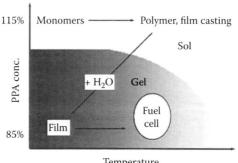

FIGURE 19.22 Sol–gel process state diagram. (Reprinted with permission from L. Xiao et al., *Chem. Mater.* 17: 5328. Copyright 2005 American Chemical Society.)

atmosphere can readily hydrolyze the PPA to PA. The sol–gel transition is attributed to PBI's excellent solubility in PPA but poor solubility in PA. Membranes prepared using the PPA process demonstrate superior mechanical properties at higher levels of acid doping than membranes prepared using the previously mentioned acid-imbibing methods [93]. Not only do membranes produced through the PPA process have better mechanical properties at higher levels of acid doping, but also this process is much less tedious and time-consuming than the conventional imbibing method and porous PBI. Figure 19.22 illustrates the sol–gel transition observed via the PPA process.

19.7.2 Impact of Chemistry on Fuel Cell Performance

Like the morphology of a polymer gel film, the chemistry of the PBI membrane greatly impacts the properties of the gel film. Conductivity, acid doping levels, and the mechanical strength of the membrane are all affected by the chemistry of the polymer. Many PBI derivatives have been studied to assess and enhance these properties. The reader is directed to more complete reviews [14,87,89,94] on how various functional groups and differences in the polymer backbone affect the properties of PBI.

19.7.2.1 Meta-PBI

The first and most thoroughly studied PBI/PA complex is poly(2,2′-*m*-phenylene-5,5′-bibenzimidazole) or *m*-PBI, which can be seen in Figure 19.23a. This research was begun in 1995 by

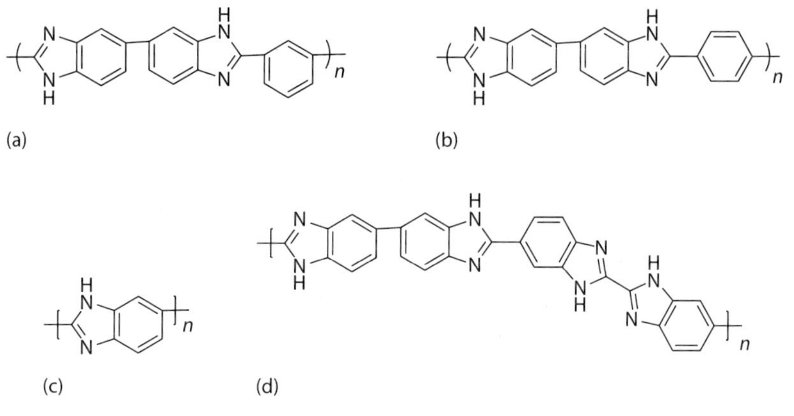

FIGURE 19.23 Chemical structures of various polybenzimidazoles: (a) *m*-PBI, (b) *p*-PBI, (c) AB-PBI, and (d) *i*-AB-PBI.

Wainright et al., which focused on various synthetic and characterization methods of PBI [96]. During this research, it was discovered that a PBI membrane could be cast from a DMAc solution, imbibed with PA, and retain conductivity at high temperatures. This research ultimately led to finding that PBI had the potential for use as a membrane in fuel cells. Since then, many researchers have worked on characterizing m-PBI as well as other PBI derivatives.

19.7.2.1.1 Acid Doping Levels

Meta-PBI has been imbibed with PA using both the conventional method and the PPA process. The method through which the membrane is processed gives rise to different properties. One of the largest differences observed when comparing these processing methods is the acid doping level [ratio of moles of PA to moles of polymer repeat units (PA/PRU)], which greatly affects conductivity. When m-PBI is prepared through conventional imbibing, the acid doping level is typically 6–10 moles PA/PRU with conductivities ranging from 0.04 to 0.08 S cm^{-1} at 150°C at varying humidities [14,89]. Table 19.16 shows the conductivity at various levels of doping for m-PBI prepared through conventional imbibing including specific testing conditions that have been reported [96–98]. When m-PBI is prepared through the PPA process, higher levels of acid doping are observed. Typical acid doping levels for m-PBI prepared through the PPA process range from 14 to 26 moles PA/PRU [14]. Conductivities of membranes prepared by the PPA process are typically higher than those prepared through conventional imbibing. In one report, it was observed that m-PBI prepared through the PPA process had a conductivity of 0.13 S cm^{-1} at 160°C under nonhumidified conditions [89]. Table 19.17 shows the conductivities at various temperatures for m-PBI prepared through the PPA process. To further compare the differences in the morphology of m-PBI prepared through conventional imbibing and the PPA process, membranes prepared through each method that had similar physical characteristics were compared. Although the acid doping levels are extremely close, the differences in conductivity and the proton diffusion coefficient can be seen in Table 19.18 [99]. These results demonstrate the differences in the proton transport architecture [89]. Not only does the membrane prepared through the PPA process have greater conductivity and proton architecture, but also the IV data show that higher-molecular-weight polymers are produced from the PPA process over the conventional method [89,93].

19.7.2.1.2 Fuel Cell Performance

Meta-PBI membranes doped through conventional imbibing have been tested many times for use in fuel cells. Li et al. [12] found that a membrane containing 6.2 moles PA/PRU reached a current

TABLE 19.16

IV, Acid Doping Levels, and Conductivity of Various m-PBI Membranes Prepared through Conventional Acid Imbibing

IV (dL g^{-1})	Acid Doping (PA/PRU)	Conductivity (S cm^{-1})	Operational Temperature (°C)	Relative Humidity	Reference
0.6	5.01	2×10^{-2}	130	NS	[96]
0.6	5	2.5×10^{-2}	150	NS	[96]
0.6	3.38	5×10^{-3}	130	NS	[96]
0.6	3.05	7×10^{-6}	30	NS	[97]
1.2	6.3	5×10^{-2}	140	30	[98]
1.2	6.3	2×10^{-2}	140	5	[98]
1.2	6.3	5.9×10^{-2}	150	30	[98]
1.2	6.3	4.7×10^{-3}	150	5	[98]

Source: W. J. Wainright et al., *J. Electrochem. Soc. 142*: L121, 1995; R. Bouchet and E. Siebert, *Solid State Ionics* 118: 287, 1999; J. A. Asensio et al., *J. Polym. Sci. Part A* 40: 3703, 2002.

Note: NS = not specified.

TABLE 19.17
Conductivity at Various Temperatures for *m*-PBI Prepared through the PPA Process

IV (dL g^{-1})	Acid Doping (PA/PRU)	Conductivity (S cm^{-1})	Temperature (°C)	Relative Humidity
1.49	14.4	5.16×10^{-2}	25	Dry
1.49	14.4	5.28×10^{-2}	40	Dry
1.49	14.4	6.23×10^{-2}	60	Dry
1.49	14.4	7.99×10^{-2}	80	Dry
1.49	14.4	9.52×10^{-2}	100	Dry
1.49	14.4	1.1×10^{-1}	120	Dry
1.49	14.4	1.2×10^{-1}	140	Dry
1.49	14.4	1.27×10^{-1}	160	Dry

Source: J. Mader et al., Polybenzimidazole/acid complexes as high-temperature membranes, *Fuel Cells II, Advances in Polymer Science* (G. G. Scherer, ed.), 216: 63, 2008.

TABLE 19.18
Comparison of *m*-PBI Prepared through Conventional Imbibing and PPA Process

Processing Method	Polymer (wt%)	Phosphoric Acid (wt%)	Water (wt%)	PA/PBI (Molar Ratio)	Conductivity (S cm^{-1})	Proton Diffusion Coefficient (cm^2 s^{-1})	IV (dL g^{-1})
Conventionally imbibed	15.6	60.7	23.7	12.2	0.048	10^{-7}	0.89
PPA process	14.4	63.3	22.3	13.8	0.13	3×10^{-6}	1.49

Source: D. C. Seel et al., High-temperature Polybenzimidazole-based Membranes, *Handbook of Fuel Cells*, pp. 300–312, 2009.

density of approximately 0.7 A cm^{-2} at 0.6 V at 190°C when hydrogen and oxygen gases were used under nonhumidified conditions. Figure 19.24 shows the fuel cell performance for *m*-PBI produced using the PPA process. Though performance was measured under high flow rate conditions, it shows its dependability at high temperatures [14]. Meta PBI has been used not only in PEM fuel cells but in direct methanol fuel cells (DMFCs) as well. When fed with 33.3 wt% methanol concentration and pure oxygen without back pressure, an open circuit voltage of 0.7 V and a maximum power density of 138.5 mW cm^{-2} at 200°C are reached, and durability tests show stable performance for over 120 h [91].

19.7.2.2 Para-PBI

Poly(2,2'-(*p*-phenylene)5,5'-bibenzimidazole) or *p*-PBI (Figure 19.23b) is one of the best performing PBI membranes for use in fuel cells [89]. Due to the rigid nature of its backbone, *p*-PBI was found to be difficult to process. In 1975, high-molecular-weight (4.2 dL g^{-1}) *p*-PBI was synthesized by the US Air Force Materials Lab; however, the method required several weeks to produce high-molecular-weight polymer, and because it could not be spun into fibers as easily as *m*-PBI, *p*-PBI was left unstudied for some time [89,100].

19.7.2.2.1 Acid Doping Levels

Kim et al. [101] produced *p*-PBI through both the PPA process and solution casting from methane sulfonic acid. When prepared through the PPA process, membranes were almost completely

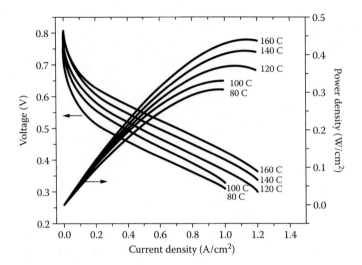

FIGURE 19.24 Performance curves of H_2/air m-PBI/PA fuel cells at various temperatures (1 atm absolute pressure). Operating conditions are as follows: constant flow rate, H_2 at 400 SCCM, air at 1300 SCCM, no humidification, 44 cm² active area, 1.0 mg cm⁻² Pt catalyst loading, Pt–C 30% on each electrode. (From J. Mader et al., Polybenzimidazole/acid complexes as high-temperature membranes, *Fuel Cells II, Advances in Polymer Science* (G. G. Scherer, ed.), 216: 63, 2008.)

amorphous and showed much higher levels of acid doping and water uptake than when produced through solution casting [87]. Yu et al. [102,103] polymerized p-PBI through the PPA process at varying monomer concentrations. They had found the acid doping levels of p-PBI to be 30–40 PA/PRU [102,103]. When p-PBI produced through the PPA process had an acid doping level of 31.8 PA/PRU, a conductivity of 0.24 S cm⁻¹ at 160°C was achieved. This shows that p-PBI has both a higher level of acid doping and a higher proton conductivity than m-PBI, which typically has 13–16 PA/PRU and a conductivity of 0.1–0.13 S cm⁻¹ at 160°C [103]. It was also found that the monomer concentration (wt%) of p-PBI greatly affected the IV. This relationship can be seen in Figure 19.25. The IV of the PBI is important because it greatly affects the mechanical properties of the membrane (Figure 19.26). Since p-PBI demonstrates a high conductivity, high acid loadings, and good mechanical properties at high acid loadings, it is an excellent candidate for use in a PEM fuel cell.

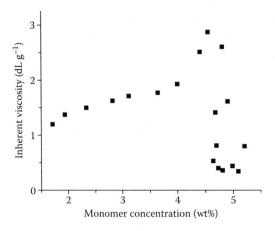

FIGURE 19.25 Effect of monomer concentration on the IV of p-PBI when polymerized at 195°C. (From S. Yu, *Fuel Cells* 9: 318, 2009.)

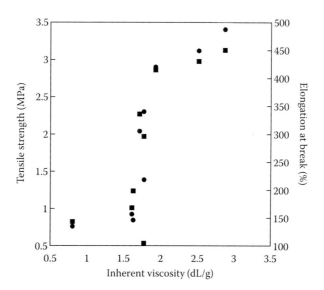

FIGURE 19.26 Effect of IV on tensile strength and elongation at break of *p*-PBI. IV measured using a Ubbelohde viscometer at a polymer concentration of 0.2 g/dL in concentrated sulfuric acid (96%) at 30°C. (Reprinted with permission from L. Xiao et al., *Chem. Mater.* 17: 5328. Copyright 2005 American Chemical Society.)

19.7.2.2.2 Fuel Cell Performance

The polarization curves for *p*-PBI membrane is shown in Figure 19.27. This fuel cell was operated under nonhumidified conditions using hydrogen and air as the fuel and oxidant. The voltage increased from 0.606 to 0.663 V when the temperature was increased from 120°C to 180°C at 0.2 A cm^{-2} [103]. This demonstrates that *p*-PBI can operate at high temperatures without the need for humidification due to the involvement of PA in the proton conduction mechanism. Fuel cell durability tests were also performed for *p*-PBI. To test long-term durability, temperature and current

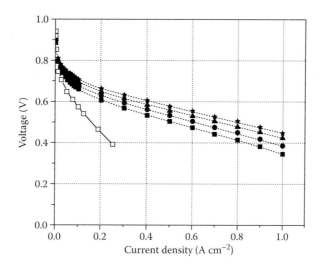

FIGURE 19.27 Polarization curves for membrane-based fuel cells with *p*-PBI membranes using 1.2–2.0 stoichiometric flows, respectively, at ambient pressure (1 atm, absolute). Squares: 120°C, circles: 140°C, triangles: 160°C, and stars: 180°C. (Open squares: m-PBI data at 150°C adapted from Wang et al., *Electrochim. Acta* 1996, 41, 193. From S. Yu, *Fuel Cells* 9: 318, 2009.)

density (0.2 A cm^{-2}) were held constant, and the voltage was recorded. Hydrogen and air were used as the fuel and oxidant at stoichiometric flows of 1.2 and 2.0, respectively [102]. The voltage was monitored at 80°C, 160°C, and 190°C to ensure durability testing over a wide range of temperatures. For the study at 80°C, the fuel cell was first broken in for 80 h and operated at 120°C. It was then run for over 900 h at 80°C and a voltage degradation rate of 45 µV h^{-1}. The performance test that had taken place at 160°C showed a voltage degradation of 4.9 µV over a 2500 h period. When tested at 190°C, the voltage degradation was found to be 60 µV h^{-1}, which shows to be significantly higher than the voltage degradation at 160°C. This could be due to multiple factors such as Pt dissolution, Pt agglomeration, and carbon support corrosion [102].

19.7.2.3 AB-PBI

The simplest known structure of PBI that can be prepared using commercial monomers is poly(2,5-benzimidazole) or AB-PBI (Figure 19.23c) [104]. This polymer is prepared using only the monomer 3,4-diaminobenzoic acid because it contains both the diamine and acid functionalities. In addition to the homopolymer, copolymers, cross-linked polymers, and substituted polymers containing AB-PBI have been studied [105]. AB-PBI has been prepared through both conventional imbibing and the PPA process.

19.7.2.3.1 Acid Doping Levels

When prepared through conventional imbibing, acid doping levels are between 2 and 10 mol PA/PRU and demonstrate a conductivity of approximately 0.05 S cm^{-1} at 180°C under anhydrous conditions [106]. It is important to note that unlike the previously mentioned *m*-PBI and *p*-PBI, AB-PBI contains one benzimidazole per repeat unit rather than two. This must be taken into consideration when comparing acid doping levels of different PBI chemistries. AB-PBI prepared using the PPA process demonstrates higher levels of acid doping (22–35 mol PA/PRU) than when prepared through conventional imbibing [14]. The optimal monomer concentration was found to be approximately 3 wt% and yielded an IV of 4.63 dL g^{-1}. These membranes, however, were found to be thermally unstable at temperatures greater than 130°C. Adjusting the PA level was done by soaking the membranes in PA baths with lower concentrations of PA in the original film [105]. These films containing lower levels of PA (14.5, 22.7, and 29.1 mol PA/PRU) were still unstable at high temperatures; therefore, conductivity and fuel cell performance were unable to be measured when prepared through the PPA process [105].

19.7.3.3.2 Isomeric AB-PBI

In 2011, a new sequence isomer of AB-PBI was reported (Figure 19.23d) [105]. This new sequence, *i*-AB-PBI, contains head-to-head, head-to-tail, and tail-to-tail linkages of the 2,5-benzimidazole groups unlike AB-PBI, which contains only head-to-tail linkages or benzimidazole-to-phenyl linkages. By inducing a new sequence in the AB polymer, two new types of bonds were added: benzimidazole-to-benzimidazole and phenyl-to-phenyl bonds. To induce this sequence change, a novel monomer, 2,2′-bisbenzimidazole-5,5′-dicarboxylic acid (Figure 19.28), was prepared and polymerized with 3,3′4,4′-tetraaminobiphenyl through the PPA process. Acid doping levels were reportedly 24–37 moles PA/PRU. The optimal monomer concentration for *i*-AB-PBI was found to be approximately 6 wt%. At this weight percentage, IVs were ~3 dL g^{-1} and showed conductivities of

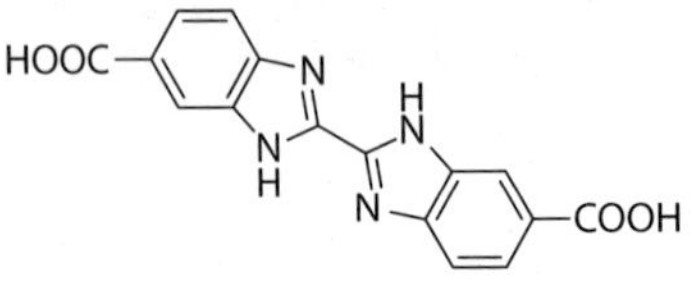

FIGURE 19.28 Chemical structure of 2,2′-bisbenzimidazole-5,5′-dicarboxylic acid.

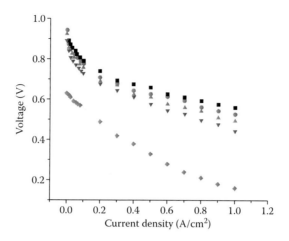

FIGURE 19.29 Polarization curves for *i*-AB-PBI membranes with stoichiometric flows of 1.2:2.0 hydrogen/oxygen, respectively. (Squares: 180°C, circles: 160°C, triangles: 140°C, inverted triangles: 120°C, diamonds: reference data S. Yu et al., *Fuel Cells* 8: 165, 2008 for AB-PBI at 130°C.) (From A. L. Gulledge et al., *J. Polym. Sci. Part A* 50: 306, 2012.)

>0.2 S cm^{-1} [105]. The conductivities are significantly higher than the conductivities of AB-PBI. Due to the high conductivity of *i*-AB-PBI, it was a good candidate for fuel cell testing. Figure 19.29 shows the polarization curve for *i*-AB-PBI and gives a comparison to AB-PBI. When fuel cell performance was evaluated using *i*-AB-PBI, the output voltage was found to be 0.65 V at 0.2 A cm^{-2} at 180°C when operating under hydrogen and air [105]. This was conducted over a period of 3500 h. These results show that *i*-AB-PBI is a promising candidate for use in high-temperature PEM fuel cells.

19.8 PBI-BASED MEMBRANES FOR PERVAPORATION SEPARATION

Because of their outstanding chemical resistance and mechanical and thermal stability, PBI membranes were used not only in high-temperature PEM fuel cells but also for many other separation applications such as RO [44–46,106–110], nanofiltration (NF) [111–114], ultrafiltration [115], forward osmosis [81,114], pervaporation [27,116–123], and gas separation [62,83,124–138], particularly for applications in aggressive environments.

Since the early 1970s, PBI membranes have been developed for RO and hemodialysis applications. Model et al. [44–46], as well as Brinegar [106], summarized most of the early development. Table 19.19 illustrates a comparison between PBI and cellulose acetate (CA) membranes for RO

TABLE 19.19
Comparative Performance of PBI and CA RO Membranes

Operating Temp. (°C)	Testing Time (h)	Final Flux of CA Membranes (gal/ft^2-day)	Final Flux of PBI Membranes (gal/ft^2-day)
70	5	12	16
120	5	20	17
167	20	5	19
194	5	0	25

Source: F. S. Model and L. A. Lee, PBI reverse osmosis membranes: An initial survey, *Reverse Osmosis Membrane Research* (H. K. Londale and H. E. Podall, eds.), Plenum Press, New York, p. 285, 1972.

application [45]. It shows that PBI has comparable RO performance as compared to CA at ambient temperatures, but outperforms CA at elevated temperatures; this is the uniqueness of PBI membranes. In the 1980s, Sawyer and Jones [107] reported the detailed morphologies of low-flux and high-flux PBI membranes, whereas Bower and Rafalko [108] as well as Sansone [109] at Hoechst Celanese invented N-substituted or ethylene-carbonated modified PBIs for RO and ultrafiltration applications. Calundann and Chung [110] developed microporous PBI membranes with a narrow pore size distribution. PBI NF membranes have also been developed recently [111–113] for the removal of heavy metal ions and pharmaceutical residuals from wastewater. Here we will only focus on PBI applications as the membrane material for pervaporation applications in this section and gas separations in Section 19.9.

Pervaporation is a membrane-based technique for liquid separation. The liquid feed solution is circulated and in direct contact with the upstream side of the membrane, while vacuum or inert gas is applied on the downstream side of the membrane. The driving force for the feed component to transport through the membrane is the chemical potential (partial vapor pressure) difference of the components between the feed and permeate sides. Thus, the separation is achieved by the differences in the diffusivities and solubilities of the components. Compared with those conventional techniques such as distillation, liquid–liquid extraction, etc., pervaporation has much higher separation efficiency, low energy consumption, simple equipment, low capital cost, and no pollution. The key factor of the pervaporation separation technology lies in the membrane. Its structure and physicochemical properties play determining factors on the pervaporation performance. In addition to achieving good selectivity and permeability, a desirable pervaporation membrane material must have good chemical and thermal stability in order to survive in the harsh environment.

As a high-performance aromatic polymeric material, many PBI membranes have been reported for pervaporation separation recently [27,116–123] for organic dehydration such as EG, acetone, acetic acid, etc., as well as some organic–organic separation. The rigid chemical structure, high hydrophilicity, and good spinnability all make PBI a promising membrane material for pervaporation dehydration. Besides, PBI is known to absorb 15 wt% water at equilibrium, and the water in PBI is mobile [48]. Therefore, water can preferentially permeate the PBI membrane due to its stronger affinity with PBI molecules and smaller molecular size relative to most organics in pervaporation dehydrations.

19.8.1 PBI-Based Membranes with Various Modifications

In spite of the many advantages of PBI, it has some weaknesses: (1) thin phase-inversed PBI membranes are brittle after drying; (2) PBI membranes can be easily swollen in aqueous solutions, resulting in a decreased membrane selectivity and unstable long-term performance; and (3) the high cost of PBI is another issue. Various modification methods including cross-linking, surface modification, blending, and thermal treatment have been proposed to overcome these issues and produce useful PBI membranes for pervaporation as follows.

19.8.1.1 Cross-Linking

P-xylene dichloride was employed as a cross-linking agent for the PBI selective layer of the PBI/P84 dual-layer hollow fibers in a recent study for acetone dehydration [119]. The cross-linking mechanism involves a reaction between the N–H functional groups of PBI molecules and the chlorine functional group of *p*-xylene dichloride to form the 1-methylimidazole groups as illustrated in Figure 19.30 [112]. Using a feed solution of 85/15 wt% acetone/water at 50°C, the pervaporation results demonstrated that the cross-linking modification can improve the separation factor significantly to 498 from 7.19 of the uncross-linked membrane, while the flux decreased to 300 $g/m^2 \cdot h$ from 1243 $g/m^2 \cdot h$. Clearly, the cross-linking modification can reduce the PBI membrane swelling and result in an enhanced pervaporation membrane for acetone–water separation.

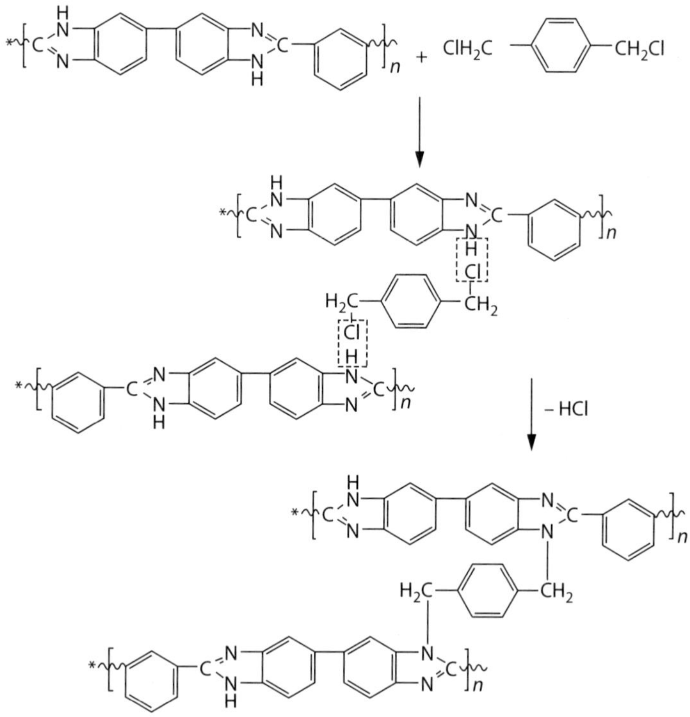

FIGURE 19.30 Proposed mechanism for the chemical cross-linking modification of PBI using *p*-xylene dichloride. (Reprinted from *Chem. Eng. Sci.*, 61, K. Y. Wang, Y. C. Xiao, and T. S. Chung, Chemically modified polybenzimidazole nanofiltration membrane for the separation of electrolytes and cephalexin, 5507, Copyright 2006, with permission from Elsevier.)

19.8.1.2 Sulfonation

Sulfonation modification of the PBI membrane surface was conducted recently by Wang et al. [116] for acetic acid dehydration. Figure 19.31 illustrates the sulfonation mechanism. Successful attachment of sulfonate groups to the imidazole ring of the PBI membrane not only enhances membrane hydrophilicity but also lowers the membrane's affinity toward acetic acid. As a result, both the flux and separation factor are enhanced significantly. The best pervaporation performance of the sulfonated PBI membrane

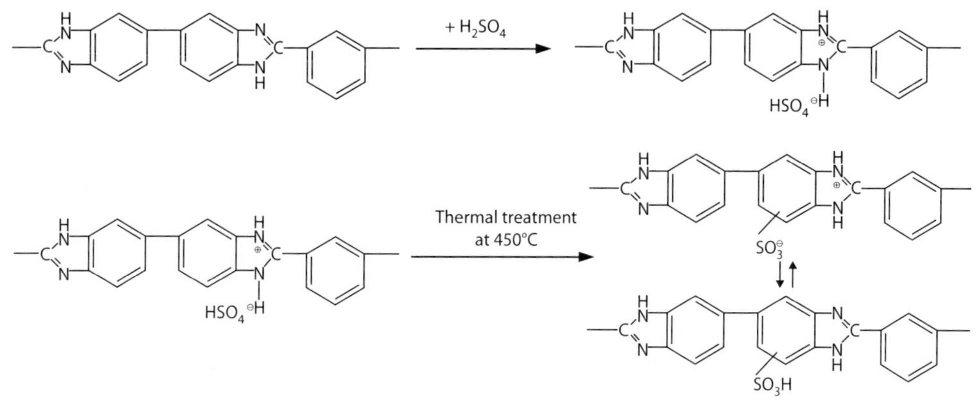

FIGURE 19.31 Sulfonation mechanism of PBI material. (Reprinted from *J. Membr. Sci.*, 415–416, Y. Wang, T. S. Chung, and M. Gruender, Sulfonated polybenzimidazole membranes for pervaporation dehydration of acetic acid, 486, Copyright 2012, with permission from Elsevier.)

has a flux of 207 g/m²·h and a separation factor of 5461 for the dehydration of a 50/50 wt% acetic acid/water feed solution at 60°C, which not only outperforms the conventional distillation process but also surpasses most other polymeric pervaporation membranes reported in the literature.

19.8.1.3 Polymer Blending

PBI/Matrimid (a commercial PI) blend membranes have been reported for the pervaporation dehydration of *tert*-butanol [122] and the separation of toluene/iso-octane mixture [27]. Because of hydrogen bonding interaction between the functional groups of these two polymers (Figure 19.32), PBI and Matrimid are completely miscible in all compositions at the molecular level. The incorporation of a small amount (up to 3.55%) of PBI into Matrimid was found to improve membrane hydrophilicity and stabilize Matrimid's chains for high-temperature pervaporation. As a consequence, the blend membrane has higher separation performance for the dehydration of *tert*-butanol/water mixtures. Similar improving mechanisms can be observed for the toluene/iso-octane separation by Kung et al. [27] using blend hollow fiber membranes. The hollow fiber comprising 10 wt% of PBI exhibits the best performance with a separation factor of 200 and a flux of 1.35 kg/m²·h for a feed solution of 50/50 wt% toluene/iso-octane at 60°C. The PBI/Matrimid blend may also have better compatibility with toluene, which favors the solubility selectivity of toluene over iso-octane.

19.8.1.4 Mixed-Matrix Membranes

Mixed-matrix membranes (MMMs) formed by incorporating as-synthesized wet-state ZIF nanoparticles into the PBI matrix have been studied by Shi et al. [120,121] for alcohol dehydration. ZIF-8 nanoparticles, with their high thermal stability, chemical resistance, and excellent compatibility with PBI, have shown their great potential to enhance PBI separation performance. Unlike most conventional MMMs, a homogeneous distribution of ZIF nanoparticles in the PBI matrix could be observed with the ZIF content not exceeding 58 wt% (as shown in Figure 19.33). This is due to the fact that the same solvent (i.e., NMP) is used, and both ZIF-8 and PBI molecules contain common imidazole functional groups. Compared with neat PBI membranes, PBI/ZIF-8 MMMs with 33.7 wt% ZIF-8 showed a fourfold increase in the water permeability (from 11.6 to 81 g/m² h) and a relative stable selectivity of about 3400 for the dehydration of 85 wt% *n*-butanol aqueous solution at 60°C [120]. The great improvement with the aid of ZIF nanoparticles in separation performance was contributed by (1) the reduction in the transport energy barrier of penetrants and (2) the suppressed solvent-induced swelling. By means of positron annihilation lifetime spectroscopy (PALS), it was found that the free volume diameter of the PBI membrane increased slightly from 4.56 to 4.78 Å together with a substantial increase in the fractional free volume (FFV) from 1.64% to 5.22% because of large cavities of ZIF-8 particles. Therefore, the MMMs have higher pervaporation permeability. In addition, sorption data confirmed that the ethanol-, methanol- and water-induced

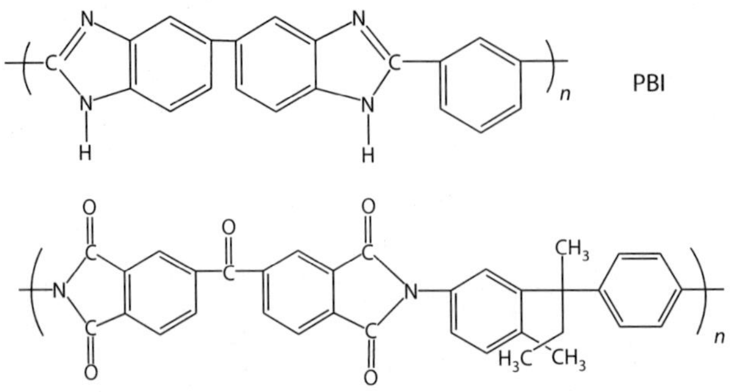

FIGURE 19.32 Hydrogen bonding interaction between PBI and Matrimid. (Reprinted from *J. Membr. Sci.*, 271, T. S. Chung, W. F. Guo, and Y. Liu, Enhanced matrimid membranes for pervaporation by homogeneous blends with polybenzimidazole (PBI), 221, Copyright 2006, with permission from Elsevier.)

Polybenzimidazoles

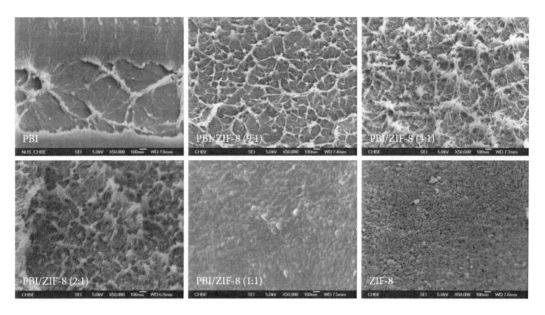

FIGURE 19.33 FESEM morphologies of PBI/ZIF-8 membranes with various ZIF-8 loadings. (Reprinted from *J. Membr. Sci.*, 415–416, G. M. Shi, T. Yang, and T. S. Chung, Polybenzimidazole (PBI)/zeoliticimidazolate frameworks (ZIF-8) mixed-matrix membranes for pervaporation dehydration of alcohols, 577, Copyright 2010, with permission from Elsevier.)

membrane swelling, especially the water-induced swelling, could be effectively suppressed in the MMMs because of the hydrophobic nature and rigid structure of ZIF-8 particles [121].

19.8.2 PBI Hollow Fiber Membranes

Compared with flat-sheet membranes, hollow fiber membranes exhibit substantial enhancement in the permeation flux because of the provision of larger surface area, less transport resistance, and lower swelling. Additionally, hollow fibers have self-contained vacuum channels where the feed can be supplied from the shell side while vacuum is applied from the lumen side, or vice versa. The porous and vacuum-dry substructure of asymmetric hollow fibers also helps reduce the swelling in the selective layer, and thus achieves a higher separation factor.

Recently, three types of PBI membranes were developed for pervaporation dehydration of EG [117,118], namely, dense flat-sheet PBI membranes and PBI single-layer and PBI/polyetherimide (PEI) dual-layer hollow fiber membranes. PBI flat-sheet dense membranes had the lowest separation performance due to the dense morphology and severe swelling. The single-layer PBI hollow fibers showed much improved separation performance in both permeation flux (1147 g/m^2·h) and separation factor [116], but they had very low tensile strains and were very fragile. It was difficult to fabricate modules from these weak fibers. The PBI/PEI dual-layer hollow fiber membranes had the best separation performance due to the unique combination of both material strengths and physicochemical properties of the PBI outer selective layer and the less swelling characteristics of the PEI supporting layer via dual-layer coextrusion. The developed PBI/PEI dual-layer hollow fiber membrane had a separation factor up to 4500 and a flux up to 186 g/m^2 h, which were far better than most other polymeric membranes [117]. Figure 19.34 shows its morphology.

PBI/P84 dual-layer hollow fiber membranes were also developed for pervaporation dehydration of tetrafluoropropanol (TFP) [123] and acetone [119] using P84 co-polyimide as the support layer. Besides good solvent resistance and thermal stability, P84 has excellent antiswelling properties and good compatibility with PBI. The PBI/P84 membrane has superior separation performance to both the PBI single-layer and P84 single-layer hollow fiber membranes. Clearly, there is great perspective

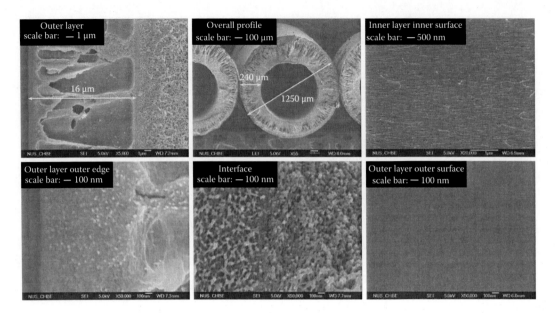

FIGURE 19.34 FESEM images of the membrane morphology of PBI/PEI dual-layer hollow fibers. (Reprinted from *J. Membr. Sci.*, 363, Y. Wang, M. Gruender, and T. S. Chung, Pervaporation dehydration of ethylene glycol through polybenzimidazole (PBI)-based membranes. 1. Membrane fabrication, 149, Copyright 2010, with permission from Elsevier.)

to develop PBI dual-layer hollow fiber membranes for pervaporation separations by using other materials as substrates. Not only can it reduce the PBI material cost, but also it can enhance the overall separation performance.

19.9 PBI-BASED MEMBRANES FOR GAS SEPARATION

PBI-based membranes have emerged as gas separation membranes for energy-related applications such as precombustion and postcombustion CO_2 capture, nitrogen/oxygen separation, acid gas removal from natural gas, separation of hydrogen from its mixtures with nitrogen, CO_2, CO, or hydrocarbons, and others.

Gas transport across polymeric membranes follows the solution-diffusion mechanism, and the gas permeability is a product of diffusion coefficient and solubility coefficient. For industrial gas separation, separation at high temperatures is advantageous since neither cooling step nor energy loss would be involved. However, most commercial polymers cannot survive at high temperatures. This makes PBI a promising membrane material for gas separation at elevated temperatures exceeding 150°C. Not only does it have remarkable thermal stability, but also it has notably high intrinsic gas-pair selectivity. PBI has a low gas permeability at room temperature, but its rigid structure makes it suitable for diffusivity-based separations at high temperatures.

In order to develop a high-performance PBI membrane for gas separation, various molecular design and modifications were proposed to improve its low gas permeability, such as monomer optimization [124,125], N-substitution [126–128], chemical cross-linking [62,129], polymer blending [62,130,131], hollow fiber engineering [129,131,132], and addition of inorganic particles [133–138].

19.9.1 Monomer-Level Optimization

By hindering chain packing and inhibiting chain flexibility simultaneously, attempts have been made to molecularly optimize the monomers for PBI syntheses [124,125]. Figure 19.35 shows the chemical structures of some selected PBI polymers, and their corresponding gas separation performance

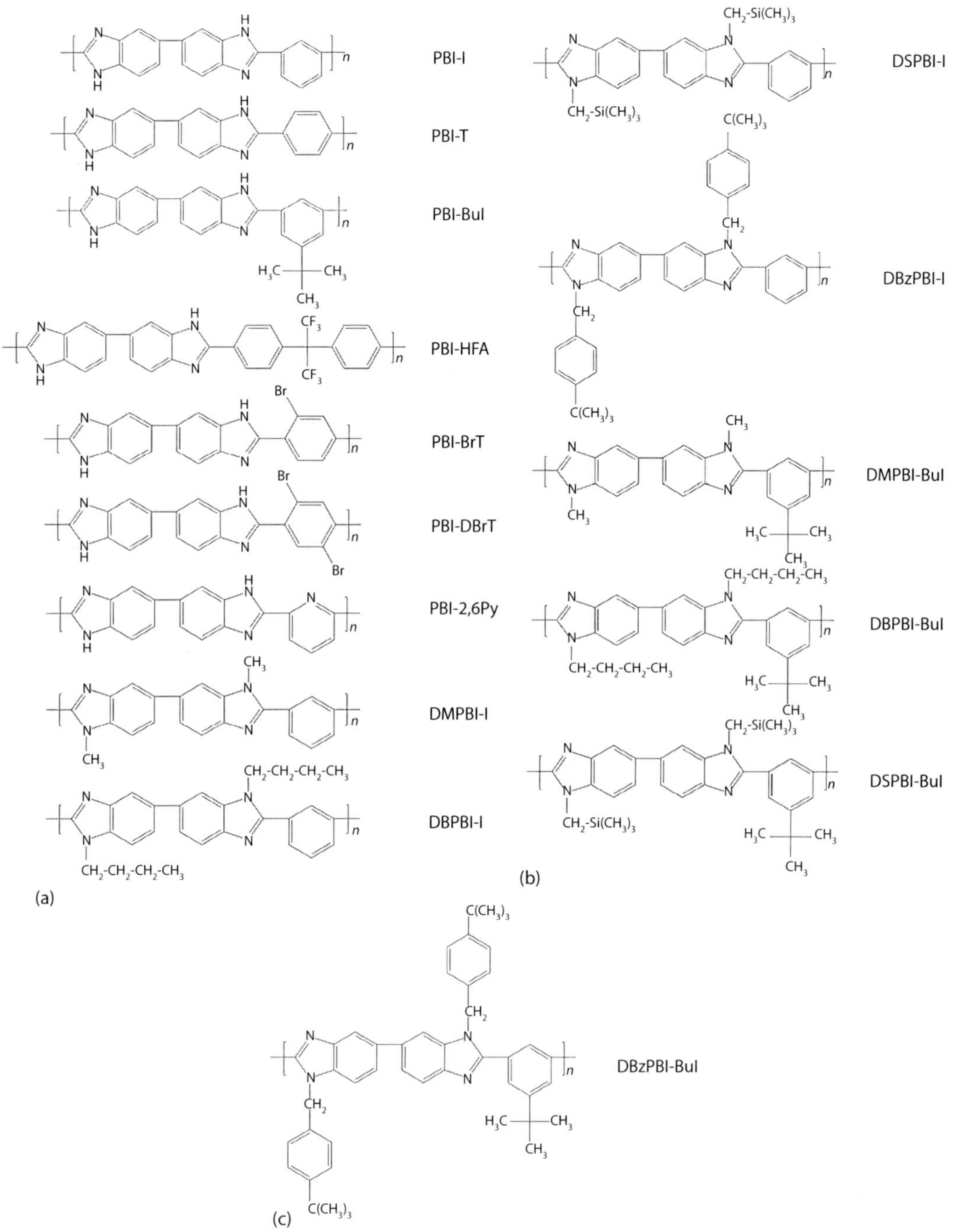

FIGURE 19.35 Chemical structures of PBIs in Table 19.20: (a) PBIs synthesized from 3,3'-diaminobenzidine (DAB) and dicarboxylic acid, (b) PBI-I with N-substitutions, and (c) PBI-BuI with N-substitutions.

is given in Table 19.20. By introducing bulky groups to some polymer chains, free volumes increase without serious loss in the chain rigidity; thus, modified PBI membranes have higher gas permeability.

Through the variation of the acid moiety during polymer syntheses, the optimized PBI polymers showed a reduced chain packing density and better solvent solubility, but slightly lower thermal stability [124,125]. For example, PBI made from 4,4'-(hexafluoroisopropylidene)bis(benzoic acid) (HFA)

TABLE 19.20
Gas Separation Performance of Pristine PBI Membranes

Material	P (atm)	Permeability (Barrer)				Selectivity				Reference
		H_2	CO_2	O_2	N_2	CH_4	H_2/CO_2	CO_2/N_2	CO_2/CH_4	H_2/O_2	
PBI-I	19.4	0.63	0.16	0.015	0.0048	0.0018	3.8	33.0	89.0	42.0	[125,126]
PBI-T	19.4	0.16	–	0.004	–	–	–	–	–	55.0	[125]
PBI-BuI	19.4	10.66	1.91	0.42	0.06	0.05	5.58	32.0	37.0	25.4	[124,125]
PBI-HFA	19.4	12.15	2.91	0.60	0.13	0.07	4.18	22.0	41.0	20.1	[124,125]
PBI-BrT	19.4	0.38	–	0.006	–	–	–	–	–	60.0	[125]
PBI-DBrT	19.4	1.89	–	0.07	–	–	–	–	–	28.0	[125]
PBI-2,6Py	19.4	1.38	–	0.045	–	–	–	–	–	31.0	[125]
DMPBI-I	19	2.18	0.19	0.052	0.005	–	11.3	38.7	–	41.9	[126,128]
DBPBI-I	19	6.54	1.79	0.36	0.09	0.06	3.7	20.0	32.4	18.2	[126,128]
DSPBI-I	19	6.81	1.62	0.38	0.07	0.04	4.2	22.1	43.0	17.9	[126,128]
DBzPBI-I	19	22.85	6.24	1.73	0.43	0.23	3.7	14.5	26.8	13.2	[126,128]
DMPBI-BuI	19	13.06	5.62	0.91	0.21	0.15	2.3	27.2	37.8	14.4	[126,128]
DBPBI-BuI	19	26.90	9.11	2.39	0.56	0.46	3.0	16.3	19.7	11.3	[126,128]
DSPBI-BuI	19	39.16	7.81	3.50	0.71	0.63	5.0	11.0	12.3	11.2	[126,128]
DBzPBI-BuI	19	46.45	21.32	6.27	1.51	1.37	2.2	14.1	15.5	7.4	[126,128]

Note: Testing temperature: 35 °C; P, transmembrane pressure; 1 Barrer = 1×10^{-10} cm^3 (STP) cm cm^{-2} s^{-1} cm Hg^{-1}.

(PBI-HFA) exhibited a H_2 permeability of 12.2 Barrer at 35°C, which was an increment of nearly 20-fold as compared to the conventional PBI [124]. However, since the glass transition temperature (T_g) of PBI-HFA dropped to 330°C (i.e., a T_g drop of 86°C from the conventional PBI of 416°C), there are modest decreases in its thermal stability and gas selectivity for industrial applications at high temperatures.

19.9.2 N-Substitution Modification

Similar to the aforementioned monomer-level optimization, the postmodification by N-substitution [126–128] is an effective modification method to improve the gas permeability of PBI membranes. By incorporating bulky side groups, the N-substitution aimed to break the intermolecular hydrogen bonding and disrupt chain packing so that the physicochemical and gas permeation properties can be enhanced. For example, both the glass transition temperature and degradation temperature of N-substituted AB-PBI dropped due to the increase in chain flexibility [127]. In spite of the decline in gas selectivity, the increase in gas permeability was observed. Gas sorption capability also increased because of the looser chain packing with the inclusion of bulky alkyl groups and weaker hydrogen bonding interactions. Solubility enhancement in some common solvents (chloroform, trichloroethylene, etc.) were also noticed.

19.9.3 Chemical Cross-Linking

Similar to previous Section 19.8.1.1, cross-linking generally leads to a higher selectivity but a lower permeability. For H_2/CO_2 gas separation [62], the *p*-xylene dichloride and *p*-xylene diamine cross-linked PBI/Matrimid blend membranes showed H_2/CO_2 selectivity of 9.43 and 26.09, respectively. Since the chemical modification may occur mainly at the PBI phase (by *p*-xylene dichloride) or the Matrimid phase (by *p*-xylene diamine), it gives us the freedom to manipulate gas pair selectivity with different chain packing density and segmental mobility.

In another study, the cross-linked PBI-HFA membranes [124] showed substantial improvement in gas selectivity but simultaneously decreased gas permeability as compared to the uncross-linked membranes.

19.9.4 Polymer Blending

Table 19.21 tabulates the gas separation performance of some selected PBI blend membranes. In the case of PBI/Matrimid blend membranes [62,130,131], the incorporation of PBI into the Matrimid matrix resulted in a higher gas selectivity but a lower gas permeability. The enhancement in selectivity is mainly attributed to a higher diffusivity selectivity since PBI has a higher molecular sieving effect, while the decline in permeability is due to the reduction in FFV and *d*-spacing. The formation of strong hydrogen bonding between blend components may also contribute to the diminishment of FFV. Delamination-free PBI/Matrimid blend dual-layer hollow fiber membranes were developed for gas separation [131]. With the aid of chemical cross-linking modification, the dual-layer hollow fiber membranes have a H_2/CO_2 selectivity of about 14.49.

Carbon molecular sieve membranes were also made from various PBI/PI blends [130]. Carbon membranes derived from PBI/Matrimid precursors exhibited better gas separation performance than other two blend precursors, namely, PBI/Torlon and PBI/P84. The newly developed PBI-based carbon membranes surpassed the trade-off lines and were of great potentials for industrial applications. The selectivity of CO_2/CH_4 and H_2/CO_2 reaches 203.95 and 33.44, respectively, while for the N_2/CH_4 separation, an unprecedented high N_2 permeability of 2.78 Barrer is achieved with a high selectivity of 7.99.

19.9.5 Mixed-Matrix Membranes

MMMs have also been proposed to overcome the low permeability of PBI-based gas membranes. Some of them and their gas separation performance are listed in Table 19.21. The key to fabricate

TABLE 19.21
Gas Separation Performance of PBI Blends and MMMs

Material	T (°C)	P (atm)	Permeability (Barrer)				Selectivity			Reference
			H_2	CO_2	N_2	CH_4	H_2/CO_2	CO_2/N_2	CO_2/CH_4	
Matrimid/PBI (75/25 wt%)	35	10	19.72	4.19	0.163	0.13	4.1	25.7	32.2	[62]
Matrimid/PBI (50/50 wt%)	35	10	13.06	2.16	0.072	0.045	6.1	30	48	[62]
Matrimid/PBI (25/75 wt%)	35	10	5.47	0.58	0.021	0.0097	9.4	27.6	59.8	[62]
PBI/PAMH (14 wt%) (proton-exchanged AMH-3, silicate)-2	35	—	1	0.025	—	—	40	—	—	[133]
PBI/PAMH (14 wt%) (proton-exchanged AMH-3, silicate)-2	200	—	18	0.82	—	—	22	—	—	[133]
PBI/SAMH (3 wt%) (swollen AMH-3, silicate)	35	—	0.9	0.03	—	—	34	—	—	[133]
PBI/SAMH (3 wt%) (swollen AMH-3, silicate)	200	—	15	0.71	—	—	21	—	—	[133]
PBI/SAMH (2 wt%) (swollen AMH-3, silicate)	35	—	1.5	0.04	—	—	34	—	—	[133]
PBI/SAMH (2 wt%) (swollen AMH-3, silicate)	200	—	18	1	—	—	18	—	—	[133]
PBI/ZIF-7 (13.7 wt%)	35	3.5	7.7	0.60	—	—	12.9	—	—	[135]
PBI/ZIF-7 (26.1 wt%)	35	3.5	15.4	1.3	—	—	11.9	—	—	[135]
PBI/ZIF-7 (26.1 wt%) (mixed gas)	180	3.5	221	26.3	—	—	8.4	—	—	[135]
PBI/ZIF-7 (42.1 wt%)	35	3.5	26.2	1.8	—	—	14.9	—	—	[135]
PBI/ZIF-7 (42.1 wt%) (mixed gas)	180	3.5	440	60.3	—	—	7.3	—	—	[135]
PBI/ZIF-8 (17.8 wt%)	35	3.5	28.5	2.2	—	—	13	—	—	[83]
PBI/ZIF-8 (20.1 wt%)	35	3.5	36.4	3	—	—	12.1	—	—	[83]
PBI/ZIF-8 (31.1 wt%)	35	3.5	82.5	6.9	—	—	12.0	—	—	[136]
PBI/ZIF-8 (31.1 wt%) (mixed gas)	230	1.0	470.5	17.9	—	—	26.3	—	—	[136]
PBI/ZIF-8 (56.1 wt%)	35	3.5	1612.8	397.6	—	—	4.1	—	—	[136]
PBI/ZIF-8 (56.1 wt%) (mixed gas)	230	1.0	2014.8	164.1	—	—	12.3	—	—	[136]
PBI/ZIF-90 (9.8 wt%)	35	3.5	12.7	0.87	—	—	14.6	—	—	[137]
PBI/ZIF-90 (24.5 wt%)	35	3.5	18.3	0.89	—	—	20.6	—	—	[137]
PBI/ZIF-90 (43.7 wt%)	35	3.5	24.5	0.98	—	—	25	—	—	[137]
PBI/ZIF-90 (43.7 wt%) (mixed gas)	180	3.5	227	16.7	—	—	13.6	—	—	[137]

Note: T, testing temperature; *P*, transmembrane pressure; 1 Barrer = 1×10^{-10} cm^3 (STP) cm cm^{-2} s^{-1} cm Hg^{-1}.

high-performance PBI-based MMMs is to disperse nanoparticles homogeneously in the MMMs via controlling the size and surface chemistry of particles plus optimizing membrane fabrication parameters [139].

The PBI-based MMM consisting of nanoporous silicate particles exhibited an extremely low H_2 permeability in spite of the enhanced H_2/CO_2 selectivity [133]. This may be due to the large particle size as well as the weak interactions between particles and the polymer. Another PBI/silica membrane was developed by Sadeghi et al. [134]. The membrane comprising 20 wt% silica particles showed simultaneous increments in CO_2 permeability and CO_2/N_2 selectivity possibly because of better interfacial interactions and smaller particle sizes.

Recently, a series of works on PBI/ZIF nanocomposite materials were reported for high-temperature hydrogen purification by Yang et al. [83,135–137]. The resultant PBI/ZIF MMMs showed very encouraging H_2/CO_2 separation performance and excellent stability under elevated temperatures due to the homogeneous microstructure and good particle–polymer interactions. The PBI/ZIF-8 (30/70 w/w%) MMM exhibited a significantly enhanced H_2 permeability of 105.4 Barrer and a high H_2/CO_2 selectivity of 12.3 at 35°C as compared with the pure PBI membrane with a permeability of 3.7 Barrer and a selectivity of 8.7 [136]. This performance was also far surpassing the well-known Robeson upper bound [140] and most other reported polymeric materials. Under the same testing condition, the 45/55 w/w% PBI/ZIF-90 MMM possessed a remarkable ideal H_2/CO_2 separation performance (a moderate H_2 permeability of 24.5 Barrer and a high H_2/CO_2 selectivity of 25.0) [137]. At an elevated test temperature of 230°C, the H_2/CO_2 selectivity of the 30/70 w/w% PBI/ZIF-8 MMM reached an impressive value of 26.3 with a H_2 permeability of around 470 Barrer. The highest H_2 permeability of 2015 Barrer was reported from the 60/40 w/w% PBI/ZIF-8 membrane. Mixed gas data also showed that the presence of CO or water vapor impurity in the feed gas stream did not significantly influence the membrane performance at 230°C. This type of PBI-based MMMs showed great potentials for hydrogen purification and CO_2 capture in industrial applications such as syngas processing, integrated gasification combined cycle (IGCC) power plant, as well as hydrogen recovery.

19.10 CURRENT PBI PRODUCTS

Not many manufacturers provide PBI products currently in the world. PBI performance Innovations Inc. is the main global manufacturer of PBI products, including powder, solution (with DMAc as the solvent), and PBI fibers in various forms. PBI fibers are available as a 1.5-denier staple fiber, and blends of PBI with other fibers, such as Nomex aramid, high-strength aramid, cotton, and rayon, are also available. In each case, PBI contributes unique thermal, flame, and chemical resistance as well as garment comfort. Some of the current applications include protective apparel, garments, and gloves for firefighters, industrial workers, race car drivers, pilots, and astronauts; it is also used in fire-blocking layers for aircraft and rocket motor insulation. Extensive flame and electric arc resistance data and test methods are provided on the PBI website [140].

Celazole PBI is a unique and highly stable linear heterocyclic polymer. In high-temperature exposure to organic chemicals, Celazole molded parts offer outstanding chemical resistance and property retention, even after extended exposures. Figure 19.36 illustrates some of these products. These Celazole stock shapes offer designers continuous temperature up to 750°F (399°C) with potential short-term exposure to 1000°F (537°C) in certain environments. Celazole is ideally suited for harsh environments ranging from oil fields to aerospace service. Some current applications include (1) rollers and pedestals for handling hot glass products; (2) seals, stem packings, and valve seals used in chemical processes; (3) insulating sprue bushings for hot runner injection molding tools; (4) fixtures for semiconductor industry; and (5) ultrasonic transducer/probe tips [141].

FIGURE 19.36 Some Celazole PBI performance parts. (From Polymer Corporation's Celazole PBI brochure, Reading, PA 19612-4235.)

ACKNOWLEDGMENTS

Professor Wang thanks Huazhong University of Science and Technology (grant no. 0124013041) and the National Science Foundation of China (grant no. 21306058). Dr. Yang and Professor Chung acknowledge also the financial support from the Singapore National Research Foundation (NRF) Competitive Research Program "New Biotechnology for Processing Metropolitan Organic Wastes into Value-Added Products" (R-279-000-311-281). Prof. Chung also thanks his former Hoechst Celanese colleagues Dr. G. W. Calundann, Dr. P. Chen, Dr. E.-W. Choe, Dr. O. R. Hughes, Dr. M. Jaffe, Dr. M. Sansone, and Dr. G. Serad for providing valuable technical data, and Daniel Conrad and Steven Quance from Polymer Corporation for providing PBI marketing information.

ABBREVIATIONS

6FCoPI	6FDA-polyimide
6FDA	4,4′-Hexafluoroisopropylidene-diphthalic anhydride
AMH-3	Three-dimensionally microporous layered material
CA	Cellulose acetate
DBPBI-BuI	PBI-BuI N-substituted by n-butyl groups
DBPBI-I	PBI-I N-substituted by n-butyl groups
DBzPBI-BuI	PBI-BuI N-substituted by 4-*tert*-butylbenzyl groups
DBzPBI-I	PBI-I N-substituted by 4-*tert*-butylbenzyl groups
DMA	Dynamic mechanical analysis
DMAc	Dimethylacetamide
DMFC	Direct methanol fuel cell
DMPBI-BuI	PBI-BuI N-substituted by methyl groups
DMPBI-I	PBI-I N-substituted by methyl groups
DMSO	Dimethylsulfoxide
DPIP	Diphenylisophthalate
DSC	Differential scanning calorimeter
DSPBI-BuI	PBI-BuI N-substituted by methylene trimethylsilyl groups
DSPBI-I	PBI-I N-substituted by methylene trimethylsilyl groups
DTA	Differential thermal analysis
EG	Ethylene glycol

FESEM	Field emission scanning electron microscopy
FFV	Fractional free volume
FTIR	Fourier transform infrared spectroscopy
HFA	Hexafluoroisopropylidene bis(benzoic acid)
HIP	Hot isostatic pressing
HMA	High-modulus polyaramide
IGCC	Integrated gasification combined cycle
IPA	Isophthalic acid
IV	Intrinsic viscosity
LCP	Liquid crystalline polymer
LOI	Limiting oxygen index
MDA	4,4'-Methylenedianiline
MEA	Membrane electrode assembly
MMM	Mixed-matrix membrane
MP	Melting point
***m*-PBI**	Poly(2,2'-*m*-phenylene-5,5'-bibenzimidazole)
NASA	National Aeronautics and Space Administration
NF	Nanofiltration
NMP	1-Methyl-2-pyrrolidinone
ODA	4,4'-Oxydianiline
OsO$_4$	Osmium tetroxide
PA	Phosphoric acid
PAI	Polyamideimide
PALS	Positron annihilation lifetime spectroscopy
PAMH	Proton-exchanged AMH-3
PAr	Polyarylate
PBA	Poly(1,4-benzamide)
PBI	Polybenzimidazole
PBI–2,6Py	PBI based on 3,3'-diaminobenzidine and 2,6-pyridinedicarboxylic acid
PBI–BrT	PBI based on 3,3'-diaminobenzidine and 2-bromoterephthalic acid
PBI–BuI	PBI based on 3,3'-diaminobenzidine and 5-*tert*-butylisophthalic acid
PBI–DBrT	PBI based on 3,3'-diaminobenzidine and 2,5-dibromoterephthalic acid
PBI–HFA	PBI based on 3,3'-diaminobenzidine and 3,3'-(hexafluoroisopropylidene) bis(benzoic acid)
PBI-I	PBI based on 3,3'-diaminobenzidine and isophthalic acid
PBI-T	PBI based on 3,3'-diaminobenzidine and terephthalic acid
PBZT	Poly(*p*-phenylenebenzobisthiazole)
PEEK	Polyetheretherketone
PEI	Polyetherimide
PEM	Polymer electrolyte membrane
PI	Polyimide
PPA	Polyphosphoric acid
PPS	Polyphenylene sulfide
PPTA	Poly(1,4-phenyleneterephthalamide)
PRU	Polymer repeat unit
PSF	Polysulfone
PVPy	Poly(4-vinyl pyridine)
RO	Reverse osmosis
SAMH	Swollen AMH-3

TAB	3,3′,4,4′-Tetraaminobiphenyl
TEM	Transmission electron microscopy
TFP	Tetrafluoropropanol
TMA	Trimetallic anhydride
ZIF	Zeolitic imidazolate framework

REFERENCES

1. H. Vogel, and C. S. Marvel, Polybenzimidazoles, I, *J. Polym. Sci.* 50: 511 (1961).
2. H. Lee, D. Stoffey, and K. Neville, Polybenzimidazoles, *New Linear Polymers*, McGraw-Hill, New York, 1967, Chap. 9.
3. A. H. Frazer, *Polymer Reviews*, Vol. 17, Interscience, New York, 1968, p. 138.
4. E. W. Neuse, Aromatic polybenzimidazoles. Syntheses, properties, and applications, *Adv. Polym. Sci.* 47: 1 (1982).
5. E. D. Powers, and G. A. Serad, History and development polybenzimidazoles, *High Performance Polymers: Their Origin and Development* (R. B. Seymour and G. S. Kirshenbaum, eds.), Elsevier, New York, 1986, p. 355.
6. A. Buckley, D. Stuetz, and G. A. Serad, Polybenzimidazoles, *Encyclopedia of Polymer Science and Engineering* (J. I. Kroschwitz, ed.), Wiley, New York, 1987, p. 572.
7. P. E. Cassidy, *Thermally Stable Polymers*, Marcel Dekker, New York, 1980, p. 168.
8. J. P. Critchley, G. J. Knight, and W. W. Wright, Polybenzimidazoles, *Heat-Resistance Polymers*, Plenum Press, New York, 1983, p. 259.
9. M. Jaffe, P. Chen, E. W. Choe, T. S. Chung, and S. Makhija, High performance polymer blends, *Adv. Polym. Sci.* 117: 297 (1994).
10. E. W. Choe, Catalysts for the preparation of polybenzimidazoles, *J. Appl. Polym. Sci.* 53: 497 (1994).
11. D. J. Jones, and J. Roziere, Recent advances in the functionalisation of polybenzimidazole and polyetherketone for fuel cell applications, *J. Membr. Sci.* 185: 41 (2001).
12. Q. Li, R. He, J. O. Jensen, and N. J. Bjerrum, PBI-based polymer membranes for high temperature fuel cells—Preparation, characterization and fuel cell demonstration, *Fuel Cells* 4: 147 (2004).
13. J. A. Asensio, and P. Gomez-Romero, Recent developments on proton conducting poly(2,5-benzimidazole) (ABPBI) membranes for high temperature polymer electrolyte membrane fuel cells, *Fuel Cells* 5: 336 (2005).
14. J. Mader, L. Xiao, T. J. Schmidt, and B. C. Benicewicz, Polybenzimidazole/acid complexes as high-temperature membranes, *Fuel Cells II, Advances in Polymer Science*, Vol. 216 (G. G. Scherer, ed.), Springer-Verlag, Berlin Heidelberg, 2008, p. 63.
15. T. Fujigaya, and N. Nakashima, Fuel cell electrocatalyst using polybenzimidazole-modified carbon nanotubes as support materials, *Adv. Mater.* 25: 1666 (2013).
16. T. S. Chung, A critical review of polybenzimidazoles: Historical development and future R & D, *J. Macromol. Sci. Rev. M.* C37: 277 (1997).
17. T. D. Dang, N. Venkat, and J. E. Mark, Rigid-rod polybenzimidazoles (PBIs): A concise review of their synthesis, properties, processing and applications, *Polyimides and Other High-Temperature Polymers: Synthesis, Characterization and Applications* (K. L. Mittal, ed.), VSP International Science Publishers, 2009, p. 145.
18. B. C. Ward, and L. P. DiSano, Polybenzimidazoles, *Engineered Material Handbook*, Vol. 2, Engineering Plastics, ASM International, (C. Dostal, ed.) Materials Park, OH, 1987, p. 147.
19. L. Leung, D. J. Williams, F. E. Karasz, and W. J. MacKnight, Miscible blends of polybenzimidazole with aromatic polyimide, *Polym. Bull.* 16: 457 (1986).
20. S. Stankovic', G. Guerra, D. J. Williams, F. E. Karasz, and W. J. MacKnight, Miscible blends of polybenzimidazole with a diisocyanate-based polyimide, *Polym. Commun.* 29: 14 (1988).
21. G. Guerra, D. J. Williams, F. E. Karasz, and W. J. MacKnight, Miscible polybenzimidazole blends with a benzophenone-based polyimide, *J. Polym. Sci. Polym. Phys.* 26: 301 (1988).
22. G. Guerra, S. Choe, D. J. Williams, F. E. Karasz, and W. J. MacKnight, Fourier transform infrared spectroscopy of some miscible polybenzimidazole/polyimide blends, *Macromolecules* 21: 231 (1988).
23. P. Musto, F. E. Karasz, and W. J. MacKnight, Hydrogen bonding in polybenzimidazole/polyimide systems: A fourier-transform infrared investigation using low-molecular-weight monofunctional probes, *Polymer* 30: 1012 (1989).

24. E. O. Stejskal, J. Schaefer, M. D. Sefcik, and R. A. McKay, Magic-angle carbon-13 nuclear magnetic resonance study of compatibility of solid polymeric blends, *Macromolecules* 14: 275 (1981).
25. S. Choe, W. J. MacKnight, and F. E. Karasz, Thermally induced phase behaviors of aromatic polybenzimidazole/polyimide blends, *Polym. Mat. Sci. Eng.* 59: 702 (1988).
26. D. L. VanderHart, G. C. Cambell, and R. M. Briber, Phase separation behavior of polybenzimidazole and polyethermide, *Macromolecules* 25: 4734 (1990).
27. G. Y. Kung, L. Y. Jiang, Y. Wang, and T. S. Chung, Asymmetric hollow fibers by polyimide and polybenzimidazole blends for toluene/iso-octane separation, *J. Membr. Sci.* 360: 303 (2010).
28. E. Foldes, E. Fekete, F. E. Karasz, and B. Pukanszky, Interaction, miscibility and phase inversion in PBI/PI blends, *Polymer* 41: 975 (2000).
29. Z. L. Xu, T. S. Chung, K. C. Loh, and B. C. Lim, Polymeric asymmetric membranes made from poly-etherimide/polybenzimidazole/poly(ethylene glycol) (PEI/PBI/PEG) for oil-surfactant-water separation, *J. Membr. Sci.* 158: 41 (1999).
30. T. S. Chung, and Z. L. Xu, Asymmetric hollow fiber membranes prepared from miscible polybenzimidazole and polyetherimide blends, *J. Membr. Sci.* 147: 35 (1998).
31. T. K. Ahn, M. Kim, and S. Choe, Hydrogen-bonding strength in the blends of polybenzimidazole with BDTA- and DSDA-based polyimides, *Macromolecules* 30: 3369 (1997).
32. Y. Wang, S. H. Goh, and T. S. Chung, Miscibility study of Torlon polyamide-imide with Matrimid® 5218 polyimide and polybenzimidazole, *Polymer* 48: 2901 (2007).
33. T. S. Chung, and P. N. Chen Sr., Polybenzimidazole (PBI) and polyarylate blends, *J. Appl. Polym. Sci.* 40: 1209 (1990).
34. T. S. Chung, and P. N. Chen Sr., Film and membrane properties of polybenzimidazole (PBI) and polyarylate blends, *Polym. Eng. Sci.* 30: 1 (1990).
35. T. S. Chung, and F. K. Herold, High-modulus polyaramide and polybenzimidazole blend fibers, *Polym. Eng. Sci.* 31: 1950 (1991).
36. T. S. Chung, F. K. Herold, and P. C. Chen Sr., Fibers from blends of PBI with polyaramide and polyarylate, *Proceedings of Fiber Producer Conf.*, Greenville, SC, May 1991, 3A-2.
37. C. Arrieta, E. David, P. Dolez, and T. Vu-Khanh, X-ray diffraction, Raman, and differential thermal analyses of the thermal aging of a Kevlar (R)-PBI blend fabric, *Polym. Compos.* 32: 362 (2011).
38. C. Arrieta, E. David, P. Dolez, and T. Vu-Khanh, Thermal aging of a blend of high-performance fibers, *J. Appl. Polym. Sci.* 5: 3031 (2010).
39. H. Chedron, M. Haubs, F. Herod, A. Schneller, O. Herrmann-Schonherr, and R. Wagener, Miscible blends of PBI and polyaramides with polyvinylpyrrolidone, *J. Appl. Polym. Sci.* 53: 507 (1994).
40. S. Makhija, E. Pearce, T. K. Kwei, and F. Liu, Miscibility studies in blends of polybenzimidazole and poly(4-vinyl pyridine), *Polym. Eng. Sci.* 30: 798 (1990).
41. T. S. Chung, M. Glick, and E. Powers, PBI and polysulfone blend fibers, *Polym. Eng. Sci.* 33: 1042 (1993).
42. T. S. Chung, The effect of LiCl on PBI and polysulfone blend fibers, *Polym. Eng. Sci.* 34: 428 (1994).
43. T. S. Chung, C. M. Tun, K. P. Pramoda, and R. Wang, Novel hollow fiber membranes with defined unit-step morphological change, *J. Membr. Sci.* 1: 123 (2001).
44. F. S. Model, and L. A. Lee, An investigation of PBI hollow fiber reverse osmosis membranes, *Org. Coat. Plast. Chem.* 32: 383 (1972).
45. F. S. Model, and L. A. Lee, PBI reverse osmosis membranes: An initial survey, *Reverse Osmosis Membrane Research* (H. K. Londale and H. E. Podall, eds.), Plenum Press, New York, 1972, p. 285.
46. F. S. Model, H. J. Davis, and P. A. Sessa, Preparation and evaluation of optimized hemodialysis membranes, Annual Report, NIH-NIAMDD-73-2200, U. S. Dept. of Health, Education and Welfare, 1973.
47. O. R. Hughes, P. N. Chen, W. M. Cooper, L. P. DiSano, E. Alvarez, and T. E. Andres, PBI powder processing to performance parts, *J. Appl. Polym. Sci.* 53: 485 (1994).
48. N. W. Brooks, R. A. Duckett, J. Rose, I. M. Ward, and J. Clements, An n.m.r. study of absorbed water in polybenzimidazole, *Polymer* 34: 4038 (1993).
49. K. C. Brinker, and I. M. Robinson, Linear polybenzimidazoles, U.S. Patent 2,895,949 (1959).
50. H. Vogel, and C. S. Marvel, Polybenzimidazoles, II, *J. Polym. Sci. A* 1: 1531 (1963).
51. A. B. Conciatori, E. C. Chenevey, T. C. Bohrer, and A. E. Prince Jr., Polymerization and spinning of PBI, *J. Appl. Polym. Sci.* 19: 49 (1967).
52. E. C. Chenevey, and A. B. Conciatori, Process for the polymerization of aromatic polybenzimidazoles, U.S. Patent 3,433,772 (1969).
53. A. E. Prince, Process for preparing polybenzimidazoles, U.S. Patent 3,551,389 (1970).
54. E. W. Neuse, and M. S. Loonat, Two-stage polybenzimidazole synthesis via poly(azomethine) intermediates, *Macromolecules* 16: 128 (1983).

55. R. A. Brand, M. Bruma, R. Kellman, and C. S. Marvel, Low-molecular-weight polybenzimidazoles from aromatic dinitriles and aromatic diamines, *J. Polym. Sci. Chem.* 16: 2275 (1978).
56. E. W. Choe, Single-stage melt polymerization process for the production of high molecular weight polybenzimidazole, U.S. Patent 4,312,976 (1982).
57. E. W. Choe, and A. B. Conciatori, Production of improved high molecular weight polybenzimidazole with tin containing catalyst, U.S. Patent 4,452,971 (1984).
58. E. W. Choe, and A. B. Conciatori, Two-stage melt production of high molecular weight polybenzimidazoles with phosphorous catalyst, U.S. Patent 4,485,232 (1984).
59. E. W. Choe, and A. B. Conciatori, Two-stage high molecular weight polybenzimidazole production with phosphorous containing catalyst, U.S. Patent 4,535,144 (1985).
60. R. W. Singleton, H. D. Noether, and J. F. Tracy, The effects of structure modifications on the critical properties of PBI fiber, *J. Polym. Sci. C* 19: 65 (1967).
61. Hoechst Celanese PBI brochure, P.O. Box 32414, Charlotte, NC 28232-9973.
62. S. S. Hosseini, M. M. Teoh, and T. S. Chung, Hydrogen separation and purification in membranes of miscible polymer blends with interpenetration networks, *Polymer* 49: 1594 (2008).
63. T. S. Chung, Z. L. Xu, and C. H. A. Huan, Halo formation in asymmetric polyetherimide and polybenzimidazole blend hollow filer membranes, *J. Polym. Sci. Pol. Phys.* 37: 1575 (1999).
64. P. Musto, F. E. Karasz, and W. J. MacKnight, Hydrogen bonding in polybenzimidazole/polyimide blends: A spectroscopic study, *Macromolecules* 24: 4762 (1991).
65. S. Choe, and T. K. Ahn, Thermally induced phase behaviors of aromatic polybenzimidazole/polyimide blends, *Polym. Mat. Sci. Eng.* 65: 331 (1991).
66. J. Grobelny, D. M. Rice, F. E. Karasz, and W. J. MacKnight, High-resolution solid state carbon-13 nuclear magnetic resonance study of polybenzimidazole/polyimide blends, *Macromolecules* 23: 2139 (1990).
67. M. Takayanagi, T. Ogata, M. Morikawa, and T. Kai, Polymer composites of rigid and flexible molecules: System of wholly aromatic and aliphatic polyamides, *J. Macromol. Sci. Phys.* B17: 591 (1980).
68. W. F. Hwang, D. R. Wiff, and C. Verschoore, Phase relationship of rigid rod polymer/flexible coil polymer/solvent ternary systems, *Polym. Eng. Sci.* 23: 784 (1983).
69. W. F. Hwang, D. R. Wiff, C. L. Benner, and T. E. Helminiak, Composites on a molecular level: Phase relationships, processing, and properties, *J. Macromol. Sci. Phys.* B22: 231 (1983).
70. W. F. Hwang, D. R. Wiff, C. Verschoore, G. E. Price, T. E. Helminiak, and W. W. Adams, Solution processing and properties of molecular composite fibers and films, *Polym. Eng. Sci.* 23: 789 (1983).
71. T. E. Helminiak, C. L. Benner, F. E. Arnold, and G. E. Husman, Aromatic heterocyclic polymer alloys and products produced therefrom, U.S. Patent 4,207,407 (1980).
72. T. E. Helminiak, Air Force rigid-rod polymer fibers and molecular composites, *Proceedings of Fiber Producer Conf.*, Greenville, SC, May 1991, 4B-2.
73. F. E. Arnold Jr., and F. E. Arnold, Rigid-rod polymers and molecular composites, *Adv. Polym. Sci.* 117: 257 (1994).
74. J. L. Kardos, and J. Raisoni, The potential mechanical response of macromolecular systems: A composite analogy, *Polym. Eng. Sci.* 15: 183 (1975).
75. G. Genc, B. Alp, D. Balkose, S. Ulku, and A. Cireli, Moisture sorption and thermal characteristics of polyaramide blend fabrics, *J. Appl. Polym. Sci.* 1: 29 (2006).
76. B. B. Ward, Effects of molecular weight and cure cycle on the properties of compress molded Celazole PBI resin, *Proceedings of the 32nd SAMPE Symposium*, Society for the Advancement of Material and Process Engineering, 1987, p. 853.
77. B. B. Ward, Molded Celazole PBI resin: Performance properties and aerospace-related applications, *Proceedings of the 33rd SAMPE Symposium*, Society for the Advancement of Material and Process Engineering, 1988, p. 146.
78. C. Delano, Polybenzimidazoles matrix resin and composites, *International Encyclopedia of Composites*, Vol. 6 (S. M. Lee, ed.), VCH, New York, p. 279.
79. Y. H. Lu, J. M. Chen, H. X. Cui, and H. D. Zhou, Doping of carbon fiber into polybenzimidazole matrix and mechanical properties of structural carbon fiber-doped polybenzimidazole composites, *Compos. Sci. Technol.* 68: 3278 (2008).
80. V. Datsyuk, S. Trotsenko, and S. Reich, Carbon-nanotube-polymer nanofibers with high thermal conductivity, *Carbon* 52: 605 (2013).
81. F. J. Fu, S. Zhang, S. P. Sun, K. Y. Wang, and T. S. Chung, POSS-containing delamination-free dual-layer hollow fiber membranes for forward osmosis and osmotic power generation, *J. Membr. Sci.* 443: 144 (2013).

82. J. S. Kim, and D. H. Reneker, Mechanical properties of composites using ultrafine electrospun fibers, *Polym. Compos.* 20: 1124 (1999).
83. T. X. Yang, G. M. Shi, and T. S. Chung, Symmetric and asymmetric zeoliticimidazolate frameworks (ZIFs)/polybenzimidazole (PBI) nanocomposite membranes for hydrogen purification at high temperatures, *Adv. Energy. Mater.* 2: 1358 (2012).
84. N. Iwamoto, A property trend study of polybenzimidazole using molecular modeling, *Polym. Eng. Sci.* 34: 434 (1994).
85. National Petroleum Council, Hard truths: Facing hard truths about energy, http://downloadcenter.connect live.com/events/npc071807/pdf-downloads/NPC-Hard_Truths-Ch2-Supply.pdf.
86. Alternative Energy, Alternative energy solutions for the 21st century, http://www.altenergy.org/.
87. H. Zhang, and P. K. Shen, Recent development of polymer electrolyte membranes for fuel cells, *Chem. Rev.* 112: 2780 (2012).
88. B. C. H. Steele, Material science and engineering: The enabling technology for the commercialisation of fuel cell systems, *J. Mater. Sci.* 36: 1053 (2001).
89. M. Molleo, T. Schmidt, and B.C. Benicewicz, Polybenzimidazole fuel cell technology: Theory, performance, and applications, *Encyclopedia of Sustainability, Science and Technology* (R. A. Meyers, ed.), Springer Science+Business Media, New York, 2012.
90. H. Ekström, B. Lafitte, J. Ihonen, H. Markusson, P. Jacobsson, A. Lundblad, P. Jannasch, and G. Lindbergh, Evaluation of a sulfophenylated polysulfone membrane in a fuel cell at 60 to 110 °C, *Solid State Ionics* 178: 959 (2007).
91. Q. Li, R. He, J. O. Jensen, and N. J. Bjerrum, Approaches and recent development of polymer electrolyte membranes for fuel cells operating above 100°C, *Chem. Mater.* 15: 4896 (2003).
92. S. R. Samms, S. Wasmus, and R. F. Savinell, Thermal stability of proton conducting acid doped polybenzimidazole in simulated fuel cell environments, *J. Electrochem. Soc.* 143: 1225 (1996).
93. L. Xiao, H. Zhang, E. Scanlon, L. S. Ramanathan, E.-W. Choe, D. Rogers, T. Apple, and B. C. Benicewicz, High-temperature polybenzimidazole fuel cell membranes via a sol–gel process, *Chem. Mater.* 17: 5328 (2005).
94. J. A. Asensio, E. M. Sanchez, and P. Gomez-Romero, Proton-conducting membranes based on benzimidazole polymers for high-temperature PEM fuel cells. A chemical quest, *Chem. Soc. Rev.* 39: 3210 (2010).
95. D. Mecerreyes, H. Grande, O. Miguel, E. Ochoteco, R. Marcilla, and I. Cantero, Porous polybenzimidazole membranes doped with phosphoric acid: Highly proton-conducting solid electrolytes, *Chem. Mater.* 16: 604 (2004).
96. W. J. Wainright, J. T. Wang, D. Weng, R. F. Savinell, and M. Litt, Acid–doped polybenzimidazoles: A new polymer electrolyte, *J. Electrochem. Soc.* 142: L121 (1995).
97. R. Bouchet, and E. Siebert, Proton conduction in acid doped polybenzimidazole, *Solid State Ionics* 118: 287 (1999).
98. J. A. Asensio, S. Borrós, and P. Gómez-Romero, Proton-conducting polymers based on benzimidazoles and sulfonated benzimidazoles, *J. Polym. Sci. Part A* 40: 3703 (2002).
99. D. C. Seel, B. C. Benicewicz, L. Xiao, and T. J. Schmidt, High-temperature polybenzimidazole-based membranes, *Handbook of Fuel Cells*, (W. Vielstich, H. Yokokawa, H.A. Gasteiger, eds.), John Wiley & Sons, Ltd. 2009, pp. 300–312.
100. C. B. Delano, R. R. Doyle, and R. J. Milligan, *United States Air Force Materials Laboratory*, AFML-TR-74-22, 1974.
101. T.-H. Kim, T.-W. Lim, and J.-C. Lee, High-temperature fuel cell membranes based on mechanically stable para-ordered polybenzimidazole prepared by direct casting, *J. Power Sources* 172: 172 (2007).
102. S. Yu, L. Xiao, and B. C. Benicewicz, Durability studies of PBI-based high temperature PEMFCs, *Fuel Cells* 8: 165 (2008).
103. S. Yu, H. Zhang, L. Xiao, E. W. Choe, and B. C. Benicewicz, Synthesis of poly(2,2′-(1,4-phenylene) 5,5′-bibenzimidazole) (para-PBI) and phosphoric acid doped membrane for fuel cells, *Fuel Cells* 9: 318 (2009).
104. J. A. Asensio, S. Borrós, and P. Gómez-Romero, Polymer electrolyte fuel cells based on phosphoric acid-impregnated poly(2,5-benzimidazole) membranes, *J. Electrochem. Soc.* 151: A304 (2004).
105. A. L. Gulledge, B. Gu, and B. C. Benicewicz, A new sequence isomer of AB-polybenzimidazole for high-temperature PEM fuel cells, *J. Polym. Sci. Part A* 50: 306 (2012).
106. W. C. Brinegar, Production of semipermeable PBI membranes, U.S. Patent 3,841,492 (1974).
107. L. C. Sawyer, and R. S. Jones, Observation on the structure of first generation PBI reverse osmosis membranes, *J. Membr. Sci.* 20: 147 (1984).

108. E. A. Bower, and J. J. Rafalko, Process for modifying polybenzimidazole polymers with ethylene carbonate, U.S. Patent 4,599,388 (1986).
109. M. J. Sansone, N-Substituted polybenzimidazole polymer, U.S. Patent 4,898,917 (1990).
110. G. W. Calundann, and T. S. Chung, Fabrication of microporous PBI membranes with narrow pore size distribution, U.S. Patent 5,091,087 (1992).
111. K. Y. Wang, and T. S. Chung, Fabrication of polybenzimidazole (PBI) nanofiltration hollow fiber membranes for removal of chromate, *J. Membr. Sci.* 281: 307 (2006).
112. K. Y. Wang, Y. C. Xiao, and T. S. Chung, Chemically modified polybenzimidazole nanofiltration membrane for the separation of electrolytes and cephalexin, *Chem. Eng. Sci.* 61: 5507 (2006).
113. K. Y. Wang, T. S. Chung, and J. J. Qin, Polybenzimidazole (PBI) nanofiltration hollow fiber membranes applied in forward osmosis process, *J. Membr. Sci.* 300: 6 (2007).
114. K. Y. Wang, Q. Yang, T. S. Chung, and R. Rajagopalan, Enhanced forward osmosis from chemically modified polybenzimidazole (PBI) nanofiltration hollow fiber membranes with a thin wall, *Chem. Eng. Sci.* 64: 1577 (2009).
115. M. J. Sansone, Process for the production of polybenzimidazole ultrafiltration membranes, U.S. Patent 4,693,824 (1987).
116. Y. Wang, T. S. Chung, and M. Gruender, Sulfonated polybenzimidazole membranes for pervaporation dehydration of acetic acid, *J. Membr. Sci.* 415–416: 486 (2012).
117. Y. Wang, M. Gruender, and T. S. Chung, Pervaporation dehydration of ethylene glycol through polybenzimidazole (PBI)-based membranes. 1. Membrane fabrication, *J. Membr. Sci.* 363: 149 (2010).
118. Y. Wang, T. S. Chung, B. Neo, and M. Gruender, Processing and engineering of pervaporation dehydration of ethylene glycol via dual-layer polybenzimidazole (PBI)/polyetherimide (PEI) membranes, *J. Membr. Sci.* 378: 339 (2011).
119. G. M. Shi, Y. Wang, and T. S. Chung, Dual-layer PBI/P84 hollow fibers for pervaporation dehydration of acetone, *AIChE J.* 58: 1133 (2012).
120. G. M. Shi, T. Yang, and T. S. Chung, Polybenzimidazole (PBI)/zeoliticimidazolate frameworks (ZIF-8) mixed matrix membranes for pervaporation dehydration of alcohols, *J. Membr. Sci.* 415–416: 577 (2010).
121. G. M. Shi, H. M. Chen, Y. C. Jean, and T. S. Chung, Sorption, swelling, and free volume of polybenzimidazole (PBI) and PBI/zeoliticimidazolate framework (ZIF-8) nano-composite membranes for pervaporation, *Polymer* 54: 774 (2013).
122. T. S. Chung, W. F. Guo, and Y. Liu, Enhanced matrimid membranes for pervaporation by homogenous blends with polybenzimidazole (PBI), *J. Membr. Sci.* 271: 221 (2006).
123. K. Y. Wang, T. S. Chung, and R. Rajagopalan, Dehydration of tetrafluoropropanol (TFP) by pervaporation via novel PBI/BTDA-TDI/MDI co-polyimide (P84) dual-layer hollow fiber membranes, *J. Membr. Sci.* 287: 60 (2007).
124. S. C. Kumbharkar, P. B. Karadkar, and U. K. Kharul, Enhancement of gas permeation properties of polybenzimidazoles by systematic structure architecture, *J. Membr. Sci.* 286: 161 (2006).
125. S. C. Kumbharkar, M. N. Islam, R. A. Potrekar, and U. K. Kharul, Variation in acid moiety of polybenzimidazoles: Investigation of physico-chemical properties towards their applicability as proton exchange and gas separation membrane materials, *Polymer* 50: 1403 (2009).
126. S. C. Kumbharkar, and U. K. Kharul, Investigation of gas permeation properties of systematically modified polybenzimidazoles by N-substitution, *J. Membr. Sci.* 357: 134 (2010).
127. S. C. Kumbharkar, and U. K. Kharul, New N-substituted ABPBI: Synthesis and evaluation of gas permeation properties, *J. Membr. Sci.* 360: 418 (2010).
128. S. C. Kumbharkar, and U. K. Kharul, N-substitution of polybenzimidazoles: Synthesis and evaluation of physical properties, *Eur. Polym. J.* 45: 3363 (2009).
129. S. C. Kumbharkar, and K. Li, Structurally modified polybenzimidazole hollow fibre membranes with enhanced gas permeation properties, *J. Membr. Sci.* 415–416: 793 (2012).
130. S. S. Hosseini, and T. S. Chung, Carbon membranes from blends of PBI and polyimides for N2/CH4 and CO_2/CH_4 separation and hydrogen purification, *J. Membr. Sci.* 328: 174 (2009).
131. S. S. Hosseini, N. Peng, and T. S. Chung, Gas separation membranes developed through integration of polymer blending and dual-layer hollow fiber spinning process for hydrogen and natural gas enrichments, *J. Membr. Sci.* 349: 156 (2010).
132. S. C. Kumbharkar, Y. Liu, and K. Li, High performance polybenzimidazole based asymmetric hollow fibre membranes for H_2/CO_2 separation, *J. Membr. Sci.* 375: 231 (2011).
133. S. Choi, J. Coronas, Z. Lai, D. Yust, F. Onorato, and M. Tsapatsis, Fabrication and gas separation properties of polybenzimidazole (PBI)/nanoporous silicates hybrid membranes, *J. Membr. Sci.* 316: 145 (2008).

134. M. Sadeghi, M. A. Semsarzadeh, and H. Moadel, Enhancement of the gas separation properties of polybenzimidazole (PBI) membrane by incorporation of silica nano particles, *J. Membr. Sci.* 331: 21 (2009).
135. T. X. Yang, Y. Xiao, and T. S. Chung, Poly-/metal-benzimidazole nano-composite membranes for hydrogen purification, *Energy Environ. Sci.* 4: 4171 (2011).
136. T. X. Yang, and T. S. Chung, High performance ZIF-8/PBI nano-composite membranes for high temperature hydrogen separation consisting of carbon monoxide and water vapor, *Int. J. Hydrogen Energy* 38: 229 (2013).
137. T. X. Yang, and T. S. Chung, Room-temperature synthesis of ZIF-90 nanocrystals and the derived nano-composite membranes for hydrogen separation, *J. Mater. Chem. A.* 1: 6081 (2013).
138. T. S. Chung, L. Y. Jiang, Y. Li, and S. Kulprathipanja, Mixed matrix membranes (MMMs) comprising organic polymers with dispersed inorganic fillers for gas separation, *Prog. Polym. Sci.* 32: 483 (2007).
139. L. M. Robeson, The upper bound revisited, *J. Membr. Sci.* 320: 390 (2008).
140. PBI Performance Products, Inc., http://www.pbiproducts.com/polymers/products/celazole-u-series.
141. Polymer Corporation' Celazole PBI brochure, Reading, PA 19612-4235.

20 Conductive Thermoplastics*

Louis M. Leung

CONTENTS

20.1 Introduction 669
    20.1.1 Development of Conductive Polymers 669
    20.1.2 Organic Semiconductors 671
20.2 Main-Chain Conductive Polymers (π-Conjugated Polymers) 672
    20.2.1 Solubilization of π-Conjugated Polymers 672
20.3 Soluble Main-Chain Conductive Polymers 674
    20.3.1 Polyethylene Dioxythiophene:Polystyrene Sulfonate Complexes 674
    20.3.2 Polyfluorenes and Their Copolymers 674
    20.3.3 Substituted Polythiophene Oligomers 675
20.4 Conductive Vinyl Polymers 676
    20.4.1 Mobility Measurements 677
    20.4.2 Hole-Transporting Vinyl Polymers 678
    20.4.3 Electron-Transporting Vinyl Polymers and Their Copolymers 680
    20.4.4 Conductive Vinyl Polymers as Emitters 682
    20.4.5 π–π Stacking Poly(Dibenzofulvene) 684
20.5 Conclusions 686
Acknowledgments 686
References 687

20.1 INTRODUCTION

20.1.1 Development of Conductive Polymers

The story regarding the discovery of conductive polymers, specifically halogen-doped polyacetylene film, is well known [1]. Its progress, however, did not end with the award of the year 2000 chemistry Nobel Prize to Alan G. MacDiarmid, Alan J. Heeger, and Hideki Shirakawa [2–9]. The discovery of traditional plastics occurred in the early twentieth century, and plastics have a distinct place from other materials of choice (such as metals, ceramic, etc.) due to their cost effectiveness, excellent processability, and dielectric properties. The contribution of plastics to the development of the modern society can best be represented by Lord Todd (Nobel laureate in chemistry 1957), then president of the Royal Society of London (1980), in answering the question "What do you think has been chemistry's biggest contribution to science, to society?" His answer was "I am inclined to think that the development of polymerization is, perhaps, the biggest thing chemistry has done, where it has had the biggest effect on everyday life. The world would be a totally different place without artificial fibers, plastics, elastomers, etc. Even in the field of electronics, what would you do without insulation? And there you come back to polymers again?" [10]. It was quite inconceivable at that time that plastic could one day replace metals such as copper, silver, and gold for its conductivity. The π-conjugated polymers, including polyacetylene, initially investigated by MacDiarmid, Heeger, and Shirakawa, however, are not truly thermoplastics [11]. These polymers

---

* Based in parts on the first-edition chapters on conducting thermoplastics and conducting thermoplastics composites.

are neither soluble nor temperature processable as their processing temperatures are well above their degradation temperatures. Methods to render some of these π-conjugated conductive polymers such as polyacetylene, poly(p-phenylene) vinylene (PPV), or polythiophene soluble will be further discussed in the following sections. The pristine (undoped) conductive polymers, however, are only semiconductive in nature. The distinction of insulator, semiconductor, and conductor can best be described by the intrinsic value of volume conductivity (σ), which has the unit siemens per meter (S/m) or siemens per centimeter (S/cm). The volume conductivity (σ) is inversely proportional to volume resistivity (ρ) (i.e., $\sigma = 1/\rho$), which has the unit ohm meter (Ω m) or ohm centimeter (Ω cm). Another relative unit, Ω/sq, is also commonly used in the industry for the comparison of surface (sheet) resistivity only. Figure 20.1 shows the relative S/m value for different metals, semiconductors, and insulators in comparison to conjugated polymers before and after doping. Although the absolute volume conductivity of most doped conductive polymers is lower than that of metals, their ratio of volume conductivity to density, however, can be one to two orders of magnitude higher due to the fact that plastics usually have lower density or specific gravity than metals. The volume resistivity (ρ) can be measured typically by employing a gold-plated four-pointed probe to measure the current (J) and voltage (V) across a bulk conductive polymer with known dimensions (width W, length L, and thickness t), in which $\rho = R (A/L) = (V/J)(W t/L)$, $R (= V/J)$ is the resistance defined by Ohm's law, and $A (= Wt)$ is the cross-section area of the sample.

The doping process for the enhancement of conductivity in π-conjugated polymers is actually a redox reaction between the small-molecule dopants and the unsaturated moieties in the π-conjugated polymers [12–15]. A typical oxidation (p-type doping) or reduction (n-type doping) reaction for polyacetylene using either iodine or sodium can be illustrated by the following redox reactions.

$$[CH]_n + 3x/2 I_2 \rightarrow [CH]_n^{x+} + x I_3^- \quad \text{(oxidative doping)}$$

$$[CH]_n + x Na \rightarrow [CH]_n^{x-} + x Na^+ \quad \text{(reductive doping)}$$

The doped polyacetylene becomes a salt in which the counters ions are either I_3^- or Na^+. The enhancement in conductivity is due to an increase in mobile charge carriers (in the format of ion–radical complexes known as polarons, bipolarons, or solitons) [16,17]. The small-molecule dopants, however, could be evaporated at elevated temperature (e.g., I_3^-) or neutralized by the environment (Na^+ by moisture in the environment). Applications of the redox-doped conductive polymers include corrosion inhibitors, antistatic coating, electromagnetic interference (EMI) shielding, electrodes for a battery or capacitor, electrochromic smart windows/displays, electroactivated actuators or sensors, etc. [2–9].

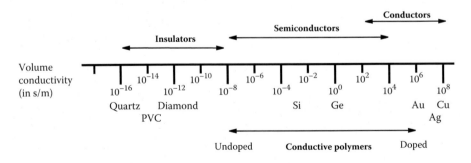

FIGURE 20.1 Conductivity (in S/m) of conductive polymers in comparison to other materials, including insulators, semiconductors, and conductors. PVC, polyvinyl chloride.

20.1.2 ORGANIC SEMICONDUCTORS

The semiconductive nature of the undoped π-conjugated polymers, however, has drawn enormous attention in the past two decades with a range of promising commercial applications [2–9]. The current progress on organic semiconductors (including the intrinsic undoped π-conjugated polymers) provided an alternative pathway to the building of integrated circuits (ICs), microelectronics, devices, and sensors, which were in the realms of traditional inorganic semiconductors such as Si, GaAs, Ge, etc. [18]. The richness of the chemistry surrounding the carbon element and its versatility, however, allow a wide range of p- and n-type organic semiconductor (including small molecules and polymers) to be designed and their properties engineered according to specific needs. In addition, organic semiconductors are compatible with high-throughput, low-temperature, and low-cost production processes, including solution-based spin/dip/silk screen coating, ink-jet printing, etc. Their structural malleability, compatibility with different substrates, and low cost make them even more suitable than traditional inorganic semiconductors to be used in applications that demand large area coverage as well as supported on flexible or foldable substrates. Some of the potential applications for organic semiconductors include back panels for active-matrix LCD (AMLCD), organic field-effect transistor (OFET) arrays, polymer light-emitting diode (PLED) displays, low-cost paper-thin illumination sources, electrophoretic display pixels, microfluidic channels, active and passive component matrices, chemical and biological sensor arrays, printable electronics, radio-frequency identification (RFID) tags and smart cards, organic photovoltaic (OPV) cells and their integrated system, etc [19–22].

Unlike conductive polymers that are characterized by either volume conductivity (σ) or volume resistivity (ρ), the critical parameter for organic semiconductor is its carrier mobility (μ) with the unit $cm^2 V^{-1} s^{-1}$. The mobility can be related to volume conductivity (σ) according to the equation $\sigma = n\mu e$, in which n is the number of charge carriers and e is the charge of an electron. Depending on the nature of the organic semiconductor, the principal charge carrier could be either holes (p-type semiconductor) or electrons (n-type semiconductor), and the mobility can be further specified as μ_h (mobility for hole carriers) or μ_e (mobility for electron carrier). In some applications such as organic light-emitting diode (OLED), a charge-balance heterojunction device required a similar order of magnitude in μ_h and μ_e for the hole (p-type) and electron (n-type) transporting layers, while some other applications such as OPV and OFET require a larger μ for effective charge separation or switching speed. Methods for the determination of mobility are more complicated than that of volume conductivity measurement and will be further elaborated in a following section.

Some of the most successful organic semiconductors have been single-crystalline small molecules due to their high charge-carrier mobility, their ability to be manipulated in the nanoscale, and the formation of uniform thin film using established vacuum vapor deposition techniques. Some of these remarkable organic semiconductors are pentacene (which has the highest mobility at 2.7 $cm^2 V^{-1} s^{-1}$) [23], oligomeric polythiophene (0.1 $cm^2 V^{-1} s^{-1}$ for α-ω-dihexyl-quinquethiophene) [24], C_{60} (0.3 $cm^2 V^{-1} s^{-1}$) [25], rubrene (30 $cm^2 V^{-1} s^{-1}$ at 200K) [26], copper phthalocyanine (CuPc, 1.0 $cm^2 V^{-1} s^{-1}$) [27], and so on. For the latest development in OPV and OFET, desirable materials are those that can be handled under ambient conditions as well as are compatible with large-throughput and large-coverage-area printing processes. Pristine conductive polymers in general have lower charge mobilities, and they have been thoroughly studied and estimated for niche applications. Traditional main-chain conjugated conductive polymers have only limited solubility. Soluble precursor methods [like ring-opening methathesis polymerization (ROMP) and sulfonium salts] allow conductive polymers with high molecular weight to be solution-processed, yet they generally require a high elimination temperature (above 150°C), and the resulting polymer can be contaminated by the elimination side products. Reducing the molecular weight (MW) (like oligomers) improved solubility, but it also reduced the mechanical strength. Attaching flexible soluble side chains or copolymerization with another monomer can improve solubility, but the long-range conductivity can be decreased due to distorted chain conformation or reduced π conjugation. Main-chain conductive polymers like polyacetylene have been attempted for semiconductor applications but were limited by their low mobility as well as poor processablility. Subsequently, progress was made on using the soluble poly(3-hexyl-thiophene) (P3HT; 0.1 cm^2 V^{-1} s^{-1}) and polythienylenevinylene (0.22 cm^2 V^{-1} s^{-1}) [28–30].

20.2 MAIN-CHAIN CONDUCTIVE POLYMERS (π-CONJUGATED POLYMERS)

20.2.1 Solubilization of π-Conjugated Polymers

Most of the π-conjugated polymers [including polyacetylene, polythiophene, polypyrrole, polyaniline, poly(phenylene vinylene), and polyphenylene, etc.] are inherently insoluble, and their melting or transition temperatures are well above their degradation temperature [31,32]. Many attempts have been made in order to render the main-chain conductive polymers soluble or thermally processable. One of the best-known methods known to conserve the characteristics of high molecular weight and maximum π conjugation along the polymer backbone is the precursor method. In this route, a high-molecular-weight yet soluble precursor polymer is synthesized using a traditional polymerization method. The precursor polymer, however, is not stable and can be transformed to the final status of the insoluble π-conjugated polymer via a temperature- or chemical-induced conversion process. The conversion is always accompanied by the elimination of small-molecular side products. The most famous Durham route or ROMP process was initially introduced by Edwards and Feast [33]. A tricyclic monomer such as 7,8-bis(trifluoromethyl)tricycle[4,2,2,0^{2,5}]deca-3,7,9-triene (CF3-TCDT) was polymerized using a 1:2 ratio of catalysis such as WCl$_6$/SnPh$_4$ or TiCl$_4$/AlEt$_3$. Uniform thin films or fibers can be prepared by dissolving the precursor polymer in solvents like acetone or chloroform. The thin films or fibers can be prestretched to align the polymer molecules so as to achieve high conductivity later upon elimination and doping. With the elimination of 1,2-bis(trifluoromethyl) benzene at a temperature from 50°C to 120°C, a high-density polyacetylene sample (mostly trans-PA due to high elimination temperature) would be obtained. Five different tricyclic monomers have been synthesized, resulting in elimination temperature that can range from 50–120°C to 175–200°C [34,35]. A "living" ROMP process has also produced soluble diblock or triblock copolymers of polyacetylene [36]. In addition, a vinyl monomer, phenyl vinyl sulfoxide, has later been introduced for the preparation of another soluble precursor to polyacetylene [37,38]. The phenyl vinyl sulfoxide is a vinyl monomer that can be polymerized using traditional addition polymerization methods, such as the free-radical or living anionic approach. The sulfoxide route produced an organic solvent soluble precursor that can be converted to polyacetylene via a zipper-like elimination process at temperatures from 80°C to 150°C. A series of polyacetylene homopolymers, random copolymers of polyacetylene and poly(phenyl vinyl sulfoxide), as well as block copolymers of polyacetylene have been synthesized, and their electrical properties as a function of the conductive fraction have been investigated and reported [37,38]. Another aqueous soluble precursor leading to the discovery of PLED is the sulfonium precursor route to PPV [39]. A water-soluble bis-sulfonium salt of *p*-xylene was polymerized under a base-catalyzed reaction under low temperature [40–43]. Uniform thin film can be casted from the polyelectrolyte precursor polymer after purification by dialysis, and the elimination temperature can reach as high as 360°C. A review on the synthesis and characterization of the poly(phenylene vinylene) can be found in the first edition of this handbook [44]. In Figure 20.2, the different precursor routes are introduced.

The other methods that can render the main-chain conjugated polymers soluble include usage of a low MW oligomer, copolymerization with another soluble monomer, or attachment of a solubilizing side chain onto the repeating unit structure [45–51]. Due to the lower molecular weight for a conductive oligomer, the oligomer has to be applied as a blend or supported on a different substrate for mechanical reinforcement. A uniform thin film from a conductive oligomer, however, can be obtained via the usage of vapor-phase vacuum evaporation. Copolymerization with another monomer may disrupt conjugation along the main chain and therefore could be detrimental to its overall electrical conductivity. But the copolymerization with another functional monomer could enhance or modify

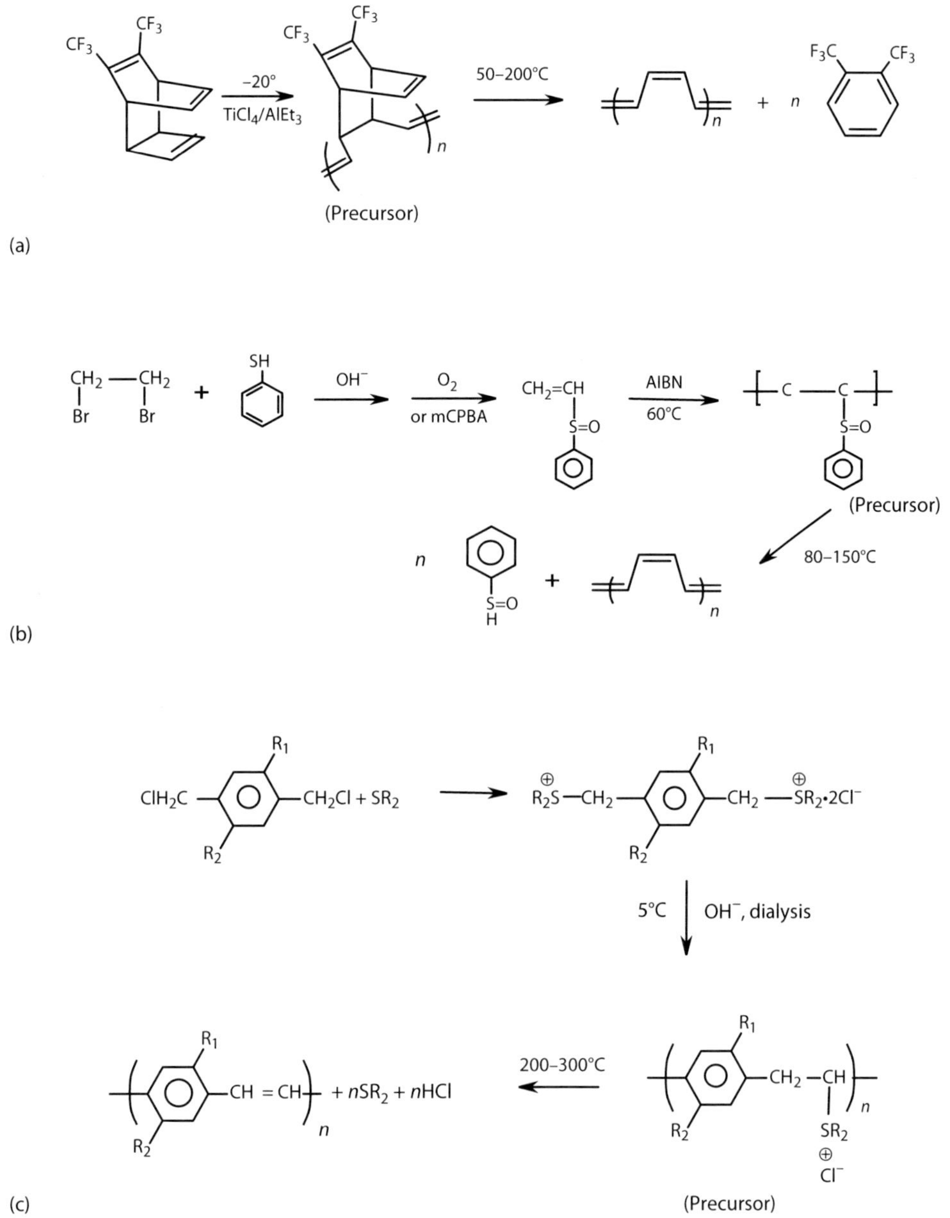

Conductive Thermoplastics

FIGURE 20.2 Precursor routes for the synthesis of (a) polyacetylene via ROMP and (b) sulfoxide approaches, and (c) the sulfonium route for the synthesis of poly(p-phenylene vinylene). mCPBA, 3-cloroperbenzoic acid.

the overall properties of the copolymers such as bandgap energy. Examples of the copolymerization of polyfluorenes or poly(3-substitued thiophene) are shown in the sections below. The usage of long solubilization side chains such as hexyl, dodecyl, or poly(ethylene-oxide) side chains has been successful in converting aromatic main-chain conductive polymers such as PPV or polythiophene to become soluble in common organic solvents [50]. Phase segregation between the aliphatic side chain and the conjugated backbone has induced some interesting transitional properties. In 3-substituted polythiophene, it was observed that the absorption maximum shifted to a lower wavelength (higher bandgap) with an increase in temperature [52]. The thermochromism phenomenon has been suggested to be the result of the stacking of the coplanar structure from the thiophene units being disrupted by increased disordering of the alkyl side chain at an increased temperature [53,54]. Only a handful of the main-chain conjugated polymers with reasonable high molecular weight remain soluble or processable, and these conductive main-chain polymers with imminent or proven commercial applications will be further discussed in the sections below.

20.3 SOLUBLE MAIN-CHAIN CONDUCTIVE POLYMERS

20.3.1 POLYETHYLENE DIOXYTHIOPHENE:POLYSTYRENE SULFONATE COMPLEXES

Polyethylene dioxythiophene (PEDOT) is a 3,4-disubstituted polythiophene. It is one of the few commercially available metal-like conductive polymer coatings. The polymer itself is not soluble, but the complex PEDOT:polystyrene sulfonate (PSS) forms a stable aqueous dispersion. The sulfonic acids can be considered as an n-type dopant for PEDOT. The polyelectrolyte is always applied in excess, with a molar ratio for PEDOT:PSS from 1:1.9 up to 1:15.2 [55,56]. A solid thin film or coating of PEDOT:PSS can be obtained by spin or dip coating, and a drying cycle up to 150°C will be applied. The conductivity of PEDOT:PSS thin film ranges from 10^{-3} to 1000 S/cm, and the conductivity is stable up to a temperature of 200°C [57]. The soluble PEDOT:PSS has found a range of commercial applications, including antistatic coating, solid electrolyte polymer capacitors, charge injection layers for OLEDs based on indium tin oxide (ITO) (ionization potential [IP] = −4.8 eV for ITO, while the work function for PEDOT:PSS = −5.0 to −5.2 eV), and transparent electrodes for a range of printable touch-screen electronic applications [58].

The monomer ethylene dioxythiophene (EDOT) can be synthesized from oxalic acid ester and thiodiacetic acid ester via a transetherification reaction. Since the monomer EDOT is only slightly soluble in water, polymerization can only be achieved in the presence of PSS. Oxidizing agents such as sodium or ammonium peroxodisulfate can be used with a small amount of catalytic Fe(III) salt for the *in situ* oxidation polymerization of EDOT in water [59]. An excellent review on the synthesis, characterization, and applications of PEDOT can be found in Ref. [60] (see Figure 20.3).

20.3.2 POLYFLUORENES AND THEIR COPOLYMERS

Poly(9,9-disubstituted-fluorene) was found to be a versatile material for PLED applications due to its excellent thermal and chemical stability, high quantum yield, as well as ability to emit color covering the complete visible spectrum [61]. The reactive 9-position carbon atom of the fluorene molecule can be modified with ease using different alkyl or alkoxy side chains for the achievement of solubility for the 9,9-disubstituted polymer [62]. The bulky side chains that also distort π–π conjugation between the adjacent coplanar fluorene moieties can be used for fine-tuning of the emission color [63].

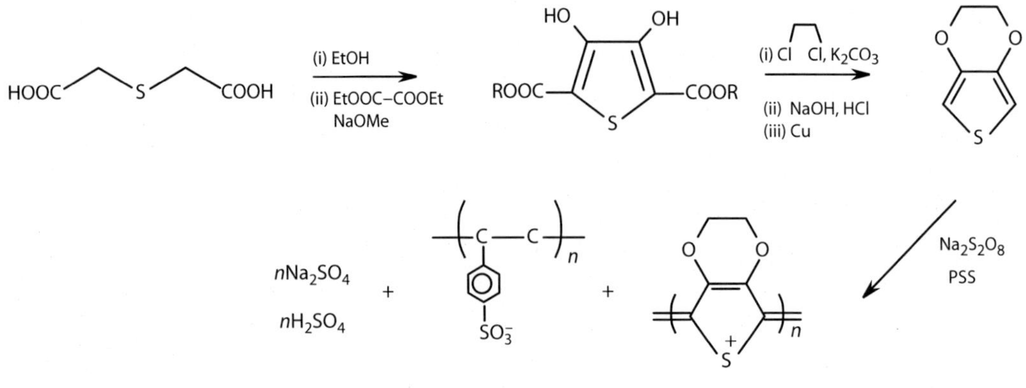

FIGURE 20.3 Synthesis of PEDOT and *in situ* oxidation of the PEDOT:PSS complexes.

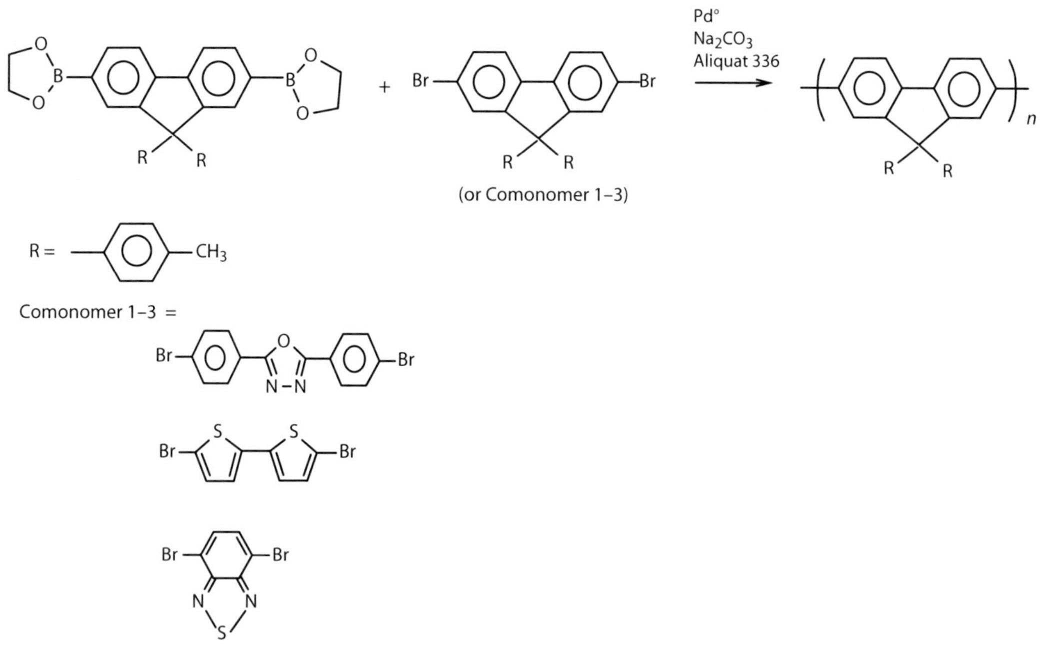

FIGURE 20.4 Modified Suzuki coupling for the synthesis of poly(9,9-substituted fluorene) and its copolymers for PLED applications.

Traditionally, polyfluorene can be prepared by the Yamamoto scheme using nickel-based catalysts [64]. Poly(9,9-dihexylfluorene) was found to be a high-efficiency blue-emitting polymer. The products, however, have limited molecular weight due to poor solubility of the intermediates in the polar reaction medium. A subsequent palladium-catalyzed Suzuki or Heck coupling reaction between a mixture of phenylboronic acids and arylbromides resulted in polymers with high molecular weight [65,66]. Copolymers with many other functional comonomers including oxadiazole (high stability) [67], bithiophene (yellow emission) [68], and benzothiadiazole (green emission) [68] have also been investigated. Upon aging, an undesired residual green emission (ca. 535 nm) can be developed for electroluminescent applications. It was suggested to be a result of excimer or aggregate formation or oxidation of disubstituted fluorene to fluorenone [69]. The addition of bulky side groups such as alkyoxyphenyl with minimum distortion of the π–π conjugation can lead to improvement in stability. Substitution of electron-transporting oxadiazole or hole-transporting triphenylamine at the C_9 methylene position has also been resulted in a stable blue emission as well as improved charge transport and injection properties [67] (see Figure 20.4).

20.3.3 Substituted Polythiophene Oligomers

The rapid development of OPV in the past decade has achieved a maximum power conversion efficiency (PCE) above 9% [70]. The PCE, η (%), can be defined as $\eta = (J_{sc}V_{oc}FF)/(P_{in})$, in which J_{sc} is the short-circuit current density, V_{oc} is the open-circuit voltage (difference between the highest occupied molecular orbital [HOMO] of the p-type and the lowest unoccupied molecular orbital [LUMO] of the n-type layers), FF is the fill factor (a normalization factor), and P_{in} is the power of the incident light [71]. The current most successful bulk heterojunction-type OPV consists of a microphase-segregated p-type thiophene-based polymer and a high-electron-affinity n-type C_{60}-based polymer (e.g., [6,6]-phenyl C_{61} butyric acid methyl ester, $PC_{60}BM$) [72,73]. The original thiophene-based polymer is a soluble regioregular and semicrystalline P3HT prepared via McCullough's route [74]. The regioregularity is essential for the development of optimum optical and electrical properties (including mobility) as a result of the side-chain crystallinity. An optimal conductivity of 1000 S/cm

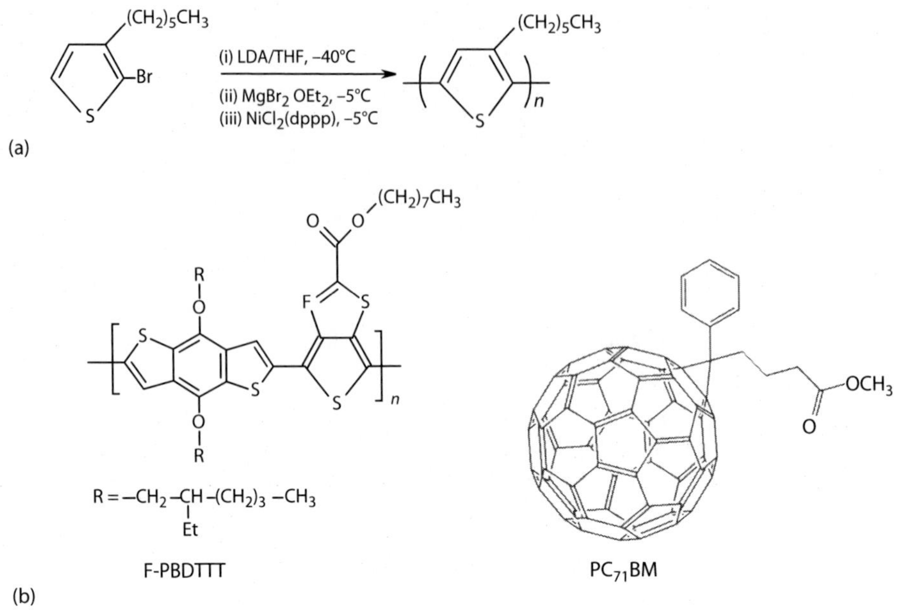

FIGURE 20.5 (a) McCullough's route to regioregular poly(3-hexylthiophene). (b) Structures of F-PBDTTT and PC$_{71}$BM for high-efficiency OPV application. LDA, lithium diisopropylamide.

has resulted in a doped polythiophene with a dodecyl side chain [75]. Oxidative coupling using FeCl$_3$ can also produce regioregular 3-substituted polythiophene [76]. Conformational changes in the polymer backbone due to an increased disordering of the aliphatic side chains at an increase in temperature are known as thermochromism [52–54].

The original P3HT-based OPV has achieved PCE of merely 3.49% [77,78]. Further improvement required a thiophene-based polymer with a low bandgap for the broadening of the absorption range, crystallinity for higher charge mobility, and a lower HOMO (together with a higher LUMO for the n-type PCBM) for maximizing the V_{oc}. For the bulk-heterojunction OPV, heterogeneity in the phase-segregated microdomain would be optimized to match the diffusion length of the degenerating excitons. A range of thiophene-based derivatives with donor-accepting repeating units with coplanar structure have been developed. As a result, a fluorinated OPV based on F-PBDTTT (Poly[(4,8-bis-(2-ethylhexyloxy)-benzo(1,2-b:4,5-b′)dithiophene)-2,6-diyl-alt-(4-(2-ethylhexanoyl)-thieno[3,4-b]thiophene-)-2-6-diyl)])– PC$_{71}$BM has achieved a maximum PCE at 9.2% [79]. There are several reviews that detail the development of conductive polymer-based printable OPV [70,71,80] (see Figure 20.5).

20.4 CONDUCTIVE VINYL POLYMERS

Vinyl polymers, including polystyrene (PS) and polyvinyl chloride, are typical insulators. They, however, can be thermally processed as well as being soluble in common organic solvents despite their high molecular weight. Vinyl polymer such as poly(N-vinyl carbazole) (PVK), however, is a well-known photoconductor [81]. In a photoconductor, free charge carriers, either holes or electrons, can be generated when the polymer is irradiated by light. Photoconductive polymers are important in the development of the photocopier and laser printer [82]. A number of the vinyl polymers with a polycyclic or heterocyclic pendant group, however, have also been described as photoconductive or semiconductive. The conductive side-chain polymers differs from the main-chain conjugated polymers as they do not have extensive or long-range π–π conjugation. Conductivity is only a result of charge hopping between the aromatic pendant side groups. For conductive vinyl polymers with coplanar pendant groups, π–π stacking between the adjacent pendant groups can result in an enhancement in charge mobility. The phenomenon can be observed by a broadening or redshift in the absorption spectrum of the conductive vinyl polymer. Other known conductive side-chain polymers include poly(1-vinylnaphthalene), poly(9-vinylanthracene), and poly(1-vinylpyrene) [83]. These vinyl polymers can be polymerized to high molecular weight using the traditional additional polymerization techniques, including free-radical or ionic processes. The applications of these polymers, however, are limited by their stability in ambient conditions (either exposure to light or oxygen). For example, photoinduced dimerization between proximate anthracene moieties can prompt gelation in poly(9-vinylanthracene) solution [83]. The mobility of these vinyl polymers is low and lie in the range of 10^{-7} to 10^{-5} cm^2 V^{-1} s^{-1}, which can only be considered for some lower-end organic semiconductor applications [84].

20.4.1 Mobility Measurements

The principal characteristic for main-chain conductive polymers is their volume conductivity (σ), while the key parameter used to quantify the transport property of organic semiconductors is their carrier mobility (μ). As defined earlier, mobility (in cm^2 V^{-1} s^{-1}) is related to the rate at which a charge carrier, either holes or electrons, moves across an organic medium under an applied electric field. The mobility not only affects the overall device efficiency but also can be used for the optimization of organic semiconductor devices. For example, the internal efficiency (η_{INT}) of OLEDs can be described as ($\eta_{INT} = \zeta\Phi\eta_{QE}\mu$), in which ζ is 0.25 for fluorescent and 0.75 for phosphorescent emitters, Φ is a charge-balance factor, η_{QE} is the quantum efficiency for the emitter, and μ is the mobility of the charge transport layers [85]. In the design for OPV, the film thickness (d) or size of heterogeneity (in bulk-heterojunction devices) is limited by the hole drift length d, where $d < \mu V \tau$, and again, μ is the mobility, V is the applied electric filed, and τ is the time constant for the dissociating excitons [72,73]. The switching speed (on/off ratio) for OFET is also limited by the mobility of the organic semiconductor [86,87].

There are several methods for the measurement of mobility. The organic semiconductor is usually spin-cast or evaporated under vacuum and sandwiched between two electrodes at a given thickness. Some of the techniques that are commonly used for measurement are summarized in Table 20.1. (1) In the time-of-flight (TOF) method, a laser is focused on the center of the organic thin film sandwiched between two electrodes (one being a transparent electrode like ITO), and the time traveled by the dissociating charges to the electrodes (the turning point for a nondispersive system) is measured and correlated to μ [88]. (2) In the organic thin-film transistor (OTFT) method, the linear region of the current-voltage (IV) characteristic of the organic transistor can be related to μ [89]. The results, however, can be affected by the gate dielectric. (3) In the J–V (current–voltage characteristic) method, a straightforward J–V fitting of the space-charge-limited current model can be used for the determination of μ; however, an ohmic contact is required for the accuracy of this method [90,91]. (4) The dark-injection space-charged-limited current (DI-SCLC) method is similar to TOF, except that a voltage pulse was used instead of the laser pulse [91,92]. It is also used for the examination of the charge injection at interface. (5) In the admittance spectroscopy (AS) method, a small alternating current (AC) signal is used instead of a laser pulse as in the TOF method, and the analysis is based on the frequency domain kinetic study [93]. (6) Lastly, charge extraction by linearly increasing voltage (CELIV) is a method similar to the TOF technique, except that the voltage between the electrodes is linearly increasing [94]. Although TOF is the most widely employed technique, it does require the deposition of a solid thin film with a larger thickness (in the order of micrometer) and minimum contamination [95]. Each of the methods has its own merits, and studies have shown that they are adapted to different classes of organic semiconductors. It is therefore desirable to compare experimental results against literature-reported data or make measurements against a known standard. Extrinsic factors such as moisture and oxygen in the environment can also affect the measurement, especially for the electron mobility measurement [95–97]. For the design of polymeric semiconductors, their chemistry is always based on the known results for some well-known small-molecule organic semiconductors. The mobility reported for some hole- and electron-transporting organic semiconductors can be found in Tables 20.2 and 20.3.

TABLE 20.1
Mobility Measurement Techniques in Organic Semiconductor

Technique	Sample geometry	Typical data	μ	Ref.
1. TOF (time of flight)	Pulsed laser	$I(t)$ vs t, τ	$\mu = \dfrac{d^2}{\tau V}$	[88]
2. OTFT (organic thin-film transistor)	I_D, V_D, V_G	I_D vs V_D (lin, sat); I_D vs V_G; $\sqrt{I_D}$ vs V_G	$\left.\dfrac{\partial I_D}{\partial V_G}\right\|_{V_D} = \dfrac{W}{L} C_i V_D \mu_{FE,lin}$ $\left.\dfrac{\partial \sqrt{I_D}}{\partial V_G}\right\|_{V_D} = \sqrt{\dfrac{W}{2L} C_i \mu_{FE,sat}}$	[89]
3. JV (current–voltage characteristics)	V_{dc}, I_{dc}	log I vs V	$J_{SCL} = \dfrac{9}{8}\mu_0\varepsilon_0\varepsilon_r \exp(0.89\beta\sqrt{F})\dfrac{F^2}{d}$	[90,91]
4. DISCLC (dark-injection space-charge-limited current)	$I(t)$	$I(t)$ vs t, τ_{DI}	$\mu_{DI} = \dfrac{0.78 d^2}{\tau_{DI} V}$	[91,92]
5. AS (admittance spectroscopy)	V_{ac}, V_{dc}, I_{ac}	$C(\omega)$, C_0; $\Delta B = \omega(C - C_0)$, $1/\tau_r$; $\omega = 2\pi f$	$\mu_{AS} = \dfrac{d^2}{0.56\tau_r V}$	[93]
6. CELIV (charge extraction by linearly increasing voltage)	Pulsed laser	j vs t, Δj, $j(0)$, t_{max}	$\mu = \dfrac{2d^2}{3At_{max}^2\left[1 + 0.36\dfrac{\Delta j}{j(0)}\right]}$	[94]

20.4.2 Hole-Transporting Vinyl Polymers

In most applications involving semiconductors, the p–n junction is the elemental building block for many electronic components like diodes, transistors, photovoltaic cells, etc. The p–n junction is defined as the interface between a p-type and an n-type semiconductor layer at which the flow of the charge carriers can be regulated, confined, recombined, or dissociated. Traditional p-type organic small molecules include N,N′-bis(3-methylphenyl)-N,N′-diphenylbenzidine (TPD) and N,N′-di(1-naphthyl)-N,N′-diphenyl-(1,1′-biphenyl)-4,4′-diamine (NPB) (see Figure 20.6b), and they are mostly arylamine compounds and their derivatives [98]. In one of the earliest inventions on high-efficiency heterojunction OLED as reported by C.W. Tang and his group at Kodak (USA) in 1987, hole-transporting and electron-transporting small molecules were evaporated in sequence on a transparent ITO glass, which functioned as an anode, and then encapsulated with a low-electron-affinity metal such as Al or Ca, which acted as the cathode [99]. The holes injected from the ITO anode and electrons from the cathode were confined at the organic p–n interface and then recombined into tris-(8-hydroxyquinoline)aluminum(III) (Alq$_3$) excitons. The excitons then degenerated and gave out photons that are characteristic of the Alq$_3$ dyes. The small-molecule hole-transporting

TABLE 20.2
Hole Mobility for Some Known Hole-Transporting Small Molecules and Polymers

	Mobility (cm^2 V^{-1} s^{-1})	Electric Field (V cm^{-1})	Ref.
TPD	~1 × 10^{-3}	1.5 × 10^5	[100]
NPB	~1 × 10^{-3}	1.5 × 10^5	[100]
m-TDATA	~3 × 10^{-5}	1 × 10^5	[101]
Spiro-TAD	~4 × 10^{-3}	1 × 10^5	[102]
PVK	~1 × 10^{-6}	/	[81]
P(H-NPA)[a]	~1 × 10^{-5}	1.6 × 10^5	[103]
P(DBF)	2.7 × 10^{-4}	7 × 10^5	[104]

Note: m-TDATA, 4,4′,4″-tri(3-methyl-phenylphenylamino) tripheny-lamine; TAD, N,N,N′,N′-tetraphenylbenzidine; PVK, polyvinyl carbazole.

[a] Measured by time-of-flight method.

compounds, including TPD and NPB, however, have low glass transition temperature (T_g = 65°C for TPD and 95°C for NPB), and they can recrystallize during operation and lead to catastrophic failure of the OLED device [105]. A soluble polymeric p-type semiconductor with improved thermal and electrical properties is therefore desired for the scaled-up large-area PLED applications.

Both arylamine-containing main-chain and side-chain conductive polymers have been reported. The main-chain arylamine-conductive polymers, however, have limited solubility and therefore are mostly low-molecular-weight oligomers [106]. Bellmann et al. have reported on the synthesis of a series of TPD-containing vinyl polymers [105]. They further designed a trimethoxyvinylsilane cross-linked variation of the TPD-containing polymers, and they were able to lower the turn-on voltage for the PLEDs as a result of improved interfacial contact with the ITO anode [107]. A facile four-step method for the preparation of a hole-transporting vinyl conductive polymer

TABLE 20.3
Electron Mobility for Some Known Electron-Transporting Small Molecules and Polymers

	Mobility (cm^2 V^{-1} s^{-1})	Electric Field (V cm^{-1})	Ref.
Alq$_3$[a]	~1 × 10^{-5}	1 × 10^6	[108]
PBD[b]	2 × 10^{-5}	1 × 10^6	[109]
PFO[a]	~1 × 10^{-3}	~7 × 10^5	[110]
OXD[c]	2 × 10^{-5}	7.5 × 10^5	[109]
PBD[b]	2 × 10^{-5}	1 × 10^6	[111]
Tri-OXD[c]	1 × 10^{-6}	7 × 10^5	[112]
Phen[a]	5 × 10^{-4}	6 × 10^5	[112]
DMPhen[b]	1 × 10^{-3}	1 × 10^6	[109]

Note: DMPhen, 2,9-dimethyl-4,7-diphenyl-1,10-phenanthroline; OXD, 1,3-bis[2-(4-tert-butylphenyl)-1,3,4-oxadiazo-5-yl]benzene; PBD, 2-(4-biphenyl)-5-(4-tert-butylphenyl)-1,3,4-oxadiazole; PFO, polyfluorene; Phen, 4,7-diphenyl-1,10-phenanthroline.

[a] Measured by time-of-flight method.
[b] Space-charge-limited current (SCLC) measurement.
[c] 50 wt.% in polycarbonate.

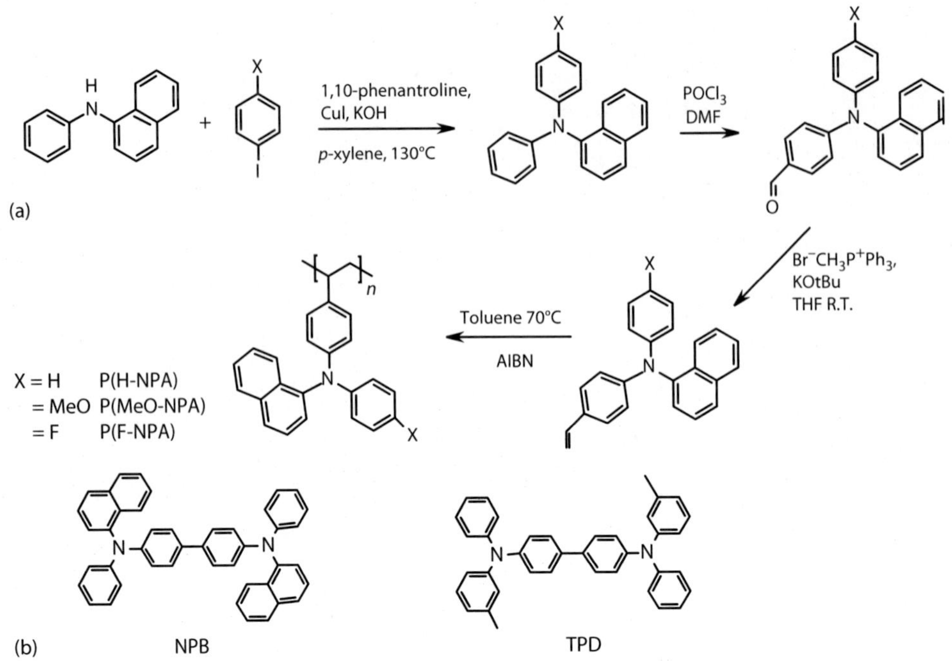

FIGURE 20.6 (a) Synthesis of the hole-transporting polymer P(X-NPA), where X = H, OMe, F. (b) Chemical structures of hole-transporting small-molecule NPB and TPD. DMF, dimethylformamide; RT, room temperature.

poly(N-(X-phenyl)-N-styryl-1-naphthylamine) [P(X-NPA)], based on the NPB chemistry has been reported [103], where X = –H, –OMe, or –F, which are different electron-withdrawing or -donating groups for the tuning of the HOMO/LUMO for the polymer. The arylamine moiety was prepared by a modified Ullmann condensation [113] followed by a Vilsmeier formylation for functionalization with an aldehyde [114,115]. A Wittig coupling reaction produced the vinyl monomer, and the polymer was prepared by a solution free-radical additional polymerization using azobisisobutyronitrile (AIBN) at 70°C. The P(X-NPA) polymers' relative number averaged molecular weight (M_n) ranged from 18,000 to 250,000 as determined by gel permeation chromatography (GPC) calibrated according to PS MW standards, yet they are soluble in common organic solvents, including tetrahydrofuran (THF), toluene, chloroform, etc. The polymers had a T_g (midpoint) above 160°C for the highest-MW polymers, and they remained stable up to 450°C under a nitrogen atmosphere. The best hole mobility determined using the TOF method for the unsubstituted poly(N-phenyl)-N-styryl-1-naphthylamine (P(H-NPA)) is 1×10^{-5} cm^2 V^{-1} s^{-1}, which is lower than that of NPB (1×10^{-3} cm^2 V^{-1} s^{-1}) but is in the same order of magnitude as the N,N,N',N'-tetraphenyl-benzidine (TPB)-containing polymers at ca. 10^{-5} cm^2 V^{-1} s^{-1}. With an increase in electron richness due to the substituted side group X = MeO > H > F, the IP of the polymers increased from −5.23 to −5.37 to −5.39 eV, respectively. The hole-transporting poly(N-(4methoxy-phenyl)-N-styryl-1-naphthylamine (P(MeO-NPA)) has a blue color fluorescence (with photoluminescent [PL] maximum at 452 nm) and a maximum electroluminescence intensity of 588 cd/m^2 at 120 mA/cm^2. For a heterojunction OLED (with Alq$_3$ being the emitter) device with the cell configuration ITO/PEDOT:PSS/P(MeO-NPA)/Alq^3/Ca/Al, a maximum luminescence of 6600 cd/m^2 at 470 mA/cm^2 was achieved [103]. For OPV application, poly(3-substituted thiophene) is the typical p-type material [47,70–73].

20.4.3 Electron-Transporting Vinyl Polymers and Their Copolymers

The electron-transporting material employed in the original heterojunction device as reported by C.W. Tang et al. was Alq$_3$ [99]. Alq$_3$ can be vacuum-evaporated to produce uniform thin film, and it is an excellent green emitter for OLED applications. It also has excellent thermal stability (T_g = 172°C) but with low electron mobility (10^{-5} to 10^{-7} cm^2 V^{-1} s^{-1}) and is susceptible to irreversible oxidation [108,116,117]. The other chemicals that can also be used as electron-transporting materials include 1,3,4-oxadiazole, fluorene-based oligomer, 4,7-diphenyl-1,10-phenanthroline, and so on [104–107].

Dailey et al. reported the synthesis of two copolymer series containing a hole-transporting triphenylamine (TPA) moiety and two electron-transporting 2,5-diphenyloxadiazole and 2,3-diphenylquinoxaline moieties [118]. The two copolymers were doped with a fluorescent dopant P580 (1,3,5,7,8-pentamethyl-2,6-di-n-butylpyrromethene-difluoroborate complexe) for PLED applications, and the maximum luminescence and quantum efficiency were found to be 1200–1557 cd/m^2 and 0.148–0.073%. A different series of soluble conductive vinyl copolymers containing a hole-transporting N-(4-methoxyphenyl)-N-phenylnaphthalen-1-amine (MeONPA) moiety and an electron-transporting (also known as hole-blocking) 2,5-diphenyl-1,3,4-oxadiazole (OXA) moiety at different composition ratios were also synthesized and characterized [119]. The copolymers were then employed in two electroluminescent applications. One was to prepare a copolymer with charge-balance properties (equal mobility for holes and electrons) for a homojunction (a single-layer device as opposed to a heterojunction device, originated from C.W. Tang) PLED application. The other would be to prepare a hole-limiting transporting layer for the optimization of a typical Alq$_3$-based heterojunction OLED. The preparation of the MeONPA monomer has been described above [103]. The OXA vinyl monomer [2-phenyl-1-5-(4-vinylphenyl)-1,3,4-oxadiazole] was prepared by a reaction between 4-formylbenzoyl chloride and 5-phenyl-2H-tetrazole to afford 4-(5-phenyl-1,3,4-oxadiazol-2-yl)benzaldehyde. The aldehyde was then reacted with methyltriphenyl phosphonium via a Wittig coupling reaction to yield the vinyl monomer (see Figure 20.7). A free-radical additional copolymerization was carried out at different feed ratios for the two monomers using AIBN in toluene at 70°C under a nitrogen atmosphere. The actual composition for the copolymers was determined by H^1 nuclear magnetic resonance (NMR), and the copolymers were found to have relative M_n from 24,000 to 80,000. The electron-transporting poly(2-phenyl-1-5-(4-vinylphenyl)-1,3,4-oxadiazole) (POXA) has T_g = 187°C and an onset of thermal degradation under nitrogen above 300°C. The thermal properties of the copolymers varied according

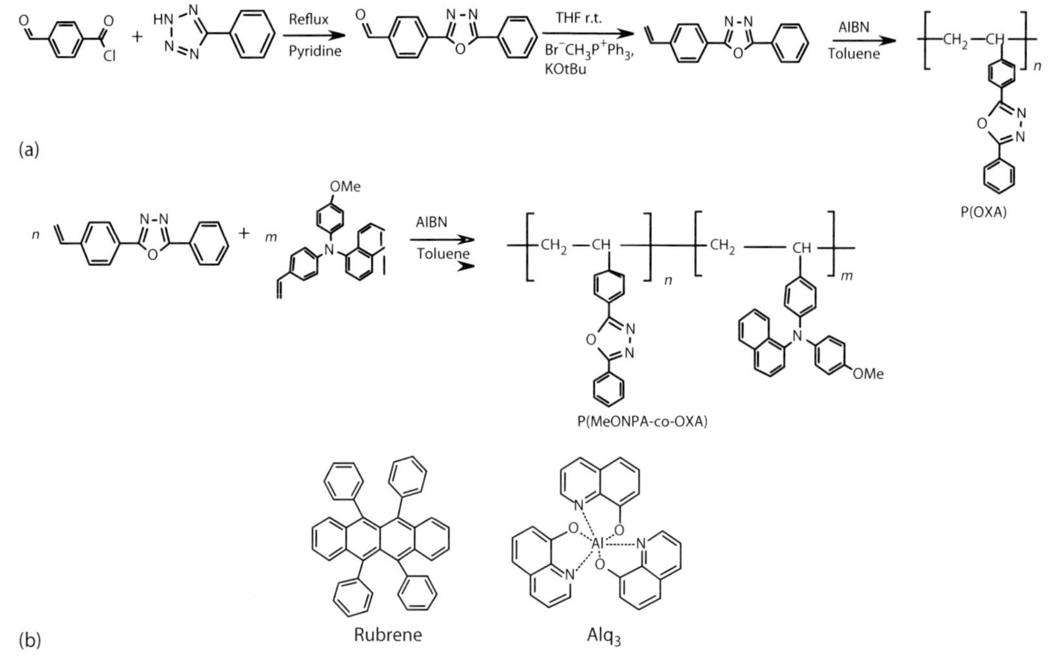

FIGURE 20.7 (a) Synthesis of the electron-transporting P(OXA) and its copolymer P(MeO-NPA-co-OXA). (b) Chemical structure of rubrene and Alq$_3$. r.t., room temperature.

to their composition, such as the T_g for the copolymers lying between that of P(MeONPA) (at 132°C) and POXA (at 187°C). Similarly, the range of onset for thermal degradation under nitrogen was between 300°C and 420°C for the copolymers. POXA has a high bandgap, and its PL emission maximum was at 365 nm. HOMO = −6.04 eV and LUMO = −2.23 eV for POXA, while the HOMO = −5.23 eV and LUMO = −2.14 eV for P(MeO-NPA), and the energy levels for the copolymers ranged within these two sets of limits. A homojunction PLED with the cell configuration ITO/PEDOT:PSS/copolymer/Ca/Al showed that the copolymer with 47 mol.% MeONPA has the best device current efficiency at 0.22 cd/A. The current efficiency for the homojunction PLED improved sevenfold with the addition of 0.1 wt.% Rubrene dopant [119].

The OXA moiety is not only electron transporting but also hole blocking due to its high HOMO level (compared to that for Alq_3). The MeONPA–OXA copolymers therefore have a range of hole-transporting properties, and their hole mobility diminished with increasing OXA composition. Although Alq_3 is an excellent electron-transporting material, it suffers from irreversible oxidation due to accumulation of excess holes at the interface for the heterojunction TPD/Alq_3 OLEDs. The performance and lifetime of the device can be optimized if the hole-transporting layer has a matching property with the electron-transporting Alq_3 layer (known as charge balance). For a heterojunction OLED with the cell configuration ITO/PEDOT:PSS/copolymer/Alq_3/Ca/Al, it was found that for a device employing the copolymer with 82 mol.% MeONPA, a maximum current efficiency and luminance at 4.2 cd/A and ca. 24,000 cd/m^2, respectively, were achieved [119]. This is compared to a typical ITO/PEDOT:PSS/NPA(1,4-bis(1-naphylphenylamino)biphenyl)/Alq3/Ca/Al device, which has a maximum current efficiency and luminescence of 3.5 cd/A and 6600 cd/m^2 respectively only.

20.4.4 Conductive Vinyl Polymers as Emitters

As mentioned above, conductive vinyl polymers can be either hole (such as P(MeO-NPA)) or electron (such as P[OXA]) transporting. Dependent on the photoluminescent properties of the polycyclic pendant side group, a conductive vinyl polymer can also function as an emitter in electroluminescent applications. Poly(9,9-disubstituted fluorene) is one of the few main-chain conductive polymers that are soluble yet electroluminescent in the full visible spectrum [61]. Conductive vinyl polymers, on the other hand, can be synthesized without expansive catalysts, and their electrical and physical properties can be tailored by copolymerization between different monomer units. The composition and ordering of the repeating units can be controlled via different but well-known additional polymerization schemes, including anionic and atom transfer radical polymerization (ATRP).

Traditionally, color tuning in PLED applications can be achieved by a guest–host approach, in which lower bandgap fluorescent or even phosphorescent dyes are doped (finely dispersed) into a semiconductive polymer matrix [120–122]. The small-molecule dyes, however, can be either phase-separated or recrystallized during operation due to ohmic heating. Vinyl copolymers containing a hole-transporting repeating unit, an electron-transporting repeating unit, and an emitting repeating unit containing a Pt- or Ir-based phosphorescent dye have been prepared [123,124]. The multilayer PLEDs based on the phosphorescent terpolymers have been recorded to have quantum efficiency above 10%. A charge-balanced terpolymer containing MeO-NPA as the hole-transporting units, OXA as the electron-transporting units, and a fluorescent dye perylene as the emitter has been prepared [125]. The charge-balance copolymer allowed the fabrication of a homojunction PLED device in which a single-layer terpolymer served multiple functions as a hole and electron transporter as well as emitter all in one. The synthesis of the vinyl emitter 3-(4-vinylphenyl)perylene and the terpolymer is outlined in Figure 20.8. The terpolymers consisted of a constant molar ratio of MeONPA/OXA of ca. 55:41, while the molar percentage of the perylene-containing repeating units varied, with values of 0.0, 0.5, 2,2, and up to 4 mol.%. The relative M_n of the terpolymer ranged from 12,000 to 22,000, and the T_g of the copolymer ranged from 162°C to 170°C. A blue to green emission for the homojunction PLED device with the simple cell configuration ITO/PEDOT:PSS/perylene terpolymer/Ca/Al has been detected due to energy transfer between the bluish NPA emitters (PL

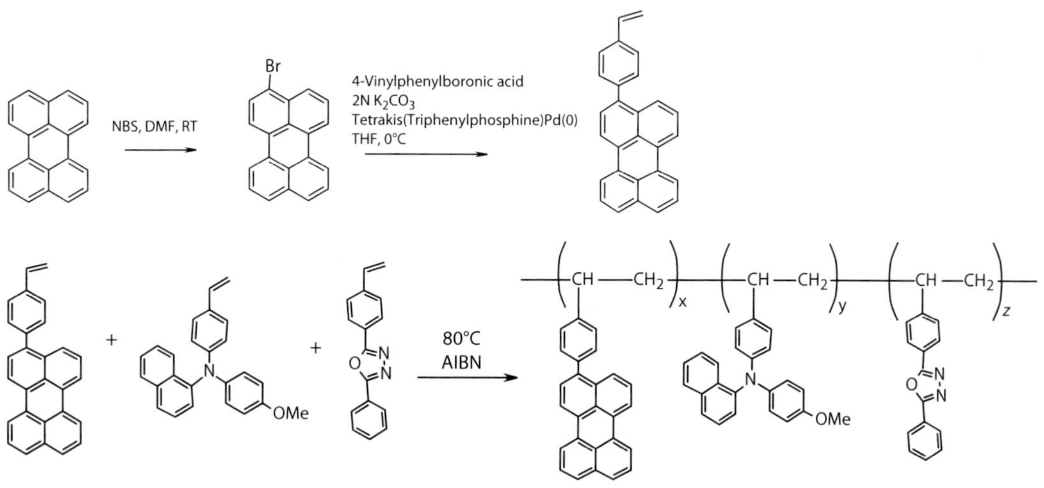

FIGURE 20.8 Synthetic pathways for 3-(4-vinylphenyl)perylene (v-peryl) and its terpolymer with vinyl–MeO-NPA and vinyl–OXA. DMF, dimethylformamide; NBS, N-Bromosuccinimede; RT, room temperature.

maximum at 460 nm) and the greenish perylene emitters (PL maximum at 493 nm). The trajectory for the Commission Internationale d'Eclairage (CIE) (1931) coordinates of the different perylene-containing terpolymers is given in Figure 20.9, showing color tuning capability by small changes in the perylene content [126]. The current efficiency and maximum luminance for the homojunction PLEDs ranged from 0.12 to 0.41 cd/A and 420 to 760 cd/m², respectively.

The design for a high-bandgap blue-emitting polymer is important as it can be tuned to emit colors with lower energy (or higher wavelength) by the simple doping process. Two novel blue fluorescent polymers, poly(9,10-di(2-naphthalenyl)-2-vinylanthracene) (P[2ADN]) and poly(9,10-di(3-quinolinyl)-2-vinylanthracene) (P[3ADQ]) have been synthesized, and their CIE coordinates at (0.15, 0.10) for P[2ADN] and (0.15, 0.13) for P[3ADQ] lie within the requirements for a true blue

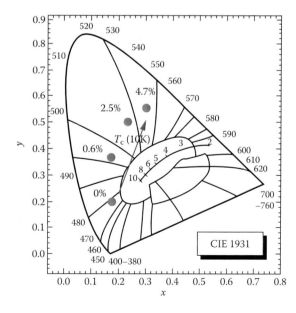

FIGURE 20.9 The CIE diagram showing color tuning for the MeO-NPA/OXA/perylene terpolymers with perylene composition varying, with values of 0, 0.6, 2.5, and up to 4.7 mol.%.

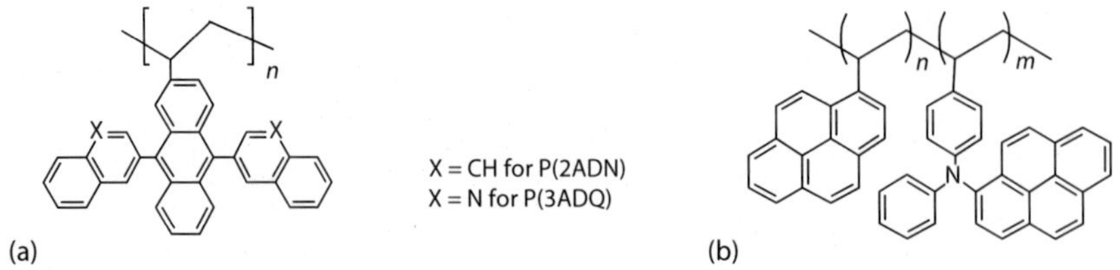

FIGURE 20.10 (a) Chemical structures of blue-emitting P(2ADN) and P(3ADQ) and (b) chemical structures for the green-emitting copolymer P(PyPA-co-VPy).

display color [127]. P(2ADN) is soluble in common organic solvent up to a relative M_n of 21,500, while P(3ADQ) is pH sensitive and can be soluble in polar solvents including ethanol and ethanol/water mixture. Their T_g values are 343°C and 298°C for P(2ADN) and P(3ADQ), respectively. The hole mobility of the two polymers, however, are relatively low at 4.0×10^{-7} and 10^{-8} cm^2 V^{-1} s^{-1} for P(2ADN) and P(3ADQ), respectively. As a result, electroluminescence pursued for their homojunction devices with the configuration ITO/MoO$_3$/AND-containing polymers/LiF/Al achieved only a meager luminance of 30–50 cd/m^2. Copolymers of P(2ADN) with styrene and 9-vinylcarbazole have also been prepared for improvement of its solubility as well as tuning for the LUMO/HOMO energy levels [128]. An aqueous soluble 2ADN-based triblock copolymer prepared via the ATRP approach has also been reported to have a concentration-dependent micellar morphology in solution [129].

Another series of soluble conductive vinyl copolymers, P(PyPA-co-VPy), based on a hole-transporting *N*-phenyl-*N*-(4-vinylphenyl)pyren-1-amine (vinyl-PyPA) and electron-transporting 1-vinyl pyrene (VPy) (see Figure 20.10b) at different molar feed ratios, have also been prepared [130]. Again, the copolymer has a high T_g, from 190°C to 201°C, and relative M_n from 7700 to 22,300. The emission maxima of these copolymers were in the range from 474.5 to 478.5 nm, which were similar to that of P(PyPA), suggesting energy transfer from PVPy to P(PyPA). The quantum yield of the copolymer, at 0.51, is higher than the homopolymers, P(PyPA) at 0.48 and PVPy at 0.13. PLEDs with the configuration ITO(CFx treated)/P(PyPA-co-VPy)/TPBi/LiF/Al achieved high luminance, from 665 to 1143 cd/m^2 (at 100 mA/cm^2), only when an electron-injecting layer 2,2′,2″-(1,3,5-Benzinetriyl)-tris(1-phenyl-1-H-benzimidazole) (TPBi) has been applied.

20.4.5 π–π Stacking Poly(Dibenzofulvene)

Conventional main-chain conjugated polymers are known to be better organic semiconductors due to their higher charge mobility. They, however, also possess numerous undesirable properties, such as insolubility and colored due to a lower bandgap, and are susceptible to oxidation. Conductive vinyl polymers, on the other hand, are soluble even with high MW and optically clear, and their overall properties can be easily tailored by copolymerization. The mobility of conductive vinyl polymers, in general, is three to four orders of magnitude lower than that of the main-chain conjugated polymers. In order to improve the mobility of conductive vinyl polymers, one can increase the size of the coplanar polycyclic or heterocyclic pendent side group, which may cause insolubility for the polymers. Another approach would be to increase the short-range interaction between the coplanar pendent side group by π–π stacking. Poly(dibenzofulvene) [Poly(DBF)] is known to adapt a unique π–π stacking conformation in which the main-chain carbon–carbon bonds exhibit an all-trans conformation and allow the fluorene moieties to stack uniformly upon one another [131]. As a result, poly(DBF) has afforded a high hole mobility at 2.7×10^{-4} cm^2 V^{-1} s^{-1} (299K) under an applied electric field of 7×10^5 V/cm [104]. The mobility is similar in magnitude to an inorganic semiconductor, selenium. Poly(DBF) has been successful polymerized by different additional polymerization techniques, including free radical, cationic, and anionic. In general, anionic polymerization produced monodisperse poly(DBF) with the highest MW yet the lowest number of conformational defects [132].

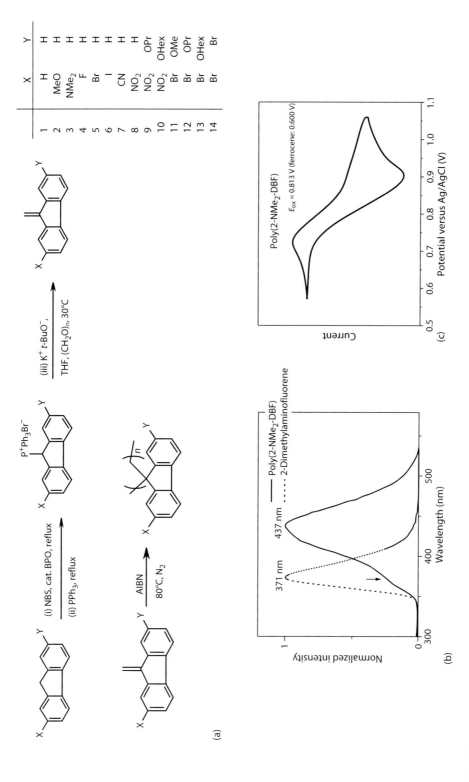

FIGURE 20.11 (a) Synthetic scheme for the 2,7-substituted DBF monomers via a stabilized ylide approach using the Wittig chemistry. (b) Emission spectrum of poly(2-NMe2-DBF) (in solid line) compared to its model compound (in dashed line); the arrow indicates the position of a nonstacked defect. (c) Cyclic voltammetry (CV) of poly(2-NMe2-DBF) in DCM showing a highly reversible oxidation state [133]. BPO, Benzoyl peroxide.

Substitution at the 2- and 7- positions of the dibenzofulvene moiety has been shown to greatly affect the solubility, MW, physical properties, and thermal stability of poly(DBF). Bulky substituents such as 2,7-di-*t*-butyldibenzofulvene have been shown to limit the MW of the disubstituted polymer to a trimer only [134]. Poly(benzofulvene), a close analogue to poly(DBF), was found to have the propensity of either thermally depolymerized or spontaneous polymerized by the steric and electronic nature of the substituent [135]. A series of 2- and 2,7-substituted dibenzofulvene monomers using Wittig chemistry based on a stabilized ylide approach has been reported with high yield [136]. Subsequently, a series of 2- and 2,7-substituted poly(DBF) has been synthesized and characterized [137] (see Figure 20.11a). It was found that the chemistry and physical properties of the polymers such as molecular weights, absorption and emission spectra, thermal stabilities, and electronic properties can be strongly affected by the substituents via weakening or enhancement of the $\pi-\pi$ stacking interaction. The presence of $\pi-\pi$ stacking conformation in the substituted poly(DBF)s was supported by an up-field shift of the fluorene protons in the ^{1}H NMR spectra due to anisotropic shielding. A bathochromic shift of about 5–13 nm in the absorption edge for the substituted poly(DBF)s in comparison with their corresponding fluorene side group (without the vinyl group) is also an indication of effective $\pi-\pi$ stacking. In general, the solubility and the molecular weight of the polymers were found to be enhanced by electron-donating substituents such as –MeO, –NMe$_2$, and –F [133]. The excimer emission for a highly polarized poly(2-NO$_2$-7-alkoxy-DBF) redshifted from 397–437 nm [which is the range for most substituted poly(DBF)] to an exception value of ca 540 nm (see Figure 20.11b). The oxidation potential of the substituted polymers was reduced by ca. 0.01–0.14 V in comparison with their corresponding fluorene side groups due to stabilization of the holes via $\pi-\pi$ stacking. Their HOMO ranged from –5.01 to –6.07 eV, while the LUMO was calculated to be –1.67 to –3.04 eV for the different substituted poly(DBF)s [137]. The bandgap energies spanned from 3.874 eV for the unsubstituted poly(DBF) to a lowest value of 2.845 eV for the highly polarized poly(2-NO$_2$-7-alkoxy-DBF). A cyclic voltammetry study on poly(2-NMe$_2$-DBF) showed its highly reversible oxidation state, rendering it a good candidate for electrochemical applications (see Figure 20.11c) Due to substituted poly(DBF) having a highly strained structure, it therefore has only moderate thermal stability, with thermogravimetric analysis (TGA) onset degradation temperature ranging from 254°C to 369°C under a nitrogen atmosphere.

20.5 CONCLUSIONS

In the recent rapid development of large-area, flexible, and printable electronics, including OPV, PLED, and OFET, organic semiconductors seem to be the materials of choice rather than the traditional rigid Si-based inorganics. Main-chain conductive polymers possess high mobility yet have limited solubility with an increase in MW. As a result, only oligomers or polymers with lengthy solubilizing side chains can be adapted for the printable applications. Their high mobility suggests that they are suitable for applications like OFET, which require high switching speed, or OPV, which require effective charge separation. At the other end, conductive vinyl polymers have low mobility but are soluble even at high MW. They are therefore suitable for self-supported large-area homojunction PLED illumination applications. The continuous growth of printable electronics demands not only the design of novel conductive polymers but also the integration of conductors with dielectrics and semiconductors. Only a good understanding of the chemistry and physics of the novel materials would allow the applications to be eventually extended into the realm of molecular electronics and machinery.

ACKNOWLEDGMENTS

The author acknowledges the contribution and discussion of Table 20.1 by Professor S.K. So of the Physics Department of Hong Kong Baptist University and also Ms. Elsa Fong of the Chemistry Department of Hong Kong Baptist University for help with the drawings.

REFERENCES

1. "The 2000 Nobel Prize in Chemistry—Popular Information," Nobelprize.org. Nobel Media AB, 2014. Available at http://www.nobelprize.org/nobel_prizes/chemistry/laureates/2000/popular.html (accessed August 12, 2014).
2. T. Skotheim, Ed., *Electroresponsive Molecular and Polymeric Systems*, Vols. 1 and 2, Marcel Dekker, New York, 1988.
3. T. Skotheim, R. Elsenbaumer, J. Reynolds, Eds., *Handbook of Conducting Polymers*, 2nd ed., Marcel Dekker, New York, 1998.
4. W. R. Salaneck, I. Lundström, B. Rånby, Eds., *Nobel Symposium in Chemistry: Conjugated Polymers and Related Materials: The Interconnection of Chemical and Electronic Structure*, Oxford University Press, Oxford, 1993.
5. H. S. Nalwa, Ed., *Handbook of Organic Conductive Molecules and Polymers*, Vols. 1–4, John Wiley & Sons, Chichester England, 1997.
6. M. Angelopoulos, "Conducting polymers in microelectronics," *IBM J. Res. Dev.*, 45(1), 57–75 (2001).
7. M. Atesa, T. Karazehir, A. S. Sarac, "Conducting polymers and their applications," *Curr. Phys. Chem.*, 2, 224–240 (2012).
8. P. J. Hesketh, D. Misra, "Conductive polymers and their applications," *The Electrochemical Society Interface*, Fall–Winter, 61 (2012).
9. T. K. Dasa, S. Prusty, "Review on conducting polymers and their applications," *Polym.-Plast. Technol Eng*, 51(14), 1487–1500 (2012).
10. Lord Todd: The state of chemistry. *Chem. Eng. News*, 58(40), 29 (1980).
11. J. C. W. Chien, *Polyacetylene: Chemistry, Physics and Materials Science*, Academic Press, New York, 1984.
12. H. Shirakawa, E. J. Louis, A. G. MacDiarmid, C. K. Chiang, A. J. Heeger, "Synthesis of electrically conducting organic polymers: Halogen derivatives of polyacetylene, (CH)x," *J. Chem. Soc. Chem. Comm.*, Issue 16, 578 (1977).
13. C. K. Chiang, C. R. Fincher, Y. W. Park, A. J. Heeger, H. Shirakawa, E. J. Louis, S. C. Gau, A. G. Macdiarmid, "Electrical conductivity in doped polyacetylene," *Phys. Rev. Lett.*, 39, 1098 (1977).
14. C. K. Chiang, M. A. Druy, S. C. Gau, A. J. Heeger, E. J. Louis, A. G. Macdiarmid, Y. W. Park, H. Shirakawa, "Synthesis of highly conducting films of derivatives of poly acetylene, (CH)x," *J. Am. Chem. Soc.*, 100, 1013 (1978).
15. C. K. Chiang, Y. W. Park, A. J. Heeger, H. Shirakawa, E. J. Louis, A. G. Macdiarmid, "Conducting polymers: Halogen-doped polyacetylene," *J. Chem. Phys.*, 69, 5098 (1978).
16. M. Winokur, Y. B. Moon, A. J. Heeger, J. Barker, D. C. Bott, H. Shirakawa, "X-ray scattering from sodium-doped polyacetylene: Incommensurate–commensurate and order–disorder transformations," *Phys. Rev. Lett.*, 58, 2329 (1987).
17. J. L. Bredas, G. B. Street, "Polarons, bipolarons, and solutions in conducting polymers," *Acc. Chem. Res.*, 18(10), 309–315 (1985).
18. M. Grundmann, *The Physics of Semiconductors*, 2nd ed., Springer-Verlag, Berlin, Heidelberg, 2010.
19. G. Hadziioannou, G. G. Malliaras, *Semiconducting Polymers: Chemistry, Physics and Engineering*, 2nd ed., Wiley-VCH, Weinheim, 2007.
20. D. A. Bernards, R. M. Owens, G. G. Malliaras, Eds., *Organic Semiconductors in Sensor Applications*, Springer Series in Materials Science, Vol. 107, Springer Verlag, Berlin, 2008.
21. F. So, Ed., *Organic Electronics: Materials, Processing, Devices, and Applications*, CRC Press, Boca Raton, FL, 2010.
22. C. Adachi, R. J. Holmes, W. Brutting, *Physics of Organic Semiconductors*, 2nd ed., Wiley-VCH Verlag, Weinheim, 2012.
23. J. H. Schon, S. Berg, C. H. Kloc, B. Batlogg, "Ambipolar pentacene field-effect transistors and inverters," *Science*, 287, 1022 (2000).
24. H. E. Katz, A. J. Lovinger, J. Johnson, C. Kloc, T. Siergist, W. Li, Y.-Y. Lin, A. Dodabalapur, "A soluble and air-stable organic semiconductor with high electron mobility," *Nature*, 404, 478 (2000).
25. R. C. Haddon, A. S. Perel, R. C. Morris, T. T. M. Palstra, A. F. Hebard, R. M. Fleming, "C_{60} thin film transistors," *Appl. Phys. Lett.*, 67, 121 (1995).
26. V. Podzorov, S. E. Sysoev, E. Loginova, V. M. Pudalov, M. E. Gerhenson, "Single-crystal organic field effect transistors with the hole mobility," *Appl. Phys. Lett.*, 83, 3504 (2003).
27. R. Zeis, T. Siegrist, C. Kloc, "Single-crystal field-effect transistors based on copper phthalocyanine," *Appl. Phys. Lett.*, 86, 22103 (2005).

28. J. H. Borroughes, C. A. Jones, R. H. Friends, "New semiconductor device physics in polymer diodes and transistors," *Nature*, 335, 137 (1988).
29. H. Sirringhaus, N. Tessler, R. H. Friend, "Integrated optoelectronic devices based on conjugated polymers," *Science*, 280, 1741 (1998).
30. H. Fuchigami, A. Tsumura, H. Koezuka, "Polythienylenevinylene thin-film transistor with high carrier mobility," *Appl. Phys. Lett.*, 63, 1372 (1993).
31. G. Wegner, "Polymers with metal-like conductivity: A review of their synthesis, structure and properties," *Angew. Chem. Int. Ed. Engl.*, 20, 361 (1981).
32. M. G. Kanatzidis, "Polymeric electrical conductors," *Chem. Eng. News*, (December), 36–54 (1990).
33. J. H. Edwards, W. J. Feast, "A new synthesis of poly acetylene," *Polymer*, 21, 595 (1980).
34. J. H. Edwards, W. J. Feast, D. C. Bott, "New routes to conjugated polymers: 1. A two-step route to polyacetylene," *Polymer*, 25, 395 (1984).
35. W. J. Feast, J. N. Winter, "An improved synthesis of polyacetylene," *J. Chem. Soc. Chem. Commun.*, 4, 202–203 (1985).
36. S. A. Krouse, R. R. Schrock, "Preparation of polyacetylene chains in low-polydispersity diblock and triblock copolymers," *Macromolecules*, 21, 1888 (1988).
37. L. M. Leung, K. H. Tan, "Synthesis and electrical properties of polyacetylene copolymers from poly(phenyl vinyl sulfoxide) and its oxidized products," *Macromolecules*, 26(17), 4426–4436 (1993).
38. L. M. Leung, K. H. Tan, "Electrical properties of anionic synthesized conducting block copolymer from the precursor polystyrene-block-poly(phenyl vinyl sulfoxide)," *Polym. Commun.*, 35(7), 1556–1560 (1994).
39. J. H. Burroughes, D. D. C. Bradley, A. R. Brown, R. N. Marks, K. Mackay, R. H. Friend, P. L. Burns, A. B. Holmes, "Light-emitting diodes based on conjugated polymers," *Nature*, 347, 539 (1990).
40. M. Knabe, M. Okawara, "Synthesis of poly(p-xylylidene) from p-xylenebis(dimethylsulfonium) terafluoroborate," *J. Polym. Sci. A-1*, 6, 1058 (1968).
41. R. A. Wessling, R. G. Zimmerman, "Polyelectrolytes from bissulfonium salts," U.S. Patent 3,401,152 (1968).
42. R. A. Wessling, "Polyxylylidene articles," U.S. Patent 3,706,677 (1972).
43. R. A. Wessling, "The polymerization of xylylene bisdialkyl sulfonium salts," *J. Polym. Sci. Polym. Symp.*, 72, 55 (1985).
44. L. M. Leung, "Polyphenylene vinylene," in *Handbook of Thermoplastics*, ed. O. Olabisi, Marcel Dekker, New York, 817–836, 1997.
45. F. S. Bates, G. L. Baker, "Soluble polyacetylene graft copolymers," *Macromolecules*, 16, 704 (1983).
46. A. Stowell, A. J. Amass, M. S. Beevers, T. R. Farren, "Synthesis of block and graft copolymers containing polyacetylene segments," *Polymer*, 30, 195 (1989).
47. J. Roncali, "Conjugated poly(thiophenes)—Synthesis, functionalization, and applications," *Chem. Rev.*, 92(4), 711–738 (1992).
48. L. M. Leung, C. G. Chik, "Phase-transfer catalyzed synthesis of disubstituted poly(phenylene vinylene)," *Polymer*, 34, 5174 (1993).
49. N. Toshima, S. Hara, "Direct synthesis of conducting polymers from simple monomers," *Prog. Polym. Sci.*, 20(1), 155–183 (1995).
50. W. J. Feast, J. Tsibouklis, K. L. Pouwer, L. Groenendaal, E. W. Meijer, "Synthesis, processing and material properties of conjugated polymers," *Polymer*, 37(22), 5017–5047 (1996).
51. L. M. Leung, "Electrically conductive polyacetylene copolymers," in *Handbook of Organic Conductive Molecules and Polymers*, Vol. 2, ed. H. S. Nalwa, John Wiley & Sons, New York, 61–95, 1997.
52. X. Zhao, "Synthesis, characterization and structure dependence of thermochromism of polythiophene derivatives," *J. Mater. Sci.*, 40(13), 3423–3428 (2005).
53. C. Roux, M. Leclerc, "Thermochromic properties of polythiophene derivatives: Formation of localized and delocalized conformational defects," *Chem. Mater.*, 6, 620–624 (1994).
54. A. Seeboth, D. Lötzsch, R. Ruhmann, O. Muehling, "Thermochromic polymers—Function by design," *Chem. Rev.*, 114(5), 3037 (2014).
55. K. E. Aasmundtveit, E. J. Samuelsen, L. A. A. Pettersson, O. Inganäs, T. Johansson, R. Feidenhans'l, "Structure of thin films of poly(3,4-ethylene dioxythiophene)," *Synth. Met.*, 101(1–3), 561–564 (1999).
56. G. Zotti, S. Zecchin, G. Schiavon, F. Louwet, L. Groenendaal, X. Crispin, W. Osikowicz, W. Salaneck, M. Fahlman, "Electrochemical and XPS studies towards the role of monomeric and polymeric sulfonate counterions in the synthesis, composition and properties of poly(3,4-ethylenedioxythiophene)," *Macromolecules*, 36, 3337–3344 (2003).

57. Clevios™: Commercial PEDOT:PSS dispersions in water and properties. © 2015 Heraeus Deutschland GmbH & Co. KG. Available at http://www.heraeus-clevios.com/en/conductivepolymers/pedot-pss-conductive-polymers.aspx.
58. S. Kirchmeyer, K. Reuter, "Scientific importance, properties and growing applications of poly(3,4-ethylenedioxythiophene)," *J. Mater. Chem.*, 15(21), 2077–2088 (2005).
59. M. Lefebvre, Z. Qi, D. Rana, P. G. Pickup, "Chemical synthesis, characterization, and electrochemical studies of poly(3,4-ethylenedioxythiophene)/poly(styrene-4-sulfonate) composites," *Chem. Mater.*, 11(2), 262–268 (1999).
60. A. Elschner, S. Kirchmeyer, W. Lovenich, U. Merker, K. Reuter, *PEDOT, Principles and Applications of an Intrinsic Conductive Polymer*, CRC Press, Boca Raton, FL, 2011.
61. U. Scherf, D. Neher (eds), *Polyfluorenes*, Advances in Polymer Science, Vol. 212, Springer-Verlag, Berlin, 2008.
62. M. T. Bernius, M. Inbasekaran, J. O'Brien, W. Wu, "Progress with light-emitting polymers," *Adv. Mater.*, 12, 1737 (2000).
63. J. L. Brédas, "Aromatic polymers: Evolution of their electronic properties as a function of bond-length alternation and torsion angle along the chains," in *Springer Series in Solid State Science*, Vol. 63, eds. H. Kuzmany, M. Mehring, S. Roth, Springer, Berlin, 166–172, 1985.
64. T. Yamamoto, A. Morita, Y. Miyazaki, T. Maruyama, H. Wakayama, Z. Zhou, Y. Nakamura, T. Kanbara, S. Sasaki, K. Kubota, "Preparation of π-conjugated poly(thiophene-2,5-diyl), poly(*p*-phenylene), and related polymers using zerovalent nickel complexes. Linear structure and properties of the π-conjugated polymers," *Macromolecules*, 25, 1214 (1992).
65. N. Miyaura, T. Yanagi, A. Suzuki, "The palladium-catalyzed cross-coupling reaction of phenylboronic acid with haloarenes in the presence of bases," *Synth. Commun.*, 11, 513 (1981).
66. M. Inbasekaran, W. Wu, E.P. Woo, "Process for preparing conjugated polymers." U.S. Patent 5,777,070 (July 7, 1998).
67. C. F. Shu, R. Dodda, F. I. Wu, M. S. Liu, A. K. Y. Jen, "Highly efficient blue-light-emitting diodes from polyfluorene containing bipolar pendant groups," *Macromolecules*, 36, 6698 (2003).
68. E. P. Woo, M. Inbasekaran, W. R. Shlang, G. R. Roof, M. T. Bernius, W. Wu, "Fluorene-containing polymers and compounds useful in the preparation thereof," U.S. Patent 6,169,163 (January 2, 2001).
69. V. N. Bliznyuk, S. A. Carter, J. C. Scott, G. Klärner, R. D. Miller, D. C. Miller, "Electrical and photo-induced degradation of polyfluorene based films and light-emitting devices," *Macromolecules*, 32, 361 (1999).
70. G. Li, R. Zhu, Y. Yang, "Polymer solar cells," *Nat. Photon.*, 6, 153–161 (2012).
71. Y.-W. Su, S.-C. Lan, K.-H. Wei, "Organic photovoltaics," *Mater. Today*, 15(12), 554–562 (2012).
72. M. C. Scharber, D. Mühlbacher, M. Koppe, P. Denk, C. Waldauf, A. J. Heeger, C. J. Brabec, "Design rules for donors in bulk-heterojunction solar cells—Towards 10% energy-conversion efficiency," *Adv. Mater.*, 18(6), 789 (2006).
73. A. J. Heeger, "25th Anniversary article: Bulk heterojunction solar cells: Understanding the mechanism of operation," *Adv. Mater.*, 26, 10–28 (2014).
74. R. D. McCullough, R. D. Lowe, "Enhanced electrical conductivity in regioselectively synthesized poly(3-alkylthiophenes)," *J. Chem. Soc., Chem. Commun.*, 1, 70–72 (1992).
75. M. Leclerc, K. Faid, "Electrical and optical properties of processable polythiophene derivatives: Structure–property relationships," *Adv. Mater.*, 9(14), 1087–1094 (1997).
76. D. Fichou, Ed., *Handbook of Oligo- and Polythiophenes*, Wiley VCH, Weinheim, 2007.
77. G. Li, V. Shrotriya, J. Huang, Y. Yao, T. Moriarty, K. Emery, Y. Yang, "High-efficiency solution processable polymer photovoltaic cells by self-organization of polymer blends," *Nat. Mater.*, 4, 864 (2005).
78. W. Ma, C. Yang, X. Gong, K. Lee, A. J. Heeger, "Thermally stable, efficient polymer solar cells with nanoscale control of the interpenetrating network morphology," *Adv. Funct. Mater.*, 15, 1617 (2005).
79. Z. He, C. Zhong, S. Su, M. Xu, H. Wu, Y. Cao, "Enhanced power-conversion efficiency in polymer solar cells using an inverted device structure," *Nat. Photon.*, 6(9), 591 (2012).
80. R. Tipnis, D. Laird, M. Mathai, "Polymer-based materials for printed electronics: Enabling high efficiency solar power and lighting," *Mater. Matters* 3(4), 92 (2008).
81. J. M. Pearson, M. Stolka, *Poly(N-vinylcarbazole)*, Polymer Monographs, Vol. 61, Gordon and Breach, New York, 1981.
82. A. Kukuta, "Chapter 12, Laser Printer Application," in *Infrared Absorbing Dyes*, ed. M. Matsuoka, Plenum Press, New York, 1990.

83. M. Stolka, D. M. Pai, "Polymers with photoconductive properties," *Adv. Polym. Sci.*, 29, 1 (1978).
84. K. S. Arora, C. G. Overberger, "Syntheses of poly[N-(1-pyrenyl)acetyl ethylenimine] [pyrene-substituted linear poly(ethylenimine)] and poly[methyl 2-(1-pyrenyl)acetamidopropenoate] [pyrene-substituted poly(dehydroalanine methyl ester)]," *J. Polym. Sci., Polym. Chem. Ed.*, 22, 1587 (1984).
85. C. Adachi, T. Tsutsui, S. Saito, "Blue light-emitting organic electroluminescent devices," *Appl. Phys. Lett.*, 56(9), 799–801 (1990).
86. H. Sirringhaus, "Device physics of solution-processed organic field-effect transistors," *Adv. Mater.*, 17(20), 2411 (2005).
87. G. Horowitz, "Organic thin film transistors: From theory to real devices," *J. Mater. Res.*, 19(7), 1946–1962 (2004).
88. P. M. Borsenberger, D. S. Weiss, *Organic Photoreceptors for Imaging Systems*, Chapter 9, Marcel Dekker, New York, 1993.
89. S. M. Sze, K. K. Ng, *Semiconductor Devices*, 3rd ed., Wiley, New York, 2006.
90. M. A. Lampert, P. Mark, *Current Injection in Solids*, Academic Press, New York, 1970.
91. P. N. Murgatroyd, "Theory of space-charge-limited current enhanced by Frenkel effect," *J. Phys. D: Appl. Phys.*, 3, 151 (1970).
92. K. C. Kao, W. Hwang, *Electrical Transport in Solids*, Pergamon, Oxford, 1981.
93. H. C. F. Martens, H. B. Brom, P. W. M. Blom, "Frequency-dependent electrical response of holes in poly(p-phenylene vinylene)," *Phys. Rev. B*, 60, R8489 (1999).
94. G. Juska, K. Arlauskas, M. Viliunas, J. Kocka, "Extraction current transients: New method of study of charge transport in microcrystalline silicon," *Phys. Rev. Lett.*, 84, 4946 (2000).
95. S. C. Tse, C. H. Cheung, S. K. So, "Charge Transport and Injection in Amorphous Organic Semiconductors," in *Organic Electronics: Materials, Processing, Devices, and Applications*, ed. F. So, CRC Press, Boca Raton, FL, 2010.
96. N. Tessler, Y. Preezant, N. Rappaport, Y. Roichman, "Charge transport in disordered organic materials and its relevance to thin-film devices: A tutorial review," *Adv. Mater.*, 21, 2741–2761 (2009).
97. V. Coropceanu, J. Cornil, D. A. da Silva Filho, Y. Olivier, R. Silbey, J.-L. Brédas, "Charge transport in organic semiconductors," *Chem. Rev.*, 107, 926–952 (2007).
98. S. A. VanSlyke, C. H. Chen, C. W. Tang, "Organic electroluminescent devices with improved stability," *Appl. Phys. Lett.*, 69, 2160 (1996).
99. C. W. Tang, S. A. VanSyke, "Organic electroluminescent diodes," *Appl. Phys. Phys. Lett.*, 51, 913 (1987).
100. M. Stolka, J. F. Janus, D. N. Pai, "Hole transport in solid solutions of a diamine in polycarbonate," *J. Phys. Chem.*, 88, 4707 (1984).
101. Y. Shirota, "Organic materials for electronic and optoelectronic devices," *J. Mater. Chem.*, 10, 1 (2000).
102. K. Okumoto, Y. Shirota, "Development of new hole-transporting amorphous molecular materials for organic electroluminescent devices and their charge-transport properties," *Mater. Sci. Eng.*, B85, 135 (2001).
103. T. H. Lee, K. L. Tong, S. K. So, L. M. Leung, "High mobility hole transporting polymers for electroluminescence applications," *JAPPL*, 44(1B), 543–545 (2005).
104. T. Nakano, T. Yade, M. Yokoyama, N. Nagayama, "Charge transport in a pi-stacked poly(dihenzofulvene) film," *Chem. Lett.*, 33, 296–297 (2004).
105. E. Bellmann, S. E. Shaheen, R. H. Grubbs, S. R. Marder, B. Kippelen, N. Peyghambarian, "Design and synthesis of stable triarylamines for hole-transport applications," *Chem. Mater.*, 11, 399 (1999).
106. Y. Kim, K. H. Bae, Y. Y. Jeong, D. K. Choi, C. S. Ha, "An electronically active molecularly doped polyimide hole injection layer for an efficient hybrid organic light-emitting device," *Chem. Mater.*, 16, 5051–5057 (2004).
107. E. Bellmann, S. E. Shaheen, R. H. Grubbs, S. R. Marder, B. Kippelen, N. Peyghambarian, "Hole transport polymers with improved interfacial contact to the anode material," *Chem. Mater.*, 12, 1349 (2000).
108. D. Ma, G. Wang, Y. Hu, Y. Zhang, L. Wang, X. Jing, C. S. Lee, "A dinuclear aluminum 8-hydroxyquinoline complex with high electron mobility for organic light-emitting diodes," *Appl. Phys. Lett.*, 82, 1296 (2003).
109. T. Yasuda, Y. Yamaguchi, D.-C. Zou, T. Tsutsui, "Carrier mobilities in organic electron transport materials determined from space charge limited current," *Jpn. J. Appl. Phys.*, 41, 5626 (2002).
110. C.-C. Wu, T.-L. Liu, W.-Y. Hung, Y.-T. Lin, K.-T. Wong, R.-T. Chen, Y.-M. Chen, Y.-Y. Chien, "Unusual nondispersive ambipolar carrier transport and high electron mobility in amorphous ter(9,9-diarylfluorene)s," *J. Am. Chem. Soc.*, 125, 3710 (2003).
111. J. Bettenhausen, P. Strohriegl, W. Brutting, H. Tokuhisa, T. Tsutsui, "Electron transport in a starburst oxadiazole," *J. Appl. Phys.*, 82, 4957 (1997).

112. S. Naka, H. Okada, H. Onnagawa, T. Tsutsui, "High electron mobility in bathophenanthroline," *Appl. Phys. Lett.*, 76, 197 (2000).
113. H. B. Goodbrand, "Low temperature arylamine processes," U.S. Patent 5,648,539 (1997).
114. G. N. Tew, M. U. Pralle, S. I. Stupp, "Supramolecular materials with electroactive chemical functions," *Angew. Chem.*, 112, 527 (2000).
115. M. Behl, E. Hattemer, M. Brehmer, R. Zentel, "Tailored semiconducting polymers: Living radical polymerization and NLO-functionalization of triphenylamines," *Macromol. Chem. Phys.*, 203, 503 (2002).
116. K. A. Higginson, X. M. Zhang, F. Papadimitrakopoulos, "Thermal and morphological effects on the hydrolytic stability of aluminum tris(8-hydroxyquinoline) (Alq3)," *Chem. Mater.*, 10, 1017 (1998).
117. Y. Hamada, "The development of chelate metal complexes as an organic electroluminescent material," *IEEE Trans. Electron Dev.*, 44(8), 1208–1217 (1997).
118. S. Dailey, W. J. Feast, R. J. Peace, I. C. Sage, S. Till, E. L. Wood, "Synthesis and device characterisation of side-chain polymer electron transport materials for organic semiconductor applications," *J. Mater. Chem.*, 11, 2238–2243 (2001).
119. T. H. Lee, K. M. Lai, L. M. Leung, "Hole-limiting conductive vinyl copolymers for AlQ3-based OLED applications," *Polymer*, 50, 4602–4611 (2009).
120. L. Leung, C. F. Kwong, C. C. Kwok, S. K. So, "Organic polymer thick film light emitting diodes (PTF-OLED)," *Displays*, 21, 199–201 (2000).
121. L. M. Leung, C. F. Kwong, S. K. So, "Effects of additives in polymer thick film organic light emitting diodes," *Displays*, 23(4), 171–175 (2002).
122. C. W. Tang, S. A. Van Slyke, C. H. Chen, "Electroluminescence of doped organic thin films," *J. Appl. Phys.*, 65, 3610 (1989).
123. M. Suzuki, S. Tokito, F. Sato, T. Igarashi, K. Kondo, T. Koyama, T. Yamaguchi, "Highly efficient polymer light-emitting devices using ambipolar phosphorescent polymers," *Appl. Phys. Lett.*, 86, 103507 (2005).
124. P. T. Furuta, L. Deng, S. Garon, M. E. Thompson, J. M. Frechet, "Platinum-functionalized random copolymers for use in solution-processible, efficient, near-white organic light-emitting diodes," *J. Am. Chem. Soc.*, 126, 15388 (2004).
125. L. M. Leung, J. Wang, M. Y. Wong, Y. C. Law, K. M. Lai, T.-H. Lee, "Charge balance conductive vinyl polymers for homojuncton organic light emitting diodes (OLEDs)," *IEEE Proc., SCET2012*, Xian, May 2012, 594–597.
126. L. T. Ho, Ph.D. Thesis, "Synthesis and characterization of electroluminescent bipolar small molecules and polymers," 2007, Hong Kong Baptist University, Hong Kong.
127. L. M. Leung, J. Wang, "Synthesis and characterization of highly soluble blue emitting poly(2-vinyl-anthracene) with 9,10-di(2-naphthalenyl) and 9,10-di(3-quinolinyl) substituents," *Dyes Pigments*, 99, 105–115 (2013).
128. J. Wang, L. M. Leung, S. K. So, C. Y. H. Chan, "Blue fluorescent conductive homopolymer poly(9,10-di(1-naphthalenyl)-2-vinylanthracene) and its highly soluble copolymers with styrene or 9-vinyl carbazole," *Polym. Int.*, 63, 363–376 (2014).
129. J. Wang, L. M. Leung, "Self-assembly and aggregation of ATRP prepared amphiphilic BAB triblock copolymers contained nonionic ethylene glycol and fluorescent 9,10-di(1-naphthalenyl)-2-vinylanthracene/1-vinyl-pyrene segments," *Eur. Polym. J.*, 49, 3722–3733 (2013).
130. J. Wang, L. M. Leung, S. K. So, C. Y. H. Chan, M. Y. Wong, "Synthesis and Characterization of Greenish-blue Emitting Vinyl Copolymer Containing Pyrene and Triarylamine Moieties," *Polymer International*, 63(10), 1797–1805 (2014).
131. T. Nakano, T. Yade, Y. Fukuda, T. Yamaguchi, S. Okumura, "Free-radical polymerization of dibenzofulvene leading to a π-stacked polymer: Structure and properties of the polymer and proposed reaction mechanism," *Macromolecules*, 38, 8140–8148 (2005).
132. T. Nakano, K. Takewaki, T. Yade, Y. Okamoto, "Dibenzofulvene, a 1,1-diphenylethylene analogue, gives a π-stacked polymer by anionic, free-radical and cationic catalysts," *J. Am. Chem. Soc.*, 123, 9182–9183 (2001).
133. W. Yin, M.Phil. Thesis, "Synthesis and characterization of substituted poly(dibenzofulvenes) and some novel fluorescent dyes," 2010, Hong Kong Baptist University, Hong Kong.
134. T. S. Chung, "The effect of LiCl on PBI and polysulfone blend fibers," *Polym. Eng. Sci.*, 34, 428 (1994).

135. T. Yade, T. Nakano, "Anionic polymerization of 2,7-di-*t*-butyldibenzofulvene: Synthesis, structure, and photophysical properties of the oligomers with a p-stacked conformation," *J. Polym. Sci. Part A: Polym. Chem.*, 44, 561–572 (2006).
136. A. Cappelli, S. Galeazzi, G. Giuliani, M. Anzini, M. Aggravi, A. Donati, L. Zetta, A. Caterina Boccia, R. Mendichi, G. Giorgi, E. Paccagnini, S. Vomero, "Anionic polymerization of a benzofulvene monomer leading to a thermoreversible π-stacked polymer. Studies in macromolecular and aggregate structure," *Macromolecules*, 41(7), 2324–2334 (2008).
137. M. Y. Wong, L. M. Leung, "A facile approach to the synthesis of substituted dibenzofulvenes—Precursors to pi-stacked poly(dibenzofulvene)s," *Tetrahedron*, 66, 3973–3977 (2010).

21 Advanced Thermoplastic Composites*

*Salvatore Iannace, Gianfranco Carotenuto,
Luigi Sorrentino, Mariano Palomba, and Luigi Nicolais*

CONTENTS

21.1 Introduction 694
21.2 Thermoplastic Matrices 695
    21.2.1 Polyarylethers 695
    21.2.2 Polyarylsulfone 696
    21.2.3 Polyetherketone 696
    21.2.4 Polyethersulfone 696
    21.2.5 Polyetherimide 696
    21.2.6 Polyetheretherketone 697
    21.2.7 Polybenzimidazole 698
    21.2.8 Polyaryletherketone 698
    21.2.9 Polyphenylene Sulfide 698
    21.2.10 Polyamideimide 699
    21.2.11 Liquid Crystalline Polymers 699
    21.2.12 Polyesters (PET, PEN) 700
21.3 Fibers 700
    21.3.1 Carbon Fibers 700
    21.3.2 Aramid Fibers 701
    21.3.3 High-Performance Glass Fibers 701
    21.3.4 Other Fibers 702
21.4 Advanced Nanocomposites 702
    21.4.1 PEEK Nanocomposites 702
    21.4.2 Polyester (PET, PEN) Nanocomposites 705
    21.4.3 Polysulfone Nanocomposites 705
    21.4.4 Thermoplastic Polyimides (PI, PAI, and PEI) 706
    21.4.5 Liquid Crystalline Polymers 706
    21.4.6 Polyphenylene Sulfide 706
21.5 Preimpregnated Composites 707
    21.5.1 Preimpregnated, True Thermoplastics 707
    21.5.2 Preimpregnated Pseudothermoplastics 707
21.6 Advanced Composites and Their Processing Technologies 708
    21.6.1 Processing of Thermoplastic Composite Material 708
        21.6.1.1 Heating 708
        21.6.1.2 Consolidation/Impregnation/Solidification 709
        21.6.1.3 Semifinished Products 709
    21.6.2 Autoclave 709
    21.6.3 Compression Molding 710

---

* Based in parts on the first-edition chapter on advanced thermoplastic composites.

21.6.4 Fiber and Tape Placement 710
    21.6.4.1 Localized Fiber Placement 710
    21.6.4.2 Automated Fiber Placement 711
21.6.5 Filament Winding 711
21.6.6 Pultrusion 712
21.6.7 Commingled Yarns 713
21.6.8 Diaphragm Forming 713
21.6.9 Thermoforming 713
21.6.10 Injection Molding 714
21.6.11 Stretch-Draw Processes 715
21.6.12 Matched-Mold Techniques 715
21.6.13 Sandwich Technology 716
    21.6.13.1 Fiber Stitching 716
    21.6.13.2 Z-Pinning 716
    21.6.13.3 Other Methodologies 716
21.7 Advanced Thermoplastic Foams 716
    21.7.1 Aromatic Polyketone-Based Foams 717
    21.7.2 Polyester-Based (PEN-Based) Foams 718
    21.7.3 Foams Based on Sulfone Polymers 718
    21.7.4 Foams Based on Thermoplastic Polyimides (PI, PAI, and PEI) 719
    21.7.5 Foamed Composites 719
21.8 Environmental Resistance 719
References 721

21.1 INTRODUCTION

Thermoplastic composites have gained importance for use as advanced materials in high-performance applications [1]. This field is still dominated by thermosetting matrix composites. Although the amount of thermoplastic-based high-performance composite materials used at present is still limited compared with thermoset-based composites, their use is growing continuously. New production techniques and the solution to the main processing issues contribute to widening the application fields of thermoplastic matrix composites. The rapid fabrication of thermoplastic composites gives significant cost reduction compared with thermosets. Thermoplastic composites: offer a reduction in initial and in-service costs; allow recycling of production scraps; do not release volatile organic compounds during curing; and since the polymer has been cured previously, do not require special storage conditions. Furthermore, thermoplastics are more resistant to environmental damage. Thermoplastic composites based on polyetheretherketone (PEEK), polyphenylene sulfide (PPS), and polyamideimide (PAI) demonstrated improvements in mechanical properties while retaining good environmental resistance. The development of impregnation technologies parallel with the development of thermoplastic polymers enables the fulfillment of the requirements of high-performance applications. Nowadays, major efforts are being expended in the development of new fabrication processes.

During the 1970s, some high-performance thermoplastic polymers (HPTPs) were developed, such as polysulfone (PSF), polyethersulfone (PES), PEEK, and PPS. All have an aromatic backbone structure and give good composites with excellent thermal resistance characteristics [2,3]. The unreinforced materials have a density, stiffness, and strength similar to those of the thermosets and, unless highly crystallized, a strain to failure well in excess of 10%. These polymers require high processing temperature and, in some cases, annealing periods to give optimum properties. In addition, their viscosity–temperature characteristics are such that until recently, it was not possible to impregnate fibers successfully.

There is a relatively long lead time in implementing a new material into aerospace structural applications. This involves property characterization, development of manufacturing processes, and subscale and full-scale evaluation, leading to in-service testing of a demonstration part prior to production [4]. Some parts have been developed recently based on PEEK and PPS thermoplastics.

21.2 THERMOPLASTIC MATRICES

An understanding of the chemical structure of the polymers and how it affects the properties is very useful, though it is not the purpose of this chapter to examine polymer chemistry in detail. The glass–rubber transition temperature (T_g) usually determines the upper service temperature of advanced composites. The upper service temperature requirements are 100°C for transport aircraft and helicopters and up to 175°C for fighter aircraft, and much higher temperatures can be encountered in specific areas, e.g., close to the engines.

All of the HPTPs are heavily aromatic in nature [5–8], usually with flexibilizing groups in the backbone [9]. PPS, PEEK, and polyetherketone (PEK) are semicrystalline [10,11]. The polymer chains are able to align themselves, forming physical linkages between adjacent chains to produce a crystalline morphology. This is possible because of the repeating sequences and the absence of bulky side groups. The semicrystalline polymers retain some stiffness and strength above T_g, and the retention depends on the level of crystallinity. Under normal conditions, PPS can develop up to 50% crystallinity, and PEEK and PEK up to 40%. Another feature of semicrystalline polymers is that it is necessary to process above the crystalline melting point (T_m).

PS, polyphenylsulfone (PPSU), and PES are all amorphous, i.e., they do not develop a defined crystalline structure. Several of the polymers contain sulfone linkages that provide a stiff unit in the polymer backbone and hence increase T_g, but these are also bulky groups, which prevent crystallization. The T_g values of the sulfone-containing polymers range from 190°C for PSF to 230°C for PES.

Polyetherimide (PEI) is the first of the polymers containing an imide group, with a T_g of 216°C. The imide link is very stiff, giving higher T_g values, and is thermally stable. It is therefore commonly found in the higher-temperature thermoplastic and thermoset polymers. In the case of PEI, the imide group is present in relatively low proportions; therefore, the polymer has a relatively low T_g and is easy to process. Where the imide is present in higher proportions, the polymers may have very high T_g values but can also be difficult to process due to the higher viscosities. For example, PAI has a very high viscosity and must be annealed to achieve the highest T_g and optimum properties. Thermoplastic polyimides (PI) are frequently impregnated as monomers or prepolymers due to the high viscosity in their fully polymerized state. The polymerization is completed during the fabrication process or during a postfabrication stage. Several different polyimides are available. Most of them are amorphous, but some have crystallinities up to 30%.

Liquid crystalline polymers have rigid backbones (frequently an aromatic polyester), which become highly aligned during processing so as to produce very ordered structures in the solid state. The polymers are therefore intrinsically anisotropic.

The characteristics required for matrices of high-performance composite materials are summarized here. Polymer matrices are discussed in some detail in view of their importance in the high-performance composite industry.

21.2.1 POLYARYLETHERS

Polyarylethers (PAEs) are characterized by high toughness and high resistance to heat and ultraviolet (UV) exposure. Of importance also are their transparency, wrap resistance, excellent flexural recovery, elevated elastic limits, and good electrical and mechanical properties, including outstanding creep resistance. In addition, they have excellent resistance to radiation and oxidation. Their thermal expansion rate is similar to that of metals. However, the material is subject to environmental stress cracking, particularly in the presence of aromatic or aliphatic hydrocarbon. PAE's properties include a heat distortion temperature of 180°C, a tensile strength of 69–165 MPa, and a flexural modulus rating of 2000–9500 MPa.

21.2.2 Polyarylsulfone

Polyarylsulfone (PAS) is a thermoplastic material with a relatively low melt-processing temperature but excellent mechanical properties [8,12]. It is hydrolytically stable and has a heat distortion temperature of about 200°C at 1900 kPa, a tensile strength over 120 MPa, and a flexural modulus of 6800 MPa. It is also highly resistant to alkali, mineral acids, and salt solutions under stress. Some of its applications include aircraft interiors, electrical and electronic items, and frozen-food packaging.

21.2.3 Polyetherketone

PEK is a high-performance, semicrystalline, aromatic ketone-based thermoplastic that is heat stable and easy to process. As a member of the ketone family, the polymer has properties like those of PEEK, i.e., good chemical resistance, exceptional rigidity, high toughness, strength, load-bearing capabilities, good radiation resistance, the best fire safety characteristics of any thermoplastic material, and the ability to be easily processed by melting. PEK is designed for advanced composites. It has a continuous service temperature of 260°C, a glass transition temperature of 200°C, and a very slow rate of crystallization that is suitable for processes with low rates of cooling from the melt.

21.2.4 Polyethersulfone

PES is a high-temperature engineering thermoplastic of the PSF family, with a T_g of 220–250°C, which begins to degrade between 450°C and 550°C. The maximum continuous working temperature is 170–190°C. It can be used for a long time at 200°C without loss of strength. Its dimensional changes at 200°C are negligible, although small variations in the electrical performance occur in the range of 0–200°C. Commercial polymers do not crystallize, because the structure, although normally linear, easily tends to link. The polymer has a very low flammability and is resistant to creep. It is recommended for load-bearing applications up to a temperature of 180°C. Even without flame retardants, it offers low flammability and has little change in its dimensions and electrical properties in the temperature range of 0–200°C. Its smoke and gas emissions are very low. At room temperature, its behavior is like a traditional engineering thermoplastic, being rough, rigid, and strong, with outstanding long-term load-bearing properties.

PES has good resistance to radiation (e.g., x-rays, β-rays, and γ-rays). Its chemical resistance is dependent on temperature and can be improved by annealing at 200°C. It has good resistance to aqueous acids, bases, and almost all inorganic solutions. All the common sterilizing solutions and anesthetics can be safely used with PES (many cleaning and degreasing solvents are based on chlorinated and fluorinated hydrocarbons). Unless PES is heavily stressed, it can be cleaned by most of these solvents.

21.2.5 Polyetherimide

PEI is an amorphous engineering thermoplastic. It is characterized by high heat resistance, high strength and high modulus, excellent electrical properties (stable over a wide range of temperatures and frequencies), and very good processability. The unmodified PEI is transparent, with an inherent flame resistance and a low smoke evolution. Its heat deflection temperature is 200°C at 1800 kPa. PEI has a T_g of 215°C; its smoke evolution remains virtually unchanged at frequencies of $60–10^9$ Hz and temperatures of 30–80°C. It has a high volume resistivity (about 6.7×10^{17} Ω cm) and an elevated dielectric strength. Its arc resistance exceeds 120 s. A key feature of PEI is its ability to maintain properties at elevated temperatures. For example, at 180°C, the tensile strength and flexural modulus of the glass-reinforced grades are high. PEI also has good creep resistance, as indicated by its apparent modulus of 24 MPa after 1000 h at 82°C under an initial load of 35 MPa.

This resin resists a broad range of chemicals under varied conditions of stress and temperature. For example, the resistance of PEI to the mineral acids is very high. However, it is attacked by such partially halogenated solvents as methylene chloride, trichloroethane, and strong acids. Its resistance to UV radiation is good, with a negligible change in tensile strength after 1000 h of xenon arc exposure. Also, the resistance to γ radiation is high, with a strength loss of less than 6% after 500 Mrad of exposure to cobalt-60 at the rate of 1 Mrad/h. Hydrolytic stability tests show that more than 85% of PEI's tensile strength is retained after 10,000 h of immersion in boiling water. This material is suitable for short-term or repeated steam exposure.

21.2.6 Polyetheretherketone

This thermoplastic polymer is a member of the ketone family that was developed in Great Britain, and composites based on it have been extensively investigated [13–15]. It offers an excellent combination of thermal and combustion characteristics. The polymer is semicrystalline [15,16] and has an exceptional resistance to chemical attack. The upper working temperature is 250–315°C because the melting point is 343°C. The T_g is 143°C. PEEK is a high-temperature crystalline thermoplastic resin suitable for many high-performance unreinforced or reinforced components [14]. The aromatic structure contributes to its high-temperature performance, and its crystalline character makes it resistant to organic solvents and dynamic fatigue, and helps it retain ductility in short-term heat aging [13]. PEEK molded compounds absorb much less water than any other thermoplastic [17]. They have good resistance to aqueous reagents, with long-term proven performance at 220°C. Their resistance to attack is over a wide pH range from 60% sulfuric acid to 40% sodium hydroxide at elevated temperatures. However, PEEK is attacked by some concentrated acids. No solvent attack has yet been observed on molded parts, although some solvents cause the crazing of highly stressed coated wire. This problem can be eliminated by orienting PEEK below its melting point. The material is molded or extruded by using conventional thermoplastic equipment, working at 350–400°C. Their mold shrinkage is about 1% for the unreinforced and 0.1–1.4% for the reinforced grades, depending on their fiber orientation. It is used in carbon fiber composites because of the excellent adhesion to carbon fibers [11,16]. Its flexural yield strength is above 315 MPa, and the flexural modulus is up to 8200 MPa.

At room temperature, PEEK is tough, strong, and rigid [10], with excellent load-bearing properties over long periods and an exceptional resistance to abrasion. On a short-term basis, this material is suitable for service temperatures in excess of 300°C. It also has excellent thermal stability for continuous operation (a lifetime of 50,000 h can be expected at 260°C). In addition, using flame-retardant additives or halogens, PEEK has an extremely low rate of smoke emission.

Tests to date show that PEEK has good resistance to radiation, but the PEEK moldings degrade by UV during outdoor weathering. When PEEK is natural or pigmented, the effect on it is minimal over a 12-month period. Paint or another applicable coating is recommended for use in prolonged or extreme weather conditions.

Fiber-impregnated PEEK feedstock can be prepared by a melt-impregnation technique. Tapes and sheets can be wound or pressed to produce manufacts. Care must be taken in cooling the PEEK products in order to obtain the correct degree of crystallinity, although more recent grades include a nucleation control agent, which makes cooling time less critical. PEEK film can be laminated to itself or to other substrates. Its bond strength depends on surface preparation and the type of adhesive used.

Welding of PEEK/carbon fiber composite laminates can be carried out by electrical heating. A prepreg tape is introduced between the laminates, and electrical power is supplied through the carbon fibers of the tape. The welding of PEEK/carbon fiber composite laminates has been modeled accounting for the electrical heating and phase-change kinetics in the polymeric matrix. The analysis highlights the relevance of processing conditions on welding, allowing the optimization of process variables, i.e., the welding time and the crystallinity content of the welded system.

21.2.7 Polybenzimidazole

Polybenzimidazole (PBI) is a thermoplastic polymer with a glass transition temperature of 430°C. Its characteristics are as follows: ultrahigh heat distortion temperature (440°C), flame retardance, and nonflammability in air. The material can withstand steady temperatures up to 427°C and short bursts up to 760°C. This material is reported to resist steam at 340°C and 15 MPa of pressure. When exposed to saturated steam, PBI absorbs only 0.4% moisture. It resists a wide range of chemicals, including harsh acids and bases.

This plastic has high mechanical and physical strength properties, including high compression strength. PBI can bear loads for short periods at temperatures up to 650°C. Its friction and thermal expansion coefficients are very low.

This wholly aromatic heterocyclic polymer is fabricated by sintering under high pressure. The low-molecular-weight PBI flows better than its high-molecular-weight (HMW) counterpart. However, HMW PBI outgases less during processing, making it really suitable to mold large parts.

PBI can be used to replace metals, ceramics, carbon, and other materials where the industry requires high heat and corrosion resistance, as in chemical and oil processing, aerospace, and transportation. The main commercial applications of PBI include its use in valve seats, seals, electrical connections, thrust washers, bearings, and other mechanical components. It can become the material of choice for parts that must resist temperatures up to 400°C.

21.2.8 Polyaryletherketone

Compared with other temperature-stable thermoset resins and fluoropolymers, polyaryletherketone (PAEK) has clear economic and processing advantages, particularly for new materials that can be highly stressed thermally and mechanically. PAEK is a leading material among the high-temperature-stable thermoplastics. This family of plastics allows continuous operating temperatures of 250°C and, depending on the type of short-term peak load, up to 350°C. The glass transition and melting temperatures are thermodynamic quantities that depend on the ratio of the ketone to the ether groups. Various complicated configurations can be obtained, such as polyetherketoneetherketoneketone (PEKEKK).

The properties of this family of plastics include a tensile strength at break of 85 MPa, an elongation at break of 56%, a tensile modulus of elasticity of 4100 MPa, a tensile stress at yield at 25°C of 104 MPa and at 160°C of 37 MPa, an elongation at yield of 6% at 25°C and of 2% at 160°C, and no break using an unnotched Izod impact test.

The processing flow behavior of PAEK does not differ fundamentally from that of other partially crystalline thermoplastics. The shear rate is similar to that of nylon 6 and PBT at 25°C above the melting point. Besides having good mechanical and rheological properties, PAEK is characterized by its favorable behavior in fire. The density of PAEK fumes in a fire is the lowest of the thermoplastics, and PAEK has exceptionally low corrosive and toxic fumes. The quantity of fire is quite low and meets aviation regulations for interior use. PAEK has high hydrolysis resistance and good resistance to many different chemicals.

21.2.9 Polyphenylene Sulfide

The crystalline, high-performance engineering thermoplastic PPS is characterized by outstanding high-temperature stability, inherent flame resistance, and a broad range of chemical resistance. PPS resins and compounds provide various combinations of high mechanical strength, impact resistance, and electrical insulation, with its high arc resistance and low arc tracking. The pigmented PPS compounds include several grades that are suitable to support current-carrying parts in electrical components. They are essentially transparent to microwave radiation.

Unreinforced PPS resins are also available for use in slurry coating and electrostatic spraying. Resin coatings are suitable for many applications, e.g., chemical processing equipment.

PPS is also available with long-fiber glass, with carbon, and in other reinforced forms [12,18]. The stampable sheet type contains fiber mat reinforcement and can be processed by compression molding. Other forms can contain predesigned reinforcement patterns for different processes, such as laminating and thermoforming. These cross-linked types of resins are more crystalline than any of the sulfones, which are generally classified as amorphous. They are quite stiff, with a flexural modulus ranging from 12,000 to 17,000 MPa. Their tensile strengths range from 69 to 172 MPa.

The polymer has excellent resistance to a wide range of chemicals, even at high temperatures. In fact, below 200°C, the resin cannot be dissolved in an organic solvent.

More recently developments have produced new linear PPSs that have a lower proportion of inorganic impurities than the conventional material. They are characterized by higher ultimate strength, elongation at break, flexural strength, and notched impact strength. However, the cross-linked product is somewhat more rigid.

The linear PPS is partially crystalline, with pronounced thermal transition ranges similar to those of polyethylenterephthalate (PET), which run from 85°C to 100°C for the T_g. Its melting point, T_m, is 280–285°C.

21.2.10 POLYAMIDEIMIDE

PAIs are engineering thermoplastics characterized by excellent dimensional stability, high tensile strength at elevated temperatures, and very good impact resistance. The polymer maintains structural integrity, during continuous use, up to a temperature of 260°C. Different grades are available, e.g., general purpose, injection moldable, poly(tetra fluoro ethylene) (PTFE)/graphite wear-resistant compounds, with 30% carbon fiber-reinforced compounds, and with 30% glass fiber-reinforced compounds. The room-temperature tensile strength of an unfilled PAI is about 200 MPa, and its compressive strength, about 220 MPa. At 232°C, its tensile strength is about 65 MPa, i.e., as strong as many engineering plastics at room temperature. Continued exposure at 260°C up to 8000 h produces no significant variation in its tensile properties. The flexural modulus in an unfilled grade is 5,000 MPa, and it can be increased up to 20,000 MPa using a carbon fiber reinforcement. The degree of retention of its modulus at temperatures to 260°C is on the order of 80%. Its creep resistance, even at high temperatures and under load, is among the best of the thermoplastics, and its dimensional stability is extremely good.

The polymer is extremely resistant to flame and has low smoke generation. The radiation resistance is good, with a tensile strength that drops to about 5% after exposure, to 10^9 rad of γ radiation. Its chemical resistance is very good, being virtually unaffected by aliphatic and aromatic hydrocarbons as well as halogenated solvents and most acid and base solutions. PAI is attacked, however, by some acids at high temperatures, by steam at high pressure and temperatures, and by strong bases. The moldings in PAI absorb moisture in humid environments or when immersed in water; however, the rate is low, and the process is reversible. For example, at 23°C with 50% relative humidity, the material absorbs about 1% by weight in 1000 h. The parts can be easily restored to the original dimensions by drying.

The material is used in structural parts that require high strength at high temperatures, such as aerospace, business equipment, industrial chemical plants, heavy-duty trucks, and underground environments. Automotive parts in which PAI is used are as follows: power and valve trains, piston skirts, tappets, piston rings, valve stems, and timing gears.

21.2.11 LIQUID CRYSTALLINE POLYMERS

Liquid crystalline polymers (LCPs; sometimes called superpolymers), which were commercially introduced in 1984, are called self-reinforcing plastics because of their densely packed fibrous polymer chains. LCPs have outstanding strength at extreme temperatures, excellent mechanical property retention after exposure to weathering and radiation, good dielectric strength as well as arc resistance, and easy processability. LCPs' ease of processing gives them the ability to fill long, narrow molds, which makes them eminently suitable for such high-performance parts. LCPs are available in both amorphous and crystalline grades. The amorphous types, with their high strength-to-weight ratios, are particularly useful for weight-sensitive items in aerospace and military parts. Most LCPs can be injection-molded, extruded, thermoformed, and blow-molded. The crystalline grades, with glass and other fibers, meet the dimensional stability at high temperatures required of products for the electrical and electronics markets. The LCPs are all exceptionally inert and resist stress cracking in the presence of most chemicals at elevated temperatures, including the aromatic and halogenated hydrocarbons as well as strong acids, bases, ketones, and other aggressive industrial products. Their hydrolytic stability in boiling water is excellent. With regard to flammability, LCPs have an oxygen index ranging from 35% to 50%.

21.2.12 Polyesters (PET, PEN)

Among the most diffuse and commercially available polyesters, poly-ethylene-naphthalate (PEN) can be considered an HPTP, thanks to its high glass transition temperature ($T_g = 125°C$) and a melting temperature higher than 265°C. PET is an engineering polymer because of its glass transition temperature ($T_g = 75°C$) and could be included within HPTPs if conveniently reinforced with nanoparticles, which are able to enhance its mechanical as well as functional characteristics.

21.3 FIBERS

Fibers are the key materials used to make composites for structural applications [19,20]. The primary function of fiber is to reduce weight while maintaining or increasing the structural strength of the composite shape. Fibers are sold in the following forms: continuous filaments for filament winding, braiding, and pultrusion; fabrics and tapes for lamination or compression molding; and papers and felts.

Carbon, aramid, and high-performance glass are the three primary fibers used in advanced thermoplastic composites. In addition, many small-volume, specialized, and expensive fibers such as boron, quartz, ceramic, and metal are also used.

The type and amount of the fiber incorporated in a composite structure is determined largely by the final properties required. Average properties for various fibers are compared in Table 21.1.

21.3.1 Carbon Fibers

The term *carbon fibers* denotes both carbon and graphite fibers. The percentage of graphite in the fiber depends primarily on the final processing temperature [21]. The higher the temperature, the greater the percentage of graphite.

TABLE 21.1
Average Properties of Various Fibers Used to Reinforce Advanced Thermoplastic Matrices

Fiber Type	Specific Gravity	Tensile Strength, MPa	Tensile Modulus, GPa
Carbon	1.75	414.3	250
Aramid	1.45	375.0	129
S-2 glass	2.49	475.0	90
Boron	2.58	364.3	414
Ceramic	2.70	178.6	157
Ultra high modulus polyethylene	0.97	267.9	121

Source: Partridge, I.K. (ed.), *Advanced Composites*, Elsevier, New York, 1989.

Carbon fibers are manufactured from one of the following three precursors: polyacrylonitrile (PAN), rayon, and pitch. In general, the manufacturing process begins by removing volatile factions through heat and tension from an organic-based precursor containing a high percentage of carbon atoms, leaving behind carbon atoms. Production yield and final fiber properties vary depending on the precursor and specific process constraints. Overall, the key properties of carbon fibers are high strength, high modulus, low density, excellent resistance to fatigue and creep, low thermal and electrical conductivity, low thermal coefficients of expansion, and good stability at high temperatures [22].

PAN, the most common precursor for carbon fibers, must contain at least 95% acrylonitrile. The initial carbon fiber is stretched to approximately five times its normal length in an oxidizing atmosphere at temperatures ranging from 200°C to 240°C. This process obtains the desired molecular orientation and reduces fiber diameter. Once the desired molecular orientation is achieved, the fibers are carbonized in an inert atmosphere at temperatures ranging from 1400°C to 3000°C. The final processing temperature influences the degree of graphitization and the resulting fiber properties.

The manufacture of rayon-based carbon fibers is a two-step process. First, rayon filaments are heat-treated in an inert atmosphere through a series of steps of increasing temperatures reaching as high as 2815°C. Second, the filaments are subjected to tensile loads and stretched while at these high temperatures. This aligns the graphite layers in a parallel direction to the axis of the filaments. At these temperatures, rayon fiber converts to a combination of carbon molecules and graphite crystals. The degree of alignment of these molecules and crystals determines the fiber's modulus and ultimate strength. Rayon-based carbon fibers do not possess the stiffness of PAN or pitch-based fibers and therefore are not used in similar applications.

The newest commercial carbon fiber is produced from pitch. Through heat treatment, the petroleum pitch is converted to a liquid crystal state called mesophase. After heating the mesophase for long periods of time at temperatures of 400–500°C to increase the asphaltene content of the pitch, the liquid is forced through a fiber-producing spinneret. These fibers are then thermoset in an oxidizing atmosphere and subsequently graphitized with hot stretching.

Since any individual carbon fiber has a very small diameter and is consequently more difficult to handle and process into a finished part, fiber manufacturers gather fibers into bundles or tows consisting of 1,000–150,000 individual fibers. Tows are then wound into spools, woven into fabric, or chopped.

21.3.2 Aramid Fibers

Kevlar 49, the most used aramid fiber in advanced composites [2], possesses greater tensile and impact strength compared to the carbon fibers but only moderate modulus and compression properties. It is less dense than carbon fiber and inherently flame resistant. One drawback to aramid fiber is its high water absorbency, a particular disadvantage for its use in aircraft.

Aramids are produced by reacting *p*-phenylenediamine and diacid chloride in a polar solvent in the presence of a catalyst such as lithium chloride (LiCl). The solution must be kept cool and well stirred. After it is washed and dried, the polymer may be stored. The fiber is produced by dissolving the polymer in concentrated sulfuric acid and passing the solution through spinnerets. Then the fiber is drawn over an air gap and quenched in cold water; once dried, the fiber can be wound on a spool.

21.3.3 High-Performance Glass Fibers

Although many different types of glass fibers exist, only S-glass, S-2 glass, R-glass, and T-glass are considered advanced fibers and have been included here.

S-glass and S-2 glass are currently the only high-performance glass fibers being widely consumed. S-glass was developed in the early 1960s for high-performance military applications and, as such, requires strict military testing and certification. S-glass was developed in the late 1960s as a lower-cost, high-performance glass. The primary differences between S-glass and S-2 glass are the sizing and the quality control procedures. Mechanical properties are similar for both S-2 glass and S-glass, with tensile strength 35–40% higher and tensile modulus 18–20% higher than E-glass. Worldwide, two competing high-performance glass materials exist: R-glass and T-glass. The properties of R-glass fall between S-2 glass and the low-performance E-glass. Newly developed T-glass is similar in performance to S-2 glass.

In general, the major advantages of glass are lower cost, high tensile and impact strengths, and high chemical resistance. Disadvantages include low modulus, high fiber abrasiveness, low fatigue resistance, and poor adhesion to matrix resins. To overcome the poor adhesion, sizing must be added to the fiber surface.

Glass fibers are produced in two basic forms: continuous filament or staple fiber. The production process for each form is identical up to the fiber-drawing stage. Raw materials, sand, limestone, alumina, and magnesia are dry-mixed and melted in a high-refractory furnace at temperatures around 1300°C. The molten glass is either drawn directly into fibers or cast into marbles for subsequent fiber production.

Producing continuous fibers requires the feeding of molten glass through a multiplicity of holes and extruding to the desired dimensions. Once a desired diameter is obtained, the filament passes through a quenching station to fix the diameter. Sizing is applied to the filaments to couple the glass fiber to the resin in the final product. Sizing also eliminates static electricity; it lubricates and protects the fiber. After sizing, the filaments are gathered into strands.

21.3.4 OTHER FIBERS

Many special fibers, such as boron, quartz, and ceramics, either are commercially available or are being developed for use in polymer composites. These products are extremely expensive compared to the fibers previously discussed and are used only when their high-performance characteristics are required.

With an excellent mix of tensile strength and high modulus, boron fibers are used primarily in applications that require high compression strength. The high price and density of boron fibers limit their applications.

Both ceramic and quartz fibers are used in applications that require high-temperature resistance and radar transparency. Silicon carbide and alumina-based fibers are the ceramic fibers most commonly used.

A new reinforcement fiber is an extended-chain polyethylene (PE) that is manufactured by first dissolving PE in a solvent and then extruding a gel fiber. After removal of the solvent, highly extended, parallel polymer chains are obtained.

21.4 ADVANCED NANOCOMPOSITES

21.4.1 PEEK NANOCOMPOSITES

A wide variety of nanofillers have been investigated to develop PEEK-based nanocomposites. Most of the works were focused on the functionalization of nanoparticles to improve their dispersion in the polymeric matrix [23–25].

MWNT/PEEK nanocomposites [26], PEEK composites reinforced with nano–zinc sulfide (ZnS) and nano–titanium dioxide (TiO_2), nano-SiO_2/PEEK, nano-SiO_2/short-carbon-fiber ternary systems [27], PEEK/Si_3N_4 [28], PEEK/AlN, PEEK/inorganic fullerene–like tungsten disulfide (WS_2), nano-TiO_2/PEEK, and Montmorillonite (MMT, Cloisite 15A)/PEEK [29] were successfully produced by conventional plastic techniques (mainly twin-screw extrusion) and evidenced good dispersion and improvements in mechanical as well as functional (thermal, dynamic-mechanical) properties. Also, the use of compatibilizer was explored to enhance the dynamic-mechanical and thermal properties of PEEK/carbon nanotube (CNT) systems.

In the literature, there are many reports related to the integration of single-wall carbon nanotubes (SWCNTs) into polymer matrices [30,31], but only a few papers deal with such composites with a PEEK matrix. In particular, the exceptional properties reported for carbon CNTs [32,33] have motivated the development of new nanotube-based composites. In addition to their unique mechanical properties [34,35], which, combined with their high aspect ratio and nanoscale dimensions, make them excellent candidates for composite reinforcement, they also possess superior thermal and electrical properties [36]. Therefore, polymer–nanotube composites are expected to have many potential applications in several fields: the addition of CNTs could improve thermal transport properties of the polymer, as they are useful as connectors, thermal interface materials, and heat skins. Alignment of CNTs under a magnetic field leads to an increase in electric conductivity, so they can be applied in electronic packaging, self-regulating heaters, and positive temperature coefficient (PTC) resistors [37]. Furthermore, incorporation of CNTs into a polymeric matrix can provide structural materials with dramatically increased modulus and strength [38]. This behavior, combined with their low density, makes them suitable for the transport industry, especially for aerospace structures, where the reduction of weight is one of the main goals in order to reduce fuel consumption. Nowadays, the use of polymeric composites in the aeronautic industry is mostly associated with secondary parts of the aircraft. Nevertheless, they are starting to be used in structural components, such as wing panels, horizontal and vertical stabilizers, and some elements of the fuselage, in order to minimize fuel and manufacturing costs. So, materials easily processed by continuous methods should be employed. PEEK CNT-reinforced composites seem to be highly suitable candidates, due to the thermal environment and mechanical requirements of those parts.

To verify the potential improvement in the CNT dispersion caused by a preprocessing step based on mechanical treatments in alcohol media, and the possible influence on the properties of the composites, PEEK/SWCNT systems were prepared by following two different procedures. As-received raw arc-grown and laser-grown SWCNTs (named ZC1 and LC1) and the same sample purified with ethanol (named ZC1p and Lc1m) were directly mixed with dried PEEK powder. All types of PEEK/SWCNT composites contained concentrations of 0.1, 0.5, and 1 wt.% SWCNTs.

To study the thermal stability of the composites, thermogravimetric analysis (TGA) characterization has been carried out under a nitrogen atmosphere. The effect of SWCNT content on the thermal behavior of the composites is shown in Table 21.2, which presents the characteristic temperatures for PEEK/SWCNT systems. Samples with different SWCNT loading begin to lose weight nearly at the same temperature, but the maximum mass loss rate is usually delayed by a few degrees. The observed stabilization may be attributed to the barrier effect of CNTs, which hinders the diffusion of the degradation products from the bulk of the polymer to the gas phase. Another plausible explanation could be the higher thermal conductivity of the composite that facilitates heat dissipation within the sample. In other words, a more uniform and fine dispersion of the SWCNTs improves the interfacial adhesion between the CNTs and the matrix and restricts the thermal motion of the PEEK chains, thereby leading to composites that are thermally more stable.

Differential scanning calorimetry (DSC) was employed to evaluate the effect of SWCNTs on the crystallization and melting behavior of different types of PEEK/SWCNT composites. The crystallization temperature, T_c, is about 309°C for pure PEEK; the incorporation of 0.1–1 wt.% SWCNTs leads to a progressive shift of the crystallization peak toward lower temperatures. However, the apparent crystallization enthalpy ΔH_c and, therefore, the crystallinity remain almost unchanged. These observations are intriguing since nanotubes usually act as nucleating agents [39,40] that increase the crystallization rate, because the nucleation starts simultaneously in multiple centers; therefore, one expects a similar if not higher T_c for the composites. Table 21.3 shows experimental DSC data and the degree of crystallinity of all composites prepared. A close analysis of this table reveals that T_c is not significantly affected by the type of SWCNT.

TABLE 21.2
Characteristic Temperatures of the PEEK/SWCNT Nanocomposites Obtained from TGA Measurements under Nitrogen Atmosphere at Heating Rate of 10°C/min

Mat. (% SWCNTs)	T_i, °C	T_{10}, °C	T_{mr}, °C
PEEK	521	544	558
PEEK/LC1m (0.1)	542	557	565
PEEK/LC1m (0.5)	544	563	573
PEEK/LC1m (1.0)	546	568	577
PEEK/LC1 (0.1)	523	547	559
PEEK/LC1 (0.5)	525	549	561
PEEK/LC1 (1.0)	526	551	564
PEEK/ZC1p (0.1)	536	550	564
PEEK/ZC1p (0.5)	540	558	574
PEEK/ZC1p (1.0)	543	564	578
PEEK/ZC1 (0.5)	520	545	560
PEEK/ZC1 (1.0)	521	547	562

Note: T_{10}, temperature for 10% weight loss; T_i, initial degradation temperature obtained at 2% weight loss; T_{mr}, temperature of maximum rate of weight loss.

TABLE 21.3
DSC Crystallization and Melting Data of PEEK/SWCNT Nanocomposites

SWCNTs	T_c, °C	ΔH_c, J/g	$(1-\lambda)_c$, %	T_m, °C	ΔH_m, J/g	$(1-\lambda)_m$, %
PEEK	309.1	55.3	42.5	344.2	58.2	44.8
PEEK/LC1m (0.1)	307.5	57.8	44.5	343.5	58.5	45.0
PEEK/LC1m (0.5)	305.1	56.5	43.7	343.1	57.3	44.3
PEEK/LC1m (1.0)	303.3	54.4	42.3	342.6	55.7	43.2
PEEK/LC1 (0.1)	307.7	56.4	43.4	343.4	58.3	44.9
PEEK/LC1 (0.5)	305.8	55.6	42.9	343.0	58.0	44.8
PEEK/LC1 (1.0)	304.9	53.9	41.8	342.8	55.8	43.3
PEEK/ZC1p (0.1)	308.1	58.2	44.8	343.9	58.6	45.0
PEEK/ZC1p (0.5)	306.6	57.0	44.1	343.2	57.5	44.4
PEEK/ZC1p (1.0)	304.3	54.6	42.3	342.6	56.6	44.0
PEEK/ZC1 (0.5)	304.0	56.4	43.6	343.9	58.5	45.2
PEEK/ZC1 (1.0)	303.1	53.2	41.3	342.3	55.4	43.0

Note: ΔH_m, apparent melting enthalpy; ΔH_c, apparent crystallization enthalpy; T_c, crystallization temperature; T_m, melting temperature; $(1-\lambda)_c$ and $(1-\lambda)_m$: crystallization and melting crystallinities.

So definitively, TGA thermograms show a substantial increase in the matrix degradation temperatures with the incorporation of SWCNTs. Higher thermal stability is found for samples with improved CNT dispersion. DSC experiments indicate a decrease in the crystallization temperature with increasing SWCNT content, whereas the melting temperature remains almost constant. This behavior can be explained by a confinement effect of the polymer chains within the CNT network, which delays the crystallization process and leads to lower values of T_c for the composites. No significant differences are found within the level of crystallinity of these composites calculated from DSC measurements. At room temperature, samples containing 0.1 wt.% SWCNTs exhibit larger crystallite size than the raw matrix. At higher concentrations, the CNT network restricts the polymer chain diffusion and hinders the formation of large-size crystals. Furthermore, over the whole concentration range, composites including purified SWCNTs present bigger crystals.

21.4.2 Polyester (PET, PEN) Nanocomposites

Several nanoparticles of different shape factors have been considered for PET reinforcement. Platelet-like nanoparticles, such as MMT or graphene, have been demonstrated to allow strong enhancement of heat deflection and glass transition temperatures, elastic modulus, and dynamic-mechanical performance.

Clay nanoparticles can be used to improve the gas barrier properties [41] or to enhance the crystallization kinetics of the polymer [42]. However, due to the thermal degradation of the conventional organic modifier, more thermally stable modifiers should be used to allow good intercalation and to limit polymer degradation during extrusion [43].

Graphene [44] as well as expanded graphite [45] have been successfully melt-blended into PET to increase the mechanical (both static and dynamic-mechanical) and functional (thermal stability, electrical conductivity) properties. CNTs have been also used as a reinforcing nano-phase into the PET matrix and have shown similar performance in the improvement of graphite nanoparticles [46–49].

SiO_2-based PEN nanocomposites have been developed since 2004 by Ahn et al. [50]. They surface-modified the silica nanoparticles to improve their dispersion in the polymer during melt blending, and they observed an increase in the elastic modulus and elongation at break at low nanoparticle content (0.4 wt.%). The use of SiO_2 nanoparticles can also lead to an improvement of the thermal stability, dynamic-mechanical behavior, and crystallization kinetics [51].

PEN nanocomposites based on CNTs have been widely investigated. Kim et al. [52] obtained a huge improvement in the viscosity by adding 2.0 wt.% of multiwalled CNT (MWCNT), but they obtained a plateau in both tensile modulus and tensile strength after the addition of just 0.5 wt.% of nanofiller [53]. Further developments also showed increases in the crystallization kinetics [51,54], thermal stability, and elastic modulus above the glass transition temperature compared to the neat polymer [51].

Layered silicate nanoparticles have also been used to prepare PEN-based nanocomposites through the direct intercalation of PEN polymer chains from the melt into the surface-treated clay. Mechanical and barrier properties measured by dynamic-mechanical and permeability analysis showed significant improvements in the storage modulus and water permeability when compared to neat PEN [55].

Exfoliated graphite (EG) has been considered by Kim and Macosko [56]. They obtained a very good dispersion of graphene sheets, by means of a twin-screw extruder, which was responsible for the significant improvements in viscosity, electrical conductivity, gas barrier, and mechanical properties. EG was used by Sorrentino et al. [124] to prepare PEN nanocomposites as a matrix for foamed materials. They reported on the improved crystallization kinetics of nanoparticles, coupled with the nucleating activity for both crystals and cells.

21.4.3 Polysulfone Nanocomposites

Sulfone polymers have been widely investigated since 2009 as a matrix for nanocomposites. Aurilia et al. investigated the effects of SiO_2 and EG nanoparticles on the solvent sorption and mechanical properties of PES nanocomposites. Both SiO_2 and EG were able to improve the solvent resistance of the amorphous matrix, acting as a barrier to the solvent diffusion. However, higher barrier properties and significant improvement of the mechanical properties were obtained in the case of EG-based nanocomposites [57].

21.4.4 Thermoplastic Polyimides (PI, PAI, and PEI)

The dispersion of nanoparticles in polyimide melts is rather difficult due to the remarkable increase in viscosity during processing. For this reason, methods such as precursor curing, for the preparation of layered silicate nanocomposites [58], or sol-gel [59] have been used in the past. Other nanoparticles have been investigated to develop polyimide (PI) nanocomposites by using thermal imidization processes, such as Al_2O_3 [60], thermally stable aromatic amine-modified montmorillonites [61], dual-intercalating-agent-system montmorillonites [62], and CNTs [63]. Yudin and Svetlichnyi [64], more recently, were able to prepare nanocomposites by using melt-blending technologies at temperatures between 350°C and 360°C. They prepared graphene sheet-based nanocomposites using specially synthesized amorphous and semicrystalline polyimide matrices, and employed silicate (natural and synthetic) and carbon nanoparticles with different morphologies (tubes, platelets, disks, and spheres) to study the influence of nanoparticle morphology on the rheological behavior of nanocomposite melts during processing [64].

PAI/MMT nanocomposites have been prepared since 2002 by means of a solvent suspension technique to intercalate or exfoliate nanoparticles, followed by thermal imidization [65]. The sol-gel technique, also followed by thermal imidization, has been used to disperse MMT [66], silica nanoparticles, and multiwalled CNTs [67]. An increase in the solvent resistance as well as in the thermal stability was observed in all the studied cases.

PEI nanocomposites have been developed since 2001, when Huang et al. prepared PEI/montmorillonite nanocomposites by melt intercalation of nanoparticles, specifically modified with thermally stable aromatic amines, which showed a dramatic decrease in solvent uptake compared to the virgin PEI [68]. Other nanofillers used in PEI nanocomposites were SWCNTs, MWCNTs [69,70], nano-TiO_2, and short carbon fiber [71]. Solution casting was also used to disperse MWCNT, $BaTiO_3$ [72], and aluminum nitride nanoparticles [73].

21.4.5 Liquid Crystalline Polymers

LCP nanocomposites based on MMT nanoparticles were studied by Bandyopadhyay et al., who experienced a significant increase in rheological properties through the addition of 3.4% of dimethyl dehydrogenated tallow quaternary ammonium-modified MMT [74,75].

MWNTs were also used as a reinforcing phase in LCP nanocomposites. In order to facilitate MWCNT dispersion, Kumar et al. developed a melt-blending technique based on twin-screw extruder equipped with an ultrasonic treatment unit. MWCNTs were also used in LCP blends with polycarbonate (PC) [76] and PEI [77] to improve the thermal, mechanical-dynamic, mechanical, and electrical properties of the composites.

21.4.6 Polyphenylene Sulfide

A nanocomposite of PPS/$CaCO_3$, prepared by using a melt-mixing process, resulted in a slight improvement in the tensile strength but in a significantly increased fracture toughness of 300%. Authors attributed this behavior to the fact that $CaCO_3$ nanoparticles, acting as stress concentration sites, promoted cavitation at the particle boundaries during loading, leading to higher plastic deformation of the matrix.

WS_2 nanoparticles also were used to prepare PPS nanocomposites via melt blending [78]. The addition of inorganic fullerene-like WS_2 (IF-WS_2) with concentrations greater than or equal to 0.5 wt.% remarkably improved the mechanical performance of PPS, with an increase in the storage modulus of 40–75% in dynamic-mechanical tests.

The preparation and characterization of PPS nanocomposites based on PEI-wrapped single-wall nanotubes and inorganic fullerene-like nanoparticles of WS_2 was reported by Diez-Pascual et al. [23]. The wrapping of SWCNTs in PEI and the addition of IF-WS_2 nanoparticles resulted an

Advanced Thermoplastic Composites

effective method for improving the dispersion of SWCNTs, leading to enhanced mechanical properties and electrical conductivity.

21.5 PREIMPREGNATED COMPOSITES

Composites are available in the form of fully impregnated tapes or prepregs [79]. Some prepregs contain a fully polymerized, true thermoplastic; others contain a prepolymer needed to complete the polymerization chemistry during the processing stage. Table 21.4 presents the principal manufacturers and trade names of thermoplastic matrices.

21.5.1 PREIMPREGNATED, TRUE THERMOPLASTICS

The processing of fully impregnated true thermoplastic requires melting the polymer and applying moderate pressure to consolidate the prepreg layers. The process typically requires a few minutes at processing temperature and pressure of about 0.5–2 MPa. The processing temperature typically varies from 200°C above T_g for semicrystalline polymers to 150°C above T_g for amorphous polymers. The heat-up rate is not a critical operative parameter for thermoplastics. On the other hand, the cooling rate for semicrystalline polymers has to be controlled to ensure the appropriate crystallinity and microstructure. In this case, the crystallinity and the crystalline morphology affect the mechanical properties of the matrix.

The controlling factor in determining toughness is the level of crystallinity rather than the processing history. For PPS matrix composites, it has been shown that a rapid quenching followed by annealing at 200°C produced the optimum properties: a tougher polymer results as a consequence of the formation of smaller crystallites.

The mechanical properties of PEEK matrix composites vary relatively little for cooling rates ranging from 5°C/min to 700°C/min as a consequence of the invariance of the crystallinity level.

21.5.2 PREIMPREGNATED PSEUDOTHERMOPLASTICS

The need to control the polymerization reactions gives less flexibility to the processing cycle of the pseudothermoplastic matrix composites. PAI (Torlon) and PI (Avimid K-III) undergo condensation polymerization, and it is necessary to remove by-products and residual solvents during the processing. The processing of pseudothermoplastics is closer to thermoset polymers, requiring longer cycles than thermoplastic processing, and in some cases, the postcuring stage is needed to optimize the properties. The long cycles and the high temperatures represent strong limitations in the use of pseudothermoplastics; the use of rapid fabrication techniques remains impractical.

TABLE 21.4
Principal Manufacturers and Trade Names of Some Thermoplastic Matrices

Manufacturer	Matrix	Trade Name
Sabic IP	PEI	Ultem
Solvay	PES, PPSU	Radel
Solvay	PAI	Torlon
Du Pont	PI	Avimid K-III
Victrex	PEEK	Victrex PEEK
Chevron Phillips	Polyphenylene sulfide	Ryton
Taijin Kasei	PEN	Teonex

21.6 ADVANCED COMPOSITES AND THEIR PROCESSING TECHNOLOGIES

There are many fabrication processes to choose from for the manufacturing of components using continuous fiber-reinforced thermoplastic systems [9,80,81]. Among these processes, some are mature (such as vacuum forming and autoclave molding), and others are at an early industrial stage or still developing (e.g., fiber or tape placement, composite expansion). Others, such as roll forming, are variations of conventional metal-forming techniques. All of these processes, with the possible exception of autoclave molding, can provide a high-rate production capability.

The process selection methodology can be combined with the material selection criteria to improve the final composite performances by exploiting the interactions between materials engineers and processing ones. In some cases, the fabrication process is determined based on the fabrication equipment on hand.

21.6.1 Processing of Thermoplastic Composite Material

During manufacturing of thermoplastic composites, the polymer is heated above its melt temperature to achieve fiber or fabric wetting (impregnation) and then solidified by proper cooling. The manufactured part is ready for postprocessing steps such as machining, trimming, and surface finishing.

The high melt viscosity, a direct consequence of the high molecular weight, hinders or restricts the use of conventional thermoset impregnation techniques to produce properly impregnated composites with a thermoplastic matrix. Different methods, such as solution impregnation, hot melt coating, film stacking, commingled yarns, powder coating, and in situ polymerization of thermoplastics precursors, have been developed to help ease the problem of high viscosity. The processing of thermoplastic composite material is divided into three main steps: heating, impregnation/consolidation, and solidification, as shown in Figure 21.1 [82].

21.6.1.1 Heating

The prepreg material must be heated to melt the matrix resin and to facilitate the bonding between the surfaces. Several modes of heating are used in practice, depending on the manufacturing setup [83] and part size. In large parts, such as in the compression-molding process, the heat is transferred through the mold to the stacked material through conduction. In high-speed processes, such as filament winding and tape placement, heating is locally conveyed to the polymer through a hot gas torch. Infrared radiation or microwave ovens can also be used to melt the material [84,85]. In all cases, the thermoplastic matrix must be melted to lower the viscosity to values adequate for the shaping and impregnation of the part.

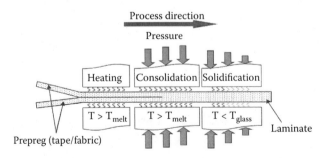

FIGURE 21.1 Melting of the thermoplastic matrix to lower the viscosity at values adequate for the shaping and impregnation of the part. (Reproduced from Mitschang, P. Manufacturing of thermoplastic fiber-reinforced polymer composites, in *Wiley Encyclopedia of Composites*, Second Edition, eds. L. Nicolais and A. Borzacchiello. John Wiley & Sons, 2012.)

21.6.1.2 Consolidation/Impregnation/Solidification

After reaching the processing temperature, voids and gaps between plies are filled by applying high pressure, which induces the transverse flow of the matrix through the laminate layers [86,87]. The duration and efficiency of the impregnation depend on the pressure applied, the melt viscosity, and the polymer–fiber interfacial compatibility.

The final composite is obtained by properly decreasing the temperature below the glass transition temperature of the polymeric matrix. During this step, an adequately high pressure should be maintained to avoid void formation, deconsolidation [88], part deformation, shrinkage, and the springback effect of the fiber network. For semicrystalline thermoplastics, the material crystallizes to a different extent depending on the cooling rate [89], so the crystallization of the polymers has to be carefully controlled to avoid internal stresses, which could reduce the structural properties of laminates [90]. These residual stresses can be removed by thermally heating the material close to its glass transition temperature or by using a controlled cooling rate.

21.6.1.3 Semifinished Products

In order to prepare composites, intermediate products can be used to reduce issues related to the impregnation processes and to speed up the process productivity. For such reasons, prepregs are prepared. They are semifinished products consisting of a single fiber reinforcement layer between two thermoplastic matrix layers. They can be already-impregnated (long fiber thermoplastic matrix reinforced thermoplastics compounded in an extrusion process, melt prepregs produced with rovings pulled through a liquid bath, made of the polymer melted through heating or dissolving in a solvent [91,92]) or unimpregnated prepregs [93] (commingled or hybrid rovings, powder-impregnated reinforcements, or polymeric film-stacked single layers). Mat-reinforced thermoplastics are plane semifinished products, whose the postprocess is limited to forming and joining the single parts.

21.6.2 Autoclave

Autoclave technology, widely used for the production of thermoset-based composites, has also been used to prepare some thermoplastic composites, due to the possibility to isotropically apply temperature (T) and pressure (p) to a preassembled semifinished product (Figure 21.2). The high viscosity of thermoplastics requires high temperatures and pressures; in turn, the process is relatively cost intensive, since autoclave structural and safety requirements are highly demanding.

After laying up the semifinished product (preimpregnated laminate, commingled yarns, stacked sequence of reinforcement and polymer films) onto a tool surface, the mold is enveloped in a bag, and vacuum is applied. This is inserted in the autoclave, which is then heated above the melting temperature of the polymer and pressurized to force the polymer to flow into the reinforcement. After this step, the internal volume of the autoclave is cooled to room temperature, and pressure is lowered to ambient

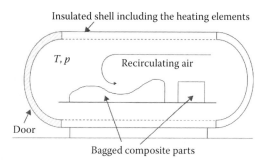

FIGURE 21.2 Schematic drawing of an autoclave. (Reproduced from Mitschang, P. Manufacturing of thermoplastic fiber-reinforced polymer composites, in *Wiley Encyclopedia of Composites*, Second Edition, eds. L. Nicolais and A. Borzacchiello. John Wiley & Sons, 2012.)

pressure. This technique can be, in principle, applied to all polymers but exhibits some disadvantages such as the high temperature needed (a factor depending on the polymer treated), the relatively low heating rate in the range of 3–4 K/min [94], the slow pressure buildup, and the high cost of acquisition.

21.6.3 Compression Molding

The compression-molding technique consists of pressure application on molten semifinished lay-ups through plates or molds. Different compression-molding arrangements are available, each characterized by a high degree of automation, short cycle times, excellent dimensional accuracy, and reproducibility. It allows the production of very different article sizes, from the centimeter to the meter scale. Every polymer can be processed in principle with the compression-molding technique, but its use is limited by the size of the part and, more importantly, by its geometry, because undercuts are only possible with a very complex mold design [95–99].

Compression molding allows the use of unidirectional fibers layers as well as fabrics (woven or nonwoven) or noncontinuous (short or long) fibers. It is the most used thermoforming technology for thermoplastic composites.

Among the different compression-molding techniques, film stacking is gaining ever more attention since it allows the consolidation of laminates and contemporary forming. It is based on the alternating of polymer films and reinforcement (unidirectional, woven or nonwoven) layers. This lay-up is often pressed between plates, to give a planar shape to the laminate, but complex curvatures can be applied. Polymer films impregnate the reinforcement after melting and fill the voids during composite consolidation. The advantages of the film-stacking method include a wide processing window, freedom of material selection, simple tools, and no expensive preproduction. The main problem with film stacking, and all the compression-molding technique variations, is that high pressures, although they can foster the impregnation of the fibers, can also induce damages to the reinforcement or compact the fibers to such an extent that no impregnation is possible. Since there is a temperature gradient throughout the thickness of laminated composites during molding, owing to their low thermal conductivity, an extended molding time is often utilized to reach temperature equilibrium in laminated composites. In order to overcome these issues, some modifications of the conventional film-stacking method have been proposed [100], in which a two-step consolidation is considered: one step is for the lamina, and the other step is for the composite. With this approach, the thin lamina allows achievement of a uniform temperature and shortening of the time required for lamina consolidation. The subsequent consolidation of the laminate can be performed as a bonding of all lamina, without worrying about the impregnation of fibers.

21.6.4 Fiber and Tape Placement

Fiber and tape placement are composite fabrication technologies in which a fiber tow or tape is placed along predefined trajectories in a composite component through a specific robotized head that heats above the melting temperature the polymer (Figure 21.3). These techniques can be used to prepare large parts and to reinforce, on the local scale, bolt-holes and cutouts [101]. The biggest advantage is that fibers or tapes can be positioned along specific directions, in order to maximize the load bearing, and very complex shapes can be realized in an automated way. The definition of the fiber trajectories is very complex, since it affects the final composite from both structural and finishing points of view. To accomplish this task, analytical methods or optimization techniques can be used, and special care should be used to reduce internal stresses or superposition of fibers. The main methods used to place fibers are localized and automated placement.

21.6.4.1 Localized Fiber Placement

Localized fiber placement consists in the realization of a dry preform by using dry fiber tows, which can be cut to the desired length and stitched onto a base structure to reinforce it in specific points and along selected directions.

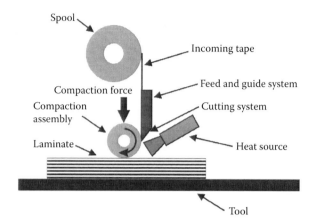

FIGURE 21.3 Automated tape placement setup. (Reproduced from Mitschang, P. Manufacturing of thermoplastic fiber-reinforced polymer composites, in *Wiley Encyclopedia of Composites*, Second Edition, eds. L. Nicolais and A. Borzacchiello. John Wiley & Sons, 2012.)

21.6.4.2 Automated Fiber Placement

In automated fiber placement (AFP), the fiber or tape placement is computer controlled. It uses prepreg tow or slit tape material form to lay up laminates on nongeodesic shapes. Fibers are laid up on the mold surface by applying low tension and compaction pressure and are bonded by locally melting the delivered thermoplastic prepreg.

The ability to control the individual tows of material coming through the delivery head is what gives fiber placement its unique process capabilities. Fiber placement will also lay up any ply orientation, and several of the machines currently available will bidirectionally lay up material. As the use of composite materials is expanding in the aircraft industry and composite structures are getting larger, AFP is ever more used, and the size of delivered materials is increasing to allow the production of large parts, such as airplane fuselage, in practical times.

AFP machines are available in different sizes, but their performance, as well as their cost, depends on the complexity and degree of freedom of the delivering head and the placement tool. The most important configurations of the AFP machines are the winding platform, the moving column, the high rail gantry, and the robotic arm, the latter being the most promising and powerful. Fiber placement processing can be used to lay up revolving tools or flat tools, which are fixed and do not rotate during placement. The cost advantages of placement technologies are related to the reduced manufacturing labor, since they can be highly automated, and the reduction of scraps. Furthermore, parts prepared by placement technologies can be postconsolidated in an autoclave, but this additional step increases the process costs compared to the out-of-autoclave process.

21.6.5 FILAMENT WINDING

Filament winding is one of the most widely used methods for the fabrication of fiber reinforced composites (FRC). In the filament winding process continuous filaments are placed on a rotating mandrel, which has a closed convex cross section. Filament winding is a technique that is used for the manufacture of pipes, tubes, cylinders, and spheres, and frequently for the construction of large tanks and pipework. By suitable design, filament-wound structures can be fabricated to withstand very high in service pressures. In general, products fabricated by filament winding have the highest strength-to-weight ratios and can have reinforcement contents of up to 80% by weight [102].

Filament winding is basically a simple process, although numerous modifications have been developed to improve the product quality. Moldings can be produced either by a wet lay-up process or from prepreg.

In recent years, filament winding has been extended to the continuous production of pipes using a continuous-steel-band mandrel. In this way, continuous lengths of pipe can be produced, with diameters ranging from about 0.3 to 3.5 m.

The fibers are drawn through a bath of molten polymer to impregnate them. The impregnated filaments are then wound under tension around a rotating mandrel. Generally, the feed head supplying the rovings to the mandrel traverses backward and forward along the mandrel.

The mandrel, which may be segmented for large-diameter pipes, is generally wrapped with a release film, such as cellophane, prior to wrapping. The mandrel may incorporate some means of heating the resin system, such as embedded electric heaters, or a provision for steam heating. Alternatively, the fully wrapped mandrel and laminate may be transferred to a curing oven to effect curing.

In order to provide a resin-rich, corrosion-resistant inner lining to the pipe, the mandrel may be wrapped with a surfacing tissue followed by one or two layers of chopped strand mat or woven tape prior to filament winding. This first layer is usually allowed to cure partially before winding commences to prevent the resin from being squeezed out into the main laminate.

Thermoplastic materials have been used in filament winding since the beginning of the 1990s. Impregnated prepregs in form of mall tows or also tapes are suitable solutions, while for cost-sensitive applications, commingled yarns are a good option. On the contrary, the impregnation of fibers directly during winding is a very complex task, due to the high viscosity of the polymeric matrix and the need to compact and impregnate fibers [103].

21.6.6 Pultrusion

Pultrusion is a technique used to produce unidirectional fiber-reinforced profiles with constant cross sections. This process is characterized by a high level of automation and high part quality, which makes it the favored choice for mass production [104,105].

The impregnation mechanism used in thermoplastic pultrusion can be performed by using reactive polymers, which polymerize to form the HMW thermoplastic polymer at the end of the pultrusion line, or by melting the thermoplastic matrix. Many types of continuous fibers, either synthetic or natural [106], can be used, even if glass and carbon fibers are the most used.

Figure 21.4 illustrates the functional principle of the pultrusion process, consisting of four main phases: the impregnation of the fibers with the matrix, the shaping of the cross section through the

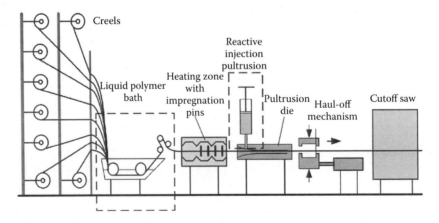

FIGURE 21.4 Schematic drawing of the pultrusion process. (Reproduced from Gutowski, T.G., *Advanced Composites Manufacturing*, John Wiley & Sons, New York, 1997.)

die, the consolidation of the laminate, and the cooling of the pultruded bars at the die exit. The maximum throughput of this process depends on the maximum tensile load of the fibers, since they are pulled through the die [107].

21.6.7 COMMINGLED YARNS

Commingled yarns, as illustrated in Figure 21.1, can represent a kind of "dry impregnation." They are essentially coweaved fibers of reinforcing fibers and polymer (Figure 21.5) [108]. The use of commingled yarns allows an easy and intimate premixing of fibers and polymers directly in the mold. This approach is very flexible because when a fabric is woven from these materials, the fabric can be designed to be highly conformable and drapeable. When heat and pressure are applied, the polymer yarn melts, thus wetting the reinforcing fibers and forming the composite laminate. The main disadvantage is the fixed polymer–fiber ratio. In fact, since any polymer can be added or drawn from the blend, the fiber volume is fixed, and some difficulties are present in producing high-fiber-content composites.

21.6.8 DIAPHRAGM FORMING

Diaphragm forming of advanced thermoplastic composites has been developed combining the analogous process developed for thermosets with the pressure forming of plasticized thermoplastic sheet preforms. The sheet is deformed plastically into the shape of a mold [109]. Unlike in thermoset-based composites, the sheet forming of fiber-reinforced thermoplastic laminates cannot be obtained at atmospheric pressure, since the continuous fibers and the polymer prevent the correct stretching of plies along the fiber's direction. In diaphragm forming, the laminate is placed between two thin plastically deformable sheets (diaphragms, made of silicone elastomers or superplastic metals) that are inflated to apply a biaxial tension on the laminate during deformation. A description of a polymeric diaphragm forming process, using an autoclave, can be found in Figure 21.6 [110]. Forming is carried out after the internal temperature of the autoclave exceeds the melting temperature of the polymeric matrix. An advantage of this technology is that during forming the laminate on the mould undergoes to hydrostatic pressure, which keep very low the forces (reactions of the reinforcement to the stretching applied by the diaphragm) that tend to distort the composite shape.

21.6.9 THERMOFORMING

Thermoforming offers vast potential for high-volume thermoplastic composite part fabrication. There are many thermoforming variations, but by basic definition, thermoforming is the heating of

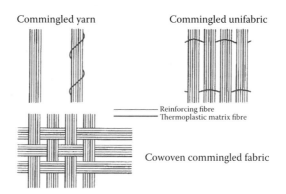

FIGURE 21.5 Commingled yarns. (Reproduced from Ye, L. et al., *Appl. Compos. Mater.*, 1, 1995.)

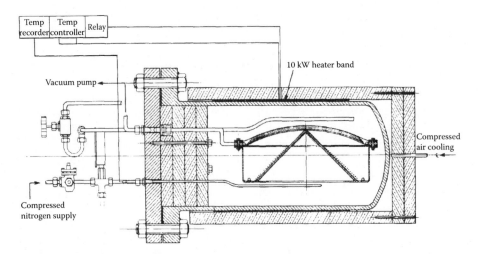

FIGURE 21.6 Schematics of the diaphragm forming process (Reproduced from Mallon, P.J. and O'Brádaigh, C.M., *Composites*, 19, 1988.)

a reinforced thermoplastic matrix sheet or kit above the softening temperature, followed by forcing the material against a contour by mechanical (e.g., matched tooling, plug) or pneumatic (e.g., differential air or hydraulic) means. The material is then held and cooled sufficiently for shape retention, and removed from the mold. Thermoforming implies only those processes applicable to thermoplastic resins, and it is often used in the same context as the term *compression molding*, which generally applies to thermosets. In this section, the process is defined as the preheating of the lay-up or reconsolidated sheet, followed by forming via a matched mold.

This technology has been applied in the fabrication of reinforced Ryton (PPS). An infrared oven is used to rapidly preheat the lay-up to 350°C in 2–3 min. The charge is quickly transferred to a preheated mold in a fast-closing press for part forming. Total cycle times of 1–3 min are feasible with this automated approach.

High production rates can be achieved using thermoforming technology. However, it is difficult to form high-quality continuous fiber-reinforced thermoplastic parts with demanding geometries due to the restricted movement of the fiber.

Long discontinuous fiber sheet products with PEEK provide easier fabrication of complex shapes. This product form is particularly attractive to helicopter manufacturers for the press forming of highly contoured secondary structure parts.

21.6.10 Injection Molding

Injection molding (IM) is a technique that is used extensively for the processing of thermoplastic materials. Due to the high mold cost, it is generally only suitable for the large-scale production of small to medium-sized components. The IM process is greatly preferred by designers because the manufacture of parts in complex shapes and three dimensions can be controlled very accurately.

During the process, the polymer moves from the hopper onto the feeding portion of the reciprocating extruder screw. The molding compound is transferred in the cold state by pressure from the material hopper into the main injection chamber. Here, it can be preheated before injection into the heated mold tool. Injection, through a special nozzle, can happen either by ram or by screw pressure. The rotating screw causes the material to move through a heated extruder barrel, where it softens so that it can be fed into the shot chamber, at the front of the screw. This motion generates a pressure, which causes the screw to retract. When the preset limit is reached, the shot size is met, and the screw stops rotating. At a preset time, the screw acts as a ram to push the melt into the mold. Injection takes place at high pressure. Adequate clamping pressure must be used to eliminate mold opening (flushing). The melt pressure within the mold is dependent on the plastic's rheology/flow behavior.

Time, pressure, and temperature controls indicate whether the performance requirements of a molded part are being met. The time factors include rate of injection, duration of ram pressure, time of cooling, time of plasticization, and screw rotational speed. Pressure factors include maximum and minimum injection pressures, back pressure on the extruder screw, and pressure loss before the plastic enters the cavity, which can be caused by a variety of restrictions in the mold. The temperature factors include mold (cavity and core), barrel, and nozzle temperatures, as well as the melt temperature from back pressure, screw speed, frictional heat, and so on.

IM has advantages over both compression and transfer molding in that the process is more automated and far higher production rates can be achieved. Although mold costs are higher than for compression molding, overall finished component costs are generally lower. IM is also better for thick parts because, with the preheating of the composites before injection into the mold, shorter molding cycles are possible.

21.6.11 STRETCH-DRAW PROCESSES

The term *stretch-draw* is applied to processes that involve a blank of material being preheated and then transferred for forming. Cooling usually takes place in situ, but the part may be removed from the forming station for cooling. The techniques normally use a blank that is at least partially consolidated. Although this is an extra step in the fabrication process, the speed of the stretch-draw processes more than offsets the cost of the extra step. Hydroforming is a metalworking technique that uses hydrostatic pressure via an oil reservoir to form the part over the tool. The blank is preheated, usually in an infrared oven, and then rapidly transferred to a male or female tool for forming. Another similar process involves pressing with a rubber block rather than an oil reservoir. A further variant is the double-action matched die, in which the preheated blank is transferred to a press with a double-action die to form the part. The tooling for the stretch-draw techniques is normally cooler than the processing temperature for the material and may be at room temperature. The type and temperature of cooling will affect the cooling rate of the part, which may, in turn, affect the properties. For example, metal tooling, oil, and the metal diaphragm used in hydroforming all have high thermal conductivity, and therefore, the part will be cooled rapidly, whereas rubber forming will give a lower cooling rate. Control of the cooling rate may be achieved by understanding the thermal properties of the composite and tooling material, and by control of the tooling temperature.

21.6.12 MATCHED-MOLD TECHNIQUES

In this technique, the material is melted and shaped between two matched dies. The dies must be carefully machined and aligned, and the prepreg must have a uniform thickness to achieve successful production of the composite with this approach. The simplest technique is to mold a flat panel using a single platen press. More complex shapes may be made using a matched male platen press. The technique requires a high-temperature press, and there is a practical limitation to the size of parts that can readily be made, owing to the temperature uniformity and clamping forces required. Heating and cooling a single press may be time-consuming. An alternative technique to the one-press technique is to melt the composite in one press and transfer it to a cooler mold for forming. Preheating may be in either another press or an oven. It is preferable to start with a sheet that is at least partially consolidated.

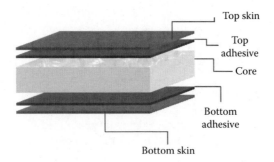

FIGURE 21.7 Schematics of the sandwich structure component stacking.

21.6.13 SANDWICH TECHNOLOGY

Among the lightweight structures, sandwich panels are the most diffuse. They are produced by bonding the stiff skins (usually made of fiber-reinforced composites) with a compliant core (a honeycomb or a foam) (Figure 21.7). Conventionally, the skins (external face sheets) are made of laminated polymer-based composite materials, such as glass or carbon fiber-reinforced polymer. They are bonded by means of adhesive layers (epoxy or vinyl-ester resins) by applying pressure and temperature profiles to the cores. Typical core materials are balsa wood, phenolic honeycombs, or foamed materials, such as PE, polyvinyl chloride (PVC), polyurethane (PU), or styrene acrylonitrile (SAN) foams.

The primary reason for the use of sandwich materials is their high strength-to-weight ratio, but the main disadvantage is the low shear properties. In order to improve them, different solutions have been developed: fiber stitching, z-pinning, and others.

21.6.13.1 Fiber Stitching

High-tensile-strength yarn (glass, carbon, or Kevlar) is sewed through the sandwich thickness before laminate consolidation. This technique increases the strength, stiffness, and fatigue performance of a laminate [111]. On the contrary, orthogonal weaving can damage the sheet reinforcement, and the planarity of the core can be loose during sewing, thus reducing the in-plane performance.

21.6.13.2 Z-Pinning

Fibrous pins/rods (metallic, carbon, or glass fibers) are inserted through the thickness of the laminate using ultrasonic insertion techniques. Z-pinning is a slow and expensive process [112], and more importantly, the insertion of very stiff rods can induce damage in the composites.

21.6.13.3 Other Methodologies

Other methods proposed to improve the sandwich performance include corrugations to improve wrinkling strength [113], new configurations of the lightweight core [114], and peel-stopper mechanisms [115].

21.7 ADVANCED THERMOPLASTIC FOAMS

HPTPs are gaining increasing attention for producing foamed materials due to their several advantages over thermosets, such as higher impact strength, recyclability, weldability, absence of volatile organic compounds, reduced processing time, and lower manpower for large-scale productions, but a deep understanding of their foaming process is still ongoing [116].

Advanced foams based on thermoplastics are difficult to produce by means of conventional thermoplastic technologies, such as the extrusion foaming process, because the high temperatures involved could be responsible for thermal degradation of the matrix, and careful control of the foaming conditions should be assured. Many studies have been conducted on processes and characterizations of foams from HPTPs. The morphology of foams is very complex to control, because it is characterized by several parameters, namely, specific density, mean cell size, cell size distribution, degree of cell interconnections (with the extremes of the range being closed cells and open cells), and cell anisotropy.

Several industrial processes are available to produce foamed polymers. The principal operations include the following: (1) solubilization of the blowing agent; (2) bubble nucleation generally obtained by a sudden modification of the local thermodynamic conditions that induce a phase separation in the matrix–gas system, which usually can be achieved by a decrease of pressure or an increase of temperature; (3) bubble growth through the diffusion of gases from the polymer–gas mixture to the nucleated bubbles; and (4) stabilization of the cellular structure through fine control of the solidification process to avoid the collapse of the structure and/or the coalescence of cells. HPTP foams can be produced by using continuous (extrusion) or discontinuous technologies (IM and batch foaming). Each technique should be used according to the final article shapes and to the number of articles to be produced.

The batch-foaming technology is based on a thermoregulated pressure vessel, whose temperature is controlled by means of an electric resistance or circulated oil bath. The polymer is placed in the vessel and tightly enclosed. The blowing agent (usually a physical one) is then injected into the vessel at the desired pressure and temperature. The physical blowing agent (PBA) diffuses into the polymer, and after a defined time interval, depending on the transport properties of the gas/polymer mixture, it saturates the sample. At the end of this stage, nucleation and growth of gas bubbles from the stable homogeneous solution can be promoted by creating a thermodynamic instability through (1) a temperature increase or (2) a pressure reduction.

Extrusion is the leading technology in thermoplastic processing, and extrusion foaming is the most important foaming process employed to produce both finished and semifinished products [117,118]. The extrusion foaming process is more critical for HPTP polymers compared to commodity polymers because all the thermal phenomena involved in the foaming process are radicalized as a result of the typical abrupt temperature drop at the die exit. It is thus fundamental to properly account for or to increase the crystallization kinetics in semicrystalline polymers to stabilize the cellular structure, or to adequately manage the glass transition temperature reduction for the plasticizing effect of the blowing agent in amorphous polymers.

Very few advanced foams are available on the market, probably due to the difficulties of manufacturing. They are often based on thermoset polymers and usually produced through complex and long processes, such as foaming from low-molecular-weight precursors or stabilizing the cellular structure by polymer cross-linking. Nanocomposite foams from high-performance thermoplastics are their most promising replacements because they can couple high heat deflection temperatures with fine control of their morphology, in turn maximizing the performance for the specific application.

Nanometric fillers gained attention due to their potential improvement in different properties when compared to the virgin polymer or conventional microcomposites [119,120]. Due to the high nucleation efficiency, nanoparticles provide a powerful way to increase cell density and reduce cell size. This is particularly beneficial for the production of microcellular foams, a class of novel lightweight and high-strength material for structural applications. Foams based on high-performance thermoplastic matrices are reported in the following sections.

21.7.1 Aromatic Polyketone-Based Foams

Aromatic polyketones (PEEK, PEK, PAEK) are widely employed in aeronautics and high-end applications due to their high stiffness and strength, high resistance to a huge number of solvents (for example, Jetfuel, Skydrol, and methyl ether ketone [MEK]), and good flame retardancy.

Microcellular PEEK foams have been prepared by a two-stage batch-foaming process using carbon dioxide as blowing agent. Two cell sizes were evident in samples exhibiting a sandwich-like structure, with nucleated cells and cell size depending on the blowing agent uptake and viscoelastic properties, respectively [121]. Microcellular foams were also produced by Behrendt et al., who expanded thin PEEK films using the batch-foaming technique with the aim to prepare electrets. They obtained a low-porosity structure, as a consequence of the crystallization during the foaming process, but the polymer was crystallized after the foaming process.

Blends based on PEEK have been used to exploit the interfaces as nucleating sites in the work of Tan et al. [122] and Nemoto et al. [123]. In particular, Nemoto et al. were able to obtain a nanoscale cell structure from PEEK/*para*-diamine polyetherimide (*p*-PEI) as well as PEEK/*meta*-diamine polyetherimide (*m*-PEI) blends by a temperature quench foaming process with CO_2 as a blowing agent.

Attempts to produce foams of PEEK-based nanocomposites have been performed using vapor-grown carbon nanofibers (CNFs) to stabilize the foaming process through an injection-molding process, in which they used commercial chemical blowing agents. This produced integral foams with both unfilled and nanocomposite systems and verified the positive effect of the nanofiller on the cellular structure.

21.7.2 Polyester-Based (PEN-Based) Foams

Among the most diffuse and commercially available polyesters, PEN is the only one that has adequate performance for inclusion in the HPTPs class, because it has a glass transition temperature of 125°C and a melting temperature higher than 265°C.

Closed-cell PEN foams were, for the first time, produced by Sorrentino et al. in 2011 [116]. They used a batch-foaming apparatus to solubilize carbon dioxide into amorphous PEN. Foams were produced within an extended temperature range (from 100°C to 240°C) and were characterized by relative densities ranging from 0.13 to 0.44 and average cell size between 5 and 15 μm. Cell nucleation strongly increased above PEN glass transition temperature, as a consequence of the higher molecular mobility.

Expanded graphite (ExG) and silica nanocomposites based on PEN were prepared by means of a twin-screw extruder by Sorrentino et al. [116]. EG nanoparticles were more effective as a nucleating agent for both crystallization and foaming compared to silica nanoparticles [124]. Microcellular nanocomposite foams with higher compressive modulus were therefore obtained in the presence of EG [125].

21.7.3 Foams Based on Sulfone Polymers

Foams from sulfone polymers have been prepared since 2001. Krause et al. investigated the foaming process of PES and polysulfone (PSF), by using carbon dioxide as a PBA, and the temperature-rise foaming technique after gas solubilization. They measured a strong glass transition temperature depression after gas sorption, and a wider temperature range for foaming was detected at high solubilization pressures. Microcellular closed-cell morphologies were obtained, but foam density was higher than 0.4 g/cm^3 [126].

Blends of PES and PEI were also investigated to exploit the combined action of (1) glass transition temperature depression after blowing agent sorption and (2) the different viscoelastic properties of the two polymeric phases. The result was the production of foams with nanocellular cells but very high foam density [127].

PPSU has also been investigated for foaming by Sorrentino et al. [116]. They detected reduced foamability compared to PES, with lower nucleated cells, higher relative foam density (0.4 and 0.2 for PPSU and PES, respectively), and average cell size (9 and 1 μm for PPSU and PES, respectively). PES nanocomposite foams based on SiO_2 and EG nanoparticles were obtained with the temperature-rise technique after solubilization in a high-pressure vessel [124]. EG-filled PES foams exhibited lower foam density, compared to SiO_2-filled ones, and allowed a two-order-of-magnitude increase of the nucleated cells compared to unfilled matrix.

21.7.4 Foams Based on Thermoplastic Polyimides (PI, PAI, and PEI)

Polyimide foams were made from powders of solid-state polyimide precursors [i.e. poly(amic acid) (PAA)] by using tetrahydrofuran (THF) as a blowing agent that complexes through hydrogen bonding to the PAA structure [128]. During the foaming process, solid powder particles of PAA precursors are heated from room temperature to produce microspheres and, ultimately, foams.

Only recently, microcellular polyimide foams were obtained by using the solid-state microcellular foaming technology with the compressed CO_2 as a PBA and THF as a coblowing agent. The presence of a coblowing agent allowed an increase of the gas sorption in PI, causing a dramatic increase in the expansion ratio of the PI foam.

PEI foams have also been prepared by solid-state batch-foaming processes using subcritical CO_2 as a blowing agent [129]. A transition from microscale cells to nanoscale cells was observed at gas concentrations in the range of 94–110 mg CO_2/g PEI, and a hierarchical structure was observed that consisted of nanocellular structures internal to microcells.

21.7.5 Foamed Composites

A method to reduce the density of a composite structure is the controlled induction of porosity into the polymeric matrix, while still ensuring load transfer to the reinforcing fibers and fiber protection from the environment. This approach can be counterintuitive since, in the advanced composite industry, voids, randomly present in the matrix and often located in fiber tows, create stress concentration and inhibit good stress transfer between the polymer matrix and reinforcing fibers. On the contrary, if very small pores are generated in a controlled way and if they are not formed directly on the reinforcement surface, then the specific mechanical properties of the composite could, in principle, be significantly improved [130].

The performance increase after the creation of submicron bubbles in a thermoplastic laminated composite using a gas foaming technique has been investigated by Sorrentino et al. [131]. Cellular thermoplastic fiber-reinforced composites (CellFRCs) were obtained by promoting and controlling the formation of microcellular bubbles within the matrix in a mold with an adjustable cavity. Through this in situ foaming process, the tuning of the cell size was achieved by controlling both the foaming process and the molding conditions. Cellular fiber-reinforced composites (CellFRCs) based on poly(ethylene-2,6-naphthalate) (PEN) reinforced with glass fiber fabrics were prepared through controlled generation of microsized gas bubbles in the thermoplastic matrix reinforced with glass fiber fabrics. A reduction of the bulk density from 1.67 g/cm^3 to less than 1 g/cm^3 was achieved. Although the specific flexural modulus of CellFRC was comparable to that of solid ones, the formation of micro or submicrocells in the matrix increases of 150% the specific value of the absorbed energy at perforation when compared to the conventional composite and sandwich structures, keeping constant the reinforcing fiber content.

21.8 ENVIRONMENTAL RESISTANCE

The effects of water, saltwater, paint stripper, jet fuel, and ethylene glycol on PEEK/carbon fiber composites have been examined [132–137]. Tests were carried out by immersing unstressed and stressed samples for 60 days. The largest effect on mechanical properties was produced after immersion in paint stripper (dichloromethane based) [135]. Only limited data are available on the effects of environment in the presence of stresses [138]. No change in mechanical properties after immersion in water has been detected for stressed PEEK/carbon composite transverse tensile specimens; only a dichloromethane environment produced relevant reduction of properties: specimens loaded at 65% of ultimate strength failed on contact with the fluid. Literature data suggest that thermoplastic composites have good resistance to typical aerospace environments; however, further information is needed.

Sensitivity of polymer matrices to the presence of moisture and solvents and to thermally induced stresses is of primary importance in determining the durability of polymer-based composite structures [134]. In fact, moisture and solvents can greatly affect not only mechanical properties of the matrix but also its adhesion properties to the reinforcing fibers [137]. Fiber-reinforced polymeric composites are currently used in structural applications when high strength combined with light weight is of critical importance. In such applications, the composites are expected to be exposed, over a wide range of temperatures, to mechanical fatigue as well as to extreme environmental conditions such as freezing and thawing at high humidity or in aggressive solvent-rich atmospheres [136].

Polymer-based materials such as fiber composites differ from other structural materials in that low-molecular-weight substances may easily migrate in them even at ambient temperature, determining, in many cases, a degradation of matrix mechanical properties [139]. In fact, although the properties in the fiber direction of these reinforced materials are mainly determined by the fiber properties, the application limits are, in many cases, governed by the final properties of the matrix. The environmental degradation of the mechanical properties of the matrix was associated with the plasticization and micromechanical damage induced by the combined effects of temperature, stress, and sorbed solvents. Environmental resistance, mainly to hygrothermal fatigue under external loading, is one of the most important matrix properties. The degradation mechanism may act on the fibers, the interfaces, and the matrix. The polymeric matrices, in fact, do not present an effective barrier to the diffusion of water. However, the combined action of moisture penetration and high temperatures strongly influences the morphology of the matrix and the sorption mechanism.

Investigations on high-performance thermoplastics, such as poly(aryletheretherketone) (PEEK), proved their high resistance to water plasticization [133]. The low amount of water sorbed in thermoplastic PEEK does not involve any significant interaction among polymer backbone and penetrant molecules. PEEK has been found to have good moisture and liquid water resistance. In fact, only a slight depression of T_g is reported for samples equilibrated with water at 60°C. The PEEK glass transition temperature is influenced only by the sterically induced structure stiffness, and although PEEK possesses several oxygens that may potentially form hydrogen bonds, it is not moisture sensitive, because these atoms are sterically shielded by the aromatic rings.

Composites for high-performance applications must have good resistance to the following different environments: organic solvents, acids, alkalis, fuels, chlorinated solvents, and hydraulic fluids. The weight uptakes of PEEK- and PI-based composites in various fluid environments are reported in Table 21.5 [140]. The weight uptake values are relatively small, except for dichloromethane in composites based on PEEK matrix.

TABLE 21.5
Weight Uptake Values of PEEK and PI Matrix Composites in Various Environments

Environment	Temp., °C	Time, days	Weight Uptake of PEEK, %	Weight Uptake of PI, %
Water	82	14	0.28	0.67
Methyl ethyl ketone	23	14	0.03	0.10
Dichloromethane	23	14	2.41	0.43
Monsanto hydraulic fluid	23	14	–	0.09
Jet fuel JP4	82	14	0.10	0.041

Source: Partridge, I.K. (ed.), *Advanced Composites*, Elsevier, New York, 1989.

**TABLE 21.6
Retention of Flexural Strength at Elevated Temperatures after Moisture Conditioning, Relative to 23°C Dry Properties**

Matrix	150°C	200°C
PEEK	0.70	0.31
PEI	0.61	–
PI	0.77	0.53
PPS	0.47	0.37

Source: Partridge, I.K. (ed.), *Advanced Composites*, Elsevier, New York, 1989.

Moisture is the most common environment. Its effect is to plasticize the polymer, reducing the T_g of the matrix. Differences in the weight uptake values in moisture environments are reported. The equilibrium weight uptake value of PEEK matrix in water is 0.67%, reducing the T_g by about 5°C. Moisture has little effect on room-temperature mechanical properties but has relevant effects at elevated temperatures. The flexural strength after moisture conditioning is shown in Table 21.6, the data having been normalized to the room-temperature unconditioned strength.

REFERENCES

1. Johnston, N. J. High performance thermoplastic: a review of neat resin and composite properties. *Proc. 32nd SAMPE Symposium*, Anaheim, CA, April 6–9, 1987, pp. 1400–1412.
2. Product literature, APC Group, ICI Fiberite, 28271 Verdugo Drive, Laguna Hills, CA 92653.
3. Product literature, Du Pont Composites Center, Chestnut Run 702, Wilmington, DE 19898.
4. Rosato, D. V., Di Mattia, D. P., and Rosato, D. V. *Designing with Plastics and Composites: A Handbook.* Van Nostrand Reinhold, New York, 1991.
5. Leach, D. C., Cogswell, F. N., and Nield, E. High temperature performance of thermoplastic aromatic polymer composites. *Proc. 31st SAMPE Symposium*, Las Vegas, NV, April 1986, pp. 434–448.
6. O'Connor, J. E., Murtha, T. P., and South A. Polyarylene sulfide high performance thermoplastic composites. *Proc. 6th International/2nd European Conference on Composite Materials*, Vol. 1, Imperial College, July 20–24, 1987, Elsevier, London, 1987, pp. 433–442.
7. Hartness, J. T. An evaluation of a high temperature thermoplastic polyimide composite. *Proc. 32nd SAMPE Symposium*, Anaheim, CA, April 6–9, 1987, pp. 154–168.
8. O'Connor, J. E., Lou, A. Y., and Brady, D. G. Polyarylene sulfide composites. *Proc. 1st Technical Conference*, American Composites Society, Dayton, OH, October 7–9, 1986, pp. 21–35.
9. Sheppard, C. H., House, E. E., and Stander, M. Advanced thermoplastic composite development. *Proc. 36th Conf Reinforced Plastic/Composites Industry*, Society of Plastics Industry, February 16–20, 1981, Paper 17-B, pp. 1–5.
10. Berglund, L. Fracture toughness of carbon fiber/PEEK composites. Linköping Studies in Science and Technology, Dissertation No. 159, Linköping University, Sweden, 1987.
11. Curtis, P. T., Davies, P., Partridge, I. K., and Sainty, J. P. Cooling rate effects in PEEK and carbon fiber–PEEK composites. *Proc. 6th International Conference on Composite Materials*, London, July 20–24, 1987, pp. 476–488.
12. O'Connor, J. E., Lou, A. Y., and Beever, W. H. Polyphenylenesulfide: A thermoplastic polymer for high performance of semi-crystalline thermoplastic composites. *Proc. 5th International Conference on Composite Materials*, San Diego, CA, 1985, pp. 963–970.
13. Jones, D. P., Leach, D. C., and Moore, D. R. Mechanical properties of poly(ether-ether-ketone) for engineering applications. *Polymer*, 26, 1385–1393 (1985).

14. Hartness, J. T. Polyetheretherketone matrix composites. *Proc. 14th SAMPE Technical Conference*, 1982, p. 26.
15. Velisaris, C. N., and Seferis, J. C. Crystallization kinetics of polyetheretherketone (PEEK) matrices. *Polym. Eng. Sci.*, 26, 1574–1581 (1986).
16. Blundell, D. J., and Osborn, B. N. Crystalline morphology of the matrix of PEEK–carbon fiber aromatic composites II. Crystallization behavior. *SAMPE Quart.*, 17(1), 1–6 (1985).
17. Del Nobile, M. A., Mensitieri, G., Netti, P. A., and Nicolais, L. Anomalous diffusion in polyether-ether-ketone (PEEK). *Chem. Eng. Sci.*, 49, 633 (1993).
18. Beever, W. H., Ryan, C. L., O'Connor, J. E., and Lou, A. Y. Ryton–PPS carbon fiber reinforced composites: The how, when and why of molding, in *Toughened Composites*, ASTM STP 937, ed. N. J. Jonston, American Society for Testing and Materials, Baltimore 1987, pp. 319–327.
19. Hoggatt, J. T., Oken, S., and House, E. E. Advanced fiber reinforced thermoplastic structures. U.S. Air Force Report AFWAL-TR-80-3023, April 1980.
20. Chung, T. S., and McMahon, P. E. Thermotropic polyester amide–carbon fiber composites. *J. Appl. Polym. Sci.*, 31, 965–977 (1986).
21. Cole, B. Torlon-C graphite composites. *Proc. 30th SAMPE Symposium*, Anaheim, CA, March 19–21, 1985, pp. 799–808.
22. Chang, I. K. Thermoplastic matrix continuous filament composites of Kevlar aramid or graphite fiber. *Compos. Sci. Technol.*, 24, 61–79 (1985).
23. Díez-Pascual, A. M., Naffakh, M., Marco, C., and Ellis, G. Mechanical and electrical properties of carbon nanotube/poly (phenylene sulphide) composites incorporating polyetherimide and inorganic fullerene-like nanoparticles. *Compos. Part A: Appl. Sci. Manuf.*, 43(4), 603–612 (2012).
24. Blundell, D. J., and Osborn, B. N. The morphology of poly (aryl-ether-ether-ketone). *Polymer*, 24(8), 953–958 (1983).
25. Díez-Pascual, A. M., Martínez, G., and Gomez, M. A. Synthesis and characterization of poly (ether ether ketone) derivatives obtained by carbonyl reduction. *Macromolecules*, 42(18), 6885–6892 (2009).
26. Deng, F., Ogasawara, T., and Takeda, N. Tensile properties at different temperature and observation of micro deformation of carbon nanotubes–poly (ether ether ketone) composites. *Compos. Sci. Technol.*, 67(14), 2959–2964 (2007).
27. Zhang, G., Chang, L., and Schlarb, A. K. The roles of nano-SiO_2 particles on the tribological behavior of short carbon fiber reinforced PEEK. *Compos. Sci. Technol.*, 69(7), 1029–1035 (2009).
28. Balaji, V., Tiwari, A. N., and Goyal, R. K. Fabrication and properties of high performance PEEK/Si3N4 nanocomposites. *J. Appl. Polym. Sci.*, 119(1), 311–318 (2011).
29. Jaafar, J., Ismail, A. F., and Matsuura, T. Effect of dispersion state of Cloisite15A® on the performance of SPEEK/Cloisite15A nanocomposite membrane for DMFC application. *J. Appl. Polym. Sci.*, 124(2), 969–977 (2012).
30. Gojny, F. H., Wichmann, M. H. G., Fieldler, B., Bauhofer, W., and Schulte, K. Influence of nano-modification on the mechanical and electrical properties of conventional fibre-reinforced composites. *Compos. Part A*, 36(11), 1525–1535 (2005).
31. Thostenston, E. T., and Chou, T. W. Carbon nanotube networks: Sensing of distributed strain and damage for life prediction and self healing. *Adv. Mater.*, 18(21), 2837–2841 (2006).
32. Thostenson, E. T., Ren, Z., and Chou, T.-W. Advances in the science and technology of carbon nanotubes and their composites: A review. *Compos. Sci. Technol.*, 61(13), 1899–1912 (2001).
33. Moniruzzaman, M., and Winey, K. I. Polymer nanocomposites containing carbon nanotubes. *Macromolecules*, 39(16), 5194–5205 (2006).
34. Enomoto, K., Kitakata, S., Yasuhara, T., Ohtake, N., Kuzumaki, T., and Mitsuda, Y. Measurement of Young's modulus of carbon nanotubes by nanoprobe manipulation in a transmission electron microscope. *Appl. Phys. Lett.*, 88(15), 15315–15317 (2006).
35. Yu, M.-F., Lourie, O., Dyer, M. J., Moloni, K., Kelly, T. F., and Ruoff, R. S. Strength and breaking mechanism of multiwalled carbon nanotubes under tensile load. *Science*, 287(5453), 637–640 (2000).
36. Pop, M., Mann, D., Wang, Q., Goodson, K., and Dai, H. Thermal conductance of an individual single-wall carbon nanotube above room temperature. *Nano Lett.*, 6(1), 96–100 (2006).
37. Valter, B., Ram, M. K., and Nicolini, C. Synthesis of multiwalled carbon nanotubes and poly(*o*-anisidine) nanocomposite material: Fabrication and characterization of its Langmuir Schaefer films. *Langmuir*, 18(5), 1535–1541 (2002).
38. Cadek, M., Coleman, J. N., Barron, V., and Hedicke, K. High-purity carbon nanotubes synthesis method by an arc discharging in magnetic field. *Appl. Phys. Lett.*, 81(4), 739–741 (2002).

39. Lee, G., Jagannathan, S., Choe, H. G., Minus, M. L., and Kumar, S. Carbon nanotube dispersion and exfoliation in polypropylene and structure and properties in the resulting composites. *Polymer*, 49(7), 1831–1840 (2008).
40. Bhattacharyya, A. R., Sreelumar, T. V., Liu, T., Kumar, S., Ericsson, L. M., and Hauge, R. H. Crystallization and orientation studies in polypropylene/single wall carbon nanotube composite. *Polymer*, 44(8), 2373–2377 (2003).
41. Hayrapetyan, S., Kelarakis, A., Estevez, L., Lin, Q., Dana, K., Chung, Y. L., and Giannelis, E. P. Nontoxic poly (ethylene terephthalate)/clay nanocomposites with enhanced barrier properties. *Polymer*, 53(2), 422–426 (2012).
42. Durmus, A., and Yalçınyuva, T. Effects of additives on non-isothermal crystallization kinetics and morphology of isotactic polypropylene. *J. Polym. Res.*, 16(5), 489–498 (2009).
43. Scamardella, A. M., Vietri, U., Sorrentino, L., Lavorgna, M., and Amendola, E. Foams based on poly (ethylene terephthalate) nanocomposites with enhanced thermal stability. *J. Cell. Plast.*, 48(6), 557–576 (2012).
44. Zhang, H. B., Zheng, W. G., Yan, Q. et al. Electrically conductive polyethylene terephthalate/graphene nanocomposites prepared by melt compounding. *Polymer*, 51(5), 1191–1196 (2010).
45. Li, M., and Jeong, Y. G. Poly (ethylene terephthalate)/exfoliated graphite nanocomposites with improved thermal stability, mechanical and electrical properties. *Compos. Part A: Appl. Sci. Manuf.*, 42(5), 560–566 (2011).
46. Lee, J. H., Shin, D. W., Makotchenko, V. G. et al. One-step exfoliation synthesis of easily soluble graphite and transparent conducting graphene sheets. *Adv. Mater.*, 21(43), 4383–4387 (2009).
47. Yesil, S., and Bayram, G. Poly (ethylene terephthalate)/carbon nanotube composites prepared with chemically treated carbon nanotubes. *Polym. Eng. Sci.*, 51(7), 1286–1300 (2011).
48. Avilés, F., Cauich-Rodríguez, J. V., Rodríguez-González, J. A., and May-Pat, A. Oxidation and silanization of MWCNTs for MWCNT/vinyl ester composites. *Expr. Polym. Lett.*, 5(9), 766–776 (2011).
49. Aurilia, M., Sorrentino, L., and Iannace, S. Modelling physical properties of highly crystallized polyester reinforced with multiwalled carbon nanotubes. *Eur. Polym. J.*, 48(1), 26–40 (2012).
50. Ahn, S. H., Kim, S. H., and Lee, S. G. Surface-modified silica nanoparticle-reinforced poly (ethylene 2,6-naphthalate). *J. Appl. Polym. Sci.*, 94(2), 812–818 (2004).
51. Kim, J. Y., Han, S. I., Kim, D. K., and Kim, S. H. Mechanical reinforcement and crystallization behavior of poly(ethylene 2,6-naphthalate) nanocomposites induced by modified carbon nanotube. *Compos. Part A: Appl. Sci. Manuf.*, 40(1), 45–53 (2009).
52. Kim, J. A., Seong, D. G., Kang, T. J., and Youn, J. R. Effects of surface modification on rheological and mechanical properties of CNT/epoxy composites. *Carbon*, 44(10), 1898–1905 (2006).
53. Kim, J. Y., and Kim, S. H. Influence of multiwall carbon nanotube on physical properties of poly(ethylene 2,6-naphthalate) nanocomposites. *J. Polym. Sci. Part B: Polym. Phys.*, 44(7), 1062–1071 (2006).
54. Kim, J. Y., Han, S. I., and Kim, S. H. Crystallization behaviors and mechanical properties of poly (ethylene 2,6-naphthalate)/multiwall carbon nanotube nanocomposites. *Polym. Eng. Sci.*, 47(11), 1715–1723 (2007).
55. Wu, T. M., and Liu, C. Y. Poly(ethylene 2,6-naphthalate)/layered silicate nanocomposites: Fabrication, crystallization behavior and properties. *Polymer*, 46(15), 5621–5629 (2005).
56. Kim, H., and Macosko, C. W. Morphology and properties of polyester/exfoliated graphite nanocomposites. *Macromolecules*, 41(9), 3317–3327 (2008).
57. Aurilia, M., Piscitelli, F., Sorrentino, L., Lavorgna, M., and Iannace, S. Detailed analysis of dynamic mechanical properties of TPU nanocomposite: The role of the interfaces. *Eur. Polym. J.*, 47(5), 925–936 (2011).
58. LeBaron, P. C., Wang, Z., and Pinnavaia, T. J. Polymer-layered silicate nanocomposites: An overview. *Appl. Clay Sci.*, 15(1), 11–29 (1999).
59. Chen, Y., and Iroh, J. O. Synthesis and characterization of polyimide/silica hybrid composites. *Chem. Mater.*, 11(5), 1218–1222 (1999).
60. Cai, K. F., Müller, E., Drašar, C., and Mrotzek, A. Preparation and thermoelectric properties of Al-doped ZnO ceramics. *Mater. Sci. Eng. B*, 104(1), 45–48 (2003).
61. Liang, Z. M., Yin, J., and Xu, H. J. Polyimide/montmorillonite nanocomposites based on thermally stable, rigid-rod aromatic amine modifiers. *Polymer*, 44(5), 1391–1399 (2003).
62. Su, H. L., Chou, C. C., Hung, D. J. et al. The disruption of bacterial membrane integrity through ROS generation induced by nanohybrids of silver and clay. *Biomaterials*, 30(30), 5979–5987 (2009).
63. Cai, H., Yan, F., and Xue, Q. Investigation of tribological properties of polyimide/carbon nanotube nanocomposites. *Mater. Sci. Eng. A*, 364(1), 94–100 (2004).

64. Yudin, V. E., and Svetlichnyi, V. M. Effect of the structure and shape of filler nanoparticles on the physical properties of polyimide composites. *Russ. J. Gen. Chem.*, 80(10), 2157–2169 (2010).
65. Ranade, A., D'Souza, N. A., and Gnade, B. Exfoliated and intercalated polyamide-imide nanocomposites with montmorillonite. *Polymer*, 43(13), 3759–3766 (2002).
66. Shantalii, T. A., Karpova, I. L., Dragan, K. S., Privalko, E. G., Karaman, V. M., and Privalko, V. P. Properties of an organosilicon nanophase generated in a poly (amide imide) matrix by the sol-gel technique. *Polym. Adv. Technol.*, 16(5), 400–404 (2005).
67. Chatterjee, S., Lee, M. W., and Woo, S. H. Adsorption of congo red by chitosan hydrogel beads impregnated with carbon nanotubes. *Bioresour. Technol.*, 101(6), 1800–1806 (2010).
68. Liang, Z. M., Yin, J., and Xu, H. J. Polyimide/montmorillonite nanocomposites based on thermally stable, rigid-rod aromatic amine modifiers. *Polymer*, 44(5), 1391–1399 (2003).
69. Liu, T., Tong, Y., and Zhang, W. D. Preparation and characterization of carbon nanotube/polyetherimide nanocomposite films. *Compos. Sci. Technol.*, 67(3), 406–412 (2007).
70. Isayev, A. I., Kumar, R., and Lewis, T. M. Ultrasound assisted twin screw extrusion of polymer-nanocomposites containing carbon nanotubes. *Polymer*, 50(1), 250–260 (2009).
71. Xian, G., Zhang, Z., and Friedrich, K. Tribological properties of micro- and nanoparticles-filled poly (etherimide) composites. *J. Appl. Polym. Sci.*, 101(3), 1678–1686 (2006).
72. Choudhury, A. Dielectric and piezoelectric properties of polyetherimide/BaTiO$_3$ nanocomposites. *Mater. Chem. Phys.*, 121(1), 280–285 (2010).
73. Peng, Y., He, J., Liu, Q. et al. Impurity concentration dependence of optical absorption for phosphorus-doped anatase TiO2. *J. Phys. Chem. C*, 115(16), 8184–8188 (2011).
74. Sinha Ray, S., Bandyopadhyay, J., and Bousmina, M. Effect of organoclay on the morphology and properties of poly (propylene)/poly [(butylene succinate)-co-adipate] blends. *Macromol. Mater. Eng.*, 292(6), 729–747 (2007).
75. Bandyopadhyay, J., Sinha Ray, S., and Bousmina, M. Nonisothermal crystallization kinetics of poly (ethylene terephthalate) nanocomposites. *J. Nanosci. Nanotechnol.*, 8(4), 1812–1822 (2008).
76. Mukherjee, M., Bose, S., Nayak, G. C., and Das, C. K. A study on the properties of PC/LCP/MWCNT with and without compatibilizers. *J. Polym. Res.*, 17(2), 265–272 (2010).
77. Nayak, G. C., Rajasekar, R., and Das, C. K. Effect of SiC coated MWCNTs on the thermal and mechanical properties of PEI/LCP blend. *Compos. Part A: Appl. Sci. Manuf.*, 41(11), 1662–1667 (2010).
78. Naffakh, M., Marco, C., Gómez, M. A., Gómez-Herrero, J., and Jiménez, I. Use of inorganic fullerene-like WS$_2$ to produce new high-performance polyphenylene sulfide nanocomposites: Role of the nanoparticle concentration. *J. Phys. Chem. B*, 113(30), 10104–10111 (2009).
79. Boyce, R. J., Gannet, T. P., Gibbs, H. H., and Wedgewood, A. R. Processing, properties and application of K-polymer composite materials based on Avimid K-III pre-preg. *Proc. 32nd SAMPE Symposium*, Anaheim, CA, April 6–9, 1987, pp. 167–184.
80. Cogswell, F. N. The processing science of thermoplastic structural composites. *Int. Polym. Proc.*, 1, 157–165 (1987).
81. Maffezzoli, A., Kenny, J. M., Torre, L., and Nicolais, L. Thermal processing of PPS and PPS-based composites. *Proc. 11th SAMPE European Chapter Conference*, Basilea, Maggio 29–31, 1990.
82. Mitschang, P. Manufacturing of thermoplastic fiber-reinforced polymer composites, in *Wiley Encyclopedia of Composites*, Second Edition, eds. L. Nicolais and A. Borzacchiello. John Wiley & Sons, Hoboken, NJ, USA, 2012.
83. Strong, A. B. *High Performance and Engineering Thermoplastic Composites*. Technomic, Lancaster, PA; Butterworth Heinemann, Oxford, 1992.
84. Lee, W. I., and Springer, G. S. Microwave curing of composites. *J. Therm. Comp. Mat.*, 18, 387–409 (1984).
85. Trende, A., Åström, B. T., Wöginger, A., Mayer, C., and Neitzel, M. Modelling of heat transfer in thermoplastic composites manufacturing: Double-belt press lamination. *Compos. Part A*, 30, 935–943 (1999).
86. Wakeman, W. D., Zingraff, L., Bourban, P. E., Månson, J.-A. E., and Blanchard, P. Stamp forming of carbon fibre/PA12 composites—A comparison of a reactive impregnation process and a commingled yarn system. *Compos. Sci. Technol.*, 66, 19–35 (2006).
87. Gutowski, T. G. *Advanced Composites Manufacturing*. John Wiley & Sons, New York, 1997.
88. Advani, S. G., and Sozer, E. M. *Process Modeling in Composite Manufacturing*. Marcel Dekker, New York, 2002.
89. Cogswell, F. N. *Thermoplastic Aromatic Polymer Composites*. Butterworth Heinemann, Oxford, 1992.
90. Parlevliet, P. P., Bersee, H. E. N., and Beukers, A. Residual stresses in thermoplastic composites—A study of the literature—Part I: Formation of residual stresses. *Compos. Part A*, 37, 1847–1857 (2006).

91. Cogswell, F. N., Hezzell, D. J., and Williams, P. J. Fibre-reinforced compositions and methods for producing such compositions. EP patent 0056703A1 (1982).
92. Van Dreumel, W. H. M. Origami-technology, creative manufacturing of advanced composites parts. *Compos. Polym.*, 3, 42–43 (1990).
93. Wöginger, A. *Prozesstechnologie zur Herstellung kontinuierlich faserverstarkter thermoplastischer Halbzeuge*. Institut für Verbundwerkstoffe GmbH, Kaiserslautern, 2004.
94. Monaghan, P. F., Brogan, M. T., and Oosthuizen, P. H. Heat transfer in an autoclave for processing thermoplastic composites. *Compos. Manuf.*, 2, 233–242 (1991).
95. Neitzel, M., and Mitschang, P. *Handbuch Verbundwerkstoffe*. Hanser Verlag, Munich, 2004.
96. Davis, B. A., Gramann, P. J., Osswald, T. A., and Rios, A. *Compression Molding*. Hanser Verlag, Munich, 2003.
97. Isayev, A. I. *Injection and Compression Molding Fundamentals*. Marcel Dekker, New York, 1987.
98. Edelmann, K. *Prozeßintegrierte Analyse des Fließverhaltens von faserverstarkten thermoplastischen Preßmassen fur die Serienfertigung*. Institut fur Verbundwerk-stoffe GmbH, Kaiserslautern, 2001.
99. Neitzel, M., and Breuer, U. *Die Verarbeitungstechnik der Faser-Kunststoff-Verbunde*. Hanser Verlag, Munich, 1997.
100. Jespersen, S. T., Wakeman, M. D., Michaud, V., Cramer, D., and Månson, J.-A. E. Film stacking impregnation model for a novel net shape thermoplastic composite preforming process. *Compos. Sci. Technol.*, 68, 1822–1830 (2008).
101. Bannister, M. Challenges for composites into the next millennium—A reinforcement perspective. *Compos. Part A Appl. Sci. Manuf.*, 31, 901–910 (2001).
102. Peters, S. T., Humphrey, W. D., and Foral, R. F. *Filament Winding Composite Structure Fabrication*. SAMPE, Covina, CA, 1991.
103. Ford, A. R. Semi-finished thermoplastic composites-realising their potential. *Mater. Des.*, 25(7), 631–636 (2004).
104. Meyer, R. W. *Handbook of Pultrusion Technology*. Chapman and Hall, New York, 1985.
105. Gutowski, T. G. *Advanced Composites Manufacturing*. John Wiley & Sons, New York, 1997.
106. Angelov, I., Wiedmer, S., Evstatiev, M., Friedrich, K., and Mennig, G. Pultrusion of a flax/polypropylene yarn. *Compos. Part A*, 38, 1431–1438 (2007).
107. Åström, B. T., Larsson, P. H., Pipes, R. B. Development of a facility for pultrusion of thermoplastic-matrix composites. *Compos. Manuf.*, 2, 114–123 (1991).
108. Ye, L., Friederich, K., and Kåstel, J. Consolidation of GF/PP commingled yarn composites. *Appl. Compos. Mater.*, 1, 415–429 (1995).
109. Mallon, P. J., O'Brádaigh, C. M., and Pipes, R. B. Polymeric diaphragm forming of complex-curvature thermoplastic composite parts. *Composites*, 20(1), 48–56 (1989).
110. Mallon, P. J., and O'Brádaigh, C. M. Development of a pilot autoclave for polymeric diaphragm forming of continuous fiber reinforced thermoplastics. *Composites*, 19(1), 37–47 (1988).
111. Mouritz, A. P., Leong, K. H., and Herzberg, I. A review of the effect of stitching on the in-plane mechanical properties of fibre-reinforced polymer composites. *Composites Part A*, 28A, 971–991 (1997).
112. Potluri, P., Kusak, E., and Reddy, T. Y. Low velocity impact damage characteristics of z-fiber reinforced sandwich panels—An experimental study. *Compos. Struct.*, 59, 251–225 (2003).
113. Grenestedt, J. L., and Reany, J. Wrinkling of corrugated skin sandwich panels. *Composites Part A*, 38, 576–589 (2007).
114. Grenestedt, J. L., and Bekisli, B. Analyses and preliminary tests of a balsa sandwich core with improved shear properties. *Int. J. Mech. Sci.*, 45, 1327–1346 (2003).
115. Jakobsen, J., Bozhevolnaya, E., and Thomsen, O. T. New peel stopper concept for sandwich structures. *Compos. Sci. Technol.*, 67, 3378–3385 (2007).
116. Sorrentino, L., Aurilia, M., and Iannace, S. Polymeric foams from high-performance thermoplastics. *Adv. Polym. Technol.*, 30(3), 234–243 (2011).
117. Lee, S. T., ed. *Foam Extrusion*. Technomic Publishing, Lancaster, PA, 2000.
118. Klempner, C., and Frish, K. C. *Handbook of Polymeric Foams and Foam Technology*. Hanser, Munich, 1992.
119. Michael, A., and Dubois, P. Polymer-layered silicate nanocomposites: Preparation, properties and uses of a new class of materials. *Mater. Sci. Eng. R: Rep.*, 28(1–2), 1–63 (2000).
120. Giannelis, E. P., Krishnamoorti, R., and Manias, E. Polymer–silicate nanocomposites: Model systems for confined polymers and polymer brushes. *Adv. Polym. Sci.*, 138, 107–147 (1999).
121. Dong, W., Gao, H., Jiang, W., and Jiang, Z. Microcellular processing and relaxation of poly(ether ether ketone). *J. Polym. Sci. Part B: Polym. Phys.*, 45(20), 2890–2898 (2007).

122. Tan, S. C., Bai, Z., Sun, H. Mark, J. E., and Lee, C. Y. C. Processing of microcellular foams Polybenzobisthiazole/polyetherketone molecular composites. *Mater. Sci.*, 38(19), 4013–4019 (2003). doi.10.1023/A:1026218817102.
123. Nemoto, T., Takagi, J., and Ohshima, M. Nanocellular foams—Cell structure difference between immiscible and miscible PEEK/PEI polymer blends. *Polym. Eng. Sci.*, 50(12), 2408–2416 (2010).
124. Sorrentino, L., Aurilia, M., Cafiero, L., and Iannace, S. Nanocomposite foams from high-performance thermoplastics. *J. Appl. Polym. Sci.*, 122(6), 3701–3710 (2011).
125. Sorrentino, L., Aurilia, M., Cafiero, L., Cioffi, S., and Iannace, S. Mechanical behavior of solid and foamed polyester/expanded graphite nanocomposites. *J. Cell. Plast.*, 48(4) 355–368 (2012).
126. Krause, B., Mettinkhof, R., van der Vegt, N. F. A., and Wessling, M. Microcellular foaming of amorphous high-T_g polymers using carbon dioxide. *Macromolecules*, 34(4), 874–884 (2001).
127. Krause, B., Sijbesma, H. J. P., Münüklü, P., van der Vegt, N. F. A., and Wessling, M. Bicontinuous nanoporous polymers by carbon dioxide foaming. *Macromolecules*, 34(25), 8792–8801 (2001).
128. Weiser, E. S., Johnson, T. F., St Clair, T. L., Echigo, Y., Kaneshiro, H., and Grimsley, B. W. Polyimide foams for aerospace vehicles. *High Perform. Polym.*, 12(1), 1–12 (2000).
129. Miller, D., Chatchaisucha, P., and Kumar, V. Microcellular and nanocellular solid-state polyetherimide (PEI) foams using sub-critical carbon dioxide I. *Process. Struct. Polym.*, 50(23), 5576–5584 (2009).
130. Rutz, B. H., and Berg, J. C. A review of the feasibility of lightening structural polymeric composites with voids without compromising mechanical properties. *Adv. Colloid Interface Sci.*, 160, 56–75 (2010).
131. Sorrentino, L., Cafiero, L., D'Auria, M., and Iannace, S. Cellular thermoplastic fibre reinforced composite (CellFRC): A new class of lightweight material with high impact properties. *Composites Part A*, 64, 223–227 (2014).
132. Kenny, J. M., Torre, L., Nicolais, L., Iannone, M., and Voto, C. Morphology and environmental resistance of PEEK/PEI systems for amorphous bonding. *Proc. 13th SAMPE European Chapter Conference*, Basilea, 1992, p. 105.
133. Mensitieri, G., Del Nobile, M. A., Apicella, A., and Nicolais, L. Time and temperature dependent sorption in poly-ether-ether-ketone (PEEK). *Polym. Eng. Sci.*, 29, 24 (1989).
134. Mensitieri, G., Del Nobile, M. A., Apicella, A., Nicolais, L., and Garbassi, F. Solvent induced crystallization in poly(aryl-ether-ether-ketone). *J. Mater. Sci.*, 25, 2963 (1990).
135. Mensitieri, G., Apicella, A., Del Nobile, M. A., and Nicolais, L. Solvent mixtures sorption in amorphous PEEK. *Polym. Bull.*, 27, 323 (1991).
136. Mensitieri, G., Apicella, A., Del Nobile, M. A., and Nicolais, L. Extreme environment resistance of PEEK matrix. *J. Reinf. Plast. Compos.*, 12, 1138 (1993).
137. Mensitieri, G., Del Nobile, M. A., Apicella, A., and Nicolais, L. Moisture–matrix interactions in polymer based composite materials. *Rev. Inst. Francais Petrole*, 4, 50 (1995).
138. Maffezzoli, A., Kenny, J. M., Torre, L., and Nicolais, L. Dynamic mechanical behaviour of thermoplastic matrix composites during crystallization, EPF 92 (V European Polymer Federation Symposium on Polymeric Materials), Baden-Baden, Germany, September 27–October 2, 1992.
139. Kenny, J. M., Torre, L., and Nicolais, L. Short- and long-term degradation of polymer-based composites. *Thermochim. Acta*, 227, 97–106 (1993).
140. Partridge, I. K. ed. *Advanced Composites*. Elsevier, New York, 1989.

22 Natural Fiber Thermoplastic Composites

Norma E. Marcovich, María M. Reboredo, and Mirta I. Aranguren

CONTENTS

22.1 Introduction 727
22.2 Matrices 728
22.3 Fibers and Surface Treatments 728
22.4 Processing 736
    22.4.1 Semifinished-Product Manufacturing 736
        22.4.1.1 Kneading 736
        22.4.1.2 Solution Mixing 737
        22.4.1.3 Granule Production 737
        22.4.1.4 Mat Production 737
    22.4.2 Direct Composite Manufacturing 738
        22.4.2.1 Compression Molding 738
        22.4.2.2 Injection Molding 738
        22.4.2.3 Extrusion 738
        22.4.2.4 Pultrusion 738
        22.4.2.5 Other Processes 739
22.5 Properties 739
    22.5.1 Agrofiber/Thermoplastic Composites 739
    22.5.2 Polyethylene Composites 740
    22.5.3 Polypropylene Composites 741
    22.5.4 Composites from Other Synthetic Thermoplastics 744
    22.5.5 Bioplastic Composites 745
22.6 Final Remarks 746
References 747

22.1 INTRODUCTION

Increasing environmental awareness throughout the world has greatly impacted materials engineering and design. Renewed interest in the utilization of natural materials addresses ecological issues such as recyclability and environmental safety (Kabir et al. 2012). Therefore, this century has witnessed remarkable achievements in green technology in the field of materials science through the development of natural fiber composites (Faruk et al. 2012).

    Natural fibers are renewable resources, abundantly available, especially in tropical areas. On the other hand, their poor dimensional stability, cracking, and degradation of mechanical properties due to moisture uptake conspire against the durability and reliability of their performance.

    Natural-reinforced plastics offer an attractive alternative for cost-effective applications; however, severe limitations are imposed due to their low strength and modulus. Traditionally, wood flour-filled molding compounds have been extensively used for products, such as panhandles, electric meter cases, and parts of cars. In addition, natural fiber-reinforced composites are particularly suited to structural and decorative applications in the transport, marine, offshore, and construction industries (Aranguren and Reborodo 2007).

The composites' shape, surface appearance, environmental tolerance, and overall durability are dominated by the matrix, while the fibrous reinforcement carries most of the structural load, thus providing macroscopic stiffness and strength (Faruk et al. 2012). As industry attempts to lessen the dependence on petroleum-based fuels and products, there is an increasing need to investigate more environmentally friendly, sustainable materials to replace the existing ones. Reinforcement of composites with plant fibers has the potential to decrease component weight relative to glass fiber reinforcement, reduce cost of materials relative to carbon fiber reinforcement, and significantly increase the fraction of composite components from a renewable source.

22.2 MATRICES

The lower thermal stability of natural fibers, with decomposition beginning around 200°C, limits the number of thermoplastics to be considered as matrix materials for natural thermoplastic composites. Unfortunately, technical thermoplastics, like polyamides, polyesters, and polycarbonates, require processing temperatures greater than 250°C and are therefore not the most adequate for this kind of composite, due to thermal degradation of the fibers (Bledzki and Gassan 1999). Thus, the only thermoplastics usable for natural fiber-reinforced composites are those whose processing temperature does not exceed 200°C.

The polymer market is dominated by commodity plastics with 80% consuming materials based on nonrenewable petroleum resources (Faruk et al. 2012). In particular, polypropylene (PP), polyethylene (PE), polystyrene (PS), and polyvinylchloride (PVC) were used as matrices for natural fiber thermoplastic composites, PP and PE being the two most commonly employed (Araujo et al. 2010; Choudhury 2008; Faruk et al. 2012; Ho et al. 2012; Kechaou et al. 2010; Li et al. 2008; Singha et al. 2010; Vilaseca et al. 2010; Zainudin et al. 2009) due to their processability, mechanical properties, stability, and durability.

However, the high durability of these polymers increases the environmental burden. Public concern about the environment, climate change, and limited fossil fuel resources are important driving forces, which motivate researchers to find alternatives to crude oil. Bio-based plastics may offer important contributions by reducing the dependence on fossil fuels and, in turn, the related environmental impacts; hence, biopolymers or "green polymers" have experienced a renaissance in recent years. Many new polymers were developed from renewable resources and used as biomatrices for natural fiber thermoplastic composites, such as starch (Liu et al. 2010; Rosa et al. 2009; Takagi 2010), which is a naturally occurring polymer that was rediscovered as a plastic material. This allows extension of the potential thermoplastic matrices, including proteins; polyesters (e.g., polyhydroxyalcanoates [PHAs]) (Faruk et al. 2012); polyhydroxybutyrate (PHB) (Barkoula et al. 2010; Hodzic et al. 2007; Zini et al. 2007); polyhydroxybutyrate-co-valerate (PHBV) (de Sousa et al. 2010; Keller 2003); polylacticacid (PLA) (Huda et al. 2008; Le Duigou et al. 2010; Lee et al. 2009); natural rubber; some polyamides; polyvinylalcohols; polyvinylacetates; polycaprolactone; and polyurethanes based on polyol derived from soybean oil, among others.

22.3 FIBERS AND SURFACE TREATMENTS

The greatest challenge in working with natural fiber-reinforced plastic composites is their large variation in properties and characteristics. The composite's properties are influenced by a number of variables, including the fiber type, environmental conditions (where the plant fibers are sourced), processing methods, and any modification of the fiber (Faruk et al. 2012).

Natural fibers are biodegradable, nonabrasive, and low-cost fibers with low density and high specific properties. They can be classified, according to their origin, into animal, vegetable, and mineral; the most used are the vegetable ones due to their wide availability and short time renewability compared to others. Vegetable fibers can also be classified according to their origin and grouped into *leaf* (abaca, cantala, curaua, date palm, henequen, pineapple, sisal, banana); *seed* (cotton); *bast* (flax, hemp, jute, ramie); or *fruit* (coir, kapok, oil palm). Some important physical and mechanical properties of the most used vegetable fibers are summarized in Table 22.1 (Bledzki and Gassan 1999; Eichhorn et al. 2001; Faruk et al. 2012; Nabi Saheb and Jog 1999; Rong et al. 2001; Shalwan and Yousif 2012).

Flax, *Linum usitatissimum*, is grown in temperate regions and is one of the oldest fiber crops in the world (Charlet et al. 2007; Faruk et al. 2012). Although traditionally it was most frequently used in the higher value-added textile markets, nowadays, it is also widely used in the composites area (Faruk et al. 2012). Another notable bast fiber crop is hemp, which belongs to the *Cannabis* family. It is an annual plant that grows in temperate climates. A considerable initiative is currently underway for its further development in Europe (Faruk et al. 2012). Jute is produced from plants of the genus *Corchorus*, which includes about 100 species. It is one of the cheapest natural fibers and is currently the bast fiber with the highest production volume. Bangladesh, India and China, provide the best conditions for the growth of jute (Faruk et al. 2012). Sisal is an agave (*Agave sisalana*) commercially produced in Brazil and East Africa (Bledzki and Gassan 1999; Faruk et al. 2012). The abaca/banana fiber, which comes from the banana plant, is durable and resistant to seawater. Abaca, the strongest of the commercially available cellulose fibers, is indigenous to the Philippines and is currently produced there and in Ecuador. It was once the preferred cordage fiber for marine applications (Faruk et al. 2012). Pineapple (*Ananas comosus*) is a tropical plant native to Brazil (Faruk et al. 2012). Ramie belongs to the family Urticaceae (*Boehmeria*), which includes about 100 species. Ramie's popularity as a textile fiber has been limited largely by regions of production and by a chemical composition that has required more extensive pretreatment than is required for the other commercially important bast fibers (Faruk et al. 2012). Coir husk fibers are located between the husk and the outer shell of the coconut. Because coir is a by-product of the production of other coconut products, its production is largely determined by demand. Abundant quantities of coconut husk imply that, given the availability of labor and other inputs, coir producers can adjust relatively

TABLE 22.1
Physical and Mechanical Properties of Several Natural Fibers

Fiber	Density (g/cm³)	ε_b (%)	TS (MPa)	E (GPa)
Abaca	1.5	3.0–10.0	400	12
Bagasse	1.25	–	290	17
Bamboo	0.6–1.1	–	140–230	
Cotton	1.5–1.6	7.0–8.0	287–597	5.5–12.6
Jute	1.3	1.5–1.8	393–773	26.5
Flax	1.5	2.7–3.2	345–1035	27.6
Hemp	1.48	1.6	690	70
Kenaf	–	1.6	930	53
Ramie	1.5	2.5–3.8	400–938	24.5–128
Sisal	1.5	2.0–2.5	511–635	9.4–22.0
Oil palm	0.7–1.55	25	248	3.2
Coir	1.2	30.0	175	4.0–6.0
Curaua	1.4	3.7–4.3	500–1150	11.8
Softwood kraft	1.5		1000	40.0
Pineapple	0.8–1.6	14.5	170–627	1.44–62.0
Sun hemp	1.07		389	35.0
Viscose (cord)		11.4	593	11.0

Note: ε_b, elongation at break; E, Young's modulus; TS, tensile strength.

rapidly to market conditions and prices (Faruk et al. 2012). Bamboo (*Bambusa* Shreb.) is a perennial plant, which grows up to 40 m in height in monsoon climates. Generally, it is used in construction, carpentry, weaving and plaiting etc. (Faruk et al. 2012). Rice is only one of the large groups of cereal grains that can be used to produce hull fibers. Nowadays wheat, corn, rye, oats, and other cereal crops are used to produce fibers for the reinforcement of composites (Faruk et al. 2012). Oil palms (*Elaeis*) comprise two species of the Arecaceae or palm family; oil palm empty fruit bunch fibers possess potential as a reinforcement fiber for plastic (Faruk et al. 2012). Bagasse is the fibrous residue that remains after sugarcane stalks are crushed to extract their juice (Faruk et al. 2012).

Climatic conditions, age, and degradation processes influence not only the structure of the fibers but also the chemical composition (Faruk et al. 2012). The major chemical component of a living tree is water. However, on a dry basis, all plant cell walls consist mainly of sugar-based polymers (cellulose, hemicellulose) that are combined with lignin with a lower amount of extractives, proteins, starch, and inorganics (Faruk et al. 2012; Nabi Saheb and Jog 1999). Their structural compositions allow moisture absorption from the environment, which leads to poor bonding with the matrix materials (Kabir et al. 2012). The chemical components are distributed throughout the cell wall, which is composed of primary and secondary wall layers (Faruk et al. 2012; Kabir et al. 2012). The chemical composition varies from plant to plant, and within different parts of the same plant. Table 22.2 (Bledzki and Gassan 1999; Eichhorn et al. 2001; Faruk et al. 2012) shows the range of the average chemical constituents for a wide variety of plant types.

The poor resistance to moisture and incompatibility with nonpolar polymers often reduce their potential as reinforcing fibers for the development of thermoplastic composites (Faruk et al. 2012; Kabir et al. 2012; Nabi Saheb and Jog 1999). Therefore, the surface properties of natural fiber are often modified to improve their adhesion with different matrices. Physical and chemical methods of different efficiency, aimed to improve the fiber–matrix adhesion by reducing the difference between hydrophilic/hydrophobic characteristics of the fiber and matrix, can be used to optimize this interface (Bledzki and Gassan 1999).

Physical methods, such as stretching, calendering, thermotreatments, electric discharge (laser, corona, plasma), vacuum ultraviolet (UV) and γ-ray bombardments, and the production of hybrid yarns, do not change the chemical composition of the fibers (Bledzki and Gassan 1999; Faruk et

TABLE 22.2
Chemical Composition of Some Common Natural Fibers

Fiber	Cellulose (%)	Hemicellulose (%)	Lignin (%)	Waxes
Bagasse	55.2	16.8	25.3	–
Bamboo	26–43	30	21–31	–
Flax	71	18.6–20.6	2.2	1.5
Kenaf	72	20.3	9	–
Jute	61–71	14–20	12–13	0.5
Hemp	68	15	10	0.8
Ramie	68.6–76.2	13–16	0.6–0.7	0.3
Abaca	56–63	20–25	7–9	3
Sisal	65	12	9.9	2
Coir	32–43	0.15–0.25	40–45	–
Oil palm	65	–	29	–
Pineapple	81	–	12.7	–
Curaua	73.6	9.9	7.5	–
Wheat straw	38–45	15–31	12–20	–
Rice husk	35–45	19–25	20	14–17
Rice straw	41–57	33	8–19	8–38

al. 2012) but do change structural and surface properties of the fiber and thereby influence the mechanical bonding to polymers (Bledzki and Gassan 1999; Faruk et al. 2012). It is noteworthy that physical treatments appear as the most eco-friendly ones (Ragoubi et al. 2012). Corona treatment is one of the most interesting techniques for surface oxidation activation. This process changed the surface energy of the cellulose fibers (Faruk et al. 2012; Ragoubi et al. 2012), resulting, for example, in a surface oxidation and an etching effect that led to an improvement of the interfacial compatibility between miscanthus fibers and polylactic acid or PP matrices (Ragoubi et al. 2012). X-ray photoelectron spectroscopy (XPS) analysis of the fibers before and after corona discharge treatment revealed an increase in the oxygen/carbon ratio as well as the generation of additional functional groups (hydroxyl, carboxyl, and carbonyl) during the treatment, while the etching effect was confirmed by scanning electron microscopy (SEM), as shown in Figure 22.1.

Plasma treatment can lead to a variety of surface modifications depending on the type and nature of the gases used. Reactive free radicals and groups can be produced, the surface energy can be increased or decreased, and surface cross-linking can be introduced (Faruk et al. 2012). For example, the hydrophilicity of cotton fabric samples increased with direct current (DC) air plasma treatment due to the formation of polar groups (Nithya et al. 2011). However, Bozaci et al. (2013) treated flax fibers by argon and air atmospheric-pressure plasma systems and demonstrated that the interfacial adhesion of argon-treated flax fiber to the high-density polyethylene (HDPE) matrix was superior to those of air-treated and untreated flax fiber, due to the lower oxygen concentration of the former fiber surface, but mainly due to the increased roughness of the argon-treated fiber surface compared to those of the air-treated or untreated ones, as shown in Figure 22.2. On the other hand, thermotreatments can be as simple as a controlled heating, as the one used by Rong et al. (2001). They noticed that just 4 h of heating of sisal fiber at 150°C in an air-circulating oven led to the removal of some aromatic lignin-like impurities but principally to an increase in crystallinity of the fiber, which was attributed to the arrangement of molecular structure at elevated temperature. According to the authors, the change in crystallinity leads to an important increase in the mechanical properties of the fiber: 36%, 3%, and 40% improvement in tensile strength, Young's modulus, and elongation at break, respectively.

Natural fibers are also amenable to chemical modification as they bear hydroxyl groups from cellulose, hemicellulose, and lignin. The hydroxyl groups may be involved in the hydrogen bonding

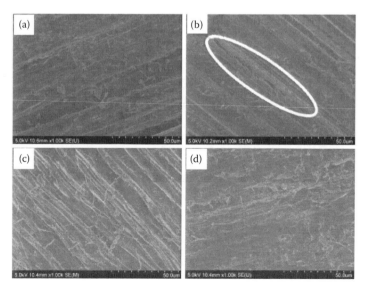

FIGURE 22.1 Effect of corona treatment on the surface of miscanthus fibers: (a) raw and after (b) 15 min, (c) 30 min, and (d) 45 min treatment. (From Ragoubi, M. et al., *Compos. Part A Appl. Sci. Manuf.*, 43, 2012.)

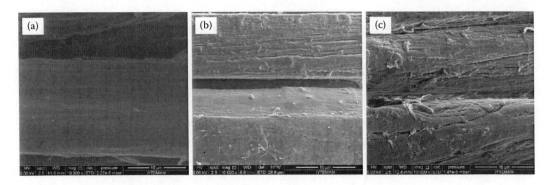

FIGURE 22.2: SEM images of flax fiber, (a) untreated flax, (b) air plasma-treated flax fiber (300 W), and (c) argon plasma-treated flax fiber (300 W). (From Bozaci, E. et al., *Compos. Part B*, 45, 2013.)

within the cellulose molecules, thereby reducing the activity towards the matrix. Chemical modifications may activate these groups or can introduce new moieties that can effectively interlock with the matrix (Faruk et al. 2012; Wang et al. 2007; Xie et al. 2010). The different surface chemical modifications of natural fibers such as mercerization, isocyanate treatment, acrylation, benzoylation, permanganate treatment, acetylation, and peroxide treatment have achieved various levels of success in improving fiber strength, fiber fitness, and fiber–matrix adhesion in natural fiber composites. Alkali treatment, also called mercerization and sometimes considered a physical treatment, involves the immersion and soaking of the fibers in alkaline medium, typically a rather concentrated NaOH aqueous solution. This treatment removes practically all noncellulose components except waxes, and thus improves the fiber–matrix adhesion due to the removal of natural and artificial impurities, having a lasting effect on the mechanical behavior of natural fibers, especially on fiber strength and stiffness (John and Anandjiwala 2008; Kabir et al. 2012; Kim and Netraval 2010). Owing to the dissolution of lignin by alkali, some pores are formed on the fiber surface. The loss of the cementing material of the vegetable fibers produces some fibrillation because of the breakage and separation of the fiber bundles. These changes increase the contact area between the fiber and the matrix, thus allowing better fiber wetting. Moreover, the number of possible reactive sites increases, and therefore, it is used quite frequently as a fiber pretreatment (Manikandan Nair et al. 2001; Marcovich et al. 1998; Sreekala and Thomas 2003; Wang et al. 2007). Scheme 22.1 gives a schematic view of the lignocellulosic fiber structure, before and after an alkali treatment. Treated fibers have lower lignin content, partial removal of wax and oil cover materials, and distension of the crystalline cellulose order.

Acetylation, benzoylation, esterification with maleic anhydride (MAN), acrylation, acrylonitrile grafting, and sodium chlorite and permanganate treatments decrease the hydrophilic nature of the

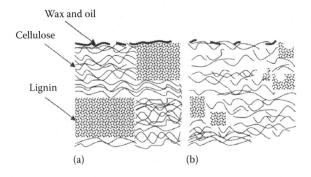

SCHEME 22.1 Typical structure of (a) untreated and (b) alkalized lignocellulosic fiber. (From Kabir, M.M. et al., *Compos. Part B*, 43, 2012.)

fibers (Frisoni et al. 2001; Kabir et al. 2012; Manikandan Nair et al. 2001; Marcovich et al. 1998; Mishra et al. 2001; Pejic et al. 2008; Sreekala et al. 2000; Wang et al. 2007) by forming covalent linkages with their accessible hydroxyl groups. Sodium chlorite treatment also removes lignin from the fiber (Pejic et al. 2008). Regarding the changes in other properties, there is no consensus in the literature. For example, Wang et al. (2007) found that benzoylation treatment led to a smooth flax fiber surface due to the substances deposited on the surface of the fiber at the same time that the fiber strength was reduced due to breakage of the bond structure. On the other hand, Manikandan Nair et al. (2001) reported that the same treatment improved the thermal stability of short sisal fibers and lead to a rougher fiber surface (compared to the untreated ones), while improving the thermodynamic compatibility between constituents and, thus, the interfacial adhesion with the PS matrix due to the presence of phenyl structures in treated fiber (similar to that of PS).

Although some researchers have reported that peroxide-induced adhesion in cellulose fiber-reinforced composites resulted in easy processability and improvement of mechanical properties (Kabir et al. 2012; Wang et al. 2007), other authors (Sreekala and Thomas 2003) found that the tensile properties of benzoyl peroxide-treated oil palm fibers decreased dramatically after treatment (Young's modulus from 6.7 to 1.1 GPa; tensile strength from 248 to 133 MPa).

Isocyanate also reacts with the hydroxyl groups of cellulose and lignin constituents of the fibers, and the urethane linkages formed block these hygroscopic sites (Gironès et al. 2007; He et al. 2012; Xie et al. 2010). Moreover, Karmarkar et al. (2007) took advantage of the reactivity of isocyanates toward hydroxyl groups of vegetable fibers and developed a novel compatibilizer (m-TMI-g-PP) with an isocyanate functional group by grafting m-isopropenyl-α,α-dimethylbenzyl-isocyanate (m-TMI) onto isotactic PP. Based on spectral evidence and previous reports (Braun and Schmitt 1998; De Roover et al. 1995), a coupling mechanism of m-TMI-g-PP in wood fiber–PP composites is proposed, as shown in Scheme 22.2.

Several compounds, known as *coupling agents*, promote adhesion by chemically coupling the fibers to the matrix. Their function is to create a chemical bridge between the reinforcement and matrix, and therefore, they possess two functions: the first one is to react with the OH groups of cellulose, and the second one, to react with functional groups of the matrix or to entangle with the polymer matrix due to their similar polarities and long chains. The m-TMI-g-PP copolymer presented previously fits into this category; however, the most extensively used coupling agents for natural fiber thermoplastic composites are copolymers containing MAN, such as maleated PP

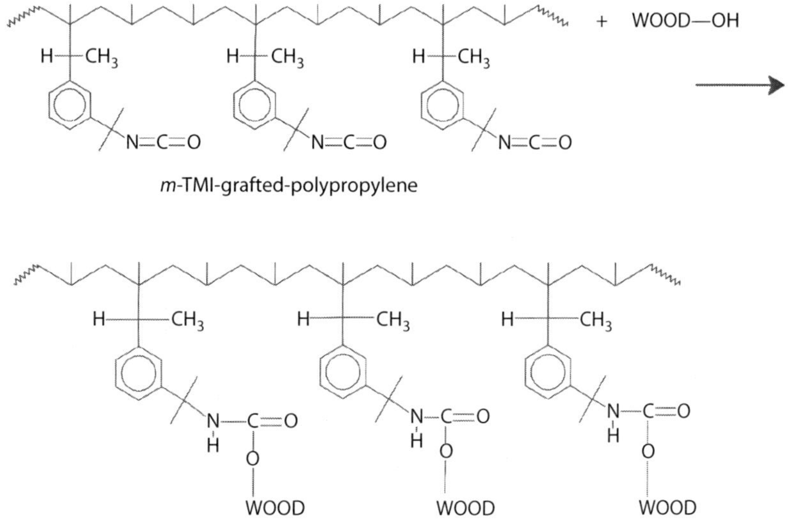

SCHEME 22.2 Coupling mechanism of TMI-grafted polypropylene. (From Karmarkar, A. et al., *Compos. Part A*, 38, 2007.)

(MAPP) or maleated PE (MAPE) (Acha et al. 2006; Keener et al. 2004; Liu et al. 2009; Saleem et al. 2008). The anhydride groups of the copolymers may react with the surface hydroxyl groups of natural fibers, forming ester bonds, while the other end of the copolymer entangles with the polymer chains of the matrix. Maleated couplers need enough acid functionality to attach to the filler and enough molecular weight to entangle or crystallize into the base polymer: a low acid number may not give the coupler enough sites for attachment to the polar filler, while a too-high acid number may hold the coupler too close to the polar surface and not allow sufficient interaction with the continuous nonpolar phase. In the same line, low molecular weight will not allow the coupler to interact and entangle sufficiently with the polyolefin phase, while too-high molecular weight may not allow the coupler to diffuse and accommodate at the interface (Keener et al. 2004). These effects are shown in Figure 22.3 (Acha et al. 2006) for jute fabric–PP composites modified by 5 wt.% of two different commercial maleated couplers (Eastman Chemical Co): Epolene E-43 wax (M_n = 3900, M_w = 9100, acid number = 4.7 mg KOH/g) and Epolene G-3003 (M_n = 27,200, M_w = 52,000, acid number = 1.4 mg KOH/g). The compatibilized composites exhibit rougher surfaces at the interfacial region of failure, showing mainly fiber breakage instead of the noticeable pullout presented by the uncompatibilized sample, added to a smother fiber surface due to the better wetting by the modified matrix. Moreover, the amount of compatibilizer used should be also carefully selected: a too-low amount of compatibilizing agent may not be enough to saturate the fiber surface (poor compatibility between fiber and matrix), while an excess of coupling agent would induce self-entanglements among compatibilizer chains rather than between compatibilizer and polymer matrix chains, resulting in slippage (Acha et al. 2006; Rana et al. 2003).

Both functional monomer and type of base resin are important factors determining the effectiveness of functionalized polyolefins in improving the properties of vegetable fiber thermoplastic matrix composites. Li and Matuana (2003) studied these effects on HDPE–wood flour composites using commercial copolymers (Uniroyal Chemical Company, Polybond line) with different base resins (HDPE, linear low density polyethylene [LLDPE], and PP) and different functional monomers (acrylic acid with 6% functionality and MAN with 1% functionality) as coupling agents. They found that maleated polyolefin coupling agents performed better than acrylic acid-functionalized polyolefins (even with higher functionality) and that PE-based maleated coupling agents were more effective than their PP-based counterparts. The former behavior was explained taking into account that MAN (a five-member ring structure) offers spatial resistance for its homopolymerization (it is more likely to copolymerize with olefins than homopolymerize during the production of maleated polyolefins), while acrylic acid tends to homopolymerize due to the lack of spatial resistance in its structure, and thus, it may be grafted as short acrylic acid chains on polyolefin backbone chains; consequently, not all carboxylic groups of acrylic acid-functionalized polyolefins are available for attachment to the surface of natural fiber. Regarding the base resin, the PE-based coupling agents were more effective than their PP-based counterparts as compatibilizers because of the compatibility between the PE backbone chains of the coupling agents and the HDPE matrix.

 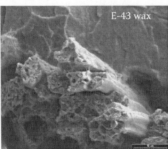 

FIGURE 22.3 Jute–PP interphase of composites modified by maleated couplers. (From Acha, B.A. et al., *Polym. Int.*, 55, 2006.)

Silane coupling agents are one of many ingredients in commercial sizing that are applied to fibers, although the structures used to couple natural fibers and thermoplastic matrices are relatively limited, the most used being the aminosilanes, especially γ-aminopropyltriethoxysilane (APS) (Kabir et al. 2012; Xie et al. 2010). The organofunctionality of the silane interacts with the polymer matrix, and the interaction modes depend on the functionality's reactivity or compatibility toward the polymer: a nonreactive alkyl group of the silane may increase the compatibility with a nonpolar matrix due to their similar polarities, while a reactive organofunctionality may covalently bond with it, as well as being physically compatible with the polymer matrices. Sreekala and Thomas (2003) modified oil palm fibers (pretreated with alkali) using a triethoxy vinyl silane coupling agent and found that the treatment increases the hydrophobic character of the fiber. Another example is the work of Rong et al. (2001), who treated sisal fibers with γ-amine propyl triethoxysilane and found that the crystallinity of the fiber decreased with the treatment because, in addition to the reaction of silanols with hydroxyls of the fiber surface, the formation of polysiloxane structures (through steps of hydrolysis, condensation, and bond formation) also took place. Such large coupling molecules would destroy the packing of cellulose chains to a certain extent, and thus, the Young's modulus and tensile stress of the treated fibers were lower compared to the untreated ones, while the elongation at break presented the opposite behavior.

Fungal treatment is another promising alternative for surface modification of natural fibers (Kabir et al. 2012). This environment-friendly and efficient biological treatment is used to remove noncellulosic components (such as wax) from the fiber surface by the action of specific enzymes. White rot fungi produce extracellular oxidase enzymes that react with lignin constituents (lignin peroxidase), which causes the removal of lignin from the fiber. It also increases hemicellulose solubility and thus reduces the hydrophobic tendency of the fiber. In addition to this, fungi produce hyphae, which create fine holes on the fiber surface and generate a rough interface for better interlocking with the matrix. Enzyme treatment of cellulosic fibers provides considerable advantages in terms of soft touch, surface luster, and improved wettability and dyeability without appreciable fiber deterioration. Currently, cellulases are widely used to alter cellulose properties for potential applications in the textile, pulp, and paper industries (Bledzki et al. 2010; Faruk et al. 2012). Bledzki et al. (2010) characterized PP composites with enzyme-treated abaca fibers. The untreated fiber surface was rough, containing waxy and protruding parts. After enzymatic treatment using both a commercial enzyme (fungamix) and natural digestion, the waxy material and cuticle of the surface of the fiber were removed, leading to a smoother surface. Fiber surface damage was also observed for naturally digested fibers, which occur in natural digestion systems.

Steam explosion is one of the examples of physicochemical treatments. With it, biomass is treated with high-pressure saturated steam, and then the pressure is suddenly reduced, so that the materials undergo an explosive decompression. It is typically initiated at a temperature of 160–260°C (corresponding pressure of 6.8–47.7 atm) for several seconds to a few minutes before the material is exposed to atmospheric pressure. Whereas the thermal effect of steam explosion causes structure variations, the mechanical effect, due to the adiabatic expansion of water present in the pores of the material, occurs in a very short time and brings about variations only at the morphological level, greatly increasing the accessibility and specific surface of the material. Hence, the reactivity of steam-exploded biopolymers is increased, resulting in a higher availability of the hydroxyl groups distributed either on the surface of the cellulose crystals or in the nonordered regions. Deepa et al. (2011) used this technique to obtain cellulose nanofibers from banana fibers. Assessment of fiber chemical composition before and after steam explosion showed evidence of the removal of noncellulosic constituents such as hemicelluloses and lignin, while surface morphological studies using SEM and atomic force microscopy (AFM) revealed that there was a reduction in fiber diameter during treatment. Moreover, the percentage yield and aspect ratio of the nanofibers obtained by this technique were found to be very high in comparison with other conventional methods.

22.4 PROCESSING

Appropriate manufacturing processes must be utilized to transform the materials into the final shape without causing any defect in the products. The way natural fibers are introduced as reinforcing material in polymer composites has to be adapted to the available production techniques. Normally, natural fiber polymer composites are fabricated by using traditional manufacturing techniques that are designed for conventional fiber-reinforced polymer composites. These techniques include compounding, compression molding, direct extrusion, and injection molding, among others. Such techniques have been well developed, and the accumulated experience has proved their success for producing composites with controllable quality. However, their suitability for natural fiber-reinforced polymer composites is still uncertain due to the chemical components and geometrical, mechanical, thermal, and structural properties of the natural fibers, which are somehow different from those of the synthetic fibers (Ho et al. 2012).

Some typical problems related to the processing of these materials are due to the hydrophilic and hygroscopic nature of the fillers, and to their poor thermal resistance.

Selection of the production technique depends on the required properties as well as the desired production rate and the product status (semi or finished). Injection molding requires prepared granules of composites, whereas pultrusion and extrusion processes can deal directly with fibers (natural or synthetic) to have a profile product. Natural fiber compounds offer numerous advantages over other injection-molding compounds, for instance, low wear of manufacturing tools, often reduced cycle time, and ease of recycling. Special attention should be paid to the choice of processing parameters; these should be carefully selected due to the limited processing window for composites, as natural fibers degrade quickly at high processing temperature.

Also, the production technique affects the fiber aspect ratio and fiber orientation, and hence, the mechanical properties are defined. For instance, in pultrusion, long and unidirectional fibers are attained within the produced profile. On the other hand, more randomness in the fiber directions is attained in extrusion and injection molding. Thus, the different fiber product forms (raw, sliver, yarn, fleece, or felts) as well as the different suitable manufacturing techniques and the production rate, fiber length, and mechanical properties should be carefully taken into account.

Besides, as has been already discussed, chemical treatments onto the fiber surface are normally required to compensate for its low bonding at the interface between the hydrophilic fiber and hydrophobic matrices. Other technical problems, such as the uniformity of fiber distributed inside the composites, thermal degradations and weathering effect of fiber and matrix, water absorption of both fibers and matrix, wettability of resin impregnated into spaces between fibrils, and breakage of fibers during mechanical stirring/mixing stages in the manufacturing processes, also limit the use of natural fibers for new composite development (Ho et al. 2012).

Drying of fibers before processing is an important factor, as water on the fiber surface acts as a separating agent in the fiber–matrix interface. Additionally, voids would appear in the composite due to the evaporation of water during processing. Both phenomena lead to a decrease in the mechanical properties of natural fiber-reinforced composites. Fiber drying can be carried out in a vacuum oven at different temperatures.

In general, natural fiber composite processing begins by either mixing or compounding, followed by shaping and finishing (Figure 22.4).

22.4.1 Semifinished-Product Manufacturing

22.4.1.1 Kneading

Otherwise called melt mixing, kneading is a type of batch production (noncontinuous) in which shear force and temperature are applied to the fibers and the molten thermoplastic matrix to ensure a homogeneous distribution. The bulky composite compound produced by kneading is granulated

Natural Fiber Thermoplastic Composites

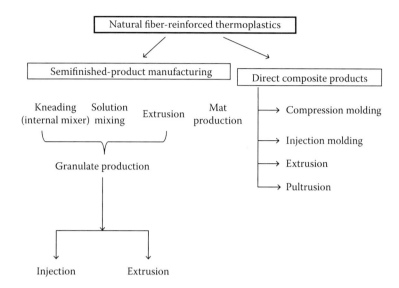

FIGURE 22.4 Natural fiber-reinforced thermoplastics processing. (Adapted from Ziegmann, G., *Flax and Hemp Fibres: A Natural Solution for the Composite Industry*, chap. 6, 100–108, Source JEC Composites, JEC Group/CELC, 2012.)

mechanically by a shredder or a mill. This process introduces fiber damage either by fiber kinking during kneading or by fiber cutting in the shredder, and hence, the efficiency of fibrous reinforcement is decreased (Ziegmann 2012).

22.4.1.2 Solution Mixing

This is another method to mix fibers and matrix: The fibers are added to a viscous solution of thermoplastics in a solvent in a stainless-steel recipient with a stainless-steel stirrer. The temperature is maintained for some time, and the mix transferred to a flat tray and kept in a vacuum oven to remove the solvent. The solution-mixing procedure avoids fiber damage that normally occurs during blending of fiber and thermoplastics by melt mixing (Li et al. 2000). Nevertheless, this process is time consuming.

22.4.1.3 Granule Production

This is the other important step in the manufacture of semifinished products. Natural fiber granules can be produced via pelleting, compounding, pultrusion, and the pull-drill process, affecting, to varying degrees, the homogeneity of the granules and the desired fiber lengths (Faruk et al. 2012). The compounding technology by extrusion has been demonstrated on a small scale (5–10 kg/h) for commodity and engineered plastics such as PP, PE, acrylonitrile butadiene styrene (ABS), PS, and for several natural fibers such as fax, hemp, jute, and kenaf. Rejected streams from pulp and paper/fiber processing industries also turned out excellent raw materials as fillers in plastics (yet2.com 2004).

22.4.1.4 Mat Production

This is one of the initial steps in semifinished-product manufacturing. Fiber mats from natural fibers could be manufactured by the carding method, aerodynamic fleece making, the fiber spreading process, and wet fleece production. The other step is the production of slivers and fiber yarns. The fiber preparation (opening, mixing, and carding) is similar to fleece production. Mats can be used directly or preimpregnated with a binder or polymeric matrix by means of a dipping technique (prepreg).

22.4.2 Direct Composite Manufacturing

22.4.2.1 Compression Molding

This is favorable for several reasons, since only two hot plates are needed to compress all components of the fiber and matrix together, and then heat is applied subsequently. However, the viscosity of the matrix during the pressing and heating processes is a concern as it is not easy to control, in particular for thick samples. The viscosity of the molten matrix should be low enough to impregnate into the space between fibers and high enough to avoid spurting out. As natural fibers are made by many small filaments, it also takes time to wet them. Therefore, control of viscosity, pressure, holding time, temperature in relation to the types of fibers and matrix, and the thickness and size of the samples are critical to produce quality composites (Ho et al. 2012). Besides, this technique has proved to be suitable for the production of profiles with any thermoplastic prepreg. (The prepregs are laid upon the mold in a desired sequence, and then, the entire laminate is placed inside the mold. The different layer orientations are thus retained after molding.) The compression molding of natural fiber thermoplastic composites is normally used in the automobile industry, especially for the inner cabin parts.

22.4.2.2 Injection Molding

This is a suitable method for processing natural fiber reinforced into sophisticated three-dimensional composite parts. It has other advantages, such as the short product cycle and excellent surface of the product. It requires a polymer with a low molecular weight to maintain low viscosity. The most important factors influencing the resulting mechanical properties are the quality of the fiber–matrix adhesion, the selection of suitable processing parameters, and the fiber orientation. In this case, the use of granular material, which already includes fibers and a coupling agent, proved to be a success. Injection molding differs from profile extrusion in that after the material is heated, it is pumped into a permanent mold, where it takes shape and cools. The mold is then opened, and the finished part, discharged. Currently, the primary application of high-thermoplastic-content lignocellulosic composites is for interior door panels and trunk liners in automobiles (Bledzki et al. 2002). The filler amount seldom exceeds 50–60 wt.%, even though sometimes, higher percentages, up to 70–80%, could be used. These conditions are relatively more extreme, and consequently, compression molding is preferred to injection molding.

22.4.2.3 Extrusion

The compounded composite material is fed into the heated barrel of the extruder using feeders and is heated so that the thermoplastic component can flow. Extrusion requires a polymer with a high molecular weight for better melt strength. The rotation of the screws inside the heating chamber plasticizes the material, which is then continually pumped and forced through a die of a given cross-section configuration. The extrudate is supported and then cooled in a water bath, after which the profile is cut to the desired length. Corotating and counter rotating twin-screw extruders can be used for the profile extrusion. They are now very widely used for long natural fiber dispersion with special feeding technology. These premixed compounds are then processed by further extrusion or injection molding (Sain and Panthapulakkal 2004).

22.4.2.4 Pultrusion

This is normally carried out either on separate rovings of natural fibers and thermoplastics or on fiber rovings from an intermingling process. The intermingled roving is produced by a simultaneous roving process on natural and thermoplastic fibers (either by extruder or from another roving). Pultrusion is carried out so that the volume of the natural fiber is defined by the number of thermoplastic rovings compared to the number of rovings of the natural fiber. The fibers are pretensioned, preheated, and then introduced into the hot mold. Finally, a cooling procedure is applied to achieve the final size profile of the pultruded yarn composite (Van de Velde and Kiekens 2001).

22.4.2.5 Other Processes

A combination of techniques is also valid. For example, composites from high-density PE filled with cork powder and coconut short fibers, in two different ratios, were prepared in a twin-screw extruder followed by a compression-molding process (Fernandes et al. 2013).

Sometimes, the PP matrix is used in the form of thin foils. The composites are manufactured by alternating stacking of fiber mats and PP foils. The individual mats are all placed with the same directional orientation, to allow this parameter to be evaluated from the composite characterization. The PP foils are used in different numbers between the fiber mats to allow a variation in volume fractions of the fibers and PP matrix. The total stack is sealed with a vacuum bag and hot-pressed in an autoclave.

The main problem with pultrusion using thermoplastic matrices lies in the full impregnation of the fiber reinforcement due to the high matrix melt viscosities. In order to overcome this problem, the distances of the polymer melt flow should be reduced. A great number of semifinished products have been developed where the matrices and the reinforcement fibers are closely mingled, for instance, in preimpregnated tapes, hybrid yarns, or powder-impregnated bundles. Nevertheless, there still remain some difficulties to achieve good mingling of the matrix with the reinforcing fibers (Angelov et al. 2007).

Rotational molding of natural fiber-reinforced thermoplastics has been also employed as an experimental technique. Different polymer–natural fiber and polymer–flour systems have been characterized. A rotomolding grade of high-density PE has been used as the polymer matrix. Fibers, including jute, sisal, and cabuya, as well as wood, pecan, and rice shell flour of different types have been used as reinforcements. The rotomolding process for those composite systems has been studied using an in-house-built two-axis rotomolding rig with variable speed. The final products have been characterized, with special emphasis on estimating the level of dispersion of the fibers and their degree of orientation, as well as the consolidation, sintering, and bubble formation processes during the different stages. Comparisons have been made with other powder processing routes, involving the sintering of fiber- (or flour-)reinforced polymer particles, including compression molding (Torres and Aguirre 2003).

22.5 PROPERTIES

22.5.1 AGROFIBER/THERMOPLASTIC COMPOSITES

Among the properties of agrofiber plastic composites, the most frequently investigated ones are the tensile and flexural properties as well as the impact properties. Additionally, some researchers have investigated the time-dependent behavior of these composites, such as creep and dynamic mechanical response, as means to understand the matrix–fiber interfacial interplay and/or predict durability.

Clearly, the use of vegetable fibers in the formulation of thermoplastic composites has attracted interest because of environmental reasons as well as the large availability, low cost, and reduced abrasive effect on processing equipment. In particular, agrofibers are interesting because they have the advantage of being derived from annual-growth crops, as compared with wood production, which requires longer growing times. But together with these economic reasons, vegetable fibers have relatively large specific mechanical properties, making them interesting reinforcing materials that add little extra weight to the final composite.

The incorporation of vegetable fibers into thermoplastics turns the material into a potential target for fungal as well as other microorganisms' attack. However, as wood replacement (one of the widespread applications of these composites), the advantage of reduced biodegradation is clear. On the other hand, if these materials are compared with the neat thermoplastics, a reduction in thermal expansion and creep is usually reported.

As already discussed, the polar nature of the fiber surfaces is an inherent drawback for the production of thermoplastic composites, since the matrices are frequently nonpolar polyolefins. The use of chemical or physical treatments of the fiber surfaces is a possibility that has been explored, although up to this moment, the use of compatibilizing agents is by far the most common applied method and certainly the one utilized in established industrial processes.

According to the nature of these materials, the composite mechanical properties depend on the two main phases, polymer and filler/reinforcement, but also on the characteristics of the interfacial region (Aranguren 2006). Some properties are more influenced by one of the phases over the other or by one given characteristic of the composite arrangement. Thus, the modulus of the material is strongly affected by the mechanical properties of the fibers, the capacity of stress transfer that occurs at the interface, the concentration of the fibers, and their orientation and aspect ratio. The strength of the material depends mostly on the matrix properties as well as the interfacial adhesion and stress concentration points.

The fracture of the material is strongly dependent on the fiber aspect ratio, and this is the reason for the better performance of fibers with longer-than-critical-length compared to shorter fibers or particles, which are stress concentrators. The impact resistance depends on a balance between interfacial adhesion and fiber debonding and other energy-absorbing mechanisms (Rowell and Stout 2007).

The role of the interfacial agent is paramount in its effect on the fracture properties of the composites. Differences in the response to notched and unnotched impact tests have also been identified. In notched tests, the crack is already present, and the fracture proceeds essentially through propagation, which is scarcely stopped by the presence of the compatibilizing agent. On the other hand, the fracture in unnotched tests involves initiation as well as propagation of the cracks, and in this case, initiation is much reduced by the action of the compatibilizer, which improves interfacial adhesion, also reducing the fiber-end negative effect in short-fiber composites.

Dynamic mechanical properties have also been studied since they can be tools to characterize interfacial interactions and also because of the direct relation to time-dependent properties, such as creep behavior, which has been specifically addressed in some cases.

The utilization of these materials requires the knowledge of their durability and stability in the conditions of use. This is particularly important in outdoor applications, due to the sensitivity of the fibers to humidity, which can also affect the interfacial adhesion, and due to the response of the polymers and fibers to UV radiation and, in general, to environmental changing conditions. The fact that the lignocellulosic fibers can be biodegraded stresses the need for studying the stability of the composites against attack by microorganisms, while it also highlights the importance of obtaining a good and complete wetting of the fibers, as well as strong interfaces that contribute to the complete encapsulation of the vegetable components. On the other hand, biodegradability in composites designed from biodegradable plastics is a searched-for characteristic, perfectly suited for applications such as disposable containers/packaging.

22.5.2 Polyethylene Composites

PE–sisal composites were studied by Joseph et al. (1996) using various fiber surface treatments (alkaline, isocyanate, and peroxide). This group has addressed the effect of processing, fiber concentration, aspect ratio, and orientation on the mechanical, viscoelastic, and dielectric properties of sisal short fibers used as reinforcement for low-density polyethylene (LDPE) (Joseph et al. 1992, 1993a,b, 1994). The fiber orientation was an important variable affecting the value of the tensile modulus of the composites (the modulus of the aligned fiber composite almost doubled the modulus of the random fiber composite). Fiber alignment also affected the dynamic mechanical properties of the materials; thus, a higher storage modulus (E') was measured for aligned fibers above a critical value of 6 mm. Chemical treatment of the fibers (particularly by means of reaction with isocyanate and peroxide) improved interfacial bonding, leading to the improvement of mechanical and dynamic mechanical properties. The treatment also contributed to improving the dimensional stability of the composites and to preserving good mechanical properties even in aged samples, which was related to the reduced sensitivity to humid environments (Joseph et al. 1995).

Silane treatment of sisal was shown to improve the interfacial shear strength in fiber-reinforced PE (Valadez-Gonzalez et al. 1999). Silane preimpregnation of the fibers for the fabrication of HDPE–hemp composites was investigated by Herrera-Franco and Valadez-Gonzalez (2004, 2005). They found that silane-treated fibers improved interfacial adhesion, which had little effect on the longitudinal tensile strength but a strong positive effect on the transversal tensile strength, which is more dependent on matrix and interphase behavior.

The use of fillers from rice husk, straw, or leaves had advantageous results when incorporated into recycled HDPE and processed by compression molding after melt compounding. The recycled HDPE composites showed improved impact strength compared to prepared virgin HDPE composites (Yao et al. 2008). The use of MAPE as a compatibilizer produced the highest improvement in the impact strength of recycled HDPE reinforced with bagasse (Lei et al. 2007).

22.5.3 Polypropylene Composites

Wambua et al. (2003) compared PP composites containing kenaf, coir, sisal, hemp, and jute fibers. The results showed similar tensile modulus (~7 GPa) for kenaf and hemp, but the hemp composite outperformed that of kenaf for tensile strength (~52 and 28 MPa, respectively), flexural modulus (5 and 2.2 GPa, respectively), and impact strength (26 and 14 kJ/m^2, respectively). The other fiber composites had lower performance, although they showed similar or better response than kenaf to impact tests.

Oksman et al. (2009) focused on the effect of the fiber microstructure of sisal, abaca, jute, and flax fibers on the impact properties PP composites.

Increased modulus and tensile strength were reported up to concentrations of 30 wt.% of sisal fibers in PP (Kuruvilla et al. 1999).

Unidirectional flax fiber–PP composites with fiber concentrations from 56 to 72 wt.% and porosities from 4% to 8% showed tensile axial modulus and strength in the range of 27–29 GPa and 251–321 MPa respectively (Madsen and Lilholt 2003). A much lower tensile modulus (3.2 GPa) was determined by Haijun and Sain (2003), who incorporated short random fibers of flax into PP and noticed that they were torn during processing, so that the actual length in the composite may be considerably reduced compared to the nominal (initial) value, which would explain why the properties of flax fibers were close to those of milled hemp composites (Haijun and Sain 2003).

Beckermann and Pickering (2008) used untreated and alkali-treated chopped hemp fibers to reinforce PP composites. The fibers were incorporated in a twin-screw extruder after careful drying at 80°C for 1 day, and the extruded material was granulated and further dried before being injected to produce tensile test specimens. For the alkaline treatment of the fibers, a solution of 5 wt.% NaOH/2 wt.% NaSO$_2$ was used, which allowed reaching the optimal reduction of the content of lignin and hemicellulose. The composite containing 40 wt.% of the treated fibers with the addition of 4 wt.% of MAPP showed the best combination of higher Young's modulus and tensile strength among the prepared composites (5.31 GPa and 50.5 MPa, respectively). The mode of failure of the composites changed from fiber debonding and tensile matrix failure in the composites made from untreated hemp fibers to local shear yielding of the matrix around the fibers for the composites prepared with the treated fibers.

The use of maleated polyolefins as compatibilizers in fiber/thermoplastic composites is widespread, and this is also the case for PP-derived composites (Bataille et al. 1989; Sanadi et al. 1994a,b, 1995). Rowell and Stout (2007) observed that uncoupled jute or kenaf fibers in PP composites showed little dependence of the tensile strength on the fiber concentration. However, the MAPP-coupled composites showed tensile strength increasing with the fiber concentration. Increasing the concentration of fibers without addition of MAPP led to lower strain to failure, which was explained by the high concentration of fiber ends in the short fiber uncoupled composite acting as points of stress concentration. Although the trend was similar, the reduction in strain was much less severe in the coupled systems. For the same reason, the impact strength was improved by the use of MAPP, especially for the case of the unnotched tests, since the coupling agent improves interfacial adhesion, increasing the resistance to crack initiation (Rowell and Stout 2007).

Comparing the performance of different fibers, Zampaloni et al. (2007) found that the highest modulus/cost ratio corresponded to the kenaf/PP/MAPP composites, as compared with composites prepared from sisal or coir or even E-glass.

Awal et al. (2011) studied the role of the interfacial modifiers by means of the fiber fragmentation test. In this test, a single fiber is centrally embedded in the polymer matrix specimen. The matrix can deform more than the fiber, and so, when the specimen is subjected to a tensile load, the fiber breaks into smaller segments because it cannot follow the matrix deformation. The fracture process continues with increasing load, up to a point that strongly depends on the fiber–matrix adhesion, and is related to the *critical fiber length*, which can be measured using optical microscopy in transparent composites (Aranguren 2006). In the work of Awal et al., the test was performed on ramie and flax composites using a PP matrix with or without the addition of MAPP (2 wt.%). The authors optimized the test (gauge length, test speed) to get more repeatable results and used Weibull analysis on the fracture results of the embedded fibers. The critical fiber lengths for the fibers used in the PP composites were 4.86 and 2.42 mm for flax, without and with MAPP, respectively. These numbers clearly show the much better interfacial adhesion obtained in the composite with a coupling agent. Analogously, ramie composites prepared with a coupling agent showed good interfacial adhesion and a critical fiber length of 0.98 mm. The interfacial fiber strength for these systems was found to be 1269 MPa and 760 MPa for ramie and flax, respectively.

Interfacial strength (IFS) and critical fiber length were also determined for a flax–PP system (PP molecular weight, M_w = 290,000) by Doan et al. (2006), finding the values of 1028.3 MPa and 1074.6 µm, respectively. The effect of adding 2 wt.% of a maleated coupling agent also showed the expected improvement of the interfacial adhesion, so that the IFS and the critical fiber length varied to 1130.1 MPa and 670.2 µm, respectively.

MAN treatment of oil palm empty fruit bunch fibers incorporated into PP allowed improvement of the impact properties. The effect of treatment appeared to reach an optimum at 5–10% weight gain (increment of the weight of the fibers due to the incorporated MAN) of the fibers; further weight gains led to marginal improvements in the impact strength for samples containing 20–60 wt.% of different-size fibers (Rozman et al. 2003).

Sim et al. (2007) used five different MAPPs to investigate the effect of molecular weight and percentage of grafted MAN of the coupling agent on the final properties of a bioflour–PP composite. They found an optimum level of grafting that allowed the coupling agent to interact with the surface; higher values resulted in molecules too closely attached to the surface of the fibers with little degree of interaction with the bulk PP. Analogously, an optimum molecular weight was found that allows interaction with the matrix, while at higher molecular weights, the mobility of the chain is reduced for diffusion toward the fiber/matrix interface.

Lignin was also used as coupling agent in PP–coconut composites, which led to improved flexural properties (Rozman et al. 2000).

Use of radiation (10 kGy) on henequen fibers incorporated into PP (40 wt.%) resulted in improved interfacial interaction, which resulted in better flexural and tensile responses, as well as improved thermal stability (Cho et al. 2007).

Dynamic mechanical studies of PP composites with different plant fillers showed that the maximum storage modulus and lowest damping properties were found for hemp composites (compounding injection), while the lowest modulus and highest damping characteristics were reported for rice hull composites. Fiber content in the composites had the effect of increasing stiffness and decreasing damping properties (Tajvidi et al. 2006).

Short-term creep of composites prepared from bidimensional jute mats and PP was carefully investigated by Acha et al. (2007), who focused the work on the effect of the interface modifiers. Figure 22.5 shows the effect of the fiber concentration on untreated jute–PP samples.

The curves show the expected result of reduced creep as the fiber concentration increases. A four-parameter Burgers model was used to fit the experimental results during creep and further used to predict the recovery step of the test with very good success. The role of the interface was

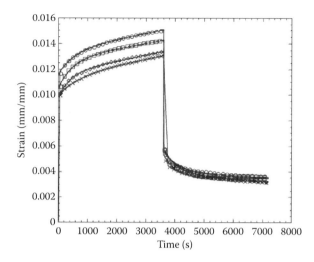

FIGURE 22.5 Experimental creep–recovery curves (symbols) and fitting model (lines) for PP–jute composites. Jute content (wt.%): [○], 9; [□], 17; [◇], 25; [×], 30. (From Acha, B.A. et al., *Compos. Part A*, 38, 2007.)

investigated by using 25 wt.% composites prepared with untreated jute, alkenyl succinic-treated jute mats, lignin addition (5 wt.% incorporated into the PP matrix) and two different types of MAPP (Epolene E-43 at 5 wt.% and Epolene G-3003 at 3, 5, 7, and 9 wt.%). The authors found that the first two modifications lead to reduced properties and increased creep deformation for those composites. On the other hand, the MAPP modifiers produced better composites with reduced creep. The results were again successfully modeled with the four-parameter model, and the authors reported that the improvement was better with the addition of Epolene G-3003, which was the MAPP of higher molecular weight. The finding was explained by the higher probability of entanglement with the PP matrix when the high-molecular-weight modifier was used and by the lubricant effect of the low-molecular-weight modifier on the PP matrix. Further results were also reported on the effect of varying the concentration of the high-molecular-weight modifier. Figure 22.6 shows the effect of the

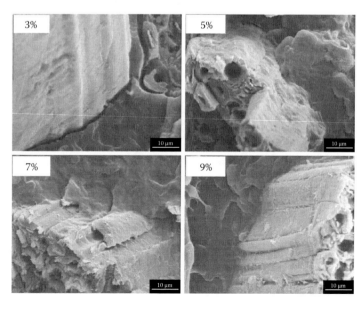

FIGURE 22.6 Scanning electronic micrographs for composites made from 25 wt.% of jute and different concentrations of Epolene G-3003. (From Acha, B.A. et al., *Compos. Part A*, 38, 2007.)

interface fiber–matrix for these composites. Apparently, 3 wt.% (based of the weight of the fibers) of the modifier was not a high enough concentration to saturate the fiber surface, and some separation was observed. However, for the composites prepared with 5 wt.% or higher concentrations of the modifier, the interfacial wetting and adhesion was very good, as can be observed in the images, and these features were well correlated with the creep results obtained.

Humidity can affect the performance of composites, as was reported in different publications. For example, rice husk particles (175–246 μm) were incorporated with PP by compounding extrusion followed by injection molding. The usual increase in tensile modulus was reported, but with decreasing ultimate properties as a consequence of poor interfacial adhesion. In that case, the 26 vol.% composite was investigated regarding its response to critical conditions in water immersion at 90°C for about 50 days (until achieving constant weight). The tensile modulus suffered a significant reduction (from 2.3 to 1.6 GPa), while the elongation at break increased from 2.8% to 3.4%. Drying of the samples did not allow recovery of the initial values, suggesting that permanent damage of the plant particles has occurred, which reduces their performance as reinforcement (Mohd Ishak et al. 2001).

22.5.4 Composites from Other Synthetic Thermoplastics

As was the case with other thermoplastics, the addition of cotton fibers or waste textile fibers (50:50 wt.% cotton–acrylic) to ABS led to the increase in the stiffness of the composite (Martins et al. 2010). The dynamic mechanical modulus of the composites showed higher values than the neat polymer, especially at temperatures above the glass transition, as was expected from the different modulus of the components. The addition of the fibers did not affect the degradation of the ABS.

PS and fibers from agave leaves were used to prepare composites (manually arranged layers of fibers and PS films, further hot-pressed in a mold) (Singha and Rana 2012).

Nontreated and methylmethacrylate- (MMA-) grafted fibers were utilized, and the mechanical properties of the composites prepared with different fiber concentrations (and fiber lengths) were analyzed. In all cases, an optimum in strength was reached at 20 wt.% of fibers, as observed for the particle fiber–PS composites, shown in Table 22.3.

The results show that the strength of the materials increased with the fiber content until reaching an optimum at 20 wt.%, probably due to the agglomeration effect at higher concentrations. Table 22.3 also shows an additional improvement when the fibers were treated before being incorporated into the composite.

Zheng et al. (2007) studied the behavior of bagasse–PVC composites, focusing on the effect of fiber concentration and fiber surface treatment. To improve the compatibility of the system, they treated the fibers with benzoic acid, which is reported to have improved the dispersion of the short fibers incorporated by dry-blending in a two-roll mill. Then, the samples were formed

TABLE 22.3
Effect of Fiber Content and Treatment on the Properties of Agave–PS Composites

Fiber (wt.%)	Tensile Strength (MPa)		Flexural Strength (MPa)		Impact Strength (kJ/mm^2)	
	Raw Fibers	MMA Graft	Raw Fibers	MMA Graft	Raw Fibers	MMA Graft
0	19.90		34.69		2.28	
10	26.42	28.48	56.10	67.20	2.62	2.63
15	27.70	28.90	66.30	78.90	2.54	3.26
20	34.16	43.76	95.10	105.30	3.09	4.02
25	29.20	33.24	80.70	96.90	2.87	3.37
30	26.54	27.74	64.20	84.30	2.72	2.77

Source: Singha, A.S. and Rana, R.K., *Mater. Des.*, 41, 2012.

by compression molding. The authors found that the tensile modulus of the composites increased with the increasing bagasse concentration, as could be expected. Additionally, the fiber treatment allowed improvement of the tensile strength and, in lower proportion, the impact strength of the composites.

On the other hand, Bakar and Hassan (2003) found that the incorporation of unmodified oil palm empty fruit bunch fibers into PVC showed only a very slight reduction in the impact strength of the composites compared to that of the neat polymer.

22.5.5 BIOPLASTIC COMPOSITES

Thermoplastic starch and starch-based polymers have received much attention because of their low cost, availability, renewability, and biodegradability.

Dufresne and Vignon (1998) reported their results on the thermomechanical behavior of glycerol-plasticized potato starch containing microfibrilated potato cellulose. Dynamic mechanical studies showed a very large increase in the rubbery modulus for these materials, which suggests improved stability. In this case, a reduction in water absorption was also registered.

Composites from unidirectional arranged flax fibers in a thermoplastic starch-based matrix (MaterBi) were prepared by hot-pressing alternating polymer and fiber layers (Romhány et al. 2003). Tensile properties were measured on 20–60 wt.% composites, and it was found that the maximum tensile strength was obtained at 40 wt.% fiber content (three times that of the unreinforced polymer) and remained almost constant for higher concentrations, while the tensile modulus increased by several orders of magnitude.

Foams prepared from starch acetate and containing 2–14 wt.% of alkali-treated corn stalks and 5 wt.% talc showed improved physical properties only up to 10 wt.% of fibers, at which point agglomeration became detrimental (Ganjyal et al. 2004; Satyanarayana et al. 2009).

Glycerol-plasticized soy flour reinforced with raw Indian grass fibers showed increased tensile and flexural properties (strength and modulus) as well as deflection temperature, when the content of fibers was increased from 0 to 30 wt.%. Alkaline treatment was shown to be effective in improving fiber dispersion and further increasing tensile, flexural, and impact strengths (60%, 40%, and 30% increase, respectively) (Liu et al. 2005).

A review on the subject of PLA–plant fiber composites was coauthored by Hassan (Hassan et al. 2012). The properties of the composites vary according to the specifics of each system, PLA utilized, type of fiber, concentration and orientation, as well as processing method. The best properties corresponded to unidirectional kenaf fiber–PLA composites (35 wt.%), which showed a tensile strength and modulus of 131 MPa and 15 GPa, respectively. The lowest performance corresponded to the randomly oriented kenaf fiber–PLA composites (30 wt.%), with 32 MPa and 4.5 GPa, for tensile strength and modulus, respectively. In both cases, the processing method was compression molding.

PLA composites with phormium (20–40 wt.%) prepared by twin-screw compounding followed by injection molding showed poor interfacial adhesion that led to pullout and debonding. However, the presence of the fibers promoted crystallinity and consequently resulted in an increase in rigidity and hardness (Santulli et al. 2011).

The tensile properties of bamboo fiber–PLA and poly(butylene succinate) (PBS) composites (Lee and Wang 2006) and sugar beet pulp–PLA composites (Liu et al. 2007a) have also been investigated with regard to the influence of the coupling agent and the structure.

Silane- and alkali-treated pineapple leaf fibers were found to perform better than untreated ones as reinforcement of PLA in laminated composites (higher modulus and deflection temperature). As observed by other authors, the surface treatment of the fibers affected the crystallization properties of the semicrystalline matrix (Huda et al. 2008).

The use of a compatibilizing agent (PLA-co-glycidyl methacrylate) in PLA–sisal composites resulted in a slight increase in tensile strength and decrease in impact strength. The effect was more obvious when PEG-plasticized PLA was used as matrix (Li et al. 2011).

Adequate selection of temperature, pressure, and mold characteristics makes it possible to obtain similar properties in grass fiber composites prepared from cellulose acetate butyrate to those obtained from analogous HDPE composites (Liu et al. 2007b).

Mohanty et al. (2004) studied the performance of cellulose acetate biocomposites reinforced with hemp fibers and prepared by extrusion–injection. The flexural strength and modulus of the composites reached 78 MPa and 5.6 GPa at 30 wt.% fiber content. The composites thus prepared outperformed those prepared by the compression-molding method.

Polybutadiene adipate-co-terephthalate (PBAT) filled with acid-hydrolyzed wheat straw showed a large modulus increase with a reduction of elongation at break (Averous and Le Digabel 2006).

Shanks et al. (2004) studied composites of polyhydroxybutyrate (PHB) with different contents of blended polyhydroxyvalerate (PHV) and flax fibers. The composites showed increased melting and glass transition temperatures, which were correlated with higher bending and storage modulus, as compared to the unfilled polymers. Flexural and impact properties of polyhydroxybutyrate-co-valerate materials were largely affected by incorporation of the kenaf fibers (flexural modulus, 1.28 and 4.02 GPa; flexural strength, 28.2 and 14.1 MPa; impact strength, 0.85 and 1.70 kJ/m^2 for the 0 and 60 wt.% composites, respectively) (Persico et al. 2011).

PHBV composites were also prepared with steam explosion-treated wheat straw (10–30 wt.%) (Avella et al. 2000). This treatment, which produces fibers with higher cellulose content, led to increased crystallization rates of the polymer (although with similar crystallinity) and higher glass transition temperatures, but lower strength and ultimate strain (8–11 MPa and 0.4–1.4%, respectively, compared to 10–19 MPa and 1–2.4% for the neat polymer).

Kenaf fiber–PBS composites compatibilized with 5 wt.% of the maleated polymer were prepared with a fiber content of 30 wt.%. The use of the maleated coupling agent resulted in reduced water uptake during water immersion tests, as well as improved flexural properties. Redrying of the composites was insufficient for recovering the original properties (Ahmad Thirmizir et al. 2013).

Benzylated China fir wood flour has been used as thermoplastic matrix of short sisal composites (Zhang et al. 2005). By chemical treatment of the wood flour (reaching 70–80 wt.% of weight gain), the wood acquires thermoplastic characteristics, showing a thermomechanical transition at about 100°C (peak in temperature scans from dynamic mechanical analysis [DMA] measurements), which allows molding of the material as plastic above this temperature. The addition of short or unidirectional sisal fibers led to materials of increased strength (tensile, flexural, and impact), with a maximum in properties at about 15 wt.% of short sisal fibers. The modulus (tensile and flexural) of the unidirectional fiber composites showed a marked increase with a maximum in properties at 30 wt.% fiber content.

22.6 FINAL REMARKS

The last three decades have witnessed an impressive growth of research, studies, as well as industrial use of natural fiber thermoplastic composites. While different fiber treatments have been considered in the literature, it appears that the use of coupling agents (or simply the untreated fibers) is the most common case in industrial processing. The low density of fibers, cost, renewability, availability, and positive view by the general public constitute the driving force for the use of natural fibers in the development of an ever-increasing number of applications.

It is most generally agreed that the addition of natural fibers leads to the increase in the thermoplastic composite stiffness (reaching an optimum value in many cases). Clearly, the characteristics of the fiber (nature, geometry, treatments, etc.) affect the results. Moreover, the role of the interface cannot be understated, since the quality of the fiber–matrix contact can affect the stress transfer, crystallinity of the matrix (when this applies), strength, fracture behavior, durability, and susceptibility to environmental conditions.

In more recent years, the industrial use of biodegradable and bio-based thermoplastics has also increased, and plant fibers are the perfect match in the formulation of biocomposites with high biocontent and biodegradability. It is to be expected that different novel applications will appear in the coming years benefiting from the combination of these materials.

REFERENCES

Acha, B.A., Reboredo, M.M., Marcovich, N.E. Effect of coupling agents on the thermal and mechanical properties of polypropylene–jute fabric composites. *Polymer International* 2006; 55: 1104–1113.

Acha, B.A., Reboredo, M.M., Marcovich, N.E. Creep and dynamic mechanical behavior of PP–jute composites: Effect of the interfacial adhesion. *Composites: Part A* 2007; 38: 1507–1516.

Ahmad Thirmizir, M.Z., Mohd Ishak, Z.A., Mat Taib, R., Rahim, S. Effect of maleated compatibiliser (PBS-g-MA) addition on the flexural properties and water absorption of poly(butylene succinate)/kenaf bast fibre composites. *Sains Malysiana* 2013; 42: 435–441.

Angelov, I., Wiedmer, S., Evstatiev, M., Friedrich, K., Mennig, G. Pultrusion of a flax/polypropylene yarn. *Composites: Part A* 2007; 38: 1431–1438.

Aranguren, M.I. Polymer composites. In *Encyclopedia of Surface and Colloid Science*, 2nd Edition, eds. P. Somasundaran and A. Hubbard, 2006, 4796–4810. New York: Taylor & Francis/CRC Press.

Aranguren, M.I., Reboredo, M.M. In *Handbook of Engineering Biopolymers: Homopolymers, Blends and Composites*, eds. D. Bhattacharyya and S. Fakirov, 2007, 193–222. Cincinnati, OH: Hanser Gardner Publications.

Araujo, J.R., Mano, B., Teixeira, G.M., Spinace, M.A.S., De Paoli, M.A. Biomicrofibrilar composites of high density polyethylene reinforced with curaua fibers: Mechanical, interfacial and morphological properties. *Composites Science and Technology* 2010; 70: 1637–1644.

Avella, M., La Rota, G., Martuscelli, E., Raimo, M., Sadocco, P., Elegir, G., Riva, R. Poly(3-hydroxybutyrate-co-3-hydroxyvalerate) and wheat straw fibre composites: Thermal and mechanical properties and biodegradation behaviour. *Journal of Materials Science* 2000; 35: 829–836.

Averous, L., Le Digabel, F. Properties of biocomposites based on lignocellulosic fillers. *Carbohydrate Polymers* 2006; 66: 480–493.

Awal, A., Cescutti, G., Ghosh, S.B., Müssig, J. Interfacial studies of natural fibre/polypropylene composites using single fibre fragmentation test (SFFT). *Composites: Part A* 2011; 42: 50–56.

Bakar, A.A., Hassan, A. Impact properties of oil palm empty fruit bunch filled impact modified unplasticised poly (vinyl chloride) composites. *Jurnal Teknologi* 2003; 39: 73–82.

Barkoula, N.M., Garkhail, S.K., Peijs, T. Biodegradable composites based on flax/polyhydroxybutyrate and its copolymer with hydroxyvalerate. *Industrial Crops and Products* 2010; 31: 34–42.

Bataille, P., Ricard, L., Sapieha, S. Effects of cellulose fibers in polypropylene composites. *Polymer Composites* 1989; 10: 103–108.

Beckermann, G.W., Pickering, K.L. Engineering and evaluation of hemp fibre reinforced polypropylene composites: Fibre treatment and matrix modification. *Composites: Part A* 2008; 39: 979–988.

Bledzki, A.K., Gassan, J. Composites reinforced with cellulose based fibres. *Progress in Polymer Science* 1999; 24: 221–274.

Bledzki, A.K., Sperber, V.E., Faruk, O. Natural and wood fiber reinforcement in polymers. *Rapra Review Reports* 2002; 13(8): 1–144.

Bledzki, A.K., Mamun, A.A., Jaszkiewicz, A., Erdmann, K. Polypropylene composites with enzyme modified abaca fibre. *Composites Science and Technology* 2010; 70: 854–860.

Bozaci, E., Sever, K., Sarikanat, M., Seki, Y., Demir, A., Ozdogan, E., Tavman, I. Effects of the atmospheric plasma treatments on surface and mechanical properties of flax fiber and adhesion between fiber–matrix for composite materials. *Composites: Part B* 2013; 45: 565–572.

Braun, D., Schmitt, M.W. Functionalization of poly(propylene) by isocyanate groups. *Polymer Bulletin* 1998; 40: 189–194.

Charlet, K., Baley, C., Morvan, C., Jernot, J.P., Gomina, M., Brard, J. Characteristics of Herme's flax fibres as a function of their location in the stem and properties of the derived unidirectional composites. *Composites Part A: Applied Science and Manufacturing* 2007; 38: 1912–1921.

Cho, D., Lee, H.S., Han, S.O., Drzal, L.T. Effects of e-beam treatment on the interfacial and mechanical properties of henequen/polypropylene composites. *Advanced Composite Materials* 2007; 16: 315–334.

Choudhury, A. Isothermal crystallization and mechanical behavior of ionomer treated sisal/HDPE composites. *Materials Science and Engineering A* 2008; 491: 492–500.

Deepa, B., Abraham, E., Cherian, B.M., Bismarck, A., Blaker, J.J., Pothan, L.A., Lopes Leao, A., Ferreira de Souza, S., Kottaisamy, M. Structure, morphology and thermal characteristics of banana nano fibers obtained by steam explosion. *Bioresource Technology* 2011; 102: 1988–1997.

De Roover, B., Sclavons, M., Carlier, V., Devaux, J., Legras, R., Momtaz, A. Molecular characterization of maleic anhydride-functionalized polypropylene. *Journal of Polymer Science Part A: Polymer Chemistry* 1995; 33: 829–842.

de Sousa, M.V., Costa, M.F., Tavares, M.I.B., Thire, R.M.S.M. Preparation and characterization of composites based on polyhydroxybutyrate and waste powder from coconut fibers processing. *Polymer Engineering and Science* 2010; 50: 1466–1475.

Doan, T.T.L., Gao, S.L., Mäder, E. Jute/polypropylene composites I. Effect of matrix modification. *Composites Science and Technology* 2006; 66: 952–963.

Dufresne, A., Vignon, M.R. Improvement of starch film performances using cellulose microfibrils. *Macromolecules* 1998; 31: 2693–2696.

Eichhorn, S.J., Baillie, C.A., Zafeiropoulos, N., Mwaikambo, L.Y., Ansell, M.P., Dufresne, A., Entwistle, K.M. et al. Review. Current international research into cellulosic fibres and composites. *Journal of Materials Science* 2001; 36: 2107–2131.

Faruk, O., Bledzki, H., Fink, H. Sain, M. Biocomposites reinforced with natural fibers: 2000–2010. *Progress in Polymer Science* 2012; 37: 1552–1596.

Fernandes, E.M., Correlo, V.M., Mano, J.F., Reis, R.L. Novel cork–polymer composites reinforced with short natural coconut fibres: Effect of fibre loading and coupling agent addition. *Composites Science and Technology* 2013; 78: 56–62.

Frisoni, G., Baiardo, M., Scandola, M. Natural cellulose fibers: Heterogeneous acetylation kinetics and biodegradation behavior. *Biomacromolecules* 2001; 2: 476–482.

Ganjyal, G.M., Reddy, N., Yang, Y.Q., Hanna, M.A. Biodegradable packaging foams of starch acetate blended with corn stalk fibers. *Journal of Applied Polymer Science* 2004; 93: 2627–2633.

Gironès, J., Pimenta, M.T.B., Vilaseca, F., de Carvalho, A.J.F., Mutjé, P., Curvelo, A.A.S. Blocked isocyanates as coupling agents for cellulose-based composites. *Carbohydrate Polymers* 2007; 68: 537–543.

Haijun, L., Sain, M. High stiffness natural fiber-reinforced hybrid polypropylene composites. *Polymer–Plastics Technology and Engineering* 2003; 42: 853–862.

Hassan, E., Wei, Y., Jiao, H., Huo, Y.M. Plant fibers reinforced poly (lactic acid) (PLA) as a green composites: Review. *International Journal of Engineering Science and Technology* 2012; 4: 4429–4439.

He, L., Li, X., Li, W., Yuan, J., Zhou, H. A method for determining reactive hydroxyl groups in natural fibers: Application to ramie fiber and its modification. *Carbohydrate Research* 2012; 348: 95–98.

Herrera-Franco, P.J., Valadez-Gonzalez, A. Mechanical properties of continuous natural fibre–reinforced polymer composites. *Composites Part A: Applied Science and Manufacturing* 2004; 35: 339–345.

Herrera-Franco, P.J., Valadez-Gonzalez, A. A study of the mechanical properties of short natural-fiber reinforced composites. *Composites Part A: Applied Science and Manufacturing* 2005; 36: 597–608.

Ho, M., Wang, H., Lee, J.H., Ho, C., Lau, K., Leng, J., Hui, D. Critical factors on manufacturing processes of natural fiber composites. *Composites: Part B* 2012; 43: 3549–3562.

Hodzic, A., Coakley, R., Curro, R., Berndt, C.C., Shanks, R.A. Design and optimization of biopolyester bagasse fiber composites. *Journal of Biobased Materials and Bioenergy* 2007; 1: 46–55.

Huda, M.S., Drzal, L.T., Mohanty, A.K., Misra, M. Effect of fiber surface treatments on the properties of laminated biocomposites from poly(lactic acid) (PLA) and kenaf fibers. *Composites Science and Technology* 2008; 68: 424–432.

John, M.J., Anandjiwala, R.D. Recent developments in chemical modification and characterization of natural fibre-reinforced composites. *Polymer Composites* 2008; 29: 187–207.

Joseph, K., Thomas, S., Pavithran, C. Viscoelastic properties of short sisal fibre filled low density polyethylene composites: Effect of fibre length and orientation. *Material Letters* 1992; 15: 224.

Joseph, K., Thomas, S., Pavithran, C. Tensile properties of short sisal fibre reinforced polyethylene composites. *Journal of Applied Polymer Science* 1993a; 47: 1731.

Joseph, K., Thomas, S., Pavithran, C. Dynamic mechanical properties of short sisal fibre reinforced low density polyethylene composites. *Journal of Reinforced Plastics and Composites* 1993b; 12: 139.

Joseph, K., Pavithran, C., Thomas, S., Baby, K., Premalatha, C.K. Melt rheological behaviour of short sisal fibre reinforced polyethylene composites. *Plastics, Rubber and Composites Processing and Applications* 1994; 21: 237.

Joseph, K., Thomas, S., Pavithran, C. Effect of ageing on the physical and mechanical properties of short sisal fibre reinforced polyethylene composites. *Composites Science and Technology* 1995; 53: 99.

Joseph, K., Thomas, S., Pavithran, C. Effect of chemical treatment on the tensile properties of short sisal fibre–reinforced polyethylene composites. *Polymer* 1996; 37: 5139–5149.

Kabir, M.M., Wang, H., Lau, K.T., Cardona, F. Chemical treatments on plant-based natural fibre reinforced polymer composites: An overview. *Composites: Part B* 2012; 43: 2883–2892.

Karmarkar, A., Chauhan, S.S., Modak, J.M., Chanda, M. Mechanical properties of wood-fiber reinforced polypropylene composites: Effect of a novel compatibilizer with isocyanate functional group. *Composites: Part A* 2007; 38: 227–233.

Kechaou, B., Salvia, M., Beaugiraud, B., Juvé, D., Fakhfakh, Z., Treheux, D. Mechanical and dielectric characterization of hemp fibre reinforced polypropylene (HFRPP) by dry impregnation process. *eXPRESS Polymer Letters* 2010; 4: 171–182.

Keener, T.J., Stuart, R.K., Brown, T.K. Maleated coupling agents for natural fibre composites. *Composites: Part A* 2004; 35: 357–362.

Keller, A. Compounding and mechanical properties of biodegradable hemp fibre composites. *Composites Science and Technology* 2003; 63: 1307–1316.

Kim, J.T., Netraval, A.N. Mercerization of sisal fibers: Effect of tension on mechanical properties of sisal fiber and fiber-reinforced composites. *Composites: Part A* 2010; 41: 1245–1252.

Kuruvilla, J., Tolêdo Filho, R.D., James, B., Thomas, S., Hecker de Carvalho, L. A review on sisal fiber reinforced polymer composites. *Revista Brasileira de Engenharia Agrícola e Ambiental*. 1999; 3: 367–379.

Le Duigou, A., Davies, P., Baley, C. Interfacial bonding of flax fibre/poly(L-lactide) bio-composites. *Composites Science and Technology* 2010; 70: 231–239.

Lee, S.H., Wang, S. Biodegradable polymers/bamboo fiber biocomposite with bio-based coupling agent. *Composites Part A: Applied Science and Manufacturing* 2006; 37: 80–91.

Lee, B.H., Kim, H.S., Lee, S., Kim, H.J., Dorgan, J.R. Bio-composites of kenaf fibers in polylactide: Role of improved interfacial adhesion in the carding process. *Composites Science and Technology* 2009; 69: 2573–2579.

Lei, Y., Wu, Q., Yao, F., Xu, Y. Preparation and properties of recycled HDPE/natural fiber composites. *Composites Part A: Applied Science and Manufacturing* 2007; 38: 1664–1674.

Li, Q., Matuana, L.M. Effectiveness of maleated and acrylic acid-functionalized polyolefin coupling agents for HDPE-wood-flour composites. *Journal of Thermoplastic Composite Materials* 2003; 16: 551–564.

Li, Y., Mai, Y., Ye, L. Sisal fibre and its composites: A review of recent developments. *Composites Science and Technology* 2000; 60: 2037–2055.

Li, Y., Hu, C., Yu, Y. Interfacial studies of sisal fiber reinforced high density polyethylene (HDPE) composites. *Composites Part A: Applied Science and Manufacturing* 2008; 39: 570–578.

Li, Z., Zhou, X., Pei, C. Effect of sisal fiber surface treatment on properties of sisal fiber reinforced polylactide composites. *International Journal of Polymer Science* 2011; doi:10.1155/2011/803428.

Liu, W., Mohanty, A.K., Drzal, L.T., Misra, M. Novel biocomposites from native grass and soy based bioplastic: Processing and properties evaluation. *Industrial and Engineering Chemistry Research* 2005; 44: 7105–7112.

Liu, L.S., Finkenstadt, V.L., Liu, C.K., Coffin, D.R., Willett, J.L., Fishman, M.L., Hicks, K.B. Green composites from sugar beet pulp and poly(lactic acid): Structural and mechanical characterization. *Journal of Biobased Materials and Bioenergy* 2007a; 1: 323–330.

Liu, W., Thayer, K., Misra, M., Mohanty, A.K., Drzal, L.T. Processing and physical properties of native grass-reinforced biocomposites. *Polymer Engineering and Science* 2007b; 47: 969–976.

Liu, H., Wu, Q., Zhang, Q. Preparation and properties of banana fiber–reinforced composites based on high density polyethylene (HDPE)/nylon-6 blends. *Bioresource Technology* 2009; 100: 6088–6097.

Liu, D., Zhong, T., Chang, P.R., Li, K., Wu, Q. Starch composites reinforced by bamboo cellulosic crystals. *Bioresource Technology* 2010; 101: 2529–2536.

Madsen, B., Lilholt, H. Physical and mechanical properties of unidirectional plant fibre composites—An evaluation of the influence of porosity. *Composites Science and Technology* 2003; 63: 1265–1272.

Manikandan Nair, K.C., Thomas, S., Groeninckx, G. Thermal and dynamic mechanical analysis of polystyrene composites reinforced with short sisal fibres. *Composites Science and Technology* 2001; 61: 2519–2529.

Marcovich, N.E., Reboredo, M.M., Aranguren, M.I. Dependence of the mechanical properties of wood flour–polymer composites with moisture content. *Journal of Applied Polymer Science* 1998; 68: 2069–2076.

Martins, J.N., Klohn, T., Bianchi, O. Fiorio, R., Freire, E. Dynamic mechanical, thermal, and morphological study of ABS/textile fiber composites. *Polymer Bulletin* 2010; 64: 497–510.

Mishra, S., Misra, M., Tripathy, S.S., Nayak, S.K., Mohanty, A.K. Potentiality of pineapple leaf fibre as reinforcement in PALF-polyester composite: Surface modification and mechanical performance. *Journal of Reinforced Plastics and Composites* 2001; 20: 321–334.

Mohanty, A.K., Wibowo, A., Misra, M., Drzal, L.T. Effect of process engineering on the performance of natural fiber reinforced cellulose acetate biocomposites. *Composites Part A: Applied Science and Manufacturing* 2004; 35: 363–370.

Mohd Ishak, Z.A., Yow, B.N., Ng, B.L., Khalil, H.P.S.A., Rozman, H.D. Hygrothermal aging and tensile behavior of injection-molded rice husk-filled polypropylene composites. *Journal of Applied Polymer Science* 2001; 81: 742–753.

Nabi Saheb, D., Jog, J.P. Natural fiber polymer composites: A review. *Advances in Polymer Technology* 1999; 18: 351–363.

Nithya, E., Radhai, R., Rajendran, R., Shalini, S., Rajendran, V., Jayakumar, S. Synergetic effect of DC air plasma and cellulase enzyme treatment on the hydrophilicity of cotton fabric. *Carbohydrate Polymers* 2011; 83: 1652–1658.

Oksman, K., Mathew, A.P., Långström, R., Nyström, B., Joseph, K. The influence of fibre microstructure on fibre breakage and mechanical properties of natural fibre reinforced polypropylene. *Composites Science and Technology* 2009; 69: 1847–1853.

Pejic, B.M., Kostic, M.M., Skundric, P.D., Praskalo, J.Z. The effects of hemicelluloses and lignin removal on water uptake behavior of hemp fibers. *Bioresource Technology* 2008; 99: 7152–7159.

Persico, P., Acierno, D., Carfagne, C., Cimino, F. Mechanical and thermal behaviour of ecofriendly composites reinforced by kenaf and caroa fibers. *International Journal of Polymer Science* 2011; doi:10.1155/2011/841812.

Ragoubi, M., George, B., Molina, S., Bienaimé, D., Merlin, A., Hiver, J.-M., Dahoun, A. Effect of corona discharge treatment on mechanical and thermal properties of composites based on miscanthus fibres and polylactic acid or polypropylene matrix. *Composites Part A: Applied Science and Manufacturing* 2012; 43: 675–685.

Rana, A.K., Mandal, A., Bandyopadhyay, S. Short jute fiber reinforced polypropylene composites: Effect of compatibiliser, impact modifier and fiber loading. *Composites Science and Technology* 2003; 62: 801–806.

Romhány, G., Karger-Kocsis, J., Czigány, T. Tensile fracture and failure behavior of thermoplastic starch with unidirectional and cross-ply flax fiber reinforcements. *Macromolecular Materials and Engineering* 2003; 288: 699–707.

Rong, M.Z., Zhang, M.Q., Liu, Y., Yang, G.C., Zeng, H.M. The effect of fiber treatment on the mechanical properties of unidirectional sisal-reinforced epoxy composites. *Composites Science and Technology* 2001; 61: 1437–1447.

Rosa, M.F., Chiou, B.S., Medeiros, E.S., Wood, D.F., Williams, T.G., Mattoso, L.H.C., Orts, W.J., Imam, S.H. Effect of fiber treatments on tensile and thermal properties of starch/ethylene vinyl alcohol copolymers/coir biocomposites. *Bioresource Technology* 2009; 100: 5196–5202.

Rowell, R.M., Stout, H.P. Jute and kenaf. In *Handbook of Fiber Chemistry*, 3rd Edition, ed. L. Menachem, 2007, 405–452. New York: Taylor & Francis/CRC Press.

Rozman, H.D., Tan, K.W., Kumar, R.N., Abubakar, A., Mohd Ishak, Z.A., Ismail, H. The effect of lignin as a compatibilizer on the physical properties of coconut fiber polypropylene composites. *European Polymer Journal* 2000; 36: 1483–1494.

Rozman, H.D., Saad, M.J., Mohd Ishak, Z.A. Flexural and impact properties of oil palm empty fruit bunch (EFB)–polypropylene composites—The effect of maleic anhydride chemical modification of EFB. *Polymer Testing* 2003; 22: 335–341.

Sain, M., Panthapulakkal, S. Chapter 9. In *Green Composites—Polymer Composites and the Environment*, ed. C. Baillie, 2004. Cambridge, UK: Woodhead Publishing.

Saleem, Z., Rennebaum, H., Pudel, F., Grimm, E. Treating bast fibres with pectinase improves mechanical characteristics of reinforced thermoplastic composites. *Composites Science and Technology* 2008; 68: 471–476.

Sanadi, A.R., Caulfield, D.F., Rowell, R.M. Reinforcing polypropylene with natural fibers. *Plastics Engineering* 1994a; 50: 27–28.

Sanadi, A.R., Young, A.A., Clemsons, C., Rowell, R.M. Recycled newspaper fibers as reinforcing fillers in thermoplastics: Part I—Analysis of tensile and impact properties in polypropylene. *Journal of Reinforced Plastics and Composites* 1994b; 13: 54–67.

Sanadi, A.R., Caulfield, D.F., Jacobson, R.E., Rowell, R.M. Renewable agricultural fibers as reinforcing fillers in plastics: Mechanical properties of kenaf fiber–polypropylene composites. *Industrial and Engineering Chemistry Research* 1995; 34: 1889–1896.

Santulli, C., Puglia, D., Sarasini, F., Kenny, J.M. Adding plant fibers improves thermomechanical properties of poly(lactic acid) matrix. *Society of Plastic Engineers. Plastics Research Online* 2011; doi:10.1002/spepro.003850.

Satyanarayana, K.G., Arizaga, G.G.C., Wypych, F. Biodegradable composites based on lignocellulosic fibers—An overview. *Progress in Polymer Science* 2009; 34: 982–1021.

Shalwan, A., Yousif, B.F. In state of art: Mechanical and tribological behaviour of polymeric composites based on natural fibres. *Journal of Materials Design* 2012; http://dx.doi.org/10.1016/j.matdes.2012.07.014.

Shanks, R.A., Hodzic, A., Wong, S. Thermoplastic biopolyester natural fiber composites. *Journal of Applied Polymer Science* 2004; 91: 2114–2121.

Sim, H., Lee, B.H., Choi, S.W., Kim, S. The effect of types of maleic anhydride-grafted polypropylene (MAPP) on the interfacial adhesion properties of bio-flour-filled polypropylene composites. *Composites Part A* 2007; 38: 1473–1482.

Singha, A.S., Rana, R.K. Natural fiber reinforced polystyrene composites: Effect of fiber loading, fiber dimensions and surface modification on mechanical properties. *Materials and Design* 2012; 41: 289–297.

Singha, A.S., Rana, R.K., Rana, A. Natural fiber reinforced polystyrene matrix based composites. *Advances in Materials Research* 2010; 123–125: 1175–1178.

Sreekala, M.S., Thomas, S. Effect of fibre surface modification on water-sorption characteristics of oil palm fibres. *Composites Science and Technology* 2003; 63: 861–869.

Sreekala, M.S., Kumaran, M.G., Joseph, S., Jacob, M., Thomas, S. Oil palm fibre reinforced phenol formaldehyde composites: Influence of fibre surface modifications on the mechanical performance. *Applied Composite Materials* 2000; 7: 295–329.

Tajvidi, M., Falk, R.H., Hermanson, J.C. Effect of natural fibers on thermal and mechanical properties of natural fiber polypropylene composites studied by dynamic mechanical analysis. *Journal of Applied Polymer Science* 2006; 101: 4341–4349.

Takagi, H. Mechanical and biodegradation behavior of natural fiber composites. *Advances in Materials Research* 2010; 123–125: 1163–1166.

Torres, F.G., Aguirre, M. Rotational moulding and powder processing of natural fibre reinforced thermoplastics. *International Polymer Processing* 2003; 2: 204–210.

Valadez-Gonzalez, A., Cervantes-Uc, J.M., Olayo, R., Herrera-Franco, P.J. Effect of fibre surface treatment on the fibre–matrix bond strength of natural fibre reinforced composites. *Composites Part B—Engineering* 1999; 30: 309–320.

Van de Velde, K., Kiekens, P. Thermoplastic pultrusion of natural fibre reinforced composites. *Composite Structures* 2001; 54: 355–360.

Vilaseca, F., Valadez-Gonzalez, A., Herrera-Franco, P.J., Pèlach, M.A., López, J.P., Mutjé, P. Biocomposites from abaca strands and polypropylene. Part I: Evaluation of the tensile properties. *Bioresource Technology* 2010; 101: 387–395.

Wambua, P., Ivens, J., Verpoest, I. Natural fibres: Can they replace glass in fibre reinforced plastics? *Composites Science and Technology* 2003; 63: 1259–1264.

Wang, B., Panigrahi, S., Tabil, L., Crerar, W. Pre-treatment of flax fibres for use in rotationally molded biocomposites. *Journal of Reinforced Plastic Composites* 2007; 26: 447–463.

Xie, Y., Hill, C.A.S., Xiao, Z., Militz, H., Mai, C. Silane coupling agents used for natural fiber/polymer composites: A review. *Composites: Part A* 2010; 41: 806–819.

Yao, F., Wu, Q., Lei, Y., Xu, Y. Rice straw fiber-reinforced high-density polyethylene composite: Effect of fiber type and loading. *Industrial Crops and Products* 2008; 28: 63–72.

yet2.com. Extrusion process for composites of polymer and natural fibers, September 2004. http://www.agrofibrecomposites.com/tech-of-the-week.pdf (accessed March 5, 2014).

Zainudin, E.S., Sapuan, S.M., Abdan, K., Mohamad, M.T.M. Thermal degradation of banana pseudo-stem filled unplasticized polyvinyl chloride (UPVC) composites. *Materials and Design* 2009; 30: 557–562.

Zampaloni, M., Pourbighart, F., Yankovich, S.A., Rodgers, B.N., Moore, J., Drzal, L.T., Mohanty, A.K., Misra, M. Kenaf natural fiber reinforced polypropylene composites: A discussion on manufacturing problems and solutions. *Composites Part A* 2007; 38: 1569–1580.

Zhang, M.Q., Rong, M.Z., Lu, X. Fully biodegradable natural fiber composites from renewable resources: All-plant fiber composites. *Composites Science and Technology* 2005; 65: 2514–2525.

Zheng, Y.-T., Cao, D.-R., Wang, D.-S., Chen, J.-J. Study on the interface modification of bagasse fibre and the mechanical properties of its composite with PVC. *Composites Part A* 2007; 38: 20–25.

Ziegmann, G. *Flax and Hemp Fibres: A Natural Solution for the Composite Industry*, Chap 6, 2012, 100–108. co-publication JEC Group/CELC.

Zini, E., Focarete, M.L., Noda, I., Scandola, M. Bio-composite of bacterial poly(3-hydroxybutyrate-co-3-hydroxyhexanoate) reinforced with vegetable fibers. *Composites Science and Technology* 2007; 67: 2085–2094.

23 Material Selection, Design, and Application

Eungkyu Kim, Patrick Lee, Joe Dooley, and Mark A. Barger

CONTENTS

23.1 Introduction 753
23.2 Product Design and Material Selection 754
    23.2.1 Product Design Process 754
    23.2.2 Material Selection Process 756
    23.2.3 Process Selection Process 757
    23.2.4 Manufacturing Cost Estimation 759
    23.2.5 Design Tools and Polymeric Material Property Database 761
23.3 Polymer Structure–Property Relations and Testing 763
    23.3.1 Polymer Structure–Property Relations 763
        23.3.1.1 Chemical Structure of Polymers 763
        23.3.1.2 Microstructure of Polymers 763
    23.3.2 Testing for Polymeric Material Selection 769
        23.3.2.1 Mechanical Properties 769
        23.3.2.2 Physical Properties 772
        23.3.2.3 Thermal Properties 772
        23.3.2.4 Electrical Properties 772
        23.3.2.5 Optical Properties 773
        23.3.2.6 Microbial, Weather, and Chemical Resistance 773
23.4 Rheological Properties of Polymer Melt for Material Selection 774
    23.4.1 Viscosity 775
        23.4.1.1 Shear Viscosity 775
        23.4.1.2 Empirical Models for Shear Viscosity–Shear Rate Relationship ... 778
        23.4.1.3 Elongational Viscosity 778
        23.4.1.4 Example: Effect of Shear Viscosity on Material Selection 780
    23.4.2 Viscoelasticity 781
        23.4.2.1 Linear Viscoelasticity 785
        23.4.2.2 Nonlinear Viscoelasticity 792
        23.4.2.3 Example: Effect of Viscoelastic Property on Material Selection ... 796
23.5 Summary and Conclusions 799
References 800

23.1 INTRODUCTION

Materials have played a fundamental and evolving role in the advancement of civilization. At present, on the order of 100,000 engineering materials exist compared to a relative handful two centuries ago. Technological advances have provided for two new classes of materials within the last century, polymers and composites, that have dramatically altered the product design landscape. This has provided the product designer with many material choices and potential design options, but it can be a challenge to effectively sort through the array of potential options and identify the best material system for a specific product design project.

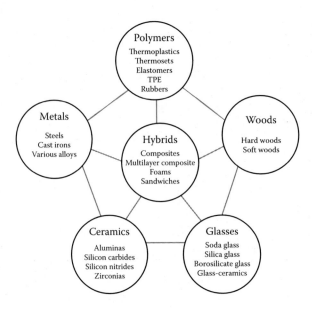

FIGURE 23.1 Six main classes of engineering materials. The basic families of polymer, metal, wood, ceramic, and glass can be combined in various configurations to create hybrids.

The choice of the right material for a design project can be critical to the success of that product [1–5]. Whether we are discussing buildings, furniture, fashion, packaging, or other products at some point, a decision will need to be made on what material to use. In some cases, this will be obvious, but in others, it may need some creative thinking. The selection of the right material can influence design on many levels. Two important considerations are performance of the end product and manufacturing costs. A balance needs to be sought among costs, manufacturing feasibility, and finding the right material for the intended application. It is not simply a case of choosing a material and then deciding on how to manufacture a product from it.

Clearly, different materials have different properties. It is helpful to classify the materials of engineering into the six broad families as shown in Figure 23.1: polymers, metals, ceramics, glasses, woods, and hybrids. The members of a family have certain features in common: similar properties, similar processing routes, and often similar applications.

Among the six engineering materials, polymers are the materials that will be the focus of this chapter. Polymers can be easily formed into an infinite array of shapes and colors, and they are displacing traditional materials in a myriad of diverse and demanding industries because of improved quality and cost competitiveness [6–10]. Today, polymers can be found in virtually every aspect of our lives. From food containers to automobiles, appliances, toys, office equipment, and life-saving medical devices, plastics affect each and every one of us. Product designers and consumers alike acknowledge that today's advanced plastics, in tandem with proper design, add to product value and versatility.

We begin with a section on basic principles of product design and materials selection (Section 23.2). We then briefly describe property–structure relationships of polymers and testing standards in Section 23.3, followed by the rheological properties of polymer melts for materials selection, which are described in Section 23.4.

23.2 PRODUCT DESIGN AND MATERIAL SELECTION

23.2.1 Product Design Process

Product design is the process of creating a new product to be sold by a business to its customers. In a very broad concept, it is essentially the efficient and effective generation and development of

Material Selection, Design, and Application

ideas through a process that leads to new products. In a systematic approach, product designers conceptualize and evaluate ideas, turning them into tangible inventions and products. The product designer's role is to combine art, science, and technology to create new products that other people can use. Their evolving role has been facilitated by digital tools that now allow designers to communicate, visualize, analyze, and actually produce tangible ideas in a way that would have taken greater manpower in the past.

Figure 23.2 shows the stages of the product design. The starting point of the product design is an assessment of market need or a new idea, while the end point is the production of the product with full specification that fills the market need or embodies the idea. The first step of the product design

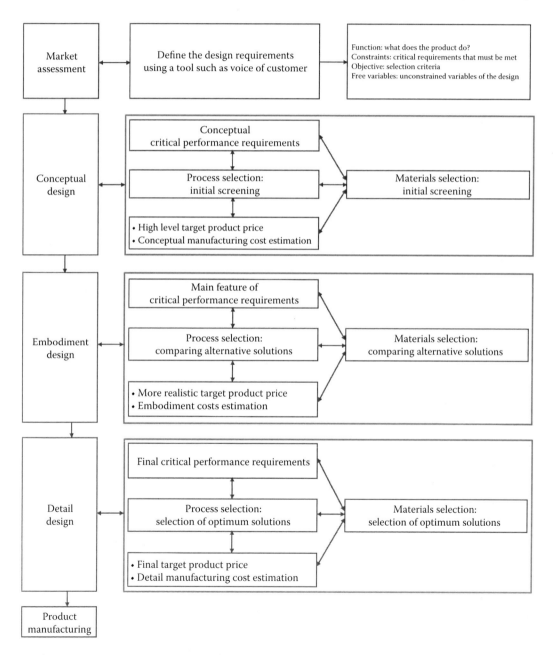

FIGURE 23.2 Stages of product design.

process is *translate*. In this step, market assessment is translated into the critical design requirements, which are (1) function (critical design requirements), (2) constraints (critical design limitations), (3) objective (selection criteria), and (4) free variables (freedom in design), which define the boundary conditions for selecting a material and a manufacturing process.

Once the decision to design a product for a specific application has been made from the assessment of the market requirement, the product design typically follows the second step of the design phase to the last stage of the product design and the related material selection stages as shown in Figure 23.2. The second stage of the product design is *conceptual design*. At this stage, all options are open. Customer critical requirements (performance requirements) and high-level target product price are developed. The design proceeds by developing concepts to perform specific functions, with each based on a working principle. Provisional material and process selections are subsequently made. In addition, an initial manufacturing cost estimate is also required. Estimates of production quantities are required to determine the manufacturing process for procuring the product's components. Lower production volume will tend to drive the design to processes that have a higher manufacturing cost, whereas higher production volume tends to drive lower manufacturing costs. A more detailed description on manufacturing cost estimation is provided in Section 23.2.4.

The third stage of the product design is *embodiment design*. In this stage, the designer takes the promising concepts and seeks to analyze their operation at an approximate level. This involves sizing the components, and selecting materials that will perform properly in the ranges of stress, temperature, and environment suggested by the design requirements, examining the implications for performance and cost. The embodiment stage ends with a feasible layout, which is then passed to the detailed design stage.

The fourth stage of the product design is *detail design*. All components in the product should be fully (or near fully) detailed during the detail product design stage. Perhaps the most important set of decisions involves the specification of the material and manufacturing process for each component. Given this information, the design team lays out the assembly, often determining the shape of the product by industrial design and the detailed dimensions by engineering analysis or manufacturability considerations. Fits and tolerances are defined by assembly guidelines and are verified with functional prototypes. Afterward, the detailed designs may undergo further refinement with the creation and commissioning of the hard tooling. Product launch usually occurs after one or more rounds of pilot production and beta testing, in which the product design and tooling may undergo revisions.

23.2.2 Material Selection Process

Material selection is a step in the process of designing a product for a specific application. In the context of product design, the main goal of material selection is to minimize cost while meeting product performance goals (see Figure 23.2). Systematic selection of the best material for a given application begins with properties, processability, and costs of the candidate materials.

Figure 23.3 shows four main steps for the materials selection process, which are *screening, ranking, supporting information,* and *testing*. A material has certain attributes such as density, strength, cost, and resistance to corrosion. A design demands a certain attribute profile such as a low density, a high strength, and a modest cost. Unbiased selection requires that all materials are considered to be candidates until shown to be otherwise. It is important to start with the full menu of materials in mind, as failure to do so may mean a missed opportunity.

The first step of the material selection process is a screening, which eliminates candidates that cannot meet the critical requirements at all because one or more of their attributes lie outside the limits set by the constraints. One essential requisite to successful materials screening process for a product design is a source of reliable and consistent data on material properties. There are many sources of information, which include government agencies, trade associations, engineering societies, textbooks, research institutes, and materials producers. This subject will be further discussed in Section 23.2.5.

Material Selection, Design, and Application

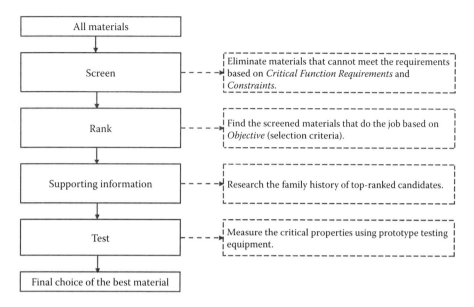

FIGURE 23.3 Material selection strategy. The four main steps are (1) screening, (2) ranking, (3) supporting information, and (4) testing.

The next step of the material selection process is rank ordering. Attribute limits do not help with ordering the candidates surviving from the screening process. Optimization criteria are required to rank the candidates. Performance is sometimes limited by a single property or sometimes by a combination of them. The property or property group that maximizes performance for a given design is called its material index. There are many such indices, each associated with maximizing some aspect of performance. They provide criteria of excellence that allow ranking of materials by their ability to perform well in the given application.

The output of the steps so far is a ranked short list of candidates that meet the constraints, and that either maximize or minimize the criterion of excellence, whichever is required. To proceed further, we seek a detailed profile of each candidate, which is deemed supporting information. Supporting information differs greatly from the structured property data used for screening. Typically, it is descriptive, graphical, or pictorial: case studies of previous uses of the material, details of its corrosion behavior in particular environments, information of availability and pricing, and experience of its environmental impact. Supporting information helps narrow the short list to a final choice, allowing for a definitive match to be made between design requirements and material attributes.

There is a significant amount of available material property information that has been generated by the resin suppliers and compounders. However, the property data are not always completely accurate or reported in a uniform format, because the data have often been developed for a particular application. The data sheets can cause considerable confusion for new customers requiring other properties for their applications. The required properties of all materials for a specific application cannot always be measured due to many reasons such as time, resources, and cost. After the previous three steps are completed, the critical properties of the candidate materials must be measured using prototype testing equipment that is designed to measure the critical properties for a specific application to lower the risks associated with the materials selection process.

23.2.3 Process Selection Process

The development of robust manufacturing process technology is fundamental to the success of any new product design and production. It is important to choose the correct processing route at an early stage in the design before the cost penalty associated with change becomes large. The manufacturing process selection depends on the design requirements, which are composed of the material type, the size and shape of the final article, required tolerances, volume requirements, and conversion cost. Note that a change in product design requirements may require a change in manufacturing process route; therefore, it is important to fully specify and finalize the design requirements prior to entering the process selection phase.

The strategy for process selection parallels that for materials as shown in Figure 23.4. Initially all processes are considered as possible candidates, then five steps follow sequentially: translation, screening, ranking, supporting information, and testing. In translation, the design requirements are expressed as function, constraints (material, shape, size, tolerance, roughness, and other process-related parameters), objective, and free variables. These constraints are used to screen out unsuitable processes, using process selection matrices (or their computer-based equivalents). These are typically displayed as simple matrices and bar charts.

The remaining processes are then ranked in accordance with economic measures. In order to do this, manufacturing cost needs to be examined. Detail discussions on manufacturing cost estimation are given in Section 23.2.4.

The top-ranked candidates are further explored, compiling as much additional supporting information as possible to enable a final choice.

In the final step, the performance of the candidate processes must be verified using a lab-scale prototype process for the selected method. Pilot plant-scale process line verification is the next step to lower the risks associated with both the materials and process selection processes before scaling up to a production-scale manufacturing process.

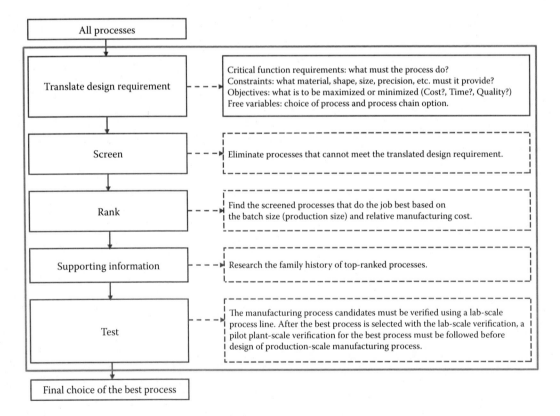

FIGURE 23.4 Process selection strategy. The five main steps are (1) translation, (2) screening, (3) ranking, (4) supporting information, and (5) testing.

23.2.4 MANUFACTURING COST ESTIMATION

The manufacture of a product consumes resources, each of which has an associated cost. Even though a new product satisfies all the performance requirements defined by the customers, if the price for the new product is too high, it will not be competitive. As shown in Figure 23.2, the manufacturing cost estimation process must be performed in parallel with both material and process selection processes.

The final manufacturing cost of the product can be divided into direct and indirect costs [11,12]. Direct costs (variable costs) are directly attributable to the manufacturing cost. Indirect costs (fixed costs) are not directly attributable to the manufacturing cost. Indirect costs are typically allocated to a cost object on some basis. Indirect costs do not vary substantially within certain production volumes or other indicators of activity, and so they may sometimes be considered to be fixed costs.

Table 23.1 lists the cost categories and how to estimate the approximate manufacturing costs. There are three major cost categories associated with the manufacturing cost: (1) operating costs, (2) maintenance costs, and (3) general costs. Both operating and maintenance costs are direct costs, and general costs are indirect costs.

The operating costs are those charges that periodically occur during production and are directly associated with the production of a product. They are the usual figures that are given as annual expenses. Figure 23.5 illustrates the inputs for the operating costs. Raw material costs and conversion costs are the two components of the operating cost. The conversion costs are those that occur between the time the raw materials enter the plant and the time the product leaves, and that are a direct result of processing. These consist of labor costs, utility costs, packaging costs, waste treatment costs, and recycling costs. There are two possible cost components that can contribute to lower total manufacturing costs: recycling and selling of by-products. Recycling raw material scrap

TABLE 23.1
Cost Categories and How to Estimate the Approximate Cost of Producing a Product

Cost Categories				Typical Method of Cost Estimation
Variable costs (direct costs)	Operating costs	Raw materials		Unit ratio material balance and material prices.
		Conversion costs	Labors	Personnel estimate and average salaries.
			Utility	Energy balance and energy costs.
			Recycling	Average recycling cost for a specific material system.
			Waste treatment	Average waste treatment cost for a specific material system.
			Packaging	Average packaging cost for disposable containers. For nondisposable containers, no separate packaging cost is listed.
	Maintenance costs	Supplies		15–20% of maintenance costs.
		Maintenance		6–7% of total sales or 5% of capital cost of constructing the plant.
		Royalties		$5/1000 lb. of product when applicable.
		Taxes		2–3% of capital cost of constructing the plant.
		Insurance		
Fixed costs (indirect costs)	General costs	Depreciation		10% of capital cost of constructing the plant for 10-year depreciation. 20% of capital cost of constructing the plant for 5-year depreciation.
		Financial charges		0–15% of capital cost of constructing the plant.
		Sales costs		5–50% of total sale (average 10%).
		Research costs		3–4% of total sales.
		General administrative costs		5% of total sales.

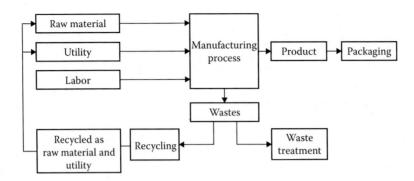

FIGURE 23.5 Inputs used for estimating the operating cost.

(i.e., recovering and reprocessing it) most likely lowers total manufacturing cost, although there is recycling cost associated with the recycling process. Some manufacturing processes generate a by-product. Selling the by-product as another form of product can also lower total manufacturing costs.

Like operating costs, maintenance costs are those charges that periodically occur during production and are directly associated with the production of a product. Service costs, supply costs, taxes, and insurance are the cost components for the maintenance costs. Service costs are those involved in keeping the plant equipment in good order, and this cost involves both equipment and labor. The supply costs are the costs associated with the supplies that may include anything from gaskets to toilet paper to lubricating oil to instrument charts. They are items that are needed in day-to-day operation of the plant. Royalties are the costs paid to the owners of patents for rights to use their inventions and know-how. Royalty costs can be incurred in the form of an up-front payment, an ongoing percentage of revenue, or a combination of the two. Taxes and insurance include property and franchise taxes and all insurance costs, which depend on the value of the physical plant. This category as presented here does not include state, local, or federal income taxes, or sales taxes. Sales taxes on raw materials are considered as raw material costs. Sales taxes on products are collected from the customer and paid directly to the correct authority. Thus, they do not need to be considered here.

General expenses are those costs that usually cannot be related directly to the cost of production. Since the corporation income is obtained from the sale of its products and services, these costs must be allocated to them. They include sales costs, research costs, and general administration costs. The sales costs are those involved with contacting customers and convincing them to buy. Included are such items as warehouses throughout the country for quick distribution, sales offices, salaries and expenses for the salespeople and their supervisors, advertising, and technical service departments. Research costs are those associated with the administration of research projects. General administrative costs are the overhead costs that it takes to operate a company. This item consists of all the administrative costs that cannot be assigned to a given project. It includes the expenses and salaries of the president, board of directors, treasurer, division managers, long-range planner, and accountants.

Two additional cost components of the general costs are finance charges and depreciation. The finance charges are the interest and other costs involved with borrowing money to finance the construction of a processing plant.

A certain amount of money must be invested if any product is to be produced. This is referred to as capital. The capital is made up of the fixed capital needed to construct the plant, the working capital needed to operate it, and the land upon which it will be constructed. None of these costs can be deducted as an expense in the year they occur. The fixed capital investment is recovered through depreciation. The working capital and land investments are eventually recovered when the plant is closed and they are sold or used for another purpose. The working capital is money that is unavailable for other purposes because it must be available for the efficient operation of the plant. It is made up of all capital items not included in the fixed capital or land. It consists of accounts receivable, raw materials inventory, work in progress, product and by-product inventories, and cash held in reserve.

Material Selection, Design, and Application

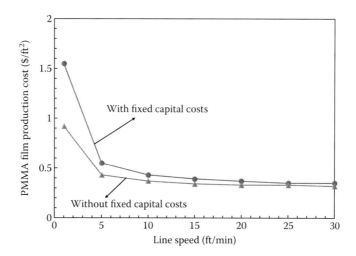

FIGURE 23.6 Polymeric film manufacturing cost as a function of line speed (production volume) with and without the fixed capital costs.

Depreciation and amortization are means of recovering an investment in property that has a useful life of more than a year, is used in a trade or business or held for the production of income, and loses its value. When a corporation constructs a new plant, the firm expects that it will last for a number of years. It is expected that when the plant begins producing, it will be worth the outlay of funds needed to construct it. However, as the plant runs, it tends to wear out and/or become obsolete. Depreciation is the means by which this loss in value can be deducted as a business expense.

Depreciation is a bookkeeping operation, and there is no physical exchange of money. This means that the money listed as a depreciation expense actually is available to the company to spend as it pleases. Depreciation is important for two reasons. First, it reduces income taxes, because the amount of depreciation occurring in any one year is considered as an expense. Second, it is a means whereby the stockholder can assess the physical value of a company. For tax purposes, it is best to depreciate property as rapidly as possible. This makes income taxes less for the first few years and greater in the later years of operation. Since the total depreciation, and hence tax deduction, is the same, this is equivalent to having money available sooner.

Fixed costs will be allocated to the volume of material produced. Spreading this cost over more volume will result in a reduction in unit cost. An example of this is shown in Figure 23.6. The manufacturing cost of producing polymethyl methacrylate (PMMA) extruded film exponentially decreases with the increase in the production mass volume.

Capital costs will normally be incurred for the internal production of a new product. In many instances, the high amount of capital cost requirement inhibits the new product development project specifically products for low-volume markets. An alternative manufacturing route that allows no capital costs is external manufacturing with paying only raw material costs and conversion costs. The conversion fee can be varied by many factors.

23.2.5 Design Tools and Polymeric Material Property Database

The conventional method in product design is to assess market needs and requirements, select the best candidates for material and process, build a prototype, and test its performance. In most development processes, there are various evolution steps and lots of prototypes necessary until all requirements are fulfilled. In fact, with the help of computer-aided engineering (CAE), simulation tools, and rapid prototyping enabled by 3D printing technology, the development time can be decreased significantly.

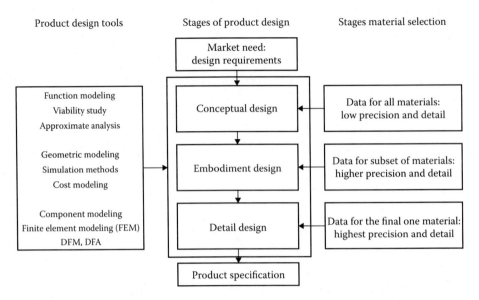

FIGURE 23.7 Design flow chart, showing how design tools and material selection enter the procedure. Information about materials is needed at each stage, but at very different levels of breadth and precision.

Product design tools are shown as inputs, attached to the left of the main backbone of the design methodology in Figure 23.7. The tools enable the modeling and optimization of a product design, easing the routine aspects of each phase. Function modelers suggest viable function structures. Configuration optimizers suggest or refine shapes. Geometric and 3D solid modeling packages allow visualization and create files that can be downloaded to numerically controlled prototyping and manufacturing systems. Optimization, design for manufacturability (DFM), design for assembly (DFA), and cost-estimation software allows manufacturing aspects to be refined. Finite element (FE) and computational fluid dynamics (CFD) packages allow precise mechanical, flow, and thermal analysis even when the geometry is complex and the deformations are large. There is a natural progression in the use of the tools as the design evolves: approximate analysis and modeling at the conceptual stage, more sophisticated modeling and optimization at the embodiment stage, and precise analysis at the detailed design stage.

TABLE 23.2
Computer-Based Polymer Material Database

Database	Website	Description
CAMPUS owned by CWFG mbH (1988)	http://www.campusplastics.com	*CAMPUS* (stands for Computer Aided Material Preselection by Uniform Standards) is a multilingual database for the properties of plastics. It is considered worldwide as a leader in regard to the level of standardization, and therefore ease of comparison, of plastics properties. It also supports diagrams to a large extent. CAMPUS is based on ISO standards 10350 for single-point value and 11403 for diagrams.
CETIM-Matériaux		Compositions and mechanical properties of materials.
CES (Cambridge Engineering Selector)	http://www.grantadesign.com	Comprehensive selection system for all classes of materials and manufacturing processes. Variety of optional reference data sources connects directly to http://www.matdata.net. Windows and web format. Modest price.

One essential requisite to successful material selection for a product design, especially for screening and ranking, is a source of reliable and consistent data on material properties. There are many sources of information, which include government agencies, trade associations, engineering societies, textbooks, research institutes, and material producers. A few hardcopy data sources span the full spectrum of materials and properties [13–28].

The number and quality of computer-based material information systems are growing rapidly. Some useful material property databases for polymeric materials are summarized in Table 23.2.

23.3 POLYMER STRUCTURE–PROPERTY RELATIONS AND TESTING

The product design and material selection processes were reviewed in Section 23.2. Both the understanding of structure–property relations of polymers and the test standards for the specific applications are vital to the material selection process. In this section, polymer structure–property relations and the international standards for bulk polymer property tests will be briefly described.

23.3.1 POLYMER STRUCTURE–PROPERTY RELATIONS

The properties of a polymer depend not only on the general chemical composition but also the more subtle differences in microstructure, which are known to exist and are many times dependent upon the fabrication process utilized [29–40]. For such a quest to be efficient, fundamental knowledge of structure–property relations is required. The problem can be examined initially on three broad planes: (1) chemical composition, (2) microstructure, and (3) morphology.

23.3.1.1 Chemical Structure of Polymers

The chemical composition deals with information on the molecular structure, namely, what type of monomer constitutes the chain and whether more than one type of monomer is used (copolymer), macromolecular tacticity and architecture, i.e., the parameters that relate ultimately to the three-dimensional aggregate structure and influence the extent of sample crystallinity and the physical properties. Chemical properties, at the nanoscale, describe how the chains interact through various physical forces. At the macroscale, they describe how the bulk polymer interacts with other chemicals and solvents.

23.3.1.1.1 Monomers and Repeat Units

A monomer is any substance that can be converted into a polymer. The identity of the monomer (repeat units) comprising a polymer is its first and most important attribute. Polymer nomenclature is generally based upon the type of monomer comprising the polymer. Polymers that contain only a single type of repeat unit are known as homopolymers, whereas polymers containing a mixture of repeat units are known as copolymers. The polymerization of a monomer often occurs in a sequential manner. Two monomer molecules first react together to form a dimer. The dimer may then react with a third monomer to yield a trimer, and so on. Dimers are usually linear molecules, but trimers, tetramers, pentamers, and so on can be linear, branched, or cyclic. Low-molecular-weight polymerization products are known as oligomers. Table 23.3 lists various types of monomers and their polymers.

23.3.1.2 Microstructure of Polymers

The microstructure of a polymer relates to the physical arrangement of monomers along the backbone of the chain. These are the elements of polymer structure that require the breaking of a covalent bond in order to change. Several kinds of isomerism or microstructural variations can be identified, and these are grouped under four main headings: architectural, orientational, configurational, and geometric. Structure has a strong influence on the other properties of a polymer. For example, two samples of natural rubber may exhibit different durability, even though their molecules comprise the same monomers.

TABLE 23.3
Various Types of Monomers and Their Polymers

Monomer		Polymer	
Ethylene	$CH_2=CH_2$	Polyethylene	$-(CH_2-CH_2)_n-$
Propylene	$CH_2=CH-CH_3$	Polypropylene	$-(CH_2-CH)_n-$ with CH_3
Styrene	$CH_2=CH-C_6H_5$	Polystyrene	$-(CH_2-CH)_n-$ with C_6H_5
Vinyl chloride	$CH_2=CH-Cl$	Polyvinyl chloride	$-(CH_2=CH)_n-$ with Cl
Vinylidene chloride	$CH_2=CCl_2$	Poly(vinylidene chloride)	$-(CH_2-C)_n-$ with Cl, Cl
Vinyl acetate	$CH_2=CH-OCOCH_3$	Poly(vinyl acetate)	$-(CH_2-CH)_n-$ with $OCOCH_3$
Tetrafluoroethylene	$CF_2=CF_2$	Poly(tetrafluoroethylene)	$-(CF_2-CF_2)_n-$
Methyl acrylate	$CH_2=CH-COOCH_3$	Poly(methyl acrylate)	$-(CH_2-CH)_n-$ with $COOCH_3$
Methyl methacrylate	$CH_2=C(CH_3)-COOCH_3$	Poly(methyl methacrylate)	$-(CH_2-C)_n-$ with $CH_3, COOCH_3$
Acrylonitrile	$CH_2=CH-CN$	Poly(acrylonitrile)	$-(CH_2-CH)_n-$ with CN
Formaldehyde	$CH_2=O$	Poly(formaldehyde)	$-(CH_2-O)_n-$
Caprolactam	$(CH_2)_5$, $O=C-NH$	Poly(caprolactam)	$[-C-(CH_2)_5-N-]_n$ with O, H
Butadiene	$CH_2=CH-CH=CH_2$	Poly(butadiene)	$(-CH_2-CH=CH-CH_2-)_n$
Chloroprene	$CH_2=C(Cl)-CH=CH_2$	Poly(chloroprene)	$(-CH_2-C=CH-CH_2-)_n$ with Cl

23.3.1.2.1 Architecture

An important microstructural feature of a polymer is its architecture and shape, which relates to the way branch points lead to a deviation from a simple linear chain. A branched polymer molecule is composed of a main chain with one or more substituent side chains or branches. Types of branched polymers include star polymers, comb polymers, brush polymers, dendronized polymers, ladders, and dendrimers. There also exist two-dimensional polymers, which are composed of topologically planar repeat units. A polymer's architecture affects many of its physical properties including, but not limited to, solution viscosity, melt viscosity, solubility in various solvents, glass transition temperature, and the size of individual polymer coils in solution (Figure 23.8).

23.3.1.2.2 Chain Length (Molecular Weights)

The physical properties of a polymer are strongly dependent on the size or length of the polymer chain. For example, as chain length is increased, melting and boiling temperatures increase quickly.

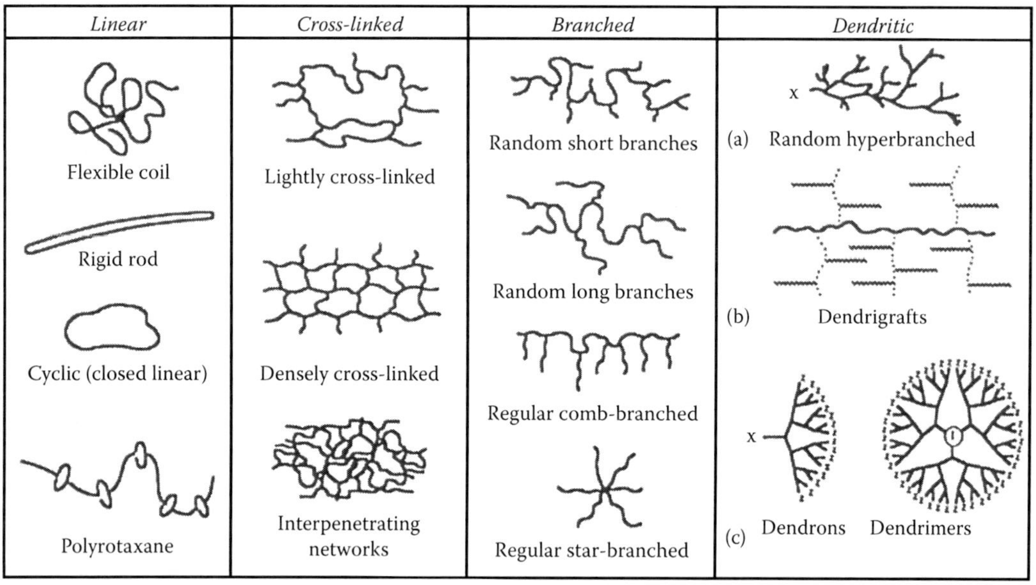

Material Selection, Design, and Application

Linear	Cross-linked	Branched	Dendritic
Flexible coil	Lightly cross-linked	Random short branches	(a) Random hyperbranched
Rigid rod		Random long branches	
Cyclic (closed linear)	Densely cross-linked	Regular comb-branched	(b) Dendrigrafts
Polyrotaxane	Interpenetrating networks	Regular star-branched	(c) Dendrons Dendrimers

FIGURE 23.8 Schematics of various polymer chain architectures.

Impact resistance also tends to increase with chain length, as does the viscosity, or resistance to flow, of the polymer in its melt state. Melt viscosity η is related to polymer chain length Z roughly as $\eta \approx Z^{3.2}$, so that a tenfold increase in polymer chain length results in a viscosity increase of over 1000 times. Increasing chain length furthermore tends to decrease chain mobility, increase strength and toughness, and increase the glass transition temperature (T_g). This is a result of the increase in chain interactions such as van der Waals attractions and entanglements that come with increased chain length. These interactions tend to fix the individual chains more strongly in position and resist deformations and matrix breakup, both at higher stresses and higher temperatures.

A common means of expressing the length of a chain is the degree of polymerization, which quantifies the number of monomers incorporated into the chain. As with other molecules, a polymer's size may also be expressed in terms of molecular weight (MW). Since synthetic polymerization techniques typically yield a polymer product including a range of MWs, the weight is often expressed statistically to describe the distribution of chain lengths present in the same.

$$M_n = \frac{\sum_i N_i M_i}{\sum_i N_i} \qquad (23.1)$$

$$M_w = \frac{\sum_i W_i M_i}{\sum_i W_i} = \frac{\sum_i N_i M_i^2}{\sum_i N_i M_i} \qquad (23.2)$$

$$M_z = \frac{\sum_i N_i M_i^3}{\sum_i N_i M_i^2} \qquad (23.3)$$

$$M_{z+1} = \frac{\sum_i N_i M_i^4}{\sum_i N_i M_i^3} \tag{23.4}$$

where N_i represents the number of molecules of MW M_i. The above MW averages are known, respectively, as the number average MW, M_n, the weight average MW, M_w, the z-average MW, M_z, and the $z+1$-average MW, M_{z+1}. The number average MW is based on averaging equally among all different polymer molecules. The weight average MW averages according to the masses W_i of individual MWs present.

If only one species of a particular MW is present, all of these MW averages M_n, M_w, M_z, and M_{z+1} are equal. If there is a distribution of MWs, we then must have

$$M_{z+1} > M_z > M_w > M_n \tag{23.5}$$

Typically MWs of commercial polymers are in the range from 10,000 to 200,000 depending upon the polymer. Breadths of MW distribution vary considerably, generally from 1.5 to 10. This depends upon the method of polymerization. However, MW distributions are often engineered for the application for which the polymer is designed.

23.3.1.2.3 Monomer Arrangement in Copolymers

In addition to homopolymers, we must consider copolymers where there are two different structural units in the backbone. Since a copolymer consists of at least two types of constituent units (also structural units), copolymers can be classified based on how these units are arranged along the chain. These may be arranged variously in random, block, or alternating manners as shown in Figure 23.9.

Copolymers may also be described in terms of the existence of branches in the polymer chain. Linear copolymers consist of a single main chain, whereas branched copolymers consist of a single main chain with one or more polymer side chains. Other special types of branched copolymers include star copolymers, brush copolymers, and comb copolymers. Terpolymer is a copolymer consisting of three different monomers.

Copolymerization is used to modify the properties of manufactured plastics to meet specific needs, for example, to reduce crystallinity, modify glass transition temperature, or improve solubility. A way of improving mechanical properties is a technique known as rubber toughening. Elastomeric phases within a rigid matrix act as crack arrestors, and so increase the energy absorption when the material is impacted, for example.

23.3.1.2.4 Tacticity

Olefin molecules that contain one unique side group, such as propylene or styrene, yield polymers that possess an asymmetric center at each monomer. Any monomer that possesses an asymmetric

Homopolymer	—A—A—A—A—A—A—A—A—A—
Alternating copolymers	—A—B—A—B—A—B—A—B—A—B—
Random copolymers	—A—B—B—B—A—B—A—B—A—A—
Block copolymers	—B—B—B—B—B—A—A—A—A—A—
Graft copolymers	—A—A—A—A—A—A—A—A—A—A— | | —B—B—B B—B—B—

FIGURE 23.9 Different types of copolymers.

Material Selection, Design, and Application

center at a skeletal atom has the capacity to form stereoregular polymers. The tacticity of a polymer describes the sequencing of these asymmetric centers along the chain. Three primary possibilities exist, called isotactic (all substituents on the same side), syndiotactic (alternating placement of substituents), and atactic (random placement of substituents) sequencing (see Figure 23.10). The tacticity of a polymer markedly affects the bulk physical properties. For example, an isotactic polymer may form a microcrystalline solid that has greater strength and rigidity than its atactic counterpart.

One further item of stereoregular nomenclature is necessary when a regular sequence of double bonds is present in the main chain as, for instance, in poly(1,4-butadiene). The configuration at each double bond may be cis or trans.

Finally, polymers can be formed that contain blocks of one type of stereoregular sequence followed by another. These are called stereoblock polymers. In general, ionic catalysts, especially anionic catalyst, favor the formation of stereoregular polymers; thus, they also favor the synthesis of microcrystalline polymers. On the other hand, free-radical catalysts tend to favor the formation of atactic polymers with a corresponding low degree of crystallinity.

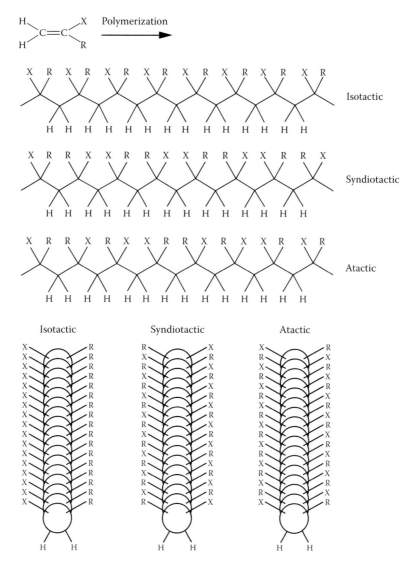

FIGURE 23.10 Tactic forms of vinyl polymer chains.

23.3.1.2.5 Chain Conformation

Conformational analysis involves the study of the ways in which molecules can alter their geometry by torsional rotations (or twisting motions) of their covalent bonds. As such, it must be distinguished from configurational analysis, which is concerned with the fixed geometric differences between closely related molecules.

Conformational changes usually result in a change in the shape of the molecule without the cleavage of bonds having occurred. Because the torsional motions of many bonds take place readily at normal temperatures, conformational changes are responsible for many of the phenomena that we associate with changes in macroscopic physical properties. Thus, the glass and melting transitions of a polymer, the absence or presence of crystallinity, the extensibility or elasticity, and the ways in which polymers raise the viscosity of small molecule solvents can be related to changes in the conformation of the macromolecules. In fact, nearly all of the useful properties of polymers can be ascribed in some way to the conformational characteristics of the component molecules.

The preferred conformation is predicted by the assumption that atoms and groups tend to avoid *collisions*. Thus, the *hard-sphere* approach predicts that, in a molecule that can undergo conformational changes by torsion of the bond that connects two tetracoordinate atoms (such as carbon), the three *staggered* minima will be accessible as shown in Figure 23.11. This leads immediately to the concept of the *threefold rotational isomeric model* that forms the basis of some of the more complex molecular structures.

23.3.1.2.6 Crystallinity

Because polymer molecules are so large, they generally pack together in a nonuniform fashion, with ordered or crystalline-like regions mixed together with disordered or amorphous domains. In some cases, the entire solid may be amorphous, composed entirely of coiled and tangled macromolecular chains. Crystallinity occurs when linear polymer chains are structurally oriented in a uniform three-dimensional matrix. Increased crystallinity is associated with an increase in rigidity, tensile strength, and opacity (due to light scattering). Amorphous polymers are usually less rigid, weaker, and more easily deformed. They are often transparent. Three factors that influence the degree of crystallinity are (1) chain length, (2) chain branching, and (3) interchain bonding.

The crystallinity of polymers is characterized by their degree of crystallinity, ranging from 0 for a completely noncrystalline polymer to 1 for a theoretical completely crystalline polymer. Polymers with microcrystalline regions are generally tougher (can be bent more without breaking) and more impact-resistant than totally amorphous polymers. Polymers with a degree of crystallinity

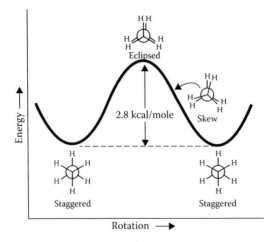

FIGURE 23.11 Potential energy profile illustrating the potential energy changes associated with rotation about the C–C bond of ethane.

approaching 0 or 1 will tend to be transparent, whereas polymers with intermediate degrees of crystallinity will tend to be opaque due to light scattering by crystalline or glassy regions. Thus, for many polymers, reduced crystallinity may also be associated with increased transparency. Crystallinity is also directly related to the density. Polymers with higher crystallinity have higher densities than those with lower crystallinity. Copolymers typically have lower crystallinities and subsequently lower densities.

23.3.2 Testing for Polymeric Material Selection

The tests commonly used in the plastics industry are established and used to describe properties and process characteristics of these polymers. This allows resin producers, product designers, processors, and end users to have a common understanding of material properties and what type of properties can be correlated to the end-use results reflecting use conditions in the application. Tests are not ends in themselves but rather a means of gathering knowledge about engineering properties and translating that to durability in-use. The real test of a material is conducted during actual service. Reliable, useful tests for making this translation should be based on an understanding of the engineering properties, processing characteristics, and the performance of these materials in similar applications.

Different geographic regions may follow different test standards. There are several different institutions that define plastics standards for different geographic regions. These are the American Society for Testing and Materials (ASTM) and the Underwriters Laboratories (UL) in the United States; International Organization for Standardization (ISO) in Geneva, Switzerland; Deutsches Institut für Normung (DIN) in Germany; British Standards Institution (BSI) in Great Britain; Association Française de Normalisation (AFNOR) in France; and Japanese Industry Standards (JIS) in Japan. Plastics testing standards developed by the Geneva-based ISO are being promoted as a universal set of testing standards that could satisfy the needs for a single global standard. Uniform global test methods and data reporting formats will enhance the efficiency and effectiveness of the plastic industry. A set of uniform broadly adopted test standards resulting in comparable data sets will improve communication among the resin suppliers, product designers, and brand owners, and in the case of multinational companies, it will improve communication between their global manufacturing sites and should improve their overall global competitiveness.

The information provided in this section allows the reader to compare the physical properties and make a preliminary selection of potential generic materials for further investigation. It also enables the reader to verify if these materials meet the application requirements, the processing characteristics for efficient production (low manufacturing cost), and optimum quality. This material investigation analysis is recommended before the final material selection process is completed.

23.3.2.1 Mechanical Properties

The mechanical properties of polymeric materials are important considerations in a product design and material selection process because almost all applications require some mechanical load-bearing capability. Various mechanical test standards have been developed to investigate tensile stress–strain behavior, flexural strength, compression stress–strain behavior, shear strength, tensile strength creep, and impact strength testing. Readers who would like to pursue further understanding on these topics and to review the property data for preliminary design selection should access some of the literature listed in the reference section [41–51].

Tensile test is the most widely used testing standard for measuring the mechanical properties of a polymer. This test determines the stress–strain relationship of the material in tension by measuring force at a constant extension rate. Tensile strength, tensile modulus, yield stress, yield strain, elongation at break, stress at break, and strain at break can be determined as schematically shown in Figure 23.12. Tensile strength is the maximum stress that a material can withstand while being stretched before failing or breaking. Tensile modulus is a measure of the stiffness of a material and

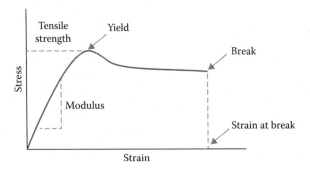

FIGURE 23.12 Tensile stress–strain curve of a thermoplastic polymer.

is defined as the ratio of stress to the corresponding strain in the range of stress in which Hooke's law holds. The yield point in a stress–strain curve is defined as a first point at which an increase in strain occurs without an increase in stress. The stress and strain at the yield point are known as yield stress and yield strain, respectively. The break point in a stress–strain test is the point at which the material ruptures under the extension. The stress, strain, and elongation at break are determined based on the break point (see Table 23.4 for ASTM test numbers).

Flexural test is the testing standard for measuring the ability of the material to withstand bending forces applied perpendicular to its longitudinal axis. A combination of compressive and tensile stresses is applied to the test specimen by the flexural loads. The flexural strength is reported using the maximum stress and strain that occur on the outside surface of the specimen. The flexural strength is defined as the maximum stress in the outer fibers at the moment of break. For materials that do not break, strain values up to 5% are recommended to calculate the flexural yield strength. Flexural modulus is a measure of the stiffness during the initial step of the bending experiment and calculated by the slope of the initial straight line portion of the flexural stress–strain curve. This property is sometimes called *modulus of elasticity in bending*. The flexural modulus is generally lower than the tensile modulus of elasticity determined by ASTM D-638.

TABLE 23.4
ASTM and ISO Test Methods for Mechanical Properties

Mechanical Tests	ASTM	Unit (SI)	ISO	Unit (SI)
Tensile modulus	D-638	MPa	527-1 and -2:1993	MPa
Yield stress	D-638	MPa	527-1 and -2:1993	MPa
Yield strain	D-638	%	527-1 and -2:1993	%
Elongation at break	D-638	%	527:1966	%
Stress at break	D-638	MPa	527-1 and -2:1993	MPa
Strain at break	D-638	%	527-1 and -2:1993	%
Flexural modulus	D-790	MPa	178:1993	MPa
Flexural strength	D-790	MPa	178:1993	MPa
Compression strength	D-695	MPa	–	–
Shear strength	D-732	MPa	–	–
Tensile strength creep	D-674	MPa	–	–
Charpy notched impact at 23°C	D-256	kJ/m^2	179:1993	kJ/m^2
Tensile impact	D-1822	kJ/m^2	8256:1990	kJ/m^2
Izod impact notched at –40°C	D-256	J/m	180:1993	kJ/m^2
Izod impact notched at 23°C	D-256	J/m	180:1993	kJ/m^2
Izod impact (unnotched) at 23°C	D-256	J/m	180:1993	kJ/m^2
Gardner falling weight impact	D-3029	J	–	–

Compression stress–strain behavior of a material when the specimen is subject to a compressive load at a low and uniform loading rate can be described by compressive properties. In general, plastic materials are stronger in compression than in tension or shear. Two principal compressive properties used in product design are compressive strength and compressive modulus. The compressive strength is defined as the maximum compressive stress either at rupture of the test specimen or at a given percentage of deformation. The compressive modulus is represented by the slope of the initial straight line portion of the stress–strain curve and calculated the same as the tensile testing procedure (Figure 23.13).

Shear strength of the plastic materials is defined as the ability to withstand the maximum load to require shearing the punching portion of the test specimen completely off from the rest of the specimen (i.e., similar to a paper punch). Shear strength data are of great importance to a designer for film and sheet products. The shear strength data in the material supplier's literature should be used with extreme caution since these shear strength values can be significantly higher than actual values. A common engineering practice is to consider the shear strength as equal to one-half of the tensile strength.

Tensile strength creep: When a thermoplastic material is subjected to a constant load over a period of time, the specimen tends to deform gradually with time. This deformation over time or creep can easily exceed beyond the initial deformation predicted by its modulus. Therefore, creep should be considered when the material is intended to support a load over a long time in order to predict the performance and useful life of a polymeric material.

Impact strength is the ability to resist loading at high strain rates. It is a critical property for plastic materials, which is directly related to the product performance, service life, product safety, and liability. Izod impact testing (ASTM D-256), Charpy impact testing (ASTM D-256), and Tensile impact testing (ASTM D-1822) are pendulum impact-type tests, whereas Gardener impact test (ASTM D-3029) is a falling weight impact test. Both Izod and Charpy impact tests use a notch in a test specimen as an artificial crack and measure the energy to propagate a crack. Sometimes, an unnotched Izod test is used for reinforced and filled materials such as fiberglass reinforced compounds, which require little energy to crack the test specimen. The tensile impact testing, which is similar to the Izod impact test with adjustment to the fixturing, was devised to overcome the limitations of Izod and Charpy tests because these tests do not break some ductile materials, especially thin films. Gardner drop weight impact testing (ASTM D-3029) uses a weight that is dropped on a test specimen positioned over a circular opening. The type of impact caused by this test is multiaxial unlike uniaxial crack initiation in Izod and Charpy tests. Therefore, this test divulges the easiest route to failure. The failure criterion is usually visual, and it is a *pass* or *fail* type of test. Instrumented versions of the drop weight impact test have been developed to better quantify impact failure when tested in this manner.

Table 23.4 summarizes the ASTM and ISO test methods for mechanical properties. The information provided in this chapter allows the readers to understand the definition and testing methods available for common mechanical properties.

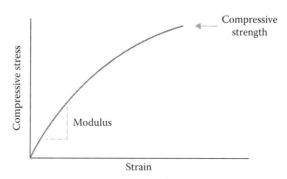

FIGURE 23.13 Compressive stress–strain curve of a thermoplastic polymer.

TABLE 23.5
ASTM and ISO Physical Test Standards

Physical Tests	ASTM/UL	Unit (SI)	ISO	Unit (SI)
Specific gravity	D-792		1183:1987	
Density	D-1505	g/cm^3	1183:1987	g/cm^3
Water absorption 24 h	D-570	%	62:1980	%
Water absorption saturation	D-570	%	62:1980	%
Rockwell hardness	D-785-60T	R,L,M,E,K	–	–
Durometer hardness	D-2240	A,D	–	–
Taber abrasion	D-1044	mg/1000 cycles	–	–
Tear resistance	D-624	N/m	–	–
Coefficient of friction	D-1894		–	–
Mold shrinkage, flow direction	D-955	%	–	%
Mold shrinkage, transverse direction	D-955	%	–	%

23.3.2.2 Physical Properties

The physical properties of polymeric materials are of major importance for certain applications [52–56]. Effects of moisture absorption on the mechanical and electrical properties and the selection of the optimum transparency, surface hardness, wear, and coefficient of friction properties for the product applications can also be determined. A brief description and the significance of the testing procedure for each physical property are provided.

Table 23.5 summarizes the ASTM and ISO test methods for the physical properties.

23.3.2.3 Thermal Properties

Thermoplastics can be sensitive to changes in temperature. One of the most important considerations while studying the performance of plastic materials at elevated temperatures is the dimensional change and decrease in the key properties such as modulus of elasticity, tensile strength, electrical properties, chemical resistance, environmental resistance, and the expected service life of the molded product [57–61]. Many factors are considered when selecting a plastic material for a high temperature application. The material must be able to support a design load under operating conditions without objectionable loss of properties caused by the high temperature, creep, expansion, or distortion. The material must not degrade or lose necessary additives that will cause a substantial reduction in the physical, electrical, and flammability properties during the expected service life of the molded product.

The molecular chain crystallinity has a number of important effects upon the thermal properties of a semicrystalline polymer. Semicrystalline thermoplastics normally have a well-defined melting point with associated thermomechanical rigidity properties. Amorphous thermoplastics, in contrast, have a gradual softening range of the polymer melt. Molecular orientation also has a significant effect on the thermal properties of thermoplastics. The melt flow orientation tends to decrease dimensional stability at higher temperatures. The MW of the polymers also affects the low-temperature flexibility and the impact strength or brittleness. The intermolecular bonding that can be achieved through cross-linking or modification through copolymerization chemicals can also have a considerable effect on the thermal properties of thermoplastic materials.

Table 23.6 summarizes the ASTM and ISO test methods for thermal properties.

23.3.2.4 Electrical Properties

The electrical properties of polymeric materials are of major importance for electrical/electronic applications [62–71]. The polymeric materials can serve as a selective insulator or signal transmitter. The two most important categories of electrical properties are the insulation or dielectric properties

TABLE 23.6
ASTM/UL Test Methods for Thermal Properties

Thermal Tests	ASTM/UL	Unit (SI)
Glass transition temperature	E-1356	°C
Melting temperature	D-2117	°C
Thermal conductivity	C-177	W/(m K)
Continuous service temperature	D-794	°C
Heat deflection temperature	D-648	°C
Vicat softening temperature	D-1525	°C
Coefficient of linear expansion	D-696	
Brittle temperature	D-746	°C
Flammability rating	UL 94	
Relative thermal index	UL746A	°C
RTI mechanical without impact	UL746A	°C
RTI mechanical with impact	UL746A	°C

TABLE 23.7
ASTM/UL Test Methods for Electrical Properties

Electrical Properties	ASTM/UL	Unit (SI)
Relative temperature index	UL746B	°C
Dielectric constant, 1 MHz and 100 Hz	D-150	–
Dissipation factor, 1 MHz and 100 Hz	D-150	–
Volume resistivity	D-257	Ω cm
Surface resistivity	D-257	Ω/Sq
Dielectric strength	D-149	kV/mm
Competitive tracking index (CTI)	UL746A	V
High ampere arc ignition (HAI)	UL746A	# of arcs to ignite
High volt arc track rate (HVTR)	UL746A	mm/min
Arc resistance	D-495	S
Hot wire ignition (HWI)	UL746A	S

and the service temperature index classifications. Table 23.7 shows the most important electrical properties of thermoplastic and thermoset resins and ASTM/UL test methods.

23.3.2.5 Optical Properties
Optical properties of polymeric materials are dependent on material selection, production process, and thermal properties [72–80]. Transparency depends on a polymer basic structure. Generally, amorphous thermoplastics are transparent. For semicrystalline thermoplastics, as the overall crystallinity and crystal size increase, the clarity of the polymer is decreased. Common optical polymeric materials include PMMA, polystyrene (PS), styrene–acrylonitrile, polycarbonate, and allyl diglycol carbonate.

Table 23.8 summarizes the ASTM and ISO test methods for optical properties.

23.3.2.6 Microbial, Weather, and Chemical Resistance
Almost all polymeric materials are affected by outdoor weather and other environmental factors such as pollution, temperature, radiation, and humidity. Most common polymers are generally not susceptible to microbial (fungi or bacteria) attack, but the chemical additives, pigments,

TABLE 23.8
ASTM Test Methods for Optical Properties

Optical Properties	ASTM
Refractive index	D-542
Luminous transmittance and haze	D-1003
Specular gloss	D-523

TABLE 23.9
ASTM Test Methods for Microbial, Weather, and Chemical Resistance Test Standards

	Test Standards	ASTM/UL	Unit (SI)
Microbial resistance test	Fungal resistance testing	ASTM G21-70	
	Bacterial resistance testing		
	Outdoor fungi and bacteria testing limitation		
Weather resistance test	Accelerated weathering testing	ASTM G53	
		ASTM D-1499	
		ASTM D-2565	
	Outdoor weathering testing		
Chemical resistance test	Immersion testing	ASTM D543	
	Solvent stress-cracking resistance testing		

and lubricants inside the polymers are vulnerable to microbial environments [81–86]. Weathering tests are also available, with some being highly specialized for the defined field of use [87–93]. Chemical resistance is the ability of a polymeric material to maintain its original properties after being exposed to a certain chemical environment for a specified time period [94–102]. The primary factors influencing the chemical resistance are the molecular structure of the polymer, the type and concentration of the chemical reagent, exposure condition, and residual and applied stresses. Polymer blends or composite structures can sometimes be used to provide improved chemical resistance relative to one of the pure components. Some of the most common test standards for microbial, weather, and chemical resistance tests are summarized in Table 23.9.

23.4 RHEOLOGICAL PROPERTIES OF POLYMER MELT FOR MATERIAL SELECTION

As previously discussed, the material selection, process selection, and cost estimation should be done in parallel for a successful product design. In the polymer industry, the typical final products are either pellet or granule forms for the downstream secondary forming processes or a final part for the end-user applications. Except for a few solid-state processing technologies, the polymeric material, which is fed into the processing equipment, has to be melted inside of the processing equipment, processed in the molten state to a final product, and solidified. There could be either reactions (reactive processing) during the melt processing or no reaction (nonreactive processing) during the processing. The fundamental understanding of the flow in processing equipment is vital to both material and process selections.

Rheology is the field of science that studies fluid behavior during flow-induced deformation [103–110]. From the variety of materials that rheologists study, polymers have been found to be the most interesting and complex. Polymer melts are often shear thinning (non-Newtonian) and viscoelastic, and their flow properties are temperature dependent.

Flow of polymer melt in processors with confined surface such as extrusion and injection molding is dictated by the viscosity of the polymer melt, whereas the flow of polymer melt in processors with nonconfined surface such as continuous profile forming and film processing is dictated by nonlinear viscoelastic properties. In this section, we first discuss shear and elongational viscosities of polymer melt followed by an example of how the shear viscosity of the polymer melt is used for both material and process selections. Second, the viscoelastic properties of polymer melt, linear and nonlinear, will be discussed followed by an example.

23.4.1 Viscosity

Viscosity is the most widely used material parameter when determining the behavior of the polymer melt during processing. Since the majority of polymer processes are shear rate dominated, the viscosity of the melt is commonly measured using shear deformation measurement devices. However, there are polymer processes, such as blow molding, thermoforming, and fiber spinning, which are dominated by either elongational deformation or a combination of shear and elongational deformation.

23.4.1.1 Shear Viscosity

There are mainly two classes of shear flows, which are drag-induced shear flow such as melt flow between parallel plates with one moving plate and coaxial cylinders, and pressure-induced shear flows such as melt flow in a tube, slit die, and annulus. The shear flow schematically shown in Figure 23.14 is the melt flow between two parallel plates with moving one plate. The velocity field in the shear flow shown in Figure 23.14 is expressed by

$$\mathbf{v} = v_1(x_2)\mathbf{e}_1 + 0\mathbf{e}_2 + 0\mathbf{e}_3 \tag{23.6}$$

where $\mathbf{e}_1$, $\mathbf{e}_2$, and $\mathbf{e}_3$ are unit vectors. The rate of deformation tensor has the form

$$\mathbf{d} = \frac{1}{2}\begin{bmatrix} 0 & \dot{\gamma} & 0 \\ \dot{\gamma} & 0 & 0 \\ 0 & 0 & 0 \end{bmatrix} \tag{23.7}$$

where $\dot{\gamma}(= dv_1/dx_2)$ is the shear rate. Our great concern for this class of motions is that internal flows in the confined surface as shown in Figure 23.14 and the flows we will consider are equivalent or similar to those occurring in screw channel in extrusion, dies, and molds.

In response to a shearing flow, the corresponding shear stresses are developed. Normal stresses arise as well. The stress field in response to shear flow has been noted by Markovitz [111] and Coleman and Noll [112,113], largely from symmetry and initial isotropy arguments, to be of the form

$$\boldsymbol{\sigma} = \begin{bmatrix} \sigma_{11} & \sigma_{12} & 0 \\ \sigma_{21} & \sigma_{22} & 0 \\ 0 & 0 & \sigma_{33} \end{bmatrix} \tag{23.8}$$

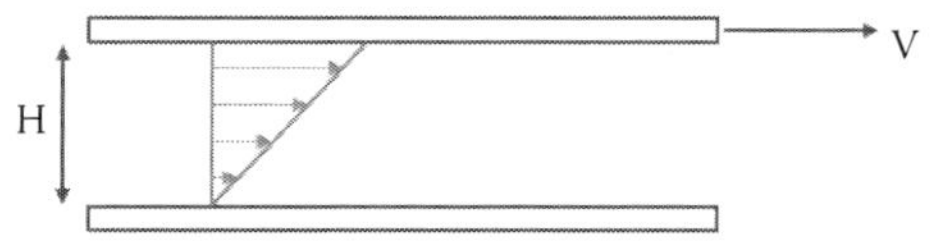

FIGURE 23.14 Schematic of shear flow field.

The sign of the shear stress varies with the sign of the shear rate, that is,

$$\sigma_{12}(\dot{\gamma}) = -\sigma_{12}(-\dot{\gamma}) \tag{23.9}$$

This suggests the formulation between σ_{12} and $\dot{\gamma}$ as

$$\sigma_{12} = \eta(\dot{\gamma})\dot{\gamma} \tag{23.10}$$

where η is the shear viscosity.

We may define three independent combinations of normal stresses. As the normal stresses are the sum of a pressure and an extra stress,

$$\sigma_{ii} = -p + P_{ii} \tag{23.11}$$

our three functions are logically a measure of this pressure p and two differences between normal stresses. These may be taken as

$$p = \frac{1}{3}(\sigma_{11} + \sigma_{22} + \sigma_{33}) \tag{23.12}$$

$$N_1 = \sigma_{11} - \sigma_{22} = P_{11} - P_{22} \tag{23.13}$$

$$N_2 = \sigma_{22} - \sigma_{33} = P_{22} - P_{33} \tag{23.14}$$

N_1 is known as the principal normal stress difference and N_2 as the second normal stress difference. N_1 and N_2 will be uniquely defined by the rheological behavior of the material, whereas p is largely determined by the externally applied hydrostatic pressure and the balance of force during flow.

The normal stresses are independent of the direction of shear. Thus,

$$N_i(\dot{\gamma}) = N_i(-\dot{\gamma}) \tag{23.15}$$

and

$$N_1 = N_1(\dot{\gamma}^2) \tag{23.16}$$

$$N_2 = N_2(\dot{\gamma}^2) \tag{23.17}$$

One may define from this normal stress coefficients Ψ_1 and Ψ_2 through

$$N_1 = \Psi_1(\dot{\gamma}^2)\dot{\gamma}^2 \tag{23.18}$$

$$N_2 = \Psi_2(\dot{\gamma}^2)\dot{\gamma}^2 \tag{23.19}$$

The purpose of shear flow rheological measurements is to determine σ_{11} and N_1 and N_2 as a function of shear rate $\dot{\gamma}$, and through them the material functions η, Ψ_1 and Ψ_2. We also seek the time-dependent or transient behavior of these functions. In addition to the rheological functions described above, it is also important to determine the boundary conditions along the metal surfaces such as slip.

The shear viscosity of most polymer melts is shear-thinning and temperature-dependent. The shear-thinning effect is the reduction in viscosity at high rates of deformation. This phenomenon occurs because at a high rate of deformation, the molecules are stretched out, enabling them to slide past each other with more ease, hence lowering the bulk viscosity of the melt. Figure 23.15 clearly shows the shear-thinning behavior and temperature dependence of the shear viscosity for polyethylene (PE) and PS.

In order to use the shear viscosity data of the material of interest in process simulation correctly, the shear viscosity data at the shear rate equivalent to the operation range of the processing equipment must be used. Table 23.10 lists rheometers, which are commonly used in polymer industry, and shear rate ranges for the rheometers.

There are many polymer fillers and additives that are mixed into a polymer to improve the performance of the polymer such as the mechanical, optical, electrical, and acoustic properties. Fillers can be divided into two categories: those that reinforce the polymer and improve its mechanical performance and those that are used to take up space and so reduce the amount of actual resin to produce a part, which are sometimes referred to as extenders. A third, less common, category of filled polymers are those where fillers are dispersed into the polymer to improve its electric conductivity. Polymers that contain fillers that improve their mechanical performance are often referred to as composites. As the level of fillers increases in the composite, the shear viscosity of the molten polymeric composite increases. Polymer melts filled with small particles exhibit an increased viscosity especially at very low shear rates. At high levels of loading, yield values Y develop [114–117]. However, some of the additives could lower the shear viscosity of the molten polymer such as a blowing agent.

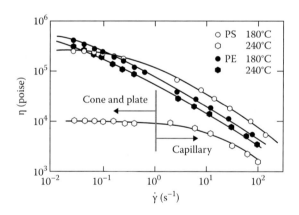

FIGURE 23.15 Influences of shear rate and temperature on shear viscosity for PE and PS melts.

TABLE 23.10
Rheometers to Measure Shear Viscosity of Polymer Melts

	Rheometers	Shear Rate Ranges (s⁻¹)
Low range of shear rate	Cone and plate rheometer	0.01–10
	Parallel plate rheometer	0.01–100
Middle range of shear rate	Slit die rheometer	1–300
High ranges of shear rate	Capillary rheometer	1–2500
	Couette rheometer	1–5000

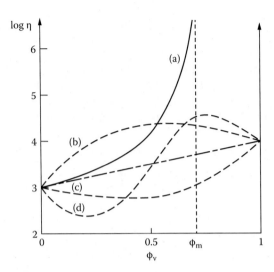

FIGURE 23.16 Concentration dependence of shear viscosity for (a) solid particle-filled liquid and (b–d) polymer blends.

Polymer blends belong to another family of polymeric materials, which are made by mixing or blending two or more polymers to enhance the physical properties of each individual component. The blending process of polymers is generally done in the molten state, particularly in shear flow. In order to understand and model the flow during the blending process, the bulk shear viscosity of a polymer blend must be understood. The rheology of multiphase systems has been investigated by many researchers [118–126]. Figure 23.16 shows the schematics of both solid particle-filled liquid (curve a) and polymer blends (curves b through d). At constant stress, the viscosity of a particulate-filled melt versus concentration (curve a in Figure 23.16) shows a monotonic increase with an asymptote at $\phi_v = \phi_m$. On the other hand, this dependence for blends may show either a positive (over 60% of blends) or negative (about 30%) deviation from the log additive rule (curves b and c in Figure 23.16). For about 10% blends, the η versus ϕ_v dependence passes through a local minimum and maximum (curve d in Figure 23.16). Blends with concentrations corresponding to these extremes are of special interest because they are easier to process and may show more interesting properties.

23.4.1.2 Empirical Models for Shear Viscosity–Shear Rate Relationship

For both material and process selection processes, various commercially available process simulation packages have been developed and used extensively in the polymer industry. Polymer processing process simulations such as extrusion, injection molding, etc., require the shear viscosity data as a function of shear rate at a minimum of three different temperatures to simulate the process of interest. A wide range of empirical viscosity–shear rate relationships have been proposed through the years. Some of the models used by polymer processors on a day-to-day basis to represent the viscosity of industrial polymers are listed in Table 23.11.

23.4.1.3 Elongational Viscosity

Elongational or extensional viscosity is a viscosity when applied stress is extensional stress. Many polymer processing technologies require either uniaxial or biaxial extensions such as fiber spinning (uniaxial extension), certain extrusion die flows, film and sheet extrusions, blow molding (biaxial extension), thermoforming, and compressing molding (Figure 23.17).

Elongational flows are motion for which the velocity field is given by

$$\mathbf{v} = v_1(x_1)\mathbf{e}_1 + v_2(x_2)\mathbf{e}_2 + v_3(x_3)\mathbf{e}_3 \tag{23.20}$$

Material Selection, Design, and Application

TABLE 23.11
Empirical Models for Viscosity–Shear Rate Relationship

Models	Parameters	Equations	References
Power law	2	$\eta(\dot{\gamma}, T) = m(T)\dot{\gamma}^{n-1}$ $m(T) = m_0 \exp\left[\frac{\Delta E}{R}\left(\frac{1}{T} - \frac{1}{T_0}\right)\right]$	[122] [123]
Ellis	3	$\frac{\eta_0}{\eta(\tau)} = 1 + \left(\frac{\tau}{\tau_{1/2}}\right)^{\alpha-1}$	[124]
Cross	4	$\frac{\eta - \eta_\infty}{\eta_0 - \eta_\infty} = \frac{1}{1 + (\lambda\dot{\gamma})^m}$	[125]
Bird–Carreau–Yasuda	5	$\frac{\eta - \eta_0}{\eta_0 - \eta_\infty} = [1 + |\lambda\dot{\gamma}|^a]^{(n-1)/a}$	[126] [127]
Modified Bird–Carreau–Yasuda	5	$\eta = \frac{k_1 a_T}{[1 + k_2 \gamma a_T]^{k_3}}$ $\ln a_T = \frac{8.86(k_4 - k_5)}{101.6 + k_4 - k_5} - \frac{8.86(T - k_5)}{101.6 + T - k_5}$	[128]
Bingham fluid	2	$\eta = \infty$ or $\dot{\gamma} = 0$ $\sigma = \sigma_Y$ $\eta = \eta_0 + \sigma_y/\dot{\gamma}$ $\sigma \geq \sigma_Y$	[129]

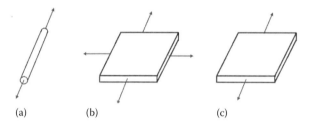

FIGURE 23.17 Classification biaxial elongational flow: (a) uniaxial, (b) equal biaxial, and (c) planar.

The rate of deformation tensor is of the form

$$\mathbf{d} = \begin{bmatrix} \dot{\gamma}_{e1} & 0 & 0 \\ 0 & \dot{\gamma}_{e2} & 0 \\ 0 & 0 & \dot{\gamma}_{e3} \end{bmatrix} \quad (23.21)$$

where

$$\dot{\gamma}_{ej} = \frac{dv_j}{dx_j} \quad (23.22)$$

For an incompressible fluid, we must have

$$\dot{\gamma}_{e1} + \dot{\gamma}_{e2} + \dot{\gamma}_{e3} = 0 \quad (23.23)$$

By *uniaxial extension*, we mean a flow for which

$$\dot{\gamma}_{e2} = \dot{\gamma}_{e3} = -\frac{1}{2}\dot{\gamma}_{e1} < 0 \tag{23.24}$$

For *equal biaxial extension*

$$\dot{\gamma}_{e1} = \dot{\gamma}_{e2} = -2\dot{\gamma}_{e3} > 0 \tag{23.25}$$

For *planar extension*

$$\dot{\gamma}_{e3} = 0, \quad \dot{\gamma}_{e2} = -\dot{\gamma}_{e1} \tag{23.26}$$

The stress response to an elongational flow will be of the form

$$\sigma = \begin{bmatrix} \sigma_{11} & 0 & 0 \\ 0 & \sigma_{22} & 0 \\ 0 & 0 & \sigma_{33} \end{bmatrix} \tag{23.27}$$

For uniaxial extension, σ_{11} is determined by an elongational viscosity, η_e, through

$$\sigma_{11} = \eta_e \dot{\gamma}_{e1}, \quad \sigma_{22} = \sigma_{33} = -p \tag{23.28}$$

For equal biaxial stretching, we have a biaxial elongational viscosity:

$$\sigma_{11} = \sigma_{22} = \eta_B \dot{\gamma}_{e1}, \quad \sigma_{33} = -p \tag{23.29}$$

For planar extension

$$\sigma_{11} = \eta_p \dot{\gamma}_{e1}, \quad \sigma_{33} = -p \tag{23.30}$$

Functions similar to η_e, η_B, and η_p may be defined for other elongational flows. One seeks to carry out long-duration constant-elongation rate or stress studies in order to obtain a well-defined rheological function equivalent to the shear viscosity. Elongational flows generally occur in post-die polymer processing operations such as melt spinning, tubular film extrusion, and blow molding.

23.4.1.4 Example: Effect of Shear Viscosity on Material Selection

An example in this section demonstrates the effect of shear viscosity of molten PS on material selection via process simulation. Three PS resins are selected as the candidates for carriers for Additive A. The minimum degradation temperature of Additive A is around 200°C. The objective of this work was to find an optimum carrier polymeric resin for Additive A without degradation during the processing. Three different PS resins are PS 1 resin (Mw ~ 295,000 g/mol), PS2 resin (Mw ~ 140,000 g/mol), and PS3 resin (Mw ~ 58,000 g/mol). Figure 23.18 shows the shear viscosity data as a function of shear rate at three different temperatures. Screw configurations used in this

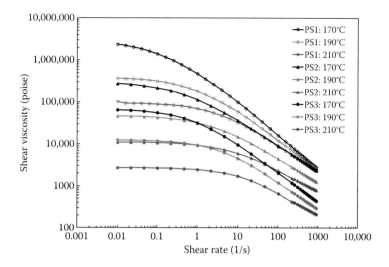

FIGURE 23.18 Dynamic viscosity data as a function of frequency at three different temperatures for three PS resins.

process simulation are shown in Figure 23.19. All simulation studies reported in this section used a feed rate of 3000 lb./h and a screw speed of 150 rpm.

Figure 23.20a shows the predicted fill factor (f), pressure (p), and melt temperature (T) profiles for PS1 resin using the screw configuration shown in Figure 23.19. A commercially available process simulation package for intermeshing corotating twin-screw extruders from The University of Akron [130] was used for this simulation study. As shown in Figure 23.20a, the calculated discharge temperature was 245°C. The discharge temperature for this resin was much higher than the target temperature. Figure 23.20b shows the predicted fill factor, pressure, and melt temperature profiles for PS2 resin. The calculated discharge temperature was 205°C. The calculated melt temperature was close to the target discharge temperature, but still slightly higher than the target discharge temperature. Figure 23.20c shows the predicted fill factor, pressure, and melt temperature profiles for PS3 resin. The calculated discharge temperature for the screw configuration shown in Figure 23.19 was 175°C. This calculated melt temperature was much lower than the target discharge temperature.

These process simulation results suggest that PS3 resin can be used as a carrier for Additive A. Formulation of the PS3 resin-based Additive A concentrate having the proper shear viscosity can reduce the melt temperature enough to prevent the Additive A concentrate melt from degradation.

23.4.2 Viscoelasticity

It is well known that a polymer, at a specific temperature and MW, may behave as a liquid or a solid depending on the speed (time scale) at which its molecules are deformed. This behavior, which ranges between liquid and solid, is generally referred to as the viscoelastic behavior or material response. Figure 23.21 shows the stress responses to a step increase in strain as a function of time for various materials. An elastic solid would show no relaxation (Figure 23.21b). If purely viscous liquid is subjected to the same deformation, the stress relaxed instantly to zero as soon as the strain becomes constant (Figure 23.21c). When a molten polymer is subjected to a step increase in strain, the stress relaxes in an exponential fashion (Figure 23.21d).

A useful parameter often used to estimate the elastic effects during flow is the Deborah number, De. The Deborah number is defined by

$$De = \frac{\lambda}{t} \tag{23.31}$$

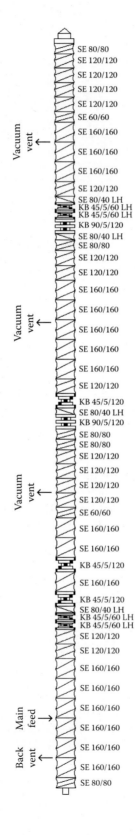

FIGURE 23.19 Screw configuration for 130-mm-diameter intermeshing corotating twin-screw extruder used in this example.

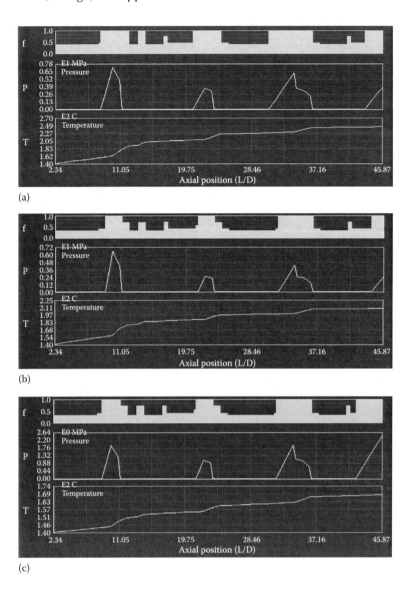

FIGURE 23.20 Simulated fill factor, pressure, and temperature profiles for (a) PS1 resin, (b) PS2 resin, and (c) PS3 resin.

where λ is the relaxation time of the polymer, and t is a characteristic process time. Figure 23.22 [131] helps visualize the relation among time scale, deformation, and applicable model. At small Deborah numbers, the polymer can be modeled as a Newtonian fluid, and at very high Deborah numbers, the material can be modeled as a Hookean solid. In between, the viscoelastic region is divided in two: the linear viscoelastic region for small deformations and the nonlinear viscoelastic region for large deformations. Second-order fluids reside in a fringe of the regime of nonlinear viscoelasticity that lies just across the border from the Newtonian domain.

The breadth of the scope of nonlinear phenomena can be grasped in part by considering the various time-dependent probes of linear viscoelasticity such as sinusoidal oscillation, creep, constrained recoil, stress relaxation after step strain, stress relaxation after steady shearing, and stress growth after start-up of steady sharing. In the linear regime, that is, at small strains or small strain rates, the experimental results of any one of these probes, for example, in simple shear, can be used to predict

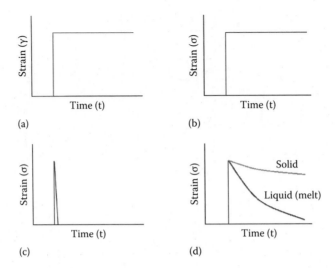

FIGURE 23.21 Stress response versus time for (a) step input in strain, (b) Hookean solid, (c) Newtonian fluid, and (d) viscoelastic melt.

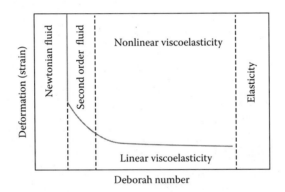

FIGURE 23.22 Schematic of Newtonian, elastic, linear, and nonlinear viscoelastic regimes as a function of deformation and Deborah number during deformation of polymer melts.

results for any of the other probes, not only for simple shearing deformations but also for any volume-preserving deformation. For strains and strain rates that are not small, each of these probes gives a nonlinear response. In principle, in the nonlinear regime, none of these probes or combinations of probes gives a response that can be used to predict the nonlinear response to any of the other probes. Nor can any amount of nonlinear data (in shear, for example) be used to generate from first principle predictions in any other type of deformation, such as uniaxial extension. Each nonlinear test gives data that, in principle, are directly relevant to that test only. Therefore, at the minimum, the task of the nonlinear rheologists is to develop empirical correlations among nonlinear measurements, or more ambitiously, it is to find constitutive equations that will allow one to predict a useful range of nonlinear behavior using as input rheological measurements over a small subset of that range.

Thus, the general nonlinear rheological behavior of a material is characterized by finding a constitutive equation appropriate for that material. The appropriateness of a constitutive equation is a balance of many factors, including the equation's accuracy in fitting and predicting various data, the soundness of its theoretical underpinnings, its simplicity, its mathematical and computational tractability, and the range of phenomena one wishes to address using the equation. The weighting of the various factors is subjective to a significant degree; hence, a rich diversity of constitutive equations has been developed over the past 60 years, many of which continue to be used.

23.4.2.1 Linear Viscoelasticity

23.4.2.1.1 Maxwell Differential Model

Some discussion of the characteristics of the one-dimensional linear formulation of viscoelasticity is really needed before we begin the consideration of the three-dimensional nonlinear theory. The theory of the flow behavior of fluids with non-Newtonian flow characteristics would seem to date to the research of James Clerk Maxwell [132] during the late 1860s. Clerk Maxwell pointed out that in an elastic material, the rate of increase in applied stress is proportional to the rate of increase in applied strain. In a viscoelastic material, the stress increases at a lesser rate than in the elastic case, and the difference might be conjectured to be proportional to the stress. Clerk Maxwell thus essentially wrote for the shear stress, σ,

$$\frac{d\sigma}{dt} = G\dot{\gamma} - C'\sigma = G\dot{\gamma} - \frac{1}{\lambda}\sigma = G\frac{d\gamma}{dt} - \frac{1}{\lambda}\sigma \qquad (23.32)$$

where G is a shear modulus, C' is a proportionality constant, and λ is the inverse of C', which is called a relaxation time. This one-dimensional expression is capable of explaining many well-known viscoelastic phenomena, such as stress relaxation following a step change in deformation and elastic recovery following a sudden release of a stressed object. Equation 23.32 predicts for steady shear flow a viscosity having a value λG, which is independent of the rate of shear.

Equation 23.32 may be recognized as a first-order ordinary nonhomogeneous differential equation, which may be solved using an integrating factor $e^{t/\tau}$. This leads to

$$G(t) = Ge^{-t/\lambda} \qquad (23.33)$$

It is traditional to express the relaxation modulus through a series of exponentials:

$$G(t) = \sum_i G_i e^{-t/\lambda_i} \qquad (23.34)$$

or through a continuous spectrum $H(\tau)$ as

$$G(t) = \int_0^\infty H(\tau) e^{-t/\lambda} \frac{d\lambda}{\lambda} \qquad (23.35)$$

Generally, the rheological behavior of polymers in the linear viscoelastic range is specified through a spectrum of relaxation times, $H(\tau)$.

23.4.2.1.2 Boltzmann Superposition Integral Model

In 1874, Ludwig Boltzmann [133] gave a three-dimensional formulation for the stresses developed in a solid with incomplete memory, which are subjected to a multistep strain history. Boltzmann's approach was to assume linearity and additivity of stress responses to the small imposed strains. He allowed for compressibility. In modern notation, Boltzmann wrote

$$\sigma = \lambda(\text{tr }\gamma)\mathbf{I} + 2G\gamma - 2\int_0^\infty [\Psi(t-s)\gamma(s) - Q(t-s)(\text{tr }\gamma(s))\mathbf{I}]\,ds \qquad (23.36)$$

where $\gamma(s)$ is the infinitesimal strain tensor, γ, λ, and G are Lame moduli, and $\Psi(t)$ and $Q(t)$ are relaxation functions.

The general one-dimensional linear theory may be derived by considering the stress to be equal to the sum of responses to a series of consecutive small strains. We presume that a one-dimensional step strain γ_0 gives a step rise in stress σ followed by relaxation through

$$\sigma(t) = G(t)\gamma_0 \tag{23.37}$$

where $G(t)$ is a relaxation modulus. A sequence of small steps at time $t_1, t_2, t_3,\ldots$ gives rise to the stress

$$\sigma(t) = \sum_i G(t-t_i)\gamma(t_i) = \int_{-\infty}^{t} G(t-s)\,d\gamma(s) = \int_{-\infty}^{t} G(t-s)\frac{d\gamma}{ds}\,ds \tag{23.38}$$

where the asymptote as the size of the strain steps becomes infinitesimally small. Equation 23.38 is called the *Boltzmann's superposition integral theory* [133]. The Boltzmann superposition principle is of extreme importance in the theory of linear viscoelasticity. The Boltzmann superposition principle states that the deformation of a polymer component is the sum or superposition of all strain results from various loads acting on the part at different times. This means that the response of a material to a specific load is independent of already existing loads. Hence, we can compute the deformation of a polymer specimen upon which several loads act at different points in time by simply adding all strain responses.

It is common following Boltzmann to express Equation 23.38 in terms of an integral over strain history rather than rate of deformation. This may be obtained from Equation 23.38 by integrating by parts to yield

$$\sigma(t) = G(0)\gamma(t) - G(\infty)\gamma(-\infty) - \int_{-\infty}^{t} \gamma(s)\,dG(t-s) \tag{23.39}$$

Noting that $G(\infty)$ is zero for fluids, Equation 23.39 may be expressed in the equivalent form:

$$\sigma(t) = G(0)\gamma(t) + \int_{0}^{\infty} \phi(z)\gamma(t-z)\,dz \tag{23.40}$$

where z is $(t-s)$ and

$$\phi(t) = -\frac{dG(t)}{dt} \tag{23.41}$$

For viscoelastic fluids, there is no preferred state, and strains are usually measured from the instantaneous configuration. Thus,

$$\gamma(t) = 0$$

$$\sigma(t) = \int_{0}^{\infty} \phi(z)\gamma(z)\,dz \tag{23.42}$$

where z is measured from time zero.

For smooth deformation, the strain $\gamma(z)$ may be written as a Taylor series in time z:

$$\gamma(z) = \sum \frac{z^n}{n!} \left(\frac{d^n\gamma}{dz^n}\right)_{z=0} \tag{23.43}$$

where $(d^n\gamma/dz^n)_0$ is the nth acceleration at time t. Substitution of Equation 23.43 into Equation 23.42 yields the linear differential equation as follows:

$$\sigma(t) = \sum_{n=1} a_n \left(\frac{d^n\gamma}{dz^n}\right)_0 (t) \tag{23.44}$$

where

$$a_n = \frac{1}{n!} \int_0^\infty z^n \phi(z)\, dz \tag{23.45}$$

23.4.2.1.3 Linear Viscoelasticity with Yield Value

Another class of rheological models of interest to industrial polymer fluids is a system with yield values. Clerk Maxwell's 1867 papers [132,134] stirred the imagination of the next generation of investigators concerned with the flow characteristics of materials with incomplete memory. The first of these investigators was the Russian Theodor Schwedoff [135], who was perhaps the first modern rheologist. In 1889–1900, he reported the steady shear flow and creep behavior of gelatin solutions. Schwedoff recognized that these systems had both memory and a viscosity, which decreases as a function of shear rate. Schwedoff expressed the response of these materials through a generalization of Maxwell's Equation 23.32:

$$\frac{d\sigma}{dt} = G\dot{\gamma} - \frac{1}{\tau}[\sigma - Y] \tag{23.46}$$

where Y is a yield value. Schwedoff was able to represent both his creep and steady shear flow data through Equation 23.46. He noted that this led to a shear stress–shear rate relationship:

$$\sigma = \left[\frac{Y}{\dot{\gamma}} + \tau G\right]\dot{\gamma} = Y + \tau G\dot{\gamma} \tag{23.47}$$

which indicated deviations from Newtonian flow due to the presence of the yield value, Y.

During the opening years of the last century, attention turned to the formulation of three-dimensional large-strain deformations for the Maxwell model, Equation 23.47. In a 1901 paper, L. Natanson [136] of the University of Krakow proposed a generalization in the form

$$\frac{D\boldsymbol{\sigma}}{Dt} = \lambda(\operatorname{tr}\mathbf{d})\mathbf{I} + 2G\mathbf{d} - \frac{1}{\tau}[\boldsymbol{\sigma} + p\mathbf{I}] \tag{23.48}$$

where

$$\frac{D\boldsymbol{\sigma}}{Dt} = \frac{\partial\boldsymbol{\sigma}}{\partial t} + (\mathbf{v}\cdot\nabla)\boldsymbol{\sigma} \tag{23.49}$$

and λ and G are equivalent to Lame's constants of the theory of elasticity.

In a paper published shortly afterward in the *Bulletin of the Academy of Krakow*, S. Zaremba [137] took exception to the formulation of Natanson largely because it did not represent the constitutive equation in a coordinate frame natural to the deforming medium. This, he argued, was a rigid coordinate system that not only translated but rotated with the fluid element. The stress components in the laboratory frame are the familiar σ_{ij}, but those from the basic constitutive law are $\pi_{\alpha\beta}$, which may be related to σ_{ij} by

$$\pi_{\alpha\beta}(t) = \sum_i \sum_j l_{\alpha i}(t) l_{\beta j}(t) \sigma_{ij}(t) \qquad (23.50)$$

The $l_{\alpha i}$ represent time-varying directional cosines relating the laboratory components to the translating and rotating coordinate system. If a simple partial time derivative of stress exists in the basic rheological model, it follows that

$$\frac{\partial \pi_{\alpha\beta}}{\partial t} = \sum_i \sum_j \left[l_{\alpha i} l_{\beta j} \frac{d\sigma_{ij}}{dt} + \left(\frac{d}{dt} l_{\alpha i}\right) l_{\beta j} \sigma_{ij} + l_{\alpha i}\left(\frac{d}{dt} l_{\beta j}\right)\right] \sigma_{ij} \qquad (23.51)$$

and

$$\frac{\partial \pi_{\alpha\beta}}{\partial t} = \sum_i \sum_j l_{\alpha i} l_{\beta j} \frac{\mathcal{D}\sigma_{ij}}{\mathcal{D}t} \qquad (23.52)$$

with

$$\frac{\mathcal{D}\sigma}{\mathcal{D}} = \frac{\mathcal{D}\sigma_{ij}}{\mathcal{D}t} + \sum_m \sigma_{im}\omega_{mj} + \sum_m \sigma_{mj}\omega_{im} \qquad (23.53)$$

where ω_{ij} is the vorticity tensor. A more complex stress derivative is thus necessary in the constitutive expression in a laboratory frame of reference. Zaremba also distinguished between distortions and volumetric relaxation writing finally

$$\frac{\mathcal{D}\sigma}{\mathcal{D}t} = \lambda(\text{tr }\mathbf{d})\mathbf{I} + 2G\mathbf{d} - \left[\frac{1}{\tau}\boldsymbol{\sigma} + \frac{1}{\tau_V}(\text{tr }\boldsymbol{\sigma})\mathbf{I}\right] \qquad (23.54)$$

where τ is a distortional relaxation time, and τ_V is a volumetric relaxation time. He applied this formulation to flow between coaxial cylinders, where he finds normal stresses arise.

Little attention was given to either Schwedoff or Zaremba by their contemporaries. A quarter century after each of their papers, their work was independently reproduced, and even a longer period passed before their views, then credited to others, were accepted. In 1913–1916, variable viscosity behavior was rediscovered [138]. Bingham [138] proposed models involving yield values. From this period, non-Newtonian viscosity was widely accepted and used to represent data. Elastic behavior was recognized only in a qualitative sense.

In 1929, Hencky [139], noting Natanson's paper, recognized its error and rederived an incompressible form of Equation 23.54. Four years later, Eisenschitz [140] recognized the validity of Hencky's theory and used it to interpret shear viscosity data on a cellulose nitrate solution. In his analysis, Eisenschitz derives expressions for normal stresses but does not recognize their significance.

Hohenemser and Prager [141] in 1932 presented the first three-dimensional theory of plastic materials with differentially viscous behavior above the yield value. In doing so, they introduced the

Material Selection, Design, and Application

use of invariants into the theory of non-Newtonian fluids. Their paper builds on von Mises theory of plastic yielding. They wrote

$$\sigma = \frac{1}{3}(\mathrm{tr}\,\sigma)\mathbf{I} + \mathbf{T}, \quad \mathbf{T}\left(1 - \frac{Y}{\sqrt{\frac{1}{2}\mathrm{tr}\,\mathbf{T}^2}}\right) = 2\eta_B \mathbf{d} \qquad (23.55)$$

where $\mathbf{T}$ is the deviatoric stress tensor and tr $\mathbf{T}^2$ its second invariant.

The simplest model would be one in which there is no deformation or a linear elastic response when the stresses are below a yield stress Y and the stresses are a constant Y above it, that is,

$$\sigma < Y \qquad (23.56)$$

$$\gamma = 0$$
$$\sigma = G_1 \gamma \qquad (23.57)$$

with γ in Equation 23.57 determined from the rest-state configuration, which must be redefined after each period of flow. When ρ reaches Y,

$$\sigma = Y, \quad \gamma = \text{indeterminate} \qquad (23.58)$$

Equations 23.56 and 23.57 are known as a *perfect plastic* or *rigid plastic* if Equation 23.56 is also valid. It is described as an *elastic plastic* if Equation 23.57 represents the behavior below the yield value.

In Bingham's theory of plastic flow, the material exhibits a viscous resistance to deformation above the yield value. This is expressed as

$$\sigma = Y + \eta_B \dot{\gamma} \qquad (23.59)$$

This defines the *Bingham plastic*. We may refer to this behavior as that of a viscous plastic. This may be further classified as a rigid viscous plastic or elastic viscous plastic, depending upon whether Equation 23.56 or 23.57 is valid below the yield value.

It is possible to generalize this formulation to include viscoelastic response above the yield value. This was first devised by Schwedoff [135] and may be considered a differential equation in which σ may be solved to give

$$\frac{d\sigma}{dt} = G\dot{\gamma} - \frac{1}{\lambda}(\sigma - Y) \qquad (23.60)$$

$$\sigma = Y + \int_{-\infty}^{t} Ge^{-(t-s)/\lambda} \gamma(s)\,ds \qquad (23.61)$$

This of course may be considered a special case of a more general expression:

$$\sigma = Y + \int_{-\infty}^{t} \phi(t-s)\gamma(s)\,ds \qquad (23.62)$$

We may classify it as rigid viscoelastic plastic and elastic viscoelastic plastic according to whether Equation 23.56 or 23.57 represents the behavior below the yield value.

23.4.2.1.4 Mechanical Models

We may present the various rheological theories described in Sections 23.4.1 and 23.4.2.1 using mechanical analogs. Three elements are included in these models: (1) a spring, representing elastic behavior; (2) a dashpot, representing viscous behavior; and (3) a slider, expressing plastic yielding. These elements are shown in Figure 23.23. The displacements in these models represent strain, and the tension deforming them represents stress.

We will only demonstrate the mechanical models representing the viscoelastic behavior in the following and recommend Refs. [103–110] to readers who are interested in the viscoplastic models.

The total strain, γ, in the model shown in Figure 23.24a has an elastic, γ_e, and a viscous, γ_v, strain contribution and can be represented as follows:

$$\gamma = \gamma_e + \gamma_v \tag{23.63}$$

Similarly, the strain rates are written as

$$\dot{\gamma} = \dot{\gamma}_e + \dot{\gamma}_v \tag{23.64}$$

Assuming that the spring follows Hooke's law, the following relation holds:

$$\dot{\gamma}_e = \frac{\dot{\sigma}}{G} \tag{23.65}$$

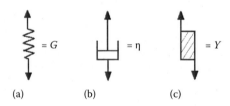

FIGURE 23.23 Mechanical models representing rheological behavior: (a) spring—linear elastic solid, (b) dashpot—linearly viscous fluid, and (c) slider—rigid perfect plastic.

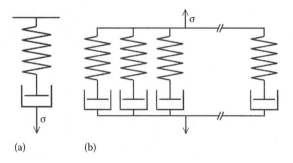

FIGURE 23.24 Mechanical models representing viscoelastic behavior: (a) Maxwell model and (b) generalized Maxwell or Wiechert model.

The viscous portion, represented by the dash pot, is written as follows:

$$\dot{\gamma}_v = \frac{\sigma}{\eta} \tag{23.66}$$

Combining Equations 23.65 and 23.66 results in

$$\dot{\gamma} = \frac{\dot{\sigma}}{G} + \frac{\sigma}{\eta} \text{ or } \frac{d\gamma}{dt} = \frac{1}{G}\frac{d\sigma}{dt} + \frac{\sigma}{\eta} \tag{23.67}$$

This is equivalent to Maxwell's equation. A series of parallel spring-dashpot models as shown in Figure 23.24b with coefficients G_i, η_i is equivalent to

$$\sigma = \sum \sigma_i$$

$$\frac{d\gamma}{dt} = \frac{1}{G_i}\frac{d\sigma_i}{dt} + \frac{\sigma_i}{\eta_i} \tag{23.68}$$

Solving these equations leads to

$$\sigma = \int_{-\infty}^{t} \sum_i G_i e^{-(t-s)/\lambda} \gamma \, ds \tag{23.69}$$

Equation 23.69 is often referred to as a *generalized Maxwell model*, which is essentially Boltzmann's superposition principle.

23.4.2.1.5 Measurement of Linear Viscoelasticity

A number of small strain experiments are used in rheology. Some of the more common techniques are stress relaxation, creep, and sinusoidal oscillations. Different experimental methods are used because they may be more convenient or better suited for a particular material or because they provide data over a particular time range. Furthermore, it is often not easy to transform results from one type of linear viscoelastic experiment to another. For example, transformation from the creep compliance $J(t)$ to the stress relaxation modulus $G(t)$ is generally difficult. Thus, both functions are often measured. In this section, we describe each linear viscoelastic experimental method.

In a stress relaxation test, a polymer test specimen is deformed a fixed amount, γ_0, and the stress required to hold that amount of deformation is recorded over time. This test is very cumbersome to perform, so the design engineer and the material scientist have tended to ignore it. In fact, several years ago, the standard relaxation test ASTM D2991 was dropped by ASTM. Rheologists and scientists, however, have been consistently using the stress relaxation test to interpret the viscoelastic behavior of polymers.

Creep experiments are particularly useful for the long time end of the relaxation spectra. The creep test, which can be performed either in shear, compression, or tension, measures the flow of a polymer component under a constant load. It is a common test that measures the strain, γ, as a function of stress, time, and temperature. Standard creep tests such as DIN 53444 and ASTM D2990 can be used.

In another important small strain experiment, the sample is deformed sinusoidally. In the sinusoidal oscillatory test, a specimen is excited with a low-frequency stress input, which is recorded along with the strain response. The shapes of the test specimen and the testing procedure vary significantly from test to test. The various tests and their corresponding specimens are described by ASTM D4065, and the terminology is described by ASTM D4092.

23.4.2.2 Nonlinear Viscoelasticity

Following Maxwell [132], Natanson [136], and Zaremba [137], we may express the constitutive equation for a viscoelastic fluid as a differential equation for the stress. Alternatively, following Boltzmann [133], the stress may be expressed in terms of integrals over the deformation history. Numerous numbers of nonlinear viscoelastic models have been developed and will be briefly introduced in the following.

23.4.2.2.1 Maxwell-Type Simple Differential Constitutive Models

The rheologists of the 1930s and early 1940s gave little attention to the papers by Hencky, Eisenschitz, and Hohenemser and Prager and used one-dimensional flow models. It is only with the papers of Reiner [142], Oldroyd [143–145], and Rivlin [146–148] in the period 1945–1950 that three-dimensional rheological theories began to be accepted. Reiner [142], seeking to explain the observation of dilation during a shear flow, used the Cayley–Hamilton theorem to derive the expression

$$\sigma_0 = a_0 \mathbf{I} + 2\eta \mathbf{d} + 4\eta_C \mathbf{d}^2 \quad (23.70)$$

where a_0, η, and η_C depend upon tr $\mathbf{d}$, tr $\mathbf{d}^2$, and tr $\mathbf{d}^3$. Here, η is the shear viscosity and η_C came to be called the cross viscosity. Reiner showed that η_C leads to normal stresses in shear flow. Rivlin [146] argued that the existence of a cross viscosity explained Weissenberg's [149] observations of normal stress effects.

Oldroyd's development followed a different route. He first independently derived [143] and generalized [144] Hohenemser and Prager's theory of Bingham plastics. In a later paper, Oldroyd [145], unaware of Zaremba's or Hencky's papers, proposed a formulation of the nonlinear viscoelastic fluid based upon a coordinate system embedded in and deforming with the medium. He called this a *convected coordinate system*. Oldroyd's paper was based on his intuition that the normal stresses observed by Weissenberg were due to the viscoelastic properties of the fluids involved. He saw that these must come from the stress derivatives. This necessitates careful consideration of the basic formulation of three-dimensional viscoelastic fluid models and leads to the idea of special embedded coordinate systems. In any case, Oldroyd's paper inspired the development of the theory of nonlinear viscoelasticity by various researchers during the next 10–20 years. Oldroyd [145] himself developed expressions based on generalizing differential models similar to Equation 23.7. He begins with models of the form

$$\sigma + \tau_1 \frac{d\sigma}{dt} = 2\eta \left[\frac{d\gamma}{dt} + \tau_2 \frac{d^2\gamma}{dt^2} \right] \quad (23.71)$$

which generalize to

$$\sigma = -p\mathbf{I} + \mathbf{P} \quad (23.72)$$

$$\mathbf{P} + \tau_1 \frac{\delta \mathbf{P}}{\delta t} = 2\eta \left[\mathbf{d} + \tau_2 \frac{\delta \mathbf{d}}{\delta t} \right] \quad (23.73)$$

where τ_1 and τ_2 have units of time and η has the unit of viscosity. The form of the derivatives was found not to be unique but depended upon whether a covariant or contravariant formulation was used. The former was called an A fluid and the latter a B fluid. Oldroyd [150] noted that as the form of Equation 23.32 is linear, any combination of second- and higher-order products of $\mathbf{P}$ and $\mathbf{d}$ could be introduced.

More general perspectives of viscoelastic fluid behavior were developed in succeeding years under the inspiration of the papers of Oldroyd. This may be seen in the papers of Rivlin [151–153], Criminale et al. [154], Coleman and Noll [155,156], and others. The stress tensor was expressed as a general function of the rates and accelerations of deformation or as a functional of deformation history. The results obtained were highly complex. This led other investigators [157–167] during the next two decades to develop a range of constitutive equations for viscoelastic fluids, which were simpler in character and could be used to correlate experimental data for polymer solutions and melts. While Zapas [168,169] made early quantitative studies on elastomers, extensive quantitative comparisons of experiments on polymer melts and theory are only found in the 1970s [170–173].

23.4.2.2.2 Lodge Integral Constitutive Model

An alternative approach is suggested by the work of Lodge [174]. The theory of a network of flexible polymer chains, that is, the kinetic theory of rubber elasticity [175], has a constitutive equation equivalent to

$$\sigma_{ij} = -p\delta_{ij} + vkTc_{ij}^{-1} \qquad (23.74)$$

where v is the cross-link density, k is Boltzmann's constant, and T is the absolute temperature. Equation 23.74, together with Equations 23.32 and 23.33, suggests that the starting point for nonlinear viscoelastic models should be a generalization of the Maxwell model, as suggested by White and Metzner [166] to be

$$\pi^{\alpha\beta} = -p\gamma^{\alpha\beta} + \pi'^{\alpha\beta}, \quad \frac{d\pi'^{\alpha\beta}}{dt} = -G\frac{d\gamma^{\alpha\beta}}{dt} - \frac{1}{\lambda}\pi'^{\alpha\beta} \qquad (23.75)$$

and the Boltzmann superposition model, as suggested by Lodge [174] to be

$$\pi^{\alpha\beta} = -p\gamma^{\alpha\beta} + \int_{-\infty}^{t} \phi(t-s)\gamma^{\alpha\beta}(s)\,ds \qquad (23.76)$$

Equations 23.75 and 23.76 form the starting point for the most useful simple nonlinear viscoelastic tensor differential and integral models.

23.4.2.2.3 Maxwell-Type Differential Constitutive Models

We may readily convert the differential Maxwell model of Equation 23.75 to a fixed-space coordinate system. The Cartesian fixed-space coordinates may be written as

$$\mathbf{P}_{ij} = 2\lambda G d_{ij} - \lambda \frac{\delta \mathbf{P}_{ij}}{\delta t} \qquad (23.77)$$

with

$$\frac{\delta \mathbf{P}_{ij}}{\delta t} = \frac{D\mathbf{P}_{ij}}{Dt} - \frac{\partial v_i}{\partial x_m}\mathbf{P}_{mj} - \frac{\partial v_j}{\partial x_m}\mathbf{P}_{im} \qquad (23.78)$$

This is the form given by White and Metzner [166]. Here, $\delta \mathbf{P}_{ij}/\delta t$ represents a convected derivative of the stress $\mathbf{P}_{ij}$. Other differential formulations have appeared in the literature. Notable are Oldroyd's formulations [145,150], and those of Giesekus [176], and Keunings and Crochet [177].

Various theories have been devised in recent years, which may be expressed in this form. Many of these follow arguments along the lines of rewriting Equations 23.77 and 23.78 as

$$\lambda\left[\frac{D\mathbf{P}}{Dt} - \mathbf{L}\cdot\mathbf{P} - \mathbf{P}\cdot\mathbf{L}\right] + \mathbf{P} = 2\eta\mathbf{d} \qquad (23.79)$$

where **L** is an effective velocity gradient associated with nonaffine effects. The theories of Johnson and Segalman [162] and Phan-Tien and Tanner [178,179] may be represented in this manner. Another formulation is the model of Leonov [180]. Both Phan-Tien [178] and Leonov [180] models are capable of predicting a non-Newtonian shear viscosity, a principal normal stress difference, a negative second normal stress difference (associated with κ), and an elongational viscosity function exhibiting a maximum.

23.4.2.2.4 Lodge-Type Integral Constitutive Models

Lodge's integral model, Equation 23.76, has served as a basis for a large literature on nonlinear viscoelastic integral models. This expression may be converted to a fixed-space coordinate system in the same manner as the differential models.

$$\boldsymbol{\sigma} = -p\mathbf{I} + \int_{-\infty}^{t} \phi(t-s)\mathbf{c}^{-1}\,ds \tag{23.80}$$

Various procedures have been proposed to generalize Equation 23.80 so that it can predict a broader range of responses. Notable are Spriggs et al. [181], Bogue [160], Bird and Carreau [159], Bogue and White [182], and Carreau [161].

A second approach to introducing nonlinearities into $\phi(t)$ is through the use of strain invariants. Such an approach was first used by Bernstein et al. [157] and later expanded upon by Zapas [168,169], White and Tokita [183], and Wagner [165,173].

23.4.2.2.5 General Constitutive Models

A more general approach has been taken by Green and Rivlin [151], Noll and Coleman [155,156,184], and others, which involves considering the stress tensor to be an arbitrary hereditary functional of the deformation history. In a hereditary functional, events in the recent past are more important than those in the distant past. They express this in terms of the Cauchy deformation tensor, which in our notation is equivalent to a covariant formation in the convected coordinate system, that is,

$$\pi_{\alpha\beta} = -p\gamma_{\alpha\beta} + F_{\alpha\beta}\left[\gamma_{\alpha\beta} \underset{0}{\overset{\infty}{(z)}}\right] \tag{23.81}$$

The alternate formulation of White [185,186] uses a Finger deformation measure that is equivalent to

$$\pi^{\alpha\beta} = -p\gamma^{\alpha\beta} + G^{\alpha\beta}\left[\gamma^{\alpha\beta} \underset{0}{\overset{\infty}{(z)}}\right] \tag{23.82}$$

This is expected to lead to a less cumbersome formulation because of the relationship of this form to Equations 23.74 and 23.76.

If we presume the stress to be a hereditary functional of the deformation history, we obtain Equations 23.81 and 23.82, depending upon the deformation measure used. Transforming these expressions to a fixed-space coordinate frame, we obtain

$$\boldsymbol{\sigma} = -p\mathbf{I} + \mathbf{F}\left[\mathbf{c} \underset{0}{\overset{\infty}{(z)}}\right] \tag{23.83}$$

Material Selection, Design, and Application

$$\sigma = -p\mathbf{I} + \mathbf{G}\left[\mathbf{c}^{-1}\,(z)\Big|_0^\infty\right] \tag{23.84}$$

There are actually a finite number of equivalent representations of the influence of deformation history on the stress field. Green and Rivlin [151] and Noll and Coleman [155,156,184] used the formulation of Equation 23.83 based upon the Cauchy deformation tensor. White [185,186] applied the formulation of Equation 23.84 because of its close relationship to Equation 23.74. White and Tokita [183] have noted that all the higher-order kernels in the integrals, such as $\psi_1(z_1, z_2)$ and $\psi_2(z_1, z_2)$, contain only a single memory function. Kaye-Bernstein, Kearsley, and Zapas (K-BKZ) and Wagner constitutive equations are of this form.

23.4.2.2.6 Second-Order Fluid Constitutive Models

The simplest constitutive model capable of predicting a first normal stress difference is the second-order fluid constitutive model. Equations 23.43 and 23.44 indicate how Boltzmann's superposition integral may be simplified for smooth flows. A similar development is possible for three-dimensional nonlinear viscoelastic fluids. Coleman and Noll [155] developed expressions for the second-order fluid as follows:

$$\sigma = -p\mathbf{I} + a_1\mathbf{B}_1 + a_2\mathbf{B}_2 + a_3\mathbf{B}_1^2 \tag{23.85}$$

The second-order fluid is the simplest of viscoelastic fluid models, and it is tempting to apply it widely. This should, however, be done only with the greatest caution. The presumption of Equation 23.85 is very severe. Viscoelastic materials respond to sudden impulses through propagation of shock waves, but such responses are not contained in Equation 23.85. It is not possible to apply Equation 23.85 to such problems.

In summary, various nonlinear viscoelastic flow models in both differential type and the integral type are summarized in Table 23.12.

TABLE 23.12
Lists of Various Nonlinear Viscoelastic Theories

Model Type	Model	References
Maxwell-type differential constitutive models	Johnson and Segalman (1977)	[162]
	White and Metzner (1963, 1977)	[166]
	Giesekus (1966, 1982)	[176]
	Leonov (1976)	[180]
	Phan Thien and Tanner (1977, 1978)	[179]
Lodge-type integral model	Sprigg et al. (1966)	[181]
	Bogue and White (1970)	[182]
	Doi and Edwards (1978)	[187]
	Papanastasiou et al. (1983)	[188]
	Wagner and Demarmels (1990)	[189]
More generalized models	Green and Rivlin (1957)	[151]
	Noll and Coleman (1960)	[155]
	White and Tokita (1967)	[183]
	Kaye (1962)–Bernstein, Kearsley, and Zapas (1963) (K-BKZ)	[157]
	Wagner et al. (1979)	[173]

23.4.2.3 Example: Effect of Viscoelastic Property on Material Selection

The effect of the viscoelastic property of a polymer melt on the material selection will be demonstrated using coextrusion processes, which have been widely used to produce multilayer sheet, blown film, tubing, wire coating, and profiles [190–195]. Feedblock coextrusion has some drawbacks since the rheology of the materials being joined together in the feedblock must be similar in order to flow uniformly through the die manifold. The resin chosen for this study was a high-impact PS resin. The rheological properties of this resin have been described previously [196]. This resin was chosen because it has exhibited significant viscoelastic properties in previous studies. The effect of the viscoelastic property on various coextruded profiles is described next using both the simulation study using a viscoelastic constitutive equation and the experimental study.

The numerical simulations of the flow of the PS resin in this study were performed using a Giesekus model with five relaxation times. Figure 23.25a shows the secondary flow patterns predicted by the simulation for flow in the square channel. Because of the symmetry of the channel shape, there are two secondary flow patterns in each quadrant resulting in a total of eight secondary flow zones. One common misconception is that the secondary flows observed in the square and triangular channels are due to the sharp corners present in these geometries. In actuality, the secondary flows are caused by second normal stress differences produced in nonradially symmetric geometries. This implies that flow through a nonradially symmetric geometry with no corners, such as an oval, would still produce secondary flows. This result is illustrated in Figure 23.25b showing the predicted secondary flow patterns in an oval channel. Note that there are only four secondary flow regions in this geometry. Figure 23.25c shows the simulation results for a rectangular channel with a 4:1 aspect ratio. These results are similar to the results from the square channel since there are eight secondary flow areas (two per quadrant). The results are different from the square channel, however, since the strength and location of the secondary flow areas have changed. Note that the secondary flow areas near the short channel wall are much smaller than those near the long channel wall. Figure 23.25d shows the simulation results for a rectangular channel in which one end has been rounded into a semicircular shape. These results show that rounding one end of the rectangular channel has eliminated two of the secondary flow areas, reducing the total number from eight to six.

Samples were produced using feedblocks with two differently pigmented layers combined with the square die channel while extruding the PS resin as shown in Figure 23.26. Figure 23.26a1 and a2 shows the feedback designs to produce 27- and 165-layer, respectively, in the square channel. Figure 23.26b1 and c1 shows cross-sectional cuts of this sample taken near the entry and the exit of this channel, respectively. Note that the interface between the black and white layers of the same PS material is fairly flat near the entry of the channel but is distorted significantly near the end of the channel. This phenomenon has been reported previously [196].

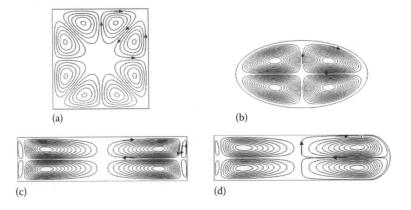

FIGURE 23.25 Flow simulation results for various channel geometries: (a) square channel, (b) oval channel, (c) rectangular channel, and (d) rounded rectangular channel.

Material Selection, Design, and Application

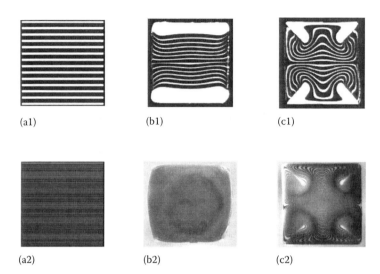

FIGURE 23.26 (a1, a2) Feedblock geometries with 27 and 165 layers, and interface location near the entry (b1, b2) and near the exit (c1, c2) of a square channel using feedblocks with two differently pigmented layers in the square die.

The shape of the final interface near the exit of the square channel shows a good correlation with the secondary flow patterns shown for a square channel in Figure 23.25a. These flow patterns would drive the material along the channel walls toward the corners and then back toward the center of the channel along the diagonal, similar to what is seen in Figure 23.26c1 and c2. Figure 23.26b1 and c1 shows the results from an experiment using the 27-layer feedblock. This experiment gives very similar results but with additional information due to the increased number of layers. Note that the central axis of symmetry is clearly visible in these results. Figure 23.26b2 and c2 shows the results from an experiment using the 165-layer feedblock. This experiment gives very similar results but with even more information due to the increased number of layers.

The experimental results of processing the PS resin through an oval channel using a 27-layer feedblock are shown in Figure 23.27a1 and b1. This increase in the number of horizontal layers produces a much better definition of the movement of the polymer in this geometry caused by the secondary flow patterns. Note how the layers are compressed horizontally along the major axis while being stretched vertically along the minor axis. Even more definition of the flow patterns in an oval

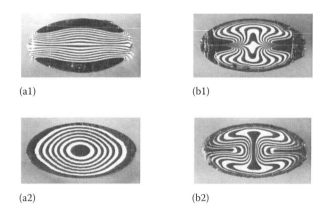

FIGURE 23.27 Interface location near the entry (a1, a2) and near the exit (b1, b2) of an oval channel using a 27-layer feedblock and a 13-concentric ring feedblock.

geometry can be realized by using the 13-concentric ring feedblock, as is shown in Figure 23.27a2 and b2. This figure shows the deformation of the ringed structure along the major and minor axes and produces even more details of the flow field in this channel compared to the other feedblocks.

Figure 23.28a1 shows the results from an experiment using the 13-ring feedblock. This experiment gives very similar results but with additional information due to the increased number of layers. Note that the symmetry of the flow field is clearly visible in these results. Figure 23.28a2 shows the results from an experiment using the 49-strand square feedblock combined with the square channel. These experimental results are very informative since the individual strands from the die entry can be followed to their final positions near the exit of the die. This type of experiment not only shows the final location of the strand but also gives some indication of the deformation it has undergone. Note that very little deformation has occurred at the center of the channel, as expected.

The experimental results of processing the PS resin through a rectangular channel using a 13-ring and a 27-layer feedblock are shown in Figure 23.29. This figure shows that the uniform white layer near the beginning of the channel gets distorted along the flow paths predicted by the simulation results shown in Figure 23.25c. Note that the layers near the short channel walls are following the flow patterns of the smaller secondary flows predicted in Figure 23.25d. Figure 23.29 shows the layer deformations in the rectangular channel using the 13-concentric ring feedblock and the 27-planar layer feedblock, respectively. In Figure 23.29a1 and b1 using the 13-concentric ring feedblock,

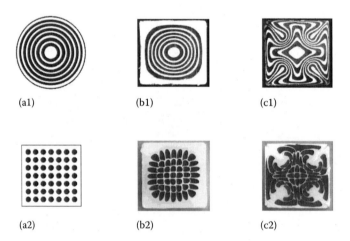

FIGURE 23.28 Interface location near the entry (b1, b2) and near the exit (c1, c2) of a square channel using a 13-ring feedblock and a 49-strand feedblock (a1, a2).

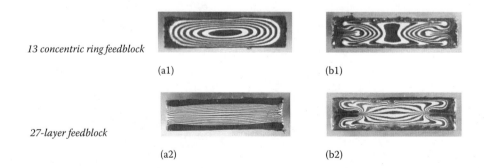

FIGURE 23.29 Interface location near the entry (a1, a2) and near the exit (b1, b2) of a rectangular channel using a two-layer feedblock.

Material Selection, Design, and Application 799

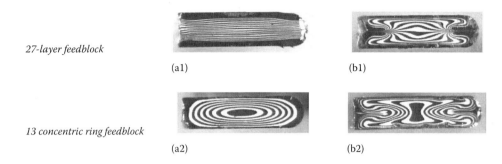

27-layer feedblock (a1) (b1)

13 concentric ring feedblock (a2) (b2)

FIGURE 23.30 Interface locations near the entry (a1, a2) and near the exit (b1, b2) of a rounded rectangular channel using a 27-planar layer feedblock and a 13-concentric ring feedblock.

the sample from near the end of the channel shows the vertical stretching of the center rings and the horizontal compression of the outer rings particularly well. These are the major deformations caused by the larger secondary flows. The smaller secondary flows near the shorter channel walls are more visible in the sample taken from near the exit of the channel using the 27-planar layer feedblock shown in Figure 23.29. This is because both the black and white layers from the 27-planar layer feedblock initially extend all the way to the short walls, but only the outside black layer from the 13-concentric ring feedblock touches the short walls.

The experimental results of processing the PS resin through the modified rectangular channel with one semicircular end in combination with the various feedblock geometries are shown in Figure 23.30. The results of the 27-planar layer feedblock are shown in Figure 23.30a1 and b1. This figure shows the major deformations in the layers caused by the larger secondary flows, which are similar to those seen in Figure 23.29, even though the geometry at one end has been altered to a semicircle. This trend also holds for the sample shown in Figure 23.30a2 and b2 produced using the 13-concentric ring feedblock when it is compared to Figure 23.29a2 and b2.

These results show how important it is to consider both the viscous and elastic effects when coextruding polymeric materials. As the differences in viscosities between layers are minimized in a coextruded structure in order to minimize viscous encapsulation effects, elastic rearrangement effects become more important in determining the final interface deformation. Appropriate resin selection, equipment design, and processing conditions are required to produce structures with good layer uniformity.

23.5 SUMMARY AND CONCLUSIONS

New product design is an iterative process. The starting point is a market need captured in a set of design requirements. Concepts for products that meet the need are devised. If initial estimates and exploration of alternatives suggest that the concept is viable, the design proceeds to the embodiment stage: working principles are selected, size and layout are decided, and initial estimates of performance and cost are made. If the outcome is successful, the designer proceeds to the detailed design stage: optimization of performance, full analysis of critical components, preparation of detailed production drawings (usually as a CAD file), specification of tolerance, precision, joining, and finishing methods, and so forth.

Material selection enters at each stage but at different levels of breadth and precision. At the conceptual stage, all materials and processes are potential candidates, requiring a procedure that allows rapid access to data for a wide range of each, though without the need for great precision. The preliminary selection passes to the embodiment stage, the calculations and optimizations of which require information at a higher level of precision and detail. They eliminate all but a small short list of candidate materials and processes for the final, detailed stage of the design. For these few, data of the highest quality are necessary.

Data exist at all these levels. Each level requires its own data-management scheme. The management is the skill: it must be design-led, yet must recognize the richness of choice and embrace the complex interaction among the material, its shape, the process by which it is given that shape, and the function it is required to perform. And it must allow rapid iteration—back-looping when a particular chain of reasoning proves to be unprofitable. Tools now exist to help with all of this.

Material selection is tackled in five steps as follows:

1. *Translation*: reinterpreting the design requirements in terms of function, constraints, objectives, and free variables
2. *Screening*: deriving attribute limits from the constraints and applying these to isolate a subset of viable materials
3. *Ranking*: ordering the viable candidates by the value of a material index, the criterion of excellence that maximizes or minimizes some measure of performance
4. *Seeking*: supporting information for the top-ranked candidates, exploring aspects of their past history, their established uses, their behavior in relevant environments, their availability, and more until a sufficiently detailed picture is built up that a final choice can be made
5. *Testing*: measuring the critical properties using a piece of prototype testing equipment using the test standards such as ASTM and ISO

The final choice among these will depend on more detailed information on their properties, considerations of manufacture, economics, and aesthetics.

REFERENCES

1. R.K. Evans and P.R. Hartley, *Materials Selector and Design Guide*, Morgan-Grampian, London, 1974.
2. N.A. Waterman and M.F. Ashby (Eds.), *The Materials Selector*, 2nd Edition, Chapman and Hall, London, 1997.
3. J.A. Jacobs and T.F. Kilduff, *Engineering Materials Technology: Structures, Processing, Properties, and Selection*, 5th Edition, Prentice Hall, Upper Saddle River, NJ, 2004.
4. K.G. Budinski and M.K. Budinski, *Engineering Materials: Properties and Selection*, 9th Edition, Prentice Hall, Upper Saddle River, NJ, 2009.
5. M.F. Ashby, *Materials Selection in Mechanical Design*, 4th Edition, Elsevier, Amsterdam, 2010.
6. R.B. Seymour, *Polymers for Engineering Applications*, ASM International, Metals Park, OH, 1987.
7. J. Murphy (Ed.), *New Horizons in Plastics, a Handbook for Design Engineers*, WEKA Publishing, London, 1990.
8. D.V. Rosato, D.V. Rosato, and M.V. Rosato, *Plastic Product Material and Process Selection Handbook*, Elsevier, New York, 2004.
9. E.A. Campo, *Selection of Polymeric Materials: How to Select Design Properties from Different Standards*, William Andrew, Norwich, NY, 2008.
10. M. Kutz (Ed.), *Applied Plastics Engineering Handbook: Processing and Materials*, Elsevier, New York, 2011.
11. W.D. Baasel, *Preliminary Chemical Engineering Plant Design*, 2nd Edition, Van Nostrand Reinhold, New York, 1990.
12. D.G. Newnan, *Engineering Economic Analysis*, 4th Edition, Engineering Press, San Jose, CA, 1991.
13. M.L. Bauccio (Ed.), *ASM Engineered Materials Reference Book*, 2nd Edition, ASM International, Metals Park, OH, 1994.
14. N.A. Waterman and M.F. Ashby (Eds.), *The Materials Selector*, 2nd Edition, Chapman and Hall, London, 1997.
15. P.P. Liu and L.D. Helsel, *Industrial Materials*, 2nd Edition, Goodheart-Willcox Publisher, Tinley Park, IL, 2007.
16. G.S. Brady, H.R. Clauser, and J.A. Vaccari (Eds.), *Materials Handbook*, 2nd Edition, McGraw-Hill, New York, 1986.

17. A. Waterman, T.E. Hirschorn, and H.J. Goldsmith, *Handbook of Thermophysical Properties of Solid Materials*, Macmillan, New York, 1961.
18. J.C. Bittence (Ed.), *Guide to Engineering Materials Producers*, ASM International, Metals Park, OH, 1994.
19. H. Saechtling (Ed.), *International Plastics Handbook*, Macmillan, London, 1983.
20. R.B. Seymour, *Polymers for Engineering Applications*, ASM International, Metals Park, OH, 1987.
21. J. Murphy (Ed.), *New Horizons in Plastics, a Handbook for Design Engineers*, WEKA Publishing, London, 1990.
22. C. Dostal (Ed.), *ASM Engineered Materials Handbook, Vol. 2: Engineering Plastics*, ASM International, Metals Park, OH, 1988.
23. C.A. Harper (Ed.), *Handbook of Plastics and Elastomers*, McGraw-Hill, New York, 1975.
24. D.M. Rady (Ed.), *Plastics: Thermoplastics and Thermosets—A Desk-Top Data Bank*, 8th Edition, International Plastics Selector, San Diego, CA, 1986.
25. H. Domininghaus, *Plastics for Engineers: Materials, Properties, Applications*, Hanser, Munich, 1992.
26. D.W. van Krevelen, *Properties of Polymers*, 3rd Edition, Elsevier, Amsterdam, 1990.
27. A.K. Bhowmick and H.L. Stephens (Eds.), *Handbook of Elastomers*, Marcel Dekker, New York, 1988.
28. T.A. Osswald, E. Baur, S. Brinkmann, K. Oberbach, and E. Schmachtenberg, *International Plastic Handbook: Resource for Plastic Engineers*, 4th Edition, Hanser, Munich, 2006.
29. R.J. Young and P.A. Lovell, *Introduction to Polymers*, 3rd Edition, CRC Press, Boca Raton, FL, 2011.
30. P.J. Flory, *Principles of Polymer Chemistry*, Cornell University Press, Ithaca, NY, 1953.
31. H.R. Allcock, E.W. Lampe, and J.E. Mark, *Contemporary Polymer Chemistry*, 3rd Edition, Prentice-Hall, Englewood Cliffs, NJ, 1990.
32. P. Hiemenz, *Polymer Chemistry: The Basic Concepts*, Marcel Dekker, New York, 1984.
33. F.W. Billmeyer Jr., *Textbook of Polymer Science*, John Wiley & Sons, New York, 1984.
34. L.H. Sperling, *Introduction to Physical Polymer Science*, 4th Edition, John Wiley & Sons, New York, 2005.
35. J.M.G. Cowie, *Polymers: Chemistry and Physics of Modern Materials*, International Textbook Company, Aylesbury, 1973.
36. M. Rubinstein and R.H. Colby, *Polymer Physics*, Oxford University Press, Oxford, 2003.
37. J.L. Halary, F. Laupretre, and L. Monnerie, *Polymer Materials*, John Wiley & Sons, New York, 2011.
38. G.R. Strobl, *The Physics of Polymers: Concepts for Understanding Their Structure and Behavior*, Springer, Berlin, 2007.
39. R.B. Seymour and C.E. Carraher, *Structure-Property Relationships in Polymers*, Plenum Press, New York, 1984.
40. A. Hiltner (Ed.), *Structure-Property Relationships of Polymeric Solids*, Plenum Press, New York, 1981.
41. G.M. Swallowe, *Mechanical Properties and Testing of Polymers*, Springer Science & Business Media, Dordrecht, 1999.
42. H. Domininghaus, *Plastics for Engineers: Materials, Properties, Applications*, Hanser, Munich, 1993.
43. R. Brown, *Handbook of Polymer Testing: Physical Methods*, Marcel Dekker, New York, 1999.
44. W. Grellmann and S. Seidler, *Polymer Testing*, Hanser Gardner Publications, Cincinnati, OH, 2007.
45. G. Wypych, *Handbook of Polymers*, ChemTec Publishing, Toronto, Canada, 2011.
46. D.W. van Krevelen, *Properties of Polymers*, Elsevier, Amsterdam, 1990.
47. L.H. Sperling, *Introduction to Physical Polymer Science*, 3rd Edition, John Wiley & Sons, New York, 2001.
48. I.M. Ward and J. Sweeney, *Mechanical Properties of Solid Polymers*, 3rd Edition, John Wiley & Sons, New York, 2012.
49. R.F. Landel and L.E. Nielsen, *Mechanical Properties of Polymers and Composites*, 2nd Edition, Marcel Dekker, New York, 1993.
50. J.E. Mark, *Physical Properties of Polymer Handbook*, Springer, Berlin, 2007.
51. E.A. Campo, *Selection of Polymeric Materials*, William Andrew, Norwich, NY, 2008.
52. J. Mark, K. Ngai, W. Graessley, L. Mandelkern, E. Samulski, J. Koenig, and G. Wignall, *Physical Properties of Polymers*, 3rd Edition, Cambridge University Press, Cambridge, 2004.
53. J.E. Mark, *Physical Properties of Polymer Handbook*, Springer, Berlin, 2007.
54. M.A. White, *Physical Properties of Materials*, 2nd Edition, CRC Press, Boca Raton, FL, 2012.
55. A.A. Askadskii, *Physical Properties of Polymers: Prediction and Control*, Gordon and Breach Publisher, Amsterdam, 1996.
56. L.A. Belfiore, *Physical Properties of Macromolecules*, John Wiley & Sons, New York, 2010.
57. Y.K. Godovsky, *Thermophysical Properties of Polymers*, Springer-Verlag, Berlin, 1992.

58. T. Hatakeyama and F.X. Quinn, *Thermal Analysis: Fundamentals and Applications to Polymer Science*, John Wiley & Sons, New York, 1999.
59. L.W. Mckeen, *Effect of Temperature and Other Factors on Plastics and Elastomers*, 2nd Edition, William Andrew, Norwich, NY, 2008.
60. G.W. Ehrenstein, G. Riedel, and P. Trawiel, *Thermal Analysis of Plastics: Theory and Practice*, Hanser, Munich, 2004.
61. G. Hartwig, *Polymer Properties at Room and Cryogenic Temperatures*, Plenum Press, New York, 1994.
62. E.A. Campo, *Selection of Polymeric Materials*, William Andrew, Norwich, NY, 2008.
63. C.J. Hilado, *Flammability Handbook for Plastics*, 5th Edition, CRC Press, Boca Raton, FL, 1998.
64. R. Bartnikas, *Electrical Properties of Solid Insulating Materials Measurement Techniques*, Vol. 2, ASTM International, West Conshohocken, PA, 1987.
65. E. Lokensgard, *Industrial Plastics: Theory and Applications*, Cengage Learning, New York, 2008.
66. M. Chanda and S.K. Roy, *Plastics Fundamentals, Properties, and Testing*, CRC Press, New York, 2010.
67. ASM International, *Characterization and Failure Analysis of Plastics*, ASM International, Metals Park, OH, 2003.
68. M. Kutz, *Applied Plastics Engineering Handbook: Processing and Materials*, William Andrew, Norwich, NY, 2011.
69. J. Troitzsch, *Plastics Flammability Handbook: Principles, Regulations, Testing, and Approval*, Hanser, Munich, 2004.
70. H. Domininghaus, *Plastics for Engineers: Materials, Properties, Applications*, Hanser, Munich, 1993.
71. A.R. Blythe, *Electrical Properties of Polymers*, Cambridge University Press, New York, 2005.
72. S. Baumer, *Handbook of Plastic Optics*, John Wiley & Sons, New York, 2011.
73. G.H. Meeten, *Optical Properties of Polymers*, Elsevier Applied Science Publishers, Amsterdam, Netherlands, 1986.
74. M.J. Weber, *Handbook of Optical Materials*, CRC Press, Boca Raton, FL, 2002.
75. M. Chanda and S.K. Roy, *Plastics Fundamentals, Properties, and Testing*, CRC Press, New York, 2010.
76. A.B. Mathur and I.S. Bhardwaj, *Testing and Evaluation of Plastics*, Allied Publishers, Mumbai, India, 2003.
77. E.A. Campo, *Selection of Polymeric Materials*, William Andrew, Norwich, NY, 2008.
78. ASM International, *Characterization and Failure Analysis of Plastics*, ASM International, Metals Park, OH, 2003.
79. J.C. Salamone, *Polymeric Materials Encyclopedia*, 12th Volume Set, Vol. 1, CRC Press, Boca Raton, FL, 1996.
80. F. Trager, *Springer Handbook of Lasers and Optics*, Springer, London, 2012.
81. V. Shah, *Handbook of Plastics Testing and Failure Analysis*, 3rd Edition, John Wiley & Sons, Hoboken, NJ, 2007.
82. C. Vasile, *Handbook of Polyolefins*, 2nd Edition, Marcel Dekker, New York, 2002.
83. G. Wypych, *Weathering of Plastics: Testing to Mirror Real Life Performance*, William Andrew, Norwich, NY, 1999.
84. D. Nichols, *Biocides in Plastics*, iSmithers Rapra Publishing, Shrewsbury, 2004.
85. G. Pritchard, *Plastics Additives: An A-Z Reference*, Vol. 1 of Polymer Science and Technology Series, Springer, London, 1998.
86. E.A. Campo, *Selection of Polymeric Materials*, William Andrew, Norwich, NY, 2008.
87. E. Lokensgard, *Industrial Plastics: Theory and Applications*, 5th Edition, Cengage Learning, Clifton Park, NY, 2008.
88. ASM International, *Characterization and Failure Analysis of Plastics*, ASM International, Metals Park, OH, 2003.
89. G. Wypych, *Weathering of Plastics: Testing to Mirror Real Life Performance*, William Andrew, Norwich, NY, 1999.
90. L.W. Mckeen, *The Effect of UV Light and Weather on Plastics and Elastomers*, 3rd Edition, William Andrew, Oxford, 2013.
91. L.K. Massey, *The Effects of UV Light and Weather on Plastics and Elastomers*, 2nd Edition, William Andrew, Norwich, NY, 2007.
92. E.A. Campo, *Selection of Polymeric Materials*, William Andrew, Norwich, NY, 2008.
93. D.A. Chasis, *Plastic Piping Systems*, 2nd Edition, Industrial Press, South Norwalk, CT, 1988.
94. W. Woishnis and S. Ebnesajjad, *Chemical Resistance of Specialty Thermoplastics: Chemical Resistance*, Vol. 3, William Andrew, Oxford, 2012.
95. C. Bonten and R. Berlich, *Aging and Chemical Resistance*, Hanser, Munich, 2001.

96. K.M. Pruett, *Chemical Resistance Guide for Plastics*, Compass Publications, La Jolla, CA, 2000.
97. D.C. Wright, *Environmental Stress Cracking of Plastics*, iSmithers Rapra Publishing, Shawbury, UK, 1996.
98. Plastic Design Library, *Chemical Resistance: Themoplastics, Thermosets, Thermoplastic Elastomers, Rubbers*, William Andrew, Norwich, NY, 2001.
99. D.J. Kemmish, *High Performance Engineering Plastics*, iSmithers Rapra Publishing, Shawbury, UK, 1995.
100. T. Sixsmith and R. Hanselka, *Handbook of Themoplastic Piping System Design*, CRC Press, Boca Roton, FL, 1997.
101. E. Lokensgard, *Industrial Plastics: Theory and Applications*, 5th Edition, Cengage Learning, Clifton Park, NY, 2008.
102. E.A. Campo, *Selection of Polymeric Materials*, William Andrew, Norwich, NY, 2008.
103. J.L. White, *Principles of Polymer Engineering Rheology*, Wiley, New York, 1990.
104. C.W. Macosko, *Rheology: Principles, Measurements and Applications*, VCH, New York, 1994.
105. J.M. Dealy and K.F. Wissbrun, *Melt Rheology and Its Role in Plastic Processing*, Van Nostrand, New York, 1990.
106. P.J. Carreau, D.C.R. De Kee, and R.P. Chhabra, *Rheology of Polymeric Systems*, Hanser, Munich, 1997.
107. R.B. Bird, R.C. Armstrong, and O. Hassager, *Dynamics of Polymeric Liquids*, 2nd Edition, Vol. 1, John Wiley & Sons, New York, 1987.
108. G.V. Gordon and M.T. Shaw, *Computer Programs for Rheologists*, Hanser, Munich, 1994.
109. R.I. Tanner, *Engineering Rheology*, Oxford University Press, New York, 1985.
110. G.V. Vinogradov and A.Y. Malkin, *Rheology of Polymers*, Mir, Moscow, 1980.
111. H. Markovitz, *Trans. Soc. Rheol.*, **1**, 37 (1957).
112. B.D. Coleman, H. Markovitz, and W. Noll, *Viscometric Flows on Non-Newtonian Fluids*, Springer, New York, 1966.
113. B.D. Coleman and W. Noll, *Arch. Rat. Mech. Anal.*, **3**, 289 (1959).
114. L.A. Utracki, *Polymer Alloys and Blends*, Hanser, Munich, 1989.
115. J. Mewis and N. Wagner, *Colloidal Suspension Rheology*, Cambridge University Press, Cambridge, 2012.
116. R.K. Gupta, Particulate suspensions. In *Flow and Rheology in Polymer Composites Manufacturing*, Ed. S.G. Advani, Elsevier, Amsterdam, Netherlands, 1994.
117. W.J. Milliken and R.L. Powell, Short-fiber suspensions. In *Flow and Rheology in Polymer Composites Manufacturing*, Ed. S.G. Avani, Elsevier, Amsterdam, Netherlands, 1994.
118. C.D. Han, *Multiphase Flow in Polymer Processing*, Academic Press, New York, 1981.
119. A.A. Collyer and D.W. Clegg (Eds.), *Rheological Measurements*, Elsevier Applied Science, London, 1988.
120. L.A. Utracki (Ed.), *Two-Phase Polymer System*, Hanser, Munich, 1991.
121. L.A. Utracki, *Commercial Polymer Blends*, Chapman and Hall, London, 1998.
122. W. Ostwald, *Kolloid-Z.*, **36**, 99 (1925).
123. A. de Waale, *Oil Color Chem. Assoc. J.*, **6**, 33 (1923).
124. S. Matsuhisa and R.B. Bird, *AIChE J.*, **11**(4), 588 (1965).
125. M.M. Cross, *J. Colloid Sci.*, **20**, 417 (1965).
126. P.J. Carreau, Ph.D Thesis, University of Wisconsin-Madison, Madison, WI, (1968).
127. K. Yasuda, R.C. Armstrong, and R.E. Cohen, *Rhel. Acta*, **20**, 163 (1981).
128. G. Menges, F. Wortberg, and W. Michaeli, *Kunststoffe*, **68**, 71 (1978).
129. E.C. Bingham, *Fluidity and Plasticity*, McGraw-Hill, New York, 1922.
130. AKRO Co-Twin Screw™ Software, Version 4 from The University of Akron, Akron, OH.
131. A.C. Pipkin, *Lectures on Viscoelasticity Theory*, Springer-Verlag, New York, 1972.
132. J.C. Maxwell, *Phil. Trans. R. Soc.*, **157**, 49 (1867).
133. L. Boltzmann, *Sitzungber Akad. Wiss.*, **10**, 275 (1874).
134. J.C. Maxwell, *Constitution of Bodies*, Encyclopedia Brittanica, London, 1867.
135. T. Schwedoff, *J. Phys.*, **9**(2), 34 (1890).
136. L. Natanson, *Phil. Mag.*, **2**(6), 342 (1901).
137. S. Zaremba, *Bull. Acad. Sci. Cracow.*, June 594 (1903).
138. E.C. Bingham, *J. Wash. Acad. Sci.*, **6**, 177 (1916); *Fluidity and Plasticity*, McGraw-Hill, New York, 1922.
139. H. Hencky, *Ann. Phys.*, **2**, 617 (1929).
140. R. Eisenschitz, *Kolloid Z.*, **54**, 184 (1933).
141. K. Hohenemser and W. Prager, *Z. Angew. Math. Mech.*, **12**, 216 (1932).
142. M. Reiner, *Am. J. Math.*, **67**, 350 (1945).
143. J.G. Oldroyd, *Proc. Camb. Phil. Soc.*, **43**, 100 (1947).
144. J.G. Oldroyd, *Proc. Camb. Phil. Soc.*, **45**, 595 (1949).
145. J.G. Oldroyd, *Proc. R. Soc.*, **A200**, 523 (1950).

146. R.S. Rivlin, *Proc. R. Soc.*, **A193**, 260 (1948).
147. R.S. Rivlin, *Phil. Trans. R. Soc.*, **A240**, 459 (1948).
148. R.S. Rivlin, *Phil. Trans. R. Soc.*, **A241**, 379 (1948).
149. K. Weissenberg, *Nature*, **159**, 310 (1947).
150. J.G. Oldroyd, *Proc. R. Soc.*, **A245**, 278 (1958).
151. A.E. Green and R.S. Rivlin, *Arch. Rat. Mech. Anal.*, **1**, 1 (1957).
152. R.S. Rivlin, *J. Rat. Mech. Anal.*, **5**, 179 (1956).
153. R.S. Rivlin and J.L. Ericksen, *J. Rat. Mech. Anal.*, **4**, 323 (1955).
154. W.O. Criminale, J.L. Ericksen, and G.L. Filbey, *Arch. Rat. Mech. Anal.*, **1**, 410 (1958).
155. B.D. Coleman and W. Noll, *Arch. Rat. Mech. Anal.*, **6**, 355 (1960); B.D. Coleman, W. Noll, and H. Markovitz, *J. Appl. Phys.*, **35**, 1 (1964).
156. B.D. Coleman and W. Noll, *Rev. Mod. Phys.*, **33**, 239 (1961).
157. B. Bernstein, E.A. Kearsley, and L.J. Zapas, *Trans. Soc. Rheol.*, **7**, 391 (1963).
158. R.B. Bird, R.C. Armstrong, and O. Hassager, *Dynamics of Polymeric Liquids*, Vol. 1, Wiley, New York, 1977.
159. R.B. Bird and P.J. Carreau, *Chem. Eng. Sci.*, **23**, 427 (1966).
160. D.C. Bogue, *IEC Fund.*, **5**, 253 (1966); D.C. Bogue and J. Doughty, *IEC Fund.*, **5**, 243 (1966).
161. P.J. Carreau, *Trans. Soc. Rheol.*, **16**, 99 (1972).
162. M.W. Johnson and D. Segalman, *J. Non-Newt. Fluid Mech.*, **14**, 279 (1984).
163. T.W. Spriggs, J.D. Huppler, and R.B. Bird, *Trans. Soc. Rheol.*, **10**(1), 191 (1966).
164. M. Takahashi, T. Masuda, and S. Onogi, *Trans. Soc. Rheol.*, **21**, 337 (1977); *J. Rheol.*, **22**, 285 (1978).
165. M.H. Wagner, *Rheol. Acta*, **15**, 136 (1976).
166. J.L. White and A.B. Metzner, *J. Appl. Polym. Sci.*, **7**, 1867 (1963).
167. M. Yamamoto, *Trans. Soc. Rheol.*, **15**, 331, 783 (1971).
168. L.J. Zapas, *J. Res. Natl. Bur. Stand.*, **70A**, 525 (1966).
169. L.J. Zapas and T. Craft, *J. Res. Natl. Bur. Stand.*, **69A**, 541 (1965).
170. I.J. Chen and D.C. Bogue, *Trans. Soc. Rheol.*, **16**, 59 (1972).
171. I.J. Chen, G.F. Hagler, L.E. Abbott, D.C. Bogue, and J.L. White, *Trans. Soc. Rheol.*, **16**, 472 (1972).
172. T. Takai and D.C. Bogue, *J. Appl. Polym. Sci.*, **19**, 419 (1975).
173. M.H. Wagner, *Rheol. Acta*, **16**, 43 (1977); **17**, 138 (1978); *J. Non-Newt. Fluid Mech.*, **4**, 39 (1978).
174. A.S. Lodge, *Trans. Faraday Soc.*, **52**, 120 (1956).
175. L.R.G. Treloar, *Trans. Faraday Soc.*, **42**, 83 (1942); *Physics of Rubber Elasticity*, 2nd Edition, Oxford University Press, Oxford, UK, 1958.
176. H. Giesekus, *J. Non-Newt. Fluid Mech.*, **14**, 47 (1984).
177. R. Keunings and M.J. Crochet, *J. Non-Newt. Fluid Mech.*, **14**, 279 (1984).
178. N. Phan-Tien, *J. Non-Newt. Fluid Mech.*, **16**, 329 (1984).
179. N. Phan-Tien and R.I. Tanner, *J. Non-Newt Fluid Mech.*, **2**, 353 (1977); N. Phan-Tien, *J. Non-Newt. Fluid Mech.*, **22**, 259 (1978).
180. A.I. Leonov, *Rheol. Acta*, **15**, 85 (1976); A.I. Leonov, E.H. Lipkina, E.D. Pashkin, and A.W. Prokunin, *Rheol. Acta*, **15**, 44 (1976).
181. T.W. Spriggs, J.D. Huppler, and R.B. Bird, *Trans. Soc. Rheol.*, **10**(1), 191 (1966).
182. D.C. Bogue and J.L. White, *Engineering Analysis of Non-Newtonian Fluids*, NATO Agardograph, 144, 1970.
183. J.L. White and N. Tokita, *J. Phys. Soc. Japan*, **22**, 719 (1967); **24**, 436 (1968).
184. C. Truesdell and W. Noll, *The Non-Linear Field Theories of Mechanics*, Handbuch der Physik, Vol. III/3, Springer, Berlin, 1965.
185. J.L. White, *J. Appl. Polym. Sci.*, **8**, 1129 (1964).
186. J.L. White, *J. Appl. Polym. Sci.*, **8**, 2339 (1964).
187. M. Doi and S.F. Edwards, *J. Chem. Soc., Faraday Trans, II*, **74**, 1789, 1978.
188. A.C. Papanastasiou, L.E. Scriven, and C.W. Macosko, *J. Rheol.*, **27**, 387 (1983).
189. M.H. Wagner and A. Demarmels, *J. Rheol.*, **34**, 943 (1990).
190. L.M. Thomka and W.J. Schrenk, *Mod. Plast.*, **49**(4), 62 (1972).
191. C.D. Han, *J. Appl. Poly. Sci.*, **19**(7), 1875 (1975).
192. W.J. Schrenk, *Plast. Eng.*, **30**(3), 65 (1974).
193. J.A. Caton, *Brit. Plast.*, **44**(3), 95 (1971).
194. L.M. Thomka, *Pack. Eng.*, **18**(2), 60 (1973).
195. C.R. Finch, *Plast. Design Forum*, **4**(6), 59 (1979).
196. J. Dooley, K.S. Hyun, and K. Hughes, *Polym. Eng. Sci.*, **38**(7), 1060 (1998).

24 Laser Processing of Thermoplastic Composites

Peter Jaeschke and Verena Wippo

CONTENTS

24.1 Introduction 805
24.2 Material Specification..807
24.3 Laser Systems 808
    24.3.1 Laser Sources and Laser Radiation Characteristics 808
    24.3.2 Laser Beam Guiding...809
    24.3.3 Focusing Systems...809
24.4 Laser Transmission Welding 810
    24.4.1 Influence of Composite Materials on the Laser Transmission Welding Process 811
    24.4.2 Process Monitoring and Control...817
    24.4.3 Joining of Composite Parts—Examples 818
    24.4.4 Laser-Based Joining of Thermoplastic Composites with Dissimilar Materials.......820
24.5 Laser Cutting 821
    24.5.1 Fundamental Process Strategy 821
    24.5.2 Cutting-Edge Characteristics..822
    24.5.3 Mechanical Properties of Laser-Processed Composites...825
    24.5.4 Influence of Moisture Content on Cutting Behavior 828
24.6 Further Processing Technologies...829
    24.6.1 Surface Activation 829
    24.6.2 Surface Ablation 829
    24.6.3 Automated Fiber Placement...830
References..831

24.1 INTRODUCTION

Due to high strength-to-weight ratios, glass fiber- and carbon fiber-reinforced composite structures offer great potential for lightweight construction in a wide variety of industrial applications. As such, reinforced composites based on fiber fabrics and noncrimp fabrics are established these days as important construction materials in many industrial fields.

The aerospace industry requires materials with high strength-to-weight ratio for the design of new components in order to contribute significantly to energy and CO_2 savings. Carbon fiber-reinforced plastic (CFRP) is then the best material for the job. Additionally, CFRP is entering applications within the automotive, railway, marine, infrastructure, as well as the sports and leisure sectors.

Although thermoset polymers (e.g., epoxy and polyurethane resins) are the predominant kind of matrix materials (Witten et al. 2014), composites based on thermoplastic polymers (TPCs) are of rising interest due to their superior producibility, outstanding impact tolerances, and recyclability. These materials are also known for their uncritical and nearly unlimited storage times, high ultimate strains, and excellent chemical resistance, as well as the possibility of quick forming processing, and mainly the weldability (Offringa 2011).

A wide variety of continuous glass fiber-reinforced thermoplastics (GFRTPs) and continuous carbon fiber-reinforced thermoplastics (CFRTPs) are available in the market. For the fabrication of TPC parts, thermoplastic matrix materials such as polypropylene (PP) and polyamide (PA) are used. For high-performance composites, polyetherimide (PEI), polyphenylene sulfide (PPS), and polyetheretherketone (PEEK) also come into operation.

In order to achieve high production rates and cycle times, fully automated process chains for the manufacturing and the assembly of TPC parts have to be developed. Trimming, drilling, and joining steps are of particular importance. Today, for cutting applications, mechanical techniques (e.g., milling, sawing, grinding or abrasive water jet cutting) and, for joining applications, adhesive bonding and welding technologies are used.

In particular, mechanical cutting shows several disadvantages, because the processing results in high tool wear or requires complex water circuit handling. In addition, all such mechanical techniques are forceful and can lead to deforming of the part or pushing the tool aside. In consequence, varying process results can appear (Sheikh-Ahmad 2009). In contrast to the fabrication of CFRP parts, laser processing is already common in the fabrication of metal parts. The major qualities that promote laser use in production are its flexibility for 3-D machining and its noncontact operation. In this case, tool wear is no issue with lasers, which also offer a process free of forces. Integration of existing laser cutting methods into the manufacturing processes of CFRTP parts and components would significantly contribute to satisfying production demands for this material class in the future. Furthermore, high-power lasers allow high feed rates; however, during laser cutting, heat-induced damage appears at the cutting edge of processed CFRTP, as it is a thermal process. This heat-affected zone (HAZ) of the workpiece is influenced by the thermal characteristics of the matrix material and the reinforcing carbon fibers. During the laser–material interaction at cutting, laser radiation is absorbed in the composite laminate and transferred to process heat. Due to the high thermal conductivity of the carbon fibers compared to the thermoplastic matrix, heat transport occurs, potentially resulting in the overheating of the local polymer matrix. Also, polymer degradation (or even evaporation) can be a result, depending on the type and orientation of the fiber reinforcement, as well as the cutting strategy. A reduced fiber–matrix adhesion close to the cutting edge could also affect the performance of the composite.

Taking into account the thermal impact during laser cutting, the influence of the laser–material interaction on the formation of a HAZ is an object of research. Of particular importance is the effect of the HAZ on the modified mechanical properties of processed CFRTP parts. Therefore, in order to determine the laser-induced influence on the mechanical properties of CFRP (e.g., static tensile strength, in-plane shear, interlaminar shear strength and fatigue), test data must be provided as an important input parameter for component construction steps.

Probably the biggest advantage of thermoplastic composite structures compared to thermoset systems is the weldability. For the manufacturing of structural elements and interior parts, components of basic geometries have to be joined to complex assemblies. For the rivetless joining of TPC, different techniques, e.g., resistance welding, induction welding, or ultrasonic welding, are used, each with advantages and disadvantages (Kagan 2003; McGrath and Cawley 2007). In particular, for the joining of high-performance polymers PPS or PEEK, adhesive bonding can rarely be performed without extensive surface preparation steps due to their chemical resistance (Chen and Zybko 2002). In addition to the manufacturing of pure TPC components, combinations of endless fiber-reinforced and unreinforced (or short fiber-reinforced) materials are in demand. As an example, such combinations are used within the aerospace or the automotive industry for the mounting of brackets, pins, and retainers.

Laser transmission welding (LTW) is regarded as an industrially established process for the joining of unreinforced thermoplastics and for thermoplastics reinforced with short glass fibers. Due to its local energy input, high flexibility, lack of melt flash, and excellent reproducibility, typical applications of this technique can be found, amongst other areas, within the automobile sector (taillights and plastic bodywork parts) and in medical technology (cardiac catheters and infusion bags) (Zweifel et al. 2009). An adaptation and the transfer of this joining technology to the welding of combinations containing TPC parts would be of great interest, especially for GFRTP–CFRTP connections.

For industrial use of LTW as a processing technology for welding composites, one must consider the aspects of laser beam guidance, shaping, and deflection. In particular, process temperatures and

Laser Processing of Thermoplastic Composites   807

optical properties have to be controlled. Finally, the resulting welding connections have to be analyzed with respect to the quality of the weld seam formation and the joint strengths for both static and dynamic testing.

This chapter mainly focuses on the laser-based welding and cutting of CFRTP and GFRTP composites. Therefore, the material specifications that are necessary for understanding the different laser–material interactions as well as the evaluation of processed parts and components are described in the following section. Afterward, fundamentals on laser systems are given, containing a description of laser sources used for processing of composites as well as system technology, such as beam delivery and beam guidance techniques. After giving detailed insight into the laser welding and cutting process, further laser-based applications in the composite field, such as repair preparation, surface activation, and automated fiber placement, are briefly discussed.

24.2 MATERIAL SPECIFICATION

For a better understanding of the respective processes and the laser–material interaction, key properties of the materials used are summarized.

The CFRTPs used for the description of the laser processes in the following sections are based on PPS, PEI, and Polyamide 6.6 (PA6.6) matrices. As is typical for organic sheets, all laminates are reinforced with crimp fabrics (FAB) in a 5 harness satin weave or twill 2/2 layup and varying fiber orientation per layer. PPS sheets consisting of noncrimp fabrics (unidirectional [UD] tapes) also come into operation. UD-reinforced laminates are used for investigating the thermal conductivity within the surface of the composite due to laser impact. Within the frame of Section 24.4, the suffixes *UD 0* and *UD 90* indicate whether the laser beam passes the surface parallel or perpendicular to the carbon fibers. The carbon fiber reinforcement consists of high-tenacity (HT) fibers.

PPS and PEI are the GFRTPs used in this study. Both matrix materials are reinforced with 5-harness satin weaves. For welding investigations, unreinforced PPS and PA6.6 also come into operation. As laser-transparent (LT) parts, both polymers are used in their natural state (*nat.*). In order to evaluate transmission welding processes based on CFRTP as a laser-absorbing (LA) part, unreinforced PPS with carbon black particles added is used. This material is represented by the suffix *c.b.* The main specifications of the materials together with the respective nomenclature are given in Table 24.1.

TABLE 24.1
Classification of Materials

Short Name	LTW Type[a]	Reinforcement/Additive	Fiber Volume Content (%)
CF-PPS FAB	LA	Carbon fiber; fabric (5-harness satin)	50
CF-PPS UD 0	LA	Carbon fiber; unidirectional (laser beam parallel to fiber orientation)	50
CF-PPS UD 90	LA	Carbon fiber; unidirectional (laser beam perpendicular to fiber orientation)	50
GF-PPS FAB	LT	Glass fiber; fabric (5-harness satin)	50
PPS nat.	LT	–	–
CF-PEI FAB	LA	Carbon fiber; fabric (5-harness satin)	50
GF-PEI FAB	LT	Glass fiber; fabric (5-harness satin)	50
CF-PA 6.6 FAB	LA	Carbon fiber; fabric (Twill 2/2)	45
PA 6.6 nat.	LT	–	–
PPS c.b.	LA	Carbon black particles; 0.5% wt.	–

[a] Type of joining partner in LTW.

24.3 LASER SYSTEMS

24.3.1 LASER SOURCES AND LASER RADIATION CHARACTERISTICS

Light amplification by stimulated emission of radiation (lasers) are used for processing materials for different industrial applications. Laser processing can be divided into four main sectors: change of material characteristics, cutting techniques, joining techniques, and additive manufacturing. A laser consists of three basic components: the active laser medium, the resonator, and the optical pumping. The laser active medium is pumped, for example, by a flashlamp or an electric current, so that the active medium reaches a higher energy level. The laser active medium is placed in between the mirrors of the resonators, which consist of a reflective and a semitransparent mirror. Due to spontaneously emitted photons, radiation is generated and reflected inside the resonator. Part of the laser radiation passes the semitransparent mirror and can be used for material processing (Bliedtner et al. 2013; Klein 2012).

The active laser medium is responsible for the emitted wavelength and can be classified as gas, liquid, semiconductor, and solid state. There are different kinds of gas lasers, which emit in different wavelengths depending on their laser active medium. Excimer lasers emit in the ultraviolet (UV) range, λ_L = 157–351 nm. These laser sources are of special interest for surface activation of CFRP. CO_2 lasers typically emit in the near-infrared (NIR) spectrum at a wavelength of λ_L = 10.6 µm. This wavelength has a low penetration depth for thermoplastic materials and thus can be used for ablation or cutting of thermoplastic materials (Bliedtner et al. 2013; Eichler and Eichler 2006; Klein 2012).

Semiconductor lasers emit in the NIR and in the visual spectrum. The laser radiation is generated at an interband transition called the p–n junction. The laser active medium can be doped with donators or acceptors. By doping the active medium with a donator, more free electrons are made available. The material can also be doped with acceptor material, which causes an excess of electron gaps. Diode lasers for material processing emit in the spectral range of λ_L = 800–1030 nm and are used, among other things, for laser welding of thermoplastic composites. This type of laser is known for high plug efficiency and reliability (Bliedtner et al. 2013; Eichler and Eichler 2006; Klein 2012).

Solid-state lasers emit laser radiation within a power range of milliwatts (mW) to kilowatts (kW). The laser active medium consists of crystals or glass doped with rare-earth or transition-metal ions, which is optically pumped. The spectral range of solid-state lasers goes from UV, over the visible (VIS) spectrum and to the infrared (IR) spectrum. A typical representative is the Neodymium Yttrium-Aluminium-Garnet (Nd:YAG) laser, which emits in the IR spectrum. Thanks to frequency doubling and tripling, there exist Nd:YAG lasers emitting in the VIS as well as the UV spectrum. Two other kinds of solid-state lasers are fiber lasers and disk lasers, which can be used for CFRP cutting applications. For fiber lasers, the laser active media are doped optical fibers consisting of quartz glass. Common wavelengths are λ_L = 1030 nm, λ_L = 1550 nm, λ_L = 1064 nm, and λ_L = 1500 nm, depending on the doping element. Disk lasers contain a laser active medium, which is a thin, disk-shaped crystal (Bliedtner et al. 2013).

The operating mode of a laser describes the kind of time-dependent energy output coupling. There are different main operation modes: continuous wave (cw), pulse wave (pw), gain-switched, Q-switched, mode-locked, and Q-switched mode-locked operation. In this chapter, only laser material processes based on cw and pw radiation are discussed. The cw mode is characterized by the emitted constant laser power. The laser power is stepless adjustable between the minimum power to the maximum power, which allows a stable and constant emitting of radiation. Cw-mode laser systems have a higher average power than pw-mode lasers. In pw mode, the radiation is emitted periodically, i.e., the radiation consists of repeating minima and maxima. The pulse duration can range from milliseconds to femtoseconds. Besides the pulse duration, other characteristics are the average pulse power and peak pulse power, which is the maximum occurring optical power. The average pulse power is defined as the quotient of pulse energy to pulse duration (Bliedtner et al. 2013; Eichler and Eichler 2006).

Another criterion for the characterization of laser radiation is the beam quality, which is an indicator of how narrowly the laser beam can be focused. The beam quality is defined by the beam parameter product M^2. $M^2 = 1$ represents an ideal Gauss beam, which can be narrowly focused. Lasers with an M^2 near 1 are called single-mode lasers. High values for M^2 represent a low beam quality, which indicates that the laser beam focusing is limited. Laser sources with high values for M^2 are called multimode lasers (Bliedtner et al. 2013).

24.3.2 LASER BEAM GUIDING

For laser material processing, the laser radiation has to be guided from the laser source to a focusing system. This is mainly conducted by mirrors or optical fibers. Using beam guidance with mirrors, the laser beam is reflected and can be shaped. In order to minimize radiation losses, mirrors are coated with dielectric substrates. For special applications, the mirrors can be made of copper or molybdenum. For solid-state and excimer lasers, highly reflective dielectric mirrors are used. Mirrors for CO_2 lasers have a metallic coating, which can consist of silver or gold layers. The mirrors are classified as spherical, aspherical, or plane mirrors, which allow shaping of the laser beam without using lenses (Bliedtner et al. 2013).

For beam guidance with an optical fiber, the radiation is transported in a cylindrical waveguide. This allows for flexible guiding of the radiation over long distances. The laser radiation, which is emitted by the laser source, is focused by lenses so that the resulting laser beam diameter is smaller than the fiber core. The length and the core diameter of the fiber affect the beam quality and the radiation losses. The optical fiber is made up of a three-layer system: outer cladding, inner cladding with a thickness of about $s = 100$ µm, and a core with a diameter $u = 20$–1000 µm. The laser radiation reflects on the interface of the core and cladding and thus propagates almost without loss in the fiber. When the radiation leaves the fiber, the angle of beam spread is limited by the numerical aperture, which is the maximum angle of an incident beam with respect to the fiber axis. Optical fibers are mainly used to guide diode laser or solid-state laser radiation. The laser radiation of CO_2 lasers cannot be guided by optical fibers, due to the absorption behavior of the fiber materials for the wavelength of CO_2 lasers (Bliedtner et al. 2013; Eichler and Eichler 2006; Klein 2012).

24.3.3 FOCUSING SYSTEMS

Independent of the respective laser process, the laser radiation, which is delivered by mirrors or fiber optics to the working head, has to be converted to a specific spot size and geometry. The size and the geometry of a laser beam at the location of the composite's surface primarily determine the kind of laser–material interaction. For welding applications, the generated radiation intensity has to be sufficient to melt the thermoplastic matrix of a TPC. In the case of cutting, much higher radiation intensities have to be absorbed by the workpiece in order to evaporate the matrix material and the reinforcing fibers simultaneously, particularly if CFRTP has to be machined. In the majority of cases, the appropriate focusing systems are based on static or dynamic lenses and mirrors. In addition, beam shaping optics can be introduced into the optical path in order to form more complex geometries.

Lenses or mirrors as optical elements are installed for laser working heads consisting of static focusing systems. In the case of NIR lasers, playing the largest role within this chapter, lenses are used for collimating and focusing the laser beam to the workpiece. Since the laser beam has a constant and invariant orientation inside the static focusing device, relative movement between the workpiece and the working head is required. This can be performed by means of robotic or multiple-axis systems. Then, a weld seam or cutting kerf can be achieved. A principle setup of beam guidance based on static focusing is given in Figure 24.1a.

Processes including such working heads are called contour welding and contour cutting. For welding applications, weld seam widths and, therefore, focus diameters in the millimeter range or

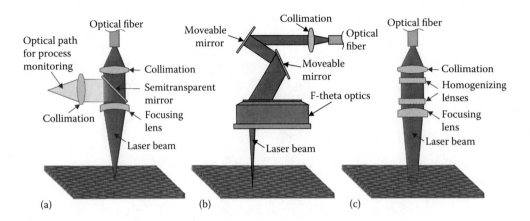

FIGURE 24.1 Principal setup of static focusing (a), dynamic focusing (b), and beam shaping (c).

higher are usually preferred (Wippo et al. 2014b). Cutting tasks require a much smaller laser beam focus ($d_f < 1$ mm). Thus, the focal lengths of focusing lenses used for laser cutting are shorter than those that are used for welding steps. In consequence, laser cutting heads have to be guided in close proximity to the surface of the composite part to be processed, in order to guarantee a constant focal plane on top of the surface. This setup allows the additional use of a cutting nozzle, providing assisting gas flow coaxial to the laser beam, which is responsible for blowing out the molten and sublimated material inside the cutting kerf.

Another possibility for the machining of composites is the use of dynamic focusing systems. Here, the focused laser beam is guided along the processing geometry while the workpiece and focusing device remain stationary. Typically, galvoscanner systems are used, deflecting the laser beam in x- and y-directions, perpendicular to each other, by the use of two galvanic-driven mirrors. After beam deflection, an optical-corrected lens system focuses the laser beam by means of specialized optics (f-theta optics). In contrast to ordinary lenses, f-theta optics are capable of focusing the laser beam within a plane area of limited size. This size is defined by the focal length of the lens system (Bliedtner et al. 2013). A schematic diagram of the optical path within a galvoscanner head is shown in Figure 24.1b. For conventional applications, so-called field sizes up to 300 mm² come into operation. Contrary to static focusing systems, this technology provides feed rates up to several meters per second, even for complex processing geometries. Instead of passing the whole geometry only once at moderate velocity, multiple cycles can be applied at relatively high speed. Processes based on this kind of dynamic focusing are called quasi-simultaneous welding or multipass cutting. If galvoscanner systems are combined with robotic or multiple-axis systems, the workable field size can be extended.

For different laser sources, the intensity distribution of a laser beam leaving the resonator varies. For the majority of processes, a Gaussian intensity distribution is sought, whereas special intensity distributions, e.g., rectangular distributions, may be advantageous for other applications. Additionally, adapted intensity profiles can be realized in terms of a line focus, a double focus, or a circular focus. Instead of generating a focal shape by deflecting the laser beam by means of galvoscanner systems, specialized beam shaping optics are available. Such beam shaping optics consist of cylindrical lenses, diffractive optics, and homogenizing optics. In Figure 24.1c, the transformation of a circular laser spot into a rectangular spot is illustrated. Beam shaping in terms of simultaneously irradiating a defined geometry is of particular importance for LTW applications (Klein 2012).

24.4 LASER TRANSMISSION WELDING

When manufacturing complex parts with high technical functions, components of basic geometries have to be joined to assemblies. Besides mechanical fastening and adhesive bonding, different welding techniques, e.g., hot plate, ultrasonic, and vibration welding, also come into operation for joining thermoplastic parts. Also, for welding CFRTP and GFRTP parts, induction and resistance welding can be applied (Grewell and Benatar 2007). In induction welding, a magnetic field is generated, which introduces heat in electrically conductive, nonmagnetic materials by resistive losses of the eddy current. For nonconductive materials, an additional electrical conduction subsector is needed. Induction welding of CFRTP is possible without any further subsector (Kagan and Nichols 2005; Rudolf et al. 2000). For resistance welding, a resistive element is needed at the interface of the joining parts. The process heat is then generated in the resistive element by an electric current based on the Joule effect. The element remains in the joining area after the welding process (Dubé 2011; Fernandez Villegas et al. 2012).

Another option is the welding of thermoplastics by laser radiation. LTW is an industrially established joining method for welding unreinforced and short GFRTP, parts. The main advantages of this welding method are high accuracy and local heat development. Also, it is contact-free and thus minimizes the amount of mechanical stress on the joining partners. Furthermore, medical and electronic devices can be welded by laser radiation due to the absence of vibration compared to, for example, ultrasonic welding (Kagan et al. 2002). The main markets for this welding technique are the automotive and medical industries due to the high cost efficiency and reliability based on in-process monitoring. For the medical market, this welding technique is especially interesting because it is lint-free and clean-room compatible (Brunnecker 2011).

24.4.1 Influence of Composite Materials on the Laser Transmission Welding Process

LTW is based on optical characteristics of the joining parts. For this joining process, both parts are placed typically in an overlap configuration. Natural thermoplastics are partially optically transparent for the NIR radiation of the laser. Thus, the laser radiation passes the upper transparent material (LT) and is absorbed by the lower material (LA). Thermoplastics highly absorb NIR radiation, when they contain carbon fibers or additives, like carbon black. The absorbing behavior can be affected by the amount and kind of additive (Chen et al. 2011; Frick 2007; Jaeschke 2012; Klein 2012). The absorbed electromagnetic radiation is transferred into heat and melts the LA part. Due to the heat conduction between the joining parts, the upper, LT part melts too, and a weld seam is generated at the interface of both parts. To support the heat conduction, the parts have to be clamped together, and a constant joining force has to be applied (Figure 24.2). The seam strength of laser-welded unreinforced thermoplastics is similar to the strength of the basic material (Frick 2007; Jaeschke 2012).

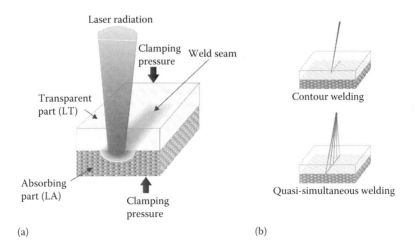

FIGURE 24.2 Principle of laser transmission welding (a) and of different welding techniques (b).

LTW can be divided into four basic techniques: simultaneous welding, mask welding, contour welding, and quasi-simultaneous welding. For simultaneous welding, the laser radiation is simultaneously applied over the entire weld seam, so that the whole weld seam material is liquefied at the same time. Therefore, the laser beam is shaped or diode laser arrays are set together to generate a laser line to meet the geometry of the weld seam. This laser beam forming makes this welding technique less flexible, and so it is mainly applied to join higher quantities of parts (Frick 2007; Jaeschke 2012). For mask welding, a mask is placed above the LT part and consists of a negative of the weld seam. The laser radiation is formed into a long line and guided over the mask. The weld seam geometry can only be changed by exchanging the mask. This technique is mainly used to weld complex contours with small dimensions (Klein 2012; Russek 2009).

For contour welding, the laser beam is guided only once over the workpiece (Chen et al. 2008). With this welding technique, the energy to melt the thermoplastic is applied all at once. Contour welding is typically utilized to generate 3-D and long weld seams. In quasi-simultaneous welding, the laser beam is guided over the workpiece several times at high speed, so that the connection area is completely plasticized (Fiegler 2007; Wilke et al. 2008). To compare these two welding methods, the reference value energy per unit length E_s is introduced:

$$E_s = \frac{nP_L}{v}. \tag{24.1}$$

This reference value consists of the number of laser beam passes over the weld seam (n), the laser output power (P_L), and welding speed (v) (Chen et al. 2008).

Other laser-based joining techniques are GLOBO® welding and transmission welding by incremental scanning technique (TWIST) laser welding, which is a combination of contour and quasi-simultaneous welding. For GLOBO welding, a glass ball is used to generate the clamping pressure as well as to focus the laser radiation. Two laser-based hybrid welding processes are IR-hybrid and ultrasonic-hybrid welding, which are a combination of a laser welding process combined with NIR radiation for heating the material and a combination with an ultrasonic welding process, respectively (Klein 2012).

Besides the laser welding technique itself, the welding process is affected by the optical characteristics of the joining members. The interaction between the radiation and the material can be divided into three main effects: transmittance (T), reflectance (R), and absorption (A). When the laser radiation hits the LT part, some of the radiation is reflected on top of the surface or in the material itself. The reflected radiation can be divided into direct and diffuse reflected radiation. The second part of the radiation is absorbed in the LT-material. The remaining transmitted radiation contributes to the welding process and generates the process heat (Bliedtner et al. 2013; Chen et al. 2011).

Furthermore, the laser radiation is affected by scatter in the LT material. Scattering depends on the crystalline phase of the matrix material and the existence of additives like pigments or reinforcements, e.g., glass fibers (Jaeschke 2012; Kagan et al. 2002; Klein 2012). Due to the scatter, less radiation is available for the actual welding process. The relationship between transmissivity and material thickness for PEI reinforced with glass fiber fabrics (GF-PEI FAB) is depicted in Figure 24.3. With increasing material thickness, the distance in the matrix material as well as the number of glass fiber fabric layers increase. Thus, the absorption and the scattering of the laser radiation rise; hence, the transmissivity decreases.

In addition to the transparent part, the absorbing part also affects the welding process. If CFRTP is used as an LA part, the carbon fiber orientation relative to the welding direction has an effect on the weld seam geometry due to the conduction of the process heat. The heat conduction in CFRP is a combination of the heat conductivity of the fibers and the matrix material. The heat conduction in the carbon fiber direction for UD CFRP can be described as

$$\lambda_p = \lambda_{pf}\varphi + \lambda_m(1 - \varphi), \tag{24.2}$$

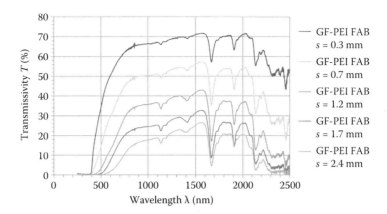

FIGURE 24.3 Transmissivity curves of GF-PEI FAB for varying material thicknesses.

where λ_p is the heat conductivity of the UD CFRP in fiber direction, λ_{pf} is the heat conductivity along the fiber, λ_m is the heat conductivity of the matrix material, and φ is the relative fiber volume content (Schuermann 2005).

Perpendicular to the fiber direction, the heat conduction can be described as

$$\lambda_n = \frac{\lambda_{nf} + \lambda_m + (\lambda_{nf} - \lambda_m)\varphi}{\lambda_{nf} + \lambda_m - (\lambda_{nf} - \lambda_m)\varphi} \lambda_m, \qquad (24.3)$$

where λ_n is the complete heat conductivity in UD CFRP perpendicular to the fiber direction and λ_{nf} is the heat conductivity perpendicular to the fiber (Schuermann 2005).

The heat conductivity of the matrix material is lower than that of the carbon fibers. Furthermore, the heat conductivity along the carbon fibers is higher than perpendicular. For welding of UD CFRTP, this leads to two main cases: the carbon fibers have the same orientation as the weld seam, so the heat stays in the weld seam and contributes to the welding process, or the carbon fibers are perpendicular to the weld seam orientation, so the heat conducts out of the laser–material interaction zone and less energy is available for the actual welding process (Figure 24.4).

If the LA part contains a carbon fiber fabric, both cases occur, depending on the roving orientation. This will lead to a fluctuation within the weld seam. The fracture patterns of two different lap shear samples are depicted in Figure 24.5. The lap shear sample with the material combination of PPS nat. and PPS c.b. shows a homogeneous fraction pattern. This is due to the homogeneous heat development in the weld seam itself arising from the all-over distribution of carbon black particles in the PPS matrix. The other lap shear sample consists of PPS nat. and CF-PPS FAB. The fraction

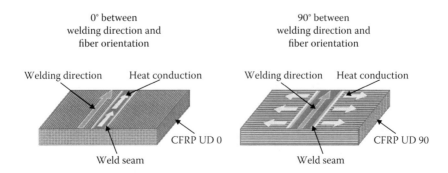

FIGURE 24.4 Heat conduction in unidirectional CFRP depending on welding direction.

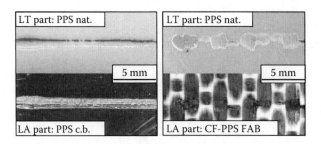

FIGURE 24.5 Fraction patterns for laser-welded combinations, based on PPS c.b. and CF-PPS FAB as laser-absorbing parts.

pattern of this sample shows a fluctuating weld seam, due to the varying carbon fiber orientations on the surface of the CF-PPS FAB and the resulting heat conduction. This can lead to a weak and leaky weld seam (Jaeschke 2012).

Furthermore, there is a dependency between the weld seam width, the welding technique and the welding parameters. The results for bead-on-plate welding with contour (C) and quasi-simultaneous (QS) welding are depicted in Figure 24.6. During bead-on-plate welding, only an absorbing part is used, which was, in this case, UD CFRTP with a PPS matrix. The fiber orientation was along (CF-PPS UD 0) as well as perpendicular (CF-PPS UD 90) to the weld seam. The welding speed was kept constant for each welding technique, and the laser output power was raised in order to increase the E_s.

The experiments were conducted with an Nd:YAG laser, which emits at a wavelength of $\lambda_L = 1064$ nm at a maximum output power of $P_L = 100$ W. The laser beam was guided over the workpiece by a galvoscanner system.

For contour welding, the difference in the average weld seam width between CF-PPS UD 0 and CF-PPS UD 90 increases for higher energies per unit length E_s. The results indicate that there are mainly two effects responsible: the first one is the change in the intensity distribution in the focal point for rising laser energies, and the second one is the heat conduction along the carbon fibers. With higher laser power, more energy is available over the whole area of the focal point to melt the matrix material, and so the width of the weld seam increases. Further, more heat is available to conduct out of the weld seam. Especially for the bead-on-plate welding on CF-PPS UD 90, the heat conduction largely effects the weld seam width.

For quasi-simultaneous welding, the heat is applied over a longer time span than in contour welding. Thus, the time in which the heat conducts out of the welding area is longer, and the welding area cools down. This leads to less heat for melting the matrix material, resulting in smaller weld seam

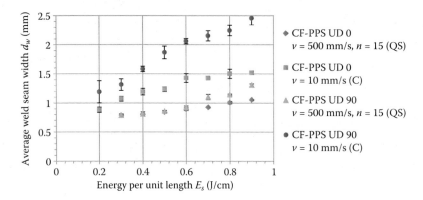

FIGURE 24.6 Average weld seam width for QS and C bead-on-plate welding on CF-PPS UD 0/90.

widths compared to contour welding. This indicates that more laser power is needed in the case of quasi-simultaneous welding to generate the same weld seam width as for contour welding (Jaeschke et al. 2014c; Wippo et al. 2012).

For welding on a CFRTP, the weld seam formation depends not only on the carbon fiber orientation relative to the weld seam orientation but also on the existence of matrix depots on the surface. To clarify this, Jaeschke et al. (2010) have performed bead-on-plate experiments with CF-PPS FAB using a diode laser emitting at a wavelength of λ_L = 940 nm. Experiments were conducted with a contour welding speed of v = 250 mm/s and with laser output power of P_L = 4.0 W and P_L = 5.5 W. Also, the surface temperature of the CF-PPS FAB was monitored by an IR camera detecting in a spectral range of λ_{IRC} = 3.4–5 µm in order to determine the influence of the fiber arrangements and the matrix depot on the temperature distribution. The upper area of Figure 24.7 shows the CF-PPS FAB surface before bead-on-plate welding and the area defined for the temperature measurement. It is possible to identify five different cases, which consist of local compositions with a complete fiber arrangement (FA1, FA2), combinations of fiber and matrix depot (FA3, FA4), as well as pure matrix depots (FA5). For all cases, different modifications of the CF-PPS FAB surface could be detected in the experiments. For the cases FA3, FA4, and FA5, the visible surface modification is limited by the matrix depot for P_L = 4.0 W.

This is due to a partial transparency of the matrix depot for the laser radiation. The radiation passed the matrix depot and was absorbed at the carbon fibers underneath the matrix depot. The generated process heat was not high enough to melt the whole matrix depot by heat conduction,

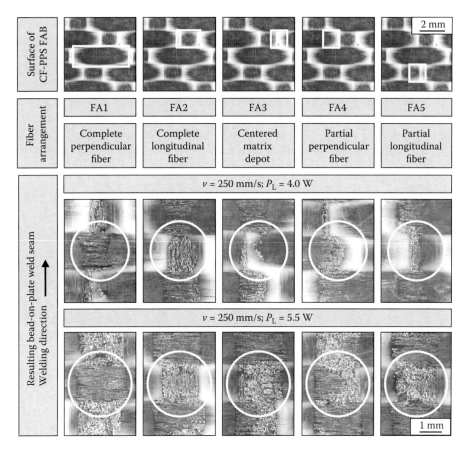

FIGURE 24.7 Bead-on-plate weld seams on CF-PPS FAB for five different fiber arrangements.

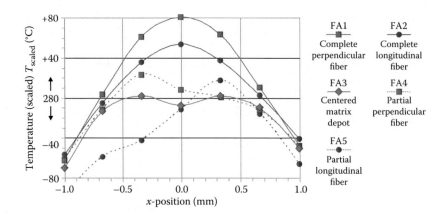

FIGURE 24.8 Temperature distribution perpendicular to the weld seam orientation while bead-on-plate welding.

and so the weld was interrupted. This effect is reflected by the detected surface temperatures in Figure 24.8, where the *x*-position (*x*-axis) is a measurement line perpendicular to the weld seam orientation. For case FA5, the temperature decreases for the middle of the matrix depot compared to the edges, where the carbon fibers are closer to the surface.

For cases FA3 and FA4, the surface temperature decreases from the edge of the carbon fibers to the matrix depot, where the temperatures are lower than the melting point of the PPS. The temperature underneath the matrix depot cannot be detected by the IR camera, due to its detection wavelength. If this bead-on-plate weld seam development is transferred to a real welding process, it would lead to gaps in the weld seam and, thus, a weak weld seam. For the bead-on-plate welding with a laser output power of $P_L = 5.5$ W, all the matrix material was molten, but the fluctuating geometry of the weld seam remained (Jaeschke 2012; Jaeschke et al. 2010).

Besides the fiber orientation, for each point in the weld seam, the main fiber orientation along the whole weld seam has to be taken into account for the realization of an LTW process. For CFRTP consisting of a satin-weave carbon fiber fabric, two main fiber directions exist: the main fiber orientation on the surface is along as well as perpendicular to the weld seam orientation. This also affects the process temperatures and, thus, the weld seam strength. Figure 24.9 shows the average weld seam strength for different energies per unit length. The material combination used is PPS nat. (LT) and CF-PPS FAB (LA) welded with a diode laser emitting at $\lambda_L = 940$ nm. For both main fiber

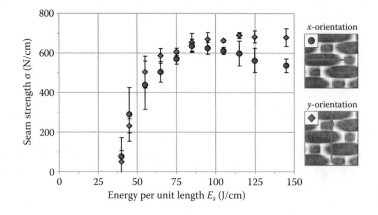

FIGURE 24.9 Seam strength depending on the main fiber orientation for different energy per unit length E_s for PPS nat. ($s = 1$ mm) and CF-PPS FAB.

Laser Processing of Thermoplastic Composites

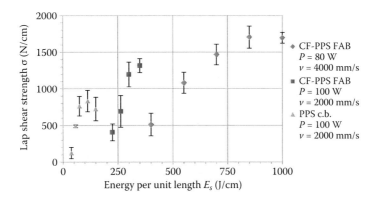

FIGURE 24.10 Lap shear strength for different welding parameters for the material combinations GF-PPS FAB–PPS c.b. and GF-PPS FAB–CF-PPS FAB.

directions, a typical lap shear curve for a laser-welded overlap configuration is achieved. The curves depict the increase of the lap shear strength until a distinct maximum is reached.

For higher energies per unit length, the process temperature exceeds the decomposition temperature of the matrix material, and so the lap shear strength decreases. The lap shear strength maximum for welding parallel to the main fiber direction (x-direction) is reached at lower energies per unit length than for the welding perpendicular to the main fiber direction (y-direction) on the surface. Furthermore, the lap shear strength for the y-direction samples remains at a higher and almost constant level toward higher energies per unit length, which results in higher process reliability (Jaeschke et al. 2009).

Apart from the fiber orientation, the welding parameters also influence the weld seam strength and the weld seam geometry. Experiments were conducted with the material combinations GF-PPS FAB–PPS c.b. and GF-PPS FAB–CF-PPS FAB (Figure 24.10). Due to the low transparency of the transparent joining part, the experiments could not be performed with contour welding and had to be conducted with quasi-simultaneous welding. The maximum seam strengths for the PPS c.b. could be clearly determined at $E_s = 112.5$ J/cm (Jaeschke 2012; Jaeschke et al. 2014c).

For the quasi-simultaneous welding of the GF-PPS FAB–PPS c.b., fewer welding cycles n are needed than for the combinations with CF-PPS FAB welded with the same laser output power and welding speed. This is due to the lower heat conductivity, which allows the process heat to remain in the joining area. To compensate for the heat loss in the welding zone, higher energy per unit length was applied by increasing the number of welding cycles. This also led to a broadening of the weld seam and, in consequence, to a higher lap shear strength. The test series for CF-PPS FAB with the same welding speed and laser output power as for the welding of PPS c.b. was aborted for energy per unit lengths of $E_s > 350$ J/cm due an overheating of the LT part. In order to avoid thermal damage, the quasi-simultaneous welding process was modified so that the process heat was generated more slowly. Therefore, the welding speed was increased to $v = 2000$ mm/s, and the laser output power was reduced to $P_L = 80$ W. These parameters led to a high number of welding cycles and therefore to a longer process time. Due to the parameter change, the heat loss in the weld seam increased, and more energy had to be applied to generate a weld seam. As a result, the strength curve is shifted to higher energy per unit length. For this case, the maximum achievable seam strengths can be clearly determined, and the LT part remained intact. It can be stated that by adapting an essential process, parameters can be optimized (Jaeschke 2012; Jaeschke et al. 2014c).

24.4.2 Process Monitoring and Control

For LTW, the weld seam temperature can be affected by impurities, e.g., foreign particles or varying absorption coefficient. These impurities can weaken the weld seam, which leads to a reduced weld seam strength or to a leaky weld seam. For in-process monitoring, different temperature detecting techniques are used, such as pyrometer, IR camera, or optical reflection monitoring methods. These techniques can be applied to detect impurities and can thus be used for quality inspection (Klein 2012).

For welding of unreinforced and short GFRTP, pyrometers are used to obtain a reference temperature in the weld seam during contour welding. Only a reference temperature can be detected, because the heat radiation emitted by the weld seam is partly shielded by the LT part, which causes a systematic error in the measurement. This error stays constant in the welding process for each material combination. For the welding procedure, the pyrometer is connected to the welding head, and its detection path is aligned to the laser radiation path by semitransparent mirrors. This allows for on-axis temperature detection directly in the laser–material interaction area. Then, the pyrometer signal can be used for in-process contour welding control, in which the laser output power is adapted, and thus, a constant welding temperature is maintained (Klein 2012; Wippo et al. 2014a). For welding of CFRTP, the weld seam temperature is affected by the carbon fiber orientation in the weld seam, which can lead to fluctuating process temperatures. If a highly transparent welding partner is used, the process temperature can also be monitored by a pyrometer. The resulting pyrometer signal can be used to control the weld seam temperature (Jaeschke 2012).

Another option is the use of IR cameras to monitor the process temperatures. An IR camera provides a 2-D pseudocolor image of the detected temperatures. The spectral detection range of the camera has to be in an area where the transparent part shows sufficient transmissivity. The detected temperatures of an IR camera are also affected by the LT part in the same way as for the pyrometer. IR cameras cannot be integrated into welding heads; they are placed beside the welding head, so that the temperature detection is conducted off-axis with a certain angle to the laser beam axis. This temperature detection method is used to observe simultaneous and quasi-simultaneous welding processes (Ackermann 2010; Klein 2012; Wippo et al. 2014a). Besides temperature-based process monitoring techniques, other methods, such as optical reflection diagnostics, digital imaging, and detection of the melt displacement, are used during quasi-simultaneous or simultaneous welding processes (Ackermann 2010; Klein 2012).

24.4.3 Joining of Composite Parts—Examples

LTW provides the possibility to join thermoplastic brackets and retainers to larger components made of CFRTP. In Figure 24.11, a pin consisting of PA6.6 attached to a CFRTP laminate consisting of a copolyamide matrix and a carbon fiber fabric is shown. The parts were welded with a diode laser emitting at a wavelength of $\lambda_L = 940$ nm. To enhance the weld seam quality, a process control was used to adapt the laser output power according to a pyrometer signal. The welded part was

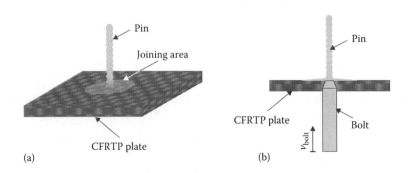

FIGURE 24.11 Schematic drawing of the joined pin with (a) CFRTP and (b) the used test setup.

tested by a push-off test. Therefore, a hole was drilled into the CFRTP plate, through which the pin was pushed away from the plate during testing.

The samples were placed into a climate chamber for at least 5 min and tested directly afterward. The ambient temperatures of the climate chamber were −50°C and 80°C. Also, adhesively bonded samples were prepared. The connection area of the bonded samples was the same as the area of the weld seam. Furthermore, in order to get a comparison between bonded and welded samples, the bonded area had the same location as the weld seam (Hansen et al. 2014).

The test results of the pin samples are illustrated in Figure 24.12. All welded samples have reached a higher strength than the adhesively bonded samples. For example, the average maximum failure load of the welded pin samples was $F_f = 919$ N and thus higher than the bonded samples, with an average failure load of $F_f = 489$ N. The maximum elongations of the adhesively bonded and welded test samples are similar at −50°C. For room temperature and for 80°C, the elongation is higher for the welded samples. These results indicate that the laser weld seam is more elastic than adhesively bonded ones.

Thus, the weld seam area could be reduced to obtain the same strength as the bonded samples. Consequentially, the area of the pin foot could be reduced, and hence, weight can be saved. Furthermore, there would be no extra weight due to the adhesive (Hansen et al. 2014).

Another laser-joined composite part is a stiffener panel manufactured by the companies AGC AeroComposites Yeovil and Laser Zentrum Hannover e.V. (Figure 24.13). This part consists of omega profiles welded to two facing sheets. All parts were made of GF-PEI FAB. The absorbing part contains the additive carbon black to become absorbing for the laser radiation. The LT part was $s = 2.4$ mm thick and had a transmissivity of about $T = 17\%$. The part was contour welded by a diode laser. The facing sheet and the omega profiles were welded together at the bottom of the omega profiles with one weld seam. Two parallel weld seams were placed at the top of the omega profile.

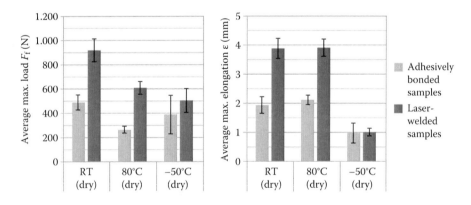

FIGURE 24.12 Average maximum load and elongation for laser-welded and adhesively bonded pin samples. RT, room temperature.

FIGURE 24.13 Laser-welded stiffener panel.

FIGURE 24.14 Insert laser welded to GF-PEI FAB.

An insert, laser welded to CF-PEI FAB and produced by AGC AeroComposites Yeovil and Laser Zentrum Hannover e.V., is shown in Figure 24.14. Both parts are less transparent for the laser radiation. To become laser absorbing, the bottom of the insert was painted with a laser-absorbing color. The joining technique used was quasi-simultaneous welding. These kinds of inserts are a mass product and need to be able to connect parts by a screwed joint.

24.4.4 LASER-BASED JOINING OF THERMOPLASTIC COMPOSITES WITH DISSIMILAR MATERIALS

In the last couple of years, the joining of thermoplastics, especially for reinforced thermoplastics, with thermoset materials or metals became of interest for mass production. Therefore, fast and reliable joining techniques are needed. Recently, research has been carried out to join these dissimilar materials by laser radiation. Studies were conducted with the thermoplastic matrix materials PA6, PA6.6, polycarbonate (PC), and polybutylene terephthalate (PBT), which were unreinforced or reinforced with short or endless fibers. Further joining partners were thermoset CFRP and glass fiber-reinforced plastic (GFRP) as well as aluminum alloy or dual-phase steel (Amend et al. 2012, 2014; Fuchs et al. 2014; Rodríguez-Vidal et al. 2014). The laser-based joining of composites with dissimilar materials can be divided into two techniques: transmission joining and heat-conduction joining (Figure 24.15). In transmission joining, the radiation passes the transparent thermoplastic part and is absorbed at the bottom part, which consists of metal or a thermoset composite. Due to heat conduction, the thermoplastic material becomes molten and wets the surface of the other joining partner.

For heat-conduction joining, the metal part is heated up directly by the laser radiation. The generated heat conducts through the metal so that the interface between the two parts is heated up and

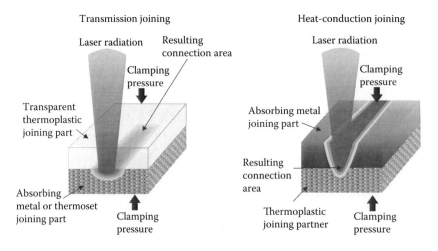

FIGURE 24.15 Schematic drawing of transmission and heat-conduction joining.

the thermoplastic material starts to melt. For both joining techniques, both parts have to be clamped together to support the heat conduction between the parts.

To enhance the joining strength, the surface of the absorbing part can be prepared by milling, grinding, blasting, or laser structuring. In the case of laser structuring, crater structures and grid structures were generated into metal parts (Amend et al. 2013). The structures consist of different structure depths and geometries, which affect the achievable connection strength. Another option is to generate microscopic pins on the metal surface. Then, the humping phenomenon is used (Fuchs et al. 2014). In the joining process, the thermoplastic melt flows into these structures and solidifies. The penetration depth of the melt depends on the geometry of the structure and the process parameters. For the surface preparation of thermoset material, the effective joining area should be enlarged by exposure of the carbon fibers. This can be used for the preparation of a thermoset absorbing partner, such as CFRP, as well as for a transparent partner, for example, GFRP, for the transmission joining technique (Amend et al. 2012).

24.5 LASER CUTTING

24.5.1 Fundamental Process Strategy

In traditional laser processing techniques, mainly based on contour cutting strategies, the geometry to be cut is passed just once by the laser beam. Essentially, the process guiding corresponds to the contour welding method described in the previous section. For this kind of machining, laser cutting heads arising from sheet metal processing are used. In order to realize a relative movement between the laser beam and the workpiece, the cutting head is guided by robots or axis systems. Regarding the overall process management, this is comparable to milling and abrasive water jet cutting.

Cutting of CFRTP parts in the thickness range of a few millimeters is typically performed by applying contour cutting techniques, using multi-kW cw lasers. The use of these lasers offers the possibility of realizing high feed rates of several meters per minute; however, since this is a thermally dominant process, heat-induced damage appears at the cutting edge of treated composites (Schneider et al. 2013; Young 2008). This is caused by major differences in the thermophysical properties of thermoplastic matrix materials and reinforcing carbon fibers.

In order to reduce the high thermal impact during the cutting process and to avoid appreciable disintegration of the fiber–matrix arrangement, multipass cutting can be applied. While using this technology, the laser radiation is guided by means of flexible galvoscanner systems. For this reason, the laser–material interaction time for a specific point along the cutting edge is reduced. During multipass cutting, the geometry is passed over several times by the laser beam at high feed rates, which differs greatly from contour cutting. Basically, this process guiding corresponds to quasi-simultaneous welding, which is described in the previous section. In this manner, instead of a whole cut, only a groove is generated inside the composite structure by laser ablation. Thus, lower energy per unit length is applied per cycle, resulting in a reduced thermal impact at the interaction zone and at the cutting edge. Hence, the extent of a HAZ decreases, and the deterioration of the fiber–matrix adhesion can be minimized. Between every two passes of the laser beam, the CFRTP cools down, resulting in lower critical temperatures inside the composite. A detailed description of the HAZ formation as a result of contour and multipass processing is given in the following section.

Not only the process strategy but also the characteristics of the laser radiation influence the achievable cutting qualities and the processing speed. As applied in micromachining of metals and nonmetals, ultrashort pulsed lasers, with pulse durations in the picosecond and nanosecond range, are capable of achieving nearly damage-free cutting, drilling, and ablation qualities. Such laser sources emit radiation in the NIR, the VIS, and the UV. Short laser–material interaction times provide the potential for almost so-called cold material ablation. Nevertheless, such laser systems are commercially available up to average output powers of a few hundred watts. This is quite low, if cutting applications in the macro range are the objective. With these kinds of lasers, feed rates in the range of some centimeters per minute can be realized (Bluemel et al. 2014a). Also, the adjustment of beam delivery and forming is time-consuming, and usually, laboratory conditions are required. With the rising need for new cutting technology in macromachining, it becomes increasingly important for composite production lines to consider using industrially established laser sources.

24.5.2 Cutting-Edge Characteristics

In general, CFRTP cut by cw lasers clearly shows an extended HAZ starting from the cutting kerf into the thermoplastic composite with changed material properties. A cross section of a CF-PPS FAB laminate processed by a fiber laser using an output power of $P_L = 6$ kW for a seven-layer arrangement with a thickness of $s = 2.2$ mm is shown in Figure 24.16. Different areas within the HAZ have been identified, in which the material damage occurs in a specific manner. The induced temperature gradient, arising from the high thermal conductivity of the carbon fibers along the fiber axis (fiber type: HT; $\lambda_c = 17$ W m^{-1} K^{-1}) and starting at the cutting kerf toward the inner bulk material, influences the composition of the thermoplastic matrix system.

In close proximity to the cutting edge, observations show damage between the single carbon fiber fabrics (interlaminar) and within the rovings themselves. In this area, the PPS matrix is vaporized, and the carbon fibers are charred (B_1). Next to this area, matrix damage becomes predominant due to the process exceeding the decomposition temperature of $T_V = 370°C$ for PPS (B_2). With increasing distance to the cutting kerf, mainly interlaminar effects, e.g., structural modification within the matrix between the carbon fiber rovings, can be seen. Porosities are detected within the PPS matrix, whereas the carbon fiber fabric remains intact (B_3). The presence of pores within CFRTP structures is known from the thermoforming procedure of laminates, if an insufficient molding pressure during the consolidation phase is applied. In this context, the pores are generated due to emission of volatile components from the matrix material after reaching the PPS melting temperature of $T_m = 285°C$ (Staehr et al. 2015). It can be assumed that the intersection between HAZ and unmodified bulk material is determined by resolidification of the polymeric matrix.

The HAZ can be quantitatively analyzed by evaluation of cross-section images with measurement of the areas B_k. All B_k are measured within the thickness of the original laminate ($s = 2.2$ mm), independent of the heat-induced burr, occurring at the cutting kerf (cf. Figure 24.16). Based on this evaluation method, the width of a heat-influenced area can be calculated, and a mean width β_k starting at the cutting edge and extending into the bulk material can be deduced by

$$\beta_k = \frac{1}{s}\sum_{i=1}^{k} B_i. \tag{24.4}$$

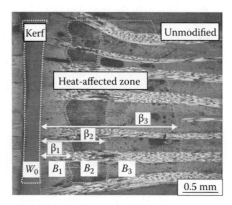

FIGURE 24.16 Determination of different regions of the HAZ due to laser impact for fiber laser–machined CF-PPS laminates.

In order to demonstrate the influence of processing parameters on the cutting quality, in Figure 24.17, the resulting width of the HAZ is shown for varying cutting velocity at constant laser output power (Jaeschke et al. 2014b). Increasing the feed rate causes an almost linear decrease in β_k for all three zones of the HAZ. The cutting kerf width W_0 likewise decreases with increasing feed rate, as long as the dimension of the focus diameter ($d_f = 0.2$ mm) at a feed rate of $v = 7.2$ m/min is reached.

However, for contour cutting, the highest temperatures inside and close to the laser–material interaction zone are generated as a consequence of applying high energy per unit length. Extensive damage of the composite structure is therefore unavoidable, even if high feed rates are applied. Furthermore, the potential of reducing the HAZ by applying higher processing speeds, in fact, is limited by the availability of laser output power, adequate movement of the cutting head, high-performance optical components, and therefore, of course, the economic efficiency (Bluemel et al. 2012).

The relationship between an increasing process speed and a decreasing HAZ directly leads to the necessity of investigating the multipass cutting method in more detail. In Figure 24.18, the extents of all three zones of the HAZ for a processed four-layer CF-PPS FAB laminate are given.

The cutting process is based on the multipass method using a cw single-mode fiber laser with an output power of $P_L = 1000$ W. For reference, the analysis of a HAZ for a contour cut is added to the graph ($n = 1$). For a constant energy per unit length of $E_s = 10$ kJ/m, multipass cutting is performed for $n = 6$ and $n = 12$ repetitions with a time gap of $\Delta t = 1$ s between two consecutive cycles. For

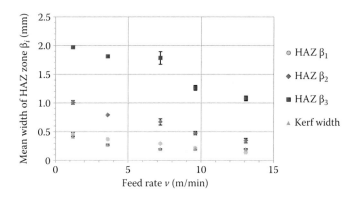

FIGURE 24.17 Mean width of different zones of the HAZ and mean kerf width for a variation in feed rate (CF-PPS FAB, seven layers; $s = 2.2$ mm; $P_L = 6$ kW).

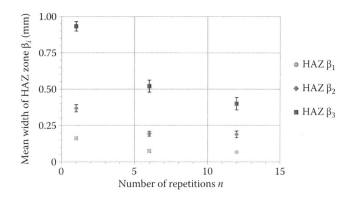

FIGURE 24.18 Mean width of HAZ as a function of the number of repetitions at constant energy per unit length (CF-PPS FAB; $s = 1.24$ mm; $P_L = 1$ kW; $E_s = 10$ kJ/m).

multipass processing, the overall energy per unit length has to be calculated taking into account the number of repetitions, which is applied. The calculation is performed analogous to the determination of E_s for quasi-simultaneous welding, described in Equation 24.1.

Figure 24.18 clearly shows a decrease in the extent of the HAZ with increasing number of repetitions. This behavior applies for all β_k. Considering the inner structure of the HAZ, it can be seen that β_1 and β_2, representing the damaged areas, are reduced to 65 and 180 µm, respectively, which is far below the values from contour cutting.

During multipass processing, the cutting geometry is passed several times by the laser beam, introducing a repetitive energy input and, therefore, heat generation, which is accumulated within the composite structure. In this context, besides the feed rate, the length of the contour and the time gap between the laser beam cycles highly influence the size of the HAZ.

Bluemel et al. (2014a) investigated the influence of the so-called loop time, which represents a coefficient of the cutting length and the applied scanning speed, on the extent of the HAZ for CF-PPS FAB. A high decrease in all β_k is observed for increased loop times and cutting lengths. This corresponds to the fact that a specific volume along the cutting line is able to give off heat during the time gap and the laser off time, respectively. This excess heat is conducted along the carbon fibers inside the bulk material. In fact, regarding the possible heat dissipation into the bulk material and the resulting process temperatures in the laser–material interaction zone, a high cutting length can be equivalent to the insertion of breaks between the single laser beam cycles. In consequence, in terms of an economic use of laser sources and process equipment, a quasi-parallel machining of thermoplastic composite laminates seems reasonable (Walter et al. 2014). In quasi-parallel machining, multipass cutting is performed for several CFRTP parts, applying the first cycle to one part and then to another part and so on. In doing so, the necessary breaks between two laser beam cycles for processing of a specific part can be transferred to efficient processing time of further parts. Since the working field of a galvoscanner system, and therefore, the size of a CFRTP part to be processed, is limited in accordance with the scanner optics used, the scanning system or the composite laminate has to be moved if larger dimensions have to be cut. Highly accurate axis systems are needed in the case of high-brightness laser sources providing small focus diameters of $d_f \approx 20$ µm and, therefore, highest intensities in the cutting kerf, as it is provided, e.g., by single-mode fiber lasers (Bluemel et al. 2013). This is of particular importance since a correct retrieval of previous positions between the different cycles has to be guaranteed. For other laser systems providing larger focus diameters, lower-accuracy robotics technology could be used (Bluemel et al. 2014b).

The quasi-parallel processing of 2-D laminates is of particular importance for the cutting and drilling of CFRTP laminates as a preliminary step for the manufacturing of thermoformed components. In a second step, after thermoforming, components have to be trimmed to their final dimensions. If the control of a multiple-axis system is linked to those of a galvoscanner system and a laser source, this final processing step could also be realized by so-called remote laser processing. In terms of productivity, the introduction of appropriate process monitoring and process control, e.g., by the use of pyrometers and thermo cameras, could lead to minimized cutting times.

Apart from cw lasers, pulsed systems are also applied for processing of composites. In contrast to cw-based machining, much higher intensities within the laser–material interaction zone can be realized (Eichler and Eichler 2006). Due to a shorter duration of radiation impact, the overall heat development close to the cutting kerf of the composite part can be reduced. By far, the largest part of today's research and development activity in this field is CFRP processing based on thermoset matrix materials (Leone et al. 2014; Negarestani et al. 2010; Weber et al. 2012). For CFRTP, investigations dealing with the influence of varying pulse lengths and repetition rates on the HAZ formation have been performed (Bluemel et al. 2014a,b). It was found that pulse lengths in the range of some tenths of a nanosecond are well suited for a gentle processing of CF-PPS FAB if the multipass method is applied. A generic example in terms of a basic contour of an aircraft clip demonstrator was published by Jaeschke et al. (2014a), wherein a diode-pumped disk laser, emitting pulse lengths in the nanosecond range, was used in conjunction with a galvoscanner system.

24.5.3 Mechanical Properties of Laser-Processed Composites

Since laser machining of CFRTP can cause critical damage at the cutting edge by disintegrating the fiber–matrix adhesion, the mechanical properties of laser-processed specimens have to be analyzed. Providing knowledge concerning the mechanical behavior of laser-treated composites under specified loads is of particular importance for design and construction steps. Detailed investigations of the influence of laser-processed CF-PPS FAB on static strength properties, representing fundamental insight, have been performed by Jaeschke et al. (2014b). Tensile tests are done in conformity with the ASTM D3039 standard. CF-PPS FAB specimens are tested with the force along the warp direction of the laminate. For benchmarking reasons, milled reference specimens were machined. It was the aim of this test series to focus on the influence of the HAZ on the mechanical properties of the composite material, not as such on the fundamental material properties. And so, a reduced sample width of $w_0 = 10$ mm was used, emphasizing the influence of the HAZ on the tensile properties.

Firstly, the static tensile strength $\sigma_{T,L}$ of laser-machined specimens and the mean widths β_k of the different zones of the HAZ are correlated. Different test samples are prepared by varying the feed rate at a constant laser output power, leading to a wide spectrum of HAZ sizes (Jaeschke et al. 2014b). In Figure 24.19, the resulting static tensile strength $\sigma_{T,L}$ is plotted against the mean widths of the HAZ for β_1 to β_3. The mean value and the corresponding standard deviation for both $\sigma_{T,L}$ and β_k are shown. A linear relationship between the maximum achievable tensile strength and the HAZ can be observed for each β_k. With decreasing β_k, $\sigma_{T,L}$ increases.

Assuming β_k would not contribute to the transfer of loads, an extrapolated $\beta_k = 0$ would constitute the tensile strength of unmodified bulk material. Doing so, the data can be fitted using linear regression, and $\sigma_{T,L}$ ($\beta_k = 0$) can be extrapolated to $\beta_k = 0$, which is equal to an absence of HAZ. It turns out that $\sigma_{T,L}$ ($\beta_1 = 0$) and $\sigma_{T,L}$ ($\beta_2 = 0$) are in the same range as the mean value of the tested milled samples, which reaches $\sigma_{T,M} = 640.3$ MPa. For $\beta_3 = 0$ mm, considerably higher strengths of $\sigma_{T,L}$ ($\beta_3 = 0$) = 683.8 MPa are found, implying a contribution of β_3 to the measured tensile strength. Decreasing the size of β_k, e.g., by applying high feed rates, will lead to higher static tensile strength $\sigma_{T,L}$ of laser-treated material. Figure 24.19 clearly shows a reduction in tensile strength with increasing HAZ. It is assumed that, due to a modified or perhaps missing fiber–matrix adhesion, the respective zones B_k cannot entirely convey the applied loads during the testing phase. Based on this assumption, with respect to the specimen's cross sections, the area able to transmit loads is scaled down by twice the

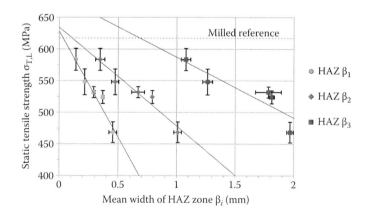

FIGURE 24.19 Static tensile strength depending on the extent of the different HAZ zones.

HAZ zone β_k on each side of the laser-cut specimen. The remaining unmodified zone is called the reduced load-bearing area. The influence of this reduced load-bearing area on the achievable tensile loads is analyzed by evaluating milled specimens, providing different widths. In Figure 24.20, the maximum tensile load F is shown as a function of the width reduction χ of laser-machined and milled reference specimens.

For laser-cut CF-PPS FAB, width reduction relates to the HAZ appearing on both cutting edges of the composite laminate, incapable of transferring tensile loads. The width reduction is given by the Equation 24.5.

$$\chi = 2\beta_k \quad (24.5)$$

Figure 24.20 also shows the tensile loads of the milled specimens, for which the width is reduced in steps of $\Delta\chi = 0.2$ mm, starting at the origin width of $w_0 = 10$ mm up to $w = 7.8$ mm, corresponding to $\chi = 2.2$ mm. The ratio of the reduced width to the original width θ is calculated according to Equation 24.6 and presented as a second x-axis in Figure 24.20.

$$\theta = \frac{w_0 - \chi}{w_0} \quad (24.6)$$

Henceforth, it becomes possible to better compare the impact of the width reduction χ as a consequence of laser-induced thermal damage, with the influence in width reduction of milled reference samples on the load applied. In contrast to β_1 and β_3, the lines of best fit for β_2 evidently run parallel and close to those of milled specimens, as presented in Figure 24.20. It can therefore be stated that β_2 accurately represents the width of laser-cut CFRTP, unable to transmit tensile loads. Figure 24.21 illustrates the formation of a reduced load-bearing area.

In a subsequent step, the ratio ϕ_T of static tensile strength of laser-machined $\sigma_{T,L}$ to milled specimens' $\sigma_{T,M}$ is calculated, taking into account the maximum tensile loads. Increasing feed rates result in higher ratios ϕ_T, up to $\phi_T = 0.91$ (Figure 24.22).

As a consequence of a reduced load-bearing area defined by β_2 (Figure 24.21), an effective ratio of the tensile strength Φ_T between laser-cut and milled specimens can be expressed as follows:

$$\Phi_T = \frac{\sigma_{A,L}}{\sigma_{T,M}}. \quad (24.7)$$

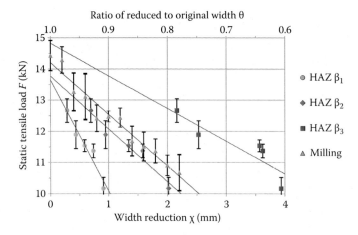

FIGURE 24.20 Maximum tensile load as a function of width reduction for both laser-processed and milled CF-PPS FAB.

Laser Processing of Thermoplastic Composites

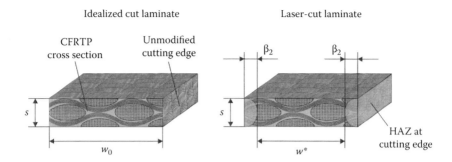

FIGURE 24.21 Reduced load-bearing area caused by thermal impact.

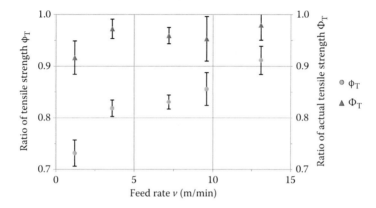

FIGURE 24.22 Ratio of tensile strength and ratio of actual tensile strength for varying feed rate.

Here, $\sigma_{A,L}$ describes the actual tensile strength of laser-processed samples, and $\sigma_{T,M}$, the tensile strength of the milled samples. Knowing the maximum load F and with s as the thickness of the laminate, $\sigma_{A,L}$ can be written as

$$\sigma_{A,L} = \frac{F}{sw^*}. \qquad (24.8)$$

According to Figure 24.21, the effective specimen width w^* can be expressed in terms of w_0 and β_2 as

$$w^* = w_0 - 2\beta_2. \qquad (24.9)$$

If w^* is considered for the calculation of Φ_T, values between $\Phi_T = 0.92$ and $\Phi_T = 0.98$ are reached (cf. Figure 24.22).

As a result, the actual tensile strength $\sigma_{A,L}$ of laser-machined CFRTP is at a level similar to those of milled laminates $\sigma_{T,M}$, if it is referenced to tensile loads, only arising from undamaged, load transmitting areas as limited by β_2.

The basic analytical model presented here is able to describe the observed reduction in tensile strength of laser-machined CFRTP laminates, and the results are in good accordance with the measuring data received while testing milled reference specimens at varying widths. A similar approach is used for modeling the behavior of laser-cut CF-PPS FAB under interlaminar shear load. Jaeschke et al. (2011) also identify a reduction of the load-bearing area, arising from a laser-induced HAZ, as the main influence factor for decreased shear strengths compared to those received for milled and abrasive water jet–cut specimens. However, since this model mainly considers intact fiber–matrix adhesion to a greater or lesser extent, contributing to the transfer of loads, this model is primarily designed for describing fiber-dominated behavior while composites load.

24.5.4 Influence of Moisture Content on Cutting Behavior

From different welding techniques such as hot-tool, vibration, or laser welding it is known that the quality of the weld seam and mechanical properties depend on the moisture content of the joining partners. In addition, the strength of welding connections can be increased if material with low moisture content is used (Kagan et al. 2005; Stokes 2001). Bearing in mind the advantages and challenges of laser cutting, detailed knowledge of the effect of the material and process conditions on the cutting behavior is required in order to offer a fully developed processing method for the growing thermoplastic composite market. Therefore, investigating the influence of moisture content on the quality of laser-cut parts and components is an essential issue. Staehr et al. (2015) studied the machinability of CF-PPS FAB and CF-PEI FAB using a single-mode fiber laser in conjunction with a galvoscanner system. By applying contour as well as multipass cutting strategies, test specimens were prepared out of CFRTP laminates at varying moisture contents, and the effect on in-plane shear properties was investigated.

Laminates with different moisture content were prepared by drying for 48 h at 120°C in an oven and stored afterward in distilled water or in ambient conditions. The moisture content ψ was calculated by dividing the weight difference between the conditioned material and the dry material by the weight of the dry material. Dried CFRTP was laser-cut after a short cool-down phase. As mechanical tests, in-plane shear measurements according to ASTM D3518M were performed. Figure 24.23 shows the average width of the HAZ of laser-machined CF-PEI FAB as a function of process strategy for all three material conditions.

For both cutting strategies separately, β_1 remains constant independent of the moisture content, whereas β_2 increases slightly for wet material. An increase of more than 100% is observed for β_3 from dry to wet material. Applying multipass cutting, β_1 and β_2 can be reduced by more than 50% compared to contour cutting. In addition, β_3 for multipass cut dry material also decreases, whereas for ambient humidity (rh) or wet material, higher values are observed. The maximum width of $\beta_3 =$ 1.63 mm is measured for a multipass cut of wet material. Much higher standard deviations arise for multipass cutting, which is supposed to be a consequence of a longer and dynamic heat insertion. The influence of the moisture content becomes primarily visible in the extent of β_3.

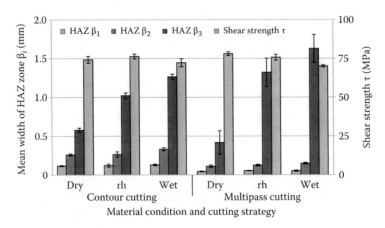

FIGURE 24.23 Mean width of HAZ and in-plane shear strength of laser-cut CF-PEI FAB using different strategies and material conditions (rh, $\psi = 0.2\%$; wet, $\psi = 0.6\%$).

Figure 24.23 also contains shear strength data for all material conditions. For contour cutting, the highest values are found for material conditioned in ambient humidity and for dry material in the case that multipass cutting is applied. Independent of the cutting strategy, the lowest shear strengths are found for wet material. If the measured shear strength is compared to the HAZ, for multipass cutting, it is found that with increasing moisture content, the achievable shear strengths decrease, while β_3 increases at the same time. The shear strength of the milled reference specimens yield τ_{ref} = 79 MPa, which is in the same range as the strength values that have been achieved for multipass cutting of CF-PEI FAB in dry and room-humidity conditions.

24.6 FURTHER PROCESSING TECHNOLOGIES

24.6.1 Surface Activation

Due to a high load transfer in the joining area and corrosion resistance, composite parts are often joined to more complex structures by adhesive bonding. In order to enhance the bond strength, the surface of composites can be prepared by plasma, chemicals, or laser radiation. These techniques are used to activate the surface and to remove release agents or other surface contaminations. As an example, UV laser radiation can be applied to break the molecular structure of the matrix material (Fischer et al. 2012a,b; Hallmann et al. 2012). It is discussed whether photothermal or photochemical effects are primarily responsible for the change in the surface characteristics. These reactions are influenced by the characteristics of the polymers, laser fluence, wavelength, and pulse duration (Sato and Nishio 2001). Srinivasan et al. (1986) investigated the ablation behavior of unreinforced polyimide (PI) and polymethylmethacrylate (PMMA) for a wavelength of λ_L = 193 nm and λ_L = 248 nm. They determined that both effects are responsible for the etching depth per laser pulse. Based on this assumption, the etching depth is the sum of the photochemical ablation and the ablation due to the thermal impact. Furthermore, they have stated that at low laser fluences, the thermal effects are negligible, but for high laser fluences these effects become significant. Investigations by Laurens et al. (1999) have shown that the surface of PEEK becomes hydrophilic after being treated with laser radiation of a wavelength of λ_L = 193 nm. This effect was related to a formation of acidic sites on the surface. Furthermore, they determined that the hydrophilic effect increases with the number of pulses. The laser-treated surface of PEEK led to improved adhesive bond strengths.

24.6.2 Surface Ablation

Removal of minor damage in fiber-reinforced thermoplastics is time-consuming and mainly done by grinding. With this manual process, a structure is ablated into the material, and the damaged area is removed completely. The pressure between the abrasive wheel and the workpiece is normally applied by the worker and varies from person to person. After the ablation process, the generated step structure can be refilled by patches consisting of CFRP or GFRP, which are adhesively bonded into the gap (Voelkermeyer et al. 2011). Other ablation techniques that can be automated are currently being researched. One technique is laser-based ablation, where NIR or UV lasers are used to generate a step structure into the composite. When machining CFRP, the heat conduction along the carbon fibers has to be taken into account for the process development. For laser-based ablation, the laser beam is guided at a high speed over the workpiece by a galvoscanner system. For the ablation of an area, hatching is applied. This means that the area is filled with uniformly spaced lines. The orientations of the hatching lines are adapted, depending on the fiber arrangement and material characteristics. Furthermore, the optimal number of ablating cycles and the hatching direction have to be determined for every material. Besides the material itself, the laser fluence, wavelength, pulse duration, feed rate, and distance between the hatching lines affect the ablation depth (Dittmar et al. 2013; Fischer et al. 2010; Voelkermeyer et al. 2011).

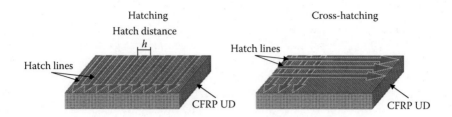

FIGURE 24.24 Hatching strategies for laser-based CFRP ablation.

Different experiments have been carried out focusing on the dependency between process parameters and ablation depth for varying materials. Voelkermeyer et al. (2011) investigated the dependency between the hatch distances and ablation depth. The experiments were conducted with a frequency-converted solid-state laser emitting at a wavelength of λ_L = 355 nm, with a pulse duration of t_p = 15 ns and a pulse frequency of f_P = 40 kHz. The laser beam was guided over the workpiece by a galvoscanner system with a feed rate of v = 1.8 m/min. The hatch distance varied, h = 5–100 µm. It was shown that the ablation depth (g) increases with shorter hatch distances and with the number of cycles. With a small hatch distance of h = 5 µm, n = 17 cycles were needed to ablate CFRP with g = 5 mm. An increase of the hatch distance to h = 20 µm leads to an increase in the number of cycles to n = 35. This is caused by the lower amount of energy applied by larger hatch distances for one point on the surface due to the lower degree of overlap of the hatch lines. Thus, more cycles are needed to ablate the same amount of material.

Wolynski et al. (2011) investigated the effect of the wavelength on the ablation behavior of CFRP. Their goal was to optimize the drilling of holes into CFRP. The experiments were conducted with lasers emitting at λ_L = 355 nm, λ_L = 532 nm, and λ_L = 1064 nm. They determined that the ablation depth rises with increasing laser output power. Furthermore, for all wavelengths, they detected similar ablation rates up to a laser output power of P_L = 10 W.

Besides the ablation depth, the surface roughness is of interest for subsequent repair steps. Voelkermeyer et al. (2013) investigated the influence of the hatch strategy on the surface quality of UD CFRP. The experiments were conducted with a pulsed fiber laser emitting at λ_L = 1067 nm. The hatching strategies included hatching perpendicular to and along the fiber orientation, as well as cross-hatching (Figure 24.24). It was observed that hatching perpendicular to the fiber orientation generates a smooth surface. When hatching along the fiber orientation, the surface became fuzzy.

The combination of both hatching strategies, so-called cross-hatching, mainly depended on the last applied hatching direction regarding the fiber orientation. When the last hatching cycle was perpendicular to the fiber orientation, the surface was relatively smooth, with exposed fibers.

24.6.3 Automated Fiber Placement

Fiber placement refers to the deposition of fiber-reinforced tapes onto a tool or a laminate. With fiber placement, 3-D or 2-D parts are built up. This technique is applied for thermoset and thermoplastic tape. With thermoplastic tapes, an out-of-autoclave processing is possible, avoiding an autoclave process step, as is needed for the processing of thermoset tapes. For thermoplastic tape placement, the laminate and tape layer are heated up and then pressed together by a compaction shoe or roller (Figure 24.25). A hot gas torch or laser radiation can be used for the heating. In laser-assisted tape placement, different kinds of laser sources, such as CO_2, Nd:YAG, and diode lasers, are used (Grouve 2012; Koelzer 2008).

Furthermore, the laser-based tape laying process is affected by the laser output power, placement speed, roller pressure, and cold or hot tooling among other things (Comer et al. 2014).

The advantages of a laser-based tape laying process are high placement velocities and the possibility to lay down complicated geometries. When benefitting from this high process speed, the

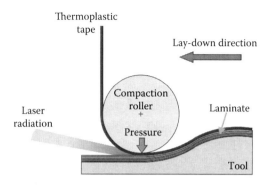

FIGURE 24.25 Schematic principle of a fiber placement process.

heating and the cooling rates of the tapes are high, and they can affect the crystallinity of semicrystalline polymers (Grouve 2012). In this manufacturing process, tapes with a fiber content of more than 60% can be used (Brecher et al. 2011).

Comer et al. (2014) investigated the interlaminar toughness and quality of carbon fiber-reinforced PEEK laminates produced by laser-assisted tape placement with different process parameters. They tested the interlaminar toughness of the tape connection by two-ply wedge-peel tests. Furthermore, they prepared microstructures and performed fracture surface analysis to evaluate the connection quality. They have, among other things, determined that samples made with an unheated nonmetallic tool resulted in the highest peel force, with $F_{peel} = 4.1$ N/mm, as well as the lowest void content of 1%. Due to fast cooling of the connection area, the PEEK matrix material exhibited only about 20% crystallinity, which affects stiffness and the chemical resistance. To increase the crystallinity, they used a metallic heated tool (280°C). However, the heated tool led to a reduction in the peel forces to $F_{peel} = 1.2$–1.5 N/mm and to an increase up to 5% of the void content.

Grouve (2012) investigated the influence of the tape placement process parameters laser output power, placement velocity, and incident laser angle on the interlaminar bond strength. The material used was carbon fiber tape with a PPS matrix. For the evaluation of the interfacial fracture toughness, mandrel peel tests were performed. Grouve determined that low fracture toughness was correlated to extremely high tape temperatures, which result in thermal degradation of the PPS. Another influencing factor is a low interface temperature, which does not provide enough energy for interdiffusion of polymer chains. Also, Grouve stated that high placement velocities and low laser output powers result in high interfacial fracture toughness. For these experiments, the laser was primarily focused on the tape to avoid the heating up of the underlying laminate, and thereby avoiding thermal degradation. Besides the production of laminates or 3-D parts, the tape placement can be used for winding tanks, e.g., pressure vessels. In this case, the tape material is pressed with a compaction roller onto a rotating mandrel (Koelzer 2008).

REFERENCES

Ackermann, J. 2010. *Prozesssicherung beim Laserdurchstrahlschweißen thermoplastischer Kunststoffe.* Bamberg: Meisenbach.

Amend, P., T. Frick and M. Schmidt. 2012. Laser-based hot melt bonding of CFRP and GFRP—A novel joining technology for multi-material-design. *31st International Congress on Applications of Lasers & Electro-Optics (ICALEO).* Anaheim, CA.

Amend, P., S. Pfindel and M. Schmidt. 2013. Thermal joining of thermoplastic metal hybrids by means of mono- and polychromatic radiation. *Physics Procedia* 41:98–105.

Amend, P., C. Mohr and S. Roth. 2014. Experimental investigations of thermal joining of polyamide aluminum hybrids using a combination of mono- and polychromatic radiation. *Physics Procedia* 56:824–834.

Bliedtner, J., H. Mueller and A. Barz. 2013. *Lasermaterialbearbeitung: Grundlagen-Verfahren-Anwendungen–Beispiele.* München: Hanser.

Bluemel, S., P. Jaeschke, V. Wippo, S. Bastick, U. Stute, D. Kracht and H. Haferkamp. 2012. Laser Machining of CFRP using a high power laser—Investigation on the heat affected zone. *15th European Conference on Composite Materials (ECCM 15)*. Venice, Italy.

Bluemel, S., R. Staehr, P. Jaeschke and U. Stute. 2013. Determination of corresponding temperature distribution within CFRP during laser cutting. *Physics Procedia* 41:401–407.

Bluemel, S., P. Jaeschke, O. Suttmann and L. Overmeyer. 2014a. Comparative study of achievable quality cutting carbon fibre reinforced thermoplastics using continuous wave and pulsed laser sources. *Physics Procedia* 56:1143–1152.

Bluemel, S., S. Brede, P. Jaeschke, O. Suttmann and L. Overmeyer. 2014b. Applying a DOE model for the determination of appropriate process windows for nanosecond laser processing of CFRP. *33rd International Congress on Applications of Lasers & Electro-Optics (ICALEO)*. San Diego, CA.

Brecher, C., M. Dubratz, J. Stimpfl and M. Emonts. 2011. Innovative manufacturing of 3D-lightweight components. *Laser Technik Journal* 5:36–40.

Brunnecker, F. 2011. Innovations powered by laser plastic welding—Laser plastic welding for modern joining technologies. *Laser Technik Journal* 8:20–24.

Chen, J. W. and J. Zybko. 2002. Laser assembly technology for planar microfluidic devices. *Annual Technical Conference of the Society of Plastics Engineers (ANTEC)*. San Francisco, CA.

Chen, M., G. Zak and P. J. Bates. 2008. Estimating contour laser transmission welding start-up conditions using a non-contact method. *Welding in the World* 52:71–76.

Chen, M., G. Zak and P. J. Bates. 2011. Effect of carbon black on light transmission in laser welding of thermoplastics. *Journal of Materials Processing Technology* 211:43–47.

Comer, A., P. Hammond, D. Ray, J. Lyons, W. Obane, D. Jones, R. O' Higgins and M. McCarthy. 2014. Wedge peel interlaminar toughness of carbon-fibre/PEEK thermoplastic laminates manufactured by laser-assisted automated-tape-placement. *SETEC 14 Tampere Conference & Table Top Exhibition*. Tampere, Finland.

Dittmar, H., F. Gaebler and U. Stute. 2013. UV-laser ablation of fibre reinforced composites with ns-pulses. *Physics Procedia* 41:266–275.

Dubé, M., P. Hubert, J. N. A. H. Gallet, D. Stavrov, H. E. N. Bersee and A. Yousefpour. 2011. Metal mesh heating element size effect in resistance welding of thermoplastic composites. *Journal of Composite Materials* 46:911–919.

Eichler, J. and H. J. Eichler. 2006. *Laser*. Berlin, Heidelberg: Springer.

Fernandez Villegas, I., L. Moser, A. Yousefpour, P. Mitschang and H. E. N. Bersee. 2012. Process and performance evaluation of ultrasonic, induction and resistance welding of advanced thermoplastic composites. *Journal of Thermoplastic Composite Materials* 26:1007–1024.

Fiegler, G. 2007. *Ein Beitrag zum Prozessverständnis des Laserdurchstrahlschweißens von Kunststoffen anhand der Verfahrensvarianten Quasi-Simultan-und Simultanschweißen*. Aachen: Shaker Verlag.

Fischer, F., L. Romoli and R. Kling. 2010. Laser-based repair of carbon fiber reinforced plastics. *CIRP Annals—Manufacturing Technology* 59:203–206.

Fischer, F., S. Kreling and K. Dilger. 2012a. Surface structuring of CFRP by using modern excimer laser sources. *Physics Procedia* 39:154–160.

Fischer, F., S. Kreling, P. Jaeschke, M. Frauenhofer, D. Kracht and K. Dilger. 2012b. Laser surface pretreatment of CFRP for adhesive bonding in consideration of the absorption behaviour. *The Journal of Adhesion* 88:350–363.

Frick, T. 2007. *Untersuchung der prozessbestimmenden Strahl-Stoff-Wechselwirkungen beim Laserschweißen von Kunststoffen*. Bamberg: Meisenbach.

Fuchs, A. N., F. X. Wirth, P. Rinck and M. F. Zaeh. 2014. Laser-generated macroscopic and microscopic surface structures for the joining of aluminum and thermoplastics using friction press joining. *Physics Procedia* 56:801–810.

Grewell, D. and A. Benatar. 2007. Welding of plastics: Fundamentals and new developments. *International Polymer Processing* 22:43–60.

Grouve, W. J. B. 2012. Weld strength of laser-assisted tape placed thermoplastic composites. PhD Thesis, University of Twente, Enschede, the Netherlands.

Hallmann, L., A. Mehl, N. Sereno and C. H. F. Haemmerle. 2012. The improvement of adhesive properties of PEEK through different pre-treatments. *Applied Surface Science* 258:7213–7218.

Hansen, P., C. Jeenjitkaew, V. Wippo and P. Jaeschke. 2014. Laser transmission welding of thermoplastic composite structures. *SAMPE Europe Conference & Exhibition—SEICO14*. Paris, France.

Jaeschke, P. 2012. Laser transmission welding of continuous carbon fiber reinforced plastics and thermoplastic polymers. PhD Thesis, Leibniz University Hanover, Germany. Hanover: PZH Produktionstechnisches Zentrum.

Jaeschke, P., D. Herzog, M. Kern, A. S. Erciyas, C. Peters, H. Purol and A. S. Herrmann. 2009. Laser transmission welding of thermoplastic composites—Fundamental investigations into the influence of the carbon fibre reinforcement and orientation on the weld formation. *Joining Plastics* 4:247–255.

Jaeschke, P., D. Herzog, H. Haferkamp, C. Peters and A. S. Herrmann. 2010. Laser transmission welding of high-performance polymers and reinforced composites—A fundamental study. *Journal of Reinforced Plastics and Composites* 29:3083–3094.

Jaeschke, P., M. Kern, U. Stute, H. Haferkamp, C. Peters and A. S. Herrmann. 2011. Investigations on interlaminar shear strength properties of disc laser machined consolidated CF-PPS laminates. *Express Polymer Letters* 5:238–245.

Jaeschke, P., K. Stolberg, S. Bastick, E. Ziolkowski, M. Roehner, O. Suttmann and L. Overmeyer. 2014a. Cutting and drilling of carbon fiber reinforced plastics (CFRP) by 70W short pulse nanosecond laser. *Photonics West, LASE Conference.* San Francisco, CA.

Jaeschke, P., M. Kern, U. Stute, D. Kracht and H. Haferkamp. 2014b. Laser processing of continuous carbon fibre reinforced polyphenylene sulfide organic sheets—Correlation of process parameters and reduction in static tensile strength properties. *Journal of Thermoplastic Composite Materials* 27:324–337.

Jaeschke, P., V. Wippo, O. Suttmann and L. Overmeyer. 2014c. Advanced laser welding of high-performance thermoplastic composites. *33rd International Congress on Applications of Lasers & Electro-Optics (ICALEO).* San Diego, CA.

Kagan, V. 2003. Innovations in laser welding of thermoplastics. This advanced technology is ready to be commercialized. *SAE World Congress Welding & Joining.* Detroit, MI.

Kagan, V. A. and R. J. Nichols. 2005. Benefits of induction welding of reinforced thermoplastics in high performance applications. *Journal of Reinforced Plastics and Composites* 24:1345–1352.

Kagan, V. A., R. G. Bray and W. P. Kuhn. 2002. Laser transmission welding of semi-crystalline thermoplastics—Part I: Optical characterization of nylon based plastics. *Journal of Reinforced Plastics and Composites* 21:1101–1122.

Kagan, V. A., S. A. Kocheny and J. E. Macur. 2005. Moisture effects on mechanical performance of laser-welded polyamide. *Journal of Reinforced Plastics and Composites* 24:1213–1224.

Klein, R. 2012. *Laser Welding of Plastics.* Weinheim: Wiley-VCH.

Koelzer, P. 2008. *Temperaturerfassungssystem und Prozessregelung des laserunterstützten Wickelns und Tapelegens.* München: Shaker.

Laurens, P., B. Sadras, F. Decobert, F. Arefi and J. Amouroux. 1999. Modifications of polyether–etherketone surface after 193 nm and 248 nm excimer laser radiation. *Applied Surface Science* 138:93–96.

Leone, C., S. Genna and V. Tagliaferri. 2014. Fibre laser cutting of CFRP thin sheets by multi-passes scan technique. *Optics and Lasers in Engineering* 53:43–50.

McGrath, G. and B. Cawley. 2007. Review of market for laser welding of plastics. *Joining Plastics* 2:175–179.

Negarestani, R., L. Li, H. K. Sezer, D. Whitehead and J. Methven. 2010. Nano-second pulsed DPSS Nd:YAG laser cutting of CFRP composites with mixed reactive and inert gases. *International Journal of Advanced Manufacturing Technology* 49:553–566.

Offringa, A. 2011. A review of continuous fiber reinforced thermoplastics in aerospace applications. *SAMPE Technical Conference Tutorial.* Long Beach, CA.

Rodríguez-Vidal, E., J. Lambarri, C. Soriano, C. Sanz and G. Verhaeghe. 2014. A combined experimental and numerical approach to the laser joining of hybrid polymer–Metal parts. *Physics Procedia* 56:835–844.

Rudolf, R., P. Mitschang and M. Neitzel. 2000. Induction heating of continuous carbon-fibre-reinforced thermoplastics. *Composites: Part A* 31:1191–1202.

Russek, U. A. 2009. *Laserschweißen von Kunststoffen.* München: Süddeutscher Verlag.

Sato, H. and S. Nishio. 2001. Polymer laser photochemistry, ablation, reconstruction, and polymerization. *Journal of Photochemistry and Photobiology C—Photochemistry Reviews* 2:139–152.

Schneider, F., N. Wolf and D. Petring. 2013. High power laser cutting of fiber reinforced thermoplastic polymers with cw- and pulsed lasers. *Physics Procedia* 41:415–420.

Schuermann, H. 2005. *Konstruieren mit Faser-Kunststoff-Verbunden.* Berlin, Heidelberg: Springer.

Sheikh-Ahmad, J. Y. 2009. *Machining of Polymer Composites.* New York: Springer.

Srinivasan, V., M. A. Smrtic and S. V. Babu. 1986. Excimer laser etching of polymers. *Journal of Applied Physics* 59:3861–3867.

Staehr, R., S. Bluemel, P. Hansen, P. Jaeschke, O. Suttmann and L. Overmeyer. 2015. The influence of moisture content on the heat affected zone and the resulting in-plane shear strength of laser cut thermoplastic CFRP. *Plastics, Rubber and Composites.* 44:111–116.

Stokes, V. K. 2001. The vibration and hot-tool welding of polyamides. *Polymer Engineering and Science* 41:1427–1439.

Voelkermeyer, F., F. Fischer, U. Stute and D. Kracht. 2011. Laser-based approach for bonded repair of carbon fiber reinforced plastics. *Physics Procedia* 12:537–542.

Voelkermeyer, F., P. Jaeschke, U. Stute and D. Kracht. 2013. Laser-based modification of wettablility for carbon fiber reinforced plastics. *Applied Physics A* 112:179–183.

Walter, J., M. Hustedt, S. Kaierle, R. Staehr, P. Jaeschke, O. Suttmann and L. Overmeyer. 2014. Laser cutting of carbon fiber reinforced plastics—Investigation of hazardous process emissions. *Physics Procedia* 56:1153–1164.

Weber, R., C. Freitag, T. Kononenko, M. Hafner, V. Onuseit and P. Berger. 2012. Short-pulse laser processing of CFRP. *Physics Procedia* 39:137–146.

Wilke, L., H. Potente and J. Schnieders. 2008. Simulation of quasi-simultaneous and simultaneous laser welding. *Welding in the Word* 52:56–66.

Wippo, V., P. Jaeschke, U. Stute, D. Kracht and H. Haferkamp. 2012. The influence of carbon fibres on the temperature distribution during the laser transmission welding process. *15th European Conference on Composite Materials (ECCM15).* Venice, Italy.

Wippo, V., R. Staehr, P. Brede, P. Jaeschke and H. Haferkamp. 2014a. Pyrometric temperature monitoring for flexible scanner-based laser transmission welding processes. *Joining Plastics* 1:39–45.

Wippo, V., P. Jaeschke, M. Brueggmann, O. Suttmann and L. Overmeyer. 2014b. Advanced laser transmission welding strategies for fibre reinforced thermoplastics. *Physics Procedia* 56:1191–1197.

Witten, E., T. Kraus and M. Kuehnel. 2014. *Composites-Marktbericht 1/2014—The Global CFRP-Market.* Frankfurt: Carbon Composites & AVK-Verlag.

Wolynski, A., T. Herrmann, P. Mucha, H. Haloui and J. L'huillier. 2011. Laser ablation of CFRP using picosecond laser pulses at different wavelengths from UV to IR. *Physics Procedia Part B* 12:292–301.

Young, T. M. 2008. Impact of Nd-YAG laser drilling on the fatigue characteristics of APC-2A/AS4 thermoplastic composite material. *Journal of Thermoplastic Composite Materials* 21:543–555.

Zweifel, H., R. D. Maier and M. Schiller. 2009. *Plastics Additives Handbook.* Cincinnati, OH: Hanser.

25 Bioplastics

Caisa Johansson

CONTENTS

25.1 Historical Development and Commercialization of Bioplastics 836
    25.1.1 Definition of Bioplastics 836
    25.1.2 Market for Bioplastics 838
    25.1.3 Sustainability Assessment of Bioplastics 838
    25.1.4 Technical Challenges for Implementation of Bioplastics 839
25.2 Polymerization and Processing Technologies of Bioplastics 840
25.3 Bioplastics from Natural Sources (Class I) 841
    25.3.1 Starch 841
        25.3.1.1 Physical and Chemical Modification of Starch—Starch Derivatives 846
        25.3.1.2 Reduction of Molecular Weight 846
        25.3.1.3 Starch Ethers 848
        25.3.1.4 Starch Esters 848
        25.3.1.5 Genetic Manipulation of Starches 848
        25.3.1.6 Thermoplastic Starch 849
    25.3.2 Lignocellulose 849
        25.3.2.1 Cellulose 849
        25.3.2.2 Hemicellulose 852
        25.3.2.3 Lignin 853
25.4 Bioplastics from Classical Chemical Synthesis (Class II) 854
    25.4.1 PE from Sugar Cane (Bio-PE) 854
    25.4.2 Polylactide 855
    25.4.3 Polycaprolactone 858
25.5 Bioplastics from Microorganisms 859
    25.5.1 Polyhydroxyalkanoates 859
25.6 Structural and Phase Characteristics and Effect on Properties of Bioplastics 861
    25.6.1 Thermoplastic Starch 861
    25.6.2 Polylactide 863
    25.6.3 Polyhydroxyalkanoates 865
25.7 Additives and Their Effects on Properties and Applications 865
    25.7.1 Thermoplastic Starch 865
    25.7.2 Polylactide 866
    25.7.3 Polyhydroxyalkanoates 866
25.8 Blends and Composites Derived from Bioplastics 866
    25.8.1 Thermoplastic Starch 867
    25.8.2 Lignocellulose 868
    25.8.3 Polylactide 869
    25.8.4 Polycaprolactone 869
    25.8.5 Polyhydroxyalkanoates 869
25.9 Recycling and Biodegradation of Bioplastics 870
25.10 Applications of Bioplastics 871
References 872

25.1 HISTORICAL DEVELOPMENT AND COMMERCIALIZATION OF BIOPLASTICS

Plastic articles have become an essential part of the everyday life in developed countries over the last 60–70 years, and it is nowadays difficult to imagine a life without plastic materials. However, conventional plastic materials are manufactured from petroleum-based feedstock, and increased concern about the continuously reducing availability of crude oil, with increased prices as a consequence, has raised the demand for economically viable alternatives. In addition, solid waste from plastics leads to increased landfill, which has severe impact on, for example, the marine life since plastic materials are not degradable in nature. Increasing environmental awareness among consumers, wholesalers, and retailers has pushed the market actors toward the development of bioplastic products for a wide variety of applications. The ambition to identify and implement more sustainable materials and processes has also been advanced by tightened legislations regarding the environmental impact of plastic materials and articles over their whole life cycle.

25.1.1 Definition of Bioplastics

The term *bioplastics* has become widely used in society to designate materials that are considered as more sustainable than conventional, synthetic plastic materials. However, a generally accepted definition does not yet exist. The term *bio* has sometimes been used to classify a biodegradable material, but more common is the designation that the material is derived from renewable resources (Johansson et al. 2012).

The European Bioplastics Association states that bioplastics are not a single kind of polymer but rather a family of materials that can vary considerably from each other. They classify bioplastics in three categories, each with its own individual characteristics (European Bioplastics 2014):

- *Biobased* or *partially biobased* but *nonbiodegradable* plastics, e.g., biobased variants of polyethylene (bio-PE), polypropylene (bio-PP), or polyethylene terephthalate (bio-PET), and biobased technical performance polymers
- Plastics that are both *biobased* and *biodegradable*, e.g., polylactides (PLAs) and polyhydroxyalkanoates (PHAs)
- Plastics that are based on *fossil resources* and are *biodegradable*, e.g., polycaprolactone (PCL)

The Sustainable Biomaterials Collaborative (SBC) was established in 2006 as a coalition of organizations that advances the introduction and use of biobased products. SBC has developed useful tools for manufacturers, purchasers, and consumers to evaluate the sustainability of biobased materials. They define biobased plastics as "plastics in which 100% of the carbon is derived from renewable agricultural and forestry resources such as corn starch, soybean protein and cellulose" (SBC 2014), but point out that not all bioplastics are biodegradable since biodegradability is directly linked to the chemical structure of the material, not to the origin of the raw materials. In addition, not all biodegradable plastics are to be defined as bioplastics.

The US Department of Agriculture defines a biobased product as a "commercial or industrial product (other than feed or food), that is composed, in whole or in significant part, of biological products or renewable domestic agricultural materials, including plant, animal or marine materials or forestry materials" (USDA 2005). The Biodegradable Products Institute (BPI), a professional association that promotes the use and recycling of biodegradable polymeric materials via composting, points out that not all biobased materials are biodegradable (BPI 2014), even if they fulfill the ASTM definition of a biobased material: "an organic material in which carbon is derived from a renewable resource via biological processes. Biobased materials include all plant and animal mass derived from CO_2 recently fixed via photosynthesis, per definition of a renewable resource."

A more detailed classification of biobased polymers (Petersen et al. 1999; van Tuil et al., 2000; Johansson et al. 2012) includes

- Polymers directly extracted from natural materials like polysaccharides (e.g., cellulose, starch, lignin, proteins, and lipids)
- Polymers produced by classical chemical synthesis from renewable biobased monomers (e.g., PLA)
- Polymers produced by microorganisms or genetically transformed by bacteria (e.g., PHAs and bacterial cellulose)

Figure 25.1 shows a schematic overview of different classes of bioplastics based on the abovementioned definitions.

Müller et al. (2014) presented different approaches to end-of-life treatment of biobased packaging materials and used the definition of bioplastics from this perspective:

- Conventional plastics made from renewable resources, which can contain a proportion of fossil resources
- Polymers made from fossil or renewable resources, which are suitable for home composting
- Polymers produced from fossil or renewable resources, which are standard for industrial composting
- Oxo-degradable plastics, made from PE and a few additives, which are degradable under atmospheric conditions under the action of ultraviolet radiation and oxygen

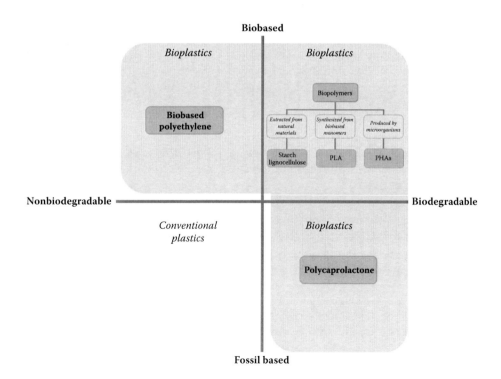

FIGURE 25.1 Schematic overview of bioplastics classified as biodegradable/nonbiodegradable and biobased/fossil based.

25.1.2 MARKET FOR BIOPLASTICS

The global production of plastic materials amounts to 280 million tons p.a. (Plastics Europe 2012; European Bioplastics 2014). Shen et al. (2009) have compiled the technical substitution potential on a material-by-material basis using information from interviews with industry experts and from these data calculated a maximum technical substitution potential. They ended up with a maximum technical substitution potential of biobased plastics and fibers replacing their petrochemical counterparts of 90% of the total consumption of plastics and fibers worldwide (2007). However, production costs and capital availability, technical challenges in scale-up, short-term availability of biobased feedstock, and the need for the plastic conversion sector to adapt to the new materials hinder the exploitation of this substitution in a short to medium term.

The global production capacities in 2012 were estimated to be about 1.4 million tons, out of which the category of biobased but nonbiodegradable materials (including bio-PE and bio-PET) equaled 56.6% and the class of biodegradable materials (including PLA, PHA, starch blends, regenerated cellulose, and others) amounted to 43.4%. The distribution of production capacities per region corresponds to 22.8% (Europe), 36.2% (Asia), and 40.7% (North and South America) (European Bioplastics 2014). European Bioplastics forecasts the production capacities to multiply by 2017 to more than 6 million tons, divided on 1 million tons for biodegradable and 5 million tons for biobased but nonbiodegradable plastics (European Bioplastics 2014).

25.1.3 SUSTAINABILITY ASSESSMENT OF BIOPLASTICS

The development of bioplastics is still in its infancy, and considerable research is still needed regarding the environmental impact of the materials throughout their life cycles (Álvarez-Chávez et al. 2012) and regarding their optimum end-of-life options (Müller et al. 2014).

To assess the sustainability of biobased plastics, a comprehensive approach must be undertaken. This includes the source from which the material is derived (preferably from renewable resources); the production processes that need to be energy efficient, safe, and healthy; and last but not least the fate of the material after its intended use, i.e., the end of life.

Halley (2005) listed a number of strategies to increase the sustainability of polymers:

- Reuse (reuse primary scrap)
- Reduce (amount of polymer in a product)
- Recycle (examine secondary uses for plastic products)
- Recover (recovering the chemicals and/or energy from plastic waste)
- Renewable (including use of renewable feedstock to produce both biodegradable and nonbiodegradable polymers)

The Business-NGO Working Group (BizNGO) is a unique collaboration of representatives from business and nongovernmental organizations (NGOs) launched in 2006 by Clean Production Action (CPA). The working group promotes "the creation and adoption of safer chemicals and sustainable materials in a way that supports market transitions to a healthy economy, healthy environment, and healthy people" (BizNGO 2014). The organization has worked out some principles for sustainable plastics that offer a systematic and holistic approach to evaluate the extent to which materials achieve the principles of (1) sustainable resources, (2) closed-loop systems, (3) energy efficiency and renewability, (4) safer chemicals, and (5) healthy workplaces and communities. They have recently released the Plastics Scorecard Version 1.0, which is a method for evaluating the chemical footprint of plastics and a guide for selecting safer alternatives. The method is a development of the previously proposed ranking from A+ to F depending on their environmental and hazard impact over the life cycle of plastic materials, where A+ refers to good life cycle performance and F to poor life cycle performance. By this approach, issues such as use of genetically modified organisms (GMOs) and hazardous pesticides, release of chemicals during processing, and toxic breakdown products are considered. This means that the very same material can get a different grade depending on the manufacturing conditions adopted in a specific case (Álvarez-Chávez et al. 2012).

Álvarez-Chávez et al. (2012) have performed an extensive analysis to define a sustainable biobased plastic by forming a list of criteria for the environmental, health, and safety impacts during the life cycle of the material. The list included, among other things, the use of GMOs, hazardous pesticides, hazardous chemicals, petroleum-based monomers, hazardous additives, and potential hazards in workplaces; disposal options; and the efficiency in the use of water, energy, and materials. Their study resulted in the formation of a Bioplastics Spectrum to be used for selection of materials. According to this spectrum, PHAs, PLA, and thermoplastic starch (TPS) are preferred over other biobased materials including cellulose and lignin from health- and safety impact perspectives as well as from an environmental perspective.

The manufacturing of biobased feedstock using industrial agricultural production methods includes use of land area, water, energy, pesticides, and fertilizers, which may lead to pollution of air, water, and soil. In addition, use of land for production of industrial polymers instead of growing crops for food production can be questioned from ethical and socioeconomical perspectives. The use of GMOs is also a concern since their long-term environmental effects are not clear, with increased resistance to herbicides and decreased genetic diversity as potential threats.

Müller et al. (2014) discussed the options of end of life of bioplastics. Three different scenarios were highlighted.

- Scenario 1: Selective collection of bioplastics together with other packaging waste and used for
  - Material recycling
  - Composting
  - Energy recovery by incineration
  - Landfill (less preferred option)
- Scenario 2: Bioplastics are collected in the residual waste container and
  - Composted
  - Incinerated
  - Go to landfill
- Scenario 3: Bioplastics are collected as organic waste and composted

New and cost-effective technologies for sorting bioplastics from conventional plastic materials and other types of waste must be developed in order to increase their usability on the market. One primary challenge is that conventional and biobased plastics are visually indistinguishable, hence making it difficult for customers to perform a first sorting step. Currently adapted technologies for industrial sorting of waste are flotation (based on different densities of materials), identification by spectrometry (visible, infrared, near-infrared), ultraviolet, x-ray, color identification, and laser sorting (Müller et al. 2014).

Álvarez-Chávez et al. (2012) concluded that none of the currently commercially used or developed biobased plastics are fully sustainable. They also pointed out the environmental and economic advantages of using by-products from the agricultural and wood industry as a way to enhance the overall sustainability of materials.

25.1.4 Technical Challenges for Implementation of Bioplastics

Going from synthetic plastic materials to bioplastic alternatives requires a continuous, secure supply of biobased raw materials for industrial use. Some potential raw materials such as starch, PLA, PHA, and lignocellulosic materials are already in production, and the production volumes are expected to increase over the coming years. A continuous use requires large stocks of raw material, which in turn requires prevention from spoilage of seasonal raw material like annual plants during storage since biobased materials are inherently sensitive to microbial attack. Bioplastics solutions also need to deal with the inevitable variety in the natural feedstock and often require purification steps in the raw material production processes. Such purification steps are not necessarily based on green chemistry.

Functionality of biobased materials compared to existing plastic films has received a lot of attention, and reliable data are readily accessible in the scientific literature. Improvements of barrier, mechanical, and various functional properties are continuously achieved on laboratory scale. However, not all modification processes are appropriate for up-scaling due to expensive processes or potential health risks.

25.2 POLYMERIZATION AND PROCESSING TECHNOLOGIES OF BIOPLASTICS

A wide variety of biobased plastics have been identified and developed over the past decades. This chapter focuses on those that have found the widest use and are the most important from a commercial perspective: TPS, lignocellulose materials, PLAs, PCL, PHAs, and PE from biobased feedstock (bio-PE). Some advantages and disadvantages of these materials are summarized in Table 25.1, and each material is described in detail next.

TABLE 25.1
Environmental, Functional, Economic, and Safety Aspects of Bioplastics

Type of Bioplastics	Environmental, Functional, and Economic Advantages	Environmental, Economic, and Safety Disadvantages
Thermoplastic starch	Renewable, biodegradable, compostable, lower CO_2 emissions[a], ~68% less energy use in production[a], low cost	Industrial agricultural production of feedstock including use of fertilizers and pesticides; GMOs sometimes used; pulverized starch can cause explosions; requires plasticizers; hydrophilic; poor mechanical properties
Lignocellulose	Renewable, slowly biodegradable, compostable, high mechanical strength	Processing requires large amounts of water and energy; strong acids, bases, and toxic sulfur substances are used; emissions to air and water
PLAs	Renewable, recyclable, compostable (>60°C), uses 30–50% less energy[a], generates 50–70% lower CO_2 emissions[a], high mechanical strength	Industrial agricultural production of feedstock including use of fertilizers and pesticides; GMOs sometimes used; brittle; potential allergic reactions
PCL	Biodegradable, water resistant	Nonrenewable
PHAs	Renewable, biodegradable, water resistant	Industrial agricultural production of feedstock including use of fertilizers and pesticides; GMOs sometimes used; requires use of toxic chemicals; high cost
Bio-PE	Renewable, recyclable, uses 70% less energy[a], lower CO_2 emissions[a], properties identical to synthetic PE	Not biodegradable

Source: Data from Álvarez-Chávez, C.R. et al., *J Cleaner Prod* 23:47–56, 2012; Halley, P.J., Thermoplastic starch biodegradable polymers. In *Biodegradable Polymers for Industrial Applications,* ed. R. Smith, 140–162, Cambridge: Woodhead Publishing, 2005; Noaves, L., Green polyethylene: Bringing renewable raw materials to the traditional plastic industry. TAPPI 2010 PLACE Conference, April 18–21, 2010, Albuquerque, NM, 2010.

[a] Compared to petroleum-based plastics.

25.3 BIOPLASTICS FROM NATURAL SOURCES (CLASS I)

25.3.1 Starch

Starch is a naturally abundant plant polymer the function of which is to serve the plant with carbohydrates. Botanical sources are seeds, roots, and tubers. In the native form, starch is organized in a semicrystalline structure called granules, which vary in size and shape depending on the plant source, with typical diameters in the range 1–100 μm. Starch is available from a variety of sources, out of which starch from cereal seeds like wheat, corn and rice, tubers (potato), and roots (tapioca) are the most important for commercial production. Starch from corn (maize) dominates the US market, whereas potato starch constitutes an important source for starch production in Europe. Tapioca is used in tropical countries like Brazil and East India (Wurzburg 1986). Other plants that have been utilized for production of starch on a smaller scale are barley, sago, sorghum, oat, rye, and pea, to mention just a few. The chemical composition and properties of starch are dependent not only on the plant source but also on the environmental conditions during plant growth and biosynthesis of starch within the plant.

Originating from a renewable resource and being biodegradable, in combination with easy accessibility for chemical modification and a relatively low price, makes starch an attractive raw material for production of a diversity of bioplastic products. Starch has been at the forefront of the renaissance of biobased plastics over the last 20 years (Shen et al. 2009) and now represents one of the most important bioplastics on the market. The properties of starch materials are thus described in detail in this chapter.

The starch polymer consists of α-D-anhydroglucose units built up of the components amylose and amylopectin (Figure 25.2). Amylose is essentially a linear polymer with the glucose units linked together by α-D(1→4)-glucosidic bonds with a few branches connected by α-D(1→6)-glucosidic linkages. Amylopectin is a highly branched molecule built up by hundreds of short chains linked by α-D(1→4)-glucosidic bonds, where the branches are interlinked to the main chain through

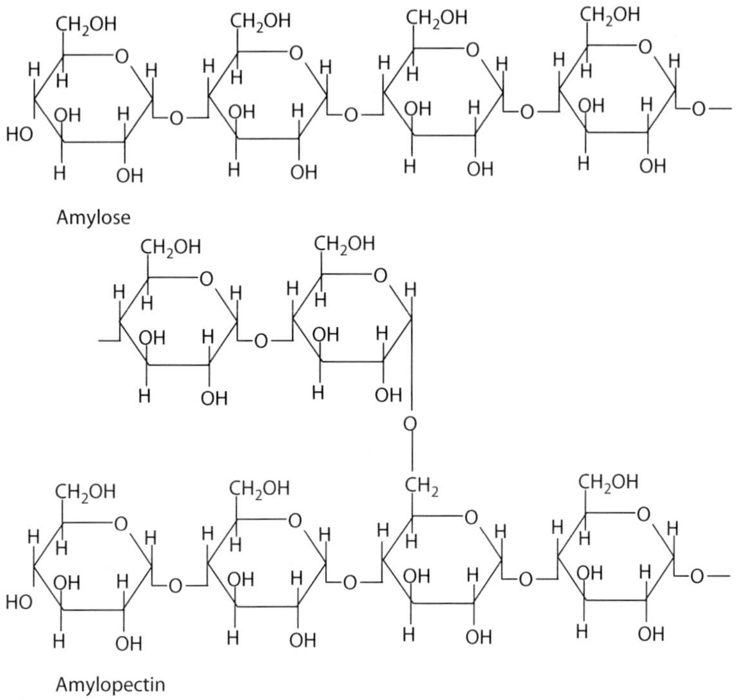

FIGURE 25.2 Structure of amylose and amylopectin.

α-D(1→6)-glucosidic linkages (Figure 25.2). The amylose fraction contains about 500–3000 anhydroglucose units, whereas amylopectin is an extremely large molecule built up of several thousand units with molecular weights, M_w, in the range 10^6–10^8 g/mol (Zobel 1990; Buléon et al. 1998; Halley 2005). The organization of the chains in amylopectin has been the subject of extensive study, and a commonly accepted model describes the arrangement of A- (outer), B- (inner), and single C-type chains in a cluster architecture, in which the C-chains contain the sole reducing terminal residue. A-chains are joined through single 1→6 bonds to the main molecule, whereas B-chains may carry one or more A- and/or B-chains on primary hydroxyl groups besides the 1→6 linkage (Zobel 1988). A comprehensive review of this complex structure can be found in Buléon et al. (1998). The amylose fraction in cereal starches typically has M_w in the range 160–250 10^6 g/mol, whereas values for potato starch amylose are much higher: $M_w \sim 1$–$1.5 \cdot 10^6$ g/mol (Zobel 1990).

The most widely accepted model for the molecular organization of the two fractions within starch granules is as alternating amorphous and semicrystalline growth rings. These semicrystalline growth rings in turn consist of amorphous and crystalline lamellae (Zobel 1988; Buléon et al. 1998).

Crystallites are formed when linear amylose and linear segments of amylopectin orient parallel to each other facilitating the formation of hydrogen bonds between adjacent segments (Wurzburg 1986). A significant portion of amylose is present in the amorphous phase because of its high mobility and ease of association (French 1984), and it is therefore primarily the amylopectin fraction that is responsible for the crystalline order in the native granule (Young 1984; Hermansson and Svegmark 1996).

The size and shape of granules, amylopectin and amylose content, percentage values of starch, and protein and moisture content in various starch sources are given in Table 25.2. Besides starch and protein, minor fractions of fibers, fat, water-soluble substances, and ash are also present.

Besides the naturally occurring starch types mentioned above, modified genotypes have been developed to provide specific properties. So-called waxy starches from, for example, corn, barley, or rice consist essentially of amylopectin, with less than 1% amylose. High-amylose variants of starch from these plant sources can contain up to 70% amylose (Buléon et al. 1998; Chiou et al. 2005).

Starch acts as a carbohydrate reserve in plants. Enzymatic activity depolymerizes the starch polymer to maltose and glucose to provide food for the plant. Since this process is dependent on the temperature and the moisture content, control of these parameters during processing and

TABLE 25.2
Granule Size and Shape, Amylopectin/Amylose Content, and Composition of Starches from Various Plant Sources

Starch Type	Granule Shape	Granule Size (μm)	Amylopectin (%)/ Amylose (%)	Starch (%)	Moisture (%)	Protein(%)	Gel Temp. (°C)
Wheat	Flat/round/elliptical	2–35	72/28	59.0	12.5	10.4	58–64
Rice	Polygonal	3–8	83/17	75.0	12.0	7.5	61–78
Corn	Polygonal/round	5–25	72/28	60.0	16.7	8.3	62–72
Potato	Oval	15–100	79/21	16.5	78.0	2.2	56–66
Tapioca	Round	5–35	83/17	25.0	68.0	3.0	52–64

Source: Data from Wurzburg, O.B., Introduction. In *Modified Starches: Properties and Uses*, ed. O.B. Wurzburg, 3–16, Boca Raton, FL: CRC Press, 1986; Wagoner, J.A., van der Burgh, L.F., Unmodified starches: Sources, composition, manufacture and properties. In *Starch and Starch Products in Paper Coating*, eds. R.L. Kearney, and H.W. Maurer, 1–14, Atlanta, GA: TAPPI Press, 1990; Zobel, H.F., Colloidal and polymeric chemistry of starches and their dispersion. In *Starch and Starch Products in Paper Coating*, eds. R.L. Kearney, and H.W. Maurer, 15–28, Atlanta, GA: TAPPI Press, 1990.

Bioplastics

storage of starch is important in order to prevent undesired depolymerization (Wagoner and van der Burgh 1990).

Manufacturing of starch utilizes various types of processes, depending on the plant source. The general production steps involve separation of the starch granules from other constituents (fiber, germ, proteins, etc.), purification (screening, washing, and centrifugation), dewatering, and drying (Wurzburg 1986). The granules are insoluble in cold water. In corn starch production, mechanical separation by steeping the grains in warm water in the presence of sulfite is often used (Wurzburg 1986). Commercial starches delivered as dry powders typically have moisture content in the range 12–20%. The drying step is a costly process requiring investment in equipment and energy. High temperature in the hot air dryers gives a higher production capacity but leads to a reduction in starch quality (Ellis et al. 1998).

The production of corn starch typically begins with a cleaning step where the corn kernels are cleaned from contaminants including broken corns, stones, dusts, and grains (Figure 25.3). The corn kernels are built up of three primary components: the *hull*, which contains mostly fiber; the *germ* consisting of oil, protein, ash, and fiber; and the *endosperm*, which contains nutrition for the lant in the form of starch. After cleaning, the corn kernels are softened by soaking in a steeping tank at about 50°C for approximately 40 h in an aqueous solution containing sulfur dioxide (Wagoner and van der Burgh 1990). Absorption of water causes the kernels to swell, and treatment with sulfur dioxide causes loosening of the protein matrix and subsequent release of soluble material. After steeping, a coarse grinding step separates the hulls and the germs. Grinding is carried out under gentle conditions to avoid leakage of oil from germs, which might cause reduced quality of the final starch product. The germs are separated, washed, dried, and utilized for production of corn oil in a separate process. A finer milling step separates the fibers from the endosperm. The fibers are collected, washed, screened, dried, and used as cattle feed. After milling, the starch slurry still

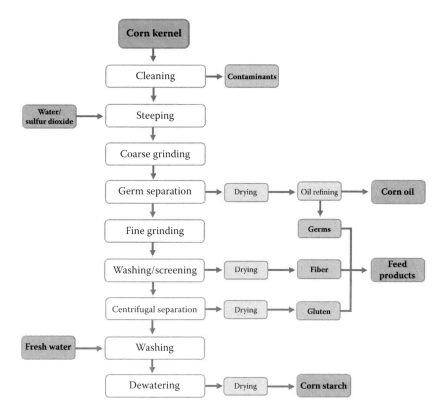

FIGURE 25.3 Schematic outline of a typical corn starch production process.

contains all dissolved proteins (gluten). The proteins are separated by centrifugal separators utilizing the density differences between starch and protein. The protein is dehydrated, dried, and used as a high protein feed to chickens.

The starch slurry contains about 2% protein and fibers after separation. By refining in a multistep cyclone plant, the protein content is reduced to below 0.3%. The refined starch slurry has a water content of about 65%; it is dewatered in centrifuges to a residual water content of about 40%. The pure starch is dried in belt driers or flash driers to a water content of about 10–14%.

In the typical production of potato starch, the potatoes are washed for removal of soil, sand, and other contaminants in a rotary washing machine (Figure 25.4). The abrasion also causes removal of most of the skin. Rasping in a drum rotated at high speed and fitted with sharp saw teeth causes almost complete disruption of the potatoes and release of the starch. Sulfur dioxide is normally added at this, and in later steps, to inhibit the action of microorganisms and oxidative enzymes that causes discoloration of the starch (Mitch 1984; Wagoner and van der Burgh 1990). The fine mash is screened in conical rotating sieves, so-called centri-sieves, allowing starch and fruit juice to pass through, while the pulp (consisting of cell fragment and coarse skin) is retarded. The pulp is dried in flash dryers and used as cattle feed with high nutritional value due to high content of protein and residual starch. In the next step, the fruit water is separated in series of hydrocyclones. This fruit water has a high content of proteins, amino acids, and mineral nutrients and is used as a fertilizer. The final washing is accomplished by addition of freshwater in a series of hydrocyclones. The remaining starch slurry consisting of about 35–40% dry matter is dewatered by rotary vacuum filtration to a moisture content below 40% and finally dried by flash or cyclone driers. The final starch powder will have an equilibrium moisture content of approximately 18–20% by weight.

Environmental concerns have been raised in recent years regarding the increased use of GMOs in crops to produce higher yields or to improve the properties of the starch (Álvarez-Chávez et al. 2012). The impact of GMOs is not well understood but may include allergic reactions, inherent toxicity of novel genes, and alterations of metabolic pathways. Additional safety precautions need to be undertaken when handling final starch products since starch in a finely pulverized form may suspend in the atmosphere and cause powerful explosions, which pose a safety hazard for workers (Álvarez-Chávez et al. 2012).

Corn and wheat starches contain about 0.3–0.4% protein and about 0.5% fat on dry basis, which, despite very low levels, still have impact on their processing and product performance. In contrast,

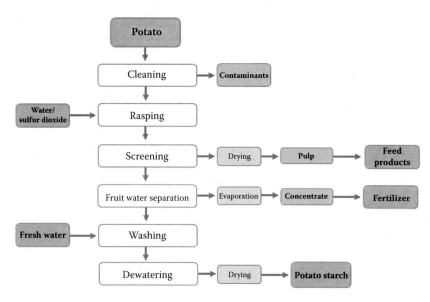

FIGURE 25.4 Schematic outline of a typical potato starch production process.

only trace amounts of such contaminating species can be found in potato starch (Wagoner and van der Burgh 1990; Chiou et al. 2005).

The minor components found in starch can be classified in three categories: particulate material (cell-wall fragments), surface components (proteins, enzymes, amino acids, and nucleic acids, removable by extraction), and internal components. Lipids constitute the most important fraction of the internal components. The presence of lipids is a characteristic of cereal starches, in contrast to tuber starches, which are free from lipids as integral components. The lipids mainly consist of lysophospholipids and free fatty acids (Buléon et al. 1998; Ellis et al. 1998). Waxy grades of barley, corn, rice, and wheat, which have very low amylose content, also have much lower lipid contents than the normal genotypes. High amylose barley, corn, and wheat grades in turn have much higher lipid content (Ellis et al. 1998). Root and tuber starches, on the other hand, contain phosphate monoesters, with an exceptionally high level in potato.

The starch molecule forms a helical structure due to intramolecular hydrogen bonding. The α-(1→4)-linkage results in a flexible molecule with the CH_2OH molecules on carbon 5 positioned on the same side of the carbohydrate chain (isotactic configuration), with a natural extended helical twist. Cellulose, in contrast, contains β-(1→4)-linked glucose units with the CH_2OH molecules positioned alternately on the sides of the carbohydrate chain forming a syndiotactic configuration, which leads to a rigid and flat structure accounting for the strength of cellulose fibers (Zobel 1990).

When studying starch granules under a microscope with polarized light, an interference pattern in the form of a Maltese cross can be observed, which originates from the crystalline nature of the granules. The intersection of this cross locates the hilum, i.e., the initial growth center of the granule (Zobel 1990). The granule is built up of concentric rings of semicrystalline shells separated by amorphous regions. X-ray diffraction analysis of native starch granules shows three basic patterns: cereal starches like corn show an A-type pattern, whereas tuber starches like potato show a B-type pattern, and root starches (tapioca) give a C-type pattern, which is a mixed form of the A- and B-types. Amylose and amylopectin can crystallize in any of these patterns, but amylose can also exhibit a V-type pattern resulting from complexation with substances such as fatty acids, phospholipids, or emulsifiers (Vázguez et al. 2011). The crystallinity is dependent on the ability of the chains to organize and form crystals, and hence on the chain mobility during crystallization. The amylose fraction crystallizes rapidly, whereas the amylopectin fraction needs several days to develop a crystalline pattern in the x-ray diffraction trace. The presence of water or the addition of a plasticizer facilitates the mobility of amylopectin chains and thus facilitates the onset of crystallization.

The strong hydrogen bonds between glucose chains that are responsible for the structural integrity of starch granules make the granules insoluble in cold water. High energy is therefore needed in order to disperse the chains and obtain a starch solution that can be used in bioplastic applications. Heating a starch solution to a certain temperature, known as the gelatinization temperature, causes the hydrogen bonds to break, allowing water to diffuse into the granule structure to hydrolyze the linear parts of the amylopectin fraction. Water entrance causes swelling of the starch granules leading to the formation of gel particles and a highly increased viscosity of the solution up to a peak value at which the granules have reached maximum hydration. At this stage, the granules rupture and amylose leaks out into the water phase. The gelatinization temperature ranges for various starches are given in Table 25.2. As the granules are fragmented, the viscosity gradually decreases. Temperature, pressure, and time are important factors in the cooking process. Simultaneous mechanical stirring facilitates the granule fragmentation and chain dispersion. The gelatinization is an irreversible process that implies a transition from an ordered to a disordered phase with dissociation of the double helices, including loss of crystallinity, loss of birefringence, uptake of heat, and hydration of the starch (Vázguez et al. 2011). Higher temperature, longer cooking time, and stirring lead to a lower viscosity of the cooked starch (Voigt 1990). A minimum cooking time of 20 min at a temperature of 93–95°C under strong agitation is required to fully swell and disperse the starch (Wagoner and van der Burgh 1990). Another way to gelatinize starch is by jet cooking in which the starch slurry is mixed with a stream of high-velocity hot steam, which leads to a rapid increase in temperature. Complete gelatinization, i.e., starch pastes free from the original granule structure, requires heating to temperatures over 121°C (Bublitz et al. 1990).

Cooked starch can undergo retrogradation, when the dissolved amylose molecules reorient themselves in a linear fashion and then associate through hydrogen bonding, which is a thermodynamically more stable form. Amorphous α-glucan chains form double helices, which are eventually aligned in crystallites (Vázguez et al. 2011). Spontaneous recrystallization into B-type crystalline structures is observed in presence of excess water (Rindlav et al. 1997). In dilute solutions, precipitates are formed that give a decrease in the viscosity (Wurzburg 1986; Voigt 1990). Precipitation can also be observed as the dispersion changes from transparent to opaque. An indication of retrogradation can be found in the viscosity changes of the dispersion. In concentrated solutions, steric hindrance prevents the molecules from orienting parallel to each other, but short segments can associate in networks forming a gel with a subsequent increase in the viscosity. The retrogradation process depends on the starch concentration, its molecular weight, the time, the temperature, the pH, and the presence of salt (Young 1984). A starch dispersion is regarded as stable over the temperature range 60–70°C (Hermansson and Svegmark 1996). Linear amylose recrystallizes rapidly and irreversibly, whereas the amylopectin fraction is more resistant to aggregation because of its branched structure (Bergh 1997). Once a gel or precipitate has formed, heating to >120°C is necessary to redissolve the amylose (Hermansson and Svegmark 1996). The amylopectin fraction has higher storage stability even at lower temperatures.

Low-molecular-weight amyloses like those found in cereals reassociate more rapidly than high-molecular-weight ones as those found in roots and tubers (Ellis et al. 1998; Chiou et al. 2005). The retrogradation process is kinetically controlled implying that the resulting product is dependent on time, temperature, and moisture content. Retrodegraded amylose is more resistant to degradation by amylolytic enzymes (Ellis et al. 1998). Control of retrogradation is necessary in applications dependent on viscosity and when a clear paste or product is required (Ellis et al. 1998).

25.3.1.1 Physical and Chemical Modification of Starch—Starch Derivatives

Native starch is highly hydrophilic and will readily disintegrate in contact with water. In addition, films from native starches are very brittle, thus limiting their application as bioplastics. A variety of physical and chemical modifications has been developed to address these weaknesses.

The hydroxyl groups on the starch glucose units can undergo substitution reactions, thus forming different types of starch ethers or esters. The number of substituent groups affects the properties of the resulting starch derivative and is reported as the degree of substitution (DS). The maximum theoretical DS value is 3, which is equal to the number of –OH groups on each glucose unit. Highly substituted starches have two or more substituent groups per glucose unit, whereas low substituted grades have an average DS of 0.2 or lower (Greif and Koval 1990). Commercial starches are typically low-substituted with one substituent group for every five glucose units.

Modifications of industrial starches are generally carried out on granular starch. The granules are suspended in water at about 40% concentration, and the required chemicals are added. The modification process takes place in large tanks fitted with impellers for agitation and heating/cooling systems for control of temperature. After completion of the modification process, the product is washed by centrifugation or on continuous vacuum filters. The final product is vacuum-filtered and dried by hot air or flash dryers, milled, and packed (Greif and Koval 1990). The reaction conditions can be varied to compensate for variations in supplies of raw material to ensure a continuous supply of uniform products meeting specific quality criteria.

25.3.1.2 Reduction of Molecular Weight

High viscosity and poor viscostability make native starches unsuitable for many applications since high viscosity limits the maximum concentration for dispersion in water, which is necessary for use in water-borne systems for film casting or coating. Mainly, there are four different approaches to reduce the viscosity by depolymerization, i.e., lowering the molecular weight: acid hydrolysis, enzymatic conversion, oxidation, and thermal–chemical treatments (Zobel 1990). These conversion/depolymerization processes are often used in combination with derivatization processes to produce starches with desired functionality.

25.3.1.2.1 Acid Hydrolysis

This involves the addition of a strong acid, e.g., hydrochloric acid, to the aqueous starch slurry that is heated and maintained at a temperature below the gelatinization temperature. The acid enters the starch granule and causes hydrolytic scission of the starch molecules (Greif and Koval 1990) while the granules are kept intact. When the desired viscosity reduction has been reached, the reaction is terminated by neutralizing the acid.

25.3.1.2.2 Oxidation

Oxidation of starch is carried out to reduce the chain length and thus lower the viscosity of the starch solution. Sodium hypochlorite (NaClO) or hydrogen peroxide (H_2O_2) is used as the oxidizing agent. Reaction with sodium hypochlorite under alkaline conditions introduces carboxylic acid and carbonyl groups on the glucose hydroxyl sites, which act to stabilize the starch by sterically hindering the retrogradation process (Rutenberg and Solarek 1984). Carboxyl group formation is favored at high pH, and the anionic charge on these groups further facilitates dispersion by electrostatic repulsion. The oxidation causes depolymerization by breakage of ether linkages, and the resulting molecular weight after oxidation determines the starch viscosity. When the target molecular weight has been achieved, the reaction system is neutralized by the addition of an acid, and sodium bisulfite is added to remove all free chlorine from the product (Greif and Koval 1990). The gelatinization temperature of the starch is reduced in proportion to the degree of oxidation.

25.3.1.2.3 Enzymatic Conversion

Enzymatic conversion utilizes the action of specific enzymes, mostly α-amylase or β-amylase, to catalyze the breakdown of long starch chains by hydrolysis, which leads to reduction of the viscosity. For commercial enzymatic conversion, α-amylase is used since it results in a minimum side production of sugars. The action of β-amylase, on the other hand, results in a maximum saccharifying reaction since β-amylase completely converts the linear macromolecule to maltose (Buléon et al. 1998). The conversion reaction is strongly time-, temperature-, and pH-dependent and requires strict control of these variables to obtain the desired viscosity of the final product. Every type of enzyme has its optimum pH level for maximum activity, and buffered systems are used to control the pH (Lovin and Wheeler 1990). The reaction is carried out below the temperature where the enzyme is inactivated.

The process starts with the preparation of a starch slurry with unmodified starch as the basis material. α-amylase is added to the starch slurry at a low dosage; typically the enzyme is fed in the range 0.01–0.1% relative to the fed amount of starch on dry basis. The breakage of glycosidic bonds starts when the starch begins to gelatinize, and the conversion process is carried out above the gelatinization temperature for a controlled time under continuous agitation. To inactivate the enzyme when the desired viscosity reduction has been achieved, the temperature is increased to 93–96°C and is held there for the required time. Another way to inactivate the enzyme is by adding chemicals that either change the pH of the system or destroy the enzyme in other ways (e.g., heavy metal salts or oxidizing agents). The use of inexpensive unmodified starch in combination with low-cost enzymes makes the process economically beneficial (Lovin and Wheeler 1990).

25.3.1.2.4 Thermal/Thermal–Chemical Conversion

Enzyme treatment of starches requires strict control over the processing parameters and consequently often results in variations between batches. A simpler technique for viscosity reduction is thermal conversion, which utilizes hot steam. An excess amount of steam is introduced in a jet cooker. This excess hot steam causes mechanical shearing, which ultimately leads to a reduction in viscosity (Bublitz et al. 1990). The amount of excess steam can be up to 500% of the amount of steam required to reach the desired temperature for complete disruption of the starch granules. The combination of thermal treatment and oxidation is commonly termed thermal–chemical conversion. As oxidizing agents, ammonium or potassium persulfate, hydrogen peroxide, or sodium hypochlorite is commonly used. The modification process takes place in the heating zone of the cooker under high temperature and pressure. Since more reaction sites are available when treating starch in its dispersed and hydrated form, the oxidation reaction is much more effective than when oxidizing starch in granular form, and the reagents can thus be fed at substantially lower levels.

25.3.1.3 Starch Ethers

25.3.1.3.1 Hydroxyalkylation

Hydroxyalkylation reactions can be carried out with the starch in its native granular form, which has the advantage that unwanted by-products can be removed by washing and filtering, and moreover, that the starch can be further modified by, for example, cross-linking. The reactions can be catalyzed by alkaline metal hydroxides, which form complexes with the starch and increase the reactivity of the hydroxyl groups. The major reagents are ethylene oxide and propylene oxide, forming hydroxyethylated and hydroxypropylated starches, respectively. To prevent solubilization of the starch during the reaction, neutral salts are added, serving as swelling inhibitors (Greif and Koval 1990). The reactivity is highest for the attachment of the –OH group in the C2 position. Consequently, the location of the substituent group is favored here over the location at the C3 or C6 positions (Tuschhoff 1986).

Hydroxypropylation improves the low-temperature stability, the stability of the dispersion viscosity, and the flexibility of the resultant films. Incorporation of the hydrophilic hydroxypropyl groups lowers the gelatinization temperature (Rutenberg and Solarek 1984), and the substitution also prevents association of the amylose chains after cooking (retrogradation). Hydroxypropylated starch has found use in both food and nonfood applications and shows excellent film-forming properties along with beneficial rheological properties. Furthermore, nonionic starch ethers are not sensitive to changes in electrolyte concentration and pH. Hydroxypropylation reduces the tensile strength of the film but increases the elongation and burst strength and the resistance to folding. Films have also been shown to be essentially impermeable to oxygen (Rutenberg and Solarek 1984).

Chemical modification of starches has also been utilized to tailor-specific properties. Replacing the hydroxyl group at the C2 starch by an *n*-butyl group causes a disruption of the intramolecular hydrogen bonding between anhydroglucose units, thus exposing more cell-binding sites for bacterial interaction. Such a modification has been shown to increase the degradation rate, which is particularly useful for single-use consumer products, which should degrade quickly after use. Creating a starch palmitate by the reaction of starch with palmitoyl chloride was, on the other hand, shown to inhibit bacterial interaction by sterically hindering cell attachment. This blocking effect is useful when slow degradation rate is preferred (Chiou et al. 2005).

25.3.1.4 Starch Esters

Reaction of starch granules with acetic anhydride results in the formation of acetylated starch esters. The dispersibility of starch granules during cooking is facilitated, and the stability of the cooked starch is enhanced due to steric hindrance to retrogradation by the acetate groups. However, acetylated starch esters are prone to undergo de-esterification reactions at either acidic or alkaline conditions (Greif and Koval 1990).

25.3.1.5 Genetic Manipulation of Starches

Besides chemical and physical modifications of natural starches, some tailored properties can also be created by manipulation of the starch biosynthesis in the plant by breeding or by the use of genetic transformation (Ellis et al. 1998). Buléon et al. (1998) listed three different strategies to modify starch biosynthesis: knocking out biosynthetic enzymes through selection of mutations, pulling down biosynthetic enzyme expression by mutations, and expressing heterologous genes—related or not—to the biosynthetic pathway. Selection of specific mutants in breeding programs can be utilized to produce starches with very small fractions of amylose or with very high amylose content, respectively. By the third strategy, it is possible to, for example, increase the degree of branching in amylopectin or to introduce novel linkages or other chemical groups on the polymer (Buléon et al. 1998). Gene modified starches should, however, be used with care due to potential environmental and health risks involved.

25.3.1.6 Thermoplastic Starch

Starch is not a true thermoplastic polymer, but in the presence of plasticizers such as water or glycerol and at higher temperatures (90–180°C) under shear, it readily melts and flows. Manufacturing of TPS requires the disruption of starch granules by breaking up the intermolecular hydrogen bonds to form a homogeneous polymer phase (gelatinization) followed by adding a plasticizer that forms hydrogen bonds with the starch molecules (Vázguez et al. 2011). One conventional method is by extrusion to produce a homogeneous molten phase by application of heat and shear in the presence of an appropriate plasticizer. Time, temperature, applied shear stress, and type and content of plasticizer control the final properties of the thermoplastic material.

Altskär et al. (2008) studied three different routes to produce films or sheets from hydroxypropylated/oxidized normal potato starch using glycerol as plasticizer: solution casting, compression molding, and film blowing. By size exclusion chromatography, it was found that both high-molecular-weight amylopectin and high-molecular-weight amylose-like molecules had degraded to some extent during processing, when compared with native starch. Solution casting was shown to have a very minor impact on the degree of degradation, whereas compression molding and film blowing resulted in degradation to a higher, but similar, extent, which was supposed to be triggered by the high temperatures used with these two techniques. Structural analysis of the resulting films also indicated phase separation into glycerol-rich and glycerol-poor regions.

25.3.2 LIGNOCELLULOSE

About half of the global amount of biomass is present as lignocellulosic materials, which are thus the most abundant organic material resources on earth (Keenan et al. 2005). Lignocellulose is derived from wood and consists of 30–50% cellulose, 20–50% hemicellulose, and 15–35% lignin, depending on the tree species and growing conditions. All three components provide properties that make them attractive for use in bioplastic applications. Historically, mainly the cellulose fraction has been used for industrial production of pulp fibers for paper manufacturing and for production of chemical derivatives. However, a paradigm shift from the traditional perspective, i.e., that the other two components are regarded as waste products, to consider them as highly valuable raw materials, has occurred over the past years. Complex biorefineries are now built up worldwide to convert a variety of renewable materials present in forest biomass to value-added products (Keenan et al. 2005).

Separation of lignocellulosic biomass in its three components can be done by various processes. One patented process (referred to by Keenan et al. 2005) utilizes a steam explosion treatment in a mix of organic solvents (water, methyl-isobutyl-ketone, ethanol, and sulfuric acid) to solubilize the lignin and hemicellulose fractions. The insoluble fraction is washed and used as cellulose pulp to produce paper products and cellulose derivatives (ethers and esters), whereas the solubilized materials are further separated in hemicellulose for microbial production of chemicals, fuels, and polymers and in lignin, which is used for synthesis of adhesives and thermosetting polymers (Keenan et al. 2005).

25.3.2.1 Cellulose

Cellulose is the most abundant natural biopolymer in the world. The cellulose polymer consists of a number of anhydroglucose units linked together by β-(1→4) glucosidic bonds arranged in a linear fashion (Figure 25.5). The degree of polymerization varies between a few hundred and several thousand glucose units for cellulose from various origins with typical M_w ~ 90,000–150,000 (Zobel 1990). The chain adopts an extended, helical conformation with the adjacent anhydroglucose units twisted around the axis, a conformation that makes the cellulose chain relatively rigid. The stiffness of the chain is dependent on steric interactions as well as on intrachain hydrogen bonding between C(3′)OH–O(5) and C(6′)OH–O(2)OH. Cellulose forms a highly crystalline structure due to intra- and intermolecular hydrogen bonds, and the energy required to break up these bonds results in a melting temperature that is higher than that at which chemical degradation takes place (Belgacem and Giandini 2011). This complex primary structure in combination with strong hydrogen bonds in secondary and tertiary configurations together with the necessity of removal of other compounds (hemicellulose, lignin, proteins, and mineral elements present in the raw stock) makes the chemical modification of cellulose a more costly and complicated process compared to modification of starch.

Chemical modification of natural cellulose involves the production of cellulose ethers (e.g., carboxymethyl cellulose [CMC] or hydroxypropyl cellulose [HPC]) and cellulose esters (e.g., cellulose acetate [CA]). The largest class of cellulose derivatives in terms of production volumes is regenerated cellulose (e.g., cellophane and man-made cellulose fibers; Shen et al. 2009). Introduction of substituent groups on the hydroxyl sites changes the hydrogen bonding but leads to new steric interactions between the substituent groups. Hence, an extended chain conformation is also retained in derivatized celluloses.

HPC is a nonionic cellulose ether formed by the base-catalyzed reaction of cellulose with propylene oxide. The addition of a strong base is necessary to break up the intermolecular hydrogen bonds on the cellulose backbone to make the –OH groups available for reaction with the propylene oxide. Substitution with hydroxypropyl groups can take place on any of the three –OH groups present on every anhydroglucose unit (marked with R in Figure 25.5). The reactivity is, in contrast to that for the starch hydroxyl groups, largest for the –OH group in the C6 position. The location and number

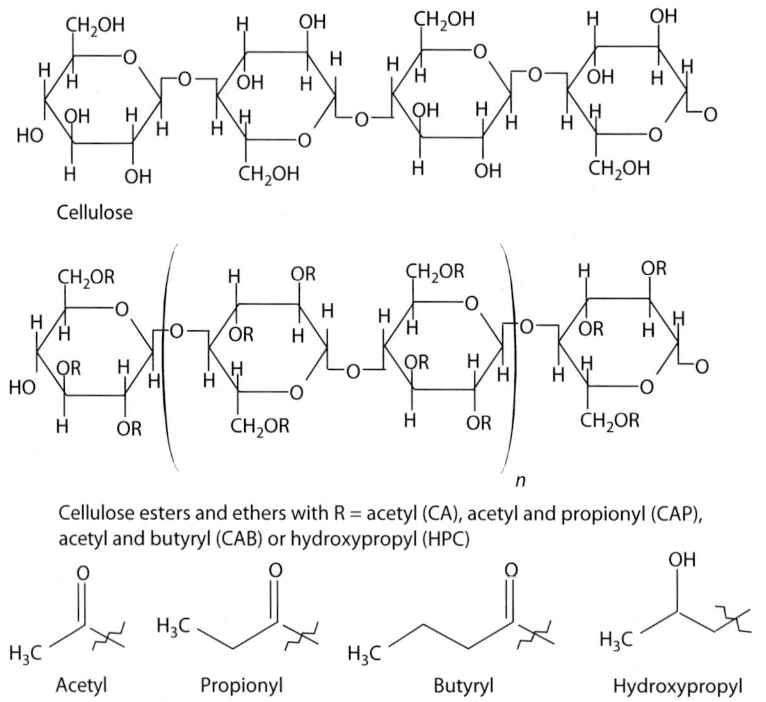

FIGURE 25.5 Cellulose polymer consisting of a number of anhydroglucose units linked together by β-(1→4) glucosidic bonds arranged in a linear fashion.

of substituent groups will affect the properties of the derivatized molecule. However, the possibility of further reaction of the –OH groups of the newly incorporated substituents leads to a theoretically unlimited number of hydroxypropyl groups per anhydroglucose unit. It is therefore more appropriate to refer to the molar substitution (MS), which is defined as the average number of hydroxyalkyl groups per anhydroglucose unit. For commercial types of cellulose ethers, MS lies in the range 3.5–4.5 with a DS of 2.2–2.8 (Nicholson and Merritt 1985). To ensure water solubility in cold water, MS should be around 4.0 (Klemm et al. 1998).

HPC has the ability to form ordered, anisotropic phases (liquid crystalline phases [LCPs]) in concentrated solutions. The rigid polymer molecules can orient parallel to each other in various layered structures, referred to as nematic, cholesteric, and smectic phases. In order to form LCPs, the intramolecular forces must be weak enough to allow the chain backbone groups to move and achieve an orientational order. The critical concentration for phase separation of derivatized celluloses is thus determined by the stiffness of the chain—the stiffer the chain, the higher the tendency to form LCPs (Larson 1999). At higher temperatures, higher critical concentration for mesophase formation is often observed simply due to the decreased chain stiffness. The critical concentration for the formation of anisotropic phases is also affected by the polymer–solvent interaction, the relative distribution of substituent groups, and, to some extent, the molecular weight of the polymer.

Since HPC contains hydroxyl groups, it is extremely difficult to completely dry the films, and the remaining, tightly bound water may affect, for example, the mechanical properties. HPC films are demonstrated to be flexible, transparent, odorless, tasteless, biodegradable, edible, and thermoplastic. Furthermore, they show moderate mechanical strength and are moderate barriers to moisture and oxygen, while they are resistant to oils and fats (Krochta and De Mulder-Johnston 1997). The use of HPC in food packaging is mentioned in the literature (Krochta and De Mulder-Johnston 1997; Petersen et al. 1999; Weber 2000).

More rigid LCP chains often display favorable mechanical and thermal properties compared with their more flexible counterparts (Lusignea and Perdikoulias 1997). In general, LCPs show high modulus and strength along the shear direction and low strength in the transverse direction as a result of the molecular orientation. In the transverse direction, the strength can be as low as one-tenth of that in the shear direction (Lusignea and Perdikoulias 1997). The problem with different mechanical properties in different directions can be overcome by producing a biaxially oriented film, for example, by using a counterrotating circular extrusion die (Lusignea 1996).

Commercial, synthetic LCPs such as Vectra (Ticona/Celanese Corp.) provide good barrier properties for both gases and water vapor even in very thin layers (Lusignea and Perdikoulias 1997). In fact, water vapor permeability of a 25-μm-thick LCP film is around 100 times lower than that for ethylene vinyl alcohol (EVOH), polyethylene terephthalate (PET), and high-density PE (HDPE) films of the same thickness (Lusignea 1998). The oxygen permeability is one, two, and three orders of magnitude lower for LCP than for the three named polymers.

Besides HPC, ethyl cellulose (EC), methyl cellulose (MC), CMC, and hydroxypropyl methyl cellulose (HPMC) are further examples of cellulose ethers that can be cast as films or coated from water and/or ethanol solutions. In contrast to HPC, these derivatives are not thermoplastic and hence do not give heat-sealable coatings (Krochta and De Mulder-Johnston 1997). All these ethers offer poor moisture barrier properties, but they have found use as oil and fat barriers.

CA is formed by reacting cellulose with acetic anhydride. The most important commercial products for plastic applications are CA and the mixed esters CA butyrate (CAB) and CA propionate (CAP) (Edgar et al. 2001; Figure 25.5). They have valuable properties in terms of clarity, stiffness, grease resistance, and moderate heat resistance and can easily be extruded and injection-molded. CA provides poor barrier properties against gases and moisture; this property is utilized in breathable films for high-moisture products since fogging is avoided (Krochta and De Mulder-Johnston 1997). However, plasticizers such as triethylcitrate are required to widen the relatively narrow window between the melt flow temperature and the decomposition temperature (Edgar et al. 2001). Biodegradable thermoplastic materials based on CA with a DS between 1.7 and 3.0 have been produced with biodegradable citric esters as plasticizers. Such plasticized CAs have shown increased elongation at break and increased biodegradation rates (Keenan et al. 2005). Commercial applications of CAB and CAP are handles for tools and toothbrushes (Mohanty et al. 2005).

Cellulosic polymers are produced mainly from wood through sulfite pulping or kraft pulping to separate cellulose from lignin and hemicellulose. Cellulose hydrate (cellophane) is commonly produced through the classical viscose process forming regenerated cellulose in a xanthate solution with subsequent formation of a gel film by immersing a slot die in the acid bath. Washing, bleaching, drying, and winding then give a clear film with typical thickness of 12–45 μm. Cellophane is highly moisture sensitive but provides good barrier properties against oxygen, flavors, fats, and oils at low relative humidity (Krochta and De Mulder-Johnston 1997). Since the barrier properties of the film are reduced at high humidity, cellophane films are often coated with polyvinylidene chloride or nitrocellulose wax to improve the moisture and temperature barrier properties (Shen et al. 2009). Coated cellophane films have found use in packaging for food and flowers not least for decorative purposes.

The extended chain conformation and microfibrillar morphology of cellulose make it effective as a load-bearing constituent in plants: this function is thus attractive to utilize for reinforcement of plastics in the form of cellulose-based nanocomposites. The highly crystalline nature of the cellulose nanofibrils, their large aspect ratio and fine diameter, as well as their high strength and modulus open wide possibilities for their use in novel composite materials (Berglund 2005).

25.3.2.2 Hemicellulose

Hemicelluloses are composed of a variety of saccharide monomers including pentoses and hexoses arranged in numerous side chains on the molecular backbone. Typical monosaccharides are D-glucose, D-galactose, D-mannose, D-xylose, and L-arabinose. Besides these, a variety of uronic acids are also present, depending on the tree species (Keenan et al. 2005). These are in turn arranged in polymers with molecular weights generally lower than for cellulose chains and form a variety of glucans, xylans, mannans, galactans, and glucuronides. Hardwood trees are characterized by a high content of an acetylated xylan (O-acetyl-4-O-methylglucuronoxylan), whereas softwood contains predominantly a partly acetylated galactoglucomannan and a smaller fraction of arabino-4-O-methylglucuronoxylan. A comprehensive overview of hemicellulose structure, its components and properties, and potential chemical modifications was given by Albertsson et al. (2011).

Hydrolysis and detoxification of the hemicellulose fraction from forest biomass result in various sugars that can be used as feedstock for production of bioplastic materials by microbial fermentation. PHA is an example of such a product, and it is further described in the following. The required pretreatment steps involve saccharification of native xylan to xylose monomers and oligomers either by chemical or physical/enzymatic methods. In general, dilute acids (HCl and H_2SO_4) are used to hydrolyze xylan to xylose at high temperatures and pressures. However, the high temperatures and low pH conditions generally result in toxic sugar decomposition products, which must be removed via detoxification methods prior to fermentation since these products are toxic and inhibitory to fermenting microorganisms (Keenan et al. 2005).

Steam explosion is a method that offers environmental advantages since it requires minimal use of chemicals and relatively low energy. Disruption of the lignocellulosic structure is promoted by high temperature and pressure followed by a sudden pressure release. The degree of hydrolysis of the hemicellulose fraction depends on the treatment conditions. Complete saccharification to xylose may require further treatment by chemical or enzymatic methods (Keenan et al. 2005).

Effective removal of xylan from cellulose fibers and/or increase in the monosaccharide content of hydrolysates can be done by addition of xylanolytic enzymes, which hydrolyzes the glycosidic bonds in xylan and increases the fraction of free xylose available for microbial fermentation (Keenan et al. 2005).

Detoxification of the lignocellulosic hydrolysates obtained by chemical and steam explosion treatments involves removal of toxic furans and furfuraldehyde compounds originating from decomposition of sugars and aromatic and phenolic compounds originating from degradation of lignin. The concentration of these toxic substances can be reduced by biological, chemical, and physical processes including oxidative enzymatic treatment, evaporation of volatile substances, precipitation and filtration, acid/base treatment, or combinations of methods (Keenan et al. 2005).

Xylan is one of the most abundant biopolymers since it is one of the major constituents of wood and plants. It is also available in large quantities from agricultural waste. Xylan consists of a β-D(1→4)-linked xylopyranosyl main chain with various side chains or substituent groups depending on its origin. Gröndahl et al. (2004) studied the properties of glucuronoxylan isolated from aspen wood. Transparent, continuous, and flexible films were formed with xylitol or sorbitol as plasticizers prepared by casting from aqueous solutions. Water-based solutions of glucuronoxylan plasticized with sorbitol were found to give films with oxygen permeability comparable with that of PVOH or EVOH plastics.

Hemicelluloses such as arabinoxylans extracted from cereal cell walls (e.g., barley husks) have also been investigated for the production of polymer films or coatings with oxygen barrier properties (Höije et al. 2006). The hydrophobicity of films from arabinoxylan extracted from maize was increased by grafting functional aliphatic acrylates (Peroval et al. 2004). The study showed a significant improvement in the water vapor barrier properties.

Albertsson et al. (2011) summarized some recent applications of hemicelluloses as packaging films and coatings. Zhang and Whistler (2004) studied the mechanical properties and water vapor permeability of plasticized films cast from corn hull arabinoxylan and found that good miscibility and moisture barrier properties could be achieved with sorbitol as plasticizer. Hartman et al. (2006) formed flexible, transparent, and highly water-resistant films from benzyl galactoglucomannan. They showed that the derivatized films possessed oxygen barrier properties that were less sensitive to moisture than unmodified acetylated galactoglucomannan.

25.3.2.3 Lignin

Lignin is a high-molecular-weight aromatic polymer that forms a complex network structure. It is readily available as a by-product in paper manufacturing and has predominantly been used as a fuel in the chemical recovery process (Kubo et al. 2005). The use of lignin as a value-added product has received increased attention over the past years since it is a renewable, nontoxic, highly abundant, and low-cost material. A wide variety of lignin and lignin derivatives with various structures and chemical properties are now commercially available. Lignins can be isolated from alkaline, acidic, or organic solvent-based pulping processes, all of which result in products with varying molecular weight, types and ratios of monomer units, types of linkages between monomers, presence of functional groups, and degree of condensation (Kubo et al. 2005). Immiscibility resulting in phase separation and the inherent brittleness of the lignin phase caused by the globular structure of lignin fragments has limited the application of lignin and its derivatives in plastics. Extensive studies have been undertaken on the incorporation of substituents in the network structure and on the formation of lignin–polymer blends to enhance the performance of lignin-based products.

Softwood kraft lignin is primarily composed of aliphatic guaiacyl units, whereas hardwood kraft lignin is built up of guaiacyl and syringyl structures. This results in a higher amount of branch points or cross-links between units in softwood lignin, which leads to a more complex network structure having, for example, a higher T_g value. These condensed linkages also results in a lower reactivity of softwood lignin (Kubo et al. 2005). Both intra- and intermolecular hydrogen bonding can be formed by the hydroxyl groups on these two different subunits, and these bonds affect the thermal processing behavior of the products.

Lignin, as being composed of phenyl propane units, is generally categorized as a hydrophobic polymer, suggesting that it should be easily compatible with other hydrophobic polymers. However, the presence of polar functional hydroxyl and carbonyl groups enables lignin to form strong electrostatic and/or acid–base interactions with hydrophilic polymers (Kubo et al. 2005).

25.4 BIOPLASTICS FROM CLASSICAL CHEMICAL SYNTHESIS (CLASS II)

25.4.1 PE FROM SUGAR CANE (BIO-PE)

Production of bio-PE from ethanol goes back to the 1970s, and the production volumes have varied over the years since then, depending on the prevailing oil prices. Large-scale production of bio-PE has become attractive over the past years due to the relatively high and anticipated future increase in the price of oil. Bio-PE is currently produced in industrial scale from biobased ethanol manufactured by fermentation of sugar cane. Other potential raw materials are sugar beet or starch-containing crops such as corn or wheat. Sugar cane has the advantage of providing the highest biomass and energy productivity compared to other sources (Noaves 2010). The process steps include cleaning, slicing, shredding, and milling the sugar cane stalks (Figure 25.6). Milling results in sugar cane juice as the major product, with sugar cane fiber (bagasse) as a by-product. Bagasse can be combusted to generate heat and/or electricity for the mill's own needs but can also be sold to external users if produced in surplus (Shen et al. 2009).

The sugar cane juice having an average sucrose content of 12–13% is anaerobically fermented to ethanol, which is further distilled to remove water and to yield an azeotropic mixture of hydrous ethanol. Another by-product, called vinasse, is generated by this distillation step. Vinasse can be further used as a fertilizer. The final step is dehydration at high temperatures over a solid catalyst to produce ethylene. Polymerization finally converts ethylene into HDPE, low-density PE (LDPE), or linear low-density PE (LLDPE). PE produced from biobased feedstock has exactly the same chemical, physical, and mechanical properties as the polymer derived from petroleum, which means that

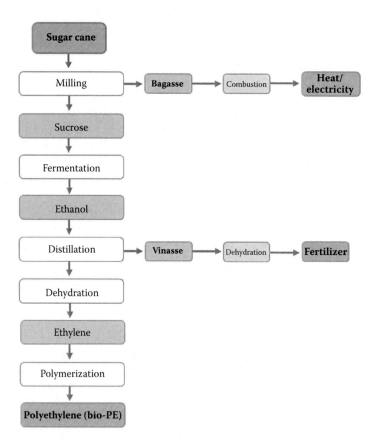

FIGURE 25.6 Schematic outline of the production process for biobased ethylene from sugar cane.

it can fully substitute conventionally produced PE. The same processing equipment can be used to produce a large variety of plastic products for flexible food packaging, cosmetics, and toys and automotive parts. However, owing to its chemical structure, bio-PE is not biodegradable but can be reused, mechanically recycled into new products alternatively being incinerated to generate energy.

Sugar cane uses CO_2 to produce sucrose by photosynthesis. When CO_2 emissions from transport and chemical processing are subtracted, the production of 1 ton bio-PE captures and fixes 2.5 tons CO_2 from the atmosphere (Noaves 2010); hence, bio-PE is being neutral regarding global warming effects. In contrast, the production of 1 ton PE from petrochemicals results in the release of 2.5 tons CO_2 to the atmosphere. In addition, the green profile of bio-PE is further enhanced by the fact that the polymerization process reduces the energy consumption by approximately 70% compared to the petrochemical production cycle (Noaves 2010).

Braskem started industrial production of bio-PE in 2010 with the goal to produce 200 kton/year of bio-PE from sugar cane (Noaves 2010). Dow and Crystalsev jointly planned to start up a plant to produce about 350 kton bio-PE/year from sugar cane in 2011, and other initiatives are on the way.

In 2014, a new cross-disciplinary project, BioInnovation, started in Sweden with the aim to produce bio-PE from forest-based cellulosic sugars. This biobased lignocellulosic PE is claimed to become competitive to bio-PE from sugar cane in terms of lower price.

25.4.2 Polylactide

Lactic acid is a natural product produced by microbial fermentation of natural carbohydrates. It is a part of the metabolic systems in mammalians and exerts a preserving effect (Södergård and Inkinen 2011). The primary starting material for lactic acid fermentation is D-glucose, which enables a wide range of naturally occurring raw materials to be used in the industrial process, e.g., sugar cane, sugar beet, rice, wheat, corn and potato starch, and last but not least agricultural waste products.

Poly(lactic acid) (PLA) is a thermoplastic aliphatic polyester, produced synthetically by polymerization of lactic acid monomers or cyclic lactide dimers. Lactic acid (2-hydroxypropionic acid) is a three-carbon chiral acid, which contains an asymmetrical carbon atom giving rise to the two optically active stereoisomers, namely, L(+)-lactic acid and D(−)-lactic acid (Figure 25.7). Lactide dimers are formed by the condensation of two lactic acid molecules, resulting in three possible stereoisomers: L-lactide, D-lactide, or LD (*meso*)-lactide (Drumright et al. 2000; van de Velde and Kiekens 2002; Auras et al. 2005; Henton et al. 2005) (Figure 25.7). The chemical structure of the lactides is 3,6-dimethyl-1,4-dioxane-2,5-dione.

Lactic acid is synthesized from pyruvate and is catalyzed by lactic acid dehydrogenase under oxygen-limiting conditions. The ratio of L- and D-lactic acid is determined by the specificity of the catalyst. The L-form is the most frequently naturally occurring of these (present in mammalian systems), whereas both forms exist in bacterial systems (Södergård and Stolt 2002). Lactic acid produced by fermentation typically results in 95.5% of the L-isomer (Jérôme and Lecomte 2005).

Production of lactic acid by fermentation usually utilizes the group of lactic acid bacteria capable of converting hexoses into lactic acid, typically *lactobacillus*, although other bacteria, fungi, or yeasts can also be used (Södergård and Stolt 2002). The industrial fermentation process is similar to the fermentation of sugars in cereal grains or grapes to produce beer and wine, respectively (Jérôme and Lecomte 2005). Temperature, atmosphere, pH, and agitation are important parameters, and both batch and continuous fermentation processes can be adopted where the latter naturally leads to higher productivity. As starting material in the microbial fermentation process, annually renewable raw materials such as maize, wheat, starch, sugar cane, or waste products from the agricultural or food industry can be used (Södergård and Inkinen 2011). The starting concentration of sugar is typically 5%, and a nitrogen-containing nutrient is added. The pH is kept constant by continuous addition of neutralizing agents like $CaCO_3$, $Ca(OH)_2$, $Mg(OH)_2$, $NaOH$, or NH_4OH since the production of the organic acid causes a shift of pH toward the acidic range as the reaction proceeds. Conversion ranging between 90% and 99% is achievable over two days (Södergård and Stolt 2002).

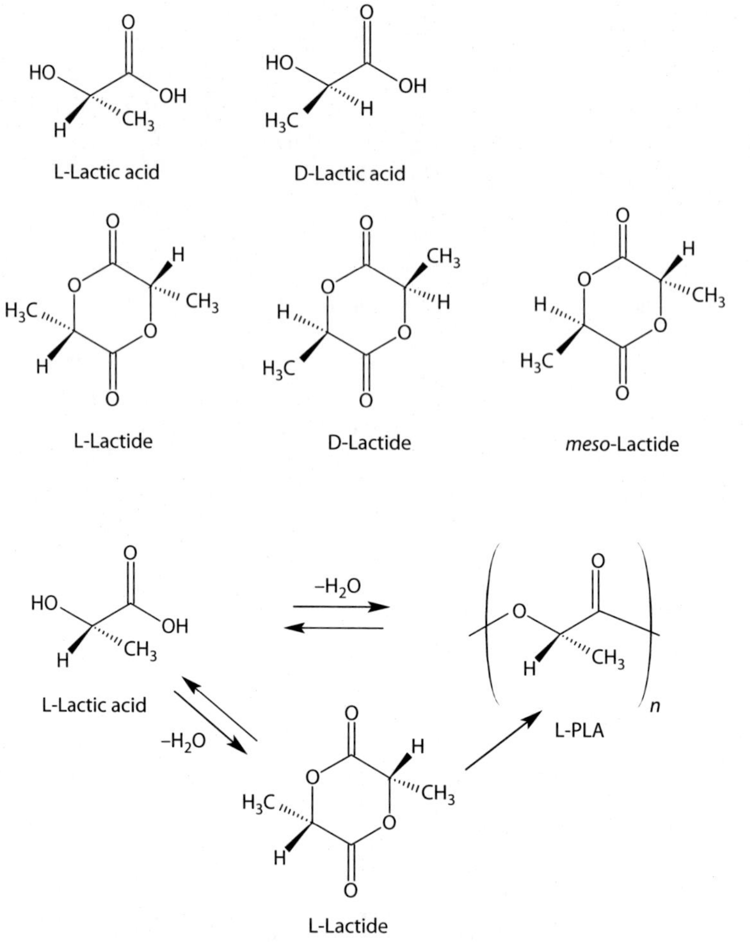

FIGURE 25.7 Stereoisomers of lactic acid and lactide. Schematic overview of the formation of L-PLA by direct polycondensation of L-lactic acid and via formation of the dehydrated, cyclic L-lactide dimer, followed by ring-opening polymerization to the final L-PLA structure. (Redrawn from Drumright, R.E. et al., *Adv Mater* 12(23):1841–1846, 2000.)

Separation of lactic acid from the fermentation broth requires a number of purification steps in which neutralization, filtration, concentration, and acidification are traditionally included. Isolation of lactic acid, which is ionized at neutral pH, and simultaneous removal of water in industrial processes require the addition of a strong mineral acid, e.g., sulfuric acid, in order to protonate the lactate salt. This treatment, however, causes a lot of waste since an equal amount on weight basis of waste salt is produced per weight of lactic acid produced, which is thus undesired from an environment perspective (Henton et al. 2005; Álvarez-Chávez et al. 2012). Further purification steps might also be utilized like ultra- or nanofiltration, electrodialysis, esterification, and ion-exchange processes (Södergård and Stolt 2002; Henton et al. 2005).

Although the terms *polylactide* and *poly(lactic acid)* are frequently used interchangeably in the literature, the use of the term *polylactide* normally refers to the polymer synthesized from the lactide dimer by the more commonly used ring-opening polymerization reaction, whereas the term *poly(lactic acid)* generally is used to describe the product prepared from polycondensation of lactic acid (Södergård and Stolt 2002). Both polymer types are, however, abbreviated PLA, and the two different routes are schematically depicted in Figure 25.7. PLA produced from the L-lactide monomer results in a polymer named poly(L-lactide), often abbreviated PLLA. The purity of lactic acid, i.e., the ratio of L- and D-enantiomers, can be determined by nuclear magnetic resonance spectroscopy, ^{13}C NMR, or by high-performance liquid chromatography (HPLC) in combination with ultraviolet detection (Södergård and Inkinen 2011).

The most important manufacturing process for production of commercial quantities of PLA is through ring-opening polymerization of lactide (Auras et al. 2005), since this route offers a more accurate control of the chemistry and properties of the resulting polymer (Södergård and Stolt 2002). The reaction also favors the formation of higher-molecular-weight PLA (Plackett et al. 2006). The final crystallinity and mechanical properties of the polymer depend on the stereochemistry of the polymer backbone, i.e., the relative amounts of the L- and D-isomers. The L-form gives a polymer with very high crystallinity, whereas copolymers of D- and L-isomers become amorphous (Weber 2000).

The ring-opening polymerization process starts with polycondensation of lactic acid, followed by depolymerization of poly(lactic acid) into a dehydrated, cyclic dimer (lactide), and ends with the ring-opening polymerization step. The racemic purity, lactide purity, and residual monomer content are crucial parameters that need to be controlled. The lactic acid molecules also have two functional groups: one hydroxyl group and one carboxyl group, which are capable of undergoing inter- and intramolecular esterification reactions during the polymerization process (Södergård and Inkinen 2011). Impurities like carboxylic acids or oligomers present may retard the polymerization or decrease the molecular weight. Removal of monomer residuals in a separate step is required to enhance the properties of the end product (Södergård and Inkinen 2011).

Production of lactide is carried out by depolymerization of poly(lactic acid) with appropriate molecular weight in either a batch or a continuous process. The depolymerization process is carried out by raising the temperature and lowering the pressure, followed by distilling off the product lactide (Södergård and Stolt 2002). The process generates a number of by-products, which have to be removed to generate a pure lactide for the next polymerization step. Purification reactions utilized involve solvent crystallization, melt crystallization, and gas-phase purification (Södergård and Inkinen 2011). The polymerization process is normally catalyzed by an organotin-based catalyst system by addition of very low concentrations of tin octanoate (Henton et al. 2005; Jérôme and Lecomte 2005; Södergård and Inkinen 2011). Tin octanoate (tin (II) bis-2-ethylhexanoic acid), whose chemical formula can be written as $Sn(O(O)CCH(C_2H_5)C_4H_9)_2$, is soluble in the molten state of lactide, has high catalytic activity, and results in low degree of racemization of the polymer and is thus preferred for bulk polymerization reactions (Drumright et al. 2000; Henton et al. 2005). The tin-based catalysts are toxic, and alternative catalysts with less toxicity have been investigated but have not yet found industrial applicability due to racemization of the PLA products (Södergård and Inkinen 2011). Hydroxyl-containing initiators such as 1-octanol are frequently used to accelerate the ring-opening polymerization step and to control the molecular weight (Drumright et al. 2000; Södergård and Stolt 2002; Henton et al. 2005; Jérôme and Lecomte 2005).

Ring-opening polymerization can be carried out in four different process routes: (1) bulk polymerization, (2) melt polymerization, (3) solution polymerization, and (4) dispersion polymerization (Södergård and Stolt 2002; Södergård and Inkinen 2011). The mechanism can, depending on the catalyst system used, be of ionic, free radical, or coordination type (Södergård and Stolt 2002). The mechanism involved in the use of tin octanoate is suggested to involve a pre-initiation step in which tin(II) 2-ethylhexanoate is converted to a tin(II) alkoxide, which acts as the actual process initiator, in reaction with an alcohol (Jérôme and Lecomte 2005). The polymerization then proceeds on the tin–oxygen bond of the alkoxide ligand (Södergård and Stolt 2002).

The generally accepted mechanism for polymerization of lactide using tin octanoate is via a coordination–insertion mechanism where ring opening of lactide occurs, followed by addition of one lactide unit (two lactyl units) to the growing end of the polymer chain (Henton et al. 2005). As initiators, hydroxyl or other nucleophilic species acts. Hydroxyl species can be present as impurities in the form of water, lactic acid, dimers, and trimers or can be added on purpose.

The patented, solvent-free production method originally implemented by Cargill Dow LLC starts with fermentation of dextrose (formed by enzymatic hydrolysis of starch) to produce lactic

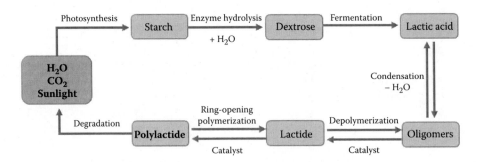

FIGURE 25.8 Schematic outline of the solvent-free process for production of PLA.

acid (Figure 25.8). The next step is a continuous condensation reaction of aqueous lactic acid to a PLA pre-polymer (oligomer) with low molecular weight (around 5000 g/mol) (Drumright et al. 2000; Vink et al. 2003; Henton et al. 2005). This pre-polymer then undergoes a depolymerization step catalyzed by a tin compound to yield a mixture of lactide stereoisomers further purified by vacuum distillation of the melt mixture. In the final step, a tin-catalyzed ring-opening polymerization in the melt results in a high-molecular-weight PLA. Monomer residuals are removed under vacuum and fed back to the initial step of the process. The typical conditions are polymerization at 180–210°C with addition of tin octanoate ranging from 100 to 1000 ppm, and a reaction time of 2–5 h, resulting n a conversion of about 95% (Drumright et al. 2000; Henton et al. 2005). The fact that the whole process is taking place in the melt has the environmental and economic benefits that the use of expensive and hazardous solvents is precluded. Final degradation of PLA results in the formation of CO_2 and $_2O$, which, in combination with sunlight, is used to produce starch via photosynthesis (Figure 25.8) and closes the loop.

PLA-based products can be either recycled or composted after use to eventually be converted to carbon dioxide and water. In addition, the CO_2 generated during biodegradation of PLA is balanced by an equal amount taken up during growth of the plant feedstock (Henton et al. 2005). Production of PLA from biomass is indeed supposed to give a net reduction in greenhouse gases over a long-term perspective (Henton et al. 2005).

Another manufacturing process of PLA, which has gained less industrial attention, is by a polycondensation process. The polymers synthesized by this self-condensation reaction are generally termed *poly(lactic acid)* (Henton et al. 2005; Södergård and Inkinen 2011). Water is released during the reaction, and removal of water becomes more difficult when the molecular weight increases due to increased viscosity of the reaction mixture (Drumright et al. 2000). The reaction is typically catalyzed by strong acids and metallic or organometallic compounds (Södergård and Inkinen 2011). Disadvantages with this production method are limited molecular weight and low yield of poly(lactic acid). The molecular weight can be enhanced by eliminating the viscosity increase and facilitating the water removal by the use of a solvent, e.g., toluene (Södergård and Inkinen 2011), a method that is less attractive from an economical and environmental perspective.

The most crucial parameter of PLA is its molecular weight, which determines the mechanical strength and thermal properties (Södergård and Inkinen 2011). Preparation of a lactic acid pre-polymer by polycondensation followed by addition of a linking molecule—typically a diisocyanate—is a route to extend the chain length.

25.4.3 Polycaprolactone

PCL is a synthetic, linear aliphatic polyester that is prepared from ring-opening polymerization of ε-caprolactone. The starting material is derived from petroleum, but despite originating from a nonrenewable resource and being produced synthetically, PCL is biodegradable.

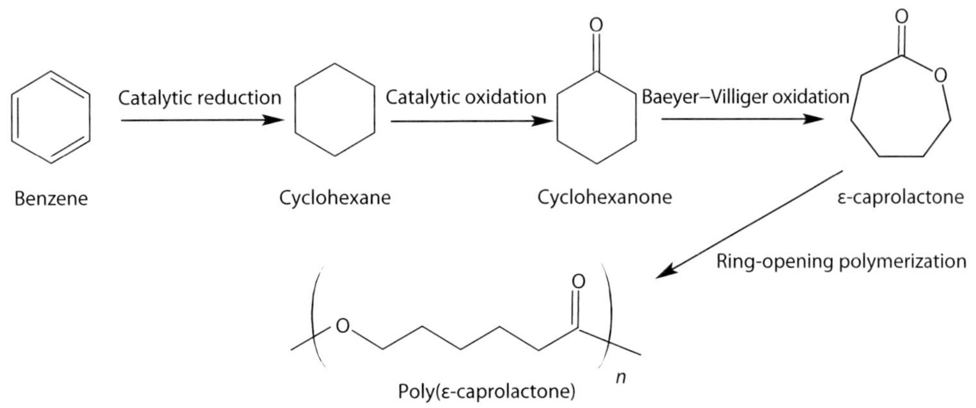

Bioplastics

FIGURE 25.9 Schematic outline of the synthesis of poly(ε-caprolactone).

A simplified description of the process starts with derivation of benzene from crude oil, followed by catalytic hydrogenation to form cyclohexane (Figure 25.9). Catalytic oxidation of cyclohexane results in the formation of cyclohexanone, which finally undergoes a Baeyer–Villiger oxidation reaction to yield ε-caprolactone (Jérôme and Lecomte 2005). Rocca et al. (2003) described a patented process (Solvay) in which a composition of antimony trifluoride and hexagonal mesoporous silica is used as a catalyst for oxidation of cyclohexanone to ε-caprolactone via a Baeyer–Villiger oxidation reaction. The catalyst is mixed with cyclohexanone; hydrogen peroxide is added as an oxidizing agent to this mixture, and the reaction proceeds under specified conditions of time and temperature.

Tin octanoate is another catalytic system used in the industrial production of PCL (Jérôme and Lecomte 2005), similarly to the ring-opening polymerization of PLA. Other strategies involve ring opening by the activity of enzymes, specifically lipases, which are considered as more environmentally friendly (Jérôme and Lecomte 2005 and references therein). The use of reactive extrusion allows the production of PCL in a continuous process with further manufacturing of films, fibers, and various shaped articles (Jérôme and Lecomte 2005). Aluminum and titanium alkoxides have been found to work well as catalysts for the polymerization of ε-caprolactone in a twin-screw extruder to give high conversion within short time.

PCL has a glass transition temperature of −60°C and a melting temperature of 60°C (Jérôme and Lecomte 2005). It has a low viscosity and can easily be processed at enhanced temperature (Gross and Kalra 2002). PCL is resistant to water, oil, solvent, and chlorine and shows relatively high thermal stability. It can readily be processed by conventional extrusion and injection molding equipment. PCL is insoluble in water, alcohol, and acetone, and hence the most commonly used solvent for solvent-based processes is chloroform (Olabarrieta et al. 2001; Sinha et al. 2004).

25.5 BIOPLASTICS FROM MICROORGANISMS

25.5.1 Polyhydroxyalkanoates

PHAs are aliphatic polyesters produced by many microorganisms for intracellular energy or carbon reserves when subjected to limited access to essential nutrients such as ammonium, iron, magnesium, phosphate, or sulfate, while carbon is available in excess. The PHAs were first visualized in the laboratory by the French microbiologist Maurice Lemoigne in the 1920s, although their true nature was not then understood, as their discovery predated the concept of macromolecules. The development of large-scale industrial fermentation processes for production of PHAs as alternative materials to petroleum-based plastics was driven by the oil crisis in the 1970s with unstable oil prices and uncertainty in the petroleum-based plastics industry. Industrially, PHAs are now produced by fermentation of biobased raw materials such as sucrose, vegetable oils, and fatty acids (Álvarez-Chávez et al. 2012) by controlled growth of certain bacteria under nutrient-limited conditions. The use of cheap carbon courses from waste products like stalks and leaves from corn plants has also been tested as feedstock and has great impact on the production economy. Bacterial biosynthesis is nowadays the most important process for production of PHAs, although chemical synthesis via ring-opening polymerization of lactones also provides a possible route.

PHAs show thermoplastic and elastomeric properties resembling those of petroleum-based plastics. Classification of PHAs is typically done into short-chain-length and medium-chain-length types based on the number of carbons in the repeating units (Pollet and Avérous 2011).

PHAs can form a variety of structures, but the most widely studied are the homopolymers poly-3-hydroxybutyrate [P(3HB)] and poly-3-hydroxyvalerate [P(3HV)]; Figure 25.10, and copolymers of 3-hydroxybutyrate (HB) and 3-hydroxyvalerate (HV), poly(3-hydroxybutyrate-co-3-hydroxyvalerate) [P(3-HB-co-3HV)]. The general chemical name of the latter structure is 3-hydroxybutanoic acid-3-hydroxypentanoic acid. The proportions between the two monomers vary, but the copolymers are generally random in sequence (Pollet and Avérous 2011). Other PHA copolymers exist where the valerate repeating unit is replaced by hexanoate, octanoate, and octadecanoate units.

Biosynthesis of PHA begins with conversion of a carbon source to acetate, followed by attachment of an enzyme cofactor (coenzyme A [CoA]) via the formation of a thioester bond. Acetyl-CoA is a basic metabolic molecule present in all PHA-producing organisms. In the next step, a reversible condensation takes place to form a dimer of acetoacetyl-CoA, which is subsequently reduced by nicotinamide adenine dinucleotide phosphate (NADPH) to (R)-3-hydroxybutyryl-CoA, the basic monomer unit for P(3HB). Polymerization of this monomer proceeds by the bonding of a second monomer unit to the free thiol group of the active site, and occurs as a continuous series of insertion reactions. The enzyme is specific for monomers with the (R) configuration, and this asymmetric center is retained in the polymer, forming completely isotactic PHAs (Pollet and Avérous 2011).

More than 250 different bacteria can produce PHAs at varying efficiency, yield, and quality, but the number of bacterial strains capable of producing PHAs at industrial scale is limited (Pollet and Avérous 2011). Some potential bacteria are *Pseudomonas oleovorans*, *Azotobacter vinelandii*, *Ralstonia eutropha*, and recombinant *Escherichia coli*, to mention just a few (Braunegg et al. 1998; Suriyamongkol et al. 2007; Pollet and Avérous 2011). Suriyamongkol et al. (2007) presented a comprehensive review to biotechnological approaches for the production of PHAs.

Recovery of the PHA product involves isolation of the polyester from biomass via mechanical, chemical, or enzymatic destruction of the cell wall, followed by dissolution of the polymer in a suitable solvent. PHAs are insoluble in water, and solubilization requires halogenated solvents like chloroform, methylene chloride, or 1,2-dichloroethane. Finally, extraction by a mixture of solvents may be required for further purification. Isolation and concentration of PHAs involves potential safety hazards by exposure to halogenated solvents and other toxic chemical substances in physical extraction and chemical digestion processes (Álvarez-Chávez et al. 2012). An alternative method is enzymatic hydrolysis and detergents to remove cell walls, proteins, and nucleic acids from bacterial cells (Suriyamongkol et al. 2007).

More recent routes to production of PHAs are by cloning of the genes responsible for PHA production in specific bacteria and transferring them to other organisms, or via genetic modification of

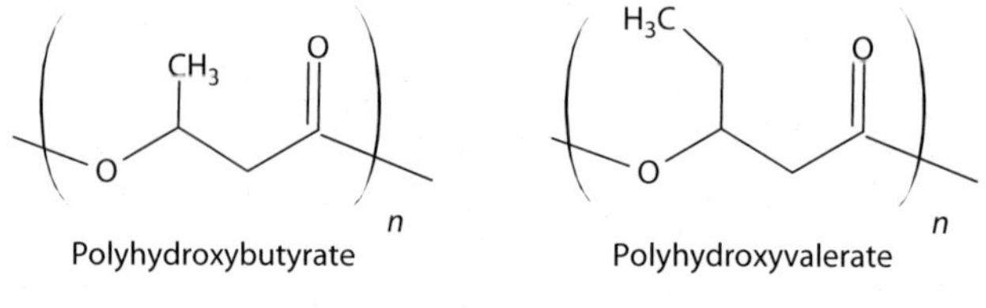

FIGURE 25.10 P(3HB) and P(3HV).

plants to produce the polymers via the same mechanisms as those used by bacteria. However, there are still challenges to produce PHAs of controlled composition at sufficiently high levels and to develop economical methods for efficient isolation of the polymers (Pollet and Avérous 2011). Even though chemical synthesis methods that produce PHAs via ring-opening polymerization of lactone monomers with the use of zinc- or aluminum-based catalysts offer a higher level of control and possibility to adjust the composition and structure of the PHA polymers, the traditional biosynthesis route provides several advantages since renewable (and less expensive) feedstock is used, and PHAs with much higher molecular weight can be produced (Pollet and Avérous 2011).

Yu and Chen (2008) presented a cradle-to-factory gate life cycle assessment for production of PHA in a biorefinery using corn stover (leaves and stalks of corn plants) as the starting material. Their analysis showed that only 0.49 kg CO_2 was emitted by the production of 1 kg PHA, which should be compared to the 2–3 kg produced for a petrochemical counterpart. In addition, only about half the fossil energy was required per kilogram bioplastics produced compared to the petrochemical counterpart.

PHAs can also be derived from the hemicellulose component in wood biomass (Keenan et al. 2005). Using detoxified hemicellulose hydrolysates as xylose-rich feedstocks for microbial production of PHA has economic advantages since this by-product from the pulping process can replace the more expensive carbon sources and separation processes traditionally adopted for production of PHA.

PHB forms highly crystalline films, which are stiff and very brittle and thus of limited practical interest. Films from copolymers containing the longer alkyl chain of HV give films with excellent toughness and strength (Chandra and Rustgi 1998). By adding propionic acid to the growth medium, the ratio of HB and HV in the copolymer can be changed. Copolymers rich in HB resemble polypropylene, while copolymers rich in HV resemble PE with respect to flexibility, tensile strength, and melting point (Krochta and De Mulder-Johnston 1997).

PHAs possess excellent film-forming and coating properties and are water-resistant due to their high hydrophobicity (Petersen et al. 1999). PHB-coated paperboard has been used for packaging of ready meals, whereas PHB/V or PLA-coated board has been used for dry products, dairy products, and beverages (Weber 2000). The drawbacks with PHAs are poor gas barrier properties and high production costs.

25.6 STRUCTURAL AND PHASE CHARACTERISTICS AND EFFECT ON PROPERTIES OF BIOPLASTICS

An overview of physical and chemical properties of bioplastics related to structural and phase effects is given below. A corresponding overview of some bioplastics (including blends and composites), manufacturers, commercial products, and application areas appears in Table 25.3.

25.6.1 THERMOPLASTIC STARCH

The linear amylose fractions in starch are able to form continuous, strong, and flexible films through hydrogen bonding, whereas the branched amylopectin structure prevents the molecules from coming close enough to form hydrogen bonds (Wurzburg 1986). Amylopectin films are thus hard and brittle (Young 1984). The film-forming behavior depends on the molecular weight of the amylose and the amylopectin fractions, the water-holding properties, and the colloidal stability of the starch dispersion. The addition of a plasticizer such as glycerol can improve the mechanical properties, but at high levels, the plasticizer can increase the mobility of amylose chains at a degree that leads to increased crystallinity (Young 1984).

Starch plastics are sensitive to water and will lose some of their properties upon contact with moisture and/or by aging effects leading to changes in crystallinity (van Soest and Vliegenthart 1997). The moisture sensitiveness makes starch films poor barriers against water vapor (Petersen et al. 1999). Increased crystallinity of starch films facilitates the barrier properties toward gases.

TABLE 25.3
Overview of Some Bioplastics (Including Blends and Composites), Manufacturers, Commercial Products, and Application Areas

Bioplastics	Manufacturer	Trade Name	Application Areas
Thermoplastic starch	Novamont	Mater-Bi	Agricultural films, packaging films, bags, foams, injection-molded items, thermoformed trays
	Plantic Technologies	Plantic	
	Biotec	Bioplast, Bioflex	
	National Starch	Eco-foam	
	Rodenburg Biopolymers	Solanyl	
	Cereplast Inc	Cereplast Compostable	
	BIOP	BioPar	
Lignocellulose	Borregard	Lignopol	Composites, fibers, adhesives, compatibilizers
Cellulose	Innovia Films	Cellophane	Breathable films, ovenable/microwaveable packaging
		Natureflex	
		Cellotherm	
PLA	NatureWorks LLC	Inego	Injection-molding items, fibers, extrusion coatings, thermoformed containers
	Cargill-Dow	Natureworks	
	Boehringer	LACEA	
	Galactic		
	Shimadzu		
	Mitsui		
PCL	Daicel	Celgreen	Compost bags, cold packaging
	Union Carbide	TONE	
	Solvay	CAPA	
PHA	Biomer	Biomer	Molded items, agricultural mulch films
	Metabolix	Biopol	
	Telles	Mirel	
	PHB Industrial	Biocycle	
	Meredian Inc.	Nodax	
Bio-PE	Braskem		Bottles for cosmetic, household products, beverages; caps, toys, shopping bags, flexible packaging
	Dow		
	Solvay		

Source: Halley, P.J., Thermoplastic starch biodegradable polymers. In *Biodegradable Polymers for Industrial Applications*, ed. R. Smith, 140–162, Cambridge: Woodhead Publishing, 2005; Jérôme, R. and Lecomte, P., New developments in the synthesis of aliphatic polyesters by ring-opening polymerization. In *Biodegradable polymers for industrial applications*, ed. R. Smith, 77–106, Cambridge: Woodhead Publishing, 2005; Shen, L. et al., Product overview and market projection of emerging bio-based plastics, PRO-BIP 2009, available at http://en.european-bioplastics.org/wp-content/uploads/2011/03/publications/PROBIP2009_Final_June_2009.pdf (accessed July 3, 2014), 2009; Plackett, D., Introductory overview. In *Biopolymers: New Materials for Sustainable Films and Coatings*, ed. D. Plackett. 2011. 3–14.

However, pores that might be present in the films have a higher impact on gas permeability and water vapor permeability than variations in crystallinity. On the other hand, starch plastics often provide moderate to good barriers to oils and grease (Krochta and De Mulder-Johnston 1997).

Reassociation in the rubbery state into more ordered structures (retrogradation) may take place by aging and results in entanglements and formation of helices and crystal structures of both the amylose and amylopectin molecules (van Soest and Vliegenthart 1997). Increased input in mechanical energy (increased extrusion screw speed) or increased time during kneading has been shown to increase the single helix-type crystallinity in amylose. In addition, the presence of lysophospholipids and other complex-forming agents may induce amylose crystallization (van Soest and Vliegenthart 1997).

Extrusion is a common processing technique to convert starch into products with thermoplastic properties. Prevention of degradation of the starch is crucial during melting and mixing at elevated temperatures. High shear stress is expected to lead to a high degree of degradation (Altskär et al. 2008). On the other hand, optimum mechanical properties of the final product require complete melting of the initial crystalline structure of the starch and subsequent homogenization of the melt. Presence of residual granules could act as weak spots in the film, thus enhancing the risk of failure.

25.6.2 POLYLACTIDE

PLA can be processed on standard converting equipment for thermoplastic materials with minimal modification, including thermoforming, injection molding, blow molding, foaming, film extrusion, and fiber extrusion (Shen et al. 2009). PLA exhibits high resistance to grease and oil and has high UV resistance (Shen et al. 2009). It is a poor barrier against oxygen and carbon dioxide but provides superior moisture barrier properties over many other bioplastics including starch.

The stereochemical composition of the PLA polymer is determined by the stereoisomer used as the monomer, i.e., L-lactide or D-lactide. PLA can be either amorphous or semicrystalline, depending on the thermal history of the product but more significantly on the stereochemistry (Henton et al. 2005). The most important commercial PLA products are copolymers of L-lactide with minor amounts of D-lactide and *meso*-lactide arranged in random structures.

PLA made of pure L-lactide is called poly(L-lactide) and has a glass transition temperature at about 60°C and an equilibrium crystalline melting point of 207°C (Drumright et al. 2000). Incorporation of D-lactide or *meso*-lactide results in kinks or defects in the crystal arrangement, which in turn leads to a reduced melting point, reduced rate of crystallization, and reduced extent of crystallization (Drumright et al. 2000 and references therein; Henton et al. 2005). The glass transition temperature increases with isomeric purity and with increased molecular weight up to a certain value, with typical T_g values in the range 55–65°C (Henton et al. 2005). The melting temperature of PLA is thus typically in the range 170–190°C because of imperfect crystals and the presence of impurities (Henton et al. 2005; Södergård and Inkinen 2011). The addition of D-lactide units in the L-PLA can be optimized to match the crystallization kinetics for specific manufacturing processes and applications (Henton et al. 2005).

For production of PLA, a high purity of the starting material is crucial. However, lactic acid (even of analytical grade) contains traces of carbohydrates and amino acids leading to coloration of the product. The presence of extremely small amounts of Na^+ causes racemization of the acid, thus resulting in a fraction of D-lactic acid (Henton et al. 2005).

Amorphous polymers with atactic structure can be formed, for example, from random copolymers of *meso*-lactide or from equimolar blends of L-lactide and D-lactide (Henton et al. 2005). Pure poly(L-lactide) and pure poly(D-lactide) are isotactic polymers (Södergård and Inkinen 2011). A number of other polymers with varying tacticity can be produced by varying the content of the monomer units.

One of the most commercially interesting applications for PLA is as packaging material (Henton et al. 2005; Södergård and Inkinen 2011). The polymer exerts excellent grease and oil resistance and heat sealability and has good barrier properties against aromas and flavors (Drumright et al. 2000). Due to their biodegradability, PLA is suitable for use, for example, in agricultural mulch films and bags (Drumright et al. 2000).

PLA shows a pseudoplastic, non-Newtonian flow behavior. It has rheological properties that make the polymer suitable for sheet extrusion, film blowing, and fiber spinning (Drumright et al. 2000; Henton et al. 2005). The rheological properties are determined by the molecular weight, temperature, shear rate, and racemic purity. Branching is a way to enhance the shear sensitivity or melt strength of PLA and is introduced by treatment with peroxide, by use of multifunctional initiators or monomers during the polymerization process. Branched PLA can be processed by extrusion coating, extrusion blow molding, and foaming (Drumright et al. 2000). However, the processing of PLA is challenging due to the hygroscopic character of the polymer, its relatively poor melt stability, and its ability to undergo hydrolytic degradation (Södergård and Inkinen 2011). Before processing, it is necessary to dry the polymer in order to minimize the risk of thermohydrolysis with subsequent molecular weight reduction during melt processing (Södergård and Inkinen 2011).

The most important parameters in melt processing of PLA involve temperature, residence time, moisture content of the polymer, and atmosphere. The major drawback with PLA is its limited thermal stability during melt processing since degradation of the ester linkages occurs when exposed to heat. Nevertheless, PLA has found use in melt processing to form blow-molded bottles, injection-molded cups, spoons and forks (Södergård and Stolt 2002 and references therein; Henton et al. 2005), paper coatings, fibers, films, and molded articles. For medical use, fracture fixation devices and sutures are examples.

One of the drawbacks of processing PLA in the molten state is its tendency to undergo thermal degradation, which is related both to the process temperature and the residence time in the extruder (Lima et al. 2008). A number of factors affect the thermal degradation of PLA: high amounts of reactive end groups and high polydispersity; presence of residual catalysts or residual monomers; and presence of moisture and impurities (Södergård and Inkinen 2011). The thermal degradation of PLA can be attributed to: (1) hydrolysis of the ester linkages by trace amounts of water, which occurs more or less randomly along the backbone of the polymer; (2) zipper-like depolymerization; (3) oxidative, random main-chain scission; (4) intermolecular transesterification to monomer and oligomeric esters; and (5) intramolecular transesterification resulting in formation of monomer and oligomer lactides of low molecular weight.

Hydrolytic degradation implies a random hydrolysis of ester linkages on the polymer backbone in the presence of water (Södergård and Inkinen 2011). At temperatures above the T_g, PLA readily undergoes hydrolysis both in the solid state and in the melt. The reaction is dependent on the molecular weight and the degree of crystallinity (Södergård and Inkinen 2011). Strategies to improve the resistance to hydrolysis involve minimizing the presence of residual monomers, impurities, and water.

Above 200°C, PLA can degrade through intra- and intermolecular ester exchange, cis-elimination, and radical and concerted nonradical reactions, resulting in the formation of CO, CO_2, acetaldehyde, and methylketene. It has also been proposed that thermal degradation of PLA can occur by way of a nonradical ester interchange reaction involving the –OH chain ends. Depending on the point in the backbone at which the reaction occurs, the product can be a lactide, an oligomeric ring structure, or acetaldehyde plus carbon monoxide. At temperatures above 270°C, homolysis of the polymer backbone can occur. The formation of acetaldehyde is expected to increase with increasing process temperature, with the highest proportion formed at 230°C. Although acetaldehyde is considered to be nontoxic and is naturally present in many foods, the acetaldehyde generated during melt processing of PLA must be minimized, especially if the converted PLA (e.g., container, bottle, and films) is to be used for food packaging. The migration of acetaldehyde into the contained food can result in off-flavors, which will impact the organoleptic properties and consumer acceptance of the product (Lima et al. 2008).

Strategies to improve the melt stability of PLA can be found in various patents as reviewed by Lima et al. (2008). Polycarbodiimide (CDI) was observed to improve the thermal stability of PLA during processing. The results showed that addition of CDI at 0.1–0.7 wt% led to stabilization of PLA at 210°C for up to 30 min. Considering the mechanism, CDI could react with the residual or newly formed moisture and lactic acid, or carboxyl and hydroxyl end groups in PLA, and thus hamper thermal degradation and hydrolysis of this polymer (Yang et al. 2008). The melt stability of PLA may vary from supplier to supplier as a result of the different processes and technologies used (Lima et al. 2008).

25.6.3 POLYHYDROXYALKANOATES

The properties of PHB resemble those of synthetic polypropylene in many respects. PHA possesses low water vapor permeability, thus making it attractive in packaging applications (Pollet and Avérous 2011). However, the most important drawbacks of PHB are very low elongation at break and brittleness of films and molded products, owing to the high crystallinity. PHAs show T_g values below room temperature. Addition of plasticizers lowers the T_g and renders the polymers less brittle at normal use temperatures (Pollet and Avérous 2011). Increasing the amount of HV results in increased impact strength, decreased crystallinity, and reduced glass transition and melting temperatures (Pollet and Avérous 2011 and references therein). For example, a decrease in melting temperature from 180°C to 137°C has been reported when going from 0% to 25% P(3HV).

The simplest of the PHAs (PHB and PHBV) are similar to PLA in terms of sensitivity to trace amounts of water, exposure to which can bring about hydrolysis of the ester linkages more or less randomly along the polymer backbone. Besides this, PHB in particular shows poor thermal stability at temperatures above the melting point, which is typically at or near 175°C (Hablot et al. 2008). Studies dedicated to thermal degradation of PHB and PHBV have revealed that thermal degradation occurs rapidly near the melting point, mainly due to a random chain scission process (Hablot et al. 2008).

25.7 ADDITIVES AND THEIR EFFECTS ON PROPERTIES AND APPLICATIONS

As with conventional thermoplastics, additives may be required to improve the resistance of bioplastic materials against hydrolysis, oxidation, discoloration, and UV light, or to enhance the flow characteristics or mechanical strength or flexibility of the products. Typical additives are plasticizers, fillers, processing aids, and colorants. In this section, some effects of various additives on the properties and applications of different types of bioplastics are discussed.

25.7.1 THERMOPLASTIC STARCH

Plasticizers are fundamental additives for the production of TPS. Plasticizers should be compatible with the material they are mixed with, and for optimum performance over the practical life time of the material, plasticizers should be nonvolatile and resistant toward migration since loss of plasticizers by these mechanisms should increase the brittleness of the polymer. In addition, for food packaging applications, it is important to prevent migration of substances from the packaging material to the food, which may adversely impact the taste and quality of the food. Of utmost concern is also the selection of a plasticizer that does not provide health risks to consumers caused by any inherent toxicity. Plasticizers should ultimately be biodegradable to form a completely biodegradable thermoplastic material.

Low-molecular-weight molecules are generally more effective as plasticizers due to the large amount of end groups present. Typical plasticizers are water, glycerol, urea, formamide, urea/formamide mixtures, glycol, xylitol, and sorbitol (van Soest and Vliegenthart 1997).

The most widely studied plasticizer is glycerol, which is capable of forming strong hydrogen bonds with starch, which increases the toughness and strength of the material (van Soest and Vliegenthart 1997). Starch is viscoelastic and its properties are strongly dependent on the plasticizer content. At a low plasticizer content, the material shows a glassy behavior, followed by a transition from glassy to rubbery at increased additions of plasticizer. The glass transition temperature of TPS strongly decreases with increased addition of glycerol (van Soest and Vliegenthart 1997). Water molecules act as plasticizer for amorphous and semicrystalline starch, and the thermal and mechanical properties of TPS are thus strongly dependent on the moisture content. At a very high plasticizer content, the interactions between starch chains in the amorphous regions are low, and the material behaves like a gel (van Soest and Vliegenthart 1997).

Higher amounts of plasticizer increase the mobility of the starch chains and lower the glass transition temperature, thereby affecting the rate of crystallization during aging. Increased strength and stiffness due to physical cross-linking caused by intermolecular crystallization has been observed (van Soest and Vliegenthart 1997). On the other hand, intramolecular crystallization in amylopectin induces internal stress and leads to deteriorated mechanical properties of starch plastics during storage and use (van Soest and Vliegenthart 1997).

Melt flow accelerators such as lecithin, glycerol monostearate, and calcium stearate are sometimes used to aid processing (van Soest and Vliegenthart 1997). Thermomechanical processing of granular starch results in complex materials consisting of residual swollen granules; partially melted, deformed, and disrupted granules; completely molten starch; and recrystallized starch (van Soest and Vliegenthart 1997), depending on the processing conditions (shear stress, viscosity, and temperature) and plasticizer content. Incomplete or intact granules lead to inferior mechanical properties of TPS materials (van Soest and Vliegenthart 1997).

25.7.2 Polylactide

PLA is readily processable by extrusion in the temperature range 60–125°C, depending on the ratio of D- and L-isomers. The brittleness of the material can be overcome by addition of plasticizers, e.g., lactic acid monomers or oligomers (Weber 2000). Other additives—both biobased and petrochemical compounds—have been developed to improve the heat resistance and impact strength of PLA products (Plackett 2011).

25.7.3 Polyhydroxyalkanoates

PHAs show T_g values below room temperature. Addition of plasticizers lowers the T_g and renders the polymers less brittle at normal use temperatures. Various studies have shown that acetyl tributyl citrate (ABTC) is an effective plasticizer for PHB and PHBV in decreasing the T_g and in increasing the elongation at break and impact strength of films (Pollet and Avérous 2011).

The influence of quaternary ammonium compounds (QACs) on the thermal and thermomechanical degradation of PHB has been studied. This is of interest because such surfactants are commonly used as organomodifiers in commercial nanoclays. Research findings suggest that QACs can have a catalytic effect on the thermal degradation of PHB, as reflected by a dramatic decrease in PHB molecular weight (Hablot et al. 2008). In another study, the thermal degradation of poly(3-hydroxybutyrate) in the presence of two plasticizers (glycerol, glycerol triacetate) was investigated as a function of the annealing time, temperature, and cooling rate after thermal treatment. The presence of glycerol leads to a significant prodegradative effect on PHB, presumably due to an alcoholysis reaction, while glycerol triacetate (triacetin) behaved as an almost inert additive in this respect (Janigová et al. 2002).

Some additives can increase the thermal stability of PHAs. For example, addition of 1 wt% of carboxyl-terminated butadiene acrylonitrile rubber (CTBN) or polyvinylpyrrolidone (PVP) modified the crystallization rate, crystallinity, melting temperature, and thermal stability of PHB (Hong et al. 2011). The degradation kinetics of PHB were changed due to the steric hindrance effects of the added PVP. The best improvement in thermal stability was obtained in the case of the PVP-modified PHB. The method of using CTBN or PVP to improve the thermal stability of PHB could be useful industrially because it does not depend on a requirement for high-purity PHB.

25.8 BLENDS AND COMPOSITES DERIVED FROM BIOPLASTICS

Blending two or more polymers is an attractive approach to reduce the costs of bioplastic products by the introduction of a cheaper material into a more expensive matrix. It is also a widely adopted route to improve some desired property (often mechanical strength) or to suppress a less desired property. To form a miscible polymer blend, it is often necessary that strong interactions between the polymers, e.g., hydrogen bonding, arise (Edgar et al. 2001).

Biocomposites have one or more of their phases derived from biological origins (e.g., plant fibers from crops such as cotton, flax, or hemp, or from recycled wood; crop processing by-products; or regenerated cellulose fibers such as viscose/rayon). The matrix phase within a biocomposite may consist of a natural polymer. The majority of biocomposites are currently used in the automotive, construction, furniture, and packaging industries, where increasing environmental awareness and the depletion of fossil fuel resources are providing the drivers for development of new renewable products.

Over the last few years, many researchers have investigated the use of natural fibers as load-bearing constituents in composite materials. The use of such composites has increased due to their relative cheapness, the possibility of recycling or composting (depending upon the polymer used) without harming the environment, and their competitive specific strength properties (Bledzki and Gassan 1999; Avérous and Digabel 2006).

The use of lignocellulosic fibers potentially presents safer handling and working conditions when compared to the use of synthetic fibers. Biofibers are nonabrasive to mixing and molding equipment, which can also contribute to significant cost reductions.

The production of 100% biobased composites as a substitute for petroleum-based products is still a challenge. Bioresins (or bioplastics) present some limitations. A viable solution would be to combine small amounts of petroleum-based resins with biobased resins to develop a cost-effective product having higher performance and a wider range of applications. Agropolymers (e.g., polysaccharides) obtained from biomass by fractionation, such as starch and cellulose, are the best known renewable resources available for making biodegradable plastics (Belgacem and Giandini 2011; Vazquéz et al. 2011).

The combination of natural fibers such as kenaf, hemp, flax, jute, henequen, pineapple leaf fiber, and sisal with polymer matrices from both finite and renewable resources to produce composite materials that are competitive with synthetic composites requires special attention (i.e., fiber–matrix interface and novel processing methods). By embedding natural fibers into biobased polymers such as cellulose, PLAs, starch, PHAs, and soy-based plastics, so-called green biocomposites are routinely being produced (Mohanty et al. 2002).

The literature describes numerous attempts to form blends and composites of bioplastic materials. Many of them are still investigated on the laboratory scale and need further development before they can be practically usable. Some of the more well-known blends and composite systems are summarized next.

25.8.1 Thermoplastic Starch

TPS shows a strong hydrophilic character; it is water-sensitive, and TPS products become brittle after aging. Furthermore, postprocessing variations in the properties of TPS products are typical. The mechanical properties of TPS films depend strongly on the amylose/amylopectin ratio and the type and amount of plasticizer (Lourdin et al. 1995; Rindlav-Westling et al. 1998). However, starch composites with mechanical properties comparable to or even higher than those of conventional polymers have been reported (Krochta and De Mulder-Johnston 1997). Modification of starches and blends of starch with other materials provides tremendous possibilities to adjust the properties of the final products to make them competitive with their synthetic counterparts.

Starch-based bioplastics can readily be produced by blending starch with synthetic polymers of various origins. The potential variations in properties are tremendous and can be controlled by selection of starch grade and by adjusting the proportions between starch and the synthetic polymer. Blends of granular (modified) starch (up to 30% vol) and some bioinert plastics made from mineral oil, e.g., PE, have been described (Röper and Koch 1990; Shogren et al. 1993; Kim and Pometto 1994; Chandra and Rustgi 1998). When the starch granules undergo enzymatic degradation, the sample loses its strength and continuity and thus disintegrates, leaving small particles of PE (Chandra and Rustgi 1998). Thus, these blends do not comply with the standards on biodegradability.

The content of starch varies generally between 30% and 80% by weight, depending on the end application. Most commercial starch blends are based on biodegradable, partially biobased feedstock (Shen et al. 2009). Examples are starch-PCL and starch-PVOH blends, whereas fully biobased and biodegradable blends have been produced from starch-PLA and starch-PHAs.

Thermodynamic immiscibility and nonwetting of starch with other polymers leads to serious deterioration of mechanical properties at starch fractions over 30 wt% in starch blends (Kalambur and Rizvi 2006). An approach to enhance the compatibility is by reactive extrusion. Blending of starch with other polymers should ideally result in the formation of covalent bonds either through existing functional groups or by introduction of new functional groups in order to form materials with high mechanical strength (Kalambur and Rizvi 2006). Addition of maleic anhydride (MA) functional groups can be done in the melt state by extrusion. The reaction is initiated by peroxide initiators, where MA becomes grafted onto, for example, PE. Further reaction of this MA-grafted PE with starch results in the formation of an ester through the reaction of free anhydride groups with starch hydroxyl groups. Similar reactions have been utilized to form maleated PLA-starch or maleated PCL-starch blends.

MA has also been used as a compatibilizer to improve the adhesion between natural fibers and the polymer matrix in composites. Lignin has been shown to act as a compatibilizer between hydrophilic fibers and hydrophobic matrix polymers, thus strengthening the fiber–matrix interface.

Blends containing more than 85% starch are used for foaming and injection molding products (Gross and Kalra 2002) to produce, for example, loose-fill material. Starch has also been blended with poly(vinyl alcohol) (PVOH) to give fully compostable products (Petersen et al. 1999). Arvanitoyannis et al. (1998) formed blends of rice or potato starch with LDPE by extrusion followed by pressing to films. Phase separation was observed due to inherent incompatibility. Films with high content of starch were found to be brittle, and the gas and water vapor permeability increased proportionally to the starch content in the blend.

Granular starch has been used as fillers in petroleum-based plastics to form partially biodegradable composites. The major challenge is to override the problems with incompatibility between hydrophilic starch and hydrophobic plastics. The granule size has a major impact on the selection of suitable starches, especially in case of thin films. Granular corn starch has been introduced as fillers in amorphous PLA to produce a more price-competitive bioplastic material (Jacobsen and Fritz 1996).

25.8.2 Lignocellulose

Lignin-based blends with PE and PP for fiber spinning have been described (Kubo et al. 2005). Good fiber spinnability was achieved by selecting lignin–polymer blends having similar molten viscosity regardless of the miscibility between the polymers. Blends between lignin and PET required high spinning temperatures due to the high spinning temperature necessary for the PET component. This resulted in poor processability since the spinning temperature of the blends was close to or higher than the thermal decomposition temperature of lignin. This high temperature also resulted in a rapid decrease in molten viscosity, presumably due to break-up of strong intermolecular interactions between lignin and PET. Although not yet confirmed, potential interactions involve hydrogen bonding between polar oxygen atoms on PET and hydroxyl groups in lignin.

Xylan isolated from bleached birch kraft pulp in combination with reinforcing nanoclays has been shown to give promising barrier properties with respect to water vapor and aromatic compounds (isoamyl acetate, limonene, cis-3-hexenol, and carvone) (Talja and Poppius-Levlin 2011). In other recent investigations on hemicellulose/clay composites, Ünlü et al. (2009) evaluated the fundamental properties of biocomposites formed from corn cob xylan and montmorillonite clay and suggested that, depending on concentration, some intercalation of xylan in the clay galleries could occur. The combination of xylans with various forms of nanocellulose could also offer a promising route to totally renewable, biodegradable films, which may find future application in fields such as food packaging (Saxena and Ragauskas 2009; Saxena et al. 2009).

Fiber-based composites from cellulose esters (CAP and CAB) reinforced with a variety of cellulose fibers (flax, sisal, and wood fibers) were formed by extrusion, compression molding, and injection molding (Toriz et al. 2005).

Dufresne et al. (2000) demonstrated that composite films formed from glycerol plasticized starch and MFC derived from homogenized potato pulp (a residual product from starch production) expressed significantly lower moisture absorption and higher modulus as compared to unfilled starch films.

25.8.3 POLYLACTIDE

Random copolymers and block copolymers of lactide and ε-caprolactone can be prepared by similar polymerization reactions catalyzed by tin octanoate as those used for polymerization of the individual monomers (Henton et al. 2005).

Physical polymer blends (also denoted alloys) of PLA and other polymers (synthetic or biobased/biodegradable) are an approach to improve the properties. EcoFlex is a petrochemically derived copolyester of terephthalic acid, adipic acid, and butanediol having a branched, long-chain structure (Mohanty et al. 2005). When blended with PLA (45/55 wt%), an aromatic–aliphatic biodegradable polyester marketed by BASF under the trade name Ecovio is formed (Shen et al. 2009) with higher impact and heat resistance compared to neat PLA. Ecovio is claimed to be biodegradable and compostable in to industrial composting facilities. The properties are similar to standard PE. This copolyester can be further blended with, for example, PCL or PHA.

Blends of PLA and other biobased polymers have attracted a lot of attention, some of which have already reached the market. Examples are starch-PLA, starch-PHA, and PLA-PCL, and improved performance with respect to thermal and mechanical properties as well as processability and potential application areas have been reported (Shen et al. 2009).

Cheng et al. (2007) described PLA biocomposites based on regenerated cellulose fiber, pure cellulose fibers, pulp fibers, or microfibrillated cellulose (MFC).

Katiyar et al. (2011) studied the dispersion of organic nanofillers in PLA by melt processing and obtained significant improvement of oxygen and water vapor barrier properties. Svagan et al. (2012) used the layer-by-layer approach to assemble layers of montmorillonite clay on extruded PLA film surfaces and achieved a substantial reduction in oxygen permeability of the films.

25.8.4 POLYCAPROLACTONE

PCL is readily miscible with many polymers, fillers, and pigments (Jérôme and Lecomte 2005) and is widely used in blends with a number of polymers, among which starch has gained considerable attention (Mohanty et al. 2005). Blending with starch to produce, for example, trash bags is a way to reduce manufacturing costs (Gross and Kalra 2002). Addition of cellulose fibers has been adopted to produce nonwovens for medical and hygiene products (Gross and Kalra 2002). PCL is highly compatible with other polymers and is used for direct coating but has also found use in composite barrier films based on starch (Myllymäki et al. 1998).

25.8.5 POLYHYDROXYALKANOATES

Blends including PHB or PHBV with a variety of other polymers have been studied with respect to mechanical properties, thermal properties, crystallinity, biodegradation, and other aspects. Examples are polysaccharides, PCL, PLA, poly(vinly acetate), and PVOH (Pollet and Avérous 2011). Increased compatibility between components has been achieved by reactive blending using, for example, peroxides, which decompose during processing in reactive melt blending to form free radicals, which cross-link the polymers in the blend (Pollet and Avérous 2011).

Owing to its polar character, PHAs show good adhesion to polar lignocellulose fibers, and incorporation of fibers is thus a way to improve the mechanical performance of PHAs. Improvements in mechanical and thermal properties have also been reported by incorporation of inorganic nanofillers (Pollet and Avérous 2011).

25.9 RECYCLING AND BIODEGRADATION OF BIOPLASTICS

Recycling or biodegradation of bioplastics requires a well-developed infrastructure for collection and handling of the material after use, especially for materials that are not water-soluble. Without efficient composting and bioconversion facilities, bioplastics might end up in landfills (Gross and Kalra 2002; Müller et al. 2014). Water-soluble biomaterials normally enter wastewater treatment facilities, which are readily available in all developed countries and in which these materials easily degrade (Gross and Kalra 2002). Optimum conditions for biodegradation of bioplastic waste include appropriate temperature and pH; sufficient amounts of water, nutrients, and oxygen; and other conditions required for efficient enzymatic activity.

Native starch and many starch derivatives fulfill the requirements of easily biodegradable materials. Completely biodegradable starch plastics should contain no nonbiodegradable polymers or additives; hence, plastic materials containing starch as a polymeric additive or a particulate filler in which only the starch fraction degrades are not considered as biodegradable plastics (van Soest and Vliegenthart 1997).

The biodegradation of cellulose esters depends on the chemical structure of the polymer, with prolonged degradation rates with increased degrees of substitution of hydrophobic ester substituents (Keenan et al. 2005). The environmental conditions also affect the degradation rate. Cellulose ethers are regarded as slowly biodegradable. The biodegradability decreases with increased DS since this makes the ether linkages more resistant to microbial attack (Chandra and Rustgi 1998).

Although interest in wider use of PLA to produce packaging materials for food products is high, apart from technical challenges and relatively high cost, recycling also presents some challenges. PLA can be separated from PET by using infrared scanning technology; however, the required infrastructure does not yet exist, and similarly, industrial-scale composting facilities are still far from common. Given that the so-called skeleton waste from tray thermoforming can represent 50% of the incoming film, there is a clear need for cost-effective PLA waste utilization. This challenge is complicated by the molecular weight reduction that inevitably occurs when melt-processing PLA. A technology for recovering monomers from PLA to be used for later PLA synthesis has been developed by the Belgian company Galactic and could be more extensively implemented if PLA volumes for recycling were sufficient.

PLA is highly resistant to attack by microorganisms at ambient temperature. Hydrolysis at elevated temperatures (>58°C) is required to start the breakdown of the high-molecular-weight polymer, which can be done in industrial composting facilities (Gross and Kalra 2002; Shen et al. 2009). Degradation by microorganisms finally converts low-molecular-weight components to carbon dioxide, water, and humus (Drumright et al. 2000).

PCL can be slowly degraded by hydrolysis, with the rate being dependent on molecular weight and crystallinity. This process can be catalyzed by enzymes or by the activity of bacteria or fungi (Fritz et al. 1995; Chandra and Rustgi 1998; Gross and Kalra 2002).

PHA biodegrades naturally under both aerobic and anaerobic environments under the action of extracellular enzymes excreted by bacteria and fungi. Degradation occurs in soil, sludge, freshwater, seawater, and compost. The water-insoluble, high-molecular-weight PHAs are first hydrolyzed into soluble oligomers and monomers. These low-molecular-weight products are then transported in the degrading microorganisms where they are further metabolized as carbon and energy sources to yield CO_2 and H_2O (Pollet and Avérous 2011). Braunegg et al. (1998) review some studies on variations in biodegradation rate for different copolymers of PHA and the dependence on various conditions.

25.10 APPLICATIONS OF BIOPLASTICS

Plastic articles are used today within numerous different areas: household, sports, car details, packaging material, furniture, interior decorations, tools, toys, etc. Biobased alternatives have to meet a number of different performance requirements in order to be competitive, not least regarding their price. Examples of application areas where bioplastics has found use up to this day are films, coatings, clothes, single-use plates and cups, toys, food containers, and fast-food tableware (Vink et al. 2003; Álvarez-Chávez et al. 2012). Biodegradable polymers have been developed and targeted for use as packaging materials including trash bags, wrappings, loose-fill foam, and food containers, and as disposable nonwovens, hygiene products, consumer goods, and agricultural mulch films and planters (Gross and Kalra 2002). Cost-effective production of bioplastics requires the use of conventional processing techniques already developed for commodity plastic materials (van Soest and Vliegenthart 1997).

The major barrier to commercialization of bioplastics has traditionally been high production costs, as in the case of PLA (Gross and Kalra 2002; Henton et al. 2005), even though the starting raw materials might be cheap as for starch- or cellulose-based materials. Cellulose-derived plastics have been available on the market for decades, and lignocellulose plastics are expected to increase rapidly due to the fast development of hemicellulose- and lignin-based derivatives.

Among the most widely used conventional plastics available is PE with its HDPE, LDPE, and LLDPE variants. A biobased alternative to this very versatile product has recently become commercially available, as bio-PE derived from sugar cane and produced by Braskem, Dow, and Solvay.

TPS has gained increased interest over the past decades due to its high natural abundance, the relatively easy process for separation of the starch from their natural sources, and their low cost. The wide variety of starch from different plants, the potential for chemical modifications, and compatibility with a broad range of other polymers, fillers, and additives are also contributing factors. Plastics from starch have been used to form coatings, garbage bags, loose-fill packaging, agricultural foils, flowerpots, and several other disposable items (van Soest and Vliegenthart 1997). TPS-based polymers or polymer blends are now commercially available, and the Italian company Novamont is a leader in this field with its line of Mater-Bi products. The relative disadvantages of starch-based plastics are their water sensitivity and inferior mechanical properties; however, technical solutions are available through blending or through various forms of chemical derivatization or graft copolymerization. Biodegradable and compostable disposable cups, plates, cutlery, and other utensils made from starch-based plastics are a growing market.

TPS is normally processed at temperatures in the range 100–200°C (van Soest and Vliegenthart 1997) by extrusion, compression molding, and injection molding. Film blowing, thermoforming, foaming, and extrusion coating are other potential converting techniques (Shen et al. 2009). Starches have also been used for coating of paper and paperboard to form fiber-based packaging material. The potential application areas for starch bioplastics are limited by their sensitivity to moisture. They are poor barriers to water vapor, especially at high relative humidity, which is an important property in, for example, food packaging applications. However, their barrier properties against oxygen and carbon dioxide are generally good, and the water sensitiveness can be reduced by chemical modification, cross-linking, or formation of starch blends and composites. The major areas of application of starch blends are packaging films, shopping bags, tableware, trays, wrap films, and biodegradable compost bags. Other areas are cups, food trays, knives and forks, agricultural mulch films, and plant pots.

Commodity products based on starch have been utilized in a variety of forms such as building materials, adhesives, foams, blends, and composites. Completely biodegradable products with physical and mechanical properties comparable to synthetic polymers have been developed including disposable containers and mulch films (Chiou et al. 2005).

Packaging products offering mechanical and thermal insulating properties, which traditionally have been made from expanded polystyrene, have recently been developed from starch-based foam composites (Chiou et al. 2005). By baking a starch dough in heated, closed molds, products with various thicknesses and shapes like cups, trays, and clamshells can be formed. Studies have shown that baked foams from wheat, corn, tapioca, or potato starch had comparable mechanical properties to containers made from expanded polystyrene and paperboard (Chiou et al. 2005). The flexural and tensile strength properties of the foam products could be improved by addition of wood fibers, whereas addition of $CaCO_3$ fillers led to a reduction in these values.

Packaging material (films, thermoformed containers, and bottles) (Gross and Kalra 2002), drinking cups and containers (Auras et al. 2005), and coatings for paper and board (Krochta and De Mulder-Johnston 1997; Kirwan and Strawbridge 2003) are typical application areas for PLA. PLA fulfills the requirements for direct food contact with aqueous, acidic, and fatty foods (de Vlieger 2000). One major drawback with PLA is its insolubility in water, thus making it less attractive for use in industrial dispersion coating for fiber-based packaging material.

PLA is now manufactured in both commodity and specialty grades worldwide. The largest share of world production is held by NatureWorks Inc. Packaging applications are the primary area of business, and PLA products are available in the form of thermoformed trays for packaging of salads and delicatessen items. PLA has also been used in blow-molding processes to manufacture noncarbonated beverage bottles. Improvements in PLA-based products are continuing, whether through the use of additives (e.g. plasticizers, impact modifiers) or through new formulations that allow the use of PLA as a coating or disposable cups for hot drinks or ready-to-eat meal trays. However, there is still a need for cost-effective methods to enhance PLA properties, especially in terms of higher gas and water vapor barrier properties, reduced brittleness, increased thermal stability, and improved fiber/matrix compatibility in natural fiber-reinforced biocomposites.

Melt spinning for production of fibers has been utilized to form clothes from PLA. The material has the advantage to quickly transport moisture away from the wearer; it is durable and has a silky feel (Gross and Kalra 2002).

PCL can be melt-extruded to sheets, bottles, and various shaped articles (Gross and Kalra 2002). It is also used in blends with other polymers in commercial packaging materials such as Mater-Bi and Bioflex. PCL is manufactured by Union Carbide Corporation and Solvay under the trade names TONE and CAPA, respectively.

Commercialization of PHAs started in the 1960s, and today, several companies worldwide are producing or in the process of producing a variety of PHA grades.

REFERENCES

Albertsson, A.-C., Edlund, U. and Varma, I.K. 2011. Synthesis, chemistry and properties of hemicelluloses. In *Biopolymers: New materials for sustainable films and coatings*, ed. D. Plackett, 133–150. Chichester: John Wiley & Sons.

Altskär, A., Andersson, R., Boldizar, A., Koch, K., Stading, M., Rigdahl, M. and Thunwall, M. 2008. Some effects of processing on the molecular structure and morphology of thermoplastic starch. *Carbohydr Polym* 71:591–597.

Álvarez-Chávez, C.R., Edwards, S., Moure-Eraso, R. and Geiser, K. 2012. Sustainability of bio-based plastics: A general comparative analysis and recommendations for improvement. *J Cleaner Prod* 23:47–56.

Arvanitoyannis, I., Biliaderis, C.G., Ogawa, H. and Kawasaki, N. 1998. Biodegradable films made from low-density polyethylene (LDPE), rice starch and potato starch for food packaging applications: Part 1. *Carbohydr Polym* 36:89–104.

Auras, R.A., Singh, P.A. and Singh, J.J. 2005. Evaluation of oriented poly(lactide) polymers vs. existing PET and oriented PS for fresh food service containers. *Packag Technol Sci* 18:207–216.

Avérous, L. and Digabel, F.L. 2006. Properties of biocomposites based on lignocellulosic fillers. *Carbohydr Polym* 66:480–493.

Belgacem, M.N. and Giandini, A. 2011. Production, chemistry and properties of cellulose-based materials. In *Biopolymers: New materials for sustainable films and coatings*, ed. D. Plackett, 152–178. Chichester: John Wiley & Sons.

Bergh, N.-O. 1997. Starches. In *Surface application of paper chemicals*, eds. J. Brander and I. Thorn, 69–108. London: Blackie Academic & Professional.

Berglund, L. 2005. Cellulose-based nanocomposites. In *Natural fibers, biopolymers and biocomposites*, eds. A.K. Mohanty, M. Misra, and L.T. Drzal, 807–832. Boca Raton, FL: CRC Press.

Biodegradable Products Institute (BPI). 2014. Confused by the terms Biodegradable & Biobased. Available at http://www.bpiworld.org/resources/Documents/PROiaelB[1].pdf (accessed July 9, 2014).

BizNGO. 2014. Business-NGO Working Group. Available at http://www.bizngo.org (accessed July 2, 2014).

Bledzki, A.K. and Gassan, J. 1999. Composites reinforced with cellulose based fibers. *Progr Polym Sci* 24:221–274.

Braunegg, G., Lefebvre, G. and Genser, K.F. 1998. Polyhydroxyalkanoates, biopolyesters from renewable resources: Physiological and engineering aspects. *J Biotechnol* 65:127–161.

Bublitz, R.H., Klem, R.E. and Craig, K.A. 1990. Thermal and thermal-chemical conversion of starch. In *Starch and starch products in paper coating*, eds. R.L. Kearney and H.W. Maurer, 109–122. Atlanta, GA: TAPPI Press.

Buléon, A., Colonna, P., Planchot, V. and Ball, S. 1998. Starch granules: Structure and biosynthesis. *Int J Biol Macromol* 23:85–112.

Chandra, R. and Rustgi, R. 1998. Biodegradable polymers. *Progr Polym Sci* 23:1273–1335.

Cheng, Q., Wang, S. and Rials, T.G. 2007. Biodegradable nanocomposites reinforced with cellulose fibrils. Technical Association of Pulp and Paper Industry (TAPPI) & Forest Product Society, *2007 International Conference on Nanotechnology for the Forest Products Industry*, June 13–15, Knoxville, TN, 61 (Poster).

Chiou, B.-S., Glenn, G.M., Imam, S.H., Inglesby, M.K., Wood, D.F. and Orts, W.J. 2005. Starch polymers: Chemistry, engineering, and novel products. In *Natural fibers, biopolymers and biocomposites*, eds. A.K. Mohanty, M. Misra, and L.T. Drzal, 639–669, Boca Raton, FL: CRC Press.

de Vlieger, J.J. 2000. Green plastic for food packaging. In *Novel food packaging techniques*, ed. R. Ahvenainen, 519–534. Cambridge: Woodhead Publishing.

Drumright, R.E., Gruber, P.R. and Henton, D.E. 2000. Polylactic acid technology. *Adv Mater* 12(23):1841–1846.

Dufresne, A., Dupeyre, D. and Vignon, M.R. 2000. Cellulose microfibrils from potato tuber cells: Processing and characterization of starch-cellulose microfibril composites. *J Appl Polym Sci* 76:2080–2092.

Edgar, K.J., Buchanan, C.M., Debenham, J.S., Rundquist, P.A., Seiler, B.D., Shelton, M.C. and Tindall, D. 2001. Advances in cellulose ester performance and application. *Progr Polym Sci* 26:1605–1688.

Ellis, R.P., Cochrane, M.P., Dale, M.F.B., Duffus, C.M., Lynn, A., Morrison, I.M., Prentice, D.M., Swanston, S. and Tiller, S.A. 1998. Starch production and industrial use. *J Sci Food Agric* 77:289–311.

European Bioplastics. 2014. Available at http://en.european-bioplastics.org (accessed May 26, 2014).

French, D. 1984. Organization of starch granules. In *Starch: Chemistry and technology*, 2nd Edition, eds. R.L. Whistler, J.N. Bemiller, and E.F. Paschall, 183–247. Orlando, FA: Academic Press.

Fritz, H.G., Aichholzer, W., Seidenstücker, T. and Widmann, B. 1995. Abbaubare Polymerwerkstoffe auf der Basis nachwachsender Rohstoffe—Möglichkeiten und Grenzen. *Starch/Stärke* 47(12):475–491.

Greif, D.S. and Koval, J.C. 1990. Modified starches: Manufacture and properties. In *Starch and starch products in paper coating*, eds. R.L. Kearney and H.W. Maurer, 29–44. Atlanta, GA: TAPPI Press.

Gröndahl, M., Eriksson, L. and Gatenholm, P. 2004. Material properties of plasticized hardwood xylans for potential application as oxygen barrier films. *Biomacromolecules* 5(4):1528–1535.

Gross, R.A. and Kalra, B. 2002. Biodegradable polymers for the environment. *Science* 297:803–807.

Hablot, E., Bordes, P., Pollet, E. and Averous, L. 2008. Thermal and thermo-mechanical degradation of poly(3-hydroxybutyrate)-based multiphase systems. *Polym Degrad Stabil* 93:413–421.

Halley, P.J. 2005. Thermoplastic starch biodegradable polymers. In *Biodegradable polymers for industrial applications*, ed. R. Smith, 140–162. Cambridge: Woodhead Publishing.

Hartman, J., Albertsson, A.-C. and Sjöberg, J. 2006. Surface- and bulk-modified galactoglucomannan hemicellulose films and film laminates for versatile oxygen barriers. *Biomacromolecules* 7:1983–1989.

Henton, D.E., Gruber, P., Lunt, J. and Randall, J. 2005. Polylactic acid technology. In *Natural fibers, biopolymers and biocomposites*, eds. A.K. Mohanty, M. Misra, and L.T. Drzal, 527–577. Boca Raton, FL: CRC Press.

Hermansson, A.-M. and Svegmark, K. 1996. Developments in the understanding of starch functionality. *Trends Food Sci Technol* 7:345–353.

Höije, A., Sandström, C., Roubroeks, J.P., Andersson, R., Gohil, S. and Gatenholm, P. 2006. Evidence of the presence of 2-O-β-D-xylopyranosyl-α-L-arabinofuranose side chains in barley husk arabinoxylan. *Carbohydr Res* 341:2959–2966.

Hong, S.-H., Gau, T.-S. and Huang, S.-C. 2011. Enhancement of the crystallization and thermal stability of polyhydroxybutyrate by polymeric additives. *J Thermal Anal Calorim* 103:967–975.

Jacobsen, S. and Fritz, H.G. 1996. Filling of poly(lactic acid) with native starch. *Polym Eng Sci* 36(22):2799–2804.

Janigová, I., Lacík, I. and Chodák, I. 2002. Thermal degradation of plasticized poly(3-hydroxybutyrate) investigated by DSC. *Polym Degrad Stabil* 77:35–41.

Jérôme, R. and Lecomte, P. 2005. New developments in the synthesis of aliphatic polyesters by ring-opening polymerization. In *Biodegradable polymers for industrial applications*, ed. R. Smith, 77–106. Cambridge: Woodhead Publishing.

Johansson, C., Bras, J., Mondragon, I., Nechita, P., Plackett, D., Simon, P., Gregor Svetec, D., Virtanen, S., Giacinti Baschetti, M., Breen, C., Clegg, F. and Aucejo, S. 2012. Renewable fibers and bio-based materials for packaging applications—A review of recent developments. *BioResources* 7(2):2506–2552.

Kalambur, S. and Rizvi, S.S.H. 2006. An overview of starch-based plastic blends form reactive extrusion. *J Plastic Film Sheeting* 22:39–58.

Katiyar, V., Gerds, N., Bender Koch, C., Risbo, J., Hansen, H.C.B. and Plackett, D. 2011. Melt processing of poly(L-lactic acid) in the presence of organomodified anionic or cationic clays. *J Appl Polym Sci* 122:112–125.

Keenan, T.M., Tanenbaum, S.W. and Nakas, J.P. 2005. Lignocellulosic biomass as a renewable and value-added feedstock for biodegradable polymer production. In *Biodegradable polymers for industrial applications*, ed. R. Smith, 219–250. Cambridge: Woodhead Publishing.

Kim, M. and Pometto III, A.L. 1994. Food packaging potential of some novel degradable starch-polyethylene plastics. *J Food Prot* 57(11):1007–1012.

Kirwan, M.J. and Strawbridge, J.W. 2003. Plastics in food packaging. In *Food packaging technology*, eds. R. Coles, D. McDowell, and M.J. Kirwan, 174–240. Boca Raton, FL: CRC Press.

Klemm, D., Philipp, B., Heinze, T., Henize, U. and Wagenknecht, W. 1998. *Comprehensive cellulose chemistry, vol 2, Functionalization of cellulose*, 207–302. Weinheim: Wiley-VCH.

Krochta, J.M. and De Mulder-Johnston, C. 1997. Edible and biodegradable polymer films: Challenges and opportunities. *Food Technol* 51(2):61–74.

Kubo, S., Gilbert, R.D. and Kadla, J.F. 2005. Lignin-based polymer blends and biocomposite materials. In *Natural fibers, biopolymers and biocomposites*, eds. A.K. Mohanty, M. Misra, and L.T. Drzal, 671–697. Boca Raton, FL: CRC Press.

Larson, J. 1999. *The structure and rheology of complex fluids*. New York: Oxford University Press.

Lima, L.T., Auras, R. and Rubino, M. 2008. Processing technologies for poly(lactic acid). *Progr Polym Sci* 33:820–852.

Lourdin, D., Della Valle, G. and Colonna, P. 1995. Influence of amylose content on starch films and foams. *Carbohydr Polym* 27:261–270.

Lovin, J.C. and Wheeler, H.R. 1990. Enzyme conversion of starch for paper coating. In *Starch and starch products in paper coating*, eds. R.L. Kearney and H.W. Maurer, 93–107. Atlanta, GA: TAPPI Press.

Lusignea, R. 1996. High barrier packaging with liquid crystal polymers. In *TAPPI 1996 Polymers, Laminations and Coatings Conference Proc.*, 115–124. Atlanta, GA: TAPPI Press.

Lusignea, R. 1998. Flexible multilayer packaging with oriented LCP barrier layer. In *TAPPI 1998 Polymers, Laminations and Coatings Conference Proc.*, 889–899. Atlanta, GA: TAPPI Press.

Lusignea, R. and Perdikoulias, J. 1997. Extrusion of oriented LCP film. In *TAPPI 1997 Polymers, Laminations and Coatings Conference Proc.*, 271–276. Atlanta, GA: TAPPI Press.

Mitch, E.L. 1984. Potato starch: Production and uses. In *Starch: Chemistry and technology*, 2nd Edition, eds. R.L. Whistler, J.N. Bemiller, and E.F. Paschall, 479–490. Orlando, FA: Academic Press.

Mohanty, A.K., Misra, M., and Drzal, L.T. 2002. Sustainable bio-composites from renewable resources: Opportunities and challenges in the Green Materials World. *J Polym Environ* 10:19–26.

Mohanty, A.K., Misra, M., Drzal, L., Selke, S.E., Harte, B.R. and Hinrichsen, G. 2005. Natural fibers, biopolymers and biocomposites: An introduction. In *Natural fibers, biopolymers and biocomposites*, eds. A.K. Mohanty, M. Misra, and L.T. Drzal, 1–36. Boca Raton, FL: CRC Press.

Müller, G., Hanecker, E., Blasius, K., Seidemann, C., Tempel, L., Sadocco, P., Ferreira Pozo, B., Boulougouris, G., Lozo, B., Jamnicki, S. and Bobu, E. 2014. End-of-life solutions for fibre and bio-based packaging materials in Europe. *Packag Technol Sci* 27:1–15.

Myllymäki, O., Myllärinen, P., Forssell, P., Suortti, T., Lähteenkorva, K., Ahvenainen, R. and Poutanen, K. 1998. Mechanical and permeability properties of biodegradable extruded starch/polycaprolactone films. *Packag Technol Sci* 11:265–274.

Nicholson, M.D. and Merritt, F.M. 1985. Cellulose ethers. In *Cellulose chemistry and its applications*, eds. T.P. Nevell and S.H. Zeronian, 363–383. Chichester: Ellis Horwood.

Noaves, L. 2010. Green polyethylene: Bringing renewable raw materials to the traditional plastic industry. In *TAPPI 2010 PLACE Conference*. Albuquerque, NM, April 18–21, 2010.

Olabarrieta, I., Forsström, D., Gedde, U.W. and Hedenqvist, M.S. 2001. Transport properties of chitosan and whey blended with poly(ε-caprolactone) assessed by standard permeability measurements and microcalorimetry. *Polymer* 42:4401–4408.

Peroval, C., Debeaufort, F., Seuvre, A.-M., Cayot, P., Chevet, B., Despré, D. and Voilley, A. 2004. Modified arabinoxylan-based films. Grafting of functional acrylates by oxygen plasma and electron beam irradiation. *J Membrane Sci* 233:129–139.

Petersen, K., Nielsen, P.V., Bertelsen, G., Lawter, M., Olsen, M.B., Nilsson, N.H. and Mortensen, G. 1999. Potential of biobased materials for food packaging. *Trends Food Sci Technol* 10:52–68.

Plackett, D. 2011. Introductory overview. In *Biopolymers: New materials for sustainable films and coatings*, ed. D. Plackett, 3–14. Chichester: John Wiley & Sons.

Plackett, D.V., Holm, V.K., Johansen, P., Ndoni, S., Vaeggermose Nielsen, P., Sipilainen-Malm, T., Södergård, A. and Verstichel, S. 2006. Characterization of L-polylactide and L-polylactide-polycaprolactone copolymer films for use in cheese-packaging applications. *Packag Technol Sci* 19:1–24.

Plastics Europe. 2012. Available at http://www.plasticseurope.org/Document/plastics-the-facts-2012.aspx?Page=DOCUMENT&FolID=2 (accessed July 3, 2014).

Pollet, P. and Avérous, L. 2011. Production, chemistry and properties of polyhydroxyalkanoates. In *Biopolymers: New materials for sustainable films and coatings*, ed. D. Plackett, 65–86. Chichester: John Wiley & Sons.

Rindlav, Å., Hulleman, S.H.D. and Gatenholm, P. 1997. Formation of starch films with varying crystallinity. *Carbohydr Polym* 34:25–30.

Rindlav-Westling, Å., Stading, M., Hermansson, A.-M. and Gatenholm, P. 1998. Structure, mechanical and barrier properties of amylose and amylopectin films. *Carbohydr Polym* 36:217–224.

Rocca, M.C., Carr, G., Lambert, A.B., MacQuarrie, D.J. and Clark, J.H. 2003. Process for the oxidation of cyclohexanone to caprolactone, US Patent 6531615.

Röper, H. and Koch, H. 1990. The role of starch in biodegradable thermoplastic materials. *Starch/Stärke* 42(4):123–130.

Rutenberg, M.W. and Solarek, D. 1984. Starch derivatives: Production and uses. In *Starch: Chemistry and technology*, 2nd Edition, eds. R.L. Whistler, J.N. Bemiller, and E.F. Paschall, 312–388. Orlando, FA: Academic Press.

Saxena, A. and Ragauskas, A.J. 2009. Water vapor transmission properties of biodegradable films based on cellulosic whiskers and xylan. *Carbohydr Polym* 78:357–360.

Saxena, A., Elder, T.J., Pan, S. and Ragauskas, A.J. 2009. Novel nanocellulosic xylan composite film. *Compos Part B: Eng* 40:727–730.

Shen, L., Haufe, J. and Patel, M.K. 2009. Product overview and market projection of emerging bio-based plastics, PRO-BIP 2009. Available at http://en.european-bioplastics.org/wp-content/uploads/2011/03/publications/PROBIP2009_Final_June_2009.pdf (accessed July 3, 2014).

Shogren, R.L., Fanta, G.F. and Doane, W.M. 1993. Development of starch based plastics—A reexamination of selected polymer systems in a historical perspective. *Starch/Stärke* 45(8):276–280.

Sinha, V.R., Bansal, K., Kaushik, R., Kumria, R. and Trehan, A. 2004. Poly-ε-caprolactone microspheres and nanospheres: An overview. *Int J Pharm* 278:1–23.

Södergård, A. and Inkinen, S. 2011. Production, chemistry and properties of polylactides. In *Biopolymers: New materials for sustainable films and coatings*, ed. D. Plackett, 43–63. Chichester: John Wiley & Sons.

Södergård, A. and Stolt, M. 2002. Properties of lactic acid based polymers and their correlation with composition. *Progr Polym Sci* 27:1123–1163.

Suriyamongkol, P., Weselake, R., Narine, S., Moloney, M. and Shah, S. 2007. Biotechnological approaches for the production of polyhydroxyalkanoates in microorganisms and plants—A review. *Biotechnol Adv* 25:148–175.

Sustainable Biomaterials Collaborative (SBC). 2014. Available at http://www.sustainablebiomaterials.org/criteria.guidelines.definitions.php (accessed July 2, 2014).

Svagan, A.J., Åkesson, A., Cárdenas, M., Bulut, S., Knudsen, J.C., Risbo, J. and Plackett, D. 2012. Transparent films based on PLA and montmorillonite with tunable oxygen barrier properties. *Biomacromolecules* 13:397–405.

Talja, R. and Poppius-Levlin, K. 2011. Xylan from wood biorefinery—A novel approach. In *FlexPakRenew Workshop*, Lyon, France, May 10, no. 5.

Toriz, G., Gatenholm, P., Seiler, B.D. and Tindall, D. 2005. Cellulose-fiber reinforced cellulose esters: Biocomposites for the future. In *Natural fibers, biopolymers and biocomposites*, eds. A.K. Mohanty, M. Misra, and L.T. Drzal, 617–638. Boca Raton, FL: CRC Press.

Tuschhoff, J.V. 1986. Hydroxypropylated starches. In *Modified starches: Properties and uses*, ed. O.B. Wurzburg, 89–96. Boca Raton, FL: CRC Press.

Ünlü, C., Gunister, E. and Atici, O. 2009. Synthesis and characterization of NaMt biocomposites with corn cob xylan in aqueous media. *Carbohydr Polym* 76:585–592.

U.S. Department of Agriculture (USDA). 2005. Biobased Products Procurement Program. Washington, DC, April 20. Available at http://www.ocio.usda.gov/sites/default/files/docs/2012/DR5023-002.htm (accessed July 8, 2014).

van de Velde, K. and Kiekens, P. 2002. Biopolymers: Overview of several properties and consequences on their applications. *Polym Testing* 21:433–442.

van Soest, J.J.G. and Vliegenthart, J.F.G. 1997. Crystallinity in starch plastics: Consequences for material properties. *Trends Biotechnol* 15:208–213.

van Tuil, R., Fowler, P., Lawther, M. and Weber, C.J. 2000. Properties of biobased packaging materials. In *Biobased packaging material for the food industry—Status and perspectives*, ed. C.J. Weber, 13–44. Copenhagen: Royal Veterinary and Agricultural University.

Vázquez, A., Foresti, M.L. and Cyras, V. 2011. Production, chemistry and degradation of starch-based polymers. In *Biopolymers: New materials for sustainable films and coatings*, ed. D. Plackett, 15–42. Chichester: John Wiley & Sons.

Vink, E.T.H., Rábago, K.R., Glassner, D.A. and Gruber, P.R. 2003. Applications of life cycle assessment to NatureWorks™ polylactide (PLA) production. *Polym Degrad Stabil* 80:403–419.

Voigt, J. 1990. Dispersion of starches. In *Starch and starch products in paper coating*, eds. R.L. Kearney and H.W. Maurer, 71–92. Atlanta, GA: TAPPI Press.

Wagoner, J.A. and van der Burgh, L.F. 1990. Unmodified starches: Sources, composition, manufacture and properties. In *Starch and starch products in paper coating*, eds. R.L. Kearney and H.W. Maurer, 1–14. Atlanta, GA: TAPPI Press.

Weber, C.J., ed. 2000. *Biobased packaging materials for the food industry—Status and perspectives*. Frederiksberg, Denmark: The Royal Veterinary and Agricultural University.

Wurzburg, O.B. 1986. Introduction. In *Modified starches: Properties and uses*, ed. O.B. Wurzburg, 3–16. Boca Raton, FL: CRC Press.

Yang, L., Chen, X. and Jing, X. 2008. Stabilization of poly(lactic acid) by polycarbodiimide. *Polym Degrad Stabil* 93:1923–1929.

Young, A.H. 1984. Fractionation of starch. In *Starch: Chemistry and technology*, 2nd Edition, eds. R.L. Whistler, J.N. Bemiller, and E.F. Paschall, 249–284. Orlando, FA: Academic Press.

Yu, J. and Chen, L.X.L. 2008. The greenhouse gas emissions and fossil energy requirement of bioplastics form cradle to gate of a biomass refinery. *Environ Sci Technol* 42:6961–6966.

Zhang, P. and Whistler, R.L. 2004. Mechanical properties and water vapour permeability of thin film from corn hull arabinoxylan. *J Appl Polym Sci* 93:2896–2902.

Zobel, H.F. 1988. Molecules to granules: A comprehensive starch review. *Starch/Stärke* 40(2):44–50.

Zobel, H.F. 1990. Colloidal and polymeric chemistry of starches and their dispersion. In *Starch and starch products in paper coating*, eds. R.L. Kearney and H.W. Maurer, 15–28. Atlanta, GA: TAPPI Press.

26 Thermoplastic Additives
Flame Retardants

Kolapo Peluola Adewale

CONTENTS

26.1 Introduction 878
26.2 Polymer Combustion and Laboratory Fire Testing 880
    26.2.1 Polymer Combustion 880
    26.2.2 Laboratory Fire Testing 881
    26.2.3 Limiting Oxygen Index 881
    26.2.4 UL94—Bunsen Burner Test 881
        26.2.4.1 Cone Calorimetry 882
26.3 Flame Retardancy, Classification, and Specification 882
    26.3.1 Sustainable Flame Retardancy—From Halogenated Chemicals to Nanocomposites ... 882
    26.3.2 Standards and Technology Life Cycle Assessment 883
    26.3.3 Classification of Fire Retardants by Mode of Action 883
        26.3.3.1 Physical Action 883
        26.3.3.2 Chemical Action 884
    26.3.4 Classification of Fire Retardants by Means of Incorporation 884
        26.3.4.1 Reactive Fire Retardants 884
        26.3.4.2 Additive Fire Retardants 884
    26.3.5 Specification for an Ideal Fire-Retardant Formulation 884
        26.3.5.1 Fire Retardancy 884
        26.3.5.2 Additive/Loading Level 885
        26.3.5.3 Compatibility 885
        26.3.5.4 Environmental Aspects and Legislation 885
26.4 Conventional Flame Retardants 885
    26.4.1 Mineral Fillers 885
    26.4.2 Thermal Effects of Mineral Fillers 886
    26.4.3 Metal Hydroxides 887
        26.4.3.1 Aluminum Trihydroxide 887
        26.4.3.2 Magnesium Dihydroxide 888
    26.4.4 Hydroxycarbonates (Hydromagnesites) 889
    26.4.5 Borates 889
26.5 Halogenated FRs 889
    26.5.1 Halogenated FR Additives 890
        26.5.1.1 Tetrabromobisphenol A 890
        26.5.1.2 Polybromodiphenylether 890
        26.5.1.3 Hexabromocyclododecane 890
        26.5.1.4 Tetrabromophthalic Anhydride 890
    26.5.2 Halogenated Monomers and Copolymers (Reactive FRs) 890
26.6 Phosphorus-Based FR 892
    26.6.1 Red Phosphorus 893
    26.6.2 Inorganic Phosphates 894

	26.6.3	Organic Phosphorus-Based Compounds	894
	26.6.4	Intumescent FR Systems	895
26.7	Nitrogen-Based FRs		895
26.8	Silicon-Based FRs		897
	26.8.1	Silicones in PC	897
	26.8.2	PC-b-PDMS in PC	897
	26.8.3	Silica in PP, PEO, and PMMA	897
26.9	Nanometric Particles as FRs		898
	26.9.1	Nanocomposite FRs—Nanoclays	898
		26.9.1.1 Polyolefin/Layered Silicate Nanocomposites	899
		26.9.1.2 Polyamide PA-6/Layered Silicate Nanocomposites	901
	26.9.2	CNTs and Sepiolite	902
	26.9.3	Nanoscale Particulate Additives, LDH, and POSS	902
		26.9.3.1 Silsesquioxane	902
		26.9.3.2 Nanometric Metallic Oxide Particles	903
26.10	FR Synergies		903
	26.10.1	Polyolefin/Layered Silicate Nanocomposites	904
	26.10.2	Polyamide/Layered Silicate Nanocomposites	905
	26.10.3	ABS/Layered Silicate Nanocomposites	905
	26.10.4	PS/Layered Silicate Nanocomposites	906
	26.10.5	Polyester/Layered Silicate Nanocomposites	906
	26.10.6	Layered Silicate Nanocomposites with Phosphorus-Based Compounds	906
	26.10.7	Synergistic Halogenated/Phosphorated Mixture	907
	26.10.8	Synergistic Nitrogenated and Phosphorous Mixtures	907
	26.10.9	Synergistic Metal Hydroxide/Metal Borate Mixtures	907
	26.10.10	Synergistic Metallic Oxide/Phosphorated Mixtures	908
26.11	Conclusions		908
References			910

26.1 INTRODUCTION

The incredibly large amount of polymeric materials in use today (hundreds of millions of metric tons annually) is made possible by an unusual combination of properties such as low cost, low weight, and ease of processing when compared with traditional materials such as wood, metals, and glass. Thermoplastic additives, which account for nearly 5% by weight of all the plastic products produced yearly, are indispensable to achieve cost effectiveness, aesthetics, and ease of processing, and to extend the useful life of the finished products. For example, plasticizers, which are mostly used in poly(vinyl chloride) (PVC) for ease of processing, alone account for over 50% of additives used by mass. Plasticizers and flame retardants (FRs) together account for over 75% of the additive market by mass in Europe [1]. Thermoplastic additives come in various forms based on functionality (end-use properties) and processability. One class of additives, namely, antioxidants, processing stabilizers, UV absorbers, and hindered amine light stabilizers (HALSs), is primarily used for protection of the thermoplastic from agents of degradation and damage, thus enhancing the value of plastic-based products. Two of these, antioxidant and light stabilizers, are used in polymer recycling. Another class, mainly used for special effects, includes FRs, antifogging agents, antistatic agents, slip agents, and antimicrobial agents. Others in the list are clarifying agents, material for extending product shell life, and optical brighteners. These additives, more than 25 distinct categories, have been described in some details in a number of publications [2–4] and on many websites including polymer-additives.specialchem.com.

Polymers in general and thermoplastics in particular are composed largely of carbon and hydrogen and are thus inherently highly combustible. This relative high flammability is often accompanied by the evolution of smoke and toxic and corrosive gases during combustion. A study by the US Federal Aviation Administration (FAA) concludes that for a thermoplastic to be inherently flame retardant, it must have heteroatoms (e.g., halogens Cl and Br, O, N, S, and P), aromatic rings, heteroatomic rings, or chemical units that can lead to cross-linking or fused aromatic rings [1,5–7].

Meeting these requirements are a number of engineered polymers such as polyphenylene sulfides, polyetheretherketones, aramids, poly(benzimidazoles), and preoxidixed polyacrylonitriles, which are inherently FR. Poly(hydroxyamide) and its derivatives represent another class of highly fire-resistant compounds. These polymers cyclize at temperatures of 250–400°C to more stable poly(benzoxazole) rings. Combustion is hindered by the endothermic cyclization reaction and the water released as by-product [5–8].

The polymers listed above often satisfy required flammability ratings without the use of additives because they are often self-extinguishing under normal conditions. However, they are all relatively more expensive and more difficult to process and are thus not suited for commodity applications [5,9].

Consequently, FRs are frequently added during processing of commodity thermoplastics primarily to increase their fire safety in applications and secondarily as a means of meeting regulations or customer specifications. There are a number of excellent publications and reviews including some that present recent developments in the use of layered silicates (clay) for designing polymer nanocomposites characterized by improved flame retardancy. The website polymer-additives.specialchem.com lists over 700 new patents and over 1500 types of current FR products from a number of manufacturers including Dow Corning, Adeka Corporation, AkzoNobel, Akrochem, Albemarle, BASF, Chemtura, Clariant, Dover (ICC Industries), Lanxess (formerly part of Bayer), ICL Industries, SABIC, RTP, and Sumitomo Chemicals. The following overview, culled from an extensive literature and from commercial products, is provided only as a quick reference [5,10–13].

Nearly all of the early FRs were halogen-based. They functioned in the gaseous phase by replacing the high-energy free radicals responsible for flame propagation with more stable species, such as Cl(-) and Br(-), and in the process suppressing the flame [14,15]. These halogenated FRs were selected for their compatibility with the host polymer and with its decomposition range. They were easy to use and straightforward to select to match the thermal stability of the target polymer in order of increasing thermal stability. The choice begins from the aliphatic bromine, aromatic bromine, and aliphatic chlorine to the aromatic chlorine for the highest temperature decomposition range. Combustion is often incomplete due to interference with the flame (chemical) reaction. Oxidation of the carbon to carbon monoxide (CO) occurs readily, but complete conversion of the CO to CO_2 is often inhibited leading to the generation of smoky and highly toxic fire effluents that are full of products of incomplete combustion. Plasticized PVC when burnt loses about 20% of its mass, generating a variety of chlorine-containing species including dichlorobenzene and other chloro-aromatic and chloro-aliphatic hydrocarbons [16]. Upward of 70 compounds including benzene, toluene, xylene, indene, and naphthalene have been identified. Many of these compounds are of significant toxicological importance including hydrogen chloride, which is even more critical than carbon monoxide [17]. Another serious concern is raised because halogenated FRs have been confirmed to leach out of polymers, and some have been proven to be endocrine disruptors [18,19].

In response to these safety and environmental concerns, there has been a concerted effort into developing alternative nonhalogenated or halogen-free FRs including metal hydroxide [20,21], and carbonate fillers [22], phosphorous compounds [23], low melt glasses [24], as well as novel materials such as nanoclay and silica nanoparticles [25,26], carbon nanotubes (CNTs) [27], expandable graphite [28], and metal chelates [29,30]. Recent research trends mostly focus on the development of new halogen-free FR systems together with combinations of traditional FR additives exhibiting enhanced efficiency or synergism when combined together.

The demand for FRs globally has recently exceeded 2.2 million metric tons/year. In 2012, global consumption of FRs was estimated at 3.9 billion lbs and at a global annualized rate of 4–5% is predicted to reach 5.2 billion lbs by 2018 [31]. As a result, demand is expected to continue to grow for both halogen-free materials and brominated FRs [32]. However, to meet safety requirements and in response to increasingly stringent regulations, various commonly used FRs such as halogenated additives are being phased out. Among halogenated FRs, polybrominated diphenyl ethers (PBDEs) have become the subject of environmental monitoring, toxicity assessment, and regulatory activity.

For many years now, the trend has been toward nonhalogenated FR materials. The most widely used FRs are the brominated FRs. These offer an excellent balance of performance, cost, properties, and processability. In response to government regulations and directives, as well as consumer and OE preferences for more environmentally friendly materials, the additive industry has intensified research and developments of alternatives to brominated and chlorinated additives especially for the electrical and electronic applications.

Halogen-free FRs as a class are rather polymer specific. Some, such as the volatile phosphorous compounds, behave like halogens, inhibiting gas-phase combustion reactions, but the majority tend to stabilize the condensed phase through barrier formation resulting from char buildup or from inorganic residue or swelling (intumescence) to create an insulating layer. They are generally less efficient compared to the halogen-based FR and require higher loading sometimes as high as 70% by weight in order to meet flammability standards. As a result, their commercial development has been limited by the availability of suitable compatibilizing agents to incorporate the additives into the polymer at suitable loadings with adequate dispersion, and polymer processing equipment such as twin screw extruders to cope with higher melt viscosities associated with higher filler loading.

Nanocomposites and their use in synergistic FR blends are just beginning. The global market for nanocomposites is reported to be worth $100–250 billion per year and is estimated by 2015 to reach ten times that amount considering the variety of market sectors where nanocomposite is expected to play. Huge investment in nanotechnology, nanocomposites, and flame retardancy is being made by national research centers and by many multinational corporations including GM, Ford, Dow and Eastman (USA); Bayer and BASF (Germany); Clariant and Basell (Switzerland) and Toyota (Japan).

Nanomaterials aside, there are more than 50 different basic types of plastics represented in more than 60,000 different plastics-based products and applications. Selecting an appropriate FR for a resin thus requires in-depth knowledge of combustion chemistry and physical chemistry. The choice is often guided by specified tests to be met and the regulatory requirements in the country of use. The mechanisms by which FRs work—the nature and chemical structure of the polymer and its decomposition mode under heat—have been well covered in many publications and will not be repeated here in any detail [2–13,33].

What follows is a brief description of the fundamentals of polymer combustion theory, followed by a summary of three simple, approved tests to describe fire behavior. FR characteristics, mechanisms of action, and examples of typical FRs are then discussed [13,14]. Toward the end of this chapter, a number of examples are provided of polymer nanocomposite-based fire retardants. Nanocomposites represent a class of current and future materials with great potential for combining physicochemical and thermomechanical performances including a significant enhancement of FR properties [9–13]. Yet, the effects of nanocomposites on flame retardancy are somewhat very complicated. Finally, synergist effects arising from combination of FR types are described.

26.2 POLYMER COMBUSTION AND LABORATORY FIRE TESTING

26.2.1 Polymer Combustion

Owing to their high carbon contents, thermoplastics are inherently flammable [2–11,13]. For combustion to occur, two ingredients are required, namely, a fuel source (any combustible material) and oxygen. The former is a reducing agent while the latter is an oxidizing agent. The process starts when a heat source increases the temperature of the thermoplastic to the point where polymer chain scissions are induced. Volatile fractions that are formed then diffuse in the air creating a combustible gaseous mixture, which subsequently ignites (exothermically) as the autoignition temperature

Thermoplastic Additives

is reached. A lower temperature ignition (at the flash point) is also possible if the mixtures come in contact with an external source of intense energy such as a spark or a flame. This has been described schematically as the combustion cycle or as the so-called fire triangle [13,14].

26.2.2 Laboratory Fire Testing

The flammability of polymers is described by their ignitability, flame spread rate, and heat release [34–36]. Numerous tests have been developed over the past decades to assess a polymer's response to fire and to measure or quantify its flame retardancy [6,11]. Comprising small-, intermediate-, or full-scale industrial flammability tests, they differ in their mode of evaluation of the heat and mass transfer phenomena, and are selected based on their suitability for material screening or specified tests for finished products. The three most commonly used laboratory tests to characterize flame retardancy for plastics, the limiting oxygen index (LOI) test, the UL94 vertical burning test, and the cone calorimetry, are described next.

26.2.3 Limiting Oxygen Index

First proposed in 1966 by Fenimore and Martin [37] and subsequently standardized as ASTM D 2863, ISO 4589, NF T 51-071 (France), the LOI—primarily used to measure a material's resistance to ignition—remains a vital screening tool and a quality control method in the plastic industry. The apparatus consists of a glass tube, in which the specimen ($80 \times 10 \times 4$ mm^3—ISO 4589) is vertically mounted. During the test, a slow stream of oxygen/nitrogen mixture is supplied at the bottom of the tube. After a 30 s purge of the column, a small candle-like flame is applied to the top of the specimen to ignite it.

The purpose of the test is to determine the minimum oxygen concentration in nitrogen that will support the combustion of the material for at least 3 min or for the consumption of 5 cm of the sample. The LOI is expressed as

$$LOI = 100 \times [O_2]/([O_2] + [N_2])$$

The more the oxygen that is required, the higher the LOI value becomes, the better the material is deemed to be flame retarded. Since air contains 21% oxygen, materials with an LOI below 21 are classified as combustible, whereas those with an LOI above 21 are classified as self-extinguishing, simply because their combustion cannot be sustained at ambient temperature without an external energy source [12,13,38–40].

26.2.4 UL94—Bunsen Burner Test

The set of UL94 tests is approved by the Underwriters' Laboratories as tests of the flammability of plastics materials for parts in devices and applications [40]. The set includes ASTM D635 (UL94HB), ASTM D3801-96 (UL94 V), ASTM D4804 (UL94 VTM), ASTM D5048-97 (UL94 5V0), and ASTM D4986-98 (UL94 HBF). The most commonly used test (the UL94 V) measures the ignitability and flame spread of vertical bulk materials exposed to a small flame.

In this test, a specimen is mounted vertically so that the lower end is located above a cotton layer (to catch any flaming drip). Flame is applied at the bottom of the specimen for 10 s, plus a subsequent 10 s application if the specimen self-extinguishes. Two sets of five specimens are tested, and the material is classified into three categories (V-0, V-1, and V-2) depending on its performance regarding the individual duration of burning for each specimen, the total burning time for all specimens, and the presence of burning drips.

During the application of the flame, the distance between the burner and the specimen must remain constant. If drops fall, the burner must be tilted through a maximum angle of 45° or slightly isolated from the specimen flame. During the test, the presence of burning drops, causing a piece of cotton under the sample to ignite, must be noted [12,40].

26.2.4.1 Cone Calorimetry

Cone calorimetry, standardized in the United States (ASTM E 1354) and internationally (ISO 5660), is one of the most effective medium-sized flame property tests for polymers. Numerous properties of the material can be determined including heat release rate (HRR), peak of heat release rate (pHRR), time to ignition (TTI), total heat released (THR), mass loss rate (MLR), peak of mass loss rate (pMLR), and specific extinction area (SEA) reflecting the smoke production. HRR and pHRR are the two most important parameters in evaluating fire safety by this test. The HRR is considered to be the driving force of the fire, whereas the pHRR represents the point in a fire where heat is apt to propagate.

The principle is based on the oxygen consumption in the combustion gases of a specimen subjected to a defined heat flux. In the experimental setup of a cone calorimeter [10], the sample (100 × 100 × 4 mm^3) is placed on a load cell in order to evaluate the evolution of mass loss during the experiment. A conical radiant electric heater uniformly irradiates the sample from above. The combustion is triggered by an electric spark. The gas flow, oxygen, CO, and CO_2 concentrations and smoke density are measured in the exhaust duct. The produced gases are collected in an exhaust duct system with centrifugal fan and a hood, and the HRR is calculated based on Huggett's observation that the quantity of heat released by most organic materials is proportional to the quantity of oxygen consumed while burning. The proportionality factor is 13.1 kJ/g of consumed oxygen [10,33,34,38,40–42].

26.3 FLAME RETARDANCY, CLASSIFICATION, AND SPECIFICATION

26.3.1 Sustainable Flame Retardancy—From Halogenated Chemicals to Nanocomposites

A FR additive can be defined in general as a compound or mixture of compounds designed to limit the risk of fire and its propagation. When carefully formulated and properly incorporated in the polymer matrix, it renders the polymer less likely to ignite, or once ignited less prone to burn effectively. It does this by increasing the TTI and improving the ability of the polymer to self-extinguish. In addition, it must act to decrease the HRR during combustion and also prevent the formation of flammable drops.

There are concerns over the potential risks to health and the environment that halogenated chemicals pose. Certain brominated organic compounds [43] are known to exhibit persistence, bioaccumulation, and toxicity (PBT), and to alleviate these problems, researchers and scientists began to develop nonhalogenated FRs. Early FRs additives that were nonhalogenated were phosphorus-based. Multiple reviews of alternatives to PBDE have been prepared by researchers in industry, in academia, and at a number of government institutes including one by the US Environmental Protection Agency (EPA) on the environmental effects of various phosphorus-based FRs [44,45]. Performance and environmental data can be found in a number of websites, including at http://www.halogen-free-flameretardants.com.

There have been many publications and patents issued covering phosphorus [46–52], aluminum trihydroxide (ATH) and magnesium dihydroxide, boron [53], and siloxane and silica [54]. Nanoparticle and nanocomposite-based FRs using naturally occurring smectite clays (layered silicates), such as montmorillonite (Mt), hectorite (Hc), or laponite (Lp), continue to be investigated and developed as well, as they hold promises of enhancement of multiple properties such as mechanical, thermal, and gas at low levels, low cost, and low risks.

Incorporation of clay in polymers has a remarkable effect on flammability, and as much as 75% decrease in the pHRR, measured in the cone calorimeter, can be realized [55]. The advantage of the nanoclays is that at small loading (a tenth of typical micrometer-sized additive), they significantly enhance material properties [56]. The clay typically would require use of organomodifier, and often it is found that the best FR results are obtained from synergism arising from combining nanoclay FR additive with one or more nonhalogenated FR additives.

26.3.2 Standards and Technology Life Cycle Assessment

Scientists and researchers at the National Institute of Standards and Technology (NIST) and the EPA continue to develop and evaluate sustainable approaches to flame retardancy with an emphasis on using nanotechnologies to limit pollution [57].

Pollution can be reduced by substituting nanoclay for PBDE FRs, but until 2008, no quantitative study had been done to confirm that such approach is sustainable. The report of a 2006 conference at the Woodrow Wilson International Center for Scholars shows that only two life cycle assessment (LCA) studies have been conducted on nanotechnology-based products as of 2005. LCAs are defined in the ISO standards (ISO14040:2006, ISO14044:2006) [58].

An early LCA study of nanocomposites used for automotive applications failed to address the issue of flame retardancy as reported by Lloyd and Lave [59]. The issues associated with performing an LCA of a product flame retarded using a nanocomposite, or another nonhalogenated FR, as compared to a halogen-based FR, are somewhat complex. Lloyd and Lave [59] provide an example of how the evaluation of a nanocomposite flame-retarded product might be conducted.

Further insight can be gained from several LCAs of various FR products (television sets, wire and cable, and sofas), performed at Swedish National Testing and Research Institute (SP) [60]. These LCAs are unique in that they include the effect of accidental fires on the LCA, something not usually included in most LCAs [61]. In the 2000 SP study of halogenated flame-retarded versus non-flame-retarded television sets made of high-impact polystyrene (HIPS), a similar approach to that taken by Lloyd and Lave was used, i.e., the incorporation of the effect of a different additive on the gases released into the environment. Additional distinctions from the SP study include the fact that the halogenated FR HIPS performed better in aging and recycling studies than the non-FR HIPS, and the FR in the HIPS did not bloom (phase separate to the surface) during tests [62,63].

Nanocomposites (which will be discussed in Section 26.9) have been found to prevent blooming [64], with potential to reduce environmental release of any additive present in the nanocomposite product. There is, however, a lack of environmental health and safety (EH&S) data on nanoparticles, which are required information for many of the inputs of an LCA. To conduct a meaningful LCA, companies need access to the results of systematic research into the mechanism of release, toxicity, and the effects on the environment that nanoparticles have on products of combustion (be it during accidental burning or planned incineration). Aside from their economics, little is known about the EH&S properties of newly engineered nanoparticles. LCAs are thus recommended not only for natural nanoparticles but also for newly developed nanoparticles with FR properties including layered double hydroxides (LDHs) [65], carbon nanofibers [66], and CNTs [67].

26.3.3 Classification of Fire Retardants by Mode of Action

The wide variety of potential fire retardants can be classified in various ways, such as mode of action (how they work) and means of incorporation, and are described briefly. Many fire retardants operate by more than one mechanism, some of which are described next. It has generally been accepted [5,6,68] that fire retardants that work by chemical action are more effective (by level of loading) than fire retardants that work by physical action.

26.3.3.1 Physical Action

A fire retardant that decomposes endothermically, such as hydrated filler, will remove heat from the material and flame, and gases evolved (H_2O, CO_2, NH_3, etc.) can dilute the flammable vapors to below the ignition limits (temperature or mass fraction) of the gas mixture. A marked endothermic reaction is known as a *heat sink*. Other additives such as CNTs increase the thermal conductivity of a material, therefore dissipating heat away from the ignition source.

Another approach is the creation of a barrier between the heat source and the underlying material, such as by use of an inert material. Carbon black and the aluminum residue resulting from the decomposition of ATH both act in this way. This layer can prevent oxygen reaching the pyrolysis zone, slowing the rate of fuel production or preventing volatile gases from reaching the flame, and can impede heat transfer by providing a radiation shield and thermal insulation. An enhanced variation of this is the formation of a swollen barrier known as an intumescent fire retardant. Finally, the incorporation of noncombustible substances, such as fillers, can inhibit burning by reducing the flammable content.

26.3.3.2 Chemical Action

Flame retardancy via a chemical modification of the burning process can occur in either the gaseous or the condensed phase. The fire retardant may decompose to evolve gases that react in the gas phase to interrupt the radical propagation mechanisms during burning. These gases typically *quench* the flame by producing unreactive radical species (Cl^* and Br^*), which are unable to propagate the free radical reaction necessary for flaming combustion, while increasing smoke, carbon monoxide, and other products of incomplete combustion. Other fire retardants work by promoting char development, therefore reducing the availability of combustible gases. This is most effective in polymers with a tendency to char, where the fire retardant promotes dehydrogenation of the polymer, leading to unsaturation, cross-linking, and cyclization. This deprives the flame of fuel, since the char requires a significantly higher temperature to burn. The char or vitrified layer acts as a physical barrier to heat and fuel transfer between the gas phase and the condensed phase.

26.3.4 Classification of Fire Retardants by Means of Incorporation

Fire retardants can be classified into two broad categories: reactive FRs and additive FRs.

26.3.4.1 Reactive Fire Retardants

Reactive fire retardants are covalently bonded to the polymer chain. They may become part of the polymer during polymerization or grafted on as side chains. As part of the modified polymer, they will not leach or migrate to the surface and remain effective for the life of the product. They are likely to profoundly affect the physical and chemical properties of the new polymer and will yield different decomposition products. By being integral to the polymer and uniformly distributed in it, they have greater efficacy as fire retardants than when used as an equivalent additive [66,68].

26.3.4.2 Additive Fire Retardants

Additive fire retardants are typically added to the polymer after polymerization, must be compatible with the polymer, and can migrate to the surface and volatilize or leach out. They do not react with the polymer but become active at higher temperatures such as at the start of a fire. They are often mineral fillers, hybrids, or organic compounds including macromolecules.

Another method of protection is the use of a fire-retardant coating, typically showing intumescent behavior [69]. Intumescent coatings, which swell or foam on heating, are used to protect the underlying materials by providing an insulating barrier to heat and gas flow to the substrate. They have also been used extensively to protect structural steelwork from excessive heat during a building fire to prevent buckling and collapse, and for the protection of flammable areas, e.g., in oil installations.

26.3.5 Specification for an Ideal Fire-Retardant Formulation

26.3.5.1 Fire Retardancy

An ideal fire-retardant additive must act to reduce flammability to a desired or specified level. Additives that decompose significantly, leach, or migrate in the working environment of the product are not acceptable because fire retardancy must be retained throughout the lifetime of the product.

26.3.5.2 Additive/Loading Level

High loading levels of additive, aside from added costs, will have adverse effects on required physical properties. Ideally, the fire retardant needs to function at low levels of addition, and should be sufficiently cheap at the effective loading.

26.3.5.3 Compatibility

An acceptable additive needs to be chemically inert in the service temperatures of the finished product. During the incorporation step, it can be added as a powder or a liquid. If added in a powder form, it must be both sufficiently small in particle size and fully compatible with the polymer to fully disperse in the polymeric matrix so as not to disrupt the physical properties such as tear strength, or melt below the mixing temperature. It must easily process in the equipment specific to the polymer.

26.3.5.4 Environmental Aspects and Legislation

A good fire retardant must satisfy health and environmental legislation in the country of use. The European Union initiated a series of directives to reduce the environmental impact of industry. Examples of these include the End of Life Vehicle (ELV) directive 2000/53/EC and the Restriction of the Use of Certain Hazardous Substances in Electrical and Electronic Equipment Regulations 2008, *the RoHS regulations*, EU Directive 2002/95. In both cases, there has been a general approach to reduce or eliminate the environmental release of toxic and bioaccumulative materials such as heavy metals like cadmium. In addition, in RoHS, polybrominated biphenyl (PBB) and PBDE fire retardants are also banned, creating a problem for the electronics industry, which has made extensive use of polymers containing brominated fire retardants [70–72].

Many international companies have published *black* and *gray* lists for suppliers of products and components. A *black* list details chemicals that are banned for supply. Many of these materials may be restricted for use in one industrial area, but the ban often covers all business areas to apply a consistent approach. A *gray* list will have materials that are not banned but may be suspected of health or environmental concerns and therefore should be avoided. The Volvo Black List STD100-0002 and Volvo Grey List STD100-0003 are typical of this approach, which include all brominated organic fire retardants, not just PBB and PBDE. The 1907/2006 EC Registration, Evaluation, Authorization and Restriction of Chemicals (REACH) regulation establishes obligations for the manufacturers, importers, and downstream users of chemical substances to ensure that chemicals in use are registered and evaluated for toxicity and environmental impact.

A list of *substances of very high concern* (SVHC) has been published by the European Chemicals Agency (ECHA), the EU agency responsible for the implementation of this directive [71]. Materials included in the list are those defined as carcinogenic, mutagenic, or toxic to reproduction; persistent; bioaccumulative; and toxic. Others in the lists are those identified from scientific evidence as causing probable serious effects to human health or the environment (e.g., endocrine disrupters). Fire-retardant chemicals such as hexabromocyclododecane (HBCD), tris (2-chloroethyl) phosphate (TCEP), and short-chain chlorinated paraffin waxes are also on the SVHC list. It is now required by law that users of materials on the SVHC list inform their customers of the presence of these chemicals in the products, and as such, it is more likely than not that the use of this class of materials will be reduced.

26.4 CONVENTIONAL FLAME RETARDANTS

26.4.1 Mineral Fillers

The flammability of any thermoplastic can be reduced by incorporation of any suitable inert or noncombustible filler. More important and most useful are the common inorganic materials such as group II or III carbonates or hydroxides, with the preferences being the metal hydroxides (especially of aluminum and magnesium), hydroxycarconates, and zinc borates. The inert fillers act to reduce the total amount of fuel and the rate of diffusion of oxygen into and fuel from the thermoplastic. Their presence also increases the thermal properties of the mixture such as the heat capacity, thermal conductivity, reflectivity, and emissivity. There are some reports of potential synergistic and even antagonistic catalytic effects [73,74] or in some cases some filler-associated surface effects, and changes to the polymer melt rheology are possible [75]. Endothermic decomposition of the inorganic materials accompanied with the release of inert gases and water vapor that occur with some inorganic materials effectively reduces flammability. The effective ones typically decompose within a narrow window range of temperature, above the polymer processing temperature but at or below their decomposition temperature.

Specifically, the following characteristics of inorganic fillers are significant for flame retardancy. Firstly, their endothermic decomposition with the resulting heat absorption cools down the polymer matrix. Secondly, they release inert effluent gases such as water vapor and carbon dioxide, which reduce the concentration of free radicals to a level below what is needed for fire to be sustained. Thirdly, they create an inert layer that shields the surface of the decomposing polymer from incoming radiation, thus preventing oxygen from reaching the fuel. They hinder flammable products of pyrolysis from reaching the gas phase and prevent radiant heat from reaching the polymer. These three effects of fire retardant have been summarized by Hull et al. [14].

The most commonly used aluminum hydroxide ($Al(OH)_3$) decomposes to form alumina (Al_2O_3) with the release of water. When used as a fire retardant, ($Al(OH)_3$) is commonly referred to as alumina trihydrate (ATH) and formulated as $Al_2O_3 \cdot 3H_2O$ (but in reality, it is neither an alumina nor a hydrate [76]). Its endothermic decomposition releases water vapor, which dilutes the free radicals in the flame. At the same time, a protective layer of alumina residue builds up.

$$180 \rightarrow 200°C$$

$$2Al(OH)_{3(s)} \rightarrow Al_2O_{3(s)} + 3H_2O_{(g)}, \Delta H = +1.3 \text{ kJ/g}$$

It is reported that the heat capacity of organic polymers varies from 0.9 to 2.1 J/K/g [74]. The decomposition enthalpy of a fire retardant mineral filler is a factor of 1000 larger. The decomposition enthalpy of 1 g $Al(OH)_3$ is equal to the heat (q) required to raise the temperature of a mass (m) of 1.5 g of low density polyethylene (LDPE) from ambient temperature (25°C) to decomposition (400°C). With constant heat capacity c = 2.3 J/K/g during heating [14] we obtain:

$$q = mc\Delta\theta, \text{ so } q = 1.5 \times 2.3 \times 375 = 1.29 \text{ kJ}$$

Of course, there are questions as to the true importance of the three contributions, from endotherm, gas, and residue heat absorption [75–77]. Khalturinskii and Berlin [78] showed significant enhancement provided by hydroxides and carbonates compared to inert fillers based on LOI test and by conducting a heat balance for processes occurring in the condensed phase and gas phase. Referring to studies of inert fillers, where loadings of 5–20% have minimal effect on the LOI, they concluded that over 80% loadings are required for effective fire. Rothon [75] reported a similar effort to quantify the three contributions to fire retardancy, but his analysis was done for two mineral fillers: $Al(OH)_3$ and nesquehonite ($MgCO_3 \cdot 3H_2O$).

26.4.2 Thermal Effects of Mineral Fillers

Hull et al. [14] quantified four physical contributions to the overall fire-retardant effects of mineral fillers. Their approach enables unexpected effects such as chemical interactions or changes of behavior resulting from different filler morphologies to be more readily identified. Estimates can be made of the decomposition endotherm, the heat capacities of the filler, its solid residue, and its vapor phase products for the temperature range over which they exist.

They made a number of simplifying assumptions as follows: (1) that the thermal conductivity of the polymer composite is unaffected by the presence of the filler; (2) that the final temperatures reached by the solid residues and the CO_2 and water in the gas phase do not vary significantly from one filler to another; (3) that the heat capacity of the filler and residue is not affected by the presence of polymer; and (4) that the decomposition endotherm of the filler is unaffected by incorporation into the polymer.

They also assumed that the only effect of the solid residue is its ability to act as a heat sink and that the only effect of the gas phase diluent is as an absorber of heat, thus neglecting any effects reducing the free radical concentration below a critical threshold. Finally, their approach does not take into account the particle size or morphology of the filler, even though these are known to be of some importance in experimental studies.

The heat capacity of any material varies as a function of temperature and can be represented by a polynomial, giving the value of the heat capacity at any temperature. The best estimate of the heat required to raise temperature over any temperature range is obtainable by integration of the Shomate equation [79,80]. The heat capacity Cp of a material is given by the Shomate equation:

$$Cp = a + bT + cT^2 + dT^3 + eT^{-2}$$

One can integrate this equation with respect to T to obtain the heat required to raise a known quantity of material over a given range of temperature, or alternatively, one can integrate the equation or a given relation numerically using a spreadsheet. Hull et al. [14] illustrated this approach to determine the heat required to raise $Mg(OH)_2$ from 25°C to 300°C (its decomposition temperature), as the area under the curve of Cp versus T. Assume Cp (T) is given as follows:

$$Cp = 84.9 + 0.0744T - 6.89 \times 10^{-5}T^2 + 2.66 \times 10^{-8}T^3 - 2.17 \times 10^6 T^{-2} \text{ in J K}^{-1} \text{mol}^{-1}$$

This gives $Cp_{298} = 77.2$ J K^{-1} mol^{-1} or $Cp_{298} = 1.32$ J K^{-1} g^{-1}, and $Cp_{573} = 1.77$ J K^{-1} g^{-1}.

The heat required to raise 1 g $Mg(OH)_2$ to its decomposition temperature at 300°C is the area under the Cp versus T curve. Hollingbery and Hull [81,82] settled on an average value of 1.44 J K^{-1} g^{-1}, allowing for some errors in the approach. The energy required to heat the resultant residue (such as MgO) from 300°C to 600°C and for the gas phase diluents can be evaluated using the same technique, recognizing the absence of phase changes. The decomposition enthalpy of the filler may be determined experimentally using differential scanning calorimetry (DSC), or obtained from validated tables in literature [14,81–85].

A number of potential fire-retardant fillers based on metal hydroxide and carbonates together with published estimates of their decomposition temperatures and endotherms have been reported by Hull et al. [14].

26.4.3 Metal Hydroxides

Metal hydroxides comprise the most important market segment for FRs. They function as polymer FRs by decomposing endothermically and releasing water at a temperature higher than the polymer processing temperature range, and around the polymer decomposition temperature [13]. ATH is the most commonly used mineral FR, followed by magnesium dihydroxide (MDH).

26.4.3.1 Aluminum Trihydroxide

On a weight basis, ATH is the most popular inorganic FR in current use. The endothermic decomposition of ATH ($Al(OH)_3$) occurs between 180°C and 200°C and leads to the release of water and the formation of alumina as described above.

As shown by differential thermal analysis (DTA), this reaction tends to occur in two stages, which correspond to two endothermic transitions. The intermediate product formed is known as boehmite AlOOH [86], which corresponds to the much lower endothermic process energy. The corresponding transition is much more perceptible with increased ATH particle size.

This reaction has several effects on the combustion of the polymer:

- It absorbs between 1050 and 1300 kJ/kg ATH, i.e., it cools down the polymer material.
- Al_2O_3 forms a thermally insulating protective coating.
- The released water vapor dilutes combustible gases and forms a protective gas layer.

However, the fire properties of ATH-filled polymers are only interesting at high loading levels at 60% or more. Such high loadings give end products with undesirable characteristics such as high density, lack of flexibility, and inferior mechanical properties. For example, the LOI can reach values higher than 50% for ethylene vinyl acetate (EVA) containing 75% (w/w) of ATH [82]. The use of ATH also decreases the HRR peak in the cone calorimeter test and considerably reduces smoke production [87]. Owing to its relatively low degradation temperature, ATH is limited to polymers with low processing temperatures, such as EVA and LDPE. Beyer [10] discussed the topic of reducing the filler content by combining ATH with layered silicates. Hull et al. [14,20] gave a more explicit role played by ATH and clay particles during the thermal degradation and combustion of EVA copolymer.

26.4.3.2 Magnesium Dihydroxide

Magnesium hydroxide (($Mg(OH)_2$) or MH) or MDH, with a higher decomposition temperature (~340°C) than ATH, is also used extensively as a stand-alone FR and in a wide range of polymers [12,14], though it is used less widely than ATH. MH acts in the same way as $Al(OH)_3$ (endothermic water release of nearly 31% of its original mass, and leaving MgO, which serves as the thermal barrier), but its endothermic degradation occurs at a higher temperature (300–340°C). The flame retardancy of MDH is recognized to be very effective up to 400°C.

This higher decomposition temperature makes it better suited to polymers such as polypropylene and polyamides [21], which are processed at temperatures (often via extrusion and injection molding) above the decomposition temperature of ATH.

$$2Mg(OH)_2 \rightarrow 2MgO + 2H_2O \ (1300 \ kJ/kg)$$

Beyond 400°C, however, the exothermic degradation predominates limiting its effectiveness. As with ATH, high loading levels of 50–60% by mass are required of magnesium hydroxide to achieve acceptable fire-retardant properties in polymers. Hornsby and Watson [83–85] proposed the following five mechanisms for the fire-retardant action of MH: (1) endothermic decomposition, which reduces the thermal decomposition of the polymer; (2) release of significant amount of water vapor, which dilutes the vapor phase; (3) the heat capacity of both the MH and the decomposition product, magnesium oxide, further reduces the thermal energy available to degrade the polymer; (4) the decomposition products promote char formation and therefore insulate the substrate from the heat source; and (5) the high loading level of MH acts as a solid-phase diluent.

During several FR tests, incandescence phenomena have been observed, and it is opined that metallic hydroxides (ATH and MDH) are potentially acting as combustion catalysts to the carbonized residues produced [86,88–90].

MDH nanoparticles may be obtained by several methods, via a sol–gel technique followed by a hypercritical drying procedure [91], a hydrothermal reaction using various precursors and solvents, [92] or by precipitation of magnesium salts with an alkaline solution [93], and have been used as FR additives. The precipitation technique using alkaline solution enables better control of the nanoparticle morphology by fine-tuning of the experimental parameters such as the chemical nature of the base used as precipitant, the type of counterion, the temperature, and the hydrothermal treatment. Qiu et al. [94] reported that by changing the base precipitant (NaOH or NH_4OH), nanometric MDHs with needle- or lamella-like morphologies, respectively, were obtained [94].

Nanometric MDH, even at lower loading levels, gives an improvement in fire retardancy. For example, when micronic MDH (2–5 mm) is replaced by nanometric Mg(OH)$_2$, the LOI obtained with EVA containing 50% (w/w) of MDH increases from 24% to 38.3% [94]. Excellent dispersion of the nanosized Mg(OH)$_2$ leads to the formation of more compact and cohesive char during the combustion test enabling the nanoparticles to improve flame retardancy of the EVA. Shen [89] and several other researchers have examined the advantages of mixtures of MH and organoclays in enhancing the flame retardancy of EVA copolymer, with a brief summary given in Ref. [14].

26.4.4 Hydroxycarbonates (Hydromagnesites)

Hydroxycarbonates are not as widely used as other conventional FRs. They nonetheless remain a viable alternative to metal hydroxides. While all carbonates release CO$_2$ at high temperatures, only magnesium and calcium carbonates release CO$_2$ below 1000°C, with magnesium carbonate presenting the lowest release temperature (550°C). Natural magnesium carbonate (magnesite) and synthetic magnesium hydroxycarbonate (hydromagnesite) both decompose endothermically at high temperature with liberation of CO$_2$, accompanied with water release.

The following reactions describe the thermal decomposition of hydromagnesite in air [95]:

$$(4MgCO_3 \, Mg(OH)_2 \cdot 4H_2O \text{ or } 5MgO \cdot 4CO_2 \cdot 5H_2O)$$

$$4MgCO_3 \cdot Mg(OH)_2 \cdot 4H_2O \rightarrow 4MgCO_3 \cdot Mg(OH)_2 + 4H_2O$$

$$4MgCO_3 \cdot Mg(OH)_2 \rightarrow 4MgCO_3 \cdot MgO + H_2O$$

$$4MgCO_3 \cdot MgO \rightarrow 5MgO + 4CO_2 \text{ (associated heat = 800 kJ/kg)}$$

While ATH and MDH have water release temperatures around 180–200°C and 300–340°C, respectively, hydromagnesite releases both water and carbon dioxide between 200°C and 550°C [96], suggesting similar or better flame retardancy activity than ATH and MDH. Hydromagnesite releases water and carbon dioxide over a wider temperature range than ATH and MDH. It has been used as a FR in polypropylene [22] and in an LDPE/EVA polymer blend (3:1 blend) [97].

26.4.5 Borates

Borates, especially zinc borates such as 2ZnO·3B$_2$O$_3$·3.5H$_2$O, are among the most frequently used inorganic additives. They decompose endothermically (503 kJ/kg) between 290°C and 450°C, releasing boric acid and boron oxide (B$_2$O$_3$). The boron oxide softens at 350°C and flows above 500°C leading to the formation of a protective vitreous layer. For polymers that contain oxygen atoms, the presence of boric acid causes dehydration, with the formation of a carbonized layer; the latter acts to protect the polymer from heat and oxygen by reducing the release of combustible gases [13,14,88].

26.5 HALOGENATED FRs

The type of halogen (fluorine, iodine, bromine, or chlorine) determines the effectiveness of halogenated FRs. Because of their low bonding energy with carbon atoms, bromine and chlorine can readily be released and take part in the combustion process, especially with the previously discussed free-radical mechanism occurring in the gas phase. Iodinated compounds are less thermally stable than most commercial polymers and therefore release halogenated species during polymer processing. Fluorine- and iodine-based compounds do not interfere with the polymer combustion process and, as a result, are not used. Fluorinated compounds are more thermally stable than most polymers and do not release halogen radicals at the same temperature range or below the decomposition temperature of the polymers [98].

26.5.1 Halogenated FR Additives

During thermally induced polymer decomposition, very reactive free-radical species such as H* and OH* are released, and they maintain combustion by a cascade-chain mechanism in the gas phase. Thus, by reacting with these species, halogenated FRs act to stop the chain decomposition and therefore prevent the thermoplastic from burning:

$$RX \rightarrow R^* + X^* \quad X \text{ may be Br or Cl}$$

$$X^* + R'H \rightarrow R'^* + HX$$

$$HX + H^* \rightarrow H_2 + X^*$$

$$HX + OH^* \rightarrow H_2O + X^*$$

The effective FR species HX is regenerated by the reaction of X* with RH, and being nonflammable, it provides a physical action on the combustion mechanism such as a protective gaseous coating and dilution of fuel gases. HX also catalyzes the oxidation of the solid phase, and the oxidation products tend to cyclize, which leads to the formation of a solid protective layer.

The four most common halogenated FR products are tetrabromobisphenol A (TBBPA), polybromodiphenylether (PBDE) compounds, HBCD, and tetrabromophthalic anhydride (TBPA) [13].

26.5.1.1 Tetrabromobisphenol A

TBBPA, the most widely used halogenated FR, is mainly incorporated as a reactive FR especially in epoxy resins for printed circuit boards (Scheme 26.1).

26.5.1.2 Polybromodiphenylether

PBDE compounds, the second most used halogenated FR family, can contain up to 10 bromine atoms attached to a diphenyl ether molecule. The polybromodiphenylethers developed as FR additives are penta-, octa-, and decabromodiphenylethers. Penta- and octabromodiphenylethers have high molecular weight and good thermal stability. They are used in styrenic polymers, polyolefins, polyesters, and nylons. They are, in composition, really mixtures of different diphenyl ethers containing various bromine atoms, with the number given 5 or 8 being the average. They have been phased out, however, since they were identified as probable dioxin precursors. Decabromodiphenylether is still finding limited use particularly outside of Europe (Scheme 26.2).

26.5.1.3 Hexabromocyclododecane

HBCD, a cycloaliphatic halogenated FR, is currently used in expanded or compact PS and textiles (Scheme 26.3).

26.5.1.4 Tetrabromophthalic Anhydride

TBPA is used both as a FR additive in unsaturated polyesters and also as a raw material for the production of other FR agents (Scheme 26.4).

26.5.2 Halogenated Monomers and Copolymers (Reactive FRs)

Reactive monomer and copolymer FRs have an advantage over the other FR additives in that they can be used in relatively low concentrations. Often used as condensation or free-radical polymerization

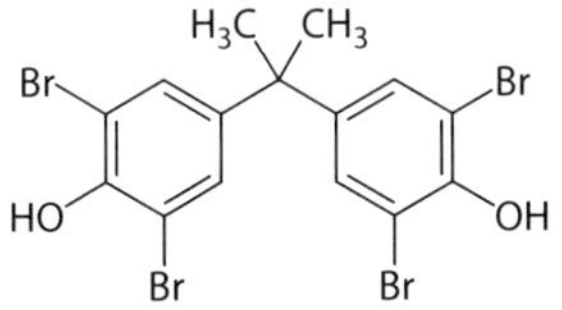

SCHEME 26.1 Chemical structure of TBBPA.

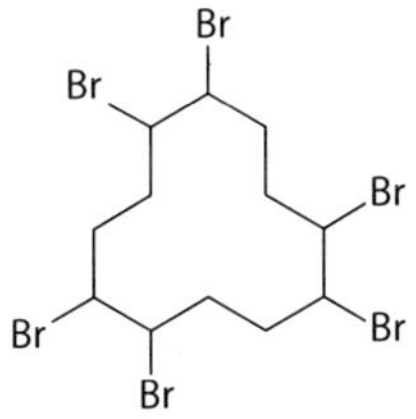

SCHEME 26.2 Chemical structure of HBCD.

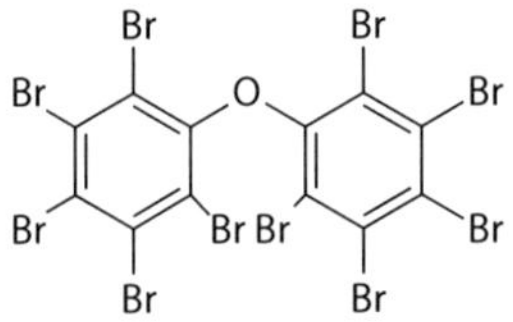

SCHEME 26.3 Chemical structure of PBDE.

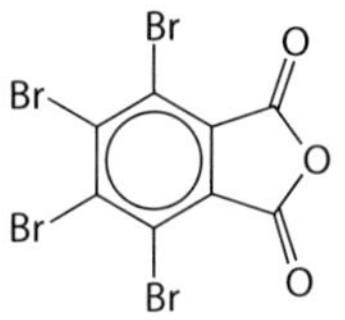

SCHEME 26.4 Chemical structure of TBPA.

monomers, copolymerized with virgin monomers or grafted onto the polymer chain, they have the disadvantage of requiring an additional synthesis step especially on the industrial scale. They increase the compatibility between the polymer and the FR agent because they are easy to incorporate directly into the polymer structure. They act to mitigate against the damage caused to the mechanical properties of the materials by heterogeneous additives. Furthermore, they limit the migration of the FR agents onto the material surface.

Monomer and copolymer halogenated FRs act in similar fashion to other halogenated FR additives by reacting with the highly reactive H_ and OH_ species and stopping the chain decomposition. It is suspected that additional mechanisms are possible [99–102]; thus, Janovic [102] investigated the thermal stability and flammability of products of copolymerization reaction of important brominated monomers with a number of known commercial vinyl monomers. They included bromine-containing monomers, vinyl bromide, brominated styrenes, aliphatic and aromatic esters of acrylic acid and methacrylic acid, brominated phenylmaleimides, and phosphorus–bromine vinyl monomers, with their results showing that copolymers of bromine-containing monomers possess FR properties that compare favorably with other brominated but low-molecular-weight compounds. They reported changes in the degradation pattern as a result of the incorporation of a low proportion of vinyl bromide (VBr) units into poly(methyl methacrylate) (PMMA) chains [94]. Janovic

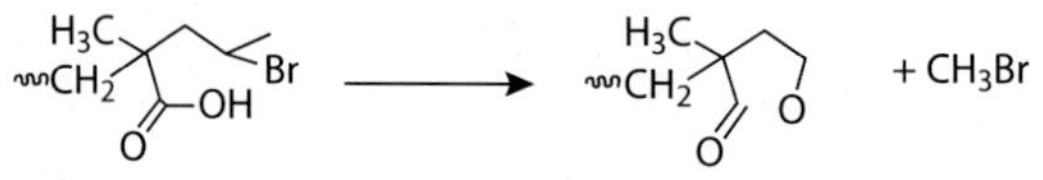

SCHEME 26.5 Intramolecular lactonization of poly(VBr-co-MMA). (From Laoutid, F. et al., *Mater. Sci. Eng. R: Reports*, 63(3):100–125, 2009; Janovic, Z., *Polym. Degrad. Stab.* 64:479, 1999.)

[102] proposed that intramolecular factorization is the mechanism for the thermal degradation of poly(VBr-co-MMA), involving adjacent bromine and ester groups in the copolymer chain with the release of alkyl bromide as a volatile product. The lactone rings formed interrupt the unzipping process, the main thermal degradation pathway for PMMA. The LOI values of the VBr-based copolymers increase with the relative VBr content in the copolymers studied (Scheme 26.5).

26.6 PHOSPHORUS-BASED FR

Phosphorus-based FR products include phosphates, phosphonates, phosphinates, phosphine oxides, phosphates, and red phosphorus (RP). They can be used as additives or incorporated into the polymer chain during polymerization, and are known to be active in the condensed and/or vapor phase.

For polymers containing oxygen (e.g., polyesters, polyamides, cellulose, etc.), the phosphorus-based FRs are very effective in the condensed phase [103,104]. With most of them, thermal decomposition results in the production of phosphoric acid, which condenses readily to produce pyrophosphate structures and liberate water.

The oxidizing gas phase is diluted by the released water. The phosphoric acid and pyrophosphoric acid can also catalyze the dehydration reaction of the terminal alcohols leading to the formation of carbocations and carbon–carbon double bonds. At high temperature, this can subsequently result in the generation of cross-linked or carbonized structures (Scheme 26.6).

At high temperature, ortho- and pyrophosphoric acids are turned into metaphosphoric acid (O)P(O)(OH) and their corresponding polymers $(PO_3H)n$. The phosphate anions (pyro- and polyphosphates) then take part, with the carbonized residues, in char formation. This carbonized layer (char) isolates and protects the polymer from the flames and

- Limits the volatilization of fuel and prevents the formation of new free-radicals
- Limits oxygen diffusion, which reduces combustion
- Insulates the polymer underneath from the heat

Phosphorus-based FRs can also volatilize into the gas phase to form active radicals such as PO_2^*, PO*, and HPO*. They also act as scavengers of H* and OH* radicals. Volatile phosphorated compounds are among the most effective combustion inhibitors since phosphorus-based radicals are, on average, 5 times more effective than bromine and 10 times more effective than chlorine radicals [13,105]. Phosphorus-based FRs are significantly more effective in oxygen- or nitrogen-containing polymers. The presence of oxygen or nitrogen atoms in the polymer chain is critical. If the case were the polymer cannot contribute to charring, a highly charring co-additive, for example, a polyol such as pentaerythritol, can be introduced in combination with the phosphorated FR (Schemes 26.7 through 26.9) [106,107].

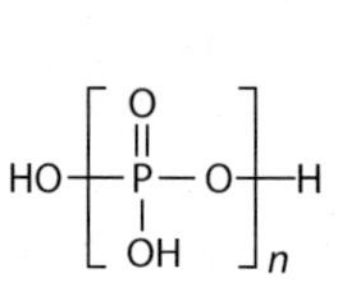

SCHEME 26.6 Polyphosphoric acid structure.

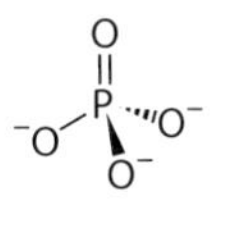

SCHEME 26.7 Chemical structure of organic phosphate.

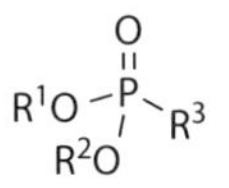

SCHEME 26.8 Chemical structure of ester of phosphonic acid–phosphonate.

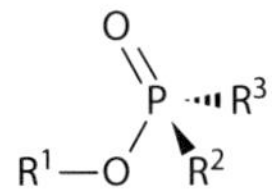

SCHEME 26.9 Chemical structure of phosphinates.

26.6.1 Red Phosphorus

The largest source of phosphorus for flame retardancy is RP. It is very effective in polymers such as polyesters, polyamides, and polyurethane when used in small quantities (<10%). For example, a glass-filled PA-6,6 containing 6–8% RP achieves a V-0 classification in the UL94 test [108]. Though Piechota [109] as far back as 1965 reported the use of RP as a FR in polyurethane, its action mechanism was not then established clearly. It is now widely accepted that in oxygen- and or nitrogen-containing polymers, RP changes, by thermal oxidation, into phosphoric acid or phosphoric anhydride, which upon heating gives polyphosphoric acid. This acid can catalyze the dehydration reaction of polymer end chains and trigger char formation [110,111].

The FR effectiveness of RP is reduced in polyesters like PET when the sample is burned in a N_2O atmosphere [111]. While a Nitrogen Oxygen Index can be defined to show whether a given FR acts mainly in the gaseous or conversely the condensed phase, this PET example indicates that some of the flame retardancy imparted by phosphorus involves gas-phase inhibition. Nonetheless, the main FR action of RP is known to occur in the condensed phase.

It is reported by Kurlya and Papa [112] that in PE, RP is active both in the gas and the condensed phase. In the gas phase, PO_ species produced from the combustion of RP quench the free-radical processes. In the condensed phase, RP substantially lowers the heat of oxidation and also traps free radicals. Thermal stability is improved accompanied with a decrease in fuel production during the burning of the material. RP is reported to be active in PE [112] and other non-oxygenated polymers [113]. As a result, a different mode of action of RP was proposed where RP depolymerizes into white phosphorus—P_4 [114,115]. P_4 can volatilize at high temperature and act in the gaseous phase, or it can diffuse from the bulk of the polymer to the burning surface, where it is oxidized to phosphoric acid derivatives that can eventually come into close contact with the flame and form phosphoric acid. This phosphoric acid could act as a char-forming agent, thus physically limiting oxygen access and fuel volatilization.

A major disadvantage of RP is that during the melting process, it can release highly toxic phosphine (PH_3) through reaction with moisture as a direct result of its poor thermostability. It is noteworthy that phosphine formation can be avoided by prior polymeric encapsulation of the RP [116], which can also further improve its effectiveness as a FR. In PE [116], for instance, the incorporation of microencapsulated RP in melamine–formaldehyde (MF) resin leads to the formation of a compact charred layer upon combustion. The RP is predominantly oxidized to various phosphoric acid derivatives. These derivatives react with the MF resin to form more stable structures containing P–O–P and P–O–C chemical bonds. Alternatively to encapsulation in MF resins, there are also systems for trapping the phosphine formed at high temperature. These systems use the capacity of phosphines to react with metallic salts [117,118]. A number of compounds known to be very efficient at trapping phosphine in polymers including $AgNO_3$, $HgCl_2$, MoS_2, HgO, PbO_2, CuO, and $(FeCl_3 \cdot H_2O)$ have been incorporated in red phosphorus-based FR formulations.

26.6.2 Inorganic Phosphates

Inorganic salts of polyphosphoric acid and ammonia [ammonium polyphosphates (APPs)] are stable, nonvolatile compounds, which when incorporated in oxygen- and/or nitrogen-containing polymers (polyesters [117], polyamides [118–120], polyurethane [121], etc.) leads to polymer charring. APP's effectiveness is loading level dependent, not being efficient at low loading in aliphatic polyamides [118–120] for example, but becoming very efficient at high loading, e.g., >10% in polyamide-6,6, >20% in polyamide-11, -12, -6, and -10, and >30% in polyamide-6.

The branched or unbranched APP has a chain length (n) that is variable and can be higher than 1000. Short, linear-chain APPs (crystalline form I: APP I) with $n < 100$ are more water sensitive and less thermally stable than longer-chain APPs (crystalline form II: APP II) with $n > 1000$. The longer-chain APPs exhibit very low water solubility (<0.1 g/100 mL). Long-chain APPs start to decompose into polyphosphoric acid and ammonia at temperatures above 300°C, whereas short-chain APPs begin to decompose at temperatures above 150°C (Scheme 26.10).

When APP decomposes, free acidic hydroxyl groups created condense by thermal dehydration to yield a cross-linked ultraphosphate and a polyphosphoric acid with a highly cross-linked structure [120]. The catalytic polyphosphoric acid reaction with the oxygen- or nitrogen-containing polymers leads to dehydration and char formation.

Additionally, APP can modify the degradation mechanism in non-self-charring polymeric materials as in the case of PMMA [120–125]. The modification occurs by the reactions of the polymethacrylate chains with polyphosphoric acid as a result of the high acidity and the reactivity of its phosphorus oxygen bonds, leading to cyclization and the formation of anhydride groups together with the elimination of ester groups. Chain depolymerization reaction in neat PMMA is inhibited by the modified structures, which in turn catalyzes a partial hydrolysis of the PMMA followed by cross-linking reactions.

26.6.3 Organic Phosphorus-Based Compounds

Organophosphorus compounds comprise mainly phosphate esters, phosphonates, and phosphinates [126,127]. Similarly to other phosphorus-based retardants, organic phosphorus derivatives can function either as additives or as reactive (co)monomers/oligomers, but their commercial use is limited by the processing temperature and the nature of the polymer to be modified.

Due to high volatility and poor fire-retardant characteristics, the (alkyl-substituted) triaryl phosphates such as triphenyl phosphate (TPP), cresyl diphenyl phosphate, isopropylphenyl diphenyl phosphate, tert-butylphenyl diphenyl phosphate, or tricresyl phosphate (TCP) are all of limited use. Oligomeric phosphates such as resorcinol bis(diphenyl phosphate) (RDP) and bisphenol A bis(diphenyl phosphate) (BDP), which have lower volatility and higher thermal stability than triaryl

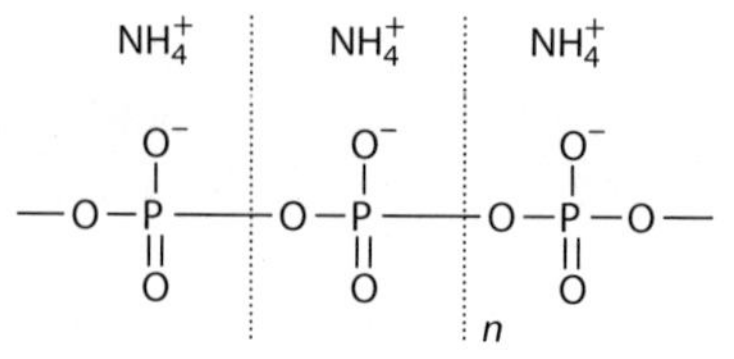

SCHEME 26.10 Chemical structure of APP I.

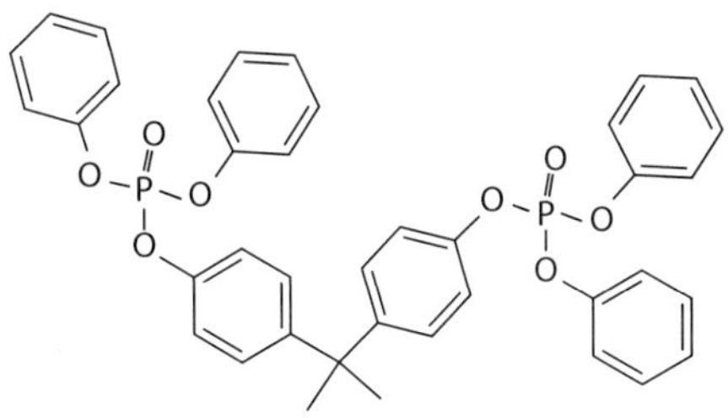

Thermoplastic Additives

SCHEME 26.11 Chemical structure of BDP.

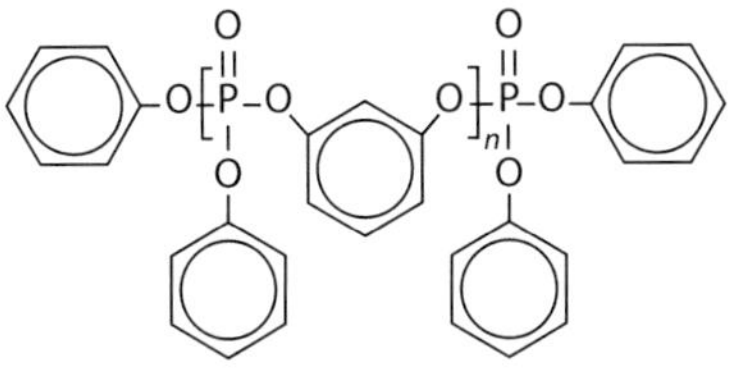

SCHEME 26.12 Chemical structure of RDP.

phosphates, are of importance in plastics engineering. Addition of BDP or RDP at 1 wt% of phosphorus in 5:1 PC/acrylonitrile–butadiene styrene (ABS) formulation yielded a V-0 rating in the UL94 test [128,129]. BDP and RDP have been shown to improve flame retardancy for poly(phenylene oxide)/high-impact PS blend (PPO/HIPS) with a V-0 rating. They have been shown to react with PC phenolic functions, inducing polymer chain cross-linking and forming char [130,131].

Using RDP and PTFE in combination shows a synergistic effect but with a lower amount of char formed compared with only BDP, an outcome that may be due in part to the partial volatilization of RDP. Combining volatile and nonvolatile phosphates has been shown also to lead to a synergistic effect [132,133]. It is also not surprising that combination of RDP or BDP with TPP or even the combination of BDP and RDP is more effective than each additive taken separately (Schemes 26.11 and 26.12).

26.6.4 Intumescent FR Systems

Intumescent systems have layers that act as insulating barriers between the heat source and the polymer surface, reducing heat transfer. They limit fuel transfer from the polymer toward the flame and hinder the diffusion of oxygen into the material. There are at least three components to the general formulation of an intumescent system: (1) an acid, inorganic acid, acid salt, or other acid source to promote the dehydration of the carbonizing agent (APP is the most widely used acid source compound); (2) a carbohydrate (or a carbonizing agent) that is dehydrated by the acid to form a char; and (3) a blowing agent, which decomposes and releases gas, causing the expansion of the polymer and producing a swollen multicellular layer (melamine, guanidine, urea, chlorinated paraffins). It is the release of the gas during the thermal decomposition of the carbonizing agent that initiates the expansion of the carbonized layer.

26.7 NITROGEN-BASED FRs

Melamine, a thermally stable crystalline product, contains 67 wt% nitrogen atoms and has a melting point as high as 345°C [98,123,124]. Upon sublimation at about 350°C, it absorbs a significant amount of energy, thus lowering temperature. Decomposition of melamine at high temperature leads to the production of such thermally stable condensates as melam, melem, and melon with release of ammonia gas. The released ammonia gas dilutes the oxygen and other combustible gases

[98,121–125]. Melam, melem, and melon are effective FR compounds. The compounds have in common that they melt or decompose at very high temperatures and that they are insoluble in any solvent, making characterization difficult [122].

Melam is a homologue of melamine; it is not a heptazine like melem and melon, but rather a fused product of 2,4-diamino-6-chloro-*s*-triazine with melamine. Melem is a heptazine. A heptazine, or tri-*s*-triazine or cyamelurine, is a type of chemical compound that consists of a planar triangular core group, C_6N_7, or three fused triazine rings, with three substituents at the corners of the triangle. Heptazines were discovered in the nineteenth century, but their study has long been hampered by their general insolubility [123].

The general form of heptazine is 1,3,4,6,7,9,9b-heptaazaphenalene. The parent compound $C_6N_7H_3$, where the three substituents are hydrogens, is called 1,3,4,6,7,9-hexaazacycl[3.3.3]azine or tri-*s*-triazine proper [124]. It is a yellow, weakly fluorescent solid with melting point over 300°C. The compound is generally stable and soluble in organic solvents such as acetonitrile but is decomposed by water. It has a peculiar crystal structure, whose cell spans 16 molecules in asymmetric positions and orientations. Melem is the specific compound with three amino (NH_2) substituents. When heptazine is polymerized with the tri-*s*-triazine units linked through an amine (NH) link, it is called melon (Schemes 26.13 through 26.15).

When melamine phosphate thermally decomposes, melamine polyphosphate is formed with the release of melamine and phosphoric acid [98,121–125]. The phosphoric acid released can phosphorylate many polymers and produce FR effects similar to phosphorus-based FR additives. Similarly, melamine polyphosphate thermally decomposes to form melam ultraphosphate and APP, releasing melamine. The melamine in the gaseous phase competes with the formation of its condensation products, such as melam ultraphosphate. The decomposition progression with the release of ammonia of melamine to melan to melem and to melon has been described by Levchik et al. [119]. Nitrogen-based compounds, in particular melamine and its derivatives, continue to find applications as alternative halogen-free FRs. Melanine cyanurate (MC), which is a salt of melamine and cyanuric acid, is used commonly in the nylon industry.

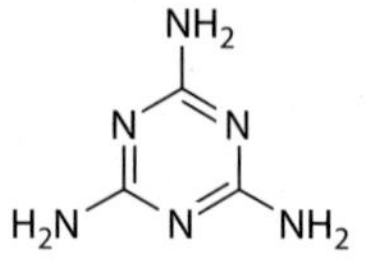

SCHEME 26.13 Chemical structure of melamine.

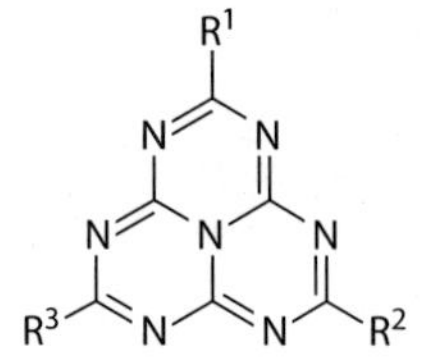

SCHEME 26.14 Chemical structure of melem (R1=R2=R3=H).

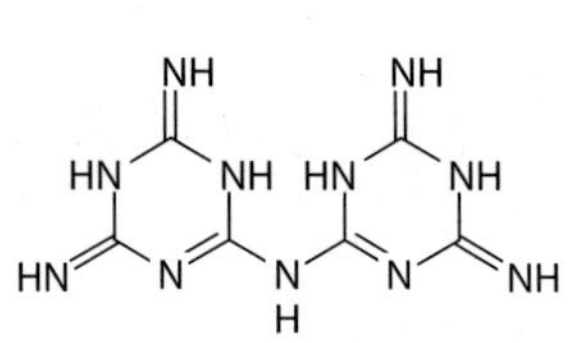

SCHEME 26.15 Chemical structure of melam.

26.8 SILICON-BASED FRs

Silicon-based compounds (silicones, silicas, organosilanes, silsesquioxanes, and silicates), when added even at low levels, can significantly improve flame retardancy. They are used as fillers incorporated in the polymer, as copolymers, or as the main polymer matrix.

26.8.1 Silicones in PC

An excellent review of flame retardancy of silicones was published by Kashiwagi and Gilman [134]. Silicones release very limited toxic material during thermal decomposition and are characterized with excellent thermal stability and high heat resistance. Their use as FR agents can be through direct blending within the polymer matrix or by synthesizing block/graft copolymers including silicone segments.

Silicone polymers as FRs in polycarbonate (PC) were investigated in depth by Iji and Serizawa [135], exploring the silicone chain topology (linear type, branched type), the pending groups along the chain (methyl, phenyl, mixture of the two), and the nature of end groups (methyl, phenyl, hydroxyl, methoxyl, vinyl). As measured by the LOI values, branched silicone polymers containing a mixture of methyl and phenyl groups along the chain and end-capped by methyl groups are the most efficient, even more so when formulated at lower molecular weights.

Silicone derivatives have excellent dispersion in PC. They migrate toward the material surface during combustion and form a highly flame-resistant char as a result of the combination of polysiloxane and condensed aromatic compounds, leading to their superior FR performance [135]. The thermal decomposition of PC containing branched-structure methyl phenyl-silicone has been investigated by Zhou and Yang [136] using TGA. It is postulated that the hydroxyl groups in the PC degradation products may be reacting with the carbon–silicon bond in the silicone to produce a cross-linked structure and a condensed aromatic structure by means of a dehydrogenation reaction between phenyl groups.

26.8.2 PC-b-PDMS in PC

Block copolymers of PC and PDMS (PC-b-PDMS with degree of polymerization [DP] 40) were incorporated as a FR agent in PC by Nodera and Kanail [137]. Compared to simple PC/PDMS blends, the copolymerization of PC and PDMS blocks leads to better FR properties in both the LOI and cone calorimeter tests. The PC/PDMS blend has a pHRR of 500 kW/m^2 compared with the PC-b-PDMS copolymer that has a pHRR of only 235 kW/m^2. Combustion of PC-b-PDMS copolymers generates large amounts of gas bubbles, promoting significantly more intensive charring compared with that of the PC/PDMS blends.

The effect of PDMS block size (DP from 15 to 350) and PDMS content (from 0 to 2.5 wt%) on LOI values recorded with PC-b-PDMS copolymers was further examined. The LOI values and the amounts of residues for each PC-b-PDMS copolymer increased rapidly until about 1.0 wt% in PDMS, and then started leveling off at higher silicone block content [137]. The PDMS block size enhances PDMS dispersion (~50 nm mean inclusion size), with the increased dispersability of the PDMS in the PC resulting in high flame retardancy. It was postulated that PC-b-PDMS block copolymers with moderate PDMS dispersion could form numerous fine bubbles through reaction of PC and PDMS during combustion, which behave as good thermal insulation domains spread throughout the material.

26.8.3 Silica in PP, PEO, and PMMA

Cone calorimeter tests were conducted by Gilman et al. [138] to study the effects of three silica gels (with different pore volume, particle size, and surface concentration) on the flammability properties of polypropylene. Silica gel pore volume has a significant effect on the HRR of PP containing 10% wt. in silica, a dramatic reduction in the HRR resulting from incorporation of high-volume silica. The improved flame retardancy at higher silica gel pore could be a result of the larger pores being able to accommodate PP chains or by increasing the molten polymer viscosity during pyrolysis, an event that is capable of inhibiting volatilization and the evolution of combustion products [138].

The flame retardancy of various types of silica, silica gel, fumed silica, and fused silica as FRs in non-char-forming thermoplastics (e.g., polypropylene) and polar char-forming thermoplastics (e.g., PE oxide) was investigated by Kashiwagi et al. [139,140]. In each case, a significant reduction in the HRR and the MLR was achieved by the incorporation of low-density, large surface area silica such as fumed silica (140 and 255 m^2/g) and silica gel (400 m^2/g) to polypropylene and PE oxide. Fused silica with lower surface area did not significantly reduce the flammability characteristics as much as the other silica samples did.

Silica has also been observed to accumulate on the surface of burned PMMA [141]. Fused silica and silica gel were incorporated in two different molecular-weight PMMA samples, with the result revealing that the specific surface area of the silica and its porous volume affected the thermal stability and FR properties of the polymer by modifying the viscosity of the system in the molten state.

26.9 NANOMETRIC PARTICLES AS FRs

Nanoparticles increase the interfacial area between the polymer and the nanofiller and can thus be used effectively at low levels. Thermal, mechanical, or fire resistance properties of thermoplastic are enhanced when nanometric particles are properly dispersed in polymer, but the contribution of each type of nanoparticle to flame retardancy varies and strictly depends on its chemical structure and geometry. As described earlier, nanometric Mg(OH)$_2$, even at lower loading levels, gives an improvement in fire retardancy. When micronic MDH (2–5 mm) is replaced by nanometric Mg(OH)$_2$, the LOI obtained with EVA containing 50% (w/w) of MDH increases from 24% to 38.3% [94]. Improved dispersion of the nanosized Mg(OH)$_2$ leads to the formation of more compact and cohesive char during the combustion test, thus enabling the nanoparticles to enhance flame retardancy of the EVA. The possibility of using nanocomposites to improve fire resistance was first suggested in 1976 by Uniika Ltd, Japan; but the extensive investigation of the flame retardancy of PA-6/clay nanocomposites by Gilman et al. [55] in 1997 and Gilman [6,7] brought the science into the mainstream.

Described briefly below are three widely investigated nanoparticles of importance for flame retardancy: (1) layered materials, such as nanoclays, e.g., montmorillonite: MMT (these are sometimes referred to as 2D nanoparticles and characterized by one nanometric dimension); (2) fibrous materials, such as CNTs and sepiolite (these are referred to as 1D nanoparticles and characterized by elongated structures with two nanometric dimensions); and (3) particulate materials, such as polyhedral oligosilsesquioxane (POSS) and spherical silica nanoparticles (referred to as 0D nanoparticles and characterized by three nanometric dimensions).

26.9.1 Nanocomposite FRs—Nanoclays

The structure, preparation, and properties of polymer/layered silicate nanocomposites (PLSNs) have been summarized in a number of publications [11–13,34,35,39]. The layered silicates are clay minerals either with a silica tetrahedral sheet fused to an aluminum octahedron with shared oxygen atoms (1:1 structures in kaolinite) or more commonly having stacked layers of two silica tetrahedra fused to an edge-shared octahedral sheet of alumina (2:1 phyllosilicates or smectites). The layer thickness is ~1 nm (hence the name nanocomposites), and the lateral dimension varies from 300 Å to several microns, giving an aspect ratio > 1000 (Figure 26.1) [55,141–143].

The exact contribution of nanoparticles to flame retardancy is rather complicated. Clay contributes to thermal stability by acting as a barrier to both heat and mass transfer during combustion. The dispersion of the clay nanolayers within the polymer matrix requires that the natural clays be modified using organic cations (alkylammonium, alkyl phosphonium, and alkyl imidazol [idin] ium cations), leading to

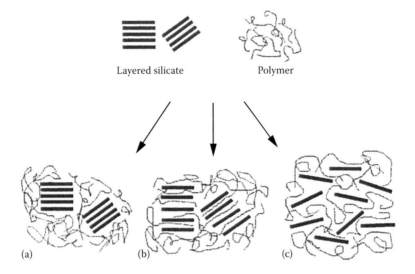

FIGURE 26.1 Probable polymer/layered silicate structures. (a) Phase separated (microcomposite). (b) Intercalated (nanocomposite). (c) Exfoliated (nanocomposite). (Reprinted from *Mater Sci Eng R-Rep*, 28. Alexandre, M., Dubois, P., Polymer-layered silicate nanocomposites: Preparation, properties and uses of a new class of materials, 1–63, Copyright 2000, with permission from Elsevier.)

the formation of organomodified nanoclays. Unfortunately, the clays and/or the decomposition products of the organic modifier can also catalyze the degradation of the polymer matrix [140,141,144,145].

Even at relatively low loading levels, the organomodified nanoclay produces a protective layer during combustion [144,146]. Without modification, the PLSN increases melt viscosity because it reduces polymer chain mobility, whereas with the organic cations, the viscosity of the molten PLSN decreases with increasing temperature and thereby facilitates the migration of the clay nanolayers to the surface.

The clay alone also hinders heat transfer and mass transfer (of the oxygen and volatiles generated from degradation of the polymer matrix); however, thermal decomposition of the organomodifier is promoted by heat transfer accompanied by the creation of strongly protonic catalytic sites onto the clay surface. The latter is assumed to be responsible for catalyzing the degradation of the polymer matrix and forming char [147,148]. The critical ingredients for flammability or degradation, namely, heat transfer into the material, evolution of flammable or combustible degradation products, and diffusion of the latter and oxygen, are all inhibited or significantly limited by the accumulation of the clay on the surface of the material. Decomposition of both the quaternary ammonium organomodifiers and the polymer chains contributes to the formation of gas bubbles that subsequently enhances the migration of the nanoclay.

26.9.1.1 Polyolefin/Layered Silicate Nanocomposites

Incorporation of nanocomposites into a polymer matrix enhances fire retardancy by a number of mechanisms. The main ones are the formation of a char layer and the migration of the nanoclay to the material surface, which constitutes a barrier against heat and diffusion of volatiles. In nanocomposites with exfoliation, increased melt viscosity often occurs. These mechanisms sometimes improve fire retardancy and sometimes make it worse depending on the base polymer and the presence of other modifiers. Synergism in PLSNs with conventional FRs has been reviewed in detail by Kiliaris and Papaspyrides [12], and these will be summarized in Section 26.10.

26.9.1.1.1 EVA/Layered Silicate Nanocomposite

There is no significant difference in the burning characteristics of unfilled EVA and that of EVA filled with conventional microcomposites, which does not have MMT exfoliation/intercalation.

However, when EVA/layered silicate nanocomposites (with high levels of clay delamination) filled with tiny amounts of organoclays (2–5%) are burned, heat release is reduced by 70% or more. A chemical mechanism to explain this has been proposed by many researchers including Camino and coworkers [106,118,120,138–140,149]. A mechanism to explain this was proposed, suggesting that the thermal decomposition of the organomodifier and the formation of acidic onium ions on the clay layers [134], facilitated by heat transfer from an external source such as the flame, catalyze the loss of the acetic acid. Costache et al. [150] studied the clay effect on the degradation pathway of EVA copolymers and their nanocomposites using a variety of means including TGA/FT-IR, TGA, GC–MS, cone calorimetry, and UV. They found that the clay accelerated the loss of acetic acid by chain stripping. They postulated that the deacetylation of the EVA produces polyene, which subsequently undergoes a multiplicity of reactions, including the formation of conjugated and cross-linked polyenes that produces the charred surface layer.

The environment can have notable effect on the FR mechanism within polymer nanocomposites as observed by Pastore et al. [149]. These researchers monitored the evolution of the (nano) structure of EVA–clay nanocomposites during thermal treatments at low temperature (under 225°C), in both nitrogen- and oxygen-rich air, and found that the interlayer distance of the organoclay decreased, whereas at higher temperature, the nanocomposite structural modification depended on the specific atmosphere. On the one hand, in nitrogen there was a reduction in polymer-layer compatibility, and segregation of the two phases occurred as the EVA deacetylation reduces the polarity of the polymer. The silicate layers became less organophilic as a result of the decomposition of the organomodifier. On the other hand, in air partial oxidation of the polymer chains resulted in an increase in matrix polarity, thereby promoting polymer–clay compatibility and enhancing the intercalation of the layered silicate.

Forming a nanocomposite, intercalated or exfoliated, is an effective way to improve FR properties. Cone calorimeter test by Bourbigot et al. [151] shows that the pHRR decreases by 50% for EVA/5% Cloisitel 30B nanocomposites and only by 25% for EVA/5% Na-MMT microcomposites when compared to the neat, unfilled polymer. Morgan et al. [152] showed that the HRR curves of intercalated and exfoliated PA-6 nanocomposites are not significantly different from each other, leaving the question as to whether simple intercalation could give similar or even better FR properties than exfoliated nanocompositions remains unresolved. The intercalated sample has an ignition time of 40 s compared to 80 s for the exfoliated nanocomposite, a result that suggests a significant difference in ignition times between the two nanomorphology types.

26.9.1.1.2 PP/PPgMA/Layered Silicate Nanocomposite

During thermal degradation of a polymer matrix with nanoclay, the char formed can significantly improve the FR behavior of the nanocomposite. However, incorporation of sodium layered montmorillonite (Na-MMT) and organomodified montmorillonite (oMMT) in polypropylene grafted maleic anhydride (PP/PPgMA) results in char layer formation, whereas incorporation of Na-MMT alone does not [146]. Irrespective of the amount of Na-MMT, when used alone, no trace of char was observed because Brønsted acid sites, which catalyze the thermal degradation of PP via a cationic mechanism, resulting in char formation through hydrogen transfer in combustion, are only formed on the oMMT surface [146]. Catalytic effect and more intensive char formation increase with organomodifier loading, proof that the organomodifier level determines the char content and cone calorimetry behavior. This has been described in detail by Song et al. [146] who investigated two organomodifier levels in PP.

There is a correlation between the change in the degradation pathway of a polymer by incorporation of nanoclay and the HRR peak as measured by cone calorimetry. As described by Jang et al. [147], polymer nanocomposites that show significant reduction in pHRR, indicating good fire retardancy, exhibit intermolecular reactions, such as interchain aminolysis/acidolysis (PA-6), radical recombination (PS), or hydrogen abstraction (EVA). On the other hand, polymers showing moderate reduction in HRR (SAN, ABS, PE, and PP) only present limited intermolecular reactions. When no significant HRR reduction was observed as in PAN and PMMA, no change in polymer degradation pathway was noticed.

Pastore et al. [149] demonstrate that the dispersion state of the clay particles in the polymer matrix has a remarkable effect on its flame retardancy characteristics. Examination of two different nanoclays (oMMT[80] and oMMT[150]) containing different amounts of ammonium salt bearing long alkyl chains (organomodifier) in PP revealed two distinct morphologies, intercalated and exfoliated, respectively, regardless of whether or not nanoclay dispersion was enhanced with PPgMA [146,149].

With PPgMA acting as a compatibilizer, improvement of the clay dispersion as evidenced by the increase in the degree of exfoliation produces a noticeable reduction in the HRR value for PP modified at two different levels, organomodified contents oMMT(80) and oMMT(150); but the reduction is more dramatic when a larger amount of organomodifier (oMMT[150]) is incorporated in the clay interlayer space [153]. This confirms that effectiveness of clay on reducing flammability depends on both clay dispersion and the amount of the organomodifier, with the organomodifier being responsible for enhancing robust formation of catalytic acid sites on the clay surface at high temperature. Increasing the nanoclay content of the polymer manifests in a sharp decrease in the HRR peak, whereas better dispersion of oMMT in the compatibilized PP matrix generates upon combustion a continuous MMT-rich char surface layer [146].

26.9.1.2 Polyamide PA-6/Layered Silicate Nanocomposites

In PA-6 clay nanocomposites, increased melt viscosity inhibits the formation of the melt state. The temperature at the surface of the nanocomposite rises more rapidly than with unfilled PA-6 because the solid state has a lower thermal conductivity than the melt state. This temperature rise increases the rate of fuel production and decreases the TTI [154]. The HRR plot versus time for PA-6 and a PA-6/clay (5 wt%) nanocomposite obtained at 35 kW/m^2 of heat flux from cone calorimetry by Gilman and coworkers [6,57,138,155] confirms that the barrier surface layer controls flame retardancy in PLSNs (Figure 26.2).

Kashiwagi et al. [140], Zanetti et al. [141], and Lewin [144] postulated similar mechanisms, arguing that clay accumulation on the material surface can result from silicate migration propelled by lower surface free energy relative to those of the carbon-based polymers or by bubbles bursting from the surface pushing clay particles to form island-like floccules in place of a continuous or carpet layer. Increasing viscosity with increasing clay loading can cause a reduction in pHRR by limiting the escape of decomposition products.

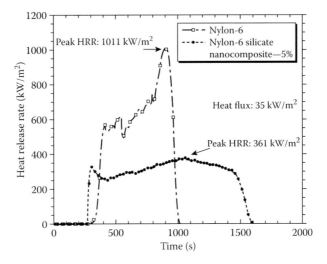

FIGURE 26.2 HRR plots of nylon-6 and nylon-6 nanocomposite (heat flux = 35 kW/m^2). (Reprinted from *Appl Clay Sci*, 15, Gilman, J.W., Flammability and thermal stability studies of polymer layered–silicate [clay] nanocomposites, 31–49, Copyright 1999, with permission from Elsevier.)

The catalytic role of clay in char formation has been confirmed by Vaia et al. [154] for PA-6/layered silicate nanocomposites, and the same is true for polymers that are not inherently char-forming including EVA copolymer [156], polypropylene [157], PE [158], PS [159], and ABS copolymer [160]. These are further discussed in Section 26.10 under FR synergism.

26.9.2 CNTs and Sepiolite

Nanofibrous materials such as CNTs and sepiolite have found application in polymer flame retardancy, the one widely studied among them being the CNTs. They represent an alternative material to the use of conventional FRs and nanoclays. CNTs are composed of carbon atoms arranged in hexagons and pentagons (graphite structure) forming cylinders. Two different types of CNTs are recognized: small-diameter (1–2 nm) single-walled nanotubes (SWNTs/SWCNTs) and larger-diameter (10–100 nm) multiwalled nanotubes (MWNTs/MWCNTs). The synthesis route could be by arc discharge, laser ablation, and thermal or plasma-enhanced chemical vapor growth deposition (CVD). Production via direct arc discharge and laser ablation requires very high temperatures and the use of a small quantity of metal catalyst. The CVD methods of production occur at lower temperatures but leave significant quantities of catalyst residues [13,161–164].

CNTs have a high aspect ratio enabling excellent dispersion, forming a network at very low loading in the polymer matrix resulting in significant improvement in mechanical [161], rheological [161–164], and FR [165–167] properties. At a low loading rate (<3 wt%), improvement in flame retardancy has been confirmed for many polymers including EVA [161], PS [164], PMMA [165], PA-6 [166], LDPE [167], and PP [168].

The effect of the nanotube dispersion state on the FR properties of PMMA nanocomposites was studied by Cipiriano et al. [164]. With proper dispersion in PMMA, addition of only 0.5 wt% of SWNT produced a substantial decrease in the HRR liberated over a much longer time range versus neat PMMA. In addition to dispersion, nanotube loading rate affects the flammability of the PMMA nanocomposites. When good nanotube dispersion is achieved, the addition of 0.1 wt% SWNT did not significantly reduce the HRR of the PMMA, but the addition of 0.2 wt% led to a 25% reduction. Further, the incorporation of 0.5 wt% of SWNT produced >50% reduction in HRR.

Using between 0.5 and 4.0 wt% of MWNT widely dispersed in PP, a significant decrease in pHRR was observed [166], with 1 wt% giving the lowest HRR (possibly due to a balance of thermal conductivity and heat shielding), but all leaving behind a residue with no holes or crack but rather a uniformly structured network.

There have been many excellent publications covering synergism in polymer-layered silicate–CNT nanocomposites including unique nanofiller synergist effects in EVA copolymer, PP, etc. [142,169–177].

26.9.3 Nanoscale Particulate Additives, LDH, and POSS

Nanotubes, nanoparticles of metal oxides, silica, LDHs and polyhedral oligomeric silsesquioxane (POSS), and spherical silica nanoparticles are defined by their isometric dimensions (they are characterized by three nanometric dimensions and sometimes called 0D nanoparticles) and exhibit significant FR characteristics.

26.9.3.1 Silsesquioxane

POSS is an inorganic silica-like nanocage ($[RSiO_{1.5}]_8$) surrounded by eight organic groups located at the corners that enhance its compatibility within organic polymers. They are sometimes called *preceramic compounds*. In combustion, POSS acts as a precursor forming thermally stable ceramic materials at high temperature.

There are two types of POSS with promising flame retardancy characteristics: one bearing eight identical R groups (R = methyl, phenyl, isobutyl, or isooctyl) and the other seven identical R groups and one functional R′ group such as an ester, silane, isocyanate, methacrylate, alcohol, epoxide, or amine. The (reactive) functionality of the R′ group improves the compatibility between the dispersed nanocages and the polymer matrix; it also enables chemical grafting of reactive polymer chains and the initiation of polymerization reactions from the POSS surface.

Kashiwagi et al. [165,168] showed the incorporation of POSS in polymers in addition to modifying both the viscosity and the mechanical properties of the molten polymer; it also affects the thermal stability and fire performances by reducing the quantity of heat released upon combustion. They reported that the incorporation of 10 wt% of methyl phenyl polysilsesquioxane in polytetramethylenylether-glycol-b-polyamide-12 (PTME-b-PA) results in a 70% decrease in pHRR during cone calorimetry tests carried out at 35 kW/m^2.

POSS can be used as a metal dispersing agent [153,171] by including a metal atom at one corner of the POSS nanocage structure. Finely dispersed metal-bearing POSS nanoparticles, at tiny concentrations (about 1 wt%), can markedly enhance the char yield in PP. This char formation is the result of catalysis of the dehydrogenation reaction by the metal-bearing POSS. Increase in flammability with metal-bearing POSS nanoparticles is possible, as revealed by lower LOI values [171,172].

Fina et al. [177,178] and Laachachi et al. [179] incorporated dimeric and oligomeric Al- and Zn-isobutyl silsesquioxane (Al- and Zn-POSS) in polypropylene and compared it with PP/octaisobutyl POSS. They found that the presence of Al-POSS led to a decrease in the HRR (decrease by 43% at 10 wt% POSS loading) as well as a reduction in CO and CO_2 production rates. Zn-POSS did not significantly affect the PP combustion behavior.

26.9.3.2 Nanometric Metallic Oxide Particles

Nanometric metallic oxide particles such as those of titanium oxide (TiO_2) and ferric oxide (Fe_2O_3) can have significant effects on the fire-retardant properties of thermoplastics. The effect of the incorporation of nanometric TiO_2 and Fe_2O_3 particles on the thermal stability and fire reaction of PMMA was examined by Laachachi et al. [180] and Fina et al. [143]. Adding just a small amount (5 wt%) of nanometric TiO_2 or Fe_2O_3 enhanced the thermal stability of PMMA nanocomposites [180]. The HRR values reported depend on the filler content and decrease at higher loadings. The incorporation of Fe_2O_3 and TiO_2 nanoparticles exhibited different behavior in cone calorimetric tests, even though the Fe_2O_3 nanoparticles had a comparable particle size and surface area (23 nm and 48 m^2/g) to TiO_2 (21 nm and 50 m^2/g). The pHRR was reduced by 50% and 37% with the incorporation of 20 wt%.

26.10 FR SYNERGIES

Several mechanisms for reducing polymer flammability are at play when nanoparticles such as organoclay, CNTs, or POSS are incorporated into a polymer matrix. These include limiting fuel transfer to the flame and formation of a barrier char layer. Unfortunately, polymer nanocomposites burn with very little reduction in total heat release and slightly improved TTI if at all. Consequently, nanoparticles have to be used in combination with other FR agents in a synergistic system in order to achieve the required fire performance levels.

The HRR of polymer nanocomposites is significantly reduced relative to the virgin polymers reflecting retardation of the flame spread in the case of developing fires [34,141,181–183]. They can as such be considered as being already flame retarded. There are, however, a number of disadvantages. Some of these are listed below:

- The total heat evolved is scarcely changed suggesting no improvement of the polymer performance in fully developed fires.
- Clay fillers, by promoting polymer degradation, may increase the probability of earlier ignition [40,183,184].
- PLSN, due to increasing polymer melt viscosity, does not perform satisfactorily in industrially important fire tests such as the LOI test or the UL94 test [40,41,185].

In order to achieve a high fire performance level and an acceptable environmental impact, it is necessary to develop a FR system based on a combination of different FR agents. One obvious approach is to combine clay fillers with conventional FRs to provide the so-called synergist systems. Many authors and researchers have adopted the concept of synergism as obvious route to optimize FR formulations and enhance the performance of mixtures of two or more additives. Simply stated, synergism is realized when the performance level due to a mixture of additives $xA + yB$ ($x + y = 1$) for a given property (P) is greater than that predicted for the linear combination ($xPA + yPB$) of the single effects of each additive (PA and PB). Conversely, antagonistic effects can be detected [13]. Synergism is used by many researchers to describe, in general, any enhanced performance of the mixture of two or more components when compared with the performance of the individual component separately in the same concentration.

In a nutshell, polymer flame retardancy can be achieved through one or more chemical and/or physical mechanisms taking place in either the gas or the condensed phase. It is nonetheless a complex process consisting of a number of individual stages with one or two dominating. The fact is that the selection of the ideal additive depends on many factors other than the mode of action. These factors include its stability, compatibility, tendency to migrate, effect on electrical performance together with toxicity, ability to cause corrosion, and ability to be colored [141,186–189]. Synergism is obtained by a combination of flame retardancy mechanisms, such as char formation by a phosphorated FR combined with a gas-phase action by a halogenated FR. It can also be realized by a combination of FR agents reinforcing the same mechanism, e.g., nanoclays and phosphorated FR agents, both acting in the condensed phase.

As an example, consider the gas-phase FR action of halogenated additives, improved by the incorporation of antimony oxide (Sb_2O_3). As described in the following reaction schemes, antimony oxide reacts with the hydracids (HCl or HBr) generated by the halogenated FRs to form antimony oxyhalides. The antimony oxyhalides are much heavier than the native hydracids, thus prolonging their residence time in the flame. These oxyhalides lead to the formation of $SbCl_3$ or $SbBr_3$, which act as scavengers of *hot* radicals such as H*:

$$SbCl_3 + H^* \rightarrow HCl + SbCl_2$$

$$SbCl_2 + H^* \rightarrow HCl + SbCl^*$$

$$SbCl + H^* \rightarrow HCl + Sb^*$$

In addition, antimony oxide and Sb* can also react by a parallel oxidation mechanism and participate in the scavenging of *hot* radicals [182]:

$$Sb^* + OH^* \rightarrow SbOH$$

$$SbOH^* + H^* \rightarrow SbO^* + H_2$$

$$SbO^* + H^+ \rightarrow SbOH$$

26.10.1 Polyolefin/Layered Silicate Nanocomposites

To reduce the ignitability of nanocomposites, decabromodiphenyl oxide (DB) was combined with antimony trioxide (AO). Halides comprise strong Lewis acid catalysts capable of promoting dehydrogenation-charring reactions operating as radical scavengers and forming a blanket that acts as a gas barrier between fuel gases and the condensed phase [46,190,191]. Zanetti et al. [192] evaluated the efficiency of the DB–AO system in poly(propylene-graft-maleic anhydride) (PP-g-MA) nanocomposites by means of calorimetric studies. Lee et al. [193] also investigated the flame-retarded PP/layered silicate nanocomposites using twin-screw extruder PP-g-MA as the compatibilizer between the matrix and the organoclay. In both studies, there was a significant drop in pHRR with the coaddition of both ingredients than otherwise. The flame-retarded polymer (PP-g-MA + 22 wt% DB + 6 wt% AO) exhibited weak and major bimodal-shaped HRR peaks (at 85 and 170 s, respectively). It was theorized that the synergy in the FR system manifests as a delayed release of the active gas-phase species evolved during combustion.

In another example, Ristolainen et al. [194] combined ATH with a nanocomposite. They improved the adhesion of the clay in the PP matrix by using either PP-g-MA or hydroxyl-functionalized PP (PP-co-OH) prepared with metallocene catalyst. At a total filler content of 30 wt%, partial substitution of ATH with 5 wt% led to the equal reduction of pHRR and THR for both compatibilizers.

26.10.2 POLYAMIDE/LAYERED SILICATE NANOCOMPOSITES

A self-extinguishing PA-6 (with a V-0 classification) was reported by Hu et al. [195]. The researchers used 5 wt% of organically modified MMT (oMMT) treated with a hexadecyl–trimethyl–ammonium salt, and obtained a partially intercalated/exfoliated structure undisturbed by the addition of the FRs. At 15 wt% DB + 5 wt% AO loading, the pHRR of the PA-6 matrix decreased from 1200 to 673 kW/m^2 in the presence of the clay filler, and to an even lower 390 kW/m^2 with the addition of the FR, confirming the synergistic effect of the mixture.

26.10.3 ABS/LAYERED SILICATE NANOCOMPOSITES

FR ABS/oMMT with the incorporation of the DB–AO system designed to pass the UL94 test was also reported by Hu et al. [195]. As for the PA-6 system, they dispersed 5 wt% of oMMT treated with a hexadecyl–trimethyl–ammonium salt in ABS and mixed it in a twin-roll mill with 15 wt% DB and 3 wt% AO. The sample with the mixture of clay filler and the FR system achieved a V-0 rating and elevated LOI value. The pHRR was also lowered by 78% compared with the pure ABS. Another example of the synergistic effect of combining nanoclay with conventional FR was provided by Ma et al. [196]. They substituted DB with a brominated epoxy resin (BER). BERs are more polar than ABS; they have greater affinity for clay nanoparticles facilitating the formation of exfoliated and more thermally stable structures. With a small amount of the halogenated compound (12 wt% BER + 4 wt% AO), the LOI of ABS containing 2 wt% clay was raised 20.5–31.4% compared to the 24 vol% required value for a V-0 rating [197–199]. The synergistic effect of clay and BER–AO and the combustion process are depicted in Scheme 26.16. As reported for the oMMT–DB–AO

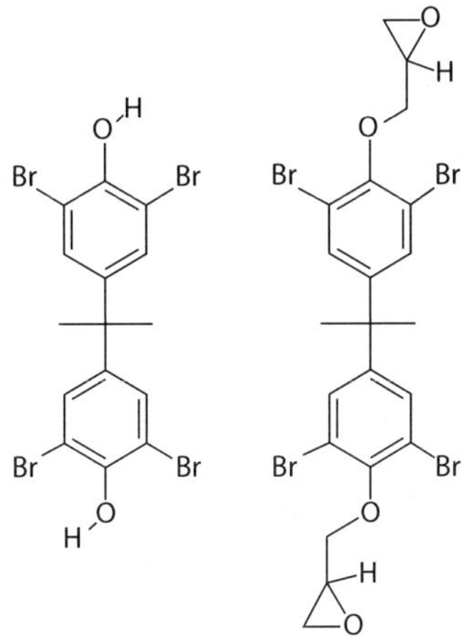

SCHEME 26.16 Chemical structure of BER. (Reprinted from *Fire Mater*, 16, Weil, E.D., Hirschler, M.M., Patel, N.G. and S. Shaki, Oxygen index: Correlations to other fire tests, 159–67, Copyright 1992, with permission from Elsevier.)

system, the oMMT–BER–AO system showed synergism as a result of the silicates forming barriers that hindered BER pyrolysis and reactions between BER–AO at lower temperatures.

26.10.4 PS/Layered Silicate Nanocomposites

Wang et al. [199] melt-blended PE, PP, and PS with 3 wt% of clay (Closite 30B, from Southern Clay Products, USA). They incorporated as FRs acrylic acid pentabromobenzyl ester (ACPB), methacrylate acid pentabromobenzyl ester (MEPB), butyric acid pentabromobenzyl ester (BUPB), and pentabromobenzyl ester polyacrylate (PBPA). In particular, they prepared flame-retarded styrene nanocomposites and nanocomposites of styrene and dibromostyrene copolymer using bulk polymerization. Two clay types, namely, fluorine-containing ammonium and diemethyl-n-hexadecyyl-4-vinylbenzyl ammonium-modified MMT, were used. Cone calorimetry results showed that with less than 6 wt% bromine, the flame retardancy of both the PS and the PP nanocomposites was significantly improved. Styrene nanocomposites bulk-polymerized in the presence of dibromostyrene showed the best result with a significant drop in the pHRR and achieved a V-0 rating.

26.10.5 Polyester/Layered Silicate Nanocomposites

Si et al. [200] melt-blended PMMA and commercial clay (Cloisite 20A) with DB–AO. Using TEM and ion mass spectroscopy, they confirmed that the incorporated clay was present in the polymer in intercalated and exfoliated form, which proved beneficial for the dispersion of the FR agents. PMMA containing 5 wt% clay, 20 wt% DB, and 5 wt% AO exhibited improved flame retardancy.

26.10.6 Layered Silicate Nanocomposites with Phosphorus-Based Compounds

There are many examples in the literature of synergy between nanocomposites and phosphorus-containing compounds. Volatile phosphorous compounds (PO*, PO_2^*, and HPO*) are considered to act in the gaseous phase to trap free radicals. They generate acids that form a thin glassy or liquid coating on the condensed phase, limiting oxygen diffusion, heat, and mass transfer. Phosphorous compounds promote the development of a carbon layer (char) on the polymer's surface. They form anhydrides of phosphoric and related acids, which act as dehydrating agents. The dehydrating reactions result in the formation of double bonds leading to cross-linked or carbonized structures [13,188,190,201].

PP nanocomposite was melt-blended with N-imidazol-O-(bicycle pentaerythritol phosphate)-O-(ethyl methacrylate phosphate) (PEBI) to determine the influence of the thermal stability and combustion characteristics of the latter. The insertion of the bulky PEBI increased the interlayer spacing of silicates into the galleries, enhancing the thermal stability of the PP matrix. Dynamic mechanical analysis (DMA) testing confirmed that, in addition to improving dispersion, PEBI acts as an internal plasticization agent and shifts the glass transition temperature (T_g) of the blend lower. PEBI enhanced the flame retardancy of the nanocomposites, improving all the usually measured fire properties.

Polyamide 66 (PA-66) was extrusion-blended with RP and an oMMT by Hao et al. [202] to prepare a FR nanocomposite. In one formulation—containing 5 wt% clay, 10 wt% RP—the LOI of the PA-66 nanocomposites increased from 25.5 to 30.4 vol% with no deleterious effect on mechanical properties. Another formulation that had 15 wt% of RP achieved a UL94/V-0 rating indicating synergism between the RP and clay particles at some loading levels. They also showed that silicates synergize well with a phosphate FR (PFR) to reduce the ignitability of a polyamide (PA-6).

Organic phosphorus-based compounds are used to enhance the flame retardancy of styrenic polymers [189]. Chigwada and Wilkie [203] examined a series of PS/clay nanocomposites including many with phosphorus-containing compounds for flame retardancy potentials. Further tests including cone calorimeter and UL94 were conducted for the promising candidates with 15 wt% FR and 3 wt% organoclay filler (Cloisite 10A, Southern Clay Products). The authors confirmed that lack of ignition was achieved only for those containing TCP, trixylyphosphate (TXP), or RDP. Reductions in pHRR and THR were proportional to the amount of added phosphate in the nanocomposite blends.

Further evidence of synergistic effects of phosphorus-based compounds with layered silicate nanocomposite is demonstrated by the works of Kim et al. [204]. Just as was done in the PS/clay nanocomposite, these authors intercalated TTP in the galleries of a commercial oMMT (Cloisite 30B, Southern Clay Products). The resulting clay was melt-blended with ABS to formulate an ABS/layered silicate nanocomposite. Shielding of the volatile TPP inside the silicates suppresses its evaporation, a scheme that permits a more efficient flame retardancy performance and an even wider range of processing conditions. The addition of the clay alone enhances the thermal stability of the ABS only slightly (the LOI remaining almost unchanged) by reason of the delayed release of the TPP. An increase in the LOI was achieved by the incorporation of an epoxy novolac system. Further improvement in thermal stability was obtained when silane agents, which promote coupling of the epoxy and the silicates, were added. An 85/9/6 wt% (ABS/epoxy/clay-TPP + silane) formulation has an LOI of 41.2 vol% versus 20.2 vol% for one 15 wt% TPP, which is just 9% above the LOI value for neat ABS polymer (18.2 vol%). As revealed by an optical micrograph of the residue, the enhancement in thermal stability is related to the formation of a more consistent char, devoid of holes and crevices. A similar synergistic effect is obtained when the TPP is replaced with a tetre-2-6-dimethylphenyl RDP (DMP-RDP).

TPP and layered silicates were solution-blended with PMMA by Kim and Wilkie [205] to produce a transparent yet FR product. XRD patterns and TEM images confirmed the occurrence of exfoliated nanocomposites. As the clay content increases, the optical transmittance measured by a UV-vis spectrophotometer diminishes. The combined use of silicates and TPP produces an optimal performance in fire properties of PMMA than when the components are incorporated individually.

26.10.7 SYNERGISTIC HALOGENATED/PHOSPHORATED MIXTURE

Synergistic effects can also be obtained by combining the gas-phase action of halogen species with the condensed-phase action of phosphorus-based compounds. Improved fire performances can thus be achieved by the use of mixtures of halogenated and phosphorated FR compounds, or by incorporation of substances containing both phosphorus and halogen groups in their molecular structure. Indeed, phosphorus halides or oxyhalides are excellent free-radical scavengers, better than hydrogen halides (HX), and can release more halogen-based radicals due to the P–X bond being weaker than the C–X bond. In addition, the phosphorus contained in phosphorus halides or oxyhalides can also act in the condensed phase to promote the formation of a protective char layer.

26.10.8 SYNERGISTIC NITROGENATED AND PHOSPHOROUS MIXTURES

For instance, the use of nitrogenated additives combined with phosphorus FRs can lead to surprising synergistic effects [206]. Formation of phosphorus–nitrogen intermediates can accelerate the *in situ* production of phosphoric acid and therefore polymer phosphorylation. It is known that P–N bonds are more reactive than P–O bonds in the phosphorylation process. By maintaining phosphorus in the condensed phase, they produce cross-linked networks that promote more intensive char formation. Phosphorus–nitrogen synergism is not a given but depends on the nature of the phosphorus and nitrogen FRs, as well as the chemical structure of the polymer matrix.

26.10.9 SYNERGISTIC METAL HYDROXIDE/METAL BORATE MIXTURES

As synergistic agents for metal hydroxides in polyolefin matrices, metal borates and particularly zinc borates have been successfully used. The endothermic effect of metallic hydroxides is improved by the use of other FR systems in order to preserve the FR effect beyond 4000°C and decrease the total filler loading rate. It was reported [207,208] that a partial substitution of 3 wt% of

MDH by zinc borate ($2ZnO \cdot 3B_2O_3 \cdot H_2O$) in EVA (8 wt% vinyl acetate at a global filler percentage of 40 wt%) increases the LOI value from 38.5% to 43% and decreases the pHRR recorded by cone calorimetry. The thermal decomposition of MDH in this system catalyzes the decomposition of zinc borate, generating boron oxide at lower temperature and forming a vitreous layer in combination with MgO at the material surface [208]. Combining both fillers triggers a physical effect equivalent to the formation of an expanded vitreous layer that proves to be more mechanically resistant than the crust formed when MDH is used alone. To improve the physical barrier effect of the protective layer formed, one can add lamellar talc particles to MDH and zinc borate in ternary systems [209]. A significant reduction of the pHRR value can be achieved also by using a combination of a small fraction of oMMT and talc together with magnesium hydroxide [210].

26.10.10 Synergistic Metallic Oxide/Phosphorated Mixtures

Laachachi et al. [179,180] combined the FR action of nanometric metallic oxides (TiO_2, Al_2O_3) with the char formation induced by phosphorated FR systems (APPs and phosphinates) in PMMA. In the case of aluminum phosphinate supplied by Clariant under the trade name Exolit OP930 (hereafter denoted phosphinate), cone calorimeter results showed that partial substitution of phosphinate by alumina nanoparticles promoted synergistic effects, with a marked decrease in pHRR, whereas no significant effect could be achieved with TiO_2 nanoparticles.

Although Al_2O_3 and TiO_2 promote positive FR effects in PMMA, their combination with phosphinate does not automatically lead to a synergistic effect. Observation of residues involving alumina nanoparticles essentially shows a continuous solid layer, as typically illustrated by the Al_2O_3 9 wt%–phosphinate 6 wt% sample. However, with TiO_2–phosphinate combinations, the char residues do not cover the entire sample surface, leading to a poor barrier effect, which explains the limited performances displayed by these compositions. It has been shown that there is no chemical reaction between Al_2O_3 (or TiO_2) and phosphinate, but only the formation of a vitreous layer, promoted by the phosphorated compound and reinforced by alumina particles. It appears that, in addition to their role in char reinforcement, alumina particles also have a positive catalytic effect on the formation of the protective layer with phosphinate (not provided by titanium oxide particles).

As mentioned earlier, depending on their nature and geometry, nanoparticles can contribute in various ways to improving the FR performances of polymer materials. In this respect, a combination of two different types of nanoparticles can be expected to promote some synergetic FR effects. For instance, Beyer [10,153,174,175] combined organomodified silicates (Nanofil 15) with MWNTs in EVA. The evolution of the HRR measured by cone calorimeter tests showed a synergistic effect when 2.5 phr organomodified silicates (MMT) were combined with 2.5 phr MWNTs compared to EVA containing either 5 phr MMT or 5 phr MWNT. Moreover, the combination of both nanoclay and MWNT has been shown to improve the cohesion of the residues recovered, which are much less cracked than those isolated.

26.11 CONCLUSIONS

Incorporation of FR additives expands the use of a small number of commodity polymers into a large number of applications and products that meet desired properties and performance. This is particularly so in a world where plastic materials are being increasingly used among many other products in transportation and buildings, and where fire prevention remains a major issue. There is a broad range of FR systems that are in current use, and many promising systems have been studied or are currently under development.

Early FRs were halogen-based (starting from the aliphatic bromine, aromatic bromine, and aliphatic chlorine to the aromatic chlorine for the highest temperature decomposition range). They functioned in the gaseous phase, suppressing the flame by replacing the high-energy free radicals responsible for flame propagation with more stable species, such as Cl(-) and Br(-). Though easy to use, they suffer from incomplete combustion and release of smoky and highly toxic fire effluents, leading to serious environmental concerns and strict regulations.

These halogen-based fire retardants are in the process of being banned due to health and environmental concerns. Nonhalogenated fire-retardant additives, such as phosphorus- and nitrogen-based compounds, are proving useful, especially in matrices containing oxygen or nitrogen atoms in their backbone. Some silicon-based additives also appear to provide efficient solutions. Relatively low amounts of nanoparticles or the synergistic effects of FRs from various families show very promising results as replacement for the halogen-based fire retardants.

The mere incorporation of nanoparticles in the polymer matrix is unfortunately insufficient in providing adequate fire resistance to meet the required standards in many cases. Using nanoparticles in synergistic combination with other FR systems such as phosphorated compounds can provide a system that can meet the stringent FR standards.

The predominant fire retardancy mechanisms in polymer/clay nanocomposites can be attributed to the formation of a heat barrier and a physical barrier against volatiles as a result of the migration of the clay nanolayer toward the material surface, accompanied by char formation and increased melt viscosity for exfoliated nanocomposites. By modifying the fire properties of the nanocomposites, these mechanisms sometimes improve flame retardancy, and at other times, they worsen them depending on the type of measurements and fire tests employed.

Considering the incredibly large thermoplastic matrices in use, flammability and flame retardancy will continue to present interesting scientific challenges and opportunities. The availability of fire-testing techniques such as cone calorimetry, TGA-EGA, or gasification studies will move research forward from empirical science into definable knowledge, profitable products, and applications.

The stability of clay's organic filler needs further studies. In a number of current investigations, an alkyl-ammonium cation is used to generate compatibility between the clay filler and the polymer matrix. Alkyl ammoniums, though successfully employed to synthesize polymer nanocomposites, decompose above 200°C to products that catalyze the degradation of the polymer matrix and promote earlier ignition. Fire resistance is negatively affected by the destruction of the polymer–clay interface and the rearrangement of the silicates to yield microstructures.

There are a number of issues that require further investigations:

- A broad quantitative relationship of cone calorimetry with UL94 and LOI test
- A theoretical foundation to describe in a quantitative manner a correlation between the reduction in pHRR and the dispersion of silicates
- A clarification of the influence of the reactive sites residing on silicates as to whether they are inert during fire or whether they react with the FR
- Stability of clay's organic modifier
- The impact of FRs and nanoclays on cost effectiveness and other key properties of thermoplastics such as the mechanical, viscoelastic, and long-term durability
- Sustainability, environmental concern, human health, and safety risks associated with nanocomposites-based and synergistic fire-retardant formulations

In order to develop sustainable nanocomposites-based FR additives, it is critical to concentrate on using nanoparticles where a significant amount of favorable EH&S data is already available. Promising nanoparticles include LDH [aluminum hydroxide, magnesium hydroxide, and carbonate (hydrotalcite)] and cellulose nanofibrils, both of which are approved by the US Food and Drug Administration for contact food.

Invariably, further research into the mechanisms by which FRs function is still needed. As a case in point, DuPont's Kevlar and Nomex (meta- and para-substituted aromatic rings) exhibit remarkable changes in heat release capacity as a result of minor variation in structure, yet we do not fully understand why.

REFERENCES

1. Beach, E.S., Weeks, B.R., Stern, R., and Anstas, P.T. 2013. Plastic additives and green chemistry. *Pure Appl Chem* 85(8):1611–1624.
2. Fink, J.K. 2010. *A Concise Introduction to Additives for Thermoplastic Polymers*. Wiley InterScience, New York City.
3. Zweifel, H. ed. 2001. *Plastic Additive Handbook*, 5th ed. Hanser, Munich.
4. Gächter, R. 1993. *Plastics Additives Handbook: Stabilizers, Processing Aids, Plasticizers, Fillers, Reinforcements, Colorants for Thermoplastic*. Hanser, Munich.
5. Troitzsch, J.H. 1990. Fire retardant plastics. In: *International Plastics Flammability Handbook: Principles, Regulations, Testing and Approval*, 2nd ed., ed. J.H. Troitzsch. Hanser, Munich, pp. 43–62.
6. Gilman, J.W. 1999. Flammability and thermal stability studies of polymer layered–silicate (clay) nanocomposites. *Appl Clay Sci* 15:31–49.
7. Zhang, H. 2004. *Fire-Safe Polymers and Polymer Composites, US Department of Transportation*. Federal Aviation Administration, Office of Aviation Research, Washington, DC.
8. Adanur, S. 1995. Applications of coated fabrics in building structures. In: *Wellington Sears Handbook of Industrial Textiles*, ed. S. Adanur. Technomic Publishing, Lancaster, PA, pp. 205–215.
9. Horrocks, A.R. and Price, D. 2001. *Fire Retardant Materials*. CRC Press, Boca Raton, FL.
10. Beyer, A.G. 2002. Nanocomposites: A new class of flame retardants for polymers. *Plast Addit Compd* 6: 22–28. Oxford University Press, Oxford, UK.
11. Pal, G. and Macskasy, H. 1991. *Plastics: Their Behavior in Fires*. Elsevier, New York.
12. Kiliaris, P. and Papaspyrides, C.D. 2010. Polymer/layered silicate (clay) nanocomposites: An overview of flame retardancy progress. *Polym Sci* 35:902–958.
13. Laoutid, F., Bonnaud, L., Alexandre, M., Lopez-Cuesta, J.-M., and Dubois, P. 2009. New prospects in flame retardant polymer materials: From fundamentals to nanocomposites. *Mater Sci Eng R: Rep*, 63(3):100–125.
14. Hull, T.R., Witkowski, A., and Hollingbery, L. 2011. Fire retardant action of mineral fillers. *Polym Degrad Stabil* 96(8):1462–1469.
15. Paul, K.T. 1989. Feasibility study to demonstrate the potential of smoke hoods in simulated aircraft fire atmospheres: Development of the fire model. *Fire Mater* 14:43–58.
16. Lebek, K., Hull, T.R., and Price, D. 2005. Products of burning rigid PVC burning under different fire conditions. In: *Fire and Polymers: Materials and Concepts for Hazard Prevention*, ACS Symposium Series No. 922. Oxford University Press, Oxford, UK, pp. 334–347.
17. Huggett, C. and Levin, B.C. 1987. Toxicity of the pyrolysis and combustion products of poly(vinyl chlorides): A literature assessment. *Fire Mater* 11:131–142.
18. de Wit, C.A. 2002. An overview of brominated flame retardants in the environment. *Chemosphere* 46(5):583–624.
19. Shaw, S.D., Blum, A., Weber, R. et al. 2010. Halogenated flame retardants: Do the fire safety benefits justify the health and environmental risks? *Rev Environ Health* 25(4):261–305.
20. Hull, T.R., Price, D., Liu, Y., Wills, C.L., and Brady, J. 2003. An investigation into the decomposition and burning behavior of ethylene-vinyl acetate copolymer nanocomposite materials. *Polym Degrad Stabil* 82(2):365–371.
21. Rothon, R.N. and Hornsby, P.R. 1996. Flame retardant effects of magnesium hydroxide. *Polym Degrad Stabil* 54(2–3):383–385.
22. Rigolo, M. and Woodhams, R.T. 1992. Basic magnesium carbonate flame retardants for polypropylene. *Polym Eng Sci* 32(5):327–334.
23. Price, D., Pyrah, K., Hull, T.R. et al. 2002. Flame retardance of poly(methyl methacrylate) modified with phosphorus containing compounds. *Polym Degrad Stabil* 77(2):227–233.
24. Bourbigot, S., Le Bras, M., Leeuwendal, R., Shen, K.K., and Schubert, D. 1999. Recent advances in the use of zinc borates in flame retardancy of EVA. *Polym Degrad Stabil* 64(3):419–425.
25. Hull, T.R., Stec, A.A., and Nazare, S. 2009. Fire retardant effects of polymer nanocomposites. *J Nanosci Nanotechnol* 9(7):4478–4486.
26. Schmaucks, G., Friede, B., Schreiner, H., and Roszinski, J.O. 2009. Amorphous silicon dioxide as additive to improve the fire retardancy of polyamides, Chapter 3. In: *Fire Retardancy of Polymers-New Strategies and Mechanisms*, eds. T.R. Hull and B.K. Kandola. Royal Society of Chemistry, Cambridge, pp. 35–48.
27. Bourbigot, S., Samyn, F., Turf, T., and Duquesne, S. 2010. Nanomorphology and reaction to fire of polyurethane and polyamide nanocomposites containing flame retardants. *Polym Degrad Stabil* 95(3):320–326.

28. Li, Z.Z., and Qu, B.J. 2003. Flammability characterization and synergistic effects of expandable graphite with magnesium hydroxide in halogen-free flame retardant EVA blends. *Polym Degrad Stabil* 81(3):401–408.
29. Wang, D.Y., Liu, Y., Wang, Y.Z. et al. 2007. Fire retardancy of a reactively extruded intumescent flame retardant polyethylene system enhanced by metal chelates. *Polym Degrad Stabil* 92(8):1592–1598.
30. Wang, D.Y., Liu, Y., Ge, X.G. et al. 2008. Effect of metal chelates on the ignition and early flaming behavior of intumescent fire-retarded polyethylene systems. *Polym Degrad Stabil* 93(5):1024–1030.
31. BBC Research LLC. 2013. *Flame Retardant Chemicals: Technologies and Global Markets (CHM014L)*. Wellesley, MA.
32. Stewart, R. 2009. Flame retardant additives and materials, *Plastics Engineering*, pp. 18–26.
33. Mack, A.G. 2004. Flame retardants, halogenated. In: *Kirk-Othmer Encyclopedia of Chemical Technology*, vol. 11, ed. A. Seidel. John Wiley & Sons Inc., New York City, pp. 454–483.
34. Kandola, B.K. 2001. Nanocomposites. In: *Fire Retardant Materials*, eds. A.R. Horrocks and D. Price. Woodhead Publishing, Cambridge, pp. 204–219.
35. Schartel, B., Bartholmai, M., and Knoll, U. 2006. Some comments on the main fire retardancy mechanisms in polymer nanocomposites. *Polym Adv Technol* 17:772–777.
36. Schartel, B. and Hull, T.R. 2007. Development of fire-retarded materials: Interpretation of cone calorimeter data. *Fire Mater* 31:327–354.
37. Fenimore, C.P. and Martin, F.J. 1966. Effect of the network structure on the flammability of cis-1,4-polyisoprene vulcanizates. *Mod Plast* 44:41–48.
38. Lyon, R.E. and Janssens, M.L. 2005. Flammability. In: *Encyclopedia of Polymer Science and Technology*, vol. 6, ed. H.F. Mark. John Wiley & Sons Inc., New York City, pp. 1–73.
39. Wilkie, C.A. 2005. An introduction to the use of fillers and nanocomposites in fire retardancy. In: *Fire Retardancy of Polymers: New Applications of Mineral Fillers*, eds. M. Le Bras, C.A. Wilkie, and S. Bourbigot. Royal Society of Chemistry, Cambridge, pp. 1–15.
40. Underwriters Laboratories. 1997. *UL94-Test for Flammability of Plastic Materials for Parts in Devices and Appliances*. Underwriters Laboratories Inc., Northbrook, IL.
41. Babrauskas, V. and Peacock, R.D. 1992. Heat release rate: The single most important variable in fire hazard. *Fire Safety J* 18:255–272.
42. Huggett, C. 1980. Estimation of rate of heat release by means of oxygen consumption measurements. *Fire Mater* 4:61–65.
43. Betts, K.S. 2008. New thinking on flame retardants. *Environ Health Perspect* 116:210.
44. Zaikov, G.E. and Lomakin, S.M. 2002. Ecological issue of polymer flame retardancy. *J Appl Polym Sci* 86:2449–2462.
45. U.S. Environmental Protection Agency. 2005. Furniture flame retardancy partnership: Environmental profiles of chemical flame-retardant alternatives for low-density polyurethane foam. *HDP Halogen Free Guideline*. Available at http://www.hdpug.org/content/publication-0 (accessed February 28, 2015).
46. Chao, C.Y.H. and Wang, J.H. 2001. Comparison of the thermal decomposition behavior of a nonfire retarded and a fire retarded flexible polyurethane foam with phosphorus and brominated additives. *J Fire Sci* 19:137–156.
47. Weil, E.D., Ravey, M., Keidar, I., and Gertner, D. 1996. Flame retardant actions of tris(1,3-dichloro-2-propyl) phosphate in flexible urethane foam. *Phosphorus Sulfur Silicon Relat Elem* 110:87–91.
48. Price, D., Pyrah, K., Hull, T.R. et al. 2000. Ignition temperatures and pyrolysis of a flame-retardant methyl methacrylate copolymer containing ethyl(methacryloyloxymethyl)-phosphonate units. *Polym Int* 49:1164–1168.
49. Green, J. 2000. Phosphorus containing flame retardants. In: *Fire Retardancy of Polymeric Materials*, eds. A. Wilkie and A.F. Grand. Marcel Dekker, New York, p. 147.
50. Ravey, M., Keidar, I., Weil, E.D., and Pearce, E.M. 1998. Flexible polyurethane foam. II. Fire retardation by tris(1,3-dichloro-2-propyl) phosphate part A. Examination of the vapor phase (the flame). *J Appl Polym Sci* 68:217–229.
51. Ravey, M. and Pearce, E.M. 1999. Flexible polyurethane foam. III. Phosphoric acid as a flame retardant. *J Appl Polym Sci* 74:1317–1319.
52. Wei, P., Li, H.X., Jiang, P.K., and Yu, H.Y. 2004. An investigation on the flammability of halogen free fire retardant PP-APP-EG systems. *J Fire Sci* 22:367–377.
53. Shen, K.K. 2001. Zinc Borates: 30 years of successful development as multifunctional fire retardants. In: *Fire and Polymers—Materials and Solutions for Hazard Prevention*, eds. G.L. Nelson and C.A. Wilkie. American Chemical Society, Washington, DC, p. 228.

54. Gilman, J.W. and Kashiwagi, K. 2000. Silicon-based flame retardants. In: *Fire Retardancy of Polymeric Materials*, eds. A.F. Grand and C.A. Wilkie. Marcel Dekker, New York, p. 353.
55. Gilman, J.W., Kashiwagi, T., and Lichtenhan, J.D. 1997. Nanocomposites: A revolutionary new flame retardant approach. *SAPME J* 33:40–46.
56. Alexandre, M. and Dubois, P. 2000. Polymer-layered silicate nanocomposites: Preparation, properties and uses of a new class of materials. *Mater Sci Eng R-Rep* 28:1–63.
57. Gilman, J.W. 2007. Flame retardant mechanism of polymer–clay nanocomposites. In: *Flame Retardant Polymer Nanocomposites*, eds. A.B. Morgan and C.A. Wilkie. John Wiley & Sons Inc, Hoboken, NJ, pp. 67–87.
58. Karn, B. and Aguar, P. 2007. Nanotechnology and life cycle assessment. In: *Synthesis of Results Obtained at a Workshop*, Washington, DC, October 2–3, 2006.
59. Lloyd, S.M. and Lave, L.B. 2003. Life cycle economic and environmental implications of using nanocomposites in automobiles. *Environ Sci Technol* 37:3458–3466.
60. Andersson, P., Simonson, M., and Stripple, H. 2003. Fire safety of upholstered furniture, A life-cycle assessment—Summary Report, SP 22.
61. Simonson, M., Boldizar, A., Tullin, C., Stripple, H. and Sundqvist, J.O. 1998. The incorporation of fire considerations in the life-cycle assessment of Polymeric composite materials, A Preparatory Study. *SP Report* 25 pp.
62. Simonson, M. Blomqvist, P., Boldizar, A., Möller, K., Rosell, L., Tullin, C., Stripple, H. and Sundqvist, J.O. 2003. Fire-LCA Model: TV case study, SP Report 13.
63. Underwriters Laboratories. 1996. UL-94: Test for flammability of plastic materials for parts in devices and applications [Includes ASTM D635-98 (UL-94 HB), ASTM D3801-96 (UL-94 V), ASTM D4804-98 (UL-94 VTM), ASTM D5048-97 (UL-94 5V), ASTM D4986-98 (UL-94 HBF)].
64. Lan, T., Psihogios, V., Bagrodia, S., Germinario, L.T., and Gilmer, J.W. 2004. Intercalates and exfoliates thereof having an improved level of extractable material, US Patent: 6,828,370 B2, Issued Date: December 12, 2004.
65. Zammarano, M., Franceschi, M., Bellayer, S., Gilman, J.W., and Meriani, S. 2005. Preparation and flame resistance properties of revolutionary self-extinguishing epoxy nanocomposites based on layered double hydroxides. *Polymer* 46(22):9314–9328.
66. Zammarano, M., Kramer, R.H., Harris, R. et al. 2008. Flammability reduction of flexible polyurethane foams via carbon nanofiber network formation. *Polym Adv Technol* 19:588–595.
67. Kashiwagi, T., Wu, M., Winey, K., Cipriano, B.H., Raghavan, S.R., and Douglas, J.F. 2008. Relation between the viscoelastic and flammability properties of polymer nanocomposites. *Polymer* 49:4358.
68. Kind, D.J. and Hull, T.R. 2012. A review of candidate fire retardants for polyisoprene. *Polym Degrad Stabil* 97:201–213.
69. Price, D., Pyrah, K., Hull, T.R. et al. 2001. Flame retarding poly (methyl methacrylate) with phosphorus-containing compounds: Comparison of an additive with a reactive approach. *Polym Degrad Stabil* 74(3):441–447.
70. National Research Council (U.S.). 1977. *Materials: State of the Art, Fire Safety Aspects of Polymeric Materials*. National Academy of Sciences, Washington, DC.
71. Cusack, P. and Perrett, T. 2006. The EU RoHS directive and its implications for the plastics industry. *Plast Addit Compd* 8(3):46–49.
72. European Chemicals Agency (ECHA). 2011. Candidate list of substances of very high concern for authorization (SVHC). Available at http://echa.europa.eu/home_en.asp.
73. Lutz, J.T. 1989. *Thermoplastic Polymer Additives: Theory and Practice*. New York, Marcel Dekker.
74. Hull, T.R., Quinn, R.E., Areri, I.G., and Purser, D.A. 2002. Combustion toxicity of fire retarded EVA. *Polym Degrad Stabil* 77(2):235–242.
75. Rothon, R.N. 2003. Effects of particulate fillers on flame retardant properties of composites, Chapter 6. In: *Particulate-Filled Polymer Composites*, 2nd ed., ed. R.N. Rothon. Rapra Technology, Shawbury, UK.
76. Horn, W.E. 2000. Inorganic hydroxides and hydroxycarbonates: Their function and use as flame retardants, Chapter 9. In: *Fire Retardancy of Polymeric Materials*, eds. A. Grand and C.A. Wilkie. Marcel Dekker, New York.
77. Stoliarov, S.I., Safronava, N., and Lyon, R.E. 2009. The effect of variation in polymer properties on the rate of burning. *Fire Mater* 33(6):257–271.
78. Khalturinskii, N.A. and Berlin, A.A. 1990. On reduction of combustibility of polymeric materials. *Int J Polym Mater* 14(1–2):109–125.
79. Shomate, C.H. 1944. High-temperature heat contents of magnesium nitrate, calcium nitrate and barium nitrate. *J Am Chem Soc* 66:928–929.

80. NIST chemistry webbook. Available at http://webbook.nist.gov/chemistry/.
81. Hollingbery, L.A. and Hull, T.R. 2010. The thermal decomposition of huntite and hydromagnesite: A review. *Thermochim Acta* 509(1–2):1–11.
82. Hollingbery, L.A. and Hull, T.R. 2010. The fire retardant behavior of huntite and hydromagnesite: A review. *Polym Degrad Stabil* 95(12):2213–2225.
83. Hornsby, P.R. and Watson, C.L. 1990. A study of the mechanism of flame retardance and smoke suppression in polymers filled with magnesium hydroxide. *Polym Degrad Stabil* 30:73–87.
84. Hornsby, P.R. and Watson, C.L. 1998. Mechanism of smoke suppression and fire retardancy in polymers containing magnesium hydroxide filler. *Plast Rub Proc Appl* 11:45–51.
85. Hornsby, P.R. and Watson, C.L. 1986. Magnesium hydroxide combined flame retardant and smoke suppressant filler for thermoplastics. *Plast Rub Proc Appl* 6:169–175.
86. Woycheshin, E.A. and Sobolev, I. 1975. Effect of particle size on the performance of alumina hydrate in glass-reinforced polyesters. *J Fire Flammability/Fire Retard Chem* 2:224–241.
87. Cross, M.S., Cusack, P.A., and Hornsby, P.R. 2003. Effects of tin additives on the flammability and smoke emission characteristics of halogen-free ethylene-vinyl acetate copolymer. *Polym Degrad Stabil* 79:309.
88. Shen, K.K. 2003. Zinc borates as multifunctional fire retardants in halogen free polyolefins. In: *Proceedings of the 14th Annual BCC Conference on Flame Retardancy*.
89. Shen, K.K. 1985. Firebrake® Zinc Borate the Unique Multifunctional Additive, *Plastics Compounding*, September/October, pp. 66–80.
90. Delfosse, L.C., Baillet, A., and Braul, D. 1989. Combustion of ethylene vinyl-acetate copolymer filled with aluminum and magnesium hydroxide. *Polym Degrad Stabil* 23:337.
91. Utamapanya, S., Klabunde, K.J., and Schlup, J.R. 1991. Nanoscale metal oxide particles/clusters as chemical reagents. Synthesis and properties of ultrahigh surface area magnesium hydroxide and magnesium oxide. *Chem Mater* 3:175.
92. Ding, Y., Zhang, G., Wu, H., Hai, B., Wang, L., and Qian, Y. 2001. Nanoscale magnesium hydroxide and magnesium oxide powders: Control over size, shape, and structure via hydrothermal synthesis. *Chem Mater* 13:435.
93. Henrist, C., Mathieu, J.-P., Vogels, C., Rulmont, A., and Cloot, R. 2003. Morphological study of magnesium hydroxide nanoparticles precipitated in dilute aqueous solution. *J Cryst Growth* 249:321.
94. Qiu, L., Xie, R., Ding, P., and Qu, B. 2003. Preparation and characterization of $Mg(OH)_2$ nanoparticles and flame-retardant property of its nanocomposites with EVA. *Compos Struct* 62:391–395.
95. Sawada, Y., Yamaguchi, J., Sakurai, O., Uematsu, K., Mizutani, N., and Kato, M. 1979. Thermal atmospheres decomposition of basic magnesium carbonates under high-pressure gas. *Thermochim Acta* 33:127.
96. Haurie, L., Fernandez, A.I., Velasco, J.I., Chimenos, J.M., Lopez-Cuesta, J.-M., and Espiell, F. 2006. Synthetic hydromagnesite as flame retardant: Evaluation of the flame behaviour in a polyethylene matrix. *Polym Degrad Stabil* 91:989.
97. Haurie, L., Fernandez, A.I., Velasco, J.I., Chimenos, J.M., Tico-Grau, J.R., and Espiell, F. 2005. Synthetic hydromagnesite as flame retardant. A study of the stearic coating process. *Macromol Symp* 221:165.
98. Levchik, S.V. 2007. Introduction to flame retardancy and polymer flammability. In: *Flame Retardant Polymer Nanocomposites*, eds. A.B. Morgan and C.A. Wilkie. John Wiley & Sons, Hoboken, NJ, pp. 1–29.
99. Price D., Anthony G., and Carty P. 2001. Introduction: Polymer combustion, condensed phase pyrolysis and smoke formation. In: Horrocks AR, Price D, editors. *Fire Retardant Materials*. Cambridge, UK: Woodhead Publishing Ltd. pp. 1–30.
100. Zutty, N.L. and Welch, F.J. 1960. The mechanism of vinyl polymerization initiated by metal alkyls: Copolymerization studies. *J Polym Sci* 43:445–452.
101. Zutty, N.L. and Welch, F. 1963. Synthesis of vinyl polymers containing α-substituted γ-butyrolactone groups in their backbones, *J Polymer Sci* A-1:2289–2297.
102. Janovic, Z. 1999. Brominated copolymers of reduced flammability. *Polym Degrad Stab* 64:479.
103. Aronson, A.M. 1992. *Phosphorus Chemistry: Developments in American Science*. American Chemical Society, Washington, DC, p. 218.
104. Weil, E.D. 1986. *Encyclopedia of Polymer Science and Technology*. Wiley Interscience, New York, p. 11.
105. Babushok, V. and Tsang, W. 2000. Inhibitor rankings for alkane combustion. *Combust Flame* 124:488.
106. Camino, G. and Frache, J. 2012. 13th International conference on fire retardant polymer materials, FRPM 11, Alessandria, Italy, June 26–30, 2011. Introduction. *Polym Degrad Stab* 97(12):2480.

107. Davis, J. and Huggard, M. 1996. The technology of halogen-free flame retardant phosphorus additives for polymeric systems. *J Vinyl Addit Technol* 2:69.
108. Piechota, H. 1965. Some correlations between raw materials, formulation, and flame-retardant properties of rigid urethane foams. *J Cell Plast* 1:186–199.
109. Granzow, A. 1978. Flame retardation by phosphorus compounds. *Acc Chem Res* 11:177.
110. Granzow, A. and Cannelongo, J.F. 1976. The effect of red phosphorus on the flammability of poly(ethylene terephthalate). *J Appl Polym Sci* 20:689.
111. van der Veen, I. and de Boer, J. 2012. Phosphorus flame retardants: Properties, production, environmental occurrence, toxicity and analysis. *Chemosphere* 88(10):1119–1153.
112. Kuryla, W.C. and Papa, A.J. eds. 1978. *Flame Retardancy of Polymeric Materials*, Vol. 4. Marcel Dekker, New York.
113. Braun, U. and Schartel, B. 2003. Fire retardancy mechanisms of red phosphorus in thermoplastics. In: *Proceeding of the Additives 2003 Conference*, San Francisco, CA.
114. Ballistreri, A., Montaudo, G., Puglisi, C., and Vitalini, D. 1983. Intumescent flame retardants for polymers. I. The poly(acrylonitrile)-ammonium polyphosphate-hexabromocyclododecane system. *J Polym Sci Polym Chem Ed* 21:679.
115. Qiang, W., Jianping, L., and Baojun, Q. 2003. Microencapsulated red phosphorus as flame retardant. *Polym Int* 52:1326.
116. Jianping, L., Qiang, W., and Baojun, Q. 2004. Thermal stability of microcapsulated chlorinated paraffin-70 with melamine-formaldehyde resin/zinc borate. *Chemistry-Peking* 67(Pt 6):456–460.
117. Shih, Y.F., Wang, Y.T., Jeng, R.J., and Wei, K.M. 2004. Expandable graphite systems for phosphorus-containing unsaturated polyesters. I. Enhanced thermal properties and flame retardancy. *Polym Degrad Stab* 86:339.
118. Levchik, S.V., Costa, L., and Camino, G. 1992. Effect of the fire-retardant ammonium polyphosphate on the thermal decomposition of aliphatic polyamides. Part I-Polyamides 11 and 12. *Polym Degrad Stab* 36:31–41.
119. Levchik, S.V., Costa, L., and Camino, G. 1992. Effect of the fire-retardant ammonium polyphosphate on the thermal decomposition of aliphatic polyamides. Part II-Polyamides 6. *Polym Degrad Stab* 36:229–237.
120. Levchik, S.V., Costa, L., and Camino, G. 1994. Effect of the fire-retardant ammonium polyphosphate on the thermal decomposition of aliphatic polyamides. Part III-Polyamides 6.6 and 6.10. *Polym Degrad Stab* 43:43–54.
121. Duquesne, S., Le Bras, M., Bourbigot, S. et al. 2001. Mechanism of fire retardancy of polyurethanes using ammonium polyphosphate. *J Appl Polym Sci* 82:3262.
122. Weil, E.D. and Levchik, S. 2004. Current practice and recent commercial developments in flame retardancy of polyamides. *J Fire Sci* 22:251–264.
123. Available at http://en.wikipedia.org/wiki/Heptazine (accessed February 14, 2015).
124. Hosmane, R.S., Rossman, M.A., and Leonard, N.J. 1982. Synthesis and structure of tri-s-triazine. *J Am Chem Soc* 104(20):5497–5499.
125. Shahbaz, M., Urano, S., LeBreton, P.R., Rossman, M.A., Hosmane, R.S., and Leonard, N.J. 1984. Tri-s-triazine: Synthesis, chemical behavior, and spectroscopic and theoretical probes of valence orbital structure. *J Am Chem Soc* 106(10):2805–2811.
126. Levchik, S.V. and Weil, D. 2005. Flame retardancy of thermoplastic polyesters: A review of the recent literature. *Polym Int* 54:11.
127. Pawlowski, K.H. and Schartel, B. 2007. Flame retardancy mechanisms of triphenyl phosphate, resorcinol bis(diphenyl phosphate) and bisphenol bis(diphenyl phosphate) in polycarbonate/acrylonitrile-butadiene-styrene blends. *Polym Int* 56:1404.
128. Levchik, S.V., Bright, D.A., Alessio, G.R., and Dashevsky, S. 2001. New halogen-free fire retardant for engineering plastic applications. *J Vinyl Addit Technol* 7(2):98.
129. Levchik, S.V., Bright, D.A., Moy, P., and Dashevsky, S. 2000. New developments in fire retardant non-halogen aromatic phosphates. *J Vinyl Addit Technol* 6(3):123.
130. Perret, B., Pawlowski, K.H., and B. Schartel. 2009. Fire retardancy mechanisms of arylphosphates in polycarbonate (PC) and PC/acrylonitrile-butadiene-styrene. *Journal of Thermal Analysis and Calorimetry* 97(3):949–958.
131. Murashko, E.A., Levchik, G.F., Levchik, S.V., Bright, D.A., and Dashevsky, S. 1999. Fire-retardant action of resorcinol bis(diphenyl phosphate) in PC-ABS blend. II. Reactions in the condensed phase. *J Appl Sci* 72:1863.
132. Levchik, S.V. and Weil, E.D. 2006. A review of recent progress in phosphorus-based flame retardants. *J Fire Sci* 24:345.

133. Bourbigot. S. and Le Bras, M. 2004. Fundamentals: Flame retardant plastics. In: *Plastics Flammability Handbook: Principles, Regulation, Testing and Approval*, ed. J. Troitzsch. Hanser Publishers/Hanser Gardner Publications Inc., Munich, Germany/Cincinnati, OH, pp. 134–148.
134. Kashiwagi, T. and Gilman, J.W. 2000. Silicon-based flame retardants. In: *Fire Retardancy of Polymeric Materials*, eds. A. F. Grand and C.A. Wilkie. Marcel Dekker, New York, pp. 353–389.
135. Iji, M. and Serizawa, S. 1998. Polym. Silicone derivatives as new flame retardants for aromatic thermoplastics used in electronic devices. *Adv Technol* 9:593.
136. Zhou, W. and Yang, H. 2007. Flame retarding mechanism of polycarbonate containing methylphenylsilicone. *Thermochim Acta* 452:43.
137. Nodera, A. and Kanail, T. 2006. Flame retardancy of a polycarbonate-polydimethylsiloxane block copolymer: The effect of the dimethylsiloxane block size. *J Appl Polym Sci* 100:565.
138. Gilman, J.W., Harris, R.H., Shields, J.R., Kashiwagi, T., and Morgan, A.B. 2006. A study of the flammability reduction mechanism of polystyrene-layered silicate nanocomposite: Layered silicate reinforced carbonaceous char. *Polym Adv Technol* 17:263–271.
139. Kashiwagi, T., Gilman, J.W., Butler, K.M., Harris, R.H., Shields, J.R., and Asano, A. 2000. Flame retardant mechanism of silica gel/silica. *Fire Mater* 24:277.
140. Kashiwagi, T., Harris, Jr., R.H., Zhang, X. et al. 2004. Flame retardant mechanism of polyamide 6-clay nanocomposites. *Polymer* 45:881–891.
141. Zanetti, M., Bracco, P., and Costa, L. 2004. Thermal degradation behavior of PE/clay nanocomposites. *Polym Degrad Stabil* 85:657–665.
142. Peeterbroeck, S., Laoutid, F., Swoboda, B., Lopez-Cuesta, J.-M., Moreau, N., Nagy, J.B., Alexandre, M., and Dubois, P. 2007. How carbon nanotube crushing can improve flame retardant behaviour in polymer nanocomposites? *Macromol B Rapid Commun* 28(3):260–264.
144. Lewin, M. 2006. Reflections on migration of clay and structural changes in nanocomposites. *Polym Adv Technol* 17:758–763.
143. Fina, A., Abbenhuis, H.C.L., Tabuani, D., Frache, A., and Camino, G. 2006. Polypropylene metal functionalised POSS nanocomposites: A study by thermogravimetric analysis. *Polym Degrad Stabil* 91:1064.
145. Lewin, M. 2003. Some comments on the modes of action of nanocomposites in the flame retardancy of polymers. *Fire Mater* 27:1–7.
146. Song, R., Wang, Z., Meng, X., Zhang, B., and Tang, T. 2007. Influences of catalysis and dispersion of organically modified montmorillonite on flame retardancy of polypropylene nanocomposites. *J Appl Polym Sci* 106:3488–3494.
147. Jang, B.N., Costache, M., and Wilkie, C.A. 2005. The relationship between thermal degradation behavior of polymer and the fire retardancy of polymer/clay nanocomposites. *Polymer* 46:10678–10687.
148. Liu, W., Varley, R.J., and Simon, G.P. 2007. Understanding the decomposition and fire performance processes in phosphorus and nanomodified high performance epoxy resins and composites. *Polymer* 48:2345–2354.
149. Pastore, H.O., Franche, A., Boccaleri, E., Marchese, L., and Camino, G. 2004. Heat induced structure modifications in polymer-layered silicate nanocomposites. *Macromol Mater Eng* 289:783.
150. Costache, M.C., Heidecker, M.J., Manias, E., and Wilkie, C.A. 2006. Preparation and characterization of poly(ethylene terephthalate)/clay nanocomposites by melt blending using thermally stable surfactants. *Polym Adv Technol* 17:764–771.
151. Bourbigot, S., Duquesne, S., Fontaine, G., Bellayer, S., Turf, T., and Samyn, F. 2008. Characterization and reaction to fire of polymer nanocomposites with and without conventional flame retardants. *Mol Cryst Liq Cryst* 486:325–339.
152. Morgan, A.B., Gilman, J.W., Kashiwagi, T., and Jackson, C.L. 2000. Fire safety developments emerging needs, product developments, non-halogen FR'S, standards and regulations, In: *Proceedings, Fire Retardant Chemicals Association*, Washington, DC, p. 25.
153. Beyer, G. 2002. Carbon nanotubes as flame retardants for polymers. *Fire Mater* 26:291–293.
154. Vaia, R.A., Price, G., Ruth, P.N., Nguyen, H.T., and Lichtenhan, J. 1999. Polymer layered silicate nanocomposites as high performance ablative materials. *Appl Clay Sci* 15:67–92.
155. Kashiwagi, T. and Gilman, J.W. 2000. Silicon-Based Flame Retardants. In: *Fire Retardancy of Polymeric Materials*, eds. A.F. Grand and C.A. Wilkie. Marcel Dekker, New York, p. 10.
156. Zanetti, M., Camino, G., Thomann, R., and Mulhaupt, R. 2001. Synthesis and thermal behaviour of layered silicate–EVA nanocomposites. *Polymer* 42:4501–4507.
157. Zanetti, M., Camino, G., Reichert, P., and Mulhaupt, R. 2001. Thermal behaviour of poly(propylene) layered silicate nanocomposites. *Macromol Rapid Commun* 22:176–180.

158. Zanetti. M., Kashiwagi. T., Falqui. L. and G. Camino. 2002. Cone calorimeter combustion and gasification studies of polymer layered silicate nanocomposites. *Chem Mater* 14:881–7.
159. Bourbigot, S., Gilman, J.W., and Wilkie, C.A. 2004. Kinetic analysis of the thermal degradation of polystyrene–montmorillonite nanocomposite. *Polym Degrad Stabil* 84:483–492.
160. Wang, S., Hu, Y., Lin, Z. et al. 2003. Flammability and thermal stability studies of ABS/montmorillonite nanocomposite. *Polym Int* 52:1045–1049.
161. Peeterbroeck, S., Laoutid, F., Taulemesse, J.-M. et al. 2007. Mechanical properties and flame-retardant behavior of ethylene vinyl acetate/high-density polyethylene coated carbon nanotube nanocomposites. *Adv Funct Mater* 17:2787.
162. Jin, Z., Pramoda, K.P., Xu, G., and Goh, S.H. 2001. Dynamic mechanical behavior of melt-processed multi-walled carbon nanotube/poly(methyl methacrylate) composites. *Chem Phys Lett* 337:43.
163. Cooper, C.A., Ravich, D., Lips, D., Mayer, J., and Wagner, H.D. 2002. Distribution and alignment of carbon nanotubes and nanofibrils in a polymer matrix. *Compos Sci Technol* 62:1105.
164. Cipiriano, B.H., Kashiwagi, T., Raghavan, T. et al. 2007. Effects of aspect ratio of MWNT on the flammability properties of polymer nanocomposites. *Polymer* 48:6086.
165. Kashiwagi, T., Du, F., Winey, K.I. et al. 2005. Flammability properties of polymer nanocomposites with single-walled carbon nanotubes: Effects of nanotube dispersion and concentration. *Polymer* 46:471.
166. Schartel, B., Potschke, P., Knoll, U., and Abdel-Goad, M. 2005. Fire behavior of polyamide 6/multiwall carbon nanotubes nanocomposites. *Eur Polym J* 41:1061–1070.
167. Bocchini, S., Frache, A., Camino, G., and Claes, M. 2007. Macromolecular nanotechnology: Polyethylene thermal oxidative stabilization in carbon nanotubes based nanocomposites. *Eur Polym J* 43:3222.
168. Kashiwagi, T., Grulke, E., Hilding, J. et al. 2004. Thermal and flammability properties of polypropylene/carbon nanotube nanocomposites. *Polymer* 45:4227.
169. Thostenson, E.T., Ren, Z., and Chou, T.W. 2001. Advances in the science and technology of carbon nanotubes and their composites: A review. *Compos Sci Technol* 61:1899–1912.
170. Dubois, P. and Alexandre, M. 2006. Performant clay/carbon nanotube polymer nanocomposites. *Adv Eng Mater* 8:147–154.
171. Fischer, J.E. 2006. Carbon nanotubes: Structures and properties. In: *Nanotubes and Nanofibers*, ed. Y. Gogotsi. CRC Press, New York, pp. 1–35.
172. Peeterbroeck, S., Alexandre, M., Nagy, J.B. et al. 2004. Polymer-layered silicate–carbon nanotube nanocomposites: Unique nanofiller synergistic effect. *Compos Sci Technol* 64:2317–2323.
173. Gao, F., Beyer, G., and Yuan, Q. 2005. A mechanistic study of fire retardancy of carbon nanotube/ethylene vinyl acetate copolymers and their clay composites. *Polym Degrad Stabil* 89:559–564.
174. Beyer, G. 2006. Flame retardancy of nanocomposites based on organoclays and carbon nanotubes with aluminium trihydrate. *Polym Adv Technol* 17:218–225.
175. Beyer, G. 2005. Filler blend of carbon nanotubes and organoclays with improved char as a new flame retardant system for polymers and cable applications. *Fire Mater* 29:61–69.
176. Bonduel, D., Mainil, M., Alexandre, M., Monteverde, F., and Dubois, P. 2005. Supported coordination polymerization: A unique way to potent polyolefin carbon nanotube nanocomposites? *Chem Commun* 6:781.
177. Fina, A., Tabuani, D., Frache, A., and Camino, G. 2005. Polypropylene–polyhedral oligomeric silsesquioxanes (POSS) nanocomposites. *Polymer* 46:7855.
178. Fina, A., Abbenhuis, H.C.L., Tabuani, D., and Camino, G. 2006. Metal functionalized POSS as fire retardants in polypropylene. *Polym Degrad Stab* 91:2275.
179. Laachachi, A., Cochez, M., Ferriol, M., Lopez Cuesta, J.M., and Leroy, E. 2005. Influence of TiO_2 and Fe_2O_3 fillers on the thermal properties of poly(methyl methacrylate) (PMMA). *Mater Lett* 59:36–39.
180. Laachachi, A., Leroy, E., Cochez, M., Ferriol, M., and Lopez Cuesta, J.M. 2005. Use of oxide nanoparticles and organoclays to improve thermal stability and fire retardancy of poly(methyl methacrylate). *Polym Degrad Stabil* 89:344–352.
181. Morgan, A.B. and Wilkie, C.A. 2009. *Fire Retardancy of Polymeric Materials*. CRC Press Taylor & Francis Group, Boca Raton, FL.
182. Lyons, J.W. 1987. *The Chemistry and Uses of Fire Retardants*. R.E. Krieger Pub. Comp., Malabar, FL.
183. Levchik, S. 2007. Introduction to Flame Retardancy and Polymer Flammability, In: *Flame Retardant Polymer Nanocomposites*, Vol. 1, eds. A.B. Morgan and C.A. Wilkie. Wiley, Hoboken, NJ.
184. Morgan, A. B. 2006. Flame retarded polymer layered silicate nanocomposites: A review of commercial and open literature systems. *Polym Adv Technol* 17:206–217.

185. Dasari, A., Yu, Z.Z., Mai, Y.W., and Liu, S. 2007. Flame retardancy of highly filled polyamide 6/clay nanocomposites. *Nanotechnology* 18(44):5602.
186. Bartholmai, M. and Schartel, B. 2004. Layered silicate polymer nanocomposites: New approach or illusion for fire retardancy? Investigations of the potentials and the tasks using a model system. *Polym Adv Technol* 15:335–364.
187. Zanetti, M. 2006. Flammability and thermal stability of polymer/layered silicate nanocomposites. In: *Polymer Nanocomposites*, eds. Y.W. Mai and Z.Z. Yu. Woodhead Publishing, Cambridge, pp. 256–272.
188. Zhao, C., Qin. H., Gong, F., Feng, M., Zhang, S., and Yang, M. 2005. Mechanical, thermal and flammability properties of polyethylene/clay nanocomposites. *Polym Degrad Stabil* 87:183–189.
189. Camino, G., Costa, L., and Luda di Cortemiglia, M.P. 1991. Overview of fire retardant mechanisms. *Polym·Degrad Stabil* 33:131–154.
190. Skinner, G.A. 1998. Flame retardancy: The approaches available. In: *Plastics Additives, an A-Z Reference*, ed. G. Pritchard. Chapman and Hall, London, UK, pp. 260–267.
191. Levchik, S.V. and Weil, E.D. 2000. Combustion and fire retardancy of aliphatic nylons. *Polym Int* 49:1033–1073.
192. Zanetti, M., Camino, G., Canavese, D., Morgan, A.B., Lamelas, F.J., and Wilkie, C.A. 2002. Fire retardant halogen–antimony–clay synergism in polypropylene layered silicate nanocomposites. *Chem Mater* 14:189–193.
193. Lee, J.H., Nam, J.H., Lee, D.H. et al. 2003. Flame retardancy of polypropylene/montmorillonite nanocomposites with halogenated flame retardants. *Polym-Korea* 27:569–575.
194. Ristolainen, N., Hippi, U., Seppälä, J., Nykänen, A., and Ruokolainen, J. 2005. Properties of polypropylene/aluminum trihydroxide composites containing nanosized organoclay. *Polym Eng Sci* 45:1568–1575.
195. Hu, Y., Wang, S., Ling, Z., Zhuang, Y., Chen, Z., and Fan, W. 2003. Preparation and combustion properties of flame retardant nylon 6/montmorillonite nanocomposite. *Macromol Mater Eng* 288:272–276.
196. Ma, H., Fang, Z., and Tong, L. 2006. Preferential melt intercalation of clay in ABS/brominated epoxy resin-antimony oxide (BER-AO) nanocomposites and its synergistic effect on thermal degradation and combustion behavior. *Polym Degrad Stabil* 91:1972–1979.
197. Wang, S., Hu, Y., Zong, R., Tang, Y., Chen, Z., and Fan, W. 2004. Preparation and characterization of flame retardant ABS/montmorillonite nanocomposite. *Appl Clay Sci* 25:49–55.
198. Weil, E.D., Hirschler, M.M., Patel, N.G., and Shaki, S. 1992. Oxygen index: Correlations to other fire tests. *Fire Mater* 16:159–167.
199. Wang, D., Echols, K., and Wilkie, C.A. 2005. Cone calorimetric and thermogravimetric analysis evaluation of halogen-containing polymer nanocomposites. *Fire Mater* 29:283–294.
200. Si, M., Zaitsev, V., Goldman, V. et al. 2007. Self-extinguishing polymer/organoclay nanocomposites. *Polym Degrad Stabil* 92:86–93.
201. Zhang, S. and Horrocks, A.R. 2003. A review of flame retardant polypropylene fibres. *Progr Polym Sci* 28:1517–1538.
202. Hao, X., Gai, G., Liu, J., Yang, Y., Zhang, Y., and Nan, C. 2006. Flame retardancy and antidripping effect of OMT/PA nanocomposites. *Mater Chem Phys* 96:34–41.
203. Chigwada, C. and Wilkie, C.A. 2003. Synergy between conventional phosphorus fire retardants and organically-modified clays can lead to fire retardancy of styrenics. *Polym Degrad Stabil* 80:551–557.
204. Kim, J., Lee, K., Lee, K., Bae, J., Yang, J., and Hong, S. 2003. Studies on the thermal stabilization enhancement of ABS; synergistic effect of triphenyl phosphate nanocomposite, epoxy resin, and silane coupling agent mixtures. *Polym Degrad Stabil* 79:201–207.
205. Kim, S. and Wilkie, C.A. 2008. Transparent and flame retardant PMMA nanocomposites. *Polym Adv Technol* 19:496–506.
206. Horacek, H. and Grabner, R. 1996. Advantages of flame retardants based on nitrogen compounds. *Polym Degrad Stab* 54:205.
207. Carpentier, F., Bourbigot, S., Le Bras, M., and Delobel, R. 2000. Charring of fire retarded ethylene vinyl acetate copolymer—Magnesium hydroxide/zinc borate formulations. *Polym Degrad Stabil* 69:83.
208. Carpentier, F., Bourbigot, S., Le Bras, M., and Delobel, R. 2000. Rheological investigations in fire retardancy: Application to ethylene-vinyl-acetate copolymer-magnesium hydroxide/zinc borate formulations. *Polym Int* 49:1216.
209. Durin-France, A., Ferry, L., Lopez-Cuesta, J.-M., and Crespy, A. 2000. Magnesium hydroxide/zinc borate/talc compositions as flame-retardants in EVA copolymer. *Polym Int* 49:1101.
210. Clerc, L., Ferry, L., Leroy, E., and Lopez-Cuesta, J.M. 2005. Influence of talc physical properties on the fire retarding behavior of (ethylene–vinyl acetate copolymer/magnesium hydroxide/talc) composites. *Polym Degrad Stabil* 88:504–511.

27 Recycling of Thermoplastics

Jesús María García-Martínez and Emilia P. Collar

CONTENTS

27.1 Introduction 919
    27.1.1 Early Polymeric Materials 920
27.2 New Polymeric Materials and Composites 920
27.3 Stability and Reactivity: Degradation and Durability of Materials 921
27.4 Wastes: Concepts and Waste Classification 923
27.5 Plastics and the Environment: Main Thermoplastic Waste Sources 924
    27.5.1 Urban Plastic Wastes 925
    27.5.2 Automotive Plastic Wastes 926
    27.5.3 Agricultural Plastic Wastes 927
    27.5.4 Plastic Wastes in Electronic Devices 927
    27.5.5 Wild Plastic Wastes: Biodegradation and Photodegradation 928
27.6 Plastic Wastes and Eco-Balance 929
    27.6.1 Primary and Secondary Recycling Routes 929
    27.6.2 Incineration and Tertiary Recycling Routes 930
    27.6.3 Economic and Ecological Dimensions 932
27.7 Conclusions and Perspectives 933
In Memoriam 934
References 934

27.1 INTRODUCTION

The recycling and disposal of organic plastic materials is of great significance and has become critical despite the fact that polymeric materials are relatively new (discovered just over a century ago) compared to traditional materials such as wood, metals, glass, and concrete. For almost a century now, organic plastic materials have been a daily presence in our everyday lives and activities. Plastic materials are ubiquitous, used in packaging, bags, film covers and containers, and building and construction. However, most thermoplastics are nondegradable and their disposal, especially in the big cities, has become a big concern and an issue characterized by negative press and environmental activism. Nonetheless, these materials are here to stay and have become extremely beneficial to society. Furthermore, there are sophisticated applications of plastic materials: as clinical implants; in greenhouses and irrigation for increasing and improving agricultural outputs; in automotive applications as bumpers; and in fuel tanks and other engineering parts in cars. They also find increasing use as composites in strong, tough, and lightweight structures and devices for the aircraft and aerospace industries.

Thermoplastic polymers constitutes nearly 80% of the total organic plastic materials used in packaging and are thus highly visible in the municipal solid waste (MWS) streams, creating sensitivity to the issue of recycling. However, from the early 1980s, the overall fraction of thermoplastic waste in the MSW in large cities has remained constant (about 10%). As polymeric materials continue to replace the traditional materials, waste from manufacturing and consumption are bound to increase, creating a backlash that can threaten the full expansion of the use of synthetic organic plastics in the marketplace.

Following the first global energy crisis of the early 1970s, there was a concerted effort against the *use-and-discard* lifestyle acknowledged to be responsible for increasing environmental pressures on limited landfills. At the beginning of the third millennium, the worldwide population exceeded 6 billion people and is estimated to be about 7 billion at present. This increases the environmental pressures and costs associated with the increase in the solid waste management by landfills as plastic waste materials are brought in from the industries and from the municipal plastic waste sources, especially in the neighborhoods of large urban or industrial areas. Therefore, and beyond the energy recovery options, recycling alternatives for thermoplastic waste materials, by mechanical or chemical means, were encouraged. The last two decades have witnessed an intensive focus on research into the recycling options for solid plastic waste as materials (primary or secondary uses) or as chemicals (tertiary). Indeed, there has been an exponential growth in published research not only dealing with the field of scientific-technical plastic materials but also concerning the social, legal, and economical dimensions of plastic waste management, with a real emphasis on the possibilities and availability of plastic recycling technologies. This has led to an increase in interdisciplinary research and collaboration between academia, corporations, and government agencies on recovery and recycling technologies, development of infrastructures, and development of a viable recycling markets. With globalization and the environmental involvement of the chemical industry from the 1980s through initiatives such as Responsible Care, Registration, Evaluation, Authorization and restriction of Chemicals (REACH), or the most recent "green chemistry" label, continuous improvement and innovations in recycling of thermoplastics are expected.

27.1.1 Early Polymeric Materials

After the industrial revolution, from the middle of the nineteenth century and the first two decades of the twentieth century, there was an exponential growth in material innovation compared to all the previous centuries of human existence. About 8 years before the Second World War broke out, polyethylene (PE) was discovered accidentally in the laboratories of Imperial Chemical Industries (ICI). Because of its lightness, nonpolarity, and almost chemical inertia, the *white dust* matched the underwater protection requirements for submarine communications cables. Ten years later, the US Army promoted the search for synthetic materials able to replace the natural silk used in parachute manufacturing, to replace materials coming from the Far East routes. Through the research work of W.H. Carothers of the DuPont Chemical Laboratories, polyamides (PAs) were synthesized and patented as nylon 6. DuPont also provided the fundamentals of the polymerization reactions, represented in a convenient physicochemical format by Flory [1]. In 1974, Flory was awarded the Nobel Prize in Chemistry. His 1953 textbook titled *Principles of Polymer Science and Technology* began with a historical introduction.

From the mid-1950s, the demand for organic synthetic materials in strategic sectors such as defense, energy, transport, and communications grew exponentially, leading to new materials for weight reduction in the aeronautical industry and specialized materials for aerospace. Communication systems profit from the development of materials suited for size reduction of components or other strategies to maximize the capabilities of storage, handling, and efficient transmission of data [2,3].

27.2 NEW POLYMERIC MATERIALS AND COMPOSITES

Materials, as a grouping of matter with sufficient structural properties to be useful in a particular application or for a final purpose, opened a broad field of nonreturn applications based on organic synthetic polymer materials and on large-scale economies, which favored the *use-and-discard* concept of the so-called consumer societies.

Before polymeric materials became commonplace, the study of material aging, durability, and degradation was limited. Moreover, the question of when a material is degraded, that is, reaches a critical level of loss of properties, rendering it useless and therefore a *solid waste*, was not relevant until recent times.

It is known from the study and practice of medicine, biology, and biochemistry that complex and bioorganic structures appeared as a collection of macromolecules, more or less complex and well dispersed in water. It is also known that small amounts of certain additives, the so-called oligoelements, metallic ions included, forming ligands and complex agents, are key in the stabilization of biomacromolecule dispersion, assuring the large efficiency and selectivity of the overall bioorganic chemical reactions that define the proper functioning of life. This knowledge on additives and stabilization is carried over into the plastics industry. In chemistry, the metallic, ionic, and covalent bonds are the primary links that constitute matter. Each one of them confers some features and properties to the material [4,5].

New advances in the materials science and technology field [6,7] reveal the relationship between the nano dimension of the interfacial domains and the successive mesoscales, microscales, and macroscales, which gives rise to the desired performance of the material as a whole in a finished part. This knowledge is exploited in the manufacturing of the most sophisticated nano-devices, up to the critical building blocks; the largest engine parts; buildings or bridges; and of course, most daily goods, such as packaging, personal computers, smart phones, car bodies, and so on. This fundamental knowledge is also critical to the disposal of these goods after use [8,9].

Another consideration is to view landfills as a potential raw material source for metals, glass, paper and board, and possibly plastic materials. These new organic and *synthetic* thermoplastic materials have now attained consumption volumes large enough that there is a great need to recycle them, just like the recycled use of "waste" or by-products of the oil refinery industries of previous decades.

Knowledge gained from the metal, glass, or paper and board industries as to the best ways to recover, sort, clean, and finally, reprocess the solid wastes from these industries is being adapted by the thermoplastic industries. Earlier attempts by industries other than polymer manufacturers to reuse and recycle organic thermoplastic waste by returning them to the material markets failed and created the false impression that these organic thermoplastic polymers are not recyclable.

27.3 STABILITY AND REACTIVITY: DEGRADATION AND DURABILITY OF MATERIALS

Figure 27.1 summarizes the different processing technologies in use in the polymer industry, including those involving physical processes, usually applied to the thermoplastic polymers, and others, which involve further chemical reactions over prepolymerized formulations to yield organic but thermosetting polymers. Unlike thermoplastics, thermosetting polymers are not suitable materials for reprocessing, due to permanent cross-links; but they can be recycled into other products. Thermosetting resins are big parts of the naval, air, and space industries, and their end-of-use recovery and treatment processes are beyond the scope of the present chapter.

Degradation can be reduced to a problem of chemical stability in atmospheric oxygen. In other words, it is a question of *reactivity*, more or less negligible in a short time scale as compared to properties of the material on a *macroscopic* scale.

The polyolefins, the largest polymers in terms of volume consumption, are probably the most sensitive to degradation processes from their production steps through their processing and even up to the finished product. The reason is their sensitivity to chemical environments involving radicals and specifically to those involving atmospheric oxygen. By combination with the thermal and shear fields imposed during the processing steps, or through exposure to sunlight, the adverse degradation reactions accelerate, not from the monomer's nature itself but because of the existence of traces of impurities on the polyolefin bulk. The impurities, although in trace levels, are always present in the material coming from the industrial polymerization processes. This explains why, almost in parallel

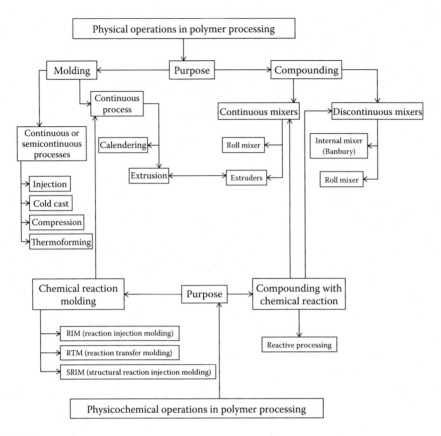

FIGURE 27.1 Polymer processing operations.

with the early stages of polyolefin applications, a new area of research in polymer science and technology was devoted to the development of additives, known as stabilizers, to promote the chemical stability of the polyolefins. Radical scavengers and UV absorbers able to avoid and/or minimize both the thermal and/or near-UV light-induced degradation processes are the main stabilizer families, on which research began several decades ago [10,11].

One may conclude that when any pristine polyolefin, alone or in composites, or polymer blends, whether for indoor or outdoor applications, are used, there always must be at least a thermo-oxidative stabilization pack for processing operations, and a design to minimize the degradation processes to extend the useful life of the end part in the use environment. The absence of a proper stabilization package would give rise to a gradual or sudden deterioration of the material properties [12,13].

Each class of thermoplastic polymer follows a unique degradation pathway, necessitating a well-defined and specific stabilization package. For instance, while polypropylene (PP) degrades by a chain scission mechanism, leading to a sharp reduction of its molecular weight, PE evolves by cross-linking mechanisms, becoming a nonsoft material by losing its thermoplastic nature.

All the other industrial thermoplastics, mostly incorporating polar groups into their backbones, require lower stabilization levels for their final applications. This is because their degradation mechanisms are less complex and catastrophic than those of polyolefins, not necessarily governed by radical reactions. Indeed, the analysis, identification, and monitoring of triggers and the generated species of the degradation processes in polar thermoplastics are easier to control and understand than those in polyolefins, a serendipitous situation for the polymer industry [14].

If a virgin polymer requires a stabilizer package in the polymer bulk to prevent the risk of thermo-oxidative degradation during processing, it seems obvious that the recycled material would need it too if it goes back to the plastic materials market, after the plastic goods it comes from reach their EOL. The well-known sensitivity of the pristine polymer to degradation processes, as *intrinsic to its molecular structure*, must be critically considered when reprocessing. If the cost of a properly formulated recycling polymer makes it uncompetitive, economic viability rather than technical feasibility becomes the critical issue.

27.4 WASTES: CONCEPTS AND WASTE CLASSIFICATION

Vian [15] compiled one of the most complete classifications of the contribution of the daily activity of human beings on our planet to the overall waste streams, and Figure 27.2 aims to illustrate all of them.

It is obvious that plastic materials appear in all of the activity areas compiled in Figure 27.2, but in very different quantities depending on the source. For instance, plastic materials hardly contribute up to 1.5% to the overall solid industrial waste fraction, while the larger part (nearly 70%) is of organic nature, with a 48:22 ratio between sludge and animal dung; followed by almost 15% building wastes. Nevertheless, in home solid wastes, the thermoplastic fraction has grown to almost 20%, averaging around 10% in the overall big urban solid waste streams, and remaining constant for several decades [16–28]. Forecasts for 2020 are around a half ton per person in the First World countries [29].

Legislative perspectives on wastes recognize their *states of aggregation: gaseous, liquid, or solid*. The first, two, denominated under the common heading of effluents, were obviously legislative matters, because of their extreme mobility and the damage they caused to the environment and living species near their points of generation and disposal.

Solid wastes, because they remained where they were discarded, did not arouse legislation until late into the twentieth century, when the uncontrolled landfills were identified as responsible sources of either organic or inorganic toxic and contaminants by rain lixiviation processes. Hence, to reduce the environmental impact of a controlled landfill requires more than just the administrative approval of its location, since disordered matter and material discharges give rise to serious health and safety problems [30]. In addition to the deterioration of the environment and the

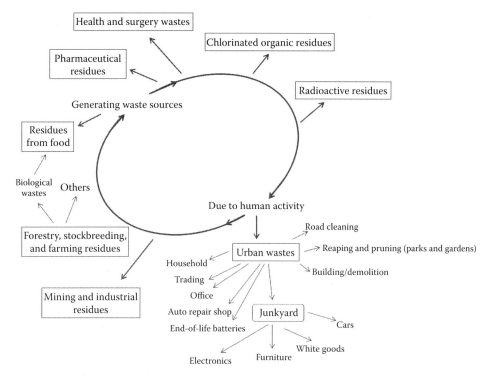

FIGURE 27.2 Waste generation due to human activity.

landscape, insects and rodents may contribute to the spread of infectious diseases, and the emergence of small fires caused by the combustion of organic matter during or following fermentation processes became a concern. In these conditions, the spread of pathogens and toxic substances is unpredictable, especially through the above-mentioned rain leaching processes, giving rise to the pollution of aquifers. According to the latest landfill engineering, the arrangement of solid waste consists in compacting thick layers, optimizing the occupied volume. Further, to favor the biological transformations of the fermentable materials present in the solid wastes, this design removes the spontaneous explosion and fire risks associated with the uncontrolled fermentation processes. To avoid lumping, a low-permeability material such as clay ground of the surroundings combined with, for instance, shredded PE is deposited between layers. It yields the most convenient hollowness level in these soils to drain the landfill liquids caused by rainwater percolation as well as to obtain the generated landfill gases that are incorporated into any energy recovery utilities nearby. Despite the well-developed landfill engineering, with almost 6 billion people and the growing floor space needs, it is more and more difficult to find available space for solid waste disposal.

According to the overall classification of wastes compiled in Figure 27.2, and from the solid waste perspective, one can sort them by generation and classes such as *industrial, agricultural, and urban wastes.*

Any solid industrial waste or scrap is an indication of a nonoptimized manufacturing process. Moreover, a significant fraction of industrial wastes implies additional costs for the company, as the waste owner must pay for the best disposal of solid wastes according to the local edicts and regulations. Consequently, producers of industrial plastic waste fractions that are generated indoors, noncontaminated, and well characterized have been promoting their recycling, giving rise to the so-called thermoplastic waste primary recycling routes.

Bioorganic matter is the main component of forestry, stockbreeding, and farming residues. Nevertheless, those earth regions that can produce crops intensively by using greenhouse technologies include as main components of their solid waste a few materials such as glass, polyvinyl chloride (PVC), and PE and ethylene copolymers—these last two being the most competitive in costs and, thus, the most frequent components of the overall fraction of agricultural synthetic solid wastes. Low-density PE (LDPE) and ethylene vinyl acetate (EVA) copolymers are indeed the most common, being characterized on one hand by the photodegradation level because of UV exposure and, on the other, by the contamination and soil residues. Besides the polymer sorting, both undesirable characteristics of the agricultural plastic wastes demand cleanness and the right treatments [31] of these kinds of plastic wastes to follow the plastic waste secondary recycling routes. The business owners, as the owners of the plastic waste, remain subjected to the legal frames in force about the management of agricultural plastic wastes in the municipalities where their greenhouses are located.

The largest flow of solid wastes is that of the so-called urban solid waste. Following the energy crisis of the 1970s, the main landfills of the largest cities became considered as alternative sources of raw materials, such as glass, paper and cardboard, metals, organic matter, and plastics. The recycling challenge for the different thermoplastics involved, mainly those coming from the packaging industries—PE, PP, polystyrene (PS), polyethylenterephtalate (PET), and PVC—then came about, and in the middle of the 1980s, the different recycling routes for thermoplastic wastes were proven to be technically feasible. The economic competitiveness of the feed forward of these recycled polymers to the material markets was another question subject to cost fluctuations of the large international markets for fresh materials [32]. As always, these issues remain of high sensitivity and concern, as evidenced by incessant news coverage.

27.5 PLASTICS AND THE ENVIRONMENT: MAIN THERMOPLASTIC WASTE SOURCES

The Greek origin of the word *plastikos* (πλαστικοσ) means "that can be shaped," and so not only the synthetic organic polymers are plastics, but also, metals, clay, glass, and ceramics are. However,

Recycling of Thermoplastics

when a person hears of *plastics*, he or she thinks only of the synthetic manmade materials, the organic polymers, and more specifically, those of the thermoplastics family.

Someone might say that in the twentieth century, mankind went into the *Plastics Age* [33]. Since the end of the nineteenth century and in a seemingly serendipitous fashion, the first synthetic polymers had been coming into the materials markets one by one: celluloid (1870), polymethylmethacrylate (PMMA, 1927), PS (1930), PVC (1931), and LDPE (1937). However, after the Second World War and because of the systematic study of those materials and the subsequent new developments and applications coming from those findings, the synthetic thermoplastics began to be referred to by families: PAs (1940), silicones (1941), aliphatic polyesters (PET, 1942), fluoride polymers (1943), polymethylenic resins (polymethylene oxide [POM], 1953), and polyimides (PIs, 1964), currently well known as engineering polymers. In the 1950s, the isotactic PP and the linear or the so-called high-density PE (HDPE) went into the so-called commodities polymer market segment, giving rise, together with PS and PVC, to the large reduction in the weight, energy, volume, and, thus, costs of the packaging industries. This translates into lowering of costs by two or three times compared to the glass, metal, and paper and board packaging industries, and this ratio remains today [34].

It is worthy of note that plastic materials themselves are the best means of disposal for the lightest fraction of oil refinery by-products, and yet the overall plastics industry represents around 5% of the overall oil refinery operations. Indeed, the largest volume, 86%, goes to energy purposes (35% thermal energy, 29% mechanical energy, and 22% electricity), while the lubricant industry represents around 10%. In other words, the problem of the synthetic thermoplastic solid wastes is not a problem of managing large amounts of material.

After an average annual growth of 8.7% from 1950 to 2012, the worldwide plastic material production was around 240 megatons by 2005, overtaking for first time that of steel, and it grew to 288 megatons by 2012 [35]. Just in Western Europe, plastic production reached a maximum of 50 megatons in 2007 and 45.9 megatons in 2012, of which almost 55% ended up in a waste stream. Around 10% appeared as postconsumer plastic wastes in big MSW streams, the other main sources of thermoplastic wastes coming from end-of-life vehicles (ELVs), agricultural applications, and electric and electronic devices. They are specifically described under the following subsections as well as the so-called wild thermoplastic wastes.

27.5.1 Urban Plastic Wastes

The integral collection of the overall MSWs gave rise to the design of integral treatment plants [36] of MSW at semi-industrial and large scales. Organic matter, metals, glass, paper and board, and the organic thermoplastic wastes are the main fractions collected, using multistage and well-understood separation technologies based on gravity, hydrodynamic, and electrostatic differences between the above-mentioned components.

Both the social dimension of the urban solid waste management and the above-mentioned reluctance of thermoplastic producers to practice plastic waste recycling led to the intervention of public authorities through legislation and specific initiatives. For instance, in the early 1980s and in some parts of the United States, the packaging industry was denied access to the use of any material that could not be recycled conveniently or in an environmentally friendly fashion at the end of service life, a decision that adversely affected PET bottles for carbonated drinks. In Europe, increasing environmental pressure regarding PVC bottles for mineral water was an important driving force to develop accurate PVC recycling procedures, such as Vinyloop [37]. From the perspective of big customers, food packaging applications based on foamed PS are forced to change if the manufacturer is unable to achieve the recycling target [21]. These experiences forced the main plastic producers to change their attitudes and positions on plastic recycling issues.

In the European Union, with an average of 524 kg per capita, municipal wastes, mainly household-generated wastes and any others similar in nature and composition, collected and managed on behalf of municipal authorities, represented an average annual generation value of 25 million tonnes from 2006 to 2012. Yet in 2006, more than a 50% went into landfill, while in, 2012, the amount fell below 36% percent. Zero plastics to landfill is the challenging but realistic goal by the year 2020 [35]. Thermoplastic wastes account for between 7% and 10% w/w of the fraction, representing a 75:25 ratio between the so-called light and heavy fractions. The light fraction is composed of about 80% film fragments of comingled polyolefins, PS foams from food packaging goods, and some other heavier-than-water films, mainly PVC. The heavy fraction is comprised of end-of-life bottles, blisters, and some rigid containers [21–25,38].

Regardless of whether they are obtained through household separation methods carried out by good citizenship, monetary rewards, or mandatory laws, both fractions of municipal thermoplastic wastes so obtained are suitable for secondary recycling [39–44]. Most common applications for the recycled light fraction are as posts, traffic barriers, garbage bags, and so on, while the separation of the heavy fraction into single polymers, polyethylene terephthalate bottles (6%), HDPE containers and caps (12%), and so on, increases the quality of the recycled pellets, which are useful in fiber applications, like carpets, fiberfill, and others [35]. Food contact applications were forbidden to avoid health risks in attending to the first live use of these thermoplastics [45]. Normative now in force accept small [46,47] very low amounts of recycling thermoplastics incorporated in the virgin polymers but preserving the technical and the health and safety standards required on food contact applications.

One of the main questions that remains open nowadays is that related to the efficiency of separation done by citizens at home. Even if they were able to do it by chemical species, variations due to different molecular weights and distribution, branching, or tacticity affect significantly the rheology of the molten polymers and, thus, the processing. Differences in crystallization behavior are also important because foreign substances, such as pigments, dies, and ink particles, may act as nucleants in a different manner in each polymer grade. It seems necessary to implement some kind of standardization activity focused on developing reasonable restrictions to the now fully open option to put new compounds in the packaging markets as well to assure the highest compositional homogeneity and a continuous supply of the raw materials from the urban solid wastes for the large MSW treatment plants.

27.5.2 Automotive Plastic Wastes

The recycling target for ELVs in Europe up to 2015 is 85% of the total weight of ELVs [48]. Technically, car recycling follows a well-defined multistage procedure that starts with depollution, by removing toxic and hazardous components, dismantling, shredding, and metal and plastic separation operations. The lightest fraction, known as the automobile shredder residue (ASR), contains the automotive plastic materials, which appear commingled with minor amounts of contaminant matter such as heavy metals, mineral oils, and hydrocarbon fractions and represent around 20% of the overall amount of solid waste from the ELV. Because of the increasing number of ELVs according to sales forecasts, research and development (R&D) activities on treatment technologies for ASR have been encouraged by public demand for the last two decades.

The average weight composition of the ASR fraction coming from ELVs determined by classical flotation separation trials [22,23] and the subsequent characterization and analysis steps shows that all the plastic materials used in car manufacturing are just above 30% and appear in this fraction [49]. Once recovered as the lightest fraction, the thermoplastic wastes coming from automotive applications are about 50% polyolefins and 30% foam rubber. The polyolefin fraction represents almost 40% of the overall plastic material used in car manufacturing and is suitable for secondary, or mechanical, recycling routes [49,50]. Several studies are available about the pyrolysis efficiency of the ASR fraction, which accounts for nearly 20% of total conversion [51]. Flotation techniques to separate the thermoplastic fractions seem to be efficient not only at facilitating the secondary recycling route of polyolefins but also because the pyrolysis conversion goes as high as 60% [49].

27.5.3 Agricultural Plastic Wastes

Almost 5% of thermoplastic products are agricultural applications utilized for the protection and enhancement of harvests (greenhouses, mulching, or high and low tunnels) and films for bale wrapping and silage applications. Ethylene and its copolymers are the main constituents of the agricultural thermoplastic wastes.

The main features of the film agricultural plastic wastes are their thickness; their pollutant level (mainly soil, sand, and pesticides); and their intrinsic degradation level because of the environmental exposure [52,53]. The film thickness is a key factor because the thinnest, around 10 μm, made the washing operations to remove soil and sand very difficult, as they adhered in film folds. Further, for mechanical recycling, the lower the film thickness, the lower the throughput of the mechanical recycling extrusion line [53]. In contrast with the household-generated plastic wastes in large cities, the selective collection of the agricultural thermoplastic wastes is feasible because of the low number of users and the insignificant amounts that each produces. To avoid cross-contamination with other thermoplastic wastes from goods that may come from polymers others than polyolefins, secondary recycling routes should be considered [53,54].

Most of the agricultural films are based on PE, sometimes incorporating up to 15% PE–vinyl acetate (EVA) copolymer. The so-called thermal films incorporate aluminum oxide or other metal oxides. As in any other commingled thermoplastic waste source, the presence of pigments or carbon black, although not difficult for mechanical recycling, restricts the market applications of the recycled thermoplastic. High photodegradation levels from extensive environment exposure may eventually favor PE cross-linking [31].

In conclusion, the characterization and analysis of each fraction of agricultural thermoplastic wastes demands the use of the best recycling route based on composition, pollutants, and degradation level [53–56].

27.5.4 Plastic Wastes in Electronic Devices

In last two decades, technological progress has extended beyond the industrial and professional environments to reach the consumer, with products in electronics such as personal computers and mobile phones. In relatively short time periods, these consumer products routinely become obsolete as their structures and capabilities are superseded by new designs and better performance of the newly introduced models. This dynamic leads to a sharp increase in solid wastes comprising obsolete electric and electronic devices, which, when mixed other materials, mainly plastic and metals, now have to be classified as hazardous solid waste. Represented by the acronym WEEE, waste electrical and electronic equipment streams are the fastest-increasing waste sources all around the world, and because they contain hazardous substances, they are harmful to the environment if not adequately treated. Cadmium and mercury are some of the components, and without a doubt, transnational legislation must be developed and implemented to assure sustainable recycling and reclamation of useful components (some components such as gold and silver can be profitably recovered).

Thermoplastic polymers represent around 20% of the overall weight fraction of the electric and electronic wastes, but a high level of integration between the different materials in the different parts is a design characteristic of the electric and electronic devices. Consequently, small amounts of thermoplastics are often embedded in thin layers within the products. After a sorting pretreatment step, the disassembly of parts from the different devices attempts to recover those useful components, making profitable the recycling processes, and to identify those parts containing hazardous components, such as PVC or brominated flame retardants, which can give rise to dioxins in downstream energy recovery processes by incineration unless special caution to avoid it is taken. After a coarse sorting by material and by using shredders [51], the wastes are reduced to homogeneous sizing ranges and undergo a separation flow for ferrous metals, nonferrous ones, copper and precious metals, and organic plastic materials, with a small fraction of rejected materials.

Because of the required presence of additives like paints and flame retardants in sensitive amounts in the thermoplastic compounds used in electric and electronic devices, these plastic wastes are not suitable for secondary recycling routes. Hence, one of the most common ways to manage them is as fuel in the smelters for recycling processes of upgraded fractions containing cooper and precious metals.

Most of the design features of the electric and electronic goods make the disassembling and recycling processes of the WEEE especially difficult. Even different types of the same goods, mobile phones, for instance, have very different assembly and disassembly routes and varying components depending on version and maker. There is a lack of standards for new developments and designs that could favor the recycling processes and treatment of the materials post-use. In Tanskanen's paper [57] and its references, there is an expanded view about the very specific problem of WEEE and the different roles of the stakeholders involved, from manufacturers to consumers.

27.5.5 Wild Plastic Wastes: Biodegradation and Photodegradation

Because of the high visibility of the plastic wastes, their *inconvenient* disposal becomes a continuous conflict between the different stakeholders involved, even stimulating restrictive legislation and regulations affecting thermoplastic uses, especially those dealing with films or bags [58]. The public remains more sensitive to the visual contamination effect of the plastic film, unlike those of other materials [59].

Responding to the large visual contamination effect of the *wild* synthetic thermoplastic wastes, several approaches claim to address the negative consequences of bad practices of garbage disposal, focusing mainly on both municipal shortcomings and the education of the citizenship. One of the approaches was to promote the so-called biodegradable polymers, those mainly based on PE–starch combinations and presumed to solve the problem and those related to the accumulation of thermoplastics in uncontrolled landfills [21].

There are noteworthy distinctions between intrinsically degradable polymers and those incorporating a degradable component such as PE–starch combinations. In the latter, the nondegradable component remains even after the starch has fully biodegraded. The PE chains just remain in the environment as small bits that become dangerous for living species [60,61]; hence, it is by no means a true or sustainable thermoplastic waste disposal route.

The truly biodegradable thermoplastics consist of polymers containing polar groups, like polylactic acid derivatives or polyalkanoates coming from bacterial synthesis. Synthetic thermoplastics characterized by structural backbones below a critical size to make them accessible to the bioorganisms are biodegradables too. Currently, the skepticism as to whether or not the degradable polymers could solve the question of overall packaging thermoplastic waste management not only remains but has grown stronger [62]. Any efficient degradation route ends with carbon dioxide evolution, which puts it parallel to incineration but without the option of energy recovery. As the degradation processes progress, the contribution of toxic residues to the environment becomes a main concern that is aggravated because the degradation rates under usual environmental conditions are very different from those in laboratories and are often too slow. Increasing amounts of the promoter degradation additives, or degradable monomers in the thermoplastic composition, almost all the time, lead to the problem that the thermoplastic degrades slowly, as it does under normal-use conditions. Further, the lack of competitiveness in cost, after several decades of experience with minor amounts of different degradable compounds in the packaging industries, has created new problems of mixing or cross-contamination. Mixing with other packaging thermoplastics that otherwise would have been suitable for secondary recycling routes leads to inefficient processing steps, machinery problems, and recycled material pellets of poor quality [63,64]. Meanwhile, more R&D activities are needed, and current trends seem to suggest that these materials will be used in very specialty applications, such as in medicine, where cost is not the key factor [65,66].

27.6 PLASTIC WASTES AND ECO-BALANCE

Several initiatives, such as the popularized 3R (*reduce, reuse, and recycle*), have been adopted to address plastic waste management. Industrially and from the point of view of *logistics*, they were launched to reduce the overall fraction of generated solid wastes from daily use and to optimize the performance/manufacturing relationship, an example being thickness or gauge reduction as much as possible in packaging items [67,68]. Other aspects promote continuous design improvements and innovation, as in the electronic devices and the automotive industries, eliminating in the new designs *unnecessary* parts of incompatible polymers that increase recycling costs (such as emerging technical problems of part disassembly or reprocessing operations of thermoplastics post-use).

The expected *loss of properties* would depend, on one hand, on the pollutant and soil levels, together with the degradation state after use, in addition to problems arising from improper reprocessing of the polymer. On the other hand, because of the existence of mixtures of polymers from different families and the thermodynamic incompatibility derived from their macromolecular nature, the loss of properties would be the expected result of bad industrial practices and wrong processing technologies applied to random mixtures of different polymers, or polymer grades, giving rise to pellets characterized by bad properties. A well-controlled morphology through appropriate processing, most often combined with the use of additives and interfacial agents, would be the key to optimize the end properties of the recycled organic thermoplastic materials.

The redefinition of thermoplastic wastes based solely on the scientific-technical frame opened up a whole field of activities in R&D, reaching also the manufacturers of polymers, who have now adopted a philosophy of promoting and encouraging proper management and treatment of the thermoplastic wastes. The management of plastic wastes in general, both thermosetting [69,70], and thermoplastic, follows a hierarchical approach in a minimization strategy beginning with reuse and followed by recycling and energy recovery routes, focusing on keeping the environment safe, preserving natural resources, and minimizing landfill usage.

27.6.1 PRIMARY AND SECONDARY RECYCLING ROUTES

According to the life cycle assessment (LCA) methodologies, the primary and secondary recycling routes assure better energy and resource proficiency, in addition to significantly reducing the carbon dioxide and nitrogen oxide emissions [39,40].

The right identification of the waste source, industrial, urban, or agricultural, has as a first step the recognition of the composition of the thermoplastic waste fraction: kind of polymer; additives; single or several polymers; and expected degradation, soil, and pollutant levels. A previous preconditioning step of milling, grinding, or shredding should prepare waste in the most convenient morphology, both size and shape, as flakes, powder, or pellets. To design the optimal washing and waste-conditioning preprocessing treatment, it is necessary to correctly characterize what and conduct an accurate qualitative and quantitative analysis of the plastic waste fraction under evaluation, soil level included. Results from this step are the keys to determine the scientific-technical viability of the thermoplastic wastes as the right raw materials for the primary or secondary recycling routes.

In compounding steps, with the use of a stabilizer package, if any, or fillers, reinforcements, or additional virgin polymer fractions, it is important to include high-value-added chemically modified thermoplastics from polymeric by-products that can be useful as interfacial modifier agents on heterogeneous systems, blends, and composites, with a view to improving the end properties of these compounds [71]. Moreover, the value of the commingled thermoplastic fractions may be enhanced through the optimal combination between reactive processing and interfacial agents able to strengthen the interfaces between the different components by, for instance, promoting grafting reactions at the nano domain of the interfaces, yielding heterogeneous but stable morphologies and assuring good properties of the recycled materials [72–77].

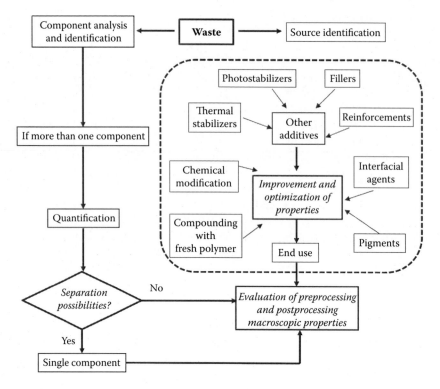

FIGURE 27.3 Technologies for primary and secondary thermoplastic recycling.

Figure 27.3 displays the general methodology to follow at the level of the primary reuse and secondary recycling routes of organic thermoplastic-based materials before resale. The message from the scientific perspective is simple: it is not possible to properly recycle a material not well characterized or known.

27.6.2 Incineration and Tertiary Recycling Routes

Thermoplastic wastes from the primary and secondary recycling routes discarded for either technical or economic reasons may pass to those described below as tertiary or quaternary thermoplastic waste disposal routes.

Aside from the primary and secondary recycling routes, because of their organic nature, thermoplastic material wastes are suitable for a third and a fourth well-differentiated way of disposal. The former, named the tertiary recycling route or chemical recycling option, deals with the conversion of the discarded plastics into high-value and feedstock chemicals, monomers included. The latter, the so-called energy recovery option, consists of converting the thermoplastic wastes into fuel feedstock for incinerators. Obviously, both options are the largest in costs, and only the petrochemical and energy industries have the expertise and implementation facilities needed.

The energy recovery options benefit from the high calorific content of the most common thermoplastics (Table 27.1), which results in similar and even higher values than those of the traditional combustibles. Average yields are around 22% of the overall energy costs of the incineration processes, and they are considered as the best means of MSW disposal, especially by countries such as Japan or Switzerland, with limited space for landfills. Those first MSW incineration experiences put into evidence either the need to reduce the waste volume to treat by incineration routes [78] or the need to enhance the burning technologies [79,80] by adapting them to the different thermoplastic burning behaviors. Besides, it must be ensured that the capturing systems for ashes and emerging

TABLE 27.1
Calorific Power of Different Thermoplastic and Organic Materials

Material	(kJ/kg)10^{-3}
Polystyrene (PS)	46
Polyethylene (PE)	46
Polyvinyl chloride (PVC)	44
Natural gas	48
Fuel oil	44
Coil	29
Lignite	20
Leather	18.9
Paperboard	16.8
Wood	16
Fats	7.8
Domestic waste	8

TABLE 27.2
EU Total Emission Limit Values for MSW Incineration Processes

Pollutant	EU Directive 2000/76/EC, mg/m^3
Total dust	10
NO$_x$	200
SO$_2$	50
HCl	10
Cd/derivates	0.05
Hg/derivates	0.05
Dioxins/furans	0.1

gases comply adequately with the local environmental quality legislation [81] (Table 27.2). The energy recovery options by waste incineration increased in the European Union from 96 kg per capita in 2005 to 102 kg in 2008, accounting for 1.3% of the overall energy production of the EU [82].

Major concerns with the thermoplastic waste burning processes come from their high viscosity in the molten states and their bulky character. These two features favor the emerging of thermal behavior that gives rise to flaming, dropping, and explosive events. To avoid these, good characterization and conditioning of the thermoplastic wastes before feeding into the ovens is needed, sometimes coupled with dilution with solvents or any other kind of profitable organic residue. The right characterization of the oven feed helps to identify the presence of hazardous pollutants, for instance, heavy and nonvolatile metals, or the presence of PVC, which can influence the fine control of the waste's residence time and temperature conditions and avoid the release of toxic dioxins. The hydrochloric acid formed must be removed from the emerging gases, for instance, by using sorbents such as calcium complexes or iron oxides, which obviously increase the costs of operation [83]. The presence of heavy metals requires additional handling of ashes, while if nonvolatiles are present, the costs of their controlled disposal may increased if the scoriae are classified as hazardous matter. Hence, good characterization and quantification of the feed composition helps to minimize the environmental impact of burning utilities and to assure environmental quality.

The chemical recycling of thermoplastic wastes means their conversion into, ideally, monomers (depolymerization), useful petrochemicals, or feedstock products. The kind of thermoplastic determines the best chemical recycling routes; for instance, the condensation polymers like linear polyesters [84–87] or PAs [88–90] are suitable for obtaining monomers, while vinyl polymers such as PS [91,92], PVC [93–95], or polyolefins [42,96,97] may act as raw materials for fuel or synthetic lubricants of base petrochemicals. Technologies from petrochemical industries, mainly, pyrolysis, gasification, viscosity breaking, thermolysis, catalytic or stream cracking, liquid–gas hydrogenation, and hydrolysis, were immediately translated to the chemical recycling processes to obtain from monomers value-added products like synthetic lubricants or fuels, mainly gasoline.

Thermochemical processing works at temperatures above the thermal stability temperatures of thermoplastics, usually between 400°C and 800°C, and benefits from the well-known thermoplastic thermal sensitivity, but thermal degradation or thermolysis processes give rise to very different ratios of liquid, gas, and energy fractions, depending on the reaction conditions. Because of the bulky nature and high viscosity of thermoplastics, both batch and continuous processes are under consideration. To take advantage of the liquid fractions, the preferred reactors are high-temperature modified extruders, while to maximize the gas fractions, the fluidized-bed reactors are the most common. The wide range of hydrocarbon products obtained put a demand on further treatment and recovery processes. The atmospheric reaction must be controlled, and depending on the chosen strategies, it is possible to obtain very different ratios of gases to liquids and to consider the energy recovery options during processing by restricting the overall gas–liquid ratios. Reaction conditions below the stoichiometric air amounts favor gasification processes and yield large amounts of hydrogen, carbon monoxide, and carbon dioxide in mutually different ratios.

Thermal cracking in the absence of oxygen by pyrolysis processes [49,98,99] defines the cleanest route and is accompanied by high calorific power from the gas fractions. In addition to being the most suitable for other organic wastes, it produces minor amounts of solid residual char and carbon fractions that are useful as active carbons or solid fuels.

It is possible to change the different thermal degradation pathways for thermoplastic wastes by using catalyst technologies, mainly based on microporous and mesoporous systems, as a way to improve the selectivity to the feedstock-desired fractions and to reduce the process temperatures [100–102]. An important question that remains open is how the deactivation of the catalysts works in the combination of thermal and catalytic cracking strategies [97].

Most of the petrochemical utility scale-up methods remain based on large production factors well associated with continuous production processes; batch or pulsating operation modes may also be considered. As in all other recycling routes, the economic competitiveness of the thermoplastic waste chemical recycling utilities sharply depends on a feasible and accurate feed supply chain. It is the right time to develop new value-added chemical recycling technologies that increase the economic incentive for this recycling route of thermoplastic wastes. The R&D laboratory-scale activities planned according to the fundamentals of chemical reaction engineering produce good results by incorporating a balance of thermodynamic species and very important design parameters such as work temperature, residence time coming from the degradation kinetics, and assessment of the energies of activation. More information on the different kinds of tertiary recycling can be found in reviews by Al-Salem et al. [42] and Aguado et al. [97] on municipal thermoplastic wastes; Briassoulis et al. [54] and Kyrikou et al. [62] on agricultural thermoplastic wastes; and Harder et al. [103] on automotive thermoplastic wastes. Additional studies about *wild* thermoplastic waste disposal and remediation have been reported as well [104,105].

27.6.3 Economic and Ecological Dimensions

The so-called eco-balance couples the economic criteria with those of technical viability and safety, health, and the environment. LCA [39,40] is the most common method to estimate the environmental impact of a product, process, or activity. There are five blocks, one accounting for each of the following: the energy and environmental inputs affecting the extraction of the original raw materials involved, their manufacturing to obtain materials, item manufacturing, the item use or consumption block, and final disposition once the item becomes residual [105].

It is not an easy task to assign the proper quantitative values to be included in both the life and cost assessment cycles, mainly, energy and primary fuels; raw materials; and solid, liquid, and gaseous emissions. The results concerning thermoplastic wastes are successively refined in the context of the worldwide problem of solid waste disposal. The household wastes span all the existing waste management routes by including collection; transport and sorting; and biological, mechanical, thermal, and chemical pretreatments, besides recycling, reuse, and landfill disposal. A review by Laurent et al. [106] analyzed more than 200 LCA works dealing with solid waste management, from all around the world, and concluded that for thermoplastic wastes, a significant number of studies show that recycling routes, both mechanical and chemical, give rise to environmental performance better than that for landfilling or the incineration processes. A majority of the works analyzed different management systems of household wastes: plastic, metal, glass, paper and organics, and mixed-waste fractions. They confirmed that paper waste landfill disposal has the worst environmental performance, and because of LCA's ability to fit into local conditions in environmental performance quantifications, it countersigns the correctness of the waste-hierarchy line of thinking, leading to better-informed decision making.

27.7 CONCLUSIONS AND PERSPECTIVES

Building on experiences of thermoplastic waste recycling technologies from the early 1980s, thermoplastic solid waste management has been recognized as a technological challenge but just one more of the many environmental issues linked to waste management in particular and humankind development in general. Because of the strategic dimension of the materials field, recycling is affected by legal, cultural, industrial, institutional, governmental, and international raw material markets, all of them largely susceptible to strong economic interests.

The widest concept of *plastic recycling* appears as a challenge for R&D activities for new and advanced thermoplastic polymers and for composites and heterogeneous materials based on them. The well-known *properties/processing/performance* relationship is employed to comply not only with those requirements dealing with applications but also with those related to the EOL of the thermoplastic goods, by minimizing their environmental impact [107,108]. Industrially and under the "green chemistry" label, the integration of environment-friendly concepts grows day by day, from synthesis and polymer production steps to those related to the final stages of manufacturing [109]. Moreover, the peripheral activities for additive package development—stabilizers, lubricants and processing aids, dyes and pigments, flame retardants, fillers and reinforcement treatments, etc.—are included in that challenge as well, and the trends are moving toward minimizing both quantitatively (i.e., amounts) and qualitatively (avoiding needless diversification or unnecessary mixtures that make recycling difficult or expensive [110]).

Polyolefins remain at the forefront of thermoplastic materials, accounting for around 60% of the total plastic wastes [111–113]. The hierarchical recycling procedures assure that after the secondary recycling operations [114,115], the progressively rejected thermoplastics, due to loss of properties, are suitable for the tertiary recycling routes by conversion to feedstock for recycling, chemicals, or monomers [116,117]. The energy recovery options remain open, but only for those hazardous solid wastes, and must be handled according to the best available technologies (BATs).

To maintain the required characteristics, such as high strength and good mechanical performance, high heat resistance, high barrier properties, etc., required in the new, good recycling performance challenges in sectors like automotive parts, household electrical appliances, electronic devices, and food packaging, it is often required to incorporate new ad hoc EOL item specifications. Among them are efforts to increase the lifetime of goods by autorepairing options from the so-called smart materials, the expected significant weight savings in the so-called advanced thermoplastic nanocomposites by the sharp decrease in the amounts of inorganic components, etc. [118–120].

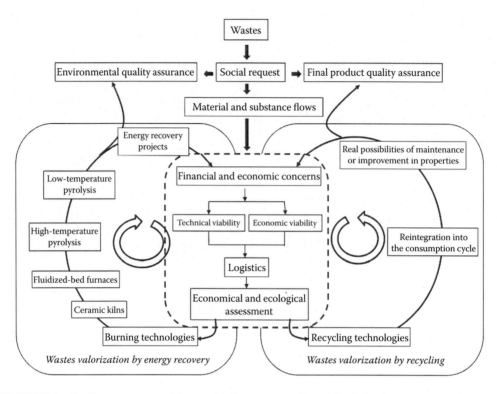

FIGURE 27.4 Quality assurance and thermoplastic waste recycling and valorization routes.

In any case, the sustainability principle demands the creation of proper information channels and the development day by day of more reliable data for production rates, realistic recycling targets, and the BATs. This is to be adapted to each kind of thermoplastic waste, attending to the source and its scientific-technical characterization results, such as composition and degradation level of the components. Research centers and universities continue to play a key role in these efforts as well as in the training of well-formed environment professionals in the overall field of waste management, cognizant but free of the strong economic pressures imposed by the competition between different materials and the raw material markets [121]. In terms of eco-balance, quality assessment (Figure 27.4) emerges as the key factor for compliance with regard to the sustainability principle.

IN MEMORIAM

This chapter is dedicated to the memory of Dr. Ovidio Laguna Castellanos (April 7, 1934–March 13, 2011), a pioneer in the study of the recycling of plastic materials and founder of the Polymer Engineering Group at the ICTP.

REFERENCES

1. Flory, P. J. *Principles of Polymer Chemistry*. Cornell University Press, Ithaca, NY, 1953.
2. Allen, G. The way forward. *Materials World (IOM)*, 2, 6, 326–331, 1994.
3. Olson, G. B. Designing a new material world. *Science*, 288, 993–998, 2000.
4. Ryan, A. J. Integrating polymer science and engineering. *Trends in Polymer Science*, 4, 138, 1996.
5. Atwood, J. L., Lehn, J. M., Davies, J. E. D., Nicol, D. D., Vogte, F. *Comprehensive Supramolecular Chemistry*. Pergamon Press, Amsterdam, 1996.
6. Vogl, O., Jaycox, G. D. Trends in polymer science. Polymer science in the 21st century. *Progress in Polymer Science*, 24, 3–6, 1999.

7. Wegner, G. The future of polymer science and technology in Europe. *European Polymer Federation Magazine*, 1, 13–18, 2001.
8. Adler, H. J., Fisher, P., Hellen, A., Jansen, I., Kuckling, D., Konber, H., Lehmann, D., Piontek, J., Pleul, D., Simon, F. Trends in polymer chemistry. *Acta Polymerica*, 50, 232–239, 1999.
9. Put, J. Business in polymeric materials, quo vadis? *European Polymer Federation Magazine*, 1, 23–27, 2001.
10. Gugumus, F. Thermolysis of polyethylene hydroperoxides in the melt, Part 10: Product yields from bimolecular and pseudo-monomolecular hydroperoxide decomposition. *Polymer Degradation and Stability*, 93, 520–532, 2008.
11. Gugumus, F. Re-examination of the thermal oxidation reactions of polymers. 2. Thermal oxidation of polyethylene. *Polymer Degradation and Stability*, 76, 329–340, 2002.
12. Zaikov, G. E. The current state of polymer aging and stabilization. *International Journal of Polymeric Materials and Polymeric Biomaterials*, 16, 1–30, 1992.
13. Geuskens, G. Fundamental aspects of the photooxidation of hydrocarbon polymers. *Journal of Polymeric Materials and Polymeric Biomaterials*, 16, 31–36, 1992.
14. Ilie, S., Senetscu, R. Polymeric materials review on oxidation, stabilization and evaluation using CL and DSC methods. TE Technical Note. CERN Publication, 2009. Available at http://cds.cern.ch/record/1201650/files/Ilie_TE_Technical_Notes.pdf?version=1.
15. Vian, A. Residue and human life. Key Lecture to introduce *"Los Residuos y sus Riesgos para la Salud,"* Jiménez, S., Ed. Spanish Royal Academy of Pharmacy, Madrid, 1998. Available at http://www.analesranf.com/index.php/mono/article/viewFile/384/405 (In Spanish).
16. Laguna, O., Collar, E. P., Taranco, J. Reuse of plastic recovered from solid wastes. Modification of properties in polyethylene blends. *Journal of Polymer Engineering*, 7, 3, 169–195, 1987.
17. Laguna, O., Collar, E. P., Taranco, J., Vigo, J. P. Reuse of urban plastic wastes. Melt behavior of HDPE/LDPE blends by using torque rheometer data. *Journal of Polymer Materials*, 4, 195–205, 1987.
18. Laguna, O., Collar, E. P. Modification of properties in polyethylene blends from recovered urban wastes. *Resources, Conservation and Recycling*, 2, 37–56, 1988.
19. Laguna, O., Collar, E. P., Taranco, J. Reuse of plastics recovered from solid wastes. Thermal and morphological studies from HDPE/LDPE blends. *Journal of Applied Polymer Science*, 38, 667–685, 1989.
20. Laguna, O., Collar, E. P., Taranco, J. Plastic film wastes reuse. A study of the mechanical behavior of blends of high and low density polyethylene. *Journal of Polymer Materials*, 7, 225–229, 1990.
21. Stein, R. S. Polymer recycling: Opportunities and limitations. *Proceedings of the National Academy of Science of the United States of America*, 89, 835–838, 1992.
22. Laguna, O., Collar, E. P., Taranco, J. In *Polymer Recycling*, La Mantia, F. P. Ed., Technomic, Toronto, 1993.
23. Bisio, A. L., Xanthos, M. *How to Manage Plastics Waste. Technology and Market Opportunities.* Hanser, Munich, 1994.
24. Collar, E. P., Taranco, J., Michelena, J., Laguna, O., García-Martinez, J. M. Post consumer plastic wastes. Polyolefin blends and modification possibilities in the molten state. *Journal of Polymer Materials*, 13, 341–349, 1996.
25. Subramanian, P. M. Plastics recycling and waste management in the US. *Resources, Conservation and Recycling*, 28, 253–263, 2000.
26. Aguado, J., Serrano, D. P., Vicente, G., Sánchez, N. Enhanced production of α-olefins by thermal degradation of high-density polyethylene (HDPE) in decalin solvent: Effect of the reaction time and temperature. *Industrial Engineering Chemical Research*, 46, 3497–3504, 2007.
27. López, A., de Marco, L., Caballero, R. M., Laresgoiti, M. F., Adrados, A. Influence of time and temperature on pyrolysis of plastic wastes in a semi-batch reactor. *Chemical Engineering Journal*, 173, 62–71, 2011.
28. López, A., de Marco, L., Caballero, R. M., Laresgoiti, M. F., Adrados, A., Torres, A. Pyrolysis of municipal plastic wastes. II: Influence of raw material composition under catalytic conditions. *Waste Management*, 31, 1973–1983, 2011.
29. Brogaard, L. K., Damgaard, A., Jensen, M. B., Barlaz, M., Christensen, T. H. Evaluation of life cycle inventory data for recycling systems. *Resources, Conservation and Recycling*, 87, 30–45, 2014.
30. Bognor, J. E., Matthews, E., Katzsbestein, A., Blake, D., Carolan, M. Greenhouse gas emissions from landfills: What we know and what we don't know. *ISAW World Congress Proceedings 2000 and Beyond-Which Choices for Wastes Management*. Paris, 421–430, July 2000.
31. PRISMA, S. L. Regeneration procedure of partially degraded polyethylene. Laguna Castellanos, O., Pérez Collar, E., Taranco González, J. Pat. Nr.: ES2051642; B29B17/00; B29B17/00; (IPC1-7): B29B17/00, 1994.

32. Recycling and Economic Development. A review of existing literature on job creation, capital investment, and tax revenues, 2009. Available at http://your.kingcounty.gov/solidwaste/linkup/index.asp.
33. Morawetz, H. *The Origins and Growth of a Science*. Wiley, New York, 1985.
34. Municipal solid waste in the United States: 2007 facts and figures. Executive Summary. Office of Solid Waste Management and Emergency response, 5306P, EPA 530-R-080-010, USEPA, November 2008. Available at http://www.epa.gov/; http://www2.epa.gov/learn-issues/learn-about-waste.
35. Association of Plastics Manufactures in Europe. Plastics the facts 2013. An analysis of European latest plastics production, demand and waste data. PlasticsEurope. Available at http://www.plasticseurope.org/Document/plastics-the-facts-2013.aspx?FolID=2.
36. Available at http://www.urbaser.es/seccion-85/Plantas-Integrales-de-Residuos.
37. Available at http://www.solvayplastics.com/sites/solvayplastics/SiteCollectionDocuments/VinyLoop/VinyLoop%20Brochure_10.2011.pdf.
38. Velghe, I., Carleer, R., Yperman, J., Schreurs, S. Study of the pyrolysis of municipal solid waste for the production of valuable products. *Journal of Analysis and Applied Pyrolysis*, 92, 366–375, 2011.
39. Rigamonti, L., Grosso, M., Sunseri, M. C. Influence of assumptions about selection and recycling efficiencies on the LCA of integrated waste management systems. *International Journal of LCA*, 14, 5, 411–419, 2009.
40. Thorneloe, S. Application of life-cycle management to evaluate integrated municipal solid waste management strategies. EPA/R-99/2006. US Environmental Protection Agency, Office of Research and Development, Washington, DC, 2006.
41. Bovea, M. D., Ibáñez-Forés, V., Gallardo, A., Colomer-Mendoza, F. J. Environmental assessment of alternative municipal solid waste management strategies. A Spanish case study. *Waste Management*, 30, 2383–2395, 2010.
42. Al-Salem, S. M., Lettieri, P., Baeyens, J. Recycling and recovery routes of plastic solid waste (PSW): A review. *Waste Management*, 29, 2625–2643, 2009.
43. Hazra, T., Goel, S. Solid waste management in Kolkata, India: Practices and challenges. *Waste Management*, 29, 470–478, 2009.
44. Guerrero, L. A., Maas, G., Hogland, W. Solid waste management challenges for cities in developing countries. *Waste Management*, 33, 220–232, 2013.
45. Misko, G. G., U. S. and EU requirements for recycled food contact materials. *Food Safety Magazine*, October/November 2013. Available at http://www.foodsafetymagazine.com/magazine-archive1/octobernovember-2013/us-and-eu-requirements-for-recycled-food-contact-materials.
46. Guidance for industry: Use of recycled plastics in food packaging: Chemistry considerations, 2011. Available at http://ec.europa.eu/food/food/chemicalsafety/foodcontact/legisl_list_en.htm.
47. Available at http://www.fda.gov/Food/IngredientsPackagingLabeling/PackagingFCS/RecycledPlastics/default.htm.
48. EC (2000) Directive 2000/53/EC of the European Parliament and of the council of 18 September 2000 on end-of-life vehicles. *Official Journal of European Communities*, 1269, 0034–0043.
49. Santini, A., Passarini, F., Vassura, I., Serrano, D., Dufour, J. Auto shredder residue recycling: Mechanical separation and pyrolysis. *Waste Management*, 32, 852–858, 2012.
50. Collar, E. P., García-Martínez, J. M., Alfagrán. Recycling of automotive plastics waste: Evaluation of properties on polypropylene matrix composites. Unpublished Results, 2000.
51. Wei, J., Realff, J. Design and optimization of drum-type electrostatic separators for plastic recycling. *Industrial and Engineering Chemical Research*, 44, 3503–3509, 2005.
52. Laguna, O., Collar, E. P., Taranco, J., Prisma. Recycling of agricultural films of polyethylene. Unpublished results, 1992.
53. Briassoulis, D., Hiskakis, M., Babou, E. Technical specifications for mechanical recycling of agricultural plastic waste. *Waste Management*, 33, 1516–1530, 2013.
54. Briassoulis, D., Hiskakis, M., Babou, E., Antiohos, S. K., Papadi, C. Experimental investigation of the quality characteristics of agricultural plastic wastes regarding their recycling and energy recovery potential. *Waste Management*, 32, 1075–1090, 2012.
55. Al-Robaidi, A. LDPE/EPDM multilayer films containing recycled LDPE for greenhouse applications. *Journal of Polymers and Environment*, 9, 1, 25–30, 2001.
56. Annual report. Evolution of plastic recycling—Spain 1999–2009. Cicloplast, 2011. Available at http://www.cicloplast.com/ingles/pdf/evolution_of_plastics_recycling.pdf.
57. Tanskanen, P. Management and recycling of electronic waste. *Acta Materialia*, 61, 1001–1011, 2013.
58. Ryan, P. G., Moore, C. J., vand Franeker, J. A., Moloney, C. L. Monitoring the abundance of plastic debris in the marine environment. *Philosophical Transactions of the Royal Society*, 364, 1999–2012, 2009.

59. Barnes, D. K. A., Galgani, F., Thompson, R. C., Barlaz, M. Accumulation and fragmentation of plastic debris in global environments. *Philosophical Transactions of the Royal Society*, 364, 1985–1998, 2009.
60. Teuten, E. L., Saquing, J. M., Knappe, D. R. U., Barlaz, M. A., Jonsson, S. et al. Transport and release of chemicals from plastics to the environment and to wildlife. *Philosophical Transactions of the Royal Society*, 364, 2027–2045, 2009.
61. Siracusa, V., Rocculi, P., Romani, S., Rosa, M. D. Biodegradable polymers for food packaging: A review. *Trends in Food Science Technology*, 19, 634–643, 2008.
62. Kyrikou, I., Briassoullis, D. Biodegradation of agricultural plastic films: A critical review. *Journal of Polymers and Environment*, 15, 125–150, 2007.
63. Narayan, R. *Biobased and Biodegradable Polymer Materials: Rationale, Drivers, and Technology Exemplars*. ACS Symposium Series 939, Chap. 18, 282–302, 2006, ACS, Washington, DC.
64. Narayan, R. Carbon footprint of bioplastics using biocarbon content analysis and life cycle assessment. *MRS Bulletin*, 36, 716–721, 2011.
65. van der Mee, M. A. J., Goossens, J. G. P., van Duin, M. Thermoreversible covalent crosslinking of maleated ethylene/propylene copolymers with diols. *Journal of Polymer Science. A Polymer Chemistry*, 46, 1810–1825, 2008.
66. Nishida, H. Development of materials and technologies for control of polymer recycling. *Polymer Journal*, 43, 435–447, 2011.
67. Bulow, J. An economic theory of planned obsolescence, 1986. Available at https://faculty-gsb.stanford.edu/bulow/articles/an%20economic%20theory%20of%20planned%20obsolescence.pdf.
68. Mycroft, N. Consumption, planned obsolescence and waste, 2006. Available at http://eprints.lincoln.ac.uk/2062/1/Obsolescence.pdf.
69. Yang, Y., Boom, R., Irion, B., van Heerden, D., Kuiper, J., de Wit, H. P. Recycling of composite materials. *Chemical Engineering and Processing*, 51, 53–68, 2012.
70. Perry, N., Bernard, A., Laroche, F., Pompidou, S. Improving design for recycling. Application to composites. *CIRP Annals. Manufacturing Technology*, 61, 151–154, 2012.
71. Utracki, L. A. *Polymer Alloys and Blends*. Hanser, Munich, 1989.
72. Pernot, H., Baumert, M., Court, F., Leibler, L. Design and properties of co-continuous nanostructures polymers by reactive blending. *Nature Materials*, 1, 54–58, 2002.
73. Collar, E. P., García-Martínez, J. M. On chemical modified polyolefins by grafting of polar monomers. A survey based on recent patents literature. *Recent Patents on Materials Science*, 3, 1, 76–91, 2010.
74. de Gennes, P. G. Conformation of polymers attached to an interface. *Macromolecules*, 13, 1069–1075, 1980.
75. García-Martínez, J. M., Areso, S., Taranco, J., Collar, E. P. Chapter 13: Heterogeneous materials based on polypropylene. In *Polyolefin Blends*, Kyu, T. and Nwabunma, D. Eds. John Wiley & Sons, Hoboken, NJ, 2008.
76. Poon, B. C., Chum, S. P., Hiltner, A., Baer, E. Modifying adhesion of linear low-density polyethylene to polypropylene by blending with a homogeneous ethylene copolymer. *Journal of Applied Polymer Science*, 92, 109–115, 2004.
77. Poon, B. C., Chum, S. P., Hiltner, A., Baer, E. Adhesion to polyethylene blends to polypropylene. *Polymer*, 45, 893–903, 2004.
78. Lea, W. R. Plastic incineration versus recycling: A comparison of energy and landfill cost savings. *Journal of Hazardous Materials*, 47, 295–302, 1996.
79. Incinerators: Myths vs Facts about "Waste to energy," 2012. Available at http://www.no-burn.org/downloads/Incinerator_Myths_vs_Facts%20Feb2012.pdf.
80. Pérez, C. A. International experiences on MSW treatments. URBASER. Available at http://www.conama10.es/conama10/download/files/SDs%202010/1335816613_ppt_CAPerez.pdf (In Spanish).
81. Directive 2000/76/EC on the incineration of waste. *Official Journal of the European Communities*, L332/91, December 28, 2000.
82. 2000 Environment Policy Review. Available at http://ec.europa.eu/environment/pdf/policy/EPR_2009.pdf.
83. COM(2000) 469 final. Green Paper. Environmental issues of PVC. Available at http://ec.europa.eu/environment/waste/pvc/pdf/en.pdf.
84. Paszun, D., Spychaj, T. Chemical recycling of poly(ethylene terephthalate). *Industrial and Engineering Chemistry Research*, 36, 1373–1383, 1997.
85. Valdya, U. R., Nadkarni, V. M. Unsaturated polyesters from PET waste: Kinetics of polycondensation. *Journal of Applied Polymer Science*, 34, 235–246, 1987.

86. Yamaye, M., Hashime, T., Yamamoto, K., Kosugi, Y., Cho, N., Ichiki, T., Kito, T. Chemical recycling of poly(ethylene terephthalate). 2. Preparation of terephthalohydroxamic acid and terephthalohydrazide. *Industrial and Engineering Chemistry Research*, 41, 3993–3998, 2002.
87. Awaja, F., Pavel, D. Review on recycling of PET. *European Polymer Journal*, 41, 1453–1477, 2005.
88. Nemade, A. M., Mishra, S., Zope, V. S. Chemical recycling of polyamide waste at various temperatures and pressures using high pressure autoclave technique. *Journal Polymers and Environmental*, 19, 110–114, 2011.
89. Kamikura, A., Yusuke, O., Tsunami, S. Supercritical secondary alcohols as useful media to convert polyamide into monomeric lactum. *Chemical Sustainable Chemistry*, 1, 82–84, 2008.
90. Kamikura, A., Yamamoto, S. A novel depolymerization of nylon in ionic liquids. *Polymer Advanced Technology*, 19, 1391–1395, 2008.
91. Wang, Z., Bai, R., Ting, Y. P. Conversion of waste polystyrene into porous and functionalized adsorbent and its application in humic acid removal. *Industrial and Engineering Chemical Research*, 47, 1861–1867, 2008.
92. Serrano, D. P., Aguado, J., Escola, J. M. Catalytic conversion of polystyrene over HMCM-41, HZSM-5 and amorphous SiO_2–Al_2O_3: Comparison with thermal cracking. *Applied Catalysis*, B25, 181–188, 2000.
93. Duangchan, A., Samart, C. Tertiary recycling of PVC-containing plastic waste by copyrolysis with cattle manure. *Waste Management*, 28, 2415–24121, 2008.
94. Bhaskar, T., Uddin, M. A., Kaneko, J., Kusaba, T., Matsui, T., Muto, A., Sakata, Y., Murata, K. Liquefaction of mixed plastics containing PVC and dechlorination by calcium bases sorbent. *Energy Fuels*, 17, 75–109, 2003.
95. Mio, H., Saeki, S., Kano, J., Saito, F. Estimation of mechanochemical dechlorination rate of poly(vinyl chloride). *Environmental Science and Technology*, 36, 1344–1350, 2002.
96. Wang, H., Chen, D., Yuan, G., Ma, X., Dai, X. Morphological characteristics of waste polyethylene/polypropylene plastics during pyrolysis and representative morphological signal characterizing pyrolysis stages. *Waste Management*, 33, 327–339, 2013.
97. Aguado, J., Serrano, D. P., Escola, J. M. Fuels from waste plastics by thermal and catalytic processes: A review. *Industrial and Engineering Chemistry Research*, 47, 7982–7992, 2008.
98. Lovett, S., Berruti, F., Behie, L. A. Ultrapyrolytic upgrading of plastic wastes and plastics/heavy oil mixtures to valuable light gas products. *Industrial and Engineering Chemical Research*, 36, 4436–4444, 1997.
99. Donaj, P., Yang, W., Blasiak, W., Forsgreen, C. Recycling of automobile shredder residue with a microwave pyrolysis combined with high temperature steam gasification. *Journal of Hazardous Materials*, 182, 1–3, 80–89, 2010.
100. Serrano, D. P., Aguado, J., Escola, J. M., Garagorri, E., Rodriguez, J. M., Morselli, L., Palazzi, G., Orsi, R. Feedstock recycling of agriculture plastic film wastes by catalytic cracking. *Applied Catalysis B: Environmental*, 49, 257–265, 2004.
101. Arandes, J. M., Torre, I., Castaño, P., Olazar, M., Bilbao, J. Catalytic cracking of waxes produced by the fast pyrolysis of polyolefins. *Energy Fuels*, 21, 561–577, 2007.
102. Harder, M. K., Tening-Forton, O. A critical review of developments in the pyrolysis of automotive shredder residue. *Journal of Analytics and Applied Pyrolysis*, 79, 387–394, 2007.
103. van Sebile, E., England, M. H. Origin, dynamics and evolution of ocean garbage patches from observed surface drifters. *Environmental Research Letters*, 7, 044040, 2012.
104. Available at http://www.theoceancleanup.com/; http://www.theoceancleanup.com/the-problem.html?gclid=COn4uYfwlL8CFenItAodbDYAbA.
105. Available at http://www.recyc-quebec.gouv.qc.ca/Upload/Publications/MICI/PPT-Valorisation08/J-Morris.pdf.
106. Laurent, A., Bakas, I., Clavreul, J., Bernstad, A., Niero, M., Gentil, E., Hauchild, M. Z., Christensen, T. H. Review of LCA studies of solid wastes management systems—Part I: Lessons learned and perspectives. *Waste Management*, 34, 573–588, 2014.
107. Ober, C. K., Cheng, S. Z. D., Hammond, P. T., Muthukumar, M., Reichmanis, E., Wooley, K. L., Lodge, P. Research in macromolecular science: Challenges and opportunities for the next decade. *Macromolecules*, 42, 465–471, 2009.
108. Scaffaro, R., Morreale, M., Mirabella, La Mantia, F. P. Preparation and recycling of plasticized PLA. *Macromolecular Materials and Engineering*, 296, 141–150, 2011.
109. REACH for Polymers. Understanding legislation. Best Practice Handbook. 2011. Available at http://www.reachforpolymers.eu.

110. Kumar, A. P., Depan, D., Tomer, N. S., Singh, R. P. Nanoscale particles for polymer degradation and stabilization—Trends and future perspectives. *Progress in Polymer Science*, 479–515, 2009.
111. Galli, P., Vecellio, G. Polyolefins: The most promising large-volume materials for the 21st century. *Journal of Polymer Science. Polymer Chemistry*, 42, 396–415, 2004.
112. Hustad, P. D. Frontiers in olefin polymerization: Reinventing the world's most common synthetic polymers. *Science*, 325, 704–707, 2009.
113. Tonzani, S. The renaissance of polyolefins. *Journal of Applied Polymer Science*, 127, 837, 2013.
114. Rosen, M. Cellulose-reinforced polypropylene: A processing study. *Plastics Engineering*, March 22–26, 2014.
115. Dicker, M. P. M., Duckworth, P. F., Baker, A. B., Francois, G., Hazzard, M. K., Weaver, P. M. Green composites: A review of material attributes and complementary applications. *Composites: Part A*, 56, 280–289, 2014.
116. Goto, T., Ashihara, S., Kato, M., Okajima, I., Sako, T. Use of single-screw extruder for continuous silane cross-linked polyethylene recycling process using supercritical alcohol. *Industrial and Engineering Chemistry Research*, 51, 6067–6971, 2012.
117. Altalhi, T., Kumeria, T., Santos, A., Losic, D. Synthesis of well-organised carbon nanotube membranes from non-degradable plastic bags with tuneable molecular transport: Towards nanotechnological recycling. *Carbon*, 423–433, 2013.
118. Shanks, R. A. Biomimetic materials: A challenge for nano-scale self-assembly. *eXPRESS Polymer Letters*, 8, 8, 543, 2014.
119. Heilgtag, F. J., Nierderberger, M. The fascinating world of nanoparticle research. *Materials Today*, 16, 7/8, 262–272, 2013.
120. Karger-Kocsis, J. Carbon dioxide "management" by polymers. *eXPRESS Polymer Letters*, 8, 142, 2014.
121. Lovelock, J. L., Margulis, L. Atmospheric homeostasis by and for the biosphere: The Gaia hypothesis. *Tellus*, 16, 1–9, 1974.

28 Environment Health and Safety
Regulatory and Legislative Issues

Emilia P. Collar and Jesús María García-Martínez

CONTENTS

28.1 Introduction 941
28.2 Sustainable Development: Legal Frameworks 942
28.3 Chemical Industries: Environmental Challenges of Plastic Materials 943
28.4 Accreditation of the Commitment: Quality Systems and Technical Competence
      of Laboratories 945
28.5 Standardization: Convergence between the Regulated and the Voluntary Field 946
28.6 Examples of Complexity of Research and Development and Legal Frameworks 948
    28.6.1 Plastic Wastes in a Standardized World 948
    28.6.2 Bioplastics in a Standardized World 950
    28.6.3 Nanotechnology and Nanometrology in a Standardized World 950
    28.6.4 The Question of Reference Materials 951
    28.6.5 Plastic Materials and Electronic Devices 952
28.7 Perspectives on Quality Infrastructure and Environmental Safety 952
References 953

28.1 INTRODUCTION

Over the years, the regulatory and legislative procedures by which governments and industries ensure that products meet the needs of citizens, health and safety included, have changed and evolved. After the industrial revolution at the early stages of the twentieth century, the first synthetic plastic materials went into the material markets almost at the same time as standardization emerged as a key activity in the expansion of transnational commercial activities. From the last quarter of the same century, and because of the strongly changing industrial and social environments, standardization as well legislative activities became more rigorous. The aim of the present chapter is to give an overview of the plastic material industry and how the overall environmental issues affect the regulatory, standardization, and legislative activities. A few examples dealing specifically with organic plastic materials are given. A number of references and electronic addresses of the main organizations responsible for regulatory, standardization, and legislative activities are provided.

Since the dawn of the twentieth century, organic polymer materials, known as *plastics*, have become a staple of daily life. The introduction of each type of plastic material into the market naturally follows that of traditional materials, metals, glass, or ceramics, by adjusting to the local legal frameworks and specific standards and regulations.

To remove technical barriers to trade, measurement standards and test procedures need to follow internationally accepted norms. Scientific metrology was required by organizations responsible for the development of the measurement standards and their maintenance. It supports, on one hand, the industrial metrology category that has to ensure the adequate functioning of measurement instruments used in the industry, production, and testing processes. On the other hand, it supports the legal metrology that provides the measurements influencing the transparency of economic transactions and particularly those where the legal verifications are key requirements of the measuring instruments [1]. Energy savings from natural gas transport or heat meters, health advances in kidney dialysis, cancer treatments, and food safety are some recent examples of how metrology and the chains of traceability contribute to these policies.

In the second decade of the third millennium, the exponential growth of the worldwide population is expected to lead to the depletion of natural resources and is a contributing factor to global climate change. Because of the finite resources of planet Earth, the feedstock costs rise, and the market consumption patterns are changing in parallel with increasing regulatory pressures. First, environmental alarm signals started in 1972 when, for the first time in an international meeting, Indira Gandhi introduced at the United Nations (UN) Conference on Human Environment, held in Stockholm, the first numerical correlation between consumerism, poverty, and environment, based on the studies by different scientists, including Erwin Schrödinger. In 1973, following the Arabian oil embargo and actions of the Organization of the Petroleum Exporting Countries (OPEC), an MIT report by Meadows et al. [2] entitled "The Limits to Growth" painted a catastrophic end scenario for the consumer society from the depletion of fossil resources. Another MIT report, entitled "Beyond the Limits" [3], reviewed the results of the former one, recommending sustainable development as a path forward to maintain the highest quality of life without jeopardizing the biosphere or the survival of future generations.

28.2　SUSTAINABLE DEVELOPMENT: LEGAL FRAMEWORKS

In adopting an Earth ecosystem philosophy, better stewardship of the resources dealing with energy and materials is required. In response to the energy crisis of the 1970s, concerted efforts were focused on better energy efficiency through the optimization of consumption, costs, and industrial waste reduction. Plastic materials came into mass production markets once their advantages of low cost, being lightweight, and processability were recognized [4]. An interesting view on the role played by the organic plastic materials in our daily lives, environmental, and health matters can be found in the 2009 theme issue of the *Philosophical Transactions of the Royal Society* [5].

The largest fraction of plastic materials today, about a third of the production, goes to disposable items, mainly found in packaging applications. For the last three decades, this fraction of organic thermoplastic materials has remained constant at about 10% of the overall fraction of the big urban solid-waste streams. In terms of industrial economy, any given process that reduces its waste fraction is improving its efficiency and thus its profitability.

A progressive approach to recycling and regulations assumes the overall concept of *from the cradle to the cave*, following methodologies and procedures based on life cycle assessment (LCA) [6]. This approach leads to a balance of ecology and economy, sometimes called eco-balance, by consideration of environmental impact and energy costs.

Competitive recycling has a number of strategic advantages, such as an overall optimization of resources, better profitability of manufacturing plants, reduction of mining and energy requirements, less dependence on imported raw materials, and a reduction of the overall waste amounts and amount of material that ends up in landfills.

Beginning from the 1970s, governments have begun to transition from a penalty to a prevention model (*better to prevent than have to repair damage, or a catastrophe*) as a way to address industrial issues. The different sovereign countries adjust the implementation of new laws and regulations, looking for harmonization between their current legislations and those that meet international harmonization policies, not necessarily by force but via voluntary commitment.

From the new perspective of globalization, it was concluded that legislation matters ought to be directed toward voluntary compliance rather than obligation, in particular, to accommodate differences in infrastructure in different countries.

The traceability chains assure the end user of the highest level of traceability as supported by the current National Metrology Institute or from secondary calibration laboratories, usually accredited as a result of different protocols or mutual recognition arrangements that are internationally recognized. The conformity assessment procedures provide a demonstration that specified requirements relating to a product, process, system, person, or body are fulfilled, from the testing, inspection, and certification of products, up to the personnel and management systems. Each national government designates the corresponding institute, or body, responsible for the specific national standards and the international representation in standardization and regulation activities. The International Organization of Legal Metrology (IOLM) is an intergovernmental treaty organization in existence since 1955, whose main aim and scope are the promotion of the global harmonization of legal metrology procedures by providing its members with metrological guidelines for the elaboration of their national and regional requirements related to all concerns of metrology.

Strong economic interests, significant differences in the level of development in different countries, and diverse cultures all combine to create delays in arriving at implementable legislation. Scientific and technical advances in the last century and the best environmental practice (BEP) and best available technologies (BAT) methodologies helped to guide legislation, such as that proposed to limit emissions of gas, vapor, or liquid effluents, or that on recycling targets from solid-waste disposal. Significant contributions to the understanding of the relationship between metrology and thermoplastic polymers and materials in general were made by many organizations, including the International Union of Pure and Applied Chemistry (IUPAC), established in 1919, and the International Union of Pure and Applied Physics (IUPAP), launched in 1923. Both organizations work to promote the exchange of information and views among the members of their respective scientific communities in their general fields of competence. Specifically, the Commission on Standards, Units, Nomenclature, Atomic Masses and Fundamental Constants of the IUPAP works from the highest scientific level in support of physical measurements, pure and applied metrology, and nomenclature and symbols for physical quantities and units. The IUPAC, as an international, nongovernmental body, is well recognized as the world authority on chemical nomenclature, terminology, standardized methods of measurement, atomic weights, and many other magnitudes to address issues involving the chemical sciences, and it assigns a specific committee devoted to organic polymer materials.

28.3 CHEMICAL INDUSTRIES: ENVIRONMENTAL CHALLENGES OF PLASTIC MATERIALS

Chemicals and polymeric materials are often the subject of intense discussion with regard to the environment, or the environmental damage risks or impact resulting from manufacturing, handling, processing, and transport of hazardous substances, whether as solids, liquids, or gases.

In 1985, the Chemistry Industry Association of Canada launched the *Responsible Care* initiative to address societal concerns on the activities of the chemical industry. Designed as a global and voluntary commitment to progress and sustainability by and for the chemical industry, the initiative looks for continuous improvement in environmental performance by the participating companies. The improvements focus particularly on performance that goes beyond the legal requirements; enhances safety and health; and encourages continuous dialog with neighbors, citizens, and all of the stakeholders, suppliers, and customers.

At the start of the Responsible Care initiative, chemicals operators in more than 50 countries (accounting for nearly 90% of the global chemical production) signed the commitment. Some of their efforts crystallized in 2006 in the Global Product Strategy (GPS), introduced by the International Council of Chemical Associations (ICCA) [7]. The main goal was to favor better management of chemical products by the development of a common language for safety communication and the strengthening of product stewardship. Currently, this initiative also works to meet the main goals compiled at the UN Strategic Approach to International Chemical Management (SAICM), developed between the chemical makers and governments to harmonize regulations and product stewardship, to ensure the safe production, use, and disposal of chemical products, based on the 2002 World Summit on Sustainability.

Plastics and all organic polymer-based materials appear to be fully included in the above-described legal and regulation scenarios, and they have been subject to standardization activities since introduction into the market.

To strengthen global and the transnational economy growth, the legislative framework on chemicals in the European Economic Area (EEA) was fully reformed after 2006, by adopting the REACH initiative. This acronym corresponds to the registration, evaluation, authorization, and restriction of chemical substances, and from 2007, it defines the new European community regulation on chemicals and their safe use. The REACH regulation gives a greater responsibility to industry to manage the risks from chemicals and to provide safety information on the substances that are registered on a central database run by the European Chemicals Agency (ECHA). The term *risk* refers to the possibility and severity of damage to human beings or ecosystems after exposition to chemicals during their production, processing, or use [8–10].

Because of the large number of polymers to register and their limited risk because of their previously well-tested nature, polymers are exempted from registration and evaluation but remain subject to authorization and restriction. There is a specific division for polymer materials established in 2008, named REACH for Polymers. Since there are blends and mixtures of the organic polymers with a lot of other chemicals and components, the plastic processing industries are concerned with the overall REACH regulations. An approach to the complexity and the major efforts that need to be matched for organic polymer-based materials to comply with the REACH legislation can be found in the best-practice documents [11–13].

During the second quarter of the twentieth century, the different industrialized countries launched standardization activities by creating national standardization bodies. The different national standardization bodies—Germany (DIN), France (AFNOR), Great Britain (BS), United States (ASTM), Austrian (ÖNORM), Spain (UNE), and Italy (UNI)—gave rise to the need for a transnational scope for such activities, leading to the formation of the International Organization for Standardization (ISO) [14], International Electrotechnical Committee (IEC) [15], and International Telecommunication Union (ITU) [16].

Currently, standardization activities continue through different organizations, both national and international. However, the ISO and the IEC remain as the main internationally recognized standardization bodies. The American National Standards Institute (ANSI) [17] and the European Committee for Standardization (CEN) [18], each one with its own well-defined organization structure, aim, and scope, represent the main regional ones, the former in the United States and the countries in its area of influence, while CEN centralizes the standardization activities within the European Union.

Further, to promote the technological cooperation initiatives, all these organizations continually review and update already published and accepted standards as well as create and adopt new ones. A material or product, tested under the corresponding standardized method and by independent organizations, upon satisfactory results, can be certified to meet the corresponding standardized requirements. In addition, these organizations remain open to any new kind of documents that have, or may have, the set of specifications and, where appropriate, the test methods to ensure the satisfactory entry of new products into the product or material markets in general.

All the standardization bodies work on the basis of the enforcement responsibilities of both producers and governments, by which the former assure that their products comply with the requirements in the documents that are in force, while the latter watch that nonconforming products are not placed on the market or put into use. Readers interested in the history and organization evolution of the different standardization bodies may find them by following their corresponding electronic addresses. Updated releases of further standards documents, technical committees, work sessions, and more might be found through Internet resources by the use of key words, acronyms, and their links, where the most recent documents in force concerning the different materials, plastics included, and the links with the regulations on industrial activity issues are kept. The Environmental Protection Agency (EPA) [19] for the United States and the EUR-Lex, "the portal to European Union law" [20], for the DGIII and DGIX, the EU directives on industry and environment, respectively, are the main proponents of the global approach, which links industrial activities and the environment by promoting the voluntary technical worldwide harmonization on these subjects.

28.4 ACCREDITATION OF THE COMMITMENT: QUALITY SYSTEMS AND TECHNICAL COMPETENCE OF LABORATORIES

From the above-mentioned Responsible Care initiative and its work principles, successively implemented by the internal establishment of *continuous improvement processes*, sometimes named the *quality culture*, affecting the entire productive value chain of the signatory companies, it is clear that a need for accreditation exists.

Accreditation, within the voluntary field, represents the formal statement of the technical competence of a body for the assessment of conformity of a given process, as defined by the corresponding standards and any other reference or normative documents that make up a given system. The ISO 17025 standard [21] is the accreditation basis document for any company to ensure that their products meet the needs of their customers. Technical laboratories, to maintain this accreditation, must undergo regular surveillance visits by the corresponding national accreditation body. The main intention is that the authorities and industry in all member countries shall accept tests and calibrations from accredited laboratories in a member country. To achieve this, the accreditation bodies subscribe internationally and regionally multilateral arrangements to both recognize and promote the equivalence of each of the other systems of certificates and test reports issued by the accredited organizations.

Most companies have implemented their environmental quality systems within the framework of the environmental management systems (EMSs), which provided the standards ISO 14001 and European Eco-Management and Audit Scheme (EMAS) [22]. EMAS is a volunteer policy of the European Union that recognizes those organizations that have implemented an EMS and have acquired a commitment to continuous improvement, verified by an independent body. The EMAS logo guarantees the reliability of the information given by the company that exhibits it.

The articulation of the various systems of the ISO 14001 series under the same approach allows the establishment of six management codes: occupational health and safety, process safety, stewardship, product tutelage, communication and emergency response, and protection of the environment. The latter—further composed of emissions into the atmosphere, discharges into water, and energy and water balances—also includes the solid-waste issues as matters of specific regulations under significantly changing frames, such as those concerning the solid plastic wastes in light of the continuous advances in the science and technology of these materials.

As occurs with the quality management systems, it is necessary to demonstrate conformity with the essential requirements of whatever directives are applicable. On the release of the technical specifications that should allow the evaluation of conformity in four ways—inspection, testing, management system certification, and product certification—the key lies in the need to harmonize the functioning of national laboratories with equal criteria of technical competence.

To attain conformity on evaluations, organizations must satisfy the requirements of independence, technical competence, and civil liability insurance. The coupling between the strong economic interests involved together with the different development levels from the different countries could make correlating with their corresponding characteristics of technical competence, and a certain risk of distrust about the recognition procedures of the conformity evaluation entities could emerge. The question is resolved by creating equivalences between the national accreditation entities based on mutual recognition agreements, by which the results issued by any accredited laboratory may be freely circulated by any of the signatory countries.

The huge petrochemical multinationals manufacturing plastic resins operate in an area separate from domestic firms, which mostly manufacture plastic products. Typically, each group has its own trade association and its own business plans and expansion. In the United States, for example, the powerful American Chemistry Council (ACC) [23] represents the polymer manufacturers, while the modest Society of Plastics Industry (SPI) [24] represents the manufacturers of products and equipment. A significant part of the current industrial management efforts goes to the convergence of the common interests of both organizations in the plastic sector under a single voice.

In the European model, the different European countries' requirements have been articulated by the standards in the EN 45000 series, under a single goal: the equivalence between certificates and the free circulation through the member states of the results issued by a laboratory accredited by the appropriate national authority. With such a scope, the acronym EAL, for the European cooperation for Accreditation of Laboratories, was created in 1994, with the integration of Germany (BMWA), Austria (BELTEST), Belgium (BBKO), Denmark (DANAK), Finland FINAS), France (COFRAC), Holland (RvA), Ireland (INAB), Italy (SINAL SIT), Norway (NA), Portugal (IPQ), Spain (ENAC), Switzerland (SAS), Sweden (SWEDAC), and the United Kingdom (UKAS).

Further information about the different worldwide national accreditation organizations and any others related to this subject is available via the electronic addresses of the former referenced acronyms. In addition, valuable information can be obtained from the regional ones: the Asia Pacific Laboratory Accreditation Co-operation (APLAC) [25]; European co-operation for Accreditation (EA) [26]; InterAmerican Accreditation Cooperation (IAAC) [27]; and Southern Africa Development Community's (SADC's) trade protocols [28] on cooperation in standardization (SADCSTAN), accreditation (SADCA), and quality assurance and metrology (SADQAM). Other subregional organizations are the East African Community (EAC) [29] and West African Economic and Monetary Union (UEMOA) [30], while from the international domain, the most representative are the International Accreditation Forum (IAF) [31] and the International Laboratory Accreditation Co-operation (ILAC) [32]. ILAC is an international cooperation network between the different laboratory accreditation schemes all around the world, launched in 1977 and operational since 1996. ILAC is indeed the main international forum for the development of laboratory accreditation practices and procedures, providing support and assistance to those developing countries in the process of developing their own laboratory accreditation systems.

28.5 STANDARDIZATION: CONVERGENCE BETWEEN THE REGULATED AND THE VOLUNTARY FIELD

The adoption of international standards places the products in contrast against large market data from different sectors. This helps suppliers and enables industrialists to compete in freer and more efficient ways and in many more markets in the world. Moreover, it would be expected that the organizations with internationally well-recognized certifications would be best positioned to face potential future fines and penalties for breach of environmental legislation and a reduction in insurance as a way of demonstrating better management of risk. In terms of reputation, public awareness of the rules can also mean a competitive advantage, favoring more and better business opportunities.

As in many other areas, all the measurements in the material field are subject to error, which means that the result of a measurement differs from the true value of the measure. Most of the error sources and the measurement errors can be identified, quantified, and even corrected by following the calibration procedures. Such procedures, once validated, are issued as the standards documents that are of customary use in the different laboratories for accurate evaluation of the properties of the different materials: plastics, concrete, ceramic, metals, glass, and paper and board.

From the Earth ecosystem perspective, the international standards contribute to the improvement of our quality of life in general, ensuring that the products, transportation, machinery, and instruments we use are safe and healthy. The international standards dealing with air, water, and soil

In each country, the national standardization body would be the entity in charge of channeling those transnational standards and the generality of those documents in force related to normalization affairs. From a global perspective, the aim of all those efforts would be a single worldwide-accepted standard or to trial it as a key factor in the development of the global markets. It would be important for the interested readers, according to their different profiles, to launch an Internet search by combining *plastic materials* with any of the above included acronyms of the standardization organizations, to access the historical backgrounds of such organizations, up to their updated mission and vision, activities, work commissions on raw materials, semifinished, and end products. Different classifications are available, devoted to the different sectors where plastic materials are significant, hot topics, such as recycling, food contact, medical, energy, chemical recycling, and the so-called bioplastics.

28.6 EXAMPLES OF COMPLEXITY OF RESEARCH AND DEVELOPMENT AND LEGAL FRAMEWORKS

28.6.1 PLASTIC WASTES IN A STANDARDIZED WORLD

In times of severe economic crisis and from sectors not always well documented and usually driven by the strong economic interests involved, organic plastic materials suffered with virulence the questioning of its environmental impact, most especially in those matters concerning solid-waste management.

Polymeric materials come from a variety of synthetic polymers, which appear in packaging, transportation, construction, electronic devices, medical applications, and almost all human activity fields. Together with their being lightweight, the higher durability of the synthetic polymeric materials compared with those traditionally used is the main reason to use them and ironically is one of their main drawbacks at the early stages of waste management strategies. Furthermore, the mixtures of plastic wastes from different sources, grades, or polymer families during reprocessing often give rise to bad properties. In the 1980s, some of the plastic materials, such as polyethylene terephthalate, were nearly banned in certain application markets, like packaging and transportation, because it was assumed to lack good properties after mechanical recycling options [35,36]. Nevertheless, the recycling of waste polymeric materials emerged as one of the key strategies to solve the twofold question of the global environment and fossil resources. By definition, recycling means that unwanted or useless polymer-based items be processed in such ways that materials return to the material cycle, giving rise to a decrease in the overall environmental contamination levels. An intelligent recycling strategy should allow the competitive recovery of valuable materials to be reused in new products. From the perspective of the LCA concepts, this gives further significant savings in energy; the recycling competitiveness is associated with the reduction in consumption of raw materials, air pollution and greenhouse gas emissions, and water pollution coming from landfill practices.

Almost 30 years later, from the business perspective, the challenge of the BEPs and the plastic waste management binomial remains open because of the strong economic interests involved. Any progress should come from the best qualification of almost all the stakeholders involved, from the specific characteristics of each plastic waste source hinting at the kinds of uses (industrial, agricultural, urban, and "wild" or noncontrolled plastic wastes) up to the expected degradation level of the plastic material (manufacture scraps, end-of-life items, and so on). Both aspects are by no means minor questions, because collaboration between industries, local authorities, waste handlers, and consumers is essential to achieve the most convenient management system in a sustainability way [37–41].

With regard to the industrial plastic waste sources and from the position of the polymer and plastic material industries, the most extended definition accepts that any statement from any organization, publicly and formally documented, on the intentions and principles of action of the organization about its environmental performance defines the policy environment. The statement usually highlights its overall objectives, including compliance with all regulatory requirements pertaining

quality, as well as gas emissions and radiation, can contribute to the effort to preserve the environment of the planet we inhabit.

For governments, the international standards provide the technological and scientific bases underpinning health, safety, and environmental quality legislation. For consumers, the conformity of products and services to international standards provides assurance of their quality, safety, and reliability. For developing countries, the international standards constitute an important source of technological expertise, by defining the expected characteristics of products and services ahead of placement in the export markets, thus providing a basis for these countries to make correct decisions to successfully invest their scarce resources and reduce wastes.

The public authorities address the legal framework to harmonize production processes and the safeguard of the rights of the consumers by launching the ISO 9000 standard series as the pilot translation document. OHSAS 18001 was similarly launched, strongly focusing on the protection of health and the work environment.

There are differences between certification and approval processes, both attesting to the fulfillment of certain characteristics but voluntary in the case of certification and mandatory in the case of type-approval systems. In the sectors of standardization and testing, homologation within the regulated field is the way to attest that an item complies with certain characteristics. In any case, the conformity assessment provides a demonstration that specified requirements relating to a product, process, system, person, or body are fulfilled, with respect to testing, inspection, certification of products, personnel, and management systems.

The ISO 9000 standard series, originally published in 1987 and first reviewed in 1994, following a procedure common to any standardization body, is continually updated, and decisions are made about whether documents need to be revised, updated, or just cancelled. Revisions are typically devoted to the elimination of internal inconsistencies detected. Reasons to delay the publication of any first revision can be attributed to the novelty of the quality management systems and the large volume of effort invested, considering that the introduction of major changes in the rules could induce a negative effect and discouragement of such efforts for the responsible committee. So it took until the year 2000 before the new release of ISO 9001 could be edited, incorporating implementation systems, process engineering, and the volume of experience generated in the more than 10 years of work invested. Five additional levels, concerning senior management commitment, were added to the revised version. They were the identification of the processes within the organization; the interfaces between them; the assurance of having the resources within the organization to operate the processes; the existence of a continuous improvement process applying to the quality management system; and the continuous tracking of the degree of customer satisfaction.

There now exist two well-consolidated reference standard series that can be used to develop any integrated quality management system, covering environment, product, processes, and/or custom management. The first one would be ISO 9001 and its updates, for the quality management systems, and the second one is the ISO 14001 series, incorporating the environment management systems.

For the accreditation of the characteristics of technical competence of laboratories, ISO 17025, in correspondence with EN 45000, is the standard. The interested reader may find a good overview of the transformation of the former ISO Guide 25, launched at the end of the 1970s, into ISO 17025 in the paper by Walsh [33]. The new ISO 17025 standard clearly differentiated for the first time between management requirements and technical requirements in laboratories. Within the frame of a voluntary system of accreditation by which a third party recognizes the quality system, the technical competence, and the impartiality of a laboratory, the ISO 17025 management requirements clearly define the responsibilities that support a robust and properly managed quality system, while the technical ones endorse the accreditation level of the ISO 17025 systems. They explain the highest level of the accredited ISO 17025 systems with respect to those just certified under the purely quality management standard requirements of ISO 9000 [34]. The most recent version of ISO 17025 was issued in 2005, and its third edition is to appear in 2020.

to the environment, and provides a framework for its action and the establishment of these objectives and goals. The feed-forward strategies aiming for closed-loop global operations are the main targets nowadays. In such a sense, the previously mentioned ISO 14000 is an internationally accepted standard series that expresses how to establish an effective EMS. The series of documents regarding environmental management, once implemented, would affect all aspects of the overall management of an organization in their environmental responsibilities, helping to address the environmental issues systematically, in order to improve the organization's environmental performance and the economic benefit opportunities.

Like any other standard that is voluntary, meaning without any legal obligation to follow it [42], the main aim and scope of this standard is to look for the balance between the entities' profitability and the progressive reduction of their impacts on the environment. It is noteworthy that ISO 14000 does not establish a set of quantitative goals regarding levels of emissions, specific methods of measuring these emissions, recycling quotes, and so on. The series focuses on each organization itself and provides a set of standards based on the procedure and guidelines from which a company can develop an efficient and competitive EMS.

The ISO 14000 series could be implemented by a wide range of organizations, whatever their current level of environmental performance. With the support of different organizations, the ISO 14000 standards provide a framework for a comprehensive and strategic approach to the policies, plans, and environmental actions of the organization. It requires a commitment to compliance with the legislation and applicable environmental regulations, together with a commitment to the continuous improvement strategies.

The main benefits claimed for the ISO 14000 implementation for any company would deal with the reduction in the cost of administration of wastes and increasing savings in the consumption of energy and materials. The reader should observe that both aspects should be coincident with the complex scenario drawn from the MIT reports [2,3] as mentioned under the corresponding sections at the beginning of the present chapter. Moreover, the company should benefit from the decrease in the costs of distribution and from an improved corporate image to the regulators, customers, and the public in general, well supported by the continuous improvement of the environmental performance of the organization. Any business willing to be sustainable in all its areas of action needs to adopt a preventive attitude that, in turn, should enable it to integrate the environmental factor in its decision mechanisms for the future.

On a general perspective, it is obvious so far that single and isolated policies could not achieve the desired recycling targets. On the contrary, integrated policies based on the commitment of the different stakeholders, mainly industry, citizens, and legal authorities, together with the development of new marketing strategies to promote the use of recycled materials in consumer products, need standards documents to be able to assure the quality of the recycling materials. It is necessary to overcome the lack of technical specifications and inconsistencies of the feedstock because of the tremendous fragmentation of the processing polymer industries downstream of the polymer production ones. The harmonization between the corresponding regulation frames and the different polymers and grades of each one of them, authorized in the packaging markets, especially those concerned with film applications, could significantly reduce the bad properties of the mixed plastics coming from the big municipal solid-waste streams.

The standards and standardization activities concerning recycled polymer materials and their corresponding manufactured products, semifinished or end products, are in a continuously changing scenario because of the lack of harmonization between certain specifications demanded by the potential customers, sometimes too complex or just too high for some technical materials to risk compromising supplies by using recycled material of uncertain quality. The interested reader on these subjects, depending on the regional areas, countries, materials, or products of interest, is kindly referred to the electronic searching tools referenced under the different sections of the present chapter. From a neutral position about the mechanical recycling options, standards documents are required to independently assess the quality of any finished compound, whether it incorporates nothing, an amount, or, as a whole, any recycled plastic coming from a specific well-controlled mechanical recycling option [43,44]. In a similar manner, the options of chemical recycling, whether or not coupled with the different energy recovery options, or just the latter, need to adjust to the applicable standards that assure that plants and utilities comply well with the product and environment quality requirements [45,46].

28.6.2 Bioplastics in a Standardized World

As described in previous sections, from the early stages of plastic materials, development to guarantee their stability, getting better durability, and long life service were among the main targets of the research and development (R&D) activities on polymeric materials. After the energy crisis of the 1970s and the subsequent one of the environmental impact of the big solid-waste streams, the degradation of plastic materials began to be required among their main characteristics, empowering in the last decade the R&D activities to produce and use biodegradable and bio-based polymers. Biodegradability means that by the action of mainly fungi or bacteria, the polymeric materials fully decompose to water, carbon dioxide, and a fraction of biomass under aerobic conditions, or generate methane in the decomposition products in anaerobic environments. Polymerization of monomers coming from renewable resources gives rise to the so-called bio-based polymers, as well to those others obtained from biomass or synthesized by microorganisms. Typically, they are aliphatic polyesters, and at this point, it is worth noting that different members of both polyamide and polyester families belonging to the classical condensation polymers and copolymers, yet coming from petrochemical sources are biodegradable materials too.

In any case, the promotion of degradation in certain plastic products characterized by a limited lifetime after they are degraded by a bio-promoted mineralization has become a desirable characteristic. Most of these environmentally degradable polymers and plastic materials, or EDPs, are present in several applications, from agriculture to medical implants, yet in strong competition with those from fossil fuel sources, which are largely more competitive in costs, processing properties, and mechanical performance. Furthermore, up to the present, and despite the variety of polymers synthesized and claimed as potentially biodegradable polymeric materials, the question of their degradation mechanisms, the safety of the degradation processes, and the kinds and distributions of the microorganisms than can degrade them remains open [47–49].

As in any other material fields, the standards elaboration devoted to EDPs has been running in parallel with R&D activities from the first moments. The corresponding documents are based on the scientific and technological state of the art [50,51], which, at any moment, defines the multifaceted and complex problem of plastic material degradability. Although polymer degradation may occur in either biotic or abiotic environments, most often, there is a combination of both. A very difficult discussion remains active about the development of controlled and comparable assessment methods of polymer biodegradability, which demands test protocols under precise and well-defined conditions that, as much as possible, mimic real scenarios. Standardized biodegradability measurements are compiled in the equivalent ASTM D6400, D6868, EN 13432, and ISO 17088 standards, while in terms of terminology, for example, the interested reader is kindly referred to Refs. [52–57], which compile the nomenclature and well-accepted definitions. The very interesting study on standardization activities of biopolymers, or bioplastics, in the work by Kran et al. [58] and the more specific one on biodegradable agricultural plastics by Briassoulis and Dejean [59] offer an interesting view on how the harmonization between the efforts and activities devoted to this subject has progressed and on its intrinsic complexity.

28.6.3 Nanotechnology and Nanometrology in a Standardized World

Nanotechnology is the understanding and control of matter at the nano domain level, where the magnitude of properties significantly differs from those at the successive mesoscopic, microscopic,

Environment Health and Safety

and macroscopic levels. The R&D activities on advanced materials based on thermoplastic matrices and nanoparticles, or the so-called nanocomposites, have been sharply encouraged from the beginning of the third millennium because of the high expectation from their enhanced properties [60–64].

Because of the expected increasing presence of nanomaterials in commercial applications, a growing public debate is open about the environmental and toxicological impact derived from direct and indirect exposure to the nanoparticles during their handling, either in the research laboratories as well in the processing industries to obtain them or during the early compounding steps. The question is far from being resolved, not only in technical terms but also as a health topic, because they escape into the biological filters of the human body, where they can be absorbed via inhalation or through the skin or ingested, causing respiratory problems [65,66]. Moreover, recent studies reveal that toxicity risks increase as the particle size decreases, and even genes might potentially be affected [67–70].

It is worth noting that nanocomposites are an excellent example of how new developments go into the legal dimensions of R&D activity. Before product launching, for health and safety legislation and environmental regulations concerning the atmosphere and workplaces on handling nanoparticles and nanomaterials in general, and nanocomposites in particular, accurate standards based on both improved and standardized characterization and measurement methods as well as well-controlled and optimized synthesis methods yielding traceable outputs and nanoparticle sizing would be required [71–73]. Furthermore, a full study of the toxicity of any polymer/nanoparticulate system ought to be extended to their durability and degradability processes in both biotic and abiotic environments to get insight into its limitations and environmental and health impact [74].

Several ISO technical committees are responsible for different standards documents under different status: published, draft, or in revision [75,76]. These documents give results relevant to different aspects of nanoparticles characterization, such as particle size analysis (TC24), surface chemical analysis (TC201), electron probe microanalysis (TC202), and fine ceramics (TC206). A more specific one, TC229, for nanotechnologies [77], is responsible for around 40 ISO standards, almost all published, including ISO 10808:2010, which specifies a protocol for nanoparticle inhalation toxicity testing. TC229 has 35 participating countries and an additional 13 observing countries [78].

28.6.4 The Question of Reference Materials

In a chapter devoted to the thermoplastic organic polymers and the legal aspects and standardization activities concerning them, it is worth noting that the issue of reference materials remains an open question [79,80]. The import–export operations between the world's trading regions concerning thermoplastic polymers imply, for instance, the specifications and analysis of the level of additives, heavy metals, and so on present in the polymeric compounds or in the manufactured products. Because of the lack of analytical measurements whose relationship to true values is well established and accepted worldwide, the effect is that analysts in different laboratories carry out extra analytical measurements comparable to one another by the use of well-characterized certified reference materials (CRMs).

As is well known, a reference material must be sufficiently homogeneous and stable with reference to specified properties to be fit for its intended use in measurement or in examination of nominal properties. Traceability should exist, then, to a reference material certified or accompanied by a certificate issued by an authoritative body that provides one or more specified property values, with associated uncertainties and demonstrated traceability established by validated or standardized procedures. ISO Guide 34 [81] compiles the requirements to comply for any reference material producer in such a way that accreditation bodies, material customers, and regulatory authorities may use it to confirm and to recognize the competence of the producers of the reference materials. It covers the production of both certified and noncertified reference materials, although as purely a guide, it is not, in any case, used for conformity assessment by the different certification bodies.

For properties related to chemical composition by spectroscopic techniques, or those others related to molecular size and distribution, size exclusion chromatography, instrument calibration, method validation, and so on, some referenced polymer grades are well accepted [82]. These referenced standards are thermoplastic polymers with a well-defined molecular weight and chemical composition and are useful in R&D activities too. Nevertheless, there are no reported certifications of those other properties of the thermoplastic polymers and the heterogeneous materials based on them, like thermal or mechanical properties, because of the key role played by the processing steps in the measured properties. The arrangement of mutual recognition between research and testing laboratories of equal levels of competence and the agreement to follow protocols, for example, ISO 17025, based on the in-force standards documents compiling the measurement procedures, are the practices followed to determine traceability chains concerning bulk properties of the thermoplastic organic polymeric materials. Repeatability of both the measuring systems and the results of the measurements and reproducibility of the results of the measurements in the absence of certified reference thermoplastic polymers require working with reference values as the basis for comparisons between magnitudes of the same kind. The reference values of the properties of the thermoplastic materials have been determined and successively reviewed through a peer-review process from early times, being compiled in well-referenced handbooks, of common use in the R&D activities on these materials [83–87]. The demand for reference materials increases because of both the increased precision of the measuring equipment and the requirements for more accurate and reliable data in the scientific and technological laboratories [88,89]. Not only reports and technical brochures from the polymer producers and compounders are needed but also statements demonstrating their competence to produce materials of the appropriate quality to match with the technical and environmental performance that the 2020 horizon demands.

28.6.5 PLASTIC MATERIALS AND ELECTRONIC DEVICES

Everybody knows that the high technological progress in the communications industry could not be possible without organic plastic materials. Several types of plastics are utilized in the manufacturing of electronic devices, and together with metals, plastics are the main materials in the solid-waste streams at their end of life. Both plastic and metals can be recovered from the electronic solid-waste streams but with very different economic competitiveness. An interesting view about the scientific-technological state of the art in this field can be found in Ref. [90]. It is noteworthy that these applications are classified under hazardous solid wastes, just because of the presence of metal.

The management of hazardous wastes appears to have been one of the three priority fields in the United Nations Environment Program (UNEP) [91] from the early 1980s. The above-mentioned differences in costs of recovery and treatment processes of solid wastes depending on what they are made of, with variations in any of the different developing countries, the Third World, from the so-called First World, gave rise to the movement of hazardous wastes all around the world and their disposal not in the most convenient ways and, in any case, far away from the most basic health and safety recommendations. The interested reader may explore the Basel Convention records [92,93] and the so-called Basel protocol [94], signed by 170 countries, that is now in force related to the liability and compensation for damage resulting from transnational movement and disposal of hazardous wastes. It is a reference document representative of the first steps to develop the desirable worldwide harmonization of the transnational solid-waste traffic.

28.7 PERSPECTIVES ON QUALITY INFRASTRUCTURE AND ENVIRONMENTAL SAFETY

Existing resources are limited. Achieving more value with less environmental impact and taking advantage of best available resources are the basis of the circular economy. Nevertheless, the change of the current economic model to a circular one is not easy. Reuse and energy and resource efficiency are some of the key aspects of this model, and innovation in sustainable chemistry should provide solutions that will help to achieve this change and to face the current challenges in society.

Thermoplastic polymers are key materials to appropriately respond to the most urgent and pressing needs facing humankind in the early twenty-first century, namely, energy, food, water, health, and the environment. Standardization activities in plastic materials helps to prevent probable barriers emerging during the development and implementation of innovations in quality and industrial safety practices as well to remove legal bottlenecks coming from obsolete techniques, laws, or policies.

Ways and means of doing things are changing in a rapid fashion, most of the time faster than the standards documents, norms, regulations, and laws do [95,96]. Leading chemical companies, valorized by-product companies, technology providers, engineering and consulting firms, universities, research organizations, technological centers, consumers' associations, governments, and nongovernmental organizations are some of the stakeholders involved in the development of those laws, norms, and standards documents that are key to getting a circular economy model, from cradle to cradle. Such a model means replacing the end-of-life concept with restoration and regeneration through the optimization of materials, products, and systems following the principle of sustainability. In addition to the responsibilities of the industrial companies, materials researchers must also help, on one hand, by ensuring that no new problems appear and, on the other, by applying their skills to solve sustainability problems, for instance, with new nature-inspired materials [97]. Strong organizations like the European Commission or the US National Science Foundation are committed to sustainability by placing it as one of their technological objectives and providing funds for research projects related to environmentally benign or green technologies, etc. [98–100].

REFERENCES

1. Metrology in short. 2008. Available at http://resource.npl.co.uk/international_office/metrologyinshort.pdf.
2. Meadows, D. H., Meadows, D. L., Randers, J., Behren III, W. W., *The Limits to Growth. A Report for the Club of Rome's Project on the Predicament of Mankind.* Potomac, London, 1972.
3. Meadows, D. L., Meadows, D. H., Randers, J., *Beyond the Limits: Confronting Global Collapse, Envisioning a Sustainable Future.* Chelsea Green Publishing, Claremont, NH, 1993.
4. Andrady, A. L., Neal, M. A., Applications and social benefits of plastics. *Philosophical Transactions of the Royal Society*, B364, 1997–1984, 2009.
5. Thompson, R. C., Swan, S. H., Moore, C. J., Saal, F. S., Our plastic age. *Philosophical Transactions of the Royal Society*, B364, 1973–1976, 2009.
6. How to know if and when it's time to commission a life cycle assessment. An Executive Guide. pp. 1–20. Available at http://www.icca-chem.org/iccadocs/acc_icc_lifecycle_2013.08.pdf.
7. Responsible care, Global Charter. 2006. Available at http://www.icca-chem.org/en/Home/ICCA-publications/.
8. National REACH/CLP Helpdesks, The European Chemical Agency. 2009. Available at http://echa.europa.eu/help/nationalhelp_en.asp.
9. Guidance Documents, The European Chemical Agency. Available at http://guidance.echa.europa.eu/guidance_en.htm#GD_PROCC_I.
10. Publications, The European Chemical Agency. Available at http://echa.europa.eu/publications_en.asp.
11. REACH for Polymers, Review of best practice and trends for complying with the REACH Legislation, 2010, Proj Nr.: LIFE08 ENV/UK/ 00020. Available at http://www.ismithers.net.
12. REACH for Polymers, Best available testing techniques and methods guide, 2011, Proj Nr.: LIFE08 ENV/UK/ 000205. Available at http://www.ascamm.com.
13. REACH for Polymers, *Best Practice Handbook*, 2011. Available at http://www.reachforpolymers.eu.
14. Available at http://www.iso.org/iso/home.html.
15. Available at http://www.iec.ch/.
16. Available at http://www.itu.int/.
17. Available at http://www.ansi.org/.
18. Available at https://www.cen.eu/.
19. Available at http://www.epa.gov/.
20. Available at http://eur-lex.europa.eu/.
21. Available at http://www.standards.org/.
22. Available at http://ec.europa.eu/environment/emas/index_en.htm.
23. Available at http://www.americanchemistry.com/.

24. Available at http://www.plasticsindustry.org/.
25. Available at https://www.aplac.org/.
26. Available at http://www.european-accreditation.org/.
27. Available at http://www.iaac.org.mx/.
28. Available at http://www.sadc.int/.
29. Available at http://www.eac.int/.
30. Available at http://www.ustr.gov/countries-regions/africa/regional-economic-communities-rec/west-african-economic-and-monetary-union-uemoa.
31. Available at http://www.iaf.nu/.
32. Available at https://www.ilac.org/.
33. Walsh, M. C., Revision of ISO Guide 25. *Accreditation and Quality Assurance*, 4 (8), 365–368, 1999.
34. Jenks, P. J., How much does quality really matter? *Spectroscopy Europe*, 25 (1), 26–27, 2013.
35. Azapagic, A., Emsley, A., Hamerton, L., *Polymers, the Environment and Sustainable Development*. John Wiley & Sons, New York, 2003.
36. Stein, R. S., Polymer recycling. Opportunities and limitations. *Proceedings of the National Academy of Sciences of the United States of America*, 89, 835–838, 1992.
37. Available at http://ec.europa.eu/environment/waste/legislation/a.htm.
38. Available at http://www.epa.gov/waste/nonhaz/municipal/.
39. Directive 2000/53/EC on end-of-life vehicles. *Official Journal of the European Union*, I. 269/34, October 21, 2000.
40. Directive 2000/96/EC on waste electrical and electronical equipment. *Official Journal of the European Union*, I. 37/34, February 13, 2003.
41. Available at http://op.bna.com/env.nsf/r?Open=aada-96fr6a.
42. Regulation N° 1221/2009/EC on the voluntary participation by organizations in a Community eco-management and audit scheme (EMAS), repealing N° 761/2001plastic materials and articles intended to come into contact with food. 10/2011/EC N° 761/2001 and Commission Decisions 2001/681/EC and 2006/193/EC. *Official Journal of the European Union*, L 342/1, December 22, 2009.
43. Yang, Y., Boom, R., Irion, B., van Heerden, D. J., Kuiper, P., de Wit, H., Recycling of composite materials. *Chemical Engineering and Processing*, 51, 53–68, 2012.
44. Perry, N., Bernard, A., Laroche, F., Pompidou, S., Improving design for recycling. Application to composites. *CRIP Annals. Manufacturing Technology*, 61, 151–154, 2012.
45. Altalhi, T., Kumeria, T., Santos, A., Losic, D., Synthesis of well-organized carbon nanotube membranes from non-degradable plastic bags with tunable molecular transport: Towards nanotechnological recycling. *Carbon*, 63, 423–433, 2013.
46. Nishida, H., Development of materials and technologies for control of polymer recycling. *Polymer Journal*, 43, 435–447, 2011.
47. Okada, M., Chemical synthesis of biodegradable polymers. *Progress in Polymer Science*, 27, 87–133, 2002.
48. Narayan, R., Biobased and biodegradable polymer materials: Rationale, drivers, and technology exemplars. *ACS Symposium Series* 939, Chapter 18, 282–292, 2006.
49. Narayan, R., Carbon footprint of bioplastics using biocarbon content analysis and life cycle assessment. *MRS Bulletin*, 36 (9), 716–721, 2011.
50. Siracusa, V., Rocculi, P., Romani, S., Rosa, M. D., Biodegradable polymers for food packaging: A review. *Trends in Food Science and Technology*, 19, 634–643, 2008.
51. La Rosa, A. D., Recca, G., Summerscales, J., Latteri, A., Cozzo, G., Cicala, G., Bio-based versus traditional polymer composites. A life cycle assessment perspective. *Journal of Cleaner Production*, 74, 135–144, 2014.
52. ASTM D883, Standard terminology relating to plastics. Philadelpia, PA, ASTM, 1991.
53. ISO 472 Plastics—Vocabulary, amendment 3, General terms and terms relating to degradable plastics. Geneva: ISO, 1993.
54. Sawada, H., ISO standard activities in standardization of biodegradability of plastics—Development of test method and definitions. *Polymer Degradation and Stability*, 59, 365–370, 1998.
55. Avella, M., Bonadies, E., Martuscelli, E., Rimedio, R., European current standardization for plastic packaging recoverable through composting and biodegradation. *Polymer Testing*, 20, 517–552, 2001.
56. CEN/TC249, CEN/TR 15351. Plastics—Guide for vocabulary in the field of degradable and biodegradable polymers and plastic items, Brussels, 2006.
57. Narayan, R., Rationale, drivers, standards, and technology for biobased materials. In *Renewable Resources and Renewable Energy*, Graziani, M. and Fornasiero, P., eds. CRC Press, Boca Raton, FL, 2011.

58. Kran, A., Hemjin, S., Miertus, S., Corti, A., Chiellini, E., Standardization and certification in the area of environmentally degradable plastics. *Polymer Degradation and Stability*, 91, 2819–2833, 2006.
59. Briassoulis, D., Dejean, C., Critical review of norms and standards for biodegradable agricultural plastics. Part I. Biodegradation in soil. *Journal of Polymers and the Environment*, 18, 384–400, 2010.
60. Breuer, O., Sundararaj, U., Big returns from small fibers: A review of polymer/carbon nanotube composites. *Polymer Composites*, 25 (6), 630–645, 2004.
61. Okada, A., Usuki, A., Twenty years of polymer–clay nanocomposites. *Macromolecular Material Engineering*, 91, 1449–1476, 2006.
62. Frontini, P. M., Pouzada, A. S. Is there any chance for polypropylene/clay nanocomposites in injection moulding? *Express Polymer Letters*, 5 (8), 661, 2011.
63. Zeng, Q. H., Yu, A. B., Lu, G. Q., Multiscale modeling and simulation of polymer nanocomposites. *Progress in Polymer Science*, 33, 191–269, 2008.
64. Pavlidou, S., Papaspyrides, C. D., A review on polymer-layered silicate nanocomposites. *Progress in Polymer Science*, 33, 1119–1198, 2008.
65. Aitken, R. J., Creely, K. S., Tran, C. L., Nanoparticles: An occupational hygiene review. Institute of Occupational Medicine, Research Report 274, HSE Books, Norwich, 2004.
66. Dreher, K. L., Health and environmental impact of nanotechnology: Toxicological assessment of manufactured nanoparticles. *Toxicology Science*, 77, 117–125, 2004.
67. Jia, G., Wang, H., Yan, L., Wang, X., Pei, R., Yan, T., Zhao, Y., Guo, X., Cytotoxicity of carbon nanomaterials: Single-wall nanotube, multi-wall nanotube, and fullerene. *Environmental Science and Technology*, 39, 1378–1383, 2005.
68. Zhang, Y., Dervishi, A. S. F., Xu, Y., Casciano, D., et al., Cytotoxicity effects of graphene and single-wall carbon nanotubes in neural phaeochromocytoma-derived PC12 cells. *ACS Nano*, 4, 3181–3186, 2010.
69. Lindberg, H. K., Falck, G. C. M., Suhonen, S., Vippola, M., et al., Genotoxicity of nanomaterials: DNA damage and micronuclei induced by carbon nanotubes and graphite nanofibres in human bronchial epithelian cells in vitro. *Toxicology Letters*, 186, 166–173, 2009.
70. Feng, L., Zhang, S., Liu, Z., Graphene based gene transfection. *Nanoscale*, 3, 1252–1257, 2011.
71. Spitalsky, Z., Tasis, D., Papagelis, K., Galiotis, C., Carbon nanotube–polymer composites: Chemistry, processing, mechanical and electrical properties. *Progress in Polymer Science*, 35, 357–401, 2010.
72. Singh, V., Joungh, D., Zhai, L., Das, S., Khondaker, S. I., Seal, S., Graphene based materials: Past, present and future. *Progress in Materials Science*, 56, 1178–1271, 2011.
73. Heiligtag, F. J., Nierderberger, M., The fascinating world of nanoparticle research. *Materials Today*, 16 (7/8), 262–271, 2013.
74. Kumar, A. P., Depan, D., Tomer, S. T., Singh, R. P., Nanoscale particles for polymer degradation and stabilization. Trends and future perspectives. *Progress in Polymer Science*, 34, 479–515, 2009.
75. Wilkening, G., Standardization efforts in nanotechnology and nanometrology—A short survey. *Euromet Lenght Workshop*, NANOtrends, Bucharest, October 2006.
76. Available at http://www.nanoforceproject.eu/nanoforce_newsletter_2_2013final.pdf.
77. Available at http://www.iso.org/iso/iso_technical_committee?commid=381983.
78. Available at http://www.iso.org/iso/home/store/catalogue_tc/.
79. Jenks, P. J., Certified reference materials and proficiency testing in an ISO 17025 accredited laboratory. Part one: Defining the role. *Spectroscopy Europe*, 24 (3), 26–27, 2012.
80. Hammond, J. P., Reference material accreditation. The way forward. *Spectroscopy Europe*, 25 (5), 24–27, 2013.
81. Available at http://www.iso.org/iso/catalogue_detail.htm?csnumber=50174.
82. Available at http://www.nist.gov/mml/msed/polymers031009.cfm.
83. Houwink, R., *Fundamentals of Synthetic Polymer Technology*. Elsevier, New York, 1949.
84. Van Krevelen, D. W., *Properties of Polymers*, 3rd. Ed. Elsevier, Amsterdam, NY, 1990.
85. Mark, J. E. Editor, *Polymer Data Handbook*. Oxford University Press, Oxford, 1999.
86. Available at http://www.polymersdatabase.com/intro/index.jsp.
87. Mark, J. E. Editor, *Physical Properties of Polymers Handbook*, 2nd Ed. Springer, New York, 2007.
88. Walker, R. F., Accreditation of reference material producers. *Accreditation and Quality Assurance*, 4, 360–365, 1999.
89. Tonzani, S., The renaissance of polyolefins. *Journal of Applied Polymer Science*, 127, 837, 2013.
90. Tanskanen, P., Management and recycling of electronic waste. *Acta Materialia*, 61, 1001–1011, 2013.
91. Available at http://www.unep.org/.
92. Available at http://www.basel.int/Home.

93. Available at http://www.basel.int/TheConvention/Overview/tabid/1271/Default.aspx.
94. Available at http://www.basel.int/Portals/4/Basel%20Convention/docs/text/BaselConventionText-e.pdf.
95. Non-Legislative Acts. Regulations. Commission Regulation on plastic materials and articles intended to come into contact with food. 10/2011/EU. *Official Journal of the European Union*, L 12/1, January 15, 2011.
96. Regulation N° 1221/2009/EC on the voluntary participation by organizations in a Community eco-management and audit scheme (EMAS), repealing N° 761/2001plastic materials and articles intended to come into contact with food. 10/2011/EC N° 761/2001 and Commission Decisions 2001/681/EC and 2006/193/EC. *Official Journal of the European Union*, L 342/1, December 22, 2009.
97. Zhou, H., Fan, T., Zhang, D., Biotemplated materials for sustainable energy and environment: Current status and challenges. *ChemSusChem*, 4, 1344–1387, 2011.
98. Editorial, Materials for sustainability. *Nature*, 419 (6907), 543, 2002.
99. Zorpas, A. A., Inglezakis, V. J., Automotive industry challenges in meeting EU 2015 environmental standard. *Technology in Society*, 34, 55–83, 2012.
100. Santini, A., Passarini, F., Vassura, I., Serrano, D., Dufour, J., Morselli, L., Auto shredder residue recycling: Mechanical separation and pyrolysis. *Waste Management*, 32, 852–858, 2012.

Index

Page numbers followed by f and t indicate figures and tables, respectively.

A

Abaca, 729
ABS, *see* Acrylonitrile–butadiene styrene (ABS)
Accreditation, 945–946; *see also* Environment health and safety
Acetaldehyde, 864
Acetal homopolymer, 193–194
Acetalization
    PVA, 93–95, 93f–94f
    PVOH, 76
        cross-linking, 76
Acetate dosing, PVB with, 103
Acetic acid, 55
Acetoacetylation, 76
Acetylene, 55
Acetyl tributyl citrate (ABTC), 866
Acid-catalyzed condensation
    of alcohols, 252
Acid doping levels
    in AB-PBI, 648
    in meta-PBI, 644, 644t, 645t
    in para-PBI, 645–646, 647f
Acid hydrolysis, 847; *see also* Molecular weight reduction, starch
Acid imbibing, conventional, 641–642
Acid-modified PCT (PCTA), 333, 334–335
Acrolein, 170
Acrylic acid (AA), 169, 170f; *see also* Poly(methyl methacrylate) (PMMA)
    historical milestones, 170
    storage, 172
    synthesis of, 170–172, 171f
Acrylic acid pentabromobenzyl ester (ACPB), 906
Acrylic fibers, 140, 152
Acrylics, 169–170; *see also* Poly(methyl methacrylate) (PMMA)
    flame-retardant efficiency, 180
    living polymerizations, 176–177
Acrylonitrile; *see also* Polyacrylonitrile (PAN)
    for acrylamide, 141
    chemical properties, 141–142
    copolymerization of, 144
    copolymers of, 149–150
        butadiene and, 149
        high-barrier application, 149–150
        medical application, 150
        optically active application, 150
    health and safety factors, 142
    homopolymerization of, 143
    manufacturing, 142
    physical properties, 141
    for synthetic rubbers, 139, 149
Acrylonitrile-butadiene copolymer, 149, 150
Acrylonitrile-butadiene-styrene (ABS), 331, 737, 895

Acrylonitrile-butadiene styrene (ABS)/layered silicate nanocomposites, 905–906
Activated monomer (AM) mechanism, 254
    vs. ACE mechanism, 255, 255f, 270, 271f
Active chain-end (ACE) mechanism, 254
    vs. AM mechanism, 255, 255f, 270, 271f
Active-matrix LCD (AMLCD), 671
Additive effects, of PVB
    antioxidants, 117
    plasticizers, 116–117, 117f
Additives
    aromatic polyamides and, 312
    fire retardants, 884
    PET and, 322–323
Additives, POM, 199–201
    antioxidants, 199–200
    antistatic agents, 201
    colorants, 201
    fillers, 201
    flame retardants, 201
    formic acid-trapping agents, 200
    heat stabilizers, 199, 200
    impact modifiers, 200
    lubricants, 200
    nucleating agents, 200–201
    stabilizers, 199–200
    UV-stabilizers, 200
Additives and bioplastics; *see also* Bioplastics
    polyhydroxyalkanoates (PHB), 866
    polylactide (PLA), 866
    thermoplastic starch (TPS), 865–866
Adhesion properties, PVOH, 109–110, 110f
Adhesives
    PVB, 129
    PVOH, 80
Administrative costs, 760
Admittance spectroscopy (AS) method, 677
Adsorbents, PAN, 152
AES, *see* Auger electron spectroscopy (AES)
AFM, *see* Atomic force microscopy (AFM)
Aggregation number, 262
Aging effect on tensile properties of hot-drawn fibers, 630t
Agricultural plastic wastes, 927; *see also* Plastics and environment
Agrofiber/thermoplastic composites, 739–740
Airline catering trolley, 448f
Air plasma-treated flax fiber, 732f
Alcohols
    acid-catalyzed condensation of, 252
Aldehydes
    PVA and, 93, 93f
    PVB with, 96, 99f
Aliphatic–aromatic polyimides, 500–501
Aliphatic (nylons) polyamides, 285–286

Aliphatic polycarbonates, 348–349, 348f, 349f, 349t
Alkaline catalysts, for hydrolysis of PVA, 56
Alkaline treatment, 732, 745
Alkoxy-substituted poly(*p*-phenylene-1,3,4-oxadiazole), 603–607; *see also* Polyhydrazides and polyoxadiazoles
Alliance for the Polyurethane Industry (API), 387
Alloys, PC, 361–362
    PC/ABS, 361–362, 361t
    PC/polyesters, 362
Allyl glycidyl ether (AGE), 256, 265
ALOAC, *see* Vinyl alcohol–alkyleneoxyacrylate (ALOAC)
Alternating current (AC) signal, 677
Alternative polymerization methods
    aromatic polyamides synthesis, 290
Alumina, 886
Alumina trihydrate (ATH), 118
    fire and, 180
Aluminum, 5
    and chemical inertness, 564
Aluminum hydroxide, 886
Aluminum isopropanol, PVB and, 129
Aluminum trihydroxide (ATH), 887–888; *see also* Metal hydroxides
American Chemistry Council (ACC), 946
American National Standards Institute (ANSI), 944
American Society for Testing and Materials (ASTM), 769, 772t–774t
Ammonium polyphosphates (APP), 894
Amoco Chemical, 326
Amoco Corporation, 421
Amorphous PC, 352–353
Amorphous PEN, 340, 341t
Amortization, 761
Amylopectin, 275, 841, 841f
    films, 861
Amylose
    about, 841, 841f
    linear, 846
    low-molecular-weight, 846
Analytical techniques
    for polycarbonates characterization, 357
Anionic polymerization
    of acrylic monomers, 176
    formaldehyde, 267–268, 267f–268f
    monomer-activated, of polyepoxides, 255–256, 255f
    polyoxetanes, 270
Anionic ring-opening polymerization (AROP)
    ethylene oxide, 259
    *lin*PG, 266
    polyepoxides, 253–254, 254f
Ansa-metallocene, 9
Anti-lock braking systems (ABS), 483
Antimony oxyhalides, 904
Antimony trioxide, 904
    PET synthesized with, 320
Antioxidants
    POM, 199–200
    PVB, 117
Antistatic agents, POM, 201
APP, *see* Ammonium polyphosphates (APP)
Aqueous dispersions
    of PTFE, 406
Arabinoxylans, 853

Aramid fibers, 701
    blends with, 439
    encapsulation of, 312
    principal properties of, 299, 300t–302t
    research and development, 297, 297t
Aramid/polyimide blends, 506
Architecture, microstructural feature of polymer, 763, 764t
Architecture, PMMA for, 184
Argon plasma-treated flax fiber, 732f
Aromatic (aramids) polyamides, 285–313
    applications, 309–312, 310t
        as biomaterials (medical applications), 311
        future developments, 312
        high-performance materials, 312
        with unique properties, 310–311, 311t
    characteristics, 286
    chemical stability, 305, 306t
    commercial fibers, 287, 288t; *see also specific entries*
        principal properties of, 299, 300t–302t
    with controlled structures, 290–297
        chain-growth polycondensation, 290–294, 291f–293f
        constitutional isomerism, 294–296, 294f, 295f
        spherical aromatic polyamides, 296–297
    cross-linked, 312
    defined, 286
    dendrimers, 305
    environmental impact and recycling, 313
    examples of uses, 310, 310t
    flammability, 304
    global market view, 308f–309f, 309
    heat resistance, 304
    hyperbranched polymers, 305
    inherent viscosity, 304
    intrinsic viscosity, 304
    LOI of, 304
    in membrane technologies, 311, 311t
    MPIA, 298; *see also* Poly(*m*-phenylene isophthalamide) (MPIA)
    ODA/PPTA, 298
    optically active (OAPs), 310–311
    overview, 285–286
    physical properties, 298t
    PPTA structure, 297f, 298; *see also* Poly(*p*-phenylene terephthalamide) (PPTA)
    processing, 305, 307–308
        dry-jet wet spinning, 308
        dry spinning, 308
        film casting of PPTA, 308
        wet spinning, 307–308
    properties, 297–305, 298t–302t
    relative viscosity, 304
    research and development, 297, 297t
    research objectives, 286
    special additives, 312
    synthesis, 287–288, 287f, 288t
        alternative polymerization methods, 290
        high-temperature solution methods, 289–290
        low-temperature solution methods, 288–289
        monomers used in, 287f
        monomer synthesis, 288
    tensile properties, 299, 299t
    in textile market, 309–310

thermal properties, 298–299, 298t
UV exposure, 312
viscosity, 303–304, 304f
Aromatic polycarbonates, 350, 350t
Aromatic poly(ether-alkyl) hydrazides based on 4,4'-dihydroxybiphenyl units
optical microscopy, 600–602
polyhydrazides to polyoxadiazoles, conversion of, 594–595, 595f
synthesis/characterization of, 588–591
thermal analysis
about, 592–593, 595–597
bulk crystallization, 597–599, 597f, 598f
thermogravimetric analysis, 591–592
x-ray diffraction analysis, 593–594, 594f, 599–600, 599t
Aromatic polyetheraroyl hydrazides based on 4-oxybenzoyl units; *see also* Polyhydrazides and polyoxadiazoles
cyclization reaction by infrared analysis, 581, 581t, 582f
optical microscopy, 584–585, 585f
polymerization procedure, 577
synthesis of monomers and polymers, 575–576, 576f, 576t
thermal analysis
about, 579–580, 582–583
bulk crystallization, 583–584, 583f, 584f
thermogravimetric (TG) analysis, 577, 577f, 578f, 578t
x-ray diffraction analysis, 580–581, 585–587
Aromatic polyhydrazides and polyoxadiazoles, *see* Polyhydrazides and polyoxadiazoles
Aromatic polyimides, 498–500, 499t
Aromatic polyketones-based foams, 717–718; *see also* Foams, thermoplastic
Association Francaise de Normalisation (AFNOR), 769
Atomic force microscopy (AFM), 735
of PTEF, 407
Atom transfer radical polymerization (ATRP), 176, 177, 375, 682
Auger electron spectroscopy (AES)
of PTEF, 407
AURUM TPI resin, 507, 508
Autoclave reactor
high-pressure, 13, 14
Autoclave technology, 709–710, 709f; *see also* Thermoplastic composites, advanced
Automated fiber placement (AFP), 711, 830–831; *see also* Fiber and tape placement
Automobile shredder residue (ASR), 926
Automotive, PMMA in, 184
Automotive applications, of PAEKs, 483
Automotive plastic wastes, 926; *see also* Plastics and environment
Average number of branches (ANB), 296–297
Avrami index, 584, 584f
A,ω-(hetero)telechelics, 261
Azobenzene-based polymers, 611; *see also* Polyoxadiazoles for high-performance membranes
Azobenzenes, photoresponsive properties, 377–378, 377f, 378f
Azobisisobutyronitrile (AIBN), 680

B

Backbiting, methacrylates, 173–174, 174f
Bacterial biosynthesis, 860
Baeyer–Villiger oxidation, 859
Bakelite polysulfone, 420
Bamboo, 730
BASF Corporation, 451, 452t
Batch-foaming technology, 717
Battery electrodes, 561
Battery separators, 561–562
Bauer, Walter, 170
B-D-glucopyranose molecules, 273
BDP, *see* Bis(diphenyl phosphate)
Bead-on-plate weld, 815f
Benzazole polymers, 371, 371f
Benzothiadiazole (green emission), 675
Benzotrichloride, 6
Benzoylation treatment, 733
Benzoyl peroxide (BPO), 685f
Benzylated China fir wood flour, 746
Benzyl galactoglucomannan, 853
Best available technologies (BAT), 943
Best environmental practice (BEP), 943
Bingham plastic, 789
Bingham's theory of plastic flow, 789
BioAmber, 326
Bio-based BDO, 325–326
Biobased feedstock, 839
Biobased materials, 836
Bio-based PBT, 326
Bio-based polycarbonates, 348–349, 348f, 349f, 349t
Bio-based succinic acid (SA), 326
Biodegradable Products Institute (BPI), 836
Biodegradable thermoplastics, 851, 928
Biodegradation
of bioplastics, 870–871
of cellulose esters, 870
BioInnovation, 855
Biomaterials
aromatic polyamides as, in medical applications, 311
Bioorganic matter, 924
Bioplastics
additives and
polyhydroxyalkanoates (PHB), 866
polylactide (PLA), 866
thermoplastic starch (TPS), 865–866
applications, 871–872
blends and composites from
about, 866–867
lignocellulose, 868–869
polycaprolactone (PCL), 869
polyhydroxyalkanoates (PHB), 869–870
polylactide (PLA), 869
thermoplastic starch (TPS), 867–868
from classical chemical synthesis (class II)
PE from sugar cane, 854–855
polycaprolactone (PCL), 858–859
polylactide, 855–858
composites, 745–746; *see also* Natural fiber thermoplastic composites
definition of, 836–837, 837f
implementation of, technical challenges for, 839–840
market for, 838

    from microorganisms
        polyhydroxyalkanoates (PHA), 859–861
    from natural sources (class I)
        lignocellulose, 849–853
        starch, 841–848
    polymerization/processing technologies of, 840, 840t
    recycling/ biodegradation of, 870–871
    in standardized world, 950; see also Environment health and safety
    structural/phase characteristics of
        polyhydroxyalkanoates (PHB), 865
        polylactide (PLA), 863–864
        thermoplastic starch (TPS), 861–863
    sustainability assessment of, 838–839
Biosynthesis of PHA, 860
Biphenyl segments, PC containing, 354–355
Birtwhistle, W. K., 320
Bis(α-diketone) and bis(o-diamine) monomers, polymerization from, 543–550, 546f, 547t–549t
Bis(α-diketones) synthesis; see also Monomer synthesis chemistry
    features of, 538–541
    via nucleophilic displacement reactions, 541f
    via oxidative reactions of bis(α-diphenylacetylene), 540f
    via oxidative route, 540f
4,4-bis[(4-amino)thiophenyl] benzophenone (BATPB), 510
3,3-bis(chloromethyl) oxetane (BCMO), 270, 270f
Bis(diphenyl phosphate) (BDP), 894, 895f
Bis-hydroxybutyl terephthalate (BHBT), 329
Bis-(2-hydroxyethyl)-terephthalate (BHET), 321, 329
3,3-bis(hydroxymethyl) oxetane (BHMO), 270, 270f
Bis(o-diamine)
    for functional PPQ, 543
    monomers, synthesis of, 541–543, 543f, 544f, 545t, 546f; see also Monomer synthesis chemistry
Bisphenol A (BPA)-containing polymer, 335
Bisphenol-A (BPA)–isosorbide copolymers, 349
Bisphenol-C PC, 360f
Bis(triphenylsilyl)chromate, 4
Bithiophene (yellow emission), 675
Black list of chemicals, 885
Blend fibers; see also Polybenzimidazoles (PBI)
    PBI/HMA, 631–635
    PBI/PAI blends, 636–637
    PBI/PAr blend, 627–631, 627f–632f, 628t–630t
    PBI/PI blends, 624–626, 625t, 626f, 627f
    PBI/PSF blends, 635–636
    PBI/PVPy blends, 637
Blends
    acrylonitrile, 150
    PMMA, 185–187, 186t
    polyolefins, 28–30, 29t
        block copolymer, 30
        mechanical, 29
        polymer–polymer phase behavior, 29
    POM, 223–226
        PCL and, 225–226
        poly(ethylene oxide) (PEO) and, 224, 224f
        polyvinylphenol (PVPh) and, 224
        PP and, 225
        terpenephenol (TPh) and, 224
    PVB, 118–120, 121t
    PVOH, 77, 78–79

Blends and composites from bioplastics; see also Bioplastics
    about, 866–867
    lignocellulose, 868–869
    polycaprolactone (PCL), 869
    polyhydroxyalkanoates (PHB), 869–870
    polylactide (PLA), 869
    thermoplastic starch (TPS), 867–868
Blends/blending
    PC, 361–362
        PC/ABS, 361–362, 361t
        PC/polyesters, 362
    PEEK, 475–477
    polyarylethersulfones, 438–439
        with aramids, 439
        compatibility, 439
        with liquid crystalline polymers, 440
        miscible, 439–440
        with other polysulfones, 440
        others, 440–441
        with polyaryletherketones and polyphenylenesulfide, 439
    polyimide, 505–507
Block copolymers, 869
    of PC and PDMS (PC-b-PDMS), 897
    of PVOH, 74–75
Block-graft copolymers, 505, 506–507
Blowing agent, 717
Blow molding, 20
    polycarbonates, 365
    for polyolefins, 44
Blown film extrusion, for polyolefins, 44
Blue light-emitting polymers, 311
Boltzmann superposition integral model, 785–787; see also Viscoelasticity
Borates, 889; see also Conventional flame retardants
Boron fibers, 702
Botanical sources, 841
BP Amoco, 336
BPA Polycarbonates, 352–353, 353t
    properties, 352–353
    structure, 350–351, 350f
BP Chemicals Ltd., 15
Bragg's law, 374, 375f
Branched polycarbonates, 355–356, 355f–356f
Branching, of PVOH, 62
British Standards Institution (BSI), 769
Brominated carbonate oligomers, 330
Brominated epoxy resin (BER), 905
Brominated PC, 330
Brominated polystyrene, 330
Bulk crystallization; see also Thermal analysis
    for aromatic poly(ether-alkyl) hydrazides based on 4,4′-dihydroxybiphenyl units, 597–599, 597f, 598f
    for aromatic polyetheraroyl hydrazides based on 4-oxybenzoyl units, 583–584, 583f, 584f
Bulk polymerization, 857
Bump-on-polymer (BOP) structure, 564
Bunsen burner test, 881–882; see also Flame retardants (FR)
Burgers model, 742
Business-NGO Working Group (BizNGO), 838

Index

Butadiene
    for BDO synthesis, 325, 325f
1,3-butadiene, 336
1,4-butanediol (BDO), 324–325
    bio-based, 325–326
    synthesis
        butadiene for, 325, 325f
        fossil-based feedstock, 326
        GENO BDO process, 325–326
        maleic anhydride for, 325, 325f
        propylene oxide for, 325, 325f
    transesterification of, 327, 327f
Butvar PVB
    dielectric properties of, 111, 111f
    mechanical properties of, 112t
Butyric acid pentabromobenzyl ester (BUPB), 906

C

Cables, PPTA fibers use in, 310
Cadmium, 927
Calico Printer's Association Limited, 320
Calorific power of thermoplastics, 931t
Capital costs, 761
Carbene catalyst complex, 39
Carbohydrate moieties, polyhydrazides/polyoxadiazoles with, 612
Carbonate-based polyols, 348
Carbon fiber-reinforced plastic (CFRP), 805
Carbon fiber-reinforced thermoplastics (CFRTP), 805, 806
Carbon fibers, 700–701
Carbon fibers, PAN for, 152–155
    carbonization, 154
    graphitization, 154
    stabilization, 153–154
Carbonization, 154
Carbon molecular sieve membranes, 657
Carbon nanofibers (CNF), 717
Carbon nanotubes (CNTs), 378, 703, 902
Carbonyl groups, PVOH, 62
Carboxyl group formation, 847
Carboxyl-terminated butadiene acrylonitrile rubber (CTBN), 866
Carboxymethyl cellulose (CMC), 850
Cartesian fixed-space coordinates, 793
Cast film extrusion process, for polyolefins, 44
Cast sheet PMMA, 182–183
Catalyst effects; *see also* Polybenzimidazoles (PBI)
    single-stage reaction, 622
    two-stage reaction, 622
Catalysts
    PET polymerization and, 320–321
Catalysts, polyolefins, 3–12
    chromium-based, 3–5
    SSC, 9–12
        metallocene, 9–11
        postmetallocene, 12
    Ziegler–Natta, 5–9
Catalytic chain-transfer polymerization (CCTP), 176, 177
Catalytic oxidation of cyclohexane, 859
Cationic copolymers of PVOH, 74
Cationic (co)polymerization, 196
Cationic ring-opening polymerization (CROP)
    polyepoxides, 254–255, 255f
    polyoxetanes, 270, 271f
    polytetrahydrofuran, 272
    trioxane, 268, 268f
Cauchy deformation tensor, 794
Cayley–Hamilton theorem, 792
Ceclon, 194, 234; *see also* Polyoxymethylene (POM)
Celanese Corporation, 234, 332
Cellular thermoplastic fiber-reinforced composites (CellFRC), 719
Celluloid, 273–274
Cellulose, 370, 849–852; *see also* Lignocellulose
    degree of polymerization (DP), 273
    polyethers from, 273–274, 273f, 274t
    structure, 253f, 273f
    vs. poly(propylene) and poly(ethylene terephthalate) (mechanical properties), 274t
Cellulose acetate (CA), 851
Cellulose acetate butyrate (CAB), 851
Cellulose acetate propionate (CAP), 851
Cellulose ethers, 870
Cellulose hydrate (cellophane), 852
Cellulosic polymers, 852
Centri-sieves, 844
Ceramic fibers, 702
Ceramic slurry, PVB for, 129
Certified reference materials (CRM), 951
Chain configuration, of PVOH, 61–62
Chain conformation, 768
    of PAN, 145
Chain depolymerization, 894
Chain-growth polycondensation methods
    aromatic polyamides, 290–294
    condensation block copolymers preparation, 292–293, 293f
    control mechanism
        based on inductive effects, 292, 292f
        based on resonance effects, 291–292, 291f
Chain length (molecular weights), 764–766
Charge extraction by linearly increasing voltage (CELIV), 677
Charpy impact testing, 771
Chauvin, Yves, 39
Chemical action, FR, 884; *see also* Flame retardants (FR)
Chemical cross-linking, 657; *see also* Gas separation, PBI-based membranes
Chemical industries, environment health and safety in, 943–945
Chemically modified PVOH, 75–77
    acetalization, 76
    cross-linking, 76
    esterification, 75–76
Chemical modification
    of natural cellulose, 850
    of starch, 846, 848
Chemical properties
    acrylonitrile, 141–142
    PAN, 148–149
Chemical resistance
    naphthalate polyesters, 339, 340t
    PAEKs, 473, 474t
    of polyarylethersulfones, 436–437, 438t

Chemical stability
    of aromatic polyamides, 305, 306t
Chemical structures
    of APP I, 894f
    of BDP, 895
    of bis(α-diketone) monomers, 542t
    of 2,2′-bisbenzimidazole-5,5′-dicarboxylic acid, 648f
    of bis(o-diamine) monomers, 545t
    of brominated epoxy resin (BER), 905f
    of ester of phosphonic acid–phosphonate, 893f
    of green-emitting copolymer, 684f
    of HBCD, 891f
    of melam, 896f
    of melamine, 896f
    of melem, 896f
    of monomers for PPQ synthesis, 539f
    of organic phosphate, 893f
    of PBDE, 891f
    of phosphinates, 893f
    of polybenzimidazoles, 643f
    of polymers, 763
    and properties of PPQ, 547t–549t
    and properties of PPQ via self-polymerizable monomers, 553t
    of RDP, 895
    of TBBPA, 891f
    of TBPA, 891f
Chemistry Industry Association of Canada, 943
Chemistry on fuel cell performance; see also Polybenzimidazoles (PBI)
    AB-PBI
        acid doping levels, 648
        isomeric, 648–649
    meta-PBI
        about, 643–644
        acid doping levels, 644, 644t, 645t
        fuel cell performance, 644–645
    para-PBI
        acid doping levels, 645–646, 647f
        fuel cell performance, 647–648, 647f
Chiral acid, 855
Chiral-bridged zirconocene, 11
Chiral polymers, see Optically active polymers (OAPs)
Chiral titanocene, 11
Chromatographic techniques
    OAPs applications, 310
Chromium-based catalysts, 3–5
    organochromium, 4–5
    Phillips catalyst, 3–4
Chromium oxide catalyst, 3–4
    SiO_2-supported, 4
Chromocene, see Bis(cyclopentadienyl)chromate
Clamping pressure, 715
Clariant GmbH, 55
Classical chemical synthesis (class II), bioplastics from; see also Bioplastics
    PE from sugar cane, 854–855
    polycaprolactone (PCL), 858–859
    polylactide, 855–858
Clay-based PC fillers, 363
Clay nanoparticles, 705
Clean Production Action (CPA), 838
3-chloroperbenzoic acid, 673f
CNT, see Carbon nanotube (CNT)

^{13}C nuclear magnetic resonance (NMR) spectra
    of PVB, 100, 102f, 102t
    of PVOH, 58, 59, 63, 64
Coatings
    PTFE processing, 406
    PVB for, 128–129
    PVOH for, 78
Cobalt-mediated radical polymerization (CMRP), 176, 177
C—O—C bond, 252, 252f
    stability of, 252
Code of Federal Regulations (CFR), 78
Coefficient of linear thermal expansion (CLTE), 363
Coefficient of thermal expansion (CTE), 551
Coir husk fibers, 729
Cold crystallization, 464
Collisions, 768
Colorants, POM, 201
Combined LCPs, 370
Combustion, 879
    polymer, 880–881
Commercial aircraft interiors
    polyarylethersulfones applications in, 448, 448f
Commercialization of bioplastics, 871
Commercial polymers, 696
Commercial starches, 843
Commercial TPIs, 507–509, 508t
Commingled yarns, 713, 713f; see also Fiber and tape placement
Commission Internationale d'Eclairage (CIE), 683, 683f
Comonomers in POM, 196, 197
    copolymer synthesis, 198t–199t
Compatible blends of polysulfones, 439
Competitive recycling, 942
Composite materials on laser transmission welding process, 811–817, 811f–817f
Composites
    PAEKs, 477–478, 478t
    PAN, 155
    polyarylethersulfones, 441, 441t
    polyolefins, 41–42
        fibers, 41, 42
        fillers, 41, 42
        interfacial agents, 41, 42
        susceptible to degradation, 42
    POM, 226, 227t
    PPQ, 555–559, 557f, 558f, 558t, 559t; see also Polyphenylquinoxalines (PPQ) processing technology
    PTFE processing, 406
    PVB, 120, 121–122
    PVOH, 77–78
    toughening of, polyarylethersulfones in, 449–450, 449f
Compression molding, 738; see also Natural fiber thermoplastic composites
    for POM processing, 221
Compression-molding technique, 710; see also Thermoplastic composites, advanced
Compression stress–strain behavior, 771, 771f
Computational fluid dynamics (CFD), 762
Conceptual design, 756
Conductive polymers
    conductivity in, 670f
    development of, 669–670
Conductive side-chain polymers, 677

Index

Conductive thermoplastics
conductive polymers, development of, 669–670, 670f
organic semiconductors, 671–672
π-conjugated polymers (main chain conductive polymers)
solubilization of, 672–674, 673f
soluble main-chain conductive polymers
polyethylene dioxythiophene (PEDOT), 674, 674f
polyfluorenes and copolymers, 674–675
substituted polythiophene oligomers, 675–676
vinyl polymers, conductive
about, 676–677
electron-transporting vinyl polymers and copolymers, 680–682, 681f
as emitters, 682–684
hole-transporting vinyl polymers, 678–680, 679t
mobility measurements, 677, 678t
π-π stacking poly(dibenzofulvene), 684–686
Cone calorimetry, 882, 897; *see also* UL94 tests, Bunsen burner test
Coniferyl alcohol, 275, 276f
Consolidation, 709
Constitutional isomerism
aromatic polyamides, 294–296, 294f, 295f
probability parameter, 294
Construction, PMMA for, 184
Continuous improvement processes, 945
Continuous wave (cw) mode laser, 808
Contour welding, 812
Controlled radical polymerization (CRP), 176–177, 177t
Controlled structures, polyamides with, 290–297
chain-growth polycondensation, 290–294, 291f–293f
constitutional isomerism, 294–296, 294f, 295f
spherical aromatic polyamides, 296–297
Convected coordinate system, 792
Conventional acid imbibing, 641–642; *see also* Polybenzimidazoles (PBI)-acid membranes, preparation of
Conventional flame retardants; *see also* Flame retardants (FR)
borates, 889
hydroxycarbonates, 889
metal hydroxides
aluminum trihydroxide (ATH), 887–888
magnesium hydroxide, 888–889
mineral fillers
about, 885–886
thermal effects of, 886–887
Conventional plastics, 837
Conversion costs, 759
Cooked starch, 846
Cooling, 715
Copolycarbonates, 348
Copolycondensation, 543
Copolyesters, CHDM-based, 333t–334t
Copolymerization, 672, 766
of acrylonitrile, 144
Copolymers
acrylonitrile, 149–150
butadiene and, 149
high-barrier application, 149–150
medical application, 150
optically active application, 150

PVOH
block copolymers, 74–75
cationic copolymers, 74
EVA, 73–74
graft copolymers, 74–75
vinyl alcohol–acrylic acid copolymers, 75
vinyl alcohol–maleic acid copolymers, 74
types of, 766f
Copper chromite catalyst, 332
Copper diffusion in polymer film, 564
Copper oxide, 564
Copper–PVOH–lactic acid complex, 77
Corn starch production process, 843f
Corona treatment, 731, 731f
Corotating extruders, 737
Cosmetics
PEG-based surfactants in, 257, 258f
Counter rotating twin-screw extruders, 738
Coupling agents, 733, 733f
Covalent protein conjugation
to PEG, 257–258
Cram, Donald J., 379
Creep–recovery curves, 743f
Critical concentration, 851
Critical fiber length, 742
Critical micelle concentration (CMC), 262
Critical micelle temperature (CMT), 263
Cross-hatching, 830
Cross-linked aromatic polyamides, 312
Cross-linked polyethylene (PEX) pipe systems, 443
Cross-linking, 650, 651f; *see also* Pervaporation separation, PBI-based membranes for
of PVOH, 76
Crown ethers, 252; *see also* Polyethers; *specific entries*
Crystalline PAN, 145
Crystalline polypropylene, 26
Crystalline POM, 208–210, 209f–210f, 211t
Crystalline PVOH, 65, 65f
Crystallinity of polymers, 768–769
Crystallites, 842
Crystallization
PBT, 329
Crystallization, degree of
PEG, 256–257
Crystallization behavior
PAEKs, 469–471, 470f
heat treatment effects on, 470–471, 471f
Crystal structures
PAEKs, 464–466, 464f, 465t
PTFE, 400–401, 400f–401f
Cu(II)–PVOH complex, 77
Cuprous iodide, 539
Current-voltage characteristics, 677, 678t
Cutting-edge characteristics, 822–825, 822f; *see also* Laser cutting
Cyclic ethers, 252; *see also* Polyethers; *specific entries*
Cyclic voltammetry (CV), 685, 686
Cyclization reaction by infrared analysis, 581, 581t, 582f; *see also* Aromatic polyetheraroyl hydrazides based on 4-oxybenzoyl units
Cyclization temperatures, 591
Cyclodehydration, 573
1,4-cyclohexanedicarboxylic acid (CHDA), 332
1,4-cyclohexanedimethanol, 349

Cyclohexanedimethanol (CHDM)-based polyesters, 331–336
  applications, 334–336, 334t
  copolyesters, 333t
  history of, 331–332
  mechanical properties, 334t
  polymerization, 332–333, 332f–333f, 333t
  properties, 334–336, 334t
  thermal properties, 334t
Cyclohexyl methacrylate, 187
Cycloolefin copolymers (COC), 40–41

D

Dark-injection space-charged-limited current (DI-SCLC) method, 677, 678t
DB, *see* Degree of branching (DB)
Deborah number, 781, 783
Decabromodiphenyl ether (decaBDE), 330
Decabromodiphenyl oxide, 904
Deformation tensor, 775, 779
Degradation in plastic products, 950
Degradation of polymeric materials, 921
Degree of branching (DB), 264, 296–297
Degree of polymerization (DP)
  cellulose, 273
  for PVOH, 59, 60t
Degree of substitution (DS), 846
Delrin®, 194
Dendrimers, 305
Density
  of polyethylenes, 22
  of polymethylene, 22
  of PVB, 106
Depolymerization process, 857
Depreciation, 761
Design for assembly (DFA), 762
Design for manufacturability (DFM), 762
Detail design, 756
Detoxification of hemicellulose, 852
Detoxification of lignocellulosic hydrolysates, 852
Deutsches Institut fur Normung (DIN), 769
Diacid chloride, 288, 289
Dialkyl magnesium compounds, 7
2,6-dialkyl napthalene, 336
1,3-diamine-4-chlorobenzene
  polymerization of, 295, 295f
1,3-diamine-4,6-dihalobenzene, 295
1,3-diamine-4-halobenzene, 295
Diamines
  polyimides synthesis from, 493–495
3,3′-diaminobenzidine, 546
Diamino diphenyl methane (DAM), 388, 389f
Dianhydrides
  polyimides synthesis from, 493–496
Diaphragm forming, 713, 714f; *see also* Fiber and tape placement
Diblock (AB)-PBI; *see also* Chemistry on fuel cell performance; Fuel cell performance, chemistry on
  acid doping levels, 648
  isomeric, 648–649
Dichroic dye (iodine) molecules, 79
Dichroism, 79
Dickson, James Tennant, 320
Dielectric properties
  of Butvar PVB, 111, 111f
  of PVB, 110–111, 111t
Dielectric strength
  of PTFE, 410
Diels–Alder reaction, 497–498
Diesters
  polyimides synthesis from, 494–495
Differential scanning calorimetry (DSC), 329, 357, 470–471, 553t, 575, 576f, 579, 703, 704t
  LCPs and, 373–374, 374f
  thermograms, 64
Differential thermal analyses (DTA), 635
Difluoro compounds, 552f
Diimides
  polyimides synthesis from, 496
Diisocyanates
  polyimides synthesis from, 495
Diluents, of PVOH, 70
Dilute acids, 852
Dimerization, 677
Dimethylacetamide (DMA), 577
Dimethyl 1,4-cyclohexanedicarboxylate (DMCD), 332
Dimethylformamide (DMF), 143, 269, 680f, 683f
2,6-dimethylphenol, 278
Dimethyl sulfoxide (DMSO), 269, 273, 379
Dimethyl terephthalate (DMT), 320, 324, 326, 332
  production of, 326–327, 326f
  transesterification of, 327, 327f
Dimethyl trans-1,4-cyclohexanedicarboxylate (DMCD), 332
Diode lasers, 808
Diphenyl carbonate (DPC), 349, 351, 352
  preparation of, 352
Dipping technique, 737
Direct composite manufacturing; *see also* Natural fiber thermoplastic composites
  compression molding, 738
  extrusion, 738
  injection molding, 738
  polypropylene (PP) matrix, 739
  pultrusion, 738
  rotomolding process, 739
Direct costs, 759
Direct methanol fuel cells (DMFC), 645
Dispersion polymerization, 857
Dissolution
  poly(vinyl acetal), 96
Dope preparation, 622, 623t; *see also* Fiber formation
Doping process, 670
Dow Chemical Company, 12, 171
Dried gel films of isotactic PAN, 146
Dry impregnation, 713
Drying
  of fibers, 736
  of polyarylethersulfones, 428
  of PTFE, 406
Dry-jet wet spinning process
  of aramid production, 307f, 308
Dry spinning, 623, 623f; *see also* Fiber formation
  aramid production, 308
  of PAN, 151–152
DSC, *see* Differential scanning calorimetry (DSC)
DuPont, 194, 286, 287, 309, 398, 415, 507
Durability of polymeric materials, 921

Durabio, 348
Durham route, 672
Dynamic focusing, 810f
Dynamic mechanical analysis (DMA), 357, 358, 559, 624, 746, 906
Dynamic-mechanical properties
 of TPUs, 391, 391f–392f, 392t
Dynamico-mechanical analysis, 604

E

Earth ecosystem perspective, 946
Eastman Chemical, 332
Eastman Kodak Co., 331
Eco-balance couples, 932
Economic aspects
 polyarylethersulfones, 451, 452t
Economic/ecological dimensions, for plastic wastes, 932–933; *see also* Recycling of thermoplastics
ECOZEN, 335
ECTFE (ethylene chlorotrifluoroethylene), 413
Edison, Thomas, 152
Ektar, 332
Elastic modulus, 508
Elastomeric phases, 766
Elastomers
 LCEs, 370, 382–384, 383f
 naphthalate-based, 340, 342t
 polyolefins, 28
Electrical and electronic components
 PAEKs applications in, 483
 polyarylethersulfones applications in, 451
Electrical properties, 772–773, 773t
 PAEKs, 475, 476t
 polyarylethersulfones, 435, 435t
 POM, 215t–218t
 PTFE, 402–403
 of PVB, 110–111, 111f, 111t
 volume resistivity, 402
Electrical resistivity
 functional polyimides, 511–512
 of PTFE, 410
Electrochromic polyamides, 310
Electrochromism (EC), 311
Electroluminescence (EL), 310
Electroluminescent materials, polyoxadiazoles as, 608–610
Electromagnetic interference (EMI) shielding, 670
Electron affinity (EA), 563
Electron conductivity, 605
Electron-deficient oxadiazole, 608
Electron-donating group (EDG), 292
Electronics, PMMA for, 185
Electron mobility, 679t
Electron-transporting vinyl polymers and copolymers, 680–682, 681f; *see also* Vinyl polymers, conductive
Electron transport material (ETM), 533, 563
Electron-withdrawing group (EWG), 292
Electro-optical applications, polyhydrazides/polyoxadiazoles for; *see also* Polyhydrazides and polyoxadiazoles
 alkoxy-substituted poly(p-phenylene-1,3,4-oxadiazole), 603–607
 polyoxadiazoles as electroluminescent materials, 608–610

Electrophilic Friedel-Crafts synthesis
 of polyarylethersulfones, 421, 425
Electrospinning, PAN, 152
Elongational or extensional viscosity, 778–780; *see also* Viscosity
El Paso Polyolefins Company, 15
Embodiment design, 756
Emitters, conductive vinyl polymers as, 682–684; *see also* Vinyl polymers, conductive
End groups, in PVOH, 62–63
 carbonyl groups, 62
 carboxyl group, 62
 hydrophobic, 62
End of Life Vehicles (ELV), 885, 925
Endosperm, 843
Endothermic decomposition, 886
Energy industry, PPQ in, 561–562, 563f; *see also* Polyphenylquinoxalines (PPQ)
Engineering materials, classes of, 754f
Engineering (thermo) plastics (ENPLA), 193
Environment, plastics and, *see* Plastics and environment
Environmental impact and recycling
 aromatic polyamides, 313
Environmental management systems (EMS), 945
Environmental resistance
 polyarylethersulfones, 437, 438t
 thermoplastic composites and, 719–721, 720t, 721t; *see also* Thermoplastic composites, advanced
Environmental stress-cracking performance (ESC), 355
Environment health and safety
 accreditation of commitment, 945–946
 bioplastics in standardized world, 950
 in chemical industries, 943–945
 nanotechnology/nanometrology in standardized world, 950–951
 overview, 941–942
 plastic materials and electronic devices, 952
 plastic wastes in standardized world, 948–950
 quality infrastructure and environmental safety, 952–953
 reference materials, 951–952
 standardization, 946–948
 sustainable development, legal frameworks, 942–943
Enzymatic conversion, 847; *see also* Molecular weight reduction, starch
Enzyme treatment, 735
Epichlorohydrine (ECH), 256
Epoxide-derived polyethers
 thermal properties, 266t
 water solubility, 266t
Epoxides, *see* Polyepoxides
Epoxy compounds, PVB with, 96, 99f
Equal biaxial extension, 780
Esterification, PVOH, 75–76
 acetoacetylation, 76
 titantrichloride, 76
Esters, polyacrylates (PAc), 187–188
ETFE (ethylene tetrafluoroethylene), 413–414
Ether bond, 252, 252f
 stability of, 252
Ethers, *see* Polyethers
Ethoxyethyl glycidyl ether (EEGE), 256, 265
Ethoxy vinyl glycidyl ether (EVGE), 260, 260f
Ethyl cellulose (EC), 851

Ethylene chlorotrifluoroethylene (ECTFE), 413
Ethylene diamine (EDA), 388, 389f
Ethylene dioxythiophene (EDOT), 610, 674
Ethylene glycol (EG), 320, 349, 634
Ethylene/norbornene copolymerization, 40–41, 40t
Ethylene oxide (EO), 255
  anionic ring-opening polymerization of, 259
Ethylene–propylene (EP) copolymer, 30–31
Ethylene tetrafluoroethylene (ETFE), 413–414
Ethylene-vinyl acetate (EVA), 888
  copolymers, 30
Ethylene vinyl acetate (EVA)/layered silicate nanocomposite, 899–900
Ethylene vinyl alcohol (EVOH), 851
Ethylene vinyl alcohol (EVA) copolymers, 73–74
3-ethyl-3-(hydroxymethyl) oxetane (EHO), 270, 270f
Ethylidenediacetate, 55
Ethyl trichloroacetate, 6
European Bioplastics Association, 836
European Chemicals Agency (ECHA), 885, 944
European Committee for Standardization (CEN), 944
European Eco-Management and Audit Scheme (EMAS), 945
European Economic Area (EEA), 944
European Medicines Agency, 275
EU total emission limit for MSW, 931t
EVA, see Ethylene vinyl acetate (EVA)
Eviva PSU, 447
Excimer lasers, 808
Exfoliated graphite (EG), 705
Expanded graphite (ExG), 718
EXTEM XH, 507
External quantum efficiency (EME), 563
External stimuli, photoresponsive LCPs, 377
Extrusion, 305, 717, 738, 863; see also Natural fiber thermoplastic composites
  of films, for PVB films, 116
  of polyarylethersulfones, 428–429
  polycarbonates, 365
  ram and paste, of PTFE, 405–406
Extrusion, of POM
  hydrostatic, 220
  melt, 219, 220t
  solid-state, 220
  supercritical carbon dioxide, 220
Extrusion forming, for polyolefins, 44
ExxonMobil, 11

F

Farnham, A. G., 421
Fatty acid esters, 265
Federal Aviation Administration (FAA), US, 879
Feedblock geometries, 797f
FEP (fluorinated ethylene propylene), 409, 411–412
Ferrocenyl glycidyl ether (fcGE), 260, 260f
Fibers; see also Thermoplastic composites, advanced
  alignment, 740
  aramid, 701
  average properties, 700t
  boron, 702
  carbon, 700–701
  ceramic, 702
  high-performance glass, 701–702
  lasers, 808
  mats, 737
  quartz, 702
Fiber and tape placement; see also Thermoplastic composites, advanced
  automated fiber placement (AFP), 711
  commingled yarns, 713, 713f
  diaphragm forming, 713, 714f
  filament winding, 711–712
  injection molding (IM), 714–715
  localized fiber placement, 710
  matched-mold techniques, 715
  pultrusion, 712–713, 712f
  sandwich technology
    fiber stitching, 716
    Z-pinning, 716
  stretch-draw process, 715
  thermoforming, 713–714
Fiber B, 286
Fiber formation; see also Polybenzimidazoles (PBI)
  dope preparation, 622, 623t
  dry-spinning process, 623, 623f
  hot drawing, 624
  sulfonation and stabilization, 624
Fiber-impregnated PEEK feedstock, 697
Fiber reinforced composites (FRC), 711
Fiber-reinforced polymeric composites, 720
Fibers
  PVOH, 79
  stitching, 716; see also Sandwich technology
  and surface treatments, 728–735; see also Natural fiber thermoplastic composites
Filament winding, 711–712; see also Fiber and tape placement
Fillers
  polycarbonates, 363–364, 363f
  POM, 201
Film casting, of PPTA, 308
Films
  PPQ, 555, 555f, 556f; see also Polyphenylquinoxalines (PPQ) processing technology
  PVB
    extrusion of, 116
    laminated safety glass, 122–126, 123f
  PVOH
    blended, 78–79
    mechanical properties, 68
    nonstretched, 69
    polarizing, 79
    preparation, 67
    properties, 67–69, 68t
    tear strength, 69
    tensile strengths, 68–69, 68f
    water-soluble, 78
  stacking, 710
Finger deformation, 794
Finite element (FE), 762
Fire and smoke properties
  of PEEK, 475, 475t
Fixed capital investment, 760
Fixed costs, 759t, 761
Flame retardants (FR)
  classification of, by means of incorporation
    additive fire retardants, 884
    reactive fire retardants, 884

Index

classification of, by mode of action
    chemical action, 884
    physical action, 883–884
conventional
    borates, 889
    hydroxycarbonates, 889
    metal hydroxides, 887–889
    mineral fillers, 885–886
    thermal effects of mineral fillers, 886–887
efficiency, acrylics, 180
halogenated FR additives
    hexabromocyclododecane (HBCD), 890
    polybromodiphenylether (PBDE), 890
    tetrabromobisphenol A (TBBPA), 890
    tetrabromophthalic anhydride (TBPA), 890
halogenated monomers and copolymers, 890–892
laboratory fire testing, 881
limiting oxygen index (LOI) test, 881
nanometric particles as
    CNT/sepiolite, 902
    nanoclays, 898–902
    nanoscale particulate additives/LDH/POSS, 902–903
nitrogen-based, 895–896
overview, 878–880
phosphorus-based
    about, 892
    inorganic phosphates, 894
    intumescent systems, 895
    organic compounds, 894–895
    red phosphorus (RP), 893–894
plasticizers and, 878
polycarbonates, 359–360, 360f
polymer combustion, 880–881
POM, 201
silicon-based
    block copolymers of PC and PDMS (PC-b-PDMS), 897
    silica in PP/PEO/PMMA, 897–898
    silicones in polycarbonate (PC), 897
specification for ideal fire-retardant formulation
    additive/loading level, 885
    compatibility, 885
    environmental aspects and legislation, 885
    fire retardancy, 884
standards and technology life cycle assessment, 883
sustainable flame retardancy-from halogenated chemicals to nanocomposites, 882
synergies
    about, 903–904
    ABS/layered silicate nanocomposites, 905–906
    layered silicate nanocomposites with phosphorus-based compounds, 906–907
    polyamide/layered silicate nanocomposites, 905
    polyester/layered silicate nanocomposites, 906
    polyolefin/layered silicate nanocomposites, 904–905
    synergistic halogenated/phosphorated mixture, 907
    synergistic metal hydroxide/metal borate mixtures, 907–908
    synergistic metallic oxide/phosph orated mixtures, 908
    synergistic nitrogenated and phosphorous mixtures, 907

UL94 tests, Bunsen burner test
    about, 881
    cone calorimetry, 882
Flammability
    aromatic polyamides, 304
    polyarylethersulfones, 435, 436t
    of polymers, 881
Flax, 729
Flax fiber, SEM images of, 732f
Flexural modulus, 770
Flexural strength, 721
Flexural test, 770
Florescence spectroscopy, 372
Flow of polymer melt, 775
Flow simulation, 796f
Fluorescence
    defined, 340
    of naphthalate polyesters, 340
Fluorinated bis(α-diketones) monomer, 543f
Fluorinated bis(o-diamine) monomer, 546f
Fluorinated compounds, 889
Fluorinated ethylene propylene (FEP), 409, 411–412
Fluorinated polyoxadiazoles, 610; *see also* Polyoxadiazoles for high-performance membranes
Fluorine-containing ancillary ligands, 12
Fluorine-containing polyimide blends, 505–506
Fluoroplastics, 397–415
    ethylene chlorotrifluoroethylene, 413
    ethylene tetrafluoroethylene, 413–414
    fluorinated ethylene propylene, 411–412
    overview, 397–398
    perfluoroalkoxy, 414–415
    polychlorotrifluoroethylene, 411
    polytetrafluoroethylene, 398–410
    polyvinyl fluoride, 412–413
    polyvinylidene fluoride, 412
Fluoro-substituted monomers, 552f
Foamed composites, 719; *see also* Foams, thermoplastic
Foams, PPQ, 559–561, 560f; *see also* Polyphenylquinoxalines (PPQ) processing technology
Foams, structural
    polyarylethersulfones applications in, 450, 450f–451f
Foams, thermoplastic; *see also* Thermoplastic composites, advanced
    about, 716–717
    aromatic polyketones-based foams, 717–718
    foamed composites, 719
    polyester-based (PEN-based) foams, 718
    sulfone polymers, foams based on, 718
    thermoplastic polyimides (PI/PAI/PEI), foams based on, 719
Focusing systems, 809–810, 810f; *see also* Laser processing of thermoplastic composites
Food and Drug Administration (FDA), 275
    PEGylated drugs, 256
Food industry, POM in, 194
Food packaging applications of PVOH, 78
Food service and food processing
    polyarylethersulfones applications in, 443
Formaldehyde
    anionic polymerization of, 267–268, 267f–268f
Formaldehyde, for POM, 194, 195–196
    excess methanol method for, 196
    health hazard, 195

inhalation of, 195–196
   methanol oxidation method for, 196
   physical properties, 195t
   polymerization, 197
Formic acid-trapping agents, 200
4-formyl-benzoic acid, 326
Forward osmosis (FO) membranes
   polyarylethersulfones applications in, 445, 446
Fossil-based feedstock
   BDO synthesis and, 326
Fossil resources, 836, 837f
Fourier transform infrared (FTIR) spectra
   of PAN homopolymer, 145, 145f
   POM, 210
   of PVB, 100, 101f
Fourier transform infrared spectroscopy (FTIR), 357, 372, 473, 575, 576f, 625
Fox, Daniel, 350
FR, *see* Flame retardants (FR)
Fracture process, 740, 742
Free fatty acids, 845
Free-radical polymerization, 2, 13–14
   of acrylonitrile, 143
   high-pressure autoclave reactor process, 13
   methacrylates
      backbiting, 173–174, 174f
      overview, 173
      Trommsdorff effect, 175
   tubular reactor process, 13–14
   of vinyl acetate (VAc), 56, 60
Friction
   PTFE and, 406–408
Friction coefficient
   of PTFE, 406
Friedel–Crafts reaction, 553t
Fuel cell performance; *see also* Chemistry on fuel cell performance
   meta-PBI and, 644–645
   para-PBI and, 647–648, 647f
Fuji, 349
Functionality of biobased materials, 840
Functional polyimides
   electrical insulating and electrolyte polyimides, 511–512
   gas permeability and permselectivity, 509–510
   NLO polyimides, 511
   photosensitive polyimides, 510
Fundamental process strategy, 821–822; *see also* Laser cutting
Fungal treatment, 735
Furan-based polymers, 277–278, 277f–278f
Furfural, 277
Furniture, PMMA for, 185

G

Gain-switched operation mode laser, 808
Gardner drop weight impact testing, 771
Gas chromatography (GC)
   OAPs applications in, 310
Gas permeability
   of naphthalate polyesters, 338–339, 339t
Gas permeability and permselectivity
   of functional polyimides, 509–510

Gas-phase technology, 17–18
   fluidized bed reactor, 18
   horizontal stirred-bed reactor, 18
   vertical stirred-bed reactor, 18
Gas separation, PBI-based membranes; *see also* Polybenzimidazoles (PBI)
   chemical cross-linking, 657
   mixed-matrix membranes (MMM), 657–659, 658t
   monomer-level optimization, 654–657
   N-substitution modification, 657
   polymer blending, 657
Gas separation membranes
   polyarylethersulfones applications in, 445–446
Gaussian intensity distribution, 810
Gelatinization, 845
Gelation, in PVOH solutions, 72–73
Gel-permeation chromatography (GPC), 356, 357
   polycarbonates, 356, 357
General constitutive models, nonlinear viscoelasticity, 794–795
General costs, 759, 759t
General Electric, 278, 350
Generalized Maxwell model, 791
Genetically modified organisms (GMO), 838, 839
Genetic manipulation of starch, 848–849
GENO BDO process, 325–326
Genomatica, 325–326
Germ, 843
Germanium
   PET polymerization and, 320
Giesekus model, 796
Glass fiber-filled PBT resins, 330, 330t
Glass fiber-reinforced PBT/PET blends, 330–331, 331t
Glass fiber-reinforced plastic (GFRP), 820
Glass fiber-reinforced thermoplastics (GFRTP), 805, 807
Glass fibers, high-performance, 701–702
Glass phase transition, PAN, 146
Glass transition temperature, 698, 765, 865
   polyarylethersulfones, 422t
   PVB, 104
Global market view
   aromatic polyamides, 308f–309f, 309
   PBT, 324
   PET, 324
Global Product Strategy (GPS), 943
GLOBO welding, 812
Glycerol, 865; *see also* Polyglycerols (PGs)
Glycerol-plasticized soy flour, 745
Glycidol, 263–264
   ROMBP of, 264
Glycidyl methacrylate, 256
Glycidyl methyl ether (GME), 256
Glycol-modified PCT (PCTG/PETG), 332, 335, 336, 362
Graft copolymers, 446–447
   of PVOH, 74–75
"Graft-on-graft" strategies, 266
Granular starch, 868
Granule production, 737; *see also* Semifinished-product manufacturing
Graphene, 705
Graphitization, 154
Gray list of chemicals, 885
Green chemistry
   PBT, 329

Index

Green polymers, 728
Grinding, 843
Group transfer polymerization (GTP)
    of acrylic monomers, 176
Grubbs, Robert H., 39

H

Haas, Otto, 170
Haehnel, W., 55
Half-time of crystallization, 583f
Halogenated aluminum alkyls, 6
Halogenated FR additives; *see also* Flame retardants (FR)
    hexabromocyclododecane (HBCD), 890
    polybromodiphenylether (PBDE), 890
    tetrabromobisphenol A (TBBPA), 890
    tetrabromophthalic anhydride (TBPA), 890
Halogenated monomers and copolymers, 890–892; *see also* Flame retardants (FR)
Halogenated/phosphorated mixture, synergistic, 907
Halogen-bonded SLCPs, 380–382, 381f
Halogen-containing flame-retardant agents, from PBT, 330
Halogen-free FR, 880
Hard and soft acid and base (HSAB) concept, 254
Hard-sphere approach, 768
Hardwood trees, 852
Hatching strategies, 830, 830f
HBCD, *see* Hexabromocyclododecane (HBCD)
*Hb*PGs, *see* Hyperbranched PGs (*hb*PGs)
Health
    PMMA for, 185
    and safety factors, acrylonitrile, 142
Health-related aspects
    polyarylethersulfones, 442
Heat-affected zone (HAZ), 806, 822–823, 822f, 825
Heat capacity of material, 887
Heat conductivity, 813, 813f
Heat deflection temperature (HDT), 334
Heat distortion temperature under load (HDTUL), 353
Heating, 708
Heat release rate (HRR), 882, 903
Heat resistance, 696
    aromatic polyamides, 304
Heat sink, 883
Heat stabilizers, POM, 199, 200
Heat treatment, PAN, 146–147, 147f
Helical (twisted zigzag) structure, of PTFE molecule, 399–400, 399f
Hemicelluloses, 852–853; *see also* Lignocellulose
Hemodialysis membranes
    polyarylethersulfones applications in, 446
Hemp, 729
Heptazine, 896
Hermogravimetry (TGA, 357
Herrmann, W. O., 55
HES (hydroxyethyl starch), 275
Heteroaromatic polymers, 541
Heterogeneous reaction
    poly(vinyl acetal), 96
Heteronuclear single quantum correlation (HSQC), 543, 546f
Heterotactic PVOH, 57

Hexabromocyclododecane (HBCD), 890; *see also* Halogenated FR additives
Hexafluoroisopropanol (HFIP), 105
Hexamethylphosphoramide (HMPA), 286, 287
High-barrier application, acrylonitrile of, 149–150
High-density PE (HDPE), 851, 924, 926
High-density polyethylenes (HDPE), 25–26, 731, 734, 741
    properties, 23t
Highest occupied molecular orbital (HOMO), 675
High-heat (Co) polycarbonates, 353–354, 354f
High-impact polystyrene (HIPS), 883
High-modulus aramide (HMA), 631–635
High-modulus polyaramides (HMA), 632, 632f, 633f
High-molecular-weight (HMW) PBI, 698
High-performance aramid materials, 313
High-performance coatings
    polyarylethersulfones applications in, 450–451
High-performance glass fibers, 701–702
High-performance liquid chromatography (HPLC), 857
High-performance thermoplastic polymers (HPTP), 694
High-temperature applications, polyhydrazides and polyoxadiazoles for, 4'-dihydroxybiphenyl units, *see* Aromatic polyetheraroyl hydrazides based on 4-oxybenzoyl units; Aromatic poly(ether–alkyl) hydrazides based on 4
High-temperature polymer electrolyte membrane fuel cells, PBI in; *see also* Polybenzimidazoles (PBI)
    impact of chemistry on fuel cell performance
        AB-PBI, 648–649
        meta-PBI, 643–645
        para-PBI, 645–648
    low *vs*. high operational temperature for PEM fuel cells
        about, 640–641
        PBI for high-temperature PEM fuel cells, 641
        preparation of PBI–acid membranes, 641–643
High-temperature solution method
    aromatic polyamides synthesis, 289–290
High-tenacity (HT) fibers, 807
Hindered amine light stabilizers (HALS), 878
HiPERTUF, 338
HMA, *see* High–modulus polyaramides (HMA)
^{1}H NMR
    of PVB, 100, 103, 103f, 103t
    of PVOH, 63–64, 63f
Hole-blocking, 681
Hole-transporting vinyl polymers, 678–680, 679t; *see also* Vinyl polymers, conductive
Hollow fiber membranes, PBI, 653–654; *see also* Pervaporation separation, PBI-based membranes for
Hollow fibers, PAN and, 152
Holographic grating, photoresponsive LCPs and, 379
Homogeneous reaction
    poly(vinyl acetal), 96
Homopolymerization of acrylonitrile, 143
Hookean solid, 784f
Hooke's law, 790
Hostaform POM copolymer, 234
Hot drawing, 624; *see also* Fiber formation
Hot radicals, 904
Hull, 843
Humidity on performance of composites, 744
Hyatt Manufacturing Company, 273
Hydra hydrate (HH), 388, 389f

Hydrazide polymers, 609
Hydroforming, 715
Hydrogels, PVOH, 76, 81–82
Hydrogen adsorption, PAN and, 154–155
Hydrogen-bonded SLCPs, 379–380, 380f
Hydrogen bonding, 845
Hydrogen bonds
    TPUs, 389t, 390
Hydrogen peroxide, 847, 859
Hydrolysis
    degree of, PVOH grades, 57–59, 57t, 58f
    at elevated temperature, 870
    of hemicellulose, 852
    of polyvinyl acetate, 54, 56
Hydrolytic degradation, 864
Hydrolytic stability
    polyarylethersulfones, 436–437
Hydrophilic–lipophilic balance (HLB), 262
Hydrophobic alkyl side chains, PMMA with, 187
Hydrostatic extrusion, of POM, 220
Hydroxyalkylation, 848; see also Natural sources (class I), bioplastics from
Hydroxycarbonates, 889; see also Conventional flame retardants
Hydroxyethyl methacrylate (HEMA)
    photopolymerization of, 178
Hydroxyethyl starch (HES), 275
Hydroxyl groups, 731
Hydroxyl-terminated POM, 213
Hydroxypropylation, 848
Hydroxypropyl cellulose (HPC), 850, 851
3-hydroxypropyldimethylvinylsilane, alcholate of, 74–75
Hydroxypropyl methyl cellulose (HPMC), 851
Hyperbranched PGs (*hb*PGs), 264–265
    applications, 265
    degree of branching (DB), 264
    history, 264
    properties, 264–265, 265f
    structure, 265f
    thermal properties, 266t
    water solubility, 266t
Hyperbranched polymers, 296, 305
Hytrel, 340

I

ICI, *see* Imperial Chemical Industries (ICI)
Ideal fire-retardant formulation, specification for; see also Flame retardants (FR)
    additive/loading level, 885
    compatibility, 885
    environmental aspects and legislation, 885
    fire retardancy, 884
I.G. Farben Company, 411
N-imidazol-*O*-(bicycle pentaerythritol phosphate)-*O*-(ethyl methacrylate phosphate), 906
Immiscibility, 853
Impact modification, polycarbonates, 360–361
Impact modifiers, POM, 200
Impact strength, 771
Imperial Chemical Industries (ICI), 2, 320, 398, 462, 920
Impregnation, 709
Incineration and tertiary recycling routes, 930–932, 931t
Incompressible fluid, 779

Indirect costs, 759
Industrial waste, 924
Infrared analysis, cyclization reaction by, 581, 581t, 582f; see also Aromatic polyetheraroyl hydrazides based on 4-oxybenzoyl units
Infrared radiation, 708
Inherent viscosity
    aromatic polyamides, 304
Injection-moldable PVOH compounds, 79
Injection molding (IM), 20, 714–715, 738; see also Fiber and tape placement; Natural fiber thermoplastic composites
    low temperatures for, 13
    of PAESs, 482–483, 482f
    of polyarylethersulfones, 428
    of polycarbonate and its resins, 364–365, 364t
    for polyolefins, 44
    POM, 214, 219, 219t
Injection-molding applications
    of PBT, 330
Injection technique, 305
Inorganic fillers, 330
Inorganic phosphates, 894; see also Phosphorus-based FR
Inorganic silica-like nanocage, 902
Institute of French Petroleum (IFP), 534
Interface location, 797f, 798f, 799f
Interfacial reaction
    polycarbonates synthesis, 351, 352f
Interfacial strength (IFS), 742
Intergovernmental treaty organization, 943
Interlayer dielectrics (ILD), 533, 564
International Council of Chemical Associations (ICCA), 943
International Organization for Standardization (ISO), 309, 353, 421, 947, 949
    in Geneva, 769
International Organization of Legal Metrology (IOLM), 943
International Technology Roadmap for Semiconductors (ITRS), 559
International Union of Pure and Applied Chemistry (IUPAC), 943
International Union of Pure and Applied Physics (IUPAP), 943
Interpenetrating polymer networks (IPNs), 505, 507
Intramolecular lactonization of poly(VBr-co-MMA), 892f
Intrinsic viscosity (IV)
    aromatic polyamides, 304
    PET, 321
Intumescent fire retardant, 884
Intumescent systems, 895; see also Phosphorus-based FR
Iodine complexes, PVOH with, 77
IOLA, 11
Ionomeric PVB, 95
IP200 PPQ films
    about, 537
    properties of, 538t
IQ VALOX, 326
IR spectrum
    of PVB, 100, 101t
    of PVOH, 64, 64f
Isobutane, 15
Isocyanates, 733
    PVB with, 96, 99f

Index

Isomeric AB-PBI, 648–649
Isomerism, constitutional
    aromatic polyamides, 294–296, 294f, 295f
    probability parameter, 294
Isophthalic acid (IPA), 287f, 322
Isophthaloyl dichloride (IPC), 287f
Isopropylidene glyceryl glycidyl ether (IGG), 259, 260f
Isosorbide, 348, 348f
    carbonate-type (Co) polymers, 348, 349t
Isotactic PAN, 146
Isotactic PP, 35–38
Izod impact testing, 771

J

Japanese Industry Standards (JIS), 769
Jarvik 7 artificial heart, 447
Johnson, R. N., 421
Joining of composite parts (examples), 818–820, 818f–820f; *see also* Laser transmission welding

K

Kasei, Asahi, 352
Kel-F 81 PCTFE, 411
Kenaf fiber–PBS composites, 746
Ketone-based thermoplastics, 461–484; *see also* Polyaryl ether ketones (PAEKs)
    applications, 483–484
    blends, 475–477, 477t
    compounds and composites, 477–478, 478t
    overview, 461–463
    processing
        injection molding, 482–483, 482f
        melt processing and, 480–483
        melt spinning, 480–481, 481f
        rubbery region processing and, 479–480
        solid-state deformation and, 478–479
    properties, 467–475
        chemical resistance, 473, 474t
        crystallization behavior, 469–471, 470f
        electrical properties, 475, 476t
        fire and smoke properties, 475, 475t
        heat treatment effect on, 470–471, 471f
        mechanical properties, 471–473, 472t
        optical, 467–469
        radiation resistance, 473, 475
        thermal, 467–469, 468f, 469t
    structure, 463–466, 463f
        crystal, 464–465, 464f, 465t
        influence of chemical architecture, 466, 466f–467f
    synthesis, 463
Kevlar fibers, 286, 299, 309, 369
    properties of, 299, 300t, 303
Kevlar yarn, 299, 302f
Klatte, Fritz, 55
Kneading, 736–737; *see also* Semifinished-product manufacturing
Kodar PETG 6763, 332
Kraft lignin, 275, 277
Kraft process, 275, 277
Kuraray Co., 55
Kwolek, S., 370–371

L

Laboratory fire testing, 881; *see also* Flame retardants (FR)
Lactic acid
    about, 855
    production of, 855
    separation of, 856
Lactobacillus, 855
Lame's constants, 787
Laminated safety glass, PVB films for, 122–126, 123f
Laminates, 828
Landfill engineering, 924
LARC-CPI, 508–509, 508t
LARC-ITPI, 492, 492f
Laser active medium, 808
Laser beam guiding, 809
Laser cutting
    cutting-edge characteristics, 822–825, 822f
    fundamental process strategy, 821–822
    mechanical properties of laser-processed composites, 825–828
    moisture content on cutting behavior, 828–829
Laser material processing, 809
Laser processing of thermoplastic composites
    automated fiber placement, 830–831
    laser cutting
        cutting-edge characteristics, 822–825, 822f
        fundamental process strategy, 821–822
        mechanical properties of laser-processed composites, 825–828
        moisture content on cutting behavior, 828–829
    laser systems
        focusing systems, 809–810, 810f
        laser beam guiding, 809
        laser sources and laser radiation characteristics, 808–809
    laser transmission welding
        about, 810
        composite materials on laser transmission welding process, 811–817, 811f–817f
        joining of composite parts (examples), 818–820, 818f–820f
        process monitoring and control, 817–818
        thermoplastic composites with dissimilar materials, laser-based joining of, 820–821, 820f
    material specification, 807, 807t
    overview, 805–807
    surface ablation, 829–830
    surface activation, 829
Laser radiation, 809, 811
Laser sources/laser radiation characteristics, 808–809
Laser transmission welding (LTW)
    about, 806, 810
    basic techniques, 812
    composite materials on laser transmission welding process, 811–817, 811f–817f
    joining of composite parts (examples), 818–820, 818f–820f
    thermoplastic composites with dissimilar materials, laser-based joining of, 820–821, 820f
Laser-transparent (LT) parts, 807
Laser-welded stiffener panel, 819f
Layered double hydroxides (LDH), 883, 902–903

Layered silicate nanocomposite
    about, 899–901
    with phosphorus-based compounds, 906–907
Layered silicate nanoparticles, 705
LCBCs (liquid crystalline block copolymers), 370, 376–377, 376f–377f
LCEs (liquid crystalline elastomers), 370, 382–384, 383f
LCP/polyimide blends, 506
LCPs, see Liquid crystalline polymers (LCPs)
LDPE, see Low-density polyethylene (LDPE)
Leather paint, PVB for, 128
LED lights, PMMA for, 184
Lehn, Jean-Marie, 379
Lewis acid catalyst adjunct, 10
Lexan EXL, 357
Lexan SLX copolymers, 355
Life cycle assessment (LCA), 883, 929, 942
    studies, 442
Ligands
    postmetallocene catalyst systems, 12
π-ligands, 9
Light amplification, 808
Lighting, PMMA for, 184
Light stabilization of PVB, 118
Lignin, 742, 853; see also Lignocellulose
    Kraft process, 275, 277
    monolignols, 275, 276f
    nanocarriers, 277
    polyethers from, 275–277, 276f
    steam-exploded, 277
    structure, 253f, 276f
Lignocellulose; see also Natural sources (class I), bioplastics from
    blends and composites on, 868–869
    cellulose, 849–852
    hemicelluloses, 852–853
    lignin, 853
Lignocellulosic fiber, structure of, 732f
Limiting oxygen index (LOI), 180, 475, 636
    aromatic polyamides, 304
    defined, 304
    for PET, 322
    test, 881, 907; see also Flame retardants (FR)
    values, of PVB, 118
Linear (linPGs), 265, 266f
    applications, 266–267
    AROP of, 266
    history, 266
    properties, 266, 266t
    synthesis of, 265, 266f
    thermal properties, 266t
    water solubility, 266t
Linear LDPE (LLDPE), 23–25, 23t
    high-pressure autoclave reactor, 14
    high-pressure tubular reactor, 15
    SSC, 33–35
Linear low density polyethylene (LLDPE), 2, 734
Linear viscoelasticity, see Viscoelasticity
LinPGs, see Linear (linPGs)
Lipids, 845
Liquid crystalline block copolymers (LCBCs), 370, 376–377, 376f–377f
Liquid crystalline elastomers (LCEs), 370, 382–384, 383f
Liquid crystalline phases (LCP), 851

Liquid crystalline polymers (LCPs), 369–384, 695, 699–700, 706; see also Nanocomposites; Thermoplastic composites, advanced
    blends with, 440
    characterization methods
        DSC, 373–374, 374f
        POM, 372–373, 373f
        XRD, 374, 375f
    classification, 370, 370f
    history, 369
    LCBCs, 376–377, 376f–377f
    LCEs, 382–384, 383f
    molecular design
        MCLCPs, 370–371, 371f
        SCLCPs, 371–372, 372f
    natural, 370
    overview, 369–370
    photoresponsive, 377–379, 377f, 378f
    physical and chemical properties
        mechanical properties, 375
        viscosity, 375
    SLCPs
        halogen-bonded, 380–382, 381f
        hydrogen-bonded, 379–380, 380f
        others, 382
    synthesis of, 375
    synthetic, 370
    textures of different phases, 373f, 373t
Liquid crystalline (LC) properties
    of polyimides, 504–505
Liquid crystals (LCs), 369
Liquid pool slurry process, 15–16
Lithium chloride (LiCl), 636
Lithium diisopropylamide (LDA), 676
Lithium hexamethyldisilazide (LiHMDS)
    phenyl-4-octylaminobenzoate polycondensation using, 292, 292f
Lithium-ion batteries
    PEG in, 259
Living polymerization, 293
Living polymerizations, of acrylic monomers, 176–177
    anionic polymerization, 176
    controlled radical polymerization (CRP), 176–177, 177t
    group transfer polymerization (GTP), 176
Load-bearing property, 697
Localized fiber placement, 710; see also Fiber and tape placement
Lodge-type integral constitutive models, 793, 794; see also Viscoelasticity
LOI, see Limiting oxygen index (LOI)
Long-chain branching (LCB), 13
    polyethylene, 22
Low-angle laser light scattering (LALLS), 105
Low-density polyethylene (LDPE), 2, 740
    branched, 22–23
        autoclave reactor, 2, 13, 15
        LCB, 22, 23
        SCB, 23
        tubular reactor, 2, 13–14, 15
    linear, 23–25, 23t
        high-pressure autoclave reactor, 14
        high-pressure tubular reactor, 15
    processing and performance properties, 23t

Index

Lower critical solution temperature (LCST)
  PEG, 257, 261
  PMMA/PC blends, 186
  PMMA/PVDF blends, 185
Lowest unoccupied molecular orbital (LUMO), 563, 675
Low-temperature solution method
  aromatic polyamides synthesis, 288–289
Luminescent converters (LUCOs), 311
Luminescent polyamides, 310
Lysophospholipids, 845

M

Magnesium chloride ($MgCl_2$), 7, 8, 8t–9t
Magnesium dihydroxide (MDH), 908
Magnesium hydroxide, 118, 888–889; *see also* Metal hydroxides
Main chain conductive polymers (π-conjugated polymers), 672–674, 673f
Main-chain LCPs (MCLCPs), 370–371, 371f
Maintenance costs, 759
Makrolon 3208
  engineering properties of, 352–353, 353t
Maleated polyethylene (MAPE), 734
Maleated polypropylene (MAPP), 734
Maleic anhydride (MA), 732, 868
  for BDO synthesis, 325, 325f
Maltese, 845
Manufacturing cost estimation, 759–761, 759t, 760f
Marketed PEGylated drugs, 258t
Market for bioplastics, 838
Mark–Houwink equation, 304
  PVB, 104
  PVOH, 60
Mask welding, 812
Mass loss rate (MLR), 882
Matched-mold techniques, 715; *see also* Fiber and tape placement
Material selection/design/application
  design tools/polymeric material property database, 761–763, 762t
  engineering materials, classes of, 754f
  manufacturing cost estimation, 759–761, 759t, 760f
  material selection process, 756–757, 757f
  overview, 753–754
  polymeric material selection, testing for
    electrical properties, 772–773, 773t
    mechanical properties, 769–771, 770t
    microbial, weather, and chemical resistance, 773–774, 774t
    optical properties, 773, 774t
    physical properties, 772, 772t
    thermal properties, 772, 773t
  polymer melt for material selection, rheological properties of
    about, 774–775
    viscoelasticity, 781–799
    viscosity, 775–781
  polymer structure-property relations
    chemical structure of polymers, 763
    microstructure of polymers, 763–769, 764t
  process selection process, 757, 758f
  product design process, 754–756, 755f
Material selection process, 756–757, 757f

Material specification, 807, 807t; *see also* Laser processing of thermoplastic composites
Mat production, 737; *see also* Semifinished-product manufacturing
Mat-reinforced thermoplastics, 709
Matrices, 728; *see also* Natural fiber thermoplastic composites
Matrices, thermoplastic
  liquid crystalline polymers (LCP), 699–700
  polyamideimide (PAI), 699
  polyaryletherketone (PAEK), 698
  polyarylethers (PAE), 695
  polyarylsulfone (PAS), 696
  polybenzimidazole (PBI), 698
  polyesters, 700
  polyetheretherketone, 697
  polyetherimide (PEI), 696–697
  polyetherketone (PEK), 696
  polyethersulfone, 696
  polyphenylene sulfide (PPS), 698–699
Matrimid, 652
Matrix resins and composites, 640; *see also* Polybenzimidazoles (PBI)
Maxwell differential model, 785; *see also* Viscoelasticity
Maxwell-type differential constitutive models, 793–794; *see also* Viscoelasticity
Maxwell-type simple differential constitutive models, 792–793
McCullough's route, 675, 676f
MCLCPs (main-chain LCPs), 370–371, 371f
Measurement of linear viscoelasticity, 791; *see also* Viscoelasticity
Mechanical models for linear viscoelasticity, 790–791; *see also* Viscoelasticity
Mechanical properties
  CHDM-based polyesters, 334t
  of laser-processed composites, 825–828; *see also* Laser cutting
  PMMA, 182
  of polymeric materials, 769–771, 770t
  POM, 215t–218t
  of PTFE, 403
  of PVB, 111–112, 112f, 112t
Mechanical properties, of TPUs
  dynamic, 391, 391f–392f, 392t
  stress-strain properties
    hard-segment content effect on, 393–394, 393f–394f
    segmented structure of TPUs effect on, 392–393, 393f
Medical applications
  acrylonitrile copolymer in, 150
  aromatic polyamides as biomaterials in, 311
  PAEKs applications in, 483–484
Medical devices
  polyarylethersulfones applications in, 447, 448f
Medical diagnostics, PMMA for, 185
Melam, 896, 896f
Melamine, 895–896, 896f
  PVB with, 96, 98f
Melamine phosphate, 896
Melanine cyanurate, 896
Melem, 896, 896f
Melt extrusion, of POM, 219, 220t

Melt flow accelerators, 866
Melting of thermoplastic matrix, 708
Melting point
   PAN, 146
   of PVOH, 66
Melting temperatures
   with different functions, of polyoxetanes, 270–271, 271t
Melt mixing, 736
Melt polymerization, 857
Melt pressure, 715
Melt processing
   of PAESs, 480–483
Melt processing, POM
   extrusion, 219, 220t
   injection molding, 214, 219, 219t
Melt-reaction
   polycarbonates synthesis, 351–352, 352f
Melt spinning, 872
   of PAESs, 480–481, 481f
   of PEEK, 480–481, 481f
Melt viscosity, 765
Melt volume rate (MVR), 353
Membrane electrode assembly (MEA), 640
Membranes
   polyarylethersulfones applications in, 444–447, 445f
   PVOH, 81–82
Membrane technologies, polyamides in, 311, 311t
Mercerization, 732
Mercury, 927
Metal complexes, of PVOH, 77
Metal hydroxide/metal borate mixtures, synergistic, 907–908
Metal hydroxides; *see also* Conventional flame retardants
   aluminum trihydroxide (ATH), 887–888
   magnesium hydroxide, 888–889
Metallic oxide/phosphorated mixtures, synergistic, 908
Metallocene SSC, 9–11
   MAO cocatalysts, 10
   significance of, 11
   on tacticity, 10–11
Metallomesogens, 382
Meta-PBI; *see also* Chemistry on fuel cell performance
   about, 643–644
   acid doping levels, 644, 644t, 645t
   fuel cell performance, 644–645
Methacrylate acid pentabromobenzyl ester (MEPB), 906
Methacrylate–butadiene–styrene (MBS), 331
Methacrylates
   free-radical polymerization, 173–176
   multifunctional, 187
   stereospecific polymerization, 177–178
   storage, 172
Methacrylic acids (MAA), 169, 170, 170f
   storage, 172
   synthesis of, 171–172, 172f
4,4′-diphenylmethane diisocyanate (MDI), 388, 389f
Methanol, 326–327
   formaldehyde, 196
   for hydrolysis of PVA, 56, 57f
   for PVB, 105, 105t
Methyl acetate, 56
Methylaluminoxane (MAO) cocatalysts, 10
Methyl ether ketone (MEK), 717

Methyl methacrylate (MMA), 170
Methylmethacrylate- (MMA-) grafted fibers, 744
Methyl vinyl sulfoxide, PVOH with, 76–77
mf-PEGs, *see* Multifunctional PEGs (*mf*-PEGs)
Michael addition, 498
Microbial, weather, and chemical resistance, 773–774, 774t; *see also* Polymeric material selection, testing for
Microcellular closed-cell morphology, 718
Microcellular polyimide foams, 719
Microelectronic applications, PPQ for, 564; *see also* Polyphenylquinoxalines (PPQ)
Microfibrillated cellulose (MFC), 869
Microfiltration (MF) membranes
   polyarylethersulfones applications in, 445, 446
Microorganisms, bioplastics from, 859–861; *see also* Bioplastics
Microstructure of polymers; *see also* Polymer structure-property relations
   architecture, 763, 764t
   chain conformation, 768
   chain length, 764–766
   crystallinity, 768–769
   monomer arrangement in copolymers, 766
   tacticity, 766–767
Microwave-assisted polycondensation, 289
Microwave radiation (MW) system, 289
Mineral fillers; *see also* Conventional flame retardants
   about, 885–886
   thermal effects of, 886–887
Miscanthus fibers, 731f
Miscible blends
   of polyarylethersulfones, 439–440
Mitsubishi Chemical, 348
Mitsui Inc., 507
Mitsui Petrochemical, 19
Mixed-matrix membranes (MMM), 652–653, 657–659, 658t; *see also* Gas separation, PBI-based membranes; Pervaporation separation, PBI-based membranes for
Mobility measurements, 677, 678t; *see also* Vinyl polymers, conductive
Modacrylic fiber, 140
Module encapsulation, PVB for, 126–128
Modulus of elasticity in bending, 770
Moisture, 721
Moisture content on cutting behavior, 828–829; *see also* Laser cutting
Molar substitution (MS), 851
Molded PBI parts, 637–638, 639t; *see also* Polybenzimidazoles (PBI)
Molded PVOH products, 79
Molding
   PPQ, 555–559, 557f, 558f, 558t, 559t; *see also* Polyphenylquinoxalines (PPQ) processing technology
   PTFE processing, 404–405
Molecular structure
   of PTFE, 399–400, 399f
Molecular structure, of PVOH
   branching, 62
   chain configuration, 61–62
   end groups, 62–63
   tacticity, 63–64

Molecular weight (MW), 671
    PAN, 147
    of PVB, 104–105
    of PVOH, 59–61, 60t
Molecular weight reduction, starch; see also Natural sources (class I), bioplastics from
    acid hydrolysis, 847
    enzymatic conversion, 847
    oxidation, 847
    thermal/thermal–chemical conversion, 847–848
Molten glass, 702
Monolignols, 275, 276f
Monomer-activated anionic polymerization polyepoxides, 255–256, 255f
Monomer-level optimization, 654–657; see also Gas separation, PBI-based membranes
Monomers; see also Polyphenylquinoxalines (PPQ) synthesis chemistry
    arrangement in copolymers, 766
    and polymers, synthesis of, 575–576, 576f, 576t
    with preformed phenylquinoxaline rings, polymerization from, 553, 554f
    and repeat units, 763, 764t
    self-polymerizable, polymerization from, 550–553, 550f–552f, 553t
Monomer synthesis
    aromatic polyamides, 288
Monomer synthesis chemistry; see also Polymerization chemistry
    about, 537–538, 539f
    bis(α-diketones), synthesis of, 538–541, 539f–541f, 542t
    bis(o-diamine) monomers, synthesis of, 541–543, 543f, 544f, 545t, 546f
Monosaccharides, 852
MonoSol, LLC, 55
Montmorillonite (MMT), 702, 706, 898
Mowiflex TC, 55
m-phenylenediamine (MPD), 287f, 288
Poly(m-phenylene isophthalamide) (MPIA), 286, 287f, 298
    chemical resistance, 305, 306t
    market breakdown for, 309f
    preparation of, 288, 288f
    in textile market, 309–310
    wet spinning of, 307
MPIA, see Poly(m-phenylene isophthalamide) (MPIA)
Multichip module (MCM), 564
Multifunctional PEGs (mf-PEGs), 253f, 259–261, 260f, 261t; see also Poly(ethylene glycol)s (PEG)
    thermal properties, 261, 261t
    water solubility, 260–261
Multipass method, 823, 824
Multiwalled carbon nanotube (MWCNT), 360, 363, 706
Multiwalled nanotubes (MWNT), 902
Municipal solid waste (MWS), 919, 925, 931t
M.W. Kellogg Company, 411

N

Nafion, 415
N-alkyl polyamides, 292
Nanocarriers, lignin, 277
Nanoclays
    about, 898–899
    polyamide PA-6/layered silicate nanocomposites, 901–902
    polymer/layered silicate nanocomposites (PLSN)
        EVA/layered silicate nanocomposite, 899–900
        PP/PPgMA/layered silicate nanocomposite, 900–901
Nanocomposites; see also Thermoplastic composites, advanced
    liquid crystalline polymers (LCP), 706
    PAN, 155
    PEEK nanocomposites, 702–705, 704t
    polyester (PET, PEN) nanocomposites, 705
    polyolefins, 42–44
        chemical reaction, 43–44
        as exfoliated, 42
        as immiscible, 42
        as intercalated, 42
        multiphase systems, 42
        performance, 42
        physical methods, 42–43
    polyphenylene sulfide (PPS), 706–707
    polysulfone nanocomposites, 705
    POM, 226, 227–232, 228t–231t
    PVB, 120, 121–122
    PVOH, 77–78
    thermoplastic polyimide (PI, PAI, PEI), 706
Nanofibers, 155
Nanofibrous materials, 902
Nanofillers, 702
Nanofiltration (NF) membranes
    polyarylethersulfones applications in, 445, 446
Nanometric fillers, 717
Nanometric metallic oxide particles, 903
Nanometric particles as FR
    CNT/sepiolite, 902
    nanoclays
        about, 898–899
        polyamide PA-6/layered silicate nanocomposites, 901–902
        polymer/layered silicate nanocomposites (PLSN), 899–901
    nanoscale particulate additives/LDH/POSS
        nanometric metallic oxide particles, 903
        silsesquioxane, 902–903
Nanometrology in standardized world, 950–951
Nanoscale particulate additives/LDH/POSS
    nanometric metallic oxide particles, 903
    silsesquioxane, 902–903
Nanotechnology in standardized world, 950–951; see also Environment health and safety
Naphthalate-based elastomers, 340, 342t
Naphthalate polyesters, 336–342
    applications, 341
    chemical resistance, 339, 340t
    fluorescence of, 340
    gas permeability, 338–339, 339t
    history, 336
    mechanical properties, 340, 341t
    naphthalate-based elastomers, 340, 342t
    optical (UV) properties, 339
    photodegradation, 340
    polymerization, 337
    thermal properties, 337–338, 337t–338t
Naphthalene dicarboxylic acid (NDA), 336
    process, 337f

2,6-naphthalene dicarboxylic acid (NDA), 335
2,6-naphthalene dicarboxylic acid dimethyl ester (NDC), 336
Native starch, 846, 870
Natta, Guilio, 5, 39; see also Ziegler–Natta catalysts
Natural fibers
    about, 727, 728
    amenable to chemical modification, 731
    chemical composition of, 730t
    physical/mechanical properties of, 729t
    surface chemical modification of, 732
Natural fiber thermoplastic composites
    fibers and surface treatments, 728–735
    matrices, 728
    overview, 727–728
    processing
        about, 736, 737f
        direct composite manufacturing, 738–739
        semifinished-product manufacturing, 736–737
    properties
        agrofiber/thermoplastic composites, 739–740
        bioplastic composites, 745–746
        polyethylene (PE) composites, 740–741
        polypropylene composites, 741–744
        synthetic thermoplastics, composites from, 744–745
    remarks on, 746–747
Natural LCPs, 370
Natural magnesium carbonate, 889
Natural (not colored) organic and conventional synthetic fibers
    vs. Teijinconex fibres, 301t
Natural-reinforced plastics, 727
Natural sources (class I), bioplastics from; see also Bioplastics
    lignocellulose
        cellulose, 849–852
        hemicelluloses, 852–853
        lignin, 853
    starch
        about, 841–846
        genetic manipulation of, 848–849
        molecular weight, reduction of, 846–848
        physical/chemical modification of, 846
        starch esters, 848
        starch ethers, 848
        thermoplastic starch (TPS), 849
NBR, see Nitrile runner (NBR)
N-Bromosuccinimede (NBS), 683f
N-Butyl perchlorocrotonate, 6
Near-infrared (NIR) spectrum, 808
Nematic optical texture of poly(ether-oxadiazole), 608f
Newtonian behavior
    polypropylene (PP), 27
Newtonian fluid, 783, 784f
NEW-TPI, 492, 493f, 508–509, 508t
Nicotinamide adenine dinucleotide phosphate (NADPH), 860
Nitrile groups, PAN and, 144
Nitrile runner (NBR), 149
4-Nitrobenzil, 540
Nitrogenated and phosphorous mixtures, synergistic, 907
Nitrogen-based FR, 895–896; see also Flame retardants (FR)
4-Nitro-2-trifluoromethylbenzil (NTFB), 541

Nitroxide-mediated polymerization (NMP), 176, 177, 375
NLO polyimides, 511
N-methyl-2-pyrrolidone (NMP), 95, 286, 287, 550
NMR spectroscopy, see Nuclear magnetic resonance (NMR) spectroscopy
Nomex 455, 303
Nomex 462, 303
Nomex fibers, 286, 302t, 303, 309
Nomex IIIA, 303
Nonbiodegradable plastics, 836
Nonlinear viscoelasticity, see Viscoelasticity
Non-Newtonian flow, 785
Nonoxinol, 257, 258f
NORYL resins, 278, 279
N-substitution modification, 657
Nuclear magnetic resonance (NMR), 541, 681
    for PAN chain, 145
    POM, 210
Nuclear magnetic resonance (NMR) spectroscopy, 295, 296, 357, 372
Nucleating agents, POM, 200–201
Nucleophilic displacement reactions, 541f
Nucleophilic displacement route
    polyarylethersulfones synthesis, 421–424
Nylon, 901f

O

4-octyloxybenzyl (OOB), 292
ODA/PPTA, 286, 287f, 298, 312
    in textile market, 309–310
    wet spinning of, 308
Ohm's law, 670
Oligoglycerols, 265
One-step synthesis of PODZ, 572–573
Operating costs, 759, 759t, 760f
Optical applications of PPQ, 563–564; see also Polyphenylquinoxalines (PPQ)
Optical fiber, 809
Optically active compounds, acrylonitrile copolymer in, 150
Optically active polymers (OAPs)
    applications, 310–311
Optical microscopy
    aromatic poly(ether-alkyl) hydrazides based on 4,4′-dihydroxybiphenyl units, 600–602
    aromatic polyetheraroyl hydrazides based on 4-oxybenzoyl units, 584–585, 585f
Optical (UV) properties
    naphthalate polyesters, 339
    PMMA, 181, 181t
        scattering studies, 181–182
    of polymeric materials, 773, 774t
OPX Biotechnologies Inc., 171
Organically modified MMT (oMMT), 905
Organic compounds, 894–895; see also Phosphorus-based FR
Organic field-effect transistor (OFET) arrays, 671
Organic light-emitting diodes (OLEDs), 310–311, 533, 563, 671
Organic photovoltaic (OPV) cells, 671
Organic semiconductor; see also Conductive thermoplastics
    about, 671–672
    mobility measurement techniques in, 678t

Index

Organic-soluble and/or melt-processable (thermoplastic) polyimides, 498
    aliphatic–aromatic, 500–501
    aromatic, 498–500, 499t
Organic thin-film transistor (OTFT) method, 677, 678t
Organochromium catalysts, 4–5
    calcined chromium oxide catalyst, 4
    esters, 4
Organomodified montmorillonite (oMMT), 900
Organophosphorus compounds, 894
Orthorhombic POM, 208–209
Oval channel geometry, 796f
Oxalic acid ester, 674
Oxetanes, *see* Polyoxetanes
Oxidation, 847; *see also* Molecular weight reduction, starch
Oxidative coupling, 676
Oxidative doping, 670
Oxiranes, 253; *see also* Polyepoxides
Oxo-degradable plastics, 837
3,4-oxydianiline (ODA), 286, 287f, 288
Oxygen
    barrier of PVOH, 78
    carbon fibers and, 153–154
    permeability, 851

P

Packaging products, 872
PAEKs, *see* Polyaryl ether ketones (PAEKs)
Parabanic polymers, 612
Para-PBI; *see also* Chemistry on fuel cell performance
    acid doping levels, 645–646, 647f
    fuel cell performance, 647–648, 647f
Paste extrusion
    of PTFE, 405–406
PBCMO, 270
    applications, 271–272
PBDE, *see* Polybromodiphenylether (PBDE)
PBI/polyimidesulfone blends, 506
PBS, *see* Poly(butylene succinate) (PBS)
PBT, *see* Poly(butylene terephthalate) (PBT)
PBT/PET blends, glass fiber-reinforced, 330–331, 331t
PC, *see* Polycarbonates (PC)
PC/ABS (polycarbonate blends with acrylonitrile–butadiene–styrene copolymers), 361–362, 361t
π-conjugated polymers; *see also* Conductive thermoplastics
    solubilization of, 672–674, 673f
p-coumaryl alcohol, 275, 276f
PC–polyarylate copolymers, 355
PC/polyesters, 362
PCTFE (polychlorotrifluoroethylene), 411
PDMS, *see* Polydimethylsiloxane (PDMS)
Pd pincer surfactant, 382
Pd–SCS pincer, 382
Peak of heat release rate (pHRR), 882
Peak of mass loss rate (pMLR), 882
Pedersen, Charles J., 379
PEEEK, *see* Polyether ether ether ketone (PEEEK)
PEEK, *see* Polyether ether ketone (PEEK)
PEEKEK, *see* Polyether ether ketone ether ketone (PEEKEK)
PEEKK, *see* Polyether ether ketone ketone (PEEKK)

PEEK/PEI blends, 476, 480
PEG, *see* Poly(ethylene glycol)s (PEG)
PEG–difatty acid ester, 257, 258f
PEG–fatty acid ester, 257, 258f
PEG–glyceryl fatty acid ester, 257, 258f
PEG–monoalkyl ethers, 257, 258f
PEG–monoalkylphenyl ethers (alkylphenolethoxylates), 257, 258f
PEGylated drugs, 257–258
    FDA-approved, 256
    marketed, 258t
PEGylation, 256, 257–258
PEK, *see* Polyether ketone (PEK)
PEKEKK, *see* Polyether ketone ether ketone ketone (PEKEKK)
PEKK, *see* Polyether ketone ketone (PEKK)
PEKKK, *see* Polyether ketone ketone ketone (PEKKK)
Pendulum impact-type tests, 771
Pentabromobenzyl ester polyacrylate (PBPA), 906
Pentaplast, 270
Penton, 270
PEO, *see* Polyethylene oxide (PEO)
Perfluoroalkoxy (PFA), 414–415
Perfluorobutenylvinylether (PBVE), 415
Perflurodimethyldioxole (PDD), 415
Performance Fibers A-701, 341
Persistence, bioaccumulation, and toxicity (PBT), 882
Pervaporation, 650
Pervaporation separation, PBI-based membranes for; *see also* Polybenzimidazoles (PBI)
    about, 649–650, 649t
    modifications
        cross-linking, 650, 651f
        mixed-matrix membranes (MMM), 652–653
        polymer blending, 652
        sulfonation, 651–652, 651f
    PBI hollow fiber membranes, 653–654
PES/PEEK blends, 439
PET, *see* Poly(ethylene terephthalate) (PET)
PETCORE Europe, 329
PETG copolyesters, 335
PET/PEN blends, 338–339
    gas barrier properties, 339
Petrochemical utility scale-up methods, 932
PFA (perfluoroalkoxy), 414–415
PFA (poly(furfuryl alcohol)), 277–278, 278f
PFA methyl vinyl ether (PMVE), 414
PGs, *see* Polyglycerols (PGs)
PHA, *see* Polyhydroxyalkanoates (PHA)
Phase segregation, 673
Phase separation, 630
    in TPUs, 390
Phenolic compounds, PVB with, 96, 97f
4-phenylethynylbenzil (PEBZ), 557
Phenylethynyl-terminated PPQ (PEPPQ)
    inherent viscosities/rheological properties of, 558t
    rheological behaviors of, 558f
    synthesis of, 557, 557f
    thermal processing parameters, 558f
Phenyl-4-octylaminobenzene
    polycondensation of, 291, 291f
Phenyl-4-octylaminobenzoate
    polycondensation of, using LiHMDS, 292, 292f
Phenyl propane, 853

Phenylquinoxaline rings, monomers with preformed, 553, 554f
Phenyl vinyl sulfoxide, 672
Phillips catalyst, 3–4
    chromium oxide, 3–4
    inorganic and organic compounds, 4
Phillips Petroleum Company, 3, 15
Phodegradation of PVB, *see* Photooxidation of PVB
Phosphate FR (PFR), 906
Phosphoric acid-doped PPQ, 563f
Phosphorus, flame-retardant efficiency of, 180
Phosphorus-based flame retardants (FR), 360; *see also* Flame retardants (FR)
    about, 892
    inorganic phosphates, 894
    intumescent systems, 895
    organic compounds, 894–895
    for PBT, 330
    red phosphorus (RP), 893–894
Photoaging of PVB, *see* Photooxidation of PVB
Photoalignment, photoresponsive LCPs, 378
Photoconductive polymers, 676
Photodegradation, 928
    naphthalate polyesters, 340
Photoluminescence (PL), 310
    of poly(ether-oxadiazole) fiber, 609f
Photooxidation of PVB, 112–114, 113f
Photoresponsive LCPs, 377–379, 377f, 378f
    external stimuli, 377
    in holographic grating, 379
    photoalignment, 378
Photosensitive polyimides (PSPIs), 510
Photosynthesis, 855
Physical action, FR, 883–884; *see also* Flame retardants (FR)
Physical modification of starch, 846
Physical properties
    acrylonitrile, 141
    formaldehyde, 195t
    PAN, 147–148
    PMMA, 178t–180t
    of polymeric materials, 772, 772t
Pineapple, 729
Planar extension, 780
Planext, 348
Plasma-enhanced chemical vapor growth deposition, 902
Plasma treatment, 731
Plasticizers, 851, 861, 865
    and FR, 878
    PVB and, 116–117, 117f
Plastic materials and electronic devices, 952; *see also* Environment health and safety
Plastics Age, 925
Plastics and environment; *see also* Recycling of thermoplastics
    about, 924–925
    agricultural plastic wastes, 927
    automotive plastic wastes, 926
    plastic wastes in electronic devices, 927–928
    urban plastic wastes, 925–926
    wild plastic wastes, 928
Plastic wastes
    in electronic devices, 927–928
    in standardized world, 948–950; *see also* Environment health and safety

Plastic wastes and eco-balance; *see also* Recycling of thermoplastics
    economic and ecological dimensions, 932–933
    incineration and tertiary recycling routes, 930–932, 931t
    primary and secondary recycling routes, 929–930, 930f
Platelet-like nanoparticles, 705
Plexiglas, 170
Plumbing components
    polyarylethersulfones applications in, 443–444, 444f
Plunkett, Roy, 397
Pluronic PPGs, 262–263, 262f, 263t
    physicochemical properties, 262, 263t
    structure, 253f
PMDA-DDE, 511
PMMA, *see* Polymethyl methacrylate (PMMA)
PODZ, *see* Poly(*p*–phenylene– 1, 3, 4–oxadiazoles)
Polarizing optical microscopy (POM)
    LCPs and, 372–373, 373f
Polarizing PVOH films
    dichroism, 79
    for LCD panels, 79
Pollution reduction, 883
Poloxamers, 257, 258f
Polyacetals, *see* Polyoxymethylene (POM)
Polyacetylene, 574
    synthesis of, 673f
Polyacrylates (PAc), *see* Poly(methyl methacrylate) (PMMA)
Polyacrylic acid (PAA)
    esters, 187–188
Polyacrylonitrile (PAN), 139, 140, 701
    adsorbents, 152
    applications, 152
    chemical properties, 148–149
    dry spinning, 151–152
    electrospinning, 152
    gel spinning, 152
    phase characteristics
        glass transition, 146
        heat treatment, 146–147, 147f
        transitions, 146
    physical properties, 147–148
    processing, 150–152
    solution fiber spinning, 151–152
    structure, 144–146
    technologies based on, 152–155
    wet spinning, 151
Polyaddition
    polyimides synthesis, 497–498
        Diels–Alder reaction, 497–498
        Michael addition, 498
Poly(amic acid) (PAA), 719
Polyamide 66 (PA-66), 906
Poly(amide-ester), 295–296
Polyamideimide (PAI), 305, 618, 636–637, 699; *see also* Thermoplastic composites, advanced
Polyamide PA-6/layered silicate nanocomposites, 901–902
Polyamide resins, 200
Polyamides, 285–286
    aliphatic (nylons), 285–286
    aromatic, *see* Aromatic (aramids) polyamides
Polyarylate (PAr), 627–631

Index

Poly(arylene ether phenylquinoxaline) synthesis via self-polymerization route, 551f
Poly(aryl ether ketone ether ketone naphthyl ketone) (PEKEKNK)
    thermal properties, 467
Polyaryl ether ketones (PAEKs), 461, 698; *see also* Ketone-based thermoplastics; Thermoplastic composites, advanced
    blends, 439, 475–477
    chemical resistance, 473, 474t
    compounds and composites, 477–478, 478t
    crystallization behavior, 469–471, 470f
    crystal structures, 464–466, 464f, 465t
    electrical properties, 475, 476t
    examples, 461, 462f
    fire and smoke properties, 475, 475t
    mechanical properties, 471–473, 472t
    PEEEK, *see* Polyether ether ether ketone (PEEEK)
    PEEK, *see* Polyether ether ketone (PEEK)
    PEEKEK, *see* Polyether ether ketone ether ketone (PEEKEK)
    PEEKK, *see* Polyether ether ketone ketone (PEEKK)
    PEK, *see* Polyether ketone (PEK)
    PEKEKK, *see* Polyether ketone ether ketone ketone (PEKEKK)
    PEKK, *see* Polyether ketone ketone (PEKK)
    PEKKK, *see* Polyether ketone ketone ketone (PEKKK)
    radiation resistance, 473, 475
    thermal properties, 467–469, 469t
Polyarylethers (PAE), 695; *see also* Thermoplastic composites, advanced
Polyaryl ether sulfones (PAESs), 420–452, 477
    applications, 443, 483–484
        commercial aircraft interiors, 448, 448f
        electrical and electronic components, 451
        food service and food processing, 443
        high-performance coatings, 450–451
        medical devices, 447, 448f
        membranes, 444–447, 445f
        plumbing components, 443–444, 444f
        structural foams, 450, 450f–451f
        toughening of advanced composites, 449–450, 449f
    blends, 438–439
        with aramids, 439
        with liquid crystalline polymers, 440
        miscible, 439–440
        with other polysulfones, 440
        others, 440–441
        with polyaryletherketones and polyphenylenesulfide, 439
    chemical resistance, 436–437, 438t
    composites, 441, 441t
    economic aspects, 451, 452t
    electrical properties, 435, 435t
    flammability behavior, 435, 436t
    glass transition temperatures, 422t
    health-related aspects, 442
    hydrolytic stability, 436–437
    mechanical properties, 430–434, 433t–434t
    overview, 420–421
    physical properties, 430–434, 430t
    polyethersulfone (PES), 420
    polyphenylsulfone (PPSU), 420
    polysulfone (PSU), 420
    processing
        drying, 428
        effect on properties, 429–430
        extrusion, 428–429
        injection molding, 428, 482–483, 482f
        melt processing and, 480–483
        melt spinning, 480–481, 481f
        rheology, 427–428
        rubbery region processing and, 479–480
        secondary fabrication operations, 429
        solid-state deformation and, 478–479
        solution processing, 429
        thermoforming, 428–429
    radiation resistance, 435–436
    safety, 442
    solubility, 437–438
    sustainability aspects, 442–443
    synthesis
        alternate routes, 424–426
        electrophilic Friedel–Crafts synthesis, 421, 425
        nucleophilic displacement route, 421–424
    thermal properties, 430–434, 430t, 431f
Polyarylsulfone (PAS), 696; *see also* Thermoplastic composites, advanced
Poly(4-benzamide) (PBA), 371
Polybenzimidazole (PBI)/polyimide blends, 506
Polybenzimidazoles (PBI), 440–441, 535, 698; *see also* Thermoplastic composites, advanced
    blend fibers
        PBI/HMA, 631–635
        PBI/PAI blends, 636–637
        PBI/PAr blend, 627–631, 627f–632f, 628t–630t
        PBI/PI blends, 624–626, 625f, 626f, 627f
        PBI/PSF blends, 635–636
        PBI/PVPy blends, 637
    catalyst effects
        single-stage reaction, 622
        two-stage reaction, 622
    fiber formation
        dope preparation, 622, 623t
        dry-spinning process, 623, 623f
        hot drawing, 624
        sulfonation and stabilization, 624
    gas separation, PBI-based membranes
        chemical cross-linking, 657
        mixed-matrix membranes (MMM), 657–659, 658t
        monomer-level optimization, 654–657
        N-substitution modification, 657
        polymer blending, 657
    in high-temperature polymer electrolyte membrane fuel cells
        impact of chemistry on fuel cell performance, 643–649
        low *vs.* high operational temperature for PEM fuel cells, 640–643, 642f, 643f
    matrix resins and composites, 640
    molded PBI parts, 637–638, 639t
    overview, 618–619
    pervaporation separation, PBI-based membranes for
        about, 649–650
        with modifications, 650–653
        PBI hollow fiber membranes, 653–654

product requirements, 621–622
products, 659, 660f
synthesis
general route to, 619, 619t, 620f
specific case for, 619–621, 620f, 621t
Polybenzimidazoles (PBI)-acid membranes, preparation of; *see also* High-temperature polymer electrolyte membrane fuel cells, PBI in
conventional acid imbibing, 641–642
porous PBI, 642
sol–gel process, 642–643
Poly(benzofulvene), 686
Polybenzoxazoles (PBO), 535
Poly(bis(3,3-hydroxymethyl)oxetane) (PBHMO), 270
Polybromodiphenylethers (PBDE), 880, 890; *see also* Halogenated FR additives
Polybutadiene adipate-co-terephthalate (PBAT), 746
Poly(butene-1) copolymers, 31
Poly(buthylene terephthalate) (PBT), 572
Poly(butylene naphthalate) (PBN), 337
Poly(butylene succinate) (PBS), 745, 746
Poly(butylene terephthalate) (PBT), 324–331, 820
applications, 329–331, 330t–331t
catalysts for production of, 328
crystallization, 329
flame-retardant ratings of, 330
global market, 324
green chemistry, 329
history, 324
injection-molding applications, 330
polycondensation, 327, 327f
polymerization process, 327–329, 327f–328f
properties, 329–331, 330t–331t
raw materials, 324–327, 324f–327f
THF by-product formation, 328–329, 328f
from TPA and BDO, 327–328, 328f
Polycaprolactone (PCL), 836
blends and composites on, 869
classical chemical synthesis (class II), bioplastics from, 858–859
Polycarbodiimide, 864
Polycarbonate blends with acrylonitrile–butadiene–styrene copolymers (PC/ABS), 361–362, 361t
Polycarbonates (PC), 348–365
aliphatic, 348–349, 348f, 349f, 349t
aromatic, 350, 350t
bio-based, 348–349, 348f, 349f, 349t
BPA, 352–353, 353t
branched, 355–356, 355f–356f
characterization of
analytical techniques, 357
relaxation studies, 358–359, 358f
rheology, 358
stability studies, 357–358
thermal analysis, 357–358
containing biphenyl segments, 354–355
copolymers with polydimethylsiloxane, 356–357
high-heat (Co)polycarbonates, 353–354, 354f
modification
blends and alloys, 361–362
fillers, 363–364, 363f
flame-retardant polycarbonates, 359–360, 360f
impact, 360–361
PC/ABS, 361–362, 361t

PC/polyesters, 362
stabilization, 359
overview, 348
PMMA and, 186
polyarylate block copolymers including resorcinol, 355
processing
blow molding, 365
extrusion, 365
injection molding, 364–365, 364t
thermoforming, 365
producers and trademarks, 350t
properties of, 352–357
synthesis of, 350–352, 350f
diphenyl carbonate preparation, 352
interfacial reaction, 351
melt-reaction, 351–352, 352f
types of, 348
UV resistance, 355
Polychlorotrifluoroethylene (PCTFE), 411
Polycondensation, 290, 589
chain-growth, aromatic polyamides, 290–294
microwave-assisted, 289
of phenyl-4-octylaminobenzene, 291, 291f
of phenyl-4-octylaminobenzoate using LiHMDS, 292, 292f
polyimides synthesis by, 493–497
Poly(cyclohexylene dimethylene cyclohexanedicarboxylate) (PCCD), 332, 335
Poly(1,4-cyclohexylenedimethylene terephthalate) (PCT), 331–332, 334
Poly(cycloolefins)
by ROMP, 39
by vinyl polymerization, 39
Poly(dibenzofulvene) (Poly(DBF)), 684
Polydicyclopentadiene, 39
Poly(9,9-dihexylfluorene), 675
Polydimethylsiloxane (PDMS), 356–357, 897
copolymers with, 356–357
Polydispersity index, 27
Poly(9,9-disubstituted fluorene), 674, 682
Poly(ε-caprolactone) (PCL)
POM and, 225–226
Polyelectrolyte, 674
Polyepichlorohydrine (PECH), 255
Polyepoxides, 253–256
anionic ring-opening polymerization, 253–254, 254f
cationic ring-opening polymerization, 254–255, 255f
monomer-activated anionic polymerization, 255–256, 255f
Polyester-based foams, 718; *see also* Foams, thermoplastic
Polyester/layered silicate nanocomposites, 906; *see also* Synergies, FR
Polyesters, 286, 700; *see also* Thermoplastic composites, advanced
CHDM-based, 331–336
applications, 334–336, 334t
history of, 331–332
polymerization, 332–333, 332f–333f, 333t
properties, 334–336, 334t
nanocomposites, 705
Polyesters, thermoplastic, 320–342
cyclohexanedimethanol-based, 331–336
naphthalate polyesters, 336–342

Index

poly(butylene terephthalate), 324–331
poly(ethylene terephthalate), 320–324
Poly(ether-alkyl) hydrazides, 4′-dihydroxybiphenyl units, *see* Aromatic poly(ether–alkyl) hydrazides based on 4
Polyetheraroyl hydrazides (PEHZ), 575, 579
Polyetheraroyloxadiazole (PEODZ)
    Avrami plots for, 584f
    codes/cyclization temperatures of, 581, 581t
    infrared spectra of, 582f
    optical micrographs, 585f, 586f
    thermal data of, 583t
Polyether ether ether ketone (PEEEK), 461–462, 462f
Polyether ether ketone (PEEK), 461, 462f, 562, 697, 702–705, 704t 720t; *see also* Nanocomposites; Thermoplastic composites, advanced
    blends, 475–477
    chemical resistance, 473, 474t
    crystallization behavior, 469–471, 470f
    crystal structures, 464–466, 464f, 465t
    fire and smoke properties, 475, 475t
    injection molding of, 482–483, 482f
    mechanical properties, 472
    melt spinning of, 480–481, 481f
    optical properties, 465t
    radiation resistance, 473, 475
    rubbery region processing, 479–480
    solid-state deformation and, 478–479
    structure, 463–464
    synthesis, 463, 463f
    thermal and optical properties, 467–469, 468t, 469t
    toxicity information, 475, 475t
Polyether ether ketone ether ketone (PEEKEK), 461, 462f
Polyether ether ketone ketone (PEEKK), 461, 462f
Polyetherimide (PEI), 475, 695, 696–697, 706; *see also* Thermoplastic composites, advanced
Polyether ketone (PEK), 461, 462, 462f, 696; *see also* Thermoplastic composites, advanced
    crystal structure, 465
    mechanical properties, 472
    synthesis, 463
Polyether ketone ether ketone ketone (PEKEKK), 461, 462, 462f
    mechanical properties, 472
    synthesis, 463
Polyether ketone ketone (PEKK), 461, 462, 462f
    crystal structure, 466, 466f
    synthesis, 463
Polyether ketone ketone ketone (PEKKK), 461, 462, 462f
Polyethers, 252–279, 253f
    accessibility, 252
    acid-catalyzed condensation, 252
    C–O–C bond, 252, 252f
    history, 252
    multifunctional poly(ethylene glycol), 259–261
    overview, 252
    polyepoxides, 253–256
    poly(ethylene glycol)s, 256–259
    polyglycerols, 263–267
    polyoxetanes, 269–272
    polyoxymethylene, 267–269
    poly(phenylene ether), 278–279
    poly(propylene glycol), 261–263

polytetrahydrofuran, 272–273
from renewable resources
    furan-based polymers, 277–278
    polysaccharides, 273–277
synthesis pathways, 252
Polyethersulfone (PES), 561, 696; *see also* Polyarylethersulfones; Thermoplastic composites, advanced
    blends of, 439
    dynamic-mechanical properties, 431–432
    electrical properties, 435, 435t
    environmental resistance, 437, 438t
    flammability behavior, 435, 436t
    overview, 420
    physical properties, 430–434, 430t
    room-temperature mechanical properties, 432–434, 433t, 434f
    thermal properties, 430–434, 430t, 431f
Polyethersulfone/polyimide blends, 506
Polyethylene dioxythiophene (PEDOT), 674, 674f; *see also* Soluble main-chain conductive polymers
Poly(ethylene glycol)s (PEG), 256–259
    applications, 257–259, 258f
    covalent protein conjugation to, 257–258
    degree of crystallization, 256–257
    history, 256
    LCST, 257, 261
    in lithium-ion batteries, 259
    marketed PEGylated drugs, 258t
    multifunctional (*mf*-PEGs), 253f, 259–261, 260f, 261t
    properties, 256–257
    structure, 253f
    as surface-active agent, 257, 258f
    thermal properties, 266t
    water solubility, 257, 266t
Poly(ethylene naphthalate) (PEN), 336, 337, 700, 705, 718; *see also* Nanocomposites
    crystal unit cell data, 338, 339t
Poly(ethylene oxide) (PEO), 374, 897–898; *see also* Poly(ethylene glycol)s (PEG)
    PMMA and, 185–186
    POM and, 224, 224f
Polyethylenes (PE), 702
    composites, 728, 740–741; *see also* Natural fiber thermoplastic composites
    density of, 22
    polymerization for, 14–18
        autoclave reactor, 15
        high-pressure processes, 14–15
        low- and medium-pressure solution processes, 16–17
        low-pressure gas-phase processes, 17–18
        low-pressure liquid slurry processes, 15–16
        tubular reactor, 15
    SSC, 33–35
Poly(ethylene terephthalate) (PET), 320–324, 700, 705, 851
    additives to, 322–323
    applications, 322–324, 323t
    crystal unit cell data, 338, 339t
    global market, 324
    grades of, 323t
    history, 320
    LOI for, 322

as packaging material, 329
polymerization process, 320–322
antimony, 320
germanium, 320
other catalysts, 320–321
solid-state, 321–322, 321f
synthesis, 321, 321f
properties, 322–324, 322t, 323t
vs. cellulose (mechanical properties), 274t
Poly(3-ethyl-3-(hydroxymethyl) oxetane) (PEHO), 270
Polyfluorenes and copolymers, 674–675; see also Soluble main-chain conductive polymers
Polyfuran
structure, 253f
Poly(furfuryl alcohol) (PFA), 277–278, 278f
Polyglycerols (PGs), 263–267
degree of branching (DB), 264
hyperbranched, 264–265
applications, 265
history, 264
properties, 264–265, 265f
linear, 265, 266f
applications, 266–267
history, 266
properties, 266, 266t
structure, 253f
Polyglycidols, see Polyglycerols (PGs)
Polyhedral oligomeric silsesquioxane (POSS), 363–364, 640, 902–903
Polyhydrazide (PHZ), 573, 575
to polyoxadiazoles, conversion of, 594–595, 595f
Polyhydrazides and polyoxadiazoles
containing carbohydrate moieties, 612
containing 1,3-imidazolidine-2,4,5-trione rings, 612–613
for electro-optical applications
alkoxy-substituted poly(p-phenylene-1,3,4-oxadiazole), 603–607
polyoxadiazoles as electroluminescent materials, 608–610
for high specialty applications, 610–613
for high-temperature applications
aromatic poly(ether-alkyl) hydrazides based on 4,4'-dihydroxybiphenyl units, 588–603
aromatic polyetheraroyl hydrazides based on 4-oxybenzoyl units, 575–588
polyoxadiazoles as electroactive polymers, 573–575
polyoxadiazoles for high-performance membranes, 610–611
with shape-persistent properties, 613, 613f
synthesis
one-step, 572–573
two-step, 573
as thermal-resistant polymers, 572
Polyhydroxyalkanoates (PHA), 728, 859–861; see also Bioplastics
additives and, 866
blends and composites on, 869–870
structural/phase characteristics on, 865
Polyhydroxybutyrate (PHB), 728, 746
Polyhydroxybutyrate-co-valerate (PHBV), 728
Poly(hydroxyethyl methacrylate) (PHEMA), 177–178, 187
Polyhydroxyvalerate (PHV), 746

Polyimide (PI), 491–512, 624–626
characterization, 501–502
functional
electrical insulating and electrolyte polyimides, 511–512
gas permeability and permselectivity, 509–510
NLO polyimides, 511
photosensitive polyimides, 510
organic-soluble and/or melt-processable (thermoplastic), 498
aliphatic–aromatic, 500–501
aromatic, 498–500, 499t
overview, 491–493
solubility parameter (SP), 502–503
structures and properties
commercial TPIs, 507–509
LC properties, 504–505
polymer alloys, 505–507
in solid state, 503–504, 503t
in solution, 502–503
synthesis
from dianhydrides and diamines, 493–494
from dianhydrides and diisocyanates or derivatives, 495
from dianhydrides and disilylated diamines, 495–496
from diesters and diamines, 494–495
from diimides and their derivatives, 496
miscellaneous, 496–497
polyaddition, 497–498
by polycondensation, 493–497
from tetracarboxylic acids and diamines, 495
thermoplastic, 706; see also Nanocomposites
Polyimide/polyimide blends, 505–506
Poly(lactic acid) (PLA), 855
composites, 728, 745
Polylactide (PLA)
additives and, 866
blends and composites on, 869
classical chemical synthesis (class II), 855–858
structural/phase characteristics on, 863–864
Polymer alloys, 505–507
Polymer blending, 652
in gas separation, PBI-based membranes, 657
in pervaporation separation, PBI-based membranes for, 657
Polymer blends, 505–507
Polymer combustion, 880–881; see also Flame retardants (FR)
Polymer electrolyte membrane (PEM) fuel cell, 640
Polymeric film manufacturing cost, 761f
Polymeric material property database, 761–763
Polymeric materials; see also Recycling of thermoplastics
early, 920
new, 920–921
Polymeric material selection, testing for; see also Material selection/design/application
electrical properties, 772–773, 773t
mechanical properties, 769–771, 770t
microbial, weather, and chemical resistance, 773–774, 774t
optical properties, 773, 774t
physical properties, 772, 772t
thermal properties, 772, 773t

Index

Polymerization
- anionic, of formaldehyde, 267–268, 267f–268f
- aromatic polyamides
  - alternative methods, 290
- cationic ring-opening (CROP)
  - of trioxane, 268, 268f
- in conversion of ethylene into HDPE, 854
- 1,3-diamine-4-chlorobenzene, 295, 295f
- living, 293
- methods, 672
- of monomer, 763
- naphthalate polyesters, 337
- poly(butylene terephthalate), 327–329, 327f–328f
- polyepoxides
  - anionic ring-opening, 253–254, 254f
  - cationic ring-opening, 254–255, 255f
  - monomer-activated anionic, 255–256, 255f
- poly(ethylene terephthalate), 320–322
  - antimony, 320
  - germanium, 320
  - other catalysts, 320–321
  - solid-state, 321–322, 321f
  - synthesis, 321, 321f
- polyoxetanes, 270, 271f
- poly(phenylene ether), 279, 279f
- polytetrafluoroethylene, 398–399
- polytetrahydrofuran, 272
- procedure, 577; see also Aromatic polyetheraroyl hydrazides based on 4-oxybenzoyl units
- ring-opening, 857
- ring-opening multibranching (ROMBP)
  - of glycidol, 264

Polymerization chemistry; see also Polyphenylquinoxalines (PPQ)
- monomer synthesis chemistry
  - about, 537–538, 539f
  - bis(α-diketones), synthesis of, 538–541, 539f–541f, 542t
  - bis(o-diamine) monomers, synthesis of, 541–543, 543f, 544f, 545t, 546f
- PPQ synthesis chemistry
  - from bis(á-diketone) and bis(o-diamine) monomers, 543–550, 546f, 547t–549t
  - from monomers with preformed phenylquinoxaline rings, 553, 554f
  - from self-polymerizable monomers, 550–553, 550f–552f, 553t

Polymer/layered silicate nanocomposites (PLSN); see also Nanometric particles as FR
- EVA/layered silicate nanocomposite, 899–900
- PP/PPgMA/layered silicate nanocomposite, 900–901

Polymer light-emitting diodes (PLEDs), 311
- displays, 671

Polymer melt for material selection, rheological properties of; see also Material selection/design/application
- about, 774–775
- viscoelasticity
  - about, 781–784
  - effect of viscoelastic property on material selection (example), 796–799
  - linear, 785–791
  - nonlinear, 792–795

viscosity
- elongational or extensional viscosity, 778–780
- shear viscosity, 775–778
- shear viscosity on material selection (example), 780–781
- shear viscosity–shear rate relationship, empirical models for, 778

Polymer polyethylene glycol (PEG), 75
Polymer processing operations, 922f
Polymer repeat units, 644
Polymer structure-property relations; see also Material selection/design/application
- chemical structure of polymers
  - monomers and repeat units, 763
- microstructure of polymers
  - architecture, 763, 764t
  - chain conformation, 768
  - chain length, 764–766
  - crystallinity, 768–769
  - monomer arrangement in copolymers, 766
  - tacticity, 766–767

Polymethylene
- density of, 22

Polymethyl methacrylate (PMMA), 169, 353, 761, 829, 897–898, 902
- applications of, 184–185
- blends, 185–187, 186t
- cast sheet, 182–183
- long hydrophobic alkyl side chains, 187
- modified, 184
- multifunctional methacrylates, 187
- PHEMA, 177–178, 187
- processing, 182–184, 183f
- properties of, 178–182
  - mechanical, 182
  - optical, 181–182, 181t
  - physical, 178t–180t
  - thermal degradation, 178, 180
- resin, 183
- spherical beads, 183–184
- stereospecific polymerization, 177–178

Poly(4-methylpentene-1) copolymers, 31–32
Poly(N-(4methoxy-phenyl)-N-styryl-1-naphthylamine (P(MeO-NPA), 680
Polynorbornene, 39
Poly(N-phenyl)-N-styryl-1-naphthylamine (P(H-NPA)), 680
Poly(N-vinyl carbazole) (PVK), 676
Poly(N-(X-phenyl)-N-styryl-1-naphthylamine) P(X-NPA), 680
Polyolefin/layered silicate nanocomposites, 904–905; see also Synergies, FR
Polyolefins, 921
- blends, 28–30, 29t
- catalysts, 3–12
  - chromium-based, 3–5
  - SSC, 9–12
  - Ziegler–Natta, 5–9
- classification and densities, 20, 20t
- composites, 41–42
  - fibers, 41, 42
  - fillers, 41, 42
  - interfacial agents, 41, 42
  - susceptible to degradation, 42

copolymers, 30–32
  EVA, 30
  poly(butene-1), 31
  poly(4-methylpentene-1), 31–32
  polypropylene (PP), 30–31
elastomers, 28
nanocomposites, 42–44
  chemical reaction, 43–44
  as exfoliated, 42
  as immiscible, 42
  as intercalated, 42
  multiphase systems, 42
  performance, 42
  physical methods, 42–43
overview, 2–3
processing methods for, 44
production technology, 13–19
properties, 20, 21f, 21t
structural characteristics, 20, 21f, 21t
Polyoxadiazoles, 609
  as electroluminescent materials, 608–610; *see also* Polyhydrazides and polyoxadiazoles
Polyoxadiazoles for high-performance membranes; *see also* Polyhydrazides and polyoxadiazoles
  azobenzene-based polymers, 611
  fluorinated polyoxadiazoles, 610
  sulfonated polyoxadiazoles, 611
Polyoxetanes, 269–272
  applications, 271–272
  history, 270
  melting temperatures with different functions, 270–271, 271t
  polymerization, 270, 271f
  properties, 270–271, 271t
  structure, 253f, 269, 270f
Polyoxybenzoate, 409
Poly(oxy-2,6-dimethyl-1,4-phenylene), *see* Poly(phenylene ether) (PPE)
Polyoxymethylene (POM), 267–269
  additives, 199–201
  anionic polymerization of formaldehyde, 267–268, 267f–268f
  applications, 232–233, 269
  blends, 223–226
  cationic ring-opening polymerization of trioxane, 268, 268f
  characteristics, 208–210
  comonomers for, 196, 197
  composites, 226, 227t
  copolymers (POM-Cs), 269t
  end-group capping, 199
  formaldehyde for, 194, 195–196
  history, 194, 267
  homopolymers (POM-Hs), 269t
  market, 233–234
  nanocomposites, 226, 227–232
  overview, 193–194
  polymerization, 201–207
  processing, 214, 219–223
  producers, 194, 195t
  properties, 210–214, 269, 269t
  reaction media for, 197
  recycling, 233
  structure, 253f

Poly(parabanic acid), 612
Poly(*p*-benzamide) (PBA), 286
Poly(phenylene ether) (PPE), 278–279
  applications, 279
  history, 278
  polymerization, 279, 279f
  properties, 279
  structure, 278f
Poly(phenylene oxide) (PPO); *see also* Poly(phenylene ether) (PPE)
  structure, 253f
Polyphenylene sulfide (PPS), 409, 698–699, 706–707; *see also* Thermoplastic composites, advanced
  blends with, 439–441
Poly(4-4′-phenylene terephthalamide) (PPTA), 371, 371f
Polyphenylquinoxalines (PPQ)
  applications
    energy, 561–562, 563f
    microelectronic, 564
    optical, 563–564
  historical development of, 535–537, 535f, 536f, 538t
  listed in SciFinder database, publications numbers, 535f
  overview, 533–535
  polymerization and processing technology of
    polymerization chemistry, 537–554, 539f–546f, 547t–549t, 550f–554f
    PPQ processing technology, 554–561, 555f–558f, 558t, 559t, 560f
Polyphenylquinoxalines (PPQ) films
  continuous production of, 556f
  in laboratory, processing steps, 555
Polyphenylquinoxalines (PPQ) processing technology
  films, 555, 555f, 556f
  foams, 559–561, 560f
  moldings/composites, 555–559, 557f, 558f, 558t, 559t
  overview, 554
Polyphenylquinoxalines (PPQ) synthesis chemistry; *see also* Polymerization chemistry
  bis(á-diketone) and bis(o-diamine) monomers, polymerization from, 543–550, 546f, 547t–549t
  monomers with preformed phenylquinoxaline rings, polymerization from, 553, 554f
  self-polymerizable monomers, polymerization from, 550–553, 550f–552f, 553t
Poly(3-phenylquinoxaline) synthesis via self-polymerization route, 550f
Polyphenylsulfone (PPSU), 420, 695, 718; *see also* Polyarylethersulfones
  dynamic-mechanical properties, 431–432
  electrical properties, 435, 435t
  environmental resistance, 437, 438t
  flammability behavior, 435, 436t
  overview, 421
  physical properties, 430–434, 430t
  room-temperature mechanical properties, 432–434, 433t, 434f
  thermal properties, 430–434, 430t, 431f
Poly(2-phenyl-1-5-(4-vinylphenyl)-1,3,4-oxadiazole) (POXA), 681, 682
Polyphosphoric acid (PPA), 553t, 572, 642
  structure, 892f
Poly(2,2′-(*p*-phenylene)5,5′-bibenzimidazole), 645

Index

Poly(*p*-phenylene- 1,3,4-oxadiazoles) (PODZ), 572
  about, 572
  synthesis of
    one-step, 572–573
    two-step, 573
Poly(*p*-phenylene terephthalamide) (PPTA), 286, 287f, 572
  chemical resistance, 305, 306t
  film casting of, 308
  global market view, 308f, 309
  market breakdown for, 309f
  in ropes and cables making, 310
  structure, 297f, 298
  in textile market, 309–310
  wet spinning of, 307
Poly(*p*-phenylene vinylene), sulfonium route for synthesis of, 673f
Polypropylene (PP)
  composites, 728, 741–744; *see also* Natural fiber thermoplastic composites
  crystalline, 26
  homopolymers, 26
  matrix, 739; *see also* Natural fiber thermoplastic composites
  Newtonian behavior, 27
  polydispersity index, 27
  polymerization processes, 18–19
    gas-phase technology, 19
    liquid pool slurry process technology, 19
  SCB of, 26
  viscosity, 27
  vs. cellulose (mechanical properties), 274t
  weight-average molecular weight, 27
Poly(propylene glycol) (PPG), 261–263, 391
  physicochemical properties, 262, 263t
  pluronic, 253f, 262–263, 262f, 263t
  properties, 262
  structure, 253f
Polypropylene grafted maleic anhydride (PP/PPgMA), 900–901, 904
  layered silicate nanocomposite, 900–901
Poly(propylene naphthalate) (PTN), 337
Poly(1,3-propylene naphthalate) (PPN), 338
Poly(propylene oxide), 559
Poly(1,3-propylene terephthalate) (PPT), 338
Poly (2-vinyl pyridine) (P2VP), 382
Polypyrrolones (PPY), 535
Polyquinoxalines (PQ), 534
Polysaccharides
  polyethers from, 273–277
    cellulose, 273–274, 273f, 274t
    hydroxyethyl starch, 275
    lignin, 275–277, 276f
    starch, 274–275, 274f
Polysorbate, 258f
Polystyrene (PS), 728
Polystyrene sulfonate (PSS), 674; *see also* Soluble main-chain conductive polymers
Polysulfone (PSF/PSU), 635–636; *see also* Polyarylethersulfones
  dynamic-mechanical properties, 431–432
  electrical properties, 435, 435t
  environmental resistance, 437, 438t
  flammability behavior, 435, 436t
  nanocomposites, 705
  overview, 420
  physical properties, 430–434, 430t
  resistance of, 434, 434f
  room-temperature mechanical properties, 432–434, 433t, 434f
  stress–strain behavior, 434, 434f
  thermal properties, 430–434, 430t, 431f
Polytetrafluoroethylene (PTFE)
  chemical properties, 402
  crystal structure, 400–401, 400f–401f
  discovery of, 397
  electrical properties, 402–403
  electronics, 410
  mechanical properties, 403
  molecular structure, 399–400, 399f
  overview, 397–398
  polymerization, 398–399
  processing
    aqueous dispersions of, 406
    coatings and composites, 406
    cooling rate and, 405
    drying, 406
    molding, 404–405
    pressure cooling, 405
    ram and paste extrusion, 405–406
    sintering, 404–405
  thermal properties, 401–402
  tribological bearing behavior, 406–410
    friction, 406–408
    radiation-modified, 409–410
    wear, 408
    wear-resistant composites, 408–409
Polytetrahydrofuran (PTHF), 272–273, 293
  applications, 272–273
  history, 272
  polymerization, 272
  properties, 272
  structure, 253f
Poly(tetramethylene ether) glycol (PTMEG), *see* Polytetrahydrofuran (PTHF)
Poly(tetramethylene glycol) (PTG), 389f
Polythiophene oligomers, substituted, 675–676; *see also* Soluble main-chain conductive polymers
Poly(trimethylene terephthalate) (PTT), 362
Polyurethanes (PUs), 387
  thermoplastic (TPUs)
    chemical structure, 387–390, 388f–389f
    dynamic-mechanical properties, 391, 391f–392f, 392t
    hard segments, 388–389
    history, 387
    hydrogen bonds, 389t, 390
    mechanical properties, 392–394
    morphology, 387–390, 388f
    overview, 387
    phase separation, 390
    stress–strain properties, 392–394
    structure, 387–390, 388f–389f
    thermoplastic (TPUs), 387
Poly(vinyl acetal), preparation of, 96
Polyvinyl acetate (PVA), 54
  acetalization of, 94, 94f
  aldehyde and, 93, 93f

hydrolysis of, 56–57, 57f, 91, 92f
hydroxyl groups, 93
vinyl acetate (VAc) to, 56
Polyvinyl alcohol (PVA/PVOH), 383, 868
  applications, 78–82
    adhesives, 80
    binders, 81
    blended films, 78–79
    fibers, 79
    hydrogels and membranes, 81–82
    molded products, 79
    packaging films and coatings, 78
    polarizing films, 79
    stabilizers/protective colloids, 80
    textile sizes, 80
    water-soluble films, 78
  blends, 77
  commercialization, 55
  historical development, 55
  hydrolysis
    degree of, 57–59, 57t
    of polyvinyl acetate (PVA), 56, 57f
  iodine complexes, 77
  metal complexes, 77
  miscibility, 77
  modification, 73–77
  molecular structure
    branching, 62
    chain configuration, 61–62
    end groups, 62–63
    tacticity, 63–64
  molecular weights, 59–61, 60t, 61f
  nanocomposites, 77–78
  overview, 54–55
  polymerization, 55–56, 56f
    degree of, 59
  production and consumption, 55
  representative properties, 54t
  solid state properties
    crystalline structure, 65, 65f
    film properties, 67–69, 68f, 68t
    surface properties, 67
    thermal properties, 66–67
  solution properties
    gelation, 72–73
    precipitation, 70
    solubility in water, 69–70, 69f
    viscosity, 71–72, 71f
Polyvinyl butyral (PVB)
  adhesion properties, 109–110, 110f
  with aldehydes, 96, 99f
  applications, 122–129
    miscellaneous, 128–129
    safety glasses, 122–126
    solar module encapsulation, 126–128
  blends, 118–120, 121t
  commercialization, 90
  composites, 120, 121–122
  density of, 106
  electrical properties, 110–111, 111f
  with epoxy compounds, 96, 99f
  historical development, 90
  with isocyanates, 96, 99f

mechanical properties, 111–112, 111t
  with melamine, 96, 98f
  monomer contents for, 94t
  nanocomposites, 120, 121–122
  with phenolic compounds, 96, 97f
  photooxidation, 112–114, 113f
  polymerization, 91–99
  processing
    extrusion of films, 116
    thermal properties, 115
    viscoelastic properties, 115, 115f
  PVOH and, 82
  recycling, 129–130
  surface tension, 106
  terpolymer structure of, 93, 94f
  thermo-oxidation, 114–115, 114f
Poly(vinyl butyral-co-ethanal), 95, 95f
Poly(vinyl butyral-co-furfural), 95, 95f
Polyvinyl chloride (PVC), 403, 728
  PMMA and, 186
Polyvinyl fluoride (PVF), 412–413
Poly(vinylidene fluoride) (PVDF), 412
  PMMA and, 185
Polyvinylphenol (PVPh)
  POM and, 224
Poly(4-vinyl pyridine) (PVPy), 637
Polyvinylpyrrolidone (PVP), 866
POM, see Polarizing optical microscopy (POM); Polyoxymethylene (POM)
POM copolymers (POM-Cs), 269t
POM homopolymers (POM-Hs), 269t
Porous PBI, 642; see also High-temperature polymer electrolyte membrane fuel cells, PBI in
Positive temperature coefficient (PTC) resistors, 703
POSS, see Polyhedral oligomeric silsesquioxane (POSS)
Postconsumer recycle (PCR) PET, 329
Potassium permanganate, 539
Potato starch production process, 844f
Power conversion efficiency (PCE), 675
PPE, see Poly(phenylene ether) (PPE)
PPG, see Poly(propylene glycol) (PPG)
P-phenylenediamine (PPD), 286, 287f, 288
PPQ, see Polyphenylquinoxalines (PPQ)
π-π stacking poly(dibenzofulvene), 684–686; see also Vinyl polymers, conductive
PPTA, see Poly(p-phenylene terephthalamide) (PPTA)
Preceramic compounds, 902
Precipitation
  poly(vinyl acetal), 96
  PVOH, 70
Preformed quinoxaline rings, 554f
Preimpregnated composites; see also Thermoplastic composites, advanced
  principal manufacturers/trade names of thermoplastic matrices, 707t
  pseudothermoplastics, 707
  true thermoplastics, 707
Prepreg material heating, 708
Pressure-cooling techniques
  PTFE and, 405
Primary recycling routes, 929–930, 930f; see also Recycling of thermoplastics
Printing inks, PVB in, 129

Index

Pristine conductive polymers, 671
Pristine (undoped) conductive polymers, 670
Processing
    PAN, 150–152
    PMMA, 182–184, 183f
    polyolefins
        blow molding, 44
        blown film extrusion, 44
        cast film extrusion, 44
        extrusion forming, 44
        injection molding, 44
        rotational molding, 44
    POM, 214, 219–223
    PVB
        extrusion of films, 116
        thermal properties, 115
        viscoelastic properties, 115, 115f
Process selection, 757, 758f; *see also* Material selection/design/application
Product design process, 754–756, 755f; *see also* Material selection/design/application
Product requirements for PBI, 621–622; *see also* Polybenzimidazoles (PBI)
Propene, 170
Propylene diamine (PDA), 389f
Propylene–ethylene copolymers, 11
Propylene oxide (PO), 254
    for BDO synthesis, 325, 325f
Protective colloid, PVOH as, 80
Proton conductivity, 562
Proton-exchange membrane (PEM), 533
    fuel cell, 640
Proton-exchange membranes fuel cell (PEMFC), 562
Protons diffuse, 640
Pseudo-living polymerization, 40
Pseudothermoplastics, 707; *see also* Preimpregnated composites
PSPIs (photosensitive polyimides), 510
PSU/PPS blends, 439
PTFE, *see* Polytetrafluoroethylene (PTFE)
PTHF, *see* Polytetrahydrofuran (PTHF)
Pulse wave (pw) mode laser, 808
Pultrusion, 712–713, 712f, 738; *see also* Fiber and tape placement; Natural fiber thermoplastic composites
PUs, *see* Polyurethanes (PUs)
PVDF (polyvinylidene fluoride), 412
PVF (polyvinyl fluoride), 412–413
PVOH–polyacrylic acid (PAA) block polymers, 74
PVOH–poly(OXZ) graft copolymers, 75
Pyrene, 609

Q

Quality infrastructure and environmental safety, 952–953
Quality systems of laboratories, 945–946; *see also* Environment health and safety
Quantum efficiency, 677
Quartz fibers, 702
Quasi-simultaneous welding, 812, 814, 817
Quaternary ammonium compounds (QAC), 866
Quaternary trimethylaminopropyl methacrylamide (QAPM), 74

R

Radel R PPSU, 421
Radiation-modified tribological behavior, of PTEF, 409–410
Radiation resistance, 699
    PAEKs, 473, 475
    polyarylethersulfones, 435–436
Radio-frequency identification (RFID) tags, 671
Rain lixiviation process, 923
Raman, spectroscopy, 357
Ram extrusion
    of PTFE, 405–406
Ramie, 729
Random copolymers, 869
Ranking, 757f
Raw materials
    costs, 759
    for PBT production, 324–327, 324f–327f
Raychem Corporation, 462
Rayon-based carbon fibers, 701
RDP, *see* Resorcinol bis(diphenyl phosphate)
REACH regulation, 944
Reactive fire retardants, 884
Reactive free radicals, 731
Reactivity, in polymer industry, 921–923
Reactor blends, 38–39
Recovery of PHA, 860
Rectangular channel geometry, 796f
Recyclability, of aromatic polyamide fibres, 313
Recycling
    of bioplastics, 870–871
    POM, 233
    PVB, 129–130
Recycling of thermoplastics
    early polymeric materials, 920
    new polymeric materials/composites, 920–921
    overview, 919–920
    plastics and environment
        about, 924–925
        agricultural plastic wastes, 927
        automotive plastic wastes, 926
        plastic wastes in electronic devices, 927–928
        urban plastic wastes, 925–926
        wild plastic wastes, biodegradation and photodegradation, 928
    plastic wastes and eco-balance
        economic and ecological dimensions, 932–933
        incineration and tertiary recycling routes, 930–932, 931t
        primary and secondary recycling routes, 929–930, 930f
        quality assurance and thermoplastic waste recycling and valorization routes, 934f
    stability and reactivity, 921–923, 922f
    wastes, 923–924, 923f
Redox reactions, 670
Red phosphorus (RP), 893–894; *see also* Phosphorus-based FR
Reductive doping, 670
Reference materials, 951–952; *see also* Environment health and safety
Refractive index, 550

Regioregularity, 675
Registration, Evaluation, Authorization and Restriction of Chemicals (REACH) regulation, 885
Reinitzer, Friedrich, 369
Relative viscosity
   aromatic polyamides, 304
Relaxation studies
   of polycarbonates, 358–359, 358f
Renewable materials, polyether-based
   furan-based polymers, 277–278
   polysaccharides, 273–277
      cellulose, 273–274, 273f, 274t
      hydroxyethyl starch, 275
      lignin, 275–277, 276f
      starch, 274–275, 274f
Reppe chemistry, 324–325, 324f
Residual stresses, 709
Resin coatings, 698
Resorcinol, 355
Resorcinol bis(diphenyl phosphate) (RDP), 894
Responsible Care, Registration, Evaluation, Authorization and restriction of Chemicals (REACH), 920
Responsible Care initiative, 943
Retrogradation process, 846
Reverse osmosis (RO) membranes
   polyarylethersulfones applications in, 445, 446
Reversible addition–fragmentation chain transfer (RAFT), 375
   polymerization
      acrylic monomers, 176, 177
      VAc, 74
R.G.C. Jenkins and Company, 15
Rheology
   of linear and branched polycarbonates, 358
   of polyarylethersulfones, 427–428
Rice, 730
Ring-opening metathesis polymerization (ROMP), 671, 672
   poly(cycloolefins) by, 39
Ring-opening multibranching polymerization (ROMBP)
   of glycidol, 264
Ring-opening polymerization (ROP), of polyepoxides
   anionic (AROP), 253–254, 254f
   cationic (CROP), 254–255, 255f
3R initiative (reduce, reuse, and recycle), in plastic wastes, 929
Ritchiethey, C. G., 320
Röhm, Otto, 170
Röhm & Haas Company, 170
ROMP, see Ring-opening metathesis polymerization (ROMP)
Room-temperature mechanical properties
   of polyarylethersulfones, 432–434, 433t, 434f
Ropes, PPTA fibers use in, 310
Rotational molding, for polyolefins, 44
Rotoforming, see Rotational molding, for polyolefins
Rotomolding, 739; see also Natural fiber thermoplastic composites; Rotational molding, for polyolefins
Rounded rectangular channel geometry, 796f
Royalty costs, 760
Rubber processing, 150
Rubber toughening, 766

Rubbery region processing
   of PAESs, 479–480
   of PEEK, 479–480
Rubrene, chemical structure of, 681f

S

SABIC Co., 507
Safety
   glass, PVB for, 95, 122–126
   polyarylethersulfones, 442
Sales taxes, 760
Sandwich technology; see also Fiber and tape placement
   fiber stitching, 716
   Z-pinning, 716
Sanitary, PMMA for, 184–185
Saponification with caustic soda, 575
SAXS, see Small-angle x-ray scattering (SAXS)
Scanning electron microscopy (SEM), 731, 743f
Scattering studies, of PMMA, 181–182
Schnell, Herman, Dr., 350
Schrock, Richard R., 39
Scientific Design, 326
SCLCPs (side-chain LCPs), 370, 371–372, 372f
Scrap, 924
Screening, 757f
Screw extrusion
   of PTFE, 405
SEC, see Size-exclusion chromatography (SEC)
Secondary fabrication operations
   of polyarylethersulfones, 429
Secondary recycling routes, 929–930, 930f; see also Recycling of thermoplastics
Second-order fluid constitutive models, 795, 795t; see also Viscoelasticity
Self-lubricating behavior, of PTEF, 406–408
Self-polymerizable monomers, polymerization from, 550–553, 550f–552f, 553t
Self-polymerization route
   poly(arylene ether phenylquinoxaline) synthesis via, 551f
   poly(3-phenylquinoxaline) synthesis via, 550f
Semiconductor lasers, 808
Semicrystalline thermoplastics, 709
Semifinished-product manufacturing; see also Natural fiber thermoplastic composites
   granule production, 737
   kneading, 736–737
   mat production, 737
   solution-mixing procedure, 737
Semifinished products, 709
Sensitivity of polymer matrices, 720
Separator for battery, 561–562
Sepiolite, 902
Shape-persistent properties, polyhydrazides with, 613, 613f
Shear rate, 698
Shear strength, 771, 817f
Shear viscosity; see also Viscosity
   about, 775–778
   on material selection (example), 780–781
Shear viscosity-shear rate relationship, empirical models for, 778
Shell's HiPERTUF PET/PEN copolymer, 341

Short-chain branching (SCB)
  as methyl groups, 22
  polyethylene, 22
Side-chain LCPs (SCLCPs), 370, 371–372, 372f
Silane
  coupling agents, 735
  organofunctionality of, 735
  treatment of sisal, 741
Silica
  nanocomposites, 718
  nanoparticles, 705
  in PP/PEO/PMMA, 897–898
Silica gel, 897
Silicone-based flame retardants, 360; *see also* Flame retardants (FR)
  block copolymers of PC and PDMS (PC-b-PDMS), 897
  silica in PP/PEO/PMMA, 897–898
  silicones in polycarbonate (PC), 897
Silicones in polycarbonate (PC), 897
Silsesquioxane, 902–903
Simultaneous welding, 812
Sinapyl alcohol, 275, 276f
Single-site catalysts (SSC), 9–12
  cycloolefin copolymers (COC) from, 40–41
  metallocene, 9–11
    MAO cocatalysts, 10
    significance of, 11
    on tacticity, 10–11
  poly(cycloolefins), 39
  polyolefins from, 32–39
  postmetallocene, 12
  process technologies for, 19
Single-stage and two-stage polymerizations, 621t
Single wall carbon nanotubes (SWNTs), 378, 703, 704t
Single-walled nanotubes (SWNT), 902
Sintering
  PTEF and, 404–405
Sisal, 729
Size-exclusion chromatography (SEC), 60, 357, 849
Sizing, 702
SK Chemical, 332
Slow monomer addition (SMA) technique, 255, 264
Slurry phase heavy-diluent loop reactor, 16
Slurry phase heavy-diluent stirred-tank reactor, 16
Slurry phase light-diluent loop reactor, 16
Slurry phase light-diluent stirred-tank reactor, 16
Slurry phase liquid pool reactor, 16
Small angle neutral scattering (SANS), 186
Small-angle x-ray scattering (SAXS), 186, 374, 471, 480, 508
Society of Plastics Industry (SPI), 946
Sodium chlorite treatment, 733
Sodium hydroxide, 76
Sodium hypochlorite, 847
Sodium layered montmorillonite (Na-MMT), 900
Softwood kraft lignin, 853
Solar module encapsulation, PVB for, 126–128
Sol-gel process, 642–643; *see also* High-temperature polymer electrolyte membrane fuel cells, PBI in
Solid crude oil, *see* Polyolefins
Solidification, 709
Solid state, polyimides in, 503–504, 503t

Solid-state deformation
  PPAESs processing and, 478–479
Solid-state extrusion, of POM, 220
Solid-state lasers, 808
Solid-state polymerization (SSP)
  of PET, 321–322
Solid state properties, of PVOH
  crystalline structure, 65, 65f
  film properties, 67–69, 68f, 68t
  surface properties, 67
  thermal properties, 66–67
Solid waste, 921, 923
Solubility
  PAN, 147
  polyarylethersulfones, 437–438
  PVB, 105–106, 105t, 107t–108t
  water-soluble PVOH, 69–70, 69f, 78
Solubility parameter (SP)
  polyimides, 502–503
Solubilization of π-conjugated polymers, 672–674, 673f; *see also* Conductive thermoplastics
Soluble main-chain conductive polymers; *see also* Conductive thermoplastics
  polyethylene dioxythiophene (PEDOT), 674, 674f
  polyfluorenes and copolymers, 674–675
  substituted polythiophene oligomers, 675–676
Solution
  casting, 849
  polyimides in, 502–503
Solution fiber spinning of PAN, 151–152
  dry spinning, 151–152
  wet spinning, 151
Solution-mixing procedure, 737; *see also* Semifinished-product manufacturing
Solution polymerization, 857
Solution polymerization processes, 16–17
  low-pressure cooled reactor, 17
  medium-pressure adiabatic reactor, 17
Solution processing
  of polyarylethersulfones, 429
Solution properties
  of PVB, 104–106, 105t, 107t–108t
  of PVOH
    gelation, 72–73
    precipitation, 70
    solubility in water, 69–70, 69f
    viscosity, 71–72, 71f
Solvay Specialty Polymers, 451, 452t
Solvent-free production method, 857, 858f
Solvents
  PVB, 104–106, 105t, 107t–108t
  PVOH, 70
Soterna thermal energy boiler
  PPSU, 444f
Specific extinction area (SEA), 882
Specific-gravity techniques, 399
Spherical aromatic polyamides, 296–297
Spherical beads, PMMA, 183–184
Spherilene gas-phase technology, 18
Spherilene process, 33
Spontaneous recrystallization, 846
Square channel geometry, 796f
SSC, *see* Single-site catalysts (SSC)
Stability, in polymer industry, 921–923

Stability studies
    of polycarbonates, 357–358
Stabilization
    PAN for carbon fibers, 153–154
    of polycarbonates, 359
Stabilizers, POM, 199–200
Standard creep tests, 791
Standardization in environment health/safety, 946–948
Standardized mortality ratios (SMR)
    acrylonitrile, 142
Standard Oil of Indiana, 3
Standards and technology life cycle assessment, 883; *see also* Flame retardants (FR)
Starch, bioplastics from, *see* Natural sources (class I)
    esters, 848
    ethers, 848
    polyethers from, 274–275, 274f
    structure, 253f, 274f
Static focusing, 810f
Staudinger, H., 55
Stealth effect, 257–258
Stealth liposomes, 258
Steam-exploded lignin, 277
Steam explosion, 735, 852
Stereoisomers of lactic acid/lactide, 856f
Stereospecific polymerization, of PMMA, 177–178
STILAN, 462
Stimuli-responsive LCPs, 370
Strategic Approach to International Chemical Management (SAICM), UN, 943
Stress relaxation test, 791
Stress–strain behavior
    of PSU, 434, 434f
Stress–strain curves, 403
Stress-strain properties, of TPUs
    hard-segment content effect on, 393–394, 393f–394f
    segmented structure of TPUs effect on, 392–393, 393f
Structural foams
    polyarylethersulfones applications in, 450, 450f–451f
Styrene acrylonitrile (SAN), 716
Styrene–ethylene–butadiene–styrene (SEBS), 331
Substances of very high concern (SVHC), 885
Substituted polythiophene oligomers, 675–676; *see also* Soluble main-chain conductive polymers
Succinic acid (SA)
    bio-based, 326
Sugar cane, PE from, 854–855; *see also* Classical chemical synthesis (class II), bioplastics from
Sulfonated polyoxadiazoles, 611; *see also* Polyoxadiazoles for high-performance membranes
Sulfonated polysulfones, 447
Sulfonated PPQ (SPPQ), 562, 563f
Sulfonation and stabilization, 624; *see also* Fiber formation
Sulfonation modification of PBI, 651–652, 651f; *see also* Pervaporation separation, PBI-based membranes for
Sulfone polymers, 705
    foams based on, 718; *see also* Foams, thermoplastic
Sulfoxide route, 672
Sulfuric acid, 561
Sumitomo Chemical Co., 451, 452t

Supramolecular cooperative motion (SMCM), 376
Supramolecular LCPs (SLCPs), 370
    halogen-bonded, 380–382, 381f
    hydrogen-bonded, 379–380, 380f
    others, 382
Surface ablation, 829–830; *see also* Laser processing of thermoplastic composites
Surface activation, 829; *see also* Laser processing of thermoplastic composites
Surface energy, 731
Surface properties
    of PVB, 106
    of PVOH, 67
Surfactant(s)
    PEG as, 257
Sustainability
    concept of, 329
Sustainability aspects
    polyarylethersulfones, 442–443
Sustainability assessment of bioplastics, 838–839
Sustainable Biomaterials Collaborative (SBC), 836
Sustainable development, legal frameworks, 942–943; *see also* Environment health and safety
Sustainable flame retardancy-from halogenated chemicals to nanocomposites, 882
Suzuki coupling, 675f
Syndiotactic PP (sPP), 11, 38
Synergies, FR; *see also* Flame retardants (FR)
    about, 903–904
    ABS/layered silicate nanocomposites, 905–906
    layered silicate nanocomposites with phosphorus-based compounds, 906–907
    polyester/layered silicate nanocomposites, 906
    polyolefin/layered silicate nanocomposites, 904–905
    synergistic halogenated/phosphorated mixture, 907
    synergistic metal hydroxide/metal borate mixtures, 907–908
    synergistic metallic oxide/phosphorated mixtures, 908
    synergistic nitrogenated and phosphorous mixtures, 907
Synthesis of PBI; *see also* Polybenzimidazoles (PBI)
    general route to, 619, 619t, 620f
    specific case for, 619–621, 620f, 621t
Synthetic LCPs, 370
Synthetic polymerization techniques, 765
Synthetic thermoplastic materials, 921
Synthetic thermoplastics, composites from, 744–745; *see also* Natural fiber thermoplastic composites
Synthofil, 79

T

Tacticity, 766–767
    of PVOH, 63–64, 63f–64f
Tape laying process, laser-based, 830–831
TBBPA, *see* Tetrabromobisphenol A (TBBPA)
TBPA, *see* Tetrabromophthalic anhydride (TBPA)
Tear strength, PVOH films, 69
Technical competence of laboratories, 945–946; *see also* Environment health and safety
Technora fibres, 286, 309
    properties of, 299–300, 301t, 302f

Index

Teflon AF amorphous fluoropolymer, 415
Teijinconex fibres
    properties of, 300, 301t, 303
    vs. natural (not colored) organic and conventional synthetic fibers, 301t
Teijin Limited, 286, 287, 309
Tennessee Eastman Co., 331, 332
Tensile impact testing, 771
Tensile load, 826f, 827
Tensile properties, 745
    aromatic polyamides, 299, 299t
Tensile strength, 827f
    creep, 771
    of HMA, 635f
    PVOH films, 68–69, 68f
Tensile test, 769
Teonex, 341
Terephthalic acid (TPA), 287f, 320, 324, 326, 349
    production of, 326, 326f
    raw material for PBT production, 327–328, 328f
Terephthalic dihydrazide (TDH), 577
Terephthaloyl dichloride (TPC), 286, 287f
Terpenephenol (TPh)
    POM and, 224
Terpolymers, 682, 683f
tert-butyl glycidyl ether (*t*BGE), 265
tert-butyl methacrylate, 187
tert-butyloxycarbonyl, 561
Tertiary recycling routes, incineration and, 930–932, 931t
Terylene, 320
Tetraaryl–BPA (TABPA), 355
Tetrabromobisphenol A (TBBPA), 890; *see also* Halogenated FR additives
Tetrabromobisphenol-A PC, 360f
Tetrabromophthalic anhydride (TBPA), 890; *see also* Halogenated FR additives
Tetracarboxylic acids
    polyimides synthesis from, 495
Tetrafluoroethylene (TFE), 397, 398
    polymerization of, 399
Tetrafluoropropanol (TFP), 653
Tetrahydrofuran (THF), 7, 680, 719; *see also* Polytetrahydrofuran (PTHF)
2,2,4,4-tetramethyl-1,3-cyclobutanediol (TMCD), 335, 348, 349, 349f
Textile industry
    MPIA fibers in, 309
Textile pigment printing, 312
Textile sizes, PVOH, 80
Thermal analysis; *see also* Polyhydrazides and polyoxadiazoles
    aromatic poly(ether-alkyl) hydrazides based on 4,4′-dihydroxybiphenyl units
        about, 592–593
        bulk crystallization, 597–599, 597f, 598f
    for aromatic polyetheraroyl hydrazides based on 4-oxybenzoyl units
        about, 579–580, 582–583
        bulk crystallization, 583–584, 583f, 584f
    of polycarbonates, 357–358
Thermal cracking, 932
Thermal cyclodehydration, 573

Thermal degradation
    of PLA, 864
    PMMA, 178, 180
    of PVB, 106
Thermal gravimetric analysis, 635
Thermally stimulated depolarization (TSD), 111
Thermal molding process, 557
Thermal properties
    aromatic polyamides, 298–299, 298t
    CHDM-based polyesters, 334t
    epoxide-derived polyethers, 266t
    mf-PEGs, 261, 261t
    naphthalate polyesters, 337–338, 337t–338t
    PAEKs, 467–469, 469t
    polyarylethersulfones, 430–434, 430t, 431f
    of polymeric materials, 772, 773t
    POM, 215t–218t
    of PTFE, 401–402
    PVB, 106
    PVOH, 66–67
Thermal-resistant polymers, polyhydrazides/polyoxadiazoles as, 572; *see also* Polyhydrazides and polyoxadiazoles
Thermal stability, 697, 704
    of poly(DBF), 686
Thermal/thermal-chemical conversion, 847–848; *see also* Molecular weight reduction, starch
Thermochemical processing, 932
Thermochromism phenomenon, 673
Thermodynamic immiscibility, 868
Thermoforming, 714
    of polyarylethersulfones, 428–429
    polycarbonates, 365
Thermogravimetric analysis (TGA), 553t, 703, 704t; *see also* Polyhydrazides and polyoxadiazoles
    for aromatic poly(ether-alkyl) hydrazides based on 4,4′-dihydroxybiphenyl units, 591–592
    for aromatic polyetheraroyl hydrazides based on 4-oxybenzoyl units, 577, 577f, 578f, 578t
Thermo-oxidation of PVB, 106
Thermo-oxidative degradation, 922
Thermo-oxidative stability, 552
    of PBI, 627f
Thermoplastic additives, *see* Flame retardants (FR)
Thermoplastic composite material processing
    consolidation/impregnation/solidification, 709
    heating, 708
    semifinished products, 709
    steps in, 708f
Thermoplastic composites, advanced
    autoclave technology, 709–710, 709f
    compression-molding technique, 710
    environmental resistance, 719–721, 720t, 721t
    fiber and tape placement
        automated fiber placement (AFP), 711
        commingled yarns, 713, 713f
        diaphragm forming, 713, 714f
        filament winding, 711–712
        injection molding (IM), 714–715
        localized fiber placement, 710
        matched-mold techniques, 715
        pultrusion, 712–713, 712f
        sandwich technology, 716

stretch-draw process, 715
thermoforming, 713–714
fibers
　aramid, 701
　average properties, 700t
　boron, 702
　carbon, 700–701
　ceramic, 702
　high-performance glass, 701–702
　quartz, 702
nanocomposites
　liquid crystalline polymers (LCP), 706
　PEEK nanocomposites, 702–705, 704t
　polyester (PET, PEN) nanocomposites, 705
　polyphenylene sulfide (PPS), 706–707
　polysulfone nanocomposites, 705
　thermoplastic polyimide (PI/PAI/PEI), 706
overview, 694–695
preimpregnated composites
　principal manufacturers/trade names of thermoplastic matrices, 707t
　pseudothermoplastics, 707
　true thermoplastics, 707
thermoplastic composite material processing
　consolidation/impregnation/solidification, 709
　heating, 708
　semifinished products, 709
　steps in, 708f
thermoplastic foams
　about, 716–717
　aromatic polyketones-based foams, 717–718
　foamed composites, 719
　polyester-based (PEN-based) foams, 718
　sulfone polymers, foams based on, 718
　thermoplastic polyimides (PI/PAI/PEI), foams based on, 719
thermoplastic matrices
　liquid crystalline polymers (LCP), 699–700
　polyamideimide (PAI), 699
　polyaryletherketone (PAEK), 698
　polyarylethers (PAE), 695
　polyarylsulfone (PAS), 696
　polybenzimidazole (PBI), 698
　polyesters (PEN/PET), 700
　polyetheretherketone, 697
　polyetherimide (PEI), 696–697
　polyetherketone (PEK), 696
　polyethersulfone, 696
　polyphenylene sulfide (PPS), 698–699
Thermoplastic composites with dissimilar materials, laser-based joining of, 820–821, 820f; see also Laser transmission welding
Thermoplastic foils, 127
Thermoplastic matrices, principal manufacturers/trade names of, 707t
Thermoplastic polyesters, 320–342
　cyclohexanedimethanol-based, 331–336
　naphthalate polyesters, 336–342
　poly(butylene terephthalate), 324–331
　poly(ethylene terephthalate), 320–324
Thermoplastic polyimides (PI/PAI/PEI), 706; see also Nanocomposites
　foams based on, 719

Thermoplastic polyimides (TPIs), 492
　commercial, 507–509, 508t
Thermoplastic polymers, 541
　development of, 273–274
Thermoplastic polyurethanes (TPUs), 387–394
　chemical structure, 387–390, 388f–389f
　cohesive energy and specific volume of polar groups, 389t, 390
　dynamic-mechanical properties, 391, 391f–392f, 392t
　hard segments, 388–389
　history, 387
　hydrogen bonds, 389t, 390
　mechanical properties
　　hard-segment content effect on stress–strain properties, 393–394, 393f–394f
　　stress-strain properties, segmented structure of TPUs effect on, 392–393, 393f
　morphology, 387–390, 388f
　overview, 387
　phase separation, 390
　structure, 387–390, 388f–389f
Thermoplastics, ketone-based, see Ketone-based thermoplastics
Thermoplastic silicones (TSI), 127
Thermoplastic starch (TPS), 745
　additives on, 865–866
　blends and composites on, 867–868
　natural sources (classI), 849
　structural/phase characteristics on, 861–863
Thermoplastic waste recycling and valorization routes, 934f
Thermotreatments, 731
Thermx, 332
Thiodiacetic acid ester, 674
Threefold rotational isomeric model, 768
Time-of-flight (TOF) method, 677, 678t
Time to ignition (TTI), 882
Tin octanoate, 859
Titanium, 5, 7
Titanium chloride ($TiCl_4$), 7
Titantrichloride, 76
p-tolualdehyde, 326
p-toluic acid, 326
Total heat released (THR), 882
Toughening, of advanced composites
　polyarylethersulfones applications in, 449–450, 449f
Toxicity information
　for PEEK, 475, 475t
TPA-based PET, 320
TPI LARC-TPI, 492, 492f
TPUs, see Thermoplastic polyurethanes (TPUs)
Traceability chains, 943
Trans double bonds, 607
Transetherification reaction, 674
Translate, product design, 756
Transmission electron microscopy (TEM), 630
Transmission welding by incremental scanning technique (TWIST) laser welding, 812
Transmissivity curves, 813f
Transportation, PMMA in, 184
Trialkyl aluminum (TMA)
　Lewis acid and, 10
　water to, 10

Index

Tribological bearing behavior, of PTFE, 406–410
    friction, 406–408
    radiation-modified, 409–410
    wear, 408
    wear-resistant composites, 408–409
Tricresyl phosphate (TCP), 894
Triethylamine, 540f
Trifluoroacetic acid (TFA), 604
Trifluoromethyl, 610
Trifunctional tris-benzil, 538
Trimellitic anhydride (TMA), 636
Trimethylsilyl glycidyl ether (TMSGE), 265
1,3,5-trioxacyclohexane, see Trioxane
Trioxane
    cationic ring-opening polymerization of, 268, 268f
    copolymerization, 197
    exposure to, 196
    polymerization, 197
    for POM, 196
    properties of, 197t
1,3,5-Trioxane, see Trioxane
Trioxymethylene, see Trioxane
Triphenylamine (TPA), 681
Triphenyl phosphate (TPP), 894
Triphenyl phosphite (TPP), 288
1,1,1-tris-p-hydroxyphenylethane (THPE), 355–356, 355f
Tritan, 335
Trommsdorff effect, 175
TSD, see Thermally stimulated depolarization (TSD)
Tubular reactor, 13–14
    high-pressure, 15
Twaron fibres, 299, 309
    properties of, 301t
Twaron yarns, 300, 303f
Two-phase polycondensation system, 289
Two-stage polymerizations, 621t
Two-step synthesis of PODZ, 573

U

UCC, see Union Carbide (UCC)
UHMWPE, see Ultrahigh-molecular-weight polyethylene (UHMWPE)
UL94 tests (Underwriters' Laboratories), Bunsen burner test; see also Flame retardants (FR)
    about, 881
    cone calorimetry, 882
Ultrafiltration (UF) membranes
    polyarylethersulfones applications in, 445, 446
Ultrahigh-molecular-weight polyethylene (UHMWPE), 19, 26
Ultralarge integrated circuit (ULSI), 537
Ultraviolet (UV)-Vis spectroscopy, 372
UL 94 V0, 330
UL 94 5VA, 330
Underwriters Laboratories (UL), 769
Uniaxial extension, 780
Uniform thin film, 672
Union Carbide (UCC), 4, 420, 421
United Nations Environment Program (UNEP), 952
Unreinforced polyphenylene sulfide, 698
Urban plastic wastes, 925–926; see also Plastics and environment
Urban solid waste, 924
Use-and-discard lifestyle, 920
US Federal Trade Commission (FTC), 286
UV absorption spectra of PVB, 100, 101f
UV light, PMMA and, 184
UV resistance
    polycarbonates, 355
UV stabilization of PVB, 118
UV-stabilizers, POM, 200

V

Vanadium, 5, 6, 40
Vapor-grown carbon fiber (VGCF), 154
Vegetable fibers, 729
Veriva PPSU, 447
Very-low-density polyethylene (VLDPE), 2, 24
Vespel TPIs, 507, 508
Vestanamer, 39
VICTREX PEEK, 462
Vinasse, 854
Vinyl acetate (VAc), 55; see also Polyvinyl alcohol (PVOH)
    to polyvinyl acetate, polymerization of, 56
Vinyl acetate–ethylene (VAE) copolymers, 30
Vinyl acetate–vinyl methanesulfonic copolymers, 75
Vinyl alcohol–acrylic acid copolymers, 75
Vinyl alcohol–alkyleneoxyacrylate (ALOAC), 75
Vinyl alcohol–maleic acid copolymers, 74
Vinyl alcohol polymers, see Polyvinyl alcohol (PVOH)
Vinyl copolymers, 682
Vinyl monomers
    stereoregularity, 5
3-(4-vinylphenyl)perylene (v-peryl), 683f
Vinyl polymerization
    poly(cycloolefins) by, 39
Vinyl polymers, conductive; see also Conductive thermoplastics
    about, 676–677
    electron-transporting vinyl polymers and copolymers, 680–682, 681f
    as emitters, 682–684
    hole-transporting vinyl polymers, 678–680, 679t
    mobility measurements, 677, 678t
    π-π stacking poly(dibenzofulvene), 684–686
Viscoelasticity; see also Polymer melt for material selection, rheological properties of
    about, 781–784
    effect on material selection (example), 796–799
    linear
        Boltzmann superposition integral model, 785–787
        Maxwell differential model, 785
        measurement of, 791
        mechanical models, 790–791
        with yield value, 787–790
    nonlinear
        general constitutive models, 794–795
        Lodge integral constitutive model, 793
        Lodge-type integral constitutive models, 794
        Maxwell-type differential constitutive models, 793–794
        Maxwell-type simple differential constitutive models, 792–793
        second-order fluid constitutive models, 795, 795t

Viscoelastic properties, of PVB, 115, 115f
Viscosity; *see also* Polymer melt for material selection, rheological properties of
   aromatic polyamides, 303–304, 304f
   elongational or extensional viscosity, 778–780
   LCPs, 375
   PVOH, 60, 60t, 71–72, 71f
      zero-shear-rate, 72
   shear viscosity, 775–778
   shear viscosity on material selection (example), 780–781
   shear viscosity-shear rate relationship, empirical models for, 778
VLDPE, *see* Very-low-density polyethylene (VLDPE)
Volume conductivity, 670

W

W. R. Grace & Co., 11
Wacker Chemie, Germany, 55
Warp sizing, PVOH for, 55
Waste generation due to human activity, 923f
Water, to trialkyl aluminum, 10
Water solubility
   epoxide-derived polyethers, 266t
   mf-PEGs, 260–261
   PEG, 257, 266t
Water-soluble biomaterials, 870
Water-soluble PVOH, 69–70, 69f
   application, 78
Water vapor permeability, 853
WAXS, *see* Wide-angle x-ray scattering (WAXS)
Waxy starches, 842
Wear, of PTFE, 408
Wear-resistant composites, of PTFE, 408–409
Weigert's effect, 378
Welding of PEEK/carbon fiber composite, 697
Wet spinning, 307
   of MPIA, 307
   of ODA/PPTA, 308
   of PAN, 151
   of PPTA, 307
Whinfield, John Rex, 320
White dust, 920
White rot fungi, 735
Wide-angle x-ray scattering (WAXS), 145, 374, 471, 480, 580, 580f, 581t, 599
   PHEMA, 181–182
   PMMA, 181
Wiechert model, 790f
Wild plastic wastes, 928; *see also* Plastics and environment
Williamson ether synthesis, 252
Wittig chemistry, 685f
Wittig coupling reaction, 680
Working capital, 760
World Health Organization (WHO), 257

X

X-ray diffraction (XRD)
   LCPs and, 374, 375f
X-ray diffraction analysis, 845; *see also* Polyhydrazides and polyoxadiazoles
   aromatic poly(ether-alkyl) hydrazides based on 4,4′-dihydroxybiphenyl units, 593–594, 594f
   aromatic polyetheraroyl hydrazides based on 4-oxybenzoyl units, 580–581, 585–587
X-ray irradiation
   of PTFE, 410
X-ray photoelectron spectroscopy (XPS), 473, 731
XRD, *see* X-ray diffraction (XRD)
Xylan
   abundant biopolymers, 853
   with nanocellulose, 869
   with reinforcing nanoclays, 868
   removal, 852
p-xylene, 336
   oxidation of, 326–327, 326f
XYLEX, 336

Y

Yamamoto scheme, 675
Yamazaki–Higashi method, 287, 289
Yield stress
   TPUs, 393–394, 393f–394f
Young's modulus
   TPUs, 393–394, 393f–394f

Z

Zeolitic imidazolate frameworks (ZIF), 640
Ziegler, Karl, 5, 39
Ziegler–Natta catalysts, 5–9
   $MgCl_2$-supported, 7, 8, 8t–9t
   titanium, 5, 7
   two components, 5, 6t
   vanadium, 5, 6
   zirconium, 5, 6
Zirconium, 5, 6
Z-pinning, 716; *see also* Sandwich technology